More Than Just a Textbook

Internet Resources

Step 1 Connect to Biology Online
biologygmh.com

Step 2 Connect to online resources by using QuickPass codes and directly access the chapter you want.

gb2843c1

Enter this code with the appropriate chapter number.

Check out the following features on your **Online Learning Center:**

Study Tools

Online StudentWorks Plus
Interactive Tutors
Personal Tutors
Vocabulary Puzzlemaker
Multilingual Science Glossary
Chapter Test Practice
Standardized Test Practice

Concepts In Motion
- Interactive Tables
- Interactive Time Lines
- Animated Illustrations
- National Geographic Visualizing Animations

(English and Spanish)
- Section Self-Check Quizzes
- e-flash cards

Extensions

Virtual Labs
Microscopy Links
Periodic Table Links
Career Links
Prescreened Web Links
WebQuest Projects
Science Fair Ideas
Internet BioLabs

For Teachers

Teacher Bulletin Board
Teaching Today and much more!

Safety Symbols

These safety symbols are used in laboratory and field investigations in this book to indicate possible hazards. Learn the meaning of each symbol and refer to this page often. *Remember to wash your hands thoroughly after completing lab procedures.*

SAFETY SYMBOLS	HAZARD	EXAMPLES	PRECAUTION	REMEDY
DISPOSAL	Special disposal procedures need to be followed.	certain chemicals, living organisms	Do not dispose of these materials in the sink or trash can.	Dispose of wastes as directed by your teacher.
BIOLOGICAL	Organisms or other biological materials that might be harmful to humans	bacteria, fungi, blood, unpreserved tissues, plant materials	Avoid skin contact with these materials. Wear mask or gloves.	Notify your teacher if you suspect contact with material. Wash hands thoroughly.
EXTREME TEMPERATURE	Objects that can burn skin by being too cold or too hot	boiling liquids, hot plates, dry ice, liquid nitrogen	Use proper protection when handling.	Go to your teacher for first aid.
SHARP OBJECT	Use of tools or glassware that can easily puncture or slice skin	razor blades, pins, scalpels, pointed tools, dissecting probes, broken glass	Practice common-sense behavior and follow guidelines for use of the tool.	Go to your teacher for first aid.
FUME	Possible danger to respiratory tract from fumes	ammonia, acetone, nail polish remover, heated sulfur, moth balls	Make sure there is good ventilation. Never smell fumes directly. Wear a mask.	Leave foul area and notify your teacher immediately.
ELECTRICAL	Possible danger from electrical shock or burn	improper grounding, liquid spills, short circuits, exposed wires	Double-check setup with teacher. Check condition of wires and apparatus. Use GFI-protected outlets.	Do not attempt to fix electrical problems. Notify your teacher immediately.
IRRITANT	Substances that can irritate the skin or mucous membranes of the respiratory tract	pollen, moth balls, steel wool, fiberglass, potassium permanganate	Wear dust mask and gloves. Practice extra care when handling these materials.	Go to your teacher for first aid.
CHEMICAL	Chemicals that can react with and destroy tissue and other materials	bleaches such as hydrogen peroxide; acids such as sulfuric acid, hydrochloric acid; bases such as ammonia, sodium hydroxide	Wear goggles, gloves, and an apron.	Immediately flush the affected area with water and notify your teacher.
TOXIC	Substance may be poisonous if touched, inhaled, or swallowed.	mercury, many metal compounds, iodine, poinsettia plant parts	Follow your teacher's instructions.	Always wash hands thoroughly after use. Go to your teacher for first aid.
FLAMMABLE	Open flame may ignite flammable chemicals, loose clothing, or hair.	alcohol, kerosene, potassium permanganate, hair, clothing	Avoid open flames and heat when using flammable chemicals.	Notify your teacher immediately. Use fire safety equipment if applicable.
OPEN FLAME	Open flame in use, may cause fire.	hair, clothing, paper, synthetic materials	Tie back hair and loose clothing. Follow teacher's instructions on lighting and extinguishing flames.	Always wash hands thoroughly after use. Go to your teacher for first aid.

Eye Safety
Proper eye protection must be worn at all times by anyone performing or observing science activities.

Clothing Protection
This symbol appears when substances could stain or burn clothing.

Animal Safety
This symbol appears when safety of animals and students must be ensured.

Radioactivity
This symbol appears when radioactive materials are used.

Handwashing
After the lab, wash hands with soap and water before removing goggles.

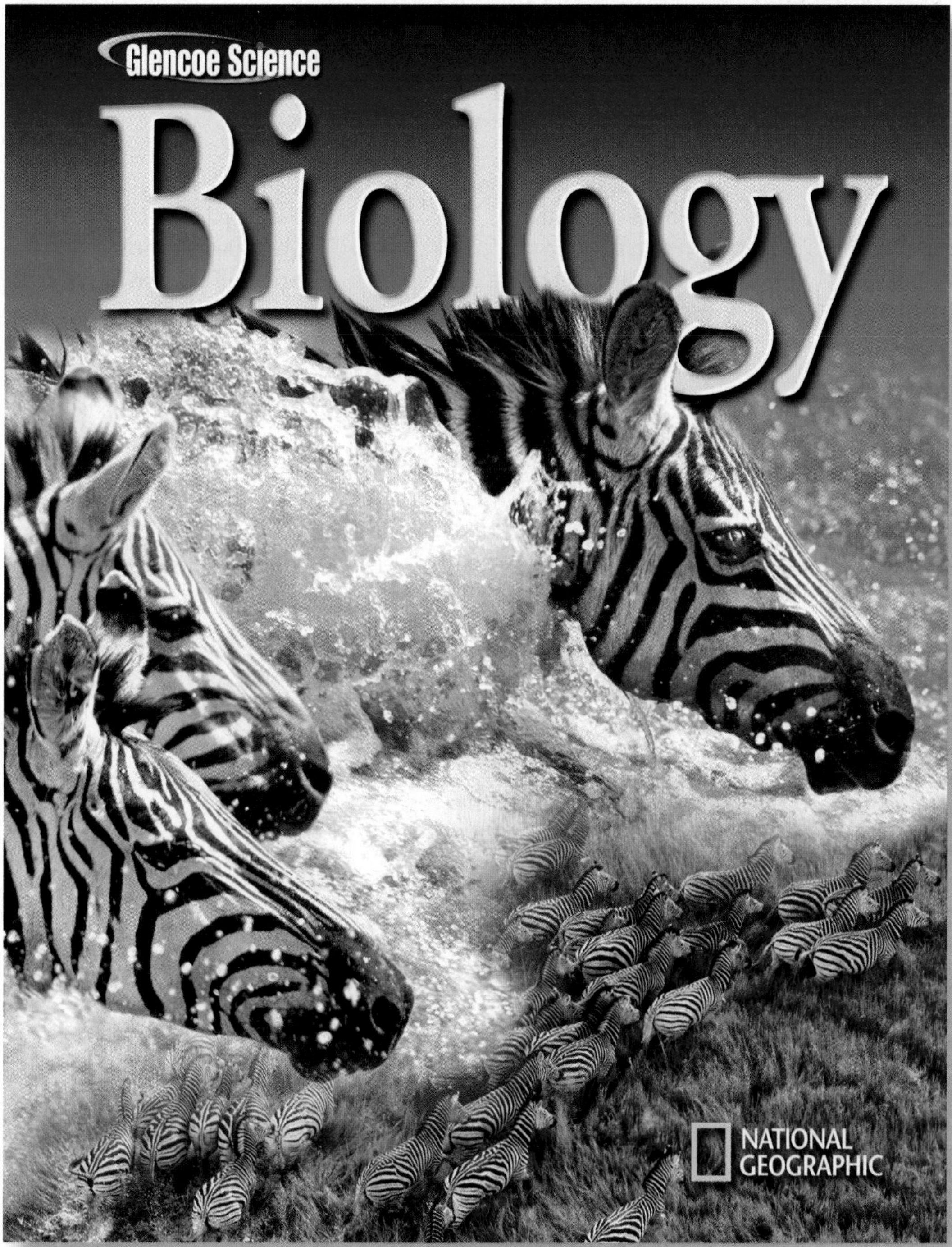

Glencoe Science

Biology

NATIONAL GEOGRAPHIC

AUTHORS

Alton Biggs • Whitney Crispen Hagins • William G. Holliday
Chris L. Kapicka • Linda Lundgren • Ann Haley MacKenzie
William D. Rogers • Marion B. Sewer • Dinah Zike
National Geographic

New York, New York Columbus, Ohio Chicago, Illinois Woodland Hills, California

biologygmh.com

Check out the following features on your **Online Learning Center:**

Study Tools

Concepts In Motion

- Interactive tables
- Interactive timelines
- Animated illustrations
- National Geographic Visualizing animations

Section Self-Check Quizzes
Chapter Test Practice
Standardized Test Practice
Vocabulary PuzzleMaker
Interactive Tutors
Personal Tutors
Multilingual Science Glossary
Study to Go—also in Spanish
Online StudentWorks™ Plus

Extensions

Virtual Labs
Microscopy Links
Periodic Table Links
Career Links
Prescreened Web Links
WebQuest Projects
Science Fair Ideas
Internet BioLabs

For Teachers

Teacher Bulletin Board
Teaching Today, and much more!

The McGraw-Hill Companies

Glencoe

Send all inquiries to:
Glencoe/McGraw-Hill
8787 Orion Place
Columbus, OH 43240-4027

ISBN-13: 978-0-07-880284-3
MHID: 0-07-880284-9

Printed in the United States of America.

2 3 4 5 6 7 8 9 10 027/055 13 12 11 10 09 08

Contents in Brief

About the Authors

Alton Biggs has been a biology educator in Texas public schools for more than 30 years. He has a BS and an MS in biology from Texas A & M University—Commerce. Mr. Biggs was the founding president of the Texas Association of Biology Teachers in 1985, received NABT's Outstanding Biology Teacher Award for Texas in 1982 and 1995, and in 1992 was the president of the National Association of Biology Teachers.

Whitney Crispen Hagins teaches biology at Lexington High School in Lexington, Massachusetts. She has a BA and an MA in biological sciences from Mount Holyoke College and an MAT from Duke University. In 1998, she received NSF funding for development of molecular biology activities. In 1999, she was a Massachusetts NABT Outstanding Biology Teacher Award recipient. In 2005, she was awarded the Siemens Foundation AP Award for Math and Science Teachers for Massachusetts. She works with the Wisconsin Fast Plant Program to develop curriculum, and she enjoys sharing ideas and activities at national meetings.

William G. Holliday is a science education professor at the University of Maryland (College Park), and before 1986, a professor at the University of Calgary (Alberta, Canada). He served as president of the National Association for Research in Science Teaching and later as an elected board member to the National Science Teachers Association. He has an MS in biological sciences and a PhD in science education. Mr. Holliday's multifaceted teaching experience totals more than 40 years.

Chris L. Kapicka is retired faculty from Northwest Nazarene University in Nampa, Idaho. She has a BS in biology from Boise State University, an MS in bacteriology and public health from Washington State University, and a PhD in cell and molecular physiology and pharmacology from the University of Nevada Medical School. In 1986, she received the Presidential Award for Science Teaching, and in 1988, she was awarded NABT's Outstanding Biology Teacher Award.

Linda Lundgren has more than 25 years of experience teaching science at the middle school, high school, and college levels, including ten years at Bear Creek High School in Lakewood, Colorado. For eight years, she was a research associate in the Department of Science and Technology at the University of Colorado at Denver. Ms. Lundgren has a BA in journalism and zoology from the University of Massachusetts and an MS in zoology from The Ohio State University. In 1991, she was named Colorado Science Teacher of the Year.

Ann Haley MacKenzie currently teaches at Miami University in Oxford, Ohio, where she works with future high school science teachers and teaches a life science inquiry course. She is the editor of *The American Biology Teacher* for the National Association of Biology Teachers. Dr. MacKenzie has a BS in biology from Purdue University, an MEd in secondary education from the University of Cincinnati, and an EdD in curriculum and instruction from the University of Cincinnati. She is a former Ohio Teacher of the Year and Presidential Award Winner for Secondary School Science.

William D. Rogers is a faculty member in the Department of Biology at Ball State University in Muncie, Indiana. He has a BA and an MA in biology from Drake University. He has a Doctor of Arts in biology from Idaho State University. He has received teaching awards for outstanding contributions to general education, and has received funding from the American Association of Colleges and Universities to study different approaches to science teaching.

Marion B. Sewer is an assistant professor at the Georgia Institute of Technology, and a Georgia Cancer Coalition Distinguished Scholar. She received a BS in biochemistry from Spelman College in 1993, and a PhD in pharmacology from Emory University in 1998. Dr. Sewer studies how the integration of various signaling pathways controls steroid hormone biosynthesis.

Dinah Zike is an international curriculum consultant and inventor who has developed educational products and three-dimensional, interactive graphic organizers for over 30 years. As president and founder of Dinah-Might Adventures, L.P., Dinah is the author of more than 100 award-winning educational publications, including *The Big Book of Science*. Dinah has a BS and an MS in educational curriculum and instruction from Texas A & M University. Dinah Zike's *Foldables* are an exclusive feature of McGraw-Hill textbooks.

National Geographic, founded in 1888 for the increase and diffusion of geographic knowledge, is the world's largest nonprofit scientific and educational organization. The Children's Books and Education Division of National Geographic supports National Geographic's mission by developing innovative educational programs. National Geographic's *Visualizing* and *In the Field* features are exclusive components of *Glencoe Biology*.

View author biographies at biologygmh.com.

Teacher Advisory Board and Reviewers

Teacher Advisory Board

The Teacher Advisory Board gave the authors, editorial staff, and design team feedback on the content and design in the Student Edition. We thank these teachers for their hard work and creative suggestions.

Chuck Cambria
Springfield North High School
Springfield, OH

Sara Haughn
Pickerington High School Central
Pickerington, OH

Michelle M. Lewis
Lancaster High School
Lancaster, OH

Daniel J. Regelski
New Albany High School
New Albany, OH

Pamela J. Schumm
Wawasee High School
Syracuse, IN

Danielle S. Scrase
Sycamore High School
Cincinnati, OH

Sharon Seckel
Canal Winchester High School
Canal Winchester, OH

Paula Weaver
Seymour High School
Seymour, IN

Reviewers

Each teacher reviewed selected chapters of *Glencoe Biology* and provided feedback and suggestions for improving the effectiveness of the instruction.

Beth Adams, EdS
Cartersville High School
Cartersville, GA

Michele Wonser Altavilla
Elk River High School
Elk River, MN

Elaine Asmus
Snake River High School
Blackfoot, ID

Shelley Barker
Danville High School
Danville, IL

Stephanie L. Baron
Patrick Henry High School
San Diego, CA

Rochelle Battersby
Sanford F. Calhoun High School
Merrick, NY

Ann Blackwell
Travelers Rest High School
Travelers Rest, SC

Fran S. Bouknight
Brookland-Cayce High School
Cayce, SC

Kimberly A. Brown
East Ridge High School
Chattanooga, TN

John H. C. Chipley
Palma High School
Salinas, CA

Melissa C. Donham
Little Rock Central High School
Little Rock, AR

William A. Donovan, Jr.
Amesbury High School
Amesbury, MA

Erica Dosch, BSN, MEd
Connetquot High School
Bohemia, NY

Julie Ertmann
University City High School
University City, MO

Mark Fife
Centerville High School
Centerville, OH

Sasha Hammond
Yakima School District
Yakima, WA

Kenneth L. Harms
Tigard High School
Tigard, OR

David Iott
Holden High School
Holden, MO

Rebecca Jackson
Summerville High School
Summerville, SC

Clinton A. Kennedy
Cascade JR/SR High School
Cascade, ID

Kristen Holly Kent
Pioneer Valley High School
Santa Maria, CA

Deborah L. Krause, EdD
West Orange Public Schools
West Orange, NJ

Kimberly Lane-Hinton
Whitney Young Magnet High School
Chicago, IL

Samantha A. Long
Carlisle High School
Carlisle, PA

Charlotte Parnell
Sullivan High School
Sullivan, MO

Cynthia A. Pousman
Davidson Fine Arts Magnet School
Augusta, GA

Rebekah R. Ravgiala, EdD
Merrimack High School
Merrimack, NH

Gail Raymond
Anchorage School District
Anchorage, AK

Kathryn A. Roberts
Lakeside Public High Schools
Hot Springs, AR

Tracy Rojo
Tucker High School
Tucker, GA

Pamela D. Sherley
Chicago Public Schools
Chicago, IL

Gary R. Smith
Pinelands Regional High School
Tuckerton, NJ

Sue Whitsett
Fond du Lac High School
Fond du Lac, WI

Karen Ann Wickersham
Troy High School
Troy, MI

Consultants

Content Consultants

Content consultants each reviewed selected chapters of *Glencoe Biology* for content accuracy and clarity.

Larry Baresi, PhD
Associate Professor of Biology
California State University, Northridge
Northridge, CA

Janice E. Bonner, PhD
Associate Professor of Biology
College of Notre Dame of Maryland
Baltimore, MD

Renea J. Brodie, PhD
Assistant Professor of Biological Sciences
University of South Carolina
Columbia, SC

Luis A Cañas, PhD
Assistant Professor
Department of Entomology/OARDC
The Ohio State University
Wooster, OH

John S. Choinski, Jr., PhD
Professor of Biology
Department of Biology
University of Central Arkansas
Conway, AR

Dr. Lewis B. Coons, PhD
Professor of Biology
The University of Memphis
Memphis, TN

Cara Lea Council-Garcia, MS
Biology Lab Coordinator
The University of New Mexico
Albuquerque, NM

Dr. Donald S. Emmeluth, PhD
Department of Biology
Armstrong Atlantic State University
Savannah, GA

Diana L. Engle, PhD
Ecology Consultant
University of California Santa Barbara
Santa Barbara, CA

John Gatz, PhD
Professor of Zoology
Ohio Wesleyan University
Delaware, OH

Alan D. Gishlick, PhD
National Center for Science Education
Oakland, CA

Yourha Kang, PhD
Assistant Professor of Biology
Iona College
New Rochelle, NY

Mark E. Lee, PhD
Assistant Professor of Biology
Spelman College
Atlanta, GA

Judy M. Nesmith, MS
Lecturer—Biology
University of Michigan—Dearborn
Dearborn, MI

Hay-Oak Park, PhD
Associate Professor
Department of Molecular Genetics
The Ohio State University
Columbus, OH

Carolyn F. Randolph, PhD
President NSTA 2001–2002
Assistant Executive Director
The SCEA
Columbia, SC

David A. Rubin, PhD
Assistant Professor of Physiology
Illinois State University
Normal, IL

Malathi Srivatsan, PhD
Assistant Professor of Biology
State University of Arkansas
Jonesboro, AR

Laura Vogel, PhD
Associate Professor of Biological Sciences
Illinois State University
Normal, IL

VivianLee Ward, MS
Director of CyberEducation; Codirector of Fellows Program; Project Director Access Excellence @ the National Health Museum
Washington, DC

Safety Consultants

Safety Consultants reviewed labs and lab materials for safety and implementation.

Jack Gerlovich
School of Education
Department of Teaching and Learning
Drake University
Des Moines, IA

Dennis McElroy
Director of Curriculum
Assistant Director for Technology
School of Education
Graceland University
Lamoni, IA

Reading Consultant

Dr. Douglas Fisher provided expert guidance on prototypes, Real-World Reading Links, and the reading strand.

Douglas Fisher, PhD
Professor of Language and Literacy Education
San Diego State University
San Diego, CA

Standardized Test Practice Consultant

Dr. Ralph Feather provided expert guidance on effective standardized test practice questions.

Ralph Feather, PhD
Assistant Professor of Education
Bloomsburg University of Pennsylvania
Bloomsburg, PA

Lab Tester

Science Kit performed and evaluated the Student Edition labs and additional Teacher Edition material, providing suggestions for improving the effectiveness of student instructions and teacher support.

Science Kit and Boreal Laboratories
Tonawanda, NY

Contents

Your book is divided into units and chapters that are organized around Themes, Big Ideas, and Main Ideas of biology.

THEMES are overarching concepts used throughout the entire book that help you tie what you learn together. They help you see the connections among major ideas and concepts.

BIG Ideas appear in each chapter and help you focus on topics within the themes. The Big Ideas are broken down even further into Main Ideas.

MAIN Ideas draw you into more specific details about biology. All the Main Ideas of a chapter add up to the chapter's Big Idea.

THEMES

Change
Diversity
Energy
Homeostasis
Scientific Inquiry

BIG Idea

One per chapter

One per section

Student Guide

Unit 1

Unit 2

Contents

Unit 9

Student Resources

Labs

LAUNCH Lab

Start off each chapter with hands-on introduction to the subject matter.

DATA ANALYSIS LAB

Build your analysis skills using actual data from real scientific sources.

DATA ANALYSIS LAB

Build your analysis skills using actual data from real scientific sources.

MiniLab

Practice scientific methods and hone your lab skills with these quick activities.

MiniLab

Practice scientific methods and hone your lab skills with these quick activities.

BIOLAB

Apply the skills you developed in Launch Labs, MiniLabs, and Data Analysis Labs in these chapter-culminating, real-world labs.

Chapter

Chapter

Real-World Biology Features

Explore today's world of biology. Discover the hot topics in biology, delve into new technologies, uncover the discoveries impacting biology, and investigate careers in biology.

BioDiscoveries

Discover pivotal advancements that have influenced the biological sciences.

CUTTING-EDGE BIOLOGY

Challenge your brain with recent cutting edge developments in biology.

Biology & Society

Examine biology in the news and sharpen your debating skills on complex issues in biology.

In the Field

Get an inside look at careers in biology.

Careers

CAREERS IN BIOLOGY

Investigate a day in the life of people working in the field of biology.

Concepts in Motion

Concepts In Motion Analyze complex science topics with animations of the National Geographic Visualizing features.

Concepts in Motion Interactive Tables

Check your understanding by viewing interactive versions of tables in your text.

Concepts In Motion **Animated Art** **Enhance and enrich your knowledge of biology concepts through simple and 3D animations of visuals.**

Concepts In Motion Interactive Time Line

Explore science and history through milestones in biology.

Concepts In Motion Plus

Concepts in Motion Plus is a library of basic and 3D animations whose depth and range extend beyond those presented in this book. These animations serve as excellent visual learning aids and can be used to enhance your understanding of biology concepts. In many cases, a number of animations on a particular concept can be found, ranging from simple to more complex explanations. "For More Help" animations are basic or serve as introductions, "On Target" animations can be used to review or strengthen understanding of a topic you are already familiar with, and "Take It Further" animations offer more detailed views of a topic and can be used as extensions. The Concepts in Motion Plus library can be found at **biologygmh.com**.

Personal Tutor For additional explanation of science topics by a science teacher, visit biologygmh.com.

Reading for Information

When you read *Glencoe Biology*, you need to read for information. Science is nonfiction writing; it describes real-life events, people, ideas, and technology. Here are some tools that *Glencoe Biology* has to help you read.

Before You Read

By reading the BIG Idea and MAIN Idea prior to reading the chapter or section, you will get a preview of the coming material.

Each unit preview lists the chapters in the unit. An overall BIG Idea is listed for each chapter. The Big Idea describes what you will learn in the chapter.

UNIT 3 Genetics

Chapter 10
Sexual Reproduction and Genetics
BIG Idea Reproductive cells, which pass on genetic traits from the parents to the child, are produced by the process of meiosis.

Chapter 11
Complex Inheritance and Human Heredity
BIG Idea Human inheritance does not always follow Mendel's laws.

Chapter 12
Molecular Genetics
BIG Idea DNA is the genetic material that contains a code for proteins.

Chapter 13
Genetics and Biotechnology
BIG Idea Genetic technology improves human health and quality of life.

CAREERS IN BIOLOGY
Geneticist
Geneticists are scientists who study heredity, genes, and variation in organisms. Geneticists, such as the ones shown here extracting genetic material from a dinosaur egg, work to uncover the building blocks of life.
WRITING in Biology
Visit biologygmh.com to learn more about geneticists. Write an account of a geneticist's contribution to the field of medicine, agriculture, biotechnology, or criminology.

266

Source: Unit 3, p.266

CHAPTER 7 Cellular Structure and Function

BIG Idea Cells are the structural and functional units of all living organisms.

Section 1
Cell Discovery and Theory
MAIN Idea The invention of the microscope led to the discovery of cells.

Section 2
The Plasma Membrane
MAIN Idea The plasma membrane helps to maintain a cell's homeostasis.

Section 3
Structures and Organelles
MAIN Idea Eukaryotic cells contain organelles that allow the specialization and the separation of functions within the cell.

Section 4
Cellular Transport
MAIN Idea Cellular transport moves substances within the cell and moves substances into and out of the cell.

BioFacts
- About ten trillion cells make up the human body.
- The largest human cells are about the diameter of a human hair.
- The 200 different types of cells in the human body come from just one cell.

HUMAN SKIN
HUMAN SKIN 2 mm
HUMAN SKIN CELLS
HUMAN SKIN CELLS

180

Source: Chapter 7, p.180

The MAIN Ideas within a chapter support the BIG Idea of the chapter. Each section of the chapter has a Main Idea that describes the focus of the section.

OTHER WAYS TO PREVIEW

- Read the chapter title to find out what the topic will be.
- Skim the photos, illustrations, captions, graphs, and tables.
- Look for key terms listed in the Reading Preview that are boldfaced and highlighted in the text.
- Create an outline using section titles and heads.

As You Read

Within each section, you will find a tool to deepen your understanding and a tool to check your understanding.

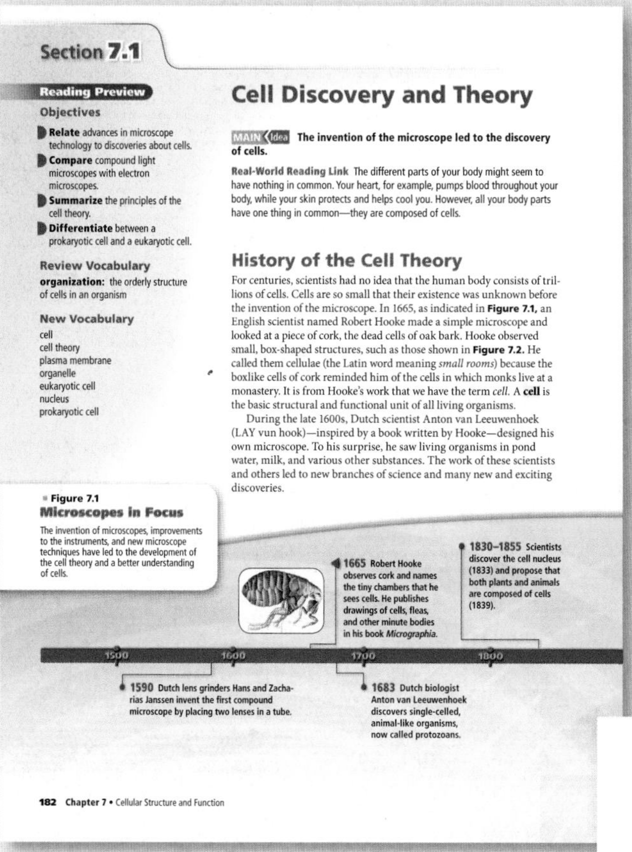

Section 7.1

Reading Preview

Objectives

- **Relate** advances in microscope technology to discoveries about cells.
- **Compare** compound light microscopes with electron microscopes.
- **Summarize** the principles of the cell theory.
- **Differentiate** between a prokaryotic cell and a eukaryotic cell.

Review Vocabulary

organization: the orderly structure of cells in an organism

New Vocabulary

cell
cell theory
plasma membrane
organelle
eukaryotic cell
nucleus
prokaryotic cell

Cell Discovery and Theory

MAIN Idea **The invention of the microscope led to the discovery of cells.**

Real-World Reading Link The different parts of your body might seem to have nothing in common. Your heart, for example, pumps blood throughout your body, while your skin protects and helps cool you. However, all your body parts have one thing in common—they are composed of cells.

History of the Cell Theory

For centuries, scientists had no idea that the human body consists of trillions of cells. Cells are so small that their existence was unknown before the invention of the microscope. In 1665, as indicated in **Figure 7.1**, an English scientist named Robert Hooke made a simple microscope and looked at a piece of cork, the dead cells of oak bark. Hooke observed small, box-shaped structures, such as those shown in **Figure 7.2.** He called them cellulae (the Latin word meaning *small rooms*) because the boxlike cells of cork reminded him of the cells in which monks live at a monastery. It is from Hooke's work that we have the term *cell*. A **cell** is the basic structural and functional unit of all living organisms.

During the late 1600s, Dutch scientist Anton van Leeuwenhoek (LAY vun hook)—inspired by a book written by Hooke—designed his own microscope. To his surprise, he saw living organisms in pond water, milk, and various other substances. The work of these scientists and others led to new branches of science and many new and exciting discoveries.

Figure 7.1
Microscopes in Focus
The invention of microscopes, improvements to the instruments, and new microscope techniques have led to the development of the cell theory and a better understanding of cells.

1590 Dutch lens grinders Hans and Zacharias Janssen invent the first compound microscope by placing two lenses in a tube.

1665 Robert Hooke observes cork and names the tiny chambers that he sees cells. He publishes drawings of cells, fleas, and other minute bodies in his book *Micrographia.*

1683 Dutch biologist Anton van Leeuwenhoek discovers single-celled, animal-like organisms, now called protozoans.

1830–1855 Scientists discover the cell nucleus (1833) and propose that both plants and animals are composed of cells (1839).

1500 1600 1700 1800

182 Chapter 7 • Cellular Structure and Function

Source: Section 7.1, p.182

The **Reading Preview** provides objectives for reading and introduces the vocabulary you will learn as you read.

The **Real-World Reading Link** describes how the section's content might relate to you.

Source: Section 7.1, p.183

The cell theory Naturalists and scientists continued observing the living microscopic world using glass lenses. In 1838, German scientist Matthias Schleiden carefully studied plant tissues and concluded that all plants are composed of cells. A year later, another German scientist, Theodor Schwann, reported that animal tissues also consisted of individual cells. Prussian physician Rudolph Virchow proposed in 1855 that all cells are produced from the division of existing cells. The observations and conclusions of these scientists and others are summarized as the cell theory. The **cell theory** is one of the fundamental ideas of modern biology and includes the following three principles:

1. All living organisms are composed of one or more cells.
2. Cells are the basic unit of structure and organization of all living organisms.
3. Cells arise only from previously existing cells, with cells passing copies of their genetic material on to their daughter cells.

Reading Check Can cells appear spontaneously without genetic material from previous cells?

Microscope Technology

The discovery of cells and the development of the cell theory would not have been possible without microscopes. Improvements made to microscopes have enabled scientists to study cells in detail, as described in **Figure 7.1.**

Turn back to the opening pages of this chapter and compare the illustrations of the skin shown there. Note that the detail increases as the magnification and resolution—the ability of the microscope to make individual components visible—increase. Hooke and van Leewenhoek would not have been able to see the individual structures within human skin cells with their microscopes. Developments in microscope technology have given scientists the ability to study cells in greater detail than early scientists ever thought possible.

LM Magnification: 100×

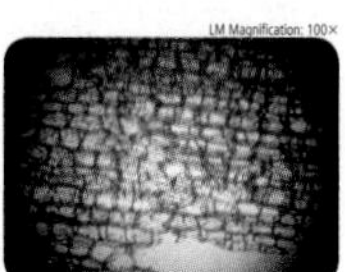

Figure 7.2 Robert Hooke used a basic light microscope to see what looked like empty chambers in a cork sample.
Infer *What do you think Hooke would have seen if these were living cells?*

LAUNCH Lab
Review Based on what you've read about cells, how would you now answer the analysis questions?

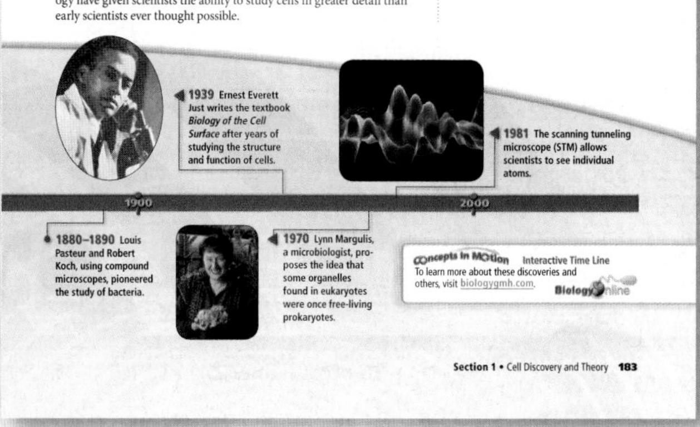

1939 Ernest Everett Just writes the textbook Biology of the Cell Surface after years of studying the structure and function of cells.
1981 The scanning tunneling microscope (STM) allows scientists to see individual atoms.
1900
2000
1880–1890 Louis Pasteur and Robert Koch, using compound microscopes, pioneered the study of bacteria.
1970 Lynn Margulis, a microbiologist, proposes the idea that some organelles found in eukaryotes were once free-living prokaryotes.

Concepts in Motion Interactive Time Line
To learn more about these discoveries and others, visit biologygmh.com. Biology Online

Section 1 • Cell Discovery and Theory 183

Reading Checks are questions that assess your understanding.

OTHER READING SKILLS

- Ask yourself what is the BIG Idea? What is the MAIN Idea?
- Think about people, places, and situations that you've encountered. Are there any similarities with those mentioned in *Glencoe Biology?*
- Relate the information in *Glencoe Biology* to other areas you have studied.
- Predict events or outcomes by using clues and information that you already know.
- Change your predictions as you read and gather new information.

After You Read

Follow-up your reading with a summary and assessment of the material to evaluate if you understood the text.

After reading, use the Personal Tutors and other study aids found at biologygmh.com to review and reinforce your learning.

Each section concludes with an assessment. The assessment contains a summary and questions. The summary reviews the section's key concepts while the questions test your understanding.

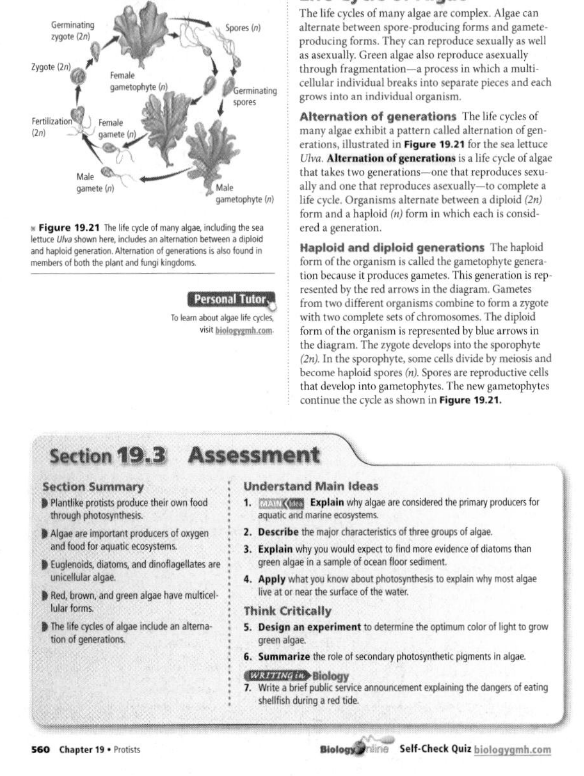

■ **Figure 19.21** The life cycle of many algae, including the sea lettuce *Ulva* shown here, includes an alternation between a diploid and haploid generation. Alternation of generations is also found in members of both the plant and fungi kingdoms.

Personal Tutor

To learn about algae life cycles, visit biologygmh.com.

Life Cycle of Algae

The life cycles of many algae are complex. Algae can alternate between spore-producing forms and gamete-producing forms. They can reproduce sexually as well as asexually. Green algae also reproduce asexually through fragmentation—a process in which a multicellular individual breaks into separate pieces and each grows into an individual organism.

Alternation of generations The life cycles of many algae exhibit a pattern called alternation of generations, illustrated in **Figure 19.21** for the sea lettuce *Ulva*. **Alternation of generations** is a life cycle of algae that takes two generations—one that reproduces sexually and one that reproduces asexually—to complete a life cycle. Organisms alternate between a diploid *(2n)* form and a haploid *(n)* form in which each is considered a generation.

Haploid and diploid generations The haploid form of the organism is called the gametophyte generation because it produces gametes. This generation is represented by the red arrows in the diagram. Gametes from two different organisms combine to form a zygote with two complete sets of chromosomes. The diploid form of the organism is represented by blue arrows in the diagram. The zygote develops into the sporophyte *(2n)*. In the sporophyte, some cells divide by meiosis and become haploid spores *(n)*. Spores are reproductive cells that develop into gametophytes. The new gametophytes continue the cycle as shown in **Figure 19.21.**

Section 19.3 Assessment

Section Summary

- Plantlike protists produce their own food through photosynthesis.
- Algae are important producers of oxygen and food for aquatic ecosystems.
- Euglenoids, diatoms, and dinoflagellates are unicellular algae.
- Red, brown, and green algae have multicellular forms.
- The life cycles of algae include an alternation of generations.

Understand Main Ideas

1. MAIN Idea **Explain** why algae are considered the primary producers for aquatic and marine ecosystems.
2. **Describe** the major characteristics of three groups of algae.
3. **Explain** why you would expect to find more evidence of diatoms than green algae in a sample of ocean floor sediment.
4. **Apply** what you know about photosynthesis to explain why most algae live at or near the surface of the water.

Think Critically

5. **Design an experiment** to determine the optimum color of light to grow green algae.
6. **Summarize** the role of secondary photosynthetic pigments in algae.

WRITING in Biology

7. Write a brief public service announcement explaining the dangers of eating shellfish during a red tide.

560 Chapter 19 • Protists

Biology Online Self-Check Quiz biologygmh.com

Source: Chapter 19, p.560

CHAPTER 7

Study Guide

STUDY TO GO Download quizzes, key terms, and flash cards from biologygmh.com.

FOLDABLES **Apply** Use what you have learned about osmosis and cellular transport to design an apparatus that would enable a freshwater fish to survive in a saltwater habitat.

Vocabulary	Key Concepts
Section 7.1 Cell Discovery and Theory	
• cell (p. 182) • cell theory (p. 183) • eukaryotic cell (p. 186) • nucleus (p. 186) • organelle (p. 186) • plasma membrane (p. 185) • prokaryotic cell (p. 186)	MAIN Idea The invention of the microscope led to the discovery of cells. • Microscopes have been used as a tool for scientific study since the late 1500s. • Scientists use different types of microscopes to study cells. • The cell theory summarizes three principles. • There are two broad groups of cell types—prokaryotic cells and eukaryotic cells. • Eukaryotic cells contain a nucleus and organelles.
Section 7.2 The Plasma Membrane	
• fluid mosaic model (p. 190) • phospholipid bilayer (p. 188) • selective permeability (p. 187) • transport protein (p. 189)	MAIN Idea The plasma membrane helps to maintain a cell's homeostasis. • Selective permeability is the property of the plasma membrane that allows it to control what enters and leaves the cell. • The plasma membrane is made up of two layers of phospholipid molecules. • Cholesterol and transport proteins aid in the function of the plasma membrane. • The fluid mosaic model describes the plasma membrane.
Section 7.3 Structures and Organelles	
• cell wall (p. 198) • centriole (p. 196) • chloroplast (p. 197) • cilium (p. 198) • cytoplasm (p. 191) • cytoskeleton (p. 191) • endoplasmic reticulum (p. 194) • flagellum (p. 198) • Golgi apparatus (p. 195) • lysosome (p. 196) • mitochondrion (p. 197) • nucleolus (p. 193) • ribosome (p. 193) • vacuole (p. 195)	MAIN Idea Eukaryotic cells contain organelles that allow the specialization and the separation of functions within the cell. • Eukaryotic cells contain membrane-bound organelles in the cytoplasm that perform cell functions. • Ribosomes are the sites of protein synthesis. • Mitochondria are the powerhouses of cells. • Plant and animal cells contain many of the same organelles, while other organelles are unique to either plant cells or animal cells.
Section 7.4 Cellular Transport	
• active transport (p. 205) • diffusion (p. 201) • dynamic equilibrium (p. 202) • endocytosis (p. 207) • exocytosis (p. 207) • facilitated diffusion (p. 202) • hypertonic solution (p. 205) • hypotonic solution (p. 204) • isotonic solution (p. 204) • osmosis (p. 203)	MAIN Idea Cellular transport moves substances within the cell and moves substances into and out of the cell. • Cells maintain homeostasis using passive and active transport. • Concentration, temperature, and pressure affect the rate of diffusion. • Cells must maintain homeostasis in all types of solutions, including isotonic, hypotonic, and hypertonic. • Some large molecules are moved into and out of the cell using endocytosis and exocytosis.

210 Chapter 7 • Study Guide

Biology Online Vocabulary PuzzleMaker biologygmh.com

Source: Chapter 7, p.210

At the end of each chapter you will find a Study Guide. The chapter's vocabulary words as well as key concepts are listed here. Use this guide for review and to check your comprehension.

OTHER WAYS TO REVIEW

- **State the BIG Idea.**
- **Relate the MAIN Idea to the BIG Idea.**
- **Use your own words to explain what you read.**
- **Apply this new information in other school subjects or at home.**
- **Identify sources you could use to find out more information about the topic.**

Scavenger Hunt

Glencoe Biology **contains a wealth of information. Complete this fun activity so you will know where to look to learn as much as you can.**

As you complete this scavenger hunt, either alone or with your teacher or family, you will learn quickly how *Glencoe Biology* is organized and how to get the most out of your reading and study time.

1. How many units are in this book? How many chapters?
2. On what page does the glossary begin? What glossary is online?
3. In what two areas can you find a listing of Laboratory Safety Symbols?
4. Suppose you want to find a list of all the MiniLabs, Data Analysis Labs, and BioLabs. Where in the front do you look?
5. How can you quickly find the pages that have information about scientist Jewell Plummer Cobb?
6. What is the name of the table that summarizes the Key Concepts of a chapter?
7. In what special feature can you find information on unit conversion? What are the page numbers?
8. On what page can you find the BIG Idea for Unit 1? On what page can you find the MAIN Idea for Chapter 2?
9. What feature at the start of each unit provides insight into biologists in action?
10. Name four activities that are found at Biology Online.
11. What study tool shown at the beginning of a chapter can you make from notebook paper?
12. Where do you go to view the Concepts In Motion?
13. CUTTING-EDGE BIOLOGY and BioDiscoveries are two types of end-of-chapter features. What are the other two types?

Investigation and Experimentation

The foundation of scientific knowledge is Investigation and Experimentation. In this section, you will read about lab safety, the proper way to take measurements, and some laboratory techniques. While not every situation you might encounter in the laboratory is covered here, you will gain practical and useful knowledge to make your investigation and experimentation a successful experience.

Laboratory Safety

Follow these safety guidelines and rules to help protect you and others during laboratory investigations.

Complete the Lab Safety Form

- Prior to each investigation your teacher will have you complete a lab safety form. This contract will inform your teacher that you have read the procedure and are prepared to perform the investigation.
- After your teacher reviews your comments, make any necessary corrections, and sign or initial the form.
- Use the lab safety form to help you prepare for each procedure and take responsibility for your own safety.

Teacher Approval Initials

Date of Approval

Lab Safety Form

Name: ______________________

Date: ______________________

Lab type (circle one) : Launch Lab MiniLab ChemLab

Lab Title: ______________________

Read carefully the entire lab and then answer the following questions. Your teacher must initial this form before you begin the lab.

1. What is the purpose of the investigation?
2. Will you be working with a partner or on a team? ______________________
3. Is this a design-your-own procedure? Circle: Yes No
4. Describe the safety procedures and additional warnings that you must follow as you perform this investigation.
5. Are there any steps in the procedure or lab safety symbols that you do not understand? Explain.

Prevent Accidents

- Always wear chemical splash safety goggles (not glasses) in the laboratory. Goggles should fit snugly against your face to prevent any liquid from entering the eyes. Put on your goggles before beginning the lab and wear them throughout the entire activity, cleanup, and hand washing.
- Wear protective aprons and the proper type of gloves as your teacher instructs.
- Keep your hands away from your face and mouth while working in the laboratory.
- Do not wear sandals or other open-toed shoes in the lab.
- Remove jewelry on hands and wrists before doing lab work. Remove loose jewelry, such as chains and long necklaces, to prevent them from getting caught in equipment.
- Do not wear clothing loose enough to catch on anything. If clothing is loose, tape or tie it down.
- Tie back long hair to keep it away from flames and equipment.
- Do not use hair spray or other flammable hair products just before or during laboratory work where an open flame is used. These products ignite easily.
- Eating, drinking, chewing gum, applying makeup, and smoking are prohibited in the laboratory.
- You are expected to behave properly in the laboratory. Practical jokes and fooling around can lead to accidents or injury.
- Notify your teacher about allergies or other health conditions that can affect participation in a lab.

Follow Lab Procedures

- Study all procedures before you begin a laboratory investigation. Ask questions if you do not understand any part of the procedures.
- Review and understand all safety symbols associated with the investigation. A table of the safety symbols is found on page xxxiii for your reference.
- Do not begin any activity until directed to do so by your teacher.
- Use all lab equipment for its intended purpose only.
- Collect and carry all equipment and materials to your work area before beginning the lab.
- When obtaining laboratory materials, dispense only the amount you will use.
- If you have materials left over after completing the investigation, check with your teacher to determine the best choice for recycling or disposing of the materials.
- Keep your work area uncluttered.
- Learn and follow procedures for using specific laboratory equipment, such as balances, microscopes, hot plates, and burners.
- When heating or rinsing a container such as a test tube or flask, point it away from yourself and others.
- Do not taste, touch or smell any chemical or substance in the lab.
- If instructed to smell a substance in a container, hold the container a short distance away and fan vapors toward your nose.
- Do not substitute other chemicals or substances for those in the materials list unless instructed to do so by your teacher.
- Do not take any chemical or material outside of the laboratory.

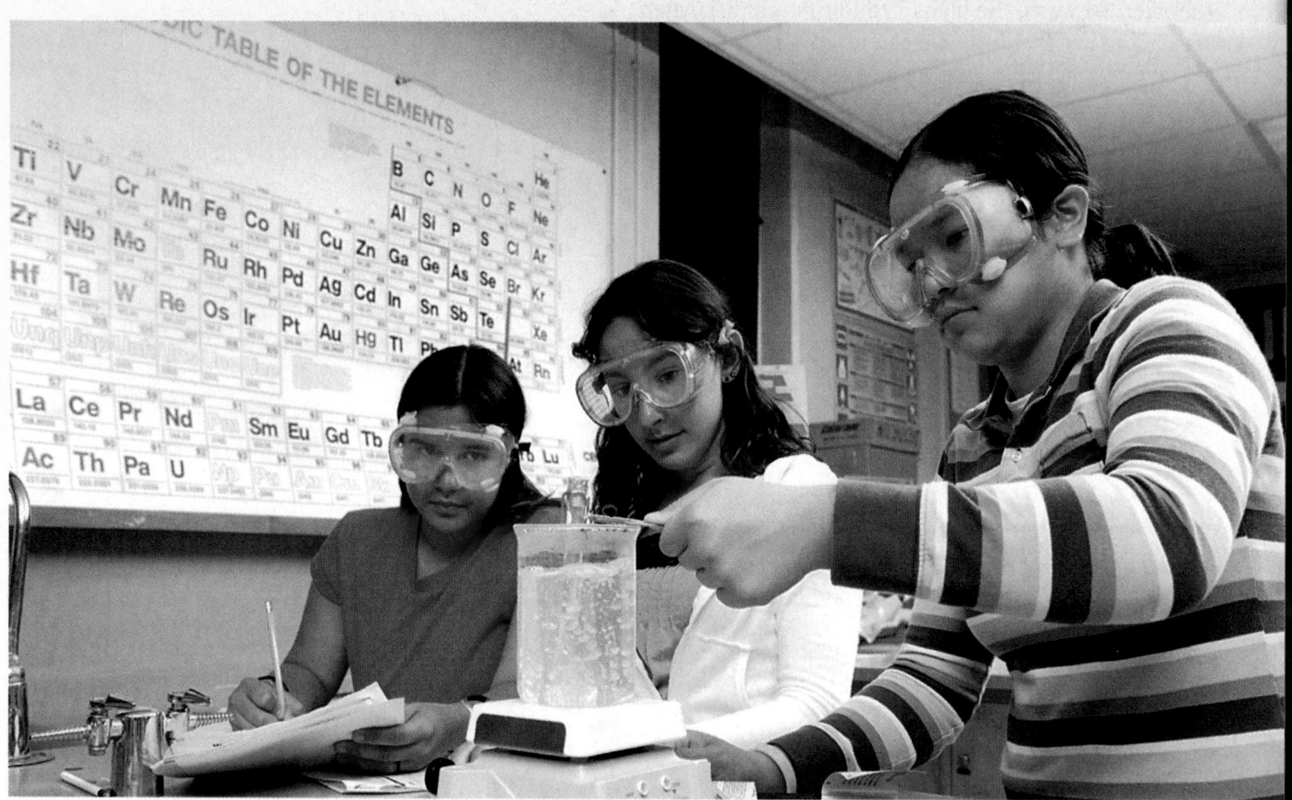

Clean up the Lab

- Turn off all burners, gas valves, and water faucets before leaving the laboratory. Disconnect all electrical devices.
- Clean all equipment as instructed by your teacher. Return everything to the proper storage place.
- Dispose of all materials properly. Place disposable items in containers specifically marked for that type of item. Do not pour liquids down the drain unless instructed to do so by your teacher.
- **Wash your hands thoroughly with soap and warm water after each activity and before removing your goggles.**

Know How to Handle Emergencies

- **Inform your teacher immediately of any mishap, such as fire, bodily injuries, burns, electrical shock, glassware breakage, and chemical or other spills.**
- Do not attempt to clean up spills unless you are given permission and instructions on how to do so. In most instances, your teacher will clean up spills.
- Know the location of the fire extinguisher, safety shower, eyewash, fire blanket, and first-aid kit. After receiving instructions, you can use the safety shower, eyewash, and fire blanket in an emergency without your teacher's permission. However, the fire extinguisher and first-aid kit should only be used by your teacher or, in an extreme emergency, with your teacher's permission.
- If chemicals come into contact with your eyes or skin, notify your teacher immediately and flush your skin or eyes with water.
- If someone is injured or becomes ill, only a professional medical provider or someone certified in first aid should perform first-aid procedures.

Be Responsible

Because your teacher cannot anticipate every safety hazard that might occur and he or she cannot be everywhere in the room at the same time, you need to take some responsibility for your own safety. The general information below should apply to nearly every science lab.

You must:

- review any safety symbols in the labs and be certain you know what they mean;
- follow all teacher instructions for safety and make certain you understand all the hazards related to the lab you are about to perform;
- be able to explain the purpose of the lab;
- be able to explain, or demonstrate, all reasonable emergency procedures, such as:
  - how to evacuate the room during emergencies;
  - how to react to any chemical emergencies;
  - how to deal with fire emergencies;
  - how to perform a scientific investigation safely;
  - how to anticipate some safety concerns and be prepared to address them;
  - how to use equipment properly and safely.
- be able to locate and use all safety equipment as directed by your teacher, such as:
  - fire extinguishers;
  - fire blankets;
  - eye protective equipment (goggles, safety glasses, face shield);
  - eyewash;
  - drench shower.
- be sure to ask questions about any safety concerns that you might have BEFORE starting any investigation.

Safety Symbols

These safety symbols are used in laboratory and field investigations in this book to indicate possible hazards. Learn the meaning of each symbol and refer to this page often. *Remember to wash your hands thoroughly after completing lab procedures.*

SAFETY SYMBOLS	HAZARD	EXAMPLES	PRECAUTION	REMEDY
DISPOSAL	Special disposal procedures need to be followed.	certain chemicals, living organisms	Do not dispose of these materials in the sink or trash can.	Dispose of wastes as directed by your teacher.
BIOLOGICAL	Organisms or other biological materials that might be harmful to humans	bacteria, fungi, blood, unpreserved tissues, plant materials	Avoid skin contact with these materials. Wear mask or gloves.	Notify your teacher if you suspect contact with material. Wash hands thoroughly.
EXTREME TEMPERATURE	Objects that can burn skin by being too cold or too hot	boiling liquids, hot plates, dry ice, liquid nitrogen	Use proper protection when handling.	Go to your teacher for first aid.
SHARP OBJECT	Use of tools or glassware that can easily puncture or slice skin	razor blades, pins, scalpels, pointed tools, dissecting probes, broken glass	Practice common-sense behavior and follow guidelines for use of the tool.	Go to your teacher for first aid.
FUME	Possible danger to respiratory tract from fumes	ammonia, acetone, nail polish remover, heated sulfur, moth balls	Make sure there is good ventilation. Never smell fumes directly. Wear a mask.	Leave foul area and notify your teacher immediately.
ELECTRICAL	Possible danger from electrical shock or burn	improper grounding, liquid spills, short circuits, exposed wires	Double-check setup with teacher. Check condition of wires and apparatus. Use GFI-protected outlets.	Do not attempt to fix electrical problems. Notify your teacher immediately.
IRRITANT	Substances that can irritate the skin or mucous membranes of the respiratory tract	pollen, moth balls, steel wool, fiberglass, potassium permanganate	Wear dust mask and gloves. Practice extra care when handling these materials.	Go to your teacher for first aid.
CHEMICAL	Chemicals that can react with and destroy tissue and other materials	bleaches such as hydrogen peroxide; acids such as sulfuric acid, hydrochloric acid; bases such as ammonia, sodium hydroxide	Wear goggles, gloves, and an apron.	Immediately flush the affected area with water and notify your teacher.
TOXIC	Substance may be poisonous if touched, inhaled, or swallowed.	mercury, many metal compounds, iodine, poinsettia plant parts	Follow your teacher's instructions.	Always wash hands thoroughly after use. Go to your teacher for first aid.
FLAMMABLE	Open flame may ignite flammable chemicals, loose clothing, or hair.	alcohol, kerosene, potassium permanganate, hair, clothing	Avoid open flames and heat when using flammable chemicals.	Notify your teacher immediately. Use fire safety equipment if applicable.
OPEN FLAME	Open flame in use, may cause fire.	hair, clothing, paper, synthetic materials	Tie back hair and loose clothing. Follow teacher's instructions on lighting and extinguishing flames.	Always wash hands thoroughly after use. Go to your teacher for first aid.

Eye Safety
Proper eye protection must be worn at all times by anyone performing or observing science activities.

Clothing Protection
This symbol appears when substances could stain or burn clothing.

Animal Safety
This symbol appears when safety of animals and students must be ensured.

Radioactivity
This symbol appears when radioactive materials are used.

Handwashing
After the lab, wash hands with soap and water before removing goggles.

Field Investigation Safety

On occasion your teacher might conduct a field investigation—an investigation on school grounds or off-campus. While many of the laboratory safety guidelines apply, the field has unique safety considerations.

Work Together

- Work with at least one other person.
- Never stray from the main group either alone or with a small group.
- Make sure each person in your group understands their task and how to perform it. Ask your teacher for clarification if necessary.
- Determine how members of your group will communicate in case of a loud environment or an emergency.
- Your teacher or chaperones should be equipped with either cell phones or walkie-talkies. They should be able to communicate with one another, the school, or emergency personnel if needed, so be sure to let your teacher know if you need help.

Dress Appropriately

- Wear your safety goggles, aprons, and gloves as indicated by the procedure.
- Protect yourself from the Sun with sunblock and a hat.
- Long pants and shirts with long sleeves will protect you from the Sun, insects, and plants such as poison ivy or poison oak.

Poison ivy

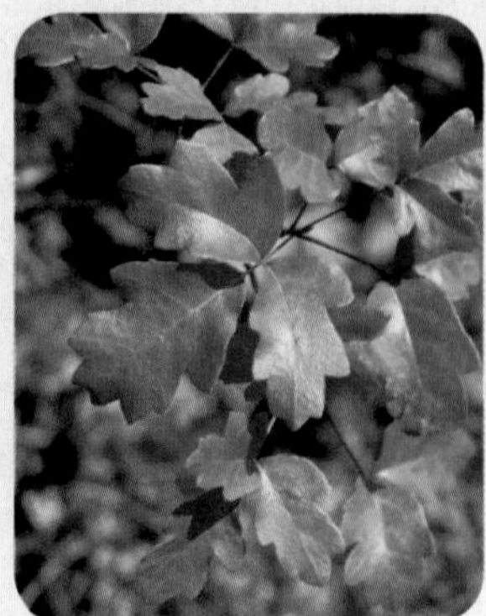

Poison oak

- Insect-repellent sprays or creams may be necessary to use.
- Be sure to wear shoes that have a closed toe and heel as well as a textured sole.
- If your investigation requires that you wade into a stream, river, lake, or other body of water, wear water-resistant clothing. Do not enter the water if you have any open sores.

Consider Your Environment

- Never approach wildlife.
- Never drink water from a stream, river, lake, or other body of water.
- Do not remove the habitat. Create a sketch of organisms you are studying.
- Stay away from power lines.
- Stay away from the edge of cliffs or ledges.
- Stay on the marked trails.

Follow General Guidelines

- Treat your field investigation like a laboratory investigation. There should be no horseplay.
- A first aid kit should be brought to the investigation site.
- Always wash your hands when you are finished. If soap and water are unavailable, use an alcohol-based hand sanitizer.

Data Collection

Biologists take measurements in many types of investigations.
- **a population biologist might count tree frogs in a rain forest survey**
- **a physical therapist might observe range of motion of an injured knee**
- **microbiologist might measure the diameters of bacteria.**

In this section, you will learn how biologists take careful measurements. When you plan and perform your biology labs, use this section as a guide.

Accuracy, Precision, and Error

In any measurement, there always will be some error—the difference between the measured value and the real or accepted value. Error comes from several sources, including the experimenter, the equipment, and even changes in experimental conditions. Errors can affect both the accuracy and precision of measurements.

- **Accuracy** refers to how close a measurement is to the real value or the accepted value.
- **Precision** refers to how close a series of measurements are to one another.

Examine the targets below while you consider a food scientist who measures the mass of a sample three times.

- Proper equipment set up and good technique—accurate and precise data
- Incorrect equipment set up but good technique—precise but inaccurate data
- Incorrect equipment set up and careless technique—inaccurate and imprecise data

Figure 1 The arrows clustered in the center represent measurements that are both accurate and precise. The arrows clustered together far from the center represent three measurements that are precise but inaccurate. These arrows are both far apart and far from the center. They represent three measurements that are inaccurate and imprecise.

Error Analysis

Imagine that an epidemiologist (eh puh dee mee AHL uh just)—a biologist who studies epidemics—tested a hypothesis about the way avian flu might spread from chickens to humans. All data have been gathered. The epidemiologist must now perform an error analysis, which is a process to identify and describe possible sources of error in measurements.

In your biology investigations, you will need to think of possible sources of measurement errors. Ask yourself questions such as:
- Did I take more than one reading of each measurement?
- Did I use the equipment properly?
- Was I objective, or did I make the results turn out as I expected they might?

Measure Mass

Triple-Beam Balance

A triple-beam balance has a pan and three beams with sliding masses called riders. At one end of the beams is a pointer that indicates whether the mass on the pan is equal to the masses shown on the beams.

To use:

1. Make sure the balance is zeroed before measuring the mass of an object. The balance is zeroed if the pointer is at zero when nothing is on the pan and riders are at their zero points.
2. Place the object to be measured on the pan.
3. Move the riders one notch at a time away from the pan. Begin with the largest rider. If moving the largest rider one notch brings the pointer below zero, begin measuring the mass with the next smaller rider.
4. Change the positions of the riders until they balance the mass on the pan and the pointer is at zero. Then add the readings from the three beams to determine the mass of the object.

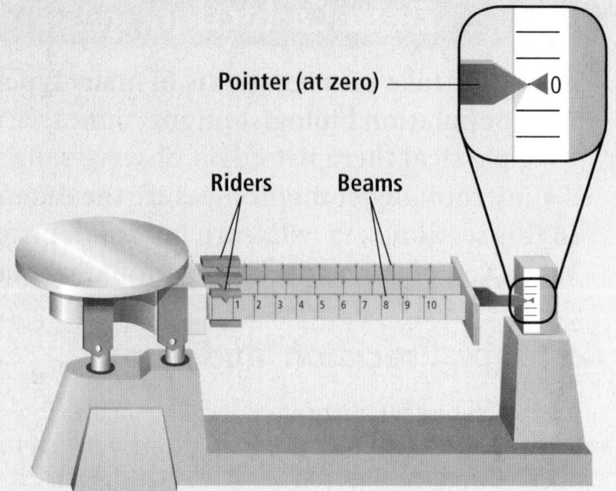

Figure 2

TIP

When using a weighing boat or weighing paper, be sure to zero the balance after you've placed the boat or paper on the pan and before you add your substance to the boat or paper.

Measure Volume

Graduated Cylinder

Use a graduated cylinder to measure the volume of a liquid.

To use:

1. Be sure to have your eyes at the level of the surface of the liquid when reading the scale on a graduated cylinder.
2. The surface of most liquids will be curved slightly down when they are held in a graduated cylinder. This curve is called the meniscus. Read the volume of the liquid at the bottom of the meniscus, as shown in **Figure 3.**
3. The volume will often be between two lines on the graduated cylinder. You should estimate the final digit in your measurement. For example, if the bottom of the meniscus appears to be exactly half way between the marks for 96 mL and 97 mL, you would read a volume of 96.5 mL.
4. To find the volume of a small solid object, record the volume of some water in a graduated cylinder. Then, measure the volume of the water after you add the object to the cylinder. The volume of the object is the difference between the first and second measurements.

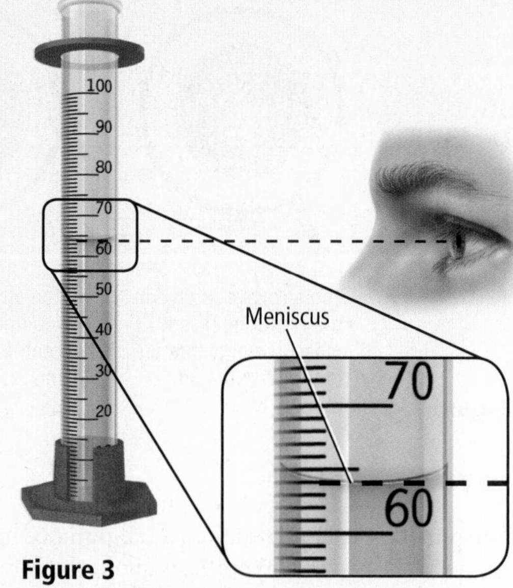

Figure 3

TIP

Do not use a beaker to measure the volume of a liquid. Beakers are used for holding and pouring liquids. To avoid overflow, be sure to use a beaker that holds roughly twice as much liquid as you need.

Measure Temperature

The thermometers you will be using measure temperature in degrees Celsius (°C). Each division on the scale represents 1°C. The average human body temperature is 37°C. A typical room temperature is between 20°C and 25°C. The freezing point of water is 0°C and the boiling point of water is 100°C.

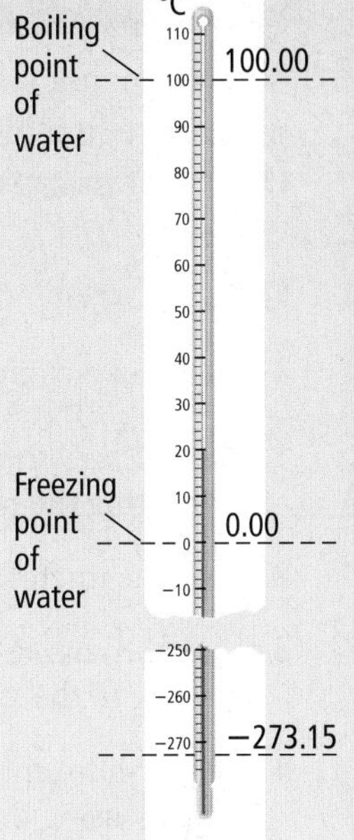

Figure 4

To use:

1. Place the thermometer in your sample and wait for 30 s before taking the reading.
2. Be sure to have your eyes at the level of the surface of the liquid when reading the scale on a thermometer.
3. The temperature will often be between two lines on the thermometer. You should estimate the final digit in your measurement. For example, if the liquid appears to be about half way between the marks for 50°C and 51°C, you would read a temperature of 50.5°C.
4. Do not touch the sides or bottom of your container with the thermometer. This can yield a false temperature.

Electronic thermometers, often called temperature probes, are used to record temperature over a range of time or to give more accurate and precise readings.

Measure Length

Use a metric ruler or meterstick to measure the length of an object. On the ruler, each marked number represents one centimeter (cm). The smaller lines between each centimeter represent millimeters (mm). There are 10 mm in one centimeter and 100 cm in one meter (m).

To use:

1. Place the metric ruler so that the 0-cm mark lines up with the end of your object.
2. Be sure to have your eyes at the level of the object when reading the scale on the ruler.
3. The accuracy of your measurement reflects the measuring tool you use and your technique. The figure below shows the estimation of the length of the same object using two different measuring tools. On the lower measuring tool, the length is between 9 and 10 cm. The measurement would be estimated to the nearest tenth of a centimeter. You would estimate the length to be 9.5 cm. On the upper measuring tool, the length is between 9.4 and 9.5 cm. The measurement would be estimated to the nearest hundredth of a centimeter. You would estimate the length to be 9.45 cm.

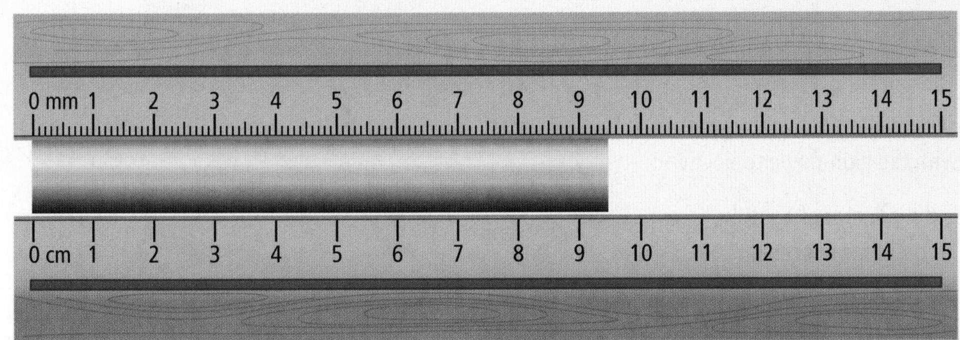

Figure 5

Laboratory Equipment and Techniques

The following six pages discuss common lab equipment and techniques that you might use in a biology lab. Refer to these pages prior to performing labs that require the use of microscopes, chromatography, gel electrophoresis, or indicators.

Use a Compound Microscope

The parts of a compound microscope are listed and diagrammed in the table below.

1. Always carry the microscope by holding the arm of the microscope with one hand and supporting the base with the other hand.
2. Place the microscope on a flat surface. The arm should be positioned toward you.
3. Look through the eyepieces. Adjust the diaphragm so that the light comes through the opening in the stage.
4. Place a slide on the stage so that the specimen is in the field of view. Hold it firmly in place by using the stage clips.
5. Always focus first with the coarse adjustment and the low-power objective lens. Once the object is in focus on low power, the high-power objective can be used. Use only the fine adjustment to focus the high-power lens.
6. Store the microscope covered.

Parts of the Compound Light Microscope

Part	Function
Base	Supports the microscope
Arm	Used to carry the microscope
Stage	Platform where the slide with specimen is placed
Stage clips	Holds the slide in place on the stage
Eyepiece	Magnifies image for the viewer
Objective lens	Low-power and high-power lenses that magnify the specimen
Coarse adjustment	Large knob used for focusing the image under low-power
Fine adjustment	Smaller knob used for focusing the image with the high-power objective
Diaphragm	Controls the amount of light that passes through the specimen
Light source	Provides light for viewing the specimen

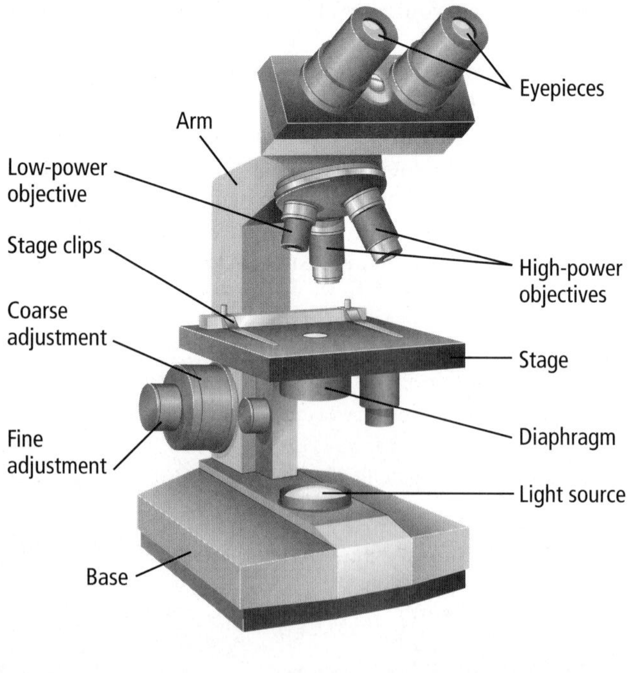

Calculate Magnification

Magnification describes how much larger an object appears when viewed through a microscope compared to the unaided eye. The numbers on the eyepiece and the objectives that are marked with the multiplication symbol (×) tell you how many times the lens of each microscope part magnifies an object.

- To calculate the total magnification of any object viewed under a microscope, multiply the number on the eyepiece by the number on the objective through which you are viewing the object.
- For example, if the eyepiece magnification is 4× and the low-power magnification is 10×, then total magnification under the low-power objective is 40×. With the same eyepiece and a high-power magnification of 40×, the total magnification under the high-power objective would be 160×.

Practice Problem 1 Calculate the low-power and the high-power magnifications of a microscope with an eyepiece magnification of 10×, a low-power objective of 40×, and a high-power objective of 60×.

Calculate the Field of View

The area you see when you look into a microscope is called the field of view. To measure the field of view of a microscope, you must use a unit called a micrometer (μm). There are 1000 micrometers in a millimeter. Use the following steps to calculate field of view and then to determine the diameters of the microscopic specimens that you are viewing.

Diameter of Low-Power Field of View Use a low-power objective to select the section of a slide that you want to examine, such as the area where pollen grains are located.

- Place the millimeter section of a clear plastic ruler over the central opening of your microscope stage.
- Use the low-power objective to locate the lines of the ruler. Center the ruler in the field of view.
- Place one of the lines representing a millimeter at the very edge of the field of view. The distance between two lines on the ruler is 1 mm, as shown in **Figure 6.**
- Estimate the diameter, in millimeters, of the field of view on low power. Use the conversion factor $\frac{1000\ \mu\text{m}}{1\ \text{mm}}$ to calculate the diameter in micrometers. For example, if you estimate the diameter to be 1.5 mm, the field of view is 1500 μm.

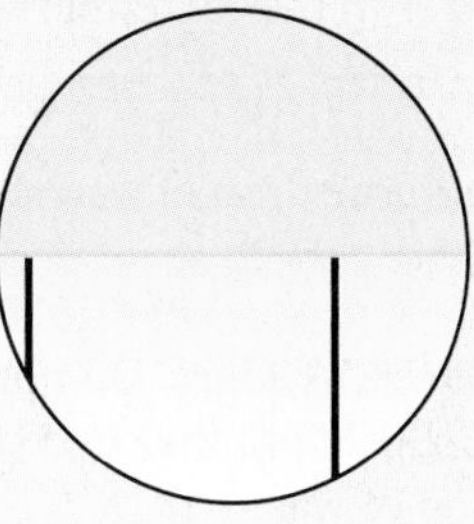

Figure 6

$$1.5\ \text{mm} \times \frac{1000\ \mu\text{m}}{1\ \text{mm}} = 1500\ \mu\text{m}$$

Diameter of High-Power Field of View After selecting a slide section on low power, use the high-power field of view to see the details on the slide, such as individual pollen grains.

- To calculate the diameter of the high-power field, divide the magnification of your high-power objective by the magnification of your low-power objective. For example, changing from a low power of 10× to a high power of 40×, you would write:

$$\frac{40\times}{10\times} = 4$$

- Then, divide the diameter of the low-power field in micrometers by this quotient. The result is the diameter of the high-power field in micrometers. For the low-power field calculated on the previous page, the diameter of the high-power field of view is

$$\frac{1500\ \mu m}{4} = 375\ \mu m$$

- To determine the diameter of a specimen in your field of view, first estimate how many of the specimens would fit end-to-end across your field of view. Then divide the diameter of the field of view by the estimated number of specimens. In the example, the diameter of the specimen is 75 μm.

$$\frac{375\ \mu m}{5} = 75\ \mu m$$

Practice Problem 2 Calculate the width of the dividing cell if the diameter of the low-power field is 720 μm, the low power is 10×, the high power is 60×, and the number of cells that fit in the field of view is 1.

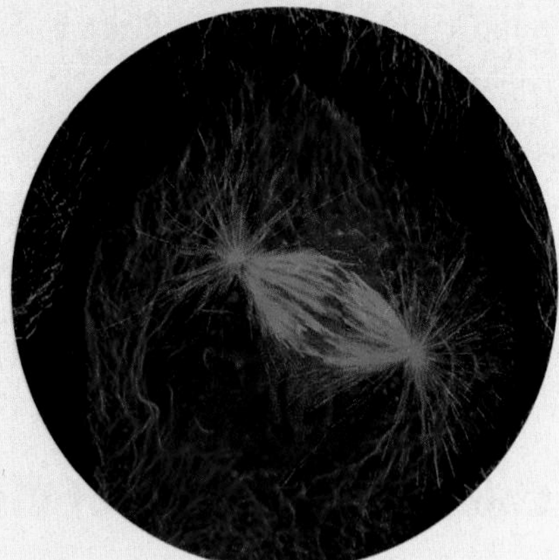

Figure 7 Dividing cell

Make a Wet Mount

Many of the slides that you will prepare for observation under the microscope will be wet mounts. Wet mounts are named as such because the object to be viewed is prepared, or mounted, in water. Follow these steps to make a wet mount.

1. Obtain a clean microscope slide and a coverslip. Add a drop or two of water to the center of the microscope slide.

2. Place the specimen in the drop of water as shown in **Figure 8.**

3. Pick up the coverslip by its edges. Do not touch the surface of the coverslip. Stand the coverslip on its edge next to the drop of water.

4. Slowly lower the coverslip over the drop of water and the specimen as shown in **Figure 9.**

5. Make sure that the object is totally covered with water. If it is not, remove the coverslip, add a little more water, and replace the coverslip.

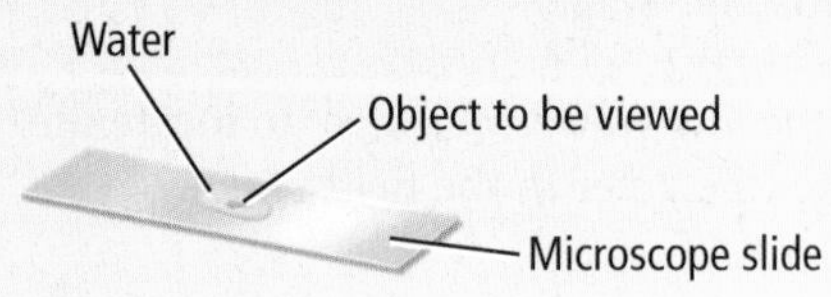

Figure 8

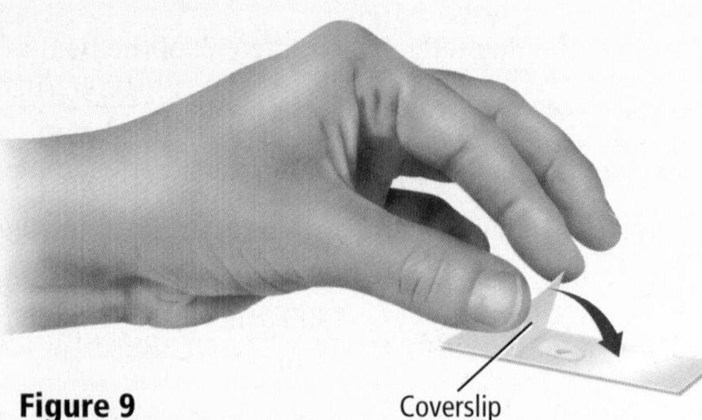

Figure 9

Stain a Slide

Staining a slide can make it easier to view a specimen. Stains enhance contrast and can call out certain features. For example, using iodine as a stain will cause carbohydrates in the specimen to become bluish-black in color. The following steps indicate one way to stain a microscope slide.

1. Prepare a wet mount, as indicated in the steps on the previous page.

2. Obtain the stain from your teacher. Using a dropper, place a drop of the stain at one end of the coverslip.

3. Place a paper towel at the end of the coverslip opposite the stain. The towel will draw the stain under the coverslip, staining the specimen.

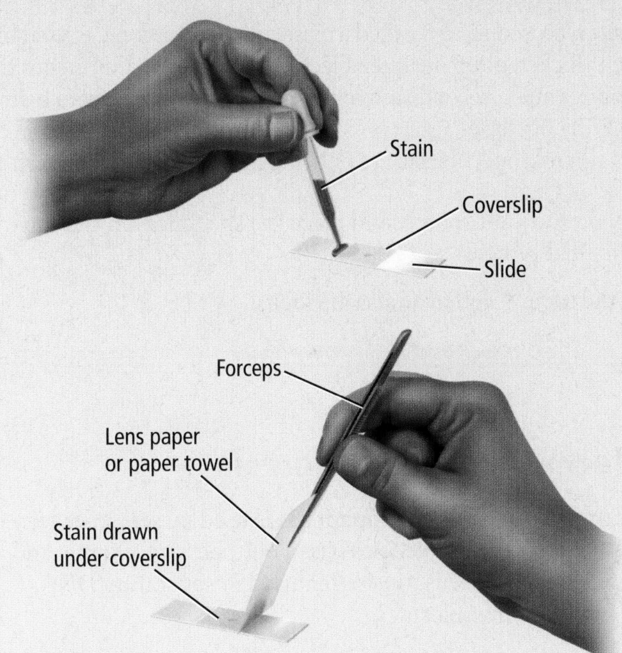

Figure 10

Make Cross Sections

When a biologist decides to the study the inner structure of a biological specimen, a basic way to expose or cut a specimen to reveal its inner structure is called a cross section. A cross-sectional exposure or cut is done at right angles to the axis of the specimen. For example, the tree trunk in **Figure 11** has been cut at right angles to the height of the trunk. Note that microscopic cross sections reveal microscopic structure, such as the bacterium's cell wall in **Figure 11.**

Think Critically Investigate cross sections by performing the following procedure using everyday materials. Then apply what you have learned to recognize more cross sections in the textbook.

1. Obtain log-shaped, rolled snack cakes that have contrasting color filling. The axis of this specimen runs through the center of one end to the center of the other end.

2. Place a snack cake on a sheet of wax paper and predict what a crosswise cut would look like.

3. Make a crosswise cut at a right angle to the axis and look at the cut ends. This view of the snack cake is a cross section.

4. Find cross-sectional diagrams in this textbook that were made in a similar way.

Figure 11

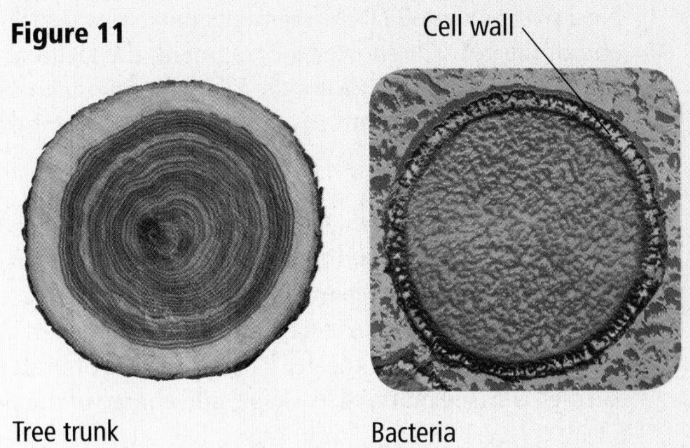

Use a Stereomicroscope

A stereomicroscope, also called a dissecting microscope, is used to observe a larger, thicker, often opaque object. A light source illuminates the object from above and a second source illuminates the object from below. The magnifying power of a stereomicroscope is much less than for a compound microscope; objects are only magnified 10–50 diameters.

- Turn on the light source, and place the specimen on the stage so it is in the field of view.
- Use the focus knob to adjust the focus.

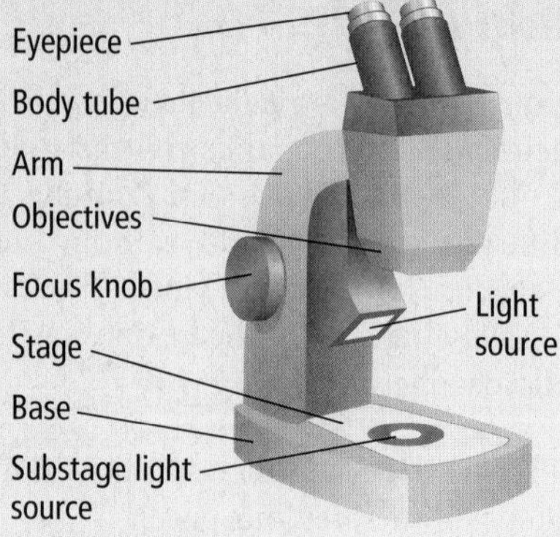

Figure 12

Perform Gel Electrophoresis

A technique called gel electrophoresis is used by scientists to separate mixtures of molecules based on their size, charge, and shape. This technique is most often used in separating DNA, RNA, and protein molecules.

Below are general guidelines for gel electrophoresis. Refer to your specific instrument's user's manual for complete instructions.

1. In the process of gel electrophoresis, scientists analyze DNA by first using special enzymes to cut a DNA sample at specific nucleotide sequences.
2. Small samples of cut DNA are prepared and placed in wells located on one end of a semisolid, gelatinlike gel, as shown in **Figure 13.**
3. The gel is placed in a buffer solution between two electrodes in a gel electrophoresis chamber. The electrodes are connected to a power supply (chamber and electrode not shown). When an electric current is applied, the buffer solution conducts the current. The current also moves through the gel. One end of the gel electrophoresis chamber becomes positively charged and the other end becomes negatively charged.

   Negatively charged DNA fragments move toward the positive end of the gel. The shorter the fragment, the farther it moves through the gel. This allows the DNA fragments to form distinct and unique patterns for study, like those shown in **Figure 13.**

This process is also used to examine protein patterns. Proteins are extracted from cells and treated with chemicals to give them a negative charge. The prepared protein samples are placed in the wells of a gel. When an electric current is applied, the protein molecules move through the gel. The separation of protein molecules is based on the size, shape, and charge of the proteins.

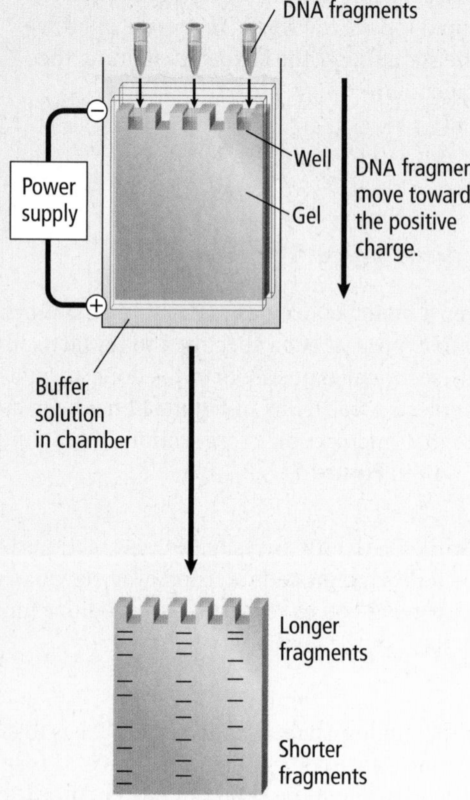

Figure 13

Perform Chromatography

Paper chromatography is a commonly used technique in the biology laboratory for separating mixtures of substances. You will perform chromatography with a special chromatography paper or with filter paper and a liquid solvent. Separation occurs based on the ability of substances in the mixture to dissolve in the solvent. The general steps for this type of chromatography are:

1. a mixture is dissolved in a liquid and placed on the paper;

2. one end of the paper is placed in a solvent;

3. the substances separate based on their tendencies to move along the surface of the paper while in the solvent.

For example, chlorophyll from leaves can be separated by paper chromatography, as shown in **Figure 14.** A dot of the chlorophyll extract is placed near one end of the strip of paper. The end of the paper nearest the dot is placed in alcohol, which acts as the solvent. The alcohol should not touch the extract to be separated, but should be just below it.

Figure 14

The alcohol moves up the paper and picks up substances in the chlorophyll extract. Substances in the extract that are tightly held to the paper will move slowly up the paper, while extract substances that are not as tightly held move quickly up the paper. This results in bands of different substances on the chromatography paper.

Use Indicators

Indicators are used to test for the presence of specific types of chemicals or substances. The table below lists commonly used indicators, what they test for, and how they react.

Indicators

Indicator	What it indicates in a solution	Reaction
Litmus paper	acid or base	• red litmus turns blue if a base • blue litmus turns red if an acid
pH paper	pH	• color change compared to a color chart to estimate the pH
Bromthymol blue	presence of carbon dioxide	• turns yellow if carbon dioxide is present • change to blue from yellow when carbon dioxide is removed
Phenolphthalein solution	presence of carbon dioxide or a basic solution	• turns from clear to a bright pink in the presence of either substance
Benedict's solution	presence of simple sugars when heated	• high sugar concentration, change from blue to red • low sugar concentration, change from blue to yellow
Biuret solution	presence of protein	• turns from light blue to purple
Lugol's solution	presence of starch	• turns from deep brown to bluish-black

CHAPTER 1

The Study of Life

BIG Idea Biology is the study of life.

Section 1
Introduction to Biology
MAIN Idea All living things share the characteristics of life.

Section 2
The Nature of Science
MAIN Idea Science is a process based on inquiry that seeks to develop explanations.

Section 3
Methods of Science
MAIN Idea Biologists use specific methods when conducting research.

BioFacts

- There are approximately 200 billion stars that make up the Milky Way galaxy.
- Humans are 1 out of an estimated 100 million species of life on Earth.
- The human brain is made up of about 100 billion neurons.

LAUNCH Lab

Why is observation important in science?

Scientists use a planned, organized approach to solving problems. A key element of this approach is gathering information through detailed observations. Scientists extend their ability to observe by using scientific tools and techniques.

Procedure

1. Read and complete the lab safety form.
2. Pick an unshelled **peanut** from the **container of peanuts.** Carefully observe the peanut using your senses and available tools. Record your observations.
3. Do not change or mark the peanut. Return your peanut to the container.
4. After the peanuts are mixed, locate your peanut based on your recorded observations.

Analysis

1. **List** the observations that were the most helpful in identifying your peanut. Which were the least helpful?
2. **Classify** your observations into groups.
3. **Justify** why it was important to record detailed observations in this lab. Infer why observations are important in biology.

Visit biologygmh.com to:

- study the entire chapter online
- explore the Interactive Time Line, Concepts in Motion, Interactive Tables, Microscopy Links, Virtual Labs, and links to virtual dissections
- access Web links for more information, projects, and activities
- review content online with Interactive Tutor and take Self-Check Quizzes

Biologists Make the following Foldable to help you organize examples of things biologists do.

STEP 1 Stack three sheets of notebook paper 2.5 cm apart as illustrated.

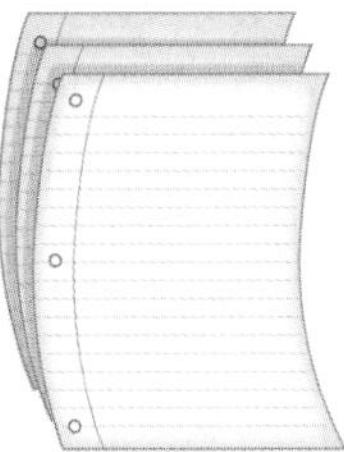

STEP 2 Bring up the bottom edges and fold to form five tabs of equal size.

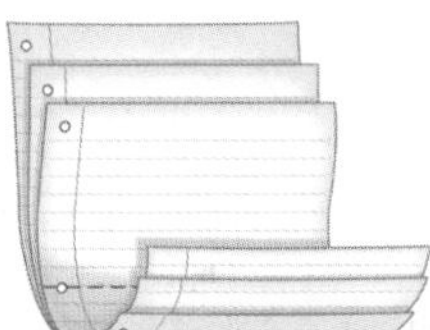

STEP 3 Rotate your Foldable 180°. Staple along the folded edge to secure all sheets. Label the tabs *Some Roles of Biologists, Study the diversity of life, Research diseases, Develop technology, Improve agriculture,* and *Preserve the environment.*

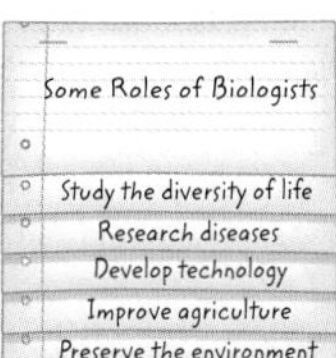

FOLDABLES **Use this Foldable with Section 1.1.** As you study the section, summarize these examples of the different roles of biologists.

Section 1.1

Reading Preview

Objectives

- **Define** biology.
- **Identify** possible benefits from studying biology.
- **Summarize** the characteristics of living things.

Review Vocabulary

environment: the living and nonliving things that surround an organism and with which the organism interacts

New Vocabulary

biology
organism
organization
growth
development
reproduction
species
stimulus
response
homeostasis
adaptation

Introduction to Biology

MAIN Idea **All living things share the characteristics of life.**

Real-World Reading Link Think about several different living or once-living things. The bacteria that live in your small intestine, the great white sharks in the ocean, a field of corn, a skateboarder, and the extinct *Tyrannosaurus rex* differ in structure and function. Who discovered what all these things have in common?

The Science of Life

Before Jane Goodall, pictured in **Figure 1.1,** arrived in Gombe Stream National Park in Tanzania, Africa, in 1960 to study chimpanzees, the world of chimpanzees was a mystery. Jane's curiosity, determination, and patience over a long period of time resulted in the chimpanzee troop's acceptance of her presence so that she was able to observe their behavior closely.

When people study living things or pose questions about how living things interact with the environment, they are learning about **biology**—the science of life. Biology comes from the Greek word *bio*, meaning *life*, and from *logos*, meaning *study*.

In biology, you will study the origins and history of life and once-living things, the structures of living things, how living things interact with one another, and how living things function. This will help you understand how humans have a vital role in preserving the natural environment and sustaining life on Earth.

Have you ever hiked in a forest and wondered why different trees have leaves with different shapes? Maybe you have watched an ant quickly cross the sidewalk toward a breadcrumb and wondered how the ant knew that the breadcrumb was there. When you ask these questions, you are observing, and you are asking questions about life.

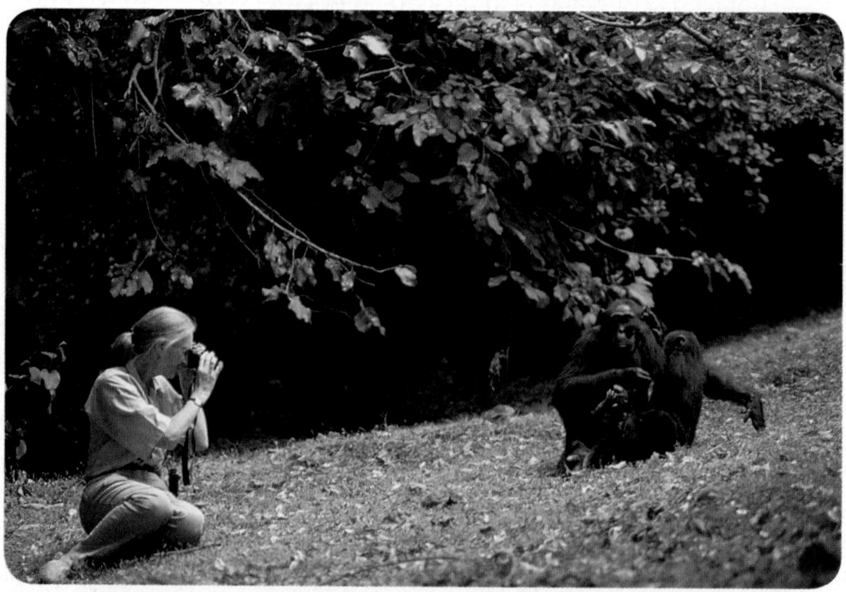

Figure 1.1 Jane Goodall conducted field research for many years to observe chimpanzee behavior.
Predict *the types of questions you would ask if you observed chimpanzee behavior.*

What do biologists do?

Imagine being the first person to look into a microscope and discover cells. What do you think it was like to find the first dinosaur fossils that indicated feathers? Who studies how organisms, including the marbled stargazer fish in **Figure 1.2,** obtain food? Will the AIDS virus be defeated? Is there life on other planets or anywhere else in the universe? The people who study biology—biologists—make discoveries and seek explanations by performing laboratory and field investigations. Throughout this textbook, you will discover what biologists in the real world do and you will learn about careers in biology.

Study the diversity of life Jane Goodall, shown in **Figure 1.1,** studied chimpanzees in their natural environments. She asked questions such as, "How do chimpanzees behave in the wild?" and "How can chimpanzee behaviors be characterized?" From her recorded and detailed observations, sketches, and maps of chimpanzees' daily travels, Goodall learned how chimpanzees grow and develop and how they gather food. She studied and recorded chimpanzee reproductive habits and their aggressive nature. She learned that they use tools. Goodall's data provided a better understanding of chimpanzees, and as a result, scientists know how to best protect them.

Research diseases Mary-Claire King also studied chimpanzees—not their behavior but their genetics. In 1973, she established that the genomes (genes) of chimpanzees and humans are 99 percent identical. Her work currently focuses on unraveling the genetic basis of breast cancer, a disease that affects one out of eight women.

Many biologists research diseases. Questions such as "What causes the disease?", "How does the body fight the disease?", and "How does the disease spread?" often guide biologists' research. Biologists have developed vaccines for smallpox, chicken pox, and diphtheria, and currently, some biologists are researching the development of a vaccine for HIV. Other biologists focus their research on diseases such as diabetes, avian flu, anorexia, and alcoholism, or on trauma such as spinal cord injuries that result in paralysis. Biologists worldwide are researching new medicines for such things as lowering cholesterol levels, fighting obesity, reducing the risk of heart attacks, and preventing Alzheimer's disease.

Develop technologies When you hear the word *technology,* you might think of high-speed computers, cell phones, and DVD players. However, technology is defined as the application of scientific knowledge to solve human needs and to extend human capabilities. **Figure 1.3** shows how new technology—a "bionic" hand—can help someone who has lost an arm.

Figure 1.2 The marbled stargazer fish lives beneath the ocean floor off the coast of Indonesia. It explodes upward from beneath the sand to grab its food.
Observe ***How does this fish hide from its food?***

FOLDABLES
Incorporate information from this section into your Foldable.

Figure 1.3 A prosthetic "bionic" hand is new technology that can help extend human capabilities.

Figure 1.4 Joanne Chory, a plant biologist, researches how plants respond to light.

Figure 1.5 *Streptococcus pyogenes* is a unicellular organism. It can infect the throat, sinuses, or middle ear.

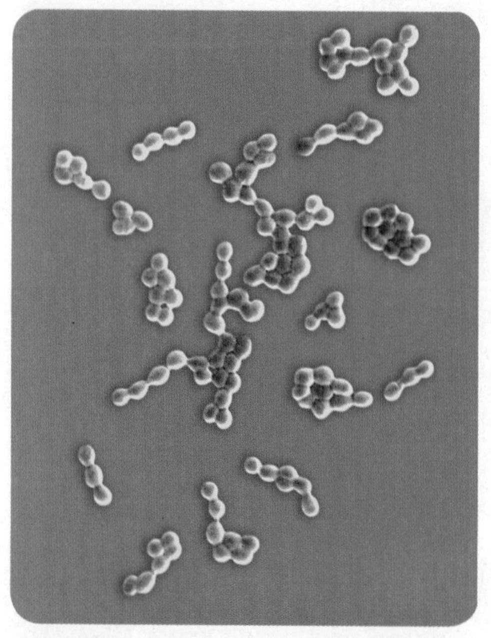

SEM Magnification: 7300×

For example, Charles Drew was a doctor who pioneered methods to separate blood plasma from blood cells and safely store and transport blood plasma for transfusions. His research led to blood banks that saved soldiers during World War II and helps countless patients today.

Biologists today continue to discover new ways to improve and save lives. For example, the field of bioengineering applies knowledge gained from studying the function of living systems to the design of mechanical devices such as artificial limbs. In addition, biologists in the field of biotechnology research cells, DNA, and living systems to discover new medicines and medical treatments.

Improve agriculture Some biologists study the possibilities of genetically engineering plants to grow in poor soils or to resist insects, fungal infections, or frost damage. Other biologists research agricultural issues to improve food production to feed the world's growing human population.

Joanne Chory, a plant biologist shown in **Figure 1.4,** studies mustard plants' sensitivity to light and their responses when exposed to different light sources, different times of exposure, and other conditions. Because of her work with plant growth hormones and light, agriculturists might be able to increase the amount of food produced from crops or to grow crops in areas where they normally would not grow.

Preserve the environment Environmental biologists seek to prevent the extinction of animals and plants by developing ways to protect them. Some biologists study the reproductive strategies of endangered species while they are in captivity. Other biologists work in nature preserves that provide safe places for endangered species to live, reproduce, and have protection against poachers.

Lee Anne Martinez is an ecologist who worked to protect the environment where outdoor toilets are common. She helped people in rural Africa construct composting toilets that use no water. The composted waste from the toilets can be added to soil to improve it for agricultural use.

The Characteristics of Life

Have you ever tried to define the word *alive?* If you were to watch a grizzly bear catch a salmon from a river, you obviously would conclude that the bear and salmon are both alive. Is fire alive? Fire moves, increases in size, has energy, and seems to reproduce, but how does fire differ from the bear and salmon?

Over time and after many observations, biologists concluded that all living things have certain characteristics, as listed in **Table 1.1.** An **organism** is anything that has or once had all these characteristics.

Made of one or more cells Have you ever had strep throat? It probably was caused by a group A streptococcal bacteria, such as the *Streptococcus pyogenes* shown in **Figure 1.5.** A bacterium is unicellular—it has just one cell—yet it displays all the characteristics of life just like a skin cell on your body or a cell in a plant's leaf. Humans and plants are multicellular—they have many cells.

Concepts In Motion
Interactive Table To explore more about the characteristics of life, visit biologygmh.com.

Table 1.1 Characteristics of Living Organisms

Characteristic of Life	Example	Description
Made of one or more cells	Magnification: 160×	All organisms are made of one or more cells. The cell is the basic unit of life. Some organisms, such as the *Paramecium sp.*, are unicellular.
Displays organization		The levels of organization in biological systems begin with atoms and molecules and increase in complexity. Each organized structure in an organism has a specific function. The structure of an anteater's snout relates to one of its functions—a container for the anteater's long tongue.
Grows and develops		Growth results in an increase in mass. Development results in different abilities. A bullfrog tadpole grows and develops into an adult bullfrog.
Reproduces		Organisms reproduce and pass along traits from one generation to the next. For a species like the koala to continue to exist, reproduction must occur.
Responds to stimuli		Reactions to internal and external stimuli are called responses. This cheetah responds to the need for food by chasing a gazelle. The gazelle responds by running away.
Requires energy		Energy is required for all life processes. Many organisms, like this squirrel, must take in food. Other organisms make their own food.
Maintains homeostasis		All organisms keep internal conditions stable by a process called homeostasis. For example, humans perspire to prevent their body temperature from rising too high.
Adaptations evolve over time		Adaptations are inherited changes that occur over time that help the species survive. Tropical orchids have roots that are adapted to life in a soil-less environment.

■ **Figure 1.6** In less than a month, these robin chicks grow and develop from helpless chicks to birds capable of flying.
Infer ***how the robins have developed in other ways.***

Cells are the basic units of structure and function in all living things. For example, each heart cell has a structure that enables it to contribute to the heart's function—continually pumping blood throughout the body. Likewise, each cell in a tree's roots has a structure that enables it to help anchor the tree in the ground and to take in water and dissolved minerals from the surrounding soil.

Displays organization Think of all the people in your high school building each day. Students, faculty, counselors, administrators, building service personnel, and food service personnel are organized based on the different tasks they perform and the characteristics they share. For example, the students are designated freshmen, sophomores, juniors, and seniors based on age and coursework.

Living things also display **organization,** which means they are arranged in an orderly way. The *Paramecium* in **Table 1.1** is made up of one cell, yet that cell is a collection of organized structures that carries on life functions. Each of those structures is composed of atoms and molecules. The many cells that make up the robin chicks in **Figure 1.6** also contain structures made of atoms and molecules. However, in multicellular organisms, specialized cells are organized into groups that work together called tissues. These tissues are organized into organs, which carry on functions such as digestion and reproduction. Organ systems work together to support an organism. You will learn in Chapter 3 how individual organisms are organized and supported by the biosphere.

MiniLab 1.1

Observe Characteristics of Life

Is it living or nonliving? In this lab, you will observe several objects to determine if they are living or nonliving.

Procedure

1. Read and complete the lab safety form.
2. Create a data table with four columns titled *Object, Prediction, Characteristic of Life,* and *Evidence.*
3. Your teacher will provide several objects for observation. List each **object** in your table. Predict whether each object is living or nonliving.
4. Carefully observe each object. Discuss with your lab partner what characteristics of life it might exhibit.
5. Use **Table 1.1** to determine whether each object is living or nonliving. List the evidence in your data table.

Analysis

1. **Compare and contrast** your predictions and observations.
2. **Explain** why it was difficult to classify some objects as living or nonliving.

Grows and develops Most organisms begin as one cell. **Growth** results in the addition of mass to an organism and, in many organisms, the formation of new cells and new structures. Even a bacterium grows. Think about how you have grown throughout your life.

Robin chicks, like those in **Figure 1.6,** cannot fly for the first few weeks of their lives. Like most organisms, robins develop structures that give them specific abilities, such as flying. **Development** is the process of natural changes that take place during the life of an organism.

Reproduces Most living things are the result of **reproduction**—the production of offspring. Reproduction is not an essential characteristic for individual organisms. Many pets are spayed or neutered to prevent unwanted births. Obviously, these pets can still live even though they cannot reproduce. However, if a species is to continue to exist, then members of that species must reproduce. A **species** is a group of organisms that can breed with one another and produce fertile offspring. If the individuals of a species do not reproduce, then when the last individual of that species dies, the species becomes extinct.

Responds to stimuli An organism's external environment includes all things that surround it, such as air, water, soil, rocks, and other organisms. An organism's internal environment is all things inside it. Anything that is part of either environment and causes some sort of reaction by the organism is called a **stimulus** (plural, stimuli). The reaction to a stimulus is a **response.** For example, if a shark smells blood in the ocean, it will respond quickly by moving toward the blood and attacking any organism present. Plants also respond to their environments, but they do so more slowly than most other organisms. If you have a houseplant and you place it near a sunny window, it will grow toward the window in response to the light. How does the Venus flytrap in **Figure 1.7** respond to stimuli?

Being able to respond to the environment is critical for an organism's safety and survival. If an organism is unable to respond to danger or to react to potential enemies, it might not live long enough to reproduce.

CAREERS IN BIOLOGY

Biology Teacher An enthusiasm for biology is one of the many reasons people become biology teachers. Other than courses in biological sciences, prospective biology teachers might take classroom management, teaching methods, and other courses needed to develop teaching skills. For more information on biology careers, visit biologygmh.com.

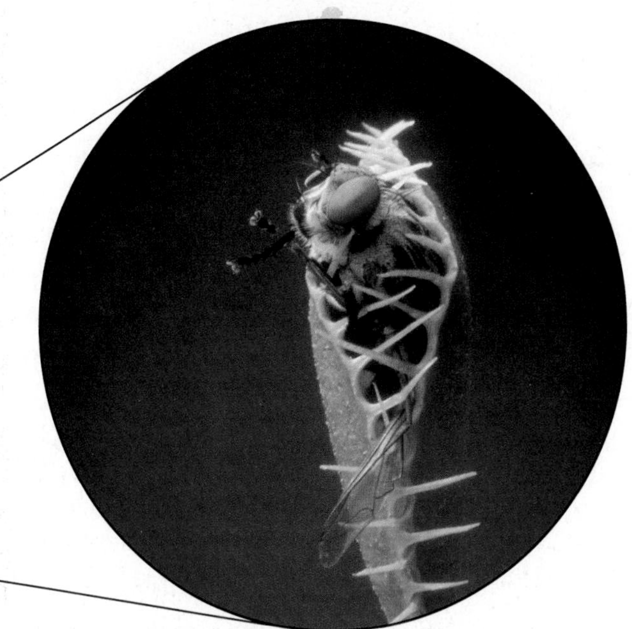

■ **Figure 1.7** In nature, this Venus flytrap grows in soils that lack certain nutrients. The plant captures and digests insects and takes in needed nutrients.
Explain *How does this plant respond to stimuli to obtain food?*

■ **Figure 1.8** The structure of a drip-tip leaf is an adaptation to rainy environments.

Requires energy Living things need sources of energy to fuel their life functions. Living things get their energy from food. Most plants and some unicellular organisms use light energy from the Sun to make their own food and fuel their activities. Other unicellular organisms can transform the energy in chemical compounds to make their food.

Organisms that cannot make their own food, such as animals and fungi, get energy by consuming other organisms. Some of the energy that an organism takes in is used for growth, development, and maintaining homeostasis. However, most of the energy is transformed into thermal energy and is radiated to the environment as heat.

Maintains homeostasis Regulation of an organism's internal conditions to maintain life is called **homeostasis** (hoh mee oh STAY sus). Homeostasis occurs in all living things. If anything happens within or to an organism that affects its normal state, processes to restore the normal state begin. If homeostasis is not restored, death might occur.

Connection to Earth Science When athletes travel to a location that is at a higher altitude than where they live, they generally arrive long before the competition so that their bodies have time to adjust to the thinner air. At higher altitudes, air has fewer molecules of gases, including oxygen, per unit of volume. Therefore, there is less oxygen available for an athlete's red blood cells to deliver to the cells and tissues, which disrupts his or her body's homeostasis. To restore homeostasis, the athlete's body produces more red blood cells. Having more red blood cells results in an adequate amount of oxygen delivered to the athlete's cells.

Adaptations evolve over time Many trees in rain forests have leaves with drip tips, like the one shown in **Figure 1.8.** Water runs off more easily and quickly from leaves with drip tips. Harmful molds and mildews will not grow on dry leaves. This means a plant with dry leaves is healthier and has a better chance to survive. Drip tips are an adaptation to the rain forest environment. An **adaptation** is any inherited characteristic that results from changes to a species over time. Adaptations like rain forest trees with drip tips enable species to survive and, therefore, they are better able to pass their genes to their offspring.

Section 1.1 Assessment

Section Summary

- Biology is the science of life.
- Biologists study the structure and function of living things, their history, their interactions with the environment, and many other aspects of life.
- All organisms have one or more cells, display organization, grow and develop, reproduce, respond to stimuli, use energy, maintain homeostasis, and have adaptations that evolve over time.

Understand Main Ideas

1. MAIN Idea **Describe** four characteristics used to identify whether something is alive.
2. **Explain** why cells are considered the basic units of living things.
3. **List** some of the benefits of studying biology.
4. **Differentiate** between response and adaptation.

Think Critically

5. MATH in Biology Survey students in your school—biology students and nonbiology students—and adults. Have participants choose characteristics of life from a list of various characteristics and rank their choices from most important to least important. Record, tabulate, average, and graph your results. Prepare a report that summarizes your findings.

Biology Online Self-Check Quiz biologygmh.com

Section 1.2

Reading Preview

Objectives

- **Explain** the characteristics of science.
- **Compare** something that is scientific with something that is pseudoscientific.
- **Describe** the importance of the metric system and SI.

Review Vocabulary

investigation: a careful search or examination to uncover facts

New Vocabulary

science
theory
peer review
metric system
SI
forensics
ethics

The Nature of Science

MAIN Idea Science is a process based on inquiry that seeks to develop explanations.

Real-World Reading Link If you see a headline that reads "Alien baby found in campsite," how do you know whether you should believe it or not? How do you know when to trust claims made in an advertisement on television or the Internet, or in a newspaper or magazine? What makes something science-based?

What is science?

Have you ever wondered how science is different from art, music, and writing? **Science** is a body of knowledge based on the study of nature. Biology is a science, as are chemistry, physics, and Earth science, which you might also study during high school. The nature, or essential characteristic, of science is scientific inquiry—the development of explanations. Scientific inquiry is both a creative process and a process rooted in unbiased observations and experimentation. Sometimes scientists go to extreme places to observe and experiment, as shown in **Figure 1.9.**

Relies on evidence Has anyone ever said to you, "I have a theory about that?" That person probably meant that he or she had a possible explanation about something. Scientific explanations combine what is already known with consistent evidence gathered from many observations and experiments.

When enough evidence from many related investigations supports an idea, scientists consider that idea a **theory**—an explanation of a natural phenomenon supported by many observations and experiments over time. For example, what happens when you throw a ball up in the air anywhere on Earth? The results are always the same. Scientists explain how the ball is attracted to Earth in the theory of universal gravitation. In biology, two of the most highly regarded theories are the cell theory and the theory of evolution. Both theories are based on countless observations and investigations, have extensive supporting evidence, and enable biologists to make accurate predictions.

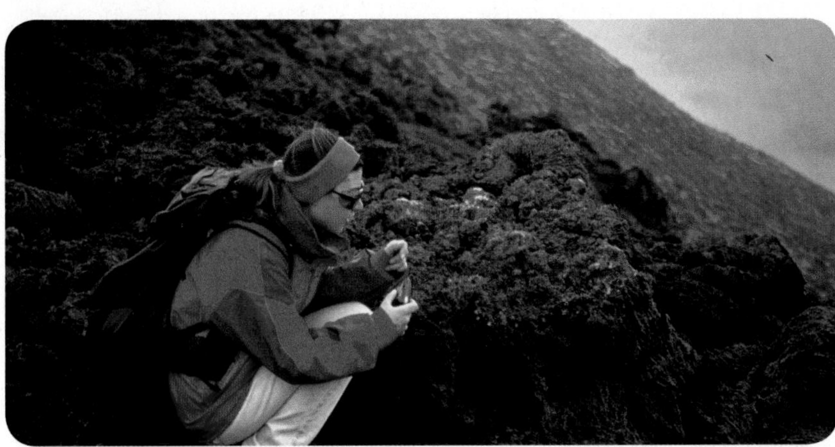

Figure 1.9 This volcanologist is collecting samples near molten lava flowing from Mount Etna. Lava temperatures can reach 750°C.

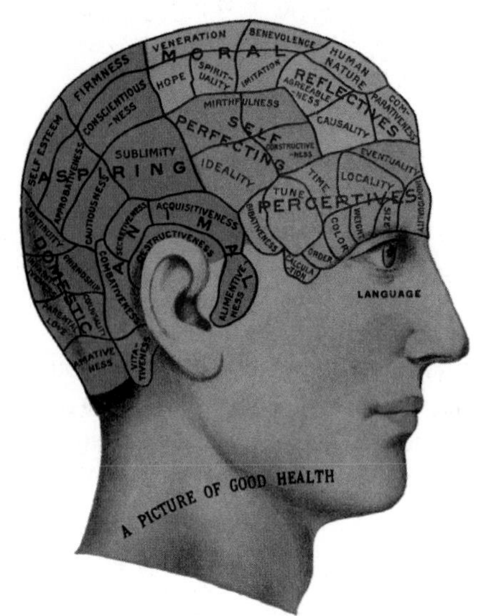

■ **Figure 1.10** Phrenology is based on observation—not scientific evidence.

Connection to History In the eighteenth and nineteenth centuries, many people practiced physiognomy (fih zee AHG nuh mee)—judging someone's character or personality based on physical features, especially facial features. Phrenology (frih NAH luh jee), the practice of reading the bumps on a person's head, illustrated in **Figure 1.10,** also is a type of physiognomy. Physiognomy often was used to determine whether individuals were appropriate for employment and other roles in society, or whether they had criminal tendencies. In fact, Charles Darwin almost did not get to take his famous voyage on the HMS *Beagle* because of the shape of his nose. Physiognomy was used and accepted even though there was no scientific evidence to support it.

Although physiognomy was based on observations and what was known at the time, it was not supported by scientific explanation. Physiognomy is considered a pseudoscience (soo doh SI uhnts). Pseudosciences are those areas of study that try to imitate science, often driven by cultural or commercial goals. Astrology, horoscopes, psychic reading, tarot card reading, face reading, and palmistry are pseudosciences. They do not provide science-based explanations about the natural world.

Reading Check **Describe** one way that science and pseudoscience differ.

Expands scientific knowledge How can you know what information is science-based? Most scientific fields are guided by research that results in a constant reevaluation of what is known. This reevaluation often leads to new knowledge that scientists then evaluate. The search for new knowledge is the driving force that moves science forward. Nearly every new finding, like the discoveries shown in **Figure 1.11,** causes scientists to ask more questions that require additional research.

With pseudoscience, little research is done. If research is done, then often it is simply to justify existing knowledge rather than to extend the knowledge base. Pseudoscientific ideas generally do not ask new questions or welcome more research.

■ **Figure 1.11**

Milestones in Biology

Major events and discoveries in the past century greatly contributed to our understanding of biology today.

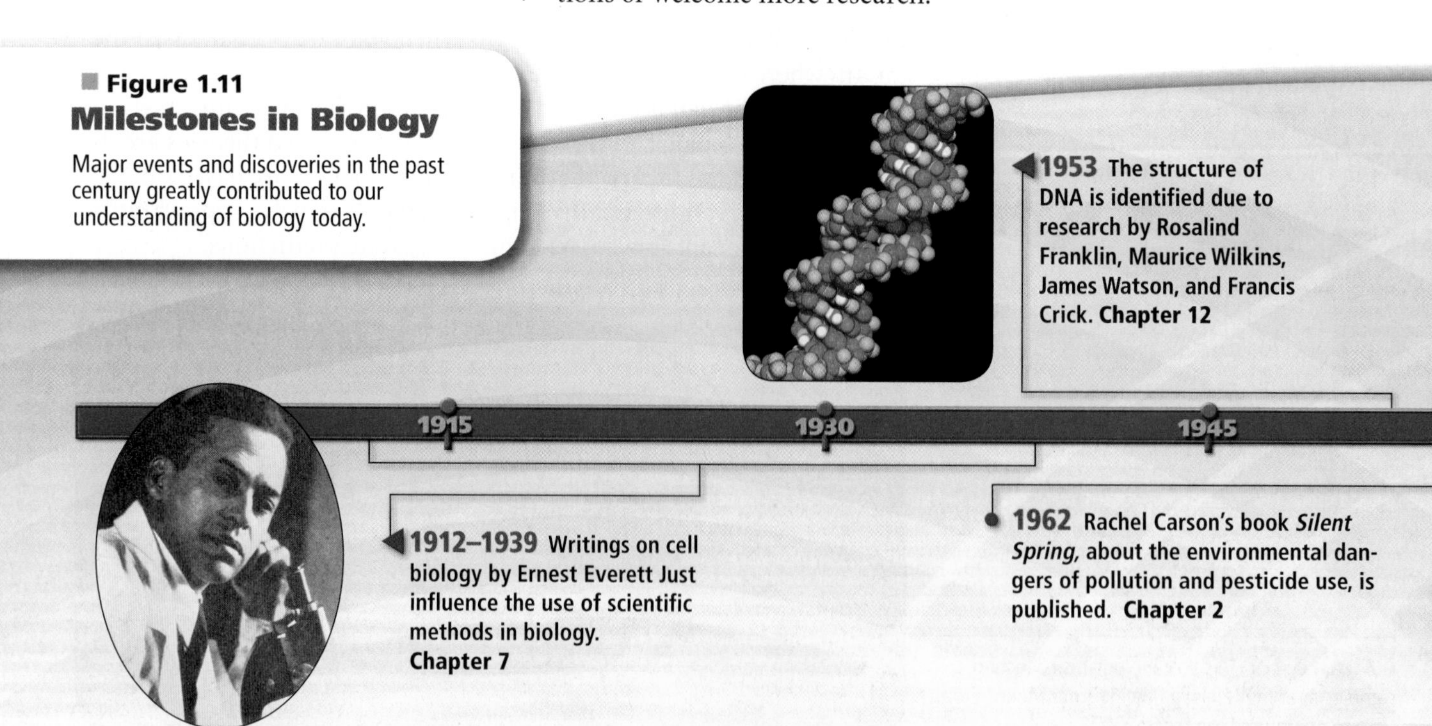

1953 The structure of DNA is identified due to research by Rosalind Franklin, Maurice Wilkins, James Watson, and Francis Crick. Chapter 12
1915
1930
1945
1912–1939 Writings on cell biology by Ernest Everett Just influence the use of scientific methods in biology. Chapter 7
1962 Rachel Carson's book Silent Spring, about the environmental dangers of pollution and pesticide use, is published. Chapter 2

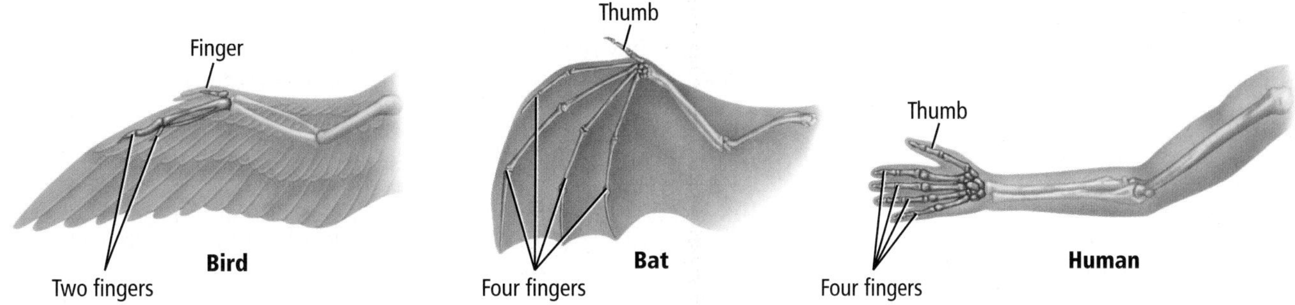

Figure 1.12 The structure of a bat's wing is more like that of a human arm than a bird's wing.

Challenges accepted theories Scientists welcome debate about one another's ideas. They regularly attend conferences and meetings where they discuss new developments and findings. Often, disagreements occur among scientists. Then additional investigations and/or experiments are done to substantiate claims.

Sciences advance by accommodating new information as it is discovered. For example, since the emergence of AIDS in the 1980s, our understanding of HIV, our ideas about how HIV is transmitted, the manner in which we treat AIDS, and the ways in which we educate people about the disease have changed dramatically due to new information from many scientific studies.

Questions results Observations or data that are not consistent with current scientific understanding are of interest to scientists. These inconsistencies often lead to further investigations. For example, early biologists grouped bats with birds because both had wings. Further study showed that bat wings are more similar to mammalian limbs than they are to bird wings, as shown in **Figure 1.12.** This led to an examination of the anatomy, genes, and proteins of rats and bats. The relationship was confirmed, and scientists established that bats were more closely related to mammals than birds. With pseudoscience, observations or data that are not consistent with beliefs are discarded or ignored.

CAREERS IN BIOLOGY

Science Writer Communicating scientific information to the public is one of the goals of a science writer. He or she might write news stories, manuals, or press releases, or edit and summarize the written materials of scientists. For more information on biology careers, visit biologygmh.com.

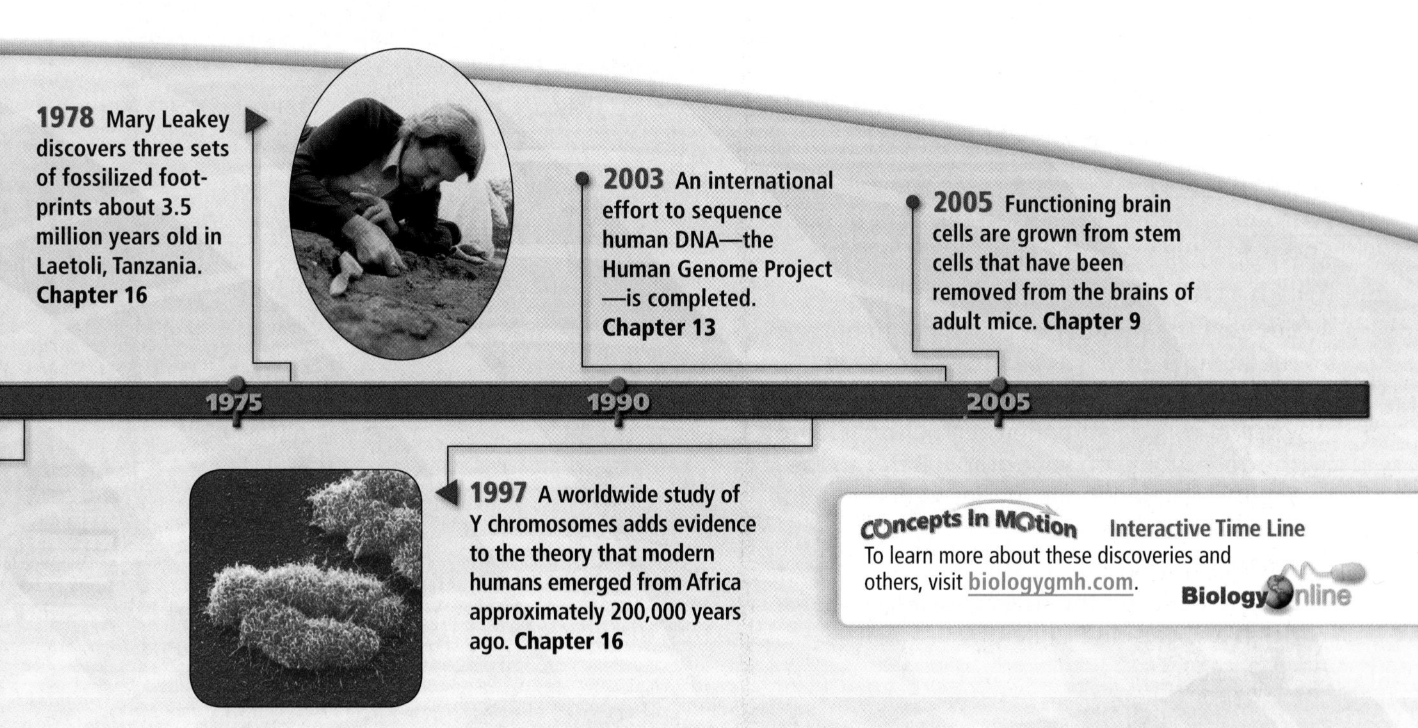

1978 Mary Leakey discovers three sets of fossilized footprints about 3.5 million years old in Laetoli, Tanzania. Chapter 16
2003 An international effort to sequence human DNA—the Human Genome Project —is completed. Chapter 13
2005 Functioning brain cells are grown from stem cells that have been removed from the brains of adult mice. Chapter 9
1975
1990
2005
1997 A worldwide study of Y chromosomes adds evidence to the theory that modern humans emerged from Africa approximately 200,000 years ago. Chapter 16

Concepts In Motion Interactive Time Line
To learn more about these discoveries and others, visit biologygmh.com.
Biology Online

VOCABULARY

ACADEMIC VOCABULARY

Unbiased:
To be objective, impartial, or fair.
The judges were unbiased in choosing the winner.

Tests claims Biologists use standard experimental procedures in their research. They make claims based on a large amount of data and observations obtained from unbiased investigations and carefully controlled experimentation. Conclusions are reached from the evidence. However, pseudoscientists often make claims that cannot be tested. These claims are often mixtures of fact and opinion.

Undergoes peer review Before it is made public, science-based information is reviewed by scientists' peers—scientists who are working in the same field of study. **Peer review** is a process by which the procedures and results of an experiment are evaluated by other scientists who are in the same field or who are conducting similar research.

Reading Check **Infer** why scientists utilize peer reviews.

Uses metric system To make communication easier, most scientists use the metric system when collecting data and performing experiments. The **metric system** uses units with divisions that are powers of ten. The General Conference of Weights and Measures established the unit standards of the metric system in 1960. The system is called the International System of Units, commonly known as **SI.** In biology, the SI units you will use most often are meter (to measure length), gram (to measure mass), liter (to measure volume), and second (to measure time).

DATA ANALYSIS LAB 1.1

Based on Real Data*

Peer Review

Can temperature be predicted by counting cricket chirps? Many outdoors enthusiasts claim that air temperature (°F) can be estimated by adding the number 40 to the number of cricket chirps counted in 15 seconds. Is there scientific evidence to support this idea?

Data and Observations

A group of students collected the data at right. They concluded that the claim is correct.

Effect of Temperature on Chirping

Temperature (°F)	Cricket Chirps (per min)
68	121
75	140
80	160
81	166
84	181
88	189
91	200
94	227

Think Critically

1. **Convert** the number of chirps per minute to the number of chirps per 15 seconds.
2. **Plot** the number of chirps per 15-second interval versus Fahrenheit temperature. Draw the best-fit line on your graph. Refer to the Skillbuilder Handbook, pages 1115–1118, for help with graphs.
3. **Write** the equation for the best-fit line.
4. **Peer review** Do the results support the students' conclusion? Explain.

*Data obtained from: Horak, V. M. 2005. Biology as a source for algebra equations: insects. *Mathematics Teacher* 99(1): 55-59.

Science in Everyday Life

There is widespread fascination with science. Popular television programs about crime are based on **forensics,** which applies science to matters of legal interest. The media is filled with information on flu epidemics, the latest medical advances, discoveries of new species, and technologies that improve or extend human lives. Clearly, science is not limited to the laboratory. The results of research go far beyond reports in scientific journals and meetings.

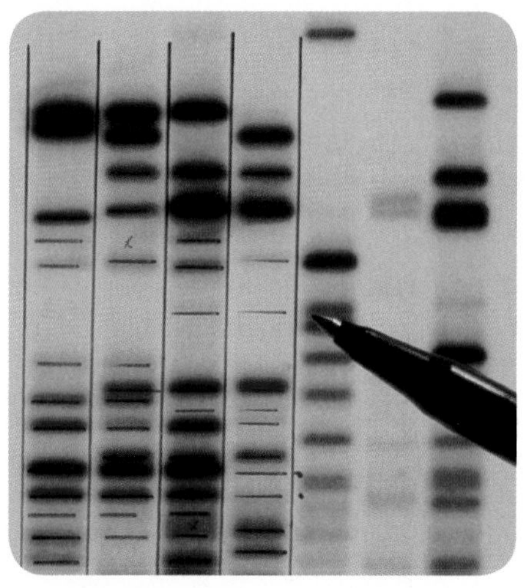

■ **Figure 1.13** DNA analysis might exclude an alleged thief because his or her DNA does not match the DNA from the crime scene.

Science literacy In order to evaluate the vast amount of information available in print, online, and on television, and to participate in the fast-paced world of the twenty-first century, each of us must be scientifically literate. A person who is scientifically literate combines a basic understanding of science and its processes with reasoning and thinking skills.

Many of the issues our world faces every day relate to biology. Drugs, alcohol, tobacco, AIDS, mental illness, cancer, heart disease, and eating disorders provide subjects for biological research worldwide. Environmental issues such as global warming, pollution, deforestation, the use of fossil fuels, nuclear power, genetically modified foods, and conserving biodiversity are issues that you and future generations will face. Also, genetic engineering, cloning—producing genetically identical individuals, genetic screening—searching for genetic disorders in people, euthanasia (yoo thuh NAY zhuh)—permitting a death for reasons of mercy, and cryonics (kri AH niks)—freezing a dead person or animal with the hope of reviving it in the future—all involve **ethics,** which is a set of moral principles or values. Ethical issues must be addressed by society based on the values it holds important.

Scientists provide information about the continued expansion of science and technology. As a scientifically literate individual, you will be an educated consumer who can participate in discussions about important issues and support policies that reflect your views. You might also serve on a jury where DNA evidence, like that shown in **Figure 1.13,** is presented. You will need to understand the evidence, comprehend its implications, and decide the outcome of the trial.

Section 1.2 Assessment

Section Summary

- Science is the study of nature and is rooted in observation and experimentation.
- Pseudoscience is not based on standard scientific research; it does not deal with testable questions, welcome critical review, or change its ideas when new discoveries are made.
- Science and ethics affect issues in health, medicine, the environment, and technology.

Understand Main Ideas

1. MAIN Idea **Describe** the characteristics of science.
2. **Define** *scientific theory.*
3. **Defend** the use of the metric system to a scientist who does not want to use it.
4. **Compare and contrast** science with pseudoscience.

Think Critically

WRITING in Biology

5. Predict what might happen to a population of people who do not understand the nature of science. Use examples of key issues facing our society.

MATH in Biology

6. One kilogram equals 1000 grams. One milligram equals 0.001 grams. How many milligrams are in one kilogram?

Reading Preview

Objectives

- **Describe** the difference between an observation and an inference.
- **Differentiate** among control, independent variable, and dependent variable.
- **Identify** the scientific methods a biologist uses for research.

Review Vocabulary

theory: an explanation of a natural phenomenon supported by many observations and experiments over time

New Vocabulary

observation
inference
scientific method
hypothesis
serendipity
experiment
control group
experimental group
independent variable
dependent variable
constant
data
safety symbol

Methods of Science

MAIN Idea **Biologists use specific methods when conducting research.**

Real-World Reading Link What do you do to find answers to questions? Do you ask other people, read, investigate, or observe? Are your methods haphazard or methodical? Over time, scientists have established standard procedures to find answers to questions.

Ask a Question

Imagine that you saw an unfamiliar bird in your neighborhood. You might develop a plan to observe the bird for a period of time. Scientific inquiry begins with **observation,** a direct method of gathering information in an orderly way. Often, observation involves recording information. In the example of your newly discovered bird, you might take photographs or draw a picture of it. You might write detailed notes about its behavior, including when and what it ate.

Science inquiry involves asking questions and processing information from a variety of reliable sources. The process of combining what you know with what you have learned to draw logical conclusions is called inferring; the conclusions themselves are called **inferences**. For instance, if you saw a photo of a bird similar to the unfamiliar bird in your neighborhood, you might infer that your bird and the bird in the photo are related. **Figure 1.14** illustrates how a field guide might be helpful in making inferences.

Scientific methods Biologists work in different places to answer their questions. For example, some biologists work in laboratories, perhaps developing new medicines, while others work outdoors in natural settings. No matter where they work, biologists all use similar methods to gather information and to answer questions. These methods sometimes are referred to as **scientific methods,** illustrated in **Figure 1.15.** Even though scientists do not use scientific methods in the same way each time they conduct an experiment, they observe and infer throughout the entire process.

Figure 1.14 Scientists might use a field guide to help them identify or draw conclusions about things they observe in nature, such as this peregrine falcon.

Visualizing Scientific Methods

Figure 1.15
The way that scientists answer questions is through an organized series of events called scientific methods. There are no wrong answers to questions, only answers that provide scientists with more information about those questions. Questions and collected information help scientists form hypotheses. As experiments are conducted, hypotheses might or might not be supported.

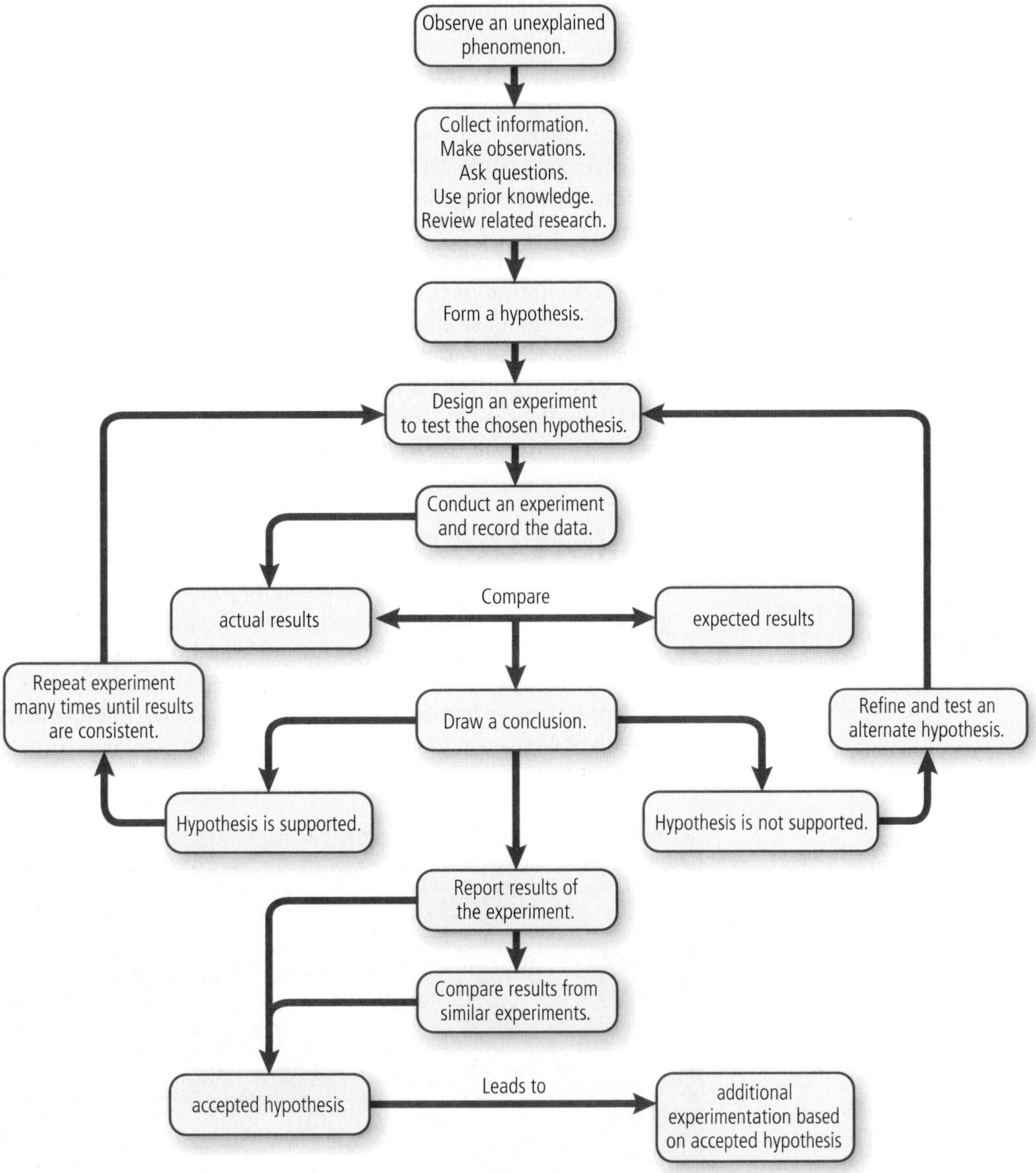

Observe an unexplained phenomenon.

Collect information.
Make observations.
Ask questions.
Use prior knowledge.
Review related research.

Form a hypothesis.

Design an experiment to test the chosen hypothesis.

Conduct an experiment and record the data.

actual results

Compare

expected results

Repeat experiment many times until results are consistent.

Draw a conclusion.

Refine and test an alternate hypothesis.

Hypothesis is supported.

Hypothesis is not supported.

Report results of the experiment.

Compare results from similar experiments.

accepted hypothesis

Leads to

additional experimentation based on accepted hypothesis

Concepts In Motion **Interactive Figure** To see an animation of scientific methods, visit biologygmh.com.

Biology Online

Form a Hypothesis

Imagination, curiosity, creativity, and logic are key elements of the way biologists approach their research. In 1969, the U.S. Air Force asked Dr. Ron Wiley to investigate how to enhance a pilot's ability to endure the effects of an increase in gravity (*g*-force) while traveling at high speed in an F-16 aircraft. It was known that isometrics, which is a form of exercise in which muscles are held in a contracted position, raised blood pressure. Wiley formed the hypothesis that the use of isometric exercise to raise blood pressure during maneuvers might increase tolerance to *g*-force and prevent blackouts. A **hypothesis** (hi PAH thuh sus) is a testable explanation of a situation.

Before Wiley formed his hypothesis, he made inferences based on his experience as a physiologist, what he read, discussions with Air Force personnel, and previous investigations. He did find that increasing a pilot's blood pressure could help the pilot withstand *g*-forces. But he also made an unexpected discovery.

During his study, Dr. Wiley discovered that isometric exercise decreased the resting blood pressure of the pilots. As a result, weight lifting and muscle-strengthening exercises are recommended today to help people lower blood pressure. **Serendipity** is the occurrence of accidental or unexpected but fortunate results. There are other examples of serendipity throughout science. For example, the discovery of penicillin was partially due to serendipity.

When a hypothesis is supported by data from additional investigations, usually it is considered valid and is accepted by the scientific community. If not, the hypothesis is revised, and additional investigations are conducted.

Study Tip

Clarification Choose a concept from the text and write its definition in the middle of a piece of paper. Circle the most important word. Around the word, write ideas related to it or examples that support it.

Collect the Data

Imagine that while in Alaska on vacation, you noticed various kinds of gulls. You saw them nesting high in the cliffs, and you wondered how they maintain their energy levels during their breeding season. A group of biologists wondered the same thing and did a controlled experiment using gulls known as black-legged kittiwakes shown in **Figure 1.16.** When a biologist conducts an **experiment,** he or she investigates a phenomenon in a controlled setting to test a hypothesis.

LAUNCH Lab

Review Based on what you've read about observing and inferring, how would you now answer the analysis questions?

■ **Figure 1.16** This colony of black-legged kittiwakes along the Alaskan coast includes nesting pairs.

Controlled experiments The biologists inferred that the kittiwakes would have more energy if they were given extra feedings while nesting. The biologists' hypothesis was that the kittiwakes would use the extra energy to lay more eggs and raise more chicks.

Biologists found nesting pairs of kittiwakes in Alaska that were similar in mass, age, size, and all other features. They set up a control group and an experimental group. A **control group** in an experiment is a group used for comparison. In this experiment, the kittiwakes not given the supplemental feedings made up the control group. The **experimental group** is the group exposed to the factor being tested. The group of kittiwakes getting the supplemental feedings made up the experimental group.

Experimental design When scientists design a controlled experiment, only one factor can change at a time. It is called the **independent variable** because it is the tested factor and it might affect the outcome of the experiment. In the kittiwakes experiment, the supplemental feeding was the independent variable. During an experiment, scientists measure a second factor. This factor is the **dependent variable.** It results from or depends on changes to the independent variable. The change in the kittiwakes' energy levels, as measured in reproductive output, was the dependent variable. A **constant** is a factor that remains fixed during an experiment while the independent and dependent variables change.

Data gathering As scientists test their hypotheses, they gather **data**—information gained from observations. The data can be quantitative or qualitative.

Data collected as numbers are called quantitative data. Numerical data can be measurements of time, temperature, length, mass, area, volume, density, or other factors. For example, when the biologists worked with the kittiwakes, they collected numerical data about the birds' energy levels.

Qualitative data are descriptions of what our senses detect. Often, qualitative data are interpreted differently because everyone does not sense things in the same way. However, many times it is the only collectible data.

Investigations Biologists conduct other kinds of scientific inquiry. They can engage in studies during which they investigate the behavior of organisms. Other biologists spend their careers discovering and identifying new species. Some biologists use computers to model the natural behavior of organisms and systems. In investigations such as these, the procedure involves observation and collection of data rather than controlled manipulation of variables.

MiniLab 1.2

Manipulate Variables

How does a biologist establish experimental conditions? In a controlled experiment, a biologist develops an experimental procedure designed to investigate a question or problem. By manipulating variables and observing results, a biologist learns about relationships among factors in the experiment.

Procedure

1. Read and complete the lab safety form.
2. Create a data table with the columns labeled *Control, Independent Variable, Constants, Hypothesis,* and *Dependent Variable.*
3. Obtain a **printed maze.** Seated at your desk, have a classmate time how long it takes you to complete the maze. Record this time on the chart. This is the control in the experiment.
4. Choose a way to alter experimental conditions while completing the same maze. Record this as the independent variable.
5. In the column labeled *Constants,* list factors that will stay the same each time the experiment is performed.
6. Form a hypothesis about how the independent variable will affect the time it takes to complete the maze.
7. After your teacher approves your plan, carry out the experiment. Record the time required to complete the maze as the dependent variable.
8. Repeat Steps 3–7 as time allows.
9. Graph the data. Use the graph to analyze the relationship between the independent and dependent variables.

Analysis

1. **Explain** the importance of the control in this experiment.
2. **Error Analysis** By completing the maze more than once, you introduced another variable, which likely affected the time required to complete the maze. Would eliminating this variable solve the problem? Explain.

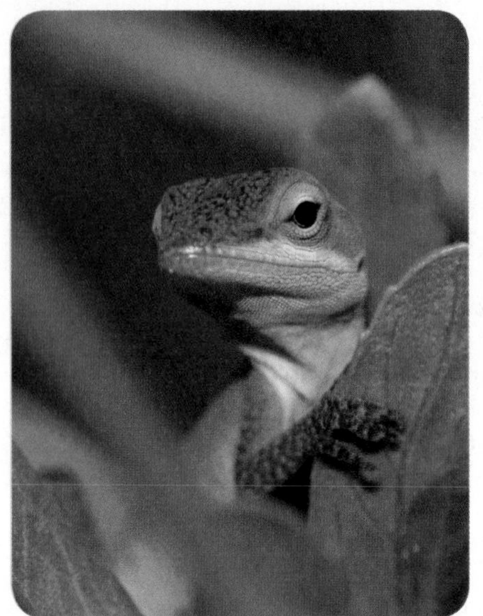

Anole

Change in Mass of Anole	
Date	Mass (g)
April 11	2.4
April 14	2.5
April 17	2.5
April 20	2.6
April 23	2.6
April 26	2.7
April 29	2.7

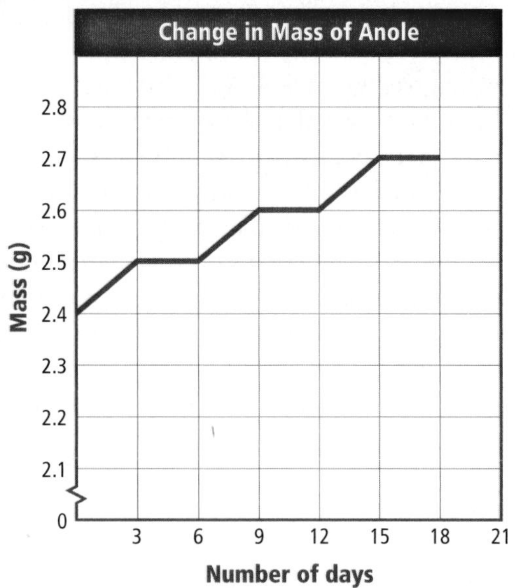

■ **Figure 1.17** After plotting the data points from the table on graph paper, draw a line that fits the pattern of the data rather than connects the dots.
Extrapolate ***What do you think the mass of the anole will be at 21 days?***

VOCABULARY
SCIENCE USAGE V. COMMON USAGE
Conclusion
Science usage: judgment, decision, or opinion formed after an investigation.
The researcher formed the conclusion that the hypothesis was not supported.

Common usage: the end or last part.
The audience left at the conclusion of the movie.

Analyze the Data

After analyzing the data from an investigation, a biologist usually asks, "Has my hypothesis been supported?" He or she then might ask, "Are more data needed?" or "Are different procedures needed?" Often, the investigation must be repeated many times to obtain consistent results.

As biologists look for explanations, patterns generally are noted that help to explain the data. A simple way to display the data is in a table or on a graph, such as the ones in **Figure 1.17,** which describe the change in mass over time of a lizard called an anole. The graph of the data makes the pattern easier to grasp. In this case, there is a regular pattern. Notice that the mass increases over a three-day period and then levels off for three days before increasing again. For more review about making graphs, refer to the Skillbuilder Handbook, pp. 1115–1118.

Because biologists often work in teams, meetings are held to discuss ongoing investigations, to analyze the data, and to interpret the results. The teams continue to examine their research plan to be certain they avoid bias, repeat their trials, and collect a large enough sample size. Analysis of the data might lead to a conclusion that the hypothesis has been supported. It also could lead to additional hypotheses, to further experimentation, or to general explanations of nature. Even when a hypothesis has not been supported, it is valuable.

Report Conclusions

Biologists report their findings and conclusions in scientific journals. Before a scientist can publish in a journal, the work is reviewed by peers. The reviewers examine the paper for originality, competence of the scientific method used, and accuracy. They might find fault with the reasoning or procedure, or suggest other explanations or conclusions. If the reviewers agree on the merit of the paper, then the paper is published for review by the public and use by other scientists.

 **Reading Check** **Infer** How does the hypothesis guide data collection and interpretation?

Ask a Question
• Have I read the entire LaunchLab, MiniLab, or BioLab?
• What question will I try to answer?
• Is background information provided for this topic? What do I know about this topic?
Form a Hypothesis
• What is my hypothesis?
• Does it give me a testable prediction?
Collect the Data
• How will I record my data?
• What is the experimental group? Control group?
• What is the independent variable? Dependent variable?
• What will I hold constant?
• What materials and equipment do I need?
• Are there additional instructions for operating the equipment?
• What safety precautions should I follow?
• Does my teacher approve this procedure?
• How will I clean up and dispose of materials?
Analyze the Data
• How will I display my data?
• How will I analyze and summarize my results?
• Are there any sources of error in my procedures?
Report a Conclusion
• Was my hypothesis supported? Why or why not?
• What is my conclusion?

■ **Figure 1.18** To ask meaningful questions, form hypotheses, and conduct careful experiments, develop research plans based on scientific methods. Use your lab report to list your procedure, record your data, and report your conclusions.

Student Scientific Inquiry

You might be given many opportunities during your study of biology to do your own investigations and experiments. You might receive a lab assignment that spells out a series of steps to follow or you might design your own procedure. Whether you are planning a lab report or an entire procedure and its lab report, be sure to ask yourself questions like those in **Figure 1.18.** For additional help with setting up experiments and using equipment, go to Investigation and Experimentation on pp. xxx–xliii of this textbook.

Lab safety During biology labs, you will be alerted of possible safety hazards by warning statements and safety symbols. A **safety symbol** is a logo designed to alert you about a specific danger. Always refer to the safety symbols chart at the front of this book before beginning any field investigation or lab activity. Carefully read the meaning of each lab's safety symbols. Also, learn the location in the classroom of all safety equipment and how and when to use it. You are responsible for being safe at all times to protect yourself and your classmates.

Section 1.3 Assessment

Section Summary

- Observations are an orderly way of gathering information.
- Inferences are based on prior experiences.
- Controlled experiments involve a control group and an experimental group.
- An independent variable is the condition being tested, and the dependent variable results from the change to the independent variable.

Understand Main Ideas

1. MAIN Idea **Describe** how a biologist's research can proceed from an idea to a published article.
2. **State** why an observation cannot be an inference.
3. **Indicate** the differences in the ways that data is collected in biological research.
4. **Differentiate** between independent variables and dependent variables.

Think Critically

5. **Design** a controlled experiment to determine whether earthworms are more attracted to perfume or to vinegar.
6. **Form a hypothesis** about one of the characteristics of life you studied in Section 1.1 and design a research project to test it. What organism would you study? What questions would you ask?

BioDiscoveries

Cancer Research

A whole new world From the first time that she peered into a microscope and saw a tiny, fascinating new world, Jewell Plummer Cobb knew that a career in biology was for her. It is no surprise that biology would fascinate her. Cobb's father was a physician, her mother was a teacher, and science often was the topic of dinner conversation. Cobb became a groundbreaking scientist, as well as a college dean, recipient of almost two dozen honorary degrees, and a champion of minority and women's rights.

Jewell Plummer Cobb has devoted her life to cancer research.

Individualized chemotherapy In 1950, Dr. Cobb joined the Harlem Hospital Cancer Research Foundation, where she pioneered chemotherapy research with Jane Cooke Wright. The two scientists determined that there should be a way to tailor therapeutic drug dosages for individuals. Cobb designed new ways to grow tissue samples so that their responses to different drug doses could be observed under a microscope and recorded using time-lapse photography. Cobb and Wright's methods of documenting cellular responses to potentially toxic drugs paved the way for further research. Their work provided scientists with another tool that could be used in the development of new, more effective chemotherapy drugs.

Skin cancer Although her research in New York was groundbreaking, Dr. Cobb did not find her niche in cancer research until 1952, when she received a grant from the National Cancer Institute. With this grant, she began her research on cancerous pigment cells and the possible role of melanin in protecting the skin from the Sun's ultraviolet rays—a cancer-causing agent. Skin cancer, called melanoma, occurs more in Caucasians than in African Americans. Because African Americans have more melanin than Caucasians, Cobb wanted to know if the melanin had protective qualities.

To determine how melanin affected the outcome of radiation therapy in cancer treatments, she designed an experiment using black and white mice that were bred to develop melanoma tumors. Dr. Cobb took samples from tumors and separated the tissue with high melanin from the tissue with low melanin. She exposed both tissues to different doses of X rays to determine if melanin protected cells against the effects of X rays. Immediately after exposure, she implanted the tissues into cancer-free mice or grew them in test tubes. The black tissues survived greater X-ray doses than the white tissues. After examination with a microscope, she concluded that melanin protected cells from X-ray damage.

Research ways to diagnose, treat, and prevent melanoma continue. For example, immunotherapy uses the body's own defenses to destroy cancer cells. A melanoma might be surgically removed from the skin or treated with chemotherapy or radiation. Immunotherapy often is combined with other forms of therapy to make them more effective or lessen side effects.

WRITING in Biology

Magazine Article For more information about the accomplishments of various scientists, visit biologygmh.com. Write an article about one individual. Include his or her contributions to science.

BIOLAB

Design Your Own

HOW CAN YOU KEEP CUT FLOWERS FRESH?

Background: **When first cut from the garden, a bouquet of flowers looks healthy and has a pleasant aroma. Over time, the flowers droop and lose their petals. Leaves and stems below the water line begin to decay.**

Question: *What steps can I take to extend the freshness of cut flowers?*

Possible Materials

Choose materials that would be appropriate for this lab.

fresh cut flowers
water
vases
scissors

Safety Precautions

Plan and Perform the Experiment

1. Read and complete the lab safety form.
2. Research strategies for extending the life of cut flowers. During your research, look for possible reasons why a specific strategy might be effective.
3. Form a hypothesis based on your research. It must be possible to test the hypothesis by gathering and analyzing specific data.
4. Design an experiment to test the hypothesis. Remember, the experiment must include an independent and dependent variable. Identify a control sample. List all factors that will be held constant.
5. Design and construct a data table.
6. Make sure your teacher approves your plan before you proceed.
7. Implement the experimental design. Organize the data you collect using a graph or chart.
8. **Cleanup and Disposal** Properly dispose of plant material. Wash hands thoroughly after handling plant material. Clean and return all lab equipment to the designated locations.

Analyze and Conclude

1. **Describe** the strategy tested by your hypothesis. Why did you choose this strategy to examine?
2. **Explain** how you established the control sample.
3. **Interpret Data** What trends or patterns do the data show?
4. **Analyze** What is the relationship between your independent and dependent variables?
5. **Draw Conclusions** Based on your data, describe one way to extend the freshness of cut flowers.
6. **Error Analysis** Critique your experimental design. Is it possible that any other variables were introduced? Explain. How could these variables be controlled?

WRITING in Biology

Brochure Compare the strategy for extending the freshness of cut flowers your group examined with strategies tested by other groups. Based on class results, create a brochure with the title "Make Cut Flowers Stay Beautiful Longer." Include tips for extending the life of cut flowers. Share the brochure with community members who might benefit from this information. To learn more about extending the freshness of cut flowers, visit BioLabs at biologygmh.com.

Study Guide

Download quizzes, key terms, and flash cards from biologygmh.com.

FOLDABLES **Brainstorm** other roles that biologists fulfill in addition to those discussed in Section 1.1. List these roles on the back of your Foldable and give examples.

Vocabulary	Key Concepts
Section 1.1 Introduction to Biology	
• adaptation (p. 10) • biology (p. 4) • development (p. 9) • growth (p. 9) • homeostasis (p. 10) • organism (p. 6) • organization (p. 8) • reproduction (p. 9) • response (p. 9) • species (p. 9) • stimulus (p. 9)	MAIN Idea All living things share the characteristics of life. • Biology is the study of life. • Biologists study the structure and function of living things, their history, their interactions with the environment, and many other aspects of life. • All organisms have one or more cells, display organization, grow and develop, reproduce, respond to stimuli, use energy, maintain homeostasis, and have adaptations that evolve over time.
Section 1.2 The Nature of Science	
• ethics (p. 15) • forensics (p. 15) • metric system (p. 14) • peer review (p. 14) • science (p. 11) • SI (p. 14) • theory (p. 11)	MAIN Idea Science is a process based on inquiry that seeks to develop explanations. • Science is the study of nature and is rooted in observation and experimentation. • Pseudoscience is not based on standard scientific research; it does not deal with testable questions, welcome critical review, or change its ideas when new discoveries are made. • Scientists worldwide use SI. • Science and ethics affect issues in health, medicine, the environment, and technology.
Section 1.3 Methods of Science	
• control group (p. 19) • constant (p. 19) • data (p. 19) • dependent variable (p. 19) • experiment (p. 18) • experimental group (p. 19) • hypothesis (p. 18) • independent variable (p. 19) • inference (p. 16) • observation (p. 16) • safety symbol (p. 21) • scientific method (p. 16) • serendipity (p. 18)	MAIN Idea Biologists use specific methods when conducting research. • Observations are an orderly way of gathering information. • Inferences are based on prior experiences. • Controlled experiments involve a control group and an experimental group. • An independent variable is the condition being tested, and the dependent variable results from the change to the independent variable.

CHAPTER 1

Assessment

Section 1.1

Vocabulary Review

Replace the underlined phrase with the correct vocabulary term from the Study Guide page.

1. The production of offspring is a characteristic of life that enables the continuation of a species.
2. The internal control of mechanisms allows for an organism's systems to remain in balance.
3. The science of life involves learning about the natural world.

Understand Key Concepts

Use the graph below to answer question 4.

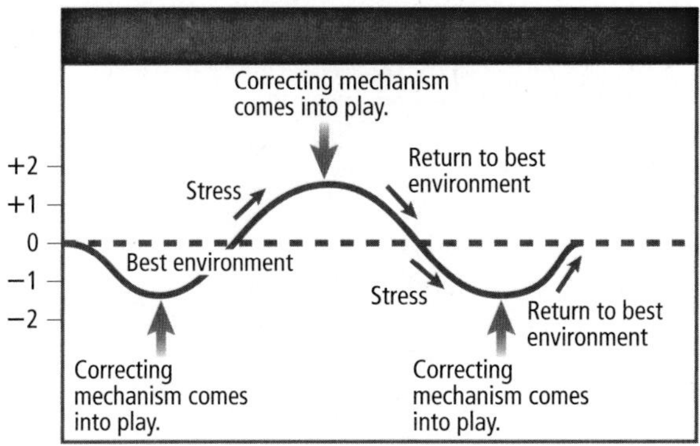

4. Which characteristic of life should be the title of this graph?
   A. Cellular Basis
   B. Growth
   C. Homeostasis
   D. Reproduction
5. Which best describes adaptation?
   A. reproducing as a species
   B. a short-term change in behavior in response to a stimuli
   C. inherited changes in response to environmental factors
   D. change in size as an organism ages

Constructed Reponse

6. **Open Ended** What is the role of energy in living organisms? Is it a more or less important role than other characteristics of life? Defend your response.

Think Critically

7. **Evaluate** how the contributions made by Goodall, Chory, and Drew reinforce our understanding of the characteristics of life.
8. **Compare and contrast** a response and an adaptation. Use examples from your everyday world in your answer.

Section 1.2

Vocabulary Review

Replace the underlined phrase with the correct vocabulary term from the Study Guide page.

9. the measurements based on powers of ten used by scientists when conducting research
10. a well-tested explanation that brings together many observations in science such as evolution, plate tectonics, biogenesis

Understand Key Concepts

Use the photo below to answer question 11.

11. Which SI base unit would be used to describe the physical characteristics of dolphins?
    A. second
    B. kilogram
    C. inches
    D. gallon
12. Which is true about scientific inquiry?
    A. It poses questions about astrology.
    B. It can be done only by one person.
    C. It is resistant to change and not open to criticism.
    D. It is testable.

Constructed Response

13. **Short Answer** Differentiate between pseudoscience and science.

Think Critically

14. **Evaluate** how technology impacts society in a positive and negative way at the same time.

Section 1.3

Vocabulary Review

Explain the differences between the terms in the following sets.

15. observation, data
16. control group, experimental group
17. independent variable, dependent variable

Understand Key Concepts

18. Which describes this statement, "The frog is 4 cm long"?
    A. quantitative data
    B. inference
    C. control group
    D. qualitative data

19. Which is a testable explanation?
    A. dependent variable
    B. independent variable
    C. hypothesis
    D. observation

Constructed Response

Use the table below to answer question 20.

Mean Body Mass and Field Metabolic Rate (FMR) of Black-Legged Kittiwakes

	Number	Mean body mass (g)	FMR
Fed females	14	426.8	2.04
Control females	14	351.1	3.08
Fed males	16	475.4	2.31
Control males	18	397.6	2.85

20. **Short Answer** Examine the data shown above. Describe the effects of feedings on the energy expenditure, FMR, of male and female kittiwakes.

Think Critically

21. **Design** a survey to investigate students' opinions about current movies. Use 10 questions and survey 50 students. Graph the data. Report the findings to the class.

Additional Assessment

22. WRITING in Biology Prepare a letter to the editor of your school newspaper that encourages citizens to be scientifically literate about topics such as cancer, the environment, ethical issues, AIDS, smoking, lung diseases, cloning, genetic diseases, and eating disorders.

DBQ Document Based Questions

Use the data below to answer questions 23 and 24.

Data obtained from: U.S. Geological Survey. *Seabirds, forage fish, and marine ecosystems.* http://www.absc.usgs.gov/research/seabird_foragefish/foragefish/index.html

Relative Fish Biomass of Three Seabird Colonies in Lower Cook Inlet

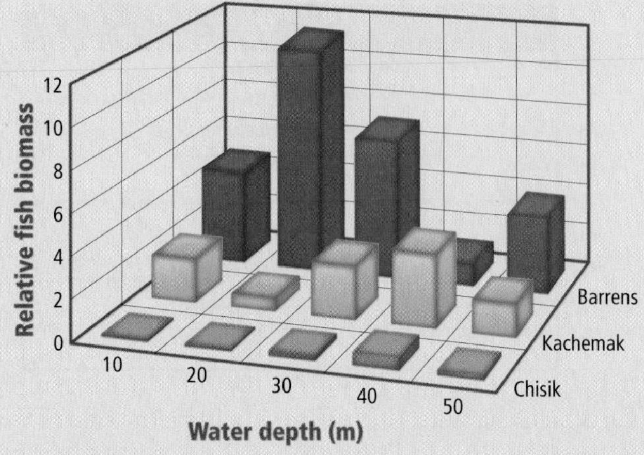

23. Identify the water depth with the highest relative fish biomass.

24. Determine which seabird colony has access to the highest fish biomass at a depth of 40 m.

CUMULATIVE REVIEW

In Chapters 2–37, Cumulative Review questions will help you review and check your understanding of concepts discussed in previous chapters.

Standardized Test Practice

Multiple Choice

1. Many scientific discoveries begin with direct observations. Which could be a direct observation?
   A. Ants communicate by airborne chemicals.
   B. Birds navigate by using magnetic fields.
   C. Butterflies eat nectar from flowers.
   D. Fish feel vibrations through special sensors.

Use this experimental description and data table to answer question 2.

A student reads that some seeds must be exposed to cold before they germinate. She wants to test seeds from one kind of plant to see if they germinate better after freezing. The student put the seeds in the freezer, took samples out at certain times, and tried to germinate them. Then she recorded her results in the table.

Germination Rate for Seeds Stored in a Freezer

Time in Freezer at –15°C	Germination Rate
30 days	48%
60 days	56%
90 days	66%
120 days	52%

2. According to the results of this experiment, how many days should seeds be stored in the freezer before planting for best germination?
   A. 30
   B. 60
   C. 90
   D. 120

Short Answer

3. Appraise one benefit to scientists of using SI units as standard units of measurement.

Extended Response

Use this drawing to answer question 4.

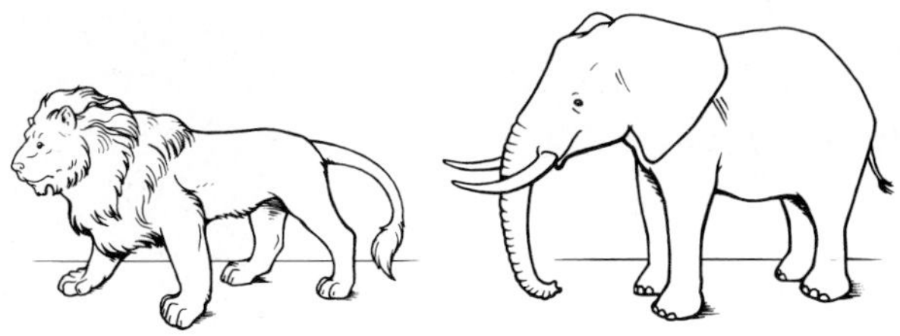

4. Look at the drawing and write five specific questions about the organisms shown that a biologist might try to investigate.

5. Compare and contrast a scientific hypothesis and a scientific theory.

Essay Question

A researcher experimented with adhesives and glues to find new and stronger adhesives. In 1968, he discovered an adhesive that was very weak rather than strong. The adhesive would stick to paper but it could be removed easily without leaving a trace of adhesive. Because he was trying to find stronger adhesives, the results of that experiment were considered a failure. Several years later, he had the idea of coating paper with the weak adhesive. This meant that notes could be stuck to paper and easily removed at a later time. Today, these removable notes are used by millions of people.

Using the information in the paragraph above, answer the following question in essay format.

6. The original adhesive experiment was considered a failure. Appraise the importance of evaluating the results of an experiment with an open mind.

NEED EXTRA HELP?						
If You Missed Question . . .	1	2	3	4	5	6
Review Section . . .	1.3	1.3	1.2	1.1	1.3	1.2

UNIT 1 Ecology

Chapter 2
Principles of Ecology

BIG Idea Energy is required to cycle materials through living and nonliving systems.

Chapter 3
Communities, Biomes, and Ecosystems

BIG Idea Limiting factors and ranges of tolerance are factors that determine where terrestrial biomes and aquatic ecosystems exist.

Chapter 4
Population Ecology

BIG Idea Population growth is a critical factor in a species' ability to maintain homeostasis within its environment.

Chapter 5
Biodiversity and Conservation

BIG Idea Community and ecosystem homeostasis depend on a complex set of interactions among biologically diverse individuals.

CAREERS IN BIOLOGY
Wildlife Biologist

As the oystercatcher researchers are doing in this photograph, **wildlife biologists** perform scientific research to study how species interact with each other and the environment. They protect and conserve wildlife species and also help maintain and increase wildlife populations.

WRITING in **Biology** Visit biologygmh.com to learn more about wildlife biology. Then write a description of the job responsibilities of wildlife biologists.

Biology Online

To read more about wildlife biologists in action, visit biologygmh.com.

CHAPTER 2

Principles of Ecology

BIG Idea **Energy is required to cycle materials through living and nonliving systems.**

Section 1
Organisms and Their Relationships

MAIN Idea Biotic and abiotic factors interact in complex ways in communities and ecosystems.

Section 2
Flow of Energy in an Ecosystem

MAIN Idea Autotrophs capture energy, making it available for all members of a food web.

Section 3
Cycling of Matter

MAIN Idea Essential nutrients are cycled through biogeochemical processes.

BioFacts

- The Pacific tree frog can change from light colored to dark colored quickly. This could be a response to changes in temperature and humidity.
- The spotted owl nests only in old growth forests and might be in danger of becoming extinct due to the loss of these forests.

LAUNCH Lab

Problems in *Drosophila* world?

As the photos on the left illustrate, what we understand to be the world is many smaller worlds combined to form one large world. Within the large world, there are populations of creatures interacting with each other and their environment. In this lab, you will observe an example of a small part of the world.

Procedure

1. Read and complete the lab safety form.
2. Prepare a data table to record your observations.
3. Your teacher has prepared a **container housing several fruit flies** *(Drosophila melanogaster)* with food for the flies in the bottom. Observe how many fruit flies are present.
4. Observe the fruit flies over a period of one week and record any changes.

Analysis

1. **Summarize** the results of your observations.
2. **Evaluate** whether or not this would be a reasonable way to study a real population.

Visit biologygmh.com to:

- study the entire chapter online
- explore the Interactive Time Line, Concepts in Motion, Microscopy Links, Virtual Labs, and links to virtual dissections
- access Web links for more information, projects, and activities
- review content online with the Interactive Tutor and take Self-Check Quizzes

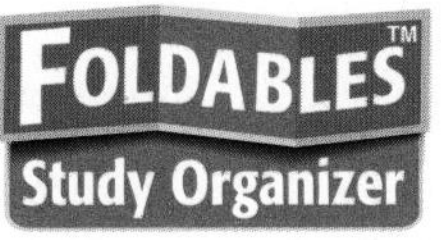

Natural Cycles Make this Foldable to help you compare and contrast the water cycle and the carbon cycle.

STEP 1 Fold a sheet of notebook paper in half lengthwise so that the side without holes is 2.5 cm shorter than the side with the holes. Then fold the paper into thirds as shown.

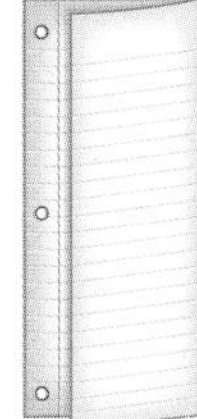

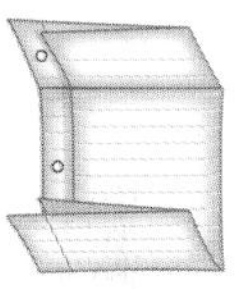

STEP 2 Unfold the paper and draw the Venn diagram. Then cut along the two fold lines of the top layer only. This makes three tabs.

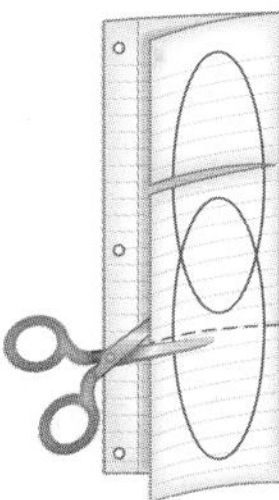

STEP 3 Label the tabs as illustrated.

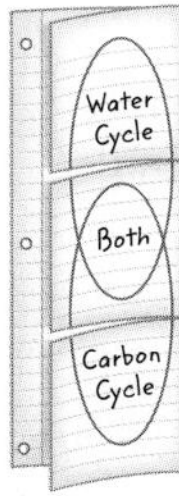

FOLDABLES **Use this Foldable with Section 2.3.** As you study the section, record what you learn about the two cycles under the appropriate tabs and determine what the cycles have in common.

Section 2.1

Reading Preview

Objectives

- **Explain** the difference between abiotic factors and biotic factors.
- **Describe** the levels of biological organization.
- **Differentiate** between an organism's habitat and its niche.

Review Vocabulary

species: group of organisms that can interbreed and produce fertile offspring in nature

New Vocabulary

ecology
biosphere
biotic factor
abiotic factor
population
biological community
ecosystem
biome
habitat
niche
predation
symbiosis
mutualism
commensalism
parasitism

Organisms and Their Relationships

MAIN Idea **Biotic and abiotic factors interact in complex ways in communities and ecosystems.**

Real-World Reading Link On whom do you depend for your basic needs such as food, shelter, and clothing? Humans are not the only organisms that depend on others for their needs. All living things are interdependent. Their relationships are important to their survival.

Ecology

Scientists can gain valuable insight about the interactions between organisms and their environments and between different species of organisms by observing them in their natural environments. Each organism, regardless of where it lives, depends on nonliving factors found in its environment and on other organisms living in the same environment for survival. For example, green plants provide a source of food for many organisms as well as a place to live. The animals that eat the plants provide a source of food for other animals. The interactions and interdependence of organisms with each other and their environments are not unique. The same type of dependency occurs whether the environment is a barren desert, a tropical rain forest, or a grassy meadow. **Ecology** is the scientific discipline in which the relationships among living organisms and the interaction the organisms have with their environments are studied.

■ **Figure 2.1**
Milestones in Ecology

Ecologists have worked to preserve and protect natural resources.

1872 Yellowstone becomes the first national park in the U.S.

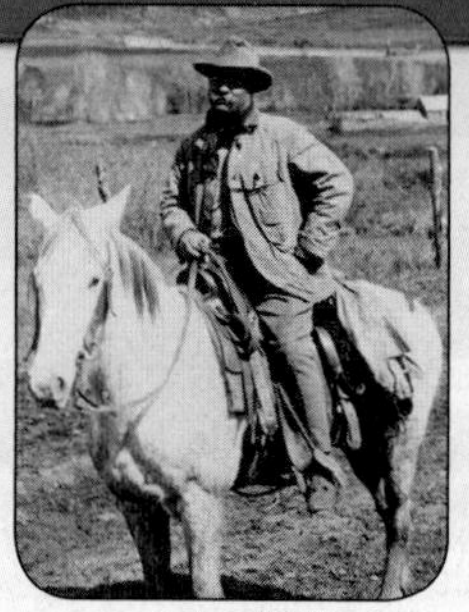

1905 Theodore Roosevelt urges the U.S. Congress to set aside over 70 million hectares of land to protect the natural resources found on them.

1900 — 1960 — 1970

1962 Rachel Carson publishes a best-selling book warning of the environmental danger of pollution and pesticides.

1967 The government of Rwanda and international conservation groups begin efforts to protect mountain gorillas, due in a large part to the work of Dian Fossey.

1971 Marjorie Carr stops the construction of the Cross Florida Barge Canal because of the environmental damage the project would cause.

■ **Figure 2.2** Ecologists work in the field and in laboratories. This ecologist is enduring harsh conditions to examine a seal.

The study of organisms and their environments is not new. The word *ecology* was first introduced in 1866 by Ernst Haeckel, a German biologist. Since that time, there have been many significant milestones in ecology, as shown in **Figure 2.1.**

Scientists who study ecology are called ecologists. Ecologists observe, experiment, and model using a variety of tools and methods. For example, ecologists, like the one shown in **Figure 2.2,** perform tests in organisms' environments. Results from these tests might give clues as to why organisms are able to survive in the water, why organisms become ill or die from drinking the water, or what organisms could live in or near the water. Ecologists also observe organisms to understand the interactions between them. Some observations and analyses must be made over long periods of time in a process called longitudinal analysis.

A model allows a scientist to represent or simulate a process or system. Studying organisms in the field can be difficult because there often are too many variables to study at one time. Models allow ecologists to control the number of variables present and to slowly introduce new variables in order to fully understand the effect of each variable.

Reading Check **Describe** a collection of organisms and their environment that an ecologist might study in your community.

VOCABULARY

WORD ORIGIN

Ecology
comes from the Greek words *oikos,* meaning *house,* and *ology,* meaning *to study.*

1990 The Indigenous Environmental Network (IEN), directed by Tom Goldtooth, is formed by Native Americans to protect their tribal lands and communities from environmental damage.

2004 Wangari Maathai wins a Nobel Prize. She began the Green Belt Movement in Africa, which hires women to plant trees to slow the process of deforestation and desertification.

1980 1990 2000

1987 The United States and other countries sign the Montreal Protocol, an agreement to phase out the use of chemical compounds that destroy atmospheric ozone.

1996 Completing a phase-out that was begun in 1973, the U.S. Environmental Protection Agency bans the sale of leaded gasoline for vehicle use.

Concepts In Motion Interactive Time Line
To learn more about these milestones and others, visit biologygmh.com. Biology Online

■ **Figure 2.3** This color-enhanced satellite photo of Earth taken from space shows a large portion of the biosphere.

The Biosphere

Because ecologists study organisms and their environments, their studies take place in the biosphere. The **biosphere** (BI uh sfihr) is the portion of Earth that supports life. The photo of Earth taken from space shown in **Figure 2.3** shows why the meaning of the term *biosphere* should be easy to remember. The term *bio* means "life," and a sphere is a geometric shape that looks like a ball. When you look at Earth from this vantage point, you can see how it is considered to be "a ball of life."

Although "ball of life" is the literal meaning of the word *biosphere,* this is somewhat misleading. The biosphere includes only the portion of Earth that includes life. The biosphere forms a thin layer around Earth. It extends several kilometers above the Earth's surface into the atmosphere and extends several kilometers below the ocean's surface to the deep-ocean vents. It includes landmasses, bodies of freshwater and saltwater, and all locations below Earth's surface that support life.

Figure 2.4 shows a satellite image of Earth's biosphere on the surface of Earth. The photo is color-coded to represent the distribution of chlorophyll. Chlorophyll is a green pigment found in green plants and algae that you will learn about in later chapters. Because most organisms depend on green plants or algae for survival, green plants are a good indicator of the distribution of living organisms in an area. In the oceans, red represents areas with the highest density of chlorophyll followed by yellow, then blue, and then pink, representing the lowest density. On land, dark green represents the area with highest chlorophyll density and pale yellow represents the area with the lowest chlorophyll density.

Reading Check **Describe** the general distribution of green plants across the United States using **Figure 2.4.**

The biosphere also includes areas such as the frozen polar regions, deserts, oceans, and rain forests. These diverse locations contain organisms that are able to survive in the unique conditions found in their particular environment. Ecologists study these organisms and the factors in their environment. These factors are divided into two large groups—the living factors and the nonliving factors.

■ **Figure 2.4** This color-coded satellite photo shows the relative distribution of life on Earth's biosphere based on the distribution of chlorophyll.

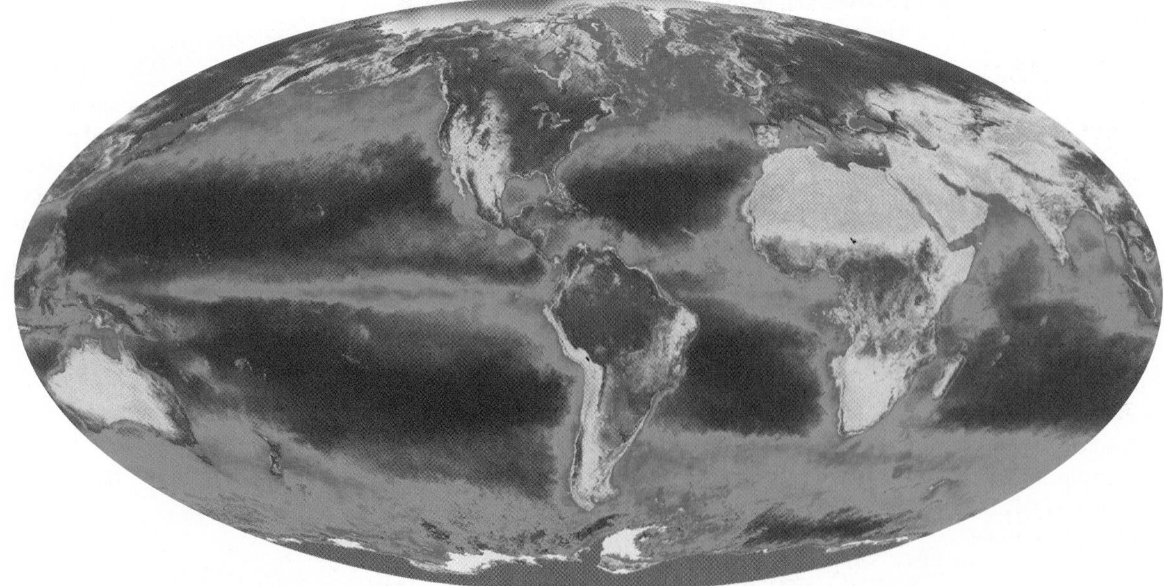

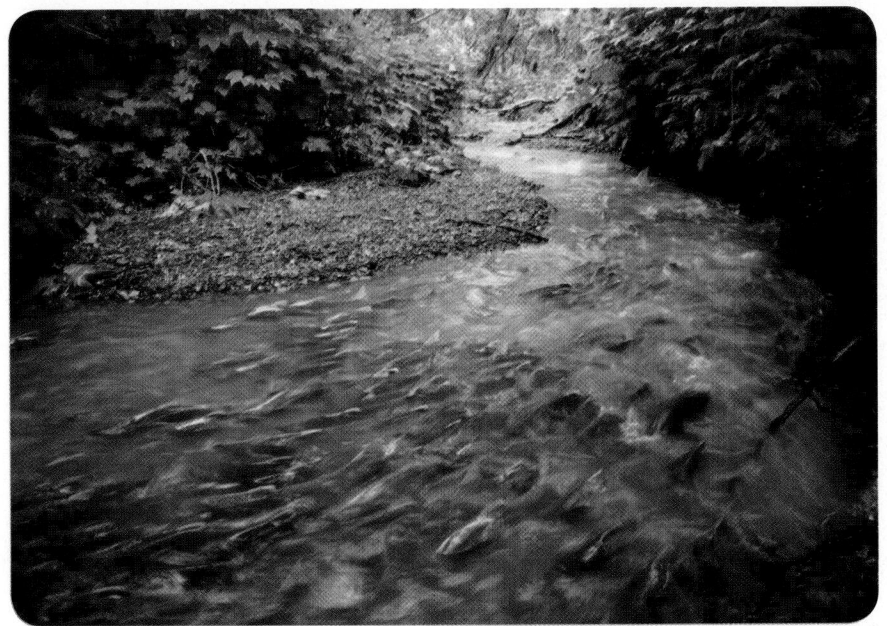

Figure 2.5 The salmon swimming upstream are biotic factors in the stream community. Other organisms in the water, such as frogs and algae, also are biotic factors.
Explain ***How are organisms dependent on other organisms?***

Biotic factors The living factors in an organism's environment are called the **biotic** (by AH tihk) **factors.** Consider the biotic factors in the habitat of salmon shown in **Figure 2.5.** These biotic factors include all of the organisms that live in the water, such as other fish, algae, frogs, and microscopic organisms. In addition, organisms that live on the land adjacent to the water might be biotic factors for the salmon. Migratory animals, such as birds that pass through the area, also are biotic factors. The interactions among organisms are necessary for the health of all species in the same geographic location. For example, the salmon need other members of their species to reproduce. Salmon also depend on other organisms for food and, in turn, are a food source for other organisms.

Abiotic factors The nonliving factors in an organism's environment are called **abiotic** (ay bi AH tihk) **factors.** The abiotic factors for different organisms vary across the biosphere, but organisms that live in the same geographic area might share the same abiotic factors. These factors might include temperature, air or water currents, sunlight, soil type, rainfall, or available nutrients. Organisms depend on abiotic factors for survival. For example, the abiotic factors important to a particular plant might be the amount of rainfall, the amount of sunlight, the type of soil, the range of temperature, and the nutrients available in the soil. The abiotic factors for the salmon in **Figure 2.5** might be the temperature range of the water, the pH of the water, and the salt concentration of the water.

Organisms are adapted to surviving in the abiotic factors that are present in their natural environments. If an organism moves to another location with a different set of abiotic factors, the organism might die if it cannot adjust quickly to its new surroundings. For example, if a lush green plant that normally grows in a swampy area is transplanted to a dry desert, the plant likely will die because it cannot adjust to abiotic factors present in the desert.

 **Reading Check** **Compare and contrast** abiotic and biotic factors for a plant or animal in your community.

CAREERS IN BIOLOGY

Ecologist The field of ecology is vast. Ecologists study the organisms in the world and the environments in which they live. Many ecologists specialize in a particular area such as marine ecology. For more information on biology careers, visit biologygmh.com.

Levels of Organization

The biosphere is too large and complex for most ecological studies. To study relationships within the biosphere, ecologists look at different levels of organization or smaller pieces of the biosphere. The levels increase in complexity as the numbers and interactions between organisms increase. The levels of organization are

- organism;
- population;
- biological community;
- ecosystem;
- biome;
- biosphere.

Refer to **Figure 2.6** as you read about each level.

Study Tip

Question Session Study the levels of organization illustrated in **Figure 2.6** with a partner. Question each other about the topic to deepen your knowledge.

Organisms, populations, and biological communities

The lowest level of organization is the individual organism itself. In **Figure 2.6,** the organism is represented by a single fish. Individual organisms of a single species that share the same geographic location at the same time make up a **population.** The school of fish represents a population of organisms. Individual organisms often compete for the same resources, and if resources are plentiful, the population can grow. However, usually there are factors that prevent populations from becoming extremely large. For example, when the population has grown beyond what the available resources can support, the population size begins to decline until it reaches the number of individuals that the available resources can support.

The next level of organization is the biological community. A **biological community** is a group of interacting populations that occupy the same geographic area at the same time. Organisms might or might not compete for the same resources in a biological community. The collection of plant and animal populations, including the school of fish, represents a biological community.

LAUNCH Lab

Review Based on what you've read about populations, how would you now answer the analysis questions?

Ecosystems, biomes, and the biosphere The next level of organization after a biological community is an ecosystem. An **ecosystem** is a biological community and all of the abiotic factors that affect it. As you can see in **Figure 2.6,** an ecosystem might contain an even larger collection of organisms than a biological community. In addition, it contains the abiotic factors present, such as water temperature and light availability. Although **Figure 2.6** represents an ecosystem as a large area, an ecosystem also can be small, such as an aquarium or tiny puddle. The boundaries of an ecosystem are somewhat flexible and can change, and ecosystems even might overlap.

The next level of organization is called the biome and is one that you will learn more about in Chapter 3. A **biome** is a large group of ecosystems that share the same climate and have similar types of communities. The biome shown in **Figure 2.6** is a marine biome. All of the biomes on Earth combine to form the highest level of organization—the biosphere.

Reading Check **Infer** what other types of biomes might be found in the biosphere if the one shown in **Figure 2.6** is called a marine biome.

Visualizing Levels of Organization

Figure 2.6
In order to study relationships within the biosphere, it is divided into smaller levels of organization. The most complex level, the biosphere, is followed by biome, ecosystem, biological community, population, and organism. Organisms are further divided into organ systems, organs, tissues, cells, molecules, and finally atoms.

Biosphere The highest level of organization is the biosphere, which is the layer of Earth—from high in the atmosphere to deep in the ocean—that supports life.

Biome A biome is formed by a group of ecosystems, such as the coral reefs off the coast of the Florida Keys, that share the same climate and have similar types of communities.

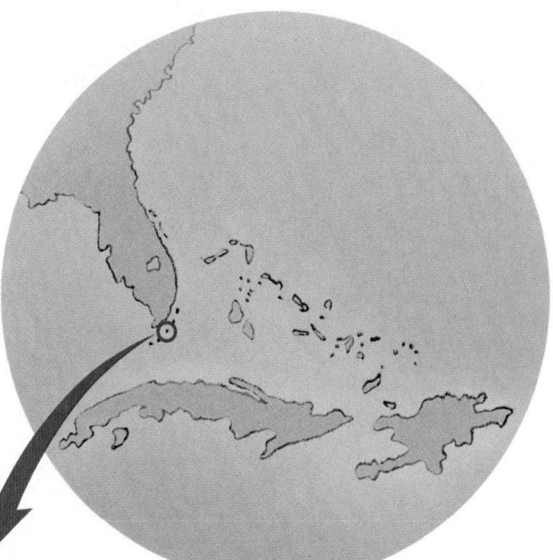

Ecosystem A biological community, such as the coral reef, and all of the abiotic factors, such as the sea water, that affect it make up an ecosystem.

Biological Community All of the populations of species—fishes, coral, and marine plants—that live in the same place at the same time make up a biological community.

Population A group of organisms of the same species that interbreed and live in the same place at the same time, such as the school of striped fish, is a population.

Organism An individual living thing, such as one striped fish, is an organism.

Organism

Population

Biological community

Ecosystem

Concepts In Motion Interactive Figure To see an animation of the levels of organization, visit biologygmh.com.

Biology Online

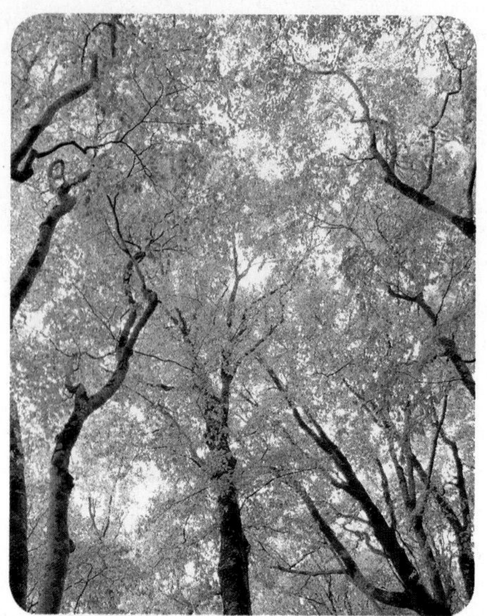

■ **Figure 2.7** These trees are the habitat for the community of organisms that live there.

Ecosystem Interactions

The interactions between organisms are important in an ecosystem. A community of organisms increases the chances for survival of any one species by using the available resources in different ways. If you look closely at a tree in the forest, like the one shown in **Figure 2.7,** you will find a community of different birds using the resources of the tree in different ways. For example, one bird species might eat insects on the leaves while another species of bird eats the ants found on the bark. The chance of survival for the birds increases because they are using different resources.

The trees shown in **Figure 2.7** also are habitats. A **habitat** is an area where an organism lives. A habitat might be a single tree for an organism that spends its life on one tree. If the organism moves from tree to tree, its habitat would be a grove of trees.

Organisms not only have a habitat—they have a niche as well. A **niche** (NIHCH) is the role or position that an organism has in its environment. An organism's niche is how it meets its needs for food, shelter, and reproduction. The niche might be described in terms of requirements for living space, temperature, moisture, or in terms of appropriate mating or reproduction conditions.

 **Reading Check** **Compare and contrast** a habitat and a niche.

Community Interactions

Organisms that live together in a biological community constantly interact. These interactions, along with the abiotic factors, shape an ecosystem. Interactions include competition for basic needs such as food, shelter, and mates, as well as relationships in which organisms depend on each other for survival.

Competition Competition occurs when more than one organism uses a resource at the same time. Resources are necessary for life and might include food, water, space, and light. For example, during a drought, as shown in **Figure 2.8,** water might be scarce for many organisms. The strong organisms directly compete with the weak organisms for survival. Usually the strong survive and the weak die. Some organisms might move to another location where water is available. At times when water is plentiful, all organisms share the resources and competition is not as fierce.

Predation Many, but not all, species get their food by eating other organisms. The act of one organism consuming another organism for food is **predation** (prih DAY shun). The organism that pursues another organism is the predator, and the organism that is pursued is the prey. If you have watched a cat catch a bird or mouse, you have witnessed a predator catch its prey.

■ **Figure 2.8** During droughts, animals compete for water; when water is plentiful, organisms share this resource.

Some insects also prey on other insects. Ladybugs and praying mantises are two examples of insects that are predators. Some insect predators also are called beneficial insects because they are used by organic gardeners for insect control. Instead of using insecticides, organic gardeners use beneficial insects to control other insect populations.

Animals are not the only organisms that are predators. The Venus flytrap, a plant native to some regions of North and South Carolina, has modified leaves that form small traps for insects and other small animals. The plant emits a sweet, sticky substance that attracts insects. When the insect lands on the leaf, the leaf trap snaps shut. Then, the plant secretes a substance that digests the insect over several days.

Symbiotic relationships Some species survive because of relationships they have developed with other species. The close relationship that exists when two or more species live together is **symbiosis** (sihm bee OH sus). There are three different kinds of symbiosis: mutualism, commensalism, and parasitism.

Mutualism The relationship between two or more organisms that live closely together and benefit from each other is **mutualism** (MYEW chuh wuh lih zum). Lichens, shown in **Figure 2.9,** display an example of a mutualistic relationship between fungi and algae. The tree merely provides a habitat for lichens, allowing it to receive ample sunlight. The algae provide food for the fungi, and the fungi provide a habitat for the algae. The close association of these two organisms provides two basic needs for the organisms—food and shelter.

Figure 2.9 Algae and fungi form lichens through a mutualistic relationship.
Explain *why lichens are an example of a mutualistic relationship.*

DATA ANALYSIS LAB 2.1

Based on Real Data*

Analyze the Data

Does temperature affect growth rates of protozoans? Researchers studied the effect of temperature on the growth rates of protozoans. They hypothesized that increasing temperature would increase the growth rate of the protozoans.

Data and Observations

The graph shows the effect of temperature on the growth rate of *Colpidium* and *Paramecium*.

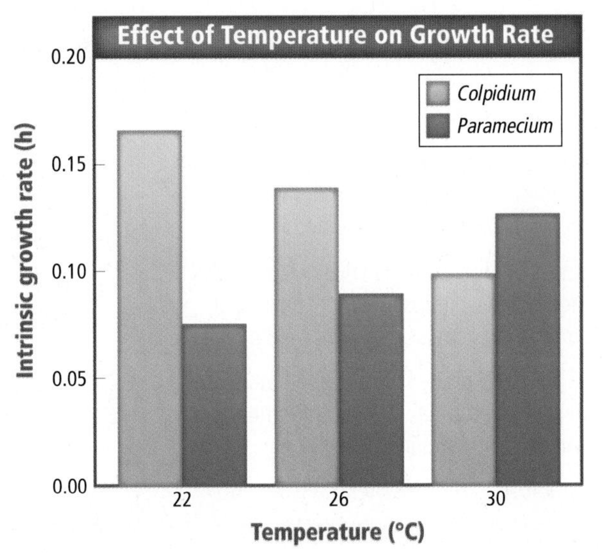

Think Critically

1. **Describe** the differences in population growth for the two species.
2. **Evaluate** What could be the next step in the researcher's investigation?

*Data obtained from: Jiang, L. and Kulczycki, A. 2004. Competition, predation, and species responses to environmental change. *Oikos* 106: 217–224.

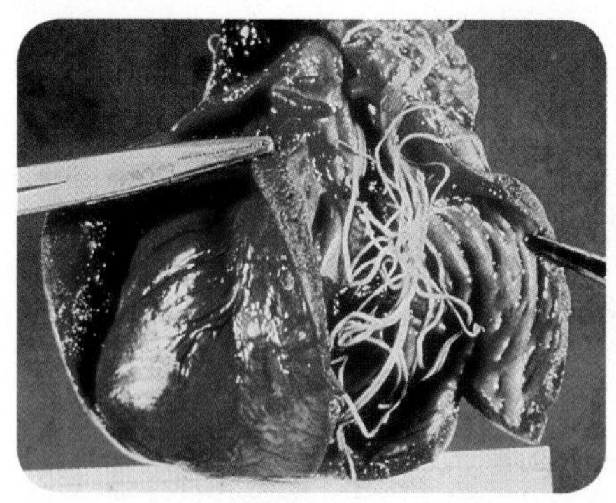

■ **Figure 2.10** This heart from a dog is infected with internal parasites called heartworms. Internal parasites depend on a host to supply their nutrients and habitat.

Commensalism Look back at **Figure 2.9.** This time, think about the relationship between the lichens and the tree. The lichens benefit from the relationship by gaining more exposure to sunlight, but they do not harm the tree. This type of relationship is commensalism. **Commensalism** (kuh MEN suh lih zum) is a relationship in which one organism benefits and the other organism is neither helped nor harmed.

The relationship between clownfish and sea anemones is another example of commensalism. Clownfish are small, tropical marine fish. Clownfish swim among the stinging tentacles of sea anemones without harm. The sea anemones protect the fish from predators while the clownfish eat bits of food missed by the sea anemones. This is a commensal relationship because the clownfish receives food and protection while the sea anemones are not harmed, nor do they benefit from this relationship.

Parasitism A symbiotic relationship in which one organism benefits at the expense of another organism is **parasitism** (PER us suh tih zum). Parasites can be external, such as ticks and fleas, or internal, such as bacteria, tapeworms, and roundworms, which are discussed in detail in Chapters 18 and 25. The heartworms in **Figure 2.10** show how destructive parasites can be. Pet dogs in many areas of the United States are treated to prevent heartworm infestation. Usually the heartworm (the parasite) does not kill the host, but it might harm or weaken it. In parasitism, if the host dies, the parasite also would die unless it quickly finds another host.

Another type of parasitism is brood parasitism. Brown-headed cowbirds demonstrate brood parasitism because they rely on other bird species to build their nests and incubate their eggs. A brown-headed cowbird lays its eggs in another bird's nest and abandons the eggs. The host bird incubates and feeds the young cowbirds. Often the baby cowbirds push the host's eggs or young from the nest, resulting in the survival of only the cowbirds. In some areas, the brown-headed cowbirds have significantly lowered the population of songbirds through this type of parasitism.

Section 2.1 Assessment

Section Summary

- Ecology is the branch of biology in which interrelationships between organisms and their environments are studied.
- Levels of organization in ecological studies include individual, population, biological community, ecosystem, biome, and biosphere.
- Abiotic and biotic factors shape an ecosystem and determine the communities that will be successful in it.
- Symbiosis is the close relationship that exists when two or more species live together.

Understand Main Ideas

1. MAIN Idea **Compare and contrast** biotic and abiotic factors.
2. **Describe** the levels of organization of an organism that lives in your biome.
3. **List** at least two populations that share your home.
4. **Differentiate** between the habitat and niche of an organism that is found in your community.

Think Critically

5. **Design an experiment** that determines the symbiotic relationship between a sloth, which is a slow-moving mammal, and a species of green algae that lives in the sloth's fur.

WRITING in **Biology**

6. Write a short story that demonstrates the dependence of all organisms on other organisms.

Section 2.2

Reading Preview

Objectives

- **Describe** the flow of energy through an ecosystem.
- **Identify** the ultimate energy source for photosynthetic producers.
- **Describe** food chains, food webs, and pyramid models.

Review Vocabulary

energy: the ability to cause change; energy cannot be created or destroyed, only transformed

New Vocabulary

autotroph
heterotroph
herbivore
carnivore
omnivore
detritivore
trophic level
food chain
food web
biomass

Flow of Energy in an Ecosystem

MAIN Idea **Autotrophs capture energy, making it available for all members of a food web.**

Real-World Reading Link When you eat a slice of pizza, you are supplying your body with energy. You might be surprised to learn that the Sun is the original source of energy for your body. How did the Sun's energy get into the pizza?

Energy in an Ecosystem

One way to study the interactions of organisms within an ecosystem is to follow the energy that flows through an ecosystem. Organisms differ in how they obtain energy, and they are classified as autotrophs or heterotrophs based on how they obtain their energy in an ecosystem.

Autotrophs All of the green plants and other organisms that produce their own food in an ecosystem are primary producers called autotrophs. An **autotroph** (AW tuh trohf) is an organism that collects energy from sunlight or inorganic substances to produce food. As you will learn in Chapter 8, organisms that have chlorophyll absorb energy during photosynthesis and use it to convert the inorganic substances carbon dioxide and water to organic molecules. In places where sunlight is unavailable, some bacteria use hydrogen sulfide and carbon dioxide to make organic molecules to use as food. Autotrophs are the foundation of all ecosystems because they make energy available for all other organisms in an ecosystem.

Heterotrophs A **heterotroph** (HE tuh roh trohf) is an organism that gets its energy requirements by consuming other organisms. Therefore, heterotrophs also are called consumers. A heterotroph that eats only plants is an **herbivore** (HUR buh vor) such as a cow, a rabbit, or grasshopper. Heterotrophs that prey on other heterotrophs, such as wolves, lions, and lynxes, shown in **Figure 2.11,** are called **carnivores** (KAR nuh vorz).

Figure 2.11 This lynx is a heterotroph that is about to consume another heterotroph.
Identify *What is an additional classification for each of these animals?*

■ **Figure 2.12** This fungus is obtaining food energy from the dead log. Fungi are decomposers that recycle materials found in dead organisms.
Explain *why decomposers are important in an ecosystem.*

In addition to herbivores and carnivores, there are organisms that eat both plants and animals, called **omnivores** (AHM nih vorz). Bears, humans, and mockingbirds are examples of omnivores.

The **detritivores** (duh TRYD uh vorz), which eat fragments of dead matter in an ecosystem, return nutrients to the soil, air, and water where the nutrients can be reused by organisms. Detritivores include worms and many aquatic insects that live on stream bottoms. They feed on small pieces of dead plants and animals. Decomposers, similar to detritivores, break down dead organisms by releasing digestive enzymes. Fungi, such as those in **Figure 2.12,** and bacteria are decomposers.

All heterotrophs, including detritivores, perform some decomposition when they consume another organism and break down its body into organic compounds. However, it is primarily the decomposers that break down organic compounds and make nutrients available to producers for reuse. Without the detritivores and decomposers, the entire biosphere would be littered with dead organisms. Their bodies would contain nutrients that would no longer be available to other organisms. The detritivores are an important part of the cycle of life because they make nutrients available for all other organisms.

Models of Energy Flow

Ecologists use food chains and food webs to model the energy flow through an ecosystem. Like any model, food chains and food webs are simplified representations of the flow of energy. Each step in a food chain or food web is called a **trophic** (TROH fihk) **level.** Autotrophs make up the first trophic level in all ecosystems. Heterotrophs make up the remaining levels. With the exception of the first trophic level, organisms at each trophic level get their energy from the trophic level before it.

MiniLab 2.1

Construct a Food Web

How is energy passed from organism to organism in an ecosystem? A food chain shows a single path for energy flow in an ecosystem. The overlapping relationships between food chains are shown in a food web.

Procedure

1. Read and complete the lab safety form.
2. Use the following information to construct a food web in a meadow ecosystem:
   - Red foxes feed on raccoons, crayfishes, grasshoppers, red clover, meadow voles, and gray squirrels.
   - Red clover is eaten by grasshoppers, muskrats, red foxes, and meadow voles.
   - Meadow voles, gray squirrels, and raccoons all eat parts of the white oak tree.
   - Crayfishes feed on green algae and detritus, and they are eaten by muskrats and red foxes.
   - Raccoons feed on muskrats, meadow voles, gray squirrels, and white oak trees.

Analysis

1. **Identify** all of the herbivores, carnivores, omnivores, and detritivores in the food web.
2. **Describe** how the muskrats would be affected if disease kills the white oak trees.

Food chains A **food chain** is a simple model that shows how energy flows through an ecosystem. **Figure 2.13** shows a typical grassland food chain. Arrows represent the one-way energy flow which typically starts with autotrophs and moves to heterotrophs. The flower uses energy from the Sun to make its own food. The grasshopper gets its energy from eating the flower. The mouse gets its energy from eating the grasshopper. Finally, the snake gets its energy from eating the mouse. Each organism uses a portion of the energy it obtains from the organism it eats for cellular processes to build new cells and tissues. The remaining energy is released into the surrounding environment and no longer is available to these organisms.

Food webs Feeding relationships usually are more complex than a single food chain because most organisms feed on more than one species. Birds, for instance, eat a variety of seeds, fruits, and insects. The model most often used to represent the feeding relationships in an ecosystem is a food web. A **food web** is a model representing the many interconnected food chains and pathways in which energy flows through a group of organisms. **Figure 2.14** shows a food web illustrating the feeding relationships in a desert community.

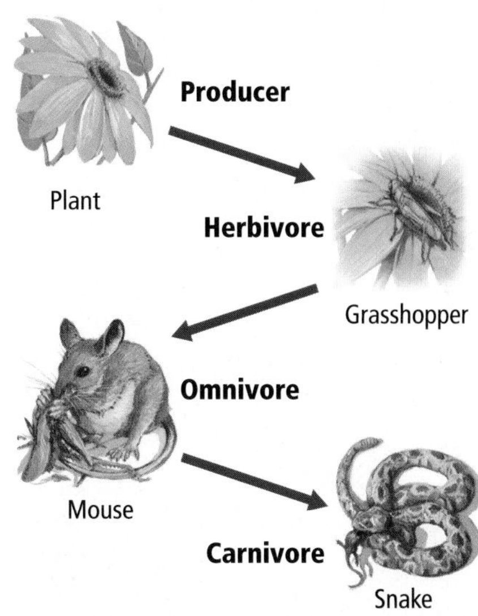

Figure 2.13 A food chain is a simplified model representing the transfer of energy from organism to organism.

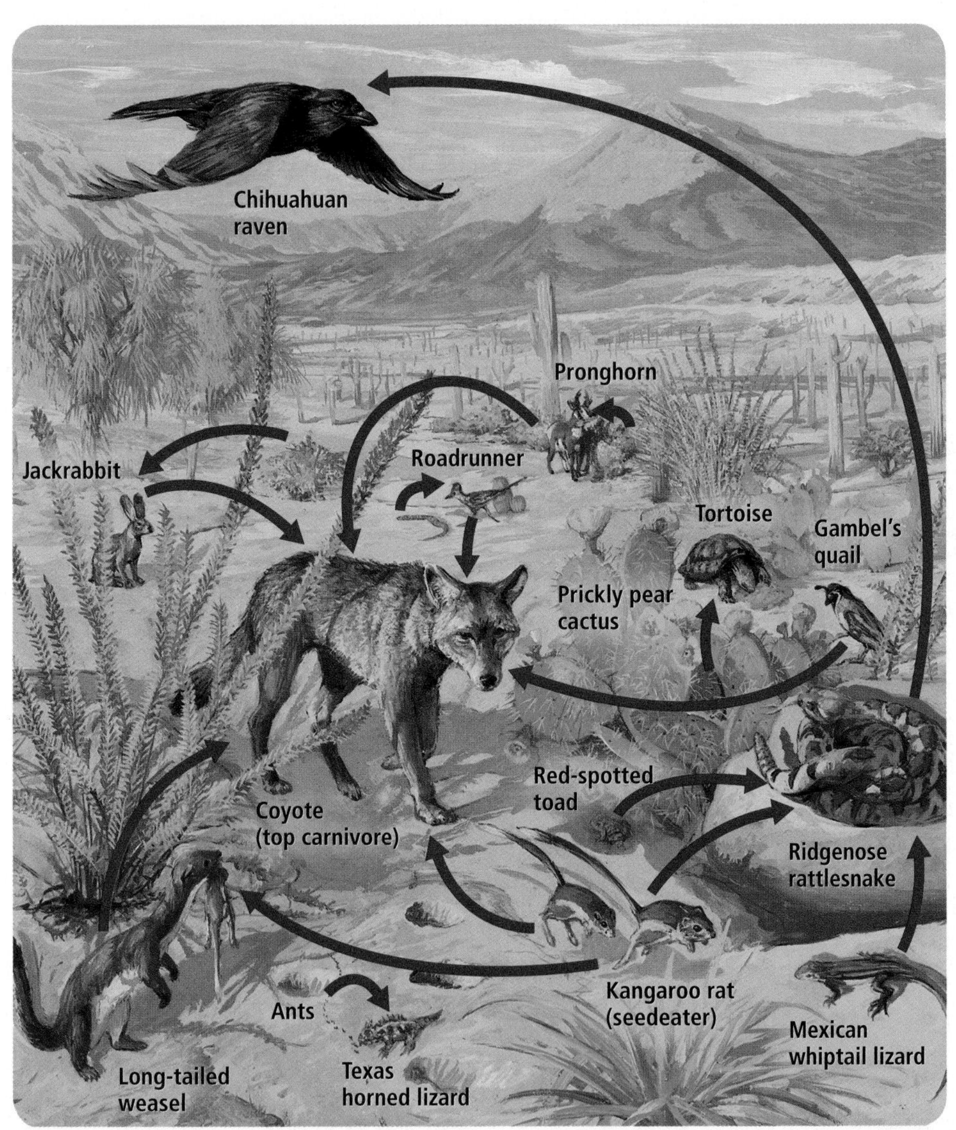

Figure 2.14 A food web is a model of the many ways in which energy flows through organisms.

Concepts In Motion

Interactive Figure To see an animation of a food web in a desert environment, visit biologygmh.com.

Pyramid of Energy
In a pyramid of energy, each level represents the amount of energy that is available to that trophic level. With each step up, there is an energy loss of 90 percent.

0.1%
Third-level consumers
Energy lost as heat
1%
Secondary consumers
10%
Primary consumers
100%
Primary producers
Available energy decreases
Parasites and decomposers feed at each level.

Pyramid of Biomass
In a pyramid of biomass, each level represents the amount of biomass consumed by the level above it.

Pyramid of Numbers
In a pyramid of numbers, each level represents the number of individual organisms consumed by the level above it.

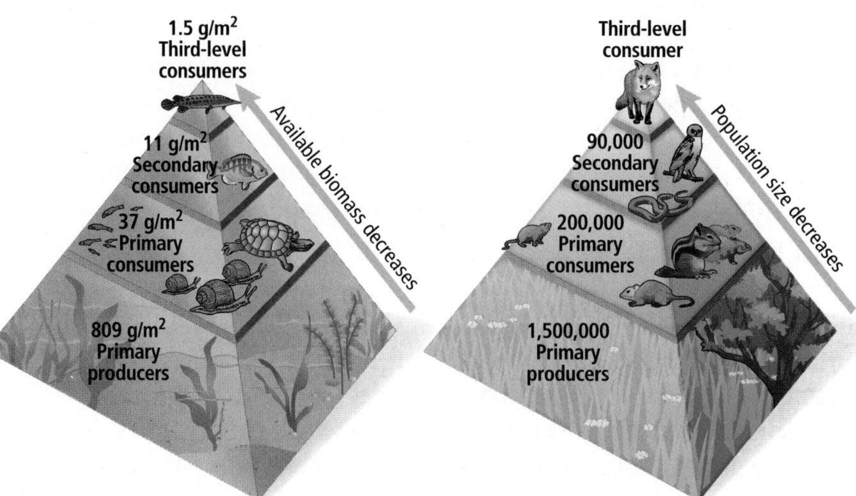

■ **Figure 2.15** Ecological pyramids are models used to represent trophic levels in ecosystems.

Ecological pyramids Another model that ecologists use to show how energy flows through ecosystems is the ecological pyramid. An ecological pyramid is a diagram that can show the relative amounts of energy, biomass, or numbers of organisms at each trophic level in an ecosystem.

Notice in **Figure 2.15** that in a pyramid of energy, approximately 90 percent of all energy is not transferred to the level above it. This occurs because most of the energy contained in the organisms at each level is consumed by cellular processes or released to the environment as heat. Usually, the amount of **biomass**—the total mass of living matter at each trophic level—decreases at each trophic level. As shown in the pyramid of numbers, the relative number of organisms at each trophic level also decreases because there is less energy available to support organisms.

Section 2.2 Assessment

Section Summary

- Autotrophs capture energy from the Sun or use energy from certain chemical substances to make food.
- Heterotrophs include herbivores, carnivores, omnivores, and detritivores.
- A trophic level is a step in a food chain or food web.
- Food chains, food webs, and ecological pyramids are models used to show how energy moves through ecosystems.

Understand Main Ideas

1. MAIN Idea **Compare and contrast** autotrophs and heterotrophs.
2. **Describe** the flow of energy through a simple food chain that ends with a lion as the final consumer.
3. **Classify** a pet dog as an autotroph or heterotroph and as an herbivore, carnivore, or omnivore. Explain.
4. **Evaluate** the impact on living organisms if the Sun began to produce less energy and then finally burned out.

Think Critically

5. **Use a Model** Create a simple food web of organisms in your community.

MATH in Biology

6. Draw an energy pyramid for a food chain made up of grass, a caterpillar, tiger beetle, lizard, snake, and a roadrunner. Assume that 100 percent of the energy is available for the grass. At each stage, show how much energy is lost and how much is available to the next trophic level.

Section 2.3

Reading Preview

Objectives

- **Describe** how nutrients move through the biotic and abiotic parts of an ecosystem.
- **Explain** the importance of nutrients to living organisms.
- **Compare** the biogeochemical cycles of nutrients.

Review Vocabulary

cycle: a series of events that occur in a regular repeating pattern

New Vocabulary

matter
nutrient
biogeochemical cycle
nitrogen fixation
denitrification

Cycling of Matter

MAIN Idea Essential nutrients are cycled through biogeochemical processes.

Real-World Reading Link Do you recycle your empty soda cans? If so, then you know that materials such as glass, aluminum, and paper can be reused. Natural processes in the environment cycle nutrients to make them available for use by other organisms.

Cycles in the Biosphere

Energy is transformed into usable forms to support the functions of an ecosystem. A constant supply of usable energy for the biosphere is needed, but this is not true of matter. The law of conservation of mass states that matter is not created or destroyed. Therefore, natural processes cycle matter through the biosphere. **Matter**—anything that takes up space and has mass—provides the nutrients needed for organisms to function. A **nutrient** is a chemical substance that an organism must obtain from its environment to sustain life and to undergo life processes. The bodies of all organisms are built from water and nutrients such as carbon, nitrogen, and phosphorus.

Connection to Chemistry In most ecosystems, plants obtain nutrients, in the form of elements and compounds, from the air, soil, or water. Plants convert some elements and compounds into organic molecules that they use. The nutrients flow through organisms in an ecosystem, such as the ecosystem shown in **Figure 2.16.** The green grass captures substances from the air, soil, and water, and then converts them into usable nutrients. The grass provides nutrients for the cow. If an organism eats the cow, the nutrients found in the cow are passed on to the next consumer. The nutrients are passed from producer—the green grass—to consumers. Decomposers return the nutrients to the cycle at every level.

The cycling of nutrients in the biosphere involves both matter in living organisms and physical processes found in the environment such as weathering. Weathering breaks down large rocks into particles that become part of the soil used by plants and other organisms. The exchange of matter through the biosphere is called the **biogeochemical cycle.** As the name suggests, these cycles involve living organisms (*bio*), geological processes (*geo*), and chemical processes (*chemical*).

Reading Check **Explain** why it is important to living organisms that nutrients cycle.

■ **Figure 2.16** Nutrients are cycled through the biosphere through organisms. In this example, the grasses are the producers and begin the cycle by capturing energy from the Sun.
Explain ***how nutrients continue to be cycled through the biosphere in this photo.***

CAREERS IN BIOLOGY

Hydrologist A hydrologist studies water processes, such as the distribution in nature, the water flow in a dam or river, or the water flow in a sewer or a city drinking-water system. For more information on biology careers, visit biologygmh.com.

The water cycle Organisms cannot live without water. Hydrologists study water found underground, in the atmosphere, and on Earth's surface in the form of lakes, streams, rivers, glaciers, ice caps, and oceans. Follow along with **Figure 2.17** to trace processes that cycle water through the biosphere.

Connection to Earth Science Water is constantly evaporating into the atmosphere from bodies of water, soil, and organisms. Water in the atmosphere is called water vapor. Water vapor rises and begins to cool in the atmosphere. Clouds form when the cooling water vapor condenses into droplets around dust particles in the atmosphere. Water falls from clouds as precipitation in the form of rain, sleet, or hail, transferring water to the Earth's surface. As shown in **Figure 2.17,** groundwater and runoff from land surfaces flow into streams, rivers, lakes, and oceans, where they evaporate into the atmosphere to continue through the water cycle. Approximately 90 percent of water vapor evaporates from oceans, lakes, and rivers; about 10 percent evaporates from the surface of plants through a process called transpiration. You will learn more about transpiration in Chapter 22.

Personal Tutor 

To learn about biogeochemical cycles, visit biologygmh.com.

All living organisms rely on freshwater. Even ocean-dwelling organisms rely on freshwater flowing to oceans to prevent high saline content and maintain ocean volume. Freshwater constitutes only about 3 percent of all water on Earth. Water available for living organisms is about 31 percent of all freshwater. About 69 percent of all freshwater is found in ice caps and glaciers, which then is unavailable for use by living organisms.

Reading Check **Identify** three processes in the water cycle.

Concepts In Motion

Interactive Figure To see an animation of the water cycle, visit biologygmh.com.

Figure 2.17 The water cycle is the natural process by which water is continuously cycled through the biosphere.

Infer ***What are the largest reservoirs of water on Earth?***

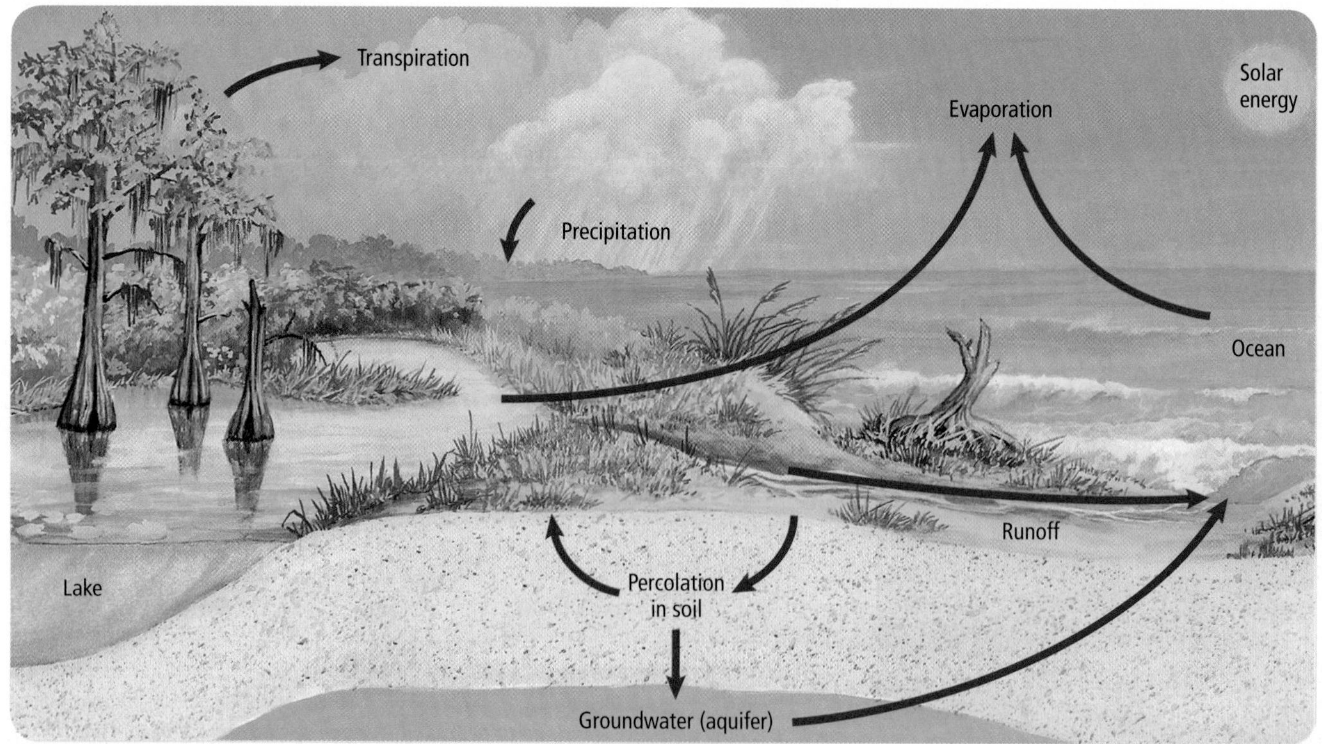

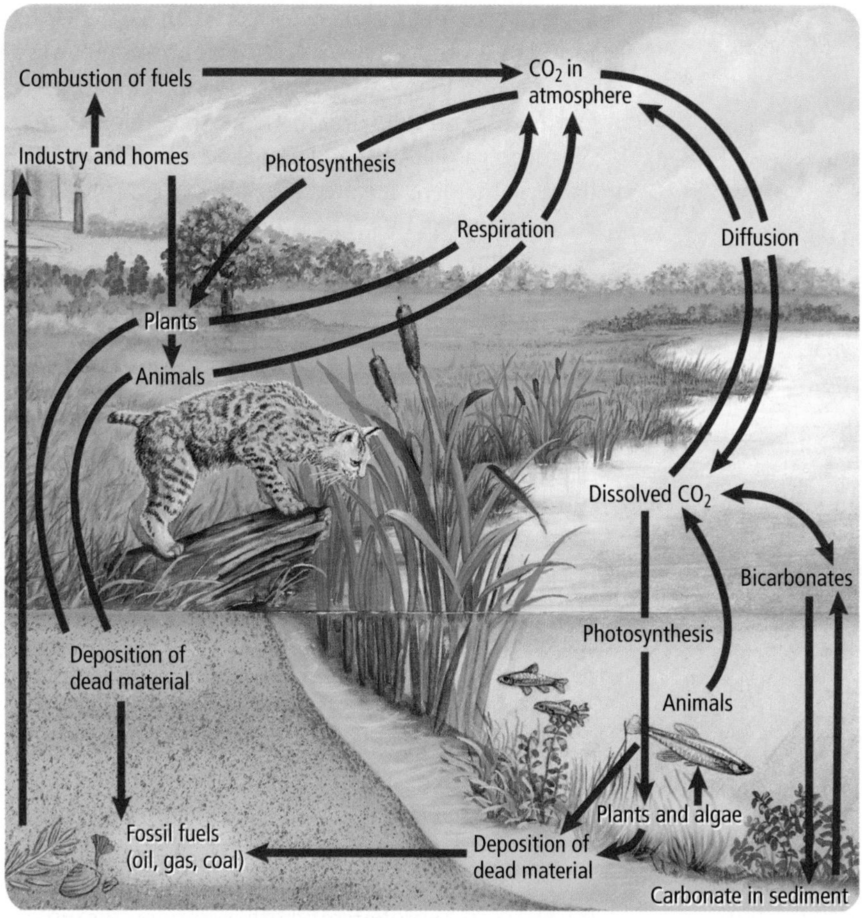

■ **Figure 2.18** The diagram shows how carbon and oxygen cycle through the environment.

Describe ***How does carbon move from the abiotic to the biotic parts of the ecosystem?***

Concepts In Motion

Interactive Figure To see an animation of the carbon cycle, visit biologygmh.com.

The carbon and oxygen cycles As you will learn in Chapter 6, all living things are composed of molecules that contain carbon. Atoms of carbon form the framework for important molecules such as proteins, carbohydrates, and fats. Oxygen is another element that is important to many life processes. Carbon and oxygen often make up molecules essential for life, including carbon dioxide and simple sugar.

Look at the cycles illustrated in **Figure 2.18.** During a process called photosynthesis, discussed in Chapter 8, green plants and algae convert carbon dioxide and water into carbohydrates and release oxygen back into the air. These carbohydrates are used as a source of energy for all organisms in the food web. Carbon dioxide is recycled when autotrophs and heterotrophs release it back into the air during cellular respiration. Carbon and oxygen recycle relatively quickly through living organisms.

Carbon enters a long-term cycle when organic matter is buried underground and converted to peat, coal, oil, or gas deposits. The carbon might remain as fossil fuel for millions of years. Carbon is released from fossil fuels when they are burned, which adds carbon dioxide to the atmosphere.

In addition to the removal of carbon from the short-term cycle by fossil fuels, carbon and oxygen can enter a long-term cycle in the form of calcium carbonate, as shown in **Figure 2.19.** Calcium carbonate is found in the shells of plankton and animals such as coral, clams, and oysters. These organisms, such as algae, fall to the bottom of the ocean floor, creating vast deposits of limestone rock. Carbon and oxygen remain trapped in these deposits until weathering and erosion release these elements to become part of the short-term cycle.

FOLDABLES
Incorporate information from this section into your Foldable.

■ **Figure 2.19** The white cliffs in Dover, England are composed almost entirely of calcium carbonate, or chalk. The calcium and oxygen found in these cliffs are in the long-term part of the cycle for calcium and oxygen.

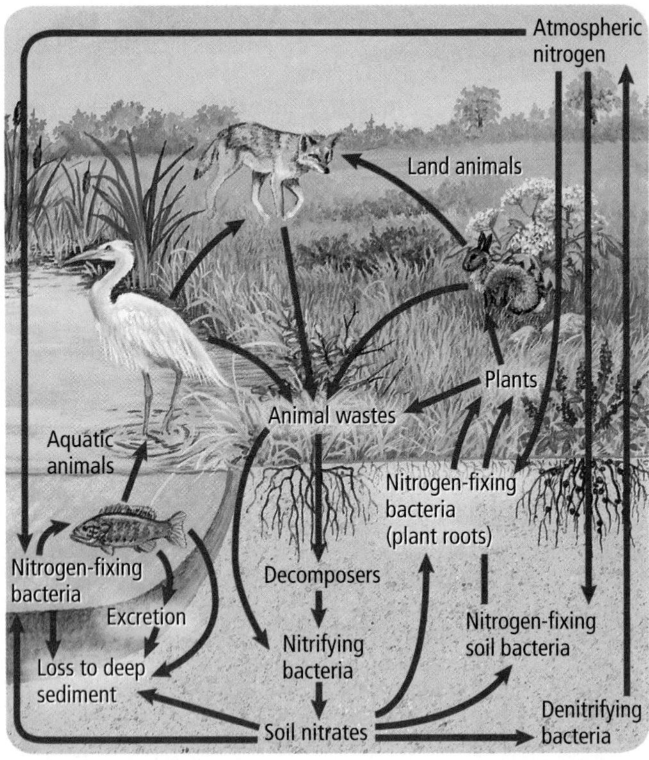

■ **Figure 2.20** Nitrogen is used and reused as it is cycled continuously through the biosphere.

Concepts In Motion

Interactive Figure To see an animation of the nitrogen cycle, visit biologygmh.com.

The nitrogen cycle Nitrogen is an element found in proteins. The largest concentration of nitrogen is found in the atmosphere. Plants and animals cannot use nitrogen directly from the atmosphere. Nitrogen gas is captured from the air by species of bacteria that live in water, the soil, or grow on the roots of some plants. The process of capture and conversion of nitrogen into a form that is useable by plants is called **nitrogen fixation.** Some nitrogen also is fixed during electrical storms when the energy from lightning bolts changes nitrogen gas to nitrates. Nitrogen also is added to soil when chemical fertilizers are applied to lawns, crops, or other areas.

Nitrogen enters the food web when plants absorb nitrogen compounds from the soil and convert them into proteins, as illustrated in **Figure 2.20.** Consumers get nitrogen by eating plants or animals that contain nitrogen. They reuse the nitrogen and make their own proteins. Because the supply of nitrogen in a food web is dependent on the amount of nitrogen that is fixed, nitrogen often is a factor that limits the growth of producers.

Nitrogen is returned to the soil in several ways, also shown in **Figure 2.20.** When an animal urinates, nitrogen returns to the water or soil and is reused by plants. When organisms die, decomposers transform the nitrogen in proteins and other compounds into ammonia. Organisms in the soil convert ammonia into nitrogen compounds that can be used by plants. Finally, in a process called **denitrification,** some soil bacteria convert fixed nitrogen compounds back into nitrogen gas, which returns it to the atmosphere.

MiniLab 2.2

Test for Nitrates

How much nitrate is found in various water sources? One ion containing nitrogen found in water can be easily tested—nitrate. Nitrate is a common form of inorganic nitrogen that is used easily by plants.

Procedure

1. Read and complete the lab safety form.
2. Prepare a data table to record your observations.
3. Obtain the **water samples** from different sources that are provided by your teacher.
4. Using a **nitrate test kit,** test the amount of nitrate in each water sample.
5. Dispose of your samples as directed by your teacher.

Analysis

1. **Determine** Did the samples contain differing amounts of nitrate? Explain.
2. **Identify** What types of human activities might increase the amount of nitrate in the water?
3. **Infer** What problems could a high nitrate level cause considering that nitrates also increase the growth rate of algae in waterways?

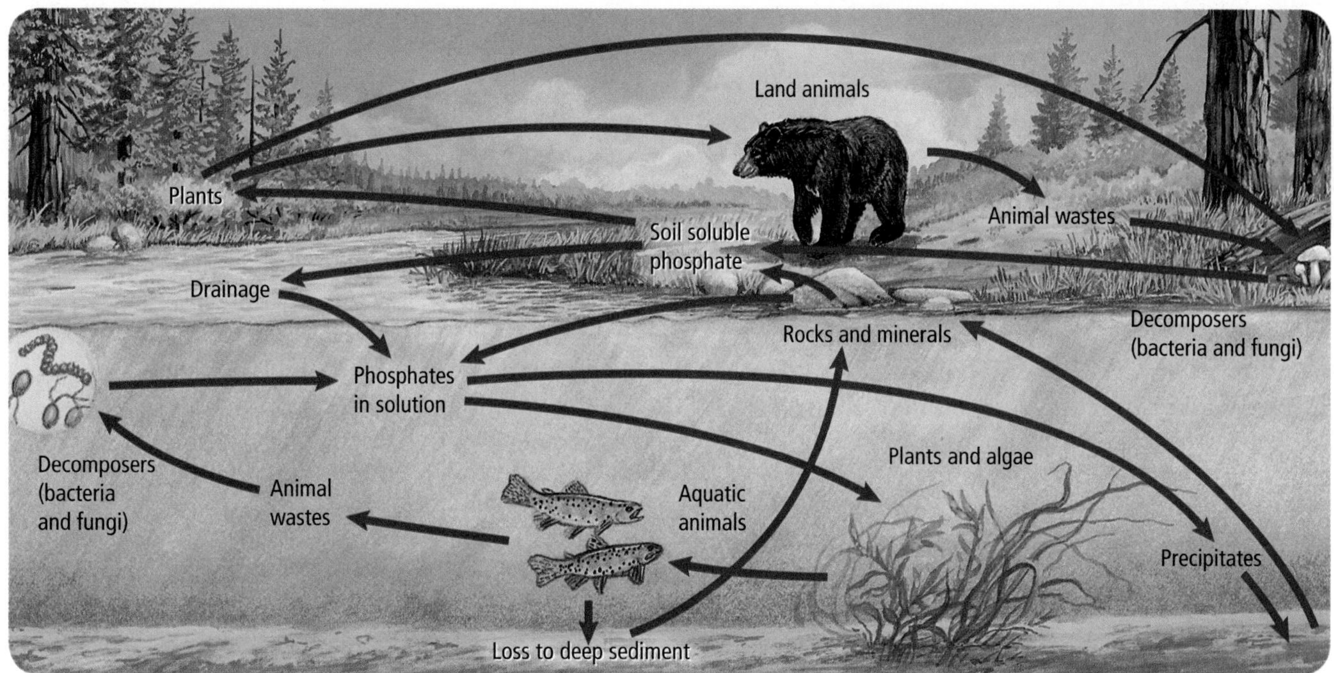

■ **Figure 2.21** The phosphorus cycle has a short-term cycle and a long-term cycle.

Concepts In Motion

Interactive Figure To see an animation of the phosphorus cycle, visit biologygmh.com.

The phosphorus cycle Phosphorus is an element that is essential for the growth and development of organisms. **Figure 2.21** illustrates the two cycles of phosphorus—a short-term and long-term cycle. In the short-term cycle, phosphorus in phosphates in solution, is cycled from the soil to producers and then from the producers to consumers. When organisms die or produce waste products, decomposers return the phosphorus to the soil where it can be used again. Phosphorus moves from the short-term cycle to the long-term cycle through precipitation and sedimentation to form rocks. In the long-term cycle, weathering or erosion of rocks that contain phosphorus slowly adds phosphorus to the cycle. Phosphorus, in the form of phosphates, might be present only in small amounts in soil and water. Therefore, phosphorus often is a factor that limits the growth of producers.

Section 2.3 Assessment

Section Summary

- Biogeochemical cycles include the exchange of important elements between the abiotic and biotic parts of an ecosystem.
- The carbon and oxygen cycles are closely intertwined.
- Nitrogen gas is limited in its ability to enter biotic portions of the environment.
- Phosphorus and carbon have short-term and long-term cycles.

Understand Main Ideas

1. MAIN Idea **List** four important biogeochemical processes that cycle nutrients.
2. **Compare and contrast** two of the cycles of matter.
3. **Explain** the importance of nutrients to an organism of your choice.
4. **Describe** how phosphorus moves through the biotic and abiotic parts of an ecosystem.

Think Critically

5. **Design an Experiment** Suppose a particular fertilizer contains nitrogen, phosphorus, and potassium. The numbers on the fertilizer's label represent the amounts of each element in the fertilizer. Design an experiment to test how much fertilizer should be added to a lawn to obtain the best results.

Biology & Society

To Dam or Not to Dam

The Glen Canyon area is a popular location for white-water rafting, fishing, hiking, and kayaking. The Glen Canyon area also is the location of a controversial dam, the Glen Canyon Dam. It was built between 1956 and 1963 in Arizona on the Colorado River. The dam holds and releases water from Lake Powell.

Economic benefits The Glen Canyon Dam provides electricity to many rural communities. It also provides water to California, New Mexico, Arizona, and Nevada. Lake Powell, which is one of the most visited tourist destinations of the southwest, provides jobs for many of the local residents. Millions of tourists visit Lake Powell each year for activities such as hiking, boating, fishing, and swimming.

The Glen Canyon Dam provides opportunities for recreation to millions of tourists every year. However, it also impacts the Colorado River ecosystem.

Impact on flora and fauna The construction of the dam has brought economic benefits to the area, but it also has negatively impacted the Colorado River ecosystem. The habitat of native fish has changed as a result of the dam. Three species of fish—the roundtail chub, the bonytail chub, and the Colorado squawfish—have become extinct.

The Lake Powell shoreline now is dominated by a non-native, semidesert scrub known as saltcedar or tamarisk. The saltcedar outcompetes native vegetation such as the sandbar willow, Gooding's willow, and fremont cottonwood. Saltcedar collects salt in its tissues over time. This salt eventually is released into the soil, making it unsuitable for many native plants.

Impact on temperature Before the dam was built, the water temperature of the Colorado River ranged from near freezing in the winter to a warm 29°C in the summer. Since the dam was built, the temperature of the water released downstream remains steady at 7–10°C. This temperature is fine for the nonnative trout that are bred for recreational activities; however, the native species do not fare as well.

The Bureau of Reclamation has proposed placing a temperature control device on the Glen Canyon Dam that would regulate the water temperature. Environmentalists suggest that this solution might not solve the problems for the native species because the native species need the fluctuating temperatures that were once part of the river system.

The Glen Canyon Dam has negatively impacted the ecosystem of the Colorado River area, but it has benefited the area economically. How do the costs weigh against the benefits? Biologists face real-world issues like these every day.

DEBATE in Biology

Collaborate Form a team to debate whether the recreational and economic opportunities outweigh the costs of damming the Colorado River. Conduct additional research at biologygmh.com prior to the debate.

BIOLAB

Design Your Own

FIELD INVESTIGATION: EXPLORE HABITAT SIZE AND SPECIES DIVERSITY

Background: **Ecologists know that a major key to maintaining not only individual species but also a robust diversity of species is preserving the proper habitat for those species.**

Question: *What effect does increasing the size of a habitat have on the species diversity within that habitat?*

Materials

Choose materials that would be appropriate for the experiment you plan.

Safety Precautions

WARNING: ***Follow all safety rules regarding travel to and from the study site. Be alert on site and avoid contact, if possible, with stinging or biting animals and poisonous plants.***

Plan and Perform the Experiment

1. Read and complete the lab safety form.
2. Form a hypothesis that you can test to answer the above question.
3. Record your procedure and list the materials you will use to test your hypothesis.
4. Make sure your experiment allows for the collection of quantitative data, which is data that can be expressed in units of measure.
5. Design and construct appropriate data tables.
6. Make sure your teacher approves your plan before you proceed.
7. Carry out the procedure at an appropriate field site.

Analyze and Conclude

1. **Graph Data** Prepare a graph of your data and the combined class data if it is available.
2. **Analyze** Do any patterns emerge as you analyze your group and/or class data and graphs? Explain.

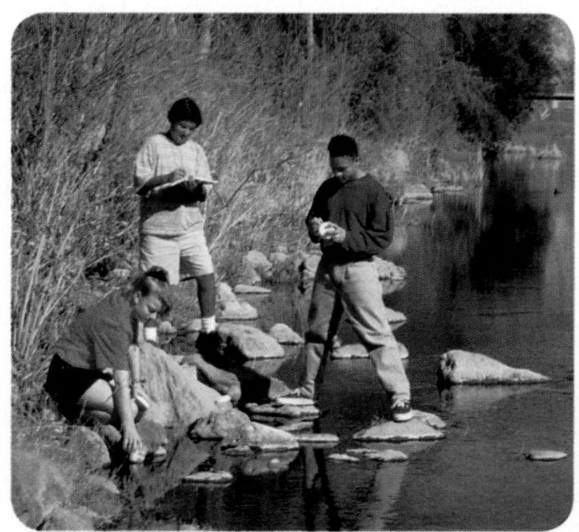

3. **Conclude** Based on your data, was your initial hypothesis correct?
4. **Error Analysis** Compare your observations and conclusions with your classmates. Did your observations and conclusions match? If not, what could explain the differences? How could you verify your results?
5. **Determine** Did the populations and diversity change proportionally as the habitat was expanded? As the habitat expanded, did it become more or less suitable for supporting life?
6. **Hypothesize** Would you expect the same results if you performed this experiment in other habitats? Explain.
7. **Think Critically** Would you expect the same results 10 years from now? 20 years from now? Explain your answer.

APPLY YOUR SKILL

Presentation Diagram and explain at least one food chain that might exist in the habitat you explored in this lab. To learn more about habitat size and species diversity, visit BioLabs at biologygmh.com.

Study Guide

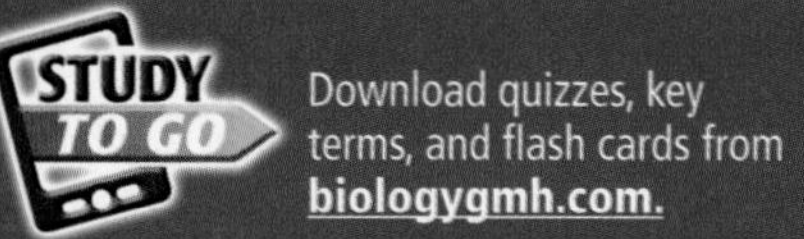

FOLDABLES **Summarize** the law of conservation of matter, and explain how it applies to the physical and chemical changes that take place in substances during natural cycles.

Vocabulary	Key Concepts
Section 2.1 Organisms and Their Relationships	
• abiotic factor (p. 35) • biological community (p. 36) • biome (p. 36) • biosphere (p. 34) • biotic factor (p. 35) • commensalism (p. 40) • ecology (p. 32) • ecosystem (p. 36) • habitat (p. 38) • mutualism (p. 39) • niche (p. 38) • parasitism (p. 40) • population (p. 36) • predation (p. 38) • symbiosis (p. 39)	MAIN Idea Biotic and abiotic factors interact in complex ways in communities and ecosystems. • Ecology is the branch of biology in which interrelationships between organisms and their environments are studied. • Levels of organization in ecological studies include individual, population, biological community, ecosystem, biome, and biosphere. • Abiotic and biotic factors shape an ecosystem and determine the communities that will be successful in it. • Symbiosis is the close relationship that exists when two or more species live together.
Section 2.2 Flow of Energy in an Ecosystem	
• autotroph (p. 41) • biomass (p. 44) • carnivore (p. 41) • detritivore (p. 42) • food chain (p. 43) • food web (p. 43) • herbivore (p. 41) • heterotroph (p. 41) • omnivore (p. 42) • trophic level (p. 42)	MAIN Idea Autotrophs capture energy, making it available for all members of a food web. • Autotrophs capture energy from the Sun or use energy from certain chemical substances to make food. • Heterotrophs include herbivores, carnivores, omnivores, and detritivores. • A trophic level is a step in a food chain or food web. • Food chains, food webs, and ecological pyramids are models used to show how energy moves through ecosystems.
Section 2.3 Cycling of Matter	
• biogeochemical cycle (p. 45) • denitrification (p. 48) • matter (p. 45) • nitrogen fixation (p. 48) • nutrient (p. 45)	MAIN Idea Essential nutrients are cycled through biogeochemical processes. • Biogeochemical cycles include the exchange of important elements between the abiotic and biotic parts of an ecosystem. • The carbon and oxygen cycles are closely intertwined. • Nitrogen gas is limited in its ability to enter biotic portions of the environment. • Phosphorus and carbon have short-term and long-term cycles.

CHAPTER 2 Assessment

Section 2.1

Vocabulary Review

Replace each underlined word with the correct vocabulary term from the Study Guide page.

1. A niche is the place in which an organism lives.
2. The presence of interbreeding individuals in one place at a given time is called a biological community.
3. A group of biological communities that interact with the physical environment is the biosphere.

Understand Key Concepts

4. Which of these levels of organization includes all the other levels?
   A. community **C.** individual
   B. ecosystem **D.** population
5. Which would be an abiotic factor for a tree in the forest?
   A. a caterpillar eating its leaves
   B. wind blowing through its branches
   C. a bird nesting in its branches
   D. fungus growing on its roots

Use the photo below to answer questions 6 and 7.

6. The insect in the photo above is gathering pollen and nectar for food, but at the same time is aiding in the plant's reproduction. What does this relationship demonstrate?
   A. predation **C.** mutualism
   B. commensalism **D.** parasitism
7. What term best describes the bee's role of gathering pollen?
   A. niche **C.** parasite
   B. predator **D.** habitat

Use the illustration below to answer question 8.

8. Which type of heterotroph best describes this snake?
   A. herbivore **C.** omnivore
   B. carnivore **D.** detritivore

Constructed Response

9. **Short Answer** Explain the difference between a habitat and niche.
10. **Open Ended** Describe two abiotic factors that affect your environment.
11. **CAREERS IN BIOLOGY** Summarize why most ecologists do not study the biosphere level of organization.

Think Critically

12. **Identify** an example of a predator-prey relationship, a competitive relationship, and a symbiotic relationship in an ecosystem near where you live.
13. **Explain** why it is advantageous for organisms such as fungi and algae to form mutualistic relationships.

Section 2.2

Vocabulary Review

Explain how the terms in each set below are related.

14. heterotroph, omnivore, carnivore
15. food chain, food web, trophic level
16. decomposer, heterotroph, carnivore
17. autotroph, food chain, heterotroph

Understand Key Concepts

18. How does energy first enter a pond ecosystem?
- **A.** through growth of algae
- **B.** through light from the Sun
- **C.** through decay of dead fish
- **D.** through runoff from fields

19. Which statement is true about energy in an ecosystem?
- **A.** Energy for most ecosystems originates from the Sun.
- **B.** Energy most often is released as light from an ecosystem.
- **C.** Energy flows from heterotrophs to autotrophs.
- **D.** Energy levels increase toward the top of the food chain.

Use the illustration below to answer questions 20 and 21.

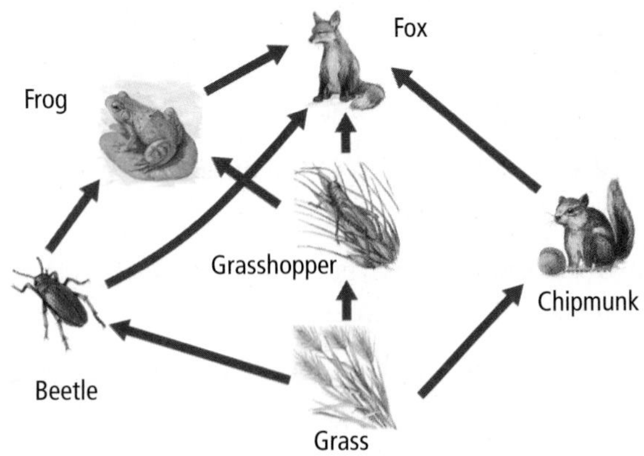

20. What does the illustration represent?
- **A.** a food web
- **B.** a food chain
- **C.** an ecological pyramid
- **D.** a pyramid of energy

21. Which organism in the illustration is an autotroph?
- **A.** frog
- **B.** grasshopper
- **C.** fox
- **D.** grass

22. Which is a detritivore?
- **A.** cat
- **B.** mouse
- **C.** sunflower
- **D.** crayfish

Constructed Response

23. Open Ended Illustrate a three-step food chain that might occur in your community. Use specific organisms.

24. Short Answer Describe why food webs usually are better models for explaining energy flow than food chains.

25. Short Answer Determine approximately how much total energy is lost from a three-step food chain if 1000 calories enter at the autotroph level.

Think Critically

26. Apply Information Create a poster of a food web that might exist in an ecosystem that differs from your community. Include as many organisms as possible in the food web.

Section 2.3

Vocabulary Review

Each of the following sentences is false. Make each sentence true by replacing the italicized word with a vocabulary term from the Study Guide page.

27. Because nitrogen is required for growth, it is considered an essential *nitrate.*

28. Converting nitrogen from a gas to a useable form by bacteria is *denitrification.*

29. The movement of chemicals on a global scale from abiotic through biotic parts of the environment is a *lithospheric process.*

Understand Key Concepts

30. What is the name of the process in which bacteria and lightning convert nitrogen into compounds that are useful to plants?
- **A.** ammonification
- **B.** denitrification
- **C.** nitrate cycling
- **D.** nitrogen fixation

Use the following diagram to answer question 31.

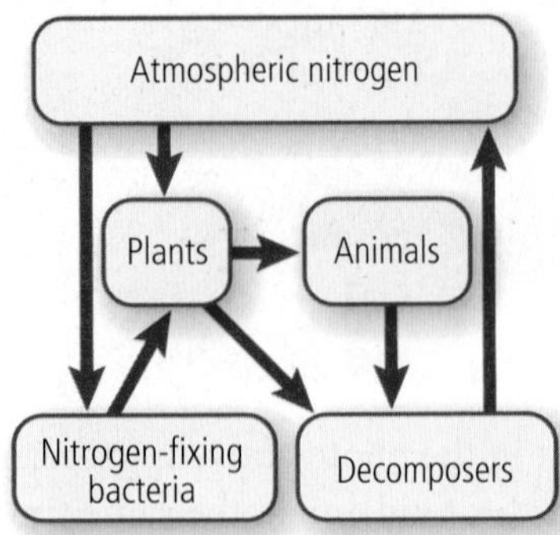

31. Where is the largest concentration of nitrogen found?
- **A.** animals
- **B.** atmosphere
- **C.** bacteria
- **D.** plants

32. What are the two major life processes that involve carbon and oxygen?
 A. coal formation and photosynthesis
 B. photosynthesis and respiration
 C. fuel combustion and open burning
 D. death and decay

33. Which process locks phosphorus in a long-term cycle?
 A. organic materials buried at the bottom of oceans
 B. phosphates released into the soil
 C. animals and plants eliminating wastes
 D. rain eroding mountains

Constructed Response

34. **Short Answer** Clarify what is meant by the following statement: Grass is just as important as mice in the diet of a carnivore such as a fox.

35. **Short Answer** The law of conservation of matter states that matter cannot be created or destroyed. How does this law relate to the cycling of carbon in an ecosystem?

36. **Short Answer** Explain the role of decomposers in the nitrogen cycle.

Think Critically

Use the illustration below to answer question 37 and 38.

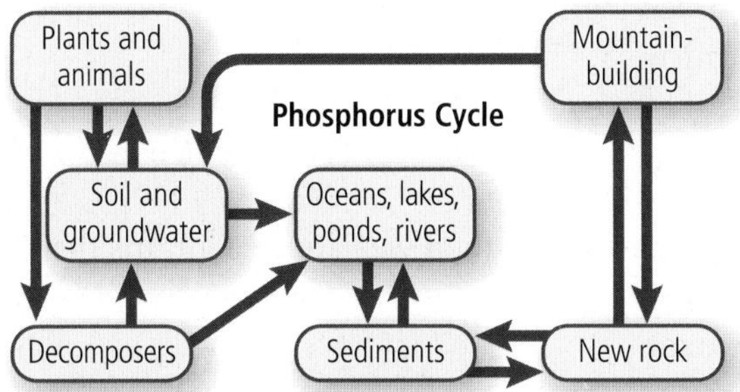

37. **Interpret Scientific Illustrations** Predict the effect of additional mountain building in the Rocky Mountains on the levels of phosphorus in the surrounding valleys.

38. **Explain** how decomposers supply phosphorus to soil, groundwater, oceans, lakes, ponds, and rivers.

Additional Assessment

39. WRITING in Biology Write a poem that includes vocabulary terms and concepts from the chapter.

Document-Based Questions

The following information pertains to an ancient sand dune in Florida that is now landlocked—Lake Wales Ridge. Read the passage and answer the following questions.

Data obtained from: Mohlenbrock, R. H. 2004–2005. Florida high. *Natural History* 113: 46–47.

The federally listed animals that live on the ridge are the blue-tailed mole skink, the Florida scrub jay, and the sand skink (which seems to "swim" through loose sand of the scrub). Other animals on the ridge are the eastern indigo snake (which can grow to more than eight feet long, making it the longest nonvenomous snake species in North America), the Florida black bear, the Florida gopher frog, the Florida mouse, the Florida pine snake, the Florida sandhill crane, the Florida scrub lizard, the gopher tortoise, Sherman's fox squirrel, and the short-tailed snake.

The gopher tortoise is particularly important because its burrows, sometimes as long as thirty feet, serve as homes for several of the rare species as well as many other more common organisms. The burrows also provide temporary havens when fires sweep through the area, or when temperatures reach high or low extremes.

40. Construct a simple food web using at least five of the organisms listed.

41. Explain how the burrows are used during fires and why they are effective.

Cumulative Review

42. Distinguish between science and pseudoscience. **(Chapter 1)**

43. Describe conditions under which a controlled experiment occurs. **(Chapter 1)**

Standardized Test Practice

Cumulative

Multiple Choice

1. Which would be considered an ecosystem?
   A. bacteria living in a deep ocean vent
   B. biotic factors in a forest
   C. living and nonliving things in a pond
   D. populations of zebras and lions

Use the illustration below to answer questions 2 and 3.

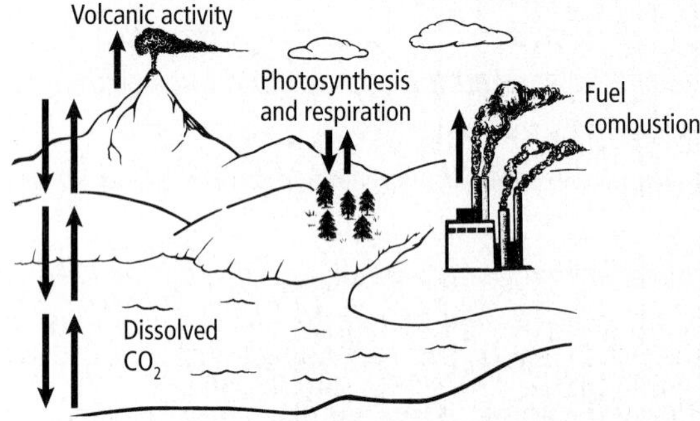

2. Which part of the diagram above relates to carbon leaving a long-term cycle?
   A. Dissolved CO_2
   B. Fuel combustion
   C. Photosynthesis and respiration
   D. Volcanic activity

3. Which part of the diagram above relates to carbon moving from an abiotic to a biotic part of the ecosystem?
   A. Dissolved CO_2
   B. Fuel combustion
   C. Photosynthesis and respiration
   D. Volcanic activity

4. Which is a scientific explanation of a natural phenomenon supported by many observations and experiments?
   A. factor
   B. hypothesis
   C. result
   D. theory

5. The mole is the SI unit for which quantity?
   A. number of particles in a substance
   B. compounds that make up a substance
   C. number of elements in a substance
   D. total mass of a substance

6. Suppose two leaf-eating species of animals live in a habitat where there is a severe drought, and many plants die as a result of the drought. Which term describes the kind of relationship the two species probably will have?
   A. commensalism
   B. competition
   C. mutualism
   D. predation

Use the illustration below to answer questions 7–9.

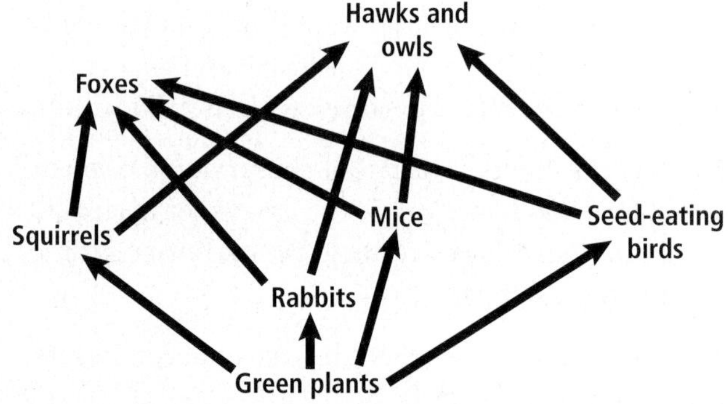

7. Which part of the food web above contains the greatest biomass?
   A. foxes
   B. green plants
   C. mice
   D. rabbits

8. Which part of the food web above contains the least biomass?
   A. foxes
   B. green plants
   C. mice
   D. rabbits

9. What happens to the energy that the fox uses for maintaining its body temperature?
   A. It is taken up by decomposers that consume the fox.
   B. It moves into the surrounding environment.
   C. It stays in the fox through the metabolism of food.
   D. It travels to the next trophic level when the fox is eaten.

Short Answer

Use the illustration below to answer questions 10 and 11.

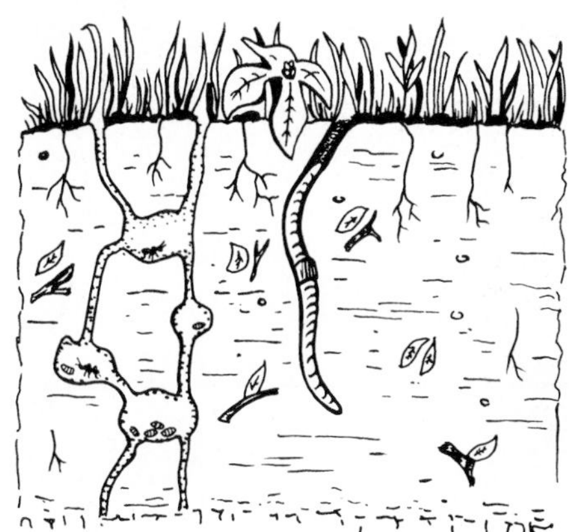

10. What are two biotic factors and two abiotic factors that affect a worm found in a situation similar to what is shown in the diagram?

11. Explain the portions of the following biogeochemical cycles that are related to the diagram above.
    A. Nitrogen cycle
    B. Oxygen cycle
    C. Carbon cycle

12. Distinguish between the everyday use of the term *theory* and its true scientific meaning.

13. Evaluate how scientific knowledge changes and how the amount of scientific knowledge grows. Suggest a reason why it probably will continue to grow.

14. Describe how a forest ecosystem might be different without the presence of decomposers and detritivores.

15. Suppose that some unknown organisms are discovered in the deep underground of Earth. Give two examples of questions that biologists might try to answer by researching these organisms.

Extended Response

Use this drawing to answer questions 16 and 17.

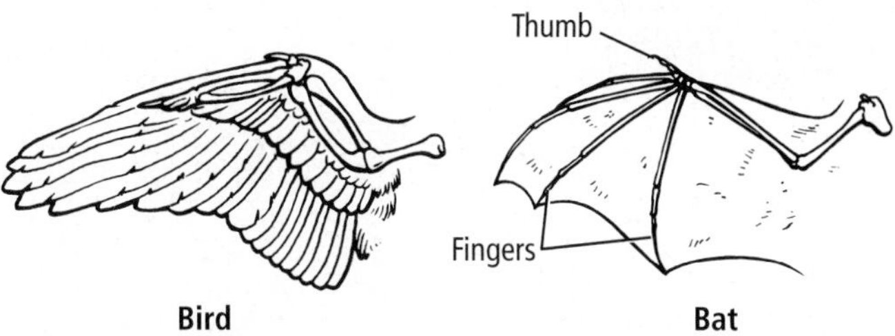

16. Someone tells you that bats and birds are closely related because they both have wings. Evaluate how this diagram could be used to critique the idea that bats and birds are not closely related.

17. Suppose you form a hypothesis that bats and birds are not closely related and you want to confirm this by comparing the way bats and birds fly. Design an experiment to test this hypothesis.

Essay Question

Various substances or elements on Earth move through long-term and short-term biogeochemical cycles as they become part of different aspects of the biosphere. The amount of a substance that is involved in a long-term cycle has an effect on the availability of that substance for use by humans and other organisms on Earth.

Using the information in the paragraph above, answer the following question in essay format.

18. Choose a substance or element that you know is involved in both long-term and short-term biogeochemical cycles. In a well-organized essay, describe how it moves through both types of cycles, and how these cycles affect its availability to humans and other organisms.

NEED EXTRA HELP?																		
If You Missed Question . . .	1	2	3	4	5	6	7	8	9	10	11	12	13	14	15	16	17	18
Review Section . . .	2.2	2.3	2.3, 2.1	1.2	1.2	2.1	2.2	2.2	2.2	2.1	2.3	1.2	1.2	2.2	1.3	1.2	1.3	2.3

Chapter 3 Communities, Biomes, and Ecosystems

BIG Idea Limiting factors and ranges of tolerance are factors that determine where terrestrial biomes and aquatic ecosystems exist.

Section 1
Community Ecology

MAIN Idea All living organisms are limited by factors in the environment.

Section 2
Terrestrial Biomes

MAIN Idea Ecosystems on land are grouped into biomes primarily based on the plant communities within them.

Section 3
Aquatic Ecosystems

MAIN Idea Aquatic ecosystems are grouped based on abiotic factors such as water flow, depth, distance from shore, salinity, and latitude.

BioFacts

- The Great Barrier Reef off the coast of northeastern Australia, shown here, is the largest living structure on Earth and is visible from space. It extends more than 2000 km in length.
- Coral reefs grow at a rate of only about 1.27 cm/y.
- Coral reefs located where the Indian and Pacific Oceans meet are the most diverse reefs; they can have as many as 700 species of coral.

Regal angel fish

Giant moray eel

Coral polyps

LAUNCH Lab

What is my biological address?

Just as you have a postal address, you also have a biological "address." As a living organism, you are part of interwoven ecological units that vary in size from as large as the whole biosphere to the place you occupy right now.

Procedure

1. Consider the following question: What do the terms *community* and *ecosystem* mean to you?
2. Describe the biological community and an ecosystem to which you belong.

Analysis

1. **Compare** Did your classmates all identify the same community and ecosystem? How would you describe, in general, the plants and animals in your area to someone from another country?
2. **Examine** Communities and ecosystems are constantly changing through a process known as succession. What changes do you think your biological community has undergone in the last 100 to 150 years?

Visit biologygmh.com to:

- study the entire chapter online
- explore Concepts in Motion, Microscopy Links, and links to virtual dissections
- access Web links for more information, projects, and activities
- review content online with the Interactive Tutor and take Self-Check Quizzes

Terrestrial Biomes Make this Foldable to help you understand primary succession and secondary succession.

STEP 1 Draw a line through the middle of a sheet of notebook paper as shown.

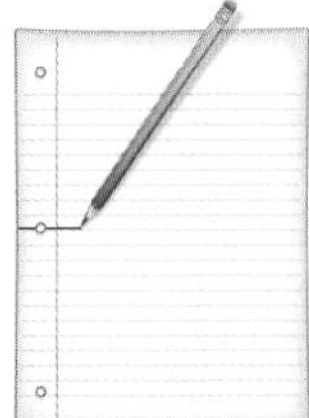

STEP 2 Fold the paper from the top and bottom so the edges meet at the center line.

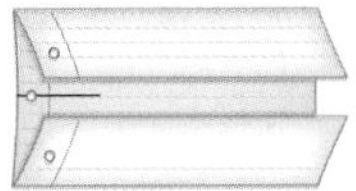

STEP 3 Label the two tabs as illustrated.

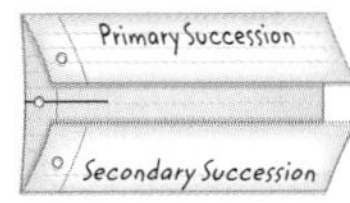

FOLDABLES **Use this Foldable with Section 3.1.** As you read the chapter, record what you learn about primary succession and secondary succession under the tabs. Use the front of the tabs to draw a visual representation of each.

Section 3.1

Reading Preview

Objectives

- **Recognize** how unfavorable abiotic and biotic factors affect a species.
- **Describe** how ranges of tolerance affect the distribution of organisms.
- **Sequence** the stages of primary and secondary succession.

Review Vocabulary

abiotic factor: the nonliving part of an organism's environment

New Vocabulary

community
limiting factor
tolerance
ecological succession
primary succession
climax community
secondary succession

Community Ecology

MAIN Idea All living organisms are limited by factors in the environment.

Real-World Reading Link Wherever you live, you probably are used to the conditions of your environment. If it is cold outdoors, you might wear a coat, hat, and gloves. Bears have adaptations such as their warm fur coat to the cold so they don't need these types of clothes.

Communities

When you describe your community, you probably include your family, the students in your school, and the people who live nearby. A biological **community** is a group of interacting populations that occupy the same area at the same time. Therefore, your community also includes plants, other animals, bacteria, and fungi. Not every community includes the same variety of organisms. An urban community is different from a rural community, and a desert community is different from an arctic community.

In Chapter 2, you learned that organisms depend on one another for survival. You also learned about abiotic factors and how they affect individual organisms. How might abiotic factors affect communities? Consider soil, which is an abiotic factor. If soil becomes too acidic, some species might die or become extinct. This might affect food sources for other organisms, resulting in a change in the community.

Organisms adapt to the conditions in which they live. For example, a wolf's heavy fur coat enables it to survive in harsh winter climates, and a cactus's ability to retain water enables it to tolerate the dry conditions of a desert. Depending on which factors are present, and in what quantities, organisms can survive in some ecosystems but not in others. For example, the plants in the desert oasis shown in **Figure 3.1** decrease in number away from the water source.

■ **Figure 3.1** Notice that populations of organisms live within a relatively small area surrounding the oasis.

Limiting factors Any abiotic factor or biotic factor that restricts the numbers, reproduction, or distribution of organisms is called a **limiting factor.** Abiotic limiting factors include sunlight, climate, temperature, water, nutrients, fire, soil chemistry, and space. Biotic limiting factors include living things, such as other plant and animal species. Factors that restrict the growth of one population might enable another to thrive. For example, in the oasis shown in **Figure 3.1,** water is a limiting factor for all of the organisms. Temperature also might be a limiting factor. Desert species must be able to withstand the heat of the Sun and the cold temperatures of desert nights.

Range of tolerance For any environmental factor, there is an upper limit and lower limit that define the conditions in which an organism can survive. For example, steelhead trout live in cool, clear coastal rivers and streams from California to Alaska. The ideal range of water temperature for steelhead trout is between 13°C and 21°C, as illustrated in **Figure 3.2.** However, steelhead trout can survive water temperatures from 9°C to 25°C. At these temperatures, steelhead trout experience physiological stress, such as inability to grow or reproduce. They will die if the water temperature goes beyond the upper and lower limits.

Have you ever had to tolerate a hot day or a boring activity? Similarly, the ability of any organism to survive when subjected to abiotic factors or biotic factors is called **tolerance.** Consider **Figure 3.2** again. Steelhead trout tolerate a specific range of temperatures. That is, the range of tolerance of water temperature for steelhead is 9°C to 25°C. Notice the greatest number of steelhead live in the optimum zone in which the temperature is best for survival. Between the optimum zone and the tolerance limits lies the zone of physiological stress. At these temperatures, there are fewer fish. Beyond the upper tolerance limit of 25°C and the lower tolerance limit of 9°C, there are no steelhead trout. Therefore, water temperature is a limiting factor for steelhead when water temperature is outside the range of tolerance.

Reading Check **Describe** the relationship between a limiting factor and a range of tolerance.

CAREERS IN BIOLOGY

Conservation Biologist Among other duties, a conservation biologist might tag and track animals in a community. Understanding the biotic and abiotic factors of the community can help explain changes in populations. For more information on biology careers, visit biologygmh.com.

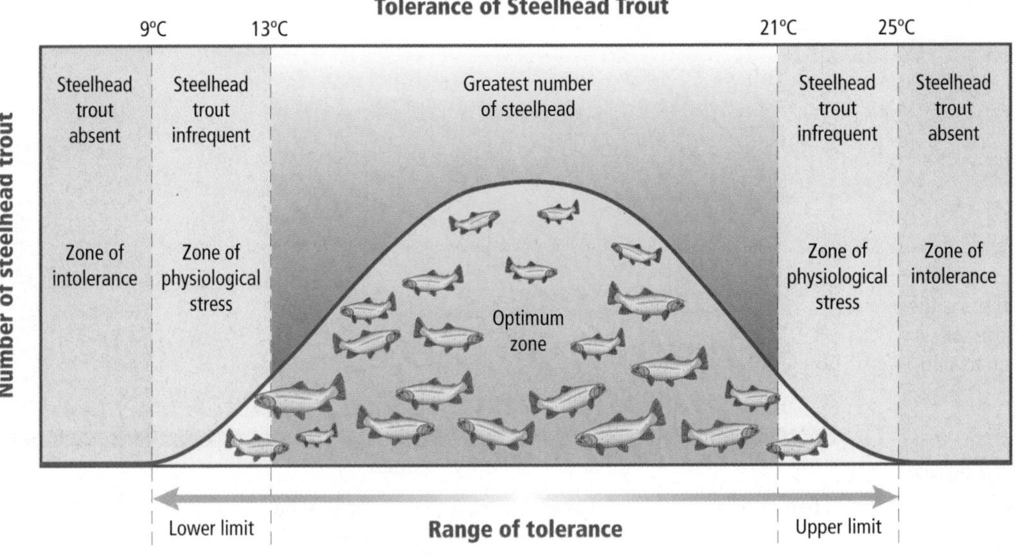

Figure 3.2 Steelhead trout are limited by the temperature of the water in which they live.
Infer *which other abiotic factors might limit the survival of steelhead trout.*

FOLDABLES Incorporate information from this section into your Foldable.

Ecological Succession

Ecosystems are constantly changing. They might be modified in small ways, such as a tree falling in the forest, or in large ways, such as a forest fire. They also might alter the communities that exist in the ecosystem. Forest fires can be good and even necessary for the forest community. Forest fires return nutrients to the soil. Some plants, such as fireweed, have seeds that will not sprout until they are heated by fire. Some ecosystems depend on fires to get rid of debris. If fires are prevented, debris builds up to the point where the next fire might burn the shrubs and trees completely. A forest fire might change the habitat so drastically that some species no longer can survive, but other species might thrive in the new, charred conditions.

The change in an ecosystem that happens when one community replaces another as a result of changing abiotic and biotic factors is **ecological succession.** There are two types of ecological succession—primary succession and secondary succession.

VOCABULARY
SCIENCE USAGE V. COMMON USAGE
Primary
Science usage: first in rank, importance, value, or order.
A doctor's primary concern should be the patient.

Common usage: the early years of formal education.
Elementary grades, up to high school, are considered to comprise a student's primary education.

Primary succession On a solidified lava flow or exposed rocks on a cliff, no soil is present. If you took samples of each and looked at them under a microscope, the only biological organisms you would observe would be bacteria and perhaps fungal spores or pollen grains that drifted there on air currents. The establishment of a community in an area of exposed rock that does not have any topsoil is **primary succession,** as illustrated in **Figure 3.3.** Primary succession usually occurs very slowly at first.

Most plants require soil for growth. How is soil formed? Usually lichens, a combination of a fungus and algae that you will learn more about in Chapter 20, begin to grow on the rock. Because lichens, along with some mosses, are among the first organisms to appear, they are called pioneer species. Pioneer species help to create soil by secreting acids that help to break down rocks.

Figure 3.3 The formation of soil is the first step in primary succession. Once soil formation starts, there is succession toward a climax community.

Concepts In Motion
Interactive Figure To see an animation of how a climax community forms, visit biologygmh.com.

As pioneer organisms die, their decaying organic materials, along with bits of sediment from the rocks, make up the first stage of soil development. At this point, small weedy plants, including ferns, and other organisms such as fungi and insects, become established. As these organisms die, additional soil is created. Seeds, brought in by animals, water, or wind, begin to grow in the newly formed soil. Eventually, enough soil is present so that shrubs and trees can grow.

A mature community can eventually develop from bare rock, as illustrated in **Figure 3.3.** The stable, mature community that results when there is little change in the composition of species is a **climax community.** Scientists today realize that disturbances, such as climate change, are ongoing in communities, thus a true climax community is unlikely to occur.

Secondary succession Disturbances such as fire, flood, or a windstorm can disrupt a community. After a disturbance, new species of plants and animals might occupy the habitat. Over time, there is a natural tendency for the species belonging to the mature community to return. **Secondary succession** is the orderly and predictable change that takes place after a community of organisms has been removed but the soil has remained intact. Pioneer species—mainly plants that begin to grow in the disturbed area—are the first species to start secondary succession.

DATA ANALYSIS LAB 3.1

Based on Real Data*

Interpret the Data

How do soil invertebrates affect secondary succession in a grassland environment? An experiment was performed by adding soil invertebrates to controlled grassland communities. The growth of various plants was measured at four months, six months, and 12 months. Growth was measured by recording shoot biomass—the mass of the grass stems.

Data and Observations

The bars on the graph indicate the change in the biomass of the plants over time.

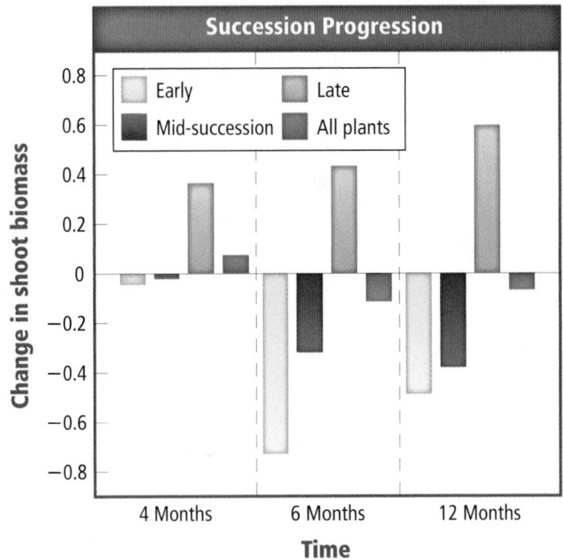

Think Critically

1. **Infer** What does a negative value of change in shoot biomass indicate?
2. **Generalize** which communities were most positively affected and which were most negatively affected by the addition of soil invertebrates.

*Data obtained from: De Deyn, G.B. et al. 2003. Soil invertebrate fauna enhances grassland succession and diversity. *Nature* 422: 711–713.

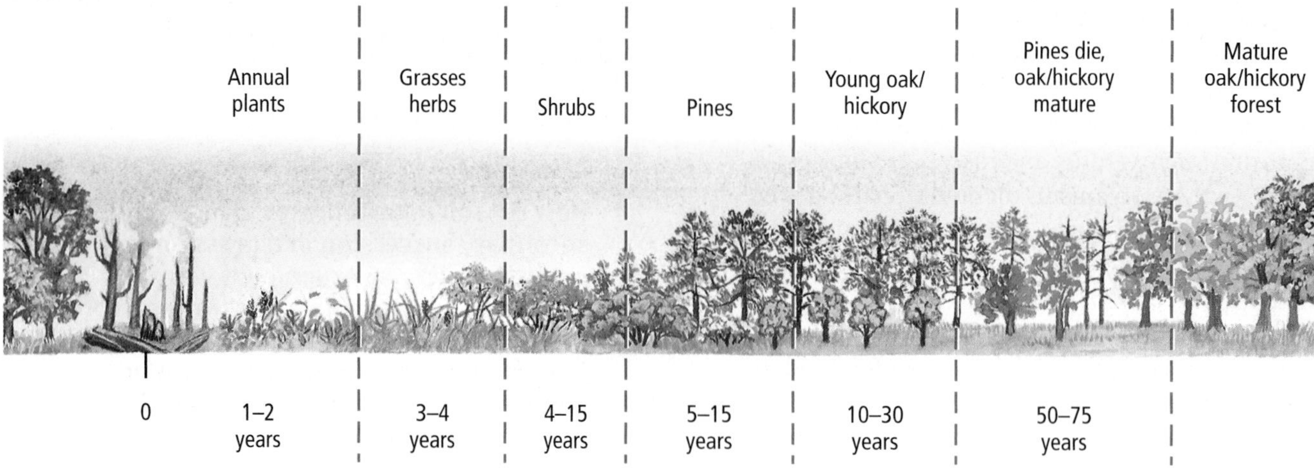

■ **Figure 3.4** After a forest fire, a forest might appear devastated. However, a series of changes ultimately leads back to a mature community.

During secondary succession, as in primary succession, the community of organisms changes over a period of time. **Figure 3.4** shows how species composition changes after a forest fire. Secondary succession usually occurs faster than primary succession because soil already exists and some species still will be present (although there might be fewer of them). Also, undisturbed areas nearby can be sources of seeds and animals.

Succession's end point Ecological succession is likely a very complex process that involves many factors. The end point of succession after a disturbance cannot be predicted. Natural communities are constantly changing at different rates, and the process of succession is very slow. Human activities also affect the species that might be present. Because of these factors, it is difficult to determine if succession has reached a climax community anywhere on Earth.

Section 3.1 Assessment

Section Summary

- Limiting factors restrict the growth of a population within a community.
- Organisms have a range of tolerance for each limiting factor that they encounter.
- Primary succession occurs on areas of exposed rock or bare sand (no soil).
- Communities progress until there is little change in the composition of species.
- Secondary succession occurs as a result of a disturbance in a mature community.

Understand Main Ideas

1. MAIN Idea **Identify** how temperature is a limiting factor for polar bears.
2. **Predict** how unfavorable abiotic and biotic factors affect a species.
3. **Describe** how ranges of tolerance affect the distribution of a species.
4. **Classify** the stage of succession of a field that is becoming overgrown with shrubs after a few years of disuse.

Think Critically

5. **Interpret the figure** Refer to **Figure 3.2** to predict the general growth trend for steelhead trout in a stream that is 22°C.

MATH in Biology

6. Graph the following data to determine the range of tolerance for catfish. The first number in each pair of data is temperature in degrees Celsius, and the second number is the number of catfish found in the stream: (0, 0); (5, 0); (10, 2); (15, 15); (20, 13); (25, 3); (30, 0); (35, 0).

Biology Online **Self-Check Quiz** biologygmh.com

Section 3.2

Reading Preview

Objectives

- **Relate** latitude and the three major climate zones.
- **Describe** the major abiotic factors that determine the location of a terrestrial biome.
- **Distinguish** among terrestrial biomes based on climate and biotic factors.

Review Vocabulary

biome: a large group of ecosystems that share the same climate and have similar types of plant communities

New Vocabulary

weather
latitude
climate
tundra
boreal forest
temperate forest
woodland
grassland
desert
tropical savanna
tropical seasonal forest
tropical rain forest

Terrestrial Biomes

MAIN Idea Ecosystems on land are grouped into biomes primarily based on the plant communities within them.

Real-World Reading Link If you live in the eastern part of the United States, you might live in an area surrounded by deciduous forests. If you live in the central part of the United States, there might be a grassy prairie nearby. Plant communities are specific to particular ecosystems.

Effects of Latitude and Climate

Regardless of where you live, you are affected by weather and climate. During a newscast, a meteorologist will make forecasts about the upcoming weather. **Weather** is the condition of the atmosphere at a specific place and time. What causes the variation in the weather patterns that you experience? What are the effects of these weather patterns on organisms that live in different areas on Earth? One of the keys to understanding communities is to be aware of latitude and climatic conditions.

Connection to Earth Science **Latitude** The distance of any point on the surface of Earth north or south from the equator is **latitude.** Latitudes range from 0° at the equator to 90° at the poles. Light from the Sun strikes Earth more directly at the equator than at the poles, as illustrated in **Figure 3.5.** As a result, Earth's surface is heated differently in different areas. Ecologists refer to these areas as polar, temperate, and tropical zones.

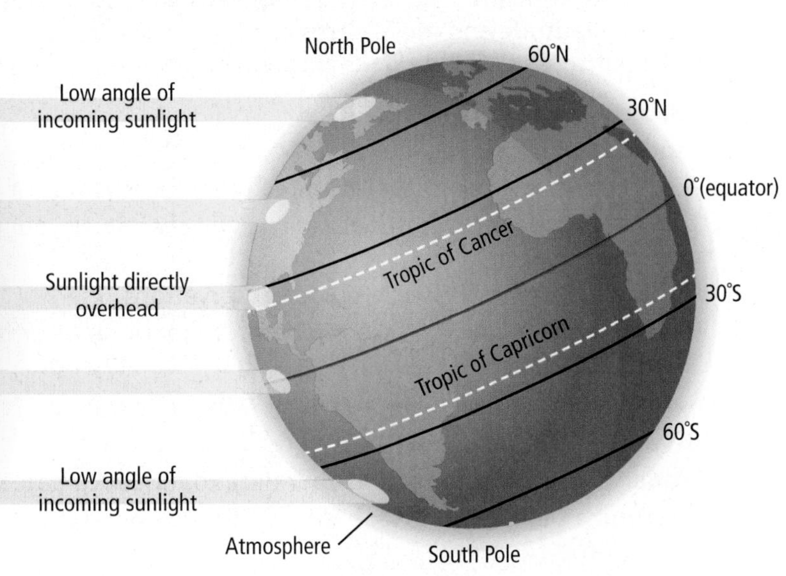

Figure 3.5 Earth's climate is determined largely by the unequal amounts of solar radiation that different areas receive.

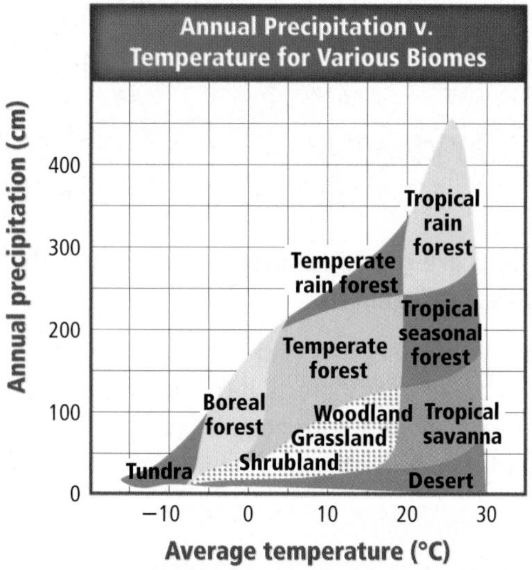

Figure 3.6 Temperature and precipitation are two major factors that influence the kind of vegetation that can exist in an area.

Analyze ***Which biome would you expect in an area that receives 200 cm of precipitation annually if the average annual temperature is 10°C?***

Personal Tutor

To learn about the greenhouse effect, visit biologygmh.com.

Climate The average weather conditions in an area, including temperature and precipitation, describe the area's **climate.** An area's latitude has a large effect on its climate. If latitude were the only abiotic factor involved in climate, biomes would be spread in equal bands encircling Earth. However, other factors such as elevation, continental landmasses, and ocean currents also affect climate. The graph in **Figure 3.6** shows how temperature and precipitation influence the communities that develop in an area. You can investigate the relationship between temperature and latitude in **Minilab 3.1.**

Recall from Chapter 2 that a biome is a large group of ecosystems that share the same climate and have similar types of communities. It is a group of plant and animal communities that have adapted to a region's climate. A biome's ecosystems occur over a large area and have similar plant communities. Even a small difference in temperature or precipitation can affect the location of a biome. Refer to **Figure 3.7** to learn how Earth's ocean currents and prevailing winds affect climate. Also illustrated in **Figure 3.7** are two ways humans have affected climate—through the hole in the ozone layer and through global warming. Global warming is in part a result of the greenhouse effect.

Major Land Biomes

Biomes are classified primarily according to the characteristics of their plants. Biomes also are characterized by temperature and precipitation. Animal species are an important characteristic of biomes as well. This section describes each of the major land biomes.

MiniLab 3.1

Formulate a Climate Model

How are temperature and latitude related? At the equator the climate is very warm. However, as you change latitude and move north or south of the equator, temperatures also change. This results in different latitudinal climate belts around the world.

Procedure

1. Read and complete the lab safety form.
2. Position a **lamp** so that it shines directly on the equator of a **globe.**
3. Predict how the temperature readings will change as you move a **thermometer** north or south away from the equator.
4. Prepare a data table to record your observations.
5. Use the thermometer to take temperature readings at different latitudes as instructed by your teacher. **WARNING:** ***The lamp and bulb will be very hot.***
6. Record temperature readings in your data table.

Analysis

1. **Model** Draw a diagram using your data to model climate belts.
2. **Recognize Cause and Effect** Why do the temperature readings change as you move north or south of the equator?

Visualizing Global Effects on Climate

Figure 3.7
Some parts of Earth receive more heat from the Sun. Earth's winds and ocean currents contribute to climate and balance the heat on Earth. Many scientists think human impacts on the atmosphere upset this balance.

Winds on Earth

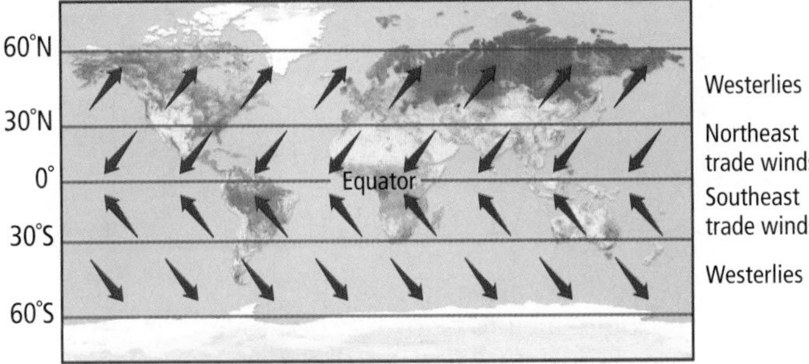

Winds are created from temperature imbalances. Distinct global wind systems transport cold air to warm areas and warm air to cold areas.

Earth's Ocean Currents

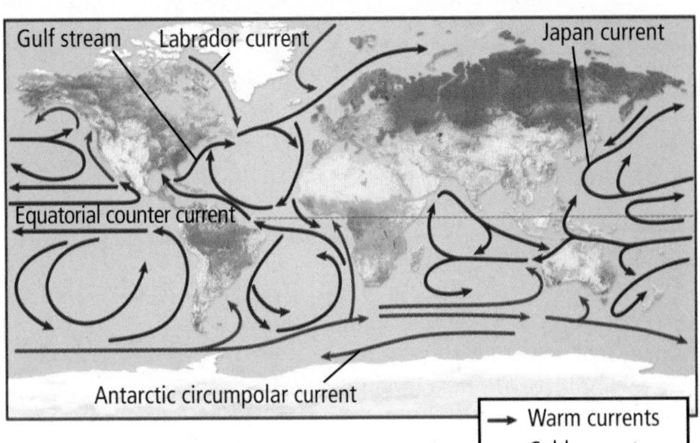

Ocean currents carry warm water toward the poles. As the water cools, it sinks toward the ocean floor and moves toward tropical regions.

Greenhouse Effect

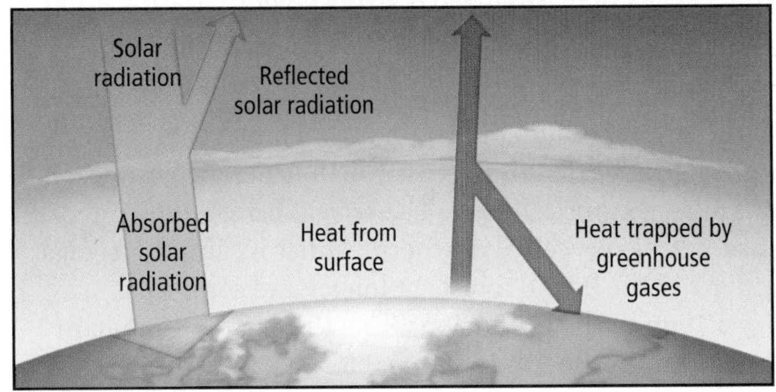

Earth's surface is warmed by the greenhouse effect. Certain gases in Earth's atmosphere, primarily water vapor, reduce the amount of energy Earth radiates into space. Other important greenhouse gases are carbon dioxide and methane.

Human Impact on the Atmosphere

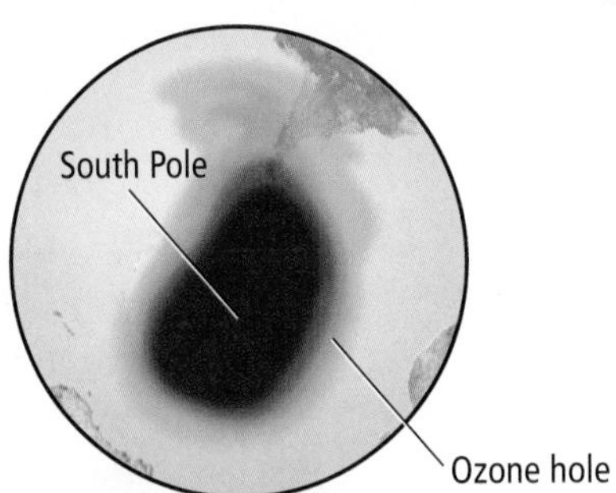

The ozone layer is a protective layer in the atmosphere that absorbs most of the harmful UV radiation from the Sun. Atmospheric studies have indicated that chlorofluorocarbons (CFCs) contribute to a seasonal reduction in ozone concentration over Antarctica, forming the Antarctic ozone hole.

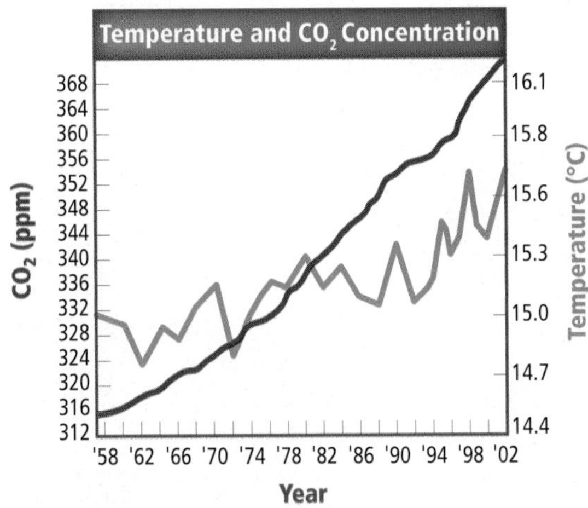

The measured increase of carbon dioxide (CO_2) in the atmosphere is mainly due to the burning of fossil fuels. As carbon dioxide levels have increased, the average global temperature has increased.

Concepts In Motion **Interactive Figure** To see an animation of the global effects on climate, visit biologygmh.com.

Biology Online

Figure 3.8 Tundra
Average precipitation: 15–25 cm per year
Temperature range: –34°C–12°C
Plant species: short grasses, shrubs
Animal species: Caribou, polar bears, birds, insects, wolves, salmon, trout
Geographic location: South of the polar ice caps in the Northern Hemisphere
Abiotic factors: soggy summers; permafrost; cold and dark much of the year

Tundra Extending in a band below the polar ice caps across northern North America, Europe, and Siberia in Asia is the tundra. The **tundra** is a treeless biome with a layer of permanently frozen soil below the surface called permafrost. Although the ground thaws to a depth of a few centimeters in the summer, its constant cycles of freezing and thawing do not allow tree roots to grow. Some animals and shallow-rooted plants that have adapted to tundra conditions are illustrated in **Figure 3.8.**

Boreal forest South of the tundra is a broad band of dense evergreen forest extending across North America, Europe, and Asia, called the boreal forest. The **boreal forest,** illustrated in **Figure 3.9,** also is called northern coniferous forest, or taiga. Summers in the boreal forest are longer and somewhat warmer than in the tundra, enabling the ground to remain warmer than in the tundra. Boreal forests, therefore, lack a permafrost layer.

Figure 3.9 Boreal forest
Average precipitation: 30–84 cm per year
Temperature range: –54°C–21°C
Plant species: spruce and fir trees, deciduous trees, small shrubs
Animal species: birds, moose, beavers, deer, wolverines, mountain lions
Geographic location: northern part of North America, Europe, and Asia
Abiotic factors: summers are short and moist; winters are long, cold, and dry

Figure 3.10 Temperate forest
Average precipitation: 75–150 cm per year
Temperature range: –30°C–30°C
Plant species: oak, beech, and maple trees, shrubs
Animal species: squirrels, rabbits, skunks, birds, deer, foxes, black bears
Geographic location: south of the boreal forests in eastern North America, eastern Asia, Australia, and Europe
Abiotic factors: well-defined seasons; summers are hot, winters are cold

Temperate forest Temperate forests cover much of southeastern Canada, the eastern United States, most of Europe, and parts of Asia and Australia. As shown in **Figure 3.10,** the **temperate forest** is composed mostly of broad-leaved, deciduous (dih SIH juh wus) trees—trees that shed their leaves in autumn. The falling red, orange, and gold leaves return nutrients to the soil. Winters are cold. In spring, warm temperature and precipitation restart the growth cycles of plants and trees. Summers are hot.

Temperate woodland and shrubland Open **woodlands** and mixed shrub communities are found in areas with less annual rainfall than in temperate forests. The woodland biome occurs in areas surrounding the Mediterranean Sea, on the western coasts of North and South America, and in South Africa and Australia. Areas that are dominated by shrubs, such as in California, are called the chaparral. **Figure 3.11** illustrates woodland and shrub communities.

Figure 3.11 Temperate woodland and shrubland
Average precipitation: 38–100 cm per year
Temperature range: 10°C–40°C
Plant species: evergreen shrubs, corn oak
Animal species: foxes, jackrabbits, birds, bobcats, coyotes, lizards, snakes, butterflies
Geographic location: surrounds the Mediterranean Sea, western coasts of North and South America, South Africa, and Australia
Abiotic factors: summers are very hot and dry; winters are cool and wet

Figure 3.12 Temperate grassland
Average precipitation: 50–89 cm per year
Temperature range: –40°C–38°C
Plant species: grasses and herbs
Animal species: gazelles, bison, horses, lions, deer, mice, coyotes, foxes, wolves, birds, quail, snakes, grasshoppers, spiders
Geographic location: North America, South America, Asia, Africa, and Australia
Abiotic factors: summers are hot, winters are cold, moderate rainfall, fires possible

Temperate grassland A biome that is characterized by fertile soils that are able to support a thick cover of grasses is called **grassland,** illustrated in **Figure 3.12.** Drought, grazing animals, and fires keep grasslands from becoming forests. Due to their underground stems and buds, perennial grasses and herbs are not eliminated by the fires that destroy most shrubs and trees. Temperate grasslands are found in North America, South America, Asia, Africa, and Australia. Grasslands are called steppes in Asia; prairies in North America; pampas, llanos, and cerrados in South America; savannas and velds in Africa; and rangelands in Australia.

Desert Deserts exist on every continent except Europe. A **desert** is any area in which the annual rate of evaporation exceeds the rate of precipitation. You might imagine a desert as a desolate place full of sand dunes, but many deserts do not match that description. As shown in **Figure 3.13,** deserts can be home to a wide variety of plants and animals.

Figure 3.13 Desert
Average precipitation: 2–26 cm per year
Temperature range: high: 20°C–49°C, low: –18°C–10°C
Plant species: cacti, Joshua trees, succulents
Animal species: lizards, bobcats, birds, tortoises, rats, antelope, desert toads
Geographic location: every continent except Europe
Abiotic factors: varying temperatures, low rainfall

Figure 3.14 Tropical savanna
Average precipitation: 50–130 cm per year
Temperature range: 20°C–30°C
Plant species: grasses and scattered trees
Animal species: lions, hyenas, cheetahs, elephants, giraffes, zebras, birds, insects
Geographic location: Africa, South America, and Australia
Abiotic factors: summers are hot and rainy, winters are cool and dry

Tropical savanna A **tropical savanna** is characterized by grasses and scattered trees in climates that receive less precipitation than some other tropical areas. Tropical savanna biomes occur in Africa, South America, and Australia. The plants and animals shown in **Figure 3.14** are common to tropical savannas.

Tropical seasonal forest **Figure 3.15** illustrates a tropical seasonal forest. **Tropical seasonal forests,** also called tropical dry forests, grow in areas of Africa, Asia, Australia, and South and Central America. In one way, the tropical seasonal forest resembles the temperate deciduous forest because during the dry season, almost all of the trees drop their leaves to conserve water.

Reading Check **Compare and contrast** tropical savannas and tropical seasonal forests.

Figure 3.15 Tropical seasonal forest
Average precipitation: >200 cm per year
Temperature range: 20°C–25°C
Plant species: deciduous and evergreen trees, orchids, mosses
Animal species: elephants, tigers, monkeys, koalas, rabbits, frogs, spiders
Geographic location: Africa, Asia, Australia, and South and Central America
Abiotic factors: rainfall is seasonal

■ **Figure 3.16** Tropical rain forest
Average precipitation: 200–1000 cm per year
Temperature range: 24°C–27°C
Plant species: broadleaf evergreens, bamboo, sugar cane
Animal species: chimpanzees, Bengal tigers, elephants, orangutans, bats, toucans, sloth, cobra snakes
Geographic location: Central and South America, southern Asia, western Africa, and northeastern Australia
Abiotic factors: humid all year, hot and wet

Tropical rain forest Warm temperatures and large amounts of rainfall throughout the year characterize the **tropical rain forest** biome illustrated in **Figure 3.16.** Tropical rain forests are found in much of Central and South America, southern Asia, western Africa, and northeastern Australia. The tropical rain forest is the most diverse of all land biomes. Tall, broad-leaved trees with branches heavy with mosses, ferns, and orchids make up the canopy of the tropical rain forest. Shorter trees, shrubs, and plants, such as ferns and creeping plants, make up another layer, or understory, of tropical rain forests.

Other Terrestrial Areas

You might have noticed that the list of terrestrial biomes does not include some important areas. Many ecologists omit mountains from the list. Mountains are found throughout the world and do not fit the definition of a biome because their climate characteristics and plant and animal life vary depending on elevation. Polar regions also are not considered true biomes because they are ice masses and not true land areas with soil.

Mountains If you go up a mountain, you might notice that abiotic conditions, such as temperature and precipitation, change with increasing elevation. These variations allow many communities to exist on a mountain. As **Figure 3.17** illustrates, biotic communities also change with increasing altitude, and the tops of tall mountains may support communities that resemble those of the tundra.

Study Tip

Summaries Review the terrestrial biomes featured in this section. Choose one or two biomes and write two sentences that summarize the information.

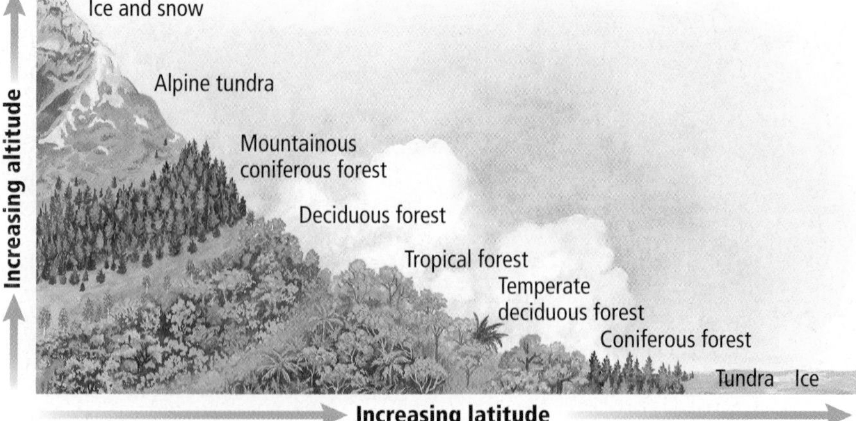

■ **Figure 3.17** As you climb a mountain or increase in latitude, the temperature drops and the climate changes.
Describe ***the relationship between altitude and latitude.***

■ **Figure 3.18** A surprising number of species inhabit the polar regions, including these penguins in Antarctica.

Polar regions Polar regions border the tundra at high latitudes. These regions are cold all year, the coldest temperature ever recorded, −89°C, was in Antarctica—the continent that lies in the southern polar region. In the northern polar region lies the ice-covered Arctic Ocean and Greenland. Covered by a thick layer of ice, the polar regions might seem incapable of sustaining life. However, as shown in **Figure 3.18,** colonies of penguins live in Antarctica. Additionally, whales and seals patrol the coasts, preying on penguins, fish, or shrimplike invertebrates called krill. The arctic polar region supports even more species, including polar bears and arctic foxes. Human societies have also inhabited this region throughout history. Although the average winter temperature is about −30°C, the arctic summer in some areas is warm enough for vegetables to be grown.

CAREERS IN BIOLOGY

Climatologist Unlike meteorologists, who study current weather conditions, climatologists study long-term climate patterns and determine how climate changes affect ecosystems. For more information on biology careers, visit biologygmh.com.

Section 3.2 Assessment

Section Summary

- Latitude affects terrestrial biomes according to the angle at which sunlight strikes Earth.
- Latitude, elevation, ocean currents, and other abiotic factors determine climate.
- Two major abiotic factors define terrestrial biomes.
- Terrestrial biomes include tundra, boreal forests, temperate forests, temperate woodlands and shrublands, temperate grasslands, deserts, tropical savannas, tropical seasonal forests, and tropical rain forests.

Understand Main Ideas

1. MAIN Idea **Describe** nine major biomes.
2. **Describe** the abiotic factors that determine a terrestrial biome.
3. **Summarize** variations in climate among three major zones as you travel south from the equator toward the South Pole.
4. **Indicate** the differences between temperate grasslands and tropical savannas.
5. **Compare and contrast** the climate and biotic factors of tropical seasonal forests and temperate forests.

Think Critically

6. **Hypothesize** why the tropical rain forests have the greatest diversity of living things.

WRITING in **Biology**

7. Tropical forests are being felled at a rate of 17 million hectares per year, which represents almost two percent of the forest area. Use this information to write a pamphlet describing how much rain forest area exists and when it might be gone.

Section 3.3

Reading Preview

Objectives

- **Identify** the major abiotic factors that determine the aquatic ecosystems.
- **Recognize** that freshwater ecosystems are characterized by depth and water flow.
- **Identify** transitional aquatic ecosystems and their importance.
- **Distinguish** the zones of marine ecosystems.

Review Vocabulary

salinity: a measure of the amount of salt in a body of water

New Vocabulary

sediment
littoral zone
limnetic zone
plankton
profundal zone
wetlands
estuary
intertidal zone
photic zone
aphotic zone
benthic zone
abyssal zone

Aquatic Ecosystems

MAIN Idea **Aquatic ecosystems are grouped based on abiotic factors such as water flow, depth, distance from shore, salinity, and latitude.**

Real-World Reading Link Think about the body of water that is closest to where you live. What are its characteristics? How deep is it? Is it freshwater or salty? For centuries, bodies of water have been central to cultures around the world.

The Water on Earth

When you think about water on Earth, you might recall a vacation at the ocean or a geography lesson in which you located Earth's oceans and seas. You probably have heard about other large bodies of water, such as the Amazon river and the Great Salt Lake. A globe of Earth is mainly blue in color because the planet is largely covered with water. Ecologists recognize the importance of water because of the biological communities that water supports. In this section, you will read about freshwater, transitional, and marine aquatic ecosystems. You also will read about the abiotic factors that affect these ecosystems.

Freshwater Ecosystems

The major freshwater ecosystems include ponds, lakes, streams, rivers, and wetlands. Plants and animals in these ecosystems are adapted to the low salt content in freshwater and are unable to survive in areas of high salt concentration. Only about 2.5 percent of the water on Earth is freshwater, as illustrated by the circle graph on the left in **Figure 3.19.** The graph on the right in **Figure 3.19** shows that of that 2.5 percent, 68.9 percent is contained in glaciers, 30.8 percent is groundwater, and only 0.3 percent is found in lakes, ponds, rivers, streams, and wetlands. Interestingly, almost all of the freshwater species live in this 0.3 percent.

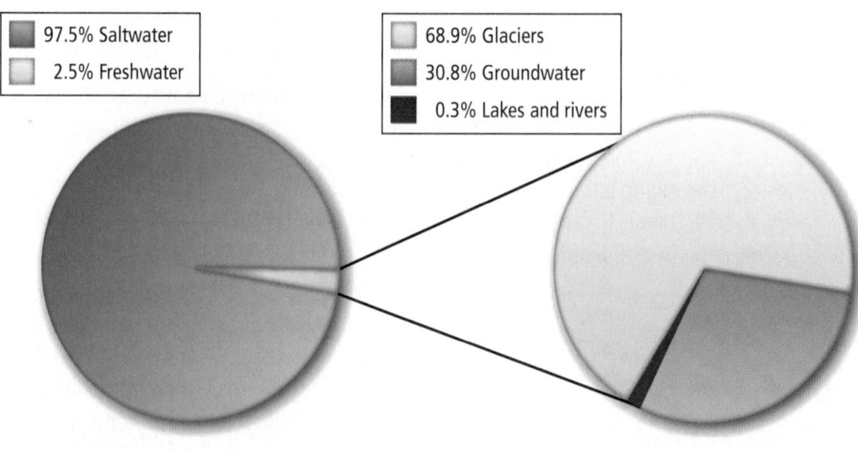

■ **Figure 3.19** The vast majority of Earth's water is salt water. Most of the freshwater supply is locked in glaciers.

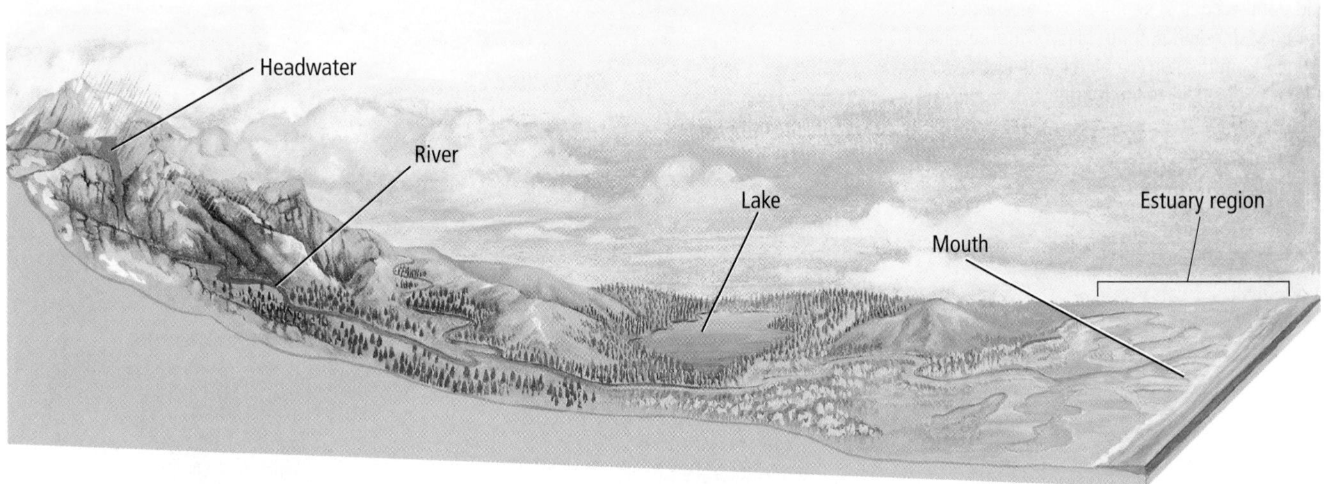

Figure 3.20 Mountain streams have clear, cold water that is highly oxygenated and supports the larvae of many insects and the coldwater fish that feed on them. Rivers become increasingly wider, deeper, and slower. At the mouth, many rivers divide into many channels where wetlands or estuaries form.

Rivers and streams The water in rivers and streams flows in one direction, beginning at a source called a headwater and traveling to the mouth, where the flowing water empties into a larger body of water, as illustrated in **Figure 3.20.** Rivers and streams also might start from underground springs or from snowmelt. The slope of the landscape determines the direction and speed of the water flow. When the slope is steep, water flows quickly, causing a lot of sediment to be picked up and carried by the water. **Sediment** is material that is deposited by water, wind, or glaciers. As the slope levels, the speed of the water flow decreases and sediments are deposited in the form of silt, mud, and sand.

The characteristics of rivers and streams change during the journey from the source to the mouth. Interactions between wind and the water stir up the water's surface, which adds a significant amount of oxygen to the water. Interactions between land and water result in erosion, in nutrient availability, and in changing the path of the river or stream.

The currents and turbulence of fast-moving rivers and streams prevent much accumulation of organic materials and sediment. For this reason, there usually are fewer species living in rapid waters similar to that in **Figure 3.21.** An important characteristic of life in rivers and streams is the ability to withstand the constant water current. Plants that root themselves into the streambed are common in areas where water is slowed by rocks or sandbars. Young fish hide in these plants and feed on the drifting microscopic organisms and aquatic insects.

In slow-moving water, insect larvae are the primary food source for many fish, including American eel, brown bullhead catfish, and trout. Other organisms, such as crabs and worms, are sometimes present in calm water. Animals that live in slow-moving water include newts, tadpoles, and frogs.

 **Reading Check** **Describe** key abiotic factors that define rivers and streams.

Figure 3.21 The turbulent churning action of fast-moving rivers and streams does not allow for many plants to take root or for other species to inhabit these waters.

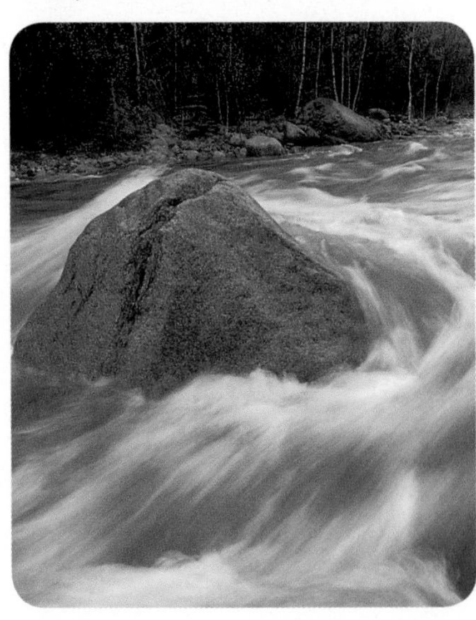

■ **Figure 3.22** The temperature of lakes and ponds varies depending on the season. During spring and autumn, deep water receives oxygen from the surface water and surface water receives inorganic nutrients from the deep water.

***Compare** the type of life that might live in a shallow lake in the tropics to one in the mid-latitudes.*

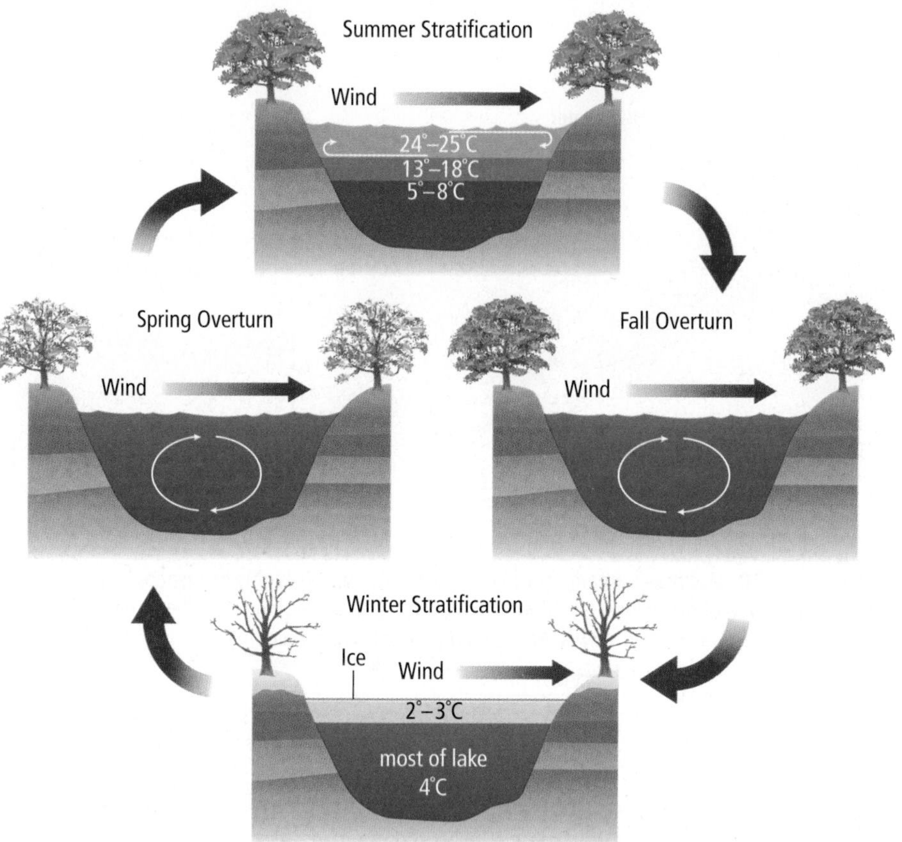

Lakes and ponds An inland body of standing water is called a lake or a pond. It can be as small as a few square meters or as large as thousands of square meters. Some ponds might be filled with water for only a few weeks or months each year, whereas some lakes have existed for thousands of years. **Figure 3.22** illustrates how in temperate regions the temperature of lakes and ponds varies depending on the season.

During the winter, most of the water in a lake or pond is the same temperature. In the summer, the warmer water on top is less dense than the colder water at the bottom. During the spring and fall, as the water warms or cools, turnover occurs. The top and bottom layers of water mix, often due to winds, and this results in a uniform water temperature. This mixing circulates oxygen and brings nutrients from the bottom to the surface.

Nutrient-poor lakes, called oligotrophic (uh lih goh TROH fihk) lakes, often are found high in the mountains. Few plant and animal species are present as a result of small amounts of organic matter and nutrients. Nutrient-rich lakes, called eutrophic (yoo TROH fihk) lakes, usually are found at lower altitudes. Many plant and animal species are present as a result of organic matter and plentiful nutrients, some of which come from agricultural and urban activities.

VOCABULARY

WORD ORIGIN

Eutrophic/oligotrophic

eu– prefix; from Greek, meaning *well*

oligo– prefix; from Greek, meaning *few*

–***trophic;*** from Greek, meaning *nourish.*

Lakes and ponds are divided into three zones based on the amount of sunlight that penetrates the water. The area closest to the shore is the **littoral** *(LIH tuh rul)* **zone.** The water in this zone is shallow, which allows sunlight to reach the bottom. Many producers, such as aquatic plants and algae, live in these shallow waters. The abundance of light and producers make the littoral zone an area of high photosynthesis. Many consumers also inhabit this zone, including frogs, turtles, worms, crustaceans, insect larvae, and fish.

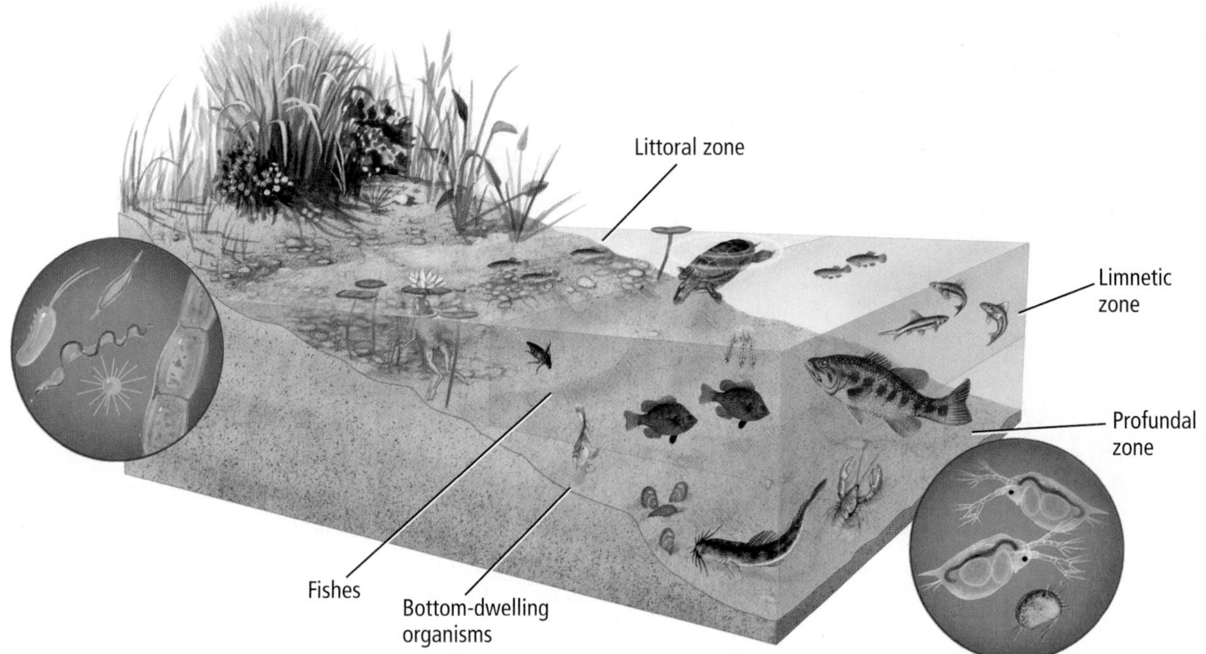

The **limnetic** (lihm NEH tihk) **zone** is the open water area that is well lit and is dominated by plankton. **Plankton** are free-floating photosynthetic autotrophs that live in freshwater or marine ecosystems. Many species of freshwater fish live in the limnetic zone because food, such as plankton, is readily available.

Minimal light is able to penetrate through the limnetic zone into the deepest areas of a large lake, which is called the **profundal** (pruh FUN dul) **zone.** The profundal zone is therefore much colder and lower in oxygen than the other two zones. A limited number of species live in this harsh environment. **Figure 3.23** illustrates the zones and biodiversity of lakes and ponds.

■ **Figure 3.23** Most of a lake's biodiversity is found in the littoral and limnetic zones. However, many species of bottom dwellers depend on materials that drift down from above.

MiniLab 3.2

Prepare a Scientific Argument

Should an environment be disturbed? One of the greatest challenges we face as a species is balancing the needs of an ever-growing human global population with the needs of wildlife and the quality of the global environment. Imagine this scenario: The county commissioners are considering a proposal to build a road through the local pond and wetlands. This road will provide much-needed access to areas of work, and will help boost the economy of a struggling town. This will mean that the pond and surrounding wetlands must be drained and filled. Many people support the proposal, while many people oppose it. How might a compromise be reached?

Procedure

1. Prepare a comparison table in which you can list pros and cons.
2. Identify the pros and cons for draining the pond and building the road, for keeping the pond and not building the road, or for building the road elsewhere.

Analysis

1. **Design** a plan to support one course of action. What steps could you take to achieve your goal? Be prepared to share and defend your plan to the rest of the class.
2. **Think Critically** Why are decisions involving the environment difficult to make?

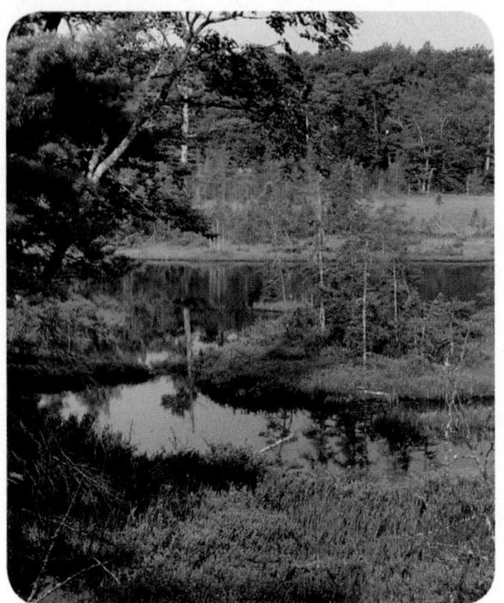

■ **Figure 3.24** Bogs are a type of wetland characterized by moist, decaying plant material and dominated by mosses.

Transitional Aquatic Ecosystems

In many areas, aquatic ecosystems do not look like a stream or a pond or even an ocean. In fact, many aquatic environments are a combination of two or more different environments. These areas, which ecologists call transitional aquatic ecosystems, can be areas where land and water or salt water and freshwater intermingle. Wetlands and estuaries are common examples of transitional aquatic ecosystems.

Wetlands Areas of land such as marshes, swamps, and bogs that are saturated with water and that support aquatic plants are called **wetlands.** Plant species that grow in the moist, humid conditions of wetlands include duckweed, pond lilies, cattails, sedges, mangroves, cypress, and willows. Bogs, like the cedar bog shown in **Figure 3.24,** are wet and spongy areas of decomposing vegetation that also support many species of organisms. Wetlands have high levels of species diversity. Many amphibians, reptiles, birds (such as ducks and herons), and mammals (such as raccoons and mink) live in wetlands.

Estuaries Another important transitional ecosystem is an estuary, shown in **Figure 3.25.** Estuaries are among the most diverse ecosystems, rivaled only by tropical rain forests and coral reefs. An **estuary** (ES chuh wer ee) is an ecosystem that is formed where freshwater from a river or stream merges with salt water from the ocean. Estuaries are places of transition—from freshwater to salt water and from land to sea—that are inhabited by a wide variety of species. Algae, seaweeds, and marsh grasses are the dominant producers. However, many animals, including a variety of worms, oysters, and crabs, depend on detritus for food. Detritus (dih TRY tus) is comprised of tiny pieces of organic material.

VOCABULARY
ACADEMIC VOCABULARY
Comprise:
to be made up of
Your community is comprised of your family, your classmates, and people who live nearby.

Mangrove trees also can be found in tropical estuaries, such as the Everglades National Park in Florida, where they sometimes form swamps. Many species of marine fishes and invertebrates, such as shrimp, use estuaries as nurseries for their young. Waterfowl, such as ducks and geese, depend on estuary ecosystems for nesting, feeding, and migration rest areas.

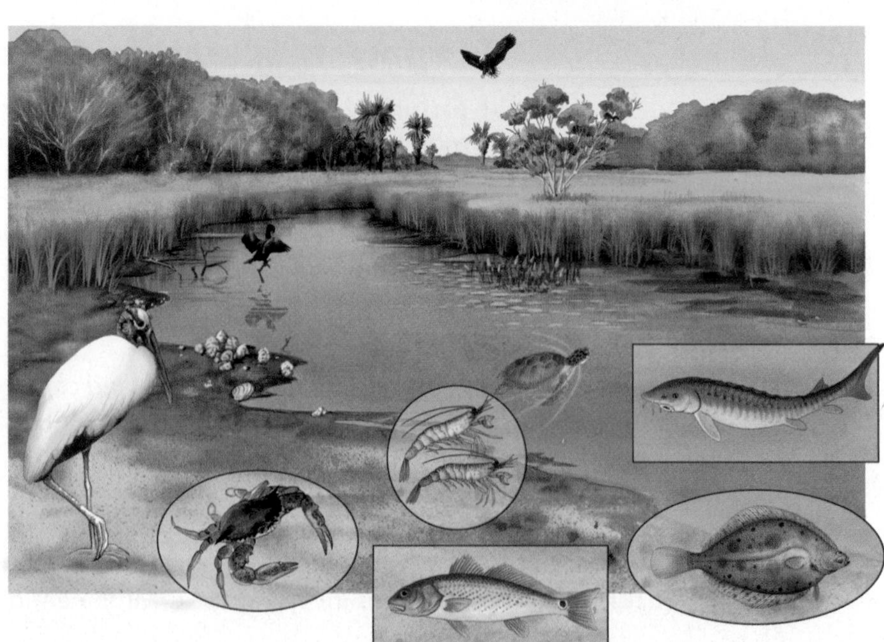

■ **Figure 3.25** Salt-tolerant plants above the low-tide line dominate estuaries formed in temperate areas.
Infer *How would an estuary differ in a tropical area?*

Salt marshes are transitional ecosystems similar to estuaries. Salt-tolerant grasses dominate above the low-tide line, and seagrasses grow in submerged areas of salt marshes. Salt marshes support different species of animals, such as shrimp and shellfish.

Marine Ecosystems

Connection to Earth Science Earth is sometimes called "the water planet." As such, marine ecosystems have a significant impact on the planet. For example, through photosynthesis, marine algae consume carbon dioxide from the atmosphere and produce over 50 percent of the atmosphere's oxygen. Additionally, the evaporation of water from oceans eventually provides the majority of precipitation—rain and snow. Like ponds and lakes, oceans are separated into distinct zones.

Intertidal zone The **intertidal** (ihn tur TY dul) **zone** is a narrow band where the ocean meets land. Organisms that live in this zone must be adapted to the constant changes that occur as daily tides and waves alternately submerge and expose the shore. The intertidal zone is further divided into vertical zones, as illustrated in **Figure 3.26.** The area of the spray zone is dry most of the time. It is only during high tides that this part of the shoreline is sprayed with salt water, and few plants and animals are able to live in this environment. The high-tide zone is under water only during high tides. However, this area receives more water than the spray zone, so more plants and animals are able to live there. The mid-tide zone undergoes severe disruption twice a day as the tides cover and uncover the shoreline with water. Organisms in this area must be adapted to long periods of air and water. The low-tide zone is covered with water unless the tide is unusually low and is the most populated area of the intertidal zone.

Reading Check **Describe** environmental variation in intertidal zones.

Figure 3.26 The intertidal zone is further divided into zones where different communities exist.
Compare and contrast *the zones illustrated in* Figures 3.23 *and* 3.26.

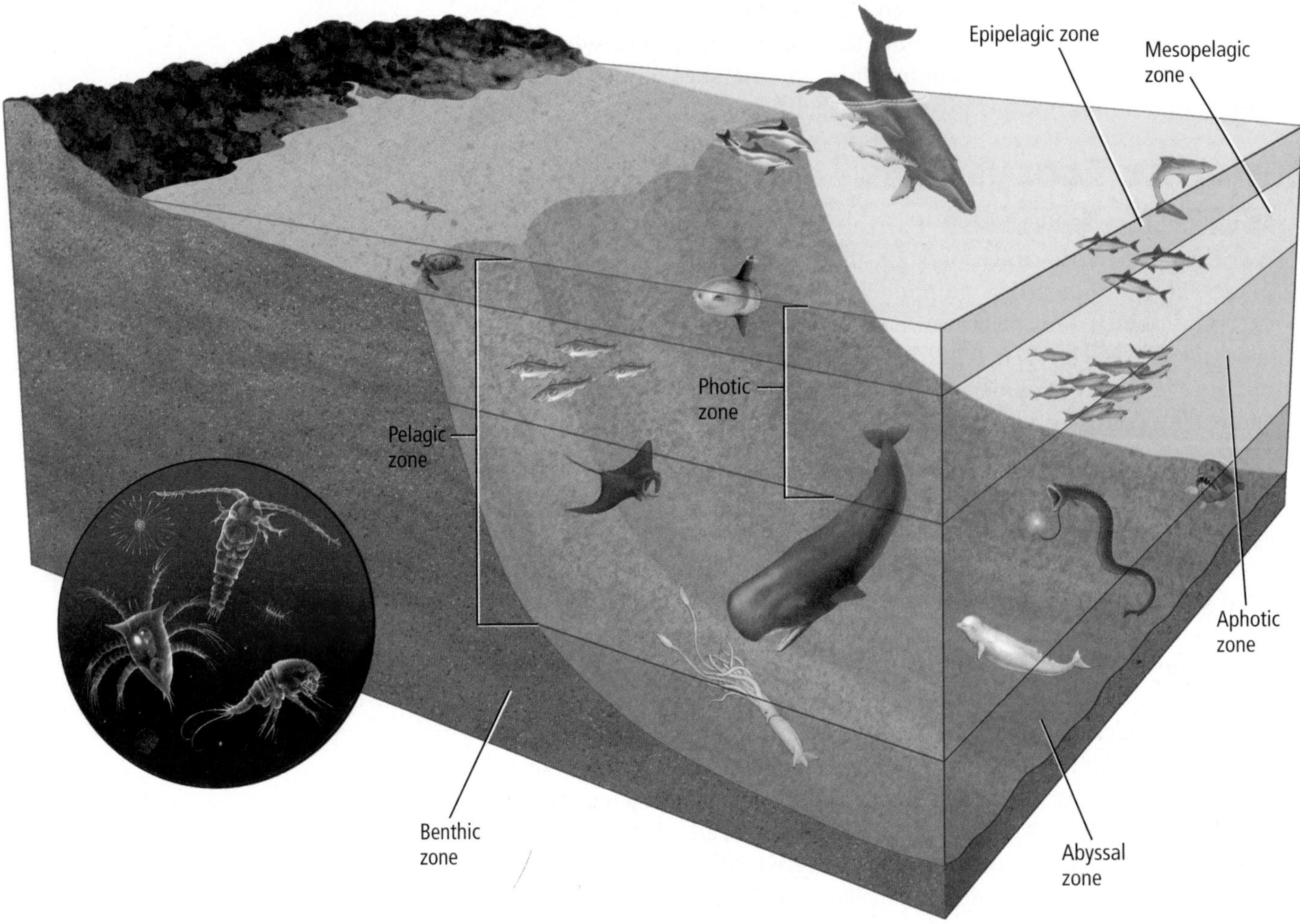

■ **Figure 3.27** Producers are found in the photic zone. Consumers live in the pelagic, abyssal, and benthic zones.

Open ocean ecosystems As illustrated in **Figure 3.27,** the zones in the open ocean include the pelagic (puh LAY jihk) zone, abyssal (uh BIH sul) zone, and benthic zone. The area to a depth of about 200 m of the pelagic (puh LAY jihk) zone is the **photic zone,** also called the euphotic zone. The photic zone is shallow enough that sunlight is able to penetrate. As depth increases, light decreases. Autotrophic organisms in the photic zone include surface seaweeds and plankton. Animals in the photic zone include many species of fish, sea turtles, jellyfish, whales, and dolphins. Many of these animals feed on plankton, but others feed on larger species.

Vocabulary

Word origin

Photic
comes from the Greek word *photos*, meaning *light*.

Below the photic zone lies the **aphotic zone**—an area where sunlight is unable to penetrate. This region of the pelagic zone remains in constant darkness and generally is cold, but there is thermal layering with a mixing of warm and cold ocean currents. Organisms that depend on light energy to survive cannot live in the aphotic zone.

The **benthic zone** is the area along the ocean floor that consists of sand, silt, and dead organisms. In shallow benthic zones, sunlight can penetrate to the bottom of the ocean floor. As depth increases, less sunlight penetrates the deeper water, and temperature decreases. Species diversity tends to decrease with depth, except in areas with hydrothermal vents, where shrimp, crabs, and many species of tubeworms are found. Many species of fishes, octopuses, and squids live in the benthic zone.

The deepest region of the ocean is called the **abyssal zone.** Water in this area is very cold. Most organisms in this zone rely on food materials that drift down from the zones above. However, on the seafloor along the boundaries of Earth's plates, hydrothermal vents spew large amounts of hot water, hydrogen sulfide, and other minerals. Scientists have found bacterial communities existing in these locations that can use the sulfide molecules for energy. These organisms are at the bottom of a food chain that includes invertebrates, such as clams and crabs, and vertebrates, such as fishes.

Coastal ocean and coral reefs One of the world's largest coral reefs is off the southern coast of Florida. Coral reefs are among the most diverse ecosystems. They are widely distributed in warm shallow marine waters. Coral reefs form natural barriers along continents that protect shorelines from erosion. The dominant organisms in coral reefs are corals. When you think of coral, you might picture a hard, stony structure, but this is only the framework secreted by tiny animal polyps. Corals are soft-bodied invertebrates that live in the stonelike structures.

Most coral polyps have a symbiotic relationship with algae called zooxanthellae (zoo uh zan THEL uh). These algae provide corals with food, and in turn, the coral provides protection and access to light for the algae. Corals also feed by extending tentacles to obtain plankton from the water. Other coral reef animals include species of microorganisms, sea slugs, octopuses, sea urchins, sea stars, and fishes. **Figure 3.28** shows only a small portion of the diversity of Florida's coral reef.

Like all ecosystems, coral reefs are sensitive to changes in the environment. Changes that are the result of naturally occurring events, such as increased sediment from a tsunami, can cause the death of a reef. Human activities, such as land development and harvesting for calcium carbonate, also can damage or kill a coral reef. Today, ecologists monitor reefs and reef environments to help protect these delicate ecosystems.

Figure 3.28 Coral reefs off the southern tip of Florida are among the world's largest and most diverse reefs.

Section 3.3 Assessment

Section Summary

- Freshwater ecosystems include ponds, lakes, streams, rivers, and wetlands.
- Wetlands and estuaries are transitional aquatic ecosystems.
- Marine ecosystems are divided into zones that are classified according to abiotic factors.
- Estuaries and coral reefs are among the most diverse of all ecosystems.

Understand Main Ideas

1. MAIN Idea **List** the abiotic factors that are used to classify aquatic ecosystems.
2. **Apply** what you know about ponds. Do you think the same organisms that would live in a seasonal pond would live in a pond that existed year-round? Explain.
3. **Describe** an ecological function of an estuary.
4. **Describe** the zones of the open ocean.

Think Critically

5. **Infer** how autotrophs in the abyssal zone of the ocean are different from those of the photic zone.

MATH in Biology

6. In November 2004, the floodgates of Glen Canyon Dam opened in an attempt to improve the Colorado River habitat. The release topped 1161 m^3/s—four times the usual daytime flow. Based on this information, about how much water normally flows through the dam on a daily basis?

In the Field

Career: Wildlife Conservation Biologist

The Last Wild Place On Earth

Imagine you are hiking through a dense forest, thick with undergrowth and trailing vines. There are no roads or even footpaths. Sound like a nightmare? To wildlife conservation biologist Dr. Michael Fay, it's paradise.

Megatransect Fay is a conservation biologist who studies how human activities affect ecosystems. While working in central Africa, Fay realized that a vast, intact forest corridor untouched by human activities ran from the center of the continent to the Atlantic Ocean. Fay envisioned walking the length of this corridor to study what he called "the last wild place on earth." He named this historic project the Megatransect.

Through the heart of Africa The Megatransect began in 1999. During the 15-month journey, Fay's team covered 3200 km on foot, traveling through the republics of Congo, Cameroon, and Gabon. Thirteen national parks have been created in Gabon as a result of Fay's work.

Megatransect data at work Megatransect data are helping to define human impact in measurable terms. Using satellite and field data, conservation biologists have designed a global map called the Human Footprint, which describes the extent of human influence in central Africa.

The Human Footprint map shown to the right indicates only limited human impact; most conservation biologists think this is changing. Fay hopes that the Megatransect will convince others of the importance of preserving untouched areas.

The success of the megatransect led to the funding of the megaflyover, a 110,000 km journey over Africa in a Cessna 182.

Fay began his eight month flight in 2004. The plane was mounted with a high-resolution digital camera that was matched with a Global Positioning Satellite (GPS) in an attempt to build on data collected during the megatransect.

The Megatransect Human Footprint

Low — High

Gradient of Human Influence

CAREERS in Biology

Oral Report Visit biologygmh.com for links to video and images from the Megatransect. Develop an oral presentation describing the skills and knowledge that made the project a success.

BIOLAB

Design Your Own

FIELD INVESTIGATION: A POND IN A JAR

Background: **Ecologists study parts of the biosphere. Each part is a unit containing many complex interactions between living things, such as food chains and food webs and the physical environment, the water cycle, and the mineral cycle. Smaller parts of the biosphere, such as communities and ecosystems, are the most practical for ecologists to explore and investigate.**

Question: *What can we learn from studying a miniaturized biological ecosystem?*

Materials

glass or clear plastic gallon jars
pond water
pond mud
appropriate cultures and select living organisms
Choose any other materials that would be appropriate to this lab.

Safety Precautions

WARNING: ***Use care when handling jars of pond water.***

Plan and Perform the Experiment

1. Read and complete the lab safety form.
2. Prepare an observation table as instructed.
3. Brainstorm and plan the step-by-step miniaturization of a pond community. Make sure your teacher has approved your plan before you proceed.
4. Decide on a particular aspect of your miniature community to evaluate and design an appropriate experiment. For example, you might test the effect of sunlight on your ecosystem.
5. Carry out your experiment.

Analyze and Conclude

1. **Explain** Why did you conduct your experiment slowly in a step-by-step manner? What might have happened if you poured everything into the jar all at once?
2. **Identify Variables** What was the independent variable? The dependent variable?
3. **Design an experiment** Did your experiment have a control? Explain.
4. **Analyze and conclude** Describe how your community differs from a pond community found in nature.
5. **Error analysis** How effective was your design? Explain possible sources of errors.

WRITING in Biology

Communicate Write a short story in which you describe what it would be like to be a protozoan (microscopic animal) living in your pond-in-a-jar. To learn more about pond ecosystems, visit BioLabs at biologygmh.com.

Study Guide

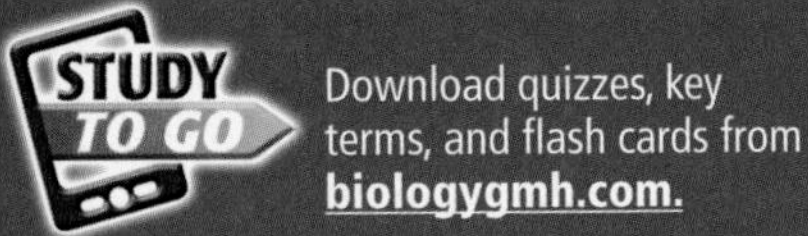

Download quizzes, key terms, and flash cards from biologygmh.com.

FOLDABLES **Research** a natural disaster that occurred twenty or more years ago. Determine what the community looked like before the disaster and what the area looks like today. Draw a "then and now" picture.

Vocabulary	Key Concepts
Section 3.1 Community Ecology	
• climax community (p. 63) • community (p. 60) • ecological succession (p. 62) • limiting factor (p. 61) • primary succession (p. 62) • secondary succession (p. 63) • tolerance (p. 61)	**MAIN Idea** All living organisms are limited by factors in the environment. • Limiting factors restrict the growth of a population within a community. • Organisms have a range of tolerance for each limiting factor that they encounter. • Primary succession occurs on areas of exposed rock or bare sand (no soil). • Communities progress until there is little change in the composition of species. • Secondary succession occurs as a result of a disturbance in a mature community.
Section 3.2 Terrestrial Biomes	
• boreal forest (p. 68) • climate (p. 66) • desert (p. 70) • grassland (p. 70) • latitude (p. 65) • temperate forest (p. 69) • tropical rain forest (p. 72) • tropical savanna (p. 71) • tropical seasonal forest (p. 71) • tundra (p. 68) • weather (p. 65) • woodland (p. 69)	**MAIN Idea** Ecosystems on land are grouped into biomes primarily based on the plant communities within them. • Latitude affects terrestrial biomes according to the angle at which sunlight strikes Earth. • Latitude, elevation, ocean currents, and other abiotic factors determine climate. • Two major abiotic factors define terrestrial biomes. • Terrestrial biomes include tundra, boreal forests, temperate forests, temperate woodlands and shrublands, temperate grasslands, deserts, tropical savannas, tropical seasonal forests, and tropical rain forests.
Section 3.3 Aquatic Ecosystems	
• abyssal zone (p. 81) • aphotic zone (p. 80) • benthic zone (p. 80) • estuary (p. 78) • intertidal zone (p. 79) • limnetic zone (p. 77) • littoral zone (p. 76) • photic zone (p. 80) • plankton (p. 77) • profundal zone (p. 77) • sediment (p. 75) • wetlands (p. 78)	**MAIN Idea** Aquatic ecosystems are grouped based on abiotic factors such as water flow, depth, distance from shore, salinity, and latitude. • Freshwater ecosystems include ponds, lakes, streams, rivers, and wetlands. • Wetlands and estuaries are transitional aquatic ecosystems. • Marine ecosystems are divided into zones that are classified according to abiotic factors. • Estuaries and coral reefs are among the most diverse of all ecosystems.

CHAPTER 3 Assessment

Section 3.1

Vocabulary Review

Choose the correct italicized term to complete each sentence.

1. An area of forest that experiences very little change in species composition is a *climax community/ primary succession.*
2. The amount of oxygen in a fish tank is a *tolerance zone/limiting factor* that affects the number of fish that can live in the tank.
3. *Ecological succession/Secondary succession* describes the events that take place on a hillside that has experienced a destructive mudslide.

Understand Key Concepts

4. Lack of iron in the photic zone of the open ocean restricts the size of plankton populations. Iron is what kind of factor for marine plankton?
   A. distribution
   B. tolerance
   C. limiting
   D. biotic

For questions 5-7, use the generalized graph below that describes an organism's tolerance to a particular factor.

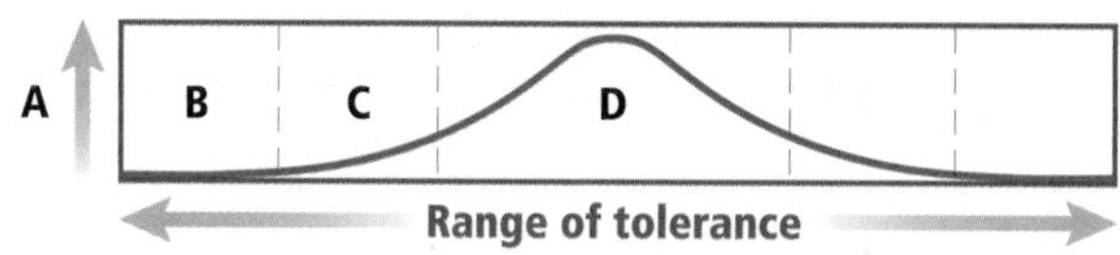

5. According to the graph, which letter represents the zone of intolerance for the factor in question?
   A. A **C.** C
   B. B **D.** D
6. What does the letter "D" in the graph represent?
   A. zone of intolerance
   B. zone of physiological stress
   C. optimum range
   D. upper limit
7. Which letter represents the zone of physiological stress?
   A. A **C.** C
   B. B **D.** D
8. Which is a place you most likely would find pioneer species growing?
   A. climax forest **C.** disturbed grassland
   B. coral reef **D.** newly formed volcano

Constructed Response

9. **CAREERS IN BIOLOGY** A state parks and wildlife department stocks several bodies of water, including rivers and lakes, with rainbow trout. The trout survive, but do not reproduce. In terms of tolerance, discuss what might be happening.

Use the image below to answer question 10.

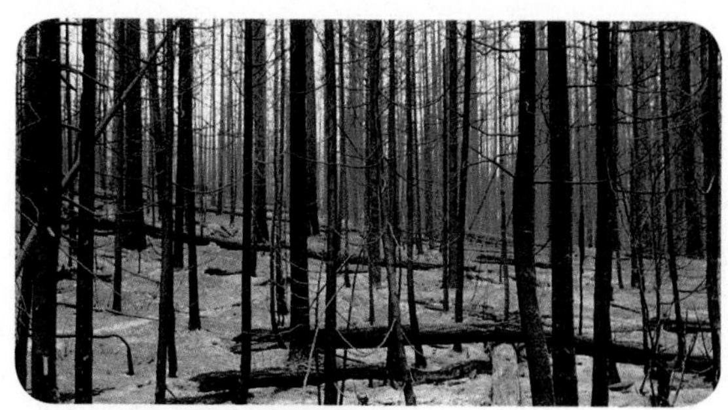

10. **Short Answer** Describe how the successional stages would differ from primary succession.
11. **Open Ended** Explain why the concepts of limiting factors and tolerance are important in ecology.

Think Critically

12. **Infer** whether species diversity increases or decreases after a fire on a grassland. Explain your response.
13. **Generalize** the difference between a successional stage and a climax community.

Section 3.2

Vocabulary Review

Choose the vocabulary term from the Study Guide page that best fits each definition below.

14. the condition of the atmosphere
15. the average conditions in an area
16. a biome characterized by evaporation exceeding precipitation

Chapter 3 Assessment

Understand Key Concepts

17. Which best describes the distribution of communities on a tall mountain?
   A. Evergreen forests exist up to the tree line and no vegetation is found above the tree line.
   B. Several communities might be stratified according to altitude and might end in an ice field at the top of the highest mountains.
   C. As altitude increases, tall trees are replaced by shorter trees, and ultimately are replaced by grasses.
   D. Tundralike communities exist at the top of the highest mountains, and deserts are found at the lower elevations.

Use the diagram below to answer question 18.

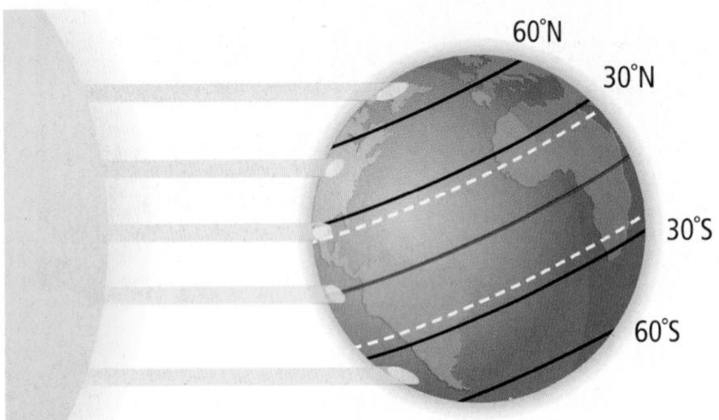

18. Which area receives the least amount of solar energy per unit of surface area?
   A. north of 60°N and south of 60°S
   B. south of 30°N and north of 30°N
   C. between the Tropic of Cancer and the Tropic of Capricorn
   D. north and south temperate zones

19. What is the name for large geographic areas with similar climax communities?
   A. assemblages
   B. communities
   C. successions
   D. biomes

20. Which biome occurs in the United States and once contained huge herds of grazing herbivores?
   A. boreal forest
   B. temperate forest
   C. grassland
   D. savanna

21. Which land biome contains the greatest species diversity?
   A. tundra
   B. grassland
   C. desert
   D. tropical rain forest

Constructed Response

Use the image below to answer question 22.

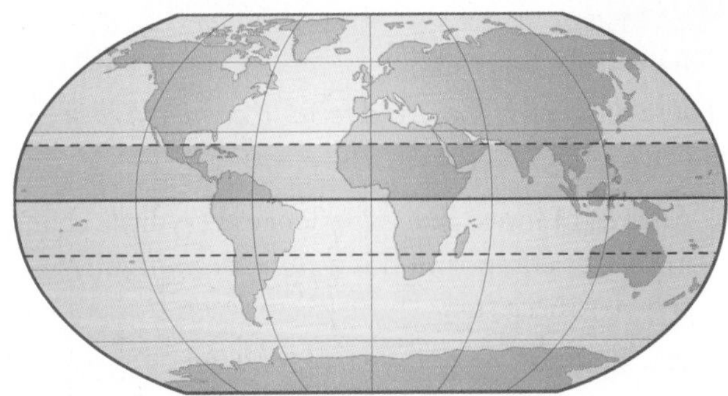

22. **Open Ended** Describe a biome that might be found in the shaded area shown above.

23. **Open Ended** In December 2004, a huge iceberg caused a large number of penguin chicks to die of starvation. Ice shelves broke apart in areas where the air temperature increased. The parents of the penguins were cut off from their food source. How is this an example of temperature as a limiting factor?

Think Critically

24. **Suggest** why land biomes are classified according to their plant characteristics rather than according to the animals that inhabit them.

25. **Classify** a biome that is warm to hot in the summer and cool or cold in the winter and that receives approximately 50–89 cm of precipitation annually.

Section 3.3

Vocabulary Review

Replace the underlined words with the correct terms from the Study Guide page.

26. A(n) area where freshwater and salt water meet provides habitat for a diversity of organisms.

27. The well-lit portion of the ocean is the area where all of the photosynthetic organisms live.

28. The shoreline of the ocean contains communities that are layered depending on how long they are submerged by tides.

Biology Online Chapter Test biologygmh.com

Understand Key Concepts

29. Where is the largest percentage of water located?
A. groundwater **C.** oceans
B. rivers **D.** glaciers

Use the diagram below to answer question 30.

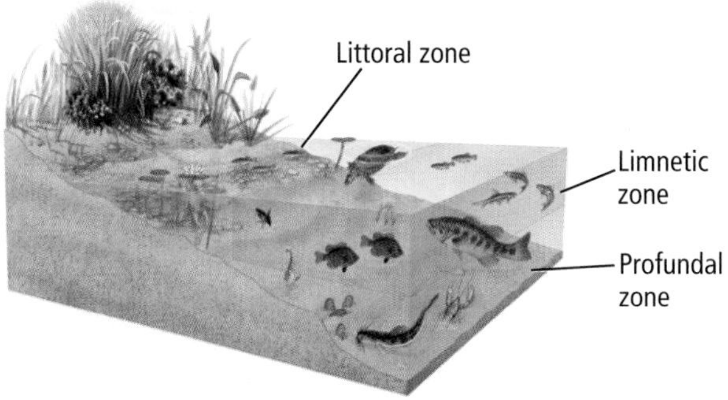

30. In which area of the lake is there likely to be the greatest diversity of plankton?
A. littoral zone **C.** profundal zone
B. limnetic zone **D.** aphotic zone

31. Which best describes the intertidal zone on a rocky shore?
A. The dominant low-energy community is likely to be an estuary.
B. The communities are adapted to shifting sands due to incoming waves.
C. The communities are stratified from the high-tide line to the low-tide line.
D. The organisms in the community constantly require dissolved oxygen.

Constructed Response

32. Short Answer How is light a limiting factor in oceans?

33. Short Answer Describe characteristics of an estuary.

34. Open Ended Describe adaptations of an organism living in the abyssal zone of the ocean.

Think Critically

35. Predict the consequences a drought would have on a river such as the Mississippi River.

36. Compare the intertidal zone with the photic zone in terms of tidal effect.

Additional Assessment

37. WRITING in Biology Choose a biome other than the one in which you live. Write an essay explaining what you think you would like and what you think you would dislike about living in your chosen biome.

DBQ Document-Based Questions

"Leaf mass per area (LMA) measures the leaf dry-mass investment per unit of light-intercepting leaf area deployed. Species with high LMA have a thicker leaf blade or denser tissue, or both."

"Plant ecologists have emphasized broad relationships between leaf traits and climate for at least a century. In particular, a general tendency for species inhabiting arid and semi-arid regions to have leathery, high-LMA leaves has been reported. Building high-LMA leaves needs more investment per unit leaf area. Construction cost per unit leaf mass varies relatively little between species: leaves with high protein content (typically low-LMA leaves) tend to have low concentrations of other expensive compounds such as lipids or lignin, and high concentrations of cheap constituents such as minerals. Leaf traits associated with high LMA (for example, thick leaf blade; small, thick-walled cells) have been interpreted as adaptations that allow continued leaf function (or at least postpone leaf death) under very dry conditions, at least in evergreen species."

Data obtained from: Wright, I.J. et al. The worldwide leaf economics spectrum. *Nature* 428:821–828.

38. From the information presented, would you expect leaves on trees in the tropical rain forest to contain large quantities of lipids? Explain your answer in terms of energy investment.

39. Hypothesize how high-LMA leaves are adapted for dry conditions.

Cumulative Review

40. Explain the difference between autotrophs and heterotrophs. **(Chapter 2)**

Standardized Test Practice

Cumulative

Multiple Choice

1. If science can be characterized as discovery, then technology can be characterized as which?
   A. application
   B. information
   C. manufacturing
   D. reasoning

Use the illustration below to answer questions 2 and 3.

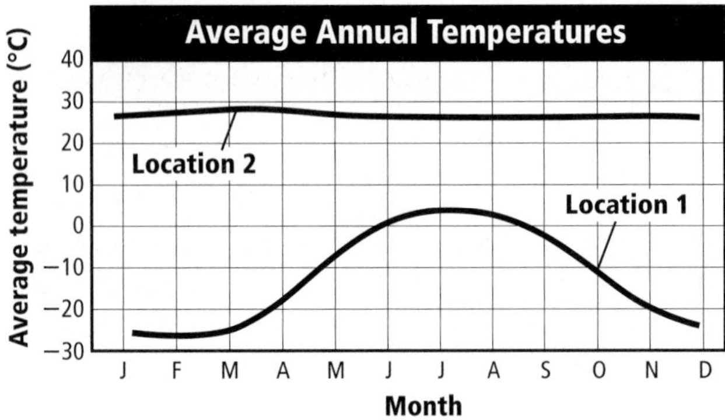

2. Based on the graph above, which term describes Location 2?
   A. oceanic
   B. polar
   C. temperate
   D. tropical

3. Suppose that in Location 2 there is very little rainfall during the year. What would be the name of that biome in this region?
   A. desert
   B. tundra
   C. temperate forest
   D. tropical rain forest

4. Which process is associated with long-term cycling of matter through the biosphere?
   A. breakdown of organic material by decomposers
   B. formation and weathering of minerals in rocks
   C. formation of compounds used for food by living organisms
   D. movement of fresh water from the land into bodies of water through run-off

Use the illustration below to answer question 5.

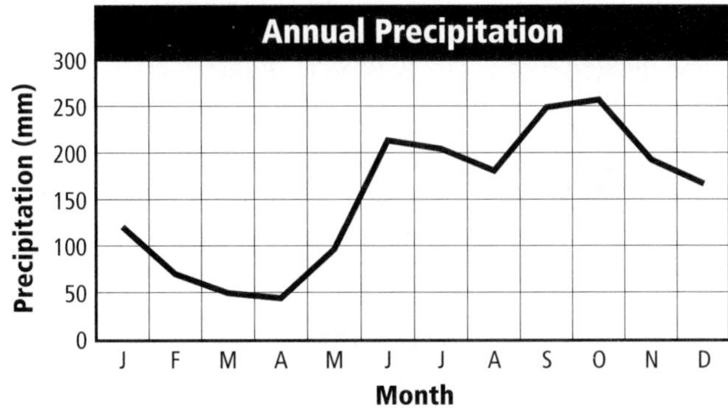

5. Look at the information in the graph. From what kind of biome are these data probably taken?
   A. desert
   B. tundra
   C. temperate forest
   D. tropical rain forest

6. Which system of measurement is the basis for many of the SI units?
   A. binary
   B. English
   C. metric
   D. number

7. Which of these organisms is a decomposer?
   A. a bacterium that makes food from inorganic compounds
   B. a clam that takes in water and filters food
   C. a fungus that gets nutrients from dead logs
   D. a plant that makes food using sunlight

8. Which distinguishes scientific ideas from popular opinions?
   A. Popular opinions are always rational and logical.
   B. Popular opinions depend on research and evidence.
   C. Scientific ideas are always testable and repeatable.
   D. Scientific ideas depend on anecdotes and hearsay.

Short Answer

9. How is a tundra similar to and different from a boreal forest? Use a Venn diagram to organize information about the similarities and differences of these biomes.

10. What is the role of a pioneer species in primary succession?

11. Give two examples of how the human body shows the living characteristic of organization.

12. Suppose a certain insect species lives only in a specific species of tree. It feeds off the sap of the tree and produces a chemical that protects the tree from certain fungi. What kind of relationship is this?

13. Why would you expect to find different animals in the photic and aphotic zones of the ocean?

14. Suppose a gardener learns that the soil in a garden has low nitrogen content. Describe two ways to increase the nitrogen available for plants in the garden.

15. Explain how the establishment of a climax community through primary succession differs from the establishment of a climax community that occurs through secondary succession.

16. Why is the ability to adapt an important characteristic of living things?

Extended Response

Use the illustration below to answer question 17.

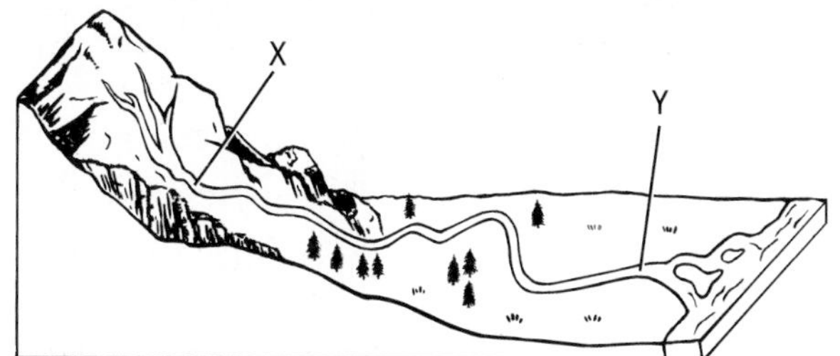

17. Based on the information in the illustration above, what can you infer about the major differences between the freshwater ecosystems at Point X and Point Y?

18. Suppose a nonnative species is introduced into an ecosystem. What is one kind of community interaction you might expect from the other organisms in that ecosystem?

Essay Question

Suppose there is a dense temperate forest where people do not live. After a few hot, dry months, forest fires have started to spread through the forest area. There is no threat of the fires reaching areas inhabited by humans. Some people are trying to get the government to intervene to control the fires, while others say the fires should be allowed to run their natural course.

Using the information in the paragraph above, answer the following question in essay format.

19. Explain which side of this debate you would support. Be sure to provide evidence based on what you know about change in ecosystems.

NEED EXTRA HELP?																			
If You Missed Question . . .	1	2	3	4	5	6	7	8	9	10	11	12	13	14	15	16	17	18	19
Review Section . . .	3.2	3.2	3.2	2.3	3.3	1.2	2.2	1.3	3.2	3.1	1.1	2.1	3.3	2.3	3.1	1.1	3.3	3.3	3.1, 3.2

Population Ecology

BIG Idea Population growth is a critical factor in a species' ability to maintain homeostasis within its environment.

Section 1
Population Dynamics
MAIN Idea Populations of species are described by density, spatial distribution, and growth rate.

Section 2
Human Population
MAIN Idea Human population growth changes over time.

BioFacts

- Deer can be found in most parts of the United States except the Southwest, Alaska, and Hawaii.
- Parasites that attack deer include fleas, ticks, lice, mites, and tapeworms.
- Diseases such as Lyme disease, chronic wasting disease, and hemorrhagic disease can kill deer.

Lyme disease bacteria
Color-Enhanced SEM
Magnification: 2850×

Deer tick
Color-Enhanced SEM
Magnification: 22×

Start-Up Activities

LAUNCH Lab

A population of one?

Ecologists study populations of living things. They also study how populations interact with each other and with the abiotic factors in the environment. But what exactly is a population? Are the deer shown on the previous page a population? Is a single deer a population?

Procedure

1. Read and complete the lab safety form.
2. In your assigned group, brainstorm and predict the meaning of the following terms: *population, population density, natality, mortality, emigration, immigration,* and *carrying capacity.*

Analysis

1. **Infer** whether it is possible to have a population of one. Explain your answer.
2. **Analyze** your definitions and determine whether a relationship exists between the terms. Explain.

Visit biologygmh.com to:

- study the entire chapter online
- explore the Interactive Time Line, Concepts in Motion, the Interactive Table, Microscopy Links, Virtual Labs, and links to virtual dissections
- access Web links for more information, projects, and activities
- review content online with the Interactive Tutor and take Self-Check Quizzes

Population Characteristics Make this Foldable to help you learn the characteristics used to describe populations.

STEP 1 Fold a sheet of paper vertically with the edges about 2 cm apart.

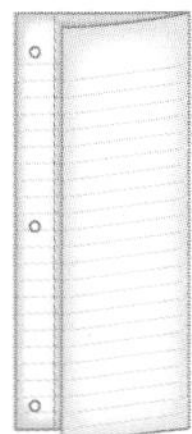

STEP 2 Fold the paper into thirds.

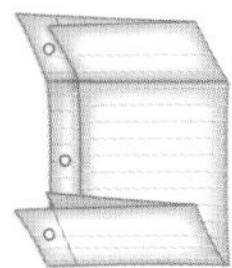

STEP 3 Unfold and cut the top layer of both folds to make three tabs.

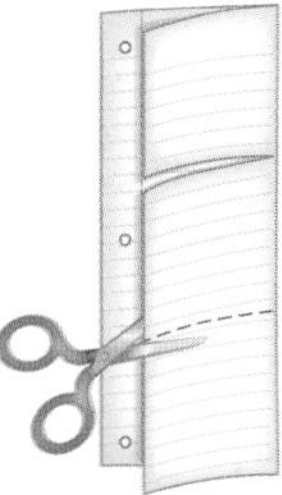

STEP 4 Label each tab as shown: *Population Density, Spatial Distribution, Growth Rate.*

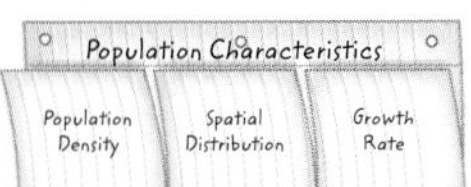

FOLDABLES Use this Foldable with Section 4.1. As you study this section, write what you learn about each characteristic under the correct tab.

Section 4.1

Reading Preview

Objectives

- **Describe** characteristics of populations.
- **Understand** the concepts of carrying capacity and limiting factors.
- **Describe** the ways in which populations are distributed.

Review Vocabulary

population: the members of a single species that share the same geographic location at the same time

New Vocabulary

population density
dispersion
density-independent factor
density-dependent factor
population growth rate
emigration
immigration
carrying capacity

Population Dynamics

MAIN Idea **Populations of species are described by density, spatial distribution, and growth rate.**

Real-World Reading Link Have you ever observed a beehive or an ant farm? The population had certain characteristics that could be used to describe it. Ecologists study population characteristics that are used to describe all populations of organisms.

Population Characteristics

All species occur in groups called populations. There are certain characteristics that all populations have, such as population density, spatial distribution, and growth rate. These characteristics are used to classify all populations of organisms, including bacteria, animals, and plants.

Population density One characteristic of a population is its **population density,** which is the number of organisms per unit area. For example, the population density of cattle egrets, shown with the water buffalo in **Figure 4.1,** is greater near the buffalo than farther away. Near the water buffalo, there might be three birds per square meter. Fifty meters from the water buffalo, the density of birds might be zero.

Spatial distribution Another characteristic of a population is called **dispersion**—the pattern of spacing of a population within an area. **Figure 4.2** shows the three main types of dispersion—uniform, clumped groups, and random. Black bears are typically dispersed in a uniform arrangement. American bison are dispersed in clumped groups or herds. White-tailed deer are dispersed randomly with unpredictable spacing. One of the primary factors in the pattern of dispersion for all organisms is the availability of resources such as food.

Figure 4.1 The population density of the cattle egrets is greater near the water buffalo.
Suggest *What type of dispersion would you expect these birds to have?*

Visualizing Population Characteristics

Figure 4.2
Population density describes how many individual organisms live in a given area. Dispersion describes how the individuals are spaced within that area. Population range describes a species' distribution.

Black Bear

Dispersion: American black bear males usually are dispersed uniformly within territories as large as several hundred square kilometers. Females have smaller territories that overlap those of males.

Density: one bear per several hundred square kilometers

Black Bear Distribution (in purple)

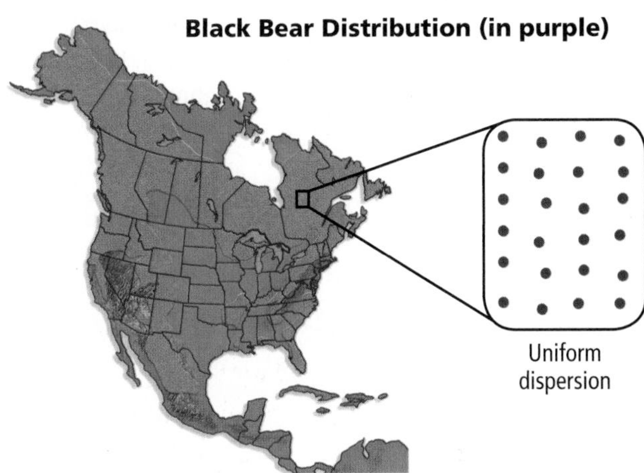

American Bison

Dispersion: American bison are found in clumped groups called herds.

Density: four bison/km^2 in Northern Yellowstone in 2000

Bison Distribution (historic range prior to 1865 in orange)

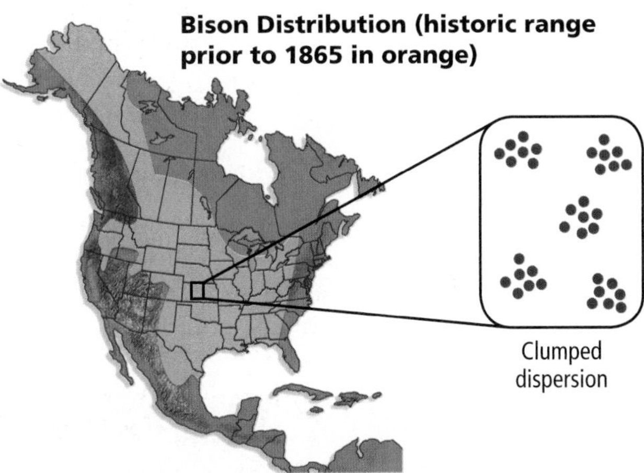

White-tailed Deer

Dispersion: White-tailed deer are dispersed randomly throughout appropriate habitats.

Density: 10 deer/km^2 in some areas of the northeastern United States

White-tailed Deer Distribution (in blue)

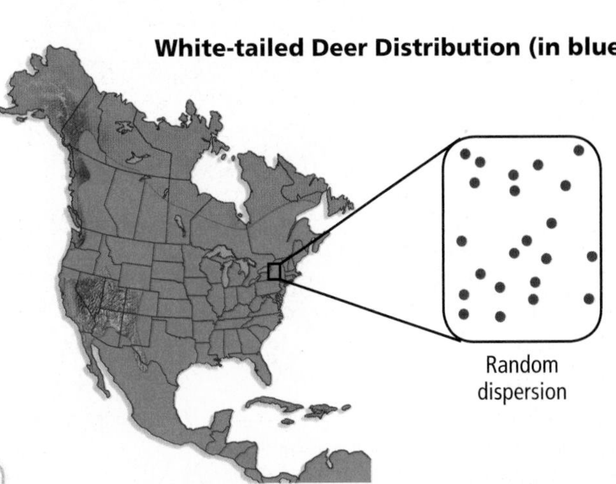

Concepts In Motion **Interactive Figure** To see an animation of population distribution, visit biologygmh.com.

VOCABULARY

SCIENCE USAGE V. COMMON USAGE

Distribution

Science usage: the area where something is located or where a species lives and reproduces

The white-tailed deer has a wide distribution that covers much of the United States.

Common usage: the handing out or delivery of items to a number of people

The distribution of report cards to students occurred today.

Population ranges No population, not even the human population, occupies all habitats in the biosphere. Some species, such as the iiwi (EE ee wee) shown in **Figure 4.3,** have a very limited population range, or distribution. This songbird is found only on some of the islands of Hawaii. Other species, such as the peregrine falcon shown in **Figure 4.3,** have a vast distribution. Peregrine falcons are found on all continents except Antarctica. Note the distribution of the animals in **Figure 4.2.**

Recall from Chapter 2 that organisms adapt to the biotic and abiotic factors in their environment. A species might not be able to expand its population range because it cannot survive the abiotic conditions found in the expanded region. A change in temperature range, humidity level, annual rainfall, or sunlight might make a new geographic area uninhabitable for the species. In addition, biotic factors, such as predators, competitors, and parasites, present threats that might make the new location difficult for survival.

Reading Check **Describe** two reasons why a species might not be able to expand its range.

Population-Limiting Factors

In Chapter 3, you learned that all species have limiting factors. Limiting factors keep a population from continuing to increase indefinitely. Decreasing a limiting factor, such as the available food supply, often changes the number of individuals that are able to survive in a given area. In other words, if the food supply increases a larger population might result, and if the food supply decreases a smaller population might result.

Density-independent factors There are two categories of limiting factors—density-independent factors and density-dependent factors. Any factor in the environment that does not depend on the number of members in a population per unit area is a **density-independent factor.** These factors usually are abiotic and include natural phenomena such as weather events. Weather events that limit populations include drought or flooding, extreme heat or cold, tornadoes, and hurricanes.

Iiwi

Peregrine falcon

Figure 4.3 The iiwi lives only on some of the Hawaiian islands. The peregrine falcon is found worldwide.

Crown fire damage

Managed ground fire damage

Figure 4.4 A crown fire is a density-independent factor that can limit population growth. However, small ground fires can promote growth of pines in a pine forest community.
Explain ***Why do these two situations involving fire have different results on the pine tree populations?***

Figure 4.4 shows an example of the effects that fire can have on a population. Fire has damaged this ponderosa pine forest community. Sometimes the extreme heat from a crown fire, which is a fire that advances to the tops of the trees, can destroy many mature ponderosa pine trees—a dominant species in forests of the western United States. In this example, the fire limits the population of ponderosa trees by killing many of the trees. However, smaller but more frequent ground fires have the opposite effect on the population. By thinning lower growing plants that use up nutrients, a healthier population of mature ponderosa pines is produced.

Populations can be limited by the unintended results of human alterations of the landscape. For example, over the last 100 years, human activities on the Colorado River, such as building dams, water diversions, and water barriers, have significantly reduced the amount of water flow and changed the water temperature of the river. In addition, the introduction of nonnative fish species altered the biotic factors in the river. Because of the changes in the river, the number of small fish called humpback chub was reduced. During the 1960s, the number of humpback chub dropped so low that they were in danger of disappearing from the Colorado River altogether.

Air, land, and water pollution are the result of human activities that also can limit populations. Pollution reduces the available resources by making some of the resources toxic.

Density-dependent factors Any factor in the environment that depends on the number of members in a population per unit area is a **density-dependent factor.** Density-dependent factors are often biotic factors such as predation, disease, parasites, and competition. A study of density-dependent factors was done on the wolf–moose populations in northern Michigan on Isle Royale, located in Lake Superior.

CAREERS IN BIOLOGY

Population Biologist A population biologist studies the characteristics of populations, such as growth, size, distribution, or genetics. For more information on biology careers, visit biologygmh.com.

■ **Figure 4.5** The long-term study of the wolf and moose populations on Isle Royale shows the relationship between the number of predators and prey over time.
Infer ***What might have caused the increase in the number of moose in 1995?***

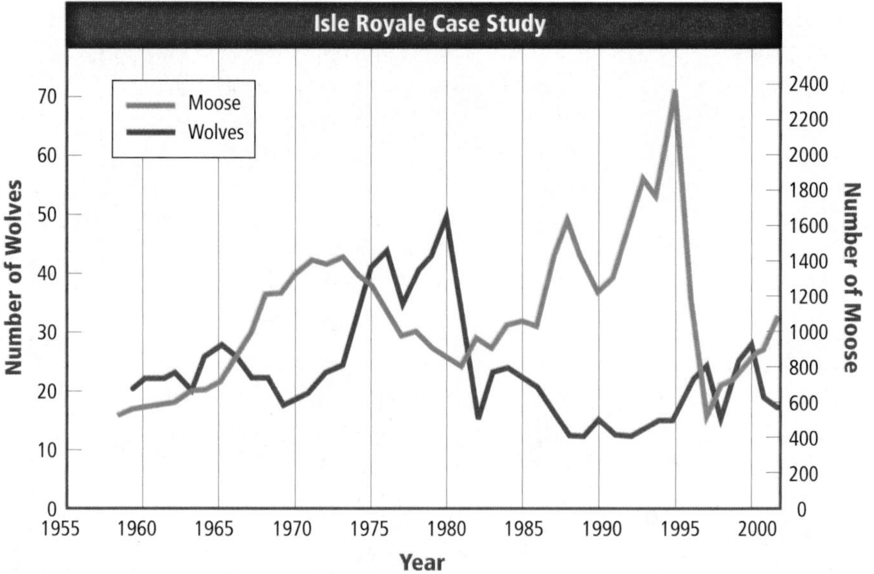

FOLDABLES
Incorporate information from this section into your Foldable.

Prior to the winter of 1947–48, apparently there were no wolves on Isle Royale. During that winter, a single pair of wolves crossed the ice on Lake Superior, reaching the island. During the next ten years, the population of wolves reached about twenty individuals. **Figure 4.5** shows some of the results from the long-term study conducted by population biologists. Notice that the rise and fall of the numbers of each group was dependent on the other group. For example, follow the wolves' line on the graph. As the number of wolves decreased, the number of moose increased.

Disease Another density-dependent factor is disease. Outbreaks of disease tend to occur when population size has increased and population density is high. When population density is high, disease is transmitted easily from one individual to another because contact between individuals is more frequent. Therefore, the disease spreads easily and quickly through a population. This is just as true for human populations as it is for populations of protists, plants, and other species of animals.

■ **Figure 4.6** Lemmings are mammals that produce offspring in large numbers when food is plentiful. When the food supply diminishes, lemmings starve and many die.

Competition Competition between organisms also increases when density increases. When the population increases to a size so that resources such as food or space become limited, individuals in the population must compete for the available resources. Competition can occur within a species or between two different species that use the same resources. Competition for insufficient resources might result in a decrease in population density in an area due to starvation or to individuals leaving the area in search of additional resources. As the population size decreases, competition becomes less severe.

The lemmings shown in **Figure 4.6** are an example of a population that often undergoes competition for resources. Lemmings are small mammals that live in the tundra biome. When food is plentiful, their population increases exponentially. As food becomes limited, many lemmings begin to starve and their population size decreases significantly.

Parasites Populations also can be limited by parasites, in a way similar to disease, as population density increases. The presence of parasites is a density-dependent factor that can negatively affect population growth at higher densities.

Population growth rate An important characteristic of any population is its growth rate. The **population growth rate** (PGR) explains how fast a given population grows. One of the characteristics of the population ecologists must know, or at least estimate, is natality. The natality of a population is the birthrate, or the number of individuals born in a given time period. Ecologists also must know the mortality—the number of deaths that occur in the population during a given time period.

The number of individuals emigrating or immigrating also is important. **Emigration** (em uh GRAY shun) is the term ecologists use to describe the number of individuals moving away from a population. **Immigration** (ih muh GRAY shun) is the term ecologists use to describe the number of individuals moving into a population. In most instances, emigration is about equal to immigration. Therefore, natality and mortality usually are most important in determining the population growth rate.

Some populations tend to remain approximately the same size from year to year. Other populations vary in size depending on conditions within their habitats. To better understand why populations grow in different ways, you should understand two mathematical models for population growth—the exponential growth model and the logistic growth model.

Exponential growth model Look at **Figure 4.7** to see how a population of mice would grow if there were no limits placed on it by the environment. Assume that two adult mice breed and produce a litter of young. Also assume the two offspring are able to reproduce in one month. If all of the offspring survive to breed, the population grows slowly at first. This slow growth period is defined as the lag phase. The rate of population growth soon begins to increase rapidly because the total number of organisms that are able to reproduce has increased. After only two years, the experimental mouse population would reach more than three million mice.

Connection to Math Notice in **Figure 4.7** that once the mice begin to reproduce rapidly, the graph becomes J-shaped. A J-shaped growth curve illustrates exponential growth. Exponential growth, also called geometric growth, occurs when the growth rate is proportional to the size of the population. All populations grow exponentially until some limiting factor slows the population's growth. It is important to recognize that even in the lag phase, the use of available resources is exponential. Because of this, the resources soon become limited and population growth slows.

Logistic growth model Many populations grow like the model shown in **Figure 4.8** rather than the model shown in **Figure 4.7.** Notice that the graphs look exactly the same through some of the time period. However, the second graph curves into an S-shape. An S-shaped curve is typical of logistic growth. Logistic growth occurs when the population's growth slows or stops following exponential growth, at the population's carrying capacity. A population stops increasing when the number of births is less than the number of deaths or when emigration exceeds immigration.

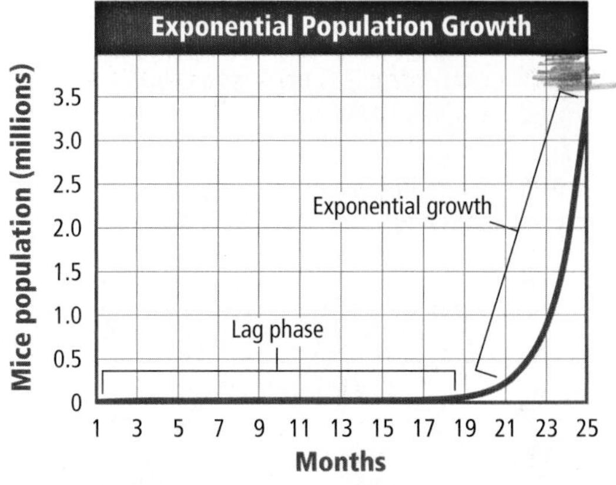

Figure 4.7 If two mice were allowed to reproduce unhindered, the population would grow slowly at first but would accelerate quickly.

Infer ***Why don't mice or other populations continue to grow exponentially?***

Figure 4.8 When a population exhibits growth that results in an S-shaped graph, it exhibits logistic growth. The population levels off at a limit called the carrying capacity.

Concepts In Motion

Interactive Figure To see an animation of population growth, visit biologygmh.com.

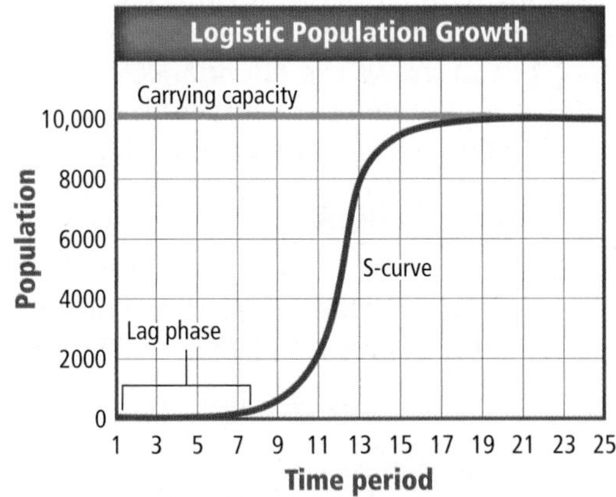

■ **Figure 4.9** Locusts, which are *r*-strategists, usually have a short life span and produce many offspring.
Infer *What specific factors might fluctuate in a locust's environment?*

Carrying capacity In **Figure 4.8** on the previous page, notice that logistic growth levels off at the line on the graph identified as the carrying capacity. The maximum number of individuals in a species that an environment can support for the long term is the **carrying capacity.** Carrying capacity is limited by the energy, water, oxygen, and nutrients available. When populations develop in an environment with plentiful resources, there are more births than deaths. The population soon reaches or passes the carrying capacity. As a population nears the carrying capacity, resources become limited. If a population exceeds the carrying capacity, deaths outnumber births because adequate resources are not available to support all of the individuals. The population then falls below the carrying capacity as individuals die. The concept of carrying capacity is used to explain why many populations tend to stabilize.

Reproductive patterns The graph in **Figure 4.8** shows the number of individuals increasing until the the carrying capacity is reached. However, there are several additional factors that must be considered for real populations. Species of organisms vary in the number of births per reproduction cycle, in the age that reproduction begins, and in the life span of the organism. Both plants and animals are placed into groups based on their reproductive factors.

Members of one of the groups are called the *r*-strategists. The rate strategy, or *r*-strategy, is an adaptation for living in an environment where fluctuation in biotic or abiotic factors occur. Fluctuating factors might be availability of food or changing temperatures. An *r*-strategist is generally a small organism such as a fruit fly, a mouse, or the locusts shown in **Figure 4.9.** *R*-strategists usually have short life spans and produce many offspring.

DATA ANALYSIS LAB 4.1

Based on Real Data*

Recognize Cause and Effect

Do parasites affect the size of a host population? In 1994, the first signs of a serious eye disease caused by the bacterium *Mycoplasma gallisepticum* were observed in house finches that were eating in backyard bird feeders. Volunteers collected data beginning three different years on the number of finches infected with the parasite and the total number of finches present. The graph shows the abundance of house finches in areas where the infection rate was at least 20 percent of the house finch population.

Data and Observations

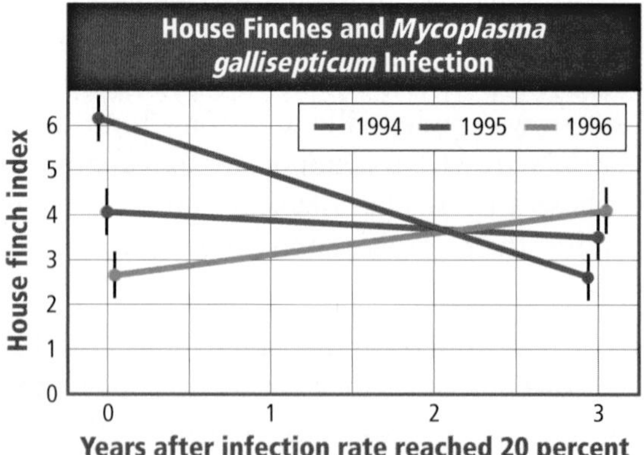

Think Critically

1. **Compare** the data from the three areas.
2. **Hypothesize** Why did the house finch abundance stabilize in 1995 and 1996?
3. **Infer** Is the parasite, *Mycoplasma gallisepthicum,* effective in limiting the size of house finch populations? Explain.

*Data obtained from: Gregory, R., et al. 2000. Parasites take control. *Nature* 406: 33–34.

■ **Figure 4.10** Elephants are *k*-strategists that produce few offspring, but they invest a lot of care in the raising of their offspring.

The reproductive strategy of an *r*-strategist is to produce as many offspring as possible in a short time period in order to take advantage of some environmental factor. They typically expend little energy or none at all in raising their young to adulthood. Populations of *r*-strategists usually are controlled by density-independent factors, and they usually do not maintain a population near the carrying capacity.

Just as some environments fluctuate, others are fairly predictable. Elephants on the savanna, shown in **Figure 4.10,** experience a carrying capacity that changes little from year to year. The carrying-capacity strategy, or *k*-strategy, is an adaptation for living in these environments. A *k*-strategist generally is a larger organism that has a long life span, produces few offspring, and whose population reaches equilibrium at the carrying capacity. The reproductive strategy of a *k*-strategist is to produce only a few offspring that have a better chance of living to reproductive age because of the energy, resources, and time invested in the care for the young. Populations of *k*-strategists usually are controlled by density-dependent factors.

VOCABULARY

ACADEMIC VOCABULARY

Fluctuate
to change from high to low levels or from one thing to another in an unpredictable way
The speed of a car fluctuates when you are driving on narrow, winding roads.

Section 4.1 Assessment

Section Summary

- There are population characteristics that are common to all populations of organisms, including plants, animals, and bacteria.
- Populations tend to be distributed randomly, uniformly, or in clumps.
- Populations tend to stabilize near the carrying capacity of their environment.
- Population limiting factors are either density-independent or density-dependent.

Understand Main Ideas

1. MAIN Idea **Compare and contrast** spatial distribution, population density, and population growth rate.
2. **Summarize** the concepts of carrying capacity and limiting factors.
3. **Sketch** diagrams showing population dispersion patterns.
4. **Analyze** the impact a nonnative species might have on a native species in terms of population dynamics.

Thinking Critically

5. **Design an experiment** that you could perform to see if a population of fruit flies—very small insects that feed on bananas—grows according to the exponential growth model or the logistic growth model.

WRITING in Biology

6. Write a newspaper article describing how a weather event, such as drought, has affected a population of animals in your community.

Reading Preview

Objectives

- **Explain** the trends in human population growth.
- **Compare** the age structure of representative nongrowing, slowly growing, and rapidly growing countries.
- **Predict** the consequences of continued population growth.

Review Vocabulary

carrying capacity: the maximum number of individuals in a species that an environment can support for the long term

New Vocabulary

demography
demographic transition
zero population growth (ZPG)
age structure

Human Population

MAIN Idea Human population growth changes over time.

Real-World Reading Link Has someone you know recently had a baby? The odds of babies surviving to adulthood are greater than ever before in most countries today.

Human Population Growth

The study of human population size, density, distribution, movement, and birth and death rates is **demography** (de MAH gra fee). The graph in **Figure 4.11** shows demographers' estimated human population on Earth for several thousand years.

Notice that the graph in **Figure 4.11** shows a relatively stable number of individuals over thousands of years—until recently. Notice also the recovery of the human population after the outbreak of the bubonic plague in the 1300s when an estimated one-third of the population of Europe died. Perhaps the most significant feature in this graph is the increase in human population in recent times. In 1804, the population of Earth was an estimated one billion people. By 1999, the human population had reached six billion people. At this growth rate, about 70 million people are added to the world population annually, and the world's population is expected to double in about 53 years.

Figure 4.11 The human population on Earth was relatively constant until recent times, when the human population began to grow at an exponential rate.

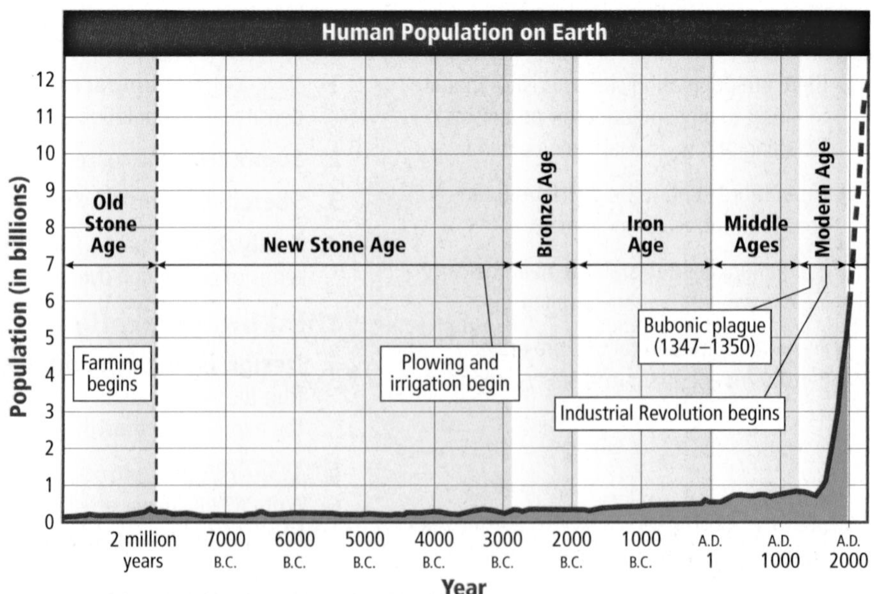

Technological advances For thousands of years, environmental conditions kept the size of the human population at a relatively constant number below the environment's carrying capacity. Humans have learned to alter the environment in ways that appear to have changed its carrying capacity. Agriculture and domestication of animals have increased the human food supply. Technological advances and medicine have improved the chances of human survival by reducing the number of deaths from parasites and disease. In addition, improvements in shelter have made humans less vulnerable to climatic impact.

Reading Check **Explain** why an improvement in shelter increased the survival rate of the human population.

Human population growth rate Although the human population is still growing, the rate of its growth has slowed. **Figure 4.12** shows the percent increase in human population from the late 1940s through 2003. The graph also includes the projected population increase through 2050. Notice the sharp dip in human population growth in the 1960s. This was due primarily to a famine in China in which about 60 million people died. The graph also shows that human population growth reached its peak at over 2.2 percent in 1962. By 2003, the percent increase in human population growth had dropped to almost 1.2 percent. Population models predict the overall population growth rate to be below 0.6 percent by 2050. The decline in human population growth is due primarily to diseases such as AIDS and voluntary population control.

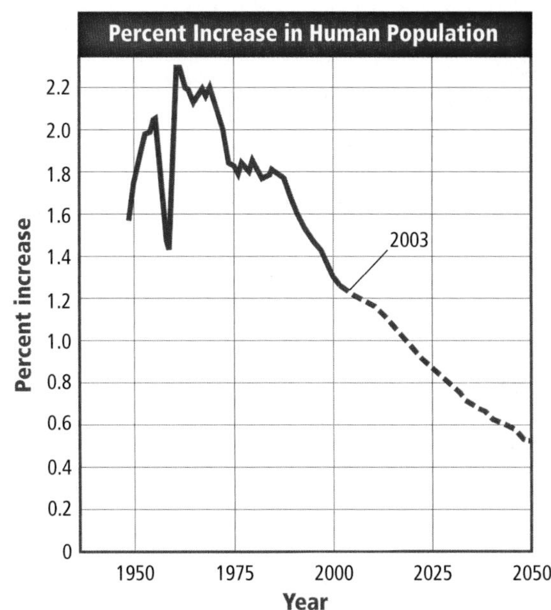

Figure 4.12 This graph shows the percent increase in the global human population using data from the late 1940s through 2003 and the projected percent increase to 2050.
Determine ***What is the approximate population increase in the year 2025?***

MiniLab 4.1

Evaluate Factors

What factors affect the growth of a human population? Technological advances have resulted in a rapid growth in human population. However, human population growth is not equal in all countries.

Procedure

1. The graph shows one factor affecting human population growth. Use the data to predict how this factor will affect the population in each country between now and the year 2050.
2. Brainstorm a list of factors, events, or conditions that affect the growth of human populations in these countries. Predict the effect of each factor on the population growth rate.

Analysis

Think Critically In your opinion, what factors or groups of factors have the greatest impact on population growth? Justify your answer.

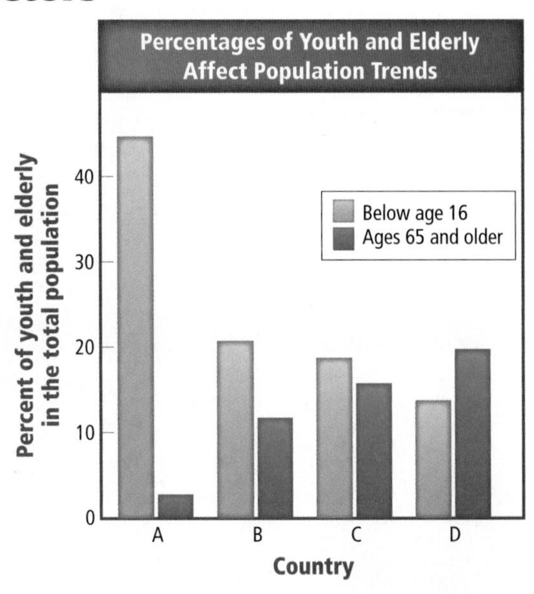

VOCABULARY

WORD ORIGIN

Demography
demo– from the Greek word *demos;* meaning *people*
–ography from the French word *graphie;* meaning *writing*

Trends in Human Population Growth

The graph in **Figure 4.12** is somewhat deceptive. Population trends can be altered by events such as disease and war. **Figure 4.13** shows a few historical events that have changed population trends. **Figure 4.12** could also easily be misinterpreted because human population growth is not the same in all countries. However, population growth trends are often similar in countries that have similar economies.

For example, one trend that has developed during the previous century is a change in the population growth rate in industrially developed countries such as the United States. An industrially-developed country is advanced in industrial and technological capabilities and has a population with a high standard of living. In its early history, the United States had a high birthrate and a high death rate. It was not uncommon for people to have large families and for individuals to die by their early forties. Many children also died before reaching adulthood. Presently, the birthrate in the United States has decreased dramatically and the life expectancy is greater than seventy years. This change in a population from high birth and death rates to low birth and death rates is called a **demographic transition.**

LAUNCH Lab

Review Based on what you've read about populations, how would you now answer the analysis questions?

Connection to Math How do population growth rates compare in industrially developed countries and developing countries—countries with a relatively low level of capabilities and a low standard of living? In the United States, the birthrate in 2005 was 14.1 births per 1000 people, the death rate was 8.2 deaths per 1000 people, and the migration rate was 3.3 people entering the country per 1000 people. The population growth rate was 0.92 percent in the United States.

In a developing country such as Honduras—location shown in **Table 4.1**—the situation is different. In 2005, Honduras had a birthrate of 30.4 births per 1000 people, a death rate of 6.9 deaths per 1000 people, and a migration rate of 1.9 people leaving the country per 1000 people. This results in a population growth rate of 2.16 percent. This is among the highest population growth rates of any country in the world.

■ **Figure 4.13**
History of Human Population Trends
Many factors have affected human population growth throughout history.

1347–1351 The bubonic plague kills one-third of Europe's population and 75 million people throughout the world.

1800 The industrial revolution leads to a dramatic population explosion.

0 — 1350 — 1800

69,000 B.C. Researchers think that as few as 15,000 to 40,000 people survived global climate changes that resulted from the eruption of the Toba supervolcano.

1798 The first essay on human population is written by Thomas Malthus, who predicted exponential population growth leading to famine, poverty, and war.

Concepts In Motion **Interactive Table** To explore more about population growth of various countries, visit biologygmh.com.

Table 4.1 Population Growth Rates of Countries

Country	Population growth rate (percent)
Afghanistan	4.77
Brazil	1.06
Bulgaria	−0.89
Germany	0.0
Honduras	2.16
India	1.40
Indonesia	1.45
Kenya	2.56
Niger	2.63
Nigeria	2.37
United States	0.92

Location

Afghanistan, Brazil, Bulgaria, Germany, Honduras, India, Indonesia, Kenya, Niger, Nigeria, United States

Developing countries will add 73 million people to the world population compared to only three million people added in the industrially developed countries. For example, between now and 2050, the developing country Niger—also shown in **Table 4.1**—is expected to be one of the most rapidly growing countries. Its population will expand from 12 to 53 million people. The industrially developed country Bulgaria is expected to have a population decline from eight to five million people in the same time period.

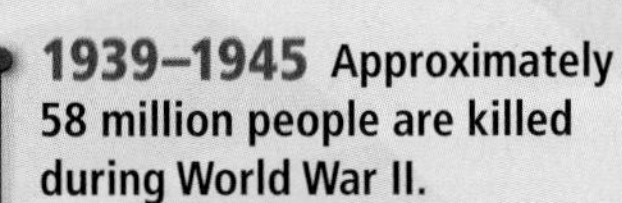

1939–1945 Approximately 58 million people are killed during World War II.

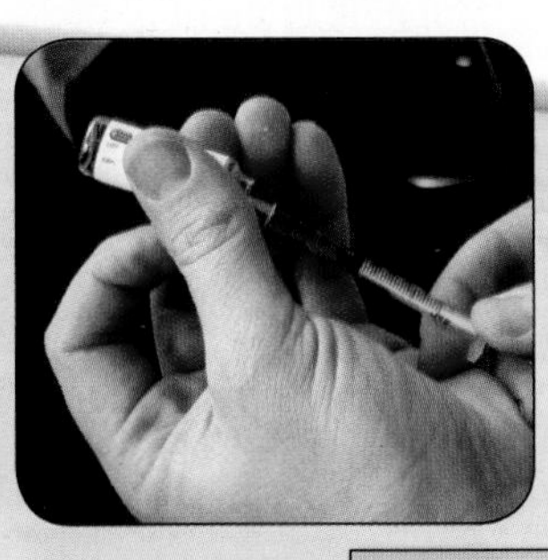

1954 Improved medical care, medicines, and sanitation leads to an increase in human population.

2006 An estimated 2.9 million people die as a result of AIDS in sub-Saharan Africa.

1900 — 1950 — 2000

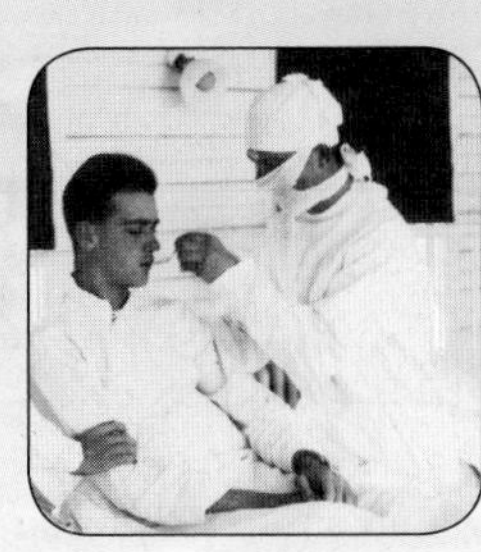

1918 The Spanish Flu kills between 20 and 40 million people.

Concepts In Motion **Interactive Time Line** To learn more about these discoveries and others, visit biologygmh.com. Biology Online

Study Tip

Interactive Reading As you read, write three questions on sticky notes about human population dynamics. The questions should begin with why, how, where, or when. Use the notes to ask a partner questions about the content in the chapter.

Zero population growth Another trend that populations can experience is zero population growth. **Zero population growth** (ZPG) occurs when births plus immigration equals deaths plus emigration. One estimate is that the world will reach ZPG between 2020 with 6.64 billion people and 2029 with 6.90 billion people. This will mean that the population has stopped growing, because births and deaths occur at the same rate. Once the world population reaches ZPG, the age structure eventually should be more balanced with numbers at pre-reproductive, reproductive, and post-reproductive ages being approximately equal.

Age structure Another important characteristic of any population is its age structure. A population's **age structure** is the number of males and females in each of three age groups: pre-reproductive stage, reproductive stage, and post-reproductive stage. Humans are considered to be pre-reproductive before age 20 even though they are capable of reproduction at an earlier age. The reproductive years are considered to be between 20 and 44, and the post-reproductive years are after age 44.

Analyze the age structure diagrams for three different representative countries in **Figure 4.14**—their locations are shown in **Table 4.1.** The age structure diagrams are typical of many countries in the world. Notice the shape of the overall diagram for a country that is rapidly growing, one that is growing slowly, and one that has reached negative growth. The age structure for the world's human population looks more like that of a rapidly growing country.

Reading Check **Compare and contrast** the age structures of the countries shown in **Figure 4.14.**

■ **Figure 4.14** The relative numbers of individuals in pre-reproductive, reproductive, and post-reproductive years are shown for three representative countries.

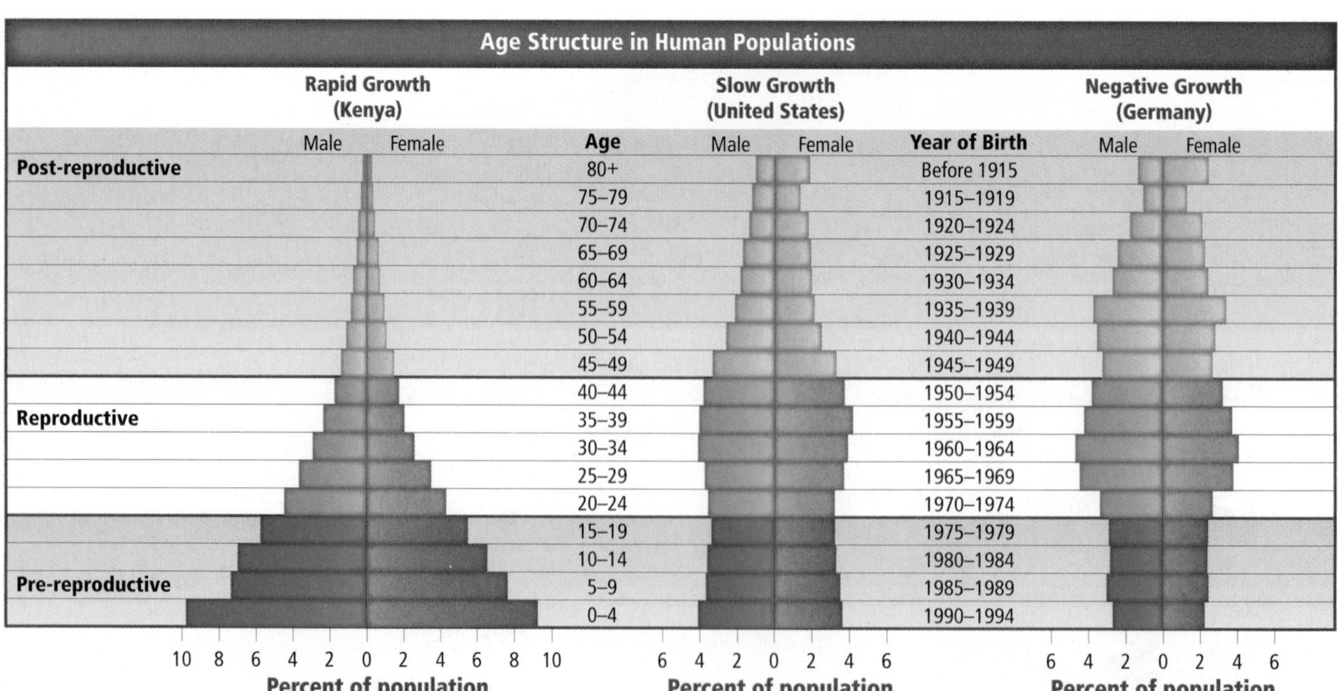

Human carrying capacity Calculating population growth rates is not just a mathematical exercise. Scientists are concerned about the human population reaching or exceeding the carrying capacity. As you learned in Section 4.1, all populations have carrying capacities, and the human population is no exception. Many scientists suggest that human population growth needs to be reduced. In many countries, voluntary population control is occurring through family planning. Unfortunately, if the human population continues to grow —as most populations do—and areas become overcrowded, disease and starvation will occur. However, technology has allowed humans to increase Earth's carrying capacity, at least temporarily. It might be possible for technology and planning to keep the human population at or below its carrying capacity.

Another important factor in keeping the human population at or below the carrying capacity is the amount of resources from the biosphere that are used by each person. Currently, individuals in industrially-developed countries use far more resources than those individuals in developing countries, as shown in **Figure 4.15**. This graph shows the estimated amount of land required to support a person through his or her life, including land used for production of food, forest products and housing, and the additional forest land required to absorb the carbon dioxide produced by the burning of fossil fuels. Countries such as India are becoming more industrialized, and they have a high growth rate. These countries are adding more people and are increasing their use of resources. At some point, the land needed to sustain each person on Earth might exceed the amount of land that is available.

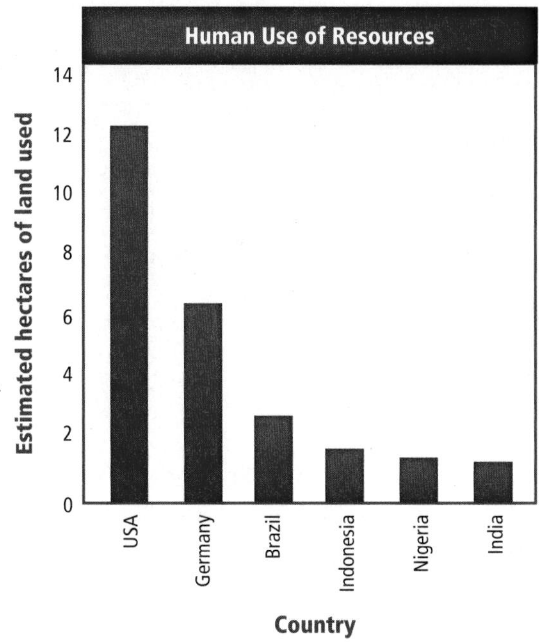

Figure 4.15 The amount of resources used per person varies around the world. Refer to **Table 4.1** for the locations of these countries.

Section 4.2 Assessment

Section Summary

- Human population growth rates vary in industrially-developing countries and industrialized countries.
- Zero population growth occurs when the birthrate of a population equals the death rate.
- The age structure of the human population is a contributing factor to population growth in some countries.
- Earth has an undefined carrying capacity for the human population.

Understand Main Ideas

1. MAIN Idea **Describe** the change in human population growth over time.
2. **Describe** the differences between the age structure graphs of nongrowing, slowly growing, and rapidly growing countries.
3. **Assess** the consequences of exponential population growth of any population.
4. **Summarize** why human population began to grow exponentially in the Modern Age.

Think Critically

5. **Predict** the short-term and long-term effects of a newly emerging disease on industrially developing and developed countries.

MATH in Biology

6. Construct an age-structure diagram using the following percentages: 0–19 years: 44.7; 20–44 years: 52.9; 45 years and over: 2.4. Which type of growth is this country experiencing?

CUTTING-EDGE BIOLOGY

Polar Bear Ecology

In late 2006, the U.S. Fish and Wildlife Service proposed that the polar bear be listed as a threatened species under the Endangered Species Act of 1973. Since then, scientists have undertaken a novel approach to studying the ecological needs of the world's largest terrestrial predator—not by tracking the bears themselves but by tracking the receding ice habitat in which the bears reside, which is vital to their survival.

Typical polar bear studies observe individual bears or track them with satellite collars, often at a considerable cost and risk to the bears and researchers alike. Now, scientists will utilize satellite and meteorological data to predict where sea ice will remain in the near future. Conservation efforts will be focused in these locations.

Bear necessities Polar bears only live in the circumpolar north, which includes the countries of the United States (Alaska), Canada, Russia, Denmark (Greenland), and Norway. The sea ice that forms each winter creates passages in which the bears travel, as well as creating an optimal environment for hunting. Polar bears rely on seasonal sea ice to stalk their favorite prey —ringed and bearded seals. As the sea ice dwindles, so does the polar bears' ability to effectively hunt these fast-swimming marine mammals.

The cold, hard facts Scientists plan to combine daily satellite and meteorological data from the past 30 years, including global climate change projections, to extrapolate where conservation efforts to save the species would be most successful. The data will also be used to create a Geographical Information Systems (GIS) map.

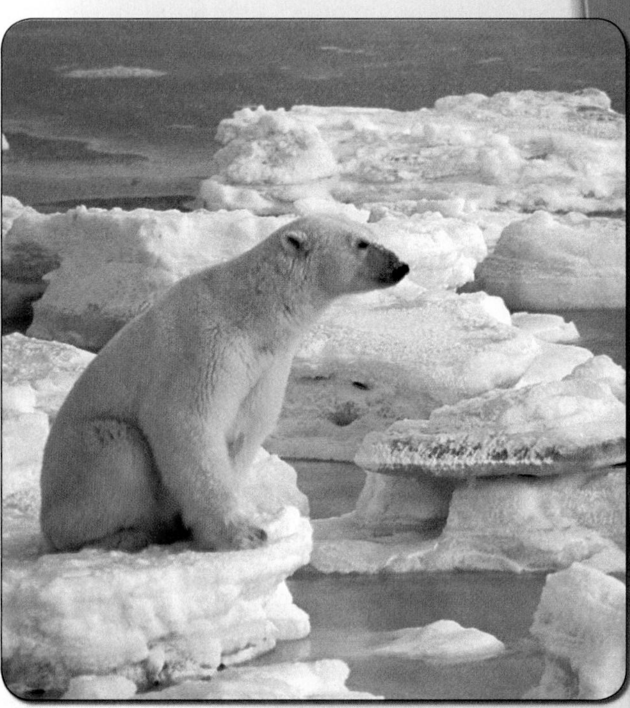

Approximately 60 percent of the polar bear population reside in Canada.

Using the GIS map, scientists think they will be able to determine short-term seasonal effects as well as large-scale phenomena (like the Arctic and North Atlantic oscillations) and their effect on the Arctic's megafauna. A scientist with the project contends, "The survival of some polar bear populations depends on the decisions we make within the next year."

WRITING in Biology

Press Release What other negative impacts could receding seasonal sea ice have on arctic ecology? Research other animal populations scientists think are being affected. Write a press release informing people of the potential effects.

BIOLAB

DO PLANTS OF THE SAME SPECIES COMPETE WITH ONE ANOTHER?

Background: Ecologists often study plant competition by comparing the biomass of individual plants in plant populations. In this lab, you will study intraspecific competition—competition among plants of the same species. As with most ecological studies, you will need to collect data for several weeks.

Question: *Do plant populations of various densities grow differently due to competition?*

Materials

marigold seeds or radish seeds
9-cm plastic pots (6)
clean potting soil
rulers
shallow tray for pots
small garden trowels
masking tape
permanent markers
balance (accurate to 0.1 g)
watering can

Safety Precautions

Procedure

1. Read and complete the lab safety form.
2. Plant seeds in several pots as instructed by your teacher. Your goal should be to have pots with the following densities of plants: 2, 4, 8, 16, 32, and 64.
3. Place the pots in a shallow tray near a sunny window or under a grow light. Continue to keep the soil moist—not drenched—throughout the course of the experiment.
4. After the seeds have sprouted, weed out any extra plants so that you have the correct density.
5. Write a hypothesis about the effect plant density will have on the average biomass of each pot's population.
6. Construct a data table. Observe the plants once each week for a 5–6 week period. Record your observations.
7. At the end of the experiment, measure the biomass of the plants in each pot by cutting each plant at soil level and quickly weighing all the plants from the same pot together. Record your measurements. Calculate the average per-plant biomass of each pot.
8. **Cleanup and Disposal** Wash and return all reusable materials. Wash your hands after watering or working with the plants. Dispose of the plants at the end of the lab as instructed by your teacher.

Analyze and Conclude

1. **Graph Data** Prepare a graph showing the relationship between the average plant biomass and the density of plants. Draw a best-fit line for your data points. What was the effect of plant density on the average biomass of each pot's population? Does this graph support your hypothesis?
2. **Infer** Draw a second graph that compares the total biomass for each population to the number of plants in each population.
3. **Think Critically** Based on your results, infer how human population growth is affected by population density.
4. **Error Analysis** What sources of error might have affected your results?

GOING FURTHER

Poster Session Create a poster using the graphs you produced as a result of your experiment. If a digital camera is available, take photos of each pot of plants to include on your poster. Add headings and legends for each graph and photograph that explain and summarize your findings. Display your poster in the classroom or a hallway of your school. To learn more about competing plants, visit BioLabs at biologygmh.com.

Chapter 4 Study Guide

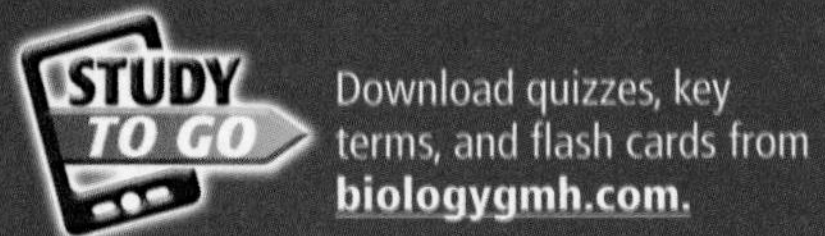

Download quizzes, key terms, and flash cards from biologygmh.com.

FOLDABLES Research Find the population density of the countries of a continent. Make a color-coded map that shows the population density of each country.

Vocabulary	Key Concepts
Section 4.1 Population Dynamics	
• carrying capacity (p. 98) • density-dependent factor (p. 95) • density-independent factor (p. 94) • dispersion (p. 92) • emigration (p. 97) • immigration (p. 97) • population density (p. 92) • population growth rate (p. 97)	**MAIN Idea** Populations of species are described by density, spatial distribution, and growth rate. • There are population characteristics that are common to all populations of organisms, including plants, animals, and bacteria. • Populations tend to be distributed randomly, uniformly, or in clumps. • Populations tend to stabilize near the carrying capacity of their environment. • Population limiting factors are either density-independent or density-dependent.
Section 4.2 Human Population	
• age structure (p. 104) • demographic transition (p. 102) • demography (p. 100) • zero population growth (ZPG) (p. 104)	**MAIN Idea** Human population growth changes over time. • Human population growth rates vary in industrially-developing countries and industrialized countries. • Zero population growth occurs when the birthrate of a population equals the death rate. • The age structure of the human population is a contributing factor to population growth in some countries. • Earth has an undefined carrying capacity for the human population.

Biology Online Vocabulary PuzzleMaker biologygmh.com

CHAPTER 4 Assessment

Section 4.1

Vocabulary Review

Replace the underlined words with the correct vocabulary term from the Study Guide page.

1. The number added to a population by movement can considerably increase a population's size.
2. Drought is a density-dependent factor.
3. Were it not for the long-term limit, a population would continue to grow exponentially.

Understand Key Concepts

Use the illustration to answer questions 4–6.

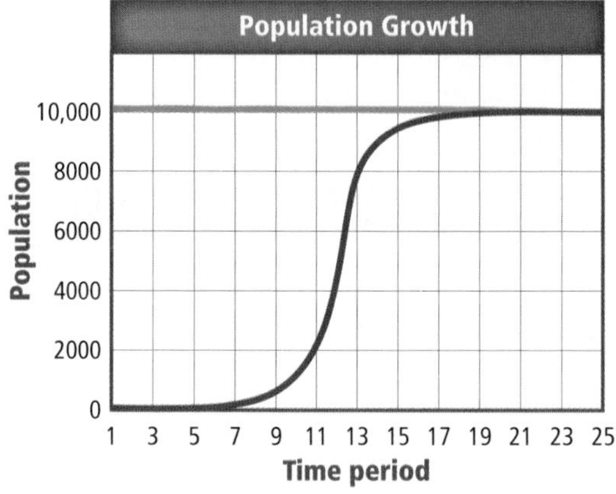

4. Which population growth model does this graph illustrate?
   A. exponential growth
   B. lag phase
   C. logistic growth
   D. straight-line growth

5. What is the horizontal line on this graph called?
   A. carrying capacity **C.** geometric growth
   B. exponential growth **D.** straight-line growth

6. What do the time periods 1–7 represent?
   A. acceleration phase **C.** exponential growth
   B. carrying capacity **D.** lag phase

7. If angelfish produce hundreds of young several times a year, which statement below is true?
   A. Angelfish have a *k*-strategy reproductive pattern.
   B. Angelfish have an *r*-strategy reproductive pattern.
   C. Angelfish probably have a low mortality rate.
   D. Angelfish provide a lot of care for their young.

8. If an aquarium holds 80 L of water and contains 170 guppies, what is the approximate density of the guppy population?
   A. 1 guppy/L **C.** 3 guppies/L
   B. 2 guppies/L **D.** 4 guppies/L

9. Which is a density-independent factor?
   A. a severe drought
   B. an intestinal parasite
   C. a fatal virus
   D. severe overcrowding

Use the photo below to answer questions 10 and 11.

10. Which is a possible reason for the relatively quick spread of the shown disease?
    A. an abiotic factor
    B. a decreased food supply
    C. increased population density
    D. increased immunity

11. Why is the life span of this finch with an eye disease most likely reduced?
    A. The bird cannot mate.
    B. The bird cannot find food or water.
    C. The bird spreads the disease to others.
    D. The bird cannot survive a temperature change.

12. What is the dispersion pattern of herding animals, birds that flock together, and fish that form schools?
    A. clumped **C.** uniform
    B. random **D.** unpredictable

Constructed Response

13. **Short Answer** Female Atlantic right whales can reproduce at ten years of age and live more than fifty years. They can produce a calf every three to five years. Assuming that a right whale begins to reproduce at age ten, produces a calf every four years, and gives birth to its last calf at age fifty, how many whales will this female produce in her lifetime?

14. **Short Answer** What is the population density of Canada and the United States if they have a combined area of approximately 12.4 million square kilometers and a combined population of approximately 524 million?

15. **Short Answer** How does the carrying capacity affect k-strategists?

16. **Open Ended** Give two examples of how two different density-independent factors can limit a specific population.

17. **Open Ended** Give two examples of how two different density-dependent factors can limit a specific population.

18. **Short Answer** Explain how competition limits a population's growth.

Think Critically

19. **Predict** the shape of a population growth curve for a game park in which a male and female rhinoceros are released.

Use the photo below to answer question 20.

20. **Infer** the reproductive strategy of the animal in the photo. Explain your answer.

21. **Generalize** Opossums are solitary animals that usually meet in nature only to mate. What is their probable dispersion pattern?

22. **Select** from the following list the species that are r-strategists: minnow, giraffe, human, beetle, bacteria, eagle, and cougar.

Section 4.2

Vocabulary Review

Using the list of vocabulary words from the Study Guide, identify the term described by the scenario.

23. A population has an equal number of births and deaths.

24. Twenty percent of a population is in pre-reproductive years, 50 percent is in the reproductive years, and 30 percent is in the post-reproductive years.

25. The size, density, and birth and death rates of a human population are studied.

Understand Key Concepts

Use the graph below of the growth of the human population through history to answer questions 26 and 27.

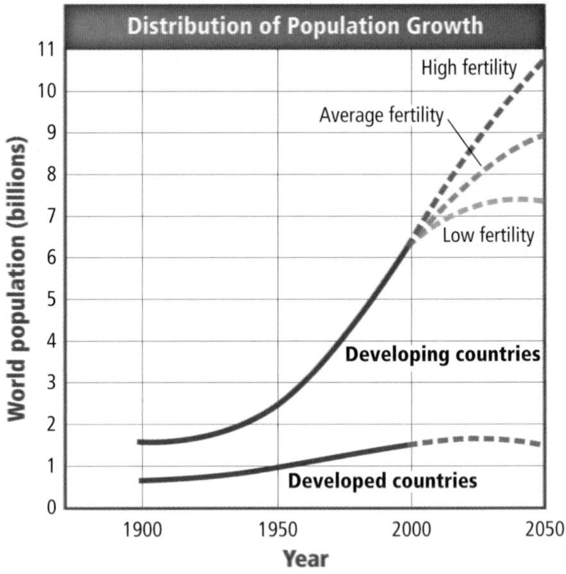

26. What is the projected population of developed countries by 2050?
    A. 1.5 billion
    B. 7.3 billion
    C. 9 billion
    D. 10.5 billion

27. What is the approximate population difference between developing countries that have low fertility rates and developing countries that have high fertility rates in 2050?
    A. 1.5 billion
    B. 1.7 billion
    C. 3.2 billion
    D. 9 billion

28. When did the human population begin to increase exponentially? Use **Figure 4.11** as a reference.
   A. 2 million years ago
   B. 6500 B.C.
   C. 1800 B.C.
   D. 1500 A.D.

29. Asia (excluding China) had a birthrate of 24 and a death rate of eight in 2004. What was the PGR?
   A. 0.16 percent
   B. 1.6 percent
   C. 16 percent
   D. 160 percent

30. Georgia, a country in Western Asia, had a birthrate of 11 and a death rate of 11 in 2004. What was the PGR of Georgia in that year?
   A. 0 percent
   B. 0.11 percent
   C. 1.1 percent
   D. 11 percent

Constructed Response

31. **Open Ended** Do you think the birthrate or the death rate is more important to human populations? Explain your answer.

32. **Short Answer** Why might a population continue to grow when the number of births equals deaths?

33. **Short Answer** Study **Figure 4.11** and identify which phase of growth occurred between the Old Stone Age and the Middle Ages.

Think Critically

34. Hypothesize the shape of the age diagram for Switzerland, a developed country in Europe.

Use the graph below to answer question 35.

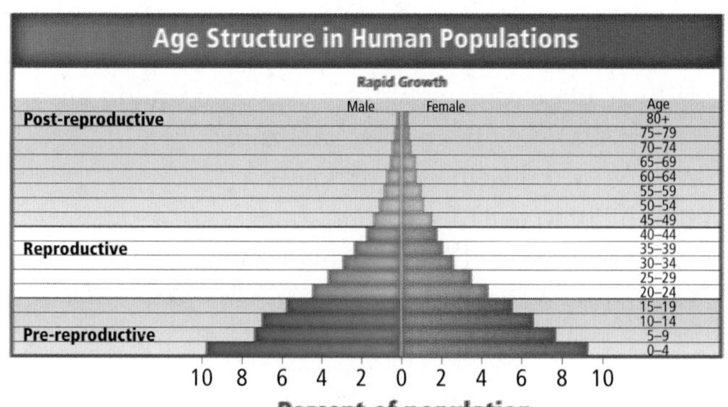

35. Describe the advantages and disadvantages of a population that has this type of age structure.

Additional Assessment

36. WRITING in Biology Write a letter to the editor of your student newspaper expressing your views on the effect of human activities on a population of animals in your area.

DBQ Document-Based Questions

Northern right whales were once abundant in the northwestern Atlantic Ocean. By 1900, their numbers were almost depleted. Today, there are an estimated 300 individuals remaining.

Use the graph below to answer the following questions.

Data obtained from: Fujiwara, M., et al. 2001. Demography of the endangered North Atlantic right whale. *Nature* 414: 537-540.

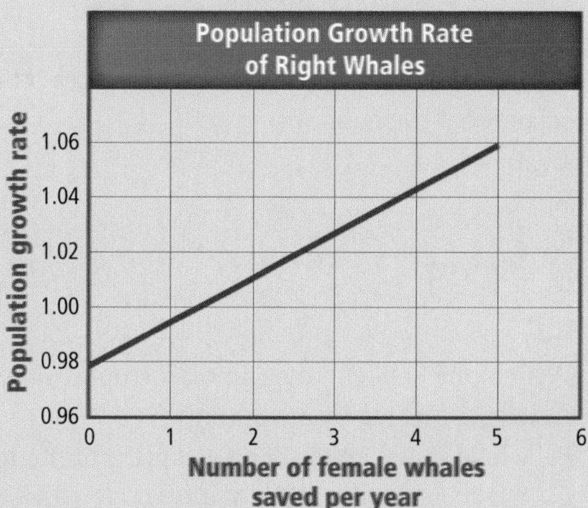

37. Predict the population growth rate if six female North Atlantic right whales were saved each year.

38. Saving females isn't the only factor to take into consideration when trying to restore the whale population. Write a hypothetical plan of action that takes into account two other factors that you think might help.

Cumulative Review

39. Predict the probable results to a community if all of the top predators were removed by hunting. **(Chapter 2)**

40. Describe three types of symbiosis. **(Chapter 2)**

Standardized Test Practice

Cumulative

Multiple Choice

1. Which is the main benefit of scientific debate for scientists?
   A. challenging accepted theories
   B. creating controversy
   C. gaining research funding
   D. publishing results

Use the graph below to answer question 2.

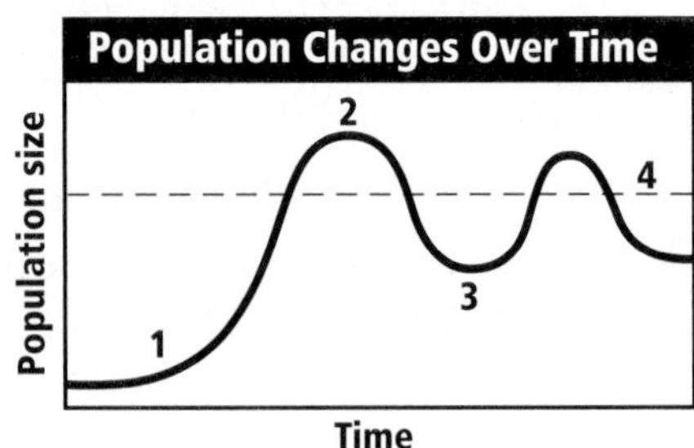

2. Which part of the graph indicates the carrying capacity of the habitat?
   A. 1
   B. 2
   C. 3
   D. 4

3. Which one is likely to be an oligotrophic lake?
   A. a lake formed by a winding river
   B. a lake in the crater of a volcanic mountain
   C. a lake near the mouth of a river
   D. a lake where algae blooms kill the fish

4. Which characteristic of a plant would NOT be studied by biologists?
   A. beauty
   B. chemical processes
   C. growth rate
   D. reproduction

5. Which statement describes the first changes in a forest that would follow a forest fire?
   A. A climax community is established.
   B. New plants grow from seeds that the wind carries to the area.
   C. New soil forms.
   D. Pioneer species are established.

Use this graph to answer question 6.

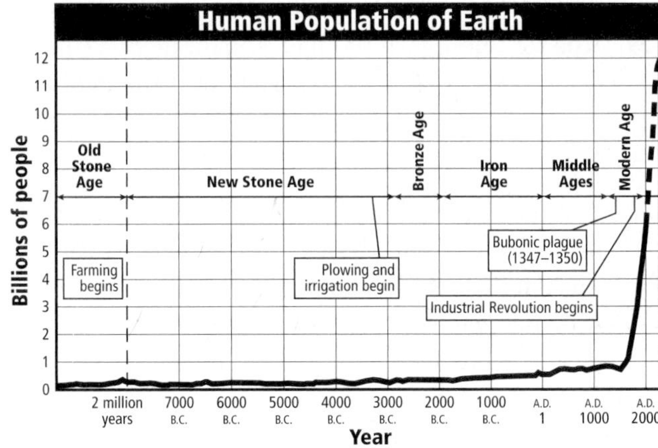

6. Which event appears to coincide with a gradual increase in human population?
   A. Bubonic plague
   B. farming
   C. Industrial Revolution
   D. plowing and irrigation

7. Suppose an organism is host to a parasitic tapeworm. Which would be beneficial to the tapeworm?
   A. death of the host from disease caused by the tapeworm
   B. absorbing enough nutrients to sustain the tapeworm without harming the host
   C. treatment of the host with antitapeworm drugs
   D. weakening of the host by the tapeworm

8. Which adaptation would you expect to find in an organism living in an intertidal zone?
   A. ability to live in total darkness
   B. ability to live in very cold water
   C. ability to survive in moving water
   D. ability to survive without water for 24 h

9. Which limiting factor is dependent on the density of the population?
   A. contagious fatal virus
   B. dumping toxic waste in a river
   C. heavy rains and flooding
   D. widespread forest fires

Biology Online Standardized Test Practice biologygmh.com

Short Answer

Use this graph to answer questions 10 and 11.

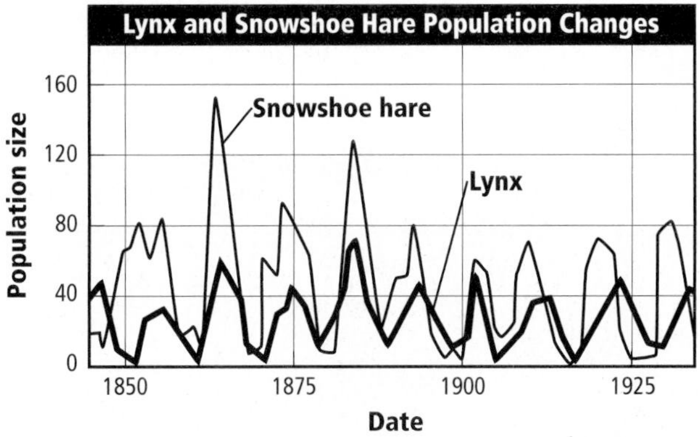

10. Assess what happened to the hare population after a sharp rise in the lynx population.

11. Lynxes hunt hares for food. Predict what would happen to the lynx population if a disease killed all of the hares.

12. Using your knowledge of current events or history, give an example of when ignorance about biology had a harmful effect on people.

13. Compare and contrast how density-dependent and density-independent factors regulate the growth of populations.

14. Describe what happens to organisms whose optimum temperature zone is between 21°C and 32°C when the temperature rises from 21°C to 50°C.

15. Give some examples of the ways that an environmental factor, such as a forest fire, can affect a population.

16. Explain how a population relates to an ecosystem.

Extended Response

Use these graphs to answer question 17.

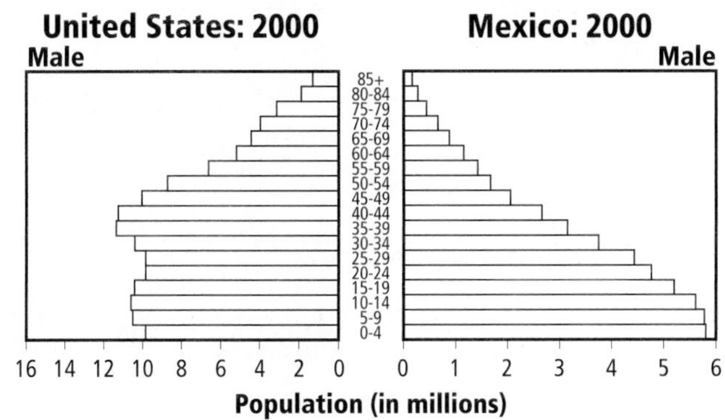

17. State what you think is the most significant area of difference between the two populations and justify your reasoning.

18. Many vertebrates that live in temperate forests hibernate in the winter. How do you think this adaptation helps with survival in this biome?

Essay Question

Author Carrie P. Snow once said, "Technology… is a queer thing. It brings you great gifts with one hand, and it stabs you in the back with the other."
C. P. Snow, New York Times, 15 March 1971

Using the information contained in the quotation above, answer the following question in essay format.

19. You are in charge of organizing a debate about whether technology is good or bad. Using your prior knowledge, choose a position and write a summary of the key points you would debate.

NEED EXTRA HELP?																			
If You Missed Question . . .	1	2	3	4	5	6	7	8	9	10	11	12	13	14	15	16	17	18	19
Review Section . . .	1.2	4.1	3.2	1.2	3.1	4.2	2.1	3.3	3.1	4.1	4.1	1.1	4.1	3.2	4.1	2.1	4.2	3.2	1.2

Biodiversity and Conservation

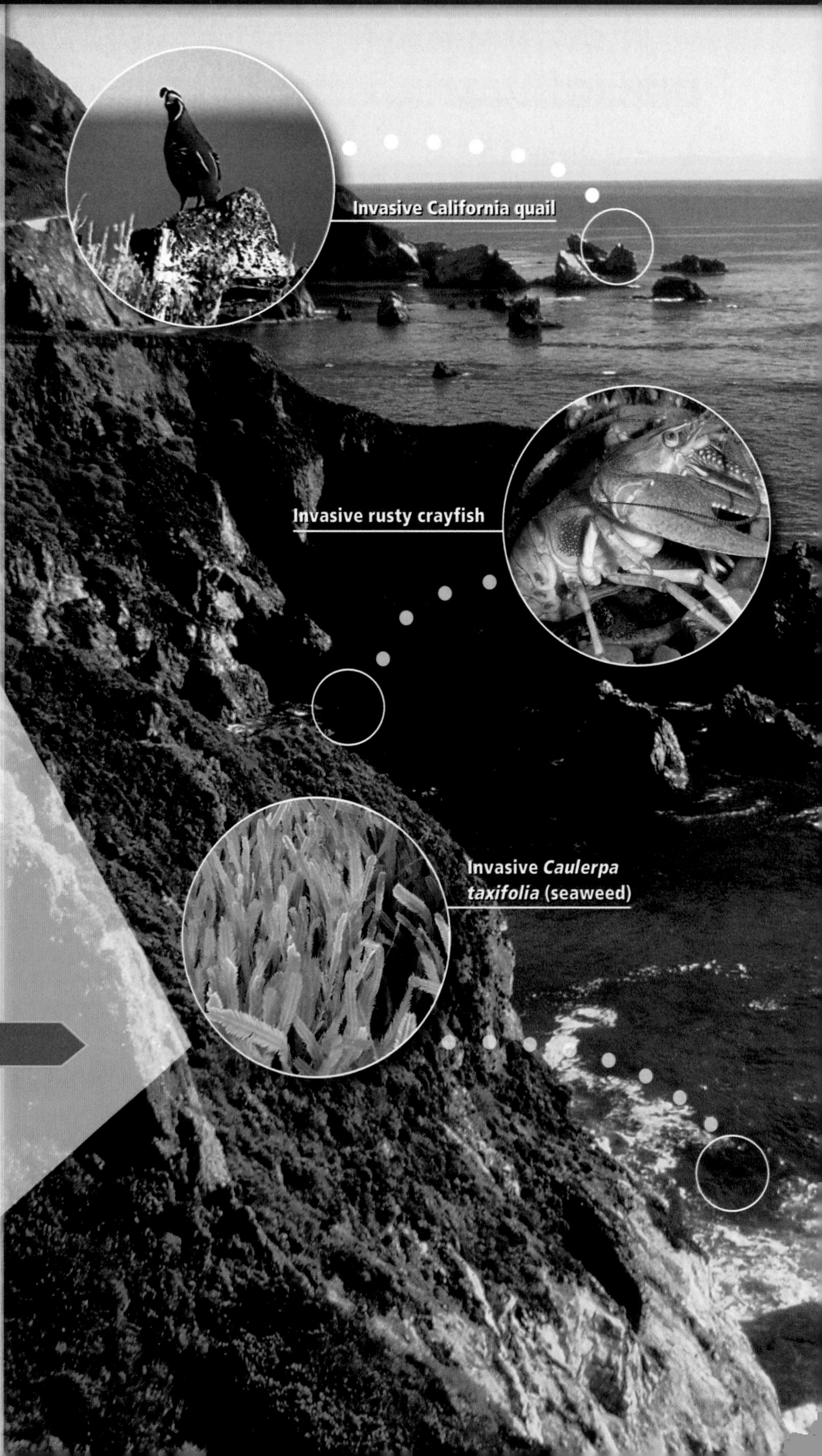

BIG Idea Community and ecosystem homeostasis depends on a complex set of interactions among biologically diverse individuals.

Section 1
Biodiversity
MAIN Idea Biodiversity maintains a healthy biosphere and provides direct and indirect value to humans.

Section 2
Threats to Biodiversity
MAIN Idea Some human activities reduce biodiversity in ecosystems, and current evidence suggests that reduced biodiversity might have serious long-term effects on the biosphere.

Section 3
Conserving Biodiversity
MAIN Idea People are using many approaches to slow the rate of extinctions and to preserve biodiversity.

BioFacts

- A hardy, cold-water strain of the tropical algae *Caulerpa taxifolia* was produced for the saltwater aquarium industry.
- *Caulerpa taxifolia* is invading the waters off the coast of California, where it can harm the local biological communities.

Invasive California quail

Invasive rusty crayfish

Invasive *Caulerpa taxifolia* (seaweed)

LAUNCH Lab

What lives here?

Some landscapes support more organisms than others. In this lab, you will infer the relative numbers of species that can be found in each environment.

Procedure

1. Read and complete the lab safety form.
2. Choose three locations in your community that are familiar to you, such as a tree, a group of trees, a drainage ditch, a field, a dumpster, a park, or a pond.
3. Rank the locations in descending order, greatest to least, according to the number of species of animals, plants, etc. you think you would find there.

Analysis

1. **Define** the term *biodiversity* in your own words.
2. **Explain** how you chose to rank the locations in order.
3. **Describe** scientific methods you could use to find out how many species live in each habitat.

Visit biologygmh.com to:

- study the entire chapter online
- explore Concepts in Motion, Interactive Tables, Microscopy Links, Virtual Labs, and links to virtual dissections
- access Web links for more information, projects, and activities
- review content online with the Interactive Tutor and take Self-Check Quizzes

Biodiversity Make the following Foldable to help you understand the three levels of biodiversity and the importance of biodiversity to the biosphere.

STEP 1 Fold a sheet of paper in half lengthwise. Make the back part about 5 cm longer than the front part.

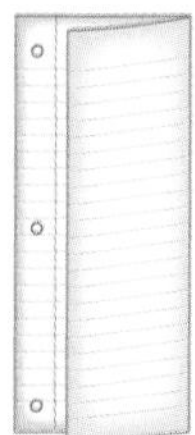

STEP 2 Turn the paper so that the fold is on the bottom, then fold it into thirds.

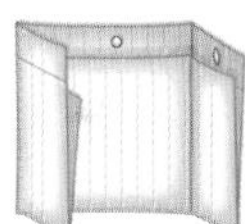

STEP 3 Unfold and cut only the top layer along each fold to make three tabs. Label the Foldable as shown.

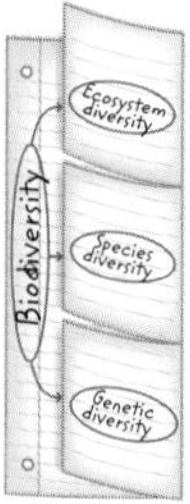

FOLDABLES **Use this Foldable with Section 5.1.** As you study the section, define *biodiversity* under the large tab and explain its importance. Describe each of the three types of biodiversity under the small tabs. Provide an example of each.

Section 5.1

Biodiversity

Reading Preview

Objectives

- **Describe** three types of biodiversity.
- **Explain** the importance of biodiversity.
- **Summarize** the direct and indirect value of biodiversity.

Review Vocabulary

gene: functional unit that controls the expression of inherited traits and is passed from generation to generation

New Vocabulary

extinction
biodiversity
genetic diversity
species diversity
ecosystem diversity

MAIN Idea Biodiversity maintains a healthy biosphere and provides direct and indirect value to humans.

Real-World Reading Link Stop for a moment and consider the effect of all the jackrabbits in a food web dying suddenly. What would happen to the other members of the food web? Is the disappearance of one species from Earth important, or will another species fill its niche?

What is biodiversity?

The loss of an entire species in a food web is not an imaginary situation. Entire species permanently disappear from the biosphere when the last member of the species dies in a process called **extinction.** As species become extinct, the variety of species in the biosphere decreases, which decreases the health of the biosphere. **Biodiversity** is the variety of life in an area that is determined by the number of different species in that area. Biodiversity increases the stability of an ecosystem and contributes to the health of the biosphere. There are three types of biodiversity to consider: genetic diversity, species diversity, and ecosystem diversity.

Genetic diversity The variety of genes or inheritable characteristics that are present in a population comprises its **genetic diversity. Figure 5.1** shows several characteristics that are shared by the ladybird beetles, such as general body structure. The variety of colors demonstrates a form of genetic diversity. The beetles have other characteristics that differ, but they are not as apparent as their color. These characteristics might include resistance to a particular disease, the ability to recover from a disease, or the ability to obtain nutrients from a new food source should the old food source disappear. The beetles with these characteristics are more likely to survive and reproduce than beetles without these characteristics.

Genetic diversity within interbreeding populations increases the chances that some species will survive during changing environmental conditions or during an outbreak of disease.

■ **Figure 5.1** These Asian ladybird beetles, *Harmonia axyridis,* demonstrate some visible genetic diversity because of their different colors. **Suggest** *some other characteristics that might vary among the beetles.*

■ **Figure 5.2** Many species gather at this watering hole, making it a habitat rich in species diversity.

Species diversity The number of different species and the relative abundance of each species in a biological community is called **species diversity.** As you look at **Figure 5.2,** notice how many different species of organisms are in this one area. This habitat represents an area with a high level of species diversity because there are so many species present in one location. However, species diversity is not evenly distributed over the biosphere. As you move geographically from the polar regions to the equator, species diversity increases. For example, **Figure 5.3** shows the number of bird species from Alaska to Central America. Use the color key to see how diversity changes as you move toward the equator.

Reading Check **Compare and contrast** genetic and species diversity.

FOLDABLES
Incorporate information from this section into your Foldable.

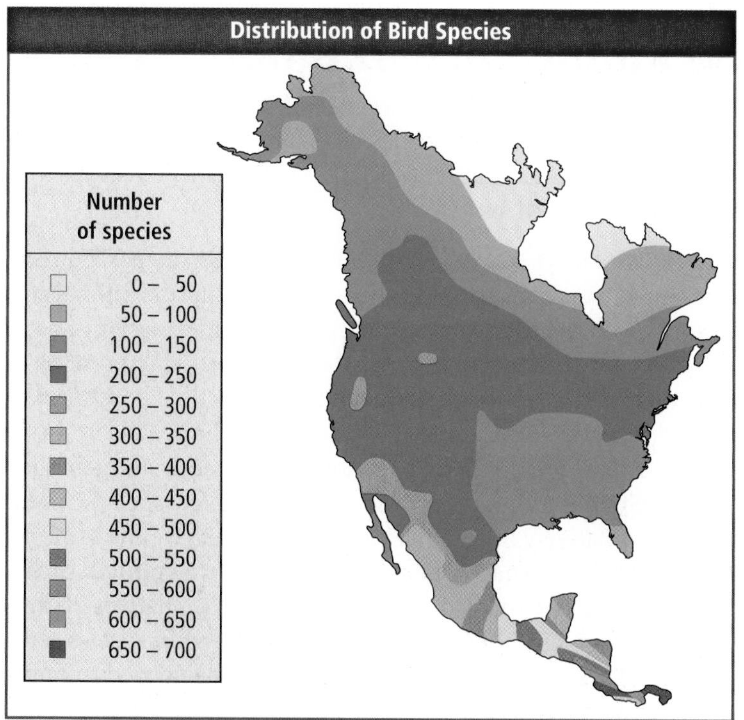

■ **Figure 5.3** This map shows the distribution of bird species in North and Central America. As you move toward the tropics, biodiversity increases.
Identify *the locations with the highest and lowest amounts of species diversity.*

Alaska

Peru

■ **Figure 5.4** The biosphere contains many ecosystems with diverse abiotic factors that support a variety of organisms.

Ecosystem diversity The variety of ecosystems that are present in the biosphere is called **ecosystem diversity.** Recall from Chapter 2 that an ecosystem is made up of interacting populations and the abiotic factors that support them. The interactions of organisms affect the development of stable ecosystems. Different locations around the world have different abiotic factors that support different types of life. For example, an ecosystem in Alaska has a set of abiotic factors that supports Dall sheep, which are shown in **Figure 5.4.** An ecosystem in South America has a different set of abiotic factors that supports tropical birds, also shown in **Figure 5.4.** All of the ecosystems on Earth support a diverse collection of organisms.

VOCABULARY

ACADEMIC VOCABULARY

Diverse:
made of different qualities
The colors and shapes of flowers are very diverse.

 **Reading Check** **Explain** why ecosystem diversity results in species diversity in a healthy biosphere.

The Importance of Biodiversity

There are several reasons to preserve biodiversity. Many humans work to preserve and protect the species on Earth for future generations. In addition, there are economic, aesthetic, and scientific reasons for preserving biodiversity.

Direct economic value Maintaining biodiversity has a direct economic value to humans. Humans depend on plants and animals to provide food, clothing, energy, medicine, and shelter. Preserving species that are used directly is important, but it also is important to preserve the genetic diversity in species that are not used directly. Those species serve as possible sources of desirable genes that might be needed in the future.

The reason there might be a future need for desirable genes is that most of the world's food crops come from just a few species. These plants have relatively little genetic diversity and share the same problems that all species share when genetic diversity is limited, such as lacking resistance to disease. In many cases, close relatives of crop species still grow wild in their native habitat. These wild species serve as reservoirs of desirable genetic traits that might be needed to improve domestic crop species.

Teosinte plant

Domestic corn plant

■ **Figure 5.5** The teosinte plant contains genes that are resistant to several viral diseases that affect domesticated corn plants. These genes have been used to produce virus-resistant domestic corn varieties.

The distant relative of corn, teosinte, shown in **Figure 5.5,** is resistant to the viral diseases that damage commercial corn crops. Using this wild species, plant pathologists developed disease-resistant corn varieties. If this wild species had not been available, this genetic diversity would have been lost, and the ability to develop disease-resistant corn varieties would also have been lost.

In addition, biologists are beginning to learn how to transfer genes that control inherited characteristics from one species to the other. This process, sometimes referred to as genetic engineering, is discussed in Chapter 13. Crops have been produced that are resistant to some insects, that have increased nutritional value, and that are more resistant to spoilage. Most wild species of plants and animals have not been evaluated for useful genetic traits. The opportunity to benefit from these genes is lost forever if wild species of plants and animals become extinct. This increases the importance of species that currently have no perceived economic value because their economic value might increase in the future.

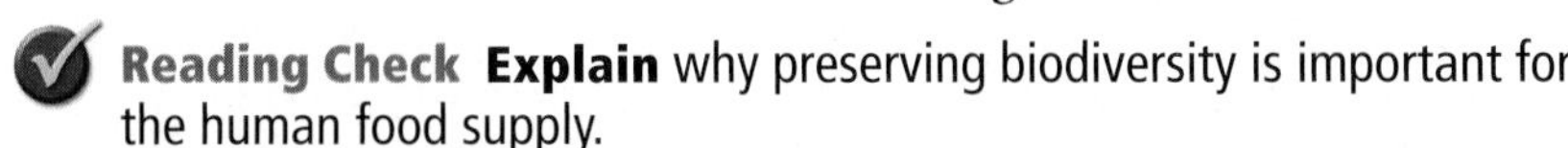

Reading Check **Explain** why preserving biodiversity is important for the human food supply.

Connection to Health Many of the medicines that are used today are derived from plants or other organisms. You probably know that penicillin, a powerful antibiotic discovered in 1928 by Alexander Fleming, is derived from bread mold. Ancient Greeks, Native Americans, and others extracted salicin, a painkiller, from the willow tree. Today, a version of this drug is synthesized in laboratories and is known as aspirin. **Figure 5.6** shows a Madagascar periwinkle flower, which recently was found to yield an extract that is useful in treating some forms of leukemia. This extract has been used to develop drugs that have increased the survival rate for some leukemia patients from 20 percent to more than 95 percent.

Scientists continue to find new extracts from plants and other organisms that help in the treatment of human diseases. However, many species of organisms are yet to be identified, especially in remote regions of Earth, so their ability to provide extracts or useful genes is unknown.

■ **Figure 5.6** Drugs developed from an extract from Madagascar periwinkle, *Catharanthus roseus,* are used to treat childhood forms of leukemia.
Summarize ***Why is it important to maintain biodiversity for medical reasons?***

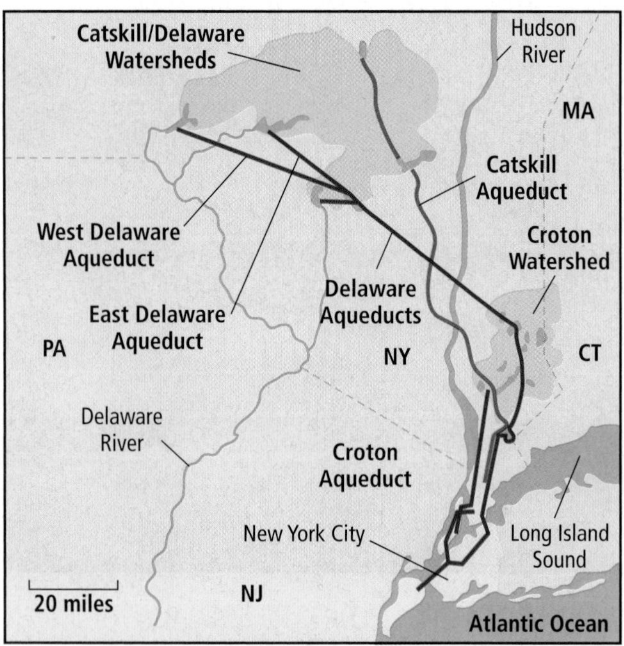

Figure 5.7 An economic study determined that restoring the biodiversity in the ecosystem that filtered the water supply for New York City was less expensive than using technology to perform the same service.
Infer ***What type of human activities could affect a watershed and lower water quality?***

Indirect economic value A healthy biosphere provides many services to humans and other organisms that live on Earth. For example, green plants provide oxygen to the atmosphere and remove carbon dioxide. Natural processes provide drinking water that is safe for human use. Substances are cycled through living organisms and nonliving processes, providing nutrients for all living organisms. As you will soon learn, healthy ecosystems provide protection against floods and drought, generate and preserve healthful fertile soils, detoxify and decompose wastes, and regulate local climates.

It is difficult to attach an economic value to the services that a healthy biosphere provides. However, some scientists and economists have attempted to do just that. In the 1990s, New York City was faced with the decision of how to improve the quality of its drinking water. A large percentage of New York City's drinking water was supplied by watersheds, shown in **Figure 5.7.** Watersheds are land areas where the water on them or the water underneath them drains to the same place. The Catskill and Delaware watersheds did not meet clean water standards and no longer could supply quality drinking water to the city.

The city was faced with two choices: build a new water filtration system for more than $6 billion or preserve and clean up the watersheds for approximately 1.5 billion dollars. The economic decision was clear in this case. A healthy ecosystem was less expensive to maintain than using technology to perform the same services.

MiniLab 5.1

Investigate Threats to Biodiversity

What are the threats to natural habitats in your local area? Investigate these threats and brainstorm possible remedies with which you can educate others.

Procedure

1. Read and complete the lab safety form.
2. With your lab group, choose one factor that is threatening the biodiversity in your community and study how it has affected the climax community.
3. Brainstorm ways that this threat could be reversed.
4. Organize this information about threats and possible solutions with your classmates.

Analysis

1. **Evaluate** What are the most important pieces of information the public needs to know about this threat?
2. **Infer** Imagine you have implemented one plan to reverse a threat you studied. Now it is 100 years later. What does the ecosystem look like? What changes have occurred? What species are there now?

■ **Figure 5.8** It is difficult to attach an economic value to the aesthetic qualities of healthy ecosystems and biodiversity.

This example shows that nature can provide services, such as water that is safe for human consumption, at less expense than using technology to provide the same service. Some scientists think the natural way should be the first choice for providing these services. Research indicates that when healthy ecosystems are preserved, the services the ecosystems provide will continue to be less expensive than performing the same services with technology.

Aesthetic and scientific value Two additional considerations for maintaining biodiversity and healthy ecosystems are the aesthetic and scientific values that they provide. It is difficult to attach a value to something that is beautiful, such as the ecosystem shown in **Figure 5.8,** or something that is interesting to study. Perhaps it would help to consider how life would be if all of Earth was a barren and desolate landscape. The value of biodiversity and healthy ecosystems might be more obvious then.

Section 5.1 Assessment

Section Summary

- Biodiversity is important to the health of the biosphere.
- There are three types of biodiversity: genetic, species, and ecosystem.
- Biodiversity has aesthetic and scientific values, and direct and indirect economic value.
- It is important to maintain biodiversity to preserve the reservoir of genes that might be needed in the future.
- Healthy ecosystems can provide some services at a lesser expense than the use of technology.

Understand Main Ideas

1. MAIN Idea **Explain** why bio-diversity is important to the biosphere.
2. **Summarize** the three types of biodiversity.
3. **Generalize** why maintaining biodiversity has a direct economic value to humans.
4. **Differentiate** between the direct and indirect economic value of biodiversity.
5. **Evaluate and discuss** the importance of maintaining biodiversity for future medical needs.

Think Critically

6. **Design a course of action** for the development of a building project in your community, such as a shopping mall, housing development, city park, or highway, that provides for the maintenance of biodiversity in the plan.

WRITING in Biology

7. Write a short report explaining the desirability of maintaining genetic diversity in domesticated animals, such as dogs, cats, pigs, cattle, and chickens. Include the advantages and disadvantages in your report.

Section 5.2

Reading Preview

Objectives

- **Describe** threats to biodiversity.
- **Compare and contrast** the background extinction rate with the current extinction rate.
- **Describe** how the decline of a single species can affect an entire ecosystem.

Review Vocabulary

food web: a model representing the many interconnected food chains and pathways in which energy and matter flow through a group of organisms

New Vocabulary

background extinction
mass extinction
natural resource
overexploitation
habitat fragmentation
edge effect
biological magnification
eutrophication
introduced species

Threats to Biodiversity

MAIN Idea **Some human activities reduce biodiversity in ecosystems, and current evidence suggests that reduced biodiversity might have serious long-term effects on the biosphere.**

Real-World Reading Link Have you ever built a structure with blocks, and then tried to remove individual blocks without causing the entire structure to collapse? Similarly, if you remove one species from a food web, the food web can collapse.

Extinction Rates

Many species have become extinct and paleontologists study fossils of those extinct species. The gradual process of species becoming extinct is known as **background extinction.** Stable ecosystems can be changed by the activity of other organisms, climate changes, or natural disasters. This natural process of extinction is not what scientists are worried about. Many worry about a recent increase in the rate of extinction. Some scientists predict that between one-third and two-thirds of all plant and animal species will become extinct during the second half of this century. Most of these extinctions will occur near the equator.

Some scientists estimate the current rate of extinction is about 1000 times the normal background extinction rate. These scientists think that we are witnessing a period of mass extinction. **Mass extinction** is an event in which a large percentage of all living species become extinct in a relatively short period of time. The last mass extinction occurred about 65 million years ago, as illustrated in **Table 5.1,** when the last of the surviving dinosaurs became extinct.

Concepts In Motion Interactive Table To explore more about mass extinctions, visit biologygmh.com.

Table 5.1 Five Most Recent Mass Extinctions

	Ordovician Period	Devonian Period	Permian Period	Triassic Period	Cretaceous Period
Time	about 444 million years ago	about 360 million years ago	about 251 million years ago	about 200 million years ago	about 65 million years ago
Example	Graptolites	*Dinichthys*	Trilobite	Cynognathus	Ammonite

Concepts In Motion

Interactive Table To explore more about mass extinctions, visit biologygmh.com.

Table 5.2 Estimated Number of Extinctions Since 1600

Group	Mainland	Island	Ocean	Total	Approximate Number of Species	Percent of Group Extinct
Mammals	30	51	4	85	4000	2.1
Birds	21	92	0	113	9000	1.3
Reptiles	1	20	0	21	6300	0.3
Amphibians*	2	0	0	2	4200	0.05
Fish	22	1	0	23	19,100	0.1
Invertebrates	49	48	1	98	1,000,000+	0.01
Flowering plants	245	139	0	384	250,000	0.2

*An alarming decrease of amphibian populations has occurred since the mid-1970s, and many species might be on the verge of extinction.

Connection to History The accelerated loss of species began several centuries ago. **Table 5.2** shows the estimated number of extinctions that have occurred by group since 1600. Many of the species' extinctions in the past have occurred on islands. For example, 60 percent of the mammals that have become extinct in the past 500 years lived on islands, and an 81 percent of bird extinctions occurred on islands.

Species on islands are particularly vulnerable to extinction because of several factors. Many of these species evolved without the presence of natural predators. As a result, when a predator, such as a dog, cat, rat, or human, is introduced to the population, the native animals do not have the ability or skills to escape. When a nonnative species is introduced to a new population, it can be a carrier of a disease to which the native population has no resistance. The native population often dies off as a result. In addition, islands typically have relatively small population sizes and individual animals rarely travel between islands, which increases the vulnerability of island species to extinction.

Reading Check Explain why organisms found on islands are more vulnerable to extinction than other organisms.

VOCABULARY

WORD ORIGIN

Native

from the Latin word *nativus;* means to be born.

Factors that Threaten Biodiversity

Scientists point out that today's high rate of extinction differs from past mass extinctions. The current high rate of extinction is due to the activities of a single species—*Homo sapiens.* After a mass extinction in the past, new species evolved, and biodiversity recovered after several million years. This time, the recovery might be different. Humans are changing conditions on Earth faster than new traits can evolve to cope with the new conditions. Evolving species might not have the natural resources they need. **Natural resources** are all materials and organisms found in the biosphere, including minerals, fossil fuels, nuclear fuels, plants, animals, soil, clean water, clean air, and solar energy.

■ **Figure 5.9** The ocelot and all species of rhinos, including the white rhinoceros, are in danger of becoming extinct, due in part to overexploitation.

Ocelot

White rhinoceros

Overexploitation One of the factors that is increasing the current rate of extinction is the **overexploitation,** or excessive use, of species that have economic value. For example, the great herds of bison that once roamed the central plains of North America were hunted to the brink of extinction because their meat and hides could be sold commercially and because they were hunted for sport. At one time, it is estimated that there were 50 million bison. By 1889, there were less than 1000 bison left.

Passenger pigeons are another example of a species that has been overexploited. At one time, there were huge flocks of these birds that would darken the skies of North America during their migration. Unfortunately, they were overhunted and forced from their habitats. By the early 1900s, they had become extinct.

The ocelot, shown in **Figure 5.9,** is found from Texas to Argentina and is in danger of becoming extinct. The increasing loss of their habitat and the commercial value of their fur are reasons for their declining numbers. The white rhinoceros, also shown in **Figure 5.9,** is one of five species of rhinos, all of which are in danger of becoming extinct. They are hunted and killed for their horns, which are then sold for medicinal purposes. Historically, overexploitation was the primary cause of species extinction. However, the number one cause of species extinction today is the loss or destruction of habitat.

Reading Check **Explain** the term overexploitation as it relates to species extinction.

Habitat loss There are several ways that species can lose their habitats. If a habitat is destroyed or disrupted, the native species might have to relocate or they will die. For example, humans are clearing areas of tropical rain forests and are replacing the native plants with agricultural crops or grazing land.

Destruction of habitat The clearing of tropical rain forests, like what is shown in **Figure 5.10,** has a direct impact on global biodiversity. As mentioned earlier, the tropical latitudes contain much of the world's biodiversity in their native populations. In fact, estimates show that more than half of all species on Earth live in the tropical rain forests. The removal of so much of the natural forest will cause many species on Earth to become extinct because of habitat loss.

■ **Figure 5.10** Cleared land often is used for agricultural crops or as grazing land for livestock. Planting large expanses of crops reduces the biodiversity of the area.

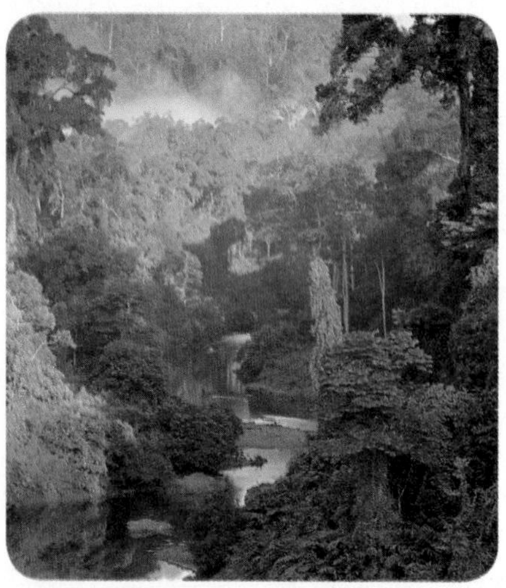

Natural tropical rain forest

Cleared tropical rain forest

Figure 5.11 A declining population of one species can affect an entire ecosystem. When the number of harbor seals and sea lions declined, killer whales ate more sea otters. The decline in sea otter population led to an increase in sea urchins, which eat kelp. This led to the ultimate decline in kelp forests.

Disruption of habitat Habitats might not be destroyed, but they can be disrupted. For example, off the coast of Alaska, a chain of events occurred in the 1970s that demonstrates how the declining numbers of one member of a food web can affect the other members. As you can see from the chain of events shown in **Figure 5.11,** the decline of one species can affect an entire ecosystem. When one species plays such a large role in an ecosystem, that species is called a keystone species. A decline in various fish populations, possibly due to overfishing, has led to a decline in sea lion and harbor seal populations. Some scientists hypothesize that global warming also played a role in the decline. This started a chain reaction within the marine ecosystem that affected many species.

 **Reading Check** **Name** the keystone species in the example in Figure 5.11.

Fragmentation of habitat The separation of an ecosystem into small pieces of land is called **habitat fragmentation.** Populations often stay within the confines of the small parcel because they are unable or unwilling to cross the human-made barriers. This causes several problems for the survival of various species.

First, the smaller the parcel of land, the fewer species it can support. Second, fragmentation reduces the opportunities for individuals in one area to reproduce with individuals from another area. For this reason, genetic diversity often decreases over time in habitat fragments. Smaller, separated, and less genetically diverse populations are less able to resist disease or respond to changing environmental conditions.

■ **Figure 5.12** The smaller the habitat size, the greater percentage of the habitat that is subject to edge effects.

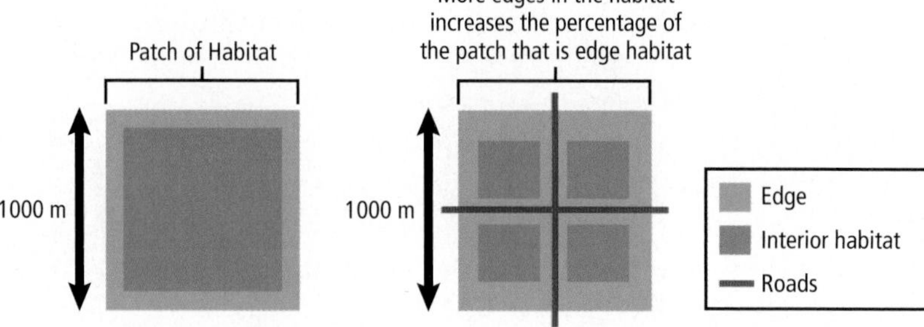

Personal Tutor

To learn about edge effects, visit biologygmh.com.

Third, carving the large ecosystem into small parcels increases the number of edges—creating edge effects, as illustrated in **Figure 5.12.** **Edge effects** are different environmental conditions that occur along the boundaries of an ecosystem. For example, edges of a forest near a road have different abiotic factors, such as temperature, wind, and humidity, than the interior of a forest. Typically, the temperature and wind will be higher and the humidity lower on the edges in a tropical forest. Species that thrive deep in the dense forest might perish on the edges of the ecosystem. Predators and parasites also thrive on the boundaries of ecosystems, which makes the species in these areas more vulnerable to attack. Edge effects do not always create a disadvantage for all species. Some species find these conditions favorable and they thrive.

Reading Check **Explain** how an increasing percentage of land is affected by edge effects when the piece of land is small.

■ **Figure 5.13** The concentration of toxic substances increases as the trophic level in a food chain increases.

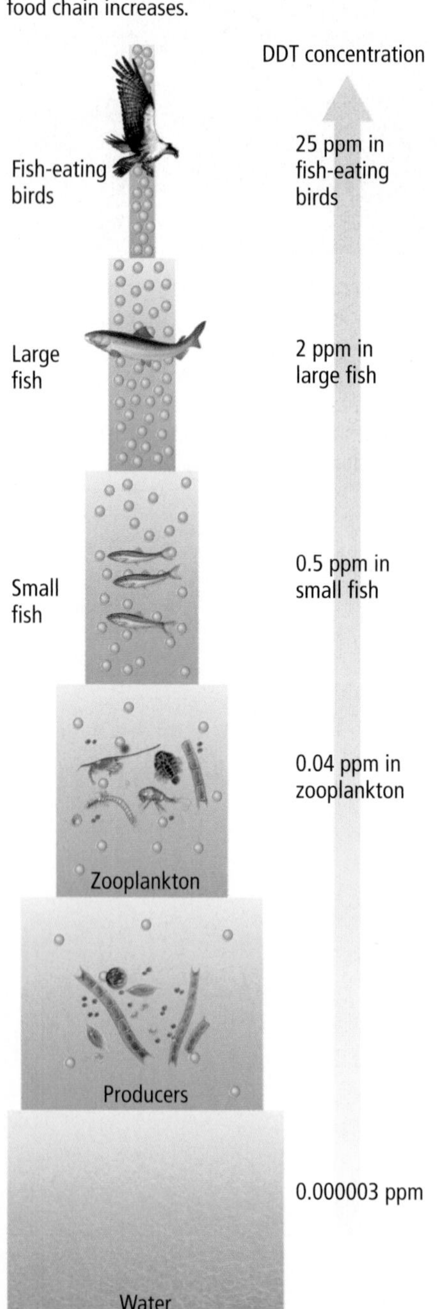

Pollution Pollution and atmospheric changes threaten biodiversity and global stability. Pollution changes the composition of air, soil, and water. There are many types of pollution. Substances—including many human-made chemicals that are not found in nature—are released into the environment. Pesticides, such as DDT (dichloro-diphenyl-trichloroethane), and industrial chemicals, such as PCBs (polychlorinated biphenyls), are examples of substances that are found in food webs. These substances are ingested by organisms when they drink water or eat other organisms that contain the toxic substance. Some substances are metabolized by the organism and excreted with other waste products. However, other substances, such as DDT and PCBs, accumulate in the tissues of organisms.

Carnivores at the higher trophic levels seem to be most affected by the accumulation because of a process called biological magnification. **Biological magnification** is the increasing concentration of toxic substances in organisms as trophic levels increase in a food chain or food web, as shown in **Figure 5.13.** The concentration of the toxic substance is relatively low when it enters the food web. The concentration of toxic substance in individual organisms increases as it spreads to higher trophic levels.

Current research implies that these substances might disrupt normal processes in some organisms. For example, DDT might have played a role in the near extinction of the American bald eagle and the peregrine falcon. DDT is a pesticide that was used from the 1940s to the 1970s to control crop-eating and disease-carrying insects. DDT proved to be a highly effective pesticide, but evidence suggested that it caused the eggshells of fish-eating birds to be fragile and thin, which led to the death of the developing birds. Once these toxic effects were discovered, the use of DDT was banned in some parts of the world.

Acid precipitation Another pollutant that is affecting biodiversity is acid precipitation. When fossil fuels are burned, sulfur dioxide is released into the atmosphere. In addition, the burning of fossil fuels in automobile engines releases nitrogen oxides into the atmosphere. These compounds react with water and other substances in the air to form sulfuric acid and nitric acid. These acids eventually fall to the surface of Earth in rain, sleet, snow, or fog. Acid precipitation removes calcium, potassium, and other nutrients from the soil, depriving plants of these nutrients. It damages plant tissues and slows their growth, as shown in **Figure 5.14.** Sometimes, the acid concentration is so high in lakes, rivers, and streams that fish and other organisms die, also as shown in **Figure 5.14.**

Eutrophication Another form of water pollution, called eutrophication, destroys underwater habitats for fish and other species. **Eutrophication** (yoo troh fih KAY shun) occurs when fertilizers, animal waste, sewage, or other substances rich in nitrogen and phosphorus flow into waterways, causing extensive algae growth. The algae use up the oxygen supply during their rapid growth and after their deaths during the decaying process. Other organisms in the water suffocate. In some cases, algae also give off toxins that poison the water supply for other organisms. Eutrophication is a natural process, but human activities have accelerated the rate at which it occurs.

Forest damage

Fish kill

Figure 5.14 Acid precipitation damages plant tissues and can kill fish if the acid concentration is high.
Infer *Which areas of the United States would most likely have acid precipitation problems?*

MiniLab 5.2

Survey Leaf Litter Samples

How do you calculate biodiversity? It is not possible to count every organism in the world, which makes calculating biodiversity difficult. Scientists use a sampling technique to do this. They calculate the biodiversity in a certain area and use that number to estimate the biodiversity in similar areas.

Procedure

1. Read and complete the lab safety form.
2. In the **leaf litter sample** your teacher has provided, count and record the species in a section that are visible to the eye. Look up any unknown species in a **field guide.**
3. Record your observations in a data table.
4. Calculate the index of diversity (IOD), using this equation (unique species is different species observed; total individual is the total of every individual observed):

$$\text{IOD} = \frac{\text{\# of unique species} \times \text{\# of samples}}{\text{\# of total individuals}}$$

Analysis

1. **Classify** which observed species are native and nonnative to your area.
2. **Infer** from your survey the effects, if any, the nonnative species have on the native species. Are these nonnative species invasive? How do you know this?
3. **Hypothesize** whether the IOD has changed in your area over the last 200 years. Explain.

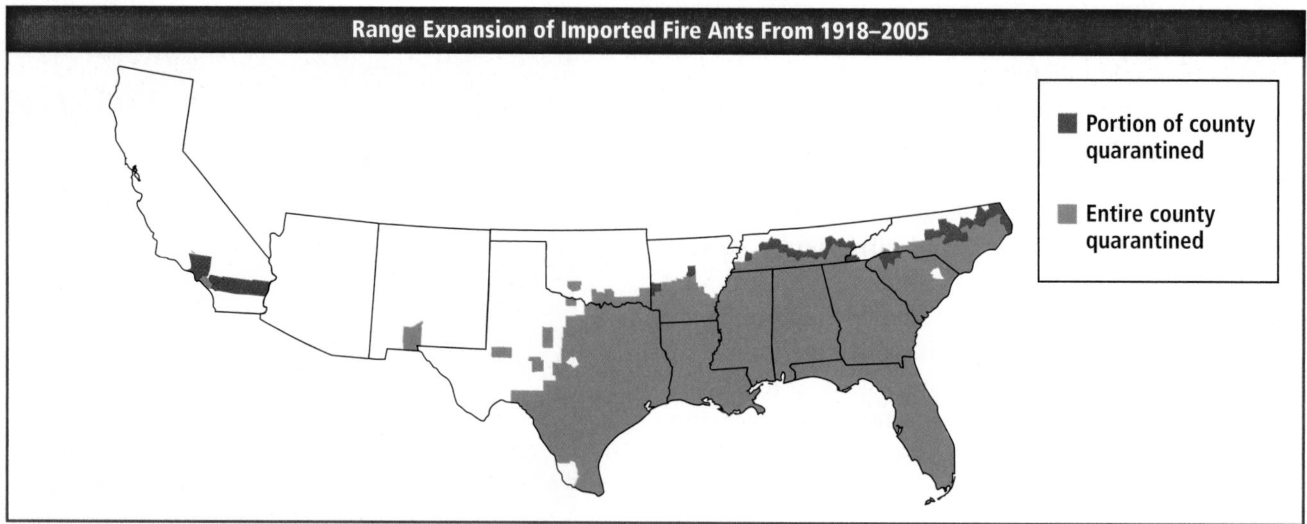

Figure 5.15 Fire ants were transported accidentally by ship to the port of Mobile in Alabama. The ants spread throughout the southern and southwestern United States.

Introduced species Nonnative species that are either intentionally or unintentionally transported to a new habitat are known as **introduced species.** These species are not a threat to biodiversity in their native habitats. Predators, parasites, and competition between species keep the native ecosystem in balance. However, when these species are introduced into a new area, these controlling factors are not in place. Introduced species often reproduce in large numbers because of a lack of predators and become invasive species in their new habitat.

Imported fire ants are a species that might have been introduced to the United States through the port of Mobile, Alabama in the 1920s by ships from South America. The fire ants spread throughout the southern and southwestern United States, as illustrated in **Figure 5.15.** Fire ants attack and feed on some wildlife, such as newborn deer and hatching or newly-hatched ground-nesting birds.

Introduced species are a worldwide environmental problem. An estimated 40 percent of the extinctions that have occurred since 1750 are due to introduced species, and billions of dollars are spent every year in an effort to clean up or control the damage caused by introduced species.

LAUNCH Lab

Review Based on what you've read about biodiversity, how would you now answer the analysis question?

Section 5.2 Assessment

Section Summary

- The current rate of species extinction is abnormally high.
- Species on islands are particularly vulnerable to extinction.
- Historically, overexploitation of some species by humans has led to their extinction.
- Human activities, such as release of pollutants, destruction of habitat, and the introduction of nonnative species, can result in a decrease in biodiversity.

Understand Main Ideas

1. MAIN Idea **Explain** three ways that humans threaten biodiversity.
2. **Summarize** why the extinction rate is greater now than in the past.
3. **Choose** one of the factors that threatens biodiversity and suggest one way in which biodiversity can be preserved in a real-life scenario.
4. **Summarize** how the overharvesting of a single species, such as a baleen whale, can affect an entire ecosystem.

Think Critically

5. **Design** a planned community that preserves biodiversity and accommodates the human population. Work in small groups to accomplish this task.
6. **Survey** your community to identify at least five threats to biodiversity and suggest ways in which biodiversity can be preserved.

BiologyOnline Self-Check Quiz biologygmh.com

Section 5.3

Reading Preview

Objectives

- **Describe** two classes of natural resources.
- **Identify** methods used to conserve biodiversity.
- **Explain** two techniques used to restore biodiversity.

Review Vocabulary

natural resources: materials and organisms found in the biosphere

New Vocabulary

renewable resource
nonrenewable resource
sustainable use
endemic
bioremediation
biological augmentation

Conserving Biodiversity

MAIN Idea People are using many approaches to slow the rate of extinctions and to preserve biodiversity.

Real-World Reading Link Have you ever broken a decorative item and repaired it? You probably carefully searched for all the pieces and then carefully glued the item together again. Repairing a damaged ecosystem is a similar process. Scientists carefully search for all the pieces of the ecosystem, repair the damages, and secure the location to protect the ecosystem from future damage.

Natural Resources

The biosphere currently supplies the basic needs for more than six billion humans in the form of natural resources. The human population continues to grow and the growth is not evenly distributed throughout the world. An increase in human population growth increases the need for natural resources to supply the basic needs of the population.

The consumption rate of natural resources is also not evenly distributed. **Figure 5.16** shows the consumption of natural resources per person for selected countries. The natural resource consumption rate is much higher for people living in developed countries than for people in developing countries. As developing countries become more industrialized and the standard of living increases, the rate of natural resource consumption also increases. Because of the rising human population growth and an increased rate of consumption of natural resources, a long-term plan for the use and conservation of natural resources is important.

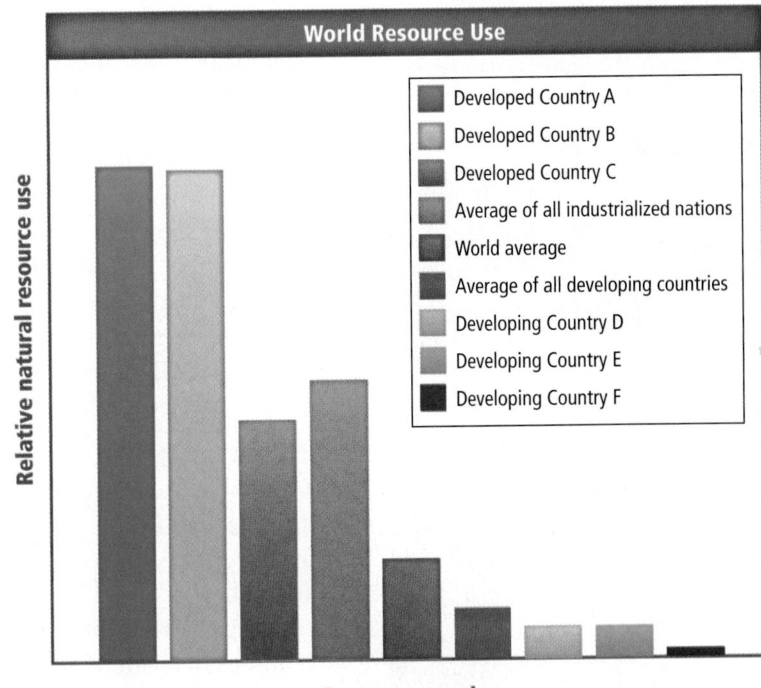

■ **Figure 5.16** This graph shows the consumption of natural resources per person for selected countries based on the equivalent kilograms of oil.
Explain *Why is the use of natural resources high for Developed Country A and Developed Country B and so low for Developing Country E and Developing Country F?*

■ **Figure 5.17** This cleared forest is considered a nonrenewable resource because there is not enough of the forest intact to provide a habitat for the organisms that live there.

Renewable resources Plans for long-term use of natural resources must take into consideration the difference between the two groups of natural resources—renewable and nonrenewable resources. Those resources that are replaced by natural processes faster than they are consumed are called **renewable resources.** Solar energy is considered a renewable resource because the supply appears to be endless. Agricultural plants, animals, clean water, and clean air are considered renewable because normally they are replaced faster than they are consumed. However, the supply of these resources is not unlimited. If the demand exceeds the supply of any resource, the resource might become depleted.

Renewable v. nonrenewable resources Those resources that are found on Earth in limited amounts or those that are replaced by natural processes over extremely long periods of time are called **nonrenewable resources.** Fossil fuels and mineral deposits, such as radioactive uranium, are considered nonrenewable resources. Species are considered renewable resources until the last of a species dies. When extinction occurs, a species is nonrenewable because it is lost forever.

The classification of a resource as renewable or nonrenewable depends on the context in which the resource is being discussed. A single tree or a small group of trees in a large forest ecosystem is renewable because replacement trees can be planted or can regrow from seeds present in the soil. Enough of the forest is still intact to serve as a habitat for the organisms that live there. However, when the entire forest is cleared, as shown in **Figure 5.17,** the forest is not considered a renewable resource. The organisms living in the forest have lost their habitat and they most likely will not survive. In this example, it is possible that more than one natural resource is nonrenewable—the forest and any species that might become extinct. If a species is found only in this forest, this species might become extinct if it loses its only habitat.

■ **Figure 5.18** Replacing resources preserves the health of the biosphere.
Explain ***Why is this process considered a sustainable use of a resource?***

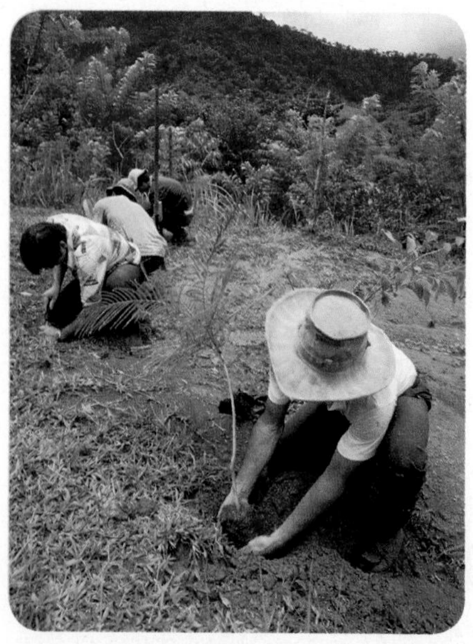

Sustainable use One approach to using natural resources called sustainable use is demonstrated in **Figure 5.18.** Just as the name implies, **sustainable use** means using resources at a rate in which they can be replaced or recycled while preserving the long-term environmental health of the biosphere. Conservation of resources includes reducing the amount of resources that are consumed, recycling resources that can be recycled, and preserving ecosystems, as well as using them in a responsible manner.

Protecting Biodiversity

In Section 2, you learned how human activities have affected many ecosystems. Many efforts are underway worldwide to slow the loss of biodiversity and to work toward sustainable use of natural resources.

Protected areas in the United States Conservation biologists recognize the importance of establishing protected areas where biodiversity can flourish. The United States established its first national park—Yellowstone National Park—in 1872 to protect the area's geological features. Many additional national parks and nature reserves have been established since 1872.

International protected areas The United States is not the only country to establish national parks or nature reserves. Currently, about seven percent of the world's land is set aside as some type of reserve. Historically, these protected areas have been small islands of habitat surrounded by areas that contain human activity. Because the reserves are small, they are impacted heavily by human activity. The United Nations supports a system of Biosphere Reserves and World Heritage sites. Costa Rica has established megareserves. These reserves contain one or more zones that are protected from human activity by buffer zones—an area in which sustainable use of natural resources is permitted. This approach creates a large managed area for preserving biodiversity while providing natural resources to the local population.

 **Reading Check** **Explain** the advantages of megareserves.

DATA ANALYSIS LAB 5.1

Based on Real Data*

Use Numbers

How is the biodiversity of perching birds distributed in the Americas? The distribution of birds, like other species, is not even. Perching birds appear to be more concentrated in some areas of the Americas than others.

Data and Observations

Use the map to answer the following questions about the biodiversity of perching birds.

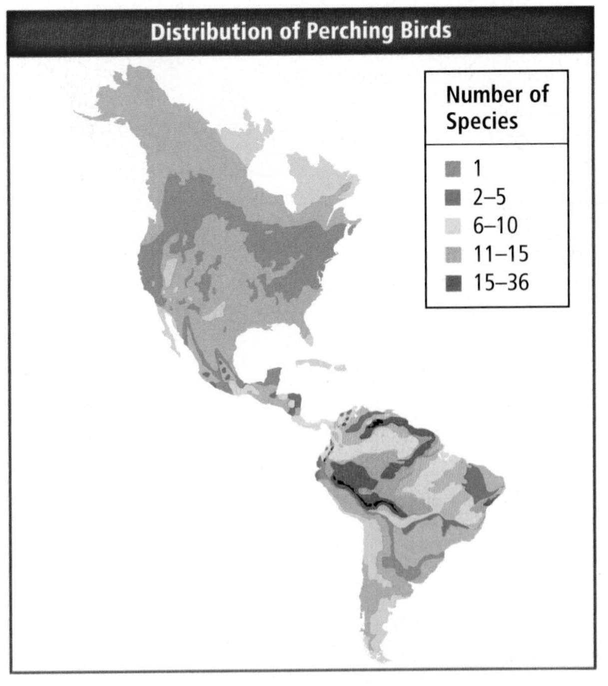

Think Critically

1. **Determine** the location of the highest concentration of perching birds.
2. **Generalize** the trend in the number of perching birds as you move from Canada to South America.
3. **Infer** Why does the number of perching birds change as you move toward the southern tip of South America?

*Data obtained from: Pimm, S.L. and Brown J.H. 2004. Domains of diversity. *Science* 304: 831–833.

Visualizing Biodiversity Hot Spots

Figure 5.19 Biodiversity hot spots, highlighted in red on the map, are ecosystems where endemic species are threatened. If these species become extinct, biodiversity will decrease.

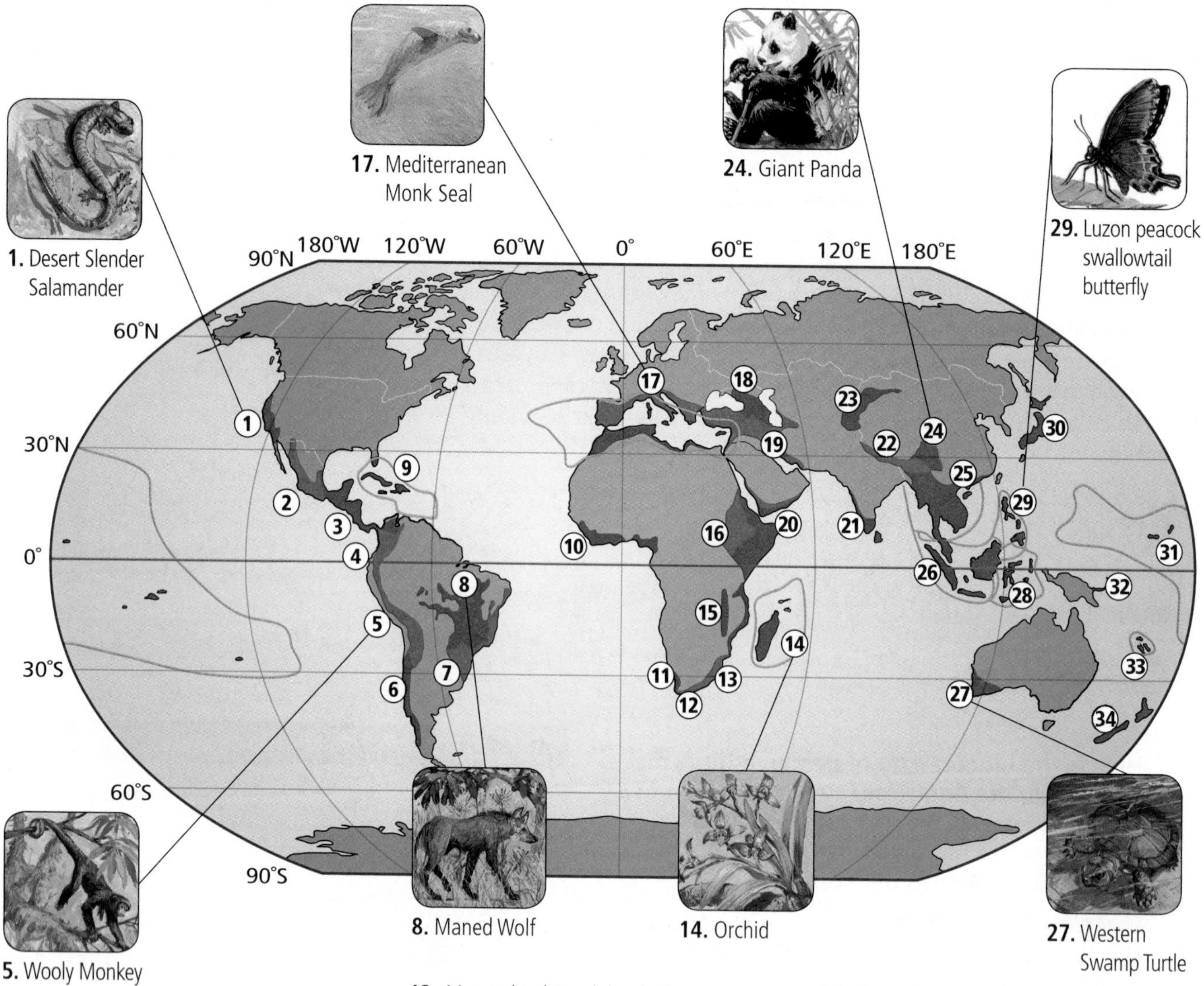

1 California Floristic Province
2 Madrean Pine-Oak Woodlands
3 Mesomerica
4 Tumbes-Chocó-Magdalena
5 Tropical Andes
6 Chilean Winter Rainfall-Valdivian Forests
7 Atlantic Forest
8 Cerrado
9 Caribbean Islands
10 Guinean Forests of West Africa
11 Succulent Karoo
12 Cape Floristic Region
13 Maputaland-Pondoland-Albany
14 Madagascar and the Indian Ocean Islands
15 Coastal Forests of Eastern Africa
16 Eastern Aforomontane
17 Mediterranean Basin
18 Caucasus
19 Irano-Anatolian
20 Horn of Africa
21 Western Ghats and Sri Lanka
22 Himalayans
23 Mountains of Central Asia
24 Mountains of Southwest China
25 Indo-Burma
26 Sundaland
27 Southwest Australia
28 Wallacea
29 Philippines
30 Japan
31 Polynesia-Micronesia
32 East Melanesian Islands
33 New Caledonia
34 New Zealand

Concepts In Motion **Interactive Figure** To see an animation of biodiversity hot spots, visit biologygmh.com.

Biodiversity hot spots Conservation biologists have identified locations around the world that are characterized by exceptional levels of **endemic** species—species that are only found in that specific geographic area—and critical levels of habitat loss. To be called a hot spot, a region must meet two criteria. First, there must be at least 1500 species of vascular plants that are endemic, and the region must have lost at least 70 percent of its original habitat. The 34 internationally recognized hot spots are shown in **Figure 5.19.**

Approximately half of all plant and animal species are found in hot spots. These hot spots originally covered 15.7 percent of Earth's surface, however, only about a tenth of that habitat remains.

Biologists in favor of recovery efforts in these areas argue that focusing on a limited area would save the greatest number of species. Other biologists argue that concentrating funding on saving species in these hot spots does not address the serious problems that are occurring elsewhere. For example, saving a wetland area might save fewer species, but the wetland provides greater services by filtering water, regulating floods, and providing a nursery for fish. These biologists think that funding should be spent in areas around the world rather than focused on the biodiversity hotspots.

Corridors between habitat fragments Conservation ecologists also are focusing on improving the survival of biodiversity by providing corridors, or passageways, between habitat fragments. Corridors, such as those shown in **Figure 5.20,** are used to connect smaller parcels of land. These corridors allow organisms from one area to move safely to the other area. This creates a larger piece of land that can sustain a wider variety of species and a wider variety of genetic variation. However, corridors do not completely solve the problem of habitat destruction. Diseases easily pass from one area to the next as infected animals move from one location to another. This approach also increases edge effect. One large habitat would have fewer edges, but often a large habitat is hard to preserve.

VOCABULARY

SCIENCE USAGE V. COMMON USAGE

Corridor

Science usage: a passageway between two habitat fragments.
The deer uses the corridor to safely travel between the two habitat fragments.

Common usage: a passageway, as in a hotel, into which rooms open.
The ice machine is in the hotel corridor by the elevators.

■ **Figure 5.20** Corridors between habitat fragments allow safe passage for animals.
Describe ***What are the advantages and disadvantages of corridors?***

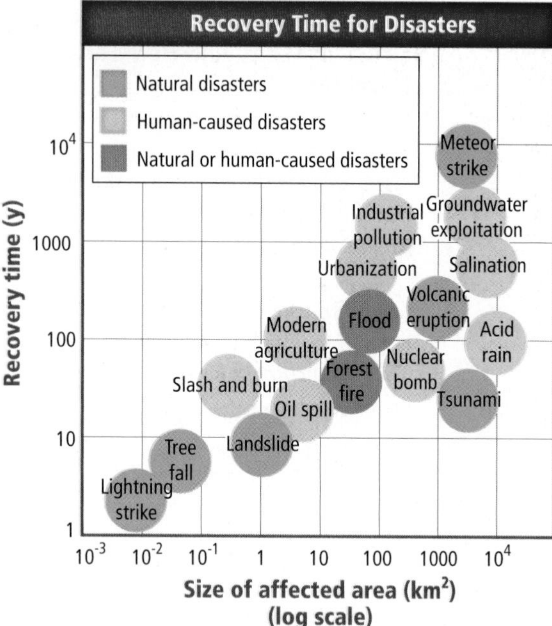

Figure 5.21 The recovery time for disasters is not dependent upon whether or not it was a natural or human-made disaster, but on the size of the area affected and on the type of disturbance.

Determine ***What is the approximate recovery time for a landslide?***

Restoring Ecosystems

Sometimes biodiversity is destroyed in an area such that it no longer provides the abiotic and biotic factors needed for a healthy ecosystem. For example, the soil from cleared tropical rain forests becomes unproductive for farming after a few years. After mining activities are completed, the land might be abandoned in a condition that does not support biodiversity. Accidental oil spills and toxic chemical spills might pollute an area to such a degree that the native species cannot live there.

Given time, biological communities can recover from natural and human-made disasters, as illustrated in **Figure 5.21.** The length of time for recovery is not related directly to whether the disaster is natural or human-made. The size of the area affected and the type of disturbance are determining factors for recovery time. In general, the larger the affected area, the longer it takes for the biological community to recover. Ecologists use two methods to speed the recovery process of these damaged ecosystems—bioremediation and biological augmentation.

Bioremediation The use of living organisms, such as prokaryotes, fungi, or plants, to detoxify a polluted area is called **bioremediation.** In 1975, a leak from a fuel-storage facility in South Carolina released about 80,000 gallons of kerosene-based jet fuel. The fuel soaked into the sandy soil and contaminated the underground water table. Microorganisms that naturally are found in the soil break down these carbon-based fuels into carbon dioxide. Scientists found that by adding additional nutrients to the soil, the rate at which the microorganisms decontaminated the area was increased. In a few years, the contamination in the area had been greatly reduced. These microorganisms can be used in other ecosystems to remove toxins from soils that are contaminated by accidental oil or fuel spills.

Some species of plants are being used to remove toxic substances, such as zinc, lead, nickel, and organic chemicals, from damaged soils, as shown in **Figure 5.22.** These plants are planted in contaminated soils, where they store the toxic metals in their tissues. The plants then are harvested, and the toxic metals are removed from the ecosystem. Bioremediation is relatively new, but there appears to be great promise in using organisms to detoxify some ecosystems that have been damaged.

Figure 5.22 Chemical waste from an industrial complex is being treated using reed beds. Bacteria and fungi in the reed beds transform a wide range of pollutants into harmless substances.

Biological augmentation Adding natural predators to a degraded ecosystem is called **biological augmentation.** For example, aphids—very small insects—eat vegetables and other plants, which can result in the destruction of farm crops. Aphids also can transmit plant diseases. Some farmers rely on ladybugs to control pests that eat their crops. Certain species of ladybugs eat aphids, as shown in **Figure 5.23,** and can be used to control aphid infestation. The ladybugs do not harm the crops, and the fields are kept free of aphids.

■ **Figure 5.23** Ladybugs can be introduced into an ecosystem to control aphid populations.

Legally Protecting Biodiversity

During the 1970s, a great deal of attention was focused on the destruction to the environment and maintaining biodiversity. Laws were enacted in countries around the world and many treaties between countries were signed in an effort to preserve the environment. In the United States, the Endangered Species Act was enacted in 1973. It was designed to legally protect the species that were becoming extinct or in danger of becoming extinct. An international treaty, the Convention on International Trade in Endangered Species of Wild Fauna and Flora (CITES), was signed in 1975. It outlawed the trade of endangered species and animal parts, such as ivory elephant tusks and rhinoceros horns. Since the 1970s, many more laws and treaties have been enacted and signed with the purpose of preserving biodiversity for future generations.

Section 5.3 Assessment

Section Summary

- There are two classes of natural resources—renewable and nonrenewable.
- One approach to using natural resources is sustainable use.
- There are many approaches used to conserve biodiversity in the world.
- Biodiversity hot spots contain a large number of endemic species that are threatened with extinction.
- Two techniques used to restore biodiversity in an ecosystem are bioremediation and biological augmentation.
- Since the 1970s, many forms of legislation have been passed to protect the environment.

Understand Main Ideas

1. MAIN Idea **Describe** three approaches used to slow down the rate of extinction or to preserve biodiversity.
2. **Identify and define** the two classes of natural resources.
3. **Choose** a human-caused disaster from **Figure 5.21.** Discuss the methods that could be used to restore biodiversity.
4. **Compare** the advantages and disadvantages of large and small nature reserves.

Think Critically

5. **Create** a script of dialogue that could occur between a conservationist and a person that lives in a biodiversity hot spot. The local person wants to use the natural resources to provide a living for his or her family. The dialogue should include a compromise in which both sides are satisfied with the use of resources.

MATH in Biology

6. If Earth has 150,100,000 km^2 of land area, how much land area is included in the biodiversity hot spots?

In the Field

Career: Conservationist

Wangari Maathai: Planting Seeds of Change

Living and working in her homeland of Kenya, Wangari Maathai was disturbed by the plight of women in rural areas of the country. Limited firewood, scarce water resources, and poor soil made it difficult for rural women to meet their families' needs. Maathai's solution? Plant trees, and teach other women to do the same.

Wangari Maathai

What began with planting trees evolved into the Green Belt Movement, with Maathai as its energetic leader. This grassroots, non-governmental organization involves Kenyans in reducing the environmental and social effects of deforestation. While tree planting is the focal activity, the movement also focuses on promoting environmental consciousness, volunteerism, conservation of local biodiversity, community development, and self-empowerment, particularly for Kenyan women and girls. Maathai was awarded the Nobel Peace Prize in 2004 for her contribution to sustainable development, democracy, and peace.

Positive change in Kenya As a leader for environmental change in Kenya, Maathai's work has helped Kenyans achieve a deeper understanding of their role in environmental conservation. Today, there are more than 600 community networks throughout Kenya that oversee about 6000 tree nurseries. These nurseries are staffed primarily by Kenyan women, and provide an income source for their families and for rural communities. Individuals working within community networks have planted more than 30 million trees throughout the country. Degraded forested areas are experiencing regrowth, resulting in areas that can support plant and animal biodiversity. Soil erosion has slowed, and both soil fertility and water-holding capacity in planted areas has increased. By promoting the planting of fruit trees and other food plants, hunger has been reduced and nutrition has improved in rural households.

The impact of the Green Belt Movement, now more than 30 years old, has been phenomenal. Expanding beyond Kenya, Green Belt methods have been adopted in other African countries, including Tanzania, Uganda, Malawi, Lesotho, Ethiopia, and Zimbabwe.

COMMUNITY SERVICE

Action Plan How can you get involved with tree planting in your community? Develop an action plan that includes contacting local groups for information, designing the project, obtaining resources, and implementing the activity. For more information about tree-planting programs across the country, visit biologygmh.com.

BIOLAB

FIELD INVESTIGATION: HOW CAN SURVEYING A PLOT OF LAND AROUND YOUR SCHOOL HELP YOU UNDERSTAND THE HEALTH OF YOUR ECOSYSTEM?

Background One of the jobs of a conservation biologist is to survey land and provide an analysis of the health of the ecosystem. Then, if problems are discovered, he or she would propose possible solutions, decide on a course of action, and implement the plan.

Question: *How can an ecosystem be restored to its natural state?*

Materials

wire coat hangers or 1-m stakes (61)
field notebook
field guide of area species (plant, animal, and fungus)
colored plastic ribbon (50 m)
string (600 m)
pencil

Safety Precautions

WARNING: *Use care in observing wildlife; do not disturb the species.*

Procedure

1. Read and complete the lab safety form.
2. Determine a site to be studied. Make sure the site owner has given permission to conduct a survey on that site.
3. With four stakes, mark off a 15 m × 15 m area within that site.
4. Further divide the area into 1 m × 1 m squares with 57 remaining stakes and string. These will be your sampling areas.
5. Using the method you used in **MiniLab 5.2,** survey your site and calculate the index of diversity.
6. Research the history of your area. How has it changed since it was first settled?
7. Research and recommend appropriate methods to care for the plot of land you surveyed in an environmentally responsible manner, perhaps by restoring it to its original state.
8. Make a plan to implement your methods. What limitations might you encounter?
9. If possible, implement part of your plan.

Analyze and Conclude

1. **Predict** how your methods of care would impact your plot of land. Why is this important?
2. **Determine** Is there a key species you expect to be affected by your plan?
3. **Analyze** What are some possible negative consequences of your plan?
4. **Defend** Is there another possible conservation biology technique that could be used? Explain.
5. **Calculate** What might the index of diversity be if you made the changes you recommended?
6. **Interpret** Was an increase in biodiversity your goal? Why or why not?

SHARE YOUR DATA

Internet Post your data at biologygmh.com. Graph the results of the current IOD and the proposed IOD of your plot and those of students analyzing other environments across the country. Describe any similarities and differences you observe in the data. To learn more about calculating biodiversity, visit BioLabs at biologygmh.com.

Chapter 5 Study Guide

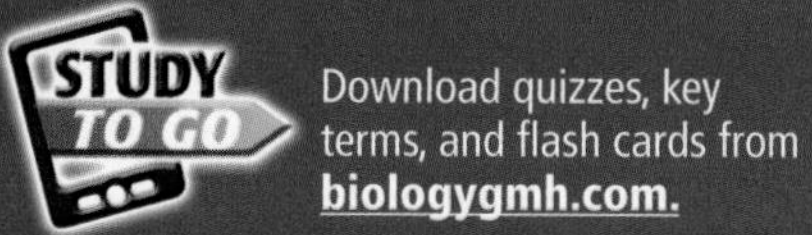

Download quizzes, key terms, and flash cards from **biologygmh.com.**

FOLDABLES Evaluate Select an endangered plant or animal and investigate what factors are contributing to its near extinction. Evaluate the organism's chances for survival, taking into consideration genetic diversity, species diversity, and ecosystem diversity.

Vocabulary	Key Concepts
Section 5.1 Biodiversity	
• biodiversity (p. 116) • ecosystem diversity (p. 118) • extinction (p. 116) • genetic diversity (p. 116) • species diversity (p. 117)	**MAIN Idea** Biodiversity maintains a healthy biosphere and provides direct and indirect value to humans. • Biodiversity is important to the health of the biosphere. • There are three types of biodiversity: genetic, species, and ecosystem. • Biodiversity has aesthetic and scientific values and direct and indirect economic value. • It is important to maintain biodiversity to preserve the reservoir of genes that might be needed in the future. • Healthy ecosystems can provide some services at a lesser expense than the use of technology.
Section 5.2 Threats to Biodiversity	
• background extinction (p. 122) • biological magnification (p. 126) • edge effect (p. 126) • eutrophication (p. 127) • habitat fragmentation (p. 125) • introduced species (p. 128) • mass extinction (p. 122) • natural resource (p. 123) • overexploitation (p. 124)	**MAIN Idea** Some human activities reduce biodiversity in ecosystems, and current evidence suggests that reduced biodiversity might have serious long-term effects on the biosphere. • The current rate of species extinction is abnormally high. • Species on islands are particularly vulnerable to extinction. • Historically, overexploitation of some species by humans has led to their extinction. • Human activities, such as release of pollutants, destruction of habitat, and the introduction of nonnative species, can result in a decrease in biodiversity.
Section 5.3 Conserving Biodiversity	
• biological augmentation (p. 135) • bioremediation (p. 134) • endemic (p. 133) • nonrenewable resource (p. 130) • renewable resource (p. 130) • sustainable use (p. 130)	**MAIN Idea** People are using many approaches to slow the rate of extinctions and to preserve biodiversity. • There are two classes of natural resources—renewable and nonrenewable. • One approach to using natural resources is sustainable use. • There are many approaches used to conserve biodiversity in the world. • Biodiversity hot spots contain a large number of endemic species that are threatened with extinction. • Two techniques used to restore biodiversity in an ecosystem are bioremediation and biological augmentation. • Since the 1970s, many forms of legislation have been passed to protect the environment.

Biology Online **Vocabulary PuzzleMaker** biologygmh.com

CHAPTER 5 Assessment

Section 5.1

Vocabulary Review

Each of these sentences is false. Make the sentence true by replacing the italicized word with a vocabulary term from the Study Guide page.

1. *Biodiversity* of a species occurs when the last member of the species dies.
2. *Genetic diversity* refers to the variety of ecosystems that are present in the biosphere.
3. *Ecosystem diversity* is the number of different species and the relative abundance of each species in a biological community.

Understand Key Concepts

4. In which location would you expect to find greater species diversity?
   A. Canada
   B. Costa Rica
   C. Mexico
   D. United States

Use the photo below to answer questions 5 and 12.

5. Which term best describes what the rabbit in each photo demonstrates?
   A. ecosystem diversity
   B. genetic diversity
   C. species richness
   D. species diversity
6. Refer to **Figure 5.3.** What is the species diversity in southern Florida?
   A. 0–50 species
   B. 50–100 species
   C. 100–150 species
   D. 150–200 species
7. Which represents an indirect economic value of biodiversity?
   A. food
   B. clothing
   C. flood protection
   D. medicines
8. Which term best describes this collection of locations: a forest, a freshwater lake, an estuary, and a prairie?
   A. ecosystem diversity
   B. extinction
   C. genetic diversity
   D. species diversity

Constructed Response

9. **Open Ended** Infer why there is more species diversity in southern Florida than there is in northern Alaska.
10. **Open Ended** Explain why increased ecosystem diversity contributes to increased biodiversity in the biosphere.
11. **Short Answer** Describe three indirect services the biosphere provides.
12. **Short Answer** Explain how a trait like the one demonstrated in the photos on the left helps the species survive.

Think Critically

13. **Explain** why it is difficult to attach a value to the aesthetic qualities of biodiversity.
14. **Describe** a service that an ecosystem provides in your community that should be protected to ensure that the quality of the service continues.

Section 5.2

Vocabulary Review

Explain the difference between each pair of terms below. Then explain how the terms are related.

15. background extinction, mass extinction
16. habitat fragmentation, edge effect
17. overexploitation, introduced species

Understand Key Concepts

18. Which group of organisms listed in **Table 5.2** has the greatest number of extinctions overall?
    A. birds
    B. flowering plants
    C. invertebrates
    D. mammals

19. Which group listed in **Table 5.2** has the greatest percentage of extinctions?
    A. birds
    B. fish
    C. mammals
    D. reptiles

Use the figure below to answer questions 20 and 21.

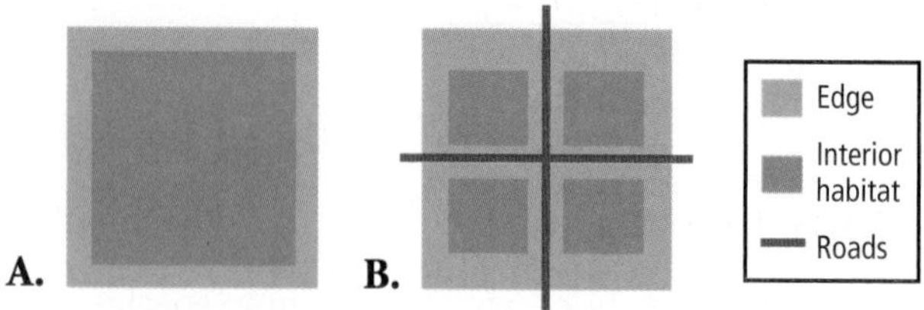

20. Which habitat has the greatest impact due to edge effects?
    A. A
    B. B
    C. A and B equally
    D. neither A nor B

21. Which habitat naturally supports the greater amount of biodiversity?
    A. A
    B. B
    C. A and B equally
    D. neither A nor B

22. Which is not a way in which species lose their habitat?
    A. background extinction
    B. destruction
    C. disruption
    D. pollution

23. Approximately how much greater is the current background extinction compared to the normal rate?
    A. 1 time
    B. 10 times
    C. 1000 times
    D. 10,000 times

24. Which condition triggered the chain of events off the coast of Alaska that caused the kelp forests to begin to disappear?
    A. a decrease in the amount of plankton
    B. an increase in the number of sea otters
    C. overharvesting of plankton-eating whales
    D. pollution caused by pesticides

Constructed Response

25. **Short Answer** Explain why rhinos are in danger of becoming extinct.

Think Critically

26. **Recommend** ways in which eutrophication can be reduced in waterways.

27. **Explain** why it is not a good idea to release exotic pets into a local ecosystem.

Section 5.3

Vocabulary Review

Answer each question with a vocabulary term from the Study Guide page.

28. What are resources called that are replaced by natural processes faster than they are consumed?

29. What are species called that are found only in one geographic location?

30. What is the process of using living organisms to detoxify a location?

31. What are resources called that are found in limited amounts or are replaced by natural processes over extremely long periods of time?

Understand Key Concepts

32. Which term is a method that is used to restore biodiversity to a polluted or damaged area?
    A. biological augmentation
    B. biological corridor
    C. renewable resource
    D. sustainable use

Use the figure below to answer question 33.

33. Which is an advantage of the habitat corridor shown above?
    A. Corridors increase the edge effect in the area.
    B. Diseases are passed easily from one area to another.
    C. Parasites are passed easily from one area to another.
    D. Members of species can move safely from one area to another.

Use the graph below to answer questions 34 and 35.

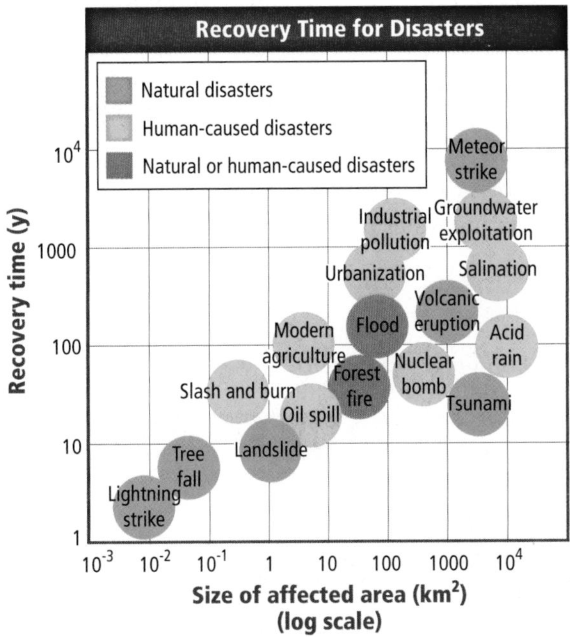

34. Which human-caused disaster requires the greatest recovery time?
A. groundwater exploitation
B. industrial pollution
C. nuclear bomb
D. oil spill

35. Which natural disaster requires the least amount of recovery time?
A. lightning strike **C.** tsunami
B. meteor strike **D.** volcanic eruption

Constructed Response

36. Short Answer Explain why reserves protect biodiversity.

37. CAREERS IN BIOLOGY Explain how an environmental microbiologist might use bioremediation to detoxify polluted areas.

Think Critically

38. Evaluate why it is important to develop a sustainable-use plan for the use of natural resources.

39. Evaluate how a sustainable-use plan for natural resources will change as the world population continues to grow, and people living in developing countries increase their standard of living.

Additional Assessment

40. *WRITING in* **Biology** Write a short essay about the importance of preserving biodiversity.

41. *WRITING in* **Biology** Choose an organism that is in danger of becoming extinct, and write a song or poem detailing the organism's situation.

DBQ Document-Based Questions

Data obtained from: Wilson, E.O. 1980. Resolutions for the 80s. *Harvard Magazine* (January–February): 20.

The quote below was obtained from one of Pulitzer Prize winner Edward O. Wilson's journal articles.

"The worst that can happen–will happen–is not energy depletion, economic collapse, limited nuclear war, or conquest by a totalitarian government. As terrible as these catastrophes would be for us, they can be repaired within a few generations. The one process ongoing in the 1980s that will take millions of years to correct is the loss of genetic and species diversity by the destruction of natural habitats. This is the folly our descendants are least likely to forgive us."

42. Describe how you think biodiversity has changed since the 1980s.

43. Why do you think Wilson compares the loss of biodiversity with energy depletion, economic collapse, nuclear war, and conquest?

44. What does Wilson mean when he says, "This is the folly our descendants are least likely to forgive us"?

Cumulative Review

45. Discuss the stages of secondary succession after a forest fire. **(Chapter 3)**

46. Describe parasitism and give an example of a parasite that is found in an ecosystem near your community. **(Chapter 2)**

47. Explain the concept of carrying capacity. **(Chapter 4)**

Standardized Test Practice

Cumulative

Multiple Choice

1. Which factor is most responsible for the lack of plants in polar regions?
   A. heavy grazing by herbivores
   B. little precipitation
   C. no soil for plants to take root
   D. not enough sunlight

Use the graph below to answer questions 2 and 3.

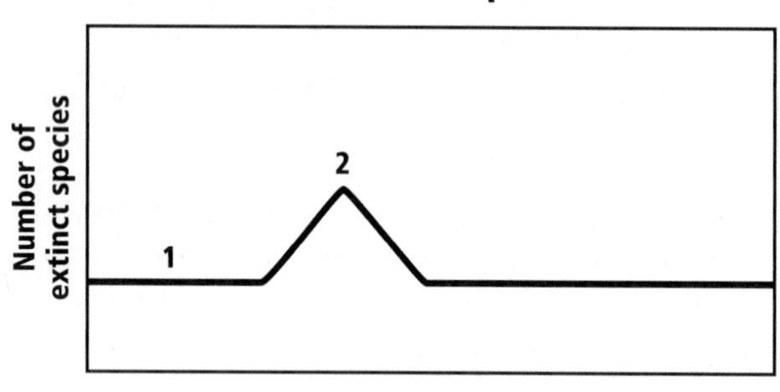

2. Which term describes the section of the graph labeled "1"?
   A. background extinction
   B. habitat destruction
   C. mass extinction
   D. species overexploitation

3. The peak labeled "2" on the graph could be related to extinctions caused by which event?
   A. destruction of a native animal's habitat as humans populate an island
   B. increasing industrialization and human influence over time
   C. introduction of a nonnative animal into an island ecosystem
   D. a fatal disease affecting a single population

4. Which factor is density-dependent?
   A. climate
   B. weather
   C. barometric pressure
   D. food competition

5. What would you expect to find in the profundal zone of a lake?
   A. algae
   B. plankton
   C. debris from dead organisms
   D. floating water plants

Use the graph below to answer questions 6 and 7.

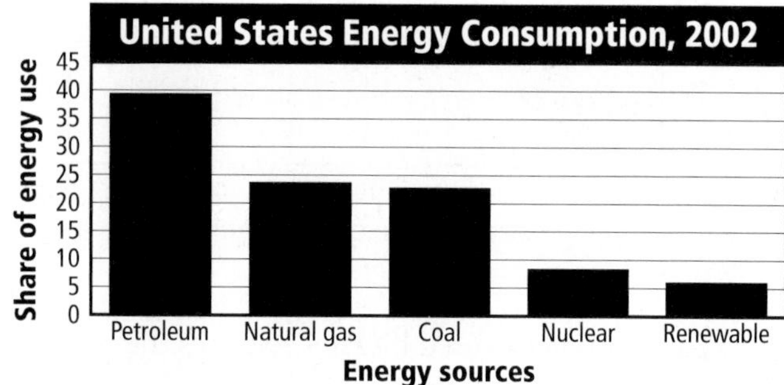

6. What percentage of the United States energy consumption in 2002 was fossil fuels?
   A. 22.7
   B. 23.6
   C. 39.3
   D. 85.6

7. What percentage of the United States energy consumption in 2002 were nonrenewable resources?
   A. 8.3
   B. 22.7
   C. 39.3
   D. 93.9

8. Based on what you know about the habitat of coral organisms, which one is an abiotic limiting factor for them?
   A. annual rainfall
   B. soil chemistry
   C. temperature throughout the year
   D. zooanthellae in the reef

Short Answer

Use the diagram below to answer questions 9 and 10.

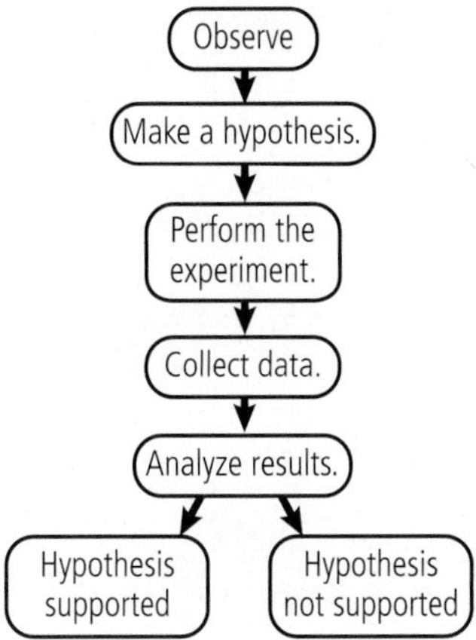

9. Based on the diagram, explain what a scientist does if the experimental data do not support his or her hypothesis.

10. Scientists do not always follow the same scientific method step-by-step. Name two steps in the scientific method shown above that often are omitted. Justify why each step is omitted.

11. If a population is experiencing a decrease in size, how do the birth and death rates compare?

12. List an example of a renewable resource and a nonrenewable resource, and analyze why they are classified as such.

13. Explain the type of information that is displayed on an age structure graph.

14. The ginger plant is considered an invasive species in Hawaii. Justify why park officials in Hawaii have to kill ginger plants.

Extended Response

Use the illustration below to answer question 15.

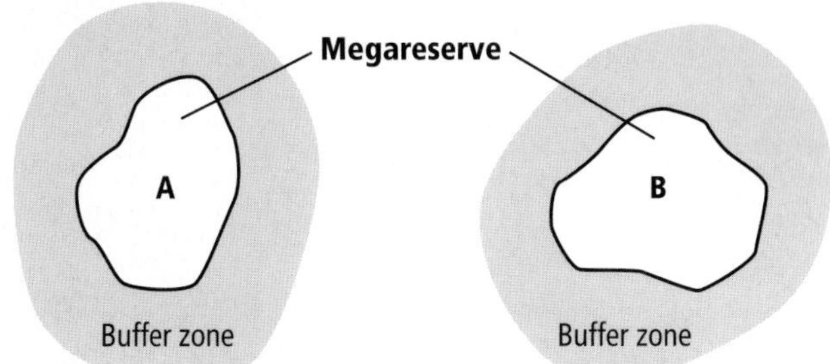

15. The map above shows two megareserves surrounded by buffer zones. Appraise a positive and negative point about these protected zones for a bird species living in Area A.

16. Explain why two species involved in a symbiotic relationship probably evolved around the same time.

Essay Question

The U.S. government takes a census of the population every ten years. The first census took place in 1790 and recorded 3.9 million people. In the last census, taken in 2000, the U.S. population was almost a quarter of a billion people. The census also shows population trends, such as people moving from rural areas to cities.

Using the information in the paragraph above, answer the following question in essay format.

17. The census provides a snapshot of the U.S. population every ten years. Many things can happen between census dates that affect the population. Compose a list of some of the factors that could contribute to a radical change in the U.S. population between each census.

NEED EXTRA HELP?

If You Missed Question . . .	1	2	3	4	5	6	7	8	9	10	11	12	13	14	15	16	17
Review Section . . .	3.2	5.2	5.2	4.1	3.3	5.3	5.3	3.2	1.3	1.3	4.2	5.3	4.2	5.2	5.3	2.1	4.2

UNIT 2 The Cell

Chapter 6
Chemistry in Biology
BIG Idea Atoms are the foundation of biological chemistry and the building blocks of all living organisms.

Chapter 7
Cellular Structure and Function
BIG Idea Cells are the structural and functional units of all living organisms.

Chapter 8
Cellular Energy
BIG Idea Photosynthesis converts the Sun's energy into chemical energy, while cellular respiration uses chemical energy to carry out life functions.

Chapter 9
Cellular Reproduction
BIG Idea Cells go through a life cycle that includes interphase, mitosis, and cytokinesis.

CAREERS IN BIOLOGY

Forensic Pathologist

Forensic pathologists are medical specialists who investigate the cause and the manner of human death. Forensic pathologists work in the field and in a laboratory to analyze medical evidence such as skulls.

WRITING in Biology Visit biologygmh.com to learn more about forensic pathology. Then write a summary of the classification of the manner of death that forensic pathologists use.

Biology Online

To read more about forensic pathologists in action, visit biologygmh.com.

CHAPTER 6

Chemistry in Biology

BIG Idea Atoms are the foundation of biological chemistry and the building blocks of all living organisms.

Section 1
Atoms, Elements, and Compounds
MAIN Idea Matter is composed of tiny particles called atoms.

Section 2
Chemical Reactions
MAIN Idea Chemical reactions allow living things to grow, develop, reproduce, and adapt.

Section 3
Water and Solutions
MAIN Idea The properties of water make it well suited to help maintain homeostasis in an organism.

Section 4
The Building Blocks of Life
MAIN Idea Organisms are made up of carbon-based molecules.

BioFacts

- Collagen is the most abundant protein in mammals.
- Collagen can be found in muscle, bone, teeth, skin, and the cornea of the eye.
- Wrinkles that become visible as people age are the result of collagen breaking down.

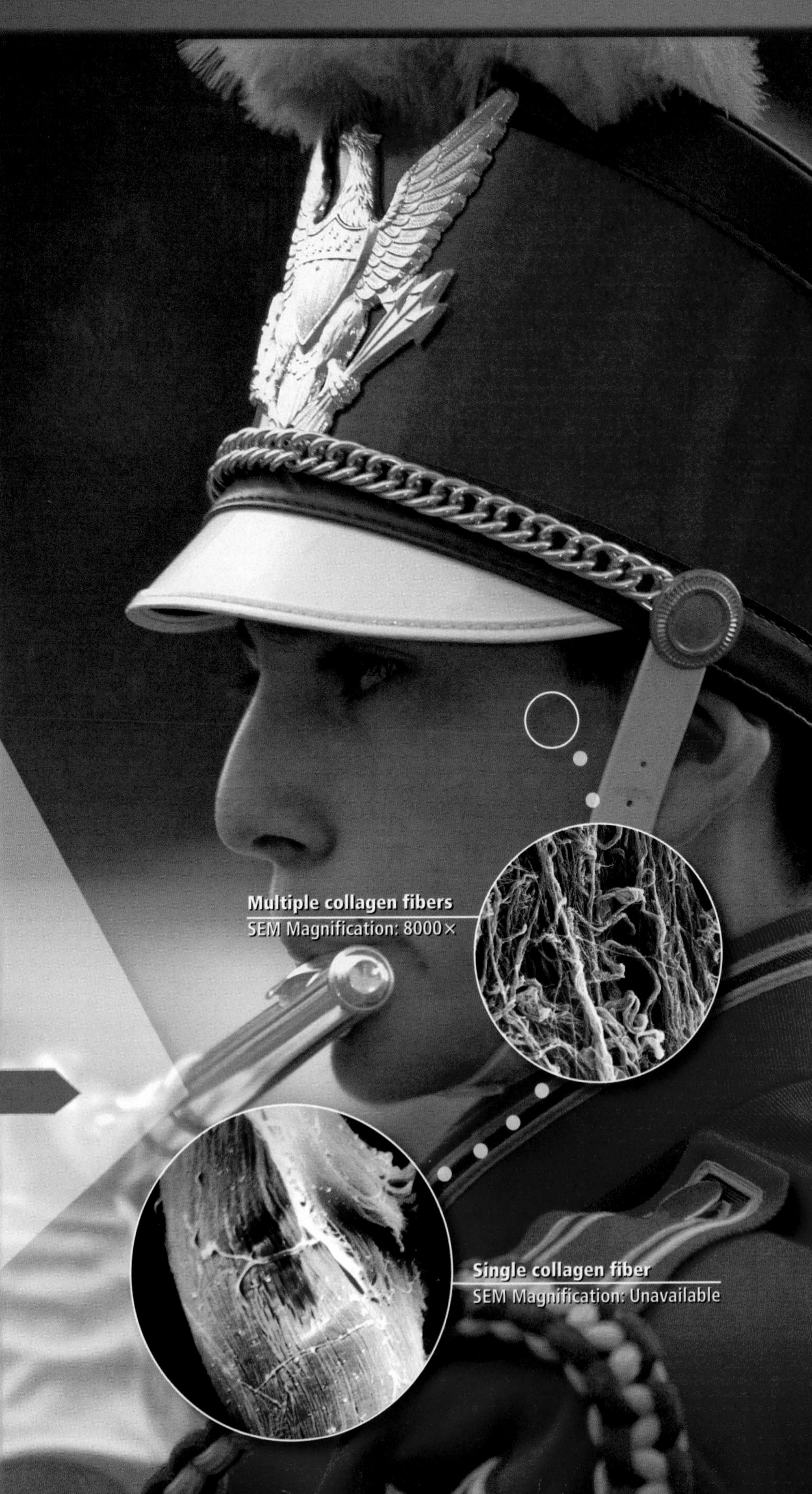

Multiple collagen fibers
SEM Magnification: 8000×

Single collagen fiber
SEM Magnification: Unavailable

LAUNCH Lab

How does the nutrient content of foods compare?

Your body's structure and function depends on chemical elements including those found in proteins, carbohydrates, fats, vitamins, minerals, and water. In this lab, you will investigate nutrients that provide those elements.

Procedure

1. Read and complete the lab safety form.
2. Construct a data chart to record grams or percent of each nutrient listed above. Include columns for Serving Size, Calories, and Calories from Fat.
3. Study and record data from the Nutrition Facts label on a **cereal box.**
4. Choose three additional **labeled food items.** Predict how the nutrients in these items compare with the nutrients in the cereal. Use the Nutrition Facts labels to record data.

Analysis

1. **Evaluate** What factors influenced your predictions of the nutrient contents? Were your predictions correct?
2. **Analyze** Which food item has the greatest amount of proteins per serving? The least?

Visit biologygmh.com to:

- study the entire chapter online
- explore Concepts in Motion, the Interactive Table, Microscopy Links, Virtual Labs, and links to virtual dissections
- access Web links for more information, projects, and activities
- review content online with the Interactive Tutor and take Self-Check Quizzes

Enzymes Make this Foldable to help you organize information about enzyme structure and function.

STEP 1 Draw a line across the middle of a piece of paper.

STEP 2 Fold the top and bottom edges to meet at the middle of the paper.

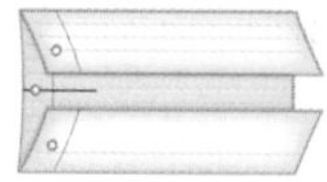

STEP 3 Fold in half to make four sections as shown.

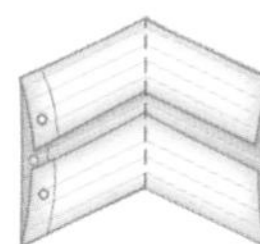

STEP 4 Cut along the fold lines of the top and bottom flaps to form four tabs of equal size. Label the tabs *A, B, C,* and *D* as shown.

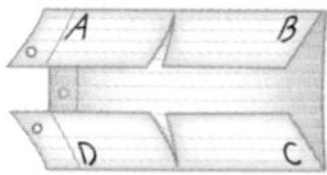

FOLDABLES **Use this Foldable with Section 6.2.** As you study the section, record what you learn about enzymes. On the front tabs, draw the four general steps of enzyme activity.

Section 6.1

Reading Preview

Objectives

- **Identify** the particles that make up atoms.
- **Diagram** the particles that make up an atom.
- **Compare** covalent bonds and ionic bonds.
- **Describe** van der Waals forces.

Review Vocabulary

substance: a form of matter that has a uniform and unchanging composition

New Vocabulary

atom
nucleus
proton
neutron
electron
element
isotope
compound
covalent bond
molecule
ion
ionic bond
van der Waals force

Atoms, Elements, and Compounds

MAIN Idea Matter is composed of tiny particles called atoms.

Real-World Reading Link Many scientists think that the universe began with a huge explosion billions of years ago. They think that the building blocks that make up the amazing diversity of life we see today are a result of that explosion. The study of those building blocks is the science of chemistry.

Atoms

Chemistry is the study of matter—its composition and properties. Matter is anything that has mass and takes up space. All of the organisms you study in biology are made up of matter. **Atoms** are the building blocks of matter.

Connection to History In the fifth century B.C., the Greek philosophers Leucippus and Democritus first proposed the idea that all matter is made up of tiny, indivisible particles. It wasn't until the 1800s that scientists began to collect experimental evidence to support the existence of atoms. As technology improved over the next two centuries, scientists proved not only that atoms exist but also that they are made up of even smaller particles.

The structure of atoms An atom is so small that billions of them fit on the head of a pin. Yet, atoms are made up of even smaller particles called neutrons, protons, and electrons, as illustrated in **Figure 6.1.** Neutrons and protons are located at the center of the atom, which is called the **nucleus. Protons** are positively charged particles (p^+), and **neutrons** are particles that have no charge (n^0). **Electrons** are negatively charged particles that are located outside the nucleus (e^-). Electrons constantly move around an atom's nucleus in energy levels. The basic structure of an atom is the result of the attraction between protons and electrons. Atoms contain an equal number of protons and electrons, so the overall charge of an atom is zero.

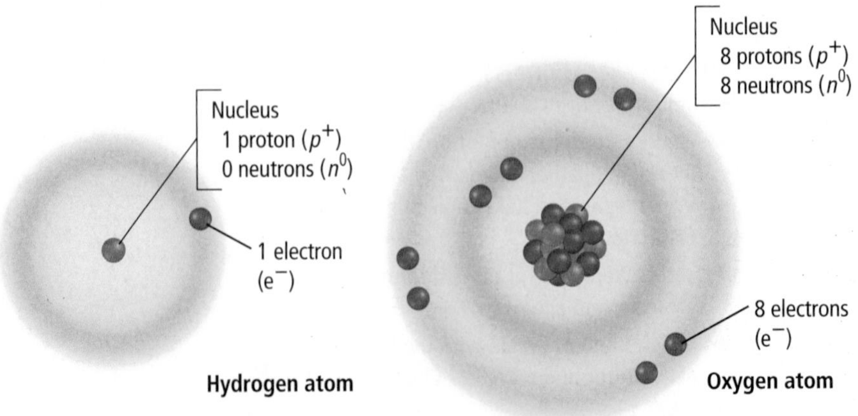

Figure 6.1 Hydrogen has only one proton and one electron. Oxygen has eight protons, eight neutrons, and eight electrons. The electrons move around the nucleus in two energy levels (shown as the darker shaded rings).
Infer *the charge of an atom if it contained more electrons than protons.*

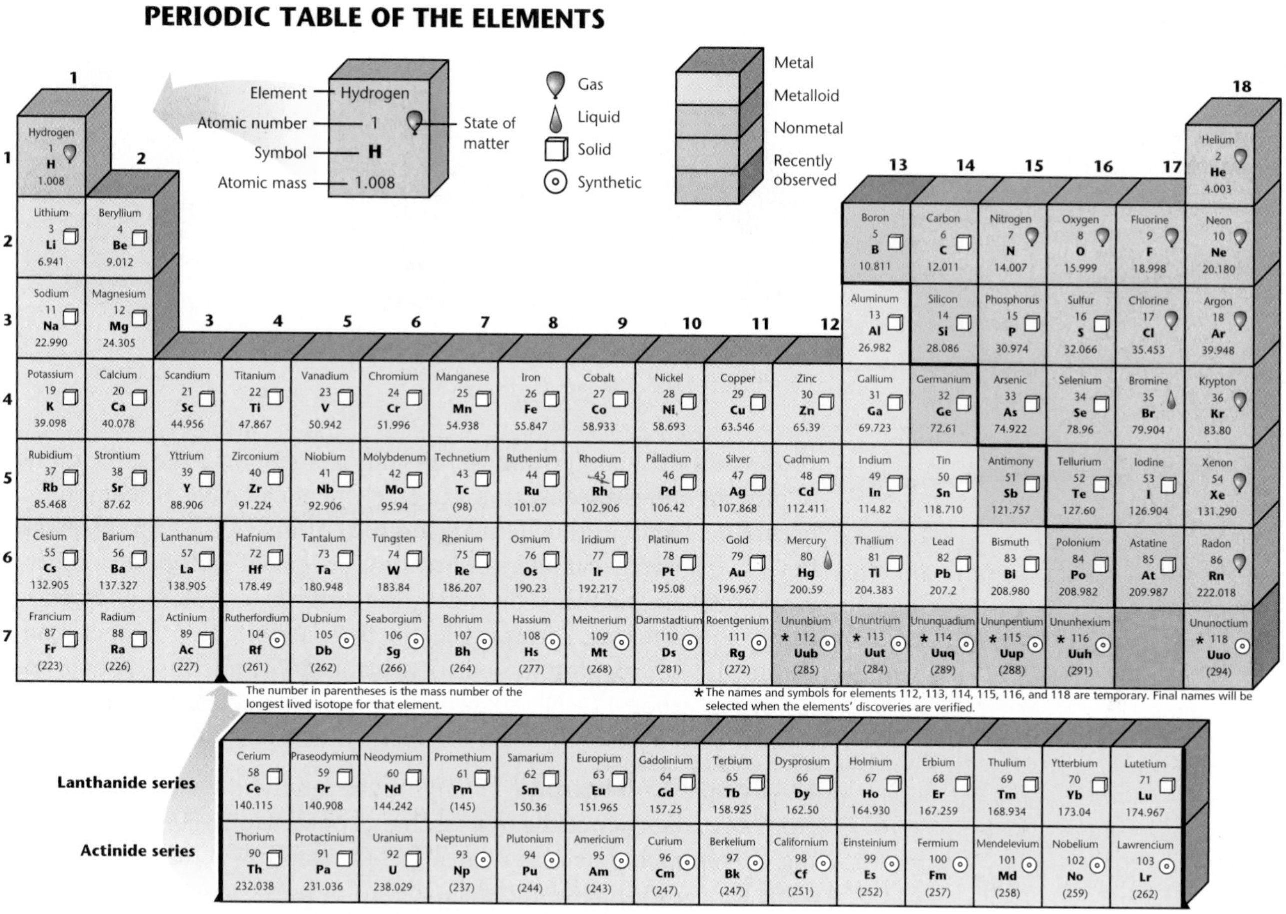

PERIODIC TABLE OF THE ELEMENTS

Element — Hydrogen; Atomic number — 1; Symbol — H; Atomic mass — 1.008; State of matter

Gas; Liquid; Solid; Synthetic

Metal; Metalloid; Nonmetal; Recently observed

Period	1	2	3	4	5	6	7	8	9	10	11	12	13	14	15	16	17	18
1	Hydrogen 1 H 1.008																	Helium 2 He 4.003
2	Lithium 3 Li 6.941	Beryllium 4 Be 9.012											Boron 5 B 10.811	Carbon 6 C 12.011	Nitrogen 7 N 14.007	Oxygen 8 O 15.999	Fluorine 9 F 18.998	Neon 10 Ne 20.180
3	Sodium 11 Na 22.990	Magnesium 12 Mg 24.305											Aluminum 13 Al 26.982	Silicon 14 Si 28.086	Phosphorus 15 P 30.974	Sulfur 16 S 32.066	Chlorine 17 Cl 35.453	Argon 18 Ar 39.948
4	Potassium 19 K 39.098	Calcium 20 Ca 40.078	Scandium 21 Sc 44.956	Titanium 22 Ti 47.867	Vanadium 23 V 50.942	Chromium 24 Cr 51.996	Manganese 25 Mn 54.938	Iron 26 Fe 55.847	Cobalt 27 Co 58.933	Nickel 28 Ni 58.693	Copper 29 Cu 63.546	Zinc 30 Zn 65.39	Gallium 31 Ga 69.723	Germanium 32 Ge 72.61	Arsenic 33 As 74.922	Selenium 34 Se 78.96	Bromine 35 Br 79.904	Krypton 36 Kr 83.80
5	Rubidium 37 Rb 85.468	Strontium 38 Sr 87.62	Yttrium 39 Y 88.906	Zirconium 40 Zr 91.224	Niobium 41 Nb 92.906	Molybdenum 42 Mo 95.94	Technetium 43 Tc (98)	Ruthenium 44 Ru 101.07	Rhodium 45 Rh 102.906	Palladium 46 Pd 106.42	Silver 47 Ag 107.868	Cadmium 48 Cd 112.411	Indium 49 In 114.82	Tin 50 Sn 118.710	Antimony 51 Sb 121.757	Tellurium 52 Te 127.60	Iodine 53 I 126.904	Xenon 54 Xe 131.290
6	Cesium 55 Cs 132.905	Barium 56 Ba 137.327	Lanthanum 57 La 138.905	Hafnium 72 Hf 178.49	Tantalum 73 Ta 180.948	Tungsten 74 W 183.84	Rhenium 75 Re 186.207	Osmium 76 Os 190.23	Iridium 77 Ir 192.217	Platinum 78 Pt 195.08	Gold 79 Au 196.967	Mercury 80 Hg 200.59	Thallium 81 Tl 204.383	Lead 82 Pb 207.2	Bismuth 83 Bi 208.980	Polonium 84 Po 208.982	Astatine 85 At 209.987	Radon 86 Rn 222.018
7	Francium 87 Fr (223)	Radium 88 Ra (226)	Actinium 89 Ac (227)	Rutherfordium 104 Rf (261)	Dubnium 105 Db (262)	Seaborgium 106 Sg (266)	Bohrium 107 Bh (264)	Hassium 108 Hs (277)	Meitnerium 109 Mt (268)	Darmstadtium 110 Ds (281)	Roentgenium 111 Rg (272)	Ununbium * 112 Uub (285)	Ununtrium * 113 Uut (284)	Ununquadium * 114 Uuq (289)	Ununpentium * 115 Uup (288)	Ununhexium * 116 Uuh (291)		Ununoctium * 118 Uuo (294)

The number in parentheses is the mass number of the longest lived isotope for that element.

* The names and symbols for elements 112, 113, 114, 115, 116, and 118 are temporary. Final names will be selected when the elements' discoveries are verified.

Lanthanide series	Cerium 58 Ce 140.115	Praseodymium 59 Pr 140.908	Neodymium 60 Nd 144.242	Promethium 61 Pm (145)	Samarium 62 Sm 150.36	Europium 63 Eu 151.965	Gadolinium 64 Gd 157.25	Terbium 65 Tb 158.925	Dysprosium 66 Dy 162.50	Holmium 67 Ho 164.930	Erbium 68 Er 167.259	Thulium 69 Tm 168.934	Ytterbium 70 Yb 173.04	Lutetium 71 Lu 174.967
Actinide series	Thorium 90 Th 232.038	Protactinium 91 Pa 231.036	Uranium 92 U 238.029	Neptunium 93 Np (237)	Plutonium 94 Pu (244)	Americium 95 Am (243)	Curium 96 Cm (247)	Berkelium 97 Bk (247)	Californium 98 Cf (251)	Einsteinium 99 Es (252)	Fermium 100 Fm (257)	Mendelevium 101 Md (258)	Nobelium 102 No (259)	Lawrencium 103 Lr (262)

■ **Figure 6.2** The periodic table of the elements organizes all of the known elements. Examine the biologists' guide to the periodic table on the back cover of this book.

Elements

An **element** is a pure substance that cannot be broken down into other substances by physical or chemical means. Elements are made of only one type of atom. There are over 100 known elements, 92 of which occur naturally. Scientists have collected a large amount of information about the elements, such as the number of protons and electrons each element has and the atomic mass of each element. Also, each element has a unique name and symbol. All of these data, and more, are collected in an organized table called the periodic table of elements.

The periodic table of elements As shown in **Figure 6.2,** the periodic table is organized into horizontal rows, called periods, and vertical columns, called groups. Each individual block in the grid represents an element. The table is called periodic because elements in the same group have similar chemical and physical properties. This organization even allows scientists to predict elements that have not yet been discovered or isolated. As shown in **Figure 6.3,** elements found in living organisms also are found in Earth's crust.

■ **Figure 6.3** The elements in Earth's crust and living organisms vary in their abundance. Living things are composed primarily of three elements—carbon, hydrogen, and oxygen.

Interpret ***What is the most abundant element that exists in living things?***

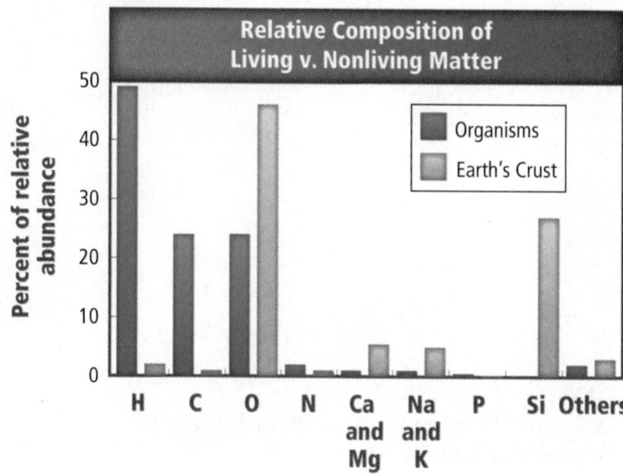

■ **Figure 6.4** Carbon-12 and carbon-13 occur naturally in living and nonliving things. All living things also contain a small amount of carbon-14.
Compare *How do the isotopes differ? How are they the same?*

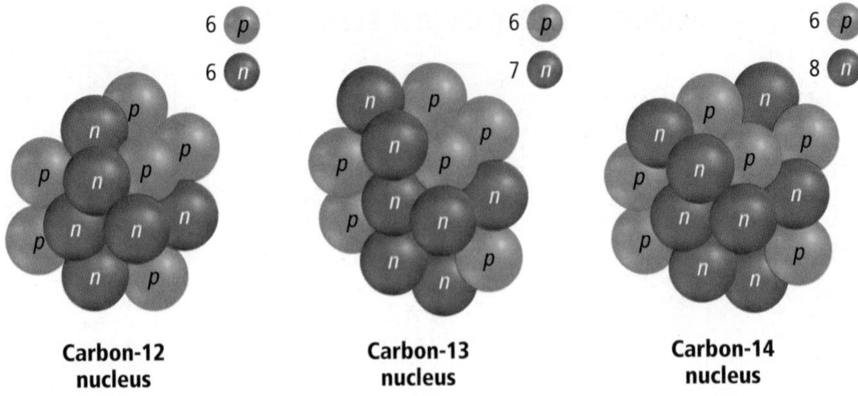

Isotopes Although atoms of the same element have the same number of protons and electrons, atoms of an element can have different numbers of neutrons, as shown in **Figure 6.4.** Atoms of the same element that have different numbers of neutrons are called **isotopes.** Isotopes of an element are identified by adding the number of protons and neutrons in the nucleus. For example, the most abundant form of carbon, carbon-12, has six protons and six neutrons in its nucleus. One carbon isotope—carbon-14—has six protons and eight neutrons. Isotopes of elements have the same chemical characteristics.

Radioactive isotopes Changing the number of neutrons in an atom does not change the overall charge of the atom. However, changing the number of neutrons can affect the stability of the nucleus, in some cases causing the nucleus to decay, or break apart. When a nucleus breaks apart, it gives off radiation that can be detected. Isotopes that give off radiation are called radioactive isotopes.

Carbon-14 is a radioactive isotope that is found in all living things. Scientists know the half-life, or the amount of time it takes for half of carbon-14 to decay, so they can calculate the age of an object by finding how much carbon-14 remains in the sample. Other radioactive isotopes have medical uses, as shown in **Figure 6.5.**

Reading Check **State** the difference between an isotope and a radioactive isotope.

■ **Figure 6.5** Radioactive isotopes are used to help doctors diagnose disease, and locate and treat certain types of cancer.

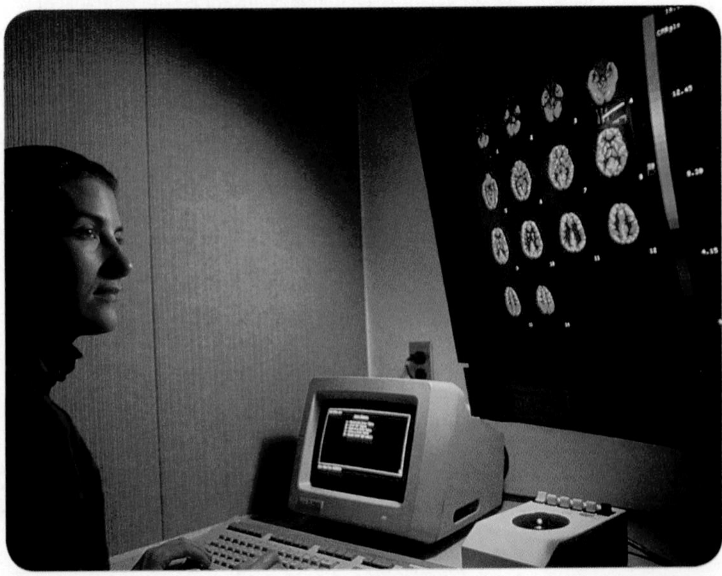

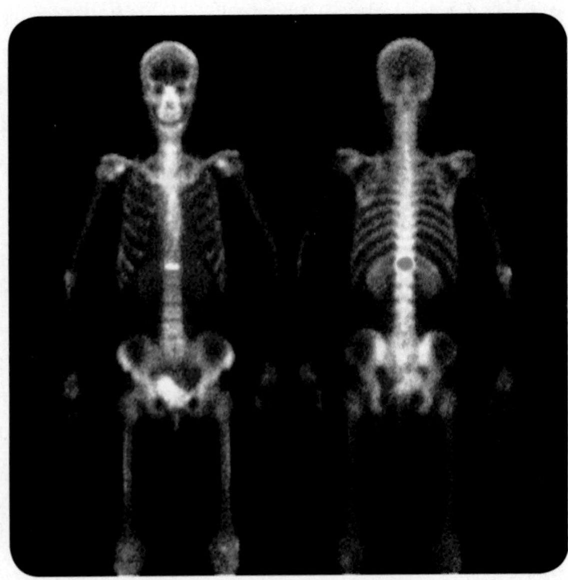

Brilliant fireworks displays depend on compounds containing the metal strontium.

Table salt is the compound NaCl.

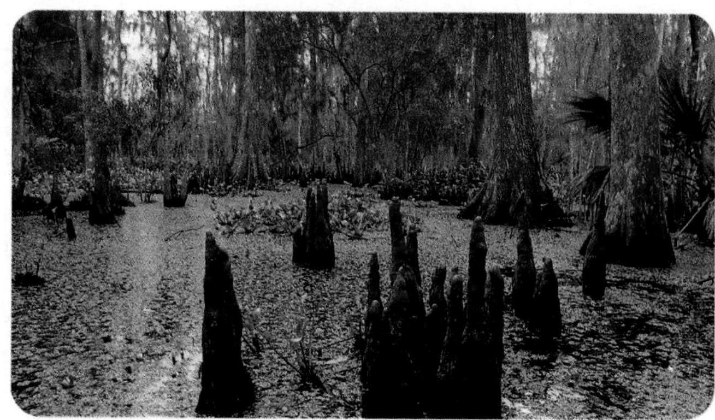
Wetlands are sources of living things made of complex compounds and the simple compound methane (CH_4).

Figure 6.6 You and your world are made of compounds.

Compounds

Elements can combine to form more complex substances. A **compound** is a pure substance formed when two or more different elements combine. There are millions of known compounds and thousands more discovered each year. **Figure 6.6** shows you a few. Each compound has a chemical formula made up of the chemical symbols from the periodic table. You might know that water is the compound H_2O. Sodium chloride (NaCl) is the compound commonly called table salt. The fuel people use in cars is a mixture of hydrocarbon compounds. Hydrocarbons only have hydrogen and carbon atoms. Methane (CH_4) is the simplest hydrocarbon. Bacteria in areas such as the wetlands shown in **Figure 6.6** release 76 percent of global methane from natural sources by decomposing plants and other organisms. They are made of compounds, too.

Compounds have several unique characteristics. First, compounds are always formed from a specific combination of elements in a fixed ratio. Water always is formed in a ratio of two hydrogen atoms and one oxygen atom, and each water molecule has the same structure. Second, compounds are chemically and physically different than the elements that comprise them. For example, water has different properties than hydrogen and oxygen.

Another characteristic of compounds is that they cannot be broken down into simpler compounds or elements by physical means, such as tearing or crushing. Compounds, however, can be broken down by chemical means into simpler compounds or into their original elements. Consider again the example of water. You cannot pass water through a filter and separate the hydrogen from the oxygen, but a process called electrolysis, illustrated in **Figure 6.7,** can break water down into hydrogen gas and oxygen gas.

Figure 6.7 Electrolysis of water produces hydrogen gas that can be used for hydrogen fuel cells.

Chemical Bonds

Compounds such as water, salt, and methane are formed when two or more substances combine. The force that holds the substances together is called a chemical bond. Think back to the protons, neutrons, and electrons that make up an atom. The nucleus determines the chemical identity of an atom, and the electrons are involved directly in forming chemical bonds. Electrons travel around the nucleus of an atom in areas called energy levels, as illustrated in **Figure 6.8.** Each energy level has a specific number of electrons that it can hold at any time. The first energy level, which is the level closest to the nucleus, can hold up to two electrons. The second can hold up to eight electrons.

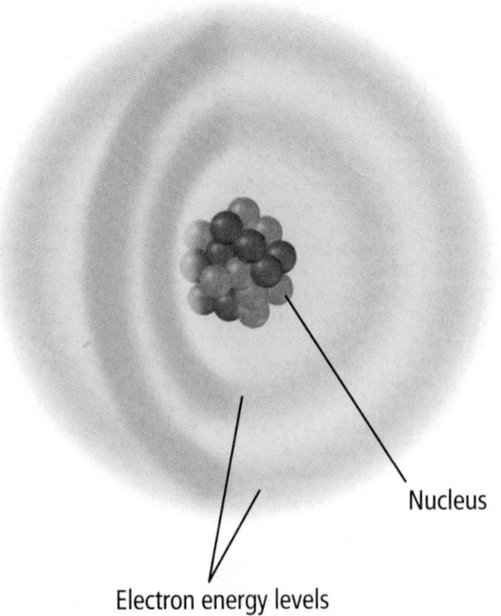

Figure 6.8 Electrons are moving constantly within the energy levels surrounding the nucleus.

A partially-filled energy level is not as stable as an energy level that is empty or completely filled. Atoms become more stable by losing electrons or attracting electrons from other atoms. This results in the formation of chemical bonds between atoms. It is the forming of chemical bonds that stores energy and the breaking of chemical bonds that provides energy for processes of growth, development, adaptation, and reproduction in living things. There are two main types of chemical bonds—covalent bonds and ionic bonds.

Covalent bonds When you were younger, you probably learned to share. If you had a book that your friend wanted to read as well, you could enjoy the story together. In this way, you both benefited from the book. Similarly, one type of chemical bond happens when atoms share electrons in their outer energy levels.

The chemical bond that forms when electrons are shared is called a **covalent bond. Figure 6.9** illustrates the covalent bonds between oxygen and hydrogen to form water. Each hydrogen (H) atom has one electron in its outermost energy level and oxygen (O) has six. Because the outermost energy level of oxygen is the second level, which can hold up to eight electrons, oxygen has a strong tendency to fill the energy level by sharing the electrons from the two nearby hydrogen atoms. Hydrogen does not completely give up the electrons, but also has a strong tendency to share electrons with oxygen to fill its outermost energy level. Two covalent bonds form, which creates water.

Most compounds in living organisms have covalent bonds holding them together. Water and other substances with covalent bonds are called molecules. A **molecule** is a compound in which the atoms are held together by covalent bonds. Depending on the number of pairs of electrons that are shared, covalent bonds can be single, double, or triple, as shown in **Figure 6.10.**

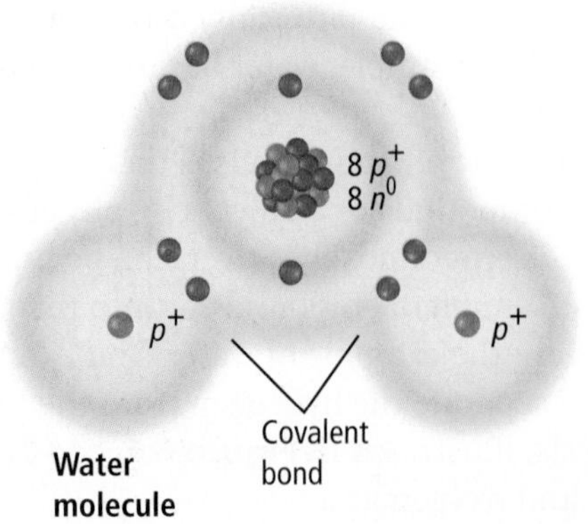

Figure 6.9 In water (H_2O), two hydrogen atoms each share one electron with one oxygen atom. Because the oxygen atom needs two electrons to fill its outer energy level, it forms two covalent bonds, one with each hydrogen atom.

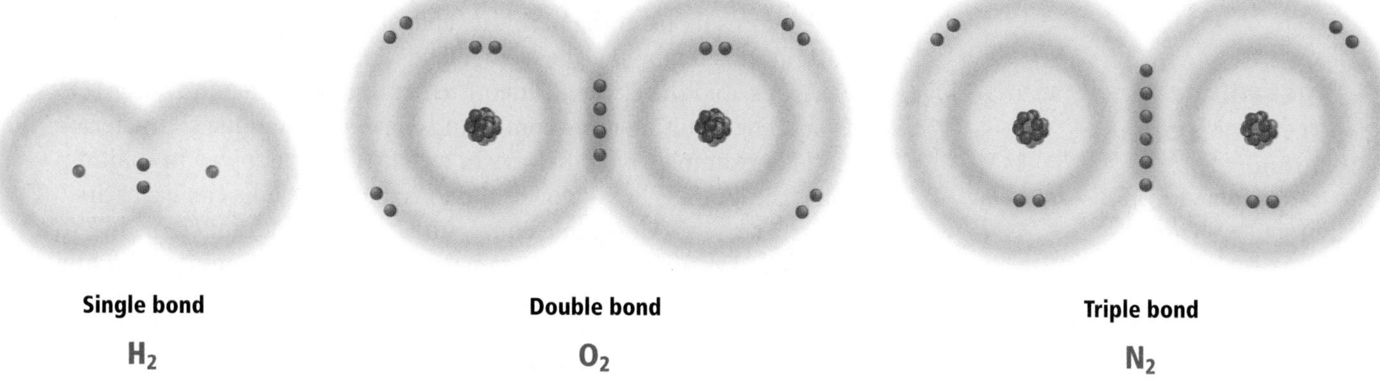

■ **Figure 6.10** A single bond has one pair of shared electrons, a double bond has two pairs, and a triple bond has three pairs.

Personal Tutor

To learn about bonds, visit biologymh.com.

Ionic bonds Recall that atoms are neutral—they do not have an electric charge. Also recall that for an atom to be most stable, the outermost energy level should be either empty or completely filled. Some atoms tend to give up (donate) or obtain (accept) electrons to empty or fill the outer energy level in order to be stable. An atom that has lost or gained one or more electrons becomes an **ion** and carries an electric charge. For example, sodium has one electron in its outermost energy level. Sodium can become more stable if it gives up this one electron, leaving its outer energy level empty. When it gives away this one negative charge, the neutral sodium atom becomes a positively charged sodium ion (Na^+). Similarly, chlorine has seven electrons in its outer energy level and needs just one electron to fill it. When chlorine accepts an electron from a donor atom, such as sodium, chlorine becomes a negatively charged ion (Cl^-).

An **ionic bond** is an electrical attraction between two oppositely charged atoms or groups of atoms called ions. **Figure 6.11** shows how an ionic bond forms as a result of the electrical attraction between Na^+ and Cl^- to produce NaCl (sodium chloride). Substances formed by ionic bonds are called ionic compounds.

Ions in living things include sodium, potassium, calcium, chloride, and carbonate ions. They help maintain homeostasis as they travel in and out of cells. In addition, ions help transmit signals among cells that allow you to see, taste, hear, feel, and smell.

■ **Figure 6.11** To form ions, sodium donates an electron and chlorine gains an electron. An ionic bond forms when the oppositely charged ions come close together.

Concepts In Motion

Interactive Figure To see an animation of how ionic bonds form, visit biologygmh.com.

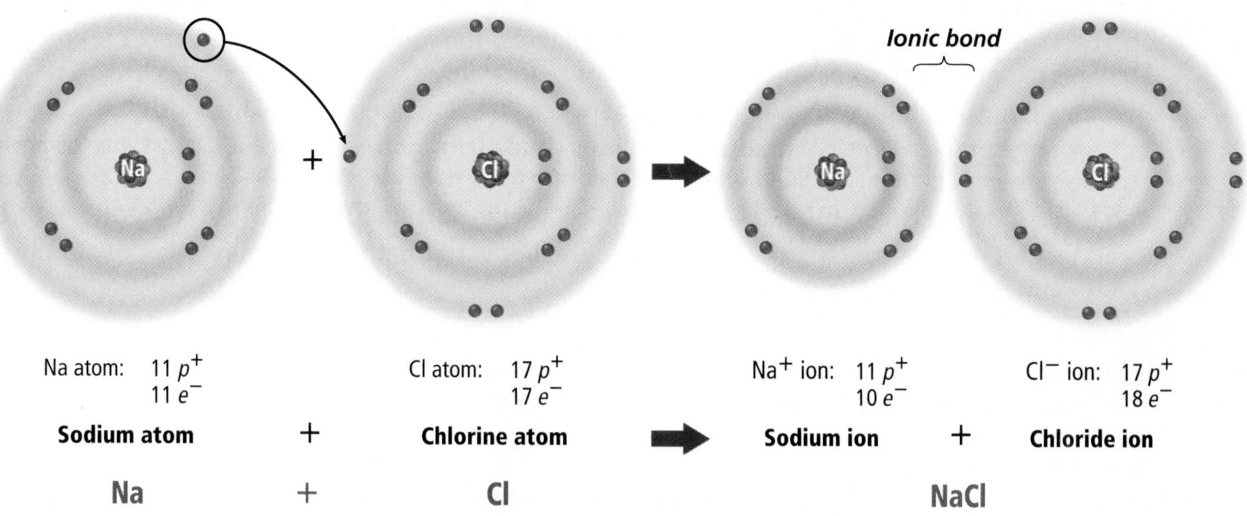

VOCABULARY

WORD ORIGIN

Atom
comes from the Greek word *atomos,* meaning *not divisible.*

Some atoms tend to donate or accept electrons more easily than other atoms. Look at the periodic table of elements inside the back cover of this textbook. The elements identified as metals tend to donate electrons, and the elements identified as nonmetals tend to accept electrons. The resulting ionic compounds have some unique characteristics. For example, most dissolve in water. When dissolved in solution, ionic compounds break down into ions and these ions can carry an electric current. Most ionic compounds, such as sodium chloride (table salt), are crystalline at room temperature. Ionic compounds generally have higher melting points than molecular compounds formed by covalent bonds.

Connection  **Earth Science** Although most ionic compounds are solid at room temperature, other ionic compounds are liquid at room temperature. Like their solid counterparts, ionic liquids are made up of positively and negatively charged ions. Ionic liquids have important potential in real-world applications as safe and environmentally friendly solvents that can possibly replace other harmful solvents. The key characteristic of ionic liquid solvents is that they typically do not evaporate and release chemicals into the atmosphere. Most ionic liquids are safe to handle and store, and they can be recycled after use. For these reasons, ionic liquids are attractive to industries that are dedicated to environmental responsibility.

Reading Check **Compare** ionic solids and liquids.

MiniLab 6.1

Test for Simple Sugars

What common foods contain glucose? Glucose is a simple sugar that provides energy for cells. In this lab, you will use an reagent called Benedict's solution, which indicates the presence of –CHO (carbon, hydrogen, oxygen) groups. A color change determines the presence of glucose and other simple sugars in common foods.

Procedure

1. Read and complete the lab safety form.
2. Create a data table with columns labeled *Food Substance, Sugar Prediction, Observations,* and *Results.*
3. Choose four **food substances** from those provided. Read the food labels and predict the presence of simple sugar in each food. Record your prediction.
4. Prepare a **hot water bath** with a temperature between 40°–50°C using a **hot plate** and **1000-mL beaker.**
5. Label four **test tubes.** Obtain a **graduated cylinder.** Add 10 mL of a different food substance to each test tube. Then add 10 mL **distilled water.** Swirl gently to mix.
6. Add 5 mL of **Benedict's solution** to each tube. Use a clean **stirring rod** to mix the contents.
7. Using **test tube holders,** warm the test tubes in the hot water bath for 2–3 min. Record your observations and results.

Analysis

1. **Interpret Data** Did any of the foods contain simple sugars? Explain.
2. **Think Critically** Could a food labeled "sugar free" test positive using Benedict's solution as an indicator? Explain.

van der Waals Forces

You have learned that positive ions and negative ions form based on the ability of an atom to attract electrons. If the nucleus of the atom has a weak attraction for the electron, it will donate the electron to an atom with a stronger attraction. Similarly, elements in a covalent bond do not always attract electrons equally. Recall also that the electrons in a molecule are in random motion around the nuclei. This movement of electrons can cause an unequal distribution of the electron cloud around the molecule, creating temporary areas of slightly positive and negative charges.

When molecules come close together, the attractive forces between these positive and negative regions pull on the molecules and hold them together. These attractions between the molecules are called **van der Waals forces,** named for the Dutch physicist Johannes van der Waals who first described the phenomenon. The strength of the attraction depends on the size of the molecule, its shape, and its ability to attract electrons. Van der Waals forces are not as strong as covalent and ionic bonds, but they play a key role in biological processes.

Scientists have determined that geckos can climb smooth surfaces due to van der Waals forces between the atoms in the hairlike structures on their toes, shown in **Figure 6.12,** and the atoms on the surface they are climbing.

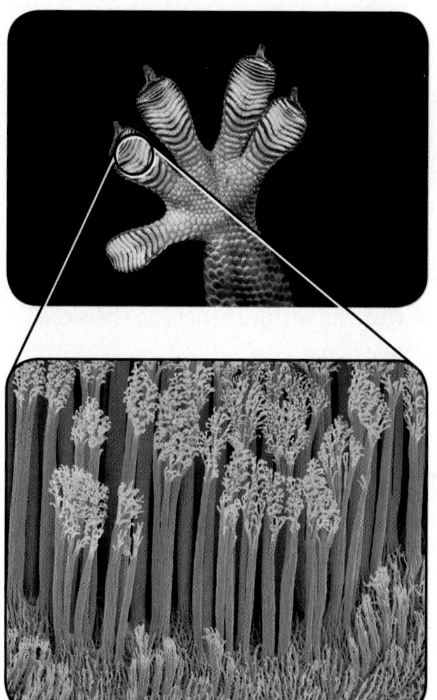

SEM Magnification: 240 ×

Figure 6.12 Geckos have millions of microscopic hairs on the bottoms of their feet that are about as long as two widths of a human hair. Each spreads into 1000 smaller pads that get close to the surface of an atom.

van der Waals forces in water Let's consider how van der Waals forces work in a common substance—water. The areas of slight positive and negative charge around the water molecule are attracted to the opposite charge of other nearby water molecules. These forces hold the water molecules together. Without van der Waals forces, water molecules would not form droplets, and droplets would not form a surface of water. It is important to understand that van der Waals forces are the attractive forces between the water molecules, not the forces between the atoms that make up water.

Section 6.1 Assessment

Section Summary

- Atoms consist of protons, neutrons, and electrons.
- Elements are pure substances made up of only one kind of atom.
- Isotopes are forms of the same element that have a different number of neutrons.
- Compounds are substances with unique properties that are formed when elements combine.
- Elements can form covalent and ionic bonds.

Understand Main Ideas

1. MAIN Idea **Diagram** Sodium has 11 protons and 11 neutrons in its nucleus. Draw a sodium atom. Be sure to label the particles.
2. **Explain** why carbon monoxide (CO) is or is not an element.
3. **Explain** Are all compounds molecules? Why or why not?
4. **Compare** van der Waals forces, ionic bonds, and covalent bonds.

Think Critically

5. **Explain** how the number of electrons in an energy level affects bond formation.

MATH in Biology

6. Beryllium has four protons in its nucleus. How many neutrons are in beryllium-9? Explain how you calculated your answer.

Section 6.2

Reading Preview

Objectives

- **Identify** the parts of a chemical reaction.
- **Relate** energy changes to chemical reactions.
- **Summarize** the importance of enzymes in living organisms.

Review Vocabulary

process: a series of steps or actions that produce an end product

New Vocabulary

chemical reaction
reactant
product
activation energy
catalyst
enzyme
substrate
active site

Chemical Reactions

MAIN Idea Chemical reactions allow living things to grow, develop, reproduce, and adapt.

Real-World Reading Link When you lie down for the night, you might think that your body is completely at rest. In fact, you are still digesting food you ate that day, the scrape on your elbow is healing, and your muscles and bones are growing and developing. All the things that happen inside your body are the result of chemical reactions.

Reactants and Products

A new car with its shining chrome and clean appearance is appealing to many drivers. Over time, however, the car might get rusty and lose some of its appeal. Rust is a result of a chemical change called a chemical reaction. A **chemical reaction** is the process by which atoms or groups of atoms in substances are reorganized into different substances. Chemical bonds are broken and formed during chemical reactions. The rust on the chain in **Figure 6.13** is a compound called iron oxide (Fe_2O_3), and it was formed when oxygen (O_2) in the air reacted with iron (Fe).

It is important to know that substances can undergo changes that do not involve chemical reactions. For example, consider the water in **Figure 6.13.** The water is undergoing a physical change. A physical change alters the substance's appearance but not its composition. It is water before and after the change.

How do you know when a chemical reaction has taken place? Although you might not be aware of all the reactions taking place inside your body, you know the surface of the chain in **Figure 6.13** has changed. What was once silver and shiny is now dull and orange-brown. Other clues that a chemical reaction has taken place include the production of heat or light, and formation of a new gas, liquid, or solid.

Figure 6.13 After a chemical change, such as rusting, a new substance is formed. During a physical change, such as ice melting or water boiling, the chemical makeup of the water is not altered.

Chemical change

Physical change

Chemical equations When scientists write chemical reactions, they express each component of the reaction in a chemical equation. When writing chemical equations, chemical formulas describe the substances in the reaction with arrows indicating the process of change.

Reactants and products A chemical equation shows the **reactants,** the starting substances, on the left side of the arrow. The **products,** the substances formed during the reaction, are on the right side of the arrow. The arrow can be read as "yields" or "react to form."

$$\text{Reactants} \rightarrow \text{Products}$$

The following chemical equation can be written to describe the reaction that provides energy in **Figure 6.14.**

$$C_6H_{12}O_6 + O_2 \rightarrow CO_2 + H_2O$$

Glucose and oxygen react to form carbon dioxide and water.

Figure 6.14 The process that provides your body with energy involves the reaction of glucose with oxygen to form carbon dioxide and water.

Balanced equations In chemical reactions, matter cannot be created or destroyed. This principle is called conservation of mass. Accordingly, all chemical equations must show this balance of mass. This means that the number of atoms of each element on the reactant side must equal the number of atoms of the same element on the product side. Coefficients are used to make the number of atoms on each side of the arrow equal.

$$C_6H_{12}O_6 + 6O_2 \rightarrow 6CO_2 + 6H_2O$$

Multiply the coefficient by the subscript for each element. You can see in this example that there are six carbon atoms, twelve hydrogen atoms, and eighteen oxygen atoms on each side of the arrow. The equation confirms that the number of atoms on each side is equal, and therefore the equation is balanced. You will study this important reaction further in Chapter 8.

VOCABULARY

ACADEMIC VOCABULARY

Coefficient:
In a chemical equation, the number written in front of a reactant or a product
The number 6 in $6Fe_2O_3$ is a coefficient.

 **Reading Check** **Explain** why chemical equations must be balanced.

Energy of Reactions

Connection to Physics A sugar cookie is made with flour, sugar, and other ingredients mixed together, but it is not a cookie until you bake it. Something must start the change from cookie dough to cookies. The key to starting a chemical reaction is energy. For the chemical reactions that transform the dough to cookies, energy in the form of heat is needed. Similarly, most compounds in living things cannot undergo chemical reactions without energy.

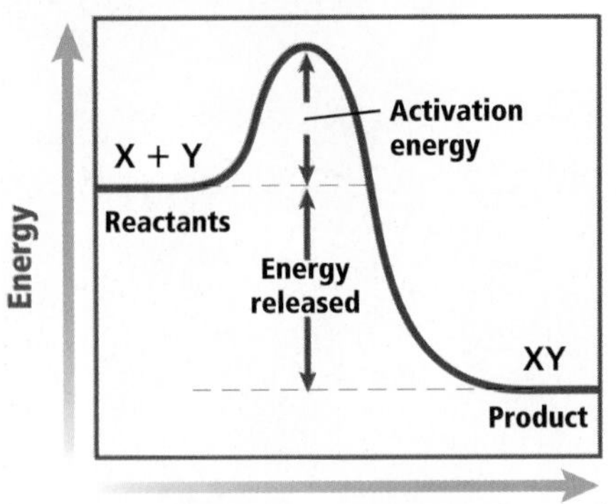

■ **Figure 6.15** The flame of the match provides activation energy—the amount of energy needed to begin a reaction. The reaction gives off energy in the form of heat and light.

Activation energy The minimum amount of energy needed for reactants to form products in a chemical reaction is called the **activation energy.** For example, you know a candle will not burn until you light its wick. The flame provides the activation energy for the reaction of the substances in the candle wick with oxygen. In this case, once the reaction begins, no further input of energy is needed and the candle continues to burn on its own. **Figure 6.15** shows that for the reactants X and Y to form product XY, energy is required to start the reaction. The peak in the graph represents the amount of energy that must be added to the system to make the reaction go. Some reactions rarely happen because they have a very high activation energy.

Energy change in chemical reactions Compare how energy changes during the reaction in **Figure 6.15** to how energy changes during the reaction in **Figure 6.16.** Both reactions require activation energy to get started. However, the reaction in **Figure 6.15** has lower energy in the product than in the reactants. This reaction is exothermic—it released energy in the form of heat. The reaction in **Figure 6.16** is endothermic—it absorbed heat energy. The energy of the products is higher than the energy of the reactant. In every chemical reaction, there is a change in energy due to the making and breaking of chemical bonds as reactants for products. Endothermic reactions keep your internal body temperature at about 37°C.

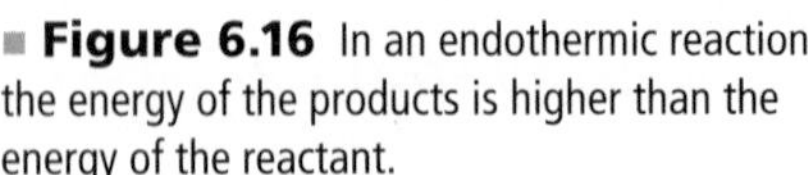

■ **Figure 6.16** In an endothermic reaction, the energy of the products is higher than the energy of the reactant.

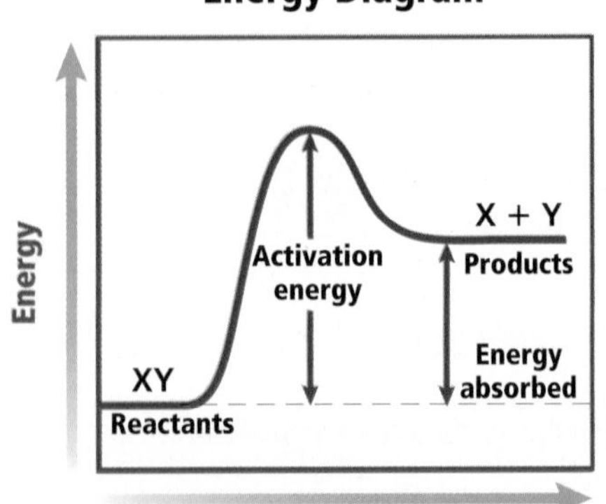

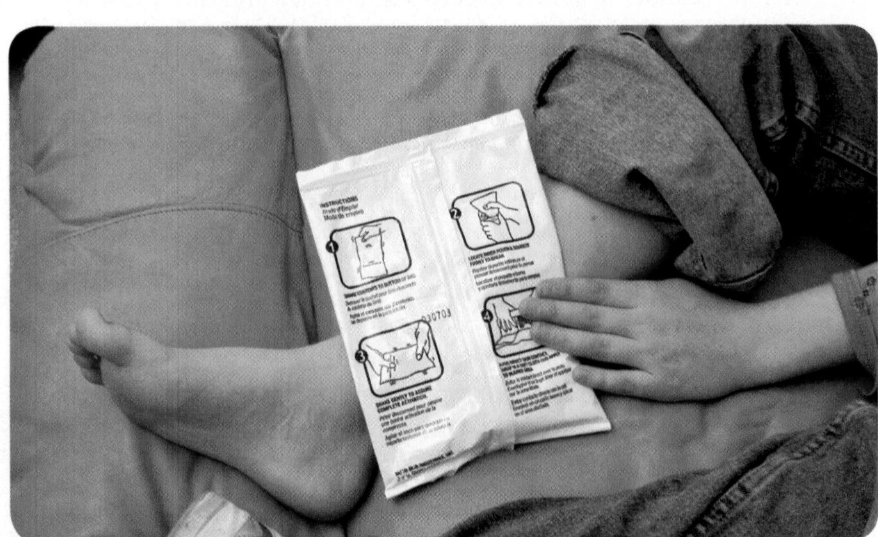

Enzymes

All living things are chemical factories driven by chemical reactions. However, these chemical reactions proceed very slowly when carried out in the laboratory because the activation energy is high. To be useful to living organisms, additional substances must be present where the chemical reactions occur to reduce the activation energy and allow the reaction to proceed quickly.

A **catalyst** is a substance that lowers the activation energy needed to start a chemical reaction. Although a catalyst is important in speeding up a chemical reaction, it does not increase how much product is made and it does not get used up in the reaction. Scientists use many types of catalysts to make reactions go thousands of times faster than the reaction would be able to go without the catalyst.

Special proteins called **enzymes** are the biological catalysts that speed up the rate of chemical reactions in biological processes. Enzymes are essential to life. Compare the progress of the reaction described in **Figure 6.17** to see the effect of an enzyme on a chemical reaction. Like all catalysts, the enzyme is not used up by the chemical reaction. Once it has participated in a chemical reaction, it can be used again.

An enzyme's name describes what it does. For example, amylase is an important enzyme found in saliva. Digestion of food begins in your mouth when amylase speeds the breakdown of amylose, one of the two components of starch. Like amylase, most enzymes are specific to one reaction.

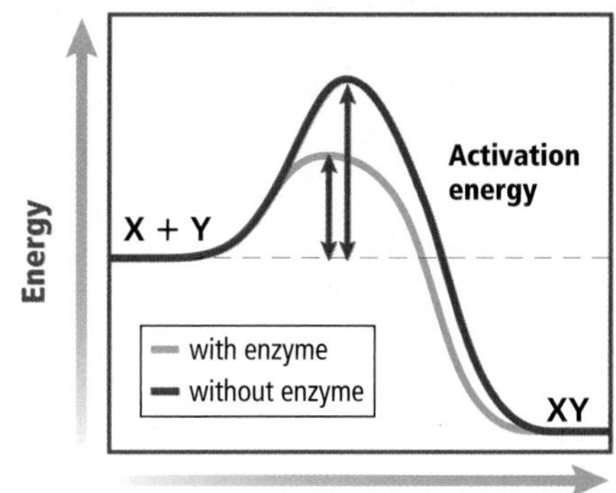

Figure 6.17 When an enzyme acts as a biological catalyst, the reaction occurs at a rate that is useful to cells.

Compare *the activation energy of the reaction without enzyme to the activation energy of the reaction with enzyme.*

MiniLab 6.2

Investigate Enzymatic Browning

What factors affect enzymatic browning? When sliced, an apple's soft tissue is exposed to oxygen, causing a chemical reaction called oxidation. Enzymes in the apple speed this reaction, producing darkened, discolored fruit. In this lab, you will investigate methods used to slow enzymatic browning.

Procedure

1. Read and complete the lab safety form.
2. Predict the relative amount of discoloration each of these apple wedges will show when exposed to air. Justify your prediction.
   Sample 1: Untreated apple wedge
   Sample 2: Apple wedge submerged in boiling water
   Sample 3: Apple wedge submerged in lemon juice
   Sample 4: Apple wedge submerged in sugar solution
3. Prepare 75 mL of each of the following: **boiling water, lemon juice,** and **sugar solution** in three **250-mL beakers.**
4. Slice an **apple** into four wedges. Immediately use **tongs** to submerge each wedge in a different liquid. Put one wedge aside.
5. Submerge the wedges for three minutes, then place on a **paper towel,** skin side down. Observe for 10 min, then record the relative amount of discoloration of each apple wedge.

Analysis

1. **Analyze** How did each treatment affect the chemical reaction that occurred on the fruit's soft tissue? Why were some of the treatments successful?
2. **Think Critically** A restaurant owner wants to serve fresh-cut fruit. What factors might be considered in choosing a recipe and preparation method?

■ **Figure 6.18** Substrates interact with enzymes at specific places called active sites. Only substrates with a specific shape can bind to the active site of an enzyme.

Concepts In Motion

Interactive Figure To see an animation of enzyme activity, visit biologygmh.com.

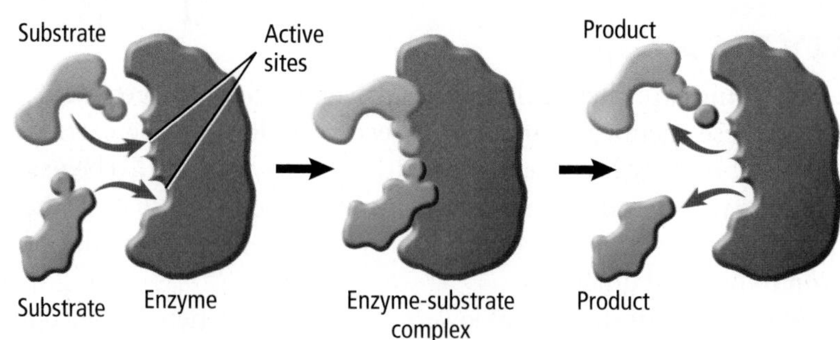

FOLDABLES Incorporate information from this section into your Foldable.

Follow **Figure 6.18** to learn how an enzyme works. The reactants that bind to the enzyme are called **substrates.** The specific location where a substrate binds on an enzyme is called the **active site.** The active site and the substrate have complementary shapes. This enables them to interact in a precise manner, similar to the way in which puzzle pieces fit together. As shown in **Figure 6.18,** only substrates with the same size and shape as the active site will bind to the enzyme.

Once the substrates bind to the active site, the active site changes shape and forms the enzyme-substrate complex. The enzyme-substrate complex helps chemical bonds in the reactants to be broken and new bonds to form—the substrates react to form products. The enzyme then releases the products.

Factors such as pH, temperature, and other substances affect enzyme activity. For example, most enzymes in human cells are most active at an optimal temperature close to 37°C. However, enzymes in other organisms, such as bacteria, can be active at other temperatures.

Enzymes affect many biological processes. When a person is bitten by a poisonous snake, enzymes in the venom break down the membranes of that person's red blood cells. Hard green apples ripen due to the action of enzymes. Photosynthesis and cellular respiration, which you will learn more about in Chapter 8, provide energy for the cell with the help of enzymes. Just as worker bees are important for the survival of a beehive, enzymes are the chemical workers in cells.

Section 6.2 Assessment

Section Summary

- Balanced chemical equations must show an equal number of atoms for each element on both sides.
- Activation energy is the energy required to begin a reaction.
- Catalysts are substances that alter chemical reactions.
- Enzymes are biological catalysts.

Understand Main Ideas

1. MAIN Idea **Identify** the parts of this chemical reaction: A+B → AB.
2. **Diagram** the energy changes that can take place in a chemical reaction.
3. **Explain** why the number of atoms of reactants must equal the number of atoms of products formed.
4. **Describe** the importance of enzymes to living organisms.

Think Critically

MATH in Biology

5. For the following chemical reaction, label the reactants and products, and then balance the chemical equation. ____H_2O_2 → ____H_2O + ____O_2

WRITING in Biology

6. Draw a diagram of a roller coaster and write a paragraph relating the ride to activation energy and a chemical reaction.

Biology Online Self-Check Quiz biologygmh.com

Section 6.3

Reading Preview

Objectives

- **Evaluate** how the structure of water makes it a good solvent.
- **Compare and contrast** solutions and suspensions.
- **Describe** the difference between acids and bases.

Review Vocabulary

physical property: characteristic of matter, such as color or melting point, that can be observed or measured without changing the composition of the substance

New Vocabulary

polar molecule
hydrogen bond
mixture
solution
solvent
solute
acid
base
pH
buffer

Water and Solutions

MAIN Idea **The properties of water make it well suited to help maintain homeostasis in an organism.**

Real-World Reading Link You probably know that the main color on a globe is blue. That's because water covers about 70 percent of Earth's surface, giving it the blue color you see from a distance. Now zoom in to a single cell of an organism on Earth. Water accounts for approximately 70 percent of that cell's mass. It is one of the most important molecules for life.

Water's Polarity

Earlier in this chapter, you discovered that water molecules are formed by covalent bonds that link two hydrogen (H) atoms to one oxygen (O) atom. Because electrons are more strongly attracted to oxygen's nucleus, the electrons in the covalent bond with hydrogen are not shared equally. In water, the electrons spend more time near the oxygen nucleus than they do near the hydrogen nuclei. **Figure 6.19** shows that there is an unequal distribution of electrons in a water molecule. This, along with the bent shape of water, results in the oxygen end of the molecule having a slightly negative charge and the hydrogen ends of the molecule a slightly positive charge. Molecules that have an unequal distribution of charges are called **polar molecules,** meaning that they have oppositely charged regions.

Polarity is the property of having two opposite poles, or ends. A magnet has polarity—there is a north pole and a south pole. When the two ends are brought close to each other, they attract each other. Similarly, when a charged region of a polar molecule comes close to the oppositely charged region of another polar molecule, a weak electrostatic attraction results. In water, the electrostatic attraction is called a hydrogen bond. A **hydrogen bond** is a weak interaction involving a hydrogen atom and a fluorine, oxygen, or nitrogen atom. Hydrogen bonding is a strong type of van der Waals force. **Figure 6.20** describes polarity and the other unique properties of water that make it important to living things.

Figure 6.19 Because water has a bent shape and electrons are not shared equally between hydrogen and oxygen, hydrogen bonds form among the molecules. Due to the attraction among the atoms that make up water, the surface of water supports a water strider.

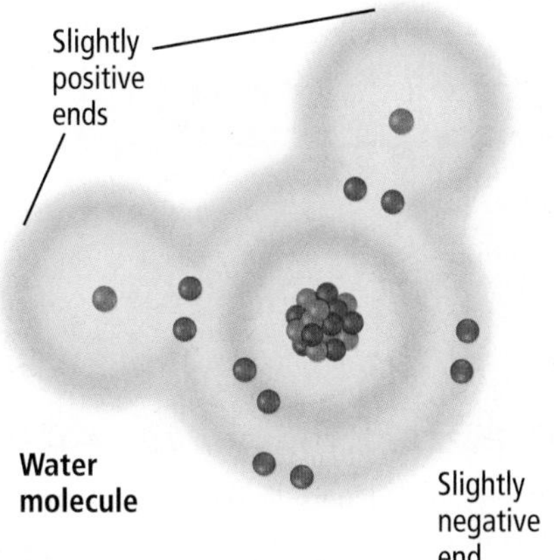

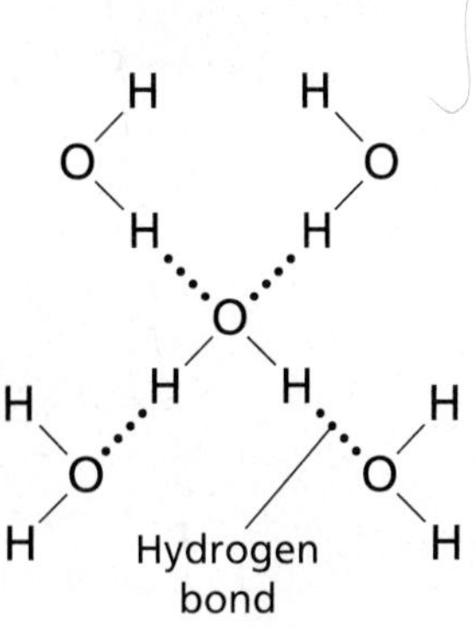

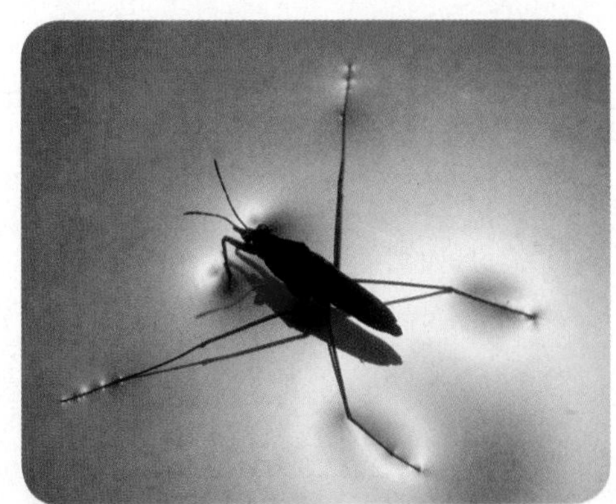

Water strider

Visualizing Properties of Water

Figure 6.20
Water is vital to life on Earth. Its properties allow it to provide environments suitable for life and to help organisms maintain homeostasis. Humans can survive many days without food, but can survive only a few days without water.

Water Molecule

- Water is made up of one oxygen atom and two hydrogen atoms.
- Water is polar. Its bent shape results in a slightly positive charge on the hydrogen atoms and a slightly negative charge on the oxygen. As a result, it forms hydrogen bonds.
- Water is called the universal solvent because many substances dissolve in it.

Hydrogen Bonding

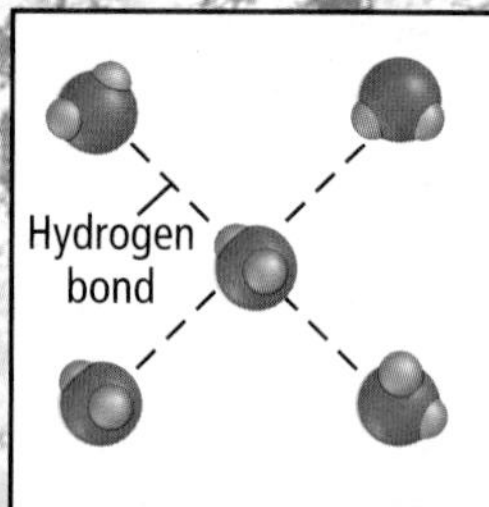

Solid

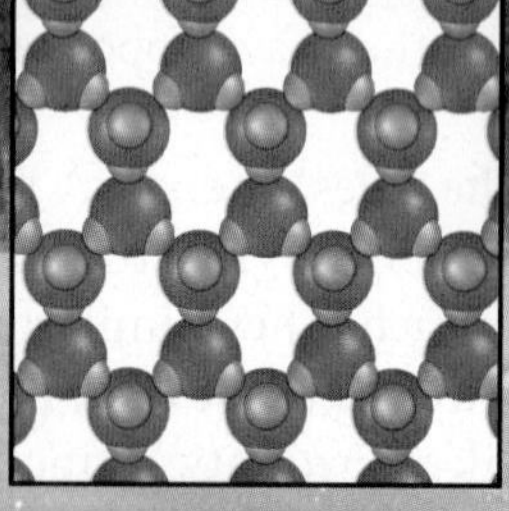

Liquid water becomes more dense as it cools to 4°C. Yet ice is less dense than liquid water. As a result, nutrients in bodies of water mix due to changes in water density during spring and fall. Also, fish can survive winter because ice floats—they continue to live and function in the water beneath the ice.

Liquid

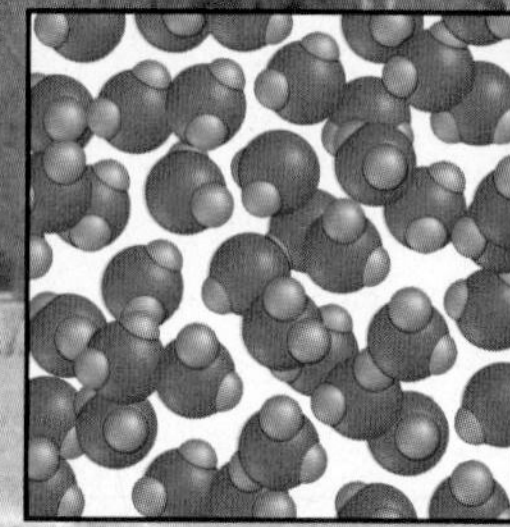

Water is adhesive—it forms hydrogen bonds with molecules on other surfaces. Capillary action is the result of adhesion. Water travels up the stem of a plant, and seeds swell and germinate by capillary action.

Water is cohesive—the molecules are attracted to each other due to hydrogen bonds. This attraction creates surface tension, which causes water to form droplets and allows insects and leaves to rest on the surface of a body of water.

Concepts In Motion **Interactive Figure** To see an animation of water, visit biologygmh.com.

Biology Online

Mixtures with water

Most students are familiar with powdered drink products that dissolve in water to form a flavored beverage. When you add a powdered substance to water, it does not react with water to form a new product. You create a mixture. A **mixture** is a combination of two or more substances in which each substance retains its individual characteristics and properties.

Homogenous mixtures When a mixture has a uniform composition throughout, it is called a homogeneous (hoh muh JEE nee us) mixture. A **solution** is another name for a homogeneous mixture. For example, in the powdered tea drink solution shown in **Figure 6.21,** tea is on top, tea is in the middle, and tea is at the bottom of the container. The water retains its properties and the drink mix retains its properties.

In a solution, there are two components: a solvent and a solute. A **solvent** is a substance in which another substance is dissolved. A **solute** is the substance that is dissolved in the solvent. In the case of the drink mix, water is the solvent and the powdered substance is the solute. A mixture of salt and water is another example of a solution because the solute (salt) dissolves completely in the solvent (water). Saliva moistens your mouth and begins the digestion of some of your food. Saliva is a solution that contains water, proteins, and salts. In addition, the air you breathe is a solution of gases.

Heterogenous mixtures Think about the last time you ate a salad. Perhaps it contained lettuce and other vegetables, croutons, and salad dressing. Your salad was a heterogeneous mixture. In a heterogeneous mixture, the components remain distinct, that is, you can tell what they are individually. Compare the mixture of sand and water to the solution of salt and water next to it in **Figure 6.22.** Sand and water form a type of heterogeneous mixture called a suspension. Over time, the particles in a suspension settle to the bottom.

A colloid is a heterogeneous mixture in which the particles do not settle out like the sand settled from the water. You are probably familiar with many colloids, including fog, smoke, butter, mayonnaise, milk, paint, and ink. Blood is a colloid made up of plasma, cells, and other substances.

Reading Check **Distinguish** between solutions and suspensions.

■ **Figure 6.21** Tea forms a homogeneous mixture in water. The particles of solute (tea) are dissolved and spread throughout the solvent (water).

VOCABULARY

ACADEMIC VOCABULARY

Suspend:
to keep from falling or sinking.
A slender thread suspended the spider from the web.

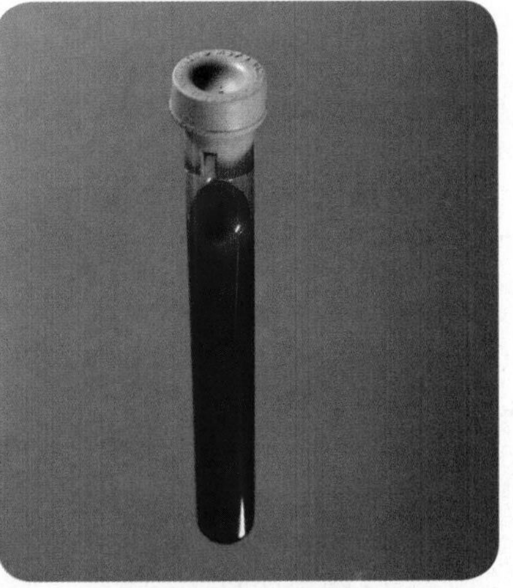

■ **Figure 6.22**
Left: Sand and water form a heterogeneous mixture—you can see both the liquid and the solid. The homogeneous mixture of salt and water is a liquid—you cannot see the salt.
Right: Blood is a heterogeneous mixture called a colloid.

Figure 6.23 Substances that release H^+ in water are acids. Substances that release OH^- in water are bases.

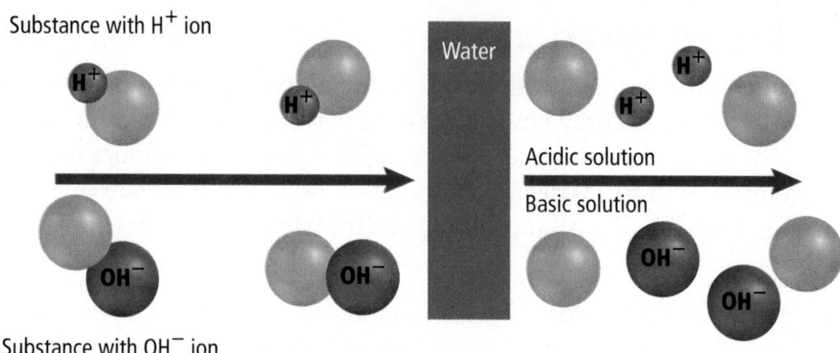

Acids and bases Many solutes readily dissolve in water due to water's polarity. This means that an organism, which might be as much as 70 percent water, can be a container for a variety of solutions. When a substance that contains hydrogen is dissolved in water, the substance might release a hydrogen ion (H^+) because it is attracted to the negatively charged oxygen atoms in water, as shown in **Figure 6.23.** Substances that release hydrogen ions when dissolved in water are called **acids.** The more hydrogen ions a substance releases, the more acidic the solution becomes.

Similarly, substances that release hydroxide ions (OH^-) when dissolved in water are called **bases.** Sodium hydroxide (NaOH) is a common base that breaks apart in water to release sodium ions (Na^+) and hydroxide ions (OH^-). The more hydroxide ions a substance releases, the more basic the solution becomes.

Acids and bases are key substances in biology. Many of the foods and beverages we eat and drink are acidic, and the substances in the stomach that break down the food, called gastric juices, are highly acidic.

DATA ANALYSIS LAB 6.1

Based on Real Data*

Recognize Cause and Effect

How do pH and temperature affect protease activity? Proteases are enzymes that break down protein. Bacterial proteases often are used in detergents to help remove stains such as egg, grass, blood, and sweat from clothes.

Data and Observations

A protease from a newly isolated strain of bacteria was studied over a range of pH values and temperatures.

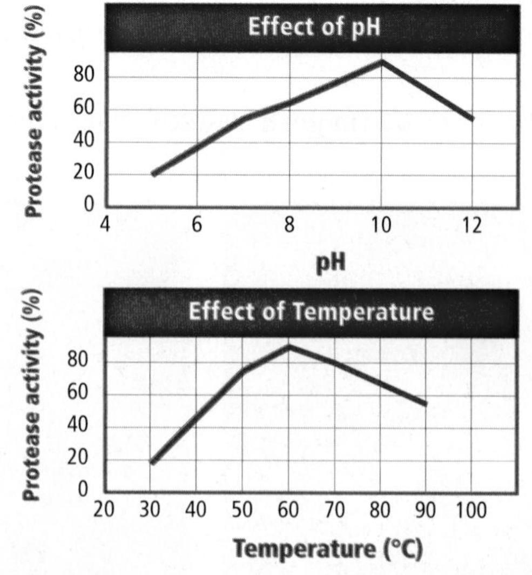

*Data obtained from: Adinarayana, et al. 2003. Purification and partial characterization of thermostable serine alkaline protease from a newly isolated *Bacillus subtilis* PE-11. *AAPS PharmSciTech 4: article 56.*

Think Critically

1. **Identify** the range of pH values and temperatures used in the experiment.
2. **Summarize** the results of the two graphs.
3. **Infer** If a laundry detergent is basic and requires hot water to be most effective, would this protease be useful? Explain.

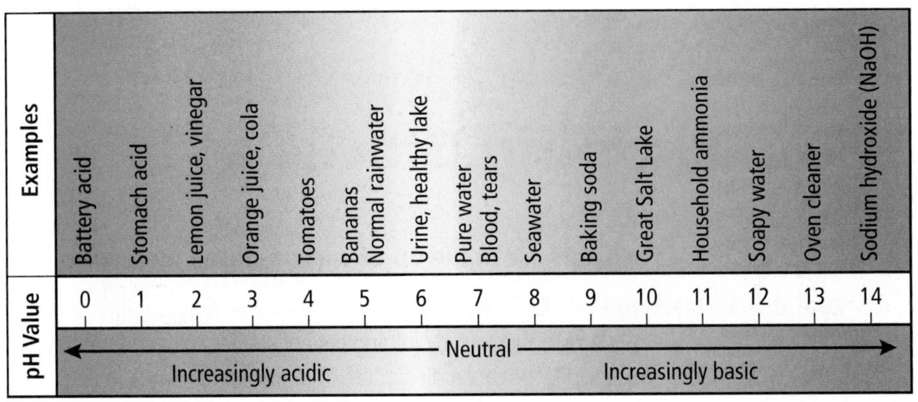

Figure 6.24 The pH scale is used to indicate the relative strength of acids and bases—in other words, the amount of hydrogen ions (H^+) in a solution.

pH and buffers The amount of hydrogen ions or hydroxide ions in a solution determines the strength of an acid or base. Scientists have devised a convenient way to measure how acidic or basic a solution is. The measure of concentration of H^+ in a solution is called **pH.** As shown in **Figure 6.24,** pure water is neutral and has a pH value of 7.0. Acidic solutions have an abundance of H^+ and have pH values lower than 7. Basic solutions have more OH^- than H^+ and have pH values higher than 7.

Connection to Health The majority of biological processes carried out by cells occur between pH 6.5 and 7.5. In order to maintain homeostasis, it is important to control H^+ levels. If you've ever had an upset stomach, you might have taken an antacid to feel better. The antacid tablet is a buffer to help neutralize the stomach acid. **Buffers** are mixtures that can react with acids or bases to keep the pH within a particular range. In cells, buffers keep the pH in a cell within the 6.5 to 7.5 pH range. Your blood, for example, contains buffers that keep the pH about 7.4.

CAREERS IN BIOLOGY

Pool Technician Every recreational body of water, such as a recreational swimming pool, training spa, or medical therapy pool, must meet strict requirements for water quality. Pool technicians make sure these requirements are met by monitoring water pH, bacteria and algae levels, and water clarity. For more information on biology careers, visit biologygmh.com.

Section 6.3 Assessment

Section Summary

- Water is a polar molecule.
- Solutions are homogeneous mixtures formed when a solute is dissolved in a solvent.
- Acids are substances that release hydrogen ions into solutions. Bases are substances that release hydroxide ions into solutions.
- pH is a measure of the concentration of hydrogen ions in a solution.

Understand Main Ideas

1. MAIN Idea **Describe** one way in which water helps maintain homeostasis in an organism.
2. **Relate** the structure of water to its ability to act as a solvent.
3. **Draw** a pH scale and label water (H_2O), hydrochloric acid (HCl), and sodium hydroxide (NaOH) in their general areas on the scale.
4. **Compare and contrast** solutions and suspensions. Give examples of each.

Think Critically

5. **Explain** how baking soda ($NaHCO_3$) is basic. Describe the effect of baking soda on the H^+ ion concentration of stomach contents with pH 4.
6. **Predict** If you add hydrochloric acid (HCl) to water, what effect would this have on the H^+ ion concentration? On the pH?

Section 6.4

Reading Preview

Objectives

- **Describe** the role of carbon in living organisms.
- **Summarize** the four major families of biological macromolecules.
- **Compare** the functions of each group of biological macromolecules.

Review Vocabulary

organic compound: carbon-based substance that is the basis of living matter

New Vocabulary

macromolecule
polymer
carbohydrate
lipid
protein
amino acid
nucleic acid
nucleotide

The Building Blocks of Life

MAIN Idea Organisms are made up of carbon-based molecules.

Real-World Reading Link Children enjoy toy trains because they can link long lines of cars together and make patterns by joining cars of similar color or function. Similarly, in biology, there are large molecules made of many smaller units joined together.

Organic Chemistry

The element carbon is a component of almost all biological molecules. For this reason, life on Earth often is considered carbon-based. Because carbon is an essential element, scientists have devoted an entire branch of chemistry, called organic chemistry, to the study of organic compounds—those compounds containing carbon.

As shown in **Figure 6.25,** carbon has four electrons in its outermost energy level. Recall that the second energy level can hold eight electrons, so one carbon atom can form four covalent bonds with other atoms. These covalent bonds enable the carbon atoms to bond to each other, which results in a variety of important organic compounds. These compounds can be in the shape of straight chains, branched chains, and rings, such as those illustrated in **Figure 6.25.** Together, carbon compounds lead to the diversity of life on Earth.

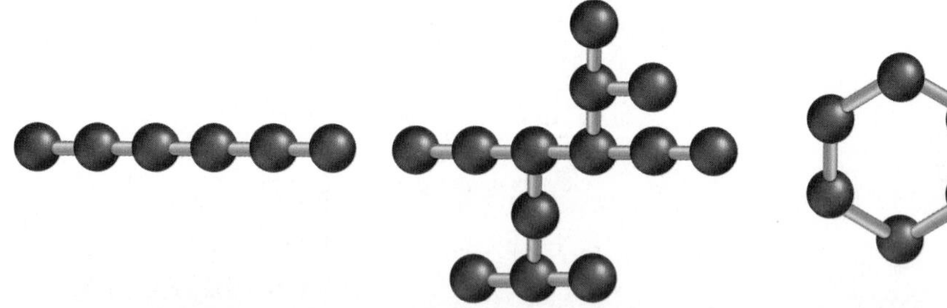

■ **Figure 6.25** The amazing diversity of life is based on the variety of carbon compounds. The half-filled outer energy level of carbon allows for the formation of straight chain, branched, and ring molecules.

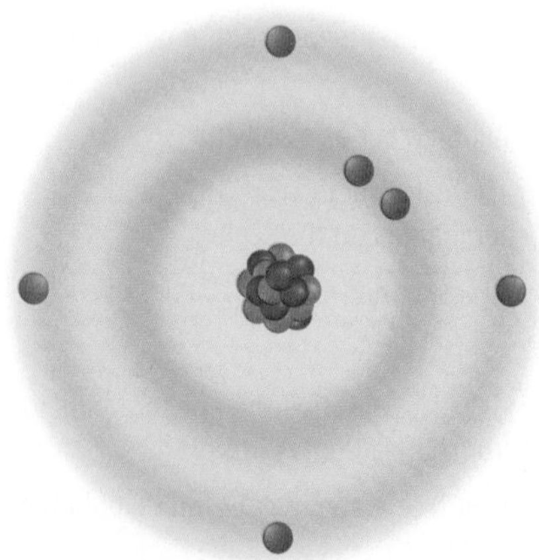

Macromolecules

Carbon atoms can be joined to form carbon molecules. Similarly, most cells store small carbon compounds that serve as building blocks for large molecules. **Macromolecules** are large molecules that are formed by joining smaller organic molecules together. These large molecules are also called polymers. **Polymers** are molecules made from repeating units of identical or nearly identical compounds called monomers that are linked together by a series of covalent bonds. As shown in **Table 6.1**, biological macromolecules are organized into four major categories: carbohydrates, lipids, proteins, and nucleic acids.

Reading Check **Use an analogy** to describe macromolecules.

VOCABULARY

WORD ORIGIN

Polymer

poly– prefix; from Greek, meaning *many.*

–***meros*** from Greek, meaning *part.*

Table 6.1 Biological Macromolecules

Concepts In Motion

Interactive Table To explore more about biological macromolecules, visit biologygmh.com.

Group	Example	Function
Carbohydrates		• Store energy • Provide structural support
Lipids		• Store energy • Provide barriers
Proteins	Hemoglobin	• Transport substances • Speed reactions • Provide structural support • Make hormones
Nucleic acids	DNA stores genetic information in the cell's nucleus.	• Store and communicate genetic information

Study Tip

Double-Entry Notes Fold a piece of paper in half lengthwise and write the boldfaced headings that appear under the *Biological Macromolecules* label on the left side. As you read the text, make a bulleted list of notes about the important ideas and terms.

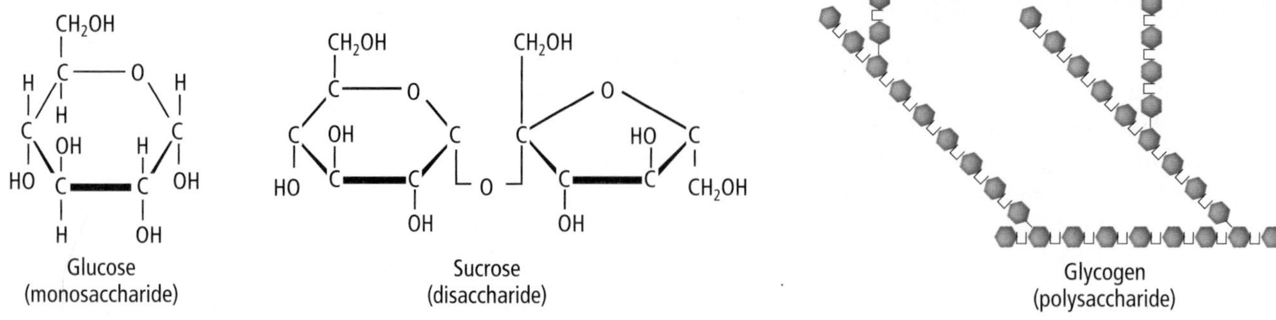

■ **Figure 6.26** Glucose is a monosaccharide. Sucrose is a disaccharide composed of glucose and fructose monosaccharides. Glycogen is a branched polysaccharide made from glucose monomers.

Carbohydrates Compounds composed of carbon, hydrogen, and oxygen in a ratio of one oxygen and two hydrogen atoms for each carbon atom are called **carbohydrates.** A general formula for carbohydrates is written as $(CH_2O)_n$. Here the subscript *n* indicates the number of CH_2O units in a chain. Biologically important carbohydrates that have values of *n* ranging from three to seven are called simple sugars, or monosaccharides (mah nuh SA kuh rid). The monosaccharide glucose, shown in **Figure 6.26,** plays a central role as an energy source for organisms.

Monosaccharides can be linked to form larger molecules. Two monosaccharides joined together form a disaccharide (di SA kuh rid). Like glucose, disaccharides serve as energy sources. Sucrose, also shown in **Figure 6.26,** which is table sugar, and lactose, which is a component of milk, are both disaccharides. Longer carbohydrate molecules are called polysaccharides. One important polysaccharide is glycogen, which is shown in **Figure 6.26.** Glycogen is an energy storage form of glucose that is found in the liver and skeletal muscle. When the body needs energy between meals or during physical activity, glycogen is broken down into glucose.

In addition to their roles as energy sources, carbohydrates have other important functions in biology. In plants, a carbohydrate called cellulose provides structural support in cell walls. As shown in **Figure 6.27,** cellulose is made of chains of glucose linked together into tough fibers that are well-suited for their structural role. Chitin (KI tun) is a nitrogen-containing polysaccharide that is the main component in the hard outer shell of shrimp, lobsters, and some insects, as well as the cell wall of some fungi.

To learn about monomers, visit biologygmh.com.

■ **Figure 6.27** The cellulose in plant cells provides the structural support for trees to stand in a forest.

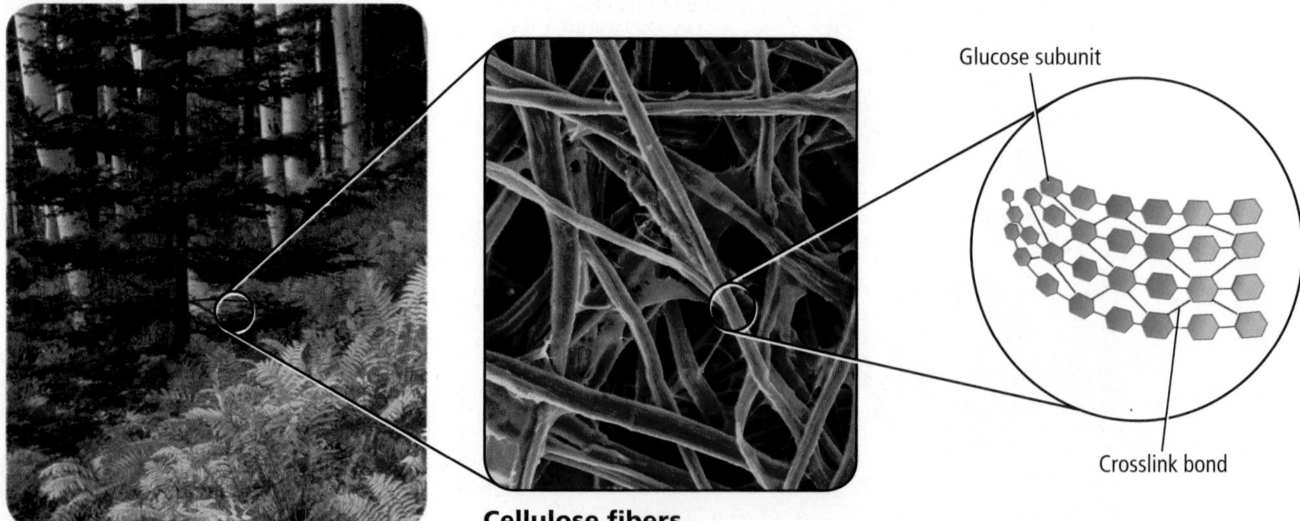

Lipids Another important group of biological macromolecules is the lipid group. **Lipids** are molecules made mostly of carbon and hydrogen that make up the fats, oils, and waxes. Lipids are composed of fatty acids, glycerol, and other components. The primary function of lipids is to store energy. A lipid called a triglyceride (tri GLIH suh rid) is a fat if it is solid at room temperature and an oil if it is liquid at room temperature. In addition, triglycerides are stored in the fat cells of your body. Plant leaves are coated with lipids called waxes to prevent water loss, and the honeycomb in a beehive is made of beeswax.

Saturated and unsaturated fats Organisms need lipids in order to function properly. The basic structure of a lipid includes fatty acid tails as shown in **Figure 6.28.** Each tail is a chain of carbon atoms bonded to hydrogen and other carbon atoms by single or double bonds. Lipids that have tail chains with only single bonds between the carbon atoms are called saturated fats because no more hydrogens can bond to the tail. Lipids that have at least one double bond between carbon atoms in the tail chain can accommodate at least one more hydrogen and are called unsaturated fats. Fats with more than one double bond in the tail are called polyunsaturated fats.

Phospholipids A special lipid shown in **Figure 6.28,** called a phospholipid, is responsible for the structure and function of the cell membrane. Lipids are hydrophobic, which means they do not dissolve in water. This characteristic is important because it allows lipids to serve as barriers in biological membranes.

Steroids Another important category of lipids is the steroid group. Steroids include substances such as cholesterol and hormones. Despite its reputation as a "bad" lipid, cholesterol provides the starting point for other necessary lipids such as vitamin D and the hormones estrogen and testosterone.

DATA ANALYSIS LAB 6.2

Based on Real Data*

Interpret the Data

Does soluble fiber affect cholesterol levels? High amounts of a steroid called cholesterol in the blood are associated with the development of heart disease. Researchers study the effects of soluble fiber in the diet on cholesterol.

Data and Observations

This experiment evaluated the effects of three soluble fibers on cholesterol levels in the blood: pectin (PE), guar gum (GG), and psyllium (PSY). Cellulose was the control (CNT).

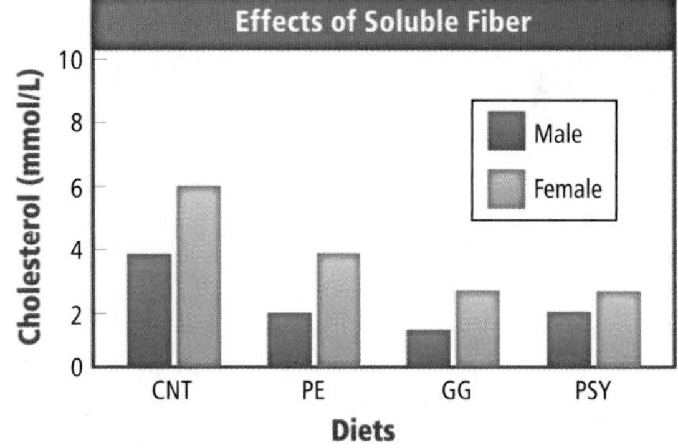

Think Critically

1. **Calculate** the percentage of change in cholesterol levels as compared to the control.
2. **Describe** the effects soluble fiber appears to have on cholesterol levels in the blood.

*Data obtained from: Shen, et al. 1998. Dietary soluble fiber lowers plasma LDL cholesterol concentrations by altering lipoprotein metabaolism in female Guinea pigs. *Journal of Nutrition* 128: 1434-1441.

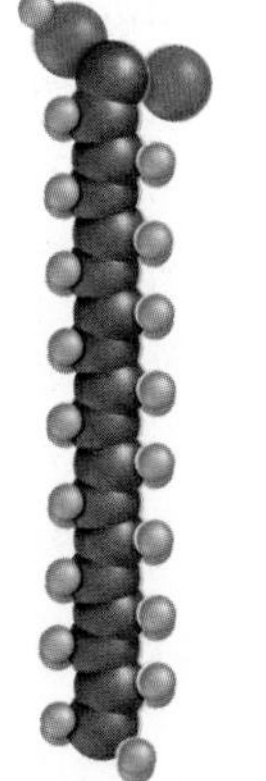

Stearic acid

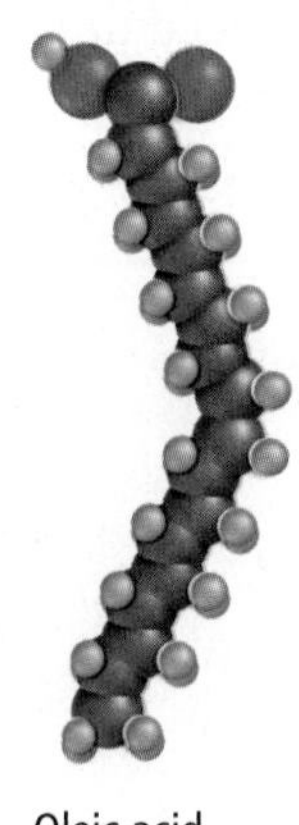

Oleic acid

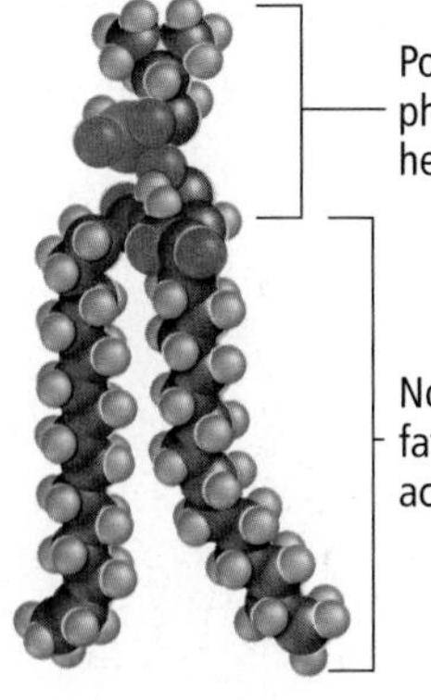
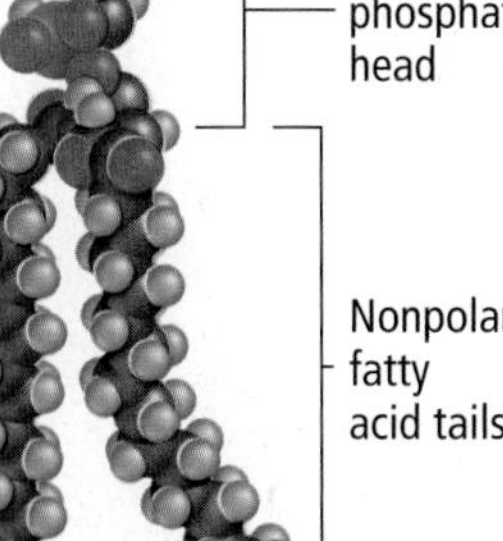

Phospholipid

■ **Figure 6.28** Stearic acid has no double bonds between carbon atoms; oleic acid has one double bond. Phospholipids have a polar head and two nonpolar tails.

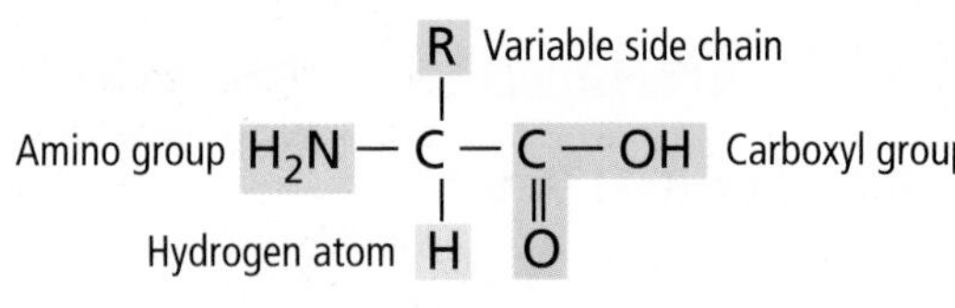

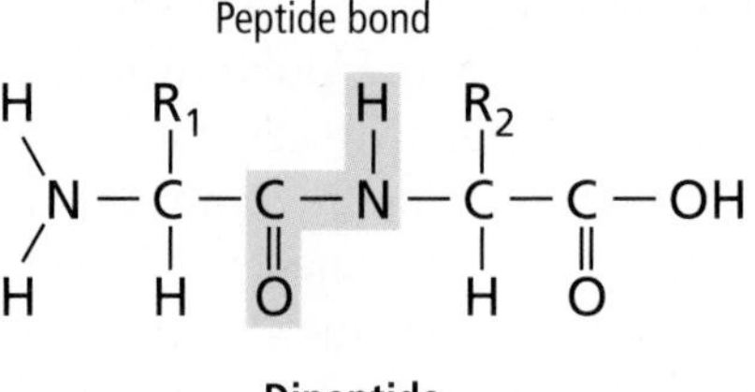

■ **Figure 6.29**
Left: The general structure of an amino acid has four groups around a central carbon.
Right: The peptide bond in a protein happens as a result of a chemical reaction.
Interpret *What other molecule is a product when a peptide bond forms?*

Concepts In Motion
Interactive Figure To see an animation of a peptide bond, visit biologygmh.com.

Proteins Another primary building block of living things is protein. A **protein** is a compound made of small carbon compounds called amino acids. **Amino acids** are small compounds that are made of carbon, nitrogen, oxygen, hydrogen, and sometimes sulfur. All amino acids share the same general structure.

Amino acid structure Amino acids have a central carbon atom like the one shown in **Figure 6.29.** Recall that carbon can form four covalent bonds. One of those bonds is with hydrogen. The other three bonds are with an amino group ($-NH_2$), a carboxyl group ($-COOH$), and a variable group ($-R$). The variable group makes each amino acid different. There are 20 different variable groups, and proteins are made of different combinations of all 20 different amino acids. Several covalent bonds called peptide bonds join amino acids together to form proteins, which is also shown in **Figure 6.29.** A peptide forms between the amino group of one amino acid and the carboxyl group of another.

Three-dimensional protein structure Based on the variable groups contained in the different amino acids, proteins can have up to four levels of structure. The number of amino acids in a chain and the order in which the amino acids are joined define the protein's primary structure. After an amino acid chain is formed, it folds into a unique three-dimensional shape, which is the protein's secondary structure. **Figure 6.30** shows two basic secondary structures—the helix and the pleat. A protein might contain many helices, pleats, and folds. The tertiary structure of many proteins is globular, such as the hemoglobin protein shown in **Table 6.1,** on page 167 but some proteins form long fibers. Some proteins form a fourth level of structure by combining with other proteins.

Protein function Proteins make up about 15 percent of your total body mass and are involved in nearly every function of your body. For example, your muscles, skin, and hair are made of proteins. Your cells contain about 10,000 different proteins that provide structural support, transport substances inside the cell and between cells, communicate signals within the cell and between cells, speed up chemical reactions, and control cell growth.

■ **Figure 6.30** The shape of a protein depends on the interactions among the amino acids. Hydrogen bonds help the protein hold its shape.

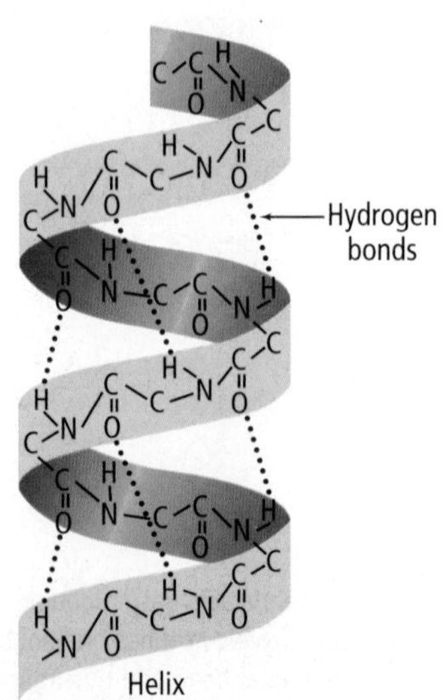

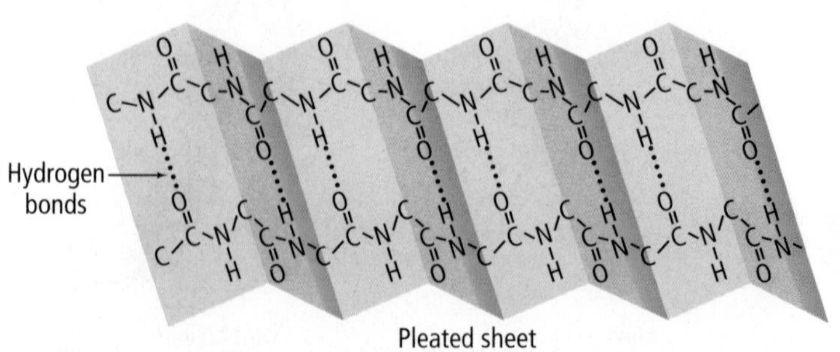

Phosphate group

Sugar

Nitrogen-containing base

Nucleotide

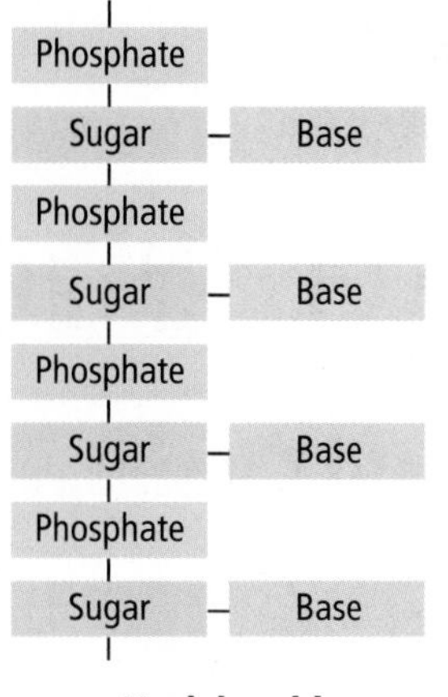

Nucleic acid

■ **Figure 6.31**
Left: DNA nucleotides contain the sugar deoxyribose. RNA nucleotides contain the sugar ribose.
Right: Nucleotides are joined together by bonds between their sugar group and phosphate group.

Nucleic acids The fourth group of biological macromolecules are nucleic acids. **Nucleic acids** are complex macromolecules that store and transmit genetic information. Nucleic acids are made of smaller repeating subunits composed of carbon, nitrogen, oxygen, phosphorus, and hydrogen atoms, called **nucleotides. Figure 6.31** shows the basic structure of a nucleotide and nucleic acid. There are six major nucleotides, all of which have three units—a phosphate, a nitrogenous base, and a ribose sugar.

There are two types of nucleic acids found in living organisms: deoxyribonucleic (dee AHK sih rib oh noo klay ihk) acid (DNA) and ribonucleic (rib oh noo KLAY ihk) acid (RNA). In nucleic acids such as DNA and RNA, the sugar of one nucleotide bonds to the phosphate of another nucleotide. The nitrogenous base that sticks out from the chain is available for hydrogen bonding with other bases in other nucleic acids. You will learn more about the structure and function of DNA and RNA in Chapter 12.

A nucleotide with three phosphate groups is adenosine triphosphate (ATP). ATP is a storehouse of chemical energy that can be used by cells in a variety of reactions. It releases energy when the bond between the second and third phosphate group is broken.

Section 6.4 Assessment

Section Summary

- Carbon compounds are the basic building blocks of living organisms.
- Biological macromolecules are formed by joining small carbon compounds into polymers.
- There are four types of biological macromolecules.
- Peptide bonds join amino acids in proteins.
- Chains of nucleotides form nucleic acids.

Understand Main Ideas

1. MAIN Idea **Explain** If an unknown substance found on a meteorite is determined to contain no trace of carbon, can scientists conclude that there is life at the metorite's origin?
2. **List** and compare the four types of biological macromolecules.
3. **Identify** the components of carbohydrates and proteins.
4. **Discuss** the importance of amino acid order to a protein's function.

Think Critically

5. **Summarize** Given the large number of proteins in the body, explain why the shape of an enzyme is important to its function.
6. **Draw** two structures (one straight chain and one ring) of a carbohydrate with the chemical formula $(CH_2O)_6$.

In the Field

Career: Field Chemist

pH and Alkalinity

Water is one of the most important abiotic factors in any ecosystem. Whether in the desert or the rainforest, the availability of water as rain, surface water, and ground water affects every living thing. It isn't surprising that the task of monitoring and testing water is an important aspect of biological field work.

Effects of pH Several factors can affect the chemistry of available water in an ecosystem, including dissolved oxygen content, salinity, and pH. Factors such as agricultural runoff and acid rain can cause the pH of a body of water to change.

Acidity Acidic conditions, which indicate high levels of H^+, can disrupt biological processes in many water-dwelling organisms, such as snails, clams, and fish. Disrupting these processes can hamper reproduction and can eventually kill the organisms. Although organisms exhibit various degrees of resistance to changes in pH, a resistant organism that is dependent on a susceptible one will feel the effects of pH through that relationship. In addition, pH also affects the solubility of certain substances. For example, the concentration dissolved aluminum in a stream or lake increases at lower pH. Aluminum in water is toxic to many living things.

Alkalinity Despite the fact that water quality is affected by pH, a body of water also has the ability to resist pH changes that are associated with increasing acidity. The ability of a body of water to neutralize acid is referred to as its alkalinity. Carbonate and bicarbonate compounds are important acid-neutralizing compounds found in lakes and streams. pH can be controlled as long as carbonates are present. If the carbonates are used up, additional acid in the water will lower the pH and possibly endanger the inhabitants.

Assessing pH Biologists who perform field testing can assess the pH and alkalinity of a stream or lake by testing the water. When monitoring water, the location of the body of water, the depth of the sample, and the speed of the current where the sample is taken are all important considerations. In general, most freshwater has a pH between 6.5 and 8.0, but there can be some variation. If the pH strays from the optimal range for the water source, local communities can take action to preserve the environment and restore the pH to normal levels.

WRITING in Biology

Research water quality and testing issues for the Gulf Coast in the wake of hurricanes Katrina and Rita. Prepare a report that explains the water-quality problems caused by the hurricanes and what solutions were developed by the crisis management teams. For more information about water quality, visit biologygmh.com.

Design Your Own

BIOLAB

WHAT FACTORS AFFECT AN ENZYME REACTION?

Background: The compound hydrogen peroxide, H_2O_2, is produced when organisms metabolize food, but hydrogen peroxide damages cell parts. Organisms combat the buildup of H_2O_2 by producing the enzyme peroxidase. Peroxidase speeds up the breakdown of hydrogen peroxide into water and oxygen.

Question: *What factors affect peroxidase activity?*

Possible Materials

400-mL beaker
kitchen knife
hot plate
test tube rack
ice
beef liver
dropper
distilled water
50-mL graduated cylinder
10-mL graduated cylinder
tongs or large forceps
square or rectangular pan
stopwatch or timer
nonmercury thermometer
3% hydrogen peroxide
potato slices
18-mm × 150-mm test tubes
buffer solutions (pH 5, pH 6, pH 7, pH 8)

Safety Precautions

CAUTION: ***Use only GFCI-protected circuits for electrical devices.***

Plan and Perform the Experiment

1. Read and complete the lab safety form.
2. Choose a factor to test. Possible factors include temperature, pH, and substrate (H_2O_2) concentration.
3. Form a hypothesis about how the factor will affect the reaction rate of peroxidase.
4. Design an experiment to test your hypothesis. Create a procedure and identify the controls and variables.
5. Create a data table for recording your observations and measurements.
6. Make sure your teacher approves your plan before you proceed.
7. Conduct your approved experiment.
8. **Cleanup and Disposal** Clean up all equipment as instructed by your teacher and return everything to its proper place. Wash your hands thoroughly with soap and water.

Analyze and Conclude

1. **Describe** how the factor you tested affected the enzyme activity of peroxidase.
2. **Graph** your data, then analyze and interpret your graph.
3. **Discuss** whether or not your data supported your hypothesis.
4. **Infer** why hydrogen peroxide is not the best choice for cleaning an open wound.
5. **Error Analysis** Identify any experimental errors or other errors in your data that might have affected the accuracy of your results.

SHARE YOUR DATA

Compare your data with the data collected by other groups in the class that are testing the same factor. Infer reasons why your group's data might have differed from the data collected by other groups. To learn more about enzymes, visit Biolabs at biologygmh.com.

Chapter 6 Study Guide

Download quizzes, key terms, and flash cards from biologygmh.com.

FOLDABLES **Examine** and report on the role of carbon in organisms and explain why so many carbon structures exist.

Vocabulary	Key Concepts
Section 6.1 Atoms, Elements, and Compounds	
• atom (p. 148) • compound (p. 151) • covalent bond (p. 152) • electron (p. 148) • element (p. 149) • ion (p. 153) • ionic bond (p. 153) • isotope (p. 150) • molecule (p. 152) • neutron (p. 148) • nucleus (p. 148) • proton (p. 148) • van der Waals force (p. 155)	MAIN Idea Matter is composed of tiny particles called atoms. • Atoms consist of protons, neutrons, and electrons. • Elements are pure substances made up of only one kind of atom. • Isotopes are forms of the same element that have a different number of neutrons. • Compounds are substances with unique properties that are formed when elements combine. • Elements can form covalent and ionic bonds.
Section 6.2 Chemical Reactions	
• activation energy (p. 158) • active site (p. 160) • catalyst (p. 159) • chemical reaction (p. 156) • enzyme (p. 159) • product (p. 157) • reactant (p. 157) • substrate (p. 160)	MAIN Idea Chemical reactions allow living things to grow, develop, reproduce, and adapt. • Balanced chemical equations must show an equal number of atoms for each element on both sides. • Activation energy is the energy required to begin a reaction. • Catalysts are substances that alter chemical reactions. • Enzymes are biological catalysts.
Section 6.3 Water and Solutions	
• acid (p. 164) • base (p. 164) • buffer (p. 165) • hydrogen bond (p. 161) • mixture (p. 163) • pH (p. 165) • polar molecule (p. 161) • solute (p. 163) • solution (p. 163) • solvent (p. 163)	MAIN Idea The properties of water make it well suited to help maintain homeostasis in an organism. • Water is a polar molecule. • Solutions are homogeneous mixtures formed when a solute is dissolved in a solvent. • Acids are substances that release hydrogen ions into solutions. Bases are substances that release hydroxide ions into solutions. • pH is a measure of the concentration of hydrogen ions in a solution.
Section 6.4 The Building Blocks of Life	
• amino acid (p. 170) • carbohydrate (p. 168) • lipid (p. 169) • macromolecule (p. 167) • nucleic acid (p. 171) • nucleotide (p. 171) • polymer (p. 167) • protein (p. 170)	MAIN Idea Organisms are made up of carbon-based molecules. • Carbon compounds are the basic building blocks of living organisms. • Biological macromolecules are formed by joining small carbon compounds into polymers. • There are four types of biological macromolecules. • Peptide bonds join amino acids in proteins. • Chains of nucleotides form nucleic acids.

Biology Online Vocabulary PuzzleMaker biologygmh.com

CHAPTER 6

Assessment

Section 6.1

Vocabulary Review

Describe the difference between the terms in each set.

1. electron—proton
2. ionic bond—covalent bond
3. isotope—element
4. atom—ion

Understand Key Concepts

Use the photo below to answer question 5.

5. What does the image above show?
   A. a covalent bond
   B. a physical property
   C. a chemical reaction
   D. van der Waals forces

6. Which process changes a chlorine atom into a chloride ion?
   A. electron gain
   B. electron loss
   C. proton gain
   D. proton loss

7. Which of the following is a pure substance that cannot be broken down by a chemical reaction?
   A. a compound
   B. a mixture
   C. an element
   D. a neutron

8. How do the isotopes of hydrogen differ?
   A. the number of protons
   B. the number of electrons
   C. the number of energy levels
   D. the number of neutrons

Constructed Response

9. **Short Answer** What is a radioactive isotope? List uses of radioactive isotopes.

10. **Short Answer** What factor determines that an oxygen atom can form two covalent bonds while a carbon atom can form four?

11. **Open Ended** Why is it important for living organisms to have both strong bonds (covalent and ionic) and weak bonds (hydrogen and van der Waals forces)?

Think Critically

Use the graph below to answer question 12.

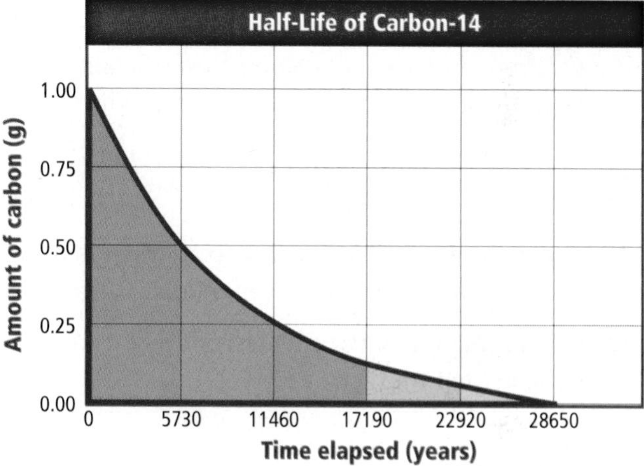

12. **Analyze** According to the data, what is the half-life of carbon-14? How can this information be used by scientists?

13. **Explain** The gecko is a reptile that climbs on smooth surfaces such as glass using van der Waals forces to adhere to the surface. How is this method of adhesion more advantageous than covalent interactions?

Section 6.2

Vocabulary Review

Match the term on the left with the correct definition on the right.

14. activation energy
15. substrate
16. enzyme
17. product

A. a protein that speeds up a reaction
B. a substance formed by a chemical reaction
C. the energy required to start a reaction
D. a substance that binds to an enzyme

Understand Key Concepts

18. Which of the following is a substance that lowers the activation energy?
A. an ion **C.** a catalyst
B. a reactant **D.** a substrate

19. In which of the following are bonds broken and new bonds are formed?
A. chemical reactions **C.** isotopes
B. elements **D.** polar molecules

20. Which statement is true of chemical equations?
A. Reactants are on the right.
B. Products are on the right.
C. Products have fewer atoms than reactants.
D. Reactants have fewer atoms than products.

Constructed Response

21. Short Answer What features do all reactions involving enzymes have in common?

22. Open Ended Identify and describe factors that can influence enzyme activity.

Think Critically

Use the graph to answer questions 23 and 24.

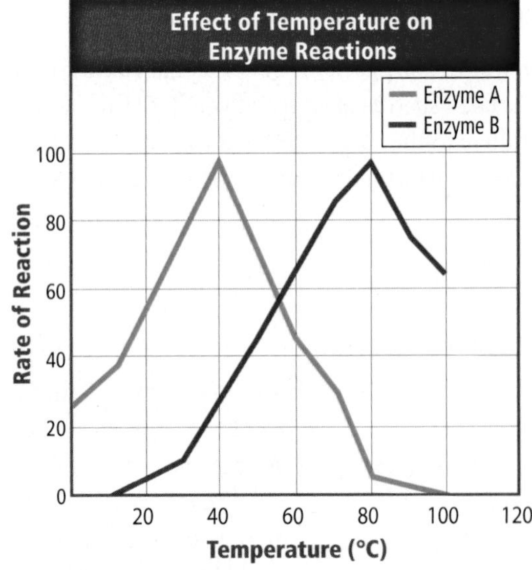

23. Describe the effect temperature has on the rate of the reactions using the graph above.

24. Infer Which enzyme is more active in a human cell? Why?

Section 6.3

Vocabulary Review

State the relationship between the terms in each set.

25. solution—mixture

26. pH—buffer

27. acid—base

28. solvent—solute

29. polar molecule—hydrogen bond

Understand Key Concepts

Use the figure below to answer question 30.

30. What does the image above show?
A. a heterogeneous mixture **C.** a solution
B. a homogeneous mixture **D.** a suspension

31. Which statement is not true about pure water?
A. It has a pH of 7.0.
B. It is composed of polar molecules.
C. It is composed of ionic bonds.
D. It is a good solvent.

32. Which is a substance that produces OH^- ions when dissolved in water?
A. a base **C.** a buffer
B. an acid **D.** salt

Constructed Response

33. Open Ended Why are hydrogen bonds so important for living organisms?

34. Short Answer Hydrochloric acid (HCl) is a strong acid. What ions are formed when HCl dissolves in water? What is the effect of HCl on the pH of water?

35. Open Ended Explain the importance of buffers to living organisms.

BiologyOnline Chapter Test biologygmh.com

Think Critically

36. **Predict** two places in the body where buffers are used to limit sharp changes in pH.

37. **Draw** a diagram of table salt (NaCl) dissolved in water.

Section 6.4

Vocabulary Review

Complete the following sentences with vocabulary terms from the Study Guide page.

38. Carbohydrates, lipids, proteins, and nucleic acids are __________.

39. Proteins are made from __________ that are joined by __________.

40. __________ make up fats, oils, and waxes.

41. DNA and RNA are examples of __________.

Understand Key Concepts

42. Which two elements are always found in amino acids?
    A. nitrogen and sulfur
    B. carbon and oxygen
    C. hydrogen and phosphorus
    D. sulfur and oxygen

43. Which joins amino acids together?
    A. peptide bonds
    B. hydrogen bonds
    C. van der Waals forces
    D. ionic bonds

44. Which substance is not part of a nucleotide?
    A. a phosphate
    B. a base
    C. a sugar
    D. water

Constructed Response

45. **Open Ended** Why do cells contain both macromolecules and small carbon compounds?

46. **Open Ended** Why can't humans digest all carbohydrates?

Think Critically

47. **Create** a table for the four main biological macromolecules that lists their components and functions.

Additional Assessment

48. WRITING in **Biology** Research and write a job description for a biochemist. Include the types of tasks biochemists perform and materials that are used in their research.

DBQ Document-Based Questions

Starch is the major carbon storehouse in plants. Experiments were performed to determine if trehalose might regulate starch production in plants. Leaf discs were incubated for three hours in sorbitol (the control), sucrose, and trelahose solutions. Then, levels of starch and sucrose in the leaves were measured. Use the data to answer the questions below.

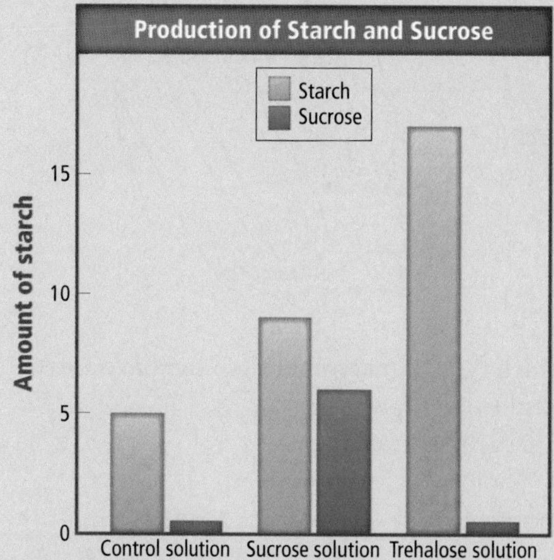

Data obtained from: Kolbe, et al. Trehalose 6-phosphate regulates starch synthesis via post translational redox activation of ADP-glucose pyrophophorylase. *Proceedings of the National Academy of Sciences of the USA* 102(31): 11118–11123.

49. Summarize the production of starch and sucrose in the three solutions.

50. What conclusion might the researchers have reached based on this data?

Cumulative Review

51. How do reproductive strategies differ? **(Chapter 4)**

52. Describe three broad categories of biodiversity value. **(Chapter 5)**

Standardized Test Practice

Cumulative

Multiple Choice

1. If a population of parrots has greater genetic diversity than a hummingbird population in the same region, which outcome could result?
   A. The parrot population could have a greater resistance to disease than the hummingbird population.
   B. Other parrot populations in different regions could become genetically similar to this one.
   C. The parrot population could have a greater variety of abiotic factors with which to interact.
   D. The parrot population could interact with a greater variety of other populations.

Use the diagram below to answer questions 2 and 3.

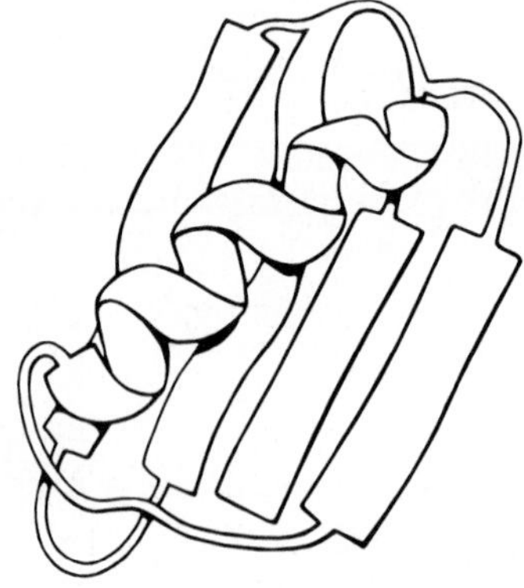

2. Which type of macromolecule can have a structure like the one shown?
   A. a carbohydrate
   B. a lipid
   C. a nucleotide
   D. a protein

3. Which molecular activity requires a folded structure?
   A. behavior as a nonpolar compound
   B. function of an active site
   C. movement through cell membranes
   D. role as energy store for the cell

4. Which describes the effects of population increase and resource depletion?
   A. increased competition
   B. increased emigration
   C. exponential population growth
   D. straight-line population growth

5. Which property of populations might be described as random, clumped, or uniform?
   A. density
   B. dispersion
   C. growth
   D. size

6. Which is an example of biodiversity with direct economic value?
   A. sparrow populations that have great genetic diversity
   B. species of a water plant that makes a useful antibiotic
   C. trees that create a barrier against hurricane winds
   D. villagers who all use the same rice species for crops

Use the illustration below to answer question 7.

7. Which term describes the part of the cycle labeled *A*?
   A. condensation
   B. evaporation
   C. run off
   D. precipitation

8. Which is a characteristic of exponential growth?
   A. the graphical representation goes up and down
   B. the graphical representation has a flat line
   C. a growth rate that increases with time
   D. a growth rate that stays constant in time

Biology Online **Standardized Test Practice** biologygmh.com

Short Answer

9. Assess what might happen if there were no buffers in human cells.

10. Choose an example of an element and a compound and then contrast them.

Use the chart below to answer question 11.

Factors Affecting Coral Survival	
Factor	**Optimal Range**
Water Temperature	23 to 25°C
Salinity	30 to 40 parts per million
Sedimentation	Little or no sedimentation
Depth	Up to 48 m

11. Using the data in the chart, describe which region of the world would be optimal for coral growth.

12. Provide a hypothesis to explain the increase in species diversity as you move from the polar regions to the tropics.

13. In a country with a very slow growth rate, predict which age groups are the largest in the population.

14. Why is it important that enzymes can bind only to specific substrates?

Extended Response

15. Suddenly, after very heavy rains, many fish in a local lake begin to die, yet algae in the water seem to be doing very well. You know that the lake receives runoff from local fields and roads. Form a hypothesis about why the fish are dying, and suggest how to stop the deaths.

16. When scientists first discovered atoms, they thought they were the smallest parts into which matter could be divided. Relate how later discoveries led scientists to revise this definition.

17. Identify and describe three types of symbiotic relationships and provide an example of each.

Essay Question

Many kinds of molecules found in living organisms are made of smaller monomers that are put together in different sequences, or in different patterns. For example, organisms use a small number of nucleotides to make nucleic acids. Thousands of different sequences of nucleotides in nucleic acids provide the basic coding for all the genetic information in living things.

Using the information in the paragraph above, answer the following question in essay format.

18. Describe how it is beneficial for organisms to use monomers to create complex macromolecules.

NEED EXTRA HELP?																		
If You Missed Question . . .	1	2	3	4	5	6	7	8	9	10	11	12	13	14	15	16	17	18
Review Section . . .	5.1	6.4	6.4	4.1	5.1	2.3	4.1	4.1	6.3	5.3	3.1, 3.3	5.1	4.2	6.2	5.2	6.1	2.1	6.4

CHAPTER 7

Cellular Structure and Function

BIG Idea Cells are the structural and functional units of all living organisms.

Section 1
Cell Discovery and Theory
MAIN Idea The invention of the microscope led to the discovery of cells.

Section 2
The Plasma Membrane
MAIN Idea The plasma membrane helps to maintain a cell's homeostasis.

Section 3
Structures and Organelles
MAIN Idea Eukaryotic cells contain organelles that allow the specialization and the separation of functions within the cell.

Section 4
Cellular Transport
MAIN Idea Cellular transport moves substances within the cell and moves substances into and out of the cell.

BioFacts

- About ten trillion cells make up the human body.
- The largest human cells are about the diameter of a human hair.
- The 200 different types of cells in the human body come from just one cell.

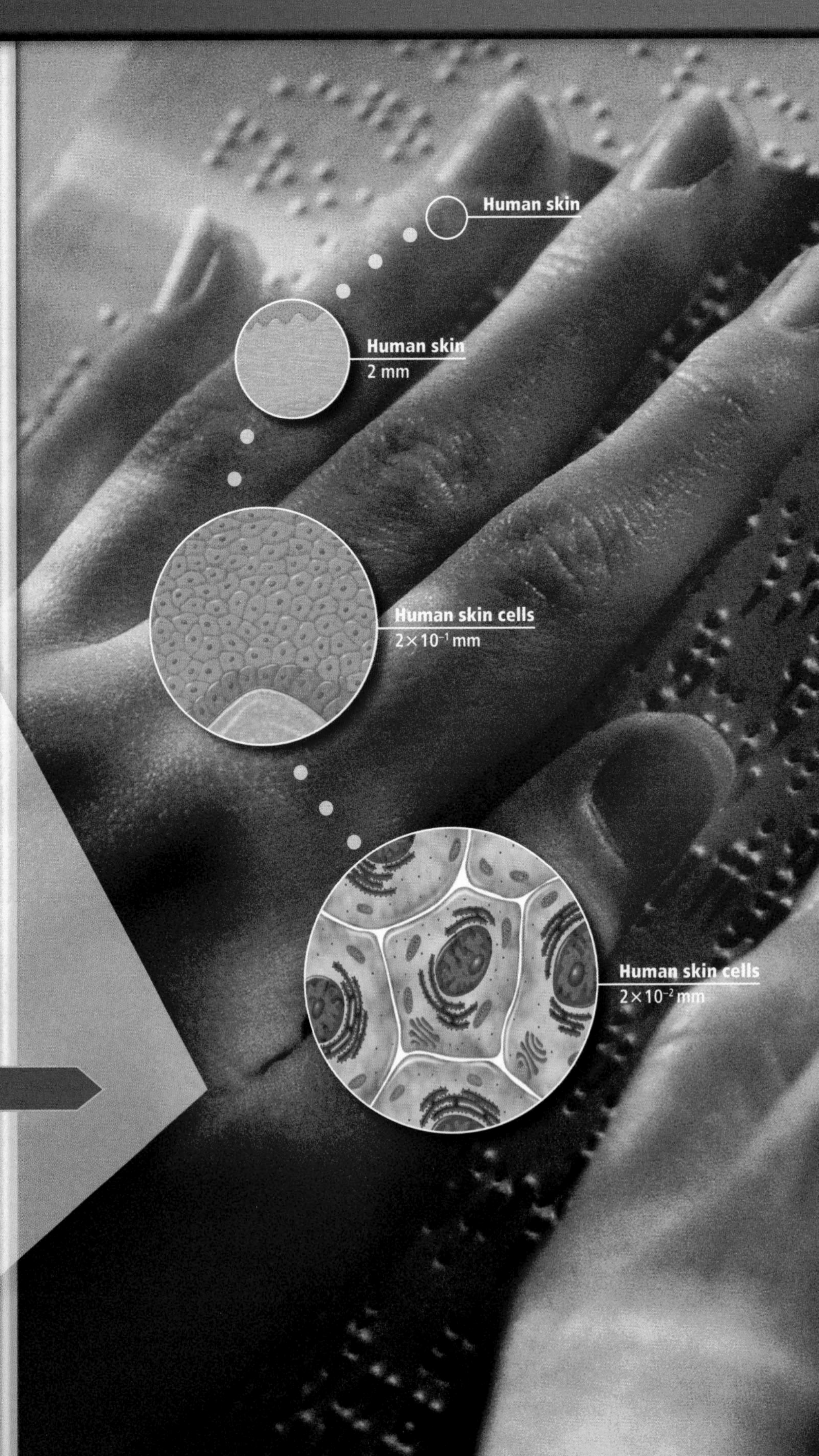

LAUNCH Lab

What is a cell?

All things are made of atoms and molecules, but only in living things are the atoms and molecules organized into cells. In this lab, you will use a compound microscope to view slides of living things and nonliving things.

Procedure

1. Read and complete the lab safety form.
2. Construct a data table for recording your observations.
3. Obtain **slides of the various specimens.**
4. View the slides through a **microscope** at the power designated by your teacher.
5. As you view the slides, fill out the data table you constructed.

Analysis

1. **Describe** some of the ways to distinguish between the living things and the nonliving things.
2. **Write** a definition of a cell based on your observations.

Visit biologygmh.com to:

- study the entire chapter online
- explore the Interactive Time Line, Concepts in Motion, the Interactive Table, Microscopy Links, Virtual Labs, and links to virtual dissections
- access Web links for more information, projects, and activities
- review content online with the Interactive Tutor, and take Self-Check Quizzes

Cellular Transport Make this Foldable to help you characterize the various methods of cellular transport.

STEP 1 Place two sheets of notebook paper 1.5 cm apart as illustrated.

STEP 2 Roll up the bottom edges making all tabs 1.5 cm in size. Crease to form four tabs of equal size.

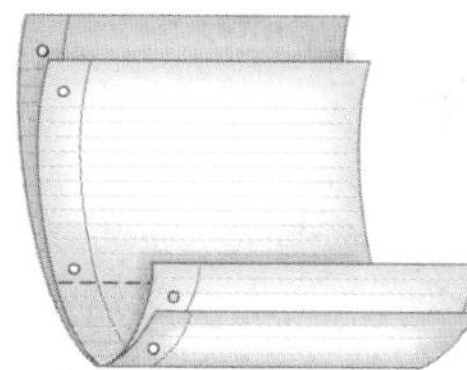

STEP 3 Staple along the folded edge to secure all sheets. Label the tabs as illustrated.

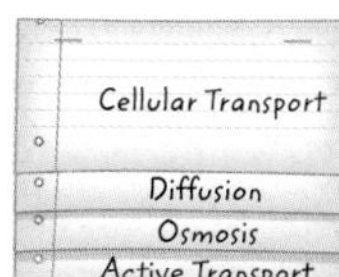

FOLDABLES **Use this Foldable with Section 7.4.** As you study the section, consider the role of energy in each of the cellular transport methods discussed.

Section 7.1

Reading Preview

Objectives

- **Relate** advances in microscope technology to discoveries about cells.
- **Compare** compound light microscopes with electron microscopes.
- **Summarize** the principles of the cell theory.
- **Differentiate** between a prokaryotic cell and a eukaryotic cell.

Review Vocabulary

organization: the orderly structure of cells in an organism

New Vocabulary

cell
cell theory
plasma membrane
organelle
eukaryotic cell
nucleus
prokaryotic cell

Cell Discovery and Theory

MAIN Idea The invention of the microscope led to the discovery of cells.

Real-World Reading Link The different parts of your body might seem to have nothing in common. Your heart, for example, pumps blood throughout your body, while your skin protects and helps cool you. However, all your body parts have one thing in common—they are composed of cells.

History of the Cell Theory

For centuries, scientists had no idea that the human body consists of trillions of cells. Cells are so small that their existence was unknown before the invention of the microscope. In 1665, as indicated in **Figure 7.1,** an English scientist named Robert Hooke made a simple microscope and looked at a piece of cork, the dead cells of oak bark. Hooke observed small, box-shaped structures, such as those shown in **Figure 7.2.** He called them cellulae (the Latin word meaning *small rooms*) because the boxlike cells of cork reminded him of the cells in which monks live at a monastery. It is from Hooke's work that we have the term *cell.* A **cell** is the basic structural and functional unit of all living organisms.

During the late 1600s, Dutch scientist Anton van Leeuwenhoek (LAY vun hook)—inspired by a book written by Hooke—designed his own microscope. To his surprise, he saw living organisms in pond water, milk, and various other substances. The work of these scientists and others led to new branches of science and many new and exciting discoveries.

■ **Figure 7.1**
Microscopes in Focus
The invention of microscopes, improvements to the instruments, and new microscope techniques have led to the development of the cell theory and a better understanding of cells.

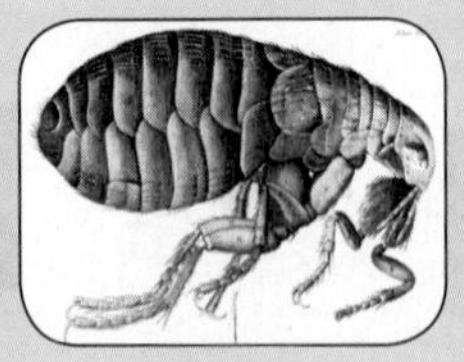

1590 Dutch lens grinders Hans and Zacharias Janssen invent the first compound microscope by placing two lenses in a tube.

1665 Robert Hooke observes cork and names the tiny chambers that he sees cells. He publishes drawings of cells, fleas, and other minute bodies in his book *Micrographia.*

1683 Dutch biologist Anton van Leeuwenhoek discovers single-celled, animal-like organisms, now called protozoans.

1830–1855 Scientists discover the cell nucleus (1833) and propose that both plants and animals are composed of cells (1839).

1500 1600 1700 1800

The cell theory Naturalists and scientists continued observing the living microscopic world using glass lenses. In 1838, German scientist Matthias Schleiden carefully studied plant tissues and concluded that all plants are composed of cells. A year later, another German scientist, Theodor Schwann, reported that animal tissues also consisted of individual cells. Prussian physician Rudolph Virchow proposed in 1855 that all cells are produced from the division of existing cells. The observations and conclusions of these scientists and others are summarized as the cell theory. The **cell theory** is one of the fundamental ideas of modern biology and includes the following three principles:

1. All living organisms are composed of one or more cells.
2. Cells are the basic unit of structure and organization of all living organisms.
3. Cells arise only from previously existing cells, with cells passing copies of their genetic material on to their daughter cells.

Reading Check Can cells appear spontaneously without genetic material from previous cells?

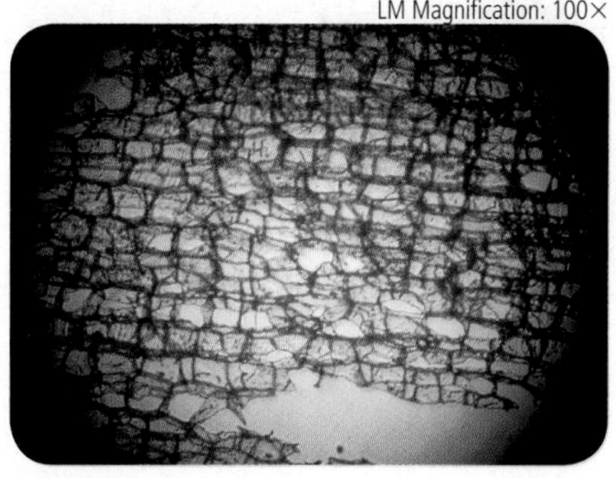

Figure 7.2 Robert Hooke used a basic light microscope to see what looked like empty chambers in a cork sample.
Infer *What do you think Hooke would have seen if these were living cells?*

Microscope Technology

The discovery of cells and the development of the cell theory would not have been possible without microscopes. Improvements made to microscopes have enabled scientists to study cells in detail, as described in **Figure 7.1.**

Turn back to the opening pages of this chapter and compare the illustrations of the skin shown there. Note that the detail increases as the magnification and resolution—the ability of the microscope to make individual components visible—increase. Hooke and van Leewenhoek would not have been able to see the individual structures within human skin cells with their microscopes. Developments in microscope technology have given scientists the ability to study cells in greater detail than early scientists ever thought possible.

LAUNCH Lab
Review Based on what you've read about cells, how would you now answer the analysis questions?

1939 Ernest Everett Just writes the textbook *Biology of the Cell Surface* after years of studying the structure and function of cells.

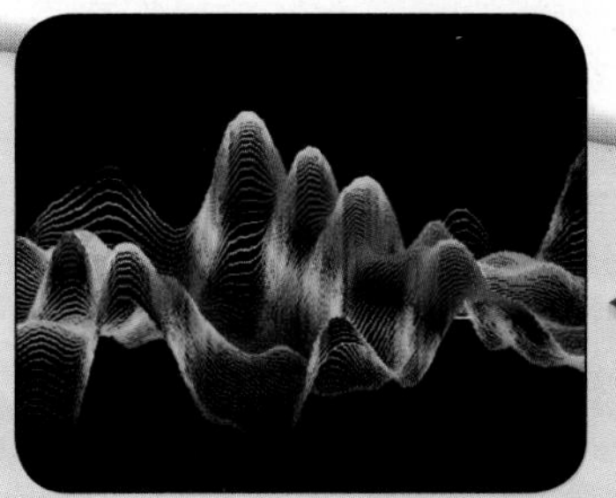

1981 The scanning tunneling microscope (STM) allows scientists to see individual atoms.

1900 — 2000

1880–1890 Louis Pasteur and Robert Koch, using compound microscopes, pioneered the study of bacteria.

1970 Lynn Margulis, a microbiologist, proposes the idea that some organelles found in eukaryotes were once free-living prokaryotes.

Concepts In Motion Interactive Time Line
To learn more about these discoveries and others, visit biologygmh.com. **Biology Online**

Compound light microscopes The modern compound light microscope consists of a series of glass lenses and uses visible light to produce a magnified image. Each lens in the series magnifies the image of the previous lens. For example, when two lenses each individually magnify 10 times, the total magnification would be 100 times (10 × 10). Scientists often stain cells with dyes to see them better when using a light microscope because cells are so tiny, thin, and translucent. Over the years, scientists have developed various techniques and modifications for light microscopes, but the properties of visible light will always limit resolution with these microscopes. Objects cause light to scatter, which blurs images. The maximum magnification without blurring is around 1000×.

Electron microscopes As they began to study cells, scientists needed greater magnification to see the details of tiny parts of the cell. During the second World War, in the 1940s, they developed the electron microscope. Instead of lenses, the electron microscope uses magnets to aim a beam of electrons at thin slices of cells. This type of electron microscope is called a transmission electron microscope (TEM) because electrons are passed, or transmitted, through a specimen to a fluorescent screen. Thick parts of the specimen absorb more electrons than thin parts, forming a black-and-white shaded image of the specimen. Transmission electron microscopes can magnify up to 500,000×, but the specimen must be dead, sliced very thin, and stained with heavy metals.

Over the past 65 years, many modifications have been made to the original electron microscopes. For example, the scanning electron microscope (SEM) is one modification that directs electrons over the surface of the specimen, producing a three-dimensional image. One disadvantage of using a TEM and an SEM is that only nonliving cells and tissues can be observed. To see photomicrographs made with electron microscopes, visit biologygmh.com and click on *Microscopy Links.*

CAREERS IN BIOLOGY

Technology Representative Companies that manufacture scientific equipment employ representatives to demonstrate and explain their products to the scientific community. A technology representative is an expert in these new technology products and brings this expertise to scientists who might use the products in the laboratory. For more information on biology careers, visit biologygmh.com.

MiniLab 7.1

Discover Cells

How can you describe a new discovery? Imagine you are a scientist looking through the eyepiece of some new-fangled instrument called a microscope and you see a field of similarly shaped objects. You might recognize that the shapes you see are not merely coincidence and random objects. Your whole idea of the nature of matter is changing as you view these objects.

Procedure

1. Read and complete the lab safety form.
2. Prepare a data table in which you will record observations and drawings for three slides.
3. View the **slide images** your teacher projects for the class.
4. Describe and draw what you see. Be sure to include enough detail in your drawings to convey the information to other scientists who have not observed cells.

Analysis

1. **Describe** What analogies or terms could explain the images in your drawings?
2. **Explain** How could you show Hooke, with twenty-first-century technology, that his findings were valid?

Another type of microscope, the scanning tunneling electron microscope (STM), involves bringing the charged tip of a probe extremely close to the specimen so that the electrons "tunnel" through the small gap between the specimen and the tip. This instrument has enabled scientists to create three-dimensional computer images of objects as small as atoms. Unlike TEM and SEM, STM can be used with live specimens. **Figure 7.3** shows DNA, the cell's genetic material, magnified with a scanning tunneling electron microscope.

The atomic force microscope (AFM) measures various forces between the tip of a probe and the cell surface. To learn more about AFM, read the *Cutting Edge Biology* feature at the end of this chapter.

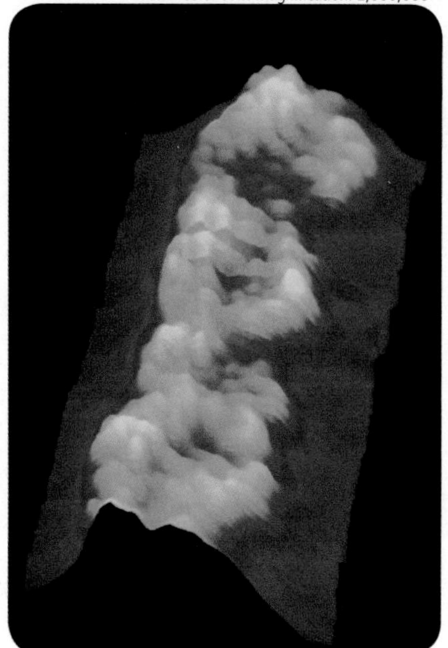

DNA

Figure 7.3 The scanning tunneling microscope (STM) provides images, such as this DNA molecule, in which cracks and depressions appear darker and raised areas appear lighter. **Name** *an application for which an STM might be used.*

Basic Cell Types

You have learned, according to the cell theory, that cells are the basic units of all living organisms. By observing your own body and the living things around you, you might infer that cells must exist in various shapes and sizes. You also might infer that cells differ based on the function they perform for the organism. If so, you are correct! However, all cells have at least one physical trait in common: they all have a structure called a plasma membrane. A **plasma membrane**, labeled in **Figure 7.4,** is a special boundary that helps control what enters and leaves the cell. Each of your skin cells has a plasma membrane, as do the cells of a rattlesnake. This critical structure is described in detail in the next section.

Cells generally have a number of functions in common. For example, most cells have genetic material in some form that provides instructions for making substances that the cell needs. Cells also break down molecules to generate energy for metabolism. Scientists have grouped cells into two broad categories. These categories are prokaryotic (pro kar ee AW tik) cells and eukaryotic (yew kar ee AW tik) cells. **Figure 7.4** shows TEM photomicrographs of these two cell types. The images of the prokayotic cell and eukaryotic cell have been enlarged so you can compare the cell structures. Eukaryotic cells generally are one to one hundred times larger than prokaryotic cells.

Reading Check **Compare** the sizes of prokaryotic cells and eukaryotic cells.

Figure 7.4 The prokaryotic cell on the left is smaller and appears less complex than the eukaryotic cell on the right. The prokaryotic cell has been enlarged for the purpose of comparing each cell's internal sturctures.

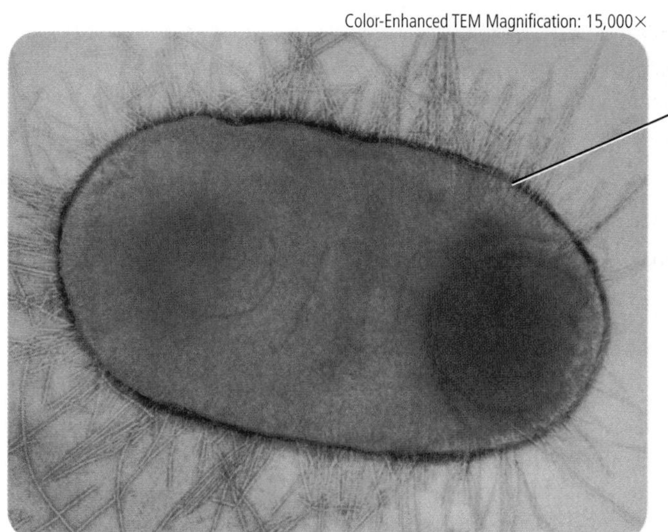

Prokaryotic cell

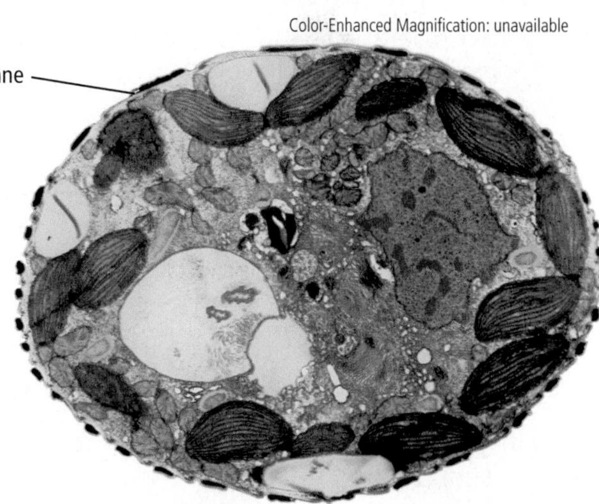

Eukaryotic cell

Vocabulary

Word origin

Eukaryote
Prokaryote
eu– prefix; from Greek, meaning *true*
pro– prefix; from Greek, meaning *before*
–kary from Greek, meaning *nucleus*

Refer again to **Figure 7.4** and compare the types of cells to see why scientists place them into two broad categories that are based on internal structures. Both have a plasma membrane, but one cell contains many distinct internal structures called **organelles**—specialized structures that carry out specific cell functions.

Eukaryotic cells contain a nucleus and other organelles that are bound by membranes, also referred to as membrane-bound organelles. The **nucleus** is a distinct central organelle that contains the cell's genetic material in the form of DNA. Organelles enable cell functions to take place in different parts of the cell at the same time. Most organisms are made up of eukaryotic cells and are called eukaryotes. However, some unicellular organisms, such as some algae and yeast, also are eukaryotes.

Prokaryotic cells are defined as cells without a nucleus or other membrane-bound organelles. Most unicellular organisms, such as bacteria, are prokaryotic cells. Thus they are called prokaryotes. Many scientists think that prokaryotes are similar to the first organisms on Earth.

Origin of cell diversity If you have ever wondered why a company makes two products that are similar, you can imagine that scientists have asked why there are two basic types of cells. The answer might be that eukaryotic cells evolved from prokaryotic cells millions of years ago. According to the endosymbiont theory, a symbiotic mutual relationship involved one prokaryotic cell living inside of another. The endosymbiont theory is discussed in greater detail in Chapter 14.

Imagine how organisms would be different if the eukaryotic form had not evolved. Because eukaryotic cells are larger and have distinct organelles, these cells have developed specific functions. Having specific functions has led to cell diversity, and thus more diverse organisms that can adapt better to their environments. Life-forms more complex than bacteria might not have evolved without eukaryotic cells.

Section 7.1 Assessment

Section Summary

- Microscopes have been used as a tool for scientific study since the late 1500s.
- Scientists use different types of microscopes to study cells.
- The cell theory summarizes three principles.
- There are two broad groups of cell types—prokaryotic cells and eukaryotic cells.
- Eukaryotic cells contain a nucleus and organelles.

Understand Main Ideas

1. MAIN Idea **Explain** how the development and improvement of microscopes changed the study of living organisms.
2. **Compare and contrast** a compound light microscope and an electron microscope.
3. **Summarize** the cell theory.
4. **Differentiate** the plasma membrane and the organelles.

Think Critically

5. **Describe** how you would determine if the cells of a newly discovered organism were prokaryotic or eukaryotic.

MATH in Biology

6. If the overall magnification of a series of two lenses is 30×, and one lens magnified 5×, what is the magnification of the other lens? Calculate the total magnification if the 5× lens is replaced by a 7× lens.

Biology Online **Self-Check Quiz** biologygmh.com

Section 7.2

The Plasma Membrane

Reading Preview

Objectives

- **Describe** how a cell's plasma membrane functions.
- **Identify** the roles of proteins, carbohydrates, and cholesterol in the plasma membrane.

Review Vocabulary

ion: an atom or group of atoms with a positive or negative electric charge

New Vocabulary

selective permeability
phospholipid bilayer
transport protein
fluid mosaic model

MAIN Idea **The plasma membrane helps to maintain a cell's homeostasis.**

Real-World Reading Link When you approach your school, you might pass through a gate in a fence that surrounds the school grounds. This fence prevents people who should not be there from entering and the gate allows students, staff, and parents to enter. Prokaryotic cells and eukaryotic cells have a structure that maintains control of their internal environments.

Function of the Plasma Membrane

Recall from Chapter 1 that the process of maintaining balance in an organism's internal environment is called homeostasis. Homeostasis is essential to the survival of a cell. One of the structures that is primarily responsible for homeostasis is the plasma membrane. The plasma membrane is a thin, flexible boundary between a cell and its environment that allows nutrients into the cell and allows waste and other products to leave the cell. All prokaryotic cells and eukaryotic cells have a plasma membrane to separate them from the watery environments in which they exist.

A key property of the plasma membrane is **selective permeability** (pur mee uh BIH luh tee), by which a membrane allows some substances to pass through while keeping others out. Consider a fish net as an analogy of selective permeability. The net shown in **Figure 7.5** has holes that allow water and other substances in the water to pass through but not the fish. Depending on the size of the holes in the net, some kinds of fish might pass through, while others are caught. The diagram in **Figure 7.5** illustrates selective permeability of the plasma membrane. The arrows show that substances enter and leave the cell through the plasma membrane. Control of how, when, and how much of these substances enter and leave a cell relies on the structure of the plasma membrane.

 **Reading Check** **Define** the term selective permeability.

Figure 7.5
Left: The fish net selectively captures fish while allowing water and other debris to pass through.
Right: Similarly, the plasma membrane selects substances entering and leaving the cell.

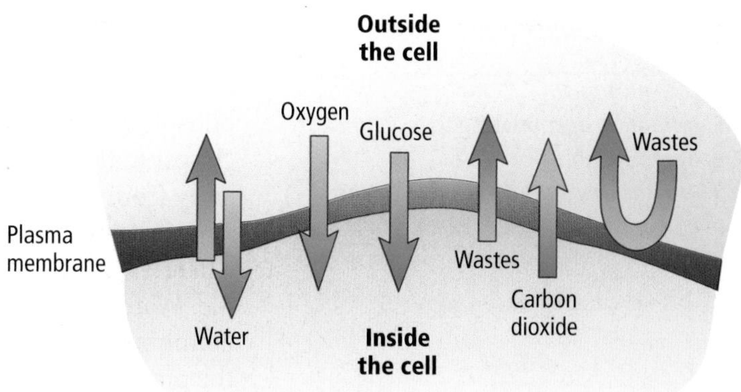

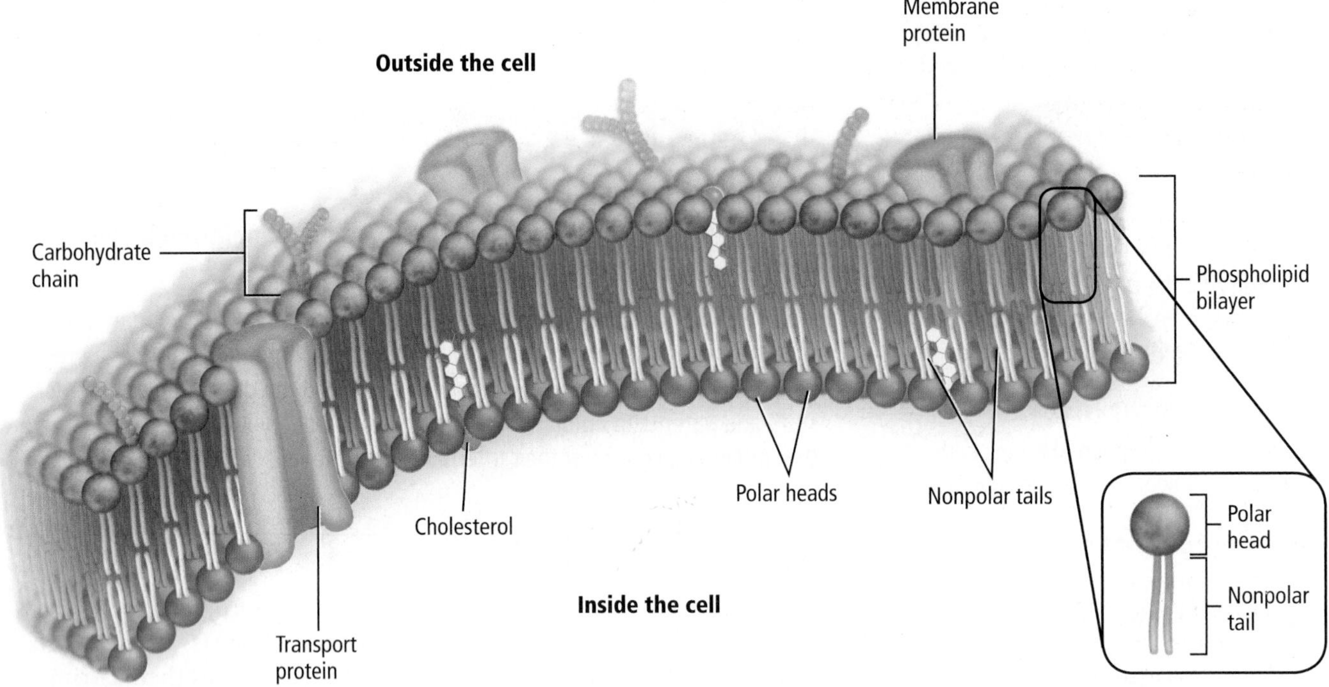

Figure 7.6 The phospholipid bilayer looks like a sandwich, with the polar heads facing the outside and the nonpolar tails facing the inside.
Infer *How do hydrophobic substances cross a plasma membrane?*

VOCABULARY

SCIENCE USAGE V. COMMON USAGE

Polar

Science usage: having an unequal distribution of charge.
The positive end of a polar molecule attracts the negative end of a polar molecule.

Common usage: relating to a geographic pole or region.
The polar ice cap in Greenland is, on average, 1.6 km thick.

Structure of the Plasma Membrane

Connection to Chemistry Most of the molecules in the plasma membrane are lipids. Recall from Chapter 6 that lipids are large molecules that are composed of glycerol and three fatty acids. If a phosphate group replaces a fatty acid, a phospholipid forms. A phospholipid (fahs foh LIH pid) is a molecule that has a glycerol backbone, two fatty acid chains, and a phosphate-containing group. The plasma membrane is composed of a **phospholipid bilayer,** in which two layers of phospholipids are arranged tail-to-tail, as shown in **Figure 7.6.** In the plasma membrane, phospholipids arrange themselves in a way that allows the plasma membrane to exist in the watery environment.

The phospholipid bilayer Notice in **Figure 7.6** that each phospholipid is diagrammed as a head with two tails. The phosphate group in each phospholipid makes the head polar. The polar head is attracted to water because water also is polar. The two fatty acid tails are nonpolar and are repelled by water.

The two layers of phospholipid molecules make a sandwich, with the fatty acid tails forming the interior of the plasma membrane and the phospholipid heads facing the watery environments found inside and outside the cell, as shown in **Figure 7.6.** This bilayer structure is critical for the formation and function of the plasma membrane. The phospholipids are arranged in such a way that the polar heads can be closest to the water molecules and the nonpolar tails can be farthest away from the water molecules.

When many phospholipid molecules come together in this manner, a barrier is created that is polar at its surfaces and nonpolar in the middle. Water-soluble substances will not move easily through the plasma membrane because they are stopped by the nonpolar middle. Therefore, the plasma membrane can separate the environment inside the cell from the environment outside the cell.

Other components of the plasma membrane Moving with and among the phospholipids in the plasma membrane are cholesterol, proteins, and carbohydrates. When found on the outer surface of the plasma membrane, proteins called receptors transmit signals to the inside of the cell. Proteins at the inner surface anchor the plasma membrane to the cell's internal support structure, giving the cell its shape. Other proteins span the entire membrane and create tunnels through which certain substances enter and leave the cell. These **transport proteins** move needed substances or waste materials through the plasma membrane, and therefore contribute to the selective permeability of the plasma membrane.

Reading Check **Describe** the benefit of a bilayer structure for the plasma membrane

Locate the cholesterol molecules in **Figure 7.6.** Nonpolar cholesterol is repelled by water and is positioned among the phospholipids. Cholesterol helps to prevent the fatty-acid tails of the phospholipid bilayer from sticking together, which contributes to the fluidity of the plasma membrane. Although avoiding a high-cholesterol diet is recommended, cholesterol plays a critical role in plasma membrane structure and it is an important substance for maintaining homeostasis in a cell.

Other substances in the membrane, such as carbohydrates attached to proteins, stick out from the plasma membrane to define the cell's characteristics and help cells identify chemical signals. For example, carbohydrates in the membrane might help disease-fighting cells recognize and attack a potentially harmful cell.

Study Tip

Question Session Work with a partner and ask each other questions about the plasma membrane. Discuss each other's answers. Ask as many questions as you think of while taking turns.

DATA ANALYSIS LAB 7.1

Based on Real Data*

Interpret the Diagram

How are protein channels involved in the death of nerve cells after a stroke? A stroke occurs when a blood clot blocks the flow of oxygen-containing blood in a portion of the brain. Nerve cells in the brain that release glutamate are sensitive to the lack of oxygen and release a flood of glutamate when oxygen is low. During the glutamate flood, the calcium pump is destroyed. This affects the movement of calcium ions into and out of nerve cells. When cells contain excess calcium, homeostasis is disrupted.

Data and Observations

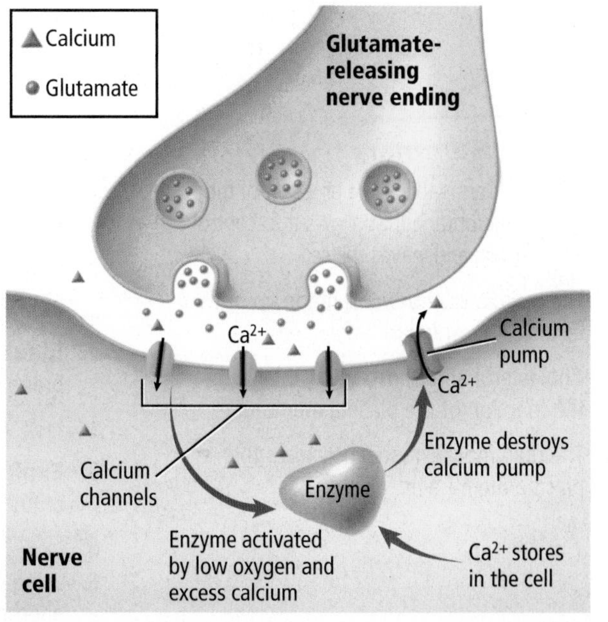

Think Critically

1. **Interpret** How does the glutamate flood destroy the calcium pump?
2. **Predict** what would happen if Ca^{2+} levels were lowered in the nerve cell during a stroke.

*Data obtained from: Choi, D.W. 2005. Neurodegeneration: cellular defences destroyed. *Nature* 433: 696–698.

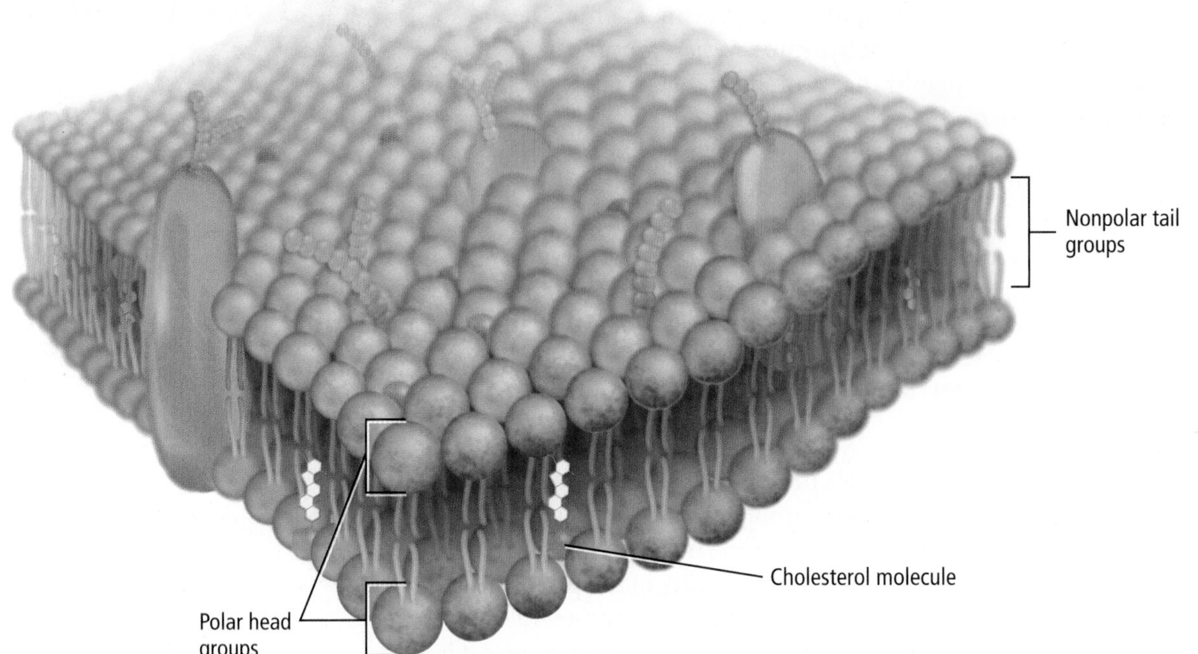

■ **Figure 7.7** The fluid mosaic model refers to a plasma membrane with substances that can move around within the membrane.

Concepts In Motion

Interactive Figure To see an animation of the fluid mosaic model, visit biologygmh.com.

Together, the phospholipids in the bilayer create a "sea" in which other molecules can float, like apples floating in a barrel of water. This "sea" concept is the basis for the **fluid mosaic model** of the plasma membrane. The phospholipids can move sideways within the membrane just as apples move around in water. At the same time, other components in the membrane, such as proteins, also move among the phospholipids. Because there are different substances in the plasma membrane, a pattern, or mosaic, is created on the surface. You can see this pattern in **Figure 7.7.** The components of the plasma membrane are in constant motion, sliding past one another.

Section 7.2 Assessment

Section Summary

- Selective permeability is a property of the plasma membrane that allows it to control what enters and leaves the cell.
- The plasma membrane is made up of two layers of phospholipid molecules.
- Cholesterol and transport proteins aid in the function of the plasma membrane.
- The fluid mosaic model describes the plasma membrane.

Understand Main Ideas

1. MAIN Idea **Describe** how the plasma membrane helps maintain homeostasis in a cell.
2. **Explain** how the inside of a cell remains separate from its environment.
3. **Diagram** the plasma membrane; label each component.
4. **Identify** the molecules in the plasma membrane that provide basic membrane structure, cell identity, and membrane fluidity.

Think Critically

5. **Explain** what effect more cholesterol in the plasma membrane will have on the membrane.

WRITING in **Biology**

6. Using what you know about the term *mosaic,* write a paragraph describing another biological mosaic.

Biology Online **Self-Check Quiz** biologygmh.com

Section 7.3

Reading Preview

Objectives

- **Identify** the structure and function of the parts of a typical eukaryotic cell.
- **Compare and contrast** structures of plant and animal cells.

Review Vocabulary

enzyme: a protein that speeds up the rate of a chemical reaction

New Vocabulary

cytoplasm
cytoskeleton
ribosome
nucleolus
endoplasmic reticulum
Golgi apparatus
vacuole
lysosome
centriole
mitochondrion
chloroplast
cell wall
cilium
flagellum

Structures and Organelles

MAIN Idea **Eukaryotic cells contain organelles that allow the specialization and the separation of functions within the cell.**

Real-World Reading Link Suppose you start a company to manufacture hiking boots. Each pair of boots could be made individually by one person, but it would be more efficient to use an assembly line. Similarly, eukaryotic cells have specialized structures that perform specific tasks, much like a factory.

Cytoplasm and Cytoskeleton

You just have investigated the part of a cell that functions as the boundary between the inside and outside environments. The environment inside the plasma membrane is a semifluid material called **cytoplasm.** In a prokaryotic cell, all of the chemical processes of the cell, such as breaking down sugar to generate the energy used for other functions, take place directly in the cytoplasm. Eukaryotic cells perform these processes within organelles in their cytoplasm. At one time, scientists thought that cell organelles floated in a sea of cytoplasm.

More recently, cell biologists have discovered that organelles do not float freely in a cell, but are supported by a structure within the cytoplasm similar to the structure shown in **Figure 7.8.** The **cytoskeleton** is a supporting network of long, thin protein fibers that form a framework for the cell and provide an anchor for the organelles inside the cells. The cytoskeleton also has a function in cell movement and other cellular activities.

The cytoskeleton is made of substructures called microtubules and microfilaments. Microtubules are long, hollow protein cylinders that form a rigid skeleton for the cell and assist in moving substances within the cell. Microfilaments are thin protein threads that help give the cell shape and enable the entire cell or parts of the cell to move. Microtubules and microfilaments rapidly assemble and disassemble and slide past one another. This allows cells and organelles to move.

■ **Figure 7.8** Microtubules and microfilaments make up the cytoskeleton.

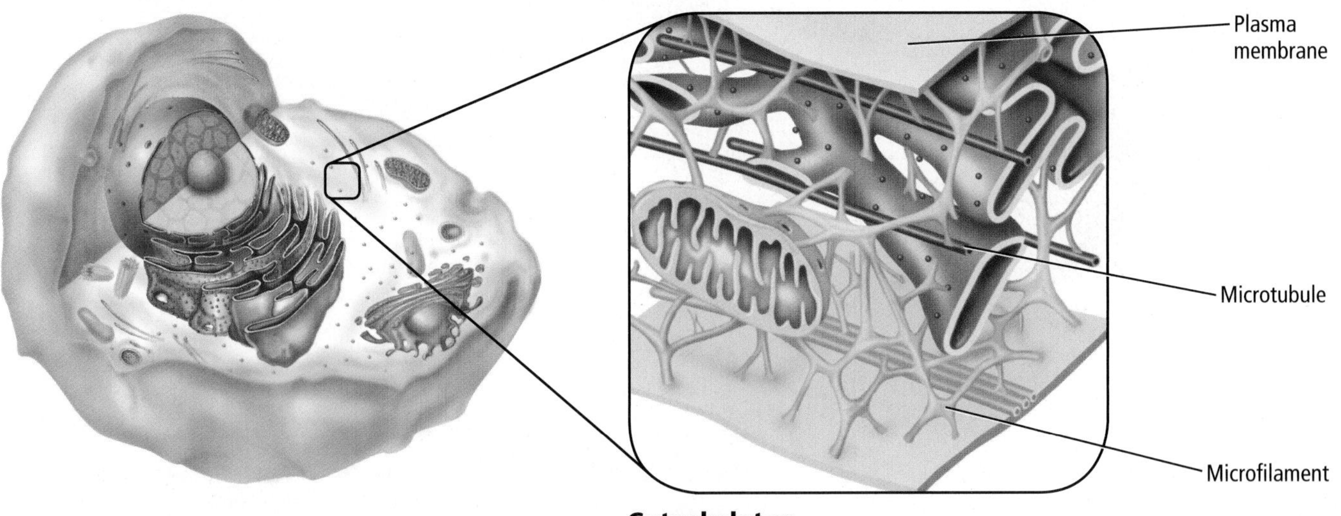

Visualizing Cells

Figure 7.9
Compare the illustrations of a plant cell, animal cell, and prokaryotic cell. Some organelles are only found in plant cells—others, only in animal cells. Prokaryotic cells do not have membrane-bound organelles.

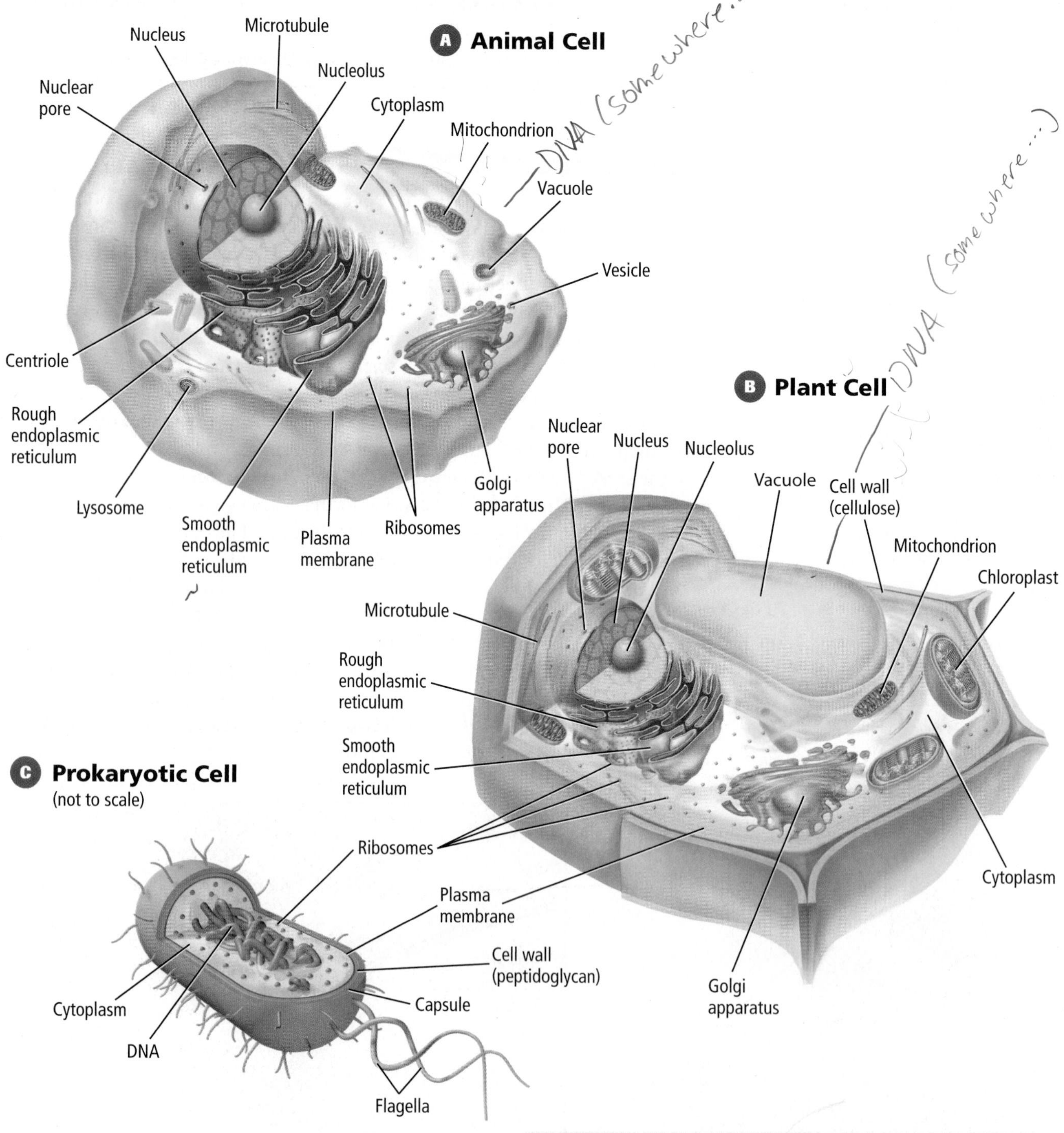

A Animal Cell

Nucleus
Microtubule
Nucleolus
Nuclear pore
Cytoplasm
Mitochondrion
Vacuole
Vesicle
Centriole
Rough endoplasmic reticulum
Lysosome
Smooth endoplasmic reticulum
Plasma membrane
Ribosomes
Golgi apparatus

B Plant Cell

Nuclear pore
Nucleus
Nucleolus
Vacuole
Cell wall (cellulose)
Mitochondrion
Chloroplast
Microtubule
Rough endoplasmic reticulum
Smooth endoplasmic reticulum
Ribosomes
Plasma membrane
Cytoplasm
Golgi apparatus

C Prokaryotic Cell
(not to scale)

Ribosomes
Plasma membrane
Cell wall (peptidoglycan)
Capsule
Cytoplasm
DNA
Flagella

Concepts in Motion **Interactive Figure** To see an animation of plant and animal cells, visit biologygmh.com.

Biology Online

Cell Structures

In a factory, there are separate areas set up for performing different tasks. Eukaryotic cells also have separate areas for tasks. Membrane-bound organelles make it possible for different chemical processes to take place at the same time in different parts of the cytoplasm. Organelles carry out essential cell processes, such as protein synthesis, energy transformation, digestion of food, excretion of wastes, and cell division. Each organelle has a unique structure and function. You can compare organelles to a factory's offices, assembly lines, and other important areas that keep the factory running. As you read about the different organelles, refer to the diagrams of plant and animal cells in **Figure 7.9** to see the organelles of each type.

The nucleus Just as a factory needs a manager, a cell needs an organelle to direct the cell processes. The nucleus, shown in **Figure 7.10,** is the cell's managing structure. It contains most of the cell's DNA, which stores information used to make proteins for cell growth, function, and reproduction.

The nucleus is surrounded by a double membrane called the nuclear envelope. The nuclear envelope is similar to the plasma membrane, except the nuclear membrane has nuclear pores that allow larger-sized substances to move in and out of the nucleus. Chromatin, which is a complex DNA attached to protein, is spread throughout the nucleus.

Reading Check **Describe** the role of the nucleus.

Ribosomes One of the functions of a cell is to produce proteins. The organelles that help manufacture proteins are called **ribosomes.** Ribosomes are made of two components—RNA and protein—and are not bound by a membrane like other organelles. Within the nucleus is the site of ribosome production called the **nucleolus,** shown in **Figure 7.10.**

Cells have many ribosomes that produce a variety of proteins that are used by the cell or are moved out and used by other cells. Some ribosomes float freely in the cytoplasm, while others are bound to another organelle called the endoplasmic reticulum. Free-floating ribosomes produce proteins for use within the cytoplasm of the cell. Bound ribosomes produce proteins that will be bound within membranes or used by other cells.

VOCABULARY

WORD ORIGIN

Cytoplasm
Cytoskeleton
cyte– prefix; from Greek, meaning *cell.*

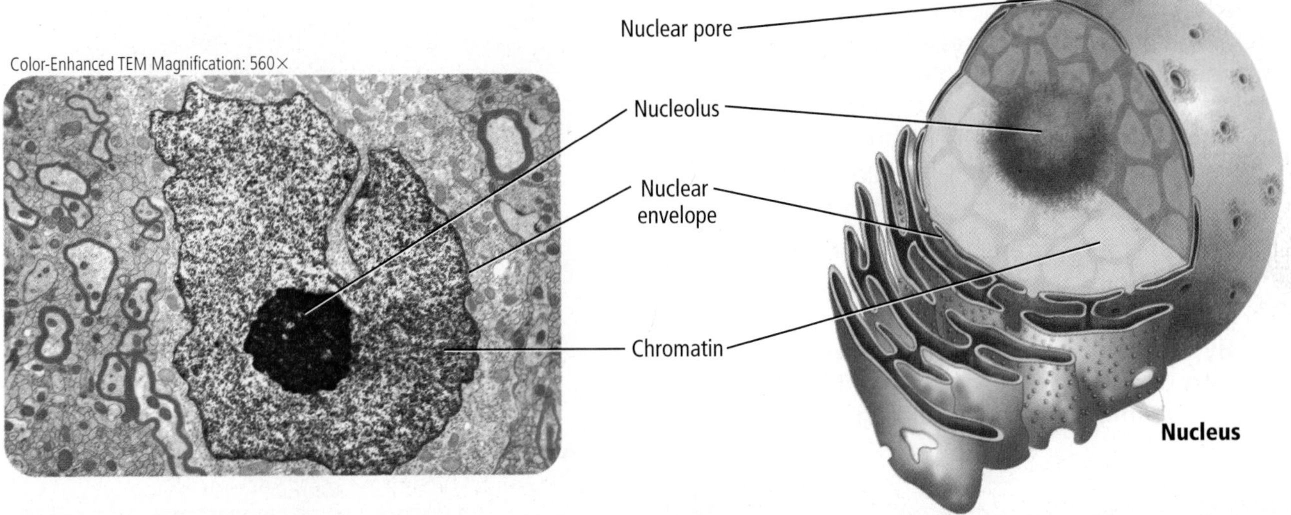

Figure 7.10 The nucleus of a cell is a three-dimensional shape. The photomicrograph shows a cross-section of a nucleus.
Infer ***Explain why all the cross-sections of a nucleus are not identical.***

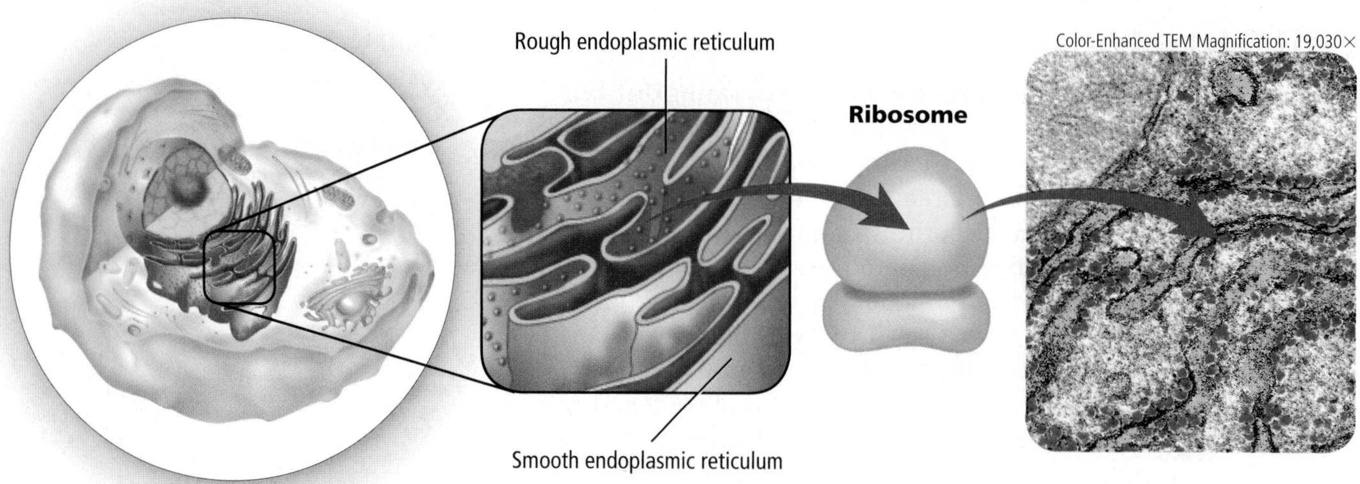

■ **Figure 7.11** Ribosomes are simple structures made of RNA and protein that may be attached to the surface of the rough endoplasmic reticulum. They look like bumps on the endoplasmic reticulum.

Endoplasmic reticulum The **endoplasmic reticulum** (en duh PLAZ mihk • rih TIHK yuh lum), also called ER, is a membrane system of folded sacs and interconnected channels that serves as the site for protein and lipid synthesis. The pleats and folds of the ER provide a large amount of surface area where cellular functions can take place. The area of ER where ribosomes are attached is called rough endoplasmic reticulum. Notice in **Figure 7.11** that the rough ER appears to have bumps on it. These bumps are the attached ribosomes that will produce proteins for export to other cells.

Figure 7.11 also shows that there are areas of the ER that do not have ribosomes attached. The area of ER where no ribosomes are attached is called smooth endoplasmic reticulum. Although the smooth ER has no ribosomes, it does perform important functions for the cell. For example, the smooth ER provides a membrane surface where a variety of complex carbohydrates and lipids, including phospholipids, are synthesized. Smooth ER in the liver detoxifies harmful substances.

DATA ANALYSIS LAB 7.2

Based on Real Lab Data*

Interpret the Data

How is vesicle traffic from the ER to the Golgi apparatus regulated? Some proteins are synthesized by ribosomes on the endoplasmic reticulum (ER). The proteins are processed in the ER, and vesicles containing these proteins pinch off and migrate to the Golgi apparatus. Scientists currently are studying the molecules that are involved in fusing these vesicles to the Golgi apparatus.

Think Critically

1. **Interpret a Diagram** Name two complexes on the Golgi apparatus that might be involved in vesicle fusion.
2. **Hypothesize** an explanation for vesicle transport based on what you have read about cytoplasm and the cytoskeleton.

Data and Observations

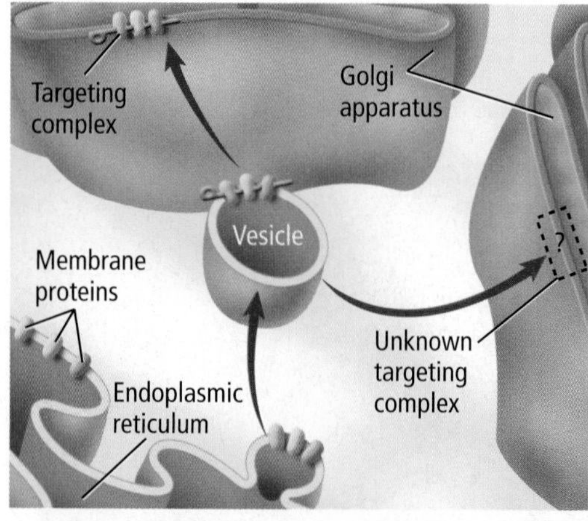

*Data obtained from: Brittle, E. E., and Waters, M. G. 2000. ER-to-golgi traffic—this bud's for you. *Science* 289: 403–404.

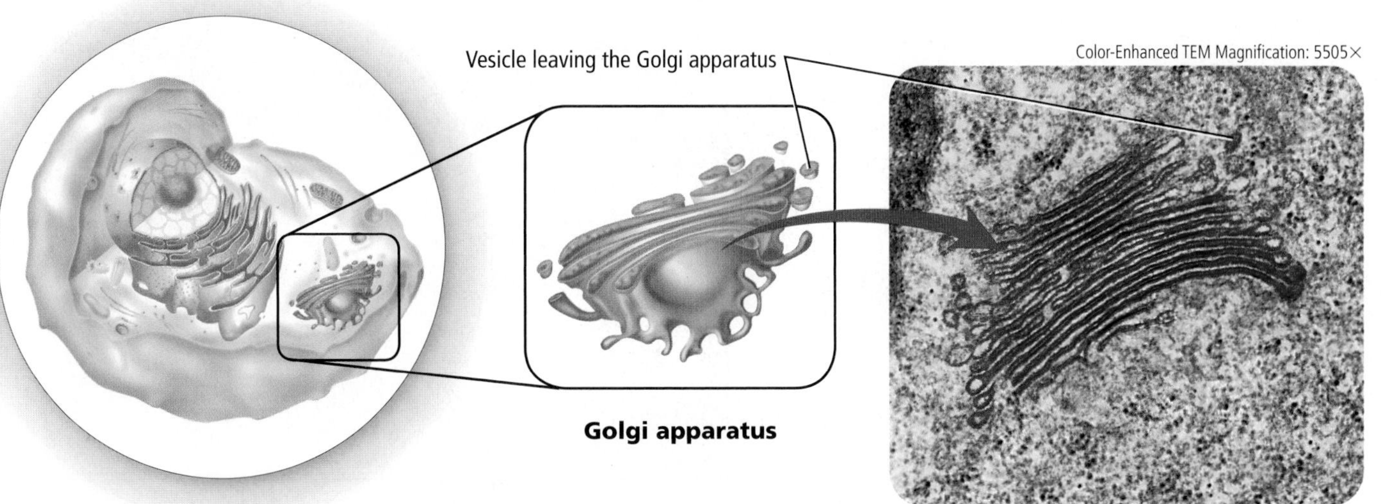

■ **Figure 7.12** Flattened stacks of membranes make up the Golgi apparatus.

Golgi apparatus After the hiking boots are made in the factory, they must be organized into pairs, boxed, and shipped. Similarly, after proteins are made in the endoplasmic reticulum, some might be transferred to the Golgi (GAWL jee) apparatus, illustrated in **Figure 7.12.** The **Golgi apparatus** is a flattened stack of membranes that modifies, sorts, and packages proteins into sacs called vesicles. Vesicles then can fuse with the cell's plasma membrane to release proteins to the environment outside the cell. Observe the vesicle in **Figure 7.12.**

Vacuoles A factory needs a place to store materials and waste products. Similarly, cells have membrane-bound vesicles called vacuoles for temporary storage of materials within the cytoplasm. A **vacuole,** such as the plant vacuole shown in **Figure 7.13,** is a sac used to store food, enzymes, and other materials needed by a cell. Some vacuoles store waste products. Interestingly, animal cells usually do not contain vacuoles. If animal cells do have vacuoles, they are much smaller than those in plant cells.

■ **Figure 7.13** Plant cells have large membrane-bound storage compartments called vacuoles.

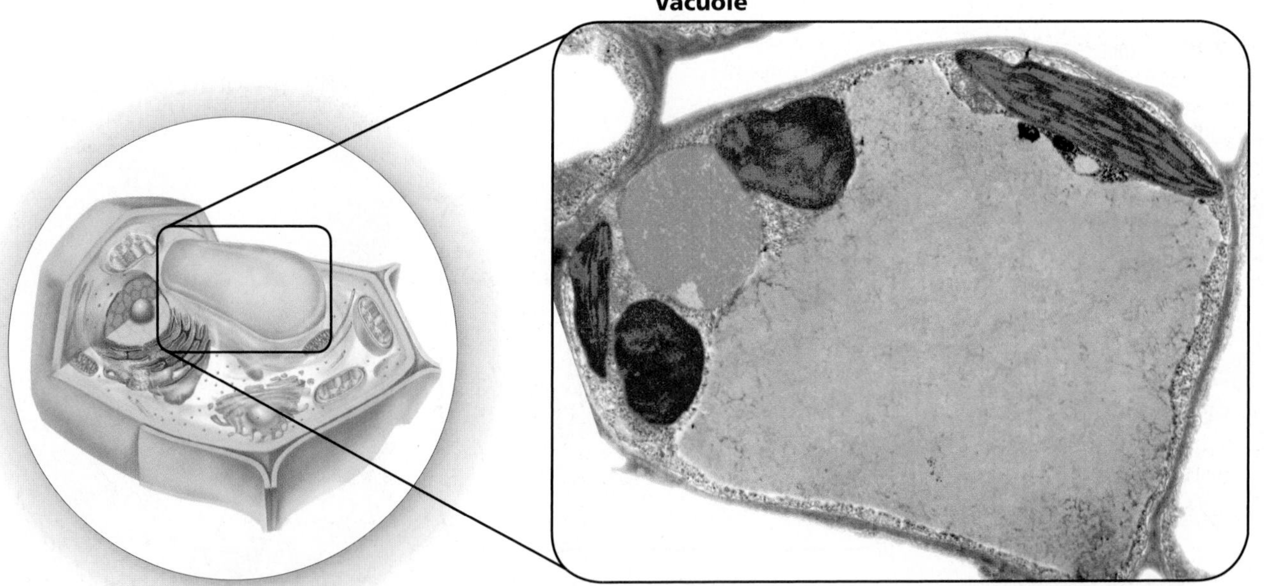

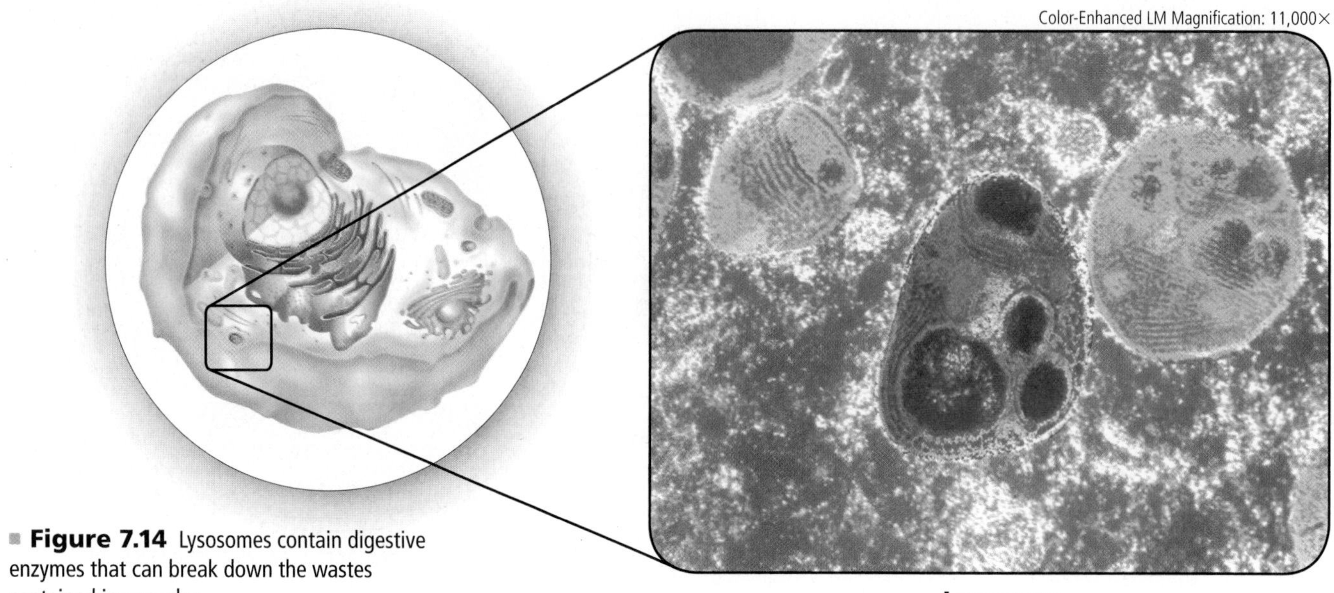

Figure 7.14 Lysosomes contain digestive enzymes that can break down the wastes contained in vacuoles.

Lysosomes Factories and cells also need clean-up crews. In the cell, **lysosomes,** shown in **Figure 7.14,** are vesicles that contain substances that digest excess or worn-out organelles and food particles. Lysosomes also digest bacteria and viruses that have entered the cell. The membrane surrounding a lysosome prevents the digestive enzymes inside from destroying the cell. Lysosomes can fuse with vacuoles and dispense their enzymes into the vacuole, digesting the wastes inside.

Centrioles Previously in this section you read about microtubules and the cytoskeleton. Groups of microtubules form another structure called a centriole (SEN tree ol). **Centrioles,** shown in **Figure 7.15,** are organelles made of microtubules that function during cell division. Centrioles are located in the cytoplasm of animal cells and most protists and usually are near the nucleus. You will learn about cell division and the role of centrioles in Chapter 9.

Figure 7.15 Centrioles are made of microtubules and play a role in cell division.

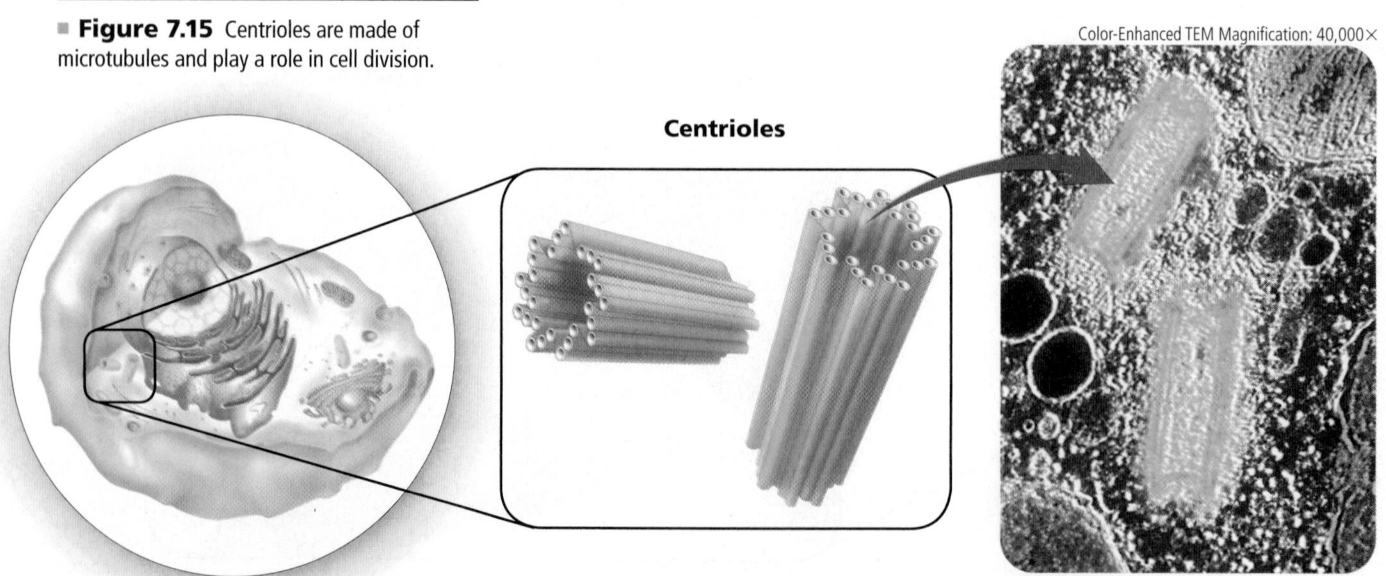

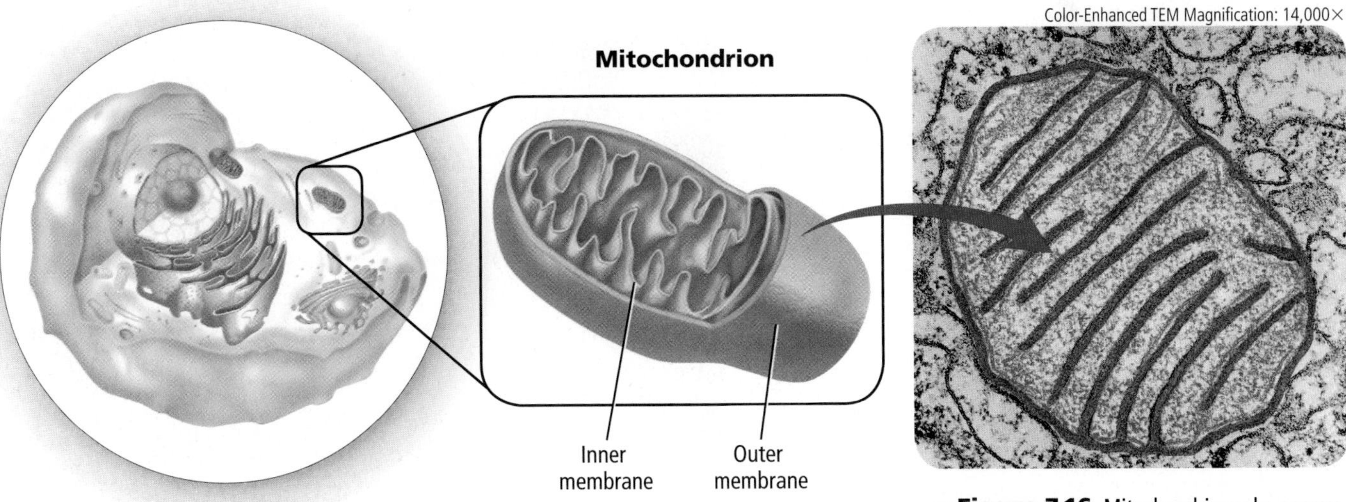

Figure 7.16 Mitochondria make energy available to the cell.
Describe *the membrane structure of a mitochondrion.*

Mitochondria Imagine now that the boot factory has its own generator that produces the electricity it needs. Cells also have energy generators called **mitochondria** (mi tuh KAHN dree uh; singular, mitochondrion) that convert fuel particles (mainly sugars), into usable energy. **Figure 7.16** shows that a mitochondrion has an outer membrane and a highly folded inner membrane that provides a large surface area for breaking the bonds in sugar molecules. The energy produced from that breakage is stored in the bonds of other molecules and later used by the cell. For this reason, mitochondria often are referred to as the "powerhouses" of cells.

Chloroplasts Factory machines need electricity that is generated by burning fossil fuels or by collecting energy from alternative sources, such as the Sun. Plant cells have their own way of using solar energy. In addition to mitochondria, plants and some other eukaryotic cells contain **chloroplasts,** which are organelles that capture light energy and convert it to chemical energy through a process called photosynthesis. Examine **Figure 7.17** and notice that inside the inner membrane are many small, disk-shaped compartments called thylakoids. It is here that the energy from sunlight is trapped by a pigment called chlorophyll. Chlorophyll gives leaves and stems their green color.

Chloroplasts belong to a group of plant organelles called plastids, some of which are used for storage. Some plastids store starches or lipids. Others, such as chromoplasts, contain red, orange, or yellow pigments that trap light energy and give color to plant structures such as flowers or leaves.

Figure 7.17 In plants, chloroplasts capture and convert light energy to chemical energy.

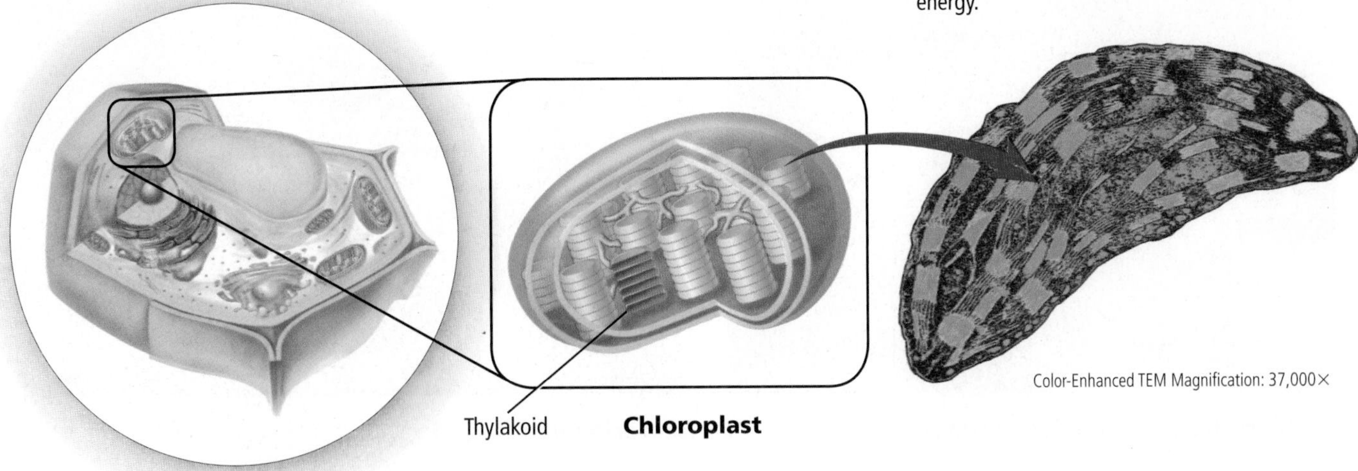

Color-Enhanced TEM Magnification: 38,000×

Plant cell 2

Plant cell walls

Plant cell 1

■ **Figure 7.18** The illustration shows plant cells and their cell walls. Compare this to the transmission electron micrograph showing the cell walls of adjacent plant cells.

Cell wall Another structure associated with plant cells is the cell wall, shown in **Figure 7.18.** The **cell wall** is a thick, rigid, mesh of fibers that surrounds the outside of the plasma membrane, protecting the cell and giving it support. Rigid cell walls allow plants to stand at various heights—from blades of grass to California redwoods. Plant cell walls are made of a carbohydrate called cellulose, which gives the wall its inflexible characteristics. **Table 7.1** lists cell walls and various other cell structures.

Cilia and flagella Some eukaryotic cell surfaces have structures called cilia and flagella that project outside the plasma membrane. As shown in **Figure 7.19, cilia** (singular, cilium) are short, numerous projections that look like hairs. The motion of cilia is similar to the motion of oars in a rowboat. **Flagella** (singular, flagellum) are longer and less numerous than cilia. These projections move with a whiplike motion. Cilia and flagella are composed of microtubules arranged in a 9 + 2 configuration, in which nine pairs of microtubules surround two single microtubules. Typically, a cell has one or two flagella.

Prokaryotic cilia and flagella contain cytoplasm and are enclosed by the plasma membrane. These structures are made of complex proteins. While both structures are used for cell movement, cilia are also found on stationary cells.

■ **Figure 7.19** The hairlike structures in the photomicrograph are cilia, and the tail-like structures are flagella. Both structures function in cell movement.
Infer ***Where in the body of an animal would you predict cilia might be found?***

Color-Enhanced SEM Magnification: 16,000×

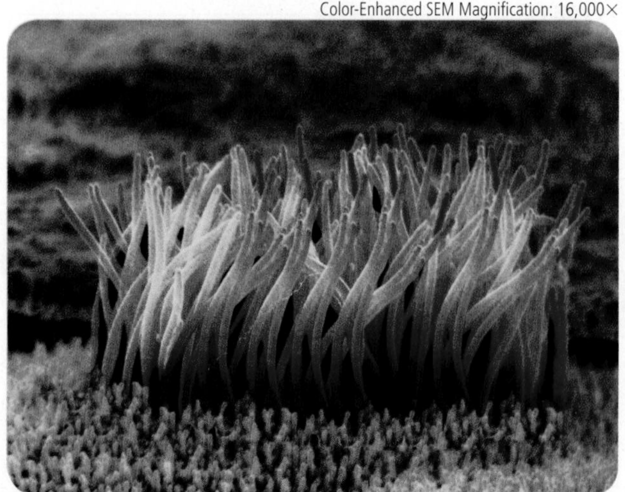

Cilia on the surface of a *Paramecium*

Color-Enhanced TEM Magnification: unavailable

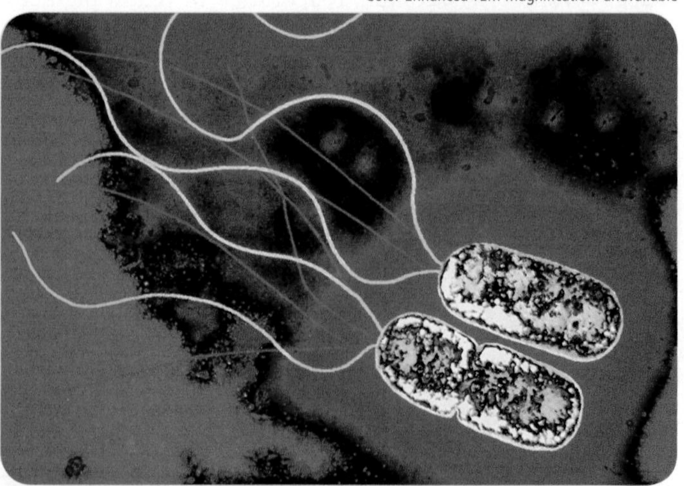

Bacteria with flagella

Concepts In Motion

Interactive Table To explore more about cell structures, visit biologygmh.com.

Table 7.1 Summary of Cell Structures

Cell Structure	Example	Function	Cell Type
Cell wall		An inflexible barrier that provides support and protects the plant cell	Plant cells, fungi cells, and some prokaryotes
Centrioles		Organelles that occur in pairs and are important for cell division	Animal cells and most protist cells
Chloroplast		A double-membrane organelle with thylakoids containing chlorophyll where photosynthesis takes place	Plant cells only
Cilia		Projections from cell surfaces that aid in locomotion and feeding; also used to sweep substances along surfaces	Some animal cells, protist cells, and prokaryotes
Cytoskeleton		A framework for the cell within the cytoplasm	All eukaryotic cells
Endoplasmic reticulum		A highly folded membrane that is the site of protein synthesis	All eukaryotic cells
Flagella		Projections that aid in locomotion and feeding	Some animal cells, prokaryotes, and some plant cells
Golgi apparatus		A flattened stack of tubular membranes that modifies proteins and packages them for distribution outside the cell	All eukaryotic cells
Lysosome		A vesicle that contains digestive enzymes for the breakdown of excess or worn-out cellular substances	Animal cells only
Mitochondrion		A membrane-bound organelle that makes energy available to the rest of the cell	All eukaryotic cells
Nucleus		Control center of the cell that contains coded directions for the production of proteins and cell division	All eukaryotic cells
Plasma membrane		A flexible boundary that controls the movement of substances into and out of the cell	All eukaryotic cells
Ribosome		Organelle that is the site of protein synthesis	All cells
Vacuole		A membrane-bound vesicle for the temporary storage of materials	Plant cells–one large; animal cells–a few small

Comparing Cells

Table 7.1 summarizes the structures of eukaryotic plant cells and animal cells. Notice that plant cells contain chlorophyll—they can capture and transform energy from the Sun into a usable form of chemical energy. This is one of the main characteristics that distinguishes plants from animals. In addition, remember that animal cells usually do not contain vacuoles. If they do, vacuoles in animal cells are much smaller than vacuoles in plant cells. Also, animal cells do not have cell walls. Cell walls give plant cells protection and support.

CAREERS IN BIOLOGY

Science Communications Specialist Many publishers of scientific material hire communications specialists to write about research and its importance to the general public. This often is accomplished through press releases, ads, pamphlets, and targeted mailings. For more information on biology careers, visit biologygmh.com.

Organelles at Work

With a basic understanding of the structures found within a cell, it becomes easier to envision how those structures work together to perform cell functions. Take, for example, the synthesis of proteins.

Protein synthesis begins in the nucleus with the information contained in the DNA. Genetic information is copied and transferred to another genetic molecule called RNA. Then RNA and ribosomes, which have been manufactured in the nucleolus, leave the nucleus through the pores of the nuclear membrane. Together, RNA and ribosomes manufacture proteins. Each protein made on the rough ER has a particular function; it might become a protein that forms a part of the plasma membrane, a protein that is released from the cell, or a protein transported to other organelles. Other ribosomes will float freely in the cytoplasm and also make proteins.

Most of the proteins made on the surface of the ER are sent to the Golgi apparatus. The Golgi apparatus packages the proteins in vesicles and transports them to other organelles or out of the cell. Other organelles use the proteins to carry out cell processes. For example, lysosomes use proteins, enzymes in particular, to digest food and waste. Mitochondria use enzymes to produce a usable form of energy for the cell.

After reading about the organelles in a cell, it becomes clearer why people equate the cell to a factory. Each organelle has its job to do, and the health of the cell depends on all of the components working together.

Section 7.3 Assessment

Section Summary

- Eukaryotic cells contain membrane-bound organelles in the cytoplasm that perform cell functions.
- Ribosomes are the sites of protein synthesis.
- Mitochondria are the powerhouses of cells.
- Plant and animal cells contain many of the same organelles, while other organelles are unique to either plant cells or animal cells.

Understand Main Ideas

1. MAIN Idea **Identify** the role of the nucleus in a eukaryotic cell.
2. **Summarize** the role of the endoplasmic reticulum.
3. **Analogy** Make a flowchart comparing the parts of a cell to an automobile production line.
4. **Infer** why some scientists do not consider ribosomes to be cell organelles.

Think Critically

5. **Hypothesize** how lysosomes would be involved in changing a caterpillar into a butterfly.

WRITING in Biology

6. Categorize the structures and organelles in **Table 7.1** into lists based on cell type, then draw a concept map illustrating your organization.

Self-Check Quiz biologygmh.com

Reading Preview

Objectives

- **Explain** the processes of diffusion, facilitated diffusion, and active transport.
- **Predict** the effect of a hypotonic, hypertonic, or isotonic solution on a cell.
- **Discuss** how large particles enter and exit cells.

Review Vocabulary

homeostasis: regulation of the internal environment of a cell or organism to maintain conditions suitable for life

New Vocabulary

diffusion
dynamic equilibrium
facilitated diffusion
osmosis
isotonic solution
hypotonic solution
hypertonic solution
active transport
endocytosis
exocytosis

Cellular Transport

MAIN Idea Cellular transport moves substances within the cell and moves substances into and out of the cell.

Real-World Reading Link Imagine studying in your room while cookies are baking in the kitchen. You probably didn't notice when the cookies were put into the oven because you couldn't smell them. But, as the cookies baked, the movement of the aroma from the kitchen to your room happened through a process called diffusion.

Diffusion

Connection to Chemistry As the aroma of baking cookies makes its way to you, the particles are moving and colliding with each other in the air. This happens because the particles in gases, liquids, and solids are in random motion. Similarly, substances dissolved in water move constantly in random motion called Brownian motion. This random motion causes **diffusion,** which is the net movement of particles from an area where there are many particles of the substance to an area where there are fewer particles of the substance. The amount of a substance in a particular area is called concentration. Therefore, substances diffuse from areas of high concentration to low concentration. **Figure 7.20** illustrates the process of diffusion. Additional energy input is not required for diffusion because the particles already are in motion.

For example, if you drop red and blue ink into a container of water at opposite ends of the container, which is similar to the watery environment of a cell, the process of diffusion begins, as shown in **Figure 7.20(A).** In a short period of time, the ink particles have mixed as a result of diffusion to the point where a purple-colored blend area is visible. **Figure 7.20(B)** shows the initial result of this diffusion.

■ **Figure 7.20** Diffusion causes the inks to move from high-ink concentration to low-ink concentration until the colors become evenly blended in the water.

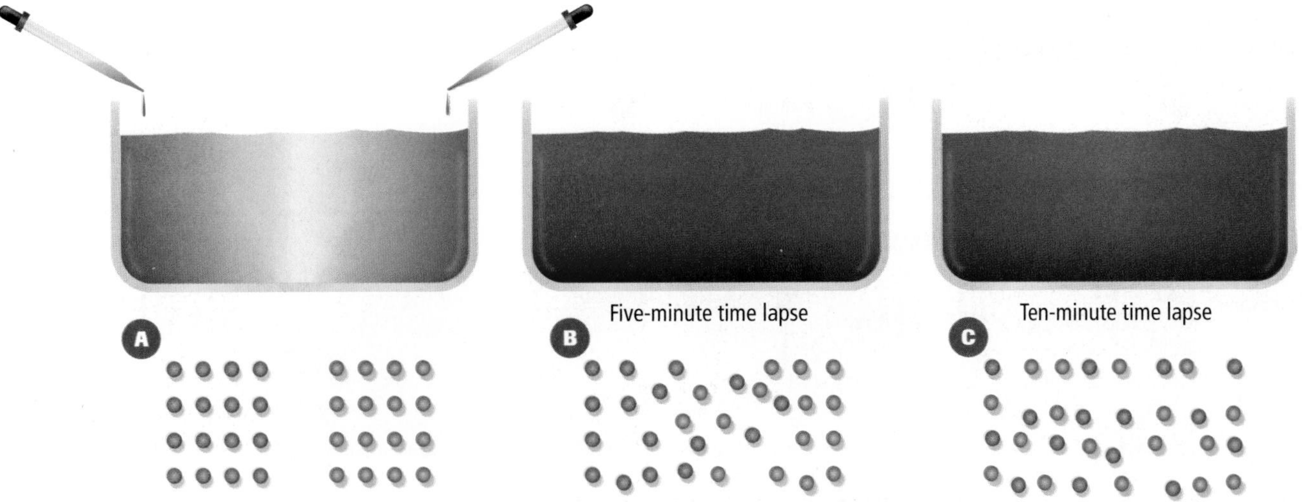

Vocabulary

Academic Vocabulary

Concentration:
the amount of component in a given area or volume.
The concentration of salt in the aquarium was too high, causing the fishes to die.

Given more time, the ink particles continue to mix and, in this case, continue to form the uniform purple mixture shown in **Figure 7.20(C).** Mixing continues until the concentrations of red ink and blue ink are the same in all areas. The final result is the purple solution. After this point, the particles continue to move randomly, but no further change in concentration will occur. This condition, in which there is continuous movement but no overall change, is called **dynamic equilibrium.**

One of the key characteristics of diffusion is the rate at which diffusion takes place. Three main factors affect the rate of diffusion: concentration, temperature, and pressure. When concentration is high, diffusion occurs more quickly because there are more particles that collide. Similarly, when the temperature or pressure increases, the number of collisions increases, thus increasing the rate of diffusion. Recall that at higher temperatures particles move faster, and at higher pressure the particles are closer together. In both cases, more collisions occur and diffusion is faster. The size and charge of a substance also affects the rate of diffusion.

Foldables
Incorporate information from this section into your Foldable.

Diffusion across the plasma membrane In addition to water, cells need certain ions and small molecules, such as chloride ions and sugars, to perform cellular functions. Water can diffuse across the plasma membrane, as shown in **Figure 7.21(A),** but most other substances cannot. Another form of transport, called **facilitated diffusion,** uses transport proteins to move other ions and small molecules across the plasma membrane. By this method, substances move into the cell through a water-filled transport protein called a channel protein that opens and closes to allow the substance to diffuse through the plasma membrane, as shown in **Figure 7.21(B).** Another type of transport protein called a carrier protein also can help substances diffuse across the plasma membrane. Carrier proteins change shape as the diffusion process continues to help move the particle through the membrane, as illustrated in **Figure 7.21(C).**

Diffusion of water and facilitated diffusion of other substances require no additional input of energy because the particles are moving from an area of high concentration to an area of lower concentration. This is also known as passive transport. You will learn later in this section about a form of cellular transport that does require energy input.

Reading Check **Describe** how sodium (Na^+) ions get into cells.

Figure 7.21 Although water moves freely through the plasma membrane, other substances cannot pass through the phospholipid bilayer on their own. Such substances enter the cell by facilitated transport.

Concepts In Motion
Interactive Figure To see an animation of how molecules can be passively transported through a plasma membrane, visit biologygmh.com.

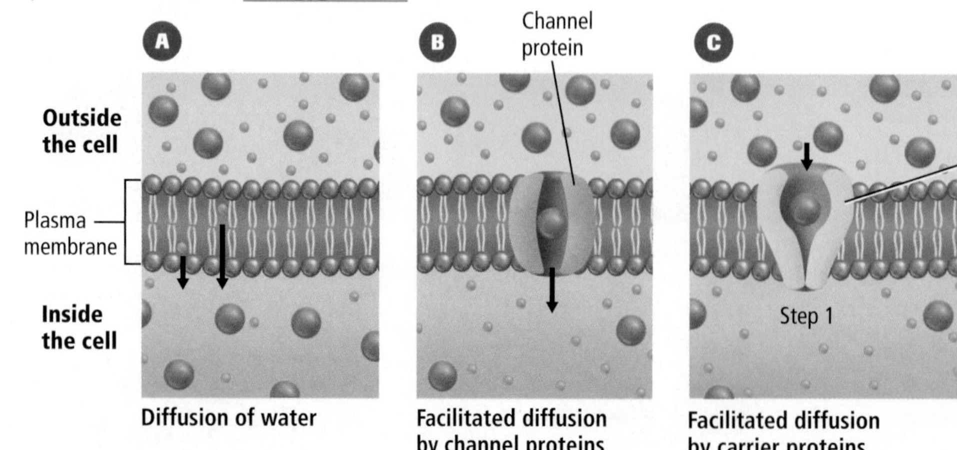

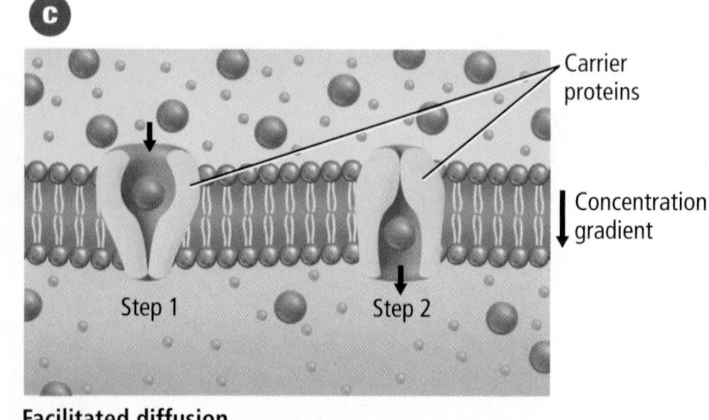

Osmosis: Diffusion of Water

Water is a substance that passes freely into and out of the cell through the plasma membrane. The diffusion of water across a selectively permeable membrane is called **osmosis** (ahs MOH sus). Regulating the movement of water across the plasma membrane is an important factor in maintaining homeostasis within the cell.

How osmosis works Recall that in a solution, a substance called the solute is dissolved in a solvent. Water is the solvent in a cell and its environment. Concentration is a measure of the amount of solute dissolved in a solvent. The concentration of a solution decreases when the amount of solvent increases.

Examine **Figure 7.22** showing a U-shaped tube containing solutions with different sugar concentrations separated by a selectively permeable membrane. What will happen if the solvent (water) can pass through the membrane but the solute (sugar) cannot?

Water molecules diffuse toward the side with the greater sugar concentration—the right side. As water moves to the right, the concentration of the sugar solution decreases. The water continues to diffuse until dynamic equilibrium occurs—the concentration of the solutions is the same on both sides. Notice in **Figure 7.22** that the result is an increase in solution level on the right side. During dynamic equilibrium, water molecules continue to diffuse back and forth across the membrane. But, the concentrations on each side no longer change.

Reading Check **Compare and contrast** diffusion and osmosis.

MiniLab 7.2

Investigate Osmosis

What will happen to cells placed in a strong salt solution? Regulating flow and amount of water into and out of the cell is critical to the survival of that cell. Osmosis is one method used to regulate a cell's water content.

Procedure

1. Read and complete the lab safety form.
2. Prepare a control **slide** using **onion epidermis, water,** and **iodine stain** as directed by your teacher.
3. Prepare a test slide using onion epidermis, **salt water,** and iodine stain as directed by your teacher.
4. Predict the effect, if any, that the salt solution will have on the onion cells in the test slide.
5. View the control slide using a **compound microscope** under low power and sketch several onion cells.
6. View the test slide under the same magnification and sketch your observations.

Analysis

1. **Analyze and Conclude** Was your prediction correct or incorrect? Explain.
2. **Explain** Use the process of osmosis to explain what you observe.

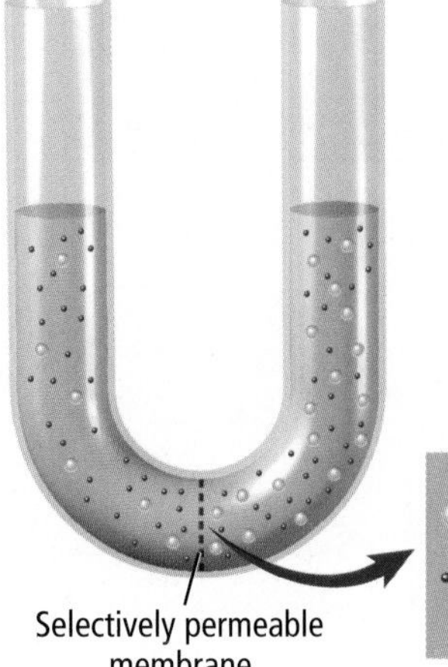

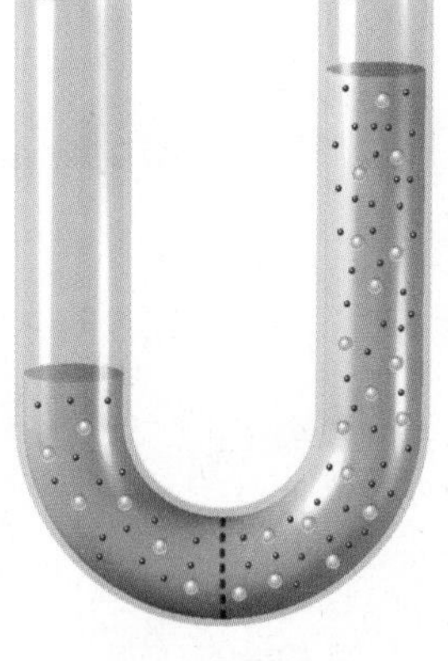

Figure 7.22 Before osmosis, the sugar concentration is greater on the right side. After osmosis, the concentrations are the same on both sides.
Name *the term for this phenomenon.*

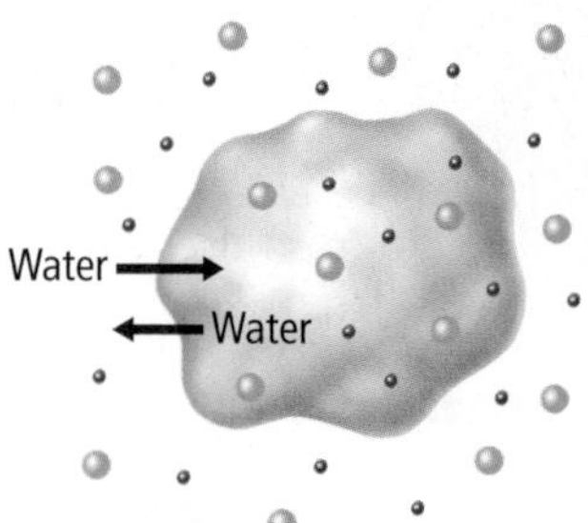

Color-Enhanced TEM Magnification: 3500×

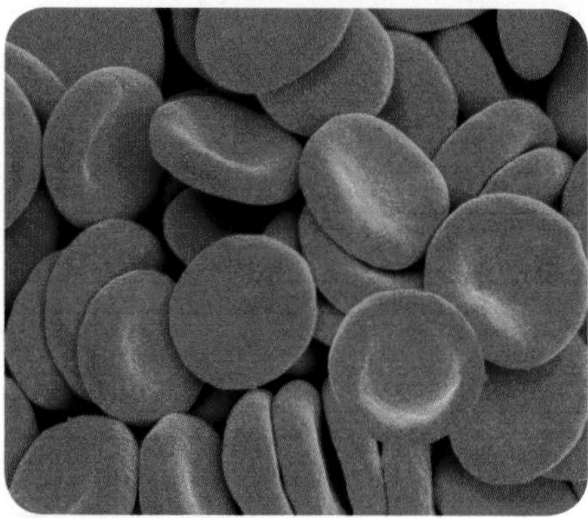

Animal cells

LM Magnification: 350×

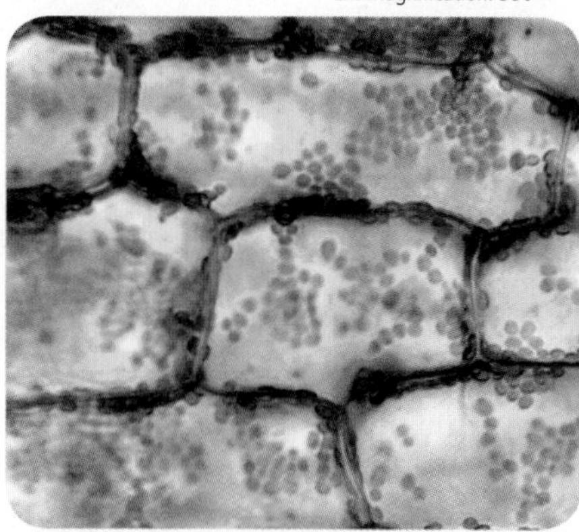

Plant cells

Figure 7.23 In an isotonic solution, water molecules move into and out of the cell at the same rate, and cells retain their normal shape. The animal cell and the plant cell have their normal shape in an isotonic solution.

Concepts In Motion

Interactive Figure To see an animation of osmosis in an isotonic, hypotonic, or hypertonic solution, visit biologygmh.com.

Cells in an isotonic solution When a cell is in a solution that has the same concentration of water and solutes—ions, sugars, proteins, and other substances—as its cytoplasm, the cell is said to be in an **isotonic solution.** *Iso-* comes from the Greek word meaning *equal.* Water still moves through the plasma membrane, but water enters and leaves the cell at the same rate. The cell is at equilibrium with the solution, and there is no net movement of water. The cells retain their normal shape, as shown in **Figure 7.23.** Most cells in organisms are in isotonic solutions, such as blood.

Cells in a hypotonic solution If a cell is in a solution that has a lower concentration of solute, the cell is said to be in a **hypotonic solution.** *Hypo-* comes from the Greek word meaning *under.* There is more water outside of the cell than inside. Due to osmosis, the net movement of water through the plasma membrane is into the cell, as illustrated in **Figure 7.24.** Pressure generated as water flows through the plasma membrane is called osmotic pressure. In an animal cell, as water moves into the cell, the pressure increases and the plasma membrane swells. If the solution is extremely hypotonic, the plasma membrane might be unable to withstand this pressure and the cell might burst.

Because they have a rigid cell wall that supports the cell, plant cells do not burst when in a hypotonic solution. As the pressure inside the cell increases, the plant's central vacuole fills with water, pushing the plasma membrane against the cell wall, shown in the plant cell in **Figure 7.24.** Instead of bursting, the plant cell becomes firmer. Grocers use this process to keep produce looking fresh by misting fruits and vegetables with water.

Figure 7.24 In a hypotonic solution, water enters a cell by osmosis, causing the cell to swell. Animal cells may continue to swell until they burst. Plant cells swell beyond their normal size as internal pressure increases.

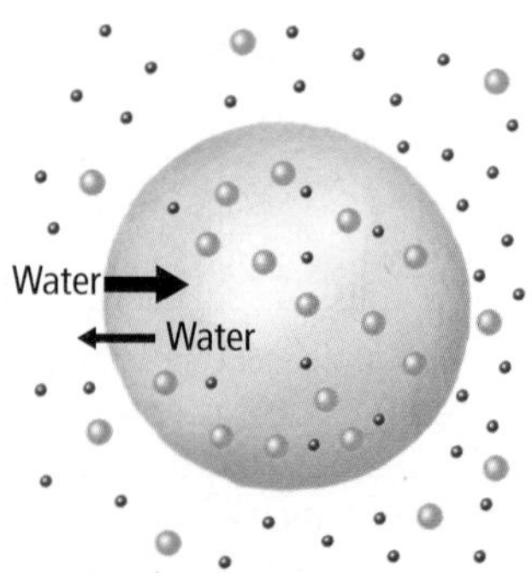

Color-Enhanced SEM Magnification: 3500×

Animal cells

LM Magnification: 400×

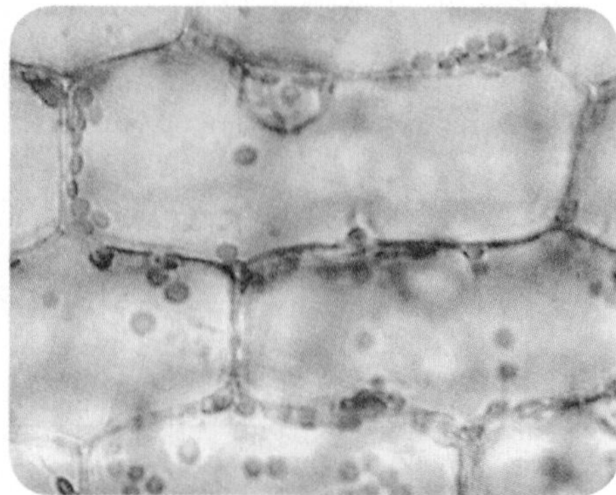

Plant cells

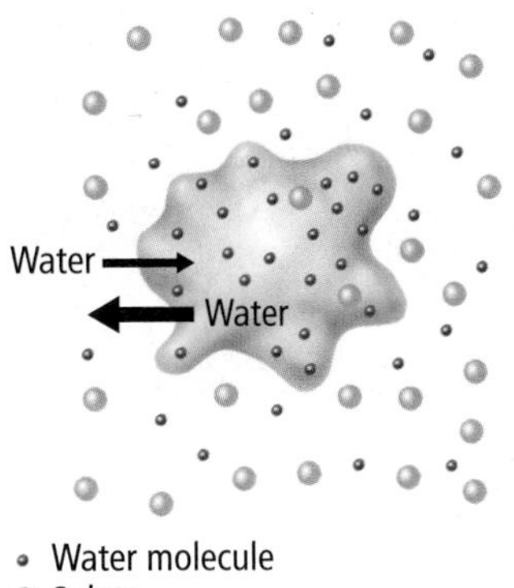

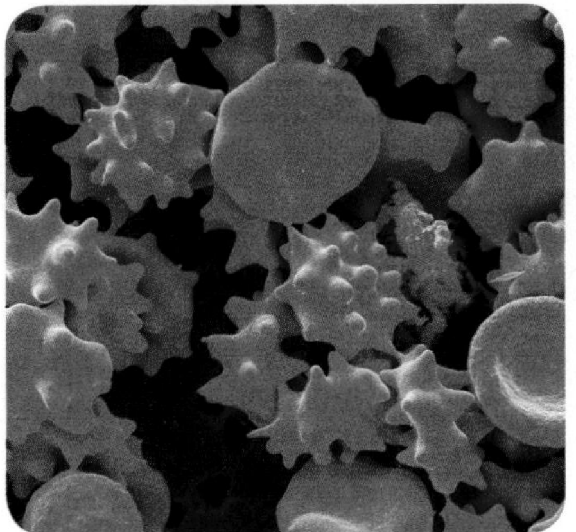

Animal cells

LM Magnification: 225×

Plant cells

■ **Figure 7.25** In a hypertonic solution, water leaves a cell by osmosis, causing the cell to shrink. Animal cells shrivel up as they lose water. As plant cells lose internal pressure, the plasma membrane shrinks away from the cell wall.

Cells in a hypertonic solution When a cell is placed in a **hypertonic solution,** the concentration of the solute outside of the cell is higher than inside. *Hyper-* comes from the Greek word meaning *above.* During osmosis, the net movement of water is out of the cell, as illustrated in **Figure 7.25.** Animal cells in a hypertonic solution shrivel because of decreased pressure in the cells. Plant cells in a hypertonic solution lose water, mainly from the central vacuole. The plasma membrane shrinks away from the cell wall. Loss of water in a plant cell causes wilting.

 **Reading Check** **Compare and contrast** the three types of solutions.

Active Transport

Sometimes substances must move from a region of lower concentration to a region of higher concentration against the passive movement from higher to lower concentration. This movement of substances across the plasma membrane against a concentration gradient requires energy, therefore, it is called **active transport. Figure 7.26** illustrates how active transport occurs with the aid of carrier proteins, commonly called pumps. Some pumps move one type of substance in only one direction, while others move two substances either across the membrane in the same direction or in opposite directions. Due to active transport, the cell maintains the proper balance of substances it needs. Active transport helps maintain homeostasis.

■ **Figure 7.26** Carrier proteins pick up and move substances across the plasma membrane against the concentration gradient and into the cell.
Explain *Why does active transport require energy?*

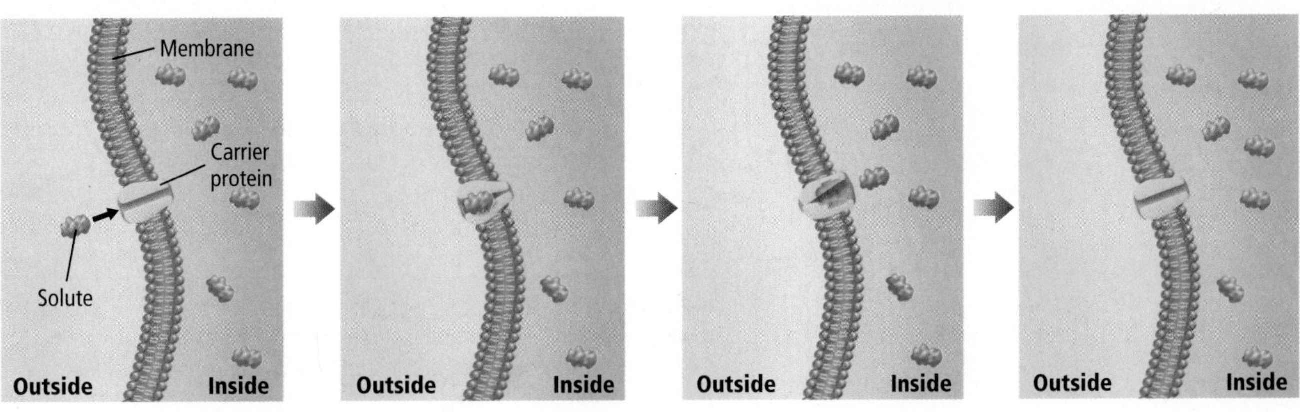

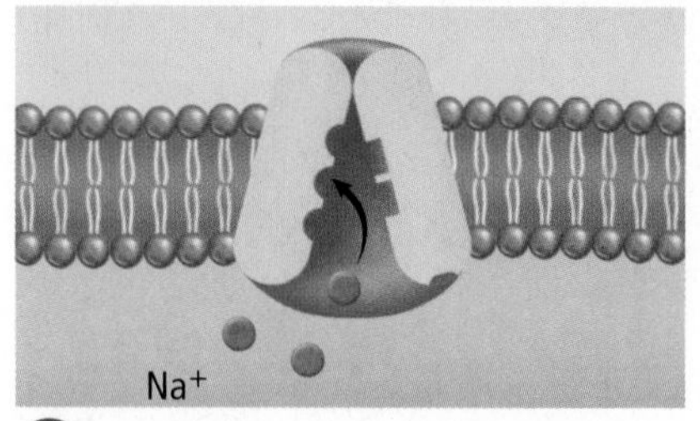

A Protein in the membrane binds intracellular sodium ions.

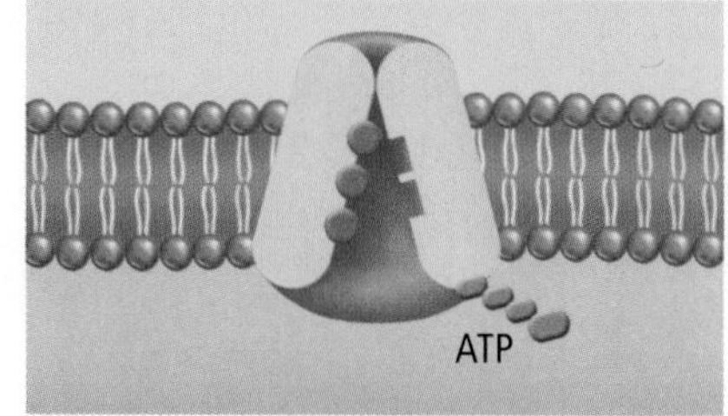

B ATP attaches to protein with bound sodium ions.

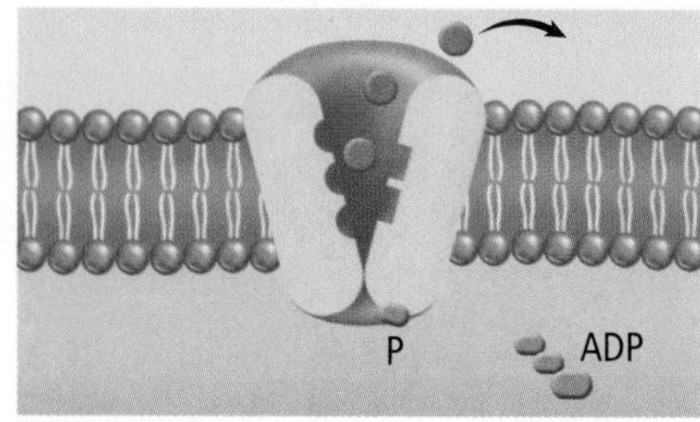

C The breakdown of ATP causes shape change in protein, allowing sodium ions to leave.

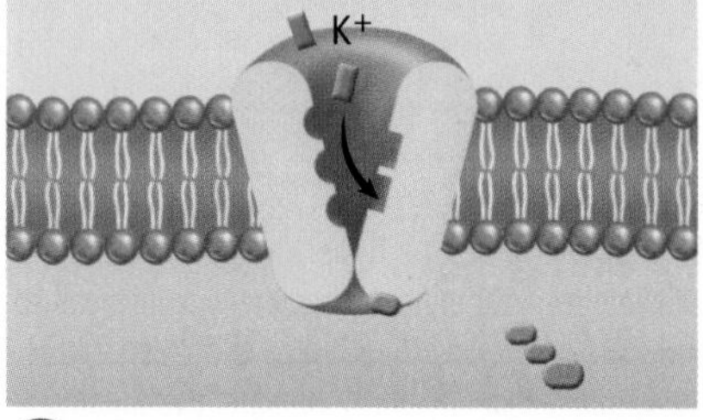

D Extracellular potassium ions bind to exposed sites.

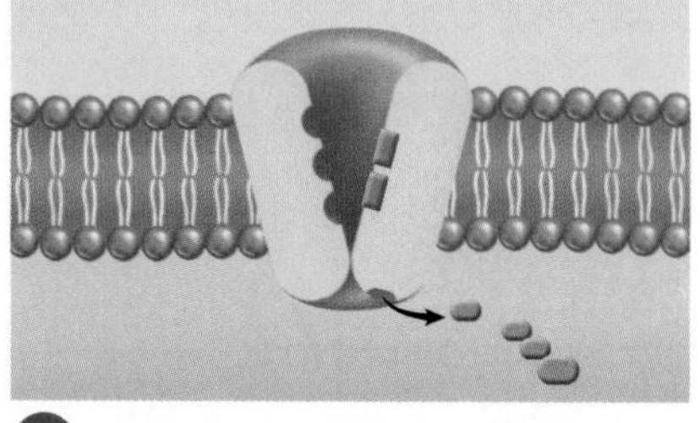

E Binding of potassium causes release of phosphate from protein.

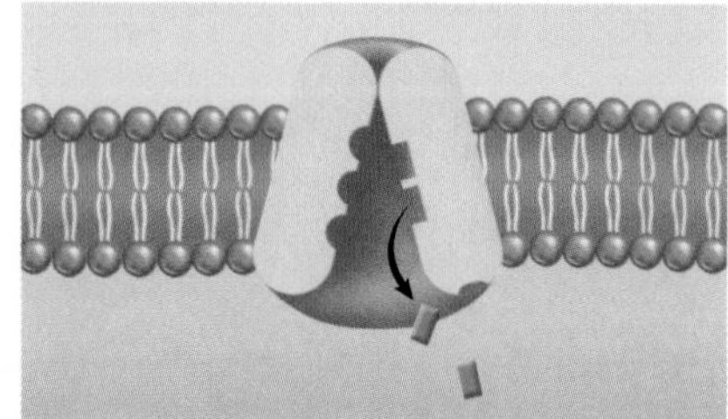

F Phospate release changes protein back to its original shape, and potassium ions move into the cell.

■ **Figure 7.27** Some cells use elaborate pumping systems, such as the Na^+/K^+ ATPase pump shown here, to help move substances through the plasma membrane.

Concepts In Motion

Interactive Figure To see an animation of how a Na^+/K^+ ATPase pump can actively transport molecules against a concentration gradient, visit biologygmh.com.

Personal Tutor

To learn about the sodium/potassium pump, visit biologygmh.com.

Na^+/K^+ ATPase pump One common active transport pump is called the sodium-potassium ATPase pump. This pump is found in the plasma membrane of animal cells. The pump maintains the level of sodium ions (Na^+) and potassium ions (K^+) inside and outside the cell. This protein pump is an enzyme that catalyzes the breakdown of an energy-storing molecule. The pump uses the energy in order to transport three sodium ions out of the cell while moving two potassium ions into the cell. The high level of sodium on the outside of the cell creates a concentration gradient. Follow the steps in **Figure 7.27** to see the action of the Na^+/K^+ ATPase pump.

The activity of the Na^+/K^+ ATPase pump can result in yet another form of cellular transport. Substances, such as sugar molecules, must come into the cell from the outside, where the concentration of the substance is lower than inside. This requires energy. Recall, however, that the Na^+/K^+ ATPase pump moves Na^+ out of the cell, which creates a low concentration of Na^+ inside the cell. In a process called coupled transport, the Na^+ ions that have been pumped out of the cell can couple with sugar molecules and be transported into the cell through a membrane protein called a coupled channel. The sugar molecule, coupled to a Na^+ ion, enters the cell by facilitated diffusion of the sodium, as shown in **Figure 7.28.** As a result, sugar enters the cell without spending any additional cellular energy.

■ **Figure 7.28** Substances "piggy-back" their way into or out of a cell by coupling with another substance that uses an active transport pump.
Compare and contrast *active and passive transport across the plasma membrane.*

Endocytosis

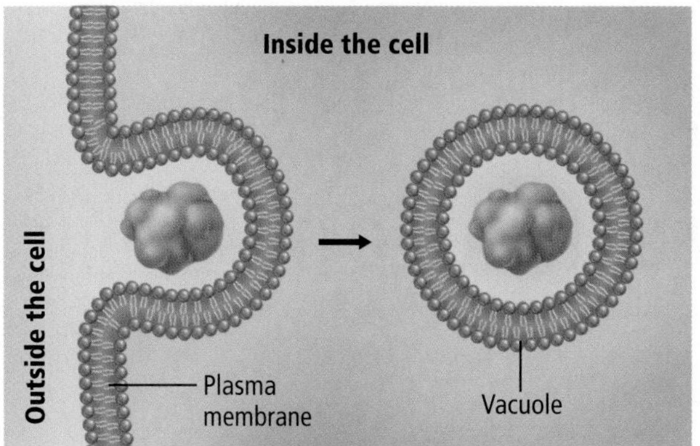

Exocytosis

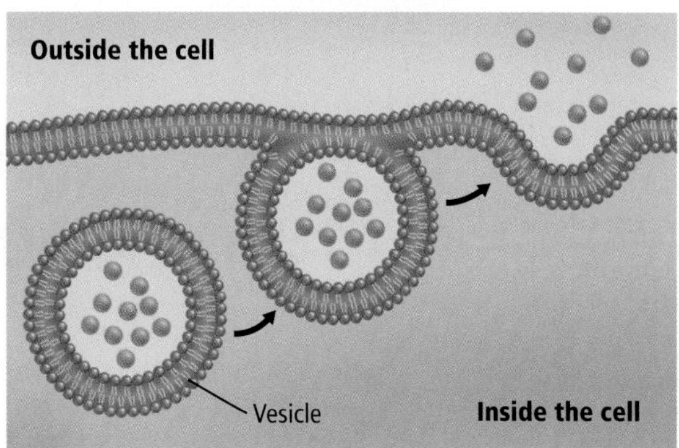

Figure 7.29
Left: Large substances can enter a cell by endocytosis.
Right: Substances can be deposited outside the cell by exocytosis.

Transport of Large Particles

Some substances are too large to move through the plasma membrane by diffusion or transport proteins and get inside the cell by a different process. **Endocytosis** is the process by which a cell surrounds a substance in the outside environment, enclosing the substance in a portion of the plasma membrane. The membrane then pinches off and leaves the substance inside the cell. The substance shown on the left in **Figure 7.29** is engulfed and enclosed by a portion of the cell's plasma membrane. The membrane then pinches off inside of the cell and the resulting vacuole, with its contents, moves to the inside of the cell.

Exocytosis is the secretion of materials at the plasma membrane. The illustration on the right in **Figure 7.29** shows that exocytosis is the reverse of endocytosis. Cells use exocytosis to expel wastes and to secrete substances, such as hormones, produced by the cell. Both endocytosis and exocytosis require the input of energy. Cells maintain homeostasis by moving substances into and out of the cell. Some transport processes require additional energy input, while others do not. Together, the different types of transport allow a cell to interact with its environment while maintaining homeostasis.

Section 7.4 Assessment

Section Summary

- Cells maintain homeostasis using passive and active transport.
- Concentration, temperature, and pressure affect the rate of diffusion.
- Cells must maintain homeostasis in all types of solutions, including isotonic, hypotonic, and hypertonic.
- Some large molecules are moved into and out of the cell using endocytosis and exocytosis.

Understand Main Ideas

1. MAIN Idea **List and describe** the types of cellular transport.
2. **Describe** how the plasma membrane controls what goes into and comes out of a cell.
3. **Sketch** a before and an after diagram of an animal cell placed in a hypotonic solution.
4. **Contrast** How is facilitated diffusion different from active transport?

Think Critically

5. **Describe** Some organisms that normally live in pond water contain water pumps. These pumps continually pump water out of the cell. Describe a scenario that might reverse the action of the pump.
6. **Summarize** the role of the phospholipid bilayer in cellular transport in living cells.

CUTTING-EDGE BIOLOGY

EXPLORING NANOTECHNOLOGY

Imagine that cancer cells could be detected and destroyed one by one, or that a new drug could be tested on a single cell to evaluate its clinical performance. Advances in technologies that allow scientists to focus on individual cells might make these scenarios a reality in the near future.

Nanotechnology (na no tek NAW luh jee) is the branch of science that deals with development and use of devices on the nanometer scale. A nanometer (nm) is one billionth of a meter (10^{-9} m). To put this scale into perspective, consider that most human cells are between 10,000 and 20,000 nm in diameter. Nanotechnology is a fast-growing branch of science that likely will leave its mark on everything from electronics to medicine.

Atomic force microscopes At the National Institute of Advanced Industrial Science and Technology in Hyogo, Japan, researchers are using nanotechnology in the form of an atomic force microscope (AFM) to operate on single cells. The microscope is actually used as a "nanoneedle." The AFM creates a visual image of a cell using a microscopic sensor that scans the cell. Then the probe of the AFM, sharpened into a needle tip that is approximately 200 nm in diameter, can be inserted into the cell without damaging the cell membrane.

Some scientists envision many applications for this technique. The nanoneedle might help scientists study how a cell responds to a new drug or how the chemistry of a diseased cell differs from that of a healthy cell. Another application for the nanoneedle might be to insert DNA strands directly into the nucleus of a cell to test new gene therapy techniques that might correct genetic disorders.

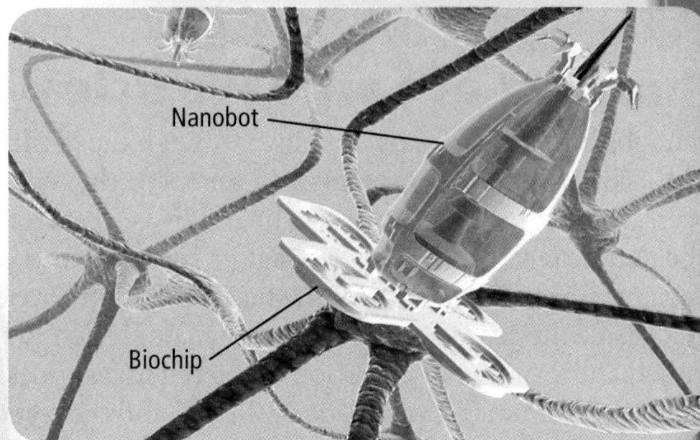

This computer-generated image shows a nanobot armed with a biochip. Someday, a biochip, which is an electronic device that contains organic materials, might repair a damaged nerve cell.

Lasers Nanotechnology applications, perhaps in the form of nanosurgery, could be used to investigate how cells work or to destroy individual cancer cells without harming nearby healthy cells. Researchers at Harvard University have developed a laser technique that allows them to manipulate a specific component of the cell's internal parts without causing damage to the cell membrane or other cell structures. Imagine having the capability to perform extremely delicate surgery on a cellular level!

In the future, nanotechnology might be our first line of defense to treat cancer. It also might become the standard technique to test new drugs or even become a favored treatment used in gene therapy.

WRITING in Biology

Review Visit biologygmh.com to learn more about nanotechnology in medicine and health care. Write an overview of one technology that you find interesting. Describe its advantages and challenges. You may include a presentation with your overview.

BIOLAB

WHICH SUBSTANCES WILL PASS THROUGH A SELECTIVELY PERMEABLE MEMBRANE?

Background: **All membranes in cells, including the plasma membrane and the membranes that surround organelles in eukaryotic cells, are selectively permeable. In this lab, you will examine the movement of some biologically important molecules through a dialysis membrane that is analogous to the plasma membrane. Because a dialysis membrane has tiny pores, it is only permeable for tiny molecules.**

Question: *Which substances pass through a dialysis membrane?*

Materials

cellulose dialysis tubing (2)
400-mL beakers (2)
string
scissors
distilled water
small plastic dishpan
starch solution
albumin solution
glucose solution
NaCl solution
iodine solution (tests starch)
anhydrous Benedict's reagent (tests glucose)
silver nitrate solution (tests NaCl)
biuret reagent (tests albumin)
10-mL graduated cylinder
test tubes (2)
test-tube rack
funnel
wax pencil
eye droppers

Safety Precautions

Procedure

1. Read and complete the lab safety form.
2. Construct a data table as instructed by your teacher.
3. Collect two lengths of dialysis tubing, two 400-mL beakers, and the two solutions that you have been assigned to test.
4. Label the beakers with the type of solution that you place in the dialysis tubing.
5. With a partner, prepare and fill one length of dialysis tubing with one solution. Rinse the outside of the bag thoroughly. Place the filled tubing bag into a beaker that contains distilled water.
6. Repeat step 5 using the second solution.
7. After 45 minutes, transfer some of the water from each beaker into separate test tubes.
8. Add a few drops of the appropriate test reagent to the water.
9. Record your results and determine whether your prediction was correct. Compare your results with other groups in your class and record the results for the two solutions that you did not test.
10. **Cleanup and Disposal** Wash and return all reusable materials. Dispose of test solutions and used dialysis tubing as directed by your teacher. Wash your hands thoroughly after using any chemical reagent.

Analyze and Conclude

1. **Evaluate** Did your test molecules pass through the dialysis tubing? Explain.
2. **Think Critically** What characteristics of a plasma membrane give it more control over the movement of molecules than the dialysis membrane?
3. **Error Analysis** How could failing to rinse the dialysis tube bags with distilled water prior to placing them in the beaker cause a false positive test for the presence of a dissolved molecule? What other sources of error might lead to inaccurate results?

POSTER SESSION

Communicate A disease called cystic fibrosis occurs when plasma membranes lack a molecule which helps transport chloride ions. Research this disease at biologygmh.com and present your finding to your class using a poster.

Study Guide

Download quizzes, key terms, and flash cards from biologygmh.com.

FOLDABLES **Apply** Use what you have learned about osmosis and cellular transport to design an apparatus that would enable a freshwater fish to survive in a saltwater habitat.

Vocabulary	Key Concepts
Section 7.1 Cell Discovery and Theory	
• cell (p. 182) • cell theory (p. 183) • eukaryotic cell (p. 186) • nucleus (p. 186) • organelle (p. 186) • plasma membrane (p. 185) • prokaryotic cell (p. 186)	**MAIN Idea** The invention of the microscope led to the discovery of cells. • Microscopes have been used as a tool for scientific study since the late 1500s. • Scientists use different types of microscopes to study cells. • The cell theory summarizes three principles. • There are two broad groups of cell types—prokaryotic cells and eukaryotic cells. • Eukaryotic cells contain a nucleus and organelles.
Section 7.2 The Plasma Membrane	
• fluid mosaic model (p. 190) • phospholipid bilayer (p. 188) • selective permeability (p. 187) • transport protein (p. 189)	**MAIN Idea** The plasma membrane helps to maintain a cell's homeostasis. • Selective permeability is the property of the plasma membrane that allows it to control what enters and leaves the cell. • The plasma membrane is made up of two layers of phospholipid molecules. • Cholesterol and transport proteins aid in the function of the plasma membrane. • The fluid mosaic model describes the plasma membrane.
Section 7.3 Structures and Organelles	
• cell wall (p. 198) • centriole (p. 196) • chloroplast (p. 197) • cilium (p. 198) • cytoplasm (p. 191) • cytoskeleton (p. 191) • endoplasmic reticulum (p. 194) • flagellum (p. 198) • Golgi apparatus (p. 195) • lysosome (p. 196) • mitochondrion (p. 197) • nucleolus (p. 193) • ribosome (p. 193) • vacuole (p. 195)	**MAIN Idea** Eukaryotic cells contain organelles that allow the specialization and the separation of functions within the cell. • Eukaryotic cells contain membrane-bound organelles in the cytoplasm that perform cell functions. • Ribosomes are the sites of protein synthesis. • Mitochondria are the powerhouses of cells. • Plant and animal cells contain many of the same organelles, while other organelles are unique to either plant cells or animal cells.
Section 7.4 Cellular Transport	
• active transport (p. 205) • diffusion (p. 201) • dynamic equilibrium (p. 202) • endocytosis (p. 207) • exocytosis (p. 207) • facilitated diffusion (p. 202) • hypertonic solution (p. 205) • hypotonic solution (p. 204) • isotonic solution (p. 204) • osmosis (p. 203)	**MAIN Idea** Cellular transport moves substances within the cell and moves substances into and out of the cell. • Cells maintain homeostasis using passive and active transport. • Concentration, temperature, and pressure affect the rate of diffusion. • Cells must maintain homeostasis in all types of solutions, including isotonic, hypotonic, and hypertonic. • Some large molecules are moved into and out of the cell using endocytosis and exocytosis.

Biology Online Vocabulary PuzzleMaker biologygmh.com

Assessment

Section 7.1

Vocabulary Review

Each of the following sentences is false. Make the sentence true by replacing the italicized word with a vocabulary term from the Study Guide page.

1. The *nucleus* is a structure that surrounds a cell and helps control what enters and exits the cell.
2. A(n) *prokaryote* has membrane-bound organelles.
3. *Organelles* are basic units of all organisms.

Understand Key Concepts

4. If a microscope has a series of three lenses that magnify individually 5×, 5×, and 7×, what is the total magnification when looking through the microscope?
   A. 25× **C.** 17×
   B. 35× **D.** 175×
5. Which is not part of the cell theory?
   A. The basic unit of life is the cell.
   B. Cells came from preexisting cells.
   C. All living organisms are composed of cells.
   D. Cells contain membrane-bound organelles.

Use the photo to answer question 6.

Color-Enhanced TEM Magnification: 15,000×

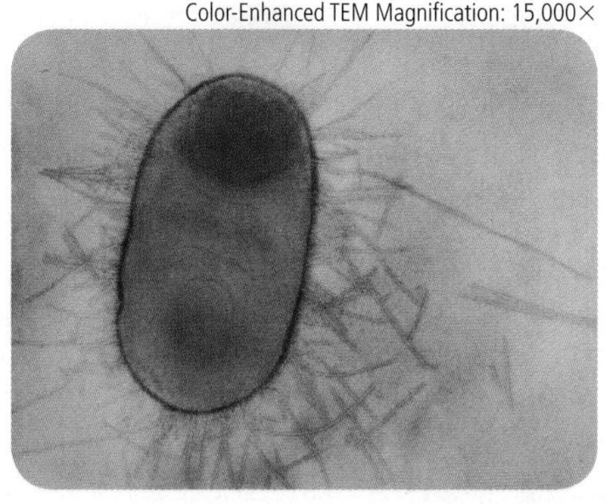

6. The photomicrograph shows which kind of cell?
   A. prokaryotic cell **C.** animal cell
   B. eukaryotic cell **D.** plant cell

Constructed Response

7. **Open Ended** Explain how the development of the microscope changed how scientists studied living organisms.
8. **Short Answer** Compare and contrast prokaryotic cells and eukaryotic cells.

Think Critically

9. **CAREERS IN BIOLOGY** Why might a microscopist, who specializes in the use of microscopes to examine specimens, use a light microscope instead of an electron microscope?
10. **Analyze** A material is found in an asteroid that might be a cell. What criteria must the material meet to be considered a cell?

Section 7.2

Vocabulary Review

Complete the sentences below using vocabulary terms from the Study Guide page.

11. A __________ is the basic structure molecule making up the plasma membrane.
12. __________ proteins move needed substances or waste materials through the plasma membrane.
13. __________ is the property that allows only some substances in and out of a cell.

Understand Key Concepts

14. Which of the following orientations of phospholipids best represents the phospholipid bilayer of the plasma membrane?

A. 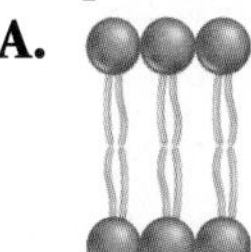

C.  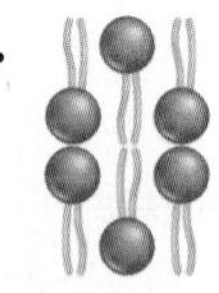

B. 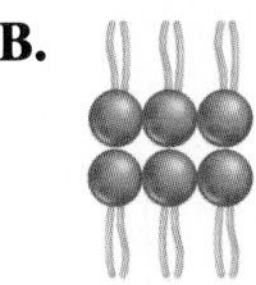

D. 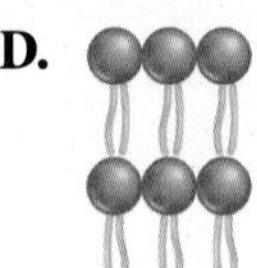

15. Which situation would increase the fluidity of a phospholipid bilayer?
    A. decreasing the temperature
    B. increasing the number of proteins
    C. increasing the number of cholesterol molecules
    D. increasing the number of unsaturated fatty acids

Constructed Response

16. **Short Answer** Explain how the plasma membrane maintains homeostasis within a cell.

17. **Open Ended** Explain what a mosaic is and then explain why the term *fluid mosaic model* is used to describe the plasma membrane.

18. **Short Answer** How does the orientation of the phospholipids in the bilayer allow a cell to interact with its internal and external environments?

Think Critically

19. **Hypothesize** how a cell would be affected if it lost the ability to be selectively permeable.

20. **Predict** What might happen to a cell if it no longer could produce cholesterol?

Section 7.3

Vocabulary Review

Fill in the blank with the vocabulary term from the Study Guide page that matches the function definition.

21. __________ stores wastes

22. __________ produces ribosomes

23. __________ generates energy for a cell

24. __________ sorts proteins into vesicles

Understand Key Concepts

Use the diagram below to answer questions 25 and 26.

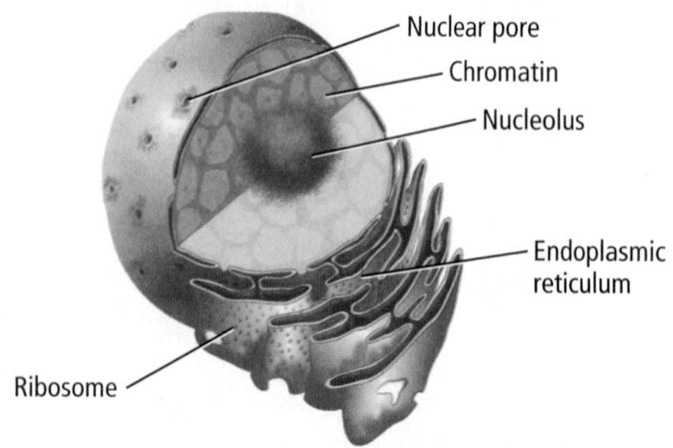

25. Which structure synthesizes proteins that will be used by the cell?
    A. chromatin
    B. nucleolus
    C. ribosome
    D. endoplasmic reticulum

26. Where are the ribosomes produced?
    A. nuclear pore
    B. nucleolus
    C. chromatin
    D. endoplasmic reticulum

27. In which structure would you expect to find a cell wall?
    A. a human skin cell
    B. a cell from an oak tree
    C. a blood cell from a cat
    D. a liver cell from a mouse

Constructed Response

28. **Short Answer** Describe why the cytoskeleton within the cytoplasm was a recent discovery.

29. **Short Answer** Compare the structures and functions of the mitochondrion and chloroplast below.

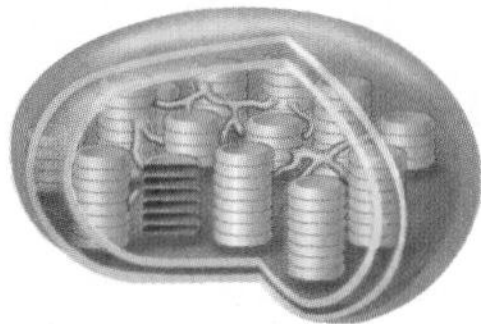

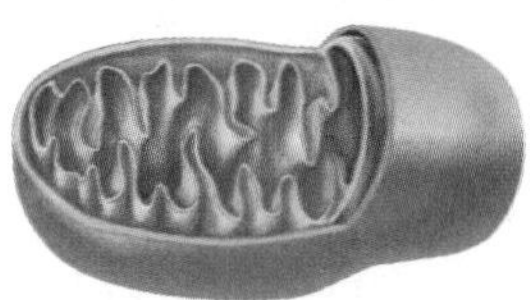

30. **Open Ended** Suggest a reason why packets of proteins collected in a vacuole might merge with lysosomes.

Think Critically

31. **Identify** a specific example where the cell wall structure has aided the survival of a plant in its natural habitat.

32. **Infer** Explain why plant cells that transport water against the force of gravity contain many more mitochondria than other plant cells.

Section 7.4

Vocabulary Review

Explain the difference in the terms given below. Then explain how the terms are related.

33. active transport, facilitated diffusion

34. endocytosis, exocytosis

35. hypertonic solution, hypotonic solution

Understand Key Concepts

36. Which is not a factor that affects the rate of diffusion?
    - A. conductivity
    - B. concentration
    - C. pressure
    - D. temperature

37. Which type of transport requires energy input from the cell?
    - A. active transport
    - B. facilitated diffusion
    - C. osmosis
    - D. simple diffusion

Constructed Response

38. **Short Answer** Why is active transport an energy-utilizing process?

39. **Short Answer** Some protists that live in a hypotonic pond environment have cell membrane adaptations that slow water uptake. What adaptations might this protist living in the hypertonic Great Salt Lake have?

LM Magnification: 75×

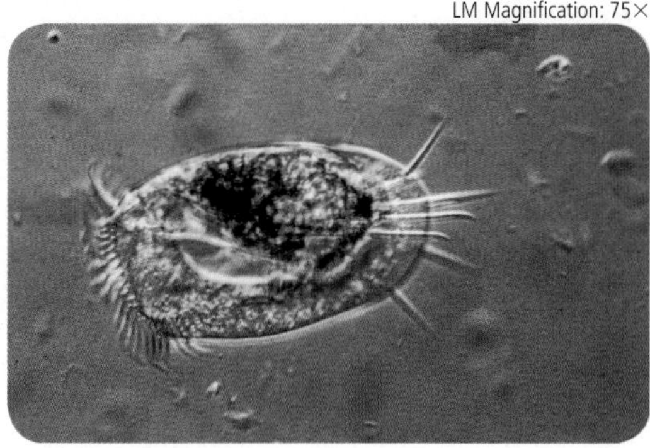

40. **Short Answer** Summarize how cellular transport helps maintain homeostasis within a cell.

Think Critically

41. **Hypothesize** how oxygen crosses the plasma membrane if the concentration of oxygen is lower inside the cell than it is outside the cell.

42. **Analyze** Farming and watering that is done in very dry regions of the world leaves salts that accumulate in the soil as water evaporates. Based on what you know about concentration gradients, why does increasing soil salinity have adverse effects on plant cells?

Additional Assessment

43. WRITING in Biology Create a poem that describes the functions of at least five cell organelles.

DBQ Document-Based Questions

The graph below describes the relationship between the amount of glucose entering a cell and the rate at which the glucose enters the cell with the help of carrier proteins. Use this graph to answer questions 44 and 45.

Data obtained from: Raven, P.H., and Johnson, G.B. 2002. *Biology*, 6th ed.: 99.

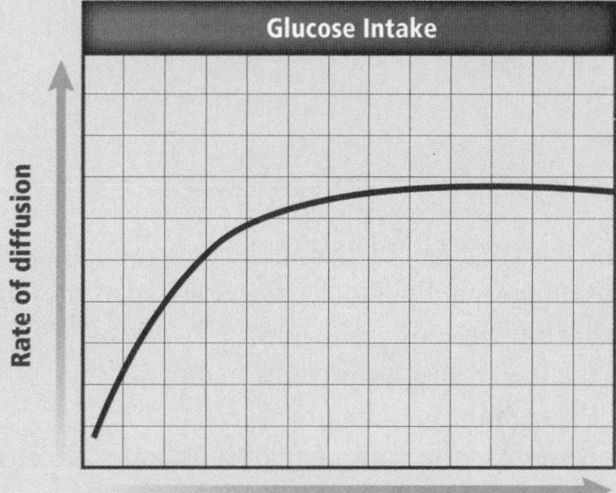

44. Summarize the relationship between the amount of glucose and the rate of diffusion.

45. Infer why the rate of diffusion tapers off with higher amounts of glucose. Make an illustration to explain your answer.

Cumulative Review

46. Rabbits were introduced into Australia in the 1800s. The population of rabbits grew unchecked. Explain why this occurred and how this could adversely affect an ecosystem. **(Chapter 3)**

47. Algae are a group of plantlike organisms. Many of these organisms produce their own food by photosynthesis. Are these organisms autotrophs or heterotrophs? Explain. **(Chapter 2)**

Standardized Test Practice

Cumulative

Multiple Choice

Use the illustration below to answer questions 1 and 2.

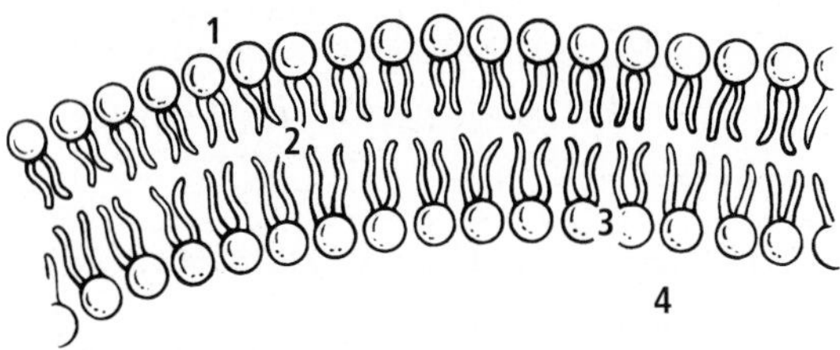

1. Which number in the illustration represents the location where you would expect to find water-insoluble substances?
   A. 1
   B. 2
   C. 3
   D. 4

2. Which is the effect of having the polar and nonpolar ends of phospholipid molecules oriented as they are in this illustration?
   A. It allows transport proteins to move easily through the membrane.
   B. It controls the movement of substances across the membrane.
   C. It helps the cell to maintain its characteristic shape.
   D. It makes more room inside the phospholipid bilayer.

3. Which of these habitats would be best suited for a population of *r*-strategists?
   A. desert
   B. grassland
   C. deciduous forest
   D. tropical rain forest

4. Which adaptation helps plants survive in a tundra biome?
   A. deciduous leaves that fall off as winter approaches
   B. leaves that store water
   C. roots that grow only a few centimeters deep
   D. underground stems that are protected from grazing animals

5. Which is a nonrenewable resource?
   A. clean water from freshwater sources
   B. energy provided by the Sun
   C. an animal species that has become extinct
   D. a type of fish that is caught in the ocean

Use this incomplete equation to answer questions 6 and 7.

$$CH_4 + 4Cl_2 \rightarrow \_\_HCl + \_\_CCl_4$$

6. The chemical equation above shows what can happen in a reaction between methane and chlorine gas. The coefficients have been left out in the product side of the equation. Which is the correct coefficient for HCl?
   A. 1
   B. 2
   C. 4
   D. 8

7. Which is the minimum number of chlorine (Cl) atoms needed for the reaction shown in the equation?
   A. 1
   B. 2
   C. 4
   D. 8

8. Why is *Caulerpa taxifolia* considered an invasive species in some coastal areas of North America?
   A. It is dangerous to humans.
   B. It is nonnative to the area.
   C. It grows slowly and invades over time.
   D. It outcompetes native species for resources.

Short Answer

9. Use a flowchart to organize information about cell organelles and protein synthesis. For each step, analyze the role of the organelle in protein synthesis.

10. Compare and contrast the functions of carbohydrates, lipids, proteins, and nucleic acids.

11. State why the polarity of water molecules makes water a good solvent.

Use the figure below to answer question 12.

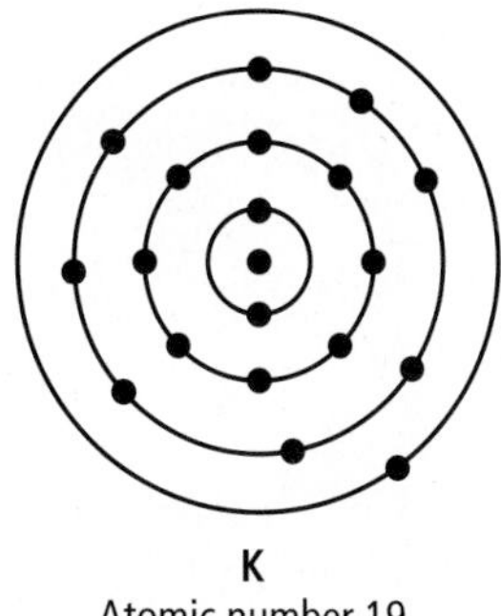

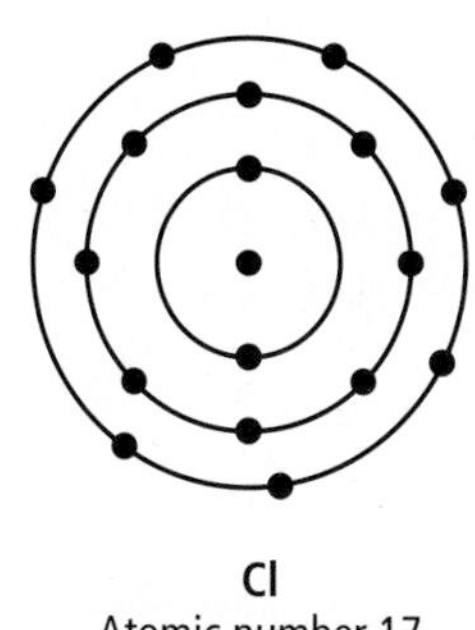

12. Use the figure to describe how the ionic compound potassium chloride (KCl) is formed.

13. What might happen if cell membranes were not selectively permeable?

14. Choose a specific natural resource and develop a plan for the sustainable use of that resource.

15. What can you infer about the evolution of bacterial cells from studying their structure?

Extended Response

The illustration below shows a single animal cell in an isotonic solution. Use the illustration to answer question 16.

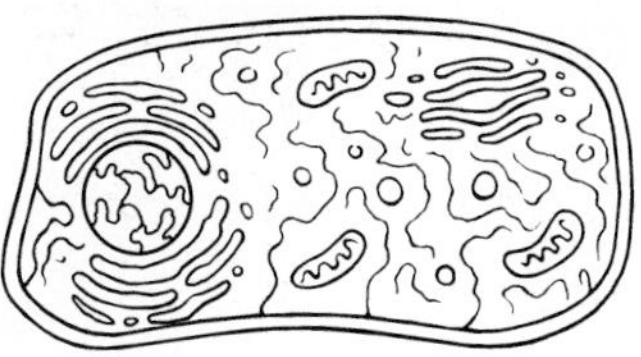

16. Describe what would happen to this cell in a hypertonic solution and in a hypotonic solution.

17. Explain why direct economic value is not the only important consideration when it comes to biodiversity.

18. Analyze why an electron microscope can produce higher magnification than a light microscope.

19. Assess why transport proteins are needed to move certain substances across a cell membrane.

Essay Question

Recently, some international trade agreements have allowed scientists and companies to patent the discoveries they make about organisms and their genetic material. For instance, it is possible to patent seeds that have genes for disease resistance or plants that can be used in medicine or industry. Owners of a patent then have greater control over the use of these organisms.

Using the information in the paragraph above, answer the following question in essay format.

20. Based on what you know about biodiversity, identify some pros and cons for a patent system. Write an essay exploring the pros and cons of patenting discoveries about organisms.

NEED EXTRA HELP?																				
If You Missed Question . . .	1	2	3	4	5	6	7	8	9	10	11	12	13	14	15	16	17	18	19	20
Review Section . . .	7.2	7.2	4.1	3.1	5.3	6.2	6.8	5.2	7.3	6.4	6.3	6.1	7.2	5.3	7.1	7.4	5.1	7.1	7.4	5.2

CHAPTER 8 Cellular Energy

BIG Idea Photosynthesis converts the Sun's energy into chemical energy, while cellular respiration uses chemical energy to carry out life functions.

Section 1
How Organisms Obtain Energy
MAIN Idea All living organisms use energy to carry out all biological processes.

Section 2
Photosynthesis
MAIN Idea Light energy is trapped and converted into chemical energy during photosynthesis.

Section 3
Cellular Respiration
MAIN Idea Living organisms obtain energy by breaking down organic molecules during cellular respiration.

BioFacts

- Sheep eat a variety of grasses to obtain glucose for energy.
- Grass is green because it contains chlorophyll, a pigment found in chloroplasts.
- A marathon runner might use 4.5 g of glucose every minute to power his or her muscles.

LAUNCH Lab

How is energy transformed?

The flow of energy in living systems is driven by a variety of chemical reactions and chemical processes. Energy is transformed from the Sun's radiant energy to chemical energy to other forms of energy along the way. In this lab, you will observe two processes in which energy is transformed.

Procedure

1. Read and complete the lab safety form.
2. Measure 100 mL of **water** using a **graduated cylinder;** pour into a **250-mL beaker.** Use a **thermometer** to record the water temperature.
3. Measure 40 g of **anhydrous calcium chloride** ($CaCl_2$). Use a **stirring rod** to dissolve $CaCl_2$ in the water. Record the solution temperature every fifteen seconds for three minutes.
4. Repeat Steps 2 and 3 using 40 g of **Epsom salts** instead of $CaCl_2$.
5. Graph your data using a different color for each process.

Analysis

1. **Describe** the graph of your data.
2. **Predict** what energy transformations occurred in the two processes.

Biology Online

Visit biologygmh.com to:

- study the entire chapter online
- explore the Interactive Time Line, Concepts in Motion, Microscopy Links, and links to virtual dissections
- access Web links for more information, projects, and activities
- review content online with the Interactive Tutor and take Self-Check Quizzes

Stages of Cellular Respiration Make the following Foldable to help you understand how all organisms get energy from food through cellular respiration.

STEP 1 Fold a 5.5 cm tab along the long edge of a sheet of 11″x 17″ paper.

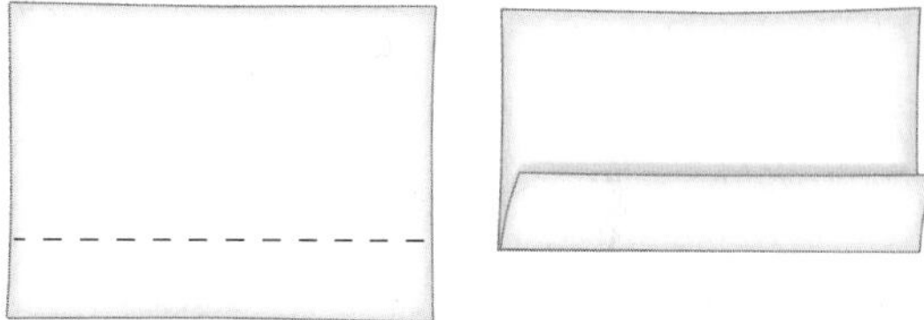

STEP 2 Fold the paper into thirds as shown.

STEP 3 Staple or glue the outer edges of the tabs to form a three-pocket book Foldable. Label the pockets as shown, and use quarter sheets of notebook paper or 3″x 5″ index cards to record information. Place them in the appropriate pockets.

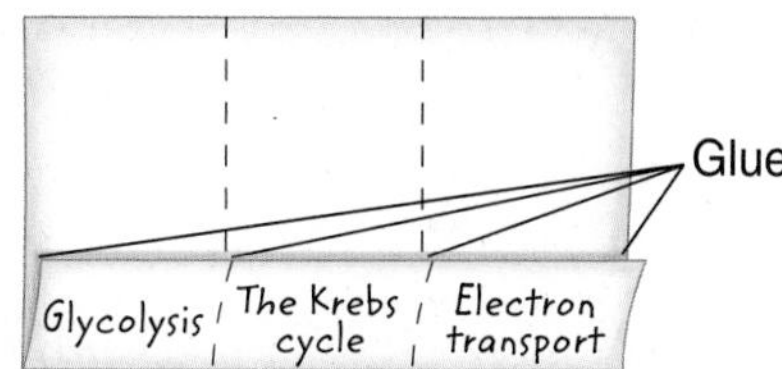

FOLDABLES Use this Foldable with Section 8.3. As you study the section, record what you learn about each stage of cellular respiration: Glycolysis, Krebs Cycle, and Electron Transport.

Section 8.1

Reading Preview

Objectives

- **Summarize** two laws of thermodynamics.
- **Compare and contrast** autotrophs and heterotrophs.
- **Describe** how ATP works in a cell.

Review Vocabulary

trophic level: each step in a food chain or a food web

New Vocabulary

energy
thermodynamics
metabolism
photosynthesis
cellular respiration
adenosine triphosphate (ATP)

How Organisms Obtain Energy

MAIN Idea **All living organisms use energy to carry out all biological processes.**

Real-World Reading Link New York City is sometimes called "the city that never sleeps." Much like the nonstop movement of a big city, living cells are sites of constant activity.

Transformation of Energy

Many chemical reactions and processes in your cells are ongoing, even when you might not think you are using any energy. Macromolecules are assembled and broken down, substances are transported across cell membranes, and genetic instructions are transmitted. All of these cellular activities require **energy**—the ability to do work. **Figure 8.1** shows some of the major advancements in the study of cellular energy. **Thermodynamics** is the study of the flow and transformation of energy in the universe.

Laws of thermodynamics The first law of thermodynamics is the law of conservation of energy, which states that energy can be converted from one form to another, but it cannot be created nor destroyed. For example, the stored energy in food is converted to chemical energy when you eat and to mechanical energy when you run or kick a ball.

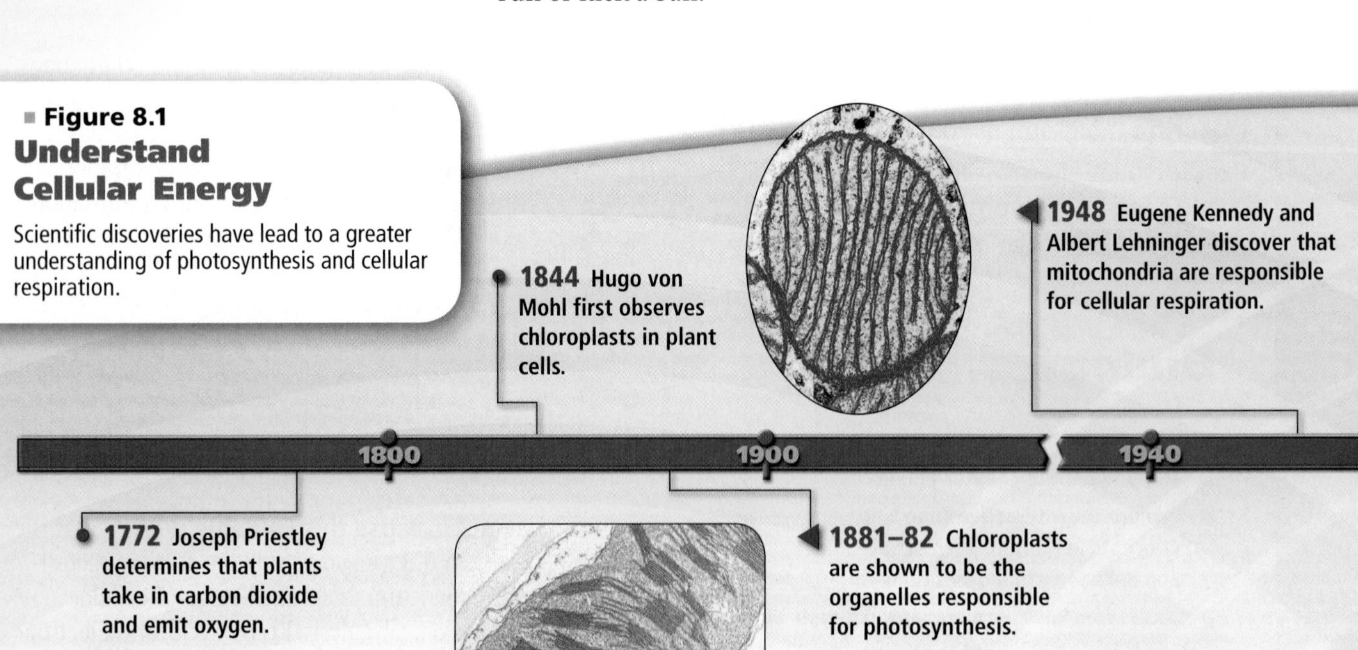

1844 Hugo von Mohl first observes chloroplasts in plant cells.
1948 Eugene Kennedy and Albert Lehninger discover that mitochondria are responsible for cellular respiration.
1800
1900
1940
1772 Joseph Priestley determines that plants take in carbon dioxide and emit oxygen.
1881–82 Chloroplasts are shown to be the organelles responsible for photosynthesis.

■ Figure 8.1
Understand Cellular Energy
Scientific discoveries have lead to a greater understanding of photosynthesis and cellular respiration.

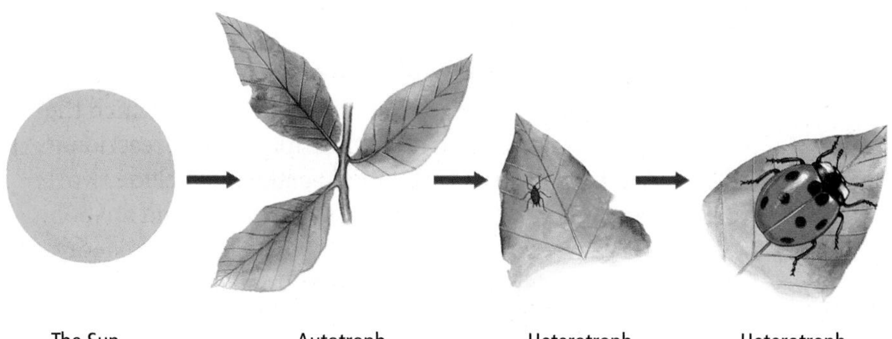

■ **Figure 8.2** Almost all the energy in living organisms originates from the Sun, and energy flows from autotrophs to heterotrophs.
Relate ***the laws of thermodynamics to the organisms in the figure.***

The second law of thermodynamics states that energy cannot be converted without the loss of usable energy. The energy that is "lost" is generally converted to thermal energy. Entropy (EN truh pee) is the measure of disorder, or unusable energy, in a system. Therefore, the second law of thermodynamics can also be stated "entropy increases." One example of the second law of thermodynamics is evident in food chains. Recall from Chapter 2 that in a food chain the amount of usable energy that is available to the next trophic level decreases.

VOCABULARY
WORD ORIGIN
Autotroph
comes from the Greek word *autotrophos,* meaning *supplying one's own food.*

Autotrophs and heterotrophs All organisms need energy to live. Directly or indirectly nearly all the energy for life comes from the Sun. Recall from Chapter 2 that some organisms make their own food, while others must obtain it from other organisms. Autotrophs are organisms that make their own food. Some autotrophs, called chemoautotrophs, use inorganic substances such as hydrogen sulfide as a source of energy. Other autotrophs, such as the plant in **Figure 8.2,** convert light energy from the Sun into chemical energy. Autotrophs that convert energy from the Sun are called photoautotrophs. Heterotrophs, such as the aphid and the ladybug in **Figure 8.2,** are organisms that need to ingest food to obtain energy.

1980 Exploring the mitochondria of fruit flies and mice, Jaime Miquel provides the first evidence that mitochondrial breakdown causes aging.

2002 Josephine S. Modica-Napolitano proposes that differences in healthy and cancerous mitochondria could lead to early cancer detection and new cancer treatments.

1960 1980 2000

1993 Fossils of the earliest known prokaryotic cells are unearthed. These cells carried out photosynthesis.

Concepts In Motion Interactive Time Line
To learn more about these discoveries and others, visit biologygmh.com. **Biology Online**

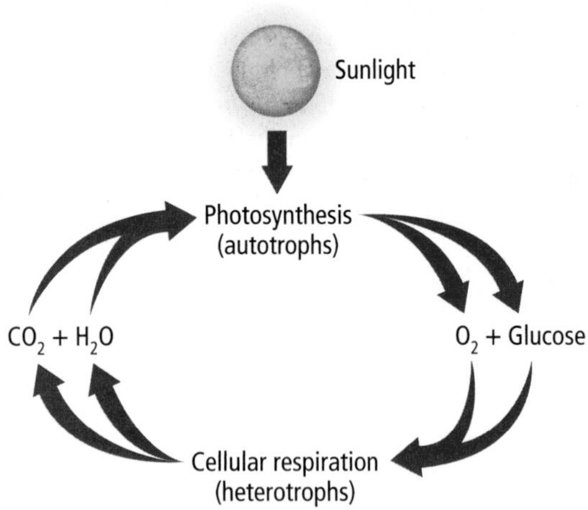

Figure 8.3 In an ecosystem, photosynthesis and cellular respiration form a cycle.
Identify *the anabolic and catabolic pathways in the figure.*

LAUNCH Lab

Review Based on what you've read about energy transformations, how would you now answer the analysis questions?

Metabolism

All of the chemical reactions in a cell are referred to as the cell's **metabolism.** A series of chemical reactions in which the product of one reaction is the substrate for the next reaction is called a metabolic pathway. Metabolic pathways include two broad types: catabolic (ka tuh BAH lik) pathways and anabolic (a nuh BAH lik) pathways. Catabolic pathways release energy by breaking down larger molecules into smaller molecules. Anabolic pathways use the energy released by catabolic pathways to build larger molecules from smaller molecules. The relationship of anabolic and catabolic pathways results in the continual flow of energy within an organism.

Energy continually flows between the metabolic reactions of organisms in an ecosystem. **Photosynthesis** is the anabolic pathway in which light energy from the Sun is converted to chemical energy for use by the cell. In this reaction, autotrophs use light energy, carbon dioxide, and water to form glucose and oxygen. As shown in **Figure 8.3,** the energy stored in the glucose produced by photosynthesis can be transferred to other organisms when the molecules are consumed as food.

Cellular respiration is the catabolic pathway in which organic molecules are broken down to release energy for use by the cell. In cellular respiration, oxygen is used to break down organic molecules, resulting in the production of carbon dioxide and water. Notice the cyclical nature of these processes in **Figure 8.3,** where the products of one reaction are the reactants for the other reaction.

MiniLab 8.1

Relate Photosynthesis to Cellular Respiration

How do photosynthesis and cellular respiration work together in an ecosystem? Use a chemical indicator to examine how carbon dioxide is transferred in photosynthesis and cellular respiration.

Procedure

1. Read and complete the lab safety form.
2. Prepare a data table to record the contents, treatment, initial color, and final color for two experimental test tubes.
3. Pour 100 mL **bromothymol blue (BTB) solution** into a **beaker.** Using a **straw,** exhale gently into the solution until it just turns yellow. **WARNING:** ***do not blow so hard that the solution bubbles over or that you get a headache. Do not suck on the straw.***
4. Fill two large **test tubes** three-quarters full with the yellow BTB solution.
5. Cover one test tube with **aluminum foil.** Place a 6-cm sprig of an **aquatic plant** into both of the tubes, tightly **stopper** the tubes, and place them in a **rack** in bright light overnight.
6. Record your observations in your data table.

Analysis

1. **Infer** the purpose of the tube covered in aluminum foil.
2. **Explain** how your results demonstrate that photosynthesis and cellular respiration depend on one another.

ATP: The Unit of Cellular Energy

Connection to Chemistry Energy exists in many forms including light energy, mechanical energy, thermal energy, and chemical energy. In living organisms, chemical energy is stored in biological molecules and can be converted to other forms of energy when needed. For example, the chemical energy in biological molecules is converted to mechanical energy when muscles contract. **Adenosine triphosphate** (uh DEN uh seen • tri FAHS fayt)—**ATP**—is the most important biological molecule that provides chemical energy.

ATP structure Recall from Chapter 6 that ATP is a multipurpose storehouse of chemical energy that can be used by cells in a variety of reactions. Although other carrier molecules transport energy within cells, ATP is the most abundant energy-carrier molecule in cells and is found in all types of organisms. As shown in **Figure 8.4,** ATP is a nucleotide made of an adenine base, a ribose sugar, and three phosphate groups.

ATP function ATP releases energy when the bond between the second and third phosphate groups is broken, forming a molecule called adenosine diphosphate (ADP) and a free phosphate group, as shown in **Figure 8.4.** Energy is stored in the phosphate bond formed when ADP receives a phosphate group and becomes ATP. As shown in **Figure 8.4,** ATP and ADP can be interchanged by the addition or removal of a phosphate group. Sometimes ADP becomes adenosine monophosphate (AMP) by losing an additional phosphate group. There is less energy released in this reaction, so most of the energy reactions in the cell involve ATP and ADP.

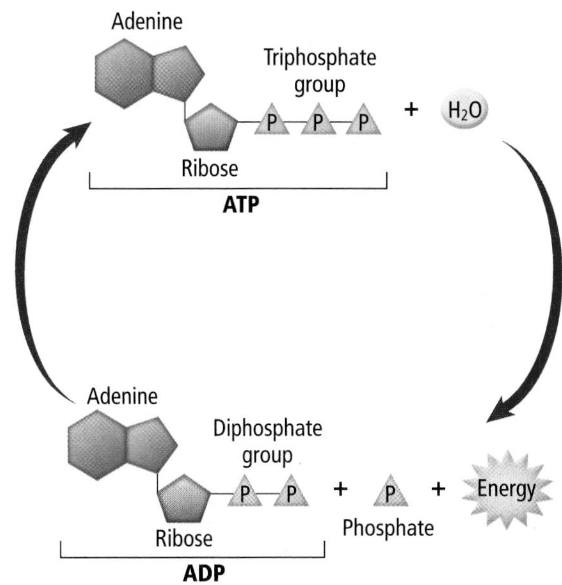

Figure 8.4 The breakdown of ATP releases energy for powering cellular activities in organisms.

Concepts In Motion

Interactive Figure To see an animation of ATP, visit biologygmh.com.

Section 8.1 Assessment

Section Summary

- The laws of thermodynamics control the flow and transformation of energy in organisms.
- Some organisms produce their own food, whereas others obtain energy from the food they ingest.
- Cells store and release energy through coupled anabolic and catabolic reactions.
- The energy released from the breakdown of ATP drives cellular activities.

Understand Main Ideas

1. MAIN Idea **Identify** the major source of energy for living organisms.
2. **Describe** an example of the first law of thermodynamics.
3. **Compare and contrast** anabolic and catabolic pathways.
4. **Explain** how ATP stores and releases energy.

Think Critically

WRITING in **Biology**

5. Write an essay describing the laws of thermodynamics. Use examples in biology to support your ideas.
6. **Create** an analogy to describe the relationship between photosynthesis and cellular respiration.

Section 8.2

Reading Preview

Objectives

- **Summarize** the two phases of photosynthesis.
- **Explain** the function of a chloroplast during the light reactions.
- **Describe** and diagram electron transport.

Review Vocabulary

carbohydrate: an organic compound containing only carbon, hydrogen, and oxygen, usually in a 1:2:1 ratio

New Vocabulary

thylakoid
granum
stroma
pigment
$NADP^+$
Calvin cycle
rubisco

Photosynthesis

MAIN Idea Light energy is trapped and converted into chemical energy during photosynthesis.

Real-World Reading Link Energy is transformed all around us every day. Batteries convert chemical energy into electric energy, and radios convert electric energy into the energy carried by sound waves. Similarly, some autotrophs convert light energy into chemical energy through photosynthesis.

Overview of Photosynthesis

Most autotrophs—including plants—make organic compounds, such as sugars, by a process called photosynthesis. Recall that photosynthesis is a process in which light energy is converted into chemical energy. The overall chemical equation for photosynthesis is shown below.

$$6CO_2 + 6H_2O \xrightarrow{\text{light}} C_6H_{12}O_6 + 6O_2$$

Photosynthesis occurs in two phases. The locations of these phases are shown in **Figure 8.5.** In phase one, the light-dependent reactions, light energy is absorbed and then converted into chemical energy in the form of ATP and NADPH. In phase two, the light-independent reactions, the ATP and NADPH that were formed in phase one are used to make glucose. Once glucose is produced, it can be joined to other simple sugars to form larger molecules. These larger molecules are complex carbohydrates, such as starch. Recall from Chapter 6 that carbohydrates are composed of repeating units of small organic molecules. The end products of photosynthesis also can be used to make other organic molecules, such as proteins, lipids, and nucleic acids.

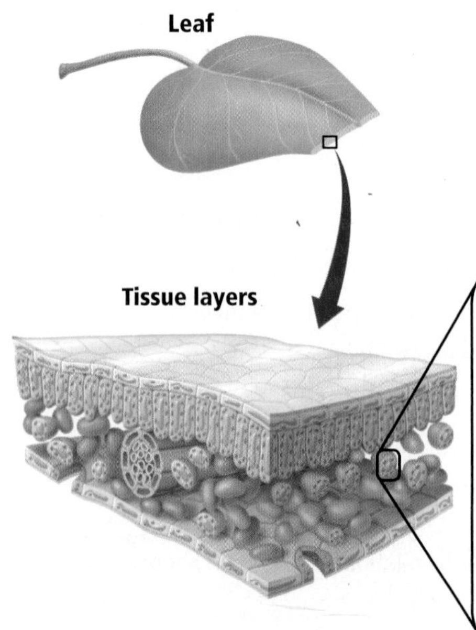

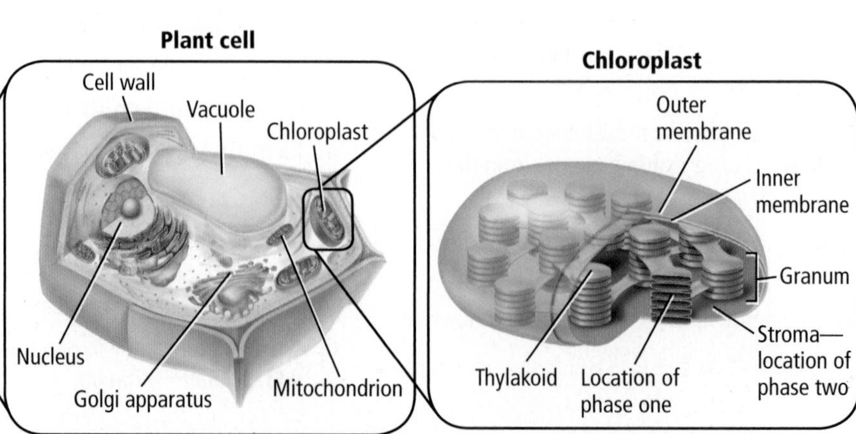

Figure 8.5 Photosynthesis occurs inside pigmented organelles called chloroplasts.

Phase One: Light Reactions

The absorption of light is the first step in photosynthesis. Plants have special organelles to capture light energy. Once the energy is captured, two energy storage molecules—NADPH and ATP—are produced to be used in the light-independent reactions.

Chloroplasts Large organelles, called chloroplasts, capture light energy in photosynthetic organisms. In plants, chloroplasts are found mainly in the cells of leaves. As shown in **Figure 8.5,** chloroplasts are disc-shaped organelles that contain two main compartments essential to photosynthesis. The first compartment is called the thylakoid (THI la koyd). **Thylakoids** are flattened saclike membranes that are arranged in stacks. These stacks are called **grana.** Light-dependent reactions take place within the thylakoids. The second important compartment is called the **stroma**—the fluid-filled space that is outside the grana. This is the location of the light-independent reactions in phase two of photosynthesis.

Pigments Light-absorbing colored molecules called **pigments** are found in the thylakoid membranes of chloroplasts. Different pigments absorb specific wavelengths of light, as illustrated in **Figure 8.6.**

The major light-absorbing pigments in plants are chlorophylls. There are several types of chlorophylls, but the most common two are chlorophyll *a* and chlorophyll *b*. The structure of chlorophyll can differ from one molecule to another, enabling distinct chlorophyll molecules to absorb light at unique areas of the visible spectrum. In general, chlorophylls absorb most strongly in the violet-blue region of the visible light spectrum and reflect light in the green region of the spectrum. This is why plant parts that contain chlorophyll appear green to the human eye.

 **Reading Check** **Distinguish** between thylakoids and stroma.

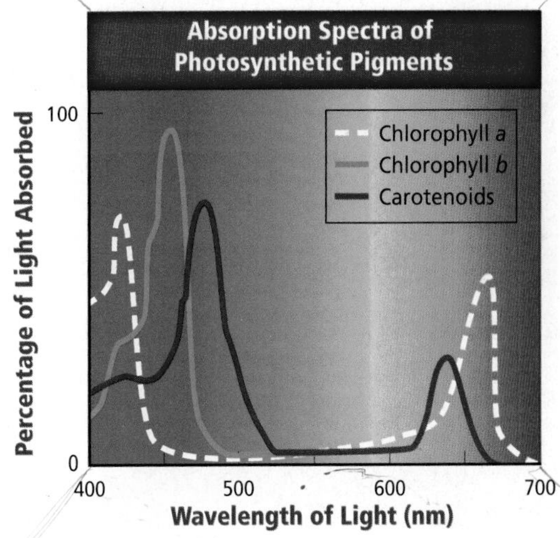

Figure 8.6 Colorful pigments found in the leaves of trees differ in their ability to absorb specific wavelengths of light.
Hypothesize *the effect on light absorption if a plant did not have chlorophyll* b.

MiniLab 8.2

Observe Chloroplasts

What do chloroplasts look like? Most ecosystems and organisms in the world depend on tiny organelles called chloroplasts. Discover what chloroplasts look like in this investigation.

Procedure

1. Read and complete the lab safety form.
2. Observe the **slides of plant and algae cells** with a **microscope.**
3. Identify the chloroplasts in the cells you observe.
4. Make a data table to record your observations and sketch the chloroplasts in the cells.

Analysis

1. **Compare and contrast** the physical features of the chloroplasts you observed in the different cells.
2. **Hypothesize** why green plant leaves vary in color.

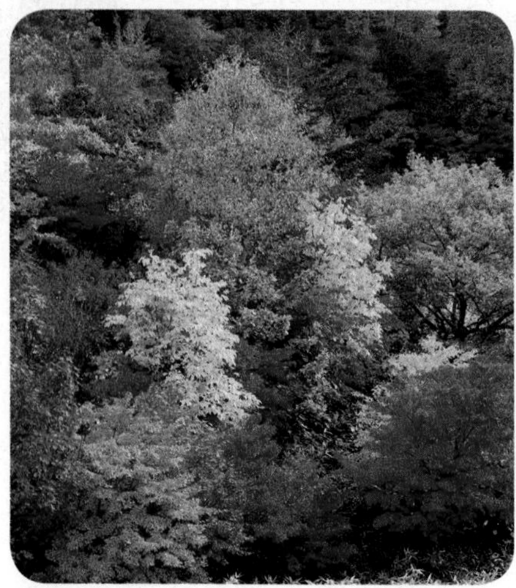

■ **Figure 8.7** When chlorophyll breaks down in the leaves of some trees, the other pigments become visible.

In addition to chlorophylls, most photosynthetic organisms contain accessory pigments that allow plants to trap additional light energy from other areas of the visible spectrum. One such group of accessory pigments is the carotenoids (kuh ROH tuh noydz). Carotenoids, such as ß-carotene (beta-carotene), absorb light mainly in the blue and green regions of the spectrum, while reflecting most light in the yellow, orange, and red regions. Carotenoids produce the colors of carrots and sweet potatoes.

Chlorophylls are more abundant than other pigments in leaves, and thus hide the colors of the other pigments. However, autumn in certain parts of the United States can bring out shades of yellow, red, and orange as the leaves turn colors, as shown in **Figure 8.7.** As trees prepare to lose their leaves before winter, the chlorophyll molecules break down, revealing the colors of the other pigments.

Electron transport The structure of the thylakoid membrane is the key to the efficient energy transfer during electron transport. Thylakoid membranes have a large surface area, which provides the space needed to hold large numbers of electron-transporting molecules and two types of protein complexes called photosystems. Photosystem I and photosystem II contain light-absorbing pigments and proteins that play important roles in the light reactions. Follow along in **Figure 8.8** as you continue to read about electron transport.

- First, the light energy excites electrons in photosystem II. The light energy also causes a water molecule to split, releasing an electron into the electron transport system, a hydrogen ion (H^+)—also called a proton—into the thylakoid space, and oxygen (O_2) as a waste product. This breakdown of water is essential for photosynthesis to occur.
- The excited electrons move from photosystem II to an electron-acceptor molecule in the thylakoid membrane.
- Next, the electron-acceptor molecule transfers the electrons along a series of electron-carriers to photosystem I.
- In the presence of light, photosystem I transfers the electrons to a protein called ferrodoxin. The electrons lost by photosystem I are replaced by electrons shuttled from photosystem II.
- Finally, ferrodoxin transfers the electrons to the electron carrier **$NADP^+$,** forming the energy-storage molecule NADPH.

VOCABULARY

ACADEMIC VOCABULARY

Transport:
to carry something from one place to another
$NADP^+$ molecules transport electrons during photosynthesis.

Chemiosmosis ATP is produced in conjunction with electron transport by the process of chemiosmosis—the mechanism by which ATP is produced as a result of the flow of electrons down a concentration gradient. The breakdown of water is not only essential for providing the electrons that initiate the electron transport chain, but also for providing the protons (H^+) necessary to drive ATP synthesis during chemiosmosis. The H^+ released during electron transport accumulate in the interior of the thylakoid. As a result of a high concentration of H^+ in the thylakoid interior and a low concentration of H^+ in the stroma, H^+ diffuse down their concentration gradient out of the thylakoid interior into the stroma through ion channels spanning the membrane, as shown in **Figure 8.8.** These channels are enzymes called ATP synthases. As H^+ moves through ATP synthase, ATP is formed in the stroma.

Reading Check Summarize the function of water during chemiosmosis in photosynthesis.

Visualizing Electron Transport

Figure 8.8
Activated electrons are passed from one molecule to another along the thylakoid membrane in a chloroplast. The energy from electrons is used to form a proton gradient. As protons move down the gradient, a phosphate is added to ADP, forming ATP.

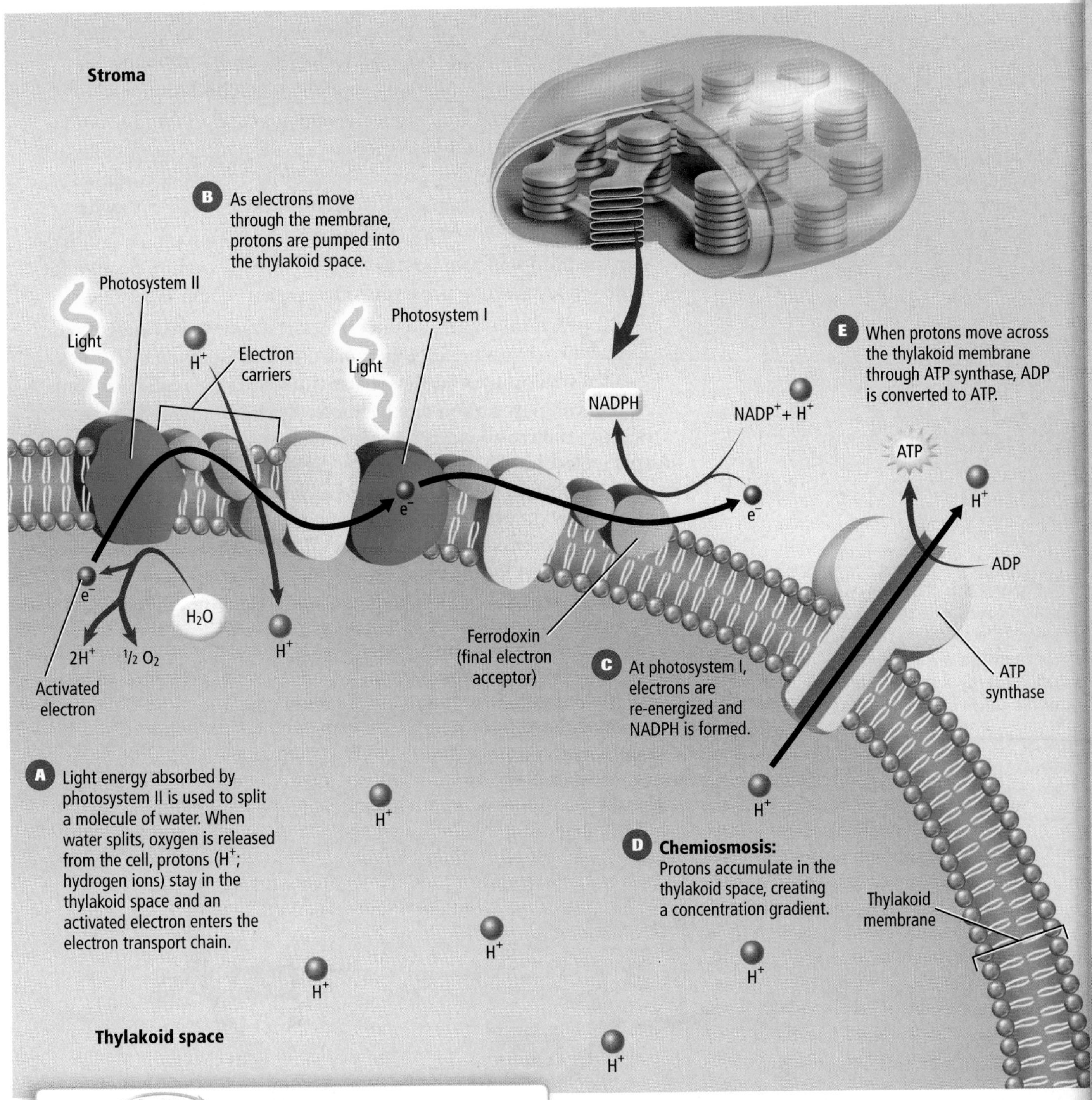

Stroma

B As electrons move through the membrane, protons are pumped into the thylakoid space.

Photosystem II

Light

H^+

Electron carriers

Light

Photosystem I

NADPH

$NADP^+ + H^+$

E When protons move across the thylakoid membrane through ATP synthase, ADP is converted to ATP.

ATP

H^+

e^-

e^-

e^-

ADP

H_2O

$2H^+$

$1/2\ O_2$

H^+

Activated electron

Ferrodoxin (final electron acceptor)

C At photosystem I, electrons are re-energized and NADPH is formed.

ATP synthase

A Light energy absorbed by photosystem II is used to split a molecule of water. When water splits, oxygen is released from the cell, protons (H^+; hydrogen ions) stay in the thylakoid space and an activated electron enters the electron transport chain.

H^+

H^+

D **Chemiosmosis:** Protons accumulate in the thylakoid space, creating a concentration gradient.

Thylakoid membrane

H^+

H^+

H^+

Thylakoid space

H^+

Concepts In Motion **Interactive Figure** To see an animation of electron transport, visit biologygmh.com.

Biology Online

Phase Two: The Calvin Cycle

Although NADPH and ATP provide cells with large amounts of energy, these molecules are not stable enough to store chemical energy for long periods of time. Thus, there is a second phase of photosynthesis called the **Calvin cycle** in which energy is stored in organic molecules such as glucose. The reactions of the Calvin cycle are also referred to as the light-independent reactions. Follow along in **Figure 8.9** as you learn the steps of the Calvin cycle.

- In the first step of the Calvin cycle called carbon fixation, six carbon dioxide (CO_2) molecules combine with six 5-carbon compounds to form twelve 3-carbon molecules called 3-phosphoglycerate (fahs foh GLI suh rayt) (3-PGA). The joining of carbon dioxide with other organic molecules is called carbon fixation.
- In the second step, the chemical energy stored in ATP and NADPH is transferred to the 3-PGA molecules to form high-energy molecules called glyceraldehyde 3-phosphates (G3P). ATP supplies the phosphate groups for forming G3P molecules, while NADPH supplies hydrogen ions and electrons.
- In the third step, two G3P molecules leave the cycle to be used for the production of glucose and other organic compounds.
- In the final step of the Calvin cycle, an enzyme called **rubisco** converts the remaining ten G3P molecules into 5-carbon molecules called ribulose 1, 5-bisphosphates (RuBP). These molecules combine with new carbon dioxide molecules to continue the cycle.

Because rubisco converts inorganic carbon dioxide molecules into organic molecules that can be used by the cell, it is considered one of the most important biological enzymes. Plants use the sugars formed during the Calvin cycle both as a source of energy and as building blocks for complex carbohydrates, including cellulose, which provides structural support for the plant.

CAREERS IN BIOLOGY

Phytochemist A biologist who studies the chemical products of plants is a phytochemist. Phytochemists might work in medical research to find new treatments for diseases. For more information on biology careers, visit biologygmh.com.

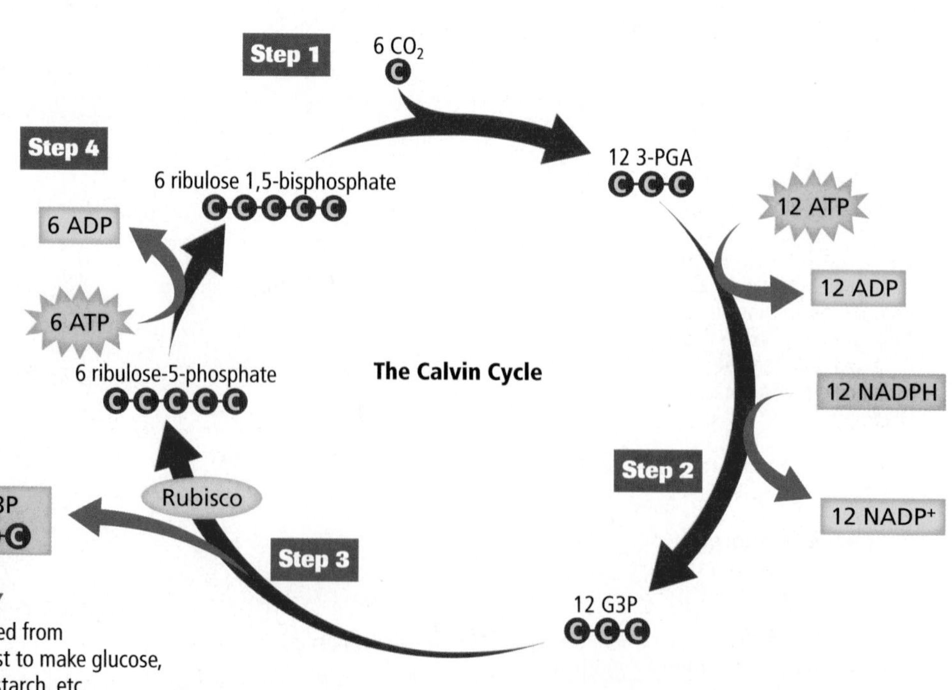

Figure 8.9 The Calvin cycle joins carbon dioxide with organic molecules inside the stroma of the chloroplast.
Determine *the compound in which energy is stored at the end of the Calvin cycle.*

Concepts In Motion
Interactive Figure To see an animation of the Calvin cycle, visit biologygmh.com.

Alternative Pathways

The environment in which an organism lives can impact the organism's ability to carry out photosynthesis. Environments in which the amount of water or carbon dioxide available is insufficient can decrease the ability of a photosynthetic organism to convert light energy into chemical energy. For example, plants in hot, dry environments are subject to excessive water loss that can lead to decreased photosynthesis. Many plants in extreme climates have alternative photosynthesis pathways to maximize energy conversion.

C_4 plants One adaptive pathway that helps plants maintain photosynthesis while minimizing water loss is called the C_4 pathway. The C_4 pathway occurs in plants such as sugarcane and corn. These plants are called C_4 plants because they fix carbon dioxide into four-carbon compounds instead of three-carbon molecules during the Calvin cycle. C_4 plants also have significant structural modifications in the arrangement of cells in the leaves. In general, C_4 plants keep their stomata (plant cell pores) closed during hot days, while the four carbon compounds are transferred to special cells where CO_2 enters the Calvin cycle. This allows for sufficient carbon dioxide uptake, while simultaneously minimizing water loss.

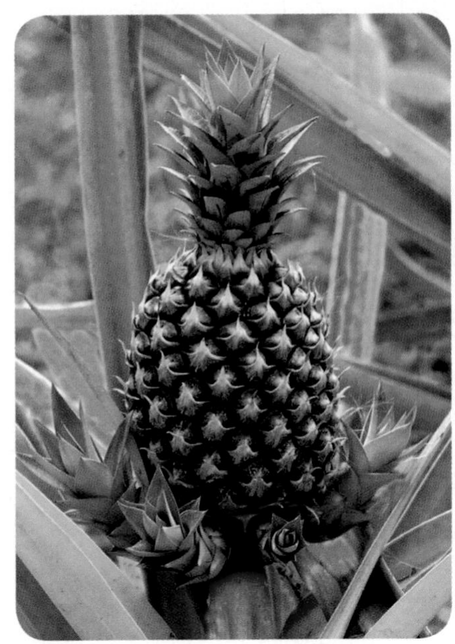

■ **Figure 8.10** This pineapple plant is an example of a CAM plant.

CAM plants Another adaptive pathway used by some plants to maximize photosynthetic activity is called crassulacean (KRAH soo lay shun) acid metabolism (CAM photosynthesis). The CAM pathway occurs in water-conserving plants that live in deserts, salt marshes, and other environments where access to water is limited. CAM plants, such as cacti, orchids, and the pineapple in **Figure 8.10,** allow carbon dioxide to enter the leaves only at night, when the atmosphere is cooler and more humid. At night, these plants fix carbon dioxide into organic compounds. During the day, carbon dioxide is released from these compounds and enters the Calvin cycle. This pathway also allows for sufficient carbon dioxide uptake, while minimizing water loss.

Section 8.2 Assessment

Section Summary

- Plants contain chloroplasts with light-absorbing pigments that convert light energy into chemical energy.
- Photosynthesis is a two-phase process that consists of light reactions and the Calvin cycle.
- In the light reactions, autotrophs trap and convert light energy into chemical energy in the form of NADPH and ATP.
- In the Calvin cycle, chemical energy in ATP and NADPH is used to synthesize carbohydrates such as glucose.

Understand Main Ideas

1. MAIN Idea **Summarize** how chemical energy is formed from light energy during photosynthesis.
2. **Relate** the structure of the chloroplast to the phases of photosynthesis.
3. **Explain** why water is essential for the light reactions.
4. **Summarize** the steps in the Calvin cycle.
5. **Diagram** and explain electron transport.

Think Critically

6. **Predict** how environmental factors such as light intensity and carbon dioxide levels can affect rates of photosynthesis.

WRITING in Biology

7. Research the effects of global warming on photosynthesis. Write an article summarizing your findings.

Section 8.3

Reading Preview

Objectives

- **Summarize** the stages of cellular respiration.
- **Identify** the role of electron carriers in each stage of cellular respiration.
- **Compare** alcoholic fermentation and lactic acid fermentation.

Review Vocabulary

cyanobacterium: a type of bacterium that is a photosynthetic autotroph

New Vocabulary

anaerobic process
aerobic respiration
aerobic process
glycolysis
Krebs cycle
fermentation

Cellular Respiration

MAIN Idea **Living organisms obtain energy by breaking down organic molecules during cellular respiration.**

Real-World Reading Link Monarch butterflies must constantly feed on nectar from flowers to provide energy to sustain themselves during their winter migration to parts of Mexico and California each year. Similarly, humans and other living organisms need reliable food sources to supply energy to survive and grow.

Overview of Cellular Respiration

Recall that organisms obtain energy in a process called cellular respiration. The function of cellular respiration is to harvest electrons from carbon compounds, such as glucose, and use that energy to make ATP. ATP is used to provide energy for cells to do work. The overall chemical equation for cellular respiration is shown below. Notice the equation for cellular respiration is the opposite of the equation for photosynthesis.

$$C_6H_{12}O_6 + 6O_2 \rightarrow 6CO_2 + 6H_2O + \text{Energy}$$

Cellular respiration occurs in two main parts: glycolysis and aerobic respiration. The first stage, glycolysis, is an anaerobic process. **Anaerobic processes** do not require oxygen. **Aerobic respiration** includes the Krebs cycle and electron transport and is an aerobic process. **Aerobic processes** require oxygen. Cellular respiration with aerobic respiration is summarized in **Figure 8.11.**

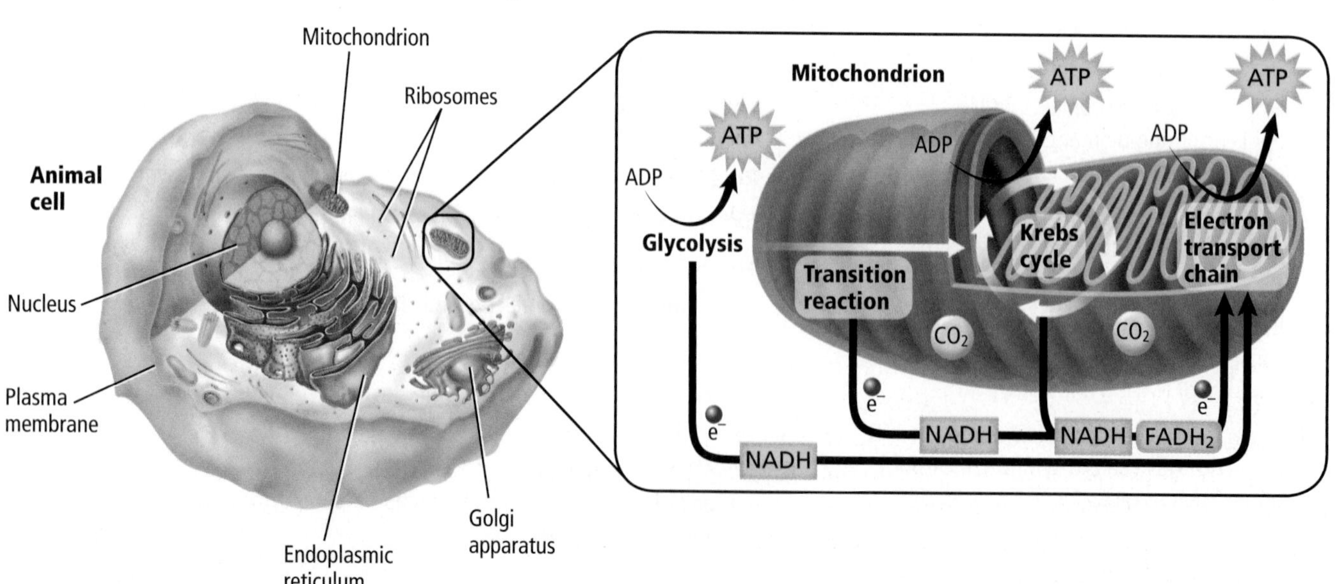

Figure 8.11 Cellular respiration occurs in the mitochondria, the energy powerhouse organelles of the cell.

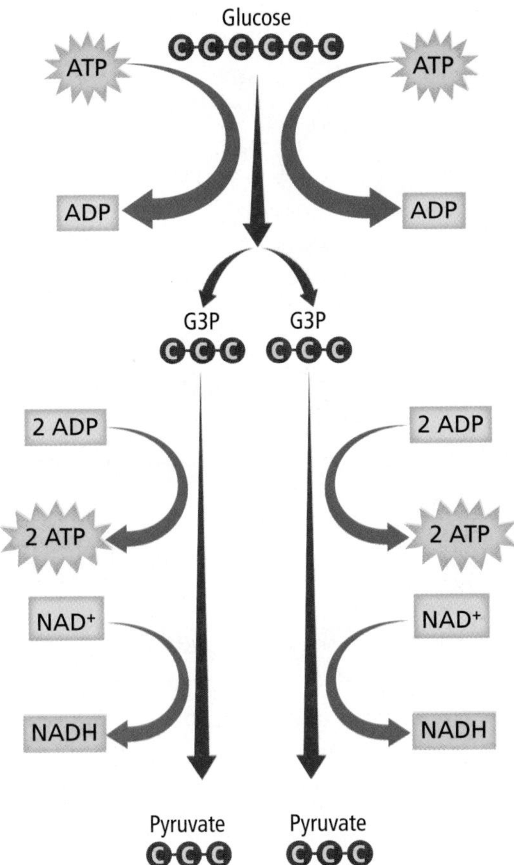

■ **Figure 8.12** Glucose is broken down during glycolysis inside the cytoplasm of cells.
Summarize ***the reactants and products of glycolysis.***

Personal Tutor

To learn about glycolysis, the Krebs cycle, and electron transport, visit biologygmh.com.

Glycolysis

Glucose is broken down in the cytoplasm through the process of **glycolysis.** Two molecules of ATP and two molecules of NADH are formed for each molecule of glucose that is broken down. Follow along with **Figure 8.12** as you read about the steps of glycolysis.

First, two phosphate groups, derived from two molecules of ATP, are joined to glucose. Notice that some energy, two ATP, is required to start the reactions that will produce energy for the cell. The 6-carbon molecule is then broken down into two 3-carbon compounds. Next, two phosphates are added and electrons and hydrogen ions (H^+) combine with two NAD^+ molecules to form two NADH molecules. NAD^+ is similar to NADP, an electron carrier used during photosynthesis. Last, the two 3-carbon compounds are converted into two molecules of pyruvate. At the same time, four molecules of ATP are produced.

Reading Check **Explain** why there is a net yield of two, not four, ATP molecules in glycolysis.

VOCABULARY

WORD ORIGIN

Glycolysis
comes from the Greek words *glykys,* meaning *sweet*
and *lysis,* meaning *to rupture or break.*

FOLDABLES
Incorporate information from this section into your Foldable.

Krebs Cycle

Glycolysis has a net result of two ATP and two pyruvate. Most of the energy from the glucose is still contained in the pyruvate. In the presence of oxygen, pyruvate is transported into the mitochondrial matrix, where it is eventually converted to carbon dioxide. The series of reactions in which pyruvate is broken down into carbon dioxide is called the **Krebs cycle** or tricarboxylic acid (TCA) cycle. This cycle also is referred to as the citric acid cycle.

■ **Figure 8.13** Pyruvate is broken down into carbon dioxide during the Krebs cycle inside the mitochondria of cells.
Trace ***Follow the path of carbon molecules that enter and leave the Krebs cycle.***

Concepts In Motion

Interactive Figure To see an animation of the Krebs cycle, visit biologygmh.com.

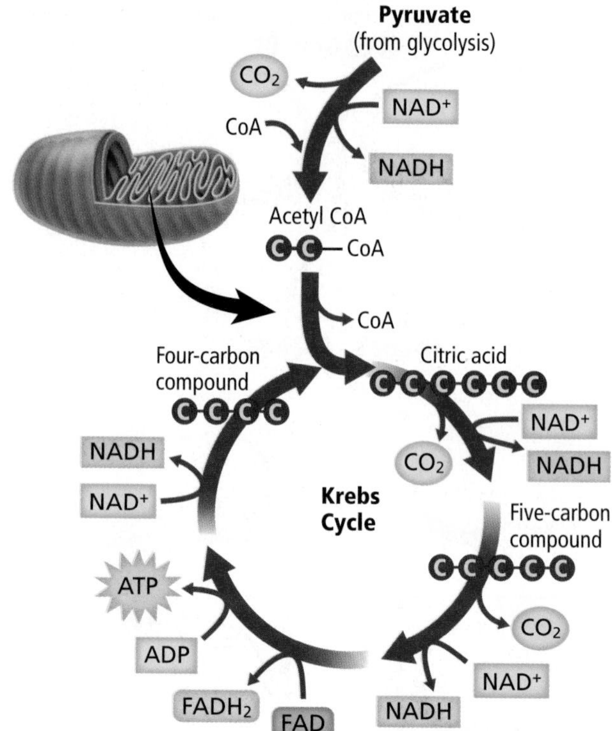

Study Tip

Clarifying Statement Work with a partner to read the text and discuss unfamiliar words and difficult concepts. Write a clarifying statement to summarize the Krebs cycle.

Steps of the Krebs cycle Prior to the Krebs cycle, pyruvate first reacts with coenzyme A (CoA) to form a 2-carbon intermediate called acetyl CoA. At the same time, carbon dioxide is released and NAD^+ is converted to NADH. Acetyl CoA then moves to the mitochondrial matrix. The reaction results in the production of two carbon dioxide molecules and two NADH. Follow along in **Figure 8.13** as you continue reading about the steps of the Krebs cycle.

- The Krebs cycle begins with acetyl CoA combining with a 4-carbon compound to form a 6-carbon compound known as citric acid.
- Citric acid is then broken down in the next series of steps, releasing two molecules of carbon dioxide and generating one ATP, three NADH, and one FADH2. FAD is another electron carrier similar to NAD+ and NADP+.
- Finally, acetyl CoA and citric acid are generated and the cycle continues.

Recall that two molecules of pyruvate are formed during glycolysis, resulting in two "turns" of the Krebs cycle for each glucose molecule. The net yield from the Krebs cycle is six carbon dioxide molecules, two ATP, eight NADH, and two $FADH_2$. NADH and $FADH_2$ move on to play a significant role in the next stage of aerobic respiration.

CAREERS IN BIOLOGY

Bioenergeticist A researcher who studies energy transfers in cells is a bioenergeticist. Some bioenergeticists study mitochondria and their relationship to aging and disease. To learn more about biology careers, visit biologygmh.com.

Electron Transport

In aerobic respiration, electron transport is the final step in the breakdown of glucose. It also is the point at which most of the ATP is produced. High-energy electrons and hydrogen ions from NADH and $FADH_2$ produced in the Krebs cycle are used to convert ADP to ATP.

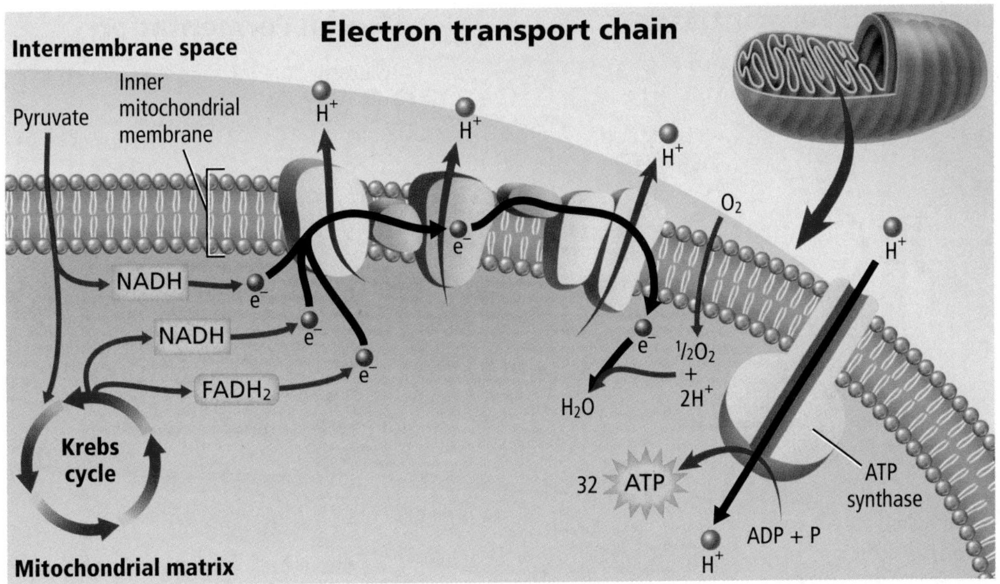

■ **Figure 8.14** Electron transport occurs along the mitochondrial membrane.
Compare and contrast ***electron transport in cellular respiration and photosynthesis.***

As shown in **Figure 8.14,** electrons move along the mitochondrial membrane from one protein to another. As NADH and $FADH_2$ release electrons, the energy carriers are converted to NAD^+ and FAD, and H^+ ions are released into the mitochondrial matrix. The H^+ ions are pumped into the mitochondrial matrix across the inner mitochondrial membrane. H^+ ions then diffuse down their concentration gradient back across the membrane and into the matrix through ATP synthase molecules in chemiosmosis. Electron transport and chemiosmosis in cellular respiration are similar to these processes in photosynthesis. Oxygen is the final electron acceptor in the electron transport system in cellular respiration. Protons and electrons are transferred to oxygen to form water.

Overall, electron transport produces 24 ATP. Each NADH molecule produces three ATP and each group of three $FADH_2$ produces two ATP. In eukaryotes, one molecule of glucose yields 36 ATP.

Prokaryotic cellular respiration Some prokaryotes also undergo aerobic respiration. Because prokaryotes do not have mitochondria, there are a few differences in the process. The main difference involves the use of the prokaryotic cellular membrane as the location of electron transport. In eukaryotic cells, pyruvate is transported to the mitochondria. In prokaryotes, this movement is unnecessary, saving the prokaryotic cell two ATP, and increasing the net total of ATP produced to 38.

VOCABULARY

SCIENCE USAGE V. COMMON USAGE

Concentration

Science usage: the relative amount of a substance dissolved in another substance.
The concentration of hydrogen ions is greater on one side of the membrane than the other.

Common usage: the directing of close, undivided attention.
The student's concentration was focused on the exam.

Anaerobic Respiration

Some cells can function for a short time when oxygen levels are low. Some prokaryotes are anaerobic organisms—they grow and reproduce without oxygen. In some cases these cells continue to produce ATP through glycolysis. However, there are problems with solely relying on glycolysis for energy. Glycolysis only provides two net ATP for each molecule of glucose, and a cell has a limited amount of NAD^+. Glycolysis will stop when all the NAD^+ is used up if there is not a process to replenish NAD^+. The anaerobic pathway that follows glycolysis is anaerobic respiration, or fermentation. **Fermentation** occurs in the cytoplasm and regenerates the cell's supply of NAD^+ while producing a small amount of ATP. The two main types of fermentation are lactic acid fermentation and alcohol fermentation.

Lactic Acid Fermentation

Glucose

GLYCOLYSIS

2 Lactic acid

2 ADP

NAD+

ATP

NADH

2 Pyruvate

Alcohol Fermentation

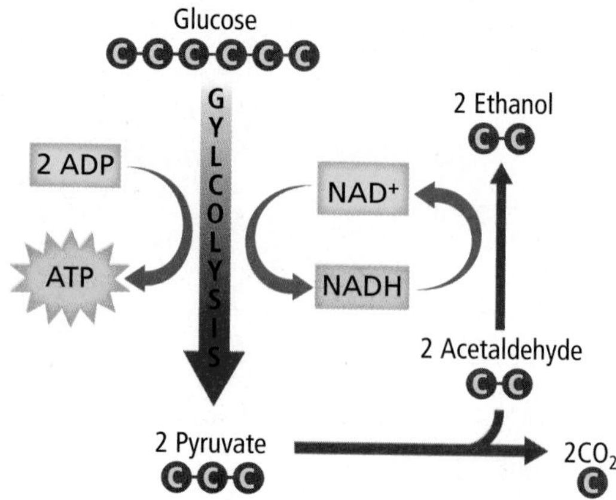

Figure 8.15 When oxygen is absent or in limited supply, fermentation can occur.
Compare and contrast *lactic acid fermentation and alcohol fermentation.*

Connection to Health **Lactic acid fermentation** In lactic acid fermentation, enzymes convert the pyruvate made during glycolysis to lactic acid, as shown in **Figure 8.15.** This involves the transfer of high-energy electrons and protons from NADH. Skeletal muscle produces lactic acid when the body cannot supply enough oxygen, such as during periods of strenuous exercise. When lactic acid builds up in muscle cells, muscles become fatigued and might feel sore. Lactic acid also is produced by several microorganisms that often are used to produce many foods, including cheese, yogurt, and sour cream.

Alcohol fermentation Alcohol fermentation occurs in yeast and some bacteria. **Figure 8.15** shows the chemical reaction that occurs during alcohol fermentation when pyruvate is converted to ethyl alcohol and carbon dioxide. Similar to lactic acid fermentation, NADH donates electrons during this reaction and NAD^+ is regenerated.

DATA ANALYSIS LAB 8.1

Based On Real Data*

Interpret the Data

How does viral infection affect cellular respiration? Infection by viruses can significantly affect cellular respiration and the ability of cells to produce ATP. To test the effect of viral infection on the stages of cellular respiration, cells were infected with a virus, and the amount of lactic acid and ATP produced were measured.

Data and Observations

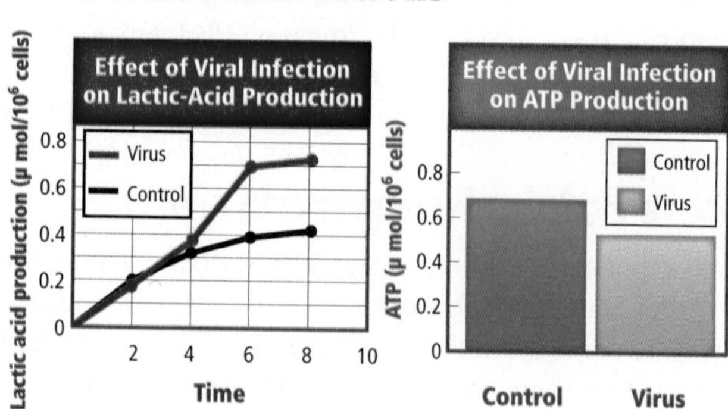

Think Critically

1. **Analyze** How did the virus affect lactic acid production in the cells?
2. **Calculate** After 8 h, by what percentage was the lactic acid higher in the virus group than in the control group? By what percentage was ATP production decreased?
3. **Infer** why having a virus like the flu might make a person feel tired.

Data obtained from: El-Bacha, T., et al. 2004. Mayaro virus infection alters glucose metabolism in cultured cells through activation of the enzyme 6-phosphofructo 1-kinase. *Molecular and Cellular Biochemistry* 266: 191–198.

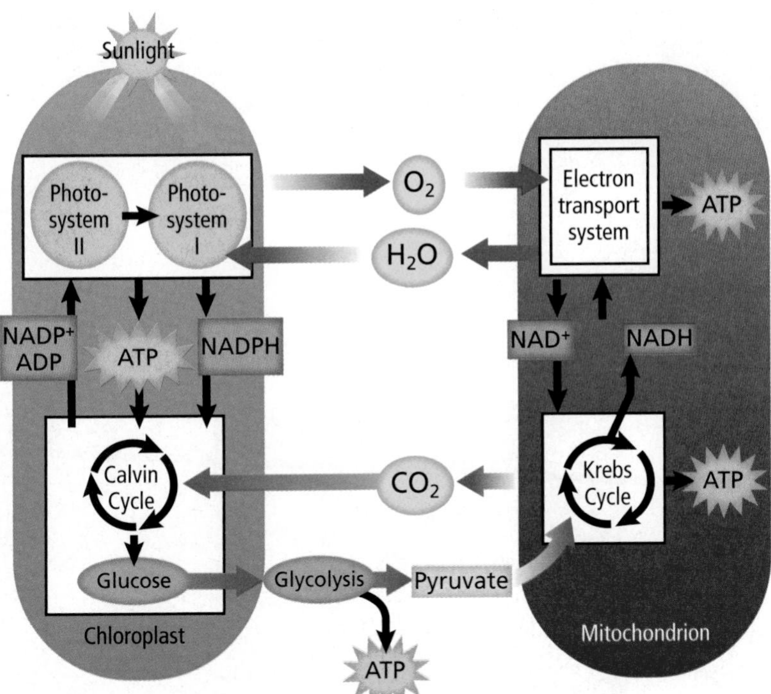

Figure 8.16 Photosynthesis and cellular respiration form a cycle in which the products of one metabolic pathway form the reactants of the other metabolic pathway.

Photosynthesis and Cellular Respiration

As you have learned, photosynthesis and cellular respiration are two important processes that cells use to obtain energy. They are metabolic pathways that produce and break down simple carbohydrates. **Figure 8.16** shows how these two processes are related. Recall that the products of photosynthesis are oxygen and glucose—the reactants needed for cellular respiration. The products of cellular respiration—carbon dioxide and water—are the reactants for photosynthesis.

Section 8.3 Assessment

Section Summary

- Many living organisms use cellular respiration to break down glucose.
- The stages of cellular respiration are glycolysis, Krebs cycle, and electron transport.
- NADH and $FADH_2$ are important electron carriers for cellular respiration.
- In the absence of oxygen, cells can sustain glycolysis by fermentation.

Understand Main Ideas

1. MAIN Idea **Name** the final form of chemical energy produced by cells during cellular respiration.
2. **Identify** How many carbons from one glucose molecule enter one round of the Krebs cycle?
3. **Explain** how high-energy electrons are used in electron transport.
4. **Describe** the role of fermentation in maintaining ATP and NAD^+ levels.

Think Critically

MATH in Biology

5. How many ATP, NADH, and $FADH_2$ are produced in each step of cellular respiration? How is the number of ATP produced different from the net ATP available?
6. **Compare and contrast** the two types of fermentation.

CUTTING-EDGE BIOLOGY

Tracking Human Evolution

DNA evidence has been used to solve mysteries that were decades, or even centuries old—but imagine trying to unravel a mystery that is millions of years old. This is exactly what scientists are doing when they use DNA analysis to track human evolution.

Mitochondrial DNA You might wonder what mitochondria have to do with DNA analysis and human evolution. Mitochondria often are called the powerhouses of the cell. They are the organelles in which cells release the energy stored in food. Mitochondria have their own DNA, which is much smaller than nuclear DNA and more abundant due to its presence outside the nucleus and the number of mitochondria in most cells. Mitochondrial DNA (mtDNA) is easier to detect and extract than nuclear DNA, making it a useful tool for unlocking some of science's toughest mysteries.

One particular characteristic of mtDNA makes it especially useful for tracking human evolution. Mitochondria are inherited through maternal lineage. When a sperm and egg combine at fertilization, the nuclear DNA of the two gametes combine but the mitochondria in the offspring are supplied solely by the egg. Therefore, mtDNA can be used as a marker to trace motherhood from generation to generation.

Tracing evolution Scientists use DNA analysis to trace the path of pre-human creatures, called hominids, as they spread around the world. The genomic DNA that is found in the nuclei of cells often is degraded or present in miniscule amounts in these ancient samples. However, scientists discovered that mtDNA is found abundantly and can be used for their analysis.

EM Magnification: 150,000×

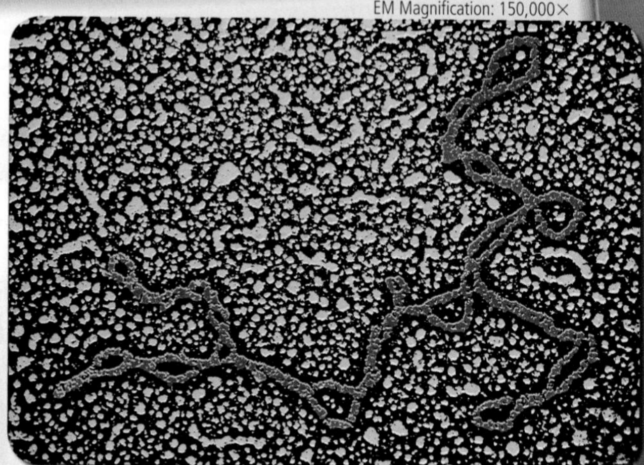

Mitochondrial DNA (red) is separate from nuclear DNA found in the nucleus of a cell.

Mutations in mtDNA occur in relatively predictable patterns, and those patterns are studied and compared by scientists. By comparing mutations in mtDNA, scientists can trace mtDNA inheritance. Based on these studies of mtDNA, scientists have determined that the most recent maternal common ancestor of people living on Earth today is "Mitochondrial Eve." Mitochondrial Eve is believed to be a woman who lived in Africa approximately 200,000 years ago.

Based on the theory of Mitochondrial Eve, an international study is being conducted to trace the migration and ancestry of early humans. The project uses mtDNA sequences in females, but uses sequences from the Y chromosome to trace ancestry in males.

WRITING in Biology

Research Paper To learn more about mtDNA, visit biologygmh.com. Choose one aspect of the research that is being done with mtDNA and write a research paper about it.

BIOLAB

Design Your Own

DO DIFFERENT WAVELENGTHS OF LIGHT AFFECT THE RATE OF PHOTOSYNTHESIS?

Background: **Photosynthesizing organisms need light to complete photosynthesis. White light is composed of the different colors of light found in the visible light spectrum, and each color of light has a specific wavelength. During this lab, you will design an experiment to test the effect of different light wavelengths on the rate of photosynthesis.**

Question: *How do different wavelengths of light affect photosynthesis rates?*

Possible Materials

Choose materials that would be appropriate for this lab.

aquatic plant material
erlenmeyer flasks
test tubes (15 mL)
graduated cylinder (10 mL)
metric ruler
colored cellophane (assorted colors)
aluminum foil
lamp with reflector and 150 W bulb
baking soda solution (0.25%)
watch with a second hand

Safety Precautions

Plan and Perform the Experiment

1. Read and complete the lab safety form.
2. Predict how different wavelengths of light will affect the rate of photosynthesis in your plant.
3. Design an experiment to test your prediction. Write a list of steps you will follow and identify the controls and variables you will use.
4. Explain how you will generate light with different wavelengths, supply the plant with carbon dioxide, and measure the oxygen production of the plants.
5. Create a data table for recording your observations and measurements.
6. Make sure your teacher approves your plan before you begin.
7. Conduct your experiment as approved.
8. **Cleanup and Disposal** Clean up all equipment as instructed by your teacher, and return everything to its proper place. Dispose of plant material as instructed by your teacher. Wash your hands thoroughly with soap and water.

Analyze and Conclude

1. **Identify** the controls and variables in your experiment.
2. **Explain** how you measured the rate of photosynthesis.
3. **Graph** your data.
4. **Describe** how the rate of photosynthesis is affected by different wavelengths of light based on your data.
5. **Discuss** whether or not your data supported your prediction.
6. **Error Analysis** Identify possible sources of error in your experimental design, procedure, and data collection.
7. **Suggest** how you would reduce these sources of error if repeating the experiment.

COMMUNICATE

Peer Review Visit biologygmh.com and post your data. Review data posted by other students. Discuss and use comments from other students in your class to improve your own methods.

Study Guide

Download quizzes, key terms, and flash cards from biologygmh.com.

FOLDABLES **Compare and Contrast** Examine the similarities and differences between the process of electron transport in mitochondria and the process of electron transport in chloroplasts.

Vocabulary	Key Concepts
Section 8.1 How Organisms Obtain Energy	
• adenosine triphosphate (ATP) (p. 221) • cellular respiration (p. 220) • energy (p. 218) • metabolism (p. 220) • photosynthesis (p. 220) • thermodynamics (p. 218)	MAIN Idea All living organisms use energy to carry out all biological processes. • The laws of thermodynamics control the flow and transformation of energy in organisms. • Some organisms produce their own food, whereas others obtain energy from the food they ingest. • Cells store and release energy through coupled anabolic and catabolic reactions. • The energy released from the breakdown of ATP drives cellular activities.
Section 8.2 Photosynthesis	
• Calvin cycle (p. 226) • granum (p. 223) • $NADP^+$ (p. 224) • pigment (p. 223) • rubisco (p. 226) • stroma (p. 223) • thylakoid (p. 223)	MAIN Idea Light energy is trapped and converted into chemical energy during photosynthesis. • Plants contain chloroplasts with light-absorbing pigments that convert light energy into chemical energy. • Photosynthesis is a two-phase process that consists of light reactions and the Calvin cycle. • In the light reactions, autotrophs trap and convert light energy into chemical energy in the form of NADPH and ATP. • In the Calvin cycle, chemical energy in ATP and NADPH is used to synthesize carbohydrates such as glucose.
Section 8.3 Cellular Respiration	
• aerobic process (p. 228) • aerobic respiration (p. 228) • anaerobic process (p. 228) • fermentation (p. 231) • glycolysis (p. 229) • Krebs cycle (p. 229)	MAIN Idea Living organisms obtain energy by breaking down organic molecules during cellular respiration. • Many living organisms use cellular respiration to break down glucose. • The stages of cellular respiration are glycolysis, Krebs cycle, and electron transport. • NADH and $FADH_2$ are important electron carriers for cellular respiration. • In the absence of oxygen, cells can sustain glycolysis by fermentation.

Assessment

Section 8.1

Vocabulary Review

Each of the following sentences is false. Make the sentence true by replacing the italicized word with a vocabulary term from the Study Guide page.

1. *Heterotrophs* are the energy currency of the cell.
2. The study of the flow and transformation of energy is called *energy.*
3. *Bioenergetics* can exist in many forms.
4. Chemical reactions that create energy within a cell are refered to as *autotrophs.*
5. Light energy is converted into chemical energy during the process of *sunlight.*

Understand Key Concepts

6. Which is not a characteristic of energy?
   A. cannot be created nor destroyed
   B. is the capacity to do work
   C. exists in forms such as chemical, light, and mechanical
   D. changes spontaneously from disorder to order
7. Which organism depends on an external source of organic compounds?
   A. autotroph
   B. heterotroph
   C. chemoautotroph
   D. photoautotroph

Use the figure to answer question 8.

8. Which part of this food chain provides energy to just one other part?
   A. chemoautotroph
   B. heterotroph
   C. the Sun
   D. photoautotroph
9. What do cells store and release as the main source of chemical energy?
   A. ATP **C.** $NADP^+$
   B. ADP **D.** NADPH

Constructed Response

10. **Short Answer** How do autotrophs and heterotrophs differ in the way they obtain energy?
11. **Open Ended** Use an analogy to describe the role of ATP in living organisms.

Think Critically

12. **Describe** how energy is released from ATP.
13. **Relate** anabolic and catabolic reactions. Create an analogy for the relationship between photosynthesis and cellular respiration.

Section 8.2

Vocabulary Review

Write the vocabulary term from the Study Guide page for each definition.

14. location of the light reactions
15. a stack of thylakoids
16. colored molecule that absorbs light
17. a process in which energy is stored in organic molecules

Understand Key Concepts

Use the equation below to answer question 18.

$$6CO_2 + 6H_2O \xrightarrow{\text{energy}} C_6H_{12}O_6 + ?$$

18. What waste product of photosynthesis is released to the environment?
    A. carbon dioxide
    B. water
    C. oxygen
    D. ammonia

19. Which is the internal membrane of the chloroplast that is organized into flattened membranous sacs?
   A. thylakoids **C.** theca
   B. mitochondria **D.** stroma

Use the figure below to answer question 20.

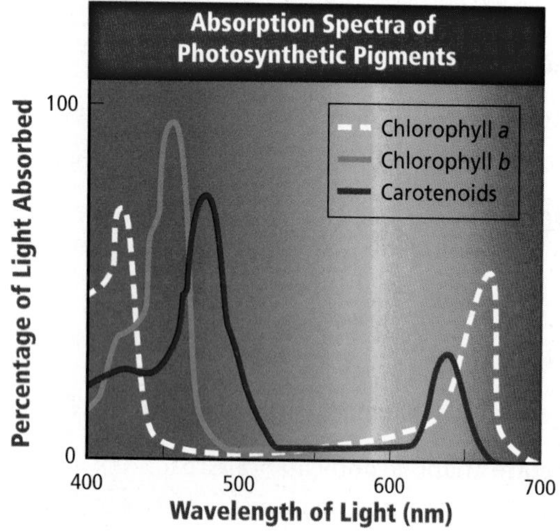

20. Of which wavelength of light do carotenoids absorb the greatest percentage?
   A. 400 **C.** 600
   B. 500 **D.** 700

21. Which supplies energy used to synthesize carbohydrates during the Calvin cycle?
   A. CO_2 and ATP
   B. ATP and NADPH
   C. NADPH and H_2O
   D. H_2O and O_2

Constructed Response

22. **Short Answer** Summarize the phases of photosynthesis and describe where each phase occurs in the chloroplast.

23. **Short Answer** Why is hydrogen ion generation essential for ATP production during photosynthesis?

24. **Short Answer** Explain why the Calvin cycle depends on light reactions.

Think Critically

25. **Explain** the following statement: The oxygen generated by photosynthesis is simply a by-product formed during the production of ATP and carbohydrates.

26. **Predict** the effect of the loss of forests on cellular respiration in other organisms.

27. **Describe** two alternative photosynthesis pathways found in plants. Suggest how these adaptations might help plants.

Section 8.3

Vocabulary Review

Define each vocabulary term in a complete sentence.

28. Krebs cycle
29. anaerobic process
30. fermentation
31. aerobic
32. glycolysis

Understand Key Concepts

Use the figure below to answer questions 33 and 34.

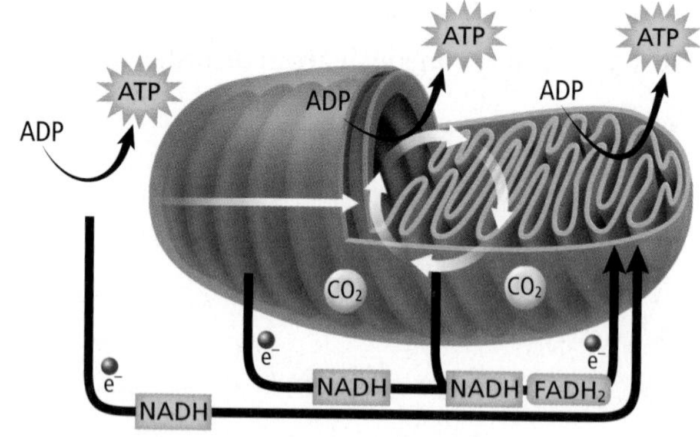

33. Which organelle is illustrated in the figure?
   A. golgi apparatus
   B. mitochondrion
   C. nucleus
   D. endoplasmic reticulum

34. Which process does not occur in the organelle illustrated above?
   A. glycolysis
   B. Krebs cycle
   C. conversion of pyruvate to acetyl CoA
   D. electron transport

Biology Online Chapter Test biologygmh.com

35. Which is not a stage of cellular respiration?
   A. glycolysis
   B. Krebs cycle
   C. electron transport chain
   D. lactic acid fermentation

36. What is produced when the electrons leave the electron transport chain in cellular respiration and bind to the final electron acceptor for the chain?
   A. H_2O
   B. O_2
   C. CO_2
   D. CO

37. In which molecule is most of the energy of glucose stored at the end of glycolysis?
   A. pyruvate
   B. acetyl CoA
   C. ATP
   D. NADH

Constructed Response

38. **Short Answer** Discuss the roles of NADH and $FADH_2$ in cellular respiration.

39. **Short Answer** In cellular respiration where do the electrons in the electron transport chain come from and what is their final destination?

40. **Short Answer** Why do your muscles hurt for some time after a large amount of strenuous exercise?

Think Critically

41. **Explain** The end products of cellular respiration are CO_2 and H_2O. Where do the oxygen atoms in the CO_2 come from? Where does the oxygen atom in H_2O come from?

42. **Infer** What is the advantage of aerobic metabolism over anaerobic metabolism in energy production in living organisms?

43. **Compare and contrast** electron transport in photosynthesis and cellular respiration.

Additional Assessment

44. *WRITING in* **Biology** Write an article using what you know about the relationship between photosynthesis and cellular respiration to explain the importance of plants in an ecosystem.

DBQ Document-Based Questions

Cadmium is a heavy metal that is toxic to humans, plants, and animals. It is often found as a contaminant in soil. Use the data below to answer questions about the effect of cadmium on photosynthesis in tomato plants.

Data obtained from: Chaffei, C., et al. 2004. Cadmium toxicity induced changes in nitrogen management in *Lycopersicon esculentum* leading to a metabolic safeguard through an amino acid storage strategy. *Plant and Cell Physiology* 45(11): 1681–1693.

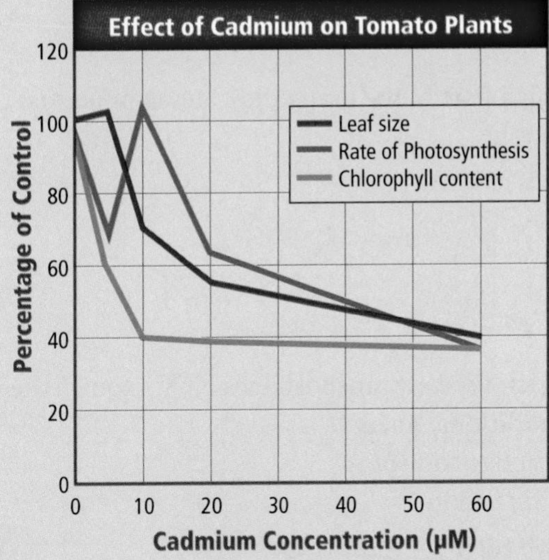

45. What was the effect of cadmium on leaf size, chlorophyll content, and photosynthesis rate?

46. At what concentration of cadmium was the largest effect on leaf size observed? On chlorophyll content? On photosynthesis rate?

47. Predict the effects on cellular respiration if an animal eats contaminated tomatoes.

Cumulative Review

48. Explain how certain toxins, such as PCBs (polychlorinated biphenyls), can increase in concentration in trophic levels. **(Chapter 5)**

Standardized Test Practice

Cumulative

Multiple Choice

1. Suppose the most common form of element X is X-97. The isotope X-99 has more of which?
   A. neutrons
   B. protons
   C. orbiting electrons
   D. overall charge

Use this graph to answer question 2.

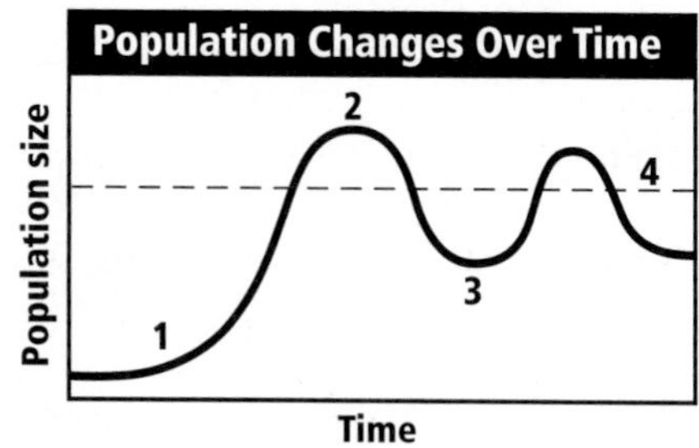

2. Which part of the graph indicates exponential growth?
   A. 1
   B. 2
   C. 3
   D. 4

3. Which type of transport does NOT require the input of additional energy?
   A. active transport
   B. diffusion
   C. endocytosis
   D. exocytosis

4. Which step occurs during the Calvin cycle?
   A. formation of ATP
   B. formation of six-carbon sugars
   C. release of oxygen gas
   D. transport of electrons by $NADP^+$

5. Which describes extinctions caused by deforestation in tropical rain forests?
   A. ecosystem pollution
   B. habitat destruction
   C. introduced species
   D. species overexploitation

Use the diagram below to answer question 6.

```
H—C=O
  |
H—C*—OH
  |
H—C*—OH
  |
H—C—H
  |
  OH
```

6. Based on the diagram, which is the correct molecular formula if the molecule shown above has 6 carbons?
   A. $C_6H_8O_4$
   B. $C_6H_{10}O_6$
   C. $C_6H_{12}O_4$
   D. $C_6H_{12}O_6$

7. Which energy transformation can occur only in autotrophs?
   A. chemical energy into mechanical energy
   B. electrical energy into thermal energy
   C. light energy into chemical energy
   D. mechanical energy into thermal energy

8. Which statement does the cell theory support?
   A. Cells can form from proteins in the environment.
   B. Cells contain membrane-bound organelles.
   C. Life-forms are made of one or more cells.
   D. Organelles are the smallest form of life.

9. Which part of the scientific method evaluates the procedures used in an experiment?
   A. form a hypothesis
   B. publish results
   C. make an observation
   D. peer review

Short Answer

Use the illustration below to answer question 10.

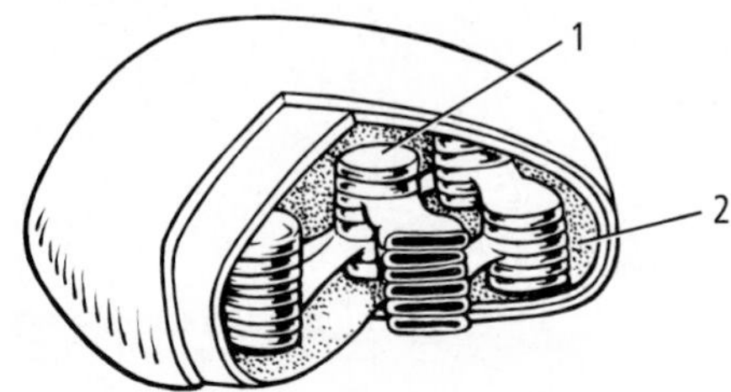

10. The diagram above shows a chloroplast. Name the two parts shown in the diagram and state which phase of photosynthesis occurs in each part.

11. Compare and contrast the structure of a cell wall and the structure of a cell membrane.

12. Relate the bonds between phosphate groups in ATP to the release of energy when a molecule of ATP is changed to ADP.

13. Name three components of a cell's plasma membrane and explain why each component is important for the function of the cell.

14. What kind of mixture is formed by stirring a small amount of table salt into water until the salt all dissolves? Identify the components of this mixture.

15. In which parts of a plant would you expect to find cells with the most chloroplasts? Explain your answer.

16. Long-distance runners often talk about training to raise their anaerobic threshold. The anaerobic threshold is the point at which certain muscles do not have enough oxygen to perform aerobic respiration and begin to perform anaerobic respiration. Hypothesize why you think it is important for competitive runners to raise their anaerobic threshold.

Extended Response

Use the graph below to answer question 17.

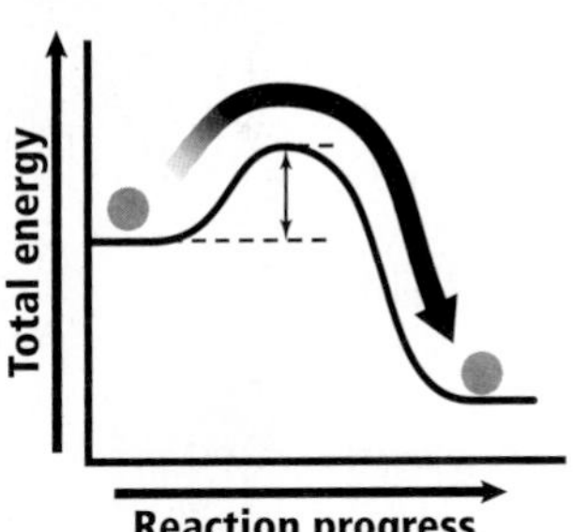

17. The graph shows the effect of an enzyme involved in the breakdown of proteins in the digestive system. Hypothesize how protein digestion would be different in a person who does not have this enzyme.

18. Which organelle would you expect to find in large numbers in cells that pump stomach acid out against a concentration gradient? Give a reason for your answer.

Essay Question

The human body constantly interacts with the environment, taking in some substances and releasing others. Many substances humans take in have a specific role in maintaining basic cellular processes such as respiration, ion transport, and synthesis of various macromolecules. Likewise, many of the substances released by the body are waste products of cellular processes.

Using the information in the paragraph above, answer the following question in essay format.

19. Write an essay that explains how humans take in substances that are important for cellular respiration, and how they release the waste products from this process.

NEED EXTRA HELP?																			
If You Missed Question . . .	1	2	3	4	5	6	7	8	9	10	11	12	13	14	15	16	17	18	19
Review Section . . .	6.1	4.1	7.4	8.2	5.2	6.4	8.1	7.1	1.3	8.2	7.3	8.1	7.2	6.3	7.3	8.3	6.2	7.2, 7.4	8.3

CHAPTER 9

Cellular Reproduction

BIG Idea **Cells go through a life cycle that includes interphase, mitosis, and cytokinesis.**

Section 1
Cellular Growth
MAIN Idea Cells grow until they reach their size limit, then they either stop growing or divide.

Section 2
Mitosis and Cytokinesis
MAIN Idea Eukaryotic cells reproduce by mitosis, the process of nuclear division, and cytokinesis, the process of cytoplasm division.

Section 3
Cell Cycle Regulation
MAIN Idea The normal cell cycle is regulated by cyclin proteins.

BioFacts

- Most animals stop growing once they reach a certain size, while most plants continue growing as long as they are alive.
- Plant roots contain regions where, at any given time, large numbers of cells undergo mitosis.
- Chemical treatments or changes in environmental conditions inhibit mitosis in onions, which prevents sprouting and extends storage times.

Root tip cells undergoing mitosis
Stained LM Magnification: 160×

Onion root tip
Stained LM Magnification: 50×

Start-Up Activities

LAUNCH Lab

From where do healthy cells come?

All living things are composed of cells. The only way an organism can grow or heal itself is by cellular reproduction. Healthy cells perform vital life functions, and they reproduce to form more cells. In this lab you will investigate the appearance of different cell types.

Procedure

1. Read and complete the lab safety form.
2. Observe prepared **slides of human cells** under high magnification using a **light microscope.**
3. Observe **onion root tip cells** under the microscope.
4. Observe other cells on the **prepared slides** your teacher will give you.
5. Draw diagrams of the sample cells you observed. Identify and label any of the structures you recognize.

Analysis

1. **Compare and contrast** the different cells you observed.
2. **Hypothesize** why the cells you observed had different appearances and structures. How could you identify diseased cells?

Visit biologygmh.com to:

- study the entire chapter online
- explore the Concepts in Motion, Microscopy Links, Virtual Labs, and links to virtual dissections
- access Web links for more information, projects, and activities
- review content online with the Interactive Tutor and take Self-Check Quizzes

Mitosis and Cytokinesis Make this Foldable to help you understand how cells reproduce by a process called mitosis, resulting in two genetically identical cells.

STEP 1 Stack three sheets of notebook paper approximately 1.5 cm apart vertically as illustrated.

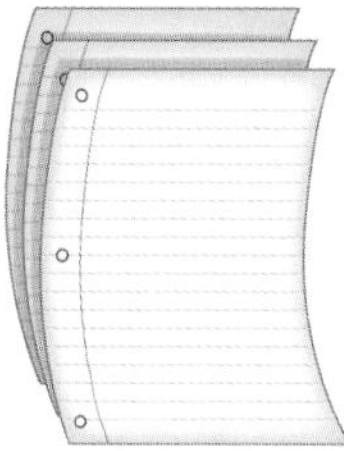

STEP 2 Roll up the bottom edges and fold to form six tabs.

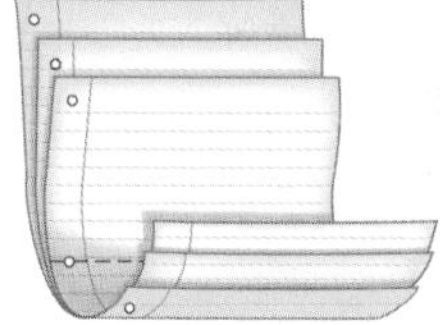

STEP 3 Staple along the folded edge to secure all sheets. Rotate the Foldable and, with the stapled end at the top, label the tabs as illustrated.

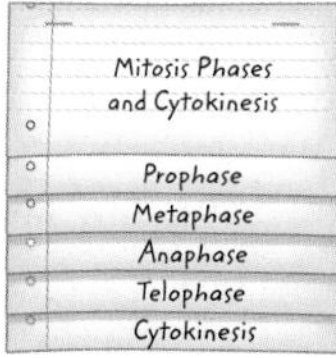

FOLDABLES **Use this Foldable with Section 9.2.** As you study the section, record what you learn about each of the four phases of mitosis. In the tab labeled *Cytokinesis,* write a brief description of cytokinesis, the division of cytoplasm.

Section 9.1

Reading Preview

Objectives

- **Explain** why cells are relatively small.
- **Summarize** the primary stages of the cell cycle.
- **Describe** the stages of interphase.

Review Vocabulary

selective permeability: process in which a membrane allows some substances to pass through while keeping others out

New Vocabulary

cell cycle
interphase
mitosis
cytokinesis
chromosome
chromatin

Cellular Growth

MAIN Idea **Cells grow until they reach their size limit, then they either stop growing or divide.**

Real-World Reading Link If you've ever played a doubles match in tennis, you probably felt that you and your partner could effectively cover your half of the court. However, if the court were much larger, perhaps you could no longer reach your shots. For the best game, the tennis court must be kept at regulation size. Cell size also must be limited to ensure that the needs of the cell are met.

Cell Size Limitations

Most cells are less than 100 μm (100×10^{-6} m) in diameter, which is smaller than the period at the end of this sentence. Why are most cells so small? This section investigates several factors that influence cell size.

Ratio of surface area to volume The key factor that limits the size of a cell is the ratio of its surface area to its volume. The surface area of the cell refers to the area covered by the plasma membrane. Recall from Chapter 7 that the plasma membrane is the structure through which all nutrients and waste products must pass. The volume refers to the space taken by the inner contents of the cell, including the organelles in the cytoplasm and the nucleus.

Connection to Math To illustrate the ratio of surface area to volume, consider the small cube in **Figure 9.1,** which has sides of one micrometer (μm) in length. This is approximately the size of a bacterial cell. To calculate the surface area of the cube, multiply length times width times the number of sides (1 μm × 1 μm × 6 sides), which equals 6 μm^2. To calculate the volume of the cell, multiply length times width times height (1 μm × 1 μm × 1 μm), which equals 1 μm^3. The ratio of surface area to volume is 6:1.

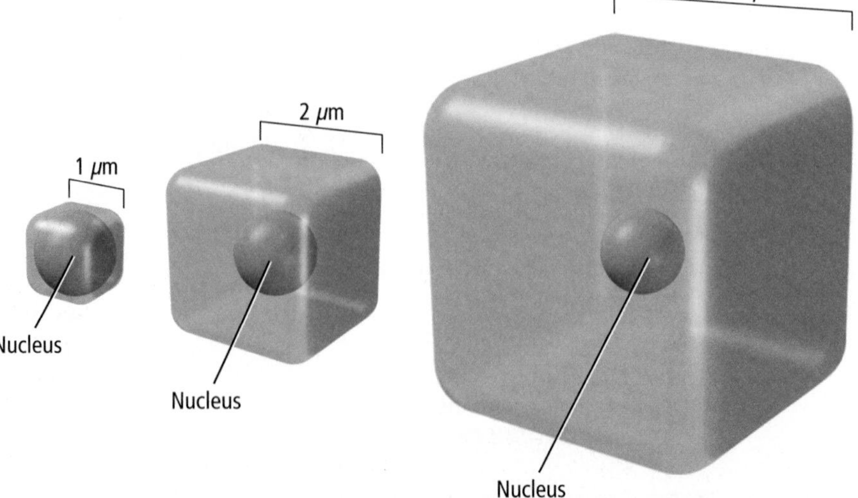

Figure 9.1 The ratio of surface area to volume decreases as a cell gets bigger. The smallest cube shown has a ratio of 6 (1 μm × 1 μm × 6 sides) to 1 (1 μm × 1 μm × 1 μm), while the largest cube has a ratio of 96 (4 μm × 4 μm × 6 sides) to 64 (4 μm × 4 μm × 4 μm) or 3:2.

If the cubic cell grows to 2 μm per side, as represented in **Figure 9.1,** the surface area becomes 24 μm^2 and the volume is 8 μm^3. The ratio of surface area to volume is now 3:1, which is less than it was when the cell was smaller. If the cell continues to grow, the ratio of surface area to volume will continue to decrease, as shown by the third cube in **Figure 9.1.** As the cell grows, its volume increases much more rapidly than the surface area. This means that the cell might have difficulty supplying nutrients and expelling enough waste products. By remaining small, cells have a higher ratio of surface area to volume and can sustain themselves more easily.

Reading Check **Explain** why a high ratio of surface area to volume benefits a cell.

Transport of substances Another task that can be managed more easily in a small cell than in a large cell is the movement of substances. Recall that the plasma membrane controls cellular transport because it is selectively permeable. Once inside the cell, substances move by diffusion or by motor proteins pulling them along the cytoskeleton. Diffusion over large distances is slow and inefficient because it relies on random movement of molecules and ions. Similarly, the cytoskeleton transportation network, shown in **Figure 9.2,** becomes less efficient for a cell if the distance to travel becomes too large. Therefore, cells remain small to maximize the ability of diffusion and motor proteins to transport nutrients and waste products. Small cells maintain more efficient transport systems.

LM Magnification: 150×

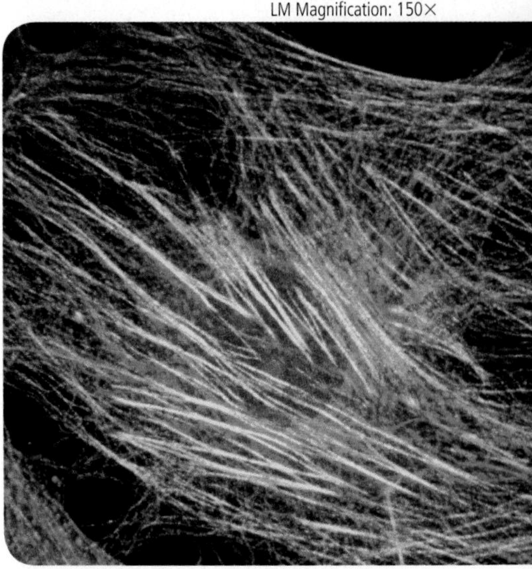

■ **Figure 9.2** In order for the cytoskeleton to be an efficient transportation railway, the distances substances must travel within a cell must be limited.

MiniLab 9.1

Investigate Cell Size

Could a cell grow large enough to engulf your school? What would happen if the size of an elephant were doubled? At the organism level, an elephant cannot grow significantly larger, because its legs would not support the increase in mass. Do the same principles and limitations apply at the cellular level? Do the math!

Procedure

1. Read and complete the lab safety form.
2. Prepare a data table for surface area and volume data calculated for five hypothetical cells. Assume the cell is a cube. (Dimensions given are for one face of a cube.)
   Cell 1: 0.00002 m (the average diameter of most eukaryotic cells)
   Cell 2: 0.001 m (the diameter of a squid's giant nerve cell)
   Cell 3: 2.5 cm
   Cell 4: 30 cm
   Cell 5: 15 m
3. Calculate the surface area for each cell using the formula: length × width × number of sides (6).
4. Calculate the volume for each cell using the formula: length × width × height.

Analysis

1. **Cause and Effect** Based on your calculations, confirm why cells don't become very large.
2. **Infer** Are large organisms, such as redwood trees and elephants, large because they contain extra large cells or just more standard-sized cells? Explain.

■ **Figure 9.3** The cell cycle involves three stages—interphase, mitosis, and cytokinesis. Interphase is divided into three substages.
Hypothesize ***Why does cytokinesis represent the smallest amount of time a cell spends in the cell cycle?***

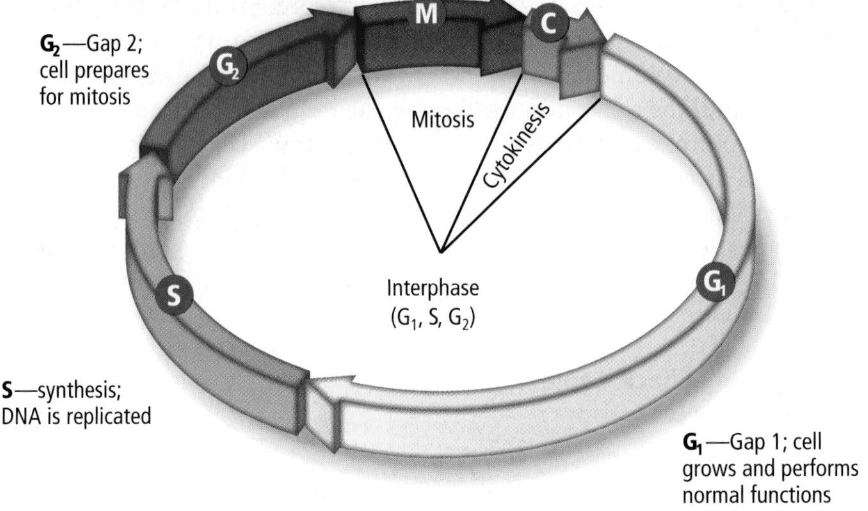

Cellular communications The need for signaling proteins to move throughout the cell also limits cell size. In other words, cell size affects the ability of the cell to communicate instructions for cellular functions. If the cell becomes too large, it becomes almost impossible for cellular communications, many of which involve movement of substances and signals to various organelles, to take place efficiently. For example, the signals that trigger protein synthesis might not reach the ribosome fast enough for protein synthesis to occur to sustain the cell.

The Cell Cycle

Once a cell reaches its size limit, something must happen—either it will stop growing or it will divide. Most cells will eventually divide. Cell division not only prevents the cell from becoming too large, but it is also the way the cell reproduces. Cellular reproduction allows you to grow and heal certain injuries. Cells reproduce by a cycle of growing and dividing called the **cell cycle.** Each time a cell goes through one complete cycle, it becomes two cells. When the cell cycle is repeated continuously, the result is a continuous production of new cells. A general overview of the cell cycle is presented in **Figure 9.3.**

There are three main stages of the cell cycle. **Interphase** is the stage during which the cell grows, carries out cellular functions, and replicates, or makes copies of its DNA in preparation for the next stage of the cycle. Interphase is divided into three substages, as indicated by the segment arrows in **Figure 9.3. Mitosis** (mi TOH sus) is the stage of the cell cycle during which the cell's nucleus and nuclear material divide. Mitosis is divided into four substages. Near the end of mitosis, a process called cytokinesis begins. **Cytokinesis** (si toh kih NEE sis) is the method by which a cell's cytoplasm divides, creating a new cell. You will read more about mitosis and cytokinesis in Section 9.2.

The duration of the cell cycle varies, depending on the cell that is dividing. Some eukaryotic cells might complete the cycle in as few as eight minutes, while other cells might take up to one year. For most normal, actively dividing animal cells, the cell cycle takes approximately 12–24 hours. When you consider all that takes place during the cell cycle, you might find it amazing that most of your cells complete the cell cycle in about a day.

VOCABULARY
WORD ORIGIN
Cytokinesis
cyto– prefix; from the Greek word *kytos,* meaning *hollow vessel*
–***kinesis*** from the Greek word *kinetikos,* meaning *putting in motion*

The stages of interphase During interphase, the cell grows, develops into a mature, functioning cell, duplicates its DNA, and prepares for division. Interphase is divided into three stages, as shown in **Figure 9.3:** G_1, S, and G_2, also called Gap 1, synthesis, and Gap 2.

The first stage of interphase, G_1, is the period immediately after a cell divides. During G_1, a cell is growing, carrying out normal cell functions, and preparing to replicate DNA. Some cells, such as muscle and nerve cells, exit the cell cycle at this point and do not divide again.

The second stage of interphase, S, is the period when a cell copies its DNA in preparation for cell division. **Chromosomes** (KROH muh sohmz) are the structures that contain the genetic material that is passed from generation to generation of cells. **Chromatin** (KROH muh tun) is the relaxed form of DNA in the cell's nucleus. As shown in **Figure 9.4,** when a specific dye is applied to a cell in interphase, the nucleus stains with a speckled appearance. This speckled appearance is due to individual strands of chromatin that are not visible under a light microscope without the dye.

The G_2 stage follows the S stage and is the period when the cell prepares for the division of its nucleus. A protein that makes microtubules for cell division is synthesized at this time. During G_2, the cell also takes inventory and makes sure it is ready to continue with mitosis. When these activities are completed, the cell begins the next stage of the cell cycle—mitosis.

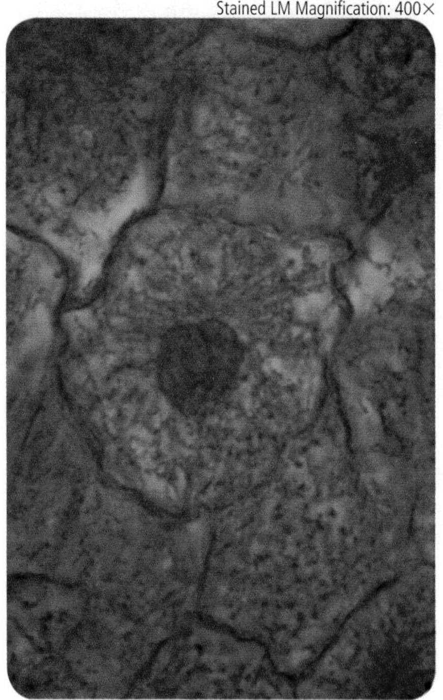

■ **Figure 9.4** The grainy appearance of this nucleus from a rat liver cell is due to chromatin, the relaxed material that condenses to form chromosomes.

Mitosis and cytokinesis The stages of mitosis and cytokinesis follow interphase. In mitosis, the cell's nuclear material divides and separates into opposite ends of the cell. In cytokinesis, the cell divides into two daughter cells with identical nuclei. These important stages of the cell cycle are described in Section 9.2.

Prokaryotic cell division The cell cycle is the method by which eukaryotic cells reproduce themselves. Prokaryotic cells, which you have learned are simpler cells, reproduce by a method called binary fission. You will learn more about binary fission in Chapter 18.

Section 9.1 Assessment

Section Summary

- The ratio of surface area to volume describes the size of the plasma membrane relative to the volume of the cell.
- Cell size is limited by the cell's ability to transport materials and communicate instructions from the nucleus.
- The cell cycle is the process of cellular reproduction.
- A cell spends the majority of its lifetime in interphase.

Understand Main Ideas

1. MAIN Idea **Relate** cell size to cell functions, and explain why cell size is limited.
2. **Summarize** the primary stages of the cell cycle.
3. **Describe** what happens to DNA during the S stage of interphase.
4. **Make a diagram** of the stages of the cell cycle and describe what happens in each.

Think Critically

5. **Hypothesize** what the result would be if a large cell managed to divide, despite the fact that it had grown beyond an optimum size.

MATH in Biology

6. If a cube representing a cell is 5 μm on a side, calculate the surface area-to-volume ratio, and explain why this is or is not a good size for a cell.

Section 9.2

Reading Preview

Section Objectives

- **Describe** the events of each stage of mitosis.
- **Explain** the process of cytokinesis.

Review Vocabulary

life cycle: the sequence of growth and development stages that an organism goes through during its life

New Vocabulary

prophase
sister chromatid
centromere
spindle apparatus
metaphase
anaphase
telophase

Mitosis and Cytokinesis

MAIN Idea **Eukaryotic cells reproduce by mitosis, the process of nuclear division, and cytokinesis, the process of cytoplasm division.**

Real-World Reading Link Many familiar events are cyclic in nature. The course of a day, the changing of seasons year after year, and the passing of comets in space are some examples of cyclic events. Cells also have a cycle of growth and reproduction.

Mitosis

You learned in the last section that cells cycle through interphase, mitosis, and cytokinesis. During mitosis, the cell's replicated genetic material separates and the cell prepares to split into two cells. The key activity of mitosis is the accurate separation of the cell's replicated DNA. This enables the cell's genetic information to pass into the new cells intact, resulting in two daughter cells that are genetically identical. In multicellular organisms, the process of mitosis increases the number of cells as a young organism grows to its adult size. Organisms also use mitosis to replace damaged cells. Recall the last time you accidently got cut. Under the scab, the existing skin cells divided by mitosis and cytokinesis to create new skin cells that filled the gap in the skin caused by the injury.

The Stages of Mitosis

Like interphase, mitosis is divided into stages: prophase, metaphase, anaphase, and telophase.

Prophase The first stage of mitosis—the longest phase—is called **prophase.** In this stage, the cell's chromatin tightens, or condenses, into chromosomes. In prophase, the chromosomes are shaped like an X, as shown in **Figure 9.5.** At this point, each chromosome is a single structure that contains the genetic material that was replicated in interphase. Each half of this X is called a sister chromatid. **Sister chromatids** are structures that contain identical copies of DNA. The structure at the center of the chromosome where the sister chromatids are attached is called the **centromere.** This structure is important because it ensures that a complete copy of the replicated DNA will become part of the daughter cells at the end of the cell cycle. Locate prophase in the cell cycle illustrated in **Figure 9.6,** and note the position of the sister chromatids. As you continue to read about the stages of mitosis, refer back to **Figure 9.6** to follow the chromatids through the cell cycle.

■ **Figure 9.5** Chromosomes in prophase are actually sister chromatids that are attached at the centromere.

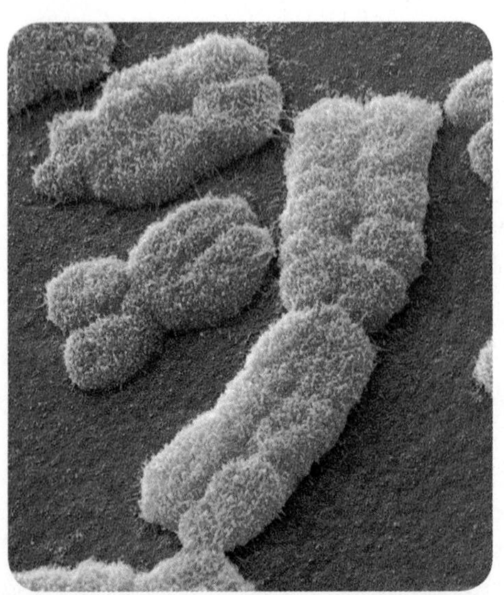

Color-Enhanced SEM magnification: 6875×

Reading Check Compare the key activity of interphase with the key activity of mitosis.

Visualizing the Cell Cycle

Figure 9.6
The cell cycle begins with interphase. Mitosis follows, occurring in four stages—prophase, metaphase, anaphase, and telophase. Mitosis is followed by cytokinesis, then the cell cycle repeats with each new cell.

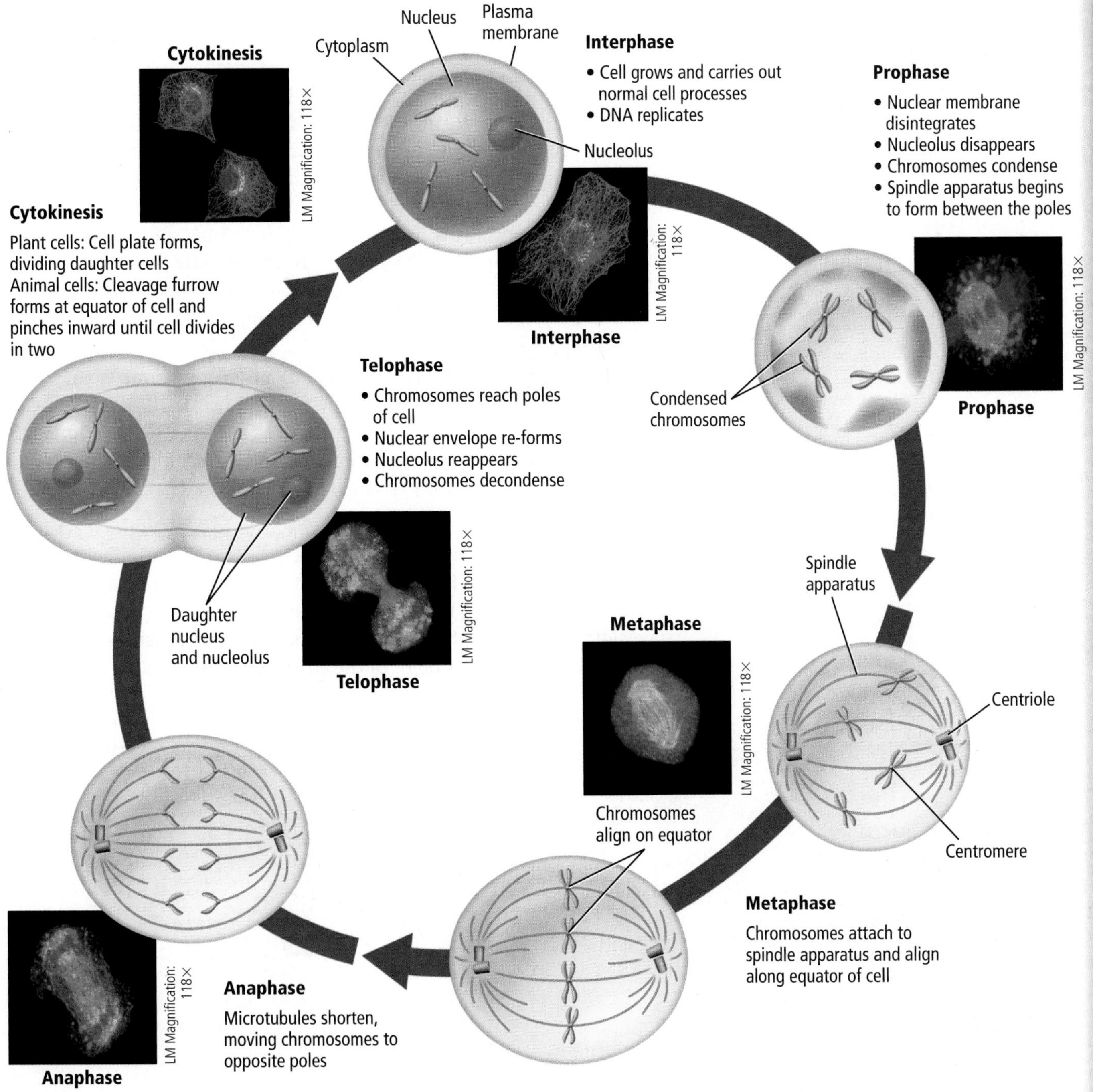

Nucleus

Plasma membrane

Cytoplasm

Nucleolus

Interphase
- Cell grows and carries out normal cell processes
- DNA replicates

LM Magnification: 118×

Interphase

Prophase
- Nuclear membrane disintegrates
- Nucleolus disappears
- Chromosomes condense
- Spindle apparatus begins to form between the poles

LM Magnification: 118×

Prophase

Condensed chromosomes

Spindle apparatus

Centriole

Centromere

Metaphase

LM Magnification: 118×

Chromosomes align on equator

Metaphase
Chromosomes attach to spindle apparatus and align along equator of cell

Anaphase
Microtubules shorten, moving chromosomes to opposite poles

LM Magnification: 118×

Anaphase

Telophase
- Chromosomes reach poles of cell
- Nuclear envelope re-forms
- Nucleolus reappears
- Chromosomes decondense

Daughter nucleus and nucleolus

LM Magnification: 118×

Telophase

Cytokinesis

LM Magnification: 118×

Cytokinesis
Plant cells: Cell plate forms, dividing daughter cells
Animal cells: Cleavage furrow forms at equator of cell and pinches inward until cell divides in two

Concepts In Motion **Interactive Figure** To see an animation of the cell cycle, visit biologygmh.com.

Biology Online

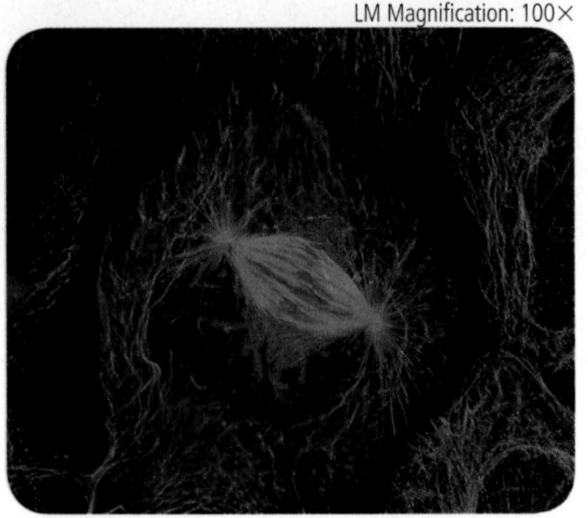

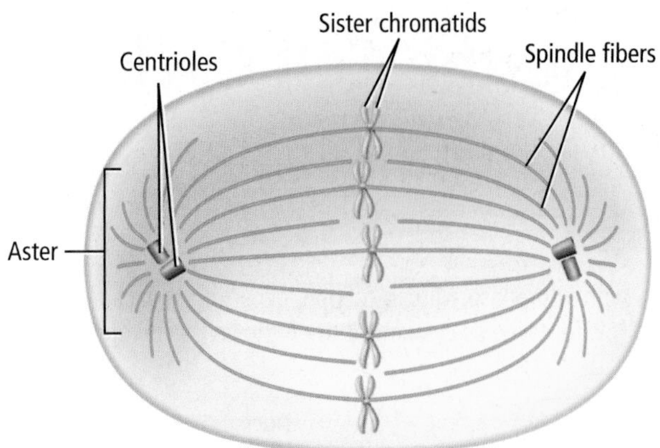

■ **Figure 9.7** In animal cells, the spindle apparatus is made of spindle fibers, centrioles, and aster fibers.

As prophase continues, the nucleolus seems to disappear. Microtubule structures called spindle fibers form in the cytoplasm. In animal cells and most protist cells, another pair of microtubule structures called centrioles migrates to the ends, or poles, of the cell. Coming out of the centrioles are yet another type of microtubule called aster fibers, which have a starlike appearance. The whole structure, including the spindle fibers, centrioles, and aster fibers, is called the **spindle apparatus** and is shown in **Figure 9.7.** The spindle apparatus is important in moving and organizing the chromosomes before cell division. Centrioles are not part of the spindle apparatus in plant cells.

Near the end of prophase, the nuclear envelope seems to disappear. The spindle fibers attach to the sister chromatids of each chromosome on both sides of the centromere and then attach to opposite poles of the cell. This arrangement ensures that each new cell receives one complete copy of the DNA.

FOLDABLES Incorporate information from this section into your Foldable.

Metaphase During the second stage of mitosis, **metaphase,** the sister chromatids are pulled by motor proteins along the spindle apparatus toward the center of the cell and line up in the middle, or equator, of the cell, as shown in **Figure 9.8.** Metaphase is one of the shortest stages of mitosis, but when completed successfully, it ensures that the new cells have accurate copies of the chromosomes.

■ **Figure 9.8** In metaphase, the chromosomes align along the equator of the cell.
Infer ***why the chromosomes align along the equator.***

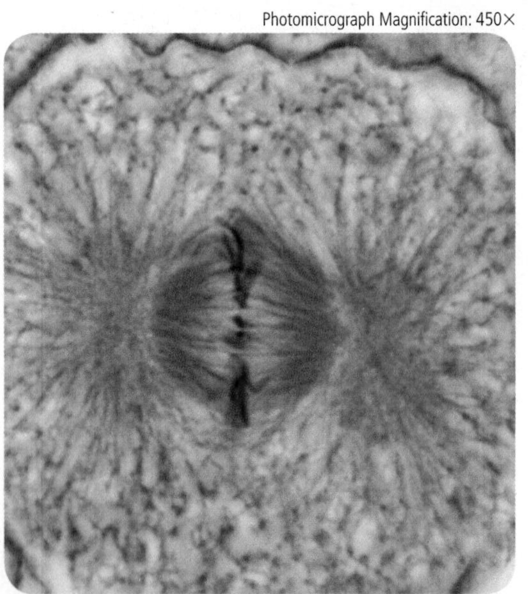

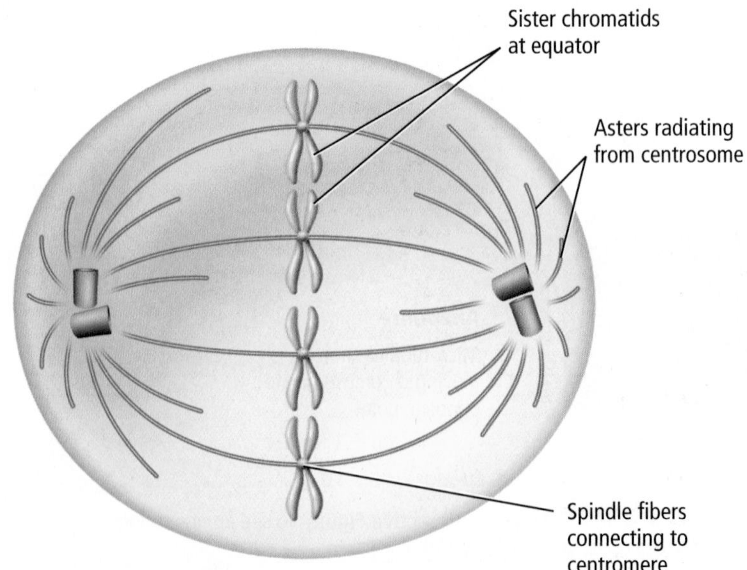

LM Magnification: 450×

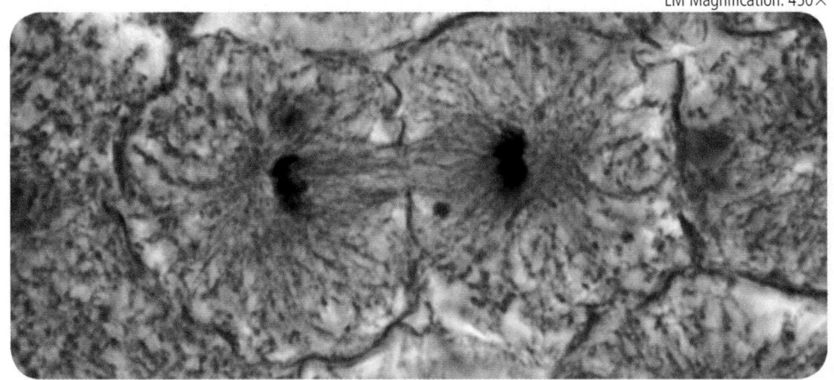

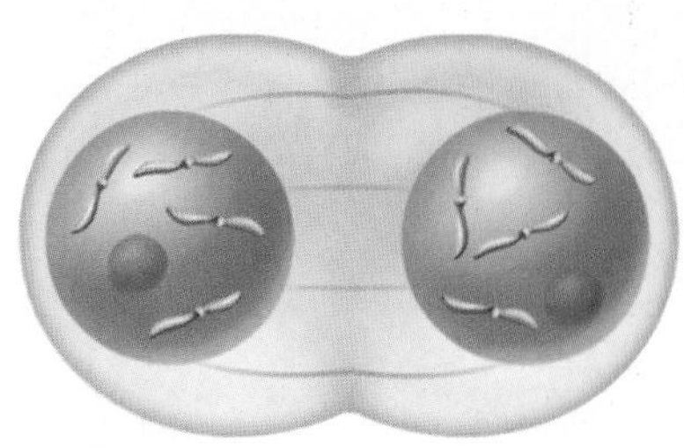

■ **Figure 9.9** By the end of telophase, the cell has completed the work of duplicating the genetic material and dividing it into two "packages," but the cell has not completely divided.

Anaphase The chromatids are pulled apart during **anaphase,** the third stage of mitosis. In anaphase, the microtubules of the spindle apparatus begin to shorten. This shortening pulls at the centromere of each sister chromatid, causing the sister chromatids to separate into two identical chromosomes. All of the sister chromatids separate simultaneously, although the exact mechanism that controls this is unknown. At the end of anaphase, the microtubules, with the help of motor proteins, move the chromosomes toward the poles of the cell.

Telophase The last stage of mitosis is called telophase. **Telophase** is the stage of mitosis during which the chromosomes arrive at the poles of the cell and begin to relax, or decondense. As shown in **Figure 9.9,** two new nuclear membranes begin to form and the nucleoli reappear. The spindle apparatus disassembles and some of the microtubules are recycled by the cell to build various parts of the cytoskeleton. Although the four stages of mitosis are now complete and the nuclear material is divided, the process of cell division is not yet complete.

DATA ANALYSIS LAB 9.1

Based on Real Data*

Predict the Results

What happens to the microtubules? Scientists performed experiments tracking chromosomes along microtubules during mitosis. They hypothesized that the microtubules are broken down, releasing microtubule subunits as the chromosomes are moved toward the poles of the cell. The microtubules were labeled with a yellow fluorescent dye, and using a laser, the microtubules were marked midway between the poles and the chromosomes by eliminating the fluorescence in the targeting region as shown in the diagram.

Think Critically

1. **Explain** What was the purpose of the fluorescent dye?
2. **Predict** Draw a diagram of how the cell might appear later in anaphase.

*Data obtained from: Maddox, P., et al. 2003. Direct observation of microtubule dynamics at kinetochores in *Xenopus* extract spindles: implications for spindle mechanics. *The Journal of Cell Biology* 162: 377-382. Maddox, et al. 2004. Controlled ablations of microtubules using picosecond laser. *Biophysics Journal* 87: 4203-4212.

Data and Observations

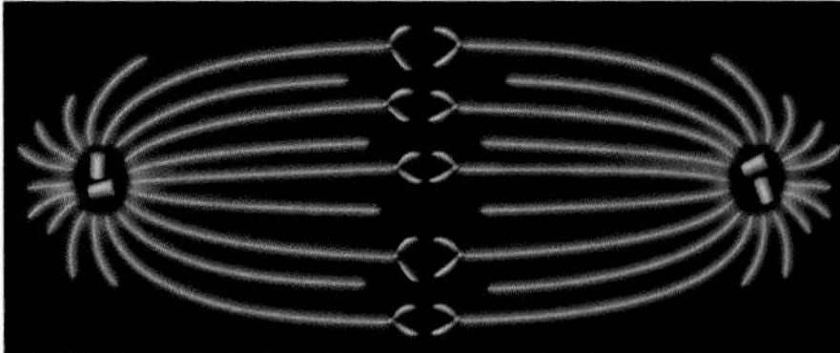

Fluorescent-labeled microtubules

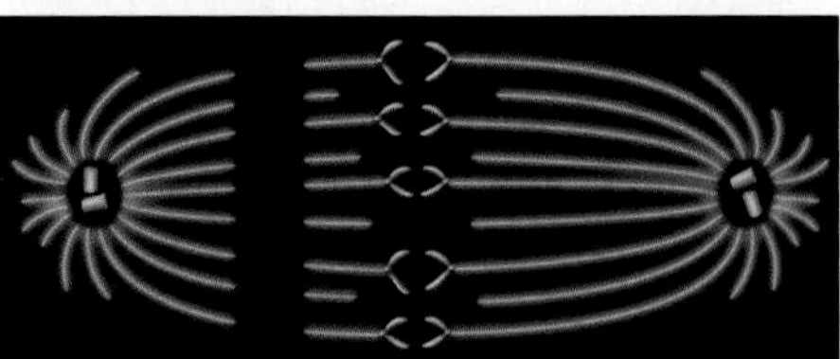

Laser-marked microtubules

Color-Enhanced SEM Magnification: 125×

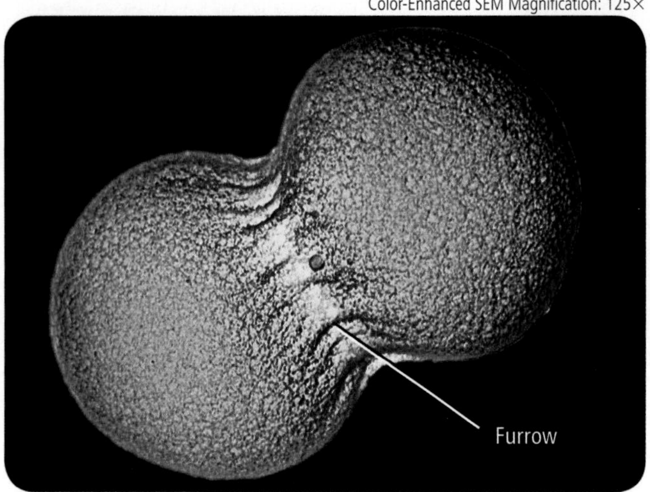

Animal cell

Stained LM Magnification: 1000×

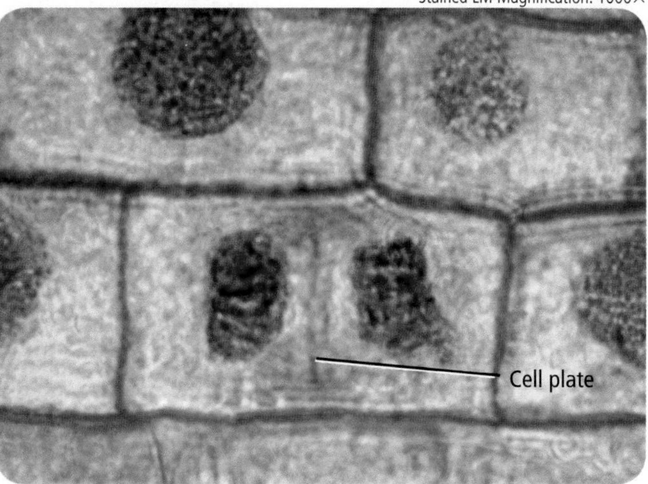

Plant cells

■ **Figure 9.10**
Left: In animal cells, cytokinesis begins with a furrow that pinches the cell and eventually splits the two cells apart.
Right: Plant cells build a cell plate that divides the cell into the two daughter cells.

Cytokinesis

Toward the end of mitosis, the cell begins another process called cytokinesis that will divide the cytoplasm. This results in two cells, each with identical nuclei. In animal cells, cytokinesis is accomplished by using microfilaments to constrict, or pinch, the cytoplasm, as shown in **Figure 9.10.** The area where constriction occurs is called the furrow.

Recall from Chapter 7 that plant cells have a rigid cell wall covering their plasma membrane. Instead of pinching in half, a new structure, called a cell plate, forms between the two daughter nuclei, as illustrated in **Figure 9.10.** Cell walls then form on either side of the cell plate. Once this new wall is complete, there are two genetically identical cells.

Prokaryotic cells, which divide by binary fission, finish cell division in a different way. When prokaryotic DNA is duplicated, both copies attach to the plasma membrane. As the plasma membrane grows, the attached DNA molecules are pulled apart. The cell completes fission, producing two new prokaryotic cells.

Section 9.2 Assessment

Section Summary

- Mitosis is the process by which the duplicated DNA is divided.
- The stages of mitosis include prophase, metaphase, anaphase, and telophase.
- Cytokinesis is the process of cytoplasm division that results in genetically identical daughter cells.

Understand Main Ideas

1. MAIN Idea **Explain** why mitosis alone does not produce daughter cells.
2. **Describe** the events of each stage of mitosis.
3. **Diagram and label** a chromosome in prophase.
4. **Identify** the stage of mitosis in which a cell spends the most time.
5. **Contrast** cytokinesis in a plant cell and an animal cell.

Think Critically

6. **Hypothesize** what might happen if a drug that stopped microtubule movement but did not affect cytokinesis was applied to a cell.

MATH in Biology

7. If a plant cell completes the cell cycle in 24 hours, how many cells will be produced in a week?

Biology Online **Self-Check Quiz** biologygmh.com

Section 9.3

Cell Cycle Regulation

Reading Preview

Objectives

- **Summarize** the role of cyclin proteins in controlling the cell cycle.
- **Explain** how cancer relates to the cell cycle.
- **Describe** the role of apoptosis.
- **Summarize** the two types of stem cells and their potential uses.

Review Vocabulary

nucleotide: subunit that makes up DNA and RNA molecules

New Vocabulary

cyclin
cyclin-dependent kinase
cancer
carcinogen
apoptosis
stem cell

MAIN Idea **The normal cell cycle is regulated by cyclin proteins.**

Real-World Reading Link No matter how many new homes a builder builds, even if building the same design, the crew always relies on blueprint instructions. Similarly, cells have specific instructions for completing the cell cycle.

Normal Cell Cycle

The timing and rate of cell division are important to the health of an organism. The rate of cell division varies depending on the type of cell. A mechanism involving proteins and enzymes controls the cell cycle.

The role of cyclins To start a car, it takes a key turning in the ignition to signal the engine to start. Similarly, the cell cycle in eukaryotic cells is driven by a combination of two substances that signal the cellular reproduction processes. Proteins called **cyclins** bind to enzymes called **cyclin-dependent kinases** (CDKs) in the stages of interphase and mitosis to start the various activities that take place in the cell cycle. Different cyclin/CDK combinations control different activities at different stages in the cell cycle. **Figure 9.11** illustrates where some of the important combinations are active.

In the G_1 stage of interphase, the combination of cyclin with CDK signals the start of the cell cycle. Different cyclin/CDK combinations signal other activities, including DNA replication, protein synthesis, and nuclear division throughout the cell cycle. The same cyclin/CDK combination also signals the end of the cell cycle.

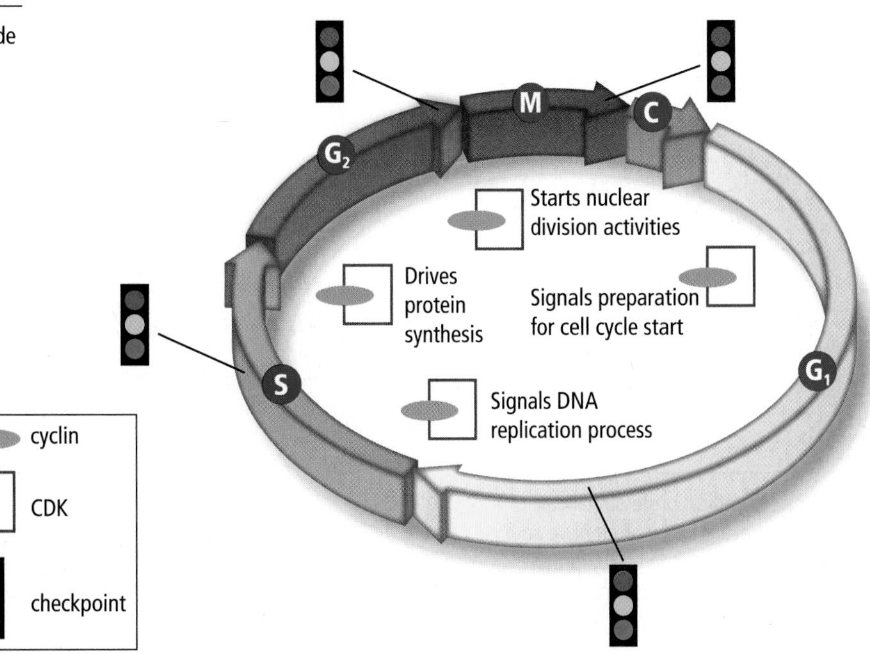

■ **Figure 9.11** Signaling molecules made of a cyclin bound to a CDK kick off the cell cycle and drive it through mitosis. Checkpoints monitor the cell cycle for errors and can stop the cycle if an error occurs.

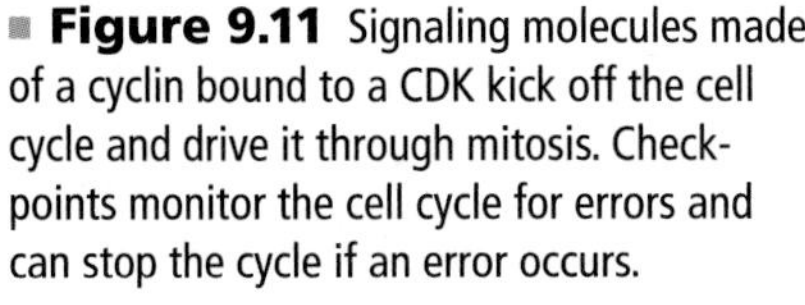

To learn about the cell cycle, visit biologygmh.com.

CAREERS IN BIOLOGY

Pharmaceutical QC Technician
Just as the cell cycle has built-in quality control checkpoints, so do biological product manufacturing processes. A QC technician in a pharmaceutical manufacturing company uses various science and math skills to monitor processes and ensure product quality. For more information on biology careers, visit biologygmh.com.

Quality control checkpoints Recall the process of starting a car. Many manufacturers use a unique microchip in the key to ensure that only a specific key will start each car. This is a checkpoint against theft. The cell cycle also has built-in checkpoints that monitor the cycle and can stop it if something goes wrong. For example, a checkpoint near the end of the G_1 stage monitors for DNA damage and can stop the cycle before entering the S stage of interphase. There are other quality control checkpoints during the S stage and after DNA replication in the G_2 stage. Spindle checkpoints also have been identified in mitosis. If a failure of the spindle fibers is detected, the cycle can be stopped before cytokinesis. **Figure 9.11** shows the location of key checkpoints in the cell cycle.

Abnormal Cell Cycle: Cancer

Connection to Health Although the cell cycle has a system of quality control checkpoints, it is a complex process that sometimes fails. When cells do not respond to the normal cell cycle control mechanisms, a condition called cancer can result. **Cancer** is the uncontrolled growth and division of cells—a failure in the regulation of the cell cycle. When unchecked, cancer cells can kill an organism by crowding out normal cells, resulting in the loss of tissue function. Cancer cells spend less time in interphase than do normal cells, which means cancer cells grow and divide unrestrained as long as they are supplied with essential nutrients. **Figure 9.12** shows how cancer cells can intrude on normal cells.

Causes of cancer Cancer does not just occur in a weak organism. In fact, cancer occurs in many healthy, active, and young organisms. The changes that occur in the regulation of cell growth and division of cancer cells are due to mutations or changes in the segments of DNA that control the production of proteins, including proteins that regulate the cell cycle. Often, the genetic change or damage that occurs is repaired by various repair systems. But if the repair systems fail, cancer can result. Various environmental factors can affect the occurrence of cancer cells. Substances and agents that are known to cause cancer are called **carcinogens** (kar SIH nuh junz).

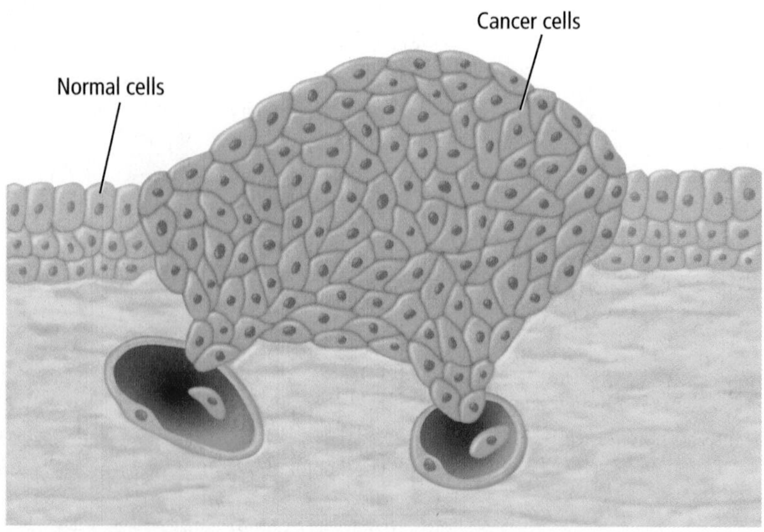

■ **Figure 9.12** A medical professional can identify cancer cells because they often have an abnormal, irregular shape compared to normal cells. If left unchecked, a cancerous tumor can grow to the point where it can kill its host organism.

Although not all cancers can be prevented, avoiding known carcinogens can help reduce the risk of cancer. A governmental agency called the Food and Drug Administration (FDA) works to make sure that the things you eat and drink are safe. The FDA also requires labels and warnings for products that might be carcinogens. Industrial laws help protect people from exposure to cancer-causing chemicals, such as asbestos, in the workplace. For example, asbestos has been removed from many old buildings to protect people living and working inside them. Avoiding tobacco of all kinds, even secondhand smoke and smokeless tobacco, can reduce the risk of cancer.

Some radiation, such as ultraviolet radiation from the Sun, is impossible to avoid completely. There is a connection between the amount of ultraviolet radiation to which a person is exposed and the risk of developing skin cancer. Therefore, sunscreen is recommended for everyone who is exposed to the Sun. Other forms of radiation, such as X rays, are used for medical purposes, such as to look at a broken bone or check for tooth cavities. To protect against exposure, you might have worn a heavy lead apron when an X ray was taken.

Cancer genetics More than one change in DNA is required to change an abnormal cell into a cancer cell. Over time, it is possible that there might be many changes in DNA. This might explain why the risk of cancer increases with age. The fact that multiple changes must occur also might explain why cancer runs in some families. An individual who inherits one or more changes from a parent is at a higher risk for developing cancer than someone who does not inherit these changes.

LAUNCH Lab

Review Based on what you've read about the abnormal cell cycle and its results, how would you now answer the analysis questions?

VOCABULARY

SCIENCE USAGE V. COMMON USAGE

Inheritance

Science usage: the passing of genetic traits from parent to offspring via DNA.

A person's body structure and facial appearance are the result of genetic inheritance.

Common usage: assets acquired from a deceased person that can be given to surviving family members.

The house was Jim's inheritance from his uncle.

MiniLab 9.2

Compare Sunscreens

Do sunscreens really block sunlight? Sunscreens contain a variety of different compounds that absorb UVB from sunlight. UVB is linked to mutations in DNA that can lead to skin cancer. Find out how effective at blocking sunlight various sunscreens are.

Procedure

1. Read and complete the lab safety form.
2. Choose one of the **sunscreen products** provided by your teacher. Record the active ingredients and the Sun protection factor (SPF) on a data sheet.
3. Obtain **two sheets of plastic wrap.** On one sheet use a **permanent marker** to draw two widely spaced circles. Place a drop of sunscreen in the middle of one circle and a drop of **zinc oxide** in the middle of the other.
4. Lay the second sheet on top of both circles. Spread the drops by pressing with a **book.**
5. Take a covered piece of **Sun-sensitive paper** and your two pieces of plastic wrap to a sunny area. Quickly uncover the paper, lay the two pieces of plastic wrap on top, and place in the sunlight.
6. After the paper is fully exposed (1–5 minutes), remove it from the sunlight and develop according to instructions.

Analysis

1. **Think Critically** Why did you compare the sunscreens to zinc oxide?
2. **Draw Conclusions** After examining the developed Sun-sensitive papers from your class, which sunscreens do you think would be most likely to prevent DNA mutations?

Apoptosis

Not every cell is destined to survive. Some cells go through a process called **apoptosis** (a pup TOH sus), or programmed cell death. Cells going through apoptosis actually shrink and shrivel in a controlled process. All animal cells appear to have a "death program" that can be activated.

One example of apoptosis occurs during the development of the human hand and foot. When the hands and feet begin to develop, cells occupy the spaces between the fingers and toes. Normally, this tissue undergoes apoptosis, with the cells shriveling and dying at the appropriate time so that the webbing is not present in the mature organism. An example of apoptosis in plants is the localized death of cells that results in leaves falling from trees during autumn. Apoptosis also occurs in cells that are damaged beyond repair, including cells with DNA damage that could lead to cancer. Apoptosis can help to protect organisms from developing cancerous growths.

VOCABULARY

ACADEMIC VOCABULARY

Mature:
to have reached full natural growth or development
After mitosis, the two new cells must mature before they divide.

Stem Cells

The majority of cells in a multicellular organism are designed for a specialized function. Some cells might be part of your skin, and other cells might be part of your heart. In 1998, scientists discovered a way to isolate a unique type of cell in humans called the stem cell. **Stem cells** are unspecialized cells that can develop into specialized cells when under the right conditions, as illustrated in **Figure 9.13.** Stem cells can remain in an organism for many years while undergoing cell division. There are two basic types of stem cells: embryonic stem cells and adult stem cells.

Embryonic stem cells After a sperm fertilizes an egg, the resulting mass of cells divides repeatedly until there are about 100–150 cells. These cells have not become specialized and are called embryonic stem cells. If separated, each of these cells has the capability of developing into a wide variety of specialized cells. If the embryo continues to divide, the cells specialize into various tissues, organs, and organ systems. Embryonic stem cell research is controversial because of ethical concerns about the source of the cells.

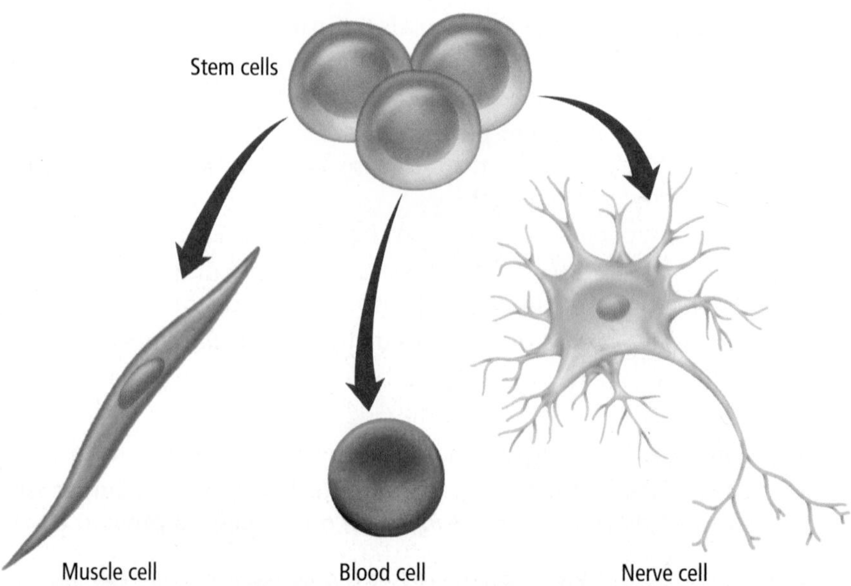

Figure 9.13 Because stem cells are not locked into becoming one particular type of cell, they might be the key to curing many medical conditions and genetic defects.
Explain *how stem cells could be used to cure nerve damage.*

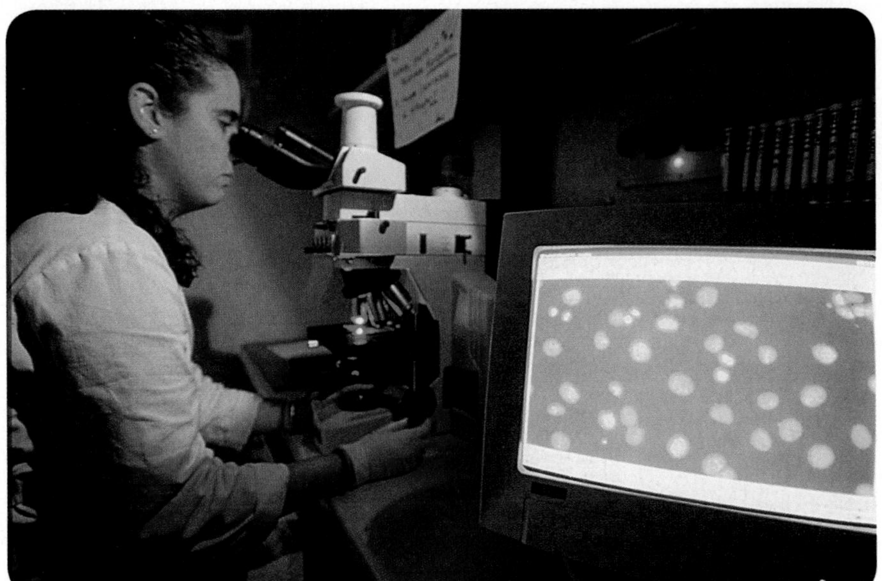

■ **Figure 9.14** Research with adult stem cells has led to advances in treatments for numerous injuries and diseases.

Adult stem cells The second type of stem cells—adult stem cells—is found in various tissues in the body and might be used to maintain and repair the same kind of tissue in which they are found. The term "adult stem cells" might be somewhat misleading because even a newborn has adult stem cells. Like embryonic stem cells, certain kinds of adult stem cells also might be able to develop into different kinds of cells, providing new treatments for many diseases and conditions. In 1999, researchers at Harvard Medical School used nervous system stem cells to restore lost brain tissue in mice. In 2000, a team of researchers at the University of Florida used pancreatic stem cells to restore pancreas function in a mouse with diabetes. Research with adult stem cells, like that shown in **Figure 9.14,** is much less controversial because the adult stem cells can be obtained with the consent of their donors.

Section 9.3 Assessment

Section Summary

- The cell cycle of eukaryotic cells is regulated by cyclins.
- Checkpoints occur during most of the stages of the cell cycle to ensure that the cell divides accurately.
- Cancer is the uncontrolled growth and division of cells.
- Apoptosis is a programmed cell death.
- Stem cells are unspecialized cells that can develop into specialized cells with the proper signals.

Understand Main Ideas

1. MAIN Idea **Describe** how cyclins control the cell cycle.
2. **Explain** how the cancer cell cycle is different from a normal cell cycle.
3. **Identify** three carcinogens.
4. **Contrast** apoptosis and cancer.
5. **Describe** a possible application for stem cells.
6. **Explain** the difference between embryonic stem cells and adult stem cells.

Think Critically

7. **Hypothesize** what might happen if apoptosis did not occur in cells that have significant DNA damage.

WRITING in Biology

8. Write a public service announcement about carcinogens. Choose a specific type of cancer, and write about the carcinogens linked to it.

Biology & Society

Stem Cells: Paralysis Cured?

A race car driver is paralyzed in a crash. A teen is paralyzed after diving into shallow water. Until recently, these individuals would have little hope of regaining the full use of their bodies, but new research on adult stem cells shows promise for reversing paralysis.

How can stem cells be used? Scientists are trying to find ways to grow adult stem cells in cell cultures and manipulate them to generate specific cell types. For example, stem cells might be used to repair cardiac tissue after a heart attack, to restore vision in diseased or injured eyes, to treat diseases such as diabetes, or to repair spinal cells to reverse paralysis. Late actor and paralysis victim Christopher Reeve was a strong proponent for stem cell research because he believed there is much potential in science to improve the condition of life for others who suffer from paralysis.

Stem cells and paralysis In Portugal, Dr. Carlos Lima and his team of researchers found that tissue taken from the nasal cavity is a rich source of adult stem cells. These stem cells become nerve cells when transplanted into the site of a spinal cord injury. The new nerve cells replace the cells that were damaged.

More than forty patients with paralysis due to accidents have undergone the Portuguese procedure. All patients have regained some sensation in paralyzed body areas. Most have regained some motor control. With intensive physical therapy, about ten percent of the patients now can walk with the aid of supportive devices, such as walkers and braces. This is promising news to the many individuals facing illnesses or injuries that have robbed them of the full use of their bodies.

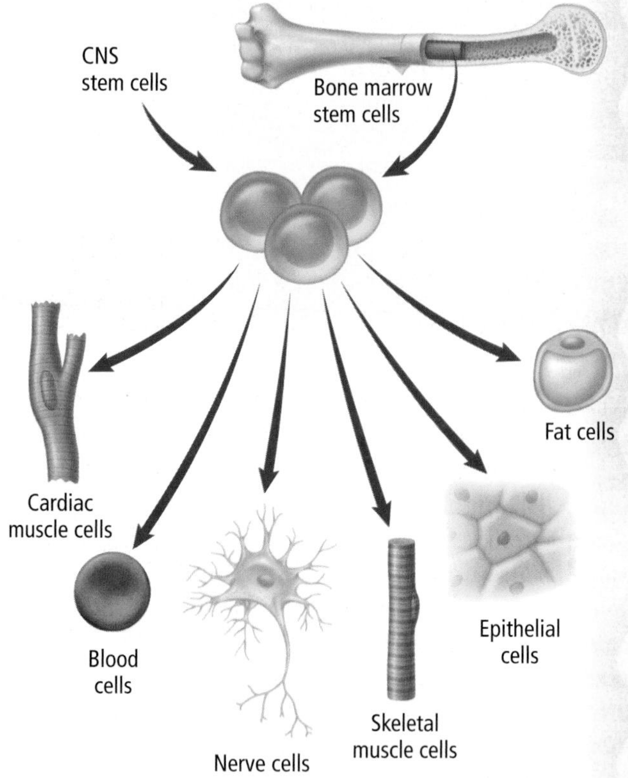

Stem cells from bone marrow or the central nervous system can be manipulated to generate many cell types that can be transplanted to treat illness or repair damage.

Stem cells and the future Scientists are eager to do the research necessary to make adult stem cell treatments a regular part of health care. Paralysis might not have to be permanent: stem cells could provide the cure.

WRITING in Biology

Pamphlet Create a pamphlet depicting the benefits of adult stem cell research. Conduct additional research on adult stem cell research at biologygmh.com in order to include the research methodology, treatment, examples, cell physiology, and history of adult stem cell research. Be sure to illustrate your pamphlet.

BIOLAB

DOES SUNLIGHT AFFECT MITOSIS IN YEAST?

Background: Ultraviolet (UV) radiation is a component of sunlight that can damage DNA and interrupt the cell cycle.

Question: *Can sunscreens prevent damage to UV-sensitive yeast?*

Materials

sterile pipettes (10)
aluminum foil
test-tube rack
sterile spreaders or sterile cotton swabs (10)
dilution of UV-sensitive yeast
yeast extract dextrose (YED) agar plates (10)
sunscreens with various amounts of SPF

Safety Precautions

Procedure

1. Read and complete the lab safety form.
2. Obtain a test tube containing a diluted broth culture of the UV-sensitive yeast.
3. Formulate a hypothesis, then choose a sunscreen and predict how it will affect the yeast when exposed to sunlight.
4. Label ten YED agar plates with your group name. Label two plates as control. The control plates will not be placed in the sunlight. Label four of the experimental plates as "no sunscreen" and four as "sunscreen."
5. Spread a 0.1 mL sample of the yeast dilution on all ten YED agar plates. Wrap the control plates in foil and give them to your teacher for incubation.
6. With direction from your teacher, decide how long to expose each of the experimental plates and label each plate accordingly. Prepare a table in which to collect your data.
7. Wrap the "no sunscreen" plates in foil. Apply sunscreen to the lids of the four sunscreen plates and wrap them in foil.
8. Remove only enough aluminum foil from each of the experimental plates to expose the dish lids. Expose the plates for the planned times. Re-cover the plates after exposure and give them to your teacher for incubation.
9. After incubation, count and record the number of yeast colonies on each plate.
10. **Cleanup and Disposal** Wash and return all reusable materials. Dispose of the YED plates as instructed by your teacher. Disinfect your work area. Wash your hands thoroughly with soap and water.

Analyze and Conclude

1. **Estimate** Assume that each yeast colony on a YED plate grew from one yeast cell in the dilution. Use the number of yeast colonies on your control plate to determine the percent of yeast that survived on each exposed plate.
2. **Graph Data** Draw a graph with the percent survival on the y-axis and the exposure time on the x-axis. Use a different color to graph the data from the plates with and without sunscreen.
3. **Evaluate** Was your hypothesis supported by your data? Explain.
4. **Error Analysis** Describe several possible sources of error.

Apply your Skill Brainstorm ideas about how UV-sensitive yeast could be used as a biological monitor to detect increases in the amounts of UV light reaching Earth's surface. To learn more about mitosis in yeast, visit Biolabs at biologygmh.com.

CHAPTER 9

Study Guide

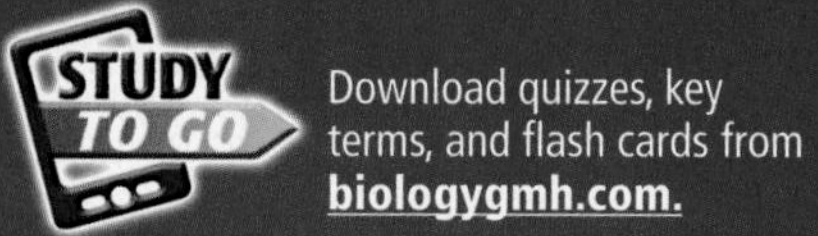

Download quizzes, key terms, and flash cards from **biologygmh.com.**

FOLDABLES **Research and sequence** key events occurring in the area of stem cell research since 1998. Include information on the discoveries of embryonic and adult stem cells and political and ethical debates over the use of embryonic stem cells in research.

Vocabulary	Key Concepts
Section 9.1 Cellular Growth	
• cell cycle (p. 246) • chromatin (p. 247) • chromosome (p. 247) • cytokinesis (p. 246) • interphase (p. 246) • mitosis (p. 246)	**MAIN Idea** Cells grow until they reach their size limit, then they either stop growing or divide. • The ratio of surface area to volume describes the size of the plasma membrane relative to the volume of the cell. • Cell size is limited by the cell's ability to transport materials and communicate instructions from the nucleus. • The cell cycle is the process of cellular reproduction. • A cell spends the majority of its lifetime in interphase.
Section 9.2 Mitosis and Cytokinesis	
• anaphase (p. 251) • centromere (p. 248) • metaphase (p. 250) • prophase (p. 248) • sister chromatid (p. 248) • spindle apparatus (p. 250) • telophase (p. 251)	**MAIN Idea** Eukaryotic cells reproduce by mitosis, the process of nuclear division, and cytokinesis, the process of cytoplasm division. • Mitosis is the process by which the duplicated DNA is divided. • The stages of mitosis include prophase, metaphase, anaphase, and telophase. • Cytokinesis is the process of cytoplasm division that results in genetically identical daughter cells.
Section 9.3 Cell Cycle Regulation	
• apoptosis (p. 256) • cancer (p. 254) • carcinogen (p. 254) • cyclin (p. 253) • cyclin-dependent kinase (p. 253) • stem cell (p. 256)	**MAIN Idea** The normal cell cycle is regulated by cyclin proteins. • The cell cycle of eukaryotic cells is regulated by cyclins. • Checkpoints occur during most of the stages of the cell cycle to ensure that the cell divides accurately. • Cancer is the uncontrolled growth and division of cells. • Apoptosis is a programmed cell death. • Stem cells are unspecialized cells that can develop into specialized cells with the proper signals.

CHAPTER 9 Assessment

Section 9.1

Vocabulary Review

Match the correct vocabulary term from the Study Guide page to the following definitions.

1. the period in which the cell is not dividing
2. the process of nuclear division
3. the sequence of events in the life of a eukaryotic cell

Understand Key Concepts

4. Which is not a reason why cells remain small?
   - **A.** Cells remain small to enable communication.
   - **B.** Large cells have difficulty diffusing nutrients rapidly enough.
   - **C.** As cells grow, their ratio of surface area to volume increases.
   - **D.** Transportation of wastes becomes a problem for large cells.

Use the hypothetical cell shown below to answer question 5.

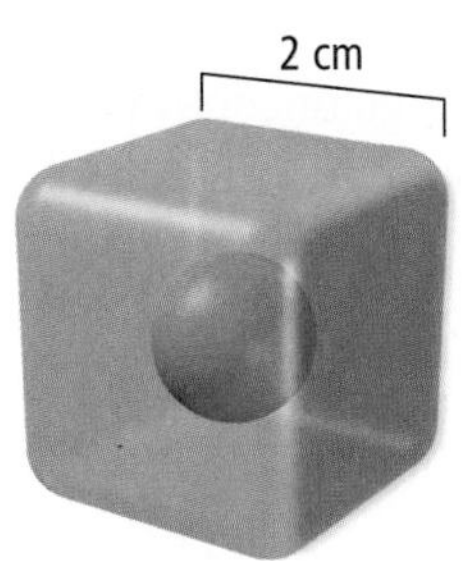

5. What is the ratio of surface area to volume?
   - **A.** 2:1 **C.** 4:1
   - **B.** 3:1 **D.** 6:1
6. Of the surface area-to-volume ratio, what does the surface area represent in a cell?
   - **A.** nucleus
   - **B.** plasma membrane
   - **C.** mitochondria
   - **D.** cytoplasm
7. Which describes the activities of a cell that include cellular growth and cell division?
   - **A.** chromatin **C.** mitosis
   - **B.** cytoplasm **D.** cell cycle
8. As a cell's volume increases, what happens to the proportional amount of surface area?
   - **A.** increases
   - **B.** decreases
   - **C.** stays the same
   - **D.** reaches its limit

Constructed Response

9. **Short Answer** Why are cellular transport and cellular communication factors that limit cell size?
10. **Short Answer** Summarize the relationship between surface area and volume as a cell grows.
11. **Short Answer** What types of activities are going on in a cell during interphase?

Think Critically

12. **Criticize** this statement: Interphase is a "resting period" for the cell before it begins mitosis.
13. **Explain** the relationship of DNA, a chromosome, and chromatin.

Section 9.2

Vocabulary Review

Complete the concept map using vocabulary terms from the Study Guide page.

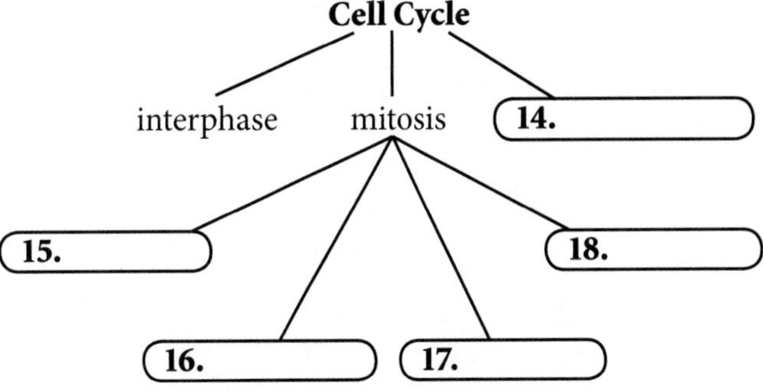

Understand Key Concepts

19. Starting with one cell that underwent six divisions, how many cells would result?
    - **A.** 13 **C.** 48
    - **B.** 32 **D.** 64

The following graph shows a cell over the course of its cell cycle. Use this graph to answer questions 20 and 21.

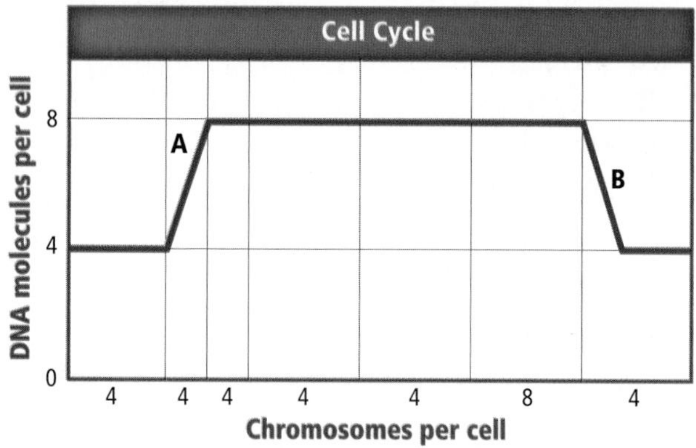

20. What stage occurred in the area labeled A?
A. prophase
B. G_1 stage
C. S stage
D. G_2 stage

21. What process occurred in the area labeled B?
A. interphase
B. cytokinesis
C. mitosis
D. metabolism

22. The cancer drug vinblastine interferes with synthesis of microtubules. In mitosis, this would interfere with what?
A. spindle formation
B. DNA replication
C. carbohydrate synthesis
D. disappearance of the nuclear envelope

Constructed Response

23. Short Answer During the cell cycle, when would a chromosome consist of two identical sister chromatids?

24. Short Answer In the following image of a section of onion root tip, identify a cell in each of the following stages: interphase, prophase, metaphase, anaphase, and telophase.

Stained LM Magnification: 130×

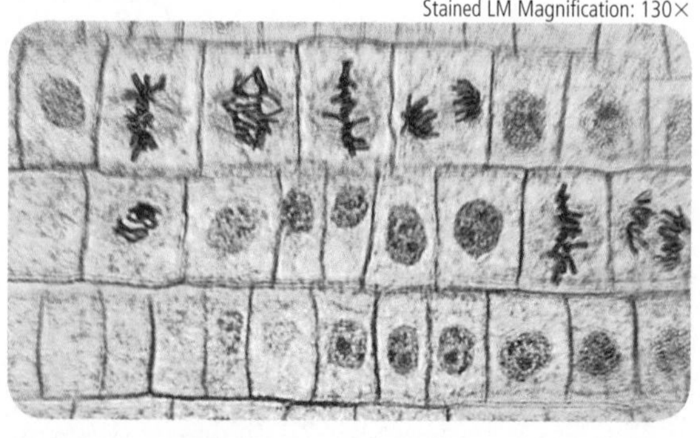

25. Short Answer Describe the events that occur in telophase.

Think Critically

26. Evaluate While looking through a microscope, you see a cell plate forming. This cell is most likely what type of cell?

27. MATH in Biology A biologist examines a series of cells and counts 90 cells in interphase, 13 cells in prophase, 12 cells in metaphase, 3 cells in anaphase, and 2 cells in telophase. If a complete cycle for this type of cell requires 24 hours, what is the average duration of mitosis?

Section 9.3

Vocabulary Review

The sentences below include term(s) that have been used incorrectly. Replace the incorrect term(s) with vocabulary terms from the Study Guide page to make the sentences true.

28. Stem cells undergo uncontrolled, unrestrained growth and division because their genes have been changed.

29. Cancer is a cell response to DNA damage that results in cell death.

30. Cyclins are substances that cause cancer.

Understand Key Concepts

31. What is the role of cyclins in a cell?
A. to control the movement of microtubules
B. to signal for the cell to divide
C. to stimulate the breakdown of the nuclear membrane
D. to cause the nucleolus to disappear

32. What substances form the cyclin-cyclin dependent kinase combinations that control the stages in the cell cycle?
A. fats and proteins
B. carbohydrates and proteins
C. proteins and enzymes
D. fats and enzymes

33. Which is a characteristic of cancer cells?
    A. controlled cell division
    B. contain multiple genetic changes
    C. cytokinesis stage is skipped
    D. cell cyclins function normally

34. Which describes apoptosis?
    A. occurs in all cells
    B. is a programmed cell death
    C. disrupts the normal development of an organism
    D. is a response to hormones

35. Why have some stem cell researchers experienced roadblocks in their studies?
    A. Stem cells cannot be found.
    B. There are ethical concerns about obtaining stem cells.
    C. There are no known uses for stem cells.
    D. Stem cells do not become specialized cells.

Constructed Response

Refer to the diagram to answer question 36.

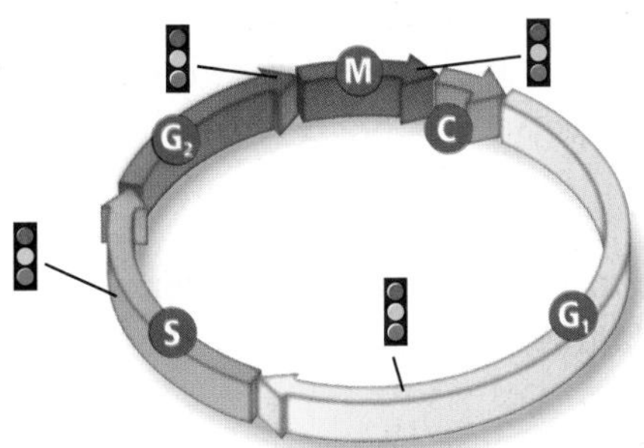

36. **Short Answer** Explain the relationship between cancer cells and the cell cycle.

37. **Short Answer** Distinguish between mitosis and apoptosis.

Think Critically

38. **Describe** how stem cells might be used to help a patient who has a damaged spinal cord.

39. **Predict** why too-frequent or too-infrequent apoptosis could endanger health.

40. **Apply** Hundreds of millions of dollars are spent annually in the U.S. on the research and treatment of cancer, with much less being spent on cancer prevention. Compose a plan that would help Americans increase cancer prevention.

Additional Assessment

41. WRITING in Biology Write a skit using props and people to demonstrate mitosis.

42. Research chemicals that are carcinogens and write about how these chemicals damage DNA.

Document-Based Questions

Dr. Chang and co-workers evaluated the risk of pancreatic cancer by studying its occurrence in a population group. Their data included age at diagnosis. The graph below shows cancer diagnosis rates for African-American men and women.

Data obtained from: Chang, K. J. et al. 2005. Risk of pancreatic adenocarcinoma. *Cancer* 103: 349-357.

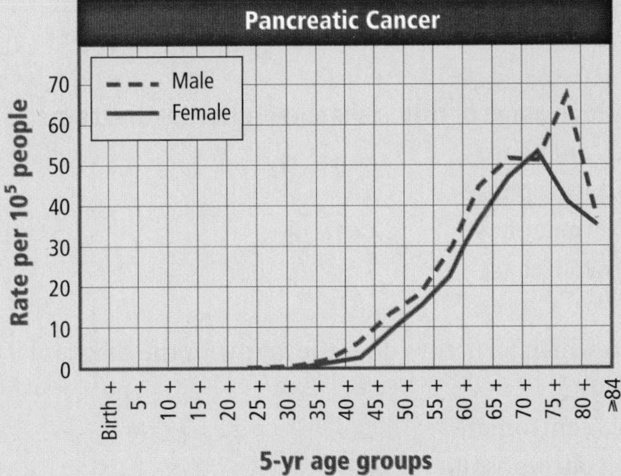

43. Summarize the relationship between the occurrence of cancer and age.

44. Considering what you know about cancer and the cell cycle, explain why incidences of cancer increase with age.

45. Compare the ages of men and women who are diagnosed with cancer.

Cumulative Review

46. Discuss the importance of enzymes in living organisms. Include the concept of catalysis in your response. **(Chapter 6)**

47. Describe the basic structure of the plasma membrane. **(Chapter 7)**

Standardized Test Practice

Cumulative

Multiple Choice

1. Carbon (C) has four electrons in its outer energy level, and fluorine (F) has seven. Which compound would carbon and fluorine most likely form?
   A. CF_2
   B. CF_3
   C. CF_4
   D. CF_5

Use the diagram below to answer questions 2 and 3.

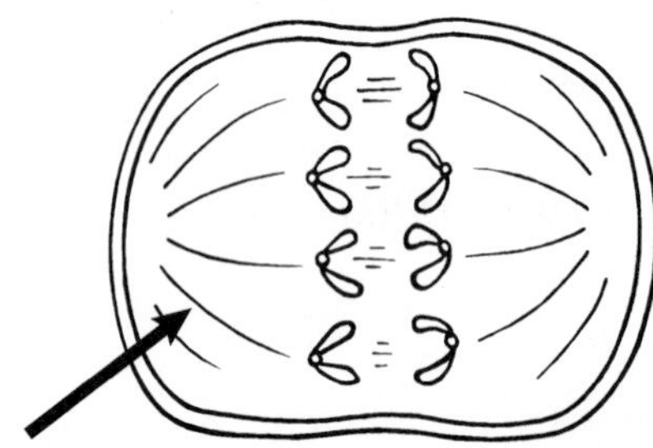

2. Which stage of mitosis is shown in this diagram?
   A. anaphase
   B. interphase
   C. metaphase
   D. telophase

3. To which structure does the arrow in the diagram point?
   A. centromere
   B. chromosome
   C. nucleolus
   D. spindle

4. Which stage of photosynthesis requires water to complete the chemical reaction?
   A. action of ATP synthase on ADP
   B. conversion of GAP molecules into RuBP
   C. conversion of NADP+ to NADPH
   D. transfer of chemical energy to form GAP molecules

5. Which carbon-containing compound is the product of glycolysis?
   A. acetyl CoA
   B. glucose
   C. lactic acid
   D. pyruvate

Use the diagram below to answer question 6.

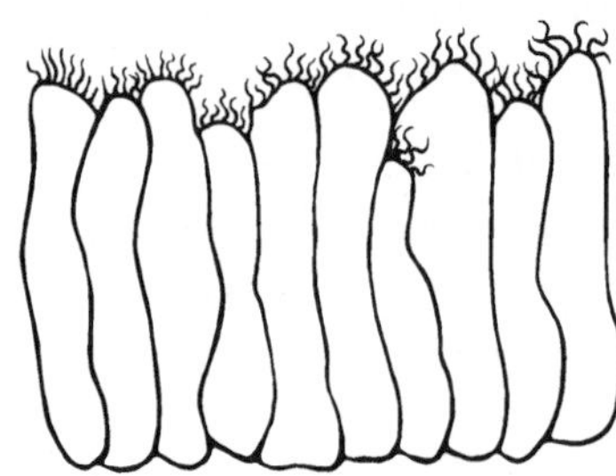

6. What are the structures projecting from the cells in the diagram?
   A. cilia
   B. flagella
   C. microfilaments
   D. villi

7. Which cellular process stores energy?
   A. the breaking of lipid chains
   B. the conversion of ADP to ATP
   C. the synthesization of proteins from RNA codons
   D. the transportation of ions across the membrane

8. Which contributes to the selective permeability of cell membranes?
   A. carbohydrates
   B. ions
   C. minerals
   D. proteins

9. If data from repeated experiments support a hypothesis, which would happen next?
   A. A conclusion would be established.
   B. The data would become a law.
   C. The hypothesis would be rejected.
   D. The hypothesis would be revised.

10. Which type of heterotroph is a mouse?
   A. carnivore
   B. detrivore
   C. herbivore
   D. omnivore

Biology Online Standardized Test Practice biologygmh.com

Short Answer

Use the diagram below to answer questions 11–13.

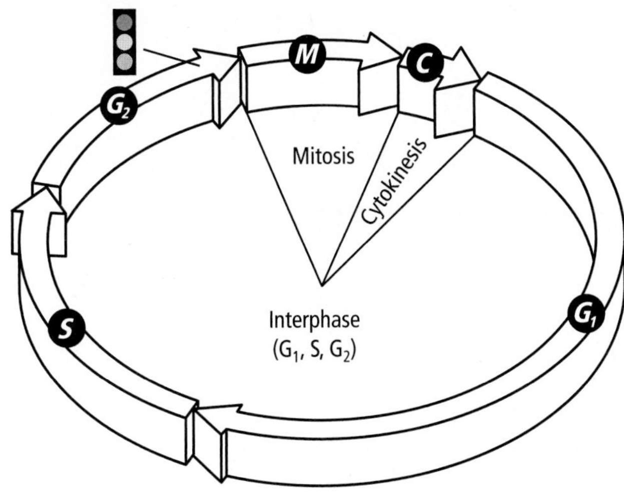

11. In the past, interphase often was called the "resting" phase of the cell cycle. Explain why this is inaccurate.

12. Explain what the cell does at the checkpoint indicated by the stoplight in the diagram.

13. Use the diagram to compare the relative rates at which mitosis and cytokinesis occur.

14. Hypothesize how an organism could be both a heterotroph and an autotroph.

15. Suppose you had ink, pebbles, and table salt. Describe what kind of mixture each one of these would make if mixed with water. Explain your answers.

16. Name two enzymes involved in photosynthesis and describe their roles.

17. Infer how the ratio of surface area to volume changes as a cell grows larger.

Extended Response

Use the diagram below to answer questions 18 and 19.

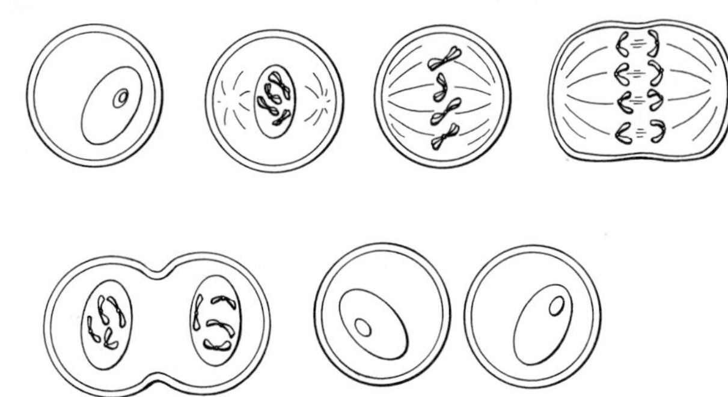

18. Analyze the diagram and describe the importance of the spindle fibers to chromatids during prophase.

19. Describe the function of the centromere and predict what might happen if cells did NOT have centromeres.

Essay Question

The same organelles are found in many different types of cells in an animal's body. However, there are differences in the number of organelles present, depending on the function of the different cells. For instance, the cells that require a great amount of energy to carry out their work would contain more mitochondria.

Using the information in the paragraph above, answer the following question in essay format.

20. How do you think two types of animal cells would differ in terms of the kinds of organelles they contain? Write a hypothesis about the cellular differences between two types of animal cells and then design an experiment to test your hypothesis.

NEED EXTRA HELP?																				
If You Missed Question . . .	1	2	3	4	5	6	7	8	9	10	11	12	13	14	15	16	17	18	19	20
Review Section . . .	6.1	9.2	9.2	8.2	8.3	7.3	8.1	7.2	1.3	2.1	9.1	9.1	8.2	8.1	6.3	8.2	9.1	9.2	9.2	7.3

UNIT 3 Genetics

Chapter 10
Sexual Reproduction and Genetics
BIG Idea Reproductive cells, which pass on genetic traits from the parents to the child, are produced by the process of meiosis.

Chapter 11
Complex Inheritance and Human Heredity
BIG Idea Human inheritance does not always follow Mendel's laws.

Chapter 12
Molecular Genetics
BIG Idea DNA is the genetic material that contains a code for proteins.

Chapter 13
Genetics and Biotechnology
BIG Idea Genetic technology improves human health and quality of life.

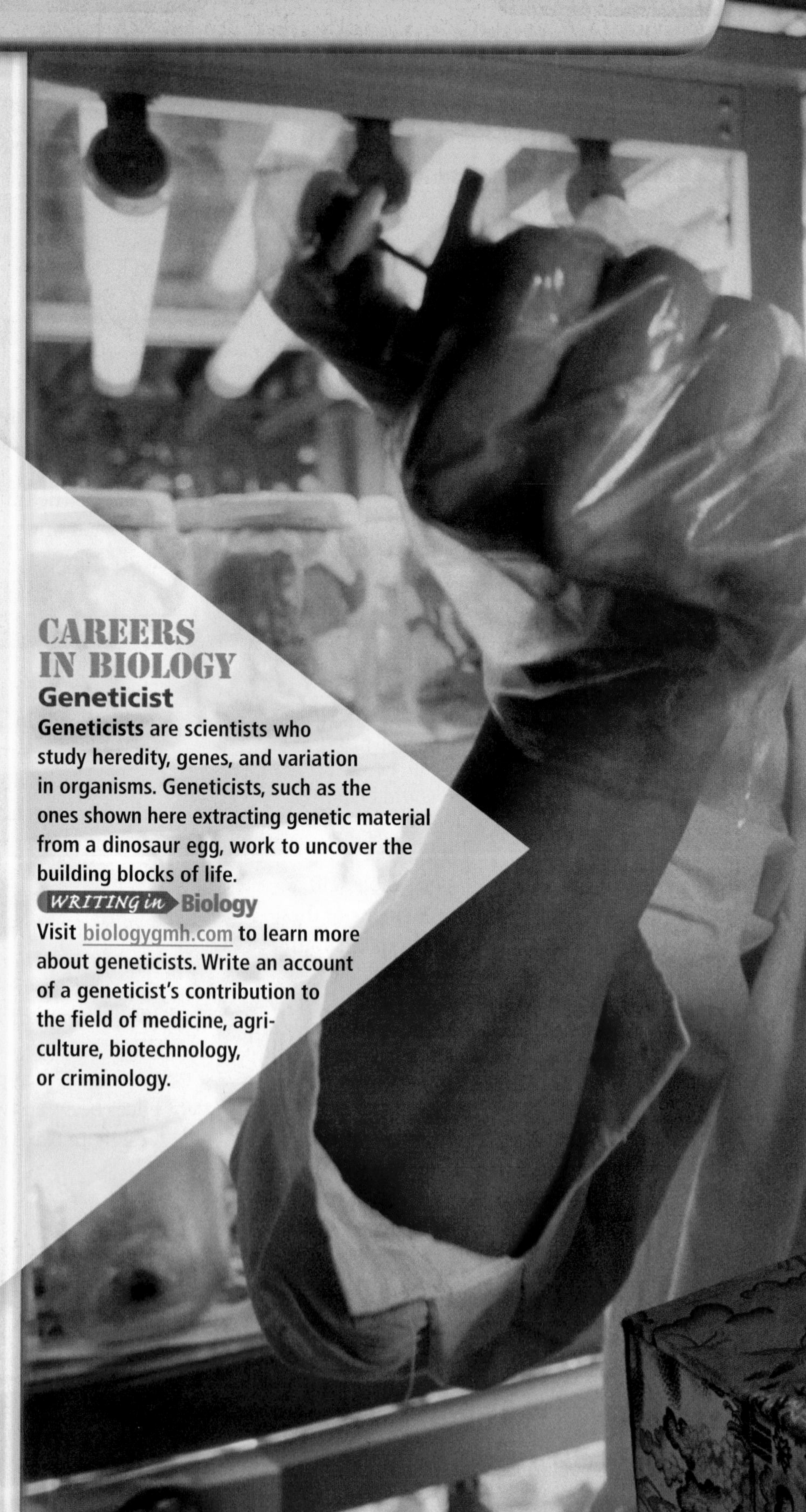

CAREERS IN BIOLOGY

Geneticist

Geneticists are scientists who study heredity, genes, and variation in organisms. Geneticists, such as the ones shown here extracting genetic material from a dinosaur egg, work to uncover the building blocks of life.

WRITING in Biology
Visit biologygmh.com to learn more about geneticists. Write an account of a geneticist's contribution to the field of medicine, agriculture, biotechnology, or criminology.

Biology Online

To read more about geneticists in action, visit biologygmh.com.

CHAPTER 10 Sexual Reproduction and Genetics

BIG Idea Reproductive cells, which pass on genetic traits from the parents to the child, are produced by the process of meiosis.

Section 1
Meiosis
MAIN Idea Meiosis produces haploid gametes.

Section 2
Mendelian Genetics
MAIN Idea Mendel explained how a dominant allele can mask the presence of a recessive allele.

Section 3
Gene Linkage and Polyploidy
MAIN Idea The crossing over of linked genes is a source of genetic variation.

BioFacts

- A female elephant gives birth after carrying her baby for 22 months.
- A baby elephant begins as a single, fertilized cell and at birth weighs about 120 kg.

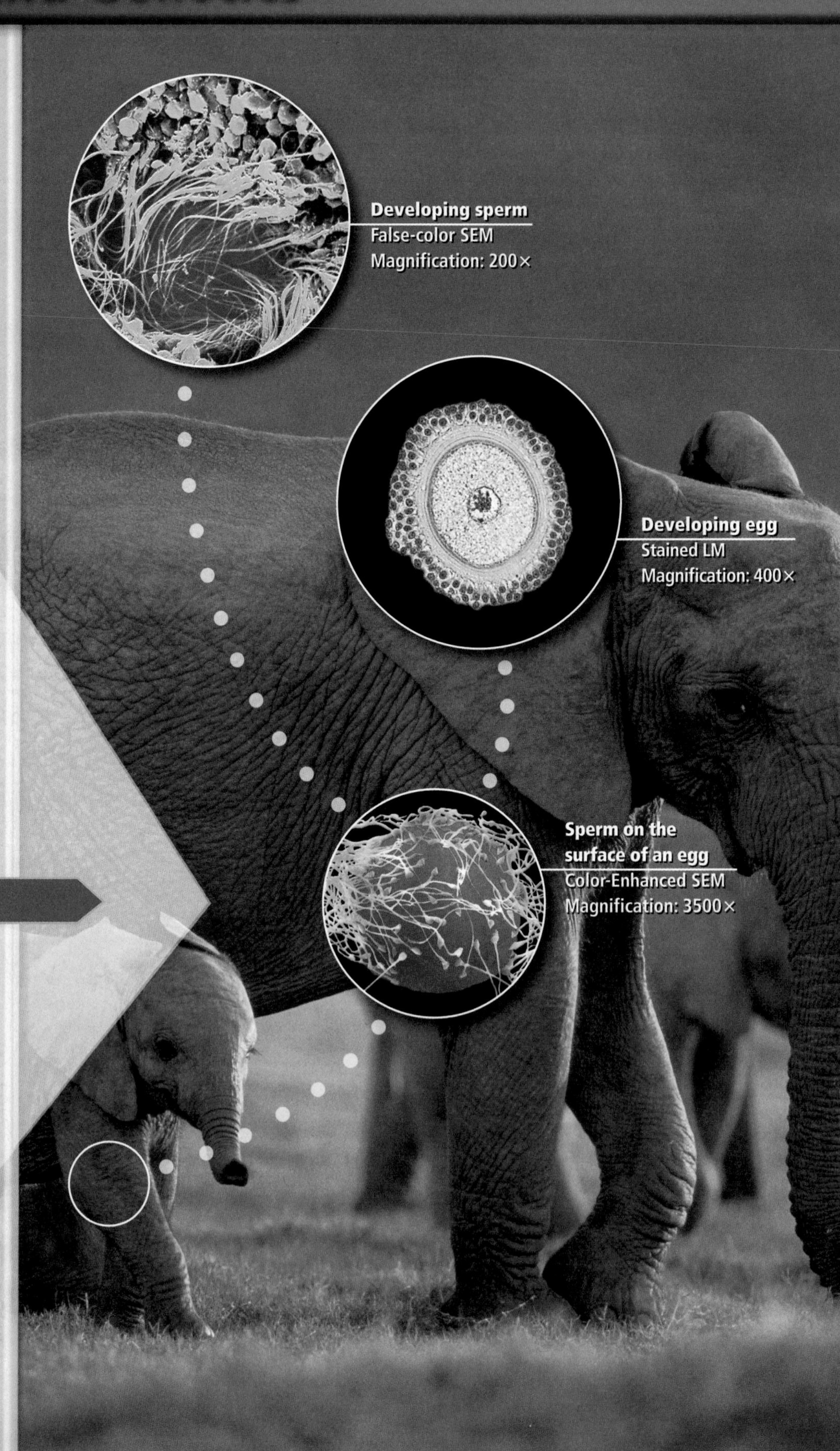

Developing sperm
False-color SEM
Magnification: 200×

Developing egg
Stained LM
Magnification: 400×

Sperm on the surface of an egg
Color-Enhanced SEM
Magnification: 3500×

Start-Up Activities

LAUNCH Lab

What would happen without meiosis?

In sexual reproduction, cells from each parent fuse; offspring have the same chromosome number as the parents. Explore what would happen to the chromosome number if mitosis were the only type of cell division.

Procedure

1. Read and complete the lab safety form.
2. Construct a data table with the headings *Cycle Number, Stage,* and *Chromosome Number.*
3. Fill in your data table for Steps 4-5.
4. Model a cell with a pair of chromosomes.
5. Demonstrate mitosis.
6. Fuse one of your cells with another student's cell.
7. Repeat Steps 4-5 two more times, recording the second and the third cycles.

Analysis

1. **Summarize** How does the chromosome number in your model change with each cycle of mitosis and fusion?
2. **Infer** What must occur when cells fuse in order for chromosome number to remain constant?

Visit biologygmh.com to:

- study the entire chapter online
- explore Concepts in Motion, the Interactive Table, Microscopy Links, Virtual Labs, and links to virtual dissections
- access Web links for more information, projects, and activities
- review content online with the Interactive Tutor and take Self-Check Quizzes

Illustrating Meiosis Make this Foldable to help you sequence, illustrate, and explain the phases of meiosis.

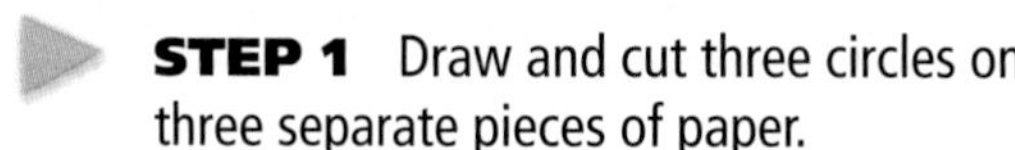

STEP 1 Draw and cut three circles on three separate pieces of paper.

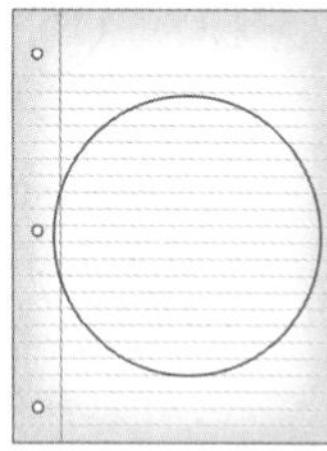
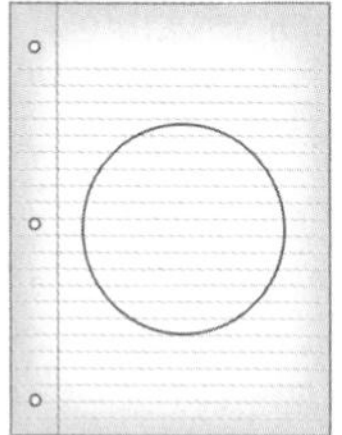
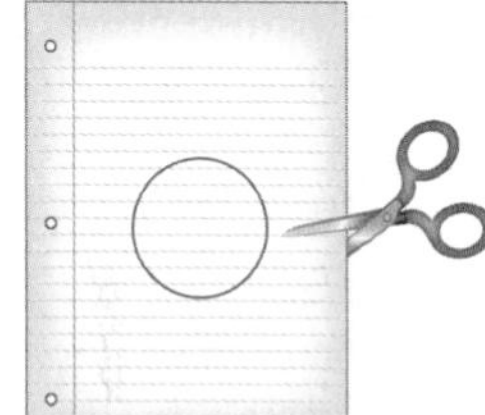

STEP 2 Fasten the circles together using a brad so that they will rotate. Label the small circle *Meiosis 1* on the top half of the circle and *Meiosis 2* on the bottom half of the circle.

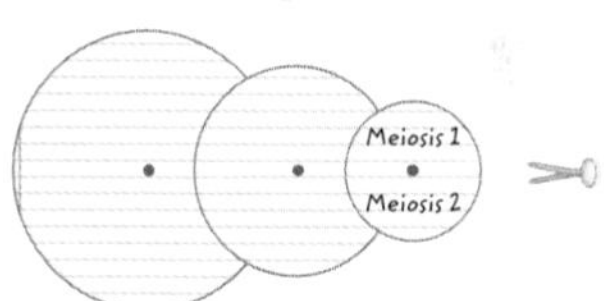

FOLDABLES **Use this Foldable with Section 10.1.** On the middle circle, write at equal intervals around the edge of the circle the following terms: *Prophase 1, Metaphase 1, Anaphase 1, Telophase 1, Prophase 2, Metaphase 2, Anaphase 2, Telophase 2.* On the largest circle, draw the phases of meiosis. Turn the circles so that both *Meiosis 1* and *Meiosis 2* align with appropriate phase names and illustrations.

Section 10.1

Reading Preview

Objectives

- **Explain** the reduction in chromosome number that occurs during meiosis.
- **Recognize and summarize** the stages of meiosis.
- **Analyze** the importance of meiosis in providing genetic variation.

Review Vocabulary

chromosome: cellular structure that contains DNA

New Vocabulary

gene
homologous chromosome
gamete
haploid
fertilization
diploid
meiosis
crossing over

Meiosis

MAIN Idea **Meiosis produces haploid gametes.**

Real-World Reading Link Look around your biology class. You might notice that the students in your class do not look the same. They might be different heights and have different eye color, hair color, and other features. This variety of characteristics is a result of two sex cells combining during sexual reproduction.

Chromosomes and Chromosome Number

Each student in your biology class has characteristics passed on to them by their parents. Each characteristic, such as hair color, height, or eye color, is called a trait. The instructions for each trait are located on chromosomes, which are found in the nucleus of cells. The DNA on chromosomes is arranged in segments called **genes** that control the production of proteins. Each chromosome consists of hundreds of genes, each gene playing an important role in determining the characteristics and functions of the cell.

Homologous chromosomes Human body cells have 46 chromosomes. Each parent contributes 23 chromosomes, resulting in 23 pairs of chromosomes. The chromosomes that make up a pair, one chromosome from each parent, are called **homologous chromosomes.** As shown in **Figure 10.1,** homologous chromosomes in body cells have the same length and the same centromere position, and they carry genes that control the same inherited traits. For instance, the gene for earlobe type will be located at the same position on both homologous chromosomes. Although these genes each code for earlobe type, they might not code for the exact same type of earlobe.

Figure 10.1 Homologous chromosomes carry genes for any given trait at the same location. The genes that code for earlobe type might not code for the exact same type of earlobe.

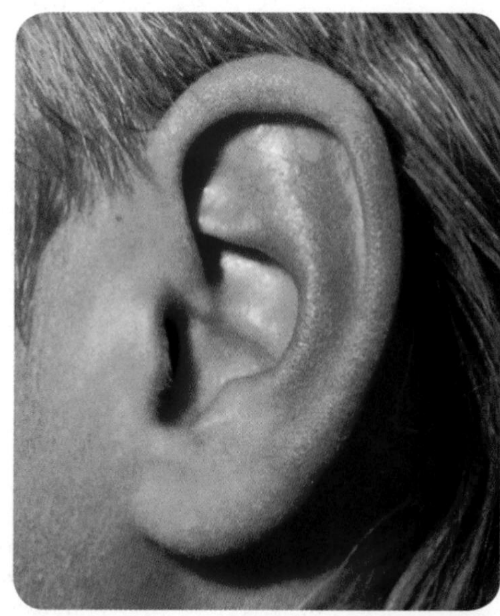

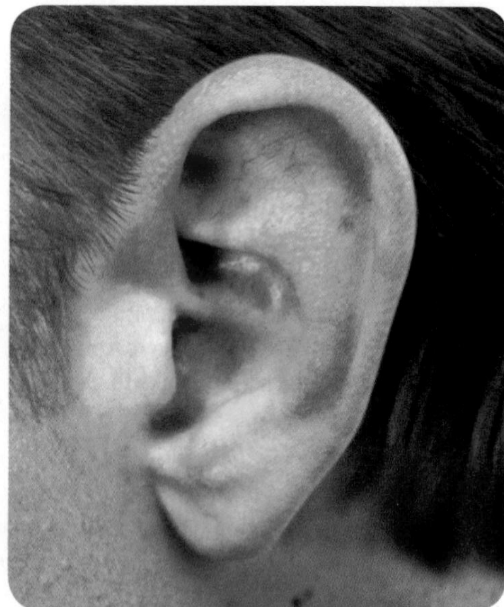

Haploid and diploid cells In order to maintain the same chromosome number from generation to generation, an organism produces **gametes,** which are sex cells that have half the number of chromosomes. Although the number of chromosomes varies from one species to another, in humans each gamete contains 23 chromosomes. The symbol n can be used to represent the number of chromosomes in a gamete. A cell with n number of chromosomes is called a **haploid** cell. Haploid comes from the Greek word *haploos,* meaning *single.*

The process by which one haploid gamete combines with another haploid gamete is called **fertilization.** As a result of fertilization, the cell now will contain a total of $2n$ chromosomes—n chromosomes from the female parent plus n chromosomes from the male parent. A cell that contains $2n$ number of chromosomes is called a **diploid** cell.

Notice that n also describes the number of pairs of chromosomes in an organism. When two human gametes combine, 23 pairs of homologous chromosomes are formed.

VOCABULARY

ACADEMIC VOCABULARY

Equator:
a circle or circular band dividing the surface of a body into two usually equal and symmetrical parts.
The chromosomes line up at the equator of the cell.

Meiosis I

Gametes are formed during a process called **meiosis,** which is a type of cell division that reduces the number of chromosomes; therefore, it is referred to as a reduction division. Meiosis occurs in the reproductive structures of organisms that reproduce sexually. While mitosis maintains the chromosome number, meiosis reduces the chromosome number by half through the separation of homologous chromosomes. A cell with $2n$ number of chromosomes will have gametes with n number of chromosomes after meiosis, as illustrated in **Figure 10.2.** Meiosis involves two consecutive cell divisions called meiosis I and meiosis II.

FOLDABLES
Incorporate information from this section into your Foldable.

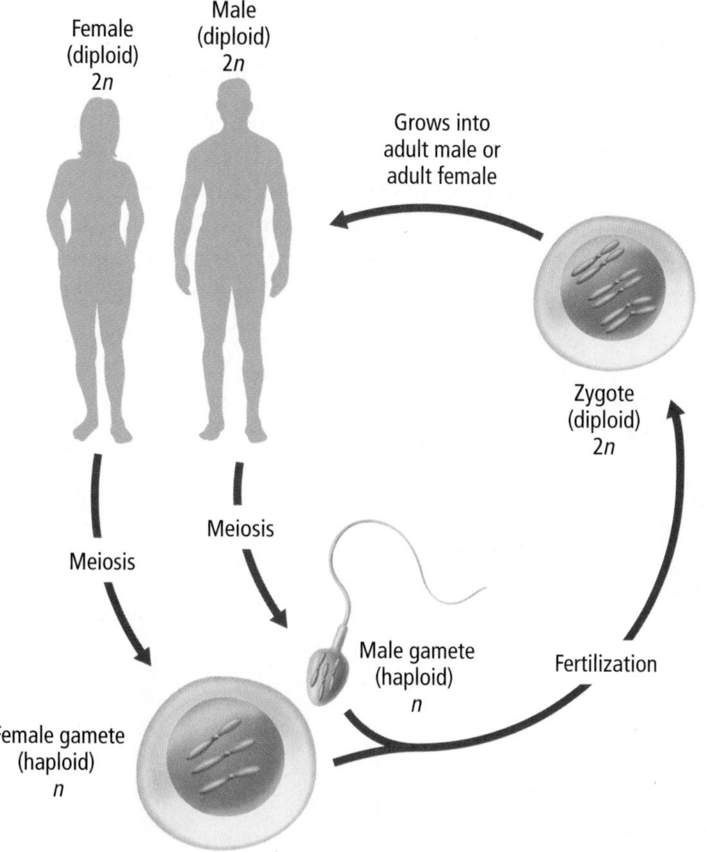

Figure 10.2 The sexual life cycle in animals involves meiosis, which produces gametes. When gametes combine in fertilization, the number of chromosomes is restored.
Describe *What happens to the number of chromosomes during meiosis?*

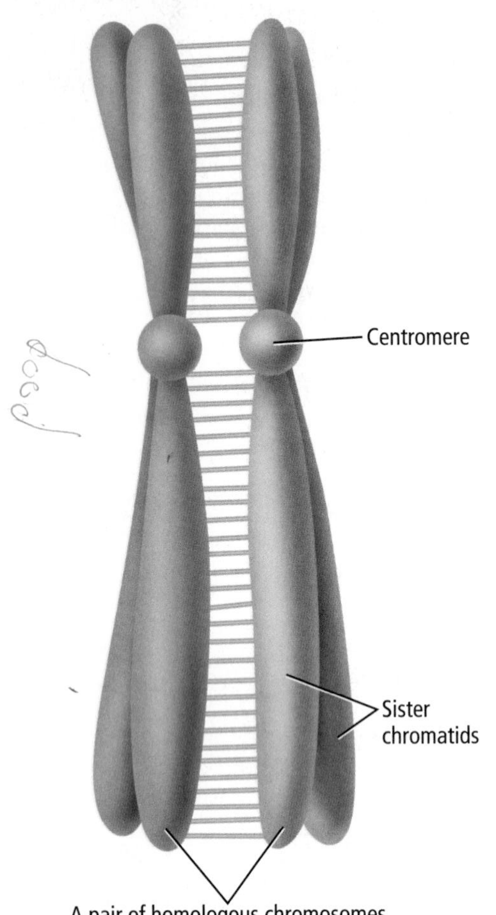

■ **Figure 10.3** The homologous chromosomes are physically bound together during synapsis in prophase I.

Interphase Recall that the cell cycle includes interphase prior to mitosis. Cells that undergo meiosis also go through interphase as part of the cell cycle. Cells in interphase carry out various metabolic processes, including the replication of DNA and the synthesis of proteins.

Prophase I As a cell enters prophase I, the replicated chromosomes become visible. As in mitosis, the replicated chromosomes consist of two sister chromatids. As the homologous chromosomes condense, they begin to form pairs in a process called synapsis. The homologous chromosomes are held tightly together along their lengths, as illustrated in **Figure 10.3.** Notice that in **Figure 10.4** the purple and green chromosomes have exchanged segments. This exchange occurs during synapsis. **Crossing over** is a process during which chromosomal segments are exchanged between a pair of homologous chromosomes.

As prophase I continues, centrioles move to the cell's opposite poles. Spindle fibers form and bind to the sister chromatids at the centromere.

Metaphase I In the next phase of meiosis, the pairs of homologous chromosomes line up at the equator of the cell, as illustrated in **Figure 10.5.** In meiosis, the spindle fibers attach to the centromere of each homologous chromosome. Recall that during metaphase in mitosis, the individual chromosomes, which consist of two sister chromatids, line up at the cell's equator. During metaphase I of meiosis, the homologous chromosomes line up as pairs at the cell's equator. This is an important distinction between mitosis and meiosis.

Anaphase I During anaphase I, the homologous chromosomes separate, which is also illustrated in **Figure 10.5.** Each member of the pair is guided by spindle fibers and moves toward opposite poles of the cell. The chromosome number is reduced from $2n$ to n when the homologous chromosomes separate. Recall that in mitosis, the sister chromatids split during anaphase. During anaphase I of meiosis, however, each homologous chromosome still consists of two sister chromatids.

Telophase I The homologous chromosomes, consisting of two sister chromatids, reach the cell's opposite poles. Each pole contains only one member of the original pair of homologous chromosomes. Notice in **Figure 10.5** that each chromosome still consists of two sister chromatids joined at the centromere. The sister chromatids might not be identical because crossing over might have occurred during synapsis in prophase I.

■ **Figure 10.4** The results of crossing over are new combinations of genes.
Determine ***Which chromatids exchanged genetic material?***

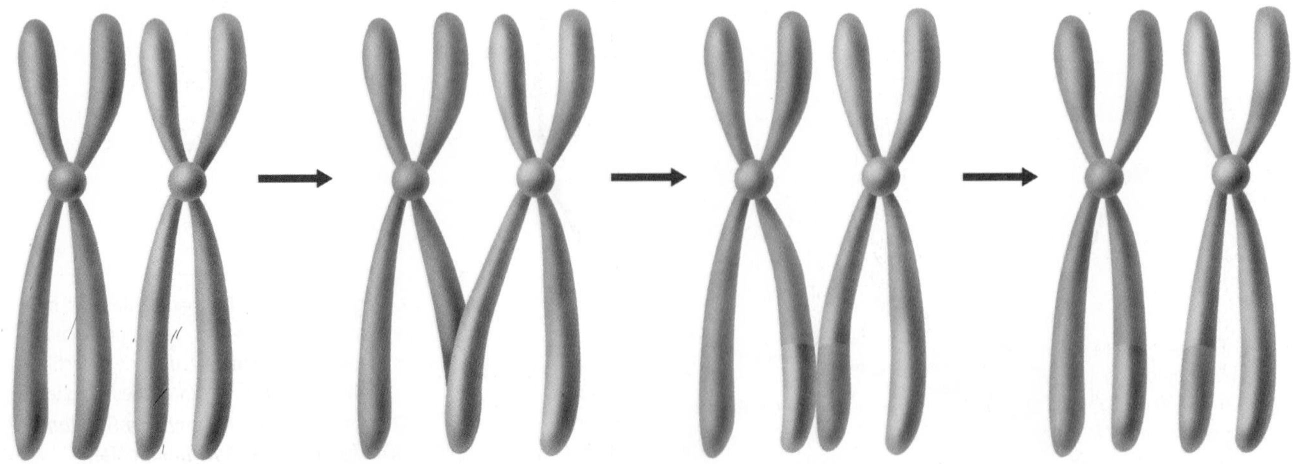

Visualizing Meiosis

Figure 10.5
Follow along the stages of meiosis I and meiosis II, beginning with interphase at the left.

1 Interphase
- Chromosomes replicate.
- Chromatin condenses.

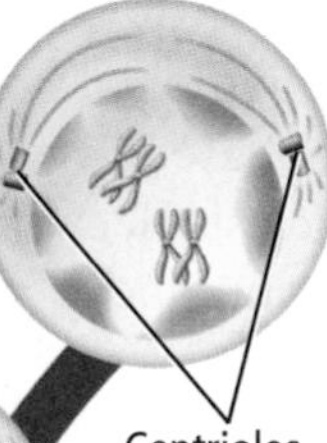

2 Prophase I
- Pairing of homologous chromosomes occurs, each chromosome consists of two chromatids.
- Crossing over produces exchange of genetic information.
- The nuclear envelope breaks down.
- Spindles form.

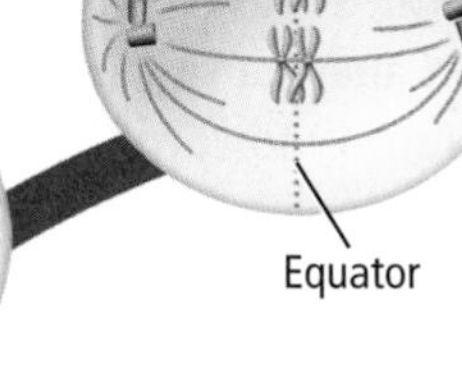

3 Metaphase I
- Chromosome centromeres attach to spindle fibers.
- Homologous chromosomes line up at the equator.

4 Anaphase I
- Homologous chromosomes separate and move to opposite poles of the cell.

5 Telophase I
- The spindles break down.
- Chromosomes uncoil and form two nuclei.
- The cell divides.

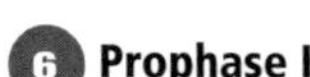

6 Prophase II
- Chromosomes condense.
- Spindles form in each new cell.
- Spindle fibers attach to chromosomes.

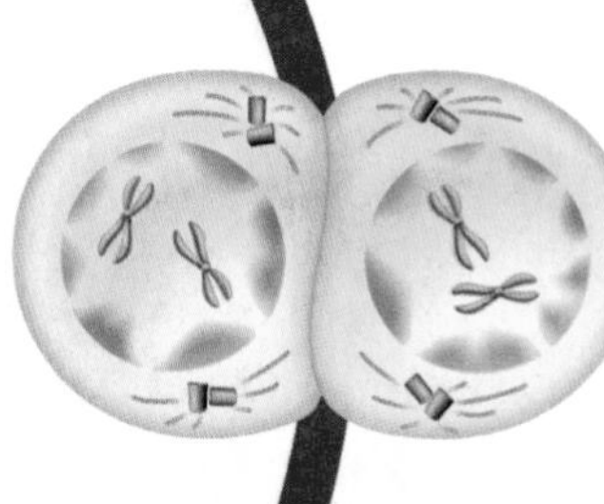

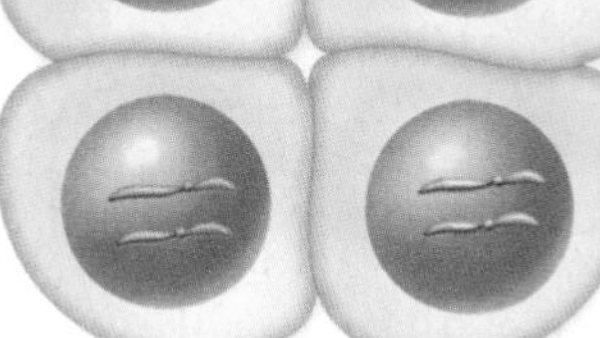

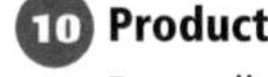

10 Products
- Four cells have formed.
- Each nucleus contains a haploid number of chromosomes.

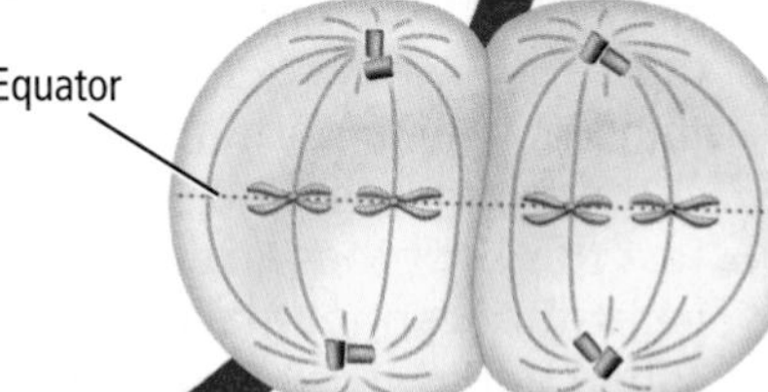

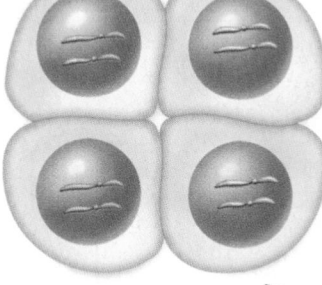

7 Metaphase II
- Centromeres of chromosomes line up randomly at the equator of each cell.

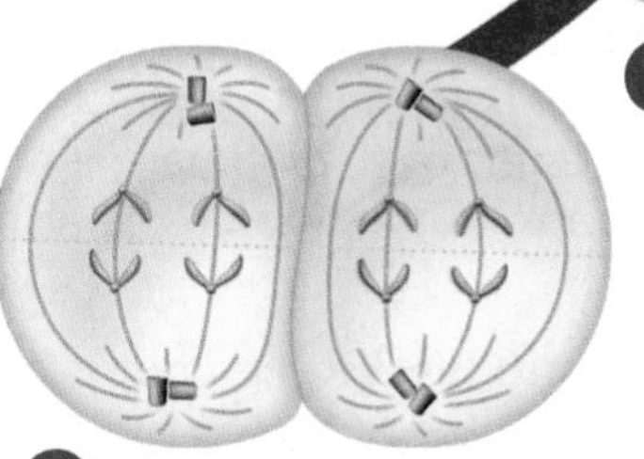

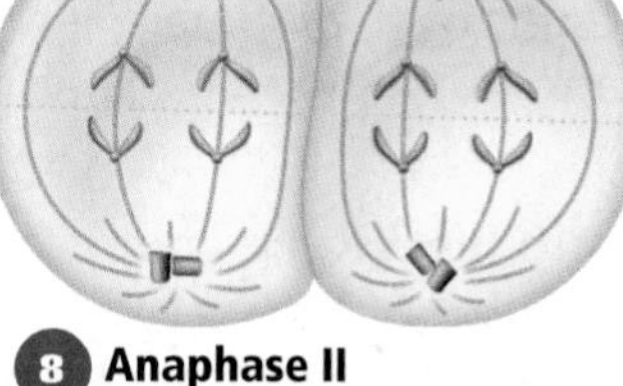

8 Anaphase II
- Centromeres split.
- Sister chromatids separate and move to opposite poles.

9 Telophase II
- Four nuclei form around chromosomes.
- Spindles break down.
- Cells divide.

Concepts In Motion **Interactive Figure** To see an animation of meiosis, visit biologygmh.com.

Biology Online

CAREERS IN BIOLOGY

Medical Geneticist A medical geneticist researches how diseases are inherited, how to diagnose genetic conditions, and treatments for genetic diseases. For more information on biology careers, visit biologygmh.com.

During telophase I, cytokinesis usually occurs, forming a furrow by pinching in animal cells and by forming a cell plate in plant cells. Following cytokinesis, the cells may go into interphase again before the second set of divisions. However, the DNA is not replicated again during this interphase. In some species, the chromosomes uncoil, the nuclear membrane reappears, and nuclei re-form during telophase I.

Meiosis II

Meiosis is only halfway completed at the end of meiosis I. During prophase II, a second set of phases begins as the spindle apparatus forms and the chromosomes condense. During metaphase II, the chromosomes are positioned at the equator by the spindle fibers, as shown in **Figure 10.5.** During metaphase of mitosis, a diploid number of chromosomes line up at the equator. During metaphase II of meiosis, however, a haploid number of chromosomes line up at the equator. During anaphase II, the sister chromatids are pulled apart at the centromere by the spindle fibers, and the sister chromatids move toward the opposite poles of the cell. The chromosomes reach the poles during telophase II, and the nuclear membrane and nuclei reform. At the end of meiosis II, cytokinesis occurs, resulting in four haploid cells, each with *n* number of chromosomes, as illustrated in **Figure 10.5.**

LAUNCH Lab

Review Based on what you have read about meiosis, how would you now answer the analysis questions?

Reading Check Infer Why are the two phases of meiosis important for gamete formation?

DATA ANALYSIS LAB 10.1

Based on Real Data*

Draw Conclusions

How do motor proteins affect cell division? Many scientists think that motor proteins play an important role in the movement of chromosomes in both mitosis and meiosis. To test this hypothesis, researchers have produced yeast that cannot make the motor protein called Kar3p. They also have produced yeast that cannot make the motor protein called Cik1p, which many think moderates the function of Kar3p. The results of their experiment are shown in the graph to the right.

Think Critically

1. **Evaluate** Does Cik1p seem to be important for yeast meiosis? Explain.
2. **Assess** Does Kar3p seem to be necessary for yeast meiosis? Explain.
3. **Conclude** Do all motor proteins seem to play a vital role in meiosis? Explain.

Data and Observations

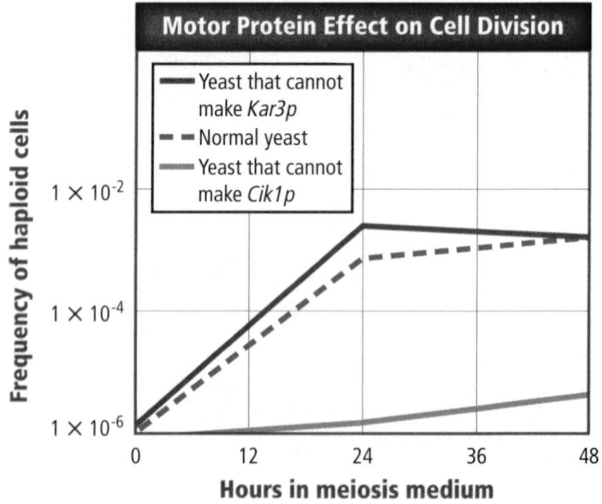

*Data obtained from: Shanks, et al. 2001. The Kar3-Interacting protein Cik1p plays a critical role in passage through meiosis I in *Saccharomyces cerevisiae. Genetics* 159: 939-951.

The Importance of Meiosis

Table 10.1 shows a comparison of mitosis and meiosis. Recall that mitosis consists of only one set of division phases and produces two identical diploid daughter cells. Meiosis, however, consists of two sets of divisions and produces four haploid daughter cells that are not identical. Meiosis is important because it results in genetic variation.

Personal Tutor

To learn about variation, visit biologygmh.com.

Concepts In Motion

Interactive Table To explore more about mitosis and meiosis, visit biologygmh.com.

Table 10.1 Mitosis and Meiosis

Mitosis	Meiosis
One division occurs during mitosis.	Two sets of divisions occur during meiosis: meiosis I and meiosis II.
DNA replication occurs during interphase.	DNA replication occurs once before meiosis I.
Synapsis of homologous chromosomes does not occur.	Synapsis of homologous chromosomes occurs during prophase I.
Two identical cells are formed per cell cycle.	Four haploid cells (*n*) are formed per cell cycle.
The daughter cells are genetically identical.	The daughter cells are not genetically identical because of crossing over.
Mitosis occurs only in body cells.	Meiosis occurs in reproductive cells.
Mitosis is involved in growth and repair.	Meiosis is involved in the production of gametes and providing genetic variation in organisms.

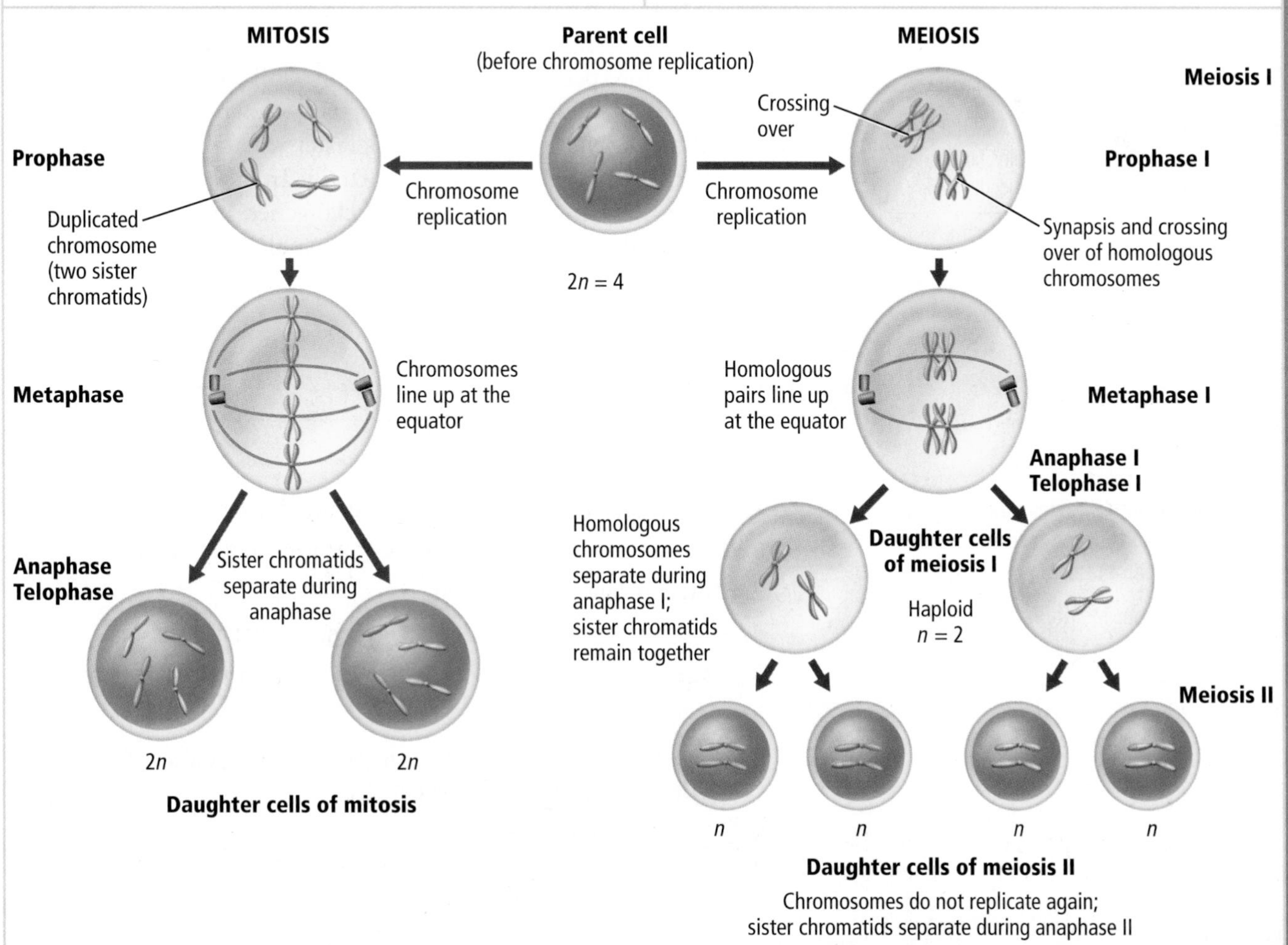

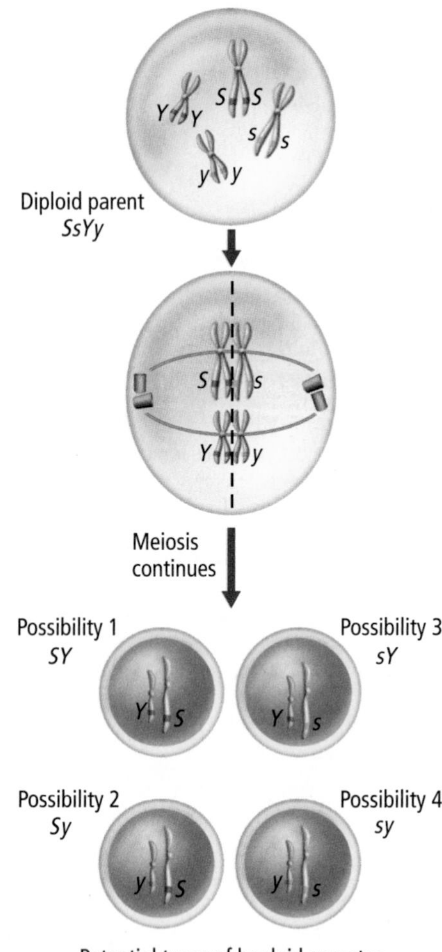

■ **Figure 10.6** The order in which the homologous pairs line up explains how a variety of sex cells can be produced.

Meiosis provides variation Recall that pairs of homologous chromosomes line up at the equator during prophase I. How the chromosomes line up at the equator is a random process that results in gametes with different combinations of chromosomes, such as the ones in **Figure 10.6.** Depending on how the chromosomes line up at the equator, four gametes with four different combinations of chromosomes can result.

Notice that the first possibility shows which chromosomes were on the same side of the equator and therefore traveled together. Different combinations of chromosomes were lined up on the same side of the equator to produce the gametes in the second possibility. Genetic variation also is produced during crossing over and during fertilization, when gametes randomly combine.

Sexual Reproduction v. Asexual Reproduction

Some organisms reproduce by asexual reproduction, while others reproduce by sexual reproduction. The life cycles of still other organisms might involve both asexual and sexual reproduction. During asexual reproduction, the organism inherits all of its chromosomes from a single parent. Therefore, the new individual is genetically identical to its parent. Bacteria reproduce asexually, whereas most protists reproduce both asexually and sexually, depending on environmental conditions. Most plants and many of the more simple animals can reproduce both asexually and sexually, compared to more advanced animals that reproduce only sexually.

Why do some species reproduce sexually while others reproduce asexually? Recent studies with fruit flies have shown that the rate of accumulation of beneficial mutations is faster when species reproduce sexually than when they reproduce asexually. In other words, when reproduction occurs sexually, the beneficial genes multiply faster over time than they do when reproduction is asexual.

Section 10.1 Assessment

Section Summary

- DNA replication takes place only once during meiosis, and it results in four haploid gametes.
- Meiosis consists of two sets of divisions.
- Meiosis produces genetic variation in gametes.

Understand Main Ideas

1. MAIN Idea **Analyze** how meiosis produces haploid gametes.
2. **Indicate** how metaphase I is different from metaphase in mitosis.
3. **Describe** how synapsis occurs.
4. **Diagram** a cell with four chromosomes going through meiosis.
5. **Assess** how meiosis contributes to genetic variation, while mitosis does not.

Think Critically

6. **Compare and contrast** mitosis and meiosis, using **Figure 10.5** and **Table 10.1,** by creating a Venn diagram.

WRITING in **Biology**

7. Write a play or activity involving your classmates, to explain the various processes that occur during meiosis.

Biology Online **Self-Check Quiz** biologygmh.com

Section 10.2

Reading Preview

Objectives

- **Explain** the significance of Mendel's experiments to the study of genetics.
- **Summarize** the law of segregation and law of independent assortment.
- **Predict** the possible offspring from a cross using a Punnett square.

Review Vocabulary

segregation: the separation of allelic genes that typically occurs during meiosis

New Vocabulary

genetics
allele
dominant
recessive
homozygous
heterozygous
genotype
phenotype
law of segregation
hybrid
law of independent assortment

■ **Figure 10.7** Gregor Mendel is known as the father of genetics.

Mendelian Genetics

MAIN Idea Mendel explained how a dominant allele can mask the presence of a recessive allele.

Real-World Reading Link There are many different breeds of dogs, such as Labrador retrievers, dachshunds, German shepherds, and poodles. You might like a certain breed of dog because of its height, coat color, and general appearance. These traits are passed from generation to generation.

How Genetics Began

In 1866, Gregor Mendel, an Austrian monk and a plant breeder, published his findings on the method of inheritance in garden pea plants. The passing of traits to the next generation is called inheritance, or heredity. Mendel, shown in **Figure 10.7,** was successful in sorting out the mystery of inheritance because of the organism he chose for his study—the pea plant. Pea plants are true-breeding, meaning that they consistently produce offspring with only one form of a trait.

Pea plants usually reproduce by self-fertilization. A common occurrence in many flowering plants, self-fertilization occurs when a male gamete within a flower combines with a female gamete in the same flower. Mendel also discovered that pea plants could easily be cross-pollinated by hand. Mendel performed cross-pollination by transferring a male gamete from the flower of one pea plant to the female reproductive organ in a flower of another pea plant.

Connection  **to History** Mendel rigorously followed various traits in the pea plants he bred. He analyzed the results of his experiments and formed hypotheses concerning how the traits were inherited. The study of **genetics,** which is the science of heredity, began with Mendel, who is regarded as the father of genetics.

Reading Check Infer why it is important that Mendel's experiments used a true-breeding plant.

The Inheritance of Traits

Mendel noticed that certain varieties of garden pea plants produced specific forms of a trait, generation after generation. For instance, he noticed that some varieties always produced green seeds and others always produced yellow seeds. In order to understand how these traits are inherited, Mendel performed cross pollination by transferring male gametes from the flower of a true-breeding green-seed plant to the female organ of a flower from a true-breeding yellow-seed plant. To prevent self-fertilization, Mendel removed the male organs from the flower of the yellow-seed plant. Mendel called the green-seed plant and the yellow-seed plant the parent generation—also known as the P generation.

Figure 10.8 The results of Mendel's cross involving true-breeding pea plants with yellow seeds and green seeds are shown here. **Explain** ***why the seeds in the F_1 generation were all yellow.***

Concepts In Motion

Interactive Figure To see an animation of the allele frequencies of three generations of flowers, visit biologygmh.com.

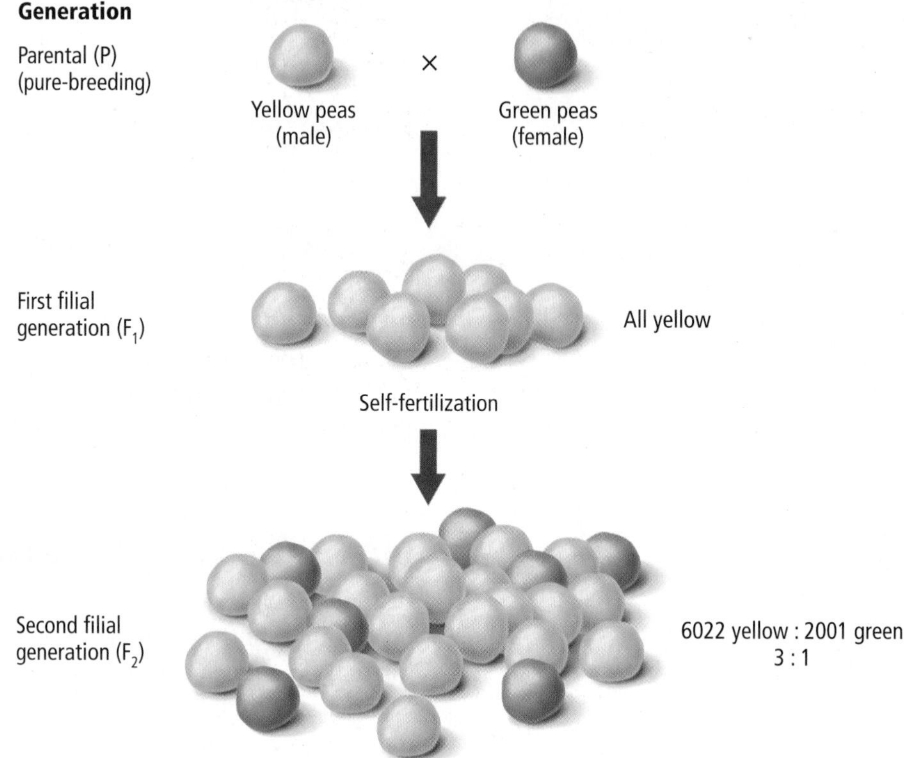

F_1 and F_2 generations When Mendel grew the seeds from the cross between the green-seed and yellow-seed plants, all of the resulting offspring had yellow seeds. The offspring of this P cross are called the first filial (F_1) generation. The green-seed trait seemed to have disappeared in the F_1 generation, and Mendel decided to investigate whether the trait was no longer present or whether it was hidden, or masked.

Mendel planted the F_1 generation of yellow seeds, allowed the plants to grow and self-fertilize, and then examined the seeds from this cross. The results of the second filial (F_2) generation—the offspring from the F_1 cross—are shown in **Figure 10.8.** Of the seeds Mendel collected, 6022 were yellow and 2001 were green, which almost is a perfect 3:1 ratio of yellow to green seeds.

Mendel studied seven different traits—seed or pea color, flower color, seed pod color, seed shape or texture, seed pod shape, stem length, and flower position—and found that the F_1 generation plants from these crosses also showed a 3:1 ratio.

Genes in pairs Mendel concluded that there must be two forms of the seed trait in the pea plants—yellow-seed and green-seed—and that each was controlled by a factor, which now is called an allele. An **allele** is defined as an alternative form of a single gene passed from generation to generation. Therefore, the gene for yellow seeds and the gene for green seeds are each different forms of a single gene.

Mendel concluded that the 3:1 ratio observed during his experiments could be explained if the alleles were paired in each of the plants. He called the form of the trait that appeared in the F_1 generation **dominant** and the form of the trait that was masked in the F_1 generation **recessive.** In the cross between yellow-seed plants and green-seed plants, the yellow seed was the dominant form of the trait and the green seed was the recessive form of the trait.

CAREERS IN BIOLOGY

Genetics Laboratory Technician
A technician in a genetics laboratory assists a researcher by conducting experiments and helping to maintain the lab. For more information on biology careers, visit biologygmh.com.

Dominance When he allowed the F_1 generation to self-fertilize, Mendel showed that the recessive allele for green seeds had not disappeared but was masked. Mendel concluded that the green-seed form of the trait did not show up in the F_1 generation because the yellow-seed form of the trait is dominant and masks the allele for the green-seed form of the trait.

When modeling inheritance, the dominant allele is represented by a capital letter, and the recessive allele is represented by a lowercase letter. An organism with two of the same alleles for a particular trait is **homozygous** (ho muh ZI gus) for that trait. Homozygous, yellow-seed plants are *YY* and green-seed plants are *yy*. An organism with two different alleles for a particular trait is **heterozygous** (heh tuh roh ZY gus) for that trait, in this case *Yy*. When alleles are present in the heterozygous state, the dominant trait will be observed.

Genotype and phenotype A yellow-seed plant could be homozygous or heterozygous for the trait form. The outward appearance of an organism does not always indicate which pair of alleles is present. The organism's allele pairs are called its **genotype.** In the case of plants with yellow seeds, their genotypes could be *YY* or *Yy*. The observable characteristic or outward expression of an allele pair is called the **phenotype.** The phenotype of pea plants with the genotype *yy* will be green seeds.

Mendel's law of segregation Mendel used homozygous yellow-seed and green-seed plants in his P cross. In **Figure 10.9(A),** the top drawing shows that each gamete from the yellow-seed plant contains one *Y*. Recall that the chromosome number is divided in half during meiosis. The resulting gametes contain only one of the pair of seed-color alleles.

The bottom drawing in **Figure 10.9(A)** shows that each gamete from the green-seed plant contains one *y* allele. Mendel's **law of segregation** states that the two alleles for each trait separate during meiosis. During fertilization, two alleles for that trait unite.

The third drawing in **Figure 10.9(B)** shows the alleles uniting to produce the genotype *Yy* during fertilization. All resulting F_1 generation plants will have the genotype *Yy* and will have yellow seeds because yellow is dominant to green. These heterozygous organisms are called **hybrids**.

VOCABULARY
WORD ORIGIN
Homozygous and **Heterozygous** come from the Greek words *homos*, meaning *the same; hetero,* meaning *other* or *different;* and *zygon,* meaning *yoke.*

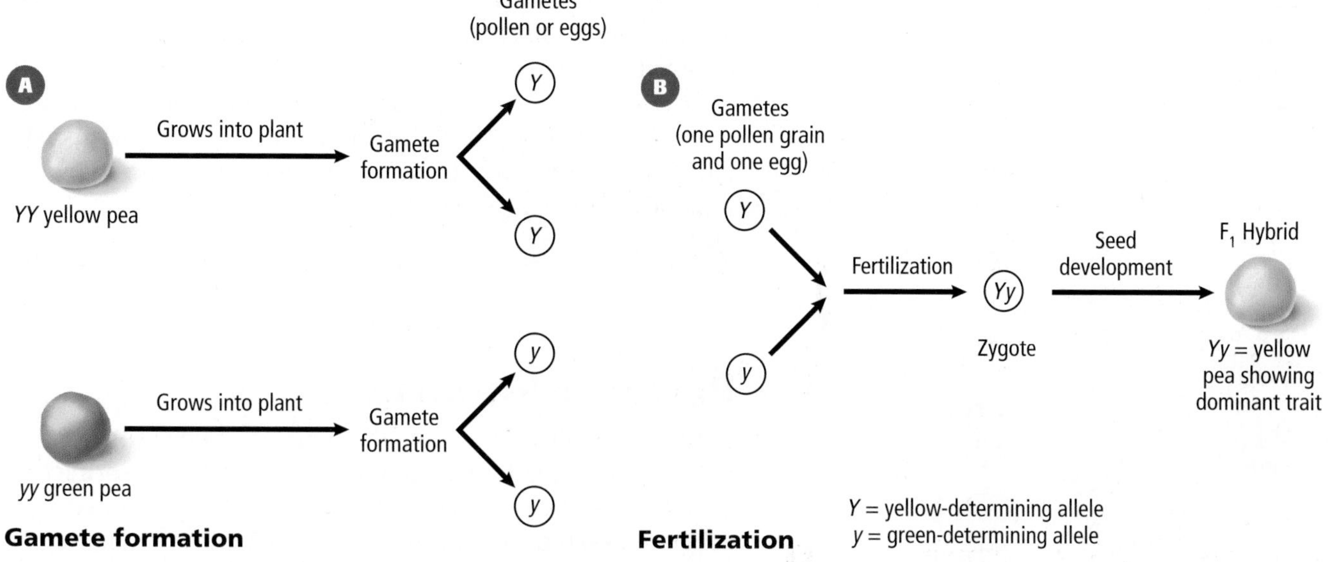

Figure 10.9 During gamete formation in the *YY* or *yy* plant, the two alleles separate, resulting in *Y* or *y* in the gametes. Gametes from each parent unite during fertilization.

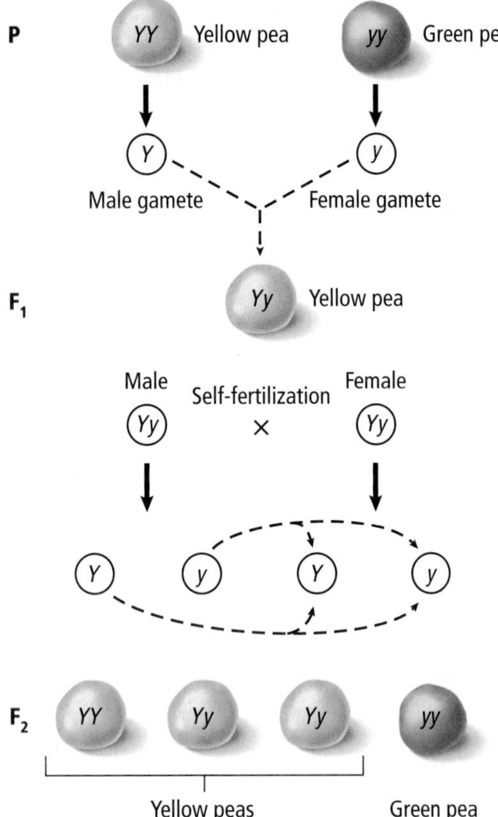

■ **Figure 10.10** During the F_1 generation self-fertilization, the male gametes randomly fertilize the female gametes.

■ **Figure 10.11** The law of independent assortment is demonstrated in the dihybrid cross by the equal chance that each pair of alleles (*Yy* and *Rr*) can randomly combine with each other.

Predict ***How many possible gamete types are produced?***

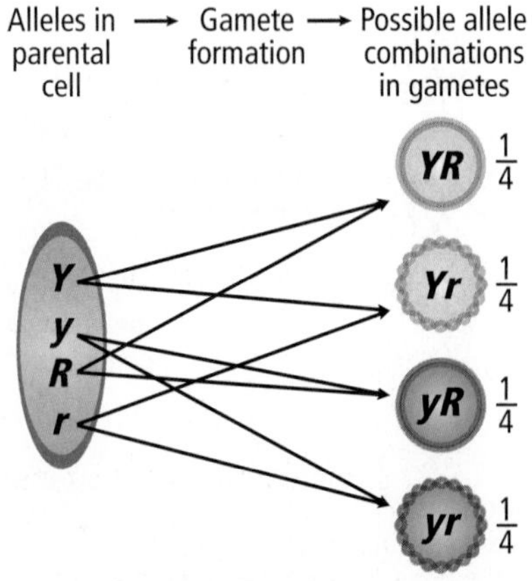

Monohybrid cross The diagram in **Figure 10.10** shows how Mendel continued his experiments by allowing the *Yy* plants to self-fertilize. A cross such as this one that involves hybrids for a single trait is called a monohybrid cross. The *Yy* plants produce two types of gametes—male and female—each with either the *Y* or *y* allele. The combining of these gametes is a random event. This random fertilization of male and female gametes results in the following genotypes—*YY, Yy, Yy,* or *yy,* as shown in **Figure 10.10.** Notice that the dominant *Y* allele is written first, whether it came from the male or female gamete. In Mendel's F_1 cross, there are three possible genotypes: *YY, Yy,* and *yy;* and the genotypic ratio is 1:2:1. The phenotypic ratio is 3:1—yellow seeds to green seeds.

Dihybrid cross Once Mendel established inheritance patterns of a single trait, he began to examine simultaneous inheritance of two or more traits in the same plant. In garden peas, round seeds *(R)* are dominant to wrinkled seeds *(r),* and yellow seeds *(Y)* are dominant to green seeds *(y).* If Mendel crossed homozygous yellow, round-seed pea plants with homozygous green, wrinkle-seed pea plants, the P cross could be represented by *YYRR* × *yyrr.* The F_1 generation genotype would be *YyRr*—yellow, round-seed plants. These F_1-generation plants are called dihybrids because they are heterozygous for both traits.

Law of independent assortment Mendel allowed F_1 pea plants with the genotype *YyRr* to self-fertilize in a dihybrid cross. Mendel calculated the genotypic and phenotypic ratios of the offspring in both the F_1 and F_2 generations. From these results, he developed the **law of independent assortment,** which states that a random distribution of alleles occurs during gamete formation. Genes on separate chromosomes sort independently during meiosis.

As shown in **Figure 10.11,** the random assortment of alleles results in four possible gametes: *YR, Yr, yR* or *yr,* each of which is equally likely to occur. When a plant self-fertilizes, any of the four allele combinations could be present in the male gamete, and any of the four combinations could be present in the female gamete. The results of Mendel's dihybrid cross included nine different genotypes: *YYRR, YYRr, YYrr, YyRR, YyRr, Yyrr, yyRR, yyRr,* and *yyrr.* He counted and recorded four different phenotypes: 315 yellow round, 108 green round, 101 yellow wrinkled, and 32 green wrinkled. These results represent a phenotypic ratio of approximately 9:3:3:1.

Reading Check **Evaluate** How can the random distribution of alleles result in a predictable ratio?

Punnett Squares

In the early 1900s, Dr. Reginald Punnett developed what is known as a Punnett square to predict the possible offspring of a cross between two known genotypes. Punnett squares make it easier to keep track of the possible genotypes involved in a cross.

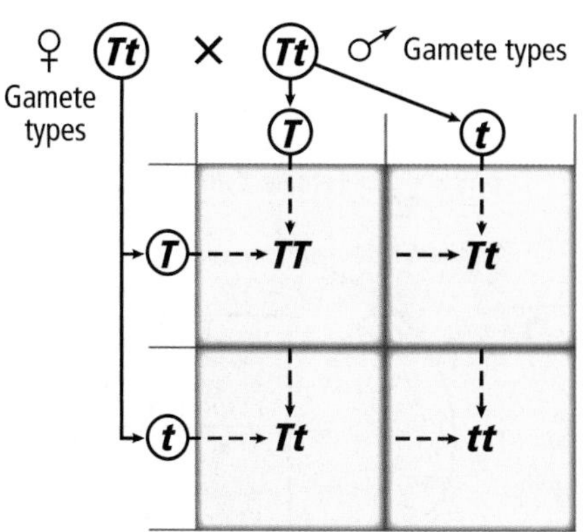

■ **Figure 10.12** The ability to roll one's tongue is a dominant trait. The Punnett square is a visual summary of the possible combinations of the alleles for the tongue-rolling trait.

Punnett square—monohybrid cross Can you roll your tongue like the person pictured in **Figure 10.12?** Tongue-rolling ability is a dominant trait, which can be represented by *T*. Suppose both parents can roll their tongues and are heterozygous (*Tt*) for the trait. What possible phenotypes could their children have?

Examine the Punnett square in **Figure 10.12.** The number of squares is determined by the number of different types of alleles—*T* or *t*—produced by each parent. In this case, the square is 2 squares × 2 squares because each parent produces two different types of gametes. Notice that the male gametes are written across the horizontal side and the female gametes are written on the vertical side of the Punnett square. The possible combinations of each male and female gamete are written on the inside of each corresponding square.

Personal Tutor

To learn about Punnett squares, visit biologygmh.com.

MiniLab 10.1

Predict Probability in Genetics

How can an offspring's traits be predicted? A Punnett square can help predict ratios of dominant traits to recessive traits in the genotype of offspring. This lab involves two parents who are both heterozygous for free earlobes (E), which is a dominant trait. The recessive trait is attached earlobes (e).

Procedure

1. Read and complete the lab safety form.
2. Determine the gamete genotype(s) for this trait that each parent contributes.
3. Draw a Punnett square that has the same number of columns and the same number of rows as the number of alleles contributed for this trait by the gametes of each parent.
4. Write the alphabetical letter for each allele from one parent just above each column, and write the alphabetical letter for each allele from the other parent just to the left of each row.
5. In the boxes within the table, write the genotype of the offspring resulting from each combination of male and female alleles.

Analysis

1. **Summarize** List the possible offspring phenotypes that could occur.
2. **Evaluate** What is the phenotypic ratio of the possible offspring? What is the genotypic ratio of the possible offspring?

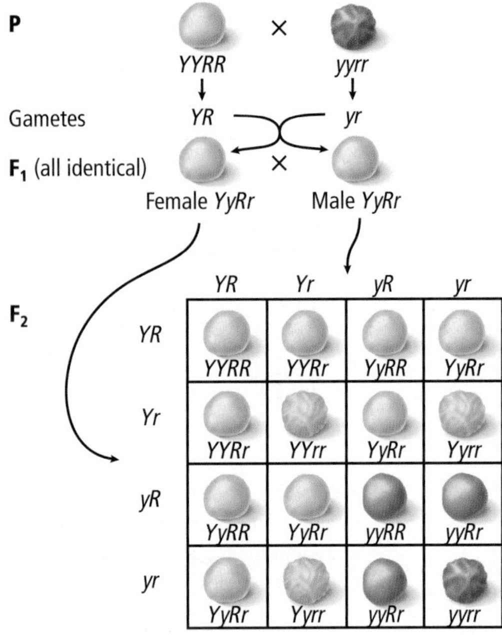

Type	Genotype	Phenotype	Number	Phenotypic Ratio
Parental	Y_R_	yellow round	315	9:16
Recombinant	yyR_	green round	108	3:16
Recombinant	Y_rr	yellow wrinkled	101	3:16
Parental	yyrr	green wrinkled	32	1:16

■ **Figure 10.13** The dihybrid Punnett square visually presents the possible combinations of the possible alleles from each parent.

How many different genotypes are found in the Punnett square? One square has *TT,* two squares have *Tt,* and one square has *tt.* Therefore, the genotypic ratio of the possible offspring is 1:2:1. The phenotypic ratio of tongue rollers to non-tongue rollers is 3:1.

Punnett square—dihybrid cross Now examine the Punnett square in **Figure 10.13.** Notice that in the P cross, only two types of alleles are produced. However, in the dihybrid cross—when the F_1 generation is crossed—four types of alleles from the male gametes and four types of alleles from the female gametes can be produced. The resulting phenotypic ratio is 9:3:3:1—9 yellow round to 3 green round to 3 yellow wrinkled to 1 green wrinkled. Mendel's data closely matched the outcome predicted by the Punnett square.

Probability

The inheritance of genes can be compared to the probability of flipping a coin. The probability of the coin landing on heads is 1 out of 2, or 1/2. If the same coin is flipped twice, the probability of it landing on heads is 1/2 each time or 1/2 × 1/2, or 1/4 both times.

Actual data might not perfectly match the predicted ratios. You know that if you flip a coin you might not get heads 1 out of 2 times. Mendel's results were not exactly a 9:3:3:1 ratio. However, the larger the number of offspring involved in a cross, the more likely it will match the results predicted by the Punnett square.

Section 10.2 Assessment

Section Summary

- The study of genetics began with Gregor Mendel, whose experiments with garden pea plants gave insight into the inheritance of traits.
- Mendel developed the law of segregation and the law of independent assortment.
- Punnett squares help predict the offspring of a cross.

Understand Main Ideas

1. MAIN Idea **Diagram** Use a Punnett square to explain how a dominant allele masks the presence of a recessive allele.
2. **Apply** the law of segregation and the law of independent assortment by giving an example of each.
3. **Use a Punnett square** In fruit flies, red eyes (*R*) are dominant to pink eyes (*r*). What is the phenotypic ratio of a cross between a heterozygous male and a pink-eyed female?

Think Critically

4. **Evaluate** the significance of Mendel's work to the field of genetics.

MATH in Biology

5. What is the probability of rolling a 2 on a six-sided die? What is the probability of rolling two 2s on two six-sided die? How is probability used in the study of genetics?

Biology Online Self-Check Quiz biologygmh.com

Section 10.3

Gene Linkage and Polyploidy

Reading Preview

Objectives

- **Summarize** how the process of meiosis produces genetic recombination.
- **Explain** how gene linkage can be used to create chromosome maps.
- **Analyze** why polyploidy is important to the field of agriculture.

Review Vocabulary

protein: large, complex polymer essential to all life that provides structure for tissues and organs and helps carry out cell metabolism

New Vocabulary

genetic recombination
polyploidy

MAIN Idea The crossing over of linked genes is a source of genetic variation.

Real-World Reading Link You might find many varieties of plants in a garden center that are not found in the wild. For example, you might have seen many varieties of roses that range in color from red to pink to white. Plant breeders use scientists' knowledge of genes to vary certain characteristics in an effort to make their roses unique.

Genetic Recombination

Connection to Math The new combination of genes produced by crossing over and independent assortment is called **genetic recombination.** The possible combinations of genes due to independent assortment can be calculated using the formula 2^n, where n is the number of chromosome pairs. For example, pea plants have seven pairs of chromosomes. For seven pairs of chromosomes, the possible combinations are 2^7, or 128 combinations. Because any possible male gamete can fertilize any possible female gamete, the number of possible combinations after fertilization is 16,384 (128×128). In humans, the possible number of combinations after fertilization would be $2^{23} \times 2^{23}$, or more than 70 trillion. This number does not include the amount of genetic recombination produced by crossing over.

Gene Linkage

Recall that chromosomes contain multiple genes that code for proteins. Genes that are located close to each other on the same chromosome are said to be linked and usually travel together during gamete formation. Study **Figure 10.14** and observe that genes A and B are located close to each other on the same chromosome and travel together during meiosis. The linkage of genes on a chromosome results in an exception to Mendel's law of independent assortment because linked genes usually do not segregate independently.

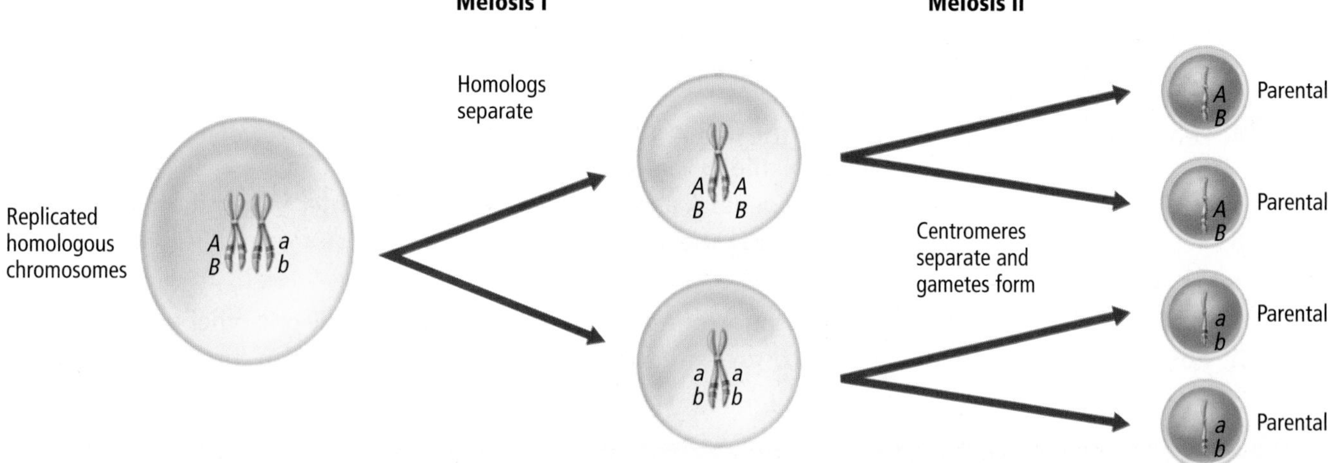

Figure 10.14 Genes that are linked together on the same chromosome usually travel together in the gamete.
Calculate *the number of possible combinations if two or three of these gametes were to combine.*

■ **Figure 10.15** This chromosome map of the X chromosome of the fruit fly *Drosophila melanogaster* was created in 1913.

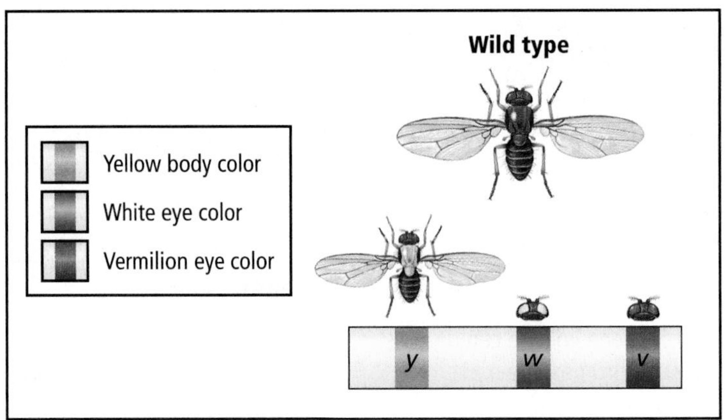

Gene linkage was first studied using the fruit fly *Drosophila melanogaster.* Thousands of crosses confirmed that linked genes usually traveled together during meiosis. However, some results revealed that linked genes do not always travel together during meiosis. Scientists concluded that linked genes can separate during crossing over.

Chromosome maps Crossing over occurs more frequently between genes that are far apart than those that are close together. A drawing called a chromosome map shows the sequence of genes on a chromosome and can be created by using crossover data. The very first chromosome maps were published in 1913 using data from thousands of fruit fly crosses. Chromosome map percentages are not actual chromosome distances, but they represent relative positions of the genes. **Figure 10.15** shows the first chromosome map created using fruit fly data. Notice that the higher the crossover frequency, the farther apart the two genes are.

MiniLab 10.2

Map Chromosomes

Where are genes located on a chromosome? The distance between two genes on a chromosome is related to the crossover frequency between them. By comparing data for several gene pairs, a gene's relative location can be determined.

Procedure

1. Read and complete the lab safety form.
2. Obtain a table of the gene-pair crossover frequencies from your teacher.
3. Draw a line on a piece of paper and make marks every 1 cm. Each mark will represent a crossover frequency of 1 percent.
4. Label one mark near the middle of the line *A.* Find the crossover frequency between Genes A and B on the table, and use this data to label *B* the correct distance from A.
5. Use the crossover frequency between genes A and C and genes B and C to infer the position of gene C.
6. Repeat steps 4–5 for each gene, marking their positions on the line.

Analysis

1. **Evaluate** Is it possible to know the location of a gene on a chromosome if only one other gene is used?
2. **Consider** Why would using more crossover frequencies result in a more accurate chromosome map?

Strawberries (8*n*)

Coffee (4*n*)

■ **Figure 10.16** Various commercial plants, such as strawberries and coffee, are polyploids.

In a cross, the exchange of genes is directly related to the crossover frequency between them. This frequency correlates with the relative distance between the two genes. One map unit between two genes is equivalent to 1 percent of the crossing over occurring between them. Genes that are farther apart would have a greater frequency of crossing over.

Polyploidy

Most species have diploid cells, but some have polyploid cells. **Polyploidy** is the occurrence of one or more extra sets of all chromosomes in an organism. A triploid organism, for instance, would be designated 3*n*, which means that it has three complete sets of chromosomes. Polyploidy rarely occurs in animals but sometimes occurs in earthworms and goldfish. In humans, polyploidy is always lethal.

Roughly one in three species of known flowering plants are polyploid. Commercially grown bread wheat *(6n)*, oats *(6n)*, and sugar cane *(8n)* are polyploidy crop plants. Polyploid plants, such as the ones shown in **Figure 10.16,** often have increased vigor and size.

Section 10.3 Assessment

Section Summary

- Genetic recombination involves both crossing over and independent assortment.
- Early chromosome maps were created based on the linkage of genes on the chromosome.
- Polyploid plants are selected by plant growers for their desirable characteristics.

Understand Main Ideas

1. MAIN Idea **Analyze** how crossing over is related to variation.
2. **Draw** Suppose genes *C* and *D* are linked on one chromosome and genes *c* and *d* are linked on another chromosome. Assuming that crossing over does not take place, sketch the daughter cells resulting from meiosis, showing the chromosomes and position of the genes.
3. **Describe** how polyploidy is used in the field of agriculture.

Think Critically

4. **Construct** a chromosome map for genes *A, B, C* and *D* using the following crossing over data: *A* to *D*=25 percent; *A* to *B*=30 percent; *C* to *D*=15 percent; *B* to *D*=5 percent; *B* to *C*=20 percent.
5. **Evaluate** what advantage polyploidy would give to a plant breeder.

WRITING in Biology

6. Write a short story describing a society with no genetic variation in humans.

NATIONAL GEOGRAPHIC

In the Field

Career: Plant Geneticist

Is it better for plants to have more chromosomes?

Compare the two flowers in the photo. What differences do you notice? Both flowers were produced by a plant known as a daylily. The larger, more robust-looking flower on the left, however, is from a polyploid plant. What makes this plant so unusual? Its cells contain more than the usual two sets of chromosomes.

Plant geneticists have been fascinated by polyploids for decades. Having multiple sets of chromosomes can dramatically affect how a plant looks, performs, and appeals to consumers.

Putting Plant Genetics to Work

Plant geneticists apply scientific methods and the principles of genetics to improve the quality and production of plants. They develop species that are more resistant to diseases, pests, and drought. Some polyploid plants, such as seedless grapes, melons, and citrus fruits, are developed to meet consumer demand. Many plant geneticists also work to make crops more nutritious.

The development of new plant varieties, including polyploid species, benefits humans in many ways. In Thailand, for example, researchers have developed polyploid rice plants with a high tolerance for salt. These plants might thrive in areas where the soil is highly salty and useless agriculturally, providing income for farmers in previously economically depressed regions.

How Does Polyploidy Occur?

Plant geneticists produce polyploids by soaking the seeds or buds of certain plants in a chemical called colchicine. This chemical interferes with cell division, causing all of the chromosomes to remain in one cell as gametes are formed. During fertilization, the number of chromosomes is doubled, producing a polyploid plant. Polyploidy occurs naturally in many flowering plants. Scientists theorize that most natural polyploids resulted from mutations during cell division.

The Benefits of Being Polyploid

Having more than one set of chromosomes can provide evolutionary advantages for plants. Polyploids often are larger and stronger, have more developed root systems, and produce larger flowers and fruits. Plant geneticists seek to understand these characteristics based on heredity and variation and to utilize them to develop plants that can thrive in specific environmental conditions.

CAREERS in Biology

Imagine that a position for a plant geneticist has become available at an arboretum, and you are assigned to write the job description. Develop a list of skills and knowledge needed for this position. To explore more about plant geneticists, visit biologygmh.com.

BIOLAB

Design Your Own

HOW CAN THE PHENOTYPE OF OFFSPRING HELP DETERMINE PARENTAL GENOTYPE?

Background: The traits of most plants have dominant and recessive alleles. Analysis of plants grown from seeds can be a good indicator of the expected genotypes of offspring as well as phenotypes and genotypes of the parent plants.

Question: *Can the phenotypes and genotypes of parent organisms be determined from the phenotype of the offspring?*

Materials

Choose materials that would be appropriate for this lab.

two groups of plant seeds
potting soil
small flowerpots or other growing containers
watering can or bottle
small gardening trowel

Safety Precautions

Plan and Perform the Experiment

1. Read and complete the lab safety form.
2. Hypothesize whether the phenotype of offspring could be used to infer the genotypes of the parents.
3. Design an experiment to test your hypothesis.
4. Decide what data you need to collect.
5. Create a data table to record your observations.
6. Make certain your teacher has approved your experiment before you proceed.
7. Conduct your experiment.
8. **Cleanup and Disposal** Properly dispose of seeds or plants considered to be invasive species in your area. Never release invasive species into the environment.

Analyze and Conclude

1. **Collect and Organize Data** Count the number of seedlings of the different phenotypes in each group of plants. Prepare a graph of your data.
2. **Calculate** the ratio of different seedlings for each of your groups of seeds.
3. **Identify** two or more possible crosses that could have resulted in your observed ratio of seedlings.
4. **Analyze** Make a Punnett square for each cross you identified in question 3. Determine whether each possible cross could have resulted in the data you collected.
5. **Evaluate** how the combined data from the two seed groups affect the ratio of seedlings.
6. **Draw Conclusions** Based on the data from your two groups of seeds, list the genotype and phenotype of the parent plants.
7. **Error Analysis** Compare your calculated ratios to those of another student. Describe any differences. Combine your data with another group's data. Infer how increasing the number of seeds analyzed affects the outcome of the experiment.

COMMUNICATE

Poster Session Prepare a poster that describes the experiment you conducted and displays the data you collected. When posters are complete, have a poster session during which you examine each others' work and compare your results. To learn more about determining genotypes, visit BioLabs at biologygmh.com.

Study Guide

Download quizzes, key terms, and flash cards from biologygmh.com.

FOLDABLES **Conclude** On the back of your Foldable, conclude how meiosis and genetic recombination work together and result in genetic diversity.

Vocabulary	Key Concepts
Section 10.1 Meiosis	
• crossing over (p. 271) • diploid (p. 271) • fertilization (p. 271) • gamete (p. 271) • gene (p. 270) • haploid (p. 271) • homologous chromosome (p. 270) • meiosis (p. 271)	MAIN Idea Meiosis produces haploid gametes. • DNA replication takes place only once during meiosis, and it results in four haploid gametes. • Meiosis consists of two sets of divisions. • Meiosis produces genetic variation in gametes.
Section 10.2 Mendelian Genetics	
• allele (p. 278) • dominant (p. 278) • genetics (p. 277) • genotype (p. 279) • heterozygous (p. 279) • homozygous (p. 279) • hybrid (p. 279) • law of independent assortment (p. 280) • law of segregation (p. 279) • phenotype (p. 279) • recessive (p. 278)	MAIN Idea Mendel explained how a dominant allele can mask the presence of a recessive allele. • The study of genetics began with Gregor Mendel, whose experiments with garden pea plants gave insight into the inheritance of traits. • Mendel developed the law of segregation and the law of independent assortment. • Punnett squares help predict the offspring of a cross.
Section 10.3 Gene Linkage and Polyploidy	
• genetic recombination (p. 283) • polyploidy (p. 285)	MAIN Idea The crossing over of linked genes is a source of genetic variation. • Genetic recombination involves both crossing over and independent assortment. • Early chromosome maps were created based on the linkage of genes on the chromosome. • Polyploid plants are selected by plant growers for their desirable characteristics.

Biology Online **Vocabulary PuzzleMaker** biologygmh.com

CHAPTER 10 Assessment

Section 10.1

Vocabulary Review

Use what you know about the terms in the Study Guide to answer the following questions.

1. When two cells with *n* number of chromosomes fuse, what type of cell results?
2. During which process are gametes formed?
3. What process results in an exchange of genes between homologous chromosomes?

Understand Key Concepts

4. How many chromosomes would a cell have during metaphase I of meiosis if it has 12 chromosomes during interphase?
   A. 6 **C.** 24
   B. 12 **D.** 36

Use the diagram below to answer questions 5 and 6.

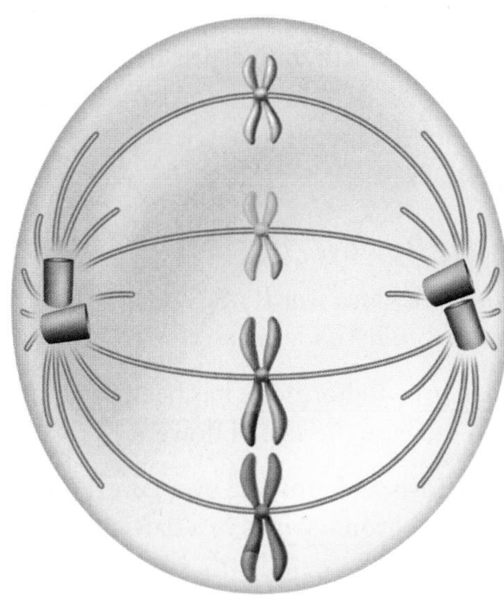

5. Which stage of meiosis is illustrated above?
   A. prophase I **C.** metaphase I
   B. prophase II **D.** metaphase II
6. What is the next step for the chromosomes illustrated above?
   A. They will experience replication.
   B. They will experience fertilization.
   C. Their number per cell will be halved.
   D. They will divide into sister chromatids.
7. Which is not a characteristic of homologous chromosomes?
   A. Homologous chromosomes have the same length.
   B. Homologous chromosomes have the same centromere position.
   C. Homologous chromosomes have the exact same type of allele at the same location.
   D. Homologous chromosomes pair up during meiosis I.

Constructed Response

8. **Short Answer** Relate the terms meiosis, gametes, and fertilization in one or two sentences.
9. **Open Ended** Plant cells do not have centrioles. Hypothesize why plant cells might not need centrioles for mitosis or meiosis.

Think Critically

10. **Analyze** A horse has 64 chromosomes and a donkey has 62. Using your knowledge of meiosis, evaluate why a cross between a horse and a donkey produces a mule, which usually is sterile.
11. **Hypothesize** In bees, the female queen bee is diploid but male bees are haploid. The fertilized eggs develop into female bees and the unfertilized eggs develop into males. How might gamete production in male bees differ from normal meiosis?

Section 10.2

Vocabulary Review

Explain the differences between the vocabulary terms in the following sets.

12. dominant, recessive
13. genotype, phenotype

Understand Key Concepts

14. If a black guinea pig (*Bb*) were crossed with a white guinea pig (*bb*) what would be the resulting phenotypic ratio?
    A. 0:1 black to white **C.** 1:1 black to white
    B. 1:0 black to white **D.** 3:1 black to white

15. In garden peas, purple flowers (*P*) are dominant to white (*p*) flowers, and tall plants (*T*) are dominant to short plants (*t*). If a purple tall plant (*PpTt*) is crossed with a white short plant (*pptt*), what is the resulting phenotypic ratio?
    A. 1:1:1:1 purple tall to purple short to white tall to white short
    B. 3:2 purple tall to purple short
    C. 9:3:3:1 purple tall to purple short to white tall to white short
    D. all purple tall

Use the figure below to answer questions 16 and 17.

16. The unusual cat shown was crossed with a cat with noncurled ears. All the kittens born from that cross had noncurled ears. Later, when these offspring were crossed with each other, the phenotypic ratio was 3:1 noncurled to curled ears. What conclusions can be made about the inheritance of curled ears?
    A. Curled ears are a result of crossing over.
    B. It is a dominant trait.
    C. It is a recessive trait.
    D. More crosses need to be done to determine how the trait is inherited.

Constructed Response

17. **Short Answer** What might occur in the F_3 generation of the curly-eared cat shown above if the F_2 generation all reproduce with cats that have non-curly ears?

18. **Short Answer** If there are five boys and no girls born into a family, does that increase the likelihood that the sixth offspring will be a girl? Explain.

Think Critically

Use the figure below to answer question 19.

19. **Predict** There are two types of American rat terrier dogs—those without hair and those with hair, as shown in the figure. The presence of hair is a genetically determined trait. Some female rat terriers with hair produce only puppies with hair, whereas other females produce rat terrier puppies without hair. Explain how this can occur.

20. **MATH in Biology** What is the probability of a couple giving birth to five girls in a row?

Section 10.3

Vocabulary Review

Replace the underlined words with the correct vocabulary term from the Study Guide page.

21. Human growth hormone has been used in agriculture to increase the size of flowers.

22. Both meiosis and crossing over contribute to the amount of chromosomes in a particular species.

Understand Key Concepts

23. Which does not contribute to genetic variation?
    A. chromosome number
    B. crossing over
    C. meiosis
    D. random mating

24. Which concept is considered an exception to Mendel's law of independent assortment?
    A. crossing over
    B. gene linkage
    C. polyploidy
    D. law of segregation

Biology Online Chapter Test biologygmh.com

Use the figure below to answer questions 25 and 26.

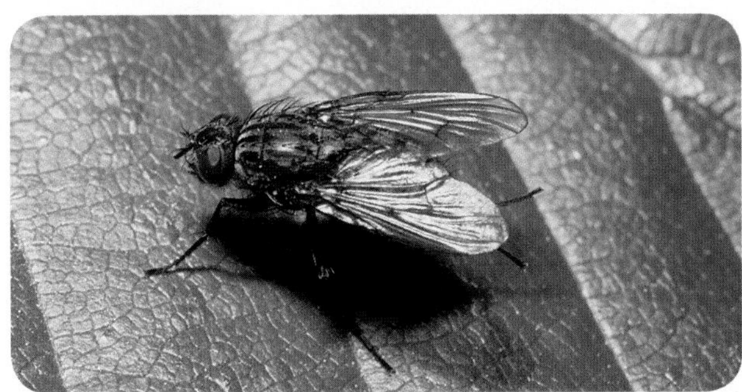

25. A housefly, shown in the photo above, has six pairs of chromosomes. If two houseflies are crossed, how many possible types of fertilized eggs could result from the random lining up of the pairs?

A. 256
B. 1024
C. 4096
D. 16,384

26. For the housefly with its six pairs of chromosomes, how many possible combinations of gametes can be produced by the random lining up of pairs in meiosis?

A. 32
B. 48
C. 64
D. 120

Constructed Response

27. Short Answer What three processes increase genetic variation?

28. Open Ended Hypothesize how a plant breeder might create a polyploid plant.

29. Short Answer How is chromosome gene linkage an exception to the law of independent assortment?

Think Critically

30. CAREERS IN BIOLOGY Horticulturists grow thousands of genetically identical plants by using cuttings. Cuttings do not involve sexual reproduction. Discuss the benefits and drawbacks of using cuttings to reproduce a certain type of plant.

31. Hypothesize Crossing over provides genetic variation, eventually changing the gene pool in a population. Yet some sexually reproducing organisms do not seem to display recombination mechanisms. Why might it be advantageous for these organisms to reduce genetic recombination?

Additional Assessment

32. WRITING in Biology In sheep, white wool is dominant and black wool is recessive. Suppose some sheep belonging to a certain flock are heterozygous for wool color. Write a plan indicating how a flock of pure-breeding white sheep could be developed.

DBQ Document-Based Questions

The paragraphs below were obtained from Mendel's publication.

Data obtained from: Mendel, Gregor. 1866. *Experiments in Plant Hybridization.* Originally translated by Bateson, William, 1901: 2.

"The hybrids of such plants must, during the flowering period, be protected from the influence of all foreign pollen, or be easily capable of such protection."

33. Mendel made the above rule for his experimental plants. Summarize why this rule was important for the success of his experiments.

Ibid: 4

"The object of the experiment was to observe these variations in the case of each pair of differentiating characters, and to deduce the law according to which they appear in successive generations. The experiment resolves itself therefore into just as many separate experiments. There are constantly differentiating characters presented in the experimental plants."

34. Describe Mendel's purpose for conducting plant breeding experiments.

Cumulative Review

35. Suggest what the consequences of biodiversity will be if globalization of species continues at its present pace. **(Chapter 5)**

36. How do prokaryotic cells differ from eukaryotic cells? **(Chapter 7)**

37. Compare and contrast the way in which plants and animals obtain energy. **(Chapter 8)**

Standardized Test Practice

Cumulative

Multiple Choice

1. A population will likely enter a long-term high growth rate when many individuals are which?
   A. below the main reproductive age
   B. just above the main reproductive age
   C. at the middle of the main reproductive age
   D. at the upper end of the main reproductive age

Use the illustration below to answer question 2.

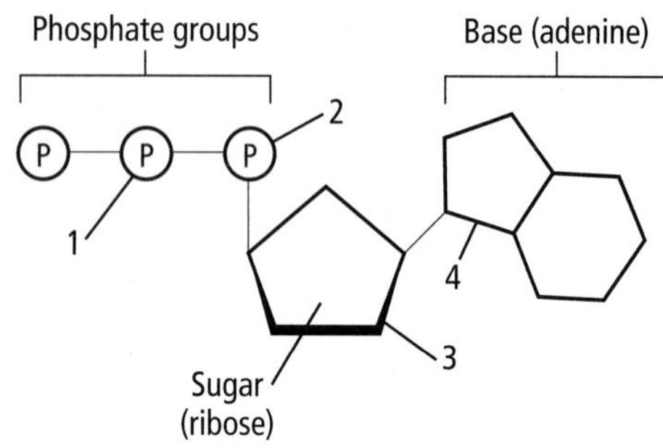

2. To release energy for use in the organism, the bond between which two groups in the ATP molecule must be broken?
   A. 1 and 2
   B. 2 and 3
   C. 2 and 4
   D. 3 and 4

3. Which process divides a cell's nucleus and nuclear material?
   A. cell cycle
   B. cytokinesis
   C. interphase
   D. mitosis

4. Which is the source of electrons in the electron transport chain stage of respiration?
   A. formation of acetyl CoA during the Krebs cycle
   B. creation of NADH and $FADH_2$ during the Krebs cycle
   C. fermentation of lactic acid
   D. breaking of bonds in glycolysis

5. Which would most likely cause lung cancer?
   A. exposure to asbestos particles
   B. exposure to fungus spores
   C. exposure to infrared radiation
   D. exposure to ultraviolet radiation

Use the illustration below to answer question 6.

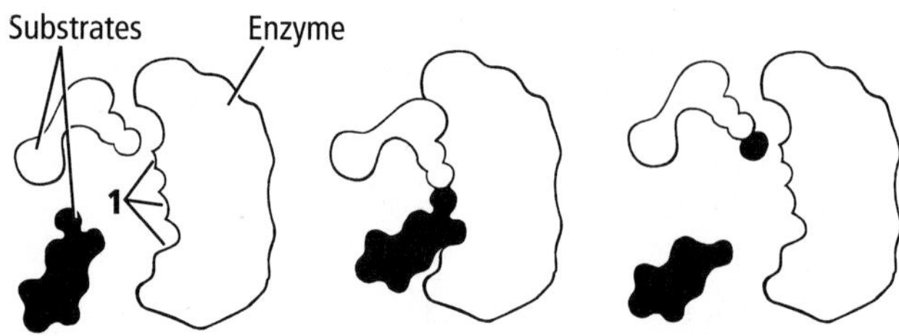

6. Which is the role of "1" in the activity of the enzyme?
   A. to make a reaction happen more slowly
   B. to make more reactants available to the substrate
   C. to provide a unique spot for substrate binding
   D. to raise the activation energy for the reaction

7. What causes the movement of calcium and sodium ions in and out of cardiac cells?
   A. charged particles in the phospholipid bilayer
   B. cholesterol molecules in the phospholipid bilayer
   C. diffusion channels in the cell membrane
   D. transport proteins in the cell membrane

8. In a cell undergoing meiosis, during which stage do the sister chromatids separate from each other?
   A. anaphase I
   B. anaphase II
   C. telophase I
   D. telophase II

9. Which is the standard SI unit for mass?
   A. candela
   B. kelvin
   C. kilogram
   D. meter

Short Answer

Use the diagram below to answer questions 10 and 11.

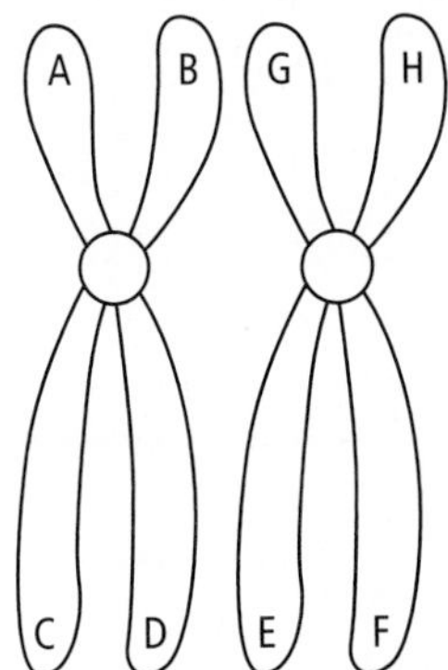

10. The diagram above shows a pair of chromosomes with different regions on the chromosomes labeled. Explain where crossing over could occur on this pair of chromosomes.

11. When is crossing over most likely to occur?

12. Suppose the concentration of CO_2 in a greenhouse decreases. Explain how the photosynthesis process could be affected by that change. Predict the overall effect on plants.

13. How does the process of meiosis promote genetic variation in a species?

14. Describe how the chromosomes change during the S phase.

15. Hypothesize why meiosis occurs in two stages—meiosis I and meiosis II.

16. Explain how factors in the environment can cause cancer to develop.

Extended Response

Use the diagram below to answer question 17.

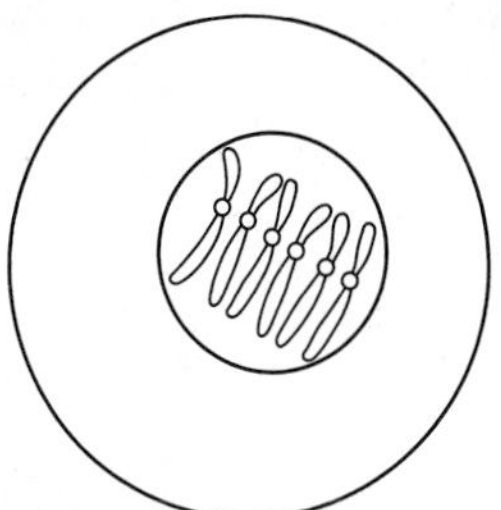

17. The diagram above shows the chromosomes found in the sex cells of a particular animal. Based on this diagram, describe what happens during fertilization in this species.

18. Assess what might happen if mitosis were NOT an extremely precise process.

Essay Question

Stem cells are cells that are not specialized for a particular function. Like other cells, stem cells contain all of the genetic material found in the organism. Stem cells, if given the correct signal, can become any type of specialized cell. There are two different types of stem cells. Embryonic stem cells are found in embryos, while adult stem cells are found in small quantities in mature tissues. The process of conducting research using stem cells, especially using embryonic stem cells, is controversial because of ethical concerns.

Using the information in the paragraph above, answer the following question in essay format.

19. Do you think medical researchers should be allowed to use stem cells as research material? Judge what you think are the benefits or risks of stem cell research.

NEED EXTRA HELP?

If You Missed Question . . .	1	2	3	4	5	6	7	8	9	10	11	12	13	14	15	16	17	18	19
Review Section . . .	4.2	8.2	9.1	8.3	9.3	6.2	7.2, 7.4	10.1	1.2	10.1	10.1	8.2	10.1, 10.3	9.2	10.2	9.3	10.1	9.2	9.2

CHAPTER 11 Complex Inheritance and Human Heredity

BIG Idea Human inheritance does not always follow Mendel's laws.

Section 1
Basic Patterns of Human Inheritance
MAIN Idea The inheritance of a trait over several generations can be shown in a pedigree.

Section 2
Complex Patterns of Inheritance
MAIN Idea Complex inheritance of traits does not follow inheritance patterns described by Mendel.

Section 3
Chromosomes and Human Heredity
MAIN Idea Chromosomes can be studied using karyotypes.

BioFacts

- Sometimes different ethnic groups can be distinguished by phenotypic traits such as skin color, hair color, and skin folds at the corner of the eyes.
- The individual genetic differences within an ethnic group can be greater than the genetic differences between individuals of two different ethnic groups.

Two X chromosomes of a human female
Colored LM Magnification: 9500×

X and Y chromosomes of a human male
Colored LM Magnification: 9500×

Start-Up Activities

LAUNCH Lab

What do you know about human inheritance?

As knowledge and understanding of human inheritance increases, long-standing ideas regarding the facts of human heredity must be reexamined. Any ideas disproven by new discoveries must be rejected.

Procedure

1. Read the statements below carefully and determine whether they are true or false.

   Statements:

   A. The father determines the gender of the child.

   B. Individuals may transmit characteristics to their offspring which they themselves do not show.

   C. Identical twins always are of the same gender.

2. Discuss your answers with your classmates and teacher.

Analysis

1. **Assess** What question was missed most often by the entire class? Discuss reasons why.
2. **Analyze** Why is it helpful to understand human heredity?

Visit biologygmh.com to:

- study the entire chapter online
- explore the Concepts in Motion, the Interactive Tables, Microscopy Links, Virtual Labs, and links to virtual dissections
- access Web links for more information, projects, and activities
- review content online with the Interactive Tutor and take Self-Check Quizzes

Genetic Disorders Make this Foldable to help you understand how variations in nucleotide base sequences are linked to genetic disorders.

STEP 1 Fold a sheet of notebook paper lengthwise, leaving a half inch between the folds as shown.

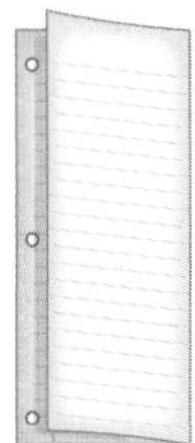

STEP 2 Rotate the paper and cut the top layer to form eight tabs of equal size, as shown.

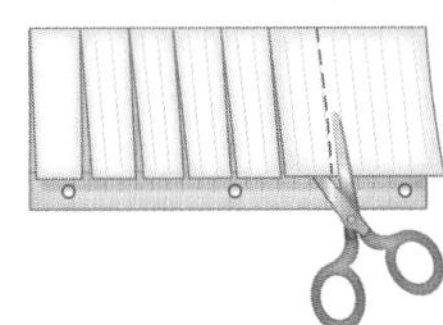

STEP 3 Label each tab with the name of a different genetic disorder. Under the tabs, write about each disorder.

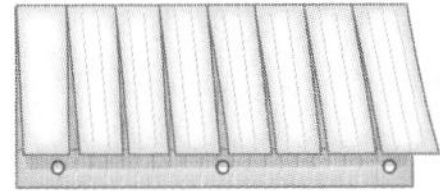

FOLDABLES **Use this Foldable with Section 11.1.** As you study the section, note how to trace genetic disorders using pedigrees.

Reading Preview

Objectives

- **Analyze** genetic patterns to determine dominant or recessive inheritance patterns.
- **Summarize** examples of dominant and recessive disorders.
- **Construct** human pedigrees from genetic information.

Review Vocabulary

genes: segments of DNA that control the production of proteins

New Vocabulary

carrier
pedigree

Basic Patterns of Human Inheritance

MAIN Idea The inheritance of a trait over several generations can be shown in a pedigree.

Real-World Reading Link Knowing a purebred dog's ancestry can help the owner know health problems that are common to that dog. Similarly, tracing human inheritance can show how a trait was passed down from one generation to the next.

Recessive Genetic Disorders

Connection to History Mendel's work was ignored for more than 30 years. During the early 1900s, scientists began to take an interest in heredity, and Mendel's work was rediscovered. About this time, Dr. Archibald Garrod, an English physician, became interested in a disorder linked to an enzyme deficiency called alkaptonuria (al kap tuh NYUR ee uh), which results in black urine. It is caused by acid excretion into the urine. Dr. Garrod observed that the condition appeared at birth and continued throughout the patient's life, ultimately affecting bones and joints. He also noted that alkaptonuria ran in families. With the help of another scientist, he determined that alkaptonuria was a recessive genetic disorder.

Today, progress continues to help us understand genetic disorders. Review **Table 11.1,** and recall that a recessive trait is expressed when the individual is homozygous recessive for that trait. Therefore, those with at least one dominant allele will not express the recessive trait. An individual who is heterozygous for a recessive disorder is called a **carrier.** Review **Table 11.2** as you read about several recessive genetic disorders.

Concepts In Motion Interactive Table To explore more about human inheritance, visit biologygmh.com.

Table 11.1 Review of Terms

Term	Example	Definition
Homozygous	True-breeding yellow-seed pea plants would be *YY,* and green-seed pea plants would be *yy.*	An organism with two of the same alleles for a particular trait is said to be homozygous for that trait.
Heterozygous	A plant that is *Yy* would be a yellow-seed pea.	An organism with two different alleles for a particular trait is said to be heterozygous for that trait. When alleles are present in the heterozygous state, the dominant trait will be observed.

Concepts In Motion
Interactive Table To explore more about recessive genetic disorders, visit biologygmh.com.

Table 11.2 Recessive Genetic Disorders in Humans

Disorder	Occurrence in the U.S.	Cause	Effect	Cure/Treatment
Cystic fibrosis	1 in 3500	The gene that codes for a membrane protein is defective.	• Excessive mucus production • Digestive and respiratory failure	• No cure • Daily cleaning of mucus from the lungs • Mucus-thinning drugs • Pancreatic enzyme supplements
Albinism	1 in 17,000	Genes do not produce normal amounts of the pigment melanin.	• No color in the skin, eyes and hair • Skin susceptible to UV damage • Vision problems	• No cure • Protect skin from the Sun and other environmental factors • Visual rehabilitation
Galactosemia	1 in 50,000 to 70,000	Absence of the gene that codes for the enzyme that breaks down galactose	• Mental disabilities • Enlarged liver • Kidney failure	• No cure • Restriction of lactose/galactose in the diet
Tay-Sachs disease	1 in 2500 (affects people of Jewish descent)	Absence of a necessary enzyme that breaks down fatty substances	• Buildup of fatty deposits in the brain • Mental disabilities	• No cure or treatment • Death by age 5

Cystic fibrosis One of the most common recessive genetic disorders among Caucasians is cystic fibrosis, which affects the mucus-producing glands, digestive enzymes, and sweat glands. Chloride ions are not absorbed into the cells of a person with cystic fibrosis but are excreted in the sweat. Without sufficient chloride ions in cells, water does not diffuse from cells. This causes a secretion of thick mucus that affects many areas of the body. The thick mucus clogs the ducts in the pancreas, interrupts digestion, and blocks the tiny respiratory pathways in the lungs. Patients with cystic fibrosis are at a higher risk of infection because of excess mucus in their lungs.

Treatment for cystic fibrosis currently includes physical therapy, medication, special diets, and the use of replacement digestive enzymes. Genetic tests are available to determine whether a person is a carrier, indicating they are carrying the recessive gene.

FOLDABLES Incorporate information from this section into your Foldable.

Albinism In humans, albinism is caused by altered genes, resulting in the absence of the skin pigment melanin in hair and eyes. You will learn more about melanin in Chapter 32. Albinism is found in other animals as well. A person with albinism has white hair, very pale skin, and pink pupils. The absence of pigment in eyes can cause problems with vision. Although we all must protect our skin from the Sun's ultraviolet radiation, those with albinism need to be especially careful.

Tay-Sachs disease Tay-Sachs (TAY saks) disease is a recessive genetic disorder. Its gene is found on chromosome 15. Often identified by a cherry-red spot on the back of the eye, Tay-Sachs disease (TSD) seems to be predominant among Jews of eastern European descent.

VOCABULARY

ACADEMIC VOCABULARY

Decline:
to gradually waste away; or a downward slope
His health declined because of the disease.

TSD is caused by the absence of the enzymes responsible for breaking down fatty acids called gangliosides. Normally, gangliosides are made and then dissolved as the brain develops. However, in a person affected by Tay-Sachs disease, the gangliosides accumulate in the brain, inflating brain nerve cells and causing mental deterioration.

Galactosemia Galactosemia (guh lak tuh SEE mee uh) is characterized by the inability of the body to digest galactose. During digestion, lactose from milk breaks down into galactose and glucose. Glucose is the sugar used by the body for energy and circulates in the blood. Galactose must be broken down into glucose by an enzyme named GALT. Persons who lack or have defective GALT cannot digest galactose. Persons with galactosemia should avoid milk products.

Dominant Genetic Disorders

Not all genetic disorders are caused by recessive inheritance. As described in **Table 11.3,** some disorders, such as the rare disorder Huntington's disease, are caused by dominant alleles. That means those who do not have the disorder are homozygous recessive for the trait.

Huntington's disease The dominant genetic disorder Huntington's disease affects the nervous system and occurs in one out of 10,000 people in the U.S. The symptoms of this disorder first appear in affected individuals between the ages of 30 and 50 years old. The symptoms include a gradual loss of brain function, uncontrollable movements, and emotional disturbances. Genetic tests are available to detect this dominant allele. However, no preventive treatment or cure for this disease exists.

Achondroplasia An individual with achondroplasia (a kahn droh PLAY zhee uh) has a small body size and limbs that are comparatively short. Achondroplasia is the most common form of dwarfism. A person with achondroplasia will have an adult height of about four feet and will have a normal life expectancy.

Interestingly, 75 percent of individuals with achondroplasia are born to parents of average size. When children with achondroplasia are born to parents of average size, the conclusion is that the condition occurred because of a new mutation or a genetic change.

Reading Check **Identify** the chances of inheriting a dominant disorder and a recessive disorder if you have one parent with the disease.

Table 11.3 **Dominant Genetic Disorders in Humans**

Concepts In Motion Interactive Table To explore more about dominant genetic disorders, visit biologygmh.com.

Disorder	Occurrence in the U.S.	Cause	Effect	Cure/Treatment
Huntington's disease	1 in 10,000	A gene affecting neurological function is defective.	• Decline of mental and neurological functions • Ability to move deteriorates	• No cure or treatment
Achondroplasia	1 in 25,000	A gene that affects bone growth is abnormal.	• Short arms and legs • Large head	• No cure or treatment

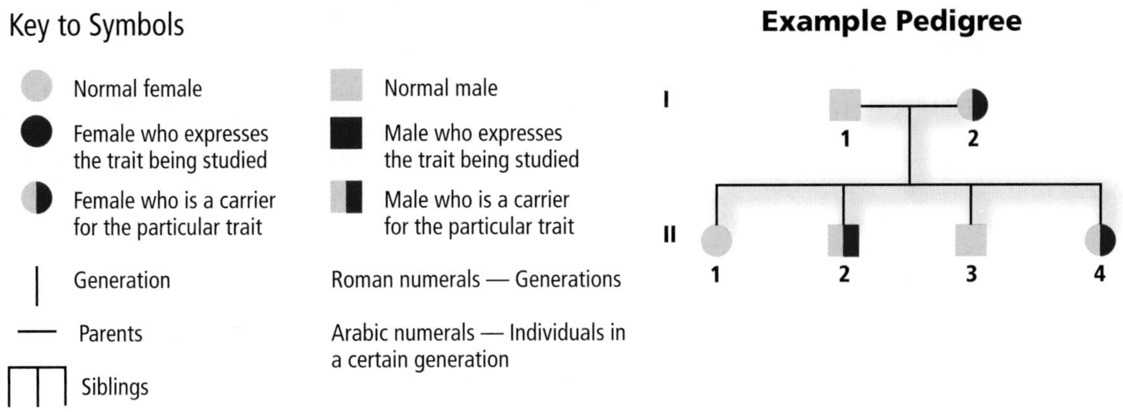

Key to Symbols

Normal female

Female who expresses the trait being studied

Female who is a carrier for the particular trait

Generation

Parents

Siblings

Normal male

Male who expresses the trait being studied

Male who is a carrier for the particular trait

Roman numerals — Generations

Arabic numerals — Individuals in a certain generation

Example Pedigree

I

1 2

II

1 2 3 4

■ **Figure 11.1** A pedigree uses standard symbols to indicate what is known about the trait being studied.

Personal Tutor

To learn about pedigrees, visit biologygmh.com.

Pedigrees

In organisms such as peas and fruit flies, scientists can perform crosses to study genetic relationships. In the case of humans, a scientist studies a family history using a **pedigree,** a diagram that traces the inheritance of a particular trait through several generations. A pedigree uses symbols to illustrate inheritance of the trait. Males are represented by squares, and females are represented by circles, as shown in **Figure 11.1.** One who expresses the trait being studied is represented by a dark, or filled, square or circle, depending on their gender. One who does not express the trait is represented by an unfilled square or circle.

A horizontal line between two symbols shows that these individuals are the parents of the offspring listed below them. Offspring are listed in descending birth order from left to right and are connected to each other and their parents.

A pedigree uses a numbering system in which Roman numerals represent generations, and individuals are numbered by birth order using Arabic numbers. For example, in **Figure 11.1,** individual II1 is a female who is the firstborn in generation II.

Analyzing Pedigrees

A pedigree illustrating Tay-Sachs disease is shown in **Figure 11.2.** Recall from **Table 11.2** that Tay-Sachs disease is a recessive genetic disorder caused by the lack of an enzyme involved in lipid metabolism. The missing enzyme causes lipids to build up in the central nervous system, which can lead to death.

Examine the pedigree in **Figure 11.2.** Note that two unaffected parents, I1 and I2, have an affected child—II3, indicating that each parent has one recessive allele—they both are heterozygous and carriers for the trait. The half-filled square and circle show that both parents are carriers.

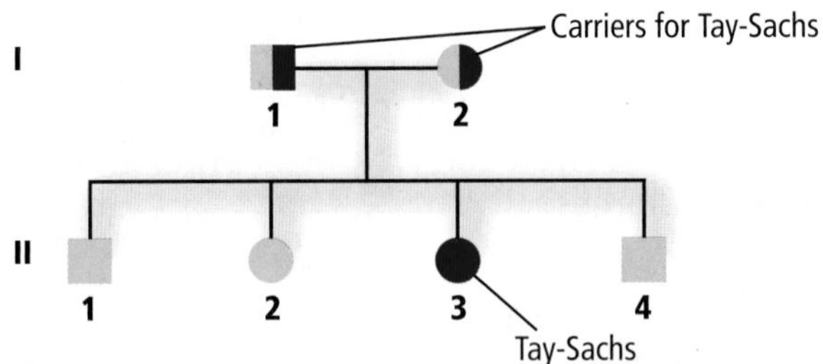

■ **Figure 11.2** This pedigree illustrates the inheritance of the recessive disorder Tay-Sachs disease. Note that two unaffected parents (I1 and I2) can have an affected child (II3).

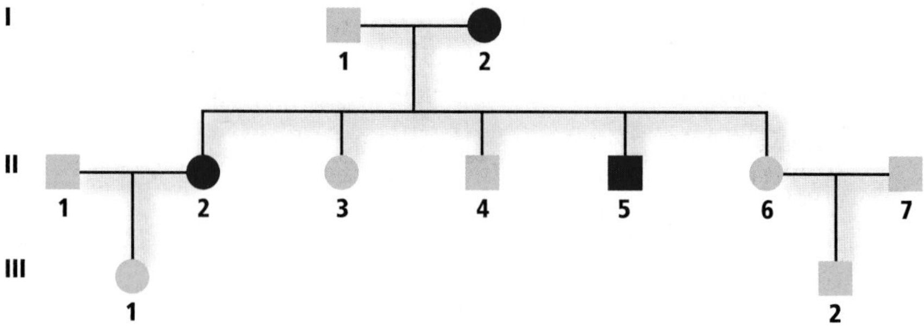

■ **Figure 11.3** This pedigree illustrates the inheritance of a dominant disorder. Note that affected parents can pass on their genes (II2, II5), but unaffected parents cannot have an affected child (III2).

The pedigree in **Figure 11.3** shows the inheritance of the dominant genetic disorder polydactyly (pah lee DAK tuh lee). People with this disorder have extra fingers and toes. Recall that with dominant inheritance the trait is expressed when at least one dominant allele is present. An individual with an unaffected parent and a parent with polydactyly could be either heterozygous or homozygous recessive for the trait. Each unaffected person would be homozygous recessive for the trait.

For example, in **Figure 11.3,** individual I2 has polydactyly, indicated by the dark circle. Because she shows the trait, she is either homozygous dominant or heterozygous. It can be inferred that she is heterozygous—having one dominant gene and one recessive gene—because offspring II3 and II4 do not have the disorder. Notice that II6 and II7, two unaffected parents, have an unaffected offspring—III2. What can be inferred about II2, based on the phenotype of her parents and her offspring?

MiniLab 11.1

Investigate Human Pedigrees

Where are the branches on the family tree? Unlike some organisms, humans reproduce slowly and produce few offspring at one time. One method used to study human traits is pedigree analysis.

Procedure

1. Read and complete the lab safety form.
2. Imagine that you are a geneticist interviewing a person about his or her family concerning the hypothetical trait of hairy earlobes.
3. From the transcript below, construct a pedigree. Use appropriate symbols and format.

   "My name is Scott. My great grandfather Walter had hairy earlobes (HEs), but great grandma Elsie did not. Walter and Elsie had three children: Lola, Leo, and Duane. Leo, the oldest, has HEs, as does the middle child, Lola; but the youngest child, Duane, does not. Duane never married and has no children. Leo married Bertie, and they have one daughter, Patty. In Leo's family, he is the only one with HEs. Lola married John, and they have two children: Carolina and Luetta. John does not have HEs, but both of his daughters do."

Analysis

1. **Assess** In what ways do pedigrees simplify the analysis of inheritance?
2. **Think Critically** Using this lab as a frame of reference, how can we put to practical use our understanding of constructing and analyzing human pedigrees?

Inferring genotypes Pedigrees are used to infer genotypes from the observation of phenotypes. By knowing physical traits, genealogists can determine what genes an individual is most likely to have. Phenotypes of entire families are analyzed in order to determine family genotypes, as symbolized in **Figure 11.3.**

Pedigrees help genetic counselors determine whether inheritance patterns are dominant or recessive. Once the inheritance pattern is determined, the genotypes of the individuals can largely be resolved through pedigree analysis. To analyze pedigrees, one particular trait is studied, and a determination is made as to whether that trait is dominant or recessive. Dominant traits are easier to recognize than recessive traits because dominant traits are exhibited in the phenotype.

A recessive trait will not be expressed unless the person is homozygous recessive for the trait. That means that a recessive allele is passed on by each parent. When recessive traits are expressed, the ancestry of the person expressing the trait is followed for several generations to determine which parents and grandparents were carriers of the recessive allele.

Predicting disorders If good records have been kept within families, disorders in future offspring can be predicted. However, more accuracy can be expected if several individuals within the family can be evaluated. The study of human genetics is difficult, because scientists are limited by time, ethics, and circumstances. For example, it takes decades for each generation to mature and then to have offspring when the study involves humans. Therefore, good record keeping, where it exists, helps scientists use pedigree analysis to study inheritance patterns, to determine phenotypes, and to ascertain genotypes within a family.

CAREERS IN BIOLOGY

Genealogist A genealogist studies or traces the descent of individuals or families. For more information on biology careers, visit biologygmh.com.

Section 11.1 Assessment

Section Summary

- Genetic disorders can be caused by dominant or recessive alleles.
- Cystic fibrosis is a genetic disorder that affects mucus and sweat secretions.
- Individuals with albinism do not have melanin in their skin, hair, and eyes.
- Huntington's disease affects the nervous system.
- Achondroplasia sometimes is called dwarfism.
- Pedigrees are used to study human inheritance patterns.

Understand Main Ideas

1. MAIN Idea **Construct** a family pedigree of two unaffected parents with a child who suffers from cystic fibrosis.
2. **Explain** the type of inheritance associated with Huntington's disease and achondroplasia.
3. **Interpret** Can two parents with albinism have an unaffected child? Explain.
4. **Diagram** Suppose both parents can roll their tongues but their son cannot. Draw a pedigree showing this trait, and label each symbol with the appropriate genotype.

Think Critically

MATH in Biology

5. Phenylketonuria (PKU) is a recessive genetic disorder. If both parents are carriers, what is the probability of this couple having a child with PKU? What is the chance of this couple having two children with PKU?
6. **Determine** When a couple requests a test for cystic fibrosis, what types of questions might the physician ask before ordering the tests?

Section 11.2

Reading Preview

Objectives

- **Distinguish** between various complex inheritance patterns.
- **Analyze** sex-linked inheritance patterns.
- **Explain** how the environment can influence the phenotype of an organism.

Review Vocabulary

gamete: a mature sex cell (sperm or egg) with a haploid number of chromosomes

New Vocabulary

incomplete dominance
codominance
multiple alleles
epistasis
sex chromosome
autosome
sex-linked trait
polygenic trait

Complex Patterns of Inheritance

MAIN Idea **Complex inheritance of traits does not follow inheritance patterns described by Mendel.**

Real-World Reading Link Imagine that you have red-green color blindness. In bright light, red lights do not stand out against surroundings. At night, green lights look like white streetlights. To help those with red-green color blindness, traffic lights always follow the same pattern. Red-green color blindness, however, does not follow the same pattern of inheritance described by Mendel.

Incomplete Dominance

Recall that when an organism is heterozygous for a trait, its phenotype will be that of the dominant trait. For example, if the genotype of a pea plant is *Tt* and *T* is the genotype for the dominant trait *tall,* then its phenotype will be tall. When red-flowered snapdragons *(RR)* are crossed with white-flowered snapdragons *(rr),* the heterozygous offspring have pink flowers *(Rr),* as shown in **Figure 11.4.** This is an example of **incomplete dominance,** in which the heterozygous phenotype is an intermediate phenotype between the two homozygous phenotypes. When the heterozygous F_1 generation snapdragon plants are allowed to self-fertilize, as in **Figure 11.4,** the flowers are red, pink, and white in a 1:2:1 ratio, respectively.

Codominance

Recall that when an organism is heterozygous for a particular trait the dominant phenotype is expressed. In a complex inheritance pattern called **codominance,** both alleles are expressed in the heterozygous condition. Sickle-cell disease provides a case study of codominant inheritance.

Personal Tutor To learn about inheritance, visit biologygmh.com.

■ **Figure 11.4** The color of snapdragon flowers is a result of incomplete dominance. When a plant with white flowers is crossed with a plant with red flowers, the offspring have pink flowers. Red, pink, and white offspring will result from self fertilization of a plant with pink flowers.
Predict *What would happen if you crossed a pink flower with a white flower?*

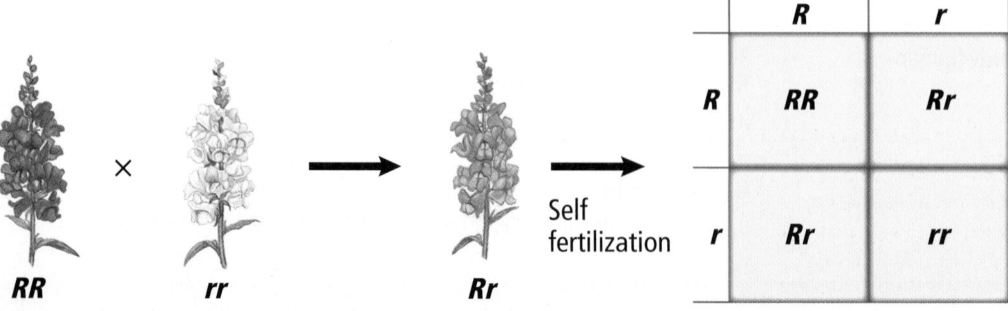

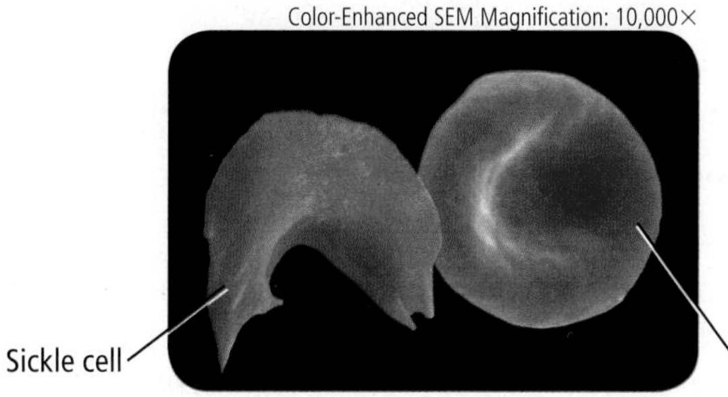

Sickle-cell disease The allele responsible for sickle-cell disease is particularly common in people of African descent, with about nine percent of African Americans having one form of the trait. Sickle-cell disease affects red blood cells and their ability to transport oxygen. The photograph in **Figure 11.5** shows the blood cells of an individual who is heterozygous for the sickle-cell trait. Changes in hemoglobin—the protein in red blood cells—cause those blood cells to change to a sickle, or "C", shape. Sickle-shaped cells do not effectively transport oxygen because they block circulation in small blood vessels. Those who are heterozygous for the trait have both normal and sickle-shaped cells. These individuals can lead relatively normal lives, as the normal blood cells compensate for the sickle-shaped cells.

Sickle-cell disease and malaria Note in **Figure 11.5** the distribution of both sickle-cell disease and malaria in Africa. Some areas with sickle-cell disease overlap areas of widespread malaria. Why might such high levels of the sickle-cell allele exist in central Africa? Scientists have discovered that those who are heterozygous for the sickle-cell trait also have a higher resistance to malaria. The death rate due to malaria is lower where the sickle-cell trait is higher. Because less malaria exists in those areas, more people live to pass on the sickle-cell trait to offspring. Consequently, sickle-cell disease continues to increase in Africa.

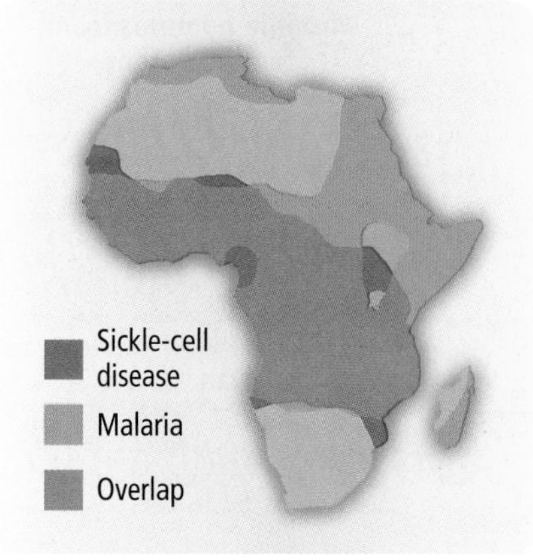

Figure 11.5
Left: Normal red blood cells are flat and disk-shaped. Sickle-shaped cells are elongated and "C" shaped. They can clump, blocking circulation in small vessels.
Right: The sickle-cell allele increases resistance to malaria.

DATA ANALYSIS LAB 11.1

Based On Real Data*

Interpret the Graph

What is the relationship between sickle-cell disease and other complications? Patients who have been diagnosed with sickle-cell disease face many symptoms, including respiratory failure and neurological problems. The graph shows the relationship between age and two different symptoms—pain and fever—during the two weeks preceding an episode of acute chest syndrome and hospitalization.

Think Critically

1. **State** which age group has the highest level of pain before being hospitalized.
2. **Describe** the relationship between age and fever before hospitalization.

Data and Observations

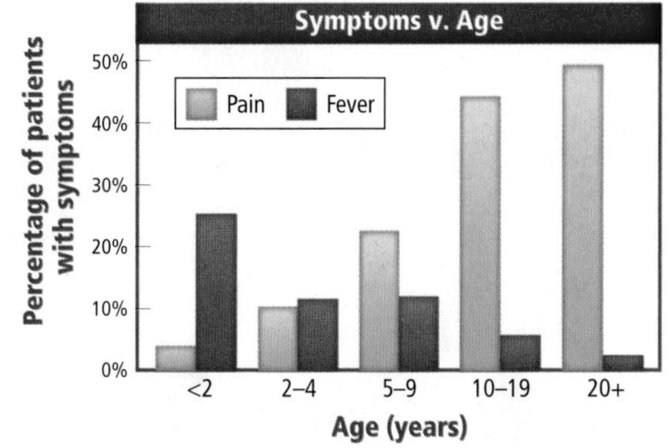

*Data obtained from: Walters, et al. 2002. Novel therapeutic approaches in sickle cell disease. *Hemotology* 17: 10-34.

Possible gametes from female parent

Possible gametes from male parent	I^A	I^B	i
I^A	I^AI^A	I^AI^B	I^Ai
I^B	I^AI^B	I^BI^B	I^Bi
i	I^Ai	I^Bi	ii

Blood types: A, AB, B, O

■ **Figure 11.6** There are three forms of alleles in the ABO blood groups—I^A, I^B, and i.

Multiple Alleles

Not all traits are determined by two alleles. Some forms of inheritance are determined by more than two alleles referred to as **multiple alleles.** An example of such a trait is human blood group.

Blood groups in humans ABO blood groups, shown in **Figure 11.6,** have three forms of alleles, sometimes called AB markers: I^A is blood type A; I^B is blood type B; and i is blood type O. Type O is the absence of AB markers. Note that allele i is recessive to I^A and I^B. However, I^A and I^B are codominant; blood type AB results from both I^A and I^B alleles. Therefore, ABO blood groups are examples of both multiple alleles and codominance.

Blood also has Rh factors, inherited from each parent. Rh factors are either positive or negative (Rh+ or Rh–); Rh+ is dominant. The Rh factor is a blood protein named after the rhesus monkey because studies of the rhesus monkey led to discovery of that blood protein.

Coat color of rabbits Multiple alleles can demonstrate a hierarchy of dominance. In rabbits, four alleles code for coat color: C, c^{ch}, c^h, and c. Allele C is dominant to the other alleles and results in a full color coat. Allele c is recessive and results in an albino phenotype when the genotype is homozygous recessive. Allele c^{ch} is dominant to c^h, and allele c^h is dominant to c and the hierarchy of dominance can be written as $C > c^{ch} > c^h > c$. **Figure 11.7** shows the genotypes and phenotypes possible for rabbit-coat color. Full color is dominant over chinchilla, which is dominant over Himalayan, which is dominant over albino.

The presence of multiple alleles increases the possible number of genotypes and phenotypes. Without multiple-allele dominance, two alleles, such as T and t, produce only three possible genotypes—in this example TT, Tt, and tt—and two possible phenotypes. However, the four alleles for rabbit-coat color produce ten possible genotypes and four phenotypes, as shown in **Figure 11.7.** More variation in rabbit coat color comes from the interaction of the color gene with other genes.

■ **Figure 11.7** Rabbits have multiple alleles for coat color. The four alleles provide four basic variations in coat color.

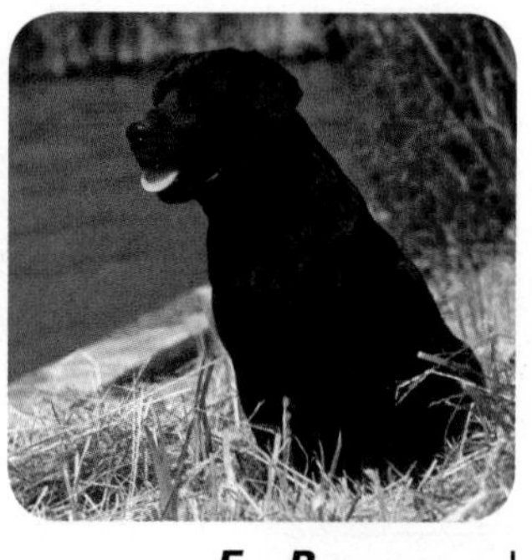

■ **Figure 11.8** The results of epistasis in coat color in Labrador retrievers show an interaction of two genes, each with two alleles. Note that an underscore in the genotype allows for either a dominant or recessive gene.

Epistasis

Coat color in Labrador retrievers can vary from yellow to black. This variety is the result of one allele hiding the effects of another allele, an interaction called **epistasis** (ih PIHS tuh sus). A Labrador's coat color is controlled by two sets of alleles. The dominant allele *E* determines whether the fur will have dark pigment. The fur of a dog with genotype *ee* will not have any pigment. The dominant B allele determines how dark the pigment will be. Study **Figure 11.8.** If the dog's genotype is *EEbb* or *Eebb,* the dog's fur will be chocolate brown. Genotypes *eebb, eeBb,* and *eeBB* will produce a yellow coat, because the *e* allele masks the effects of the dominant *B* allele.

Sex Determination

Each cell in your body, except for gametes, contains 46 chromosomes, or 23 pairs of chromosomes. One pair of these chromosomes, the **sex chromosomes,** determines an individual's gender. There are two types of sex chromosomes—X and Y. Individuals with two X chromosomes are female, and individuals with an X and a Y chromosome are male. The other 22 pairs of chromosomes are called **autosomes.** The offspring's gender is determined by the combination of sex chromosomes in the egg and sperm cell, as shown in **Figure 11.9.**

■ **Figure 11.9**
Left: The size and shape of the Y chromosome and the X chromosome are quite different from one another.
Right: The segregation of the sex chromosomes into gametes and the random combination of sperm and egg cells result in an approximately 1:1 ratio of males to females.

Color-Enhanced SEM Magnification: unavailable

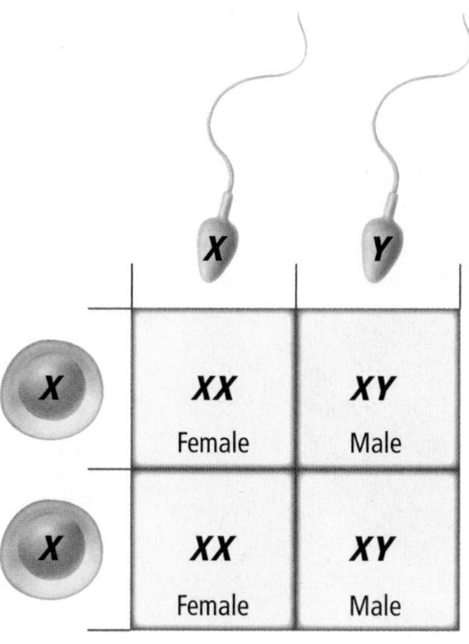

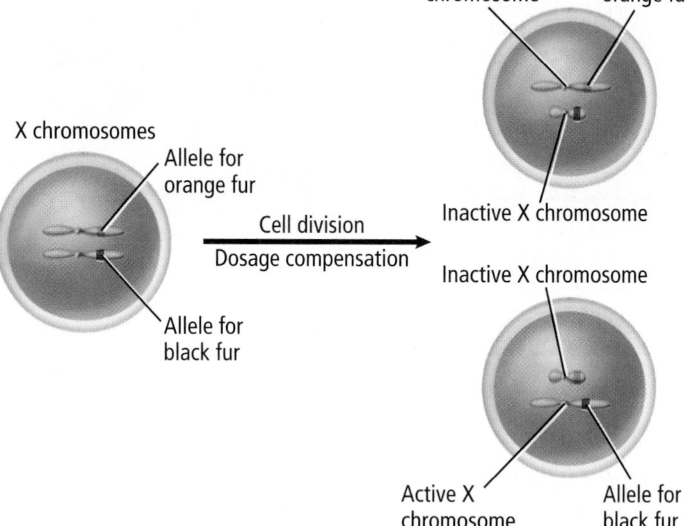

Figure 11.10 The calico coat of this cat results from the random inactivation of the X chromosomes. One X chromosome codes for orange fur, and one X chromosome codes for black fur, as illustrated on the right.

Dosage Compensation

Human females have 22 pairs of autosomes and one pair of X chromosomes. Males have 22 pairs of autosomes along with one X and one Y chromosome. If you examine the X and Y chromosomes in **Figure 11.9,** you will notice that the X chromosome is larger than the Y chromosome. The X chromosome carries a variety of genes that are necessary for the development of both females and males. The Y chromosome mainly has genes that relate to the development of male characteristics.

Because females have two X chromosomes, it seems as though females get two doses of the X chromosome and males get only one dose. To balance the difference in the dose of X-related genes, one of the X chromosomes stops working in each of the female's body cells. This often is called dosage compensation or X-inactivation. Which X chromosome stops working in each body cell is a completely random event. Dosage compensation occurs in all mammals.

As a result of the Human Genome Project, the National Institutes of Health (NIH) has released new information on the sequence of the human X chromosome. Researchers now believe that some genes on the inactivated X chromosome are more active than previously thought.

Chromosome inactivation The coat colors of the calico cat shown in **Figure 11.10** are caused by the random inactivation of a particular X chromosome. The resulting colors depend on the X chromosome that is activated. The orange patches are formed by the inactivation of the X chromosome carrying the allele for black coat color. Similarly, the black patches are a result of the inactivation of the X chromosome carrying the allele for orange coat color.

Barr bodies The inactivated X chromosomes can be observed in cells. In 1949, Canadian scientist Murray Barr observed inactivated X chromosomes in female calico cats. He noticed a condensed, darkly stained structure in the nucleus. The darkly stained, inactivated X chromosomes, such as the one shown in **Figure 11.11,** are called Barr bodies. It was discovered later that only females, including human females, have Barr bodies in their cell nuclei.

Figure 11.11 Inactivated X chromosomes in female body cells are called Barr bodies, a dark body usually found near the nucleus.

Phase contrast LM Magnification: 1000×

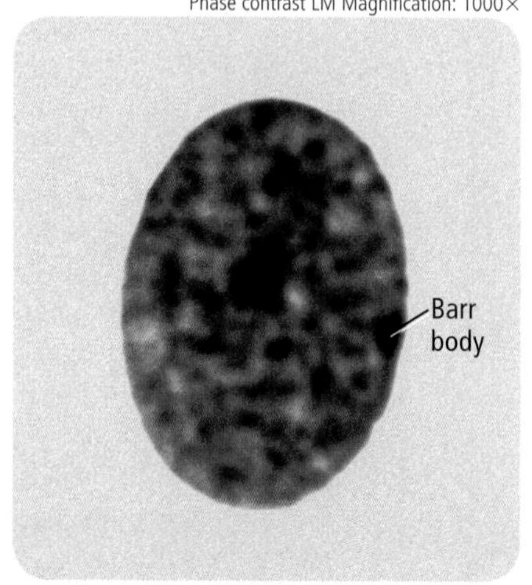

Sex-Linked Traits

Traits controlled by genes located on the X chromosome are called **sex-linked traits**—or X-linked traits. Because males have only one X chromosome, they are affected by recessive X-linked traits more often than are females. Females are less likely to express a recessive X-linked trait because the other X chromosome may mask the effect of the trait.

Some traits that are located on autosomes may appear to be sex-linked even though they are not. This occurs when an allele appears to be dominant in one gender but recessive in the other. For example, the allele for baldness is recessive in females but dominant in males, causing hair loss that follows a typical pattern called male-pattern baldness. A male would be bald if he were heterozygous for the trait, while the female would be bald only if she were homozygous recessive.

Red-green color blindness The trait for red-green color blindness is a recessive X-linked trait. About 8 percent of males in the United States have red-green color blindness. The photo in **Figure 11.12** shows how a person with red-green color blindness might view colors compared to a person who does not have red-green color blindness.

Study the Punnett square shown in **Figure 11.12.** The mother is a carrier for color blindness because she has the recessive allele for color blindness on one of her X chromosomes. The father is not color blind because he does not have the recessive allele. The sex-linked trait is represented by writing the allele on the X chromosome. Notice that the only child that can possibly have red-green color blindness is a male offspring. As a result of it being an X-linked trait, red-green color blindness is very rare in females.

Figure 11.12 People with red-green color blindness view red and green as shades of gray.

Explain ***Why are there fewer females who have red-green color blindness than males?***

X^B = Normal
X^b = Red-green color blind
Y = Y chromosome

	X^B	Y
X^B	X^BX^B	X^BY
X^b	X^BX^b	X^bY

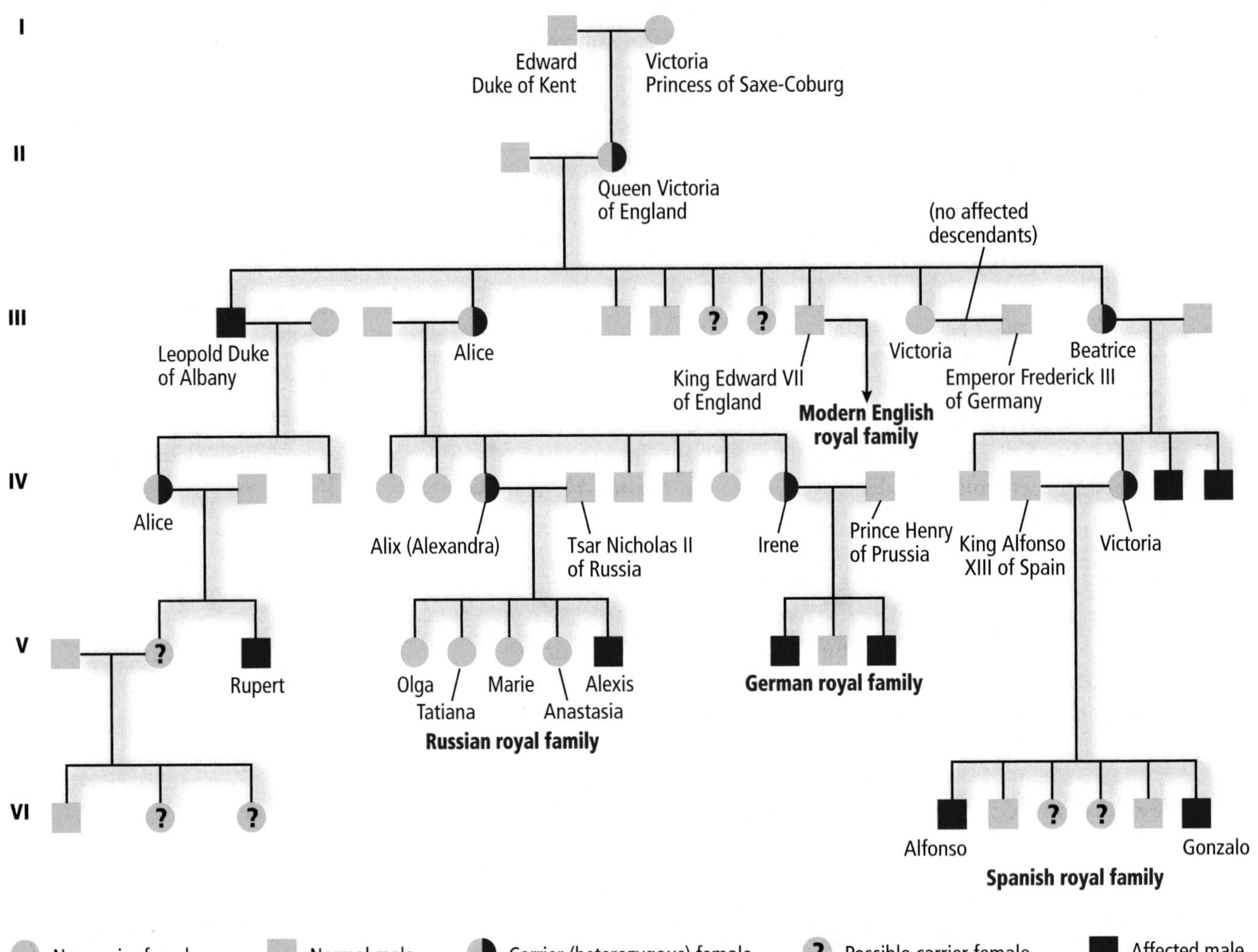

■ **Figure 11.13** The pedigree above shows the inheritance of hemophilia in the royal families of England, Germany, Spain, and Russia, starting with the children of Queen Victoria.

Determine ***Which of Alexandra's children inherited the disorder?***

Hemophilia Hemophilia, another recessive sex-linked disorder, is characterized by delayed clotting of the blood. Like red-green color blindness, this disorder is more common in males than in females.

A famous pedigree of hemophilia is one that arose in the family of Queen Victoria of England (1819-1901). Her son Leopold died of hemophilia, and her daughters Alice and Beatrice, illustrated in the pedigree in **Figure 11.13,** were carriers for the disease. Alice and Beatrice passed on the hemophilia trait to the Russian, German, and Spanish royal families. Follow the generations in this pedigree to see how this trait was passed through Queen Victoria's family. Queen Victoria's granddaughter Alexandra, who was a carrier for this trait, married Tsar Nicholas II of Russia. Irene, another granddaughter, passed the trait on to the German royal family. Hemophilia was passed to the Spanish royal family through a third granddaughter, whose name also was Victoria.

Men with hemophilia usually died at an early age until the twentieth century when clotting factors were discovered and given to hemophiliacs. However, blood-borne viruses such as Hepatitis C and HIV were often contracted by hemophiliacs until the 1990s, when safer methods of blood transfusion were discovered.

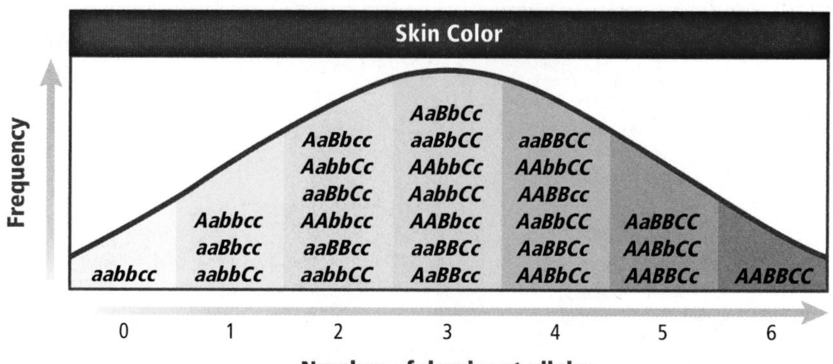

■ **Figure 11.14** This graph shows possible shades of skin color from three sets of alleles, although the trait is thought to involve more than three sets of alleles.

Predict ***Would more gene pairs increase or decrease the number of possible phenotypes?***

Polygenic Traits

You have examined traits determined by a pair of genes. Many phenotypic traits, however, arise from the interaction of multiple pairs of genes. Such traits are called **polygenic traits.** Traits such as skin color, height, eye color, and fingerprint pattern are polygenic traits. One characteristic of polygenic traits is that, when the frequency of the number of dominant alleles is graphed, as shown in **Figure 11.14,** the result is a bell-shaped curve. This shows that more of the intermediate phenotypes exist than do the extreme phenotypes.

 **Reading Check** **Infer** Why would a graph showing the frequency of the number of dominant alleles for polygenic traits be a bell-shaped curve?

Environmental Influences

The environment also has an effect on phenotype. For example, the tendency to develop heart disease can be inherited. However, environmental factors such as diet and exercise also can contribute to the occurrence and seriousness of the disease. Other ways in which environment influences phenotype are very familiar to you. You may not have thought of them in terms of phenotype, however. Sunlight, water, and temperature are environmental influences that commonly affect an organism's phenotype.

Sunlight and water Without enough sunlight, most flowering plants do not bear flowers. Many plants lose their leaves in response to water deficiency.

Temperature Most organisms experience phenotypic changes from extreme temperature changes. In extreme heat, for example, many plants suffer. Their leaves droop, flower buds shrivel, chlorophyll disappears, and roots stop growing. These are examples that probably do not surprise you, though you might never have thought of them as phenotypic changes. What other environmental factors affect the phenotypes of organisms? Temperature also influences the expression of genes. Notice the fur of the Siamese cat shown in **Figure 11.15.** The cat's tail, feet, ears, and nose are dark. These areas of the cat's body are cooler than the rest. The gene that codes for production of the color pigment in the Siamese cat's body functions only under cooler conditions. Therefore, the cooler regions are darker; and the warmer regions, where pigment production is inhibited by temperature, are lighter.

■ **Figure 11.15** Temperature affects the expression of color pigment in the fur of Siamese cats.

■ **Figure 11.16** When a trait is found more often in both members of identical twins than in fraternal twins, the trait is presumed to have a significant inherited component.

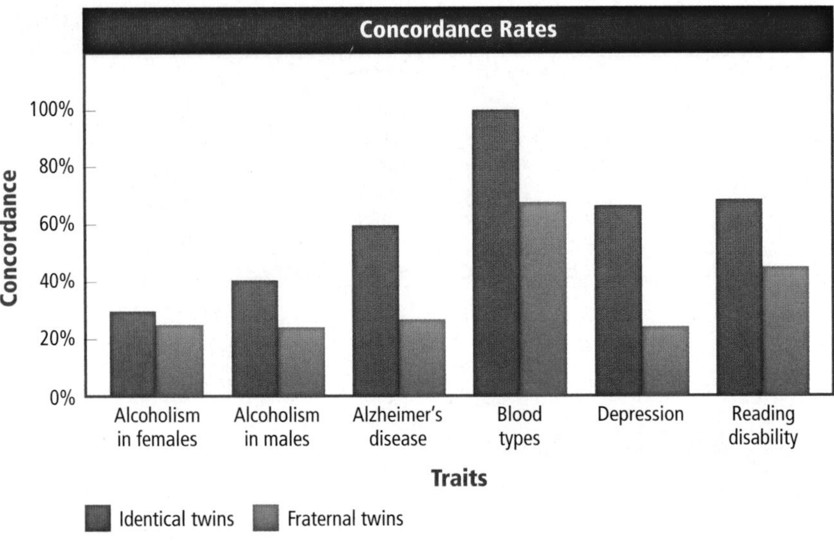

LAUNCH Lab

Review Based on what you've read about human inheritance, how would you now answer the analysis questions?

Twin Studies

Another way to study inheritance patterns is to focus on identical twins, which helps scientists separate genetic contributions from environmental contributions. Identical twins are genetically the same. If a trait is inherited, both identical twins will have the trait. Scientists conclude that traits that appear frequently in identical twins are at least partially controlled by heredity. Also, scientists presume that traits expressed differently in identical twins are strongly influenced by environment. The percentage of twins who both express a given trait is called a concordance rate. Examine **Figure 11.16** for some traits and their concordance rates. A large difference between fraternal twins and identical twins shows a strong genetic influence.

Section 11.2 Assessment

Section Summary

- Some traits are inherited through complex inheritance patterns, such as incomplete dominance, codominance, and multiple alleles.
- Gender is determined by X and Y chromosomes. Some traits are linked to the X chromosome.
- Polygenic traits involve more than one pair of alleles.
- Both genes and environment influence an organism's phenotype.
- Studies of inheritance patterns of large families and twins give insight into complex human inheritance.

Understand Main Ideas

1. MAIN Idea **Distinguish** between complex inheritance and inheritance patterns described in Chapter 10.
2. **Explain** What is epistasis, and how is it different from dominance?
3. **Determine** the genotypes of the parents if the father is blood type A, the mother is blood type B, the daughter is blood type O, one son is blood type AB, and the other son is blood type B.
4. **Analyze** how twin studies help to differentiate the effects of genetic and environmental influences.

Think Critically

5. **Evaluate** whether having sickle-cell disease would be advantageous or disadvantageous to a person living in central Africa.

MATH in Biology

6. What is the chance of producing a son with normal vision if the father is color-blind and the mother is homozygous normal for the trait? Explain.

Biology Online Self-Check Quiz biologygmh.com

Section 11.3

Reading Preview

Objectives

- **Distinguish** normal karyotypes from those with abnormal numbers of chromosomes.
- **Define and describe** the role of telomeres.
- **Relate** the effect of nondisjunction to Down syndrome and other abnormal chromosome numbers.
- **Assess** the benefits and risks of diagnostic fetal testing.

Review Vocabulary

mitosis: a process in the nucleus of a dividing cell, including prophase, metaphase, anaphase, and telophase

New Vocabulary

karyotype
telomere
nondisjunction

Chromosomes and Human Heredity

MAIN Idea **Chromosomes can be studied using karyotypes.**

Real-World Reading Link Have you ever lost one of the playing pieces belonging to a game? You might not have been able to play the game because the missing piece was important. Just as a misplaced game piece affects a game, a missing chromosome has a significant impact on the organism.

Karyotype Studies

The study of genetic material does not involve the study of genes alone. Scientists also study whole chromosomes by using images of chromosomes stained during metaphase. The staining bands identify or mark identical places on homologous chromosomes. Recall from Chapter 9 that during metaphase of mitosis, each chromosome has condensed greatly and consists of two sister chromatids. The pairs of homologous chromosomes are arranged in decreasing size to produce a micrograph called a **karyotype** (KER ee uh tipe). Karyotypes of a human male and a human female, each with 23 pairs of chromosomes, are shown in **Figure 11.17.** Notice that the 22 autosomes are matched together with one pair of nonmatching sex chromosomes.

Telomeres

Scientists have found that chromosomes end in protective caps called **telomeres.** Telomere caps consist of DNA associated with proteins. The cap serves a protective function for the structure of the chromosome. Scientists have discovered that telomeres also might be involved in both aging and cancer.

Figure 11.17 Karyotypes arrange the pairs of homologous chromosomes from increasing to decreasing size.
Distinguish ***Which two chromosomes are arranged separately from the other pairs?***

False-Color LM Magnification: 1400×

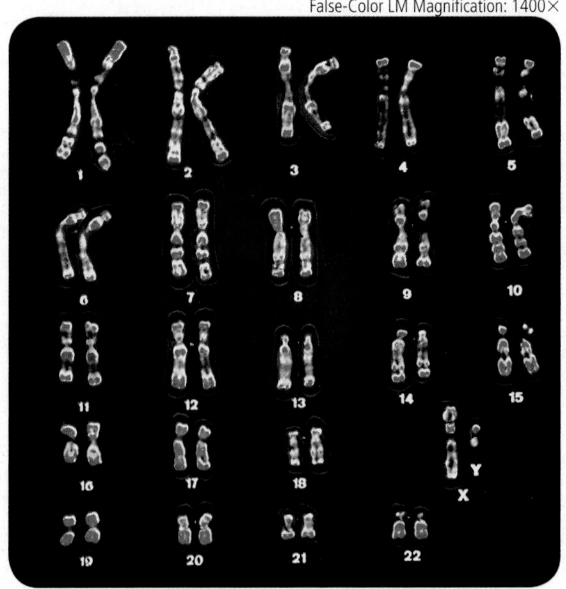

False-Color LM Magnification: 1400×

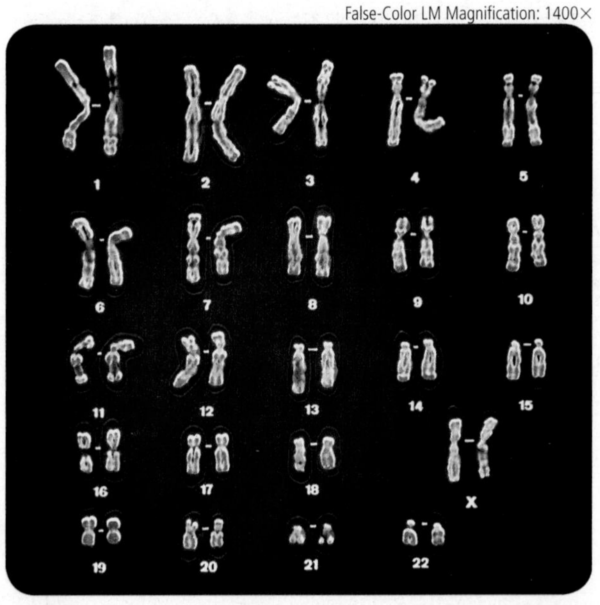

Visualizing Nondisjunction

Figure 11.18
Gametes with abnormal numbers of chromosomes can result from nondisjunction during meiosis. The orange chromosomes come from one parent, and the blue chromosomes come from the other parent.

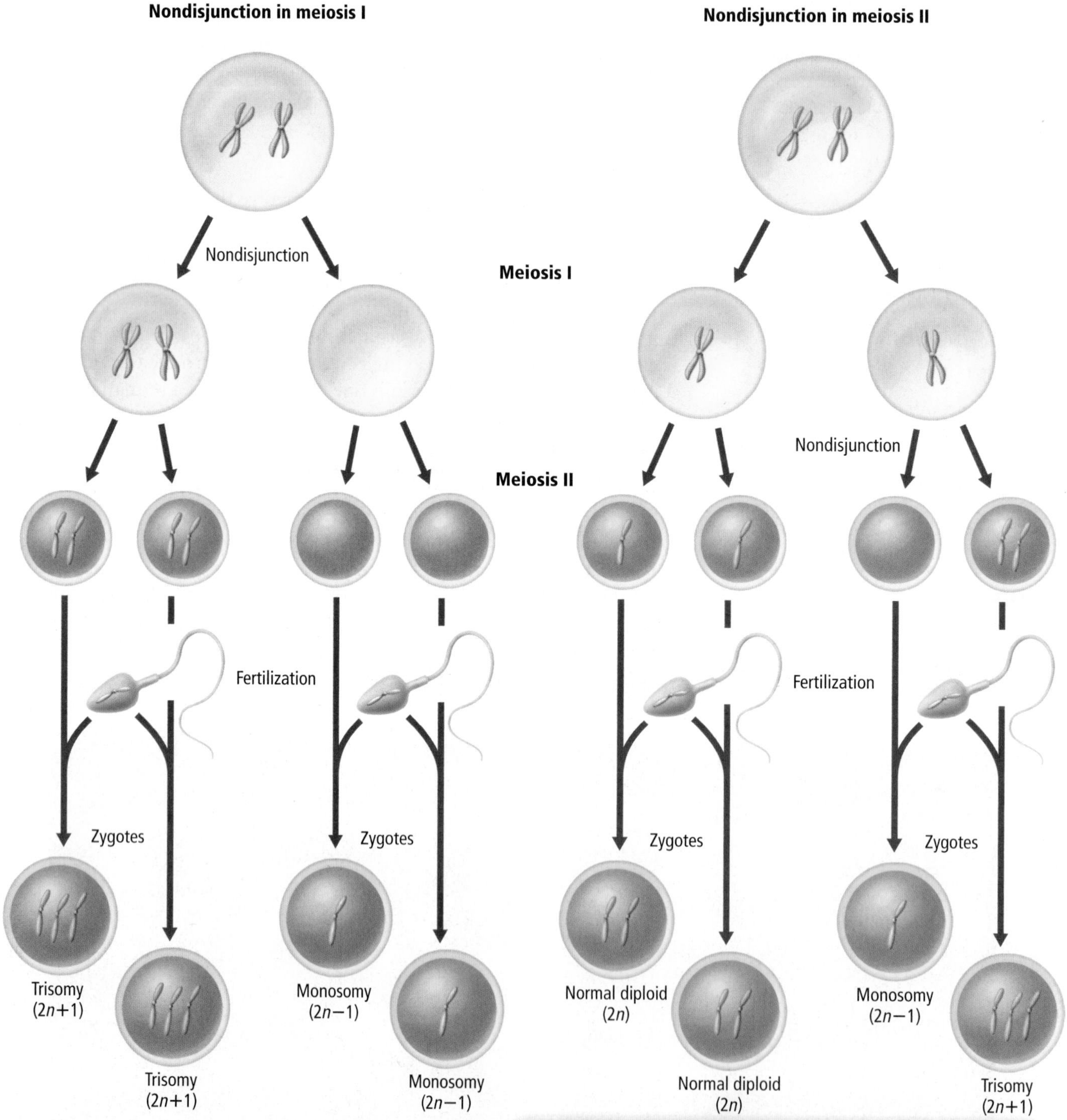

Concepts In Motion **Interactive Figure** To see an animation of nondisjunction, visit biologygmh.com.

Biology Online

Nondisjunction

During cell division, the chromosomes separate, with one of each of the sister chromatids going to opposite poles of the cell. Therefore, each new cell has the correct number of chromosomes. Cell division during which sister chromatids fail to separate properly, which does happen occasionally, is called **nondisjunction.**

If nondisjunction occurs during meiosis I or meiosis II, as shown in **Figure 11.18,** the resulting gametes will not have the correct number of chromosomes. When one of these gametes fertilizes another gamete, the resulting offspring will not have the correct number of chromosomes. Notice that nondisjunction can result in extra copies of a certain chromosome or only one copy of a particular chromosome in the offspring. Having a set of three chromosomes of one kind is called trisomy (TRI so me). Having only one of a particular type of chromosome is called monosomy (MAH nuh so me). Nondisjunction can occur in any organism in which gametes are produced through meiosis. In humans, alterations of chromosome number are associated with serious human disorders, which are often are fatal.

Down syndrome One of the earliest known human chromosomal disorders is Down syndrome. It usually is the result of an extra chromosome 21. Therefore, Down syndrome often is called trisomy 21. Examine the karyotype of a child with Down syndrome, shown in **Figure 11.19.** Notice that she has three copies of chromosome 21. The characteristics of Down syndrome include distinctive facial features as shown in **Figure 11.19,** short stature, heart defects, and mental disability. The frequency of children born with Down syndrome in the United States is approximately one out of 800. The frequency of Down syndrome increases with the age of the mother. Studies have shown that the risk of having a child with Down syndrome is about 6 percent in mothers who are 45 and older.

CAREERS IN BIOLOGY

Research Scientist Research scientists know and research a particular field of science, such as genetic disorders. For more information on biology careers, visit biologygmh.com.

■ **Figure 11.19** A person with Down syndrome has distinctive features and will have a karyotype that shows three copies of chromosome number 21.

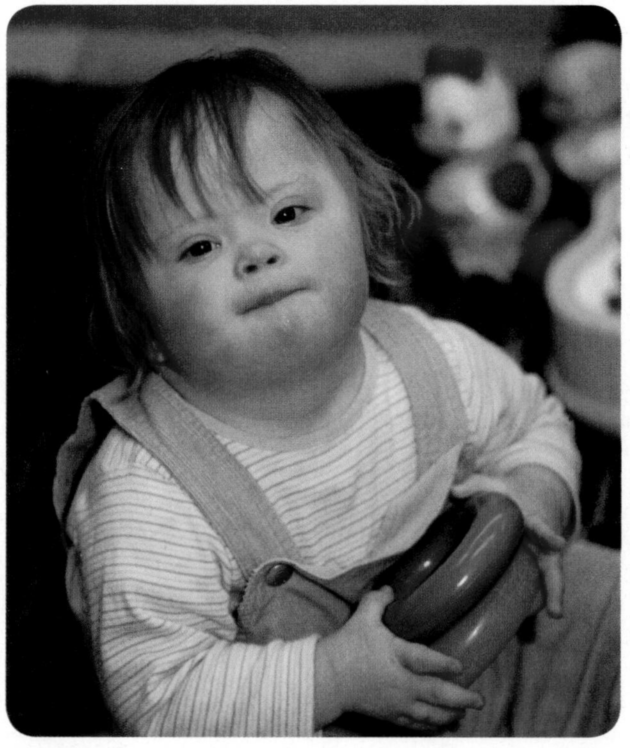

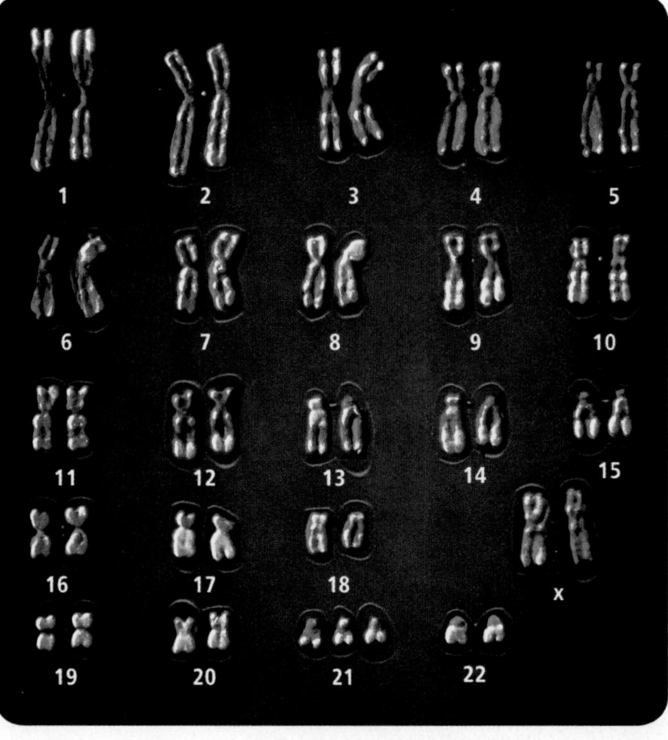

Concepts In Motion

Interactive Table To explore more about nondisjunction in human sex chromosomes, visit biologygmh.com.

Table 11.4 Nondisjunction in Sex Chromosomes

Genotype	XX	XO	XXX	XY	XXY	XYY	OY
Example							
Phenotype	Normal female	Female with Turner's syndrome	Nearly normal female	Normal male	Male with Klinefelter's syndrome	Normal or nearly normal male	Results in death

Sex chromosomes Nondisjunction occurs in both autosomes and sex chromosomes. Some of the results of nondisjunction in human sex chromosomes are listed in **Table 11.4.** Note that an individual with Turner's syndrome has only one sex chromosome. This condition results from fertilization with a gamete that had no sex chromosome.

Fetal Testing

Couples who suspect they might be carriers for certain genetic disorders might want to have a fetal test performed. Older couples also might wish to know the chromosomal status of their developing baby, known as the fetus. Various types of tests for observing both the mother and the baby are available.

MiniLab 11.2

Explore the Methods of the Geneticist

How do geneticists learn about human heredity? Traditional methods used to investigate the genetics of plants, animals, and microbes are not suitable or possible to use on humans. A pedigree is one useful tool for investigating human inheritance. In this lab, you will explore yet another tool of the geneticist—population sampling.

Procedure

1. Read and complete the lab safety form.
2. Construct a data table as instructed by your teacher.
3. Survey your group for the hitchhiker's thumb trait.
4. Survey your group for other traits determined by your teacher.
5. Compile the class data, and analyze the traits you investigated in the survey population. Determine which of the traits are dominant and which are recessive.

Analysis

1. **Interpret Data** What numerical clue did you look for to determine whether each trait surveyed was dominant or recessive?
2. **Think Critically** How could you check to see if you correctly identified dominant and recessive traits? Explain why you might have misidentified a trait.

Concepts In Motion
Interactive Table To explore more about fetal testing, visit biologygmh.com.

Table 11.5 Fetal Tests

Test	Benefit	Risk
Amniocentesis	• Diagnosis of chromosome abnormalities • Diagnosis of other defects	• Discomfort for expectant mother • Slight risk of infection • Risk of miscarriage
Chorionic villus sampling	• Diagnosis of chromosome abnormality • Diagnosis of certain genetic defects	• Risk of miscarriage • Risk of infection • Risk of newborn limb defects
Fetal blood sampling	• Diagnosis of genetic or chromosome abnormality • Checks for fetal blood problems and oxygen levels • Medications can be given to the fetus before birth	• Risk of bleeding from sample site • Risk of infection • Amniotic fluid might leak • Risk of fetal death

Connection to Health Many fetal tests can provide important information to the parents and the physician. **Table 11.5** describes the risks and benefits of some of the fetal tests that are available. Physicians must consider many factors when advising such examinations. At least a small degree of risk usually is possible in any test or procedure. The physician would not want to advise tests that would endanger the mother or the fetus; therefore, when considering whether to recommend fetal testing, the physician would need to consider previous health problems of the mother and also the health of the fetus. If the physician and parents determine that any fetal test is needed, the health of both the mother and the fetus need to be closely monitored throughout the testing.

Section 11.3 Assessment

Section Summary

- Karyotypes are micrographs of chromosomes.
- Chromosomes terminate in a cap called a telomere.
- Nondisjunction results in gametes with an abnormal number of chromosomes.
- Down syndrome is a result of nondisjunction.
- Tests for assessing the possibility of genetic and chromosomal disorders are available.

Understand Main Ideas

1. MAIN Idea **Explain** how a scientist might use a karyotype to study genetic disorders.
2. **Summarize** the role of telomeres.
3. **Illustrate** Draw a sketch to show how nondisjunction occurs during meiosis.
4. **Analyze** Why might missing sections of the X or Y chromosome be a bigger problem in males than deletions would be in one of the X chromosomes in females?

Think Critically

5. **Create** a karyotype of a female organism in which $2n = 8$ showing trisomy of chromosome 3.
6. **Infer** What might be the benefits of fetal testing? What might be the risks?

MATH in Biology

7. Conduct research on the consequences of nondisjunction other than trisomy 21. Write a paragraph about your findings.

In the Field

Career: Genetic Counselor

Genetic Testing and Support

Have you ever looked at your family tree? Do you know of any disorders or diseases that "run" in families? Genetic counselors specialize in uncovering, interpreting, and explaining this information.

Genetic counselors Genetic counselors apply their knowledge of genetics to provide information and support to people who are affected by genetic disorders. They specialize in evaluating genetic tests and indicating prevention, monitoring, and treatment options related to specific genetic conditions. Genetic counselors are also trained to deal with the emotional aspects associated with learning the results of a genetic test. They serve as patient advocates, referring individuals to community or state support services.

What does genetic testing involve? Tests are done to determine if any abnormalities are present in a particular gene or chromosome. Testing usually involves a sample of blood or tissue. In the case of prenatal genetic testing, a sample of amniotic fluid or tissue from around a fetus is taken.

It can be helpful to provide medical details about other people in your family, usually going back to your grandparents' generation, prior to meeting with a genetic counselor. Sometimes a family history gives doctors enough information to diagnose a genetic condition.

Who gets genetic testing? Sometimes a doctor recommends genetic testing. Other times, individuals seek it for themselves.

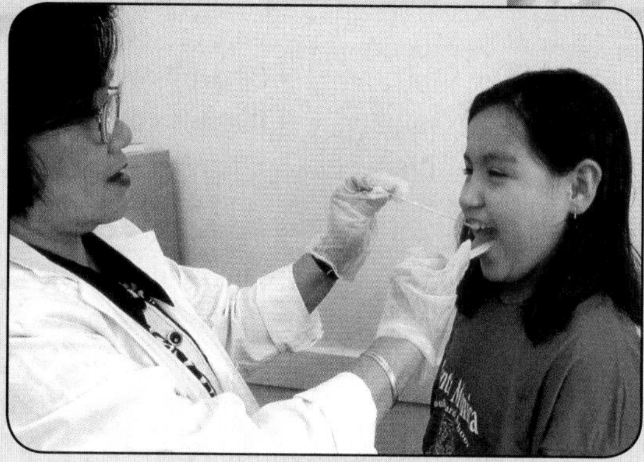

Sometimes a simple mouth swab is all that is needed to extract a genetic sample for testing.

Possible reasons for genetic testing include:

- a family history of genetic disorders;
- an unusual occurrence of certain types of cancer;
- having a child with learning difficulties or health problems, which might have a genetic cause;
- couples planning pregnancy who wish to determine if their child is at risk for a genetic condition.

Several hundred genetic tests are currently in use, with more being developed. While a doctor or health care specialist can order a genetic test, they often refer patients to genetic counselors who have received special training to interpret such tests, suggest available options, and provide supportive counseling.

WRITING in Biology

Debate Use the Skillbuilder Handbook, p. 1113, to organize a debate about the use and potential implications of genetic testing. Write a summary of your notes and your argument before participating in the debate. For more information on genetic testing, visit biologygmh.com.

BIOLAB

WHAT'S IN A FACE? INVESTIGATE INHERITED HUMAN FACIAL CHARACTERISTICS

Background: Most people know that they inherit their hair color and their eye color from their parents. However, there are many other head and facial traits that humans inherit. In this lab, you will investigate a number of different inherited facial structures that combine to compose a human face.

Question: *What structures that comprise the human face are actually determined genetically?*

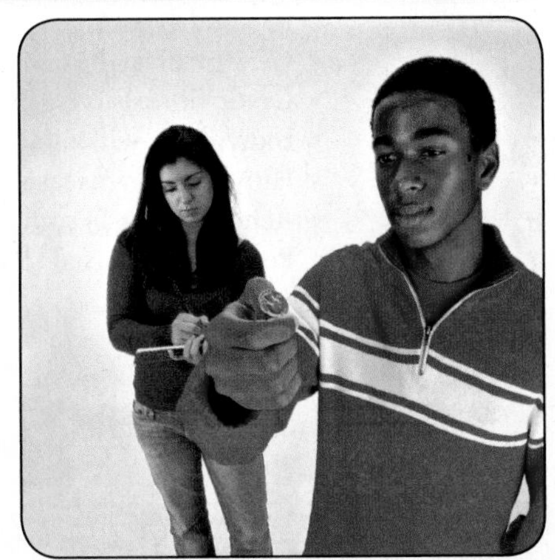

Materials

coins, 2 per team: heads=dominant trait, tails=recessive trait
table of inherited human facial characteristics

Procedure

1. Read and complete the lab safety form.
2. Partner with a classmate.
3. One member of the team will represent the father, and one member will represent the mother. Decide which partner will represent the father and who will represent the mother.
4. Have the person representing the father flip a coin. If the coin lands *heads* facing up, the offspring is a female; if the coin lands *tails* facing up, the offspring is a male. Record the gender of the offspring.
5. Flip your coin at the same time as your partner. Flip the coins only once for each trait.
6. Continue to flip coins for each trait shown in the table. After each coin flip, record the trait of your offspring by placing a check in the appropriate box in the table.
7. Once the traits are determined, draw the offspring's facial features, give him/her a name, and be prepared to introduce the offspring to the rest of the class.

Analyze and Conclude

1. **Think Critically** Why did the partner representing the father flip the coin initially to determine the gender of the offspring?
2. **Calculate** What percent chance was there of producing male offspring? Female offspring? Explain.
3. **Recognize Cause and Effect** What are the possible genotypes of parents of the following three children: a boy with straight hair (hh), a daughter with wavy hair (Hh), and a son with curly hair (HH)?
4. **Observe and Infer** Which traits show codominance?
5. **Analyze and Conclude** Would you expect other student pairs in the class to have offspring exactly like yours? Explain.

WRITING in Biology

Research Imagine that you write a science column for a large newspaper. A reader has written to you asking for a job description for a genetic counselor. Research this question; then write a short newspaper column answering the question. To learn more about inherited characteristics, visit BioLabs at biologygmh.com.

Study Guide

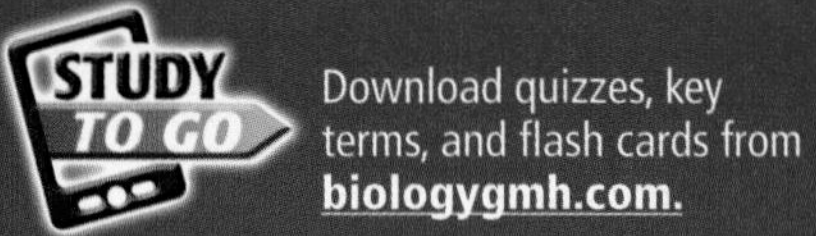

Download quizzes, key terms, and flash cards from **biologygmh.com.**

FOLDABLES **Research** Find additional information on how variations in nucleotide base sequences are linked to genetic disorders. Use the information you gathered in your Foldable and other information you learned in the chapter to describe the scientific methods you used.

Vocabulary	Key Concepts
Section 11.1 Basic Patterns of Human Inheritance	
• carrier (p. 296) • pedigree (p. 299)	MAIN Idea The inheritance of a trait over several generations can be shown in a pedigree. • Genetic disorders can be caused by dominant or recessive alleles. • Cystic fibrosis is a genetic disorder that affects mucus and sweat secretions. • Individuals with albinism do not have melanin in their skin, hair, and eyes. • Huntington's disease affects the nervous system. • Achondroplasia sometimes is called dwarfism. • Pedigrees are used to study human inheritance patterns.
Section 11.2 Complex Patterns of Inheritance	
• autosome (p. 305) • codominance (p. 302) • epistasis (p. 305) • incomplete dominance (p. 302) • multiple alleles (p. 304) • polygenic trait (p. 309) • sex chromosome (p. 305) • sex-linked trait (p. 307)	MAIN Idea Complex inheritance of traits does not follow inheritance patterns described by Mendel. • Some traits are inherited through complex inheritance patterns, such as incomplete dominance, codominance, and multiple alleles. • Gender is determined by X and Y chromosomes. Some traits are linked to the X chromosome. • Polygenic traits involve more than one pair of alleles. • Both genes and environment influence an organism's phenotype. • Studies of inheritance patterns of large families and twins give insight into complex human inheritance.
Section 11.3 Chromosomes and Human Heredity	
• karyotype (p. 311) • nondisjunction (p. 313) • telomere (p. 311)	MAIN Idea Chromosomes can be studied using karyotypes. • Karyotypes are micrographs of chromosomes. • Chromosomes terminate in a cap called a telomere. • Nondisjunction results in gametes with an abnormal number of chromosomes. • Down syndrome is a result of nondisjunction. • Tests for assessing the possibility of genetic and chromosomal disorders are available.

CHAPTER 11

Assessment

Section 11.1

Vocabulary Review

Use what you know about the vocabulary terms from the Study Guide page to answer the questions.

1. Which term describes a person who is heterozygous for a recessive disorder?
2. How is the inheritance pattern between parents and offspring represented diagrammatically?

Understand Key Concepts

3. Which condition is inherited as a dominant allele?
   A. albinism
   B. cystic fibrosis
   C. Tay-Sachs disease
   D. Huntington's disease
4. Which is not a characteristic of a person with cystic fibrosis?
   A. chloride channel defect
   B. digestive problems
   C. lack of skin pigment
   D. recurrent lung infections

Use the diagram below to answer questions 5 and 6.

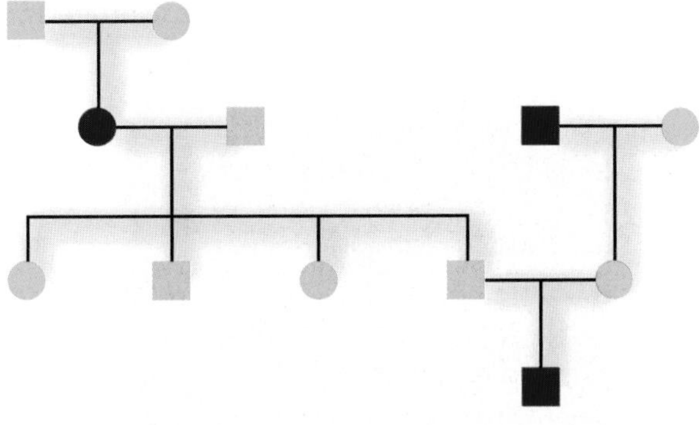

5. Which disorder could not follow the inheritance pattern shown?
   A. cystic fibrosis
   B. albinism
   C. Tay-Sachs disease
   D. Huntington's disease
6. How many affected males and females are in the pedigree?
   A. 1 male, 2 females **C.** 1 male, 1 female
   B. 2 males, 1 female **D.** 2 males, 2 females

Constructed Response

Use the photo below to answer question 7.

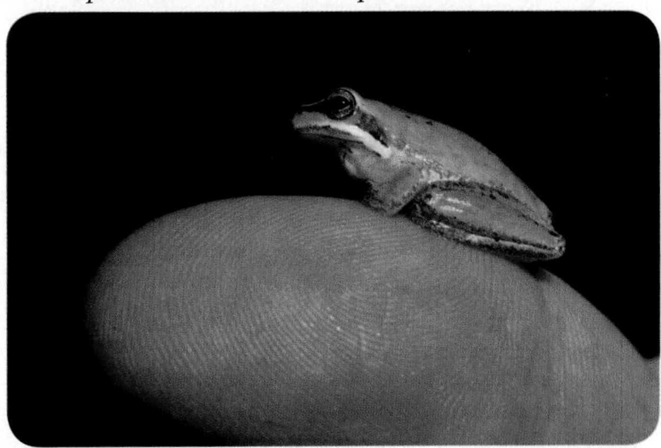

7. **Open Ended** Imagine that all animals have the same genetic disorders that humans have. What is the biological name of the genetic disorder that this dwarf tree frog would have? Describe the inheritance pattern of the genetic disorder.
8. **Short Answer** Predict the genotypes of the children of a father with Huntington's disease and an unaffected mother.

Think Critically

9. **Draw a conclusion** about the relationship of chloride ions to the excessively thick mucus in a patient suffering from cystic fibrosis.

Section 11.2

Vocabulary Review

Replace each underlined word with the correct vocabulary term from the Study Guide page.

10. Codominance is an inheritance pattern in which the heterozygous genotype results in an intermediate phenotype between the dominant and recessive phenotype.
11. A characteristic that has more than one pair of possible traits is said to be a(n) epistasis.
12. Genes found on the sex chromosomes are associated with multiple alleles.

Chapter 11 Assessment

Understand Key Concepts

13. What determines gender in humans?
    - **A.** the X and Y chromosome
    - **B.** chromosome 21
    - **C.** codominance
    - **D.** epistasis

14. Which two terms best describe the inheritance of human blood types?
    - **A.** incomplete dominance and codominance
    - **B.** codominance and multiple alleles
    - **C.** incomplete dominance and multiple alleles
    - **D.** codominance and epistasis

Use the photos below to answer question 15.

15. In radishes, color is controlled by incomplete dominance. The figure above shows the phenotype for each color. What phenotypic ratios would you expect from crossing two heterozygous plants?
    - **A.** 2: 2 red: white
    - **B.** 1: 1: 1 red: purple: white
    - **C.** 1: 2: 1 red: purple: white
    - **D.** 3: 1 red: white

Constructed Response

16. **Short Answer** How does epistasis explain the differences in coat color in Labrador retrievers?

17. **Short Answer** Explain whether a male could be heterozygous for red-green color blindness.

18. **Short Answer** What types of phenotypes would one look for if a phenotype were due to polygenic inheritance?

Think Critically

19. **Evaluate** why it might be difficult to perform genetic analysis in humans.

20. **Summarize** the meaning of the following information regarding trait inheritance: For a certain trait, identical twins have a concordance rate of 54 percent and fraternal twins have a rate of less than five percent.

Section 11.3

Vocabulary Review

Identify the vocabulary term from the Study Guide page described by each definition.

21. the protective ends of the chromosome

22. an error that occurs during cell division

23. a micrograph of stained chromosomes

Understand Key Concepts

24. What could explain a human karyotype showing 47 chromosomes?
    - **A.** monosomy
    - **B.** trisomy
    - **C.** codominance
    - **D.** dominant traits

25. Why does nondisjunction occur?
    - **A.** Cytokinesis does not occur properly.
    - **B.** The nucleoli do not disappear.
    - **C.** The sister chromatids do not separate.
    - **D.** The chromosomes do not condense properly.

Use the photo below to answer question 26.

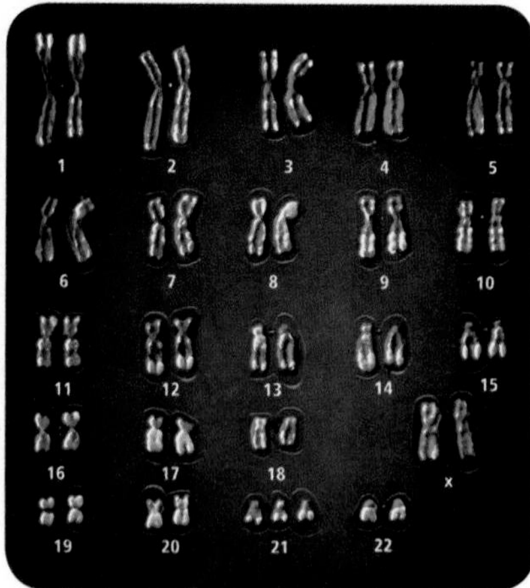

26. What disorder can be identified in the karyotype?
    - **A.** Turner's syndrome
    - **B.** Klinefelter's syndrome
    - **C.** Down syndrome
    - **D.** The karyotype shows no disorder.

27. Which statement concerning telomeres is not true?
    - **A.** They are found on the ends of chromosomes.
    - **B.** They consist of DNA and sugars.
    - **C.** They protect chromosomes.
    - **D.** They are involved with aging.

Constructed Response

Use the photo below to answer question 28.

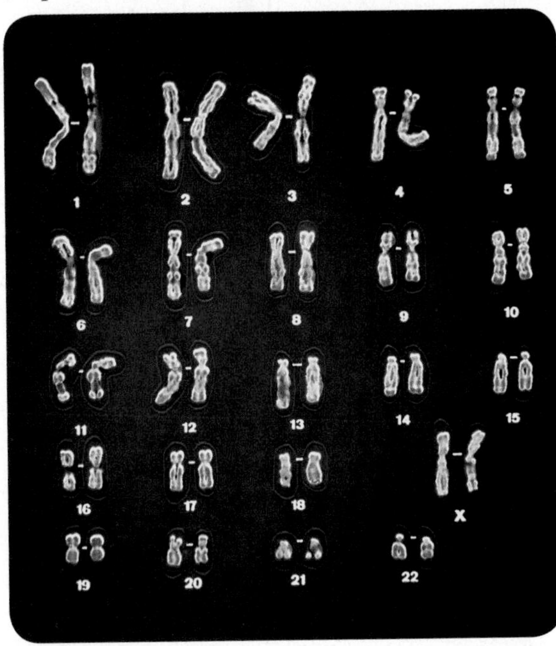

28. **Short Answer** Describe a fetal test that results in the karyotype shown above.

29. **Short Answer** What characteristics are associated with Down syndrome?

30. **Open Ended** Most cases of trisomy and monosomy in humans are fatal. Why might this be?

Think Critically

31. **Hypothesize** why chromosomes need telomeres.

32. **Explain** why a girl who has Turner's syndrome has red-green color blindness even though both of her parents have normal vision.

33. **Illustrate** what might have occurred to result in an extra chromosome in the following example: A technician is constructing a karyotype from male fetal cells. The technician discovers that the cells have one extra X chromosome.

Additional Assessment

34. **WRITING in Biology** Write a scenario for one of the genetic disorders described in **Table 11.2.** Then create a pedigree illustrating the scenario.

DBQ Document-Based Questions

Answer the questions below concerning the effect of environment on phenotype.

Data obtained from: Harnly, M.H. 1936. Genetics. *Journal of Experimental Zoology* 56: 363-379.

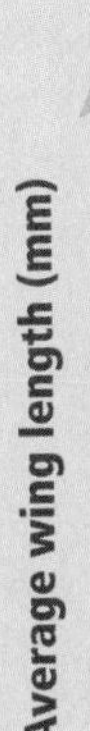

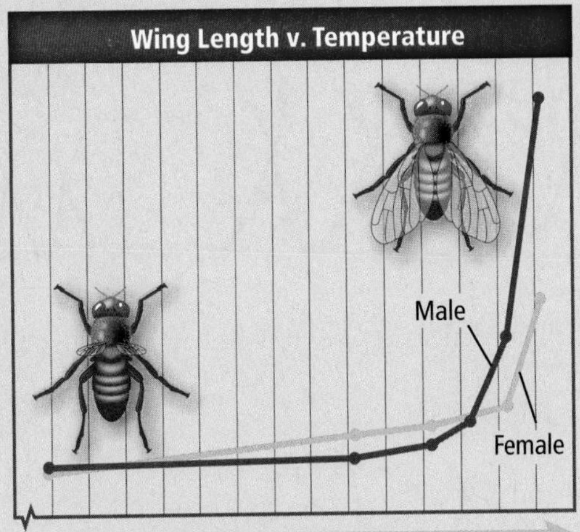

0 18 19 20 21 22 23 24 25 26 27 28 29 30 31

Environmental temperature during development (°C)

35. At which temperature is wing length the greatest?

36. Is male or female wing length more influenced by temperature? Explain.

37. Summarize the relationship between temperature and wing length for all flies.

Cumulative Review

38. Describe the structure of an atom. Elaborate on the organization of protons, neutrons, and electrons. **(Chapter 6)**

39. Compare photosynthesis to cellular respiration, relating both to the body's energy needs. **(Chapter 8)**

Standardized Test Practice

Cumulative

Multiple Choice

1. Which is affected when a cell has a low surface-area-to-volume ratio?
   A. the ability of oxygen to diffuse into the cell
   B. the amount of energy produced in the cell
   C. the diffusion of proteins through the cells
   D. the rate of protein synthesis in the cell

Use the diagram below to answer questions 2 to 4.

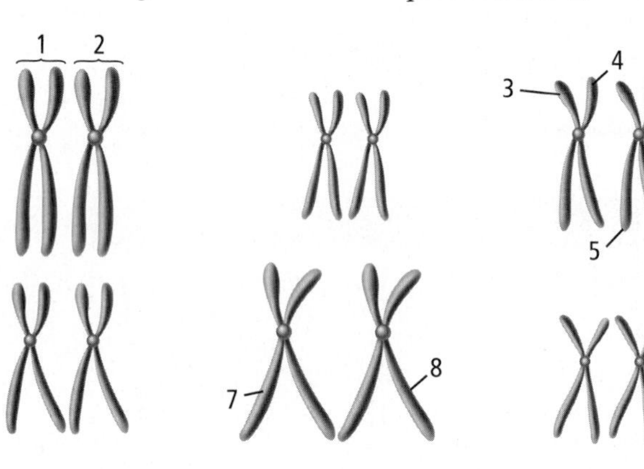

2. Which labeled structures represent a homologous pair?
   A. 1 and 2
   B. 3 and 4
   C. 3 and 6
   D. 7 and 8

3. Which parts of the chromosomes shown could appear together in a gamete of this organism?
   A. 1 and 2
   B. 3 and 6
   C. 3 and 7
   D. 5 and 6

4. If the diagram shows all the chromosomes from a body cell, how many chromosomes would be in a gamete of this organism at the end of meiosis I?
   A. 3
   B. 6
   C. 9
   D. 12

5. Which represents a polyploid organism?
   A. 1/2 n
   B. 1 1/2 n
   C. 2 n
   D. 3 n

Use the pedigree below to answer questions 6 and 7.

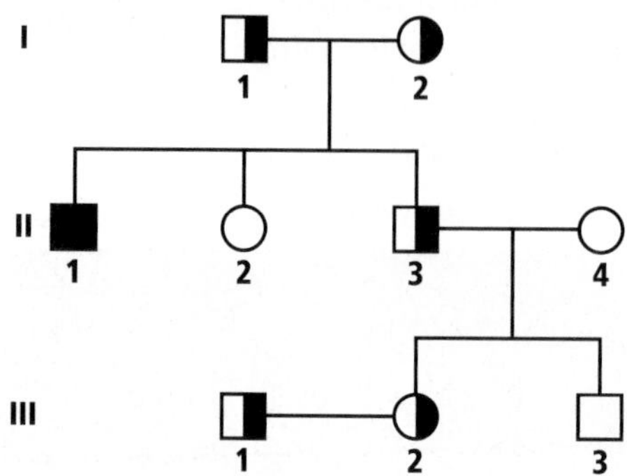

6. Which person could develop symptoms of the disease that is tracked in the pedigree?
   A. I1
   B. II1
   C. II2
   D. III2

7. According to the pedigree, who is a carrier and cannot have children with the disease?
   A. I1
   B. II1
   C. II3
   D. III1

8. Which condition would trigger mitosis?
   A. Cells touch each other.
   B. Cyclin builds up.
   C. Environmental conditions are poor.
   D. Growth factors are absent.

9. Shivering when you are cold raises your body temperature. This is an example of which characteristic of life?
   A. Your body adapts over time.
   B. Your body grows and develops.
   C. Your body has one or more cells.
   D. Your body maintains homeostasis.

Short Answer

10. In pea plants, yellow seed color is the dominant trait, and green seed color is the recessive trait. Use a Punnett square to show the results of a cross between a heterozygous yellow-seed plant and a green-seed plant.

11. Based on your Punnett square from question 10, what percentage of the offspring would have a homozygous genotype? Explain your answer.

12. Because Huntington's disease is a dominant genetic disorder, it might seem that it would be selected out of a population naturally. Write a hypothesis that states why the disease continues to occur.

13. Explain how a cancerous tumor results from a disruption of the cell cycle.

14. Write, in order, the steps that must occur for cell division to result in an organism with trisomy.

15. Which function in metabolism is performed by both the thylakoid membrane and the mitochondrial membrane? Give a reason why this function might or might not be important.

16. Suppose two parents have a mild form of a genetic disease, but their child is born with a very severe form of the same disease. What kind of inheritance pattern took place for this disease?

17. Describe an example of each of the following: species diversity, genetic diversity, and ecosystem diversity.

Extended Response

Use the diagram below to answer question 18.

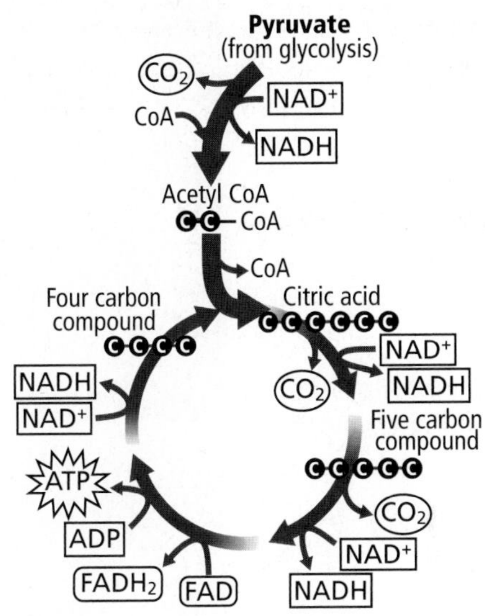

18. Identify the cycle in the figure and summarize the steps of the cycle.

19. Describe the function of microtubules, and predict what might happen if cells did NOT have microtubules.

Essay Question

The type of pea plants that Mendel investigated had either purple flowers or white flowers. One flower-color trait is dominant, and the other is recessive.

Using the information in the paragraph above, answer the following question in essay format.

20. Explain what crosses Mendel would have performed to determine which color is the dominant trait.

NEED EXTRA HELP?																				
If You Missed Question . . .	1	2	3	4	5	6	7	8	9	10	11	12	13	14	15	16	17	18	19	20
Review Section . . .	9.1	10.1	10.1	10.1	10.1	11.1	11.1	9.2	1.1	10.2	10.2	11.1	9.3	11.3	8.2, 8.3	11.2	2.2	8.3	7.3	10.2

CHAPTER 12 Molecular Genetics

BIG Idea DNA is the genetic material that contains a code for proteins.

Section 1
DNA: The Genetic Material
MAIN Idea The discovery that DNA is the genetic code involved many experiments.

Section 2
Replication of DNA
MAIN Idea DNA replicates by making a strand that is complementary to each original strand.

Section 3
DNA, RNA, and Protein
MAIN Idea DNA codes for RNA, which guides protein synthesis.

Section 4
Gene Regulation and Mutation
MAIN Idea Gene expression is regulated by the cell, and mutations can affect this expression.

BioFacts

- The human body has about 100 trillion cells that contain the 46 chromosomes in which DNA is stored.
- If all of the DNA in a human cell were stretched end to end, it would form a line about 1.8 m long.

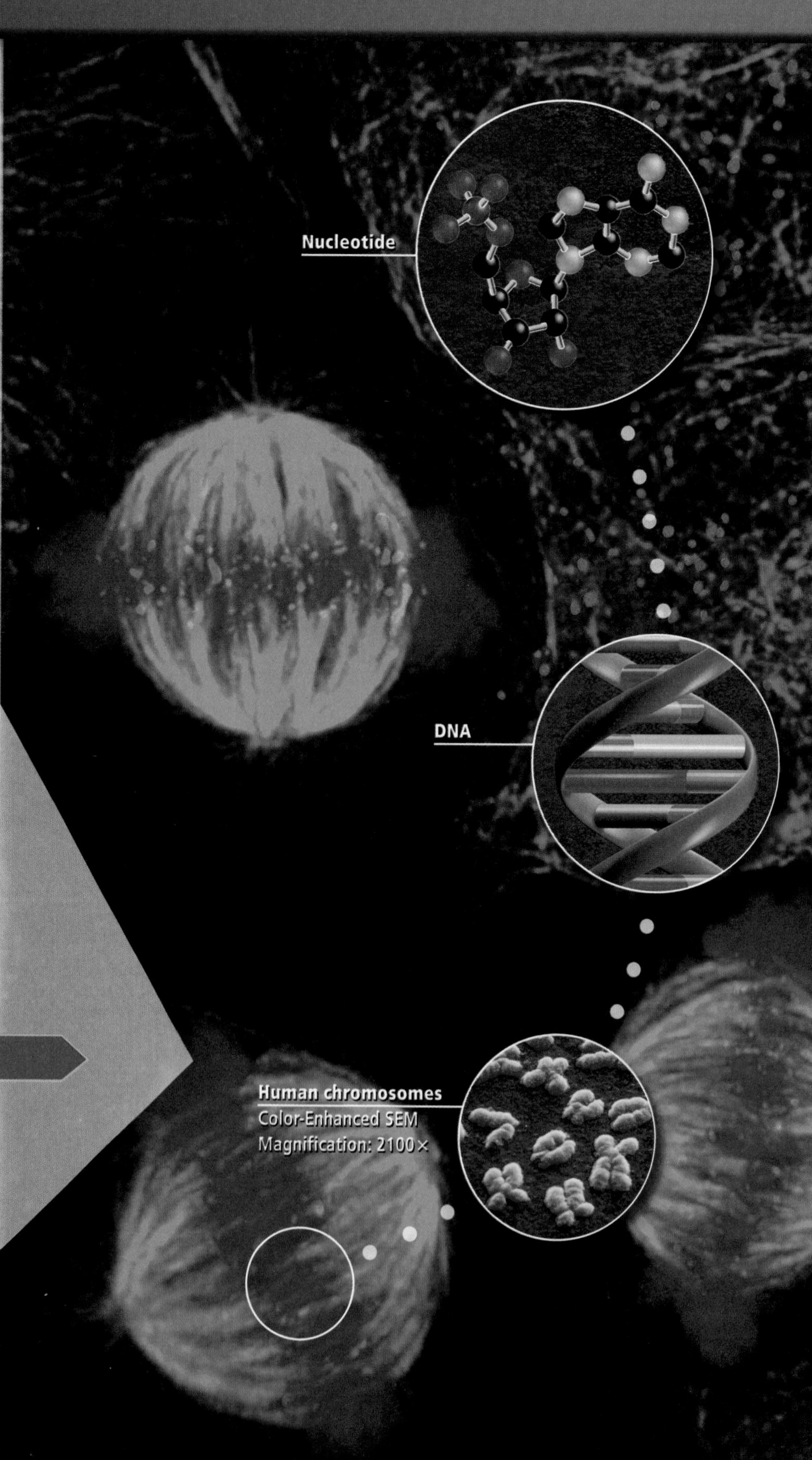

LAUNCH Lab

Who discovered DNA?

The body of knowledge concerning genetics, DNA, and biotechnology has been accumulating for nearly one and a half centuries. In this lab you will make a time line of the discovery of DNA.

Procedure

1. Work in groups of 3-4 to identify scientists and experiments that made important contributions to the understanding of genetics and DNA.
2. Preview the chapter in this **textbook.**
3. Make a time line showing when each important discovery mentioned in the text was made.

Analysis

1. **Compare and contrast** your group's time line with other time lines in the class.
2. **Infer** how the results of past experiments are important for each scientist that follows.

Visit biologygmh.com to:

- study the entire chapter online
- explore Concepts in Motion, the Interactive Table, Microscopy Links, and links to virtual dissections
- access Web links for more information, projects, and activities
- review content online with the Interactive Tutor and take Self-Check Quizzes

Comparing Transcription and Translation Use this Foldable to compare the processes of transcription and translation.

STEP 1 Fold a sheet of paper in half horizontally.

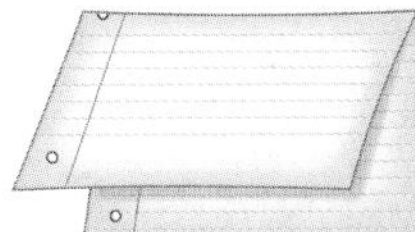

STEP 2 Fold the paper in half again as shown.

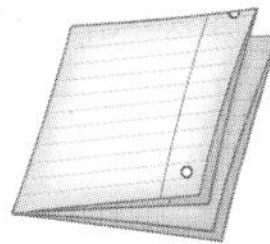

STEP 3 Cut along the fold lines in the top layer only. This will make two tabs. Label the tabs as illustrated.

FOLDABLES **Use this Foldable with Section 12.3.** Diagram and explain the processes of translation and transcription under each tab.

Section 12.1

Reading Preview

Objectives

- **Summarize** the experiments leading to the discovery of DNA as the genetic material.
- **Diagram and label** the basic structure of DNA.
- **Describe** the basic structure of the eukaryotic chromosome.

Review Vocabulary

nucleic acid: complex biomolecule that stores cellular information in the form of a code

New Vocabulary

double helix
nucleosome

DNA: The Genetic Material

MAIN Idea **The discovery that DNA is the genetic code involved many experiments.**

Real-World Reading Link Do you like to read mystery novels or watch people on television solve crimes? Detectives search for clues that will help them solve the mystery. Geneticists are detectives looking for clues in the mystery of inheritance.

Discovery of the Genetic Material

Once Mendel's work was rediscovered in the 1900s, scientists began to search for the molecule involved in inheritance. Scientists knew that genetic information was carried on the chromosomes in eukaryotic cells, and that the two main components of chromosomes are DNA and protein. For many years, scientists tried to determine which of these macromolecules—nucleic acid (DNA) or proteins—was the source of genetic information.

Griffith The first major experiment that led to the discovery of DNA as the genetic material was performed by Fredrick Griffith in 1928. Griffith studied two strains of the bacteria *Streptococcus pneumoniae,* which causes pneumonia. He found that one strain could be transformed, or changed, into the other form.

Of the two strains he studied, one had a sugar coat and one did not. Both strains are shown in **Figure 12.1.** The coated strain causes pneumonia and is called the smooth (S) strain. The noncoated strain does not cause pneumonia and is called the rough (R) strain because, without the coat, the bacteria colonies have rough edges.

Follow Griffith's study described in **Figure 12.2.** Notice the live S cells killed the mouse in the study. The live R cells did not kill the mouse, and the killed S cells did not kill the mouse. However, when Griffith made a mixture of live R cells and killed S cells and injected the mixture into a mouse, the mouse died. Griffith isolated live bacteria from the dead mouse. When these isolated bacteria were cultured, the smooth trait was visible, suggesting that a disease-causing factor was passed from the killed S bacteria to the live R bacteria. Griffith concluded that there had been a transformation from live R bacteria to live S bacteria. This experiment set the stage for the search to identify the transforming substance.

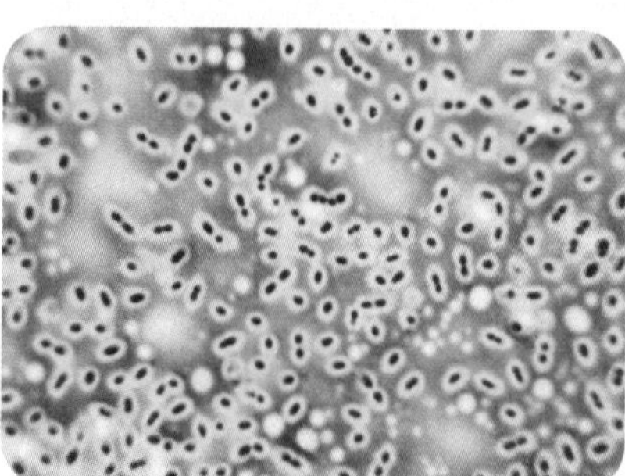

Smooth strain—*S. pneumoniae*

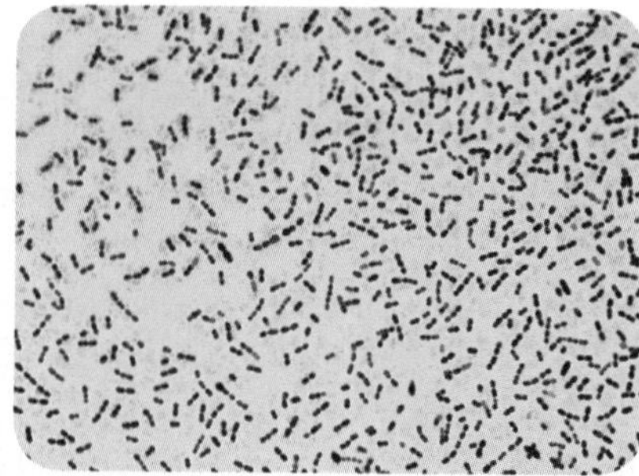

Rough strain—*S. pneumoniae*

■ **Figure 12.1** The smooth (S) strain of *S. pneumoniae* can cause pneumonia, though the rough (R) strain is not disease causing. The strains can be identified by the appearance of the colonies.

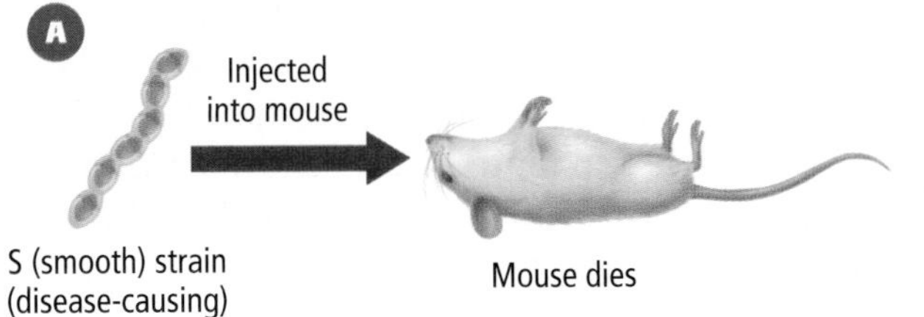

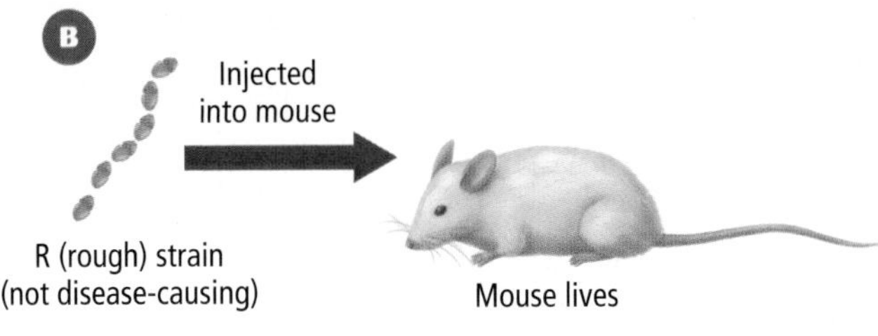

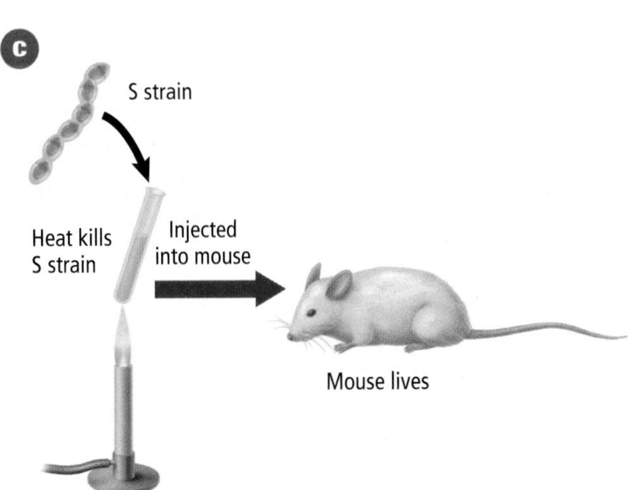

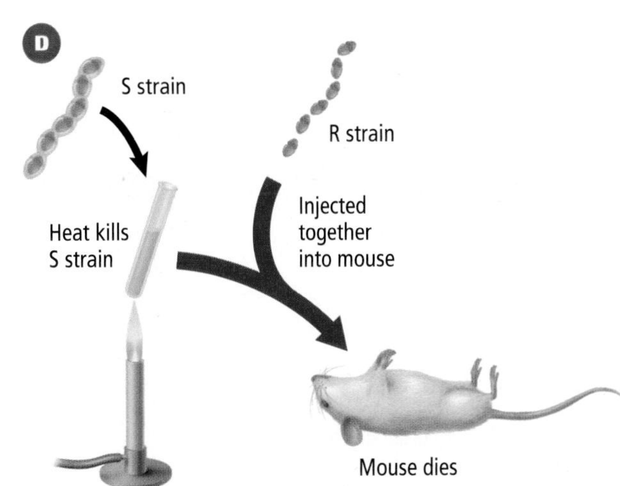

■ **Figure 12.2** Griffith's transformation experiment demonstrates the change of rough bacteria into smooth bacteria.
Explain ***why Griffith concluded there had been a change from live R bacteria to live S bacteria.***

Avery In 1944, Oswald Avery and his colleagues identified the molecule that transformed the R strain of bacteria into the S strain. Avery isolated different macromolecules, such as DNA, protein, and lipids, from killed S cells. Then he exposed live R cells to the macromolecules separately. When the live R cells were exposed to the S strain DNA, they were transformed into S cells. Avery concluded that when the S cells in Griffith's experiments were killed, DNA was released. Some of the R bacteria incorporated this DNA into their cells, and this changed the bacteria into S cells. Avery's conclusions were not widely accepted by the scientific community, and many biologists continued to question and experiment to determine whether proteins or DNA were responsible for the transfer of genetic material.

Reading Check **Explain** how Avery discovered the transforming factor.

Hershey and Chase In 1952, Alfred Hershey and Martha Chase published results of experiments that provided definitive evidence that DNA is the transforming factor. These experiments involved a bacteriophage (bak TIHR ee uh fayj), a type of virus that attacks bacteria. Two components made the experiment ideal for confirming that DNA is the genetic material. First, the bacteriophage used in the experiment was made of DNA and protein. Second, viruses cannot replicate themselves. They must inject their genetic material into a living cell to reproduce. Hershey and Chase labeled both parts of the virus to determine which part was injected into the bacteria and, thus, which part was the genetic material.

VOCABULARY
ACADEMIC VOCABULARY
Transform:
to cause a change in type or kind.
Avery used DNA to transform bacteria.

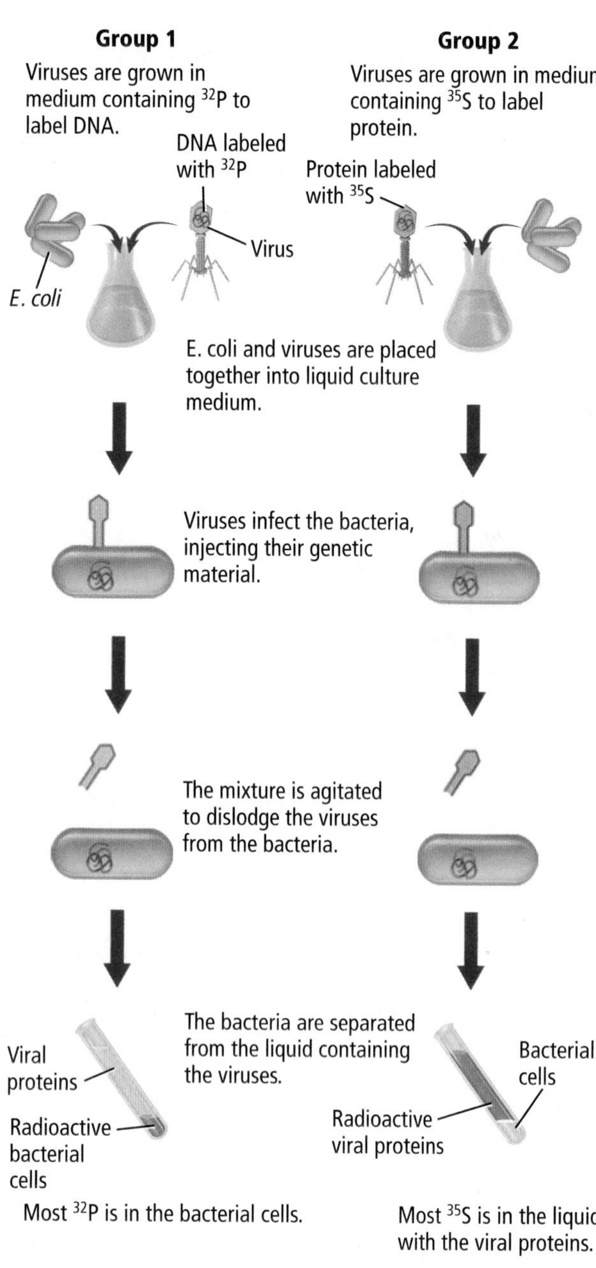
Group 1
Viruses are grown in medium containing ^{32}P to label DNA.
DNA labeled with ^{32}P
Group 2
Viruses are grown in medium containing ^{35}S to label protein.
Protein labeled with ^{35}S
Virus
E. coli
E. coli and viruses are placed together into liquid culture medium.
Viruses infect the bacteria, injecting their genetic material.
The mixture is agitated to dislodge the viruses from the bacteria.
The bacteria are separated from the liquid containing the viruses.
Viral proteins
Radioactive bacterial cells
Bacterial cells
Radioactive viral proteins
Most ^{32}P is in the bacterial cells.
Most ^{35}S is in the liquid with the viral proteins.

■ **Figure 12.3** Hershey and Chase used radioactive labeling techniques to demonstrate that DNA is the genetic material in viruses.

Radioactive labeling Hershey and Chase used a technique called radioactive labeling to trace the fate of the DNA and protein as the bacteriophages infected bacteria and reproduced. Follow along in **Figure 12.3** as you continue learning about the Hershey-Chase experiment. They labeled one set of bacteriophages with radioactive phosphorus (^{32}P). Proteins do not contain phosphorous, so DNA and not protein in these viruses would be radioactive. Hershey and Chase labeled another set of bacteriophages with radioactive sulfur (^{35}S). Because proteins contain sulfur and DNA does not, proteins and not DNA would be radioactive.

Hershey and Chase infected bacteria with viruses from the two groups. When viruses infect bacteria, they attach to the outside of the bacteria and inject their genetic material. The infected bacteria then were separated from the viruses.

Tracking DNA Hershey and Chase examined Group 1 labeled with ^{32}P and found that the labeled viral DNA had been injected into the bacteria. Viruses later released from the infected bacteria contained ^{32}P, further indicating that DNA was the carrier of genetic information.

When examining Group 2 labeled with ^{35}S, Hershey and Chase observed that the labeled proteins were found outside of the bacterial cells. Viral replication had occurred in the bacterial cells, indicating that the viruses' genetic material had entered the bacteria, but no label (^{35}S) was found. **Table 12.1** summarizes the results of the Hershey-Chase experiment.

Based on their results, Hershey and Chase concluded that the viral DNA was injected into the cell and provided the genetic information needed to produce new viruses. This experiment provided powerful evidence that DNA, not protein, was the genetic material that could be passed from generation to generation in viruses.

Reading Check Explain why it is important that new viruses were produced in the bacteria.

Table 12.1 Summary of Hershey-Chase Results

Concepts In Motion
Interactive Table To explore more about Hershey and Chase, visit biologygmh.com.

Group 1 (Viruses labeled with ^{32}P)		Group 2 (Viruses labeled with ^{35}S)	
Infected Bacteria	**Liquid with Viruses**	**Infected Bacteria**	**Liquid with Viruses**
• Labeled viral DNA (^{32}P) found in the bacteria • Viral replication occurred • New viruses contained ^{32}P	• No labeled DNA • No viral replication	• No labeled viral proteins (^{35}S) • Viral replication occurred • New viruses did not have a label	• Labeled proteins found • No viral replication

Nucleotide structure

Phosphate

Sugar

Base

Purine Bases

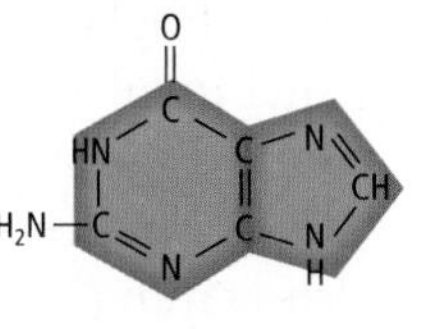

Adenine (A)

Guanine (G)

Pyrimidine Bases

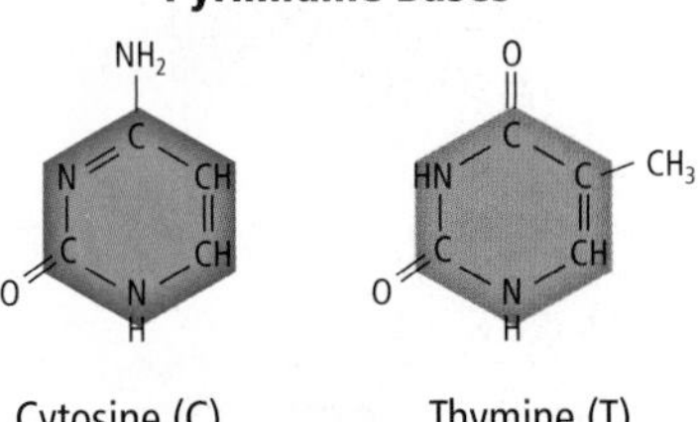

Cytosine (C)

Thymine (T) (DNA only)

Uracil (U) (RNA only)

■ **Figure 12.4** Nucleotides are made of a phosphate, sugar, and a base. There are five different bases found in nucleotide subunits that make up DNA and RNA.
Identify ***What is the structural difference between purine and pyrimidine bases?***

DNA Structure

After the Hershey-Chase experiment, scientists were more confident that DNA was the genetic material. The clues had led to the identification of the genetic material, but the questions of how nucleotides came together to form DNA and how DNA could communicate information remained.

Nucleotides In the 1920s, the biochemist P. A. Levene determined the basic structure of nucleotides that make up DNA. Nucleotides are the subunits of nucleic acids and consist of a five-carbon sugar, a phosphate group, and a nitrogenous base. The two nucleic acids found in living cells are DNA and RNA, which you learned about in Chapter 6. DNA nucleotides contain the sugar deoxyribose (dee ahk sih RI bos), a phosphate, and one of four nitrogenous bases: adenine (A duh neen), guanine (GWAH neen), cytosine (SI tuh seen), or thymine (THI meen). RNA nucleotides contain the sugar ribose, a phosphate, and one of four nitrogenous bases: adenine, guanine, cytosine, or uracil (YOO ruh sihl). Notice in **Figure 12.4** that guanine (G) and adenine (A) are double-ringed bases. This type of base is called a purine base. Thymine (T), cytosine (C), and uracil (U) are single-ringed bases called pyrimidine bases.

Chargaff Erwin Chargaff analyzed the amount of adenine, guanine, thymine, and cytosine in the DNA of various species. A portion of Chargaff's data, published in 1950, is shown in **Figure 12.5.** Chargaff found that the amount of guanine nearly equals the amount of cytosine, and the amount of adenine nearly equals the amount of thymine within a species. This finding is known as Chargaff's rule: C = G and T = A.

The structure question When four scientists joined the search for the DNA structure, the meaning and importance of Chargaff's data became clear. Rosalind Franklin, a British chemist; Maurice Wilkins, a British physicist; Francis Crick, a British physicist; and James Watson, an American biologist, provided information that was pivotal in answering the DNA structure question.

■ **Figure 12.5** Chargaff's data showed that though base composition varies from species to species, within a species C = G and A = T.

Chargaff's Data

	Base Composition (Mole Percent)			
Organism	A	T	G	C
Escherichia coli	26.0	23.9	24.9	25.2
Yeast	31.3	32.9	18.7	17.1
Herring	27.8	27.5	22.2	22.6
Rat	28.6	28.4	21.4	21.5
Human	30.9	29.4	19.9	19.8

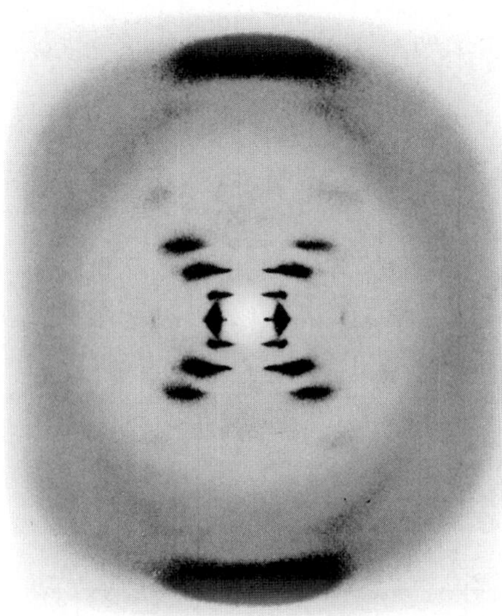

■ **Figure 12.6** Rosalind Franklin's Photo 51 and X-ray diffraction data helped Watson and Crick solve the structure of DNA. When analyzed and measured carefully, the pattern shows the characteristics of helix structure.

X-ray diffraction Wilkins was working at King's College in London, England, with a technique called X-ray diffraction, a technique that involved aiming X rays at the DNA molecule. In 1951, Franklin joined the staff at King's College. There she took the now famous Photo 51 and collected data eventually used by Watson and Crick. Photo 51, shown in **Figure 12.6,** indicated that DNA was a **double helix,** or twisted ladder shape, formed by two strands of nucleotides twisted around each other. The specific structure of the DNA double helix was determined later by Watson and Crick when they used Franklin's data and other mathematical data. DNA is the genetic material of all organisms, composed of two complementary, precisely paired strands of nucleotides wound in a double helix.

Watson and Crick Watson and Crick were working at Cambridge University in Cambridge, England, when they saw Franklin's X-ray diffraction picture. Using Chargaff's data and Franklin's data, Watson and Crick measured the width of the helix and the spacing of the bases. Together, they built a model of the double helix that conformed to the others' research. The model they built is shown in **Figure 12.7.** Some important features of their proposed molecule include the following:

1. two outside strands consist of alternating deoxyribose and phosphate
2. cytosine and guanine bases pair to each other by three hydrogen bonds
3. thymine and adenine bases pair to each other by two hydrogen bonds

LAUNCH Lab

Review Based on what you've read about the history of DNA experiments, how would you now answer the analysis questions?

DNA structure DNA often is compared to a twisted ladder, with the rails of the ladder represented by the alternating deoxyribose and phosphate. The pairs of bases (cytosine–guanine or thymine–adenine) form the steps, or rungs, of the ladder. A purine base always binds to a pyrimidine base, ensuring a consistent distance between the two rails of the ladder. This proposed bonding of the bases also explains Chargaff's data, which suggested that the number of purine bases equaled the number of pyrimidine bases in a sample of DNA. Remember, cytosine and thymine are pyrimidine bases, adenine and guanine are purines, and C = G and A = T. Therefore, C + T = G + A, or purine bases equal pyrimidine bases. Complementary base pairing is used to describe the precise pairing of purine and pyrimidine bases between strands of nucleic acids. It is the characteristic of DNA replication through which the parent strand can determine the sequence of a new strand.

Reading Check **Explain** why Chargaff's data was an important clue for putting together the structure of DNA.

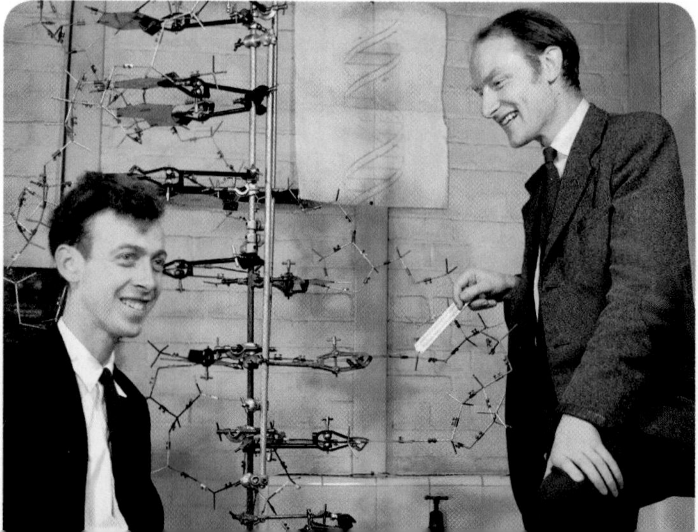

■ **Figure 12.7** Using Chargaff's and Franklin's data, Watson and Crick solved the puzzle of the structure of DNA.

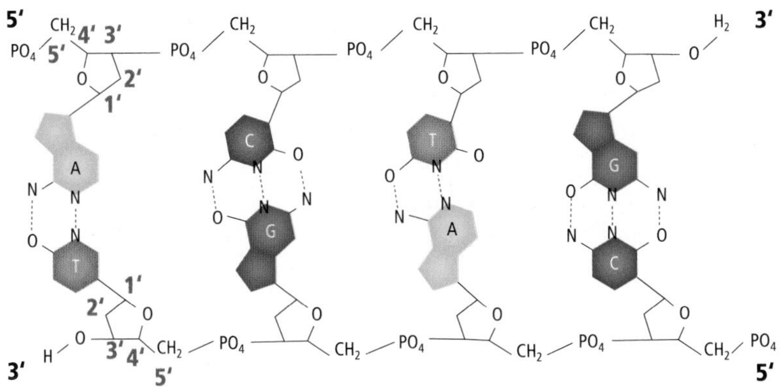

■ **Figure 12.8** Two strands of DNA running antiparallel make up the DNA helix.
Explain ***Why are the ends of the DNA strands labeled 3′ and 5′?***

Concepts In Motion

Interactive Figure To see an animation of the structure of DNA, visit biologygmh.com.

Orientation Another unique feature of the DNA molecule is the direction, or orientation, of the two strands. Carbon molecules can be numbered in organic molecules. **Figure 12.8** shows the orientation of the numbered carbons in the sugar molecules on each strand of DNA. On the top rail, the orientation of the sugar has the 5' (read "five-prime") carbon on the left, and on the end of that rail, the 3' (read "three-prime") carbon is on the right of the sugar-phosphate chain. The strand is said to be oriented 5' to 3'. The strand on the bottom runs in the opposite direction and is oriented 3' to 5'. This orientation of the two strands is called antiparallel. Another way to visualize antiparallel orientation is to take two pencils and position them so that the point of one pencil is next to the eraser of the other and vice versa.

The announcement In 1953, Watson and Crick surprised the scientific community by publishing a one-page letter in the journal *Nature* that suggested a structure for DNA and hypothesized a method of replication for the molecule deduced from the structure. In articles individually published in the same issue, Wilkins and Franklin presented evidence that supported the structure proposed by Watson and Crick. Still, the mysteries of how to prove DNA's replication and how it worked as a genetic code remained.

VOCABULARY

SCIENCE USAGE V. COMMON USAGE

Prime

Science usage: a mark located above and to the right of a character, used to identify a number or variable.
Carbon molecules in organic molecules are numbered and labeled with a prime.

Common usage: first in value, excellence, or quality.
The student found the prime seats in the stadium for watching the game.

MiniLab 12.1

Model DNA Structure

What is the structure of the DNA molecule? Construct a model to better understand the structure of the DNA molecule.

Procedure

1. Read and complete the lab safety form.
2. Construct a model of a short segment of DNA using the materials provided by your teacher.
3. Identify which parts of the model correspond to the different parts of a DNA molecule.

Analysis

1. **Describe** the structure of your DNA molecule.
2. **Identify** the characteristics of DNA that you focused on when constructing your model.
3. **Infer** In what way is your model different from your classmates' models? How does this relate to differences in DNA among organisms?

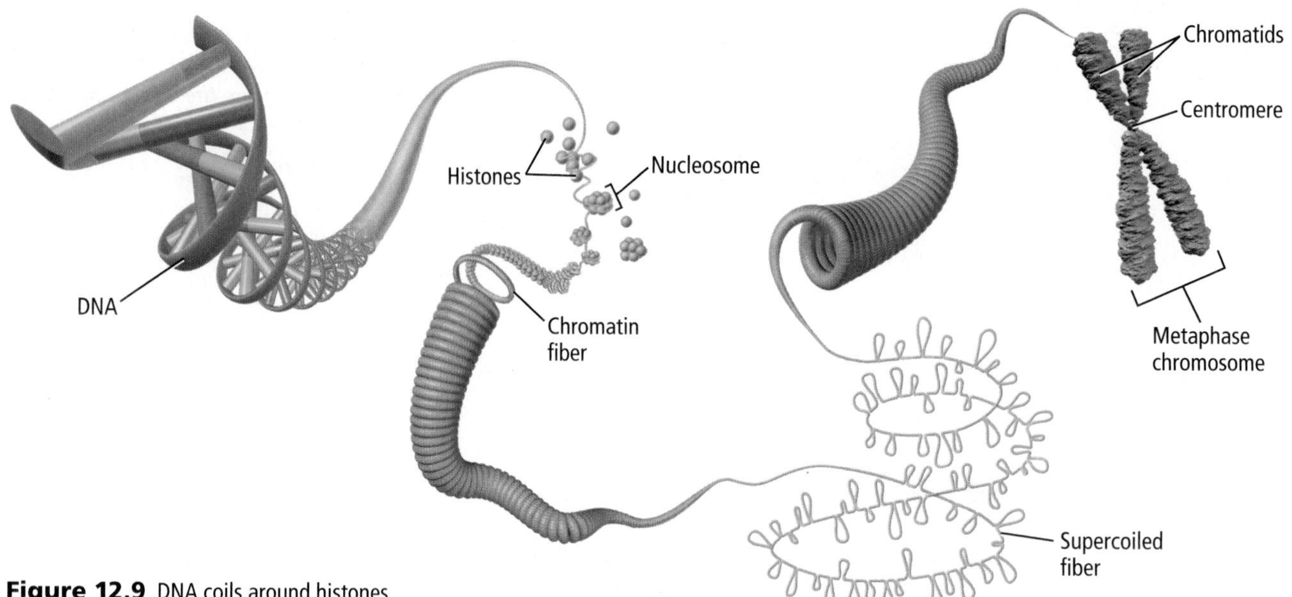

■ **Figure 12.9** DNA coils around histones to form nucleosomes, which coil to form chromatin fibers. The chromatin fibers supercoil to form chromosomes that are visible in the metaphase stage of mitosis.

Chromosome Structure

In prokaryotes, the DNA molecule is contained in the cytoplasm and consists mainly of a ring of DNA and associated proteins. Eukaryotic DNA is organized into individual chromosomes. The length of a human chromosome ranges from 51 million to 245 million base pairs. If a DNA strand 140 million nucleotides long was laid out in a straight line, it would be about five centimeters long. How does all of this DNA fit into a microscopic cell? In order to fit into the nucleus of a eukaryotic cell, the DNA tightly coils around a group of beadlike proteins called histones, as shown in **Figure 12.9.** The phosphate groups in DNA create a negative charge, which attracts the DNA to the positively charged histone proteins and forms a **nucleosome.** The nucleosomes then group together into chromatin fibers, which supercoil to make up the DNA structure recognized as a chromosome.

Section 12.1 Assessment

Section Summary

- Griffith's bacterial experiment and Avery's explanation first indicated that DNA is the genetic material.
- The Hershey-Chase experiment provided evidence that DNA is the genetic material of viruses.
- Chargaff's rule states that, in DNA, the amount of cytosine equals the amount of guanine and the amount of thymine equals the amount of adenine.
- The work of Watson, Crick, Franklin, and Wilkins provided evidence of the double-helix structure of DNA.

Understand Main Ideas

1. MAIN Idea **Summarize** the experiments of Griffith and Avery that indicated that DNA is the genetic material.
2. **Describe** the data used by Watson and Crick to determine the structure of DNA.
3. **Draw and label** a segment of DNA showing its helix and complementary base pairing.
4. **Describe** the structure of eukaryotic chromosomes.

Think Critically

5. **Describe** two characteristics that DNA needs to fulfill its role as a genetic material.
6. **Evaluate** Hershey and Chase's decision to use radioactive phosphorus and sulfur for their experiments. Could they have used carbon or oxygen instead? Why or why not?

Biology Online Self-Check Quiz biologygmh.com

Section 12.2

Reading Preview

Objectives

- **Summarize** the role of the enzymes involved in the replication of DNA.
- **Explain** how leading and lagging strands are synthesized differently.

Review Vocabulary

template: a molecule of DNA that is a pattern for synthesis of a new DNA molecule

New Vocabulary

semiconservative replication
DNA polymerase
Okazaki fragment

Replication of DNA

MAIN Idea **DNA replicates by making a strand that is complementary to each original strand.**

Real-World Reading Link When copies are made using a photocopy machine, they are expected to be exact copies of the original. Making a copy would not be very efficient if it contained errors that were not in the original. Think about how your body might make copies of DNA.

Semiconservative Replication

When Watson and Crick presented their model of DNA to the science community, they also suggested a possible method of replication—semiconservative replication. During **semiconservative replication,** parental strands of DNA separate, serve as templates, and produce DNA molecules that have one strand of parental DNA and one strand of new DNA. Recall from Chapters 9 and 10 that DNA replication occurs during interphase of mitosis and meiosis. An overview of semiconservative replication is in **Figure 12.10.** The process of semiconservative replication occurs in three main stages: unwinding, base pairing, and joining.

Unwinding DNA helicase, an enzyme, is responsible for unwinding and unzipping the double helix. When the double helix is unzipped, the hydrogen bonds between the bases are broken, leaving single strands of DNA. Then, proteins called single-stranded binding proteins associate with the DNA to keep the strands separate during replication. As the helix unwinds, another enzyme, RNA primase, adds a short segment of RNA, called an RNA primer, on each DNA strand.

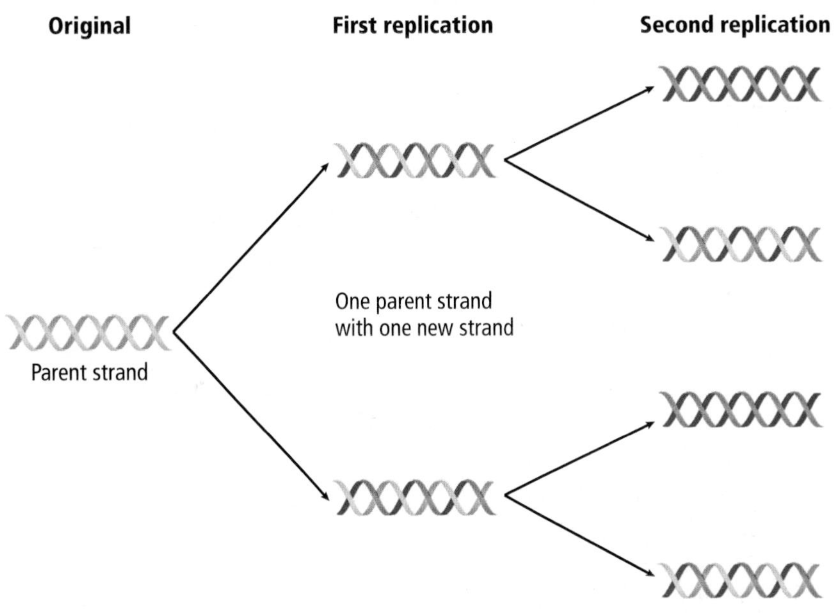

Figure 12.10 In semiconservative replication, the parental DNA separates and serves as templates to produce two daughter DNA, which then can separate to produce four DNA.

MiniLab 12.2

Model DNA Replication

How does the DNA molecule replicate? Use a model to better understand the replication of the DNA molecule.

Procedure

1. Read and complete the lab safety form.
2. Use your DNA model from **MiniLab 12.1** and extra pieces to model the replication of your segment of DNA.
3. Use your model to demonstrate DNA replication for a classmate, and identify the enzymes involved in each step.

Analysis

1. **Explain** how your model of DNA replication shows semiconservative replication.
2. **Infer** How would DNA replication in a cell be affected by an absence of DNA ligase?
3. **Identify** Where in the replication process could errors occur?

Base pairing The enzyme **DNA polymerase** catalyzes the addition of appropriate nucleotides to the new DNA strand. The nucleotides are added to the 3' end of the new strand, as illustrated in **Figure 12.11.** DNA polymerase continues adding new DNA nucleotides to the chain by adding to the 3' end of the new DNA strand. Recall that each base binds only to its complement—A binds to T and C binds to G. In this way, the templates allow identical copies of the original double-stranded DNA to be produced.

Notice in **Figure 12.11** that the two strands are made in a slightly different manner. One strand is called the leading strand and is elongated as the DNA unwinds. This strand is built continuously by the addition of nucleotides to the 3' end.

The other strand of DNA, called the lagging strand, elongates away from the replication fork. It is synthesized discontinuously into small segments, called **Okazaki fragments,** by the DNA polymerase in the 3' to 5' direction. These fragments are later connected by the enzyme DNA ligase. Each Okazaki fragment is about 100–200 nucleotides long in eukaryotes. Because one strand is synthesized continuously and the other is synthesized discontinuously, DNA replication is said to be semidiscontinuous as well as semiconservative.

Reading Check Explain How does base pairing during replication ensure that the strands produced are identical to the original strand?

Concepts In Motion

Interactive Figure To see an animation of DNA replication, visit biologygmh.com.

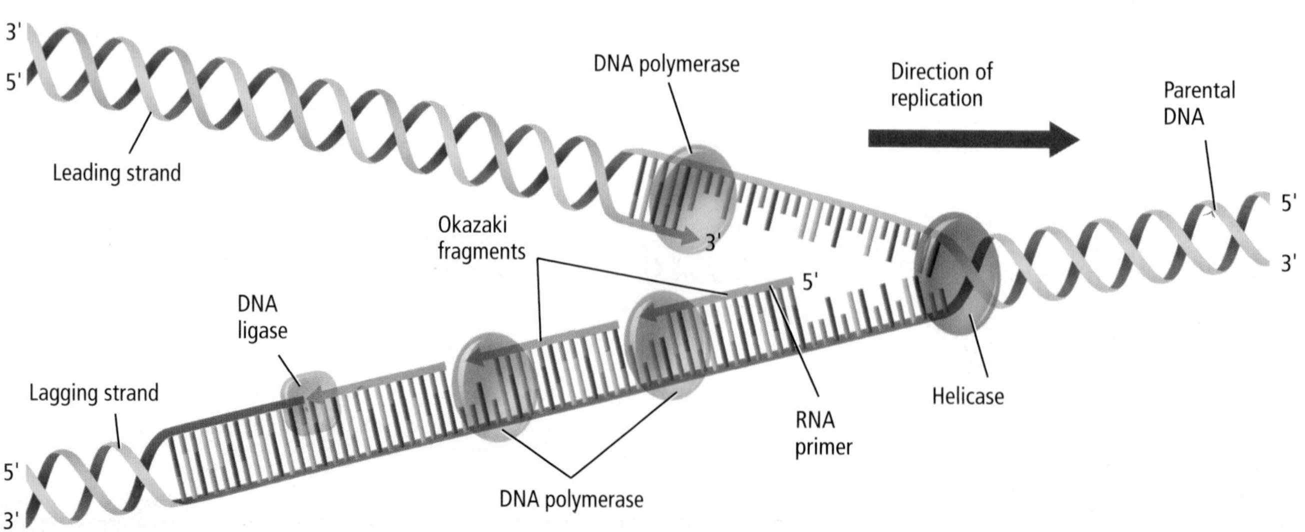

■ **Figure 12.11** The DNA strands are separated during replication as each parent strand serves as a template for new strands.
Infer *why the lagging strand produces fragments instead of being synthesized continuously.*

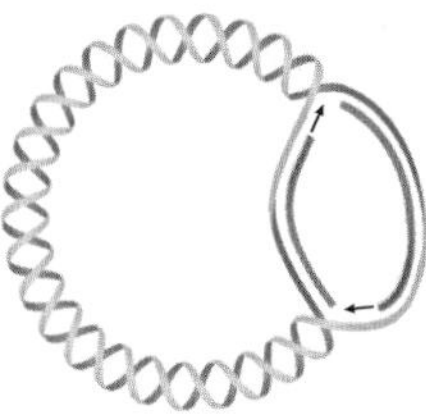

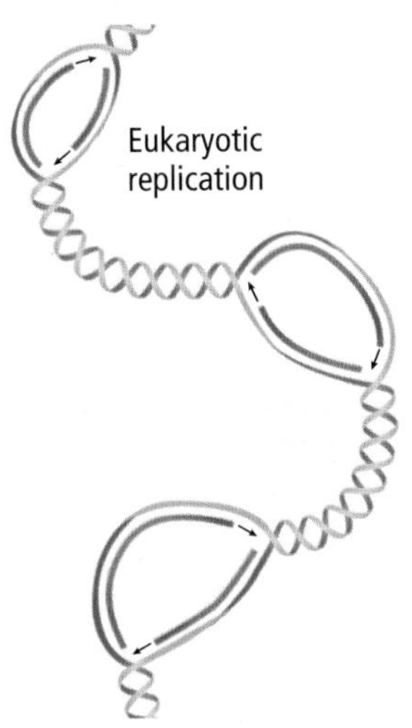

Figure 12.12 Eukaryotes have many origins of replication. Bacteria have one origin of replication, with the DNA replicating in both directions when it unzips.

Joining Even though the leading strand is synthesized continuously, in eukaryotic DNA replication there often are many areas along the chromosome where replication begins. When the DNA polymerase comes to an RNA primer on the DNA, it removes the primer and fills in the place with DNA nucleotides. When the RNA primer has been replaced, DNA ligase links the two sections.

Comparing DNA Replication in Eukaryotes and Prokaryotes

Eukaryotic DNA unwinds in multiple areas as DNA is replicated. Each individual area of a chromosome replicates as a section, which can vary in length from 10,000 to one million base pairs. As a result, multiple areas of replication are occurring along the large eukaryotic chromosome at the same time. Multiple replication origins look like bubbles in the DNA strand, as shown in **Figure 12.12.**

In prokaryotes, the circular DNA strand is opened at one origin of replication, as shown in **Figure 12.12.** Notice in the figure that DNA replication occurs in two directions, just as it does in eukaryotes. Recall from Chapter 7 that prokaryotic DNA typically is shorter than eukaryotic DNA and remains in the cytoplasm—not packaged in a nucleus.

Section 12.2 Assessment

Section Summary

- The enzymes DNA helicase, RNA primase, DNA polymerase, and DNA ligase are involved in DNA replication.
- The leading strand is synthesized continuously, but the lagging strand is synthesized discontinuously, forming Okazaki fragments.
- Prokaryotic DNA opens at a single origin of replication, whereas eukaryotic DNA has multiple areas of replication.

Understand Main Ideas

1. MAIN Idea **Indicate** the sequence of the template strand if a nontemplate strand has the sequence 5' ATGGGGCGC 3'.
2. **Describe** the role of DNA helicase, DNA polymerase, and DNA ligase.
3. **Diagram** the way leading and lagging strands are synthesized.
4. **Explain** why DNA replication is more complex in eukaryotes than in bacteria.

Think Critically

MATH in Biology

5. If the bacteria *E. coli* synthesize DNA at a rate of 100,000 nucleotides per min and it takes 30 min to replicate the DNA, how many base pairs are in an *E. coli* chromosome?

Section 12.3

Reading Preview

Objectives

- **Explain** how messenger RNA, ribosomal RNA, and transfer RNA are involved in the transcription and translation of genes.
- **Summarize** the role of RNA polymerase in the synthesis of messenger RNA.
- **Describe** how the code of DNA is translated into messenger RNA and is utilized to synthesize a protein.

Review Vocabulary

synthesis: the composition or combination of parts to form a whole

New Vocabulary

RNA
messenger RNA
ribosomal RNA
transfer RNA
transcription
RNA polymerase
intron
exon
codon
translation

DNA, RNA, and Protein

MAIN Idea **DNA codes for RNA, which guides protein synthesis.**

Real-World Reading Link Computer programmers write their programs in a particular language, or code. The computer is designed to read the code and perform a function. Like the programming code, DNA contains a code that signals the cell to perform a function.

Central Dogma

One of the important features of DNA that remained unresolved beyond the work of Watson and Crick was how DNA served as a genetic code for the synthesis of proteins. Recall from Chapter 6 that proteins function as structural building blocks for the cells and as enzymes.

Geneticists now accept that the basic mechanism of reading and expressing genes is from DNA to RNA to protein. This chain of events occurs in all living things—from bacteria to humans. Scientists refer to this mechanism as the central dogma of biology: DNA codes for RNA, which guides the synthesis of proteins.

RNA RNA is a nucleic acid that is similar to DNA. However, **RNA** contains the sugar ribose, the base uracil replaces thymine, and usually is single stranded. Three major types of RNA are found in living cells. **Messenger RNA** (mRNA) molecules are long strands of RNA nucleotides that are formed complementary to one strand of DNA. They travel from the nucleus to the ribosome to direct the synthesis of a specific protein. **Ribosomal RNA** (rRNA) is the type of RNA that associates with proteins to form ribosomes in the cytoplasm. The third type of RNA, **transfer RNA** (tRNA) are smaller segments of RNA nucleotides that transport amino acids to the ribosome. **Table 12.2** compares the structure and function of the three types of RNA.

Table 12.2 **Comparison of Three Types of RNA**

Concepts In Motion
Interactive Table To explore more about the types of RNA, visit biologygmh.com.

Name	mRNA	rRNA	tRNA
Function	Carries genetic information from DNA in the nucleus to direct protein synthesis in the cytoplasm	Associates with protein to form the ribosome	Transports amino acids to the ribosome
Example			

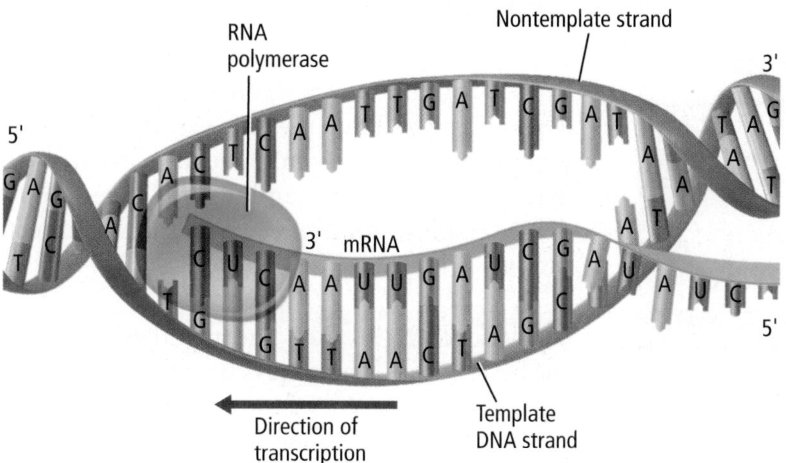

Figure 12.13 RNA is grown in the 5′ to 3′ direction.
Identify ***which enzyme adds nucleotides to the growing RNA.***

Transcription The first step of the central dogma involves the synthesis of mRNA from DNA in a process called **transcription** (trans KRIHP shun). Through transcription, the DNA code is transferred to mRNA in the nucleus. The mRNA then can take the code into the cytoplasm for protein synthesis. Follow along with the process of transcription in **Figure 12.13.** The DNA is unzipped in the nucleus and **RNA polymerase,** an enzyme that regulates RNA synthesis, binds to a specific section where an mRNA will be synthesized. As the DNA strand unwinds, the RNA polymerase initiates mRNA synthesis and moves along one of the DNA strands in the 3' to 5' direction. The strand of DNA that is read by RNA polymerase is called the template strand, and mRNA is synthesized as a complement to the DNA nucleotides. The DNA strand not used as the template strand is called the nontemplate strand. The mRNA transcript is manufactured in a 5' to 3' direction, adding each new RNA nucleotide to the 3' end. Uracil is incorporated instead of thymine as the mRNA molecule is made. Eventually, the mRNA is released, and the RNA polymerase detaches from the DNA. The new mRNA then moves out of the nucleus through nuclear pores into the cytoplasm.

FOLDABLES
Incorporate information from this section into your Foldable.

 **Reading check** **Explain** the direction in which the mRNA transcript is manufactured.

RNA processing When scientists compared the coding region of the DNA with mRNA that ultimately coded for a protein, they found that the mRNA code is significantly shorter than the DNA code. Upon closer examination, they discovered that the code on the DNA is interrupted periodically by sequences that are not in the final mRNA. These sequences are called intervening sequences, or **introns.** The coding sequences that remain in the final mRNA are called **exons.** In eukaryotes, the original mRNA made in the nucleus is sometimes called pre-mRNA and contains all of the DNA code. Before the pre-mRNA leaves the nucleus, the introns are removed from it. Other processing of the pre-mRNA includes adding a protective cap on the 5' end and adding a tail of many adenine nucleotides, called the poly-A tail, to the 3' end of the mRNA. Research shows that the cap aids in ribosome recognition, though the significance of the poly-A tail remains unknown. The mRNA that reaches the ribosome has been processed.

First Base	Second Base: U	Second Base: C	Second Base: A	Second Base: G	Third Base
U	UUU phenylalanine	UCU serine	UAU tyrosine	UGU cysteine	U
U	UUC phenylalanine	UCC serine	UAC tyrosine	UGC cysteine	C
U	UUA leucine	UCA serine	UAA *stop*	UGA *stop*	A
U	UUG leucine	UCG serine	UAG *stop*	UGG tryptophan	G
C	CUU leucine	CCU proline	CAU histidine	CGU arginine	U
C	CUC leucine	CCC proline	CAC histidine	CGC arginine	C
C	CUA leucine	CCA proline	CAA glutamine	CGA arginine	A
C	CUG leucine	CCG proline	CAG glutamine	CGG arginine	G
A	AUU isoleucine	ACU threonine	AAU asparagine	AGU serine	U
A	AUC isoleucine	ACC threonine	AAC asparagine	AGC serine	C
A	AUA isoleucine	ACA threonine	AAA lysine	AGA arginine	A
A	AUG *(start)* methionine	ACG threonine	AAG lysine	AGG arginine	G
G	GUU valine	GCU alanine	GAU aspartate	GGU glycine	U
G	GUC valine	GCC alanine	GAC aspartate	GGC glycine	C
G	GUA valine	GCA alanine	GAA glutamate	GGA glycine	A
G	GUG valine	GCG alanine	GAG glutamate	GGG glycine	G

■ **Figure 12.14** This "dictionary" of the genetic code is helpful for knowing which codons code for which amino acids.
Determine ***the possible sequences that would produce the amino acid chain: start—serine—histidine—tryptophan—stop.***

The Code

Biologists began to hypothesize that the instructions for protein synthesis are encoded in the DNA. They recognized that the only way the DNA varied among organisms was in the sequence of the bases. Scientists knew that 20 amino acids were used to make proteins, so they knew that the DNA must provide at least 20 different codes.

Connection to Math The hypothesis for how the bases formed the code is based on math and logic. If each base coded for one amino acid, then the four bases could code for four amino acids. If each pair of bases coded for one amino acid, then the four bases could only code for 16 (4×4 or 4^2) amino acids. However, if a group of three bases coded for one amino acid, there would be 64 (4^3) possible codes. This provides more than the 20 codes needed for the 20 amino acids, but is the smallest possible combination of bases to provide enough codes for the amino acids.

This reasoning meant that the code was not contained in the base pairs themselves, but must run along a single strand of the DNA. Experiments during the 1960s demonstrated that the DNA code was indeed a three-base code. The three-base code in DNA or mRNA is called a **codon.** Each of the three bases of a codon in the DNA is transcribed into the mRNA code. **Figure 12.14** shows a "dictionary" of the genetic code. Notice that all but three codons are specific for an amino acid—these three are stop codons. Codon AUG codes for the amino acid methionine and also functions as the start codon.

Translation Once the mRNA is synthesized and processed, it moves to the ribosome. In eukaryotes, this means the mRNA must leave the nucleus and enter the cytoplasm. Once in the cytoplasm, the 5' end of the mRNA connects to the ribosome. This is where the code is read and translated to make a protein through a process called **translation.** Follow along in **Figure 12.15** as you learn about translation.

In translation, tRNA molecules act as the interpreters of the mRNA codon sequence. The tRNA is folded into a cloverleaf shape and is activated by an enzyme that attaches a specific amino acid to the 3' end. At the middle of the folded strand, there is a three-base coding sequence called the anticodon. Each anticodon is complementary to a codon on the mRNA. Though the code in DNA and RNA is read 5' to 3', the anticodon is read 3' to 5'.

Visualizing Transcription and Translation

Figure 12.15
Transcription takes place in the nucleus. Translation occurs in the cytoplasm and results in the formation of polypeptides.

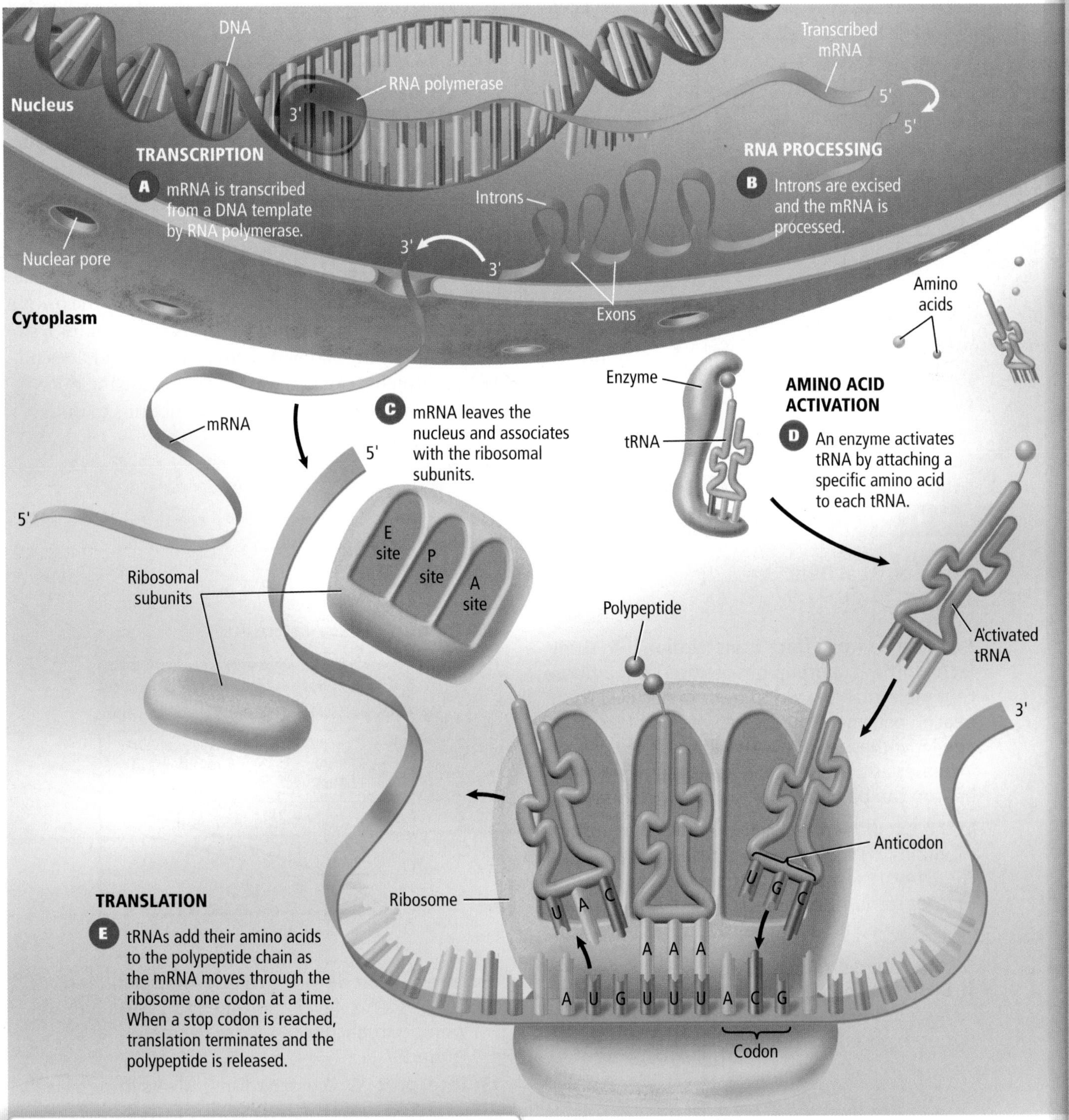

DNA
Transcribed mRNA
Nucleus
RNA polymerase
3'
5'
5'

TRANSCRIPTION

A mRNA is transcribed from a DNA template by RNA polymerase.

RNA PROCESSING

B Introns are excised and the mRNA is processed.

Introns
Nuclear pore
3'
3'
Exons
Cytoplasm
Amino acids
Enzyme
tRNA

AMINO ACID ACTIVATION

D An enzyme activates tRNA by attaching a specific amino acid to each tRNA.

mRNA
5'

C mRNA leaves the nucleus and associates with the ribosomal subunits.

5'
E site
P site
A site
Ribosomal subunits
Polypeptide
Activated tRNA
3'
Anticodon
U G C
U A C
A A A
A U G U U U A C G
Codon
Ribosome

TRANSLATION

E tRNAs add their amino acids to the polypeptide chain as the mRNA moves through the ribosome one codon at a time. When a stop codon is reached, translation terminates and the polypeptide is released.

Concepts In Motion Interactive Figure To see an animation of transcription and translation, visit biologygmh.com.

Biology Online

The role of the ribosome The ribosome consists of two subunits, as shown in **Figure 12.15.** These subunits are not associated when they are not involved in protein translation. When the mRNA leaves the nucleus, the two parts of the ribosome come together and attach to the mRNA to complete the ribosome. Once the mRNA is associated with the ribosome, a tRNA with the anticodon CAU carrying a methionine will move in and bind to the mRNA start codon—AUG—on the 5' end of the mRNA. The ribosome structure has a groove, called the P site, where the tRNA that is complementary to the mRNA moves in.

A second tRNA moves into a second groove in the ribosome, called the A site, and corresponds to the next codon of the mRNA. The next codon is UUU, so a tRNA with the anticodon AAA moves in, carrying the amino acid phenylalanine.

Part of the rRNA in the ribosome now acts as an enzyme catalyzing the formation of a bond between the new amino acid in the A site and the amino acid in the P site. As the two amino acids join, the tRNA in the P site is released to the third site, called the E site, where it exits the ribosome. The ribosome then moves so the tRNA found in Groove A is shifted to Site P, as shown in **Figure 12.15.** Now a new tRNA will enter the A site, complementing the next codon on the mRNA. This process will continue adding and linking amino acids in the sequence determined by the mRNA.

The ribosome continues to move along until the A site contains a stop codon. The stop codon signals the end of protein synthesis and does not complement any tRNA. Proteins called release factors cause the mRNA to be released from the last tRNA and the ribosome subunits to disassemble, ending protein synthesis.

Study Tip

Flowchart Draw a flowchart that connects the processes of DNA replication, transcription, and translation.

DATA ANALYSIS LAB 12.1

Based on Real Data*

Interpret the Data

How can a virus affect transcription? To study RNA synthesis, a group of scientists used a fluorescent molecular beacon to trace molecules. This beacon becomes flourescent when it binds to newly synthesized RNA. The fluorescence increases as the RNA chain lengthens. Thus, the beacon can be used to follow RNA synthesis.

In this experiment, scientists added the antibiotic rifampin (rif) to RNA polymerase from a virus (T7 RNAP), *Escherchia coli (E. coli* RNAP*)*, and *Mycobacterium smegmatis (M. smegmatis* RNAP*)* and followed RNA synthesis.

Data and Observations

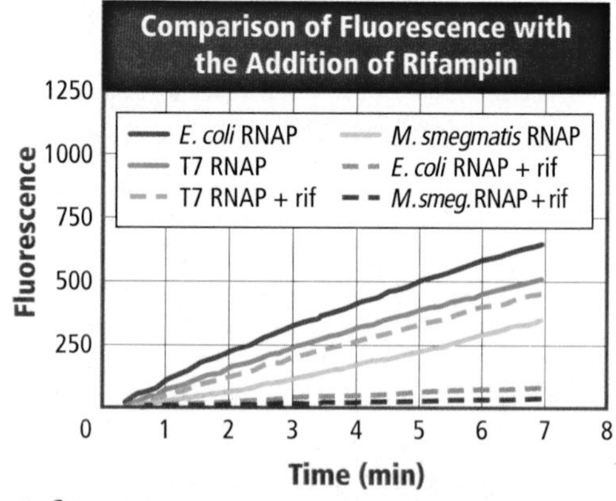

Think Critically

1. **Describe** the relationship between the fluorescence level and time in each experiment not exposed to rifampin.
2. **Infer** What does the relationship between fluorescence level and time indicate is happening in each case where rifampin was added?
3. **Interpret** Which organism's RNA synthesis is affected most by the antibiotic rifampin?

*Data obtained from: Marras, Salvatore A.E., et al. 2004. Real-time measurement of *in vitro* transcription. *Nucleic Acids Research* 32.9.e: 72.

One Gene—One Enzyme

Once scientists learned how DNA works as a code, they needed to learn the relationships between the genes and the proteins for which they coded. Experiments on the mold *Neurospora* were the first to demonstrate the relationship between genes and enzymes. In the 1940s, George Beadle and Edward Tatum provided evidence that a gene can code for an enzyme. They studied mold spores that were mutated by exposure to X rays. Examine **Figure 12.16** to follow along with their experiment.

Normally, *Neurospora* can grow on an artificial medium that provides no amino acids. This type of medium is called minimal medium. Complete medium provides all the amino acids that *Neurospora* needs to function. In Beadle and Tatum's experiment, the spores were exposed to X rays and grown on a complete medium. To test for a mutated spore, the scientists grew spores on a minimal medium. When a spore was unable to grow on the minimal medium, the mutant was tested to see what amino acid it lacked. When the mold-spore type grew on a minimal medium with a supplement such as arginine, Beadle and Tatum hypothesized that the mutant was missing the enzyme needed to synthesize arginine.

Beadle and Tatum came up with what is known as the "one gene—one enzyme" hypothesis. Today, because we know that polypeptides make up enzymes, their hypothesis has been modified slightly to refer to the fact that one gene codes for one polypeptide.

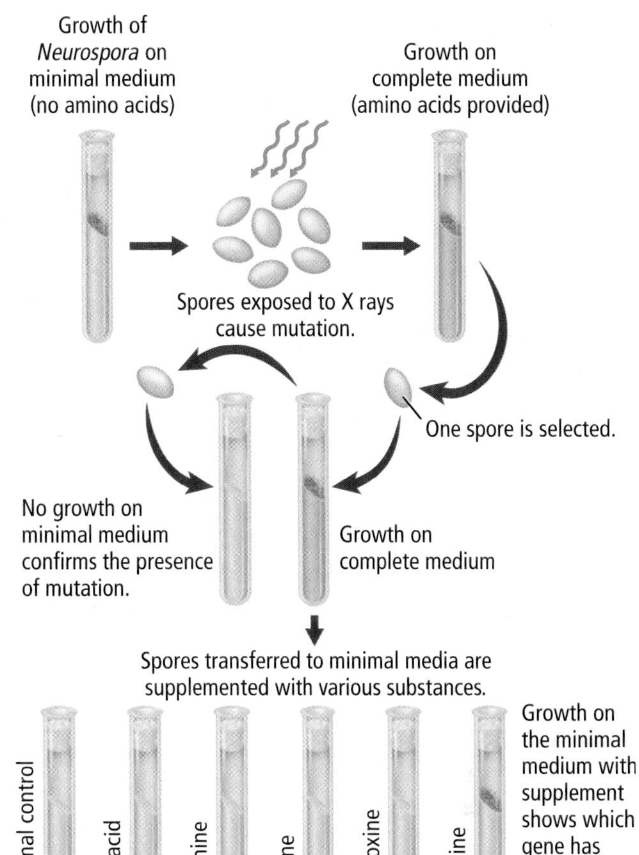

Figure 12.16 The Beadle and Tatum experiment showed that a gene codes for an enzyme. We now know that a gene codes for a polypeptide.

Section 12.3 Assessment

Section Summary

- Three major types of RNA are involved in protein synthesis: mRNA, tRNA, and rRNA.
- The synthesis of the mRNA from the template DNA is called transcription.
- Translation is the process through which the mRNA attaches to the ribosome and a protein is assembled.
- In eukaryotes, the mRNA contains introns that are excised before leaving the nucleus. A cap and poly-A tail also are added to the mRNA.

Understand Main Ideas

1. MAIN Idea **Summarize** the process by which the DNA code is made into a protein.
2. **Describe** the function of each of the following in protein synthesis: rRNA, mRNA, and tRNA.
3. **Differentiate** between codons and anticodons.
4. **Explain** the role of RNA polymerase in mRNA sythesis.
5. **Draw a Conclusion** Why has Beadle and Tatum's "one gene, one enzyme" hypothesis been modified since they presented it in the 1940s?

Think Critically

MATH in Biology

6. If the genetic code used four bases as a code instead of three, how many code units could be encoded?

Section 12.4

Reading Preview

Objectives

- **Describe** how bacteria are able to regulate their genes by two types of operons.
- **Discuss** how eukaryotes regulate transcription of genes.
- **Summarize** the various types of mutations.

Review Vocabulary

prokaryote: organism that does not have membrane-bound organelles and DNA that is organized in chromosomes

New Vocabulary

gene regulation
operon
mutation
mutagen

Gene Regulation and Mutation

MAIN Idea Gene expression is regulated by the cell, and mutations can affect this expression.

Real-World Reading Link When you type a sentence on a keyboard, it is important that each letter is typed correctly. The sentence "The fat cat ate the rat" is quite different from "The fat cat ate the hat." Though there is a difference of only one letter between the two sentences, the meaning is changed.

Prokaryote Gene Regulation

How do prokaryotic cells regulate which genes will be transcribed at particular times in the lifetime of an organism? **Gene regulation** is the ability of an organism to control which genes are transcribed in response to the environment. In prokaryotes, an operon often controls the transcription of genes in response to changes in the environment. An **operon** is a section of DNA that contains the genes for the proteins needed for a specific metabolic pathway. The parts of an operon include an operator, promoter, regulatory gene, and the genes coding for proteins. The operator is a segment of DNA that acts as an on/off switch for transcription. A second segment of DNA, called the promoter, is where the RNA polymerase first binds to the DNA. The bacteria *Escherichia coli (E. coli)* respond to tryptophan, an amino acid, and lactose, a sugar, through two operons.

The *trp* operon In bacteria, tryptophan synthesis occurs in a series of five steps, and each step is catalyzed by a specific enzyme. The five genes coding for these enzymes are clustered together on the bacterial chromosome with a group of DNA that controls whether or not they are transcribed. This cluster of DNA is called the tryptophan *(trp)* operon and is illustrated in **Figure 12.17.**

■ **Figure 12.17** The *trp* operon is an example of the gene expression of repressible enzymes.

Concepts In Motion

Interactive Figure To see an animation of the *trp* operon, visit biologygmh.com.

***Trp* operon "on"**

Repressor gene
Genes for tryptophan enzymes
DNA
RNA polymerase
Promoter
5'
mRNA
3'
5'
3'
mRNA coding for tryptophan enzymes
Protein
Repressor (inactive)
Enzymes for tryptophan synthesis

***Trp* operon "off"**

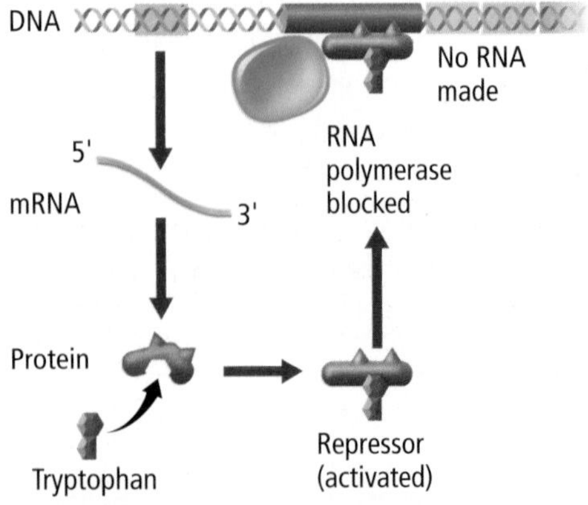

The *trp* operon is referred to as a repressible operon because transcription of the five enzyme genes normally is repressed, or turned off. When tryptophan is present in the cell's environment, the cell has no need to synthesize it and the *trp* repressor gene turns off, or represses, the transcription process by making a repressor protein. Tryptophan in *E. coli* combines with an inactive repressor protein to activate it, and the complex binds to the operator in the promoter sequence. If the repressor is bound to the operator, RNA polymerase cannot bind to it, which prevents the transcription of the enzyme genes. This prohibits the synthesis of tryptophan by the cell.

When tryptophan levels are low, the repressor is not bound to tryptophan and is inactive—it does not bind to the operator. The RNA polymerase is able to bind to the operator, turning on transcription of the five enzyme genes. This transcription enables the synthesis of tryptophan by the cell. Notice the location of the repressor protein in **Figure 12.17** when the operon is turned both off and on.

Reading Check **Summarize** the effect of tryptophan on the *trp* operon.

The *lac* operon When lactose is present in the cell, *E. coli* makes enzymes that enable it to use lactose as an energy source. The lactose (*lac*) operon, illustrated in **Figure 12.18,** contains a promoter, an operator, a regulatory gene, and three enzyme genes that control *lac* digestion. In the *lac* operon, the regulatory gene makes a repressor protein that binds to the operator in the promoter sequence and prevents the transcription of the enzyme genes.

When a molecule called an inducer is present, the inducer binds to the repressor and inactivates it. In the *lac* operon, the inducer is allolactose, a molecule that is present in food that contains lactose. Thus, when lactose is present, the allolactose binds to the repressor and inactivates it. With the repressor inactivated, RNA polymerase then can bind to the promoter and begin transcription. The *lac* operon is called an inducible operon because transcription is turned on by an inducer.

CAREERS IN BIOLOGY

Microbiologist Scientists who study microbes, primarily prokaryotes, are called microbiologists. They might research to learn about which genes control the production of particular proteins or how a protein affects the life of a cell. To learn more about biology careers, visit biologygmh.com.

Figure 12.18 The *lac* operon is an example of the gene expression of inducible enzymes.
Identify ***What is the repressor bound to when the lac operon is turned off?***

Concepts In Motion
Interactive Figure To see an animation of the lac operon, visit biologygmh.com.

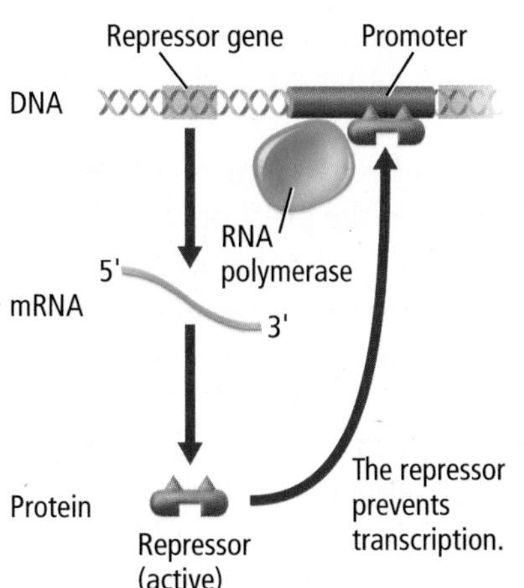

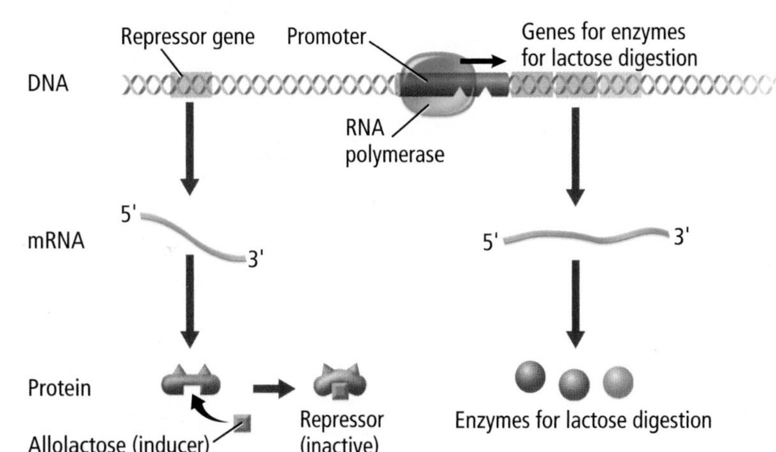

Adult *Drosophila*

Drosophila embryo

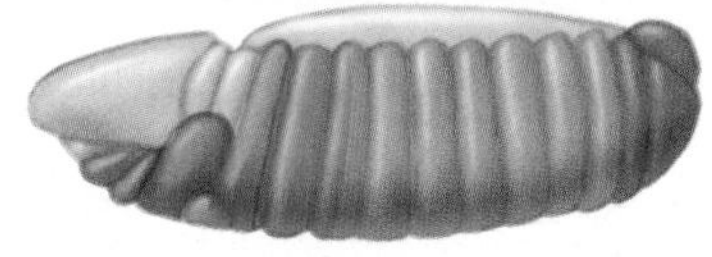

Drosophila Hox genes

■ **Figure 12.19** Hox genes are responsible for the general body pattern of most animals. Notice that the order of the genes is the same as the order of the body sections the genes control.

Eukaryote Gene Regulation

Eukaryotic cells also must control what genes are expressed at different times in the organism's lifetime. In eukaryotic cells, many genes interact with one another, requiring more elements than a single promoter and operator for a set of genes. The organization and structure of eukaryotic cells is more complex than in prokaryotic cells, increasing the complexity of the control system.

Controlling transcription One way that eukaryotes control gene expression is through proteins called transcription factors. Transcription factors ensure that a gene is used at the right time and that proteins are made in the right amounts. There are two main sets of transcription factors. One set of transcription factors forms complexes that guide and stabilize the binding of the RNA polymerase to a promoter. The other set includes regulatory proteins that help control the rate of transcription. For instance, proteins called activators fold DNA so that enhancer sites are close to the complex and increase the rate of gene transcription. Repressor proteins also bind to specific sites on the DNA and prevent the binding of activators.

The complex structure of eukaryotic DNA also regulates transcription. Recall that eukaryotic DNA is wrapped around histones to form nucleosomes. This structure provides some inhibition of transcription, although regulatory proteins and RNA polymerase still can activate specific genes even when they are packaged in the nucleosome.

Hox genes Gene regulation is crucial during development. Recall that multicellular eukaryotes develop from a single cell called a zygote. The zygote undergoes mitosis, producing all the different kinds of cells needed by the organism. Differentiation is the process through which the cells become specialized in structure and function. One group of genes that controls differentiation has been discovered. These genes are called homeobox (Hox) genes. Hox genes are important for determining the body plan of an organism. They code for transcription factors and are active in zones of the embryo that are in the same order as the genes on the chromosome. For example, the colored regions of the fly and fly embryo in **Figure 12.19** correspond to the colored genes on the piece of DNA in the figure. These genes, transcribed at specific times, and located in specific places on the genome, control what body part will develop in a given location. One mutation in the Hox genes of fruit flies has yielded flies with legs growing where their antennae should be. Studying these flies has helped scientists understand more about how genes control the body plan of an organism. Similar clusters of Hox genes that control body plans have been found in all animals.

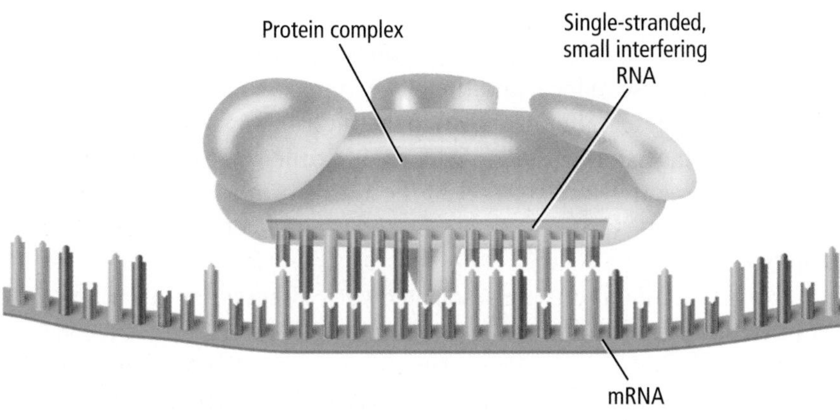

Figure 12.20 RNA interference can stop the mRNA from translating its message.
Describe *how the RNA-protein complex prevents the translation of the mRNA.*

RNA interference Another method of eukaryotic gene regulation is RNA interference (RNAi). Small pieces of double-stranded RNA in the cytoplasm of the cell are cut by an enzyme called dicer. The resulting double-stranded segments are called small interfering RNA. They bind to a protein complex that degrades one strand of the RNA. The resulting single-stranded small interfering RNA and protein complex bind to sequence-specific sections of mRNA in the cytoplasm, causing the mRNA in this region to be cut and thus preventing its translation. **Figure 12.20** shows the single-stranded small interfering RNA and protein complex binding to the mRNA. Research and clinical trials are being conducted to investigate the possibility of using RNAi to treat cancer, diabetes, and other diseases.

 **Reading Check** **Explain** how RNA interference can regulate eukaryotic gene expression.

Mutations

Do you ever make mistakes when you are typing an assignment? When you type, sometimes you might strike the wrong key. Just as you might make a mistake when typing, cells sometimes make mistakes during replication. However, these mistakes are rare, and the cell has repair mechanisms that can repair some damage. Sometimes a permanent change occurs in a cell's DNA and this is called a **mutation.** Recall that one inheritance pattern that Mendel studied was round and wrinkled pea seeds. It is now known that the wrinkled phenotype is associated with the absence of an enzyme that influences the shape of starch molecules in the seeds. Because the mutation in the gene causes a change in the protein that is made, the enzyme is nonfunctional.

Types of mutations Mutations can range from changes in a single base pair in the coding sequence of DNA to the deletions of large pieces of chromosomes. Point mutations involve a chemical change in just one base pair and can be enough to cause a genetic disorder. A point mutation in which one base is exchanged for another is called a substitution. Most substitutions are missense mutations, where the DNA code is altered so that it codes for the wrong amino acid. Other substitutions, called nonsense mutations, change the codon for an amino acid to a stop codon. Nonsense mutations cause translation to terminate early. Nearly all nonsense mutations lead to proteins that cannot function normally.

VOCABULARY

ACADEMIC VOCABULARY

Substitution:
the act of replacing one thing with another
The substitution of adenine for guanine in the DNA caused a dysfunctional protein.

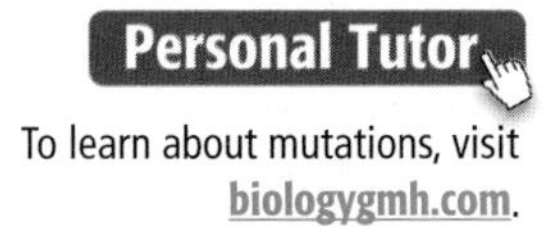

To learn about mutations, visit biologygmh.com.

Another type of mutation that can occur involves the gain or loss of a nucleotide in the DNA sequence. Insertions are additions of a nucleotide to the DNA sequence, and the loss of a nucleotide is called a deletion. Both of these mutations change the multiples of three, from the point of the insertion or deletion. These are called frameshift mutations because they change the "frame" of the amino acid sequence. **Table 12.3** illustrates various types of mutations and their effect on the DNA sequence.

Sometimes mutations are associated with diseases and disorders. One example is Garrod's alkaptonuria, which was described in Chapter 11. Patients with this disorder have a mutation in their DNA coding for an enzyme involved in digesting the amino acid phenylalanine. This mutation results in the black-colored homogentisic acid that discolors the urine. Studies have shown that patients with alkaptonuria have a high occurrence of frameshift and missense mutations in a specific region of their DNA. **Table 12.3** lists some more examples of diseases associated with types of mutation.

Concepts In Motion

Interactive Table To explore more about types of mutations, visit biologygmh.com.

Table 12.3 Mutations

Mutation Type	Analogy Sentence	Example of Associated Disease
Normal	THE BIG FAT CAT ATE THE WET RAT	
Missense (substitution)	THE BIZ FAT CAT ATE THE WET RAT	Achondroplasia: improper development of cartilage on the ends of the long bones of arms and legs resulting in a form of dwarfism
Nonsense (substitution)	THE BIG RAT	Muscular dystrophy: progressive muscle disorder characterized by the progressive weakening of many muscles in the body
Deletion (causing frameshift)	THB IGF ATC ATA TET HEW ETR AT	Cystic fibrosis: characterized by abnormally thick mucus in the lungs, intestines, and pancreas
Insertion (causing frameshift)	THE BIG ZFA TCA TAT ETH EWE TRA	Crohn's disease: chronic inflammation of the intestinal tract, producing frequent diarrhea, abdominal pain, nausea, fever, and weight loss
Duplication	THE BIG FAT FAT CAT ATE THE WET RAT	Charcot-Marie-Tooth disease (type 1A): damage to peripheral nerves leading to weakness and atrophy of muscles in hands and lower legs
Expanding mutation (tandem repeats) **Generation 1** **Generation 2** **Generation 3**	THE BIG FAT CAT ATE THE WET RAT THE BIG FAT CAT CAT CAT ATE THE WET RAT THE BIG FAT CAT CAT CAT CAT CAT CAT ATE THE WET RAT	Huntington's disease: a progressive disease in which brain cells waste away, producing uncontrolled movements, emotional disturbances, and mental deterioration

Large portions of DNA also can be involved in a mutation. A piece of an individual chromosome containing one or more genes can be deleted or moved to a different location on the chromosome, or even to a different chromosome. Such rearrangements of the chromosome often have drastic effects on the expression of these genes.

Connection to Health In 1991, a new kind of mutation was discovered that involves an increase in the number of copies of repeated codons, called tandem repeats. The increase in repeated sequences seems to be involved in a number of inherited disorders. The first known example was fragile X syndrome—a syndrome that results in a number of mental and behavioral impairments. Near the end of a normal X chromosome, there is a section of CGG codons that repeat about 30 times. Individuals with fragile X have CGG codons that repeat hundreds of times. The syndrome received its name because the repeated area on the tip of the X chromosomes appears as a fragile piece hanging off the X chromosome, as illustrated in **Figure 12.21.** Currently, the mechanism by which the repeats expand from generation to generation is not known.

Figure 12.21 Fragile X syndrome is due to many extra repeated CGG units near the end of the X chromosome, making the lower tip of the X chromosome appear fragile.

Reading Check **List and describe** three types of mutations.

Protein folding and stability You might expect that large changes in the DNA code, such as frameshift mutations or changes in position, lead to genetic disorders. However, small changes like substitutions also can lead to genetic disorders. The change of one amino acid for another can change the sequence of amino acids in a protein enough to change both the folding and stability of the protein, as illustrated in **Figure 12.22.**

In Chapter 11, you learned about a genetic disorder caused by a single point mutation called sickle-cell disease. In the case of sickle-cell disease, the codon for a glutamic acid (GAA) has been changed to a valine (GUA) in the protein. This change in composition changes the structure of hemoglobin and is the cause of this disorder.

Figure 12.22 A single amino acid substitution can cause the genetic disorder sickle-cell disease.
Recall *What happens to the protein with the substituted amino acid?*

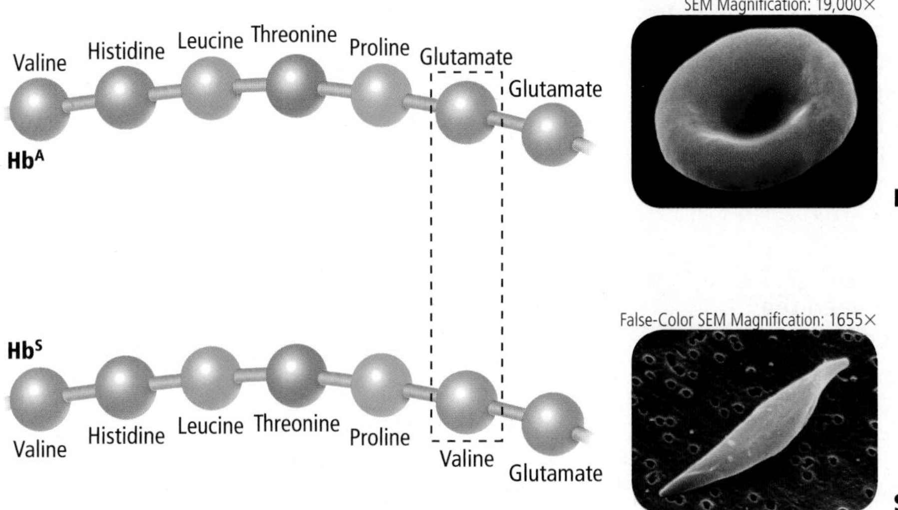

Normal shape of red blood cell

Sickle shape of red blood cell

Hemoglobin is made of four polypeptide chains—two sets of two identical chains. The molecule also contains a large carbon-ring structure that binds iron called the heme group. The substituted glutamic acid is located near the start of one set of chains, as shown in **Figure 12.22.** Glutamic acid is a polar amino acid, but the valine that substitutes for it in sickle-cell disease is nonpolar. Because of the charge difference, the sickle-cell hemoglobin folds differently than normal hemoglobin. The abnormal folding of the protein caused by the mutation results in a change to the sickle shape of the red blood cell. Numerous other diseases involve problems with protein folding, including Alzheimer's disease, cystic fibrosis, diabetes, and cancer.

Causes of mutation Some mutations, especially point mutations, can occur spontaneously. During replication, DNA polymerase sometimes adds the wrong nucleotides. Because the DNA polymerase has a proofreading function, the wrong nucleotide gets added only for one in one hundred thousand bases; it goes unfixed in less than one in one billion.

Certain chemicals and radiation also can damage DNA. Substances which cause mutations are called **mutagens** (MYEW tuh junz). Many different chemicals have been classified as mutagens. Some of these chemicals affect DNA by changing the chemical structure of the bases. Often these changes cause bases to mispair, or bond, with the wrong base. Other chemical mutagens have chemical structures that resemble nucleotides so closely that they can substitute for them. Once these imposter bases are incorporated into the DNA, it can not replicate properly. This type of chemical has become useful medically, especially in the treatment of HIV—the virus that causes AIDS. Many drugs used to treat HIV and other viral infections mimic various nucleotides. Once the drug is incorporated in the viral DNA, the DNA cannot copy itself properly.

VOCABULARY

WORD ORIGIN

Mutagen
comes from the Latin word *mutare,* meaning *to change* and from the Greek word *genes,* meaning *born.*

DATA ANALYSIS LAB 12.2

Based on Real Data*

Interpret the Graph

How can we know if a compound is a mutagen? The Ames test is used to identify mutagens. The test uses a strain of bacteria that cannot make the amino acid histidine. The bacteria are exposed to a suspected mutagen and grow on a medium without histidine. The bacteria that grow have a mutation called a reversion because they reverted to the natural condition of making histidine. The compounds in the graph were Ames tested.

Think Critically

1. **Describe** the relationship between the amount of the compound and the mutation.
2. **Analyze** Which compound is the strongest mutagenic compound?

Data and Observations

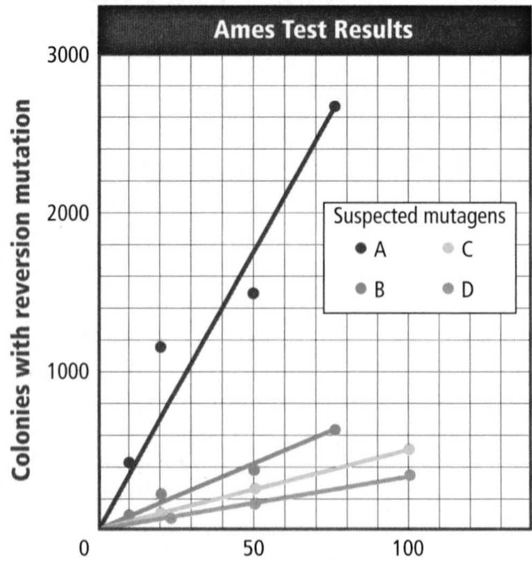

*Data obtained from: Ames, B.N. 1979. Identifying environmental chemicals causing mutations and cancer. *Science* 204: 587-593.

High-energy forms of radiation, such as X rays and gamma rays, are highly mutagenic. When the radiation reaches the DNA, electrons absorb the energy. The electrons can escape their atom, leaving behind a free radical. Free radicals are charged atoms with unpaired electrons that react violently with other molecules, including DNA. Ultraviolet (UV) radiation from the Sun contains less energy than X-ray radiation and does not cause electrons to be ejected from the atoms. However, UV radiation can cause adjacent thymine bases to bind to each other, disrupting the structure of DNA, as shown in **Figure 12.23.** DNA with this structure disruption, or kink, are unable to replicate properly unless repaired.

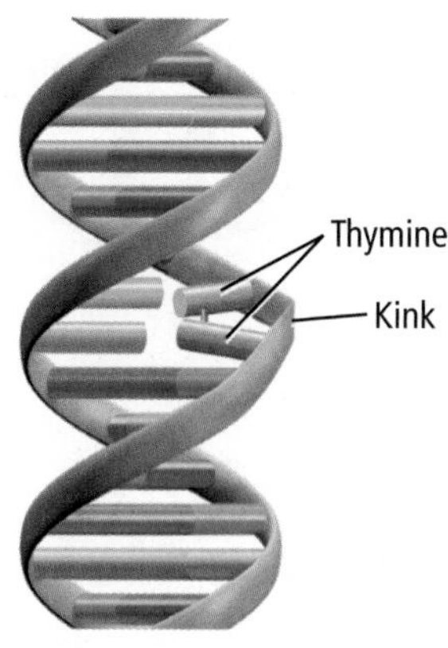

Figure 12.23 Ultraviolet radiation can cause adjacent thymines to bind to each other instead of to their complementary bases, making the DNA "kink" and preventing replication.

Body-cell v. sex-cell mutation When a mutation in a body cell, also called a somatic cell, escapes the repair mechanism, it becomes part of the genetic sequence in that cell and in future daughter cells. Somatic cell mutations are not passed on to the next generation. In some cases, the mutations do not cause problems for the cell. They could be sequences not used by the adult cell when the mutation occurred, the mutation might have occurred in an exon, or the mutation might not have changed the amino acid for which it was coded. These mutations are called neutral mutations. When the mutation results in the production of an abnormal protein, the cell might not be able to perform its normal function, and cell death might occur. In Chapter 9, you learned that mutations in body cells that cause the cell cycle to be unregulated can lead to cancer. All of these effects are contained within the cells of the organism as long as only body cells are affected.

When mutations occur in sex cells, also called germ-line cells, the mutations are passed on to the organism's offspring and will be present in every cell of the offspring. In many cases, these mutations do not affect the function of cells in the organism, though they might affect the offspring drastically. When the mutations result in an abnormal protein in the sex cell, the offspring is impacted. However, the offspring is not impacted when an abnormal protein is produced in an isolated body cell.

Section 12.4 Assessment

Section Summary

- Prokaryotic cells regulate their protein synthesis through a set of genes called operons.
- Eukaryotic cells regulate their protein synthesis using various transcription factors, eukaryotic nucleosome structures, and RNA interference.
- Mutations range from point mutations to the deletion or movement of large sections of the chromosome.
- Mutagens, such as chemicals and radiation, can cause mutations.

Understand Main Ideas

1. MAIN Idea **Relate** gene regulation and mutations.
2. **Identify** the two main types of mutagens.
3. **Diagram** how adding lactose to a culture affects the *lac* operon of *E. coli.*
4. **Analyze** how a point mutation can affect the overall protein shape and function, using hemoglobin as an example.
5. **Compare and contrast** prokaryotic and eukaryotic gene regulation.

Think Critically

6. **Explain** why most mutations in eukaryotes are recessive.
7. **Hypothesize** why DNA replication has such accuracy.

WRITING in **Biology**

8. Write an article describing how Hox genes regulate development in animals.

BioDiscoveries

Unraveling the Double Helix

Moving from the science of death and the atomic bomb, the post-World War II scientific community was eager to explore the science of life—mainly the cell and genetics. An atmosphere of intense competition arose—everyone wanted to be the first to solve the mystery of DNA structure.

Rosalind Franklin took the X-ray diffraction photos used to determine the double-helix structure of DNA.

Building on the past Rosalind Franklin moved to France after the war and learned X-ray diffraction, a technique that uses X rays to produce images of crystalline substances. Though typically used for single-element crystals, Franklin used this technology to take pictures of biological molecules. In January of 1951, Franklin went to King's College to decipher the structure of DNA.

Adding data In the fall of 1951, Franklin discovered that DNA had two forms—dry and wet. She also pioneered a microfocus X-ray camera and a technique to orient the DNA in the beam. She figured out how to extract single DNA strands. Finally, she used long X-ray exposures, some up to 100 hours, to take pictures that revealed keys to DNA structure.

Photo 51 One of Franklin's pictures of the wet DNA was an obvious "X," a characteristic helix diffraction pattern. Franklin thought the dry form would reveal DNA structure, so she put the picture, labeled Photo 51, aside. Early in 1953, Franklin decided to leave King's College to study viral structures. Around this time, James Watson and Francis Crick saw Photo 51 and Franklin's unpublished data. Her co-worker, Maurice Wilkins, was working independently with Watson and Crick, both of whom had been unsuccessful in modeling DNA structure.

The structure solved In March 1953, Watson and Crick published their model of DNA, which was based largely on Franklin's data. In the same issue, Franklin published her findings which supported Watson and Crick's theory. Franklin went on to have a successful career in virology, paving the way for structural virology, the study of the molecular structure of viruses. In 1958, she died of ovarian cancer.

The Nobel Prize In 1962, Watson, Crick, and Wilkins received the Nobel Prize for their discovery of the double-helix structure of DNA. Franklin was ineligible for the prize because she had died. In 1968, Watson admitted in his book *The Double Helix* that they had used her data without her knowledge. Since then, Franklin has been acknowledged as an important contributor to the discovery of DNA structure.

WRITING in Biology

News Article Imagine that you are a reporter in 1953 when the discovery of the double helix is made. Research and write a news article covering the "race to decipher DNA structure" as well as the discovery's implication for science. For more information about the double helix shape of DNA, visit www.biologygmh.com.

BIOLAB

FORENSICS: HOW IS DNA EXTRACTED?

Background: **DNA tests are important for biologists, doctors, and even detectives. Imagine that you are working in a lab where someone has brought a sample of corn from a crime scene to be analyzed. You decide to test the DNA of the corn to look for genes to identify the type of corn. Before the DNA sequence can be examined, the DNA must be extracted.**

Question: *How can DNA be extracted?*

Materials

corn kernels (50 g)
beakers (2)
blender
cheesecloth (4 squares—30 cm on each edge)
rubber band
glass spooling hook
homogenization medium (100–150 mL)
plastic centrifuge tube (30–50 mL)
contact lens cleaning tablet (containing papain)
95% ethanol (12 mL)
distilled water (3 mL)
test tube
container of ice
water bath at 60°C
stirring rod
timer or clock

Safety Precautions

Procedure

1. Read and complete the lab safety form.
2. Carefully weigh out 50 g of corn kernels.
3. Place the corn kernels into a beaker and cover with homogenization medium that has been warmed to 60°C. Place the beaker in a 60°C water bath for 10 min. Gently stir every 45 s.
4. Remove the beaker from the water bath and chill quickly in an ice bath for 5 min.
5. Pour the mixture into a blender and homogenize, or blend, to achieve a consistent texture.
6. Filter the homogenized mixture through four layers of cheesecloth into a clean large beaker on ice.
7. Pour 15 mL of the filtrate into a 30–50 mL plastic centrifuge tube.
8. Dissolve one contact lens cleaning tablet in 3 mL of distilled water in a test tube. Add this to the filtrate tube and mix gently.
9. Hold the filtrate tube at an angle and slowly pour 12 mL of cold 95% ethanol down the side of the tube.
10. Observe the DNA rising into the alcohol layer as a cloudy suspension of white strings. Use a hooked glass rod to spool the DNA, and allow it to dry.
11. **Cleanup and Disposal** Clean your lab area, disposing of chemicals and materials as directed by your teacher. Be sure to wash your hands when you are finished.

Analyze and Conclude

1. **Describe** the appearance of the DNA in suspension and once it has dried.
2. **Explain** why you put the corn kernels into the blender.
3. **Think Critically** Why is it important not to contaminate a sample of DNA that is to be sequenced? How would you know if you had contaminated your sample?

WRITING in **Biology**

Report Imagine you are the first researcher to extract DNA from corn. Write a report detailing your methods and possible applications of your discovery. To learn more about DNA extraction, visit BioLabs at biologygmh.com.

CHAPTER 12

Study Guide

Download quizzes, key terms, and flash cards from biologygmh.com.

FOLDABLES **Assess** the importance of transcription and translation in the central dogma about genes and proteins.

Vocabulary	Key Concepts
Section 12.1 DNA: The Genetic Material	
• double helix (p. 330) • nucleosome (p. 332)	**MAIN Idea** The discovery that DNA is the genetic code involved many experiments. • Griffith's bacterial experiment and Avery's explanation first indicated that DNA is the genetic material. • The Hershey-Chase experiment provided evidence that DNA is the genetic material of viruses. • Chargaff's rule states that, in DNA, the amount of cytosine equals the amount of guanine and the amount of thymine equals the amount of adenine. • The work of Watson, Crick, Franklin, and Wilkins provided evidence of the double-helix structure of DNA.
Section 12.2 Replication of DNA	
• DNA polymerase (p. 334) • Okazaki fragment (p. 334) • semiconservative replication (p. 333)	**MAIN Idea** DNA replicates by making a strand that is complementary to each original strand. • The enzymes DNA helicase, RNA primase, DNA polymerase, and DNA ligase are involved in DNA replication. • The leading strand is synthesized continuously, but the lagging strand is synthesized discontinously, forming Okazaki fragments. • Prokaryotic DNA opens at a single origin of replication, whereas eukaryotic DNA has multiple areas of replication.
Section 12.3 DNA, RNA, and Protein	
• codon (p. 338) • exon (p. 337) • intron (p. 337) • messenger RNA (p. 338) • ribosomal RNA (p. 336) • RNA (p. 336) • RNA polymerase (p. 337) • transcription (p. 337) • transfer RNA (p. 336) • translation (p. 338)	**MAIN Idea** DNA codes for RNA, which guides protein synthesis. • Three major types of RNA are involved in protein synthesis: mRNA, tRNA, and rRNA. • The synthesis of the mRNA from the template DNA is called transcription. • Translation is the process through which the mRNA attaches to the ribosome and a protein is assembled. • In eukaryotes, the mRNA contains introns that are excised before leaving the nucleus. A cap and poly-A tail also are added to the mRNA.
Section 12.4 Gene Regulation and Mutation	
• gene regulation (p. 342) • mutagen (p. 348) • mutation (p. 345) • operon (p. 342)	**MAIN Idea** Gene expression is regulated by the cell, and mutations can affect this expression. • Prokaryotic cells regulate their protein synthesis through a set of genes called operons. • Eukaryotic cells regulate their protein synthesis using various transcription factors, eukaryotic nucleosome structures, and RNA interference. • Mutations range from point mutations to the deletion or movement of large sections of the chromosome. • Mutagens, such as chemicals and radiation, can cause mutations.

CHAPTER 12 Assessment

Section 12.1

Vocabulary Review

Each of the following sentences is false. Make the sentence true by replacing the underlined word with the correct vocabulary term from the Study Guide page.

1. The twisted ladder shape of DNA is called a <u>nucleotide</u>.

2. A <u>double helix</u> consists of DNA wrapped around the histone proteins.

Understand Key Concepts

3. What are the basic building blocks of DNA and RNA?
   A. ribose **C.** nucleotides
   B. purines **D.** phosphorus

4. If a section of DNA has 27 percent thymine, how much cytosine will it have?
   A. 23 percent **C.** 46 percent
   B. 27 percent **D** 54 percent

5. Which was a conclusion of Griffith's work with *Streptococcus pneumoniae?*
   A. DNA is the genetic material in viruses.
   B. The structure of DNA is a double helix.
   C. Bacteria exposed to DNA can incorporate the DNA and change phenotype.
   D. The amount of thymine equals the amount of adenine in DNA.

Refer to the figure below to answer questions 6 and 7.

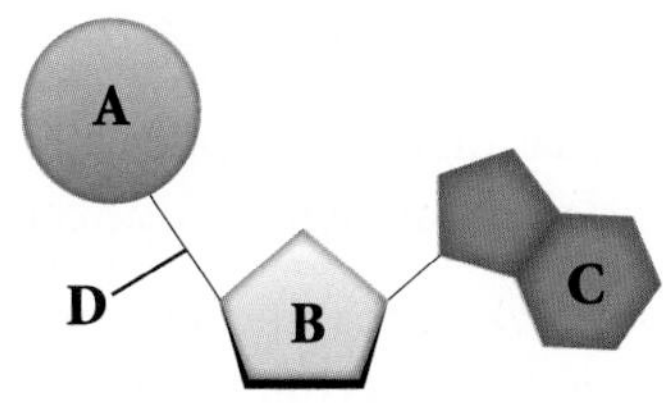

6. What is the entire labeled structure called?
   A. nucleotide **C.** base
   B. RNA **D.** phosphate

7. Which label represents the coding part of DNA?
   A. A **C.** C
   B. B **D.** D

Constructed Response

8. **Short Answer** Explain how DNA forms chromosomes in eukaryotic cells.

Use the figure below to answer question 9.

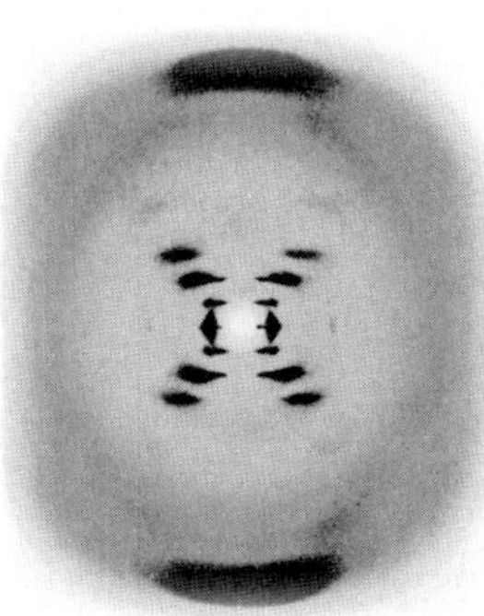

9. **Short Answer** Summarize the experiments and data shown in the photo that led to the discovery of DNA.

Think Critically

10. **Design** How might you use radioactive phosphorus to demonstrate that the transforming compound of bacteria in Griffith's experiment was DNA?

11. **Analyze** How would the results of the Hershey-Chase experiment have been different if protein were the genetic material?

Section 12.2

Vocabulary Review

Write a sentence defining each of the following vocabulary terms.

12. DNA polymerase

13. semiconservative replication

14. Okazaki fragment

Understand Key Concepts

15. With what does the synthesis of a new strand of DNA begin?
    A. RNA primer
    B. nucleotide unit
    C. messenger RNA
    D. transfer RNA

16. Which is true about the elongation of the lagging strand?
   A. does not require a template strand
   B. produces Okazaki fragments
   C. requires the action of RNA ligase
   D. proceeds by continually adding nucleotides to the 3' end

Constructed Response

17. **Short Answer** List the enzymes involved in replication and describe their function.

18. **Short Answer** Summarize the process of DNA replication in a diagram. Add labels to explain what is happening.

Think Critically

Use the figure below to answer questions 19 and 20.

19. **Determine** Imagine you are a scientist looking at a cell through a microscope. You see DNA replicating in several areas. Determine what type of cell you are looking at based on the origins of replication.

20. **Hypothesize** why it is important for the DNA in the figure to have multiple origins of replication.

21. **Infer** how complementary base pairing is responsible for semiconservative replication.

Section 12.3

Vocabulary Review

Write a sentence that connects the vocabulary terms in each pair.

22. mRNA — tRNA

23. codon — RNA polymerase

24. intron — exon

Understand Key Concepts

25. Which correctly lists the changes to eukaryotic pre-mRNA to form mRNA?
   A. cap added, introns excised, and poly T tail added
   B. cap added, exons excised, and poly T tail added
   C. cap added, introns excised, and poly A tail added
   D. cap added, exons excised, and poly A tail added

Use the figure below to answer questions 26 and 27.

26. What is the mRNA sequence for the template strand DNA sequence in the figure?
   A. 5' ATGTTTGATCTT 3'
   B. 5' AUGUUUGAUCUU 3'
   C. 5' TACAAACTAGAA 3'
   D. 5' UACAAACUAGAA 3'

27. What is the sequence for the nontemplate strand of the DNA in the figure?
   A. 5' ATGTTTGATCTT 3'
   B. 5' AUGUUUGAUCUU 3'
   C. 5' TACAAACTAGAA 3'
   D. 5' UACAAACUAGAA 3'

Constructed Response

28. **Short Answer** Compare and contrast transcription and translation and indicate where they occur in prokaryotic cells and eukaryotic cells.

29. **Short Answer** Describe the experiment that led to the One Gene–One Enzyme hypothesis.

Think Critically

30. **Identify** the mRNA sequence and orientation if the nontemplate strand has the sequence 5' ATGCCAGTCATC 3'. Use **Figure 12.14** to determine the amino acid sequence coded by the mRNA.

Section 12.4

Vocabulary Review

Write the vocabulary term from the Study Guide page that describes each of the following processes.

31. regulation of a prokaryotic genome

32. control of the functional units of DNA

33. changes in DNA sequence

Understand Key Concepts

34. Which demonstrates an insertion mutation of the sequence 5' GGGCCCAAA 3'?
- **A.** 5' GGGGCCAAA 3'
- **B.** 5' GGGCCAAA 3'
- **C.** 5' GGGAAACCC 3'
- **D.** 5' GGGCCCAAAAAA 3'

35. Which is true about eukaryotic gene regulation?
- **A.** Eukaryotic gene regulation is exactly like prokaryotic gene regulation.
- **B.** Replication factors guide the binding of eukaryotic RNA polymerase to the promoter.
- **C.** Activator proteins fold DNA to enhancer sites that increase the rate of gene transmission.
- **D.** Repressor proteins bind to activators, preventing them from binding to the DNA.

36. Which is not a type of mutation?
- **A.** base substitutions
- **B.** insertions
- **C.** RNA interference
- **D.** translocation

Constructed Response

37. Short Answer Illustrate the effect of adding tryptophan to a culture of *E. coli*.

38. Short Answer Describe RNA interference.

Think Critically

39. Infer why base substitutions in the third position are least likely to cause a change in the amino acid for which it coded.

40. Hypothesize how it might be possible for bacteria to respond to environmental stress by increasing the rate of mutations during cell division.

Additional Assessment

41. *WRITING in* **Biology** The book *Jurassic Park* by Michael Crichton presents the idea of isolating DNA from extinct organisms and "resurrecting" them. If this were possible, should this be done? Defend your opinion in an essay.

DBQ Document-Based Questions

Data obtained from: Watson, J.D. and Crick, F.H.. 1953. Molecular Structure of Nucleic Acids. *Nature* 171: 737-738.

The following excerpts are from Watson and Crick's description of the structure of DNA.

"The novel feature of the structure is the manner in which the two chains are held together by the purine and pyrimidine bases. The planes of the bases are perpendicular to the fibre axis. They are joined together in pairs, a single base from one chain being hydrogen-bonded to a single base from the other chain so that the two lie side by side with identical z-co-ordinates. One of the pair must be a purine and the other a pyrimidine for bonding to occur."

"It has not escaped our notice that the specific pairing we have postulated immediately suggests a possible copying mechanism for the genetic material."

42. Draw a diagram of DNA structure based on the description above.

43. According to the description, how are the bases joined together?

44. What did Watson and Crick see as a possible copying mechanism?

Cumulative Review

45. Explain why species diversity is so great in estuaries and coral reefs. **(Chapter 3)**

46. Under what conditions does the exponential phase of logistic growth occur? **(Chapter 4)**

47. Describe the process by which gametes are produced. **(Chapter 10)**

Standardized Test Practice

Cumulative

Multiple Choice

1. Which macromolecule can be formed using the sugars produced by plants during photosynthesis?
   A. cellulose
   B. DNA
   C. lipid
   D. protein

Use the diagram below to answer questions 2 and 3.

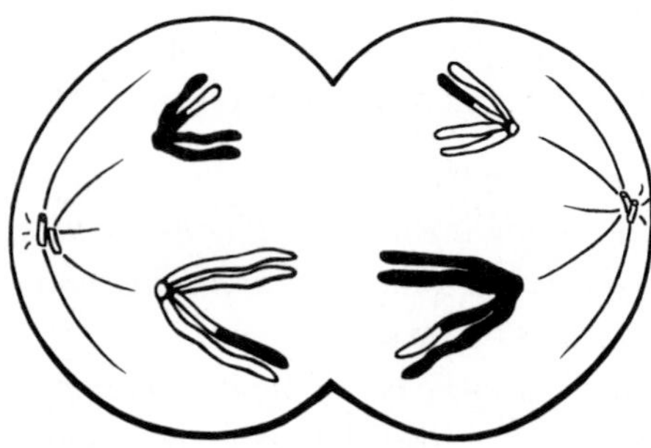

2. Which stage of meiosis is represented in the diagram?
   A. anaphase I
   B. anaphase II
   C. metaphase I
   D. metaphase II

3. Which process can take place during the stage of meiosis that follows the stage in the diagram?
   A. change to diploid
   B. crossing over
   C. cytokinesis
   D. DNA replication

4. What enzyme is responsible for "unzipping" the DNA strand during replication?
   A. DNA helicase
   B. DNA ligase
   C. DNA polymerase
   D. RNA primase

Use the illustration below to answer question 5.

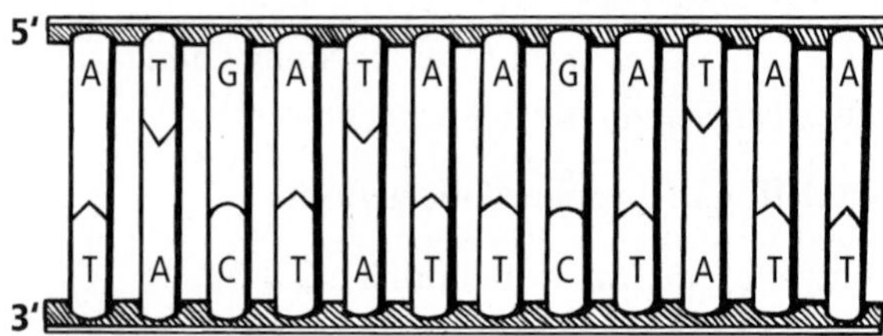

5. Which sequence is possible for mRNA formed from the DNA strand shown in the illustration?
   A. 5'AATAGAATAGTA3'
   B. 5'AAUAGAAUAGUA3'
   C. 5'ATGATAAGATAA3'
   D. 5'AUGAUAAGAUAA3'

6. Which cells would likely undergo apoptosis?
   A. cells between fingers
   B. cells reproducing normally
   C. cells reproducing slowly
   D. cells surrounding the heart

7. Which genotype could be the one of a person whose blood type is A?
   A. I^BI^B
   B. ii
   C. I^Ai
   D. I^AI^B

8. Which sex chromosomes are present in a person with Kleinfelter Syndrome?
   A. OY
   B. XO
   C. XXY
   D. XYY

Short Answer

9. Using the law of independent assortment, describe a dihybrid cross of heterozygous yellow, round-seed pea plants (YyRr). Include a Punnett square and phenotype ratios in your response.

10. Give an example of a technological development, and explain how it contributed to scientists' understanding of the structure of DNA.

11. Which probably causes the coat color variations that occur only in the females of a certain animal? Give a reason to support your conclusion.

12. Suppose you perform a dihybrid cross between two organisms with the genotype RrYy. What percentage of the offspring would be homozygous for both traits? Explain how you determined the answer.

13. Why do you think Mendel's work preceded the search for molecules involved in inheritance?

14. Suppose an organism (with a chromosome number of $2n = 6$) has monosomy of chromosome 3. How many chromosomes are in the organism's karyotype? Explain your answer.

15. Explain why the number of bases in a strand of mRNA can be different from the number in the DNA from which it was synthesized.

16. Explain why a hypothesis must be testable.

Extended Response

Use the figure below to answer question 17.

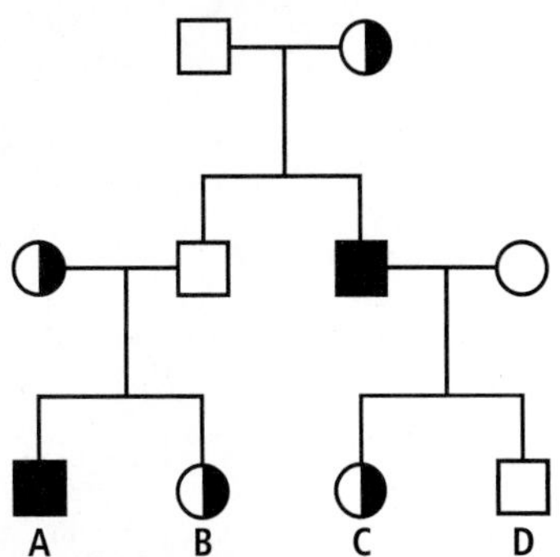

17. Describe the pattern of inheritance of the disease tracked in the pedigree above.

18. Human nerve cells seldom divide after they are formed. Evaluate how this might affect a person with a spinal cord injury.

19. Explain the role that publication of findings had in the discovery of DNA's structure.

Essay Question

For certain kinds of research studies, scientists recruit pairs of twins to be participants or subjects of the research. They might recruit identical or fraternal twins, depending on the focus of the study. Twins can be particularly helpful in studies about genetics and heredity.

Using the information in the paragraph above, answer the following question in essay format.

20. Imagine you are a research scientist. Write a plan for a research study that would require participants to be twins. Explain what you are trying to learn, whether you are looking for identical or fraternal twins, and why it is important to have twins as subjects for your study.

NEED EXTRA HELP?																				
If You Missed Question . . .	1	2	3	4	5	6	7	8	9	10	11	12	13	14	15	16	17	18	19	20
Review Section . . .	8.2	10.1	10.1	12.2	12.3	9.3	11.2	11.3	11.2	12.2	11.2	10.2	12.2	11.3	11.2	12.3	11.1	9.1	12.1	11.2

CHAPTER 13 Genetics and Biotechnology

BIG Idea Genetic technology improves human health and quality of life.

Section 1
Applied Genetics
MAIN Idea Selective breeding is used to produce organisms with desired traits.

Section 2
DNA Technology
MAIN Idea Researchers use genetic engineering to manipulate DNA.

Section 3
The Human Genome
MAIN Idea Genomes contain all of the information needed for an organism to grow and survive.

BioFacts

- The human genome consists of approximately 20,000–25,000 genes.
- Biotechnology enables scientists to study individual genes as well as the entire genome of an organism.
- Mutated genes might be repaired with the use of genetic engineering.

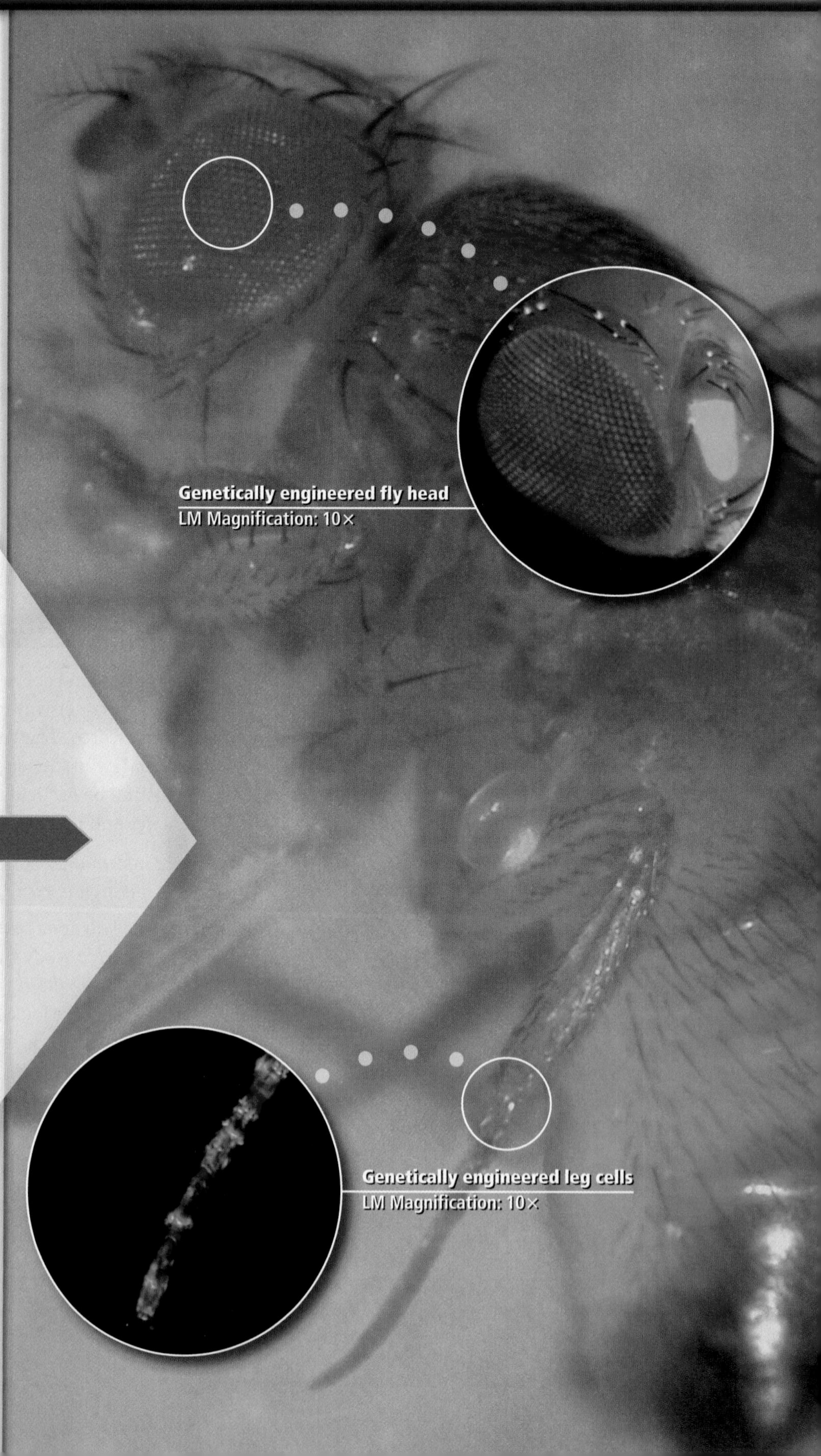

Genetically engineered fly head
LM Magnification: 10×

Genetically engineered leg cells
LM Magnification: 10×

LAUNCH Lab

How does selective breeding work?

A deck of cards can represent the genome of a population of organisms. In this lab, you will model selective breeding to create a population of cards with similar suits.

Procedure

1. Read and complete the lab safety form.
2. Shuffle a **deck of cards.** Choose one suit to represent the gene you wish to select.
3. Lay the entire deck face up in 26 pairs.
4. Select the pairs that contain at least one card from your chosen suit.
5. Record the number of cards remaining and calculate the percentage of cards not selected from the starting pile.
6. Shuffle the remaining cards and repeat Steps 2–4 until all of your cards are of the suit you selected.

Analysis

1. **Infer** why the cards were laid out in pairs.
2. **Relate** changes in the percentage of cards discarded after each round to how the percentage of genes might change in a population.

Visit biologygmh.com to:

- study the entire chapter online
- explore the Interactive Time Line, Concepts in Motion, Interactive Tables, Microscopy Links, Virtual Labs, and links to virtual dissections
- access Web links for more information, projects, and activities
- review content online with the Interactive Tutor, and take Self-Check Quizzes

Recombinant DNA Make this Foldable to help you sequence and describe DNA tools.

STEP 1 Fold a sheet of notebook paper in half lengthwise so that the side without the holes is 2.5 cm shorter than the side with holes.

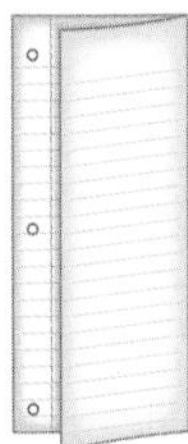

STEP 2 Fold the folded paper into thirds.

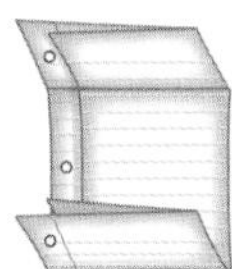

STEP 3 Unfold once and cut along the two fold lines of the top layer only. This will make three tabs.

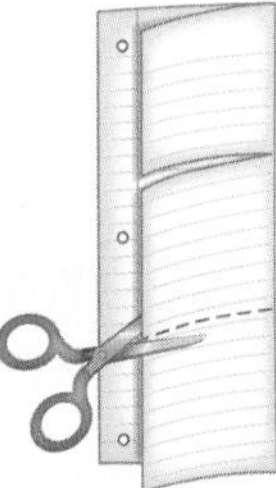

STEP 4 Label the tabs as illustrated.

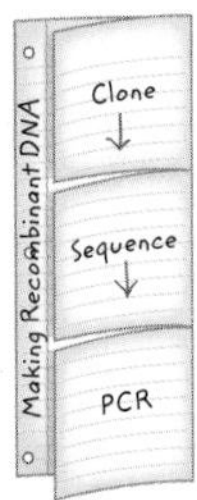

FOLDABLES **Use this Foldable with Section 13.2.** As you study the section, summarize what happens when using DNA tools under the appropriate tabs.

Section 13.1

Reading Preview

Objectives

- **Describe** how selective breeding is used to produce organisms with desired traits.
- **Compare** inbreeding and hybridization.
- **Assess** the genotypes of organisms using a Punnett square test cross.

Review Vocabulary

hybrid: organism that is heterozygous for a particular trait

New Vocabulary

selective breeding
inbreeding
test cross

Applied Genetics

MAIN Idea **Selective breeding is used to produce organisms with desired traits.**

Real-World Reading Link Coin collectors separate rare coins from all other coins because the rare ones are more valuable. Just as certain coins are selected for their value, certain plants and animals have been selected and bred to produce organisms with traits that are valuable to humans.

Selective Breeding

You might be familiar with different breeds of dogs, such as Saint Bernards, huskies, and German shepherds. Observe some of the phenotypic traits of these breeds in **Figure 13.1.** All three have strong, muscular bodies. Saint Bernards have traits such as a keen sense of smell that make them good rescue dogs. Huskies are endurance runners and pull sleds long distances. German shepherds are highly trainable for special services.

Since ancient times, humans have bred animals with certain traits to obtain offspring that have desired traits. As a result, these traits become more common. Breeding for desired traits is not restricted to animals alone. Plants also are bred to produce desired traits, such as larger fruits and shorter growing times. The process by which desired traits of certain plants and animals are selected and passed on to their future generations is called **selective breeding.** Through the processes of hybridization and inbreeding, desired traits can be passed on to future generations.

Figure 13.1 Dogs have traits that make them suited for different tasks: Saint Bernard—keen sense of smell; husky—endurance to run long distances; and German shepherd—high trainability.

Saint Bernard
Rescue dog

Husky
Sled dog

German shepherd
Service dog

Hybridization Recall from Chapter 10 that crossing parent organisms with different forms of a trait to produce offspring with specific traits results in hybrids. Farmers, animal breeders, scientists, and gardeners widely use the production of hybrids, also known as hybridization. They select traits that will give hybrid organisms a competitive edge. These hybrid organisms can be bred to be more disease-resistant, to produce more offspring, or to grow faster. For example, plant breeders might choose to cross two different varieties of tomato plants in order to produce a hybrid that has both the disease resistance of one parent and the fast growth rate of the other parent.

Care must be taken to identify organisms with desired traits and successfully cross them to yield the right combination of traits from both parents. A disadvantage of hybridization is that it is time consuming and expensive. For example, it took rice breeders three decades to produce hybrid rice varieties that can produce higher yields than nonhybrid varieties. Because hybrids can be bred to be more nutritious, to have the ability to adapt to a wide range of changes in the environment, and to produce greater numbers of offspring, the advantages of hybridization sometimes outweigh the disadvantages.

Inbreeding Once a breeder observes a desired trait in an organism, a process is needed to ensure that the trait is passed on to future generations. This process, in which two closely related organisms are bred to have the desired traits and to eliminate the undesired ones in future generations, is called **inbreeding.**

Pure breeds are maintained by inbreeding. Clydesdale horses, Angus cattle, and German shepherd dogs are all examples of organisms produced by inbreeding. You might have seen Clydesdale horses at parades and petting zoos. Horse breeders first bred the Clydesdale horse in Scotland hundreds of years ago for use as a farm horse. Because of their strong build, agility, and obedient nature, Clydesdales originally were inbred and used extensively for pulling heavy loads.

A disadvantage of inbreeding is that harmful recessive traits also can be passed on to future generations. Inbreeding increases the chance of homozygous recessive offspring. If both parents carry the recessive allele, the harmful trait likely will not be eliminated.

Reading Check **Explain** the difference between hybridization and inbreeding.

MiniLab 13.1

Model Hybridization

How are hybrid lilies produced? In this lab, you will examine techniques used by both professional plant breeders and amateur gardeners to produce the wide variety of lilies you might see growing in landscaped areas.

Procedure

1. Read and complete the lab safety form.
2. Obtain a **labeled drawing of a lily flower** and a **fresh open lily flower.** Examine the flower with a **hand lens** and identify the male anthers and the female pistil.
3. Use a **cotton swab** to gently rub an anther to pick up pollen.
4. Trade flowers with another lab group and, using the cotton swab, gently apply the pollen from your flower to the stigma of the pistil of the new flower.

Analysis

1. **Infer** When breeders hybridize lilies, they transfer pollen to the stigma of an unopened lily flower and then cover the stigma with a foil cap. Why do you think this would be necessary?
2. **Think Critically** A breeder produces a hybrid lily which then is allowed to grow and produce seeds naturally. When these seeds are planted, the new lily plants do not have the same characterisitics as the hybrid parent. Hypothesize why this would occur.

LAUNCH Lab

Review Based on what you've read about selective breeding, how would you now answer the analysis questions?

Homozygous white grapefruit

Homozygous red grapefruit	W	W
w	Ww	Ww
w	Ww	Ww

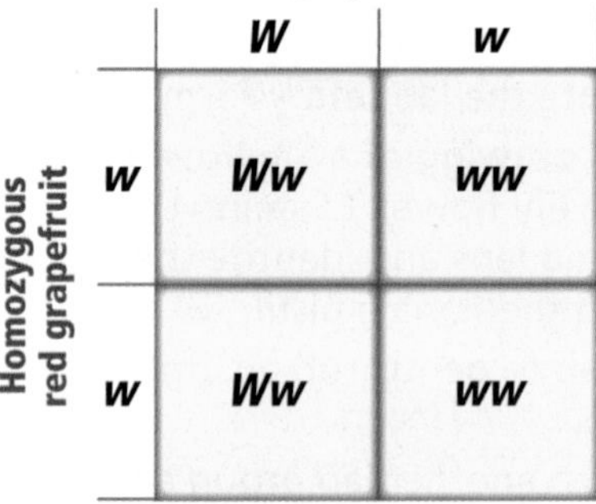

■ **Figure 13.2** The genotype of a white grapefruit tree can be determined by the results of a test cross with a homozygous red grapefruit.

Test Cross

An important thing a breeder has to determine when producing a hybrid is the genotype of the hybrid. Once a breeder observes the desired trait, if the trait is dominant, then the genotype of the organism could be homozygous dominant or heterozygous. The exact genotype is determined by performing a test cross. A **test cross** involves breeding an organism that has the unknown genotype with one that is homozygous recessive for the desired trait. If the parent's genotype is homozygous dominant, all the offspring will have the dominant phenotype; if it is heterozygous, the offspring will show a 1:1 phenotypic ratio.

Performing a test cross Suppose a breeder wants to produce hybrid white grapefruits. In grapefruit trees, white color is the dominant trait, while red is recessive. Therefore, the red grapefruit trees in the orchard must be homozygous recessive (*ww*). The genotype of the hybrid white grapefruit tree obtained by the breeder can be homozygous dominant (*WW*) or heterozygous (*Ww*) for the white color. Therefore, the breeder has to perform a test cross to determine the genotype of the white grapefruit tree. Recall from Chapter 10 that when performing a cross, pollen from the flower of one plant is transferred to the female organ in a flower of another plant.

Results As shown in the top Punnett square in **Figure 13.2,** if the white grapefruit tree is homozygous dominant (*WW*) and is crossed with a red grapefruit tree (*ww*), then all the offspring will be heterozygous (*Ww*) and white in color. In this case, all of the offspring will have the dominant phenotype. However, as shown in the second Punnett square in **Figure 13.2,** if the white grapefruit tree is heterozygous (*Ww*), then half the number of offspring will be white and half will be red, and the phenotypic ratio will be 1:1. Review the results in the Punnett squares in **Figure 13.2.** If the white grapefruit tree is homozygous, all offspring will be heterozygous—white in color. If the tree is heterozygous, half of the test-cross offspring will be white and half will be red.

Section 13.1 Assessment

Section Summary

- Selective breeding is used to produce organisms with traits that are beneficial to humans.
- Hybridization produces organisms with desired traits from parent organisms with different traits.
- Inbreeding creates pure breeds.
- A test cross can be used to determine an organism's genotype.

Understand Main Ideas

1. MAIN Idea **Assess** the effect of selective breeding on food crops.
2. **Describe** three traits that might be desired in sheep. How can these traits be passed on to the next generation? Explain.
3. **Compare and Contrast** inbreeding and hybridization.
4. **Predict** the phenotype of offspring from a test cross between a seedless orange (*ss*) and an orange with seeds (*Ss*).

Think Critically

5. **Evaluate** Should a cow and a bull that both carry recessive alleles for a mutation that causes decreased milk production be bred? Why or why not?

MATH in Biology

6. A breeder performs a test cross to determine the genotype of a black cat. He crosses the black cat (*BB* or *Bb*) with a white cat (*bb*). If 50 percent of the offspring are black, what is the genotype of the black cat?

Biology Online Self-Check Quiz biologygmh.com

Section 13.2

DNA Technology

Reading Preview

Objectives

- **Describe** how genetic engineering manipulates recombinant DNA.
- **Compare** selective breeding to genetic engineering.
- **Summarize** how genetic engineering can be used to improve human health.

Review Vocabulary

DNA: the genetic material of all organisms, composed of two complementary chains of nucleotides wound in a double helix

New Vocabulary

genetic engineering
genome
restriction enzyme
gel electrophoresis
recombinant DNA
plasmid
DNA ligase
transformation
cloning
polymerase chain reaction
transgenic organism

MAIN Idea **Researchers use genetic engineering to manipulate DNA.**

Real-World Reading Link Have you seen a handmade patchwork quilt? Patchwork quilts are created by combining different pieces of fabric. Scientists use a similar process and combine DNA from different sources to create an organism with unique traits.

Genetic Engineering

By about 1970, researchers had discovered the structure of DNA and had determined the central dogma that information flowed from DNA to RNA and from RNA to proteins. However, scientists did not know much about the function of individual genes. Suppose your friend told you the final score of a high school football game but did not tell you how each player contributed to the game. Your curiosity about the details of the game is similar to the curiosity scientists experienced because they did not know how each gene contributed to a cell's function.

The situation changed when scientists began using **genetic engineering,** technology that involves manipulating the DNA of one organism in order to insert exogenous DNA (the DNA of another organism). For example, researchers have inserted a gene for a bioluminescent protein called green fluorescent protein (GFP) into various organisms. GFP, which is a substance naturally found in jellyfishes that live in the north Pacific Ocean, emits a green light when it is exposed to ultraviolet light. Organisms that have been genetically engineered to synthesize the DNA for GFP, such as the mosquito larvae shown in **Figure 13.3,** can be easily identified in the presence of ultraviolet light. The GFP DNA is attached to exogenous DNA to verify that the DNA has been inserted into the organism. These genetically engineered organisms are used in various processes, such as studying the expression of a particular gene, investigating cellular processes, studying the development of a certain disease, and selecting traits that might be beneficial to humans.

Magnification: unavailable

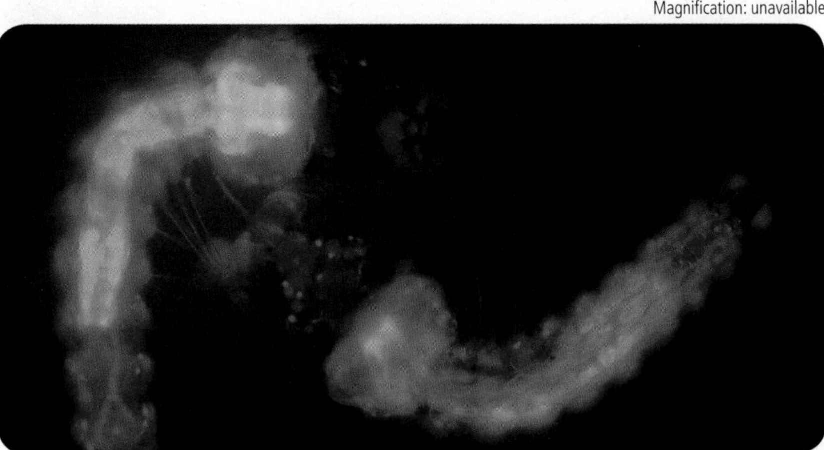

Genetically engineered mosquito larvae

Figure 13.3 The gene for green fluorescent protein (GFP) was introduced into mosquito larvae so that researchers could verify that exogenous DNA was inserted.
Predict *how genetic engineering might be used in the future by the medical field.*

VOCABULARY

ACADEMIC VOCABULARY

Manipulate:
to manage or utilize skillfully
Scientists use technology to manipulate genetic information in order to test scientific hypotheses.

DNA Tools

You have learned that selective breeding is used to produce plants and animals with desired traits. Genetic engineering can be used to increase or decrease the expression of specific genes in selected organisms. It has many applications from human health to agriculture.

An organism's **genome** is the total DNA present in the nucleus of each cell. As you will learn in the next section, genomes, such as the human genome, can contain millions and millions of nucleotides. In order to study a specific gene, DNA tools can be used to manipulate DNA and to isolate genes from the rest of the genome.

Restriction enzymes Some types of bacteria contain powerful defenses against viruses. These cells contain proteins called **restriction enzymes** that recognize and bind to specific DNA sequences and cleave the DNA within that sequence. A restriction enzyme, also called an endonuclease (en doh NEW klee ayz), cuts the viral DNA into fragments after it enters the bacteria. Since their discovery in the late 1960s, scientists have identified and isolated hundreds of restriction enzymes. Restriction enzymes are used as powerful tools for isolating specific genes or regions of the genome. When the restriction enzyme cleaves genomic DNA, it creates fragments of different sizes that are unique to every individual.

EcoRI One restriction enzyme that is used widely by scientists is known as *EcoRI.* As illustrated in **Figure 13.4,** *EcoRI* specifically cuts DNA containing the sequence GAATTC. The ends of the DNA fragments created by *EcoRI* are called sticky ends because they contain single-stranded DNA that is complementary. The ability of some restriction enzymes to create fragments with sticky ends is important because these sticky ends can be joined together with other DNA fragments that have complementary sticky ends.

 **Reading Check** **Generalize** how restriction enzymes are used.

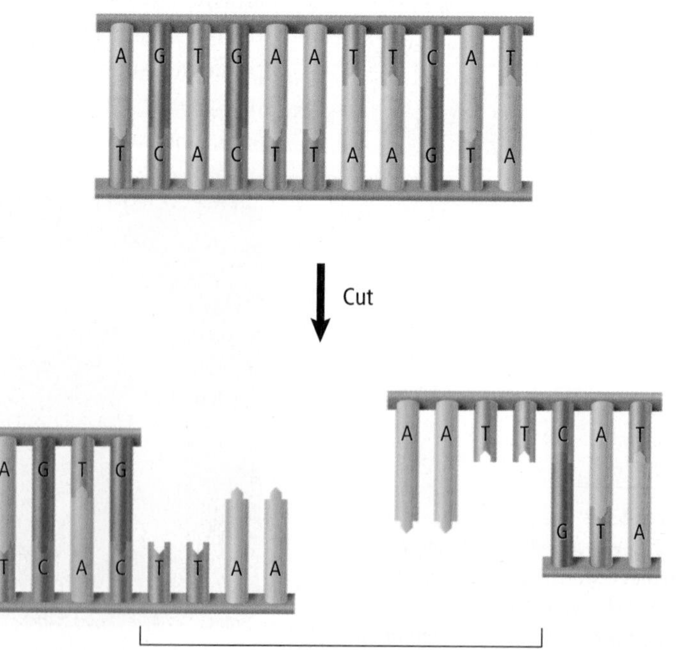

Figure 13.4 DNA containing the sequence GAATTC can be cut by the restriction enzyme *EcoRI* to produce sticky ends.

Concepts In Motion

Interactive Figure To see an animation of restriction enzymes cleaving strands of DNA between specific nucleotides, visit biologygmh.com.

Loading the gel Solution containing DNA is dropped into holes at one end of the gel with a pipette.

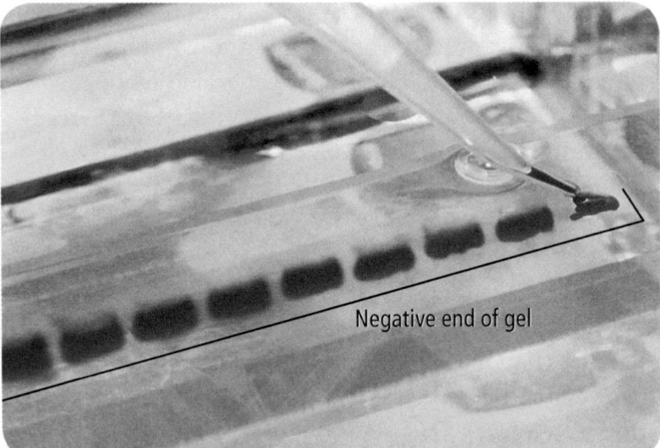

Fragment pattern A staining solution binds to the separated DNA fragments in the gel, making them visible under ultraviolet light.

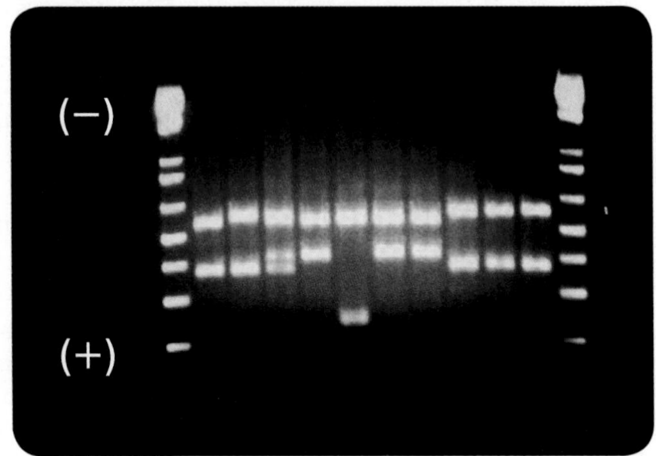

Figure 13.5 When the loaded gel is placed in an electrophoresis tank and the electric current is turned on, the DNA fragments separate.

However, not all restriction enzymes create sticky ends. Some enzymes produce fragments containing blunt ends—created when the restriction enzyme cuts straight across both strands. Blunt ends do not have regions of single-stranded DNA and can join to any other DNA fragment with blunt ends.

Connection to Physics **Gel electrophoresis** An electric current is used to separate the DNA fragments according to the size of the fragments in a process called **gel electrophoresis. Figure 13.5** shows how the DNA fragments are loaded on the negatively charged end of a gel. When an electric current is applied, the DNA fragments move toward the positive end of the gel. The smaller fragments move farther faster than the larger ones. The unique pattern created based on the size of the DNA fragment can be compared to known DNA fragments for identification. Also, portions of the gel containing each band can be removed for further study.

MiniLab 13.2

Model Restriction Enzymes

How are sticky ends modeled? Use scissors and tape to produce paper DNA fragments with sticky ends and a recombinant DNA plasmid.

Procedure

1. Read and complete the lab safety form.
2. Obtain one **straight paper DNA sequence,** which will represent genomic DNA, and one **circular paper DNA sequence,** which will represent a plasmid.
3. Find each GAATTC sequence recognized by the restriction enzyme *EcoRI* and cleave the genome and plasmid DNA using **scissors.**
4. Use **tape** to make a recombinant DNA plasmid.

Analysis

1. **Analyze and Conclude** Compare your plasmid to those made by other lab groups. How many different recombinant plasmids could be made using this particular genomic sequence? Explain.
2. **Infer** What enzyme did the scissors represent? Explain.

Recombinant DNA Technology

When DNA fragments have been separated by gel electrophoresis, fragments of a specific size can be removed from the gel and combined with DNA fragments from another source. This newly generated DNA molecule, with DNA from different sources, is called **recombinant DNA.** Recombinant DNA technology has revolutionized the way scientists study DNA because it enables individual genes to be studied.

Large quantities of recombinant DNA molecules are needed in order to study them. A carrier, called a vector, transfers the recombinant DNA into a bacterial cell called the host cell. Plasmids and viruses are commonly used vectors. **Plasmids**—small, circular, double-stranded DNA molecules that occur naturally in bacteria and yeast cells—can be used as vectors because they can be cut with restriction enzymes. If a plasmid and a DNA fragment obtained from another genome have been cleaved by the same restriction enzyme, the ends of each DNA fragment will be complementary and can be combined, as shown in **Figure 13.6.** An enzyme normally used by cells in DNA repair and replication, called **DNA ligase,** joins the two DNA fragments chemically. Ligase joins DNA fragments that have sticky ends as well as those that have blunt ends.

Examine **Figure 13.6** again. Notice that the resulting circular DNA molecule contains the plasmid DNA and the DNA fragment isolated from another genome. This recombinant plasmid DNA molecule now can be inserted into a host cell so that large quantities of this type of recombinant DNA can be made.

 **Reading Check** **Relate** restriction enzymes to recombinant DNA.

Figure 13.6 Recombinant DNA is created by joining together DNA from two different sources.

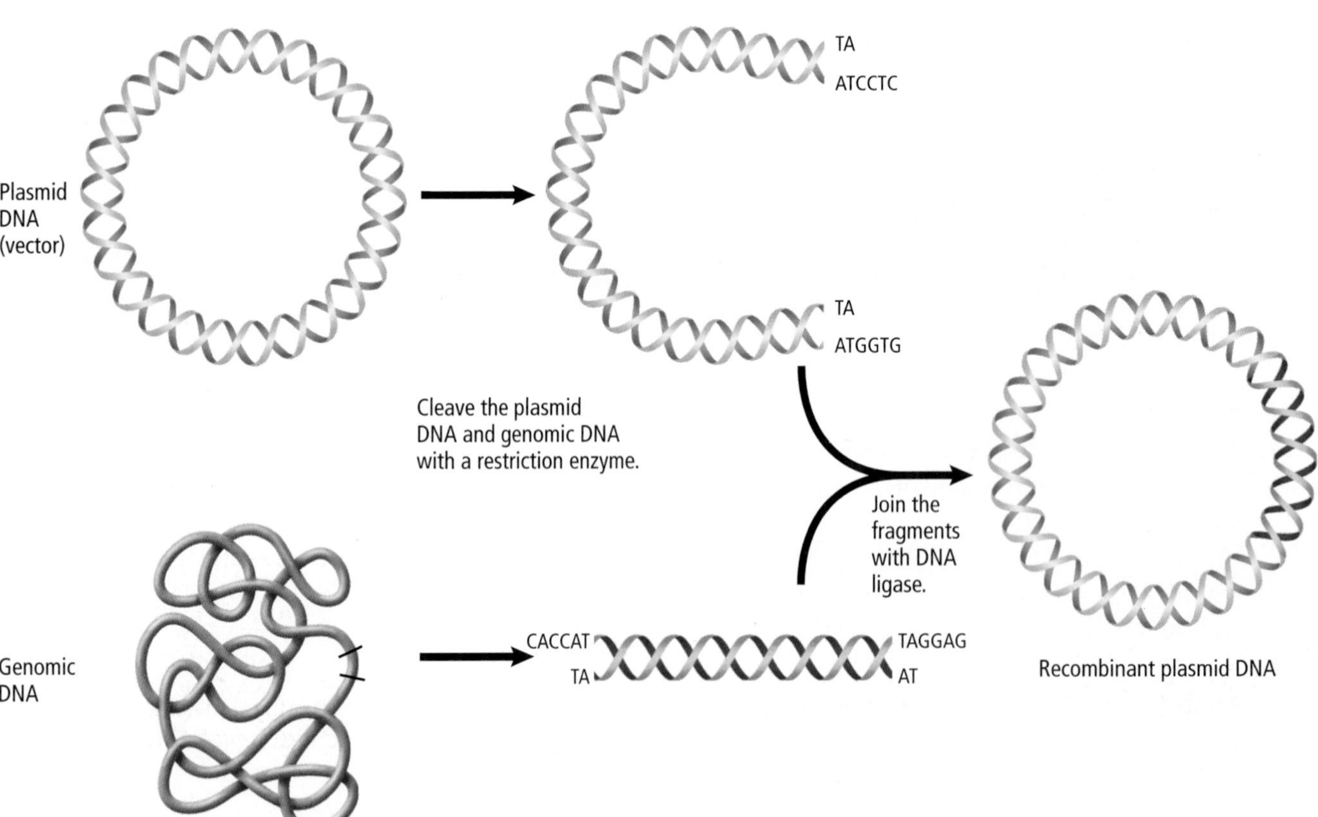

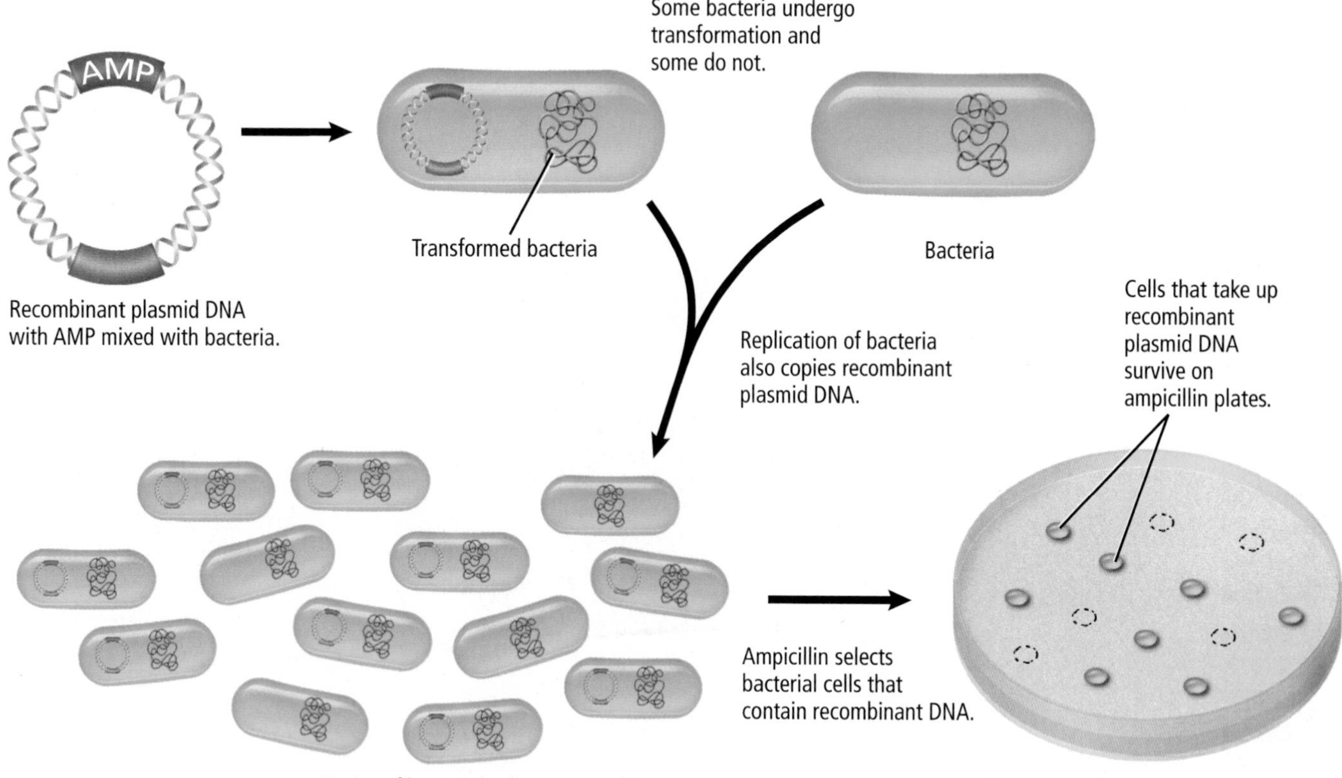
AMP
Some bacteria undergo transformation and some do not.
Transformed bacteria
Bacteria
Recombinant plasmid DNA with AMP mixed with bacteria.
Replication of bacteria also copies recombinant plasmid DNA.
Cells that take up recombinant plasmid DNA survive on ampicillin plates.
Ampicillin selects bacterial cells that contain recombinant DNA.
Copies of bacterial cells

■ **Figure 13.7** Clones containing copies of the recombinant DNA can be identified and used for further study when the bacterial cells that do not contain recombinant DNA die.

Gene cloning To make a large quantity of recombinant plasmid DNA, bacterial cells are mixed with recombinant plasmid DNA. Some of the bacterial cells take up the recombinant plasmid DNA through a process called **transformation,** as shown in **Figure 13.7.** Bacterial cells can be transformed using electric pulsation or heat. Recall that all cells, including bacterial cells, have plasma membranes. A short electric pulse or a brief rise in temperature temporarily creates openings in the plasma membrane of the bacteria. These temporary openings allow small molecules, such as the recombinant plasmid DNA, to enter the bacterial cell. The bacterial cells make copies of the recombinant plasmid DNA during cell replication. Large numbers of identical bacteria, each containing the inserted DNA molecules, can be produced through this process called **cloning.**

Recombinant plasmid DNA contains a gene that codes for resistance to an antibiotic such as ampicillin (AMP). Researchers use this gene to distinguish between bacterial cells that have taken up the recombinant plasmid DNA and those that have not. Notice in **Figure 13.7** that when the transformed bacterial cells are exposed to the specific antibiotic, only the bacterial cells that have the plasmid survive.

DNA sequencing The sequence of the DNA nucleotides of most organisms is unknown. Knowing the sequence of an organism's DNA or of a cloned DNA fragment provides scientists with valuable information for further study. The sequence of a gene can be used to predict the function of the gene, to compare genes with similar sequences from other organisms, and to identify mutations or errors in the DNA sequence. Because the genomes of most organisms are made up of millions of nucleotides, the DNA molecules used for sequencing reactions first must be cut into smaller fragments using restriction enzymes.

VOCABULARY

SCIENCE USAGE V. COMMON USAGE

Transformation

Science usage: process by which one type of bacterium takes up the DNA from another source. *Transformation of bacteria involves the uptake of plasmid DNA.*

Common usage: the act of change *The transformation of the room was complete with the addition of new drapes.*

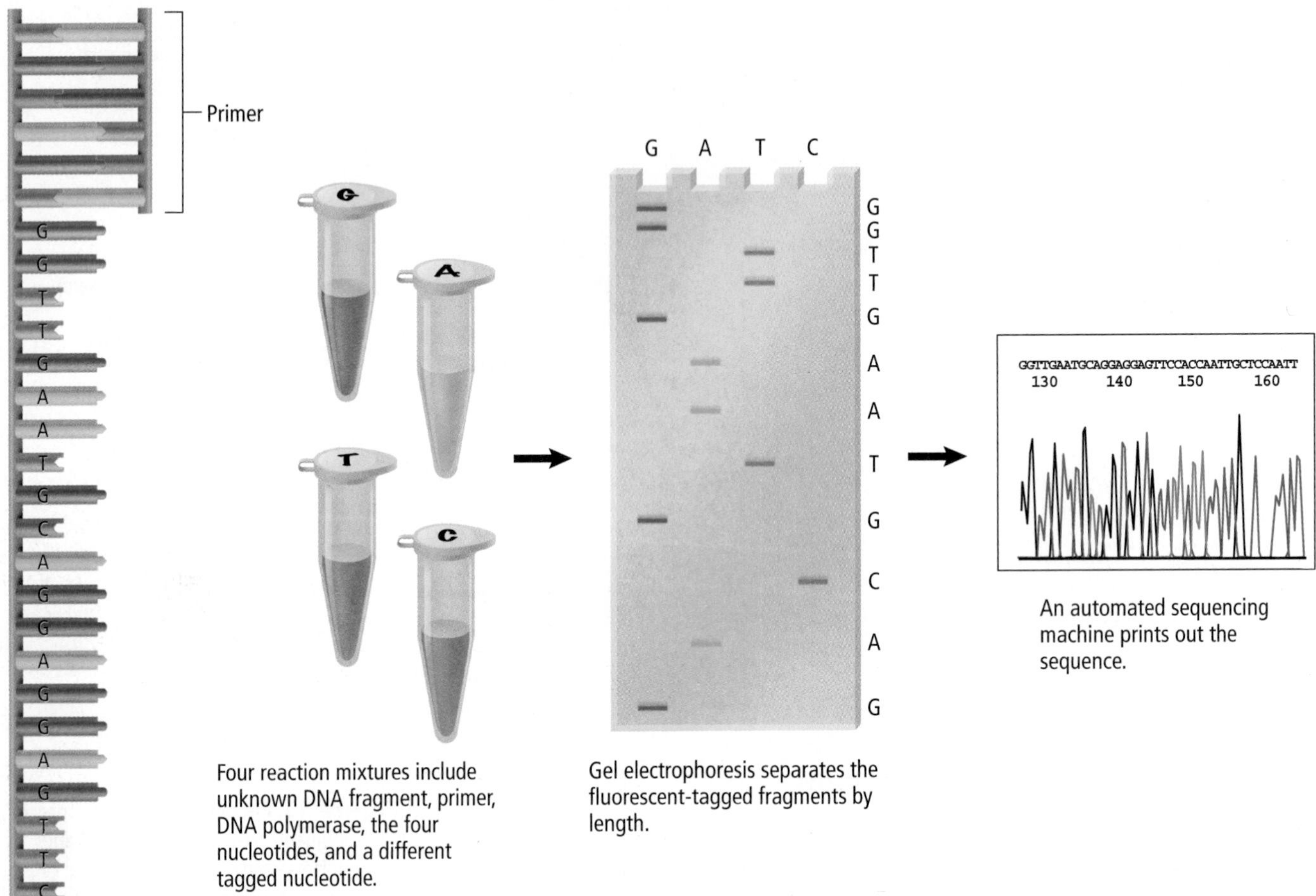

■ **Figure 13.8** DNA can be sequenced using fluorescent-tagged nucleotides.
Describe ***how the sequence of the original DNA template is determined.***

Follow **Figure 13.8** to understand how DNA is sequenced. Scientists mix an unknown DNA fragment, DNA polymerase, and the four nucleotides—A, C, G, T in a tube. A small amount of each nucleotide is tagged with a different color of fluorescent dye, which also modifies the structure of the nucleotide. Every time a modified fluorescent-tagged nucleotide is incorporated into the newly synthesized strand, the reaction stops. This produces DNA strands of different lengths. The sequencing reaction is complete when the tagged DNA fragments are separated by gel electrophoresis. The gel is then analyzed in an automated DNA sequencing machine that detects the color of each tagged nucleotide. The sequence of the original DNA template is determined from the order of the tagged fragments.

Polymerase chain reaction Once the sequence of a DNA fragment is known, a technique called the **polymerase chain reaction** (PCR) can be used to make millions of copies of a specific region of a DNA fragment. PCR is extremely sensitive and can detect a single DNA molecule in a sample. PCR is useful because this single DNA molecule then can be copied, or amplified, numerous times to be used for DNA analysis. Follow **Figure 13.9** as you read about the steps of PCR.

Step 1 PCR is performed by placing the DNA fragment to be copied, DNA polymerase, the four DNA nucleotides, and two short single-stranded pieces of DNA called primers in a tube. The primers are complementary to the ends of the DNA fragment that will be copied and used as starting points for DNA synthesis. PCR begins when the tube is heated.

Step 2 The heat separates the two strands of the template DNA fragment. When the tube is cooled, the primers can bind to each strand of the template DNA. An automated machine called a thermocycler is used to cycle the tube containing all of the components involved in PCR through various hot and cool temperatures.

Step 3 As shown in **Figure 13.9,** each primer is made to bind to one strand of the DNA fragment. Once the primers are bound, DNA polymerase incorporates the correct nucleotides between the two primers as in DNA replication. This process of heating, cooling, and nucleotide incorporation is repeated 20 to 40 times, resulting in millions of copies of the original fragment. Because the separation of DNA strands requires heat, the DNA polymerase used in PCR has to be able to withstand high temperatures. This special DNA polymerase was isolated from a thermophilic, or heat-loving, bacterium such as those found living in the hot springs of Yellowstone National Park.

Because PCR can detect a single DNA molecule in a sample, it has become one of the most powerful tools used by scientists. PCR is not used only by researchers in laboratories, but also by forensic scientists to identify suspects and victims in crime investigations, and by doctors to detect infectious diseases, such as AIDS.

Reading Check **Describe** the polymerase chain reaction using an analogy.

■ **Figure 13.9** PCR is a biological version of a copy machine. During each PCR cycle, the reaction mixture is heated to separate the DNA strands and then cooled to allow primers to bind to complementary sequences. The DNA polymerase then adds nucleotides to form new DNA molecules.

Concepts In Motion

Interactive Figure To see an animation of how PCR works, visit biologygmh.com.

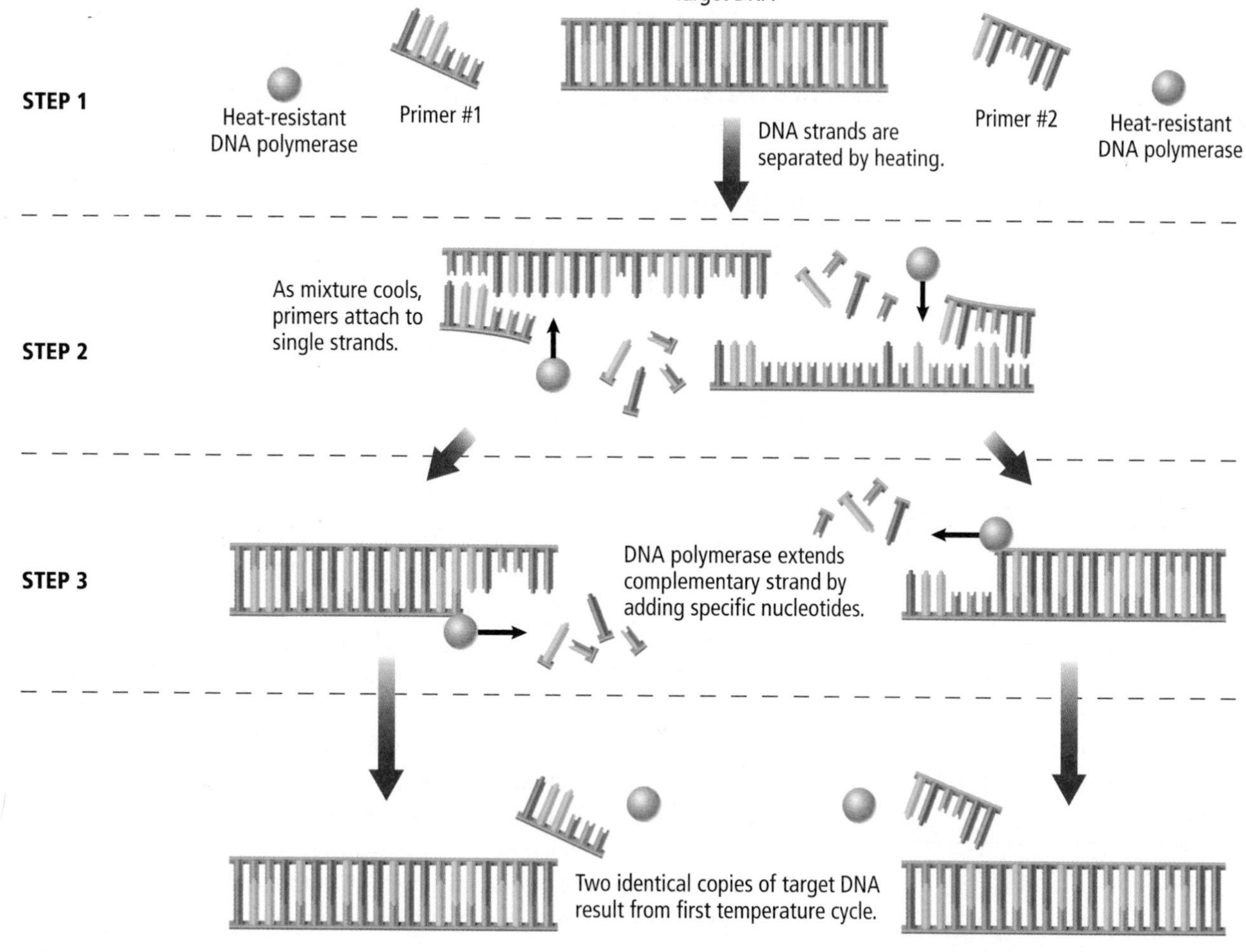
Target DNA
STEP 1
Heat-resistant DNA polymerase
Primer #1
DNA strands are separated by heating.
Primer #2
Heat-resistant DNA polymerase
STEP 2
As mixture cools, primers attach to single strands.
STEP 3
DNA polymerase extends complementary strand by adding specific nucleotides.
Two identical copies of target DNA result from first temperature cycle.

Concepts In Motion

Interactive Table To explore more about genetic engineering, visit biologygmh.com.

Table 13.1	Genetic Engineering	
Tool/Process	Function	Applications
Restriction enzymes Ex: *EcoRI*	Cut DNA strands into fragments	Used to create DNA fragments with sticky ends or blunt ends that can join with other DNA fragments
Gel electrophoresis	Separates DNA fragments by size	Used to study DNA fragments of various sizes
Recombinant DNA technology	Combines a DNA fragment with DNA from another source (exogenous DNA)	Used to create recombinant DNA to be used to study individual genes and genetically engineered organisms, and in the treatment of certain diseases
Gene cloning	Produces large numbers of identical recombinant DNA molecules	Used to create large amounts of recombinant DNA to be used in genetically engineered organisms
DNA sequencing	Identifies the DNA sequence of cloned recombinant DNA molecules for further study	Used to identify errors in the DNA sequence, to predict the function of a particular gene, and to compare to other genes with similar sequences from different organisms
Polymerase chain reaction (PCR)	Makes copies of specific regions of sequenced DNA	Used to copy DNA for any scientific investigation including forensic analysis, and medical testing

Genetic engineering uses powerful tools, summarized in **Table 13.1,** to study and manipulate DNA. Although researchers investigate many different problems, their experimental procedures often include cleavage by a restriction enzyme, isolation of fragments, combination with exogenous DNA, cloning or PCR, and identification of sequences.

Biotechnology

Biotechnology—the use of genetic engineering to find solutions to problems—makes it possible to produce organisms that contain individual genes from another organism. Recall that organisms such as the mosquito larvae shown in **Figure 13.3** have a gene from another organism. Such organisms, genetically engineered by inserting a gene from another organism, are called **transgenic organisms.** Transgenic animals, plants, and bacteria are used not only for research, but also for medical and agricultural purposes.

Transgenic animals Currently, scientists produce most transgenic animals in laboratories for biological research. Mice, fruit flies, and the roundworm *Caenorhabditis elegans,* also called *C. elegans,* are widely used in research laboratories around the world to study diseases and develop ways to treat them. Some transgenic organisms, such as transgenic livestock, have been produced to improve the food supply and human health. Transgenic goats have been engineered to secrete a protein called antithrombin III, which is used to prevent human blood from forming clots during surgery. Researchers are working to produce transgenic chickens and turkeys that are resistant to diseases. Several species of fishes also have been genetically engineered to grow faster. In the future, transgenic organisms might be used as a source of organs for organ transplants.

CAREERS IN BIOLOGY

Geneticist Using many of the DNA tools, a geneticist might research genes, inheritance, and the variations of organisms. Some geneticists are medical doctors who diagnose and treat genetic conditions. For more information on biology careers, visit biologygmh.com.

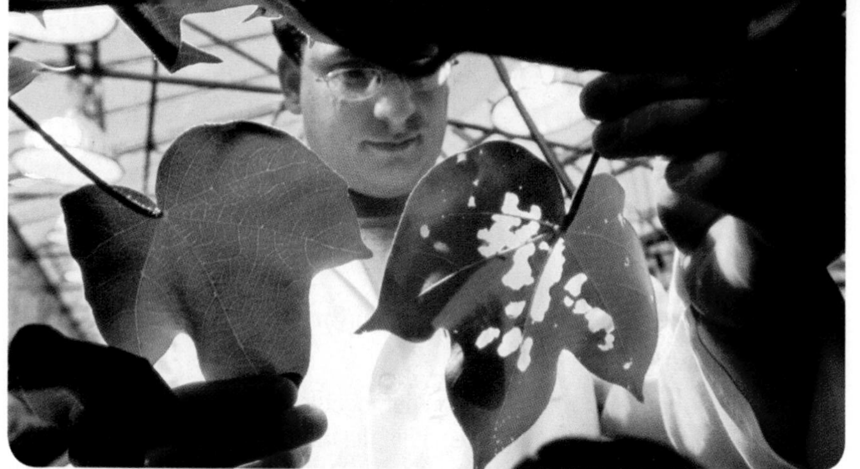

■ **Figure 13.10** This researcher is examining cotton plant leaves. The leaf on the left has been genetically engineered to resist insect infestation.

Transgenic plants Many species of plants have been genetically engineered to be more resistant to insect or viral pests. In 2006, about 69.9 million hectares grown by 7 million farmers in 18 countries were planted with transgenic crops. These crops included herbicide- and insecticide-resistant soybeans, corn, cotton, and canola. Scientists now are producing genetically engineered cotton, as shown in **Figure 13.10,** that resists insect infestation of the bolls. Researchers also are developing peanuts and soybeans that do not cause allergic reactions.

Other crops are being grown commercially and being field-tested. These crops include sweet-potato plants that are resistant to a virus that could kill most of the African harvest, rice plants with increased iron and vitamins that could decrease malnutrition in Asian countries, and a variety of plants able to survive extreme weather conditions. Prospective crops include bananas that produce vaccines for infectious diseases, such as hepatitis B, and plants that produce biodegradable plastics.

Transgenic bacteria Insulin, growth hormones, and substances that dissolve blood clots are made by transgenic bacteria. Transgenic bacteria also slow the formation of ice crystals on crops to protect them from frost damage, clean up oil spills more efficiently, and decompose garbage.

Section 13.2 Assessment

Section Summary

- Genetic engineering is used to produce organisms that are useful to humans.
- Recombinant DNA technology is used to study individual genes.
- DNA fragments can be separated using gel electrophoresis.
- Clones can be produced by transforming bacteria with recombinant DNA.
- The polymerase chain reaction is used to make copies of small DNA sequences.

Understand Main Ideas

1. MAIN Idea **Sequence** how recombinant DNA is made and manipulated.
2. **Explain** why some plasmids contain a gene for resistance to an antibiotic.
3. **Apply** How can genetic engineering improve human health?
4. **Contrast** What is one major difference between selective breeding and genetic engineering?

Think Critically

5. **Hypothesize** Restriction enzymes play an essential role in recombinant DNA technology. How can a bacterium produce restriction enzymes that do not cleave its DNA?

WRITING in **Biology**

6. Why would a business synthesize and sell DNA? Who would their customers be? Write a list of possible uses for DNA that is synthesized in a laboratory.

Section 13.3

The Human Genome

Reading Preview

Objectives

- **Describe** components of the human genome.
- **Describe** how forensic scientists use DNA fingerprinting.
- **Explain** how information from the human genome can be used to diagnose human diseases.

Review Vocabulary

codon: the triplet of bases in the DNA or mRNA

New Vocabulary

DNA fingerprinting
bioinformatics
DNA microarray
single nucleotide polymorphism
haplotype
pharmacogenomics
gene therapy
genomics
proteomics

MAIN Idea **Genomes contain all the information needed for an organism to grow and survive.**

Real-World Reading Link When you put together a jigsaw puzzle, you might first find all the border pieces and then fill in the other pieces. Sequencing the human genome can be compared to putting together a jigsaw puzzle. Just as you have to figure out which puzzle pieces fit together, scientists had to determine the sequence of the base pairs along the length of a human chromosome.

The Human Genome Project

The Human Genome Project (HGP) was an international project that was completed in 2003. A genome is the complete genetic information in a cell. The goal of the HGP was to determine the sequence of the approximately three billion nucleotides that make up human DNA and to identify all the 23,299 human genes. If all the nucleotides in the human genome were the size of the type on this page and fused together in one continuous line, the line would extend from Los Angeles, California, to Panama, as illustrated in **Figure 13.11.**

Though the HGP is finished, analysis of the data generated from this project will continue for many decades. To complete this huge task, researchers also have studied the genomes of several other organisms, including the fruit fly, the mouse, and *Escherichia coli*—the bacterium present in the human gut. Studies in nonhuman organisms help to develop the technology required to handle the large amounts of data produced by the Human Genome Project. These technologies help to interpret the function of newly identified human genes.

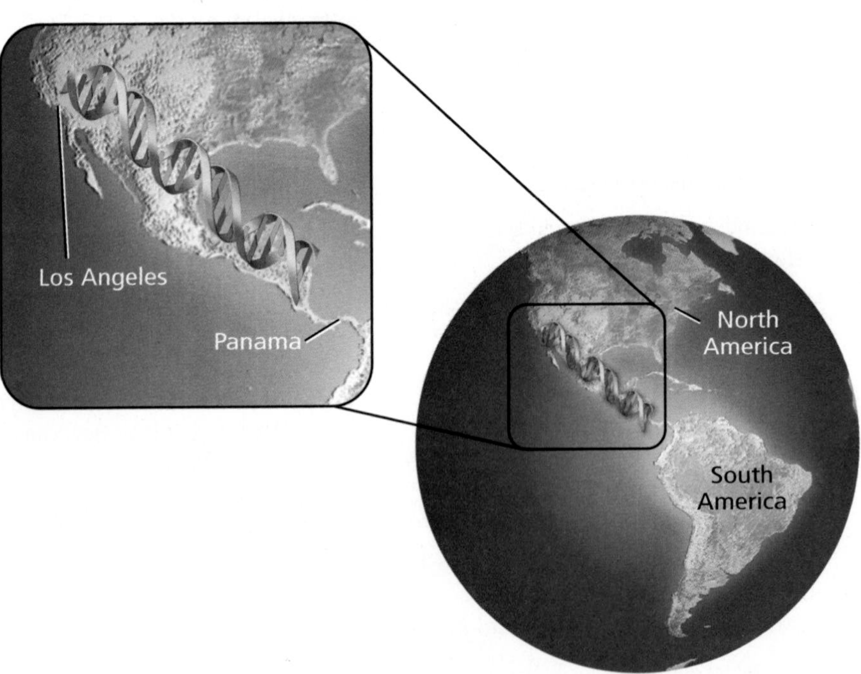

■ **Figure 13.11** If all the DNA in the human genome were fused together in one continuous line, it would stretch from California to Panama.

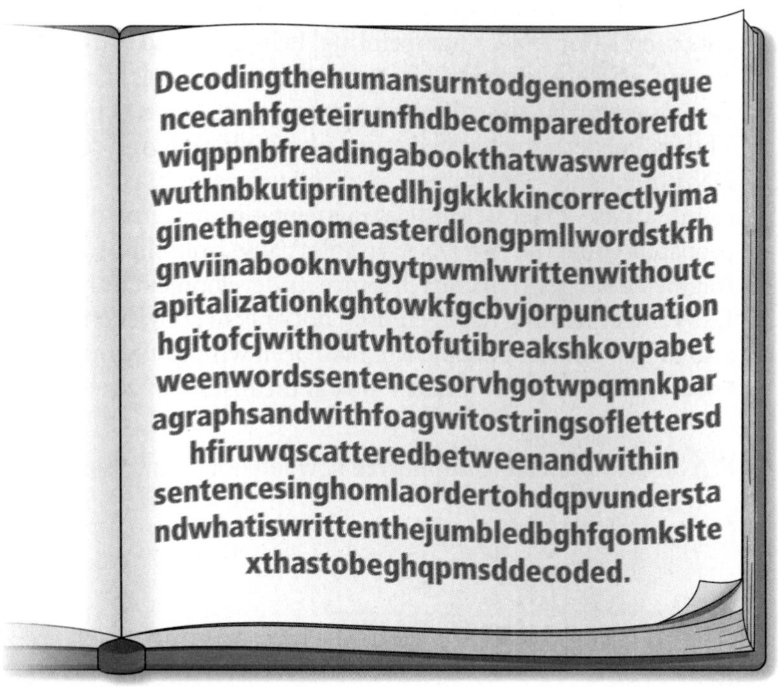

Decodingthehumansurntodgenomeseque
ncecanhfgeteirunfhdbecomparedtorefdt
wiqppnbfreadingabookthatwaswregdfst
wuthnbkutiprintedlhjgkkkkincorrectlyima
ginethegenomeasterdlongpmllwordstkfh
gnviinabooknvhgytpwmlwrittenwithoutc
apitalizationkghtowkfgcbvjorpunctuation
hgitofcjwithoutvhtofutibreakshkovpabet
weenwordssentencesorvhgotwpqmnkpar
agraphsandwithfoagwitostringsoflettersd
hfiruwqscatteredbetweenandwithin
sentencesinghomlaordertohdqpundersta
ndwhatiswrittenthejumbledbghfqomkslte
xthastobeghqpmsddecoded.

■ **Figure 13.12** The genetic information contained within the human genome has to be decoded in order to uncover important sequences.
***Interpret** the text by decoding the jumbled sentences.*

Sequencing the genome Recall from Chapter 10 that human DNA is organized into 46 chromosomes. In order to determine one continuous human genome sequence, each of the 46 human chromosomes was cleaved. Several different restriction enzymes were used in order to produce fragments with overlapping sequences. These fragments were combined with vectors to create recombinant DNA, cloned to make many copies, and sequenced using automated sequencing machines. Computers analyzed the overlapping regions to generate one continuous sequence.

Decoding the sequence of the human genome can be compared to reading a book that was printed in code. Imagine the genome as words in a book written without capitalization, punctuation, or breaks between words, sentences, or paragraphs. Suppose there are strings of letters scattered between and within sentences. **Figure 13.12** illustrates how a page from such a book might look. In order to understand what is written, you have to decode the jumbled text. Similarly, scientists had to decode the genetic code in the human genome.

After sequencing the entire human genome, scientists observed that less than two percent of all of the nucleotides in the human genome code for all the proteins in the body. That is, the genome is filled with long stretches of repeated sequences that have no direct function. These regions are called noncoding sequences.

DNA fingerprinting Unlike the protein-coding regions of DNA that are almost identical among individuals, the long stretches of noncoding regions of DNA are unique to each individual. When these regions are cut by restriction enzymes, as described earlier in this chapter, the set of DNA fragments produced is unique to every individual. **DNA fingerprinting** involves separating these DNA fragments using gel electrophoresis in order to observe the distinct banding patterns that are unique to every individual. Forensic scientists use DNA fingerprinting to identify suspects and victims in criminal cases, to determine paternity, and to identify soldiers killed in war.

VOCABULARY

ACADEMIC VOCABULARY

Sequence (SEE kwens):
a continuous series.
The sequence of colors formed a beautiful pattern.

CAREERS IN BIOLOGY

Forensic Scientist Genetic engineering is a technology used widely by forensic scientists. They use the various tools and processes, such as DNA fingerprinting, in criminal and archaeological investigations. For more information on biology careers, visit biologygmh.com.

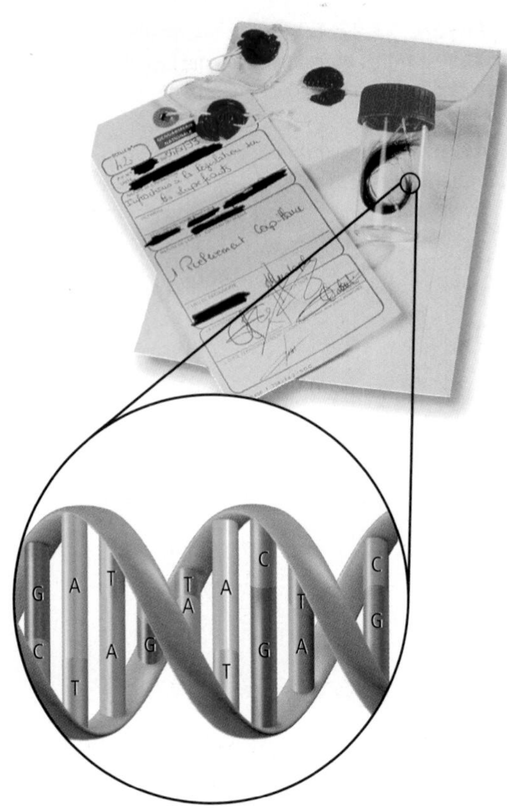

■ **Figure 13.13** People can be identified using the genetic information contained in blood, hair, semen, or skin.

Figure 13.13 shows a sample obtained from hair that forensic scientists can use for DNA fingerprinting. PCR is used to copy this small amount of DNA to create a larger sample for analysis. The amplified DNA then is cut using different combinations of restriction enzymes. The fragments are separated by gel electrophoresis and compared to DNA fragments from known sources, such as victims and suspects in a criminal case, to locate similar fragmentation patterns. There is a high probability that the two DNA samples came from the same person if two fragmentation patterns match. Since its development in England in 1985, DNA fingerprinting has been used not only to convict criminals but also to free innocent people who had been wrongfully imprisoned. **Figure 13.14** provides a closer look at the history of genetic technology.

 **Reading Check** **Summarize** how forensic scientists use DNA fingerprinting.

Identifying Genes

Once the genome has been sequenced, the next step in the process is to identify the genes and determine their functions. The functions of many of the genes in the human genome are still unknown. Researchers use techniques that integrate computer analysis and recombinant DNA technology to determine the function of these genes.

For organisms such as bacteria and yeast, whose genomes do not have large regions of noncoding DNA, researchers have identified genes by scanning the sequence for open reading frames (or ORFs, pronounced "orphs"). ORFs are stretches of DNA containing at least 100 codons that begin with a start codon and end with a stop codon. While these sequences might indicate a gene, they will be tested to determine if these sequences produce functioning proteins.

■ **Figure 13.14**
Discoveries in Genetics

Many studies in genetics have led to advances in biotechnology.

1960 — 1970 — 1980

1959 Down syndrome is the first chromosomal abnormality identified in humans.

1972 Paul Berg creates the first recombinant DNA molecules.

1973 Herbert Boyer, Annie Chang, Stanley Cohen, and Robert Helling discover that recombinant DNA reproduce if inserted into bacteria.

1983 Kary Mullis invents the polymerase chain reaction, for which he will be awarded the Nobel Prize in 1993.

Recall from Chapter 12 that a codon is a group of three nucleotides that code for an amino acid. Researchers look for the start codon AUG and a stop codon such as UAA, UGA, or UAG. ORF analysis has been used to identify correctly over 90 percent of genes in yeast and bacteria. However, the identification of genes in more complex organisms such as humans requires more sophisticated computer programs called algorithms. These algorithms use information, such as the sequence of the genomes of other organisms, to identify human genes.

Bioinformatics

The completion of the HGP and the sequencing of the genomes of other organisms have resulted in large amounts of data. Not only has this enormous amount of data required careful storage, organization, and indexing of sequence information, but it also has created a new field of study. This field of study, called **bioinformatics,** involves creating and maintaining databases of biological information. The analysis of sequence information involves finding genes in DNA sequences of various organisms and developing methods to predict the structure and function of newly discovered proteins. Scientists also study the evolution of genes by grouping protein sequences into families of related sequences and comparing similar proteins from different organisms.

Study Tip

BioJournal As you read about the human genome, list several beneficial uses of this information.

DNA Microarrays

Analyzing all the expressed genes from a given organism or a specific cell type can be useful. This analysis can be done using **DNA microarrays,** which are tiny microscope slides or silicon chips that are spotted with DNA fragments. DNA microarrays can contain a few genes, such as the genes that control the cell cycle, or all of the genes of the human genome. Therefore, a large amount of information can be stored in one small slide or chip. DNA microarrays help researchers determine whether the expression of certain genes is caused by genetic factors or environmental factors.

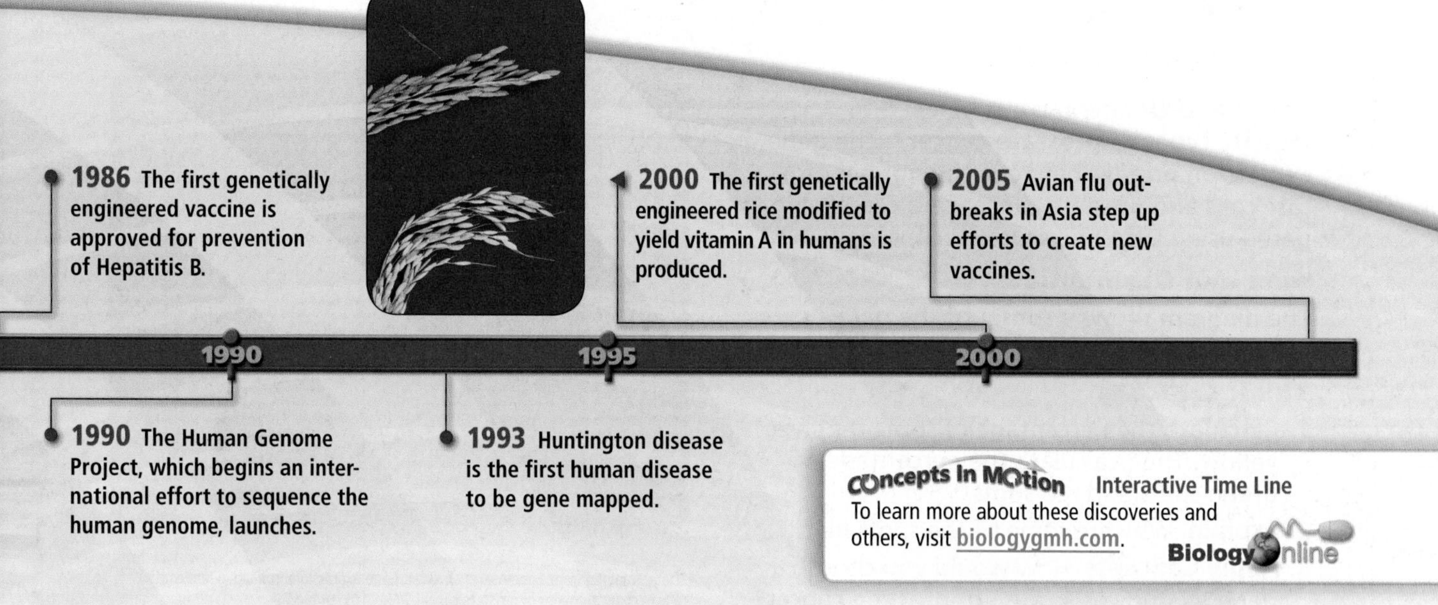

1986 The first genetically engineered vaccine is approved for prevention of Hepatitis B.

1990 The Human Genome Project, which begins an international effort to sequence the human genome, launches.

1993 Huntington disease is the first human disease to be gene mapped.

2000 The first genetically engineered rice modified to yield vitamin A in humans is produced.

2005 Avian flu outbreaks in Asia step up efforts to create new vaccines.

Concepts In Motion Interactive Time Line
To learn more about these discoveries and others, visit biologygmh.com. Biology Online

Follow the steps involved in doing the DNA microarray experiment shown in **Figure 13.15.** mRNA from two different populations of cells is isolated and converted into complementary DNA (cDNA) strands using an enzyme called reverse transcriptase. The complementary DNA from each cell population is labeled with a specific fluorescent dye—for example, red for cancer cells and green for normal cells. Both pools of complementary DNA are combined on the microarray slide and incubated.

Figure 13.15 shows the fluorescent signals that are produced when the microarray slide is analyzed. When the expression of a gene is the same in both the normal and cancer cells, a yellow spot is produced on the chip. If the expression of a gene is higher in cancer cells, then the spot formed is red. However, if the expression is higher in normal cells, then the spot formed is green.

Because one DNA microarray slide can contain thousands of genes, researchers can examine changes in the expression patterns of multiple genes at the same time. Scientists also are using DNA microarrays to identify new genes and to study changes in the expression of proteins under different growth conditions.

The Genome and Genetic Disorders

Although more than 99 percent of all nucleotide base sequences are exactly the same in all people, sometimes there are variations that are linked to human diseases. These variations in the DNA sequence that occur when a single nucleotide in the genome is altered are called **single nucleotide polymorphisms** or SNPs (SNIHPS). For a variation to be considered an SNP, it must occur in at least one percent of the population. Many SNPs have no effect on cell function, but scientists hypothesize that SNP maps will help identify many genes associated with many different types of genetic disorders.

DATA ANALYSIS LAB 13.1

Based on Real Data*

Apply Concepts

How can DNA microarrays be used to classify types of prostate cancer? The gene expression profiles between the normal prostate cells and prostate cancer cells can be compared using DNA microarray technology.

Data and Observations

The diagram shows a subset of the data obtained.

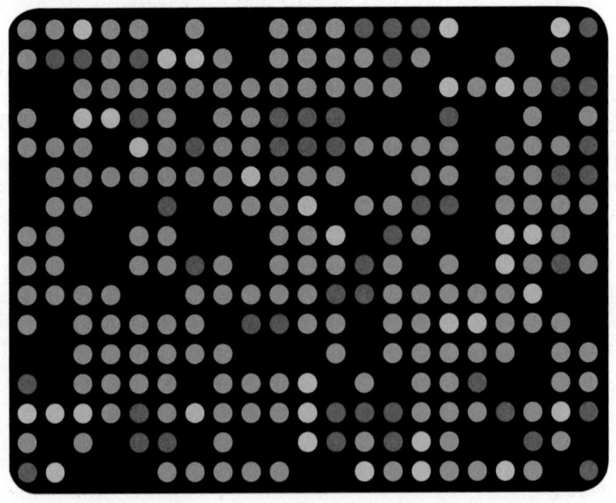

Think Critically

1. **Calculate** the percentage of spots that are yellow. Then calculate the percentage of green spots and red spots.
2. **Explain** Why are some of the spots black?
3. **Apply Concepts** How would you choose a gene to study as a cause of prostate cancer?

*Data obtained from: Lapointe, et al. 2004. Gene expression profiling identifies clinically relevant subtypes of prostate cancer. *PNAS* 101: 811–816.

Visualizing Microarray Analysis

Figure 13.15
In this experiment, the expression of thousands of human genes was detected by DNA microarray analysis. Each spot on the microarray chip represents a gene. A red spot indicates expression of a gene is higher in cancer cells compared to normal cells. A green spot indicates an expression in normal cells is higher, and yellow spots indicate no difference in expression between cancer cells and normal cells.

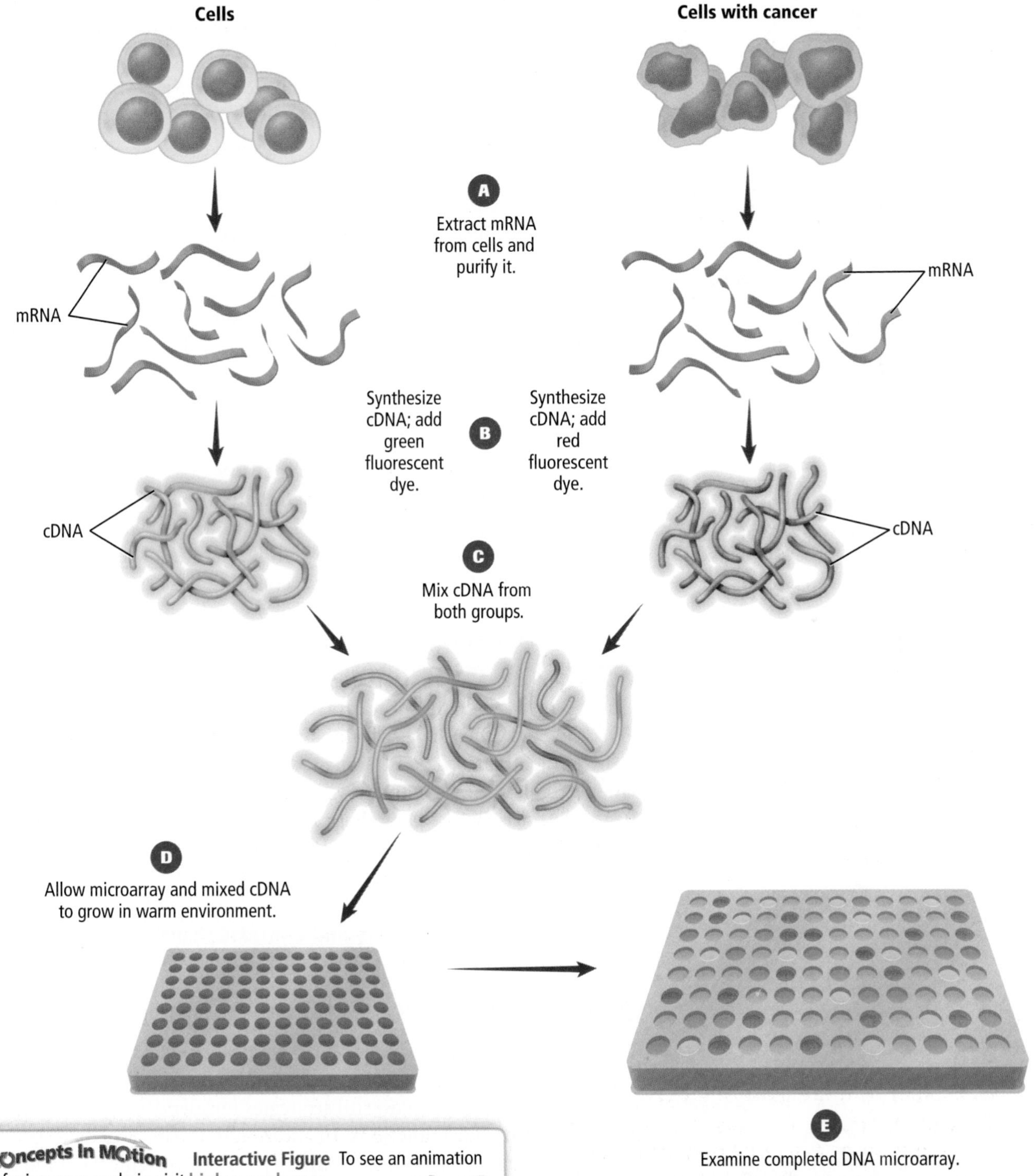

Cells

Cells with cancer

A Extract mRNA from cells and purify it.

mRNA

mRNA

B Synthesize cDNA; add green fluorescent dye.

Synthesize cDNA; add red fluorescent dye.

cDNA

cDNA

C Mix cDNA from both groups.

D Allow microarray and mixed cDNA to grow in warm environment.

E Examine completed DNA microarray.

Concepts In Motion **Interactive Figure** To see an animation of microarray analysis, visit biologygmh.com.

Biology Online

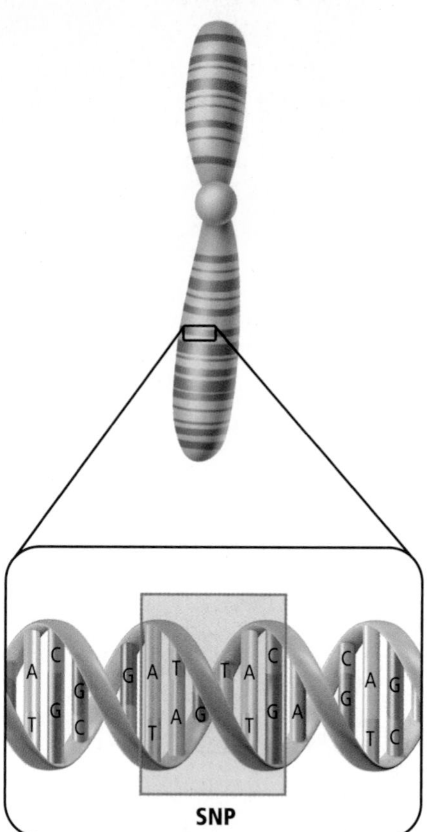

■ **Figure 13.16** The HapMap project involves grouping all adjacent SNPs that are inherited together into haplotypes.

The HapMap project An internationl group of scientists is creating a catalog of common genetic variations that occur in humans. Recall from Chapter 10 that linked genes are inherited together. Similarly, genetic variations located close together also tend to be inherited together. Therefore, regions of linked variations in the human genome, known as **haplotypes,** can be located. The project to create this catalog is called the haplotype map, or HapMap project. Assembling the HapMap involves identifying groups of SNPs in a specific region of DNA.

Figure 13.16 shows how the genome is divided into haplotypes. Once completed, the HapMap will describe what these variations are, where they occur in our DNA, and how they are distributed among people within populations and among populations in different parts of the world. This information will help researchers find genes that cause disease and affect an individual's response to drugs.

Pharmacogenomics Sequencing the human genome combines the knowledge of genes, proteins, and SNPs with other areas of science. The study of how genetic inheritance affects the body's response to drugs is called **pharmacogenomics** (far muh koh jeh NAW mihks). The benefits of pharmacogenomics include more accurate dosing of drugs that are safer and more specific. Researchers hope that pharmacogenomics will allow for drugs to be custom-made for individuals based on their genetic makeups. Prescribing drugs based on an individual's genetic makeup will increase safety, speed recovery, and reduce side effects. Perhaps one day when you are sick, your doctor will read your genetic code and prescribe medicine tailor-made for you.

Gene therapy A technique aimed at correcting mutated genes that cause human diseases is called **gene therapy.** Scientists insert a normal gene into a chromosome to replace a dysfunctional gene. In most gene therapy studies, inserting a normal gene into a viral vector, like the one in **Figure 13.17,** produces recombinant DNA. Target cells in the patient are infected with the virus and the recombinant DNA material is released into the affected cells. Once deposited into cells, the normal gene inserts itself into the genome and begins functioning.

■ **Figure 13.17** DNA can be encapsulated in a virus and delivered into a patient to replace a defective gene. Once the virus enters the cells, the genetic information is released into the nucleus and inserted into the genome.

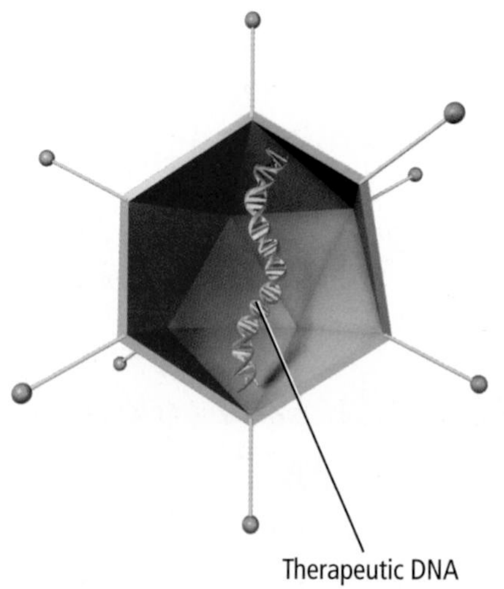

Connection to Health In the 1990s, many hospitals around the United States conducted clinical trials using gene therapy. However, in 2003, the Food and Drug Administration halted all gene therapy trials in the United States. The FDA made this decision after the death of a patient undergoing gene therapy. The death was caused by a reaction to the viral vector. Before gene therapy becomes an effective treatment for genetic disorders, viral vectors that are nontoxic and do not activate a body's defense reaction need to be engineered.

Reading Check **Compare and Contrast** pharmacogenomics to gene therapy.

Genomics and Proteomics

Sequencing the human genome began what researchers call "the genomic era." **Genomics** is the study of an organism's genome. Genomics has become one of the most powerful strategies for identifying human genes and interpreting their functions. In addition to the mass of data obtained from sequencing the genomes of humans, rice, mice, fruit flies, and corn, scientists also are investigating the proteins produced by these genes.

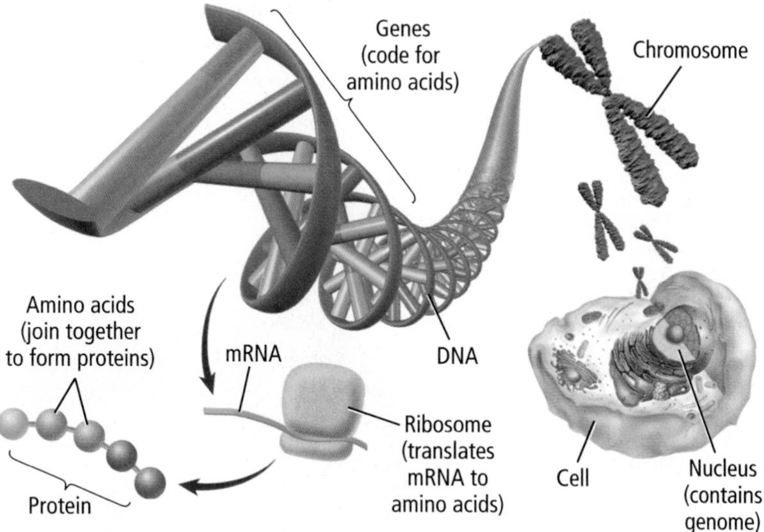

Figure 13.18 The central dogma is that the information in genes eventually flows to the synthesis of proteins.

Genes are the primary information storage units, whereas proteins are the machines of a cell. Recall that when a gene is expressed, a protein is produced, as illustrated in **Figure 13.18.** Therefore, an understanding of how proteins function also is important. For instance, if the genome represents the words in a dictionary, the proteome, which represents all the proteins found in a cell, provides the definition of these words and how to use these words in a sentence. The large-scale study and cataloging of the structure and function of proteins in the human body is called **proteomics.** Proteomics allows researchers to look at hundreds or thousands of proteins at the same time. This type of broad analysis will better define both normal and disease states. Scientists anticipate that proteomics will revolutionize the development of new drugs to treat diseases such as Type II diabetes, obesity, and atherosclerosis.

Section 13.3 Assessment

Section Summary

- Researchers who worked on the HGP sequenced all nucleotides in the human genome.
- DNA fingerprinting can be used to identify individuals.
- DNA microarrays allow researchers to study all the genes in the genome simultaneously.
- Gene therapy might be used in the future to correct genetic disorders.
- Genomics is the study of an organism's genome and proteomics is the study of the proteins in the human body.

Understand Main Ideas

1. MAIN Idea **Relate** the human genome to blueprints for a house.
2. **Analyze** the role of DNA fingerprinting in criminal investigations.
3. **Indicate** why the HapMap project is useful in diagnosing human disease.
4. **Explain** the goal of gene therapy. What is one of the obstacles that this technology faces?

Think Critically

5. **Hypothesize** Most of the human genome consists of noncoding DNA. From where did all of this noncoding DNA come?

MATH in Biology

6. If 1.5 percent of the human genome consists of protein-coding sequences, and the entire genome has 3.2×10^9 nucleotides, how many codons are in the human genome? Remember that a codon is three nucleotides in length.

In the Field

Career: Biomedical Research

Illuminating Medical Research

Have you ever watched fireflies glow on a summer evening? A chemical reaction in firefly cells produces light through a process called bioluminescence. Among bioluminescent marine organisms, a tiny jellyfish named *Aequorea victoria* has emerged as a hero to biomedical researchers. This jellyfish produces a substance called green fluorescent protein (GFP), which makes parts of its body shine with an emerald green light.

Incorporating GFP into tumor cells might allow for the separation of tumor cells from healthy cells.

Shining Light on Cell Functions Found off the west coast of North America, the diminutive *Aequorea victoria* is only five to ten centimeters in diameter. Its cells contain aequorin, a bioluminescent protein that emits a deep blue light. GFP absorbs this light and converts it into a glowing emerald green. In the early 1990's, scientists removed the GFP gene from *Aequorea victoria* and cloned it. Today, biomedical researchers can fuse GFP to other proteins inside cells of living organisms. When illuminated with light of a specific frequency, these marked proteins glow, making it possible to observe their behavior during cell processes.

Biological Marking at Work GFP allows scientists to determine where proteins are located during different stages of a cell's life, and to observe how proteins interact to produce disease. Researchers can attach GFP to a virus and observe the spread of the virus throughout the host.

By injecting tumor cells marked with GFP, scientists can analyze how they develop, spread, and destroy healthy cells over time. The photo above shows an animal tumor marked with GFP. The black areas on the image are blood vessels on the tumor's surface. Bioluminescent imaging can be used to evaluate the effectiveness of various treatments on these types of tumors. Ultimately, scientists hope to incorporate GFP directly into human tumor cells, then use bioluminescence to identify the mass as a separate cell population within the body. Easily differentiated from healthy cells and tissues, the glowing cancerous cells would be marked for treatment.

WRITING in Biology

Research and Communicate GFP is used to investigate the effectiveness of gene therapy, vaccines, and cancer treatments. Research how GFP was used in cancer studies and share your findings with classmates. For more information about GFP, visit biologygmh.com.

BIOLAB

FORENSICS: HOW CAN GENETIC ENGINEERING BE USED TO SOLVE A CRIME?

Background: Although all humans are similar genetically, variations do occur in certain segments of DNA. When cut with restriction enzymes, the variety of sizes of these fragments can be used to determine the source of a sample of DNA. In this lab, DNA from suspects will be analyzed.

Question: *Based on the DNA samples, were any of the suspects at the scene?*

Materials

various DNA samples
electrophoresis chamber
power source
micropipette and tips
prepared agarose gels
restriction enzyme
microcentrifuge tubes and rack
sample-loading dye
nontoxic dye
staining and destaining containers
DNA fragments of known size (control)
ruler
ice in foam container
water bath at 37°C

Safety Precautions

Procedure

1. Read and complete the lab safety form.
2. Read the entire procedure.
3. Label your DNA samples.
4. Design and construct a data table you can use to record your observations when you perform gel electrophoresis of your samples.
5. Your teacher will instruct you how to prepare your samples, set up the gel electrophoresis equipment, load your samples, and run the electrophoresis.
6. Use the gel-staining dye to detect the location of DNA fragments in the gel for each of your samples.
7. Use a ruler to measure (in mm) the distance of each migrated DNA band from the wells. Record this information in your data table.
8. **Cleanup and Disposal** Wash and return all reusable materials. Dispose of gels and other reagents in properly labeled containers. Wash your hands thoroughly.

Analyze and Conclude

1. **Interpret Data** Based on your observations, predict which suspect is incriminated by the DNA evidence.
2. **Think Critically** While the amount of DNA needed for electrophoresis is not large, the amount that can be extracted from a few hairs might not be enough. How might a forensic scientist solve this problem?
3. **Error Analysis** DNA fingerprints have a very high level of accuracy if they are run correctly. What are some sources of error that could lead to inaccurate results?

WRITING in Biology

Plan a procedure. Find a news article describing the use of DNA fingerprinting in investigations such as a criminal investigation or identifying a bacterium involved in a disease outbreak. Write a mock lab that explains the techniques and steps that might be taken in the situation described by the article. To learn more about DNA fingerprinting, visit Biolabs at biologygmh.com.

Study Guide

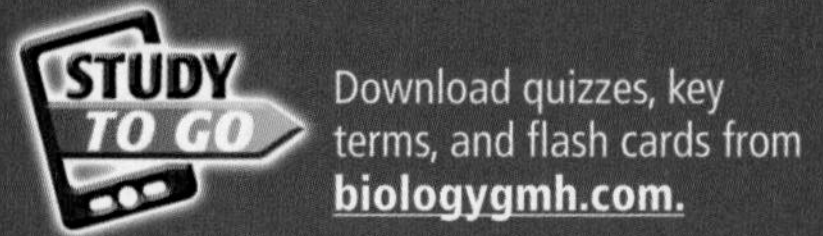

Download quizzes, key terms, and flash cards from biologygmh.com.

FOLDABLES **Summarize and Predict** Use what you have learned and the information recorded on your Foldable to summarize the current uses of recombinant DNA technology and to predict future uses.

Vocabulary	Key Concepts
Section 13.1 Applied Genetics	
• inbreeding (p. 361) • selective breeding (p. 360) • test cross (p. 362)	MAIN Idea Selective breeding is used to produce organisms with desired traits. • Selective breeding is used to produce organisms with traits that are beneficial to humans. • Hybridization produces organisms with the desired traits from parent organisms with different traits. • Inbreeding creates pure breeds. • A test cross can be used to determine an organism's genotype.
Section 13.2 DNA Technology	
• cloning (p. 367) • DNA ligase (p. 366) • gel electrophoresis (p. 365) • genetic engineering (p. 363) • genome (p. 364) • plasmid (p. 366) • polymerase chain reaction (p. 368) • recombinant DNA (p. 366) • restriction enzyme (p. 364) • transformation (p. 367) • transgenic organism (p. 370)	MAIN Idea Researchers use genetic engineering to manipulate DNA. • Genetic engineering is used to produce organisms that are useful to humans. • Recombinant DNA technology is used to study individual genes. • DNA fragments can be separated using gel electrophoresis. • Clones can be produced by transforming bacteria with recombinant DNA. • The polymerase chain reaction is used to make copies of small DNA sequences.
Section 13.3 The Human Genome	
• bioinformatics (p. 375) • DNA fingerprinting (p. 373) • DNA microarray (p. 375) • gene therapy (p. 378) • genomics (p. 378) • haplotype (p. 378) • pharmacogenomics (p. 378) • proteomics (p. 379) • single nucleotide polymorphism (p. 376)	MAIN Idea Genomes contain all of the information needed for an organism to grow and survive. • Researchers who worked on the HGP sequenced all nucleotides in the human genome. • DNA fingerprinting can be used to identify individuals. • DNA microarrays allow researchers to study all the genes in the genome simultaneously. • Gene therapy might be used in the future to correct genetic disorders. • Genomics is the study of an organism's genome and proteomics is the study of the proteins in the human body.

CHAPTER 13 Assessment

Section 13.1

Vocabulary Review

Fill in the blanks with the correct term from the Study Guide page.

1. A __________ is used to determine the genotype of a plant or animal.
2. The offspring produced by __________ are homozygous for most traits.

Understand Key Concepts

Use the illustration below to answer questions 3 and 4.

Heterozygous white grapefruit

Homozygous red grapefruit	*W*	*w*
w	*Ww*	*ww*
w	*Ww*	*ww*

3. What is the genotypic ratio of the offspring in the cross above?
   A. 1:2:1
   B. 1:1
   C. All are homozygous recessive.
   D. All are heterozygous.
4. The cross above could be used to determine the genotype of a parent with a dominant phenotype. What is this type of cross called?
   A. a homozygous cross
   B. a heterozygous cross
   C. a test cross
   D. a parental cross

Constructed Response

5. **Short Answer** Predict the phenotype of the parent plants of hybrid tomato plants that grow fast and are resistant to pesticides. Explain.
6. **Short Answer** How do polygenic traits affect selective breeding? *Hint: Review Chapter 10.*
7. **Short Answer** Discuss the advantages and disadvantages of selective breeding.

Think Critically

8. **Infer** Why don't purebred animals exist in the wild?
9. **Determine** Suppose a phenotype is controlled by more than one gene. Can a test cross be used to determine the genotype? Why or why not?

Section 13.2

Vocabulary Review

Fill in the blank with the correct vocabulary term from the Study Guide page.

10. Transgenic animals are produced by __________.
11. Biologists use __________ to join two DNA molecules together.
12. During __________, a cell takes in DNA from outside the cell.
13. Small, circular DNA molecules that are found in bacterial cells are called __________.

Understand Key Concepts

Use the illustration below to answer question 14.

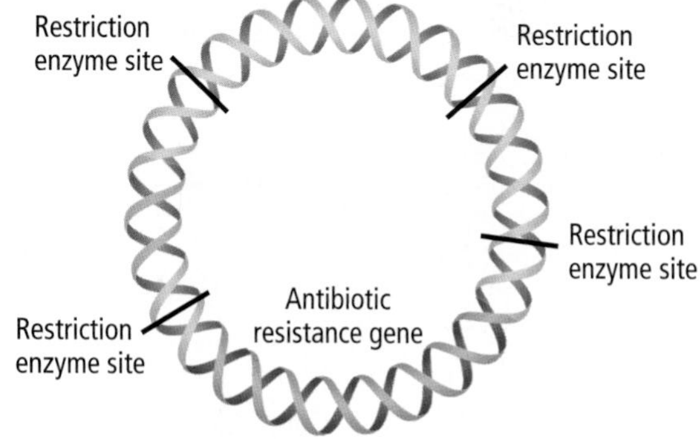

14. What is the role of the molecule above in DNA cloning?
    A. to carry the foreign DNA into the host cell
    B. to identify the source of DNA as foreign
    C. to identify the host cell that has taken up the gene of interest
    D. to make the foreign DNA susceptible to digestion with enzymes

Chapter 13 Assessment

15. Which of the following enzymes will produce a blunt end? *Hint: The cut site is indicated by the *.*

A. *Eag*I	C*GGCC G G CCGG*C
B. *Eco*RV	GAT*ATC CTA*TAG
C. *Nsi*I	A TGCA*T T*ACGT A
D. *Taq*I	T*CG A A GC*T

16. Why is the polymerase chain reaction used?
    A. to amplify DNA **C.** to ligate DNA
    B. to cut DNA **D.** to separate DNA

Constructed Response

17. **Open Ended** Predict what effect genetic engineering will have on the evolution of a species.

18. **Short Answer** Suppose you transform bacteria with a recombinant DNA plasmid and by mistake grow the transformed cells without an antibiotic. What result would you observe? Why?

19. **Interpret the Figure** Refer to **Figure 13.9** to make a flowchart diagramming the steps in the PCR.

Think Critically

20. **Conclude** A recombinant DNA molecule was created by joining a plasmid vector and a DNA fragment. Gel electrophoresis was done to verify that the plasmid and the DNA fragment ligated.

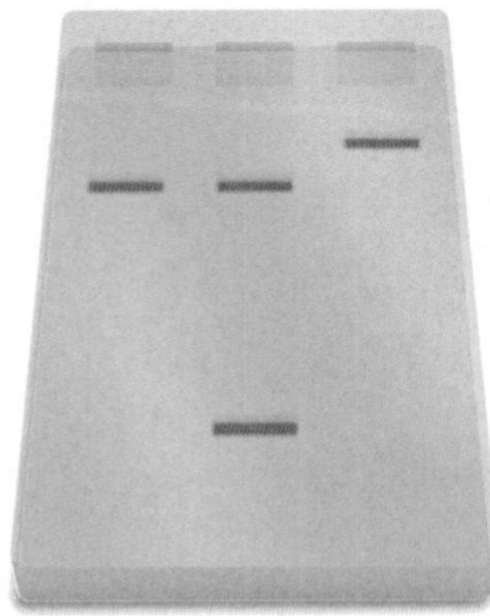

a. Which lane in the gel corresponds to the recombinant DNA?
b. Which lane corresponds to the plasmid?
c. Which lane represents cleaving using a restriction enzyme of the recombinant DNA molecule?

21. **Differentiate** The plasmid below was cut to produce the five fragments shown in the diagram. The fragments then were separated by gel electrophoresis. Draw a diagram of a gel and the location of each fragment. Label ends as positive or negative.

1633 bp

257 bp

1108 bp

1400 bp

601 bp

22. **Assess** A small DNA molecule was cleaved with several different restriction enzymes, and the size of each fragment was determined by gel electrophoresis. The following data were obtained.

DNA Fragmentation Patterns Created by *Eco*RI and *Hind*III

Enzyme	Number of Fragments	Fragment Size (kilobases)
*Eco*RI	2	1.5 kb 1.5 kb
*Hind*III	1	3.0 kb
*Eco*RI + *Hind*III	3	0.8 kb 0.7 kb 1.5 kb

a. Is the original DNA linear or circular?
b. Draw a restriction-site map showing distances consistent with the data.

Section 13.3

Vocabulary Review

Fill in the blanks with the correct vocabulary term from the Study Guide page.

23. The field of __________ uses computers to index and organize information created by sequencing the human genome.

24. Genetic variations that are located close together are called __________.

Understand Key Concepts

25. Which statement about the human genome is false?
    A. The human genome contains approximately 25,000 genes.
    B. The human genome contains long stretches of DNA with no known function.
    C. The human genome was sequenced by scientists from around the world.
    D. The human genome contains nucleotide sequences that all code for proteins.

26. What are variations in specific nucleotides that are linked to human diseases called?
    A. proteomes
    B. haplotypes
    C. single nucleotide polymorphisms
    D. genomes

27. For what purpose is DNA fingerprinting used?
    A. to sequence DNA from bacteria
    B. to separate DNA fragments
    C. to identify individuals who have committed crimes
    D. to identify single nucleotide polymorphisms

Constructed Response

28. **Short Answer** Discuss the advantages and disadvantages of using DNA microarrays.

29. **Short Answer** List three ways patients will benefit from pharmacogenomics.

30. **Open Ended** What impact does sequencing the human genome have on diagnosing and treating diseases?

Think Critically

31. **Describe** how DNA microarrays and DNA sequencing can be used to identify the defective gene.

32. **CAREERS IN BIOLOGY** A forensic scientist finds a strand of hair at a crime scene. Draw a flow chart and explain the steps that the forensic scientist has to take to determine the identity of the person to whom the hair belongs.

Additional Assessment

33. **WRITING in Biology** Write a paragraph discussing the approach you would take to create a transgenic organism and the drawbacks to creating it.

DBQ Document-Based Questions

The data below were obtained during a study on mosquito biting patterns. DNA fingerprints were obtained from individuals A, B, and C who were bitten by mosquitoes. In order to determine which mosquitoes bit each individual, a group of mosquitoes was collected and their DNA fingerprints were obtained. The mosquitoes were numbered 1–8.

Use the data to answer the questions below.

Data obtained from: Michael, et al. 2001. Quantifying mosquito biting patterns on humans by DNA fingerprinting of blood meals. *American Journal of Tropical Medicine and Hygiene* 65(6): 722–728.

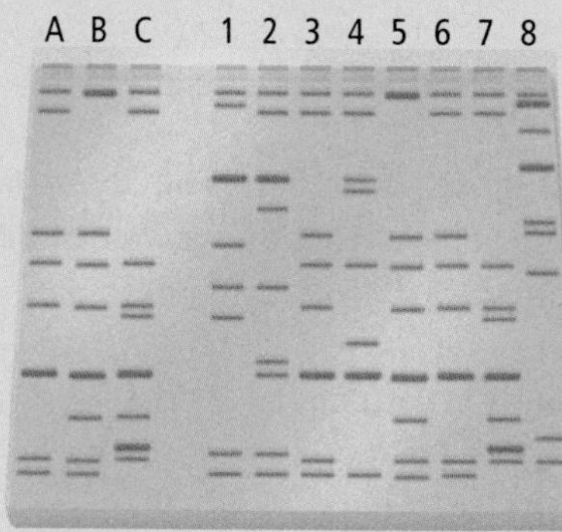

34. Examine the banding patterns and match each individual with the mosquito(es) that bit him or her.

35. What can researchers gain by knowing which mosquito bit which individual?

36. Based on your answer to question 34, what is a disadvantage of using this DNA fingerprinting to identify disease-carrying mosquitoes in the environment?

Cumulative Review

37. Explain the goal of sustainable use in biodiversity. **(Chapter 5)**

38. Identify and describe a population, community, and ecosystem in your area. List four abiotic factors that affect the biotic factors. **(Chapter 2)**

Standardized Test Practice

Cumulative

Multiple Choice

1. Which describes the process of cytokinesis?
   A. chromosomes duplicate
   B. spindle disintegrates
   C. nucleus disappears
   D. cytoplasm divides

Use the illustration below to answer questions 2 and 3.

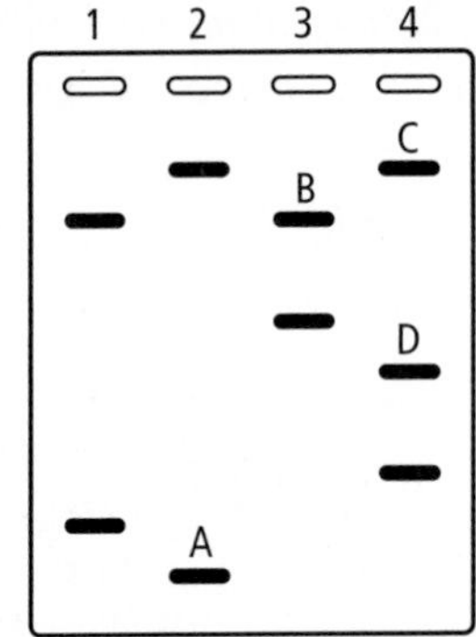

2. The figure above shows bands of DNA that were separated using gel electrophoresis. Which band contains the smallest DNA fragments?
   A. Band A
   B. Band B
   C. Band C
   D. Band D

3. What could the results of this gel electrophoresis show to a scientist?
   A. the amount of noncoding DNA present
   B. the fingerprint of a person's DNA
   C. the number of genes in a piece of DNA
   D. the random patterns of DNA

4. Which process plays a part in genetic recombination?
   A. asexual reproduction
   B. cytokinesis
   C. independent assortment
   D. mitotic division

5. Which correctly lists the following terms in order from smallest to largest: DNA, chromatin, chromosomes, nucleosomes?
   A. chromatin, chromosomes, DNA, nucleosomes
   B. chromosomes, DNA, chromatin, nucleosomes
   C. DNA, nucleosomes, chromatin, chromosomes
   D. nucleosomes, DNA, chromatin, chromosomes

Use the figure below to answer question 6.

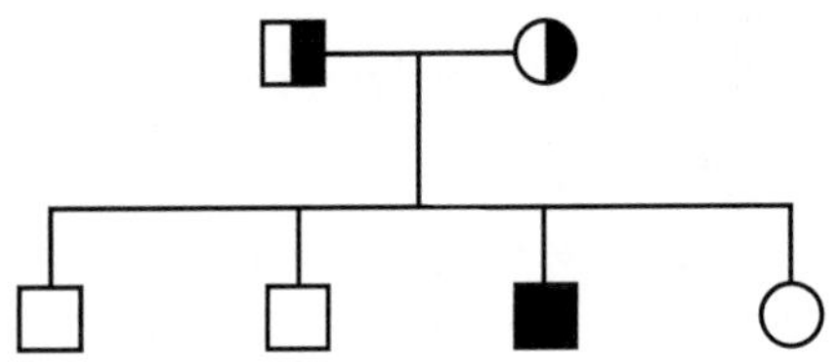

6. In a particular family, one child out of four is born with Tay-Sachs disease. Which pair of symbols represents the parents of these offspring?
   A.
   B.
   C.
   D.

7. Which is a stop codon in mRNA?
   A. AUG
   B. AUU
   C. CAU
   D. UAA

8. In a triploid organism, how many alleles are present for each gene per cell?
   A. 1
   B. 3
   C. 6
   D. 9

Short Answer

Use the figure below to answer question 9.

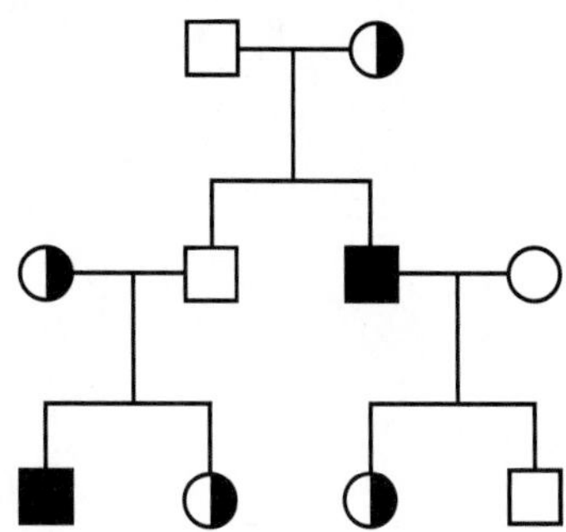

9. The pedigree in the figure tracks a dominant genetic disease. Explain the meaning of the symbols in the last generation.

10. Why are the protein-coding regions of most human genomes identical?

11. If hemophilia is a sex-linked recessive gene, what is the chance that a father with hemophilia and a mother who is a carrier for hemophilia will have a boy with hemophilia? Explain.

12. Compare and contrast the two major processes in protein synthesis.

13. List three genetic disorders; classify them as dominant or recessive; and name the affected organ systems.

14. Why might it take many generations to develop a purebred animal?

15. List the purine bases and the pyrimidine bases in DNA; explain their importance in DNA structure.

Extended Response

16. Give the names of two DNA mutations, and illustrate how each one would change the following DNA sequence.

    CGATTGACGTTTTAGGAT

17. Chemosynthetic autotrophs might have evolved long before the photosynthetic ones that currently are more common on Earth. Propose an explanation for this difference in evolution.

18. Explain how the noncoding sequences in the human genome make it difficult to interpret the DNA code.

19. Even though chloroplasts and mitochondria perform different functions, their structures are similar. Relate the similarity of their structures to their functions.

Essay Question

Suppose a scientist uses gel electrophoresis to separate the DNA extracted from a cell line. After performing the experiment, the scientist observes that several bands are missing and that other bands have traveled to the far end of the gel.

Using the information in the paragraph above, answer the following question in essay format.

20. Using what you know about DNA separation and gel electrophoresis, explain what might have gone wrong with the experiment. Then, describe how to adjust the experimental procedures to test your explanation.

NEED EXTRA HELP?																				
If You Missed Question . . .	1	2	3	4	5	6	7	8	9	10	11	12	13	14	15	16	17	18	19	20
Review Section . . .	9.2	13.2	13.3	10.3	12.1	11.1	12.3	10.3	11.1	13.3	11.3	12.3	11.2	13.1	12.1	12.4	2.2	13.3	7.3	13.2

UNIT 4 History of Biological Diversity

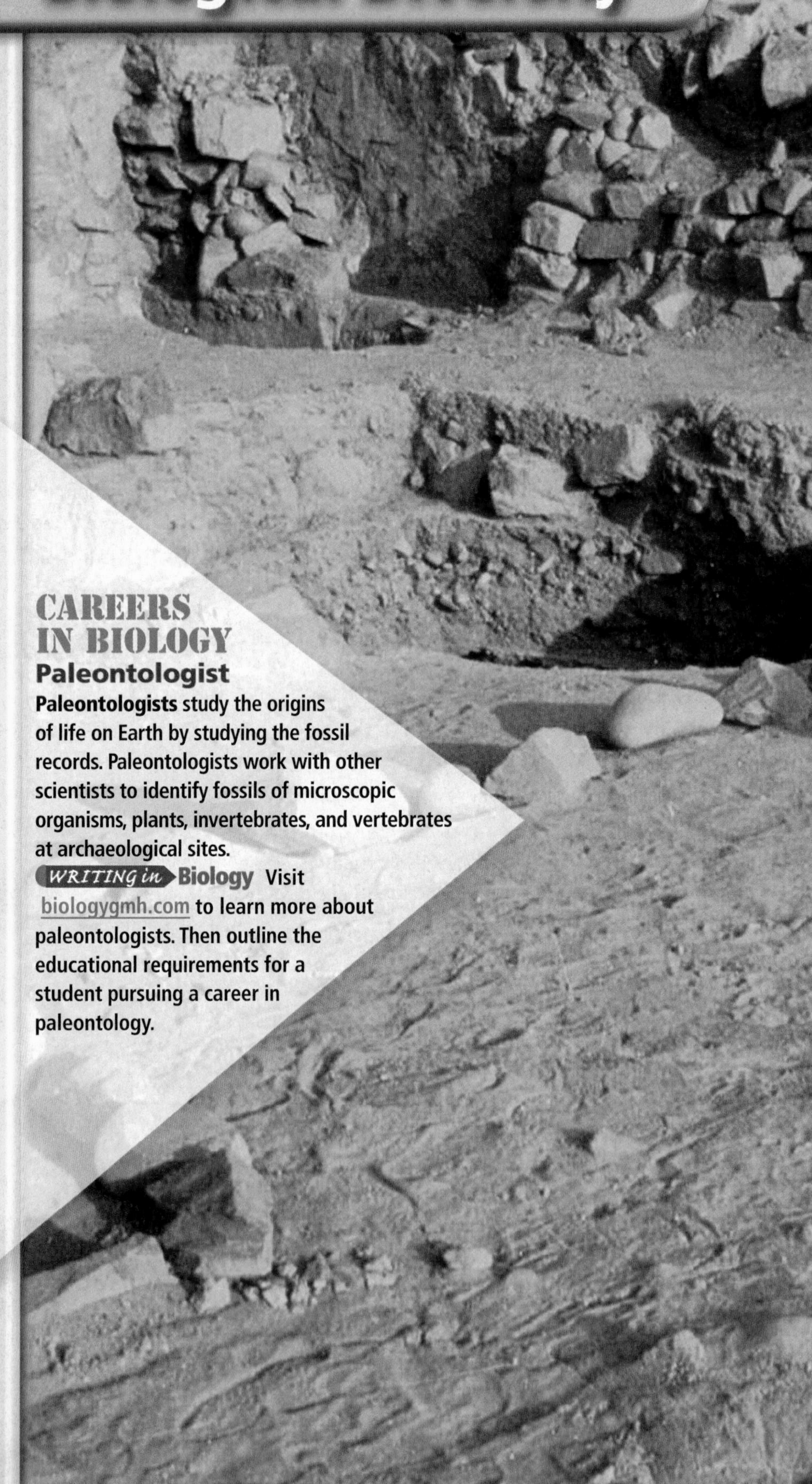

Chapter 14
The History of Life
BIG Idea Fossils provide key evidence for understanding the origin and the history of life on Earth.

Chapter 15
Evolution
BIG Idea The theory of natural selection explains evolution and the diversity of life.

Chapter 16
Primate Evolution
BIG Idea Evolutionary change in a group of small, tree-living mammals eventually led to a diversity of species that includes modern humans.

Chapter 17
Organizing Life's Diversity
BIG Idea Evolution underlies the classification of life's diversity.

CAREERS IN BIOLOGY
Paleontologist

Paleontologists study the origins of life on Earth by studying the fossil records. Paleontologists work with other scientists to identify fossils of microscopic organisms, plants, invertebrates, and vertebrates at archaeological sites.

WRITING in Biology Visit biologygmh.com to learn more about paleontologists. Then outline the educational requirements for a student pursuing a career in paleontology.

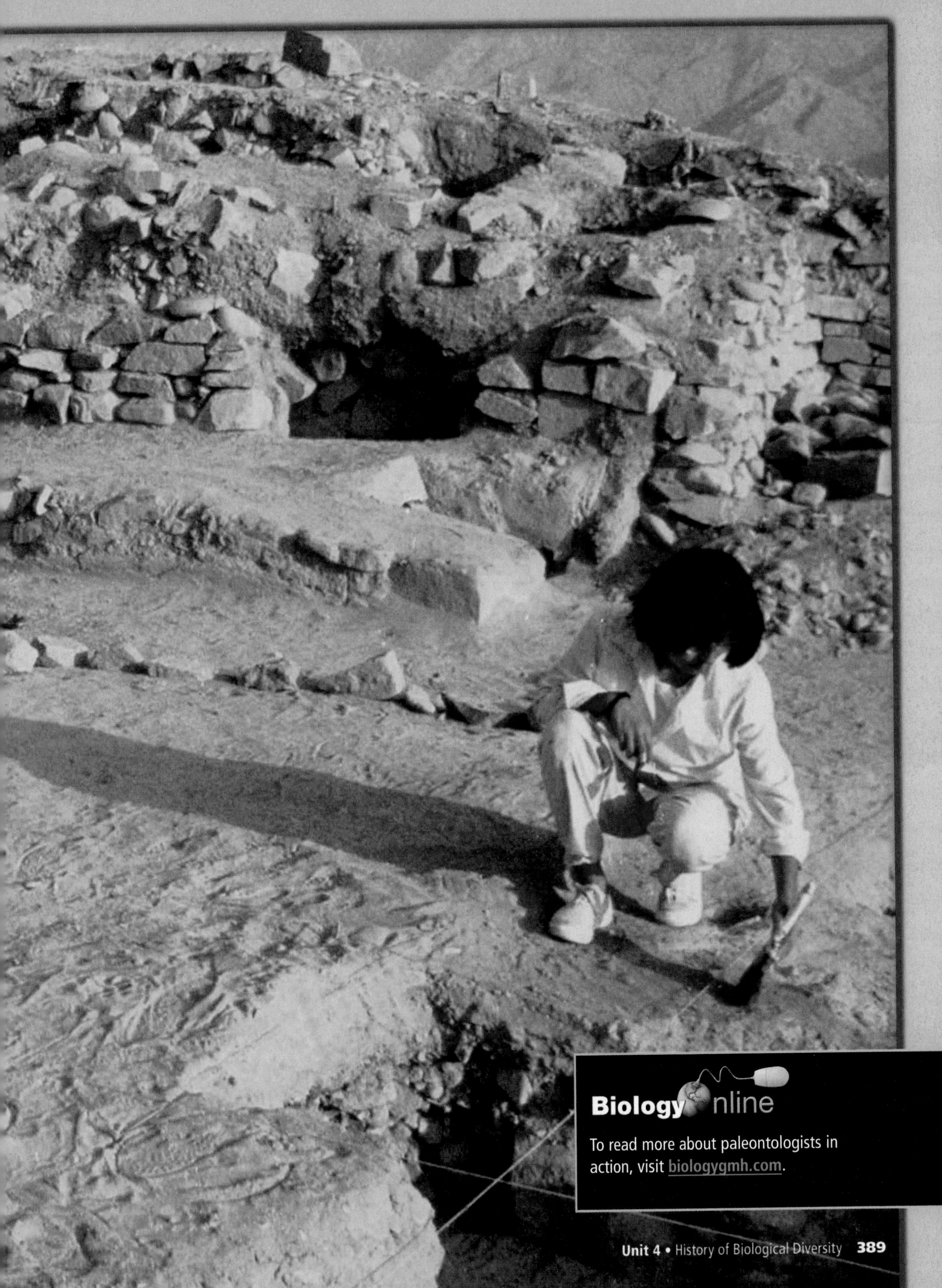

Biology Online

To read more about paleontologists in action, visit biologygmh.com.

CHAPTER 14

The History of Life

BIG Idea **Fossils provide key evidence for understanding the origin and the history of life on Earth.**

Section 1
Fossil Evidence of Change

MAIN Idea Fossils provide evidence of the change in organisms over time.

Section 2
The Origin of Life

MAIN Idea Evidence indicates that a sequence of chemical events preceded the origin of life on Earth and that life has evolved continuously since that time.

BioFacts

- After laying their eggs, many dinosaurs remained on their nests to protect their young.
- A dinosaur fossil found in the 1920s atop an egg-containing nest was named *Oviraptor*, which means "egg thief."
- In the 1990s, similar *Oviraptor* fossils were found that contained fossilized *Oviraptor* embryos.

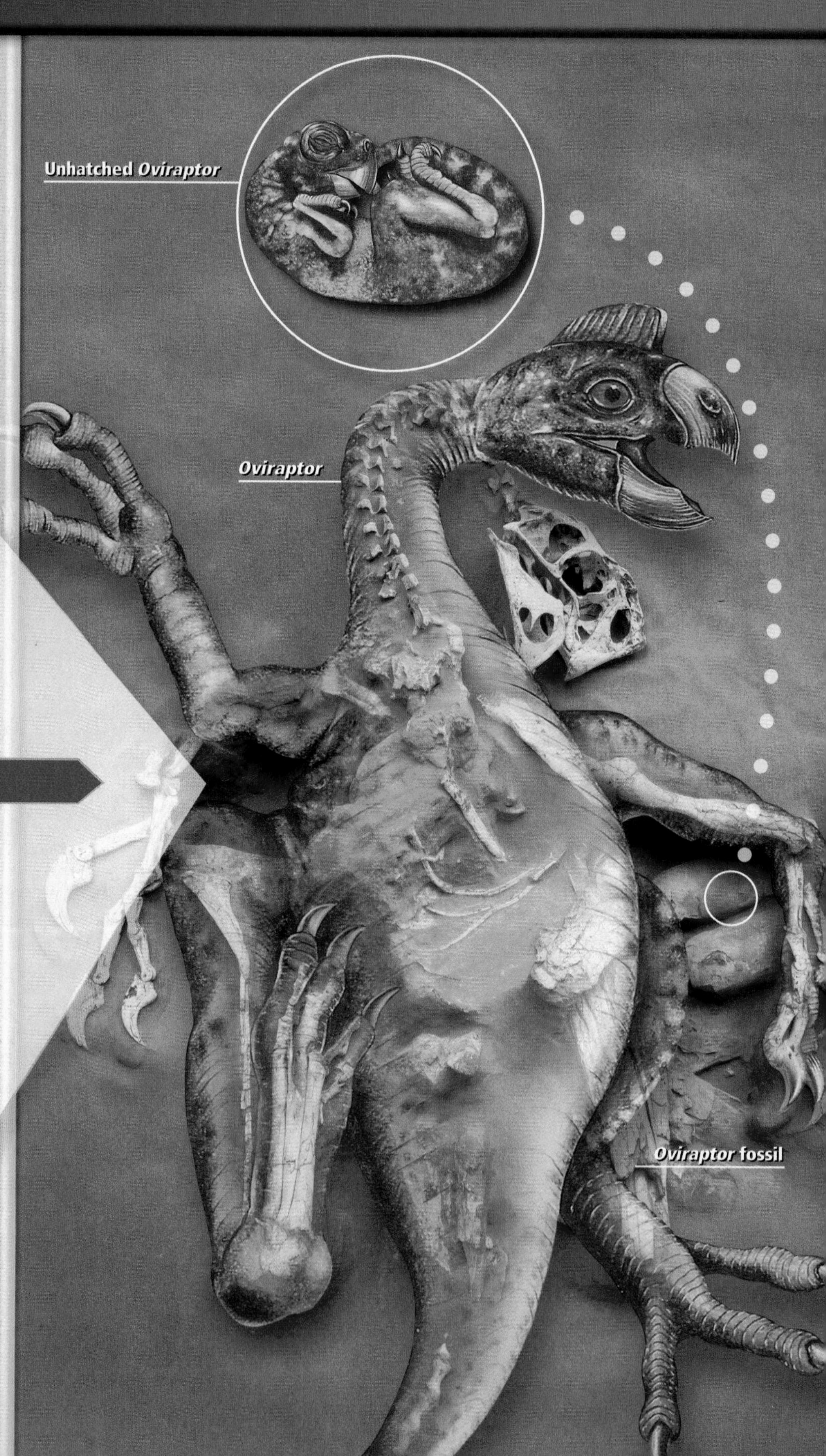

Start-Up Activities

LAUNCH Lab

What can skeletal remains reveal?

Fossils are all that remain of extinct organisms. Paleontologists study fossils to understand how organisms looked and behaved when they lived on Earth. In this lab, you will infer an organism's characteristics based on skeletal remains.

Procedure

1. Read and complete the lab safety form.
2. Choose an unidentified animal from the list provided by your teacher.
3. Imagine that the animal you selected has been extinct for millions of years. Study **skeletal parts, teeth, diagrams, and photos** provided.
4. Based on skeletal remains alone, list the animal's physical and behavioral characteristics.
5. Learn the identity of your animal from your teacher. Now make a new list of characteristics.

Analysis

1. **Compare** the two lists. Do fossils limit what paleontologists can infer about an extinct organism? Explain.
2. **Conclude** Based on your observations, what general characteristics can be inferred about most animals based on fossilized remains?

Visit biologygmh.com to:

- study the entire chapter online
- explore Concepts in Motion, the Interactive Table, Microscopy Links, and links to virtual dissections
- access Web links for more information, projects, and activities
- review content online with the Interactive Tutor and take Self-Check Quizzes

Origin of Life Make this Foldable to help you understand some of the early experiments related to the origin of life.

STEP 1 Fold a sheet of notebook paper in thirds lengthwise as shown.

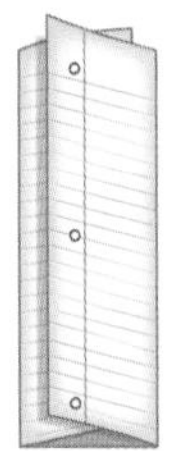

STEP 2 Unfold the paper and make a fold a quarter of the way down the page.

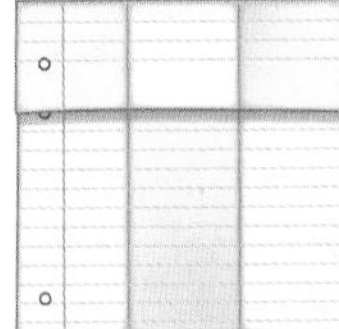

STEP 3 With a pencil or pen, trace the fold lines to make a three-column chart.

STEP 4 Label the columns: *Redi, Pasteur,* and *Miller and Urey.*

FOLDABLES Use this Foldable with Section 14.2. As you study the chapter, record what you learn about each scientist and list the steps that helped him investigate spontaneous generation and biogenesis.

Section 14.1

Reading Preview

Objectives

- **Describe** a typical sequence of events in fossilization.
- **Compare** techniques for dating fossils.
- **Identify** and describe major events using the geologic time scale.

Review Vocabulary

extinction: the death of all individuals of a species

New Vocabulary

fossil
paleontologist
relative dating
law of superposition
radiometric dating
half-life
geologic time scale
epoch
era
period
Cambrian explosion
K-T boundary
plate tectonics

Fossil Evidence of Change

MAIN Idea **Fossils provide evidence of the change in organisms over time.**

Real-World Reading Link Did you know that when you look at the stars at night you are looking into the past? The stars are so far away that the light you see left the stars thousands and sometimes millions of years ago. You also are looking into the past when you look at rocks. The rocks formed thousands or even millions of years ago. Rocks can tell us what Earth was like in the distant past, and sometimes they can tell us what lived during that time.

Earth's Early History

What were the conditions on Earth as it formed, and how did life arise on a lifeless planet? Because there were no people to witness Earth's earliest history, it might seem that this is a mystery. Like any good mystery, however, it left clues behind. Each clue to Earth's history and life's origin is open to investigation by the scientists who study the history of the Earth.

Land environments By studying other planets in the solar system and rocks on Earth, scientists conclude that Earth was a molten body when it formed about 4.6 billion years ago. Gravity pulled the densest elements to the center of the planet. After about 500 million years, a solid crust formed on the surface, much like the crust that forms on the top of lava, as shown in **Figure 14.1.** The surface was rich in lighter elements, such as silicon. From the oldest rocks remaining today, scientists infer that Earth's young surface included a number of volcanic features. In addition, the cooling interior radiated much more heat to the surface than it does today. Meteorites would have caused additional heating as they crashed into Earth's surface. If there had been any life on Earth, it most likely would have been consumed by the intense heat.

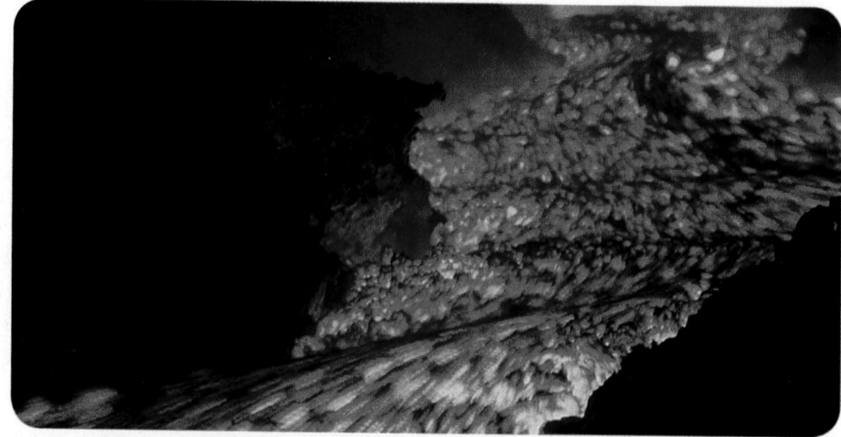

Molten iron in lava flow

■ **Figure 14.1** Just as a crust forms on top of cooling lava, a crust formed atop Earth's early surface.
Infer ***the importance of the crust to the origin of life on Earth.***

Atmosphere Because of its gravitational field, Earth is a planet that is able to maintain an atmosphere. However, no one can be certain about the exact composition of Earth's early atmosphere. The gases that likely made up the atmosphere are those that were expelled by volcanoes. Volcanic gases today include water vapor (H_2O), carbon dioxide (CO_2), sulfur dioxide (SO_2), carbon monoxide (CO), hydrogen sulfide (H_2S), hydrogen cyanide (HCN), nitrogen (N_2), and hydrogen (H_2). Scientists infer that the same gases would have been present in Earth's early atmosphere. The minerals in the oldest known rocks suggest that the early atmosphere, unlike today's atmosphere, had little or no free oxygen.

Vocabulary
Word origin
Fossil
from the Latin word *fossilis,* meaning *dug up.*

Clues in Rocks

Earth eventually cooled to the point where liquid water formed on its surface, which became the first oceans. It was a very short time after this—maybe as little as 500 million years—that life first appeared. The earliest clues about life on Earth date to about 3.5 billion years ago.

The fossil record A **fossil** is any preserved evidence of an organism. Six categories of fossils are shown in **Table 14.1.** Plants, animals, and even bacteria can form fossils. Although there is a rich diversity of fossils, the fossil record is like a book with many missing pages. Perhaps more than 99 percent of the species that ever have lived are now extinct, but only a tiny percentage of these organisms are preserved as fossils.

Most organisms decompose before they have a chance to become fossilized. Only those organisms that are buried rapidly in sediment are readily preserved. This occurs more frequently with organisms living in water because the sediment in aquatic environments is constantly settling, covering, and preserving the remains of organisms.

Concepts In Motion
Interactive Table To explore more about categories of fossil types, visit biologygmh.com.

Table 14.1 Categories of Fossil Types

Category	Trace fossil	Molds and casts	Replacement	Petrified or permineralized	Amber	Original material
Example						
Formation	A trace fossil is any indirect evidence left by an organism. Footprints, burrows, and fossilized feces are trace fossils.	A mold is an impression of an organism. A cast is a mold filled with sediment.	The original material of an organism is replaced with mineral crystals that can leave detailed replicas of hard or soft parts.	Empty pore spaces are filled in by minerals, such as in petrified wood.	Preserved tree sap traps an entire organism. The sap hardens into amber and preserves the trapped organism.	Mummification or freezing preserves original organisms.

Figure 14.2 **(A)** Organisms usually become fossilized after they die and are buried by sediment. **(B)** Sediments build up in layers, eventually encasing the remains in sedimentary rock. **(C)** Minerals replace, or fill in the pore space of, the bones and hard parts of the organism. **(D)** Erosion can expose the fossils.

Fossil formation Fossils do not form in igneous (IHG nee us) or metamorphic (meh tuh MOR fihk) rocks. Igneous rocks form when magma from Earth's interior cools. Metamorphic rocks form when rocks are exposed to extreme heat and pressure. Fossils usually do not survive the heat or pressure involved in the formation of either of these kinds of rocks.

Nearly all fossils are formed in sedimentary rock through the process described in **Figure 14.2.** The organism dies and is buried in sediments. The sediments build up until they cover the organism's remains. In some cases, minerals replace the organic matter or fill the empty pore spaces of the organism. In other cases, the organism decays, leaving behind an impression of its body. The sediments eventually harden into rock.

A **paleontologist** (pay lee ahn TAH luh jist) is a scientist who studies fossils. He or she attempts to read the record of life left in rocks. From fossil evidence, paleontologists infer the diet of an organism and the environment in which it lived. In fact, paleontologists often can create images of extinct communities.

Study Tip

Background Knowledge Check Based on what you know, predict the meaning of each new vocabulary term before reading the section. As you read, check the actual meaning compared to your prediction.

Connection to Earth Science When geologists began to study rock layers, or strata, in different areas, they noticed that layers of the same age tended to have the same kinds of fossils no matter where the rocks were found. The geologists inferred that all strata of the same age contained similar collections of fossils. This led to the establishment of a relative age scale for rocks all over the world.

Dating fossils **Relative dating** is a method used to determine the age of rocks by comparing them with those in other layers. Relative dating is based on the **law of superposition,** illustrated in **Figure 14.3,** which states that younger layers of rock are deposited on top of older layers. The process is similar to stacking newspapers in a pile as you read them each day. Unless you disturb the newspapers, the oldest ones will be on the bottom.

A

B

C

D

E

F

■ **Figure 14.3** According to the law of superposition, rock layers are deposited with the youngest undisturbed layers on top.
Infer ***Which layer shows that an aquatic ecosystem replaced a land ecosystem?***

Radiometric dating uses the decay of radioactive isotopes to measure the age of a rock. Recall from Chapter 6 that an isotope is a form of an element that has the same atomic number but a different mass number. The method requires that the **half-life** of the isotope, which is the amount of time it takes for half of the original isotope to decay, is known. The relative amounts of the radioactive isotope and its decay product must also be known.

One radioactive isotope that is commonly used to determine the age of rocks is Uranium 238. Uranium 238 (U^{238}) decays to Lead 206 (Pb^{206}) with a half life of 4510 million years. When testing a rock sample, scientists calculate the ratio of the parent isotope to the daughter isotope to determine the age of the sample.

Radioactive isotopes that can be used for radiometric dating are found only in igneous or metamorphic rocks, not in sedimentary rocks, so isotopes cannot be used to date rocks that contain fossils. Igneous rocks that are found in layers closely associated with fossil-bearing sedimentary rocks often can be used for assigning relative dates to fossils.

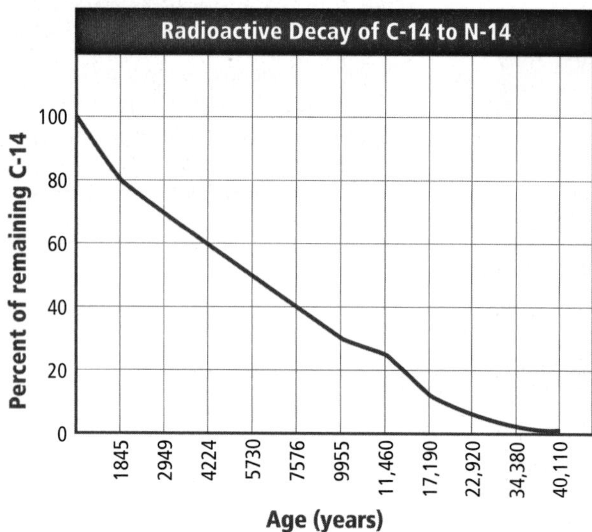

■ **Figure 14.4** The graph shows how the percent of carbon-14 indicates age.

Interpret the graph ***What would the age of a rock be if it contained only 10 percent of C-14?***

Materials, such as mummies, bones, and tissues, can be dated directly using carbon-14 (C-14). Given the half-life of carbon-14, shown in **Figure 14.4,** only materials less than 60,000 years old can be dated accurately with this isotope.

The Geologic Time Scale

Think of geologic time as a ribbon that is 4.6 m long. If each meter represents one billion years, each millimeter represents one million years. Earth was formed at one end of the ribbon, and humans appear at the very tip of the other end.

The **geologic time scale,** shown in **Figure 14.5,** is a record of Earth's history. All the major geological and biological events in Earth's history can be identified within the geologic time scale. Because geologic time spans more than 4 billion years, a subdivision of time is usually identified by how many millions of years ago (mya) it occurred. The geologic time scale is divided into two distinct segments—Precambrian time and the Phanerozoic (fan eh roh ZOH ihk) eon. **Epochs** are the smallest unit of geologic time lasting several million years. **Periods** are divisions of geologic time lasting tens of millions of years. An **Era** is a unit of geologic time consisting of two or more periods that lasts hundreds of millions of years.

In 2004, geologists worldwide agreed on a revision of the names and dates in the geologic time scale based on a project coordinated by the International Commission on Stratigraphy. As in all fields of science, continuing research and discoveries might result in future revisions.

Reading Check **Explain** why C-14 would not be useful for dating something from the Precambrian.

MiniLab 14.1

Correlate Rock Layers Using Fossils

How can paleontologists establish relative age? Scientists use fossils from many locations to piece together the sequence of Earth's rock layers. This is the process of correlation.

Procedure 

1. Read and complete the lab safety form.
2. Your teacher will assign you to a group and will give your group a **container** with layers of material embedded with fossils.
3. Carefully remove each layer, noting any embedded materials.
4. Make a sketch of the cross section, and label each layer and any materials contained within it.
5. Collect copies of sketches from the other groups and use them to determine the sequence of all the layers the class has studied.

Analysis

1. **Describe** the materials in each cross section. What patterns did you observe?
2. **Explain** how your analysis would be different if different layers contained the same materials. What if some of the layers didn't overlap? Suggest a way to gather additional data that might resolve these issues.

Visualizing the Geologic Time Scale

Figure 14.5
Eras, periods, and epochs are shown on this geologic time scale that begins with Earth's formation 4.6 billion years ago. Though not to scale, this diagram illustrates the approximate appearance of various organisms over time.

Cenozoic
Mesozoic
Paleozoic
Precambrian

Eon	Era	Period	Epoch	MYA	Biological events
Phanerozoic	Cenozoic	Neogene	Holocene		• Humans form civilizations
			Pleistocene	0.01	• Ice ages occur • Modern humans appear
			Pliocene	1.8	• Hominins appear • Flowering plants are dominant
			Miocene	5.3	• Apes appear • Climate is cooler
		Paleogene	Oligocene	23.0	• Monkeys appear • Climate is mild
			Eocene	33.9	• Flowering plants scattered • Most mammal orders exist
			Paleocene	55.8	• Mammals, birds, and insects scatter • Climate is tropical
		Mass extinction			
	Mesozoic	Cretaceous		65.5	• Flowering plants appear • Dinosaur population peaks
		Jurassic		145.5	• First birds appear • Dinosaurs scatter • Forests are lush
		Triassic		199.6	• Gymnosperms are dominant • Dinosaurs appear • First mammals appear
		Mass extinction			
	Paleozoic	Permian		251.0	• Reptiles scatter
		Carboniferous		299.0	• Ferns and evergreens make up forests • Amphibians appear • Insects scatter
		Devonian	Mass extinction	359.2	• Sharks and bony fishes appear
		Silurian		416.0	• Coral and other invertebrates are dominant • Land plants and insects appear
		Mass extinction			
		Ordovician		443.7	• First vertebrates appear • First plants appear
		Cambrian		488.3	• Cambrian explosion • All body plans arise
Precambrian — Proterozoic	Neoproterozoic	Ediacaran		542	• Soft-bodied organisms appear
		Cryogenian		630	• Invertebrates and algae diversify
		Tonian		850	• Multicellular organisms appear
				1500	• Oldest fossils of eukaryotes
				2500	• Oldest fossils of prokaryotes
				4600	• Earth forms

Concepts In Motion **Interactive Figure** To see an animation of the geologic time scale, visit biologygmh.com.

Biology Online

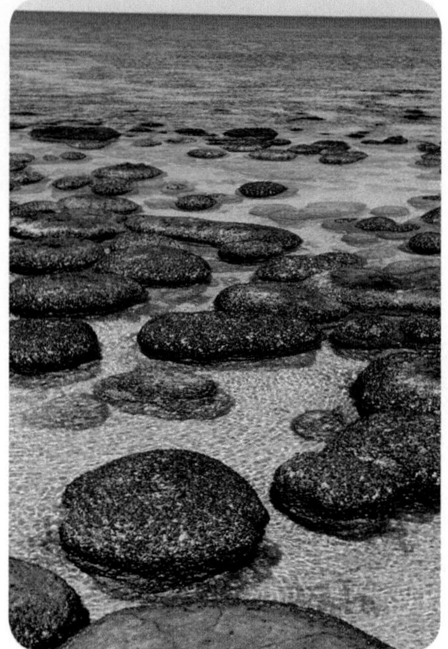

Figure 14.6 Fossils much like these stromatolites are found in rocks almost 3.5 billion years old. Modern day stromatolites are formed by cyanobacteria.
Explain *the importance of the organisms that left these stromatolites.*

Precambrian The first 4 billion years, as shown in **Figure 14.5,** make up the Precambrian. This is nearly 90 percent of Earth's entire history, stretching from the formation of Earth to the beginning of the Paleozoic era about 542 million years ago. The Precambrian was an important time. Earth formed and life first appeared. Eventually, autotrophic prokaryotes, much like the cyanobacteria that made the stromatolites in **Figure 14.6,** enriched the atmosphere with oxygen. Eukaryotic cells also emerged, and by the end of the Precambrian, life was flourishing and the first animals had appeared.

Extensive glaciation marked the second half of the Precambrian. This might have delayed the further evolution of life until the ice receded at the beginning of the Ediacaran (ee dee UH kur uhn) period. The Ediacaran period was added to the time scale in 2004. It is the first new period added to the time scale since 1891 and reflects new knowledge of Earth's history. The Ediacaran period lasted from about 630 million years ago to about 542 million years ago, representing about three quarters of a meter on the time ribbon at the end of the Precambrian. Simple organisms, such as the fossil in **Figure 14.7,** inhabited Ediacaran marine ecosystems. Food chains probably were short, and were dominated by animals that consumed tiny particles suspended in the water and by animals that ate debris on the bottom of the sea.

 **Reading Check** **Infer** the process by which early autotrophic prokaryotes produced oxygen.

The Paleozoic era A drastic change in the history of animal life on Earth marked the start of the Paleozoic (pay lee uh ZOH ihk) era. In the space of just a few million years, the ancestors of most major animal groups diversified in what scientists call the **Cambrian explosion.** Not all major groups of organisms evolved rapidly at this time, and paleontologists still do not know when the rapid changes started or ended.

Major changes in ocean life occurred during the Paleozoic. More importantly, it seems the first life on land emerged during this era. Life in the oceans continued to evolve through the Cambrian period. Fish, land plants, and insects appeared during the Ordovician and Silurian periods. Organisms of many kinds, including huge insects, soon flourished in swampy forests that dominated the land, as shown in **Figure 14.8.** Tetrapods, the first land vertebrates (animals with backbones), emerged in the Devonian period. By the end of the Carboniferous period, the first reptiles were roaming the forests.

Figure 14.7 Paleontologists disagree about scarce Ediacaran fossils such as this one. Some paleontologists suggest that they are relatives of today's living invertebrates such as segmented worms, while others think they represent an evolutionary dead end of giant protists or simple metazoans.

■ **Figure 14.8** During the Carboniferous period, swamp forests covered much of Earth's land surface. Insects dominated the air, and tetrapods flourished in freshwater pools.
Infer ***How were the plants of the Paleozoic era different from those of today?***

A mass extinction ended the Paleozoic era at the end of the Permian period. Recall from Chapter 5 that a mass extinction is an event in which many species become extinct in a short time. Mass extinctions have occurred every 26 to 30 million years on average. Between 60 and 75 percent of the species alive went extinct in each of these events. During the Permian mass extinction, 90 percent of marine organisms disappeared. Geologists disagree about the cause of the Permian extinction, but most agree that geological forces, including increased volcanic activity, would have disrupted ecosystems or changed the climate.

The Mesozoic era At the beginning of the Triassic period, the ancestors of early mammals were the dominant land animals. Mammals and dinosaurs first appeared late in the Triassic period, and flowering plants evolved from nonflowering plants. Birds evolved from a group of predatory dinosaurs in the middle Jurassic period. For the rest of the Mesozoic, reptiles, such as the dinosaurs illustrated in **Figure 14.9,** were the dominant organisms on the planet. Then, about 65 million years ago, a meteorite struck Earth.

The primary evidence for this meteorite impact is found in a layer of material between the rocks of the Cretaceous (krih TAY shus) period and the rocks of the Paleogene period, the first period of the Cenozoic era. Paleontologists call this layer the **K-T boundary.** Within this layer, scientists find unusually high levels of an element called iridium. Iridium is rare on Earth, but relatively common in meteorites. Therefore, the presence of iridium on Earth indicates a meteorite impact.

Many scientists think that this impact is related to the mass extinction at the end of the Mesozoic era, which eliminated all dinosaurs except birds, most marine reptiles, many marine invertebrates, and numerous plant species. The meteorite itself did not wipe out all of these species, but the debris from the impact probably stayed in the atmosphere for months or even years, affecting global climate. Those species that could not adjust to the changing climate disappeared.

Reading Check **Recall** What were the dominant land animals in the Triassic and Jurassic periods?

■ **Figure 14.9** The dominant organisms during the Mesozoic era were dinosaurs. A mass extinction occurred at the end of the Mesozoic era that eliminated all dinosaurs, with the exception of their avian and reptilian descendants.

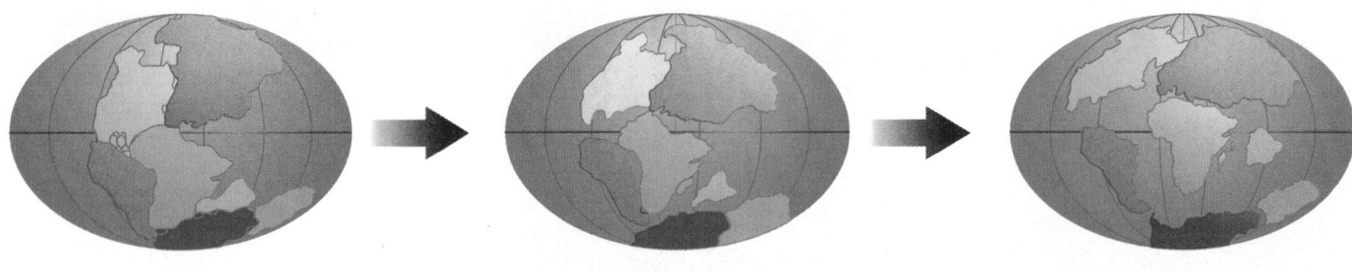

■ **Figure 14.10** These illustrations show the movement of Earth's major tectonic plates from about 225 million years ago to 65 million years ago, when all of the continents were joined into one landmass called Pangaea.

Concepts In Motion

Interactive Figure To see an animation of continental drift, visit biologygmh.com.

Scientists also think that the course of evolution in the Cenozoic era was shaped by the massive geological changes shown in **Figure 14.10** that characterized the Mesozoic era. While it might appear to us that continents are immobile, they have been moving since they formed. Alfred Wegener, a German scientist, presented the first evidence for continental drift in the 1920s. Continental drift has since become part of the theory of plate tectonics. **Plate tectonics** describes the movement of several large plates that make up the surface of Earth. These plates, some of which contain continents, move atop a partially molten layer of rock underneath them.

The Cenozoic era The most recent era is the one in which mammals became the dominant land animals. At the beginning of the Cenozoic (sen uh ZOH ihk) era, which means "recent life," most mammals were small and resembled shrews. After the mass extinction at the end of the Mesozoic era, mammals began to diversify into distinct groups, including primates—the group to which you belong. Humans appeared very recently, near the end of the geologic time scale, in the current Neogene period. Humans survived the last ice age, but many species of mammals did not. To get an idea of how recently modern humans have appeared, you need to remove about two threads at the end of your geologic time ribbon. These threads represent the time that humans have existed on Earth.

Section 14.1 Assessment

Section Summary

- Early Earth was lifeless for several hundred million years.
- Fossils provide evidence of past life.
- Relative dating and radiometric dating are two methods used to determine the age of fossils.
- The geologic time scale is divided into eras and periods.
- Major events in the geological time scale include both biological and geological changes.

Understand Main Ideas

1. MAIN Idea **Discuss** how fossils provide evidence of change from the earliest life-forms to those alive today.
2. **Diagram** a typical sequence of events in fossilization.
3. **Discuss** two ways that radiometric dating can be used to establish the age of a fossil.
4. **Explain** major events in three periods of the geologic time scale.

Think Critically

5. **Infer** what changes you might observe in the fossil record that would indicate the occurrence of a mass extinction.

MATH in Biology

6. Out of the total of Earth's history (approximately 4.6 billion years), modern humans have existed for only 100,000 years. To put this in perspective, calculate the percentage of Earth's history that modern humans have existed.

Biology Online **Self-Check Quiz** biologygmh.com

Section 14.2

The Origin of Life

Reading Preview

Objectives

- **Differentiate** between spontaneous generation and biogenesis.
- **Sequence** the events that might have led to cellular life.
- **Describe** the endosymbiont theory.

Review Vocabulary

amino acid: building blocks for proteins

New Vocabulary

spontaneous generation
theory of biogenesis
endosymbiont theory

MAIN Idea **Evidence indicates that a sequence of chemical events preceded the origin of life on Earth and that life has evolved continuously since that time.**

Real-World Reading Link In a recipe, some steps can be out of order, but some steps have to occur earlier than others or the end result will be different from what was intended. In the same way, to arrive at the pattern of life that is seen today, events leading to the emergence of life had to occur in specific ways.

Origins: Early Ideas

Perhaps one of the oldest ideas about the origin of life is spontaneous generation. **Spontaneous generation** is the idea that life arises from nonlife. For example, at one time people thought that mice could be created by placing damp hay and corn in a dark corner, or that mud could give rise to worms, insects, and fish. These ideas might seem humorous to us today, but before much was known about reproduction, it is easy to see how someone might form these conclusions.

One of the first recorded investigations of spontaneous generation came in 1668. Francesco Redi, an Italian scientist, tested the idea that flies arose spontaneously from rotting meat. He hypothesized that flies—not meat—produced other flies. In his experiment, illustrated using present-day equipment in **Figure 14.11,** Redi observed that maggots, the larvae of flies, appeared only in flasks that were open to flies. Closed flasks had no flies and no maggots. The results of his experiments failed to convince everyone, however. Although people were beginning to use the microscope during Redi's time and knew that organisms invisible to the naked eye could be found almost everywhere, some thought that these tiny organisms must arise spontaneously, even if flies did not.

Figure 14.11 Francesco Redi showed that flies and maggots did not arise spontaneously from rotting meat.
Infer *the purpose of the covered flask in Redi's experiment.*

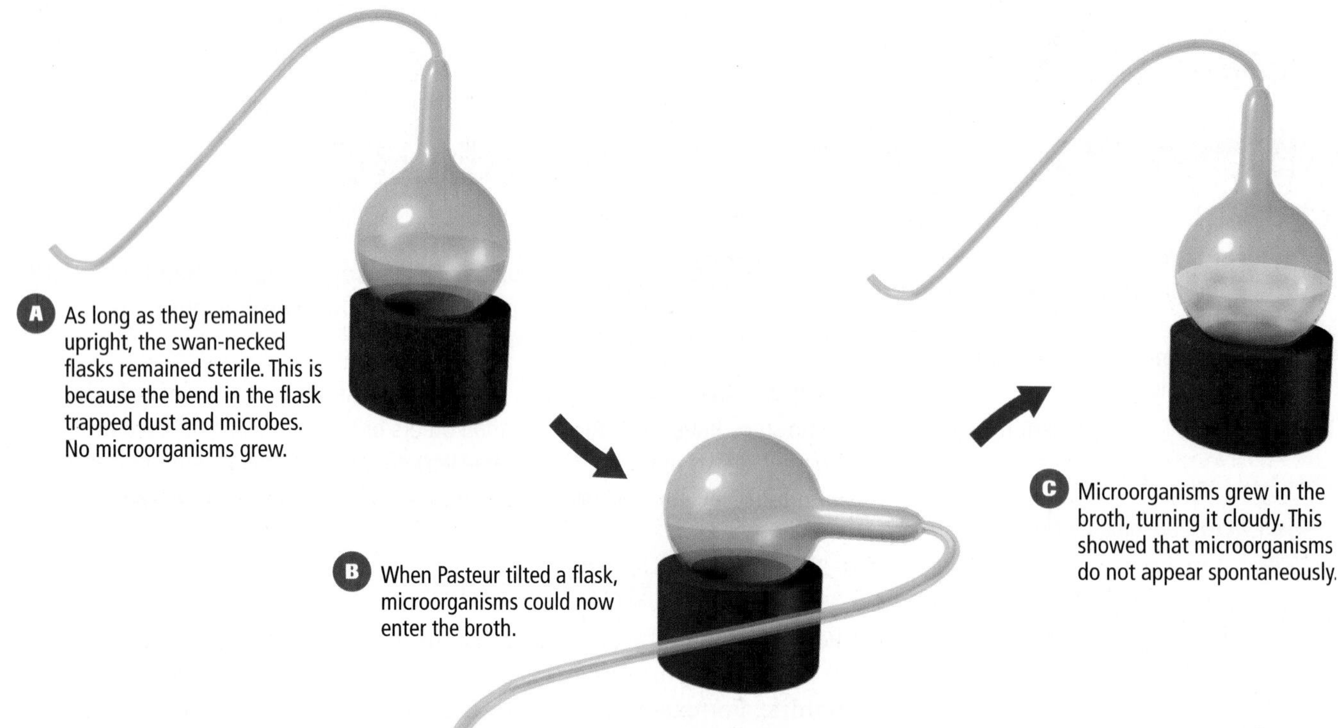

A As long as they remained upright, the swan-necked flasks remained sterile. This is because the bend in the flask trapped dust and microbes. No microorganisms grew.
B When Pasteur tilted a flask, microorganisms could now enter the broth.
C Microorganisms grew in the broth, turning it cloudy. This showed that microorganisms do not appear spontaneously.

Figure 14.12 Pasteur's experiment showed that sterile broth remained free of microorganisms until exposed to air.

FOLDABLES
Incorporate information from this section into your Foldable.

The idea of spontaneous generation was not completely rejected until the mid-1800s. It was replaced by the **theory of biogenesis** (bi oh JEN uh sus), which states that only living organisms can produce other living organisms. Louis Pasteur designed an experiment to show that biogenesis was true even for microorganisms. Pasteur's experiment is illustrated in **Figure 14.12.** In one flask, only air was allowed to contact a sterile nutrient broth. Nutrient broth supports the growth of microorganisms. In another flask, both air and microorganisms were allowed to contact the broth. No microorganisms grew in the first container. They did, however, grow in the second container.

Origins: Modern Ideas

If life can arise only from pre-existing life, then how did the first life-form appear? Most biologists agree that life originated through a series of chemical events early in Earth's history. During these events, complex organic molecules were generated from simpler ones. Eventually, simple metabolic pathways developed. Such pathways allowed molecules to be synthesized or broken down more efficiently. These pathways might have led to the emergence of life as we know it. How this happened is a topic of ongoing research among scientists today.

Simple organic molecule formation The primordial soup hypothesis was an early hypothesis about the origin of life. Scientists Alexander Oparin and John Haldane suggested this hypothesis in the 1920s. They thought that if Earth's early atmosphere had a mix of certain gases, organic molecules could have been synthesized from simple reactions involving those gases in the early oceans. UV light from the Sun and electric discharge in lightning might have been the primary energy sources. They thought that these organic molecules would have eventually supplied the precursors to life.

Connection to Chemistry In 1953, American scientists Stanley Miller and Harold Urey were the first to show that simple organic molecules could be made from inorganic compounds, as proposed by Oparin and Haldane. Miller and Urey built a glass apparatus, illustrated in **Figure 14.13,** to simulate the early Earth conditions hypothesized by Oparin. They filled the apparatus with water and the gases that they thought had made up the early atmosphere. The water was boiled and electric discharges were used to simulate lightning as an energy source. Upon examination, the resulting mixture contained a variety of organic compounds including amino acids. Because amino acids are the building blocks of proteins, this discovery supported the primordial soup hypothesis.

Later, other scientists found that hydrogen cyanide could be formed from even simpler molecules in simulated early Earth environments. Hydrogen cyanide can react with itself to eventually form adenine, one of the nucleotide bases in the genetic code. Many other experiments have since been carried out under conditions that probably reflect the atmosphere of early Earth more accurately. The final reaction products in these experiments were amino acids and sugars as well as nucleotides.

Some scientists suggest that the organic reactions that preceded life's emergence began in the hydrothermal volcanic vents of the deep sea, where sulfur forms the base of a unique food chain. Still others think that meteorites brought the first organic molecules to Earth.

CAREERS IN BIOLOGY

Evolutionary Biochemist Scientists who study chemistry and how it relates to life are biochemists. Evolution biochemists specifically study the structure and function of molecules from Earth's early history. For more information on biology careers, visit biologygmh.com.

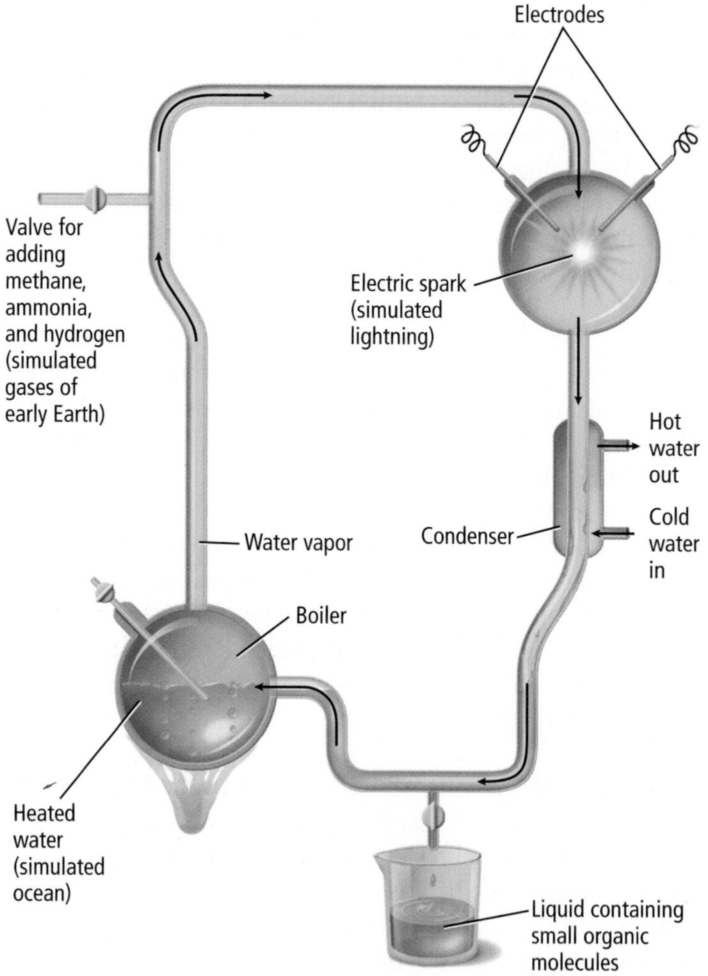

Concepts In Motion

Interactive Figure To see an animation of the Miller-Urey experiment, visit biologygmh.com.

Figure 14.13 The Miller-Urey experiment showed for the first time that organic molecules could be produced from gases proposed to have made up the atmosphere of early Earth.

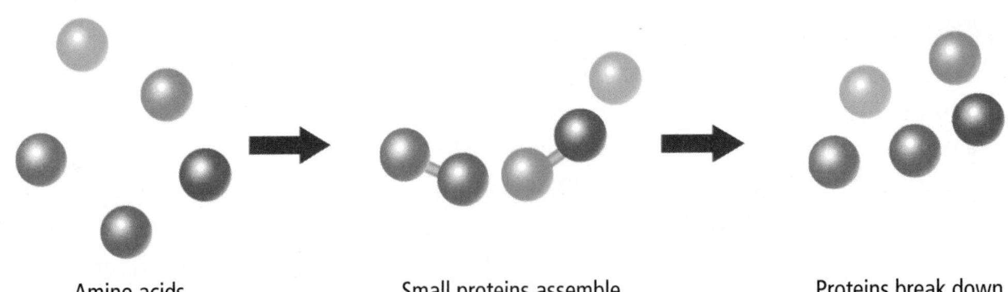

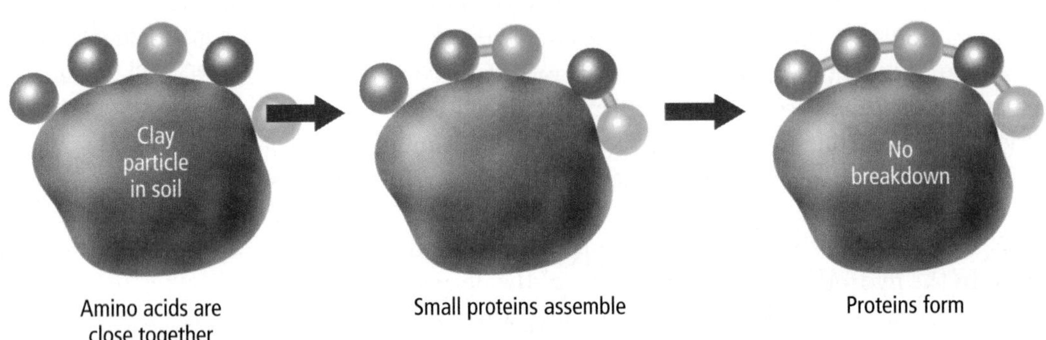

Figure 14.14 Without clay, amino acids could have formed small, unstable proteins. In the presence of clay, amino acids might have come together in a more stable manner.

Personal Tutor

To learn about protein formation, visit biologygmh.com.

Making proteins Wherever the first organic molecules originated, it is clear that the next critical step was the formation of proteins. Amino acids alone are not sufficient for life. Life requires proteins, which, as you might recall from Chapter 6, are chains of amino acids. In the Miller-Urey experiment, amino acids could bond to one another, but they could separate just as quickly, as illustrated in **Figure 14.14.** One possible mechanism for the formation of proteins would be if amino acids were bound to a clay particle. Clay would have been a common sediment in early oceans, and it could have provided a framework for protein assembly.

Genetic code Another requirement for life is a coding system for protein production. All modern life has such a system, based on either RNA or DNA. Because all DNA-based life-forms also contain RNA, and because some RNA sequences appear to have changed very little through time, many biologists consider RNA to have been life's first coding system. Researchers have been able to demonstrate that RNA systems are capable of evolution by natural selection. Some RNAs also can behave like enzymes. These RNA molecules, called ribozymes, could have carried out some early life processes. Other researchers have proposed that clay crystals could have provided an initial template for RNA replication, and that eventually the resulting molecules developed their own replication mechanism.

Molecules to cells Another important step in the evolution of life was the formation of membranes. Researchers have tested ways of enclosing molecules in membranes, allowing early metabolic and replication pathways to develop. In this work, as in other origin-of-life research, the connection between the various chemical events and the overall path from molecules to cells remains unresolved.

VOCABULARY

ACADEMIC VOCABULARY

Mechanism:
an instrument or process by which something is done or comes into being.
The mechanism for protein synthesis was unknown for a long time.

Cellular Evolution

What were the earliest cells like? Scientists don't know because the first life left no fossils. The earliest fossils are 3.5 billion years old. Chemical markings in rocks as old as 3.8 billion years suggest that life was present at that time even though no fossils remain. In 2004, scientists announced the discovery of what appeared to be fossilized microbes in volcanic rock that is 3.5 billion years old. This suggests that cellular activity had become established very early in Earth's history. It also suggests that early life might have been linked to volcanic environments.

The first cells Scientists hypothesize that the first cells were prokaryotes. Recall from Chapter 7 that prokaryotic cells are much smaller than eukaryotic cells, and they lack a defined nucleus and most other organelles. Many scientists think that modern prokaryotes called archaea (ar KEE uh) are the closest relatives of Earth's first cells. These organisms often live in extreme environments, such as the hot springs of Yellowstone Park or the volcanic vents in the deep sea, such as the one shown in **Figure 14.15.** These are environments similar to the environment that might have existed on early Earth.

■ **Figure 14.15** Some archaea live near deep-sea hydrothermal vents. They use energy from inorganic molecules to form the base of the vent food web.
Infer ***Why do some scientists think these microorganisms most resemble the first cells?***

Photosynthesizing prokaryotes Although archaea are autotrophic, they do not obtain their energy from the Sun. Instead, they extract energy from inorganic compounds such as sulfur. Archaea also do not need or produce oxygen.

Scientists think that oxygen was absent from Earth's earliest atmosphere until about 1.8 billion years ago. Any oxygen that appeared earlier than 1.8 billion years ago likely bonded with free ions of iron as oxygen does today. Evidence that iron oxide was formed by oxygen generated by early life is found in unique sedimentary rock formations, such as those shown in **Figure 14.16,** that are between about 1.8 billion and 2.5 billion years old. Scientists hypothesize that after 1.8 billion years ago, the early Earth's free iron was saturated with oxygen, and oxygen instead began accumulating in the atmosphere.

Many scientists think that photosynthesizing prokaryotes evolved not long after the archaea—very early in life's history. Fossil evidence of these primitive prokaryotes, called cyanobacteria, has been found in rocks as old as 3.5 billion years. Cyanobacteria eventually produced enough oxygen to support the formation of an ozone layer. Once an ozone shield was established, conditions would be right for the appearance of eukaryotic cells.

■ **Figure 14.16** These sedimentary rock formations appear as banded layers. Scientists believe that banding is a result of cyclic peaks in oxygen production.

The endosymbiont theory Eukaryotic cells appeared in the fossil record about 1.8 billion years ago, around two billion years after life first formed. Eukaryotic cells have complex internal membranes, which enclose various organelles, including mitochondria and, in plant cells, chloroplasts. Mitochondria metabolize food through cellular respiration, and chloroplasts are the site of photosynthesis. Both mitochondria and chloroplasts are about the size of prokaryotic cells and contain similar prokaryote features. This lead some scientists to speculate that prokaryotic cells were involved in the evolution of eukaryotic cells.

In 1966, biologist Lynn Margulis proposed the endosymbiont theory. According to the **endosymbiont theory,** the ancestors of eukaryotic cells lived in association with prokaryotic cells. In some cases, prokaryotes even might have lived inside eukaryotes. Prokaryotes could have entered a host cell as undigested prey, or they could have been internal parasites. Eventually, the relationship between the cells became mutually beneficial, and the prokaryotic symbionts became organelles in eukaryotic cells. This theory explains the origin of chloroplasts and mitochondria, as illustrated in **Figure 14.17.**

Evidence for the endosymbiont theory When Margulis first proposed the endosymbiont theory, many scientists were hesitant to accept it. There is evidence, however, that at least mitochondria and chloroplasts formed by endosymbiosis. For example, mitochondria and chloroplasts contain their own DNA. It is arranged in a circular pattern, just as it is in prokaryotic cells. Mitochondria and chloroplasts also have ribosomes that more closely resemble those in prokaryotic cells than those in eukaryotic cells. Finally, like prokaryotic cells, mitochondria and chloroplasts reproduce by fission, independent from the rest of the cell.

DATA ANALYSIS LAB 14.1

Based on Real Data*

Analyze Scientific Illustrations

How did plastids evolve?
Chloroplasts belong to a group of organelles called plastids, which are found in plants and algae. Chloroplasts perform photosynthesis. Other plastids store starch and make substances needed as cellular building blocks or for plant function.

Think Critically

1. **Summarize** the process described in the diagram. Include the definition of phagocytosis in your description.
2. **Compare** secondary endosymbiosis to the endosymbiont theory described in **Figure 14.17.**

Data and Observations

The illustration shows a way these plastids might have evolved.

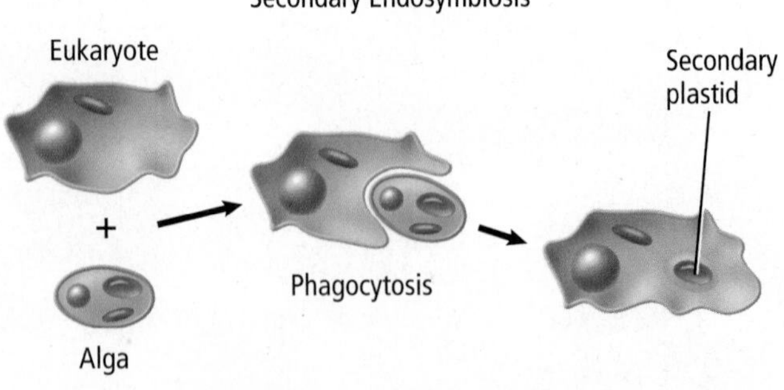

*Data obtained from: Dyall, S.D., et al. 2004. Ancient invasions: from endosymbionts to organelles. *Science* 304: 253–257.

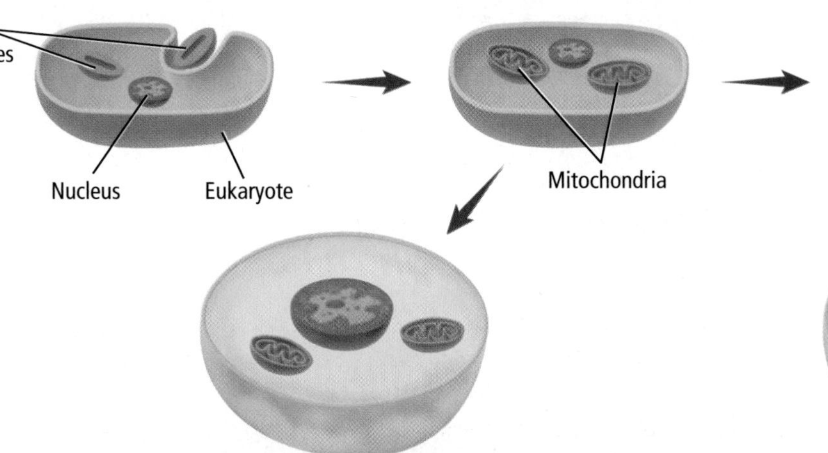
An early eukaryote was parasitized by or ingested some aerobic prokaryotes. The cells were protected and produced energy for the eukaryote.
Over millions of years, the aerobic prokaryotes became mitochondria, no longer able to live on their own.
Some eukaryotes also formed symbiotic relationships with photosynthetic bacteria, which contain photosynthetic pigments.
Aerobic prokaryotes
Nucleus
Eukaryote
Mitochondria
Photosynthetic prokaryotes
Chloroplasts
The aerobic prokaryotes became mitochondria in all eukaryotic cells.
The photosynthetic bacteria became chloroplasts in protist or plant cells.

Though the endosymbiont theory is widely endorsed, it is important to understand that scientists do not know the early steps that led to the emergence of life or to its early evolution. It is unlikely that any traces of the first life will ever be found. What scientists do know is that the conditions on Earth shortly after it took shape allowed the precursors of life to form.

The evolution of life is better understood than how the first life appeared. Fossil, geologic, and biochemical evidence supports many of the proposed steps in life's subsequent evolution. However, future discoveries might alter any or all of these steps. Scientists will continue to evaluate new evidence and test new theories in years to come.

Figure 14.17 This illustration shows how Margulis hypothesized that eukaryotic cells and their organelles evolved.

Concepts In Motion

Interactive Figure To see an animation about the endosymbiont hypothesis, visit biologygmh.com.

Section 14.2 Assessment

Section Summary

- Spontaneous generation was disproved in favor of biogenesis.
- The origin of life is hypothesized to be a series of chemical events.
- Organic molecules, such as amino acids, might have been formed from simpler molecules on early Earth.
- The first cells probably were autotrophic and prokaryotic.
- The endosymbiont theory explains how eukaryotic cells might have evolved from prokaryotic cells.

Understand Main Ideas

1. MAIN Idea **Infer** why scientists hypothesize that chemical events preceded the origin of life on Earth.
2. **Compare and contrast** spontaneous generation and biogenesis.
3. **Discuss** why prokaryotic cells probably appeared before eukaryotic cells.
4. **Hypothesize** whether prokaryotic cells might have been symbiotic before the evolution of eukaryotic cells.

Think Critically

5. **Sequence** Describe the hypothesized sequence of chemical and biological events that preceded the origin of eukaryotic cells.

WRITING in Biology

6. Write a persuasive paragraph that explains why many scientists accept the endosymbiont theory.

In the Field

Career: Paleontologist

Paleontologists Debate the Evolution of Birds

Along lakeshores in northeastern China 130 million years ago, volcanic eruptions sealed the fate of millions of organisms. Ash rains buried dinosaurs, mammals, fish, insects, and amphibians. Entombed for tens of millions of years, their bodies fossilized, sometimes leaving impressions of feathers, fur, and even stomach contents! Today, in the fossil-rich area of the Laioning Province in China, paleontologists are making important discoveries about life in the early Cretaceous period.

Caudipteryx zoui is an important fossil that shows that some dinosaurs had feathers.

A feathered dinosaur Organisms like the fossil specimen *Caudipteryx zoui* in the figure cause excitement in the paleontology community. In the fossil of *C. zoui,* there are clear traces of feathers from head to tail on the roughly one-meter long dinosaur. These feathers were not used for flight, but might have provided more stability for bipedal running.

An early bird A 130 million year old fossil of a new bird species, *Confuciusornis dui,* was discovered in the same general area as *C. zoui. C. dui* appears to have been a well-developed, tree-dwelling bird, not a feathered dinosaur that lived on the ground. *C. dui* and *C. zoui* lived during roughly the same time in history—between 120–150 mya. The coexistence of *C. dui* and *C. zoui* in this region provides an example of ancestral and derived species living together.

Link to the past Paleontologists often interpret fossil evidence to make evolutionary connections between organisms. Paleontologists agree that an evolutionary link exists between birds and dinosaurs. They share many anatomical features, including hollow, thin-walled bones, flexible wrists, clawed hands, and a fused collarbone that forms a wishbone. Paleontologists think that birds came from dinosaurs, but they continue to debate about when the divergence took place. Fossil finds like those in China help to provide evidence and insight into the evolution of birds.

CAREERS IN BIOLOGY

Interview a Paleontologist
Work with a team to create a list of questions you would like to ask a paleontologist. Conduct an interview with a paleontologist at a local college or university. Use the information you gather to write an article which describes what you learned from the conversation. Post your writing at biologygmh.com.

BIOLAB

IS SPONTANEOUS GENERATION POSSIBLE?

Background: **In the mid-1800s, Louis Pasteur conducted an experiment that showed that living organisms come from other living organisms—not from nonliving material. Pasteur's classic experiment, which disproved the notion of spontaneous generation, laid an essential foundation for modern biology by supporting the concept of biogenesis. In this lab, you will carry out an experiment based on Pasteur's work.**

Question: *How can the idea of spontaneous generation be disproved?*

Materials

beef broth	string
graduated cylinder	rubber stopper (2)
Erlenmeyer flask (2)	bunsen burner (2)
ring stand (2)	5 cm of plastic tubing
wire gauze (2)	30 cm of plastic tubing

Safety Precautions

Procedure

1. Read and complete the lab safety form.
2. Study the description of Louis Pasteur's classic experiment that disproved spontaneous generation.
3. Design and construct a data table to record changes in color, smell, and the presence of sediments.
4. Label the flasks "A" and "B." Flask A will be capped with a stopper holding a 5-cm piece of tubing. Flask B will be capped with a stopper holding a 30-cm piece of tubing.
5. Place 50 mL of beef broth in each flask. Cap each flask with the appropriate stopper.
6. Put each flask on a wire gauze on a ring stand over a bunsen burner.
7. Bend the tubing on Flask B until it forms a U-shape. The bottom of the U should be near the base of the flask. Tie the end of the tubing to the ring stand to hold the U-shape.
8. Boil the broth in each flask for 30 min.
9. After the equipment and broth cool, move the apparatuses to an area of the lab where they will not be disturbed.
10. Observe the flasks over the next two weeks. Record your observations in your data table.
11. **Cleanup and Disposal** Dispose of beef broth according to your teacher's instructions. Clean and return all equipment to the appropriate location.

Analyze and Conclude

1. **Describe** the experimental procedure you followed. How does it compare to the steps followed by Louis Pasteur?
2. **Compare** your findings to Pasteur's findings.
3. **Describe** why it is important for scientists to verify one another's data.
4. **Think Critically** Explain how Pasteur's findings disprove spontaneous generation.
5. **Error Analysis** If your results did not match Pasteur's results, explain a possible reason for the difference.

WRITING in Biology

Pasteur's experiment resulted in wide acceptance of biogenesis by the scientific community. Write an essay explaining how Pasteur's work contributed to some of the central ideas of biology. To learn more about biogenesis, visit BioLabs at biologygmh.com.

Chapter 14 Study Guide

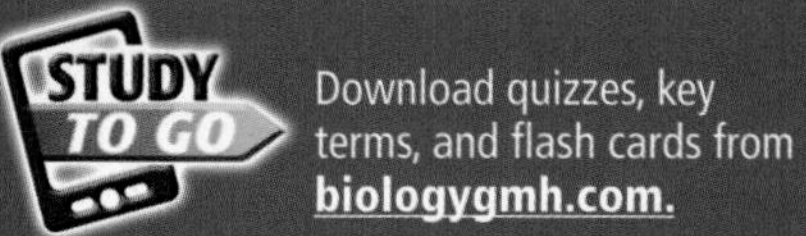

Download quizzes, key terms, and flash cards from biologygmh.com.

FOLDABLES **Categorize** Make a list of requirements for the existence of life. Put them in the sequence in which you think that they had to occur in order for life to appear successfully.

Vocabulary	Key Concepts
Section 14.1 Fossil Evidence of Change	
• Cambrian explosion (p. 398) • epoch (p. 396) • era (p. 396) • fossil (p. 393) • geologic time scale (p. 396) • half-life (p. 395) • K-T boundary (p. 399) • law of superposition (p. 394) • paleontologist (p. 394) • period (p. 396) • plate tectonics (p. 400) • radiometric dating (p. 395) • relative dating (p. 394)	**MAIN Idea** Fossils provide evidence of the change in organisms over time. • Early Earth was lifeless for several hundred million years. • Fossils provide evidence of past life. • Relative dating and radiometric dating are two methods used to determine the age of fossils. • The geologic time scale is divided into eras and periods. • Major events in the geologic time scale include both biological and geological changes. 
Section 14.2 The Origin of Life	
• endosymbiont theory (p. 406) • spontaneous generation (p. 401) • theory of biogenesis (p. 402)	**MAIN Idea** Evidence indicates that a sequence of chemical events preceded the origin of life on Earth and that life has evolved continuously since that time. • Spontaneous generation was disproved in favor of biogenesis. • The origin of life is hypothesized to be a series of chemical events. • Organic molecules, such as amino acids, might have been formed from simpler molecules on early Earth. • The first cells probably were autotrophic and prokaryotic. • The endosymbiont theory explains how eukaryotic cells might have evolved from prokaryotic cells. 

Biology Online Vocabulary PuzzleMaker biologygmh.com

CHAPTER 14 Assessment

Section 14.1

Vocabulary Review

Choose the vocabulary term from the Study Guide page that best describes each of the following phrases.

1. determining the age of a fossil by radioactive elements
2. the remains or evidence of an organism
3. scientist who studies fossils

Understand Key Concepts

Use the table below to answer questions 4 and 5.

Radioactive Isotope	Product of Decay	Half-Life (Years)
Carbon-14	Nitrogen-14	5730
Chlorine-36	Argon-36	300,000
Beryllium-10	Boron-10	1.52 million
Uranium-235	Lead-207	700 million

4. According to the table above, if one-fourth of the original radioactive carbon is present in a fossil, what is the fossil's age?
   A. 2857.5 years old
   B. 5730 years old
   C. 11,460 years old
   D. 17,145 years old

5. Which isotope would be best for measuring the age of a rock layer estimated to be about one million years old?
   A. beryllium-10
   B. carbon-14
   C. chlorine-36
   D. uranium-235

6. Which fossil type provides the most anatomical information to paleontologists?
   A. trace
   B. molds
   C. replacement
   D. amber

Use the graph below to answer questions 7 and 8.

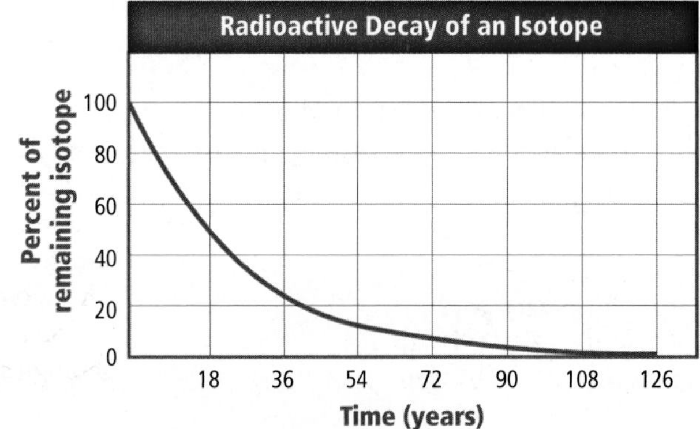

7. Which is the half-life of the radioactive isotope shown in the graph?
   A. 18 years **C.** 54 years
   B. 36 years **D.** 72 years

8. Assuming that you can only date material that has at least one percent of the radioisotope remaining, which age would be too old to date with this isotope?
   A. 35 years **C.** 75 years
   B. 50 years **D.** 125 years

9. What is the name of the period that followed extensive glaciation in the Precambrian?
   A. Cambrian
   B. Ediacaran
   C. Precambrian
   D. Neogene

10. Nearly all fossils occur in what kind of rocks?
    A. batholithic
    B. igneous
    C. metamorphic
    D. sedimentary

Constructed Response

11. **Short Answer** How does the law of superposition help paleontologists?

12. **Open Ended** Explain the geologic time scale using an analogy other than a ribbon of time.

13. **Short Answer** Calculate the percentage of Earth's existence occupied by the Cenozoic era (65 million years). Show your work.

Think Critically

14. **Infer** Imagine that you found a piece of amber in a sedimentary rock layer. What environment likely was present at the time of the fossil's formation?

15. **Describe** a fossil type and how it helps paleontologists understand an organism's anatomy.

Use the photo below to answer question 16.

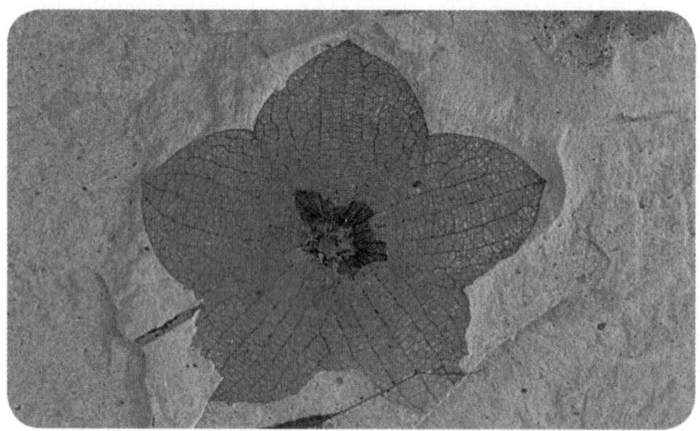

16. **Infer** If you found the above fossil of a flowering plant in a layer of rock, what would you conclude about the age of the layer? Would you look in layers above or below the layer with the flower to learn about the Permian mass extinction?

Section 14.2

Vocabulary Review

Replace the underlined words with the correct vocabulary term from the Study Guide page.

17. The belief that organisms originate from nonliving matter was disproven by Redi and Pasteur.

18. The explanation that bacteria might have lived inside prokaryotes and eventually became organelles was proposed by Lynn Margulis.

Understand Key Concepts

19. Pasteur's experiments led to which theory?
    A. biogenesis theory
    B. endosymbiont theory
    C. evolution theory
    D. spontaneous generation theory

Use the illustration below to answer questions 20 and 21.

20. The organisms represented in the photo above had which effect on early Earth?
    A. produced the first amino acids
    B. increased oxygen in the atmosphere
    C. became the first mitochondria
    D. consumed the first heterotrophs

21. When did the fossils of organisms like those in the photo first appear in the fossil record?
    A. 1.0 million years ago
    B. 2.0 million years ago
    C. 3.5 billion years ago
    D. 4.5 billion years ago

22. Clay most likely was involved in which process?
    A. producing the first oxygen in the atmosphere
    B. forming the first plasma membranes
    C. providing a framework for amino acid chains
    D. capturing prokaryotes for chloroplast evolution

23. Scientists have fossil evidence for which idea for the origin of life?
    A. first amino acids
    B. first RNA
    C. first cells
    D. first autotrophs

24. Banded iron formations are important evidence for which idea in the early evolution of life?
    A. photosynthetic autotrophs
    B. endosymbiont organelles
    C. heterotrophic prokaryotes
    D. heterotrophic eukaryotes

Constructed Response

25. **Open Ended** What would you expect the first step to be in the emergence of life from nonliving matter?
26. **Open Ended** Explain the significance of the Miller-Urey experiment for understanding the origin of cells.
27. **Open Ended** Which evidence do you think is most important for the endosymbiont hypothesis? Why?

Think Critically

28. **Sequence** the hypothesized events that led from a lifeless Earth to the presence of eukaryotic cells.
29. **Compare** the contributions of Redi and Pasteur in disproving spontaneous generation.

Use the photo below to answer question 30.

30. **Infer** How is the hot spring shown above similar to conditions on early Earth? What kind of organisms can survive in this type of environment?
31. **Infer** How was evolution affected by the increase in oxygen caused by the first photosynthetic organisms?
32. **CAREERS IN BIOLOGY** How could a biochemist studying DNA sequences provide evidence for the endosymbiont theory?
33. **Analyze and critique** the endosymbiont theory. What are its strengths and weaknesses?

Additional Assessment

34. *WRITING in* **Biology** Assume that you are a scientist searching for the cause of a mass extinction. Several causes have been hypothesized. Write a paragraph that explains how you could use dating methods to accept or reject them.

DBQ Document-Based Questions

"Probably all of the organic beings which have ever lived on this Earth have descended from some one primordial form."
Charles Darwin in *The Origin of Species,* 1859.

35. If Darwin was alive today, do you think he would include proteins among "organic beings"? Why or why not?
36. Use the quote above to support why you think Darwin would or would not have supported the endosymbiont theory.
37. Discuss what Darwin might have meant by the phrase, "...descended from some one primordial form."

Cumulative Review

Use the diagram below to answer question 38.

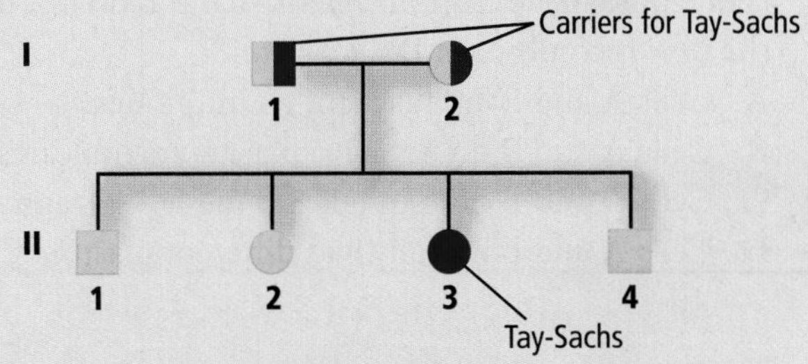

38. Which type of inheritance is represented in this pedigree? **(Chapter 11)**
39. Compare mitosis in animal and plant cells. **(Chapter 9)**
40. How does diffusion differ from active transport? **(Chapter 7)**

Standardized Test Practice

Cumulative

Multiple Choice

1. Which is associated with gene regulation in prokaryotic cells?
   A. DNA pairing
   B. repressor proteins
   C. RNA interference
   D. transcription factor

Use the illustration below to answer questions 2 and 3.

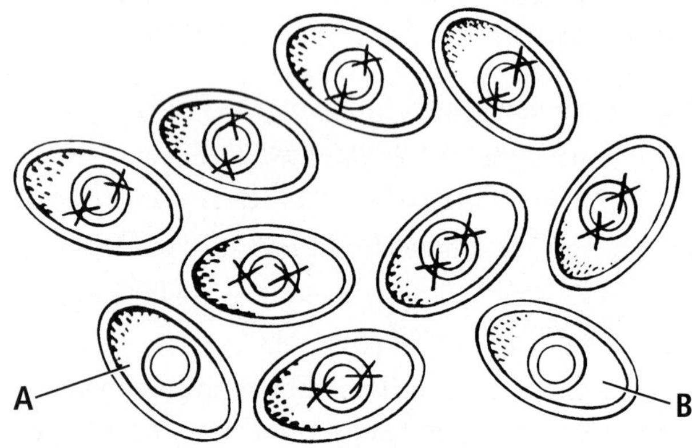

2. The bacterial cells in the figure above were transformed after they were mixed with recombinant DNA—represented by "XX" in the diagram. Which is one possible reason that Cells A and B do not have the new recombinant DNA plasmid?
   A. Cells A and B are resistant to antibiotics.
   B. Cells A and B do not have plasma membranes.
   C. Cells A and B did not take up the DNA fragment.
   D. Cells A and B initially had different plasmids.

3. In the figure, which step is likely to happen after the transformation of bacterial cells?
   A. Cells with the new plasmid will die after exposure to an antibiotic.
   B. Cells with the new plasmid will replicate quicker.
   C. Cells without the new plasmid will die after exposure to an antibiotic.
   D. Cells without the new plasmid will replicate more quickly.

4. A piece of DNA has the following sequence: CCCCGAATT. Suppose a mutation causes the following change: CCTCGAATT. Which term describes this mutation?
   A. chromosomal
   B. deletion
   C. duplication
   D. missense

Use the graph below to answer question 6.

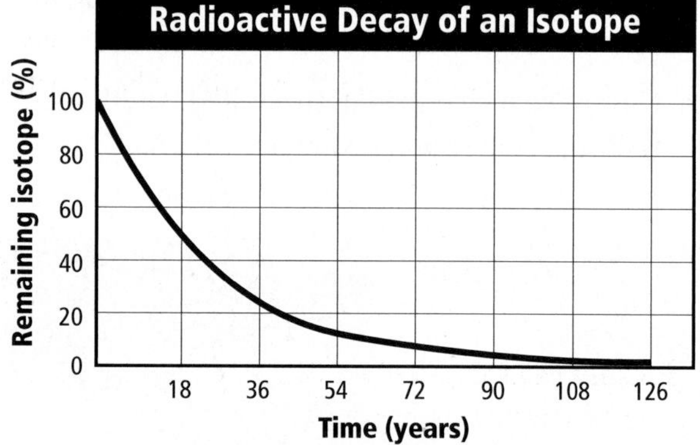

5. What is the half-life of the isotope represented in the graph?
   A. 9 years
   B. 18 years
   C. 54 years
   D. 90 years

6. Which causes DNA fragments to separate during gel electrophoresis?
   A. charge on the fragments
   B. DNA extraction of chemicals
   C. gel medium components
   D. source of the DNA

7. Where can Barr bodies be found?
   A. female body cells
   B. female sex cells
   C. male body cells
   D. male sex cells

Short Answer

Use the illustration below to answer questions 8 and 9.

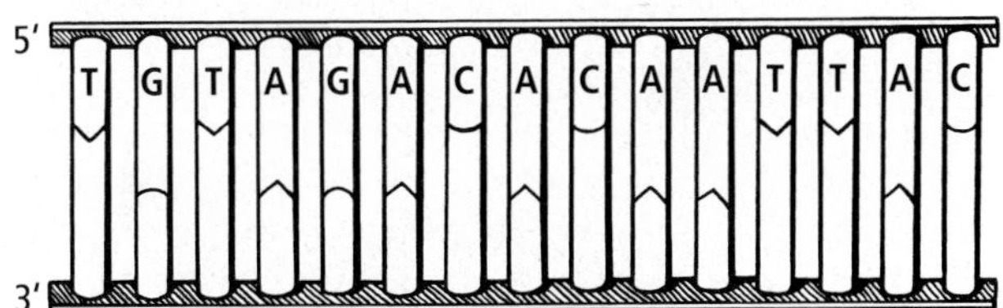

8. The diagram shows a molecule of DNA. What is the complementary DNA strand base code? Be sure to indicate the orientation of the strand.

9. Suppose the adjacent thymine bases in the figure formed a dimer after being exposed to ultraviolet radiation. How would the dimer affect the structure of the DNA molecule?

10. Describe the difference between petrified and replacement fossils.

11. Explain the three steps that take place in a polymerase chain reaction (PCR).

12. Describe why scientists infer that oxygen was absent from the early atmosphere on Earth.

13. Use a chart to show the role that different enzymes play in the replication of DNA. Be sure to put the steps in the correct order.

14. What are restriction enzymes? Assess why they are an important tool for genetic engineering.

15. How does a paleontologist use geologic principles for the relative dating of fossils?

Extended Response

16. How is selective breeding related to genetic engineering?

17. Appraise how your body temperature is related to homeostasis.

Essay Question

Some genes contain instructions for controlling when our cells grow, divide, and die. Certain genes that promote cell division are called oncogenes. Others that slow down cell division, or cause cells to die at the right time, are called tumor suppressor genes. It is known that cancers can be caused by DNA mutations (changes) that "turn on" oncogenes or "turn off" tumor suppressor genes.

The BRCA genes (BRCA1 and BRCA2) are tumor suppressor genes. When they are mutated, they no longer function to suppress abnormal growth and breast cancer is more likely to develop. Certain inherited DNA changes can result in a high risk for the development of breast cancer in people who carry these genes and are responsible for the cancers that run in some families.

Using the information in the paragraph above, answer the following question in essay format.

18. How could oncogenes and tumor suppressor genes play a part in the development of breast cancer? Use what you know about molecular genetics to write an essay explaining how these genes might contribute to the formation of tumors.

NEED EXTRA HELP?																		
If You Missed Question . . .	1	2	3	4	5	6	7	8	9	10	11	12	13	14	15	16	17	18
Review Section . . .	12.4	13.2	13.2	12.4	14.1	13.2	11.2	12.2	12.4	14.1	13.2	14.2	12.3	13.2	14.1	13.2, 12.2	1.1	12.3, 12.4

CHAPTER 15

Evolution

BIG Idea **The theory of natural selection explains evolution and the diversity of life.**

Section 1
Darwin's Theory of Evolution by Natural Selection

MAIN Idea Charles Darwin developed a theory of evolution based on natural selection.

Section 2
Evidence of Evolution

MAIN Idea Multiple lines of evidence support the theory of evolution.

Section 3
Shaping Evolutionary Theory

MAIN Idea The theory of evolution continues to be refined as scientists learn new information.

BioFacts

- Darwin was fascinated by the evolutionary adaptations of orchids and their pollinators.
- Orchids use scent, color, or shape to attract insects.
- The strong-smelling orchid *Stanhopea wardii* mimics the shape of the female euglossine bee.

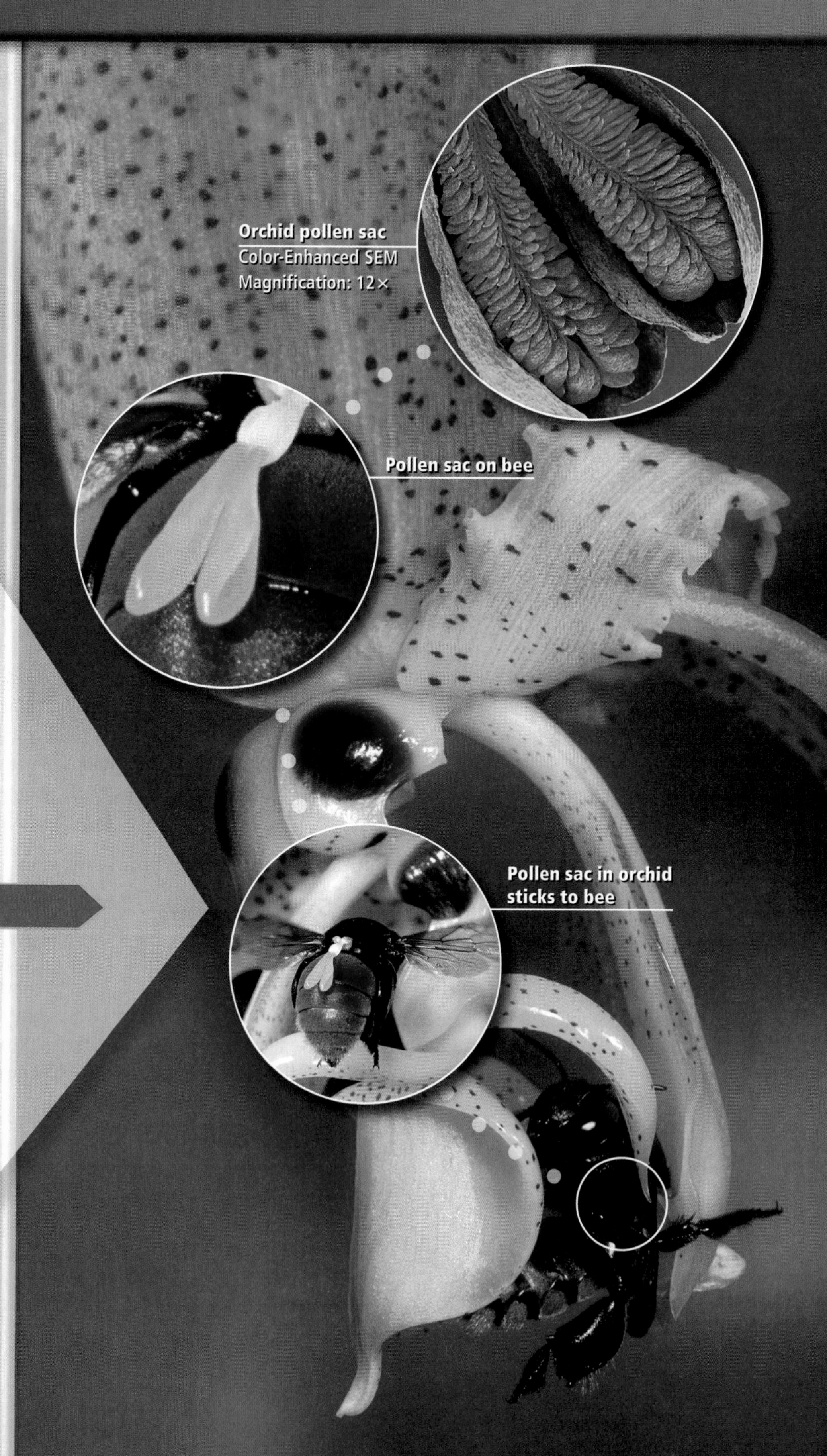

LAUNCH Lab

How does selection work?

Predators can cause changes in populations by choosing certain organisms as prey. In this lab, you will look at how prey populations might respond to a predator.

Procedure

1. Read and complete the lab safety form.
2. Work in groups of two to cut ten 3-cm-by-3-cm squares out of a piece of **black paper** and a piece of **red paper.**
3. Make two groups of ten squares: one with two red squares and the other with eight red squares.
4. Number the squares in each group, making sure that #1 is always red.
5. Place squares numbered side down, then choose a red square and record its number.
6. Repeat Step 5 ten times.

Analysis

1. **Compare** the number of times you chose Square 1 in the group with two red squares versus the group with eight red squares.
2. **Infer** A predator prefers red squares. In which group is Square 1 less likely to be eaten? Explain.

Visit biologygmh.com to:

- study the entire chapter online
- explore the Concepts in Motion, Interactive Tables, Microscopy Links Virtual Labs, and links to virtual dissections
- access Web links for more information, projects, and activities
- review content online with the Interactive Tutor and take Self-Check Quizzes

Populations Change Over Time Make this Foldable to help you organize what you learn about the steps of natural selection.

STEP 1 Fold the top and bottom edges of a sheet of notebook paper to meet in the middle.

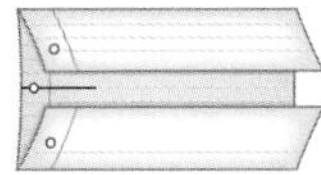

STEP 2 Fold in half horizontally.

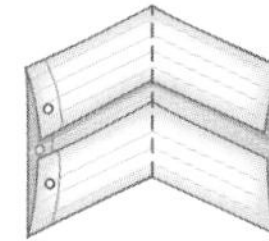

STEP 3 Open and cut along the inside fold lines to form four tabs.

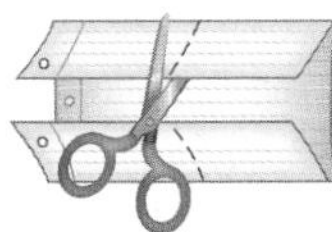

STEP 4 Label the tabs as follows:

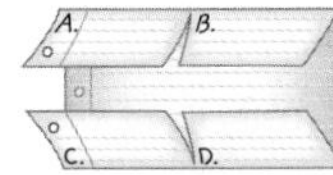

A. *In any population, individuals have variations.*
B. *Variations are inherited.*
C. *Organisms usually produce more offspring than can survive.*
D. *Inherited variations that increase reproductive success will eventually predominate in a population.*

FOLDABLES **Use this Foldable with Section 15.1.** As you study the section, record what you learn about the four principal ideas of natural selection.

Section 15.1

Reading Preview

Objectives

- **Discuss** the evidence that convinced Darwin that species could change over time.
- **List** the four principles of natural selection.
- **Show** how natural selection could change a population.

Review Vocabulary

selective breeding: process by which a breeder develops a plant or animal to have certain traits

New Vocabulary

artificial selection
natural selection
evolution

■ **Figure 15.1** Charles Darwin (1809-1882) posed for this portrait shortly after he returned from his voyage aboard the HMS *Beagle*.

Darwin's Theory of Evolution by Natural Selection

MAIN Idea **Charles Darwin developed a theory of evolution based on natural selection.**

Real-World Reading Link Today, a jet can travel from London to New York in hours. Imagine how different things were when it took almost five years for Charles Darwin to circle the globe aboard a small, cramped ship.

Developing the Theory of Evolution

When Charles Darwin, shown in **Figure 15.1,** boarded the HMS *Beagle* in 1831, the average person believed that the world was about 6000 years old. Almost everyone, including the young Darwin, thought that animals and plants were unchanging. The concept of gradual change over time was still years away.

Darwin on the HMS *Beagle* The primary mission of the *Beagle* was to survey the coast of South America. In 1831, the *Beagle* set sail from England for Maderia and then proceeded to South America, as shown on the map in **Figure 15.2.** Darwin's role on the ship was as naturalist and companion to the captain. His job was to collect biological and geological specimens during the ship's travels. Darwin had a degree in theology from Christ's College, Cambridge, though he previously had studied medicine and the sciences.

Over the course of the ship's five-year voyage, Darwin made extensive collections of rocks, fossils, plants, and animals. He also read a copy of Charles Lyell's *Principles of Geology*—a book proposing that Earth was millions of years old. This book influenced his thinking as he observed fossils of marine life at high elevation in the Andes, unearthed giant fossil versions of smaller living mammals, and saw how earthquakes could lift rocks great distances very quickly.

The Galápagos Islands In 1835, the *Beagle* arrived in the Galápagos (guh LAH puh gus) Islands off the coast of South America. Darwin was initially disappointed by the stark barrenness of these volcanic islands. But as he began to collect mockingbirds, finches, and other animals on the four islands he visited, he noticed that the different islands seemed to have their own, slightly different varieties of animals. These differences, however, only sparked a mere curiosity. He took little notice of the comment from the colony's vice governor that the island origins of the giant tortoises could be identified solely by the appearance of the tortoises' shells.

Reading Check **Summarize** some of the experiences or observations that influenced Darwin during his voyage on the *Beagle*.

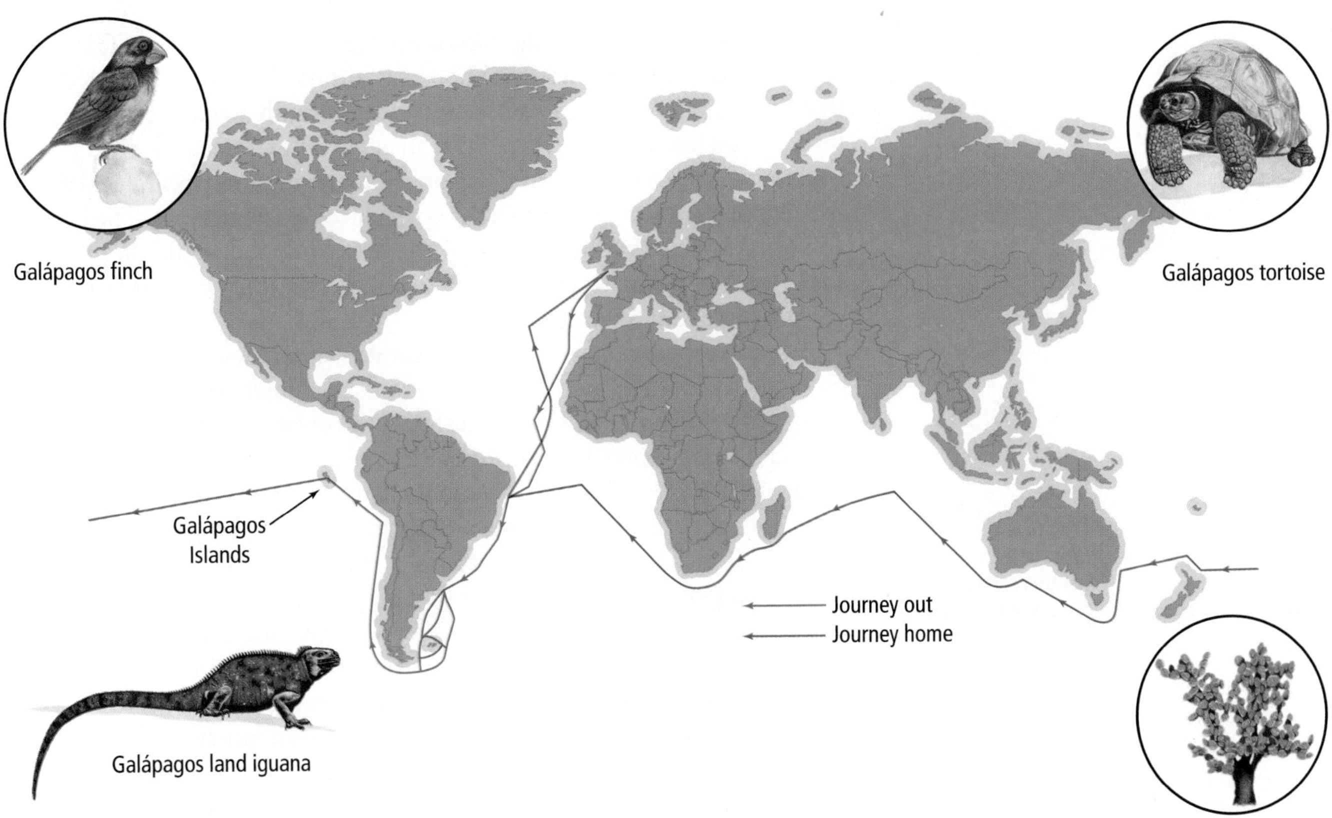

■ **Figure 15.2** The map shows the route of the *Beagle's* voyage. The species shown are all unique to the Galápagos Islands.
Infer ***How did the first organisms reach the Galápagos?***

A few years after Darwin returned to England, he began reconsidering his observations. He took note of the work of John Gould, an ornithologist who was classifying the birds Darwin brought back from the Galápagos. Gould discovered that the Galápagos finches were separate species and determined that the finches of the Galápagos did not live anywhere else in South America. In fact, almost every specimen that Darwin had collected on the islands was new to European scientists. These new species most closely resembled species from mainland South America, although the Galápagos and the mainland had different environments. Island and mainland species should not have resembled one another so closely unless, as Darwin began to suspect, populations from the mainland changed after reaching the Galápagos.

Darwin continued his studies Darwin hypothesized that new species could appear gradually through small changes in ancestral species, but he could not see how such a process would work. To understand it better, he turned to animal breeders—pigeon breeders in particular.

Different breeds of pigeons have certain distinctive traits that also are present in that breed's offspring. A breeder can promote these traits by selecting and breeding pigeons that have the most exaggerated expressions of those traits. For example, to produce pigeons with fan-shaped tails, the breeder will breed pigeons with the most fan-shaped tails. The process of directed breeding to produce offspring with desired traits, referred to as selective breeding in Chapter 13, was called **artificial selection** by Darwin.

Artificial selection also occurs when developing new breeds of dogs or new strains of crop plants. Darwin inferred that if humans could change species by artificial selection, then perhaps the same process could work in nature. Further, Darwin thought that, given enough time, perhaps this process could produce new species.

Natural selection While thinking about artificial selection, Darwin read an essay by the economist Thomas Malthus. The essay suggested that the human population, if unchecked, eventually would outgrow its food supply, leading to a competitive struggle for existence. Darwin realized that Malthus's ideas could be applied to the natural world. He reasoned that some competitors in the struggle for existence would be better equipped for survival than others. Those less equipped would die. This is the process of **natural selection.** Here, finally, was the framework for a new theory about the origin of species.

Darwin's theory of evolution by natural selection has four basic principles that explain how traits of a population can change over time. First, individuals in a population show differences, or variations. Second, variations can be inherited, meaning that they are passed down from parent to offspring. Third, organisms have more offspring than can survive on available resources. The average cardinal, for example, lays nine eggs each summer. If each baby cardinal survived and reproduced just once, it would take only seven years for the first pair to have produced one million birds. Finally, variations that increase reproductive success will have a greater chance of being passed on than those that do not increase reproductive success. If having a fantail helps a pigeon reproduce successfully, future generations would include more pigeons with fan-shaped tails.

FOLDABLES Incorporate information from this section into your Foldable.

Given enough time, natural selection could modify a population enough to produce a new species. Natural selection is now considered the mechanism by which evolution takes place. **Figure 15.3** shows how natural selection might modify a population of sunflowers.

Reading Check **Explain** the four principles of natural selection.

DATA ANALYSIS LAB 15.1

Based on Real Data*

Interpret the Data

How did artificial selection change corn? Plant breeders have made many changes to crops. In one of the longest experiments ever conducted, scientists selected maize (corn) for oil content in kernels.

Data and Observations

Look at the graph and compare the selection in the different plant lines.

Line IHO was selected for high oil content, and line ILO was selected for low oil content. The direction of selection was reversed in lines RHO (started from IHO) and RLO (started from ILO) at generation 48. In line SHO (derived from RHO), selection was switched back to high oil content at generation 55.

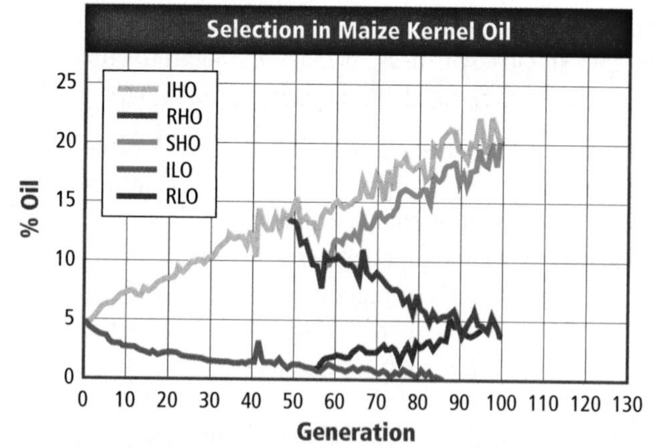

Think Critically

1. **Measure** What were the highest and lowest percentages of oil seen in the experiment?
2. **Predict** If the trend continues for line RHO, approximately how many generations will it take until the oil content reaches zero percent?

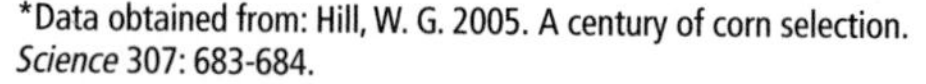
*Data obtained from: Hill, W. G. 2005. A century of corn selection. *Science* 307: 683-684.

Visualizing Natural Selection

Figure 15.3
Natural selection is the mechanism by which, if given enough time, a population—in this case, a population of sunflowers—could be modified to produce a new species. There are four principles of natural selection that explain how this can occur—variation, heritability, overproduction, and reproductive advantage.

Variation Individuals in a population differ from one another. For example, some sunflowers are taller than others.

Heritability Variations are inherited from parents. Tall sunflowers produce tall sunflowers, and short sunflowers produce short sunflowers.

Overproduction Populations produce more offspring than can survive. Each sunflower has hundreds of seeds, most of which will not germinate.

Reproductive Advantage Some variations allow the organism that possesses them to have more offspring than the organism that does not possess them. For example, in this habitat, shorter sunflowers reproduce more successfully.

Over time, the average height of the sunflower population is short if the short sunflowers continue to reproduce more successfully. After many generations, the short sunflowers might become a new species if they are unable to breed with the original sunflowers.

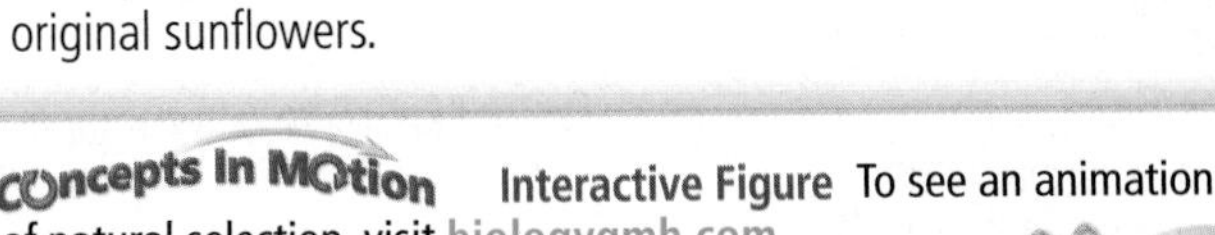
Concepts In Motion Interactive Figure To see an animation of natural selection, visit biologygmh.com.

Biology Online

Concepts In Motion Interactive Table To explore more about the principles of natural selection, visit biologygmh.com.

Table 15.1 Basic Principles of Natural Selection

Principle	Example
Individuals in a population show variations among others of the same species.	The students in a classroom all look different.
Variations are inherited.	You look similar to your parents.
Animals have more young than can survive on the available resources.	The average cardinal lays nine eggs per summer. If each cardinal lived only one year, in seven years there would be a million cardinals if all offspring survived.
Variations that increase reproductive success will be more common in the next generation.	If having a fan-shaped tail increases reproductive success of pigeons, then more pigeons in the next generation will have fan-shaped tails.

The Origin of Species

Darwin had likely formulated his theory of evolution by natural selection by about 1840. Soon after, he began writing a multi-volume book compiling evidence for evolution and explaining how natural selection might provide a mechanism for the origin of species. **Table 15.1** summarizes the principles of natural selection described in Darwin's work. He continued to compile evidence in support of his theory for many years. For example, he spent eight years studying relationships among barnacles.

In 1858, Alfred Russel Wallace, another English naturalist, proposed a theory that was almost identical to Darwin's theory. Both men's ideas were presented to the Linnean Society of London. One year later, Darwin published *On the Origin of Species by Means of Natural Selection*—a condensed version of the book he had started many years before.

In his book, Darwin used the term *evolution* only on the last page. Today, biologists use the term **evolution** to define cumulative changes in groups of organisms through time. Natural selection is not synonymous with evolution; it is a mechanism by which evolution occurs.

VOCABULARY
WORD ORIGIN
Evolve
comes from the Latin word *evolvere*, meaning *unroll* or *unfold*.

Section 15.1 Assessment

Section Summary

- Darwin drew from his observations on the HMS *Beagle* and later studies to develop his theory of evolution by natural selection.
- Natural selection is based on ideas of excess reproduction, variation, inheritance, and advantages of certain traits in certain environments.
- Darwin reasoned that the process of natural selection eventually could result in the appearance of new species.

Understand Main Ideas

1. MAIN Idea **Describe** the evidence Charles Darwin gathered that led to his theory of evolution.
2. **Explain** how the idea of artificial selection contributed to Darwin's ideas on natural selection.
3. **Identify** the four principles of natural selection and provide examples not used in the section.
4. **Discuss** Wallace's contribution to the theory of evolution by natural selection.

Think Critically

5. **Infer** the consequences for evolution if species did not vary.

WRITING in Biology

6. Write a short story about what it might have been like to visit the Galápagos Islands with Darwin.

Biology Online Self-Check Quiz biologygmh.com

Section 15.2

Evidence of Evolution

Reading Preview

Objectives

- **Describe** how fossils provide evidence of evolution.
- **Discuss** morphological evidence of evolution.
- **Explain** how biochemistry provides evidence of evolution.

Review Vocabulary

fossil: remains of an organism or its activities

New Vocabulary

derived trait
ancestral trait
homologous structure
vestigial structure
analogous structure
embryo
biogeography
fitness
camouflage
mimicry

MAIN Idea **Multiple lines of evidence support the theory of evolution.**

Real-World Reading Link The evidence for evolution is like a set of building blocks. Just as you cannot build something with only one building block, one piece of evidence does not make a theory. The evidence for evolution is more convincing when supported by many pieces of evidence, just as a structure is more sturdy when built with many blocks.

Support for Evolution

Darwin's book *On the Origin of Species* demonstrated how evolution might happen. The book also provided evidence that evolution has occurred on our planet. The concepts of natural selection and evolution are different, though related. Darwin's theory of evolution by natural selection is part of the larger theory of evolution. Recall from Chapter 1 that a theory provides an explanation for a natural phenomenon based on observations. Theories explain available data and suggest further areas for experimentation. The theory of evolution states that all organisms on Earth have descended from a common ancestor.

The fossil record Fossils provide a record of species that lived long ago—supplying some of the most significant evidence of evolutionary change. This record can show ancient species that are similar to current ones, as illustrated in **Figure 15.4.** Fossils also show some species, such as the horseshoe crab, have remained unchanged for millions of years. The fossil record is an important source of information for determining the ancestry of organisms and patterns of evolution.

Figure 15.4 The giant armadillo-like glyptodont, *Glyptodon*, is an extinct animal that Darwin thought must be related to the living armadillos.
Observe ***What features of the 2000-kg glyptodont are similar to those of the 4-kg armadillo?***

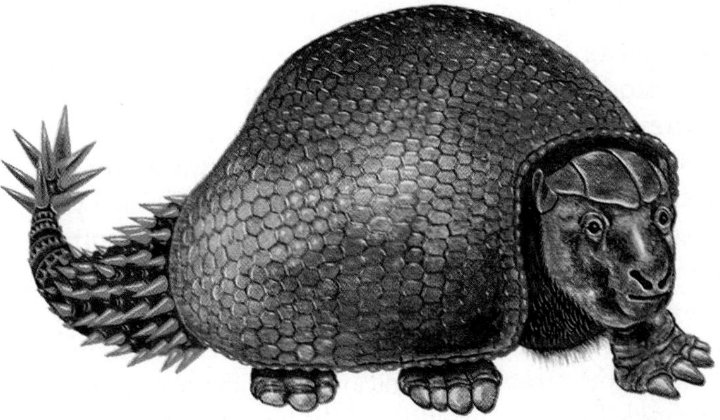

Glyptodont

Armadillo

■ **Figure 15.5** This artist's rendering of *Archaeopteryx* shows that it shares many features with modern birds while retaining ancestral dinosaur features.

Connection to Earth Science Though Darwin recognized the limitations of the fossil record, he predicted the existence of fossils intermediate in form between species. Today, scientists studying evolutionary relationships have found hundreds of thousands of transitional fossils that contain features shared by different species. For example, certain dinosaur fossils show feathers of modern birds and the teeth and bony tails of reptiles. **Figure 15.5** shows an artist's rendering of *Archaeopteryx*, one of the first birds. *Archaeopteryx* fossils provide evidence of characteristics that classify it as a bird, and also show that the bird retained several distinct dinosaur features.

Researchers consider two major classes of traits when studying transitional fossils: derived traits and ancestral traits. **Derived traits** are newly evolved features, such as feathers, that do not appear in the fossils of common ancestors. **Ancestral traits,** on the other hand, are more primitive features, such as teeth and tails, that do appear in ancestral forms. Transitional fossils provide detailed patterns of evolutionary change for the ancestors of many modern animals, including mollusks, horses, whales, and humans.

VOCABULARY
WORD ORIGIN
Homologous
from the Greek words *homos*, meaning *same*, and *logos*, meaning *relation* or *reasoning*.

Comparative anatomy Why do the vertebrate forelimbs shown in **Figure 15.6** have different functions but appear to be constructed of similar bones in similar ways? Evolutionary theory suggests that the answer lies in shared ancestry.

Homologous structures Anatomically similar structures inherited from a common ancestor are called **homologous structures.** Evolution predicts that an organism's body parts are more likely to be modifications of ancestral body parts than they are to be entirely new features. The limbs illustrated in **Figure 15.6** move animals in different ways, yet they share similar construction. Bird wings and reptile limbs are another example. Though birds use their wings to fly and reptiles use their limbs to walk, bird wings and reptile forelimbs are similar in shape and construction, which indicates that they were inherited from a common ancestor. While homologous structures alone are not evidence of evolution, they are an example for which evolution is the best available explanation for the biological data.

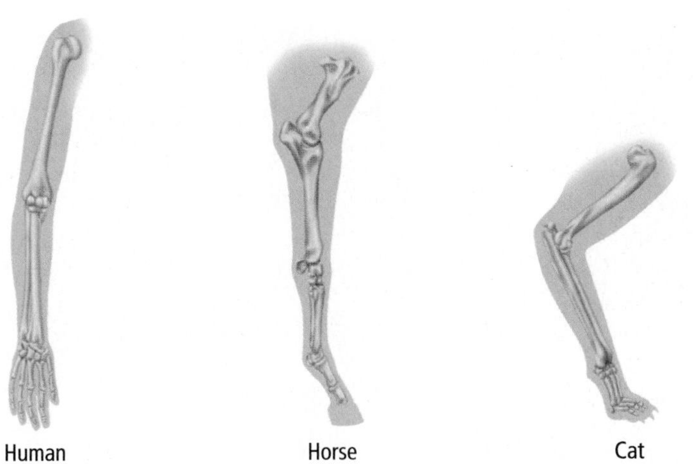

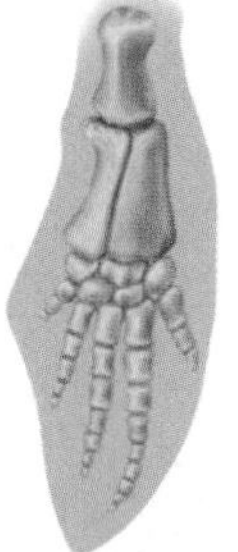

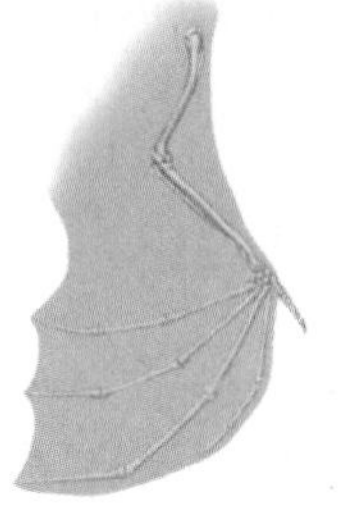

Figure 15.6 The forelimbs of vertebrates illustrate homologous structures. Each limb is adapted for different uses, but they all have similar bones.

Infer ***Which of the forelimbs shown would most likely resemble a whale's fluke?***

Vestigial structures In some cases, a functioning structure in one species is smaller or less functional in a closely related species. For example, most birds have wings developed for flight. Kiwis, however, have very small wings that cannot be used for flying. The kiwi wing is a kind of homologous structure called a vestigial structure. **Vestigial structures** are structures that are the reduced forms of functional structures in other organisms. **Table 15.2** illustrates some vestigial structures in different species. Evolutionary theory predicts that features of ancestors that no longer have a function for that species will become smaller over time until they are lost.

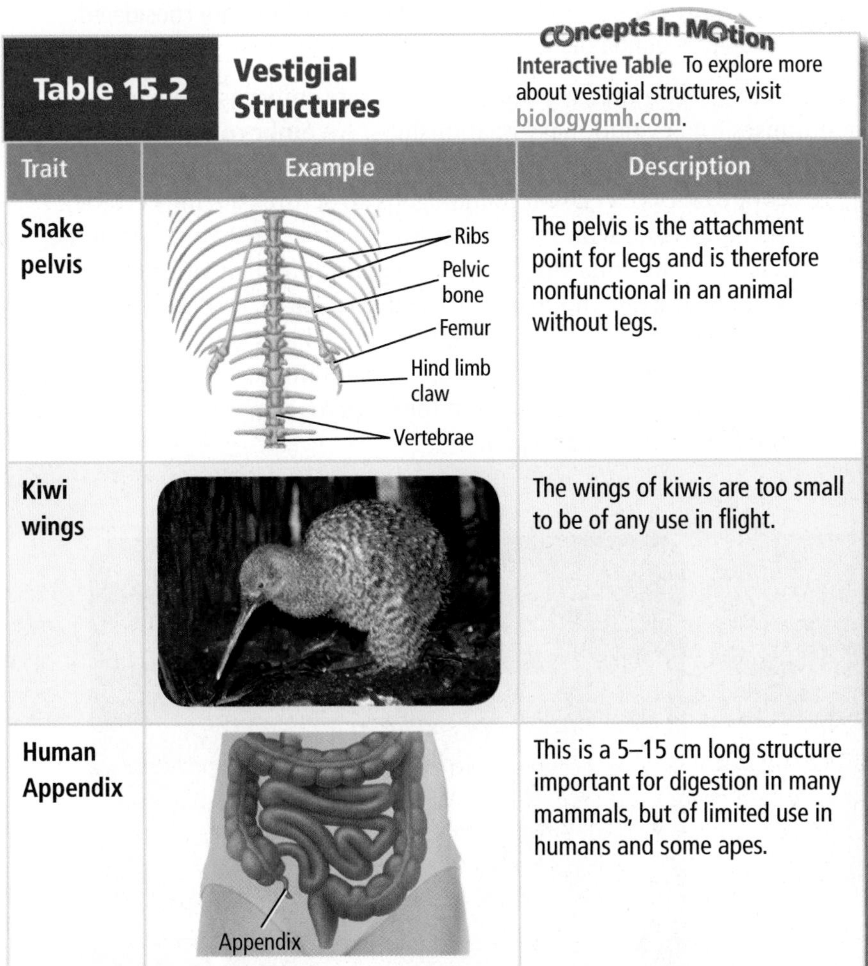

Table 15.2 **Vestigial Structures**

Concepts In Motion
Interactive Table To explore more about vestigial structures, visit biologygmh.com.

Trait	Example	Description
Snake pelvis		The pelvis is the attachment point for legs and is therefore nonfunctional in an animal without legs.
Kiwi wings		The wings of kiwis are too small to be of any use in flight.
Human Appendix		This is a 5–15 cm long structure important for digestion in many mammals, but of limited use in humans and some apes.

Figure 15.7 Eagles and beetles use their wings to fly, but their wing structures are different.

Explain ***how scientists know that the wings of eagles and beetles are analogous structures.***

Bald Eagle

May Beetle

Not all anatomically similar features are evidence of common ancestry. **Analogous structures** can be used for the same purpose and can be superficially similar in construction but are not inherited from a common ancestor. As shown in **Figure 15.7,** the wings of an eagle and the wings of a beetle have the same function—they both enable the organism to fly. But the wings are constructed in different ways from different materials. While analogous structures do not indicate close evolutionary relationships, they do show that functionally similar features can evolve independently in similar environments.

Reading Check **Explain** why vestigial structures are considered examples of homologous structures.

Comparative embryology Vertebrate embryos provide more glimpses into evolutionary relationships. An **embryo** is an early, pre-birth stage of an organism's development. Scientists have found that vertebrate embryos exhibit homologous structures during certain phases of development but become totally different structures in the adult forms. The embryos shown in **Figure 15.8,** like all vertebrate embryos, have a tail and paired structures called pharyngeal pouches. In fish the pouches develop into gills. In reptiles, birds, and mammals, these structures become parts of the ears, jaws, and throats. Though the adult forms differ, the shared features in the embryos suggest that vertebrates evolved from a shared ancestor.

Figure 15.8 Embryos reveal evolutionary history. Bird and mammal embryos share several developmental features.

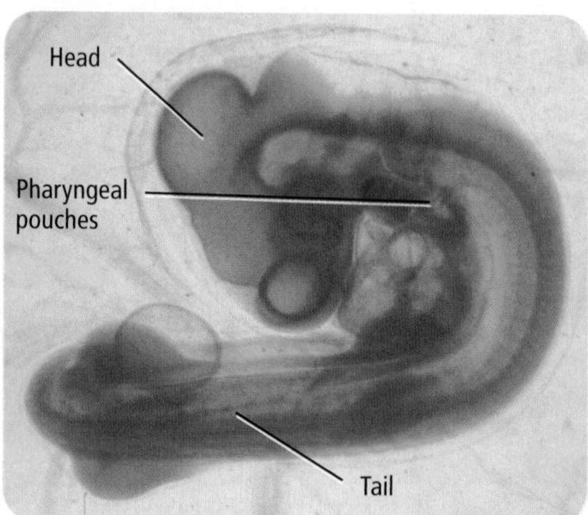

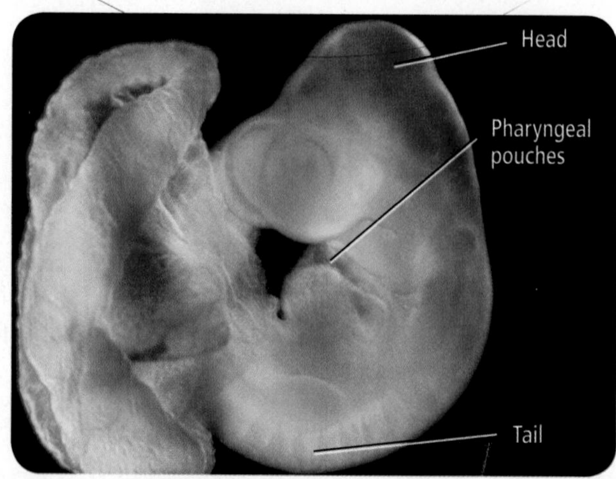

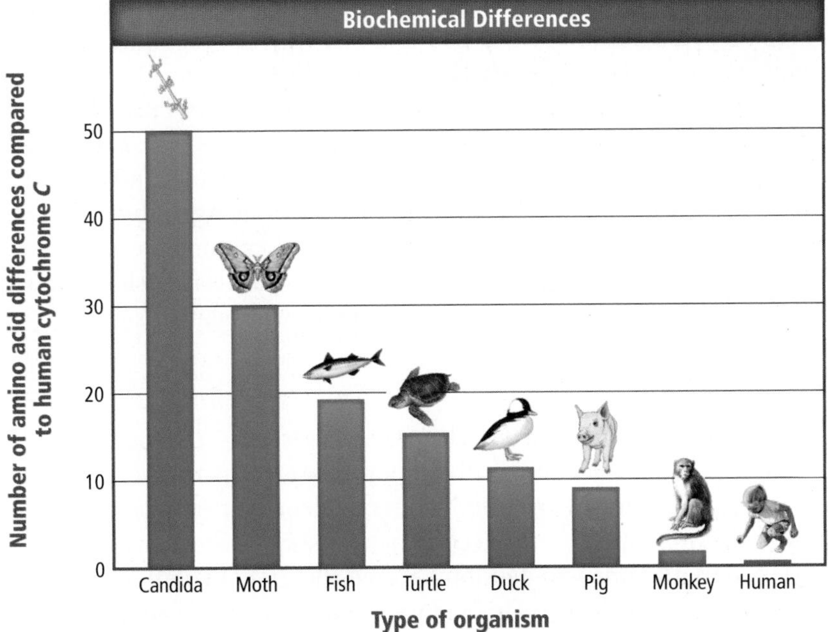

■ **Figure 15.9** This illustration compares amino acid sequences of cytochrome c in humans and other organisms.
Infer ***Would the cytochrome*** c ***of a reptile or a duck be expected to have more amino acid differences when compared with that of a human? Explain.***

Comparative biochemistry Scientific data also shows that common ancestry can be seen in the complex metabolic molecules that many different organisms share. Cytochrome *c* is an enzyme that is essential for respiration and is highly conserved in animals. This means that despite slight variations in its amino acid sequence, the molecule has changed very little over time.

Evolutionary theory predicts that molecules in species with a recent common ancestor should share certain ancient amino acid sequences. The more closely related the species are, the greater number of sequences will be shared. This predicted pattern is what scientists find to be true in cytochrome *c*. For example, as illustrated in **Figure 15.9,** the cytochrome *c* in the pig and in the monkey share more amino acid sequences with humans than the cytochrome *c* in birds shares with humans.

Connection to Chemistry Scientists have found similar biochemical patterns in other proteins, as well as in DNA and RNA. DNA and RNA form the molecular basis of heredity in all living organisms. The fact that many organisms have the same complex molecules suggests that these molecules evolved early in the history of life and were passed on through the life-forms that have lived on Earth. Comparisons of the similarities in these molecules across species reflect evolutionary patterns seen in comparative anatomy and in the fossil record. Organisms with closely related morphological features have more closely related molecular features.

Geographic distribution The distribution of plants and animals that Darwin saw during his South American travels first suggested evolution to Darwin. He observed that animals on the South American mainland were more similar to other South American animals than they were to animals living in similar environments in Europe. The South American mara, for example, inhabited a niche that was occupied by the rabbit in Europe. You can compare a mara and a rabbit in **Figure 15.10.** Darwin realized that the mara was more similar to other South American species than it was to the rabbit because it shared a closer ancestor with the South American animals.

■ **Figure 15.10** The mara *(Dolichotis patagonum)* exists in a niche similar to that of the English rabbit *(Oryctolagus cuniculus).*

Mara

English Rabbit

Patterns of migration were critical to Darwin when he was developing his theory. Migration patterns explained why, for example, islands often have more plant diversity than animal diversity: the plants are more able to migrate from the closest mainland as seeds, either by wind or on the backs of birds. Since Darwin's time, scientists have confirmed and expanded Darwin's study of the distribution of plants and animals around the world in a field of study now called **biogeography.** Evolution is intimately linked with climate and geological forces, especially plate tectonics, which helps explain many ancestral relationships and geographic distributions seen in fossils and living organisms today.

Adaptation

The five categories discussed in the previous section—the fossil record, comparative anatomy, comparative embryology, comparative biochemistry, and geographic distribution—offer evidence for evolution. Darwin drew on all of these except biochemistry—which was not well developed in his time—to develop his own theory of evolution by natural selection. At the heart of his theory lies the concept of adaptation.

Types of adaptation An adaptation is a trait shaped by natural selection that increases an organism's reproductive success. One way to determine how effectively a trait contributes to reproductive success is to measure fitness. **Fitness** is a measure of the relative contribution an individual trait makes to the next generation. It often is measured as the number of reproductively viable offspring that an organism produces in the next generation.

The better an organism is adapted to its environment, the greater its chances of survival and reproductive success. This concept explains the variations Darwin observed in the animals on the Galápagos Islands. Because the finches were each adapted to their individual islands, they had variations in their beaks.

VOCABULARY

SCIENCE USAGE V. COMMON USAGE

Adaptation

Science usage: a trait shaped by natural selection to increase the survival or reproductive success of an organism.
The prehensile tail of monkeys is an adaptation for life in the trees.

Common usage: adjustment or change.
The movie script is an adaptation of the original play.

Camouflage Some species have evolved morphological adaptations that allow them to blend in with their environments. This is called **camouflage** (KA muh flahj). Camouflage allows organisms to become almost invisible to predators, as shown in **Figure 15.11.** As a result, more of the camouflaged individuals survive and reproduce.

■ **Figure 15.11** It would be easy for a predator to overlook a leafy sea dragon, *Phycodurus eques*, in a sea grass habitat because of the animal's effective camouflage.

California Kingsnake

Western Coral Snake

■ **Figure 15.12** Predators avoid the harmless California kingsnake because it has color patterns similar to those of the poisonous western coral snake.

Mimicry Another type of morphological adaptation is mimicry. In **mimicry,** one species evolves to resemble another species. You might expect that mimicry would make it difficult for individuals in one species to find and breed with other members of their species, thus decreasing reproductive success. However, mimicry often increases an organism's fitness. Mimicry can occur in a harmless species that has evolved to resemble a harmful species, such as the example shown in **Figure 15.12.** Sometimes mimicry benefits two harmful species. In both cases, the mimics are protected because predators can't always tell the mimic from the animal it is mimicking, so they learn to avoid them both.

 **Reading Check** **Compare** mimicry and camouflage.

Antimicrobial resistance Species of bacteria that originally were killed by penicillin and other antibiotics have developed drug resistance. For almost every antibiotic, at least one species of resistant bacteria exists. One unintended consequence of the continued development of antibiotics is that some diseases, which were once thought to be contained, such as tuberculosis, have reemerged in more harmful forms.

MiniLab 15.1

Investigate Mimicry

Why do some species mimic the features of other species? Mimicry is the process of natural selection shaping one species of organism to look similar to another species. Natural selection has shaped the nontoxic viceroy butterfly to look like the toxic monarch butterfly. Investigate the mimicry displayed during this lab.

Procedure

1. Read and complete the lab safety form.
2. Create a data table for recording your observations and measurements of the **monarch** and **viceroy butterflies.**
3. Observe the physical characteristics of both butterfly species and record your observations in your data table.

Analysis

1. **Compare and contrast** the physical characteristics of the two butterfly species.
2. **Hypothesize** why the viceroy butterflies have bright colors that are highly visible.

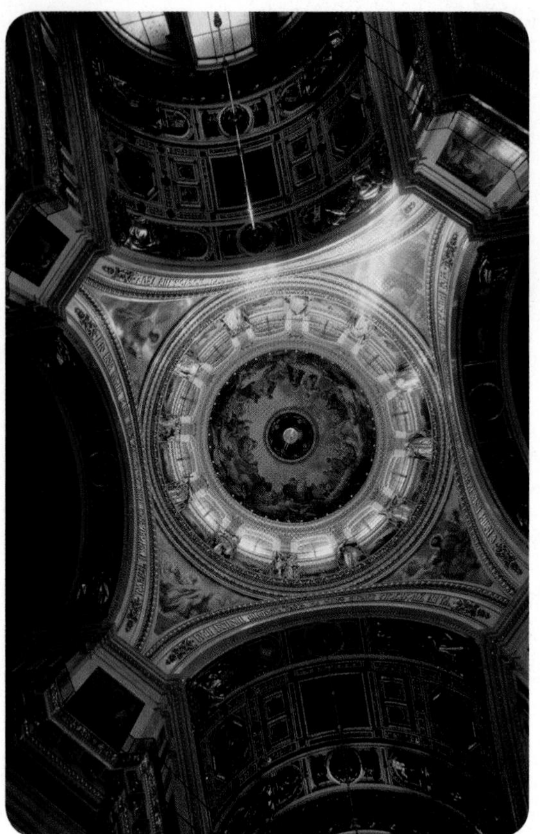

■ **Figure 15.13** Spaces between arches set in a square to support a dome are called spandrels and are often decorative. Some features might be like spandrels—a consequence of another adaptation.

Consequences of adaptations Not all features of an organism are necessarily adaptive. Some features might be consequences of other evolved characteristics. Biologists Stephen Jay Gould and Richard Lewontin made this point in 1979 in a paper claiming that biologists tended to overemphasize the importance of adaptations in evolution.

Spandrel example To illustrate this concept, they used an example from architecture. Building a set of four arches in a square to support a dome means that spaces called spandrels will appear between the arches, as illustrated in **Figure 15.13.** Because spandrels are often decorative, one might think that spandrels exist for decoration. In reality, they are an unavoidable consequence of arch construction. Gould and Lewontin argued that some features in organisms are like spandrels because even though they are prominent, they do not increase reproductive success. Instead, they likely arose as an unavoidable consequence of prior evolutionary change.

Human example A biological example of a spandrel is the helplessness of human babies. Humans give birth at a much earlier developmental stage than other primates. This causes them to need increased care early in their life. Many scientists think that the helplessness of human babies is a consequence of the evolution of big brains and upright posture. To walk upright, humans need narrow pelvises, which means that babies' heads must be small enough to fit through the pelvic opening at birth. In contrast, scientists previously thought that the helplessness of human infants provided an adaptive advantage, such as increased attention from parents and more learning.

Section 15.2 Assessment

Section Summary

- Fossils provide strong direct evidence to support evolution.
- Homologous and vestigial structures indicate shared ancestry.
- Examples of embryological and biochemical traits provide insight into the evolution of species.
- Biogeography can explain why certain species live in certain locations.
- Natural selection gives rise to features that increase reproductive success.

Understand Main Ideas

1. MAIN Idea **Describe** how fossils provide evidence of evolution.
2. **Explain** what natural selection predicts about mimicry, camouflage, homologous structures, and vestigial structures.
3. **Indicate** how biochemistry provides evidence of evolution.
4. **Compare** the morphological evidence and the biochemical evidence supporting evolution.

Think Critically

5. **Hypothesize** Evidence suggests that the bones in bird wings share a number of features with the bones of dinosaur arms. Based on this evidence, what hypothesis could you make about the evolutionary relationship between birds and dinosaurs?
6. **Apply** Research has shown that if a prescribed dose of antibiotic is not taken completely, some bacteria might not be killed and the disease might return. How does natural selection explain this phenomenon?

Biology Online **Self-Check Quiz** biologygmh.com

Section 15.3

Reading Preview

Objectives

- **Discuss** patterns observed in evolution.
- **Describe** factors that influence speciation.
- **Compare** gradualism with punctuated equilibrium.

Review Vocabulary

allele: alternate forms of a character trait that can be inherited

New Vocabulary

Hardy-Weinberg Principle
genetic drift
founder effect
bottleneck
stabilizing selection
directional selection
disruptive selection
sexual selection
prezygotic isolating mechanism
postzygotic isolating mechanism
allopatric speciation
sympatric speciation
adaptive radiation
gradualism
punctuated equilibrium

Shaping Evolutionary Theory

MAIN Idea The theory of evolution continues to be refined as scientists learn new information.

Real-World Reading Link The longer you operate a complicated piece of electronics, the better you understand how it works. The device does not change, but you become more familiar with its functions. Scientists have been studying evolution for almost 150 years, yet they are still learning new ways in which evolution leads to changes in species.

Mechanisms of Evolution

Darwin's theory of natural selection remains a central theme in evolution. It explains how organisms adapt to their environments and how variations can give rise to adaptations within species. Scientists now know, however, that natural selection is not the only mechanism of evolution. Studies from population genetics and molecular biology have led to the development of evolutionary theory. At the center of this is the understanding that evolution occurs at the population level, with genes as the raw material.

Population genetics At the turn of the twentieth century, genes had not been discovered. However, the allele was understood to be one form of an inherited character trait, such as eye color, that gets passed down from parent to offspring. Scientists didn't understand why dominant alleles wouldn't simply swamp recessive alleles in a population.

In 1908, English mathematician Godfrey Hardy and German physician Wilhelm Weinberg independently came up with the same solution to this problem. They showed mathematically that evolution will not occur in a population unless allelic frequencies are acted upon by forces that cause change. In the absence of these forces, the allelic frequency remains the same and evolution doesn't occur. According to this idea, which is now known as the **Hardy-Weinberg principle,** when allelic frequencies remain constant, a population is in genetic equilibrium. This concept is illustrated in **Figure 15.14.**

Figure 15.14 According to the Hardy-Weinberg principle, even though the number of owls doubled, the ratio of gray to red owls remained the same.

Connection to Math To illustrate the Hardy-Weinberg principle, consider a population of 100 humans. Forty people are homozygous dominant for earlobe attachment *(EE).* Another 40 people are heterozygous *(Ee).* Twenty people are homozygous recessive *(ee).* In the 40 homozygous dominant people, there are 80 *E* alleles (2 *E* alleles × 40), and in the 20 homozygous recessive people there are 40 *e* alleles (2 *e* alleles × 20). The heterozygous people have 40 *E* alleles and 40 *e* alleles. Summing the alleles, we have 120 *E* alleles and 80 *e* alleles for a total of 200 alleles. The *E* allele frequency is 120/200, or 0.6. The *e* allele frequency is 80/200, or 0.4.

The Hardy-Weinberg principle states that the allele frequencies in populations should be constant. This often is expressed as $p + q = 1$. For our example, p can represent the *E* allele frequency and q can represent the *e* allele frequency.

Squaring both sides of the equation yields the new equation $p^2 + 2pq + q^2 = 1$. This equation allows us to determine the equilibrium frequency of each genotype in the population: homozygous dominant (p^2), heterozygous ($2pq$), and homozygous recessive (q^2). From the above example, $p = 0.6$, and $q = 0.4$, so $(0.6)(0.6) + 2(0.6)(0.4) + (0.4)(0.4) = 1$. In the example population, the equilibrium frequency for homozygous dominant will be 0.36, the equilibrium frequency of heterozygous will be 0.48, and the equilibrium frequency of homozygous recessive will be 0.16. Note that the sum of these frequencies equals one.

CAREERS IN BIOLOGY

Biometrician Almost all scientific research papers include some statistics. Many researchers consult biometricians—people who specialize in statistics related to biology—to help design studies and analyze study results. For more information on biology careers, visit biologygmh.com.

Reading Check Determine when a population is in equilibrium.

Conditions According to the Hardy-Weinberg principle, a population in genetic equilibrium must meet five conditions—there must be no genetic drift, no gene flow, no mutation, mating must be random, and there must be no natural selection. Populations in nature might meet some of these requirements, but hardly any population meets all five conditions for long periods of time. If a population is not in genetic equilibrium, at least one of the five conditions has been violated. These five conditions listed in **Table 15.3** are known mechanisms of evolutionary change.

Concepts in Motion

Interactive Table To explore more about the Hardy-Weinberg principle, visit biologygmh.com.

Table 15.3 The Hardy-Weinberg Principle

Condition	Violation	Consequence
The population is very large.	Many populations are small.	Chance events can lead to changes in population traits.
There is no immigration or emigration.	Organisms move in and out of the population.	The population can lose or gain traits with movement of organisms.
Mating is random.	Mating is not random.	New traits do not pass as quickly to the rest of the population.
Mutations do not occur.	Mutations occur.	New variations appear in the population with each new generation.
Natural selection does not occur.	Natural selection occurs.	Traits in a population change from one generation to the next.

Genetic drift Any change in the allelic frequencies in a population that is due to chance is called **genetic drift.** Recall from Chapter 10 that for simple traits only one of a parent's two alleles passes to the offspring, and that this allele is selected randomly through independent assortment. In large populations, enough alleles "drift" to ensure that the allelic frequency of the entire population remains relatively constant from one generation to the next. In smaller populations, however, the effects of genetic drift become more pronounced, and the chance of losing an allele becomes greater.

Founder effect The founder effect is an extreme example of genetic drift. The **founder effect** can occur when a small sample of a population settles in a location separated from the rest of the population. Because this sample is a random subset of the original population, the sample population carries a random subset of the population's genes. Alleles that were uncommon in the original population might be common in the new population, and the offspring in the new population will carry those alleles. Such an event can result in large genetic variations in the separated populations.

The founder effect is evident in the Amish and Mennonite communities in the United States in which the people rarely marry outside their own communities. The Old Order Amish have a high frequency of six-finger dwarfism. All affected individuals can trace their ancestry back to one of the founders of the Order.

Bottleneck Another extreme example of genetic drift is a **bottleneck,** which occurs when a population declines to a very low number and then rebounds. The gene pool of the rebound population often is genetically similar to that of the population at its lowest level, that is, it has reduced diversity. Researchers think that cheetahs in Africa experienced a bottleneck 10,000 years ago, and then another one about 100 years ago. Throughout their current range, shown in **Figure 15.15,** cheetahs are so genetically similar that they appear inbred. Inbreeding decreases fertility, and might be a factor in the potential extinction of this endangered species.

Reading Check **Explain** how genetic drift affects populations.

Study Tip

Concept Map Make a concept map, placing the term *evolution* in the top oval. The second row of ovals should contain the following terms: *genetic drift, gene flow, nonrandom mating, mutation,* and *natural selection.* As you read the chapter, fill in definitions and write examples that illustrate each term.

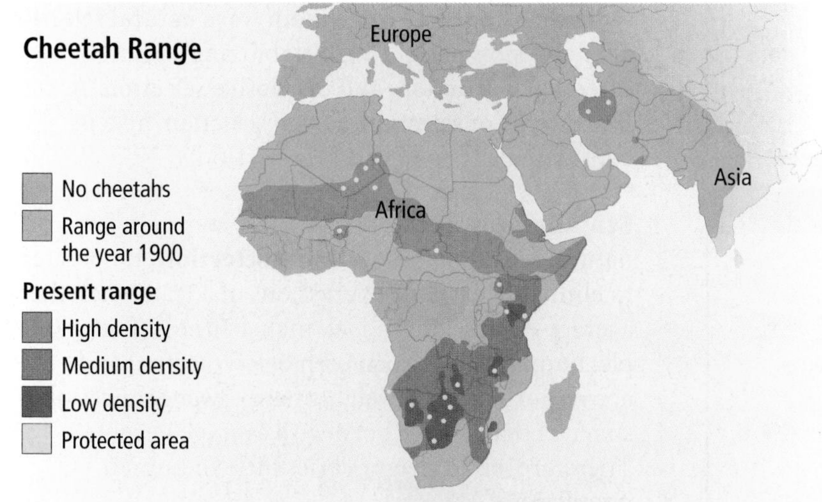

Figure 15.15 The map shows the present range of cheetahs in Africa. It is believed that cheetahs had a much larger population until a bottleneck occurred.
Apply Concepts ***What effect has the bottleneck had on the reproductive rate of cheetahs?***

Gene flow A population in genetic equilibrium experiences no gene flow. It is a closed system, with no new genes entering the population and no genes leaving the population. In reality, few populations are isolated. The random movement of individuals between populations, or migration, increases genetic variation within a population and reduces differences between populations.

Nonrandom mating Rarely is mating completely random in a population. Usually, organisms mate with individuals in close proximity. This promotes inbreeding and could lead to a change in allelic proportions favoring individuals that are homozygous for particular traits.

Personal Tutor

To learn about types of selection, visit biologygmh.com.

Mutation Recall from Chapter 12 that a mutation is a random change in genetic material. The cumulative effect of mutations in a population might cause a change in allelic frequencies, and thus violate genetic equilibrium. Though many mutations cause harm or are lethal, occasionally a mutation provides an advantage to an organism. This mutation will then be selected for and become more common in subsequent generations. In this way, mutations provide the raw material upon which natural selection works.

Reading Check **Summarize** how mutation violates the Hardy-Weinberg principle.

■ **Figure 15.16** Natural selection can alter allele frequencies of a population in three ways. The bell-shaped curve shown as a dotted line in each graph indicates the trait's original variation in a population. The solid line indicates the outcome of each type of selection pressure.

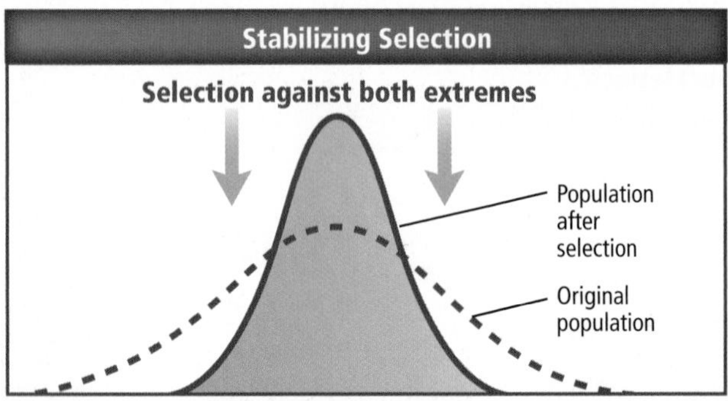

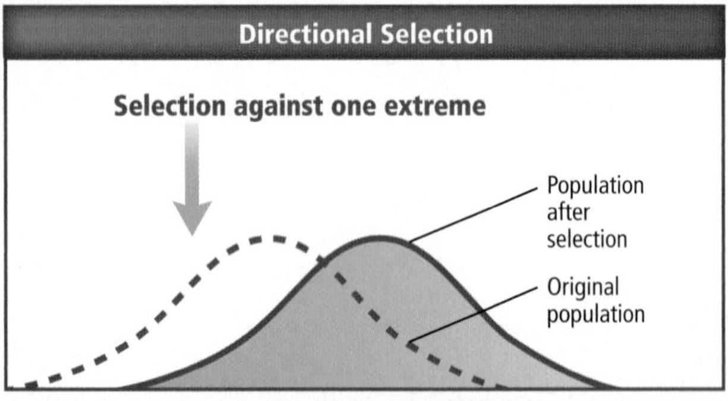

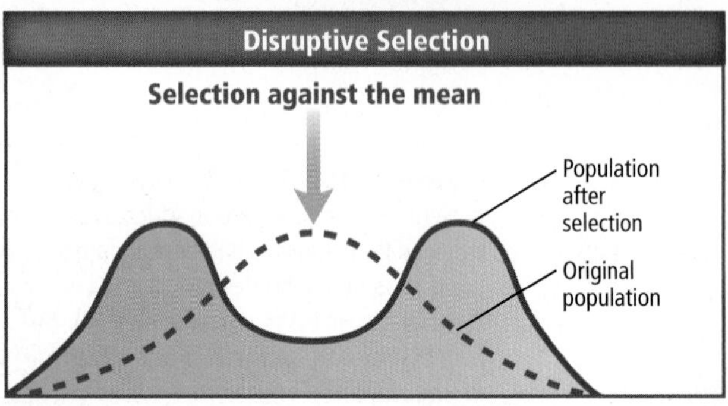

Natural selection The Hardy-Weinberg principle requires that all individuals in a population be equally adapted to their environment and thus contribute equally to the next generation. As you have learned, this rarely happens. Natural selection acts to select the individuals that are best adapted for survival and reproduction. Natural selection acts on an organism's phenotype and changes allelic frequencies. **Figure 15.16** shows three main ways natural selection alters phenotypes: through stabilizing selection, directional selection, and disruptive selection. A fourth type of selection, sexual selection, also is considered a type of natural selection.

Stabilizing selection The most common form of natural selection is **stabilizing selection.** It operates to eliminate extreme expressions of a trait when the average expression leads to higher fitness. For example, human babies born with below-normal and above-normal birth weights have lower chances of survival than babies born with average weights. Therefore, birth weight varies little in human populations.

Directional selection If an extreme version of a trait makes an organism more fit, **directional selection** might occur. This form of selection increases the expression of the extreme versions of a trait in a population. One example is the evolution of moths in industrial England. The peppered moth has two color forms, or morphs, as shown in **Figure 15.17.** Until the mid 1850s, nearly all peppered moths in England had light-colored bodies and wings. Beginning around 1850, however, dark moths began appearing. By the early 1900s, nearly all peppered moths were dark. Why? Industrial pollution favored the dark-colored moths at the expense of the light-colored moths. The darker the moth, the more it matched the sooty background of its tree habitat, and the harder it was for predators to see. Thus, more dark moths survived, adding more genes for dark color to the population. This conclusion was reinforced in the mid-1900s when the passage of air pollution laws led to the resurgence of light-colored moths. This phenomenon is called industrial melanism.

Directional selection also can be seen in Galápagos finches. For three decades in the latter part of the twentieth century, Peter and Rosemary Grant studied populations of these finches. The Grants found that during drought years, food supplies dwindled and the birds had to eat the hard seeds they normally ignored. Birds with the largest beaks were more successful in cracking the tough seed coating than were birds with smaller beaks. As a result, over the duration of the drought, birds with larger beaks came to dominate the population. In rainy years, however, the directional trend was reversed, and the population's average beak size decreased.

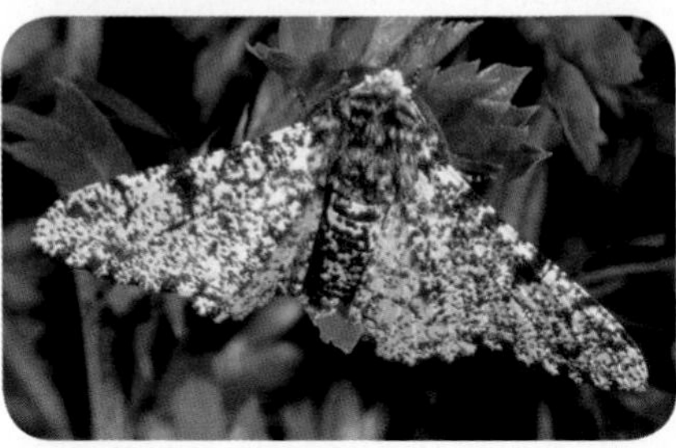

Figure 15.17 The peppered moth exists in two forms— light colored and dark colored.
Infer ***How might natural selection have caused a change in the frequencies of the two forms?***

DATA ANALYSIS LAB 15.2

Based on Real Data*

Interpret the Graph

How does pollution affect melanism in moths?
The changing frequencies of light-colored and dark-colored moths have been studied for decades in the United States. The percentage of the melanic, or dark, form of the moth was low prior to the industrial revolution. It increased until it made up nearly the entire population in the early 1900s. After antipollution laws were passed, the percentage of melanic moths declined, as shown in the graph.

Think Critically

1. **Interpret** What was the percent decrease in Pennsylvania melanic moth population?
2. **Hypothesize** Why might the percentage of melanic moths have remained at a relatively low level in Virginia?

*Data obtained from: Grant, B. S. and L. L. Wiseman. 2002. Recent history of melanism in American peppered moths. *Journal of Heredity* 93: 86-90.

Data and Observations

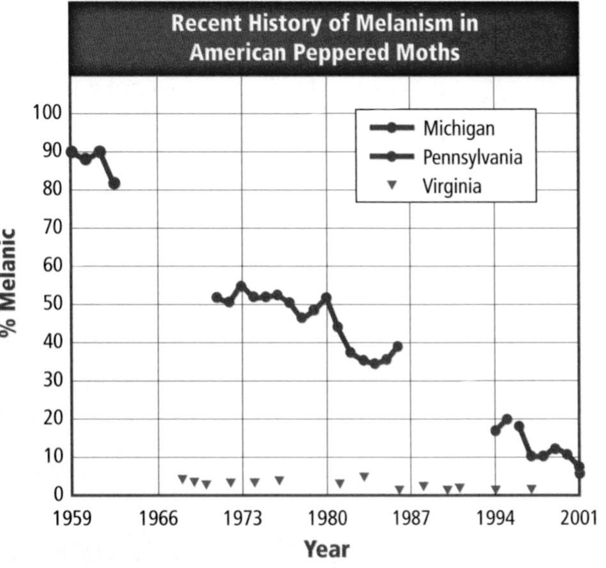

■ **Figure 15.18** Northern water snakes have two different color patterns depending on their habitat. Intermediate color patterns would make them more visible to predators.

Disruptive selection Another type of natural selection, **disruptive selection,** is a process that splits a population into two groups. It tends to remove individuals with average traits but retain individuals expressing extreme traits at both ends of a continuum. Northern water snakes, illustrated in **Figure 15.18,** are an example. Snakes living on the mainland shores inhabit grasslands and have mottled brown skin. Snakes inhabiting rocky island shores have gray skin. Each is adapted to its particular environment. A snake with intermediate coloring would be disadvantaged because it would be more visible to predators.

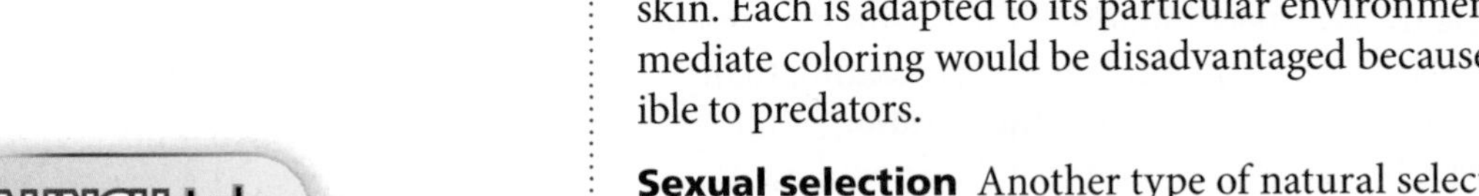

LAUNCH Lab

Review Based on what you have learned about adaptation, how would you now answer the analysis questions?

Sexual selection Another type of natural selection, in which change in frequency of a trait is based on the ability to attract a mate is called **sexual selection.** This type of selection often operates in populations where males and females differ significantly in appearance. Usually in these populations, males are the largest and most colorful of the group. The bigger the tail of a male peacock, as shown in **Figure 15.19,** the more attractive the bird is to females. Males also evolve threatening characteristics that intimidate other males; this is common in species, such as elk or deer, where the male keeps a harem of females.

Darwin wondered why some qualities of sexual attractiveness appeared to be the opposite of qualities that might enhance survival. For example, the peacock's tail, while attracting females, is large and cumbersome, and it might make the peacock a more likely target for predators. Though some modern scientists think that sexual selection is not a form of natural selection, others think that sexual selection follows the same general principle: brighter colors and bigger bodies enhance reproductive success, whatever the chances are for long-term survival.

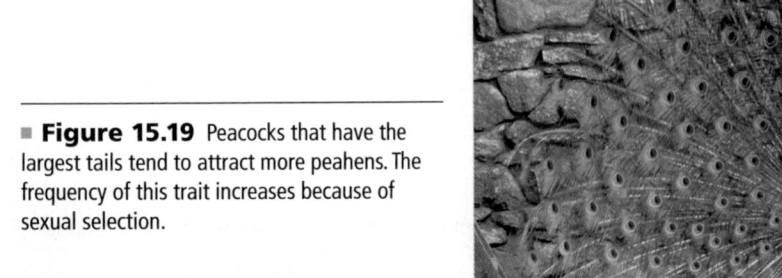

■ **Figure 15.19** Peacocks that have the largest tails tend to attract more peahens. The frequency of this trait increases because of sexual selection.

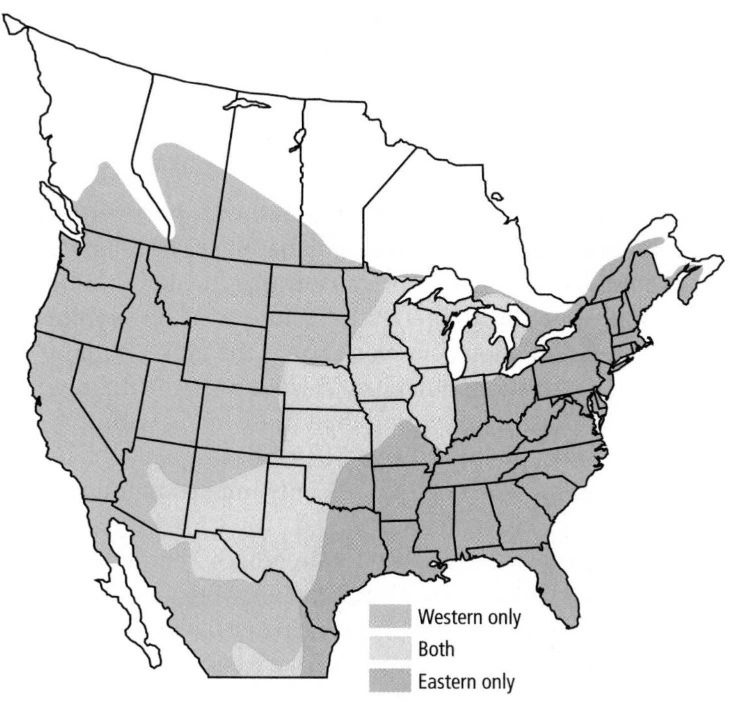

■ **Figure 15.20** The map shows the overlapping ranges of the Eastern meadowlark and Western meadowlark. While the two are similar in appearance, their songs separate them behaviorally.
Infer ***how different songs prevent the meadowlarks from breeding.***

Reproductive Isolation

Mechanisms of evolution—genetic drift, gene flow, nonrandom mating, mutation, and natural selection—violate the Hardy-Weinberg principle. To what extent each mechanism contributes to the origin of new species is a major topic of debate in evolutionary science today. Most scientists define speciation as the process whereby some members of a sexually reproducing population change so much that they can no longer produce fertile offspring with members of the original population. Two types of reproductive isolating mechanisms prevent gene flow among populations. **Prezygotic isolating mechanisms** operate before fertilization occurs. **Postzygotic isolating mechanisms** operate after fertilization has occurred to ensure that the resulting hybrid remains infertile.

Prezygotic isolation Prezygotic isolating mechanisms prevent reproduction by making fertilization unlikely. These mechanisms prevent genotypes from entering a population's gene pool through geographic, ecological, behavioral, or other differences. For example, the eastern meadowlark and the western meadowlark, pictured in **Figure 15.20,** have overlapping ranges and are similar in appearance. These two species, however, use different mating songs and do not interbreed. Time is another factor in maintaining a reproductive barrier. Closely related species of fireflies mate at different times of night, just as different species of trout live in the same stream but breed at different times of the year.

Postzygotic isolation When fertilization has occurred but a hybrid offspring cannot develop or reproduce, postzygotic isolation has occurred. Postzygotic isolating mechanisms prevent offspring survival or reproduction. A lion and a tiger are considered separate species because even though they can mate, the offspring—a liger, shown in **Figure 15.21**—is sterile.

■ **Figure 15.21** The offspring of a male lion and a female tiger is a liger. Ligers are sterile.

Speciation

For speciation to occur, a population must diverge and then be reproductively isolated. Biologists usually recognize two types of speciation: allopatric and sympatric.

Allopatric speciation In **allopatric speciation,** a physical barrier divides one population into two or more populations. The separate populations eventually will contain organisms that, if enough time has passed, will no longer be able to breed successfully with one another. Most scientists think that allopatric speciation is the most common form of speciation. Small subpopulations isolated from the main population have a better chance of diverging than those living within it. This was the conclusion of the biologist Ernst Mayr, who argued as early as the 1940s that geographic isolation was not only important but required for speciation.

Vocabulary

Academic Vocabulary

Isolation:
the condition of being separated from others.
After infection, a patient is kept in isolation from other patients to prevent the infection from spreading.

Geographic barriers can include mountain ranges, channels between islands, wide rivers, and lava flows. The Grand Canyon, pictured in **Figure 15.22,** is an example of a geographic barrier. The Kaibab squirrel is found on the canyon's north rim, while the Abert squirrel lives on the south rim. Scientists think that the two types of squirrels diverged from an ancestral species and today are reproductively isolated by the width of the canyon. While these animals officially belong to the same species, they demonstrate distinct differences and, in time, they might diverge enough to be classified as separate species.

Sympatric speciation In **sympatric speciation,** a species evolves into a new species without a physical barrier. The ancestor species and the new species live side by side during the speciation process. Evidence of sympatric evolution can be seen in several insect species, including apple maggot flies, which appear to be diverging based on the type of fruit they eat. Scientists think that sympatric speciation happens fairly frequently in plants, especially through polyploidy. Recall from Chapter 10 that polyploidy is a mutation that increases a plant's chromosome number. As a result, the plant is no longer able to interbreed with the main population.

Figure 15.22 The Grand Canyon is a geographic barrier separating the Abert and Kaibab squirrels.

■ **Figure 15.23** More than 300 species of cichlid fishes once lived in Lake Victoria. Their adaptive radiation is remarkable because it is thought to have occurred in less than 14,000 years.

Patterns of evolution

Many details of the speciation process remain unresolved. Relative to the human life span, speciation is a long process, and first-hand accounts of speciation are expected to be rare. However, evidence of speciation is visible in patterns of evolution.

Adaptive radiation More than 300 species of cichlid fish, six of which are illustrated in **Figure 15.23,** once lived in Africa's Lake Victoria. Data shows that these species diverged from a single ancestor within the last 14,000 years. This is a dramatic example of a type of speciation called **adaptive radiation.** Adaptive radiation, also called divergent evolution, can occur in a relatively short time when one species gives rise to many species in response to the creation of new habitat or another ecological opportunity. Likely, a combination of factors caused the explosive radiation of the cichlids, including the appearance of a unique double jaw, which allowed these fish to exploit various food sources. Adaptive radiation often follows large-scale extinctions. Adaptive radiation of mammals at the beginning of the Cenozoic following the extinction of dinosaurs likely produced the diversity of mammals visible today.

Coevolution Many species evolve in close relationship with other species. The relationship might be so close that the evolution of one species affects the evolution of other species. This is called coevolution. Mutualism is one form of coevolution. Recall from Chapter 2 that mutualism occurs when two species benefit each other. For example, comet orchids and the moths that pollinate them have coevolved an intimate dependency: the foot-long flowers of this plant perfectly match the foot-long tongue of the moth, shown in **Figure 15.24.**

In another form of coevolution, one species can evolve a parasitic dependency on another species. This type of relationship is often called a coevolutionary arms race. The classic example is a plant and an insect pathogen that is dependent on the plant for food. The plant population evolves a chemical defense against the insect population. The insects, in turn, evolve the biochemistry to resist the defense. The plant then steps up the race by evolving new defenses, the insect escalates its response, and the race goes on. Complex coevolutionary relationships like these might reflect thousands of years of evolutionary interaction.

■ **Figure 15.24** By coevolving, this moth and the comet orchid it pollinates exist in a mutualistic relationship.

Table 15.4	Convergent Evolution	Concepts In Motion — Interactive Table To explore more about convergent evolution, visit biologygmh.com.
Niche	Placental Mammals	Australian Marsupials
Burrower	Mole	Marsupial mole
Anteater	Lesser anteater	Numbat (anteater)
Mouse	Mouse	Marsupial mouse
Glider	Flying squirrel	Flying phalanger
Wolf	Wolf	Tasmanian wolf

Convergent evolution Sometimes unrelated species evolve similar traits even though they live in different parts of the world. This is called convergent evolution. Convergent evolution occurs in environments that are geographically far apart but that have similar ecology and climate. The mara and rabbit discussed in Section 15.2 provide an example of convergent evolution. The mara and the rabbit are unrelated, but because they inhabit similar niches, they have evolved similarities in morphology, physiology, and behavior. **Table 15.4** shows examples of convergent evolution between Australian marsupials and the placental mammals on other continents.

Rate of speciation Evolution is a dynamic process. In some cases, as in a coevolutionary arms race, traits might change rapidly. In other cases, traits might remain unchanged for millions of years. Most scientists think that evolution proceeds in small, gradual steps. This is a theory called **gradualism.** A great deal of evidence favors this theory. However, the fossil record contains instances of abrupt transitions. For example, certain species of fossil snails looked the same for millions of years, then the shell shape changed dramatically in only a few thousand years. The theory of **punctuated equilibrium** attempts to explain such abrupt transitions in the fossil record. According to this theory, rapid spurts of genetic change cause species to diverge quickly; these periods punctuate much longer periods when the species exhibit little change.

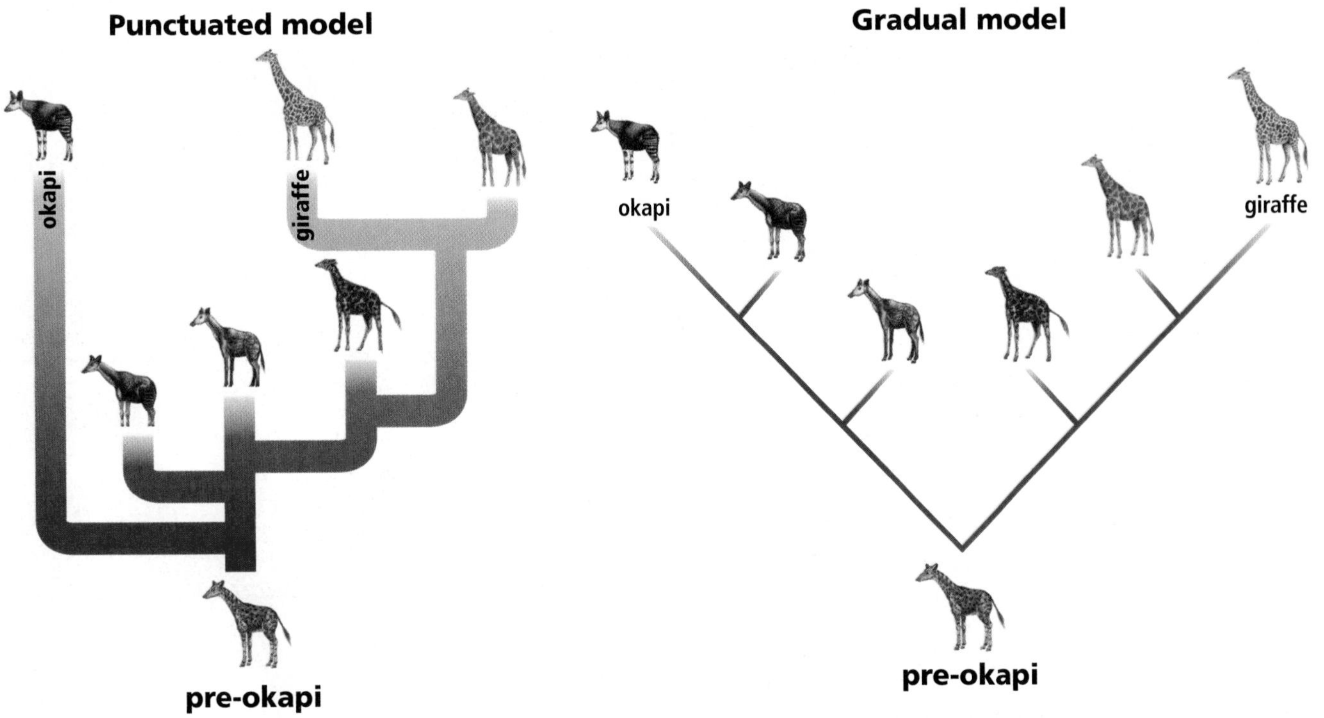

Punctuated model

Gradual model

okapi

giraffe

okapi

giraffe

pre-okapi

pre-okapi

■ **Figure 15.25** Gradualism and punctuated equilibrium are two competing models describing the tempo of evolution.

Concepts In Motion

Interactive Figure To see an animation of gradualism and punctuated equilibrium, visit biologygmh.com.

The two theories for the tempo of evolution are illustrated in **Figure 15.25.** The tempo of evolution is an active area of research in evolutionary theory today. Does most evolution occur gradually or in short bursts? Fossils can show only morphological structures. Changes in internal anatomy and function go unnoticed. How, then, does one examine the past for evidence?

The question of the tempo of evolution is an excellent illustration of how science works. Solving this puzzle requires insights from a variety of disciplines using a variety of methods. Like many areas of scientific endeavor, evolution offers a complex collection of evidence, and it does not yield easily to simple analysis.

Section 15.3 Assessment

Section Summary

- The Hardy-Weinberg principle describes the conditions within which evolution does not occur.
- Speciation often begins in small, isolated populations.
- Selection can operate by favoring average or extreme traits.
- Punctuated equilibrium and gradualism are two models that explain the tempo of evolution.

Understand Main Ideas

1. MAIN Idea **Describe** one new line of evidence supporting evolution that scientists learned after Darwin's book was published.
2. **List** three of the conditions of the Hardy-Weinberg principle.
3. **Discuss** factors that can lead to speciation.
4. **Indicate** which pattern of evolution is shown by the many species of finches on the Galápagos Islands.

Think Critically

5. **Design an Experiment** Biologists discovered two populations of frogs separated by the Amazon River. What experiment could be designed to test whether the two populations are one species or two?

MATH in Biology

6. What type of mathematical results would you expect from the experiment you designed above if the two populations diverged only recently?

CUTTING-EDGE BIOLOGY

T. Rex was a chicken?

After excavating a *T. Rex* fossil in 2003, scientists found it was too big to transport by helicopter. The scientists carefully broke the thighbone in half to ship the bone. Later tests on the broken bone led to an incredible surprise—the bone held preserved soft tissues! These tissues, shown in Figure 1, included the connective tissue, blood vessels, and possibly even blood cells.

Soft tissue In 2007, the fossil of the 68-million-year-old *T. Rex* was tested for the first time to see if dinosaurs could be shown to share genetic markers with modern animals. The examination of dinosaur fossils allows scientists to understand how life on Earth has changed over time. The discovery of soft tissue allows new tests to be performed. It is possible that many more dinosaur bones contain soft tissue samples.

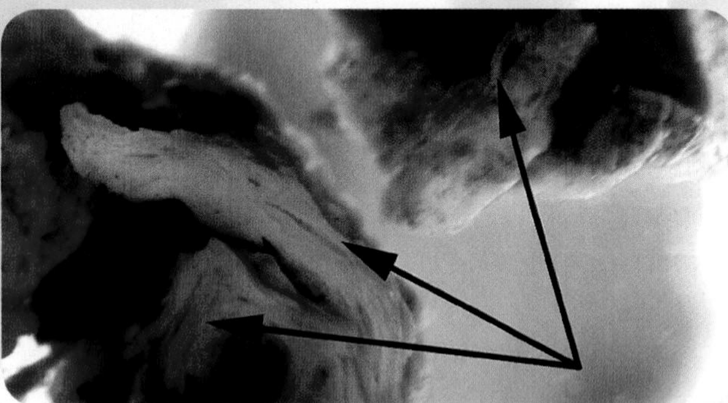

Figure 1 The soft tissue from the *T. Rex* discovered in 2003 was almost perfectly preserved.

Two independent tests on the soft tissue found in the fossil indicate that the *T. Rex* is likely related to the present-day chicken. This new research provides molecular evidence that supports hypotheses that a common ancestor linked birds and dinosaurs. In previous studies, physical similarities between early bird fossils and dinosaur fossils supported this link.

Figure 2 Molecular evidence suggests the *Tyrannosaurus Rex* is an ancestor of the chicken.

Figure 2 shows a chicken and an artist's rendering of a *T. Rex.* For example, some fossils showed the earliest birds had feet very similar to dinosaurs. In addition, several dinosaur fossils show evidence of feathers.

The test A group of scientists at North Carolina State University introduced a protein to chicken and the *T. Rex* soft tissue. The protein reacted strongly in the presence of the collagen found in chickens. A similar reaction was observed when the protein was administered to the dinosaur tissue. This indicates a molecular similarity between chicken tissues and dinosaur tissues

In another study performed by a team of researchers from Harvard Medical School, scientists obtained protein sequences from the T. Rex. The amino acid sequence in the proteins was similar to the amino acid sequence in chickens, showing clear support for an ancestral link between chickens and dinosaurs.

WRITING in Biology

Design an experiment that uses existing technology to test the tissue samples to answer a question you have about the link between birds and dinosaurs. For more information about *T. Rex* soft tissues visit biologygmh.com.

BIOLAB

CAN SCIENTISTS MODEL NATURAL SELECTION?

Background: **Natural selection is the mechanism Darwin proposed to explain evolution. Through natural selection, traits that allow individuals to have the most offspring in a given environment tend to increase in the population over time.**

Question: *How can natural selection be modeled in a laboratory setting?*

Materials

small, medium, and large beads
forceps
short-nosed pliers
tray or pan
stopwatch

Safety Precautions

Procedure

1. Read and complete the lab safety form.
2. Divide into groups of three. One student will use tweezers to represent one adult member of a predator population, one will use pliers to represent another adult member of the predator population, and the third will keep time and score.
3. Mix prey items (beads) on a tray or pan.
4. In 20 seconds, try to pick up all possible beads using forceps or pliers.
5. After the 20 seconds, assign three points for each large bead, two points for each medium bead, and one point for each small bead.
6. Add up the points and use the following rules: survival requires 18 points, and the ability to produce a new offspring requires an additional 10 points.
7. Determine the number of survivors and the number of offspring.
8. Repeat the procedure 10 times and combine your data with the other groups.

Analyze and Conclude

1. **Calculate** Combining all of the trials of all of the groups, determine the percentage of tweezers and pliers that survived.
2. **Evaluate** Using data from the entire class, determine the total number of offspring produced by the tweezer adult and the plier adult.
3. **Summarize** The original population was divided evenly between the tweezer adult and the plier adult. If all of the adults left, what would be the new population ratio? Use the results from the entire class.
4. **Infer** Given the survival and reproduction data, predict what will happen to the two organisms in the study. Which adult—the tweezer or the pliers—is better adapted to produce more offspring?
5. **Conclude** Using the principles of natural selection, how is this population changing?

APPLY YOUR SKILL

Make Inferences Given the results of the experiment, how will the prey populations (beads) change as the predator population changes? Explain your inference. To learn more about natural selection, visit BioLabs at biologygmh.com.

Study Guide

Download quizzes, key terms, and flash cards from biologygmh.com.

FOLDABLES **Research** Consider how natural selection has changed the plants and animals in your community. Select a plant or animal that is native to your community. Research and diagram some of the evolutionary changes the plant or animal went through from ancestors to the present organism.

Vocabulary	Key Concepts
Section 15.1 Darwin's Theory of Evolution by Natural Selection	
• artificial selection (p. 419) • evolution (p. 422) • natural selection (p. 420)	MAIN Idea Charles Darwin developed a theory of evolution based on natural selection. • Darwin drew from his observations on the HMS *Beagle* and later studies to develop his theory of evolution by natural selection. • Natural selection is based on ideas of excess reproduction, variation, inheritance, and advantages of certain traits in certain environments. • Darwin reasoned that the process of natural selection eventually could result in the appearance of new species.
Section 15.2 Evidence of Evolution	
• analogous structure (p. 426) • ancestral trait (p. 424) • biogeography (p. 428) • camouflage (p. 428) • derived trait (p. 424) • embryo (p. 426) • fitness (p. 428) • homologous structure (p. 424) • mimicry (p. 429) • vestigial structure (p. 425)	MAIN Idea Multiple lines of evidence support the theory of evolution. • Fossils provide strong direct evidence to support evolution. • Homologous and vestigial structures indicate shared ancestry. • Examples of embryological and biochemical traits provide insight into the evolution of species. • Biogeography can explain why certain species live in certain locations. • Natural selection gives rise to features that increase reproductive success.
Section 15.3 Shaping Evolutionary Theory	
• adaptive radiation (p. 439) • allopatric speciation (p. 438) • bottleneck (p. 433) • directional selection (p. 435) • disruptive selection (p. 436) • founder effect (p. 433) • genetic drift (p. 433) • gradualism (p. 440) • Hardy-Weinberg principle (p. 431) • postzygotic isolating mechanism (p. 437) • prezygotic isolating mechanism (p. 437) • punctuated equilibrium (p. 440) • sexual selection (p. 436) • stabilizing selection (p. 434) • sympatric speciation (p. 438)	MAIN Idea The theory of evolution continues to be refined as scientists learn new information. • The Hardy-Weinberg principle describes conditions within which evolution does not occur. • Speciation often begins in small, isolated populations. • Selection can operate by favoring average or extreme traits. • Punctuated equilibrium and gradualism are two models that explain the tempo of evolution.

CHAPTER 15

Assessment

Section 15.1

Vocabulary Review

Replace the underlined portions of the sentences below with words from the Study Guide to make each sentence correct.

1. Natural selection is a mechanism for species change over time.
2. Selective breeding was used to produce purebred Chihuahuas and cocker spaniels.
3. Differential survival by members of a population with favorable adaptations is a mechanism for a theory developed by Charles Darwin.

Understand Key Concepts

4. Which best describes the prevailing view about the age of Earth and evolution before Darwin's voyage on the HMS *Beagle?*
   A. Earth and life are recent and have remained unchanged.
   B. Species evolved rapidly during the first six thousand to a few hundred thousand years.
   C. Earth is billions of years old, but species have not evolved.
   D. Species have evolved on Earth for billions of years.

Use the photo below to answer question 5.

5. Which statement about the tortoise above would be part of an explanation for tortoise evolution based on natural selection?
   A. All tortoises look like the above tortoise.
   B. Tortoises with domed shells have more young than tortoises with flat shells.
   C. All the tortoises born on the island survive.
   D. The tortoise shell looks nothing like the shell of either parent.

Constructed Response

6. **Open Ended** Summarize Darwin's theory of evolution by using an example.
7. **Short Answer** How is artificial selection similar to natural selection?

Think Critically

8. **Sequence** Sequence events leading to evolution by natural selection.
9. **Recognize Cause and Effect** What is the likely evolutionary effect on a species of an increase in global temperatures over time?

Section 15.2

Vocabulary Review

The sentences below include terms that have been used incorrectly. Make the sentences true by replacing the italicized word with a vocabulary term from the Study Guide page.

10. Anatomical parts that have a reduced function in an organism are *analogous structures.*
11. *Biogeography* is a measure of the relative contribution an individual trait makes to the next generation.
12. *Camouflage* occurs when two or more species evolve adaptations to resemble each other.

Understand Key Concepts

Use the photos below to answer question 13.

13. These organisms have similar features that are considered what kind of structures?
   A. vestigial
   B. homologous
   C. analogous
   D. comparative

Use the photo below to answer question 14.

14. The photo of the bird above shows what kind of morphological adaptation?
    A. vestigial organ
    B. camouflage
    C. mimicry
    D. analogous structure

15. Which is not an example of a morphological adaptation?
    A. Cytochrome *c* is similar in monkeys and humans.
    B. Butterflies evolve similar color patterns.
    C. A harmless species of snake resembles a harmful species.
    D. Young birds have adaptations for blending into the environment.

16. Industrial melanism could be considered a special case of which of the following?
    A. embryological adaptation
    B. mimicry
    C. physiological adaptation
    D. structural adaptation

17. Which sets of structures are homologous?
    A. a butterfly's wing and a bat's wing
    B. a moth's eyes and a cow's eyes
    C. a beetle's leg and a horse's leg
    D. a whale's flipper and a bird's wing

Constructed Response

18. **Short Answer** Describe how cytochrome *c* provides evidence of evolution.

19. **Short Answer** What can be concluded from the fact that many insects are resistant to certain pesticides?

20. **Short Answer** Why are fossils considered to provide the strongest evidence supporting evolution?

Think Critically

21. **Design an Experiment** How could you design an experiment to show that a species of small fish has the ability to evolve a camouflage color pattern?

22. **CAREERS IN BIOLOGY** An evolutionary biologist is studying several species of closely related lizards found on Cuba and surrounding islands. Each species occupies a somewhat different niche, but in some ways they all look similar to the green anole lizard found in Florida. Suggest the pattern of lizard evolution.

Section 15.3

Vocabulary Review

Choose the vocabulary term from the Study Guide page that best matches each of the following descriptions.

23. one species evolves over millions of years to become two different but closely related species

24. a species evolves into a new species without a physical barrier

25. the random changes in gene frequency found in small populations

Understand Key Concepts

Use the figure below to answer question 26.

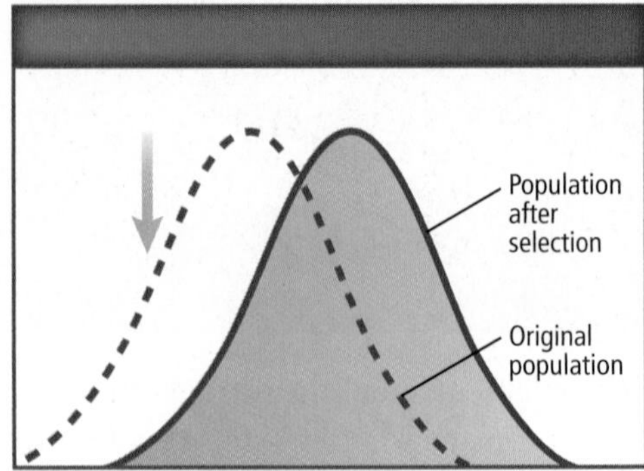

26. The graph above best represents which kind of selection?
    A. directional
    B. disruptive
    C. sexual
    D. stabilizing

Use the photo below to answer question 27.

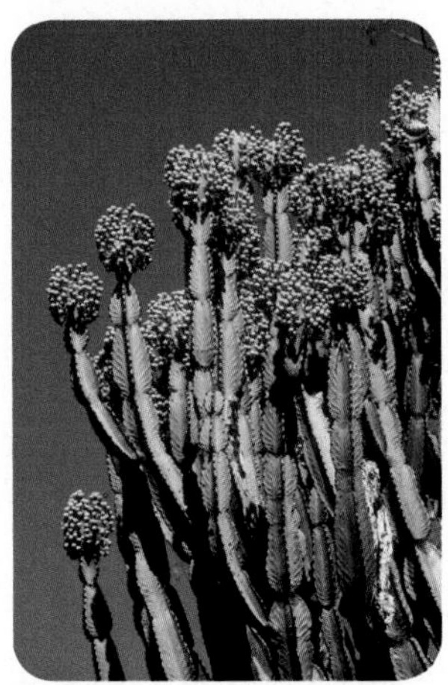

27. The plant in the above illustration looks like a cactus but is classified in a completely separate group of plants. This would be an example of which mechanism?
    A. adaptive radiation
    B disruptive selection
    C. convergent evolution
    D. punctuated equilibrium

Constructed Response

28. **Open Ended** Discuss why the Hardy-Weinberg principle is often violated in real populations.

29. **Open Ended** Sea stars eat clams by pulling apart the two halves of a clam's shell. Discuss how this could result in directional selection of clam muscle size.

30. **Short Answer** Compare and contrast genetic drift and natural selection as mechanisms of evolution.

Think Critically

31. **Make and Use Graphs** Draw a graph that would illustrate a population that has a wide variation of color from light to dark brown. Then draw on the same graph what that population would look like after several years of stabilizing selection. Label your graph.

32. **Drawing a Conclusion** What would you conclude about the evolutionary process that produces two unrelated species that share similar niches on different continents?

Additional Assessment

33. **WRITING in Biology** Imagine that you are Charles Darwin and write a letter to your father detailing your observations aboard the *Beagle.*

34. **WRITING in Biology** Write a paragraph that explains why a genetic bottleneck can be an important evolutionary factor for a species.

DBQ Document-Based Questions

Darwin,Charles. 1859. *On the Origin of Species by Means of Natural Selection, or the Preservation of Favoured Races in the Struggle for Life.*

Naturalists continually refer to external conditions, such as climate, food, etc., as the only possible cause of variation. In one very limited sense, as we shall hereafter see, this may be true; but it is preposterous to attribute to mere external conditions, the structure, for instance, of the woodpecker, with its feet, tail, beak, and tongue, so admirably adapted to catch insects under the bark of trees.

35. In Darwin's time, most naturalists considered only external conditions as causes of variation. What nonexternal mechanism did Darwin propose as a cause of variation?

36. How would modern scientists explain the nonexternal mechanisms that Darwin proposed?

37. Consider Darwin's example of the woodpecker. Explain the role of natural selection in producing a bird species with a woodpeckerlike beak.

Cumulative Review

38. Explain the importance of radiometric dating to paleontologists. **(Chapter 14)**

39. Discuss two ways in which losses in biodiversity could affect humans. **(Chapter 5)**

40. In this chapter, you learned that mutations provide the new variations that are involved in natural selection. Explain how mutations occur, and discuss the consequences of a point mutation and a frameshift mutation. **(Chapter 12)**

Standardized Test Practice

Cumulative

Multiple Choice

1. Which experimental setup did Francesco Redi use to test the idea of spontaneous generation?
   A. a flask filled with all the chemicals present on early Earth
   B. mice sealed in jars with lit candles and jars with unlit candles
   C. rotten meat in covered jars and uncovered jars
   D. special flasks that were filled with broth

Use the illustration below of tortoises on two different islands to answer questions 2 and 3.

Large Island

Small Island

2. The above illustrates which principle of natural selection?
   A. inheritance
   B. variation
   C. differential reproduction
   D. overproduction of offspring

3. Tortoises that have shells with higher openings can eat taller plants. Others can only reach vegetation close to the ground. Judging from the differences in the tortoises' shells, what kind of vegetation would you expect to find on the large and small islands?
   A. Both islands have a dense ground cover of low-growing plants.
   B. Both islands have similar plants, but vegetation is more spread out on large islands.
   C. On the large island, the land is mostly dry, and only tall trees grow.
   D. The small island is less grassy, and plants grow with their leaves farther above ground.

4. A dinosaur footprint in rocks would be which kind of fossil?
   A. cast fossil
   B. petrified fossil
   C. replacement fossil
   D. trace fossil

5. Which concept is essential for the process of DNA fingerprinting?
   A. location of genes for related traits on different chromosomes
   B. organization of human DNA into 46 chromosomes
   C. provision by DNA of the codes for proteins in the body
   D. uniqueness of each person's pattern of noncoding DNA

6. Chargaff's rules led to the understanding of which aspect of DNA structure?
   A. base pairing
   B. helix formation
   C. alternation of deoxyribose and phosphate
   D. placement of 3' and 5' carbons

Use the Punnett square below to answer question 7.

	B	?
b		
b		

7. A test cross, shown in the Punnett square above, is used to determine the genotype of an animal that is expressing a dominant gene (B) for a particular characteristic. If the animal is homozygous for the dominant trait, which percentage of its offspring will have the dominant gene?
   A. 25%
   B. 50%
   C. 75%
   D. 100%

8. What prevents the two strands of DNA from immediately coming back together after they unzip?
   A. addition of binding proteins
   B. connection of Okazaki fragments
   C. parting of leading and lagging strands
   D. use of multiple areas of replication

Short Answer

Use the diagram of Miller and Urey's experiment below to answer questions 9 and 10.

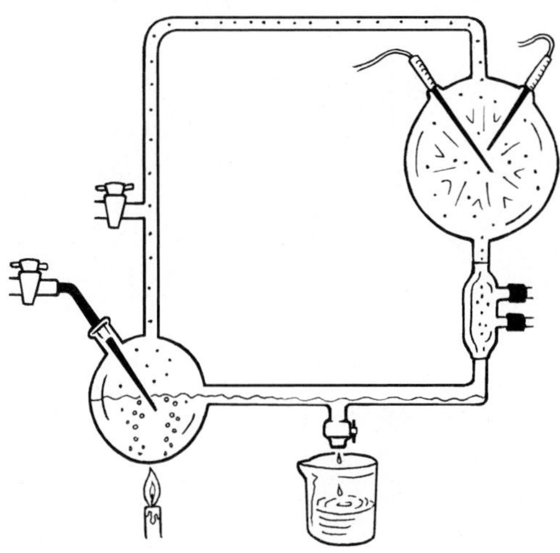

9. What are the possible consequences of a different mix of gases in the apparatus?

10. Some scientists think that lightning might not have been present on Earth in the past. What other energy sources might have caused these reactions?

11. Describe briefly how scientists could use a particular kind of bacteria to synthesize a specific protein.

12. Predict two positive outcomes and two negative outcomes of using transgenic plants for agricultural purposes.

13. Explain the connection between excess reproduction and the concept of natural selection as formulated by Darwin.

14. How would a primitive cell benefit from a symbiotic relationship with a mitochondrion?

Extended Response

Use the diagram below to answer questions 15 and 16.

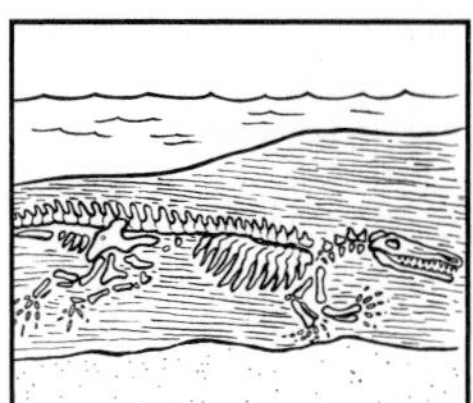

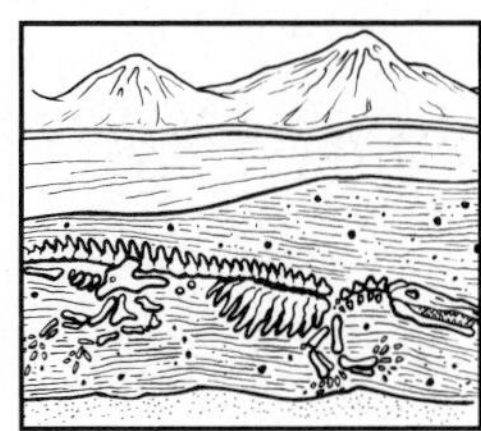

15. Describe the process illustrated in the figure.

16. Explain why a fossil is more likely to form in a wet environment than in a dry environment.

Essay Question

Scientists think that archaea living today are similar to ancient archaea. Many archaea today are found in places such as hot springs, deep ocean hydrothermal vents, polar ice, and in other extreme environments. The organisms living in these environments might be similar to organisms that existed in the distant past.

Using the information in the paragraph above, answer the following question in essay format.

17. Scientists also study organisms in extreme environments to help identify where life might exist on other planets. Why would understanding the origins of life on Earth help with discovering life on other planets?

NEED EXTRA HELP?																	
If You Missed Question . . .	1	2	3	4	5	6	7	8	9	10	11	12	13	14	15	16	17
Review Section . . .	14.2	15.1, 15.3	15.1	14.1	12.1	13.1	10.2	12.2	14.2	14.2	13.2	13.2	15.1	14.2	14.1	14.1	14.2

CHAPTER 16

Primate Evolution

BIG Idea **Evolutionary change in a group of small, tree-living mammals eventually led to a diversity of species that includes modern humans.**

Section 1
Primates
MAIN Idea Primates share several behavioral and biological characteristics, which indicates that they evolved from a common ancestor.

Section 2
Hominoids to Hominins
MAIN Idea Hominins, a subgroup of the hominoids, likely evolved in response to climate changes of the Miocene.

Section 3
Human Ancestry
MAIN Idea Tracing the evolution of the genus *Homo* is important for understanding the ancestry of humans, the only living species of *Homo*.

BioFacts

- The prehensile tail of a red howler monkey can be 50–75 cm long, and can support the mass of its entire body.
- As a rainstorm begins, red howler monkeys howl together as if in a chorus.

LAUNCH Lab

What are the characteristics of primates?

If you've been to a zoo or seen pictures of African wildlife, you have probably observed monkeys, chimpanzees, and gorillas. Maybe you've even seen pictures of lemurs. What makes these animals primates? What makes you a primate? In this lab, you will investigate the features that you share with these other primates.

Procedure

1. Read and complete the lab safety form.
2. Scan through the pictures of primates in Section 1 of Chapter 16. Do not read the text.
3. Explain what physical features you see in primates that appear in humans.
4. Create a data table in which to record your observations.

Analysis

1. **Compare** the human and ape characteristics in your table. Which features are similar?
2. **Contrast** How are primates different from the cats and dogs and other mammals around you?

Visit biologygmh.com to:

- study the entire chapter online
- explore the Interactive Time Line, Concepts in Motion, the Interactive Table, Microscopy Links, and links to virtual dissections
- access Web links for more information, projects, and activities
- review content online with the Interactive Tutor and take Self-Check Quizzes

Organize Make this Foldable to help you organize Old World monkeys and New World monkeys.

STEP 1 Fold a sheet of notebook paper in half lengthwise so that the side without holes is 2.5 cm shorter than the side with holes.

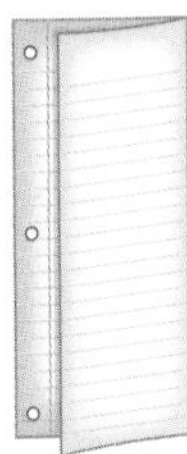

STEP 2 Fold the paper in thirds.

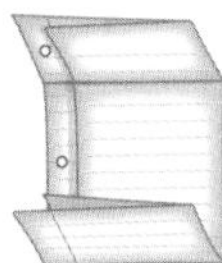

STEP 3 Unfold and cut along the two fold lines of the top layer only. This will make three tabs.

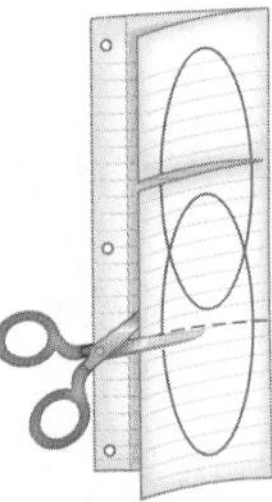

STEP 4 Label as illustrated.

FOLDABLES **Use this Foldable with Section 16.1.** As you read this section, take notes under the tabs, list characteristics, and give examples to compare and contrast Old World monkeys and New World monkeys.

Section 16.1

Reading Preview

Objectives

- **Describe** characteristics of primates.
- **Compare** major primate groups.
- **Trace** the evolution of primates.

Review Vocabulary

extinction: the disappearance of a species when the last of its members dies

New Vocabulary

opposable first digit
binocular vision
diurnal
nocturnal
arboreal
anthropoid
prehensile tail
hominin

Primates

MAIN Idea Primates share several behavioral and biological characteristics, which indicates that they evolved from a common ancestor.

Real-World Reading Link You can often tell that your aunts, uncles, or cousins are related to you. Perhaps they have the same color hair, similar features, or they are as tall as you. Just as you can tell that you are related to your biological family, characteristics of primates show that they are also a related family.

Characteristics of Primates

Humans, apes, monkeys, and lemurs belong to a group of mammals called primates. Though primates are highly diverse, they share some general features. Some primates have a high level of manual dexterity, which is the ability to manipulate or grasp objects with their hands. They usually also have keen eyesight and long, highly movable arms. Compared to other animals, they have large brains. The primates with the largest brains, which includes humans, have the capacity to reason.

Manual dexterity Primates are distinguished by their flexible hands and feet. All primates typically have five digits on each hand and foot; as you know, humans have fingers and toes. Most have flat nails and sensitive areas on the ends of their digits. The first digits on most primates' hands are opposable, and the first digit on many primates' feet are opposable. An **opposable first digit,** either a thumb or a great toe, is set apart from the other digits. This digit can be brought across the palm or foot so that it touches or nearly touches the other digits. This action allows the primate to grasp an object in a powerful grip. Some primates also have lengthened first digits that provide added dexterity. **Figure 16.1** shows a monkey using its opposable thumbs to grasp its food.

Figure 16.1 This squirrel monkey is using its opposable digits to hold its dinner—a lantern fly.
Infer *other ways a primate might use an opposable digit.*

Senses Though there are exceptions, primates rely more on vision and less on their sense of smell than other mammals do. Their eyes, protected by a bony eye socket, are on the front of their face. This creates overlapping fields of vision, often called **binocular vision.** Forward-looking eyes allow for a greater field of depth perception, and enable primates to judge relative distance and movement of an object.

Most primates are **diurnal** (di YUR nul), which means they are active during the day. Because these primates are active in daylight, most also have color vision. Primates that are **nocturnal** (nahk TUR nul) are active at night. They only see in black and white. An increased sense of vision is generally accompanied by a decreased sense of smell. Their snouts are smaller and their faces tend to be flattened, which increases the degree of binocular vision. Their teeth are reduced in size and usually are unspecialized, meaning that they are suitable for many different types of diets.

Locomotion Another characteristic of primates is their flexible bodies. Primates have limber shoulders and hips, and primarily rely on hind limbs for locomotion. Most primates live in trees and have developed an extraordinary ability to move easily from branch to branch. When on the ground, all primates except humans walk on all four limbs. Many primates can walk upright for short distances and many have a more upright posture compared to four-legged animals.

Complex brain and behaviors Primates tend to have large brains in relation to their body size. Their brains have fewer areas devoted to smell and more areas devoted to vision. They also tend to have larger areas devoted to memory and coordinating arm and leg movement. Along with larger brains, many primates have problem-solving abilities and well-developed social behaviors, such as grooming and communicating. Most diurnal primates spend a great deal of time socializing by spending time grooming each other. In addition, many primates have complex ways of communicating to each other, which include a wide range of facial expressions.

Reproductive rate Most primates have fewer offspring than other animals. Usually, primates give birth to one offspring at a time. Compared to other mammals, pregnancy is long, and newborns are dependent on their mothers for an extended period of time. For many primates, this time period allows for the increased learning of complex social interactions. A low reproductive rate, the loss of tropical habitats, and human predation has threatened some primate populations. Many are endangered. **Figure 16.2** illustrates the tropical areas of the world, such as Africa and Southeast Asia, where primates live.

LAUNCH Lab

Review Based on what you have read about primate characteristics, how would you now answer the analysis questions?

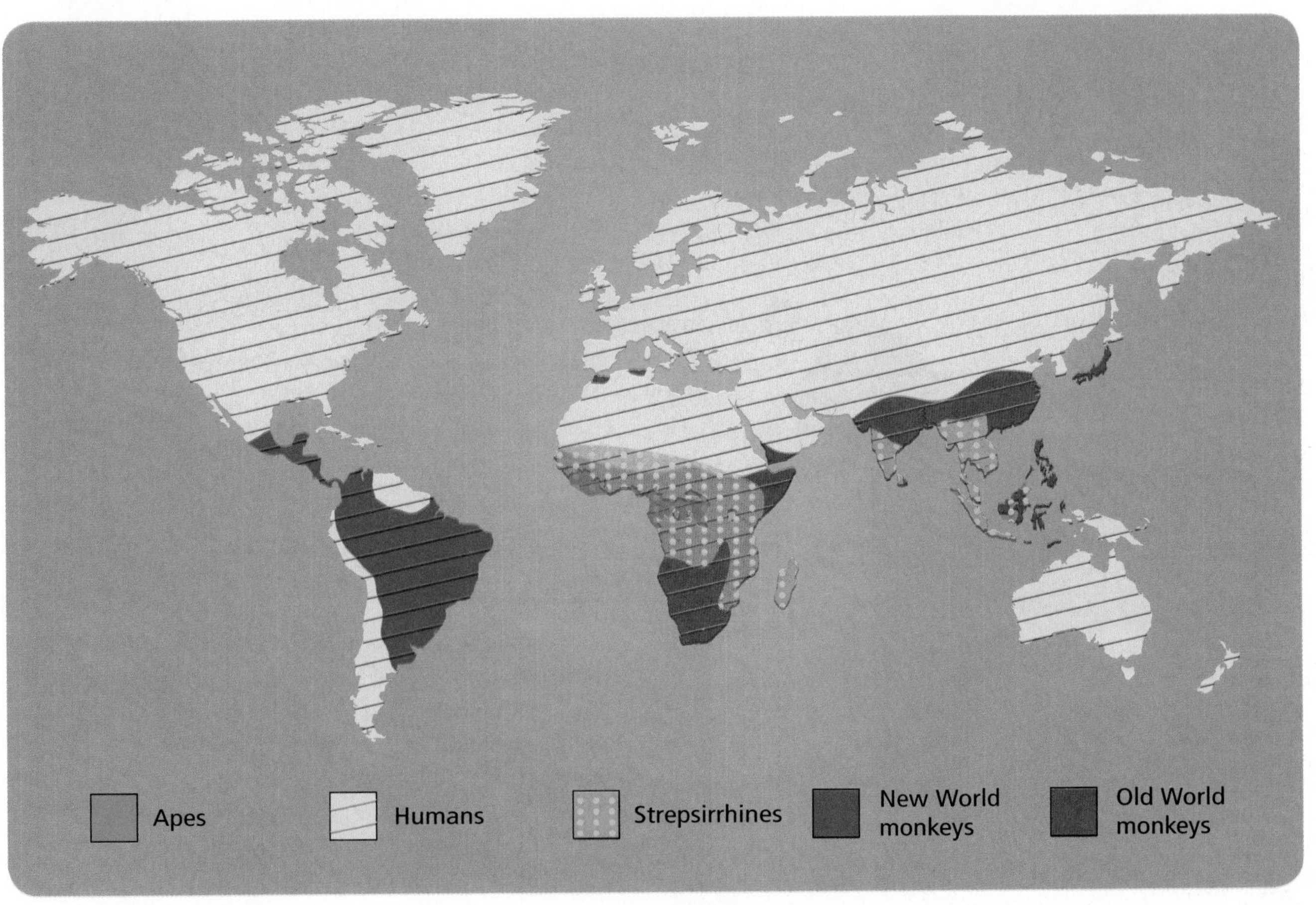

■ **Figure 16.2** Non-human primates live in a broad area spanning most of the world's tropical regions. Use this map as you read about the different primates.

Visualizing Primates

Figure 16.3
Primates are members of a highly diverse order of mammals. Most primates share common features such as binocular vision and opposable digits.

A The strepsirrhines are relatively small, have large eyes, and are nocturnal. They resemble the earliest primates.

B New World monkeys are characterized by a relatively long tail. Many have prehensile tails.

C Old World monkeys resemble New World monkeys, but lack a prehensile tail. Some have reduced tails.

D Asian apes are long-armed and inhabit tropical rain forests. Apes lack tails.

E African apes live in family groups or small bands and display complex social behavior.

F Humans, *Homo sapiens,* are the only living species in the hominin group. Hominins are unique because they possess the ability to walk for long distances on two legs.

Concepts In Motion **Interactive Figure** To explore more about primates, visit biologygmh.com.

Biology Online

Primate Groups

Primates are a large, diverse group of more than 200 living species. Examine **Figure 16.3** as you read about this diverse group. Most primates are **arboreal** (ar BOHR ee uhl), or tree-dwelling. Arboreal primates live in the world's tropical and subtropical forests. Primates that live on the ground are considered terrestrial primates.

Primates are classified into two subgroups based on characteristics of their nose, eyes, and teeth. The most basic subgroup is the strepsirrhines (STREP sihr ines), (also called "wet-nosed primates"), such as the lemur. The second subgroup consists of the haplorhines (HAP lohr ines), also called "dry-nosed" primates. The haplorhines include the **anthropoids** (AN thruh poydz), a group of large-brained, diurnal monkeys and hominoids.

Reading Check **Differentiate** between strepsirrhines and haplorhines.

Strepsirrhines

Strepsirrhines can be identified by their large eyes and ears. However, they are the only primates that rely predominantly on smell for hunting and social interaction. Some members of this primate group can be found in tropical Africa and Asia. Most are found in Madagascar and nearby islands. Madagascar drifted away from the African mainland as these animals evolved leaving them reproductively isolated. This isolation resulted in their diversification. **Table 16.1** lists characteristics of some strepsirrhine groups.

VOCABULARY

WORD ORIGIN

Lemur
comes from Latin, meaning *spirit of the night.*

Concepts In Motion

Interactive Table To explore more about strepsirrhines, visit biologygmh.com.

Table 16.1 Characteristics of Strepsirrhines

Group	Lemurs	Aye-Ayes	Lorises	Galagos
Example				
Active Period	Large—diurnal Small—nocturnal	Nocturnal	Nocturnal	Mostly nocturnal
Range	Madagascar	Madagascar	Africa and Southeast Asia	Africa
Characteristics	• Vertical leaper • Uses long bushy tail for balance • Herbivores and omnivores	• Taps bark, listens, fishes out grubs with long third finger	• Small and slow climber, solitary • Lack tails • Some have toxic secretions	• Small and fast leaper • No opposable digit • Long tail

■ **Figure 16.4** Lemurs vary in their size and color. Some lemurs, like this sifaka, spend time on the ground.

Most small lemurs are nocturnal and solitary. Only a few large species, such as the sifaka shown in **Figure 16.4,** are diurnal and social. The indri is unique because it does not have a tail, unlike most lemurs that use their bushy tails for balance as they jump from branch to branch. Lorises are similar to lemurs but are found primarily in India and Southeast Asia. Galagos (ga LAY gohs), also called bushbabies, are found only in Africa.

Haplorhines

The second group of primates is a much larger group. The haplorhines include tarsiers, monkeys, and apes. The apes, in turn, include gibbons, orangutans, gorillas, chimpanzees, and humans.

The tarsier is found only on Borneo and the Philippines. It is a small, nocturnal creature with large eyes. It has the ability to rotate its head 180 degrees like an owl. It lives in trees, where it climbs and leaps among the branches. The tarsier shares characteristics with both lemurs and monkeys. Scientists once classified it with the lemurs, but new evidence suggests that it is more closely related to anthropoids, which makes it part of the haplorhine group.

Anthropoids are generally larger than strepsirrhines, and they have large brains relative to their body size. They are more likely to be diurnal, with eyes adapted to daylight and sometimes to color. Anthropoids also have more complex social interactions. They tend to live longer than lemurs and other strepsirrhines. The anthropoids are split into the New World monkeys and the Old World monkeys. "New World" refers to the Americas; "Old World" refers to Africa, Asia, and Europe. New World monkeys are the only monkeys that live in the Americas.

FOLDABLES
Incorporate information from this section into your Foldable.

New World monkeys The New World monkeys are a group of about 60 species of arboreal monkeys that inhabit the tropical forests of Mexico, Central America, and South America. New World monkeys include the marmosets and tamarins. These are among the smallest and most unique primates. Neither species has fingernails or opposable digits.

The New World monkeys also include the squirrel monkeys, spider monkeys, and capuchin monkeys. Some of these monkeys have opposable digits and most are diurnal and live together in social bands. Most are also distinguished by their prehensile (pree HEN sul) tails. A **prehensile tail** functions like a fifth limb. It can grasp tree branches or other objects and support a monkey's weight, like that shown in **Figure 16.5.**

■ **Figure 16.5** This spider monkey uses its prehensile tail as a fifth limb.

Old World monkeys Old World monkeys live in a wide variety of habitats throughout Asia and Africa, from snow-covered mountains in Japan to arid grasslands in Africa. Some Old World monkeys live in Gibraltar, which is located at the southern tip of Spain. There are about 80 species in this group, including macaques and baboons in one subgroup, and colobus and proboscis monkeys in another. Old World monkeys are similar to New World monkeys in many ways. They are diurnal and live in social groups. However, their noses tend to be narrower and their bodies are usually larger. They also spend more time on the ground. None have prehensile tails, and some have no tails. Most Old World monkeys have opposable digits.

Apes Only a handful of ape species exist today. Apes generally have larger brains in proportion to their body size than monkeys. They also have longer arms than legs, barrel-shaped chests, no tails, and flexible wrists. They are often highly social and have complex vocalizations. They are classified into two subcategories: the lesser apes, which include the gibbons and siamangs, and the great apes, which include orangutans, gorillas, chimpanzees, and humans.

Lesser apes The Asian gibbons and their close relatives, the larger siamangs, are the arboreal gymnasts of the ape family. Though they have the ability to walk on either two or four legs like all great apes, they generally move from branch to branch using a hand-over-hand swinging motion called brachiation. This motion, as shown in **Figure 16.6,** enables an adult gibbon to move almost 3 m in one swing.

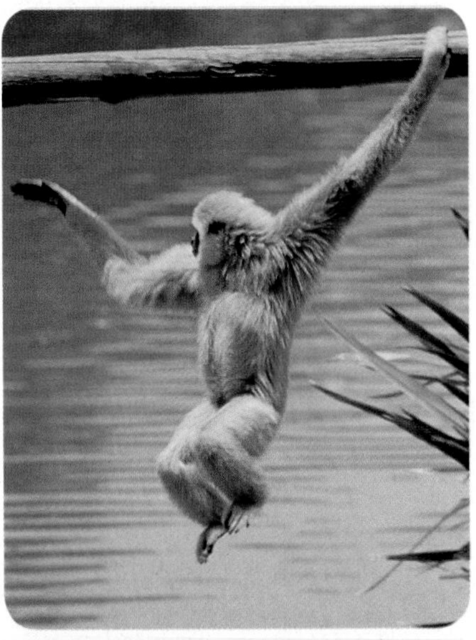

■ **Figure 16.6** Lesser apes, such as this gibbon, move through trees primarily by brachiation—a hand-over-hand swinging motion.

Great apes Orangutans are the largest arboreal primates and the only great ape species that lives exclusively in Asia. Orangutans are large enough that the males are often more comfortable on the ground, though they are not efficient walkers. Female orangutans give birth once every eight years and nurse their young for up to six years. A male orangutan with prominent cheek pads and female orangutan with her offspring are shown in **Figure 16.7.**

The gorillas are the largest of the primates. Like all great apes, they are predominantly terrestrial animals. They walk on all four limbs, supporting themselves by their front knuckles. Also, like other great apes, they use sticks as simple tools in the wild, and some living in captivity have been taught to recognize characters and numbers.

■ **Figure 16.7** Male orangutans are much larger and more solitary than females. The females spend most of their time raising their offspring.

Figure 16.8 The bonobo is slightly smaller than the chimpanzee. Like the chimpanzee, it is structurally and behaviorally similar to humans.

Chimpanzees and their close relatives, the bonobos, are also knuckle-walkers. They have well-developed communication systems, such as body positions and gestures, and social behavior, and they live in a wide variety of habitats. They are more like humans in their physical structure and behavior than any other primate. The bonobo, shown in **Figure 16.8,** is slightly smaller than the chimpanzee. It was once called the "pygmy chimpanzee," but it now is considered a separate species.

Humans are included in the great ape family. They are then classified in a separate subcategory of hominids called hominins. **Hominins** are humanlike primates that appear to be more closely related to present-day humans than they are to present-day chimpanzees and bonobos. Though many species of hominins have existed on Earth, only one species—the group to which you belong—survives today. The diagram in **Figure 16.9** illustrates evolutionary relationships among primates.

Primate Evolution

Most primates today are arboreal. Prehensile tails, long limbs, binocular vision, brachiation, and opposable digits are traits that help them take full advantage of their forest environments.

Arboreal adaptation Some scientists suggest that primates evolved from ground-dwelling animals that searched for food in the top branches of forest shrubbery. They then evolved into additional food-gathering niches in trees. For example, the flexible hand with its opposable digits evolved not to grasp tree branches but to catch insects. Other scientists suggest that the rise of flowering plants provided new niche opportunities, and that arboreal adaptations allowed primates to take advantage of the fruits and flowers of trees.

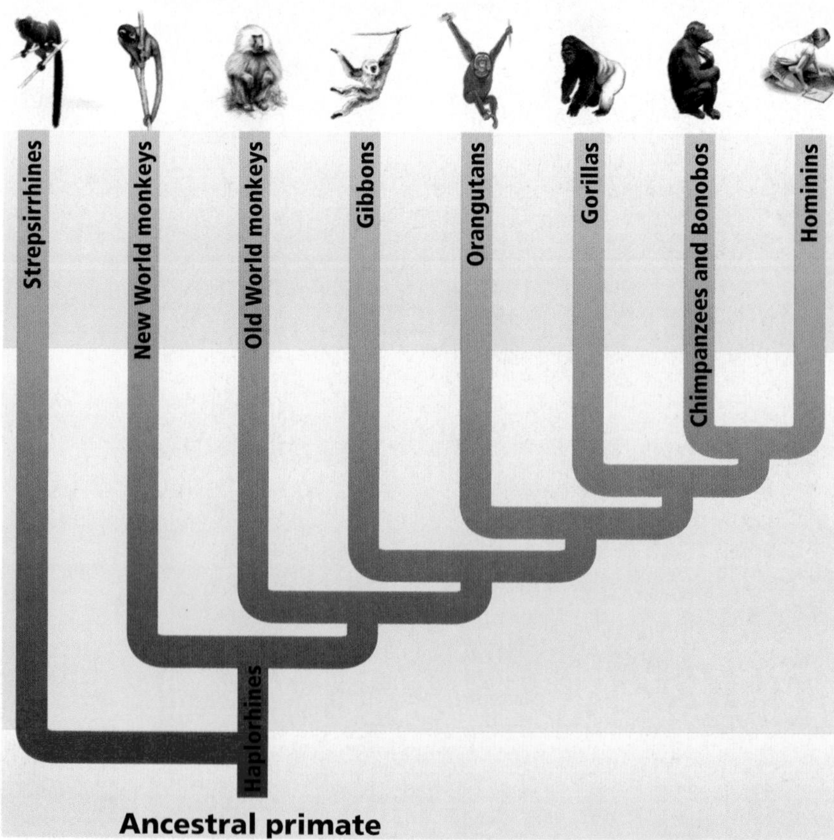

Figure 16.9 This branching diagram illustrates the diverging pattern of primate evolution.

Trace *Which primate was the earliest to diverge?*

Primate ancestors Genetic data suggests that the first primates probably lived about 85 mya, when dinosaurs still roamed Earth. However, the earliest primate fossils do not appear in the fossil record until the beginning of the Eocene, about 60 mya. One of the earliest fossil primates, called *Altiatlasius* (al tee aht lah SEE us), was a small, nocturnal animal that ate insects and fruits using its hands and feet for grasping. It might have resembled the tiny tree shrew in **Figure 16.10,** but it had some features similar to those of lemurs today. Learn more about the early evolution of primates in **Data Analysis Lab 16.1.**

Diverging primates Lemurlike primates were widespread by about 50 mya, and many species existed on all continents except Australia and Antarctica. Sometime around 50 mya, and possibly earlier, the anthropoids diverged from the tarsiers; this might have occurred in Asia, where the tarsiers are found today. The earliest anthropoids leaped less and walked more than the strepsirrhines and tarsiers, but they were still tiny and their brains were still small. By the end of the Eocene, 30–35 mya, the anthropoids had diverged and spread widely.

Displacement Many early strepsirrhines appear to have become extinct by the end of the Eocene. This might have been caused by a change in climate. Many major geological events took place at the end of the Eocene and temperatures became cooler. Or it could have been caused by the divergence of the anthropoids. The anthropoids of this time generally were larger and had bigger brains than the strepsirrhines. Thus, the anthropoids might have outcompeted some of the strepsirrhines for resources. This idea is supported by the observation that today the nocturnal strepsirrhines do not interact with the diurnal anthropoids when the habitats of these two groups overlap.

Figure 16.10 The earliest primate ancestor might have looked like this tree shrew.

DATA ANALYSIS LAB 16.1

Based on Real Data*

Interpret Scientific Illustrations

When did early primate lineages diverge?

The fossil record for primate evolution is sparse. In the simplified primate evolutionary tree at right, the green diagram shows the present divergence according to known fossils. The red diagram shows the time line with presumed fossils filling the gaps. Use the diagrams to answer the following questions.

Think Critically

1. **Summarize** Why are lemurs, lorises, and bushbabies considered descendants of the earliest primates?
2. **Extrapolate** How far back might the divergence of the lemurs have occurred?
3. **Infer** Are the tarsiers more closely related to apes or the lemurs?

Data and Observations

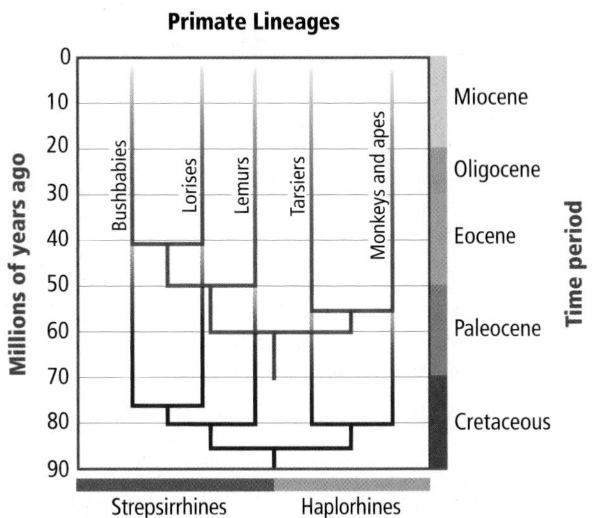

*Data obtained from: Martin, Robert D. 2003. Paleontology: combing the primate record. *Nature* 422: 388–391.

Monkeys The end of the Eocene also saw the appearance of the monkeys. Early monkeys had larger brains than their anthropoid ancestors, and their eyes were more forward-looking. Their snouts were less pointed and they relied less on smell. Scientists hypothesize that the New World monkeys diverged from the line that gave rise to the Old World monkeys sometime between 35 and 25 mya in Africa. While the Old World monkeys continued to evolve in Africa, the New World monkeys developed distinct characteristics in South America. By this time, Africa and South America had separated into two continents. How, then, did the New World monkeys arrive in South America?

Journey to South America Many scientists hypothesize that the New World monkeys evolved from an isolated group of ancestral anthropoids that somehow drifted to South America from Africa, perhaps on rafts of vegetation and soil, much like how the ancestors of lemurs might have drifted to Madagascar from the African mainland. Some scientists suggest that the New World monkeys might have diverged from the anthropoid lineage and made their journey millions of years earlier when sea levels were lower and the continents were closer.

VOCABULARY

ACADEMIC VOCABULARY

Diverge:
to become different in character or form
Their ideas diverged so much that they could not come to an agreement.

Aegyptopithecus In Africa and Asia, anthropoids continued to evolve. Many anthropoid fossils have been found at a site in present-day Egypt called the Fayum basin. Now a desert, the Fayum was predominantly tropical when dozens of anthropoid species lived there 36–31 mya. The largest among them was *Aegyptopithecus* (ee gypt oh PIH thuh kus), often called the dawn ape. Some scientists hypothesize that this arboreal animal, which was about the size of a domestic cat, was ancestral to the apes. It might have been part of the anthropoid line that split from the Old World monkeys and might have given rise to orangutans, gorillas, chimpanzees, and humans.

Section 16.1 Assessment

Section Summary

- All primates share certain anatomical and behavioral characteristics.
- Primates include lemurs, New World monkeys, Old World monkeys, apes, and humans.
- Strepsirrhines are the most primitive living lineages of primates to evolve. They diverged from haplorhines before 55 mya.
- Anthropoids diverged from tarsiers by 50 mya.
- New World monkeys are the only nonhuman primates in the Americas.

Understand Main Ideas

1. MAIN Idea **List** four characteristics that are representative of most primates and lead paleoanthropologists to conclude that primates share a common ancestry.
2. **Describe** how the characteristics of primates make them well-adapted for an arboreal lifestyle.
3. **Diagram** the evolutionary relationships of primates.
4. **Compare and contrast** major primate groups.

Think Critically

5. **Hypothesize** how the breakup of Pangaea might have contributed to the evolutionary history of primates.

MATH in Biology

6. Assume that life on Earth began 3.5 billion years ago. To the nearest percent, how much of this time have anthropoids been living?

BiologyOnline Self-Check Quiz biologygmh.com

Section 16.2

Reading Preview

Objectives

- **Describe** hominoid and hominin features.
- **Trace** hominoid evolution from *Proconsul* to *Homo.*
- **Compare** various australopithecine species.

Review Vocabulary

savanna: a flat grassland of tropical or subtropical regions

New Vocabulary

hominoid
bipedal
australopithecine

Hominoids to Hominins

MAIN Idea Hominins, a subgroup of the hominoids, likely evolved in response to climate changes of the Miocene.

Real-World Reading Link Have you ever tried to put together a puzzle that is missing most of its pieces? Human evolution is like that puzzle. Scientists who try to understand how humans evolved are slowed by the holes in the fossil record. Recent advances in genetics and molecular biology have helped, but the puzzle that is human evolution remains only partially assembled.

Hominoids

Hominoids (HAH mih noydz) include all nonmonkey anthropoids—the living and extinct gibbons, orangutans, chimpanzees, gorillas, and humans. The fossil transition from early anthropoid to ape is not clear; very few fossils from the late Oligocene exist. The earliest hominoid fossils appear in the fossil record only about 25 mya at the beginning of the Miocene. These hominoids retained some ancestral primate features. For example, most had bodies adapted for brachiation. There is evidence that they had relatively large brains and had shoulders and hips that moved freely, and some might even have had the ability to stand on two legs.

Connection to Chemistry Scientists use fossils to help them determine when ancestral hominoids diverged into the hominoids that exist today. But because the fossil record for hominoids is so sparse, scientists also turn to biochemical data to help them in this task. By comparing the DNA of living hominoid species, researchers conclude that gibbons likely diverged first from an ancestral anthropoid, followed by orangutans, gorillas, chimpanzees and bonobos, and finally, humans. **Figure 16.11** shows the potential divergence of these species. Chimpanzees and bonobos are the closest living relatives to humans. All three share at least 96 percent of their DNA sequences.

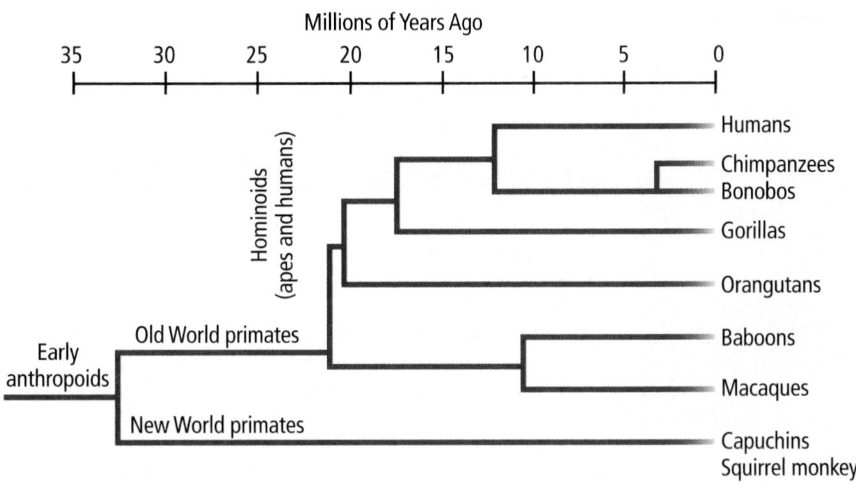

Figure 16.11 Orangutans, gorillas, bonobos, and chimpanzees all diverged from an ancestral anthropoid.

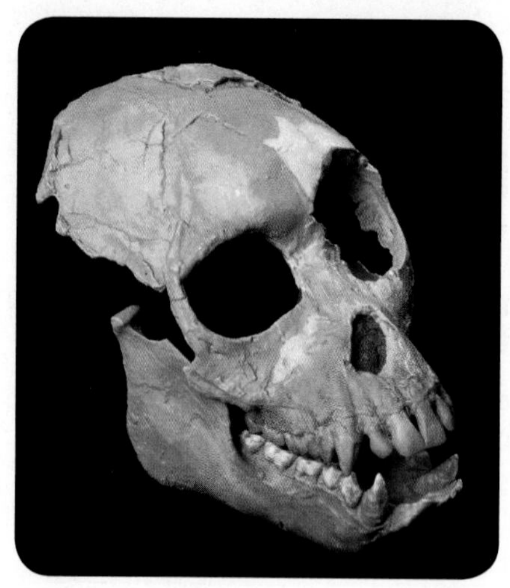

Figure 16.12 *Proconsul* was an early, small-brained hominoid that might have been a human ancestor.

Hominoid characteristics Hominoids are the largest of the primates, and they have the largest brain size in relation to their body size. They tend to have broad pelvises, long fingers, no tail, and flexible arm and shoulder joints. They also have semi-upright or upright posture, and, except for hominins, their arms are longer than their legs. Their teeth are less specialized than those of other animals, and their molars have a distinctive pattern that scientists use to distinguish hominoid fossils from other primate fossils.

Hominoid biogeography During the Miocene (24–5 mya), the world's climate became warmer and drier. As a result, tropical rain forests in Africa began to shrink. Many new animals, including new hominoids, evolved as they adapted to the changing environments. Between about 23 and 14 mya, perhaps as many as 100 hominoid species existed. Early hominoids were more diverse than the modern apes, and they migrated from Africa to Europe and Asia.

Proconsul The best-known hominoid fossils, and some of the oldest, are those from the genus *Proconsul*. **Figure 16.12** shows a fossil skull of one *Proconsul* species discovered by Mary Leakey in Kenya in 1948. This *Proconsul* species generally had the smallest brains of the hominoids. Most had freely moving arms and legs, and while they lived predominantly in trees, some might have had the ability to walk upright. Some scientists think that this *Proconsul* species is a human ancestor, but others suggest that one of the European hominoids—whose fossils are in some ways more humanlike than *Proconsul*—might have returned to Africa at the end of the Miocene and given rise to the human line.

Hominins

The lineage that most likely led to humans split off from the other African apes sometime between 8 and 5 mya. The hominins include humans and all their extinct relatives. These extinct relatives are more closely related to humans than to chimpanzees. The time line in **Figure 16.13** highlights some important hominin discoveries.

Figure 16.13
Hominin Evolution

Discoveries have shaped our understanding of how *Homo sapiens* evolved from hominoids.

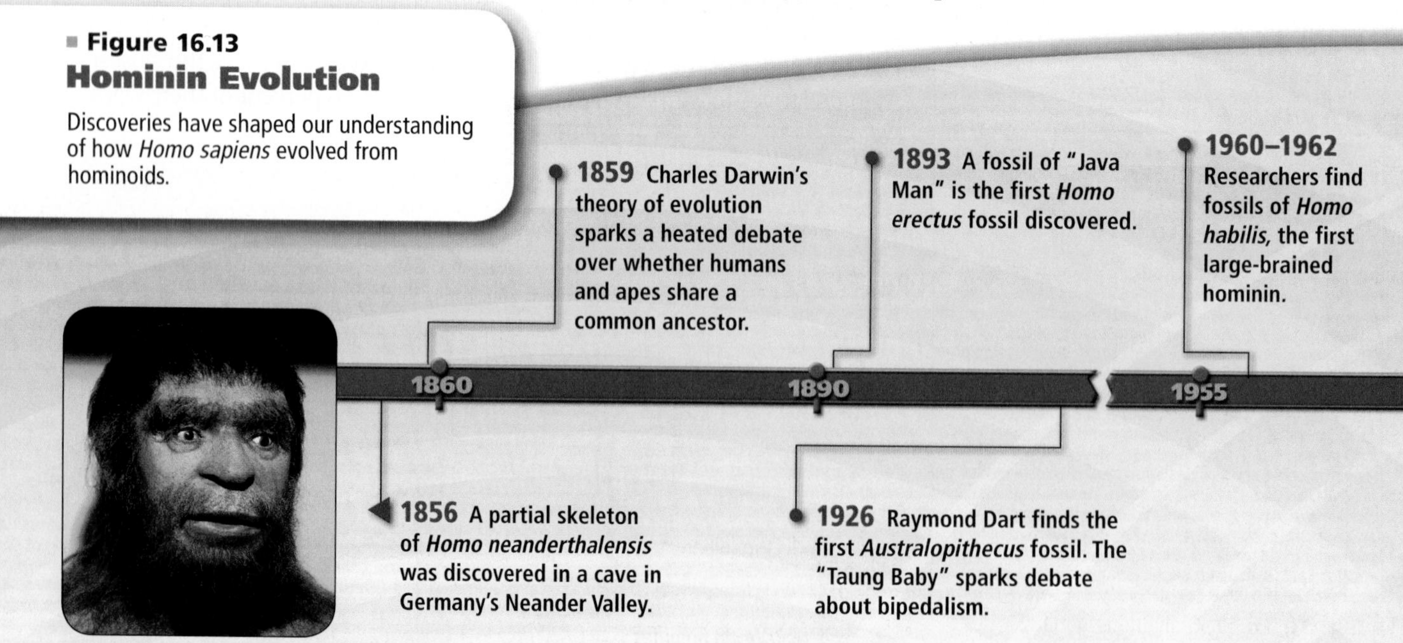

1859 Charles Darwin's theory of evolution sparks a heated debate over whether humans and apes share a common ancestor.
1893 A fossil of "Java Man" is the first *Homo erectus* fossil discovered.
1960–1962 Researchers find fossils of *Homo habilis,* the first large-brained hominin.
1860
1890
1955
1856 A partial skeleton of *Homo neanderthalensis* was discovered in a cave in Germany's Neander Valley.
1926 Raymond Dart finds the first *Australopithecus* fossil. The "Taung Baby" sparks debate about bipedalism.

Chimpanzee	Hominin
Skull attaches posteriorly	Skull attaches inferiorly
Spine slightly curved	S-shaped spine
Arms longer than legs and used for walking	Arms shorter than legs and not used for walking
Long, narrow pelvis	Bowl-shaped pelvis
Femur angled outward	Femur angled inward

■ **Figure 16.14** A comparison between chimpanzee and hominin skeletons illustrates evolutionary changes leading to bipedalism.
Observe and Infer ***What differences in the lengths of the arms and the legs do you detect?***

Hominin characteristics Hominins have bigger brains than other hominoids, with more complexity in parts of the brain where high-level thought occurs. The hominin face is thinner and flatter than that of other hominoids. Hominin teeth are also smaller. With lengthened thumbs and more flexible wrists, hominins have high manual dexterity. Hominins are also **bipedal,** which means that they can walk upright on two legs.

Examine **Figure 16.14,** which illustrates anatomical differences in a quadruped and a biped. When becoming bipedal, hominins developed a fully upright stance, shortened arms, restructured pelvic bones and foot bones, and a change in the position of the head on the spinal cord. In quadrupedal animals, or those that walk on all four limbs, the foramen magnum—the hole in the skull where the spine extends from the brain—is located at the back of the skull. In hominins, it is positioned at the base of the skull.

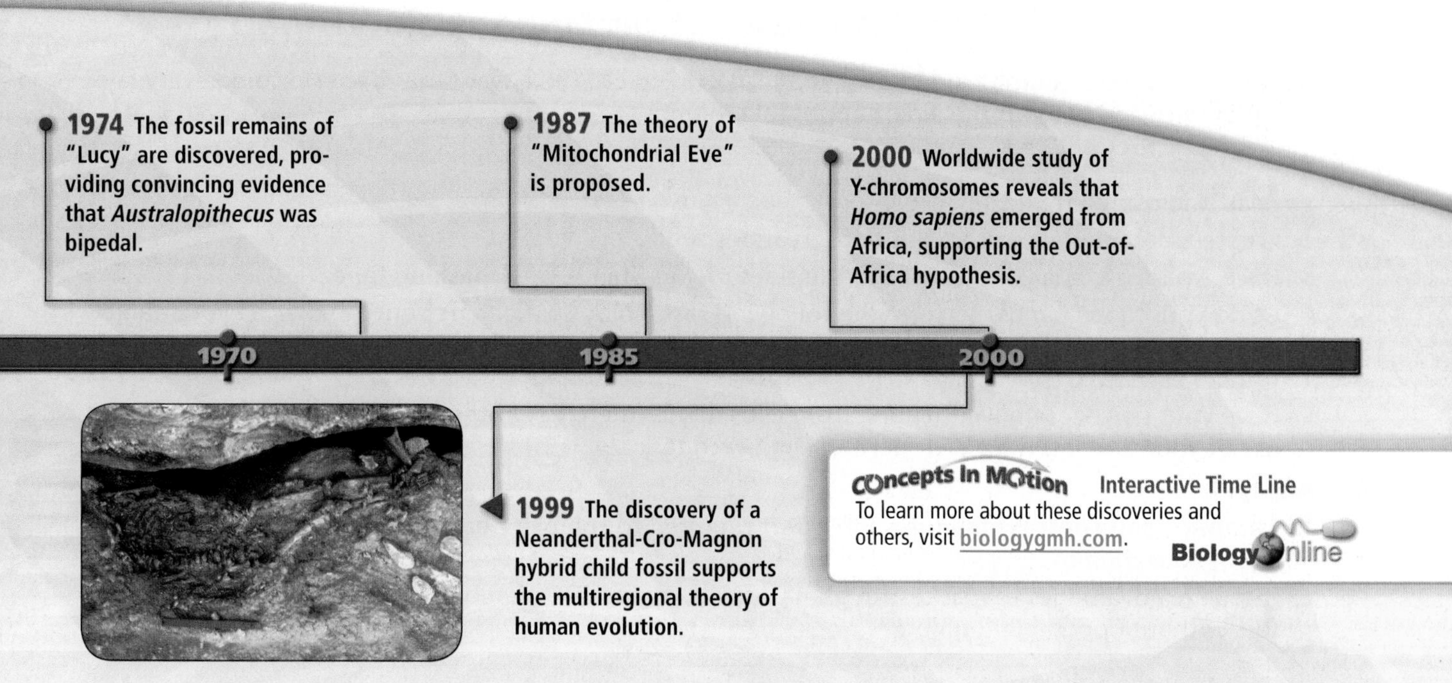

1974 The fossil remains of "Lucy" are discovered, providing convincing evidence that *Australopithecus* was bipedal.
1987 The theory of "Mitochondrial Eve" is proposed.
2000 Worldwide study of Y-chromosomes reveals that *Homo sapiens* emerged from Africa, supporting the Out-of-Africa hypothesis.
1970
1985
2000
1999 The discovery of a Neanderthal-Cro-Magnon hybrid child fossil supports the multiregional theory of human evolution.
Concepts In Motion Interactive Time Line
To learn more about these discoveries and others, visit biologygmh.com.
Biology Online

Disadvantages of bipedalism Bipedalism is not necessarily more efficient than quadrupedalism. Bipedal individuals are easier for predators to see, they might not run as fast, and bipedalism puts greater strain on the hips and back. Also, standing upright defies gravity and therefore requires more energy. Why, then, did hominins become bipedal when their ancestors were so well adapted to life in the trees?

Advantages of bipedalism The African landscape was changing during the period when hominins evolved, and many scientists suggest that bipedalism was an adaptation to the new savanna environment. In this scenario, food resources were sparse and far apart. Bipedalism could have been selected for because it uses less energy than walking on all fours over long distances. Also, standing upright could have made it easier to see food sources. Walking upright for long distances might also have reduced the total area of the body exposed to sunlight and increased the area exposed to cooling winds.

Another hypothesis suggests that hominins were better able to carry objects and use their hands for grasping, and therefore developed upright posture to keep their hands free. Another hypothesis suggests that an upright posture would have helped hominins reach fruit on low branches as they traveled from tree to tree.

There is no single answer to the question of why bipedalism developed. A changing environment might have played only a minor role. Some scientists suggest that the most successful hominins might have been those that evolved on the edge of the forest and savanna. The earliest hominins could have slept in trees and ventured out to the savanna in search of food. In this view, adaptability to a range of environments might have its origin in the hominin use of forest and savanna.

Reading Check **Summarize** the advantages and disadvantages of bipedalism.

MiniLab 16.1

Observe the Functions of an Opposable Thumb

How do opposable thumbs aid in everyday tasks? Explore the advantages of performing everyday activities with and without the aid of opposable thumbs.

Procedure 

1. Read and complete the lab safety form.
2. Create a data table to record your observations.
3. Have a partner tape your thumbs to the sides of your hands with **masking tape.**
4. Using your taped hands, perform the following tasks: pick up a **pen or pencil** and write your name on a **piece of paper,** tie your **shoelaces,** and open a **closed door.** Have your partner use a **stopwatch** to time each task.
5. Have your partner remove the tape from your hands, then repeat the activities in Step 4 with the use of your thumbs. Have your partner time each task.

Analysis:

1. **Compare and contrast** the time and effort required to complete each task with and without the aid of your thumbs.
2. **Infer** the advantages that ancestral primates with opposable thumbs would have had over competitors without opposable thumbs.

Hominin fossils Bipedalism evolved before many other hominin traits, and it is often used to identify hominin fossils. The earliest fossils of species that show some degree of bipedalism are 6–7 million years old. The first hominins that were truly bipedal, however, were the australopithecines (aw stray loh PIH thuh seens).

Australopithecines lived in the east-central and southern part of Africa between 4.2 and 1 mya. They were small—the males were only about 1.5 m tall—and they had apelike brains and jaws. However, their teeth and limb joints were humanlike.

Taung baby The anthropologist Raymond Dart (1893-1988) identified the first australopithecine fossil, the "Taung baby," in Africa in 1926. He called the species *Australopithecus africanus,* meaning "southern ape from Africa." *A. africanus* likely lived between 3.3 and 2.3 mya. The placement of the foramen magnum in the skull of the Taung baby, shown in **Figure 16.15,** convinced Dart that *A. africanus* was bipedal. Not everyone agreed because *A. africanus* had a small brain. Some scientists thought that larger brains evolved before bipedalism. The question continued to be debated for many years, even after the discovery of other African australopithecine fossils such as *A. bosei* and *A. robustus* that indicated bipedalism and small brains.

Figure 16.15 The Taung baby skull convinced Raymond Dart that *A. africanus* walked upright.

Lucy In 1974 in Kenya, the anthropologist Donald Johanson discovered an australopithecine skeleton that helped resolve the debate. Lucy is one of the most complete australopithecine fossils ever found. She was a member of the species *A. afarensis,* which lived between 4 and 2.9 mya.

Lucy was about the size of a chimpanzee. She had the typical australopithecine skull and small brain, and her arms were still somewhat long in proportion to her legs. She also had finger bones that were more curved than those of modern humans, which indicates that she was capable of arboreal activity. But her hip and knee joints were humanlike, and it was clear that she walked upright. A few years later, Mary Leakey uncovered further evidence that australopithecines were bipedal when she discovered fossilized australopithecine footprints. Lucy's skeleton and the footprints of her relatives are illustrated in **Figure 16.16.**

Figure 16.16 Fossilized footprints indicate that Lucy was bipedal. Though incomplete, this skeleton of Lucy indicates that *A. afarensis* had a small brain but also had the ability to walk upright.
Infer ***what bones scientists would examine to determine if Lucy walked upright.***

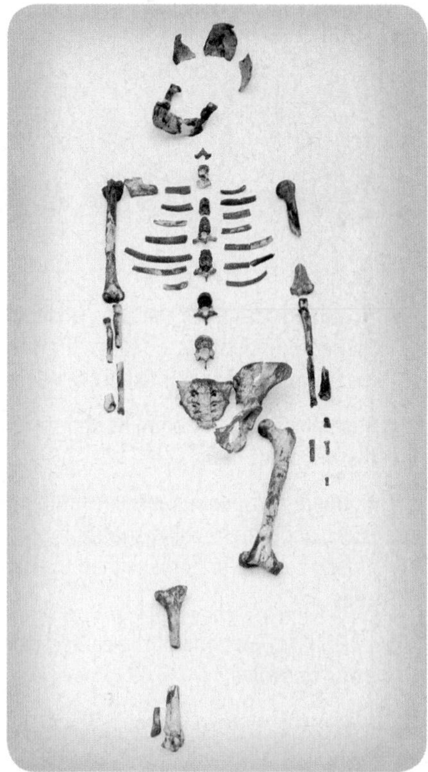

Mosaic pattern Like other hominin fossils, Lucy and her relatives show a patchwork of human and apelike traits. In this way, they follow a mosaic pattern of evolution. Mosaic evolution occurs when different body parts or behaviors evolve at different rates. For example, hominins developed the ability to walk upright nearly two million years before they developed modern flat faces and larger brains.

VOCABULARY

WORD ORIGIN

Australopithecine
from the Latin word *australis,* meaning *southern,* and the Greek word *pithekos,* meaning *ape*

Hominin evolution Within the last 30 years, scientists have discovered many more early hominin fossils. Some defy characterization and have led to new genus designations. Scientists have estimated that *Kenyanthropus platyops* (ken yan THROH pus • PLAT ee ops), for example, lived between 3.5 and 3.2 mya. Some scientists thought that *K. platyops,* which means "flat-faced man," represents a completely new hominin genus.

Paranthropus There is also confusion about where *A. bosei* and *A. robustus* fit in the classification of hominins. Traditionally, these two species have been classified as robust forms of australopithecines, distinguished from the smaller, more slender forms by their size and muscular jaws. Today, many scientists prefer to put these primates in a separate genus called *Paranthropus.* Paranthropoids, which thrived between 2 and 1.2 mya, were an offshoot of the human line that lived alongside human ancestors but were not directly related.

Overlapping hominins However they are classified, these robust hominins appear to have lived alongside some of the slender australopithecines. They might have overlapped, for example, with *A. garhi,* an African australopithecine that was discovered in 1999. The illustration of the evolution of hominins is more like a bush than a tree. Many species lived successfully for years, often overlapping with earlier species and then—for unknown reasons—became extinct. By 1 mya, all australopithecines had disappeared from the fossil record. The only hominin fossils found after that time belong to the genus *Homo.*

Section 16.2 Assessment

Section Summary

- Hominoids are all of the apes, including gibbons, orangutans, gorillas, chimpanzees, and humans and their extinct relatives.
- Several species of hominins appear in the fossil record.
- Hominins include humans, australopithecines, and other extinct species more closely related to humans than to chimpanzees.
- Bipedalism was one of the earliest hominin traits to evolve.

Understand Main Ideas

1. MAIN Idea **Summarize** how the climate of the Miocene spurred the evolution of hominins.
2. **Describe** characteristics unique to hominoids.
3. **Describe** characteristics unique to hominins.
4. **Outline** the hominin family "bush."
5. **Compare** australopithecine species.

Think Critically

6. **Discuss** Do you think hominins would have evolved if the climate had not changed during the Miocene? Why?
7. **Classify** If you found a primate skeleton with arms shorter than legs, what general category would you put it in?

Biology Online Self-Check Quiz biologygmh.com

Section 16.3

Human Ancestry

Reading Preview

Objectives

- **Describe** species in the genus *Homo.*
- **Explain** the Out-of-Africa hypothesis.
- **Compare** Neanderthals and modern humans.

Review Vocabulary

mitochondrion: an organelle found in eukaryotic cells containing genetic material and responsible for cellular energy

New Vocabulary

Homo
Neanderthal
Cro-Magnon

MAIN Idea Tracing the evolution of the genus *Homo* is important for understanding the ancestry of humans, the only living species of *Homo*.

Real-World Reading Link Have you ever heard anyone use the term "cave man" in an insulting way? Unfortunately, this term is used sometimes to indicate brutish behavior. But the people who lived in caves 40,000 years ago were very much like modern humans. Their art was beautiful and their tools were sophisticated.

The Genus *Homo*

The African environment became considerably cooler between 3 and 2.5 mya. Forests became smaller in size, and the range of grasslands was extended. The genus ***Homo,*** which includes living and extinct humans, first appeared during these years and although the fossil record is lacking fossils, many scientists infer that they evolved from an ancestor of the australopithecines.

Homo species had bigger brains, lighter skeletons, flatter faces, and smaller teeth than their australopithecine ancestors. They are also the first species known to control fire and to modify stones for tool use. As they evolved, they developed language and culture.

***Homo habilis* used stone tools** The earliest known species that is generally accepted as a member of the genus *Homo* is *Homo habilis,* called "handy man" because of its association with primitive stone tools. This species lived in Africa between about 2.4 and 1.4 mya. **Figure 16.17** shows a scientific illustrator's idea of what *H. habilis* might have looked like.

H. habilis possessed a brain averaging 650 cm^3, about 20 percent larger than that of the australopithecines. It also had other *Homo* species traits, including a smaller brow, reduced jaw, flatter face, and more humanlike teeth. Like australopithecines, it was small, long-armed, and it seems to have retained the ability to climb trees. Other *Homo* species might have coexisted with *H. habilis,* among them a species called *Homo rudolfensis.* Because few fossils of *H. rudolfensis* have been found, its exact relationship to the rest of the *Homo* line is uncertain.

Figure 16.17 Scientific illustrators use fossils and their knowledge of anatomy to create drawings of what *H. habilis* might have looked like.

■ **Figure 16.18** Models of nonliving species also can be created from fossil remains. *H. ergaster* appeared in the fossil record about 1.8–1.3 mya.

***Homo ergaster* migrated** Within about 500,000 years of the appearance of *H. habilis,* another *Homo* species, *Homo ergaster,* emerged with an even larger brain. *H. ergaster,* illustrated in **Figure 16.18,** appeared only briefly in the fossil record, from about 1.8 to 1.3 mya. *H. ergaster* was taller and lighter than *H. habilis,* and had longer legs and shorter arms. Its brain averaged 1000 cm^3, and it had a rounded skull, reduced teeth, and what many scientists think is the first human nose (with the nostrils facing downward).

Tools Carefully made hand axes and other tools associated with *H. ergaster* fossils suggest to some scientists that *H. ergaster* was a hunter, but others think that *H. ergaster* was primarily a scavenger and used the tools to scrape the meat off of scavenged bones.

MiniLab 16.2

Explore Hominin Migration

Where did early hominins live? Scientists carefully record the locations where fossils are found. The latitude and longitude coordinates represent the known geographic points of each *Homo* species' range.

Procedure

1. Read and complete the lab safety form.
2. Plot the following fossil sites on the map your teacher gives you. Use a different color for each species. When you are finished, lightly shade in the approximate boundaries.

   H. habilis (2.4–1.4 million years ago): 37°E: 4°S, 36°E: 3°N, 36°E: 7°N, 43°E: 8°N

   H. erectus (2 million–400,000 years ago): 112°E: 38°N, 13°E: 47°N, 7°W: 34°N, 112°E: 8°S

   H. neanderthalensis (300,000–200,000 years ago): 8°E: 53°N, 66°E: 39°N, 5°W: 37°N, 36°E: 33°N

   H. sapiens (195,000 years ago–present): 70°E: 62°N, 24°E: 30°S, 138°E: 34°S, 112°E: 38°N, 99°W: 19°N, 102°W: 32°N

Analysis

1. **Hypothesize** According to the map you made, when was the earliest that hominins could have migrated out of Africa? Where did they go?
2. **Determine** what sets of fossils overlapped in geographic ranges. What does this suggest?

■ **Figure 16.19** *H. erectus* might have lived in caves, made tools, and used fire.
Explain ***some of the advantages*** **H. erectus** ***would have over*** **H. ergaster.**

Migration Both scavenging and hunting are associated with a migratory lifestyle, and *H. ergaster* appears to have been the first African *Homo* species to migrate in large numbers to Asia and possibly Europe, perhaps following the trail of migrating animals. The later Eurasian forms of *H. ergaster* are called *Homo erectus.* Because *H. ergaster* shares features with modern humans, scientists hypothesize that *H. ergaster* is an ancestor of modern humans.

***Homo erectus* used fire** *H. erectus,* illustrated in **Figure 16.19,** lived between 1.8 million and 400,000 years ago and appears to have evolved from *H. ergaster* as it migrated out of Africa. While some scientists consider *H. ergaster* and *H. erectus* a single species, *H. erectus* appears to have evolved traits that the early African *H. ergaster* species did not have. Members of this species seem to have been more versatile than their predecessors, and they adapted successfully to a variety of environments. *H. erectus* includes "Java Man," discovered in Indonesia in the 1890s, and "Peking Man," discovered in China in the 1920s.

In general, *H. erectus* was larger than *H. habilis* and had a bigger brain. It also had teeth that were more humanlike. Brain capacity ranged from about 900 cm^3 in early specimens to about 1100 cm^3 in later ones. It was as tall as *H. sapiens* but it had a longer skull, lower forehead, and thicker facial bones than either *H. ergaster* or *H. sapiens.* It also had a more prominent browridge. Evidence indicates that *H. erectus* made sophisticated tools, used fire, and sometimes lived in caves.

***Homo floresiensis*—"The Hobbit"** In 2004 a curious set of fossils were discovered on the Indonesian island of Flores. These fossils, which are about 18,000 years old, are heavily debated in the scientific community. Some scientists think they might represent a species called *Homo floresiensis* (flor eh see EN sus). Others think that the fossils belong to early human dwarfs and do not warrant classification as a separate species. *H. floresiensis,* nicknamed "The Hobbit," was only about 1 m tall when full grown. While it had brain and body proportions like all the australopithecines, primitive stone tools were found with its fossils. In 2007 a study showed that *H. floresiensis* had apelike wrist bones—further support for its status as a separate species. You can compare *H. floresiensis* and *H. sapiens* skulls in **Figure 16.20.**

Reading Check What are the evolutionary relationships among *H. habilis, H. ergaster,* and *H. erectus?*

■ **Figure 16.20** Scientists are debating whether *H. floresiensis* is a new species. The *H. floresiensis* skull on the left is smaller than the human skull on the right.
Infer ***what this skull comparison might predict about the evolutionary relationship between*** **H. floresiensis** ***and*** **H. sapiens.**

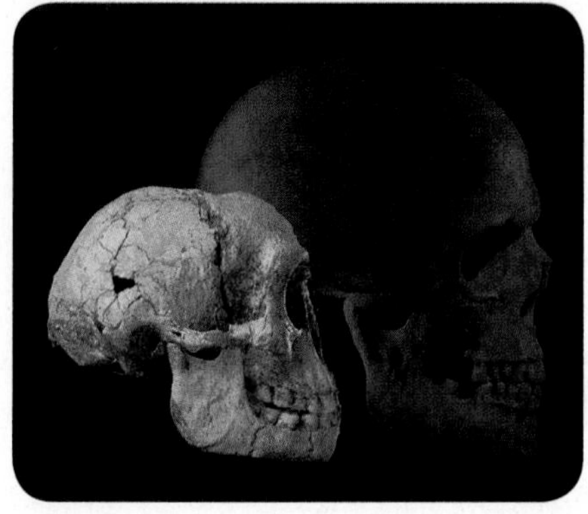

Homo heidelbergensis—traits The transition from *H. ergaster* to modern humans appears to have occurred gradually. Numerous transitional fossils have been found that display a mixture of *H. ergaster* and *H. sapiens* traits. These fossils are often categorized as *Homo heidelbergensis,* but others put them in the category *Homo sapiens.* These humans generally had larger brains and thinner bones than *H. ergaster,* but they still had browridges and receding chins.

Reading Check **Relate** *H. heidelbergensis* to *H. sapiens.*

Homo neanderthalensis built shelter A distinct human species called *Homo neanderthalensis,* or the **Neanderthals,** evolved exclusively in Europe and Asia about 200,000 years ago, likely from *H. erectus* or a *Homo* intermediary. Neanderthals were shorter but had more muscle mass than most modern humans. Their brains were sometimes even larger than the brains of modern humans, though the brains might have been organized in different ways. Neanderthals had thick skulls, bony browridges, and large noses. They also had a heavily muscled, robust stature, as illustrated in **Figure 16.21.** Evidence of heavy musculature appears in the extremely large muscle attachments and the bowing of the long bones.

Neanderthals lived near the end of the Pleistocene ice age, a time of bitter cold. Their skeletons reflect lives of hardship; bone fractures and arthritis seem to have been common. There is evidence that they used fire and constructed complex shelters. They hunted and skinned animals, and it is possible that they had basic language. There is also some evidence that they cared for their sick and buried their dead.

Are Neanderthals our ancestors? In some areas of their range, particularly in the Middle East and southern Europe, Neanderthals and modern humans overlapped for as long as 10,000 years. Some scientists suggest that the two species interbred. However, DNA tests on fossil bones suggest that Neanderthals were a distinct species that did not contribute to the modern human gene pool. Neanderthals went extinct about 30,000 years ago.

Figure 16.21 *H. neanderthalensis* had much thicker bones than modern humans and a pronounced browridge. Neanderthals were hunters who used fire and tools.

Emergence of Modern Humans

The species that displaced the Neanderthals, *Homo sapiens*, is characterized by a more slender appearance than all other *Homo* species. They have thinner skeletons, rounder skulls, and smaller faces with prominent chins. Their brain capacity averages 1350 cm^3. *H. sapiens* first appeared in the fossil record, in what is now Ethiopia, about 195,000 years ago. These early *H. sapiens* made chipped hand axes and other sophisticated stone tools. They appear to have had the ability to use a range of resources and environments, and at some point they began migrating out of Africa. **Table 16.2** compares modern humans with other *Homo* species.

Concepts In Motion
Interactive Table To explore more about the *Homo* species, visit biologygmh.com.

Table 16.2 Characteristics of the *Homo* species

Species	Skull	Time in fossil record	Characteristics
Homo habilis		2.4–1.4 million years ago	• Average brain had a capacity of 650 cm^3 • Used tools
Homo ergaster		1.8–1.2 million years ago	• Average brain had a capacity of 1000 cm^3 • Had thinner skull bones • Had humanlike nose
Homo erectus		1.8 million–400,000 years ago	• Average brain had a capacity of 1000 cm^3 • Had thinner skull bones • Used fire
Homo neanderthalensis		300,000–200,000 years ago	• Average brain had a capacity of 1500 cm^3 • Buried their dead • Possibly had a language
Homo sapiens		195,000 years ago to present	• Average brain has a capacity of 1350 cm^3 • Does not have browridge • Has a small chin • Has language and culture

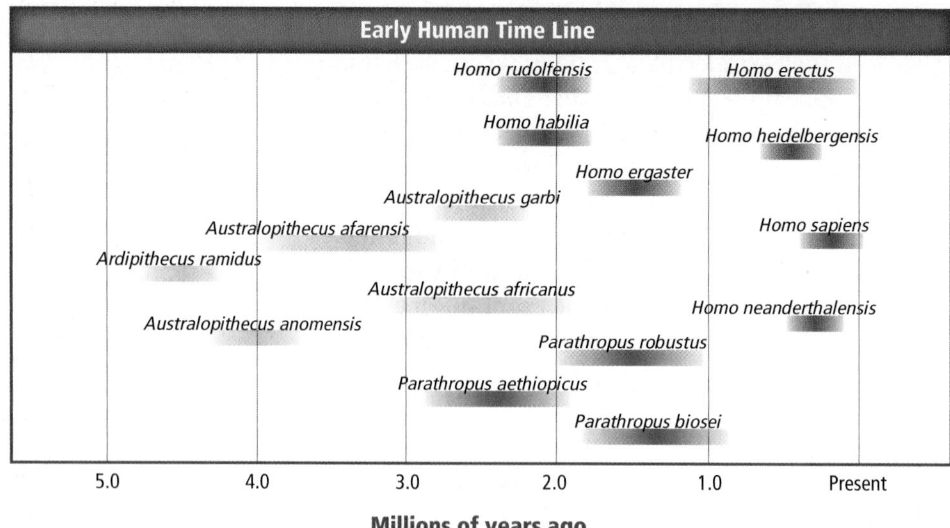

Figure 16.22 The period of existence of several early hominins overlapped until about 30,000 years ago.

Out-of-Africa hypothesis The world's population 200,000 years ago looked significantly different than it does today. It was inhabited by a morphologically diverse genus of hominins, including primitive humans, Neanderthals, and modern humans, as illustrated in **Figure 16.22.** By 30,000 years ago, however, only modern humans remained. Some scientists propose that these modern humans evolved from several dispersed populations of early *Homo* species at the same time in different areas of the world. According to this multiregional evolution model, modern races of humans arose in isolated populations by convergent evolution.

Most scientists explain the global dominance of modern humans with the African Replacement model or, more commonly, the Out-of-Africa hypothesis. According to this hypothesis, which was first proposed by Christopher Stringer and Peter Andrews of the British Museum of Natural History in 1988, modern humans evolved only once, in Africa, and then migrated to all parts of the world, eventually displacing other hominins.

"Mitochondrial Eve" The Out-of-Africa hypothesis was supported by mitochondrial DNA analysis of contemporary humans in the early 1990s. Mitochondrial DNA changes very little over time, and humans living today have nearly identical mitochondrial DNA. Researchers Allan Wilson and Rebecca Cann of the University of California, Berkeley, reasoned that the population with the most variation should be the population that has had the longest time to accumulate diversity. This was exactly what they found in the mitochondrial DNA of Africans. Because mitochondrial DNA is inherited only from the mother, this analysis suggested that *H. sapiens* emerged in Africa about 200,000 years ago from a hypothetical "Mitochondrial Eve."

Later, work by other scientists studying DNA sequences in the male Y chromosome yielded similar results. While some scientists think that a single movement of only a few hundred modern humans ultimately gave rise to the world's current population, others think the process occurred in phases, with some interbreeding among the species that humans displaced.

Reading Check **Describe** evidence in support of the Out-of-Africa hypothesis.

Study Tip

Discussion Group Discuss with your classmates what you've learned about human evolution. What characteristics of early hominins have surprised you or your classmates?

■ **Figure 16.23** Cro-Magnons were known for their sophisticated cave paintings, tools, and weapons. This painting was found in Lascaux Cave in France.

The beginning of culture The first evidence of complex human culture appeared in Europe only about 40,000 years ago, shortly before the Neanderthals disappeared. Unlike the Neanderthals, early modern humans expressed themselves symbolically and artistically in decorative artifacts and cave drawings, as illustrated in **Figure 16.23.** They developed sophisticated tools and weapons, including spears and bows and arrows. They were the first to fish, the first to tailor clothing, and the first to domesticate animals. These and many other cultural expressions marked the appearance of fully modern humans, the subspecies *Homo sapiens.* Some people call them **Cro-Magnons.** They represent the beginning of historic hunter-gatherer societies.

Connection to History Humans continued their migration throughout Europe and Asia. They probably reached Australia by boat and traveled to North America via a land bridge from Asia. From North America, they spread to South America. They adapted to new challenges along the way, leaving behind a trail of artifacts that we study today.

Section 16.3 Assessment

Section Summary

- The genus *Homo* is thought to have evolved from genus *Australopithecus.*
- Of the many species that have existed in the hominin group, only one species survives today.
- The first member of the genus *Homo* was *H. habilis.*
- The Out-of-Africa hypothesis suggests that humans evolved in Africa and migrated to Europe and Asia.
- *H. neanderthalensis* went extinct about 30,000 years ago and *H. sapiens* moved into those areas inhabited by *H. neanderthalensis* at about the same time.

Understand Main Ideas

1. MAIN Idea **Hypothesize** why only one genus and species remains in the hominin group.
2. **Describe** how *H. habilis* might have lived.
3. **Apply** what you have learned about the Out-of-Africa hypothesis to what you know about the arrival of *H. sapiens* in North America.
4. **Compare and contrast** *H. neanderthalensis* and *H. sapiens* fossils.

Think Critically

5. **Classify** How would you classify a fossil that was found in France and dated to about 150,000 years if the skull had a thick browridge, but in most other ways appeared human?

WRITING in Biology

6. **Hypothesize** the importance of language to the early modern humans, and how it might have contributed to their success.

BioDiscoveries

One Family, One Amazing Contribution to Science

Out of Africa Growing up in Africa, Louis Leakey (1903–1972) thought, like Darwin, that humans evolved in Africa. After all, that is where our closest primate relatives—chimpanzees and gorillas—live. When Louis finished his education in England, he went to the Olduvai Gorge in Tanzania, Africa. Hints of early man, such as stone tools, had been found in the gorge, and Louis was determined to find hominin bones.

Louis and Mary Leakey on a dig.

Proconsul In 1948, while Louis and his wife, Mary (1913–1996), were living on an island in Africa's Lake Victoria, Mary found a hominoid skull that she and Louis named *Proconsul africanus.* By comparing the skull to other objects nearby, they determined that the skull was 20 million years old.

"Zinj" Shortly after discovering *Proconsul,* Mary and Louis returned to Olduvai Gorge. There, while walking her dogs one day in 1959, Mary saw a skull poking out of a rock. Further examination yielded a hominin skull with teeth still intact. The skull, which Mary named *Zinjanthropus boisei* (zihn JAN thruh pus • BOY see) (now called *Paranthropus boisei*), was 1.75 million years old—the oldest humanlike ancestor then known.

Renewed vigor The Leakeys began spending more time at Olduvai and hired workers to help them find fossils. Louis, though often given credit, never found many hominin fossils. Mary, their three children, and their skilled fossil-hunting staff were the actual discoverers. The renewed vigor at Olduvai soon yielded many more discoveries, including the first *Homo habilis* fossil and, in 1976, fossilized australopithecine footprints.

More skeletons By the 1970s, the Leakeys' son Richard and his wife, Maeve, were making their own important discoveries at a camp in nearby Kenya. In 1984, Richard discovered "Turkana Boy," one of the most complete and oldest *H. ergaster* fossils yet found. In 1999, Maeve and her daughter Louise discovered *Kenyanthropus platyops* (ken yen THROH pus • plat ee ops), which lived about 3.5 mya. Some scientists think that *K. platyops* represents a new hominin genus.

Enduring legacy Over the course of six decades, three generations of the Leakey family have discovered pieces of the puzzle of the evolution of the *Homo* genus. Today, Louise heads the Koobi Fora Research Project on the shores of Lake Turkana.

E-COMMUNICATION

Travelogue To learn more about the Leakey family, visit biologygmh.com. Imagine that you are accompanying the Leakeys on one of their digs. Write a travelogue to detail an amazing fossil they have just found, including potential implications it might have for human evolution. Share your travelogue with your classmates.

BIOLAB

WHAT CAN YOU LEARN ABOUT BIPEDALISM FROM COMPARING BONES?

Background: **Humans and chimpanzees have the same number of bones in the same places, but humans walk upright and chimpanzees do not. Can you identify the skeletal features that enable humans to walk upright on two legs? Assume that you are a paleontologist and have been given chimpanzee and human bones to identify and assemble. Then, you receive a third set. How is the mystery skeleton related to the human and chimpanzee skeletons?**

Question: *What unique skeletal features did humans evolve to become bipedal?*

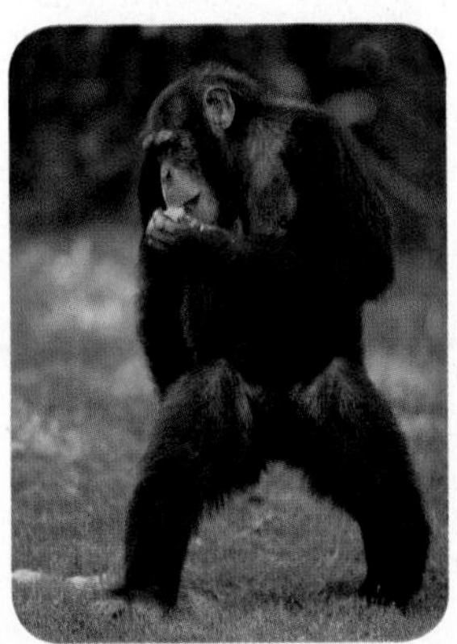

Materials

envelopes containing paper bones and clues (2)
paper, pencil, and ruler

Safety Precautions

Procedure

1. Read and complete the lab safety form.
2. Make a data table to help you compare the following characteristics of each of the three fossil sets you will examine: skull, rib cage, pelvis, arms, legs, and feet.
3. Make sure your teacher approves your table.
4. Open envelope #1.
5. Using the clues in your envelope, identify the bones, determine to which species they belong, and write down at least one distinguishing characteristic of each on your table.
6. Open envelope #2.
7. Using the new set of clues, classify each new bone as: chimpanzee, human, similar to both, similar to chimpanzee, or similar to human. Record this data in your table.

Analyze and Conclude

1. **List** features that a scientist might use to determine if a fossil organism was bipedal.
2. **Think Critically** Based on your knowledge, do you think the mystery fossil is bipedal? Why?
3. **Conclude** What organism do you think your mystery bones represent?
4. **Compare** your table with those of other students in the class. Did you arrive at the same conclusions? If not, discuss the differences.
5. **Experiment** Chimpanzees cannot completely straighten—or lock—their knees as humans can and must use more muscles when standing upright. Try standing for 10 s with your knees locked and for 10 s with your knees bent. Describe how your legs feel at the end.
6. **Reason,** from your mystery fossil bones, what it means to say that humans evolved in a mixed, or mosaic, pattern.

WRITING in Biology

Research and discuss why bipedalism is often thought of as an evolutionary compromise. List skeletal injuries that humans suffer as a result of walking upright. To learn more about bipedalism, visit BioLabs at biologygmh.com.

CHAPTER 16

Study Guide

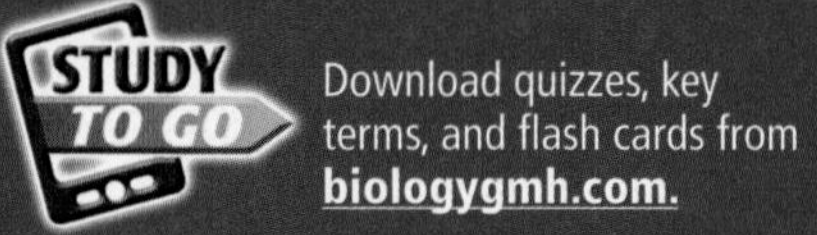

Download quizzes, key terms, and flash cards from biologygmh.com.

FOLDABLES **Summarize** the characteristics of Old World monkeys and New World monkeys. Explain why they are primates. Compare and contrast them to another type of primate.

Vocabulary	Key Concepts
Section 16.1 Primates	
• anthropoid (p. 455) • arboreal (p. 455) • binocular vision (p. 452) • diurnal (p. 452) • hominin (p. 458) • nocturnal (p. 452) • opposable first digit (p. 452) • prehensile tail (p. 456)	**MAIN Idea** Primates share several behavioral and biological characteristics, which indicates that they evolved from a common ancestor. • All primates share certain anatomical and behavioral characteristics. • Primates include lemurs, New World monkeys, Old World monkeys, apes, and humans. • Strepsirrhines are the most primitive living lineages of primates to evolve. They diverged from haplorhines before 55 mya. • Anthropoids diverged from tarsiers by 50 mya. • New World monkeys are the only nonhuman primates in the Americas.
Section 16.2 Hominoids to Hominins	
• australopithecine (p. 465) • bipedal (p. 463) • hominoid (p. 461)	**MAIN Idea** Hominins, a subgroup of the hominoids, likely evolved in response to climate changes of the Miocene. • Hominoids are all of the apes, including gibbons, orangutans, gorillas, chimpanzees, and humans and their extinct relatives. • Several species of hominins appear in the fossil record. • Hominins include humans, australopithecines, and other extinct species more closely related to humans than to chimpanzees. • Bipedalism was one of the earliest hominin traits to evolve.
Section 16.3 Human Ancestry	
• Cro-Magnon (p. 473) • *Homo* (p. 467) • Neanderthal (p. 470)	**MAIN Idea** Tracing the evolution of the genus *Homo* is important for understanding the ancestry of humans, the only living species of *Homo*. • The genus *Homo* is thought to have evolved from the genus *Australopithecus*. • Of the many species that have existed in the hominin group, only one species survives today. • The first member of the genus *Homo* was *H. habilis*. • The Out-of-Africa hypothesis suggests that humans evolved in Africa and migrated to Europe and Asia. • *H. neanderthalensis* went extinct about 30,000 years ago and *H. sapiens* moved into those areas inhabited by *H. neanderthalensis* about the same time.

CHAPTER 16 Assessment

Section 16.1

Vocabulary Review

Replace the underlined words with the correct vocabulary term from the Study Guide page.

1. A fifth limb might be used by a primate to grip a limb while engaged in reaching for and eating food.
2. Primates that are active at night are "wet-nosed" primates.
3. Depth perception evolved as the faces of primates became flattened.

Understand Key Concepts

Use the figure below to answer question 4.

4. Which is the term for the movement demonstrated by this gibbon?
   A. brachiation
   B. knuckle-walking
   C. quadruped movement
   D. upright locomotion
5. Which group was the first to evolve?
   A. African apes **C.** New World monkeys
   B. hominins **D.** Old World monkeys
6. Which adaptation results in a better gripping ability?
   A. complex brain
   B. flexible forelimbs
   C. opposable digits
   D. prehensile tail
7. The first primates most resembled which animal?
   A. gibbon
   B. gorilla
   C. tamarin
   D. lemur

Constructed Response

8. **Open Ended** Describe the usefulness of binocular and color vision.
9. **Short Answer** Which groups of primates make up the anthropoids?

Think Critically

10. **Hypothesize** Why do you think primate fossils haven't been found on Antarctica?
11. **Classify** Suppose on a trip to Brazil you found a fossil of a primate that closely resembles a squirrel monkey. Into which group of anthropoids would the specimen be placed?

Section 16.2

Vocabulary Review

Define the following vocabulary terms in complete sentences.

12. australopithecine
13. bipedal
14. hominoid

Understand Key Concepts

15. Which hominin species made the fossilized footprints shown in **Figure 16.16?**
    A. *A. afarensis* **C.** *Paranthropus*
    B. *A. africanus* **D.** *Proconsul*
16. Which hominoid might be ancestral to apes and humans?
    A. *A. afarensis*
    B. *A. africanus*
    C. *Paranthropus*
    D. *Proconsul*

17. Which is the correct sequence of fossils as evidenced by the fossil record?
    A. *A. africanus, A. afarensis, Paranthropus, Proconsul*
    B. *Proconsul, A. afarensis, A. africanus, Paranthropus*
    C. *Proconsul, Paranthropus, A. afarensis, A. africanus*
    D. *Paranthropus, Proconsul, A. africanus, A. afarensis*

18. *A. afarensis* was bipedal, but she exhibited apelike traits. What type of evolutionary pattern might account for this?
    A. convergence
    B. mosaic
    C. divergence
    D. coevolution

Constructed Response

19. **Open Ended** Discuss the debate regarding the classification of *Paranthropus*.

Use the figure below to answer question 20.

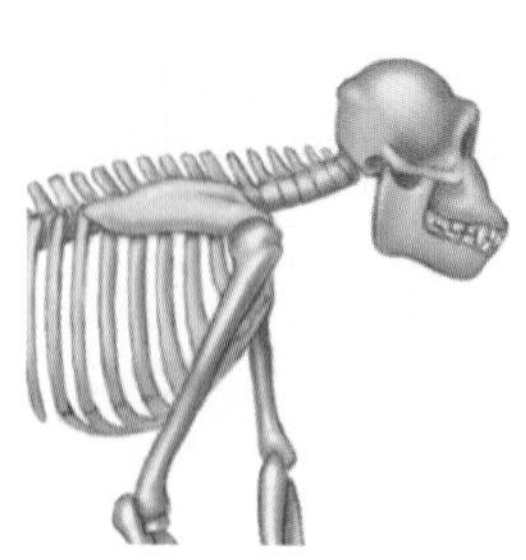
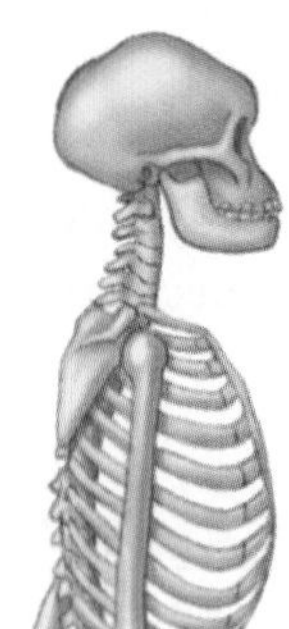

20. **Short Answer** Describe the relevance of the foramen magnum's location to bipedalism.

Think Critically

21. **Explain** how climate change might have contributed to the evolution of bipedalism.

22. **Hypothesize** Why is biochemical evidence important in helping scientists learn about the divergence of primate groups?

Section 16.3

Vocabulary Review

Each of the following sentences is false. Make the sentence true by replacing the underlined word with a vocabulary term from the Study Guide page.

23. The genus *Australopithecus* is thought to be ancestral to the genus *Proconsul*.

24. Cro-Magnons were adapted to cold climates. They eventually were replaced by modern humans.

25. *H. neanderthalensis* is the scientific name for modern humans.

Understand Key Concepts

Use the figure below to answer question 26.

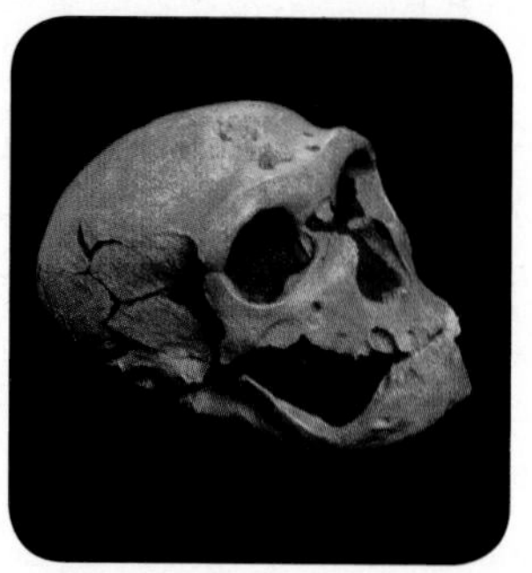

26. The large brain and thickened browridges illustrated by the skull above are characteristic of which species?
    A. Cro-magnon humans
    B. modern *H. sapiens*
    C. Neanderthal humans
    D. *Proconsul*

27. The first undisputed member of the hominin group was which of the following?
    A. *A. africanus*
    B. *H. antecessor*
    C. *H. ergaster*
    D. *H. habilis*

28. Which hominin was likely the first to migrate long distances?
    A. *H. ergaster*
    B. *H. antecessor*
    C. *H. neanderthalensis*
    D. *H. sapiens*

29. Which hominin likely first used fire, lived in caves, and made tools?
    A. *H. ergaster*
    B. *H. erectus*
    C. *H. neanderthalensis*
    D. *H. sapiens*

30. *H. heidelbergensis* is generally considered part of which group?
    A. Neanderthals
    B. *H. sapiens*
    C. Cro-Magnons
    D. australopithecines

Use the figure below to answer questions 31 and 32.

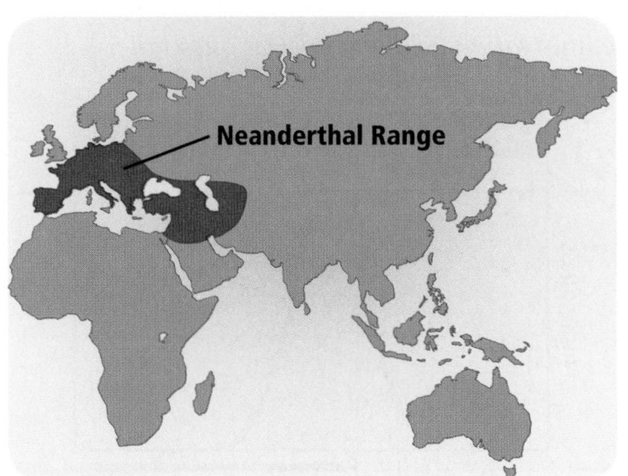

31. The map above represents the geographic range of which species?
A. *Homo erectus*
B. *Homo sapiens*
C. *Homo neanderthalensis*
D. *Homo heidelbergensis*

32. During what time did the species represented on the map live?
A. 300,000–200,000 years ago
B. 100,000–12,000 years ago
C. 2.4–1.4 million years ago
D. 1.8 and 1.2 million years ago

Constructed Response

33. Open Ended Describe the importance of *H. habilis* in human evolution.

34. Short Answer Describe the importance of fire to the migration of early *Homo* species.

35. Open Ended From what you have learned about the evolution of primates, do you think *Homo sapiens,* our species, will continue to evolve? Why?

Think Critically

36. Apply Concepts Explain why mitochondrial DNA instead of nuclear DNA is used to study the evolution of modern humans.

37. Predict If modern humans had not arrived in Europe, do you think Neanderthals would have persisted?

38. Hypothesize How might *H. floresiensis* have coexisted with modern humans?

Additional Assessment

39. WRITING in Biology Write a paragraph to describe what you imagine a day in the life of *A. afarensis* to have been like.

DBQ Document-Based Questions

Scientists generally consider walking, but not running, to be a key trait in the evolution of humans. Like apes, humans are poor sprinters when compared to quadruped animals such as horses and dogs. Unlike apes, but like some quadrupeds, humans are capable of endurance running (ER), running long distances over extended time periods. The graph below compares speed during ER to length of an organism's stride (two steps for a human).

Data obtained from: Bramble, D. and Lieberman, D. 2004. Endurance running and the evolution of *Homo. Nature* 432: 345–352.

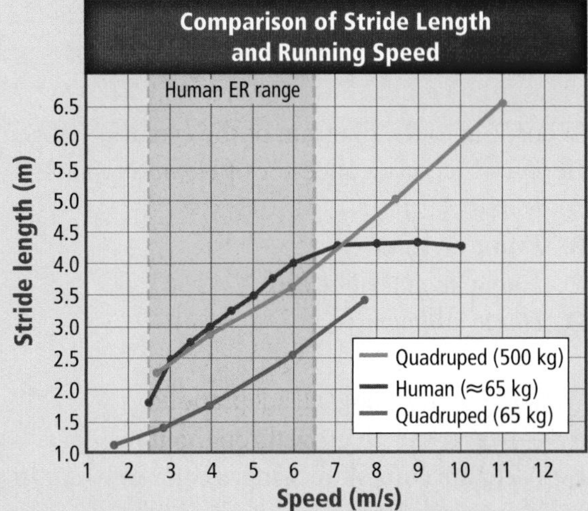

40. During ER, is the stride length of a human more like that of a 65-kg quadruped or a 500-kg quadruped?

41. Is a human more efficient at endurance running than a similar-size quadruped such as a cheetah or a leopard? Explain.

Cumulative Review

42. Describe the role of mitochondria in eukaryotic cells. **(Chapter 7)**

43. Under what conditions would a population experience zero population growth? **(Chapter 4)**

Standardized Test Practice

Cumulative

Multiple Choice

1. A scientific understanding of which natural process helped Darwin formulate the concept of natural selection?
   A. artificial selection
   B. continental drift
   C. group selection
   D. plant genetics

Use the diagram below to answer question 2.

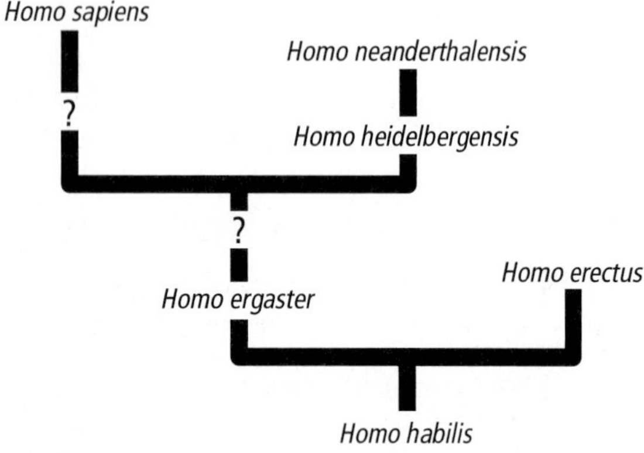

2. According to the diagram of the evolution of genus *Homo*, which is an ancestor of *Homo sapiens*?
   A. *Homo erectus*
   B. *Homo ergaster*
   C. *Homo neanderthalensis*
   D. *Homo rudolfensis*

3. Which is a physiological adaptation?
   A. A beaver's teeth grow throughout its life.
   B. A chameleon's skin changes color to blend in with its surroundings.
   C. A human sleeps during the day in order to work at night.
   D. An insect does not respond to a chemical used as an insecticide.

4. Which process can include the use of selective breeding?
   A. curing a tree of a disease
   B. finding the gene that makes a type of tree susceptible to disease
   C. mapping the genome of a fungus that causes disease in trees
   D. producing trees that resist certain diseases

Use the illustration below to answer question 5.

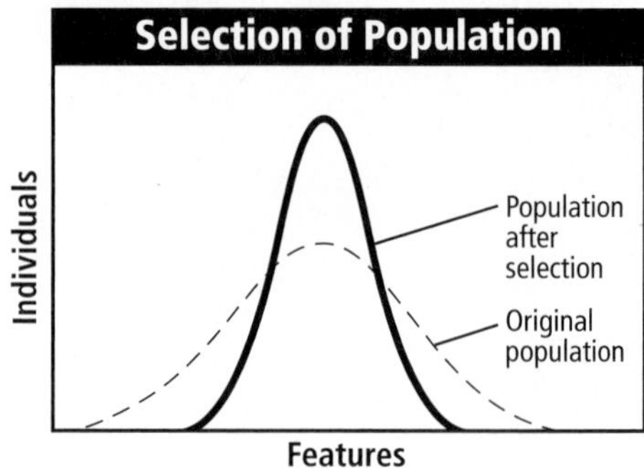

5. Which description fits the graph above?
   A. Average-sized features are selected for in population X.
   B. Larger features are selected for in population X.
   C. Smaller features are selected for in population X.
   D. Average-sized features are selected against in population X.

6. Which sequence correctly traces the order of hominin evolution?
   A. *Australopithecus afarensis* → *Australopithecus africanus* → *Proconsul* → *Homo*
   B. *Australopithecus africanus* → *Australopithecus afarensis* → *Proconsul* → *Homo*
   C. *Homo* → *Australopithecus africanus* → *Australopithecus afarensis* → *Proconsul*
   D. *Proconsul* → *Australopithecus afarensis* → *Australopithecus africanus* → *Homo*

7. Which of the following is NOT a reason scientists support the endosymbiont theory?
   A. Mitochondria and chloroplasts are found living outside eukaryotic cells.
   B. Mitochondria and chloroplasts reproduce by fission.
   C. The size and structure of mitochondria and chloroplasts is similar to prokaryotic cells.
   D. The genetic material in mitochondria and chloroplasts is circular.

Short Answer

8. The gene that controls the fur color of guinea pigs codes for either dominant black fur (B) or recessive white fur (b). Suppose you want to find the genotype of a black guinea pig. Explain how you would do a test cross. Then use one or both Punnett squares below to show possible testcross results.

9. A species of bird has a chemical in its tissue that is poisonous to many potential predators. Suppose you find another bird with a coloring pattern similar to the feathers of the first bird. What is this adaptation? Explain its importance.

10. Contrast the multiregional hypothesis and the "Out of Africa" hypothesis for human evolution.

11. Malathion is a pesticide used to control mosquitoes. Suppose a population of mosquitoes develops an ability to survive malathion spraying. How does this phenomenon fit with the ideas of variation and heritability in natural selection?

Use the figure below to answer question 12.

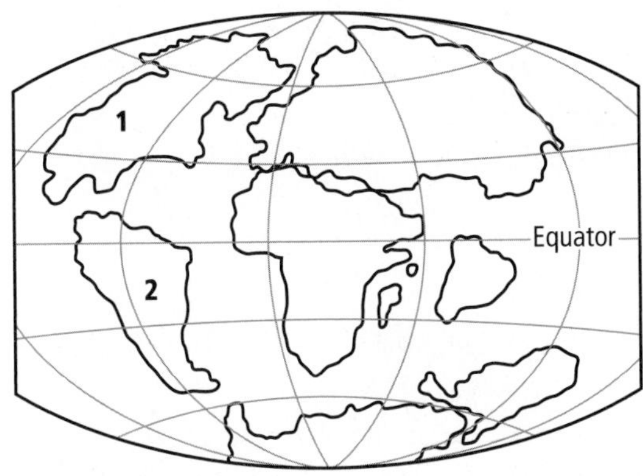

12. Discuss how land species and environments might change if the two numbered continents in the figure collided.

Extended Response

13. Suppose you are explaining human evolution to someone who is unfamiliar with the topic. Hypothesize why *Homo sapiens* is the only surviving member of the human family.

14. Some aggressive bacterial infections are treated with combinations of antibiotics. How would such a treatment affect drug resistance?

Essay Question

"If evolution almost always occurs by rapid speciation in small, peripheral isolates, then what should the fossil record look like? We are not likely to detect the event of speciation itself. It happens too fast, in too small a group, isolated too far from the ancestral range. Only after its successful origin will we first meet the new species as a fossil—when it reinvades the ancestral range and becomes a large central population in its own right. During its recorded history in the fossil record, we should expect no major change."

Gould, Stephen Jay. "Ladders, Bushes, and Human Evolution" *Natural History* 85 (April 1976): 30–31.

Using the information in the paragraph above, answer the following question in essay format.

15. Gould's research in evolution was devoted, in part, to explaining his theory of punctuated equilibrium. In an essay, explain why the fossil record is incomplete.

NEED EXTRA HELP?															
If You Missed Question . . .	1	2	3	4	5	6	7	8	9	10	11	12	13	14	15
Review Section . . .	15.1	16.3	15.2	13.1	15.3	16.2	14.2	13.1	15.3	16.3	15.1	14.1	16.3	15.2	15.3

Organizing Life's Diversity

BIG Idea **Evolution underlies the classification of life's diversity.**

Section 1
The History of Classification
MAIN Idea Biologists use a system of classification to organize information about the diversity of living things.

Section 2
Modern Classification
MAIN Idea Classification systems have changed over time as information has increased.

Section 3
Domains and Kingdoms
MAIN Idea The most widely used biological classification system has six kingdoms within three domains.

BioFacts

- The Sonoran Desert is the most biologically diverse desert in North America.
- The cardon cactus, found in the Sonoran Desert, can be up to 21 m tall.
- Scientists withheld water from a barrel cactus for six years and it still survived.

Ground squirrel

Cardon cactus

Ocotillo plant

Desert orangetip

Start-Up Activities

LAUNCH Lab

How can desert organisms be grouped?

You might think of a desert as a place without much biodiversity, but a wide variety of species have adaptations for desert life. Some adaptations are useful for grouping these organisms. In this lab, you will develop a system for grouping desert organisms.

Procedure

1. Read and complete the lab safety form.
2. List the desert organisms in the **photo.**
3. Identify physical characteristics, behaviors, or other factors that vary among the organisms in your list. Choose one factor you can use to sort them into groups.
4. Sort the list based on the factor you selected.
5. Brainstorm a list of desert organisms not shown in the photo. Add each to the appropriate group.

Analysis

1. **Compare and contrast** your grouping strategy with those developed by other students.
2. **Determine** What modifications would make your system more useful?

Visit biologygmh.com to:

- study the entire chapter online
- explore the Concepts in Motion, the Interactive Table, Microscopy Links, Virtual Labs, and links to virtual dissections
- access Web links for more information, projects, and activities
- review content online with the Interactive Tutor and take Self-Check Quizzes

The Six Kingdoms Make the following Foldable to help you organize information about the six kingdoms.

STEP 1 Place three sheets of paper 1.5 cm apart as illustrated.

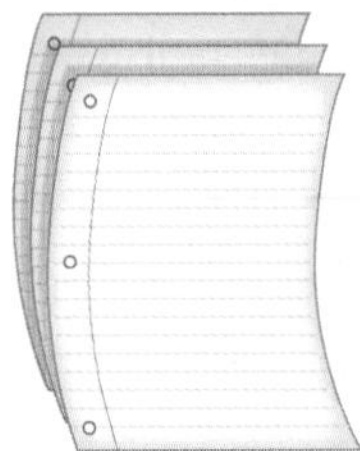

STEP 2 Fold all three sheets to make six tabs 1.5 cm in size, as shown.

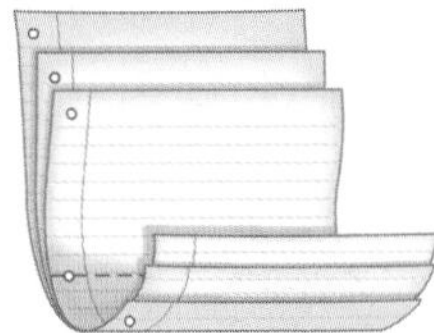

STEP 3 Rotate your Foldable 180°. Staple along the folded edge to secure all sheets. With the stapled end on top, label the tabs *The Six Kingdoms, Bacteria, Archaea, Protista, Fungi, Plantae,* and *Animalia.*

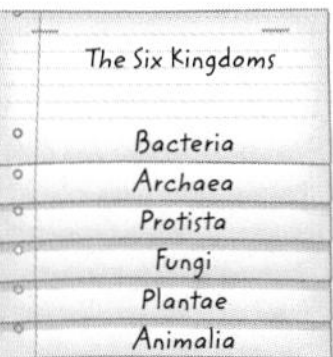

FOLDABLES **Use this Foldable with Section 17.3.** As you study the chapter, write the characteristics and list examples of each kingdom under each tab.

Section 17.1

Reading Preview

Objectives

- **Compare** Aristotle's and Linnaeus's methods of classifying organisms.
- **Explain** how to write a scientific name using binomial nomenclature.
- **Summarize** the categories used in biological classification.

Review Vocabulary

morphology: the structure and form of an organism or one of its parts

New Vocabulary

classification
taxonomy
binomial nomenclature
taxon
genus
family
order
class
phylum
division
kingdom
domain

The History of Classification

MAIN Idea Biologists use a system of classification to organize information about the diversity of living things.

Real-World Reading Link Think about how frustrating it would be if you went into a music store and all the CDs were in one big pile. You might need to go through all of them to find the one you want. Just as stores group CDs according to type of music and artist, biologists group living things by their characteristics and evolutionary relationships.

Early Systems of Classification

Has anyone ever told you to get organized? You are probably expected to keep your room in order. Your teachers might have asked you to organize your notes or homework. Keeping items or information in order makes them easier to find and understand. Biologists find it easier to communicate and retain information about organisms when the organisms are organized into groups. One of the principal tools for this is biological classification. **Classification** is the grouping of objects or organisms based on a set of criteria.

Aristotle's system More than two thousand years ago, the Greek philosopher Aristotle (394–322 B.C.) developed the first widely accepted system of biological classification. Aristotle classified organisms as either animals or plants. Animals were classified according to the presence or absence of "red blood." Aristotle's "bloodless" and "red-blooded" animals nearly match the modern distinction of invertebrates and vertebrates. Animals were further grouped according to their habitats and morphology. Plants were classified by average size and structure as trees, shrubs, or herbs. **Table 17.1** shows how Aristotle might have divided some of his groups.

Table 17.1 Aristotle's Classification System

Concepts In Motion Interactive Table To explore more about classification systems, visit biologygmh.com.

Plants		
Herbs	**Shrubs**	**Trees**
Violets Rosemary Onions	Blackberry bush Honeysuckle Flannelbush	Apple Oak Maple
Animals with red blood		
Land	**Water**	**Air**
Wolf Cat Bear	Dolphin Eel Sea bass	Owl Bat Crow

Aristotle's system was useful for organizing, but it had many limitations. Aristotle's system was based on his view that species are distinct, separate, and unchanging. The idea that species are unchanging was common until Darwin presented his theory of evolution. Because of his understanding of species, Aristotle's classification did not account for evolutionary relationships. Additionally, many organisms do not fit easily into Aristotle's system, such as birds that don't fly or frogs that live both on land and in water. Nevertheless, many centuries passed before Aristotle's system was replaced by a new system that was better suited to the increased knowledge of the natural world.

Linnaeus's system In the eighteenth century, Swedish naturalist Carolus Linnaeus (1707–1778) broadened Aristotle's classification method and formalized it into a scientific system. Like Aristotle, he based his system on observational studies of the morphology and the behavior of organisms. For example, he organized birds into three major groups depending on their behavior and habitat. The birds in **Figure 17.1** illustrate these categories. The eagle is classified as a bird of prey, the heron as a wading bird, and the cedar waxwing is grouped with the perching birds.

Linnaeus's system of classification was the first formal system of taxonomic organization. **Taxonomy** (tak SAH nuh mee) is a discipline of biology primarily concerned with identifying, naming, and classifying species based on natural relationships. Taxonomy is part of the larger branch of biology called systematics. Systematics is the study of biological diversity with an emphasis on evolutionary history.

Binomial nomenclature Linnaeus's method of naming organisms, called binomial nomenclature, set his system apart from Aristotle's system and remains valid today. **Binomial nomenclature** (bi NOH mee ul • NOH mun klay chur) gives each species a scientific name that has two parts. The first part is the genus (JEE nus) name, and the second part is the specific epithet (EP uh thet), or specific name, that identifies the species. Latin is the basis for binomial nomenclature because Latin is an unchanging language, and, historically, it has been the language of science and education.

■ **Figure 17.1** Linnaeus would have classified these birds based on their morphological and behavioral differences.
Infer *In what group might Linnaeus have placed a robin?*

American bald eagle
Bird of prey

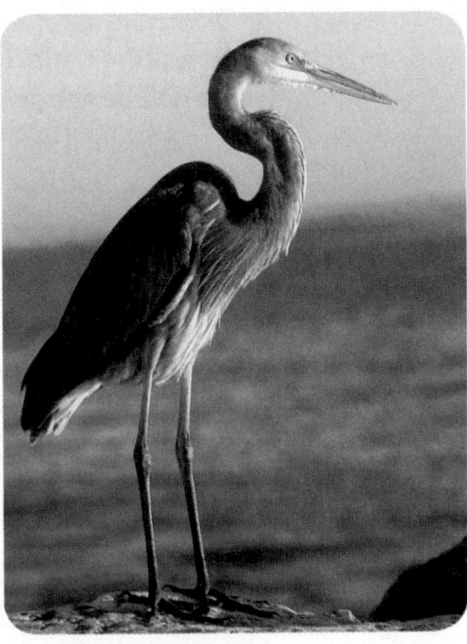

Great blue heron
Wading bird

Cedar waxwing
Perching bird

■ **Figure 17.2** *Cardinalis cardinalis* is a bird with many common names and is seen throughout much of the United States. It is the state bird of Illinois, Indiana, Kentucky, North Carolina, and Ohio.
Identify ***some other animals that have multiple common names.***

Biologists use scientific names for species because common names vary in their use. Many times the bird shown in **Figure 17.2** is called a redbird, sometimes it is called a cardinal, and other times it is called a Northern cardinal. In 1758, Linnaeus gave this bird its scientific name, *Cardinalis cardinalis.* The use of scientific names avoids the confusion that can be created with common names. Binomial nomenclature also is useful because common names can be misleading. If you were doing a scientific study on fish, you would not include starfish in your studies. Starfish are not fish. In the same way, great horned owls do not have horns and sea cucumbers are not plants.

VOCABULARY
WORD ORIGIN
Binomial nomenclature
comes from the Latin words *bi,* meaning *two; nomen,* meaning *name;* and *calatus,* meaning *list.*

When writing a scientific name, scientists follow these rules.

- The first letter of the genus name always is capitalized, but the rest of the genus name and all letters of the specific epithet are lowercase.
- If a scientific name is written in a printed book or magazine, it should be italicized.
- When a scientific name is written by hand, both parts of the name should be underlined.
- After the scientific name has been written completely, the genus name often will be abbreviated to the first letter in later appearances. For example, the scientific name of *Cardinalis cardinalis* can be written *C. cardinalis.*

 **Reading Check** **Explain** why Latin is the basis for many scientific names.

Modern classification systems The study of evolution in the 1800s added a new dimension to Linnaeus's classification system. Many scientists at that time, including Charles Darwin, Jean-Baptiste Lamarck, and Ernst Haekel, began to classify organisms not only on the basis of morphological and behavioral characteristics. They also included evolutionary relationships in their classification systems. Today, while modern classification systems remain rooted in the Linnaeus tradition, they have been modified to reflect new knowledge about evolutionary ancestry.

Taxonomic Categories

Think about how things are grouped in your favorite video store. How are the DVDs arranged on the shelves? They might be arranged according to genre—action, drama, or comedy—and then by title and year. Although taxonomists group organisms instead of DVDs, they also subdivide groups based on more specific criteria. The taxonomic categories used by scientists are part of a nested-hierarchal system—each category is contained within another, and they are arranged from broadest to most specific.

Species and genus A named group of organisms is called a **taxon** (plural, taxa). Taxa range from having broad diagnostic characteristics to having specific characteristics. The broader the characteristics, the more species the taxon contains. One way to think of taxa is to imagine nesting boxes—one fitting inside the other. You already have learned about two taxa used by Linnaeus—genus and species. Today, a **genus** (plural, genera) is defined as a group of species that are closely related and share a common ancestor.

Note the similarities and differences among the three species of bears in **Figure 17.3.** The scientific names of the American black bear *(Ursus americanus)* and Asiatic black bear *(Ursus thibetanus)* indicate that they belong to the same genus, *Ursus.* All species in the genus *Ursus* have massive skulls and similar tooth structures. Sloth bears *(Melursus ursinus),* despite their similarity to members of the genus *Ursus,* usually are classified in a different genus, *Melursus,* because they are smaller, have a different skull shape and size, and have two fewer incisor teeth than bears of the genus *Ursus.*

Family All bears, both living and extinct species, belong to the same family, Ursidae. A **family** is the next higher taxon, consisting of similar, related genera. In addition to the three species shown in **Figure 17.3,** the Ursidae family contains six other species: brown bears, polar bears, giant pandas, Sun bears, and Andean bears. All members of the bear family share certain characteristics. For example, they all walk flatfooted and have forearms that can rotate to grasp prey closely.

CAREERS IN BIOLOGY

Wildlife Biologist A scientist who studies organisms in the wild is called a wildlife biologist. Wildlife biologists might study populations of bears or work to educate the public about nature. For more information on biology careers, visit careers at biologygmh.com.

Figure 17.3 All species in the genus *Ursus* have large body size and massive skulls. Sloth bears are classified in the genus *Melursus.*

Ursus americanus
American black bear

Ursus thibetanus
Asiatic black bear

Melursus ursinus
Sloth bear

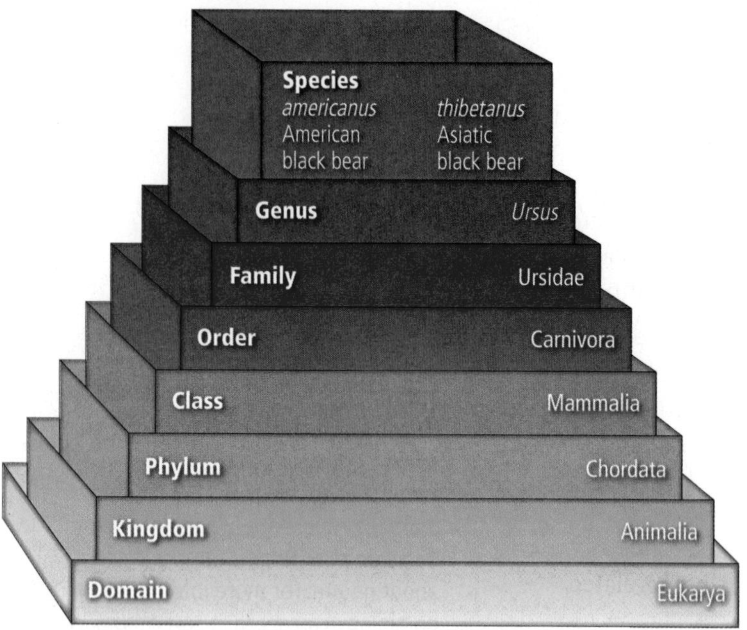

■ **Figure 17.4** Taxonomic categories are contained within one another like nesting boxes. Notice that the American black bear and Asiatic black bear are different species; however, their classification is the same for all other categories.

Higher taxa An **order** contains related families. A **class** contains related orders. The bears in **Figure 17.3** belong to the order Carnivora and class Mammalia. A **phylum** (FI lum) (plural, phyla) or **division** contains related classes. The term *division* is used instead of *phylum* for the classification of bacteria and plants. Sometimes scientists break the commonly used taxa into subcategories, such as subspecies, subfamilies, infraorders, and subphyla.

The taxon composed of related phyla or divisions is a **kingdom**. Bears are classified in phylum Chordata, Kingdom Animalia, and Domain Eukarya. The **domain** is the broadest of all the taxa and contains one or more kingdoms. The basic characteristics of the three domains and six kingdoms are described later in this chapter.

Figure 17.4 shows how the taxa are organized into a hierarchical system. The figure also shows the complete classification from domain to species for the American black bear and the Asiatic black bear. Notice that although these bears are classified as different species, the rest of their classification is the same.

MiniLab 17.1

Develop a Dichotomous Key

How can you classify items? Scientists group organisms based on their characteristics. These groups are the basis for classification tools called dichotomous keys. A dichotomous key consists of a series of choices that lead the user to the correct identification of an organism. In this lab, you will develop a dichotomous key as you group familiar objects.

Procedure

1. Read and complete the lab safety form.
2. Remove one **shoe** and make a shoe pile with other shoes from your group.
3. Write a question in your dichotomous key regarding whether the shoe has a characteristic of your choice. Divide the shoes into two groups based on that distinguishing characteristic.
4. Write another question for a different characteristic in your dichotomous key. Divide one of the subgroups into two smaller groups based on this distinguishing characteristic.
5. Continue dividing shoes into subgroups and adding questions to your key until there is only one shoe in each group. Make a branching diagram to identify each shoe with a distinctive name.
6. Use your diagram to classify your teacher's shoe.

Analysis

1. **Relate** taxa to the other groups you used to classify shoes. Which group relates to kingdom, phyla, and so on?
2. **Explain** how you were able to classify your teacher's shoe in Step 6.
3. **Critique** How could your classification system be modified to be more effective?

■ **Figure 17.5** This systematist is studying a cactus to determine if it is a new species.

Systematics Applications

Scientists who study classification provide detailed guides that help people identify organisms. Many times a field guide will contain a dichotomous key, which is a key based on a series of choices between alternate characteristics. You can find out whether a plant or animal is poisonous by using a dichotomous (di KAHT uh mus) key to identify it.

CAREERS IN BIOLOGY Systematists, like the one shown in **Figure 17.5** also work to identify new species and relationships among known species. They incorporate information from taxonomy, paleontology, molecular biology, and comparative anatomy in their studies. While the discovery of new species is exciting and important, learning a new connection between species also impacts science and society. For example, if a biologist knows that a certain plant, such as the Madagascar periwinkle *Catharanthus roseus,* produces a chemical that can be used to treat cancer, he or she knows that it is possible related plants also might produce the same or similar chemicals.

Section 17.1 Assessment

Section Summary

- Aristotle developed the first widely accepted biological classification system.
- Linnaeus used morphology and behavior to classify plants and animals.
- Binomial nomenclature uses the Latin genus and specific epithet to give an organism a scientific name.
- Organisms are classified according to a nested hierarchical system.

Understand Main Ideas

1. MAIN Idea **Explain** why a biological classification system is important.
2. **Summarize** the rules for using binomial nomenclature.
3. **Compare and contrast** how modern classification systems differ from those used by Aristotle and Linnaeus.
4. **Classify** a giant panda, *Ailuropoda melanoleuca,* completely from domain to species level by referring to **Figure 17.4.**

Think Critically

WRITING in Biology

5. Write a short story describing an application of biological classification.
6. **Consider** Would you expect to see more biodiversity among members of a phyla or among members of a class? Why?
7. **Differentiate** between taxonomy and systematics.

Section 17.2

Reading Preview

Objectives

- **Compare and contrast** species concepts.
- **Describe** methods used to reveal phylogeny.
- **Explain** how a cladogram is constructed.

Review Vocabulary

evolution: the historical development of a group of organisms

New Vocabulary

phylogeny
character
molecular clock
cladistics
cladogram

Modern Classification

MAIN Idea Classification systems have changed over time as information has increased.

Real-World Reading Link Did you ever try a new way of organizing your school notes? Just as you sometimes make changes in the way you do something based on a new idea or new information, scientists adjust systems and theories in science when new information becomes available.

Determining Species

It isn't always easy to define a species. Organisms that are different species by one definition might be the same species by a different definition. As knowledge increases, definitions change. The concept of a species today is much different than it was 100 years ago.

Typological species concept Aristotle and Linnaeus thought of each species as a distinctly different group of organisms based on physical similarities. This definition of species is called the typological species concept. It is based on the idea that species are unchanging, distinct, and natural types, as defined earlier by Aristotle. The type specimen was an individual of the species that best displayed the characteristics of that species. When another specimen was found that varied significantly from the type specimen, it was classified as a different species. For example, in **Figure 17.6** the color patterns on the butterflies' wings are all slightly different. At one time, they would have been classified as three different species because of these differences, but now they are classified as the same species.

Because we now know that species change over time, and because we know that members of some species exhibit tremendous variation, the typological species concept has been replaced. However, some of its traditions, such as reference to type specimens, remain.

Figure 17.6 Although these tropical butterflies vary in their color patterns, they are classified as different varieties of the same species, *Heliconius erato*.
Describe *Why might early taxonomists have classified them as separate species?*

Biological species concept Theodosius Dobzhansky and Ernst Mayr, two evolutionary biologists, redefined the term species in the 1930s and 1940s. They defined a species as a group of organisms that is able to interbreed and produce fertile offspring in a natural setting. This is called the biological species concept, and it is the definition for species used throughout this textbook. Though the butterflies in **Figure 17.6** have variable color patterns, they can interbreed to produce fertile offspring and therefore are classified as the same species.

There are limitations to the biological species concept. For example, wolves and dogs, as well as many plant species, are known to interbreed and produce fertile offspring even though they are classified as different species. The biological species concept also does not account for extinct species or species that reproduce asexually. However, because the biological species concept works in most everyday experiences of classification, it is used often.

Phylogenetic species concept In the 1940s, the evolutionary species concept was proposed as a companion to the biological species concept. The evolutionary species concept defines species in terms of populations and ancestry. According to this concept, two or more groups that evolve independently from an ancestral population are classified as different species. More recently, this concept has developed into the phylogenetic species concept. **Phylogeny** (fi LAH juh nee) is the evolutionary history of a species. The phylogenetic species concept defines a species as a cluster of organisms that is distinct from other clusters and shows evidence of a pattern of ancestry and descent. When a phylogenetic species branches, it becomes two different phylogenetic species. For example, recall from Chapter 15 that when organisms become isolated—geographically or otherwise—they often evolve different adaptations. Eventually they might become different enough to be classified as a new species.

This definition of a species solves some of the problems of earlier concepts because it applies to extinct species and species that reproduce asexually. It also incorporates molecular data. **Table 17.2** summarizes the three main species concepts.

Study Tip

Note Discussions While you read, use self-adhesive notes to mark passages that you do not understand. In addition, mark passages you do understand and can explain to others with your own explanations, examples, and ideas. Then, discuss them with your classmates.

LAUNCH Lab

Review Based on what you've read about classification systems, how would you now answer the analysis questions?

Table 17.2 Species Concepts

Concepts In Motion
Interactive Table To explore more about species concepts, visit biologygmh.com.

Species Concept	Description	Limitation	Benefit
Typological species concept	Classification is determined by the comparison of physical characteristics with a type specimen.	Alleles produce a wide variety of features within a species.	Descriptions of type specimens provide detailed records of the physical characteristics of many organisms.
Biological species concept	Classification is determined by similar characteristics and the ability to interbreed and produce fertile offspring.	Some organisms, such as wolves and dogs that are different species, interbreed occasionally. It does not account for extinct species.	The working definition applies in most cases, so it is still used frequently.
Phylogenetic species concept	Classification is determined by evolutionary history.	Evolutionary histories are not known for all species.	Accounts for extinct species and considers molecular data.

VOCABULARY

SCIENCE USAGE V. COMMON USAGE

Character

Science usage: a feature that varies among species

Organisms are compared based on similar characters

Common usage: imaginary person in a work of fiction—a play, novel, or film.

The queen was my favorite character in the book.

Characters

To classify a species, scientists often construct patterns of descent, or phylogenies, by using **characters**—inherited features that vary among species. Characters can be morphological or biochemical. Shared morphological characters suggest that species are related closely and evolved from a recent common ancestor. For example, because hawks and eagles share many morphological characters that they do not share with other bird species, such as keen eyesight, hooked beaks, and taloned feet, they should share a more recent common ancestor with each other than with other bird groups.

Morphological characters When comparing morphological characters, it is important to remember that analogous characters do not indicate a close evolutionary relationship. Recall from Chapter 15 that analogous structures are those that have the same function but different underlying construction. Homologous characters, however, might perform different functions, but show an anatomical similarity inherited from a common ancestor.

Birds and dinosaurs Consider the oviraptor and the sparrow shown in **Figure 17.7.** At first you might think that dinosaurs and birds do not have much in common and do not share a close evolutionary relationship. A closer look at dinosaur fossils shows that they share many features with birds. Some fossil dinosaur bones, like those of the large, carnivorous theropod dinosaurs, show that their bones had large hollow spaces. Birds have bones with hollow spaces. In this respect, they are more like birds than most living reptiles, such as alligators, lizards, and turtles, which have dense bones. Also, theropods have hip, leg, wrist, and shoulder structures that are more similar to birds than to other reptiles. Recently, scientists have discovered some fossil dinosaur bones that suggest some theropods had feathers. The evidence provided by these morphological characters indicates that modern birds are related more closely to theropod dinosaurs than they are to other reptiles.

Reading Check **Explain** how morphological characters have influenced the classification of dinosaurs and birds.

■ **Figure 17.7** This artist's conception of *Oviraptor philoceratops* might not appear to be related to the sparrow *Zonotrichia leucophrys*, but these animals share many characteristics that indicate a shared evolutionary history.
Deduce ***which similarities might prompt you to think that these species are more closely related than was commonly thought.***

Oviraptor philoceratops

Zonotrichia leucophrys

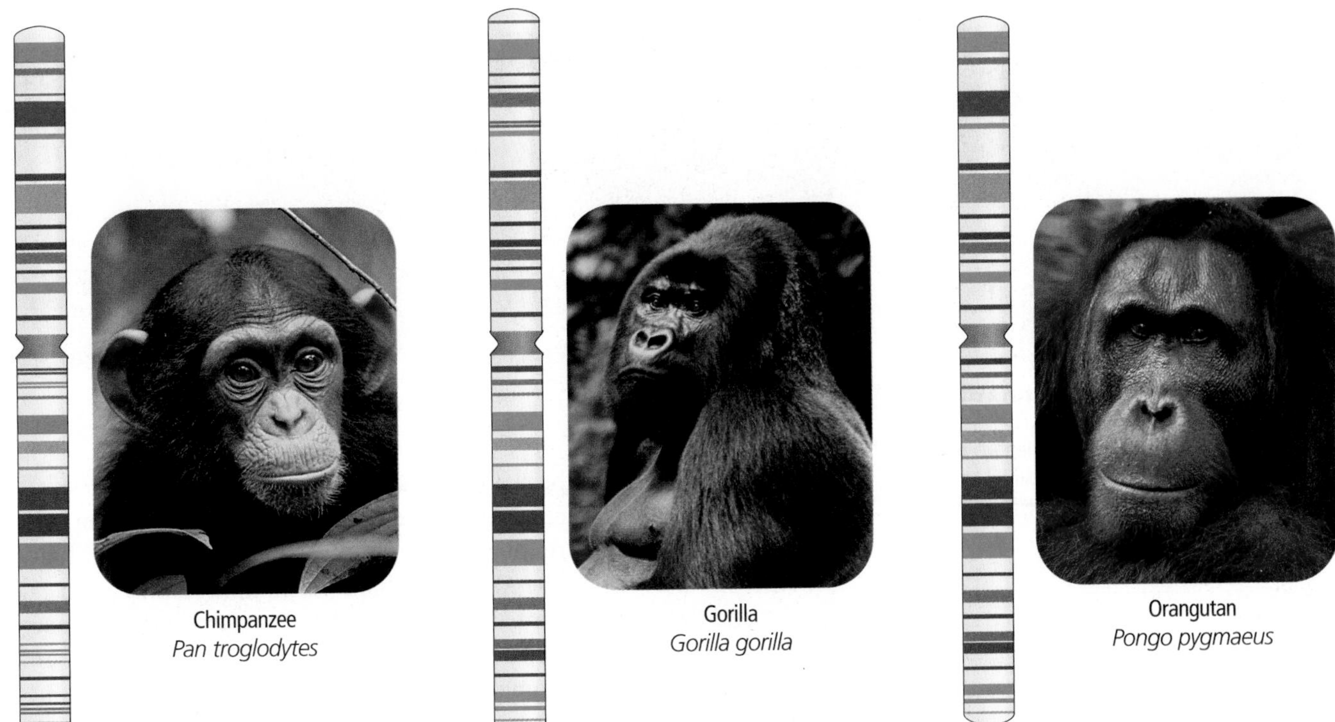

■ **Figure 17.8** The representation of chromosome-banding patterns for these homologous chromosomes illustrates the evidence of a close evolutionary relationship among the chimpanzee, gorilla, and orangutan.

Biochemical characters Scientists use biochemical characters, such as amino acids and nucleotides, to help them determine evolutionary relationships among species. Chromosome structure and number is also a powerful clue for determining species similarities. For example, members of the mustard family (Cruciferae)—including broccoli, cauliflower, and kale—all look different in the garden, but these plants have almost identical chromosome structures. This is strong evidence that they share a recent common ancestor. Likewise, the similar appearance of chromosomes among chimpanzees, gorillas, and orangutans suggests a shared ancestry. **Figure 17.8** shows the similar appearance of a chromosome-banding pattern in these three primates.

DNA and RNA analyses are powerful tools for reconstructing phylogenies. Remember that DNA and RNA are made up of four nucleotides. The nucleotide sequences in DNA define the genes that direct RNA to make proteins. The greater the number of shared DNA sequences between species, the greater the number of shared genes—and the greater the evidence that the species share a recent common ancestor.

Scientists use a variety of techniques to compare DNA sequences when assessing evolutionary relationships. They can sequence and compare whole genomes of different organisms. They can compare genome maps made by using restriction enzymes, like those you learned about in Chapter 13. They also use a technique called DNA-DNA hybridization, during which single strands of DNA from different species are melted together. The success of the hybridization depends on the similarity of the sequences—complementary sequences will bind to each other, while dissimilar sequences will not bind. Comparing the DNA sequences of different species is an objective, quantitative way to measure evolutionary relationships.

VOCABULARY

ACADEMIC VOCABULARY

Corresponding:
Being similar or equivalent in character, quantity, origin, structure, or function.
The corresponding sequences matched perfectly.

African elephant (savannah)

African elephant (forest)

Asiatic elephant

■ **Figure 17.9** The two populations of African elephants have been classified as the same species; however, DNA analysis shows that they might be separate species. The Asiatic elephant belongs to a separate genus.

A species example The classification of elephants is one example of how molecular data has changed traditional taxonomic organization. **Figure 17.9** shows pictures of elephants that live in the world today. Taxonomists have classified the Asiatic elephant *(Elephas maximus)* as one species and the African elephant *(Loxodonta africana)* as another for over 100 years. However, they have classified the two types of African elephant as the same species, even though the two populations look different. The forest-dwelling elephants are much smaller and have longer tusks and smaller ears than the savanna-dwelling elephants. Even so, scientists thought that the elephants interbred freely at the margins of their ranges. Recent DNA studies, however, show that the African elephants diverged from a common ancestor about 2.5 million years ago. Scientists have proposed renaming the forest-dwelling elephant *Loxodonta cyclotis.* Use **Data Analysis Lab 17.1** to explore molecular evidence for renaming the forest-dwelling elephant.

DATA ANALYSIS LAB 17.1

Based on Real Data*

Draw a Conclusion

Are African elephants a separate species? Efforts to count and protect elephant populations in Africa were based on the assumption that all African elephants belong to the same species. Evidence from a project originally designed to trace ivory samples changed that assumption.

A group of scientists studied the DNA variation among 195 African elephants from 21 populations in 11 of the 37 nations in which African elephants range and from seven Asian elephants. They used biopsy darts to obtain plugs of skin from the African elephants. The researchers focused on a total of 1732 nucleotides from four nuclear genes that are not subject to natural selection. The following paragraph shows the results of the samples.

*Data obtained from: Roca, A.L., et al. 2001. Genetic evidence for two species of elephant in Africa. *Science* 293(5534): 1473-1477.

Data and Observations

"Phylogenetic distinctions between African forest elephant and savannah elephant population corresponded to 58% of the difference in the same genes between elephant genera Loxodonta (African) and Elephas (Asian)."

Think Critically

1. **Describe** the type of evidence used in the study.
2. **Explain** the evidence that there are two species of elephants in Africa.
3. **Propose** other kinds of data that could be used to support three different scientific names for elephants.
4. **Infer** Currently *Loxodonta africana* is protected from being hunted. How might reclassification affect the conservation of forest elephants?

Molecular clocks You know that mutations occur randomly in DNA. As time passes, mutations accumulate, or build up, in the chromosomes. Some of these mutations do not affect the way cells function, and they are passed down from parent to offspring. Systematists can use these mutations to help them determine the degree of relationship among species. A **molecular clock** is a model that is used to compare DNA sequences from two different species to estimate how long the species have been evolving since they diverged from a common ancestor. **Figure 17.10** illustrates how a molecular clock works.

Scientists use molecular clocks to compare the DNA sequences or amino acid sequences of genes that are shared by different species. The differences between the genes indicate the presence of mutations. The more mutations that have accumulated, the more time that has passed since divergence. When the molecular clock technique was first introduced in the 1960s, scientists thought the rate of mutation within specific genes was constant. Hence, they used the clock as an analogy. However, scientists now know that the speed by which mutations occur is not always the same in a single gene or amino acid sequence.

The rate of mutation is affected by many factors, including the type of mutation, where it is in the genome, the type of protein that the mutation affects, and the population in which the mutation occurs. In a single organism, different genes might mutate, or "tick," at different speeds. This inconsistency makes molecular clocks difficult to read. Researchers try to compare genes that accumulate mutations at a relatively constant rate in a wide range of organisms. One such gene is the gene for cytochrome *c* oxidase, which is found in the mitochondrial DNA of most organisms.

Despite their limitations, molecular clocks can be valuable tools for determining a relative time of divergence of a species. They are especially useful when used in conjunction with other data, such as the fossil record.

Reading Check **Explain** What does the molecular clock use to compare DNA?

Phylogenetic Reconstruction

The most common systems of classification today are based on a method of analysis called cladistics. **Cladistics** (kla DIHS tiks) is a method that classifies organisms according to the order that they diverged from a common ancestor.

Character types Scientists consider two main types of characters when doing cladistic analyses. An ancestral character is found within the entire line of descent of a group of organisms. Derived characters are present members of one group of the line but not in the common ancestor. For example, when considering the relationship between birds and mammals, a backbone is an ancestral character because both birds and mammals have a backbone and so did their shared ancestor. However, birds have feathers and mammals have hair. Therefore, having hair is a derived character for mammals because only mammals have an ancestor with hair. Likewise, having feathers is a derived character for birds.

CAREERS IN BIOLOGY

Molecular Systematist An evolutionary geneticist uses genetic analysis to establish evolutionary relationships. Often, they work at colleges and universities where they teach and perform research. For more information on biology careers, visit biologygmh.com.

Figure 17.10 This molecular clock diagram shows how mutations might accumulate over time.
Infer *Why is a clock not a good analogy for this process?*

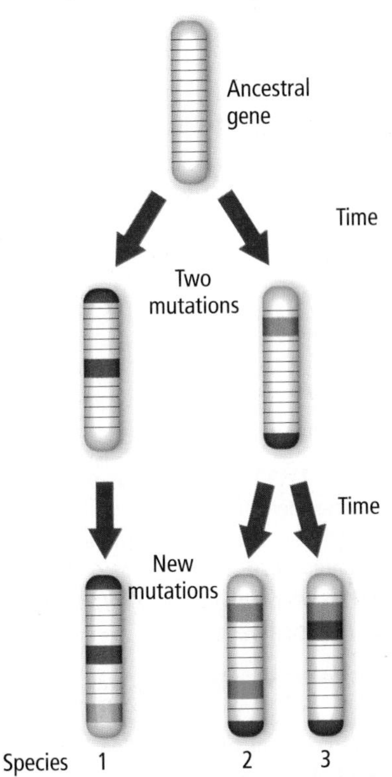

VOCABULARY

WORD ORIGIN

Cladistics
comes from the Greek word *klados,* meaning *sprout* or *branch.*

Cladograms Sytematists use shared derived characters to make a cladogram. A **cladogram** (KLAD uh gram) is a branching diagram that represents the proposed phylogeny or evolutionary history of a species or group. A cladogram is a model similar to the pedigrees you studied in Chapter 11. Just as a pedigree's branches show direct ancestry, a cladogram's branches indicate phylogeny. The groups used in cladograms are called clades. A clade is one branch of the cladogram.

Constructing a cladogram **Figure 17.11** is a simplified cladogram for some major plant groups. This cladogram was constructed in the following way. First, two species were identified, conifers and ferns, to compare with the lily species. Then, another species was identified that is ancestral to conifers and ferns. This species is called the outgroup. The outgroup is the species or group of species on a cladogram that has more ancestral characters with respect to the other organisms being compared. In the diagram below, the outgroup is moss. Mosses are more distantly related to ferns, conifers, and lilies.

The cladogram is then constructed by sequencing the order in which derived characters evolved with respect to the outgroup. The closeness of clades in the cladogram indicate the number of characters shared. The group that is closest to the lily shares the most derived characters with lilies and thus shares a more recent common ancestor with lilies than with the groups farther away. The nodes where the branches originate represent a common ancestor. This common ancestor generally is not a known organism, species, or fossil. Scientists hypothesize its characters based on the traits of its descendants.

The primary assumption The primary assumption that systematists make when constructing cladograms is that the greater the number of derived characters shared by groups, the more recently the groups share a common ancestor. Thus, as shown in **Figure 17.11,** lilies and conifers have three derived characters in common and are presumed to share a more recent common ancestor than lilies and ferns, which share only two characters.

A cladogram also is called a phylogenetic tree. Detailed phylogenetic trees show relationships among many species and groups of organisms. **Figure 17.12** illustrates a phylogenetic tree that shows the relationships among the domains and kingdoms of the most commonly used classification system today.

Personal Tutor

To learn about cladograms, visit biologygmh.com.

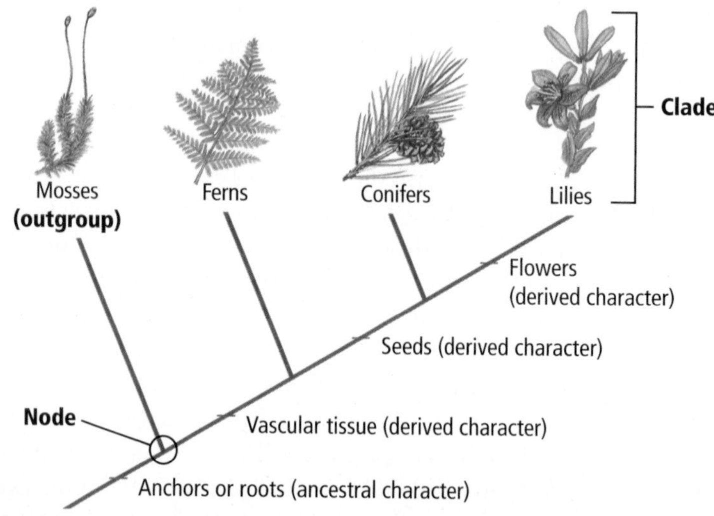

Figure 17.11 This cladogram uses the derived characters of plant taxa to model its phylogeny. Groups that are closer to the lily on the cladogram share a recent common ancestor. **Identify** ***which clades have chloroplasts but do not produce seeds.***

Concepts In Motion

Interactive Figure To see an animation of the cladistic method of classification, visit biologygmh.com.

Visualizing the Tree of Life

Figure 17.12
This phylogenetic tree shows the main branches in the "tree of life." Notice the three domains and the four kingdoms of Domain Eukarya. All of the branches are connected at the trunk, which is labeled *Common Ancestor.*

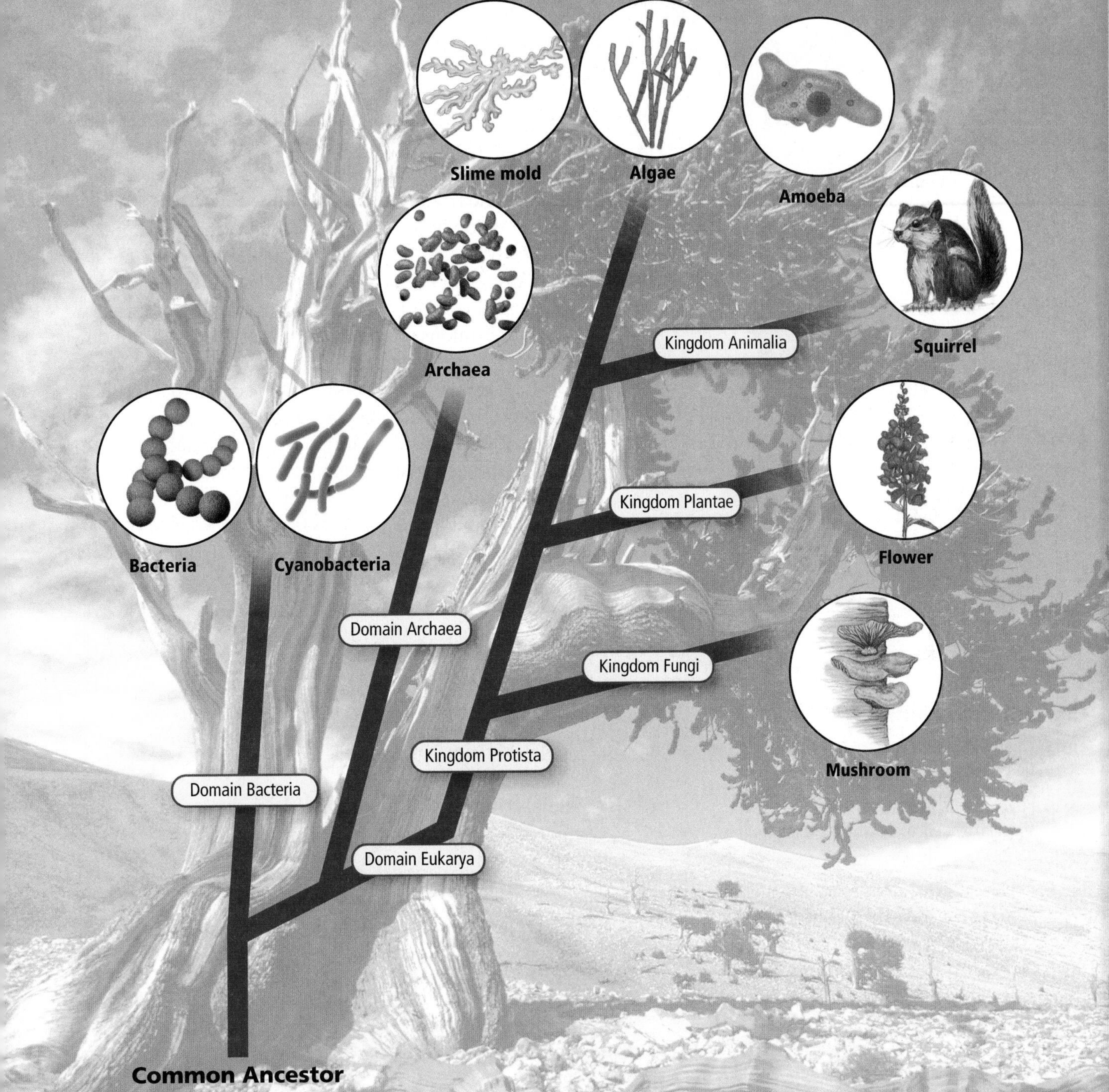

Slime mold
Algae
Amoeba
Squirrel
Archaea
Kingdom Animalia
Bacteria
Cyanobacteria
Kingdom Plantae
Flower
Domain Archaea
Kingdom Fungi
Mushroom
Kingdom Protista
Domain Bacteria
Domain Eukarya
Common Ancestor

Concepts In Motion **Interactive Figure** To see an animation of the tree of life, visit biologygmh.com.

Biology Online

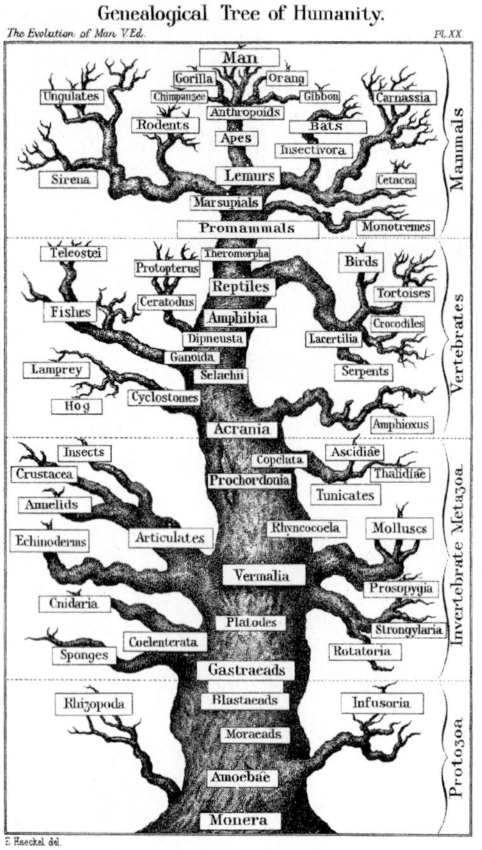

Figure 17.13 This illustration, made by Ernst Haeckel in the nineteenth century, was one of the first graphic depictions of evolutionary relationships.

Concepts In Motion

Interactive Figure To see an animation of evolutionary trees, visit biologygmh.com.

Connection to History **The tree of life** In his book *On the Origin of Species*, Charles Darwin used the analogy of a tree to suggest that all of the species developed from one or a few species. He imagined the tree's trunk to represent ancestral groups and each of the branches to have similar species. From each branch, smaller and smaller branches grew. Finally, at the tips of the twigs of these branches were the leaves, consisting of individual living species. This concept was developed further, and the term *tree of life* was coined by German biologist Ernst Haeckel (1834–1919). **Figure 17.13** shows Haeckel's Genealogical Tree of Humanity. Haeckel was the first to represent phylogenies in the form of a tree, and while his phylogenies are no longer completely accurate, they represent the first step in the reconstruction of phylogenies.

The tree of life diagram in **Figure 17.13** is a representation of the diversity of living organisms. A tree of life that incorporates all known organisms is almost unimaginably large. Scientists have discovered and described nearly 1.75 million species, and they estimate that millions more remain unclassified. Assembling a comprehensive tree of life requires a convergence of data from phylogenetic and molecular analysis. It also requires collaboration among many scientists representing many disciplines, from molecular biology to Earth science to computer science. Many scientists believe that the construction of a comprehensive tree of life, though an enormous task, is an important goal. Knowing how all organisms are related would benefit industry, agriculture, medicine, and conservation.

Section 17.2 Assessment

Section Summary

- The definition of species has changed over time.
- Phylogeny is the evolutionary history of a species, evidence for which comes from a variety of studies.
- A molecular clock uses comparisons of DNA sequences to estimate phylogeny and rate of evolutionary change.
- Cladistic analysis models evolutionary relationships based on sequencing derived characters.

Understand Main Ideas

1. MAIN Idea **Describe** how the changing species concept has affected classification systems.
2. **List and describe** the different concepts of a species.
3. **Describe** some methods used to determine phylogeny.
4. **Organize** the following derived characters on a cladogram in order of ascending complexity: multicellular, hair, backbone, unicellular, and four appendages.

Think Critically

MATH in Biology

5. **Describe** the mathematical challenges of counting the "ticks" of a molecular clock.
6. **Evaluate** the analogy of a tree for the organization of species based on phylogeny.
7. **Indicate** the hypothetical evolutionary relationship between two species if their DNA sequences share a 98 percent similarity.

Section 17.3

Domains and Kingdoms

Reading Preview

Objectives

- **Compare** major characteristics of the three domains.
- **Differentiate** among the six kingdoms.
- **Classify** organisms to the kingdom level.

Review Vocabulary

eukaryote: an organism composed of one or more cells containing a nucleus and membrane-bound organelles

New Vocabulary

archaea
protist
fungus

MAIN Idea The most widely used biological classification system has six kingdoms within three domains.

Real-World Reading Link You know that one of the classification categories is kingdom. Why would scientists use that term? Think about how a kingdom of medieval times could relate to groups of organisms.

Grouping Species

The broadest category in the classification system used by most biologists is the domain. There are three domains: Bacteria, Archaea, and Eukarya. Within these domains are six kingdoms: Bacteria, Archaea, Protists, Fungi, Plantae, and Animalia. Organisms are classified into domains according to cell type and structure, and into kingdoms according to cell type, structure, and nutrition.

This three-domain, six-kingdom classification system has been in use for less than three decades. It was modified from a system that did not have domains but had five kingdoms after scientists discovered an entirely new kind of organism in the 1970s. These new organisms are unicellular prokaryotes that scientists named archaea (ar KEE uh). Subsequent biochemical studies found that archaea are significantly different from the only other prokaryotes then known—the bacteria—and, in 1990, they were renamed and a new classification scheme was proposed to accommodate them. The Archaea are now members of their own domain.

Domain Bacteria

Connection to Chemistry Bacteria, members of Domain and Kingdom Bacteria, are prokaryotes whose cell walls contain peptidoglycan (pep tih doh GLY kan). Peptidoglycan is a polymer that contains two kinds of sugars that alternate in the chain. The amino acids of one sugar are linked to the amino acids in other chains, creating a netlike structure that is simple and porous, yet strong. Two examples of bacteria are shown in **Figure 17.14.**

Figure 17.14 Bacteria vary in their habitats and their methods of obtaining nourishment. The bacteria *Mycobacterium tuberculosis* that cause tuberculosis are heterotrophs. Cyanobacteria, such as *Anabaena,* are autotrophs.

Color-Enhanced SEM Magnification: 15,000×

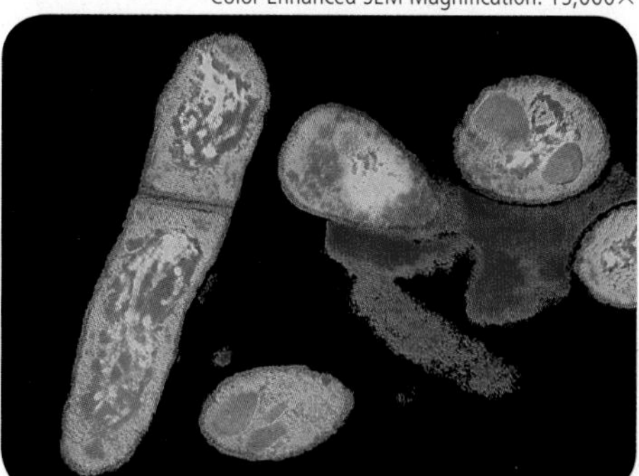

Mycobacterium tuberculosis

LM Magnification: 450×

Anabaena

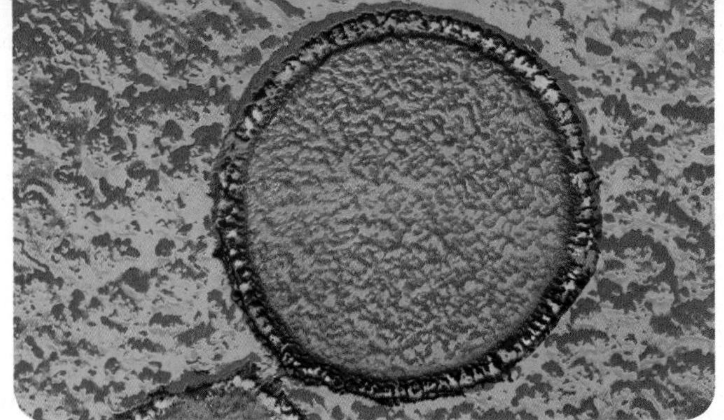

■ **Figure 17.15** This electron microscope image of *Staphylothermus marinus* shows the cell wall (green) and cell contents (pink). *S. marinus* is an extremophile found in deep ocean thermal vents.

Bacteria are a diverse group that can survive in many different environments. Some are aerobic organisms that need oxygen to survive, while others are anaerobic organisms that die in the presence of oxygen. Some bacteria are autotrophic and produce their own food, but most are heterotrophic and get their nutrition from other organisms. Bacteria are more abundant than any other organism. There are probably more bacteria in your body than there are people in the world. You can view some different types of bacteria in **MiniLab 17.2.**

VOCABULARY

WORD ORIGIN

Archaea
comes from the Greek word *archaios,* meaning *ancient* or *primitive.*

Domain Archaea

Archaea (ar KEE uh), the species classified in Domain Archaea, are thought to be more ancient than bacteria and yet more closely related to eukaryote ancestors. Their cell walls do not contain peptidoglycan, and they have some of the same proteins that eukaryotes do. They are diverse in shape and nutrition requirements. Some are autotrophic, but most are heterotrophic. Archaea are called extremophiles because they can live in extreme environments. They have been found in boiling hot springs, salty lakes, thermal vents on the ocean's floor, and in the mud of marshes where there is no oxygen. The archaea *Staphylothermus marinus,* shown in **Figure 17.15,** is found in deep ocean thermal vents and can live in water temperatures up to 98°C.

MiniLab 17.2

Compare Bacteria

How do the physical characteristics of various types of bacteria compare? Investigate the different features of bacteria by viewing prepared bacteria slides under the microscope during this lab.

Procedure

1. Read and complete the lab safety form.
2. Observe the prepared **slides of bacteria** with a **compound light microscope.**
3. Create a data table to compare the shapes and features of the bacteria you observe.
4. Compare and contrast the bacteria from the prepared slides and record your observations and comparisons in your data table.

Analysis

1. **Compare and contrast** the shapes of the individual bacteria cells you observed.
2. **Describe** Did any of your bacteria samples form colonies? What does a colony look like?
3. **Design** a classification system for the bacteria you observed based on the data you collected.

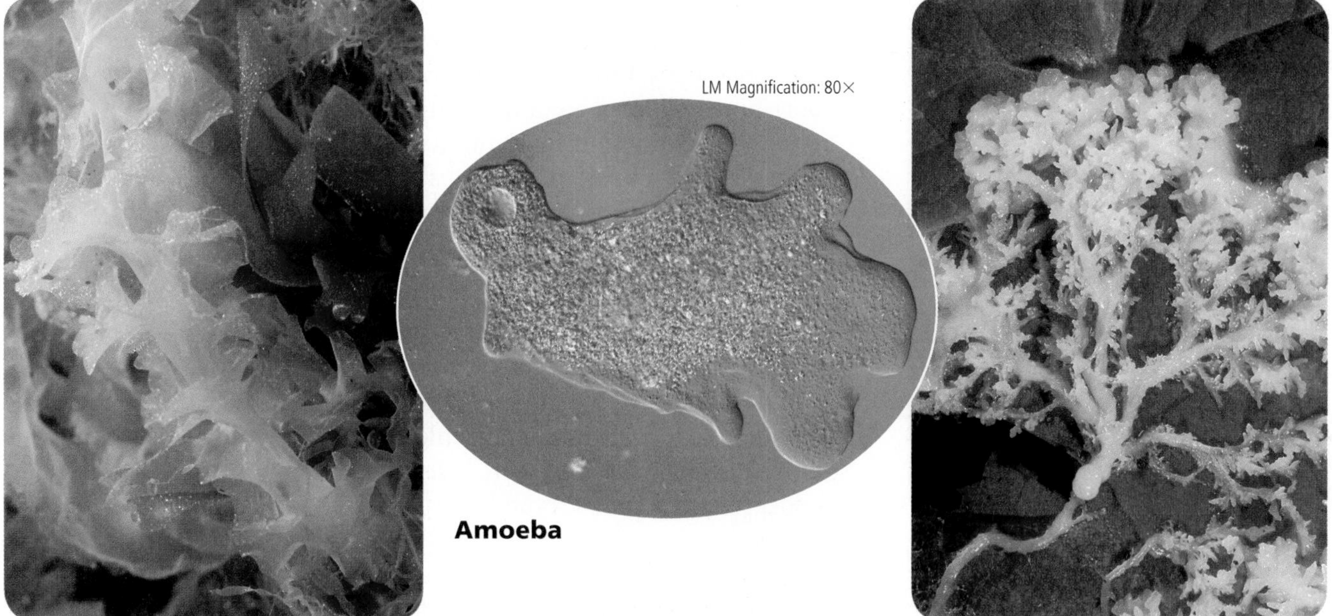

■ **Figure 17.16** These protists look different, but they all are eukaryotes, live in moist environments, and do not have organs.
Infer ***Which of these protists are plant-like? Animal-like? Funguslike?***

Domain Eukarya

Recall from Chapter 7 that cells with a membrane-bound nucleus and other membrane-bound organelles are called eukaryotic cells. All organisms with these cells are called eukaryotes and are classified in Domain Eukarya. Domain Eukarya contains Kingdom Protista, Kingdom Fungi, Kingdom Plantae, and Kingdom Animalia.

Kingdom Protista The wide variety of species shown in **Figure 17.16** belong to Kingdom Protista. Members of Kingdom Protista are called protists. **Protists** are eukaryotic organisms that can be unicellular, colonial, or multicellular. Unlike plants or animals, protists do not have organs. Though protists are not necessarily similar to each other, they don't fit in any other kingdoms. They are classified into three broad groups.

The plantlike protists are called algae. All algae, such as kelp, are autotrophs that perform photosynthesis. Animal-like protists are called protozoans. Protozoans, such as amoebas, are heterotrophs. Funguslike protists are slime molds and mildews, and they comprise the third group of protists. Euglenoids (yoo GLEE noyds) are a type of protist that have both plantlike and animal-like characteristics. They usually are grouped with the plantlike protists because they have chloroplasts and can perform photosynthesis.

FOLDABLES
Incorporate information from this section into your Foldable.

Kingdom Fungi A **fungus** is a unicellular or multicellular eukaryote that absorbs nutrients from organic materials in its environment. Members of Kingdom Fungi are heterotrophic, lack motility—the ability to move—and have cell walls. Their cell walls contain a substance called chitin, which is a rigid polymer that provides structural support. A fungus consists of a mass of threadlike filaments called hyphae (HI fee). Hyphae are threadlike filaments that are responsible for the fungus's growth, feeding, and reproduction. Fungi fossils exist that are over 400 million years old, and there are more than 70,000 known species.

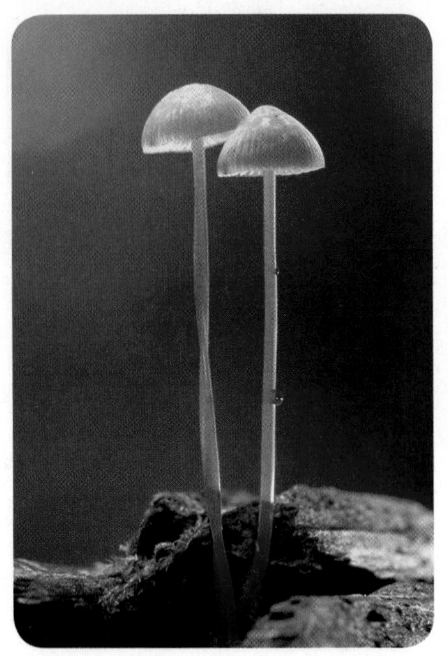

■ **Figure 17.17** Fungi come in a variety of sizes, from microscopic yeasts to multicellular forms, such as the mushrooms shown here.

Fungi, such as the mushrooms in **Figure 17.17,** are heterotrophic organisms. Some fungi are parasites—organisms that grow and feed on other organisms. Other fungi are saprobes—organisms that get their nourishment from dead or decaying organic matter. Unlike heterotrophs that digest their food internally, fungi secrete digestive enzymes into their food source and then absorb digested materials directly into their cells. Fungi that live in a mutualistic relationship with algae are called lichens. Lichens get their food from the algae that live among their hyphae.

Kingdom Plantae There are more than 250,000 species of plants in Kingdom Plantae (PLAN tuh). These organisms form the base of all terrestrial habitats. All plants are multicellular and have cell walls composed of cellulose. Most plants contain chloroplasts, where photosynthesis is carried out, but a few plants are heterotrophic. For example, the parisitic dodder has no green parts and extracts its food from host plants through suckers.

All plants possess cells that are organized into tissues, and many plants also possess organs such as roots, stems, and leaves. Like the fungi, plants lack motility. However, some plants do have reproductive cells that have flagella, which propel them through water. The characteristics of plants and members of the other five kingdoms are summarized in **Table 17.3.**

Reading Check **Describe** three characteristics of plants.

Table 17.3 Kingdom Characteristics

Concepts In Motion Interactive Table To explore more about the six kingdoms, visit biologygmh.com.

Domain	Bacteria	Archaea	Eukarya			
Kingdom	Bacteria	Archaea	Protista	Fungi	Plantae	Animalia
Example	*Pseudomonas* SEM Magnification: 5500×	*Methanopyrus* TEM Magnification: 25,000×	*Paramecium* LM Magnification: 150×	Mushroom	Moss	Earthworm
Cell type	Prokaryote		Eukaryote			
Cell walls	Cell walls with peptidoglycan	Cell walls without peptidoglycan	Cell walls with cellulose in some	Cell walls with chitin	Cell walls with cellulose	No cell walls
Number of cells	Unicellular		Unicellular and multicellular	Most multicellular	Multicellular	
Nutrition	Autotroph or heterotroph			Heterotroph	Autotroph	Heterotroph

Coral

Fish

Rabbit

■ **Figure 17.18** Members of Kingdom Animalia can look very different from each other even though they are in the same kingdom.

Kingdom Animalia Members of Kingdom Animalia are commonly called animals. More than one million animal species have been identified. All animals are heterotrophic, multicellular eukaryotes. Animal cells do not have cell walls. All animal cells are organized into tissues, and most tissues are organized into organs, such as skin, a stomach, and a brain. Animal organs often are organized into complex organ systems, like digestive, circulatory, or nervous systems. Animals range in size from a few millimeters to many meters. They live in the water, on land, and in the air. **Figure 17.18** shows some of the variety of organisms classified in Kingdom Animalia. Most animals are motile, although some, such as coral, lack motility as adults.

Viruses—an exception Have you ever experienced a cold or the flu? If so, you've had a close encounter with a virus. A virus is a nucleic acid surrounded by a protein coat. Viruses don't possess cells, nor are they cells, and are not considered to be living. Because they are nonliving, they usually are not placed in the biological classification system. Virologists, scientists who study viruses, have created special classification systems to group viruses. You will learn more about viruses in Chapter 18.

Section 17.3 Assessment

Section Summary

- Domains Bacteria and Archaea contain prokaryotes.
- Organisms are classified at the kingdom level based on cell type, structures, and nutrition.
- Domain Eukarya contains four kingdoms of eukaryotes.
- Because viruses are not living, they are not included in the biological classification system.

Understand Main Ideas

1. MAIN Idea **State** the three domains and the kingdoms in each.
2. **Compare and contrast** characteristics of the three domains.
3. **Explain** the difference between Kingdom Protista and Kingdom Fungi based on substances in their cell walls.
4. **Classify** to the kingdom level an organism that has organ systems, lacks cell walls, and ingests food.

Think Critically

5. **Summarize** the reason why systematists separated Domain Bacteria from Domain Archaea.

WRITING in **Biology**

6. Write an essay for or against including viruses in a classification system.

CUTTING-EDGE BIOLOGY

DNA Bar Codes

Most people would find it odd if their friend collected vials containing muscles from 940 different species of fish—but then again most people haven't undertaken a project as ambitious as this one.

DNA UPC Paul Herbert, a geneticist at the University of Guelph in Ontario, Canada, is trying to gather cell samples from all of the world's organisms. With small pieces of tissue no larger than the head of a pin, Herbert and his colleagues are working to assign DNA bar codes to every living species.

Herbert has shown that the segment of mitochondrial DNA, called cytochrome *c* oxidase I, or COI, can be used as a diagnostic tool to tell animal species apart. The COI gene is simple to isolate and allows for identification of an animal. A different gene would need to be used for plants. Just like UPC codes, the DNA segment sequence could be stored in a master database that would allow for easy access to the material. A hand scanner, when supplied with a small piece of tissue, such as a scale, a hair, or a feather, could identify the species almost instantly.

Potential benefits This technology has several potential benefits. A doctor might use it to pinpoint disease-causing organisms quickly to prevent epidemics or to determine what antivenom to give a snakebite victim. Health inspectors could scan foods for plant and animal contaminants. People who are curious about their surroundings could learn what lives around them. Farmers would be able to identify pests and use species-specific methods for their removal.

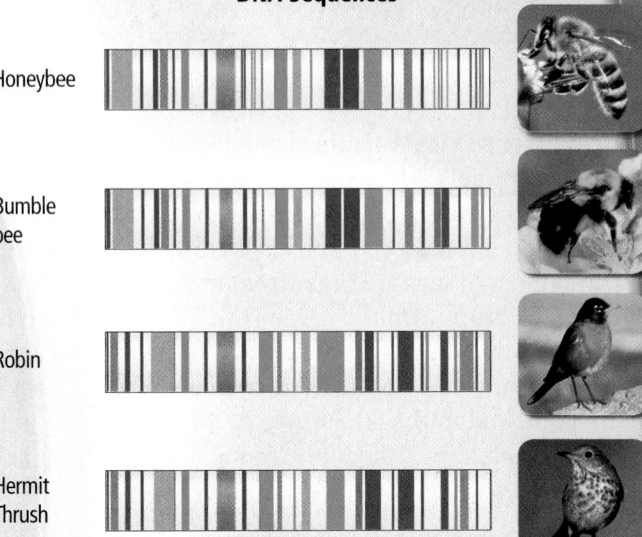

This representation of DNA barcodes shows that more closely related species would have more similar barcodes.

A new way to classify Using bioinformatics—a field of science in which biology, computer science, and information technology merge—to create a database of DNA barcodes allows taxonomists to classify more organisms quickly.

Currently, taxonomists have identified approximately two million species. Scientists estimate that anywhere between 10 and 100 million species exist. Historically, species have been classified using morphology, genetics, phylogeny, habitat, and behavior. While the bar codes would not replace classic taxonomic methods, they could supplement them by giving scientists another tool to use.

E-COMMUNICATION

Fact Finder Think of at least three questions you have about DNA bar coding. Research to find answers to your questions. Then, share your questions and answers with your class by e-mailing them to your teacher. For more information on DNA bar coding, visit biologygmh.com.

BIOLAB

HOW CAN ORGANISMS BE GROUPED ON A CLADOGRAM?

Background: When a cladogram is made, derived characters are used to divide the organisms into groups called clades. In this exercise, you will use simulated data to learn how to make a simple cladogram and then make your own cladogram.

Question: *How can you use organisms' characteristics to construct a cladogram?*

Materials

paper and pencil
examples of cladograms
photographs of various organisms
books describing characteristics of organisms

Procedure

1. Read and complete the lab safety form.
2. Examine the data table provided.
3. Compare the shared derived characteristics of the sample organisms. Assume that all the characteristics of your outgroup are ancestral. To make the data easier to compare, note that a "0" has been assigned to each ancestral character and a "1" to all derived characters.
4. Use the information to develop a cladogram that best shows the relationships of the organisms.
5. Make sure your teacher approves your cladogram before you proceed.
6. Choose four organisms from one of the domains you have studied that you believe are closely related.
7. Develop a table of derived characteristics of these organisms similar to the table you used in Step 2. Use your table to develop a cladogram that groups the organisms based on their shared derived characters.

Data Table for Cladistic Analysis

Organisms	Characters			
	1	2	3	4
A	b(1)	a(0)	a(0)	b(1)
B	b(1)	b(1)	b(1)	a(0)
C	b(1)	a(0)	b(1)	a(0)

Data obtained from: Lipscomb, D. 1998. Basics of cladistic analysis. George Washington University. http://www.gwu.edu/~clade/faculty/lipscomb/Cladistics.pdf

Analyze and Conclude

1. **Think Critically** How did you determine which were the ancestral and which were the derived characters of the organisms you examined?
2. **Explain** how you determined which characteristics to use to separate the clades.
3. **Explain** Which organism is the outgroup on your cladogram? Why?
4. **Critique** Trade data tables with another lab group. Use their data to draw a cladogram. Compare the two cladograms and explain any differences.
5. **Error Analysis** What type of error would mistaking analogous structures as homologous introduce into a cladogram? Examine your second cladogram and determine if you have made this error.

APPLY YOUR SKILL

Construct Molecular data, such as the amino acid sequences of shared proteins, can be used to make cladograms. Research cytochrome *c*, a protein important in aerobic respiration, and decide how it could be used to construct a cladogram. To learn more about cytochrome *c* and cladograms, visit BioLabs at biologygmh.com.

Study Guide

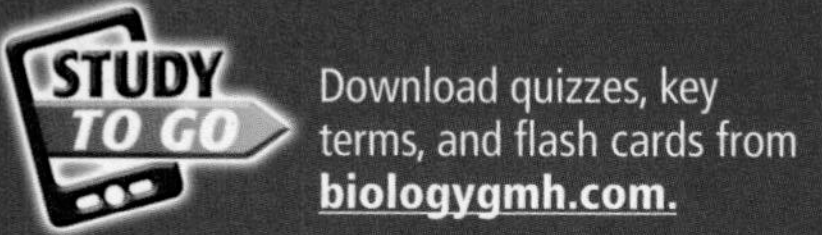

Download quizzes, key terms, and flash cards from biologygmh.com.

FOLDABLES Interpret On the back of your Foldable, draw a cladogram or phylogenetic tree that displays the order of evolution of the six kingdoms. Explain the reasoning for your interpretation.

Vocabulary	Key Concepts
Section 17.1 The History of Classification	
• binomial nomenclature (p. 485) • class (p. 488) • classification (p. 484) • division (p. 488) • domain (p. 488) • family (p. 487) • genus (p. 487) • kingdom (p. 488) • order (p. 488) • phylum (p. 488) • taxon (p. 487) • taxonomy (p. 485)	MAIN Idea Biologists use a system of classification to organize information about the diversity of living things. • Aristotle developed the first widely accepted biological classification system. • Linnaeus used morphology and behavior to classify plants and animals. • Binomial nomenclature uses the Latin genus and specific epithet to give an organism a scientific name. • Organisms are classified according to a nested hierarchical system.
Section 17.2 Modern Classification	
• character (p. 492) • cladistics (p. 495) • cladogram (p. 496) • molecular clock (p. 495) • phylogeny (p. 491)	MAIN Idea Classification systems have changed over time as information has increased. • The definition of species has changed over time. • Phylogeny is the evolutionary history of a species, evidence for which comes from a variety of studies. • A molecular clock uses comparisons of DNA sequences to estimate phylogeny and rate of evolutionary change. • Cladistic analysis models evolutionary relationships based on sequencing derived characters.
Section 17.3 Domains and Kingdoms	
• archaea (p. 500) • fungus (p. 501) • protist (p. 501)	MAIN Idea The most widely used biological classification system has six kingdoms within three domains. • Domains Bacteria and Archaea contain prokaryotes. • Organisms are classified at the kingdom level based on cell type, structures, and nutrition. • Domain Eukarya contains four kingdoms of eukaryotes. • Because viruses are not living, they are not included in the biological classification system.

Biology Online Vocabulary PuzzleMaker biologygmh.com

CHAPTER 17

Assessment

Section 17.1

Vocabulary Review

Match each definition with the correct term from the Study Guide page.

1. system of naming species using two words

2. taxon of closely related species that share a recent common ancestor

3. branch of biology that groups and names species based on studies of their different characteristics

Understand Key Concepts

4. On what did Linnaeus base his classification?
A. derived characters
B. binomial nomenclature
C. morphology and habitat
D. evolutionary relationship

Use the table to answer questions 5 and 6.

Classification of Selected Mammals				
Kingdom	Animalia	Animalia	Animalia	Animalia
Phylum	Chordata	Chordata	Chordata	Chordata
Class	Mammalia	Mammalia	Mammalia	Mammalia
Order	Cetacea	Carnivora	Carnivora	Carnivora
Family	Mysticeti	Felidae	Canidae	Canidae
Genus	*Balenopora*	*Felis*	*Canis*	*Canis*
Species	*B. physalis*	*F. catus*	*C. latrans*	*C. lupus*
Common name	Blue whale	Domestic cat	Coyote	Wolf

5. Which animal is the most distant relative to the others?
A. wolf
B. coyote
C. domestic cat
D. blue whale

6. At which level does the domestic cat diverge from the coyote?
A. family
B. class
C. order
D. genus

Constructed Response

7. Short Answer Explain the rules and uses of binomial nomenclature.

8. Short Answer Why is 'seahorse' not a good scientific name?

Think Critically

9. Recognize Relationships How does the system of classification relate to the diversity of species?

Section 17.2

Vocabulary Review

Differentiate between the following pairs.

10. phylogeny, character

11. cladogram, molecular clock

Understand Key Concepts

Use the figure below to answer questions 12 and 13.

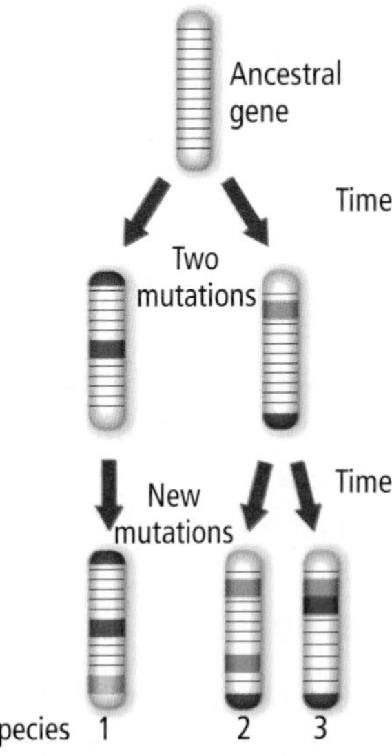

12. What does this figure represent?
A. pedigree
B. cladogram
C. molecular clock
D. phylogenetic tree

13. What do the colored bands in the figure represent?
A. mutations
B. derived characters
C. ancestral characters
D. genomes

14. Which species concept defines a species as a group of organisms that are able to reproduce successfully in the wild?
    A. typological species concept
    B. biological species concept
    C. evolutionary species concept
    D. phylogenetic species concept

Use the figure below to answer questions 15 and 16.

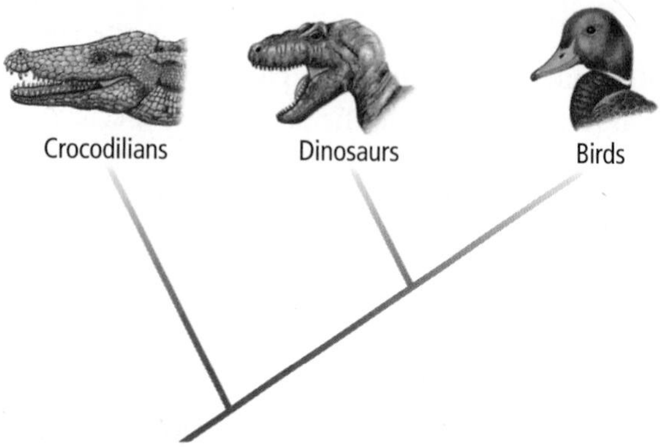

15. According to the figure, which organism diverged last?
    A. alligators
    B. birds
    C. crocodiles
    D. dinosaurs

16. Which is represented by the figure?
    A. pedigree
    B. cladogram
    C. molecular clock
    D. character

17. Which does not affect the rate of mutation in a molecular clock?
    A. type of mutation
    B. location of gene in genome
    C. the protein affected
    D. the time of divergence

Constructed Response

18. **Open Ended** Two scientists produce two different cladograms for the same groups of organisms. Explain how the differences are possible.

19. **Short Answer** Describe how to make a cladogram. Include the types of characters that are used and the judgments you must make about the characters.

20. **Short Answer** Summarize how biochemical characters can be used to determine phylogeny.

Think Critically

21. **Differentiate** between the typological species concept and the phylogenetic species concept.

22. **Decide** How should molecular clocks be used if not all mutations occur at the same rate? Should they be considered reliable evidence of phylogeny? Explain your answer.

Use the figure below to answer question 23.

23. **Evaluate** evidence that suggests that the two organisms in the figure are closely related.

Section 17.3

Vocabulary Review

Replace the italicized words with the correct vocabulary terms from the Study Guide page.

24. Algae are a type of *archaea.*

25. *Bacteria* are called extremophiles because they grow in extreme environments.

26. Some types of *protists* are used to make food products like bread and cheese.

Understand Key Concepts

27. Which taxon contains one or more kingdoms?
    A. genus
    B. phylum
    C. family
    D. domain

28. In which would prokaryotes found living in acid run-off or sulfur vents of volcanoes likely be classified?
    A. Bacteria
    B. Archaea
    C. Eubacteria
    D. Protista

BiologyOnline Chapter Test biologygmh.com

Use the photograph below to answer question 29.

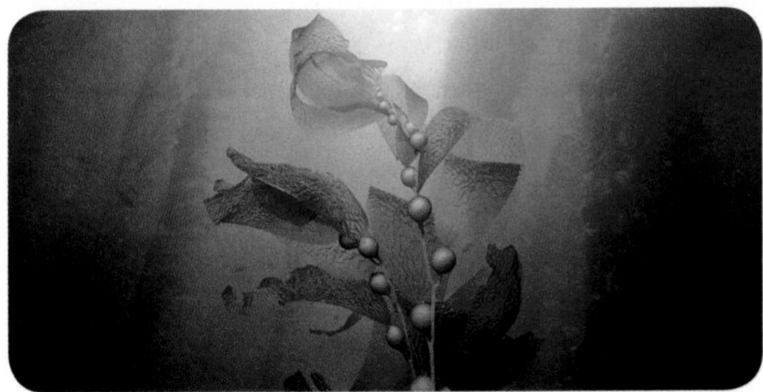

29. In which kingdom would this organism, which has chloroplasts, cell walls, but no organs, be classified?
    A. Plantae **C.** Protista
    B. Animalia **D.** Fungi

30. Which substance would most likely be in the cell walls of an organism with chloroplasts and tissues?
    A. peptidoglycan **C.** hyphae
    B. chitin **D.** cellulose

Constructed Response

31. **Open Ended** Indicate the relationship between domains and kingdoms.

32. **Short Answer** Predict in which domain a taxonomist would place a newly discovered photosynthetic organism that has cells without membrane-bound organelles and no peptidoglycan.

33. **Open Ended** Write an argument for or against including Bacteria and Archaea in the same domain. How would this affect the phylogenetic tree of life?

Think Critically

34. **Analyze** Using the model in **Figure 17.13,** decide which three of the kingdoms in Domain Eukarya evolved from the fourth.

35. **CAREERS IN BIOLOGY** A biologist studied two groups of frogs in the laboratory. The groups looked identical and produced fertile offspring when interbred. However in nature they don't interbreed because their reproductive calls are different and their territories do not overlap. Use your knowledge of species concepts and speciation to decide why they should or should not be placed in the same species.

Additional Assessment

36. **WRITING in Biology** Suppose you found a cricket near your home. After a biologist from a local university studies your find, you learn that the cricket is a new species. Write a paragraph to explain how the biologist might have determined that the cricket is a new species.

DBQ Document-Based Questions

Data obtained from: Blaxter, M. 2001. Sum of the arthropod parts. *Science* 413: 121-122.

Scientists continue to debate about evolutionary relationships among organisms. Groups of arthropods were thought to be related in the way shown on the left, but new molecular evidence suggests that the grouping on the right is more accurate.

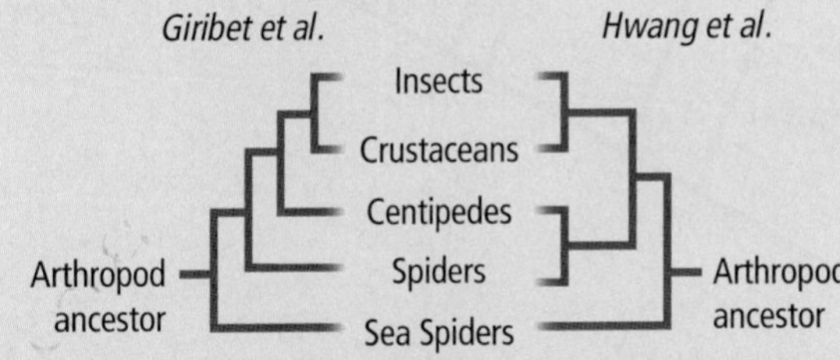

37. Compare and contrast the two cladograms. How did the molecular evidence change the relationship between centipedes and spiders?

38. To which group are crustaceans most closely related?

39. Which group in the cladogram appears to be the most ancestral?

Cumulative Review

40. Describe how hemophilia is inherited. **(Chapter 11)**

41. Choose three lines of evidence that support evolution. Give an example of each. **(Chapter 15)**

Standardized Test Practice

Cumulative

Multiple Choice

1. Which data shows that Neanderthals are not the ancestors of modern humans?
   A. differences in Neanderthal and human DNA
   B. evidence from Neanderthal burial grounds
   C. muscular build of Neanderthals, as compared to humans
   D. patterns of Neanderthal extinction

Use the illustration below to answer questions 2 and 3.

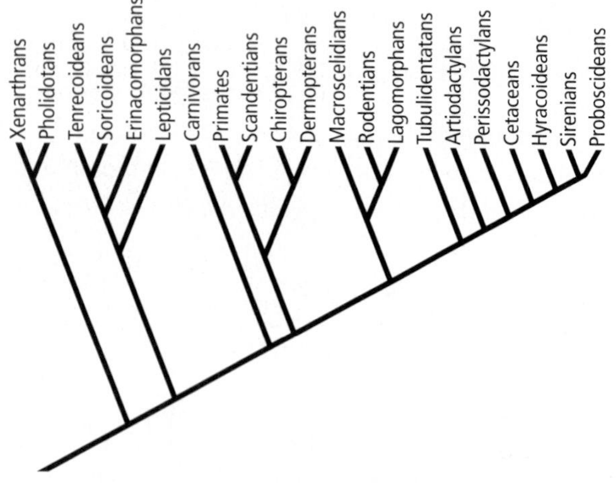

2. According to the cladogram of mammals, which two groups of animals have a more recent common ancestor?
   A. carnivorans and chiropterans
   B. cetaceans and hyracoideans
   C. dermopterans and carnivorans
   D. rodentians and lagomorphans

3. Which mammal is most closely related to bats (chiropterans)?
   A. carnivorans
   B. xenarthrans
   C. primates
   D. rodentians

4. Which radioactive isotope would be used to determine the specific age of a Paleozoic rock formation?
   A. Beryllium-10 (1.5 million years)
   B. Carbon-14 (5715 years)
   C. Thorium-232 (14 billion years)
   D. Uranium-235 (704 million years)

5. According to the Hardy-Weinberg principle, which situation would disrupt genetic equilibrium?
   A. A large population of deer inhabits a forest region.
   B. A particular population of flies mates randomly.
   C. A population of flowering plants always has the same group of natural predators.
   D. A small population of birds colonizes a new island.

Use the diagram below to answer question 6.

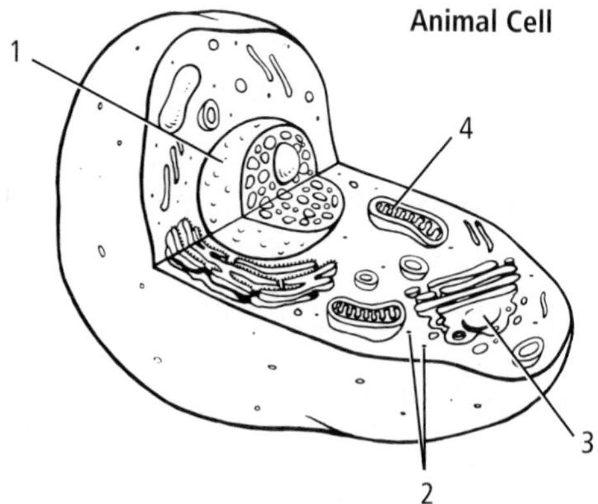

6. Which labeled structure contains the cell's genetic information?
   A. 1
   B. 2
   C. 3
   D. 4

7. Which structure is a vestigial structure?
   A. human appendix
   B. deer horns
   C. multiple cow stomachs
   D. snake tail

8. According to the endosymbiont theory, which part of the eukaryotic cell evolved from a prokaryotic cell?
   A. chloroplast
   B. golgi apparatus
   C. nucleus
   D. ribosome

Short Answer

9. List three primate adaptations found in humans, and explain how each one relates to a tree-dwelling habitat.

10. Assess how molecular clocks are useful in investigating phylogeny in ways that morphological characteristics are not.

11. In terms of their evolution, how are homologous structures and analogous structures different?

12. Assess the advantage of bipedalism.

13. Infer why Aristotle only used two kingdoms to classify living things.

14. Assess the significance of the discovery of the Lucy fossil.

15. Contrast one of the characteristics of living things with the characteristics of nonliving things such as rocks.

Use the figure below to answer question 16.

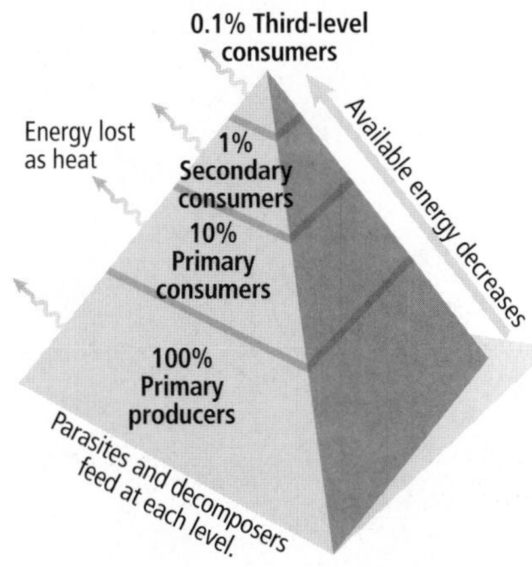

16. How much energy from one trophic level is available to organisms at the next higher trophic level?

Extended Response

17. How could a mutagen cause a change in the protein for which a DNA strand is coding? Trace the effect of a specific mutation through the process of protein synthesis.

18. Assess the value of the binomial system of naming organisms.

19. Name two animals that you would expect to have similar chromosomal characters. Design an experiment to test whether they are similar.

Essay Question

Scientists often use multiple types and sources of data in order to determine when different groups of organisms evolved. Taken together, the data can help construct an evolutionary history.

Using the information in the paragraph above, answer the following question in essay format.

20. What kind of evidence could help scientists determine whether bacteria or archaea evolved earlier on Earth? Write an essay that justifies what specific kinds of data would need to be collected to make this judgment.

NEED EXTRA HELP?																				
If You Missed Question . . .	1	2	3	4	5	6	7	8	9	10	11	12	13	14	15	16	17	18	19	20
Review Section . . .	16.3	17.2	17.2	14.1	15.3	16.2	15.2	14.2	16.1	7.2	15.2	16.2	17.1	16.2	1.1	2.2	12.3	17.1	17.2	17.2, 17.3

UNIT 5

Bacteria, Viruses, Protists, and Fungi

CAREERS IN BIOLOGY

Microbiologist

Microbiologists study the growth and characteristics of microscopic organisms, including bacteria, viruses, protists, and fungi. Environmental microbiologists, like the one shown here, focus their research on biological and chemical pollutants in the environment.

WRITING in Biology Visit biologygmh.com to learn more about microbiologists. Compile a list of microbiology specialties associated with the food, agricultural, and pharmaceutical industries.

Biology Online

To read more about microbiologists in action, visit biologygmh.com.

CHAPTER 18

Bacteria and Viruses

BIG Idea **Bacteria are microscopic organisms, and viruses are nonliving microscopic agents that invade cells.**

Section 1
Bacteria

MAIN Idea Prokaryotes are diverse organisms that live in nearly all environments.

Section 2
Viruses and Prions

MAIN Idea Viruses and prions are smaller and less complex than bacteria; they invade cells and can alter cellular functions.

BioFacts

- One spoonful of soil contains more than 100 million bacteria.
- A human has ten times more bacterial cells living on the body than body cells.
- More than 300 different viruses are known to infect humans.

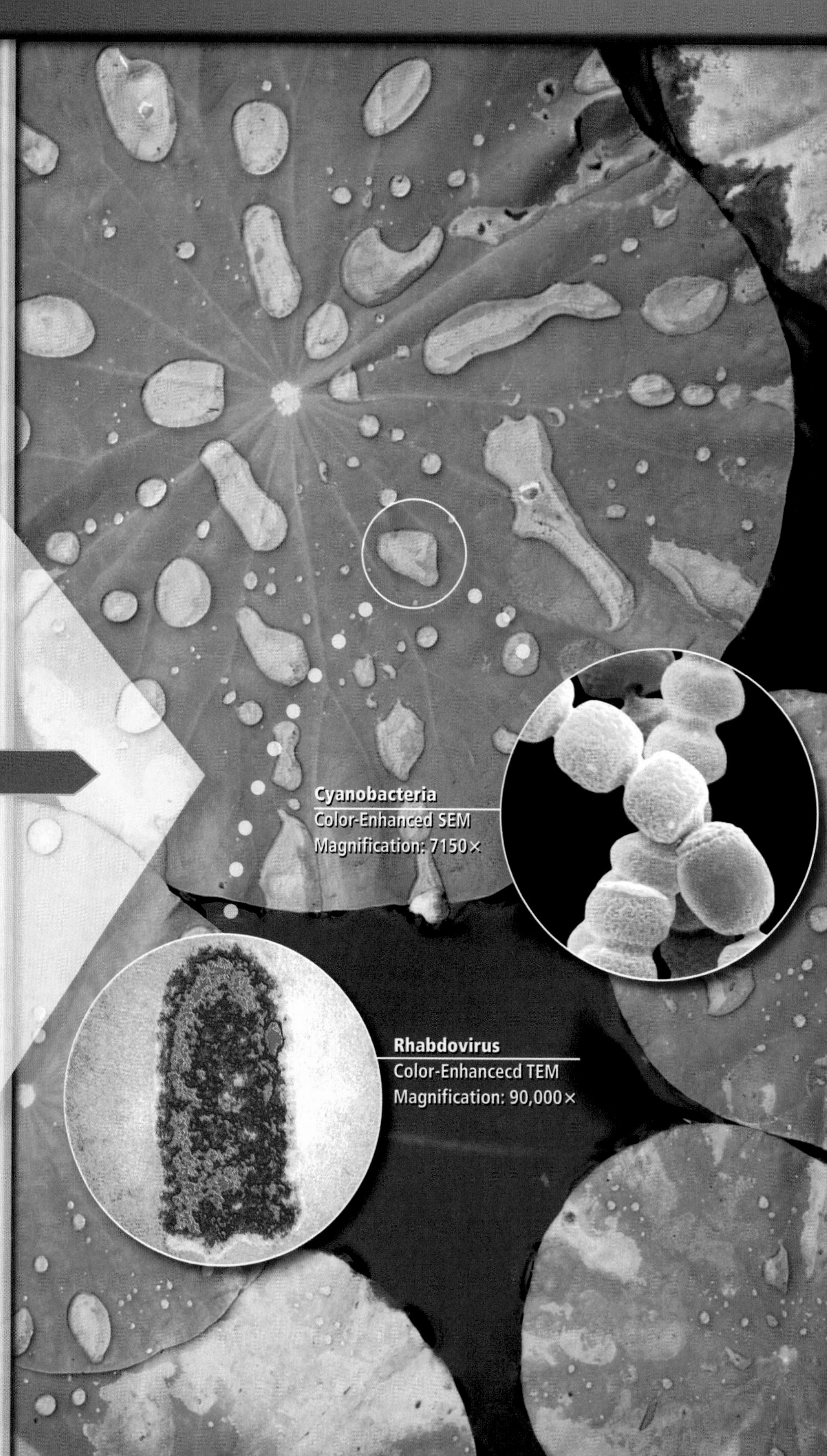

Cyanobacteria
Color-Enhanced SEM
Magnification: 7150×

Rhabdovirus
Color-Enhancecd TEM
Magnification: 90,000×

Start-Up Activities

LAUNCH Lab

What are the differences between animal cells and bacterial cells?

You are already familiar with animal cells. How do animal cells compare to the cells of bacteria? Bacteria are the most common organisms in your environment. In fact, billions of bacteria live on and in your body. Many species of bacteria can cause diseases. What makes bacteria different from your own cells?

Procedure

1. Read and complete the lab safety form.
2. Use a **compound light microscope** to observe the slides of **animal and bacterial cells.**
3. Complete a data table listing the similarities and differences between the two types of cells.

Analysis

1. **Describe** the different cells you observed. What did you notice about each?
2. **Infer** whether they are living things. What leads you to these conclusions?

Visit biologygmh.com to:

- study the entire chapter online
- explore Concepts in Motion, Interactive Tables, Microscopy Links, links to virtual dissections, and the Interactive Time Line
- access Web links for more information, projects, and activities
- review content online with the Interactive Tutor and take Self-Check Quizzes

Viral Replication Make the following Foldable to help you organize the cycles of viral replication.

STEP 1 Fold a sheet of paper in half vertically.

STEP 2 Fold it in half again as shown.

STEP 3 Cut along the middle fold of the top layer only.

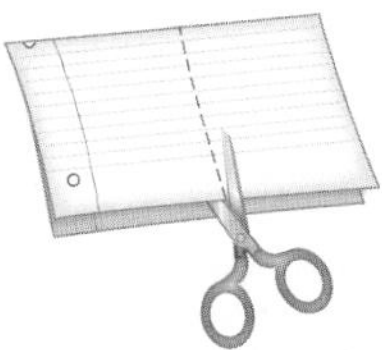

STEP 4 Label the tabs as illustrated.

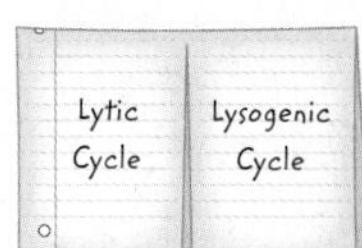

FOLDABLES **Use this Foldable as you study viral infection in Section 18.2.** Draw the stages of the two cycles under the flaps.

Section 18.1

Reading Preview

Objectives

- **Differentiate** among archaea and bacteria and their subcategories.
- **Describe** survival mechanisms of bacteria at both the individual and population levels.
- **Describe** ways that bacteria are beneficial to humans.

Review Vocabulary

prokaryotic cell: cell that does not contain any membrane-bound organelles

New Vocabulary

bacteria
nucleoid
capsule
pilus
binary fission
conjugation
endospore

Bacteria

MAIN Idea Prokaryotes are diverse organisms that live in nearly all environments.

Real-World Reading Link What do yogurt, cheese, and strep throat have in common? You might wonder what food and disease have in common, but they each are the result of microscopic organisms called bacteria.

Diversity of Prokaryotes

Many scientists think that the first organisms on Earth were microscopic, unicellular organisms called prokaryotes. Today, prokaryotes are the most numerous organisms on Earth. They are found everywhere from the depths of the oceans to the highest mountaintops. Some prokaryotes are the only organisms able to survive in hostile environments, such as the water in hot sulfur springs or the Great Salt Lake. The word *prokaryote* is a Greek word that means *before a nucleus.*

All prokaryotes were once classified into one group—Kingdom Monera—based on their lack of a nucleus and membrane-bound organelles. However, modern research has shown that great differences exist among prokaryotes. They are now divided into two domains—Domain Bacteria and Domain Archaea. **Bacteria** (sometimes called eubacteria) are prokaryotic organisms that belong to Domain Bacteria. Bacteria live in nearly every environment on Earth and are important in the human body, industry, and food production. Archaea (previously called archaeabacteria) live in extreme environments and are sometimes called extremophiles. Archaea have been found to have some similarities with eukaryotic cells, such as cytoplasm proteins and histones. **Figure 18.1** shows representatives of these two domains.

Figure 18.1 Archaea are similar to the first life-forms on Earth. The middle photo shows cells of bacteria. The right photo shows cyanobacteria.

Color-Enhanced SEM Magnification: unavailable

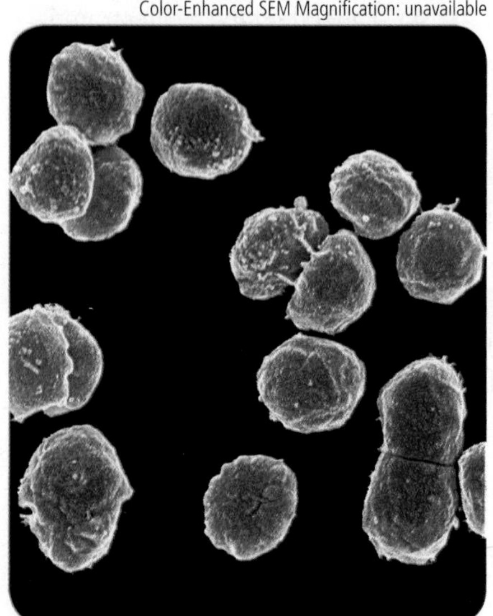

Archaea

Color-Enhanced SEM Magnification: 23,000×

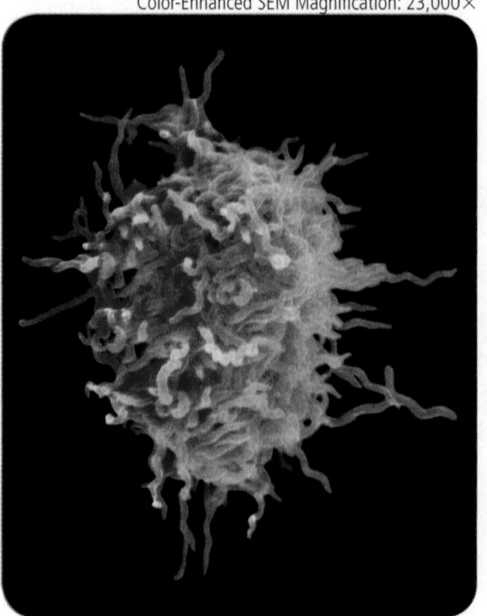

Bacteria

Color-Enhanced SEM Magnification: 260×

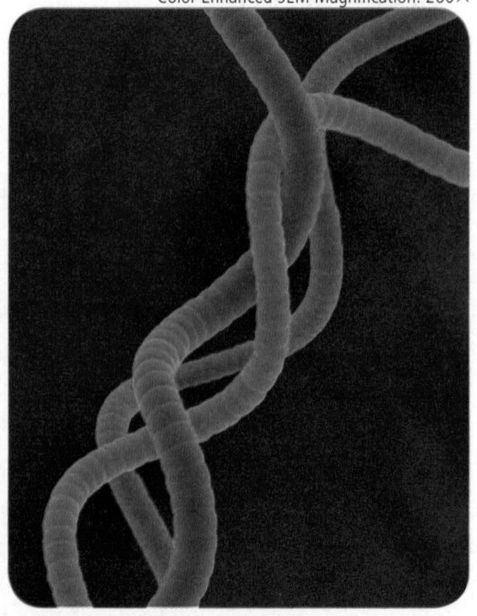

Photosynthetic Bacteria

Hot springs

Great Salt Lake

■ **Figure 18.2** Some members of the Domain Archaea can live in hostile environments, such as the sulfur hot springs in Yellowstone National Park and the Great Salt Lake in Utah.
Hypothesize ***What other hostile places might you find archaebacteria?***

Bacteria Bacteria are the most-studied organisms and are found almost everywhere except in extreme environments where mostly archaea are found. Bacteria have strong cell walls that contain peptidoglycan. Some bacteria have a second cell wall, a property which can be used to classify them. Additionally, some bacteria, such as the cyanobacteria in **Figure 18.1,** are photosynthetic.

Archaea In extreme environments that are hostile to most other forms of life, archaea predominate. Some archaea called thermoacidophiles (thur muh uh SIH duh filz) live in hot, acidic environments including sulfur hot springs shown in **Figure 18.2,** thermal vents on the ocean floor, and around volcanoes. These archaea thrive in temperatures above 80°C and a pH of 1–2. Some of these archaea cannot survive temperatures as low as 55°C. Many are strict anaerobes, which means that they die in the presence of oxygen.

Other archaea called halophiles (HA luh filz) live in very salty environments. The salt concentration in your cells is 0.9 percent, oceans average 3.5 percent salt, and the salt concentrations in the Great Salt Lake, shown in **Figure 18.2,** and the Dead Sea can be greater than 15 percent. Halophiles have several adaptions that allow them to live in salty environments. Halophiles usually are aerobic, and some carry out a unique form of photosynthesis using a protein instead of the pigment chlorophyll.

The methanogens (meh THAHN oh jenz) are the third group of archaea. These organisms are obligate anaerobes, which means they cannot live in the presence of oxygen. They use carbon dioxide during respiration and give off methane as a waste product. Methanogens are found in sewage treatment plants, swamps, bogs, and near volcanic vents. Methanogens even thrive in the gastrointestinal tract of humans and other animals and are responsible for the gases that are released from the lower digestive tract.

Differences between bacteria and archaea Bacteria and archaea have many differences that have led them to be classified in different domains. Recall from Chapter 17 that there are three domains. Based on their classification, we understand that bacteria and archaea are as different from each other as they are from eukaryotic cells. Some differences include: bacterial cell walls contain peptidoglycan, but archaea do not; different lipids in their plasma membranes; and different ribosomal proteins and RNA. The ribosomal proteins in archaea are similar to those of eukaryotic cells.

VOCABULARY
WORD ORIGIN
Halophile
halo- from the Greek word *hals,* meaning *salt.*
-phile from the Greek word *phileo,* meaning *like.*

■ **Figure 18.3** Prokaryotic cells have structures that are necessary for carrying out life processes.
Compare and Contrast ***How does a bacterial cell differ structurally from a eukaryotic cell?***

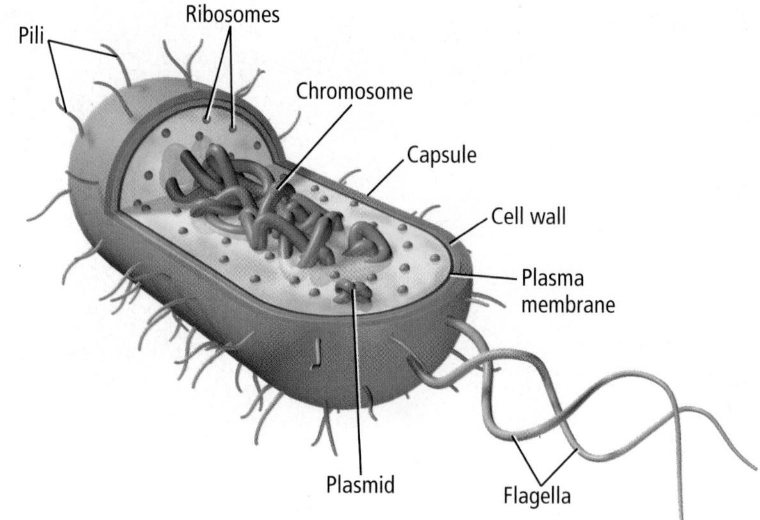

Prokaryote Structure

Prokaryotes are microscopic, unicellular organisms. They have some characteristics of all cells, such as DNA and ribosomes, but they lack a nuclear membrane and other membrane-bound organelles, such as mitochondria and chloroplasts. Although a prokaryotic cell is very small and doesn't have membrane-bound organelles, it has all it needs to carry out life functions. Examine **Figure 18.3** as you read about the structure of prokaryotic cells.

LAUNCH Lab
Review **Based on what you've read about bacterial cells, how would you now answer the analysis questions?**

Chromosomes The chromosomes in prokaryotes are arranged differently than the chromosomes found in eukaryotic cells. Their genes are found on a large, circular chromosome in an area of the cell called the **nucleoid.** Many prokaryotes also have at least one smaller piece of DNA, called a *plasmid,* which also has a circular arrangement.

Capsule Some prokaryotes secrete a layer of polysaccharides around the cell wall, forming a **capsule,** illustrated in **Figure 18.3.** The capsule has several important functions, including preventing the cell from drying out and helping the cell attach to surfaces in its environment. The capsule also helps prevent the bacteria from being engulfed by white blood cells and shelters the cell from the effects of antibiotics.

Pili Structures called pili are found on the outer surface of some bacteria. **Pili** (singular, pilus) are submicroscopic, hairlike structures that are made of protein. Pili help bacterial cells attach to surfaces. Pili also can serve as a bridge between cells. Copies of plasmids can be sent across the bridge, thus providing some prokaryotes with new genetic characteristics. This is one way of transferring the resistance to antibiotics.

■ **Figure 18.4** A size comparison shows how a human cheek cell is much larger than bacteria found in a human mouth.

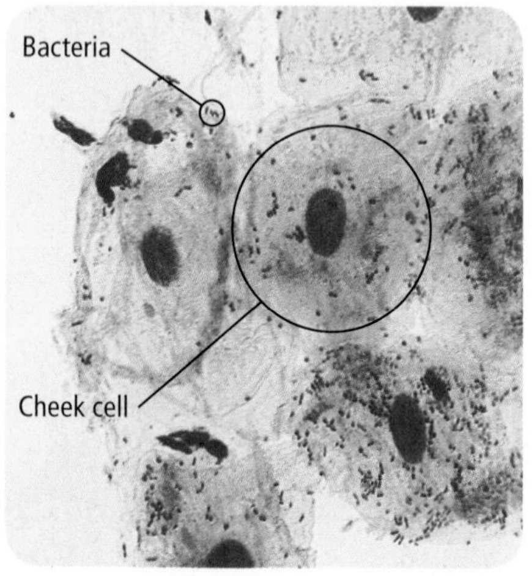

Stained LM Magnification: 400×

Size Even when using a typical light microscope, prokaryotes are small when magnified 400 times. Prokaryotes are typically only 1 to 10 micrometers long and 0.7 to 1.5 micrometers wide. Study **Figure 18.4,** which shows a bacterial cell and a human cell. Notice the relative size of bacterial cells found adjacent to a cheek cell.

Recall from Chapter 9 that small cells have a larger, more favorable surface area-to-volume ratio than large cells. Because prokaryotes are so small, nutrients and other substances the cells need can diffuse to all parts of the cell easily.

Prokaryote Characteristics

As with other types of organisms, prokaryotes now can be identified using molecular techniques. By comparing DNA, evolutionary relationships can be determined. Historically, scientists identified prokaryotes using criteria such as shape, cell wall, and movement.

Shape There are three general shapes of prokaryotes, as shown in **Figure 18.5.** Spherical or round prokaryotes are called cocci (KAHK ki) (singular, coccus), rod-shaped prokaryotes are called bacilli (buh SIH li) (singular, bacillus), and spiral-shaped prokaryotes, or spirilli (spi RIH li) (singular, spirillium), are called spirochetes (SPI ruh keets).

Cell walls Scientists also classify bacteria according to the composition of their cell walls. All bacterial cells have peptidoglycan in their cell walls. Peptidoglycan is made of disaccarides and peptide fragments. Biologists add dyes to the bacteria to identify the two major types of bacteria—those with and those without an outer layer of lipid, in a technique called a Gram stain.

Bacteria with a large amount of peptidoglycan appear dark purple once they are stained, and are called gram positive. Bacteria with the lipid layer have less peptidoglycan and appear a light pink after staining. These bacteria are called gram negative. Because some antibiotics work by attacking the cell wall of bacteria, physicians need to know the type of cell wall that is present in the bacteria they suspect is causing illness in order to prescribe the proper antibiotic.

Movement Although some prokaryotes are stationary, others use flagella for movement. Prokaryotic flagella are made of filaments, unlike the flagella of eukaryotes, which are made of microtubules . Flagella help prokaryotes to move toward light, higher oxygen concentration, or chemicals such as sugar or amino acids that they need to survive. Other prokaryotes move by gliding over a layer of secreted slime.

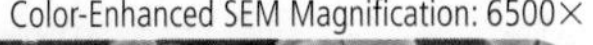
Color-Enhanced SEM Magnification: 6500×

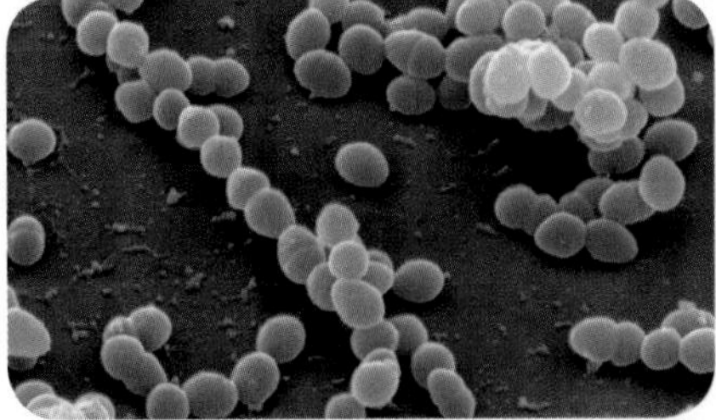

Cocci

Color-Enhanced SEM Magnification: 50,000×

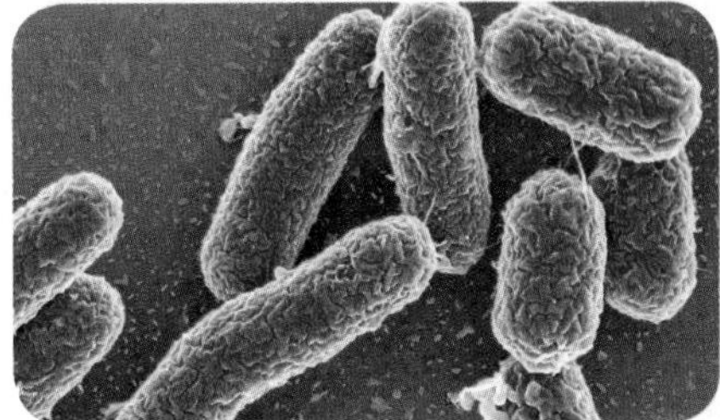

Bacilli

Color-Enhanced SEM Magnification: 2000×

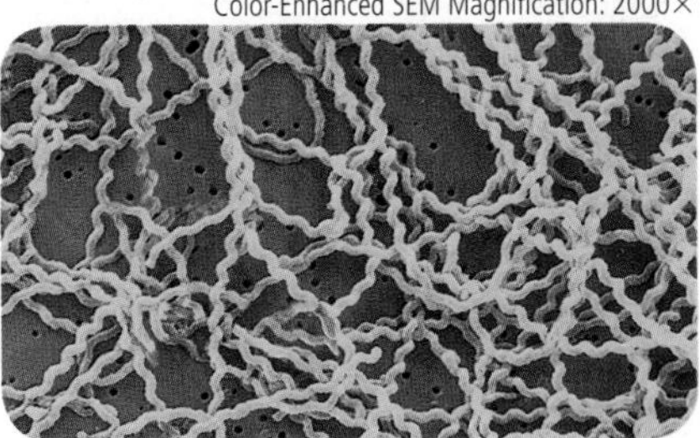

Spirochetes

■ **Figure 18.5** There are three shapes of prokaryotes: cocci, bacilli, and spirochetes.

MiniLab 18.1

Classify Bacteria

What types of characteristics are used to divide bacteria into groups? Bacteria can be stained to show the differences in peptidoglycan (PG) in their cell walls. Based on this difference in their cell walls, bacteria are divided into two main groups.

Procedure

1. Read and complete the lab safety form.
2. Choose four different **slides of bacteria** that have been stained to show cell wall differences. The slides will be labeled with the names of the bacteria and marked either thick PG layer or thin PG layer.
3. Use the oil immersion lens of your **microscope** to observe the four slides.
4. Record all of your observations, including those about the cell color, in a table.

Analysis

1. **Interpret Data** Based on your observations, make a hypothesis about how to differentiate between the two groups of bacteria.
2. **Describe** two different cell shapes you saw on the slides you observed.

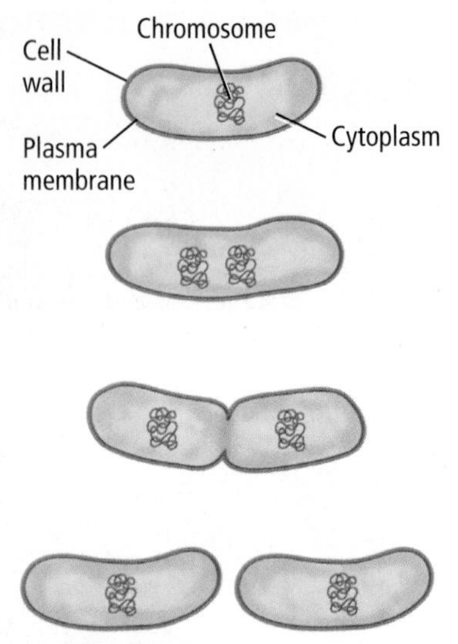

Binary fission

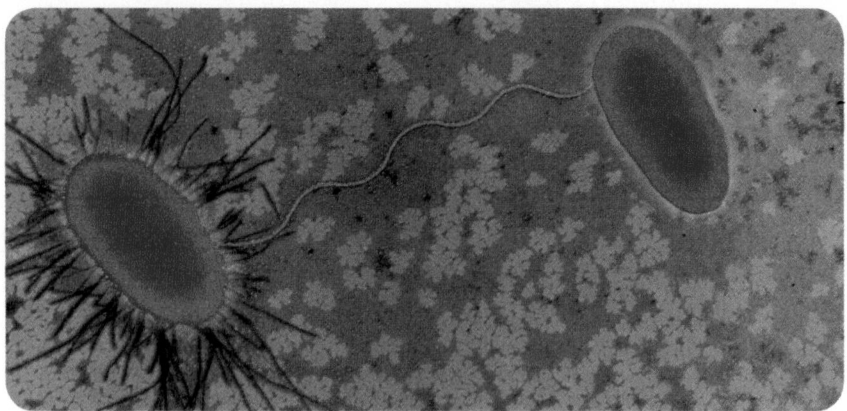

Conjugation

Figure 18.6 Binary fission is an asexual form of reproduction used by some prokaryotes. Conjugation is a method of exchanging genetic material used by some prokaryotes.
Analyze *Which means of reproducing shown here exchanges genetic information?*

Reproduction of Prokaryotes

Most prokaryotes reproduce by an asexual process called binary fission, illustrated in **Figure 18.6. Binary fission** is the division of a cell into two genetically identical cells. In this process, the prokaryotic chromosome replicates, and the original chromosome and the new copy separate. As this occurs, the cell gets larger by elongating. A new piece of plasma membrane and cell wall forms and separates the cell into two identical cells. Under ideal environmental conditions, this can occur quickly—as often as every 20 minutes. If conditions are just right, one bacterium could become one billion bacteria through binary fission in just ten hours.

Some prokaryotes exhibit a form of reproduction called **conjugation,** in which two prokaryotes attach to each other and exchange genetic information. As shown in **Figure 18.6,** the pilus is important for the attachment of the two cells so that there can be a transfer of genetic material from one cell to the other. In this way, new gene combinations are created and diversity of prokaryote populations is increased.

Metabolism of Prokaryotes

Anaerobic prokaryotes do not use oxygen for growth or metabolism. Obligate anaerobes cannot live or grow in the presence of oxygen. They obtain energy through fermentation. Facultative anaerobes can grow either in the presence of oxygen or without it. Obligate aerobes require oxygen to grow. Besides being classified on how they use oxygen, prokaryotes can also be classified by how they obtain energy for cellular respiration or fermentation, as shown in **Figure 18.7.**

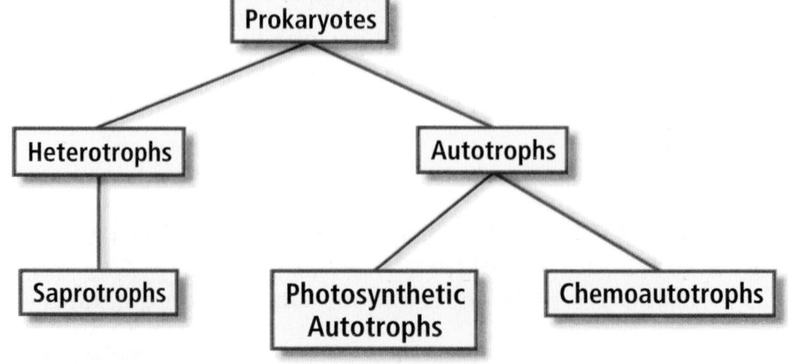

Figure 18.7 Prokaryotes are grouped according to how they obtain nutrients for energy. Heterotrophic bacteria can also be saprotrophs; autotrophs can be photosynthetic or chemoautotrophic.

Heterotrophs Some prokaryotes are heterotrophs, meaning they cannot synthesize their own food and must take in nutrients. Many heterotrophic bacteria are saprotrophs, or saprobes. They obtain their energy by decomposing organic molecules associated with dead organisms or organic waste.

Photoautotrophs Some bacteria are photosynthetic autotrophs (AW tuh trohfs)—they carry out photosynthesis in a similar manner as plants. These bacteria must live in areas where there is light, such as shallow ponds and streams, in order to synthesize organic molecules to use as food.

Scientists once thought that these organisms were eukaryotes and called them blue-green algae. Later, it was discovered that they were prokaryotes and they were renamed cyanobacteria. These bacteria, like plants, are ecologically important because they are at the base of some food chains and release oxygen into the environment. Cyanobacteria are thought to have been the first group of organisms to release oxygen into Earth's early atmosphere, approximately three billion years ago.

Chemoautotrophs A second type of bacteria are autotrophs that do not require light for energy. These organisms are called chemoautotrophs. They break down and release inorganic compounds that contain nitrogen or sulfur, such as ammonia and hydrogen sulfide, in a process called chemosynthesis. Some chemoautotrophs are important ecologically because they keep nitrogen and other inorganic compounds cycling through ecosystems.

Study Tip

Summarization Write a summary paragraph that addresses the diversity of prokaryotes, how they reproduce, and the importance of prokaryotes.

Survival of Bacteria

How can bacteria survive if their environment becomes unfavorable? They have several mechanisms that help them survive such environmental challenges as lack of water, extreme temperature change, and lack of nutrients.

Endospores When environmental conditions are harsh, some types of bacteria produce a structure called an **endospore.** The bacteria that cause anthrax, botulism, and tetanus are examples of endospore producers. An endospore can be thought of as a dormant cell. Endospores are resistant to harsh environments and might be able to survive extreme heat, extreme cold, dehydration, and large amounts of ultraviolet radiation. Any of these conditions would kill a typical bacterial cell.

As illustrated in **Figure 18.8,** when a bacterium is exposed to harsh environments, a spore coat surrounds a copy of the bacterial cell's chromosome and a small part of the cytoplasm. The bacterium itself might die, but the endospore remains. When environmental conditions become favorable again, the endospore grows, or germinates, into a new bacterial cell. Endospores are able to survive for long periods of time. Because a bacterial cell usually only produces one endospore, this is considered a survival mechanism rather than a type of reproduction.

Figure 18.8 Endospores can survive extreme environmental conditions.

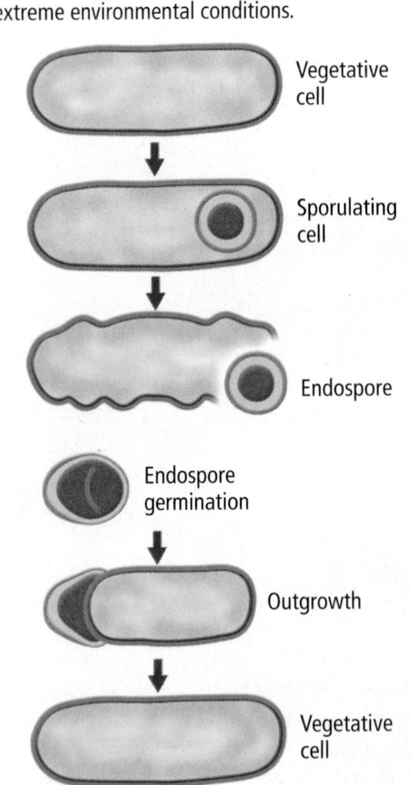

Mutations If the environment changes and bacteria are not well adapted to the new conditions, extinction of the bacteria is a possibility. Because bacteria reproduce quickly and their population grows rapidly, genetic mutations can help bacteria survive in changing environments. Mutations, which are changes or random errors in a DNA sequence, lead to new forms of genes, new gene combinations, new characteristics, and genetic diversity. If the environment happens to change, some bacteria in a population might have the right combination of genes to allow them to survive and reproduce. From the human point of view, this can lead to problems, such as antibiotic-resistant bacteria, as you will learn more about in Chapter 37.

CAREERS IN BIOLOGY

Food Scientist Food scientists help protect the flavor, color, texture, nutritional quality, and safety of our food. They test for amounts of nutrients and presence of harmful organisms such as bacteria. For more information on biology careers, visit biologygmh.com.

Ecology of Bacteria

When many people think of bacteria, they immediately think of germs or disease. Most bacteria do not cause disease, and many are beneficial. In fact, it has been said that humans owe their lives to bacteria because they help fertilize fields, recycle nutrients, protect the body, and produce foods and medicines.

Nutrient cycling and nitrogen fixation In Chapter 2, you learned how nutrients are cycled in an ecosystem. Some organisms get their energy from the cells and tissues of dead organisms and are called decomposers or detrivores. Bacteria are decomposers, returning vital nutrients to the environment. Without nutrient recycling, all raw materials necessary for life would be used up. Without nitrogen fixation, far more fertilizer would be needed for growing plants.

Connection to Chemistry All forms of life require nitrogen. Nitrogen is a key component of amino acids, the building blocks of proteins. Nitrogen also is needed to make DNA and RNA. Most of Earth's nitrogen is found in the atmosphere in the form of nitrogen gas (N_2). Certain types of bacteria can use nitrogen gas directly. These bacteria have enzymes that can convert nitrogen gas into nitrogen compounds by a process called nitrogen fixation. Some of these bacteria live in the soil.

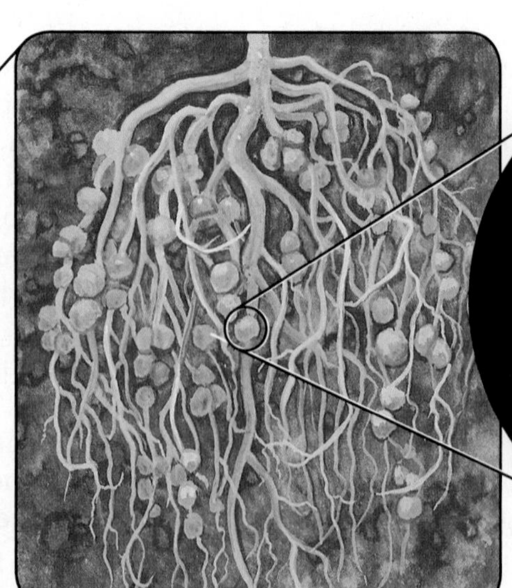

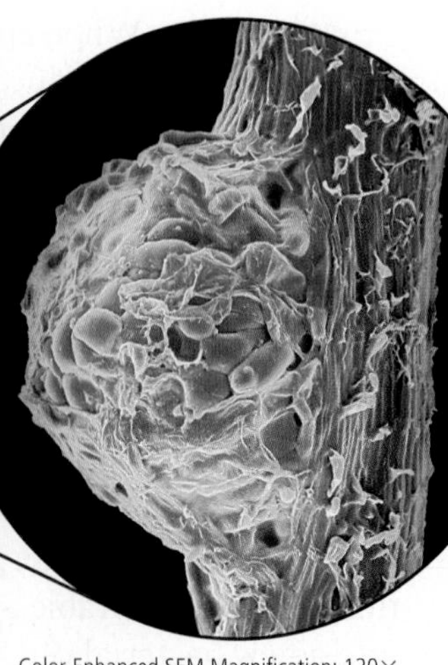

Color-Enhanced SEM Magnification: 120×

Figure 18.9 Nitrogen-fixing bacteria on a plant root nodule are able to remove nitrogen from the air and convert it into a form the plant can use.

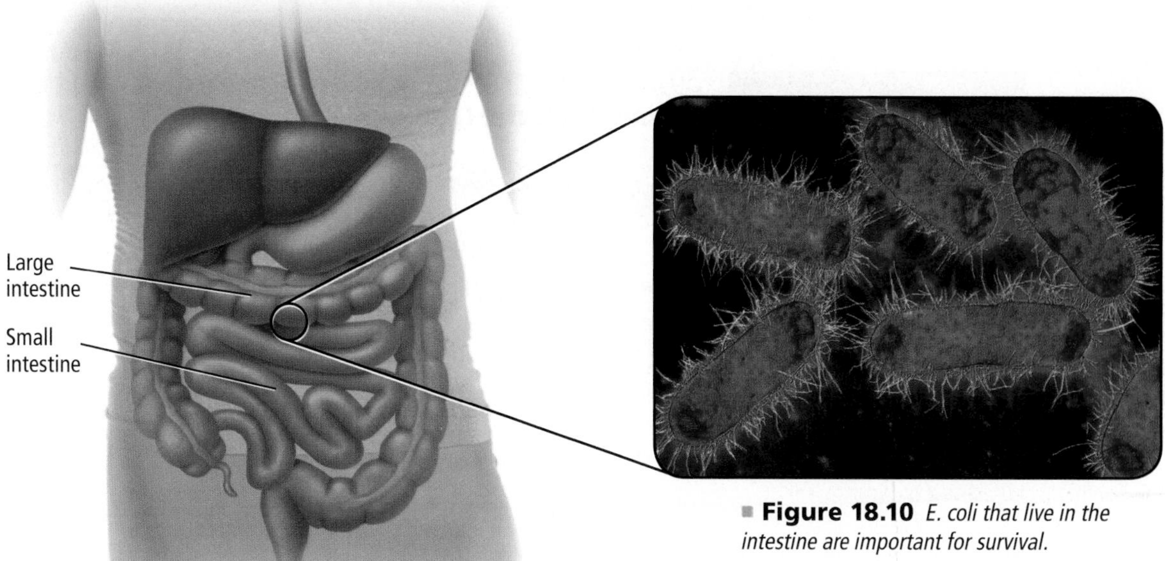

■ **Figure 18.10** *E. coli that live in the intestine are important for survival.*

Some nitrogen-fixing bacteria live in a symbiotic relationship in the root nodules of plants such as soybeans, clover, and alfalfa. The bacteria use the nitrogen in the atmosphere to produce forms of nitrogen the plant can use. The plants then are able to take up ammonia (NH_3) and other forms of nitrogen from the soil. These plants are at the base of a food chain and the nitrogen is passed along to organisms that eat them. **Figure 18.9** shows where nitrogen-fixing bacteria live on root nodules.

Normal flora Your body is covered with bacteria inside and out. Most of the bacteria that live in or on you are harmless. These are called normal flora. Normal flora are of great importance to the body. By living and replicating on the body, they compete with harmful bacteria and prevent them from taking hold and causing disease.

A certain type of bacteria called *Escherichia coli (E. coli)* lives inside your intestines, and is illustrated in **Figure 18.10.** Some *E. coli* strains can cause food poisoning. The type that lives in the digestive tracts of humans and other mammals is harmless and important for survival. The *E. coli* that live in humans make vitamin K, which humans absorb and use in blood clotting. In this symbiotic relationship, *E. coli* are provided with a warm place with food to live. In return, the bacteria provide the body with an essential nutrient.

Foods and medicines Think about what you have eaten in the last few days. Have you had pizza? How about a cheeseburger? Cheese, yogurt, buttermilk, and pickles, as well as other foods, are made with the aid of bacteria.

Bacteria are even used in the production of chocolate. Although bacteria are not found in the chocolate products you eat, bacteria are used to break down the covering of cocoa beans during the production of cocoa. Bacteria also are responsible for commercial production of vitamins, such as vitamin B12 and riboflavin.

Bacteria also are important in the fields of medicine and research. Although some bacteria cause disease, others are useful in fighting disease. Streptomycin, bacitracin, tetracycline, and vancomycin are commonly prescribed antibiotics that were originally made by bacteria.

 **Reading Check** **Describe** ways that bacteria are beneficial.

Concepts In Motion
Interactive Table To explore more about bacterial disease, visit biologygmh.com.

Table 18.1 Human Bacterial Diseases

Category	Disease
Sexually transmitted diseases	Syphilis, gonorrhea, chlamydia
Respiratory diseases	Strep throat, pneumonia, whooping cough, tuberculosis, anthrax
Skin diseases	Acne, boils, infections of wounds or burns
Digestive tract diseases	Gastroenteritis, many types of food poisoning, cholera
Nervous system diseases	Botulism, tetanus, bacterial meningitis
Other diseases	Lyme disease, typhoid fever

Disease-causing bacteria Only a small percentage of bacteria cause disease. Some of the diseases caused by bacteria are listed in **Table 18.1.** The small percentage of bacteria that cause disease do so in two ways. Some bacteria multiply quickly at the site of infection before the body's defense systems can destroy them. In cases of serious infections, bacteria then might spread to other parts of the body.

Other bacteria secrete a toxin or other substance that might cause harm. The bacteria that cause botulism secrete a toxin that paralyzes cells in the nervous system. Bacteria that cause cavities in teeth use sugar in the mouth for energy, and in turn secrete acids that erode the teeth.

Bacteria also can cause disease in plants, and most plants can become infected. Such infections can destroy entire crops and have long-ranging consequences on local ecosystems. For example, citrus canker, a bacterial disease that kills orange trees, has severely impacted the Florida citrus crop and prompted eradication programs.

Section 18.1 Assessment

Section Summary

- Many scientists think that prokaryotes were the first organisms on Earth.
- Prokaryotes belong to two domains.
- Most prokaryotes are beneficial.
- Prokaryotes have a variety of survival mechanisms.
- Some bacteria cause disease.

Understand Main Ideas

1. MAIN Idea **Diagram** a bacterium.
2. **Discuss** possible rationales that taxonomists might have used when deciding to group prokaryotes into two distinct domains instead of in one group.
3. **Explain** survival mechanisms of bacteria, at the individual and population levels.
4. **List** three examples of how bacteria are beneficial to humans.

Think Critically

5. **Analyze** why it is more difficult for biologists to understand the diversity in prokaryotes as compared to plants or animals.

MATH in Biology

6. Imagine that today at 1 P.M. a single *Salmonella* bacterial cell landed on potato salad sitting on your kitchen counter. Assuming your kitchen provides an optimal environment for bacterial growth, how many bacterial cells will be present at 3 P.M. today?

Biology Online Self-Check Quiz biologygmh.com

Section 18.2

Reading Preview

Objectives

- **Illustrate** the general structure of viruses.
- **Compare and contrast** the sequence of events in viral replication by the lytic cycle, the lysogenic cycle, and retroviral replication.
- **Discuss** the structure, replication, and action of prions in relationship to causing disease.

Review Vocabulary

protein: large, complex polymer composed of carbon, hydrogen, oxygen, nitrogen, and sometimes sulfur

New Vocabulary

virus
capsid
lytic cycle
lysogenic cycle
retrovirus
prion

Viruses and Prions

MAIN Idea **Viruses and prions are smaller and less complex than bacteria; they invade cells and can alter cellular functions.**

Real-World Reading Link "It's Cold and Flu Season," "1918 Spanish Flu Epidemic Kills Millions," "New Cases of SARS Reported," "Human Cases of Bird Flu Reported"—headlines tell many stories about diseases that spread worldwide. What do colds, severe acute respiratory syndrome (SARS), and types of flu have in common? They all are caused by viruses.

Viruses

Although some viruses are not harmful, other viruses are known to infect and harm all types of living organisms. A **virus** is a nonliving strand of genetic material within a protein coat. Most biologists don't consider viruses to be living because they do not exhibit all of the characteristics of life. Viruses have no organelles to take in nutrients or use energy, they cannot make proteins, they cannot move, and they cannot replicate on their own. In humans, some diseases, such as those listed in **Table 18.2,** are caused by viruses. Just as there are some bacteria that cause sexually transmitted disease, some viruses can cause sexually transmitted diseases—such as genital herpes and AIDS. These viruses can be spread through sexual contact. Diseases caused by these viruses have no cure or vaccine to prevent them.

Virus size Viruses are some of the smallest disease-causing structures that are known. They are so small that powerful electron microscopes are needed to study them. Most viruses range in size from 5 to 300 nanometers (a nanometer is one billionth of a meter). It would take about 10,000 cold viruses to span the period at the end of this sentence.

Table 18.2 Human Viral Diseases

Concepts In Motion
Interactive Table To explore more about viral diseases, visit biologygmh.com.

Category	Disease
Sexually transmitted diseases	AIDS (HIV), genital herpes
Childhood diseases	Measles, mumps, chicken pox
Respiratory diseases	Common cold, influenza
Skin diseases	Warts, shingles
Digestive tract diseases	Gastroenteritis
Nervous system diseases	Polio, viral meningitis, rabies
Other diseases	Smallpox, hepatitis

■ **Figure 18.11** Viruses have several different types of arrangements, but all viruses have at least two parts: an outer capsid portion made of proteins, and genetic material.

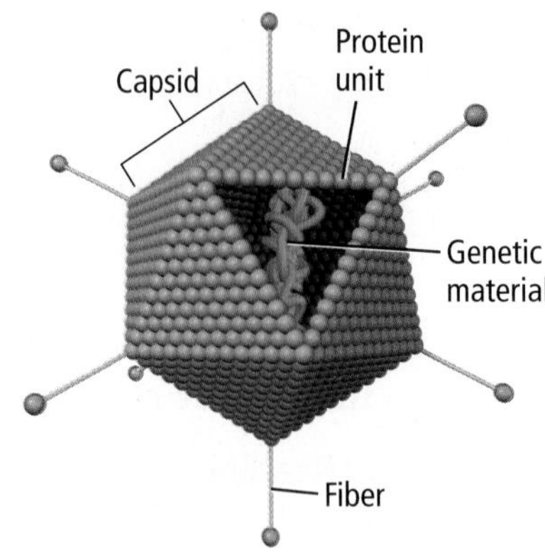

Adenovirus

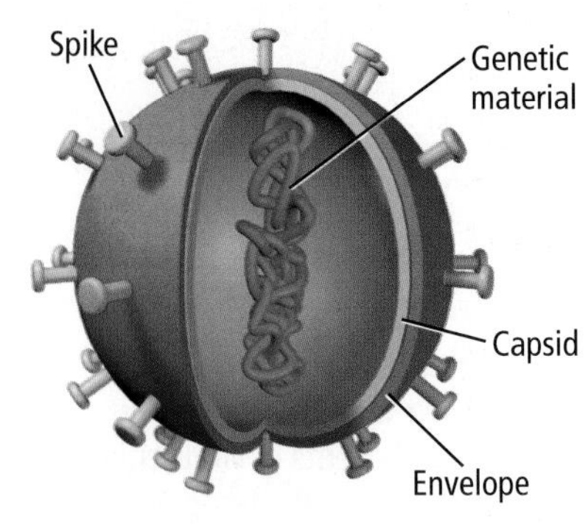

Influenza virus

CAREERS IN BIOLOGY

Virologist Virologists study the natural history of viruses and the diseases they cause. Most virologists spend many hours in the laboratory conducting experiments. For more information on biology careers, visit biologygmh.com.

Virus origin Although the origin of viruses is not known, scientists have several theories about how viruses evolved. One theory, now considered to be most likely, is that viruses came from parts of cells. Scientists have found that the genetic material of viruses is similar to cellular genes. These genes somehow developed the ability to exist outside of the cell.

Virus structure **Figure 18.11** shows the structures of adenovirus, influenza virus, bacteriophage, and tobacco mosaic virus. Adenovirus infection causes the common cold, and influenza virus is responsible for causing the flu. A virus that infects bacteria is called a bacteriophage (bak TIHR ee uh fayj). Tobacco mosaic virus causes disease in tobacco leaves. The outer layer of all viruses is made of proteins and is called a **capsid.** Inside the capsid is the genetic material, which could be DNA or RNA, never both. Viruses generally are classified by the type of nucleic acid they contain.

Reading Check **Sketch** the general structure of a virus.

■ **Figure 18.12**

The History of Smallpox

Though it has been eradicated, smallpox has been an important and deadly disease throughout history.

1157 B.C. Smallpox kills Egyptian Pharaoh Ramses V. Two centuries earlier, Egyptian prisoners caused the first known smallpox epidemic when they were captured by the Hittites in Syria.

243 B.C. A terrible epidemic ravages China. Invading Huns bring smallpox to China where the disease is called "Hun-pox."

0 — 1000 — 1500

1017 A hermit in China introduces mild cases of smallpox into humans to build immunity (variolation).

1519 Hernando Cortes and his crew spread smallpox to Mexico, which ends up decimating the Aztec population.

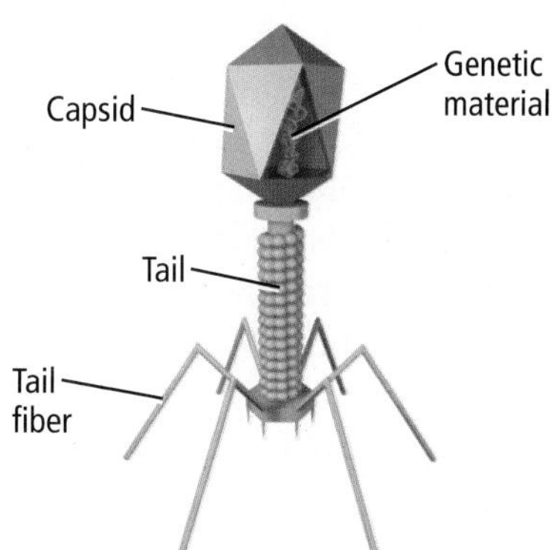

Bacteriophage

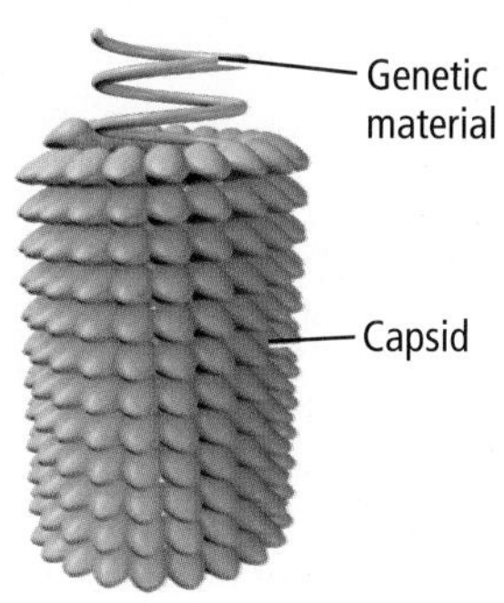

Tobacco mosaic virus

Connection to History The virus that causes smallpox is a DNA virus. Outbreaks of smallpox have occurred in the human population for thousands of years. A successful program of worldwide vaccination eliminated the disease and routine vaccination was stopped. For a closer look at the history of the discovery of the virus that causes smallpox and smallpox vaccination, examine **Figure 18.12.**

Viral Infection

In order to replicate, a virus must enter a host cell. The virus attaches to the host cell using specific receptors on the plasma membrane of the host. Different types of organisms have receptors for different types of viruses, which explains why many viruses cannot be transmitted between different species.

Once the virus successfully attaches to a host cell, the genetic material of the virus enters the cytoplasm of the host. In some cases, the entire virus enters the cell and the capsid is broken down quickly, exposing the genetic material. The virus now uses the host cell to replicate by either the lytic cycle or the lysogenic cycle.

VOCABULARY

ACADEMIC VOCABULARY

Widespread:
Widely diffused or prevalent.
Finding a cure for HIV is of widespread interest in the world.

1796 Edward Jenner develops a smallpox vaccine from cowpox pustules.

1959 World Health Organization adopts a plan to eradicate smallpox. Eight years later, freeze-dried vaccines become available.

1977 The last case of smallpox occurs in Somalia.

1700 1800 1960

1717 Mary Wortley Montagu introduces variolation to England after observing the technique in Turkey.

Concepts In Motion Interactive Time Line
To learn more about these discoveries and others, visit biologygmh.com.
Biology Online

Lytic cycle In the **lytic cycle,** illustrated in **Figure 18.13,** the host cell makes many copies of the viral RNA or DNA. The viral genes instruct the host cell to make more viral protein capsids and enzymes needed for viral replication. The protein coat forms around the nucleic acid of new viruses. These new viruses leave the cell by exocytosis or by causing the cell to burst, or lyse, releasing new viruses that are free to infect other cells. Viruses that replicate by the lytic cycle often produce active infections. Active infections usually are immediate, meaning symptoms of the illness caused by the virus start to appear one to four days after exposure. The common cold and influenza are two examples of widespread viral diseases that are active infections.

FOLDABLES
Incorporate information from this section into your Foldable.

Lysogenic cycle In some cases, the viral DNA might enter the nucleus of the host cell. In the **lysogenic cycle,** also illustrated in **Figure 18.13,** the viral DNA inserts, or integrates into a chromosome in a host cell. Once integrated, the infected cell will have the viral genes permanently. The viral genes might remain dormant for months or years. Then at some future time, the viral genes might be activated by many different factors. Activation results in the lytic cycle. The viral genes instruct the host cell to manufacture more viruses. The new viruses will leave the cell by exocytosis or by causing the cell to lyse.

Many disease-causing viruses have lysogenic cycles. Herpes simplex I is an example of a virus that causes a latent infection. This virus is transmitted orally, and a symptom of this infection is cold sores. When the viral DNA enters the nucleus, it is inactive. It is thought that during times of stress, whether physical, emotional, or environmental, the herpes genes become activated and the production of viruses occurs.

DATA ANALYSIS LAB 18.1

Based on Real Data*

Model Viral Infection

Is protein or DNA the genetic material?
In 1952, Alfred Hershey and Martha Chase designed experiments to find out whether protein or DNA provides genetic information. Hershey and Chase labeled the DNA of bacteriophages—viruses that infect bacteria—with a phosphorus isotope and the protein in the capsid with a sulfur isotope. The bacteriophages were allowed to infect the bacteria *E. coli.*

Data and Observations

- At least 80 percent of the sulfur-containing proteins stayed on the surface of the host cell.
- Most of the viral DNA entered the host cell on infection.
- After replication inside the host cell, 30 percent or more of the copies of the virus contained radioactive phosphorus.

Think Critically

1. **Analyze and Conclude** Do the results of these experiments support the idea that proteins are the genetic material or DNA is the genetic material? Explain.
2. **Infer** If proteins and DNA had entered the cell, would this data be useful to answer Hershey and Chase's question?

*Data obtained from: Hershey, A.D. and Chase, M. 1952. Independent functions of viral protein and nucleic acid in growth of bacteriophage. *Journal of General Physiology* 36: 39–56.

Visualizing Viral Replication

Figure 18.13
In the lytic cycle, the entire replication process occurs in the cytoplasm. The viruses' genetic material enters the cell; the cell replicates the viral RNA or DNA. The viral genes instruct the host cell to manufacture capsids and assemble new viral particles. The new viruses then leave the cells.

In the lysogenic cycle, the viral DNA inserts into a chromosome of the host cell. Many times the genes are not activated until later. Then the viral DNA instructs the host cell to make more viruses.

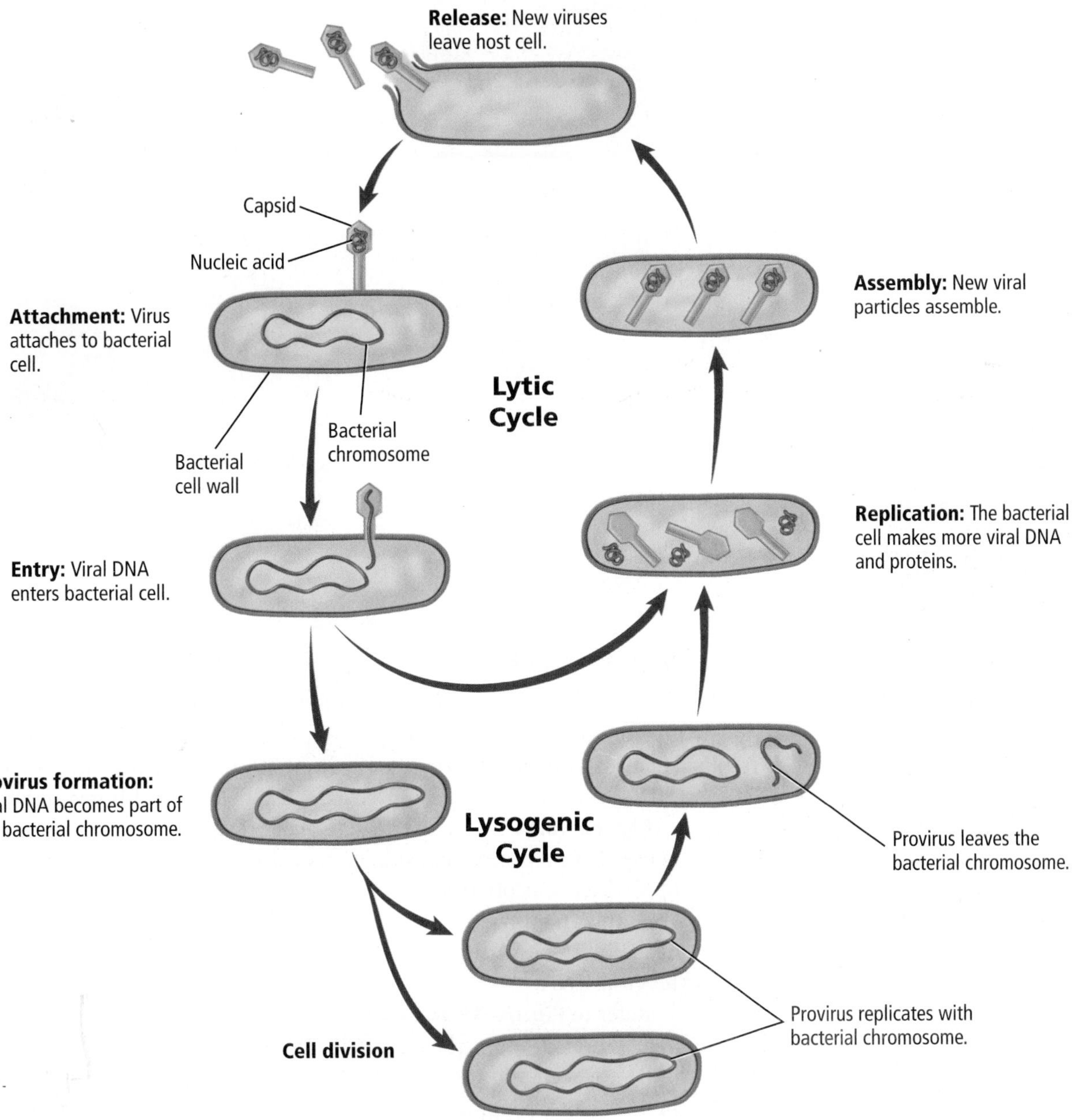

Release: New viruses leave host cell.

Capsid

Nucleic acid

Attachment: Virus attaches to bacterial cell.

Assembly: New viral particles assemble.

Lytic Cycle

Bacterial chromosome

Bacterial cell wall

Entry: Viral DNA enters bacterial cell.

Replication: The bacterial cell makes more viral DNA and proteins.

Provirus formation: Viral DNA becomes part of the bacterial chromosome.

Lysogenic Cycle

Provirus leaves the bacterial chromosome.

Provirus replicates with bacterial chromosome.

Cell division

Concepts In Motion **Interactive Figure** To see an animation of viral replication, visit biologygmh.com.

Biology Online

Concepts In Motion

Interactive Figure To see an animation of how retroviruses replicate, visit biologygmh.com.

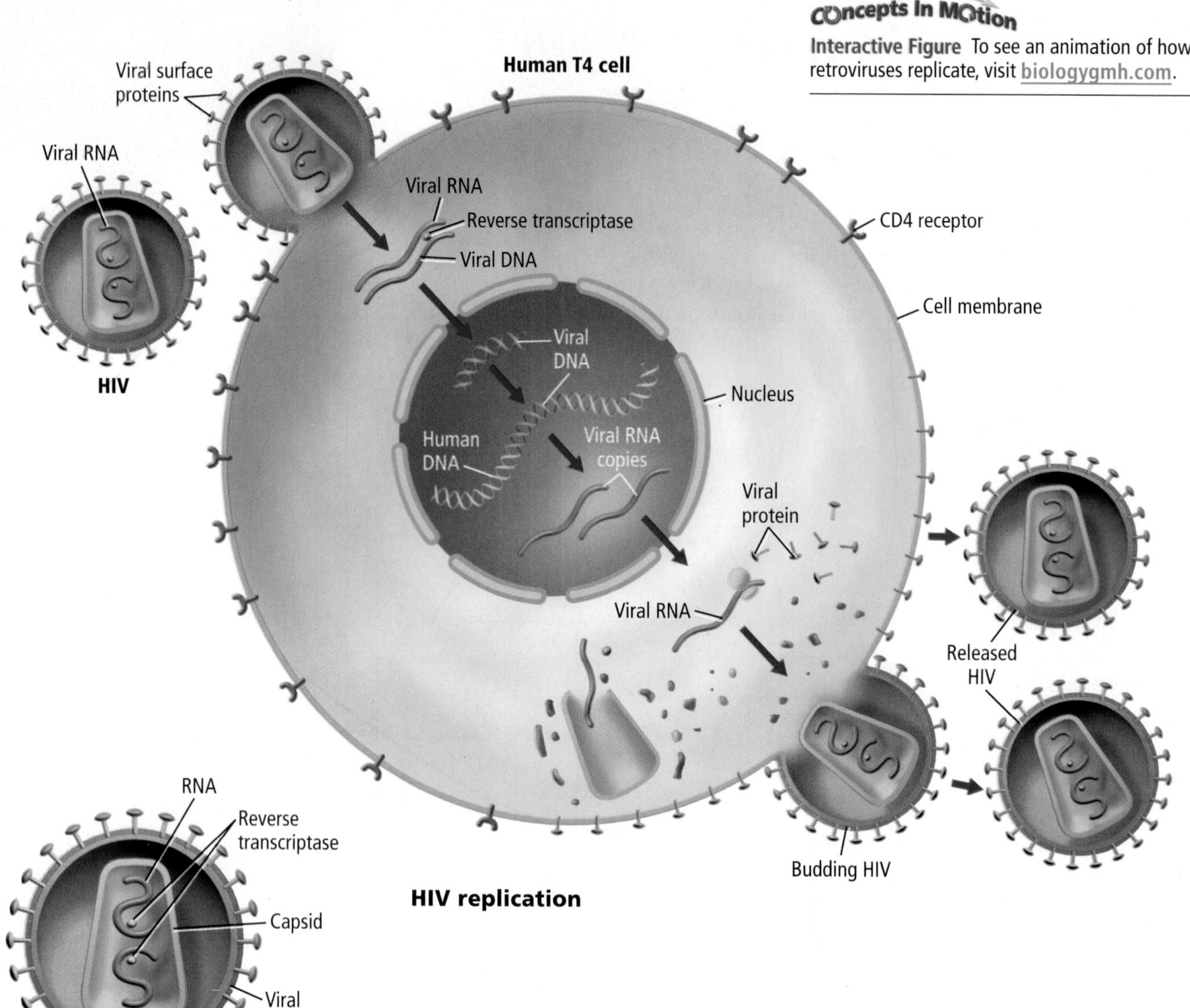

■ **Figure 18.14** The genetic material and replication cycle of a retrovirus, such as HIV, is different from that of DNA viruses.

Infer ***What is unique about the function of reverse transcriptase?***

Retroviruses

Some viruses have RNA instead of DNA for their genetic material. This type of virus is called a **retrovirus** and has a complex replication cycle. The best-known retrovirus is the human immunodeficiency virus (HIV). Some cancer-causing viruses also belong to this group.

Figure 18.14 shows the structure of HIV. Like all viruses, retroviruses have a protein capsid. Surrounding the capsid is a lipid envelope, which was obtained from the plasma membrane of a host cell. RNA and an enzyme called reverse transcriptase are in the core of the virus. Reverse transcriptase is the enzyme that transcribes DNA from the viral RNA.

Refer to **Figure 18.14** as you learn about the replication cycle of HIV. When HIV attaches to a cell, the virus moves into the cytoplasm of the host cell and the viral RNA is released. Reverse transcriptase synthesizes DNA using the viral RNA as a template. Then, the DNA moves into the nucleus of the host cell and integrates into a chromosome. The viral DNA might lie inactive for a period of years before it is activated. Once it is activated, RNA is transcribed from the viral DNA, and the host cell manufactures and assembles new HIV particles.

Prions

A protein that can cause infection or disease is called a proteinaceous (pro te NAY shuhs) infectious particle, or a **prion** (PREE ahn). Although diseases now believed to be caused by prions have been studied for decades, they were not well understood until 1982, when Stanley B. Prusiner first identified that the infectious particle was a protein.

Prions normally exist in cells, although their function is not well understood. Normal prions are shaped like a coil. Mutations in the genes that code for these proteins occur, causing the proteins to be misfolded. Mutated prions are shaped like a piece of paper folded many times. Mutated prions are associated with diseases known as transmissible spongiform encephalopathies (SPUN gee form • in SEH fuh la pah thees) (TSE). Examples of diseases caused by prions include mad cow disease in cattle, Creutzfeldt-Jakob disease (CJD) in humans, scrapie (SKRAY pee) in sheep, and chronic wasting disease in deer and elk.

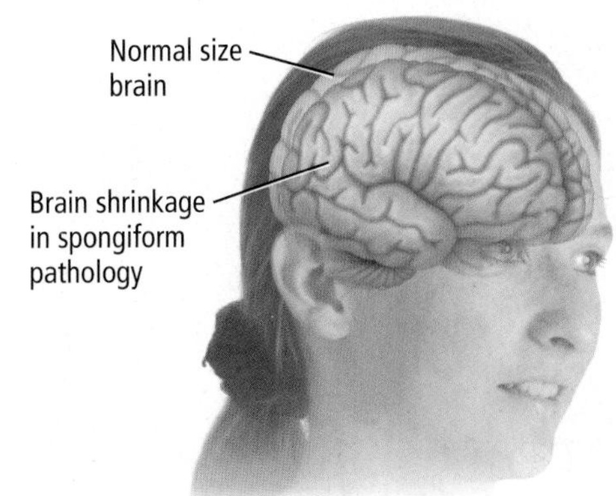

■ **Figure 18.15** A normal brain compared with the brain of a patient with Creutzfeldt-Jakob disease is pictured here.

Prion infection **Figure 18.15** shows a normal brain compared with a brain infected with prions. What scientists find fascinating about these misfolded proteins is that these prions can cause normal proteins to mutate. These prions infect nerve cells in the brain, causing them to burst. This results in spaces in the brain, hence the description of spongiform (spongelike) encephalopathy (brain disease).

In the mid-1980s, a new variant of CJD, or nvCJD, was discovered in England. Scientists do not fully agree on the origin of nvCJD, but a leading hypothesis is that the prions are transmitted from cattle. Abnormal prions can be found in the brains and spinal cords of cattle. The hypothesis is that if the spinal cord is cut in the butchering process, the prions might contaminate the beef and then be transmitted to humans that eat the beef. Although this mode of transmission is not agreed upon, the United States government has strict regulations concerning the importation of cattle and beef from other countries.

Section 18.2 Assessment

Section Summary

- Viruses have a nucleic acid core and a protein-containing coat.
- Viruses are classified by their genetic material.
- Viruses have three different patterns of replication.
- Many viruses cause disease.
- Proteins called prions also might cause disease.

Understand Main Ideas

1. MAIN Idea **Describe** how viruses and prions can alter cell functions.
2. **Compare and contrast** similarities and differences in the replication of a herpes simplex virus with a human immunodeficiency virus.
3. **Draw** a diagram of a virus and label the parts.
4. **Sequence** the steps in the process of how prions might be transmitted from cattle to humans.

Think Critically

5. **Propose** ideas for the development of drugs that could stop viral replication cycles.

WRITING in Biology

6. Write a paragraph explaining why it is difficult to make drugs or vaccines against HIV, given the fact that each time reverse transcriptase works, it makes a slight miscopy.

CUTTING-EDGE BIOLOGY

Innovations in the Fight Against Viral Infections

TEM Magnification: 100,000×

Drugs are being developed to help fight infection by viruses, like this herpes virus.

Perhaps stress or lack of sleep is wearing down your immune system. You start to feel feverish, an indication that your immune system is responding to a new threat. You have contracted a viral infection.

Viruses can cause illnesses that range in severity from mild to life-threatening. Viruses are not alive; they must hijack a host's cells in order to replicate and spread. Because they use a host's cells, inhibiting the replication of the virus to prevent illness can damage the host as well as the virus. In addition, some viruses mutate easily, making previous treatments ineffective. But now there is a new player on the field, one that might make the development of antiviral drugs as direct as following a recipe.

Bioinformatics As the genomes of viruses are decoded, researchers identify proteins that can be targeted and destroyed with the help of bioinformatics—a branch of science that melds biology and computer science to organize and analyze large amounts of scientific data. Researchers enter a virus genome sequence into the database, which then sorts through the many drugs already in existence to find the best one to fight the invader. If there are not any drugs that will help defeat this particular strain of virus, scientists can design a candidate for development and testing using a computer program.

Antiviral Arsenal One promising drug prevents the contact of two proteins necessary for the replication of the viral invader herpes simplex virus (HSV). A molecule called BP5 slips into the binding site of these two proteins. Without this connection, HSV is unable to replicate its DNA. Because it cannot replicate, it cannot spread and the host will be spared from further infection.

Another relatively new drug is oseltamivir (oh sel TAM ih veer), used extensively in Asia in 2005 to treat the large number of H5N1 bird flu cases there. H5N1 bird flu is a subset of the *Influenza A* virus. Oseltamivir blocks a specific protein which prevents the H5N1 virus from budding and breaking off from the host cell.

Current research is giving us a better understanding of viruses and the drugs that may subdue them. While prevention through vaccination remains the prevailing method for heading off viral outbreaks, new drugs like oseltamivir and an ever-increasing bioinformatics database are offering new options in the fight against viruses. Perhaps in the future, treating viruses with medication will be as common as treating bacterial infections with antibiotics.

WRITING in Biology

Pamphlet The HIV/AIDS epidemic has reached epic proportions worldwide. Research the life cycle of HIV/AIDS and create a pamphlet detailing how it is spread, its life cycle, and the treatment options available. For more information about HIV/AIDS, visit biologygmh.com.

BIOLAB

Design Your Own

HOW CAN THE MOST EFFECTIVE ANTIBIOTICS BE DETERMINED?

Background: A patient is suffering from a serious bacterial infection, and as the doctor you must choose from several new antibiotics to treat the infection.

Question: *How can the effectiveness of antibiotics be tested?*

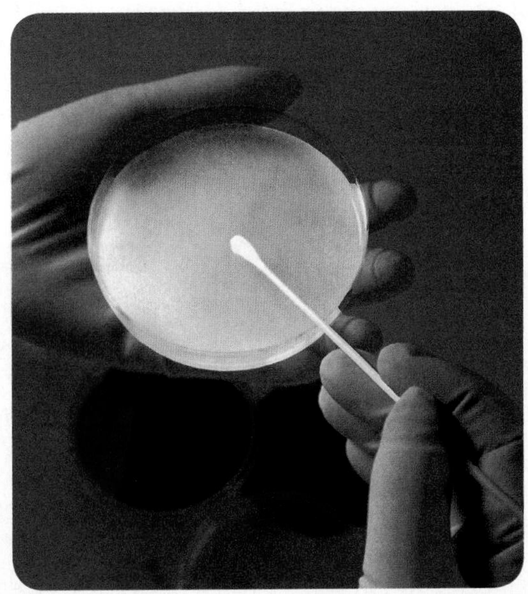

Materials

bacteria cultures
sterile nutrient agar
petri dishes
antibiotic disks
control disks
forceps
Bunsen burner
marking pen
long-handle cotton swabs
70% ethanol
thermometer
container
disinfectant
autoclave disposal bag

Safety Precautions

WARNING: ***Clean your work area with disinfectant after you finish.***

Plan and Perform the Experiment

1. Read and complete the lab safety form.
2. Design an experiment to test the effectiveness of different antibiotics. Identify the controls and variables in your experiment.
3. Create a data table for recording your observations and measurements.
4. Make sure your teacher approves your plan before you proceed.
5. Conduct your experiment.
6. **Cleanup and Disposal** Dispose of all materials according to your teacher's instructions. Disinfect your area.

Analyze and Conclude

1. **Compare and contrast** What are the effects of the different antibiotics for the bacteria species you tested?
2. **Hypothesize** Why would a doctor instruct you to take all of your prescribed antibiotics for a bacterial infection even if you start feeling better before the pills run out?
3. **Explain** What were the limitations of your experimental design?
4. **Error Analysis** Compare and contrast the observations and measurements collected by your group with the data from the experiments designed by other groups. Identify possible sources of error in your experimental data.

COMMUNITY INVOLVEMENT

Create a Poster Misuse of antibiotic prescriptions and use of antibacterial household items are contributing to antibiotic-resistant bacteria. Research the causes of bacterial resistance to drugs and the steps people in your community can take to help solve this problem. Create a poster display to educate the people in your community about this issue. To learn more about antibiotic-resistant bacteria, visit BioLabs at biologygmh.com.

Study Guide

Download quizzes, key terms, and flash cards from biologygmh.com.

FOLDABLES **Point out** the differences between viruses and prions. Research what is known about normal and mutated prions. Use current knowledge to help you develop a program to prevent the spread of any transmissible spongiform encephalopathy, such as chronic wasting disease in elk and deer.

Vocabulary	Key Concepts
Section 18.1 Bacteria	
• bacteria (p. 516) • binary fission (p. 520) • capsule (p. 518) • conjugation (p. 520) • endospore (p. 521) • nucleoid (p. 518) • pilus (p. 518)	**MAIN Idea** **Prokaryotes are diverse organisms that live in nearly all environments.** • Many scientists think that prokaryotes were the first organisms on Earth. • Prokaryotes belong to two domains. • Most prokaryotes are beneficial. • Prokaryotes have a variety of survival mechanisms. • Some bacteria cause disease.
Section 18.2 Viruses and Prions	
• capsid (p. 526) • lysogenic cycle (p. 528) • lytic cycle (p. 528) • prion (p. 531) • retrovirus (p. 530) • virus (p. 525)	**MAIN Idea** **Viruses and prions are smaller and less complex than bacteria; they invade cells and can alter cellular functions.** • Viruses have a nucleic acid core and a protein-containing coat. • Viruses are classified by their genetic material. • Viruses have three different patterns of replication. • Many viruses cause disease. • Proteins called prions also might cause disease.

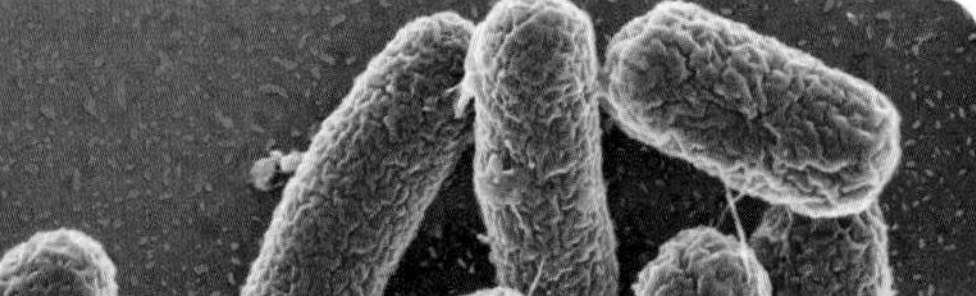

Color-Enhanced SEM Magnification: 50,000×

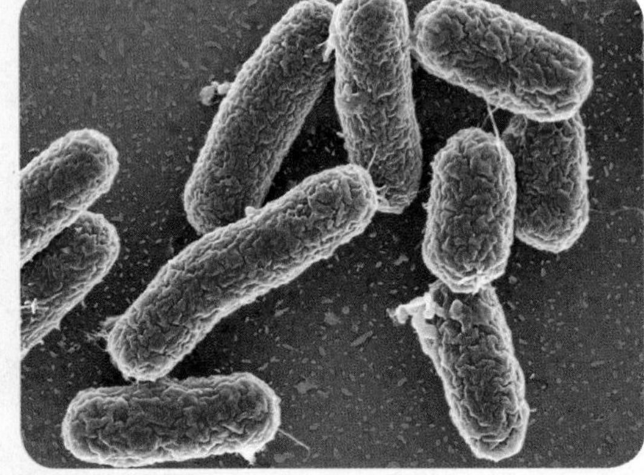

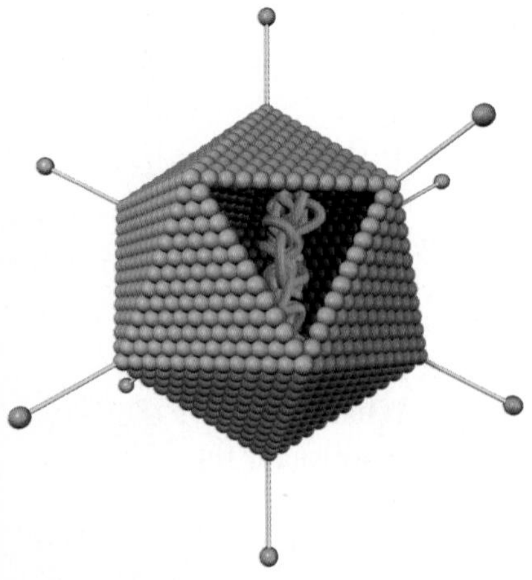

Biology Online Vocabulary PuzzleMaker biologygmh.com

CHAPTER 18

Assessment

Section 18.1

Vocabulary Review

For each set of terms below, choose the one that does not belong and explain why it does not belong.

1. capsule—pilus—endospore
2. binary fission—nitrogen fixation—conjugation
3. endospore—nucleoid—nitrogen fixation

Understand Key Concepts

4. Which organism is not included in Domain Archaea?
   A. cyanobacteria
   B. methanogens
   C. halophiles
   D. thermoacidophiles

5. Why is an electron microscope useful when studying bacteria?
   A. Electrons can penetrate through the capsule surrounding bacteria.
   B. Bacteria are tiny.
   C. Bacteria move quickly; the electrons stun the bacteria.
   D. Bacteria organelles are small and tightly packed together.

Use the figure below to answer questions 6 and 7.

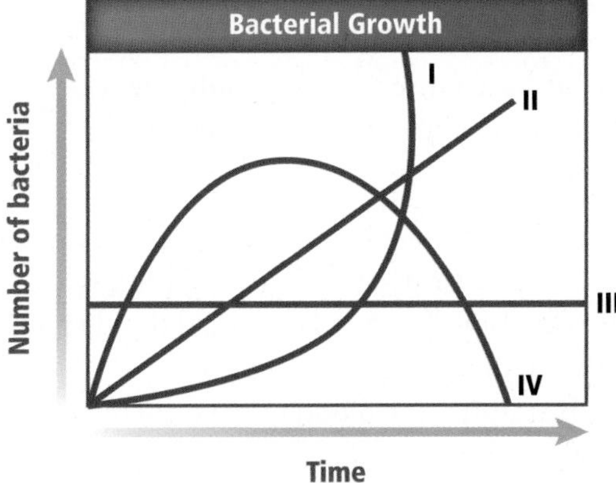

6. Which line on the graph best indicates the growth rate of a population of bacteria living in ideal conditions?
   A. line I **C.** line III
   B. line II **D.** line IV

7. Which line on the graph best indicates the growth rate of a population of bacteria exposed to an effective antibiotic?
   A. line I **C.** line III
   B. line II **D.** line IV

8. You have just been named a contestant on a reality show. Your first challenge is to swallow microbes. Which would be the most dangerous to swallow?
   A. thermoacidophiles
   B. halophiles
   C. *Escherichia coli*
   D. a bacteriophage

Use the photos below to answer question 9.

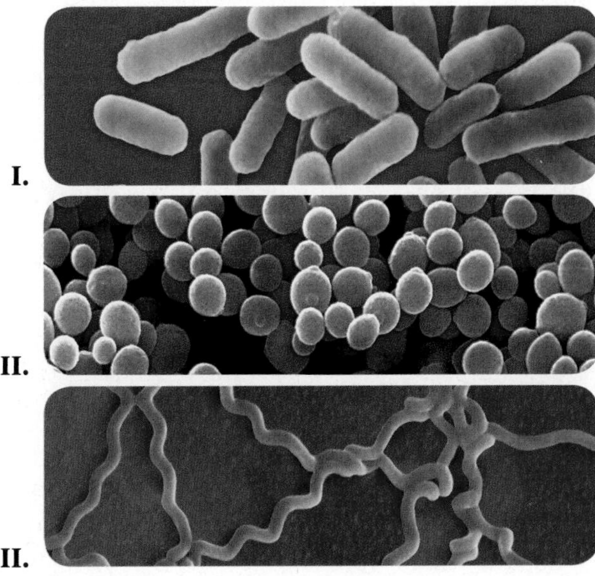

9. Which is the correct identification for the bacteria shown above?
   A. I—cocci, II—bacilli, III—spirochetes
   B. I—bacilli, II—cocci, III—spirochetes
   C. I—spirochetes, II—cocci, III—bacilli
   D. I—bacilli, II—spirochetes, III—cocci

10. What is the likely cause of tooth decay?
   A. a lysogenic virus infecting the living cells of the tooth
   B. bacteria feeding on the sugar in the mouth and producing acid
   C. an excess of vitamin K production by mouth bacteria
   D. nitrogen-fixing bacteria releasing ammonia that is eroding the tooth enamel

Constructed Response

11. **Open Ended** Make an argument for or against the following statement: Living organisms on Earth owe their lives to bacteria.

12. **Short Answer** Describe characteristics of bacteria (both at the individual and population level) that make them tough to destroy.

13. **Open Ended** What types of arguments do you think biologists use when they say prokaryotes were the first organisms on Earth?

Think Critically

14. **Speculate** what life on Earth might be like if cyanobacteria had never evolved.

15. **Predict** any ecological consequences that would result if all types of nitrogen-fixing bacteria suddenly went extinct.

16. **Describe** some of the diverse characteristics of prokaryotes.

Section 18.2

Vocabulary Review

Use what you know about the vocabulary terms on the Study Guide page to describe what the terms in each pair below have in common.

17. lytic cycle—lysogenic cycle

18. prion—virus

19. capsid—prion

20. virus—retrovirus

Understand Key Concepts

21. Viruses contain which substances?
    - **A.** genetic material and a capsid
    - **B.** a nucleus, genetic material, and a capsid
    - **C.** a nucleus, genetic material, a capsid, and ribosomes
    - **D.** a nucleus, genetic material, a capsid, ribosomes, and a plasma membrane

Use the figure below to answer questions 22 and 23.

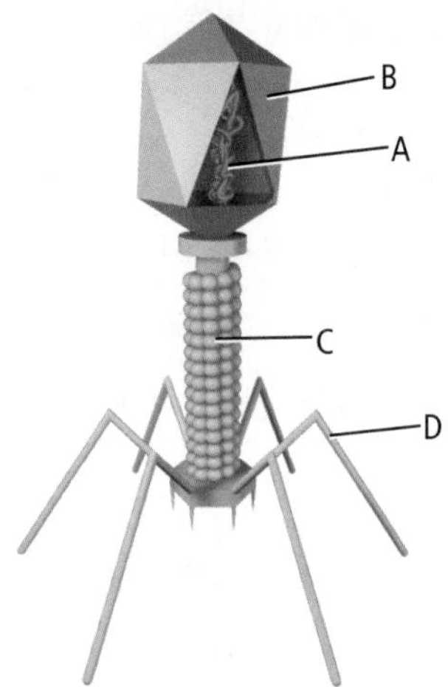

22. Which labeled structure represents the genetic material of a virus?
    - **A.** A
    - **B.** B
    - **C.** C
    - **D.** D

23. Which structure represents the capsid of a virus?
    - **A.** A
    - **B.** B
    - **C.** C
    - **D.** D

24. HIV is a retrovirus. What does this mean?
    - **A.** Viral RNA is used to make DNA.
    - **B.** Viral DNA is used to make RNA.
    - **C.** Protein is made directly from viral RNA.
    - **D.** Protein is made directly from viral DNA.

25. Which statement about prions is true?
    - **A.** Prions are renegade pieces of RNA that infect cells.
    - **B.** Prions are infectious proteins.
    - **C.** Prion-based diseases affect only cows.
    - **D.** Prions are a newly discovered type of genetic material.

26. Imagine that a patient in a hospital has died mysteriously. A doctor suspects the cause of death is Creutzfeldt-Jacob disease. How could this diagnosis be confirmed?
    - **A.** by examining the blood to see if there is a high viral count
    - **B.** by asking the patient's family and friends if the patient consumed a lot of meat
    - **C.** by examining the brain to see if there are a lot of spaces in the tissue
    - **D.** by examining nerve cells to see if they have been affected by a bacterial neurotoxin

Use the figure below to answer question 27.

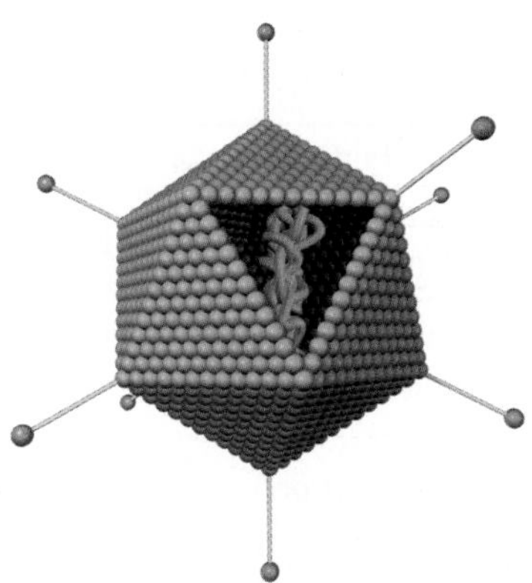

27. Which organisms does this virus infect?
    A. humans
    B. bacteria
    C. plants
    D. fungi

Constructed Response

28. **Open Ended** Make an argument for or against the following statement: Viruses are living organisms.

29. **Open Ended** Should people with highly contagious, potentially deadly viruses be quarantined? Defend your response.

30. **Open Ended** Make an argument for or against the following statement: Prions are just viruses that lack a capsid.

Think Critically

31. **Infer** why it is more difficult to make an antiviral drug that fights a virus that replicates through the lysogenic cycle than it is to make one that fights a virus that replicates through the lytic cycle.

32. **Evaluate** why it is easier to make drugs that fight bacteria than drugs that fight viruses, even though viruses are structurally less complex than bacteria.

33. **Hypothesize** and develop a technique to slow down or stop a viral replication cycle.

34. **Develop** a list of different careers that are associated with bacteria, viruses, and prions.

Additional Assessment

35.  WRITING in Biology Prepare a newspaper article that clearly explains the differences between disease-causing bacteria and viruses.

36. WRITING in Biology **Compose** a sentence that explains each step in the sequence of events in the replication of HIV.

DBQ Document-Based Questions

U.S. Data: Centers for Disease Control http://www.cdc.gov/flu/avian/pdf/avianflufacts.pdf.
Global Data: Scotland Government http://www.scotland.gov.uk/library5/health/pfle-00.asp

There were three worldwide influenza epidemics during the twentieth century. The number of deaths is presented in the table below.

	Spanish Flu	Asian Flu	Hong Kong Flu
Years	1918–1919	1957–1958	1968–1969
U.S. deaths	500,000	70,000	34,000
Global deaths	20–40 million	1 million	1–4 million

37. Which epidemic was the most deadly?

38. Why were deaths not as high in the United States with the Hong Kong flu compared to the Asian flu, but were higher worldwide?

39. Hypothesize why a flu epidemic eventually stops instead of eliminating all human life.

Cumulative Review

40. Explain how the concepts of observation, inference, and skepticism differ. **(Chapter 1)**

41. Summarize the overall reactions of photosynthesis and cellular respiration. **(Chapter 8)**

42. Summarize how a cancer cell cycle is different than a normal cell cycle. **(Chapter 9)**

43. Describe the primate groups that comprise the anthropoids. **(Chapter 16)**

Standardized Test Practice

Cumulative

Multiple Choice

1. Which primate is an Asian ape?
   A. baboon
   B. gorilla
   C. lemur
   D. orangutan

Use the chart below to answer questions 2 and 3.

Common Name	Scientific Name
Grey wolf	*Canis lupus*
Red wolf	*Canis rufus*
African hunting dog	*Lycaon pictus*
Pampas fox	*Pseudalopex gymnocercus*

2. Which animal is related most closely to the Sechura fox *Pseudalopex sechurae*?
   A. African hunting dog
   B. grey wolf
   C. pampas fox
   D. red wolf

3. Which kind of difference is a valid reason to classify the red wolf and pampas fox in separate genera?
   A. different prey
   B. different region of habitation
   C. different structure of skulls
   D. different age of evolutionary origin

4. Which describes the role of an endospore in bacteria?
   A. a dormant state of bacteria that can survive in unfavorable conditions
   B. a form of sexual reproduction in bacteria during which genetic information is exchanged
   C. a protective covering that bacteria secrete to protect them against harsh environments
   D. a tiny hairlike structure made of proteins that attaches the bacteria to a surface

5. Which information constitutes a scientific hypothesis?
   A. defined data
   B. proven explanation
   C. published conclusion
   D. reasonable guess

Use the table below to answer questions 6 and 7.

Identifying Bacteria			
Bacterial Strain	Gram Staining	Morphology	Related Disease
Bacillus cereus	Gram-positive	Rods; arranged in chains	Meningitis
Escherichia coli	Gram-negative	Cocci	Traveler's diarrhea
Pseudomonas aeruginosa	Gram-negative	Rod-like; occur in pairs or short chains	Pneumonia
Serratia mercescens	Gram-negative	Rod-like	Pneumonia

6. Which kind of bacteria stains Gram-negative and appears rodlike in short chains?
   A. *Bacillus cereus*
   B. *Escherichia coli*
   C. *Pseudomonas aeruginosa*
   D. *Serratia marcescens*

7. Which related disease would be associated with a bacterium that is Gram-negative and in paired rods?
   A. meningitis
   B. pneumonia
   C. cystic fibrosis
   D. traveler's diarrhea

8. Which taxon gives you the most general information about an organism?
   A. class
   B. domain
   C. family
   D. phylum

9. A population of rodents on an island makes up a distinct species that is similar to a species found on the mainland. Which process caused this speciation?
   A. behavioral isolation
   B. geographic isolation
   C. reproductive isolation
   D. temporal isolation

Short Answer

10. Suppose that two mosquitoes are classified as different species using the typological species concept. What data could scientists use, under the biological species concept, to show that they are the same species?

11. Compare the basic shapes of bacteria.

12. Contrast the typological species concept and the phylogenetic species concept.

13. Hypothesize how the evolution of bipedalism made it possible for hominoids to survive better in the drier African environment of the Miocene Epoch.

Use the illustration below to answer questions 14 and 15.

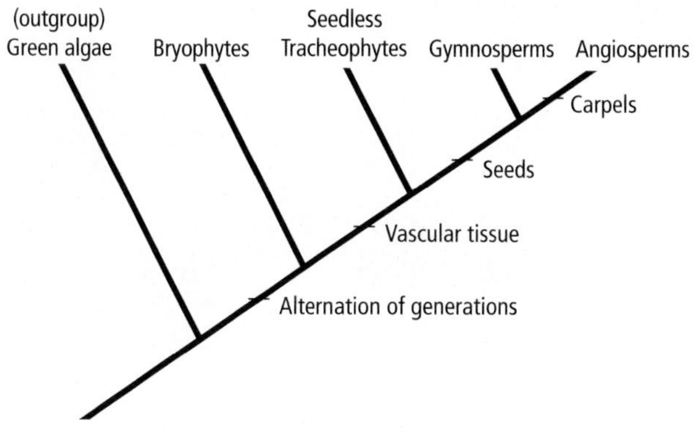

14. According to the plant cladogram, what characteristic separates plants from earlier organisms?

15. Specify an example of an ancestral character and a derived character among angiosperms.

Extended Response

16. Certain bacteria fix nitrogen in the root nodules of a bean plant. Assess how the location of those bacteria in nodules is beneficial to the bacteria and the plants.

17. Give one justification for why a farmer might plant beans in the fields when not growing other crops.

18. Compare and contrast Domain Bacteria and Domain Archaea.

19. Justify why a doctor would not prescribe an antibiotic to treat the flu.

Essay Question

Although scientists have made many discoveries to piece together the steps in human and primate evolution, there are still areas of disagreement and gaps in the evidence. For instance, not all scientists agree about the naming of different species in the genus *Homo*, or about the ways to depict the human evolutionary tree.

Using the information in the paragraph above, answer the following question in essay format.

20. Write an essay that describes an area of debate in human evolution that interests you. What are some aspects of the debate or disagreement that you would want to find out more about? What kind of research would you be able to do if you were going to investigate this debate further?

NEED EXTRA HELP?

If You Missed Question . . .	1	2	3	4	5	6	7	8	9	10	11	12	13	14	15	16	17	18	19	20
Review Section . . .	16.1	17.1	17.1	18.1	1.2	18.1	18.1	17.1	15.3	17.2	18.1	17.1	16.2	17.2	17.2	18.1	18.1	17.3	18.2	16.2, 16.3

CHAPTER 19 Protists

BIG Idea Protists are a diverse group of unicellular and multicellular organisms that do not necessarily share the same evolutionary history.

Section 1
Introduction to Protists
MAIN Idea Protists form a diverse group of organisms that are subdivided based on their method of obtaining nutrition.

Section 2
Protozoans—Animal-like Protists
MAIN Idea Protozoans are animal-like, heterotrophic protists.

Section 3
Algae—Plantlike Protists
MAIN Idea Algae are plantlike, autotrophic protists that are the producers for aquatic ecosystems.

Section 4
Funguslike Protists
MAIN Idea Funguslike protists obtain their nutrition by absorbing nutrients from dead or decaying organisms.

BioFacts

- A protist that lives symbiotically in the gut of termites helps it digest cellulose found in wood.
- An estimated five million protists can live in one teaspoon of soil.

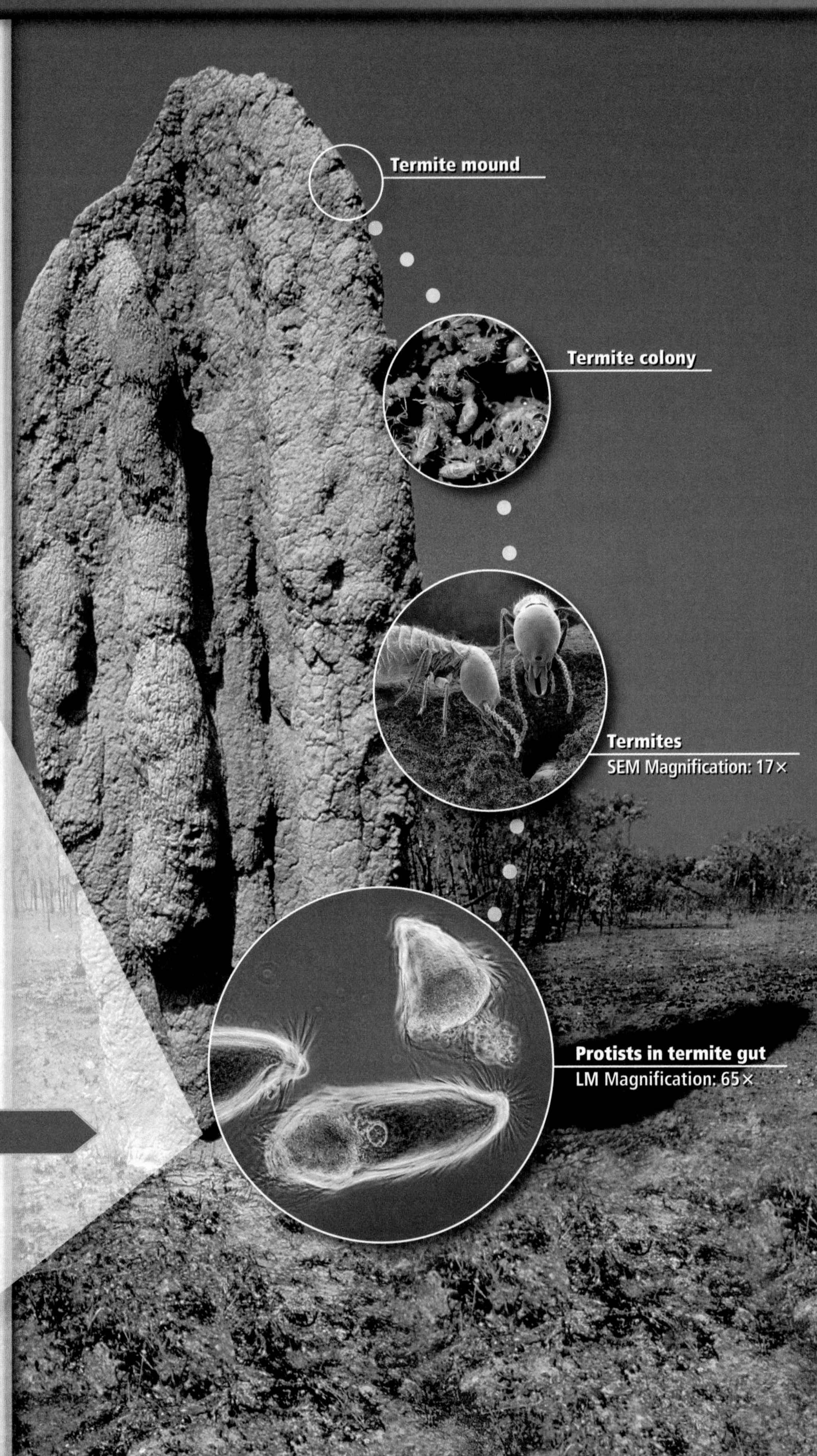

LAUNCH Lab

What is a protist?

The Kingdom Protista is similar to a drawer or closet in which you keep odds and ends that do not seem to fit any other place. The Kingdom Protista is composed of three groups of organisms that do not fit in any other kingdom. In this lab, you will observe the three groups of protists.

Procedure

1. Read and complete the lab safety form.
2. Construct a data table to record your observations.
3. Observe **different types of protists** with a **microscope,** noting their similarities and differences. Record your observations, notes, and illustrations in your data table.

Analysis

1. **Organize** the protists with similar characteristics into groups using the data that you collected.
2. **Infer** which of your groups are animal-like, plantlike, or funguslike.

Visit biologygmh.com to:

- study the entire chapter online
- explore Concepts in Motion, Interactive Tables, Microscopy Links, and links to virtual dissections
- access Web links for more information, projects, and activities
- review content online with the Interactive Tutor, and take Self-Check Quizzes

Classify Protists Make this Foldable to help you organize the characteristics of protists.

STEP 1 Fold a sheet of notebook paper in half vertically. Fold the sheet into thirds.

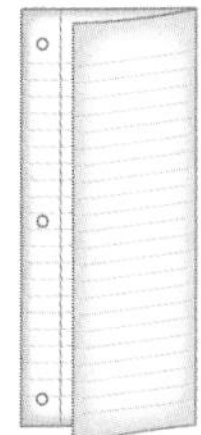

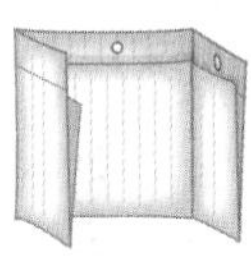

STEP 2 Cut along the creases of the top layer to form three tabs.

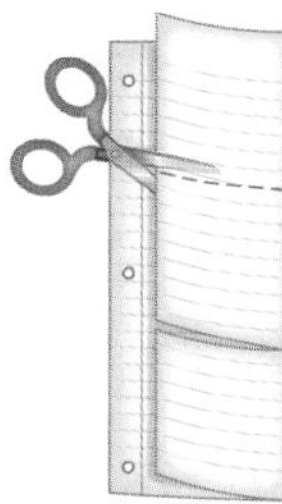

STEP 3 Label the edge with holes *Protists.* Label the top tab *Animal-like Protists,* the middle tab *Plantlike Protists,* and the bottom tab *Funguslike Protists.*

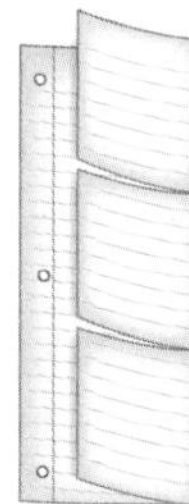

FOLDABLES **Use this Foldable with Section 19.1.** As you study the section, summarize the characteristics of each group under the appropriate tab.

Section 19.1

Reading Preview

Objectives

- **Explain** how protists are classified.
- **Describe** how some protists with mitochondria might have evolved.
- **Describe** how some protists with chloroplasts might have evolved.
- **Explain** why the organization of Kingdom Protista might change.

Review Vocabulary

heterotroph: organism that cannot make its own food and must get its energy and nutrients from other organisms

New Vocabulary

protozoan
microsporidium

Introduction to Protists

MAIN Idea **Protists form a diverse group of organisms that are subdivided based on their method of obtaining nutrition.**

Real-World Reading Link Hurricanes, such as Katrina in 2005, bring winds and water surges that leave destruction and devastation. Contaminated flood waters, damaged sewage systems, and crowded shelters provide breeding grounds for infectious bacteria, viruses, and microorganisms called protists.

Protists

Protists are classified more easily by what they are not than by what they are. Protists are not animals, plants, or fungi because they do not have all of the characteristics necessary to place them in any of these kingdoms. The Kingdom Protista was created to include this diverse group of more than 200,000 organisms.

All protists share one important trait—they are eukaryotes. You learned in Chapter 7 that eukaryotic cells contain membrane-bound organelles. Like all eukaryotes, the DNA of protists is found within the membrane-bound nucleus. Although protists have a cellular structure similar to other eukaryotes, there are remarkable differences in their reproductive methods. Some reproduce asexually by mitosis while others exchange genetic material during meiosis.

Classifying protists Because they are such a diverse group of organisms, some scientists classify protists by their method of obtaining nutrition. Protists are divided into three groups using this method: animal-like protists, plantlike protists, and funguslike protists. The **protozoan** (proh tuh ZOH un) (plural, protozoa or protozoans), shown in **Figure 19.1,** is an example of an animal-like protist because it is a heterotroph—it ingests food. Additional examples of protists and a summary of characteristics are shown in **Table 19.1.**

Color-Enhanced SEM Magnification: 1000×

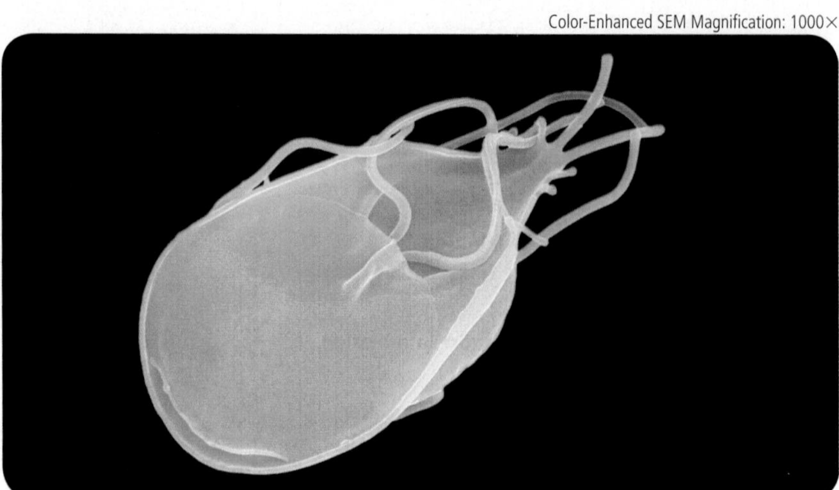

Giardia lamblia

■ **Figure 19.1** This animal-like protist is a parasite that might be found in the intestinal tract of a person who has consumed contaminated water.
Infer *how this protist obtains its nutrients.*

Animal-like protists The amoeba is an example of a unicellular, animal-like protist or protozoan. Protozoans are heterotrophs and usually ingest bacteria, algae, or other protozoans. The amoeba shown in **Table 19.1** is in the process of capturing and ingesting another unicellular protozoan—a paramecium.

Plantlike protists The giant kelp, shown in **Table 19.1,** is an example of a plantlike protist that makes its own food through photosynthesis. Plantlike protists commonly are referred to as algae (AL jee) (singular, alga). Some algae are microscopic. The unicellular algae *Micromonas* are about 10^{-6} m in diameter. Other forms of algae are multicellular and are quite large. The giant kelp, *Macrocystis pyrifera,* can grow up to 65 m long.

Funguslike protists The water mold in **Table 19.1** is an example of a funguslike protist that is absorbing nutrients from a dead salamander. Funguslike protists are similar to fungi because they absorb their nutrients from other organisms. These organisms are not classified as fungi because funguslike protists contain centrioles—small, cylindrical organelles that are involved in mitosis and usually are not found in the cells of fungi. Fungus and funguslike protists also differ in the composition of their cell walls.

Reading Check **Compare and contrast** the three groups of protists.

VOCABULARY
WORD ORIGIN
Protist
comes from the Greek word *protistos,* meaning *the very first.*

FOLDABLES
Incorporate information from this section into your Foldable.

Concepts In Motion
Interactive Table To explore more about protists, visit biologygmh.com.

Table 19.1 The Protists

	Animal-like protists (Protozoans)	Plantlike protists (Algae)	Funguslike protists
Group	Ciliates, amoebas, apicomplexans, and zooflagellates	Euglenoids, diatoms, dinoflagellates, green algae, red algae, brown algae, yellow-green algae, and golden-brown algae	Slime molds, water molds, and downy mildews
Example	**Amoeba**	**Giant kelp**	**Water mold**
Distinguishing Characteristics	• Considered animal-like because they consume other organisms for food • Some are parasites.	• Considered plantlike because they make their own food through photosynthesis • Some consume other organisms or are parasites when light is unavailable for photosynthesis.	• Considered funguslike because they feed on decaying organic matter and absorb nutrients through their cell walls • Some slime molds consume other organisms and a few slime molds are parasites.

Figure 19.2 The protists, green algae, live in the fur of this tree sloth, forming a symbiotic relationship.

Infer ***What type of symbiotic relationship do these organisms have?***

Habitats Protists typically are found in damp or aquatic environments such as decaying leaves, damp soil, ponds, streams, and oceans. Protists also live in symbiotic relationships. **Microsporidia** (MI kroh spo rih dee uh) are microscopic protozoans that cause disease in insects. Some species of microsporidia can be used as insecticides. New technology might allow these microsporidia to be used to control insects that destroy crops.

One beneficial protist lives in the hair of a sloth, shown in **Figure 19.2.** A sloth is a large, slow-moving mammal that lives in the uppermost branches of trees in tropical rain forests. The sloth spends most of its life hanging upside down. Green algae help the brown sloth blend into the leaves on the tree, providing camouflage for the sloth.

LAUNCH Lab

Review Based on what you've read about protists, how would you now answer the analysis questions?

DATA ANALYSIS LAB 19.1

Based on Real Data*

Interpret Scientific Illustrations

What is the relationship between green alga and *Ginkgo biloba* cells? In 2002, scientists in France reported the first confirmed symbiotic relationship between plantlike protists called green algae and a land plant's cells. The figure at the right represents an alga inside a cell from the *Ginkgo biloba* tree.

Think Critically

1. **Examine** the figure and estimate the size of the algal cell.
2. **Explain** why the term endophytic (en duh FIT ihk) is appropriate to describe these algae. The prefix *endo* means "within" and the suffix *-phyte* means "plant."

Data and Observations

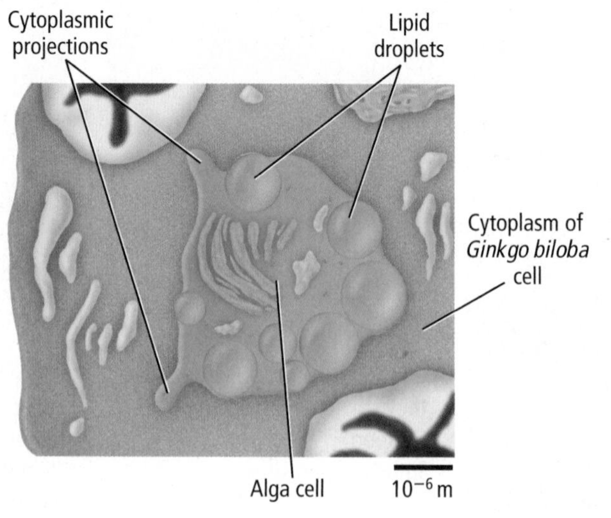

*Data obtained from: Tremoullaux-Guiller, et al. 2002. Discovery of an endophytic alga in *Ginkgo biloba*. *American Journal of Botany* 89(5): 727–733.

Origin of Protists

In Chapter 14, you read about the theory of endosymbiosis, which was proposed by Lynn Margulis. This theory suggests that eukaryotes, including protists, formed when a large prokaryote engulfed a smaller prokaryote. The two organisms lived symbiotically. Eventually, the organisms evolved into a single, more highly-developed organism. Some scientists think that the mitochondria and chloroplasts found in some eukaryotes, including protists, were once individual organisms. Protists might have been the first eukaryotes to appear billions of years ago.

Grouping protists by how they obtain nutrition is a convenient method of classifying them. However, this method does not consider an organism's evolutionary history. Scientists are still trying to sort out the evolutionary relationships between protists and the other kingdoms. As scientists learn more information, the organization of Kingdom Protista most likely will change.

The diagram in **Figure 19.3** shows the current understanding of the evolutionary history of protists based on the theory of endosymbiosis. Notice in the diagram that all of the protists have a common ancestral eukaryotic cell. Examine the diagram and find where mitochondria entered into the evolutionary process. Mitochondria became part of protist cells early in the evolutionary process. Now, locate where chloroplasts entered cells. Follow the path of the arrow and you can see that algae are the only protists with chloroplasts and that undergo photosynthesis.

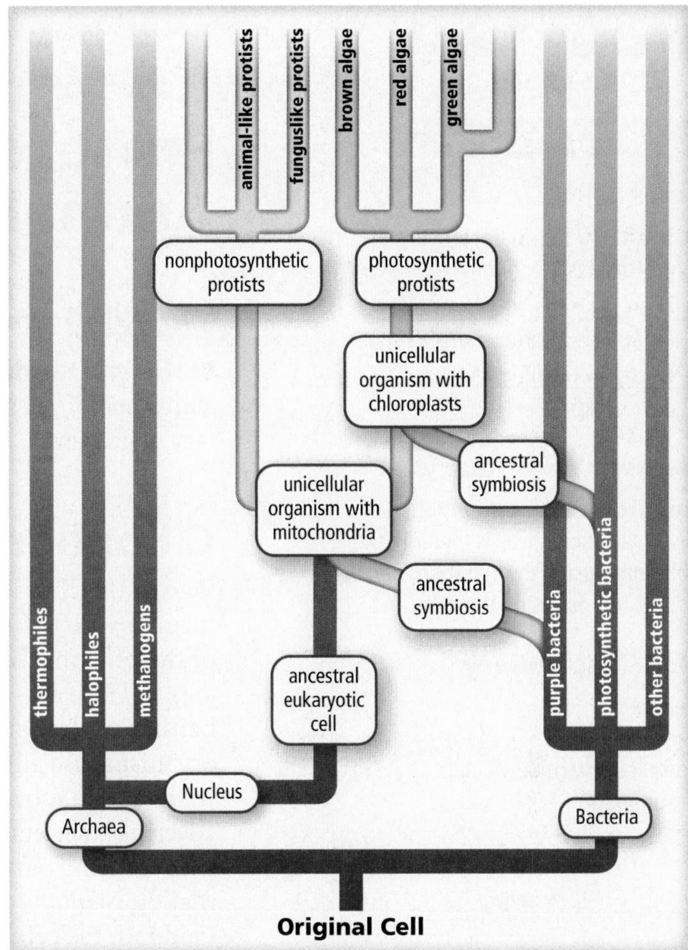

Figure 19.3 This diagram shows how the theory of endosymbiosis explains the evolution of the protist kingdom.

Section 19.1 Assessment

Section Summary

- Protists include unicellular and multicellular eukaryotes.
- Protists are classified by their methods of obtaining food.
- The first protists might have formed through endosymbiosis.
- Protists might have been the first eukaryotic cells with chloroplasts and mitochondria, evolving billions of years ago.

Understand Main Ideas

1. MAIN Idea **Explain** why some scientists use nutrition to classify organisms in the Kingdom Protista.
2. **Sketch** a diagram that illustrates how the first protists might have formed from prokaryotes.
3. **Explain** why scientists have classified protists in one kingdom when they are such a diverse group.

Think Critically

4. **Apply Concepts** What if you discovered a new protist? What characteristics would help you decide the group in which it belongs?
5. **Compare and contrast** using nutrition methods and evolutionary relationships to classify protists.

Section 19.2

Reading Preview

Objectives

- **Identify** the characteristics of protozoans.
- **Describe** the structures and organelles of protozoans.
- **Explain** the life cycles of protozoans.

Review Vocabulary

hypotonic: the concentration of dissolved substances is lower in the solution outside the cell than the concentration inside the cell

New Vocabulary

pellicle
trichocyst
contractile vacuole
pseudopod
test

Protozoans—Animal-like Protists

MAIN Idea **Protozoans are animal-like, heterotrophic protists.**

Real-World Reading Link Have you ever looked at pond water under a microscope? If you saw tiny organisms darting around, then you most likely have seen protozoans.

Ciliophora

One of the characteristics that biologists use to further classify protozoans into different phyla is their method of movement. Members of the phylum Ciliophora (sih lee AH fuh ruh), also known as ciliates (SIH lee ayts), are animal-like protists that have numerous short, hairlike projections. Recall from Chapter 7 that some unicellular organisms use cilia (singular, cilium) to propel themselves through water and to move food particles into the cell. Some ciliates have cilia covering their entire plasma membrane, while others have groups of cilia covering parts of their membrane, as shown in **Figure 19.4.** Note that the *Stentor's* cilia are located on the anterior end; they help propel food into the cell. The ciliate *Trichodina pediculus* has two visible sets of cilia. The outer ring is used for movement and the inner ring is used for feeding.

There are more than 7000 species of ciliates. They are abundant in most aquatic environments—ocean waters, lakes, and rivers. They also are found in mud, and it is estimated that as many as 20 million ciliates can inhabit one square meter in some mud flats.

Figure 19.4 *Stentor* and *Trichodina pediculus* are protozoans that have cilia.

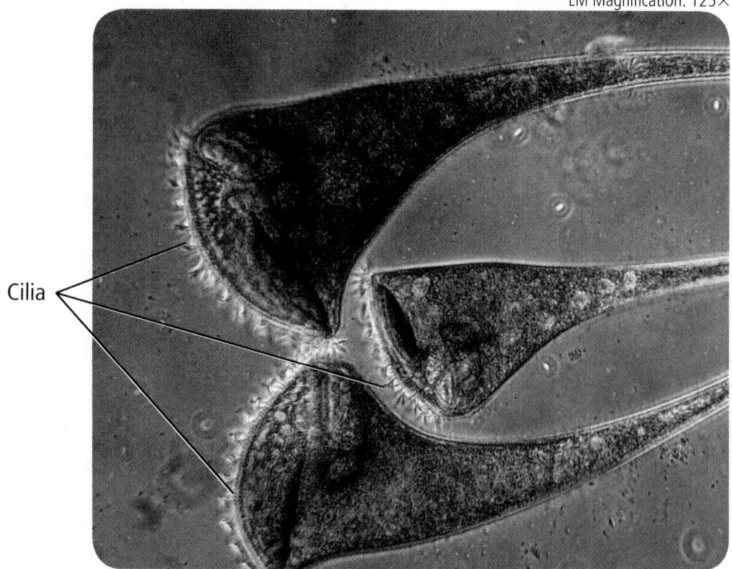

***Stentor*—use cilia for feeding**

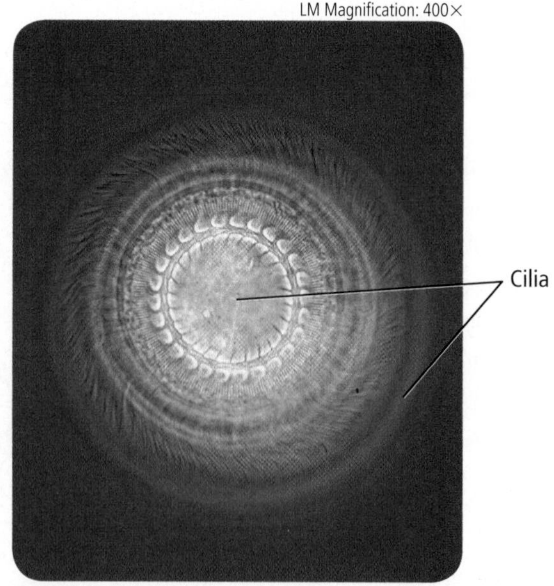

***Trichodina pediculus*—use cilia for feeding and movement**

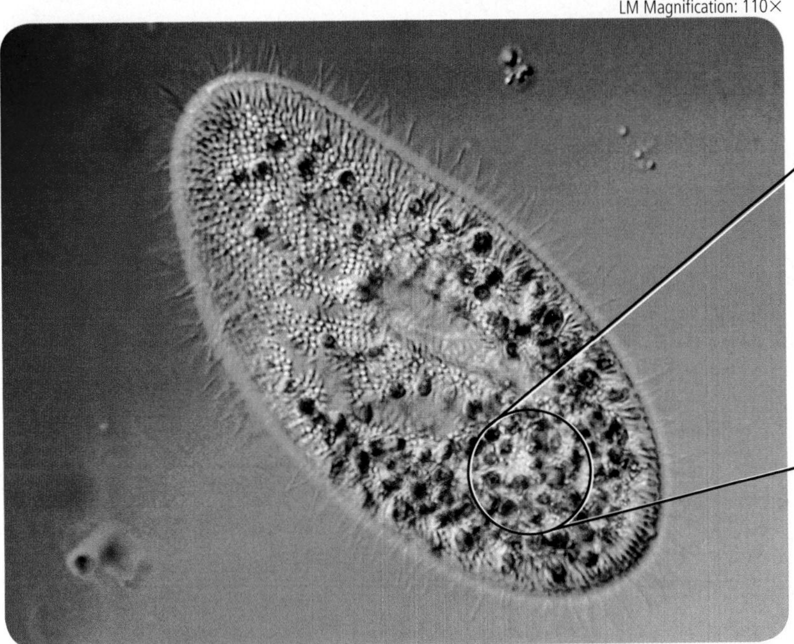

Paramecium bursaria

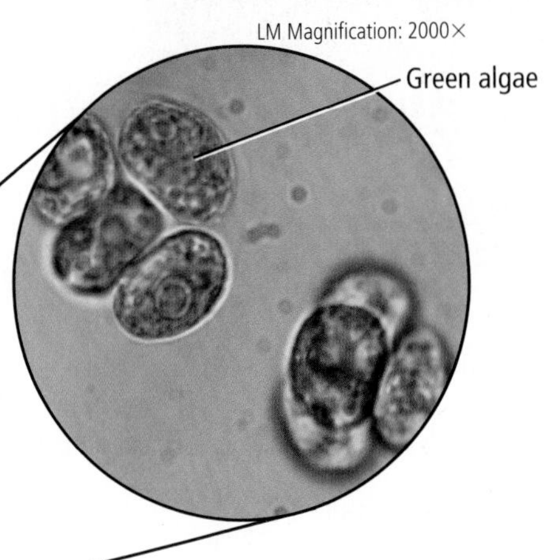

■ **Figure 19.5** *Paramecium bursaria* provides a home for green algae that enter the paramecium during the feeding process, but the green algae are not digested.
Infer ***What type of symbiotic relationship does this represent?***

Paramecia Some of the most commonly studied ciliates are found in the genus *Paramecium* (per uh MEE see um) (plural, paramecia). The paramecium in **Figure 19.5** lives symbiotically with green algae. The green algae undergoes photosynthesis, providing nutrients to the paramecium.

A paramecium is a unicellular protozoan. It is enclosed by a layer of membrane called a **pellicle.** Directly beneath the pellicle is a layer of cytoplasm called ectoplasm. Embedded in the ectoplasm are the **trichocysts** (TRIH kuh sihsts), which are elongated, cylindrical bodies that can discharge a spinelike structure. The function of trichocysts is not completely understood, but they might be used for defense, as a reaction to injury, as an anchoring device, or to capture prey.

Cilia Notice the cilia on the paramecium in **Figure 19.5,** which are used for movement and feeding. Cilia completely cover the organism—including the oral groove. Locate the oral groove on the paramecium in **Figure 19.6.** The cilia covering the wall of the oral groove are used to guide food, primarily bacteria, into the gullet. Once the food reaches the end of the gullet, it is enclosed in a food vacuole. Enzymes within the food vacuole break down the food into nutrients that can diffuse into the cytoplasm of the paramecium. Waste products from the paramecium are excreted through the anal pore.

Contractile vacuoles Because freshwater paramecia live in a hypotonic environment, water constantly enters the cell by osmosis. Recall from Chapter 7 that a hypotonic solution is one in which the concentration of dissolved substances is lower in the solution outside the cell than the concentration inside the cell. The **contractile vacuoles,** shown in **Figure 19.6,** collect the excess water from the cytoplasm and expel it from the cell. The expelled water might contain waste products, which is another way paramecia can excrete waste. Paramecia often have two or three contractile vacuoles that help to maintain homeostasis in the cell.

Reading Check **Explain** why the contractile vacuoles are necessary in hypotonic environments to maintain homeostasis.

CAREERS IN BIOLOGY

Microbiologist A microbiologist studies organisms that usually are seen only with a microscope. One area of study in which a microbiologist can specialize is the study of protists. For more information on biology careers, visit biologygmh.com.

VOCABULARY

SCIENCE USAGE V. COMMON USAGE

Expel

Science usage: to force out.
Contractile vacuoles expel water from cells.

Common usage: to force to leave.
The principal will expel students for breaking school rules.

Visualizing Paramecia

Figure 19.6
Paramecia are unicellular organisms with membrane-bound organelles. They undergo a process called conjugation in which a pair of paramecia will exchange genetic information as shown in the diagram at the bottom of the page. This is not considered sexual reproduction because new individuals are not formed.

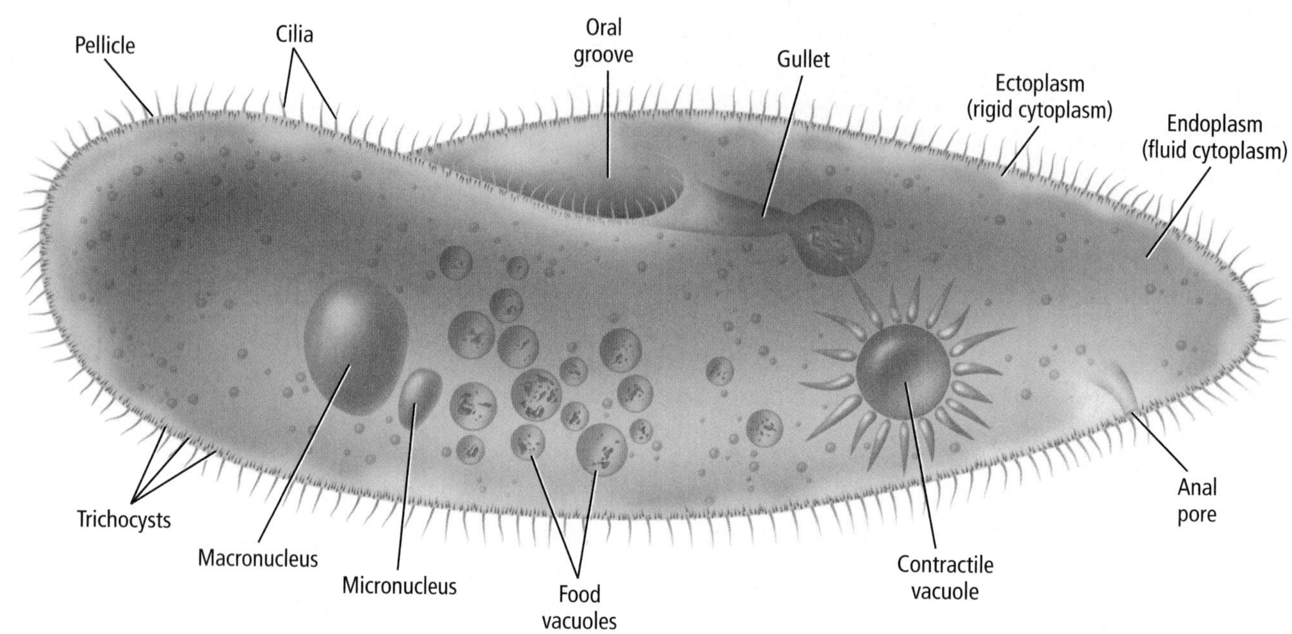

Conjugation

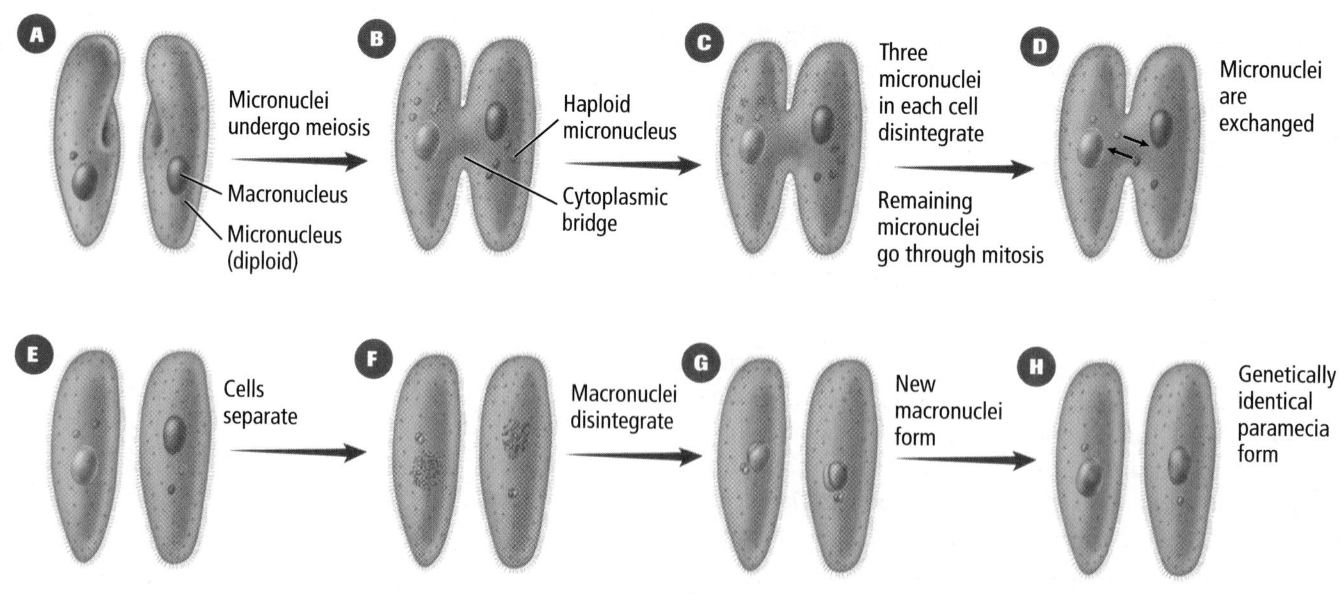

Concepts In Motion **Interactive Figure** To see an animation of paramecia, visit biologygmh.com.

Biology Online

Reproduction in ciliates All known ciliates have two kinds of nuclei—the macronucleus and a smaller micronucleus. A cell might contain more than one of each of these nuclei. Both nuclei contain the genetic information for the cell. The macronuclei contain multiple copies of the cell's genome, which controls the everyday functions of the cell such as feeding, waste elimination, and maintaining water balance within the cell. The micronucleus is used for reproduction.

Ciliates reproduce asexually by binary fission. During this process, the macronucleus elongates and splits rather than undergoing mitotic division. Most ciliates maintain genetic variation by undergoing conjugation—a sexual process in which genetic information is exchanged. Conjugation is considered a sexual process, but it is not considered sexual reproduction because new organisms are not formed.

The process of conjugation for *Paramecium caudatum* is typical of most ciliates and is illustrated in **Figure 19.6.** During conjugation, two paramecia form a cytoplasmic bridge and their diploid micronuclei undergo meiosis. After three of the newly formed haploid micronuclei dissolve, the remaining micronucleus undergoes mitosis. One micronucleus from each connected cell is exchanged, and the two paramecia separate. The macronucleus disintegrates in each paramecium, and the micronuclei combine and form a new, diploid macronucleus. Each cell now contains a macronucleus, micronuclei, and a new combination of genetic information.

Reading Check **Explain** the purpose of the cytoplasmic bridge, shown in **Figure 19.6,** during conjugation.

VOCABULARY

SCIENCE USAGE V. COMMON USAGE

Conjugation

Science usage: asexual reproduction in which there is an exchange of genetic information.

Paramecia reproduce in a process called conjugation.

Common usage: in grammar an arrangement of the correct form of a verb.

Victoria is practicing the conjugation of Spanish verbs.

DATA ANALYSIS LAB 19.2

Based on Real Data*

Recognize Cause and Effect

How does solution concentration affect the contractile vacuole? The contractile vacuole moves water from inside a paramecium back into its freshwater environment. Researchers have studied the effects of solution concentrations on paramecia.

Data and Observations

Paramecia were allowed to adapt to various solutions for 12 h. Then, they were placed into hypertonic and hypotonic solutions. The graphs show the change in rate of water flow out of the contractile vacuole over time.

Paramecium A

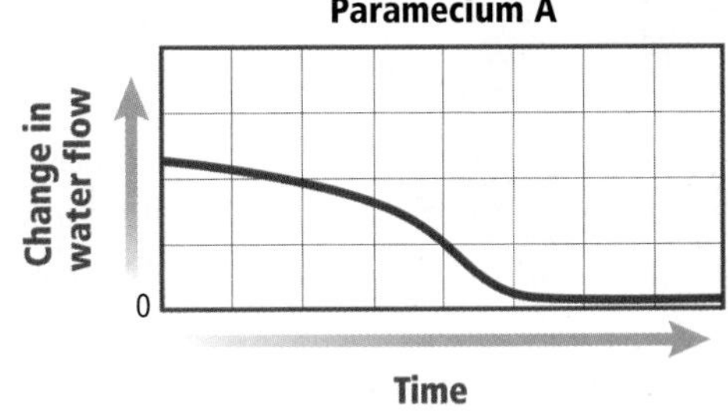

Paramecium B

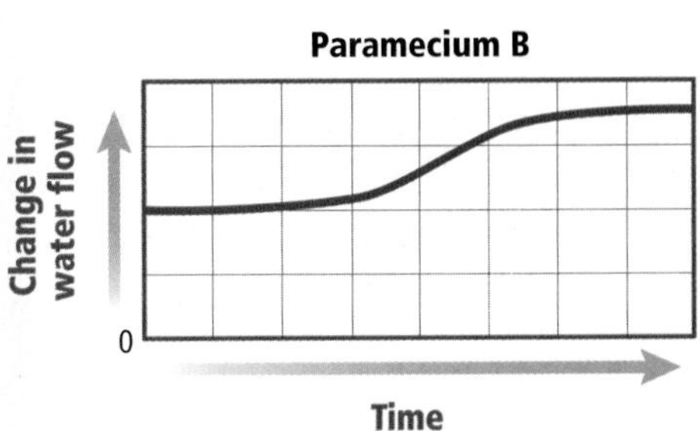

Think Critically

1. **Analyze** What do the downward and upward slopes in the graphs indicate about the contractile vacuole?
2. **Infer** which paramecium was placed into a hypertonic solution. Explain.

*Data obtained from: Stock, et al. 2001. How external osmolarity affects the activity of the contractile vacuole complex, the cytosolic osmolarity and the water permeability of the plasma membrane in *Paramecium Multimicronucleatum. The Journal of Experimental Biology* 204: 291–304.

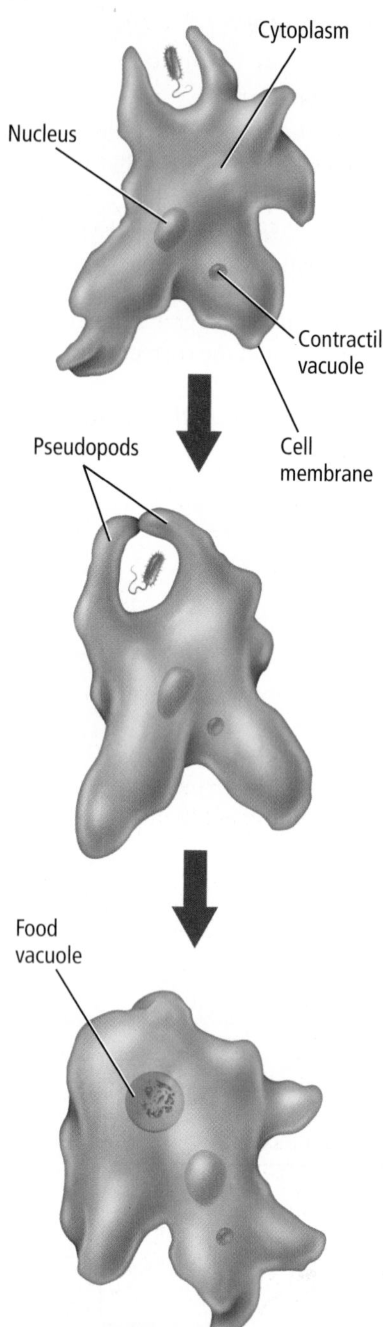

■ **Figure 19.7** Chemical stimuli from smaller organisms can cause the amoeba to form pseudopods from their cell membrane.

Sarcodina

Members of the phylum Sarcodina (sar kuh DI nuh), also called sarcodines (SAR kuh dinez), are animal-like protists that use pseudopods for feeding and locomotion. A **pseudopod** (SEW duh pahd) is a temporary extension of cytoplasm and is shown in **Figure 19.7.** These extensions surround and envelop a smaller organism, forming a food vacuole. Digestive enzymes are secreted and break down the captured organism.

Some of the most commonly studied sarcodines are found in the genus *Amoeba.* Most amoebas are found in saltwater, although some freshwater species live in streams, in the muddy bottoms of ponds, and in damp patches of moss and leaves. Some amoebas are parasites that live inside an animal host.

Amoeba structure The structure of an amoeba is simple, as shown in **Figure 19.7.** Amoebas are enveloped in an outer cell membrane and an inner thickened cytoplasm called ectoplasm. Inside the ectoplasm, the cytoplasm contains a nucleus, food vacuoles, and occasionally a contractile vacuole. Notice that an amoeba does not have an anal pore like the paramecium. Waste products and undigested food particles are excreted by diffusion through the outer membrane into the surrounding water. The oxygen needed for cellular processes also diffuses into the cell from the surrounding water.

Foraminiferans (fuh rah muh NIH fur unz) and radiolarians (ray dee oh LER ee unz) are types of amoebas that have tests. A **test** is a hard, porous covering similar to a shell, which surrounds the cell membrane. Most of these amoebas live in marine environments, although there are some freshwater species.

Connection to Earth Science Foraminiferans have tests made of calcium carbonate ($CaCO_3$), grains of sand, and other particles cemented together. Geologists use the fossilized remains of foraminiferans to determine the age of some rocks and sediments, and to identify possible sites for oil drilling. Radiolarians, another amoeba with tests shown in **Figure 19.8,** have tests made mostly of silica (SiO_2).

Amoeba reproduction Amoebas reproduce by asexual reproduction during which a parent cell divides into two identical offspring. During harsh environmental conditions, some amoebas become cysts that help them survive until environmental conditions improve and survival is more likely.

SEM Magnification: 190×

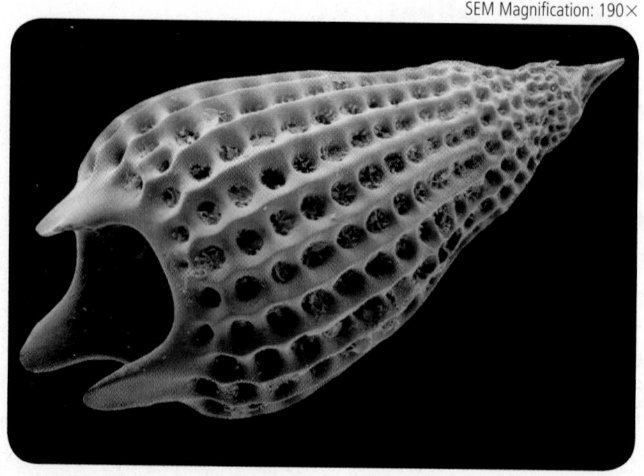

■ **Figure 19.8** Radiolarians have tests made of silica. Foraminiferans and radiolarians extend their pseudopods through openings in their tests.

Apicomplexa

Animal-like protists that belong to the phylum Apicomplexa (ay puh KOM pleks uh) are also known as sporozoans (spo ruh ZOH unz). They are called sporozoans because they produce spores at some point in their life cycle. Spores are reproductive cells that form without fertilization. Sporozoans lack contractile vacuoles and methods for locomotion. Respiration and excretion occur by diffusion through the plasma membrane.

All sporozoans are parasitic. Recall from Chapter 2 that parasites get their nutritional requirements from a host organism. Sporozoans infect vertebrates and invertebrates by living as internal parasites. Organelles at one end of the organism are specialized for penetrating host cells and tissues, allowing them to get their nutrients from their host.

The life cycle of sporozoans has both sexual and asexual stages. Often two or more hosts are required for an organism to complete a life cycle. The life cycle of *Plasmodium,* which causes malaria, is shown in **Figure 19.9.**

Sporozoans cause a variety of illnesses in humans, some of which are fatal. The sporozoans responsible for the greatest number of human deaths are found in the genus *Plasmodium.* These parasites cause malaria in humans and are transmitted to humans by female *Anopheles* mosquitoes. Malaria causes fever, chills, and other flu-like symptoms. Its greatest impact is in tropical and subtropical regions where factors such as high temperature, humidity, and rainfall favor the growth of mosquitoes and sporozoans, and preventative measures are too costly.

VOCABULARY

WORD ORIGIN

Apicomplexa

apicalis from Latin; meaning *uppermost point or tip*

complexus from Latin; meaning *comprised of multiple objects*

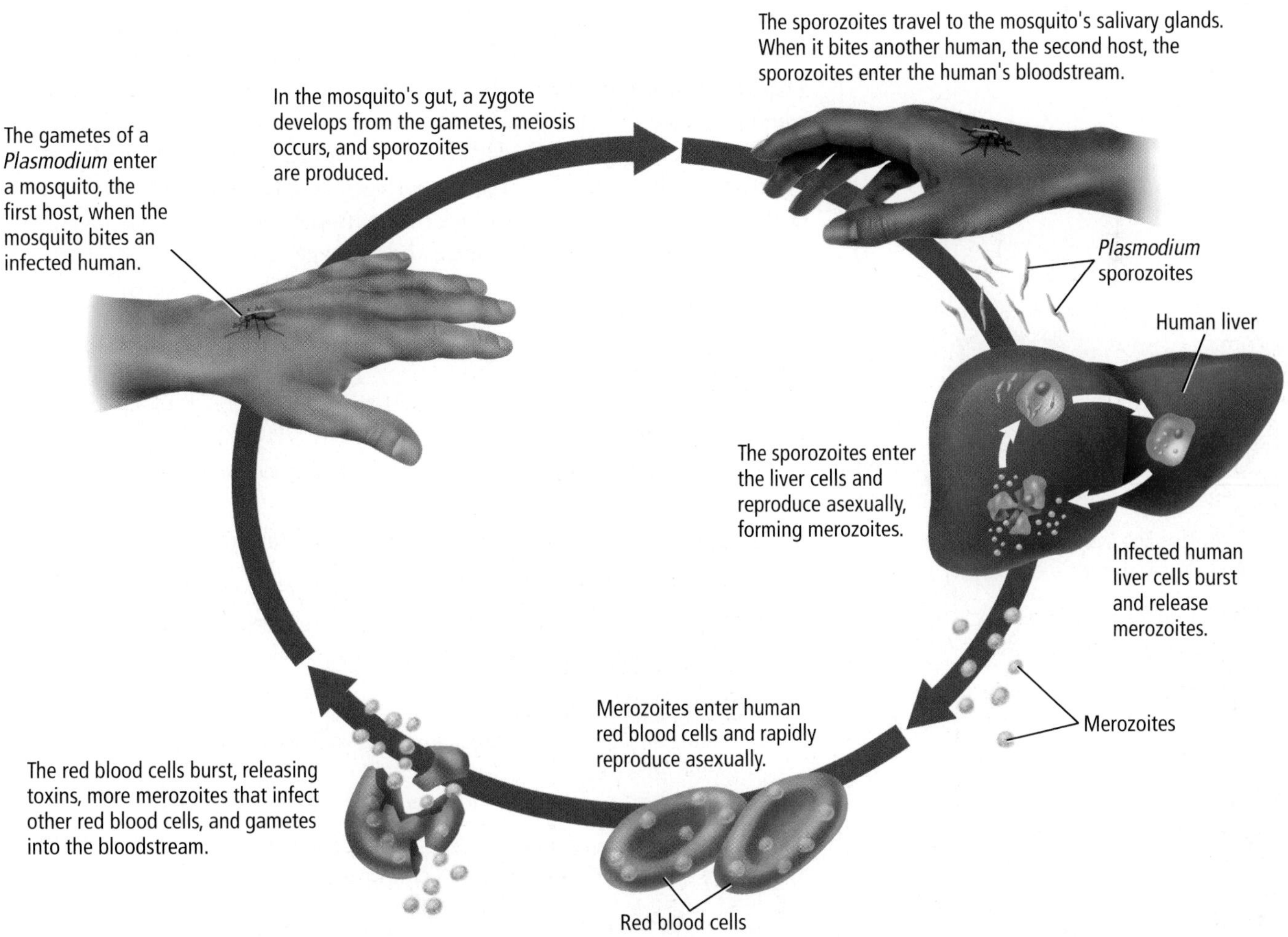

■ **Figure 19.9** Malaria is caused by the sporozoan *Plasmodium*, which is transmitted by mosquito.

Identify ***What are the two hosts that are required for this sporozoan to be successful?***

The gametes of a *Plasmodium* enter a mosquito, the first host, when the mosquito bites an infected human.

In the mosquito's gut, a zygote develops from the gametes, meiosis occurs, and sporozoites are produced.

The sporozoites travel to the mosquito's salivary glands. When it bites another human, the second host, the sporozoites enter the human's bloodstream.

Plasmodium sporozoites

Human liver

The sporozoites enter the liver cells and reproduce asexually, forming merozoites.

Infected human liver cells burst and release merozoites.

Merozoites

Merozoites enter human red blood cells and rapidly reproduce asexually.

Red blood cells

The red blood cells burst, releasing toxins, more merozoites that infect other red blood cells, and gametes into the bloodstream.

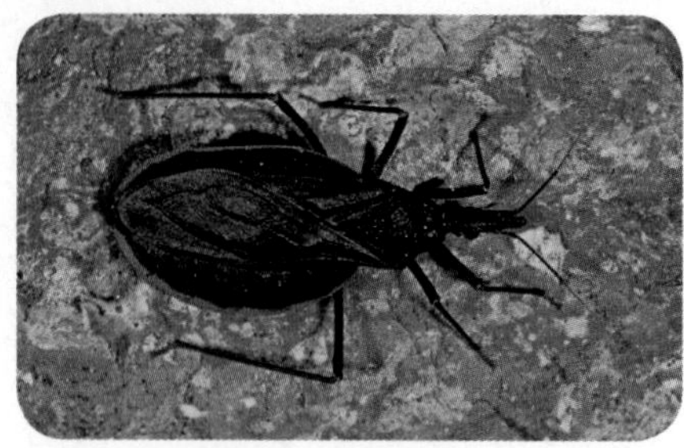

Reduviid bug

Tsetse fly

■ **Figure 19.10** The insects that carry protozoans from person to person are controlled by insecticides.

Zoomastigina

Protozoans in the phylum Zoomastigina (zoh oh mast tuh JI nuh) are called zooflagellates. Zooflagellates (zoh oh FLA juh layts) are animal-like protozoans that use flagella for movement. Flagella are long whip-like projections that protrude from the cell and are used for movement. Some zooflagellates are free living, but many are parasites.

At least three species of zooflagellates from the genus *Trypanosoma* (TRY pan uh zohm uh) cause infectious diseases in humans that often are fatal because of limited treatment options. One species found in Central and South America causes Chagas disease, sometimes called American sleeping sickness. The second species causes East African sleeping sickness. The third species causes West African sleeping sickness.

American sleeping sickness The zooflagellates that cause Chagas' disease are similar to the sporozoans that cause malaria because they have two hosts in their life cycle and insects spread the diseases through the human population. The reduviid bug (rih DEW vee id) bug, shown in **Figure 19.10,** serves as one host for the protist in Central and South America. The parasitic zooflagellates reproduce in the gut of this insect. The reduviid bug gets its nutrients by sucking blood from a human host. During the feeding process, the zooflagellates pass out of the reduviid body through its feces. The zooflagellates enter the human body through the wound site or mucus membranes. Once the zooflagellate enters the body, it multiplies in the bloodstream and can damage the heart, liver, and spleen.

African sleeping sickness The life cycles of the zooflagellates that cause both African sleeping sicknesses are similar to the one that causes American sleeping sickness. The insect host is the tsetse (SEET see) fly, shown in **Figure 19.10.** The blood-sucking tsetse fly becomes infected when it feeds on an infected human or other mammal. The zooflagellate reproduces in the gut of the fly and then migrates to its salivary glands. When the fly bites the human, the zooflagellate is transferred to the human host. The zooflagellates reproduce in the human host and cause fever, inflammation of the lymph nodes, and damage to the nervous system.

Section 19.2 Assessment

Section Summary

- Protozoans are unicellular protists that feed on other organisms to obtain nutrients.
- Protozoans live in a variety of aquatic environments.
- Protozoans reproduce in a variety of ways, including sexually and asexually.
- Protozoans have specialized methods for movement, feeding, and maintaining homeostasis.

Understand Main Ideas

1. MAIN Idea **Compare** the methods of feeding, locomotion, and reproduction of three groups of protozoa.
2. **Explain** the function of three organelles found in protozoans.
3. **Diagram** and explain the life cycle of a member of the genus *Plasmodium.*
4. **Explain** why paramecium conjugation is not considered sexual reproduction.

Think Critically

WRITING in Biology

5. Create an informational brochure about zooflagellates for people living in South America.

MATH in Biology

6. There are approximately 50,000 species of protozoa, of which about 7000 are ciliates. What percentage of protozoans are ciliates?

Biology Online Self-Check Quiz biologygmh.com

Section 19.3

Reading Preview

Objectives

- **Describe** the characteristics of several phyla of algae.
- **Identify** secondary photosynthetic pigments that are characteristic of some algae.
- **Explain** how diatoms differ from most other types of algae.

Review Vocabulary

chloroplasts: chlorophyll-containing organelles found in the cells of green plants and some protists that capture light energy and convert it to chemical energy

New Vocabulary

bioluminescent
colony
alternation of generations

Algae—Plantlike Protists

MAIN Idea **Algae are plantlike, autotrophic protists that are the producers for aquatic ecosystems.**

Real-World Reading Link Have you ever looked at a group of people and wondered what they had in common? You might discover that they all like the same type of music or they like the same type of sports. Most plantlike protists have something in common—they make their own food.

Characteristics of Algae

The group of protists called algae (singular, alga) is considered plant-like because the members contain photosynthetic pigments. Recall from Chapter 8 that photosynthetic pigments enable organisms to produce their own food using energy from the Sun in a process called photosynthesis. Algae differ from plants because they do not have roots, leaves, or other structures typical of plants.

The light-absorbing pigments of algae are found in chloroplasts. In many algae, the primary pigment is chlorophyll—the same pigment that gives plants their characteristic green color. Many algae also have secondary pigments that allow them to absorb light energy in deep water. As water depth increases, much of the sunlight's energy is absorbed by the water. These secondary pigments allow algae to absorb light energy from wavelengths that are not absorbed by water. Because these secondary pigments reflect light at different wavelengths, algae are found in a variety of colors, as shown in **Figure 19.11.**

Reading Check **Explain** the function of chloroplasts and photosynthetic pigments in algae.

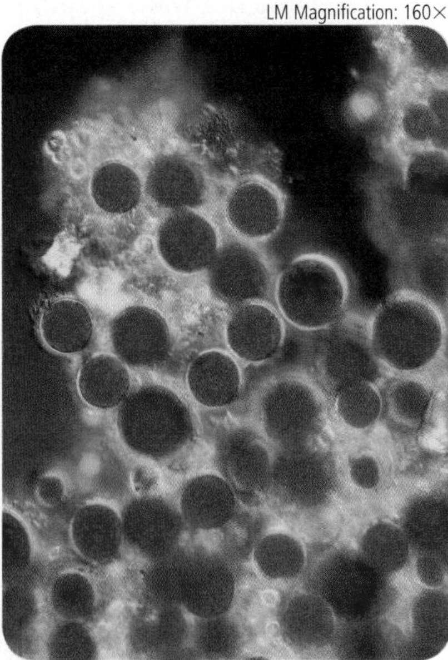

LM Magnification: 160×

Red algae

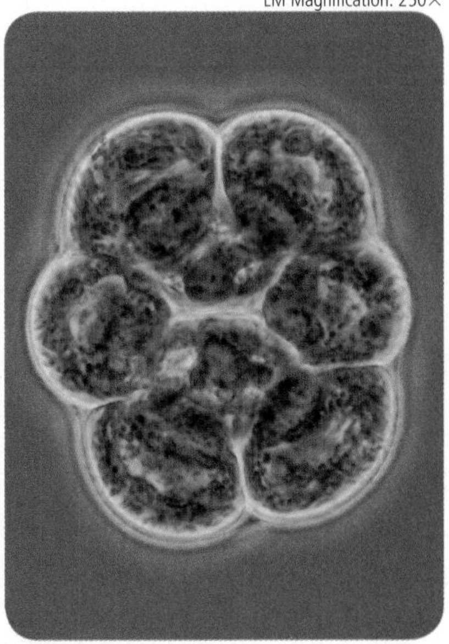

LM Magnification: 250×

Green algae

■ **Figure 19.11** Algae vary in color because they contain different light-absorbing pigments.

LM Magnification: 30×

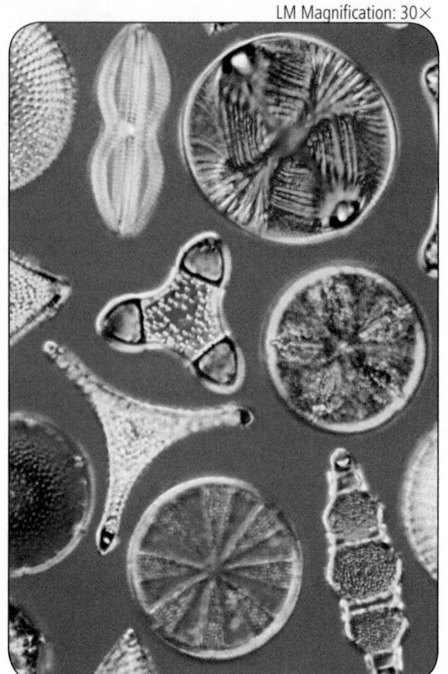

■ **Figure 19.12** The various species of diatoms have different shapes and sizes.

Diversity of Algae

Algae have more differences than their color. For example, many algae exist as single cells, whereas others are huge multicellular organisms reaching 65 m in length. Some unicellular algae are referred to as phytoplankton—meaning "plant plankton." Phytoplankton is vital in aquatic ecosystems because it provides the base of the food web in these environments. As a by-product of photosynthesis, they also produce much of the oxygen found in Earth's atmosphere.

The great diversity of algae makes them a challenge to classify. Algologists usually use three criteria to classify algae: the type of chlorophyll and secondary pigments, the method of food storage, and the composition of the cell wall.

Diatoms The unicellular algae, shown in **Figure 19.12,** are members of the phylum Bacillariophyta (BAH sih LAYR ee oh FI tuh). These intricately shaped organisms are called diatoms. Look at **Figure 19.13** and notice that the diatom consists of two unequal halves—one fits neatly inside the other, forming a small box with a lid.

Connection to Physics Diatoms are photosynthetic autotrophs. They produce food by photosynthesis using chlorophyll and secondary pigments called carotenoids, which give diatoms their golden-yellow color. Diatoms store their food as oil instead of as a carbohydrate. The oil not only makes diatoms a nutritious food source for many marine animals, but it also provides buoyancy. Oil is less dense than water, so diatoms float closer to the surface of the water, where they can absorb energy from the Sun for photosynthesis.

Diatoms reproduce both sexually and asexually, as illustrated in **Figure 19.14.** Asexual reproduction occurs when the two separated halves each create a new half that can fit inside the old one. This process produces increasingly smaller diatoms. When a diatom is about one-quarter of the original size, sexual reproduction is triggered and gametes are produced. The gametes fuse to form a zygote that develops into a full-sized diatom. The reproduction cycle then repeats.

The hard silica walls of the diatom last long after the diatom has died. The silica walls accumulate on the ocean floor to form sediment known as diatomaceous earth. This sediment is collected and used as an abrasive and a filtering agent. The gritty texture of many tooth polishes and metal polishes is due to the presence of diatom shells.

CAREERS IN BIOLOGY

Algologist A biologist who specializes in the study of algae is an algologist. Algologists might work at fish hatcheries or conduct marine research. For more information on biology careers, visit biologygmh.com.

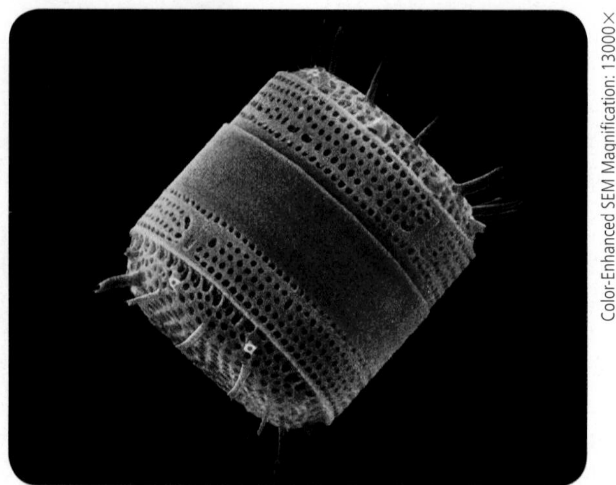

Color-Enhanced SEM Magnification: 13000×

■ **Figure 19.13** Diatoms are found in both marine and freshwater environments. A unique feature of the diatom is its cell wall made of silica.

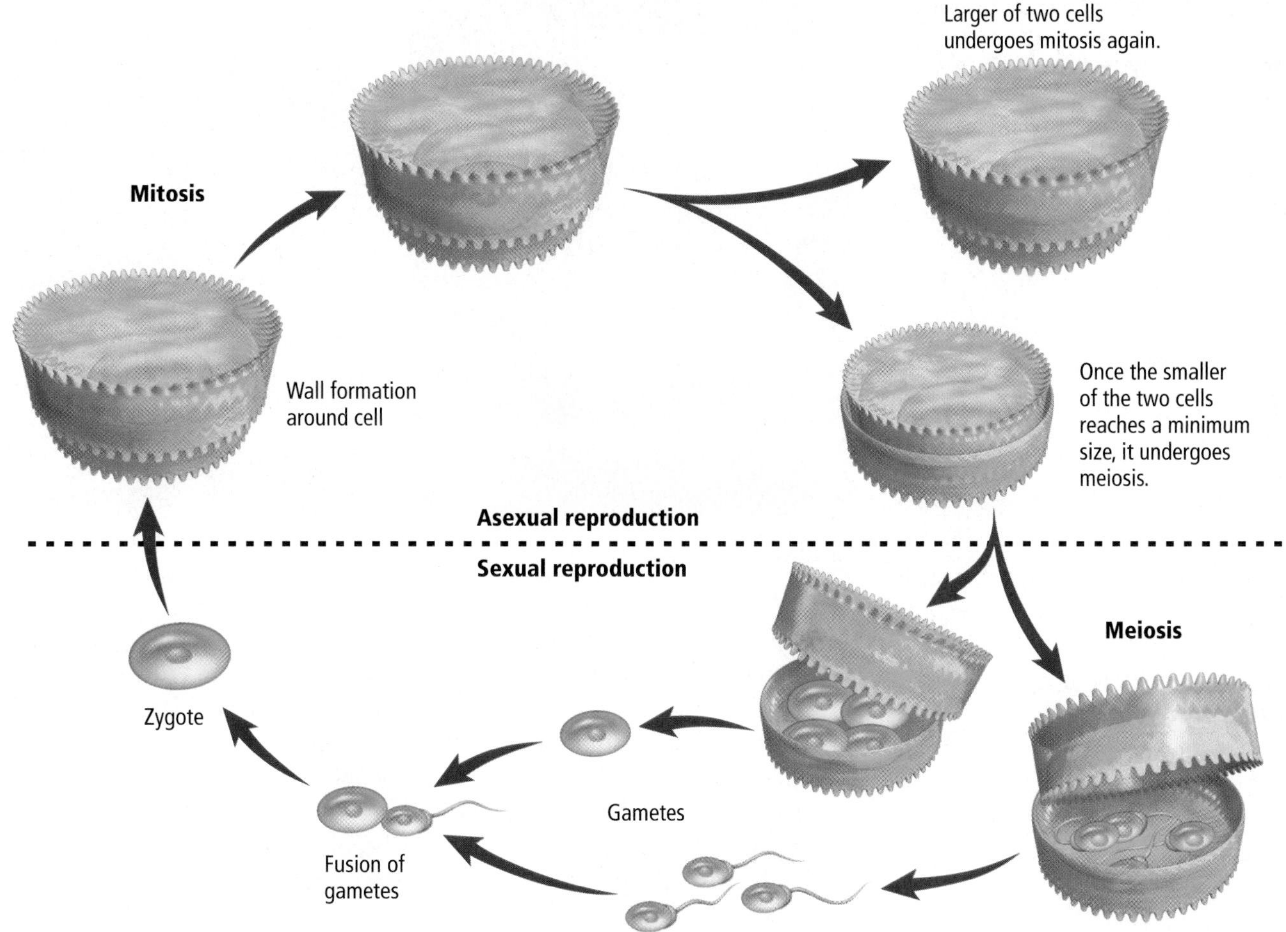

■ **Figure 19.14** Diatoms reproduce asexually for several generations before undergoing sexual reproduction.

Dinoflagellates Plantlike protists that are members of the phylum Pyrrophyta (puh RAH fuh tuh) are called dinoflagellates (di nuh FLA juh layts). Most members of the phylum are unicellular and have two flagella at right angles to one another. As these flagella beat, a spinning motion is created, so dinoflagellates spin as they move through the water. Some members in this group have cell walls made of thick cellulose plates that resemble helmets or suits of armor. Other members of this group are **bioluminescent,** which means they emit light. Although there are a few freshwater dinoflagellates, most are found in saltwater. Like diatoms, photosynthetic dinoflagellates are a major component of phytoplankton.

Dinoflagellates vary in how they get their nutritional requirements. Some dinoflagellates are photosynthetic autotrophs, and other species are heterotrophs. The heterotrophic dinoflagellates can be carnivorous, parasitic, or mutualistic. Mutualistic dinoflagellates have relationships with organisms such as jellyfishes, mollusks, and coral.

Algal blooms When food is plentiful and environmental conditions are favorable, dinoflagellates reproduce in great numbers. These population explosions are called blooms. Algal blooms can be harmful when they deplete the nutrients in the water. When the food supply diminishes, the dinoflagellates die in large numbers. As the dead algae decompose, the oxygen supply in the water is depleted, suffocating fish and other marine organisms. Additional fish suffocate when their gills become clogged with the dinoflagellates.

VOCABULARY
WORD ORIGIN
Pyrrophyta
pyro– prefix; from Greek; meaning *fire*
–phyton from Greek word *phyton,* meaning *plant.*

■ **Figure 19.15** The microscopic organism *Gonyaulax catanella* is one species of dinoflagellates that causes red tides. During red tides, many marine organisms die and shellfish can be too toxic for humans to eat.

Color-Enhanced SEM Magnification: 12000×

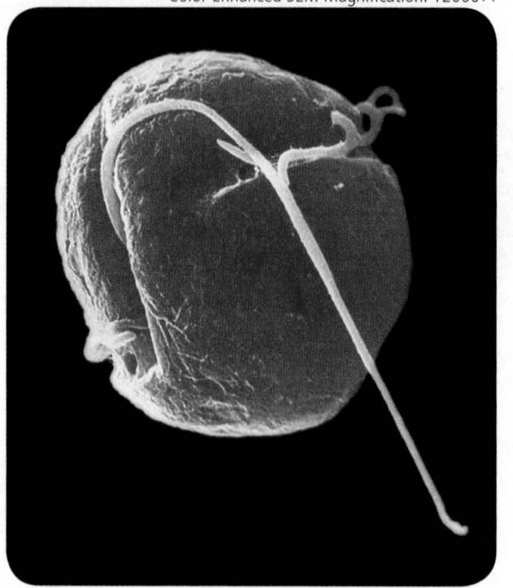

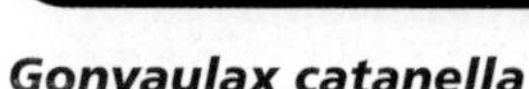

Gonyaulax catanella

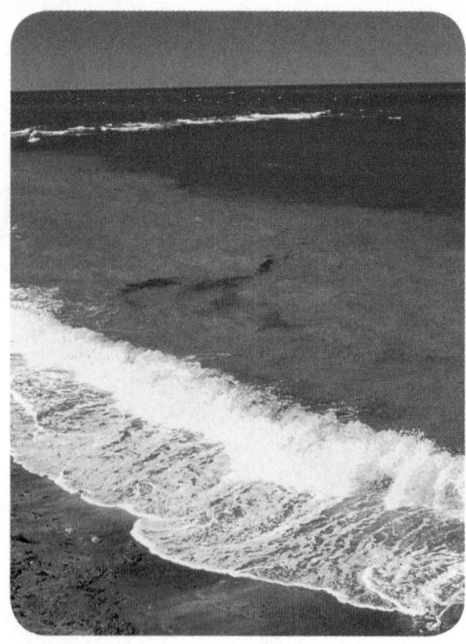

Red tide

Red tides Some dinoflagellates have red photosynthetic pigments, and when they bloom, the ocean is tinged red, as shown in **Figure 19.15.** These blooms are called red tides. Red tides can be a serious threat to humans because some species of dinoflagellates produce a potentially lethal nerve toxin. The toxins affect people primarily when people eat shellfish. Shellfish that feed by filtering particles ingest the toxic dinoflagellates from the water. The toxins become concentrated in tissues of the shellfish. People and other organisms can become seriously ill or die from consuming these toxic shellfish.

Red tides must be closely monitored. One method scientists use to track red tides is reviewing satellite images. However, floating robots are being developed that can constantly measure the concentration of red tide algae. If the concentration becomes too high, scientists can issue a warning to stop shellfish harvesting.

Euglenoids Members of the phylum Euglenophyta are unicellular, plantlike protists called euglenoids (yoo GLEE noydz). Most euglenoids are found in shallow freshwater, although some live in saltwater. Euglenoids are challenging to classify because they have characteristics of both plants and animals. Most euglenoids contain chloroplasts and photosynthesize, which is characteristic of plants, yet they lack a cell wall. Euglenoids also can be heterotrophs. When light is not available for photosynthesis, some can absorb dissolved nutrients from their environment. Others can ingest other organisms such as smaller euglenoids, which is a characteristic of animals. There even are a few species of euglenoids that are animal parasites.

The structure of a typical euglenoid is shown in **Figure 19.16.** Notice that instead of a cell wall, a flexible, tough outer membrane, called a pellicle, surrounds the cell membrane, which is similar to a paramecium. The pellicle allows euglenoids to crawl through mud when the water level is too low to swim. Note the flagella that are used to propel the euglenoid toward food or light. The eyespot is a light-sensitive receptor that helps orient the euglenoid toward light for photosynthesis. The contractile vacuole serves the same purpose in the euglenoid as it does in paramecia. It expels excess water from the cell to maintain homeostasis inside the cell.

■ **Figure 19.16** *Euglena gracilis* are unicellular, plantlike algae that have characteristics of both plants and animals.

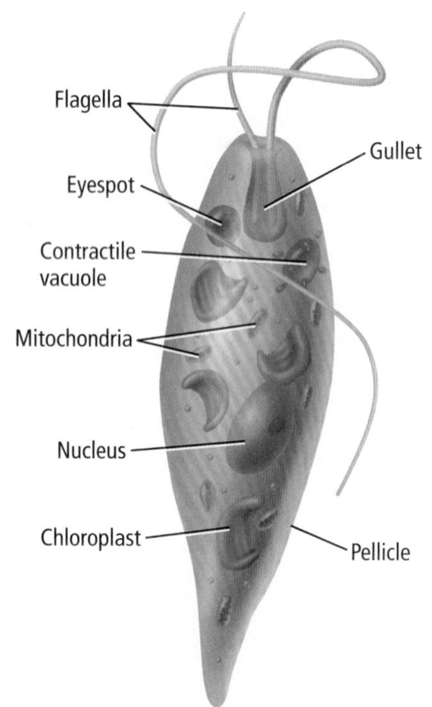

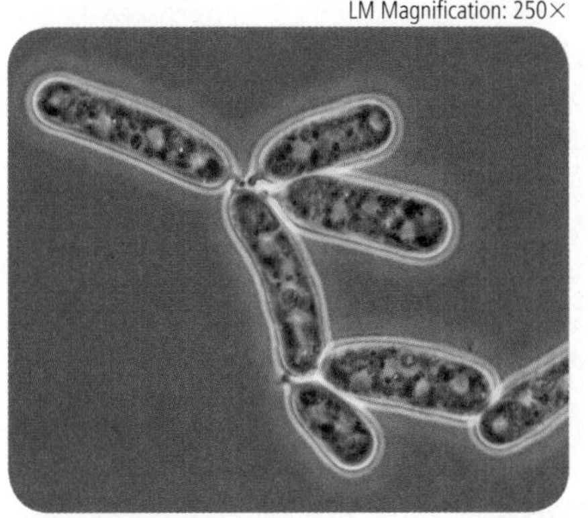

Yellow-green algae

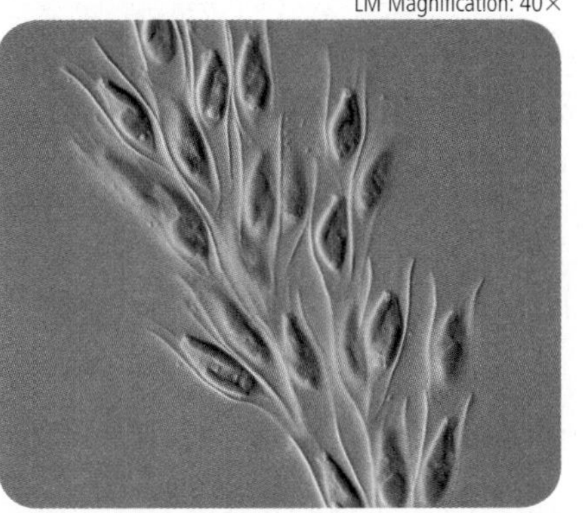

Golden-brown algae

■ **Figure 19.17** Chrysophytes, like yellow-green and golden-brown algae, have carotenoids—secondary pigments used in photosynthesis.

Chrysophytes Yellow-green algae and golden-brown algae are in the phylum Chrysophyta (KRIS oh fyt uh) and are called chrysophytes (KRIS oh fytz). Like diatoms, these algae have yellow and brown carotenoids that give them their golden brown color. The algae in **Figure 19.17** are two examples of organisms from this phylum. Most members of this phylum are unicellular, but some species form colonies. A **colony** is a group of cells that join together to form a close association. The cells of chrysophytes usually contain two flagella attached at one end of the cell. All chrysophytes are photosynthetic, but some species also can absorb dissolved organic compounds through their cell walls or ingest food particles and prokaryotes. They reproduce both asexually and sexually, although sexual reproduction is rare. Chrysophytes are components of both freshwater and marine plankton.

 **Reading Check** **Identify** the substance that gives chrysophytes their golden-brown color.

Study Tip

Shared Reading Have a partner read two paragraphs aloud. Then, you summarize the key ideas in the paragraphs. Then, you read aloud and have your partner summarize the key ideas in your paragraphs.

Brown algae Brown algae are members of the phylum Phaeophyta (FAY oh FI tuh) and are some of the largest multicellular plantlike algae. These algae get their brown color from a secondary carotenoid pigment called fucoxanthin (fyew ko ZAN thun). Most of the 1500 species of brown algae live along rocky coasts in cool areas of the world. Look back at **Table 19.1** to see kelp, an example of a brown alga. The body of a kelp is called the thallus, as shown in **Figure 19.18.** The blades are the flattened portions, the stipe is the stalklike part, the holdfast is the rootlike structure, and the bladder is the bulging portion of the alga. The bladder is filled with air and keeps the alga floating near the surface of the water where light is available for photosynthesis.

Green algae The diverse group of algae from the phylum Chlorophyta (kloh RAH fy tuh) contains more than 7000 species. Green algae have several characteristics in common with plants. Green algae and plants both contain chlorophyll as a primary photosynthetic pigment, which gives both groups a green color. Both green algae and plant cells have cell walls, and both groups store their food as carbohydrates. These shared characteristics lead some scientists to think there is an evolutionary link between these two kingdoms. You will learn more about the plant kingdom in Chapter 21.

Most species of green algae are found in freshwater, but about ten percent are marine species. Green algae also are found on damp ground, tree trunks, and in snow. Green algae even are found in the fur of some animals, such as the sloth shown in **Figure 19.2.**

■ **Figure 19.18** Underwater kelp forests provide a habitat for many marine organisms, as well as provide algin—an additive used in many products.
Explain *What is the function of the bladder in kelp?*

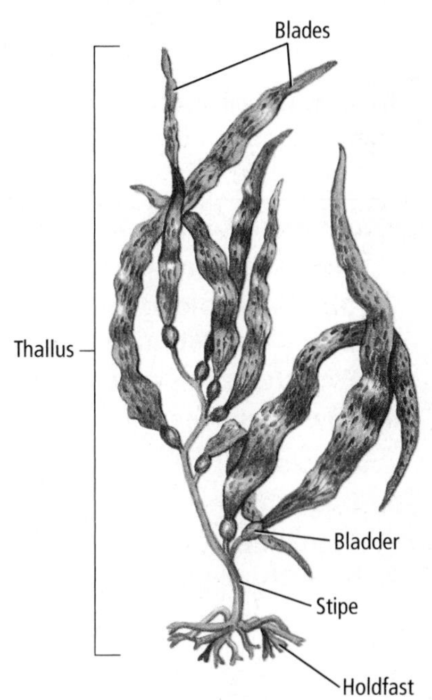

LM Magnification: 125×

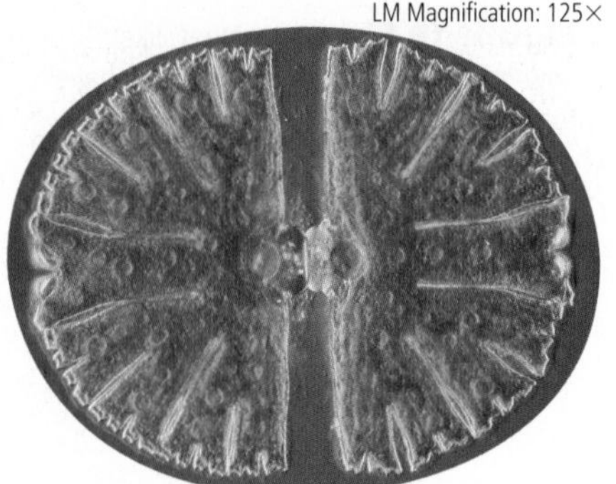

Desmids

LM Magnification: 300×

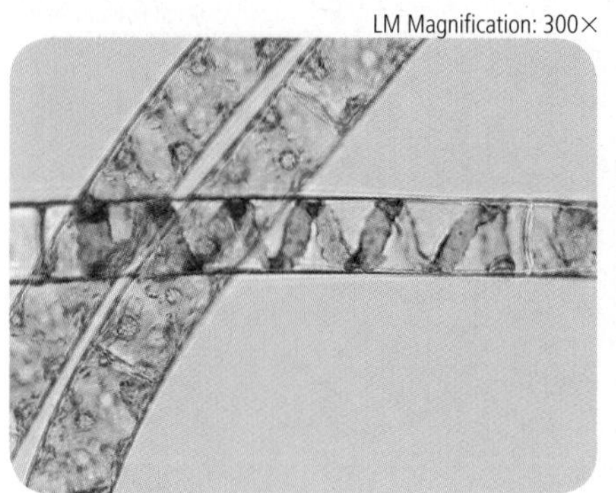

Spirogyra

LM Magnification: 15×

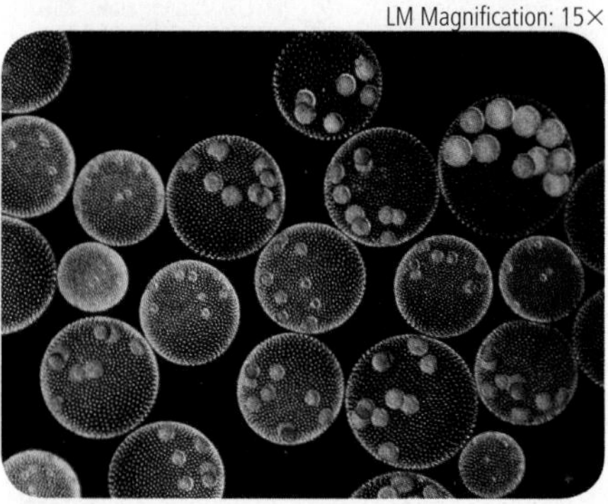

Volvox

■ **Figure 19.19** *Desmids* are unicellular green algae that have elaborate cell walls. The green alga *Spirogyra* is named for its spiraling chloroplasts. Many cells that make up the *Volvox* colony have daughter colonies within the larger colony.

There are a variety of growth patterns exhibited by green algae. The unicellular algae *Desmids,* shown in **Figure 19.19,** are characterized by their symmetrically divided cells. Notice how the cells have two identical sides that are connected by a bridge. Another growth pattern is found in Spirogyra, shown in **Figure 19.19.** *Spirogyra* is a multicellular green algae characterized by its long, thin filaments. The name Spirogyra comes from the spiral pattern of the chloroplasts. *Volvox,* shown in **Figure 19.19,** is an example of an alga that has a colonial growth pattern.

The single cells of the *Volvox* colony are held together by a gelatinlike secretion called cytoplasmic strands. Each cell has flagella that beat in unison to move the colony. *Volvox* colonies might include hundreds or even thousands of cells that form a hollow ball. Smaller colonies, called daughter colonies, form balls inside the larger colony. When the daughter cells have matured, they digest the parental cell and become free-swimming.

 **Reading Check** **Identify** the growth patterns for the algae above.

MiniLab 19.1

Investigate Photosynthesis in Algae

How much sunlight does green alga need to undergo photosynthesis? Algae contain photosynthetic pigments that allow them to produce food by using energy from the Sun. Observe green algae to determine whether the amount of light affects photosynthesis.

Procedure:

1. Read and complete the lab safety form.
2. Obtain samples of **green algae** from your teacher. Place the sample of each type of algae in different locations in the classroom. Be sure one location is completely dark.
3. Hypothesize what will happen to the algae in each location.
4. Check each specimen every other day for a week. Record your observations.

Analysis

1. **Describe** the evidence you used to determine whether photosynthesis was occurring.
2. **Conclude** Was your hypothesis supported? Explain.
3. **Identify** What organelles would you expect to see if you looked at each type of algae under a microscope?

Coralline

■ **Figure 19.20** The red photosynthetic pigments allow the red algae to live in deep water and still use sunlight to photosynthesize.
Explain ***How do the red photosynthetic pigments make this possible?***

Red Algae Most red algae in phylum Rhodophyta (roh dah FI duh) are multicellular. Look at **Figure 19.20** to see how red algae got their name. These organisms contain red photosynthetic pigments called phycobilins that give them a red color. These pigments enable the red algae to absorb green, violet, and blue light that can penetrate water to a depth of 100 m or more. This allows red algae to live and photosynthesize in deeper water than other algae.

Some red algae also contribute to the formation of coral reefs. The cell walls of the red alga *Coralline* contain calcium carbonate. The calcium carbonate binds together the bodies of other organisms called stony coral to form coral reefs. You will learn more about the formation of coral reefs in Chapter 27.

VOCABULARY

ACADEMIC VOCABULARY

Supplement:
something that completes or makes an addition.
Vitamins are taken to supplement one's diet.

Uses for Algae

Algae are used as a source of food for animals and people worldwide. In coastal areas of North America and Europe, algae are fed to farm animals as a food supplement. Algae are found in many dishes and processed foods, as described in **Table 19.2.** Algae are nutritious because of their high protein content and because they contain minerals, trace elements, and vitamins. Some of the substances found in algae also are used to stabilize or improve the texture of processed foods.

Concepts In Motion
Interactive Table To explore more about the uses for algae, visit biologygmh.com.

Table 19.2 Some Uses for Algae

Type of Algae	Uses
Red algae	A species of red alga, *Porphyra,* is called nori, which is dried, pressed into sheets, and used in soups, sauces, sushi, and condiments. Some species of red algae provide agar and carrageenan, which are used in the preparation of scientific gels and cultures. Agar also is used in pie fillings and to preserve canned meat and fish. Carrageenan is used to thicken and stabilize puddings, syrups, and shampoos.
Brown algae	Brown algae are used to stabilize products, such as syrups, ice creams, and paints. The genus *Laminaria* is harvested and eaten with meat or fish and in soups.
Green algae	Species from the genera *Monostroma* and *Ulva,* also called sea lettuce, are eaten in salads, soups, relishes, and in meat or fish dishes.
Diatoms	Diatoms are used as a filtering material for processes such as the production of beverages, chemicals, industrial oils, cooking oils, sugars, water supplies, and the separation of wastes. They also are used as abrasives.

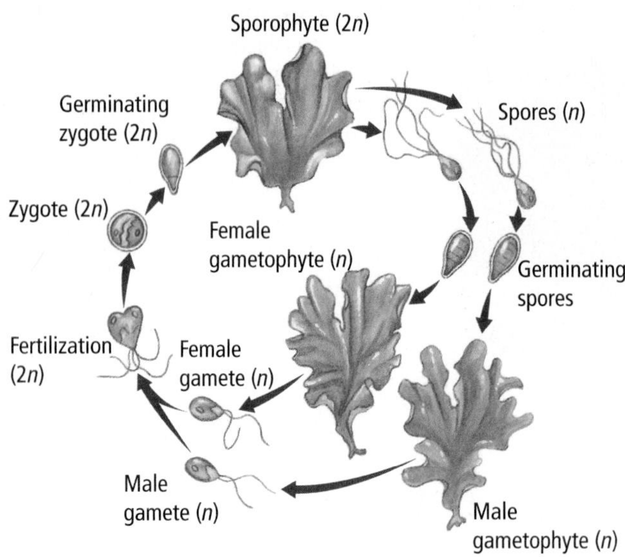

■ **Figure 19.21** The life cycle of many algae, including the sea lettuce *Ulva* shown here, includes an alternation between a diploid and haploid generation. Alternation of generations is also found in members of both the plant and fungi kingdoms.

Personal Tutor

To learn about algae life cycles, visit biologygmh.com.

Life Cycle of Algae

The life cycles of many algae are complex. Algae can alternate between spore-producing forms and gamete-producing forms. They can reproduce sexually as well as asexually. Green algae also reproduce asexually through fragmentation—a process in which a multicellular individual breaks into separate pieces and each grows into an individual organism.

Alternation of generations The life cycles of many algae exhibit a pattern called alternation of generations, illustrated in **Figure 19.21** for the sea lettuce *Ulva*. **Alternation of generations** is a life cycle of algae that takes two generations—one that reproduces sexually and one that reproduces asexually—to complete a life cycle. Organisms alternate between a diploid *(2n)* form and a haploid *(n)* form in which each is considered a generation.

Haploid and diploid generations The haploid form of the organism is called the gametophyte generation because it produces gametes. This generation is represented by the red arrows in the diagram. Gametes from two different organisms combine to form a zygote with two complete sets of chromosomes. The diploid form of the organism is represented by blue arrows in the diagram. The zygote develops into the sporophyte *(2n)*. In the sporophyte, some cells divide by meiosis and become haploid spores *(n)*. Spores are reproductive cells that develop into gametophytes. The new gametophytes continue the cycle as shown in **Figure 19.21.**

Section 19.3 Assessment

Section Summary

- Plantlike protists produce their own food through photosynthesis.
- Algae are important producers of oxygen and food for aquatic ecosystems.
- Euglenoids, diatoms, and dinoflagellates are unicellular algae.
- Red, brown, and green algae have multicellular forms.
- The life cycles of algae include an alternation of generations.

Understand Main Ideas

1. MAIN Idea **Explain** why algae are considered the primary producers for aquatic and marine ecosystems.
2. **Describe** the major characteristics of three groups of algae.
3. **Explain** why you would expect to find more evidence of diatoms than green algae in a sample of ocean floor sediment.
4. **Apply** what you know about photosynthesis to explain why most algae live at or near the surface of the water.

Think Critically

5. **Design an experiment** to determine the optimum color of light to grow green algae.
6. **Summarize** the role of secondary photosynthetic pigments in algae.

WRITING in **Biology**

7. Write a brief public service announcement explaining the dangers of eating shellfish during a red tide.

Section 19.4

Reading Preview

Objectives

- **Describe** the characteristics of cellular and acellular slime molds.
- **Compare** the life cycle of cellular and acellular slime molds.
- **Explain** how water molds obtain their nutrition.

Review Vocabulary

cellulose: a glucose polymer that forms the cell walls of plants and some funguslike protists

New Vocabulary

plasmodium
acrasin

Funguslike Protists

MAIN Idea **Funguslike protists obtain their nutrition by absorbing nutrients from dead or decaying organisms.**

Real-World Reading Link Have you ever heard the saying, "don't judge a book by its cover"? The same could be said of funguslike protists. Although at first glance they look like fungi, when they are examined more closely, many traits are revealed that are not true of fungi.

Slime Molds

As you can imagine, funguslike protists are protists that have some characteristics of fungi. Fungi and slime molds use spores to reproduce. Like fungi, slime molds feed on decaying organic matter and absorb nutrients through their cell walls. However, fungi and slime mold differ in the composition of their cell walls. Fungi cell walls are composed of a substance called chitin (KI tun). Chitin is a complex carbohydrate that is found in the cell walls of fungi, and in the external skeletons of insects, crabs, and centipedes. The cell walls of funguslike protists do not contain chitin as a true fungus does. The cell walls of these protists contain cellulose or celluloselike compounds.

Slime molds are found in a variety of colors, ranging from yellows and oranges to blue, black, and red as shown in **Figure 19.22.** They usually are found in damp, shady places where decaying organic matter is located, such as on a pile of decaying leaves or on rotting logs. Slime molds are divided into two groups—acellular slime molds and cellular slime molds.

 **Reading Check** **Compare and contrast** fungi and slime molds.

Figure 19.22 Slime molds have a variety of colors and shapes, but they all have funguslike characteristics.
Infer *Where might these slime molds be obtaining their nutrition?*

Myxamoebae slime mold

Red raspberry slime mold

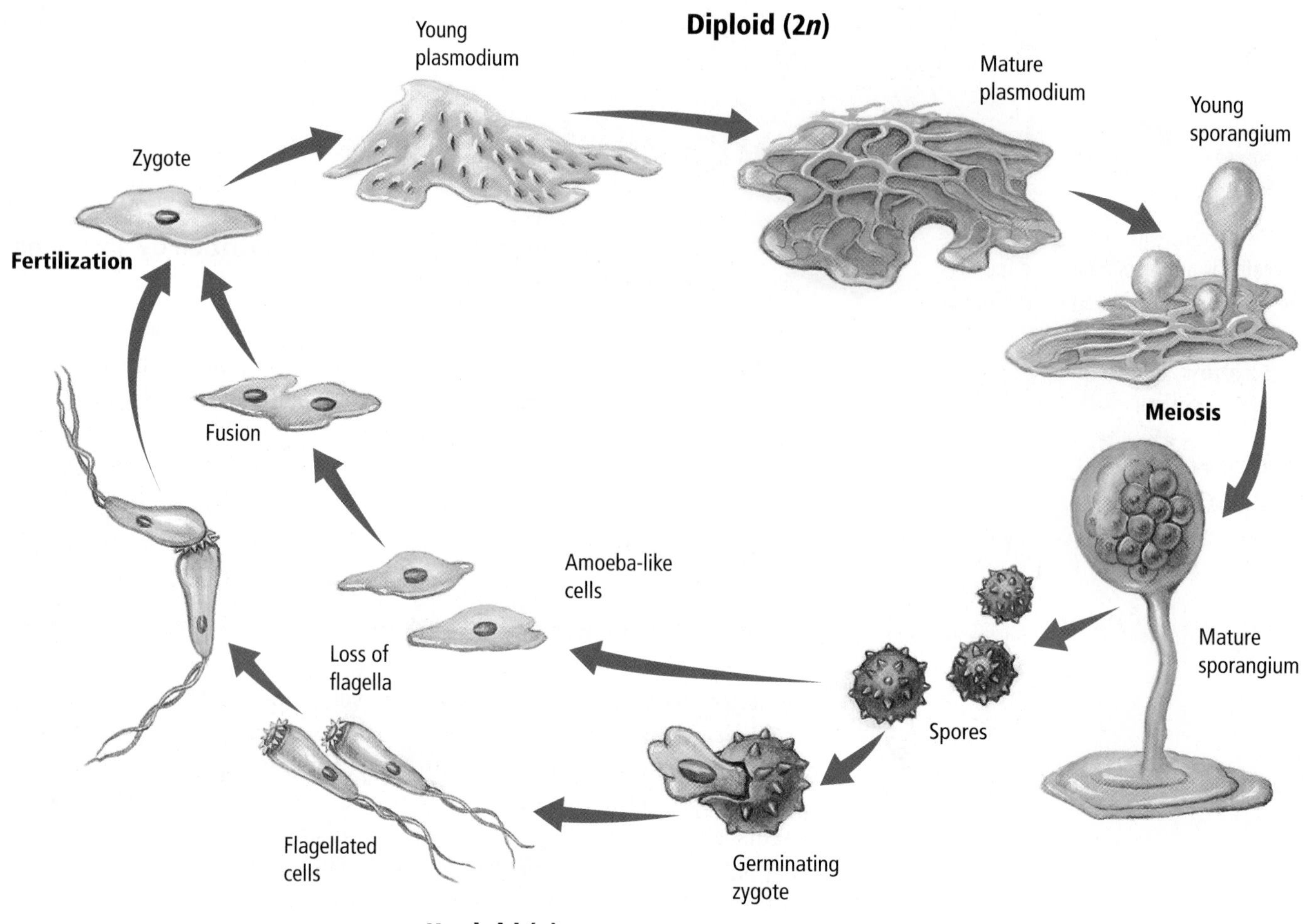

■ **Figure 19.23** Acellular slime mold goes through haploid and diploid phases during its life cycle.

Acellular slime molds Funguslike protists called acellular slime molds are found in the phylum Myxomycota (mihk soh mi COH tuh). They are acellular because they go through a phase in their life cycle in which the nucleus divides but no internal cell walls form, resulting in a mass of cytoplasm with multiple nuclei.

Follow the life cycle of a typical acellular slime mold shown in **Figure 19.23.** Acellular slime molds begin life as spores, usually when conditions are harsh—such as during a drought. In the presence of water, the spore produces a small mass of cytoplasm, or an amoeboid cell, or a cell with a flagella. The cell is propelled by the flagella until it comes in contact with a favorable surface. Then, the flagella permanently retract and the cell produces pseudopods that allow it to move like an amoeba. Both the flagellated cell and the amoeba-like cell are gametes and are haploid *(n).*

When two gametes unite, the next phase of the life cycle begins. The fertilized cells undergo repeated divisions of the nuclei, forming a plasmodium. A **plasmodium** (plaz MOH dee um)is a mobile mass of cytoplasm that contains many diploid nuclei but no separate cells. This is the feeding stage of the organism. It creeps over the surface of decaying leaves or wood like an amoeba and can grow as large as 30 cm in diameter. When food or moisture becomes limited, the slime mold develops spore-producing structures. Spores are produced through meiosis and dispersed by the wind. Once the spores are in the presence of water, the cycle repeats.

VOCABULARY

ACADEMIC VOCABULARY

Phase: a particular state in a regular cycle of changes.
The phases of the moon are based on the positions of Earth, the Sun, and the moon.

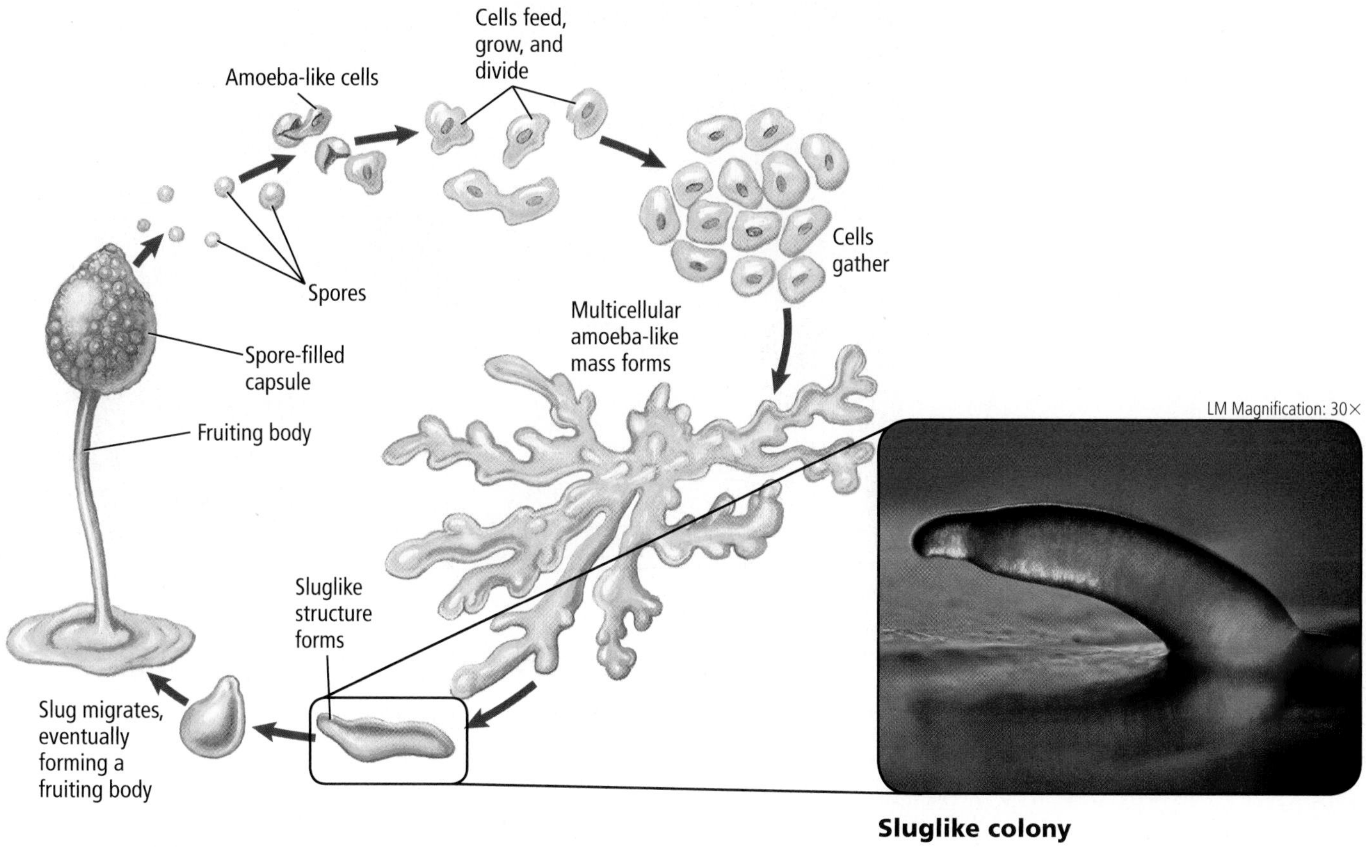

Cells feed, grow, and divide

Amoeba-like cells

Spores

Spore-filled capsule

Fruiting body

Cells gather

Multicellular amoeba-like mass forms

Sluglike structure forms

Slug migrates, eventually forming a fruiting body

LM Magnification: 30×

Sluglike colony

■ **Figure 19.24** Cellular slime molds reproduce both sexually and asexually. Amoeba-like cells congregate during asexual reproduction, shown above, to form a sluglike colony, which functions like a single organism.
Explain *why the sluglike stage is considered a colony.*

Cellular slime molds Cellular slime molds are found in the phylum Acrasiomycota (uh kray see oh my COH tuh). These funguslike protists creep over rich, moist soil and engulf bacteria. Unlike acellular slime molds, they spend most of their life cycle as single amoeba-like cells and they have no flagella.

The life cycle of cellular slime molds is shown in **Figure 19.24.** When food is plentiful, the single amoeba-like cells reproduce rapidly by sexual reproduction. During sexual reproduction, two haploid amoebas unite and form a zygote. The zygote develops into a giant cell and undergoes meiosis followed by several divisions by mitosis. Eventually, the giant cell ruptures, releasing new haploid amoebas.

When food is scarce, the single amoeba-like cells reproduce asexually. The starving amoeba-like cells give off a chemical called **acrasin** (uh KRA sun). The amoeba-like cells begin to congregate in response to the chemical signal, forming a sluglike colony that begins to function like a single organism. The colony migrates for a while, eventually forming a fruiting body, like the one shown in **Figure 19.25.** The fruiting body produces spores. Once the spores are fully developed, they are released. The spores germinate, forming amoeba-like cells, and the cycle repeats.

Reading Check **Infer** why the stages in the life cycle of cellular slime molds contribute to their long-term survival.

■ **Figure 19.25** Cellular slime molds produce fruiting bodies that contain spores during part of their life cycle.

SEM Magnification: 2700×

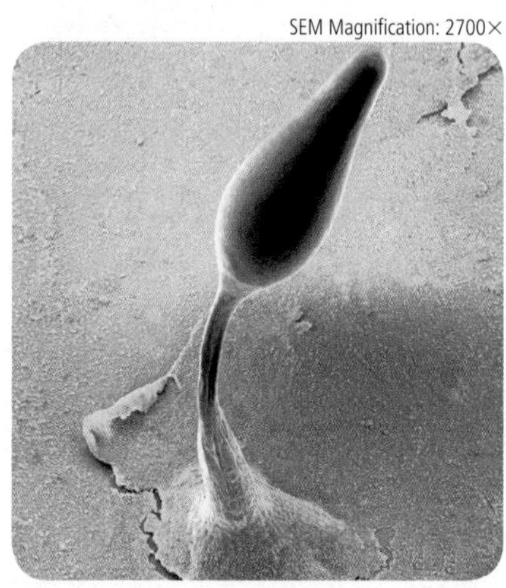

■ **Figure 19.26** This water mold is absorbing nutrients found in this dead insect.
Explain *What funguslike characteristic do water molds have?*

LM Magnification: 10×

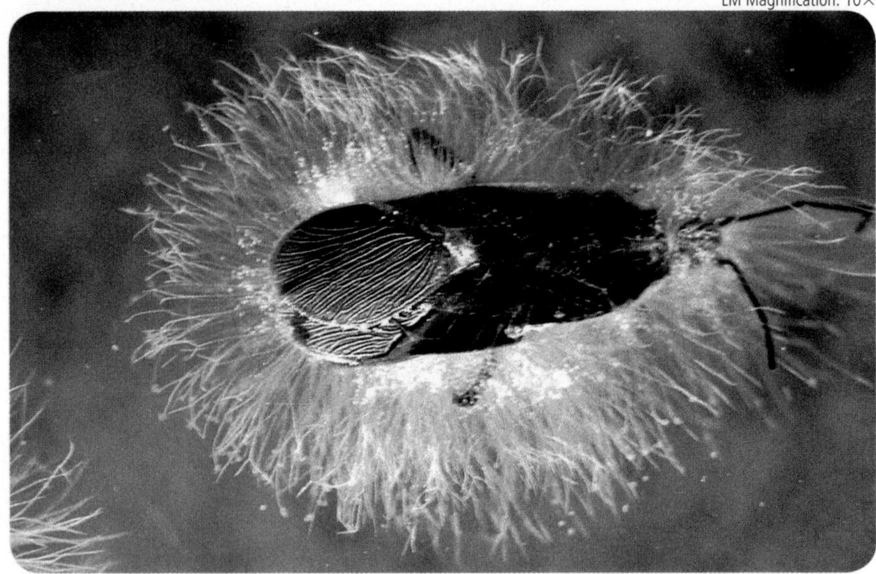

Water mold

Water Molds and Downy Mildew

There are more than 500 species of water molds and downy mildews in the phylum Oomycota (oo oh my COH tuh). Most members of this group of funguslike protists live in water or damp places. Some absorb their nutrients from the surrounding water or soil, while others obtain their nutrients from other organisms, as shown in **Figure 19.26.**

Originally, water molds were considered fungi because of their method of obtaining nutrients. Like fungi, water molds envelope their food source with a mass of threads; they break down the tissue, and absorb the nutrients through their cell walls. Although this is characteristic of fungi, water molds differ from fungi in the composition of their cell walls and they produce flagellated reproductive cells. Recall that the cell walls of funguslike protists are composed of cellulose and celluloselike compounds.

 **Reading Check** **Compare and contrast** water molds and fungi.

MiniLab 19.2

Investigate Slime Molds

What is a slime mold? In a kingdom of interesting creatures, slime molds perhaps are the most interesting. Observe different types of slime molds and observe the unusual nature of their bodies.

Procedure

1. Read and complete the lab safety form.
2. Obtain **slides of different specimens of slime molds.** Examine the slides under a **microscope.**
3. Create a data table to record your information. Sketch and describe each specimen.

Analysis

1. **Compare and contrast** the specimens.
2. **Identify** specimens that have similar characteristics. Explain why the specimens are similar.
3. **Think Critically** How would you classify each specimen that you examined? Explain.

Infected potato

Healthy potato

■ **Figure 19.27** Compare the infected potato on the left with the normal one on the right. *Phytophthora infestans* will destroy the harvested potato in a matter of weeks.

Connection to History One member of phylum Oomycota has had a far-reaching impact. The downy mildew *Phytophthora infestans* (FI toh fah thor uh • in FEST unz) infects potato plants and destroys the potato, as shown in **Figure 19.27.** This organism devastated the potato crop of Ireland in the 19th century. Because the potato was their primary food source, about one million people died of starvation or famine-related diseases in Ireland. Ironically, during this time many other agricultural products were produced in Ireland. The Irish peasants could not afford to purchase the agricultural products, so the products were exported to Britain. The British government did provide some assistance to the peasants, but it was too little to prevent the widespread famine. During this time, a large number of people emigrated from Ireland to the United States to escape the terrible famine.

Section 19.4 Assessment

Section Summary

- The cell walls of funguslike protists do not contain chitin.
- Slime molds, water molds, and downy mildew grow in aquatic or damp places.
- Acellular slime molds form a plasmodium that contains many nuclei but no separate cells.
- Cellular slime molds form colonies of cells to reproduce.
- Water molds envelop their food source with a mass of threads.

Understand Main Ideas

1. MAIN Idea **Explain** how funguslike protists obtain their nutrition.
2. **Explain** the life cycle of a cellular slime mold.
3. **Describe** how amoeba-like cells move.
4. **Outline** the life cycles of cellular and acellular slime molds.
5. **Classify** an organism that has cell walls made of cellulose and absorbs its nutrients from dead organisms.

Think Critically

6. **Design an experiment** to determine the moisture requirements of an acellular slime mold.
7. **Recommend** a procedure a garden shop owner should follow in order to prevent slime molds from growing on his or her wooden benches.

WRITING in Biology

8. Write a short newspaper article about the Irish potato famine.

NATIONAL GEOGRAPHIC

In the Field

Career: Nanotechnologist

Diatoms: Living Silicon Chips

Diatoms have recently gained the attention of nanotechnologists—scientists who engineer devices on the atomic level. Diatoms build intricate shells with incredible precision and regularity. Nanotechnologists think these organisms could be used to build useful structures from silicon on the atomic level.

Nature's nanotechnologists Humans still have a lot to learn from diatoms about constructing materials on the nanoscale. Currently, nanotechnologists etch features on to silicon and other materials to produce components. The process is costly, time-consuming, and generates chemical waste.

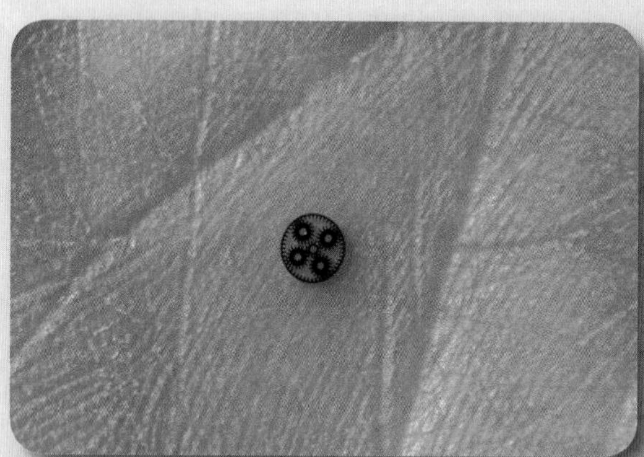

Silicon dioxide shell in a hand

Living silicon chips Diatoms have been described as living silicon chips because they construct their shells atom by atom. Silicon derived from sea water is processed into intricate microstructures to form a rigid silica shell, such as the one shown in the photo. Each diatom species forms a unique and potentially useful shell structure.

Diatoms

To create nanomaterials from diatoms, scientists prepare feeding solutions containing silicon and other elements they wish to test. The diatoms take these elements in and use them to build shells. When diatoms replace silicon atoms in their shell with elements like magnesium or titanium, a structurally intact unit with a desired shape and chemical makeup is produced. Scientists are working to use diatom shell patterns, many of which cannot currently be duplicated by nanotechnologists, as templates to build components with desired specifications.

Future Applications Diatoms might prove to be an important tool in the evolving science of nanotechnology with potential applications in biomedicine, telecommunications, and energy storage and production.

WRITING in Biology

Newspaper Article The worldwide need for nanotechnology workers could reach two million by the year 2015. Visit biologygmh.com to find more information about the field of nanotechnology. Write a want ad for a specific career in nanotechnology.

BIOLAB

Design Your Own

INVESTIGATE: HOW DO PROTOZOA BEHAVE?

Background: Animals respond and react to the world around them. One such type of reaction is known as *taxis,* in which an animal orients itself toward (positive) or away (negative) from a stimulus. Some of the things animals respond to are: light (phototaxis), temperature (thermotaxis), chemicals (chemotaxis), and gravity (gravitaxis).

Question: *How do simple unicellular, animal-like protozoa respond to stimuli?*

Materials

cultures of live protozoa
compound microscope
glass slides and coverslips
materials needed to produce stimuli

Safety Precautions

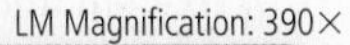

WARNING: ***Use care when handling slides. Dispose of any broken glass in a container provided by your teacher.***

LM Magnification: 390×

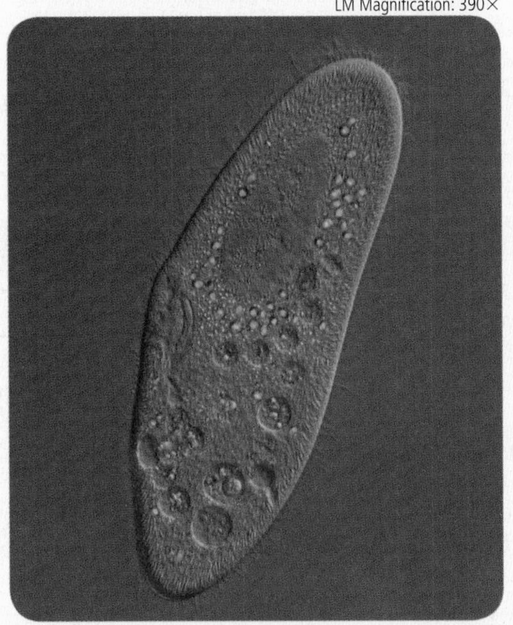

Plan and Perform the Experiment

1. Read and complete the lab safety form.
2. Design an experiment to answer the question to the left. Reword the original question to include the taxis you plan to investigate.
3. Make sure your teacher approves your plan before you proceed.
4. Collect the materials and supplies needed and begin conducting your experiment.
5. Dispose of your protozoan cultures as instructed by your teacher.

Analyze and Conclude

1. **Observe and Infer** Some protozoa often are described as animal-like. What animal-like characteristics did you observe?
2. **State the Problem** What stimuli were you trying to test with your experimental design?
3. **Hypothesize** What was your hypothesis for the question to be solved?
4. **Summarize** What data did you collect during the experiment?
5. **Analyze and Conclude** Did your data support your hypothesis? What is your conclusion?
6. **Error Analysis** Compare your data and conclusions with other students in your class. Explain the differences in data.

WRITING in Biology

Report In this lab, you tested the response of an organism to a stimuli. To find out more about how scientists test an organism's response to a stimulus, visit Biolabs at biologygmh.com. Write a short report critiquing your methods. Include ways in which you can improve your techniques.

Study Guide

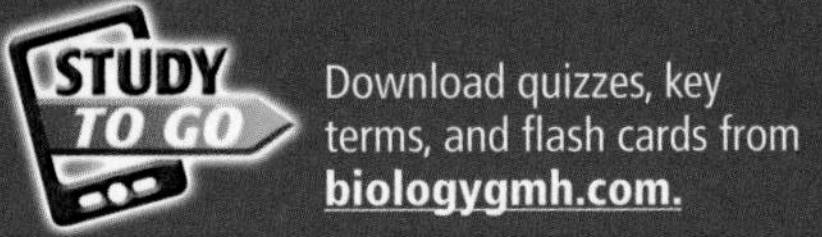

Download quizzes, key terms, and flash cards from biologygmh.com.

FOLDABLES **Hypothesize** Is it possible to describe the typical protist? Hypothesize why the organisms in the Kingdom Protista are more diverse than the organisms in any of the other kingdoms.

Vocabulary	Key Concepts
Section 19.1 Introduction to Protists	
• microsporidium (p. 544) • protozoan (p. 542)	MAIN Idea Protists form a diverse group of organisms that are subdivided based on their method of obtaining nutrition. • Protists include unicellular and multicellular eukaryotes. • Protists are classified by their methods of obtaining food. • The first protists might have formed through endosymbiosis. • Protists might have been the first eukaryotic cells with chloroplasts and mitochondria, evolving billions of years ago.
Section 19.2 Protozoans—Animal-like Protists	
• contractile vacuole (p. 547) • pellicle (p. 547) • pseudopod (p. 550) • test (p. 550) • trichocyst (p. 547)	MAIN Idea Protozoans are animal-like, heterotrophic protists. • Protozoans are single-celled protists that feed on other organisms to obtain nutrients. • Protozoans live in a variety of aquatic environments. • Protozoans reproduce in a variety of ways, including sexually and asexually. • Protozoans have specialized methods for movement, feeding, and maintaining homeostasis.
Section 19.3 Algae—Plantlike Protists	
• alternation of generations (p. 560) • bioluminescent (p. 555) • colony (p. 557)	MAIN Idea Algae are plantlike, autotrophic protists that are the producers for aquatic ecosystems. • Plantlike protists produce their own food through photosynthesis. • Algae are important producers of oxygen and food for aquatic ecosystems. • Euglenoids, diatoms, and dinoflagellates are unicellular algae. • Red, brown, and green algae have multicellular forms. • The life cycles of algae include an alternation of generations.
Section 19.4 Funguslike Protists	
• acrasin (p. 563) • plasmodium (p. 562)	MAIN Idea Funguslike protists obtain their nutrition by absorbing nutrients from dead or decaying organisms. • The cell walls of funguslike protists do not contain chitin. • Slime molds, water molds, and downy mildew grow in aquatic or damp places. • Acellular slime molds form a plasmodium that contains many nuclei but no separate cells. • Cellular slime molds form colonies of cells to reproduce. • Water molds envelop their food source with a mass of threads.

Biology Online Vocabulary PuzzleMaker biologygmh.com

CHAPTER 19 Assessment

Section 19.1

Vocabulary Review

Answer the following questions with complete sentences.

1. What is another name for animal-like protists?
2. What are microscopic protozoans that are found in the gut of insects?

Understand Key Concepts

3. Which process is most likely the way in which the first protists formed?
   A. aerobic respiration **C.** endosymbiosis
   B. decomposition **D.** photosynthesis
4. Which method below is used to divide protists into three groups?
   A. method of getting food
   B. method of movement
   C. type of reproduction
   D. type of respiration
5. Which is least likely to be a suitable environment for protists?
   A. decaying leaves **C.** damp soil
   B. the ocean **D.** dry sand

Use the photo below to answer questions 6 and 7.

LM Magnification: 125×

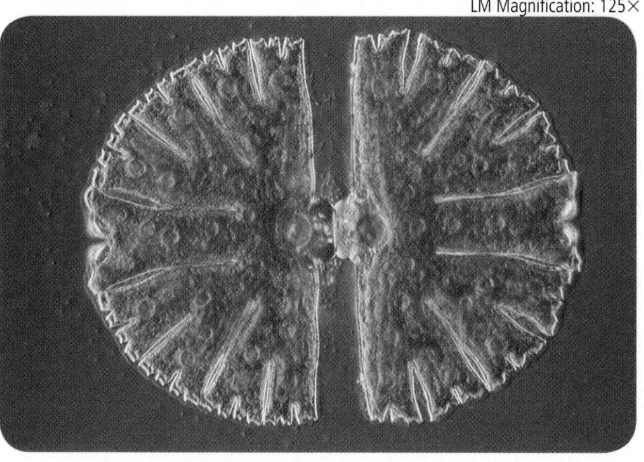

6. To which group does the protist belong?
   A. algae **C.** funguslike
   B. animal-like **D.** protozoan
7. Which term best describes this protist?
   A. acellular **C.** multicellular
   B. eukaryotic **D.** prokaryotic

Constructed Response

8. **Open Ended** Describe three locations near your home or school where you might be able to find protists.
9. **CAREERS IN BIOLOGY** If you were a taxonomist given the task of organizing protists into groups, would you use the same method described in this book? Explain your answer.

Think Critically

10. **Predict** changes in protist populations if an area had an above-average amount of rainfall.

Section 19.2

Vocabulary Review

Define each of the structures below and provide an example of an organism where it could be found.

11. pseudopod
12. contractile vacuole
13. test

Understand Key Concepts

Use the diagram below to answer question 14.

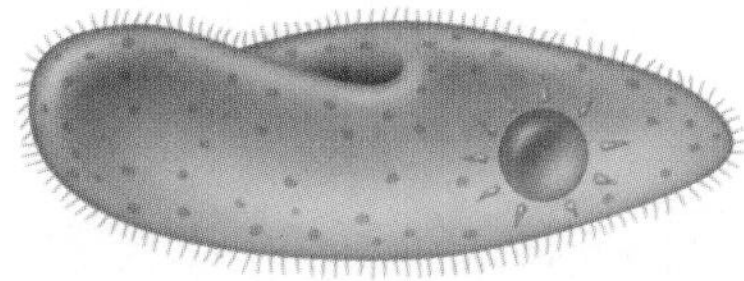

14. Which structure does this organism use for movement?
    A. cilia
    B. contractile vacuole
    C. flagella
    D. pseudopodia
15. What does the paramecium's contractile vacuole help regulate inside the cell?
    A. amount of food **C.** movement
    B. amount of water **D.** reproduction
16. Which are most likely to form fossils?
    A. apicomplexans **C.** foraminifera
    B. flagellates **D.** paramecia

Constructed Response

17. **Open Ended** Explain why termites might die if their symbiotic flagellates died.

18. **Short Answer** Describe the process of conjugation in paramecia.

Think Critically

19. **Apply Concepts** Recommend several options a village might consider to slow down the spread of malaria.

20. **Research Information** Research other diseases that are caused by protozoans. Use a map and plot locations where the diseases occur.

Section 19.3

Vocabulary Review

Match each definition below with the correct vocabulary term from the Study Guide page.

21. a life cycle of algae that requires two generations

22. a group of cells living together in close association

23. gives off light

Understand Key Concepts

Use the photo below to answer question 24.

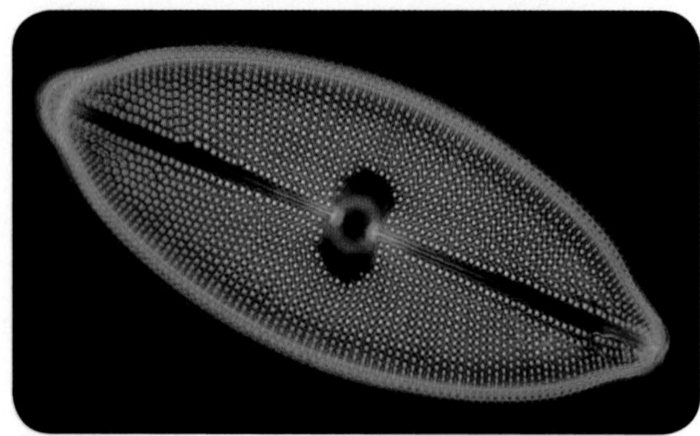

24. Where does this organism store its excess food?
    A. cellulose **C.** protein
    B. oil **D.** carbohydrate

25. Which are used in the human food supply?
    A. dinoflagellates **C.** protozoans
    B. euglenoids **D.** red algae

26. Which organism has silica walls?
    A. brown alga **C.** dinoflagellate
    B. diatom **D.** euglenoid

Use the illustration below to answer questions 27 and 28.

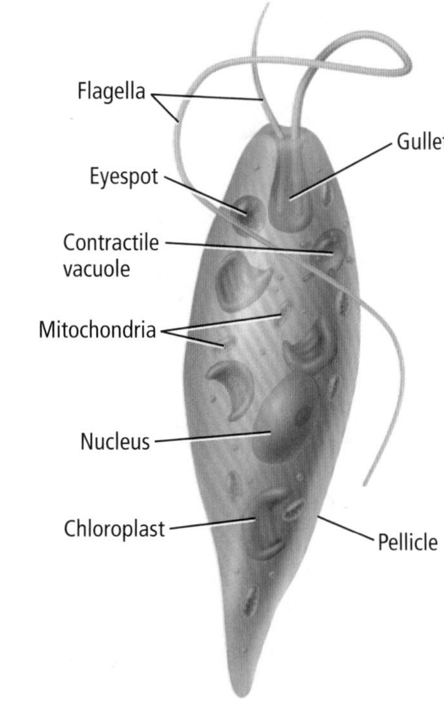

27. Which structure is used by the organism above for movement?
    A. cilia **C.** flagella
    B. contractile vacuole **D.** pseudopod

28. Which structure is used to sense light?
    A. chloroplast **C.** nucleus
    B. eyespot **D.** pellicle

Constructed Response

29. **Open Ended** Why are there more fossils of diatoms, foraminiferans, and radiolarians than of other algae?

30. **Short Answer** Explain why diatoms must sometimes reproduce sexually.

31. **Short Answer** Explain the relationship between the sporophyte and gametophyte in alternation of generations.

Think Critically

32. **Analyze** the difference between freshwater algae and marine algae.

33. **Recognize Cause and Effect** Explain the effects of a marine parasite that kills all phytoplankton.

Section 19.4

Vocabulary Review

Replace the underlined words with the correct vocabulary term from the Study Guide page.

34. A motile organism that consists of many diploid nuclei but no separate cells is a protoplasm.

35. Starving amoeboid cells give off a chemical called arsenic.

Understand Key Concepts

36. Acellular slime molds have many nuclei, but what structure do they not have?
    - **A.** chromosomes
    - **B.** spores
    - **C.** separate cells
    - **D.** cilia

37. Which is present in the life cycle of water molds in a flagellated form?
    - **A.** nuclei
    - **B.** plasmodia
    - **C.** pseudopods
    - **D.** reproductive cells

Constructed Response

38. **Short Answer** Compare and contrast a water mold and a cellular slime mold.

39. **Open Ended** Describe some environmental conditions that might lead to the production of spores by an acellular slime mold.

Think Critically

40. **Analyze and Conclude** During the multinucleated plasmodial stage, could acellular slime molds be classified as multicellular organisms? Explain your reasoning.

Additional Assessment

41. *WRITING in* **Biology** Choose one protist and help it "evolve" by determining a new organelle or structure that is going to develop. How will this new condition affect the protist? Will this change increase or decrease the chance of survival?

DBQ Document-Based Questions

The text below describes a new detection method for finding microscopic organisms in water sources.

The protozoans *Giardia lamblia* and *Cryptosporidium parvom* are major causes of waterborne intestinal diseases throughout the world. A very sensitive detection method was developed using the DNA amplification procedure—polymerase chain reaction. This procedure can detect the presence of incredibly small amounts of these pathogens—as little as a single cell in two liters of water.

Data obtained from: Guy, et al. 2003. Real-time PCR for quantification of *Giarida* and *Cryptosporidium* in environmental water samples and sewage. *Applications of Environmental Biology* 2003 69(9): 5178-5185.

42. Explain how this detection method might be used by municipal water departments.

43. Analyze the significance of this research for global human health concerns especially in remote regions of the world.

44. Predict how this detection method might be used to monitor the level of organisms that cause red tides.

Cumulative Review

45. Point out how meiosis provides genetic variety. **(Chapter 10)**

46. Sketch a branching diagram that explains evolution of hominoids from genus *Proconsul* to genus *Homo.* **(Chapter 16)**

47. Pick the traits you would use to make a key for classifying the kingdoms. Describe why you chose the characteristics on the list. **(Chapter 17)**

Standardized Test Practice

Cumulative

Multiple Choice

1. Which environment would likely have chemosynthetic autotrophic eubacteria?
   A. coral reef
   B. deep-ocean volcanic vent
   C. lake in the mountains
   D. soil near a spring

Use the diagram below to answer questions 2 and 3.

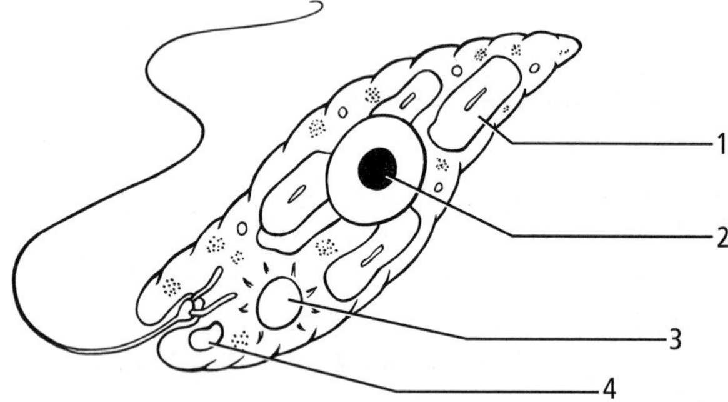

2. Which number represents the eyespot of the *Euglena?*
   A. 1
   B. 2
   C. 3
   D. 4

3. Which number represents an organelle that captures energy for the cell from sunlight?
   A. 1
   B. 2
   C. 3
   D. 4

4. Which do the two bats *Craseonycteris thonglongyai* and *Noctilio leporinus* have in common?
   A. division
   B. genus
   C. phylum
   D. species

5. Suppose you are investigating bone characteristics of two birds to determine how closely they are related in terms of phylogeny. Which type of evidence are you using?
   A. biochemical characters
   B. cellular characters
   C. chromosomal characters
   D. morphological characters

Use the diagram below to answer question 6.

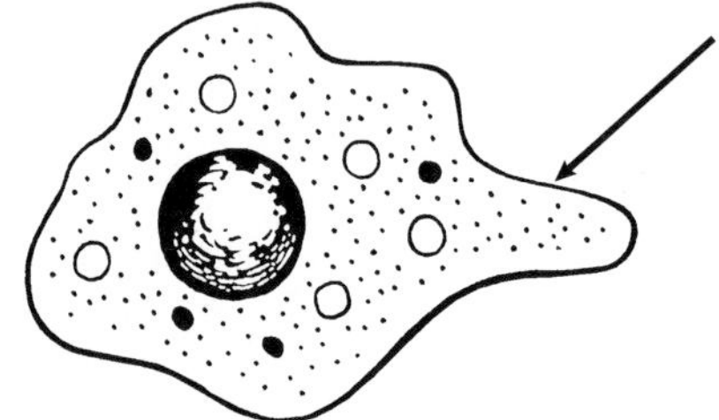

6. Members of the phylum Sarcodina use this structure for locomotion and which other activity?
   A. conjugation
   B. feeding
   C. protection
   D. reproduction

7. How do prions harm their host?
   A. by activating synthesis of viral RNA
   B. by causing normal proteins to mutate
   C. by deactivating part of the host's DNA
   D. by disrupting the way cells reproduce

8. Which could be a derived, rather than ancestral, character in one group of vertebrates?
   A. nervous system
   B. organized systems of tissues
   C. role of ATP in mitochondria
   D. wings used for flight

Short Answer

Use the diagram below to answer questions 9 and 10.

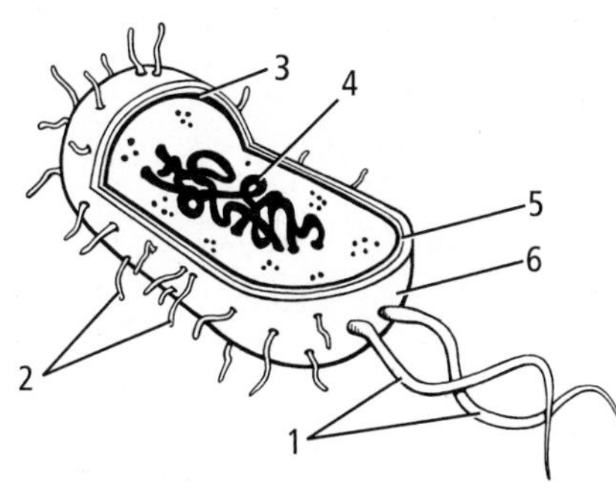

9. Name the parts of this bacterium and classify each part according to its function.

10. From the diagram, deduce how the structure of a typical bacterium enables it to survive in a harsh environment that frequently changes.

11. Imagine that you have been asked to do an experiment in which you boil different leaves and flower petals in different solutions to extract the pigments. State what safety equipment would be appropriate for your experiments and give reasoning for your choices.

12. Organisms in Kingdom Fungi and Kingdom Plantae used to be classified in the same kingdom. State a reason to classify them in different kingdoms.

13. Write a hypothesis about how the life cycle of a retrovirus, such as HIV, might be disrupted to slow or stop the reproduction of the virus.

Extended Response

Use the illustration below to answer question 14.

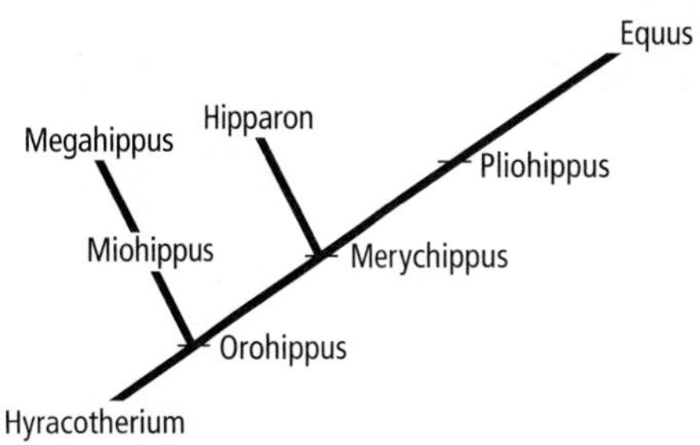

14. The figure above shows the evolution of horses, including the modern-day horse, *Equus*. Does this diagram support the idea of gradualism or of punctuated equilibrium? Explain your answer.

Essay Question

One challenge people face is the presence of antibiotic-resistant bacteria. Antibiotics are used to treat many diseases. Generally, they improve the quality of life of people. However, the widespread use and misuse of antibiotics has created antibiotic-resistant bacteria. This means that some diseases caused by bacteria no longer can be cured with the same antibiotics. Doctors must use new and stronger antibiotics to cure the disease. This gives bacteria an opportunity to develop a resistance to the new antibiotics. Unfortunately, antibiotic resistance in bacteria is spreading faster than new antibiotics are being developed.

Using the information in the paragraph above, answer the following question in essay format.

15. Evaluate how the characteristics of bacteria contribute to the rapid development of antibiotic-resistant bacteria.

NEED EXTRA HELP?															
If You Missed Question . . .	1	2	3	4	5	6	7	8	9	10	11	12	13	14	15
Review Section . . .	18.1	19.3	19.3	17.1	17.2	19.2	18.2	17.2	18.1	18.1	1.3	17.3	18.2	15.3	18.1

CHAPTER 20 Fungi

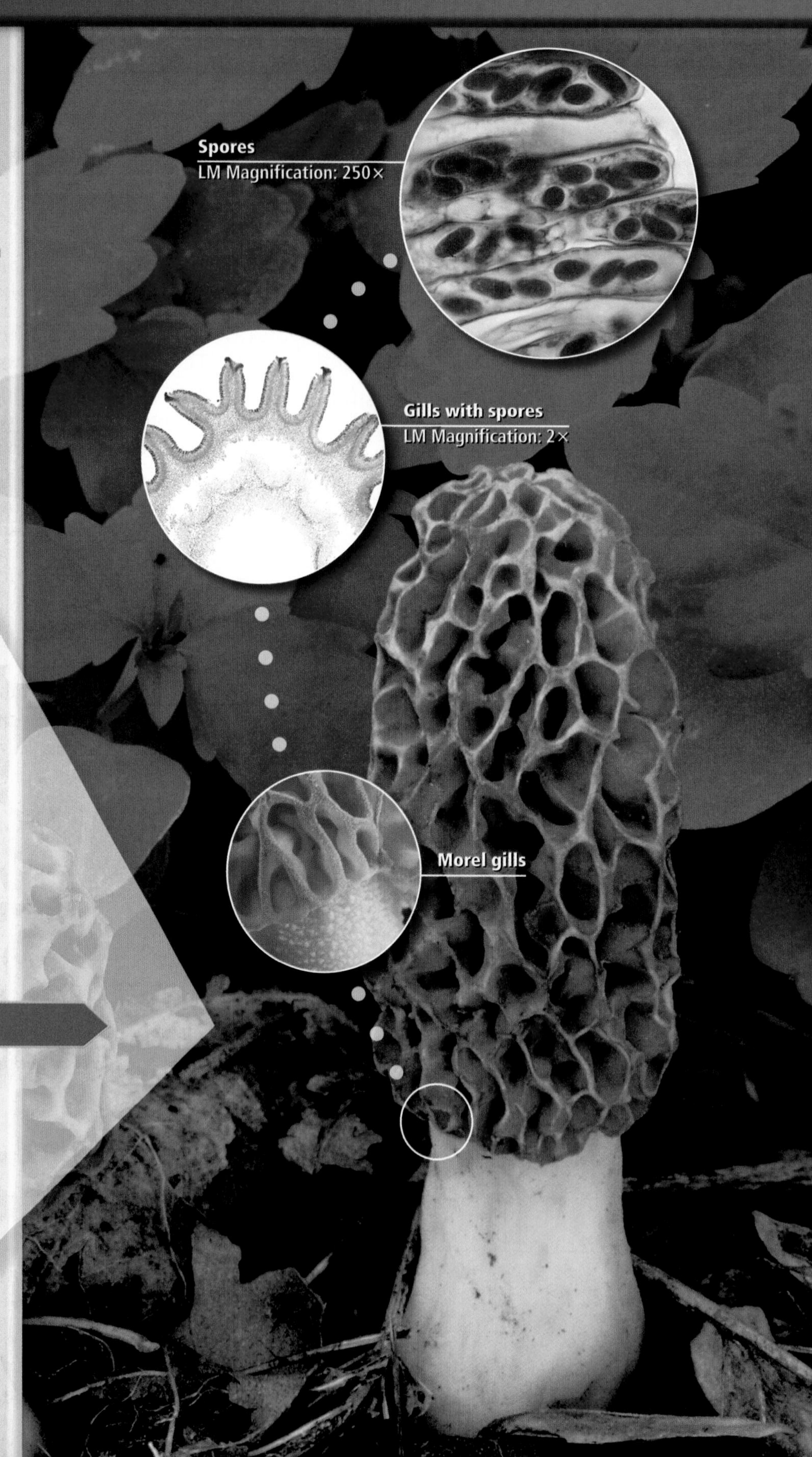

BIG Idea The Kingdom Fungi is made up of four phyla based on unique structures, methods of nutrition, and methods of reproduction.

Section 1
Introduction to Fungi
MAIN Idea Fungi are unicellular or multicellular eukaryotic heterotrophs that are decomposers.

Section 2
Diversity of Fungi
MAIN Idea Fungi exhibit a broad range of diversity and are classified into four major phyla.

Section 3
Ecology of Fungi
MAIN Idea Lichens and mycorrhizae demonstrate important symbiotic relationships between fungi and other organisms.

BioFacts

- Throughout history, people have used fungi to make disease-treating drugs like antibiotics.
- Fungi provide us with delicious food products such as soy sauce and blue cheese.
- One portobello mushroom has more potassium than a banana.

LAUNCH Lab

What differences exist among fungi?

Fungi display enormous diversity. The organisms in this kingdom vary in size from a single cell to a fungus found in the Malheur National Forest that is 5.6 km wide! In this lab, you will observe some of the differences among fungi.

Procedure

1. Read and complete the lab safety form.
2. Create a data table to record your observations of the fungi samples provided by your teacher.
3. Study each fungus carefully. Wash your hands thoroughly after handling fungi.
4. Describe each fungus sample as completely as you can. Include properties like color, shape, size, and growth medium.
5. Dispose of fungi and clean your work station according to your teacher's instructions.

Analysis

1. **Contrast** What physical characteristics varied most among your samples?
2. **Compare** Summarize any similarities you observed or can infer among the fungi you examined.

Visit biologygmh.com to:

- study the entire chapter online
- explore the Concepts in Motion, the Interactive Table, Microscopy Links, and links to virtual dissections
- access Web links for more information, projects, and activities
- review content online with the Interactive Tutor, and take Self-Check Quizzes

Obtaining Nutrients Make the following Foldable to help you identify three types of fungi that differ in how they obtain food.

STEP 1 Fold a sheet of paper into thirds.

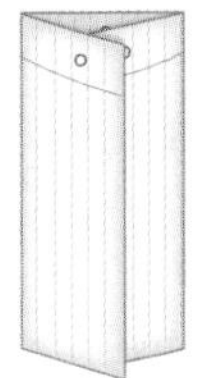

STEP 2 With a pencil or pen, trace along the fold lines to form a three-column chart.

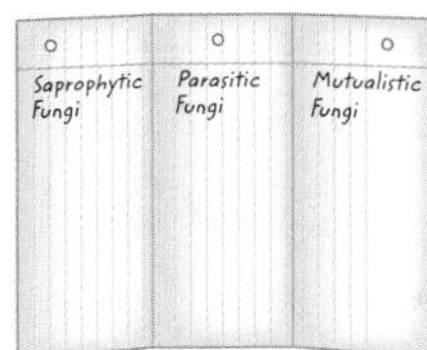

STEP 3 Label the columns *Saprophytic Fungi, Parasitic Fungi,* and *Mutualistic Fungi.*

FOLDABLES **Use this Foldable with Section 20.1.** As you study the section, summarize how the three types of fungi obtain nutrients.

Section 20.1

Introduction to Fungi

Reading Preview

Objectives

- **Identify** the major characteristics of organisms in the Kingdom Fungi.
- **Explain** how fungi obtain nutrients, including their role as decomposers.
- **Identify** three types of asexual reproduction in fungi.

Review Vocabulary

decomposer: organism that feeds on and breaks down dead organisms, recycling nutrients back into food webs

New Vocabulary

chitin
hypha
mycelium
fruiting body
septum
haustorium
spore
sporangium

MAIN Idea **Fungi are unicellular or multicellular eukaryotic heterotrophs that are decomposers.**

Real-World Reading Link When you listen to the radio, how is it that you can always identify your favorite group? Maybe it is by the common characteristics of the band, such as their instruments or the lead singer's voice. Organisms in the same kingdom also share common, identifying characteristics.

Characteristics of Fungi

Some of the largest and oldest organisms on Earth belong to the Kingdom Fungi. When you see the word *fungi* (FUN ji) (singular, fungus), you might envision the mushrooms found in grocery stores or ones that grow in your backyard. In eastern Oregon, there is a honey mushroom that is so big that it is called the "Humongous Fungus." The honey mushroom, similar to the one shown in **Figure 20.1,** is estimated to be at least 2400 years old. All fungi are eukaryotic heterotrophs. More than 100,000 species of fungi have been identified.

Multicellular fungi Most members of the Kingdom Fungi, such as the honey mushroom, are multicellular. At first glance, you might think these multicellular fungi look like plants. Although they do not contain chloroplasts, at one time fungi were classified as plants because they appeared to have some characteristics similar to plants. However, after careful study, scientists decided that fungi are different enough to be placed in their own kingdom.

Unicellular fungi Yeasts are unicellular fungi. They are found throughout the world in soils, on plant surfaces, and even in the human body. While there are hundreds of different kinds of yeasts, the most familiar yeasts are used commercially to produce breads, beer, and wine. The yeast *Candida albicans,* shown in **Figure 20.1,** can cause a yeast infection in humans.

■ **Figure 20.1** Most fungi are multicellular like this honey mushroom growing on a tree. Some fungi are unicellular, such as this yeast colony called *Candida albicans.*

Honey mushroom

Color-Enhanced SEM Magnification: 1250×

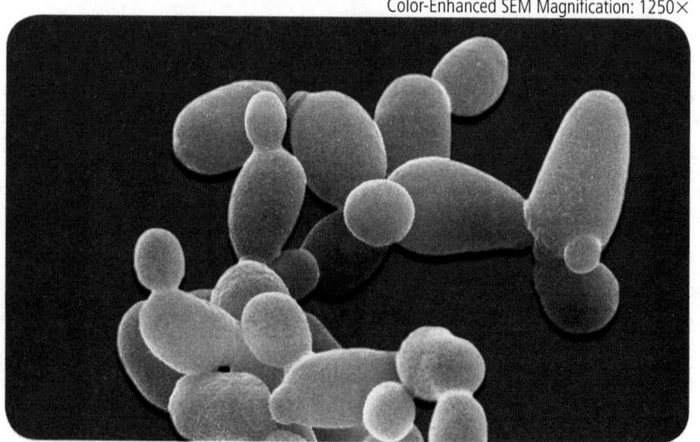

Colony of *Candida albicans*

Major Features in Fungi

Some features in fungi that distinguish them from plants include their cell walls, their hyphae, and their cross-walls.

Cell walls One significant difference between plants and fungi is the composition of their cell walls. Plants have a cell wall composed of cellulose, while fungi have a cell wall composed of chitin. **Chitin** (KI tun)is a strong, flexible polysaccharide that is found in the cell walls of all fungi and in the exoskeletons of insects and crustaceans. Recall from Chapter 6 that polysaccharides are carbohydrate polymers that are composed of many simple sugar subunits. Chitin is one of the most abundant organic compounds on Earth.

Hyphae The physical structure of fungi also differs from plants. Look at the magnified image of the fungus in **Figure 20.2** and notice that it is composed of long chains of cells. Without a microscope, hyphae appear to be threadlike filaments. These filaments are the basic structural units that make up the body of a multicellular fungus and are called **hyphae** (HI fee) (singular, hypha). Hyphae grow at their tips and branch repeatedly to form a netlike mass called a **mycelium** (mi SEE lee um) (plural, mycelia). While the mycelium is visible in some fungi, it is packed so tightly in mushrooms that it is almost impossible to distinguish the individual hyphae. The fungus that you see above ground, illustrated in **Figure 20.2,** is a reproductive structure called the **fruiting body.** The hyphae form all parts of the mushroom, including the fruiting body above ground and the mycelium below ground. The extensive hyphae of fungi give them an advantage in obtaining nutrients by providing a large surface area for nutrient absorption.

Reading Check **Describe** the structural unit of a mushroom.

Connection to History Fungal hyphae are found in the work of many medieval painters. Victorian illustrators often link fairies and toadstools (another name for mushrooms). Today, the colorful spotted cap of a *Fly Agaric* mushroom often is associated with a gnome or sprite in children's stories.

VOCABULARY
WORD ORIGIN
Hypha
comes from the Greek word *hyphos,* meaning *web.*

Figure 20.2
Left: The visible body and underground structure of a multicellular fungus is made up of long chains of cells called hyphae.
Right: A multicellular fungus consists of an above-ground fruiting body.
Infer ***What do you think are the advantages of having a filamentous body?***

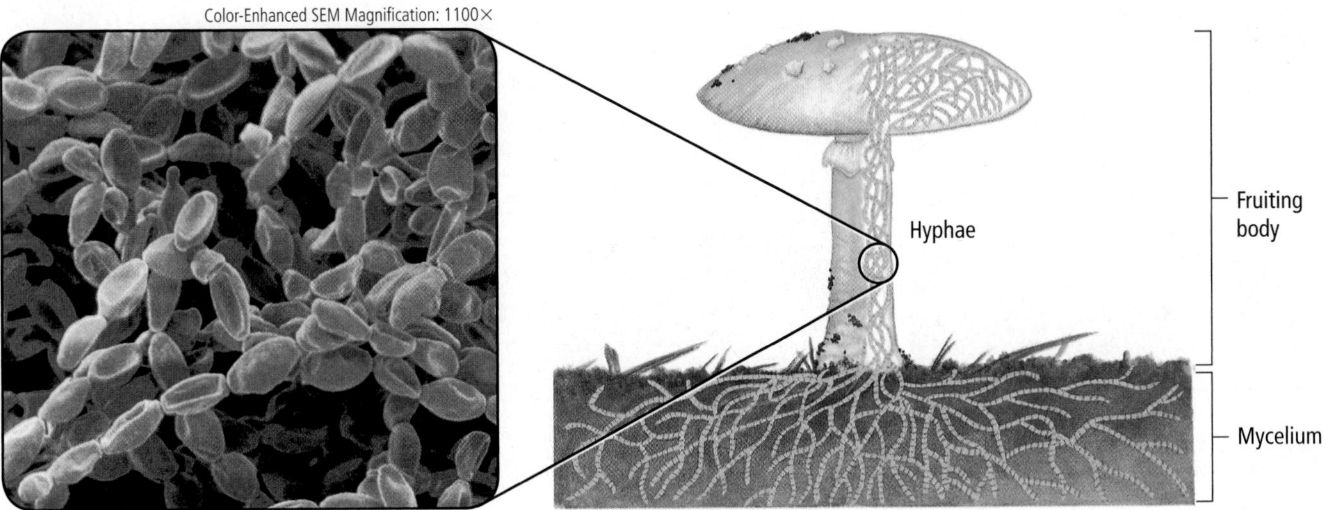

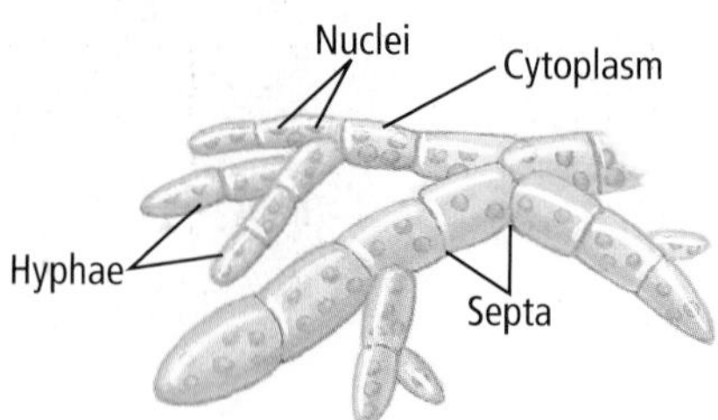

Septate Hyphae

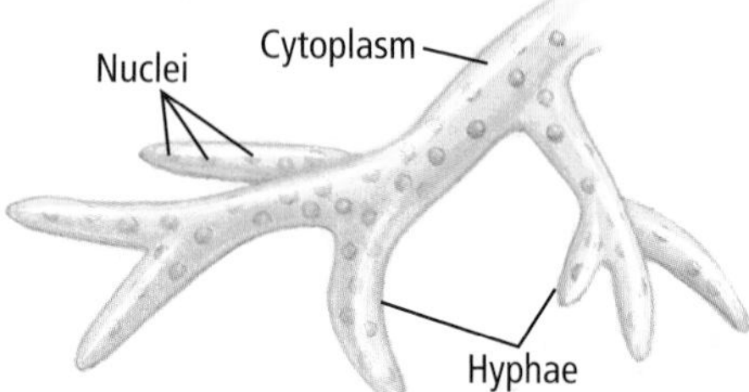

Aseptate Hyphae

■ **Figure 20.3**
Top: Some fungi have hyphae that are divided by cross-walls called septa.
Bottom: Other fungi do not have hyphae with septa.

Cross-walls In many fungi, hyphae are divided into cells by cross-walls called **septa** (singular, septum), as shown in **Figure 20.3.** The septa have large pores that allow nutrients, cytoplasm, organelles, and, in some cases, nuclei to flow between cells.

Some fungi are aseptate, meaning they have no septa. The cytoplasm, containing hundreds or thousands of nuclei, flows freely through the hyphae. This condition is a result of repeated mitosis without cytokinesis. Nutrients and other materials flow very quickly through aseptate hyphae.

Nutrition in Fungi

Unlike humans who ingest their food and then digest it, fungi digest their food before they ingest it. Many fungi produce enzymes that break down organic material, allowing the nutrients to be absorbed through their thin cell walls. All fungi are heterotrophs, but there are three types of fungi that differ in how they obtain their nutrients.

Saprophytic fungi A saprobe is an organism that feeds on dead organisms or organic wastes. Saprophytic fungi, such as the bracket fungus shown in **Figure 20.4,** are decomposers and recycle nutrients from dead organisms back into food webs. The fungus in **Figure 20.5** also is a saprobe.

Parasitic fungi Parasitic fungi absorb nutrients from the living cells of another organism, called a host. Many parasitic fungi produce specialized hyphae called **haustoria** (haws TOH ree ah), which grow into the host's tissues and absorb their nutrients. *Arthrobotrys* is a group of parasitic soil fungi that trap prey with rings of hyphae.

Mutualistic fungi Some fungi live in a mutualistic relationship with another organism, such as a plant or an alga. The mycelia of a particular fungus cover the root of a soybean plant. The fungus receives sugar from the host plant. The mycelia increase water uptake and mineral absorption for the host plant.

■ **Figure 20.4** There are three different ways that fungi can obtain food—through decomposition, through parasitism, and through mutualism.

Bracket fungus feeding on a log

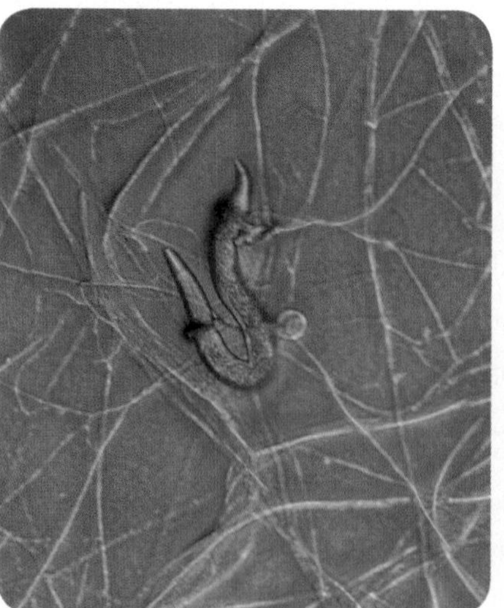

***Arthrobotrys* hyphae trapping a nematode**

Color-Enhanced SEM Magnification: 150×

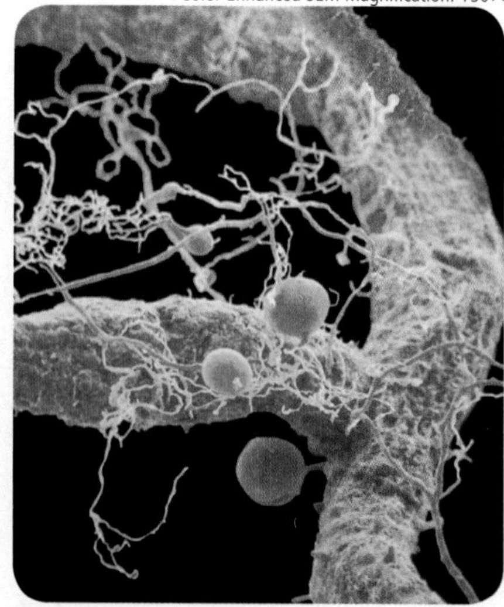

Mycelia on a root

Visualizing a Fairy Ring

Figure 20.5
Fungi produce spores in reproductive structures called *fruiting bodies*. The fruiting body is made up of hyphae that grow outward by extending their lengths, growing into new areas where a fresh supply of nutrients can be found in the soil. This creates a ring of mushrooms called a fairy ring.

Nuclei

Septa

Cap

Gills

Stalk

Mycelium

Hyphae

Fruiting body

Fairy ring In this fairy ring, what you recognize as a mushroom is the fruiting body of the fungus *Marasmius oreades*.

A fairy ring forms because the fruiting bodies all share the same underground hyphae.

Fruiting bodies

The mycelium produces fruiting bodies.

Mycelium

Concepts In Motion Interactive Figure To see an animation of a fairy ring in fungi, visit biologygmh.com.

Biology Online

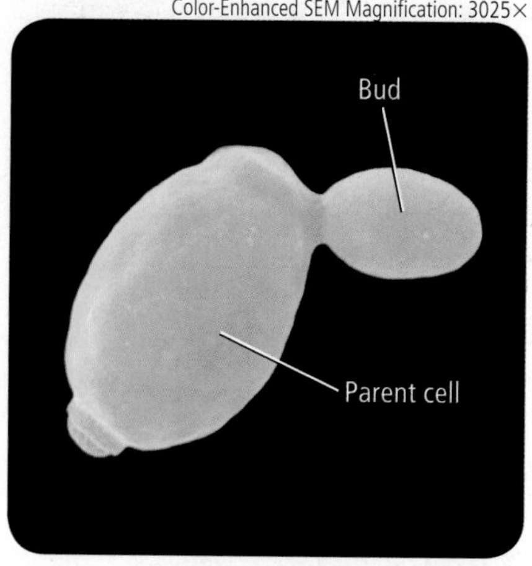

■ **Figure 20.6** Notice how the plasma membrane is begining to separate the bud from the parent cell.
Infer *Is this an example of sexual or asexual reproduction?*

Reproduction in Fungi

Fungi are classified by their structure and patterns of reproduction. Some fungi only can reproduce asexually through mitosis. Asexual reproduction in fungi also includes fragmentation, budding, and spore production. Many fungi can reproduce both asexually and sexually. Sexually reproducing fungi produce spores by the process of meiosis.

Budding Unicellular yeast cells reproduce asexually by budding. As shown in **Figure 20.6,** the new cell develops while attached to the parent cell. The plasma membrane pinches off to partially separate the new individual from the parent cell.

Fragmentation Fragmentation is a form of asexual reproduction that occurs when the mycelium of a fungus is physically broken apart or fragmented. This can occur in a number of different ways. One example is by an animal digging in the soil where a fungus is growing. If the fragments of mycelia land in a location with suitable growing conditions, the hyphae will grow into new mycelia.

Spore production The asexual and sexual life cycles of most fungi include the production of spores. A **spore** is a reproductive haploid cell with a hard outer coat that develops into a new organism without the fusion of gametes. These spores produce new hyphae that eventually form a mycelium. Some fungal spores are thin-walled and germinate quickly; others are thick-walled and take longer periods of time to germinate. In sexual reproduction, the fungi's diploid, reproductive structure produces haploid spores by meiosis. These spores form the next generation, which grows into new mycelia.

MiniLab 20.1

Examine Yeast Growth

What is the relationship between yeast reproduction and the availability of food? Yeasts are unicellular fungi. These organisms feed on sugars, producing carbon dioxide and ethyl alcohol in the process. Yeasts reproduce asexually and multiply quickly under optimal growth conditions.

Procedure

1. Read and complete the lab safety form.
2. Label four **250-mL Erlenmeyer flasks** 1–4.
3. Create a data table to record your results.
4. Add **100 mL warm water** to each flask and do not cover the flasks.
5. Add 0.0 g, 0.5 g, 1.0 g, or 1.5 g of **table sugar** to each one of the flasks.
6. Add one packet of **dry yeast** to each flask. Stir flasks with a **glass rod** until contents are thoroughly mixed.
7. Observe and record the changes in the flasks every five minutes for 20 minutes.
8. Clean up your work station according to your teacher's instructions.

Analysis

1. **Conclude** What is the relationship between yeast reproduction and the availability of sugar?
2. **Analyze** How might your results have changed if the flasks had been covered during your experiment?

Adaptations for survival Most fungi, like the puffball shown in **Figure 20.7,** produce trillions of spores. Producing such large quantities of spores is an adaptation for survival. This adaptation ensures that at least a small percentage of the spores will land in suitable locations and begin to grow, producing the next generation.

Additional adaptations for survival include the physical traits of the spores. They are so small and lightweight that wind and even the smallest animals, such as insects, can disperse them. A spore also is protected by its cell wall. This wall often is tough and waterproof, which allows the spore to survive extremes of temperature and moisture.

Examine **Figure 20.7** again to see the cloud of spores being released. When spores are dispersed by wind, the spores can travel hundreds of miles across land and water. In fact, fungal spores can be found almost everywhere.

Sporophores The fruiting body of a spore-forming fungus is called a sporophore (SPOH ruh for) and is characteristic of that species of fungus. The classification of a fungus is based primarily on the type of sporophore it produces. For example, in some primitive fungi, such as black bread mold, specialized hyphae called sporangiophores (spuh RAN jee uh forz) have a spore-containing structure called a sporangium (plural, sporangia) on each of their tips. A **sporangium** is a sac or case in which spores are produced. The sporangia provide protection for the spores, preventing them from drying out prematurely.

In Section 20.2, you will learn that some fungi have common names such as sac fungi and club fungi. These names are descriptive for the type of sporophores these fungi produce. Section 20.2 also contains information about the life cycles and the types of spores and sporophores produced by members of each of the major phyla of fungi.

■ **Figure 20.7** Puffball fungi can produce trillions of spores. The slight touch of an animal brushing against the fungus or a falling raindrop can trigger the release of spores.

Section 20.1 Assessment

Section Summary

- Fungi produce hyphae that form a netlike mass called a mycelium.
- There are three different methods by which fungi obtain food.
- Fungi can reproduce asexually by budding, fragmentation, or producing spores.
- Most fungi can reproduce sexually.

Understand Main Ideas

1. MAIN Idea **Name** three major characteristics of the Kingdom Fungi.
2. **Diagram** the difference between septate and aseptate hyphae.
3. **State** how fungi feeding differs from animal feeding.
4. **Contrast** the methods that parasitic, saprophytic, and mutualistic fungi use to obtain food.
5. **Describe** three methods of asexual reproduction in fungi.

Think Critically

6. **Predict** how a slice of bread that is left out on the table for a few weeks can become covered with bread mold. From where does the mold come?

WRITING in **Biology**

7. Fungi can be used as a biocontrol to control common insect pests. Research and write an article for a gardening magazine about the value of fungi in your garden. Include several examples of fungi used in gardens.

Biology Online Self-Check Quiz biologygmh.com

Section 20.2

Reading Preview

Objectives

- **Identify** four major phyla of fungi.
- **Summarize** the distinguishing traits of each fungus phylum.
- **Describe** the reproductive strategies of each fungus phylum.

Review Vocabulary

flagellated: having long projections that propel organisms with a whiplike motion

New Vocabulary

stolon
rhizoid
gametangium
conidiophore
ascocarp
ascus
ascospore
basidiocarp
basidium
basidiospore

Diversity of Fungi

MAIN Idea **Fungi exhibit a broad range of diversity and are classified into four major phyla.**

Real-World Reading Link Think of the many sizes, shapes, and colors of insects you might have seen. While they are all insects, they are very diverse. Within the Kingdom Fungi, there is also much diversity in structures and life cycles.

Classification of Fungi

Biologists use fungal structure and methods of reproduction to divide fungi into four major phyla—Chytridiomycota, Zygomycota, Ascomycota, and Basidiomycota. The cladogram shown in **Figure 20.8** shows the evolutionary relationships between the phyla of fungi as they currently are understood.

Fungi are likely to have colonized the land with plants more than 450 million years ago, possibly due to mutualistic associations with plants. Yet, molecular evidence supports the view that fungi are more closely related to animals than plants. Evidence suggests that fungi and animals diverged from a common protist ancestor.

Chytrids

The fungi in the phylum Chytridiomycota (ki TRIHD ee oh mi koh tuh) often are referred to as chytrids or chytridiomycetes. Some chytrids are saprophytes, whereas others parasitize protists, plants, and animals. Most chytrids are aquatic, and they are unique among fungi because they produce flagellated spores. For this reason, scientists originally grouped chytrids with protists.

Chytrids (KI trihdz), like the one in **Figure 20.8,** have been reclassified as new information about them became available. Recent molecular evidence suggests chytrids are related more closely to fungi than to protists because of similar protein and DNA sequences. Another characteristic that indicates a close relationship with fungi is chitin-containing cell walls. There is evidence that chytrids were perhaps the first fungi and are the evolutionary link between the funguslike protists and fungi.

Figure 20.8 This phylogenetic tree shows the evolutionary relationships of the phyla of fungi. Chytrids are more similar to fungi than protists.

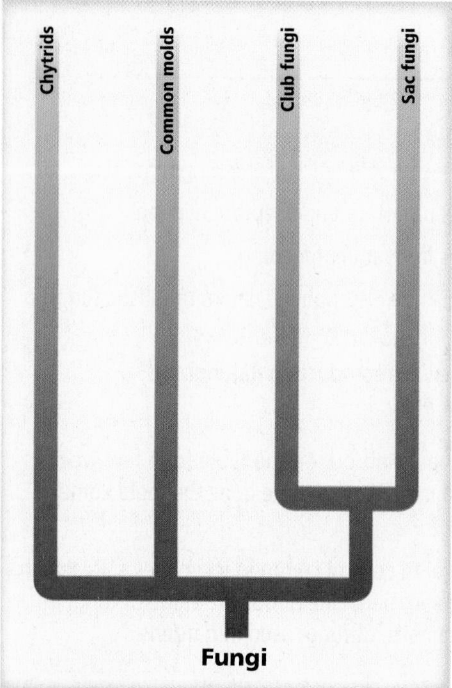

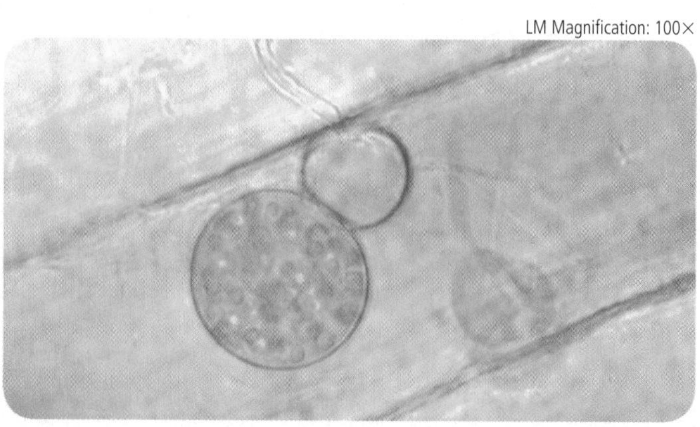

Common Molds

The most familiar member of the phylum Zygomycota (zi goh mi KOH tuh) is a common mold that grows on bread and other foods called *Rhizopus stolonifer.* Common molds are mostly terrestrial and some live in mutualistic relationships with plants. Molds form a type of hyphae called **stolons** (STOH lunz) that spread across the surface of food. Another type of hyphae, called **rhizoids** (RIH zoydz), penetrates the food and absorbs nutrients, as shown in **Figure 20.9.** Other functions of rhizoids include anchoring the mycelium and producing digestive enzymes. Zygomycetes also can be found on decaying plant and animal material.

Life cycle Zygomycetes reproduce both asexually and sexually, as illustrated in **Figure 20.9.** Asexual reproduction occurs when sporangia form at the tips of upright hyphae called sporangiophores. Within each sporangium are thousands of haploid spores that can be spread by wind or other air movements. When released, spores will produce new hyphae if they land in a favorable environment.

If conditions in the environment are no longer favorable to sustain life, zygomycetes can reproduce sexually. There are no defined male and female fungi, but rather plus (+) and minus (–) mating strains. Haploid hyphae from two compatible mating strains—one plus and one minus—fuse. Each hypha produces a **gametangium** (ga muh TAN jee um) (plural, gametangia), which is a reproductive structure that contains a haploid nucleus. As shown in **Figure 20.9,** the two haploid nuclei from each gametangium fuse to form a diploid zygote. The zygote develops a thick wall and becomes a dormant zygospore (ZI guh spor), sometimes called a zygosporangium.

Investigate Mold Growth

How does salt affect mold growth? Chemical preservatives, including salt (sodium chloride), often are used to influence mold growth on a variety of foods.

Procedure

1. Read and complete the lab safety form.
2. Obtain two slices of **bread.** Touch one object in the room with both sides of both slices.
3. Using a **spray bottle** filled with **water,** lightly moisten both sides of both slices of bread evenly.
4. Place one bread slice into a **self-sealing bag.** Seal the bag and label it with your name, date, and the object wiped with the bread.
5. Sprinkle **salt** on both sides of the second slice. Place the slice into another bag and seal it. Label this bag as you did the first, but note that salt was added.
6. Create a table to record your observations.
7. Record observations daily for ten days. Your table should include descriptions, as well as measurements of any mold that has formed.

Analysis

1. **Identify** Which slice grew more mold?
2. **Conclude** Did the salt affect mold growth?
3. **Analyze** Why did the salt affect the mold?

Figure 20.9 *Rhizopus stolonifer,* a common bread mold, is an example of a zygomycete that undergoes both asexual and sexual reproduction.

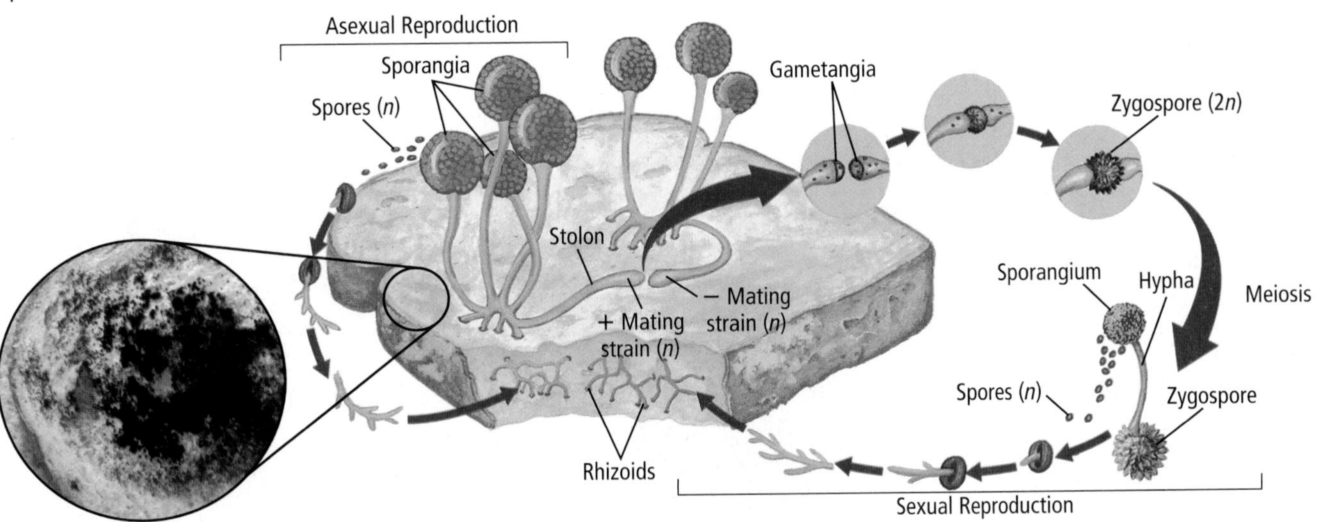

CAREERS IN BIOLOGY

Mycologist A mycologist studies various aspects about fungi such as their biochemistry, genetics, evolution, classification, and economic uses. For more information on biology careers, visit biologygmh.com.

The zygospore can remain dormant for months until environmental conditions improve. Then, the zygospore germinates and undergoes meiosis to produce hyphae with a sporangium. Each haploid spore formed in the new sporangium can grow into a new mycelium. This process of sexual reproduction provides greater genetic diversity that helps ensure the survival of the species by allowing zygomycetes to survive in changing environments.

Sac Fungi

There are more than 60,000 species of sac fungi found in the phylum Ascomycota (AS koh mi koh tuh). It contains more species than any other phylum of fungi. Species found in this phylum are referred to as ascomycetes or sac fungi. Although the most well-known ascomycete—yeast—is unicellular and microscopic, most members of this group are multicellular.

Life Cycle Sac fungi can reproduce both sexually and asexually. During asexual reproduction, spores are formed at the tips of the hyphae. These spore-producing hyphae are called **conidiophores** (koh NIH dee uh forz), and the spores they generate are called conidia. Instead of forming inside sporangia, conidia form externally at the tips of the conidiophore. These spores are dispersed easily by wind, water, and animals.

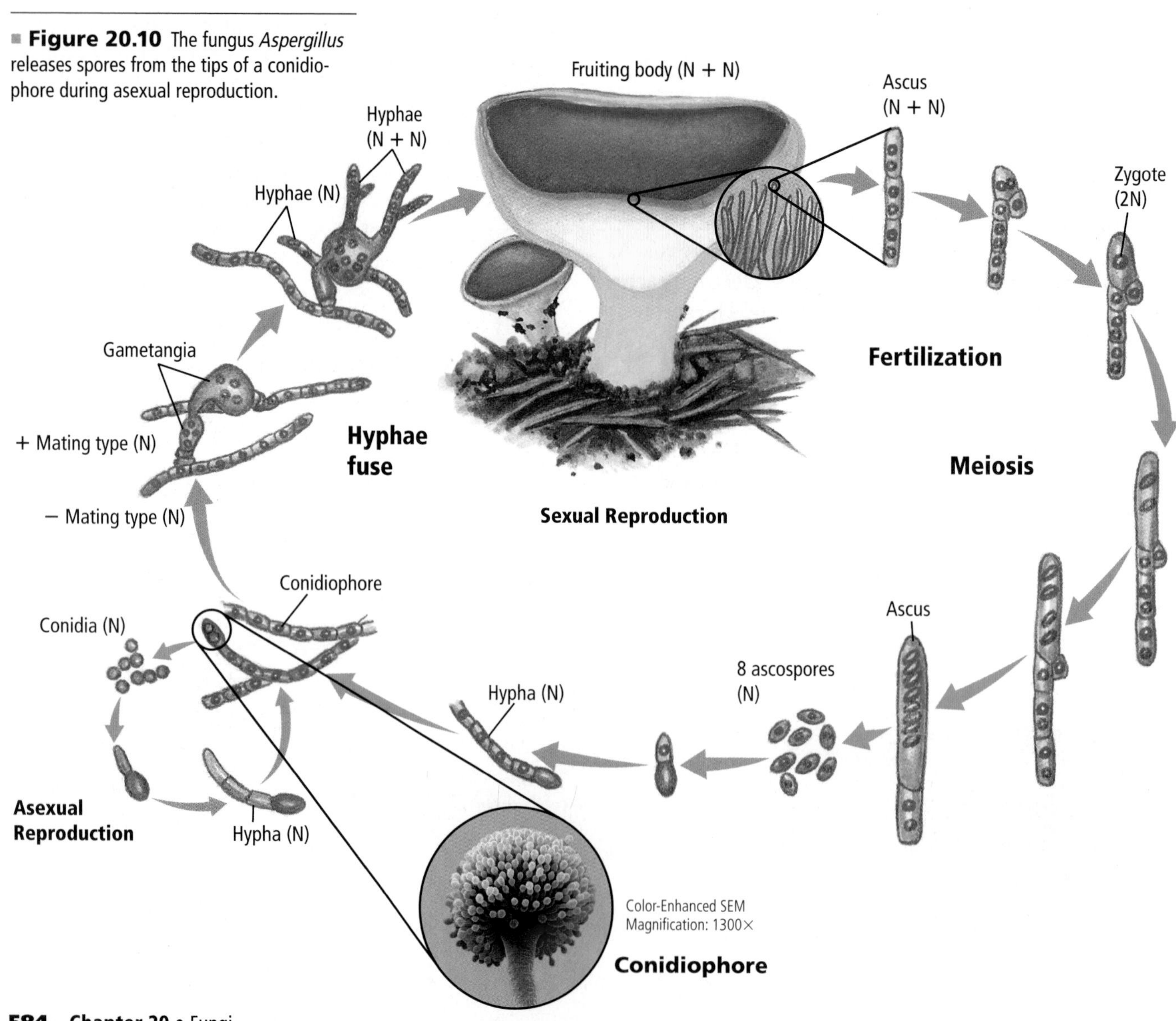

■ **Figure 20.10** The fungus *Aspergillus* releases spores from the tips of a conidiophore during asexual reproduction.

Ascomycete sexual reproduction is a complicated process, illustrated in **Figure 20.10.** It takes place when hyphae from opposite mating types fuse and one nucleus from each type pairs off in separate cells. The hyphae that continue to grow are septate. Each cell contains two haploid nuclei–one from each mating type. Eventually the hyphae will develop a specialized reproductive structure called an **ascocarp.** Within the ascocarp, the haploid nuclei fuse to form a zygote. The zygote divides by meiosis, producing four haploid nuclei. These nuclei then divide by mitosis, forming a total of eight haploid nuclei. The nuclei develop into spores in the **ascus,** a saclike structure. Spores produced by an ascus are called **ascospores.** Just like other spores, when growing conditions are favorable, each ascospore can develop into a haploid mycelium.

Study Tip

Tables Write a short paragraph using the table below to compare the number of phyla and the approximate number of species in each phylum. Predict how these figures will compare to the figures found in upcoming chapters on plants and animals.

Club Fungi

Table 20.1 compares the characteristics of the fungi in phylum Basidiomycota with the fungi in other phyla. Among the 25,000 members of the Basidiomycota (buh SIH dee oh mi koh tuh) phylum are the mushrooms, perhaps the most commonly recognized type of fungus.

Concepts In Motion

Interactive Table To explore more about the fungi phyla, visit biologygmh.com.

Table 20.1 Fungi Phyla

Phylum (Common Name)	Example	Number of Species	Characteristics
Chytridiomycota (chytrids)	Magnification: unavailable	1300+	• Unicellular • Most are aquatic • Some are saprophytic, while others are parasitic • Produce flagellated spores
Zygomycota (common molds)	LM Magnification: 30×	800	• Multicellular • Most are terrestrial • Many form mutualistic relationships with plants • Reproduce sexually and asexually
Ascomycota (sac fungi)		60,000+	• Most are multicellular, but some are unicellular • Variety of habitats • Saprophytic, parasitic or mutualistic • Reproduce sexually and asexually
Basidiomycota (club fungi)		25,000	• Most are multicellular • Most are terrestrial • Saprophytic, parasitic, or mutualistic • Rarely reproduce asexually
Deuteromycota (imperfect fungi)		25,000	• No sexual stage observed • Very diverse group • Might not be considered a true phylum

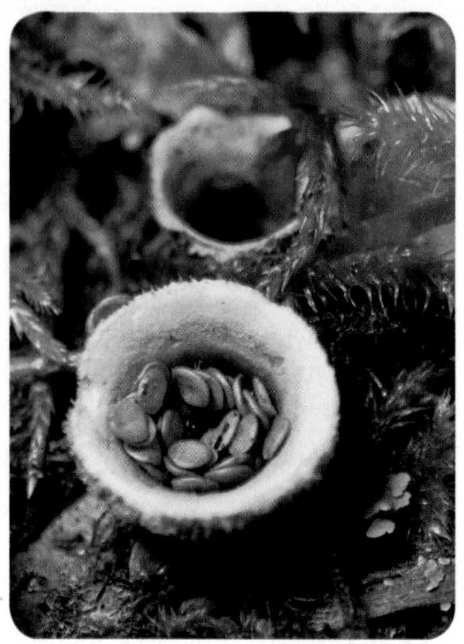

■ **Figure 20.11** This club fungus, called Bird's Nest Fungus or *Crucibulum vulgare,* has basidiocarps that resemble eggs in a bird's nest.

Species found in phylum Basidiomycota also are called basidiomycetes (buh SIH dee ah mi see teez) or club fungi. They can be saprophytic, parasitic, or mutualistic. The saprophytic basidiomycetes are major decomposers of wood. They produce enzymes that can break down complex polymers found in wood, such as lignin.

Life Cycle Basidiomycetes rarely produce asexual spores. They spend most of their life cycle as dikaryotic mycelia, meaning that each cell has two nuclei. The mycelium periodically will reproduce sexually by forming a **basidiocarp** (buh SIH dee oh karp), or a fruiting body, as shown in **Figure 20.11.** The mushrooms that you see growing in the woods or that you add to your salad are basidiocarps.

Basidiocarps can grow quickly, sometimes appearing full-grown in a few hours or overnight, with rapid growth due to cell enlargement rather than cell division. The underside of the cap is composed of **basidia**—club-shaped hyphae that produce spores. Within the basidia, the two nuclei fuse to form a diploid nucleus. This nucleus divides meiotically into four haploid nuclei that will develop into **basidiospores**—the haploid spores released by basidia during reproduction. The basidiospores can be dispersed by wind, water, or animals. It is estimated that some mushrooms will produce as many as a billion basidiospores.

Other Fungi

Organisms in phylum Deuteromycota share only one unique trait—sexual reproduction in these fungi never has been observed. Because these fungi appear to lack a sexual stage, they are referred to as the imperfect fungi. Scientists currently use modern genetic techniques, such as DNA and protein comparisons, to reassign some of these fungi in one of the other four phyla.

Section 20.2 Assessment

Section Summary

- The four major phyla of fungi are Chytridiomycota, Zygomycota, Ascomycota, and Basidiomycota.
- Zygomycetes reproduce sexually by forming zygospores.
- Ascomycetes produce ascospores within a saclike structure called an ascus during sexual reproduction.
- Basidiomycetes produce basidiospores during sexual reproduction.
- Sexual reproduction in the phylum Deuteromycota has never been observed.

Understand Main Ideas

1. MAIN Idea **Identify** two characteristics of each of the four major phyla of fungi.
2. **Explain** why fungi produce so many spores.
3. **Diagram** the life cycle of ascomycetes.
4. **Describe** What are the imperfect fungi?
5. **Compare** sexual reproduction in ascomycetes and basidiomycetes.

Think Critically

6. **Predict** what might happen to an ecosystem if a virus destroyed all the basidiomycetes. What effect would that have on the recycling of nutrients in a forest?

WRITING in Biology

7. Write a news story detailing how a scientist reclassified a species of imperfect fungi once sexual reproduction has been identified.

BiologyOnline Self-Check Quiz biologygmh.com

Section 20.3

Reading Preview

Objectives

- **Identify** the characteristics of lichens.
- **Describe** the characteristics of mycorrhizal relationships.
- **List** some beneficial and harmful effects fungi can have on humans.

Review Vocabulary

bioremediation: the use of organisms to detoxify a polluted area

New Vocabulary

lichen
bioindicator
mycorrhiza

Ecology of Fungi

MAIN Idea Lichens and mycorrhizae demonstrate important symbiotic relationships between fungi and other organisms.

Real-World Reading Link You might think that the only time you encounter fungi is when you order pizza or when you take a nature walk. But you might be surprised to know that some of the antibiotics that you take are derived from fungi, and that athlete's foot is caused by fungi.

Fungi and Photosynthesizers

Lichens and mycorrhizae are two examples of mutualistic relationships between fungi and other organisms. As you learned in Chapter 2, mutualism is a type of symbiosis in which both organisms benefit from the relationship.

Lichens A symbiotic relationship between a fungus and an alga or a photosynthetic partner is called a **lichen** (LI ken). The fungus usually is an ascomycete, but lichens also may contain basidiomycetes. The photosynthetic partner is either a green alga or cyanobacterium which provides food for both organisms. The fungus provides a dense web of hyphae in which the alga or cyanobacterium can grow. Examine **Figure 20.12** to see the structure of a lichen. Notice that fungal tissues account for most of the mass of a lichen.

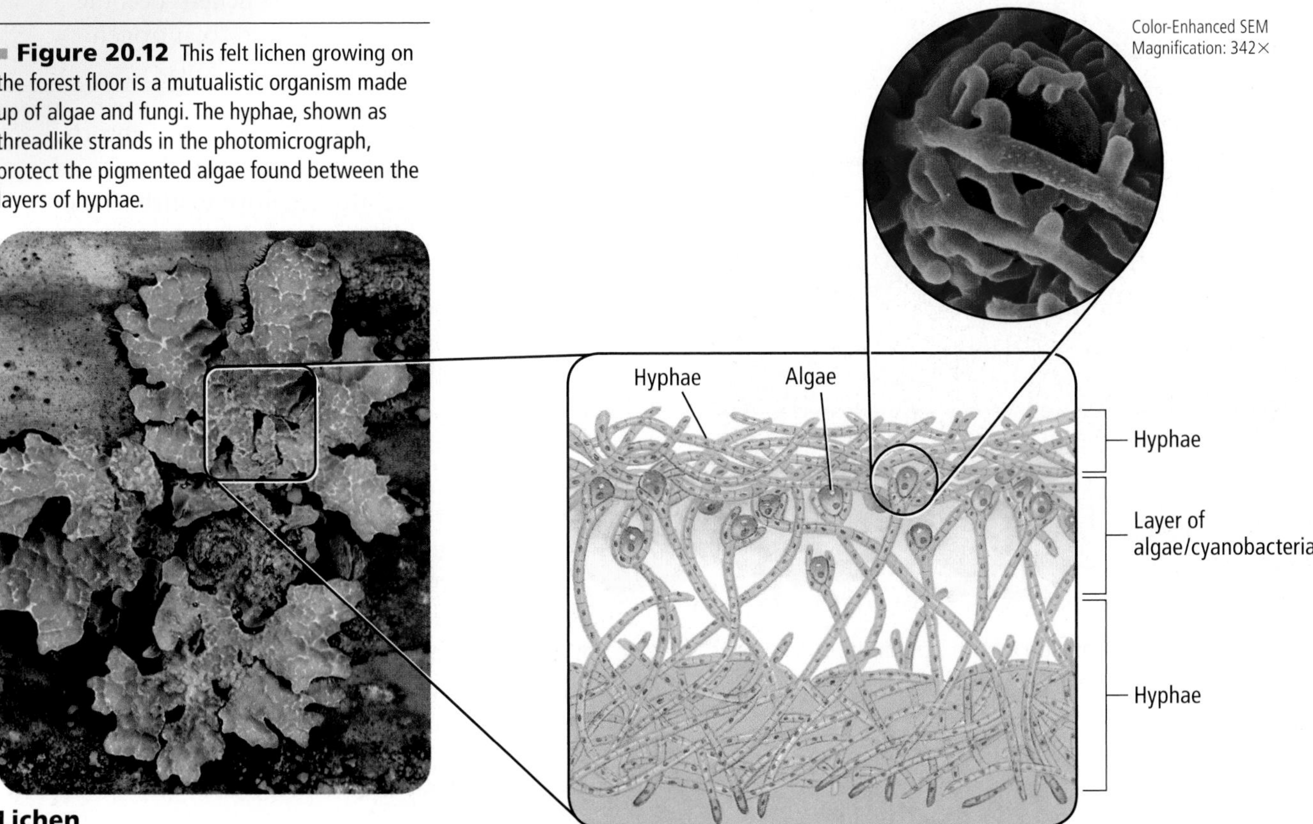

■ **Figure 20.12** This felt lichen growing on the forest floor is a mutualistic organism made up of algae and fungi. The hyphae, shown as threadlike strands in the photomicrograph, protect the pigmented algae found between the layers of hyphae.

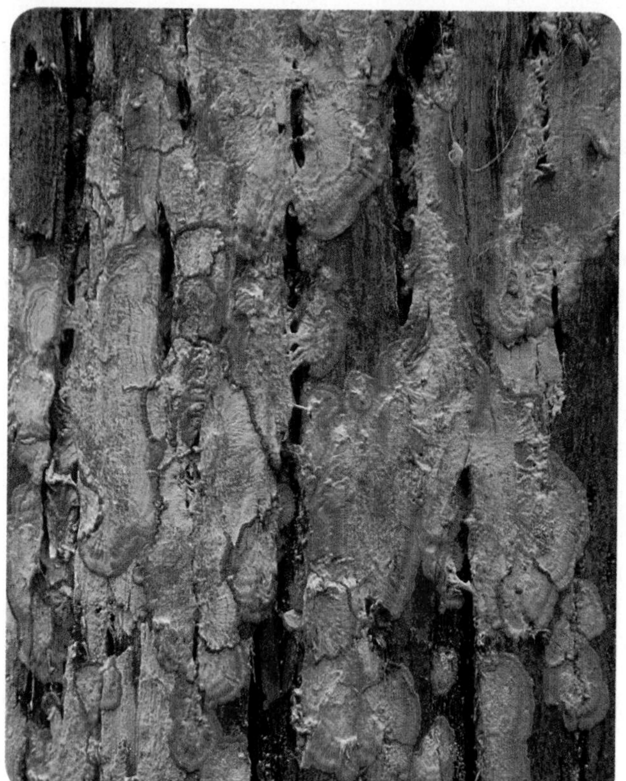

Red blanket lichen on cypress bark

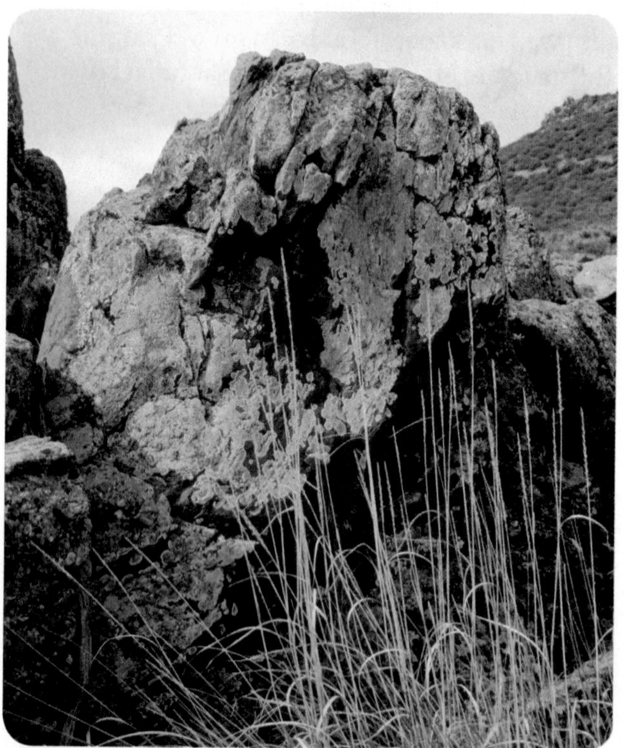

Lepraria on basalt rock

■ **Figure 20.13** Lichens grow in harsh enviroments, such as the surfaces of trees or bare rocks, where fungi, algae, and cyanobacteria would not grow alone.
Explain *Why lichens often are pioneer species.*

Diversity of lichens There are over 25,000 species of lichens. To see two of the different forms, examine **Figure 20.13.** Because they need only light, air, and minerals to grow, lichens are found in some of the harshest environments. Lichens absorb needed minerals from rainwater or from dust in the air. The fungus traps rainwater, but it also can absorb moisture from the air. Some fungi produce toxic compounds that keep animals, including insects, from eating the lichen. Fungi also inhibit moss and bacterial growth.

Although a few types of lichens can be found in deserts or the tropics, the majority of these organisms grow in temperate or arctic areas. They form the primary ground cover on the tundra, providing food for grazing animals. Caribou utilize an enzyme called lichenase (LI kun ayz) to help them digest the lichen.

Lichens also survive severe drought. They dry out, stop photosynthesis, and become brittle. Pieces can break off, blow away, and reestablish in another location—a form of asexual reproduction. When water is available again, the lichens rapidly absorb large quantities of the water and begin the process of photosynthesis again.

Recall from Chapter 3 that a pioneer species is a species that can grow with little soil or on rocks. Lichens often are the pioneer species in an area of newly cleared soil or rock following natural disasters, such as fire or volcanic activity. Acids produced by the fungal portion of the lichen help penetrate and break down rocks to help form soil. As lichens become established, they help trap soil and fix nitrogen, which helps in the colonization of plants.

Lichens as bioindicators Because they absorb much of their water and minerals from the air and rain, lichens are especially sensitive to airborne pollutants. When air pollution levels rise in an area, lichens often will die. In addition, lichens typically do not grow in or near cities where air pollution is high. They usually are found in rural areas where there is little or no air pollution. Lichens absorb pollutants that are dissolved in rain and dew. Because of their sensitivity to pollution, lichens are important bioindicators.

A **bioindicator** is a living organism that is sensitive to changes in environmental conditions and is one of the first organisms to respond to changing conditions. The levels of air pollutants in an area can be correlated to the changes in lichen growth. As the level of pollution decreases, the populations of lichens increase.

Reading Check **Explain** why lichens are bioindicators.

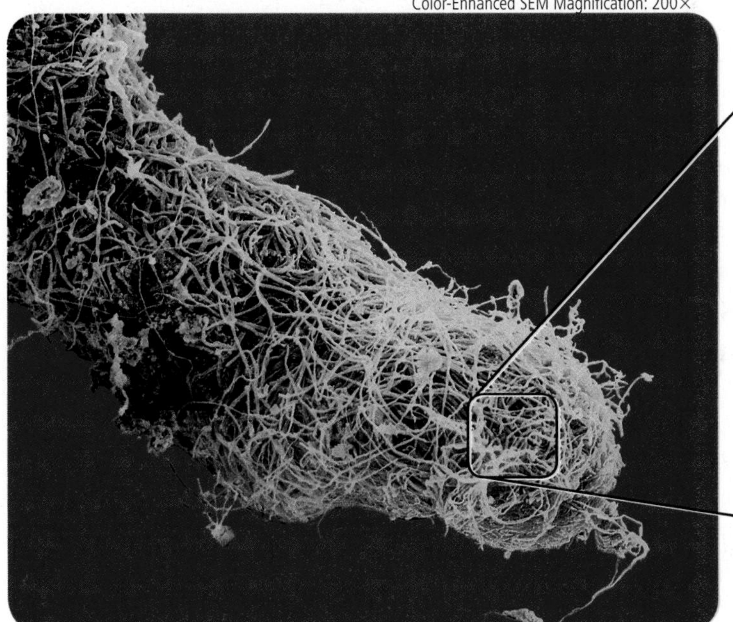

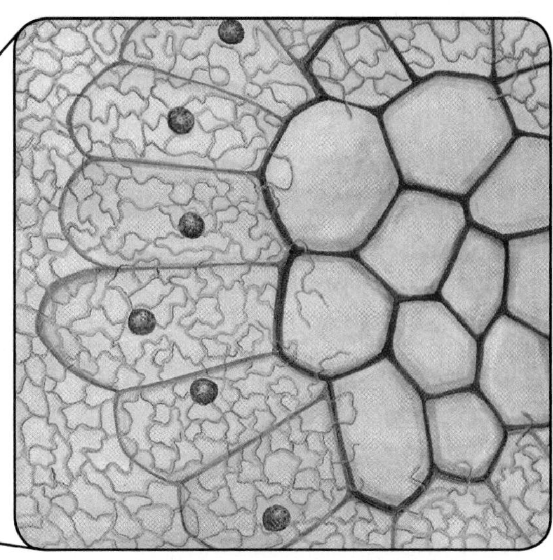

■ **Figure 20.14** The mycelium of the *Scleroderma geaster* fungus growing on the *Eucalyptus* tree root increases the surface area in which water and nutrients can be absorbed by the root.
Explain ***How does the fungus benefit from this relationship?***

Mycorrhizae Another mutalistic relationship involving a fungus is **mycorrhiza** (my kuh RHY zuh) (plural, mycorrhizae)—a symbiotic relationship between a specialized fungus and plant roots. Plants with mycorrhizae are healthier and more vigorous than similar plants lacking mycorrhizae. Other plant species, such as orchids, cannot survive without mutualistic partners. Orchid seeds will not germinate unless they are infected by a fungal partner or provided with the fungal carbohydrate trehalose.

Figure 20.14 shows a mycorrhizal relationship between a *Schleroderma geaster* fungus and a *Eucalyptus* tree. The fungus absorbs and concentrates various minerals for the plant. The hyphae of the fungus also increase the plant's root surface area for water and mineral absorption. In return, the fungus receives carbohydrates and amino acids from the plant.

Between 80 and 90 percent of plants, including primitive plants, have mycorrhizae. Mycorrhizae are extremely important in natural habitats and for agricultural crops. Crops associated with mycorrhizae include corn, carrots, potatoes, tomatoes, and strawberries.

VOCABULARY
ACADEMIC VOCABULARY

Cooperate:
To work or act together toward a common end or purpose.
Organisms that cooperate with members of their own or different species might be more successful than those organisms that do not cooperate.

Fungi and Humans

For the most part, fungi have a positive effect on the lives of humans. Their most important role is as decomposers, assisting with the recycling of nutrients found in dead organisms. Decomposing organic matter makes nutrients available for other organisms and prevents dead organisms and their wastes from littering the surface of Earth.

Beneficial fungi Fungi have many medical uses. The ascomycete, *Penicillium notatum*, can be used as a source of penicillin. This antibiotic has saved countless lives. Chemical compounds found in the fungus *Claviceps purpurea* are used to reduce high blood pressure, to control excessive bleeding, to treat migraine headaches, and to promote contractions during childbirth. The Norwegian deuteromycete *Tolypocladium inflatum* is the source for cyclosporine. Cyclosporine is an immune suppressant drug given to organ transplant patients to keep their bodies from rejecting the new organ.

CAREERS IN BIOLOGY

Food Technologist A food technologist can choose from a range of jobs, including the development of new food products and setting standards for producing, packaging, and marketing food technologies. They also might test products for nutritional and microbiological quality. For more information on biology careers, visit biologygmh.com.

Foods Many of the foods we eat are made from fungi or fungal products. The most obvious are the many different mushrooms we eat. Yeast makes bread rise by releasing carbon dioxide gas during fermentation, as discussed in Chapter 8. Another product of fermentation is the alcohol found in beer and wine. Truffles are fungi and are one of the most expensive food items. Many other fungi are enjoyed similarly as delicacies. The flavors of some cheeses such as Brie, Camembert, and Roquefort are the result of fungi. The citrus flavor found in colas is created by the fungus *Aspergillus.* This fungus also is used to make soy sauce.

Bioremediation Fungi also can be used for cleaning the environment of pollutants that are threatening some ecosystems. The fungi are mixed with water or soil where they decompose organic materials in the pollutants. During this process, called bioremediation, the pollutants are broken down into harmless substances. The rate at which microorganisms, such as fungi and bacteria, remove environmental pollutants can be increased if additional nutrients are added to the water or soil. Bioremediation is a relatively new scientific field, and new discoveries and processes are being developed to be used in environmental clean-up projects.

Connection to Chemistry Researchers are using white-rot fungi to break down top priority pollutants, such as dyes and polycyclic aromatic hydrocarbons (PAHs). PAHs are carcinogenic (cancer-causing) molecules. Researchers also are taking advantage of the fact that these fungi contain enzymes that degrade lignin, a molecule found in wood fiber that hardens and strengthens the cell walls of plants, enabling them to recycle wood. These enzymes also can attach to structurally similar chemicals, including many man-made pollutants.

DATA ANALYSIS LAB 20.1

Based on Real Data*

Interpret The Data

Does the addition of salt to the soil affect asparagus production? *Fusarium oxysporum* is a disease-causing organism of many crops, including asparagus. The fungus penetrates the roots and spreads up through the plant, often reducing the flow of water to the stem and leaves. Infected plants produce fewer and smaller spears than healthy plants. The fungus stays in the soil year after year.

Data and Observations

Salt (sodium chloride) treatment is a common method for suppressing disease in plants. The table shows data collected after an asparagus field was treated with a dusting of salt.

Asparagus Production		
	Spear number	Spear mass
Before treatment with salt	78.2	1843.2
After treatment with salt	89.1	2266.1

Think Critically

1. **Calculate** What is the percentage change in spear number and spear mass?
2. **Interpret** How does the salt treatment affect the asparagus crop?
3. **Hypothesize** Why might salt have this effect on the plants? How would you test your hypothesis?

*Data obtained from: Elmer, W.H. 2002. Influence of formononetin and NaCl on mycorrhizal colonization and fusarium crown and root rot of asparagus. *Plant Disease* 86(12): 1318-1324.

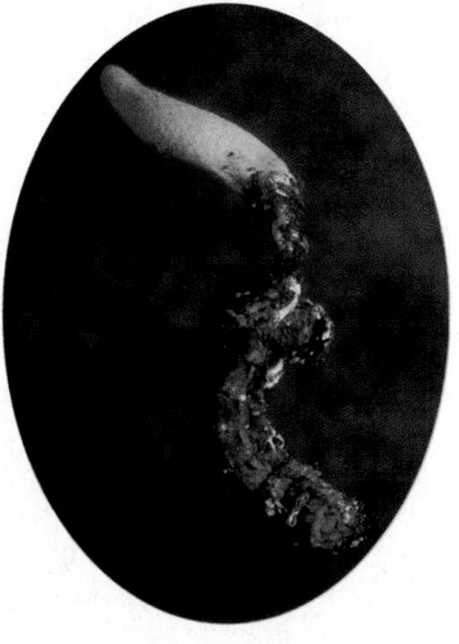

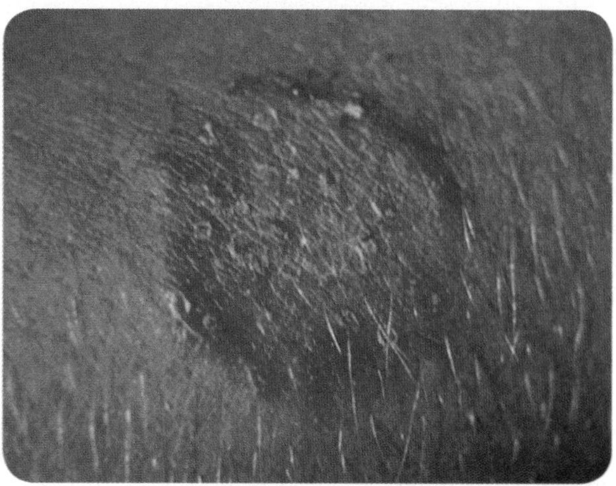

Figure 20.15 Parasitic fungi can be harmful to humans as well as other organisms.
Left: Grapes are infected by a fungus that causes noble rot.
Middle: Scarlet caterpillar fungus can kill caterpillars.
Right: Ringworm is caused by a fungus.

Harmful fungi Some fungi can be harmful to other organisms. For example, American elm trees are killed by the fungus *Ceratocystis ulmi* and American chestnut trees are killed by the fungus *Endothia parasitica.* The fungi quickly spread from tree to tree, and they have killed many trees in North America. Agricultural crops are also damaged by fungi. The fungal parasite *Leptoterochila medicaginis* causes leaf blotch in alfalfa plants and can diminish crop production by as much as 80 percent. The ripe grapes shown in **Figure 20.15** have been infected with a parasitic fungus *Botrytis cinerea,* causing what is known as noble rot. The fungus attacks the grapes and causes an increase in their sugar content. Certain wines are produced from such grapes in France.

Fungi can also parasitize humans and other animals. The parasitic fungus *Cordyceps militaris* can infect larvae and pupae of butterflies and moths, as shown in **Figure 20.15.** Athlete's foot, ringworm, yeast infections, and oral thrush are infections in humans that are caused by fungi. **Figure 20.15** shows skin infected with ringworm, which can be caused by several species of fungi.

Section 20.3 Assessment

Section Summary

- Lichens are examples of mutualistic relationships between a fungus and an alga or a cyanobacterium.
- Mycorrhizae help plants obtain water and minerals by increasing the surface area of their roots.
- Compounds obtained from fungi are used for a variety of medicines.
- Many foods eaten by people are made from fungi.
- Fungi can have an adverse effect on humans and plants.

Understand Main Ideas

1. MAIN Idea **Identify** the characteristics of the mutualistic relationship between fungi and algae.
2. **Explain** why lichens are important for the environment.
3. **Apply** what you know about enzymes to design a lichenase enzyme for lichen-eating animals.
4. **Construct** a table to show the beneficial and harmful effects of fungi on humans.

Think Critically

5. **Infer** the effect on world food production if a fungicide was discovered that destroys all the fungi in agricultural settings.

MATH in Biology

6. Lichens grow an average of one centimeter per year. How long would it take for a lichen to grow the width of your hand?

Biology & Society

Fungi Superheroes

The Iceman, whose mummified corpse was discovered in 1991, provided a picture of what life was like 5000 years ago in the Stone Age. In his belt were two walnut-sized lumps of birch fungus (known as the chaga mushroom). This fungus can both cause diarrhea and serve as an antibiotic. The chaga mushroom helped alleviate the effects of the parasites that were living in the Iceman's colon by helping his body eliminate the parasite's eggs.

A folk hero The chaga mushroom has been used as a traditional folk remedy in Eastern European countries since the sixteenth century and has been mentioned in Chinese texts dating back 4600 years. Chaga mushrooms have been used as a treatment for tuberculosis, various cancers, and intestinal ailments, usually crushed up and ingested as an herbal tea.

Chaga mushroom Chaga is a parasitic fungus that grows from the trunks of birch trees. In Russia, chaga is known as the "birch-killer" because it leads to the tree's death within five to seven years. It is estimated that as few as 1 in 15,000 birch trees have chaga. Chaga which grow on birch trees in Siberia are particularly prized among herbalists. Their high concentration of beneficial compounds is attributed to their ability to persist in the relatively harsh environment.

Making better cancer drugs Current scientific research is backing up the claims that folk medicine has maintained for generations. Chaga transforms the compound "betulin" found in the bark of the birch tree into a form which can be ingested. Betulin has been shown to exhibit anti-malarial, anti-inflammatory, and anti-HIV activity, in addition to being toxic toward some tumor cells. In 1998 a study showed that betulin, in the form of betulinic acid, triggered apoptosis (programmed cell death) when injected into tumor cells.

The chaga mushroom is a parasitic fungus that feeds off of birch, alder, and beech trees.

Chaga mushrooms contain high levels of antioxidants, known to inhibit the cell-damaging effects of free radicals (highly reactive, unpaired electrons). A 2005 study revealed that extracts from the mushroom protected DNA in human lymphocytes against oxidative damage. In addition, active polysaccharides present in the mushroom have been found to stimulate the immune system.

Scientists estimate that more than one million species of fungi exist and many are unidentified. The National Cancer Institute is collecting 1000 fungi samples yearly from the tropical rain forests to see if they might contain disease-fighting compounds. In the meantime, researchers continue to study existing herbal remedies in order to reinforce or dispel the claims made of these natural curatives.

MAKE A BOARD GAME

Work with a Team Create a board game depicting the development of a cancer treatment based on the discovery of a fungus found living on a plant in the rain forest. Conduct additional research at biologygmh.com regarding drug development, drugs developed from fungi, and drug treatment in cancer research.

BIOLAB

Design Your Own

HOW DO ENVIRONMENTAL FACTORS AFFECT MOLD GROWTH?

Molds can grow under a wide range of conditions. Consider the differences in your kitchen alone. Molds can grow in a cool refrigerator or in a dark bread box on the counter. They grow on foods that contain varying amounts of sugar, protein, and moisture.

Question: *How does a specific environmental factor change the rate of mold growth?*

Materials

Choose materials that would be appropriate for this lab. Possible materials include:

mold from a food source
plain powdered gelatin (contains protein only)
bread
sugar
prepared gelatin in a small cup
cotton swab
aluminum foil or plastic wrap
small cup
thermometer
graduated cylinder
spray bottle

Safety Precautions

WARNING: *Never eat food used in the lab.*

Plan and Perform the Experiment

1. Read and complete the lab safety form.
2. Make a list of environmental factors that might affect mold growth. Based on this list, develop a question to investigate.
3. Design an experiment that will help you answer this question. Remember, only one environmental factor should vary in your experimental conditions.
4. Write your hypothesis and design a data table.
5. Make sure your teacher has approved your experiment before you proceed.
6. Use cotton swabs to transfer mold from the food source to your trial cups.
7. Record observations for 5–7 days.
8. **Cleanup and Disposal** Place trial cups in the area designated by your teacher. Clean and return all equipment used in the lab. Wash your hands thoroughly.

Analyze and Conclude

1. **Identify** What is the independent variable and dependent variable in your experiment? Explain how the independent variable was changed.
2. **Compare** Describe differences you noticed among trial samples.
3. **Describe** What steps did you take to limit variables in this experiment? Make a list of constants.
4. **Interpret the Data** How did the environmental factor you changed affect the rate of mold growth?
5. **Conclude** Was your hypothesis supported? Explain.
6. **Error Analysis** Is it possible that more than one variable was introduced in your experiment? How would you change your experimental plans?

WRITING in Biology

Communicate Share your results with other groups. Develop a class list of environmental factors tested and results obtained. Based on these results, create a list of environmental factors that lead to optimal growth of the mold utilized in this experiment. To learn more about molds, visit BioLabs at biologygmh.com.

CHAPTER 20

Study Guide

Download quizzes, key terms, and flash cards from biologygmh.com.

FOLDABLES **Research** additional information on the different methods by which fungi obtain nutrients. Use what you learn and the information you gathered in the Foldable to generate a questionnaire for classifying fungi.

Vocabulary	Key Concepts
Section 20.1 Introduction to Fungi	
• chitin (p. 577) • fruiting body (p. 577) • haustorium (p. 578) • hypha (p. 577) • mycelium (p. 577) • septum (p. 578) • sporangium (p. 581) • spore (p. 580)	**MAIN Idea** Fungi are unicelluar or multicellular eukaryotic heterotrophs that are decomposers. • Fungi produce hyphae that form a netlike mass called a mycelium. • There are three different methods by which fungi obtain food. • Fungi can reproduce asexually by budding, fragmentation, or producing spores. • Most fungi can reproduce sexually.
Section 20.2 Diversity of Fungi	
• ascocarp (p. 585) • ascospore (p. 585) • ascus (p. 585) • basidiocarp (p. 586) • basidiospore (p. 586) • basidium (p. 586) • conidiophore (p. 584) • gametangium (p. 583) • rhizoid (p. 583) • stolon (p. 583)	**MAIN Idea** Fungi exhibit a broad range of diversity and are classified into four major phyla. • The four major phyla of fungi are Chytridomycota, Zygomycota, Ascomycota, and Basidiomycota. • Zygomycetes reproduce sexually by forming zygospores. • Ascomycetes produce ascospores within a saclike structure called an ascus during sexual reproduction. • Basidiomycetes produce basidiospores during sexual reproduction. • Sexual reproduction in the phylum Deuteromycota has never been observed.
Section 20.3 Ecology of Fungi	
• bioindicator (p. 588) • lichen (p. 587) • mycorrhiza (p. 589)	**MAIN Idea** Lichens and mycorrhizae demonstrate important symbiotic relationships between fungi and other organisms. • Lichens are examples of mutualistic relationships between a fungus and an alga or cyanobacterium. • Mycorrhizae help plants obtain water and minerals by increasing the surface area of their roots. • Compounds obtained from fungi are used for a variety of medicines. • Many foods eaten by people are made from fungi. • Fungi can have an adverse effect on humans and plants.

CHAPTER 20 Assessment

Section 20.1

Vocabulary Review

Each of the following sentences is false. Make the sentence true by replacing the italicized word with a vocabulary term found on the Study Guide page.

1. *Hyphae* is/are the cross-walls between fungal cells.
2. *Chitin* is/are the threadlike filaments found in certain fungi.
3. A tough, flexible polysaccahride is called a *septa.*

Understand Key Concepts

4. Which does not describe a method by which fungi obtain food?
   A. parasitism **C.** photosynthesis
   B. decomposition **D.** mutualism
5. Which structure of fungi is different from plants?
   A. composition of cytoplasm
   B. composition of cell walls
   C. exoskeletons
   D. cellulose

Use the image to answer question 6.

Color-Enhanced SEM Magnification: 1100×

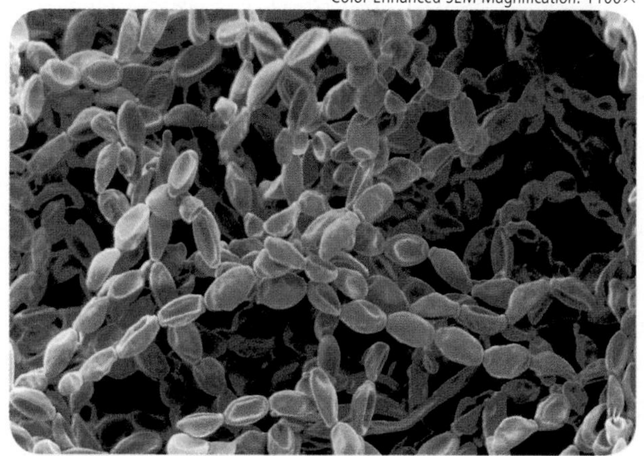

6. What is the structure shown above?
   A. hyphae **C.** chitin
   B. septae **D.** spores
7. Which can be used for asexual and sexual reproduction?
   A. gametes **C.** fragmentation
   B. budding **D.** spores

Use the diagram to answer question 8.

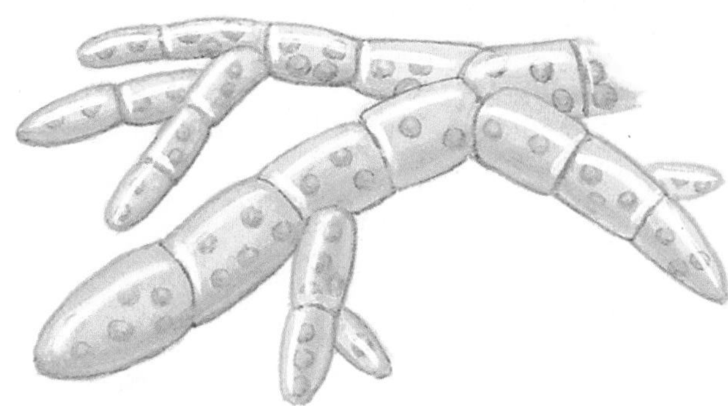

8. What is the structure shown above?
   A. mycelium
   B. spore
   C. septate hyphae
   D. aseptate hyphae

Constructed Response

9. **Short Answer** Distinguish between parasitic fungi and saprophytic fungi.
10. **Short Answer** Distinguish between hypha and mycelium.
11. **Open Ended** Hypothesize the best method of reducing the number of mold spores in your classroom. How would you test your hypothesis?

Think Critically

12. **Infer** how the structure of the aseptate hyphae allows for more rapid growth.
13. **Assess** the ability of fungi to disperse their spores.

Section 20.2

Vocabulary Review

Explain the differences between the vocabulary terms in the following sets.

14. stolon, rhizoid
15. ascospore, ascus
16. basidiocarp, basidia

Understand Key Concepts

17. Which fungi have flagellated spores?
- **A.** basidiomycetes
- **B.** zygomycetes
- **C.** ascomycetes
- **D.** chytridiomycetes

18. What is the function of stolons?
- **A.** to penetrate the food
- **B.** to spread across the surface of food
- **C.** to digest the food
- **D.** to reproduce

19. Which is a unicellular fungi?
- **A.** bread mold
- **B.** yeast
- **C.** mushroom
- **D.** Bird's Nest fungus

Use the diagram to answer question 20.

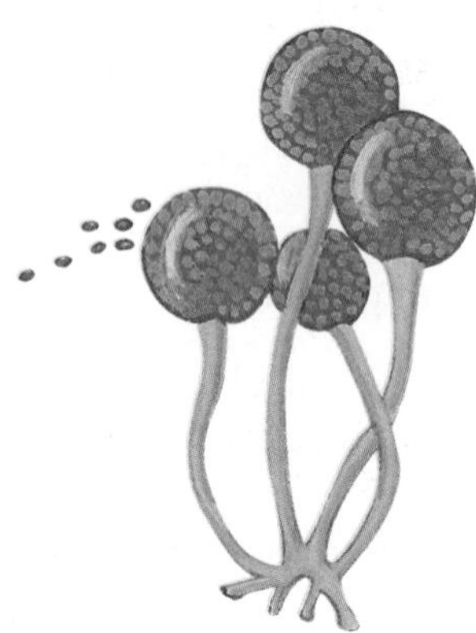

20. Within which structure do the fungi in the diagram form their spores?
- **A.** ascocarp
- **B.** sporangium
- **C.** ascus
- **D.** ascophore

21. What word best describes the structure you identified in question 20?
- **A.** bud
- **B.** haploid
- **C.** diploid
- **D.** fragmented

Constructed Response

22. Short Answer Choose one type of fungus that reproduces asexually and describe the process.

23. Open Ended Research the different size spores produced by basidiomycetes and prepare a graphic organizer for the class.

24. Open Ended Compose an argument to defend the placement of chytrids in the Kingdom Fungi instead of the Kingdom Protista.

Think Critically

25. Design an experiment to determine if homemade bread is more or less likely to grow fungus than commercially produced bread.

26. Collect and interpret data on how many of your classmates have mold allergies. Calculate the percentage of allergic classmates.

Section 20.3

Vocabulary Review

Use what you know about the vocabulary terms found on the Study Guide page to answer the following questions.

27. What term describes a symbiotic relationship between a fungus and an alga?

28. What term describes a symbiotic relationship between a fungus and a plant root?

29. What is the name of a living organism that is sensitive to environmental pollutants?

Understand Key Concepts

Use the image to answer question 30.

30. In an area recovering from a forest fire, what is this lichen's main function?
- **A.** absorbing water
- **B.** bioindicator
- **C.** pioneer species
- **D.** keeping insects away

BiologyOnline Chapter Test biologygmh.com

31. Why are lichens important bioindicators?
   A. They are susceptible to drought.
   B. They are unicellular.
   C. They are mutualistic.
   D. They are susceptible to air pollutants.

Use the image below to answer question 32.

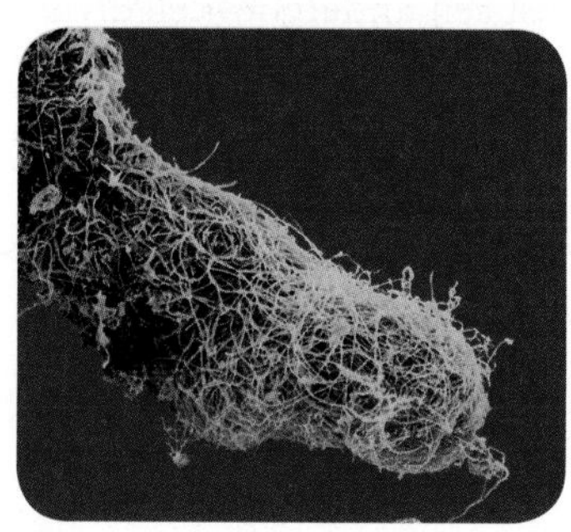

32. How is this mycorrhizae benefiting the plant?
   A. increases the surface area for gathering light
   B. decreases the need for water
   C. decreases the surface area of the roots
   D. decreases the temperature

Constructed Response

33. **Short Answer** In what ways are fungi beneficial to humans?

34. **Short Answer** Evaluate the role of lichens in the arctic environments.

Think Critically

35. **Predict** how the availability of the antibiotic penicillin during World War II impacted the soldiers.

36. **Design an experiment** that will allow you to test the antibiotic effects of two or three common fungi.

37. **CAREERS IN BIOLOGY** Write a want ad for a mycologist in a research labratory.

38. **Design** an organism that cultivates its own food production using fungi. How might the fungi benefit from this relationship?

39. **Hypothesize** why mycorrhizae might have been important for the colonization of land by plants. What kind of evidence would you look for to support your hypothesis?

Additional Assessment

40. *WRITING in* **Biology** Imagine yourself as a fungal spore landing near your home or school. Evaluate your chances of survival.

DBQ Document-Based Questions

Data obtained from: Stokstad, E. 2004. Plant pathologists gear up for battle with dread fungus. *Science* 306: 1672–1673.

This map shows where Asian soybean rust Phakopsora pachyrhizi *is found in the United States. It is a recent arrival from Brazil and other parts of South America. Its presence in each state was officially diagnosed by the USDA. Soybean rust is a disease caused by the fungus* Phakopsora pachyrhiz *that recently has become a problem for soybean farmers in the United States. Losses from this infection can amount to 80 percent of the crop.*

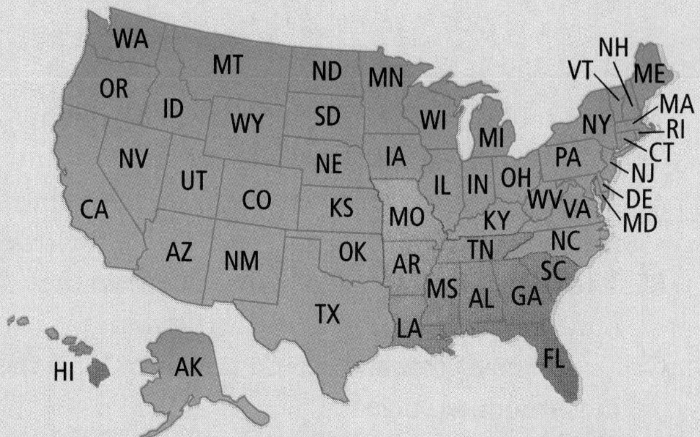

41. Evaluate the map and speculate about the factors affecting the distribution of soybean rust in the U.S. In which states is soybean rust most prevalent?

42. Apply what you know about fungi to recommend a course of action to eradicate this fungus.

43. Estimate the impact of this fungus on the future of soybean production in the U.S.

Cumulative Review

44. Suppose the molecular clock technique indicates that two organisms have begun to evolve into separate species. Indicate the kinds of data or evidence you would expect to be able to find in the organisms. **(Chapter 17)**

Standardized Test Practice

Cumulative

Multiple Choice

1. Which are autotrophic protists commonly referred to as?
   A. algae
   B. protozoans
   C. slime molds
   D. water molds

Use the diagram below to answer questions 2 and 3.

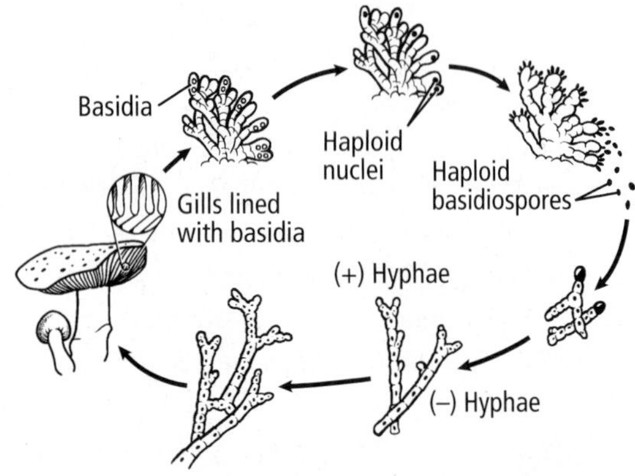

2. What part of this life cycle can be used to explain why many mushrooms grow quickly?
   A. The cap pulls in organic matter from the soil to fill the mushroom.
   B. The gills in the cap rapidly divide to form the mushroom.
   C. The hyphae grow and extend rapidly to form the mushroom.
   D. The basidia grow and lengthen the mushroom.

3. Which occurrence pictured in the diagram allows the mating types to fuse?
   A. basidia form
   B. hyphae unite
   C. mushroom forms
   D. spores release

4. A certain tree-dwelling primate has a prehensile tail and nails on its digits. To which group of primates would you expect this animal to belong?
   A. Asian apes
   B. New World monkeys
   C. Old World monkeys
   D. prosimians

5. Which occurs during the lytic cycle of a viral infection?
   A. The host cell becomes a factory that continually makes more copies of the virus.
   B. The host cell undergoes cell division that makes more copies of the virus.
   C. The virus incorporates its nucleic acid into the DNA of the host cell and lies dormant.
   D. The virus takes over the cell, makes copies of itself, and usually kills the host cell.

Use the figure below to answer question 6.

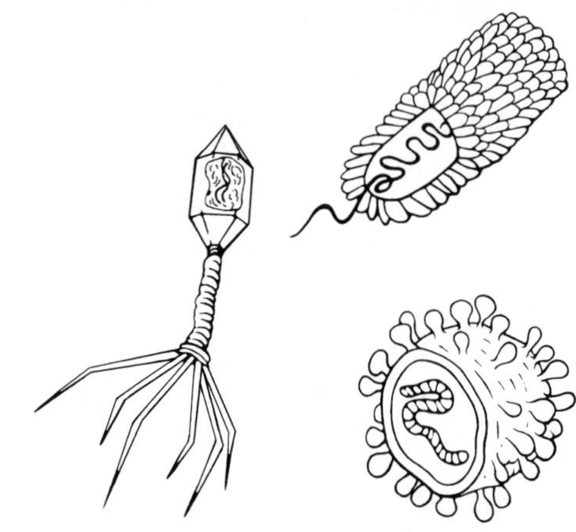

6. On what property do scientists base their classification of viruses?
   A. capsid proteins
   B. chromosome number
   C. host resistance
   D. type of genetic material

7. Which characteristic distinguishes australopithecines from earlier hominoids?
   A. binocular vision
   B. bipedalism
   C. fingernails
   D. opposable thumb

8. Which is a characteristic of an acellular slime mold?
   A. cytoplasm with many cells
   B. locomotion by means of cilia
   C. plasmodium with many nuclei
   D. reproduction by fragmentation

Short Answer

Use the diagram below to answer question 9.

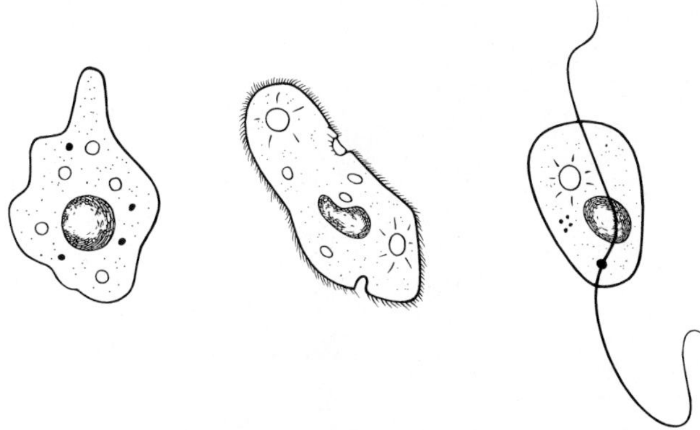

9. Identify the structure used for locomotion in each of these organisms and briefly describe how each structure functions.

10. Imagine that you found a unicellular organism living in the mud at the bottom of a pond. Write a plan to determine how you would classify it.

11. Some people think that technology can solve all human problems. Name and critique an example of a problem that technology might not be able to solve.

12. What characteristics are used to classify protists into three groups? Explain your answer.

13. Describe how sexual reproduction begins in Ascomycetes and assess its significance.

14. Give examples of three ways that fungi are important for human foods.

Extended Response

15. Create a flowchart to show how isolation of a small population can lead to speciation.

16. Assess the value of mycorrhizae for plants.

17. Evaluate how viruses benefit from their small size and simple composition.

18. Imagine that you've noticed that mushrooms grow in one corner of a field every time it rains. Give a reason why picking the mushrooms immediately following a rainshower will not stop them from growing back.

Essay Question

Light is needed for photosynthesis to take place. The algae depend on the energy from light to carry out photosynthesis. The main photosynthetic pigment of green algae is chlorophyll. Sunlight is made up of all the different wavelengths of visible light, but only blue and red are absorbed by chlorophyll. Other algae contain larger amounts of other pigments such as carotenoids. Carotenoids absorb energy from green light. Since algae live in water, this becomes important because water absorbs the different colors of light at different rates.

Using the information in the paragraph above, answer the following question in essay format.

19. Red light does not penetrate into water. Algae in water must be able to use light energy that is available underwater. Write an essay about why carotenoids are better than chlorophyll for algae living well below the surface.

NEED EXTRA HELP?																			
If You Missed Question . . .	1	2	3	4	5	6	7	8	9	10	11	12	13	14	15	16	17	18	19
Review Section . . .	19.3	20.2	20.2	16.1	18.2	18.2	16.2	19.4	19.2	19.2	18.1	1.2	20.2	20.3	15.3	20.2	18.2	20.2	19.3

UNIT 6 Plants

Chapter 21
Introduction to Plants
BIG Idea Plants have changed over time and are now a diverse group of organisms.

Chapter 22
Plant Structure and Function
BIG Idea The diverse nature of plants is due to the variety of their structures.

Chapter 23
Reproduction in Plants
BIG Idea The life cycles of plants include various methods of reproduction.

CAREERS IN BIOLOGY

Botanist

Botanists are scientists who study plants. Botanists might specialize in many disciplines from bryology (the study of mosses and simple plants) to dendrology (the study of trees and woody plants). This giant sequoia researcher is a dendrologist.

WRITING in Biology Visit biologygmh.com to learn more about botany careers. Write a paragraph to briefly explain why the demand for botanists is increasing as the human population grows around the world.

Biology Online

To read more about botanists in action, visit biologygmh.com.

CHAPTER 21

Introduction to Plants

BIG Idea Plants have changed over time and are now a diverse group of organisms.

Section 1
Plant Evolution and Adaptations

MAIN Idea Adaptations to environmental changes on Earth contributed to the evolution of plants.

Section 2
Nonvascular Plants

MAIN Idea Nonvascular plants are small and usually grow in damp environments.

Section 3
Seedless Vascular Plants

MAIN Idea Because they have vascular tissues, seedless vascular plants generally are larger and better adapted to drier environments than nonvascular plants.

Section 4
Vascular Seed Plants

MAIN Idea Vascular seed plants are the most widely distributed plants on Earth.

BioFacts

- The number of plant species is three times greater than the number of animal species.
- Nearly 98 percent of Earth's biomass consists of plants and plant products.

Alpine forest
Appalachian Mountains

Agave plants
Chihuahuan Desert

Giant water lilies
Amazon River

LAUNCH Lab

What characteristics differ among plants?

Scientists use specific characteristics to group plants within the plant kingdom. In this lab, you will examine some of the characteristics of plants.

Procedure

1. Read and complete the lab safety form.
2. Label five plant specimens using letters *A, B, C, D,* and *E.*
3. Study each plant carefully. Wash your hands thoroughly after handling plant material.
4. Based on your observations, list characteristics that describe the differences and similarities among these plants.
5. Rank your list of characteristics based on what you consider the most and least important.

Analysis

1. Compare your list to your classmates' lists.
2. Describe the diversity among the plants you studied.
3. List plant characteristics that you could not observe that might be useful in organizing these plants into groups.

Visit biologygmh.com to:

- study the entire chapter online
- explore Concepts in Motion, Microscopy Links, Virtual Labs, and links to virtual dissections
- access Web links for more information, projects, and activities
- review content online with the Interactive Tutor, and take Self-Check Quizzes

Plant Adaptations Make this Foldable to help you understand some adaptations that enabled plants to inhabit different land environments.

STEP 1 Stack three sheets of notebook paper so that the top edges are 1.5 cm apart.

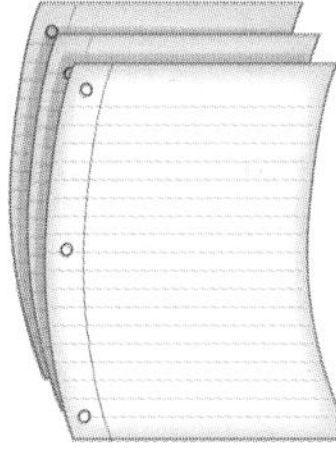

STEP 2 Fold up the bottom edges to form five tabs of equal size.

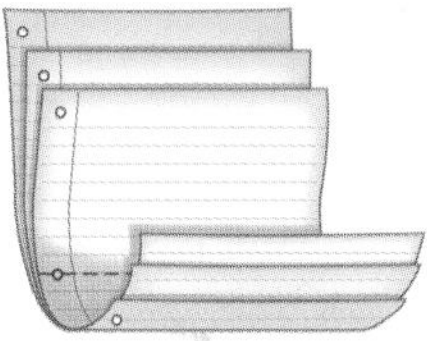

STEP 3 Staple along the folded edge to secure all sheets, place the stapled edge at the top, and then label the tabs as shown.

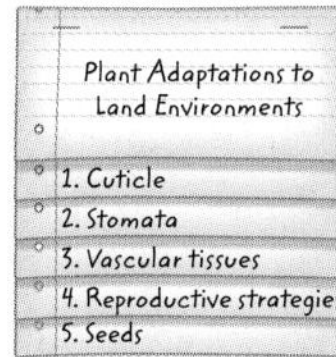

FOLDABLES **Use the Foldable with Section 21.1.** As you study the section, record on your Foldable what you learn about the importance of each adaptation.

Reading Preview

Objectives

- **Compare** the characteristics of plants and green algae.
- **Identify and evaluate** adaptations of plants to land environments.
- **Assess** the importance of vascular tissue to plant life on land.
- **Explain** alternation of generations of plants.
- **List** the divisions of the plant kingdom.

Review Vocabulary

limiting factor: any abiotic or biotic factor that restricts the existence, numbers, reproduction, or distribution of organisms

New Vocabulary

stomata
vascular tissue
vascular plant
nonvascular plant
seed

Plant Evolution and Adaptations

MAIN Idea Adaptations to environmental changes on Earth contributed to the evolution of plants.

Real-World Reading Link Perhaps you have seen a photo of your ancestors and noticed that some of your living relatives resemble people in the photo. In a similar way, scientists who study evolution notice common characteristics between ancient groups of organisms and present-day groups.

Plant Evolution

Plants are vital to our survival. The oxygen we breathe, the food we eat, and many of the things that make our lives comfortable, such as clothing, furniture, and our homes, come from or are parts of plants. If you were asked to describe a plant, would you describe a tree, a garden flower, or a houseplant? Biologists describe plants as multicellular eukaryotes with tissues and organs that have specialized structures and functions. For example, most plants have photosynthetic tissues, and organs that anchor them in soil or to an object or another plant. However, does this description apply to ancient plants?

Connection to Earth Science You read in Chapter 14 that Earth is about 4.6 billion years old. Can you imagine ancient Earth without land plants? That was the case until about 400 million years ago when primitive land plants appeared. However, fossil evidence from about 500 million years ago indicates that the shallow waters of ancient Earth were filled with a variety of organisms—archaea, bacteria, algae and other protists, and animals, such as sponges, corals, and worms.

There is strong evidence, including biochemical and fossil evidence, that multicellular land plants and present-day green algae share a common ancestor, as diagrammed in the evolutionary tree in **Figure 21.1.** This common ancestor might have been able to survive periods of drought. Through natural selection, drought-resistant adaptations in that ancestor, such as protected embryos and other survival characteristics, might have passed to future generations. When scientists compare present-day plants and present-day green algae, they find the following common characteristics:

- cell walls composed of cellulose
- cell division that includes the formation of a cell plate
- the same type of chlorophyll used in photosynthesis
- similar genes for ribosomal RNA
- food stored as starch
- the same types of enzymes in cellular vesicles

Figure 21.1 This evolutionary tree shows the relationship of ancient freshwater green algae to present-day plants.

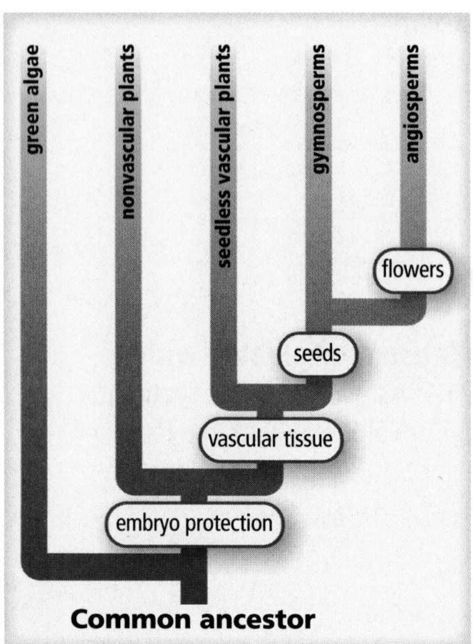

■ **Figure 21.2** The cuticle is produced by the outer layer of cells. Plants in dry environments often have a thick waxy layer over the cuticle.
Infer ***what advantage this waxy layer provides to plants in dry environments.***

Plant Adaptations to Land Environments

While living on land might seem advantageous for many organisms, there are challenges for land organisms that aquatic organisms do not face. Over time, plants that inhabited land developed adaptations that helped them survive limited water resources as well as other environmental factors.

Cuticle Have you ever noticed that some plant leaves appear shinier than others, or that some leaves have a grayish appearance, such as those of the agave in **Figure 21.2?** An adaptation found on most above-ground plant parts is a fatty coating called the cuticle on the outer surface of their cells. Wax can also be a component of the cuticle, giving it a grayish appearance. Fats and waxes are lipids and are insoluble in water, as you learned in Chapter 6. Because of this, the cuticle helps prevent the evaporation of water from plant tissues and can also act as a barrier to invading microorganisms.

VOCABULARY
WORD ORIGIN
Cuticle
from the Latin diminutive *cuticula*, meaning *skin*.

MiniLab 21.1

Compare Plant Cuticles

Does the cuticle vary among different types of plants? Plant leaves are covered with a cuticle that reduces water loss. The thickness of cuticle material varies among plants.

Procedure

1. Read and complete the lab safety form.
2. Observe the **plant leaves** provided by your teacher. Write a description of each leaf type.
3. Pile each type of leaf on separate but identical **plastic plates.** Measure the mass and then adjust the number of leaves on each plate until they are of equal mass. Record the masses.
4. The next day, examine each plate of leaves. Record your observations.
5. Measure the mass of each plate of leaves and record the data.

Analysis

1. **Interpret Data** Which leaves appeared to have lost more water? Do the data support your observation?
2. **Infer** which leaves might have a thicker cuticle.

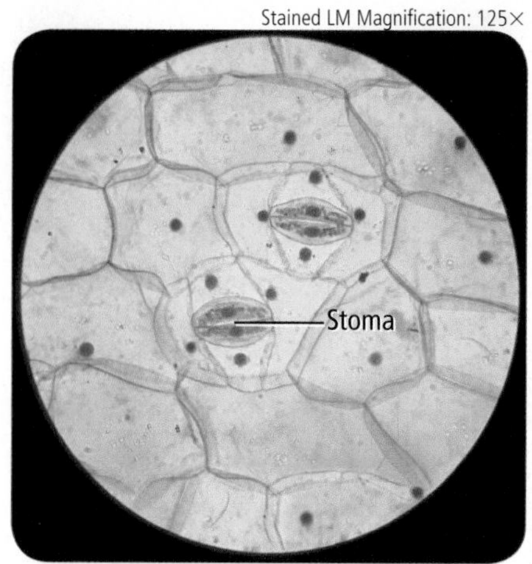

Figure 21.3 Stomata are common on the lower surfaces of most plants' leaves.

FOLDABLES
Incorporate information from this section into your Foldable.

Stomata Like algae, most plants carry on photosynthesis, which produces glucose and oxygen from carbon dioxide and water. The exchange of gases between plant tissues and the environment is necessary for photosynthesis to occur. If the cuticle reduces water loss, it also might prevent the exchange of gases between a plant and its environment. **Stomata** (singular, stoma) are adaptations that enable the exchange of gases even with the presence of a cuticle on a plant. Stomata are openings in the outer cell layer of leaves and some stems, as shown in **Figure 21.3.**

Although photosynthesis can occur in some green stems, plant leaves usually are the sites of photosynthesis and are where most stomata are found. You will read more about the structures and functions of stomata and leaves in Chapter 22.

Vascular tissues Another plant adaptation to land environments is **vascular tissue**—specialized transport tissues. Recall from Chapter 7 that many substances slowly move into and out of cells and from cell to cell by osmosis or diffusion. However, vascular tissue enables faster movement of substances than by osmosis and diffusion, and over greater distances. Plants with vascular tissues are called **vascular plants,** like those in **Figure 21.4.** In some plants, substances slowly move from cell to cell by osmosis and diffusion. They are the **nonvascular plants** and lack specialized transport tissues.

Vascular tissues also provide structure and support. The presence of thickened cell walls in some vascular tissue provides additional support. Therefore, vascular plants can grow larger than nonvascular plants. You will read more about vascular tissues in Chapter 22.

Reproductive strategies You learned in Chapter 19 that a spore is a haploid cell capable of producing an organism. Some land plants reproduce by spores that have waterproof protective coverings. However, the gametophytes of those land plants must have a film of water covering them for sperm to swim to eggs. Water is a limiting factor in the environments of these plants. In Chapter 23 you will read about adaptations of seed plants that enable a sperm to reach an egg without the presence of water.

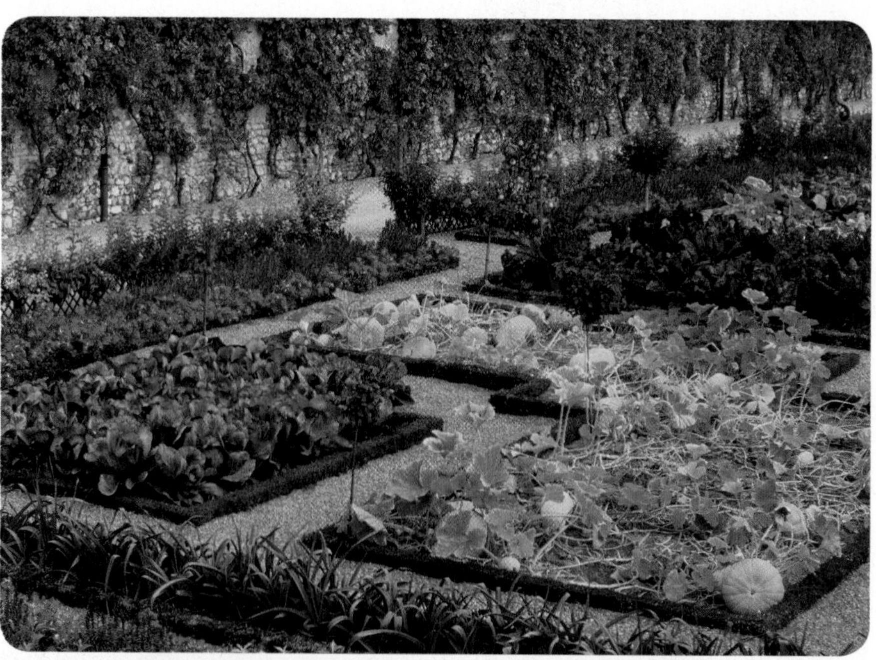

Figure 21.4 Vascular plants have many shapes and sizes.
List *the plants that you recognize.*

Seeds The evolution of the seed was another important adaptation that helped ensure the success of some vascular plants. A **seed,** as shown in **Figure 21.5,** is a plant structure that contains an embryo, contains nutrients for the embryo, and is covered with a protective coat. These features enable seeds to survive harsh environmental conditions and then sprout when favorable conditions exist. Seeds also can have different structural adaptations that help scatter them. You will read more about these structural adaptations of seeds in Section 4 of this chapter.

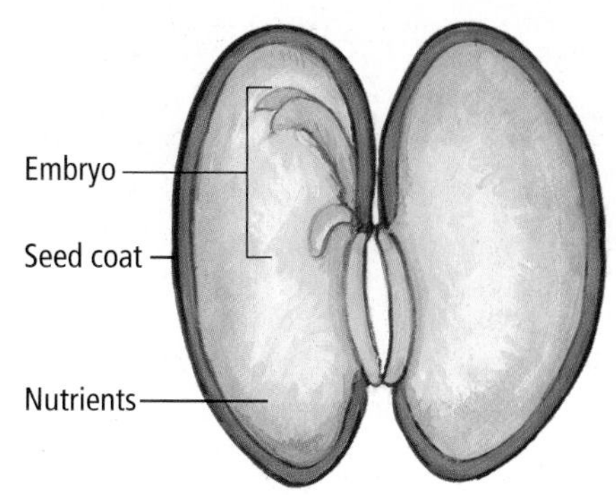

Figure 21.5 The seed coat protects the embryo—the new sporophyte generation.

Alternation of Generations

You read in Chapter 19 that the life cycles of some organisms include an alternation of generations—a haploid gametophyte generation and a diploid sporophyte generation. The gametophyte generation produces gametes—sperm and eggs. Some plants produce sperm and eggs on separate gametophytes while others produce them on one gametophyte. When a sperm fertilizes an egg, a diploid zygote forms that can undergo countless mitotic cell divisions to form a multicellular sporophyte. The sporophyte generation produces spores that can grow to form the next gametophyte generation.

Depending on the type of plant, one generation is dominant over the other. This means it is larger and, therefore, more noticeable and lasts longer. Most of the plants you see—houseplants, grasses, garden plants, and trees—are the diploid sporophyte generation for those plants. During plant evolution, the trend was from dominant gametophytes to dominant sporophytes that contain vascular tissue. In land plants, the gametophyte generation of vascular plants is microscopic, as shown in **Figure 21.6,** but is larger in nonvascular plants and can be observed without using a magnifying device. You will see more examples of gametophytes and sporophytes later in this chapter.

Vocabulary

Academic Vocabulary

Dominant (DAH muh nunt): most immediately noticeable.
Oaks are the dominant trees in the forest.

 **Reading Check** **Identify** the generation of a plant's life cycle that produces sperm and eggs.

Figure 21.6 The sporophyte of a maple tree—the roots, trunk, and branches—is larger than the tiny male gametophyte found in its pollen. The maple sporophyte also lives longer than the pollen.

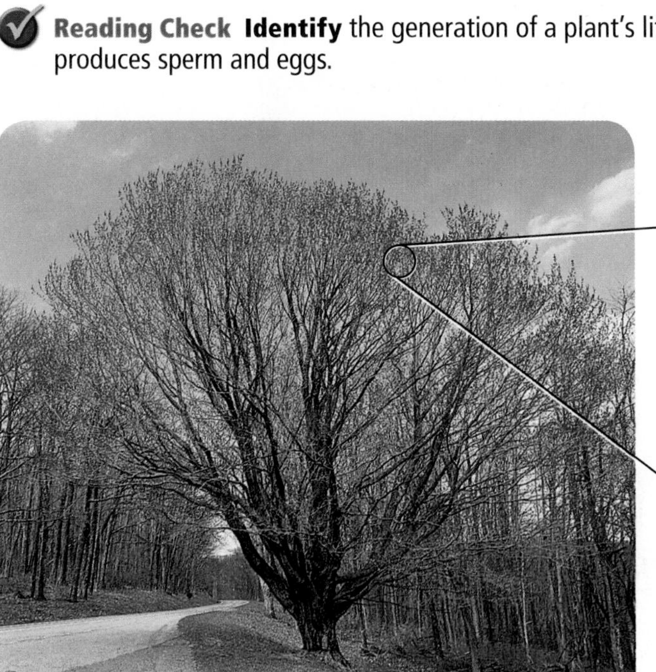

Sporophyte generation—maple tree

Color-Enhanced SEM Magnification: 155×

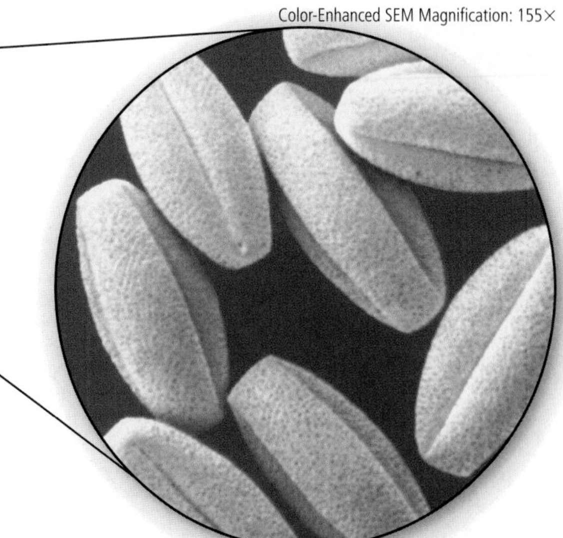

Gametophyte generation—maple pollen

Visualizing the Plant Kingdom

Figure 21.7
One way to classify the divisions of the plant kingdom is as either nonvascular or vascular plants. In addition, vascular plants can be classified as nonseed or seed plants.

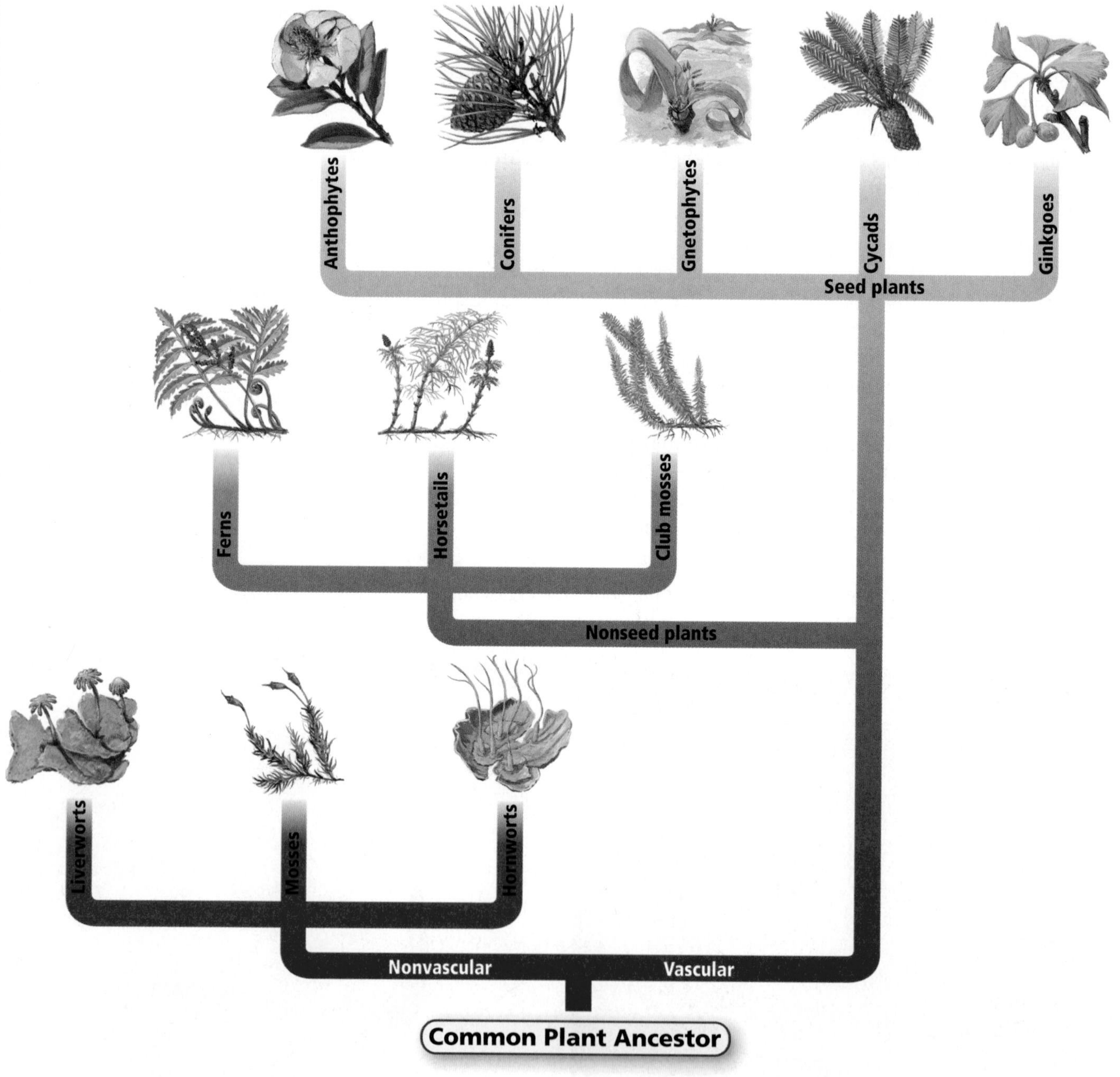

Anthophytes
Conifers
Gnetophytes
Cycads
Ginkgoes
Seed plants
Ferns
Horsetails
Club mosses
Nonseed plants
Liverworts
Mosses
Hornworts
Nonvascular
Vascular
Common Plant Ancestor

Concepts In Motion **Interactive Figure** To see an animation of plant classification, visit biologygmh.com.

Biology Online

Plant Classification

Over time, plant adaptations resulted in a diversity of plant characteristics. Botanists use these characteristics to classify all plants of Kingdom Plantae into twelve divisions. Recall from Chapter 17 that other kingdoms, except for bacteria, are divided into phyla not divisions. When referring to members of a division, it is common practice to drop the *-a* from the division name and add *-es*. Therefore, members of Division Bryophyta are called bryophytes (BRI uh fites).

The twelve plant divisions can be placed into two groups—the nonvascular plants and the vascular plants, eleven of which are illustrated in **Figure 21.7.** As you learned in this section, nonvascular plants lack specialized transport tissues. They include

- bryophytes—mosses;
- anthocerophytes (an tho SAIR uh fites)—hornworts; and
- hepaticophytes (hih PA tih koh fites)—liverworts.

You also learned that vascular plants have specialized transport tissues. Vascular plants are divided into two smaller groups—plants that do not produce seeds and plants that produce seeds. Two seedless vascular plants discussed later in this chapter are

- lycophytes (LI kuh fites)—club mosses; and
- pterophytes (TER uh fites)—ferns and horsetails.

Also discussed are five seed-producing vascular plants

- cycadophytes (si KAH duh fites)—cycads or sago palms;
- gnetophytes (NEE tuh fites)—joint firs;
- ginkgophytes (GIHN koh fites)—ginkgoes;
- coniferophytes (kuh NIHF uh ruh fites)—pines and similar plants; and
- anthophytes (AN thuh fites)—flowering plants.

CAREERS IN BIOLOGY

Botanist Individuals who study plant life are botanists. They study the interactions of plants and the environment, and the structure and functions of plants. They might identify new plant species. A botanist can teach, conduct research, or work at a botanical garden, plant nursery, or greenhouse, or for a governmental agency. For more information on biology careers, visit biologygmh.com.

LAUNCH Lab

Review Based on what you've read about plant characteristics, how would you now answer the analysis questions?

Section 21.1 Assessment

Section Summary

- Plants are multicellular organisms and most are photosynthetic.
- Evidence indicates that ancient, unicellular, freshwater green algae were the ancestors of present-day plants.
- Present-day plants and green algae have many common characteristics.
- Over time, plants developed several adaptations for living on land.
- Plants alternate between a sporophyte and a gametophyte generation.

Understand Main Ideas

1. MAIN Idea **Identify** adaptations that make it possible for plants to survive on land.
2. **Explain** why scientists hypothesize that green algae and plants share a common ancestor.
3. **Name** the plant divisions. Which ones are seedless vascular plants?
4. **Differentiate** between a gametophyte and a sporophyte.

Think Critically

5. **Apply** what you know about lipids to explain why the cuticle helps prevent water loss in plants.
6. **Assess** the importance of a plant's vascular tissue to its ability to live on land.

WRITING in **Biology**

7. Find a poem about any plant and then analyze its scientific accuracy.

Section 21.2

Nonvascular Plants

Reading Preview

Objectives

- **Identify** the structures of nonvascular plants.
- **Compare and contrast** the characteristics of the nonvascular plant divisions.

Review Vocabulary

symbiosis: a relationship in which two organisms live together in a close association

New Vocabulary

thallose

MAIN Idea **Nonvascular plants are small and usually grow in damp environments.**

Real-World Reading Link Have you ever used a garden hose to water a lawn or wash a car? Why didn't you carry water from the faucet in a bucket? As you probably realize, using a garden hose to transport water is more efficient than using a bucket. You learned in the previous section that nonvascular plants lack structures that can move water and other substances. However, because of their small size, moving substances by diffusion and osmosis is sufficient for them.

Diversity of Nonvascular Plants

As shown in the evolutionary tree in **Figure 21.8,** nonvascular plants make up one of the four major groups of plants that evolved along with green algae from a common ancestor. In general, nonvascular plants usually are small, which enables most materials to move within them easily. These plants often are found growing in damp, shady areas—an environment that provides the water needed by nonvascular plants for nutrient transport and reproduction.

Division Bryophyta The most familiar bryophytes are the mosses. You might have seen these small, nonvascular plants growing on a damp log or along a stream. Although they do not have true leaves, mosses have structures that are similar to leaves. Their photosynthetic, leaflike structures usually consist of a layer of cells that is only one cell thick.

Mosses produce rootlike, multicellular rhizoids that anchor them to soil or another surface, as shown in **Figure 21.8.** Water and dissolved minerals can diffuse into a moss's rhizoids. Although mosses have some tissue that transports water and food, these plants do not have true vascular tissues. Water and other substances move throughout a moss by osmosis and diffusion, processes explained in Chapter 7.

■ **Figure 21.8** Embryo protection is a characteristic of nonvascular and vascular plants. The dense carpet of moss—a nonvascular plant—consists of hundreds of moss plants, each with leafy stems and rhizoids.

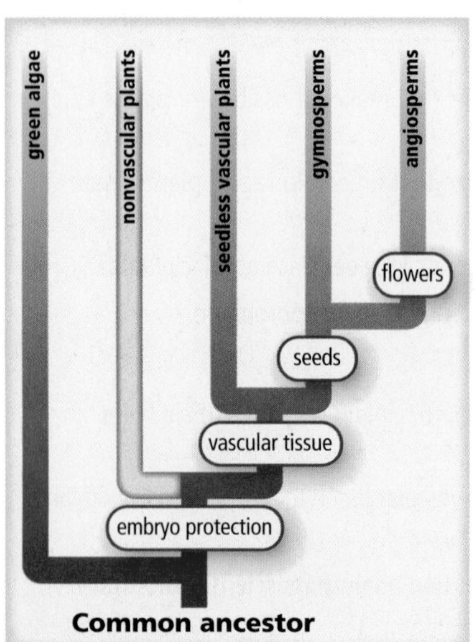

Carpet of moss

Mosses exhibit variety in structure and growth. Some mosses have stems that grow upright and others have trailing vinelike stems. Other mosses form extensive mats that help slow erosion on rocky slopes. Over time, *Sphagnum* (a type of moss) and other plant matter accumulated, decayed, and formed deep deposits called peat. Peat can be cut into blocks and burned as a fuel. Gardeners and florists often add peat moss to soil to help it retain moisture.

Scientists estimate that as much as one percent of Earth's surface might be covered by bryophytes. Many mosses, like those in **Figure 21.8,** grow in temperate regions and freeze and thaw without damage. Others can survive an extreme loss of water and then resume growth when moisture returns.

 **Reading Check** **Explain** how peat is formed.

Division Anthocerophyta The smallest division of nonvascular plants is division Anthocerophyta. Anthocerophytes are called hornworts because of their hornlike sporophytes, as shown in **Figure 21.9.** Water and nutrients move in hornworts by osmosis and diffusion. About 100 hornwort species have been identified.

An identifying feature of these plants is the presence of one large chloroplast in each cell of the gametophyte and sporophyte. This feature can be observed under a microscope. However, the hornwort sporophyte produces much of the food used by its sporophyte and gametophyte generations.

While examining hornwort tissue under a microscope, besides the large chloroplast in each cell, you also might observe that the spaces around cells are filled with mucilage, or slime, rather than air. Cyanobacteria in the genus *Nostoc* often grow in this slime. The cyanobacteria and hornwort exhibit mutualism.

DATA ANALYSIS LAB 21.1

Based on Real Data*

Form a Hypothesis

How does *Nostoc* benefit a hornwort? Cyanobacteria, usually species of *Nostoc,* form mutualistic relationships with a few liverworts and the majority of hornworts.

Data and Observations

Nostoc colonies appear as dark spots within gametophyte tissue, as shown in the photo.

Think Critically

1. **Form a hypothesis** about the benefit(s) the cyanobacteria receive from the hornwort.
2. **Design an experiment** to test your hypothesis.

*Data obtained from: Costa, J-L., et al. 2001. Genetic diversity of *Nostoc* symbionts endophytically associated with two bryophyte species. *Appl. Envir. Microbiol.* 67: 4393–4396.

Figure 21.9 The hornlike sporophyte of a hornwort is anchored to the gametophyte.

Figure 21.10 A thallose liverwort's shape resembles lobes of a liver. Leafy liverworts have leaflike structures but not true leaves.

Thallose liverwort

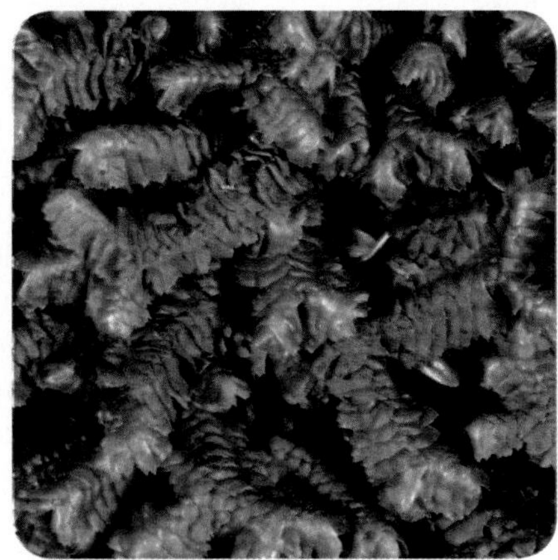

Leafy liverwort

VOCABULARY

SCIENCE USAGE V. COMMON USAGE

Fleshy

Science usage: having a juicy or pulpy texture.
Peaches and plums are fleshy fruits.

Common usage: relating to, consisting of, or resembling flesh.
The piece of beef was fleshy—not bony.

Division Hepaticophyta Because of their appearance and use as a medicine to treat liver ailments during medieval times, hepaticophytes are referred to as liverworts. This division of nonvascular plants contains more than 6000 species. They are found in a variety of habitats ranging from the tropics to the arctic. Liverworts tend to grow close to the ground and in areas where moisture is plentiful, such as damp soil, near water, or on damp decaying logs. A few species even can survive in relatively dry areas. Like other nonvascular plants, water, nutrients, and other substances are transported throughout liverworts by osmosis and diffusion.

Liverworts are classified as either **thallose** (THAL lohs) or leafy, as shown in **Figure 21.10.** A thallose liverwort has a body that resembles a fleshy, lobed structure. Leafy liverworts have stems with flat, thin leaflike structures arranged in three rows—a row on each side of the stem and a row of smaller leaves on the undersurface. Liverworts have unicellular rhizoids, unlike mosses that have multicellular rhizoids.

DNA analysis has shown that liverworts lack DNA sequences that most other land plants contain. This suggests that liverworts are the most primitive of land plants.

Section 21.2 Assessment

Section Summary

- Distribution of nonvascular plants is limited by the plants' ability to transport water and other substances.
- Mosses are small plants that can grow in different environments.
- Like other nonvascular plants, hornworts rely on osmosis and diffusion to transport substances.
- The two types of liverworts are described as thallose and leafy.

Understand Main Ideas

1. MAIN Idea **Summarize** the characteristics of a moss.
2. **Identify** environmental changes that might have influenced the evolution of nonvascular plant structures.
3. **Distinguish** between a liverwort and a hornwort.
4. **Generalize** the economic value of bryophytes.

Think Critically

5. **Apply** what you know about osmosis and diffusion to suggest why nonvascular plants usually are small.
6. **Predict** the changes that would occur at the cellular level when a moss dries out.
7. **Compare and contrast** the habitats of mosses, hornworts, and liverworts.

Biology Online Self-Check Quiz biologygmh.com

Section 21.3

Reading Preview

Objectives

- **Identify and analyze** the characteristics of seedless vascular plants.
- **Compare and contrast** the characteristics of club mosses and ferns.

Review Vocabulary

spore: a reproductive haploid cell with a hard outer coat that can develop into a new organism without the fusion of gametes

New Vocabulary

strobilus
epiphyte
rhizome
sporangium
sorus

Seedless Vascular Plants

MAIN Idea Because they have vascular tissues, seedless vascular plants generally are larger and better adapted to drier environments than nonvascular plants.

Real-World Reading Link Whether it is to brush your teeth, get a drink of water, or to wash something, when you turn on a faucet, water flows out. The plumbing in your home carries water to and from different places. The vascular tissue in plants can be thought of as a plant's plumbing because it carries water and dissolved substances throughout the plant.

Diversity of Seedless Vascular Plants

Club mosses, also known as spike mosses, and the fern group make up the seedless vascular plant group. Keep in mind that though the common name for a club moss identifies it as a moss, it is not like the mosses described in the previous section. As indicated in **Figure 21.11,** this plant group is one of the three plant groups with vascular tissues. Seedless vascular plants exhibit a great diversity of form and size.

Regardless of their size, an adaptation seen in some seedless vascular plant sporophytes is the strobilus (STROH bih lus) (plural, *strobili*). A **strobilus** is a compact cluster of spore-bearing structures. The tiny spores produced in the strobilus often are carried by the wind. If a spore lands in a favorable environment, it can grow to form the gametophyte.

Division Lycophyta Present-day lycophytes or club mosses are descendants of the oldest group of vascular plants. Fossil evidence suggests that ancient lycophytes were tree-sized plants—some as tall as 30 m. They formed a large part of the vegetation of Paleozoic forests. After this vegetation died, its remains changed over time and eventually became part of the coal that humans mine for fuel.

Unlike true mosses, the sporophyte generation of lycophytes is dominant. They resemble moss gametophytes, and their reproductive structures that produce spores are club-shaped or spike-shaped, as shown in **Figure 21.11.**

Figure 21.11 Seedless vascular plants, such as the club moss called wolf's claw, produce spores in strobili instead of seeds.

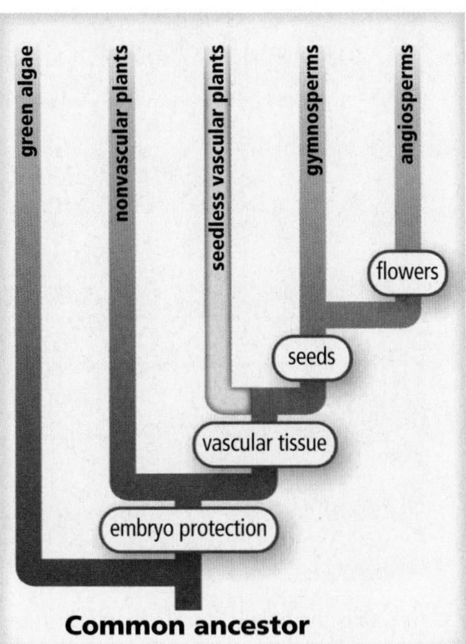

***Lycopodium* sp.—wolf's claw**

***Selanginella* sp.**

■ **Figure 21.12** This club moss belongs to the genus *Selaginella.*

Lycophytes have roots, stems, and small, scaly, leaflike structures. Another name for some lycophytes is ground pines because they resemble miniature pine trees. Stems are either branched or unbranched and grow either upright or creep along the soil's surface. Roots grow from the base of a stem. Extending down the middle of each scaly leaflike structure is a vein of vascular tissue.

Most of the club mosses belong to two genera—*Lycopodium* and *Selanginella*—like the examples shown in **Figure 21.11** and **Figure 21.12.** Many tropical lycophyte species are epiphytes. An **epiphyte** is a plant that lives anchored to an object or another plant. When anchored in treetops, they create another habitat for insects and other small animals in the forest canopy.

Reading Check **Identify** the contribution of ancient lycophytes to present-day economies.

Division Pterophyta This plant division includes ferns and horsetails. The horsetails once were in their own plant division. However, recent biochemical studies reveal that they are closely related to ferns and should be grouped with them.

Connection to Earth Science During the Carboniferous Period, about 300–359 million years ago, ferns were the most abundant land plants. Vast forests of treelike ferns existed, and some of them produced seedlike structures. Ferns grow in many different environments. Although ferns are most common in moist environments, they can survive dry conditions. When water is scarce, the life processes of some ferns slow so much that the fern appears to be lifeless. When water becomes available, the fern resumes growth. Examples of ferns growing in diverse habitats are shown in **Figure 21.13.**

■ **Figure 21.13** Ferns are a diverse group of plants that occupy a variety of habitats.

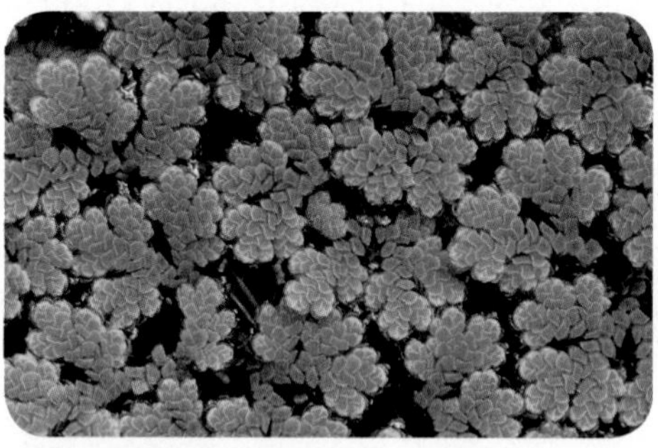

The aquatic fern *Azolla* is mutualistic with a cyanobacterium.

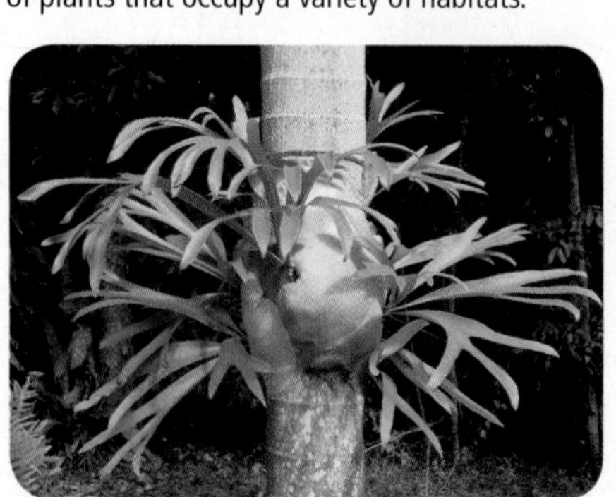

The staghorn fern grows as an epiphyte.

Dryopteris grows best in shady, dry environments.

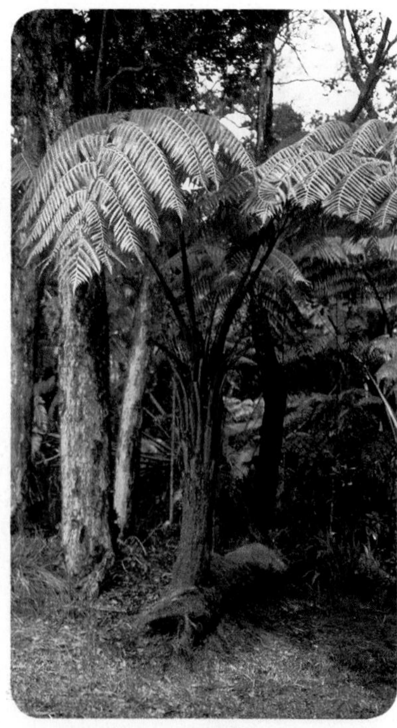

Hawaii is the only U.S. state to which tree ferns are native in tropical forests.

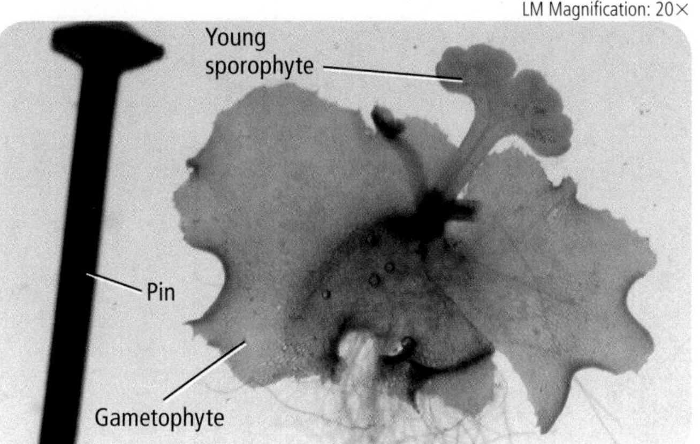

Fern gametophyte and sporophyte

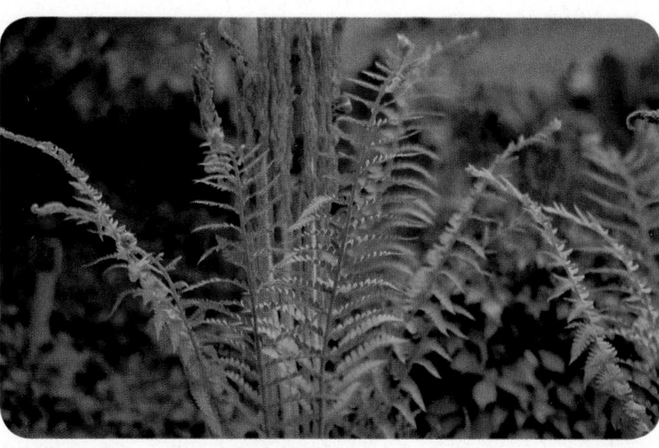

Mature fern sporophyte

■ **Figure 21.14** Fern gametophytes and sporophytes differ greatly in size and appearance. A mature fern sporophyte is many times larger than the gametophyte.

It is unlikely that you have seen a fern gametophyte like the one in **Figure 21.14.** This tiny, thin structure is smaller than a pin. It grows from a spore and has male and female reproductive structures. Following fertilization, the sporophyte grows from and is briefly dependent on the gametophyte. One adaptation of some ferns that live in dry areas is that they can produce sporophytes without fertilization. Eventually, the sporophyte produces roots and a thick underground stem called a **rhizome.** The rhizome is a food-storage organ. The aboveground structures of some ferns die at the end of a growing season. The breakdown of the rhizome's stored food releases energy when growth resumes.

The familiar parts of a fern are its photosynthetic leafy structures, or fronds, shown in **Figure 21.14.** The frond is part of the sporophyte generation of ferns. Fronds have branched vascular tissue and vary greatly in size.

DATA ANALYSIS LAB 21.2

Based on Real Data*

Analyze Models

When did the diversity of modern ferns evolve? Researchers analyzed fossil evidence and DNA sequence data of ferns. They found that ferns have shown greater diversity in more recent evolutionary history. They concluded that the diversity of modern ferns evolved after angiosperms dominated terrestrial ecosystems.

Data and Observations

Observe the two models showing the evolution of the diversity of organisms.

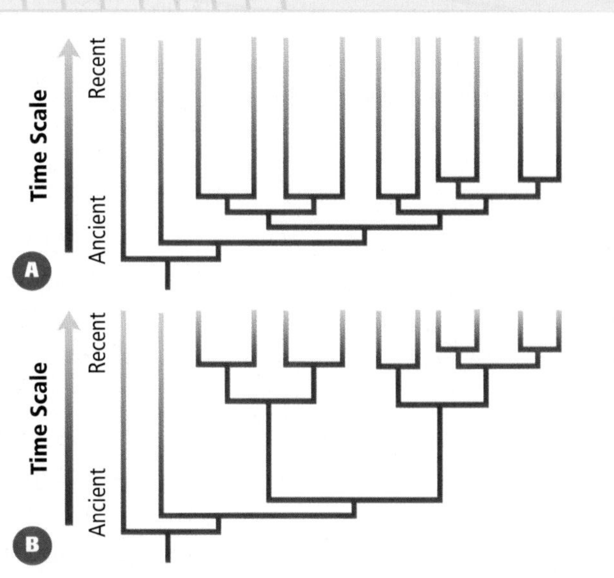

Think Critically

1. **Select** the model that best fits the researchers' conclusion described above.
2. **Infer** Angiosperms are flowering plants and trees. How might angiosperms have influenced fern diversity?

*Data obtained from: Eriksson, Torsten. 2004. Evolutionary biology: Ferns reawakened. *Nature* 428: 480–481.

Bird's nest fern

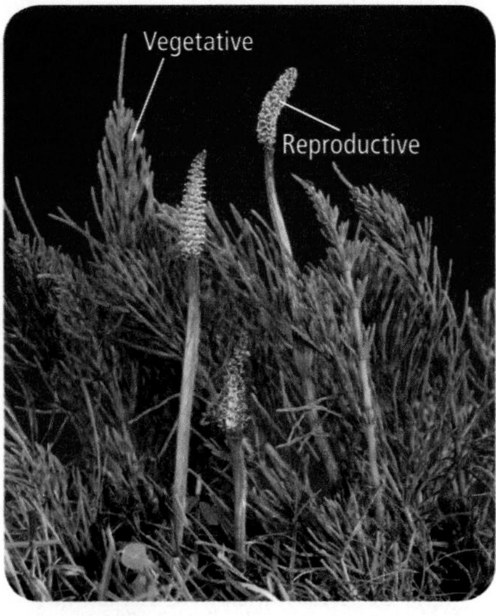

Horsetails

■ **Figure 21.15** The sori of the bird's nest fern contain spores and form lines on the underside of a frond. Some horsetails produce two different sporophyte stalks—vegetative and reproductive.

Fern spores form in a structure called a **sporangium** (plural, sporangia), and clusters of sporangia form a **sorus** (plural, sori). Sori usually are located on the undersides of fronds, as shown in **Figure 21.15.**

Figure 21.15 also shows the typical structure of horsetails—ribbed, hollow stems with circles of scalelike leaves. Like lycophytes, horsetails produce spores in strobili at the tips of reproductive stems. When released into a favorable environment, horsetail spores can develop into gametophytes.

Another common name for horsetails is scouring rushes because in colonial days, they often were used to scrub pots and pans. Horsetails contain a scratchy substance called silica. You can feel it when you rub your finger along a horsetail stem.

Present-day horsetail species are much smaller than their ancient ancestors. Most horsetails grow in wet areas, such as marshes, swamps, and stream banks. Some species grow in the drier soil of fields and roadsides only because their roots grow into underlying, water-saturated soil.

Section 21.3 Assessment

Section Summary

- Seedless vascular plants have specialized transport tissues and reproduce by spores.
- The sporophyte is the dominant generation in vascular plants.
- Lycophytes and pterophytes are seedless vascular plants.

Understand Main Ideas

1. MAIN Idea **Make a table** that lists the characteristics of seedless vascular plant groups.
2. **Compare** the sporophyte and gametophyte generations of vascular and nonvascular plants.
3. **Infer** the advantages of the fern sporophyte's initial dependency upon the gametophyte.

Think Critically

4. **Design an experiment** that would test the ability of fern gametophytes to grow on different soils.
5. **Evaluate** the advantage of branching vascular tissue in fern fronds.
6. **Construct a Venn diagram** showing characteristics of club mosses and ferns.

Biology Online **Self-Check Quiz** biologygmh.com

Section 21.4

Reading Preview

- **Compare and contrast** the characteristics of the seed plants.
- **Identify** the divisions of gymnosperms.
- **Summarize** the life spans of anthophytes.

Review Vocabulary

adaptation: inherited characteristic that results from response to an environmental factor

New Vocabulary

cotyledon
cone
annual
biennial
perennial

Vascular Seed Plants

MAIN Idea **Vascular seed plants are the most widely distributed plants on Earth.**

Real-World Reading Link You put a letter in an envelope to mail it because you are hoping that the envelope will protect your letter. In a similar way, a new seed plant is protected within the seed until environmental conditions are favorable for growth.

Diversity of Seed Plants

Vascular seed plants produce seeds. Each seed usually contains a tiny sporophyte surrounded by protective tissue. Seeds have one or more **cotyledons** (kah tuh LEE dunz)—structures that either store food or help absorb food for the tiny sporophyte. Plants whose seeds are part of fruits are called angiosperms. Those plants whose seeds are not part of fruits are called gymnosperms. The word *gymnosperm* comes from two Greek words that together mean *naked seed.*

Seed plants have a variety of adaptations for the dispersal or scattering of their seeds throughout their environments, like those shown in **Figure 21.16.** Dispersal is important because it limits competition between the new plant and its parent and other offspring.

The sporophyte is dominant in seed plants and produces spores. These spores divide by meiosis to form male gametophytes (pollen grains) and female gametophytes. Each female gametophyte consists of one or more eggs surrounded by protective tissues. Both gametophytes are dependent on the sporophyte generation for their survival.

Figure 21.16 Examine these structural adaptations for seed dispersal.

The cocklebur has hooks that can attach to an animal's fur or a human's clothing.

The dry fruit of a witch hazel plant can eject its two seeds more than 12 m from the plant.

These pine seeds have winglike structures that enable them to move with the wind.

The coconut, with its seed inside, can float great distances on ocean currents.

Parachutelike structures help disperse milkweed seeds.

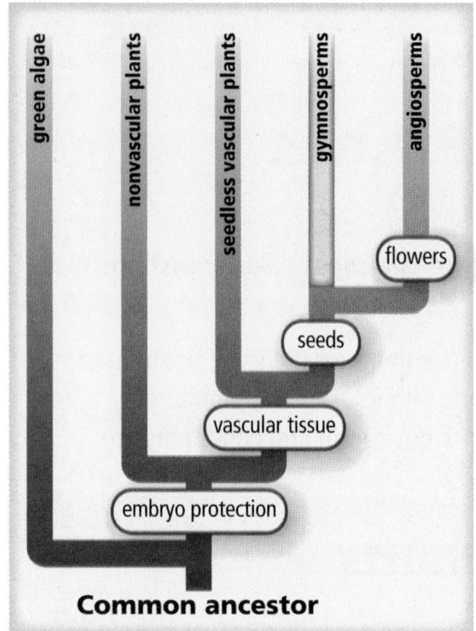

■ **Figure 21.17** The evolutionary tree above shows that the development of cones was an earlier evolutionary adaptation than flowers.

Earlier in this chapter you read that water must be present for a sperm to reach an egg in both nonvascular plants and seedless vascular plants. Most seed plants do not require a film of water for this process. This is an important difference between seed plants and other plants. This adaptation enables seed plants to thrive in different environments, including areas where water is scarce.

Division Cycadophyta As shown in the evolutionary tree in **Figure 21.17,** plants with cones—the gymnosperms—evolved before plants with flowers—the angiosperms. A **cone** is a structure that contains the male or female reproductive structures of cycads and other gymnosperm plants. A male cone produces clouds of pollen grains that produce male gametophytes. Female cones contain female gametophytes. Cycad cones can be as long as 1 m and weigh as much as 35 kg. Male and female cones grow on separate cycad plants.

Because cycads have large divided leaves and some grow more than 18 m tall, people often think that cycads are related to palm trees. However, cycads have structural differences and different reproductive strategies than do palms. While cycads might resemble woody trees, they actually have a soft stem or trunk consisting mostly of storage tissue.

The natural habitats for cycads are the tropics or subtropics. There is only one species native to the United States. Its native habitat is southern Florida. Cycads grew in abundance 200 million years ago, but today there are only about 11 genera and 250 species.

Reading Check **Compare** a cone with a strobilus.

Division Gnetophyta Plants in division Gnetophyta can live as long as 1500–2000 years. There are just three genera of gnetophytes and each exhibits unusual structural adaptations to its environment.

The genus *Ephedra* is the only gnetophyte genus that grows in the United States. The genus *Gnetum* includes about 30 species of tropical trees and climbing vines. The remaining genus, *Welwitschia,* has only one species—a bizarre-looking plant shown in **Figure 21.18**—found exclusively in the deserts of southwest Africa. It has a large storage root, and two continuously growing leaves that eventually can exceed 6 m in length. *Welwitschia* takes in available moisture from fog, dew, or rain through its two leaves.

CAREERS IN BIOLOGY

Wood Scientist An individual who is involved in one or more aspects of converting wood to wood products is a wood scientist. A wood scientist can conduct research or work in manufacturing as a product or process developer, quality or production controller, engineer, or manager. For more information on biology careers, visit biologygmh.com.

■ **Figure 21.18** *Welwitschia* leaves are blown about by the wind. This causes them to split many times and makes the two leaves appear as many leaves.

Division Ginkgophyta Only one living species, *Ginkgo biloba*, represents division Ginkgophyta. Early in the 19th century, fossil remains of *Ginkgo biloba* were discovered in the state of Washington. The ginkgo disappeared from North America during the Ice Age. However, it survived in China where it was grown for its seeds—a food delicacy only eaten at weddings and during holidays.

This distinctive tree has small, fan-shaped leaves. Like cycads, male and female reproductive systems are on separate plants. The male tree produces pollen grains in strobiluslike cones growing from the bases of leaf clusters, as shown in **Figure 21.19.** The female tree produces cones, also shown in **Figure 21.19,** which, when fertilized, develop foul-smelling, fleshy seed coats. Because they tolerate smog and pollution, ginkgoes are popular with gardeners and urban landscapers. However, male trees usually are favored because they do not produce foul-smelling fleshy cones.

Male reproductive structures

Female reproductive structures

■ **Figure 21.19** Both male and female ginkgo reproductive structures grow from the bases of leaf clusters but on separate trees.
Predict *how pollen travels to the female reproductive structure.*

Division Coniferophyta Conifers range in size from low-growing shrubs that are several centimeters tall to towering trees over 50 m in height. Pines, firs, cypresses, and redwoods are examples of conifers. Conifers are the most economically important gymnosperms. They are sources of lumber, paper pulp, and the resins used to make turpentine, rosin, and other products.

Reproductive structures of most conifers develop in cones. Most conifers have male and female cones on different branches of the same tree or shrub. The small male cones produce pollen. Larger female cones remain on the plants until the seeds have matured. The characteristics of female cones, such as those shown in **Figure 21.20,** can be used to identify conifers.

Conifers like all plants exhibit adaptations to their environments. What connection can you make between the facts that most conifers have drooping branches and that many conifers grow in snowy climates? Another adaptation is a waxlike coating called cutin that covers conifer needlelike or scalelike leaves and reduces water loss.

■ **Figure 21.20** Female cones of conifers can be described as woody, berrylike, or fleshy.

Douglas fir—woody cones

Juniper—berrylike cones

Pacific yew—fleshy cones

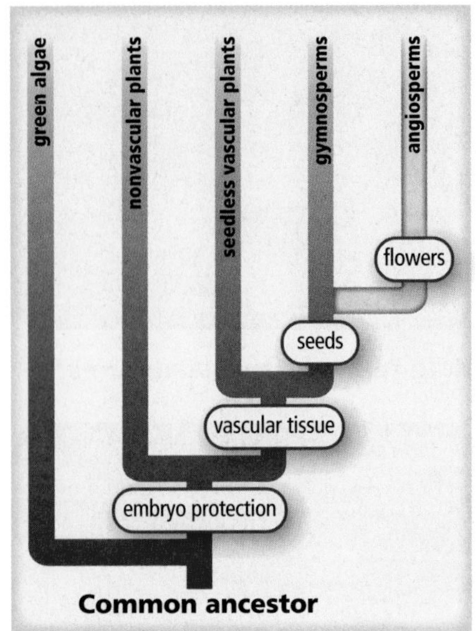

■ **Figure 21.21** The flowering plants were the most recently evolved of the plant kingdom.

When you hear the word *evergreen,* do you think of a pine or another conifer? Most plants in northern temperate regions called evergreen are conifers. However, in subtropical and tropical regions, other plants, such as palms, also are evergreen. Botanists define an evergreen plant as one with some green leaves throughout the year. This adaptation enables it to undergo photosynthesis whenever conditions are favorable. A plant that loses its leaves at the end of the growing season or when moisture is scarce is called deciduous. Some conifers are deciduous, such as larches and bald cypresses. Whether deciduous or evergreen, you can identify a conifer species by its leaves, as demonstrated in **MiniLab 21.2.**

Reading Check **Explain** why some trees are called evergreens.

Division Anthophyta Flowering plants, or anthophytes, are the most widely distributed plants because of adaptations that enable them to grow in terrestrial and aquatic environments. Anthophytes, also known as angiosperms, first appeared in the fossil record about 130 million years ago after the appearance of gymnosperms, as shown in **Figure 21.21.** Today, flowering plants make up more than 75 percent of the plant kingdom.

Traditionally, botanists classified anthophytes as monocots or dicots. The names refer to the number of seed leaves called cotyledons: monocot—one seed leaf, dicot—two seed leaves. However, botanists now classify dicots as eudicots, based on the structure of their pollen. You will read more about monocots and eudicots in Chapters 22 and 23.

Life spans A few weeks or years describe the life spans of anthophytes. An **annual** plant completes its life span—sprouts from a seed, grows, produces new seeds, and dies—in one growing season or less. This group includes many garden plants and most weeds.

MiniLab 21.2

Investigate Conifer Leaves

What similarities and differences exist among conifer leaves? Some conifer trees are among the tallest and oldest organisms on Earth. Most conifers have needlelike leaves that differ in a variety of ways. Leaf characteristics are important in conifer identification.

Procedure

1. Read and complete the lab safety form.
2. Obtain one of each of the **conifer samples** your teacher has identified. Label each sample by name.
3. Design a data table to record your observations.
4. Compare and contrast the leaves. Make a list of characteristics that you determine are important for describing each sample. Record these characteristics for each conifer sample.
5. Develop a system for grouping the conifer samples. Be prepared to justify your system.
6. Wash your hands thoroughly after handling plant samples.

Analysis

1. **Explain** the reasoning for your classification system.
2. **Compare** your classification system to those created by other students. Explain why your system is an efficient way to classify the conifer samples that you studied.

First-year growth

Second-year growth

■ **Figure 21.22** An evening primrose (a biennial) produces leaves, an underground stem, and roots the first growing season. It flowers in the second year of growth.

A **biennial** plant's life spans two years. During the first year, it produces leaves and a strong root system. Refer to **Figure 21.22.** Some biennials, like carrots, beets, and turnips, develop fleshy storage roots that are harvested after the first growing season. If the biennial is not harvested, the aboveground tissues die. However, roots and other underground parts remain alive for biennials that are adapted to their environments. In the second year, stems, leaves, flowers, and seeds grow. The plant's life ends the second year.

Perennial plants can live for several years and usually produce flowers and seeds yearly. Some perennials respond to harsh conditions by dropping leaves, and others completely die back so only their roots remain alive. They resume growth when favorable growing conditions return. Fruit and shade trees, shrubs, irises, peonies, roses, and many types of berries are perennial plants.

The life spans of all plants are determined genetically and reflect adaptations for surviving harsh conditions. However, all plant life spans are affected by environmental conditions.

Section 21.4 Assessment

Section Summary

- Vascular seed plants produce seeds containing the sporophyte generation.
- Vascular seed plants exhibit numerous adaptations for living in varied environments.
- There are five divisions of vascular seed plants. Each division has distinct characteristics.
- Flowering plants are annuals, biennials, or perennials.

Understand Main Ideas

1. MAIN Idea **Describe** the advantages of a plant that produces seeds.
2. **Compare and contrast** a gymnosperm and an angiosperm.
3. **Distinguish** between male and female cones of gymnosperms.
4. **Identify** the divisions of gymnosperms.
5. **Differentiate** between a monocot and a eudicot.
6. **Compare and contrast** the three types of anthophyte life spans.

Think Critically

7. **Consider** A Christmas tree farmer saw an advertisement that read, "Bald cypresses—the way to quick profits. Plant these fast-growing trees and harvest them in just five years." Would these trees be profitable for the farmer? Explain.

MATH in Biology

8. The smallest flowering plant is 1 mm tall and the largest conifer can be 90 m tall. How many times taller is the largest conifer than the smallest flowering plant?

NATIONAL GEOGRAPHIC

In the Field

Career: Forensic Palynology

The Proof Is in the Pollen

Forensic palynology (pah luh NAW luh gee), a relatively new science, uses pollen and spore evidence in legal cases to help police solve crimes. A jogger was attacked, dragged to a nearby wooded area, and murdered. The police questioned a key suspect who admitted that he was in the area, but claimed he did not see the jogger. He also said that he never had been in the wooded area where the body was found. Was he telling the truth?

Incriminating evidence Soil from the crime scene contained large amounts of pine pollen and fern spores. A survey revealed that no other nearby locations contained both pine trees and ferns. When police searched the suspect's apartment, they found a sweater and pants that they believed he was wearing during the attack. When a forensic palynologist examined the clothes, she found the same pine pollen and fern spores as that of the crime scene. The suspect eventually was tried and convicted of the murder.

Palynologists at a crime scene Detectives collect many types of evidence from a crime scene, including fingerprints. Can palynologists collect fingerprints? In a way, yes. Each seed-plant species produces unique pollen grains. They can be thought of as a species' "fingerprint" and can be used for identification. Also, dirt and dust often contain large amounts of pollen and spores. Fibers in woven fabrics can act as filters and trap pollen and spores. Blown by the wind, pollen can become trapped between strands of hair.

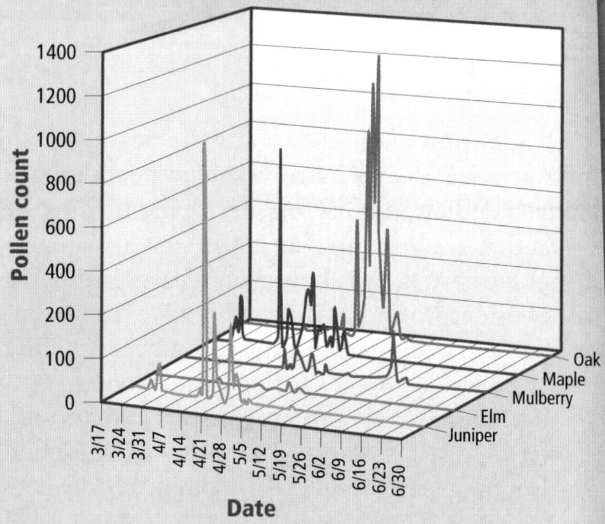

Forensic palynology A pollen study can help investigators narrow the list of suspects, making this a valuable investigative tool. Because it requires extensive background knowledge and training in collecting and preserving samples without contamination, forensic palynology is a specialized science.

MATH in Biology

Interpret the Graph Examine the graph of tree pollen counts. What types of pollen might you expect to find if the crime occurred on April 14, May 12, or June 2? To find out more about the types of pollen found in your community and to learn more about palynology, visit biologygmh.com.

BIOLAB

FIELD INVESTIGATION: HOW CAN YOU IDENTIFY AND CLASSIFY TREES?

Background: Botanists and others interested in plants often use field guides and dichotomous keys to identify plants. In this BioLab, you will use a field guide to identify trees in a given area. Then, you will create your own dichotomous key to identify the trees in your area.

Question: *What characteristics can be used to identify trees and to create a dichotomous key for them?*

Materials

field guide of trees (for your area)
metric ruler
magnifying lens

Safety Precautions

WARNING: ***Stay within the area of study and be alert for plants, insects, or other organisms that might pose a hazard.***

Procedure

1. Read and complete the lab safety form.
2. Study the field guide provided by your teacher to determine how it is organized.
3. Based on your examination of the field guide and what you learned about plant characteristics in this chapter, make a list of characteristics that will help you identify the trees in your area.
4. Create a data table based on the list you made in Step 3.
5. Use a field guide to identify a tree in the area designated by your teacher. Confirm your identification with your teacher.
6. Record in your data table the characteristics of your identified tree.
7. Repeat Steps 5 and 6 until you have identified all trees required for this lab.
8. Review your data table. Choose the characteristics most helpful in identifying trees. These characteristics will form the basis of your dichotomous key.
9. Determine in what rank the characteristics should appear in the dichotomous key. Create a written description for each characteristic.
10. Create your dichotomous key. The traits described at each step of a dichotomous key usually are pairs of contrasting characteristics. For example, the first step might compare needlelike or scalelike leaves to broad leaves.

Analyze and Conclude

1. **Interpret Data** Based on the data you collected, describe plant diversity in the area you studied.
2. **Critique** Exchange your dichotomous key with a classmate's dichotomous key. Use the key to identify trees in the study area. Give your classmate suggestions to improve his or her key.
3. **Predict** How effective would your dichotomous key be for someone trying to identify trees in the study area? Explain.
4. **Error Analysis** What changes could you make to improve the effectiveness of your dichotomous key?

SHARE YOUR DATA

Compare and Contrast Post your data at biologygmh.com and compare it to other data found there. What plants are common to all of the posted dichotomous keys? To learn more about dichotomous keys, visit Biolabs at biologygmh.com.

CHAPTER 21

Study Guide

Download quizzes, key terms, and flash cards from biologygmh.com.

FOLDABLES **Design** the perfect plant for a given land environment using information from the Foldable at the beginning of the chapter. Explain and justify each feature of your plant. For example, design a plant that could survive in the Atacama Desert of Chile.

Vocabulary	Key Concepts
Section 21.1 Plant Evolution and Adaptations	
• nonvascular plant (p. 606) • seed (p. 607) • stomata (p. 606) • vascular plant (p. 606) • vascular tissue (p. 606)	MAIN Idea Adaptations to environmental changes on Earth contributed to the evolution of plants. • Plants are multicellular organisms and most are photosynthetic. • Evidence indicates that ancient, unicellular, freshwater green algae were the ancestors of present-day plants. • Present-day plants and green algae have common characteristics. • Over time, plants developed several adaptations for living on land. • Plants alternate between a sporophyte and a gametophyte generation.
Section 21.2 Nonvascular Plants	
• thallose (p. 612)	MAIN Idea Nonvascular plants are small and usually grow in damp environments. • Distribution of nonvascular plants is limited by the plants' ability to transport water and other substances. • Mosses are small plants that can grow in different environments. • Like other nonvascular plants, hornworts rely on osmosis and diffusion to transport substances. • The two types of liverworts are described as thallose and leafy.
Section 21.3 Seedless Vascular Plants	
• epiphyte (p. 614) • rhizome (p. 615) • sorus (p. 616) • sporangium (p. 616) • strobilus (p. 613)	MAIN Idea Because they have vascular tissues, seedless vascular plants generally are larger and better adapted to drier environments than nonvascular plants. • Seedless vascular plants have specialized transport tissues and reproduce by spores. • The sporophyte is the dominant generation in vascular plants. • Lycophytes and pterophytes are seedless vascular plants.
Section 21.4 Vascular Seed Plants	
• annual (p. 620) • biennial (p. 621) • cone (p. 618) • cotyledon (p. 617) • perennial (p. 621)	MAIN Idea Vascular seed plants are the most widely distributed plants on Earth. • Vascular seed plants produce seeds containing the sporophyte generation. • Vascular seed plants exhibit numerous adaptations for living in varied environments. • There are five divisions of vascular seed plants. Each division has distinct characteristics. • Flowering plants are annuals, biennials, or perennials.

CHAPTER 21

Assessment

Section 21.1

Vocabulary Review

For questions 1–3, match each phrase with a vocabulary term from the Study Guide page.

1. plant structure that contains the embryo
2. transport tissue
3. enable exchange of gases

Understand Key Concepts

4. Approximately when did primitive land plants appear?
   A. 30 mya
   B. 400 mya
   C. 500 mya
   D. 2000 mya

5. Which is not a trait shared by freshwater green algae and plants?
   A. cellulose cell walls
   B. chlorophyll
   C. food stored as starch
   D. contain vascular tissue

6. Which is likely to ensure the survival of the embryo?

A.

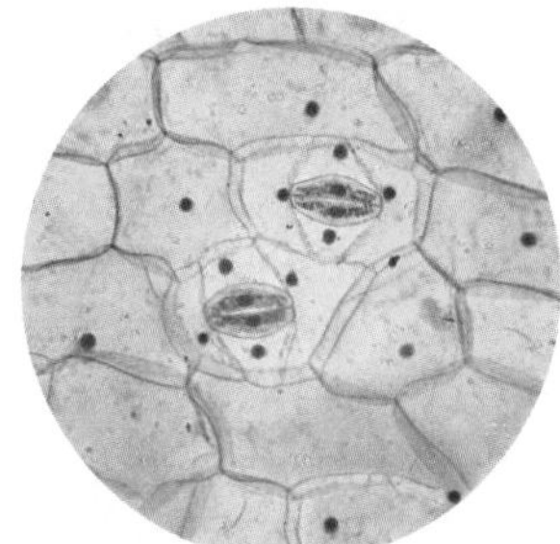

C.

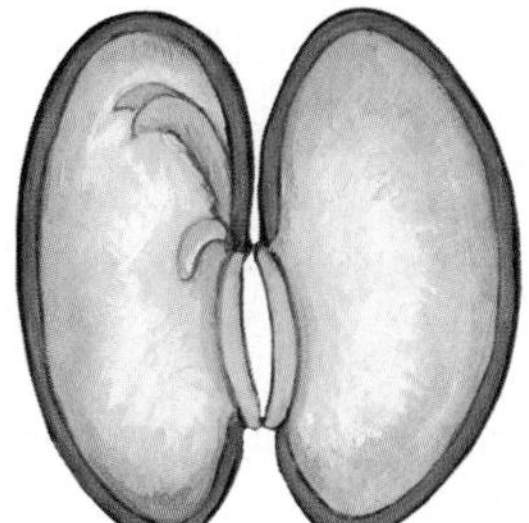

B.

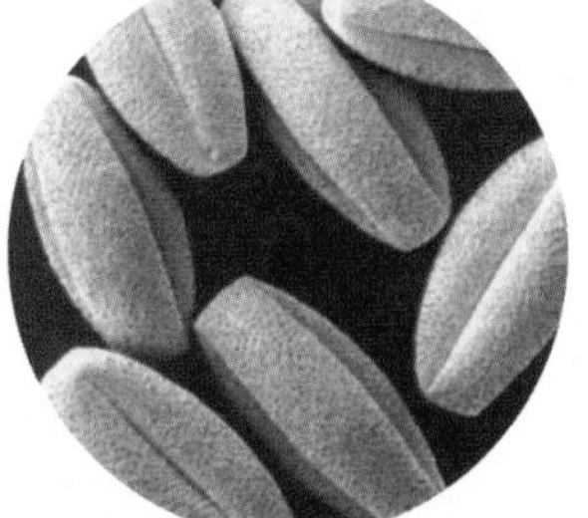

D.

7. Which was a major obstacle for plants to live on land?
   A. obtaining enough light
   B. obtaining enough soil
   C. obtaining enough water
   D. obtaining enough oxygen

Constructed Response

8. **Short Answer** Describe the adaptations that you would expect to find in an aquatic plant.

9. **Open Ended** Of the adaptations discussed in Section 21.1, which one do you predict would be most important to a plant living in the desert?

Think Critically

10. **Organize** the adaptations to life on land from the most important to the least important. Defend your decisions.

Section 21.2

Vocabulary Review

Write a sentence using the following vocabulary term correctly.

11. thallose

Understand Key Concepts

Use the photo below to answer question 12.

12. Which word does not describe the plant shown above?
    A. multicellular
    B. nonvascular
    C. seedless
    D. thallose

13. Which is a characteristic of mosses?
   A. vascular tissue
   B. flowers
   C. seeds
   D. rhizoids

Constructed Response

14. **Short Answer** Refer to **Figure 21.9** and analyze the need for a nonvascular sporophyte to remain dependent on the gametophyte generation.

15. **Open Ended** Describe a habitat in your community that would support nonvascular plants.

Think Critically

16. **Research** nonvascular plants at biologygmh.com and make a list of those that grow in your state.

Section 21.3

Vocabulary Review

For questions 17–19, match each definition with a vocabulary term from the Study Guide page.

17. spore-bearing structures that form a compact cluster

18. thick underground stem

19. plant that lives anchored to another plant or object

Understand Key Concepts

Use the concept map below to answer question 20.

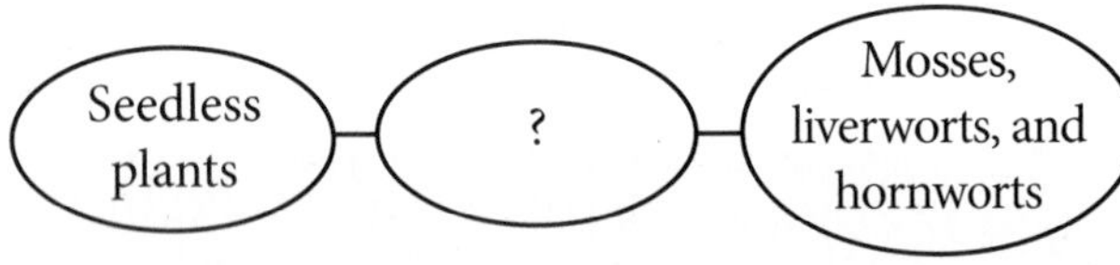

20. Which term correctly completes the concept map shown above?
   A. nonvascular
   B. flowering
   C. vascular
   D. seed-producing

21. What structure contains clusters of sporangia?
   A. sorus
   B. frond
   C. stem
   D. blade

22. Which is not part of the fern sporophyte generation?
   A. rhizome
   B. sorus
   C. frond
   D. rhizoid

23. Which photo shows sori?

Constructed Response

24. **Short Answer** Summarize the characteristics of ferns.

25. **Short Answer** Differentiate between Division Pterophyta and Division Lycophyta.

Think Critically

26. **Infer** the advantage of fern sori being on the under surface of fronds rather than on the upper.

Section 21.4

Vocabulary Review

For questions 27–29, replace each underlined word with the correct vocabulary term from the Study Guide page.

27. A <u>root</u> of a seed provide nutrients when it sprouts.

28. A plant that lives for several growing seasons is a <u>rhizome</u>.

29. A <u>flower</u> contains the male or female reproduction structures of gymnosperms.

Understand Key Concepts

30. Which plant division has plants with needlelike or scaly leaves?
    A. Gnetophyta
    B. Anthophyta
    C. Coniferophyta
    D. Cycadophyta

Use the photo below to answer question 31.

31. Which plant division has plants that produce female reproductive structures like those shown above?
    A. Coniferophyta
    B. Anthophyta
    C. Gnetophyta
    D. Ginkgophyta

32. Which describes the importance of seed dispersal?
    A. ensures more favorable environments for growth
    B. creates greater biodiversity
    C. limits competition with parent plants and other offspring
    D. provides greater resources

Constructed Response

33. **Open Ended** What might be the adaptive advantage of having a gametophyte dependent on a sporophyte?

34. **Short Answer** Make a list of the traits you would use to differentiate between coniferophytes and anthophytes.

Think Critically

35. **Compare and contrast** cones and strobili.

36. **Infer** why there are more conifers than flowering plants in colder environments such as those in northern Canada and Alaska.

Additional Assessment

37. *WRITING in* **Biology** Imagine yourself as one of the first plants that survived living on land. What stories could you tell your grandchildren about the difficulties you faced?

DBQ Document-Based Questions

Data obtained from: Qiu, Yin-Long, et al. 1998. The gain of three mitochondrial introns identifies liverworts as the earliest land plants. *Nature* 394: 671.

Here we survey 352 diverse land plants and find that three mitochondrial Group II introns are present . . . in mosses, hornworts and all major lineages of vascular plants, but are entirely absent from liverworts, green algae and all other eukaryotes. These results indicate that liverworts are the earliest land plants, with the three introns having been acquired in a common ancestor of all other land plants, and have important implications concerning early plant evolution.

38. Evaluate the research above by making a cladogram.

39. Explain how this research lead scientists to suggest that liverworts are the ancestors of all other plants.

40. Apply what you read in Chapter 13 about polymerase chain reactions to predict how the scientists determined which plants contained these introns.

Cumulative Review

41. Describe the cause of Down syndrome. **(Chapter 11)**

42. Discuss how plate tectonics explains why similar organisms can be associated on distant continents. **(Chapter 14)**

43. Describe some of the things Darwin saw that caused him to hypothesize that species evolve. **(Chapter 15)**

44. Compare and contrast the characteristics of prokaryotic and eukaryotic cells. **(Chapter 18)**

Standardized Test Practice

Cumulative

Multiple Choice

1. Which substance do yeasts produce that causes bread to rise?
   A. carbon dioxide
   B. ethanol
   C. oxygen
   D. simple sugars

2. Which must a virus have in order to attack a host cell?
   A. a DNA or RNA sequence that is recognized by the ribosomes of the host cell
   B. the enzymes to burst the host cell so that the host cell can be used as raw materials
   C. a particular shape that matches the proteins on the surface of the host cell
   D. the proper enzyme to puncture the membrane of the host cell

3. Which group of protists is characterized by parasitic behavior?
   A. chrysophytes
   B. dinoflagellates
   C. sarcodines
   D. sporozoans

Use the following illustration to answer question 4.

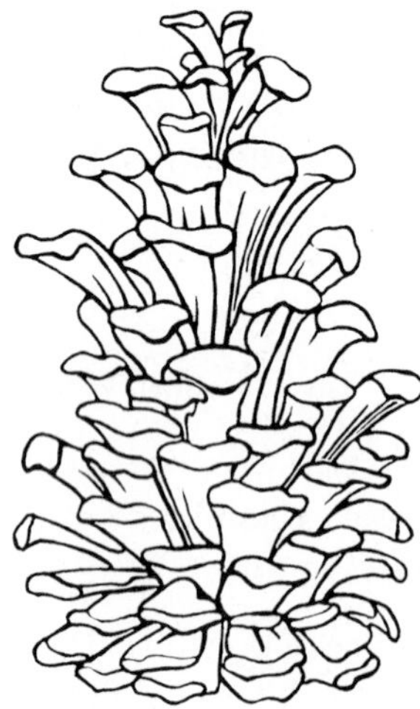

4. In which division of seed plants would you expect to find the structure in the above illustration?
   A. Anthophyta
   B. Coniferophyta
   C. Cycadophyta
   D. Ginkgophyta

5. Suppose a cell from the frond of a fern contains 24 chromosomes. How many chromosomes would you expect to find in the spores?
   A. 6
   B. 12
   C. 24
   D. 48

Use the diagram below to answer questions 6 and 7.

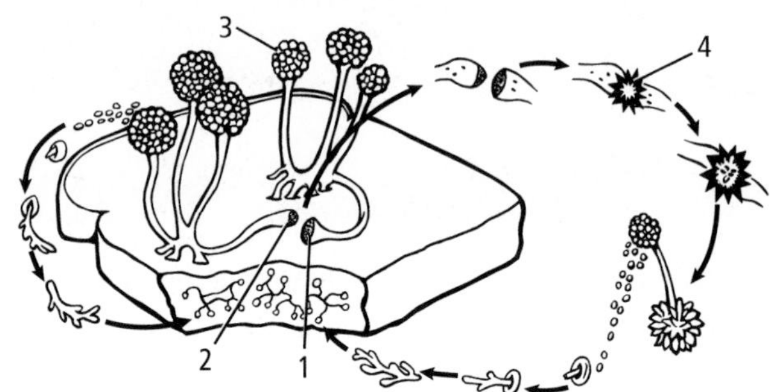

6. Which phylum of fungi has these kinds of reproductive structures?
   A. Ascomycota
   B. Basidiomycota
   C. Deuteromycota
   D. Zygomycota

7. Which of these structures is involved in asexual reproduction?
   A. 1
   B. 2
   C. 3
   D. 4

8. Why is conjugation important for protists?
   A. It expands the habitat.
   B. It improves locomotion speed.
   C. It increases genetic variation.
   D. It restores injured parts.

Short Answer

9. Compare the sporophyte generation in nonvascular plants to the sporophyte generation in seedless vascular plants.

10. Describe the two membranes that make up an amoeba and suggest why it is beneficial for the amoeba to have two membranes.

11. What is the relationship between bat wings and monkey arms? Explain the importance of this relationship for the classification of organisms.

12. Describe how a multicellular fungus obtains nutrients from its environment and assess how that affects its role in the environment.

Use the diagram of the lichen below to answer questions 13 and 14.

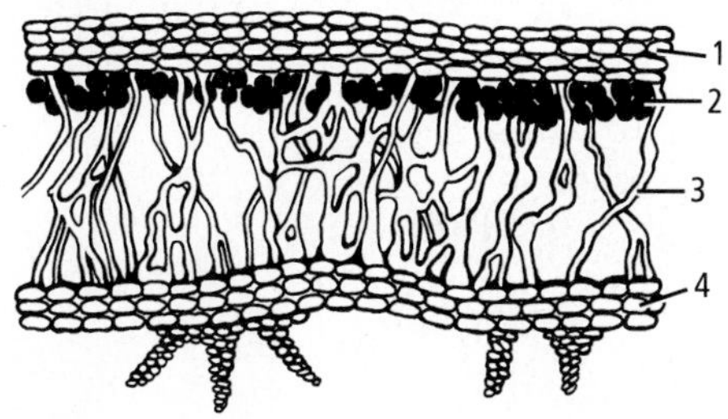

13. Identify and evaluate the importance of the layer of the lichen where photosynthesis takes place.

14. Analyze how the photosynthesizer and fungus benefit from being part of a lichen.

15. Evaluate how spore production gives fungi an advantage in an ecosystem.

Extended Response

Use the diagram below to answer question 16.

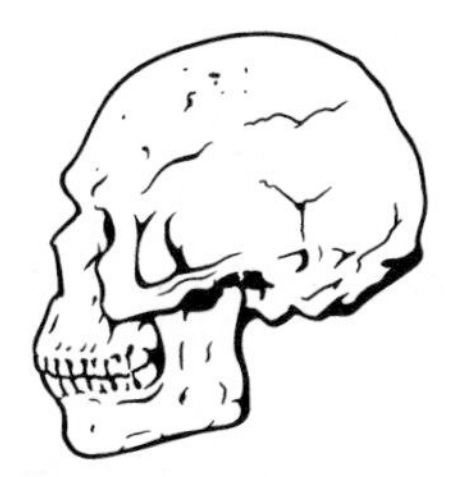

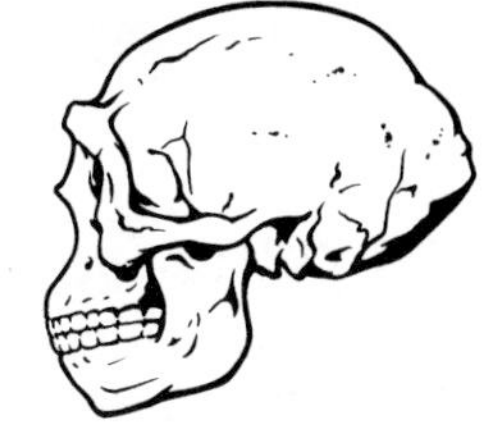

16. Look at the two skulls in the diagram. Infer which one you think is more closely related to *Homo sapiens.* Explain your inference.

17. Compare and contrast reproduction in paramecia and amoebas.

Essay Question

During the 1840s, the potato was an extremely popular crop plant in Ireland. Many people in rural Ireland were completely dependent on potatoes for food. From 1845 to 1847, the potato blight—a funguslike disease—wiped out potato crops. The blight produces spores on the leaves of the potato plant. The spores can be transmitted by water or wind. They are carried into the soil by water, where they infect the potato tubers, and can survive through winter on the potatoes left buried in the fields. Close to one million people died from starvation and nearly as many left Ireland for America and other countries.

Using the information in the paragraph above, answer the following question in essay format.

18. Write an essay that indicates why potato blight spread so quickly through Ireland and how the spread of the fungus might have been slowed by different farming practices.

NEED EXTRA HELP?

If You Missed Question . . .	1	2	3	4	5	6	7	8	9	10	11	12	13	14	15	16	17	18
Review Section . . .	20.1	18.2	19.1	21.4	21.3	20.2	20.3	19.2	21.3	19.2	15.2	20.1	20.3	20.3	20.2	16.3	19.2	20.1

CHAPTER 22 Plant Structure and Function

Cross section of coleus stem
Stained LM Magnification: 47×

Cross section of coleus leaf
Stained LM Magnification: 75×

BIG Idea The diverse nature of plants is due to the variety of their structures.

Section 1
Plant Cells and Tissues
MAIN Idea Different types of plant cells make up plant tissues.

Section 2
Roots, Stems, and Leaves
MAIN Idea The structures of plants are related to their functions.

Section 3
Plant Hormones and Responses
MAIN Idea Hormones can affect a plant's responses to its environment.

BioFacts

- The pigment in coleus leaves that give them their reddish color serves as a sunblock—it protects the plant from harmful UV rays from the Sun.
- For more than 2000 years, humans have grown plants for their stem fibers that are woven to make linen fabrics.
- With a few exceptions, 80–90 percent of a plant's roots grow in the top 30 cm of the soil.

Start-Up Activities

LAUNCH Lab

What structures do plants have?

Most plants have structures that absorb light and others that take in water and nutrients. In this lab, you will examine a plant and observe and describe structures that help the plant survive.

Procedure

1. Read and complete the lab safety form.
2. Carefully examine a **potted plant** provided by your teacher. Use a **magnifying lens** to get a closer look. Make a list of each type of structure you observe.
3. Gently remove the plant from the pot and observe the plant structures in the soil. Do not break up the soil. Record your observations and place the plant back into the pot.
4. Sketch your plant and label each part.

Analysis

1. **Compare** your list with those of other students. What structures were common to all plants?
2. **Infer** how each structure might be related to a function of the plant.
3. **Predict** the type of structural adaptations of plants living in dry environments.

Visit biologygmh.com to:

- study the entire chapter online
- explore the Concepts in Motion, Interactive Tables, Virtual Labs, Microscopy Links, and links to virtual dissections
- access Web links for more information, projects, and activities
- review content online with the Interactive Tutor, and take Self-Check Quizzes

Leaf Structure and Function Make this Foldable to help you investigate the structure and function of a typical leaf.

STEP 1 Stack three sheets of paper, keeping all edges aligned.

STEP 2 Fold the stack in half. Crease and staple the fold to make a six-page booklet.

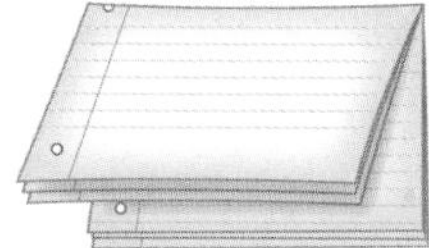

STEP 3 Draw the outline of a large leaf on the front page and label the page *Cuticle.*

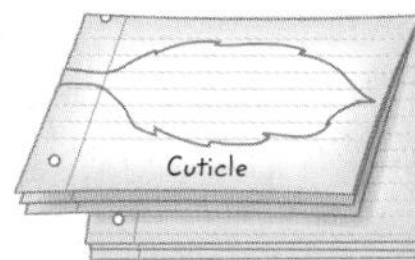

STEP 4 Label the remaining five pages in the following order: *Upper epidermis, Palisade mesophyll, Spongy mesophyll, Lower epidermis,* and *Cuticle.*

FOLDABLES Use this Foldable with Section 22.2. As you read the section, write a description of each layer's structure and function on its page.

Section 22.1

Reading Preview

Objectives

- **Describe** the major types of plant cells.
- **Identify** the major types of plant tissues.
- **Distinguish** among the functions of plant cells and tissues.

Review Vocabulary

vacuole: membrane-bound vessicle used for storage or transport

New Vocabulary

parenchyma cell
collenchyma cell
sclerenchyma cell
meristem
vascular cambium
cork cambium
epidermis
guard cell
xylem
vessel element
tracheid
phloem
sieve tube member
companion cell
ground tissue

Plant Cells and Tissues

MAIN Idea **Different types of plant cells make up plant tissues.**

Real-World Reading Link Buildings are made of a variety of materials. Different materials are used for stairways, plumbing, doors, and the electrical system because each of these has a different function. Similarly, different plant structures have cells and tissues that function efficiently for specific tasks.

Plant Cells

Recall from Chapters 7 and 21 that you can identify a typical plant cell, like the one in **Figure 22.1,** by the presence of a cell wall and large central vacuole. Also, plant cells can have chloroplasts. However, there are many different types of plant cells—each with one or more adaptations that enable it to carry out a specific function. Three types of plant cells form most plant tissues. Together they provide storage and food production, strength, flexibility, and support.

Parenchyma cells Most flexible, thin-walled cells found throughout a plant are **parenchyma** (puh RENG kuh muh) **cells.** They are the basis for many plant structures and are capable of a wide range of functions, including storage, photosynthesis, gas exchange, and protection. These cells are spherical in shape and their cell walls flatten when they are packed tightly together, as shown in **Table 22.1.** An important trait of parenchyma cells is that they can undergo cell division when mature. When a plant is damaged, parenchyma cells divide to help repair it.

Depending on their function, parenchyma cells can have special features. Some parenchyma cells have many chloroplasts, also shown in **Table 22.1.** These cells often are found in leaves and green stems, and can carry on photosynthesis, producing glucose. Some parenchyma cells, such as those found in roots and fruits, have large central vacuoles that can store substances, such as starch, water, or oils.

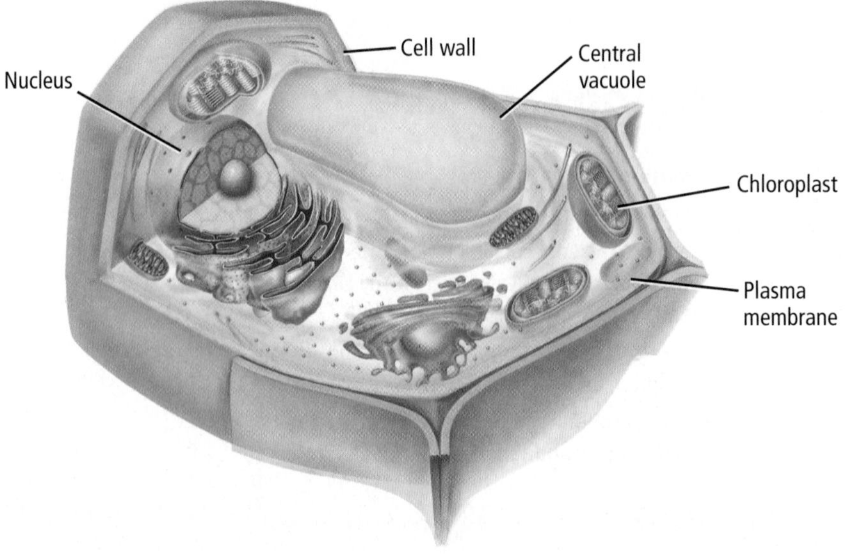

Figure 22.1 Features unique to a plant cell include a cell wall and a large central vacuole. Plant cells also can contain chloroplasts where photosynthesis occurs.
Infer *why chloroplasts are not part of all plant cells.*

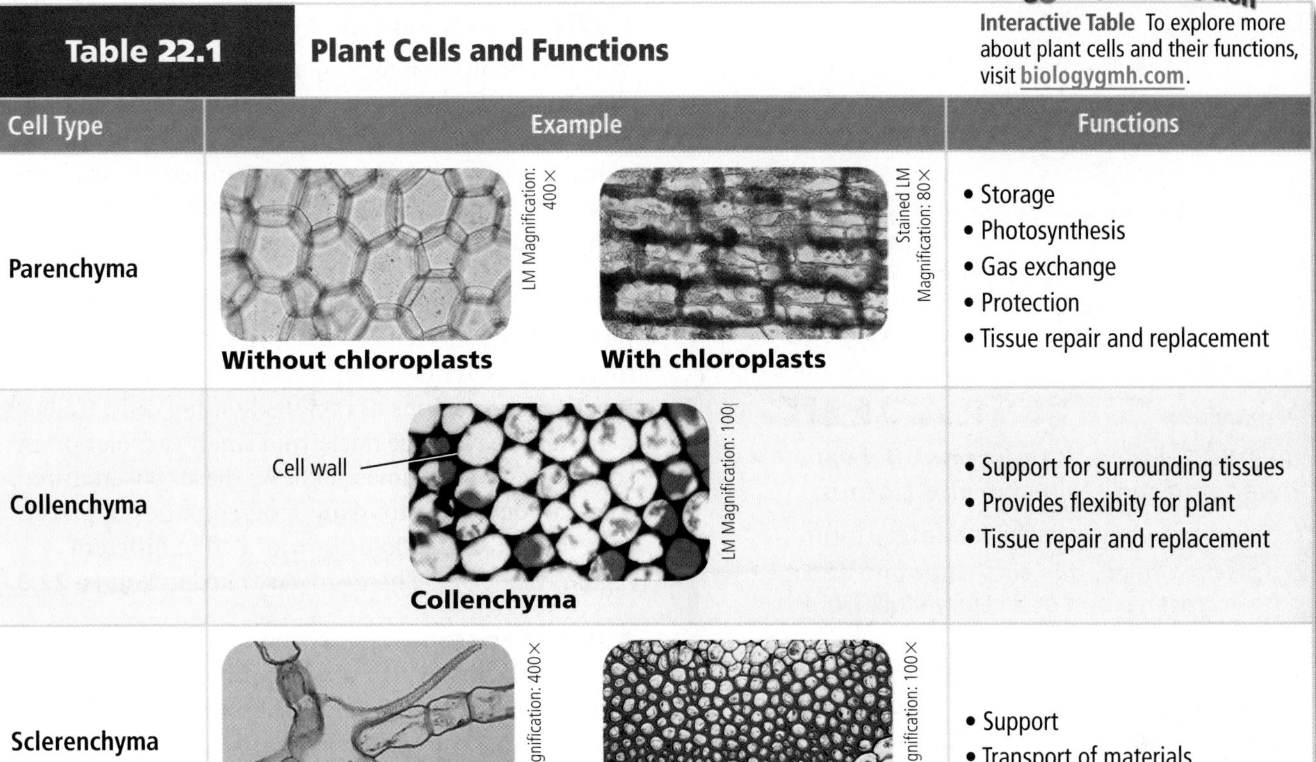

Table 22.1 **Plant Cells and Functions**

Concepts In Motion

Interactive Table To explore more about plant cells and their functions, visit biologygmh.com.

Cell Type	Example	Functions
Parenchyma	Without chloroplasts (LM Magnification: 400×); With chloroplasts (Stained LM Magnification: 80×)	• Storage • Photosynthesis • Gas exchange • Protection • Tissue repair and replacement
Collenchyma	Cell wall; Collenchyma (LM Magnification: 100×)	• Support for surrounding tissues • Provides flexibity for plant • Tissue repair and replacement
Sclerenchyma	Sclereid (LM Magnification: 400×); Fibers (LM Magnification: 100×)	• Support • Transport of materials

Collenchyma cells If you have eaten celery, you might be familiar with collenchyma (coh LENG kuh muh) cells. These cells make up those long strings that you can pull from a celery stalk. **Collenchyma cells** are plant cells that often are elongated and occur in long strands or cylinders that provide support for the surrounding cells. As shown in **Table 22.1,** collenchyma cells can have unevenly thickened cell walls. As a collenchyma cell grows, the thinner portions of its cell wall can expand. Because of this growth pattern, collenchyma cells are flexible and can stretch, which enables plants to bend without breaking. Like parenchyma cells, collenchyma cells retain the ability to undergo cell division when mature.

Sclerenchyma cells Unlike parenchyma and collenchyma cells, **sclerenchyma** (skle RENG kuh muh) **cells** are plant cells that lack cytoplasm and other living components when they mature, but their thick, rigid cell walls remain. These cells provide support for a plant, and some are used for transporting materials within the plant. Sclerenchyma cells make up most of the wood we use for shelter, fuel, and paper products.

There are two types of sclerenchyma cells—sclereids and fibers—shown in **Table 22.1.** You might have eaten sclereids—they create the gritty texture of pears. Sclereids are also called stone cells and can be distributed randomly throughout a plant. They are shorter than fibers and are somewhat irregularly shaped. The toughness of seed coats and nut shells results from the presence of sclereids. Sclereids also function in transport. A fiber cell is needle-shaped, has a thick cell wall, and has a small interior space. When stacked end-to-end, fibers form a tough, elastic tissue. Humans have used these fibers for making ropes and linen, canvas, and other textiles for centuries, as shown in **Figure 22.2.**

Figure 22.2 Fiber cells in plants have been used to make textiles such as these ancient Egyptian sandals.

MiniLab 22.1

Observe Plant Cells

How can a microscope be used to distinguish plant cell types? Investigate the three different types of plant cells by making and observing slides of some common plant parts.

Procedure

WARNING: ***Iodine is poisonous if swallowed and can stain skin and clothes.***

1. Read and complete the lab safety form.
2. Obtain a small, thin **slice of potato** and a thin **cross section of a celery stalk** from your teacher.
3. Place the potato slice on a **slide,** add a drop of **iodine,** and cover with a **coverslip.** Use a **microscope** to observe the potato slice. Record your observations.
4. Place the celery slice on a slide, add a drop of **water,** and cover with a coverslip.
5. Put a drop of **dye** at one end of the coverslip, and then touch a **paper towel** to the other end to draw the dye under the coverslip. Use a microscope to observe the celery slice. Record your observations.
6. Obtain a small amount of **pear tissue,** place it on a slide, and add a coverslip.
7. Using a **pencil eraser,** press gently but firmly on the coverslip until the pear tissue is a thin even layer. Use a microscope to observe the pear tissue. Record your observations.

Analysis

1. **Identify** the type of specialized plant cell observed on each slide.
2. **Infer** why there are different cell types in a potato, a celery stalk, and pear tissue.

Plant Tissues

You learned in Chapter 9 that a tissue is a group of cells that work together to perform a function. Depending on its function, a plant tissue can be composed of one or many types of cells. There are four different tissue types found in plants—meristematic (mer uh stem AH tihk), dermal, vascular, and ground.

Meristematic tissue Throughout their lives, plants can continue to produce new cells in their meristematic tissues. Meristematic tissues make up **meristems**—regions of rapidly dividing cells. Cells in meristems have large nuclei and small vacuoles or, in some cases, no vacuoles at all. As these cells mature, they can develop into many different kinds of plant cells. Meristematic tissues are located in different regions of a plant. These are illustrated in **Figure 22.3.**

Apical meristems Meristematic tissues at the tips of roots and stems, which produce cells that result in an increase in length, are apical (AY pih kul) meristems, as shown in **Figure 22.3.** This growth is called primary growth. Because plants are usually stationary, stems and roots enter different environments or different areas of the same environments.

Intercalary meristems Another type of meristem, called intercalary (in TUR kuh LAYR ee) meristem, is related to a summer job you might have had—mowing grass. This meristem is found in one or more locations along the stems of many monocots. Intercalary meristem produces new cells that result in an increase in stem or leaf length. If grasses only had apical meristems, they would stop growing after the first mowing. They continue to grow because they have more than one type of meristematic tissue.

Lateral meristems Increases in root and stem diameters result from secondary growth produced by two types of lateral meristems. Only nonflowering seed plants, eudicots, and a few monocots have secondary growth.

The **vascular cambium,** also shown in **Figure 22.3,** is a thin cylinder of meristematic tissue that can run the entire length of roots and stems. It produces new transport cells in some roots and stems.

In some plants, another lateral meristem, the **cork cambium,** produces cells that develop tough cell walls. These cells form a protective outside layer on stems and roots. Cork tissues make up the outer bark on a woody plant like an oak tree. Recall that cells of cork tissue are what Robert Hooke observed when he looked through his microscope.

Visualizing Meristematic Tissues

Figure 22.3
Most plant growth results from the production of cells by meristematic tissues. Stems and roots increase in length mostly due to the production of cells by apical meristems. A plant's vascular cambium produces cells that increase root and stem diameters.

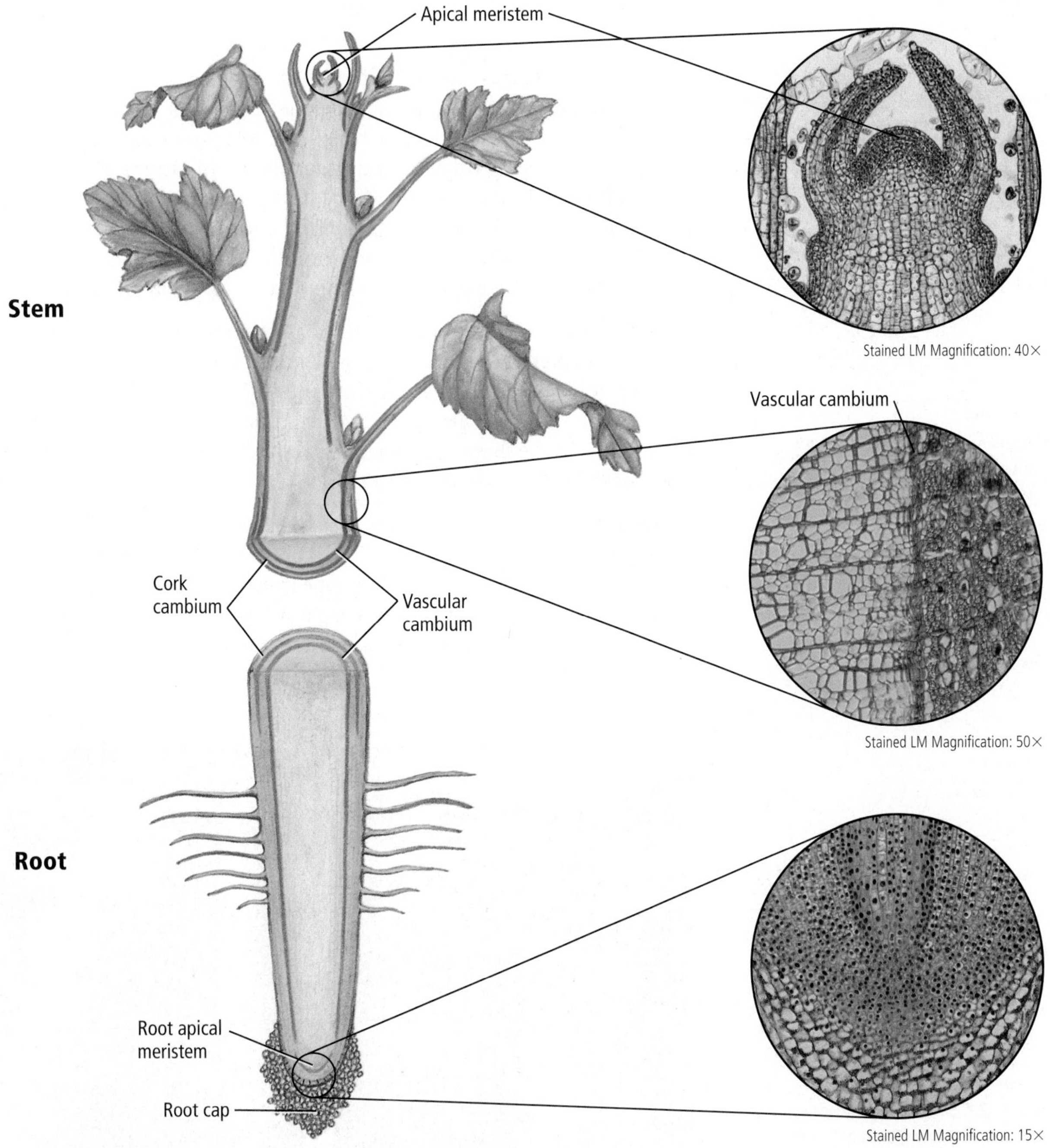

Apical meristem

Stem

Stained LM Magnification: 40×

Vascular cambium

Cork cambium

Vascular cambium

Stained LM Magnification: 50×

Root

Root apical meristem

Root cap

Stained LM Magnification: 15×

Concepts In Motion **Interactive Figure** To see an animation of plant tissues, visit biologygmh.com.

Biology Online

■ **Figure 22.4** The surface of a leaf is composed of tightly-packed epidermal cells that help protect the plant and prevent water loss. Stomata open and close to allow gases in and out.

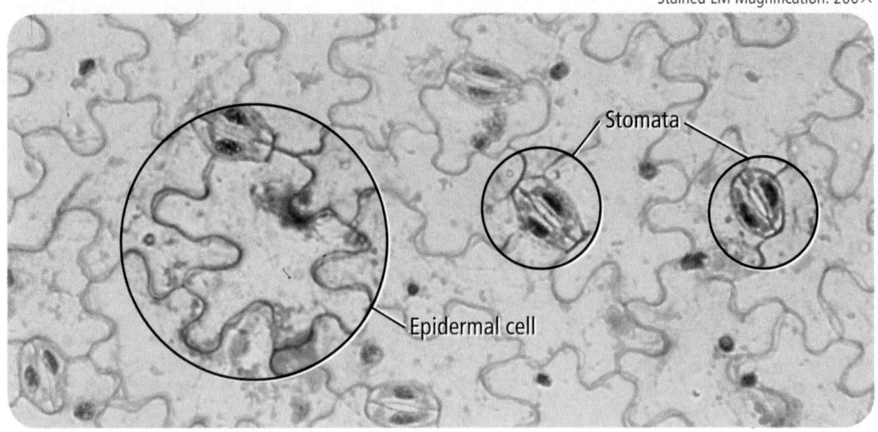

Dermal tissue—the epidermis The layer of cells that makes up the outer covering on a plant is dermal tissue, also called the **epidermis.** Cells of the epidermis resemble pieces of a jigsaw puzzle with interlocking ridges and dips, as shown in **Figure 22.4.** Most epidermal cells can secrete a fatty substance that forms the cuticle. You might recall from Chapter 21 that the cuticle helps reduce water loss from plants by slowing evaporation. The cuticle also can help prevent bacteria and other disease-causing organisms from entering a plant.

Vocabulary

Word origin

Trichome
from the Greek word *trickhma,* meaning *growth of hair.*

Stomata Plants can have several adaptations of their epidermis. Recall from Chapter 21 that the epidermis of most leaves and some green stems have stomata—small openings through which carbon dioxide, water, oxygen, and other gases pass. The two cells that form a stoma are **guard cells.** Changes in the shapes of guard cells result in the opening and closing of stomata, as shown in **Figure 22.4.**

Trichomes Some epidermal cells on leaves and stems produce hairlike projections called trichomes (TRI kohmz), shown in **Figure 22.5.** Trichomes can give leaves a fuzzy appearance and can help protect the plant from insect and animal predators. Some trichomes even release toxic substances when touched. Trichomes help keep some plants cool by reflecting light.

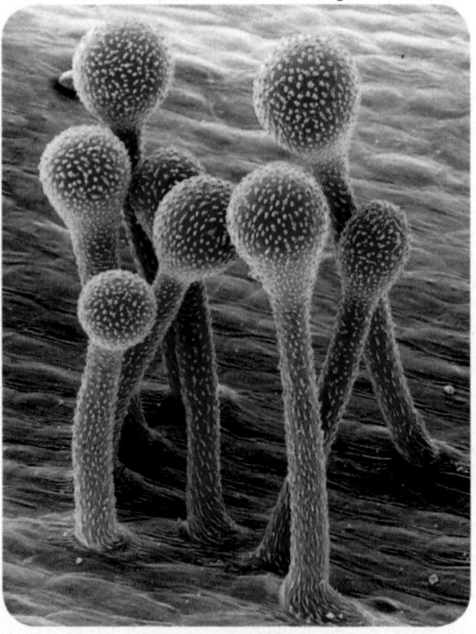

Trichomes on a leaf

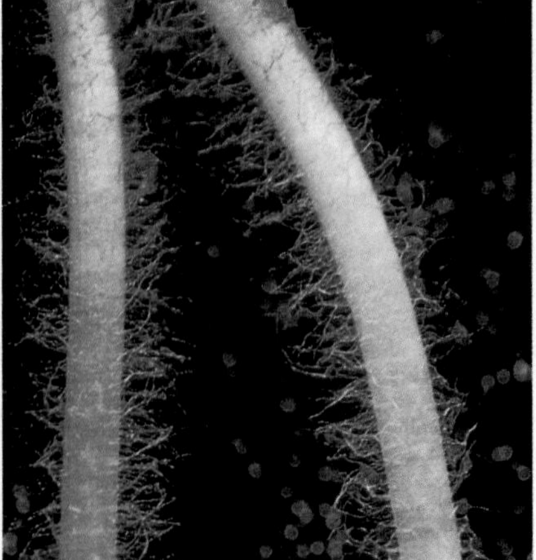

Root hairs

■ **Figure 22.5** Epidermal adaptations help plants survive. The tiny glands at the tip of a trichome can contain toxic substances. Root hairs increase the root's surface area.
Infer ***why it is important to water recently replanted plants.***

Root hairs Some roots have root hairs—fragile extensions of root epidermal cells. Root hairs, as shown in **Figure 22.5,** increase a root's surface area and enable the root to take in a greater volume of materials than it can without root hairs.

Vascular tissues Food, water, and other substances are carried throughout your body in your blood vessels. In a plant, the transportation of water, food, and dissolved substances is the main function of two types of vascular tissue—xylem and phloem.

Xylem Water that contains dissolved minerals enters a plant through its roots. Some of the water is used in photosynthesis. The dissolved minerals have many functions in cells. This water with dissolved minerals is transported throughout a plant within a system of xylem that flows continuously from the roots to the leaves. **Xylem** (ZI lum) is the water-carrying vascular tissue composed of specialized cells—vessel elements and tracheids (tray KEY ihdz). When mature, each vessel element and tracheid consists of just its cell wall. This lack of cytoplasm at maturity allows water to flow freely through these cells.

Vessel elements are tubular cells that are stacked end-to-end, forming strands of xylem called vessels. Vessel elements are open at each end with barlike strips across the openings. In some plants, mature vessel elements lose their end walls. This enables the free movement of water and dissolved substances from one vessel element to another.

Tracheids (tray KEY ihdz) are long, cylindrical cells with pitted ends. The cells are found end-to-end and form a tubelike strand. Unlike some mature vessel elements, mature tracheids have end walls. For this reason, tracheids are less efficient than vessel elements at transporting materials. Compare the structure of tracheids to vessel elements in **Figure 22.6.**

In gymnosperms or nonflowering seed plants, xylem is composed almost entirely of tracheids. However, in flowering seed plants, xylem consists of tracheids and vessels. Because vessels are more efficient at transporting water and materials, scientists propose that this might explain why flowering plants inhabit many different environments.

CAREERS IN BIOLOGY

Turf Scientist Parks, golf courses, and sports stadiums require the skills of a turf scientist to help maintain the grasses that grow there. He or she has an educational background that includes science, managerial, and business courses. For more information on biology careers, visit biologygmh.com.

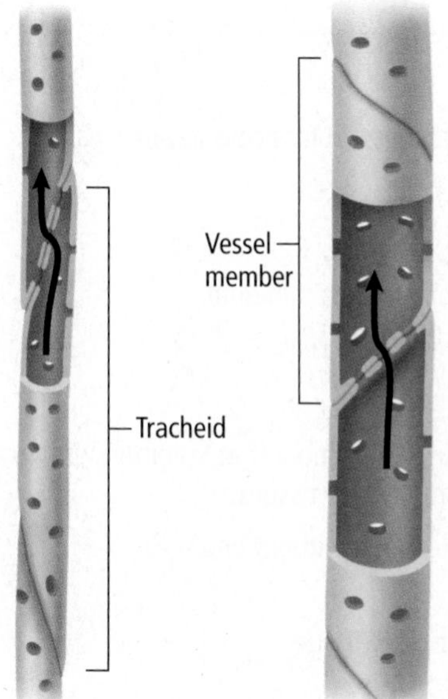

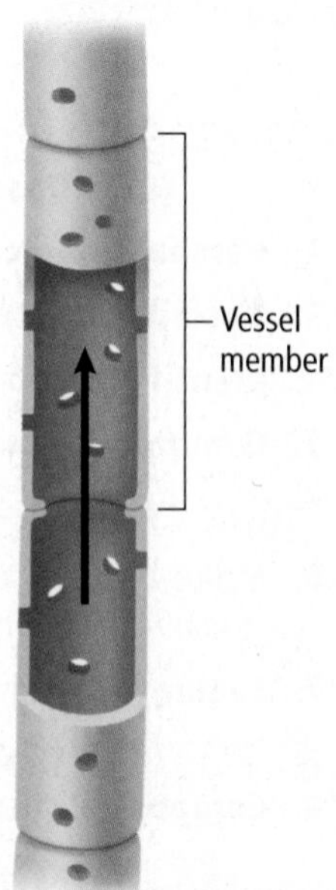

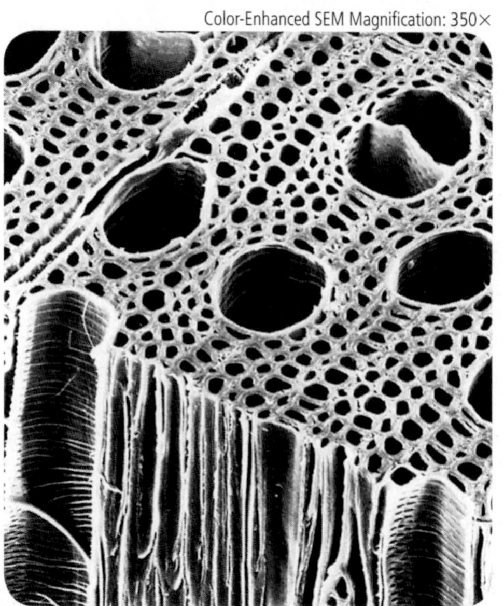

■ **Figure 22.6** Tracheids and vessel elements are the conducting cells of the xylem.

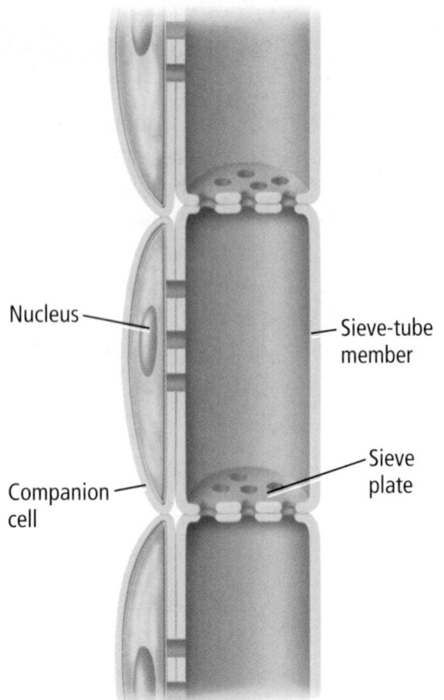

■ **Figure 22.7** Notice the openings in the sieve plates between the sieve-tube members.

Phloem The main food-carrying tissue is **phloem** (FLOH em). It transports dissolved sugars and other organic compounds throughout a plant. Recall that xylem only transports materials away from the roots. Phloem, however, transports substances from the leaves and stems to the roots and from the roots to the leaves and stems. Although not used for transport, there are sclereids and fibers associated with the phloem. These sturdy sclerenchyma cells provide support for the plant.

Phloem consists of two types of cells—sieve tube members and companion cells, shown in **Figure 22.7.** Each **sieve tube member** contains cytoplasm but lacks a nucleus and ribosomes when it is mature. Next to sieve tube members are **companion cells,** each with a nucleus. Scientists hypothesize that this nucleus functions for both the companion cell and the mature sieve tube member. In flowering plants, structures called cell plates are at the end of the sieve tube members. The cell plates have large pores through which dissolved substances can flow.

Some of the glucose produced in leaves and other photosynthetic tissue is metabolized by the plant. However, some is converted to other carbohydrates and transported and stored in regions of the plant called sinks. Examples of sinks are the parenchyma storage cells in the root cortex, which are described in the next section of this chapter. The transport in phloem of dissolved carbohydrates from sources to sinks and other dissolved substances is translocation.

Ground tissue The category for plant tissues that are not meristematic tissues, dermal tissues, or vascular tissues is ground tissue. **Ground tissues** consist of parenchyma, collenchyma, and sclerenchyma cells and have diverse functions, including photosynthesis, storage, and support. Most of a plant consists of ground tissue. The ground tissue of leaves and green stems contains cells with numerous chloroplasts that produce glucose for the plant. In some stems, roots, and seeds, cells of ground tissue have large vacuoles that store sugars, starch, oils, or other substances. Ground tissues also provide support when they grow between other types of tissue.

Section 22.1 Assessment

Section Summary

- There are three types of plant cells—parenchyma, collenchyma, and sclerenchyma cells.
- The structure of a plant cell is related to its function.
- There are several different types of plant tissues—meristematic, dermal, vascular, and ground tissues.
- Xylem and phloem are vascular tissues.

Understand Main Ideas

1. MAIN Idea **Describe** the different types of plant cells in plant tissues.
2. **Compare and contrast** the types of plant cells.
3. **Describe** a root hair and explain its function.
4. **Identify** the location and function of vascular cambium.
5. **Compare** the two types of specialized xylem cells.

Think Critically

6. **Make a table,** using information in this section, that summarizes the structures and functions of the different plant tissues.
7. **Evaluate** the advantage of vessel elements without end walls.

WRITING in **Biology**

8. **Compose** a limerick about a type of plant tissue.

Biology Online Self-Check Quiz biologygmh.com

Section 22.2

Reading Preview

Objectives

- **Relate** the structures of roots, stems, and leaves to their functions.
- **Compare and contrast** the structure and function of roots, stems, and leaves.

Review Vocabulary

apical meristem: tissue at the tips of roots and stems that produces cells, which results in an increase in length

New Vocabulary

root cap
cortex
endodermis
pericycle
petiole
palisade mesophyll
spongy mesophyll
transpiration

Roots, Stems, and Leaves

MAIN Idea **The structures of plants are related to their functions.**

Real-World Reading Link Using a fork to eat a lettuce salad usually is more effective than using a spoon. However, if you were eating tomato soup, a spoon would be more useful than a fork. These are examples of the common expression "the right tool for the right job." The same applies in nature. The variety of plant structures relates to the diversity of plant functions.

Roots

If you ever have eaten a carrot, a radish, or a sweet potato, then you have eaten part of a plant root. The root usually is the first structure to grow out of the seed when it sprouts. For most plants, roots take in water and dissolved minerals that are transported to the rest of the plant. If you have tried to pull a weed, you experienced another function of roots—they anchor a plant in soil or to some other plant or object. Roots also support a plant against the effects of gravity, extreme wind, and moving water.

In some plants, the root system is so vast that it makes up more than half of the plant's mass. The roots of most plants grow 0.5 to 5 m down into the soil. However, some plants, such as the mesquite (mes KEET) that grows in the dry southwestern part of the United States, have roots that grow downward as deep as 50 m toward available water. Other plants, such as some cacti, have many, relatively shallow branching roots that grow out from the stem in all directions as far as 15 m. Both root types are adaptations to limited water resources.

Root structure and growth The tip of a root is covered by the **root cap,** as shown in **Figure 22.8.** It consists of parenchyma cells that help protect root tissues as the root grows. The cells of the root cap produce a slimy substance that, together with the outside layer of cells, form a lubricant that reduces friction as the root grows through the soil, a crack in a sidewalk, or some other material. Cells of the root cap that are rubbed off as the root grows are replaced by new cells produced in the root's apical meristem. Recall from Section 22.1 that the root's apical meristem also produces cells that increase the root's length. These cells develop into the numerous types of root tissues that perform different functions.

You also learned in Section 22.1 that an epidermal layer covers the root. Some root epidermal cells produce root hairs that absorb water and dissolved minerals. The layer below this epidermal layer is the **cortex.** It is composed of ground tissues made of parenchyma cells that are involved in transport and storage of plant substances. The cortex is between the epidermis and the vascular tissues of the root. To reach vascular tissues, all water and nutrients that are taken in by the epidermal cells must move through the cortex.

 **Reading Check** **List** three functions of roots.

Figure 22.8 The root cap covers the root tip and loses cells as the root grows through soil.

Color-Enhanced SEM Magnification: 220×

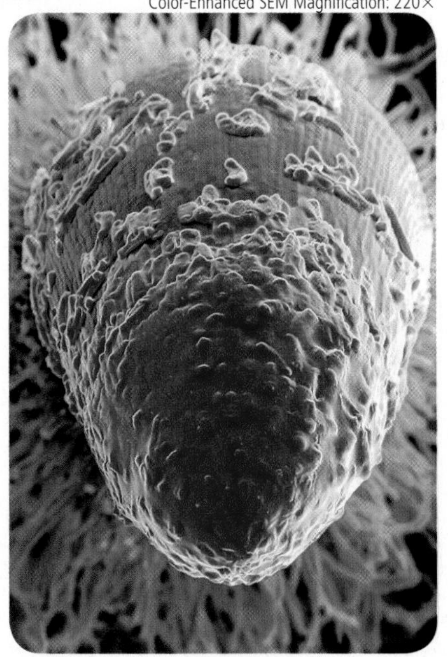

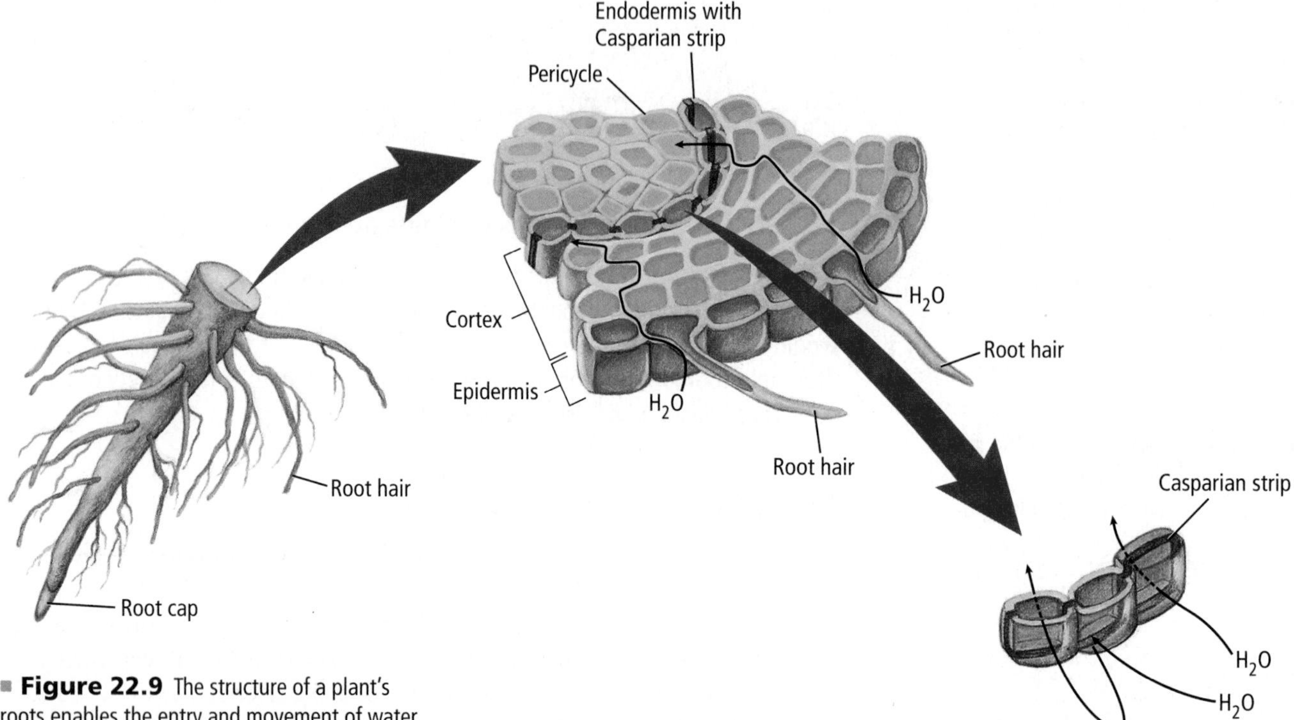

■ **Figure 22.9** The structure of a plant's roots enables the entry and movement of water and dissolved minerals into the plant.
Sequence ***the tissues through which water passes as it moves from a root hair to xylem tissue of a root.***

Concepts In Motion
Interactive Figure To see an animation showing ways that nutrients can enter the cells of roots, visit biologygmh.com.

At the inner boundary of the cortex is a layer of cells called the **endodermis,** as illustrated in **Figure 22.9.** Encircling each cell of the endodermis as part of the cell wall is a waterproof strip called a Casparian strip. Its location is similar to that of mortar that surrounds bricks in a wall. The Casparian strip creates a barrier that forces water and dissolved minerals to pass through endodermal cells rather than around them. Therefore, the plasma membranes of endodermal cells regulate the material that enters the vascular tissues.

The layer of cells directly next to the endodermis toward the center of the root is called the **pericycle.** It is the tissue that produces lateral roots. In most eudicots, and some monocots, a vascular cambium develops from part of the pericycle. Recall that the vascular cambium produces vascular tissues that contribute to an increase in the root's diameter. The vascular tissues—xylem and phloem—are in the center of a root. Monocots and eudicots can be distinguished by the pattern of the xylem and phloem in their roots, as shown in **Figure 22.10.**

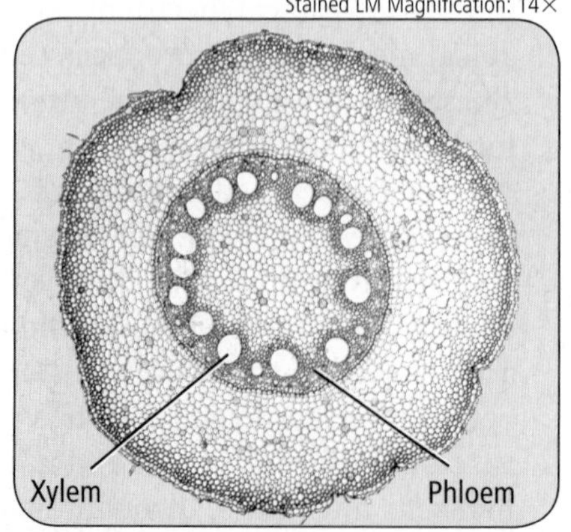

Monocot

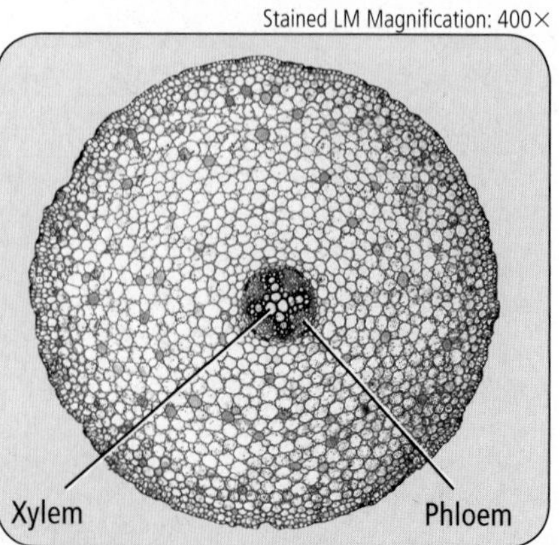

Eudicot

■ **Figure 22.10** In monocots, strands of xylem and phloem cells alternate, usually surrounding a central core of cells called pith. The xylem in eudicot roots is in the center and forms an X shape. Phloem cells are between the arms of the X.

Types of roots The two major types of root systems are taproots and fibrous roots. A taproot system consists of a thick root with few smaller, lateral-branching roots. Some plants, such as radishes, beets, and carrots, as shown in **Table 22.2,** store food in the parenchyma cells of a taproot. Other taproots, like those of poison ivy plants, grow deep into the soil toward available water.

Fibrous root systems, also shown in **Table 22.2,** have numerous branching roots that are about the same size and grow from a central point, similar to the way that the spokes of a bicycle wheel are arranged. Plants also can store food in fibrous roots systems. For example, sweet potatoes develop on fibrous roots.

Other root types, also shown in **Table 22.2,** are adapted to diverse environments. In arid regions, some plants produce huge water-storage roots. Cypress, mangrove, and some other trees that live in water develop modified roots that help supply oxygen to the roots called pneumatophores (new MA toh forz). Adventitious (ad vehn TIH shus) roots form where roots normally do not grow and can have different functions. For example, some tropical trees have adventitious roots that help support their branches. As these roots develop, they resemble trunks.

LAUNCH Lab

Review Based on what you've read about plant structures, how would you now answer the analysis questions?

Concepts In Motion

Interactive Table To explore more about root systems and adaptations, visit biologygmh.com.

Table 22.2 Root Systems and Adaptations

Type	Taproot system	Fibrous root system	Modified root
Example			
Function	• Anchors plant • Food and water storage	• Anchors plant • Rapid water storage	Water storage

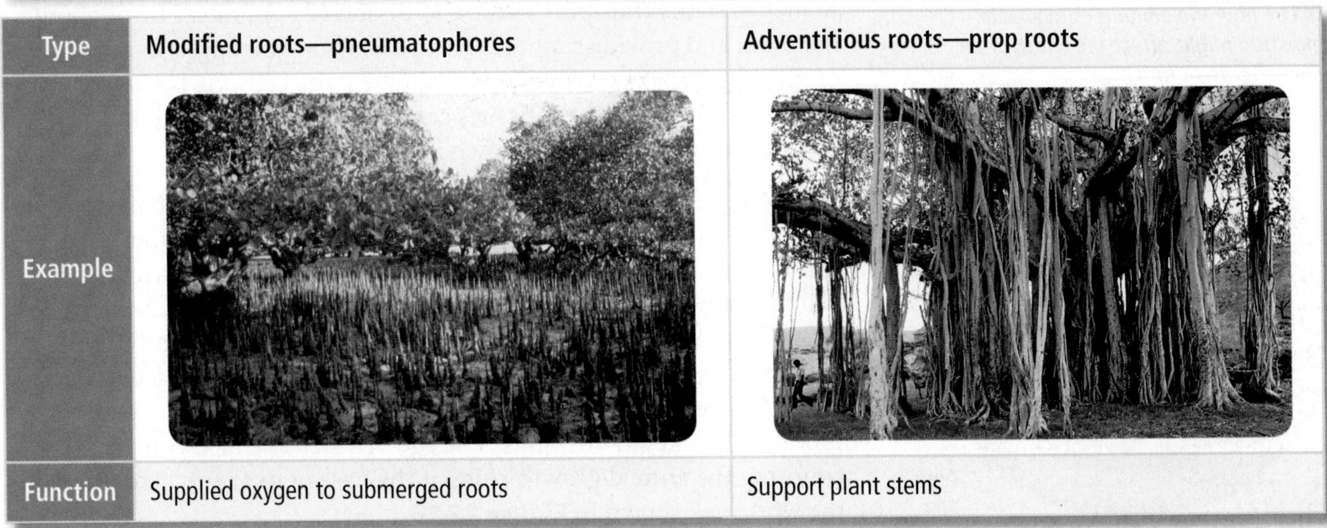

Type	Modified roots—pneumatophores	Adventitious roots—prop roots
Example		
Function	Supplied oxygen to submerged roots	Support plant stems

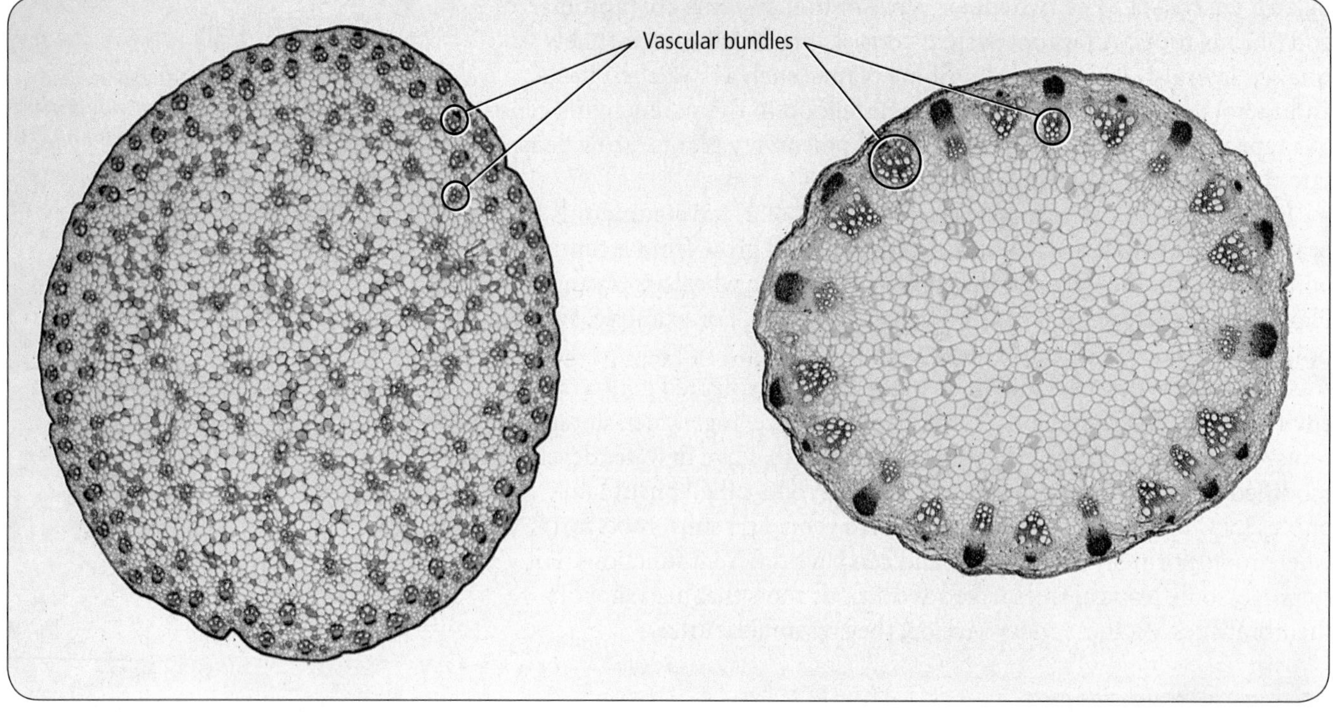

■ **Figure 22.11** The xylem and phloem of stems are grouped together in vascular bundles. The vascular bundles in a monocot stem are scattered. Eudicot stems have one ring or concentric rings of vascular bundles.

Stems

While you might know that asparagus spears are stems, you might be surprised to learn that there are many types of plant stems. Some stems, like asparagus, are soft, flexible, and green due to the presence of chloroplasts, and therefore can perform photosynthesis. These stems are called herbaceous (hur BAY shus) and most annual plants have this type of stem. Palms and bamboos have rigid, fibrous stems. Trees, shrubs, and many perennials have sturdy, woody stems that do not carry on photosynthesis. Some older plants have stems that are covered with bark. This tough, corky tissue can protect the stem from physical damage and insect invasion. Some trees even have survived a forest fire with minimal damage because of the bark that covers their trunks.

Stem structure and function The main function of a plant's stem is support of a plant's leaves and reproductive structures. Vascular tissues in stems transport water and dissolved substances throughout the plant and provide support. These tissues are arranged in bundles, or groups, that are surrounded by parenchyma cells. As is true for roots, the pattern of these tissues can be used to distinguish between monocots and eudicots, as shown in **Figure 22.11.**

Growth of a stem Cells produced by the apical meristem result in an increase in the length of the stem. As the plant grows taller, an increase in stem diameter provides additional support. In annual plants, an increase in stem diameter mostly is due to an increase in cell size. The increase in stem diameter in plants, such as perennial eudicots and conifers, is due to the production of cells by the vascular cambium. The production of xylem and phloem throughout the year can produce annual growth rings. The age of a tree can be estimated by counting the annual growth rings at the base of its trunk, like those of the white oak shown in **Figure 22.12.**

■ **Figure 22.12** An annual growth ring forms in the stem of a woody plant when growth resumes after a period of little or no growth.
Infer ***how the amount of available moisture might affect the width of an annual growth ring.***

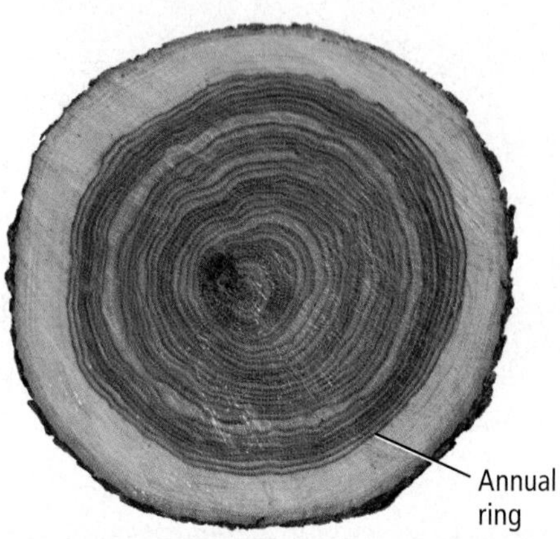

Types of stems All stems have adaptations that help plants survive. In some plants, these adaptations enable stems to store excess food, and in other plants, they help withstand drought, cold, or heat. While you easily might identify stems of tomatoes and oaks, other plants have stems that do not resemble typical stems.

For example, a white potato is a type of stem called a tuber—a swollen, underground stem with buds from which new potato plants can grow. The stem of an onion, a tulip, or a tiger lily is part of a bulb. A bulb is a shortened, compressed stem surrounded by fleshy leaves. Irises and some ferns have rhizomes—underground horizontal stems. Some rhizomes store food. Runners, or stolons, are horizontal stems that grow along the soil's surface in nature, like those of strawberry plants and some grasses. Crocuses and gladiolas are examples of plants that form corms. A corm is composed almost entirely of stem tissue with some scaly leaves at its top. Examples of some of these stem types are shown in **Table 22.3.**

Concepts In Motion
Interactive Table To explore more about types of stems, visit biologygmh.com.

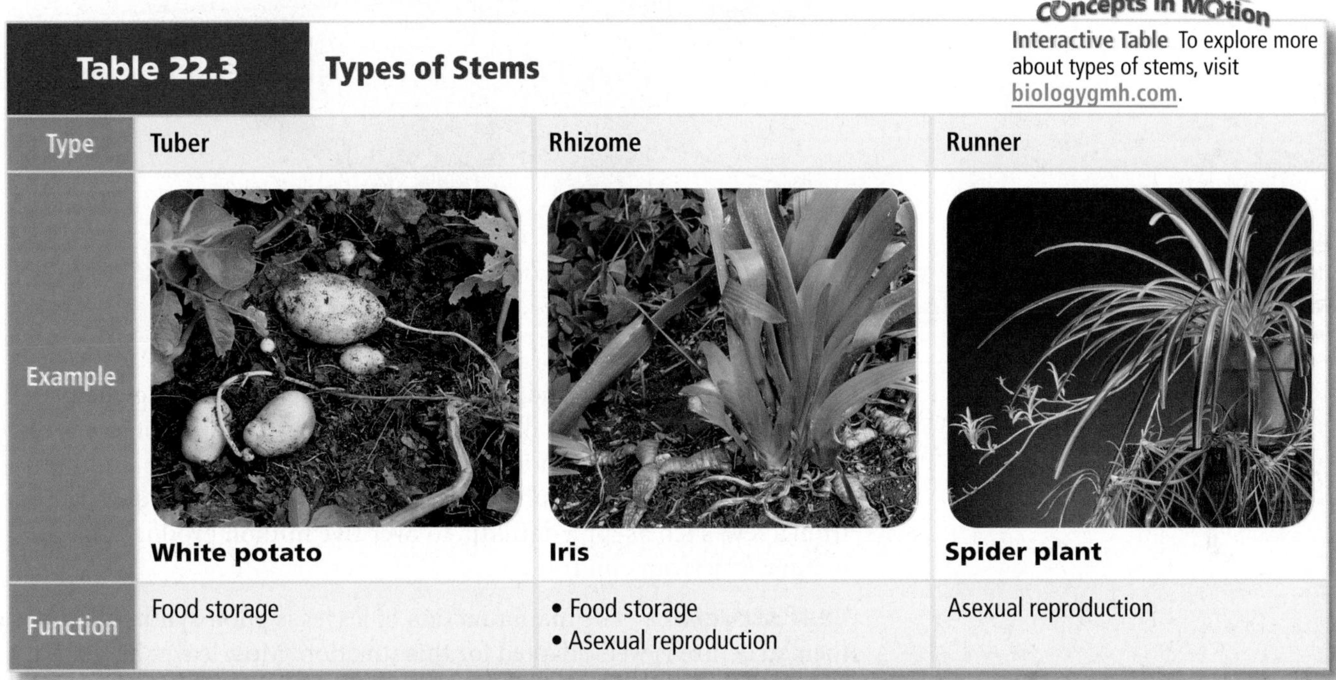

Table 22.3 Types of Stems

Type	Tuber	Rhizome	Runner
Example	White potato	Iris	Spider plant
Function	Food storage	• Food storage • Asexual reproduction	Asexual reproduction

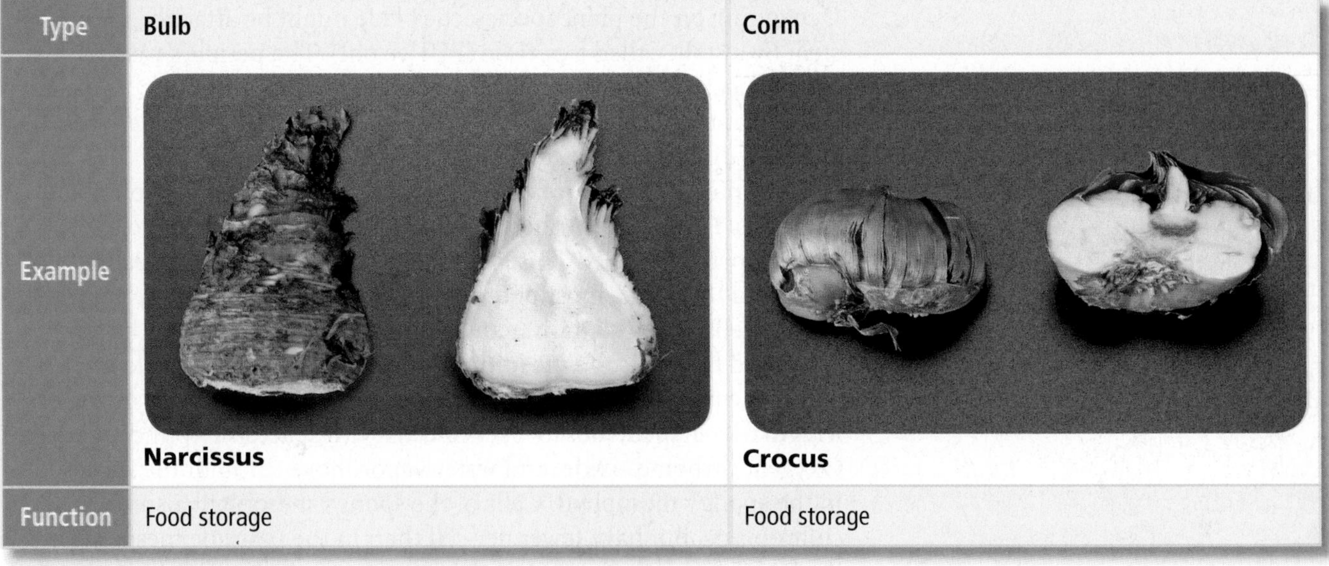

Type	Bulb	Corm
Example	Narcissus	Crocus
Function	Food storage	Food storage

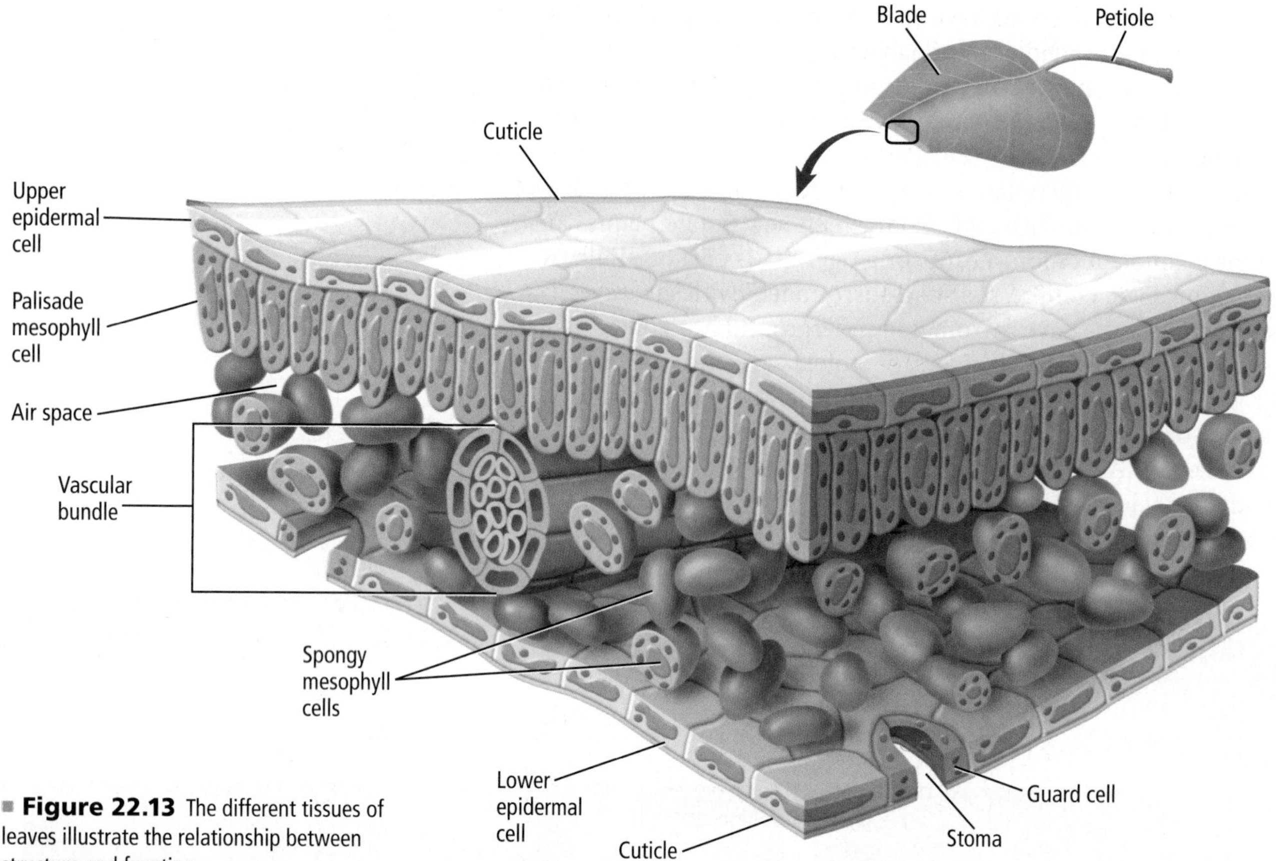

■ **Figure 22.13** The different tissues of leaves illustrate the relationship between structure and function.
Infer *why having a transparent cuticle is important to a plant.*

Leaves

There are many shapes and colors of leaves, and their arrangements on plants are different for different species. Also, the sizes of leaves can range from as large as 2 m in diameter to less than 1 mm in length. In a growing season, the number of leaves that a plant can produce varies from a few, such as for a daffodil, to over five million produced by a mature American elm tree.

Leaf structure The main function of leaves is photosynthesis, and their structure is well-adapted for this function. Most leaves have a flattened portion called the blade that has a relatively large surface area. Depending on the plant species, the blade might be attached to the stem by a stalk called a **petiole** (PET ee ohl). The petiole's vascular tissue connects the stem's vascular tissues to the leaf's vascular tissue or veins. Plants such as grasses lack petioles, and their leaf blades are attached directly to the stem.

The internal structure of most leaves is well-adapted for photosynthesis. **Figure 22.13** shows tightly packed cells directly below a leaf's upper epidermis. This location has the maximum exposure to light, and therefore, most photosynthesis takes place in these column-shaped cells. They contain many chloroplasts and make up the tissue called the **palisade mesophyll** (mehz uh fihl), or palisade layer. Below the palisade mesophyll is the **spongy mesophyll,** consisting of irregularly-shaped, loosely packed cells with spaces surrounding them. Oxygen, carbon dioxide, and water vapor move through the spaces in the spongy mesophyll. Cells of the spongy mesophyll also contain chloroplasts, but have fewer per cell than in the palisade mesophyll.

VOCABULARY

WORD ORIGIN

Mesophyll

meso- comes from the Greek word *mesos,* meaning *middle*

-phyll comes from the Greek word *phyllon,* meaning *leaf.*

Gas exchange and transpiration The epidermis covers a leaf. Except for submerged leaves of aquatic plants, the epidermis contains stomata. There usually are more stomata on the underside of leaves than on the upper side. Recall that two guard cells border a stoma. When more water diffuses into the guard cells than out of them, their shapes change in such a way that the stoma opens. Conversely, when more water diffuses out of the guard cells than into them, their shapes change in such a way that the stoma closes. In Chapter 8, you learned that carbon dioxide is used in photosynthesis and that oxygen gas is a by-product of photosynthesis. The diffusion of these and other gases into and out of a plant also occurs through stomata.

In most plants, water travels from the roots up through the stems and into the leaves, replacing the water used in photosynthesis and lost from the plant by evaporation. Water evaporates from the inside of a leaf to the outside through stomata in a process called **transpiration** that helps pull the water column upward.

Characteristics of leaves Can you identify a maple tree by looking at its leaves? Some people can use differences in the size, shape, color, and texture of leaves to help them identify types of plants. Some leaves are simple, which means the leaf blade is not divided into smaller parts. Compound leaves have leaf blades that are divided into two or more smaller parts called leaflets, as shown in **Figure 22.14.**

The arrangement of leaves on the stem, also shown in **Figure 22.14,** can also be used to distinguish between types of plants. If two leaves are directly opposite of each other on a stem, the growth arrangement is called opposite. An alternate growth arrangement is when the positions of leaves alternate on opposite sides of the stem. A third arrangement, called whorled, is when three or more leaves are evenly spaced around a stem at the same position.

The arrangement of veins in a leaf, or the venation pattern, can also be used to identify leaves. Monocots usually have parallel venation and eudicots usually have branched or netlike venation.

FOLDABLES
Incorporate information from this section into your Foldable.

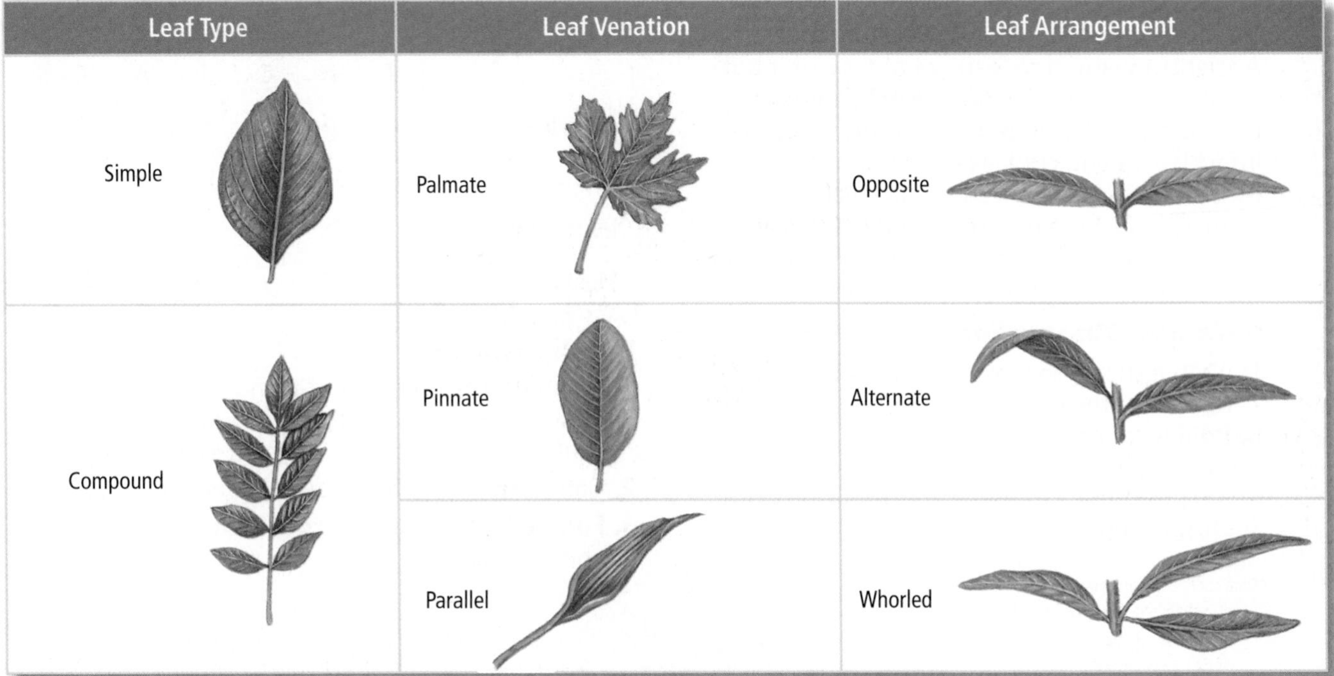

Leaf Type
Leaf Venation
Leaf Arrangement
Simple
Palmate
Opposite
Compound
Pinnate
Alternate
Parallel
Whorled

■ **Figure 22.14** Each species of seed plants has leaves with a unique set of characteristics, some of which are shown here.

■ **Figure 22.15** Cactus spines grow in clusters from small, raised areas on the stem called areoles. The leaves of a jade plant are water-storage organs.

Cactus

Jade plant

VOCABULARY

SCIENCE USAGE V. COMMON USAGE

Spine

Science usage: thin, pointed modified leaf of a cactus or other succulent.
The cactus's spines pierced the animal's flesh.

Common usage: the backbone of an animal.
The motorcyclist's spine was injured as a result of the accident.

Leaf modifications Although the primary function of leaves is photosynthesis, there are many chemical and structural leaf modifications related to other functions. Many succulents, like the cacti in **Figure 22.15,** have modified leaves called spines. In addition to reducing water loss, the spines help protect cacti from being eaten by animals. Other succulents have leaves used as water storage sites. The leaves swell with water when water is available, and when water is scarce, these reserves can help ensure the long-term survival of plants.

DATA ANALYSIS LAB 22.1

Based on Real Data*

Form a Hypothesis

Do *Pieris* caterpillars prefer certain plants?
A scientist wanted to learn what type of input—smell, taste, or touch—helps *Pieris* caterpillars choose food. She used four Petri dishes each of intact leaves and cut leaves. Each set of leaves consisted of a nonmustard family plant (the control) and three different mustard family plants. A caterpillar was added to each dish and its behavior was observed and recorded.

Data and Observations

The table shows the results of the experiment. *T* represents the caterpillar touched the plant but did not bite it. *A* represents the caterpillar took a bite but then abandoned the leaf. *C* represents that the caterpillar chose the leaf and ate it for a time.

Plants offered	Intact leaves			Cut leaves		
	T	A	C	T	A	C
Control	8	0	0	8	0	0
Mustard 1	14	16	14	17	18	13
Mustard 2	16	18	19	22	24	25
Mustard 3	8	10	9	15	19	23

Think Critically

1. **Examine** the data. What trend do you observe about caterpillars choosing mustard-family plants and control plants?
2. **Compare** the data from intact and cut leaves.
3. **Form a hypothesis** to explain the caterpillars' choice of leaves.

*Data obtained from: Chew, F. S. 1980. Foodplant preferences of *Pieris* caterpillars. *Oecologia* 46: 347–353.

Poinsettias

Pitcher plant

■ **Figure 22.16** Leaf modifications relate to different functions. In poinsettias, bracts grow below flowers and attract pollinators. The inside of the pitcher plant's modified leaf has hairs that grow downward. This prevents a trapped animal from crawling out.

In plants like poinsettias, leaves called bracts at the tips of stems change from green to another color in response to the number of hours of darkness in their environments. These plants usually have tiny flowers at the center of the colored leaves, shown in **Figure 22.16.** The leaves look like flower petals and attract pollinators.

The leaves of the sundew plant produce a sticky substance that traps insects. The pitcher plant, also shown in **Figure 22.16,** has cylinderlike modified leaves that fill with water and can trap and drown insects and small animals. Both of these adaptations enable the plants to get nutrients, especially nitrogen, from the insects they capture.

You might be familiar with poison ivy or poison oak that can cause severe skin irritation for some people. These are examples of leaves that contain toxic chemicals that deter organisms from touching them. Some leaves have modifications that deter herbivores from eating them. For example, the epidermis of tomato and squash leaves and stems have tiny hairs with glands at their tips called "trichomes." The glands contain substances that repel insects and other herbivores.

When you read about stems, you learned that bulbs were shortened stems with leaves. A bulb's leaves are modified food storage structures. They provide the dormant bulb with necessary energy resources when favorable growth conditions exist.

Section 22.2 Assessment

Section Summary

- Roots anchor plants and absorb water and nutrients.
- Stems support the plant and hold the leaves.
- Leaves are the sites of photosynthesis and transpiration.
- There are many different modifications of roots, stems, and leaves.
- Modifications help plants survive in different environments.

Understand Main Ideas

1. MAIN Idea **Summarize** the functions of the root cap, cortex, and endodermis.
2. **Compare** a leaf's palisade mesophyll to its spongy mesophyll.
3. **Describe** two leaf modifications and their functions.
4. **Draw and label** the arrangement of vascular tissue in a monocot stem and root and in a eudicot stem and root.

Think Critically

5. **Evaluate** why the role of stomata in a plant is important.
6. MATH in Biology A forest produces approximately 970 kg of oxygen for every metric ton of wood produced. If the average person breathes about 165 kg of oxygen per year, how many people does this forest support?

Section 22.3

Plant Hormones and Responses

Reading Preview

Objectives

- **Identify** the major types of plant hormones.
- **Explain** how hormones affect the growth of plants.
- **Describe and analyze** the different types of plant responses.

Review Vocabulary

active transport: the movement of materials across the plasma membrane against a concentration gradient

New Vocabulary

auxin
gibberellins
ethylene
cytokinin
nastic response
tropism

MAIN Idea **Hormones can affect a plant's responses to its environment.**

Real-World Reading Link As you might have learned in health class or another science course, various responses of your body are controlled by hormones. When you eat, hormones signal cells of your digestive system to release digestive enzymes. Although plants don't have digestive systems with enzymes, hormones do control many aspects of their growth and development.

Plant Hormones

You read in Chapter 6 that hormones are organic compounds that are made in one part of an organism, and then are transported to another part where they have an effect. It takes only a tiny amount of a hormone to cause a change in an organism. Were you surprised to read that plants produce hormones? Plant hormones can affect cell division, growth, or differentiation. Research results indicate that plant hormones work by chemically binding to the plasma membrane at specific sites called receptor proteins. These receptors can affect the expression of a gene, the activity of enzymes, or the permeability of the plasma membrane. You will learn more about human hormones in Chapter 35.

Auxin One of the first plant hormones to be identified was **auxin.** There are different kinds of auxins, but indoleacetic (IHN doh luh see tihk) acid (IAA) is the most widely studied. IAA is produced in apical meristems, buds, young leaves, and other rapidly growing tissues. It moves throughout a plant from one parenchyma cell to the next by a type of active transport. The rate of this movement has been measured at 1 cm per hour. Some auxins also move in the phloem. Also, an auxin moves in only one direction—away from where it was produced.

Connection to Chemistry Auxin usually stimulates the lengthening, or elongation, of cells. Research indicates that in young cells this is an indirect process. Auxin promotes a flow of hydrogen ions through proton pumps from the cytoplasm into the cell wall. This creates a more acidic environment, which weakens the connections between the cellulose fibers in the cell wall. It also activates certain enzymes that help to break down the cell wall. Due to the loss of hydrogen ions in the cytoplasm, water enters the cell, as shown in **Figure 22.17.** The combination of weakened cell walls and increased internal pressure results in cell elongation.

The effect of auxin in a plant varies greatly depending on its concentration and location. For example, in some plants the concentration of auxin that promotes stem growth can inhibit root growth. Low concentrations of auxin usually stimulate cell elongation. However, at higher concentrations, auxin can have the reverse effect. The presence of other hormones can modify the effects of an auxin.

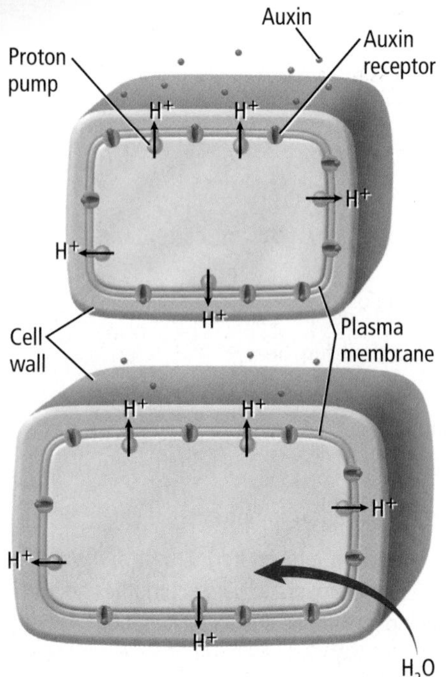

Figure 22.17 Auxin promotes the flow of hydrogen ions into the cell wall, which weakens it. Water enters the cell and it lengthens.

The presence of auxin also creates a phenomenon called apical dominance, which is when plant growth is mostly upward with few or no side branches. The auxin produced by an apical meristem inhibits the growth of side or lateral branches. Removing a plant's apical meristem, however, decreases the amount of auxin present. This promotes the growth of side branches. **Figure 22.18** shows the difference this makes.

Auxins affect fruit formation and inhibit the dropping of fruit. Research results show that the production of auxin slows as cells mature. At the end of the growing season, the decreased amount of auxin in some trees and shrubs causes ripened fruits to fall to the ground and leaves to fall before winter.

Reading Check **Compare and contrast** how different concentrations of auxin can affect a plant.

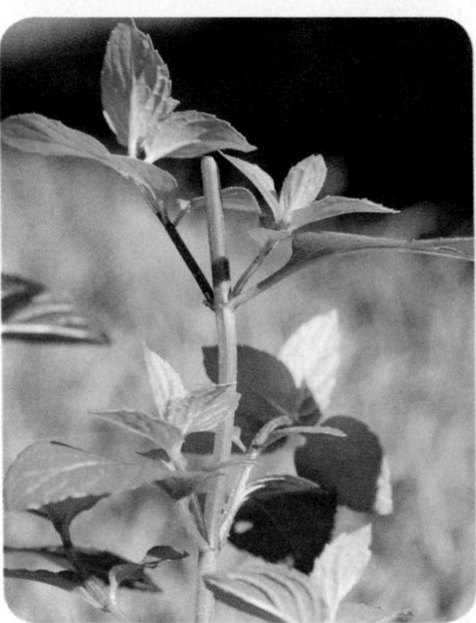

Figure 22.18
Top: Auxin inhibits the growth of side or lateral branches.
Bottom: Removing the apical meristem decreases the amount of auxin and the side branches grow.

Gibberellins The group of plant hormones called **gibberellins** causes cell elongation, stimulates cell division, and affects seed growth. Gibberellins are transported in vascular tissue. Dwarf plants often lack either the genes for gibberellins production or the genes for gibberellins protein receptors. When treated with gibberellins, plants that lack the genes for gibberellins but have gibberellins receptors grow taller. Applying gibberellins to a plant can cause an increase in height, as shown in **Figure 22.19.**

Ethylene The only known gaseous hormone is **ethylene.** It is a simple compound composed of two carbon and four hydrogen atoms. Ethylene is found in plant tissues such as ripening fruits, dying leaves, and flowers. Since ethylene is a gas, it can diffuse through the spaces between cells. It also is transported within the phloem.

Although ethylene can affect other parts of plants, it primarily affects the ripening of fruits. Ethylene causes cell walls of unripe fruit to weaken and complex carbohydrates to break down into simple sugars. The results of ethylene exposure are fruits that are softer and sweeter than unripe fruits.

Because ripe fruits and vegetables are bruised easily during shipping, growers often pick and ship unripe fruits and vegetables. Once they reach their destinations, a treatment with ethylene speeds up the ripening process.

Figure 22.19 These plants do not have genes for gibberellins production. However, the plant on the right grew when treated with gibberellins.

CAREERS IN BIOLOGY

Plant Physiologist The chemistry of plants, including how plant hormones work, is one of many topics studied by plant physiologists. Many plant physiologists teach and conduct research at universities. For more information on biology careers, visit biologygmh.com.

Cytokinins Growth-inducing **cytokinins** (si tuh KI nihnz) are produced in rapidly dividing cells. They travel to other parts of the plant within xylem. Cytokinins promote cell division by stimulating the production of the proteins needed for mitosis and cytokinesis. Because cytokinins increase the rate of growth, they often are added to the growth media used for plant tissue culture—a laboratory technique for growing plants from pieces of plant tissues. The presence of other hormones, especially auxins, influences the effects of cytokinins. For example, IAA alone stimulates cell elongation, but combined with a cytokinin, it promotes rapid cell division and results in rapid growth.

 **Reading Check** **Describe** two ways hormones can affect plants.

Plant Responses

Have you ever wondered why the leaves of a houseplant grow toward a window, or how a vine can climb a pole? These and other events—roots growing down, stems growing upward, trees dropping their leaves, and leaves of some plants trapping an insect—are due to responses of plants to their environment.

Nastic responses A plant response that causes movement that is not dependent on the direction of the stimulus is a **nastic response.** It is not a growth response, is reversible, and can be repeated. An example of a nastic response is the opening of leaves during the day and the closing of leaves at night to conserve heat, or the movement of *Mimosa pudica* leaflets when they are touched. Nastic movements are caused by a change in water pressure in the leaf cells. Once the stimulus ends, the leaves return to their original positions.

MiniLab 22.2

Investigate a Plant Response

What stimulus causes a Venus flytrap to shut its leaves? A Venus flytrap has specialized leaves that trap and digest insects. In this lab, you will learn what type of stimulus is necessary to trigger the trapping response.

Procedure

1. Read and complete the lab safety form.
2. Examine a **Venus flytrap plant** with open leaves.
3. Using a **small paintbrush,** carefully touch one of the trigger hairs on the inner surface of a leaf.
4. Wait 60 seconds. Now use your paintbrush and touch two different trigger hairs. Alternatively, touch one trigger hair and then touch it again in about ten seconds.
5. After you have stimulated the leaves to snap shut, whenever possible, observe your plant to determine how long it takes the trap to open again.

Analysis

1. **Identify** the type of stimulus necessary to trigger the plant leaf to shut. How long did it take the leaf to reopen?
2. **Think Critically** If you drop a dead insect onto a leaf, the leaf might close. However, it will not close tightly and will reopen later without digesting the insect. Based on this lab, hypothesize how the plant might distinguish between a living insect and a dead one.

Another example of a nastic response is the closing of a Venus flytrap's leaves. Recent research shows that this results from a movement of water within each half of the leaf trap. The movement results in uneven expansion until the leaf's curved shape suddenly changes and snaps the trap shut.

Tropic responses What do you notice about the plants in **Table 22.4?** These are examples of tropic responses, or tropisms. A **tropism** (TROH pih zum) is a plant's growth response to an external stimulus. If resulting plant growth is toward the stimulus, it is called a positive tropism. If the resulting plant growth is away from the stimulus, it is called a negative tropism. There are several different types of tropisms, including phototropism, gravitropism, and thigmotropism.

Phototropism is a plant growth response to light caused by an unequal distribution of auxin. There is less auxin on the side of the plant toward the light source and more auxin on the side away from the light source. Because auxin can cause cell elongation, the cells on the side away from the light elongate, making that side of the stem longer. This results in the stem curving toward the direction of the light.

Gravitropism is a plant growth response to gravity. Roots generally show a positive gravitropism. The downward growth of roots into soil helps to anchor the plant and brings roots in contact with water and minerals. However, a stem exhibits a negative gravitropism when it grows upward, away from gravity. This growth positions leaves for maximum exposure to light.

Another tropism found in some plants is thigmotropism. This is a growth response to mechanical stimuli, such as contact with an object, another organism, or even wind. Thigmotropism is evident in vining plants that twist around a nearby structure such as a fence or tree.

Table 22.4 Plant Tropisms

Concepts In Motion — Interactive Table To explore more about phototropism, visit biologygmh.com.

Tropism	Stimulus/ Response	Example
Phototropism	Light • Growth toward light source	
Gravitropism	Gravity • Positive: downward growth • Negative: upward growth	
Thigmotropism	Mechanical • Growth toward point of contact	

Personal Tutor

To learn about tropisms, visit biologygmh.com.

Section 22.3 Assessment

Section Summary

- Plant hormones are produced in very small amounts.
- Hormones can affect cell division, growth, and differentiation.
- Nastic responses are not dependent on the direction of the stimulus.
- Tropisms are responses to stimuli from a specific direction.

Understand Main Ideas

1. MAIN Idea **Identify** plant hormones and classify them according to the effects that they have on a plant.
2. **Name and describe** three tropisms.
3. **Compare and contrast** tropisms and nastic responses.

Think Critically

4. **Construct** a model to show how auxin can move from one cell to another.
5. **Judge** the scientific basis of the saying, "One rotten apple spoils the whole barrel."

Biology Online **Self-Check Quiz** biologygmh.com

BioDiscoveries

Plants and Their Defenses

When you think of a food chain, you might picture a predator stalking and capturing prey. However, plants are sessile—they cannot move away from herbivores. How do plants defend themselves against predators? Understanding plant chemical defenses helps humans devise strategies to protect crops and other vegetation.

Defend or die Some plants evolved adaptations, such as hairs, spines, prickles, or thorns on the epidermis, to repel predators. Others have silica inside their leaves, which makes them tough to eat and wears down the predator's teeth.

Many plants produce secondary plant compounds not needed for plant metabolism. These substances might be bitter to taste or toxic to the predator. Some interfere with the predator's digestion, growth, or reproduction. In 2005, researchers discovered that the roots of a type of cabbage produce substances that protect the plant by killing a wide variety of bacteria in the soil.

Insect or not It is known that plants can distinguish between an insect attack and other types of damage, such as pruning. Scientists have learned that some plants respond to certain chemicals in insect saliva. For example, a team of biochemists determined that when an insect nibbles on the plant's leaves, a chemical signal spreads throughout the plant. This signal stimulates increased toxin production by all the leaves—not just the attacked leaves.

Calling for help When some plants are damaged by herbivores, the plants release chemical signals that attract natural enemies of the herbivores. For example, the tobacco plant in the photos guides the parasitic wasp to the caterpillar eating the tobacco leaves.

As a caterpillar feeds on a tobacco plant, the caterpillar's saliva causes the plant to release chemicals into the air, which attract a parasitic wasp—a predator of the caterpillar.

Chemical labeling studies confirmed that the signaling chemicals are not stored in the undamaged plant. Plants develop and release the signals soon after damage begins, and release them most strongly during the time when the natural enemies are most active. Also, different herbivores elicit different signals. Although advances in chemical technology and biotechnology speed the discovery of natural plant signals that might aid in protecting crops, evidence shows that the signals might also help herbivores locate food.

WRITING in Biology

Advertisement Imagine that you developed a remarkable new pesticide using natural plant defenses. For more information about natural plant pesticides, visit biologygmh.com. Write an advertisement describing your product, why it is different from other available products, and how it can prevent pest resistance.

BIOLAB

Design Your Own

INTERNET: HOW DO DWARF PLANTS RESPOND TO GIBBERELLINS?

Background: **Some dwarf plants lack a gene for gibberellin production and some lack gibberellin receptors. In this lab, you will design an experiment to determine if you can change the growth pattern of dwarf pea-plant seedlings by applying gibberellic acid (a form of gibberellins) to them.**

Question: *Can you use gibberellins to change the growth of dwarf pea plants?*

Materials

gibberellic acid in varying concentrations
sheets of poster board or cardboard
dishwashing liquid (wetting agent)
potted dwarf pea-plant seedlings
spray bottles
cotton swabs
light source
large plastic bags
plant fertilizer
distilled water
metric rulers
graph paper
Choose materials that would be appropriate for this lab.

Safety Precautions

Plan and Perform the Experiment

1. Read and complete the lab safety form.
2. Form a hypothesis that explains how gibberellins will affect the growth of dwarf pea plants.
3. Design an experiment to test your hypothesis. Be sure that your experiment has a control group.
4. Make a list of factors that must be constant for your experimental and control groups. Be sure to test only one variable.
5. Determine a way to apply gibberellins to the plants and decide how often you will apply it.
6. Design and construct a data table to record data from your experiment.
7. Make sure your teacher approves your plan before you proceed.
8. Collect the supplies you need and set up your experimental and control plants.
9. Complete the approved experiment.
10. Record measurements and observations of the plants in your data table.
11. Graph the data from your experimental and control groups.
12. **Cleanup and Disposal** Return unused gibberellic acid to your teacher for disposal. Empty spray bottles and thoroughly rinse. Dispose of used cotton swabs in the trash. Dispose of plants as directed by your teacher.

Analyze and Conclude

1. **Analyze** your graph and determine the effect of gibberellic acid on the dwarf pea plants.
2. **Hypothesize** Based on your results, explain why the pea plants are dwarfs.
3. **Think Critically** Why might a genetic change, such as one that causes a plant not to produce gibberellins, be a problem for plants in a natural environment?
4. **Error Analysis** What might have occurred in your experimental setup that could have caused your data to be inaccurate? How would you change your procedure?

SHARE YOUR DATA

Peer Review Visit BioLabs at biologygmh.com and post your data. Compare and contrast your graph to those of other students who completed this lab.

CHAPTER 22

Study Guide

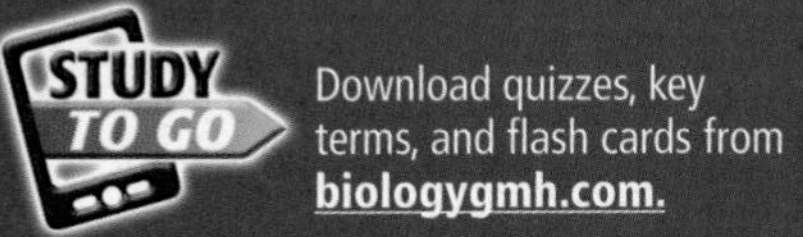

FOLDABLES **Justify** On the back of your Foldable, rank the leaf layers in order of importance to a plant. Justify your ranking.

Vocabulary	Key Concepts
Section 22.1 Plant Cells and Tissues	
• collenchyma cell (p. 633) • companion cell (p. 638) • cork cambium (p. 634) • epidermis (p. 636) • ground tissue (p. 638) • guard cell (p. 636) • meristem (p. 634) • parenchyma cell (p. 632) • phloem (p. 638) • sclerenchyma cell (p. 633) • sieve tube member (p. 638) • tracheid (p. 637) • vascular cambium (p. 634) • vessel element (p. 637) • xylem (p. 637)	MAIN Idea Different types of plant cells make up plant tissues. • There are three types of plant cells—parenchyma, collenchyma, and sclerenchyma cells. • The structure of a plant cell is related to its function. • There are several different types of plant tissues—meristematic, dermal, vascular, and ground tissues. • Xylem and phloem are vascular tissues.
Section 22.2 Roots, Stems, and Leaves	
• cortex (p. 639) • endodermis (p. 640) • palisade mesophyll (p. 644) • pericycle (p. 640) • petiole (p. 644) • root cap (p. 639) • spongy mesophyll (p. 644) • transpiration (p. 645)	MAIN Idea The structures of plants are related to their functions. • Roots anchor plants and absorb water and nutrients. • Stems support the plant and hold the leaves. • Leaves are the sites of photosynthesis and transpiration. • There are many different modifications of roots, stems, and leaves. • Modifications help plants survive in different environments.
Section 22.3 Plant Hormones and Responses	
• auxin (p. 648) • cytokinin (p. 650) • ethylene (p. 649) • gibberellins (p. 649) • nastic response (p. 650) • tropism (p. 651)	MAIN Idea Hormones can affect a plant's responses to its environment. • Plant hormones are produced in very small amounts. • Hormones can affect cell division, growth, and differentiation. • Nastic responses are not dependent on the direction of the stimulus. • Tropisms are responses to stimuli from a specific direction.

Biology Online Vocabulary PuzzleMaker biologygmh.com

CHAPTER 22 Assessment

Section 22.1

Vocabulary Review

Distinguish between the words in each pair.

1. sclerenchyma, collenchyma
2. xylem, phloem
3. epidermis, guard cell

Understand Key Concepts

4. Which is the vascular tissue that transports water and dissolved minerals from roots to leaves?
   A. epidermis **C.** xylem
   B. parenchyma **D.** phloem

5. Which is the region of actively dividing cells at the tip of the stem?
   A. apical meristem **C.** dermal tissue
   B. vascular tissue **D.** lateral meristem

Use the photos below to answer questions 6 and 7.

6. Which image shows a trichome?

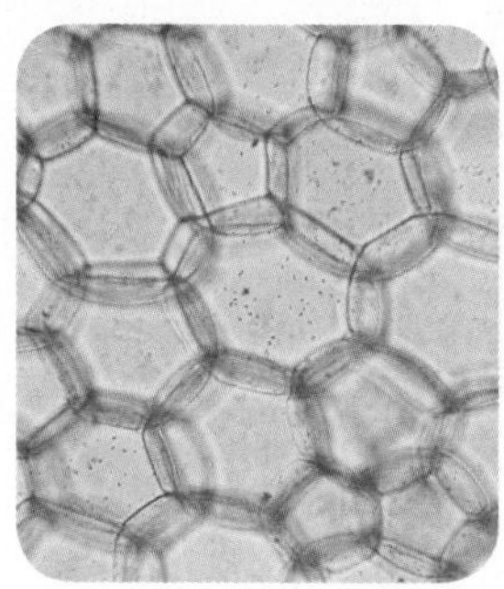

A.

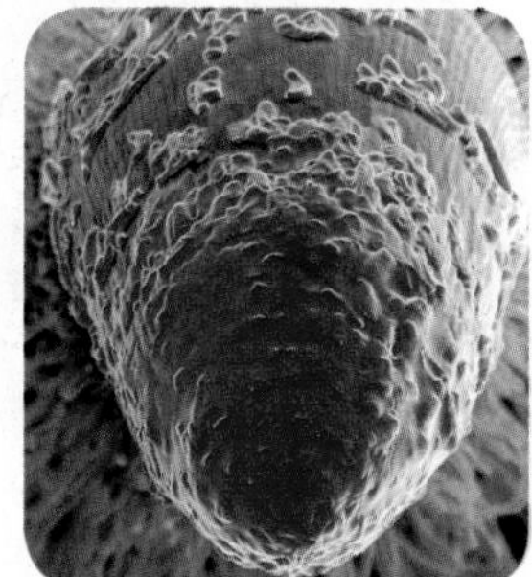

B.

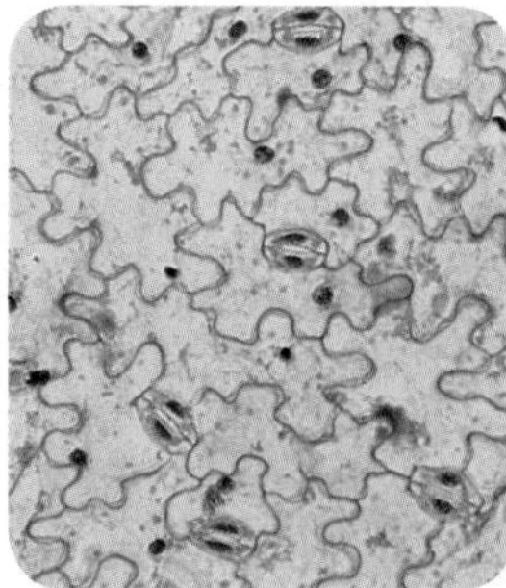

C.

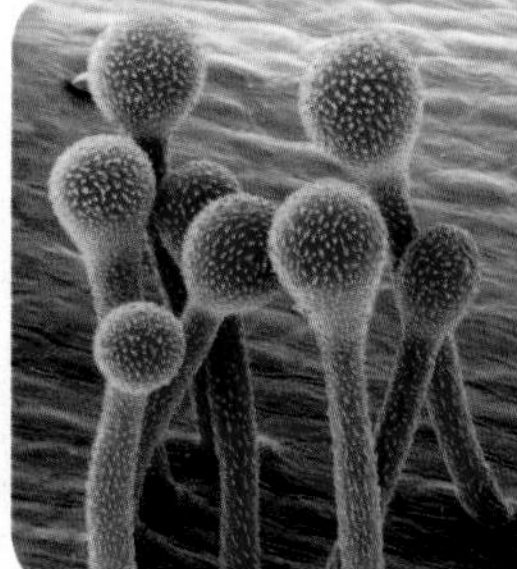

D.

7. Which image shows parenchyma cells?
   A. A **C.** C
   B. B **D.** D

8. Which is one of the differences between nonflowering seed plants and flowering seed plants?
   A. presence of stomata in the roots
   B. amount of sugar stored in the roots
   C. presence of tracheids and vessels
   D. structure of parenchyma cells

Constructed Response

Use the image below to answer question 9.

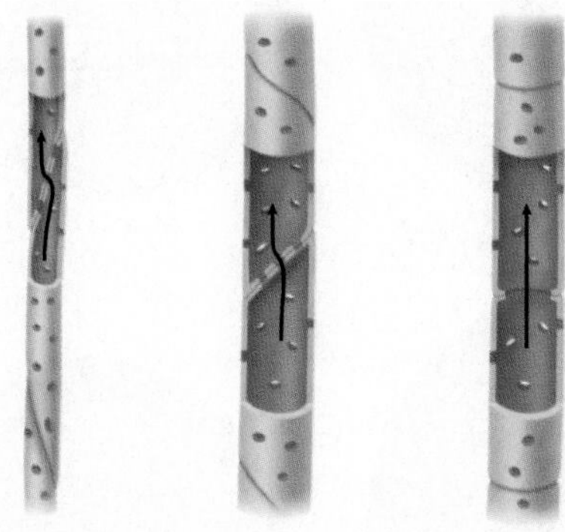

9. **Short Answer** Explain one advantage of these vessels.
10. **Short Answer** Compare and contrast root hairs and trichomes.
11. **Open Ended** Do you think plants could survive without ground tissue? Defend your answer.

Think Critically

12. **Construct** a graphic organizer that lists each of the four different types of tissue, the function of each, and the types of cells it contains.
13. **Compare** the dermal tissue of plants to your skin. What are some characteristics that make it more efficient than your skin? What are some characteristics that make your skin more efficient than the plant's epidermis?

Section 22.2

Vocabulary Review

Correctly use each set of words in a sentence.

14. endodermis, pericycle
15. petiole, transpiration
16. spongy mesophyll, palisade mesophyll

Understand Key Concepts

17. Which fill(s) the space between spongy mesophyll cells?
    - **A.** chlorophyll
    - **B.** gases
    - **C.** cells
    - **D.** vascular tissue

18. Which image shows a eudicot stem?

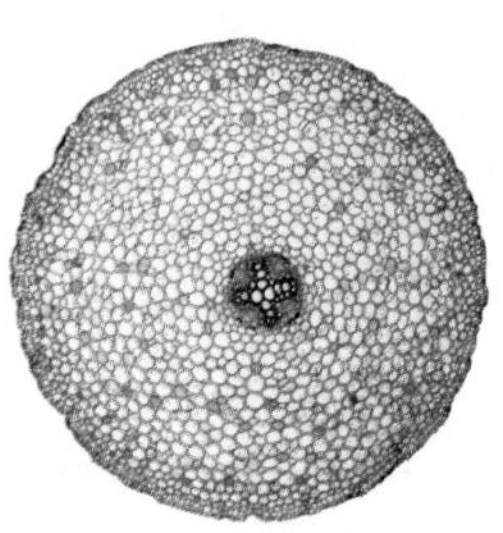
A.

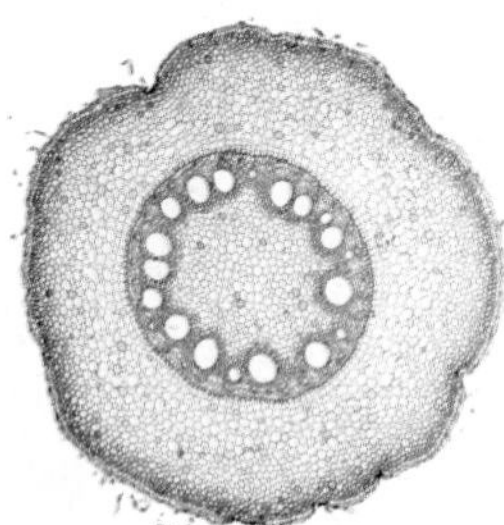

B.

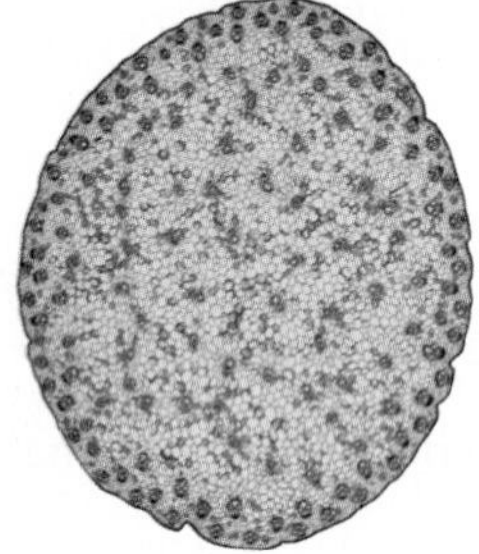
C.

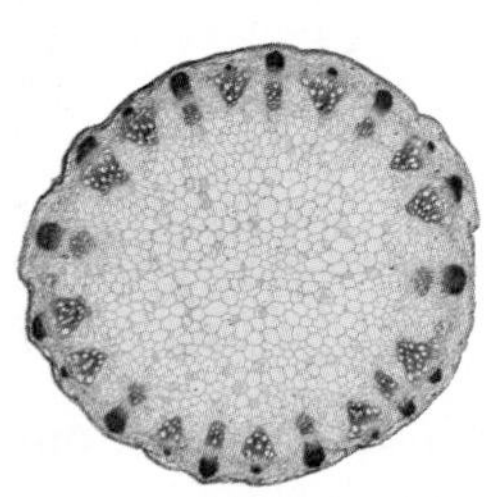
D.

19. Which image above shows one ring of vascular bundles?
    - **A.** A
    - **B.** B
    - **C.** C
    - **D.** D

20. Which plant structure is not part of a root?
    - **A.** endodermis
    - **B.** root cap
    - **C.** pericycle
    - **D.** stomata

21. Which control(s) the movement of water vapor through the stomata?
    - **A.** bark
    - **B.** pericycle
    - **C.** guard cells
    - **D.** vascular tissues

Constructed Response

22. **Open Ended** List some environmental factors that might affect transpiration.

23. **Short Answer** Describe the control of materials as they are transported from soil to a root's vascular tissue.

Think Critically

24. **Evaluate** some leaf modifications in terms of their functions.

25. **Summarize** the reasons why eudicot stems can have a greater increase in diameter than most monocot stems.

Section 22.3

Vocabulary Review

Explain the difference between the terms in each pair below. Then explain how they are related.

26. hormone, auxin

27. ethylene, gibberellins

28. tropic response, nastic response

Understand Key Concepts

Use these photos to answer questions 29 and 30.

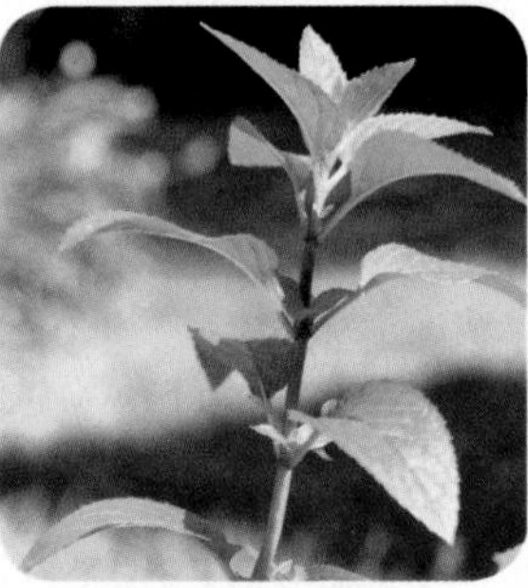

29. What plant condition do these photos show?
    - **A.** apical dominance
    - **B.** dwarfism
    - **C.** leaf drop
    - **D.** nastic movements

30. Which hormone controls this plant condition?
    - **A.** auxin
    - **B.** gibberellin
    - **C.** ethylene
    - **D.** cytokinin

31. Which describes a positive phototropism?
    - **A.** The plant grows away from the light.
    - **B.** The plant grows toward the light.
    - **C.** The plant grows toward gravity.
    - **D.** The plant grows away from gravity.

32. Which is involved in the transport of gibberellins throughout the plant?
    - **A.** cork cambium
    - **B.** guard cells
    - **C.** vascular tissue
    - **D.** apical meristem

Chapter Test biologygmh.com

Use the images below to answer question 33.

33. Which stem shown above is exhibiting negative gravitropism?

A. A
B. B
C. C
D. D

Constructed Response

34. Open Ended Discuss the pros and cons of the transport of auxin from one parenchyma cell to another instead of in the vascular tissue.

35. Short Answer Refer to **Figure 22.17** and explain how auxin can cause cell elongation.

36. Short Answer Explain why tropic responses are permanent while nastic responses are reversible.

Think Critically

37. Design an experiment to determine if bean plants show apical dominance.

38. Evaluate the following statement: "Seeds soaked in gibberellins will germinate faster than seeds not soaked in gibberellins."

39. CAREERS IN BIOLOGY Farmers must evaluate the use of plant hormones to increase crop production. Do you think it is a good idea? Compare it to the use of growth hormones that are used to increase the milk production of cows.

Additional Assessment

40. WRITING in Biology What if you could develop a new plant hormone? What would you have it do? How would it work and what would you name it?

Document-Based Questions

A team of biologists studied the effect of temperature and carbon dioxide on ponderosa pines. The graph below represents the amounts of tracheids with various diameters grown at different temperatures.

Use the graph to answer questions 41–42.

Data obtained from: Maherali, H., and DeLucia, E. H. 2000. Interactive effects of elevated CO_2 and temperature on water transport in ponderosa pine. *Amer. Journal of Botany* 87: 243-249.

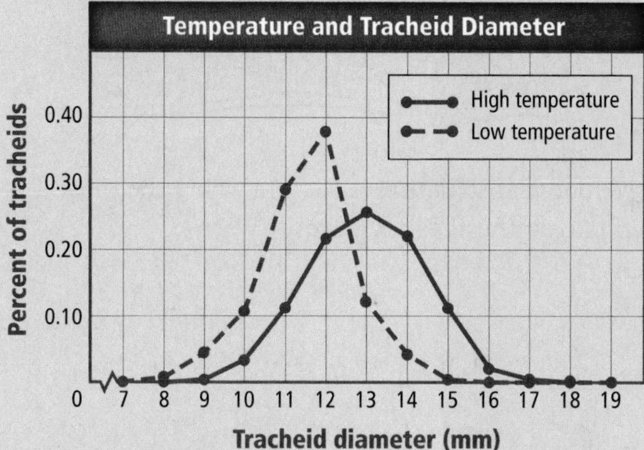

41. How does the temperature affect the diameter of developing tracheid cells?

42. How does the relationship between temperature and diameter relate to the tracheid function?

Cumulative Review

43. In pigeons, the checker pattern of feathers (*P*) is dominant to the nonchecker pattern (*p*). Suppose a checker pigeon with the genotype *Pp* mates with a nonchecker pigeon. Use a Punnett square to predict the genotypic ratio of their offspring. **(Chapter 10)**

44. Create an analogy that illustrates why two species in the same family of organisms also must be in the same order. **(Chapter 17)**

Standardized Test Practice

Cumulative

Multiple Choice

1. The Miller-Urey experiment tested which hypothesis?
   A. Margulis's endosymbiont theory
   B. Miller's amino acid origin
   C. Oparin's primordial soup idea
   D. Pasteur's biogenesis theory

Use the diagram below to answer question 2.

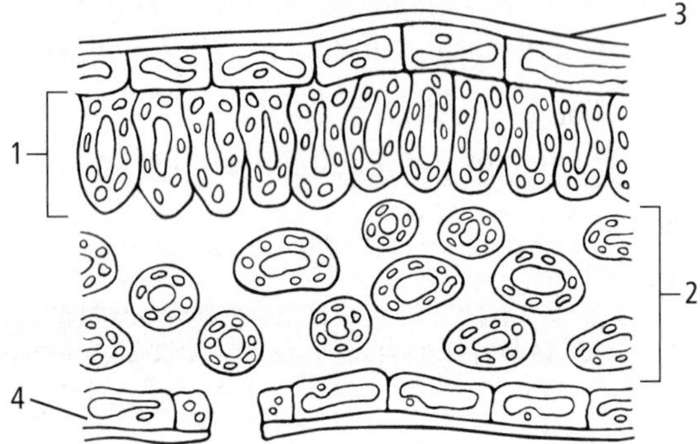

2. Which leaf structure is the site where the most photosynthesis takes place?
   A. 1
   B. 2
   C. 3
   D. 4

3. Lichens can be an indicator of environmental quality. If a coal-fired electric plant was built and then the lichens in the area decreased, which would be the most likely cause?
   A. air quality decreased
   B. annual temperatures decreased
   C. humidity patterns changed
   D. rainfall patterns changed

4. Which is one method of asexual reproduction that can occur in fungi?
   A. conjugation
   B. fragmentation
   C. segmentation
   D. transformation

5. Which development in plants contributed most to the evolution of large trees?
   A. alternation of generations
   B. flowers
   C. seeds
   D. vascular tissue

6. Which describes how funguslike protists obtain food?
   A. They absorb nutrients from decaying organisms.
   B. They obtain nutrients by feeding on unicellular organisms.
   C. They have a symbiotic relationship with an animal host, obtaining nutrients from it.
   D. They produce sugars as a nutrient source by using energy from sunlight.

7. Which is the function of a plant's root cap?
   A. generate new cells for root growth
   B. help the root tissues absorb water
   C. protect root tissue as the root grows
   D. provide support for the root tissues

Use the diagram below to answer question 8.

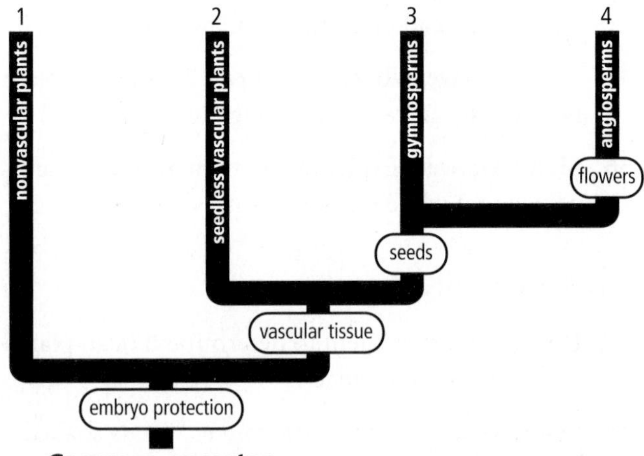

8. Which number represents where you would expect to find cycadophytes on this evolutionary tree?
   A. 1
   B. 2
   C. 3
   D. 4

Short Answer

Use the illustration below to answer question 9.

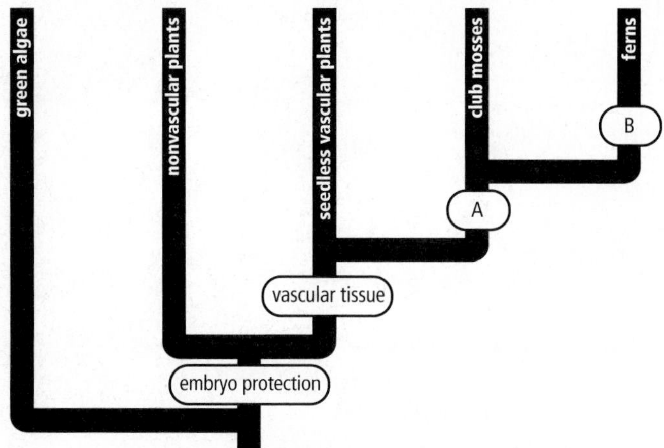

9. Look at the evolutionary tree in the diagram above. What word or phrase would best describe branching points A and B in the diagram?

10. Compare and contrast septate and aseptate hyphae.

11. Write a hypothesis about the benefit of the stem adaptations that allow some plants to store excess food.

12. Use a chart to organize information about how annuals, biennials, and perennials are similar and different.

13. Name and describe the function of the two types of vascular tissue found in plants.

14. What are three characteristics of ancient algae that enabled them to survive and can be found in all plants today?

15. Describe the function of the vascular tissue in a leaf.

Extended Response

Use the illustration below to answer question 16.

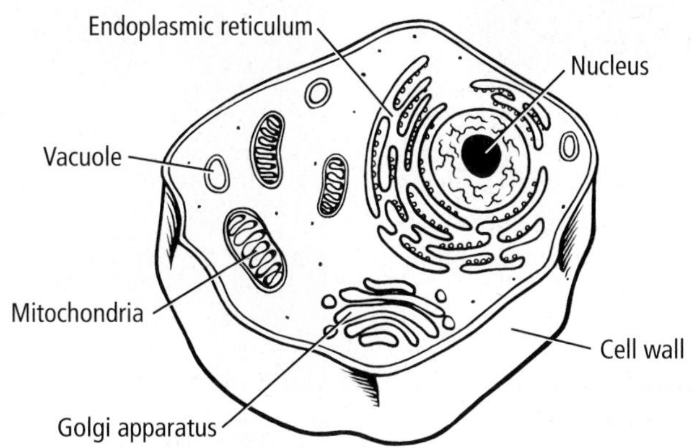

16. Based on the characteristics of the cell above, how would you classify the organism from which it was taken? Justify your method for classifying the organism.

17. Evaluate why the structure of the thylakoid in a chloroplast is well suited for its function.

Essay Question

Imagine that you are planning to turn an area of land near your school into a small garden. You can order seeds to plant, or you can transplant small plants to the site. Your main goal is to have some plants growing in your garden every season of the year.

Using the information in the paragraph above, answer the following question in essay format.

18. Based on what you know about plants and the climate where your school is located, what type of plants would be best to grow? Describe your plan in a well-organized essay, and be sure to explain how the different types of plants you plan to use will meet the criteria for the garden.

NEED EXTRA HELP?

If You Missed Question . . .	1	2	3	4	5	6	7	8	9	10	11	12	13	14	15	16	17	18
Review Section . . .	14.2	22.2	20.3	20.1	21.1, 21.3	19.4	22.2	21.1	21.2	20.1	22.2	21.4	22.1	21.1	22.2	17.3	8.2	21.2, 21.4

CHAPTER 23

Reproduction in Plants

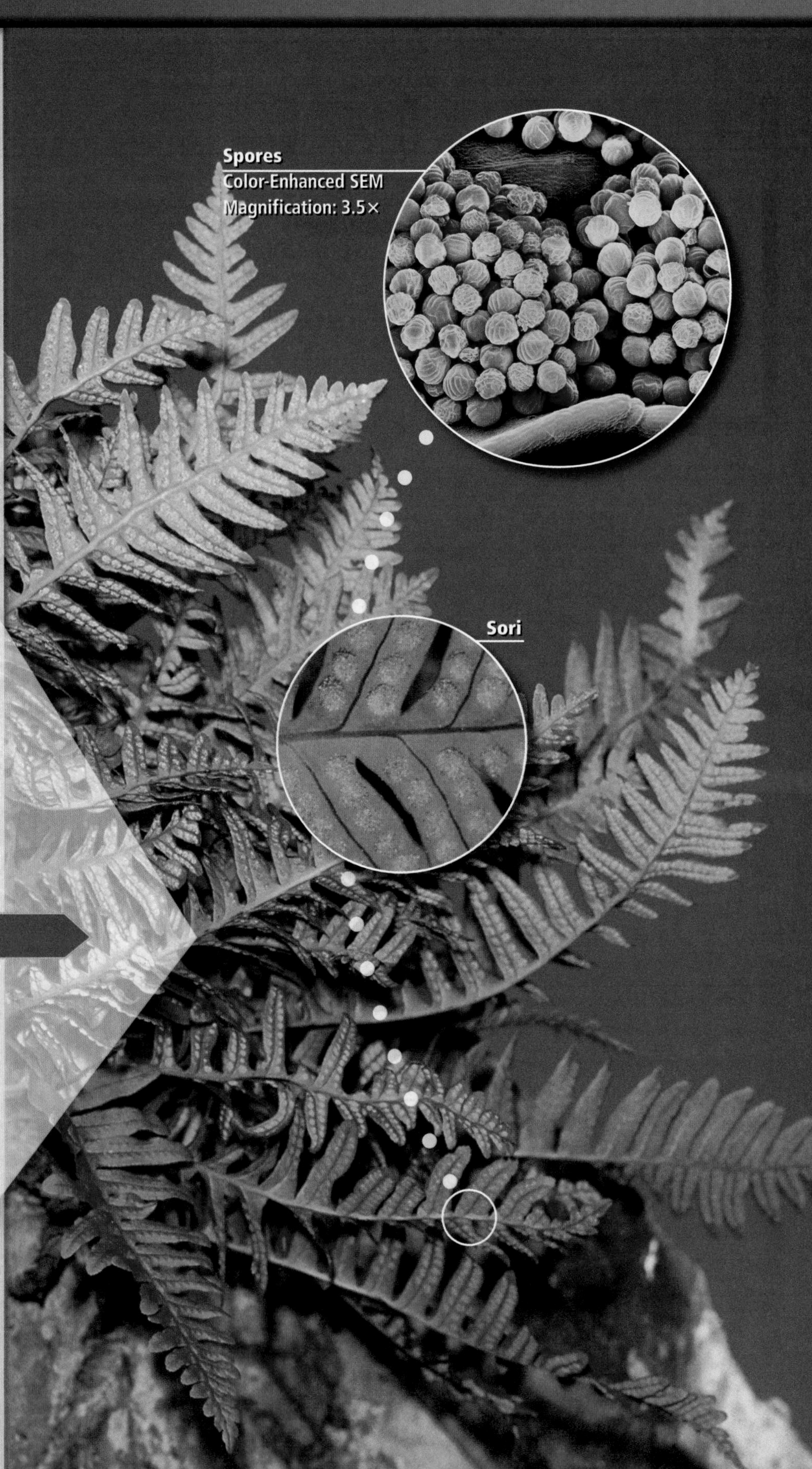

BIG Idea **The life cycles of plants include various methods of reproduction.**

Section 1
Introduction to Plant Reproduction

MAIN Idea Like all plants, the life cycles of mosses, ferns, and conifers include alternation of generations.

Section 2
Flowers

MAIN Idea Flowers are the reproductive structures of anthophytes.

Section 3
Flowering Plants

MAIN Idea In anthophytes, seeds and fruits can develop from flowers after fertilization.

BioFacts

- A moss or fern can produce millions of spores.
- Some cones can only open and release their seeds when heated by fire.
- The world's largest flower grows on the tropical plant *Rafflesia arnoldii,* and it smells like rotting meat.
- The largest seed is that of the coco de mer, *Lodoicea maldivica.* It can weigh more than 20 kg at maturity.

LAUNCH Lab

What are plant reproductive structures?

Have you ever noticed that sometimes flowers seem to appear suddenly on trees, shrubs, and other plants in the spring? Have you picked up a cone while walking under pine trees and wondered why these trees have cones? Like many organisms, plants have reproductive structures and reproduce sexually. Mosses, ferns, conifers, and flowering plants have unique reproductive structures. Investigate these structures during this lab.

Procedure

1. Read and complete the lab safety form.
2. Create a data table to record your observations and measurements of the plant reproductive structures your teacher gives you.
3. Observe the reproductive structures of a moss, fern, conifer, and flowering plant. Record your observations in your data table.

Analysis

1. **Identify** the similarities and differences in the reproductive structures of the plants.
2. **Describe** how flowering plants might use flowers to reproduce based on what you already know about plants.

Visit biologygmh.com to:

- study the entire chapter online
- explore Concepts in Motion, the Interactive Table, Microscopy Links, and links to virtual dissections
- access Web links for more information, projects, and activities
- review content online with the Interactive Tutor and take Self-Check Quizzes

Flowering Plant Life Cycle Make this Foldable to help you organize what you learn about the life cycle of a flowering plant.

STEP 1 Mark the center of a piece of notebook paper. Fold the top and bottom edges so that they meet at the center and form two equal tabs.

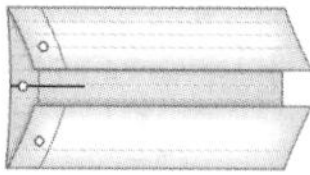

STEP 2 Fold the paper in half from side to side.

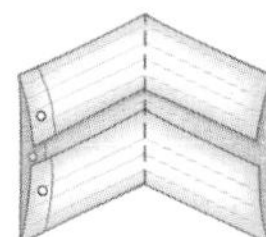

STEP 3 Open the fold and cut along the fold lines to form four tabs.

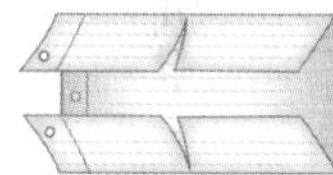

STEP 4 Using a colored pencil or pen, sketch and label the stages of the sporophyte generation for flowering plants on three of the tabs. Then, using a different colored pencil or pen, sketch and label gametophyte generation on the remaining tab.

FOLDABLES **Use this Foldable with Section 23.3.** As you study the section, diagram and record what you learn about alternation of generations in flowering plants.

Section 23.1

Reading Preview

Objectives

- **Summarize** forms of vegetative reproduction.
- **Review** the stages of alternation of generations.
- **Compare** reproduction of mosses, ferns, and conifers.

Review Vocabulary

flagellated: having one or more flagellum that propel a cell by whiplike motion

New Vocabulary

vegetative reproduction
chemotaxis
protonema
prothallus
heterosporous
megaspore
microspore
micropyle

Introduction to Plant Reproduction

MAIN Idea Like all plants, the life cycles of mosses, ferns, and conifers include alternation of generations.

Real-World Reading Link Have you ever seen photos of your friends when they were younger? Were you able to recognize most of them? Some plants differ greatly in appearances throughout their life stages. Recognizing that the different life stages of a plant are the same plant is not as easy as recognizing your friends from their old photos.

Vegetative Reproduction

Recall from Chapter 9 that reproduction without the joining of an egg and a sperm is called asexual reproduction. **Vegetative reproduction** is a form of asexual reproduction in which new plants grow from parts of an existing plant. The new plants are clones of the original plant because their genetic makeups are identical to the original plant.

There are several advantages of vegetative reproduction. It usually is a faster way to grow plants than from a spore or a seed. In Chapter 10 you read that an organism produced sexually will have a combination of features from its parents. However, plants produced vegetatively are more uniform than those that result from sexual reproduction. Also, some fruits do not produce seeds, and vegetative reproduction is the only way to reproduce them.

Naturally occurring vegetative reproduction There are many examples of natural vegetative reproduction. When conditions are dry, some mosses dry out, become brittle, and easily are broken and scattered by animals or wind. When conditions improve and water is available, some of these fragments can resume growth. Liverworts reproduce asexually by producing small, cuplike structures on the gametophyte thallus, as shown in **Figure 23.1.** Strawberry plants produce horizontal stems called stolons. A new strawberry plant can grow at the end of a stolon, and if the stolon is cut, the plant can continue to grow.

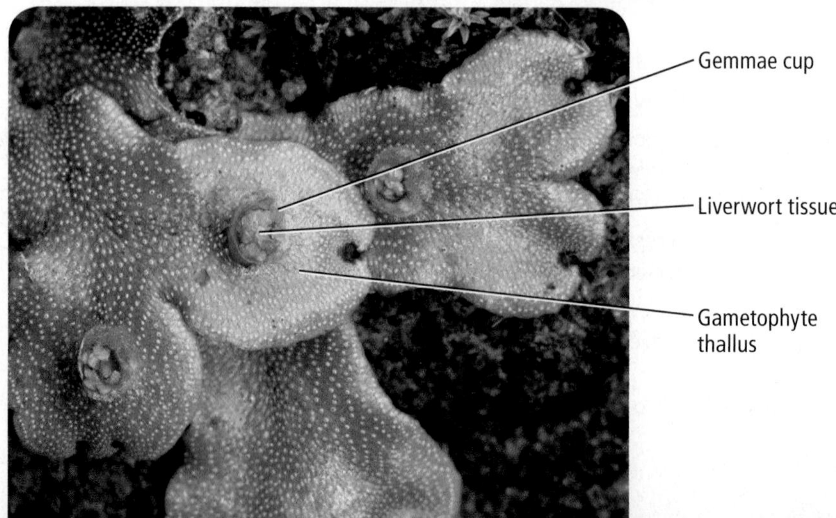

■ **Figure 23.1** Gemmae (JE mee) cups or splash cups contain small pieces of liverwort tissue. If knocked from or splashed out of the cup, they can grow into plants.
Infer *the genetic makeup of the new liverworts.*

Humans use vegetative reproduction Farmers, horticulturists, and scientists have been using vegetative reproduction for years. Leaves, roots, or stems, when cut from certain plants, can grow and become new plants if kept under proper environmental conditions. For example, white potatoes can be cut into sections. As long as each section contains an eye or bud and is planted in a favorable environment, a new plant can grow from the section and produce new potatoes. Some plants can be grown from a few cells of plant tissue using a technique called tissue culture. The plant tissue is grown on nutrient agar in sterile conditions, as shown in **Figure 23.2.** Eventually, hundreds of identical plants can be produced.

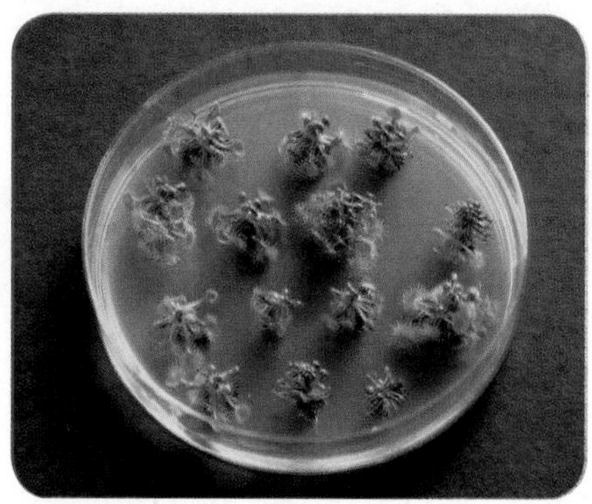

Figure 23.2 These cactus clones were produced using tissue culture techniques.

Alternation of Generations

As you read in Chapter 21, the life cycle of a plant includes an alternation of generations that has a diploid ($2n$) sporophyte stage, and a haploid (n) gametophyte stage. As shown in **Figure 23.3,** the sporophyte stage produces haploid spores that divide by mitosis and cell division and form the gametophyte generation. Depending on the plant species, the size of a gametophyte can be tiny or a larger structure. In the plant kingdom, there is an evolutionary trend for smaller gametophytes as plants become more complex.

The gametophyte stage produces gametes—eggs and sperm. One distinguishing characteristic among plants is how a sperm gets to an egg. Sperm of nonvascular plants and some of the vascular plants must have at least a film of water to reach an egg. Sperm of flowering plants do not need water for sperm to reach the egg.

Fertilization of an egg by a sperm forms a zygote that is the first cell of the sporophyte stage. As plants evolved and became more complex, sporophytes became larger. In addition to the size of the sporophyte, another distinguishing feature among plants is the growth pattern of the sporophyte. Flowering plants and other vascular plants have sporophytes that live completely independent of the gametophyte. Most nonvascular plants have sporophytes that depend on the gametophyte for support and food.

CAREERS IN BIOLOGY

Tissue-Culture Technician
Working with plant tissue while maintaining sterile conditions is one of many tasks performed by a technician in a tissue-culture lab. Besides knowledge of plants, a tissue-culture technician needs excellent eye-hand coordination, good concentration skills, and the ability to keep accurate records. For more information on biology careers, visit biologygmh.com.

Figure 23.3 The form of the sporophyte (blue) and gametophyte (yellow) is different for different plant species.

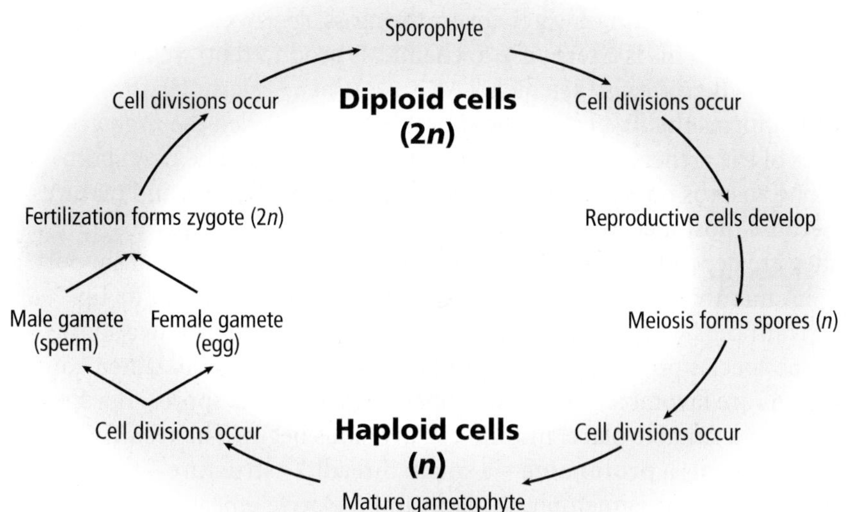

Concepts In Motion

Interactive Figure To see an animation of alternation of generations, visit biologygmh.com.

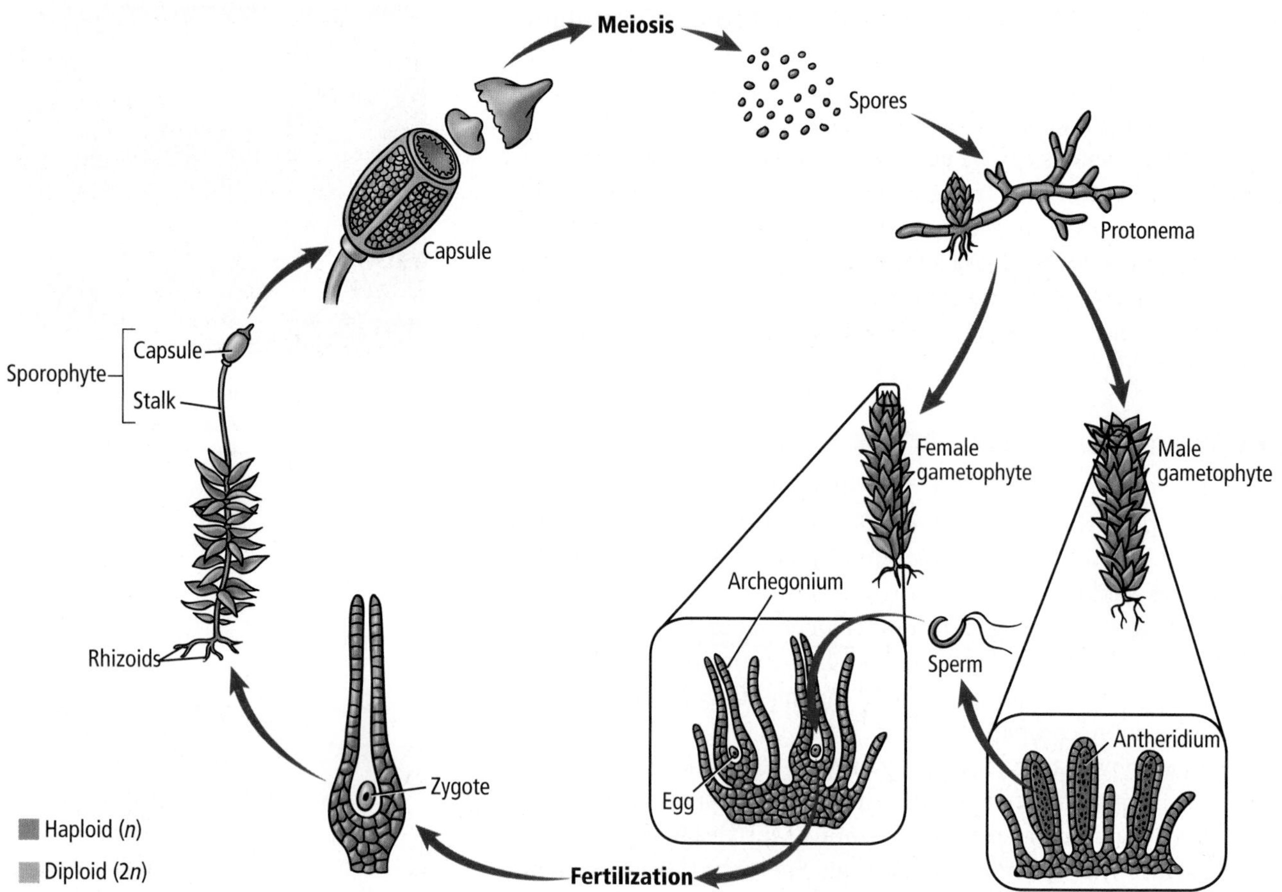

■ **Figure 23.4** Spores produced by a moss sporophyte grow into moss gametophyte plants. Following fertilization, the sporophyte develops while attached to the gametophyte and eventually will release spores, continuing the cycle.

Concepts In Motion

Interactive Figure To see an animation of a moss's life cycle, visit biologygmh.com.

VOCABULARY

WORD ORIGIN

Chemotaxis

chemo– comes from the late Greek word *chemeia,* meaning *alchemy*

–taxis comes from the Greek word *taxis,* meaning *responsive movement*

Moss Reproduction and Life Cycle

The reproduction and life cycle of mosses, as shown in **Figure 23.4,** exhibits alternation of generations and is typical of most nonvascular plants. The dominant stage is the gametophyte stage that you might see growing in damp shady places or on rocks along a stream. Gametophytes produce archegonia and antheridia. These structures can be on the same moss plant or, as is often the case, on separate plants. Depending on the moss species, an archegonium produces one or more eggs. The tissues of the archegonium surround the egg or eggs with a protective layer.

Antheridia produce flagellated sperm that need water to get to the archegonia. If a film of water covers the moss, sperm can move toward archegonia. This is a response to chemicals produced by archegonia and is called **chemotaxis** (kee moh TAK sus). When a sperm fertilizes an egg, it forms the first cell of the sporophyte stage called the zygote. Tissues of the archegonium protect the new sporophyte. The new sporophyte absorbs nutrients from the archegonium as it grows and matures. Because most mature moss sporophytes cannot undergo photosynthesis, they are dependent upon their gametophytes for nutrition and support.

A mature sporophyte consists of a stalk with a capsule at its tip. Certain cells within the capsule undergo meiosis and produce spores. Some species produce up to 50 million spores per capsule. When conditions are favorable, the capsule opens, releasing the spores. If a spore lands in a suitable place, mitotic cell divisions begin. The resulting growth forms a **protonema**—a small, threadlike structure—that can develop into the gametophyte plant, and the cycle repeats.

Fern Reproduction and Life Cycle

When you visit a forest or a plant conservatory, you might see the lacy fronds of ferns. Fronds are part of a fern's sporophyte stage. If you look closely at a frond, you might find spore-producing structures called sori on it. Each sorus consists of sporangia. Certain cells in a sporangium undergo meiosis and the resulting spores are the beginning of the new gametophyte generation.

If a fern spore lands on damp, rich soil, it can grow and form a tiny heart-shaped gametophyte called a **prothallus** (pro THA lus) *(plural, prothalli)*, as shown in **Figure 23.5.** Cells of the prothallus contain chloroplasts; therefore, photosynthesis can occur. Most prothalli develop both antheridia and archegonia. Antheridia produce flagellated sperm that need water to move to archegonia. Each archegonium contains one egg. If fertilization occurs, the resulting zygote is the first cell of the sporophyte generation. Chemical reactions between sperm and eggs of the same prothallus can prevent fertilization.

The zygote undergoes mitotic cell divisions and forms a photosynthetic, multicellular sporophyte. Initially, the sporophyte grows on the prothallus and receives support and nutrition. Later, the prothallus disintegrates and the sporophyte develops fronds and a rhizome—a thick underground stem that produces roots and supports the photosynthetic fronds.

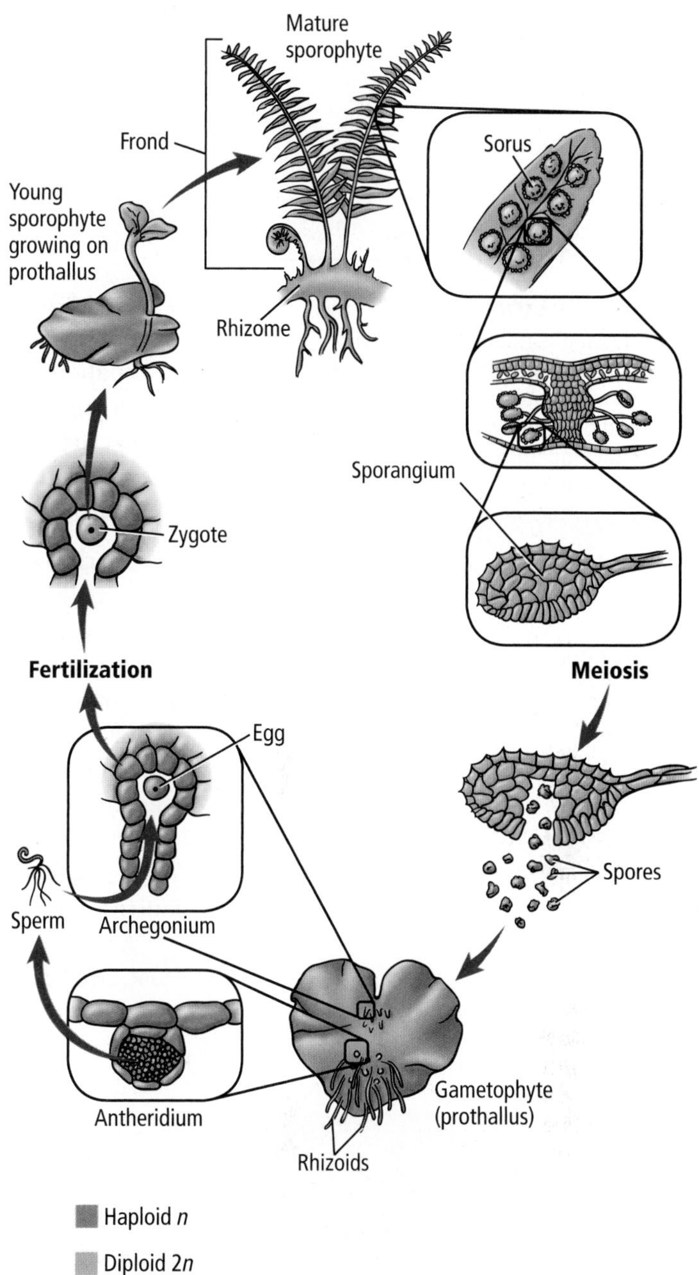

Figure 23.5 There usually is a substantial size difference between the sporophyte and gametophyte stages of ferns.

Conifer Reproduction and Life Cycle

Have you ever seen the surface of a car or a pond covered with fine yellow dust? It's possible that this dust came from one or several plants called conifers. The tree or shrub that you might recognize as a pine or other conifer is that plant's sporophyte generation. Conifers, like a few lycophytes and pterophytes, are **heterosporous** (he tuh roh SPOR us)—they produce two types of spores that develop into male or female gametophytes.

Female cones Each female cone is composed of many scales. At the base of each scale are two ovules. Within each ovule, meiosis of a cell in the megasporangium produces four **megaspores.** Three of these megaspores disintegrate. The remaining megaspore undergoes mitotic cell divisions and becomes the female gametophyte. When fully developed, the female gametophyte consists of hundreds of cells and contains two to six archegonia. Each archegonium eventually contains an egg.

Male cones The pollen-producing cone, commonly called the male cone, consists of small reproductive scales that have hundreds of sporangia. Certain cells in these sporangia undergo meiosis and form **microspores.** A pollen grain—the male gametophyte— consists of four cells and develops from a microspore. Pollen grains are transported on air currents.

 **Reading Check** **Compare** the sizes of a moss gametophyte and a pine gametophyte.

Study Tip

Group Study With two other classmates, study the life cycles of mosses, ferns, and conifers. Each of you should choose one of these life cycles and prepare a summary of it. When you study together, teach each other your life cycle summaries.

Pollination When a pollen grain from one species of seed plants lands on the female reproductive structure of a plant of the same species, pollination occurs. If a conifer pollen grain lands near the **micropyle,** or the opening of the ovule, it can be trapped in a sticky substance called a pollen drop. As the pollen drop slowly evaporates or is absorbed into the ovule, the pollen grain is pulled closer to the micropyle. Over the next year, the pollen grain will continue to develop.

Seed development Following pollination, the pollen grain generates a pollen tube. It grows through the micropyle, and into the ovule. This process can take a year or longer. One of the four cells in the pollen grain undergoes mitosis, forming two nonflagellated sperm. The sperm travel in the pollen tube to an egg, as shown in **Figure 23.6.** Fertilization occurs when an egg and a sperm join to form the zygote. The remaining sperm and the pollen tube disintegrate. The zygote is dependent on the female gametophyte for nutrition as it undergoes mitotic cell divisions that result in the formation of an embryo with one or more cotyledons. These undergo photosynthesis and provide nutrition for the embryo when the seed sprouts.

As the embryo develops, the outside layer of the ovule forms a seed coat. Seed development can take as long as three years. When seeds mature, the female cone opens and releases them.

MiniLab 23.1

Compare Conifer Cones

How do cones from the different conifers compare? Have you ever noticed the many different types of cones that fall from conifers? Investigate the types of cones during this lab.

Procedure

1. Read and complete the lab safety form.
2. Create a data table for recording your observations, measurements, and comparisons of cones.
3. Obtain **cones** from your teacher.
4. Observe the physical characteristics of your cones and record your observations and measurements in your data table. Do not damage the cones in any way.
5. Identify the conifer species of your cones by using a **tree identification guidebook.** Record this data.
6. Return the cones to your teacher.

Analysis

1. **Compare and contrast** the cones.
2. **Describe** Were there any seeds present? How do you think seeds form in conifers?

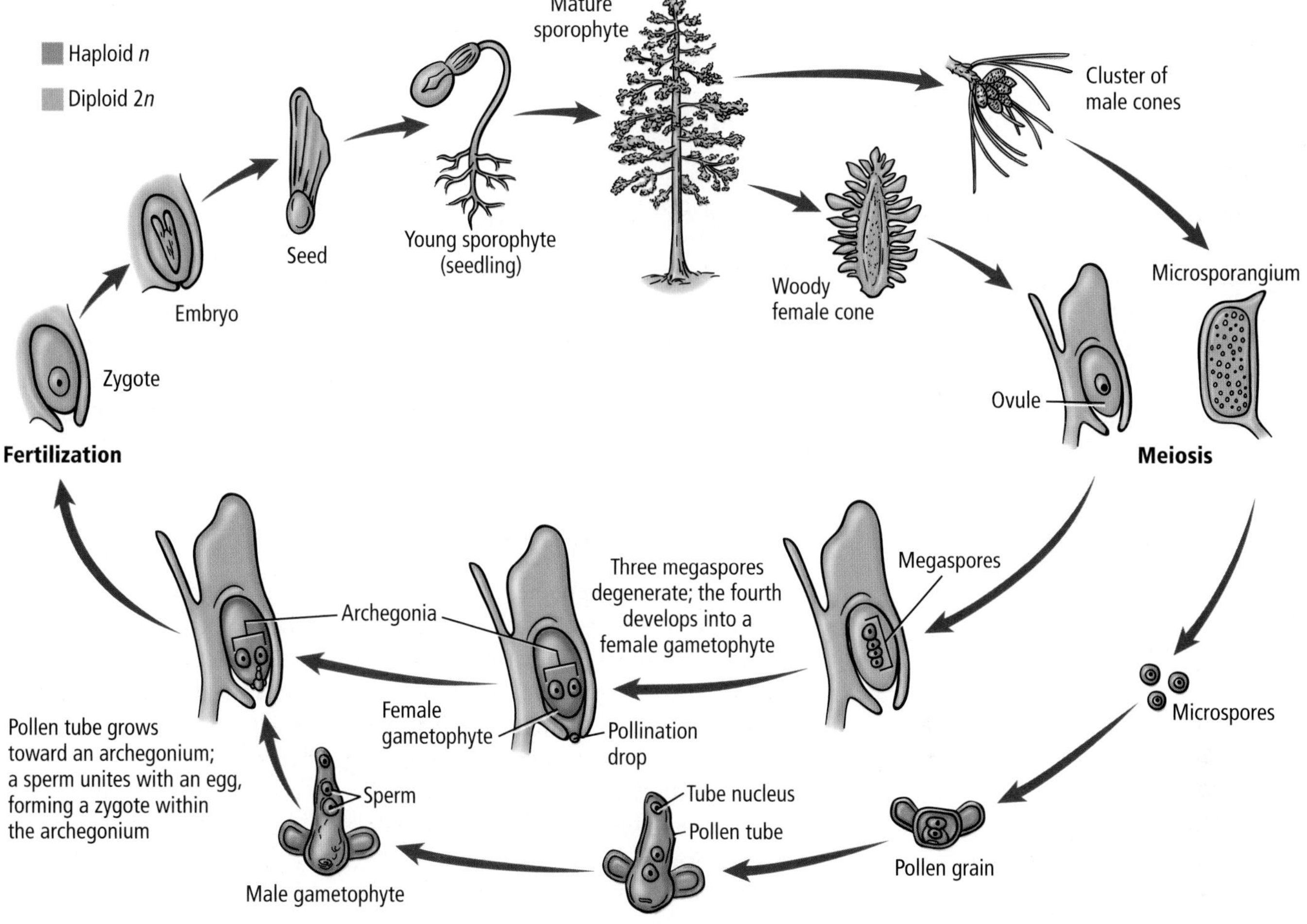

Figure 23.6 The sporophyte generation is dominant in the life of a conifer.

Concepts In Motion

Interactive Figure To see an animation of a conifer's life cycle, visit biologygmh.com.

Reproduction in conifers is diverse. The time for a conifer life cycle varies from species to species. Not all conifers produce cones. For example, yews produce ovules covered by fleshy tissue. Juniper seed cones look like berries. Regardless of differences, conifer reproduction ensures the survival of this plant division.

Section 23.1 Assessment

Section Summary

- Vegetative reproduction produces new plants without sexual reproduction.
- The moss sporophyte depends on the gametophyte.
- A fern sporophyte can live independently of the gametophyte.
- Conifer gametophytes develop within sporophyte tissues.

Understand Main Ideas

1. MAIN Idea **Describe** the stages of alternation of generations.
2. **List** advantages of vegetative reproduction.
3. **Explain** how the fern sporophyte is dependent upon the gametophyte.
4. **Compare and contrast** the life cycles of mosses and conifers.

Think Critically

5. **Determine** how the distribution of conifers might be affected if water was needed for reproduction.

MATH in Biology

6. Calculate the number of spores that could be released in three square meters if the density of moss plants is 100 plants per square meter and the average number of spores released per plant is 10,000.

Section 23.2

Flowers

Reading Preview

Objectives

- **Identify** the parts of a flower and their functions.
- **Describe** complete, incomplete, perfect, and imperfect flowers.
- **Distinguish** between monocot and eudicot flowers.
- **Relate** the pollination mechanism of a flower to its structure.
- **Explain** photoperiodism.

Review Vocabulary

nocturnal: active only at night

New Vocabulary

sepal
petal
stamen
pistil
photoperiodism
short-day plant
long-day plant
intermediate-day plant
day-neutral plant

MAIN Idea **Flowers are the reproductive structures of anthophytes.**

Real-World Reading Link Have you ever worn a corsage or boutonniere to a dance? Perhaps you have given a flower to someone to let him or her know that he or she is special to you. You probably can think of many other instances when flowers were important to you. However, from a scientific viewpoint, the most important role of flowers is in anthophyte sexual reproduction.

Flower Organs

Vivid orange, deep purple, ghostly white, fragrant, rancid, spectacular, and inconspicuous—these all are terms that can be used to describe flowers. The colors, shapes, and sizes of flowers are determined by each species' genetic makeup. It is important to remember that flowers can vary in structure and form from species to species.

Flowers have several organs. Some organs provide support or protection, while others can be involved directly in reproduction. In general, flowers have four organs—sepals, petals, stamens, and one or more pistils, illustrated in **Figure 23.7.** **Sepals** protect the flower bud and can look like small leaves or even resemble the flower's petals. **Petals** usually are colorful structures that can both attract pollinators and provide them with a landing platform. Sepals and petals, if present, are attached to a flower stalk, called a peduncle.

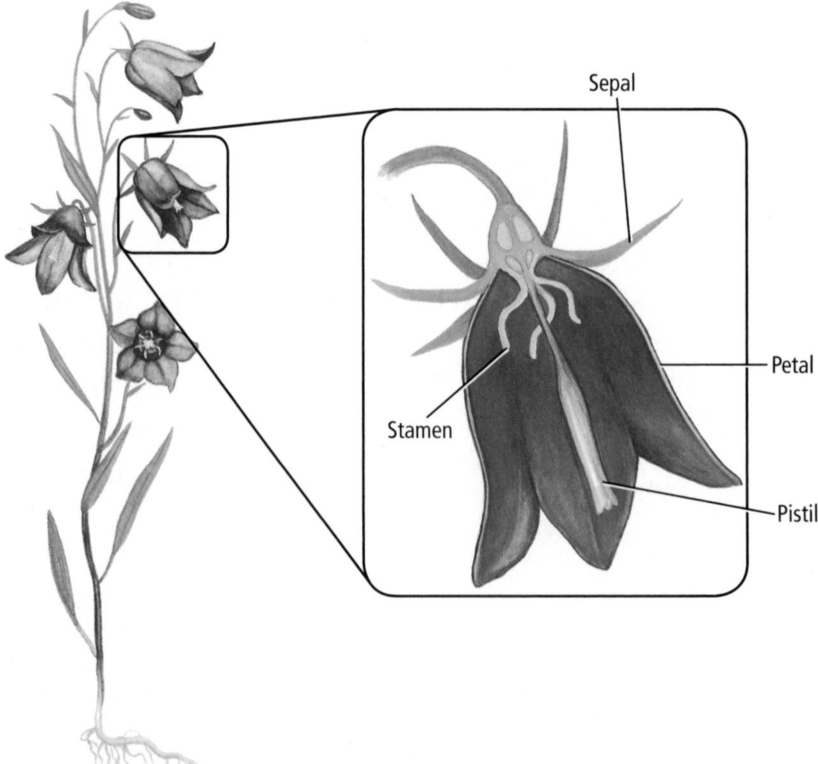

Figure 23.7 The typical flower has four organs—sepals, petals, stamens, and one or more pistils.

Concepts In Motion

Interactive Figure To see an animation of the organs of a flower, visit biologygmh.com.

Most flowers have several **stamens**—the male reproductive organs. A stamen is composed of two parts—the filament and the anther. The filament, or stalk, supports the anther. Inside the anther are cells that undergo meiosis and then mitotic cell divisions, forming pollen grains. Two sperm eventually form inside each pollen grain.

The female reproductive structure of a flower is the **pistil.** In the center of a flower is one or more pistils. A pistil usually has three parts—the stigma, the style, and the ovary. The stigma is the tip of the pistil and is where pollination occurs. The style is the part that connects the stigma to the ovary that contains one or more ovules. A female gametophyte develops in each ovule, and an egg forms inside each female gametophyte.

VOCABULARY

SCIENCE USAGE V. COMMON USAGE

Stigma

Science usage: the tip of a flower's pistil where pollination occurs

The stigma of an iris's pistil has three parts.

Common usage: a mark of shame or discredit.

A criminal record often is a stigma for an individual trying to reenter society

Flower Adaptations

The flower organs described in the previous paragraphs are typical of most flowers. However, many flowers can have modifications to one or more organs. Scientists categorize flowers using these modifications.

Structural differences Flowers that have sepals, petals, stamens, and one or more pistils are called complete flowers. If a flower is missing one or more of these organs, it is an incomplete flower. For example, wild ginger flowers are called incomplete because they have no petals. Other descriptive terms relating to flower organs are perfect and imperfect. Flowers that have both stamens and pistils are called perfect flowers. Some plants, such as cucumbers and squash, have imperfect flowers. An imperfect flower has either functional stamens or pistils. The stamen-containing, or male, flowers release pollen grains. Following fertilization, a fruit forms from the pistil-containing, or female, flowers.

The number of each flower organ varies from species to species. However, the number of flower organs distinguishes eudicots from monocots. When the petal number for a flower is a multiple of four or five, the plant usually is a eudicot. The number of other organs—the sepals, pistils, and stamens—is often the same multiple of four or five. For example, the members of the mustard family of plants have flowers with four sepals and petals, as shown in **Figure 23.8.** Monocots generally have flower organs in multiples of three. The daylily, also shown in **Figure 23.8,** has three sepals and petals and six stamens.

Figure 23.8 Some plants can be identified as either a monocot or a eudicot by their flowers.

This plant is related to those whose seeds are used to make canola oil.

At a glance, this daylily's petals and sepals are indistinguishable.

Visualizing Pollination

Figure 23.9
Flowers have several adaptations that ensure pollination. Pollen might be carried by the wind or by animals. While feeding, an animal can become covered with pollen and can transfer the pollen to the next flower it visits.

Wind disperses lightweight oak pollen that can cause allergic reactions for many people. Tassels hang down and can wave in the wind.

Hummingbirds are attracted to red flowers. The hummingbird's long beak reaches nectar at the base of this flower. Some yellow and orange pigments reflect light in ranges invisible to the human eye. Even so, the markings are highly visible to bees and other insects.

As night falls, heavy scents and pale colors make it easier for moths to locate certain flowers.

The carrion flower has a rancid odor that attracts fly and beetle pollinators.

Nectar producing flowers often attract insect pollinators as they seek food.

Concepts In Motion **Interactive Figure** To see an animation of pollination, visit biologygmh.com.

Biology Online

Pollination mechanisms Different anthophyte species have flowers of distinctive sizes, shapes, colors, and petal arrangements. Many of these adaptations relate to pollination. Some of these adaptations are shown in **Figure 23.9.**

Self pollination and cross pollination Connection to History
Recall from Chapter 10 that Mendel knew that pea flowers tend to self-pollinate, but also can be cross-pollinated. Self-pollinating flowers can pollinate themselves or another flower on the same plant. Cross-pollinated flowers receive pollen from another plant. Some flowers must be cross-pollinated. This is one reason that pollinators play important roles in anthophyte reproduction. Pollinators provide a way to transfer pollen for flowers that must be cross-pollinated. Pollinators also ensure that reproduction can occur for imperfect flowers, like squash blossoms, as shown in **Figure 23.10.**

Animal pollination As shown in **Figure 23.9,** many animal-pollinated flowers are brightly colored, have strong scents, or produce a sweet liquid called nectar. When insects and other small animals move from flower to flower searching for nectar, they can carry pollen from one flower to another flower. Other insects collect pollen for food. The bright colors and sweet scents of peonies, roses, and lilacs attract insects such as bees, butterflies, beetles, and wasps. White or pale yellow flowers are more visible at dusk and at night, and attract nocturnal animals, such as moths and bats. The fruity smell of some flowers attracts fruit-eating bats that act as the flowers' pollinators. On the first page of this chapter, you read about *Rafflesia,* a flower. It gives off the odor of rotting meat. Flowers with this trait attract fly pollinators. Bird-pollinated flowers often give off little or no aroma. A bird generally has a poor sense of smell, so it usually locates flowers by sight.

Wind pollination Flowers that generally lack showy or fragrant floral parts, also shown in **Figure 23.9,** usually are wind-pollinated. They produce huge amounts of lightweight pollen. This helps to ensure that some pollen grains will land on the stigma of a flower of the same species. Also, the stamens of wind-pollinated flowers often hang below the petals, exposing them to the wind. The stigma of a wind-pollinated flower often is large, which helps to ensure that a pollen grain might land on it. Wind-pollinated plants include most trees and grasses.

CAREERS IN BIOLOGY

Plant Breeder Knowledge of flower structures, pollination mechanisms, and genetics are essential for a plant breeder. A plant breeder conducts selective breeding by choosing plants with desirable traits, breeding them together, and then recording results. For more information on biology careers, visit biologygmh.com.

Figure 23.10 Honeybees or other insects must transfer pollen from the male squash flower to the female squash flower for the fruit—a squash—to form.
Determine ***Are squash flowers perfect or imperfect? Explain.***

LAUNCH Lab

Review Based on what you've read about plant reproduction, how would you now answer the analysis questions?

Photoperiodism After noticing that certain plants only flowered at certain times of the year, plant biologists conducted experiments to explain this observation. The research initially focused on the number of hours of daylight to which the plants were exposed. However, researchers discovered that the critical factor that influenced flowering was the number of hours of uninterrupted darkness, not the number of hours of daylight. This flowering response is known as **photoperiodism** (foh toh PIHR ee uh dih zum). Scientists also learned that the beginning of flower development for each plant species was a response to a range in the number of hours of darkness. This range of hours is called the plant's critical period.

Botanists classify flowering plants into one of four different groups—short-day plants, long-day plants, intermediate-day plants, or day-neutral plants. This classification is based on the critical period. The names reflect the researchers' original focus—the number of hours of daylight. It is important to remember that a more accurate term for a short-day plant, for example, would be a long-night plant. As you read the descriptions of these plants, refer to **Figure 23.11.**

Short-day photoperiodism A **short-day plant** flowers when exposed daily to a number of hours of darkness that is greater than its critical period. For example, a short-day plant could flower when exposed to 16 hours of darkness. Short-day plants flower during the winter, spring, or fall, when the number of hours of darkness is greater than the number of hours of light. Some short-day plants you might recognize are pansies, poinsettias, tulips, and chrysanthemums.

Long-day photoperiodism A **long-day plant** flowers when the number of hours of darkness is less than its critical period. These plants flower during the summer. Examples of long-day plants are lettuce, asters, coneflowers, spinach, and potatoes.

MiniLab 23.2

Compare Flower Structures

How do the structures of flowers vary? Just a quick browse through a flower garden or florist's shop reveals that there is great diversity among flowers. Investigate how flowers differ from species to species.

Procedure

1. Read and complete the lab safety form.
2. Create a data table to record your observations and measurements.
3. Obtain the **flowers** for this lab from your teacher.
4. Observe the differences in structure, color, size, and odor of the flowers. Do not damage the flowers in any way.
5. Make a sketch of each flower and record other observations in your data table.
6. Return the flowers to your teacher.

Analysis

1. **Compare and contrast** the flower structures you observed.
2. **Infer** why the flower petals that you observed were different colors.
3. **Propose** an explanation for the different sizes and shapes of flower structures.

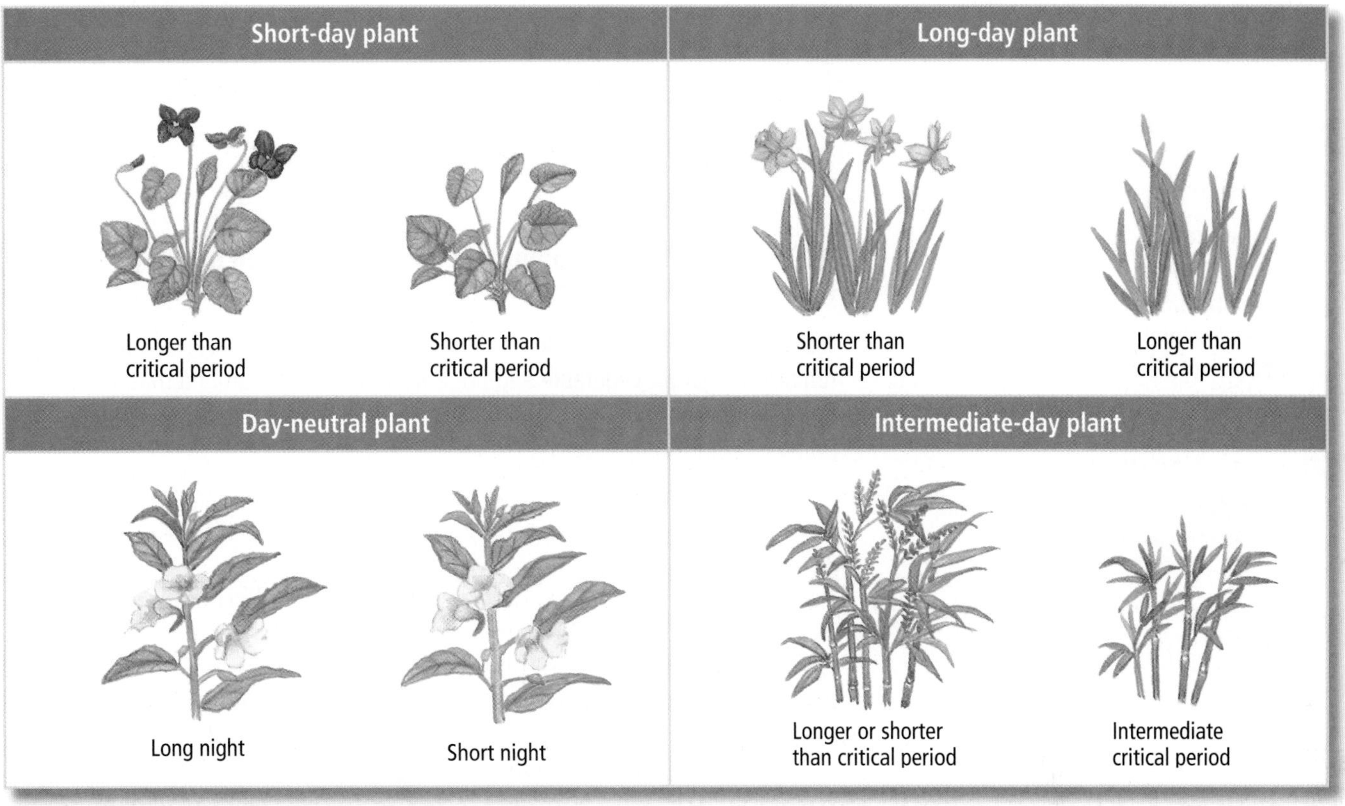

Short-day plant

Longer than critical period

Shorter than critical period

Long-day plant

Shorter than critical period

Longer than critical period

Day-neutral plant

Long night

Short night

Intermediate-day plant

Longer or shorter than critical period

Intermediate critical period

Intermediate-day photoperiodism Many plants that are native to tropical regions are **intermediate-day plants.** This means that they will flower as long as the number of hours of darkness is neither too great nor too few. Sugarcane and some grasses are examples of intermediate-day plants.

Day-neutral photoperiodism Some plants will flower regardless of the number of hours of darkness as long as they receive enough light for photosynthesis that supports growth. A plant that flowers over a range in the number of hours of darkness is a **day-neutral plant.** Buckwheat, corn, cotton, tomatoes, and roses are examples of day-neutral plants.

■ **Figure 23.11** A plant's critical period determines when the plant will flower.

Section 23.2 Assessment

Section Summary

- A typical flower has sepals, petals, stamens, and one or more pistils.
- Flower form differs from species to species.
- Some flower modifications distinguish monocots from eudicots.
- Modifications make flowers more attractive to pollinators.
- Photoperiodism can influence when a plant flowers.

Understand Main Ideas

1. MAIN Idea **Compare and contrast** the function of each of the four organs of a typical flower.
2. **Describe** traits of a typical monocot flower and a typical eudicot flower.
3. **Compare and contrast** complete and incomplete flowers.
4. **Predict** which type of photoperiodism should produce blooms at this time of the year.

Think Critically

5. **Design a plan** to develop flowers on long-day plants during the winter.
6. **Assess** the importance of pollinators for imperfect flowers.

WRITING in **Biology**

7. Write a description, from the point of view of a pollinator, of a visit to a flower.

Section 23.3

Flowering Plants

Reading Preview

Objectives

- **Sequence** the life cycle of a flowering plant.
- **Describe** the process of fertilization and seed formation in flowering plants.
- **Summarize** seed germination.

Review Vocabulary

cytoskeleton: the long, thin protein fibers that form a cell's framework

New Vocabulary

polar nuclei
endosperm
seed coat
germination
radicle
hypocotyl
dormancy

MAIN Idea In anthophytes, seeds and fruits can develop from flowers after fertilization.

Real-World Reading Link In 1893, the U.S. Supreme Court ruled that a tomato is legally a vegetable and not a fruit. The justices argued that a tomato is not a fruit because it is not sweet. As you read this section, decide whether this ruling is scientifically accurate.

Life Cycle

Anthophytes are the most diverse and widespread group of plants. They are unique because they have flowers. Anthophytes have distinctive life cycles and, like all plants, exhibit an alternation of generations. Like conifers, the sporophyte generation of anthophytes is dominant and supports the gametophyte generation. However, there are many variations of the anthophyte reproductive process.

Gametophyte development In anthophytes, the development of male and female gametophytes begins in an undeveloped flower. Anthophytes are heterosporous—pistils produce megaspores, and stamens produce microspores. A specialized cell in the ovule of a pistil's ovary undergoes meiosis, producing four megaspores. Usually, three of these megaspores disintegrate and disappear. The nucleus of the functional megaspore undergoes mitosis. Mitotic division continues and the megaspore grows until there is one large cell with eight nuclei. As shown in **Figure 23.12,** two nuclei migrate toward the center and membranes form around the other six nuclei. The result is three nuclei at each end of the cell and two nuclei in the center called **polar nuclei.** One of the three nuclei at the end closest to the micropyle becomes the egg. The cell that contains the egg and seven nuclei is the female gametophyte.

The development of the female gametophyte and the male gametophyte might or might not occur at the same time. Within the anther, specialized cells undergo meiosis and produce microspores. As shown in **Figure 23.13,** the nucleus in each microspore undergoes mitosis that forms two nuclei called the tube nucleus and the generative nucleus. A thick, protective cell wall forms around a microspore. At this point, the microspore is an immature male gametophyte, or pollen grain.

■ **Figure 23.12** The megaspore results from meiosis, and the egg results from mitosis. This plant has 12 chromosomes.
Infer *the chromosome number of the egg.*

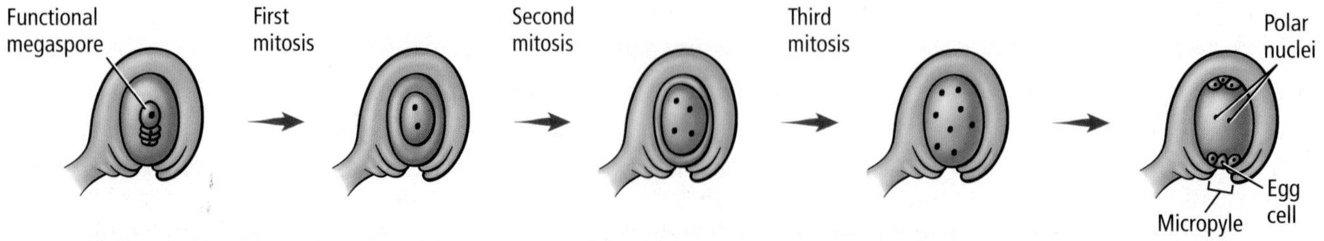

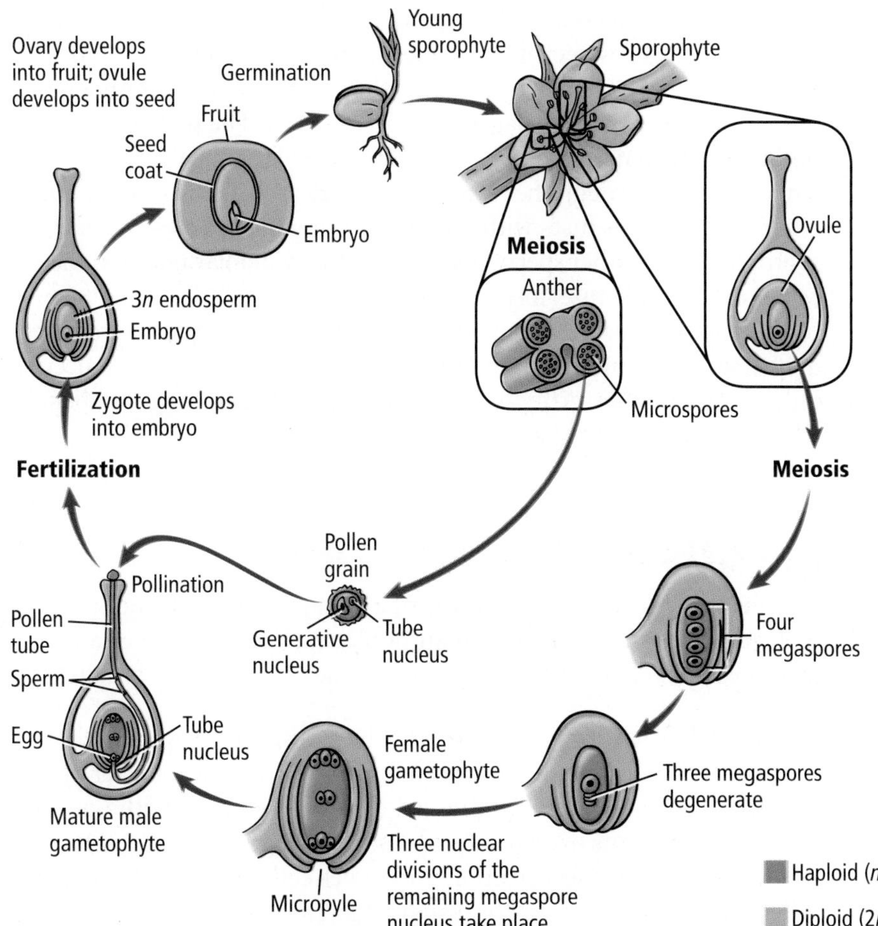

■ **Figure 23.13** The life cycle of a flowering plant, like a peach, includes gametophyte and sporophyte generations. The male and female gametophytes are surrounded by sporophyte tissue.

Scientists can identify the family or genus of a pollen grain by the distinctive outer layer of its cell wall called the exine. This characteristic is useful to paleontologists and forensic investigators. Paleontologists can trace the agricultural history of certain regions using pollen fossils. For over 50 years, forensic scientists have used pollen evidence to help determine where and when some crimes were committed.

Pollination and fertilization Earlier in this chapter, you learned that various flower adaptations help to ensure the successful transfer of pollen from the anther to the stigma of the pistil. Once pollination occurs, the pollen grain can form a pollen tube—an extension of the pollen grain. Usually, the pollen tube grows down through the style to the ovary and the two nuclei travel in the pollen tube toward the ovule.

Connection to Chemistry The pollen grain's exine can contain compounds that react with compounds of the pistil's stigma. These reactions can stimulate or inhibit the growth of the pollen tube. For example, in some poppies, a chemical reaction disrupts the formation of the pollen grain's cytoskeleton. This inhibits the pollen tube's growth. Different mechanisms prevent incompatible pollen from producing a functional pollen tube.

When a compatible pollen grain lands on a stigma, the pollen grain absorbs substances from the stigma and a pollen tube starts to form, also shown in **Figure 23.13.** The tube nucleus directs the growth of the pollen tube. However, recent research suggests that the growth of the pollen tube toward the ovule is a chemotaxic response. In some plants, it has been found that calcium affects the direction of the pollen tube's growth.

FOLDABLES
Incorporate information from this section into your Foldable.

VOCABULARY

ACADEMIC VOCABULARY

Compatible:
Capable of functioning together.
Because agricultural corn's pollen is compatible with sweet corn's pollen, the two crops must be planted some distance apart to prevent contamination of the sweet corn.

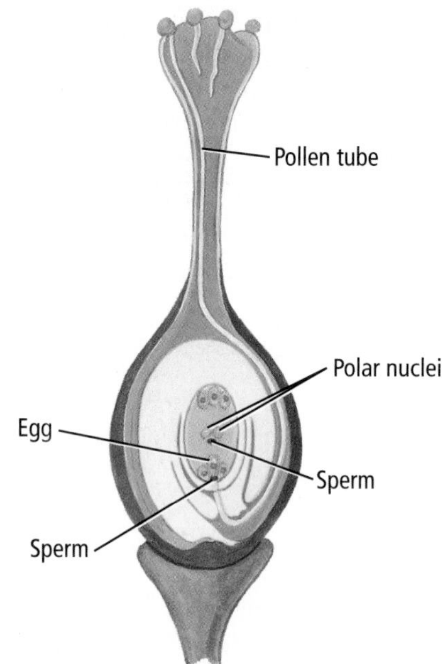

■ **Figure 23.14** Double fertilization results in the formation of diploid and triploid tissues.

Concepts In Motion

Interactive Figure To see an animation of double fertilization in flowering plants, visit biologygmh.com.

The length of a pollen tube depends on the length of the pistil, and can vary from a few centimeters or less to more than 50 cm in some corn plants. As the pollen tube grows, the generative nucleus undergoes mitosis, forming two nonflagellated sperm nuclei. The pollen grain is now a mature male gametophyte. When the pollen tube reaches the ovule, it grows through the micropyle and releases the two sperm nuclei. One sperm nucleus fuses with the egg, forming the zygote—the new sporophyte. The other sperm nucleus and the two polar nuclei in the center of the ovule fuse, forming a triploid or $3n$ cell.

Because two fertilizations occur in an anthophyte egg, this is called double fertilization, shown in **Figure 23.14.** Double ferilization occurs only in anthophytes. After fertilization, the ovule and the ovary begin to develop into the seed and fruit, respectively.

Results of Reproduction

Fertilization is only the beginning of a long process that finally ends with the formation of a seed. In anthophytes, a seed is part of a fruit that develops from the ovary and sometimes other flower organs.

Seed and fruit development The sporophyte begins as a zygote—or a $2n$ cell. Numerous cell divisions produce a cluster of cells that eventually develops into an elongated embryo with one cotyledon in monocots or two cotyledons in eudicots. The $3n$ cell formed as a result of double fertilization undergoes cell divisions. A tissue called the **endosperm** (EN duh spurm) forms as a result of these divisions and provides nourishment for the embryo. Initially, these cell divisions occur rapidly without cell wall formation. As the endosperm matures, cell walls form. In some monocots, the endosperm is the major component of the seed and makes up most of the seed's mass. For example, the coconut palm is a monocot. The liquid inside a fresh coconut is liquid endosperm—cells without cell walls. In eudicots, the cotyledons absorb most of the endosperm tissue as the seed matures. Therefore, the cotyledons of eudicot seeds provide much of the nourishment for the embryo. Examples of eudicot and monocot seeds are shown in **Figure 23.15.**

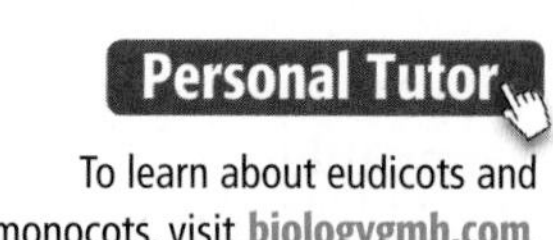

To learn about eudicots and monocots, visit biologygmh.com.

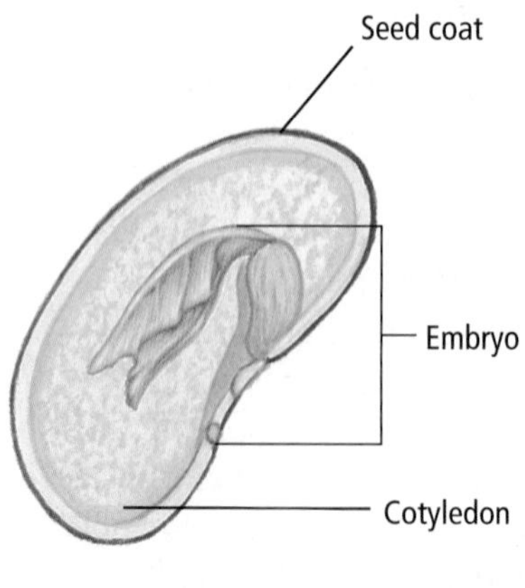

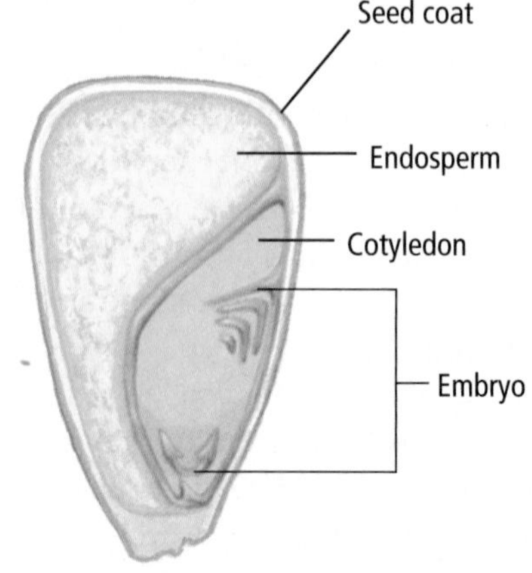

■ **Figure 23.15** Seeds of monocots differ from those of eudicots.
Identify ***the embryo's food source in each seed.***

As the endosperm matures, the outside layers of the ovule harden and form a protective tissue called the **seed coat.** You might notice the seed coats of beans or peas when you eat them. The seed coat is the thin, outer covering that often comes off or loosens as seeds are cooked.

Have you ever eaten a tomato or cucumber and noticed the number of seeds inside? Depending on the plant, the ovary can contain one ovule or hundreds. As the ovule develops into a seed, changes occur in the ovary that lead to the formation of a fruit.

Fruits form primarily from the ovary wall. In some cases, the fruit consists of the ovary wall and other flower organs. For example, the seeds of the apple are within the core that develops from the ovary. The juicy tissue that we eat develops from other flower parts.

Besides the apple, other fruits, such as peaches and oranges, are fleshy, while some are dry and hard, such as walnuts and grains. Study **Table 23.1** to learn about types of fruit.

Reading Check **Compare and contrast** the formation of a seed and a fruit.

CONCEPTS In MOtion

Interactive Table To explore more about types of fruit, visit Tables at biologygmh.com.

Table 23.1 Types of Fruit

Fruit Type	Example of Flower and Fruit	Description
Simple fleshy fruits	**Peach**	Simple fleshy fruits can contain one or more seeds. Apples, peaches, grapes, oranges, tomatoes, and pumpkins are simple fleshy fruits.
Aggregate fruits	**Raspberry**	Aggregate fruits form from flowers with multiple female organs that fuse as the fruits ripen. Strawberries, raspberries, and blackberries are examples of aggregate fruits.
Multiple fruits	**Pineapple**	Multiple fruits form from many flowers that fuse as the fruits ripen. Figs, pineapples, mulberries, and osage oranges are examples of multiple fruits.
Dry fruits	**Redbud**	When mature, these fruits are dry. Examples of dry fruits include pods, nuts, and grains.

Seed dispersal In addition to providing some protection for seeds, fruits also help disperse seeds. Dispersal of seeds away from the parent plant increases the survival rate of offspring. For example, when many plants are growing in one area, there is competition for light, water, and soil nutrients. Seeds sprouting next to parent plants and with other offspring compete for these resources.

Fruits that are attractive to animals can be transported great distances away from the parent plant. Animals that gather and bury or store fruits usually do not recover all of them, so the seeds might sprout. Some of the animals, such as deer, bears, and birds, consume fruits. The seeds pass through their digestive tracts undamaged and then are deposited on the ground along with the animals' wastes. Some seeds have structural modifications that enable them to be transported by water, animals, or wind. You can review seed dispersal in Chapter 21.

Seed germination When the embryo in a seed starts to grow, the process is called **germination.** There are a number of factors that affect germination, including the presence of water and/or oxygen, temperature, and those described in **Data Analysis Lab 23.1.** Most seeds have an optimum temperature for germination. For example, some seeds can germinate when soil is cool, but others need the warmer soils.

Germination begins when a seed absorbs water, either as a liquid or gas. As cells take in water, the seed swells; this can break the seed coat. Water also transports materials to the growing regions of the seed.

Within the seed, digestive enzymes help start the breakdown of stored food. This broken-down food and oxygen are the raw materials for cellular respiration, which results in the release of energy for growth.

DATA ANALYSIS LAB 23.1

Based on Real Data*

Recognize Cause and Effect

What is allelopathy? In nature, some plants produce chemicals that affect nearby plants. This is called allelopathy (uh LEEL luh pa thee). Some scientists studied the connection between allelopathy and the spread of nonnative plants, such as garlic mustard *Alliaria petiolata*. They investigated the effect of garlic mustard on the seed germination of native plants *Geum urbanum* and *Geum laciniatum.*

Think Critically

1. **Describe** the effect of garlic mustard on seed germination.
2. **Design an experiment** Alfalfa is known to allelopathically inhibit germination of some seeds. Use alfalfa sprouts to investigate their effect on seeds of your choice.

Data and Observations

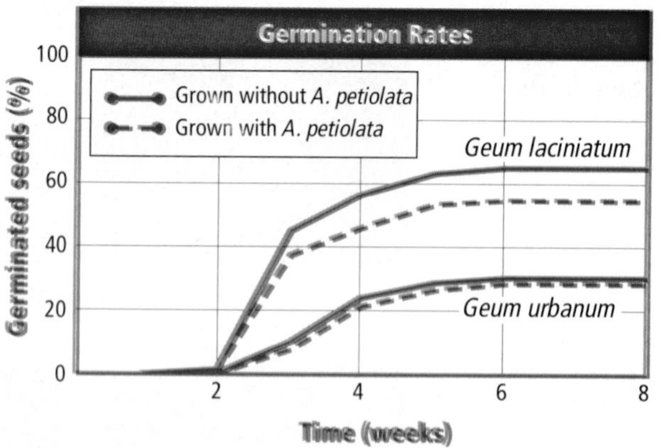

*Data obtained from: Prati, D. and O Bossdorf. 2004. Allelopathic inhibition of germination by *Alliaria petiolata* (Brassicaceae). *Amer. Journal of Bot.* 91(2): 285–288.

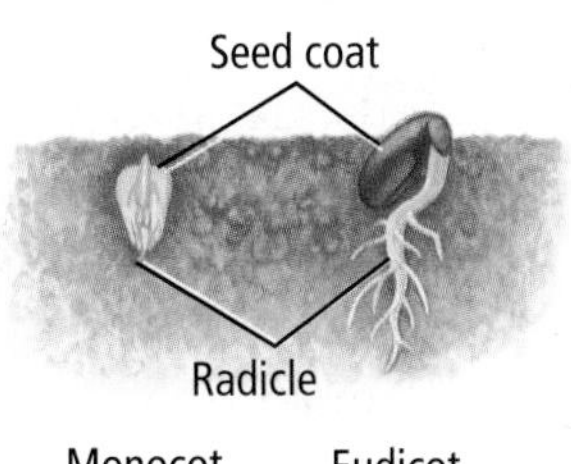

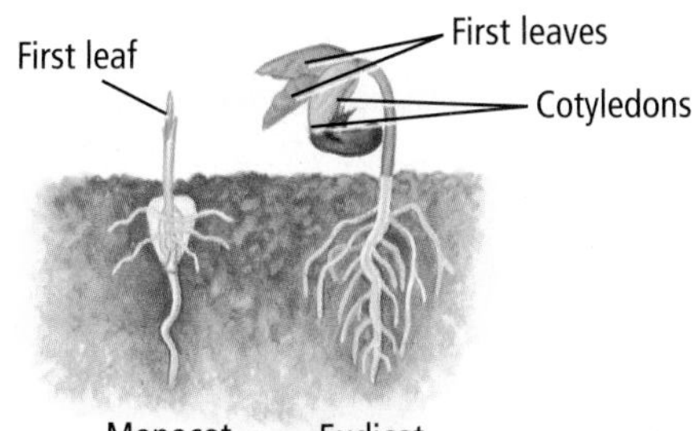

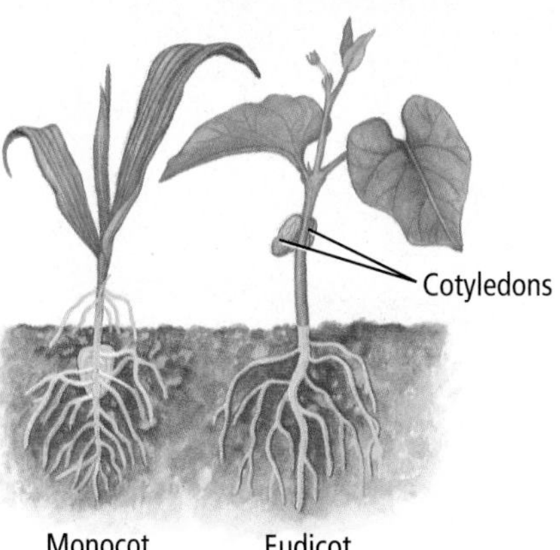

The first part of the embryo to appear outside the seed is a structure called the **radicle** that starts absorbing water and nutrients from its environment. The radicle, as shown in **Figure 23.16,** will develop into the plant's root. The **hypocotyl** is the region of the stem nearest the seed and, in many plants, it is the first part of the seedling to appear above the soil. In some eudicots, as the hypocotyl grows, it pulls the cotyledons and the embryonic leaves out of the soil. Photosynthesis begins as soon as the seedling's cells that contain chloroplasts are above ground and exposed to light. In monocots, seedling growth is slightly different because the cotyledon usually stays in the ground when the stem emerges from the soil.

Some seeds can survive harsh environmental conditions, such as drought and cold. Other seeds germinate soon after dispersal and others can germinate after long periods. Some maple seeds must germinate within two weeks after dispersal or they will not germinate at all. Most seeds produced at the end of a growing season enter **dormancy**—a period of little or no growth. Dormancy is an adaptation that increases the survival rate of seeds exposed to harsh conditions. The length of dormancy varies from species to species.

Figure 23.16 Seed germination differs in monocots and eudicots.

Concepts In Motion

Interactive Figure To see an animation of seed germination in flowering plants, visit biologygmh.com.

Section 23.3 Assessment

Section Summary

- The life cycle of anthophytes includes alternation of generations.
- The development of gametophytes occurs in the flower.
- Double fertilization is unique to anthophytes.
- Seeds provide nutrition and protection for the embryonic sporophyte.
- Fruits help protect and disperse seeds.
- Environmental conditions affect seed germination.

Understand Main Ideas

1. MAIN Idea **Diagram** the steps of the flowering-plant life cycle.
2. **Summarize** the development of the male gametophyte.
3. **Illustrate** the internal structure of a eudicot seed.
4. **Discuss** the importance of double fertilization.
5. **Compose** an argument for the 1893 court ruling that tomatoes are legally a vegetable, not a fruit.

Think Critically

6. **Evaluate** the mechanism that prevents incompatible pollen from producing a pollen tube.
7. **Compare and contrast** the germination of monocot and eudicot seeds.

MATH in Biology

8. As many as three million seeds can form inside an orchid pod. What is the percentage of germination, if all three million seeds are planted and 1,860,000 germinate?

Biology & Society

Genetically Modified Plants

Did you have cornflakes, orange juice, or wheat toast for breakfast? If they were purchased from a large grocery store, then there is a good chance you ate genetically modified foods. People have been altering the genetics of plants for centuries through selective breeding. Only recently have scientists modified the genetic makeup of plants.

What are genetically modified plants? Before genetic engineering, there was selective breeding. For example, if a fungus infected a corn crop, then a farmer would collect seeds from those plants with little or no signs of infection. If the farmer continued to select seeds from fungus-free plants, fungus-resistant corn could be developed over time.

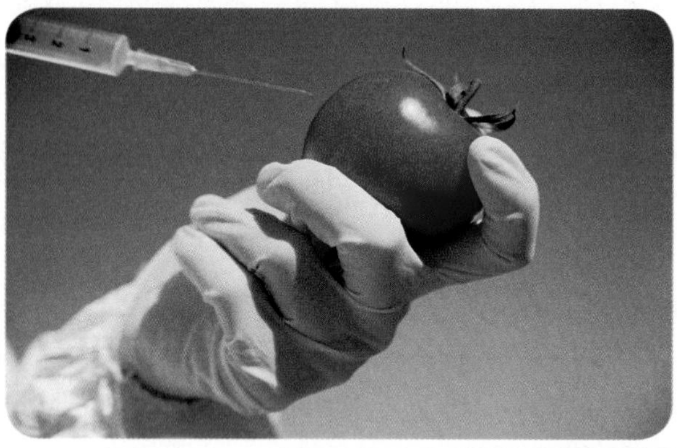

This tomato doesn't appear different, but has been modified to not soften prematurely.

In recent years, scientists have performed intraspecies gene transfers to alter plants. Genes for resistance to insects or disease are transferred from one variety of plant into another variety of the same species. Generally, plants that result from intraspecies gene transfer are considered safe to eat. In 1994, the first genetically modified food—a tomato that would not soften prematurely, shown above—became available to the public.

What are the benefits of genetically modified plants? Besides tomatoes that do not soften prematurely, other genetic modifications have produced seeds that have improved nutritional quality and that can be used for industrial products. Plants with herbicide, virus, and disease resistance, and plant products with longer shelf life have been produced. Also, plants have been developed that withstand environmental stresses. Farmers have better crop yields, and can use their land more efficiently. Currently, genetically engineered plants are being tested that produce drugs against HIV, tuberculosis, diabetes, and rabies.

What are the drawbacks of genetically modified plants? The main concern about genetically modified plants is the potential long-term risks. There also is the possibility that some of the genetically modified genes could enter wild populations of organisms. In fact, scientists show that the transgenic plants are 20 times more likely to cross-pollinate with other plants than mutated plants do.

One of the most controversial genetic modifications is the terminator gene. Plants with this gene produce seeds that cannot germinate. This means that farmers cannot gather seeds from their current crop for future planting. For many farmers in developing countries, gathered seeds are their only seed source for the next season. The company that purchased the patent for this gene has stopped development but has the option to resume development in the future.

DEBATE in Biology

Debate Should interspecies genetic plant modification continue without any controls? Conduct additional research at biologygmh.com. Prepare arguments that support your side and refute the other side.

BIOLAB

Design Your Own

HOW DO MONOCOT AND EUDICOT FLOWERS COMPARE?

Background: Flowers are the reproductive structures of flowering plants, and there is great diversity in flower form. Botanists classify flowering plants into two groups—monocots and eudicots—based on the structure of their seeds. However, their flower structures also differ. Explore the differences between these two groups of plants by completing this lab.

Question: *What are the structural differences between monocot and eudicot flowers?*

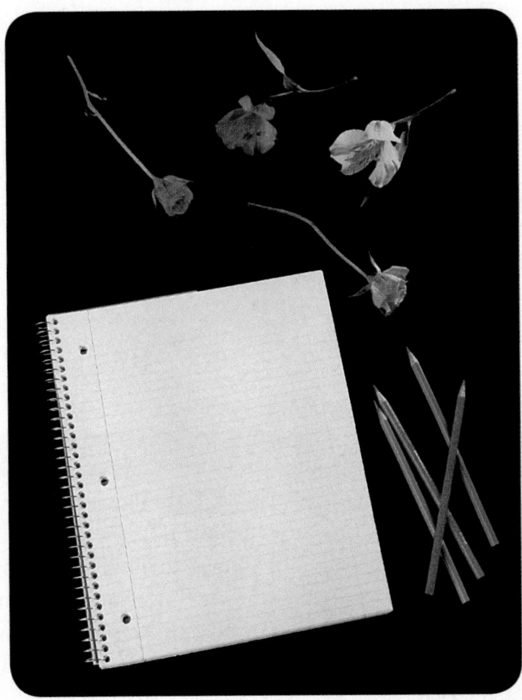

Materials

monocot flowers
eudicot flowers
colored pencils
Choose other materials that would be appropriate for this lab.

Safety Precautions

WARNING: ***Use dissecting tools with extreme caution.***

Plan and Perform the Experiment

1. Read and complete the lab safety form.
2. Choose several features of monocot and eudicot flowers to observe and compare.
3. Create a data table to record your observations of flowers—monocots and eudicots. Include sketches of each flower type.
4. Make sure your teacher approves your plan before you proceed.
5. Make observations as you planned.
6. Label and color-code the female and male reproductive structures and other flower parts of one of your monocot flower sketches.
7. Repeat Step 6 using one of the eudicot flower sketches.
8. **Cleanup and Disposal** Properly dispose of the flower parts. Clean all equipment as instructed by your teacher and return everything to its proper storage location.

Analyze and Conclude

1. **Compare and contrast** the characteristics of monocot and eudicot flowers.
2. **Conclude** Which of the flowers that you examined were monocots? Eudicots?
3. **Error Analysis** Compare your data with the data collected by your classmates. Explain any differences.

APPLY YOUR SKILL

Field Investigation Visit a local florist, greenhouse, or plant conservatory on your own or with a friend. Make a list of monocot and eudicot plants, based on their flower structures, that you observe at the location. Ask permission before touching any plants. To learn more about monocot and eudicot flowers, visit BioLabs at biologygmh.com.

CHAPTER 23

Study Guide

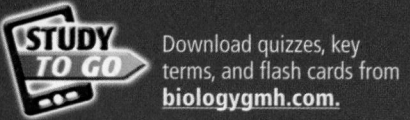

STUDY TO GO Download quizzes, key terms, and flash cards from biologygmh.com.

FOLDABLES **Infer** Consider why many fruits are commonly referred to as vegetables. For example, squash are treated as vegetables in cookbooks, yet botanically they are fruits. Analyze the common use of the term *fruit*.

Vocabulary	Key Concepts
Section 23.1 Introduction to Plant Reproduction	
• chemotaxis (p. 664) • heterosporous (p. 665) • megaspore (p. 665) • micropyle (p. 666) • microspore (p. 666) • prothallus (p. 665) • protonema (p. 664) • vegetative reproduction (p. 662)	MAIN Idea Like all plants, the life cycles of mosses, ferns, and conifers include alternation of generations. • Vegetative reproduction produces new plants without sexual reproduction. • The moss sporophyte depends on the gametophyte. • A fern sporophyte can live independently of the gametophyte. • Conifer gametophytes develop within sporophyte tissues.
Section 23.2 Flowers	
• day-neutral plant (p. 673) • intermediate-day plant (p. 673) • long-day plant (p. 672) • petal (p. 668) • photoperiodism (p. 672) • pistil (p. 669) • sepal (p. 668) • short-day plant (p. 672) • stamen (p. 669)	MAIN Idea Flowers are the reproductive structures of anthophytes. • A typical flower has sepals, petals, stamens, and one or more pistils. • Flower form differs from species to species. • Some flower modifications distinguish monocots from eudicots. • Modifications make flowers more attractive to pollinators. • Photoperiodism can influence when a plant flowers.
Section 23.3 Flowering Plants	
• dormancy (p. 679) • endosperm (p. 676) • germination (p. 678) • hypocotyl (p. 679) • polar nuclei (p. 674) • radicle (p. 679) • seed coat (p. 677)	MAIN Idea In anthophytes, seeds and fruits can develop from flowers after fertilization. • The life cycle of anthophytes includes alternation of generations. • The development of gametophytes occurs in the flower. • Double fertilization is unique to anthophytes. • Seeds provide nutrition and protection for the embryonic sporophyte. • Fruits help protect and disperse seeds. • Environmental conditions affect seed germination.

CHAPTER 23

Assessment

Section 23.1

Vocabulary Review

The sentences below are incorrect. Make each sentence correct by replacing the italicized word with a vocabulary term on the Study Guide page.

1. The *megaspore* of a conifer develops into the pollen grain.
2. A *protonema* is the gametophyte of a fern.
3. *Chemotaxis* is the growth of a new plant from a piece of the old plant.

Understand Key Concepts

4. Which is a fern prothallus?

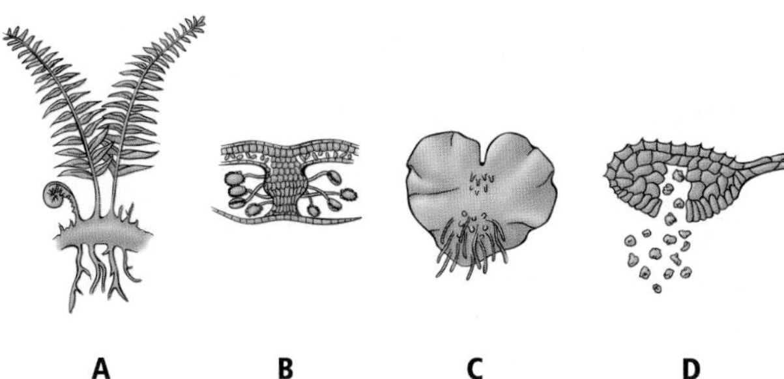

5. Which phrase accurately compares a fern sporophyte to the fern gametophyte?
   A. smaller than
   B. larger than
   C. always independent of
   D. always dependent on
6. From which structure does a conifer female gametophyte develop?
   A. prothallus
   B. fertilized egg
   C. microspore
   D. megaspore
7. Which is not an advantage of vegetative reproduction?
   A. uniform plant features
   B. genetically identical plants
   C. faster reproduction
   D. greater genetic variation

Constructed Response

8. **Short Answer** Explain the benefits of female gametophyte development within a conifer cone.
9. **Short Answer** What are some advantages and disadvantages of the moss sporophyte growing on the gametophyte?

Use the image below to answer question 10.

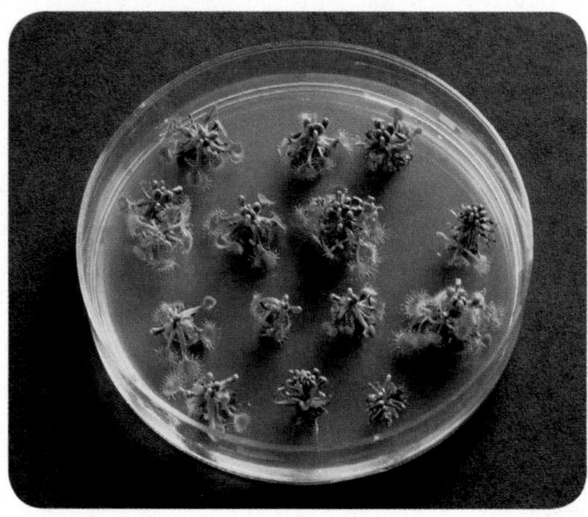

10. **Short Answer** Explain the genetic relationship among the offspring shown above.

Think Critically

11. **Discuss** the advantages or disadvantages of heterosporous plants.
12. **Suggest** a possible mechanism for the development of independent sporophyte generations as seen in conifers.

Section 23.2

Vocabulary Review

Distinguish between the vocabulary terms in each set.

13. pistil, stamen
14. long-day plant, short-day plant
15. petal, sepal

Understand Key Concepts

16. Which flower organ produces pollen?
   A. stamen
   B. pistil
   C. petal
   D. sepal

17. What dark/light conditions produce flowers in a short-day plant?
    A. hours of darkness are greater than the hours of light
    B. hours of darkness are less than hours of light
    C. hours of darkness are equal to hours of light
    D. hours of darkness and light are not factors

Use the image below to answer question 18.

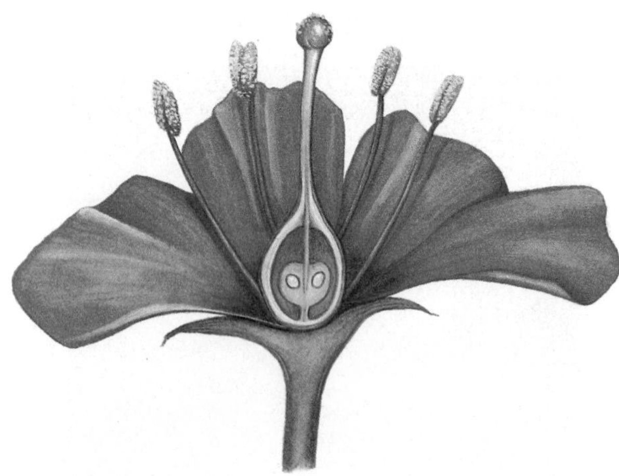

18. Which terms describe the flower above?
    A. perfect, complete
    B. perfect, incomplete
    C. imperfect, incomplete
    D. imperfect, complete

19. Which best describes pollen production in wind-pollinated flowers?
    A. small amounts of pollen
    B. larger pollen grains
    C. large amounts of pollen
    D. large quantities of nectar

20. Which terms could describe a monocot flower?
    A. four sepals, four petals
    B. five sepals, ten petals
    C. twelve sepals, twelve petals
    D. four sepals, eight petals

Constructed Response

21. **Short Answer** Explain why *short-day* and *long-day* are not the best descriptive terms for these types of flowering plants.

22. **Open Ended** Suggest a flower modification that would make water necessary for pollination. Justify your suggestion.

23. **Short Answer** Explain how modifications in flower structure make pollination more successful.

Think Critically

24. **Design an experiment** to test the ability of butterflies to distinguish between a real flower and an artificial flower.

25. **Assess** the benefits of photoperiodism.

Section 23.3

Vocabulary Review

Explain the relationship between the vocabulary terms in each pair below.

26. dormancy, germination

27. hypocotyl, radicle

28. polar nuclei, endosperm

Understand Key Concepts

29. Which is not part of a seed?
    A. cotyledon **C.** endosperm
    B. embryo **D.** pollen

30. Which describes the embryo of an anthophyte?
    A. diploid **C.** monoploid
    B. haploid **D.** triploid

31. From what structure does a pollen grain develop?
    A. egg **C.** endosperm
    B. embryo **D.** microspore

Use the image below to answer question 32.

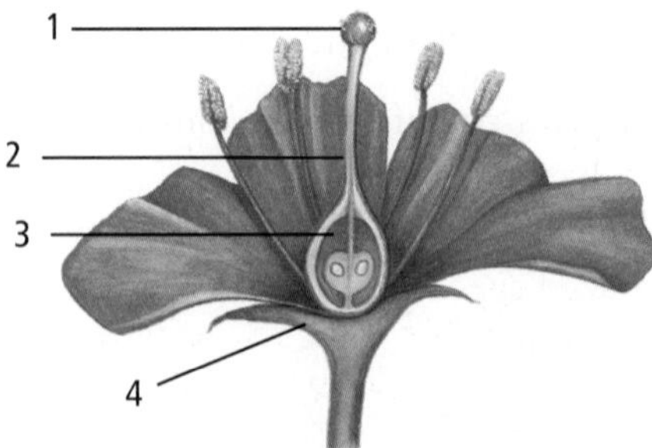

32. From which structure is a fruit usually formed?
    A. 1
    B. 2
    C. 3
    D. 4

Biology Online Chapter Test biologygmh.com

33. What is the inactive period of a seed?
   A. alternation of generations
   B. dormancy
   C. fertilization
   D. photoperiodism

Constructed Response

34. **Short Answer** Explain why fruit and/or seed dispersal is so important.

35. **Open Ended** Hypothesize why an anthophyte's female gametophyte produces so many nuclei when only two are involved in fertilization.

36. **Open Ended** When a seed germinates, as shown in **Figure 23.16,** the radicle usually is the first structure to break through the seed coat. Why is this beneficial for the embryo?

Think Critically

Use the graph below to answer questions 37–38.

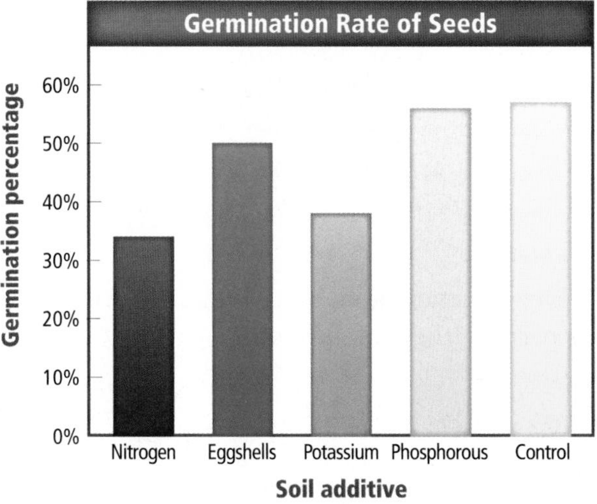

37. **Compare** the effects of each soil additive on the rate of germination to the control's rate of germination.

38. **Design** an experiment to test the effect on the rate of germination for various amounts of a soil additive. Choose one of the soil additives listed in the graph above.

39. **Analyze** the reduction in size of the gametophyte from mosses, to ferns, to anthophytes. What are the advantages or disadvantages of this trend?

Additional Assessment

40. WRITING in Biology Write a short story about the life of a pollen grain.

 Document-Based Questions

Data obtained from: Lang, A. et al. 1977. Promotion and inhibition of flower formation in a day-neutral plant in grafts with a short-day plant and a long-day plant. *Proc. Natl. Acad. Sci.* 74 (6): 2412-2416.

The day-neutral plant flowered sooner when it was grafted to the short-day plant that was exposed to its critical period. The flowering of another day-neutral plant also was accelerated when it was grafted to a long-day plant that was exposed to its critical period.

41. Examine the drawings. Form a hypothesis about why the grafted day-neutral plants flowered before the day-neutral plant that was not grafted.

42. Predict what might happen if a long-day plant was grafted to a short-day plant and they were exposed to the critical period of the short-day plant.

43. Design an experiment to determine the "longest day" under which a long-day plant flowers.

Cumulative Review

44. Relate genetic engineering to agriculture. **(Chapter 13)**

45. Choose three lines of evidence that support evolution. Give an example of each. **(Chapter 15)**

46. Describe the types of environments where you would expect to find protists. **(Chapter 19)**

Standardized Test Practice

Cumulative

Multiple Choice

1. Which vascular tissue is composed of living tubular cells that carry sugars from the leaves to other parts of the plant?
   A. cambium
   B. parenchyma
   C. phloem
   D. xylem

Use the diagram below to answer question 2.

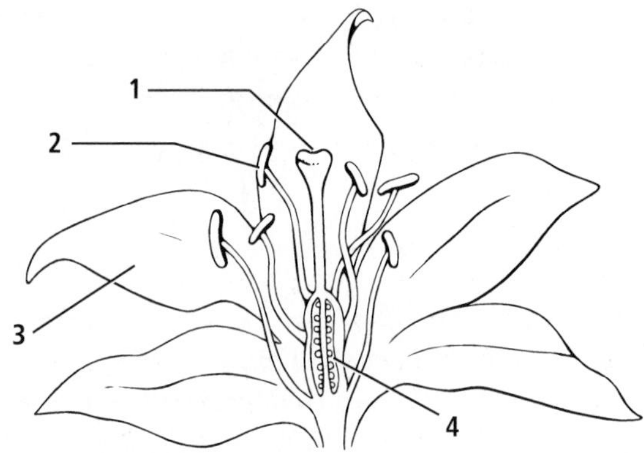

2. Which labeled structure is part of a flower's male reproductive organ?
   A. 1
   B. 2
   C. 3
   D. 4

3. Which statement provides evidence that anthophytes evolved after other seed plants?
   A. About 75 percent of all plants are anthophytes.
   B. Anthophytes do not require water to facilitate the fertilization of an egg.
   C. Prehistoric tree-like ferns were the main coal-forming plants.
   D. The seeds of anthophytes are more advanced than those of other seed plants.

4. Which precedes the haploid generation in seedless vascular plants?
   A. epiphytes
   B. gametophytes
   C. rhizomes
   D. spores

5. Which is the primary pollinator for conifers?
   A. birds
   B. insects
   C. water
   D. wind

Use the diagram below to answer question 6.

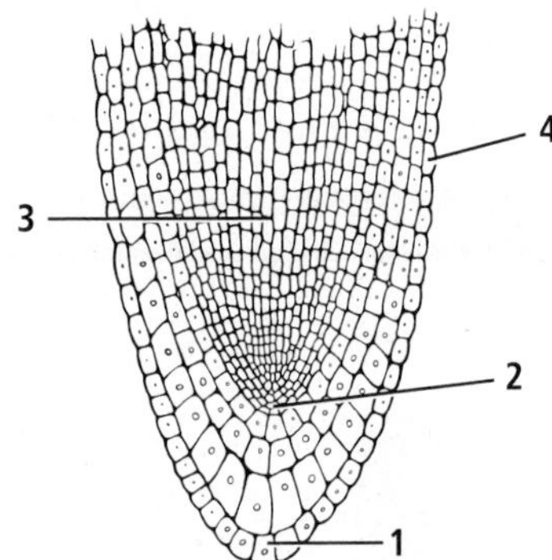

6. Which structure produces cells that result in an increase in length of the root?
   A. 1
   B. 2
   C. 3
   D. 4

7. Which statement is true of an aseptate fungus?
   A. Cell walls are made of cellulose.
   B. Cell walls are made of thin membranes.
   C. Hyphae are not divided by cross walls.
   D. Hyphae are not present except during reproduction.

8. A tuber is an adaptation of which structure?
   A. cell
   B. leaf
   C. root
   D. stem

Short Answer

Use the diagram below to answer question 9.

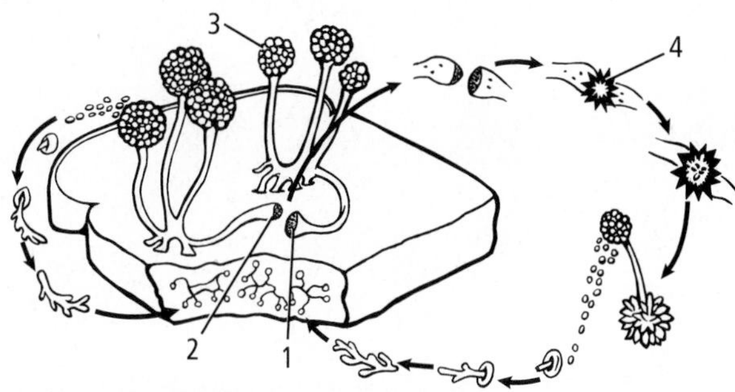

9. Describe two ways that bread mold could spread in a kitchen.

10. List two characteristics of nonvascular plants that compensate for their lack of transport tissues.

11. A certain type of fern has a chromosome number of 14. What would be the chromosome number of the prothallus? Explain why.

12. Explain the benefit to nonvascular plants of having very thin rhizoids and leaflike structures.

13. Name and describe the three types of plant cells and their functions.

14. Interpret how the actions of plate tectonics affected the evolution of primates.

15. Imagine that a friend who lives in Montana gives you some seeds from a plant. You plant the seeds in Florida but they do not grow. Predict why the seeds do not germinate in Florida.

Extended Response

16. Infer how collenchyma cells support surrounding plant tissues.

17. Critique the idea that roots in the ground do not need oxygen to survive.

18. A forest near a city provides drainage for rainfall runoff. A group of citizens is protesting new housing developments in the forest because they believe flooding and property destruction will result. Analyze the value of biodiversity that describes their concern.

19. Suppose that a couple wants to have children and neither the man nor the woman has cystic fibrosis. However, some distant family members have cystic fibrosis. Could their child have the disease? Write an explanation summarizing the risk for this couple.

Essay Question

Water is important for functions in plants. For example, it is one of the reactants in the chemical reactions of photosynthesis. Water enters a plant by diffusion. Most of the water that enters a plant diffuses into roots. Therefore, water must be in a higher concentration in the soil than in the roots. After water enters the roots, it moves through vascular tissue to tissues that contain chloroplasts. The water also diffuses into the plants' cells, making them rigid.

Using the information in the paragraph above, answer the following question in essay format.

20. When more water leaves a plant than enters it, the plant begins to wilt. Explain the role of guard cells in regulating the amount of water in a plant.

NEED EXTRA HELP?

If You Missed Question . . .	1	2	3	4	5	6	7	8	9	10	11	12	13	14	15	16	17	18	19	20
Review Section . . .	22.1	23.2	21.4	21.3	23.1	22.1	20.1	22.2	20.3	21.2	23.1	21.2	22.1	16.1	23.3	22.1	22.2	5.1	11.1	22.2

UNIT 7

Invertebrates

Chapter 24
Introduction to Animals
BIG Idea Animal phylogeny is determined in part by animal body plans and adaptations.

Chapter 25
Worms and Mollusks
BIG Idea Worms and mollusks have evolved to have a variety of adaptations for living as parasites or for living in water or soil.

Chapter 26
Arthropods
BIG Idea Arthropods have evolved to have a variety of adaptations for successful diversity, population, and persistence.

Chapter 27
Echinoderms and Invertebrate Chordates
BIG Idea Echinoderms and invertebrate chordates have features that connect them to the chordates that evolved after them.

CAREERS IN BIOLOGY

Entomologist

Entomologists are scientists who study insects and their behavioral patterns. Entomologists who conduct research, such as this field entomologist is doing, contribute to a better understanding of the ecosystem's function as a whole.

WRITING in Biology Visit biologygmh.com to learn more about careers in entomology. Write two scenarios in which entomologists study insects that are directly beneficial to humans.

Biology Online

To read more about entomologists in action, visit biologygmh.com.

CHAPTER 24

Introduction to Animals

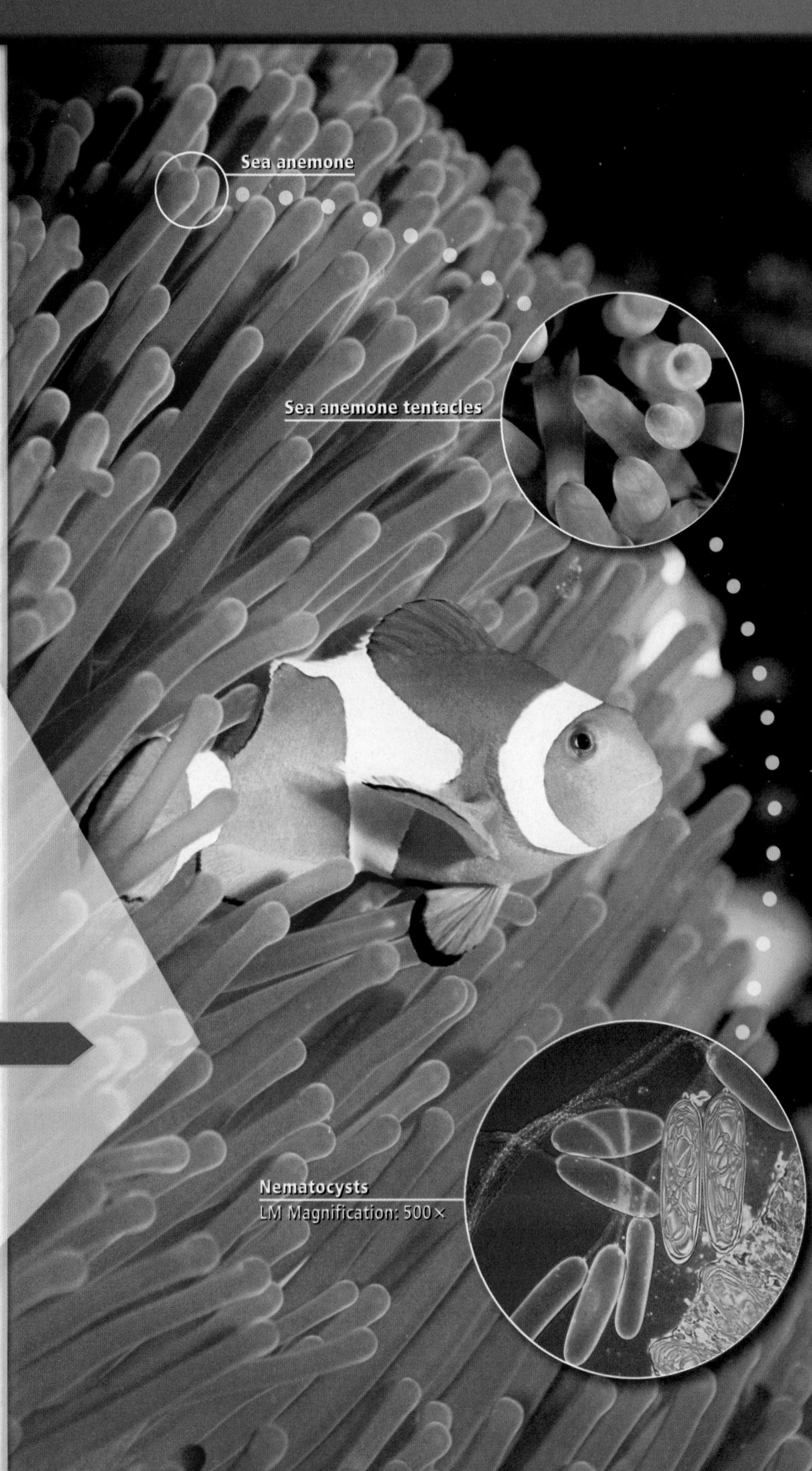

BIG Idea Animal phylogeny is determined in part by animal body plans and adaptations.

Section 1
Animal Characteristics
MAIN Idea Animals are multicellular, eukaryotic heterotrophs that have evolved to live in many different habitats.

Section 2
Animal Body Plans
MAIN Idea Animal phylogeny can be determined, in part, by body plans and the ways animals develop.

Section 3
Sponges and Cnidarians
MAIN Idea Sponges and cnidarians were the first animals to evolve from a multicellular ancestor.

BioFacts

- Sea anemones protect clown fishes from predators, and clown fishes attract bigger fish prey to anemones.
- Sea anemone tentacles have stinging structures called nematocysts for stunning prey.
- A layer of mucus on their scales protects clown fish from anemone stings.

Start-Up Activities

LAUNCH Lab

What is an animal?

Although animals share some characteristics with all other living organisms, they also have unique characteristics. In this lab, you will compare and contrast two organisms and determine which one is an animal.

Procedure

1. Read and complete the lab safety form.
2. Observe the **two organisms** you are given.
3. Compare and contrast the organisms using a **hand lens or stereomicroscope** if available.
4. Describe any specialized structures that you observe.
5. Based on your observations, predict how the form of each organism might be an adaptation to its habitat.

Analysis

1. **Identify** any structures that might be specific to animals.
2. **Predict** Based on your observations, can you predict which one of these organisms is more likely an animal? Explain.

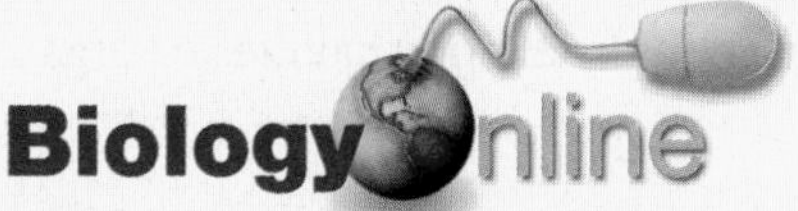

Visit biologygmh.com to:

- study the entire chapter online
- explore the Interactive Time Line, Concepts in Motion, Interactive Tables, Microscopy Links, and links to virtual dissections
- access Web links for more information, projects, and activities
- review content online with the Interactive Tutor, and take Self-Check Quizzes

Animal Body Plans Make the following Foldable to help you identify the characteristics of a coelomate, pseudocoelomate, and acoelomate body plans.

STEP 1 Stack two sheets of paper about 1.5 cm apart vertically as illustrated.

STEP 2 Fold up the bottom edges of the paper to form four equal tabs.

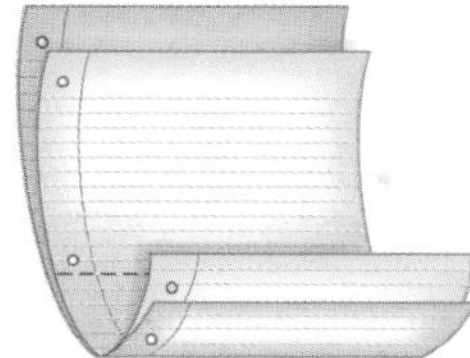

STEP 3 Staple along the folded edge to secure all sheets. With the stapled edge at the top, label each tab as shown below.

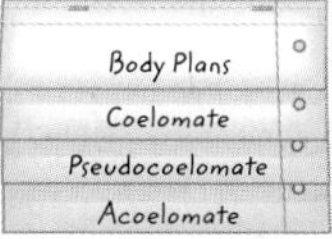

FOLDABLES **Use this Foldable with Section 24.2.** As you read the section, record information about each body plan under the tabs, and use what you learn to identify the body plans of animals around you.

Section 24.1

Reading Preview

Objectives

- **Examine** adaptations that enable animals to live in different habitats.
- **Compare and contrast** animal structure and function.
- **Distinguish** among the stages of embryonic development in animals.

Review Vocabulary

protist: diverse group of unicellular or multicellular eukaryotes that lack complex organ systems and live in moist environments

New Vocabulary

invertebrate
exoskeleton
endoskeleton
vertebrate
hermaphrodite
zygote
internal fertilization
external fertilization
blastula
gastrula
endoderm
ectoderm
mesoderm

Animal Characteristics

MAIN Idea Animals are multicellular, eukaryotic heterotrophs that have evolved to live in many different habitats.

Real-World Reading Link When you think of animals, you might think of creatures that are furry and fuzzy. However, animals can have other outer coverings, such as feathers on birds and scales on fishes. Some animals even might be mistaken for plants.

General Animal Features

Recall from Chapter 17 that biologists have created an evolutionary tree to organize the great diversity of living things. The ancestral animals at the beginning of the evolutionary tree are eukaryotic and multicellular—they are made up of many cells. The tiger in **Figure 24.1** and all other present-day animals might have evolved from choanoflagellates (KOH uh noh FLA juh layts), which are protists that formed colonies in the sea 570 million years ago. Choanoflagellates, such as the ones shown in **Figure 24.1,** might have been the earliest true animals. As animals evolved from this multicellular ancestor, they developed adaptations in structure that enabled them to function in numerous habitats. These features mark the branching points of the evolutionary tree and are discussed in the next section. In this section, you will learn about the characteristics that all animals have in common.

Feeding and Digestion

Animals are heterotrophic, so they must feed on other organisms to obtain nutrients. A sea star obtains its food from a clam it has pried open, and a butterfly feeds on nectar from a flower. The structure or form of an animal's mouth parts determines how its mouth functions. You can investigate how some animals obtain food by performing **MiniLab 24.1.** After obtaining their food, animals must digest it. Some animals, such as sponges, digest their food inside specific cells. Others, such as earthworms and humans, digest their food in internal body cavities or organs.

Figure 24.1 Present-day animals, such as this Bengal tiger, might have evolved from choanoflagellates such as this colony of *Zoothamnium*.

Bengal tiger

LM Magnification: 50×

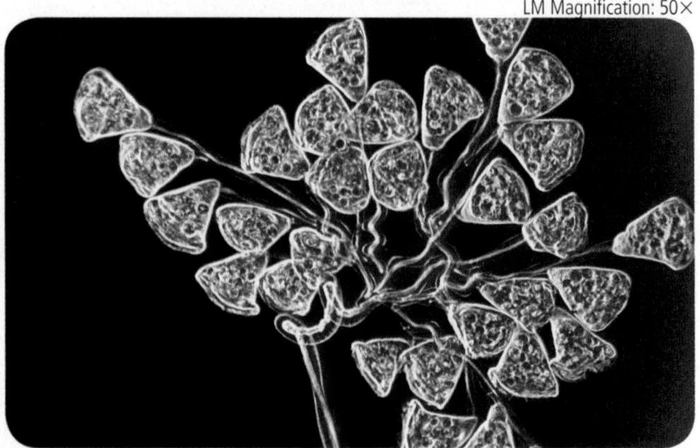

Colony of *Zoothamnium*

Support

Just as animals digest their food in different ways, they support their bodies in different ways. Between 95 and 99 percent of animal species are **invertebrates**—animals without backbones. The bodies of many invertebrates are covered with **exoskeletons,** which are hard or tough outer coverings that provide a framework of support. Exoskeletons also protect soft body tissues, prevent water loss, and provide protection from predators. As the animal grows, like the cicada in **Figure 24.2,** it must shed the old exoskeleton and make a new one.

Some invertebrates, such as sea urchins and sea stars, have internal skeletons called **endoskeletons.** If an animal has an endoskeleton and a backbone, it is called a **vertebrate.** An endoskeleton grows with the animal, like the squirrel in **Figure 24.2.** The material making up the endoskeleton varies. Sea urchins and sea stars have endoskeletons made of calcium carbonate, sharks have endoskeletons made of cartilage, and fishes, amphibians, reptiles, birds, and mammals have endoskeletons made of bone. An endoskeleton protects internal organs, provides support for the body, and can provide an internal brace for muscles to pull against. You will learn more about the human skeletal and muscular systems in Unit 9.

 **Reading Check** **Distinguish** between vertebrates and invertebrates.

Habitats

Animal bodies have a variety of adaptations, such as those for feeding, digestion, and support. These body variations enable animals to live in numerous habitats. Vertebrates and invertebrates live in oceans, in freshwater, and on land. They can be found in deserts, grasslands, rain forests, polar regions, and all other land biomes and aquatic ecosystems.

Cicada

Squirrel

Figure 24.2 A cicada must shed its old exoskeleton (outlined in white) in order to grow. A squirrel has an endoskeleton that grows as the squirrel grows.
Infer ***How might an exoskeleton be a disadvantage for animals?***

MiniLab 24.1

Investigate Feeding in Animals

How do animals obtain food? Small aquatic animals called hydras consume brine shrimp to obtain food.

Procedure

1. Read and complete the lab safety form.
2. Obtain several **hydras** in a **plastic Petri dish** containing **water.**
3. Add several **brine shrimp** to the dish. Using a **hand lens or stereomicroscope,** observe the activity of the hydras.
4. Record your observations.

Analysis

1. **Draw Conclusions** Based on your observations, how do the hydras react to the food?
2. **Infer** What factors in their environment might influence how the hydras find food?

Animal Cell Structure

No matter where an animal lives or what adaptations it has, its cells do not have cell walls. Recall from Chapter 7 that plants also are multicellular organisms, but their cells have cell walls. The cells of all animals, except sponges, are organized into structural and functional units called tissues. Recall from Chapter 9 that a tissue is a group of cells that is specialized to perform a specific function. For example, nerve tissue is involved in the transmission of nerve impulses throughout the body and muscle tissue enables the body to move.

CAREERS IN BIOLOGY

Systematist Using observation, inference, and the latest technology, a systematist classifies new species based on evolutionary relationships. For more information on biology careers, visit biologygmh.com.

Connection to History Beginning with Aristotle in the fourth century B.C. and continuing into the nineteenth century, living organisms were classified into two kingdoms—Animalia (animals) and Plantae (plants). In 1866, Ernst Haeckel, a German scientist, proposed adding a third kingdom called Protista. The organisms in this kingdom are mainly unicellular eukaryotes. Some protists have cell walls, while others do not, making them neither plant nor animal. During the 1960s, as more was learned about cell structure, bacteria and fungi were placed into their own kingdoms. **Figure 24.3** illustrates how the classification of living things continues to develop.

Movement

The evolution of nerve and muscle tissues enables animals to move in ways that are more complex and faster than organisms in other kingdoms. This is one notable characteristic of the animal kingdom. A gecko running across a ceiling, a mosquito buzzing around your ear, and a school of minnows swimming against the current are all exhibiting movements unique to animals. Some animals are stationary as adults, yet most have a body form that can move during some stage of development.

■ **Figure 24.3**
History of Classification

The process of scientifically classifying organisms began in 350 B.C. when Aristotle, a Greek philosopher, placed organisms into two large groups—plant and animal. Advances in scientific knowledge and technology helped develop the classification system we use today.

1735 Biologist Carolus Linnaeus devises a classification system for all organisms using Latin binomial nomenclature.

1500 1600 1700 1800

1555 The book *L'Histoire de la Nature des Oyseaux* (Natural History of Birds) uses body form and structures to classify species.

1682 Naturalist John Ray establishes the use of species as the basic unit of classification.

1859 Naturalist Charles Darwin proposes the classification of organisms based on their shared ancestry.

Reproduction

Most animals reproduce sexually, although some species can reproduce asexually. Most commonly in sexual reproduction, male animals produce sperm and female animals produce eggs. Some animals, such as earthworms, are **hermaphrodites** (hur MAF ruh dites), which produce both eggs and sperm in the same animal body. In general, hermaphrodites produce eggs and sperm at different times, so another individual of the same species still is needed for sexual reproduction.

Fertilization occurs when the sperm penetrates the egg to form a fertilized egg cell called the **zygote** (ZI goht). Fertilization can be internal or external. **Internal fertilization** occurs when the sperm and egg combine inside the animal's body. For example, male turtles fertilize the eggs of the female internally. **External fertilization** occurs when egg and sperm combine outside the animal's body. This process requires an aquatic environment for the sperm to swim to the egg. In many fishes, the female lays eggs in the water and the male sheds sperm over the eggs, as shown in **Figure 24.4.**

Figure 24.4 Fertilization is external in some fishes. In the photo, strands of sperm are being shed over eggs laid in the water.
Infer ***Why do animals lay a large number of eggs when fertilization is external?***

Recall that asexual reproduction means that a single parent produces offspring that are genetically identical to itself. Although few animal species reproduce asexually, when they do, they use one or more methods to do so. Some of the common methods of asexual reproduction include:

- budding—an offspring develops as a growth on the body of the parent
- fragmentation—the parent breaks into pieces and each piece can develop into an adult animal
- regeneration—a new organism can regenerate, or regrow, from the lost body part if the part contains enough genetic information
- parthenogenesis (par thuh noh JE nuh sus)—a female animal produces eggs that develop without being fertilized

Reading Check **Infer** the advantages and disadvantages of asexual reproduction in animals.

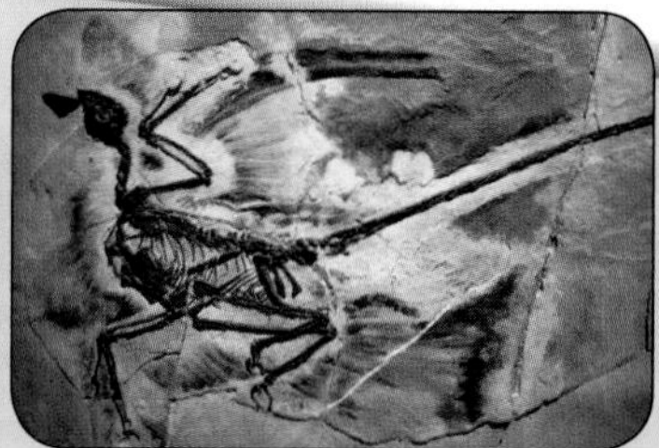

1977 Microbiologist Carl Woese uses ribosomal RNA to show the evolutionary relationships among organisms.

2003 Paleontologists find feathered dinosaur fossils that might alter the classification of some species.

1900

2000

1891 Marine zoologist Mary Jane Rathbun begins establishing the basic taxonomic information on crustaceans.

1982 Biologist Lynn Margulis is instrumental in reorganizing and improving the classification of organisms into the current five kingdoms.

Concepts In Motion Interactive Time Line
To learn more about these discoveries and others, visit biologygmh.com. **Biology Online**

Early development In most animals, the zygote undergoes mitosis and a series of cell divisions to form new cells. After the first cell division, in which the zygote forms two cells, the developing animal is called an embryo. The embryo continues to undergo mitosis and cell division, forming a solid ball of cells. These cells continue to divide, forming a fluid-filled ball of cells called the **blastula** (BLAS chuh luh), as shown in **Figure 24.5.** During these early stages of development, the number of cells increases, but the total amount of cytoplasm in the embryo remains the same as that in the original cell. Therefore, the total size of the embryo does not increase during early development.

In animals such as lancelets, the outer blastula is a single layer of cells, while in animals such as frogs, there might be several layers of cells surrounding the fluid. The blastula continues to undergo cell division. Some cells move inward to form a **gastrula** (GAS truh luh)—a two-cell-layer sac with an opening at one end. A gastrula looks like a double bubble—one bubble inside another bubble.

Look again at **Figure 24.5.** Notice how the diagrams of the two-cell stage, the 16-cell stage, and the blastula differ from the photographs of these same stages. The diagrams illustrate early development in embryos that develop inside the adult animal. The photographs illustrate early development in embryos that develop outside of the adult animal. The large ball that does not divide is the yolk sac. It provides food for the developing embryo.

Reading Check **Explain** the differences between the blastula and the gastrula.

VOCABULARY

WORD ORIGIN

Gastrula

gastr– prefix; from Greek; meaning *stomach* or *belly.*

–ula suffix; from Latin; meaning *resembling.*

Figure 24.5 The fertilized eggs of most animals follow a similar pattern of development. Beginning with one fertilized egg cell, cell division occurs and a gastrula is formed.

Concepts In Motion

Interactive Figure To see the development of a zygote into specialized cells, visit biologygmh.com.

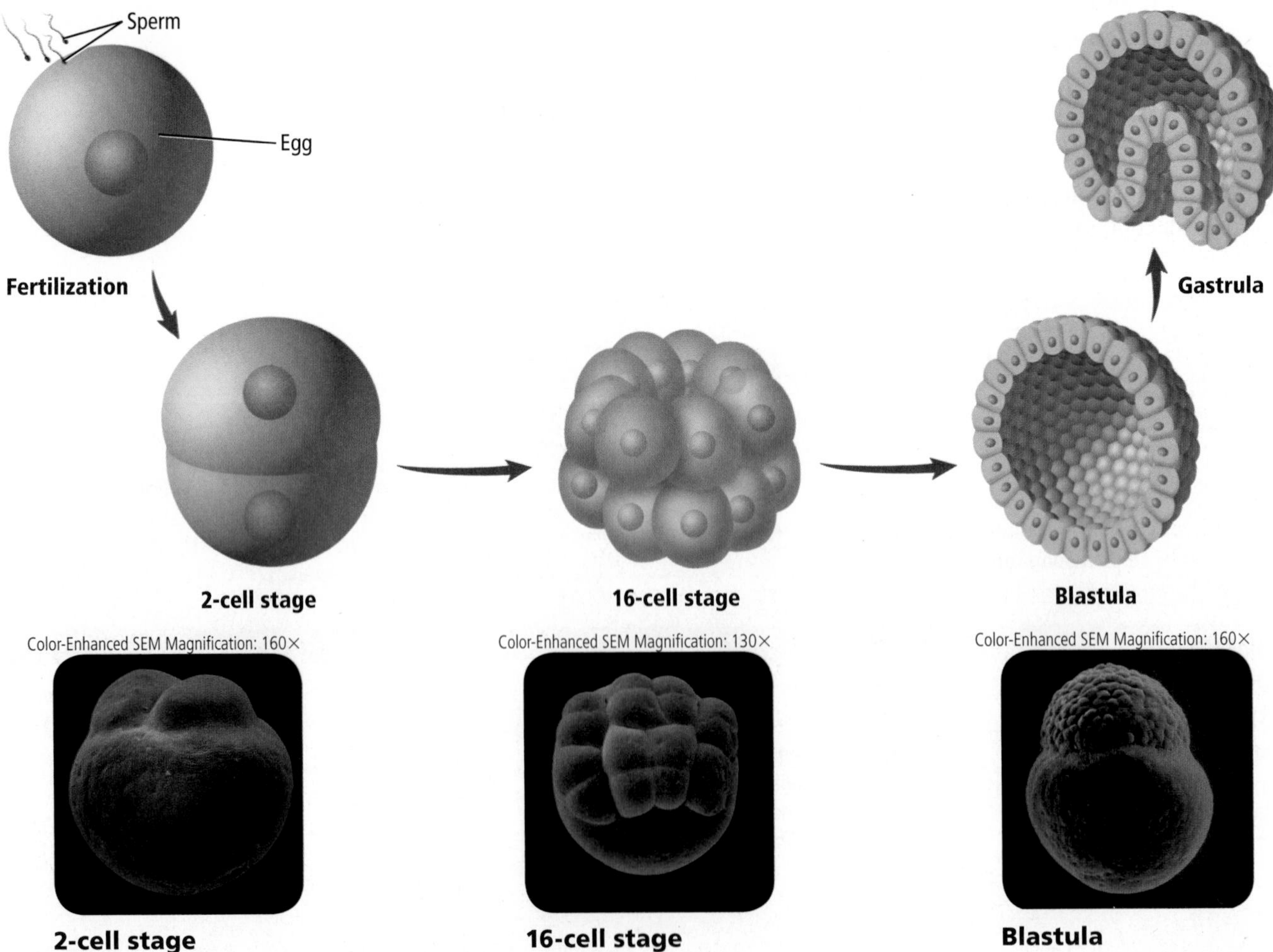

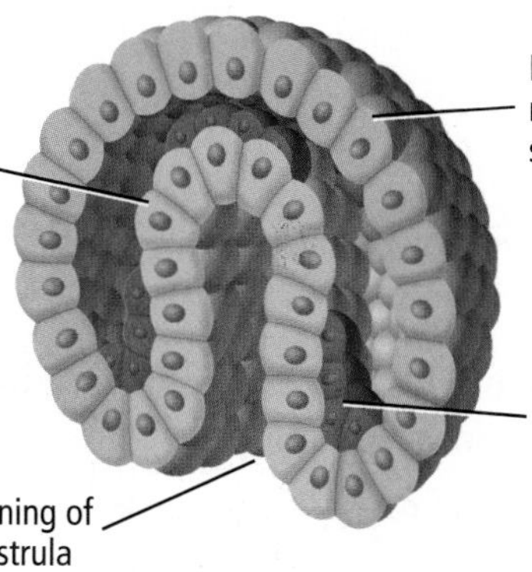

Endoderm becomes digestive organs and digestive tract lining.

Ectoderm becomes nervous tissue and skin.

Mesoderm becomes muscle tissue and the circulatory, excretory, and respiratory systems.

Opening of gastrula

■ **Figure 24.6** As development continues, each cell layer differentiates into specialized tissues.

Tissue development Notice in **Figure 24.6** that the inner layer of cells in the gastrula is called the **endoderm.** The endoderm cells develop into the digestive organs and the lining of the digestive tract. The outer layer of cells in the gastrula is called the **ectoderm.** The ectoderm cells in the gastrula continue to grow and become the nervous tissue and skin.

Cell division in some animals continues in the gastrula until another layer of cells, called the **mesoderm,** forms between the endoderm and the ectoderm. In some animals, the mesoderm forms from cells that break away from the endoderm near the opening of the gastrula. In more highly evolved animals, the mesoderm forms from pouches of endoderm cells on the inside of the gastrula. As development continues, mesoderm cells become muscle tissue, the circulatory system, the excretory system, and, in some species of animals, the respiratory system.

Recall from Chapter 12 that Hox genes might be expressed in ways that give proteins new properties that cause variations in animals. Much of the variation in animal bodies is the result of changes in location, number, or time of expression of Hox developmental genes during the course of tissue development.

LAUNCH Lab

Review Based on what you've read about animal characteristics, how would you now answer the analysis questions?

Section 24.1 Assessment

Section Summary

- Animals must get their nutrients from other organisms.
- Animals have diverse means of support and live in diverse habitats.
- Animal cells do not have cell walls, and most animals have cells that are organized into tissues.
- Most animals undergo sexual reproduction, and most animals can move.
- During embryonic development, animal cells become tissue layers, which become organs and systems.

Understand Main Ideas

1. MAIN Idea **Infer** why colonial organisms that lived grouped together might have been one of the first steps toward multicellular organisms in the course of evolution.
2. **Infer** how an exoskeleton enables invertebrates to live in a variety of habitats.
3. **Describe** how the evolution of nerve and muscle tissue is related to one of the main characteristics of animals.
4. **Diagram** how an animal zygote becomes a gastrula.

Think Critically

5. **Model** the stages of cell differentiation by comparing them to pushing in the end of a balloon. Draw a diagram of this process and label it with the stages of cell differentiation.

MATH in Biology

6. Biologists have observed that it is common for an animal that doubles its mass to increase its length 1.26 times. Suppose an animal has a mass of 2.5 kg and is 30 cm long. If this animal grows to a mass of 5 kg, how long will it be?

Section 24.2

Reading Preview

Objectives

- **Analyze** how animal body plans are related to phylogeny.
- **Demonstrate** how body cavities are related to animal phylogeny.
- **Distinguish** between the two types of coelomate development.

Review Vocabulary

phylogeny: evolutionary history of a species based on comparative relationships of structures and comparisons of modern life-forms with fossils

New Vocabulary

symmetry
radial symmetry
bilateral symmetry
anterior
posterior
cephalization
dorsal
ventral
coelom
pseudocoelom
acoelomate
protostome
deuterostome

Animal Body Plans

MAIN Idea **Animal phylogeny can be determined, in part, by body plans and the ways animals develop.**

Real-World Reading Link People often classify or group things based on what they have in common. If you want to rent an action movie, you would look in the action movie section at the store. You would not find comedies or dramas in this section. In biology, animals generally are classified into groups because they have some of the same features.

Evolution of Animal Body Plans

Recall that the evolutionary tree is organized like a family tree, and the phylogeny of animals is represented by the branches. For example, all of the mammals in **Figure 24.7** belong on the chordate branch of the tree. The trunk represents the earliest animals and the branches represent the probable evolution of the major phyla of animals from a common ancestor, as shown in **Figure 24.8.**

Anatomical features in animals' body plans mark the branching points on the evolutionary tree. For example, animals without tissues are grouped separately from animals with tissues, and animals without segments are grouped separately from animals with segments. Recall from Chapter 17 that the relationships among animals on this tree are inferred by studying similarities in embryological development and shared anatomical features. This traditional phylogeny, with animals classified into 35 phyla, is still used by most taxonomists. However, molecular data suggest other relationships among animals. Recent molecular findings, based on comparisons of DNA, ribosomal RNA, and proteins, indicate that the relationships between arthropods and nematodes and between flatworms and rotifers might be closer than anatomical features suggest.

Reading Check **Summarize** the structure of an evolutionary tree.

■ **Figure 24.7** Although these animals look very different from each other, they all have features that place them on the chordate branch of the evolutionary tree.

Mouse

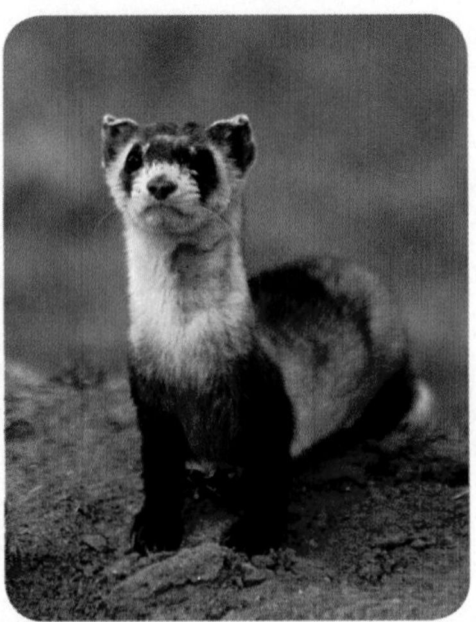
Ferret

Chimpanzee

Development of Tissues

As animals evolved from the first multicellular forms, the first anatomical feature to indicate a major change in body plan was the development of tissues. Therefore, tissues mark the first branching point on the evolutionary tree. Notice in **Figure 24.8** that the only animals without tissues are sponges. These animals descended from a common ancestor that lacked tissues, and they are on the no-true-tissue branch of the evolutionary tree. Follow the tissue branch of the evolutionary tree, and you will see that all other phyla have tissues.

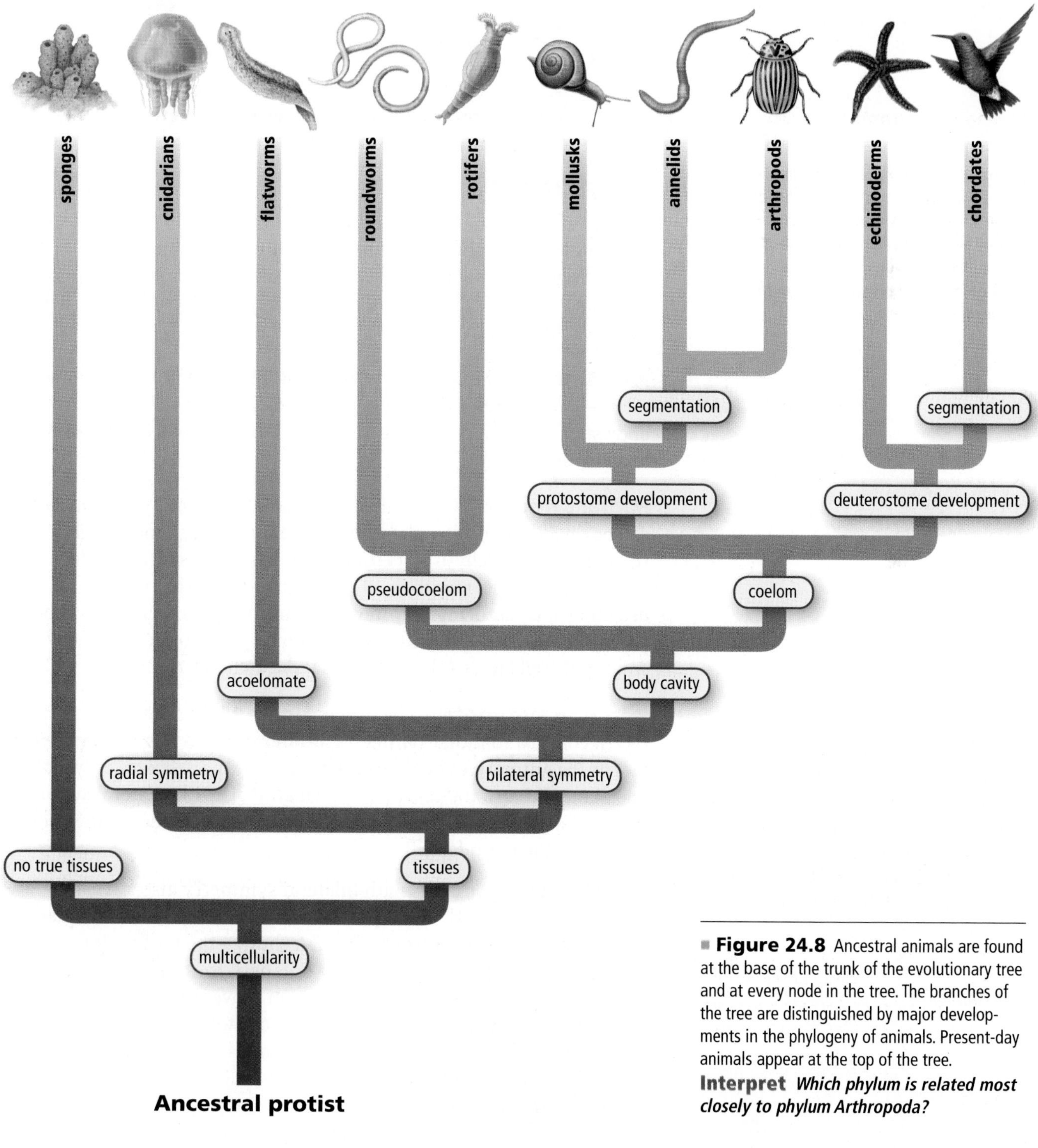

Figure 24.8 Ancestral animals are found at the base of the trunk of the evolutionary tree and at every node in the tree. The branches of the tree are distinguished by major developments in the phylogeny of animals. Present-day animals appear at the top of the tree.
Interpret ***Which phylum is related most closely to phylum Arthropoda?***

Sponge—asymmetry

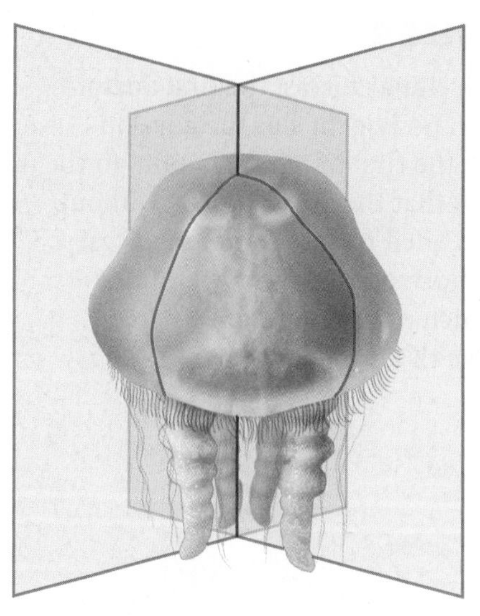

Jellyfish—radial symmetry

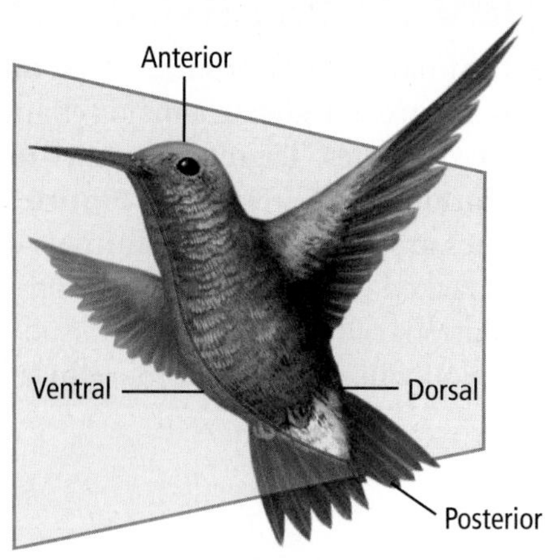

Hummingbird—bilateral symmetry

Figure 24.9 Animals have different arrangements of body structures. The sponge has an irregular shape and is asymmetrical. The jellyfish has radial symmetry, and the hummingbird has bilateral symmetry.
List ***objects you see in the room that have bilateral symmetry.***

Personal Tutor

To learn about symmetry, visit biologygmh.com.

Symmetry

Move along the tissue branch on the evolutionary tree in **Figure 24.8** and you will find the next branching point to be symmetry. **Symmetry** (SIH muh tree) describes the similarity or balance among body structures of organisms. The type of symmetry an animal has enables it to move in certain ways.

Asymmetry The sponge in **Figure 24.9** has no tissue and has asymmetry—it is irregular in shape and has no symmetry or balance in its body structures. In contrast, animals with tissues have either radial or bilateral symmetry.

Radial symmetry An animal with **radial** (RAY dee uhl) **symmetry** can be divided along any plane, through a central axis, into roughly equal halves. The jellyfish in **Figure 24.9** has radial symmetry. Its tentacles radiate from its mouth in all directions—a body plan adapted to detecting and capturing prey moving in from any direction. Jellyfishes and most other animals with radial symmetry develop from only two embryonic cell layers—the ectoderm and the endoderm.

Bilateral symmetry The bird in **Figure 24.9** has bilateral symmetry. In contrast to radial symmetry, **bilateral** (bi LA tuh rul) **symmetry** means the animal can be divided into mirror image halves only along one plane through the central axis. All animals with bilateral symmetry develop from three embryonic cell layers—the ectoderm, the endoderm, and the mesoderm.

Vocabulary

Science usage v. Common usage

Plane

Science usage: an imaginary line that divides a body form into two parts.
The dog can be divided into its ventral and dorsal parts by a plane.

Common usage: an aircraft.
The pilot flew the plane from Cleveland to Chicago.

Cephalization Animals with bilateral symmetry also have an **anterior,** or head end, and a **posterior,** or tail end. This body plan is called **cephalization** (sef uh luh ZA shun)—the tendency to concentrate nervous tissue and sensory organs at the anterior end of the animal. Most animals with cephalization move through their environments with the anterior end first, encountering food and other stimuli. In addition to cephalization, animals with bilateral symmetry have a **dorsal** (DOR sul) surface, also called the backside, and a **ventral** (VEN trul) surface, also called the underside or belly.

Body Cavities

In order to understand the next branching point on the evolutionary tree, it is important to know about certain features of animals with bilateral symmetry. Body plans of animals with bilateral symmetry include the gut, which is either a sac inside the body or a tube that runs through the body, where food is digested. A saclike gut has one opening—a mouth—for taking in food and disposing of wastes. A tube-like gut has an opening at both ends—a mouth and an anus—and is a complete digestive system that digests, absorbs, stores, and disposes of unused food.

Coelomates Between the gut and the outside body wall of most animals with bilateral symmetry is a fluid-filled body cavity. One type of fluid-filled cavity, the **coelom** (SEE lum), shown in **Figure 24.10,** has tissue formed from mesoderm that lines and encloses the organs in the coelom. You have a coelom, as do insects, fishes, and many other animals, therefore, you are a coelomate. The coelom was a key adaptation in the evolution of larger and more specialized body structures. Specialized organs and body systems that formed from mesoderm developed in the coelom. As more efficient organ systems evolved, such as the circulatory system and muscular system, animals could increase in size and become more active.

Pseudocoelomates Follow the body cavity branch on the evolutionary tree in **Figure 24.8** until you come to the pseudocoelomates—animals with pseudocoeloms. A **pseudocoelom** (soo duh SEE lum) is a fluid-filled body cavity that develops between the mesoderm and the endoderm rather than developing entirely within the mesoderm as in coelomates. Therefore, the pseudocoelom, as shown in **Figure 24.10,** is lined only partially with mesoderm. The body cavity of pseudocoelomates separates mesoderm and endoderm, which limits tissue, organ, and system development.

Acoelomates Before the body cavity branch on the evolutionary tree in **Figure 24.8,** notice that the branch to the left takes you to the acoelomate animals. **Acoelomates** (ay SEE lum ayts), such as the flatworm in **Figure 24.10,** are animals that do not have a coelom. The body plan of acoelomates is derived from ectoderm, endoderm, and mesoderm—the same as in coelomates and pseudocoelomates. However, acoelomates have solid bodies without a fluid-filled body cavity between the gut and the body wall. Nutrients and wastes diffuse from one cell to another because there is no circulatory system.

Personal Tutor

To learn about body cavities, visit biologygmh.com.

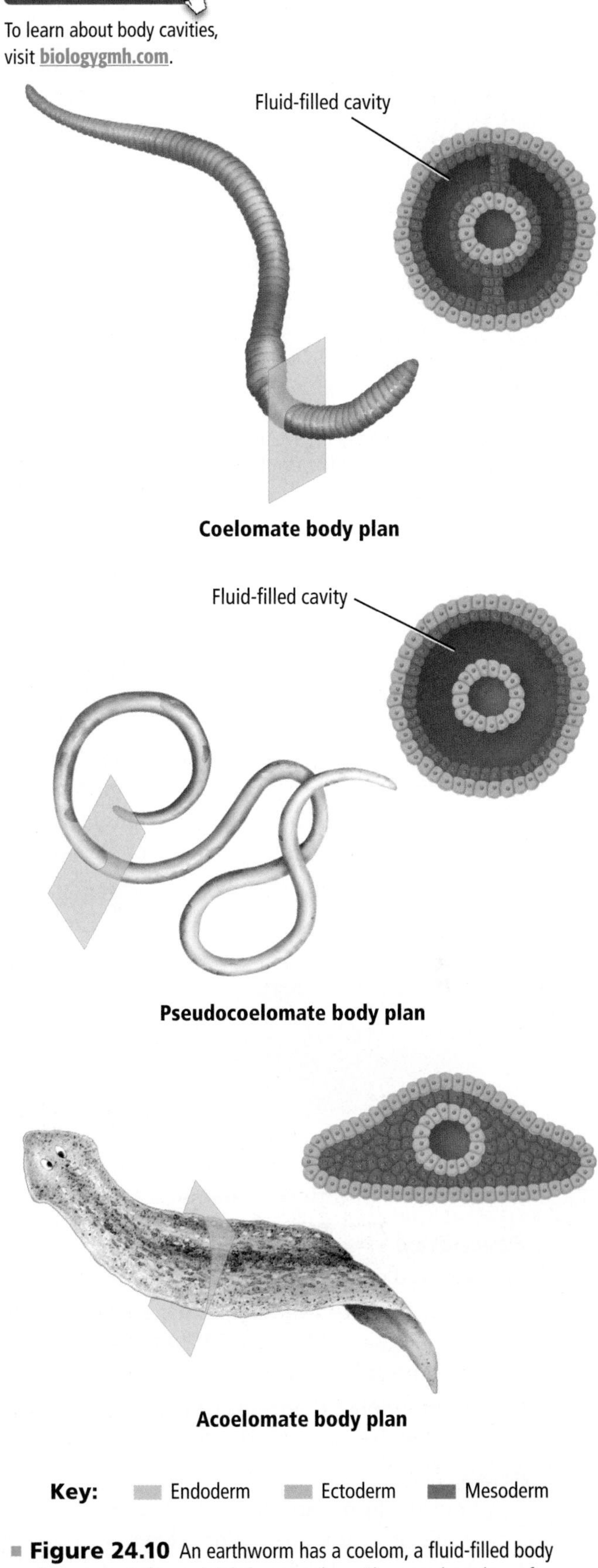

Figure 24.10 An earthworm has a coelom, a fluid-filled body cavity surrounded completely by mesoderm. The pseudocoelom of a roundworm develops between the mesoderm and endoderm. A flatworm has a solid body without a fluid-filled cavity.

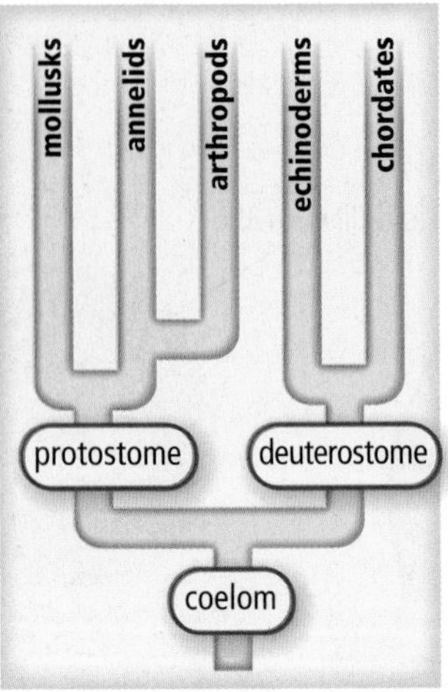

■ **Figure 24.11** This part of the evolutionary tree shows that protostomes and deuterostomes are branches of coelomate animals.

Development in Coelomate Animals

The evolutionary tree in **Figure 24.11** begins at the coelomate branch. Notice that two major lines of development have been identified in coelomate animals. One is protostome development, which occurs in animals such as snails, earthworms, and spiders. The other is deuterostome development, which occurs in animals such as sea urchins, dogs, and birds. Biologists can tell if animals are closely related based on their patterns of embryonic development.

Protostomes In organisms that are **protostomes** (PROH tuh stohms), the mouth develops from the first opening in the gastrula. As protostomes develop, the final outcome for each cell in the embryo cannot be altered. If one cell of the embryo is removed, the embryo will not develop into a normal larva, as shown in **Figure 24.12.** In addition, in the eight-cell stage of embryonic development, the top four cells are offset from the bottom four cells, giving the embryo a spiral appearance. As the embryo continues to develop, the mesoderm splits down the middle. The cavity between the two pieces of mesoderm becomes the coelom.

Deuterostomes In organisms that are **deuterostomes** (DEW tihr uh stohms), the anus develops from the first opening in the gastrula. The mouth develops later from another opening of the gastrula. During the development of deuterostomes, the final outcome for each cell in the embryo can be altered. In fact, each cell in the early embryo, if removed, can form a new embryo, as shown in **Figure 24.12.** In contrast to protostome development, in the eight-cell stage of embryonic deuterostome development the top four cells are directly aligned on the bottom four cells. As the embryo develops, the coelom forms from two pouches of mesoderm.

Reading Check **Determine** Are you a protosome or a deuterostome? Explain.

MiniLab 24.2

Examine Body Plans

What is the importance of a body plan? One way to classify animals is by body plan. Looking at cross sections of different animals can help you distinguish between the different body plans.

Procedure

1. Read and complete the lab safety form.
2. Obtain **prepared slides of cross sections of an earthworm and a hydra.** Using a **microscope,** observe each slide under low-power magnification.
3. Sketch each cross section.
4. Obtain **labeled diagrams of cross sections of each animal** from your teacher. Make a list of how your sketches are like the diagrams and another list of how they are different.

Analysis

1. **Compare and Contrast** What type of body cavity does each of these animals have? Are they acoelomate or coelomate? What do your observations tell you about the phylogeny of these animals?
2. **Infer** how the body plan of each animal is related to how each of these animals obtains food.

Visualizing Protostome and Deuterostome Development

Figure 24.12
Developmental differences characterize protostome and deuterostome development.

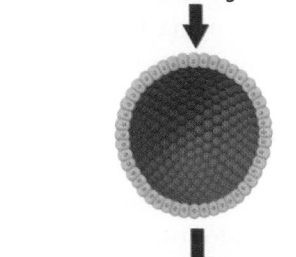

Protostome Development

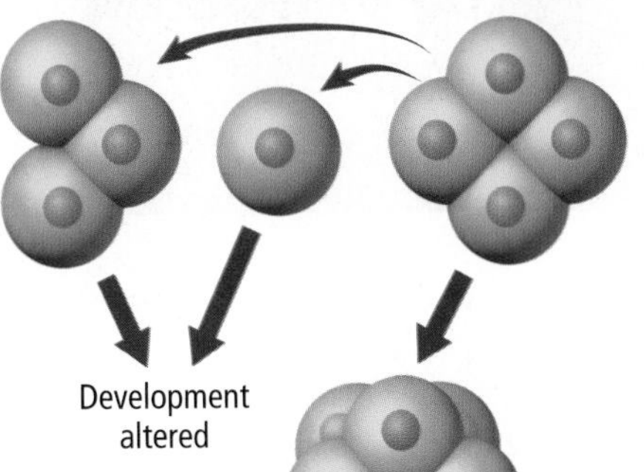

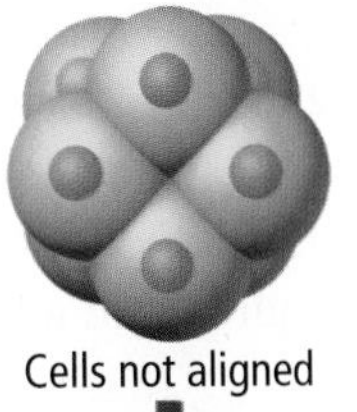

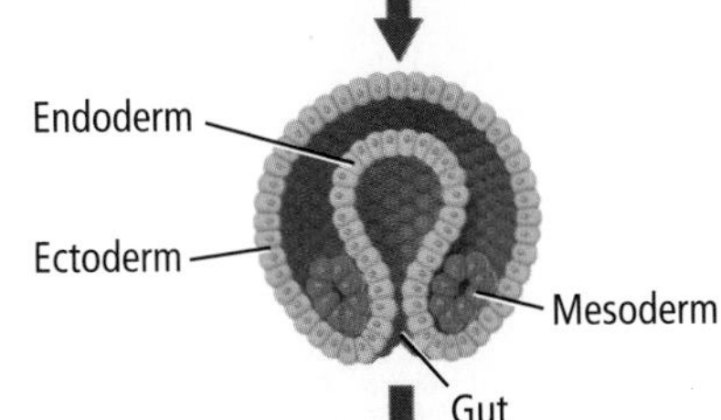

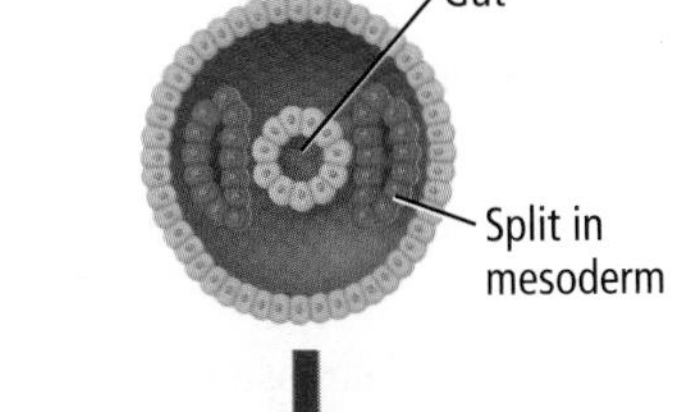

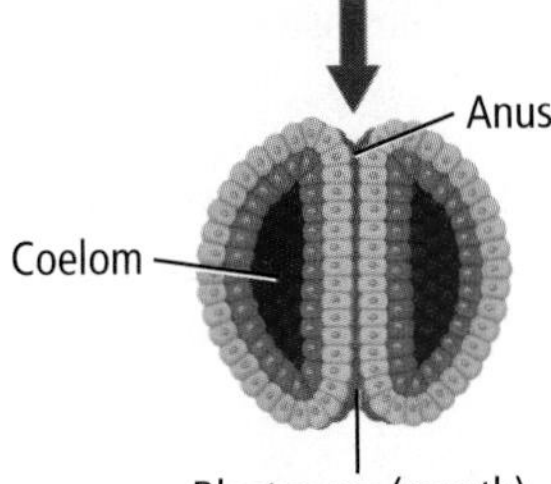

A If one cell is removed from a protostome at the four-cell stage, the development of all embryos is altered. If a cell is removed in a deuterostome at this stage, each cell or group of cells is not altered and will develop into a normal embryo.

B Another difference is apparent at the eight-cell stage. In protostomes, the four cells are between the other four cells. In deuterostomes, the cells align.

C A blastula forms in both types of development.

D Note the location of mesoderm as the gastrula forms.

E As the embryo continues to develop, the mesoderm splits in protostomes to form the coelom. In deuterostomes, the coelom is formed from pouches of mesoderm that separate from the gut.

F The opening in the gastrula, called a blastopore, becomes the mouth in protostomes and the anus in deuterostomes.

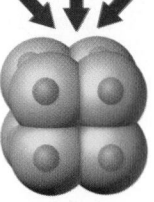

Deuterostome Development

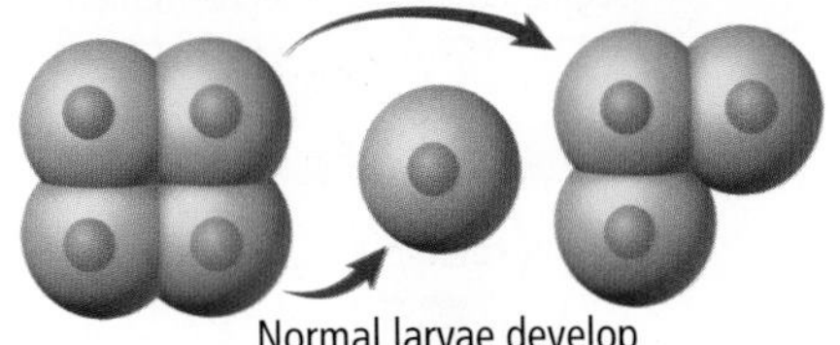

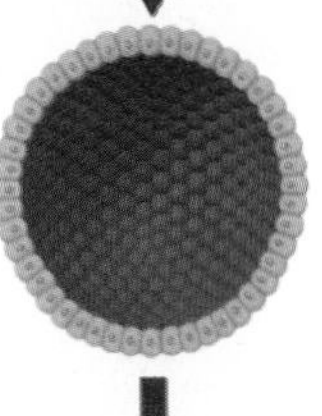

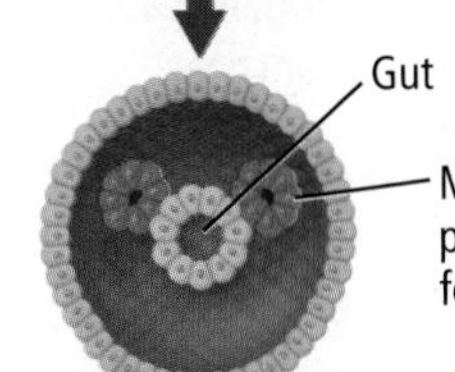

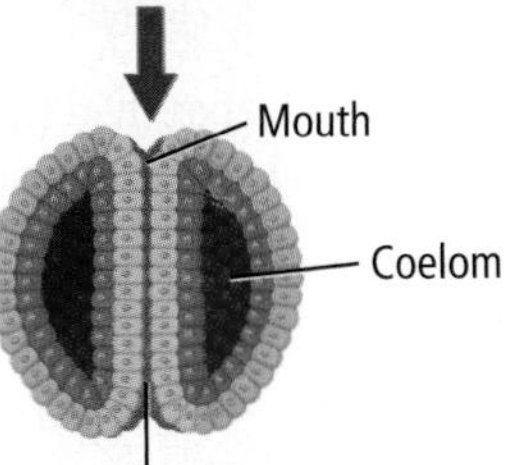

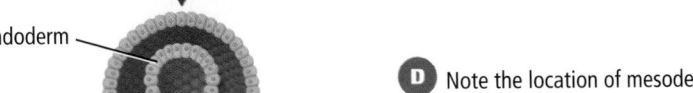

Concepts In Motion **Interactive Figure** To see an animation of protostomes and deuterostomes, visit biologygmh.com. **Biology Online**

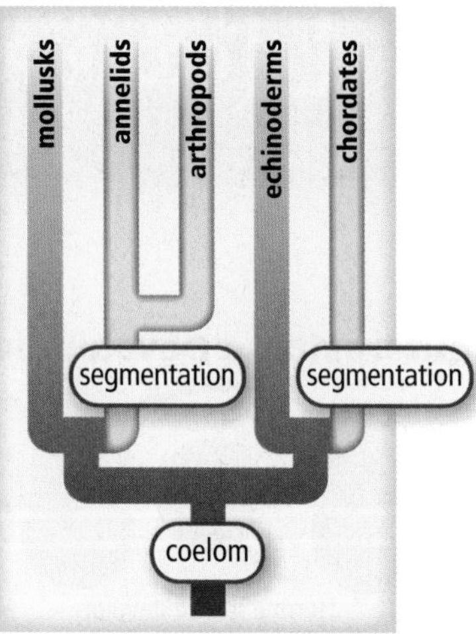

Scorpion

■ **Figure 24.13** Segmentation enables a scorpion to move its stinger in different directions to attack prey or for defense.

Segmentation

Examine the next branching point on the evolutionary tree in **Figure 24.13.** Segmentation is an important feature in the evolution of coelomate animals. Just as a chain is constructed from a series of links, segmented animals can be "put together" from a succession of similar parts.

The segmentation, such as that seen in scorpions, has two advantages. First, segmented animals can survive damage to one segment because other segments might be able to carry out the damaged section's function. Second, movement is more effective because segments can move independently. Therefore, the scorpion in **Figure 24.13** has more flexibility and can move in ways that are very complex. Segments allow the scorpion to arch its tail over its back to sting prey.

Section 24.2 Assessment

Section Summary

- Animal phylogeny can be compared to a tree with branches.
- The branches of a phylogenetic evolutionary tree show the relationships among animals.
- Animal phylogeny can be determined, in part, by the animal's type of body cavity or lack of a body cavity.
- After gastrulation, two types of development can occur in coelomate animals.
- Segmentation is an important feature in some coelomate animals.

Understand Main Ideas

1. MAIN Idea **Explain** how body symmetry is related to the phylogeny of animals.
2. **Name** the features marking the main branching points on the evolutionary tree of animals.
3. **Illustrate** how body cavities distinguish branches of development of animals with bilateral symmetry.
4. **Compare and contrast** deuterostome and protostome development.

Think Critically

5. **Diagram** Make diagrams of animals not shown in **Figure 24.9** that have radial and bilateral symmetry. Indicate the type of symmetry by showing planes passing through the animals. Label each animal as having either radial or bilateral symmetry.

WRITING in **Biology**

6. Write a paragraph summarizing the differences among coelomates, pseudocoelomates, and acoelomates.

Biology Online **Self-Check Quiz** biologygmh.com

Section 24.3

Sponges and Cidarians

Reading Preview

Objectives

- **Distinguish** structure and function in sponges and cnidarians.
- **Describe** the diversity of sponges and cnidarians.
- **Evaluate** the ecology and importance of sponges and cnidarians.

Review Vocabulary

diploid: cell with two of each kind of chromosome

New Vocabulary

filter feeder
sessile
cnidocytes
nematocyst
gastrovascular cavity
nerve net
polyp
medusa

MAIN Idea **Sponges and cnidarians were the first animals to evolve from a multicellular ancestor.**

Real-World Reading Link Have you ever double-bagged your groceries? If so, you have an idea of what a sponge is like—a layer, or sac, of cells within another sac of cells. These sacs of cells are among the first animals to evolve from the common ancestor of all animals.

Sponges

If you examine a living sponge, you might wonder how these animals do so much with so little. They have no tissues, no organs, and most have no symmetry. You can break apart a sponge into its individual cells and those cells will come together again to form a sponge. Other animals cannot do this.

Locate sponges on the evolutionary tree in **Figure 24.14.** They are in the phylum Porifera (po RIF uh ruh), which contains between 5000 and 10,000 members. Most live in marine environments. Biologists hypothesize that sponges evolved from the colonial choanoflagellates because sponges have cells that look similar to these protist cells.

Body structure Notice the asymmetrical appearance and bright colors of the sponge in **Figure 24.14.** It is difficult to think that these are animals, especially if you see one washed up on a beach where it might appear as a black blob. Recall that tissues form from ectoderm, endoderm, and mesoderm in a developing embryo. Sponge embryos do not develop endoderm or mesoderm, and, therefore, sponges do not develop tissues. How does a sponge's body function without tissues?

■ **Figure 24.14** It might be hard to believe that the sponges on the right are animals that take in and digest food, grow, and reproduce.

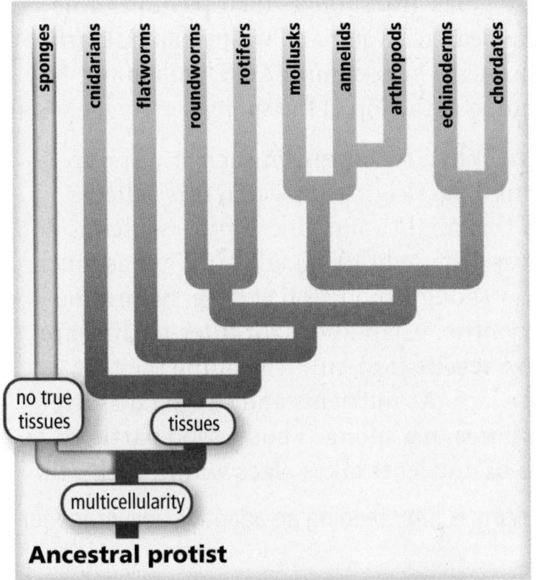

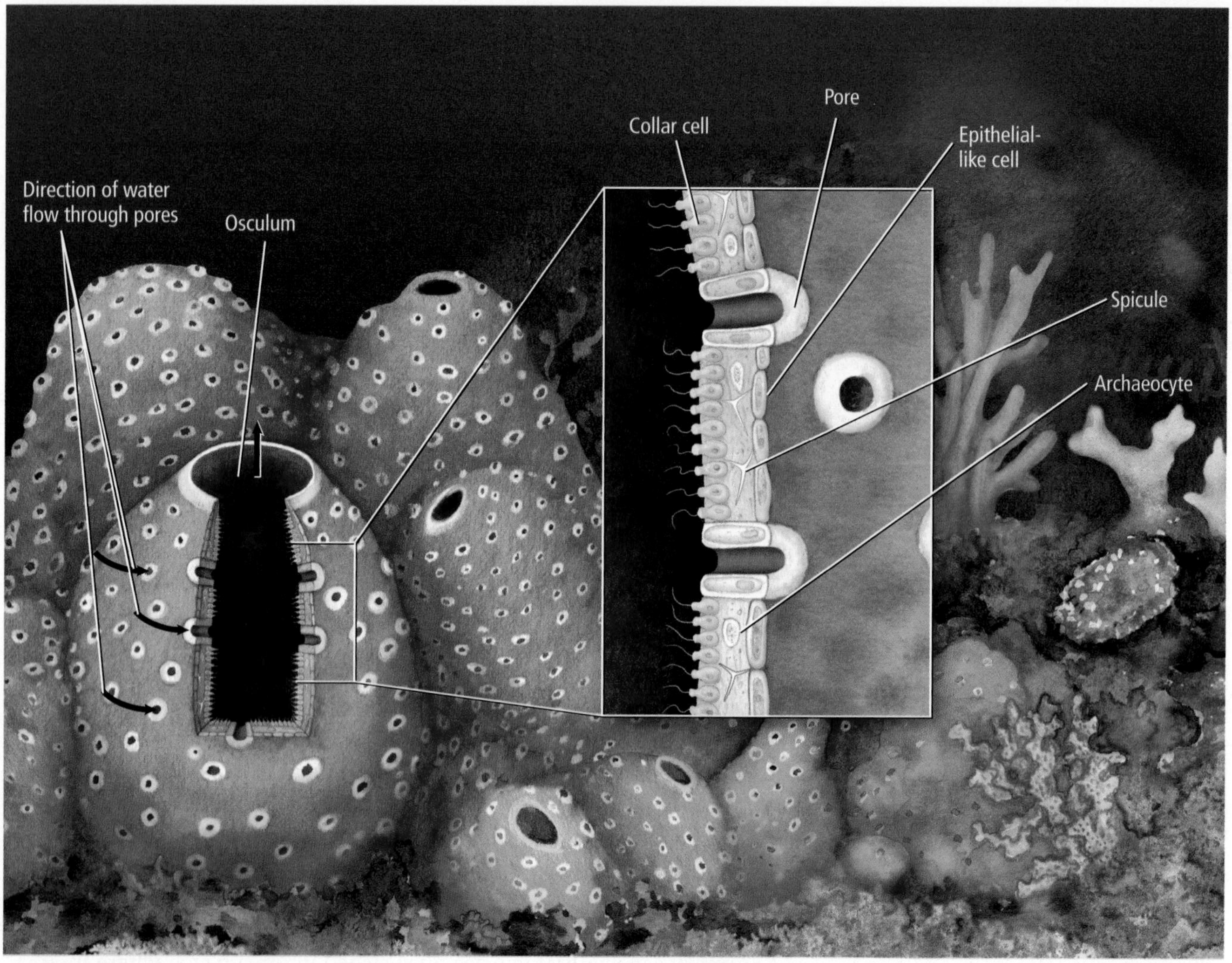

Figure 24.15 Sponges have no tissues or organs and have a body made of two layers of cells.

Concepts In Motion

Interactive Figure To see details of the anatomy of a sponge, visit biologygmh.com.

Two layers of independent cells with a jellylike substance between the layers accomplish all of the life functions of sponges. As illustrated in **Figure 24.15,** epithelial-like cells cover the sponge and protect it. Collar cells with flagella line the inside of the sponge. As collar-cell flagella whip back and forth, water is drawn into the body of the sponge through pores. These pores give sponges their phylum name Porifera, which means "pore-bearer." Water and waste materials are expelled from the sponge through the osculum (AHS kyuh lum), which is the mouthlike opening at the top of the sponge.

Feeding and digestion When an organism such as a sponge gets its food by filtering small particles from water, it is called a **filter feeder.** Even though this might sound like a process that is not very active, consider that a sponge only 10 cm tall can filter as much as 100 L of water each day. Although sponges have free-swimming larvae, the adults move very little. Adaptations for filter-feeding are common in animals that are **sessile** (SES sul)—meaning they are attached to and stay in one place. As nutrients and oxygen dissolved in water enter through the pores in a sponge's body, food particles cling to the cells. Digestion of nutrients takes place within each cell.

Reading Check Infer Why is filter feeding an adaptive advantage for sponges?

Study Tip

Think Aloud Read the text and captions aloud. As you read, say aloud your questions and comments. For instance, when you come to the mention of **Figure 24.15,** look at the figure to say how it relates to the text.

Demosponge

■ **Figure 24.16** Bath sponges are harvested from the sea and processed for human use.

Support Within the jellylike material that lies between the two cell layers of a sponge are amoeba-like cells—cells that can move and change shape. These amoeba-like cells are called archaeocytes (ar kee OH sites) and are illustrated in **Figure 24.15.** These cells are involved in digestion, production of eggs and sperm, and excretion. Archaeocytes also can become specialized cells that secrete spicules (SPIH kyuhls)—the support structures of sponges. Spicules are small, needlelike structures made of calcium carbonate, silica, or a tough fibrous protein called spongin.

Sponge diversity Biologists place sponges into three classes based on the type of support system they have. Most sponges belong to class Demospongiae (deh muh SPUN jee uh), the demosponges, and have spicules composed of spongin fibers, silica, or both. Natural bath sponges, like the ones in **Figure 24.16,** have spongin support. Class Calcarea (kal KER ee uh) consists of sponges with spicules composed of calcium carbonate. Calcareous sponges, like the one in **Figure 24.17,** often have a rough texture because the calcium carbonate spicules can extend through the outer covering of the sponge. The sponges in class Hexactinellida (heks AK tuh nuh LEE duh) are called glass sponges and have spicules composed of silica. These spicules join together to form a netlike skeleton that often looks like spun glass, as illustrated in **Figure 24.17.**

■ **Figure 24.17** Calcareous sponges are small and have a rough texture. The skeletons of glass sponges look like brittle spun glass.

Calcareous sponge

Glass sponge skeleton

A Sperm are released into the water and float on water currents to other sponges.
B Sperm are caught by the collar cells of another sponge, and eggs are fertilized internally. Free-swimming larvae are released.
C The larvae swim using tiny cilia.
D A larva eventually settles on a surface.
E A sessile larva develops into an adult that can reproduce.

Figure 24.18 Sexual reproduction in sponges requires water currents to carry sperm from one sponge to another.
Evaluate ***Is fertilization internal or external in sponge sexual reproduction?***

Response to stimuli A sponge does not have a nervous system. They do have epithelial-like cells that detect external stimuli, such as touch or chemical signals, and respond by closing their pores to stop water flow.

Reproduction Sponges can reproduce asexually by fragmentation, through budding, or by producing gemmules (JEM yewlz). In fragmentation, a piece of sponge that is broken off due to a storm or other event develops into a new adult sponge. In budding, a small growth, called a bud, forms on a sponge, drops off, and settles in a spot where it grows into a new sponge. Some freshwater sponges form seedlike particles called gemmules during adverse conditions like droughts or freezing temperatures. Gemmules contain sponge cells protected by spicules that will survive and grow again when favorable conditions occur.

Most sponges reproduce sexually, illustrated in **Figure 24.18.** Some sponges have separate sexes, but most sponges are hermaphrodites. Recall that a hermaphrodite is an animal that can produce both eggs and sperm. During reproduction, eggs remain within a sponge, while sperm are released into the water. Sperm released from one sponge can be carried by water currents to the collar cells of another sponge. The collar cells then change into specialized cells that carry the sperm to an egg. After fertilization occurs, the zygote develops into a larva that is free-swimming and has flagella. The larva eventually attaches to a surface, then develops into an adult.

VOCABULARY
ACADEMIC VOCABULARY
Survive:
to remain alive.
Sponge gemmules survive despite adverse conditions.

 **Reading Check** **Describe** the methods by which sponges reproduce.

Sponge ecology Although spicules and toxic or distasteful compounds in sponges discourage most potential predators, sponges are food for some tropical fishes and turtles. Sponges also are common habitats for a variety of worms, fishes, shrimp, and colonies of symbiotic green algae. Some sponges even live on and provide camouflage for mollusks, as shown in **Figure 24.19.**

Sponges also are beneficial to humans. Sponges with spicules made of spongin fibers often are used for household scrubbing purposes. Medical research is focusing on sponge chemicals that appear to discourage prey and prevent infection. Ongoing studies of these sponge chemicals as possible pharmaceutical agents have shown that they might have antibiotic, anti-inflamatory, or antitumor possibilities. They also might have potential importance as respiratory, cardiovascular, and gastrointestinal medicines.

Connection to Health For example, researchers discovered a powerful antitumor substance in the deep water sponge shown in **Figure 24.20.** This substance, discodermolide (disk uh DER muh lide), stops cancer cells from dividing by breaking down the nucleus and rearranging the microtubule network. Recall from Chapter 7 that microtubules are part of a cell's skeleton and help the cell maintain its shape. Note the differences in the nuclei and microtubules between the untreated and treated cancer cells in **Figure 24.20.**

Figure 24.19 This crab hides from predators by carrying a living sponge on its back. The crab uses two pairs of legs to hold the sponge in place.

Figure 24.20 Discodermolide, a substance taken from the sponge *Discodermia dissoluta,* breaks down the nucleus in a cancer cell and rearranges its microtublules.

Discodermia dissoluta

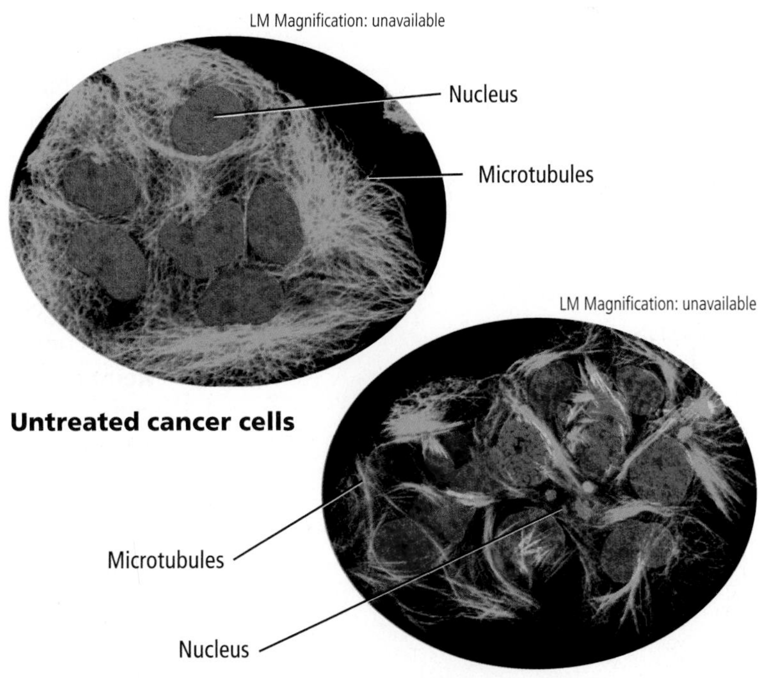

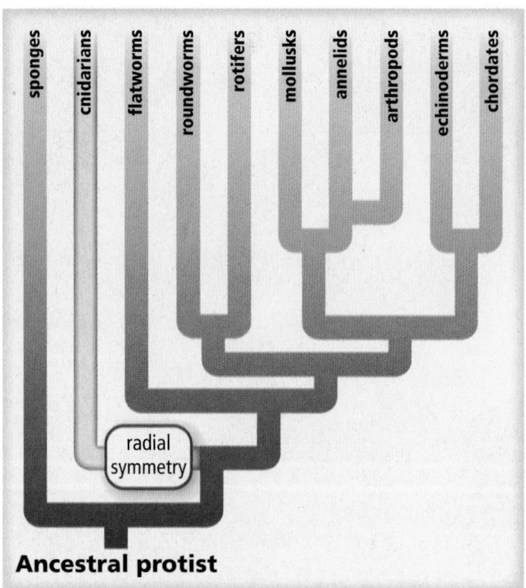

Jellyfish—free floating

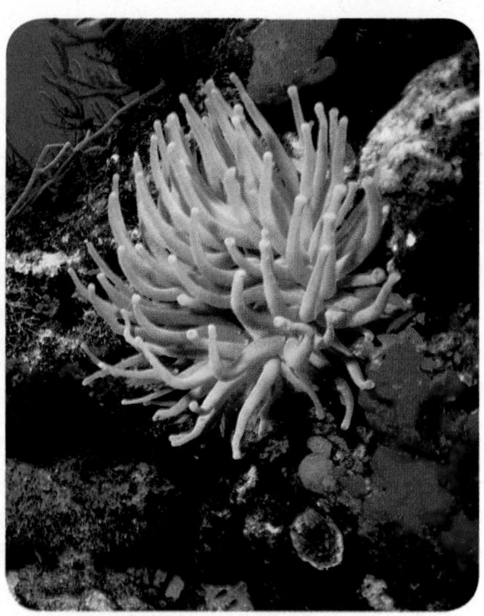
Sea anemone—sessile

■ **Figure 24.21** Cnidarians have radial symmetry and can be free floating or sessile.
Explain how radial symmetry helps a cnidarian obtain food.

Cnidarians

Imagine that you go snorkeling around a coral reef, and you wear a bodysuit to protect yourself from the stings of jellyfishes that float on the water. Later, when you go ashore to visit a tidepool, you might see colorful sea anemones that look somewhat like flowers. The jellyfish and sea anemone in **Figure 24.21** belong to phylum Cnidaria (ni DARE ee uh). This phylum consists of about 10,000 species, most of which are marine.

Body structure Like sponges, cnidarians (ni DARE ee uns) have one body opening and most have two layers of cells. However, in cnidarians, the two cell layers are organized into tissues with specific functions. The outer layer functions in protecting the internal body, while the inner layer functions mainly in digestion. Because cnidarians have tissues, they also have symmetry. As shown in **Figure 24.21,** cnidarian bodies have radial symmetry. Recall that radial symmetry enables slow moving or sessile animals to detect and capture prey from any direction. Cnidarians are adapted to aquatic floating or sessile attachment to surfaces under the water.

Feeding and digestion Cnidarian tentacles are armed with stinging cells called **cnidocytes** (NI duh sites). Cnidarians get their name from these stinging cells. Cnidocytes contain nematocysts, as shown in **Figure 24.22.** A **nematocyst** (nih MA tuh sihst) is a capsule that holds a coiled, threadlike tube containing poison and barbs.

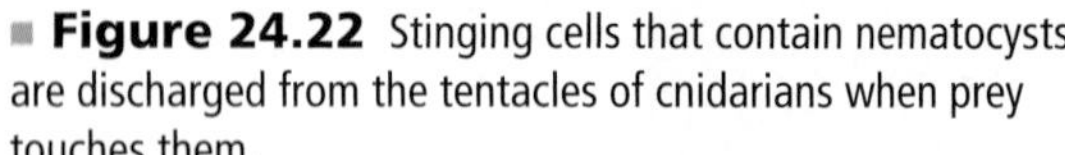

■ **Figure 24.22** Stinging cells that contain nematocysts are discharged from the tentacles of cnidarians when prey touches them.

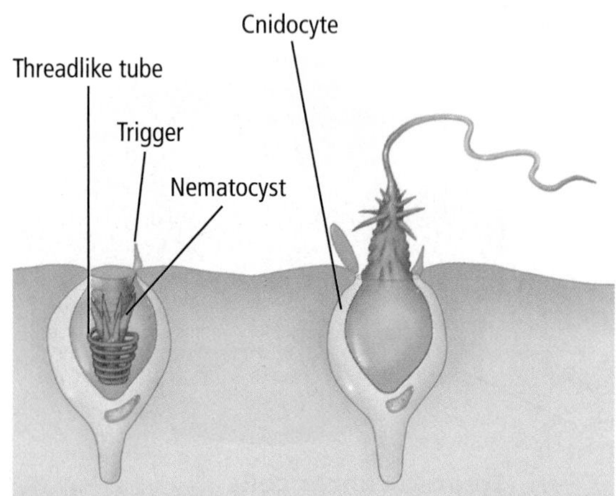

Connection to Physics A nematocyst works like a tiny but very powerful harpoon. Recall from Chapter 7 that osmosis is the diffusion of water through a selectively permeable membrane. The pressure provided by this flow of water is called osmotic pressure. The water inside an undischarged nematocyst is under an osmotic pressure of more than 150 atmospheres. This pressure is about 20 times the pressure in an inflated bicycle tire.

In response to being touched or to a chemical stimulus, the permeability of the nematocyst membrane increases, allowing more water to rush in. As the osmotic pressure increases, the nematocyst discharges forcefully. A barb is capable of penetrating a crab shell.

Nematocyst discharge is one of the fastest cellular processes in nature. It happens so quickly—in just 3/1000ths of a second—that it is impossible to escape after touching these cells. After capture by nematocysts and tentacles, the prey is brought to the mouth of the cnidarian.

The inner cell layer of cnidarians surrounds a space called the **gastrovascular** (gas troh VAS kyuh lur) **cavity,** illustrated in **Figure 24.23.** Cells lining the gastrovascular cavity release digestive enzymes over captured prey. Undigested materials are ejected through the mouth. Recall that digestion occurs within each cell of a sponge. However, in cnidarians, digestion takes place in the gut cavity—a major evolutionary adaptation.

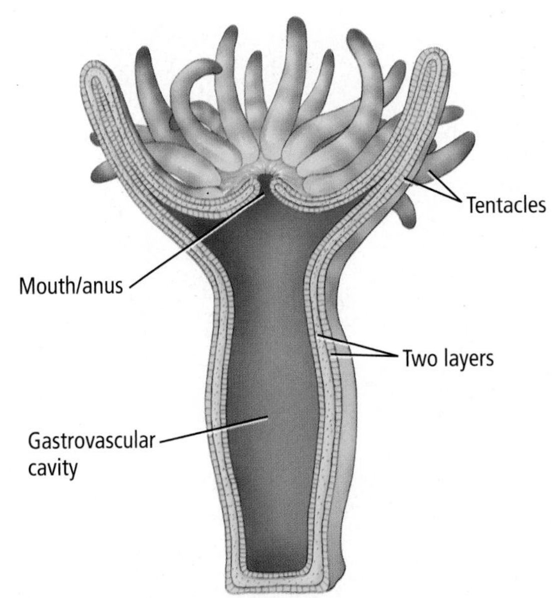

Figure 24.23 A cnidarian's mouth leads directly into its gastrovascular cavity. Because the digestive tract has only one opening, wastes are expelled through the mouth.

Response to stimuli In addition to cells adapted for digestion, cnidarians have a nervous system consisting of a **nerve net** that conducts impulses to and from all parts of the body. The impulses from the nerve net cause contractions of musclelike cells in the two cell layers. The movement of tentacles during prey capture is the result of contractions of these musclelike cells. Cnidarians have no blood vessels, respiratory systems, or excretory organs. Look at **Table 24.1** to compare the structures and functions of sponges and cnidarians.

Reading Check Compare How does a cnidarian's response to stimuli differ from a sponge's response?

Concepts In Motion Interactive Table To explore more about sponges and cnidarians, visit biologygmh.com.

Table 24.1 Comparison of Sponges and Cnidarians

	Sponges	Cnidarians
Example		
Body Plan	• Generally has asymmetry	• Has radial symmetry
Feeding and digestion	• Filter feed • Digestion takes place within individual cells	• Capture prey with nematocysts and tentacles • Digestion takes place in gastrovascular cavity
Movement	• Sessile	• Aquatic floating or sessile
Response to stimuli	• No nervous system • Cells react to stimuli	• Simple nervous system consisting of a nerve net
Reproduction	• Hermaphrodites reproduce sexually • Asexual reproduction by fragmentation, budding, or gemmule production	• Separate sexes reproduce sexually • Polyp stage reproduces asexually by budding

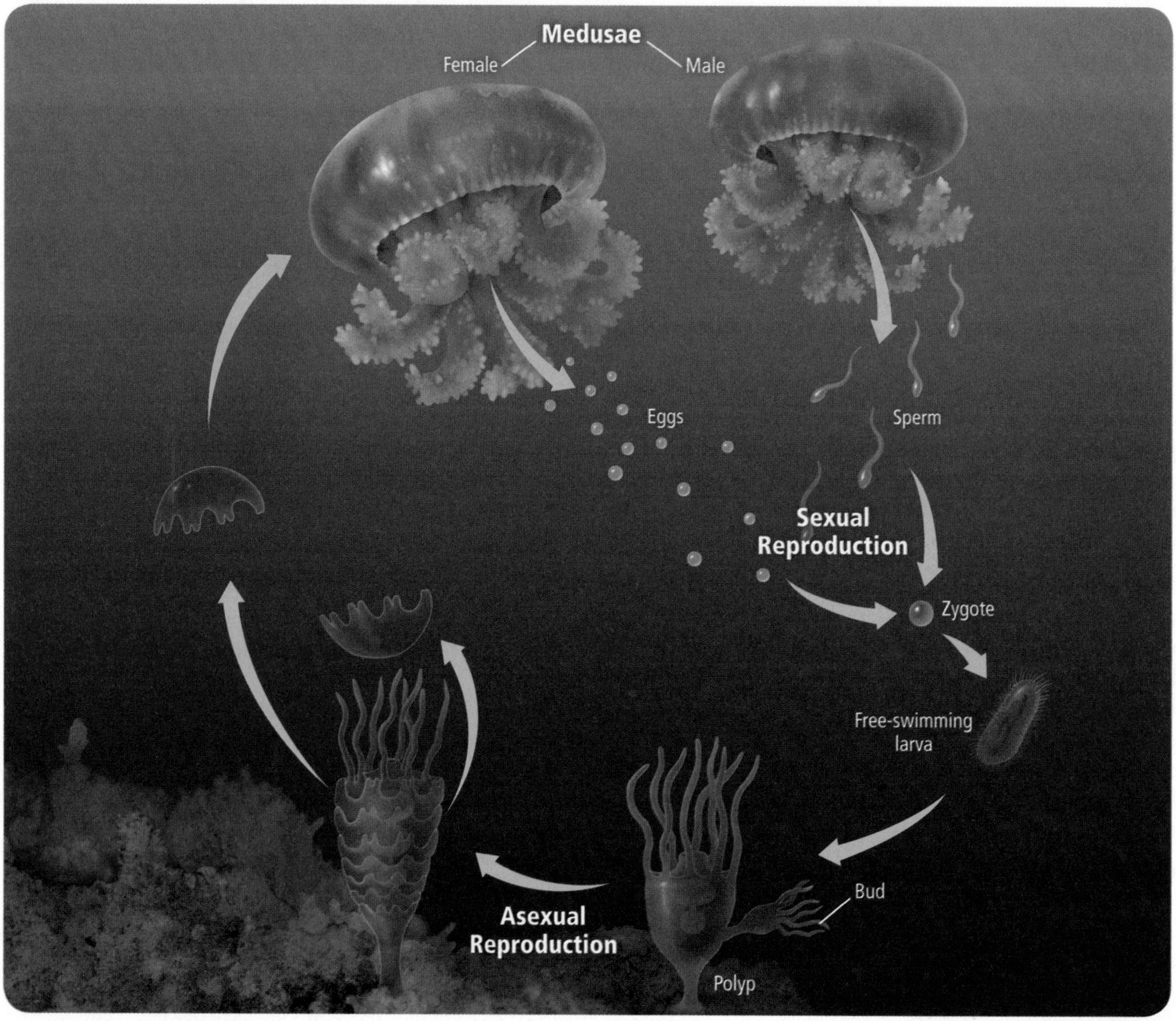

■ **Figure 24.24** Jellyfishes reproduce by alternating sexual and asexual stages of their life cycle.

Concepts In Motion

Interactive Figure To see an animation of the reproductive cycle in cnidarians, visit biologygmh.com.

CAREERS IN BIOLOGY

Marine Ecologist Using submersibles and deep-sea robots, a marine ecologist studies the relationships between marine animals and their environments. For more information on biology careers, visit biologygmh.com.

Reproduction In addition to stinging cells, cnidarians have another adaptation not seen in most animals of recent origin. Most cnidarians have two body forms: a **polyp** (PAH lup) with a tube-shaped body and a mouth surrounded by tentacles, and a **medusa** (mih DEW suh) (plural, medusae) with an umbrella-shaped body and tentacles that hang down. The mouth of a medusa is on the ventral surface between the tentacles.

The two body forms of cnidarians can be observed in the life cycle of jellyfishes, illustrated in **Figure 24.24.** To reproduce, jellyfishes in the medusa stage release eggs and sperm into the water where fertilization occurs. The resulting zygotes eventually develop into free-swimming larvae that settle and grow into polyps. These polyps reproduce asexually to form new medusae. It would be easy to confuse the life cycle of cnidarians with the alternation of generations you studied in plants. However, in plants, one generation is diploid and the other is haploid. In cnidarians, both the medusae and polyps are diploid animals.

Reading Check **Compare** How are the larvae of sponges and cnidarians similar?

Cnidarian diversity There are four main classes of cnidarians: Hydrozoa, the hydroids; two classes of jellyfishes—Scyphozoa and Cubozoa (the box jellyfishes); and Anthozoa, the sea anemones and corals.

Hydroids Most of the approximately 2700 known species of hydroids have both polyp and medusa stages in their life cycles. Most hydroids form colonies, such as the Portuguese man-of-war in **Figure 24.25.** Another well-known hydroid is the freshwater hydra, which is unusual because it has only a polyp stage.

Jellyfishes There are about 200 known species of jellyfishes. They are transparent or translucent in appearance and float near the water's surface. The medusa is the dominant body form, although a polyp stage does exist. They are called jellyfishes because the substance between the outer body covering and the inner body wall is jellylike. The structure of the inner and outer body layers with the jellylike structure between can be compared to a jelly sandwich. The box jellyfishes take their name from the boxlike medusae that are their dominant form. The stings of some box jellyfish species can be fatal to humans.

Sea anemones and corals Generally colorful and inviting, sea anemones and corals still possess stinging cells like all cnidarians. The 6200 known species of sea anemones and corals are different from the jellyfishes because the polyp stage is the dominant stage of their life cycles. Recent research indicates that these anthozoans might have bilateral symmetry. This would alter the evolutionary tree because this adaptation usually is seen only in animal groups that evolved later than cnidarians.

Sea anemones live as individual animals, while corals live in colonies of polyps. Corals secrete protective calcium carbonate shelters around their soft bodies. The living portion of a coral reef is a thin, fragile layer growing on top of the shelters left behind by previous generations. Coral reefs form from these shelters over thousands of years.

Coral polyps extend their tentacles to feed, as shown in **Figure 24.26.** They also harbor symbiotic photosynthetic protists called zooxanthellae (zoh oh zan THEH lee). The zooxanthellae produce oxygen and food that corals use, while using carbon dioxide and waste materials produced by the corals. These protists are primarily responsible for the bright colors found on healthy coral reefs.

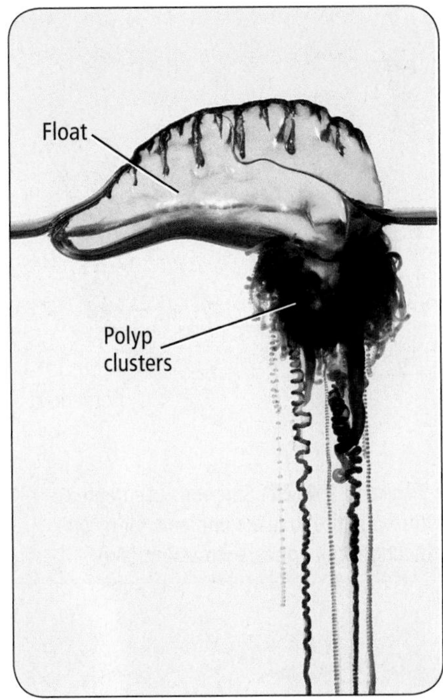

Figure 24.25 This Portuguese man-of-war is composed of a colony of hydroids. One hydroid polyp forms the large float, while other hydroid polyps cluster beneath the float.

Figure 24.26 Coral polyps capture food by extending their tentacles.

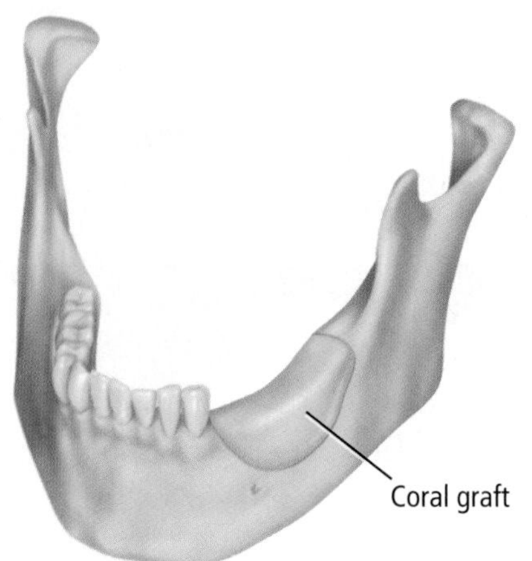

■ **Figure 24.27** Surgeons use treated hydroxyapatite to make implants for reconstructing facial bones, such as this jaw.

The health of a coral reef depends on proper water temperature, adequate light, and appropriate water depth. If these environmental conditions deteriorate in areas where there are reefs, the health of the reef might also deteriorate. You can examine this problem in **Data Analysis Lab 24.1.**

Cnidarian ecology Mutualism, a relationship in which both organisms benefit, is common in cnidarians. One species of sea anemone wraps itself around hermit crabs' shells; the anemones obtain food scraps and the crabs are protected. Some sea slugs feed on cnidarians and incorporate the unfired nematocysts into their bodies for their own defense. As shown in the photo at the beginning of the chapter, clown fishes are protected by the tentacles of anemones. One theory as to how clown fishes are protected from the tentacles of anemones is that the fish incorporates mucus from an anemone into its own mucous coating, which prevents the nematocysts from discharging.

People benefit from cnidarians in many ways. Some people enjoy visiting a coral reef. In the medical field, some stony coral species are used in surgical procedures. A calcium phosphate mineral in coral called hydroxyapatite (hi DROX ee ap uh TITE) can be treated so that it has the same structure and chemical composition as human bone. Small pieces of coral are implanted as bone grafts, especially in face and jaw reconstruction and in arm and leg surgery. The grafts anchor to the adjacent bone, as shown in **Figure 24.27,** and eventually are replaced by new human bone growth.

DATA ANALYSIS LAB 24.1

Based on Real Data*

Interpret Data

Where are coral reefs being damaged?
Some corals have ejected their symbiotic algae and become bleached, or lost their coloring. Coral reef bleaching is a common response to reef ecosystem damage. However, some corals appear to be recovering from bleaching.

Data and Observations

The graph indicates the percentage of the damage that has occurred to specific reefs.

Degradation of Various Coral Reef Ecosystems

Percent degradation: 0, 10, 20, 30, 40, 50, 60, 70, 80

Coral reef areas: Outer Great Barrier Reef, Inner Great Barrier Reef, Torres Strait Islands, Southern Red Sea, Northern Red Sea, Belize, Bermuda, Cayman Islands, Bahamas, Eastern Panama, Moreton Bay, U.S. Virgin Islands, Western Panama, Jamaica

Think Critically

1. **Interpret** What part of the world has suffered the most damage to its coral reefs? What part of the world has suffered the least damage to its reefs?
2. **Model** On a world map, locate the coral reefs noted in the graph. Color code the map based on the percent of degradation.

*Data obtained from: Pandolfi, J.M. et al., 2003. Global trajectories of the long-term decline of coral reef ecosystems. *Science* 301 (5635): 955–958.

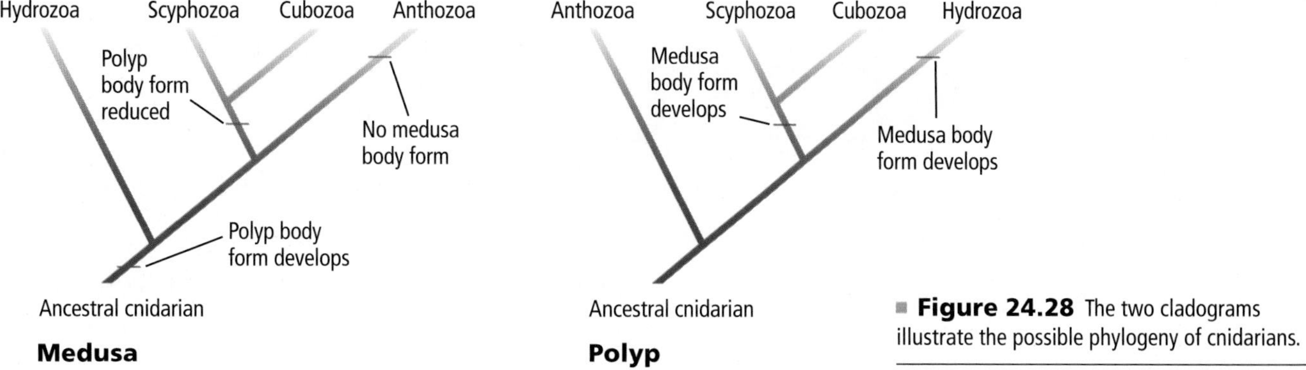

Figure 24.28 The two cladograms illustrate the possible phylogeny of cnidarians.

Evolution of cnidarians There are two major interpretations of the phylogeny of cnidarians. The fact that cnidarians have two body forms—medusa and polyp—raises the question of whether the ancestral cnidarian had a medusa or a polyp body form. The cladograms in **Figure 24.28** present both interpretations.

In the cladogram on the left, the ancestral cnidarian has a medusa body form. As cnidarians evolved, a polyp stage developed. The life cycles of hydrozoans have both polyp and medusa stages. As the scyphozoans and cubozoans developed, the medusa stage became the dominant stage in their life cycles. The most highly evolved cnidarians, the anthozoans, have no medusa stage.

In the cladogram above, the ancestral cnidarian has a polyp body form. The anthozoans evolved first, and the polyp stage is the dominant stage of their life cycles. The medusa stage evolved independently in hydrozoans and in scyphozoans and cubozoans. As you examine the cladograms, notice how the classes of cnidarians are arranged in each.

Section 24.3 Assessment

Section Summary

- Sponges can be described according to animal features they do not have and according to features they do have.
- Sponges do not have tissues, but carry out the same life functions as other animals.
- Cnidarians have unique features that other animals do not have.
- Cnidarians have more highly evolved body forms and structures than sponges.
- Sponges and cnidarians are important to the ecology of their habitats and to humans.

Understand Main Ideas

1. MAIN Idea **Explain** why sponges and cnidarians were the first animals to evolve.
2. **Describe** the differences between the body plans of sponges and cnidarians.
3. **List** two characteristics that are unique to sponges and two characteristics that are unique to cnidarians.
4. **Demonstrate** your knowledge of cnidarians by describing how they affect other marine organisms.

Think Critically

5. **Hypothesize** How are nematocysts an adaptive advantage for cnidarians?

MATH in Biology

6. Review the text under the heading *Cnidarian diversity.* Make a circle graph that shows the proportions of each of the three groups of cnidarians to the total numbers of cnidarians. In addition to the groups in this section, there are 900 species of other cnidarians. Analyze this information and hypothesize why one group is so much smaller than the others.

BioDiscoveries

New Species Everywhere

When Wildlife Conservation Society researcher Rob Timmins went to the market, he didn't find a great sale—he found a new species. While in a food market in Ben Lak, Laos, Timmins saw some unusual black- and brown-striped rabbits. DNA analysis of tissue samples confirmed that the rabbit was a new species, now named the Annamite rabbit.

Discovering new species A species is a group of genetically distinct organisms that share common characteristics and are capable of interbreeding. New species—animals that were previously unknown to scientists—are being discovered all the time. Recent finds include the Christmas tree coral—a new species of coral found off the coast of southern California—and a new species of honeyeater bird found on the island of New Guinea.

This white Christmas tree coral was discovered by researchers Milton Love (UC Santa Barbara) and Mary Yoklavich (NOAA NMFS Santa Cruz) from the mini-submersible *Delta* during surveys on rocky banks off the coast of southern California in 150-m-deep water.

Cataloging species Whether in the Amazon forest or through deep-sea exploration, in-depth cataloging of species is revealing that there are still many species left to identify. The Census of Marine Life (CoML), a multinational project to catalog sea life, found 106 new species of marine fish in their 2004 survey of the world's oceans. That's an average of more than two new species per week.

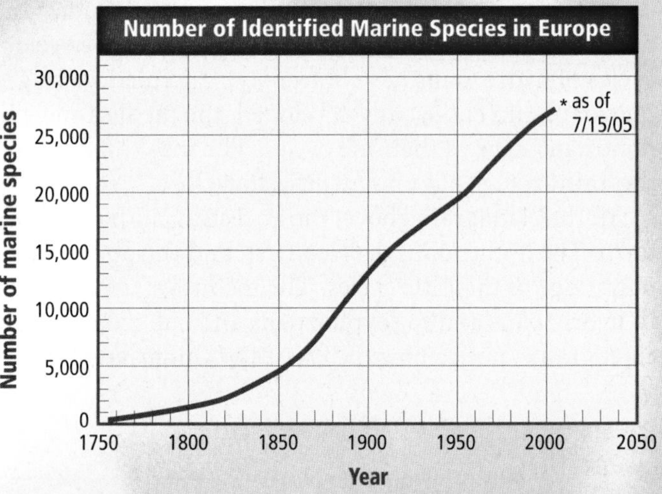

Future finds? The graph above shows how the number of known marine species in Europe increased in 255 years. CoML plans to continue surveying the world's sea life through 2010, so the trend observed in Europe also should be seen globally. The continuing discovery of new species shows one way science is constantly changing.

WRITING in Biology

Interpret Data Based on the data in the graph above, estimate the number of marine species in Europe that might be identified by 2050. Explain your answer. Then infer why the rate of identifying new marine species might be higher in other parts of the world than in Europe. Visit biologygmh.com to learn more about finding new species.

BIOLAB

Design Your Own

FIELD INVESTIGATION: WHAT CHARACTERISTICS DO ANIMALS HAVE?

Background: A small pond is an ecosystem in which organisms interact to accomplish essential life functions. They exhibit a wide variety of body plans, obtain food in different ways, and use various methods of movement.

Question: *What kinds of animals live in ponds?*

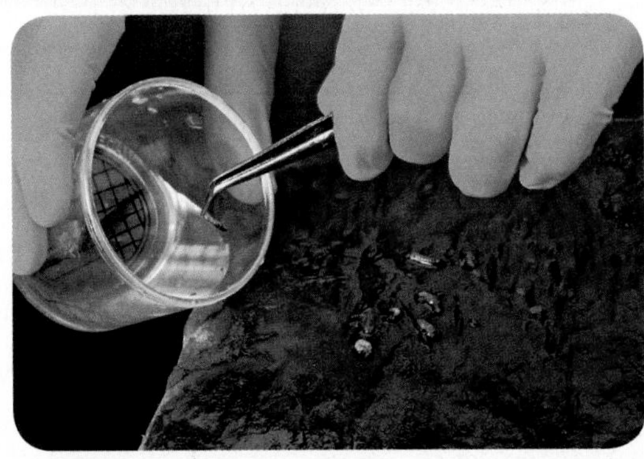

Materials

wading boots
tweezers
aquarium
Petri dishes
dissecting microscopes
Choose other materials that would be appropriate for this lab.

Safety Precautions

WARNING: ***Handle living animals with care.***

Plan and Perform the Experiment

1. Read and complete the lab safety form.
2. Locate a pond to use for your observations and collections. Make sure you have permission to use the pond.
3. Determine methods to observe and record animals you see at the pond that you do not collect.
4. Design and construct a data table to record your observations.
5. Make sure your teacher approves your plan before you proceed.
6. **Cleanup and Disposal** Wash your hands after handling any live organisms. Return the animals and any pond water to the pond. Wash and return all reusable lab materials and correctly dispose of other materials used in the lab as directed by your teacher.

Analyze and Conclude

1. **Use Scientific Explanations** How were you able to determine if the organisms you observed were animals?
2. **Summarize** the adaptations you observed used for obtaining food. Were any of the adaptations similar to those you observed in **MiniLab 24.1?**
3. **Compare and contrast** the methods of movement used by each of the animals you observed.
4. **Interpret Data** Look at drawings or photographs of the animals you observed. What do these illustrations tell you about the body plan of each organism? What gut type does each animal have?
5. **Error Analysis** What other types of observations could you make to verify your conclusions about each organism?

WRITING in Biology

Make a Booklet Choose one of the animals you observed in your pond study. Develop an illustrated booklet that shows how this animal obtains food, how it reproduces, its body plan, and its stages of development. Share the information with your class. To learn more about animals, visit biologygmh.com.

Study Guide

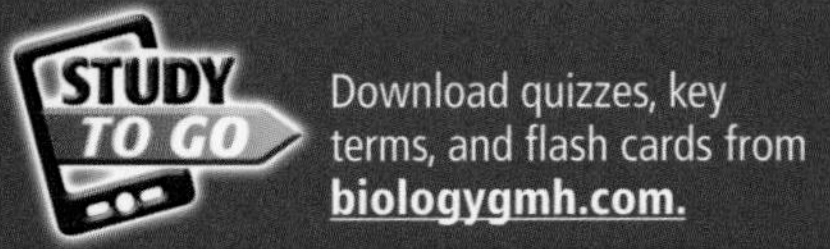

Download quizzes, key terms, and flash cards from biologygmh.com.

FOLDABLES **Compare** all three body plans of animals with bilateral symmetry, and consider why the greatest diversity is found among the coelomates. Determine the levels of diversity of the other two body plans and rank them as second and third, respectively. Explain your reasoning.

Vocabulary	Key Concepts
Section 24.1 Animal Characteristics	
• blastula (p. 696) • ectoderm (p. 697) • endoderm (p. 697) • endoskeleton (p. 693) • exoskeleton (p. 693) • external fertilization (p. 695) • gastrula (p. 696) • hermaphrodite (p. 695) • internal fertilization (p. 695) • invertebrate (p. 693) • mesoderm (p. 697) • vertebrate (p. 693) • zygote (p. 695)	MAIN Idea Animals are multicellular, eukaryotic heterotrophs that have evolved to live in many different habitats. • Animals must get their nutrients from other organisms. • Animals have diverse means of support and live in diverse habitats. • Animal cells do not have cell walls, and most animals have cells that are organized into tissues. • Most animals undergo sexual reproduction, and most animals can move. • During embryonic development, animal cells become tissue layers, which become organs and systems.
Section 24.2 Animal Body Plans	
• acoelomate (p. 701) • anterior (p. 700) • bilateral symmetry (p. 700) • cephalization (p. 700) • coelom (p. 701) • deuterostome (p. 702) • dorsal (p. 700) • posterior (p. 700) • protostome (p. 702) • pseudocoelom (p. 701) • radial symmetry (p. 700) • symmetry (p. 700) • ventral (p. 700)	MAIN Idea Animal phylogeny can be determined, in part, by body plans and the ways animals develop. • Animal phylogeny can be compared to a tree with branches. • The branches of a phylogenetic evolutionary tree show the relationships among animals. • Animal phylogeny can be determined, in part, by the animal's type of body cavity or lack of a body cavity. • After gastrulation, two types of development can occur in coelomate animals. • Segmentation is an important feature in some coelomate animals.
Section 24.3 Sponges and Cnidarians	
• cnidocyte (p. 710) • filter feeder (p. 706) • gastrovascular cavity (p. 711) • medusa (p. 712) • nematocyst (p. 710) • nerve net (p. 711) • polyp (p. 712) • sessile (p. 706)	MAIN Idea Sponges and cnidarians were the first animals to evolve from a multicellular ancestor. • Sponges can be described according to animal features they do not have and according to features they do have. • Sponges do not have tissues, but carry out the same life functions as other animals. • Cnidarians have unique features that other animals do not have. • Cnidarians have more highly-evolved body forms and structures than sponges. • Sponges and cnidarians are important to the ecology of their habitats and to humans.

CHAPTER 24 Assessment

Section 24.1

Vocabulary Review

Match the definitions below with the correct vocabulary terms from the Study Guide page.

1. a hard outer covering that provides support
2. a two-layer sac with an opening at one end formed in embryonic development
3. an animal that produces both eggs and sperm

Understand Key Concepts

Use the diagram below to answer question 4.

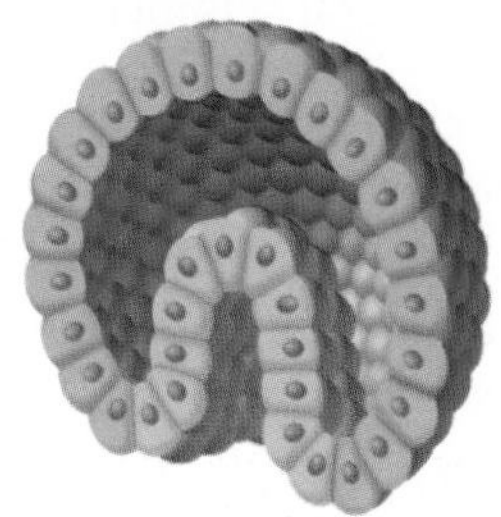

4. The embryo is in which stage of development?
   A. gastrula **C.** egg cell
   B. blastula **D.** zygote
5. Which material is not found in endoskeletons?
   A. calcium carbonate **C.** silica
   B. bone **D.** cartilage
6. Hox genes are active during which process?
   A. cell differentiation **C.** digestion
   B. movement **D.** neural stimulation

Constructed Response

7. **Open Ended** How are animals different from plants?
8. **Open Ended** Describe the advantages and disadvantages of internal and external fertilization.

Think Critically

9. **Interpret** this statement by Hans Spemann, a biologist who studied embryonic development: "We are standing and walking with parts of our body which could have been used for thinking had they developed in another part of the embryo."
10. **Hypothesize** what might happen to an embryo that suffers damage to some mesoderm cells.

Section 24.2

Vocabulary Review

Distinguish between the vocabulary terms in each pair.

11. bilateral symmetry and radial symmetry
12. ventral and dorsal
13. coelom and pseudocoelom

Understand Key Concepts

Use the diagram below to answer questions 14 and 15.

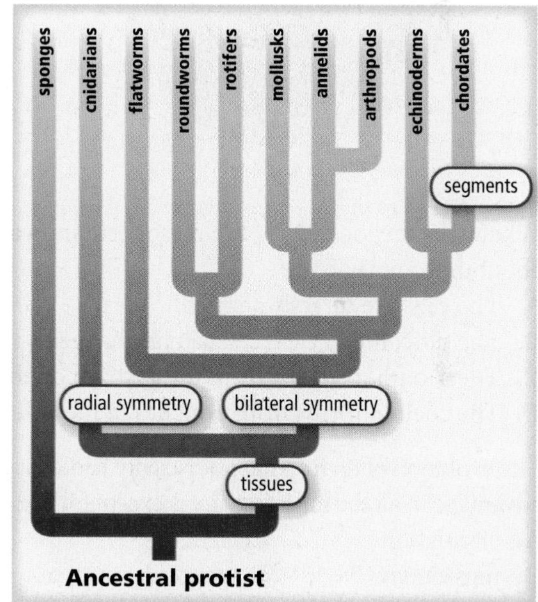

14. Based on the evolutionary tree above, which statement is true?
    A. True tissues evolved after bilateral symmetry.
    B. Segments evolved after bilateral symmetry.
    C. The common animal ancestor was a sponge.
    D. Most animals have radial symmetry.
15. On the evolutionary tree, which animals are related most closely?
    A. an earthworm and a snail
    B. a flatworm and an earthworm
    C. a roundworm and an earthworm
    D. an earthworm and a sea star

16. **CAREERS IN BIOLOGY** An embryologist, a scientist who studies embryos, discovers a new marine animal. When one cell is removed during its early development, this cell develops into a complete animal. This animal is which of the following?
   A. acoelomate
   B. deuterostome
   C. protostome
   D. pseudocoelomate

Use the diagram below to answer question 17.

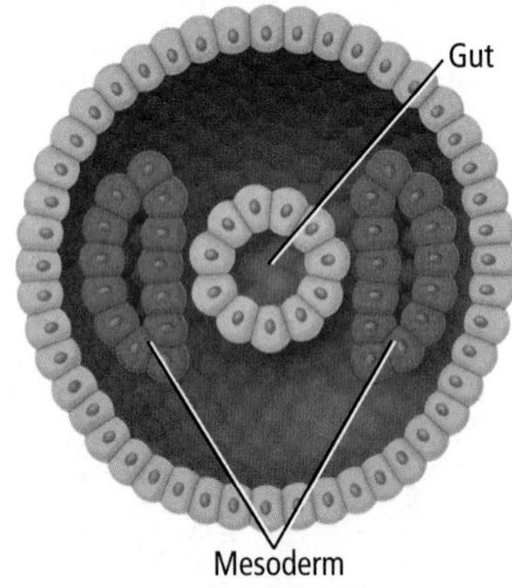

17. What does the location of the mesoderm indicate about this embryo?
   A. The cells are directly aligned.
   B. The outcome of each cell can be changed.
   C. The mouth develops from the gastrula opening.
   D. The coelom forms from pouches of mesoderm.

18. The evolution of an internal body cavity had adaptive advantage in all the following areas except for which?
   A. circulation
   B. movement
   C. feeding
   D. muscular system

19. Based on the evolutionary tree in **Figure 24.8,** what characteristics does an earthworm have that a flatworm does not?
   A. a coelom, a body cavity, bilateral symmetry, and no tissues
   B. a coelom and segmentation
   C. a coelom, protostome development, and segmentation
   D. a pseudocoelom, a body cavity, and bilateral symmetry

20. What is the lighter undersurface of a frog called?
   A. dorsal surface
   B. ventral surface
   C. anterior surface
   D. posterior surface

Constructed Response

21. **Open Ended** Construct a working model of cell differentiation using clay, salt dough, or other materials. Make the first stage, then make that stage into the next, and that stage into the next until you have completed the steps.

22. **Open Ended** Describe how you would choreograph a dance or skit that would illustrate symmetry for elementary school children.

Think Critically

23. **Hypothesize** Biologists have recently determined that some sea anemones seem to possess bilateral symmetry. Hypothesize how this changes ideas for how and when bilateral symmetry evolved.

24. **Recognize Cause and Effect** Explain how segmentation and exoskeletons gave some animals an adaptive advantage over those that were not segmented and did not have exoskeletons.

Section 24.3

Vocabulary Review

For each set of terms below, choose the term that does not belong and explain why it does not belong.

25. cnidocyte, nematocyst, cnidarian, spicule

26. pores, gemmule, filter feeder, nematocyst

27. alternation of generations, polyp, spongin, medusa

Understand Key Concepts

Use the diagram below to answer question 28.

28. The animal in the diagram above possesses which characteristic?
   A. cephalization
   B. cnidocytes
   C. bilateral symmetry
   D. asymmetry

 Chapter Test biologygmh.com

29. Cidarians evolved directly from which group?
   A. sponges
   B. multicellular choanoflagellates
   C. flatworms
   D. animals with bilateral symmetry

Use the diagram below to answer question 30.

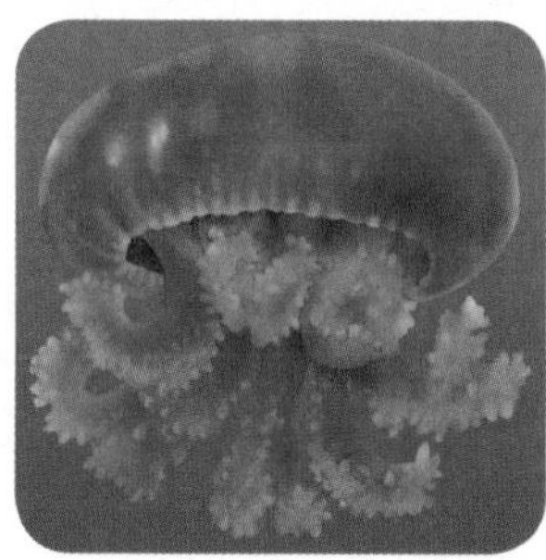

30. How does the animal shown in the diagram reproduce?
   A. fragmentation
   B. external fertilization
   C. internal fertilization
   D. regeneration

31. Which is not a characteristic of sponges?
   A. filter feeding
   B. digestion inside cells
   C. asymmetry
   D. tissues

32. Which pair of words is mismatched?
   A. sponges—filter feeding
   B. cnidarians—nematocysts
   C. sponges—free swimming larva
   D. cnidarians—spicules

Constructed Response

33. **Open Ended** Examine want ads in the paper to see how they are organized, and then use your knowledge of cnidarians to write a want ad that describes an ideal jellyfish homesite.

Think Critically

34. **Calculate** Assume that a sponge filters 1.8 mL of water per second. How much water is filtered in one hour? In 12 hours?

35. **Create** Make a concept map using the following words: coral, polyp, cnidocyte, reef, calcium carbonate, zooxanthellae.

Additional Assessment

36. WRITING in Biology Write an editorial for a newspaper advocating protection for coral reefs. Explain the dangers that corals are facing and make suggestions about what could be done to preserve and protect reefs.

DBQ Document-Based Questions

Transplantation experiments with early embryos of newts show that when tissue responsible for tail development was added into a different fluid-filled gastrula, it caused the effects shown below.

Data obtained from: Niehrs, C. 2003. A tale of tails. *Nature* 424: 375–376.

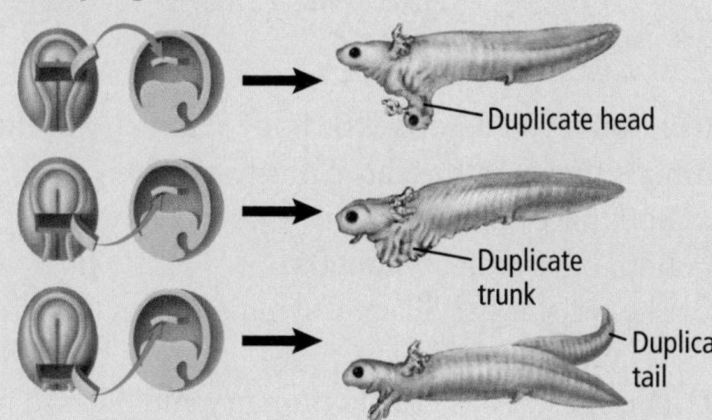

37. When a section from the top of the area was transplanted, where did the new tissue grow?

38. When a section from the bottom of the area was transplanted, where did the new tissue grow?

39. Make a summary statement that describes where new tissue grew when portions of the embryo responsible for tail development were transferred to fluid in the gastrula.

Cumulative Review

40. Review what you learned about microscopic agents that cause disease. Which of these are considered living and which are not? Explain. **(Chapter 18)**

Standardized Test Practice

Cumulative

Multiple Choice

1. Which color of flower is most likely to attract nocturnal pollinators such as bats and moths?
   A. blue
   B. red
   C. violet
   D. white

Use the illustration below to answer questions 2 and 3.

2. How would you describe the body symmetry of the animals shown in the above illustration?
   A. Both have bilateral symmetry.
   B. Both have radial symmetry.
   C. The sea star has bilateral symmetry and the bird has radial symmetry.
   D. The sea star has radial symmetry and the bird has bilateral symmetry.

3. How does the body shape of the sea star help with its survival?
   A. It enables the sea star to capture many kinds of prey.
   B. It enables the sea star to capture prey from many directions.
   C. It enables the sea star to move through the water quickly.
   D. It enables the sea star to move through the water feebly.

4. Which structure in nonvascular plants is similar to roots in vascular plants?
   A. chloroplast
   B. mucilage
   C. rhizoid
   D. sporophyte

5. Which hormone stimulates the ripening of fruit?
   A. auxin
   B. cytokinins
   C. ethylene
   D. gibberellins

Use the diagram below to answer question 6.

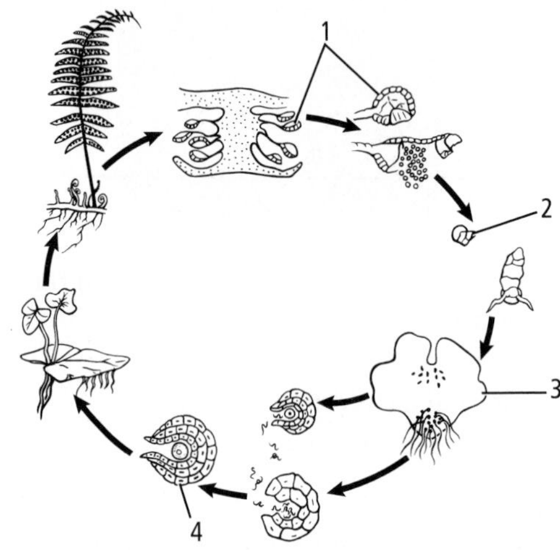

6. At which stage of the fern life cycle does the chromosome number change from haploid to diploid?
   A. 1
   B. 2
   C. 3
   D. 4

7. Which is the role of sclerenchyma cells in plants?
   A. gas exchange
   B. photosynthesis
   C. food storage
   D. support

8. What evidence would help scientists determine that colonial organisms were an early step in the evolution of multicellularity?
   A. similarities in DNA or RNA of early multicellular organisms and colonial unicellular organisms
   B. differences in DNA or RNA of early multicellular organisms and colonial unicellular organisms
   C. similarities of early multicellular organisms and present-day multicellular organisms
   D. differences between early multicellular organisms and present-day multicellular organisms

Short Answer

Use the diagram below to answer question 9.

9. A student conducted an experiment using the above set up. Explain the purpose of this experiment.

10. Which type of fossils would tell a paleontologist the most about the soft tissues of an animal?

11. Explain why most spore-producing plants live in moist areas.

Use the diagram below to answer question 12.

12. How is the age of a tree estimated? What is the approximate age of this tree?

13. Name four flower adaptations that attract insects.

14. What are three kinds of evidence that can be used to confirm whether animals with different body structures are related closely?

Extended Response

15. *Plasmodium* is a sporozoan that causes the disease malaria. Identify the different stages of the sporozoan that occur in mosquitoes. Assess the importance of the stages of the life cycle that occur in mosquitoes.

16. Why does Mendel's law of segregation only apply to organisms that reproduce sexually?

17. Summarize egg development and fertilization in flowering plants.

Essay Question

Pollen analysis, or palynology, is an important tool used in archaeology. Palynologists take samples of soil from archaeological sites and analyze the pollen from different soil layers. By examining the changes in pollen types over time, palynologists can learn about historical land use. The pollen in the soil indicates how the land was used—whether it was cultivated, a forest was cleared, or if it was abandoned.

Using the information in the paragraph above, answer the following question in essay format.

18. Scientists have been trying to find the origin of corn. They know that corn was domesticated from a plant called *teosinte* that grew somewhere in the central valley of Mexico between 12,000 and 6,000 years ago. It often is hard to find intact corncobs because they do not fossilize well. How could a palynologist help determine the origin of corn?

NEED EXTRA HELP?

If You Missed Question . . .	1	2	3	4	5	6	7	8	9	10	11	12	13	14	15	16	17	18
Review Section . . .	23.1	24.2	24.2	21.2	22.3	23.1	22.1	24.1, 24.3	22.3	14.1	23.1	22.2	23.2	24.1	19.2	10.2	23.2	23.3

CHAPTER 25 Worms and Mollusks

BIG Idea Worms and mollusks have evolved to have a variety of adaptations for living as parasites or for living in water or soil.

Section 1
Flatworms
MAIN Idea Flatworms are thin, flat, acoelomate animals that can be free-living or parasitic.

Section 2
Roundworms and Rotifers
MAIN Idea Roundworms and rotifers have a more highly evolved gut than flatworms.

Section 3
Mollusks
MAIN Idea Mollusks are coelomates with a muscular foot, a mantle, and a digestive tract with two openings.

Section 4
Segmented Worms
MAIN Idea Segmented worms have segments that allow for specialization of tissues and for efficiency of movement.

BioFacts

- One hectare of soil can hold more than 2.5 million earthworms.
- An earthworm's setae can hold it so firmly to soil that a bird cannot pull the worm from its burrow.

LAUNCH Lab

What do earthworms feel like?

In this lab, you will examine a familiar worm—an earthworm like the ones on the facing page.

Procedure

1. Read and complete the lab safety form.
2. Obtain an **earthworm** from your teacher. **WARNING:** ***Treat the earthworm in a humane manner at all times.***
3. Run your finger along the ventral side, or underside, of the worm. Repeat in the opposite direction. Record your observations.
4. Examine the ventral side of the worm with a **magnifying glass.** Record your observations.
5. Wash your hands and return the earthworm to your teacher.

Analysis

1. **Compare** the way the earthworm felt to you when you brushed it in each direction.
2. **Infer** how any differences you observed might be important adaptations.
3. **Interpret** What did you see on the worm's ventral side that might explain how the worm felt to you?

Biology Online

Visit biologygmh.com to:
- study the entire chapter online
- explore Concepts in Motion, the Interactive Table, Microscopy Links, Virtual Labs, and links to virtual dissections
- access Web links for more information, projects, and activities
- review content online with the Interactive Tutor, and take Self-Check Quizzes

Segmented Worms Make the following Foldable to help you describe the three main classes of segmented worms.

STEP 1 Fold a piece of paper into thirds vertically.

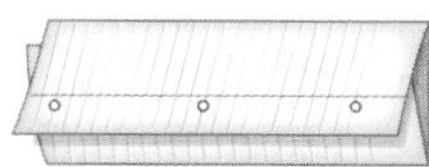

STEP 2 Fold the paper down 2.5 cm from the top.

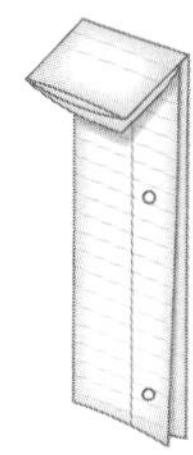

STEP 3 Unfold and draw lines along the 2.5-cm fold. Label the tabs *Earthworms, Bristleworms,* and *Leeches* as shown.

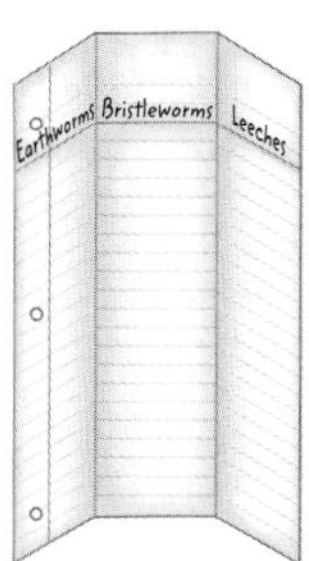

FOLDABLES **Use this Foldable with Section 25.4.** As you read the section, describe the features and unique characteristics of each class in the appropriate column.

Section 25.1

Flatworms

Reading Preview

Objectives

- **Compare** the adaptations of free-living flatworms to parasitic flatworms.
- **Explain** how flatworms maintain homeostasis.
- **Compare** the three classes of flatworms.

Review Vocabulary

acoelomate: an animal without any body cavities

New Vocabulary

pharynx
flame cell
ganglion
regeneration
scolex
proglottid

MAIN Idea Flatworms are thin, flat, acoelomate animals that can be free-living or parasitic.

Real-World Reading Link Think about a time you were caught in an unexpected rain shower without rain gear. If you were wearing layers of clothing, the rain might not have soaked through to your skin. As you read about worms, think about how it is easier for the rain to move through one thin layer than through multiple heavy layers.

Body Structure

The evolutionary tree in **Figure 25.1,** shows that flatworms are on the acoelomate branch of the tree, while roundworms are on the pseudocoelomate branch. However, flatworms and roundworms both have bilateral symmetry—they can be divided along only one plane into mirror-image halves. Bilateral symmetry is a major evolutionary step that allows parts of the body to evolve different organs. Animals that have bilateral symmetry also have more efficient movement than animals with radial symmetry.

Phylum Platyhelminthes (pla tee HEL min theez)—flatworms—consists of about 20,000 species. **Figure 25.1** shows some of the variety seen in this phylum. Flatworms range in length from many meters to 1 mm or less. They have thin, flat bodies that resemble a ribbon. Unlike sponges and cnidarians, flatworms have a definite head region and body organs. Recall from Chapter 24 that flatworms are acoelomates and therefore lack a coelom. Their bodies have no cavities.

Most flatworms are parasites living in the bodies of a variety of animals, while others are free-living in marine, freshwater, or moist land habitats. Freshwater planarians are often seen on the underside of rocks in swiftly flowing streams.

■ **Figure 25.1** Notice on the evolutionary tree that flatworms, such as flukes and tapeworms, were among the first animals to show bilateral symmetry.
Explain *how the symmetry of flatworms is different from that of cnidarians.*

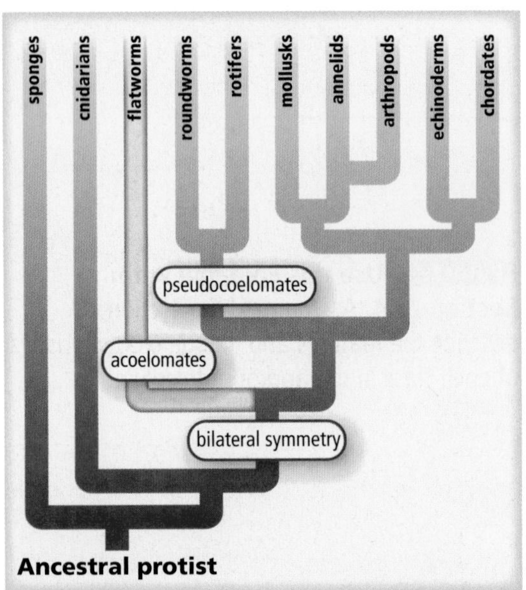

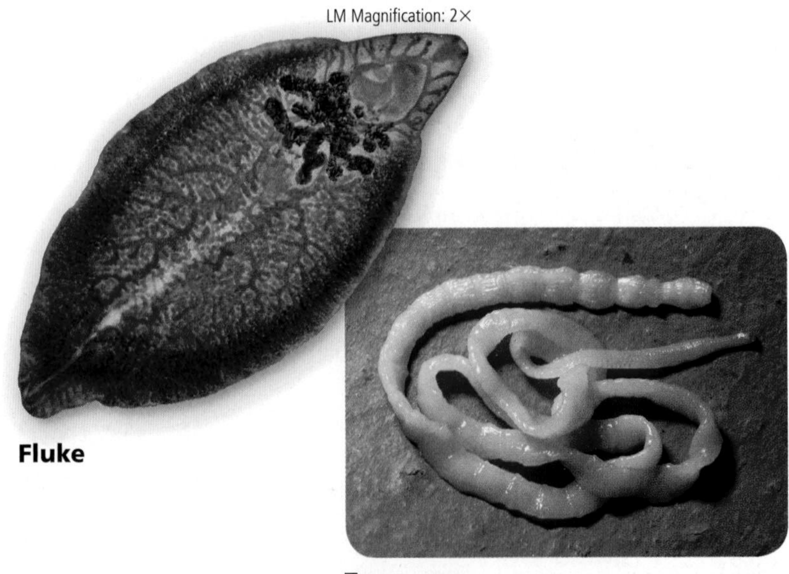

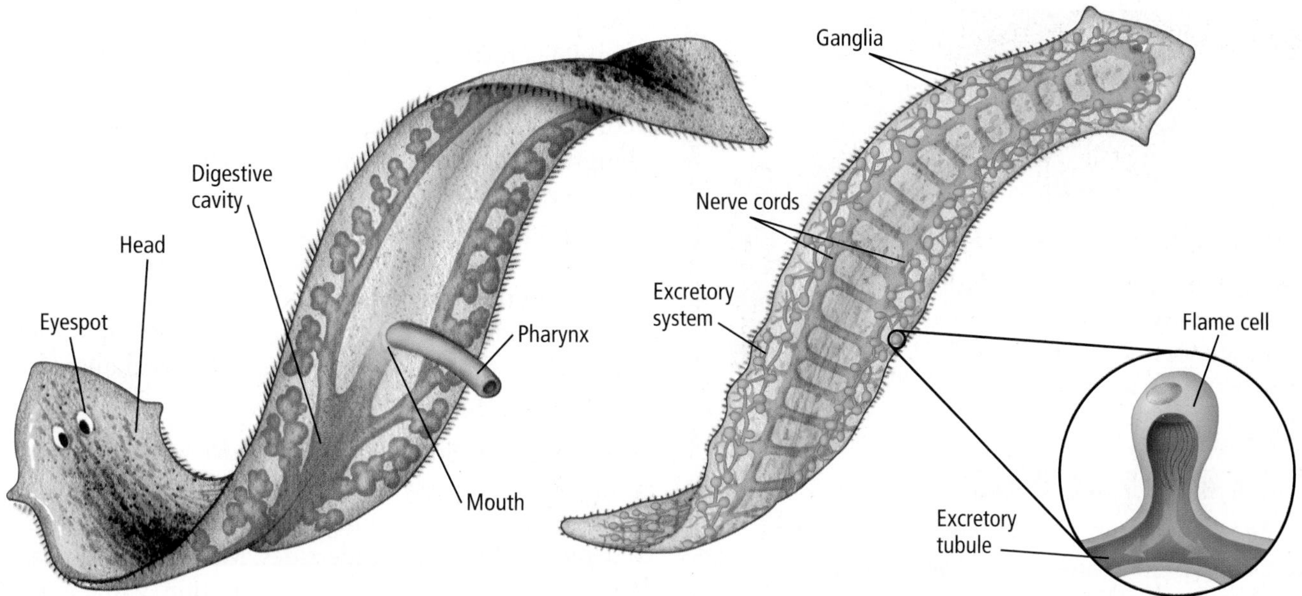

■ **Figure 25.2** Simple organ systems, such as the excretory and nervous systems, are found in flatworms.

Concepts In Motion

Interactive Figure To see an animation of the basic anatomy of a planarian, visit biologygmh.com.

Feeding and digestion Free-living flatworms feed on dead or slow-moving organisms. They extend a tubelike muscular organ, called the **pharynx** (FAHR ingks), out of their mouths. The pharynx, shown in **Figure 25.2,** releases enzymes that begin the digestion of prey. Then food particles are sucked into the digestive tract where digestion continues. Because flatworms have only one body opening, wastes are ejected through the mouth.

Parasitic flatworms have modified feeding structures called hooks and suckers, which enable them to stay attached to their hosts. Some parasitic flatworms have a reduced digestive system and feed on blood and other body tissues. Other parasitic flatworms lack a digestive system. Because they are so thin, like a single layer of cloth, and are surrounded by nutrients in their host's intestines, these parasites can absorb directly through their body walls partially or completely digested food eaten by the host.

 **Reading Check** **Compare** feeding and digestion in free-living flatworms and parasitic flatworms.

Respiration, circulation, and excretion Like sponges and cnidarians, flatworms do not have circulatory organs or respiratory organs. Because flatworms are so thin, their cells can use the process of diffusion to move dissolved oxygen and nutrients to all parts of their bodies. Carbon dioxide and other wastes also are removed from flatworm cells by diffusion.

Unlike sponges and cnidarians, flatworms have an excretory system that consists of a network of small tubes that run through the body. On side branches of the tubes, as shown in **Figure 25.2,** bulblike **flame cells** lined with cilia sweep water and excretory substances into tubules. These substances then exit through pores to the outside of the body. Flame cells were named because the flickering movements of the cilia inside the cells look like the light of a candle flame. Because flame cells move water out of the body, they keep flatworm cells from becoming waterlogged. In addition to the action of flame cells, flatworms also excrete waste products and maintain homeostatic water balance through their mouths.

VOCABULARY

SCIENCE USAGE V. COMMON USAGE

Host

Science usage: an animal or plant on which or in which a parasite lives.
Some parasitic worms live in the intestines of their hosts.

Common usage: a person who entertains guests.
Tyler's dad was the host for the football party.

MiniLab 25.1

LM Magnification: 10×

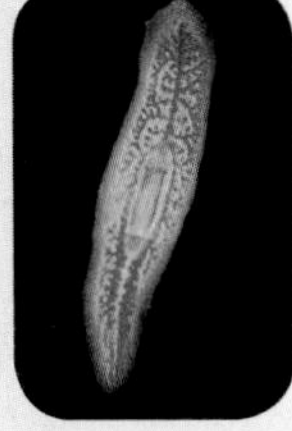

Planarian

Observe a Planarian

How does a planarian behave? Investigate the physical features and behavior of a planarian by observing this common flatworm.

Procedure

1. Read and complete the lab safety form.
2. Observe the **planarian** in a **water-filled observation dish** by using a **magnifying glass.**
3. Create a data table to record your observations.
4. Record the physical characteristics and behaviors of the flatworm.
5. Place a small piece of **cooked egg white** into the dish, and observe the feeding behavior of the planarian.

Analysis

1. **Compare and contrast** the physical features of the planarian with the features of the earthworm you observed in the Launch Lab.
2. **Analyze** how the body shape and movement of a planarian enables it to live in its environment.
3. **Infer** why scientists classify planaria into a group separate from other worms.

Response to stimuli The nervous system regulates the body's response to stimuli. In most flatworms, the nervous system consists of two nerve cords with connecting nerve tissue that run the length of the body. In most flatworms, the connecting nerve tissue looks like the rungs of a ladder, as illustrated in **Figure 25.2.** At the anterior end of the nerve cords is a small swelling composed of ganglia that send nerve signals to and from the rest of the body. A **ganglion** (plural, ganglia) is a group of nerve cell bodies that coordinates incoming and outgoing nerve signals.

Movement Some flatworms move by contracting muscles in the body wall. To escape predators and to find food, most free-living flatworms glide by using cilia located on their undersides. Mucus lubricates the worms and improves the gliding motion, while muscular action lets the animals twist and turn. If you ever have tried to loosen planaria worms from the bottoms of rocks, you know that their outer mucus covering enables them to stick tightly—an important adaptation in a swiftly moving stream. You can observe the features and behavior of a flatworm in **MiniLab 25.1.**

Reproduction Flatworms are hermaphrodites because they produce both eggs and sperm. During sexual reproduction, two different flatworms exchange sperm, and the eggs are fertilized internally. In marine flatworms, zygotes in cocoons are released into the water where they hatch within a few weeks.

Free-living flatworms can reproduce asexually by **regeneration**—a process in which body parts that are missing due to damage or predation can be regrown. A planarian that is cut in half horizontally can grow a new head on the tail end and a new tail on the head end, forming two new organisms, as shown in **Figure 25.3.**

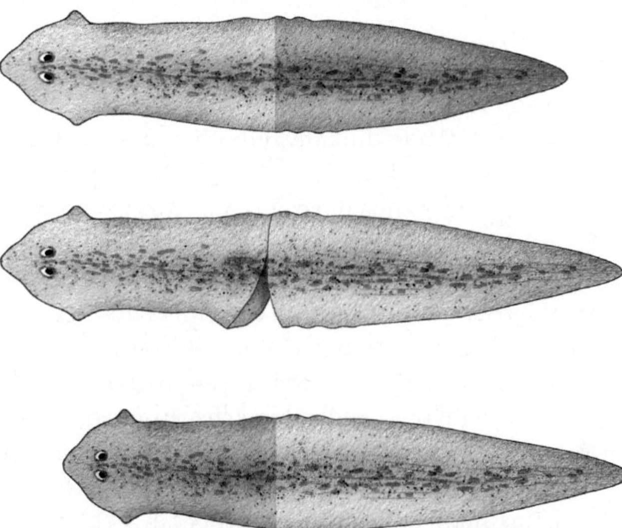

Figure 25.3 Two new planaria form when one planarian is cut in half horizontally. Some planaria can regenerate from almost any piece of their bodies.

Diversity of Flatworms

There are three main classes of flatworms: Turbellaria (tur buh LER ee uh), Trematoda (trem uh TOH duh), and Cestoda (ses TOH duh). Class Turbellaria consists of the free-living flatworms. Class Trematoda and class Cestoda consist of parasitic flatworms.

Turbellarians Members of the class Turbellaria are called turbellarians. Most turbellarians, like planarians, live in marine or freshwater habitats, while some live in moist soils. They vary in size, color, and body shape. As shown in **Figure 25.4,** turbellarians have eyespots that can detect the presence or absence of light. They also have sensory cells that help them identify chemicals and water movement.

The cells sensitive to chemicals are concentrated on small projections called auricles (OR ih kulz) at the anterior end of the worm. When a planarian hunts, it might wave its head back and forth as it crawls forward, exposing the auricles to chemical stimuli coming from food. At the same time, its eyespots might help it perceive light conditions that would protect it from predators.

Trematodes Flukes belong to class Trematoda—the trematodes. They are parasites that infect the blood or body organs of their hosts. The life cycle of the parasitic fluke *Schistosoma* is shown in **Figure 25.5.** Notice that this parasite requires two hosts to complete its life cycle.

When humans contract schistosomiasis (shihst tuh soh MI uh sis), the fluke eggs clog blood vessels, causing swelling and eventual tissue damage. Schistosomiasis can be prevented by proper sewage treatment and by wearing protective clothing when wading or swimming in infested water. Schistosomiasis infections are not common in the United States.

LM Magnification: 1×

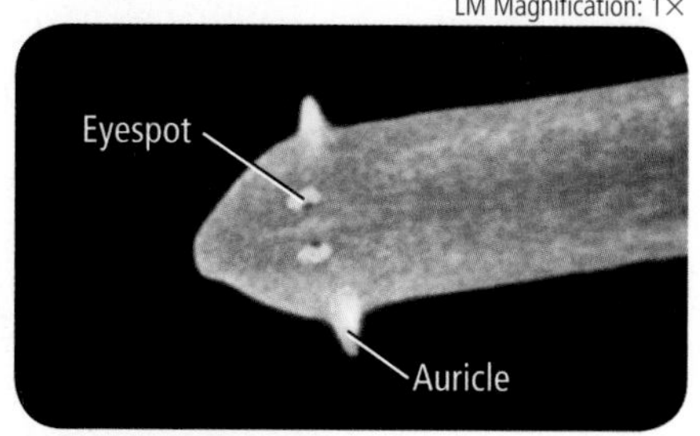

Figure 25.4 Dark clusters of light-sensitive cells form the eyespots on this planarian. Note the auricles projecting from the same area.

Figure 25.5 Two hosts—humans and snails—are needed to complete the life cycle of the fluke *Schistosoma.*

Infer ***Why are the two larval forms of the fluke different shapes?***

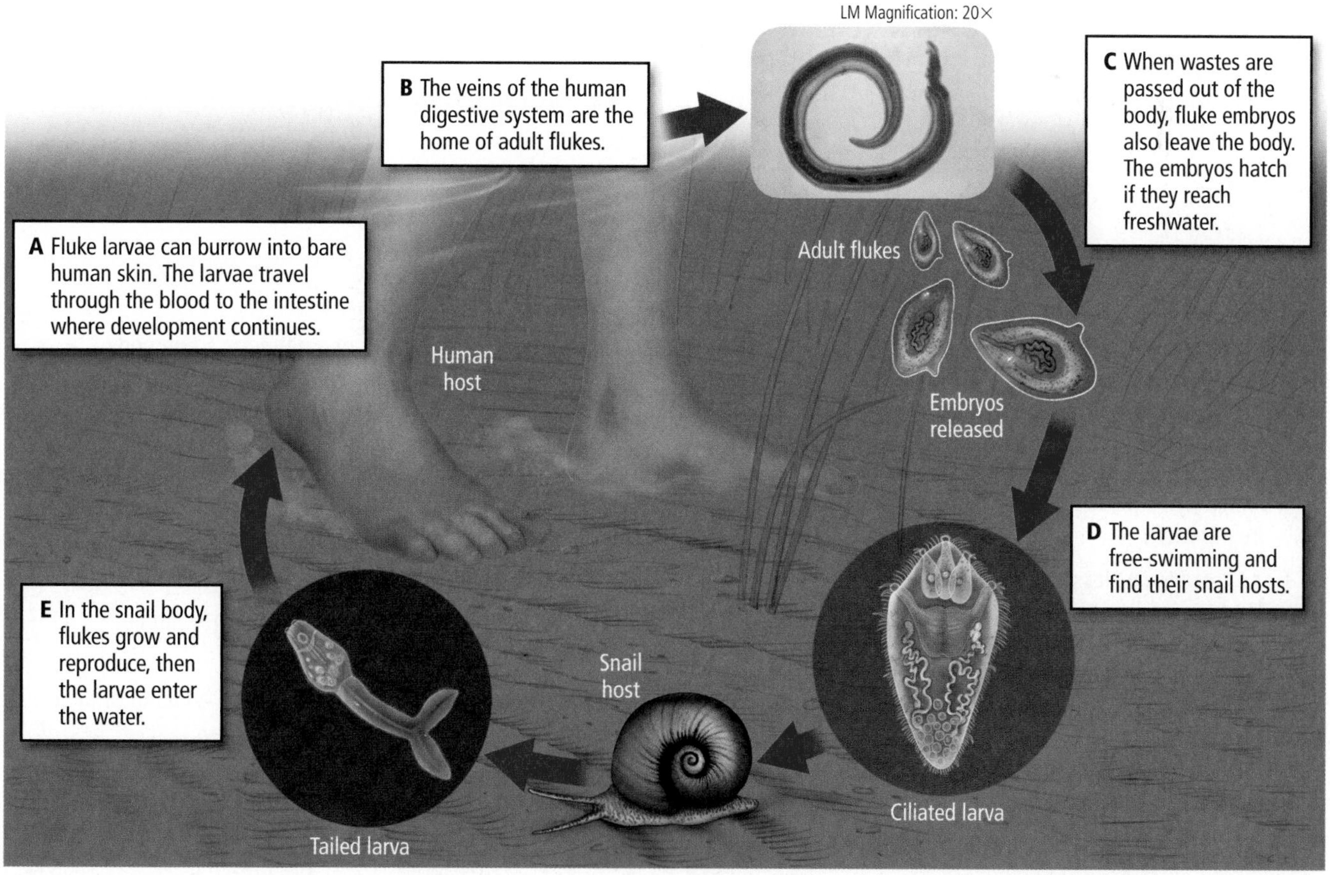

LM Magnification: 20×

A Fluke larvae can burrow into bare human skin. The larvae travel through the blood to the intestine where development continues.

B The veins of the human digestive system are the home of adult flukes.

C When wastes are passed out of the body, fluke embryos also leave the body. The embryos hatch if they reach freshwater.

D The larvae are free-swimming and find their snail hosts.

E In the snail body, flukes grow and reproduce, then the larvae enter the water.

Human host

Adult flukes

Embryos released

Ciliated larva

Snail host

Tailed larva

■ **Figure 25.6** As the proglottids behind the scolex mature, new proglottids form.

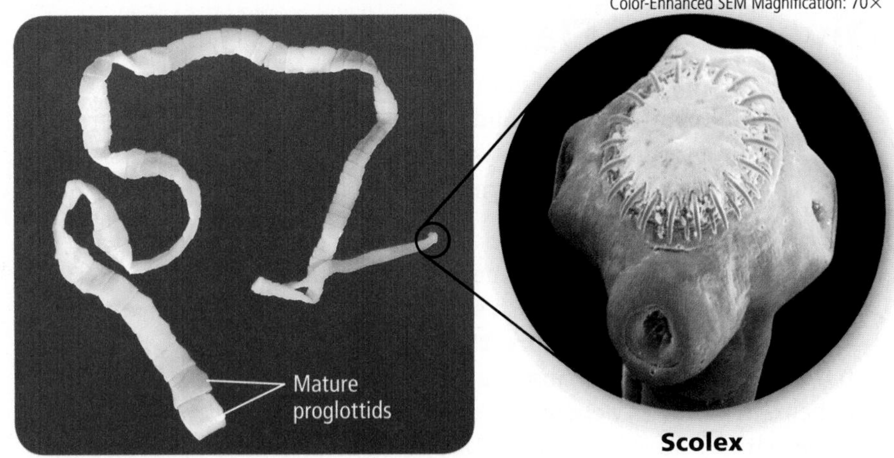

Cestodes All tapeworms are members of class Cestoda—the cestodes. They are parasites adapted to life in the intestines of their hosts. Look at the anterior end, or head, of the tapeworm in **Figure 25.6.** This is the **scolex** (SKOH leks), a knob-shaped structure with hooks and suckers that attach to the intestinal lining of a host such as a cow or a human.

Behind the scolex of the worm are a series of individual sections called **proglottids** (proh GLAH tihdz), each of which contains muscles, nerves, flame cells, and male and female reproductive organs. Proglottids form continuously; as new ones form near the scolex, older proglottids move farther back and mature. After eggs in the mature proglottids are fertilized, the last segments with developing embryos break off and pass out of the intestines of their hosts. Animals such as cattle might feed on vegetation or drink water contaminated by the tapeworm proglottids, and then the cycle of tapeworm growth is repeated.

When eaten by cattle, tapeworms burrow through intestinal walls, entering blood and muscle. If undercooked infected beef is eaten, human infection by tapeworms is likely. Tapeworm infections are uncommon in industrialized nations because of beef inspections.

Section 25.1 Assessment

Section Summary

- Flatworms were among the first animals to exhibit bilateral symmetry.
- Flatworms are acoelomates with limited numbers of organs and systems.
- Some flatworms are free-living, and others are parasitic.
- The three main classes of flatworms are Turbellaria, Trematoda, and Cestoda.
- Flatworms that are parasitic have specialized adaptations for parasitic life.

Understand Main Ideas

1. MAIN Idea **Evaluate** the advantages of a flatworm's thin body.
2. **Compare and contrast** the adaptations of free-living flatworms and parasitic flatworms.
3. **Prepare** a chart that compares digestion, respiration, movement, and reproduction in the free-living and parasitic flatworms.
4. **Analyze** the importance of flame cells in a flatworm.

Think Critically

5. **Design an experiment** to determine what habitat conditions planarians prefer.
6. **Evaluate** how the two classes of parasitic worms are adapted to their habitats.
7. **Diagram** bilateral symmetry using a planarian as an example. Explain the adaptive advantage of bilateral symmetry to a planarian.

Biology Online Self-Check Quiz biologygmh.com

Section 25.2

Reading Preview

Objectives

- **Compare** the features of roundworms to the features of flatworms.
- **Identify** roundworms based on movement.
- **Evaluate** the risk of contracting roundworm parasites.

Review Vocabulary

cilia: short, numerous projections that look like hairs

New Vocabulary

hydrostatic skeleton
trichinosis

Roundworms and Rotifers

MAIN Idea **Roundworms and rotifers have a more highly evolved gut than flatworms.**

Real-World Reading Link If you were to guess what animal is one of the most common in the world, what animal would you choose? Would you guess a roundworm? With 20,000 species of roundworms known, scientists estimate that there might be 100 times as many more kinds of roundworms still undiscovered.

Body Structure of Roundworms

Roundworms are in phylum Nematoda (ne muh TOH duh) and often are called nematodes. Locate roundworms on the evolutionary tree in **Figure 25.7.** Notice that they are pseudocoelomates. Recall that pseudocoelomates have a fluid-filled body cavity that is partially lined with mesoderm. Roundworms have bilateral symmetry and are cylindrical, unsegmented worms that are tapered at both ends. Roundworms come in many sizes, as shown in **Figure 25.7.** Most are less than 1 mm long. However, the longest known roundworm, living in certain whales, can grow to 9 m in length.

Roundworms are found in both marine and freshwater habitats and on land. Some are parasites on plants and animals. A spadeful of garden soil might contain one million roundworms. One study revealed that a rotting apple contained 1074 roundworms! Dogs and cats can be plagued by roundworms if they are not wormed when they are young and at regular intervals during adulthood. Roundworms have adaptations that enable them to live in many places.

Figure 25.7 Roundworms are pseudocoelomates with bilateral symmetry.

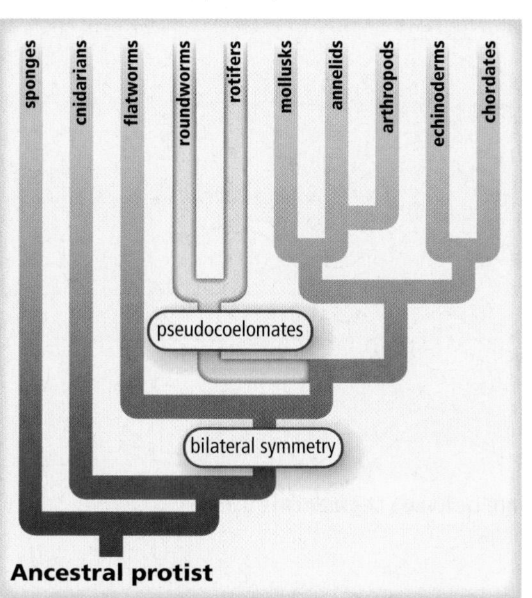

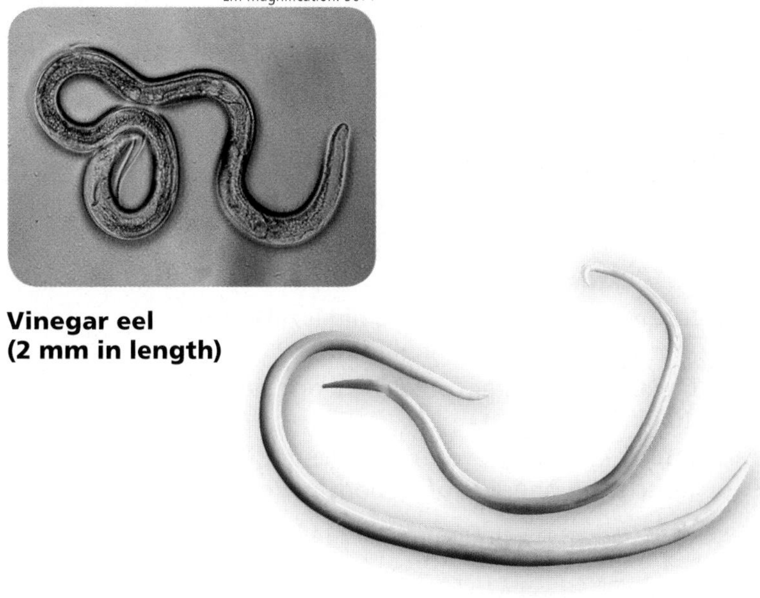

Vinegar eel (2 mm in length)

Ascarid worms (10-35 cm in length)

CAREERS IN BIOLOGY

Veterinary Parasitologist As scientists who keep parasites out of the food supply, veterinary parasitologists are responsible for the health of farm and domesticated animals. They determine ways to control and prevent parasites in animals such as cows, pigs, and chickens. For more information on biology careers, visit biologygmh.com.

Feeding and digestion Most roundworms are free-living, but some are parasites. Some free-living roundworms are predators of other tiny invertebrates, while others feed on decaying plant and animal matter. Free-living forms have a key evolutionary adaptation in their digestive systems. Recall from Chapter 24 that in the course of evolution, pseudocoelomate animals were the first to have a body cavity. The pseudocoelom of a nematode separates the endoderm-lined gut from the rest of the body. The movement of food through the gut, or digestive tract, is one-way—food enters through the mouth, and undigested food leaves through an opening at the end of the digestive tract called the anus.

Respiration, circulation, excretion, and response to stimuli Like flatworms, roundworms have no circulatory organs or respiratory organs, and they depend on diffusion to move nutrients and gases throughout their bodies. Most roundworms exchange gases and excrete metabolic wastes through their moist outer body coverings. More complex forms have excretory ducts that enable them to conserve water for living on land, while others have flame cells.

Ganglia and associated nerve cords coordinate nematode responses. Nematodes are sensitive to touch and to chemicals. Some have structures that might detect differences between light and dark.

Movement Roundworms have muscles that run the length of their bodies. These muscles cause their bodies to move in a thrashing manner as one muscle contracts and another relaxes. These muscles also pull against the outside body wall and the pseudocoelom. The pseudocoelom acts as a **hydrostatic skeleton**—fluid within a closed space that provides rigid support for muscles to work against. If you were to observe a roundworm moving, it might resemble a tiny piece of wriggling thread. Learn more about worm movement in **Data Analysis Lab 25.1.**

DATA ANALYSIS LAB 25.1

Based on Real Data*

Interpret the Diagram

How does a nematode move? A nematode alternately contracts and relaxes muscles running lengthwise on each side of its body, which moves it forward in successive stages.

Data and Observations

Look at the diagram and observe how a nematode moves.

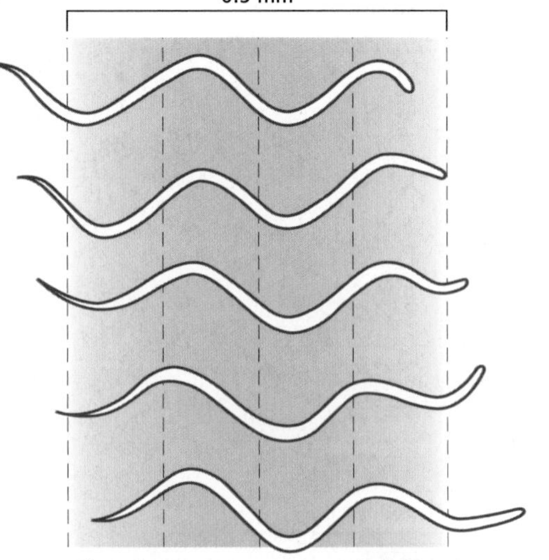

Time between segments is 0.33 s

Think Critically

1. **Interpret** About how long did it take the worm to move to its final location?
2. **Calculate** How far could the worm move in 10 min?
3. **Infer** How might worm movement differ if muscles on one side of its body were damaged?

*Data obtained from: Gray, J. and H.W. Lissmann. 1964. The locomotion of nematodes. *Journal of Experimental Biology* 41:135–154.

Reproduction Roundworms reproduce sexually. The females produce eggs, and the males, which often are smaller than the females, produce sperm. Fertilization is internal. In free-living roundworms, larvae hatch from the fertilized eggs, then grow into adults. In parasitic roundworms, development often is more complicated, involving one or more hosts or different locations in the host's body.

The adult roundworm *Caenorhabditis elegans (C. elegans),* shown in **Figure 25.8,** contains only 959 cells; zygotes mature to adults in just three days. These characteristics make it an extremely important subject of research on development, aging, and genetics. *C. elegans* was the first multicellular organism to have its entire genome sequenced. The *C. elegans* genome contains 97 million DNA bases encoding over 19,000 different genes. See the BioDiscoveries feature at the end of this chapter to learn more about *C. elegans.*

 **Reading Check Explain** why the features of *C. elegans* make it a good subject for research.

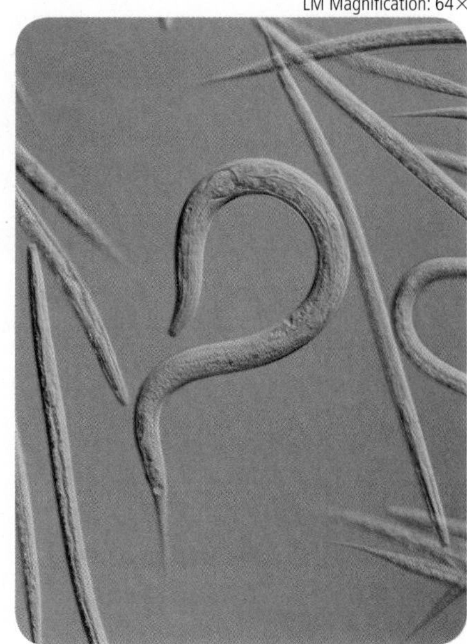

Figure 25.8 *C. elegans* is the subject of much genetic research. With comparatively few cells and rapid development, scientists can easily research developmental changes.

Diversity of Roundworms

Of the 20,000 known roundworm species, approximately half are parasites. These parasitic roundworms cause a variety of diseases in plants and animals, including humans. Many of these diseases in humans are the result of carelessness, a lack of personal hygiene, or poor sanitation.

Trichinella worms A disease called **trichinosis** (trih keh NOH sis) can be contracted by eating raw or undercooked pork and pork products, or wild game infected with the larvae of *Trichinella,* like the one shown in **Figure 25.9.** After ingestion by a host organism, the worms mature in one to two days. Female worms with fertilized eggs burrow into the intestinal walls of humans, pigs, and other mammals. After the eggs hatch, the larvae burrow into muscles where they form cysts, causing muscle pain. Trichinosis can be prevented by cooking meat properly.

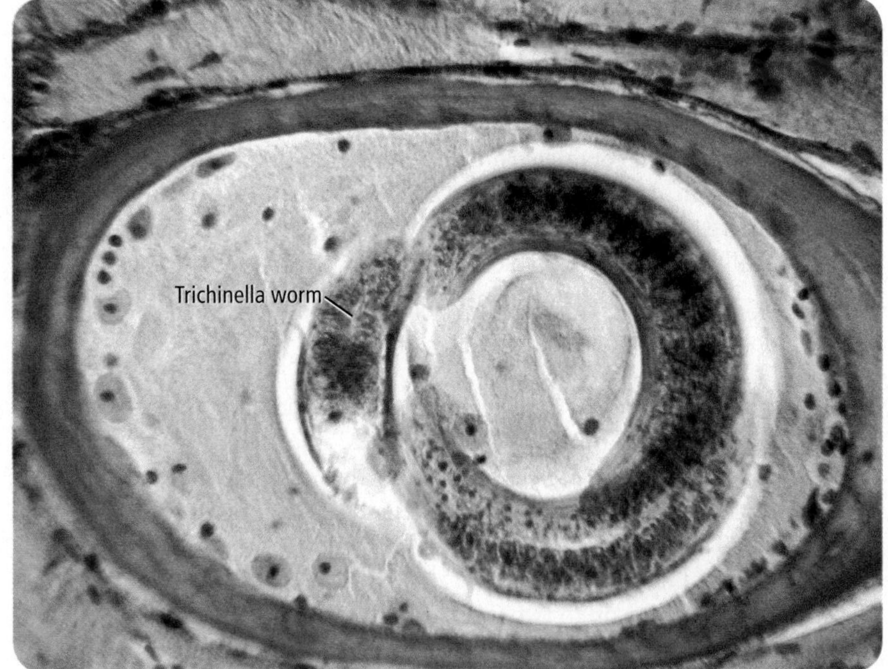

Figure 25.9 A trichinella worm larva is seen curled up inside of a cyst (purple) in pig muscle.
Infer *A person with trichinosis might have what kind of physical symptoms?*

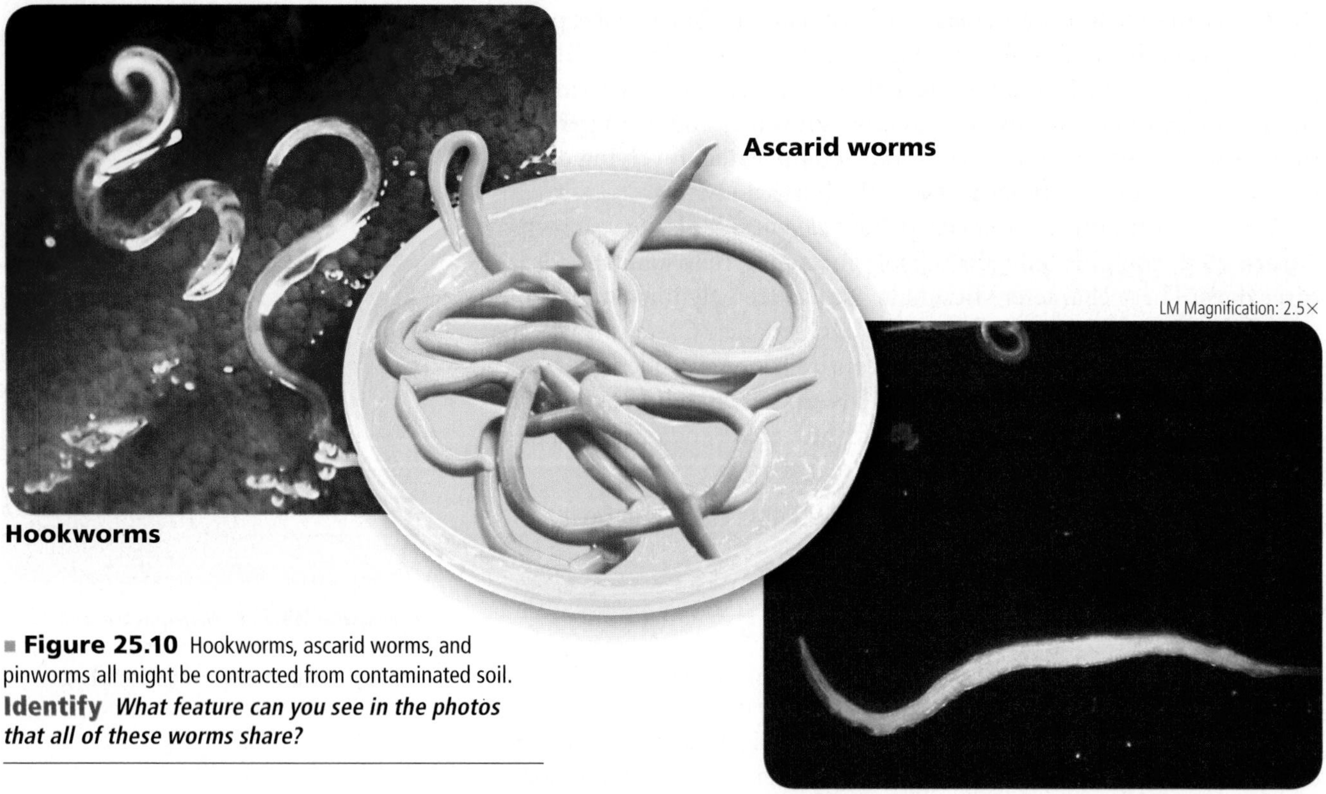

Ascarid worms

LM Magnification: 2.5×

Hookworms

Pinworm

■ **Figure 25.10** Hookworms, ascarid worms, and pinworms all might be contracted from contaminated soil.
Identify ***What feature can you see in the photos that all of these worms share?***

Hookworms Hookworm infections are common in warm climates when people go barefoot on contaminated soil. When a hookworm, shown in **Figure 25.10,** contacts bare human skin, it cuts its way inside, travels in the bloodstream to the lungs, and then to the windpipe, or esophagus, where it is coughed up and swallowed. The parasite then moves to the small intestine, where it attaches to intestinal walls and feeds on blood and other tissue. Hookworm infection can be prevented by wearing shoes.

Ascarid worms The most common worm infection in humans is ascariais (AS kuh RI uh sus), which is caused by ascarid (AS kuh rid) worms like the ones shown in **Figure 25.10.** Eggs of ascarid worms are found in soil in subtropical and tropical areas. They enter the human body through the mouth and live in the intestine. Infection can result when unwashed vegetables from contaminated soil are eaten or when hands contaminated with infected soil are put in the mouth, as young children are likely to do. Infection by ascarid worms can be controlled by carefully washing vegetables and hands.

Reading Check **Explain** how humans can prevent infection from hookworms and ascarid worms.

Vocabulary
Word origin
Nematode
nemato– prefix; from Greek– meaning *thread*
–ode suffix; from the Greek word *oeides,* meaning *similar to.*

Pinworms **Figure 25.10** shows a pinworm—the most common nematode parasite in humans in the United States. The highest incidence of infection occurs in children. At night, female pinworms living in the intestine move out of the anus and lay eggs on nearby skin. When the skin is scratched because of the itching caused by pinworm activity, the eggs are transferred to hands and then to any surface that is touched. These eggs can survive for up to two weeks on surfaces and are ready to hatch if another person ingests them. This infection can spread quickly among children who put toys and other objects in their mouths.

Filarial worms Elephantiasis (el uh fun TI uh sus) is a disease caused by filarial (fuh LER ee uhl) worms—roundworm parasites that live in tropical areas. A mosquito is the intermediate host of filarial worms. When a mosquito sucks blood from a person who is infected with this roundworm, worm embryos are passed into the insect's bloodstream. The embryos grow into larvae, which then are passed to another person when the mosquito bites again. Adult worms accumulate in the lymphatic system and obstruct the flow of lymph—the tissue fluid in the spaces between cells. This fluid builds up in tissue, causing legs and other body regions to enlarge. Controlling mosquitoes and using mosquito netting at night can aid in preventing this disease.

Another disease caused by a filarial worm—heartworm—is found in dogs and cats throughout the United States. Heartworms are transmitted to dogs and cats through mosquito bites. Once in the bloodstream, the worms travel to the heart and block the flow of blood. Regular doses of oral medications prevent heartworm in dogs and cats.

Reading Check **Identify** In what parts of the human body do pinworms and filarial worms live?

Nematodes in plants Some species of roundworms cause diseases in plants. Nematodes can infect and kill pine trees, soybean crops, and food plants such as tomatoes. When they infect plant roots, as shown in **Figure 25.11,** they damage the plant.

Most species of nematodes are either harmless or beneficial to plants. Certain nematodes are used to control the spread of cabbage worm caterpillars, Japanese beetle grubs, and many other pests of crop plants. Spraying a solution of nematodes and water on areas that are infested with crop pests is most effective when the targeted pest is at the stage in its life cycle when it lives in the soil.

Connection to Health In addition to treating plant pests, nematodes are used to control pests of humans and animals. Nematodes eat flea larvae, controlling the flea population in yards. This reduces or eliminates exposure of humans and animals to traditional chemicals used to treat flea infestations.

Figure 25.11 The growth of the vascular system of plants can be slowed down when nematodes move into the roots and form cysts.

Potato plant without nematodes

Potato plant with nematodes

Nematode cysts on roots

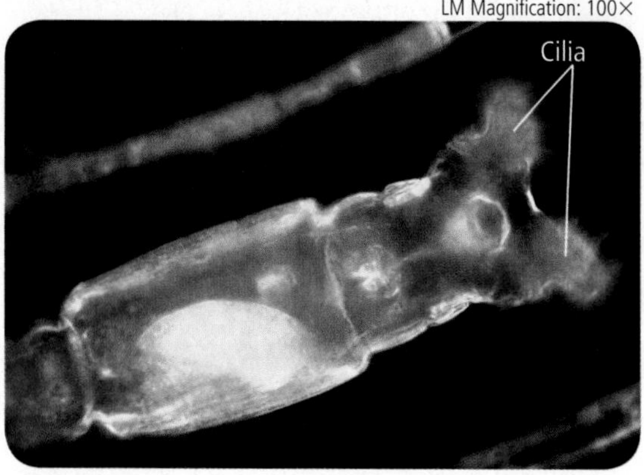

■ **Figure 25.12** Rotifers have two rings of cilia at their anterior end.

Rotifers

Rotifers are tiny animals only about 0.1 mm to 0.5 mm in length. The phylum Rotifera, meaning *wheel-bearer,* gets its name from the rings of cilia around the mouths of the animals. About 1800 species of rotifers have been studied, mostly from freshwater habitats including ponds, streams, and lakes. It would not be unusual to find 40 to 500 rotifers in a liter of pond water. A few species are marine. Refer back to the evolutionary tree in **Figure 25.7** and see that, although rotifers and roundworms occupy separate branches, both are pseudocoelomates.

Rotifer features and movement Rotifers are similar to roundworms because they have bilateral symmetry and are pseudocoelomates with a gut open at both ends. Unlike roundworms, rotifers move through the water by means of their ciliated wheel-like structures, which are shown in **Figure 25.12.** The posterior end of a rotifer generally has "toes" and glands that secrete an adhesive material that enables a rotifer to attach itself to a surface in the water.

Organ systems of rotifers Rotifers feed by using cilia to gather protists and organic materials into a complete digestive tract, which includes a mouth and an anus. Like other pseudocoelomates, rotifers exchange gases and excrete metabolic wastes by diffusion through body walls. Sensory structures include sensory bristles and eyespots on the head. Some rotifers reproduce sexually, while others have complex life cycles involving diploid eggs producing diploid females and haploid eggs producing haploid males. These life cycles are dependent on environmental conditions, such as spring rains, frost, or the presence of stagnant water.

Section 25.2 Assessment

Section Summary

- Roundworms are closely related to flatworms, but roundworms have an evolutionary adaptation related to their gut.
- Roundworms, like flatworms, have a limited number of organs and systems.
- Roundworms are either free-living or parasitic.
- Roundworms cause many human and plant diseases.
- Rotifers are pseudocoelomates that appear on a different branch of the evolutionary tree than roundworms.

Understand Main Ideas

1. MAIN Idea **Describe** the evolutionary adaptation of the digestive tract of roundworms.
2. **Compare and contrast** the features of flatworms and roundworms.
3. **Explain** how roundworms make their distinctive thrashing movements.
4. **Compare and contrast** the various ways humans might risk contracting roundworm parasites.

Think Critically

5. **Hypothesize** Imagine that you are digging in your garden and find some tiny threadlike animals making thrashing movements. Make a hypothesis about what these animals might be. Explain your answer.

MATH in Biology

6. Make a circle graph that shows the number of roundworm species known compared to the estimated number of roundworm species that might exist.

Biology Online Self-Check Quiz biologygmh.com

Section 25.3

Mollusks

Reading Preview

Objectives

- **Evaluate** the importance of the coelom to mollusks.
- **Interpret** the function of the mantle and its adaptive advantage to mollusks.
- **Analyze** the importance of mucus and the muscular foot to mollusks.

Review Vocabulary

herbivore: an organism that eats only plants

New Vocabulary

mantle
radula
gill
open circulatory system
closed circulatory system
nephridium
siphon

MAIN Idea Mollusks are coelomates with a muscular foot, a mantle, and a digestive tract with two openings.

Real-World Reading Link Have you ever watched a rocket blast off into space? The rocket is powered by jet propulsion—a stream of heated gas is forced out of the engine, pushing the rocket in the opposite direction. Some animals, such as octopuses, also move by jet propulsion, forcefully expelling streams of water to push them away from danger.

Body Structure

Mollusks are members of the phylum Mollusca. They range from the slow-moving slug to the jet-propelled squid, from scallops and cuttlefish to chitons and nudibranchs. Mollusks range in size from almost microscopic snails to giant squids, which can grow to be 21 m long.

Look at the evolutionary tree in **Figure 25.13.** Mollusks, such as the nudibranch and the octopus in **Figure 25.13,** undergo protostome development and might have been the first animals in the course of evolution to have a coelom, which allowed for the development of more complex tissues and organs. There are more than 110,000 species of mollusks. Many are marine, some live in freshwater, and others live in moist land environments.

Mollusks are coelomate animals with bilateral symmetry, a soft internal body, a digestive tract with two openings, a muscular foot, and a mantle. The **mantle** (MAN tuhl) is a membrane that surrounds the internal organs of the mollusk. In mollusks with shells, the mantle secretes calcium carbonate to form the shell. Other mollusks, including slugs and squids, are adapted to life without a hard outer covering.

Figure 25.13 Mollusks, like the nudibranch and octopus, have coeloms.
Infer *What is the main difference between mollusks and roundworms based on the evolutionary tree?*

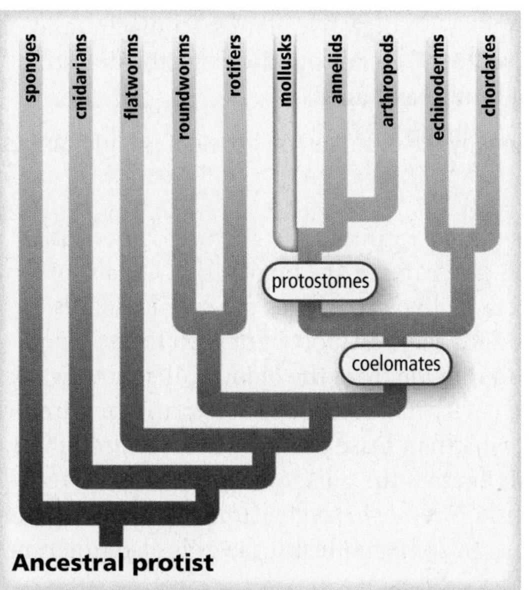

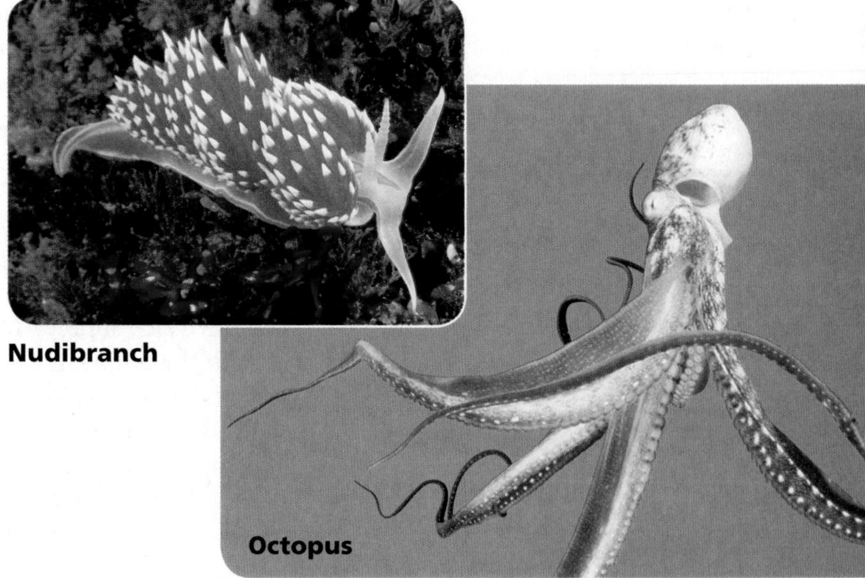

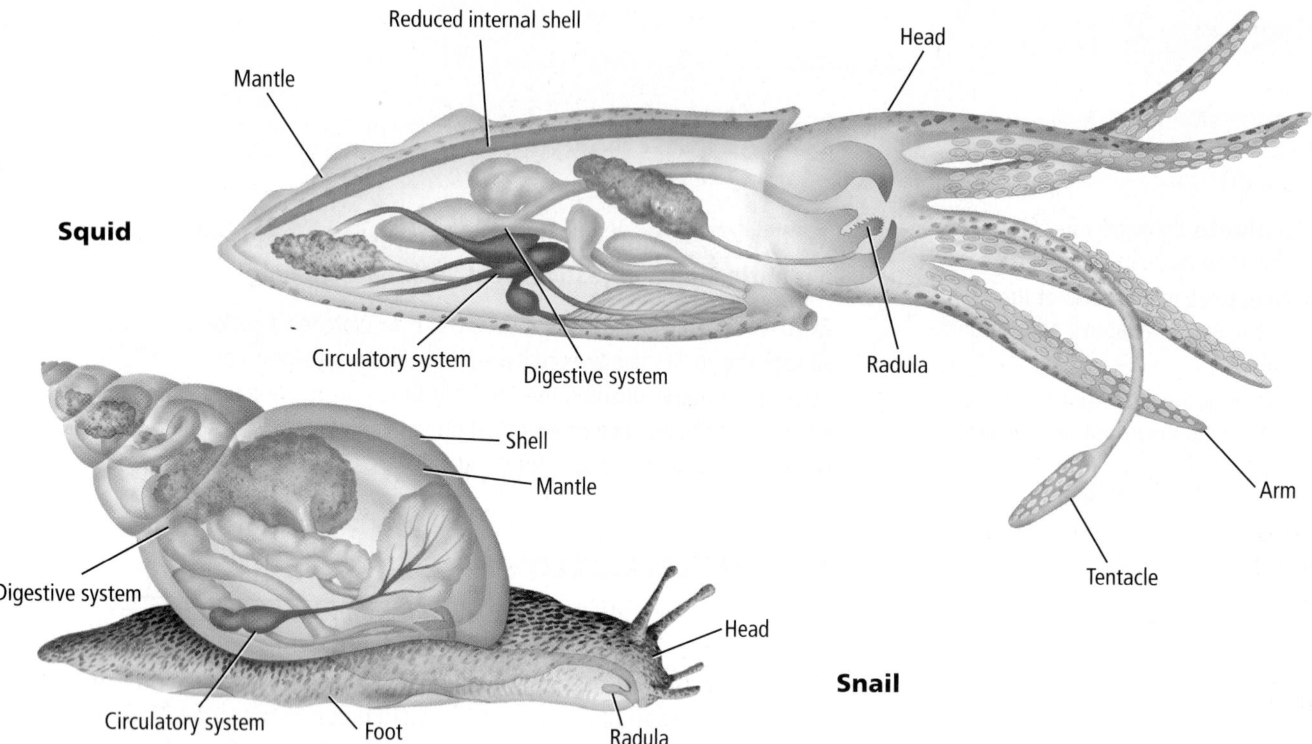

■ **Figure 25.14** Many mollusks have shells. Inside the shell is a soft body consisting of a foot, organs, and a mantle.
Compare and contrast ***the bodies of the snail and the squid.***

Compare the bodies of the snail and the squid in **Figure 25.14.** Their external features are very different from each other. However, both have coelomate body plans and highly evolved body systems, such as the digestive system, respiratory system, circulatory system, and nervous system.

Feeding and digestion Many mollusks use a rasping structure called a radula to scrape food into their mouths. Located in a mollusk's mouth, a **radula** (RA juh luh) is a tonguelike organ with rows of teeth, as shown in **Figure 25.15.** Herbivorous mollusks use their radulas to scrape algae off rocks. Carnivorous mollusks use their radulas to drill into other mollusks and feed on their internal body parts. Some of these predators, such as octopuses and squids, use their radulas to tear up the food they capture with their tentacles. Other mollusks, such as clams, are filter feeders and do not have radulas.

Mollusks have complete guts with digestive glands, stomachs, and intestines, as shown in **Figure 25.16.** As in roundworms, the digestive system has two openings—a mouth and anus.

Reading Check **Explain** why the evolution of a coelom is important to mollusks.

■ **Figure 25.15** Many mollusks feed using a radula. At top, the radula is at rest. At bottom, you can see the toothlike scraping structures on the radula as it is extended to feed.

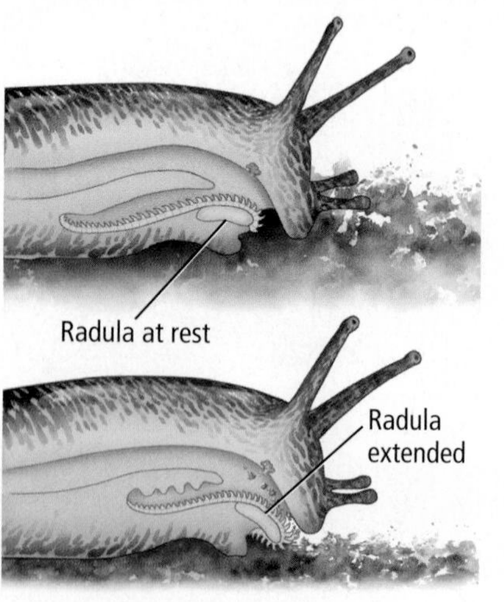

Respiration Most mollusks have respiratory structures called gills. **Gills,** shown in **Figure 25.16,** are parts of the mantle that consist of a system of filamentous projections like the fringes of a blanket. Gills contain a rich supply of blood for the transport of oxygen to the blood and for the removal of carbon dioxide from the blood. Gills move water into and through the mantle cavity in a continuous stream. They are highly branched structures, which increase the surface area through which gases can diffuse. This enables the gills to take in more oxygen from water. Land snails and slugs remove oxygen from the air using the lining of their mantle cavities. In some mollusks, the gills also function in filter feeding.

Circulation Mollusks have a well-developed circulatory system that includes a chambered heart. Most mollusks have an **open circulatory system,** in which the blood is pumped out of vessels into open spaces surrounding the body organs. This adaptation enables animals to diffuse oxygen and nutrients into tissues that are bathed in blood and also to move carbon dioxide from tissues into the blood. Slow-moving animals, such as snails and clams, utilize this system effectively because they do not need rapid delivery of oxygen and nutrients for quick movements.

Some mollusks, such as squids, move nutrients and oxygen through a closed circulatory system, which was a major adaptation in the evolution of animals. In a **closed circulatory system,** blood is confined to vessels as it moves through the body. A closed system efficiently transports oxygen and nutrients to cells where they are converted to usable forms of energy. Mollusks that move quickly, such as the octopus and squid, need more energy than slow-moving mollusks, and the closed circulatory system quickly delivers nutrients and oxygen. A closed circulatory system is like the heating ducts in some houses. A furnace is efficient at delivering warm air to the rooms in a house because the air travels through a series of ducts or pipes. Rooms of a house would not be evenly heated if the furnace did not have a delivery system.

Excretion Most mollusks get rid of metabolic wastes from cellular processes through structures called **nephridia** (nih FRIH dee uh), shown in **Figure 25.16.** After nephridia filter the blood, waste is passed out through the mantle cavity. Nephridia are an evolutionary adaptation enabling mollusks to efficiently maintain homeostasis in their body fluids.

Response to stimuli Mollusks have nervous systems that coordinate their movements and behavior. Mollusks that are more highly evolved, such as octopuses, have a brain. In addition, octopuses have complex eyes similar to human eyes with irises, pupils, and retinas. Most mollusks have simple structures in the eyes that reflect light.

CAREERS IN BIOLOGY

Laboratory Assistant Marine biologists depend on laboratory assistants to help collect specimens such as mollusks and to maintain databases. These entry-level workers also set up equipment and prepare samples for testing. For more information on biology careers, visit biologygmh.com.

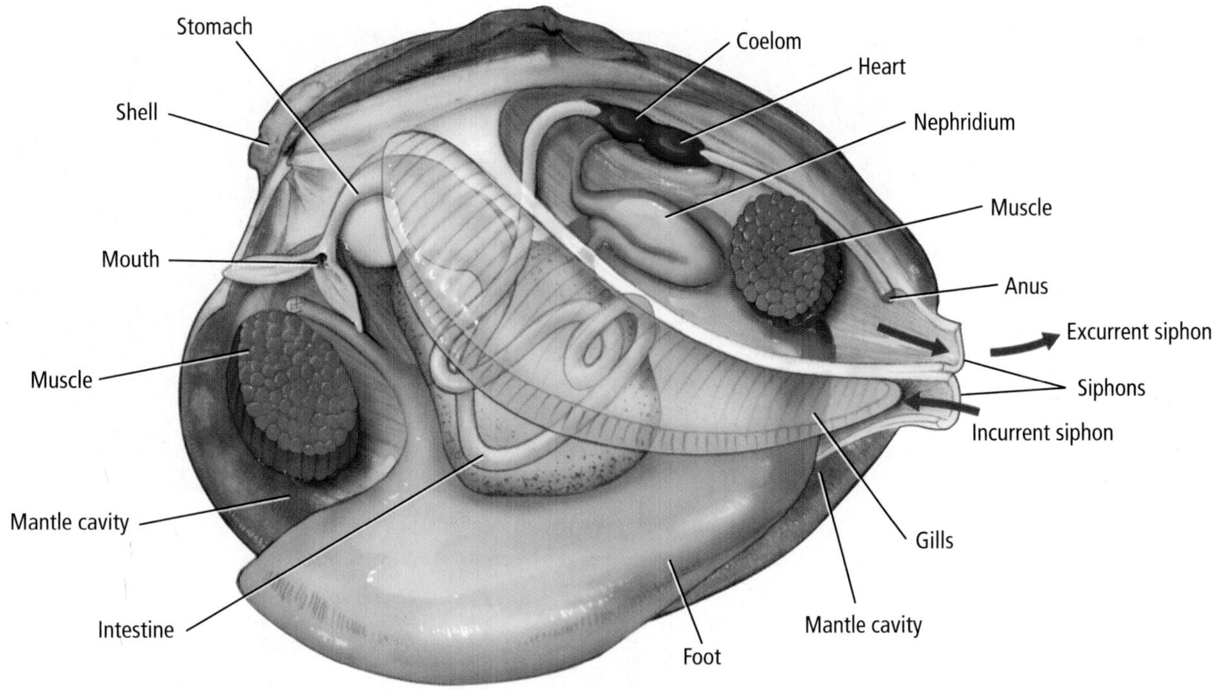

■ **Figure 25.16** The internal anatomy of a clam illustrates the well-developed organ systems in mollusks.

Visualizing Movement in Mollusks

Figure 25.17
Mollusks move in a variety of ways. The type of movement used often depends on a mollusk's unique adaptations.

Gastropods

A gastropod moves by sending waves of contractions along its muscular foot. A film of mucus lubricates the foot and helps propel the animal forward.

Note the waves of muscle contractions as the snail moves along its mucous trail.

Bivalves

Most bivalves don't move much, unless they are threatened by a predator. Then, a bivalve either uses its muscular foot to burrow into sediment, as shown on the left, or uses jet propulsion to flee, as shown at right.

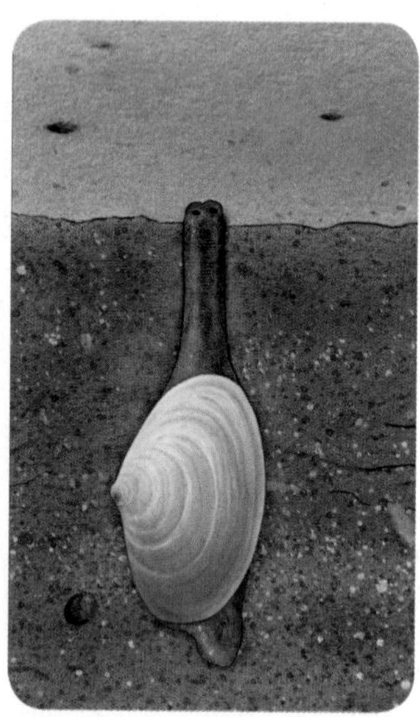

A clam can rapidly bury itself in sand using its muscular foot.

A scallop pulls its shells together, forcing jets of water toward the shell hinge. The force of the water pushes the scallop in the direction of the shell opening.

Cephalopods

Members of class Cephalopods, such as octopuses and squids, move by jet propulsion. To avoid predators, a cephalopod draws in water through slits in the body wall. Then the water is pumped rapidly through the siphon, jet-propelling the cephalopod away from danger.

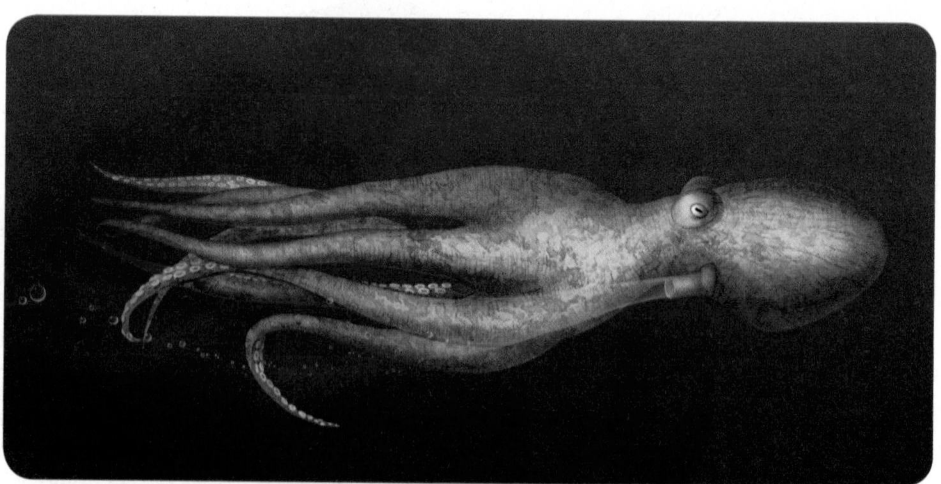

An octopus changes the direction it moves by alternating the direction of its siphon.

Concepts In Motion **Interactive Figure** To see an animation of mollusk movement, visit biologygmh.com.

Biology Online

Movement The muscular foot of a clam enables it to burrow into wet sand. Mollusks with two shells can clap their shells together for short bursts of rapid swimming. Most slugs and snails creep along moist areas on a slime trail of mucus secreted by glands in the foot. Octopuses and squids take water into the mantle cavity and expel it through a tube called a **siphon.** When threatened, they can eject the water so rapidly that they appear to be jet-propelled. **Figure 25.17** illustrates the ways mollusks move.

Reading Check **Compare** movement in two-shelled mollusks, snails, and squids.

Reproduction Mollusks reproduce sexually, as illustrated in **Figure 25.18.** The males and females of most aquatic species release their eggs and sperm into the water at the same time, and fertilization is external. A few bivalves and many gastropods that live on land are hermaphrodites, in which fertilization takes place internally.

All members of the phylum Mollusca share similar developmental patterns, even though their adult forms vary widely. One larval stage of most mollusks—the trochophore (TRAH kuh for)—looks very similar to the larval stage of the next group of animals you will study—the segmented worms. Because the larval forms are similar in both segmented worms and mollusks, scientists hypothesize that segmented worms and mollusks are closely related.

Figure 25.18 The life cycle of a clam illustrates the characteristic developmental stages of all mollusks.

A A female clam releases eggs into the water, where they are then fertilized by sperm released by a male clam.

B After fertilization, the trochophore larvae change into veliger larvae. Both forms are free swimming.

C The veliger larvae shed their velums—the ciliated "sails" that enable them to swim—and settle on a surface.

D The final larval stage, the pediveligers, develop into adult clams.

Abalone

Scallop

■ **Figure 25.19** Most gastropods, such as the abalone, have single shells for protection. Bivalves, such as the scallop, have two shells.

Diversity of Mollusks

Animals in the three major classes of mollusks—gastropods, bivalves, and cephalopods—are grouped based on differences in their shell and foot structures.

Gastropods The largest class of mollusks is Gastropoda, the stomach-footed mollusks. The name comes from the way the animal's large foot is positioned under the stomach on the ventral surface. Most species of gastropods have a single shell, like the abalone in **Figure 25.19.** Single-shelled gastropods also include snails, conches, periwinkles, limpets, cowries, whelks, and cones. They can be found in aquatic habitats and in moist terrestrial habitats, and they can quickly draw their bodies into their shells for protection when threatened.

Slugs and nudibranchs do not have shells, but secrete a thick mucus that covers their bodies. To protect themselves, land slugs hide in dark locations under forest or garden litter. Nudibranchs incorporate into their own tissues the poisonous nematocysts of the jellyfishes they eat. The presence of nematocysts is advertised to predator fishes by the bright colors of the nudibranchs.

VOCABULARY

WORD ORIGIN

Gastropod

gastro– prefix; from the Greek word *gaster,* meaning *belly*

–pod suffix; from Greek, meaning *foot.*

Bivalves One word—slow—best describes most behavior of the class Bivalvia—the two-shelled mollusks. Bivalves, such as clams, mussels, oysters, and the scallop shown in **Figure 25.19,** are all aquatic animals. Most are marine, but some are found in freshwater habitats. Bivalves might seem to be inactive even though they are continuously filter feeding and carrying on all bodily functions.

If you have ever been clamming or have seen people clamming, you know that you might have to dig deeply to find the clams because they use a muscular foot to burrow far down into wet sand. Mussels attach to rocks with a sticky, gluelike substance called byssal threads. Scallops are more active than other bivalves because they can clap their shells together to move more quickly through water.

 **Reading Check** **Compare** the foot and shell of a snail with those of a clam.

Cephalopods Quick is a word that best describes some behaviors of the class Cephalopoda. Cephalopods are the head-footed mollusks (from the Greek word *cephalo,* meaning *head,* and from *pod,* meaning *foot*), which includes the squid, octopus, chambered nautilus, and the cuttlefish in **Figure 25.20.** The chambered nautilus is the only cephalopod with an external shell. Squids and cuttlefishes have an internal shell, while octopuses do not have a shell. The foot of a cephalopod is divided into arms and tentacles with suckers, which are used to capture prey.

Cuttlefish

■ **Figure 25.20** Cuttlefish have eight arms and two tentacles. The tentacles often are not visible because they are withdrawn into pouches under the eyes.
Compare ***What other differences do you see between cephalopods and gastropods?***

Protection Although most cephalopods don't have a hard external shell, they have evolved other protective mechanisms. Octopuses forcefully expel water to propel themselves away from threat. They hide in crevices or caves in the daytime. At night, they creep about in search of prey.

When threatened, an octopus shoots out an inky substance that forms a cloud. Scientists hypothesize that the ink visually confuses predators, and it also might act as a narcotic. Octopuses can change color to blend in with their surroundings. Squids and cuttlefishes also use ink and camouflage to escape predators. A chambered nautilus can pull into its shell for protection. It also uses its shell as camouflage. The dark top of the shell blends in with the ocean bottom when seen from above, while the white bottom of the shell blends in with the water above when seen from below.

Learning Octopuses are considered to be the most intelligent mollusks. They are capable of complex learning, such as being trained to select an object of a certain shape, color, or texture. See **Data Analysis Lab 25.2** to study this phenomenon.

DATA ANALYSIS LAB 25.2

Based on Real Data*

Interpret the Data

Can untrained octopuses learn to select certain objects? Two groups of octopuses were trained to select either a red ball or a white ball. Each trained group was observed by different groups of octopuses that were not trained.

Data and Observations

The graphs show the results of untrained octopus selection of white or red balls.

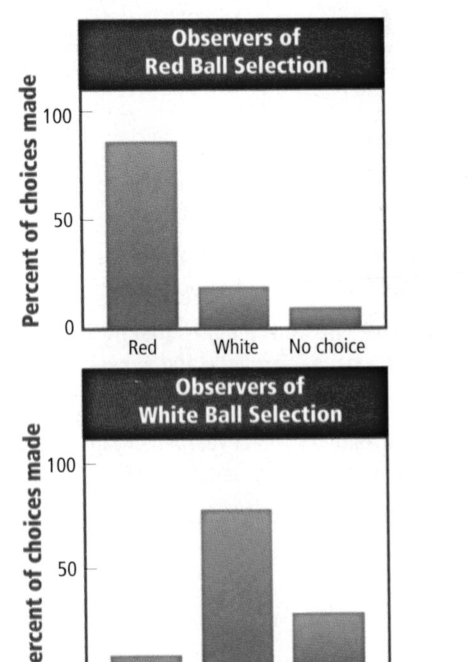

Think Critically

1. **Analyze the Data** How many octopuses selected the red ball or the white ball after observing the red ball being selected?
2. **Analyze the Data** How many octopuses selected the red ball or the white ball after observing the white ball being selected?
3. **Draw Conclusions** Can untrained octopuses learn by observation? Explain.

*Data obtained from: Fiorito, G. and P. Scotto. 1992. Observational learning in *Octopus vulgaris. Science* 256: 545–547.

■ **Figure 25.21** Cone snails are prized for their beauty.

Ecology of Mollusks

Mollusks play important roles in aquatic and terrestrial food chains as herbivores, predators, scavengers, and filter feeders. In many areas, certain mollusks are considered keystone species. A keystone is the stone at the top of an arch that holds the arch together, so a keystone species is one whose health influences the health of the entire ecosystem. For example, the hard clam is a keystone species for the Great South Bay in Long Island, New York. These clams filter water, which cleans the ecosystem. If the hard clam population declines, the water isn't filtered. This disrupts the food web, causing algal blooms and a decline in water quality.

The ability of some mussels to accumulate toxins in their body tissues can be useful to scientists who are monitoring water quality. By examining these mollusks, scientists can find out more about water quality than they could by testing the water alone.

Cone snails, as shown in **Figure 25.21,** are highly prized by collectors for the beauty of their shells and, as a result, might be close to extinction.

Connection to Health Certain cone snails produce powerful venom to kill prey. These venoms are being studied as potential treatments for pain, heart disease, clinical depression, and brain diseases, such as Alzheimer's disease, Parkinson's disease, and epilepsy.

Some mollusks cause damage, while others benefit humans. Some marine bivalve species, such as the shipworm, burrow into wood, causing much damage to wooden marinas and boats. On the other hand, people enjoy beautiful pearls that come from oysters. Pearls result when a grain of sand or a tiny parasite becomes trapped in an oyster. The mantle of the mollusk secretes a coating around the object to protect the mollusk, resulting in a pearl. Pearl producers implant pieces of shell or tiny plastic spheres in oysters and harvest cultured pearls in about five to seven years.

Section 25.3 Assessment

Section Summary

- Mollusks were the first animals in the course of evolution to develop a coelom.
- Mollusks are divided into three main classes based on different characteristics.
- Mollusks have two body features that no other animals have—a mantle and a muscular foot.
- Mollusks have more well-developed organ systems than roundworms and flatworms.
- Mollusks play important roles in the ecosystems in which they live.

Understand Main Ideas

1. MAIN Idea **Summarize** the main features of the three classes of mollusks.
2. **Evaluate** the ways in which the development of the coelom allowed for adaptations in mollusks that were not possible in earlier animals.
3. **Draw** a diagram of a representative mollusk and show the main evolutionary adaptations common to mollusks.
4. **Analyze** the importance to mollusks of the following adaptations: the mantle, mucus, and the muscular foot.

Think Critically

5. **Design an experiment** A species of bivalves on one beach is a pale color compared to the same species that is a much darker color on a beach 1100 km to the north. Design an experiment that might explain the differences in shell color.
6. **Classify** Make a dichotomous key that would distinguish the differences among the three classes of mollusks.

Section 25.4

Reading Preview

Objectives

- **Compare** segmented worms to flatworms and roundworms.
- **Evaluate** the importance of segmentation as an adaptation to survival in segmented worms.
- **Differentiate** the features of the three main classes of annelids that make them well-suited for their habitats.

Review Vocabulary

protostome: an animal with a mouth that develops from the opening in the gastrula

New Vocabulary

crop
gizzard
seta
clitellum

Segmented Worms

MAIN Idea Segmented worms have segments that allow for specialization of tissues and for efficiency of movement.

Real-World Reading Link You watch a train as it roars around a curve. The train follows the curve of the track because it is made up of individual cars that are linked together. The links give the train the flexibility it needs to stay on the track. In the same way, the individual segments that make up a segmented worm enable it to be flexible.

Body Structure

Earthworms are annelids and belong to phylum Annelida, which is characterized by animals with a body plan consisting of segments. As shown on the evolutionary tree in **Figure 25.22,** both mollusks and annelids undergo protostome development and, therefore, are considered close relatives.

There are more than 11,000 species of annelids, most of which live in the sea. Most of the remaining species are earthworms. Annelids live almost everywhere except in the frozen soil of the polar regions and in the sand of dry deserts.

Annelids include earthworms, marine worms, such as the ones shown in **Figure 25.22,** and parasitic leeches. These worms are different from flatworms and roundworms because they are segmented and have a coelom. Most annelids also have a larval stage that is similar to that of certain mollusks, suggesting a common ancestor. Annelids have bilateral symmetry like flatworms and roundworms, and have two body openings like roundworms.

Reading Check Describe two important ways segmented worms are different from flatworms and roundworms.

Figure 25.22 Annelids, like these marine worms, show protostome development, have coeloms, and are segmented.

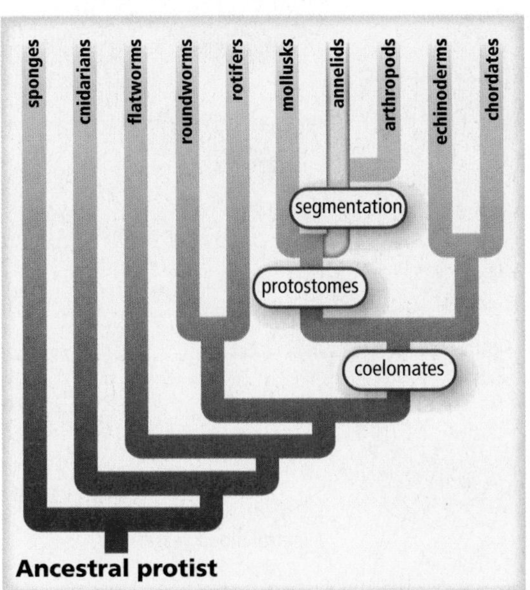

Even though annelids have the same cylindrical body shape as roundworms, annelid bodies are divided into segments. Externally, the segments look like a stack of thick coins or a stack of donuts. Inside the worm, the segments are divided almost completely from each other by walls of tissue, similar to the way walls separate the segments of a submarine. Each segment contains structures for digestion, excretion, and locomotion. The fluid within the coelom of each segment makes a rigid support system for the worm similar to the rigidity of a filled water balloon.

This rigidity in annelid segments creates a hydrostatic skeleton that muscles can push against. Segmentation also permits segments to move independently of each other and enables a worm to survive damage to a segment because other segments with the same functions exist.

Segments can be specialized, and groups of segments might be adapted to a particular function. For example, some segments might be adapted to sensing, while others are adapted to reproduction. As you continue to study annelids, earthworms will be used to show examples of typical annelid features.

Reading Check **Explain** how segments relate to a hydrostatic skeleton.

Feeding and digestion Running through all earthworm segments from the mouth to the anus is the digestive tract, a tube within a tube. Locate the digestive tract in the earthworm in **Figure 25.23.** Food and soil taken in by the mouth pass through the pharynx into the **crop,** where they are stored until they pass to the gizzard. The **gizzard** is a muscular sac containing hard particles that help grind soil and food before they pass into the intestine. Nutrients are absorbed from the intestine, then undigested material passes out of the worm's body through the anus. Parasitic annelids have pouches along the digestive tract that hold enough food to last for months.

Figure 25.23 As an earthworm pushes through the soil, it takes soil into its mouth. Nutrients are absorbed from the organic matter in the soil as it passes through the intestine.

Concepts In Motion

Interactive Figure To see an animation of the many systems in an earthworm, visit biologygmh.com.

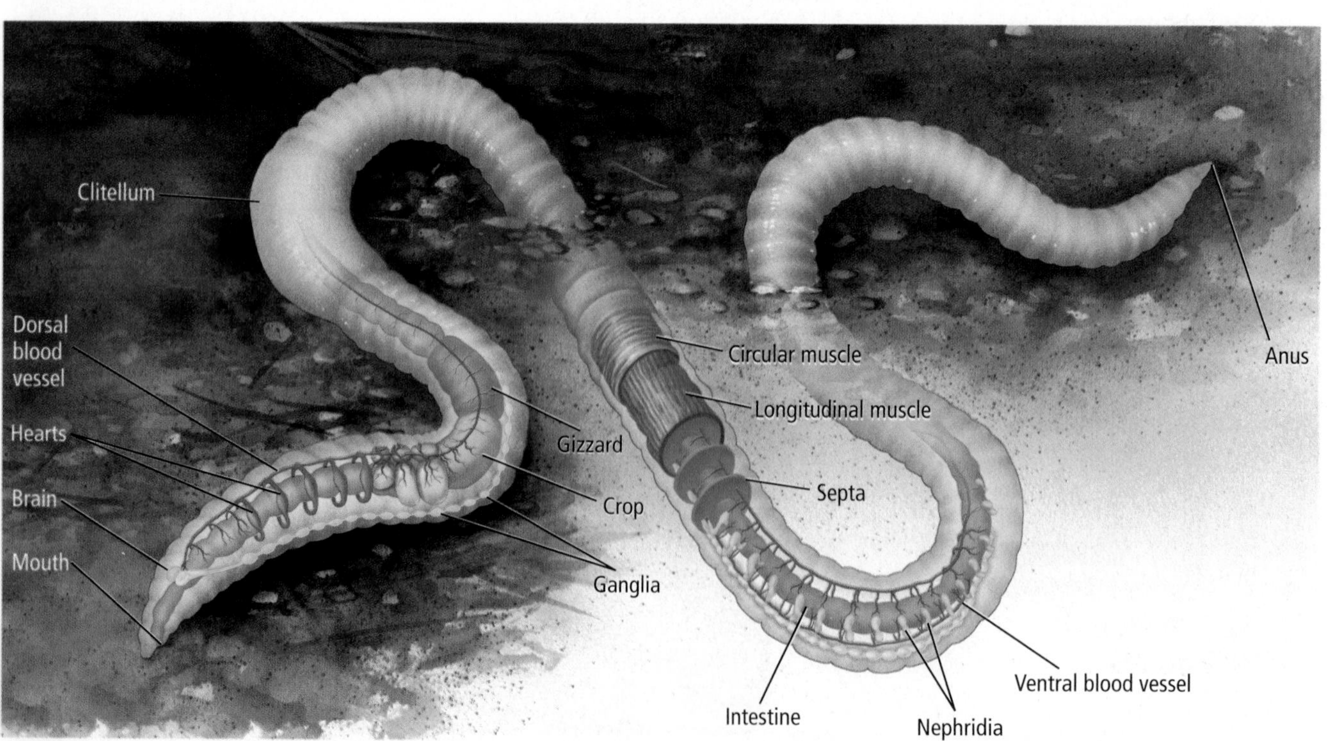

Circulation Unlike most mollusks, most annelids have a closed circulatory system. Oxygen and nutrients move to various parts of their bodies through their blood vessels. At the same time carbon dioxide and metabolic wastes are removed from the blood and excreted. Some of the vessels at the anterior end, or head, are large and muscular, as shown in **Figure 25.24,** and serve as hearts that pump the blood. The blood moves toward the anterior end of the worm in the dorsal blood vessel and toward the posterior end in the ventral blood vessel.

Respiration and excretion Earthworms take in oxygen and give off carbon dioxide through their moist skin. Some aquatic annelids have gills for the exchange of gases in the water. Segmented worms have two nephridia—similar to those in mollusks—in almost every segment. Cellular waste products are collected in the nephridia and are transported in tubes through the coelom and out of the body. Nephridia also function in maintaining homeostasis of the body fluids of annelids, ensuring that the volume and composition of body fluids are kept constant.

Response to stimuli In most annelids, such as the earthworm, the anterior segments are modified for sensing the environment. The brain and nerve cords composed of ganglia are shown in the earthworm in **Figure 25.23.** You might have seen an earthworm quickly withdraw into its burrow when you shine a flashlight on it or step close to it. These observations show that earthworms can detect both light and vibrations.

Movement When an earthworm moves, it contracts circular muscles running around each segment. This squeezes the segment and causes the fluid in the coelom to press outward like paste in a tube of toothpaste being squeezed. Because the fluid in the coelom is confined by the tissues between segments, the fluid pressure causes the segment to get longer and thinner. Next, the earthworm contracts the longitudinal muscles that run the length of its body. This causes the segment to shorten and return to its original shape, pulling its posterior end forward and resulting in movement.

Many annelids have setae on each segment. **Setae** (SEE tee) (singular, seta), as shown in **Figure 25.25,** are tiny bristles that push into the soil and anchor the worm during movement. By anchoring some segments and retracting others, earthworms can move their bodies forward and backward segment by segment.

Reading Check **Describe** how longitudinal and circular muscles work together to enable an earthworm to move.

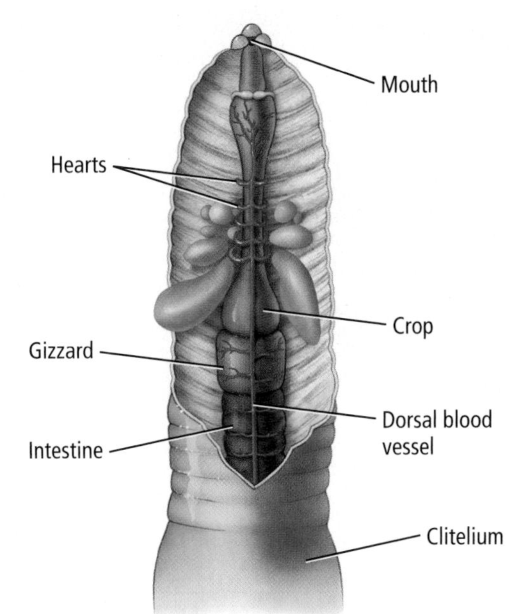

Figure 25.24 An earthworm has five hearts that pump blood through its circulatory system.

LAUNCH Lab

Review Based on what you've read about earthworm movement, how would you now answer the analysis questions?

Figure 25.25 This earthworm cross section shows how the setae extend from the body. Setae dig into soil and anchor the worm as it pushes forward.

Evaluate ***whether an earthworm would move faster on a rough or a smooth surface.***

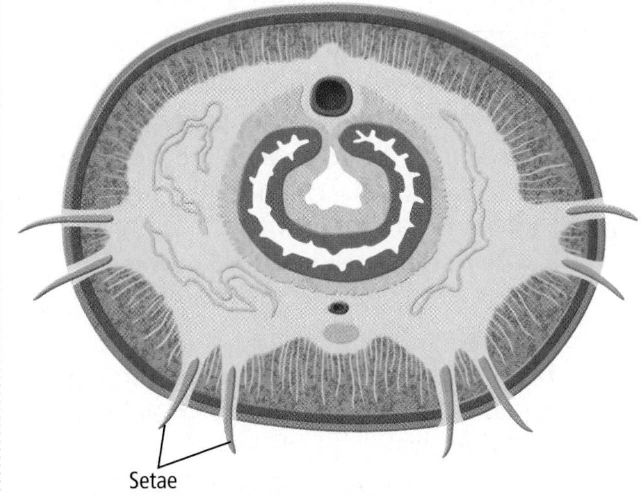

■ **Figure 25.26** After developing in the cocoon for two to three weeks, a young earthworm hatches.

Reproduction Annelids can reproduce both sexually and asexually. Most annelids have separate sexes, but some, such as earthworms and leeches, are hermaphrodites. Sperm are passed between two worms near segments called the clitellum. Refer back to **Figure 25.23** and notice that the **clitellum** is a thickened band of segments. It produces a cocoon from which young earthworms hatch, as shown in **Figure 25.26.** Sperm and eggs pass into the cocoon as it slips forward off the body of the worm. After fertilization, the young are protected in the cocoon as they develop. Some annelids reproduce asexually by fragmentation. If a worm breaks apart, the missing parts can be regenerated.

Diversity of Annelids

The phylum Annelida is divided into three classes: class Oligochaeta (ohl ih goh KEE tuh)—the earthworms and their relatives, class Polychaeta (pah lih KEE tuh)—the bristleworms and their relatives, and class Hirudinea (hur uh DIN ee uh)—the leeches.

FOLDABLES Incorporate information from this section into your Foldable.

Earthworms and their relatives Earthworms probably are the best-known annelids. They are used as bait for fishing and are found in garden soil. An earthworm can eat its own mass in soil every day. Earthworms ingest soil to extract nutrients. In this way, earthworms aerate the soil—they break up the soil to allow air and water to move through it.

In addition to earthworms, class Oligochaeta—the oligochaetes (AH lee goh keetz)—includes tubifex worms and lumbriculid worms. Tubifex worms are small, threadlike aquatic annelids that are common in areas of high pollution. Lumbriculid (lum BRIH kyuh lid) worms are freshwater oligochaetes that are about 6 cm long and live at the edges of lakes and ponds. You can observe a feature common to oligochaetes in **MiniLab 25.2.**

MiniLab 25.2

Observe Blood Flow in a Segmented Worm

How does blood flow in a segmented worm? The California blackworm has a closed circulatory system and a transparent body. Its blood can be viewed as it flows along the dorsal blood vessel.

Procedure

1. Read and complete the lab safety form.
2. Moisten a piece of **filter paper** with **spring water** and place it in a **Petri dish.**
3. Examine a **blackworm** on the moist paper using a **stereomicroscope.**
4. Locate the dorsal blood vessel in a segment near the midpoint of the worm. Observe how blood flows in each segment.
5. Use a **stopwatch** to record how many pulses of blood occur per minute. Repeat this for two more segments, one near the head and one near the tail of the worm. Record your data in a table.

Analysis

1. **Summarize** how blood moves through each segment, including the direction of blood flow.
2. **Compare and contrast** the rate of blood flow near the head, at the midpoint, and near the tail of the worm.

Bristleworm

Fan worms

■ **Figure 25.27** Note the paddlelike parapodia on the bristleworm, which are used for swimming and crawling. Fan worms withdraw quickly into their tubes when there is a disturbance in the water.

Marine annelids Polychaetes (PAH lee keetz), which belong to class Polychaeta, mainly are marine animals. They include bristleworms and fan worms, shown in **Figure 25.27.** Polychaetes have head regions with well-developed sense organs, including eyes. Most body segments of polychaetes have many setae. Most body segments also have a pair of appendages called parapodia, shown in **Figure 25.27,** that are used for swimming and crawling. Fan worms are sessile—they stay in one place—and are filter feeders. They trap food in the mucus on their fan-shaped structures. If there is a threat nearby, fan worms retreat into their tubes.

Leeches As shown in **Figure 25.28,** leeches in class Hirudinae are external parasites with flattened bodies and usually have no setae. Most leeches live in freshwater streams or rivers where they attach to the bodies of their hosts—including fishes, turtles, and humans. Leeches attach to their hosts using front and rear suckers. When a leech bites, its saliva contains chemicals that act as an anesthetic. Other chemicals in the saliva reduce swelling and prevent the host's blood from clotting.

 **Reading Check** **Describe** the habitats of the three classes of annelids.

Leech

■ **Figure 25.28** A leech uses its suckers to attach to its host and feeds by drawing blood into its muscular pharynx.
Compare and contrast *the feeding methods of leeches and tapeworms.*

Ecology of Annelids

Segmented worms play important roles in the ecology of ecosystems. Some are beneficial to plants and animals, while others benefit humans.

Earthworms Many different animals, including frogs and birds, eat earthworms as a part of their diets. Earthworms also mix leaf litter into soil, aerating it so that roots can grow easily and water can move through the soil efficiently. Both functions are important for a healthy ecosystem. However, nonnative earthworms are moving into areas, especially northern forests, where they are altering ecosystems in harmful ways. The disappearance of leaf litter on forest floors due to nonnative earthworms removes the shelter and moisture needed by many native plants and animals. Earthworms can be introduced into new habitats inadvertently by fishers and gardeners when the earthworms left from a day's fishing are dumped on the ground, or when a nursery plant from another part of the country is planted in a home garden.

VOCABULARY

ACADEMIC VOCABULARY

Convert:
to change from one form to another.
Through photosynthesis, plants convert water and carbon dioxide to sugar and oxygen.

Polychaetes Marine polychaetes help convert the organic debris of the ocean floor into carbon dioxide. Marine plant plankton take in carbon dioxide and use it during photosynthesis. Marine polychaetes are an important part of the diet of many marine predators.

Connection to History Leeches have been used as medical treatments for centuries. They were used to suck blood out of patients who were believed to be ill because of an excess of blood. Today, leeches are used after microsurgical procedures to prevent blood from accumulating in the surgical area. However, there are drawbacks to using leeches. They have a small feeding capacity and must be replaced often. They can cause bacterial infections, and many people cringe at the thought of having a leech attached to them. To solve these problems, a nonliving mechanical leech has been designed to do the work of the animal. **Table 25.1** summarizes the ecological benefits of earthworms, polychaetes, and leeches.

Reading Check **Evaluate** the effects of removing polychaetes from oceans.

Table 25.1 **Ecological Importance of Annelids**

Concepts In Motion **Interactive Table** To explore more about annelid ecology, visit biologygmh.com.

Type of Annelid	Example	Characteristics	Habitat	Ecological Benefit
Earthworms		• Few setae on most body segments	Terrestrial	• They aerate soil so roots can grow more easily and water can move efficiently. • They are food for many different animals.
Polychaetes		• Well-developed sense organs • Many setae on most body segments • Parapodia	Mainly marine	• They convert organic debris in oceans into carbon dioxide, which is used by marine plankton for photosynthesis.
Leeches		• Usually no setae on body segments • Front and rear suckers	Mainly freshwater	• They maintain blood flow after microsurgery.

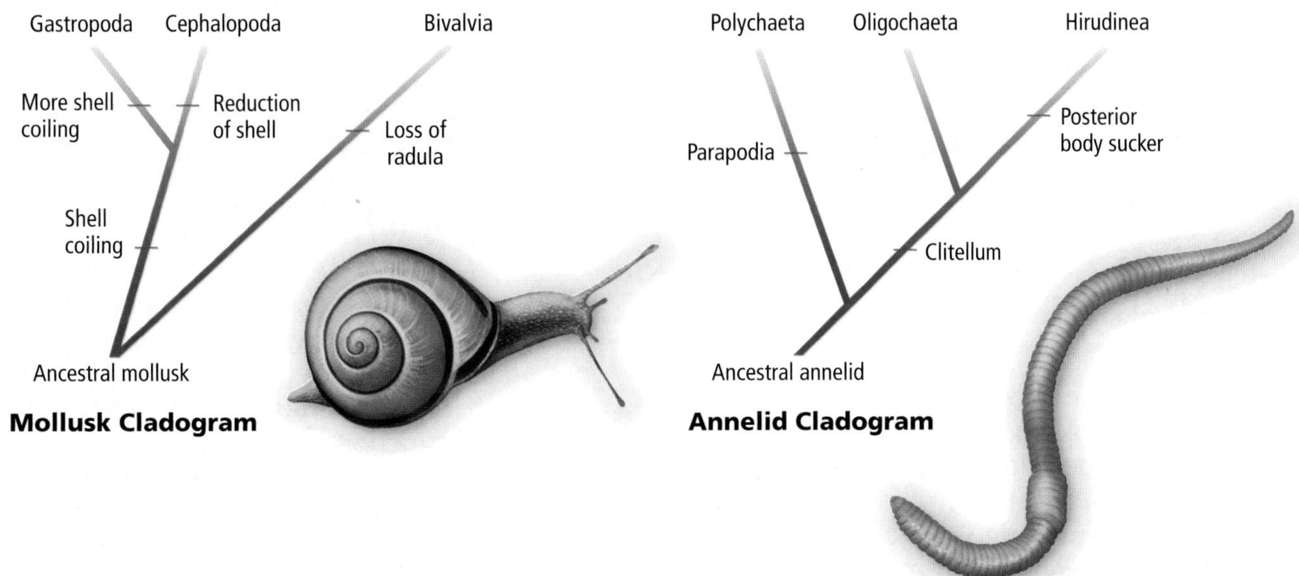

■ **Figure 25.29** These cladograms show how mollusks and annelids might have evolved.

Evolution of Mollusks and Annelids

In **Figure 25.29,** the cladogram on the left is one interpretation of the evolution of mollusks. As shown, gastropods have more shell coiling than cephalopods. In shell coiling, the shell grows in a circular manner, making it more compact and more stable than an uncoiled shell. Cephalopods have a reduced shell. Bivalves are considered to have evolved later than gastropods and cephalopods because they lack a radula.

The cladogram on the right in **Figure 25.29** is one interpretation of the evolution of annelids. In this interpretation, early segmented worms—the polychaetes—developed parapodia in the course of their evolution. Later annelids—the oligochaetes and leeches—developed clitella. Leeches developed posterior body suckers even later.

Section 25.4 Assessment

Section Summary

- Two main body features characterize annelids and distinguish them from roundworms and flatworms.
- There are three main classes of annelids based on distinctive features.
- Segmentation is an adaptation important to evolution.
- Annelids are important parts of terrestrial and marine habitats.

Understand Main Ideas

1. MAIN Idea **Summarize** how segmentation was an important evolutionary milestone.
2. **Compare and contrast** earthworms to flatworms and roundworms.
3. **Model** Using clay or salt dough, make models of typical examples from the three classes of annelids. Describe the adaptations that enable each annelid to live in its environment.
4. **Summarize** how earthworm muscles interact to cause movement.

Think Critically

5. **Hypothesize** Form a hypothesis about what might happen to a farm field if earthworms suddenly disappeared.
6. **Compare and contrast** circulation in annelids and mollusks.

WRITING in Biology

7. Write a paragraph explaining why leeches might be used after microsurgical procedures based on what you know about leech saliva.

BioDiscoveries

Fountain of Youth?

In 1513, Juan Ponce de Leon discovered Florida while looking for the "fountain of youth." Today, people are getting closer to finding it. The answer might be in the genome of a nematode. The roundworm *Caenorhabditis elegans* just might hold the key to aging.

Cynthia Kenyon's research involving *C. elegans* might lead to humans living longer and healthier lives.

Old worm, new insight Although *C. elegans* was determined to be an excellent model organism for developmental biology in 1963, it was not until thirty years later that it took center stage for aging studies. A strain of *C. elegans* was discovered in 1993 by Cynthia Kenyon, a professor at the University of California, San Francisco, to have a life span more than twice as long as normal. Mutations in a single gene caused the increase in life span. The worms also aged more slowly than normal. These findings revitalized the study of aging and have contributed to advances in that field.

Small worm, big uses In December 1998, *C. elegans* wormed its way into the history books once again because scientists decoded its entire genome. It is the first multicellular organism to hold that distinction. With a relatively small number of genes—97 million base pairs as opposed to 3 billion in humans—*C. elegans* is easy to observe and manipulate genetically.

The genome of *C. elegans* is surprisingly similar to that of humans (40 percent similarity). This little worm also carries out some of the same processes as humans. From a single fertilized cell, *C. elegans* develops into an adult with complex tissues and organ systems through mitosis. This makes it particularly useful for studying aging or other genetic puzzles such as cancer and insulin production.

From worms to humans It is evident that this little worm still holds many mysteries for us to discover. Cynthia Kenyon is amazed at all the progress science has made on aging. "There wasn't a field when we started," she says. "Now, there are a lot of people working on aging. We've learned a huge amount." Her work with *C. elegans* has led to similar findings in fruit flies and mice. These findings might lead to the extension of human life spans and to the postponement of age-related diseases. If we can unlock this mystery, it would be the key to the "fountain of youth."

CAREERS IN BIOLOGY

Model Imagine that you are a geneticist working on the *C. elegans* genome. You are asked by a local school system to talk about your work with this roundworm. For more information about *C. elegans,* visit biologygmh.com. Then, using papier-mâché, clay, or another similar medium, create a three-dimensional model of the worm to show to the students. Use different colors to highlight the different internal body parts.

BIOLAB

HOW DO WORMS AND MOLLUSKS MOVE?

Background: **The worm and mollusk phyla display wide diversity in behavior and physical characteristics. Throughout this chapter, you have been introduced to some of the various species that make up these phyla. In this lab, you will compare the form of movement used by a flatworm (a planarian), a roundworm (a vinegar eel), a mollusk (a land snail), and a segmented worm (a blackworm).**

Question: *What kind of motion do worms and mollusks display?*

Materials

plastic droppers (2)
petri dish (1 or 2)
microscope slide (1 or 2)
coverslip (1 or 2)
500-mL beaker
magnifying glass
dissecting microscope
light microscope
spring water or aged tap water (500 mL)
live cultures of planaria, vinegar eels, land snails, and blackworms

Safety Precautions

WARNING: ***Be sure to treat live animals in a humane manner at all times. Use caution when working with a microscope, glass slides, and coverslips.***

Procedure

1. Read and complete the lab safety form.
2. Create a data table to record your observations.
3. Observe the movement of a flatworm by placing it in a drop of water in a petri dish or on a slide with no coverslip.
4. Make a wet mount of a vinegar eel and observe its movement under low-power magnification.
5. Place a land snail on a petri dish. Gently tip the dish to observe the snail's movement from underneath.
6. Place a blackworm on a moist paper towel and observe it with a magnifying glass.
7. Place the blackworm in a beaker of aged tapwater and observe its movement.
8. Record your observations in your data table.
9. **Cleanup and Disposal** Wash reusable materials and place them where your teacher directs. Return all live specimens to the cultures provided by your teacher.

Analyze and Conclude

1. **Compare and contrast** the movements of the flatworm, roundworm, land snail, and segmented worm.
2. **Infer** how the forms of the flatworm, roundworm, land snail, and segmented worm are designed to move the animals.
3. **Describe** what happens to each segment of the blackworm as it crawls on land.
4. **Compare** the forward and backward motion of the blackworm on land. How might this be an adaptation for survival?
5. **Infer** how the blackworm might be able to escape from predators in the water.

APPLY YOUR SKILL

Experiment Design an experiment that you could perform to investigate how temperature affects worm and mollusk movement. If you have all the materials you will need, you might want to conduct the experiment. To learn more about worms and mollusks, visit biologygmh.com.

CHAPTER 25

Study Guide

Download quizzes, key terms, and flash cards from biologygmh.com.

FOLDABLES **Formulate** a question concerning the number of earthworms in a given area, such as "How many earthworms are in the first 30 cm of soil depth on the football field?" Develop a procedure to answer the question.

Vocabulary	Key Concepts
Section 25.1 Flatworms	
• flame cell (p. 727) • ganglion (p. 728) • pharynx (p. 727) • proglottid (p. 730) • regeneration (p. 728) • scolex (p. 730)	MAIN Idea Flatworms are thin, flat, acoelomate animals that can be free-living or parasitic. • Flatworms were among the first animals to exhibit bilateral symmetry. • Flatworms are acoelomates with limited numbers of organs and systems. • Some flatworms are free-living, and others are parasitic. • The three main classes of flatworms are Turbellaria, Trematoda, and Cestoda. • Flatworms that are parasitic have specialized adaptations for parasitic life.
Section 25.2 Roundworms and Rotifers	
• hydrostatic skeleton (p. 732) • trichinosis (p. 733)	MAIN Idea Roundworms and rotifers have a more highly evolved gut than flatworms. • Roundworms are closely related to flatworms, but roundworms have an evolutionary adaptation related to their gut. • Roundworms, like flatworms, have a limited number of organs and systems. • Roundworms are either free-living or parasitic. • Roundworms cause many human and plant diseases. • Rotifers are pseudocoelomates that appear on a different branch of the evolutionary tree than roundworms.
Section 25.3 Mollusks	
• closed circulatory system (p. 739) • gill (p. 738) • mantle (p. 737) • nephridium (p. 739) • open circulatory system (p. 739) • radula (p. 738) • siphon (p. 741)	MAIN Idea Mollusks are coelomates with a muscular foot, a mantle, and a digestive tract with two openings. • Mollusks were the first animals in the course of evolution to develop a coelom. • Mollusks are divided into three main classes based on different characteristics. • Mollusks have two body features that no other animals have—a mantle and a muscular foot. • Mollusks have more well-developed organ systems than roundworms and flatworms. • Mollusks play important roles in the ecosystems in which they live.
Section 25.4 Segmented Worms	
• clitellum (p. 748) • crop (p. 746) • gizzard (p. 746) • setae (p. 747)	MAIN Idea Segmented worms have segments that allow for specialization of tissues and for efficiency of movement. • Two main body features characterize annelids and distinguish them from roundworms and flatworms. • There are three main classes of annelids based on distinctive features. • Segmentation is an adaptation important to evolution. • Annelids are important parts of terrestrial and marine habitats.

CHAPTER 25

Assessment

Section 25.1

Vocabulary Review

Use what you know about the vocabulary terms found on the Study Guide page to answer the following questions.

1. What is a group of nerve cell bodies that coordinates ingoing and outgoing messages?
2. What is a tubelike muscular organ that releases digestive enzymes?
3. What structure attaches to the intestinal lining of a host with hooks and suckers?

Understand Key Concepts

Use the diagram below to answer question 4.

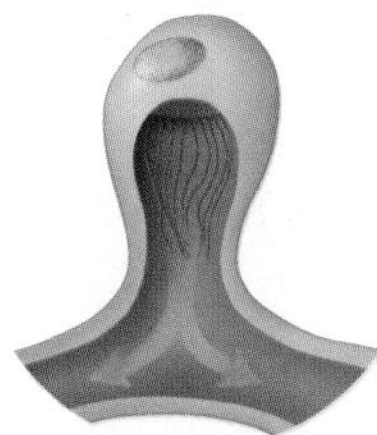

4. What function does the structure in the diagram perform?
   A. digestion **C.** maintains homeostasis
   B. movement **D.** provides support
5. Which animals have proglottids?
   A. flukes **C.** tapeworms
   B. planarians **D.** roundworms
6. Which classification fits a flatworm that is free-living?
   A. Turbellaria **C.** Trematoda
   B. Cestoda **D.** Nematoda
7. Which is not involved in planarian movement?
   A. cilia **C.** mucus
   B. muscles **D.** flame cells

Constructed Response

8. **Open Ended** A certain tapeworm secretes a chemical that slows the intestinal pulsations of its host. This helps to ensure that the tapeworm is not expelled from the body of the host. Explain how adding this chemical to a drug for humans might increase the drug's effectiveness.
9. **Open Ended** Design a parasitic worm for an animal that lives in the desert. Remember to consider how the animal will contract the parasite and how the parasite will stay within its host.

Think Critically

10. **Design an Experiment** Design an experiment that would determine what a planarian prefers to eat.

Section 25.2

Vocabulary Review

Each of the following sentences is false. Make each sentence true by replacing the italicized word with a vocabulary term from the Study Guide page.

11. Roundworms are bilaterally symmetrical, cylindrical, *segmented,* and are tapered at both ends.
12. *Trichinosis* can be contracted by walking barefoot on contaminated soil.
13. Roundworms have *crosswise* muscles that cause their bodies to move in a thrashing manner.

Understand Key Concepts

Use the diagram below to answer questions 14 and 15.

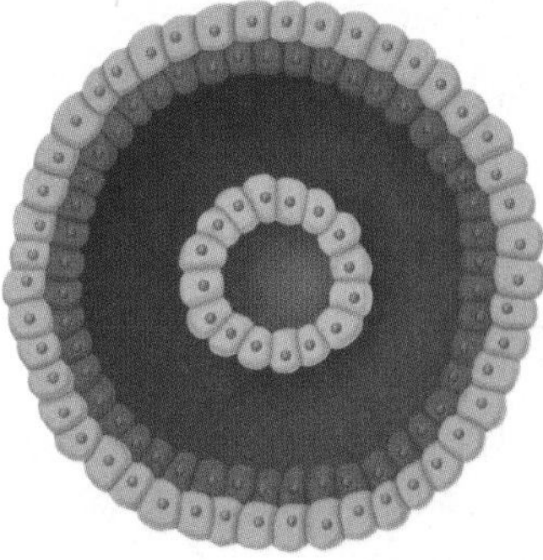

14. Which feature of roundworms is illustrated in the diagram above?
    A. pseudocoelom **C.** circulatory system
    B. scolex **D.** nervous system
15. The feature in the diagram led to which adaptation in roundworms?
    A. a coelom **C.** a mantle
    B. a gut **D.** segments

Constructed Response

16. **Short Answer** Make a diagram that shows the life cycle of a beef tapeworm.

17. **Open Ended** Select a human parasite and indicate with a key on a map of the world where the parasite is most common. Visit biologygmh.com to find information.

Think Critically

18. **Concept Mapping** Make a concept map that uses the following words: nematode, pseudocoelomate, digestive tract with two openings, parasitic, free-living, lengthwise muscles, host.

19. **Design an Experiment** Imagine that you find a tiny worm in the garden. How could you determine whether it is a flatworm or a roundworm?

Section 25.3

Vocabulary Review

An analogy is a comparison relationship between two pairs of words and can be written in the following manner: A is to B as C is to D. In the analogies that follow, one of the words is missing. Complete each analogy with a vocabulary term from the Study Guide page.

20. Kidney is to metabolic waste as __________ is to cellular waste.

21. Tongue is to candy as __________ is to algae.

22. Legs are to running as __________ is to jet-propelled swimming.

Understand Key Concepts

23. If the mantle of a bivalve was damaged, the bivalve would not be able to do which function?
    - **A.** maintain its shell
    - **B.** digest food
    - **C.** circulate blood
    - **D.** excrete wastes

24. Which word pair is related most closely?
    - **A.** shell—circulation
    - **B.** radula—feeding
    - **C.** jet-propelled swimming—bivalve
    - **D.** open circulatory system—octopus

Use the diagram below to answer questions 25 and 26.

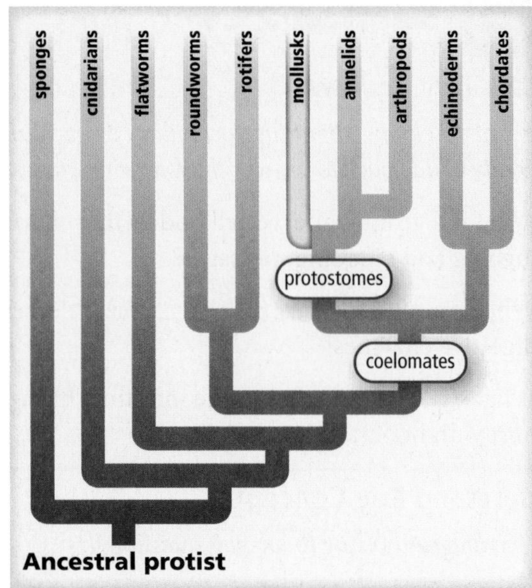

25. The phylogenetic tree of animals shows that mollusks have which feature?
    - **A.** a pseudocoelom
    - **B.** a coelom
    - **C.** solid bodies
    - **D.** shells

26. Which group is related most closely to mollusks?
    - **A.** nematodes
    - **B.** enchinoderms
    - **C.** annelids
    - **D.** chordates

Constructed Response

27. **Open Ended** Make a dichotomous key to identify mollusk shells that you find in pictures in books about animals, shells that you have collected, or shells that your teacher supplies.

Think Critically

28. **CAREERS IN BIOLOGY** Marine biologists know that zebra mussels are pests in many aquatic ecosystems. They form ultra dense colonies and can clog pipes leading to water-treatment and industrial plants. However, some marine biologists hypothesize that because the mussels live in such dense groups, they might serve as water-purification systems in places such as zoo ponds and other park ponds that have excessive algae blooms in the summer. Design an experiment that would determine if zebra mussels could be used to purify water.

Section 25.4

Vocabulary Review

An analogy is a comparison relationship between two pairs of words and can be written in the following manner: A is to B as C is to D. In the analogies that follow, one of the words is missing. Complete each analogy with a vocabulary term from the Study Guide page.

29. Teeth are to human as __________ is to earthworm.

30. Cocoon is to butterfly as __________ is to earthworm.

31. Vacuole is to protist as __________ is to earthworm.

Understand Key Concepts

Use the diagram below to answer questions 32 and 33.

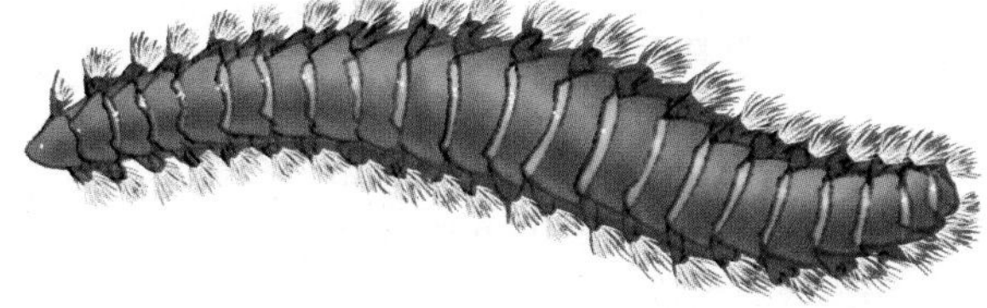

32. Which animal is illustrated in the diagram?
A. roundworm
B. leech
C. polychaete
D. earthworm

33. What feature is characteristic of this animal?
A. foot
B. parapodia
C. sucker
D. shell

Constructed Response

34. Open Ended If global warming continues, predict how earthworms might change as a result of natural selection.

Think Critically

35. CAREERS IN BIOLOGY Rheumatologists, doctors who treat arthritis, have observed that when leeches are applied for a short time to the skin near joints of people affected with arthritis, pain is relieved for up to six months. Design an experiment that would explain this phenomenon.

Additional Assessment

36. WRITING in Biology Research mollusks that live in areas of hydrothermal vents. Write a report emphasizing the differences between hydrothermal vent mollusks and those that live in the habitats you studied in this chapter.

DBQ Document-Based Questions

The data below represent the percentages of the three main classes of flatworms.

Data obtained from: Pechenik, J. 2005. *Biology of the Invertebrates.* New York: McGraw-Hill.

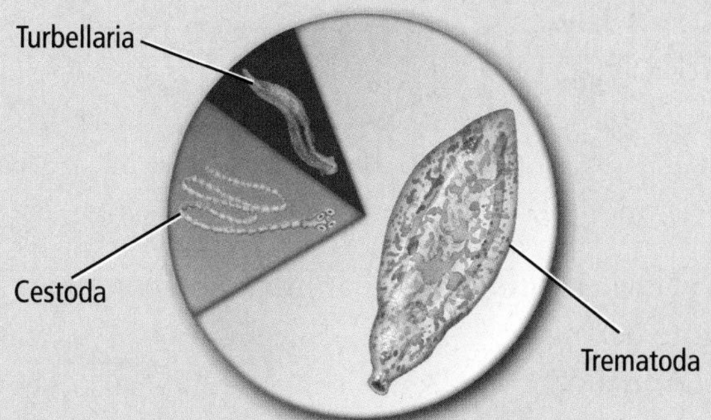

37. Approximately what percentage of flatworms are flukes?

38. Which group of flatworms has the least number of species?

39. Infer why there might be so many more of one kind of flatworm than any other kind.

Cumulative Review

40. Place the following steps of DNA translation in the correct order. **(Chapter 12)**
1. tRNA carrying a methionine moves into the P site.
2. The mRNA attaches to the ribosome.
3. A tRNA brings the appropriate amino acid to the A site.
4. The tRNA is released to the E site.
A. 2, 1, 3, 4
B. 4, 3, 1, 2
C. 3, 1, 2, 4
D. 3, 2, 4, 1

Standardized Test Practice

Cumulative

Multiple Choice

1. During dry weather, pieces of a moss might be scattered by the wind. When it rains, these pieces can grow into new plants. Which process does this display?
   A. alternation of generations
   B. gametophyte reproduction
   C. sporophyte generation
   D. vegetative reproduction

Use the diagram below to answer questions 2 and 3.

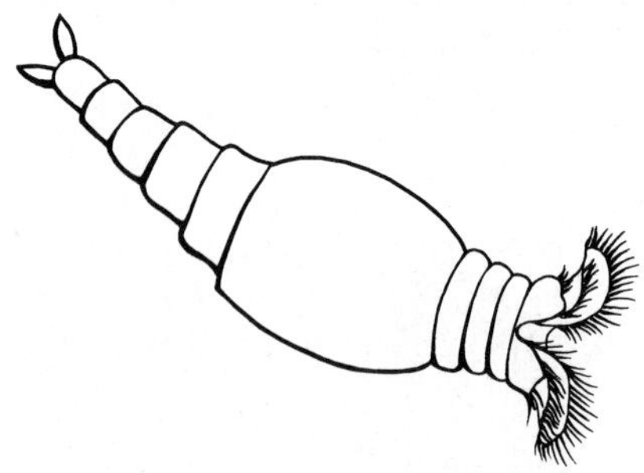

2. In which phylum does the animal shown in the figure belong?
   A. Annelida
   B. Nematoda
   C. Platyhelminthes
   D. Rotifera

3. Roundworms differ from the organism shown above because roundworms have which characteristic?
   A. a complete digestive tract
   B. inability to live in fresh water
   C. a smaller body size
   D. a body surface covered with cilia

4. Which is one characteristic of all cnidarians?
   A. Their tentacles contain cnidocytes.
   B. Their tentacles contain fibroblasts.
   C. They only live in freshwater environments.
   D. They spend some time as sessile animals.

5. Which is an example of a nastic response?
   A. bamboo plants growing toward a light
   B. corn plant roots growing downward
   C. sunflowers tracking the Sun
   D. vines growing up a tree

Use the diagram below to answer question 6.

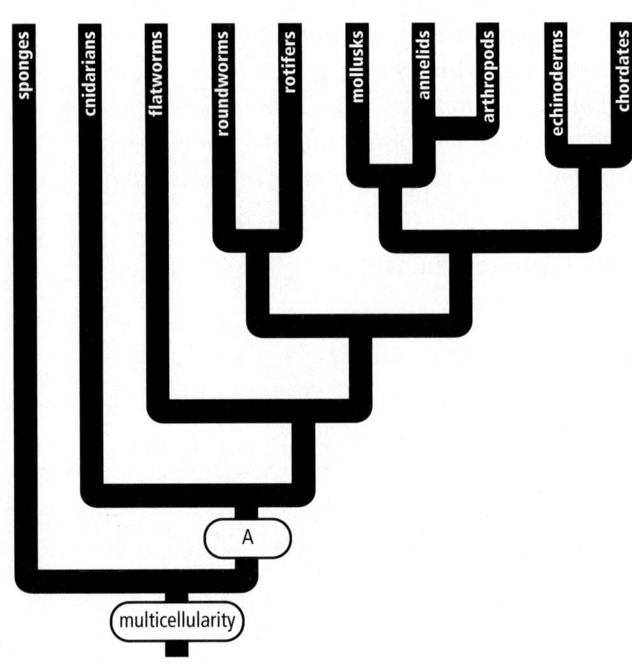

6. Which body structures are typical of all the animals above Point A on the evolutionary tree?
   A. cell walls
   B. coeloms
   C. tentacles
   D. tissues

7. How does an aggregate fruit, such as a blackberry or strawberry, form?
   A. when a flower has multiple female organs that fuse together
   B. when a fruit has multiple seeds that fuse together
   C. when multiple flowers from the same plant fuse together
   D. when multiple simple fruits fuse together

8. How do hornworts differ from other nonvascular plants?
   A. Their cells allow nutrients and water to move by diffusion and osmosis.
   B. Their cells can contain a type of cyanobacteria.
   C. They can be classified as either thallose or leafy.
   D. They have chloroplasts in some of their cells.

Biology Online Standardized Test Practice biologygmh.com

Short Answer

Use the diagram below to answer questions 9 and 10.

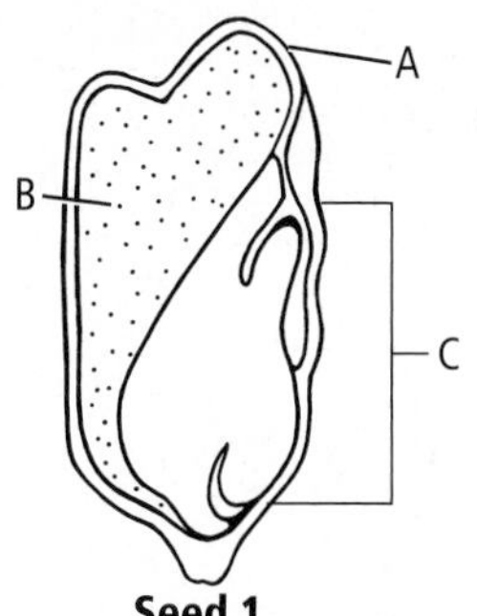

Seed 1

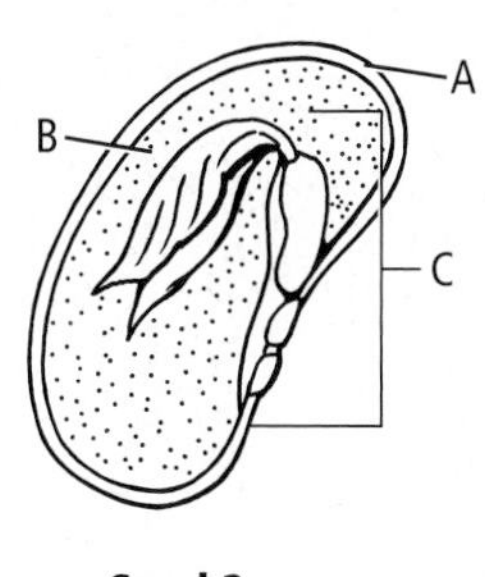

Seed 2

9. Name the parts of these seeds.

10. Which seed is a monocot and which is a eudicot? How do you know?

11. Explain why squids and clams are both included in Phylum Mollusca even though they appear to be very different kinds of animals.

12. What is one thing that humans can do to preserve reefs? Explain your reasoning.

13. Specify an example of an ancestral character and a derived character that angiosperms have.

14. Describe the alternation of generations in plants.

15. Describe how cellular slime molds reproduce and identify whether this process is sexual or asexual. Assess how this form of reproduction is beneficial for cellular slime molds.

Extended Response

16. List two reasons why animals benefit from segmentation. Assess the importance of these benefits.

17. Suppose you are a scientist trying to determine the water quality of a river where mussels live. What data could you collect from the mussels in order to determine the quality of the river water?

Essay Question

Schistosomiasis is caused by flukes, which have complex life cycles involving specific freshwater snail species as intermediate hosts. Infected snails release large numbers of minute, free-swimming larvae (cercariae) that are capable of penetrating the unbroken skin of a human host. Even brief exposure to contaminated freshwater, such as wading, swimming, or bathing, can result in infection. Human schistosomiasis cannot be acquired by wading or swimming in salt water (oceans or seas). The cercariae of birds and aquatic mammals can penetrate the skin of humans who enter infested fresh or salt water in many parts of the world, including cool temperate areas. The cercariae die in the skin but may elicit a puritic rash ("swimmer's itch" or "clam-digger's itch").

Using the information in the paragraph above, answer the following question in essay format.

18. Schistosomiasis is a disease that is most common in sub-Saharan Africa, the Philippines, southern China, and Brazil. Propose a plan to control this disease in a specific area. What steps would need to be taken to keep people from getting the disease? Develop a plan and explain it in a well-organized essay.

NEED EXTRA HELP?																		
If You Missed Question . . .	1	2	3	4	5	6	7	8	9	10	11	12	13	14	15	16	17	18
Review Section . . .	23.1	25.3	25.2	24.3	22.3	24.3	23.3	21.2	23.3	23.3	25.3	24.3	18.2	23.1	19.4	24.2	15.2	25.1

CHAPTER 26

Arthropods

BIG Idea Arthropods have evolved to have a variety of adaptations for successful diversity, population, and persistence.

Section 1
Arthropod Characteristics
MAIN Idea Arthropods have segmented bodies and tough exoskeletons with jointed appendages.

Section 2
Arthropod Diversity
MAIN Idea Arthropods are classified based on the structure of their segments, types of appendages, and mouthparts.

Section 3
Insects and Their Relatives
MAIN Idea Insects have structural and functional adaptations that have enabled them to become the most abundant and diverse group of arthropods.

BioFacts

- Copepods are tiny, but they exist in such large numbers that they are a major source of protein in the oceans.
- A single copepod might eat 200,000 microscopic diatoms in one day.
- Copepod eggs can lie dormant for months or years until conditions are right for hatching.

Copepods
LM Magnification: 20×

Individual copepod
LM Magnification: unavailable

Jointed copepod antenna
LM Magnification: 100×

LAUNCH Lab

What structures do arthropods have?

Arthropods form a group of animals that includes all bees, flies, crabs, millipedes, centipedes, spiders, and ticks. Discover the features arthropods share by observing two different arthropods.

Procedure

1. Read and complete the lab safety form.
2. Create a data table to record your observations.
3. Observe the physical characteristics of live or preserved **specimens of a crayfish** and a **pill bug.** Record your observations in your data table. **Warning: *Treat live animals in a humane manner at all times.***
4. Observe the movements of the two animals, if possible, and record your observations.

Analysis

1. **Describe** the structures of the two animals that are similar.
2. **Identify** the defensive structure that the two animals have in common. How does this feature allow them to protect themselves from predators?

Visit biologygmh.com to:

- study the entire chapter online
- explore Concepts in Motion, the Interactive Table, Microscopy Links, Virtual Labs, and links to virtual dissections
- access Web links for more information, projects, and activities
- review content online with the Interactive Tutor and take Self-Check Quizzes

Arthropod Adaptations Make the following Foldable to help you understand and compare arthropod adaptations to terrestrial and aquatic habitats.

STEP 1 Fold one sheet of paper into thirds lengthwise. Then fold the paper into fourths widthwise.

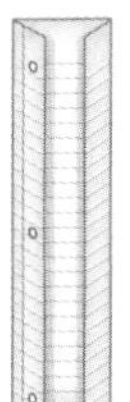

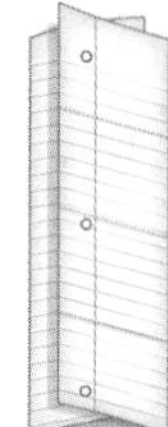

STEP 2 Unfold, lay the paper lengthwise, and draw lines along the folds.

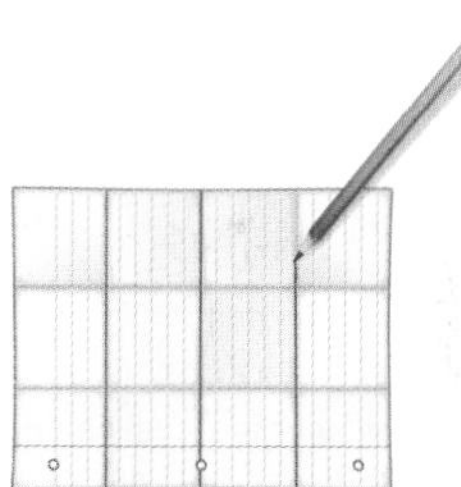

STEP 3 Add the following labels to your table as shown: *Respiration, Circulation/Excretion, Movement, Aquatic Arthropods,* and *Terrestrial Arthropods.*

	Respiration	Circulation/ Excretion	Movement
Aquatic Arthropods			
Terrestrial Arthropods			

FOLDABLES **Use this Foldable with Section 26.1.** As you read the section, record what you learn about the differences between terrestrial and aquatic arthropods.

Section 26.1

Arthropod Characteristics

Reading Preview

Objectives

- **Evaluate** the importance of exoskeletons, jointed appendages, and segmentation to arthropods.
- **Compare** organ system adaptations in arthropods.
- **Differentiate** arthropod organs that enable them to maintain homeostasis.

Review Vocabulary

ganglion: a group of nerve cell bodies that coordinates incoming and outgoing messages

New Vocabulary

thorax
abdomen
cephalothorax
appendage
molting
mandible
tracheal tube
book lung
spiracle
Malpighian tubule
pheromone

MAIN Idea **Arthropods have segmented bodies and tough exoskeletons with jointed appendages.**

Real-World Reading Link Think about what animal group might have more individuals than any other group. Did copepods come to mind? Even though copepods are numerous, most people have never seen one. The copepods in the opening photo are tiny arthropods that float in the open ocean and feed on even smaller protists. They can be found almost anywhere there is water.

Arthropod Features

Copepods belong to phylum Arthropoda (ar THRAH puh duh). Between 70 and 85 percent of all named animal species are arthropods (AR thruh pahdz). As shown in the circle graph in **Figure 26.1,** the majority of arthropods are insects, which includes beetles, butterflies, moths, flies, bees, and wasps.

Find arthropods on the evolutionary tree shown in **Figure 26.1.** Follow the branches and you will see that, like annelids, arthropods are segmented invertebrates with bilateral symmetry, coelomate body cavities, and protostome development. Unlike annelids, arthropods have exoskeletons with jointed appendages that enable them to move in complex ways. All three of these features—segmentation, exoskeletons, and jointed appendages—are important keys to their success.

Reading Check **Compare and contrast** arthropods and annelids.

■ **Figure 26.1** Most arthropods are insects, as shown by the blue segments on the graph. Arthropods are coelomates and show protostome development.
Interpret ***Crustaceans and spiders make up what percentage of arthropods?***

Percentages of Arthropod Species

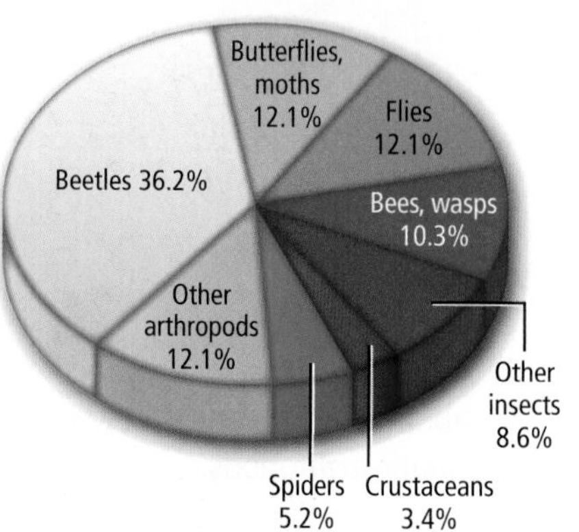

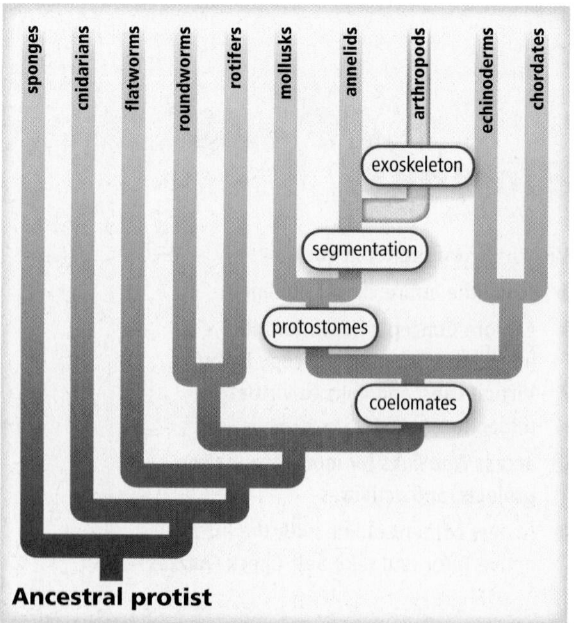

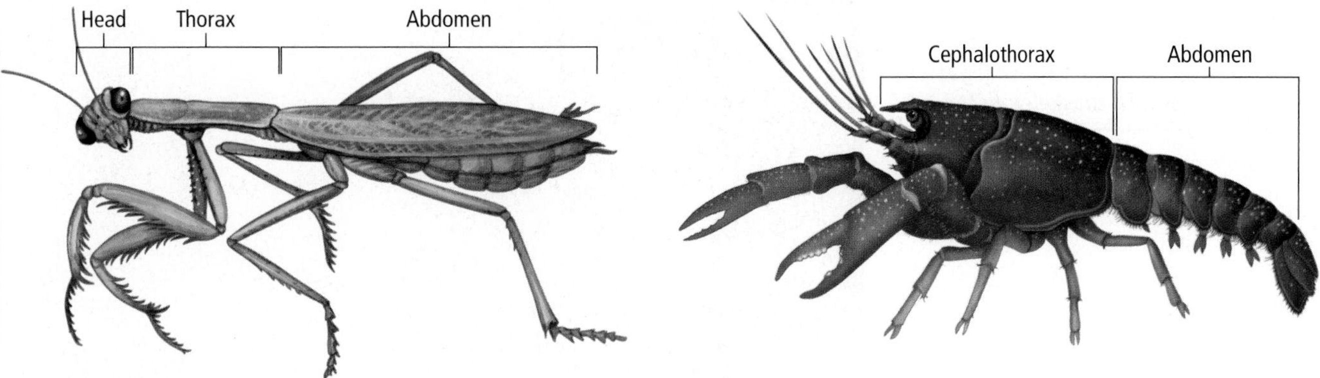

Figure 26.2 Some segments in arthropods are fused. The praying mantis shows fusion of segments into its head, thorax, and abdomen. The crayfish shows a different fusion of segments into its cephalothorax and abdomen.

Personal Tutor

To learn about arthropods, visit biologygmh.com.

Segmentation Arthropods are segmented allowing for efficient and complex movements. Notice in **Figure 26.2** that the praying mantis's segments are fused into three main body regions—a head, a thorax, and an abdomen. The heads of arthropods have mouthparts for feeding and various types of eyes. Many have antennae. Antennae are long sensory structures that contain receptors for smell and touch. The **thorax** is the middle body region, consisting of three fused main segments to which, in many arthropods, the legs and wings are attached. The **abdomen,** which also contains fused segments and is at the posterior end of the arthropod, bears additional legs and contains digestive structures and the reproductive organs. Some arthropods, such as the crayfish in **Figure 26.2,** have the thorax region fused with the head into a single structure called a **cephalothorax** (sef uh luh THOR aks).

In some groups of arthropods, segmentation is more obvious during early development. For example, a caterpillar has many obvious segments, while the adult butterfly has only three body segments.

 **Reading Check** **Summarize** the main body regions in arthropods.

Exoskeleton Arthropods have hard exoskeletons on the outside of their bodies, similar to a lightweight suit of armor. Recall from Chapter 24 that the exoskeleton provides a framework for support, protects soft body tissues, and slows water loss in animals that live on land. It also provides a place for muscle attachment.

Connection to Chemistry The exoskeleton of an arthropod is made of chitin—a nitrogen-containing polysaccharide bound with protein. While the exoskeleton of a grasshopper is leathery, the exoskeletons of some crustaceans, such as lobsters, incorporate calcium salts that harden them to such an extent that a hammer would be needed to crush them. An arthropod's exoskeleton can be hard in some places and thin and flexible in others, providing for movable joints between body segments and within appendages.

There is a limit to how hard and thick an exoskeleton can be. It is thin in small arthropods, such as the copepod, because tiny muscles pull against it; it is thicker in larger arthropods, such as crabs and lobsters, to bear the pull of larger muscles. Imagine a fly as large as a bird. The fly's exoskeleton would have to be so thick to withstand the pull of the large muscles that the fly would not be able to move under the weight of the exoskeleton.

■ **Figure 26.3** Like a door hinge, the joint in this fly's leg can bend in only one direction.
Explain ***How do jointed appendages benefit animals with exoskeletons?***

Jointed appendages Arthropods have paired appendages. **Appendages** (uh PEN dih juz) are structures, such as legs and antennae, that grow and extend from an animal's body. Appendages of arthropods are adapted for a variety of functions, such as feeding, mating, sensing, walking, and swimming. Notice in **Figure 26.3** that the appendages of arthropods have joints. To understand how important jointed appendages are, imagine yourself without joints—no finger joints, no wrist, elbow, knee, hip, or ankle joints. Without jointed appendages, you could not play a computer game, sit in a movie theater, shoot a basketball, or even walk. Jointed appendages enable arthropods to have flexible movements and to perform other life functions, such as getting food and mating, that would be impossible without joints.

LAUNCH Lab

Review **Based on what you've read about arthropod features, how would you now answer the analysis questions?**

Molting Because the exoskeleton of arthropods is made of nonliving material and cannot grow, arthropods must shed their outer coverings in order to grow. This process of shedding the exoskeleton is called **molting.** Arthropods make their own new exoskeletons. Glands in the skin make a fluid that softens the old exoskeleton while the new exoskeleton forms underneath. As the fluid increases in volume, the pressure increases and eventually cracks the old exoskeleton. This process is similar to freezing water in a closed glass container—as the water expands, the glass cracks. **Figure 26.4** shows a tarantula next to its shed exoskeleton. Before the new exoskeleton hardens, blood circulation increases to all parts of the body and the animal puffs up. Some arthropods also take in air, which assists in making the hardening exoskeleton a little larger for "growing room."

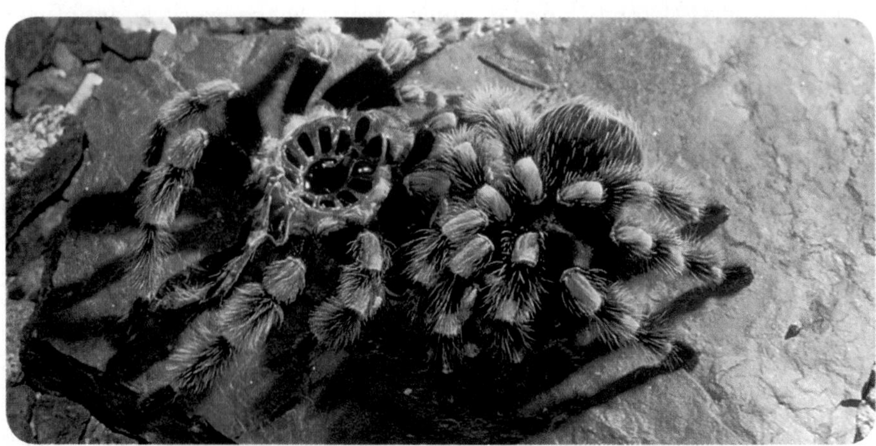

■ **Figure 26.4** Arthropods must molt so that their bodies can continue to grow. This tarantula has just come out of its outgrown exoskeleton.

■ **Figure 26.5** This leafcutter ant uses its mandibles to cut a leaf from a tree. Once fungus grows on the leaf cutting, the ant will feed the fungus to its larvae.

Body Structure of Arthropods

Arthropods have complex organ systems that enable them to live in many diverse habitats. Adaptations in several organ systems, such as the respiratory system and the nervous system, have contributed to the success of these animals.

Feeding and digestion The great diversity of arthropods is reflected in their enormous variety of feeding habits and structures. The mouthparts of most arthropods include a pair of appendages called **mandibles** (MAN duh bulz) that can be adapted for biting and chewing, as shown in **Figure 26.5.** Depending on their feeding habits, other arthropods have mouthparts modified like feathery strainers, stabbing needles, cutting swords, or sucking straws. Observe the structure of arthropod mouthparts in **MiniLab 26.1.** Arthropods can be herbivores, carnivores, filter feeders, omnivores, or parasites. To digest food, arthropods have a complete, one-way digestive system with a mouth, gut, and an anus, along with various glands that produce digestive enzymes.

Study Tip

Key Ideas Work with another student to determine this section's key ideas. Notice that the headings often are clues to key ideas. Also, many paragraphs have topic sentences that state the key idea.

MiniLab 26.1

Compare Arthropod Mouthparts

How do the mouthparts of arthropods differ? Arthropods eat a wide variety of foods, from nectar and plants to fish and small birds. Explore how the mouthparts of different types of arthropods are designed for their specific diets.

Procedure

1. Read and complete the lab safety form.
2. Create a data table to record your observations about the mouthparts of the arthropods and your inferences about the function of each type of mouth.
3. Using a **magnifying lens** or a **stereomicroscope,** observe the **mouthparts of preserved specimens of different arthropods**. Record your observations in your data table.
4. Infer the specific function of each type of mouth based on the structure of its parts.

Analysis

1. **Compare and contrast** the different mouthparts you observed.
2. **Infer** the type of diet each arthropod might eat based upon your observations of their mouthparts.

Visualizing Respiratory Structures

Figure 26.6
Arthropods take in oxygen by using one of three basic structures—gills, tracheal tubes, or book lungs.

Gills

A crayfish lives in an aquatic environment and uses gills to obtain oxygen. The cross section illustrates how the gills are divided. This provides a large surface area in a small space for the exchange of gases.

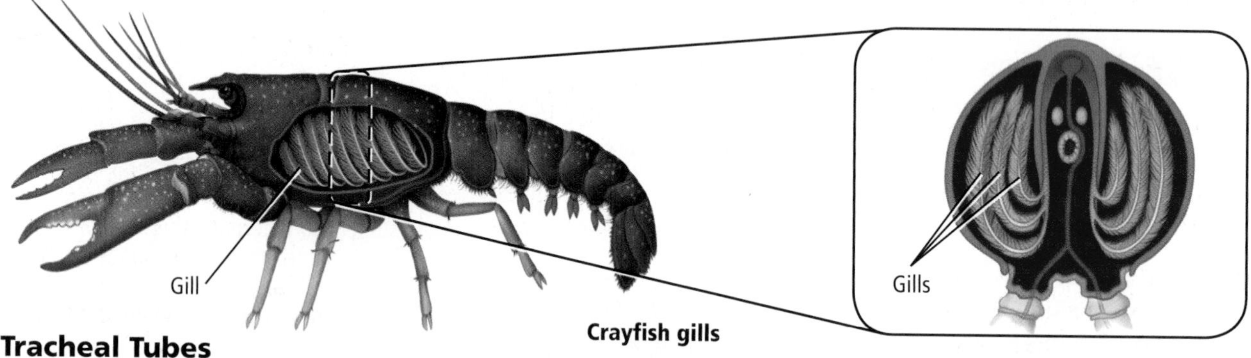

Crayfish gills

Tracheal Tubes

Insects such as this beetle have tracheal tubes that branch into smaller and smaller tubules to carry oxygen throughout the body. Air enters the respiratory system through spiracles, then travels from the tracheal tubes to tracheal tubules until it reaches muscle.

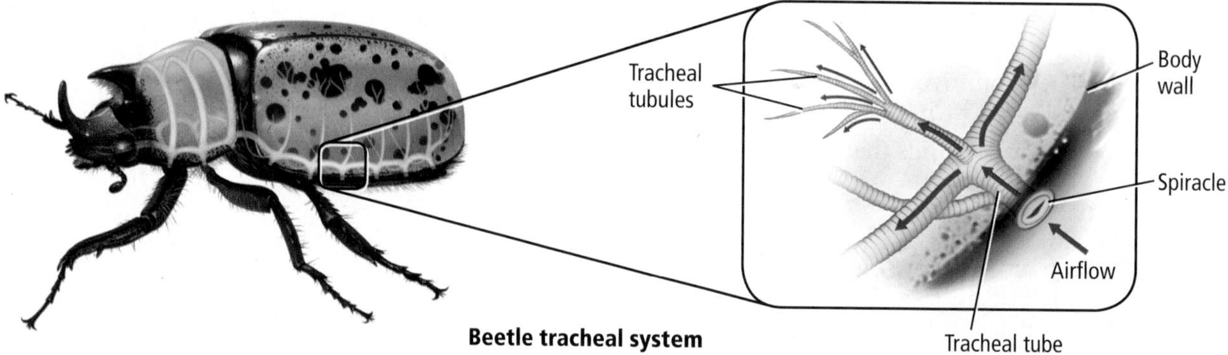

Beetle tracheal system

Book Lungs

This spider uses book lungs to draw in oxygen. As in arthropods with tracheal tubes, air enters the book lungs through spiracles.

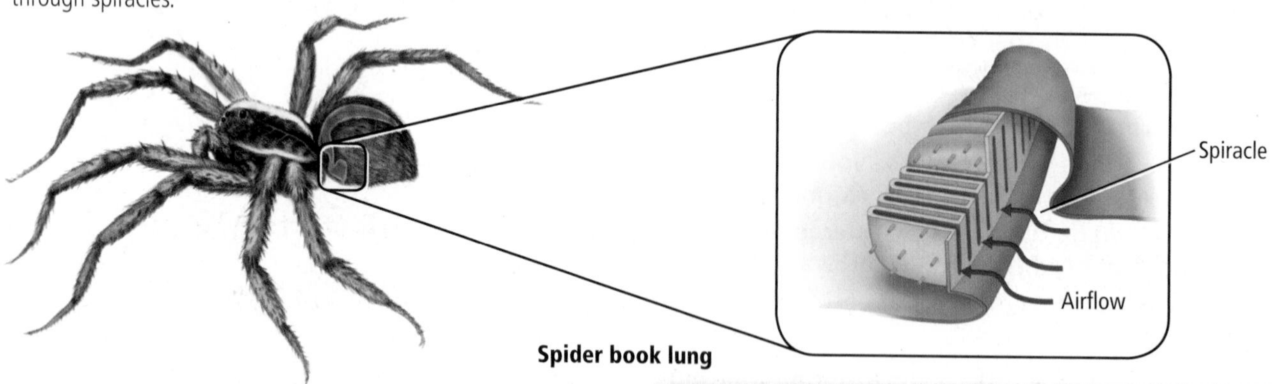

Spider book lung

Concepts In Motion **Interactive Figure** To see an animation of arthropod respiratory structures, visit biologygmh.com.

Biology Online

Respiration Arthropods obtain oxygen by using one of three structures—gills, tracheal tubes, or book lungs. Recall from Chapter 25 that maintaining a certain homeostatic balance of oxygen in body tissues enables animals to have energy for a variety of functions. Most aquatic arthropods have gills, like those shown in **Figure 26.6,** that function in the same way as the gills in mollusks. All terrestrial arthropod body tissues need to be near airways to obtain oxygen.

Terrestrial arthropods depend on respiratory systems rather than circulatory systems to carry oxygen to cells. Most terrestrial arthropods have a system of branching tubes called **tracheal** (TRAY kee ul) **tubes,** shown in **Figure 26.6,** that branch into smaller and smaller tubules. These tubules carry oxygen throughout the body.

Some arthropods, including spiders, have **book lungs,** saclike pockets with highly folded walls for respiration. In **Figure 26.6,** notice how the membranes in book lungs are like the pages in a book. The folded walls increase the surface area of the lungs and allow an efficient exchange of gases. You also can see how both tracheae and book lungs open to the outside of the body of the arthropod in openings called **spiracles** (SPIHR ih kulz).

FOLDABLES
Incorporate information from this section into your Foldable.

Circulation Even though most arthropods do not rely on their circulatory systems to deliver oxygen, they do rely on their circulatory systems to transport nutrients and remove wastes. Arthropod blood is pumped by a heart into vessels that carry the blood to body tissues. The tissues are flooded with blood, which returns to the heart through open body spaces. The blood maintains homeostasis in tissues by delivering nutrients and removing wastes.

VOCABULARY

ACADEMIC VOCABULARY

Transport:
To transfer from one place to another. *Blood transports nutrients to cells throughout the body.*

Excretion In most arthropods, cellular wastes are removed from the blood through **Malpighian** (mal PIH gee un) **tubules.** These tubules also help terrestrial arthropods preserve water in their bodies to maintain homeostatic water balance. In insects, the tubules, as shown in **Figure 26.7,** are located in the abdomen, unlike in segmented worms, where nephridia exist in each segment. Malpighian tubules are attached to and empty into the gut, which contains the undigested food wastes to be eliminated from the body. Crustaceans and some other arthropods do not have Malpighian tubules. They have modified nephridia, similar to those in annelids, to remove cellular wastes.

Figure 26.7 Most arthropods get rid of cellular wastes through Malpighian tubules.
Describe *another function of Malpighian tubules.*

■ **Figure 26.8** Compound eyes enable flying arthropods to see things in motion easily. The image the fly sees might not be as clear as that seen by a vertebrate. That blurry image is all the fly requires for its way of life.
Infer ***If a fly has blurry vision, how does it stay safe from predators?***

Response to stimuli Most arthropods have a double chain of ganglia throughout their bodies, on the ventral surface. Fused pairs of ganglia in the head make up the brain. Although most behaviors, such as feeding and locomotion, are controlled by the ganglia in each segment, the brain can inhibit these actions.

Vision Have you ever tried to swat a fly with a flyswatter? The fly's accurate vision allows the fly to spot even the slightest movement, and the fly often escapes. Most arthropods have one pair of large compound eyes. A compound eye, as shown in **Figure 26.8,** has many facets, which are hexagonal in shape. Each facet sees part of an image. The brain combines the images into a mosaic. The compound eyes of flying arthropods, such as dragonflies, enable them to analyze a fast-changing landscape during flight. Compound eyes can detect the movements of prey, mates, or predators, and also can detect colors. In addition, many arthropods have three to eight simple eyes. A simple eye has one lens and functions by distinguishing light from dark. In locusts and some other flying insects, simple eyes act as horizon detectors that help stabilize flight.

Hearing In addition to having eyes that detect movement and distinguish light from dark, many arthropods also have another sense organ called a tympanum (tihm PA num). A tympanum is a flat membrane used for hearing. It vibrates in response to sound waves. Arthropod tympanums can be located on the forelegs as in crickets, on the abdomen as in some grasshoppers, or on the thorax as in some moths.

Chemicals Imagine ants carrying off potato chip pieces, following each other like soldiers marching in formation. Ants communicate with each other by **pheromones** (FER uh mohnz), chemicals secreted by many animal species that influence the behavior of other animals of the same species. The ants use their antennae to sense the odor of pheromones and to follow the scent trail. Arthropods give off a variety of pheromones that signal behaviors such as mating and feeding.

CAREERS IN BIOLOGY

Biochemist As a scientist who determines how biological processes work, a biochemist might study the chemicals in pheromones to develop effective pest-management treatments. For more information on biology careers, visit biologygmh.com.

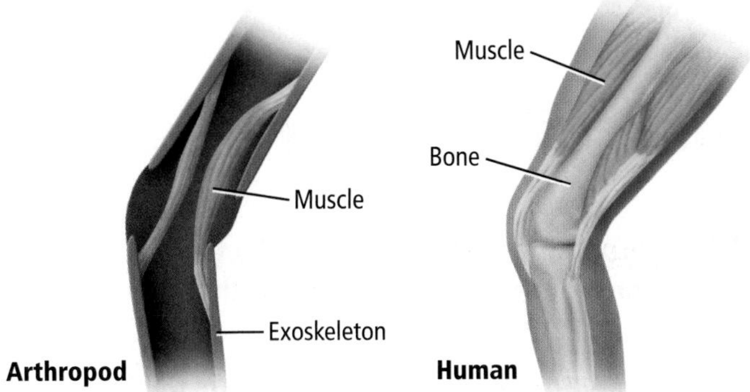

■ **Figure 26.9** The muscles of an arthropod attach inside of the exoskeleton to each side of the joint. The muscles in a human limb attach to the outer surfaces of the bones.

Movement Think again about the ants carrying the potato chip pieces and how fast they were moving. Arthropods generally are quick, active animals. They are able to crawl, run, climb, dig, swim, and fly because of their well-developed muscular systems. Refer to **Figure 26.9** to compare muscle attachment in human and arthropod limbs. The muscles in a human leg are attached to the outer surfaces of the bones. The muscles in an arthropod limb are attached to the inner surface of the exoskeleton on both sides of the joint. The strength of muscle contraction in arthropods depends on the rate at which nerve impulses stimulate muscles. In contrast, in vertebrates, the strength of muscle contraction depends on the number of muscle fibers contracting.

Reproduction Most arthropods reproduce sexually and have a variety of adaptations for reproduction. Most arthropods have separate sexes, but a few, such as barnacles, are hermaphrodites and undergo cross-fertilization. Most crustaceans brood, or incubate, their eggs in some way, but they do not care for their hatched offspring. Some spiders and insects also incubate their eggs, and some, such as bees, care for their young.

Section 26.1 Assessment

Section Summary

- Arthropods can be identified by three main structural features.
- Arthropods have adaptations that make them the most successful animals on Earth.
- Arthropod mouthparts are adapted to a wide variety of food materials.
- In order to grow, arthropods must molt.
- Arthropods have organ system modifications that have enabled them to live in all types of habitats and to increase in variety and numbers.

Understand Main Ideas

1. MAIN Idea **Evaluate** the three main features of arthropods that have enabled them to be successful.
2. **Explain** why jointed appendages are important to an animal with an exoskeleton.
3. **Summarize** the three main methods of respiration in arthropods.
4. **Infer** what might happen to an arthropod that had malformed Malpighian tubules. Be specific.

Think Critically

5. **Formulate Models** Design an arthropod adapted to conditions on a cold and windy mountaintop with low-growing grasses and arthropod-eating birds.

WRITING in **Biology**

6. Write a paragraph describing how an arthropod might protect itself while waiting for its new exoskeleton to harden after molting.

Section 26.2

Reading Preview

Objectives

- **Distinguish** structure and function in the major groups of arthropods.
- **Compare** adaptations in the major groups of arthropods.
- **Identify** characteristics of crustaceans and arachnids.

Review Vocabulary

sessile: an organism that is attached to and stays in one place

New Vocabulary

cheliped
swimmeret
chelicera
pedipalp
spinneret

Arthropod Diversity

MAIN Idea **Arthropods are classified based on the structure of their segments, types of appendages, and mouthparts.**

Real-World Reading Link Imagine turning over a rock on the forest floor. The ground beneath the rock suddenly seems to come alive with small animals creeping, crawling, and scurrying every which way. A spider darts under a leaf, a pill bug inches its way out of the light, and ants pour out of a tiny hole. All of these animals are arthropods.

Arthropod Groups

Spiders, pill bugs, and ants are arthropods. In the previous section, you learned why they all are considered arthropods. In the next two sections, you will learn how they differ from one another. Arthropods are classified into groups based on shared similarities, such as the structure of their body segments, appendages, and mouthparts. Taxonomists continue to debate the classification of arthropods. In this section, you will learn about two of the major groups—the crustaceans (krus TAY shunz), such as crabs and lobsters, and the arachnids (uh RAK nids), such as spiders and their relatives. In the next section, you will learn about the third major group—the insects and their relatives. **Table 26.1** summarizes the common characteristics of the three main groups of arthropods.

Concepts In Motion
Interactive Table To explore more about arthropod characteristics, visit biologygmh.com.

Table 26.1 Arthropod Characteristics

Group	Crustaceans	Spiders and Their Relatives	Insects and Their Relatives
Example		 	
Characteristics	Two pairs of antennae, two compound eyes, mandibles, five pairs of legs (chelipeds and walking legs), and swimmerets	No antennae, two body sections (cephalothorax and abdomen), and six pairs of jointed appendages (chelicerae, pedipalps, and four pairs of walking legs)	Antennae, compound eyes, simple eyes, three body sections (head, thorax, and abdomen), three pairs of legs, and generally two pairs of wings on the thorax

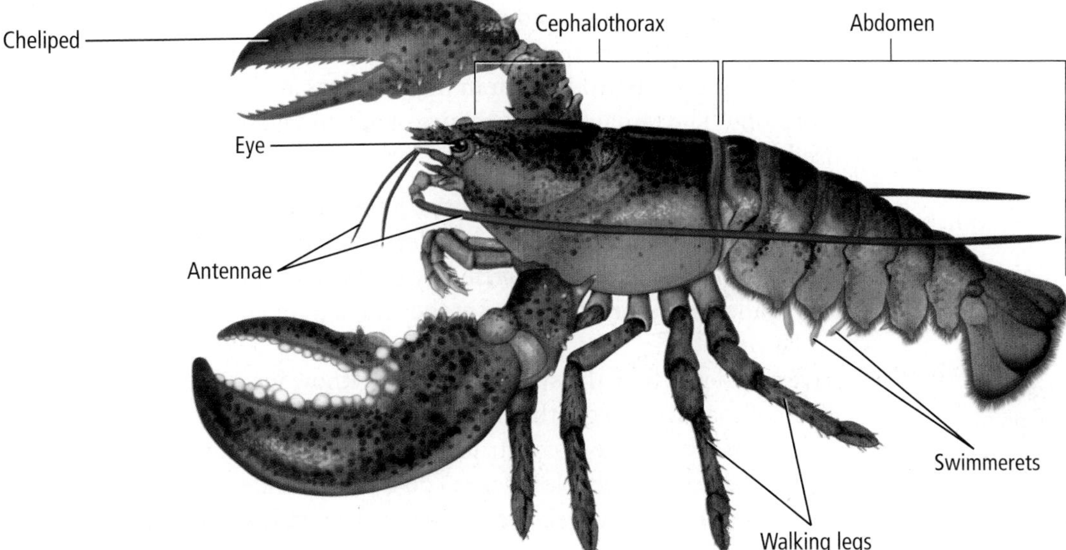

■ **Figure 26.10** Lobsters are aquatic crustaceans. Note the chelipeds for catching and crushing food, the thick cephalothorax with attached walking legs, the antennae, and the abdomen with attached swimmerets.
Consider ***how else a lobster might use its chelipeds.***

Crustaceans

Crabs, shrimps, lobsters, crayfishes, barnacles, water fleas, and pill bugs are crustaceans, and they live in marine, freshwater, and terrestrial habitats. Class Crustacea consists of about 35,000 named species. Most are aquatic and have two pairs of antennae, two compound eyes that are often on the tips of slender movable stalks, and mandibles for chewing. Crustacean mandibles open and close from side to side, instead of in an up-and-down movement like human jaws. Crustaceans possess branched appendages and have a free-swimming larval stage called a nauplius (NAW plee us) larva. A larva is an immature form of an animal that is markedly different in form and appearance from the adult.

Most crustaceans, such as crayfishes, lobsters, and crabs, have five pairs of legs. The first pair of legs—the **chelipeds,** shown in **Figure 26.10** —has large claws adapted to catch and crush food. Behind the chelipeds are four pairs of walking legs used primarily for locomotion. **Swimmerets** are the short legs behind the walking legs. They are used for reproduction and as flippers during swimming. If you have ever seen a lobster swim, you might have been surprised at how it can snap its tail beneath its body and move backward quickly. Some crustaceans, such as barnacles, are sessile and use their legs to kick food into their mouths.

Sow bugs and pill bugs are terrestrial crustaceans that live in damp places, such as under logs. They have seven pairs of legs.

Reading Check **Summarize** the functions of a crustacean's appendages.

Spiders and Their Relatives

Spiders belong to class Arachnida (uh RAK nuh duh) in which there are about 57,000 named species. Arachnids include spiders, ticks, mites, and scorpions.

Most arachnids have two body sections—a cephalothorax and an abdomen—and six pairs of jointed appendages. They do not have antennae. An arachnid's most anterior pair of appendages is modified into mouthparts called **chelicerae** (kih LIH suh ree) (singular, chelicera). Chelicerae are adapted to function as fangs or pincers and often are connected to a poison gland. Most spiders in the United States are not poisonous to humans. Exceptions include the black widow and the brown recluse shown in **Figure 26.11.**

■ **Figure 26.11** The inconspicuous brown recluse spider has a violin-shaped mark on its cephalothorax. If a person is bitten by this spider, he or she will require medical treatment because the venom is poisonous.

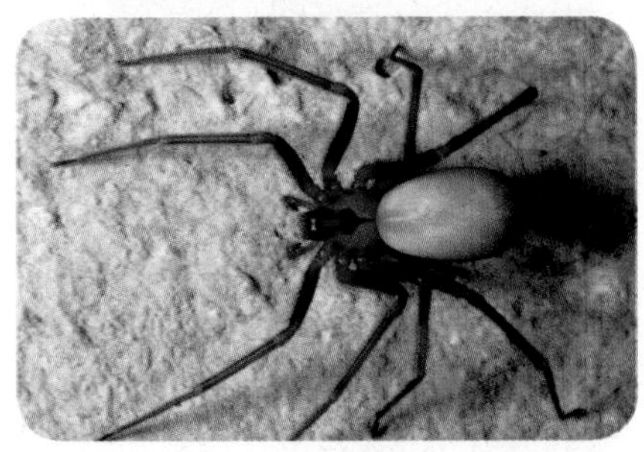

In arachnids, the second pair of appendages is called the **pedipalps.** These appendages are used for sensing and holding prey. The pedipalps also are used for reproduction in male spiders and as large pincers in scorpions. The remaining four pairs of appendages in arachnids are used for locomotion.

Spiders All spiders are carnivores. Some spiders, such as wolf spiders and tarantulas, hunt prey. Other spiders catch prey in silken webs. Silk is made from a fluid protein secreted by glands and spun into silk by structures called **spinnerets,** located at the end of a spider's abdomen.

Have you ever watched a spider weave a web? If you have, you might have wondered how the spider seemed to know just what to do and where it managed to get the training to do such intricate work.

Spiders are capable of constructing only specific kinds of webs. This instinctive behavior enables them to do this efficiently and effectively time after time. **Figure 26.12** shows the stages of construction of an orb web.

After catching an insect or other arthropod in their webs, many spiders wrap their prey in a silken cocoon until they are ready to feed. Digestion begins externally, when a spider secretes digestive enzymes onto its prey. After liquification occurs, the spider ingests the softened food. The remaining nutrients are digested internally.

To reproduce, a male spider deposits sperm on a small web he has built, picks up the sperm, and stores it in a cavity on his pedipalps. After a courtship ritual, the male inserts the sperm into the female. The female lays her eggs in a cocoon spun of spider silk. There can be as many as 100 eggs in one cocoon. The young hatch after about two weeks, then molt between five and ten times before reaching their adult size.

Vocabulary

Science usage v. Common usage

Weave

Science usage: to spin a web.
Spiders weave webs to catch prey.

Common usage: to construct by interlacing strips of material.
Artists weave baskets out of reeds and other natural fibers.

Reading Check **Compare and contrast** the appendages that crustaceans and arachnids use to capture prey.

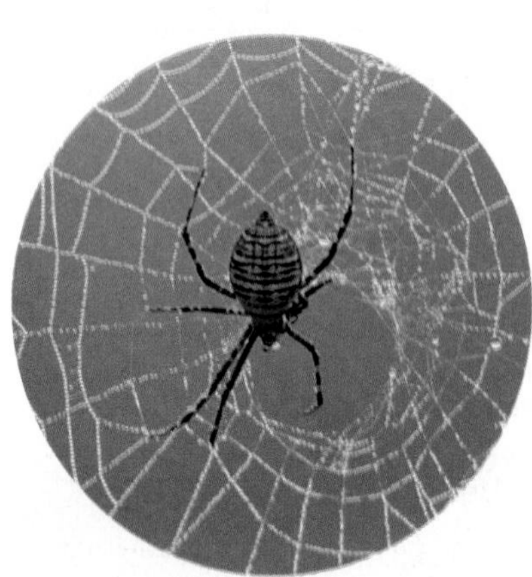

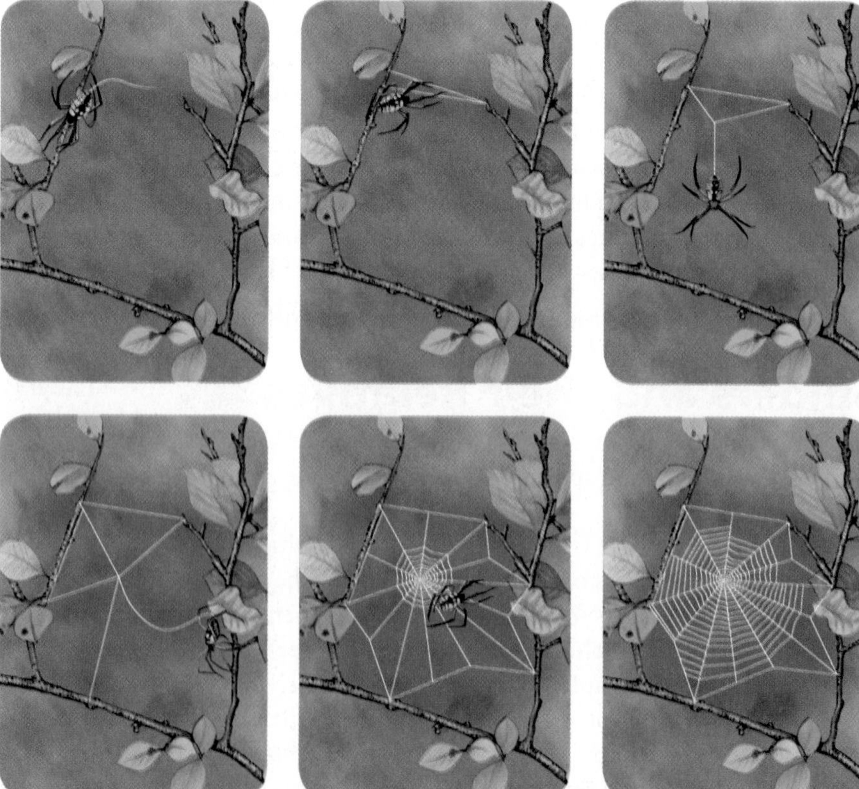

Figure 26.12 Orb-weaving spiders usually attach their webs to vegetation. An area of the web that is not sticky enables the spider to pass from one side of the web to the other.

Color-Enhanced SEM Magnification: 75×

Tick

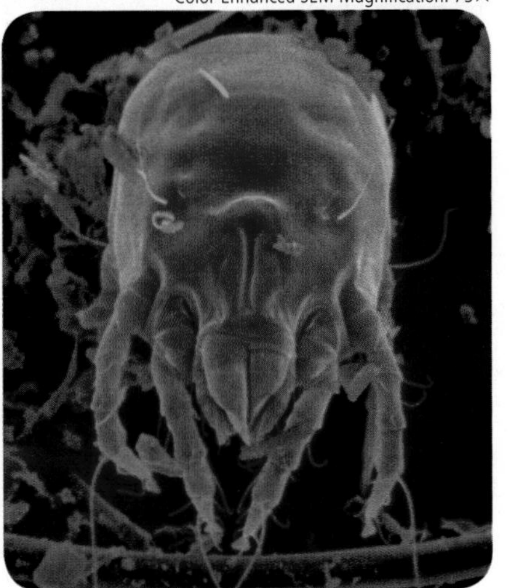

Mite

Scorpion

■ **Figure 26.13** Ticks, mites, and scorpions are in the same class as spiders.
Describe ***What characteristics of this class can you see in the photos?***

Ticks, mites, and scorpions Other members of class Arachnida—ticks, mites, and scorpions—are shown in **Figure 26.13.** Most mites are less than 1 mm long, with the cephalothorax and abdomen fused into one oval-shaped body section. They can be predators or parasites of other animals. Ticks are parasites that feed on blood after attaching themselves to the surface of their hosts. Ticks also frequently harbor disease-causing agents, such as viruses, bacteria, and protozoa, and introduce them to their hosts when they bite. Some of these diseases, such as Lyme disease and Rocky Mountain spotted fever, affect humans.

Scorpions feed on insects, spiders, and small vertebrates that they capture with their pedipalps and tear apart with their chelicerae. They generally are nocturnal, hiding under logs or in burrows during the day. When you think of a scorpion, you might think of the stinger at the end of the abdomen. Most scorpions that live in the United States do not have venom in their stinger but their sting can be quite painful. Compare different arthropod groups in **MiniLab 26.2.**

MiniLab 26.2

Compare Arthropod Characteristics

How do the physical characteristics of arthropods differ? Classify arthropods by observing specimens from the three major groups of arthropods.

Procedure

1. Read and complete the lab safety form.
2. Create a data table to record your observations of **live or preserved arthropod specimens. WARNING:** ***Treat live specimens in a humane manner at all times.***
3. Observe the arthropod specimens and record your observations about their physical characteristics in your data table.

Analysis

1. **Identify** the physical characteristics your arthropod specimens have in common.
2. **Classify** the arthropods into different taxonomic groups.

Figure 26.14 Horseshoe crabs come to shore to lay eggs in the sand.

Horseshoe crabs Horseshoe crabs are an ancient group of marine animals, related to the arachnids, that have remained basically unchanged since the Triassic Period more than 200 million years ago. They have unsegmented heavy exoskeletons in the shape of a horseshoe. The chelicerae, pedipalps, and the next three pairs of legs are used for walking and getting food from the bottom of the sea. The animals feed on annelids, mollusks, and other invertebrates, which they capture with their chelicerae. The posterior appendages are modified with leaflike plates at their tips and can be used for digging or swimming.

Horseshoe crabs, shown in **Figure 26.14,** come to shore to reproduce at high tide. The female burrows into the sand to lay her eggs. A male follows behind and adds sperm before the female covers the eggs with sand. Young larvae hatch after a period of being warmed by the Sun and then return to the ocean during another high tide.

Section 26.2 Assessment

Section Summary

- Arthropods are divided into three major groups.
- Crustaceans have modified appendages for getting food, walking, and swimming.
- The first two pairs of arachnid appendages are modified as mouthparts, as reproductive structures, or as pincers.
- Spiders are carnivores that either hunt prey or trap it in webs that they spin out of silk.
- Horseshoe crabs are ancient arthropods that have remained unchanged for more than 200 million years.

Understand Main Ideas

1. MAIN Idea **Classify** a small, quickly moving arthropod with two pairs of antennae, a segmented body, and mandibles that move from side to side.
2. **Compare and contrast** the ways of life of crustaceans and arachnids, and explain how their body forms are adapted to their environments.
3. **Summarize** the differences in function among the various appendages of spiders.
4. **Identify** the common characteristics among ticks, scorpions, and horseshoe crabs.

Think Critically

5. **Make a Hypothesis** Caribbean spiny lobsters have a navigation system that enables them to return to their original habitat after being moved to an unfamiliar location. Make a hypothesis about what signals the lobsters might use to orient themselves in the direction of their original habitat.
6. **Design an Experiment** A biologist wants to find out what brown recluse spiders eat. After some observation, she hypothesizes that the spiders prefer dead prey to live prey. Design an experiment that would test this hypothesis.

Biology Online Self-Check Quiz biologygmh.com

Section 26.3

Reading Preview

Objectives

- **Identify** characteristics of insects.
- **Analyze** how structure determines function in insects.
- **Compare and contrast** complete and incomplete metamorphosis.

Review Vocabulary

pollen: a fine powder produced by certain plants when they reproduce

New Vocabulary

metamorphosis
pupa
nymph
caste

Insects and Their Relatives

MAIN Idea Insects have structural and functional adaptations that have enabled them to become the most abundant and diverse group of arthropods.

Real-World Reading Link Think about a time you were stung by a bee, admired a bright butterfly flitting from flower to flower, or heard a cricket chirp. Insects are everywhere, and they affect your life in many ways.

Diversity of Insects

Scientists estimate that there are as many as 30 million insect species, which is more species than all other animals combined. Recall that arthropods make up about three-fourths of all named animal species. About 80 percent of arthropods are insects. They are the most abundant and widespread of all terrestrial animals. You can find insects in soil, in forests and deserts, on mountaintops, and even in polar regions.

Insects live in many habitats because of their ability to fly and their ability to adapt. Their small size enables them to be moved easily by wind or water. Diversity of insects also is enhanced by the hard exoskeleton that protects them and keeps them from drying out in deserts and other dry areas. In addition, the reproductive capacity of insects ensures that they are successful in any areas they inhabit. Insects produce a large number of eggs, most of the eggs hatch, and the offspring have short life cycles, all of which can lead to huge insect populations.

External Features

Insects have three body areas—the head, thorax, and abdomen, shown in **Figure 26.15.** Head structures include antennae, compound eyes, simple eyes, and mouthparts. Insects have three pairs of legs and generally two pairs of wings on the thorax. Some only have one pair of wings, and others do not have wings at all.

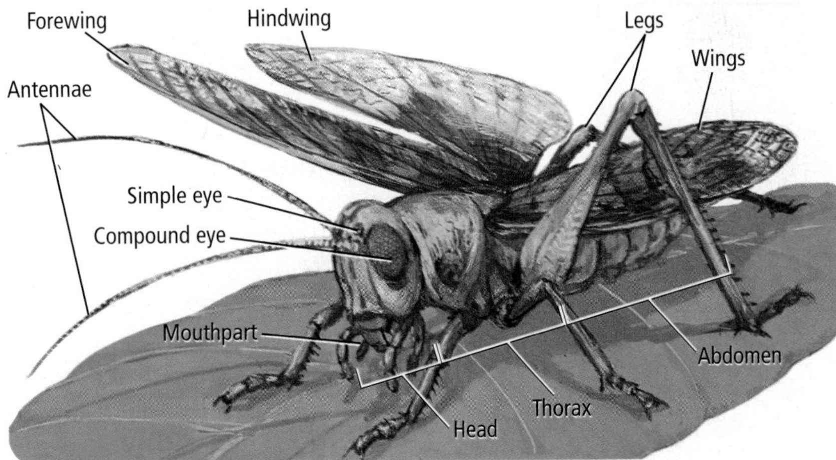

Figure 26.15 The head, thorax, and abdomen regions of this cricket are characteristic of insects.
Compare *How do the body regions of insects differ from those of crustaceans?*

Concepts In Motion
Interactive Figure To see an animation of the basic anatomy of a grasshopper, visit biologygmh.com.

Insect Adaptations

Structural adaptations to legs, mouthparts, wings, and sense organs have led to increased diversity in insects. These adaptations enable insects to utilize all kinds of food and to live in many different types of environments. Taking advantage of a variety of food sources, insects might be parasites, predators, or plant-sap suckers.

Legs Insect legs are adapted to a variety of functions. Beetles have walking legs with claws that enable them to dig in soil or crawl under bark. Flies have walking legs with sticky pads on the ends that enable them to walk upside down. Honeybee legs have adaptations for collecting pollen, while the hind legs of grasshoppers and crickets are adapted to jumping. Water striders have legs adapted to skimming over the surface of water. On its footpads, a water strider has water-repellent hairs that do not break the surface tension of the water. As it skates over the water, this insect propels itself with its back legs and steers with its front legs, like a rear-wheel-drive car.

Mouthparts Insects' mouthparts are adapted to the food they eat, as shown in **Table 26.2.** Butterflies and moths have a long tube through which they draw nectar from flowers in a motion similar to sipping through a straw. Different types of flies, such as houseflies and fruit flies, have sponging and lapping mouthparts that take up liquids. Some insects, such as leafhoppers and mosquitoes, have piercing mouthparts for feeding on plant juices or prey. Insects such as beetles and ants cut animal skin or plant tissue with their mandibles to reach the nutrients inside.

Table 26.2 Insect Mouthparts

Concepts In Motion **Interactive Table** To explore more about insect mouthparts, visit biologygmh.com.

Type of mouthpart	Siphoning	Sponging	Piercing/Sucking	Chewing
Example				
Function	Feeding tube is uncoiled and extended to suck liquids into the mouth.	Fleshy end of mouthpart acts like a sponge to mop up food.	A thin, needlelike tube pierces the skin or plant wall to suck liquids into the mouth.	Mandible pierces or cuts animal or plant tissue, and other mouthparts bring food to the mouth.
Insects with adaptation	Butterflies, moths	Houseflies, fruit flies	Mosquitoes, leafhoppers, stink bugs, fleas	Grasshoppers, beetles, ants, bees, earwigs

Wings Insects are the only invertebrates that can fly. Unlike bird and mammal wings that are modified limbs, insect wings are outgrowths of the body wall. Wings are formed of a thin double membrane of chitin, which is the same material that makes up the exoskeleton, and they have rigid veins that give them strength. Wings can be thin, as in flies, or thick, as in beetles. The wings of butterflies and moths are covered with fine scales, as shown in **Figure 26.16.** Investigate how butterflies might use their wing scales to attract mates in **Data Analysis Lab 26.1.** Flying requires complex movements of the wings. Forward thrust, upward lift, balance, and steering are all important. Most insects rotate their wings in a figure-eight pattern, as shown in **Figure 26.16.**

Reading Check **Compare** How are wings like an exoskeleton?

Sense organs Along with leg, mouthpart, and wing adaptations, insects have a variety of adaptations in their sense organs. Recall how arthropods use their antennae and eyes to sense their environment. Insects also have hairlike structures that are sensitive to touch, pressure, vibration, and odor. In addition to visually detecting motion, a fly detects changes in airflow using the hundreds of hairs that cover its body. It's no wonder that a fly often is long gone before the flyswatter can strike.

Some insects detect airborne sounds with their tympanic organs, while others can detect vibrations coming from the ground. These sensory cells often are located on the legs.

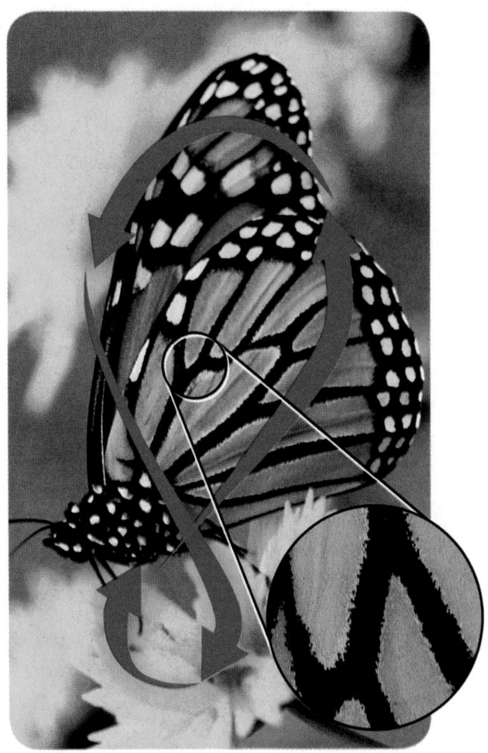

Figure 26.16 Butterfly wings are covered with fine scales. Notice how the up-and-down strokes of insect wings make a figure-eight pattern.

DATA ANALYSIS LAB 26.1

Based on Real Data*

Interpret the Graph

Do butterflies use polarized light for mate attraction? Light waves with electric fields vibrating in the same direction are said to be polarized. Scientists hypothesized that the iridescent wing scales in some butterflies, such as the one shown at right, create polarized light to attract certain males to females. The graph shows the response of males to polarized light versus nonpolarized light from female iridescent butterfly wings.

Think Critically

1. **Interpret the Graph** To which view of wings does the male butterfly respond more often?
2. **Infer** Researchers have noted that forest-dwelling butterflies tend to have iridescent wings, while meadow-dwelling butterflies do not. What might explain this difference?

Data and Observations

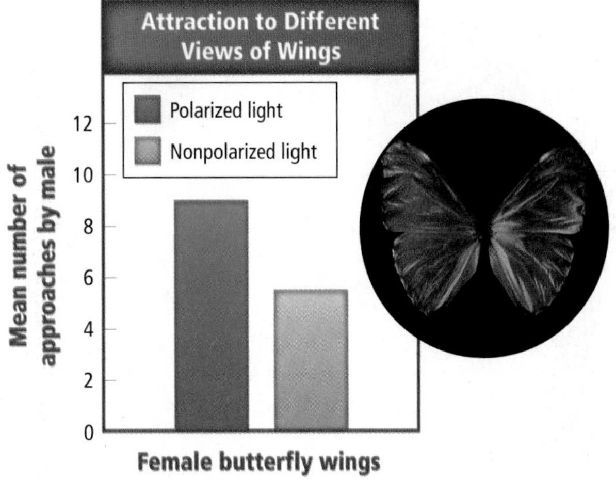

*Data obtained from: Sweeney, A., et al. 2003. Insect communication: polarized light as a butterfly mating signal. *Nature* 423: 31-32.

Most insects have keen chemical senses. Chemical receptors, or chemoreceptors, for taste and smell are located on mouthparts, antennae, or legs. Some insects, such as moths, can detect odors several kilometers away. Chemical signals in the form of pheromones enable insects to communicate with one another to attract mates or to gather members in large colonies to migrate or survive periods of cold weather.

Vocabulary

Word origin

Metamorphosis
from the Greek word *metamorphoun,* meaning *to transform.*

Metamorphosis Most insects lay their eggs in a specific habitat where the young can survive. For example, a monarch butterfly lays its eggs on milkweed plants, which the young feed on after they hatch. After hatching, most insects undergo **metamorphosis,** a series of major changes from a larval form to an adult form.

Complete metamorphosis Most insects develop through the four stages of complete metamorphosis—egg, larva, pupa, and adult. As shown in **Figure 26.17,** when the egg of a butterfly hatches, the wormlike larva that appears commonly is called a caterpillar. At this stage, the larva usually has chewing mouthparts and behaves like a feeding machine. The larva molts several times as it grows. A **pupa** (PYEW puh) is a nonfeeding stage of metamorphosis in which the animal changes from the larval form into the adult form. Adult insects are generally specialized for reproduction. Some adult insects do not live long enough to feed—a female adult mayfly only lives for five minutes. If adults feed, they generally do not compete with larvae for food.

Reading Check **Summarize** the life cycle of an insect that undergoes complete metamorphosis.

Incomplete metamorphosis Insects that undergo incomplete metamorphosis, as shown in **Figure 26.17,** hatch from eggs as **nymphs** (NIHMFS)—the immature form of insects that look like small adults without fully developed wings. After several molts, nymphs become adults.

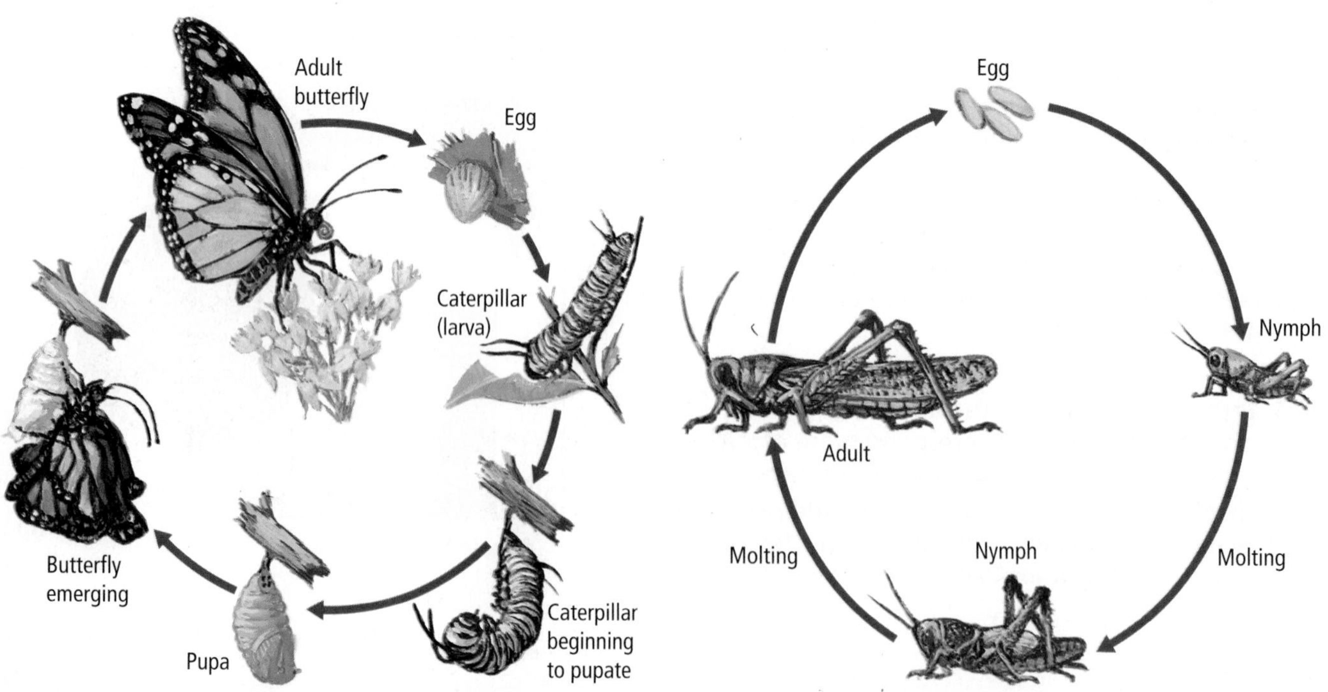

Figure 26.17 Insects that undergo complete metamorphosis have a resting stage called a pupa. This stage is absent in insects that undergo incomplete metamorphosis.

Concepts In Motion

Interactive Figure To see an animation of the stages in the complete metamorphosis of a butterfly, visit biologygmh.com.

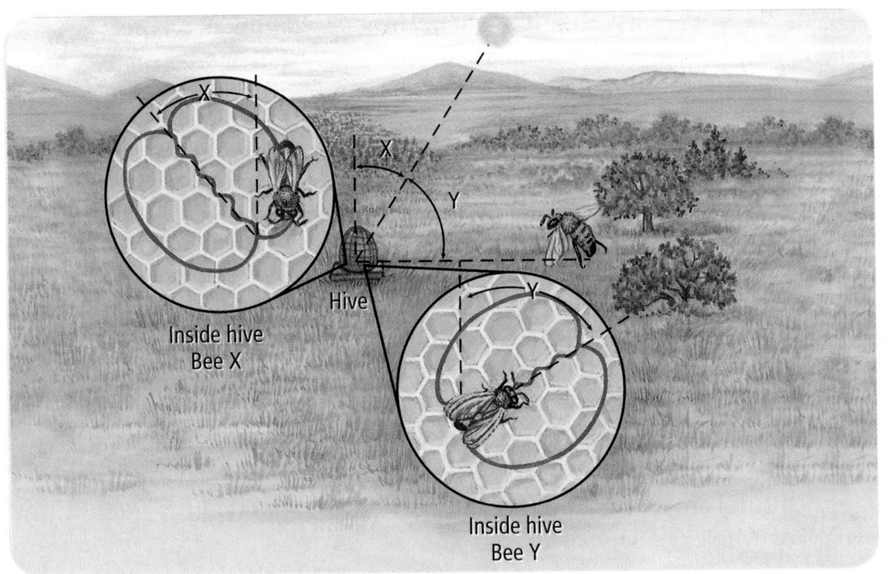

Figure 26.18 The straight run of the honeybee's figure-eight waggle dance indicates the direction of the food in relation to the angle of the Sun.
Interperet ***Where has Bee X found food?***

Concepts In Motion

Interactive Figure To see an animation of how honeybees use movement to communicate, visit biologygmh.com.

Insect societies The players on a basketball team work together to win the game. Insects such as honeybees, ants, and termites organize into social groups and cooperate in activities necessary for their survival. Honeybees have a complex society, with as many as 70,000 bees in one hive. There are three castes in a hive. A **caste** is a group of individuals within a society that perform specific tasks. Workers are females that do not reproduce. They gather nectar and pollen, build the honeycomb, manufacture honey, care for young, and guard the hive. Drones are the reproductive males. The queen is the only reproductive female.

Communication methods Honeybees have evolved an efficient system of communication, using bodily movements to indicate the location of food sources. One of the movements by which honeybees communicate is called the waggle dance. This dance is performed when a bee returns to the hive from a faraway food source. First, the returning bee makes a circle with a diameter about three times the bee's length. The bee then moves in a straight line while waggling its abdomen from side to side. The orientation of the line indicates the direction to the food source. Finally, the bee makes another circle in the opposite direction from the first circle. It traces this figure-eight pattern many times. The duration of the dance indicates the distance to the food source.

Connection to Math The most significant part of the waggle dance is the straight line because it tells the other bees where the food is in relation to the hive. The direction of the line relative to the vertical indicates the direction of the food relative to the Sun, as shown in **Figure 26.18.** If food is located 70 degrees to the right of the Sun, the straight line of the dance will be 70 degrees to the right of vertical.

Round dances also convey information about food sources and are used only if the food is close to the hive. In a round dance, the bee traces a clockwise circle followed by a counterclockwise circle and repeats this dance many times. The dance does not indicate distance or direction.

Ants also have evolved various societal behaviors for living in colonies. Females that do not reproduce gather food, care for young, and protect the colony from predators. Like honeybees, the males die after mating with the queen, whose sole function is to lay eggs.

CAREERS IN BIOLOGY

Entomologist Scientists who study insects are entomologists. They might study insect life cycles and behaviors, research insect pests and how to control them, or work with beneficial insects like honeybees. A beekeeper cares for bee colonies that are used for crop pollination and honey production. For more information on biology careers, visit biologygmh.com.

■ **Figure 26.19** Not only are insects important in pollinating flowers, some are important in feeding on harmful insects. This ladybird beetle feeds on plant pests.
Explain ***how insects maintain stability in ecosystems.***

Insects and humans It might be difficult to think of insects as beneficial when a mosquito buzzes around your head or when a bee stings you, but insects are an integral part of all ecosystems on Earth. Most insect species are not harmful to humans. Insects pollinate most flowering plants, including almost ten billion dollars' worth of food crops in the United States. They produce honey and silk used by humans and serve as food for many birds, fishes, and other animals. Insect predators, such as praying mantises and ladybird beetles, feed on plant pests such as aphids and mites, as shown in **Figure 26.19.**

Insects can also can be harmful to humans. Lice and bloodsucking flies are human parasites. Fleas can carry plague, houseflies can carry typhoid fever, and mosquitoes can carry malaria, yellow fever, and the West Nile virus. Weevils, cockroaches, ants, and termites cause much property destruction. Grasshoppers, corn borers, and boll weevils destroy agricultural crops. Bark beetles, spruce budworms, and gypsy moths can destroy whole portions of forests.

How is all this insect damage kept in check? In the past, chemicals were used indiscriminately to control insects. However, the overuse of chemicals disrupted food chains, reduced numbers of beneficial insects, and insects developed resistance to the insecticides. Use of biological controls has become increasingly important. Integrated pest management, a technique used by many farmers today, offers long-term control of pests. This strategy employs resistant plant varieties, crop rotation, and critical timing of planting and other agricultural practices along with small amounts of chemicals at critical times to control insect pests.

Centipedes and Millipedes

The centipedes of class Chilopoda and the millipedes of class Diplopoda are close relatives of insects. Centipedes move quickly and live in moist places under logs, bark, and stones. They have long, segmented bodies, and each segment has one pair of jointed legs. The first pair of appendages is modified to form poison claws, which a centipede uses to kill prey. Most species of centipedes are not harmful to humans.

Millipedes have two pairs of appendages on their abdominal segments and one pair on their thorax segments. Millipedes are herbivorous and live, as centipedes do, in moist places under logs or stones. Unlike centipedes, they do not wriggle quickly, but walk with a slow, graceful motion. Millipedes do not have poison claws and feed primarily on damp and decaying vegetation. Compare the centipede and millipede in **Figure 26.20.**

■ **Figure 26.20** Centipedes have one pair of appendages on each segment and poison claws on the first segment. Millipedes have two pairs of appendages on each abdominal segment, while the thorax has one pair of appendages on each segment.

Centipede

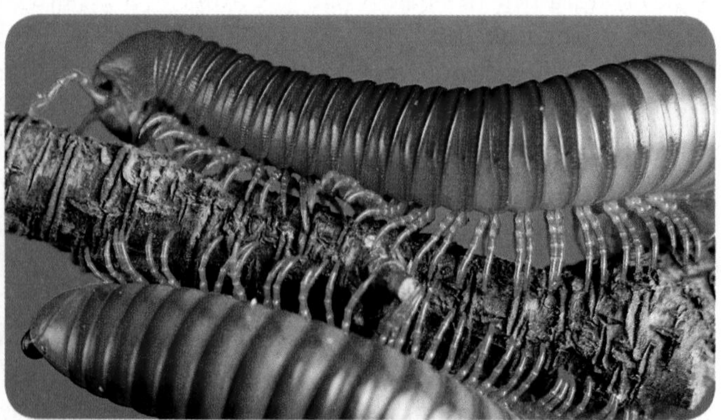

Millipede

Trilobite fossil

Color-Enhanced SEM Magnification: 750×

Tardigrade

■ **Figure 26.21** Extinct trilobites are considered to be some of the first arthropods. They were abundant in Cambrian times. Tardigrades, belonging to a phylum that might be related to annelids and arthropods, are called water bears and can live in areas that are alternately wet and dry.

Evolution of Arthropods

The relationships of tardigrades, trilobites, and arthropods have been under close scrutiny as new evidence is discovered. Fossil records show that trilobites, abundant in the mid-Cambrian but now extinct, were early arthropods. Trilobites, like the one shown in **Figure 26.21,** were oval, flattened, and divided into three body sections like some modern arthropods. The large number of identical segments of these ancestral arthropods evolved to more specialized appendages and fewer segments in modern arthropods.

Tardigrades also are related to arthropods, but they appear to be related less closely to arthropods than trilobites are. The tardigrade shown in **Figure 26.21** illustrates why these tiny animals are known commonly as water bears. The largest are 1.5 mm long with four pairs of stubby legs. They feed on algae, decaying matter, nematodes, and other soil animals. They inhabit freshwater, marine, and land habitats. During temperature extremes and drought, tardigrades can survive for years in a completely dry state with reduced metabolism until favorable conditions return.

Section 26.3 Assessment

Section Summary

- Insects make up approximately 80 percent of all arthropod species.
- A variety of adaptations have enabled insects to live in almost all habitats on Earth.
- Insect mouthparts reflect their diets.
- Most insects undergo metamorphosis.
- In some insects, social structure, including individual specializations, is necessary for the survival of the colony.

Understand Main Ideas

1. MAIN Idea **Evaluate** three adaptations of insects in terms of the role they played in enabling insects to become so diverse and abundant.
2. **Identify** features common to all insects.
3. **List** adaptations of the mouthparts of insects that feed on three different food sources and explain each one.
4. **Identify** one reason most insects undergo complete metamorphosis.

Think Critically

5. **Design an Experiment** Different species of firefly beetles flash their light in different sequences of short and long flashes. Design an experiment that would explain why fireflies flash their lights.

MATH in Biology

6. There are approximately 1.75 million named animal species. About three-fourths of all known animal species are arthropods, and 80 percent of arthropods are insects. Approximately how many named species are insects?

NATIONAL GEOGRAPHIC

In the Field

Career: Forensic Entomologist

Insect Evidence

Insects often are the first to arrive at a crime scene. Blowflies can arrive within minutes. Over time, other insects arrive. As the insects feed, grow, and lay eggs, they follow predictable developmental cycles. For forensic entomologists—scientists who apply their knowledge of insects to help solve crimes—these cycles reveal information about the time and location of death.

Time of death Forensic entomologists use two methods to determine time of death. The first method is used when the victim has been dead for at least one month. While blowflies and houseflies arrive almost immediately, other species arrive later in the decomposition process. Some species arrive to feed on other insects already at the scene. The succession of insects provides information about the time that passed since death occurred.

When death has occurred within a few weeks, a second method used involves the developmental cycle of blowflies. Within a couple of days, the blowflies lay eggs. The next stages of development are determined in part by temperature, as shown in the graph. Based on the stage of insect development and area temperatures, entomologists can determine a range of days in which the first insects laid eggs in the body, establishing a time of death.

Location of death Insects help determine if a body was relocated after death. If insects found on the body are not native to the habitat where the body is found, investigators can assume that the body was moved. The species that are present also provide clues about the area where death took place.

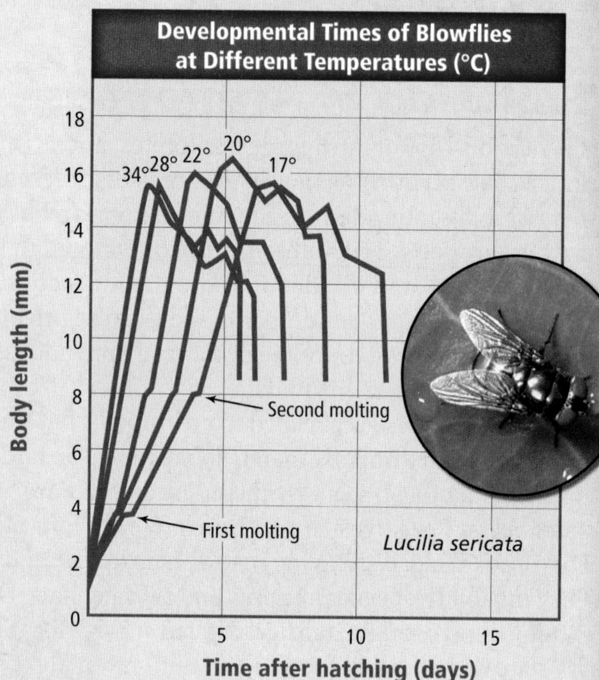

Limitations In many locations, forensic entomology is less useful in winter, when insects are less active and less abundant. In addition, insects might be prevented from invading a body if it is frozen, buried deeply, or wrapped tightly. In many cases, however, insects can give crucial testimony about the details of a crime.

MATH in Biology

Study the graph to solve this problem: Blowfly larvae with a body length of about 6 mm are found on a corpse with a temperature of 22°C. How much time has passed since death? For more information about careers in biology, visit biologygmh.com.

BIOLAB

INTERNET: WHERE ARE MICROARTHROPODS FOUND?

Background: Microarthropods range from 0.1 to 5 mm in size—barely visible to human eyes. Dozens of microarthropod species can be unearthed in one shovelful of soil. Discover these hidden animals during this investigation.

Question: *What types of microarthropods can be found in your local environment?*

Materials

soil sample
clear funnel
ring stand
gooseneck lamp
wire mesh
beaker
95% ethanol
plastic collection vials
magnifying lens
arthropod field guide
metric ruler

Safety Precautions

Procedure

1. Read and complete the lab safety form.
2. Obtain a sample of leaf litter and soil from your teacher.
3. Create a data table to record your observations.
4. Place the funnel in the ring stand.
5. Cut the mesh screen in a circle so it rests inside the funnel.
6. Pour ethanol into the beaker until the beaker is two-thirds full. Set the beaker under the funnel.
7. Remove your soil sample from the bag and place it carefully on the mesh screen in the funnel.
8. Place the lamp at least 10 cm above the sample. Switch on the light and leave it on for several hours. The heat from the lamp dries the soil. This forces the microarthropods downward until they fall through the screen and into the alcohol.
9. Use a magnifying lens to observe the physical characteristics of the microarthropods you collected.
10. **Cleanup and Disposal** Be certain to properly dispose of the alcohol and specimens you collected by following your teacher's instructions.

Analyze and Conclude

1. **Classify** Place the microarthropods you collected into the three major groups of arthropods. Place unidentified specimens into a separate group.
2. **Graph** Use the data you collected to graph the abundances of each type of arthropod.
3. **Describe** Write a description of the physical characteristics of the microarthropod specimens that you could not classify into any of the three major groups.
4. **Hypothesize** How do microarthropods help create a healthy soil ecosystem?
5. **Error Analysis** Check your findings against those for the microarthropods collected by other classmates. Did you classify the microarthropods into the same group? If not, explain why.

SHARE YOUR DATA

Report Use a field guide or dichotomous key to identify the microarthropods you collected. Visit Biolabs at biologygmh.com and post your findings in the table provided for this activity. Write a report comparing your findings to those of students in another area of the country.

CHAPTER 26

Study Guide

STUDY TO GO Download quizzes, key terms, and flash cards from biologygmh.com.

FOLDABLES **Create** a scenario in which a species of terrestrial arthropod has been transferred from its native habitat to a nonnative habitat. Describe the possible short-term and long-term effects on the arthropod and on the habitat.

Vocabulary	Key Concepts
Section 26.1 Arthropod Characteristics	
• abdomen (p. 763) • appendage (p. 764) • book lung (p. 767) • cephalothorax (p. 763) • Malpighian tubule (p. 767) • mandible (p. 765) • molting (p. 764) • pheromone (p. 768) • spiracle (p. 767) • thorax (p. 763) • tracheal tube (p. 767)	MAIN Idea Arthropods have segmented bodies and tough exoskeletons with jointed appendages. • Arthropods can be identified by three main structural features. • Arthropods have adaptations that make them the most successful animals on Earth. • Arthropod mouthparts are adapted to a wide variety of food materials. • In order to grow, arthropods must molt. • Arthropods have organ system modifications that have enabled them to live in all types of habitats and to increase in variety and numbers.
Section 26.2 Arthropod Diversity	
• chelicera (p. 771) • cheliped (p. 771) • pedipalp (p. 772) • spinneret (p. 772) • swimmeret (p. 771)	MAIN Idea Arthropods are classified based on the structure of their segments, types of appendages, and mouthparts. • Arthropods are divided into three major groups. • Crustaceans have modified appendages for getting food, walking, and swimming. • The first two pairs of arachnid appendages are modified as mouthparts, as reproductive structures, or as pincers. • Spiders are carnivores that either hunt prey or trap it in webs that they spin out of silk. • Horseshoe crabs are ancient arthropods that have remained unchanged for more than 200 million years.
Section 26.3 Insects and Their Relatives	
• caste (p. 779) • metamorphosis (p. 778) • nymph (p. 778) • pupa (p. 778)	MAIN Idea Insects have structural and functional adaptations that have enabled them to become the most abundant and diverse group of arthropods. • Insects make up approximately 80 percent of all arthropod species. • A variety of adaptations have enabled insects to live in almost all habitats on Earth. • Insect mouthparts reflect their diets. • Most insects undergo metamorphosis. • In some insects, social structure, including specializations, is necessary for the survival of the colony.

CHAPTER 26 Assessment

Section 26.1

Vocabulary Review

An analogy is a relationship between two pairs of words and can be written in the following manner: A is to B as C is to D. *Complete each analogy by providing the missing vocabulary term from the Study Guide page.*

1. Spiracles are to breathing as __________ are to excreting wastes.
2. Compound eye is to sense organ as mandible is to __________.
3. Head is to thorax as __________ is to abdomen.

Understand Key Concepts

Use the diagram below to answer questions 4 and 5.

4. Which labeled structure helps terrestrial arthropods maintain water balance?
   A. 1
   B. 2
   C. 3
   D. 4
5. Which labeled structure would an arthropod use to sense odors in its environment?
   A. 1
   B. 2
   C. 3
   D. 4
6. Which group of words has one that does not belong?
   A. exoskeleton, chitin, molting, growth
   B. mandible, antennae, appendage, leg
   C. cephalothorax, thorax, head, abdomen
   D. simple eye, compound eye, tympanum, thorax
7. The relationship between muscle size and exoskeleton thickness limits which in an arthropod?
   A. diet
   B. habitat
   C. motion
   D. size

Constructed Response

8. **Open Ended** Make a table that lists arthropod structures, their functions, and an analogy of what each structure is like in a world of human-made devices. For example, a particular bird's bill that pulls insects out of bark might be compared to tweezers that can pull a sliver out of skin. Use the following structures in your table: antennae, exoskeleton, mandibles, tracheal tubes, and tympanum.
9. **Open Ended** Katydids are members of the grasshopper family. Most katydids are green, but occasionally both pink and yellow katydids appear. Make a hypothesis to explain why pink and yellow katydids sometimes appear.

Think Critically

Use the diagram below to answer question 10.

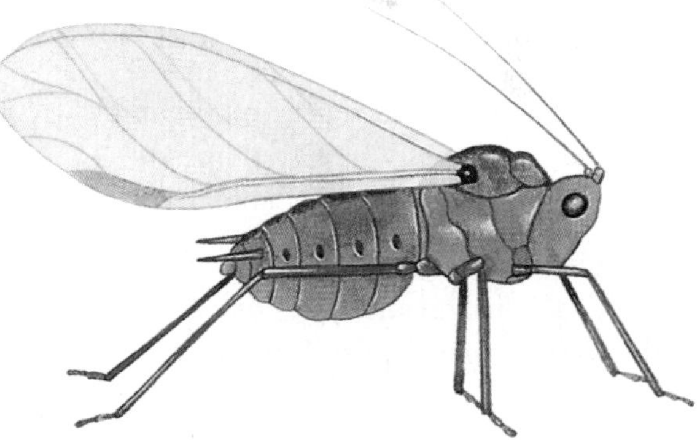

10. **CAREERS IN BIOLOGY** Arborists, people who specialize in caring for trees, sometimes spray horticultural oils on fruit trees to control aphids, the plant pest shown in the diagram. Based on your knowledge of insect anatomy, analyze why oils are an effective treatment to control plant pests.
11. **Infer** Some species of flowers produce heat that attracts certain beetles to live inside the bloom. Infer how the plant and the beetle both benefit from this relationship.

Section 26.2

Vocabulary Review

For each set of vocabulary terms, explain the relationship that exists.

12. cheliped, swimmeret

13. chelicera, pedipalp

14. cheliped, chelicera

Understand Key Concepts

Use the diagram below to answer question 15.

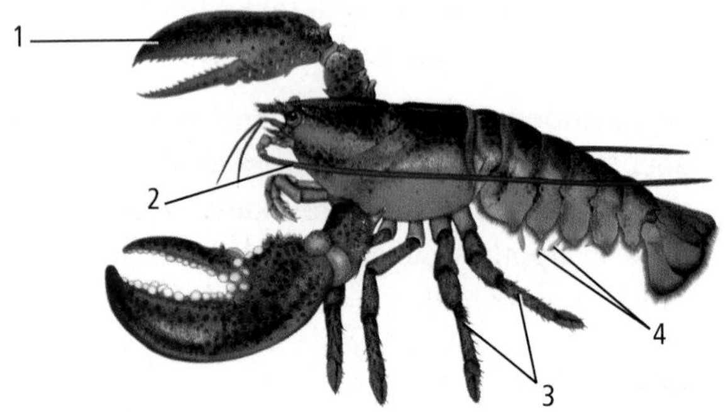

15. Which structure would a lobster use to catch and crush food?
A. 1 **C.** 3
B. 2 **D.** 4

16. Which is not a characteristic of arachnids?
A. chelicerae **C.** spinnerets
B. pedipalps **D.** antennae

17. An animal you found on the forest soil has two body sections, no antennae, and large pincers as the second pair of appendages. What type of animal is it?
A. tick **C.** spider
B. scorpion **D.** lobster

18. In spiders, the spinnerets are involved in which activity?
A. defense **C.** circulation
B. getting rid of waste **D.** spinning silk

19. Which is not a characteristic of mites?
A. one oval-shaped body section
B. carry lyme disease bacteria
C. less than 1 mm long
D. animal parasite

Constructed Response

20. Short Answer Compare the body forms of aquatic crustaceans to those of terrestrial arachnids, showing how each is adapted to its environment.

21. Open Ended What would happen if crustaceans could not molt?

Think Critically

22. Formulate Models Draw and describe a model of a spider that would be adapted to conditions in a hot, dry attic with only crawling insects as a food source.

23. Interpret Scientific Illustrations Based on the lobster diagram in **Figure 26.10** and your knowledge of crustaceans, what adaptations enable a lobster to survive in its aquatic enviroment?

Section 26.3

Vocabulary Review

For each set of vocabulary terms, choose the one term that does not belong and explain why it does not belong.

24. incomplete metamorphosis, pupa, larva, adult

25. complete metamorphosis, nymph, adult, molt

26. pupa, larva, nymph, caste, adult

Understand Key Concepts

Use the diagram below to answer question 27.

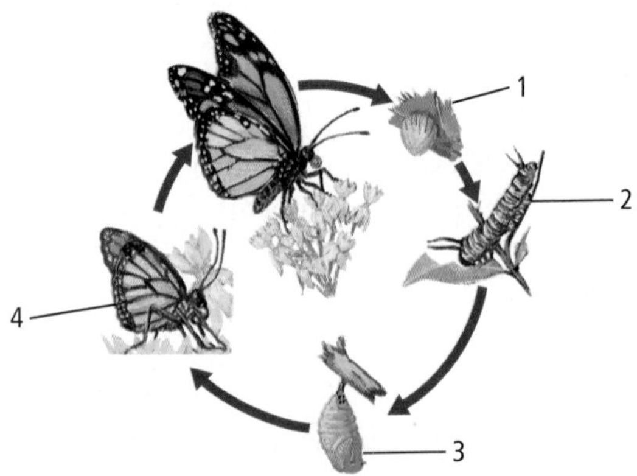

27. Which stage does not belong in the diagram of complete metamorphosis?
A. 1 **C.** 3
B. 2 **D.** 4

28. If the food is 40 degrees to the right of the Sun, what will be the angle of the straight line of the figure-eight waggle dance?
   A. 60 degrees to the right of vertical
   B. 40 degrees to the right of vertical
   C. 60 degrees to the right of horizontal
   D. 40 degrees to the right of horizontal

29. If a farm field has an infestation of insects, which method would the farmer use to manage it for the long-term?
   A. genetic engineering
   B. insecticides
   C. integrated pest management
   D. pesticide resistance

Constructed Response

Use the diagram below to answer questions 30 and 31.

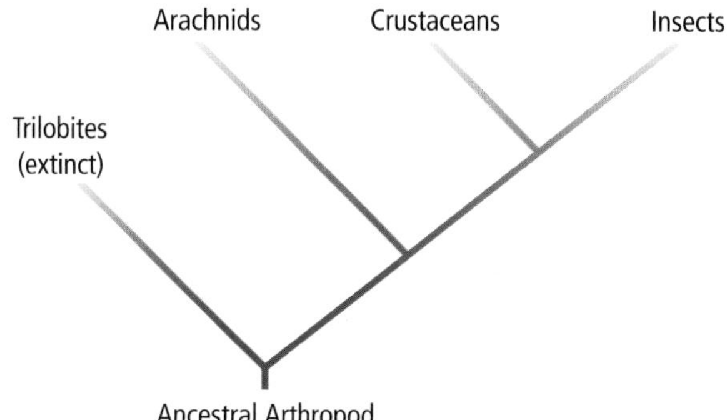

30. **Open Ended** Based on this interpretation of the phylogeny of arthropods, which group developed the earliest? Which group developed most recently?

31. **Open Ended** Examine the cladogram and sequence the order of appearance, from oldest to most modern, of the following features in the evolution of insects: chelicerae, mandibles, body divided into two regions, segmentation. Explain your reasoning.

Think Critically

32. **Hypothesize** A certain species of beetle looks very much like an ant. Make a hypothesis about the advantage to the beetle of looking like a particular ant.

33. **Design an experiment** that would answer this question: Why do crickets chirp?

Additional Assessment

34. WRITING in Biology Malaria is spread by mosquitos and is one of the world's worst diseases in terms of numbers of people affected and the difficulties in treating and preventing it. Research and write an essay on how scientists are using fungi to prevent this disease.

DBQ Document-Based Questions

Desert locusts have two distinct phases in their lives: the solitary insect that stays in one area and the social phase in which locusts band together in swarms of billions and move kilometers in search of food. Biologists found that exposing individual insects to jostling by small paper balls induced swarming. Examine the locust below. Each color indicates the percentage of social behavior induced by touching the locust on various parts of the body.

Data obtained from: Enserink, M. 2004. Can the war on locusts be won? *Science* 306 (5703): 1880–1882.

35. What percentage of social behavior resulted from touching the insect's thorax?

36. What part of the insect's body is the most sensitive for generating social activity when touched?

37. Draw a conclusion about what physical trigger causes locusts to swarm.

Cumulative Review

38. Compare alternation of generations in plants and alternation of generations in jellyfishes. **(Chapter 24)**

Standardized Test Practice

Cumulative

Multiple Choice

1. Which common function do both the endoskeletons and exoskeletons of animals perform?
   A. growing along with the animal
   B. preventing water loss
   C. supporting the body
   D. providing protection from predators

Use the diagram below to answer questions 2 and 3.

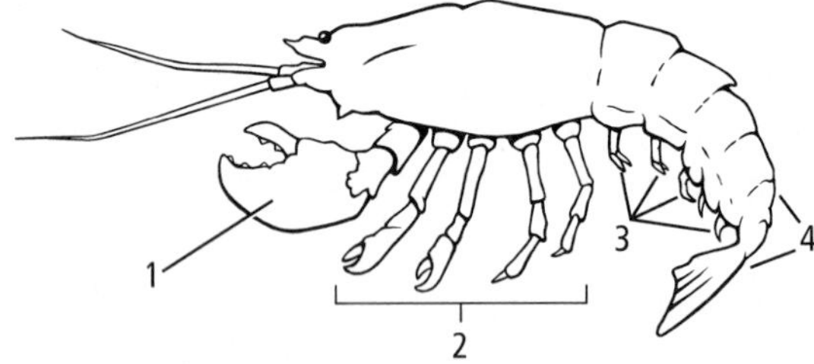

2. In which group does this animal belong?
   A. copepods
   B. crustaceans
   C. insects
   D. spiders

3. Which part of the body does this animal use for reproduction?
   A. 1
   B. 2
   C. 3
   D. 4

4. How are the organisms in Kingdom Protista different from animals?
   A. Some are multicellular.
   B. Some are prokaryotes.
   C. Some have cell walls.
   D. Some have tissues.

5. Which kind of asexual reproduction is possible in flatworms?
   A. budding
   B. fertilization
   C. parthenogenesis
   D. regeneration

Use the drawing below to answer question 6.

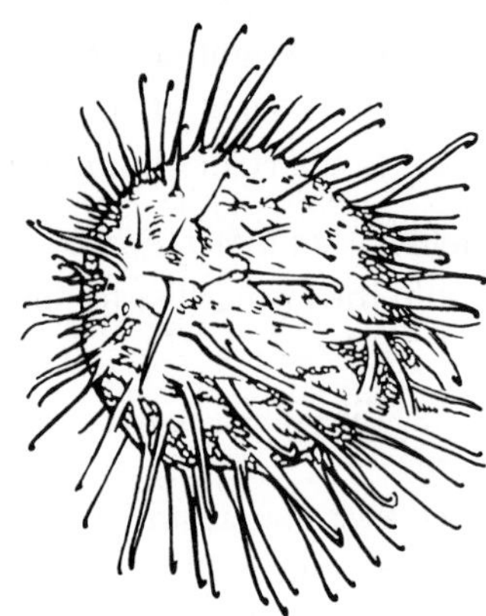

6. Which is the method of seed dispersal for this seed?
   A. animals
   B. gravity
   C. water
   D. wind

7. Which process is related to sexual reproduction in animals?
   A. budding
   B. fertilization
   C. fragmentation
   D. parthogenesis

8. Which is the role of an earthworm's clitellum in reproduction?
   A. It breaks off, allowing fragmentation to occur.
   B. It indicates whether or not an earthworm is hermaphroditic.
   C. It leaves the earthworm's body and forms a cocoon for developing earthworms.
   D. It produces sperm and eggs.

9. Which is used to classify protists?
   A. feeding
   B. habitat
   C. structure
   D. reproduction

Short Answer

Use the diagram below to answer question 10.

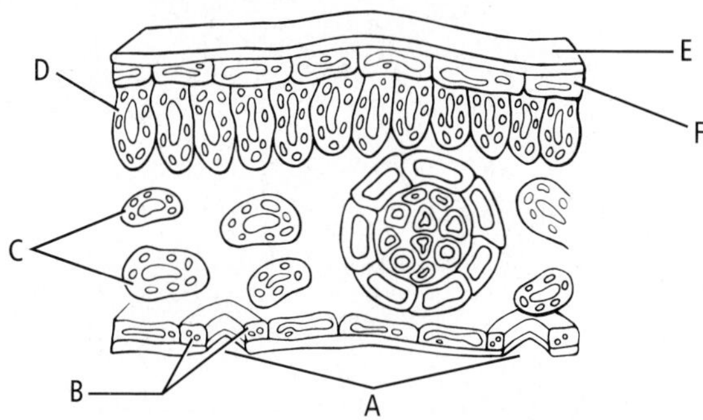

10. Identify the labeled parts of this leaf and state a function for each part.

11. Which characteristics differentiate arthropods from other invertebrates?

12. Describe embryonic development from a zygote to a gastrula. Provide the name of each stage, and explain how it is unique.

13. What characteristics do all mollusks share?

14. Compare and contrast how blood circulates through an insect with the circulation of blood in another kind of animal.

15. Explain the theory of endosymbiosis as it applies to protists. Assess the possible connection between certain organelles in eukaryotic protists and the structures of prokaryotic organisms.

16. Assess the importance of algae to all living things.

Extended Response

Use the illustrations below to answer question 17.

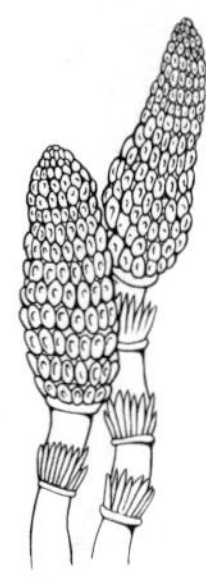

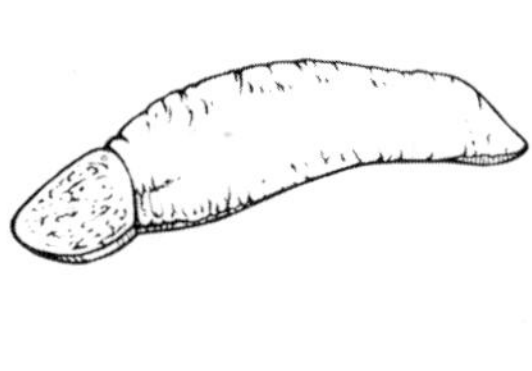

17. The figures above show spores and seeds from different kinds of plants. Explain why one of these structures would have an advantage and would be more likely to be naturally selected.

18. Evaluate the advantages and disadvantages of an exoskeleton.

Essay Question

The world's coral reefs and associated ecosystems are threatened by an increasing array of pollution, habitat destruction, invasive species, disease, bleaching, and global climate change. The rapid decline of these complex and biologically diverse marine ecosystems has significant social, economic, and environmental impacts in the U.S. and around the world. The U.S. Coral Reef Task Force identified two basic themes for national action:

- understand coral reef ecosystems and the processes that determine their health and viability
- reduce the adverse impacts of human activities on coral reefs and associated ecosystems

Using the information in the paragraph above, answer the following question in essay format.

19. What steps do you think the U.S. should take to preserve coral reef ecosystems?

NEED EXTRA HELP?																			
If You Missed Question . . .	1	2	3	4	5	6	7	8	9	10	11	12	13	14	15	16	17	18	19
Review Section . . .	24.1	26.2	26.2	24.1	25.1	23.3	24.1	25.4	19.1	22.2	26.1	24.1	25.3	26.1	19.1	19.3	21.3	26.1	24.3

CHAPTER 27

Echinoderms and Invertebrate Chordates

BIG Idea **Echinoderms and invertebrate chordates have features that connect them to the chordates that evolved after them.**

Section 1

Echinoderm Characteristics

MAIN Idea Echinoderms are marine animals with spiny endoskeletons, water-vascular systems, and tube feet; they have radial symmetry as adults.

Section 2

Invertebrate Chordates

MAIN Idea Invertebrate chordates have features linking them to vertebrate chordates.

BioFacts

- A single crown-of-thorns sea star eats 2–6 m^2 of coral per year.
- Crown-of-thorns sea stars have spines that are covered with poison-filled skin.
- Another echinoderm, the sea cucumber, protects itself by changing the consistency of its skin from near liquid to solid and back again.

LAUNCH Lab

Why are tube feet important?

Like all echinoderms, the crown-of-thorns sea star in the opening photo has structures called tube feet. In this lab, you will observe tube feet and determine their function.

Procedure

1. Read and complete the lab safety form.
2. Place a **live sea star** in a **petri dish** filled with **water from a saltwater aquarium. WARNING:** ***Treat the sea star in a humane manner at all times.***
3. Observe the ventral side of the sea star under a **dissecting microscope.** Look for the rows of tube feet that run down the middle of each arm, and draw a diagram of the structures.
4. Gently touch the end of a tube foot with a **glass probe.** Record your observations.
5. Return the sea star and water to the aquarium.

Analysis

1. **Describe** the structure of the sea star's tube feet.
2. **Infer** Based on your observations, what is the function of an echinoderm's tube feet?

Visit biologygmh.com to:

- study the entire chapter online
- explore Concepts in Motion, the Interactive Table, Microscopy Links, and links to virtual dissections
- access Web links for more information, projects, and activities
- review content online with the Interactive Tutor and take Self-Check Quizzes

Describing Invertebrate Chordates Make the following Foldable to help you understand the physical features that link invertebrate chordates to vertebrate chordates.

STEP 1 Collect three sheets of paper and layer them about 1.5 cm apart vertically. Keep the edges level.

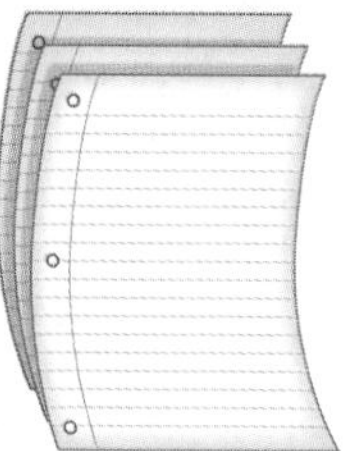

STEP 2 Fold up the bottom edges of the paper to form six tabs.

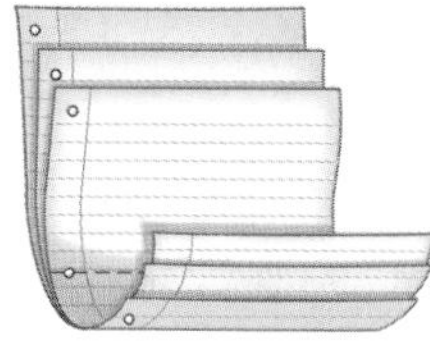

STEP 3 Crease well along the fold to hold the tabs in place. Staple along the fold. Rotate the paper so the fold is at the top, and label each tab as shown.

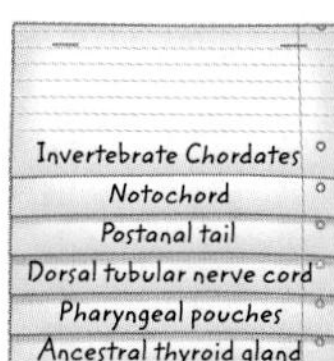

FOLDABLES Use this Foldable with Section 27.2. As you read the section, record information about the physical features of invertebrate chordates that link them to vertebrate chordates.

Section 27.1

Reading Preview

Objectives

- **Summarize** the characteristics common to echinoderms.
- **Evaluate** how the water-vascular system and tube feet are adaptations that enable echinoderms to be successful.
- **Distinguish** between the classes of echinoderms.

Review Vocabulary

endoskeleton: an internal skeleton that provides support and protection and can act as a brace for muscles to pull against

New Vocabulary

pedicellaria
water-vascular system
madreporite
tube foot
ampulla

Echinoderm Characteristics

MAIN Idea Echinoderms are marine animals with spiny endoskeletons, water-vascular systems, and tube feet; they have radial symmetry as adults.

Real-World Reading Link To take a blood-pressure reading, a health care professional squeezes a bulb that forces air through a tube and into the blood-pressure cuff around your arm. The cuff remains tight around your arm until the pressure is released when the air is let out. Some animals use this same kind of system to obtain food and move.

Echinoderms Are Deuterostomes

As shown in the evolutionary tree in **Figure 27.1,** echinoderms (ih KI nuh durmz) are deuterostomes—a major transition in the phylogeny of animals. Notice how the evolutionary tree branches at deuterostome development.

The mollusks, annelids, and arthropods you studied in previous chapters are protostomes. Recall that during development, a protostome's mouth develops from the opening on the gastrula, while a deuterostome's mouth develops from elsewhere on the gastrula. This might not seem important, but consider that only echinoderms and the chordates that evolved after echinoderms have this kind of development. Echinoderms and chordates are related more closely than groups that do not develop in this way. Animals with spinal cords, including humans, are chordates.

The approximately 6000 living species of echinoderms are marine animals and include sea stars, sea urchins and sand dollars, sea cucumbers, brittle stars, sea lilies and feather stars, and sea daisies. Two echinoderms are shown in **Figure 27.1.**

Figure 27.1 Echinoderms are marine animals and are the first animals in evolutionary history to have deuterostome development and an endoskeleton.

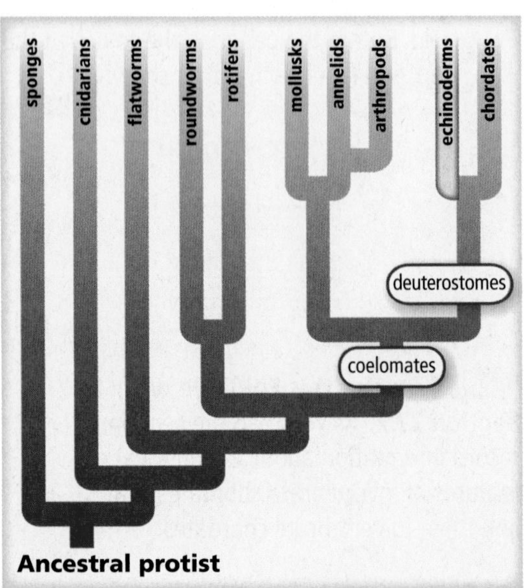

Body Structure

The brittle star is an example of an echinoderm with the spiny endoskeleton that is characteristic of the organisms in this phylum. Echinoderms are the first group of animals in evolutionary history to have endoskeletons. In echinoderms, the endoskeleton consists of calcium carbonate plates, often with spines attached, and is covered by a thin layer of skin. On the skin are **pedicellariae** (PEH dih sih LAH ree ee) (singular, pedicellaria) small pincers that aid in catching food and in removing foreign materials from the skin.

All echinoderms have radial symmetry as adults. In **Figure 27.2,** you can see this feature in the five arms of the brittle star radiating out from a central disk. However, echinoderm larvae have bilateral symmetry, as shown in **Figure 27.2.** In the next chapter, you will learn how bilateral symmetry shows an embryonic link to the vertebrate animals that evolved later.

No other animals with the complex organ systems of echinoderms have radial symmetry. Scientists theorize that the ancestors of echinoderms did not have radial symmetry. Primitive echinoderms might have been sessile, and radial symmetry developed, to enable them to carry on a successful stationary existence. Free-moving echinoderms might have evolved from the sessile animals. Investigate the features of echinoderms in **MiniLab 27.1.**

Reading Check **Infer** how radial symmetry is important to animals that cannot move quickly.

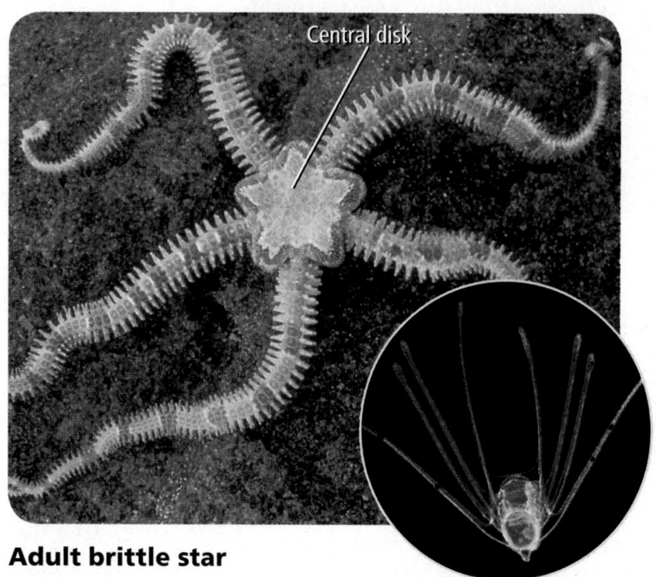

■ **Figure 27.2** Brittle star larvae have bilateral symmetry and can be divided along only one plane into mirror-image halves. Adult brittle stars have radial symmetry and can be divided through a central axis, along any plane, into equal halves.

MiniLab 27.1

Observe Echinoderm Anatomy

What are the characteristics of echinoderms? Although they have many shapes and sizes, all echinoderms have some features in common.

Procedure

1. Read and complete the lab safety form.
2. Study preserved specimens of **a sand dollar, a sea cucumber, a sea star,** and **a sea urchin.**
3. Create a data table to record your observations. Complete the table by describing the major features of each specimen. Include a sketch of each specimen.
4. Label any external features you can identify.
5. Clean all equipment and return it to the appropriate place. Wash your hands thoroughly after handling preserved specimens.

Analysis

1. **Compare** the external features of the echinoderms you studied. Can your observations completely justify why these four organisms are classified in the same phylum? Explain.
2. **Observe and Infer** What features are most important in helping echinoderms avoid being eaten by predators?

Visualizing an Echinoderm

Figure 27.3
Sea urchins can be found in tidal areas of the sea. They burrow into crevices in rocks to hide, and they scrape algae with a hard five-plated structure, called Aristotle's lantern, in their mouths. Imagine that these plates are like teeth that move.

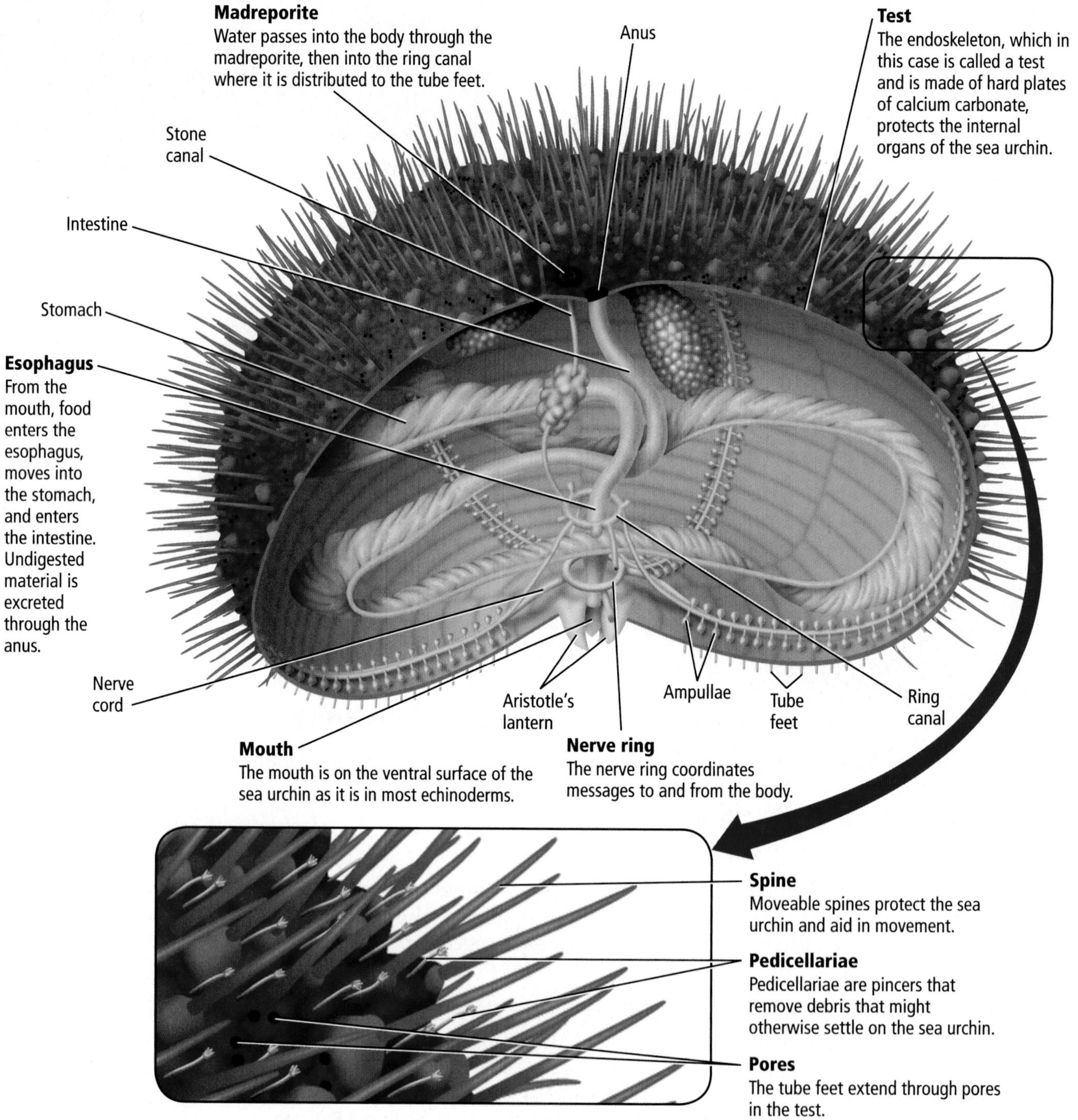

Madreporite
Water passes into the body through the madreporite, then into the ring canal where it is distributed to the tube feet.

Anus

Test
The endoskeleton, which in this case is called a test and is made of hard plates of calcium carbonate, protects the internal organs of the sea urchin.

Stone canal

Intestine

Stomach

Esophagus
From the mouth, food enters the esophagus, moves into the stomach, and enters the intestine. Undigested material is excreted through the anus.

Nerve cord

Aristotle's lantern

Ampullae

Tube feet

Ring canal

Mouth
The mouth is on the ventral surface of the sea urchin as it is in most echinoderms.

Nerve ring
The nerve ring coordinates messages to and from the body.

Spine
Moveable spines protect the sea urchin and aid in movement.

Pedicellariae
Pedicellariae are pincers that remove debris that might otherwise settle on the sea urchin.

Pores
The tube feet extend through pores in the test.

Concepts In Motion **Interactive Figure** To see an animation of echinoderm features, visit biologygmh.com.

Biology Online

Water-vascular system Another feature of echinoderms is their **water-vascular system**—a system of fluid-filled, closed tubes that work together to enable echinoderms to move and get food. The strainerlike opening to the water-vascular system, shown in **Figure 27.3,** is called the **madreporite** (MA druh pohr it). Water is drawn into the madreporite, then moves through the stone canal to the ring canal. From there, the water moves to the radial canals and eventually to the tube feet.

Tube feet are small, muscular, fluid-filled tubes that end in suction-cuplike structures and are used in movement, food collection, and respiration. The opposite end of the tube foot is a muscular sac, called the **ampulla** (AM pyew luh). When muscles contract in the ampulla, water is forced into the tube foot and it extends. Imagine holding a small, partly inflated balloon in your hand and squeezing it. The balloon will extend from between your thumb and forefinger, which is similar to the way the tube foot extends. The suction-cuplike structure on the end of the tube foot attaches it to the surface. This hydraulic suction enables all echinoderms to move and some, such as sea stars, to apply a force strong enough to open the shells of mollusks, as illustrated in **Figure 27.4.**

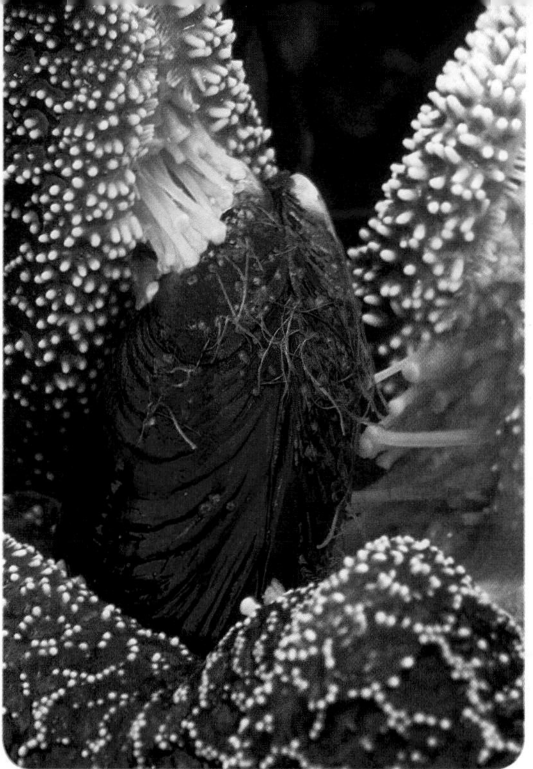

Figure 27.4 A sea star uses its tube feet to open the two shells of a clam.
Describe *the sea star's feeding method.*

Feeding and digestion Echinoderms use a great variety of feeding strategies in addition to tube feet. Sea lilies and feather stars extend their arms and trap food. Sea stars prey on a variety of mollusks, coral, and other invertebrates. Many species of sea stars can push their stomachs out of their mouths and onto their prey. They then spread digestive enzymes over the food and use cilia to bring the digested material to their mouths. Brittle stars can be active predators or scavengers, and they can trap organic materials in mucus on their arms. Most sea urchins use teethlike plates, shown in **Figure 27.3,** to scrape algae off surfaces or feed on other animals. Many sea cucumbers extend their branched, mucous-covered tentacles to trap floating food.

Respiration, circulation, and excretion Echinoderms also use their tube feet in respiration. Oxygen diffuses from the water through the thin membranes of the tube feet. Some echinoderms carry out diffusion of oxygen through all thin body membranes in contact with water. Others have thin-walled skin gills that are small pouches extending from the body. Many sea cucumbers have branched tubes, called respiratory trees, through which water passes and oxygen moves into the body.

Circulation takes place in the body coelom and the water-vascular system, while excretion of cellular wastes occurs by diffusion through thin body membranes. Cilia move water and body fluids throughout these systems aided by pumping action in some echinoderms. In spite of the simplicity of these organs and systems, echinoderms maintain homeostasis effectively with adaptations that are suited to their way of life.

 **Reading Check** **Summarize** the functions of an echinoderm's tube feet.

LAUNCH Lab

Review Based on what you've read about the water-vascular system, how would you now answer the analysis questions?

■ **Figure 27.5** A sea star lifts the end of an arm to sense light and movement.

Response to stimuli Echinoderms have both sensory and motor neurons with varying degrees of complexity in different species. In general, a nerve ring surrounds the mouth with branching nerve cords connecting to other body areas.

Sensory neurons respond to touch, chemicals dissolved in the water, water currents, and light. At the tips of the arms of sea stars are eyespots, clusters of light-sensitive cells, illustrated in **Figure 27.5.** Many echinoderms also sense the direction of gravity. For example, a sea star will return to an upright position after being overturned by a wave or current.

Movement Echinoderm locomotion is as varied as echinoderm body shapes. The structure of the endoskeleton is important for determining the type of movement an echinoderm can undertake. The movable bony plates in the endoskeletons of echinoderms enable them to move easily. Feather stars move by grasping the soft sediments of the ocean bottom with their cirri—long, thin appendages on their ventral sides—or by swimming with up-and-down movements of their arms. Brittle stars use their tube feet and their arms in snakelike movements for locomotion. Sea stars use their arms and tube feet for crawling. Sea urchins move by using tube feet and burrowing with their movable spines. Sea cucumbers crawl using their tube feet and body wall muscles.

Reading Check **Summarize** In addition to using their tube feet, in what other ways do echinoderms move?

VOCABULARY

SCIENCE USAGE V. COMMON USAGE

Structure

Science usage: the arrangement of parts of an organism.

The structure of an insect's mouth determines how it functions.

Common usage: something that is constructed, such as a building.

The workers built the structure in three months.

Reproduction and development Most echinoderms reproduce sexually. The females shed eggs and the males shed sperm into the water where fertilization takes place. The fertilized eggs develop into free-swimming larvae with bilateral symmetry. After going through a series of changes, the larvae develop into adults with radial symmetry. Recall that echinoderms have deuterostome development, making them an important evolutionary connection to vertebrates.

The sea star in **Figure 27.6** illustrates an echinoderm regenerating a lost body part. Many echinoderms can drop off an arm when they are attacked, enabling them to flee while the predator is distracted. Others can expel part of their internal organ systems when threatened, an action that might surprise and deter predators. All of these body parts can be regenerated.

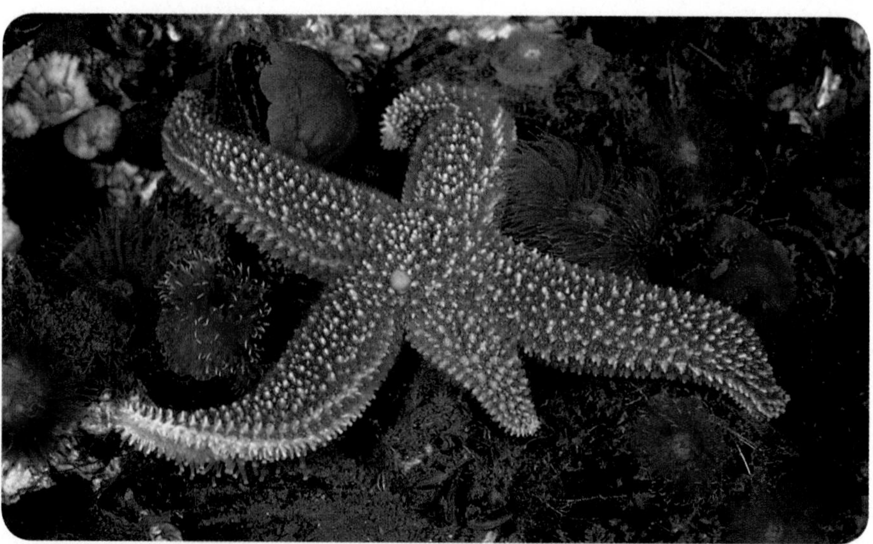

■ **Figure 27.6** This sea star is regenerating one of its arms, a process that can take up to one year.

Explain ***how regenerating body parts helps echinoderms survive.***

Concepts In Motion
Interactive Table To explore more about echinoderms, visit biologygmh.com.

Table 27.1 Classes of Echinoderms

Class	Asteroidea	Ophiuroidea	Echinoidea	Crinoidea	Holothuroidea	Concentricycloidea
Examples						
Class Members	Sea stars	Brittle stars	Sea urchins, Sand dollars	Sea lilies, Feather stars	Sea cucumbers	Sea daisies
Distinctive Features	• Often five-armed • Tube feet used for feeding and movement	• Often five-armed • Arms break off easily and can be regenerated • Move by arm movement • Tube feet have no suction cups	• Body encased in a test with spines • Sea urchins burrow in rocky areas. • Sand dollars burrow in sand.	• Sessile for some part of life • Sea lilies have long stalks. • Feather stars have long branching arms.	• Cucumber shape • Leathery outer body • Tube feet modified to tentacles near mouth	• Less than 1 cm in diameter • No arms • Tube feet located around a central disk

Echinoderm Diversity

The major classes of living echinoderms include Asteroidea (AS tuh ROY dee uh), the sea stars; Ophiuroidea (OH fee uh ROY dee uh), the brittle stars; Echinoidea (ih kihn OY dee uh), the sea urchins and sand dollars; Crinoidea (kri NOY dee uh), the sea lilies and feather stars; Holothuroidea (HOH loh thuh ROY dee uh), the sea cucumbers; and Concentricycloidea (kahn sen tri sy CLOY dee uh), the sea daisies. Recall that they all are marine animals with radial symmetry as adults, a water-vascular system with tube feet, endoskeletons often bearing spines, and larvae with bilateral symmetry. The classes of echinoderms are summarized in **Table 27.1.**

Sea stars If you ever have seen an echinoderm, it probably was a sea star. Most species of sea stars have five arms arranged around a central disk. Some, such as the one in **Figure 27.7,** have more than five arms. Sea stars can be found in shallow water near the shore and in tide pools when the tide recedes. They can be found in groups clinging to rocks by means of their tube feet. A single tube foot can exert a pull of 0.25–0.30 N. This is equal to the force required to lift 25–30 large paper clips. Because a sea star might have as many as 2000 tube feet, it can exert quite a large force as it crawls or opens mollusks for food. Sea stars are important predators in marine ecosystems, feeding on clams and other bivalves. Because of their spiny skin, sea stars usually are not food for other marine predators.

Figure 27.7 Sunflower stars can have twenty or more arms.

■ **Figure 27.8** One type of brittle star, the basket star, extends its branched arms into the current to filter feed.

Analyze ***How are brittle stars different from sea stars?***

Brittle stars Like sea stars, most brittle stars have five arms, but the brittle star's arms are thin and very flexible, as shown in **Figure 27.8.** They do not have suckers on their tube feet, so they cannot use them for movement as sea stars do. Brittle stars move by rowing themselves quickly over the bottom rocks and sediments or by snakelike movements of their arms. When attacked by a predator, a brittle star can release an arm and make a quick getaway. The missing arm will be regenerated later. Brittle stars hide in the crevices of rocks by day and feed at night. They feed on small particles suspended in the water or catch suspended materials on mucous strands between their spines.

Some brittle stars respond to light. The spherical structures covering the body of these brittle stars might function as light-gathering lenses. Brittle stars are more abundant and have more numbers of species than any other class of echinoderm.

 **Reading Check Compare and contrast** the locomotion of sea stars and brittle stars.

Sea urchins and sand dollars Burrowing is a key characteristic of sea urchins and sand dollars. Sand dollars can be found in shallow water burrowing into the sand, while sea urchins burrow into rocky areas. These echinoderms each have a compact body enclosed in a hard endoskeleton, called a test, that looks like a shell. The tube feet extend through pores in the test. Closely fitting plates of calcium carbonate make up the test. Sea urchins and sand dollars lack arms, but their tests reflect the five-part pattern of arms in sea stars and brittle stars. Spines also are an important feature of this class, as seen in **Figure 27.9.** Some sea urchin spines and pedicellariae contain venom and are used for fending off predators. The poison in pedicellariae can paralyze prey. Sea urchins also can be herbivorous grazers, scraping algae from rocks, while sand dollars filter organic particles from the sand in which they are partially buried.

■ **Figure 27.9** Sea urchins burrow themselves into rocky crevices with their sharp, movable spines. Sand dollars burrow themselves into the sand, where they filter out small food particles.

Sand dollar

Sea urchins

Aristotle's lantern

Five-sided lantern

■ **Figure 27.10** Aristotle's lantern is a five-sided mouthpart similar in shape to a five-sided lantern. The force of a sea urchin's chewing plates is so strong that it has been known to chew through concrete.

Connection to History Most sea urchins have a chewing apparatus inside their mouths consisting of five hard plates, similar to teeth. This structure, shown in **Figure 27.10,** is called Aristotle's lantern. It was named after a description written by Aristotle, a Greek philosopher, in his book *Historia Animalium* (The History of Animals). In the fourth century B.C., which is when Aristotle lived, people used five-sided lanterns with side panels made of thin, translucent horn, called horn lanterns. Aristotle thought the mouth of a sea urchin looked like a horn lantern without the panels.

Sea lilies and feather stars Fossil records show that sea lilies and feather stars are the most ancient of the echinoderms and were abundant before other echinoderms evolved. They are different from other echinoderms in that they are sessile for part of their lives. As shown in **Figure 27.11,** sea lilies have a flower-shaped body at the top of a long stalk, while the long-branched arms of feather stars radiate upward from a central area. Though they might stay in one place for a long time, sea lilies and feather stars can detach themselves and move elsewhere. Both sea lilies and feather stars capture food by extending their tube feet and arms into the water, where they catch suspended organic materials.

Reading Check **Compare** How are feather stars and sea lilies similar?

■ **Figure 27.11** This fossil illustrates how a sea lily's body is flower-shaped at the tip of a long stalk. A feather star extends its arms from a central point where they are attached.
Infer ***How is the shape of the arms of feather stars adapted to a lifestyle that includes little movement?***

Sea lily

Feather star

■ **Figure 27.12** Some of the sea cucumber's tube feet are modified into tentacles that trap food particles from the water.
Identify ***What substance coats the tentacles and helps trap food particles?***

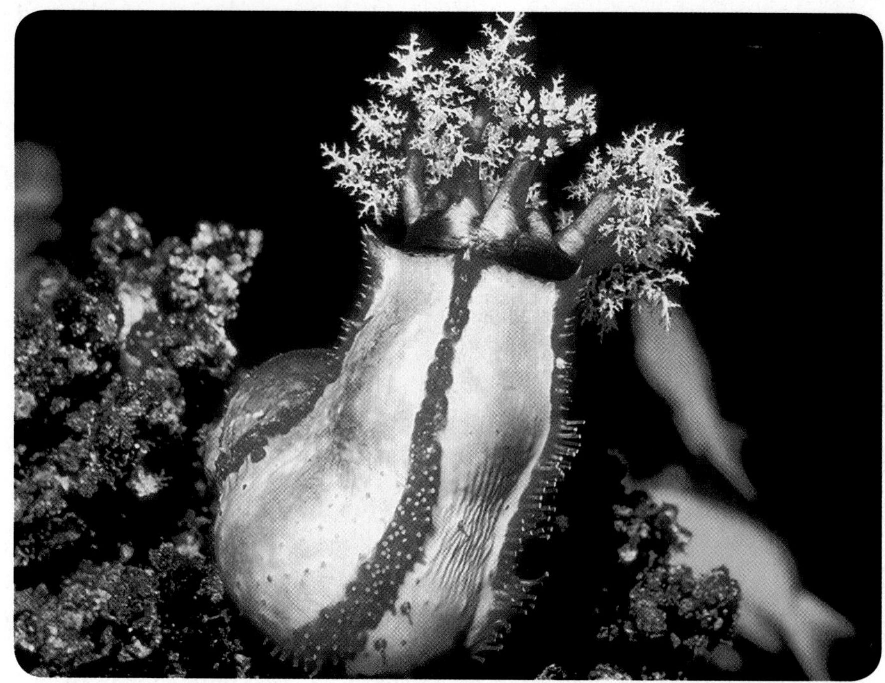

CAREERS IN BIOLOGY

Marine Biologist Scientists in this field study plants and animals, such as echinoderms, that live in the ocean. They also study how pollution affects the marine environment. For more information on biology careers, visit biologygmh.com.

Sea cucumbers Sea cucumbers don't look like other echinoderms. Some might say they don't even look like animals. Can you guess why they are called sea cucumbers? They look like cartoon cucumbers creeping over the ocean floor. Their elongated bodies move sluggishly by means of tube feet assisted by contractions in their muscular body wall. Their calcium carbonate plates are reduced in size and do not connect as they do in other echinoderms, so their outer bodies generally appear leathery. Some of their tube feet are modified to form tentacles which extend from around their mouths to trap suspended food particles, as shown in **Figure 27.12.** The tentacles are covered with mucus, which increases their ability to trap food. Once food has been trapped on a tentacle, it is drawn into the mouth where the food is sucked off. This process is similar to licking your finger after putting it in a bowl of pudding.

Sea cucumbers are the only echinoderms to have respiratory organs in the form of respiratory trees. These many-branched tubes pump in seawater through the anus for oxygen extraction. The respiratory tree also functions in excretion by removing cellular wastes.

When a sea cucumber is threatened, it can cast out some of its internal organs through its anus. A potential predator might be confused by this action and move on. The sea cucumber can regenerate its lost parts. Even though the sea cucumber's adaptations might seem odd, it is important to remember that this animal maintains homeostasis with adaptations that fit its way of life in its particular habitat.

■ **Figure 27.13** Sea daisies are tiny disc-shaped echinoderms.

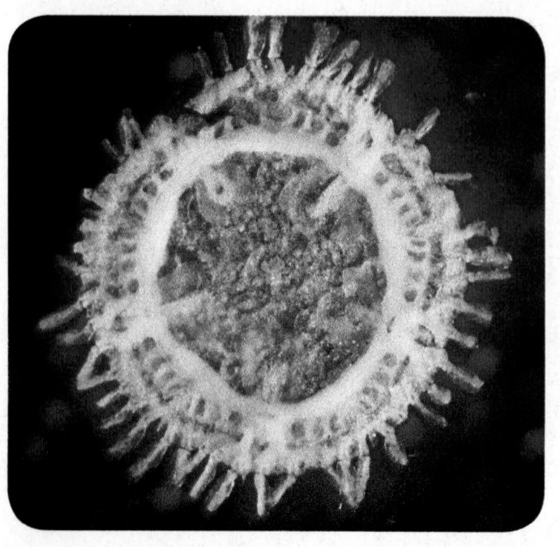

Sea daisies Discovered in 1986 off the coast of New Zealand, sea daisies have been difficult to classify and study, because so few have been found. They are less than 1 cm in diameter and are disc-shaped with no arms. Their tube feet are located around the edge of the disc. **Figure 27.13** shows that they have five-part radial symmetry, as do other echinoderms. Notice the daisy pattern of petals, or tube feet, around the edges of the disc.

 **Reading Check** **Infer** What characteristics place sea daisies in the phylum Echinodermata?

Ecology of Echinoderms

Sea cucumbers and sea urchins are sources of food for people in some Asian countries. The muscles of certain sea cucumbers are eaten as sushi, and dried sea cucumbers are added to flavor soups, vegetables, and meat. The egg masses of sea urchins are eaten raw or slightly cooked.

Commensal relationships exist between some echinoderms and other marine animals. Recall from Chapter 2 that commensalism is a relationship in which one organism benefits and the other organism is neither helped nor harmed. For example, some species of brittle stars live inside sponges. The brittle star leaves the protective interior of the sponge and feeds on materials that have settled on the sponge.

Echinoderm benefits Marine ecosystems also depend on some echinoderms. When populations of echinoderms decline, a change in the ecosystem is often noted. For example, a sea urchin species that lives in the Caribbean and the Florida keys declined in numbers by more than 95 percent in 1983 due to disease. After this, algae increased greatly on the coral reefs and virtually destroyed the reefs in many areas. Sea urchins and sea cucumbers are bioturbinators—organisms that stir up sediment on the ocean floor. This action is important to the entire marine ecosystem, as it makes nutrients in the seafloor available to other organisms.

Echinoderm harm Some echinoderms can harm marine ecosystems. The crown-of-thorns sea star shown in the photo at the beginning of this chapter feeds on coral polyps. When these sea stars increase in numbers, coral reefs are destroyed. Although the causes of the population explosions of these sea stars continue to be debated, the numbers seem to decline on their own. At a later time, they might increase again with no apparent explanation. Sea urchins are a favorite food of sea otters, as shown in **Figure 27.14.** The number of sea otters in California has declined in recent years, leading to an increase in the number of sea urchins. The sea urchins are eating the kelp forests, destroying the habitat of fish, snails, and crabs.

Figure 27.14 Without enough sea otters to keep the sea urchin population under control, sea urchins will continue to increase in number, threatening the kelp forests on which they feed.

Section 27.1 Assessment

Section Summary

- Adult echinoderms can be identified by four main structural features.
- Larval echinoderms have features that link them to relatives that evolved after them.
- Echinoderms have a water-vascular system and tube feet.
- Echinoderms have a variety of adaptations for feeding and movement.
- There are six major classes of living echinoderms.

Understand Main Ideas

1. MAIN Idea **Identify** the four main features that distinguish adult echinoderms.
2. **Explain** how a water-vascular system works.
3. **Sketch** line drawings that represent each of the six classes of echinoderms.
4. **Suggest** how feeding and movement are related to each other in echinoderms.

Think Critically

5. **Hypothesize** A certain species of red-and-white-striped shrimp is often found on a species of colorful brittle stars. Form a hypothesis about the relationship between the shrimp and the brittle stars.

MATH in Biology

6. If it takes a force of 20 N to pull apart a bivalve's shells, how many tube feet will it take to pull apart a bivalve if each tube foot has a pull of 0.25 N?

Section 27.2

Reading Preview

Objectives

- **Interpret** the features of invertebrate chordates that place them in the phylum Chordata.
- **Analyze** the features of invertebrate chordates that place them with invertebrates.
- **Compare** the adaptations of lancelets with those of sea squirts.

Review Vocabulary

deuterostome: an animal whose mouth develops from cells other than those at the opening of the gastrula

New Vocabulary

chordate
invertebrate chordate
notochord
postanal tail
dorsal tubular nerve cord
pharyngeal pouch

Invertebrate Chordates

MAIN Idea Invertebrate chordates have features linking them to vertebrate chordates.

Real-World Reading Link Worms, snails, bees, fishes, birds, and dogs are all animals because they share common characteristics. Think about the features these animals have in common and the features that make them different from each other. The animals that share the most features are related more closely than the animals that share only a few features.

Invertebrate Chordate Features

Look at **Figure 27.15,** the evolutionary tree of animal phylogeny. Notice that invertebrate chordates, such as lancelets and tunicates, are deuterostomes like echinoderms but they have additional chordate features that echinoderms do not have.

Fossil evidence suggests animals such as the lancelet, shown in **Figure 27.15,** separated from the echinoderms during the Cambrian. This group of animals is known as the invertebrate chordates. It has become a diverse group over the last 500 million years. The lancelet is also called an amphioxus (am fee AHK sus).

The lancelet is a small eel-like animal that spends most of its life buried in the sand filtering the water for food. They are tiny, headless creatures with translucent, fish-shaped bodies. Fossil and molecular data show that the lancelet is one of the closest living relatives of vertebrates. In 2007, genetic studies showed that tunicates, shown in **Figure 27.15,** are the closest invertebrate relative to vertebrates. You are more closely related to the lancelet and tunicate than you are to any other invertebrate—yet related very distantly compared to other vertebrates.

Figure 27.15 Like echinoderms, invertebrate chordates, such as lancelets and tunicates, show deuterostome development.

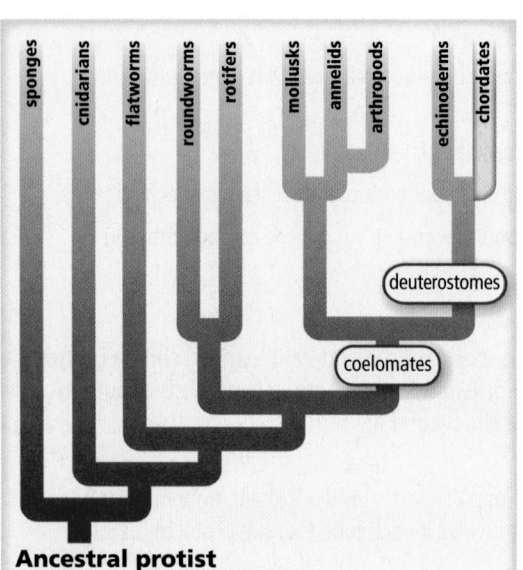

sponges
cnidarians
flatworms
roundworms
rotifers
mollusks
annelids
arthropods
echinoderms
chordates
deuterostomes
coelomates
Ancestral protist

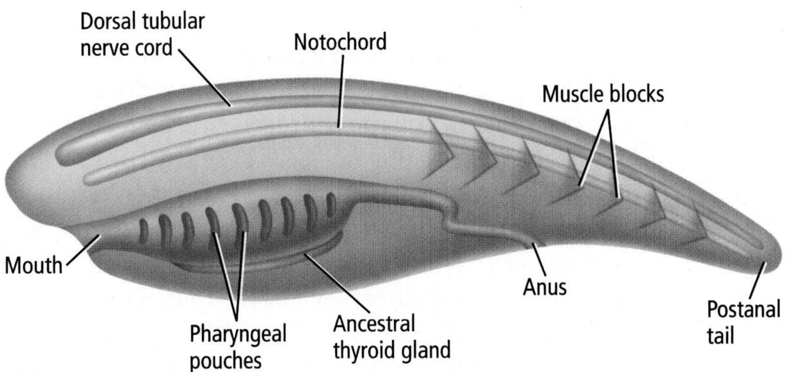

Figure 27.16 Chordates have a dorsal tubular nerve cord, a notochord, pharyngeal pouches, a postanal tail, and, possibly, some form of a thyroid gland.
Infer *Which of these features did you have when you were an embryo?*

Chordates are animals belonging to the phylum Chordata (kor DAH tuh) that have four distinctive features—a dorsal tubular nerve cord, a notochord, pharyngeal pouches, and a postanal tail—at some point during their development. Recent evidence suggests that all chordates also might have some form of a thyroid gland. In addition, they have a coelom and segmentation. Study **Figure 27.16** to see the main features of chordates. Recall that vertebrates are animals with backbones. Most chordates are vertebrates. **Invertebrate chordates,** which belong to two of the subphyla of chordates—Cephalochordata and Urochordata, also have a dorsal tubular nerve cord, a notochord, pharyngeal pouches, a postanal tail, and, possibly, an ancestral thyroid gland. They have no backbone, however.

Notochord The **notochord** (NOH tuh kord) is a flexible, rodlike structure that extends the length of the body. It is located just below the dorsal tubular nerve cord. In most vertebrates, the notochord eventually is replaced by bone or cartilage. In invertebrate chordates, the notochord remains. The flexibility of the notochord enables the body to bend, rather than shorten, during contractions of the muscle segments. An animal with a notochord can make side-to-side movements of the body and tail, the first time in the course of evolution that fishlike swimming is made possible.

Postanal tail A free-swimming animal moves efficiently by using a postanal tail. A **postanal tail** is a structure used primarily for locomotion and is located behind the digestive system and anus. In most chordates, the postanal tail extends beyond the anus. Tails in nonchordates have parts of the digestive system inside and the anus is located at the end of the tail. The postanal tail with its muscle segments can propel an animal with more powerful movements than the body structure of invertebrates without a postanal tail.

Dorsal tubular nerve cord In nonchordates, the nerve cords are ventral to, or below, the digestive system and are solid. Chordates have a **dorsal tubular nerve cord** that is located dorsal to, or above, the digestive organs and is a tube shape. The anterior end of this cord becomes the brain and the posterior end becomes the spinal cord during development of most chordates.

 **Reading Check Analyze** How is a notochord important to invertebrate chordates?

VOCABULARY
WORD ORIGIN
Notochord
noto- prefix; from Greek, meaning *back.*
-chord from Greek, meaning *chord* or *string.*

FOLDABLES
Incorporate information from this section into your Foldable.

Study Tip

Cooperation Form a group with four other students. Each group member should pick one of the five boldfaced headings under the heading *Invertebrate Chordate Features.* Each group member should read the selected section and then teach the rest of the group the material presented in that section of text.

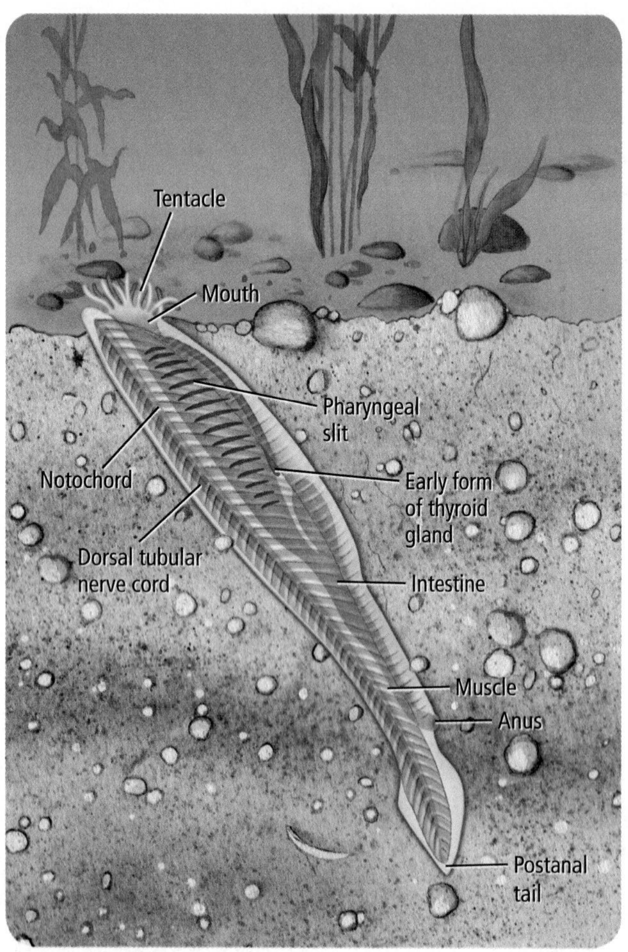

■ **Figure 27.17** The lancelet is an invertebrate chordate that has the main features of chordates.
Infer ***How might the short tentacles surrounding the lancelet's mouth function?***

Pharyngeal pouches In all embryos, paired structures called **pharyngeal pouches** connect the muscular tube that links the mouth cavity and the esophagus. In aquatic chordates, the pouches contain slits that lead to the outside. These structures were used first for filter feeding and later evolved into gills for gas exchange in water. In terrestrial chordates, the pharyngeal pouches do not contain slits and develop into other structures, such as the tonsils and the thymus gland. Pharyngeal pouches are thought to be evidence of the aquatic ancestry of all vertebrates.

Ancestral thyroid gland The thyroid gland is a structure that regulates metabolism, growth, and development. An early form of a thyroid gland had its origins in cells of early chordates that secreted mucus as an aid in filter feeding. Invertebrate chordates have an endostyle—cells in this same area that secrete proteins similar to those secreted by the thyroid gland. Only vertebrate chordates have a thyroid gland.

Connection  **Health** Iodine is concentrated in the endostyle and plays an important role in thyroid gland function. It is essential for the production of thyroid hormones. In the United States, iodine is added to salt to prevent iodine deficiency. Other sources of iodine include fish, dairy products, and vegetables grown in iodine-rich soil.

Reading Check **Explain** why an endostyle might be considered an early form of a thyroid gland.

Diversity of Invertebrate Chordates

Like echinoderms, all invertebrate chordates are marine animals. There are about 23 species of lancelets belonging to subphylum Cephalochordata. Tunicates, in subphylum Urochordata, consist of about 1250 species.

Lancelets Recall the amphioxus at the beginning of this section. Most lancelets belong to the genus *Branchiostoma* (formerly *Amphioxus).* They are small, fishlike animals without scales. As shown in **Figure 27.17,** lancelets burrow their bodies into the sand in shallow seas.

Lancelets lack color in their skin, which is only one cell layer deep, enabling an observer to view some body functions and structures. Water flowing through the body can be observed as a lancelet filter feeds. To get food, water enters the mouth of the lancelet and passes through pharyngeal gill slits. Food is trapped and passed on to a stomachlike structure to be digested. Water exits through the gill slits.

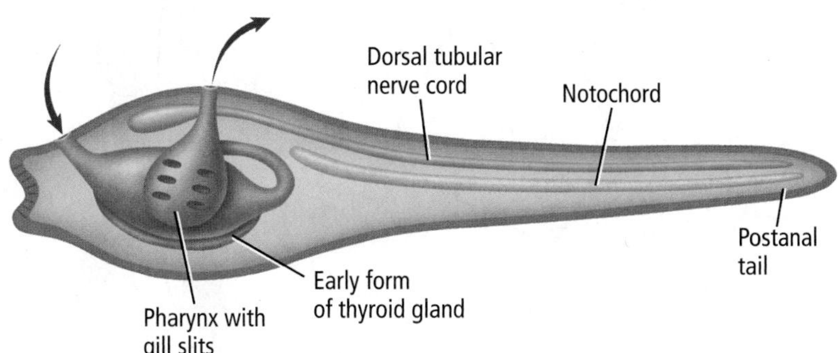

■ **Figure 27.18** Larval tunicates look like tadpoles and have all of the chordate features. The arrows indicate where water flows into and out of the body.

Just as filter feeding can be observed in the lancelet, its muscles can also be seen. Observe the internal structures of a lancelet in **Figure 27.17.** The arrangement of segmented muscle blocks is similar to that in vertebrates, enabling lancelets to swim with a fishlike motion. Unlike vertebrates, they have no heads or sensory structures other than light receptors and small sensory tentacles near the mouth. The nervous system consists of main branching nerves and a simple brain at the anterior end of the animal. Blood passes through the body by the action of pumping blood vessels, as there is no true heart. Lancelets have separate sexes, and fertilization is external.

Tunicates Often called sea squirts, tunicates (TEW nuh kayts) are named for the thick outer covering, called a tunic, that covers their small, saclike bodies. Most tunicates live in shallow water; some live in masses on the ocean floor. In general, tunicates are sessile, and only in the larval stages do they show typical chordate features. Locate the notochord, postanal tail, dorsal tubular nerve cord, pharyngeal pouches, and ancestral thyroid gland on the tunicate larva in **Figure 27.18.**

Water is drawn into the saclike body of an adult tunicate through the incurrent siphon, as shown in **Figure 27.19,** by the action of beating cilia. Food particles are trapped in a mucous net and moved into the stomach where digestion takes place. In the meantime, water leaves the body, first through gill slits in the pharynx and then out through the excurrent siphon.

Circulation in the body of the tunicate is performed by a heart and blood vessels that deliver nutrients and oxygen to body organs. The nervous system consists of a main nerve complex and branching neurons. Tunicates are hermaphrodites—they produce both eggs and sperm—with external fertilization.

Why are tunicates called sea squirts? When they are threatened, they can eject a stream of water with force through the excurrent siphon, possibly distracting a potential predator.

Reading Check Compare tunicates and lancelets.

■ **Figure 27.19** Adult tunicates look like sacs. The only chordate features that remains in the adult is pharyngeal gill slits and the thyroid gland. The arrows indicate water flow in and out of the body.
Compare ***What other invertebrates have you studied that are filter feeders?***

Concepts In Motion

Interactive Figure To see an animation of a tunicate's anatomy, visit biologygmh.com.

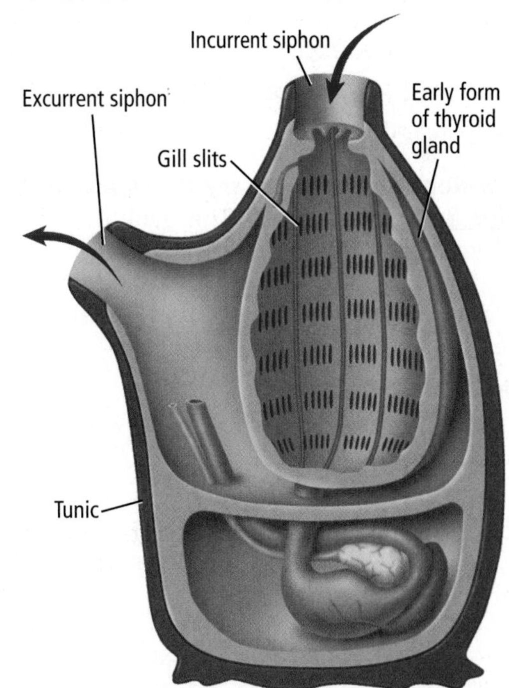

Figure 27.20 This cladogram shows the phylogeny of echinoderms.
Interpret *Which is the most recent class of echinoderms to evolve?*

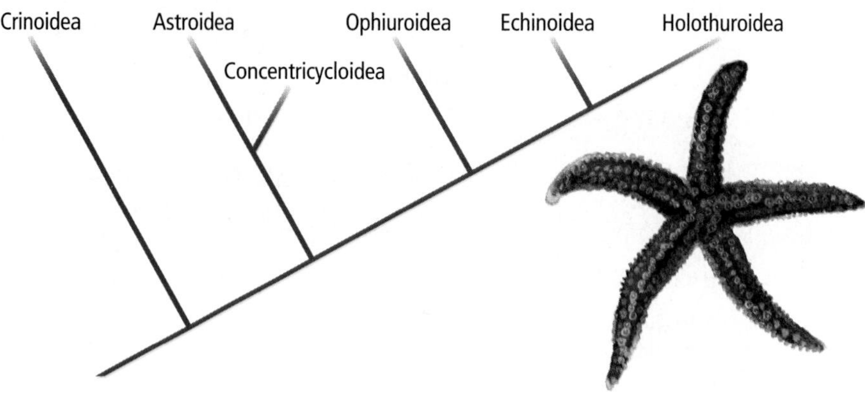

CAREERS IN BIOLOGY

Evolutionary Biologist Scientists in this field study how organisms have changed over time. For more information on biology careers, visit biologygmh.com.

Evolution of Echinoderms and Invertebrate Chordates

Biologists are studying fossil and molecular evidence to learn how echinoderms and invertebrate chordates are related to the vertebrates that evolved later.

Phylogeny of echinoderms The fossil record of echinoderms extends back to the Cambrian. Scientists think that they evolved from ancestors with bilateral symmetry because echinoderms have bilaterally symmetrical larvae. Their radial symmetry develops later in the adult stage. Many biologists think that ancient echinoderms were sessile and attached to the ocean floor by a long stalk, just as the sea lily does today.

Echinoderms also undergo deuterostome development. Recall that this type of development links them phylogenetically to chordates, which also have deuterostome development. The cladogram in **Figure 27.20** shows one interpretation of the evolution of echinoderms.

DATA ANALYSIS LAB 27.1

Based on Real Data*

Interpret Scientific Illustrations

How does an evolutionary tree show relationships among sea stars? This evolutionary tree is a representation of various species of sea stars and their phylogenetic history based on molecular data. Each letter represents a specific sea star species.

Think Critically

1. **Identify** Which sea star is most closely related to sea star A?
2. **Interpret** Which is the oldest sea star?
3. **Analyze** Which group of sea stars has the most diversity—C,G,N or L,K,M? How did you decide?

Data and Observations

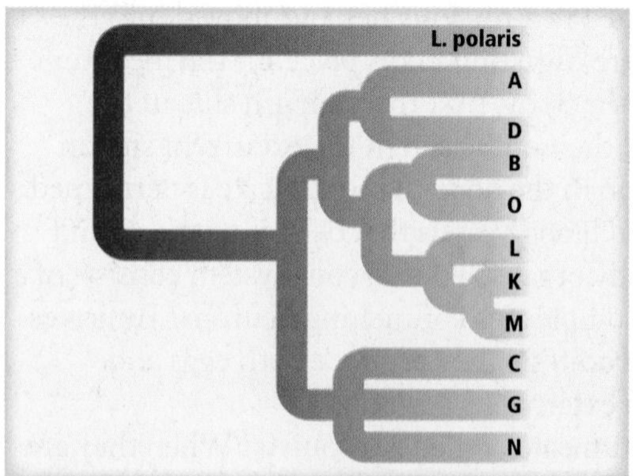

*Data obtained from: Hrincevich, A.W., et al. 2000. Phylogenetic analysis of molecular lineages in a species-rich subgenus of sea stars (*Leptasterias* subgenus *Hexasterias*). *American Zoologist* 40: 365-374.

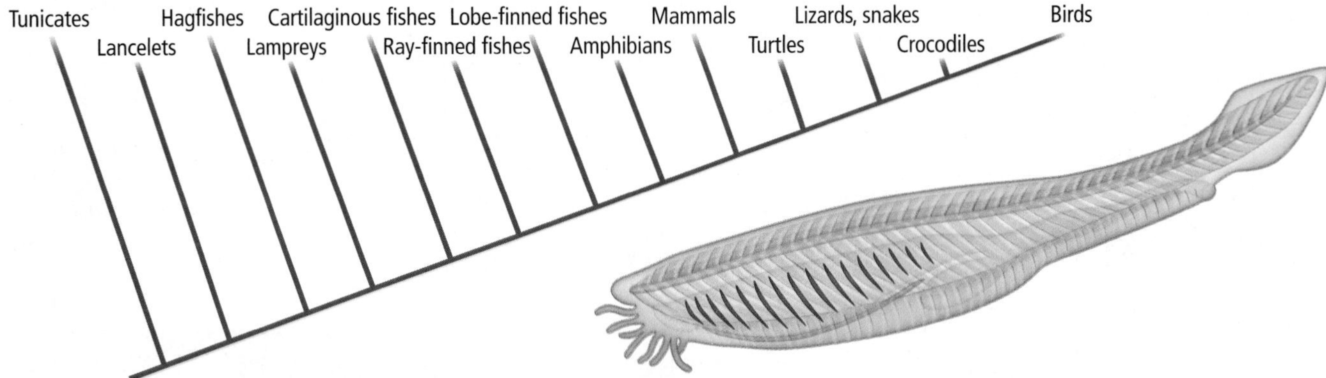

Figure 27.21 This cladogram shows a possible phylogeny of invertebrate chordates and other chordates that evolved later.

Phylogeny of invertebrate chordates How could sleek, burrowing, fishlike lancelets be close relatives of saclike, sessile sea squirts? They are related because of their dorsal tubular nerve cords, notochords, pharyngeal gill slits, and postanal tails, even if sea squirts have all these features only in their larvae. Beyond this answer, scientists debate about the evolution of these animals and raise still unanswered questions. For example, from which invertebrate chordate did the fishlike tadpole larvae arise? What was the original form of that first fishlike animal?

One thing is certain: the notochord, that flexible tough rod, provided support for the animal, and it also provided a place for muscles to attach. With this arrangement, chordates could swing their backs from side to side and swim through the water, a key development in the evolution of chordates. This advance also led to the first large animals. Examine the cladogram in **Figure 27.21** to see one interpretation of how chordates are related.

Section 27.2 Assessment

Section Summary

- Chordates have four main features that make them different from animals that are not chordates.
- Invertebrate chordates have all the features of chordates, except they do not have the main feature of vertebrate chordates.
- The notochord is an adaptation that enabled animals to move in ways they had never moved before.
- Lancelets are fish-shaped invertebrate chordates that, as adults, have all the main features of chordates.
- Tunicates are sac-shaped invertebrate chordates that have chordate features as larvae.

Understand Main Ideas

1. MAIN Idea **Summarize** the main features of invertebrate chordates that show their close relationship to vertebrate chordates.
2. **Describe** the characteristic of invertebrate chordates that places them with other invertebrates rather than with vertebrates.
3. **Model** Make models of a lancelet and a sea squirt from clay or salt dough. Identify features that place these animals in the phylum Chordata.
4. **Compare** the adaptations of sea squirts with those of lancelets that enable them to live in their environments.

Think Critically

5. **Design an experiment** to determine if lancelets prefer a light environment or a dark environment.
6. **Interpret** Use **Figure 27.21** to determine which subphylum of chordates evolved next after the cephalochordates.

WRITING in **Biology**

7. Write a paragraph describing how sponges and tunicates are alike. Write another paragraph describing how they are different from each other.

CUTTING-EDGE BIOLOGY

Echinoderms Aid Medical Research

How did the comic book character the Incredible Hulk increase his body size without ripping his body to pieces?

Believe it or not, producers consulted an expert on echinoderms before creating a film about this character because they wondered if any living creature could perform such feats. Sea cucumbers, specifically, can stretch and then shrink back to their normal size, much like the Incredible Hulk does in the film.

Because the collagen in a sea cucumber's connective tissue is not fixed, its body can change from the consistency of liquid gelatin to a rigid structure and back again in seconds.

Connective tissue When Greg Szulgit was a graduate student in biology, he discovered the amazing power of sea cucumbers to increase their body size and then shrink back to their normal size. How do sea cucumbers change their body size? It's all due to their connective tissue—the tissue that connects, supports, and surrounds other tissues and organs in the body.

A sea cucumber's connective tissue is similar to a human's connective tissue. Connective tissue fibers contain a protein called collagen. In humans, collagen is a fixed part of the tissue. Szulgit and other researchers found that the collagen in the connective tissue of echinoderms is not fixed, but instead slides back and forth. When the collagen particles in the endoskeleton are sliding past each other, a sea cucumber's body is soft and flexible. A sea cucumber's cells can release a substance that locks the collagen and stops it from sliding. This stiffens its endoskeleton, making it immobile.

Connective-tissue disorders Szulgit studies the sea cucumber's body-stretching abilities with the hope of someday being able to treat connective-tissue disorders in humans. These disorders include Ehlers-Danlos syndrome, osteogenesis imperfecta, and Marfan syndrome.

People with Ehlers-Danlos syndrome have abnormally fragile connective tissue, resulting in joint problems and weakened internal organs. In osteogenesis imperfecta, the body doesn't produce enough collagen or it produces poor quality collagen, leading to fragile bones that break easily. People with Marfan syndrome have connective tissue that isn't as stiff as it should be, causing skeletal abnormalities and weakened blood vessels.

By studying the connective tissue in echinoderms such as the sea cucumber, researchers are moving closer to successfully treating debilitating illnesses that prevent people from having freedom of movement in their joints due to connective tissue diseases.

WRITING in Biology

Journal Visit biologygmh.com to learn more about scientific research involving echinoderms. Create a biologist's research journal describing his or her work with an echinoderm. The journal should include thorough descriptions, charts, graphs, and sketches of echinoderms.

BIOLAB

INTERNET: HOW DO ECHINODERMS SURVIVE WITHOUT A HEAD, EYES, OR A BRAIN?

Background: Echinoderms have evolved unlike any other animals on Earth. Lacking eyes and a brain, they also have no heart, and pump seawater through their bodies rather than blood. Some echinoderms can change their endoskeletons from rock hard to nearly liquid within seconds. Some can break off an arm to distract a predator. Sound unusual? Not for echinoderms.

Question: *How do echinoderms survive in the competitive marine environment?*

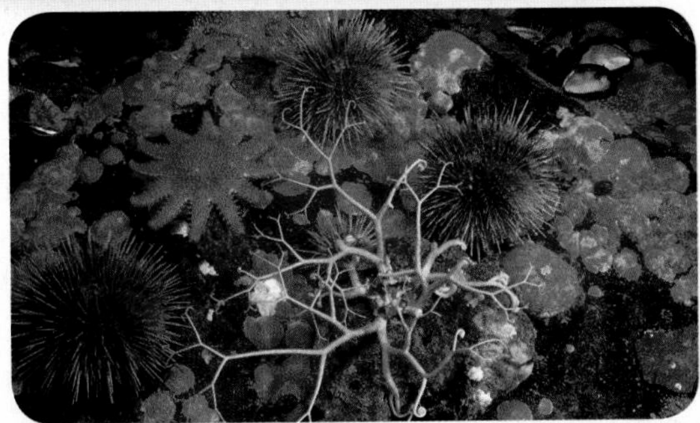

These sea stars, basket stars, and sea urchins are from the Gulf of Maine.

Materials

internet access
echinoderm reference book
field journal

Procedure

1. Read and complete the lab safety form.
2. Design and construct a data table for recording the species; physical characteristics; food sources/strategies for obtaining food; predators; defense strategies; reproduction and development; and other interesting facts about six animals.
3. Choose one species from each of the six major classes of echinoderms to study. List the species in your data table.
4. Research the species you chose and fill in information in your data table. Observe the echinoderms in their natural habitat by visiting a local zoo or aquarium. If you cannot observe the animals in their natural habitats, obtain information about the echinoderms from a reference book or visit biologygmh.com.
5. Record your observations in your field journal. Transfer the information to your data table.
6. Post your results at biologygmh.com. Use data posted by other students to complete missing portions of your table.

Analyze and Conclude

1. **Describe** some basic physical characteristics shared by echinoderms.
2. **Compare** sexual and asexual reproductive strategies used by echinoderm species.
3. **Think Critically** Echinoderm larvae and mature echinoderms differ in several important ways. Describe the differences, and infer the advantages they provide.
4. **Interpret Data** What are the major food sources of the echinoderms you studied?
5. **Draw Conclusions** Are echinoderms well-adapted to survive in the marine environment? Justify your answer.
6. **Error Analysis** Describe advantages and disadvantages of obtaining information about echinoderms from the Internet.

WRITING in Biology

Resource Book Use the data you gathered to create a fact sheet including photos and interesting information about each echinoderm you studied. Combine your fact sheets with those developed by other students to create an echinoderm resource book for your school's media center. To find out more about echinoderms, visit Biolabs at biologygmh.com.

CHAPTER 27

Study Guide

Download quizzes, key terms, and flash cards from **biologygmh.com.**

FOLDABLES **Analyze** Use what you have learned in this chapter to debate the placement of invertebrate chordates in the phylum Chordata.

Vocabulary	Key Concepts
Section 27.1 Echinoderm Characteristics	
• ampulla (p. 795) • madreporite (p. 795) • pedicellaria (p. 793) • tube foot (p. 795) • water-vascular system (p. 795)	MAIN Idea Echinoderms are marine animals with spiny endoskeletons, water-vascular systems, and tube feet; they have radial symmetry as adults. • Adult echinoderms can be identified by four main structural features. • Larval echinoderms have features that link them to relatives that evolved after them. • Echinoderms have a water-vascular system and tube feet. • Echinoderms have a variety of adaptations for feeding and movement. • There are six major classes of living echinoderms. 
Section 27.2 Invertebrate Chordates	
• chordate (p. 803) • dorsal tubular nerve cord (p. 803) • invertebrate chordate (p. 803) • notochord (p. 803) • pharyngeal pouches (p. 804) • postanal tail (p. 803)	MAIN Idea Invertebrate chordates have features linking them to vertebrate chordates. • Chordates have four main features that make them different from animals that are not chordates. • Invertebrate chordates have all the features of chordates, except they do not have the main feature of vertebrate chordates. • The notochord is an adaptation that enabled animals to move in ways they had never moved before. • Lancelets are fish-shaped invertebrate chordates that, as adults, have all the main features of chordates. • Tunicates are sac-shaped invertebrate chordates that have chordate features as larvae. 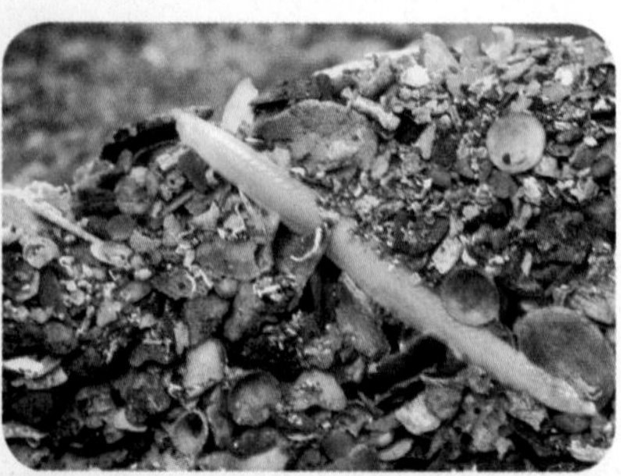

Biology Online **Vocabulary PuzzleMaker** **biologygmh.com**

CHAPTER 27 Assessment

Section 27.1

Vocabulary Review

Distinguish between the terms in each of the following pairs.

1. tube foot, ampulla

2. madreporite, water-vascular system

Understand Key Concepts

3. Which is not an echinoderm?

A. 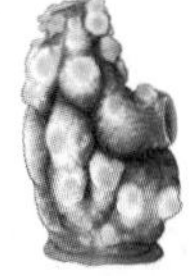

C. 

B. 

D. 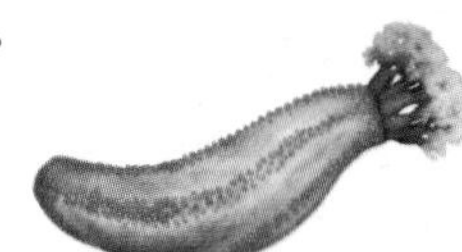

A. A **C.** C
B. B **D.** D

4. Which echinoderm is sessile for part of its life?
A. sea cucumber **C.** brittle star
B. sea lily **D.** sea urchin

5. What three functions do tube feet perform?
A. reproduction, feeding, respiration
B. feeding, respiration, neural control
C. feeding, respiration, movement
D. development, reproduction, respiration

6. Which is not associated with deuterostomes?
A. a pattern of development
B. mouth develops from somewhere on the gastrula away from the opening
C. echinoderms
D. arthropods

7. Which are involved in protecting an echinoderm?
A. endoskeleton, pedicellariae, spines
B. madreporite, tentacles, endoskeleton
C. water-vascular system, ampulla, pedicellariae
D. exoskeleton, pedicellariae, spines

8. What is the main difference between echinoderm larvae and adults?
A. Larvae are protostomes and adults are deuterostomes.
B. Larvae are deuterostomes and adults are protostomes.
C. Larvae have bilateral symmetry and adults have radial symmetry.
D. Larvae have radial symmetry and adults have bilateral symmetry.

9. Which group of echinoderms has respiratory trees with many branches?
A. sea cucumbers
B. sea stars
C. sea lilies and feather stars
D. sea urchins and sand dollars

Constructed Response

Use the diagram below to answer questions 10 and 11.

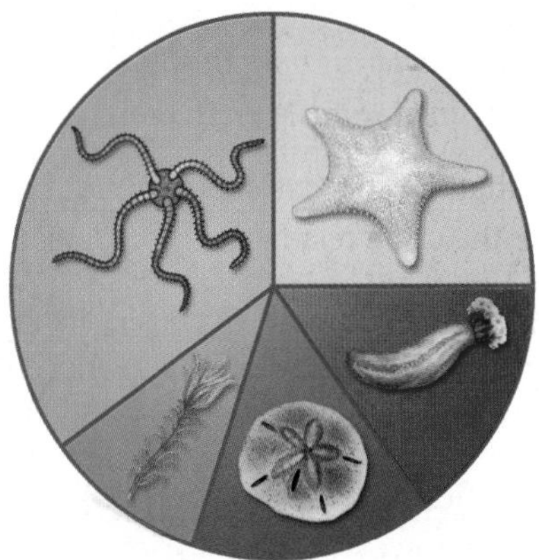

10. **Short Answer** Examine the circle graph and estimate the percentage of echinoderms that are sea cucumbers.

11. **Open Ended** Examine the circle graph and explain why class Concentricycloidea does not appear with the other classes of living echinoderms.

12. **Open Ended** Scientists have discovered a fossil that has the following characteristics: an endoskeleton similar to that of echinoderms, a tail-like structure with an anus at the end of the tail, a structure that might be a gill, and symmetry similar to echinoderms. How might scientists explain this animal in terms of echinoderm classification?

13. **Open Ended** Tidal animals suffer when water and air temperatures rise beyond the limits of tolerance of the animals. The temperature of sea stars remain about 18 degrees cooler than those of the surrounding mussels on a hot day. Make a hypothesis about why sea stars have a lower body temperature.

Think Critically

14. **Observe and Infer** You are walking on the beach and find an animal that has many feathery arms and tube feet. What kind of animal might this be?

15. **Hypothesize** Some sea urchins seem to have relatively long lifespans. Make a hypothesis about why they live so long.

Section 27.2

Vocabulary Review

Using the vocabulary terms from the Study Guide page, replace the underlined words with the correct term.

16. Animals that are chordates, but do not have backbones are the close relatives of chordates.

17. Located just below the nerve cord is a structure in chordates that enables invertebrate chordates to swim by moving their tails back and forth.

18. The connections between the muscular tube that links the mouth cavity and the esophagus develop slits and are used for filter feeding in some invertebrate chordates.

Understand Key Concepts

19. Chordates have which features at some time in their lives?
    - **A.** water-vascular system, notochord, pharyngeal pouches, postanal tail
    - **B.** tunic, pharyngeal pouches, dorsal tubular nerve cord, postanal tail
    - **C.** tube feet, notochord, pharyngeal pouches, postanal tail
    - **D.** dorsal tubular nerve cord, notochord, pharyngeal pouches, postanal tail

20. Which is the main function of a postanal tail?
    - **A.** circulation
    - **B.** digestion
    - **C.** flexibility
    - **D.** locomotion

Use the diagram below to answer questions 21 and 22.

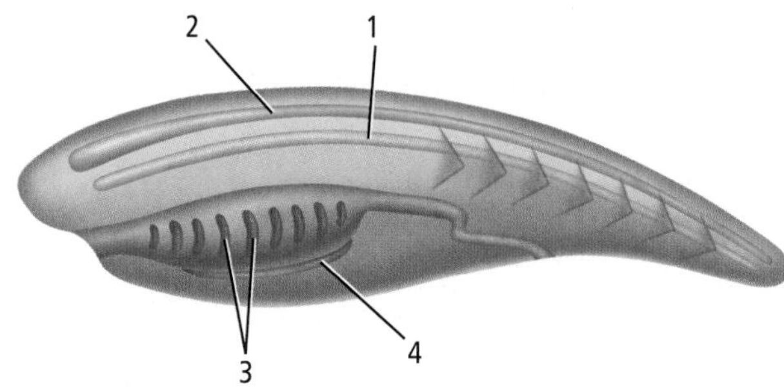

21. Fishlike swimming is made possible by which labeled structure above?
    - **A.** 1
    - **B.** 2
    - **C.** 3
    - **D.** 4

22. Which structure develops into the brain and spinal cord in most chordates?
    - **A.** 1
    - **B.** 2
    - **C.** 3
    - **D.** 4

23. Which describes adult sea squirts?
    - **A.** They are bilaterally symmetrical.
    - **B.** They have the same adult features as lancelets.
    - **C.** As adults, they have only two chordate features.
    - **D.** They are actively swimming predators.

24. In invertebrate chordates, what does the endostyle secrete?
    - **A.** proteins similar to thyroid hormone
    - **B.** mucus
    - **C.** the notochord
    - **D.** pharyngeal pouches

25. The phylogeny of echinoderms indicates that echinoderms are related to chordates because they both have which feature?
    - **A.** pharyngeal pouches
    - **B.** deuterostome development
    - **C.** protostome development
    - **D.** pseudocoeloms

Biology Online Chapter Test biologygmh.com

26. Which structure might be an early form of the thyroid gland?
 A. dorsal tubular nerve cord
 B. endostyle
 C. notochord
 D. pharyngeal pouches

27. Which chordate feature enabled large animals to develop?
 A. dorsal tubular nerve cord
 B. notochord
 C. pharyngeal pouches
 D. postanal tail

Constructed Response

28. **Open Ended** Infer why there are no freshwater invertebrate chordates.

29. **Open Ended** What would happen if all lancelets disappeared?

Use the diagram below to answer questions 30 and 31.

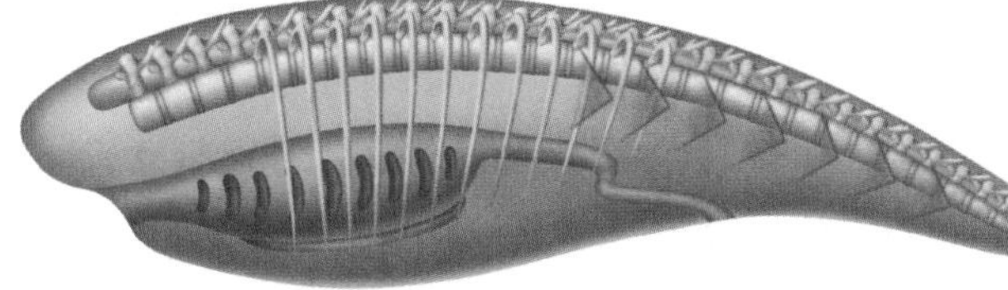

30. **Short Answer** Examine the diagram and explain why this animal could not be an invertebrate chordate.

31. **Short Answer** What features does this animal share with invertebrate chordates?

Think Critically

32. **Analyze** How do the larvae of organisms help scientists classify and determine the phylogeny of animals?

33. **Use the Internet** Make a visual report of the newest information, both molecular and fossil evidence, gathered by scientists on the origins of chordates.

Additional Assessment

34. *WRITING in* **Biology** **Create** a poem that describes your favorite echinoderm. Make sure you point out the actual features of the echinoderm.

DBQ Document-Based Questions

Study the illustration of the progression of development of arms in a specific sea star.

Diagram based on examples from: Sumrall, Colin D., 2005. Unpublished research on the growth stages of *Neoisorophusella lanei*. The University of Tennessee. http://web.eps.utk.edu/Faculty/sumrall/research2.htm

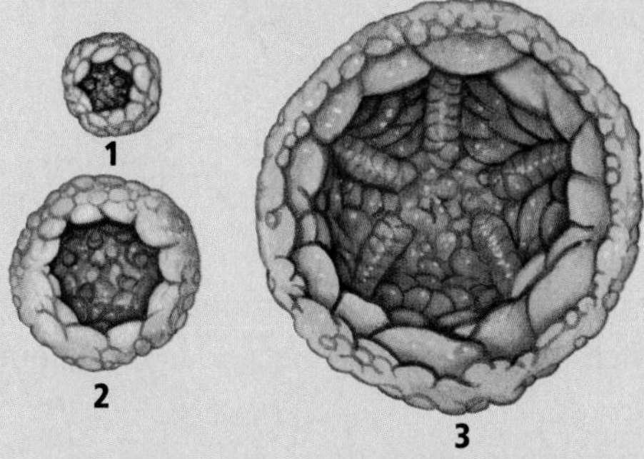

35. What kind of symmetry is shown in the diagram labeled 1?

36. Infer how additional arms might develop.

37. How does the number of arms in diagram 3 reflect the characteristics of all echinoderms?

Cumulative Review

38. Compare Neanderthals and modern humans. **(Chapter 16)**

39. Compare and contrast the animal-like, plantlike, and funguslike protists. **(Chapter 19)**

40. Prepare a list of vocabulary words that describe general fungal structures, and sketch illustrations of each one. **(Chapter 20)**

41. Name three hormones and the effects they can have on plants. **(Chapter 22)**

42. Sequence the steps involved in the production of the pollen grain and egg in anthophytes. **(Chapter 23)**

Standardized Test Practice

Cumulative

Multiple Choice

1. In which structure of a flowering plant do eggs develop?
   A. anther
   B. ovule
   C. seed
   D. stigma

Use the diagram below to answer question 2.

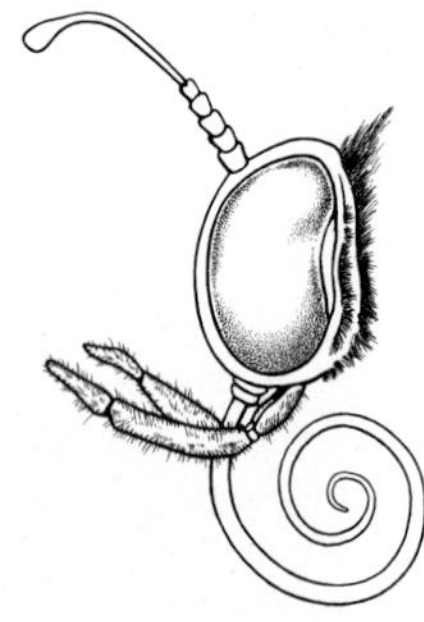

2. Arthropods have specialized mouthparts for feeding. For which type of feeding method is this mouthpart specialized?
   A. getting nectar from flowers
   B. sponging liquids from a surface
   C. sucking blood from a host
   D. tearing and shredding leaves

3. Which statement about a group of invertebrates is correct?
   A. Cnidarians have collar cells.
   B. Flatworms have flame cells.
   C. Flatworms have nematocysts.
   D. Sponges have a nervous system.

4. Echinoderms have which characteristic that is an evolutionary connection to vertebrates?
   A. bilateral symmetry as adults
   B. free-swimming larvae
   C. deuterostome development
   D. radial symmetry as larvae

5. Which special adaptation would be essential for an insect that swims in water?
   A. compound eyes
   B. modified legs
   C. sticky foot pads
   D. sharp mouth parts

Use the diagram below to answer questions 6 and 7.

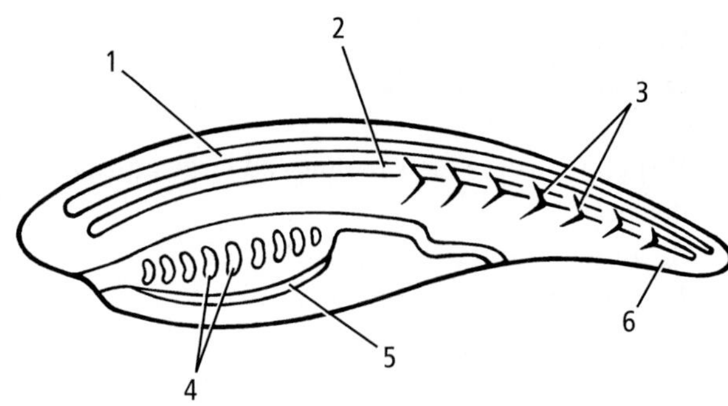

6. Which structure is replaced by bone or cartilage in vertebrate chordates?
   A. 1
   B. 2
   C. 4
   D. 5

7. Which structure is a bundle of nerves protected by fluid?
   A. 1
   B. 3
   C. 5
   D. 6

8. What kind of body organization or body structure first appeared with the evolution of flatworms?
   A. bilateral symmetry
   B. coelomic cavity
   C. nervous system
   D. radial symmetry

9. Suppose a cell from the frond of a fern contains 24 chromosomes. How many chromosomes would you expect to find in the spores?
   A. 6
   B. 12
   C. 24
   D. 48

Short Answer

10. Use what you know about the body structure of a sponge to explain how it obtains food.

11. Sea stars are echinoderms that feed on oysters. Justify why oyster farmers should not cut up sea stars and toss the parts back into the water.

12. Evaluate the defense adaptations of the two groups of invertebrate chordates.

13. Contrast the main characteristics of echinoderms with the characteristics of the organisms in another phylum that you already know.

Use the diagram below to answer questions 14 and 15.

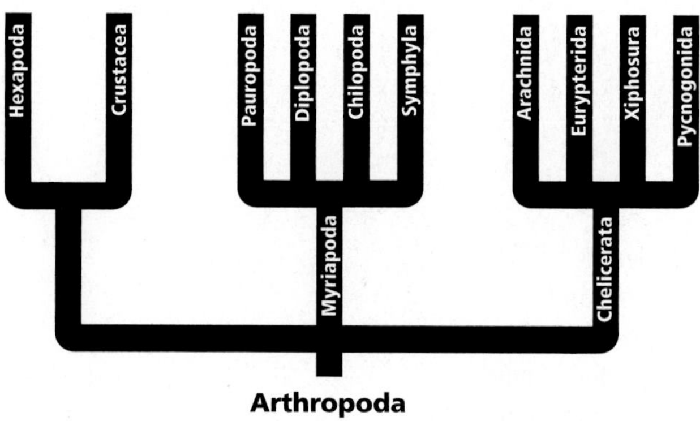

14. Write a hypothesis about why horseshoe crabs (in class Xiphosura) are more closely related to spiders than to regular crabs and lobsters.

15. Write a hypothesis about where trilobites would fit into this phylogenetic tree.

Extended Response

16. Explain how echinoderms and annelids are similar, and how they are different.

17. In animals, how are mitosis and meiosis different?

18. Evaluate the idea that it was not a large evolutionary jump for aquatic arthropods to move onto land.

19. Suppose that one crow in an area's population is born with longer claws on its feet than other crows in the same population. According to Darwin's theory of natural selection, under what circumstances would this trait become common in the area's crow population?

Essay Question

In the past, many horror movies have featured giant arthropods attacking major cities. These giant arthropods have included ants, grasshoppers, crabs, and spiders. Actually, the largest living insects are not very big. The longest insect, a walking stick, is about 40 cm long. Some marine arthropods grow larger. The largest arthropod is the Japanese spider crab that can grow up to 4 m wide. Some fossil marine arthropods are even larger. However, none of these are nearly as large as the size of the giant arthropod villains in the movies.

Using the information in the paragraph above, answer the following question in essay format.

20. Write an essay about why real-life arthropods cannot become as large as the giant arthropods shown in horror movies.

NEED EXTRA HELP?																				
If You Missed Question . . .	1	2	3	4	5	6	7	8	9	10	11	12	13	14	15	16	17	18	19	20
Review Section . . .	23.3	26.1	25.1	27.1	26.3	27.2	27.2	25.1	21.3	24.3	27.1	27.2	27.2	26.2	26.3	24.2	10.1	26.2	15.1	26.1

UNIT 8 Vertebrates

Chapter 28
Fishes and Amphibians
BIG Idea Fishes have adaptations for living in aquatic environments. Most amphibians have adaptations for living part of their lives on land.

Chapter 29
Reptiles and Birds
BIG Idea Reptile and bird adaptations enable them to live and reproduce successfully in terrestrial habitats.

Chapter 30
Mammals
BIG Idea Mammals have evolved to have a variety of adaptations for maintaining homeostasis and living in a variety of habitats.

Chapter 31
Animal Behavior
BIG Idea Many animal behaviors are influenced by both genetics and environmental experiences.

CAREERS IN BIOLOGY

Veterinarian

Veterinarians are medical specialists trained to prevent, diagnose, and treat medical conditions in domestic, wildlife, zoo, and laboratory animals. Some veterinarians conduct research to expand knowledge of a particular species, much like this giant panda bear researcher is doing.

WRITING in Biology Visit biologygmh.com to learn more about veterinarians. Write a paragraph to compare and contrast the duties of veterinarians in private practices and in public institutions.

Biology Online

To read more about veterinarians in action, visit biologygmh.com.

CHAPTER 28 Fishes and Amphibians

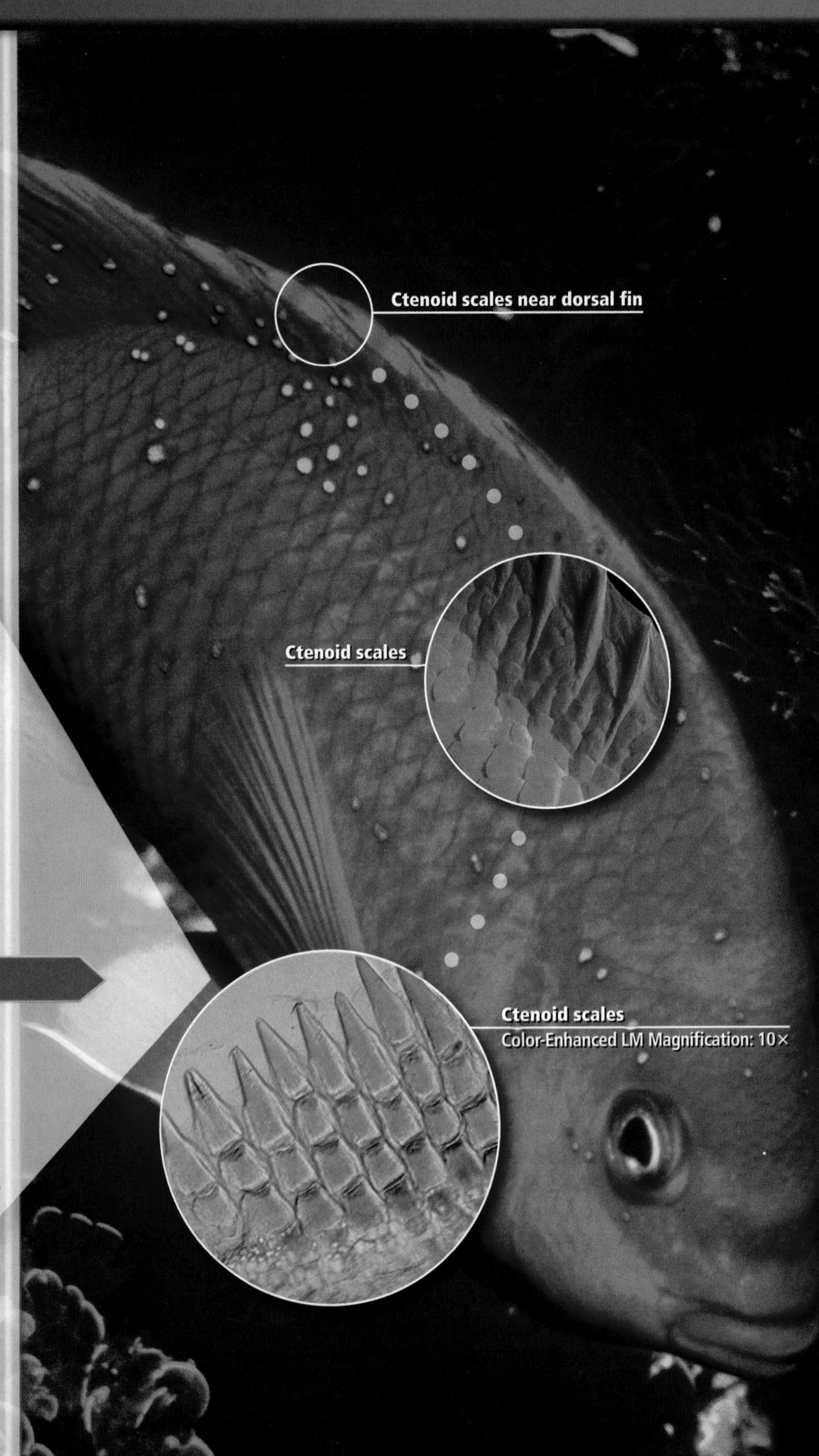

BIG Idea Fishes have adaptations for living in aquatic environments. Most amphibians have adapted to living part of their lives on land.

Section 1
Fishes
MAIN Idea Fishes are vertebrates that have characteristics allowing them to live and reproduce in water.

Section 2
Diversity of Today's Fishes
MAIN Idea Scientists classify fishes into three groups based on body structure.

Section 3
Amphibians
MAIN Idea Most amphibians begin life as aquatic organisms then live on land as adults.

BioFacts

- Fish scales have growth rings similar to those of a tree trunk.
- Some types of scales contain enamel, the same material that makes up teeth.
- Fish scales do not have color. The apparent color comes from the skin just beneath the scales.

Start-Up Activities

LAUNCH Lab

What are the characteristics of fishes in different groups?

Fishes are classified into three main groups—jawless fishes, cartilaginous fishes, and bony fishes. They are classified based on external and internal characteristics. In this lab, you will compare the external characteristics of fishes in the three groups.

Procedure

1. Read and complete the lab safety form.
2. Examine **photos** of representatives from each of the three groups of fishes. Look at features such as skin/scales, fin position, fin shape, eyes, mouth shape and teeth, body shape, and tail shape.
3. Construct a table and record information about the external characteristics of the different groups of fishes.

Analysis

1. **Summarize** What are the main external differences between these groups of fishes?
2. **Infer** Why is it important to examine and compare the internal structures and characteristics of organisms when trying to classify them?

Visit biologygmh.com to:

- study the entire chapter online
- explore Concepts in Motion, the Interactive Table, Microscopy Links, Virtual Labs, and links to virtual dissections
- access Web links for more information, projects, and activities
- review content online with the Interactive Tutor and take Self-Check Quizzes

Fishes and Amphibians Make the following Foldable to help you identify the characteristics of fishes, early tetrapods, and amphibians.

STEP 1 Lay two sheets of paper about 1.5 cm apart vertically. Keep the edges level.

STEP 2 Fold up the bottom edges of the paper to form four equal tabs.

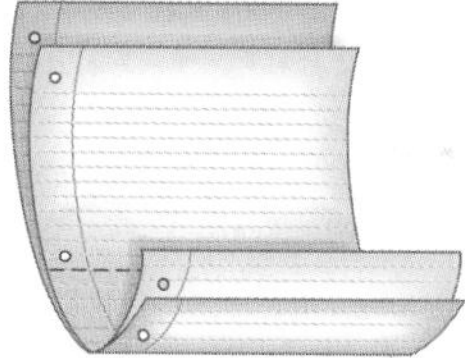

STEP 3 Crease well to hold the tabs in place. Staple along the fold. Label each tab as shown.

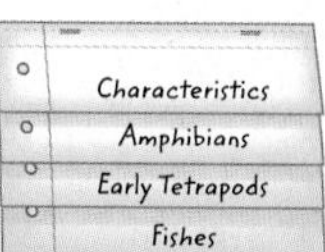

FOLDABLES **Use this Foldable with Section 28.3.** As you study the section, record and sketch what you learn about the characteristics of each group.

Section 28.1

Reading Preview

Section Objectives

- **Identify** the features of vertebrates that make them different from invertebrates.
- **Describe** the characteristics that most fishes have in common.
- **Summarize** how the characteristics of fishes are adapted to aquatic life.

Review Vocabulary

notochord: a flexible, rodlike structure that extends the length of the body

New Vocabulary

cartilage
neural crest
fin
scale
operculum
atrium
ventricle
nephron
lateral line system
spawning
swim bladder

Fishes

MAIN Idea Fishes are vertebrates that have characteristics allowing them to live and reproduce in water.

Real-World Reading Link You might have seen an aquarium full of colorful fishes similar to the one in the photo at the beginning of the chapter. What adaptations do fishes have for living in water? Fishes have unique characteristics that allow them to live and reproduce in water.

Characteristics of Vertebrates

Until now you have been studying sponges, worms, and sea stars, which are all invertebrates. Recall that the four main characteristics of chordates are that they have a dorsal nerve cord, a notochord, pharyngeal pouches, and a postanal tail. Animals belonging to subphylum Vertebrata are called vertebrates. Vertebrates have a vertebral column and specialized cells that develop from the nerve cord. The vertebral column, also called a spinal column, is the hallmark feature of vertebrates. Classes of vertebrates include fishes, amphibians, reptiles, birds, and mammals.

Vertebral column In most vertebrates, the notochord is replaced by a vertebral column that surrounds and protects the dorsal nerve cord. The replacement of the notochord happens during embryonic development. Cartilage or bone is the building material of most vertebrate endoskeletons. **Cartilage** (KAR tuh lihj) is a tough, flexible material making up the skeletons or parts of skeletons of vertebrates.

The vertebral columns, shown in **Figure 28.1,** are important structures in terms of the evolution of animals. The vertebral column functions as a strong, flexible rod that muscles can pull against during swimming or running. Separate vertebrae enhance an animal's ability to move quickly and easily. Bones enable forceful contraction of muscles, improving the strength of an animal.

Figure 28.1 The vertebral column is present in most vertebrates, including bony fishes and reptiles as shown by the art.

Triggerfish

Sidewinder

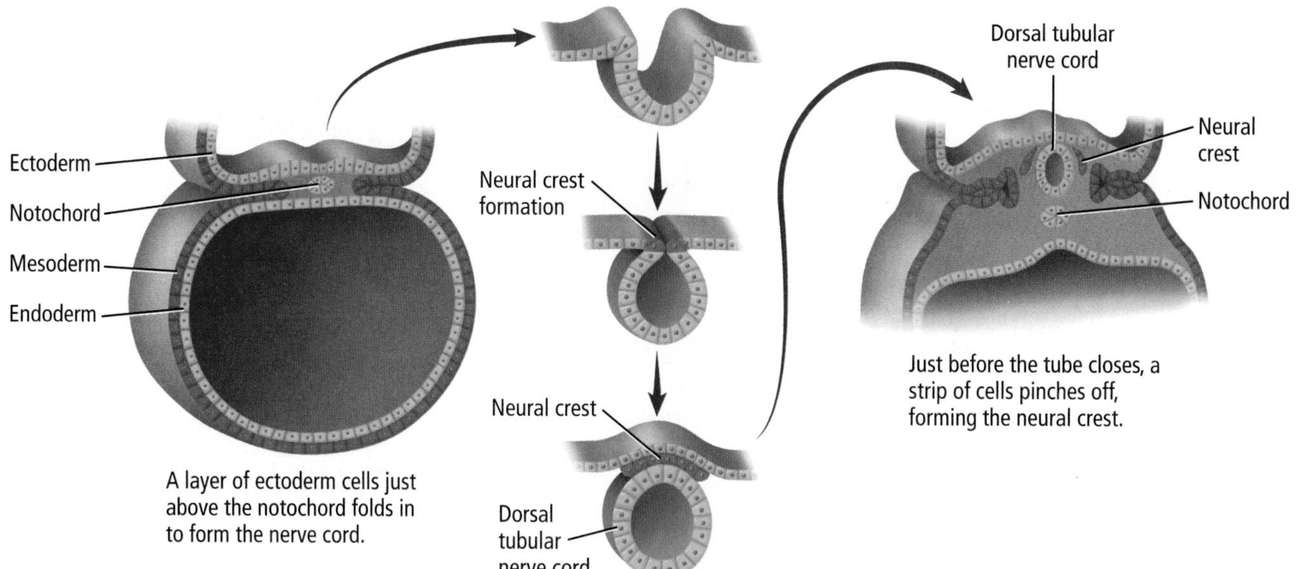

Figure 28.2 The neural crest of vertebrates develops from the ectoderm of the embryo.

Personal Tutor

To learn about neural crest formation, visit biologygmh.com.

Neural crest As the nerve cord forms during embryonic development in vertebrates, another important process occurs: a neural (NOOR ul) crest forms. A **neural crest** is a group of cells that develop from the nerve cord in vertebrates. The process of neural crest formation is shown in **Figure 28.2.** Even though this group of cells is small, it is significant in the development of vertebrates because many important vertebrate features develop from the neural crest. These features include portions of the brain and skull, certain sense organs, parts of pharyngeal pouches, some nerve fibers, insulation for nerve fibers, and certain gland cells.

Other features that are characteristic of vertebrates include internal organs, such as kidneys and a liver. A heart and closed circulatory system also are features of all vertebrates.

Reading Check **Explain** why the neural crest is an important vertebrate feature.

Characteristics of Fishes

Fishes live in most aquatic habitats on Earth—seas, lakes, ponds, streams, and marshes. Some fishes live in complete darkness at the bottom of the deep ocean. Others live in the freezing waters of the polar regions and have special proteins in their blood to keep the blood from freezing. There are about 24,600 species of living fishes, more than all other vertebrates combined. They range in size from whale sharks that can be 18 m long to tiny cichlids that are the size of a human fingernail.

The features of fishes provided the structural basis for the development of land animals during the course of evolution. Important characteristics of fishes include the development of jaws and, in some fishes, lungs. As shown in the evolutionary tree in **Figure 28.3,** there are three groups of fishes, all of which are vertebrates. Although fishes' body shapes and structures vary a great deal, they all have several characteristics in common. Most fishes have vertebral columns, jaws, paired fins, scales, gills, and single-loop blood circulation, and they are not able to synthesize certain amino acids.

Figure 28.3 The branches of the different groups of fishes are highlighted in this evolutionary tree.

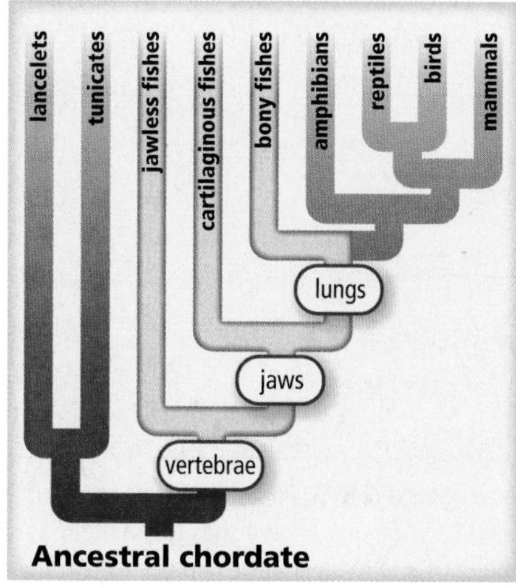

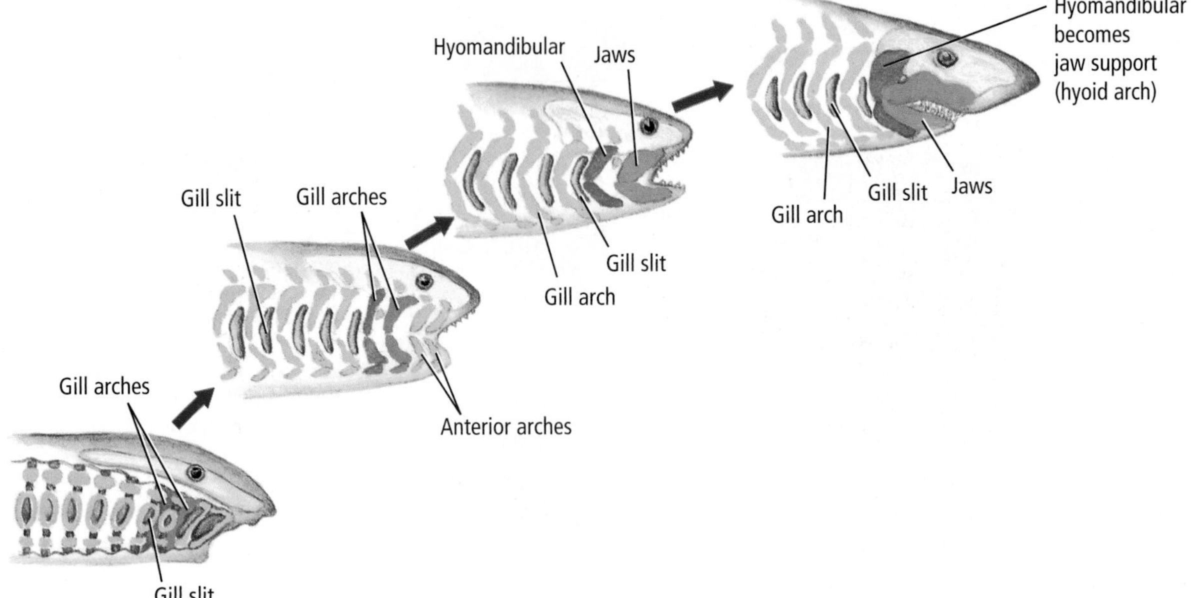

■ **Figure 28.4** Anterior gill arches evolved into jaws in ancient fishes.

Jaws Most fishes have jaws. The evolution of jaws is shown in **Figure 28.4,** where you can see that the anterior gill arches evolved to form jaws in ancient fishes. The development of jaws allowed ancient fishes to prey on a larger range of animals. This included being able to prey on fishes that were larger in size and more active. Fishes grasp prey with their teeth and quickly crush them using powerful jaw muscles. Jaws also allow for a biting defense against predators.

Reading Check Describe why the evolution of jaws in fishes was important.

Paired fins At the same time jaws were evolving, paired fins were also appearing in fishes. A **fin** is a paddle-shaped structure on a fish or other aquatic animal that is used for balance, steering, and propulsion. Pelvic fins and pectoral fins, like the ones shown in **Figure 28.5,** give fishes more stability. Most fishes have paired fins. Paired fins reduce the chance of rolling to the side and allow for better steering during swimming.

While fishes in ancient seas moved with precision and skill, they also were able to use their jaws in new ways. Both jaws and paired fins contributed to the evolution of a predatory way of life for some fishes and also enabled them to live in new habitats and produce more offspring.

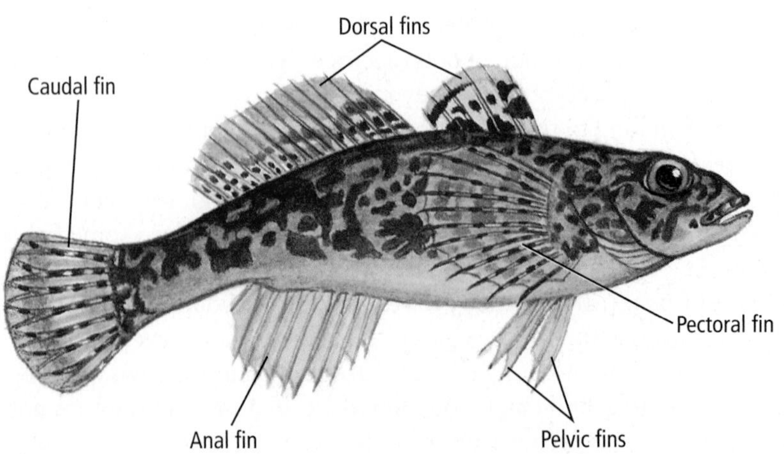

■ **Figure 28.5** Paired fins, such as pelvic and pectoral fins, allow fishes to maintain balance and to steer in the water.

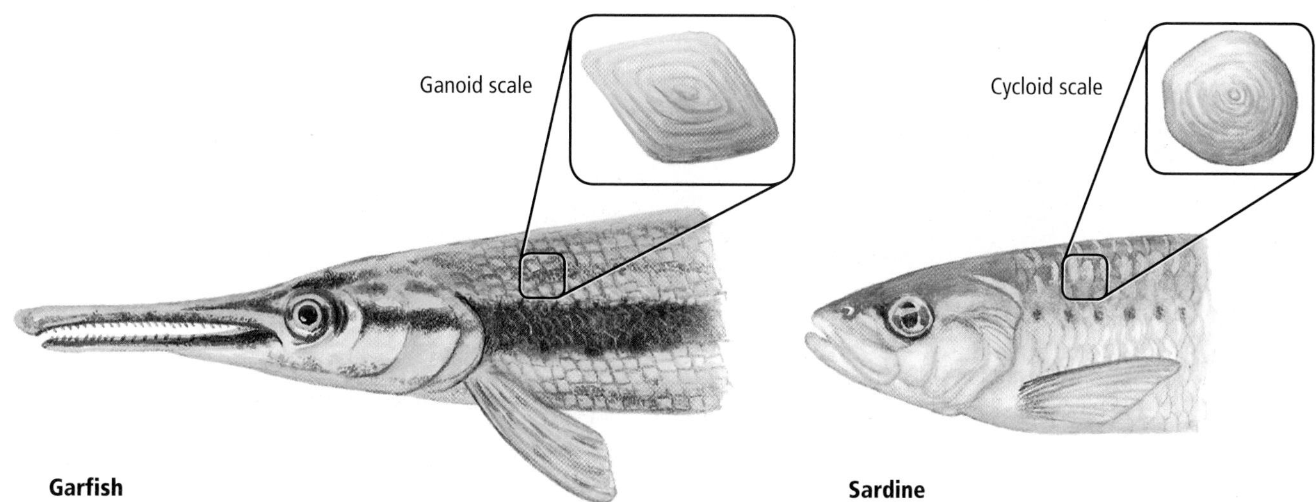

■ **Figure 28.6** Two types of fishes' scales are shown here—ganoid scales and cycloid scales.
Describe ***the difference in the appearance of cycloid and ganoid scales.***

Scales Most fishes have at least one of four different types of scales. A **scale** is small, flat, platelike structure near the surface of the skin. There are four types of fish scales. Ctenoid (TEH noyd) scales, shown in the photos at the beginning of the chapter, and cycloid (SY kloyd) scales are made of bone and skin and are thin and flexible. Cycloid scales are shown in **Figure 28.6.** Placoid (PLA koyd) scales, which can be seen in **Figure 28.15** in Section 28.2, are made of toothlike materials and are rough and heavy. Ganoid (GAN oyd) scales, shown in **Figure 28.6,** are diamond-shaped and made of both enamel and bone.

Reading Check **Infer** why different fishes have different kinds of scales.

MiniLab 28.1

Observe a Fish

What inferences can you make about characteristics of fishes through observation? In this lab, you will observe a fish in its aquatic environment.

Procedure

1. Read and complete the lab safety form.
2. Observe the **fish(es)** in an **aquarium.**
3. Make a diagram of a fish and label the following applicable structures: dorsal fin, caudal fin, anal fin, pectoral fins, pelvic fins, scales, mouth, eye, and gill covering.
4. Observe how the fish moves through the water. Illustrate how the fish moves its body and its fins as it moves forward in the water.

Analysis

1. **Infer** A fish's body is divided into three regions: head, trunk, and tail. Label these regions on your diagram of the fish you observed.
2. **Apply** Suppose a fish lost one of its pectoral fins when fighting off a predator. How might this affect its ability to move through the water?

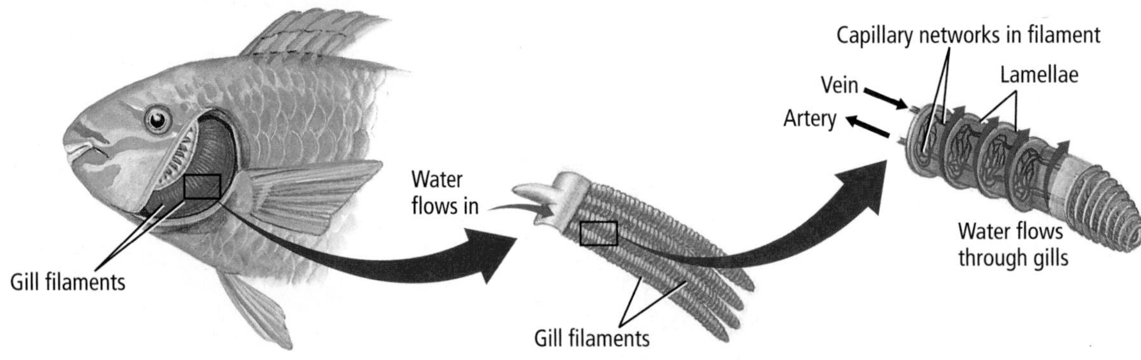

■ **Figure 28.7** The lamellae in a fish's gills have many blood vessels.
Infer ***Why are the gills of fishes made up of very thin tissue?***

Gills Another adaptation that allows fishes to live in aquatic environments is their ability to get oxygen from the water. Fishes get oxygen when water that enters their mouths flows across their gills, where oxygen from the water diffuses into the blood. Gills are composed of thin filaments that are covered with highly-folded, platelike lamellae (luh MEH lee). The gill structure of most fishes is shown in **Figure 28.7.** The lamellae have many blood vessels that can take in oxygen and give off carbon dioxide.

The flow of blood in the gill is opposite to the flow of water over the gill surface. This countercurrent flow is an efficient mechanism by which oxygen can be removed from water. Up to 85 percent of oxygen dissolved in water is removed as water flows over the gills in one direction and blood in the other. Some fishes have an **operculum** (oh PUR kyuh lum), a movable flap that covers the gills and protects them. An operculum also aids in pumping water coming in the mouth and over the gills. Some fishes, such as lungfishes, can live out of water for short times by using structures resembling lungs. Eels can breathe through their moist skin when they are out of water.

VOCABULARY
WORD ORIGIN
Atrium
from the Latin word *atrium*, meaning *central hallway.*

Circulation Vertebrates have a closed circulatory system in which the heart pumps blood through blood vessels. The circulatory system of fishes is shown in **Figure 28.8.** In most fishes, the blood is passed through the heart in a one-way loop. From the heart, the blood goes to the gills, and then through the body, delivering oxygenated blood to tissues. The blood then returns to the heart. From the heart, blood is pumped back to the gills and then to the body again. Because this system is a complete and uninterrupted circuit, it is called a single-loop circulatory system.

In most fishes, the heart consists of two main chambers that are analogous to parts of your own heart—an atrium and a ventricle. The **atrium** is the chamber of the heart that receives blood from the body. From there, blood is passed to the **ventricle,** a chamber of the heart that pumps blood from the heart to the gills. Once the blood passes over the gills, it travels to the rest of the body.

■ **Figure 28.8** A fish's heart pumps blood through a closed circulatory system.

Concepts In Motion
Interactive Figure To see an animation of how blood circulates through a fish, visit biologygmh.com.

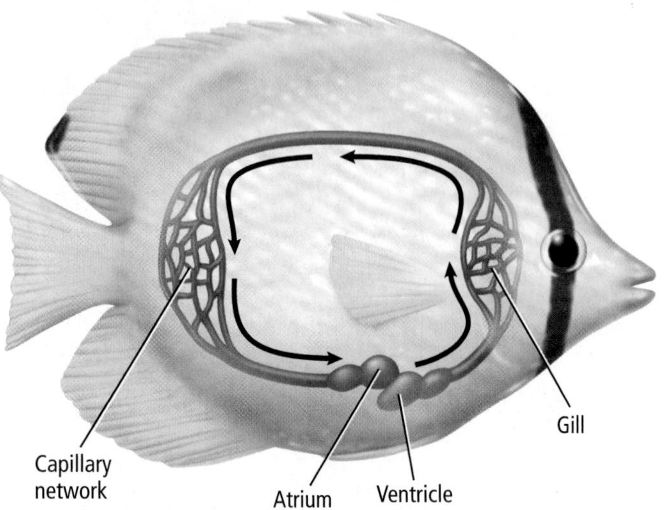

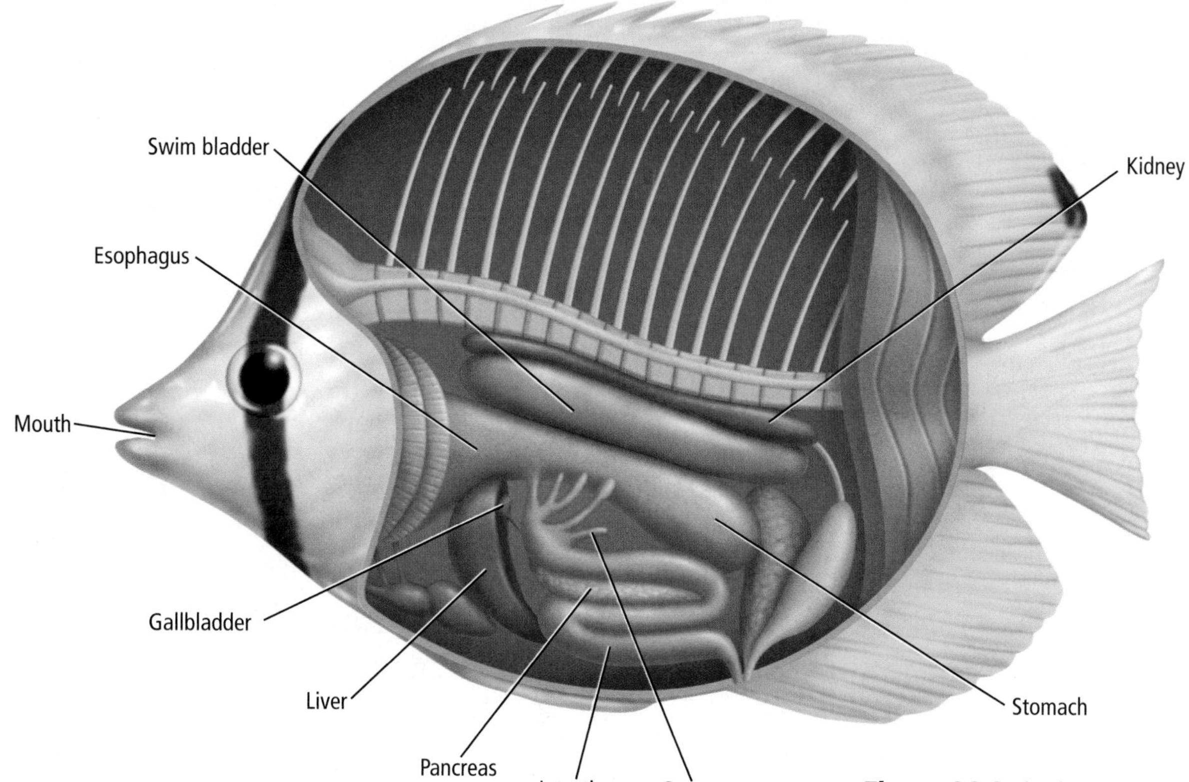

■ **Figure 28.9** The digestive organs of a fish are similar to those of other vertebrates.
List ***the structures food passes through as it is being digested.***

Feeding and digestion Ancient fishes most likely were filter feeders or scavengers, sucking up organic debris on the ocean floor. With the evolution of jaws, fishes became efficient predators, and the diets of fishes changed dramatically. The digestive tract of fishes, illustrated in **Figure 28.9,** consists of organs similar to those of other vertebrates.

Most fishes swallow their food whole, passing it through a tube called the esophagus (ih SAH fuh gus) to the stomach, where digestion begins. Food then passes to the intestine, where most digestion occurs. Some fishes have pyloric (pi LOR ihk) ceca (SEE kuh) (singular, cecum), which are small pouches at the junction of the stomach and the intestine that secrete enzymes for digestion and absorb nutrients into the bloodstream. The liver, pancreas, and gallbladder add digestive juices that complete digestion.

Fishes are described not only by structures and their functions, but also by one important thing that they cannot do. They are not able to synthesize certain amino acids. Therefore, not only fishes, but all the vertebrates that evolved from them must get these same amino acids from the foods they eat.

Excretion Cellular wastes are filtered from fishes' blood by organs called kidneys. The main functional unit of the kidney is the nephron. A **nephron** is a filtering unit within the kidney that helps to maintain the salt and water balance of the body and to remove cellular waste products from the blood. Some cellular wastes are excreted by the gills.

Connection to Chemistry The bodies of freshwater fishes take in water by osmosis because the surrounding water is hypotonic—the water contains more water molecules than the fishes' tissues. The opposite occurs in saltwater bony fishes. Because the surrounding water is hypertonic—the water contains fewer water molecules than their tissues—their bodies tend to lose water. Kidneys, gills, and other internal mechanisms adjust the water and salt balance in the bodies of freshwater and saltwater fishes.

CAREERS IN BIOLOGY

Ichthyologist Researching information about the behavior, ecology, and anatomy and physiology of fishes, both in the field and in the lab, is just one of the jobs an ichthyologist (ihk thee AHL uh jist) might have. Ichthyologists also manage aquariums, organize museum collections, teach at universities, and work to conserve fish populations. For more information on biology careers, visit biologygmh.com.

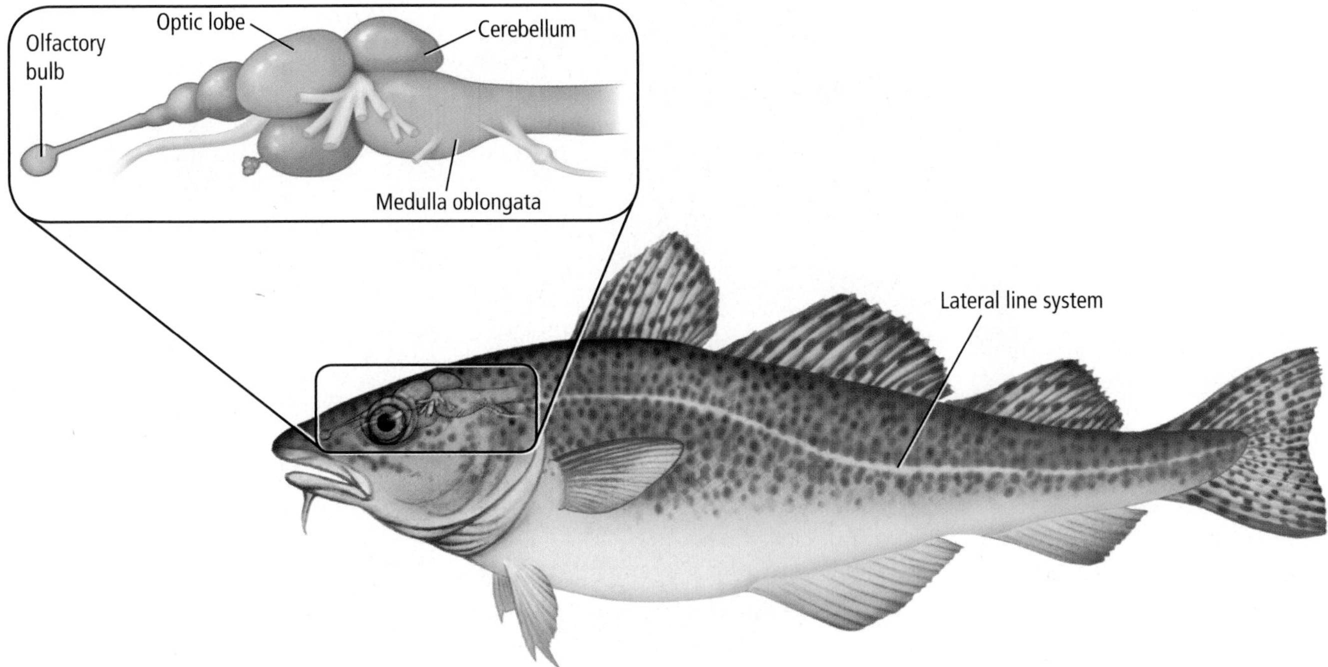

■ **Figure 28.10** Fishes have a brain that enables them to carry out their life functions.
Infer ***In what way would the brain of a fish that lived passively on the bottom of a pond feeding on organic debris be different from a predatory fish that had to swim swiftly after prey?***

The brain and senses As in other vertebrates, the nervous system of fishes consists of a spinal cord and a brain. A fish brain is shown in **Figure 28.10.** The cerebellum is involved in coordinating movement and controlling balance. Fishes have receptors for the sense of smell that enable them to detect chemicals in the water. The olfactory (ohl FAK tree) bulbs record and respond to incoming chemical input. Fishes also have color vision. The optic lobes are responsible for visual input. The cerebrum coordinates input from other parts of the brain. Internal organs are under the control of the medulla oblongata.

If you have spent any time fishing, you know that fishes can detect the slightest movement in the water. Fishes can do this because they have special receptors called the lateral line system. The **lateral line system** enables fishes to detect movement in the water and also helps to keep them upright and balanced. You can see the lateral line system of a fish in **Figure 28.10.**

Reproduction The majority of fishes reproduce through external fertilization. Male and female fishes release their gametes near each other in the water in a process called **spawning.** Developing embryos get nutrition from food stored in the yolk of the egg. Some fishes, such as sharks, reproduce through internal fertilization. Although fertilization takes place internally, development of the embryo of some fish species might occur outside of the female's body when fertilized eggs are laid. Some species of fishes have internal fertilization as well as internal development of offspring. In this case, the developing embryos get nutrition from the female's body.

Fishes that reproduce through external fertilization can produce millions of eggs in a single season. Most fishes do not protect or care for their eggs or offspring. As a result, many eggs and juvenile fishes are prey to other animals. The production of large numbers of eggs ensures that some offspring develop and survive to reproduce. One exception is the male sergeant major fish, shown in **Figure 28.11.** The male fish guards the fertilized eggs from predators until they hatch.

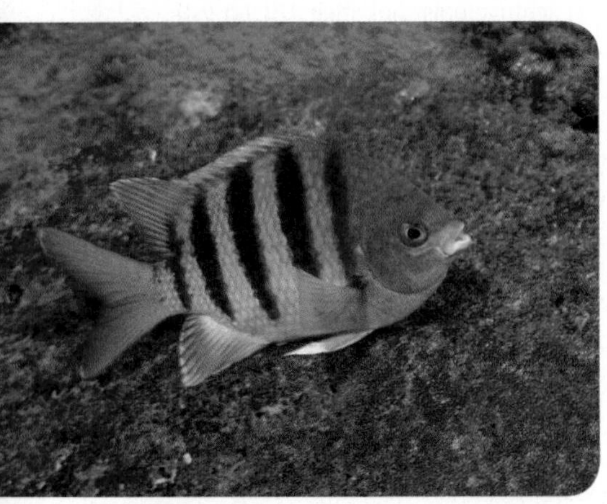

■ **Figure 28.11** Most fishes do not care for their young; however, male sergeant majors guard the eggs as the embryos develop.

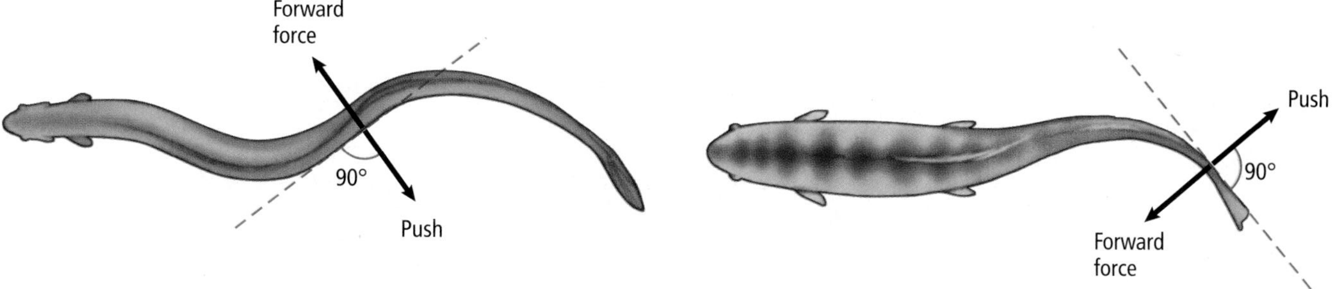

■ **Figure 28.12** An eel moves its whole body in an S-shaped pattern. Other, faster-moving fishes, such as trout, move only their tails as they push forward through the water.

Movement Fishes are well adapted to swimming in the water. Most fishes have a streamlined shape. Most also have mucus that lubricates their body surface and reduces friction between the fish and the water. Fins enable fishes to steer and maneuver in a variety of ways. The buoyant force of water reduces the effect of gravity on fishes. In addition, the **swim bladder,** which is shown in **Figure 28.9,** is a gas-filled space, like a balloon, found in bony fishes that allows a fish to control its depth. When gases diffuse out of the swim bladder, a fish can sink. When gases from the blood diffuse into the swim bladder, a fish can rise in the water column.

Connection to Physics Examine **Figure 28.12.** Fishes move through the water by contracting muscle groups on either side of their bodies. The arrangement of muscle in a fish allows muscle contraction to bend a large portion of the fish's body. As the body of a fish bends, it pushes against the water, creating an opposing force that moves the fish forward, but at an angle. Alternate contraction of muscles, first on one side of the body and then on the other, keep the fish moving in an S-shaped pattern.

Section 28.1 Assessment

Section Summary

- Vertebrates include fishes, amphibians, reptiles, birds, and mammals.
- All vertebrates have a notochord. In most vertebrates, the notochord is replaced by a vertebral column during embryonic development.
- Fishes share certain characteristics and, therefore, are classified together.
- The bodies of fishes have unique adaptations that enable them to live their entire lives in water.

Understand Main Ideas

1. MAIN Idea **Describe** the characteristics fishes have that allow them to live and reproduce in water.
2. **Summarize** the features of vertebrates that make them different from invertebrates.
3. **Evaluate** the importance of the evolution of jaws in fishes.
4. **Identify** the characteristics most fishes have in common.
5. **Explain** why freshwater and saltwater bony fishes have to adjust the balance of salt and water in their bodies.

Think Critically

6. **Hypothesize** Male three-spined stickleback fishes build nests using bright, shiny materials that are in limited supply and are chosen more frequently by females. Form a hypothesis about why this might ensure that a female is choosing a male that has strong traits of his species.
7. **Infer** How might an injury to a fish's lateral line system affect that fish's ability to escape predation?

Section 28.2

Diversity of Today's Fishes

Reading Preview

Section Objectives

- **Identify** the characteristics of different groups of fishes.
- **Compare** the key features of various types of fishes.
- **Explain** the evolution of fishes.

Review Vocabulary

adaptive radiation: the process of evolution that produces many species from an ancestral species

New Vocabulary

tetrapod

MAIN Idea **Scientists classify fishes into three groups based on body structure.**

Real-World Reading Link You already know that the basic structures and their functions in fishes are similar. Now think about all the different types of fishes you have seen in aquariums, photos, or on television.

Classes of Fishes

You have read about jellyfish, crayfish, and various shellfish, but none of these are true fishes. True fishes belong to three groups based on their body structure. Hagfishes and lampreys are jawless fishes; sharks, skates, and rays are cartilaginous (kar tuh LAJ uh nus) fishes; and bony fishes include both ray-finned and lobe-finned fishes.

Jawless fishes Hagfishes, as shown in **Figure 28.13,** are jawless, eel-shaped fishes that do not have scales, paired fins, or a bony skeleton. Members of class Myxini (mik SEE nee), hagfishes have a notochord throughout life. Although they do not develop a vertebral column, they do have gills and many other characteristics of fishes. They live on the seafloor and feed on soft-bodied invertebrates and dead or dying fishes. Even though they are almost blind, their keen chemical sense enables them to locate food. They either enter the body of the fish through the mouth or they scrape an opening into the fish with toothlike structures on their tongues. After eating the internal parts of the fish, the hagfish leaves only a sac of skin and bones.

Hagfishes are known for their ability to produce slime. If threatened, they secrete fluid from glands in their skin. The fluid, when in contact with seawater, forms a slime that is slippery enough to prevent the hagfishes from being caught by predators.

Figure 28.13 Hagfishes are jawless fishes that have toothlike structures on their tongues. Lampreys are parasites on other living fishes.
Describe ***What adaptations for life on the seafloor can you see in this hagfish photo?***

Hagfish

Lamprey

Lampreys, like hagfishes, also are jawless, eel-shaped fishes that lack scales, paired fins, or a bony skeleton. Lampreys, members of class Cephalaspidomorphi (ceh fah las pe doh MOR fee), retain a notochord throughout life, as do hagfishes. Lampreys have gills and other characteristics of fishes. Adult lamprey, shown in **Figure 28.13,** are parasites that feed by attaching themselves to other fishes. They use their sucker-like mouth and tongue with toothlike structures to feed on the blood and bodily fluids of their hosts.

Reading Check **List** the characteristics of jawless fishes.

Cartilaginous fishes When you hear the word *shark*, the first thing that might come to mind is a large fish with many sharp teeth. In spite of being famous for teeth, a shark's main distinguishing feature is its skeleton. All cartilaginous fishes have skeletons made of cartilage. The skeleton of a shark also is made of cartilage, which gives the skeleton flexibility, and calcium carbonate, which gives it strength.

Sharks belong to class Chondrichthyes (kon DRIK thees). Some species of sharks have several rows of sharp teeth as shown in **Figure 28.14.** As teeth are broken or lost, new ones move forward to replace them. Most sharks also have a streamlined shape, with a pointed head and a tail that turns up at the end, as shown in **Figure 28.15.**

These streamlining features, along with strong swimming muscles and sharp teeth, make sharks one of the top predators in the sea. They can sense chemicals in the water, allowing them to detect prey from a distance of one kilometer. As they move in closer, their lateral line systems can detect vibrations in the water. Finally, when they are in the last stages of pursuit, they use their vision and other receptors to detect the bioelectrical fields given off by all animals. An additional adaptation to a predatory life includes tough skin with placoid scales, shown in **Figure 28.15.**

Not all sharks have rows of teeth. Whale sharks, the largest living sharks, are filter-feeders with specialized straining structures in their mouths. Other sharks have mouths adapted to feeding on shelled mollusks.

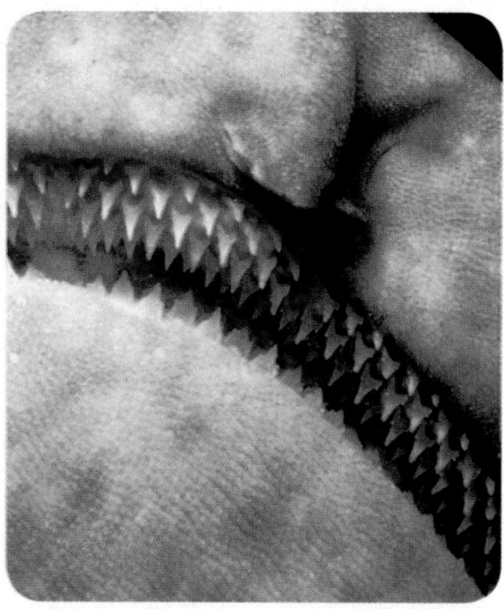

Figure 28.14 Some sharks have several rows of teeth. As teeth in the front row fall or are pulled out, teeth from the row behind them move up into their place.

Great white shark

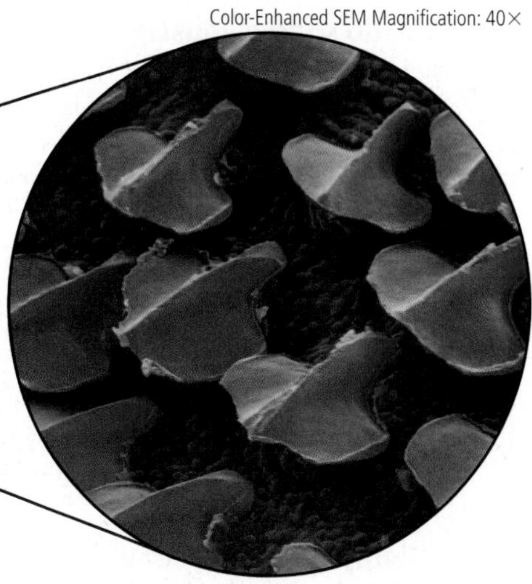

Placoid scales

Figure 28.15 Great white sharks have streamlined bodies and are covered with tough placoid scales.
Infer ***What would a shark's skin feel like if you touched it?***

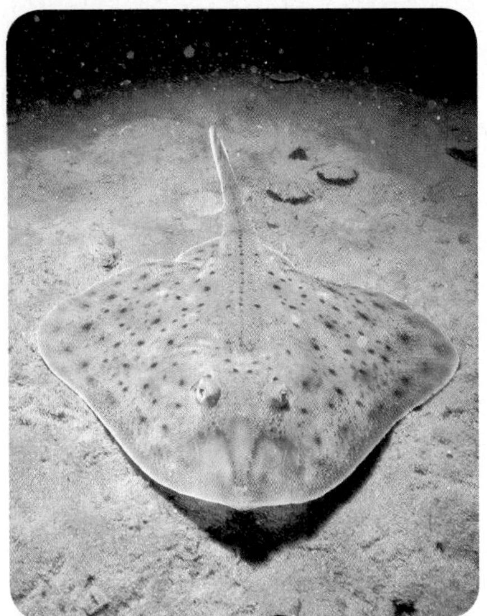

■ **Figure 28.16** Skates have flattened bodies that are adapted for living on the ocean floor.

Skates and rays are cartilaginous fishes adapted to life on the bottom of the sea. In addition to their flattened bodies, their pectoral fins, shown in **Figure 28.16,** are enlarged and attached to their heads. Their winglike fins flap slowly as they swim along the seafloor in search of mollusks and crustaceans, which they crush with their teeth.

Bony fishes Class Osteichthyes (ahs tee IHK theez) contains two groups of bony fishes: the ray-finned fishes, belonging to subclass Actinopterygii (AK tihn ahp TUR ee jee i), and the lobe-finned fishes, belonging to the subclass Sarcopterygii (SAR kahp TUR ee jee i). Modern ray-finned fishes have a bony skeleton, ctenoid or cycloid scales, an operculum covering the gills, and a swim bladder. The most distinguishing feature of ray-finned fishes is in their name. The thin membranes of these fishes' fins are supported by thin, spinelike rays, which are shown in **Figure 28.17.** Most fishes alive today, including salmon and trout, are ray-finned fishes.

There are only eight species of lobe-finned fishes living today. Their fins, shown in **Figure 28.17,** have muscular lobes and joints similar to those of land vertebrates. This makes the fins more flexible than those of ray-finned fishes. Lobe-finned fishes, such as the lungfish, usually have lungs for gas exchange. When drought occurs, a lungfish can burrow with its fleshy fins into the mud and breathe air. When rain returns, lungfishes come out of their burrows.

Coelacanths (SEE luh kanths) are another small group of lobe-finned fishes that many people thought had become extinct about 70 million years ago. However, in 1938, some people fishing off the coast of South Africa caught a coelacanth. Since that time, other coelacanths have been caught. A third group of lobe-finned fishes, now extinct, is thought to be the ancestor of tetrapods. A **tetrapod,** shown in **Figure 28.17,** is a four-footed animal with legs that have feet and toes that have joints.

DATA ANALYSIS LAB 28.1

Based on Real Data*

Analyze Data

How do sharks' muscles function? Lamnid sharks have two types of muscles. Red muscle tissue does not tire easily and is used more during cruising. White muscle tissue is used more during short bursts of speed. Both muscles, however, are always used at the same time.

Data and Observations

The peaks of the graph represent when each muscle type contracts.

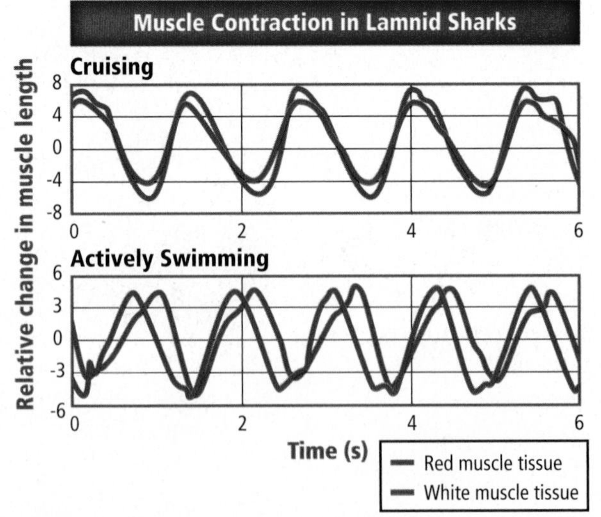

*Data obtained from: Donley, J., et al. 2004 Convergent evolution in mechanical design of lamnid sharks and tunas. *Nature* 429: 61-65.

Think Critically

1. **Evaluate** Does the timing of the contractions of the two types of muscle differ when the sharks are cruising?
2. **Compare** How does the timing of the contractions between the two muscle types change when the sharks are actively swimming?

Visualizing Bony Fishes

Figure 28.17
Class Osteichthyes consists of the bony fishes, and it can be divided into two subclasses—ray-finned fishes and lobe-finned fishes. An extinct lobe-finned fish is thought to be the ancestor of modern tetrapods.

Ray-finned fish

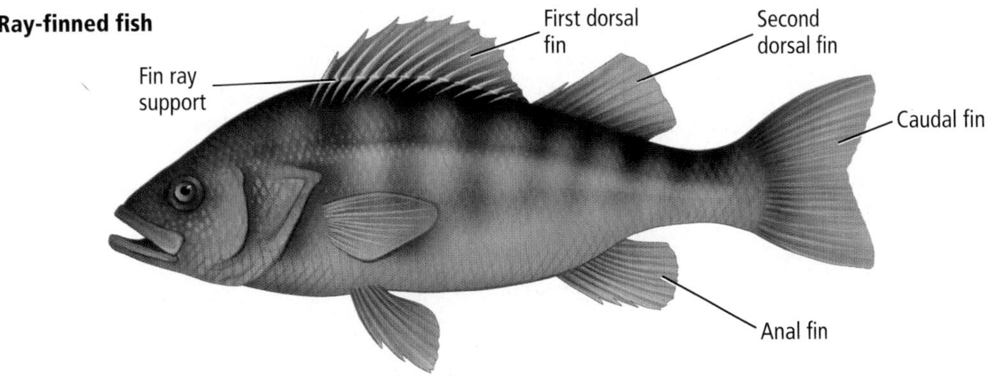

Ray-finned fishes have thin, spinelike rays that support the membranes of their fins.

Lobe-finned fish

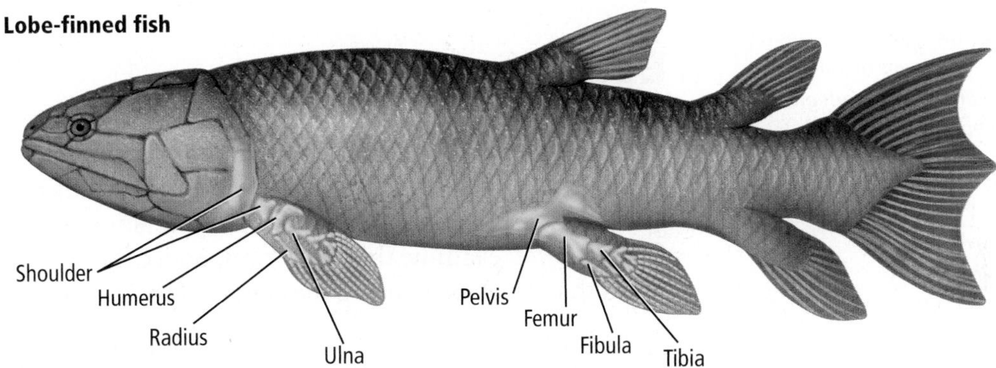

Lobe-finned fishes have muscular lobes and joints similar to those of tetrapods.

Early tetrapod

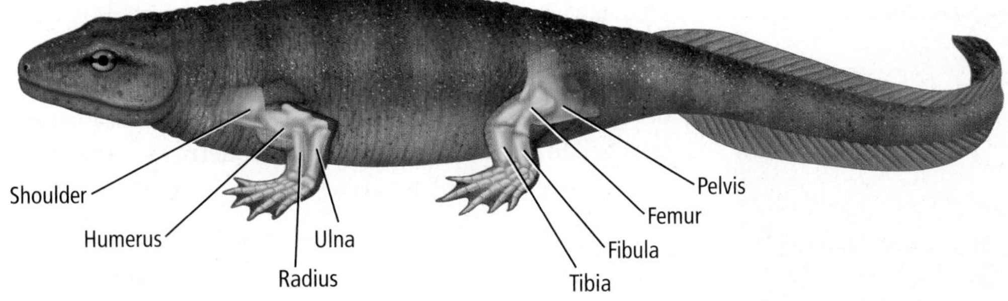

The limbs of tetrapods evolved from the fins of lobe-finned fishes. *Ichthyostega* was a tetrapod that lived about 325 million years ago, had fully formed limbs, and walked on land.

Concepts In Motion **Interactive Figure** To see an animation of bony fishes, visit biologygmh.com.

Biology Online

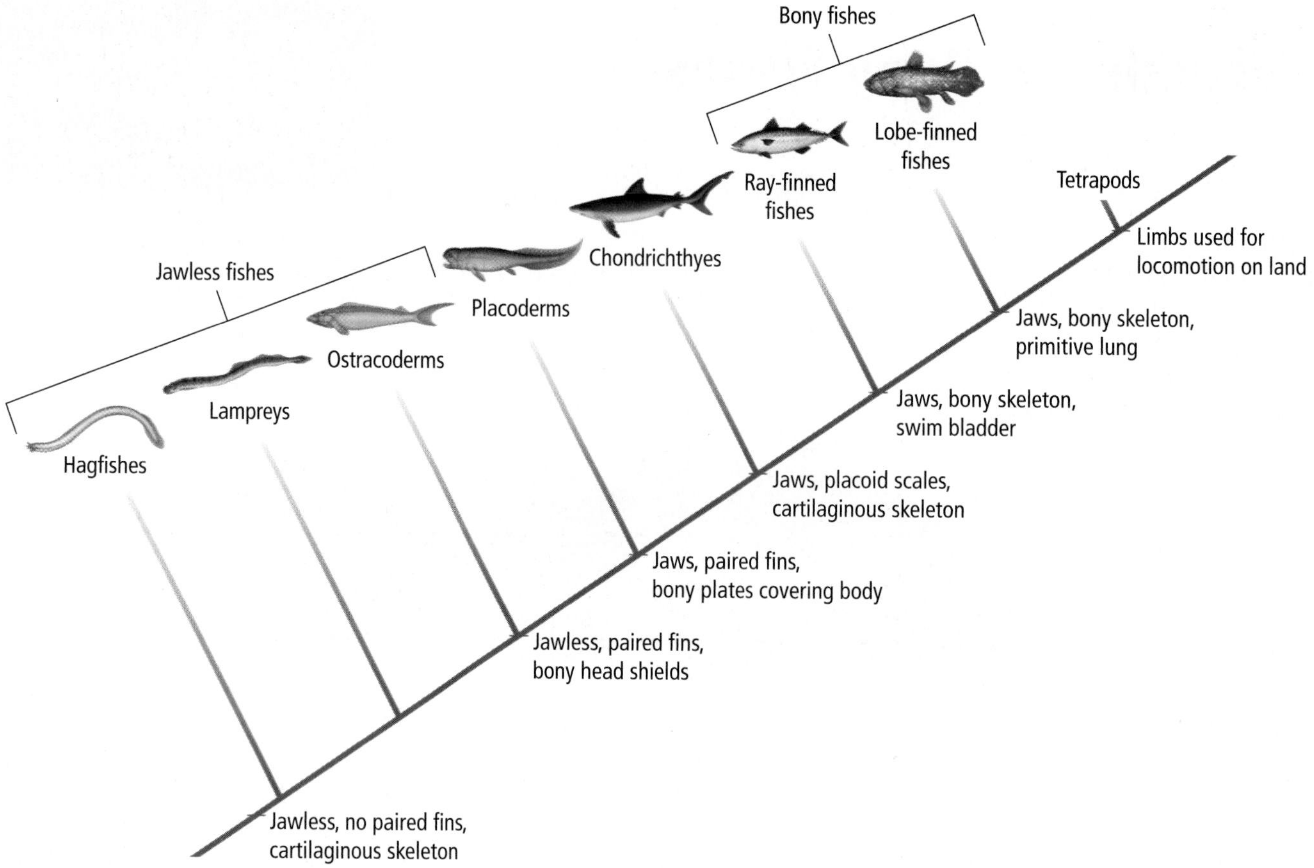

Bony fishes

Lobe-finned fishes

Ray-finned fishes

Tetrapods

Limbs used for locomotion on land

Chondrichthyes

Jawless fishes

Placoderms

Jaws, bony skeleton, primitive lung

Ostracoderms

Jaws, bony skeleton, swim bladder

Lampreys

Hagfishes

Jaws, placoid scales, cartilaginous skeleton

Jaws, paired fins, bony plates covering body

Jawless, paired fins, bony head shields

Jawless, no paired fins, cartilaginous skeleton

■ **Figure 28.18** The cladogram shows one interpretation of the phylogeny of fishes.
Identify ***According to the cladogram, which fishes did not have jaws?***

■ **Figure 28.19** *Dinichthys,* also called *Dunkleosteus,* was a placoderm that had armor plating around its head.

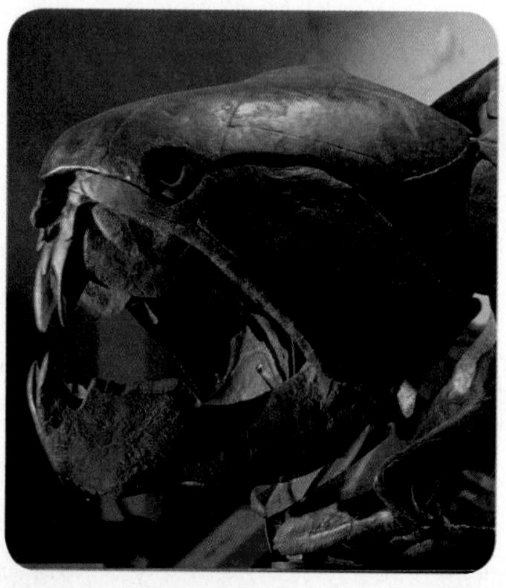

Evolution of Fishes

In the ancient seas of the Cambrian, the first vertebrates wriggled through the water. They were jawless and toothless, but they did have gills, heads, and tails that moved them through the water. The cladogram shown in **Figure 28.18** is one interpretation of the phylogeny of fishes. As you examine the cladogram, notice the characteristics of fishes that developed during the course of their evolution.

First fishes These first jawless, toothless fishes sucked up organic materials from the ocean floor as if they were miniature vacuum cleaners. Ostracoderms (OS tra koh dermz) were the next group of fishes to appear in the fossil record of the Ordovician Period. Ostracoderms had head shields made of bone, a bony outer covering, and paired fins. They were jawless filter-feeders, many of which rested on the bottom of ancient seas.

The bony armor of ostracoderms was an evolutionary milestone in the development of vertebrates. Stronger movement is possible when muscle is attached to bone. Even though ostracoderms became extinct, scientists hypothesize that modern fishes share an ancestry with ostracoderms.

Age of fishes During the Devonian Period, modern fishes had their beginnings. Some were jawless, while some, such as the placoderms, had jaws, a covering of bony plates, paired fins, and an internal skeleton. A fossil of a placoderm fish is shown in **Figure 28.19.** Recall that three of these features are characteristics of the fishes that eventually replaced the placoderms as they became extinct. The Devonian Period often is referred to as the Age of Fishes because of the adaptive radiation of fishes that occurred at that time.

Ecology of Fishes

Fishes are an important source of food in all aquatic ecosystems. Yet, their freshwater habitats and saltwater habitats are being changed by human activities, such as damming rivers or pollution. Fishes are good bioindicators of the environmental health of an aquatic system. When noncommercial fish populations decrease, the main cause often is habitat alteration. When fishes decline in numbers, not only are there negative human economic impacts, ecosystems also can become unbalanced as well.

Habitat alteration Some fishes, such as salmon, migrate. Salmon spend their adult lives in the ocean but return to freshwater to spawn in the streams where they hatched. In the Pacific Northwest, river and stream habitats have been changed by the construction of dams. Dams interfere with the upstream and downstream migration of salmon, as shown in **Figure 28.20.** The end result in the Pacific Northwest, for example, is that the number of salmon swimming upstream now is only about three percent of the 10 to 16 million salmon that swam up the rivers 150 years ago.

Figure 28.20 Not all salmon are able to get over the dams used to generate hydroelectricity. In order to spawn, salmon must return to the streams where they hatched.

Pollution The habitats of fishes can be changed by pollution which can reduce the quality of water in lakes, rivers, and streams. This can result in a decline in both the number and diversity of fishes in an area. In some cases, when the cause of habitat alteration is stopped and suitable conditions return, fishes also return. For example, Atlantic salmon were not observed in the Penobscot River in Maine for ten years during a time when intense pollution altered water quality. When the pollution was stopped, the salmon returned.

LAUNCH Lab

Review Based on what you've read about different fishes, how would you now answer the analysis questions?

Section 28.2 Assessment

Section Summary

- Fishes can be placed into one of three main groups—jawless fishes, cartilaginous fishes, and bony fishes.
- Hagfishes and lampreys are examples of jawless fishes.
- Sharks, rays, and skates are examples of cartilaginous fishes.
- Bony fishes consist of two subclasses of fishes—ray-finned fishes and lobe-finned fishes.
- Ancient extinct fishes had features that enabled them to evolve into modern fishes.
- Habitat alteration and pollution can negatively affect fish populations.

Understand Main Ideas

1. MAIN Idea **Compare and contrast** the structures of jawless fishes, cartilaginous fishes, and bony fishes.
2. **Identify** the characteristics of the two subclasses of bony fishes.
3. **Sketch** the basic shape that represents each of the three main groups of fishes.
4. **Describe** the evolutionary sequence of the different groups of fishes.
5. **Hypothesize** Bony fishes have either cycloid or ctenoid scales. Form a hypothesis that explains how scale type is related to diversity.

Think Critically

MATH in Biology

6. The number of fish species often decreases with latitude. In fact, the number of fish species in tropical lakes is much greater than the number of fish species in temperate lakes. Suggest a hypothesis that accounts for this mathematical phenomenon.

Section 28.3

Reading Preview

Section Objectives

- **Analyze** the kinds of adaptations that were important as animals moved to the land.
- **Summarize** the characteristics of amphibians.
- **Distinguish** among the orders of amphibians.

Review Vocabulary

metamorphosis: a series of developmental changes in the form or structure of an organism

New Vocabulary

cloaca
nictitating membrane
tympanic membrane
ectotherm

Amphibians

MAIN Idea **Most amphibians begin life as aquatic organisms then live on land as adults.**

Real-World Reading Link Think about the last time you went swimming. How is moving in water different from moving on land? Just as fishes have adaptations for living in water, tetrapods have adaptations for living on land.

Evolution of Tetrapods

Tetrapods are four-legged vertebrates that first appeared on Earth 360 million years ago. Modern amphibians are descendants of these early tetrapods. Examine the evolutionary tree in **Figure 28.21** to see how amphibians are related to other vertebrates. As millions of years passed, animals adapted to the conditions of life on land.

The move to land Animals faced several physical challenges in the move from water to land. **Table 28.1** lists some of the differences between conditions of life in the water and life on land. These differences include buoyancy, oxygen concentration, and temperature. **Table 28.1** also gives examples of how terrestrial vertebrates adapted to life on land.

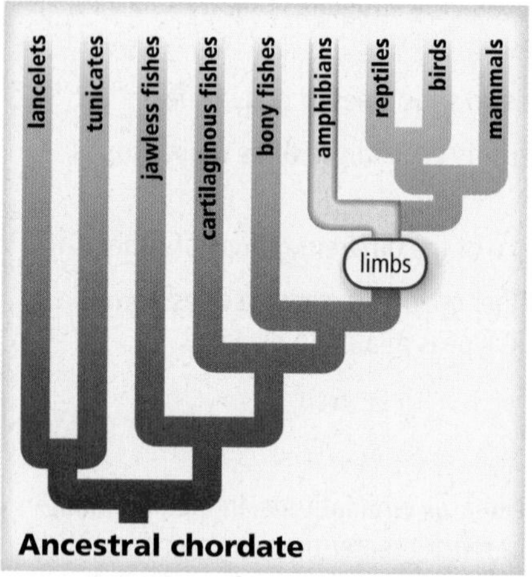

Figure 28.21 The evolutionary tree shows how amphibians are related to other vertebrates.

Table 28.1 Adaptations to Land

Concepts In Motion
Interactive Table To explore more about adaptations for life on land, visit biologygmh.com.

Conditions in Water	Conditions on Land	Terrestrial Vertebrate Adaptations
Water exerts a buoyant force that counters the force of gravity.	• Air is about 1000 times less buoyant than water. • Animals must move against gravity.	Limbs develop and the skeletal and muscular systems of terrestrial animals become stronger.
Oxygen is dissolved in water and must be removed by gills through countercurrent circulation.	• Oxygen is at least 20 times more available in air than in water.	With lungs, terrestrial animals can get oxygen from air more efficiently than from water.
Water retains heat, so the temperature of water does not change quickly.	• Air temperature changes more easily than water temperature. • Daily temperatures can change by 10°C between day and night.	Terrestrial animals develop behavioral and physical adaptations to protect themselves from extreme temperatures.

In addition to the differences listed in **Table 28.1,** another difference between conditions in water and on land is that sound travels more quickly through water. Fishes use lateral line systems to sense vibrations, or sound waves, in water. A lateral line system is not effective in air. The ears of terrestrial vertebrates developed to sense sound waves traveling through air.

Terrestrial habitats In spite of the challenges associated with terrestrial life, there are many habitats available to animals on land. The different biomes on land, including tropical rain forests, temperate forests, grasslands, deserts, taiga, and tundra, provide suitable habitats for animals with appropriate adaptations.

Characteristics of Amphibians

Have you ever watched a tadpole in a jar of pond water? Examine and describe the tadpole in **Figure 28.22.** A tadpole is the limbless, gill-breathing, fishlike larva of a frog. Day by day, the tadpole undergoes a metamorphosis (me tuh MOR fuh sihs)—hind legs form and grow longer, the tail shortens, gills are replaced by lungs, and forelimbs sprout. In just a few weeks or months, depending on the species, the tadpole becomes an adult frog. Most amphibians begin life as aquatic organisms. After metamorphosis, they are equipped to live life on land.

Modern amphibians include frogs, toads, salamanders, newts, and legless caecilians. Most amphibians are characterized by having four legs, moist skin with no scales, gas exchange through skin, lungs, a double-loop circulatory system, and aquatic larvae.

Feeding and digestion Most frog larvae are herbivores, whereas salamander larvae are carnivores. However, as adults their diets are similar as both groups become predators and feed on a variety of invertebrates and small vertebrates. Some salamanders and legless amphibians use just their jaws to catch prey. Others, such as frogs and toads, can flick out their long, sticky tongues with great speed and accuracy to catch flying prey.

Food moves from the mouth through the esophagus to the stomach, where digestion begins. From the stomach, food moves to the small intestine, which receives enzymes from the pancreas to digest food. From the intestine, food is absorbed into the bloodstream and delivered to body cells. Food moves from the small intestine into the large intestine before waste material is eliminated. At the end of the intestine is a chamber called the cloaca. The **cloaca** (kloh AY kuh) is a chamber that receives the digestive wastes, urinary waste, and eggs or sperm before they leave the body.

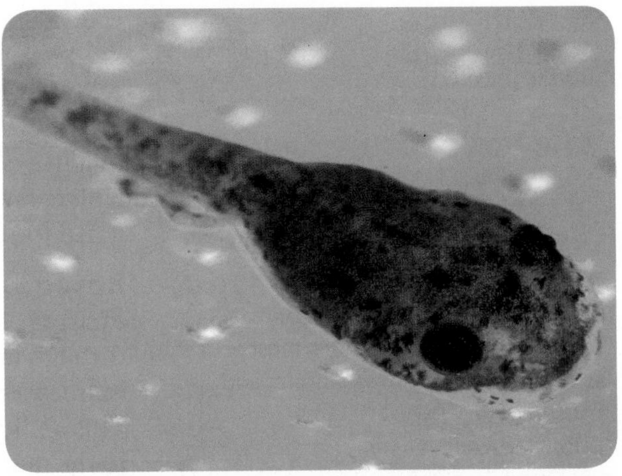

Figure 28.22 The cayenne slender-legged tree frog is found in South America.
Top: A tadpole is limbless.
Middle: The frog is undergoing metamorphosis to become an adult frog. Notice the development of limbs.
Bottom: An adult frog has fully developed limbs and lacks a tail.

Concepts In Motion

Interactive Figure To see an animation of a frog's life cycle, visit biologygmh.com.

Excretion Amphibians filter wastes from the blood through their kidneys, and excrete either ammonia or urea as the waste product of cellular metabolism. Ammonia is the end product of protein metabolism and is excreted by amphibians that live in the water. Amphibians that live on land excrete urea that is made from ammonia in the liver. Unlike ammonia, urea is stored in the urinary bladder until it is eliminated from the body through the cloaca.

Respiration and circulation As larvae, most amphibians exchange gases through their skin and gills. As adults, most breathe through lungs, their thin, moist skin, and the lining of the mouth cavities. Frogs can breathe through their skin either in or out of water. This ability enables them to spend the winter protected from the cold in the mud at the bottom of a pond.

The circulatory system of amphibians is shown in **Figure 28.23.** It consists of a double loop instead of the single loop you learned about in fishes. The first loop moves oxygen-poor blood from the heart to pick up oxygen in the lungs and skin, and then moves the oxygen-filled blood back to the heart. During circulation in the second loop, blood filled with oxygen moves from the heart through vessels to the body, where the oxygen diffuses into cells.

Amphibians have three-chambered hearts. The atrium is completely separated into two atria by tissue. The right atrium receives deoxygenated blood from the body, while the left atrium receives oxygenated blood from the lungs. The ventricle of amphibians remains undivided.

Reading Check **Describe** how the amphibian circulatory system is adapted to life on land.

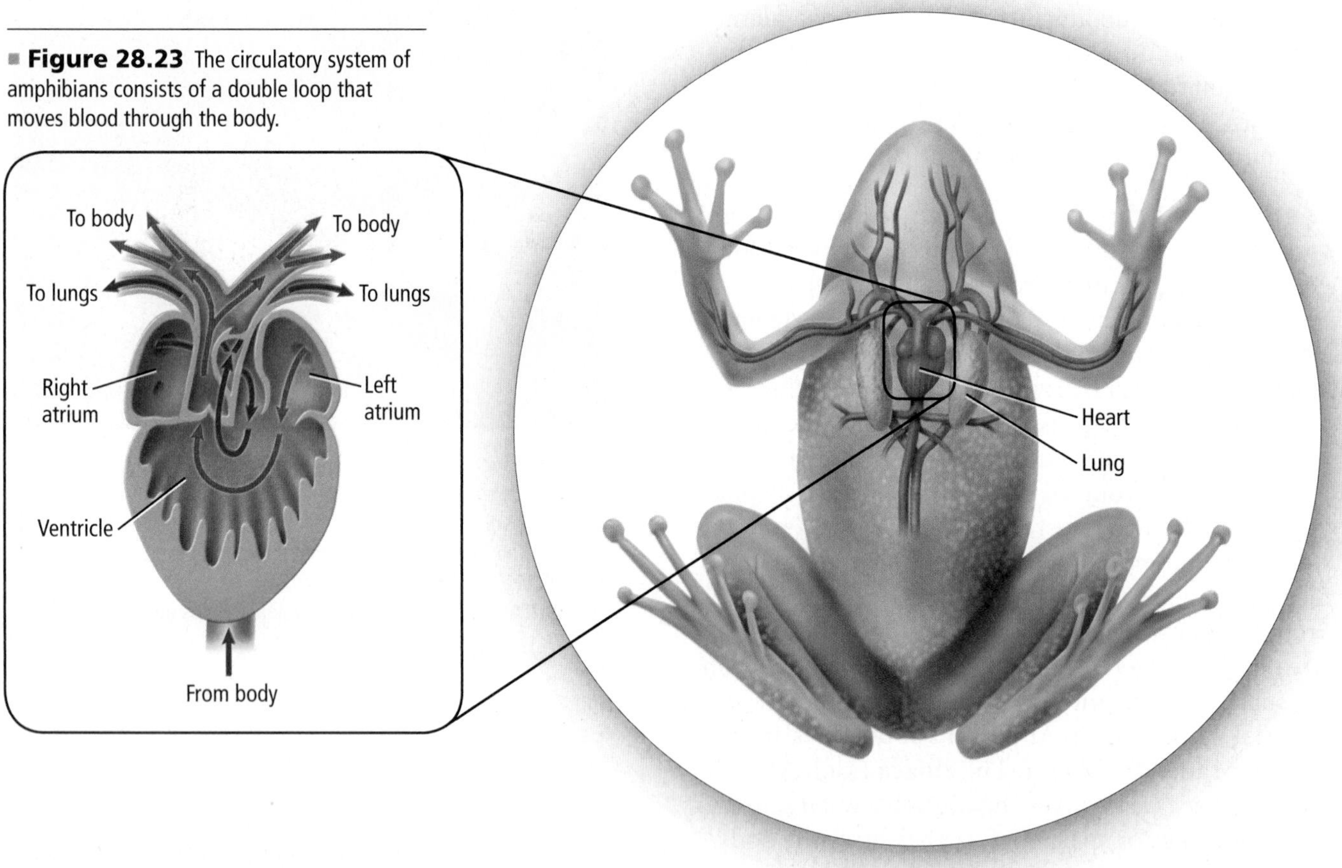

VOCABULARY

SCIENCE USAGE V. COMMON USAGE

Amphibian

Science usage: organisms that are members of class Amphibia; most spend part of their lives in water and part on land.
A frog is an amphibian.

Common usage: an airplane designed to take off from and land on either land or water.
The amphibian landed smoothly in the lake water.

Figure 28.23 The circulatory system of amphibians consists of a double loop that moves blood through the body.

The brain and senses Like fishes, the nervous systems of amphibians are well developed. The differences in conditions between life in the water and life on land are reflected in the differences between the brains of fishes and those of amphibians. For example, the forebrain of frogs contains an area that is involved with the detection of odors in the air. The cerebellum, which is important in maintaining balance in fishes, is not as well developed in terrestrial amphibians that stay close to the ground.

Vision is an important sense for most amphibians. They use sight to locate and capture prey that fly at high speeds and to escape predators. Frogs' eyes have structures called nictitating (NIK tuh tayt ing) membranes. The **nictitating membrane** is a transparent eyelid that can move across the eye to protect it underwater and keep it from drying out on land.

The amphibian ear also shows adaptation to life on land. The **tympanic** (tihm PA nihk) **membrane** is an eardrum. In frogs, it is a thin external membrane on the side of the head, as shown in **Figure 28.24.** Frogs use their tympanic membrane to hear high-pitched sounds and to amplify sounds from the vocal cords. Other senses in amphibians include touch, chemical receptors in skin, taste buds on the tongue, and sense of smell in the nasal cavity.

It is important for amphibians to sense the temperature of their environment because they are ectotherms. **Ectotherms** are animals that obtain their body heat from the external environment. Ectotherms cannot regulate their body temperatures through their metabolism, so they must be able to sense where they can go to get warm or to cool down. For example, if it is cold, a toad can find a warm, moist rock on which to bask and warm itself.

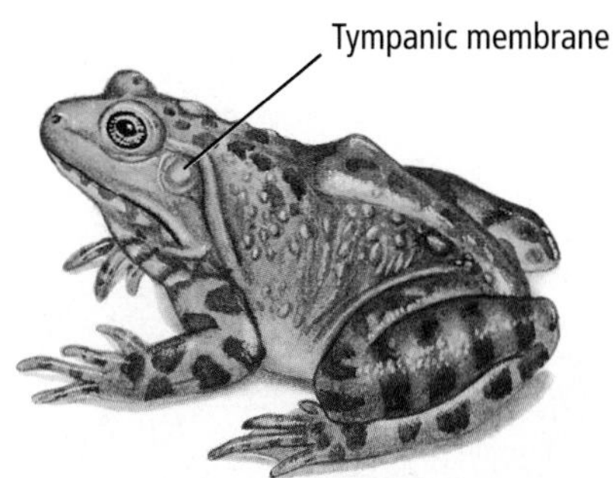

Figure 28.24 The tympanic membrane is an adaptation for life on land.

Concepts In Motion

Interactive Figure To see an animation of the adaptations of a frog, visit biologygmh.com.

DATA ANALYSIS LAB 28.2

Based on Real Data*

Interpret a Graph

How does temperature affect the pulse rate of calling in tree frogs? Male tree frogs make calls that females can identify easily based on the rate of the sound pulses in the call.

Data and Observations

The graph shows the pulse rate of two species of frogs versus temperature.

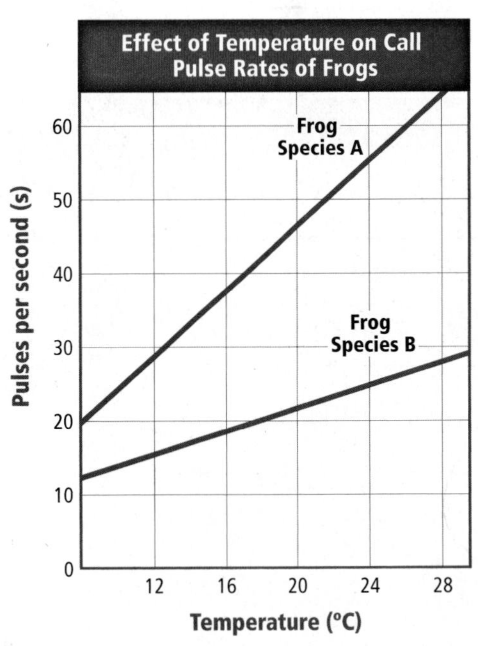

Think Critically

1. **Interpret the Data** What is the relationship between sound pulses and temperature?
2. **Compare** How did temperature affect the rate of pulses in species A and in species B?
3. **Infer** Why is it important that the two species of frogs do not have the same pulse rate in their calls at the same temperature?

*Data obtained from: Gerhardt, H.C. 1978. Temperature coupling in the vocal communication system in the grey treefrog *Hyla versicolor*. *Science* 199: 992–994.

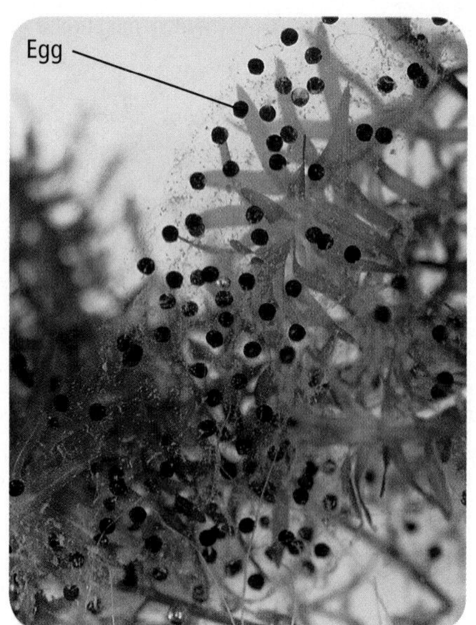

■ **Figure 28.25** Amphibian eggs do not have shells that would protect them from drying out.
Infer ***What adaptation ensures that amphibian eggs do not dry out?***

Reproduction and development Like many amphibians, female frogs lay eggs to be fertilized by males in the water. The eggs do not have shells or protective coverings to keep them from drying out. The eggs, like the ones shown in **Figure 28.25,** are covered with a sticky, jelly-like substance that helps them stay anchored to vegetation in the water. After fertilization, the developing embryo uses the yolk in the egg for nourishment until it hatches into a tadpole. A tadpole, like the one shown in **Figure 28.22,** changes from a gill-breathing, legless herbivore with fins and a two-chambered heart into a lung-breathing, four-legged carnivore with a three-chambered heart. The stages of metamorphosis are primarily under the control of chemicals released within the tadpole's body.

Amphibian Diversity

Biologists classify modern amphibians into three orders. Order Anura (a NOOR ah) contains 4200 species of frogs and toads. Order Caudata (kaw DAY tah) has about 400 species of salamanders and newts. One hundred and fifty species of wormlike caecilians make up order Gymnophiona (JIHM noh fee oh nah). Frogs, toads, and salamanders live in moist areas in a variety of habitats, while newts are aquatic. Caecilians are tropical burrowing animals.

Frogs and toads Frogs and toads, shown in **Figure 28.26,** lack tails and have long legs enabling them to jump. Frogs have longer and more powerful legs than toads and are able to make more powerful jumps than the small hops of toads. Frogs have moist, smooth skin, while the skin of toads tends to be bumpy and dry. Though both need to be near water to carry out reproduction, toads generally live farther away from water than do frogs. Another difference between frogs and toads is that toads have kidney-bean-shaped glands near the back of their heads that release a bad-tasting poison. The poison discourages predators from eating them.

Reading Check **Compare and contrast** the characteristics of frogs and toads.

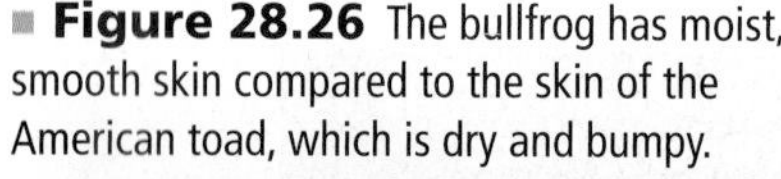

■ **Figure 28.26** The bullfrog has moist, smooth skin compared to the skin of the American toad, which is dry and bumpy.

Bullfrog

American toad

Red salamander

Warty newt

■ **Figure 28.27** The red salamander is found in the eastern United States. The warty newt breeds in deep ponds that contain aquatic vegetation.

Salamanders and Newts Unlike frogs and toads, salamanders and newts have long, slim bodies with necks and tails, as shown in **Figure 28.27.** Most salamanders have four legs and thin, moist skin and cannot live far from water. Like frogs, most salamanders lay their eggs in water. The larvae look like miniature salamanders, except that they have gills. Newts, like the one in **Figure 28.27,** generally are aquatic throughout their lives, while most salamanders, as adults, live in moist areas such as under logs or in leaf litter. Salamanders range in size from about 15 cm long to the giant salamander that is 1.5 m long. An adult salamander's diet consists of worms, frog eggs, insects, and other invertebrates.

Caecilians Caecilians (si SILH yenz) are different from other amphibians because they are legless and wormlike, as shown in **Figure 28.28.** They burrow in the soil and feed on worms and other invertebrates. Skin covers the eyes of many caecilians, so they might be nearly blind. All caecilians have internal fertilization. They lay their eggs in moist soil located near water. Caecilians can be found in tropical forests of South America, Africa, and Asia.

CAREERS IN BIOLOGY

Animal Curator An animal curator works at a zoo, managing some part of the zoo's animal collection, such as the amphibian exhibit. Curators work together to decide the best way to maintain an animal's habitat in the zoo. For more information on biology careers, visit biologygmh.com.

■ **Figure 28.28** Caecilians do not have ear openings. It is not known if, or how, they can hear sounds.

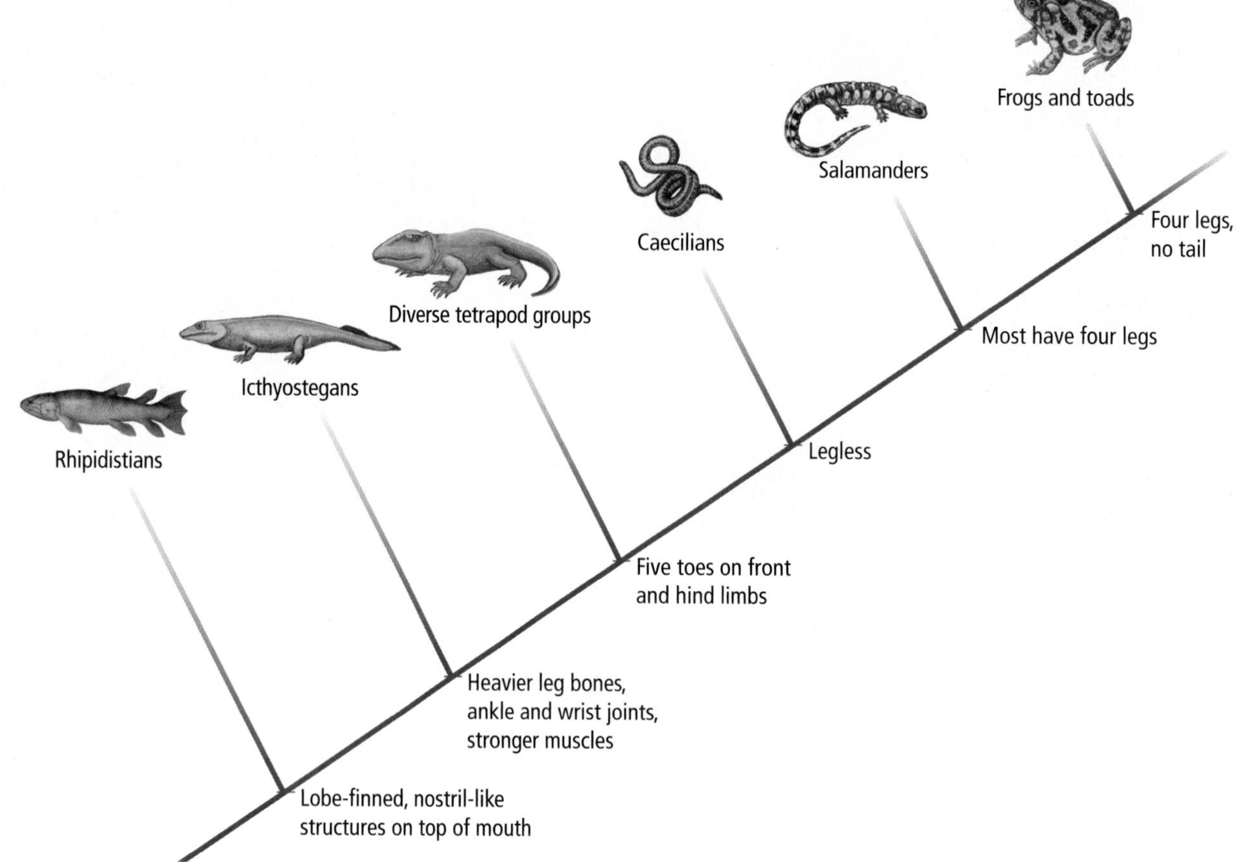

■ **Figure 28.29** The cladogram shows one interpretation of amphibian evolution.
Interpret ***Which modern amphibians are more closely related to the first tetrapods?***

Evolution of Amphibians

Fossil evidence shows that the first tetrapods evolved limbs in water before they moved to land. Many adaptations that are useful on land first evolved in water. For example, legs with feet and toes could be helpful in moving through bottom vegetation. Ankles and wrists might have increased maneuverability. The attachment of hip bones to the vertebral column might have helped predators attack prey more easily.

The cladogram shown in **Figure 28.29** is one interpretation of the evolution of amphibians. Many scientists think that early tetrapods are most closely related to a group of now extinct, lobe-finned fishes called rhipidistians (RI pih dihs tee unz). Characteristics that both early tetrapods and rhipidistians share include similar bone structure in the skull and limbs, nostril-like openings in the tops of their mouths, and a similar tooth structure.

Early tetrapods had defined legs with feet, but the construction of the legs was too weak for these animals to walk easily on land. Ichthyostegans, shown in **Figure 28.17,** had more support in the shoulder bones, heavier leg bones, and more muscular features that enabled them to pull themselves onto land and move a little more easily. The skull had the same general shape as the skulls of lobe-finned fishes. Tetrapod groups branched out to produce the three major groups of amphibians alive today, as well as reptiles, birds, and mammals.

VOCABULARY

ACADEMIC VOCABULARY

Diversify:
to produce variety.
The bakery diversified the flavors of doughnuts it made, giving customers more choices.

 **Reading Check Explain** why scientists think that tetrapods evolved from rhipidistians.

Ecology of Amphibians

In recent decades, amphibian populations have been declining worldwide. Scientists have been collecting data to determine possible causes for the decline. The results have varied. In some cases, the cause can be isolated to a local condition. In other cases, the cause might be the result of several factors occurring on a large scale.

Local factors In some cases, such as that of the California red-legged frog, the decline is due to habitat destruction. When wetlands are drained and buildings are built in the areas instead, these areas of water are no longer available to amphibians that must lay eggs in or near water to reproduce successfully. In other areas, the introduction of exotic species—species that are not found in that area naturally—has affected amphibian populations. The exotic species compete with the amphibians for food and habitat space, or they are predators of amphibians. The introduction of trout, which prey on tadpoles, into the high-altitude lakes of California's Sierra Nevada Mountains is thought to have contributed to the near extinction of the mountain yellow-legged frog found in that region.

Global factors In addition to local factors, various global factors might be at work causing amphibian decline. Aspects of global climate change, such as increased temperature, decreased soil moisture, increased length of the dry season, and changes in rainfall can cause either death or stress to the bodies of amphibians, making them more susceptible to disease.

Figure 28.30 compares healthy toad eggs versus those infected by a fungus. Some scientists think that global climate changes that have led to a decreased amount of rainfall leave developing amphibians' eggs in shallow pond water. Because the depth of the water is reduced, the eggs are exposed to more ultraviolet light. Laboratory experiments have shown that increased exposure to UV light leads to an increased risk of fungal infection in amphibian eggs.

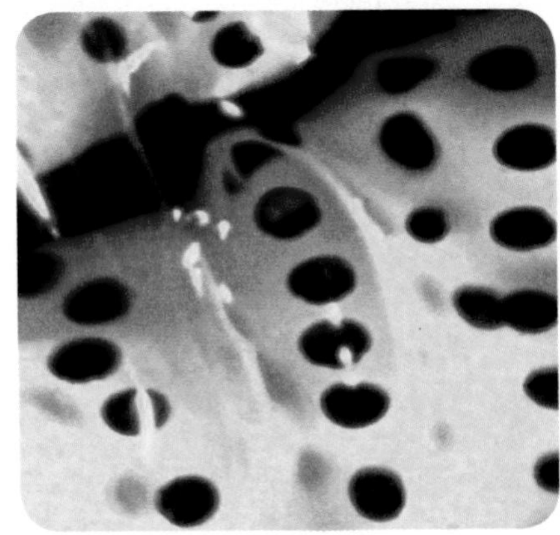

Healthy toad eggs

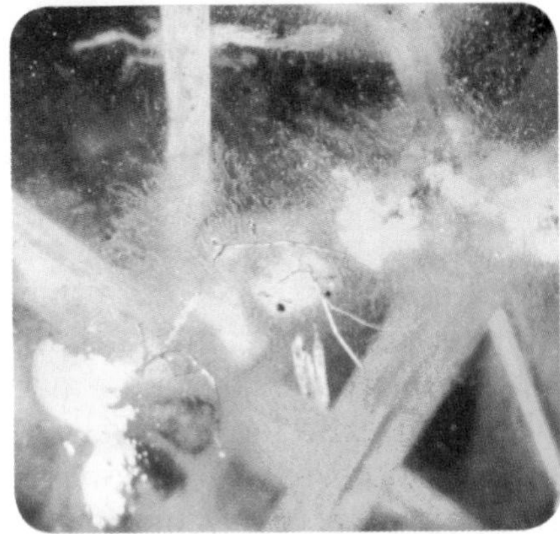

Fungus-infected toad eggs

■ **Figure 28.30** Healthy toad eggs are laid in single file in the water. Infected toad eggs are covered by fungus; fungus infection might account for a decrease in some toad populations.

Section 28.3 Assessment

Section Summary

- The transition of animals to land required a variety of adaptations.
- The bodies of amphibians have unique adaptations that enable them to live on land.
- Amphibians belong to three orders based on structural similarities.
- Ancient tetrapods evolved aquatic adaptations that eventually were used on land.
- Amphibian populations are declining worldwide for a variety of reasons.

Understand Main Ideas

1. MAIN Idea **Summarize** the adaptations of amphibians that make them adapted to life on land.
2. **Compare** the conditions of a land environment to that of an aquatic environment.
3. **Analyze** the kinds of adaptations that were important as animals moved to land.
4. **Summarize** the characteristics of each order of amphibians.

Think Critically

5. **Interpret Scientific Illustrations** Examine **Figure 28.29** and explain which of the three groups of amphibians is most recent and which is the most ancient.

WRITING in **Biology**

6. On a hike in a marshy area near your home, you find a dead frog with deformed limbs. Hypothesize possible reasons why these deformities might have occurred.

BioDiscoveries

What is causing frog malformations?

From classroom to newsroom What began as a class field trip ended up alerting environmentalists to a potentially important problem. In August 1995, a class of students in LeSeuer, Minnesota, took a field trip to a wetland to learn about the ecosystem. While they were there, they noticed a large population of frogs—more than 50 percent of their catch—had malformations. In 1996, reports of frogs with malformations, including missing or extra legs, partially formed limbs, and missing eyes, were showing up elsewhere, including all across Minnesota and in several other states and countries.

Leaping into the laboratory Several studies are being conducted to determine the cause of the malformations. The results of one study indicate that poor water quality could be the cause. When frogs were grown in the laboratory in different samples of water, as many as 75 percent of the frogs were malformed when grown in the water from the Minnesota sites, compared to 0 percent when grown in purified water. The problem contaminant has not been identified yet. Scientists are testing other hypotheses as well, including that tadpoles are being infected by a parasitic worm or a fungus that could be causing the malformations. Another hypothesis being tested is whether or not increased exposure of frog eggs to UV light is causing the malformations.

Each of the studies has yielded data that support the hypotheses tested, but since the types and frequency of malformations are not the same from site to site, as shown in the graphs above, the real-world cause probably is a combination of factors.

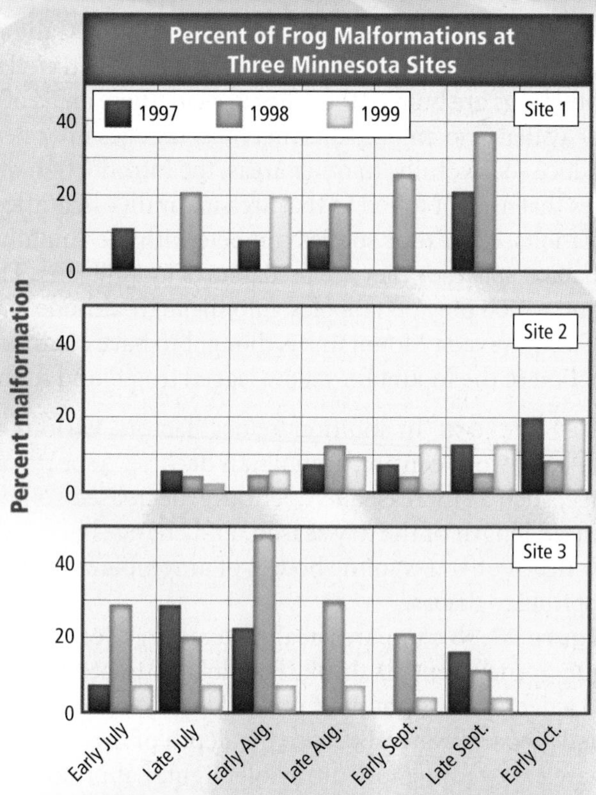

Malformation rates peak at different times from year to year and from site to site.

For instance, increased phosphorus and nitrogen in the water, caused by chemical use, can create an algal bloom. That algal bloom, in turn, increases the population of snails that carry the parasite that can cause malformations. Or, chemical mixtures, while harmless individually, might become toxic when mixed, or might change when exposed to sunlight.

MATH in Biology

Analyze Data The graphs above show the differences in the percent of malformations in frogs at three sites in Minnesota over a three-year period. Find the average percent of malformations at each site for the three-year period. Which site had the largest percent of malformations? For more information about frog malformations, visit biologygmh.com.

BIOLAB

HOW DO SOME ECTOTHERMS REGULATE BODY TEMPERATURE?

Background: Recall that amphibians are ectotherms. Many ectotherms live in habitats where the temperature may fluctuate by 10 or 15°C throughout each day. In this lab, you will investigate the strategies some ectothermic animals use to maintain a suitable body temperature.

Question: *How do ectotherms maintain their body temperature within a specific range?*

Materials

thermometers (2)
plastic containers (2)
metric ruler
room temperature water
paper towels
sand
soil
high-wattage lightbulb and lamp

Safety Precautions

WARNING: *Lamps may become hot when lightbulb is lit.*

Procedure

1. Read and complete the lab safety form.
2. Obtain two thermometers. These will be a model of an ectotherm animal. Record the temperature of each thermometer. Place one thermometer in a plastic container. Place the other thermometer in another container and fill the container so that the thermometer is covered by at least 5 cm of water.
3. Place each container under a lit lightbulb. Monitor the temperature of the thermometers. You must maintain the temperature of each thermometer within a range of 36–39°C for the next 15 min. Decide how often you will measure the temperature of the thermometers and record the data in a table. Record what actions you took to maintain the temperature of the thermometers within the given range.

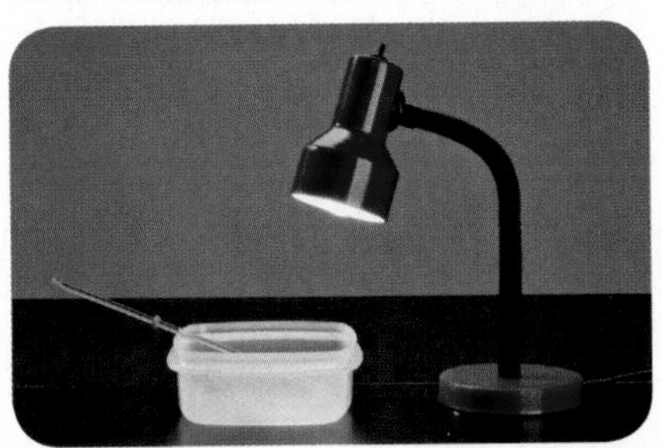

4. Pour the water out of the container and dry the container thoroughly. Allow the thermometers to return to room temperature.
5. Place one thermometer in a container and fill the container with soil so that the thermometer is covered by at least 5 cm of soil. Place the other thermometer in a container and cover it with at least 5 cm of sand.
6. Repeat Step 3.

Analyze and Conclude

1. **Summarize** Did you successfully maintain the temperature within a given range for all steps of the experiment? How did you do this?
2. **Analyze** Were there differences in how you maintained the temperature of the thermometers in water, soil, or sand? In which substance was it easiest to maintain the temperature range? Why?
3. **Draw Conclusions** What are the challenges associated with being an ectothermic animal? Explain.
4. **Think Critically** How do real ectotherms, such as amphibians and reptiles, keep their body temperatures within a specific range?

APPLY YOUR SKILL

Poster Research ectotherms and make a poster that describes the adaptations they require to survive in cold temperatures. To learn more about ectotherms, visit BioLabs at biologygmh.com.

Study Guide

Download quizzes, key terms, and flash cards from biologygmh.com.

FOLDABLES **Analyze Cause and Effect** On the back of your Foldable, explain the cause and effect relationship between the method of movement of an animal and its circulatory system. For example, how did walking change the oxygen needs of an amphibian?

Vocabulary	Key Concepts
Section 28.1 Fishes	
• atrium (p. 824) • cartilage (p. 820) • fin (p. 822) • lateral line system (p. 826) • nephron (p. 825) • neural crest (p. 821) • operculum (p. 824) • scale (p. 823) • spawning (p. 826) • swim bladder (p. 827) • ventricle (p. 824)	MAIN Idea Fishes are vertebrates that have characteristics allowing them to live and reproduce in water. • Vertebrates include fishes, amphibians, reptiles, birds, and mammals. • All vertebrates have a notochord. In most vertebrates, the notochord is replaced by a vertebral column during embryonic development. • Fishes share certain characteristics and, therefore, are classified together. • The bodies of fishes have unique adaptations that enable them to live their entire lives in water.
Section 28.2 Diversity of Today's Fishes	
• tetrapod (p. 830)	MAIN Idea Scientists classify fishes into three groups based on body structure. • Fishes can be placed into one of three main groups—jawless fishes, cartilaginous fishes, and bony fishes. • Hagfishes and lampreys are examples of jawless fishes. • Sharks, rays, and skates are examples of cartilaginous fishes. • Bony fishes consist of two subclasses of fishes—ray-finned fishes and lobe-finned fishes. • Ancient extinct fishes had features that enabled them to evolve into modern fishes. • Habitat alteration and pollution can negatively affect fish populations.
Section 28.3 Amphibians	
• cloaca (p. 835) • ectotherm (p. 837) • nictitating membrane (p. 837) • tympanic membrane (p. 837)	MAIN Idea Most amphibians begin life as aquatic organisms then live on land as adults. • The transition of animals to land required a variety of adaptations. • The bodies of amphibians have unique adaptations that enable them to live on land. • Amphibians belong to three orders based on structural similarities. • Ancient tetrapods evolved aquatic adaptations that eventually were used on land. • Amphibian populations are declining worldwide for a variety of reasons.

CHAPTER 28 Assessment

Section 28.1

Vocabulary Review

Complete each sentence by providing the missing vocabulary term from the Study Guide page.

1. The process by which male and female fishes release their gametes near each other in the water is called __________.

2. The __________ is the chamber of the heart that receives blood from the body.

3. A group of cells that develop from the nerve cord in vertebrates is called a __________.

Understand Key Concepts

Use the diagram below to answer questions 4 and 5.

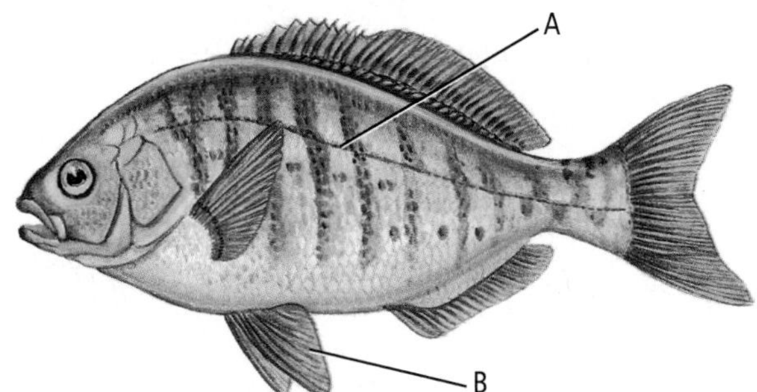

4. Which is the structure labeled A?
   A. ctenoid scales **C.** neural crest
   B. lateral line system **D.** operculum

5. Which is the structure labeled B?
   A. gills
   B. swim bladder
   C. ventricle
   D. pelvic fins

6. Which structure allows fishes to control their depth in an aquatic environment?
   A. operculum
   B. swim bladder
   C. lateral line
   D. jaws

7. Which adaptation allows fishes to be predators?
   A. paired fins
   B. placoid scales
   C. jaws
   D. gills

Constructed Response

8. **Open Ended** There are more species of vertebrates living in the ocean than there are on land. Form a hypothesis to explain why this is true.

9. **CAREERS IN BIOLOGY** After ichthyologists discovered a new species of deep-sea predatory dragonfish, they were curious about the function of a long, thin luminescent protrusion called a barbel that was attached under its chin and trailed below its body. Design an experiment that would determine the function of the dragonfish's barbel.

Think Critically

10. **Draw Conclusions** Bluegill males make a nest and protect the eggs and newly hatched offspring. Sometimes intruding males are able to fertilize some of the eggs. The bluegill fathers can identify their biological offspring and will care only for them and not others that might hatch from the same nest. Why is it important for male bluegills to identify their own offspring and care only for them?

Section 28.2

Vocabulary Review

Complete the sentence by providing the missing vocabulary term from the Study Guide page.

11. A __________ is a four-footed animal with legs that have feet and toes that have joints.

Understand Key Concepts

12. Which illustration shows an external parasite?

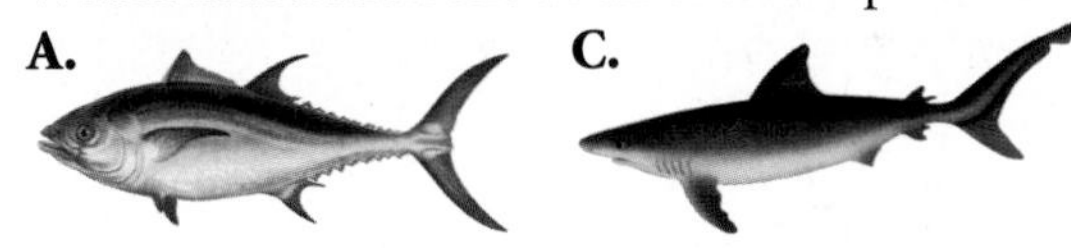

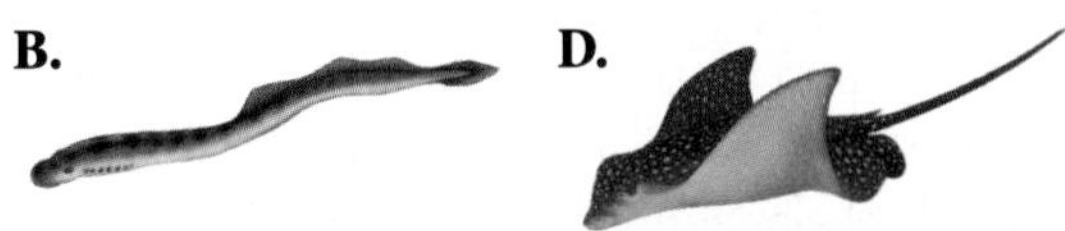

13. Which ancient extinct fishes are ancestors to modern fishes?
    A. ostracoderms
    B. ray-finned fishes
    C. coelacanths
    D. jawless fishes

14. Which are characteristics of sharks?
    A. jawless, cartilaginous skeleton, lateral line
    B. jawless, cartilaginous skeleton, ray-finned
    C. jaws, bony skeleton, swim bladder
    D. jaws, cartilaginous skeleton, lateral line

Constructed Response

15. **Open Ended** Sketch the body forms of each of the main groups of fishes. Include and explain external adaptations of fishes to their environment.

Think Critically

Use the diagram below to answer questions 16 and 17.

Biologists have studied muscle activity in trout swimming in a free stream flow compared to trout behind a barrier during the same stream movement. The red dots indicate the most intense muscle activity, orange indicates moderate muscle activity, while a white circle indicates no muscle activity.

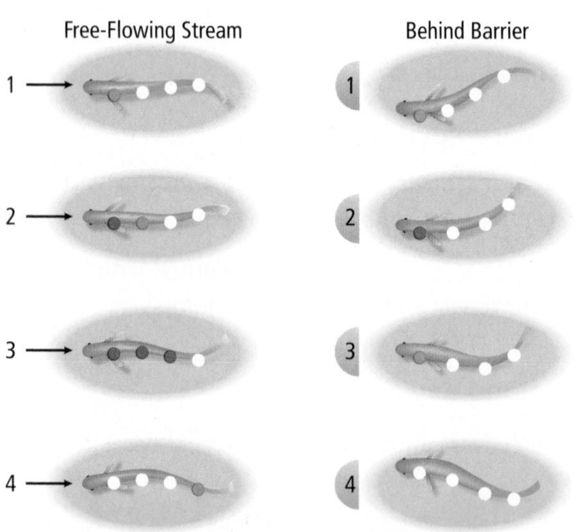

16. **Evaluate** In which situation does the trout use the most energy?

17. **Infer** Based on this experiment, if trout are trying to conserve energy, in which part of the stream would they be found?

Section 28.3

Vocabulary Review

Each of the following sentences is false. Make the sentence true by replacing the italicized word with a vocabulary term from the Study Guide page.

18. The *atrium* is a chamber that receives the digestive wastes, urinary waste, and eggs or sperm before they leave the body.

19. The *nictitating membrane* enables amphibians to hear sounds.

20. Amphibians have *tympanic membranes* to protect their eyes from drying out.

Understand Key Concepts

21. Which is a caecilian?

A.

C.

B.

D.

22. What does the phylogeny of tetrapods indicate about lobe-finned fishes?
    A. Lobe-finned fishes are the ancestors of amphibians.
    B. Lobe-finned fishes are similar to amphibians because they both have fins.
    C. Lobe-finned fishes are most closely related to hagfishes.
    D. Lobe-finned fishes are similar to amphibians because they both have lateral lines as adults.

23. What structures do amphibians use to maintain homeostatic water balance?
    A. nictitating membranes
    B. tympanic membranes
    C. kidneys
    D. swim bladders

Chapter Test biologygmh.com

24. Which is not associated with a tadpole?
   A. lungs
   B. tail
   C. gills
   D. herbivorous feeding

Constructed Response

25. **Open Ended** Draw a picture that would illustrate how amphibians are affected by increased exposure to ultraviolet light.

26. **Open Ended** Describe how the structure and physiology of amphibians, presently adapted to temperate and tropical climates, might be modified to enable them to live in colder climates.

27. **Open Ended** Describe how the senses of amphibians are adapted to life on land.

Think Critically

28. **Design an Experiment** Tadpole larvae of certain frogs gather in clusters so close together that the group looks like a moving football in the water. Design an experiment that would test a hypothesis about why the tadpoles exhibit this behavior.

29. **Create** Read the advertisements for homes in the newspaper to see how they are written, and write an ad for an amphibian home site based on what you know about the habitat, nutrition, and other needs of frogs.

Use the diagram below to answer question 30.

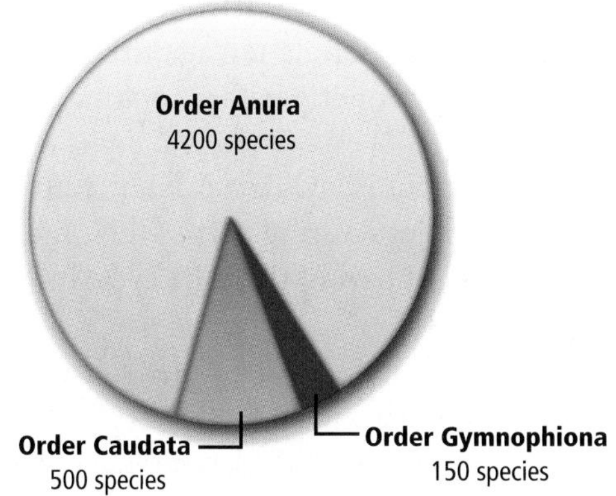

30. **Calculate** Determine the percent that each order of amphibians contributes to the total amount of amphibians.

Additional Assessment

31. WRITING in Biology Research what efforts are being made by scientists to preserve amphibians. Write a newspaper article summarizing what you learned.

DBQ Document-Based Questions

Scientists are trying to determine the cause or causes for the decline in amphibian populations over the past few decades. The graph below shows the results of one study in which the survival rate of amphibian embryos was measured against the depth of the water in which they developed.

Data obtained from: Kiesecker, J., et al. 2001. Complex causes of amphibian population declines. *Nature* 410: 681-683.

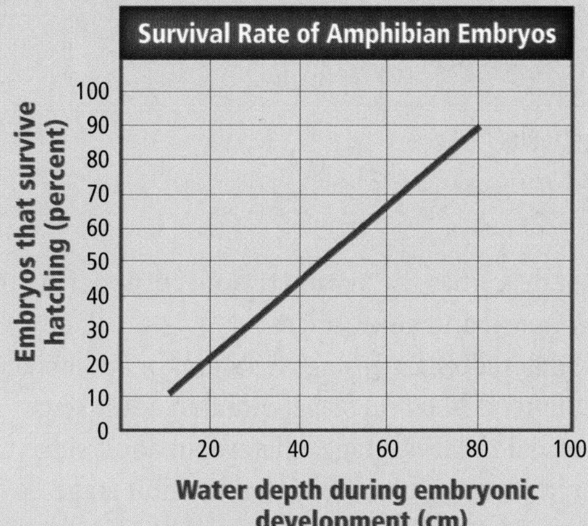

32. Describe the relationship between water depth during development and the survival rate of embryos.

33. Form a hypothesis about the decline of amphibian populations in relation to changes in climate.

Cumulative Review

34. Explain the theory of evolution in terms of violating Hardy-Weinberg conditions. **(Chapter 15)**

35. Describe the adaptations that helped plants survive on land. **(Chapter 21)**

Standardized Test Practice

Cumulative

Multiple Choice

Use the diagram below to answer question 1.

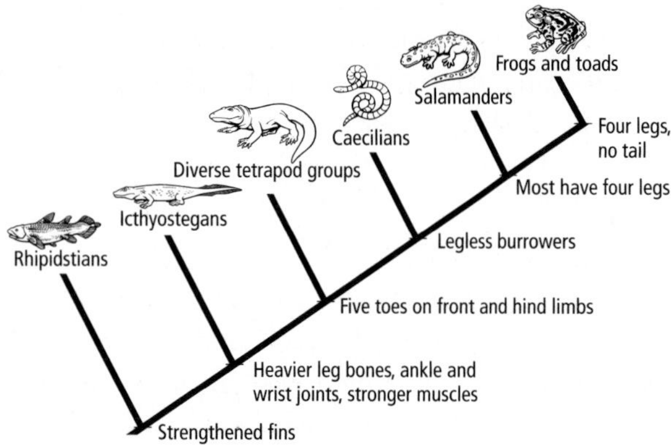

1. According to the cladogram, which is the earliest relative of the amphibians?
   A. ichthyostegans
   B. rhipidstians
   C. rays
   D. sharks

2. What describes the symmetry of echinoderms in their larval and adult stages?
   A. bilateral in larval stage, bilateral in adult stage
   B. bilateral in larval stage, radial in adult stage
   C. radial in larval stage, bilateral in adult stage
   D. radial in larval stage, radial in adult stage

3. Which is the role of the Malpighian tubules in arthropods?
   A. adding digestive enzymes to the intestines
   B. allowing oxygen into the body
   C. maintaining homeostatic water balance
   D. transporting blood to body tissues

4. Which method of communication does a honeybee use to tell others in the hive about the location of food?
   A. chemical pheromones
   B. complex dances
   C. quiet buzzing sounds
   D. rapid wing beating

5. What is the function of the simple eye in arthropods?
   A. to analyze landscapes during flight
   B. to detect colors
   C. to distinguish light from dark
   D. to see movement

Use the chart below to answer question 6.

Row	Group	Characteristics
1	Invertebrate chordates	Lack a backbone
2	Jawless fishes	Lack a notochord
3	Bony fishes	Have a skeleton made of bone
4	Cartilaginous fishes	Have a skeleton made of cartilage

6. Which row in the chart contains incorrect information?
   A. 1
   B. 2
   C. 3
   D. 4

7. Which statement describes the most reasonable way to prevent the disease of trichinosis in humans?
   A. Cook pork thoroughly before eating it.
   B. Treat infected pigs for trichinosis worm.
   C. Vaccinate the population against trichinosis.
   D. Wash pork properly before cooking it.

8. Which statement is NOT true about amphibians?
   A. Many lack legs during part of their life cycle.
   B. Many spend part of their life cycle in the water and part on land.
   C. Most depend on outside water sources to keep their bodies moist.
   D. Most have developed a lateral line system.

9. Which mutation is often caused by the addition or deletion of a single base pair?
   A. frame shift
   B. missense
   C. substitution
   D. tandem repeat

Short Answer

Use the diagrams below to answer question 10.

10. Describe a body structure from each of the mollusks shown above, and explain how these structures are related.

11. Sequence the energy transitions that have to take place in order for the Sun's energy to be used by a heterotroph.

12. Evaluate why the notochord is considered an evolutionary advancement.

13. Analyze which characteristics of a shark enabled it to be a fast swimmer. Explain your answer.

14. Name two groups of invertebrate chordates and describe how they feed. Relate their feeding patterns to their way of life.

15. Compare three characteristics of fishes to three characteristics of another group of animals that you already know.

Extended Response

16. Create a Venn diagram to organize information about endoderm and ectoderm tissues that form during embryonic development. Then explain how endoderm and ectoderm tissues are similar and different.

17. Explain how the first photosynthetic prokaryotes changed life on Earth.

18. Contrast the circulatory systems of lancelets and tunicates and justify the classification of both kinds of organisms as invertebrate chordates.

19. Hypothesize whether incomplete metamorphosis or complete metamorphosis in insects is more primitive. Explain your reasoning.

Essay Question

Most of the invertebrate animal groups living today trace their evolutionary history back more than 500 million years. Every other group of invertebrates living today expanded from the oceans to fresh water and, in many cases, to land. The fossil record for echinoderms shows that they have changed radically throughout their evolutionary history. Today, there are many diverse forms of echinoderms yet none have ever left the ocean.

Using the information in the paragraph above, answer the following question in essay format.

20. Hypothesize why echinoderms have continued living only in the ocean while other invertebrate groups have migrated to fresh water and land.

NEED EXTRA HELP?																				
If You Missed Question . . .	1	2	3	4	5	6	7	8	9	10	11	12	13	14	15	16	17	18	19	20
Review Section . . .	28.3	27.1	26.1	26.3	26.1	28.2	25.2	28.3	12.4	25.4	8.2, 8.3,	27.2	28.2	27.2	28.1	24.1	14.2	27.2	26.3	27.1

CHAPTER 29

Reptiles and Birds

BIG (Idea Reptile and bird adaptations enable them to live and reproduce successfully in terrestrial habitats.

Section 1
Reptiles

MAIN (Idea Reptiles are fully adapted to life on land.

Section 2
Birds

MAIN (Idea Birds have feathers, wings, lightweight bones, and other adaptations that allow for flight.

BioFacts

- The fangs of a rattlesnake lie flat on the roof of its mouth when its mouth is closed.
- When a rattlesnake's mouth is opened during a strike, its fangs rotate forward, ready to inject venom from the venom gland in the jaw through openings in the fangs.
- The speed of a rattlesnake strike is an amazing 2.4 m/s.

LAUNCH Lab

Are cultural symbols of reptiles and birds scientifically accurate?

Throughout history, reptiles and birds have been feared, revered, and symbolized. In this lab, you will review examples of symbolized reptiles and birds and determine if the representations are scientifically accurate.

Procedure

1. Read and complete the lab safety form.
2. Research symbols, stories, or legends about reptiles or birds from different cultures.
3. Analyze the information in the materials you find from Step 2 for scientific accuracy. Hypothesize as to why a reptile or bird was used as a symbol or legend in each situation.

Analysis

1. **Evaluate** How much of the information you analyzed was scientifically accurate? Why do you think some information was inaccurate?
2. **Synthesize** Choose one symbol or legend that contained inaccurate information and modify it so that it is scientifically accurate.

Visit biologygmh.com to:

- study the entire chapter online
- explore the Concepts in Motion, the Interactive Table, Virtual Labs, Microscopy Links, and links to virtual dissections
- access Web links for more information, projects, and activities
- review content online with the Interactive Tutor, and take Self-Check Quizzes

Characteristics of Reptiles and Birds Make the following Foldable to help you compare and contrast the characteristics of reptiles and birds.

STEP 1 Fold one sheet of paper lengthwise, leaving the holes uncovered.

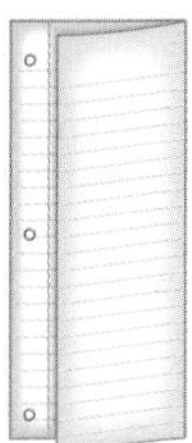

STEP 2 Fold into thirds.

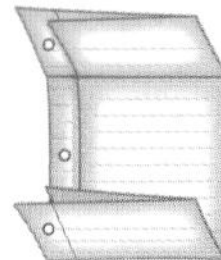

STEP 3 Unfold and draw overlapping ovals. Cut the top sheet along the folds.

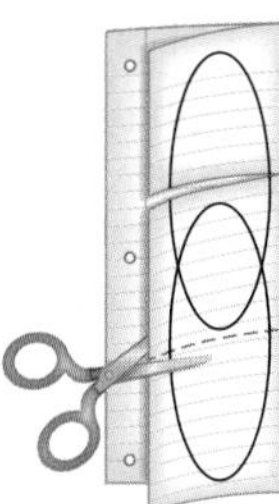

STEP 4 Label the Venn diagram as shown.

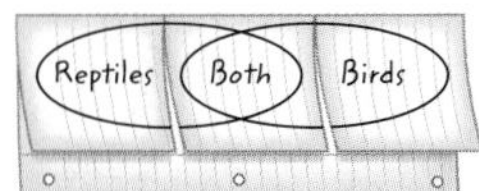

FOLDABLES **Use this Foldable with Sections 29.1 and 29.2.** As you read each section, record characteristics that are unique to reptiles and birds and those they have in common.

Section 29.1

Reading Preview

Objectives

- **Explain** the importance of the amniotic egg in the transition to life on land.
- **Summarize** the characteristics of reptiles.
- **Distinguish** between the orders of reptiles.

Review Vocabulary

embryo: the earliest stage of development of plants and animals after an egg has been fertilized

New Vocabulary

amnion
amniotic egg
Jacobson's organ
plastron
carapace

Reptiles

MAIN Idea **Reptiles are fully adapted to life on land.**

Real-World Reading Link Think about the last time you saw a movie in which a reptile was a main character. Maybe it was a giant anaconda or a ferocious *Tyrannosaurus rex.* Maybe it was an animated character that was funny. As you read this section, think about whether the characteristics of the movie reptile were scientifically accurate.

Characteristics of Reptiles

In Chapter 28, you learned that vertebrates with well-developed limbs, circulatory and respiratory systems, and other adaptations moved from water to land. However, amphibians were left vulnerable to the drying effects of life on land with their shell-less eggs and larvae that breathed through gills. In contrast, reptiles, like the Western fence lizard shown in **Figure 29.1,** are fully adapted to life on land and were the first completely terrestrial vertebrates. Characteristics that allow reptiles to succeed on land include a shelled egg, scaly skin, and more efficient circulatory and respiratory systems.

Amniotic eggs As you can see in the evolutionary tree in **Figure 29.1,** reptiles have characteristics in common with other groups that have an amnion and other membranes that surround the embryo as it develops. An **amnion** (AM nee ahn) is a membrane that surrounds a developing embryo. It is filled with fluid that protects the embryo during development. Animals that undergo this type of development are called amniotes and include reptiles, birds, and mammals.

Figure 29.1
Right: This Western fence lizard is one of 7000 species of reptiles belonging to class Reptilia. Reptiles live in a variety of terrestrial and aquatic habitats.
Left: The phylogenetic tree shows that reptiles, along with birds and mammals, have an amnion.

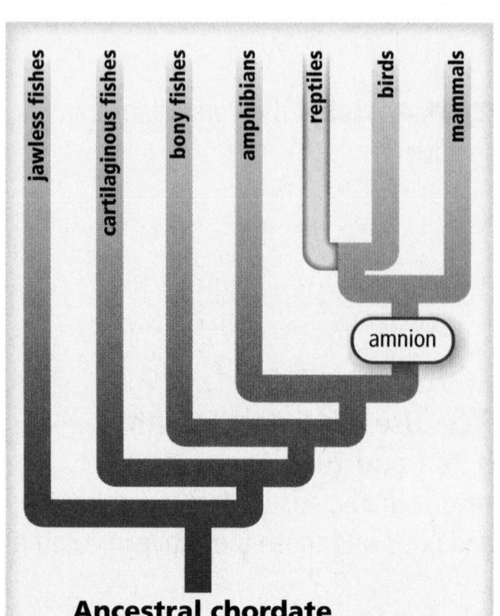

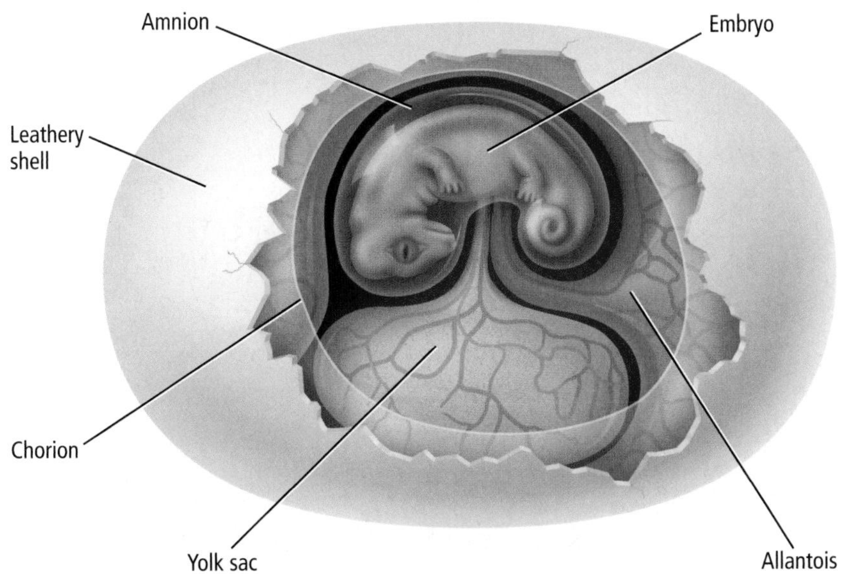

■ **Figure 29.2** The amniotic egg is protected by a shell and membranes with fluid that help to protect the embryo and keep it from drying out during development.

Concepts In Motion

Interactive Figure To see an animation of the form and function of an amniotic egg, visit biologygmh.com.

An **amniotic egg,** like the one shown in **Figure 29.2,** is covered with a protective shell and has several internal membranes with fluids contained between the membranes. Inside the egg, the embryo is self-sufficient because it gets its nutrition from food in the yolk sac inside the egg. Bathing the embryo within the amnion is amniotic fluid. Amniotic fluid mimics the aquatic environments of fish and amphibian embryos. The allantois (uh LAN tuh wus) is a membrane that forms a sac that contains wastes produced by the embryo. The outermost membrane of the egg is the chorion (KOR ee ahn), which allows oxygen to enter and keeps fluid inside the egg. In reptiles, the leathery shell protects the internal fluids and embryo, and prevents the egg from drying out on land. In birds, the shell is hard instead of leathery.

Dry, scaly skin In addition to keeping fluid in their eggs, reptiles also must keep fluids in their bodies. The dry skin of reptiles keeps them from losing internal fluids to the air. A layer of scales on the exterior of many reptiles also keeps them from drying out. However, one problem with having a tough outer covering is that an organism could have difficulty growing larger. In order to grow, some reptiles, like the snake in **Figure 29.3,** periodically must shed their skins in a process called molting. You might have seen the molt of a snake's skin while hiking a nature trail.

■ **Figure 29.3** Some reptiles molt as they grow larger.
Compare ***molting in reptiles to molting in arthropods.***

Respiration Most reptiles, except for some aquatic turtles, depend primarily on lungs for gas exchange. Recall that when amphibians breathe, they squeeze their throats to force air into their lungs. Reptiles are able to suck air into their lungs, or inhale, by contracting muscles of the rib cage and body wall to expand the upper part of the body cavity in which the lungs are held. They exhale by relaxing these same muscle groups. Reptiles exchange gases in lungs that have larger surface areas for gas exchange than the lungs of amphibians. With more oxygen, more energy can be released through metabolic reactions and made available for more complex movements.

Reading Check **Evaluate** why the amniotic egg is important for an animal to be able to live exclusively on land.

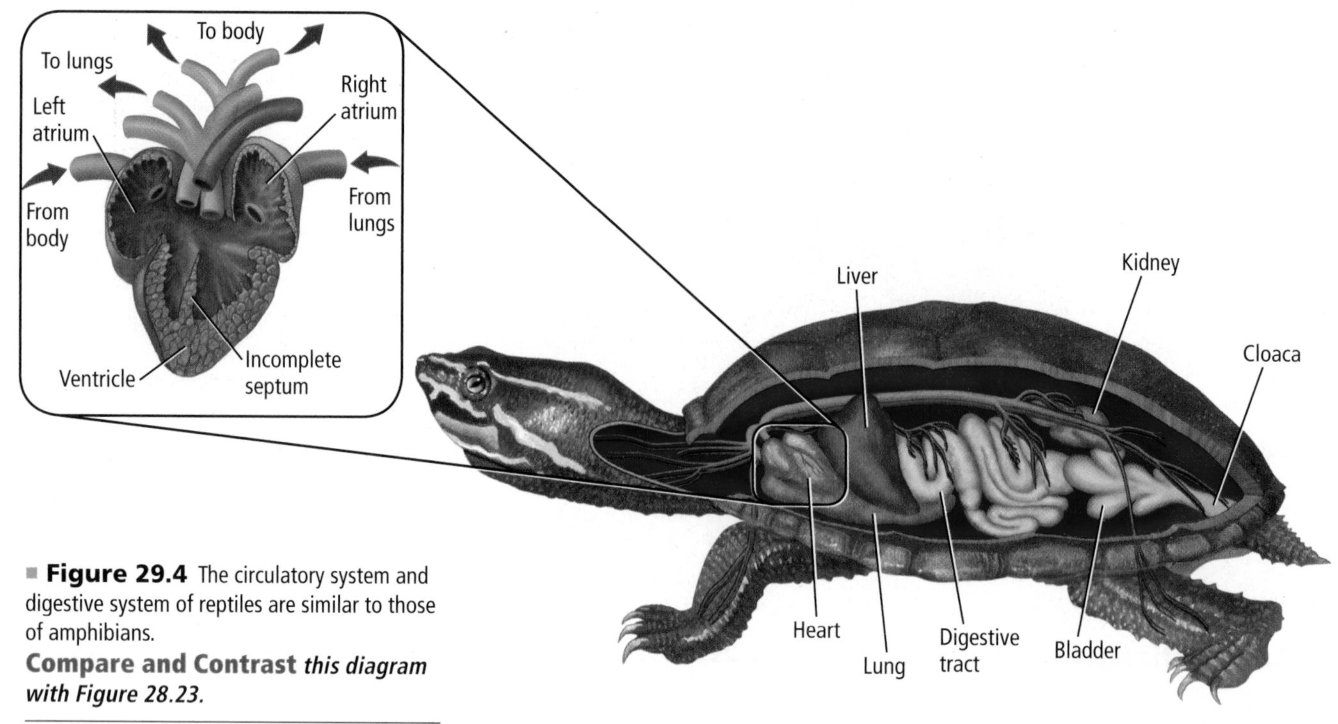

■ **Figure 29.4** The circulatory system and digestive system of reptiles are similar to those of amphibians.
Compare and Contrast *this diagram with Figure 28.23.*

Circulation In most reptiles, oxygen from the lungs enters into a circulatory system that is similar to that of amphibians. Most reptiles have two separate atria and one ventricle that is partially divided by an incomplete septum, as shown in **Figure 29.4.** In crocodiles, however, the septum in the ventricles is complete, thereby resulting in a four-chambered heart. The separation into two ventricles keeps oxygen-rich blood separate from the oxygen-poor blood throughout the heart.

Because reptiles generally are larger than amphibians, they need to pump blood forcefully enough to reach parts of the body far away from the heart. In an example from the past, the dinosaur *Brachiosaurus* had to pump blood more than 6 m from the heart to the head!

Feeding and digestion The organs of the digestive system of reptiles, shown in **Figure 29.4,** are similar to those of fish and amphibians. Reptiles have a variety of feeding methods and diets. Most reptiles are carnivores, but some, such as iguanas and tortoises, are herbivores that feed on plants, and some turtles are omnivores. Turtles and crocodiles have tongues that help them swallow. Some lizards, such as the chameleon, have long, sticky tongues for catching insects.

Snakes have the ability to ingest prey much larger than themselves. The bones of the skull and jaws of snakes are joined loosely so that they can spread apart when taking in large food materials, as shown in **Figure 29.5.** To swallow, the opposite sides of the upper and lower jaws can alternately thrust forward and retract to draw in the food. Some snakes have venom that can paralyze and begin digestion of their prey.

■ **Figure 29.5** Snakes can consume a meal that is larger than their mouths because their jaws are loosely jointed, and upper and lower jaws can move independently of each other.

Excretion The excretory system of reptiles is adapted to life on land. The kidneys, such as the one shown in **Figure 29.4,** filter the blood to remove waste products. When urine enters the cloaca, water is reabsorbed to form uric acid, which is a semisolid excretion. This method of water reabsorption enables reptiles to conserve water and maintain homeostasis of water and minerals in their bodies.

The brain and senses Reptile brains are similar to amphibian brains, except that the cerebrum of reptiles is larger. Because vision and muscle function are more complex, the optic lobes and cerebellum portions in the brain of reptiles are larger than those of amphibians. Vision is the most sensitive sense for most reptiles, and some reptiles even have color vision. Hearing varies in reptiles. Some reptiles have tympanic membranes similar to those of amphibians, while others, such as snakes, detect vibrations through their jaw bones.

The sense of smell is more highly evolved in reptiles than it is in amphibians. You might have seen a snake rapidly flicking its forked tongue. When a snake sticks its tongue out, odor molecules stick to it. The snake then brings the tongue and the odor molecules into its mouth. Inside the mouth, the odor molecules transfer to a pair of saclike structures that sense odors called **Jacobson's organs**. **Figure 29.6** shows one of these structures. Without Jacobson organs, snakes would not be able to find prey or mates.

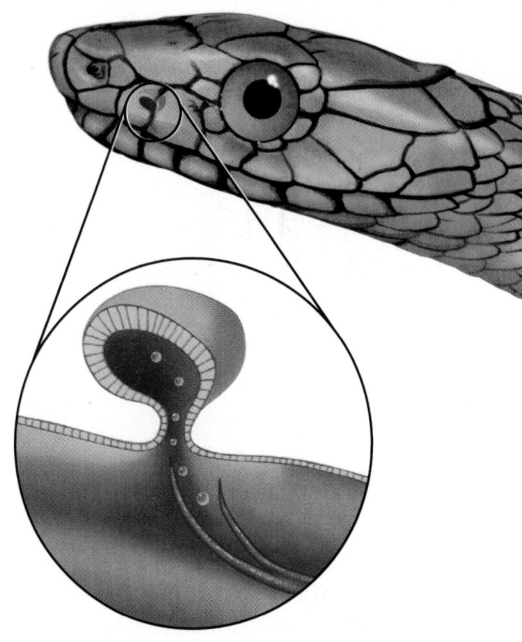

Figure 29.6 In snakes, Jacobson's organs in the mouth are used to sense odors.

Temperature control Like amphibians, reptiles are ectotherms that cannot generate their own body temperatures. Because they cannot regulate their body temperatures internally, they must regulate it behaviorally. You might have seen a turtle basking on a rock on a sunny day. Heat from the Sun and the rock raise the turtle's body temperature. Body temperature can be lowered by moving into the shade or a cool burrow. Some reptiles in temperate regions survive winter by burrowing or going into a state of inactivity with lower body metabolism and lower body temperature. Others, such as some snakes, gather together in masses of hundreds during the winter. Heat loss is reduced when the snakes are covering each other.

Movement Compare the leg position of the salamander to the leg position of the crocodile shown in **Figure 29.7.** Note that the salamander's belly is on the ground, while the crocodile's belly is above the ground. Like amphibians, some reptiles move with limbs sprawled to their sides and push against the ground while swinging their bodies from side to side. Crocodiles, however, have their limbs rotated farther under the body and, as a result, can bear more weight and move faster. To bear more body weight on land, reptiles' skeletons are stronger with heavier bone structure. Reptiles also have claws on their toes which aid in digging, climbing, and gripping the ground for traction.

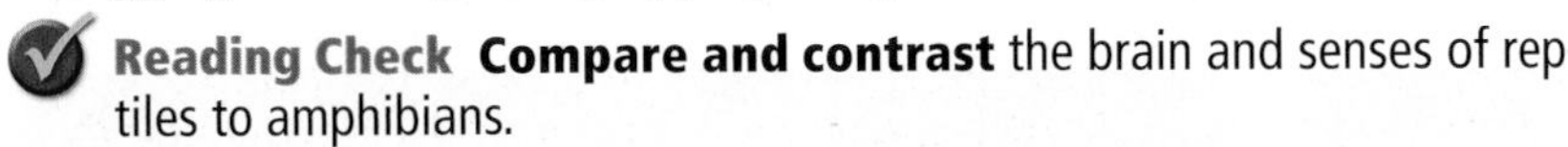

Reading Check **Compare and contrast** the brain and senses of reptiles to amphibians.

Figure 29.7 Salamanders move with splayed legs pushing against the ground as their bodies drag along. Crocodiles have legs that are rotated underneath their bodies, which holds their bodies off the ground.

Salamander

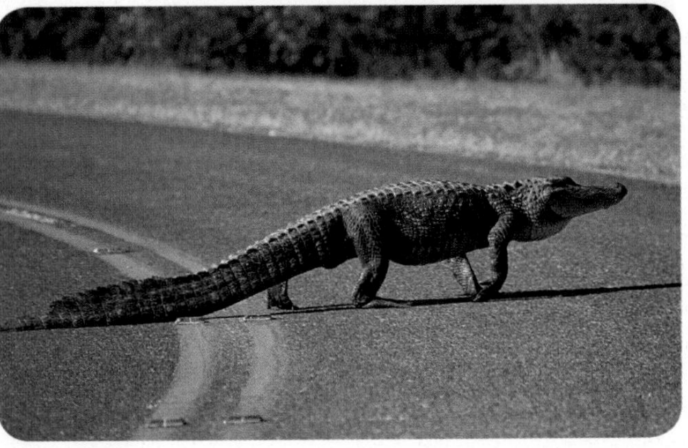

Crocodile

Reproduction Recall from Chapter 28, amphibian females lay eggs that later get fertilized. Reptile reproduction is significantly different, mainly because reptiles have internal fertilization. After fertilization, an amniotic egg and embryo develop. The yolk of the egg nourishes the embryo. The female reproductive system then produces a leathery shell around the egg. The female usually digs a hole and lays her eggs in the ground or in plant debris. After laying the eggs, most female reptiles leave them unattended to hatch. Alligators and crocodiles build a nest in which to lay eggs and tend to young after they hatch. Some snakes and lizards keep their eggs in their bodies until they hatch. In this way, the eggs are protected in the mother's body until they are fully developed young.

Diversity of Modern Reptiles

There are currently four living orders of reptiles—snakes and lizards belonging to order Squamata (skwuh MAHD uh), crocodiles and alligators belonging to order Crocodilia, turtles and tortoises belonging to order Testudinata, and tuataras belonging to order Sphenodonta (sfee nuh DAHN tuh).

VOCABULARY

WORD ORIGIN

Squamata
squama- from Latin, meaning *scale.*
-ata from Latin, meaning *to bear.*

Lizards and snakes Lizards commonly have legs with clawed toes. Also, lizards usually have movable eyelids, a lower jaw with a movable hinge joint allowing for flexibility in jaw movement, and tympanic membranes. Common lizards include iguanas, chameleons, geckos, and anoles. An iguana is shown in **Figure 29.8.**

Snakes are legless and have shorter tails than lizards. Snakes lack movable eyelids and tympanic membranes. Like lizards, however, snakes have joints in their jaws enabling them to eat prey larger than their heads. Some snakes, such as the rattlesnake shown at the beginning of the chapter, have venom that can slow down or even kill their prey. Other snakes, such as the python shown in **Figure 29.8,** anacondas, and boas, are constrictors. Constrictors generally are very large snakes. They suffocate their prey by wrapping around the prey's body and tightening until the prey dies because it no longer can breathe.

Reading Check **Describe** the different methods by which snakes capture prey.

Figure 29.8 The green iguana and the green tree python are both members of order Squamata.

Green iguana

Green tree python

Eastern box turtle

American alligator

■ **Figure 29.9** The shell of a turtle helps protect it from predators. An alligator has a broad snout and thick scales covering its body.

Turtles Turtles are unique because they are encased by a protective shell, as shown in **Figure 29.9.** A turtle can hide from predators by pulling its head and legs inside this hard shell. The ventral part of the shell is called the **plastron** (PLAS trahn) and the dorsal part of the shell is called the **carapace** (KAR ah pays). The vertebrae and the ribs of most turtles are fused to the inside of the carapace. Another unique aspect of turtles is that they do not have teeth. Instead, they have a sharp beak that can deliver a powerful bite. Like other reptiles, there are aquatic turtles and terrestrial turtles. Turtles that live on the land are called tortoises.

Crocodiles and alligators Order Crocodilia includes crocodiles, alligators, and caimans. Unlike most reptiles, Crocodilians have a four-chambered heart. Because a four-chambered heart can deliver oxygen more efficiently to their powerful muscles, crocodilians move quickly and aggressively, both in and out of the water. These quick movements help in capturing large prey.

Crocodiles have a long snout, sharp teeth, and powerful jaws. Alligators, like the one in **Figure 29.9,** generally have a broader snout than crocodiles. The upper jaw of an alligator is wider than the lower jaw. When an alligator closes its mouth, the upper jaw overlaps the lower jaw and its teeth are almost completely covered. The upper and lower jaws of a crocodile are about the same width. So when a crocodile closes its mouth, some teeth in the lower jaw are easily visible. Caimans are closely related to alligators but lack a bony separation between their nostrils. The teeth of crocodiles are similar to those of dinosaurs and the earliest birds.

Tuataras Tuataras (tyew ah TAR ahz) look like large lizards, as shown in **Figure 29.10.** Tuataras have a spiny crest that runs down the back and a "third eye" on top of the head. This structure is covered with scales but can sense sunlight. Biologists think that it might keep the tuatara from overheating in the Sun. One distinguishing feature of tuataras is that they have unique teeth compared to those of other reptiles. Two rows of teeth on the upper jaw shear against one row in the lower jaw, making them effective predators of small vertebrates. The only two living species of tuataras are found exclusively on islands off the coast of New Zealand.

■ **Figure 29.10** Tuataras reach a length of about 2 m and can live up to 80 years in the wild.

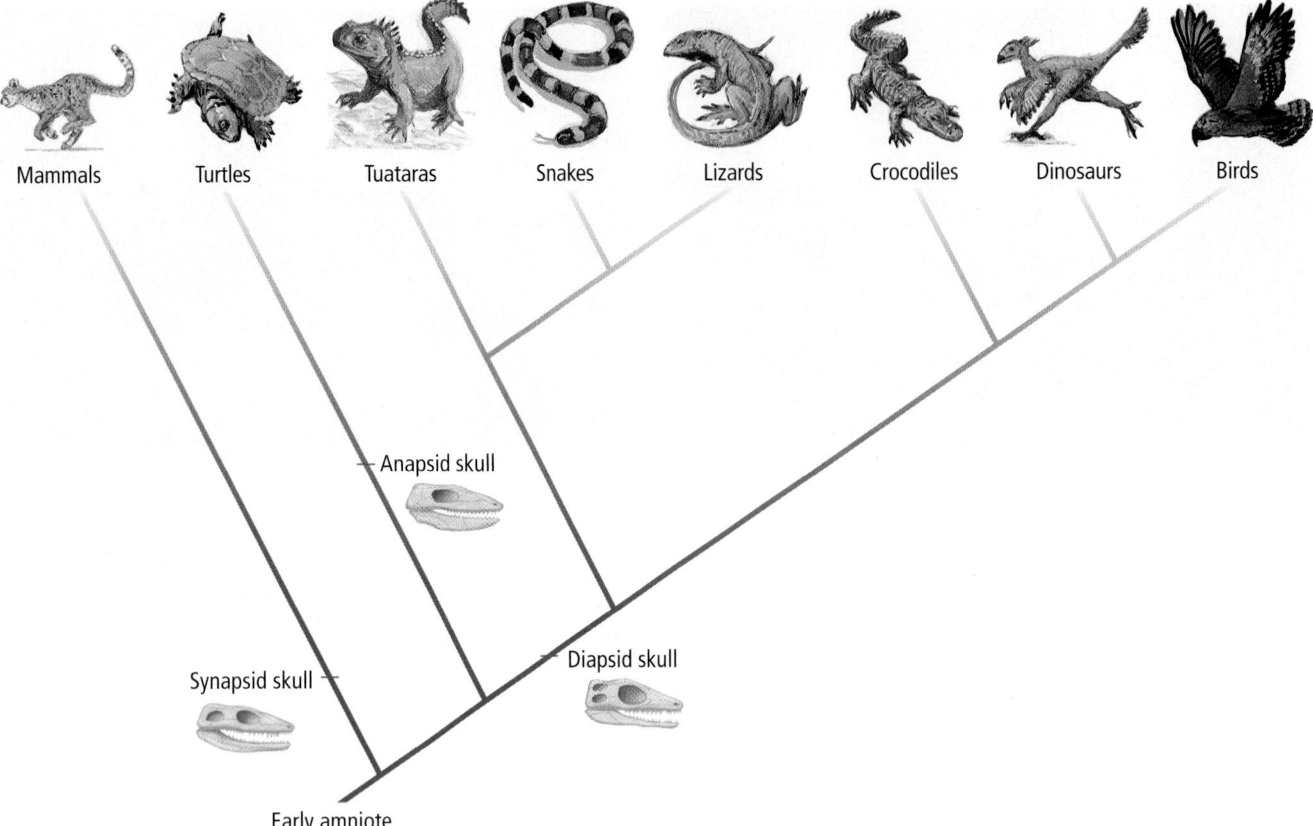

■ **Figure 29.11** The cladogram shows one interpretation of the relationships between amniotes.
Interpret ***Which modern reptiles evolved first? Which evolved most recently?***

Vocabulary
Academic Vocabulary
Interpretation :
A particular adaptation or version of a work, method, or style.
Scientists might make several interpretations of fossil evidence.

Evolution of Reptiles

The cladogram in **Figure 29.11** shows one interpretation of how early amniotes underwent adaptive radiation, giving rise to reptiles as well as modern birds and mammals. Recall that amniotes are vertebrates in which the embryo is encased in an amniotic membrane. As shown in the cladogram, early amniotes separated into three lines, each having a different skull structure. Anapsids, which might have given rise to turtles, have a skull that has no openings behind the eye sockets. Diapsids, which gave rise to crocodiles, dinosaurs, modern birds, tuataras, snakes, and lizards, have a skull with two pairs of openings behind each eye socket. Synapsids, which gave rise to modern mammals, have one opening behind each eye socket.

Reading Check **Identify** which part of a reptile fossil would be a major indicator in classifying it as a lizard or a dinosaur.

Dinosaurs For 165 million years, dinosaurs dominated Earth. Some, such as *Tyrannosaurus rex,* stood almost 6 m high, were 14.5 m long, weighed more than 7 tonnes, and were predatory. Others, such as *Triceratops,* had massive horns and were herbivores. Despite their diversity, dinosaurs can be divided into two groups based on the structure of their hips. A comparison of the two groups is shown in **Figure 29.12.** Saurischians (saw RISK ee unz) had hip bones that radiated out from the center of the hip area. In Ornithischians, some bones projected back toward the tail.

Like birds and crocodiles, some dinosaurs built nests and cared for eggs and young. Some dinosaurs might have had the ability to regulate their body temperatures. Fossil evidence shows that one group of dinosaurs had feathers and evolved into today's birds.

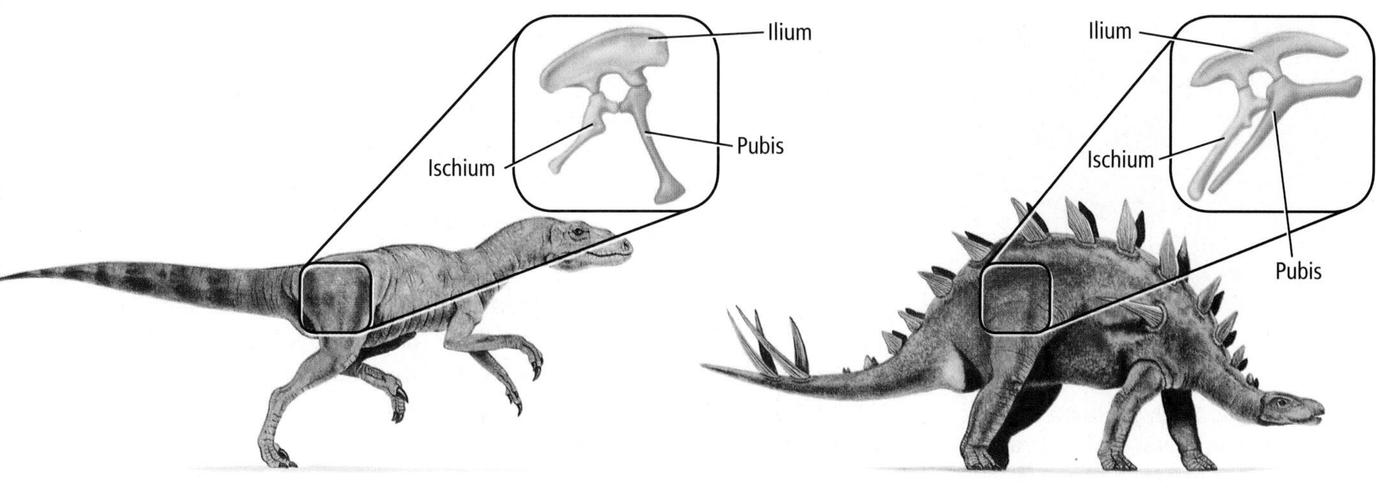

■ **Figure 29.12** Saurischians had a hip bone that pointed forward. Ornithischians had the same bone pointing back toward the tail end of the animal.

Connection to Earth Science The Cretaceous Period is known for worldwide mass extinction of many species, including all dinosaurs. Some scientists hypothesize that a meteorite crashed to Earth and caused this extinction. Clouds of dust might have blocked the Sun, causing a much cooler climate to develop. This change, along with fires, toxic dust, and gases, could have caused the death of many plants and animals at this time. When dinosaurs disappeared, the niches they had occupied were made available for other vertebrates to evolve.

DATA ANALYSIS LAB 29.1

Based on Real Data*

Interpret the Data

How fast did dinosaurs grow? Scientists study thin sections of fossilized bone tissue to determine how rapidly bone grew. By studying how quickly dinosaurs grew, scientists can learn about their populations and ecology.

Think Critically

1. **Compare** During what age span did the dinosaurs experience the greatest growth? Explain.
2. **Analyze Data** Which dinosaur grew at the slowest rate? The fastest rate?
3. **Infer** Fast-growing bones have many blood vessels. How would the bones of *Tyrannosaurus* compare to those of *Daspletosaurus*?

Data and Observations

The graph shows bone-based growth curves comparing several dinosaurs.

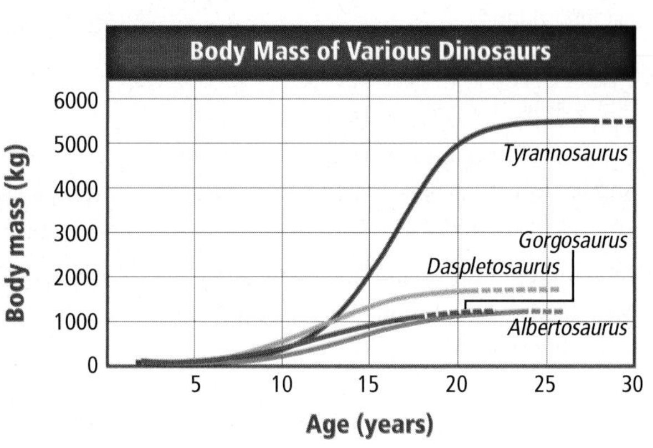

*Data obtained from: Stokstad, E. 2004. Dinosaurs under the knife. *Science* 306: 962-965.

■ **Figure 29.13** The San Francisco garter snake *(Thamnophis sirtalis tetrataenia)* lives in wetlands or grasslands near ponds and marshes.

Ecology of Reptiles

Reptiles are important parts of food chains both as prey and as predators. The balance of an ecosystem can be disrupted when a reptile species is removed. For example, when certain snakes are removed from an environment, rodent populations can increase. Loss of habitat and the introduction of exotic species are factors that contribute to the decline in population of some reptile species.

Habitat loss Both the American alligator *(Alligator mississippiensis)* and the American crocodile *(Crocodylus acutus)* have been affected by habitat loss in the Florida Everglades. The destruction and fragmentation of wetlands for building development has led to reduced numbers of these reptiles. The American crocodile remains endangered, with only 500-1200 remaining in Florida. With the passage of laws to protect wetlands in certain areas, the American alligator population has rebounded enough so that its status has been changed from endangered to threatened.

Introduction of exotic species An exotic species is a species that is not naturally found in an area, and when introduced, the local animals might suffer due to predation or competition for resources. For example, when the mongoose, a small mammal, was introduced into Jamaica to kill rats in sugarcane fields, the mongoose fed mostly on several lizard species which are now endangered. This includes the Jamaican iguana, which was thought to have been extinct due to the introduction of the mongoose. In 1990, a small population was discovered in a remote area of Jamaica.

Some species, such as the San Francisco garter snake shown in **Figure 29.13,** have suffered a population decline due to both habitat loss and the introduction of exotic species. The use of land for building houses, other buildings, and agriculture has led to habitat loss for this snake. The American bullfrog, which is not native to California, eats both the garter snake as well as the red-legged frog, a food source of the garter snake.

Section 29.1 Assessment

Section Summary

- Reptiles have several types of adaptations for life on land.
- Eggs of reptiles are adapted to development on land.
- Reptiles belong to four living orders: snakes and lizards, crocodiles and alligators, turtles and tortoises, and tuataras.
- Modern reptiles evolved from early amniotes. Many ancient reptiles, including dinosaurs, became extinct.

Understand Main Ideas

1. MAIN Idea **Identify** features that allow reptiles to live on land successfully.
2. **Describe** the parts of an amniotic egg. How did this structure allow the move to land?
3. **Compare and contrast** members of order Squamata with members of order Sphenodonta.
4. **Explain** the difference between anapsids, diapsids, and synapsids. Which gave rise to groups of reptiles?

Think Critically

5. **Formulate Models** Make a model of the amniotic egg shown in **Figure 29.2.** Relate the function of each membrane.

MATH in Biology

6. The biting force of alligators is directly proportional to their lengths. An alligator that is 1 m long has a biting force of 268 kg. What is the biting force of a larger alligator that is 3.6 m long?

Biology Online Self-Check Quiz biologygmh.com

Section 29.2

Reading Preview

Objectives

- **Summarize** the characteristics of birds.
- **Relate** the adaptations of birds to their ability to fly.
- **Describe** different orders of birds.

Review Vocabulary

terrestrial: living on or in land

New Vocabulary

endotherm
feather
contour feather
preen gland
down feather
sternum
air sac
incubate

Birds

MAIN Idea **Birds have feathers, wings, lightweight bones, and other adaptations that allow for flight.**

Real-World Reading Link You probably have heard the sayings: "Free as a bird," "Birds of a feather flock together," or "Light as a feather." As people talk, listen for "bird words." As you read, see if these sayings refer to real science.

Characteristics of Birds

Suppose your teacher asked you to describe a bird. You might respond that birds have feathers and that they fly. Birds belong to class Aves and include about 8600 species, making them the most diverse of all terrestrial vertebrates. Birds range in size from tiny hummingbirds hovering over bright flowers to large flightless ostriches running across the African plains. Birds are found in deserts, forests, mountains, prairies, and on all seas.

As shown on the evolutionary tree in **Figure 29.14,** birds and reptiles have a common ancestor. Birds have many characteristics that demonstrate their reptilian roots. For example, birds lay amniotic eggs. In addition, scales similar to those of reptiles cover the legs of birds.

You can think of a bird as a collection of adaptations to a lifestyle that includes flight. The adaptations include being able to generate their own body heat internally, feathers, and lightweight bones. The respiratory and circulatory systems of birds are also adapted to provide more oxygen to working muscles to support flight.

Endotherms Unlike reptiles, birds are endotherms. An **endotherm** is an organism that generates its body heat internally by its own metabolism. The high metabolic rate associated with endothermy generates a large amount of ATP that can be used to power flight muscles or for other purposes. Endotherms generate heat due to their normal body metabolism. The body temperature of a bird is about 41°C. Your body temperature is about 37°C. A high body temperature enables the cells in a bird's flight muscles to use the large amounts of ATP needed for rapid muscle contraction during flight.

Reading Check **Explain** why endothermy is an important adaptation for flight.

Feathers Birds are the only living animals to have feathers. **Feathers** are specialized outgrowths of the skin of birds. They are made of keratin (KER ah tihn), a protein in the skin that also makes up hair, nails, and horns of other animals. Feathers have two main functions: flight and insulation. Feathers keep heat generated during metabolism from escaping from the body of the bird. When a bird fluffs its feathers, it creates a dead air space that traps the heat. Similarly, if you are covered with a quilt while you sleep, the quilt creates dead air space between you and the cool air in the room so that you do not lose body heat.

Figure 29.14 The evolutionary tree shows that feathers are a unique characteristic of birds.

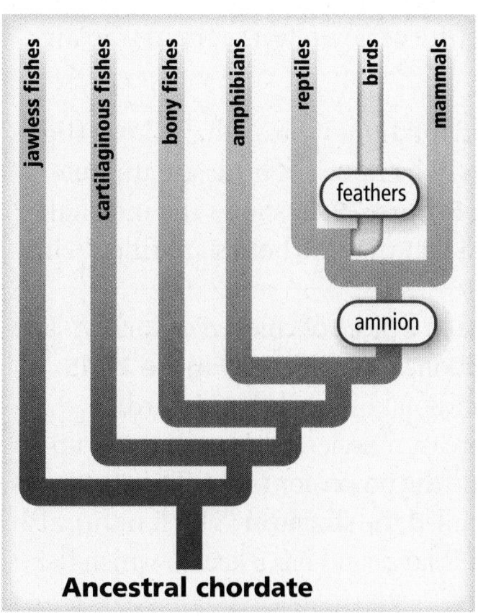

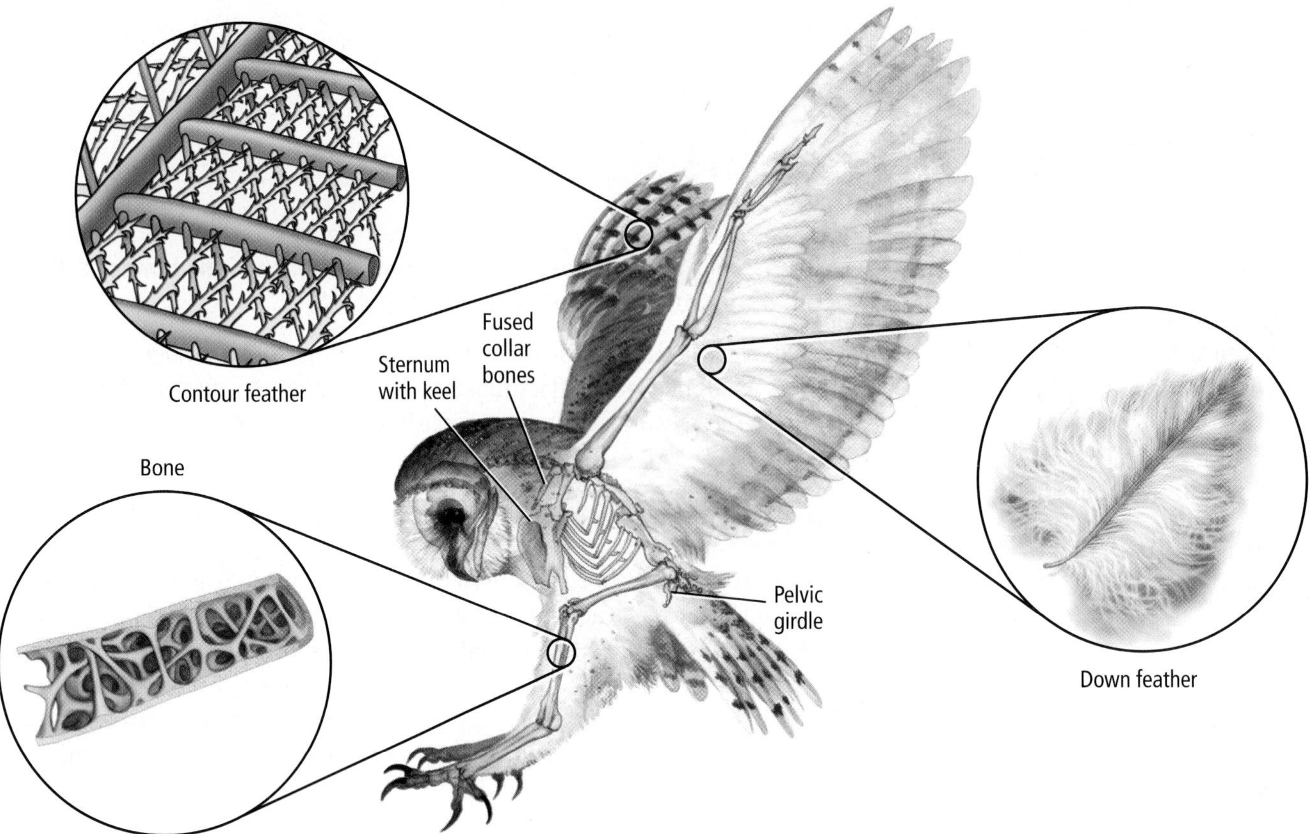

■ **Figure 29.15** Birds have contour feathers, down feathers, and lightweight bones.

Concepts In Motion

Interactive Figure To see an animation of the adaptations of a bird, visit biologygmh.com.

VOCABULARY

SCIENCE USAGE V. COMMON USAGE

Preen

Science usage: to maintain or repair using a bill (of a bird).

The bluejay was preening its feathers before it flew away.

Common usage: to gloat or congratulate oneself on an achievement.

Jim was preening over his victory at the track meet.

Feathers that cover the body, wings, and tail of a bird are called **contour feathers.** Examine the contour feathers shown in **Figure 29.15.** Contour feathers consist of a shaft with barbs that branch off. Barbules branch off barbs and are held together by hooks. If two adjoining barbs become separated, they can be rejoined like the teeth of a zipper. Birds repair broken links when they preen their feathers. They use their bills to preen their feathers, drawing the length of the feather through the bill to zip up broken links. Birds spend a large amount of time maintaining their feathers. Many birds have a **preen gland,** a gland located near the base of the tail that secretes oil. During preening, birds spread oil from the preen gland over their feathers, thereby adding a waterproofing coating. **Down feathers,** shown in **Figure 29.15,** are soft feathers located beneath contour feathers. Down feathers do not have hooks to hold barbs together. As a result, the looser structure of down feathers can trap air that acts as insulation.

Lightweight bones Another adaptation of birds that allows flight is their strong, lightweight skeletons. The bones of birds are unique because they contain cavities of air. **Figure 29.15** shows the internal structure of a bird bone. Despite the fact that the bones are filled with air, they are still strong.

Have you ever found the wishbone in a piece of chicken or turkey? The wishbone is formed from fused collarbones, as shown in **Figure 29.15.** Fusion of bones in the skeleton of a bird makes the skeleton sturdier, another adaptation for flight. Large breast muscles, which can make up 30 percent of a bird's total weight, provide the power for flight. These muscles connect the wing to the breastbone, called the **sternum** (STUR num), also shown in **Figure 29.15.** The sternum is large and has a keel to which the muscles attach.

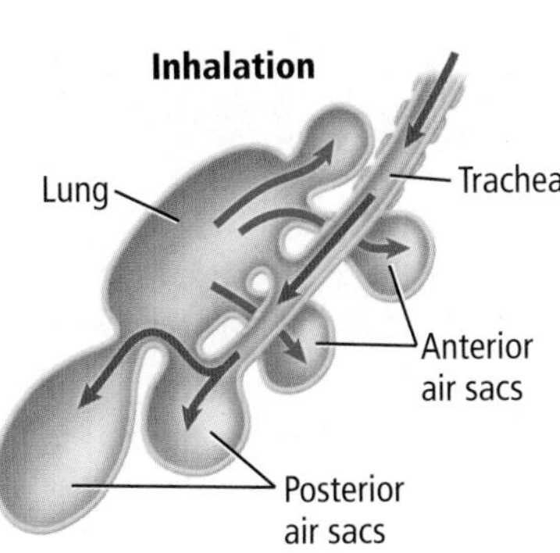

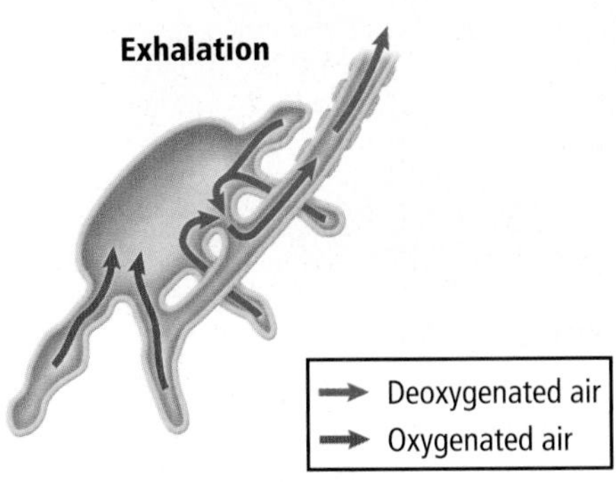

Figure 29.16 When a bird breathes, air always flows in a single direction, and highly efficient gas exchange can be achieved.

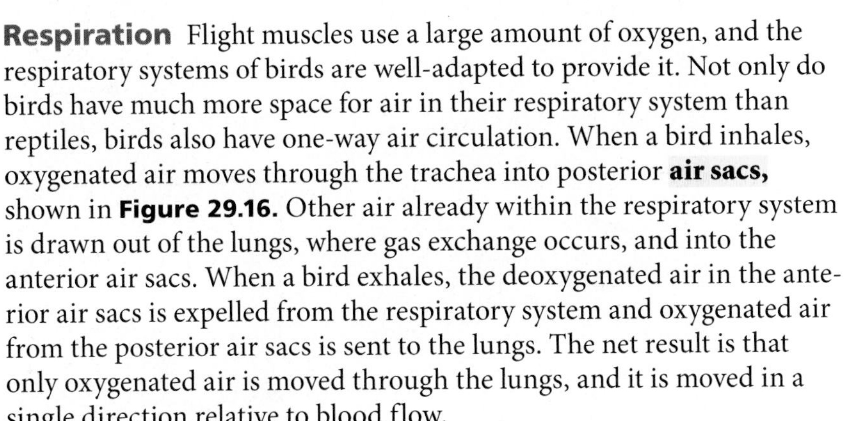

Respiration Flight muscles use a large amount of oxygen, and the respiratory systems of birds are well-adapted to provide it. Not only do birds have much more space for air in their respiratory system than reptiles, birds also have one-way air circulation. When a bird inhales, oxygenated air moves through the trachea into posterior **air sacs,** shown in **Figure 29.16.** Other air already within the respiratory system is drawn out of the lungs, where gas exchange occurs, and into the anterior air sacs. When a bird exhales, the deoxygenated air in the anterior air sacs is expelled from the respiratory system and oxygenated air from the posterior air sacs is sent to the lungs. The net result is that only oxygenated air is moved through the lungs, and it is moved in a single direction relative to blood flow.

Circulation A bird's circulatory system also helps it maintain high levels of energy by efficient delivery of oxygenated blood to the body. Recall that crocodiles are the only reptiles to have a heart ventricle completely divided by a septum. Birds also have a four-chambered heart, as shown in **Figure 29.17.** Having two ventricles keeps the oxygenated and deoxygenated blood separated and makes delivery of oxygenated blood more efficient. The left atrium receives blood from the lungs. This blood is pumped into the left ventricle and out to the body. Blood returning from the body is delivered to the right atrium, then moves into the right ventricle and on to the lungs where it will pick up more oxygen.

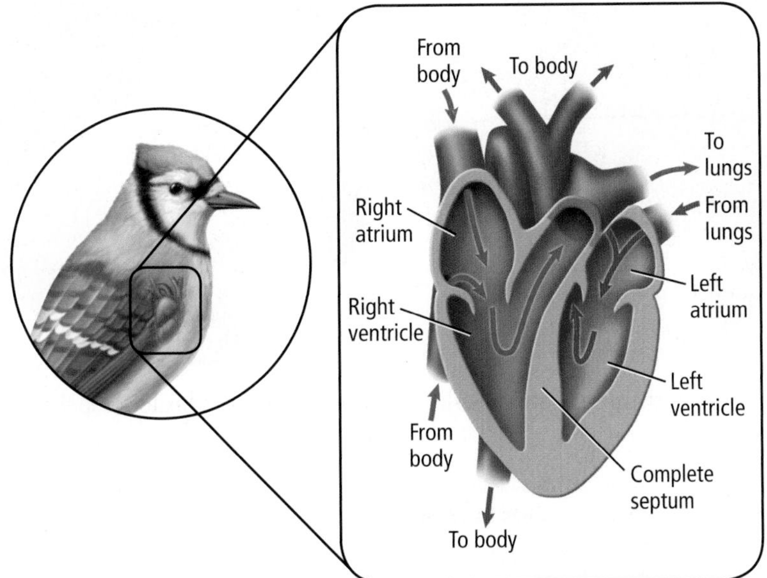

Figure 29.17 Birds have a four-chambered heart that keeps oxygenated and deoxygenated blood separate.
Compare *the heart of a bird to that of the reptile shown in Figure 29.4.*

Visualizing Feeding and Digestion

Figure 29.18
Examine the organs in the digestive system of a bird. Aside from having unique adaptations to their digestive systems, birds have beaks that are adapted to the type of food they eat.

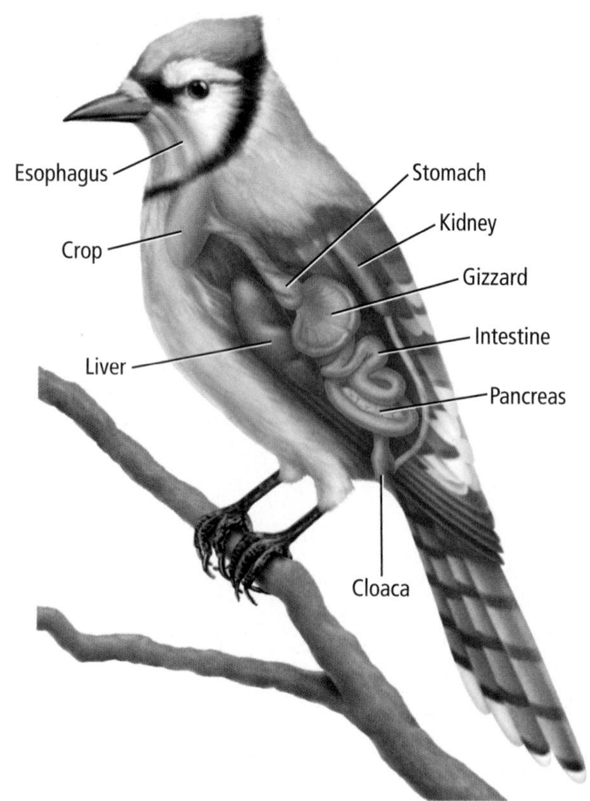

Herons use their long, thin, sharp bills to stab and capture fish and small amphibians as prey.

Hummingbirds have long, thin beaks shaped for drinking nectar from flowers.

An eagle uses its sharp beak to tear flesh from its prey.

A pelican uses its beak to scoop fish out of the water.

Concepts In Motion **Interactive Figure** To see an animation of feeding and digestion in birds, visit biologygmh.com.

Biology Online

Feeding and digestion Birds require large amounts of food to maintain their high metabolic rate. Once they have taken in food, birds process it with unique adaptations of their digestive systems, shown in **Figure 29.18.** Many birds have a storage chamber, called the crop, at the base of their esophagus. The crop stores food that the bird is ingesting. From the crop, food moves to the stomach. The posterior end of the stomach is a thick, muscular sac called the gizzard (GIH zurd). The gizzard often contains small stones that, together with the muscular action of the gizzard, crush food the birds have swallowed. The smaller food particles that result are easier to digest. Birds have no teeth and cannot chew their food. Digestion and absorption of food occurs primarily in the small intestine where secretions from the pancreas and liver aid the digestive process.

Excretion As in reptiles, bird kidneys filter wastes from the blood and convert it to uric acid. Birds also have a cloaca, shown in **Figure 29.18,** where the water is reabsorbed from the uric acid. Birds do not have urinary bladders where urine is stored. Stored urine would add weight during flight, so having no urinary bladder can be considered an adaptation for flight. Birds excrete uric acid in the form of a white, pasty substance.

The brain and senses The brains of birds, shown in **Figure 29.19,** are large compared to the body size of the bird. The cerebellum is large because birds need to coordinate movement and balance during flight. The optic lobes coordinate visual input. The core of the cerebrum also is large because it is the primary integrating center of the brain. This area of the brain controls eating, singing, flying, and instinctive behavior. The medulla oblongata controls automatic functions such as respiration and heartbeat.

Birds generally have excellent vision. Birds of prey, like the hawk shown in **Figure 29.19,** have a focusing system that instantaneously enables them to stay focused on moving prey as they make a dive for their food. The position of a bird's eyes on its head relates to its life habits. Birds of prey have eyes that are at the front of the head. This enables them to recognize the distance of an object because both eyes can focus on the same object. A pigeon has eyes on the sides of its head. This enables the bird to see nearly 360 degrees of the space nearby, with each eye focusing on different areas. A pigeon eats grain and seeds and does not pursue prey. Its eyesight is adapted to scout out predators that might be nearby. Birds also have a good sense of hearing. Owls can hear the faintest sound of a scurrying mouse in the night. Even as the mouse runs for cover, the owl can catch it by following just the sound.

FOLDABLES
Incorporate information from this section into your Foldable.

LAUNCH Lab
Review Based on what you've read about reptiles and birds, how would you now answer the analysis questions?

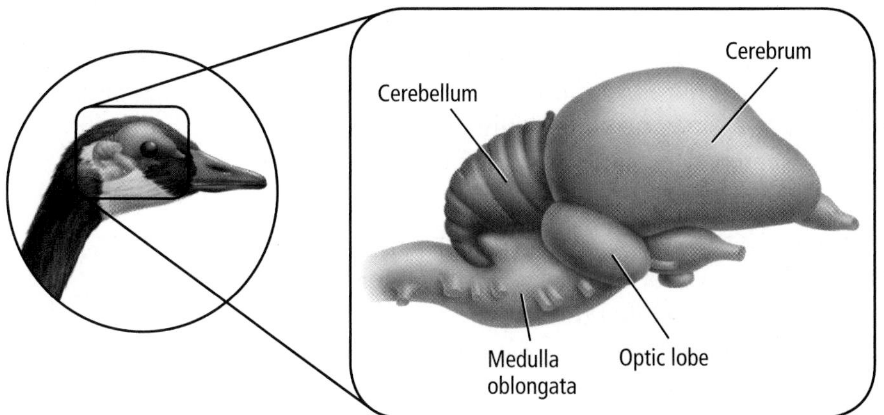

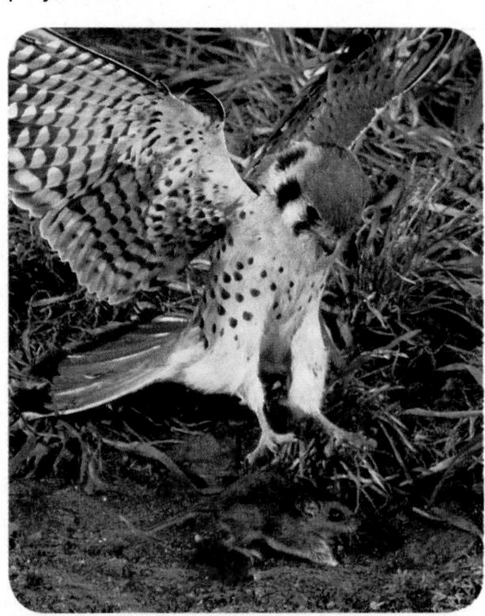

Figure 29.19
Left: Birds have large cerebellums that enable them to balance and coordinate movements. The medulla oblongata controls automatic processes.
Right: A hawk's eyes stay focused on moving prey as it dives.

Reproduction The reproductive activities of birds are complex. They include establishing territories, locating mates, courtship behavior, mating, constructing nests, incubating eggs, and feeding young. During breeding season, many birds gather in large colonies where they breed and take care of young. All birds have internal fertilization. Generally, after fertilization, the amniotic egg develops and is encased within a hard shell while still within the body of the female. After the shell forms, the egg or eggs are released through the cloaca to a nest, where the male or female or both birds incubate the egg or eggs and feed the young after hatching. To **incubate** means to maintain favorable conditions for hatching. Birds sit on their eggs to incubate them.

Diversity of Modern Birds

Modern birds are divided into about 27 living orders, depending on the classification system used. Anatomical differences, specific behaviors, songs, and the habitats occupied distinguish the orders. In **Table 29.1,** you will study the most common orders of birds and their adaptations. The largest order of birds is the Passeriformes, which often are called perching birds or songbirds. There are more than 5000 species in order Passeriformes. Flightless birds, including ostriches, emus, and kiwis, have reduced or no wings. The kiwi, a bird about the size of a chicken found in New Zealand, lays an egg that is extremely large given the size of the bird. Some birds, such as penguins, geese, and ducks, have adaptations that allow them to swim. Penguins use their wings as paddles to swim through the water. Ducks and geese have webbed feet.

MiniLab 29.1

Survey Local Birds

What birds live in your local area? A variety of birds can be found in almost any environment. Explore the area around your school to survey the different birds that live there.

Procedure

1. Read and complete the lab safety form.
2. Predict the number of different kinds of birds you can observe in the area around your school. Make a data table to keep track of birds you observe.
3. Go for a 10-min walk in the area near your school. Be sure to follow your teacher's instructions about where you are allowed to go. Record information about the birds you observe. Use **binoculars** if necessary. If you cannot identify a bird, use a **field guide** for local birds.
4. Compile your findings as a class. Research information about the birds you observed.

Analysis

1. **Count** the number of bird species you observed. List the types of birds you observed.
2. **Identify** Were the birds you saw native to your area or have they been introduced?
3. **Analyze** Did any patterns emerge as you compiled the data?
4. **Predict** Would this list differ if you surveyed the area around your house? If so, how?

Concepts In Motion **Interactive Table** To explore more about bird orders, visit biologygmh.com.

Table 29.1 Diversity of Bird Orders

Order	Example	Members	Distinguishing Characteristics
Passeriformes **Perching songbirds; about 5000 species**		Thrushes, warblers, mockingbirds, crows, blue jays, nuthatches, finches	Members of this order have feet that are adapted for perching on thin stems and twigs. Many birds in this order sing. The vocal organ, called the syrinx, is well-developed in these birds. Other species, such as crows and ravens, do not sing.
Piciformes **Cavity-nesters; about 380 species**		Woodpeckers, toucans, honeyguides, jacamars, puffbirds	Members of this order have highly specialized bills that are related to their feeding habits. They all build nests in cavities—for example, a hole in a dead tree. The feet have two toes that extend forward and two toes that extend backward, allowing them to cling to tree trunks.
Ciconiiformes **Wading birds and vultures; about 90 species**		Herons, egrets, bitterns, storks, flamingoes, ibises, vultures	Members of this order are medium- to large-sized birds that have long necks and long legs. Most are wading birds that live in large colonies in wetlands. Vultures are closely related to storks but are detritovores.
Procellariiformes **Marine birds; about 100 species**		Albatrosses, petrels, shearwaters, storm-petrels	All members of this order are marine birds. They have hooked beaks that aid in feeding on fish, squid, and small crustaceans. They all have tube-shaped nostrils located on the top of their beaks. Many have webbed feet.
Sphenisciformes **Penguins; about 17 species**		Penguins	Penguins are marine birds that use their wings as flippers to swim through the water rather than fly. The bones of penguins are solid, lacking the air spaces of other birds. All species are found in the southern hemisphere.
Strigiformes **Owls; about 135 species**		Owls	Owls are nocturnal birds with large eyes, strong, hooked beaks, and large, sharp talons on their feet. All of these adaptations aid in capturing prey. Many species have feathers on their legs. Owls are found worldwide except for Antarctica.
Struthioniformes **Flightless birds; 10 species**		Ostriches, kiwis, cassowaries, emus, rheas	All members of this order have reduced wings and are flightless birds. The ostrich is the largest living bird, reaching a height of more than 2 m and a weight of 130 kg. All species are found in the southern hemisphere.
Anseriformes **Waterfowl; about 150 species**		Swans, geese, ducks	Members of this order live in aquatic environments. They have webbed feet that aid in moving them through the water. Many have broad, round beaks. They feed on aquatic plants and sometimes crustaceans or small fish using broad, round beaks.

■ **Figure 29.20** *Archaeopteryx* had a long, reptilelike tail, clawed fingers, teeth, and feathers. This artist's rendering of *Caudipteryx* shows long feathers that might have been used for insulation and balance.

Evolution of Birds

Fossil evidence shows that birds descended from archosaurs, the same line from which crocodiles and dinosaurs evolved, as you saw in **Figure 29.11.** Similarities between birds and reptiles are apparent. They have similar skeletal features, kidney and liver function, amniotic eggs, and behaviors such as nesting and caring for young.

Feathered dinosaurs Three different species of birdlike dinosaurs from Chinese fossil beds have been carefully studied. *Sinosauropteryx* had a coat of downy, featherlike fibers. *Protoarchaeopteryx* and *Caudipteryx,* illustrated in **Figure 29.20,** had long feathers on their front appendages and on their tails. The downy dinosaur feathers might have functioned as insulation, and the front appendage feathers might have served as balancing devices as the dinosaurs ran along the ground.

Connection to History In 1861, in southern Germany, paleontologist Hermann von Meyer discovered what is now known to be the oldest bird fossil—*Archaeopteryx. Archaeopteryx,* illustrated in **Figure 29.20,** lived about 150 million years ago. This ancient bird had a long reptilelike tail, clawed fingers in the wings, and teeth. These are features that modern birds do not have. Yet, like modern birds, its body was covered with feathers. The feathers were asymmetrical, like those found only in modern birds that fly. Recent fossil evidence also shows that the brain of *Archeopteryx* was much like that of modern birds.

Recent discoveries In 2006, a new fossil bird, *Ganus youmenensis,* was discovered. It lived 110 mya and might be the link between *Archaeopteryx* and modern birds. *Ganus* had webbed feet, leading some scientists to think modern birds evolved from aquatic birds. Molecular evidence from *Tyrannosaurus rex* soft tissue has also strengthened the link between dinosaurs and birds.

CAREERS IN BIOLOGY

Paleontologist Scientists who learn more about the history of life on Earth by studying fossils are paleontologists. Many paleontologists teach at colleges and conduct research in the field by uncovering fossils. Some work at museums, managing fossil exhibits. For more information on biology careers, visit biologygmh.com.

Ecology of Birds

Birds are important parts of food chains as predators of small mammals, arthropods, and other invertebrates. For example, you probably have seen a robin pulling a worm out of the ground. Birds are also important parts of food chains and food webs as prey of larger birds and mammals.

Birds play an important role in the dispersal of seeds. Birds eat seeds or fruits and berries, and, after digestion, eliminate them in a different location. Seeds also get caught on the feathers and drop off as birds move from one location to another. Some birds, such as hummingbirds, feed on the nectar of flowers, and pollinate the flowers as they feed.

Habitat destruction Many birds are threatened with extinction as the habitat they require either disappears or is degraded by pesticides and other chemical pollutants. Waterfowl populations depend on wetlands—a habitat that is disappearing rapidly as wetlands are drained for development. Deforestation of tropical rain forests has also led to some species of birds being endangered.

Illegal trade Illegal pet-bird trade is increasing. Many pet birds are raised in captivity, but other exotic birds are taken from the wild in a multibillion dollar industry. In some cases illegal capture has led to the disappearance of rare birds in the wild. The little blue macaw, shown in **Figure 29.21,** only exists in captivity. An international trade agreement was enacted in 1975 to ensure that the buying and selling of wild animals does not endanger their survival. Currently, 160 countries participate in the agreement; however, illegal wildlife trade continues.

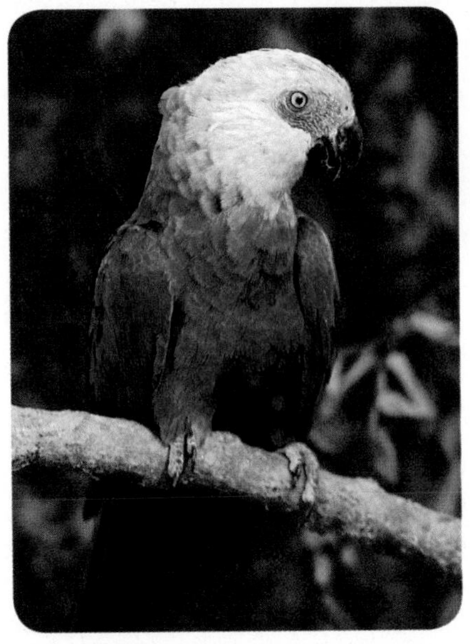

Figure 29.21 There are no little blue macaws left in the wild. Only about 70 of these birds remain in captivity.

Section 29.2 Assessment

Section Summary

- Birds have characteristics that make them well-adapted for flight.
- Birds can generate their body heat internally.
- Birds have lightweight bones.
- The shape of a bird's beak is related to the type of food it eats.
- Birds generally have excellent vision.
- Birds belong to about 27 living orders.
- Modern birds evolved from dinosaurs.
- Habitat destruction and illegal trade can negatively affect some species of birds.

Understand Main Ideas

1. MAIN Idea **Identify** the characteristics of birds that make them adapted for flight.
2. **Compare and contrast** contour feathers and down feathers.
3. **Explain** how respiration and circulation in birds are adapted for flight.
4. **Compare and contrast** reproduction in birds and reptiles.
5. **Describe** how the characteristics of birds in order Strigiformes are different from those of birds in order Anseriformes.
6. **Describe** how scientists have been able to conclude that birds evolved from dinosaurs.

Think Critically

7. **Scientific Illustrations** Draw and label the parts of a bird's brain. Explain the function of the different parts of the brain.

WRITING in **Biology**

8. Most small land birds that feed their young lay from two to 12 eggs in their nests. Some larger birds, such as waterfowl, have young that are able to care for themselves after hatching and are not fed by parents. These birds lay up to 20 eggs in their nests. Write a detailed hypothesis that explains why some bird species might lay fewer eggs than other species.

Biology & Society

Invasive Species Run Wild

What happens when pet owners buy infant Burmese pythons and then decide when the pythons grow to be 4–5 m long that they no longer can care for them? Scientists have discovered that pet owners are dumping these large snakes in the Everglades. In the Everglades, Burmese pythons now are considered an invasive species that is causing increasing problems in the area. Other invasive species are causing similar problems for their host environments around the country.

What is an invasive species? Invasive species are organisms that are introduced, by human action, to an area where they do not naturally live, and in which they do not naturally breed. They successfully breed, become a pest in the new area, and threaten biodiversity. The Burmese python is just one of thousands of nonnative species that are now in the United States. The rooting of feral pigs, which have spread throughout Florida, is destructive to native vegetation and the nests of sea turtles. Officials actively have been removing feral pigs from Florida state parks in an effort to reduce their impact. The graph shows the number of feral pigs that have been removed each year since 1993.

What are the costs of invasive species? Invasive species can cause billions of dollars of damage annually to crops, rangelands, and waterways throughout the United States. The presence of invasive species is the second leading cause of species endangerment and extinction. Invasive plant species can threaten bird populations by causing habitat loss at their breeding or wintering grounds. Invasive animal species prey on animals native to an area. Competing for space and prey is another key way invasive species overrun the native species.

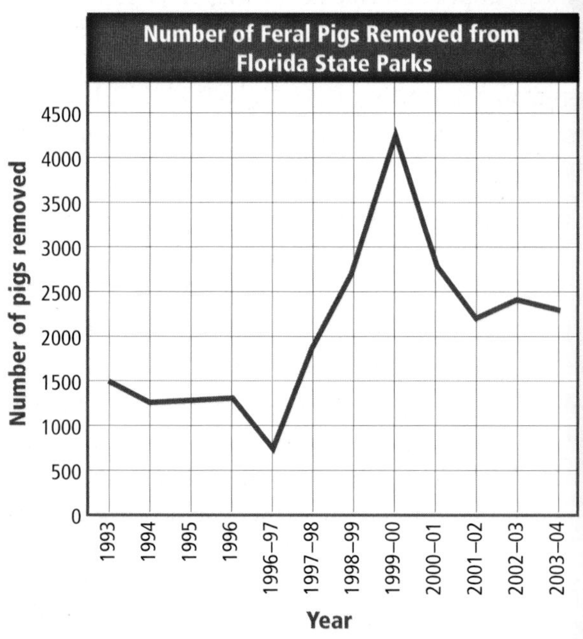

Feral pigs have been in the process of being removed from Florida State Parks since 1993.

Solutions Invasive species can be controlled through various methods, including legislation such as the National Invasive Species Act established in 1996. Research is ongoing by scientists who are constantly examining invasive species in order to understand methods to control their spread, life cycle, and behavior. Public education can provide people with knowledge to make informed decisions to vote on important policies. Legislation related to environmental issues also can help improve situations involving invasive species.

COMMUNITY INVOLVEMENT

Lesson Plan Develop a lesson plan about an invasive species of your choice that is affecting your state. The lesson plan should be geared toward the elementary school children in your district. Be sure you involve the elementary students in an activity. For more information about invasive species, visit biologygmh.com.

BIOLAB

Design Your Own

HOW CAN YOU MODEL A HABITAT FOR REPTILES AND BIRDS?

Background: Your class has been asked to help plan a new zoo exhibit about adaptations in birds and reptiles. In this lab, you will research a variety of birds and reptiles to understand how their body structures are adapted to habitats and food sources. You will use this information to help make a model habitat in which reptiles and birds would live in the zoo.

Question: *How can you make a model habitat based on what you know about an organism's adaptations to its environment?*

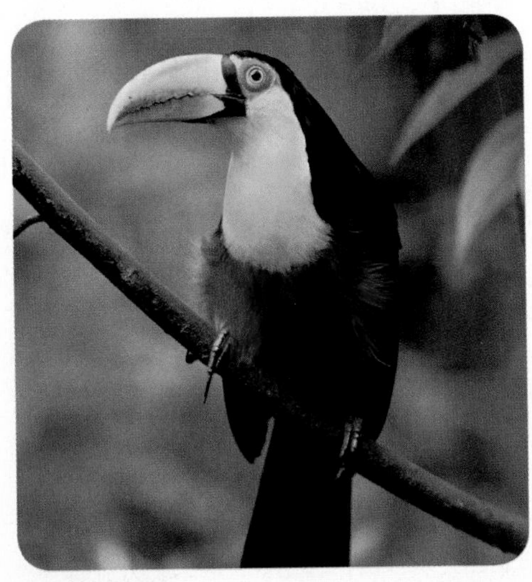

Materials

field guides for birds and reptiles
sand
soil
cardboard pieces
cardboard box
wooden sticks
toothpicks
glue
scissors
colored markers
dried beans
rocks/pebbles
felt pieces

Safety Precautions

Plan and Perform the Experiment

1. Read and complete the lab safety form.
2. Choose one reptile species and one bird species. Research the adaptations of each species. Find out information about the habitat in which they live, the food they eat, and their behavior. Examine how their body structures and behaviors give them a competitive advantage in the habitats in which they live.
3. Use the information you collected to make a detailed description of the habitat that should be set up in the exhibit for each reptile and bird you investigated.
4. Make sure your teacher approves your plan before your proceed.
5. Use the materials available to make a model habitat for the reptile species and a model habitat for the bird species that they would live in at the zoo.
6. Present and explain your models to the class.

Analyze and Conclude

1. **Describe** How did the differences between reptiles and birds lead to differences in the models you made for each habitat?
2. **Identify** any weaknesses in your model. Would your model habitats support the needs of each species? What changes would you make to your model?
3. **Describe** how the structures and behaviors of the organisms give them a competitive advantage in their habitat.

WRITING in Biology

Take-Home Pamphlet Write and illustrate a pamphlet that people visiting your exhibit could take home. Include informational text about the animals in the exhibit as well as illustrations of their natural habitats. To learn more about zoo exhibits, visit BioLabs at biologygmh.com.

Study Guide

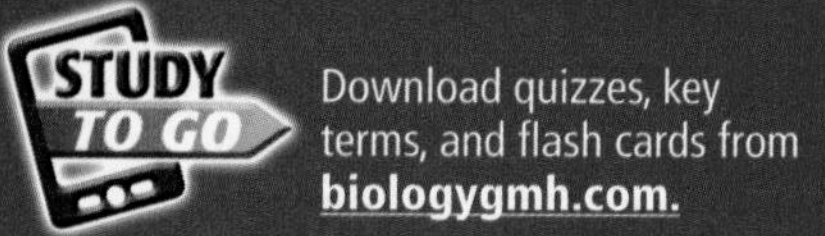

Download quizzes, key terms, and flash cards from biologygmh.com.

FOLDABLES **Infer** why fishes and amphibians do not have amniotic eggs, and describe your reasoning on the back of your Foldable.

Vocabulary	Key Concepts
Section 29.1 Reptiles	
• amnion (p. 852) • amniotic egg (p. 853) • carapace (p. 857) • Jacobson's organ (p. 855) • plastron (p. 857)	MAIN Idea Reptiles are fully adapted to life on land. • Reptiles have several types of adaptations for life on land. • Eggs of reptiles are adapted to development on land. • Reptiles belong to four living orders: snakes and lizards, crocodiles and alligators, turtles and tortoises, and tuataras. • Modern reptiles evolved from early amniotes. Many ancient reptiles, including dinosaurs, became extinct. 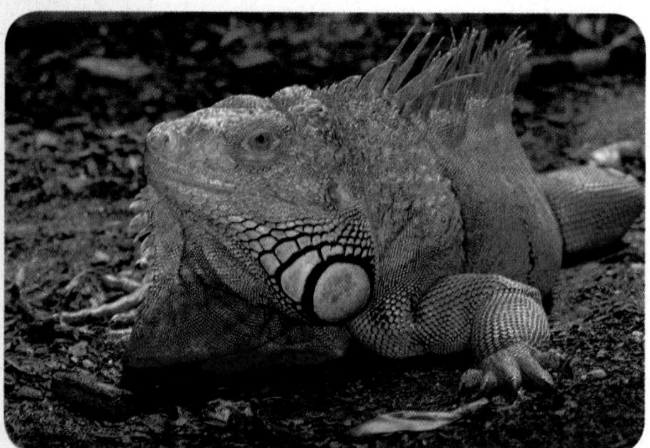
Section 29.2 Birds	
• air sac (p. 863) • contour feather (p. 862) • down feather (p. 862) • endotherm (p. 861) • feather (p. 861) • incubate (p. 866) • preen gland (p. 862) • sternum (p. 862)	MAIN Idea Birds have feathers, wings, lightweight bones, and other adaptations that allow for flight. • Birds have characteristics that make them well-adapted for flight. • Birds can generate their body heat internally. • Birds have lightweight bones. • The shape of a bird's beak is related to the type of food it eats. • Birds generally have excellent vision. • Birds belong to about 27 living orders. • Modern birds evolved from dinosaurs. • Habitat destruction and illegal trade can negatively affect some species of birds. 

Biology Online Vocabulary PuzzleMaker biologygmh.com

CHAPTER 29 Assessment

Section 29.1

Vocabulary Review

Each of the sentences below is false. Make the sentence true by replacing the italicized word with the correct vocabulary term from the Study Guide page.

1. Several membranes are inside a(an) *carapace.*
2. The ventral part of a turtle's shell is called the *Jacobson's organ.*
3. The *plastron* is responsible for the sense of smell in snakes.
4. The dorsal part of a turtle's shell is the *amniotic egg.*

Understand Key Concepts

5. Which is not a reptile?

A.

C.

B.

D.

6. Which is not true about respiration in reptiles?
   A. Most reptiles use lungs for gas exchange.
   B. As reptiles inhale, the muscles of the rib cage relax.
   C. As reptiles exhale, the muscle of the body wall relax.
   D. The lungs of reptiles have a larger surface area than those of amphibians.

7. In which structure in reptiles can uric acid be found?
   A. the lungs
   B. the cloaca
   C. the heart
   D. the stomach

8. Which statement best represents scientists' understanding of early reptiles?
   A. Dinosaurs evolved into modern-day reptiles such as lizards, snakes, and turtles.
   B. Birds and crocodiles are the closest relatives of dinosaurs.
   C. The earliest reptiles did not have amniotic eggs.
   D. Dinosaurs became extinct because they were too big.

Constructed Response

9. **Open Ended** Make a table that lists the following structures, their functions, and an analogy of what that structure is like in the world of human-made devices: amnion, ventricle, bladder, Jacobson's organ, carapace and plastron, kidney.

10. **Open Ended** Make a dichotomous key that would allow a person to determine which order of reptile they are examining.

Think Critically

11. **Apply Concepts** The feet of a gecko are covered by billions of tiny hairlike structures that stick to surfaces. When the hairs contact a surface, attractions between molecules bond the gecko's foot to the surface. These structures can support up to 400 times the body weight of the gecko. How could scientists use the way in which a gecko's foot sticks to surfaces to make a tool that would be useful to people?

Use the graph below to answer questions 12 and 13. The brown four-fingered skink was introduced to the Pacific island of Guam in the early 1950s.

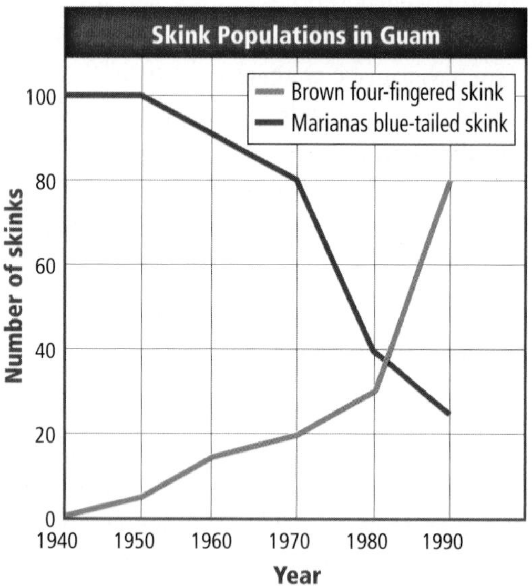

12. **Analyze Data** How have the populations of the brown four-fingered skink and the Marianas blue-tailed skink changed since the 1950s?

13. **Hypothesize** Form a detailed hypothesis that might explain the decline in population of the Marianas blue-tailed skink.

14. **Compare** How does circulation in reptiles compare to circulation in amphibians?

15. **Illustrate** Make a diagram, flowchart, concept map, or illustration that shows how the loss of habitat and the introduction of exotic species has affected the population of the San Francisco garter snake.

Section 29.2

Vocabulary Review

Explain the relationship that exists between the vocabulary terms in each set.

16. endotherm, down feather

17. contour feather, down feather

18. preen gland, contour feather

19. sternum, air sac

Understand Key Concepts

20. Which group of words has a word that does not belong?
    A. ventricle, atrium, oxygenated blood, deoxygenated blood
    B. kidney, nitrogenous waste, uric acid, cloaca
    C. cerebellum, cerebrum, optic lobes, medulla
    D. amniotic egg, cloaca, kidney, amnion

Use the figure below to answer question 21.

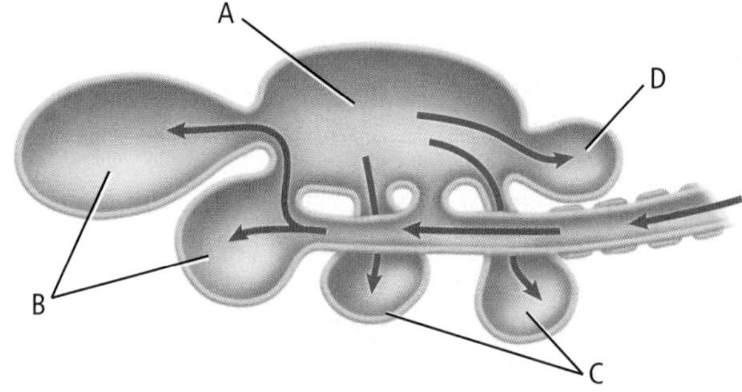

21. When a bird breathes in, oxygenated air goes into which structure(s)?
    A. Structure A
    B. Structure B
    C. Structure C
    D. Structure D

22. To which system do the kidney and cloaca belong?
    A. excretory
    B. nervous
    C. digestive
    D. reproductive

23. What type of beak would a bird need if it feeds on aquatic plants?
    A. broad and flat
    B. large and scooped
    C. sharp and hooked
    D. long, thin, and pointed

24. What does fossil evidence in dinosaurs show?
    A. Dinosaurs were not related to birds.
    B. Dinosaurs were not warm-blooded.
    C. Dinosaurs were not flying animals.
    D. Dinosaurs were not feathered.

Biology Online Chapter Test biologygmh.com

Constructed Response

25. **CAREERS IN BIOLOGY** Ornithologists hypothesized that the long-term memory of certain migratory birds would be better than that of nonmigrants. To test this hypothesis, two rooms were decorated—one with ivy and one with geraniums. Food was placed in only one room. Both migrant and nonmigrant birds were allowed to explore both of the rooms. One year later, the same birds were allowed to explore the rooms. Migrant birds spent significantly more time exploring the room that had contained food than the nonmigrants. Draw a conclusion about the long-term memories of these birds.

Think Critically

26. **Hypothesize** Birds often sing at dawn. Biologists think that the birds are announcing their territories or letting potential mates know where they are. Biologists also have discovered that the larger a bird's eyes are, the earlier in the day it sings. Form a hypothesis about why eye size might be correlated to how early birds sing.

27. **Infer** Biologists know that the young of modern birds curl up their bodies in their nests to conserve body heat. Recently, fossils of dinosaurs' young have been found in a curled position in their nests. This particular line of dinosaurs is one with a direct lineage to birds. Infer what this curled-up position might mean about the bodies of these dinosaurs.

Use the figure below to answer question 28.

28. **Infer** What type of food does this bird eat? How does it use its beak during feeding?

Additional Assessment

29. **WRITING in Biology** Write a summary for a yearbook page about the Ornithology Club, in which students went bird watching, recorded species, and conducted species counts.

DBQ Document-Based Questions

Sea snakes have highly toxic venom that they inject into prey. In many cases, the toxin paralyzes the muscles that pump water across the gills of fishes. The graph shows the rate of mortality of five species of fish when given different doses of venom from an olive sea snake.

Data obtained from: Zimmerman, K.D., et al. 1992. Survival times and resistance to sea snake *(Aipysurus laevis)* venom by five species of prey fish. *Toxicon* 30: 259–264.

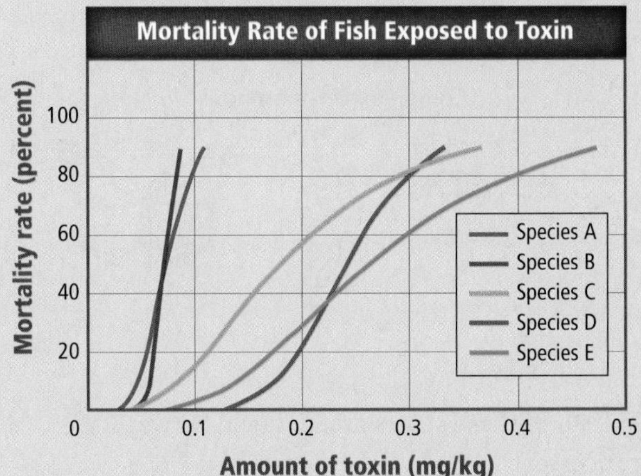

30. Which fish species is most affected by the venom? Which species is least affected? Explain how you know this.

31. The species of fish least affected by the venom have the ability to respire through their skin as well as through their gills. Why would this ability be important to surviving a bite by a sea snake?

Cumulative Review

32. Sketch the four phases of mitosis in a cell with two chromosomes. **(Chapter 9)**

33. Explain the meaning of alternation of generations as it applies to plants. **(Chapter 21)**

Standardized Test Practice

Cumulative

Multiple Choice

1. The word *echinoderm* means "spiny skin." Which describes the skin of an echinoderm?
   - **A.** Calcium carbonate plates with spines covered with a thin skin.
   - **B.** Calcium carbonate spines that protrude through the skin.
   - **C.** Silicon plates with spines covering the entire surface.
   - **D.** Silicon spines that protrude through the skin.

Use the figure below to answer questions 2 and 3.

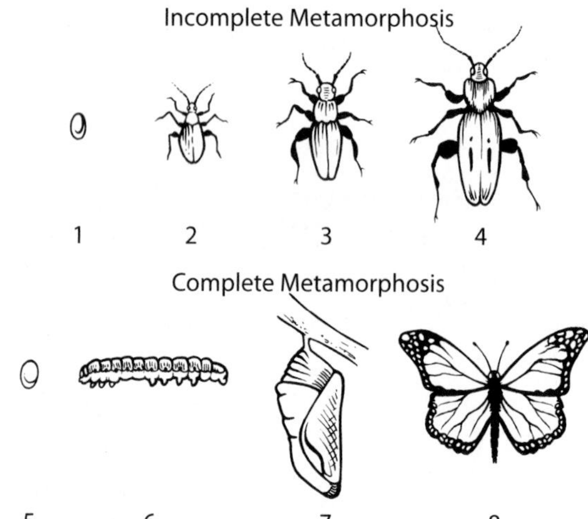

2. Which of these stages are identical between the processes?
   - **A.** 1, 5
   - **B.** 2, 7
   - **C.** 3, 8
   - **D.** 4, 7

3. During which stages do immature insects feed?
   - **A.** 1, 5
   - **B.** 1, 7
   - **C.** 2, 6
   - **D.** 4, 8

4. How do pseudocoelomates take in gases and excrete metabolic wastes?
   - **A.** Their digestive tract is used for gas exchange.
   - **B.** Gas exchange occurs through the endoderm tissue.
   - **C.** Materials diffuse through their body walls.
   - **D.** Materials exchange in the primitive respiratory system.

5. Which is a function of the lateral line system in fishes?
   - **A.** detecting chemicals in the water
   - **B.** detecting water pressure changes
   - **C.** keeping a fish upright and balanced
   - **D.** sending signals between fishes

Use the diagram below to answer questions 6 and 7.

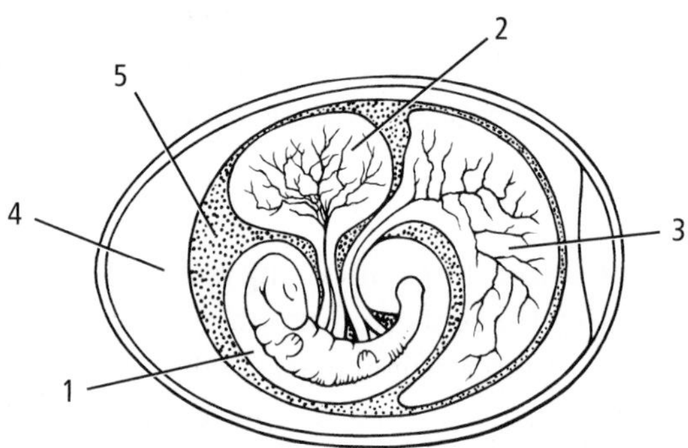

6. Which number represents the fluid-filled membrane that prevents dehydration and cushions the pictured embryo?
   - **A.** 1
   - **B.** 2
   - **C.** 4
   - **D.** 5

7. Which number represents the main food supply for the pictured developing reptile embryo?
   - **A.** 1
   - **B.** 2
   - **C.** 3
   - **D.** 4

8. Which structures are used in most adult amphibians to take in oxygen and transport it to body cells?
   - **A.** gills and a closed circulatory system
   - **B.** gills and an open circulatory system
   - **C.** lungs and a closed circulatory system
   - **D.** lungs and an open circulatory system

Short Answer

Use the diagram to answer questions 9 and 10.

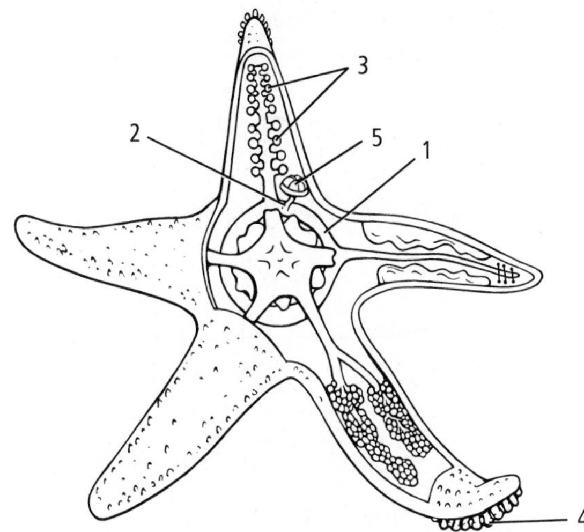

9. Name each of the numbered structures and describe how they enable a sea star to move.
10. Analyze how a sea star opens an oyster by relating the process to the above numbered structures.
11. Give a reason why most viruses are limited to attacking only a few types of cells and hypothesize why this might be important information for a medical researcher.
12. Relate evidence scientists use to propose that the lobe-finned fishes are the ancestors of the amphibians.
13. Describe how reptiles regulate their body temperature.
14. Explain why birds need an efficient respiratory system.
15. Generalize changes a tadpole goes through before becoming a frog.

Extended Response

16. Contrast the circulatory system of a frog with the circulatory system of a fish and assess the importance of those differences.
17. Select a technology that has changed the way in which scientists learn about genetics, and describe how that technology has brought about the change.

Essay Question

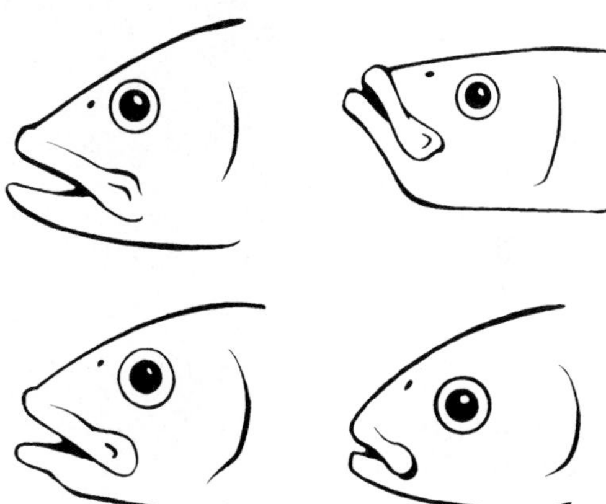

The evolution of a jaw was an important advancement in fish structure. The evolution of the jaw was a specialized adaptation for feeding. The jaw of fishes continued to evolve as fishes became more specialized in their feeding behaviors. The shape of the jaw gives important information about how a fish feeds and, in some cases, what it feeds on. By studying the different shapes of the jaw, scientists can understand how different species became adapted for their particular environments.

Using the information in the paragraph above, answer the following question in essay format.

18. Justify how each of these four types of jaws is suited for the food fishes eat.

NEED EXTRA HELP?

If You Missed Question . . .	1	2	3	4	5	6	7	8	9	10	11	12	13	14	15	16	17	18
Review Section . . .	27.1	26.3	26.3	25.1, 25.2	28.1	29.1	29.1	28.3	27.1	27.1	18.2	28.2	29.1	29.2	28.3	28.3	13.1	28.1

CHAPTER 30

Mammals

BIG Idea **Mammals have evolved to have a variety of adaptations for maintaining homeostasis and living in a variety of habitats.**

Section 1
Mammalian Characteristics

MAIN Idea Mammals have two distinct characteristics: hair and mammary glands.

Section 2
Diversity of Mammals

MAIN Idea Class Mammalia is divided into three subgroups based on reproductive methods.

BioFacts

- The hairs in a deer's winter coat are hollow. This helps insulate the deer against the cold and keep it afloat when moving through water.
- The hairs of a polar bear are transparent and have no color. The bears look white because the hollow hairs reflect and scatter light.
- Some red foxes actually have black fur, silver fur, or, in rare cases, both black and silver fur.

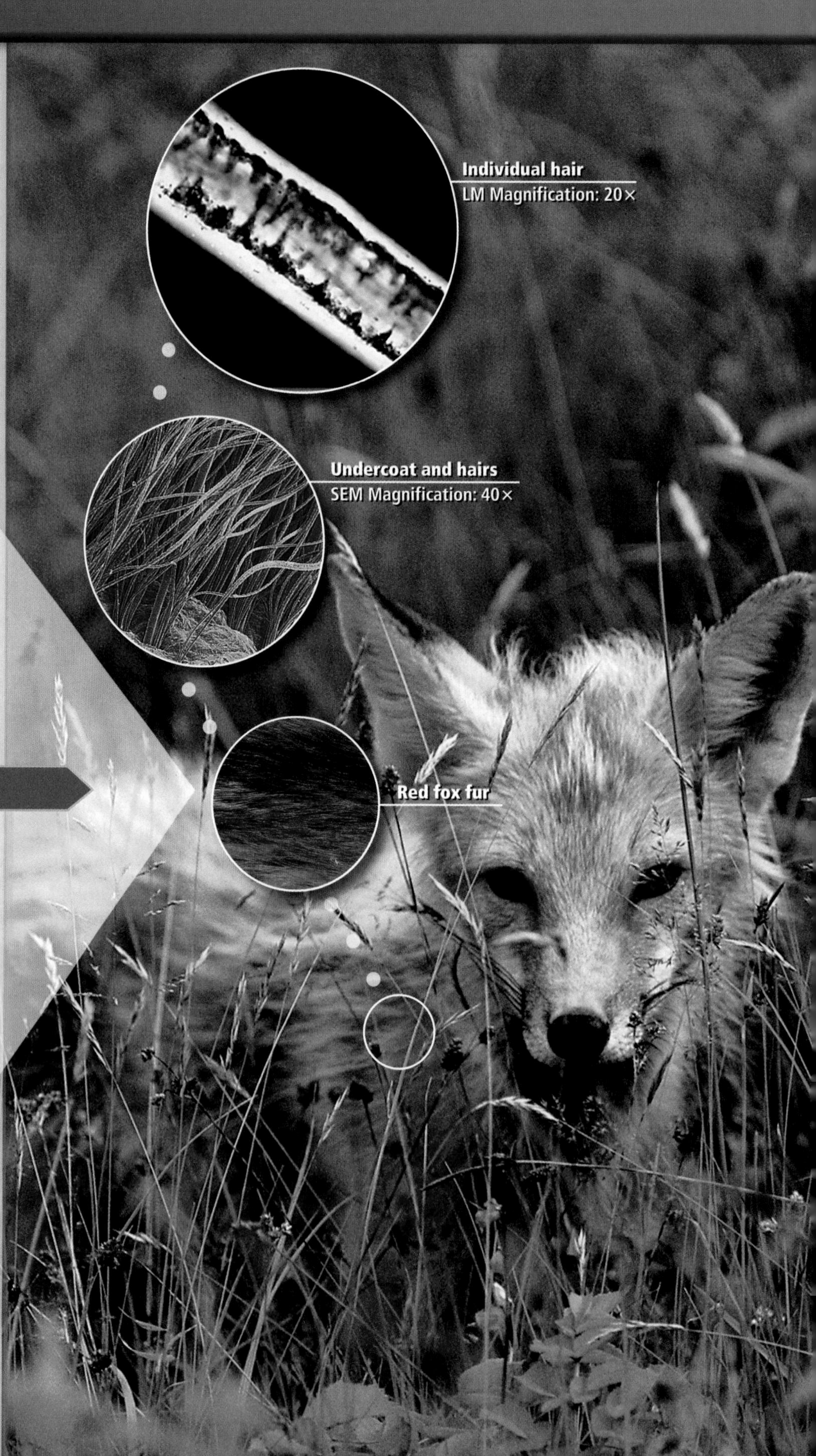

Start-Up Activities

LAUNCH Lab

What is a mammal?

You see mammals every day—the neighborhood dog, a squirrel scampering across the grass, the people with whom you live. What characteristics do mammals share?

Procedure

1. Read and complete the lab safety form.
2. Examine **specimens or photographs of mammals,** including the red fox shown on the opposite page.
3. Identify characteristics that the mammals in the photographs share.
4. Create a data table to record your observations.

Analysis

1. **Infer** the function of each physical characteristic shared by mammals.
2. **Describe** the wide diversity of mammalian characteristics and behaviors using the photographs and your experiences with other mammals.
3. **Infer** how scientists would use different characteristics to classify mammals into specific groups.

Visit biologygmh.com to:

- study the entire chapter online
- explore Concepts in Motion, Interactive Tables, Microscopy Links, and links to virtual dissections
- access Web links for more information, projects, and activities
- review content online with the Interactive Tutor, and take Self-Check Quizzes

Mammalian Subclasses Make the following Foldable to help you compare the characteristics of mammals in each subgroup.

STEP 1 Fold a sheet of notebook paper into thirds.

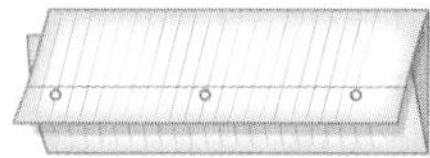

STEP 2 Fold the paper down 2.5 cm from the top.

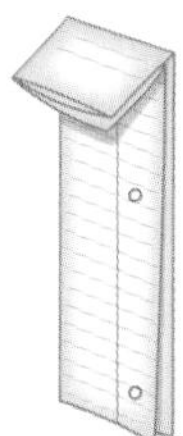

STEP 3 Open and draw lines along the 2.5-cm fold. Write the following labels on the tabs: *Monotremes, Marsupials,* and *Placentals.*

FOLDABLES **Use this Foldable with Section 30.2.** As you read the section, record what you learn about the characteristics of mammals in each subgroup, and use the information to compare and contrast each.

Section 30.1

Mammalian Characteristics

Reading Preview

Objectives

- **Identify** characteristics of mammals.
- **Describe** how mammals maintain a constant temperature to achieve homeostasis.
- **Distinguish** how respiration in mammals differs from that of other vertebrates.

Review Vocabulary

metabolic rate: the rate at which all the chemical reactions that occur within an organism take place

New Vocabulary

mammary gland
diaphragm
cerebral cortex
cerebellum
gland
uterus
placenta
gestation

MAIN Idea **Mammals have two distinct characteristics: hair and mammary glands.**

Real-World Reading Link Think about the characteristics of the other classes of vertebrates you have studied. Think about how you are different from the animals in the other classes. The characteristics you have as a mammal allow you to carry out your daily life functions and activities.

Hair and Mammary Glands

Two characteristics that distinguish members of class Mammalia from other vertebrate animals are hair and mammary glands. **Mammary glands** produce and secrete milk that nourishes developing young. Recall that if an animal has feathers, it is a bird. In a similar way, if an animal has hair, it is a mammal. As you can see on the evolutionary tree in **Figure 30.1,** mammals have their own branch labeled *hair and mammary glands.*

Functions of hair Mammals' hair has several functions, including:

1. Insulation—One of the most important functions of hair is to insulate against the cold. Mammals benefit from having fur or hair that traps their body heat and prevents it from escaping.
2. Camouflage—The striped coat of a Bengal tiger allows it to blend into its natural habitat—the jungle.
3. Sensory devices—In some cases, hair has been modified into sensitive whiskers. Seals use the whiskers on their snouts to track prey in murky water by sensing changes in water movements when a fish is nearby.
4. Waterproofing—You might know how cool it feels when you come out of a swimming pool on a hot day. As the water evaporates from your skin, your body loses heat. Many aquatic mammals, such as the sea otter shown in **Figure 30.2,** have hair that keeps water from reaching their skin. This helps them maintain their body temperature.

Figure 30.1 Hair and mammary glands are two characteristics that distinguish mammals from other vertebrates.

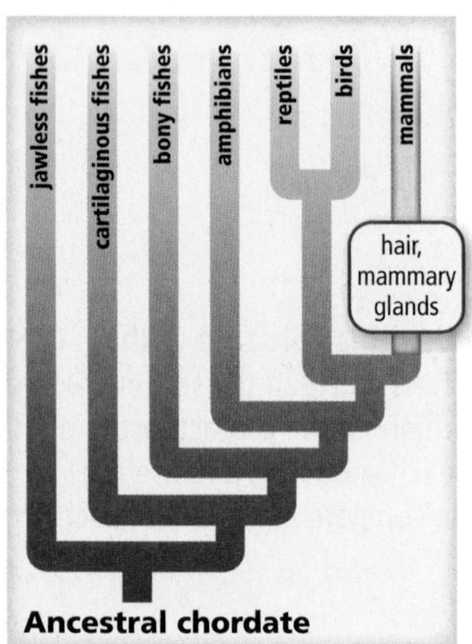

Figure 30.2 The hair covering this sea otter helps keep water from reaching its skin.

Signaling

Defense

■ **Figure 30.3**
Left: The white hair on the tails of these deer signal them to follow each other as they run from predators.
Right: The quills of a porcupine are modified hairs that are used as protection from predation.

5. Signaling—Hair can function as a signaling device. The white-tailed deer in **Figure 30.3** raise their tails, the undersides of which are white, when they run so that other deer can follow.
6. Defense—Hair also can function as a defense against predators. The porcupine in **Figure 30.3** has sharp quills, which are modified hairs, that are easily detached when the animal is threatened by a predator. The quills stick to and stab predators that touch the porcupine.

Structure of hair The hair in a bushy fox tail and the hair on your head contains a tough, fibrous protein called keratin, which is a protein that also makes up nails, claws, and hooves. A coat of hair usually consists of two kinds of hair—long guard hairs that protect a dense layer of shorter insulating underhair. The air trapped in the thick underhair layer provides insulation against the cold and retains body heat.

 **Reading Check** **Explain** why hair is important to mammals.

Other Characteristics

In addition to having hair and mammary glands, mammals share other characteristics. These include a high metabolic rate, which supports endothermy, specialized teeth and digestive systems, a diaphragm to aid in respiration, a four-chambered heart, and a highly developed brain.

Endothermy Mammals are endotherms, which means they produce their body heat internally. The source of body heat is internal—the result of heat produced by a high metabolic rate. Body temperature is regulated by internal feedback mechanisms that send signals between the brain and sensors throughout the body.

For example, when some mammals become warm as a result of exertion or because the air is warm, sweat glands in the skin are stimulated to secrete sweat that evaporates from the skin. As the sweat evaporates, it draws heat away from the body and cools it. When body temperature lowers, sweating stops. For other mammals that do not sweat, panting cools the body. You might have seen a dog panting in the summer's heat. During panting, water evaporates from the mouth and nose. Because mammals can regulate their body temperatures internally to maintain homeostasis, they can live in a range of ecosystems from the frigid temperatures of polar regions to the sweltering heat of deserts and the tropics.

Study Tip

Prediction Preview this section by looking at the bold titles and the photos. Predict the distinguishing characteristics of mammals. Use the titles and photos to help anticipate how to take notes on this section.

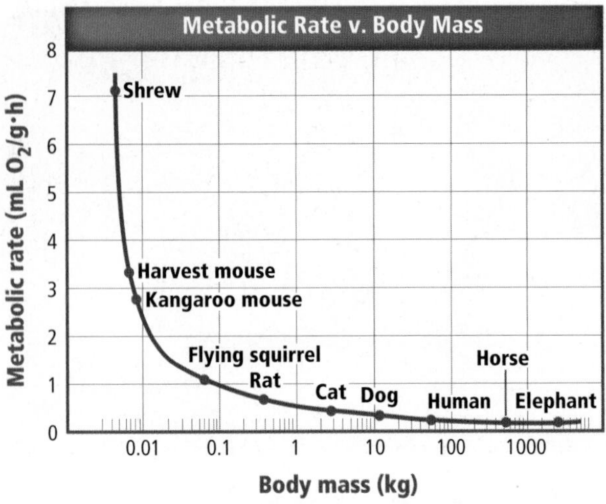

■ **Figure 30.4** Due to their high metabolic rates, some small mammals, such as mice and shrews, must eat food masses that are equivalent to their body masses each day in order to maintain temperature homeostasis.

Analyze ***Approximately how much food (in kg) does a shrew have to eat each day to survive?***

Feeding and digestion Maintaining an endothermic metabolism requires large amounts of energy. Mammals get the energy they need from the breakdown of food. Much of an endotherm's daily intake of food is used to generate heat to maintain a constant body temperature.

Examine the graph in **Figure 30.4.** The graph shows the relationship of a mammal's metabolic rate to its body mass. Small mammals, including shrews, bats, and mice, have a high metabolic rate relative to their body mass. As a result, these small mammals must hunt and eat food almost constantly in order to fuel their metabolisms.

Trophic categories Mammalogists divide mammals into four trophic categories based on what they eat:

1. Insectivores, such as moles and shrews, eat insects and other small invertebrates.
2. Herbivores, such as rabbits and deer, feed on vegetation.
3. Carnivores, such as foxes and lions, mostly feed on herbivores.
4. Omnivores, such as raccoons and most primates, feed on both plants and animals.

A mammal's adaptations for finding, capturing, chewing, swallowing, and digesting food all influence the mammal's structure and life habits. The fibers of plants are more difficult to digest and take longer than the digestion of meat. As a result, mammals that eat plant material have a larger cecum and longer digestive tracts than those that eat meat, as shown in **Figure 30.5.**

Ruminant herbivores Cellulose, a component of the cell walls of plants, can be a source of nutrition and energy. However, the enzymes in the digestive system of mammals cannot digest cellulose. Instead, some herbivores have bacteria in the cecum—a pouch where the small intestine meets the large intestine. Other herbivores have bacteria in their stomachs that break down the cellulose and release nutrients the animals can use. These mammals, called ruminants, have large, four-chambered stomachs. Cattle, sheep, and buffalo are all ruminants. As a ruminant feeds, plant material passes into the first and second stomach chambers. Plants are partially digested by bacteria into a material called cud. The ruminant brings the cud back up into the mouth and chews the cud for a long period of time. This further crushes the grass fibers. Once the cud is swallowed, it eventually reaches the fourth chamber of the stomach where digestion continues.

CAREERS IN BIOLOGY

Mammalogist The branch of biology that focuses on the study of mammals is mammalogy. A mammalogist might research the behavior, anatomy, or ecology of one or more species of mammal, or he or she can compare characteristics, such as digestion, in many species of mammals. For more information on biology careers, visit biologygmh.com.

Reading Check **Infer** the type of relationship that exists between a ruminant and the bacteria in its stomach.

Visualizing the Digestive Systems of Mammals

Figure 30.5
The digestive systems of mammals are adapted to maximize the digestion and absorption of food. The protein consumed by carnivores and insectivores is readily digestible. Plant materials contain cellulose, which resists digestion, water, and some carbohydrates. Compare the structure of each digestive system below.

Insectivore Digestive System
The diet of an insectivore is easily digested and absorbed in a relatively short digestive system.

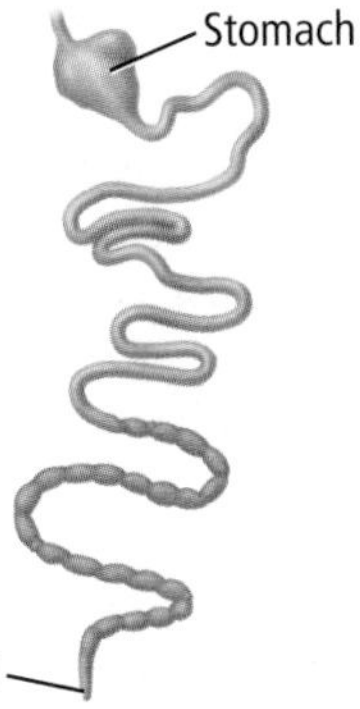

Short-tailed shrew

Eastern cottontail rabbit

Nonruminant Herbivore Digestive System
Digestion and absorption of nutrients begins in the stomach. Cellulose is broken down by bacteria in the cecum.

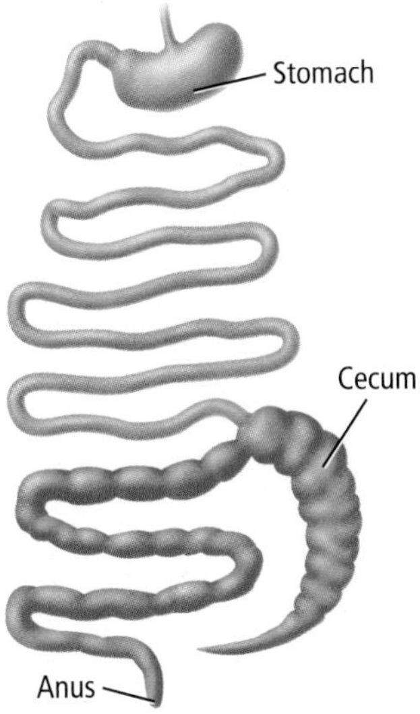

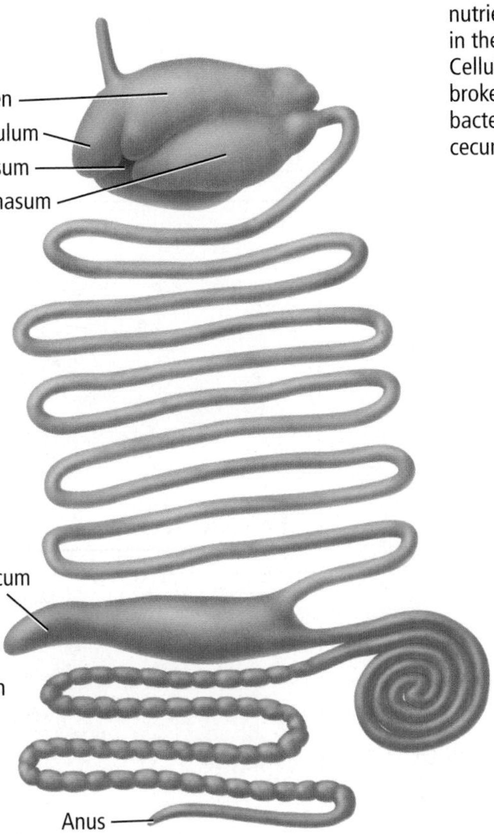

Deer

Ruminant Herbivore Digestive System
A multi-chambered stomach helps break down plant materials before they enter the intestines. Longer intestines and the cecum increase nutrient absorption.

Red fox

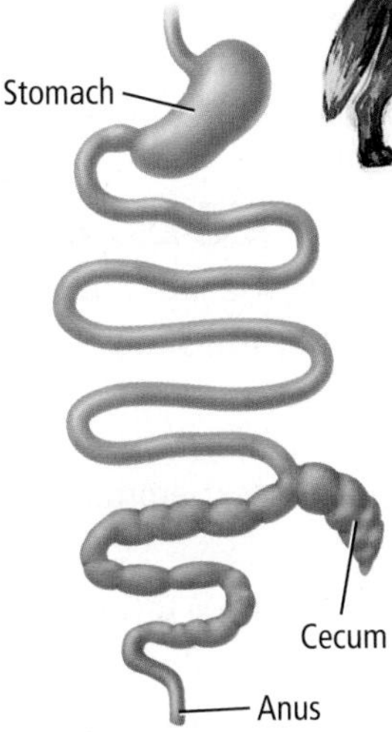

Carnivore Digestive System
The digestive system of a carnivore is similar to that of an insectivore. Unlike the herbivore, the cecum serves no vital function for carnivore digestion.

Concepts In Motion Interactive Figure To see an animation of mammal digestion, visit biologygmh.com.

Biology Online

VOCABULARY

ACADEMIC VOCABULARY

Retain:
to keep in possession or use.
You can retain your teeth throughout adulthood by brushing and flossing.

Teeth In addition to adaptations of the digestive system, teeth, perhaps more than any other physical characteristic, reveal the life habits of a mammal. Generally, in fish and reptile species, all teeth in the mouth look very much alike. This is because these animals use all their teeth in similar ways—for seizing prey or for tearing prey apart before swallowing. In contrast, mammals have different types of teeth that are specialized for various functions. Examine the four types of mammalian teeth—canines, incisors, premolars, and molars—illustrated in **MiniLab 30.1** below.

A fox's canines are long and sharp. Carnivores use canines to stab and pierce their prey. The canines of herbivores often are reduced in size. This is illustrated in the cow skull shown in **MiniLab 30.1.** The premolars and molars of carnivores are used to slice and shear meat from the bones of their prey, while crushing and grinding are the functions of premolars and molars in herbivores. The incisors of insectivores are long and curved, functioning as pincers in seizing insect prey. The chisel-like incisors of beavers are modified for gnawing. Because the teeth of mammals reflect their feeding habits, biologists can determine what a mammal eats by examining its teeth. Complete **MiniLab 30.1** to see what inferences you can make about a mammal's diet based on its teeth.

Excretion The kidneys of mammals excrete metabolic wastes and maintain the homeostatic balance of body fluids. Kidneys filter urea, an end product of cellular metabolism, from the blood. Mammalian kidneys excrete or retain the proper amount of water in body fluids as well. Kidneys enable mammals to live in extreme environments, such as deserts, because they can control the amount of water in body fluids and cells.

MiniLab 30.1

Compare Mammalian Teeth

How are the teeth of mammals specialized? Explore how the teeth of different mammal species are related to their diets.

Procedure

1. Read and complete the lab safety form.
2. Observe **teeth from the skulls of different mammal species.**
3. List the similarities and differences among the teeth of the different mammal species.

Analysis

1. **Infer** the function of each type of tooth based on its shape.
2. **Identify** the type of tooth common to all of the mammals you studied.
3. **Describe** how each mammal you studied uses its teeth to obtain and ingest food.
4. **Explain** how scientists might use the differences in mammalian teeth to classify mammals into different groups.

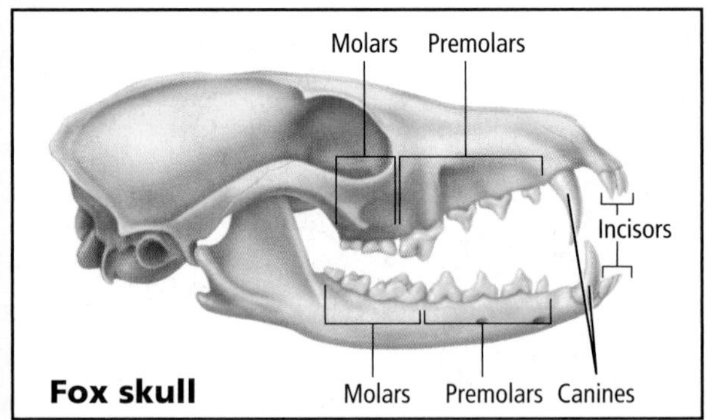

Fox skull

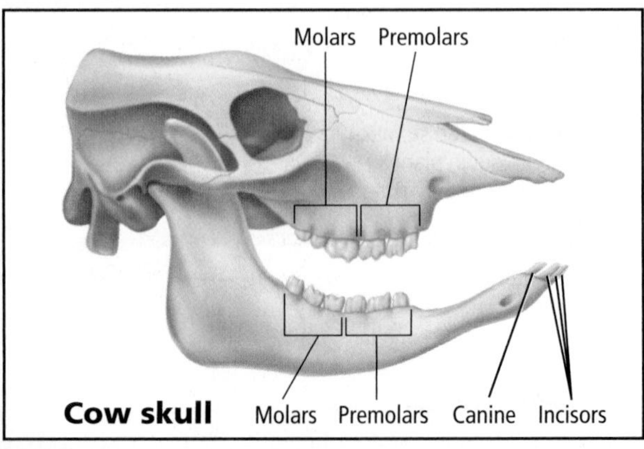

Cow skull

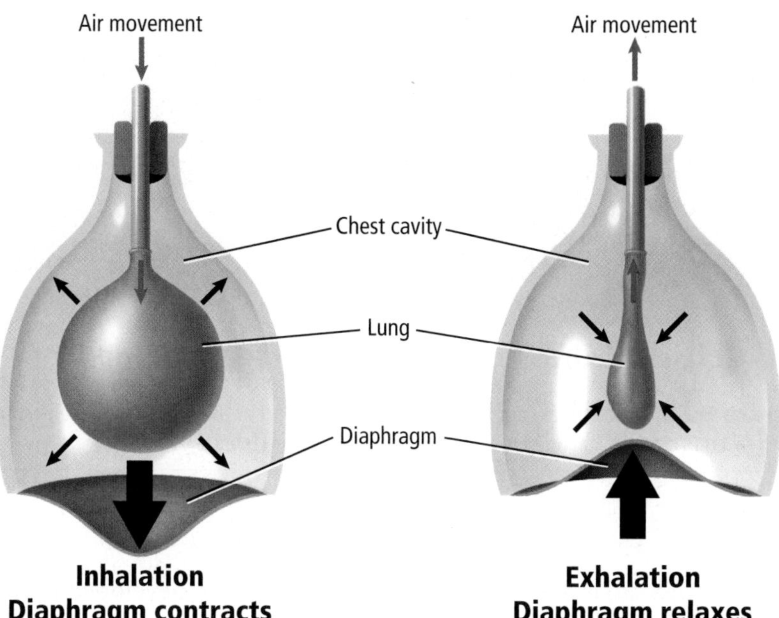

Figure 30.6 The flask with the balloon is an analogy of how the diaphragm aids breathing in mammals.

Describe ***What happens to the chest cavity as the diaphragm contracts and relaxes?***

Respiration The food a mammal eats is used to maintain high energy levels. High levels of oxygen also are required to maintain a high level of metabolism. Oxygen is taken into the lungs of mammals during respiration. Although other animals, such as birds and reptiles, have lungs, mammals are the only animals that have a diaphragm. A **diaphragm** is a sheet of muscle located beneath the lungs that separates the chest cavity from the abdominal cavity where other organs are located. As the diaphragm contracts, it flattens, causing the chest cavity to enlarge, as shown in **Figure 30.6.** Once air enters the lungs, oxygen in the air moves by diffusion into blood vessels. When the diaphragm relaxes, the chest cavity becomes smaller and air is exhaled.

Circulation Once oxygen is in the blood, vessels carry it to the heart, which pumps it out to the body. Like birds, mammals have a four-chambered heart. Also as in birds, oxygenated blood is kept entirely separate from deoxygenated blood in mammals. This is illustrated in **Figure 30.7.** Because most mammals are physically active and all are endotherms—they require a consistent supply of nutrients and oxygen to maintain homeostasis. Keeping oxygenated and deoxygenated blood separate makes the delivery of nutrients and oxygen more efficient.

Connection to Physics The circulatory system of a mammal also functions to help maintain a constant internal temperature. When body temperature increases, the blood vessels near the surface of the skin dilate, or expand, and deliver more blood than usual. Heat moves from the blood to the surface of the skin by conduction. At the skin's surface, heat is lost from the body by radiation and the evaporation of sweat. When body temperature decreases, blood vessels near the surface of the skin contract and do not deliver as much blood as usual. This action reduces the loss of body heat.

Reading Check **Describe** how the respiratory system of mammals is different from other animals.

Personal Tutor

To learn about four-chambered hearts, visit biologygmh.com.

Figure 30.7 Mammals have a four-chambered heart in which the atria and ventricles are separated by the septum.

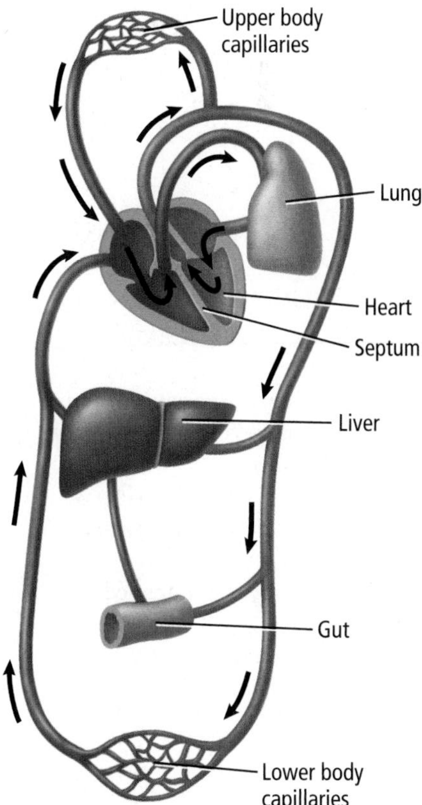

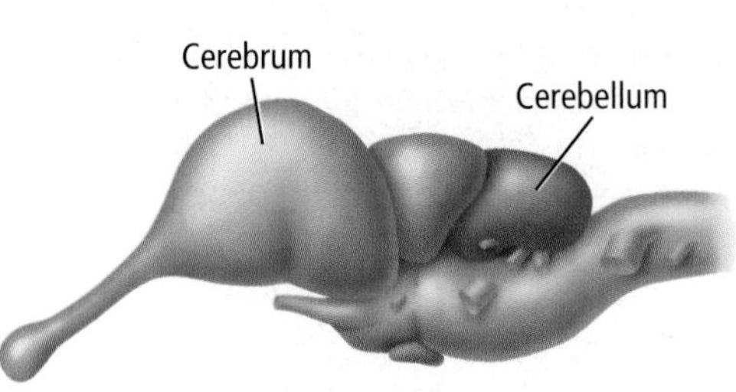

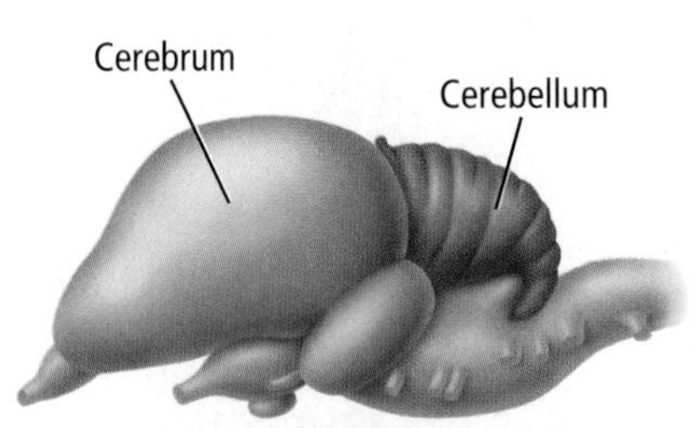

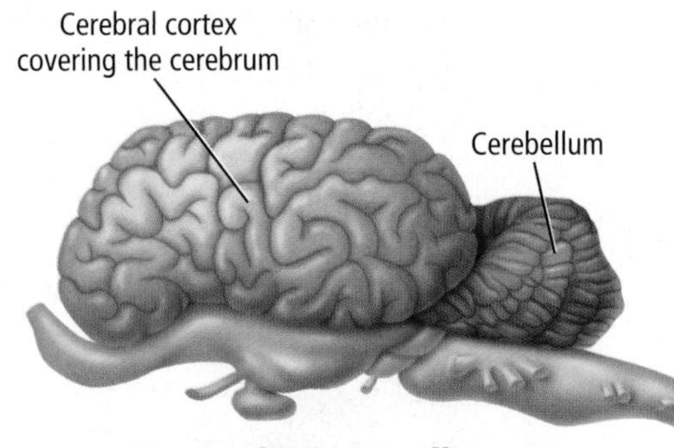

■ **Figure 30.8** The cerebral cortex is the most complex part of the brain and is the part that has increased in size and changed most during the course of vertebrate evolution.

The brain and senses Mammals have highly developed brains, especially the cerebrum. The **cerebral cortex,** shown in **Figure 30.8,** is the highly folded outer layer of the cerebrum. The foldings allow the brain to have a larger surface area for nerve connections while allowing it to still fit inside the skull. The cerebral cortex is responsible for coordinating conscious activities, memory, and the ability to learn.

Another area of the mammalian brain that is well-developed is the cerebellum. The **cerebellum** is responsible for balance and coordinating movement. Compare the size and structure of the cerebellums of a reptile, a bird, and a mammal that are shown in **Figure 30.8.** A well-developed cerebellum allows an animal to have precise motor movements and to make complex movements in three dimensional space.

Complex behavior The mother fox will teach her young to hunt. Because mammals can learn and teach their young survival skills, they have an increased chance of survival. Mammals can carry out complex behaviors, such as learning and remembering what they have learned. Many mammals can get information about their environment and retain it. This information can then be used later. For example, mice that have had a chance to explore a habitat are able to avoid predators better than mice that have not had a chance to explore and learn about the same habitat.

VOCABULARY
SCIENCE USAGE V. COMMON USAGE

Sense

Science Usage: a specialized animal function that involves a sense organ and a stimulus.
Dogs use their sense of smell to get information about their environments.

Common Usage: an intended meaning.
As I read the paragraph, I tried to get a sense of the main idea.

Senses The importance of the senses varies from one group of mammals to the next. In some mammals, such as humans, vision is extremely important, while hearing is most important to mammals such as bats. Bats produce high-frequency sounds that bounce off objects and return to them. In this way a bat can detect objects in its path. This process is called echolocation. If you have seen how a dog sniffs people and objects in its surroundings, you recognize the importance of the sense of smell to this mammal. In some cases, a dog's sense of smell is one million times more sensitive than a human's sense of smell.

 **Reading Check** **Use an analogy** to describe the advantage of having folds in the outer layer of the cerebral cortex.

Glands A system of glands secretes a variety of fluids that helps to regulate a mammal's internal environment. A **gland** is a group of cells that secrete fluid to be used elsewhere in the body. Sweat glands help maintain body temperature. Mammary glands produce and secrete milk that nourishes developing young. Milk contains water, carbohydrates in the form of the sugar lactose, fat, and protein. The proportion of these nutrients differs according to species.

Examine **Table 30.1** to see the proportions of nutrients in the milk of various mammals. The proportion of nutrients is highly variable among different species of mammals. For example, fat amounts can range from one percent to 50 percent. Aquatic mammals, which use a layer of fat to help keep warm, usually have the highest percent of fat in their milk.

Scent glands produce substances that mammals use to mark their territories or attract mates. Oil glands in the skin maintain the quality of the animal's hair and skin. Other glands produce hormones that regulate internal processes, such as growth and release of eggs from ovaries.

Reading Check **Explain** why the fat content in milk would be higher in aquatic mammals.

Reproduction In mammals, the egg is fertilized internally. In most mammals, development of the embryo takes place in the female uterus. The **uterus** is a saclike muscular organ in which embryos develop. In most mammals, the developing embryo is nourished by the **placenta,** an organ that provides food and oxygen to and removes waste from the developing young. The amount of time the young stay in the uterus before they are born is called **gestation.** Gestation periods in mammals vary by species—the shortest being that of the Virginian opossum which can be only 12 days. The longest gestation period occurs in the African elephant, which is an average of 660 days and can be as long as 760 days. In general, the larger the mammal, the longer the gestation period. After birth, the offspring of mammals drink milk for nourishment from the mother's mammary glands.

VOCABULARY
WORD ORIGIN
Gestation
gest– from the Latin word *gestare,* meaning *to bear*
–ation suffix; from Latin meaning *action* or *process.*

Table 30.1 **Proportion of Nutrients in the Milk of Mammals**

Concepts In Motion
Interactive Table To explore more about nutrients in milk, visit biologygmh.com.

Nutrient	Dog	Dolphin	Harp Seal	Rabbit	Zebra
Water	76.3	44.9	43.8	71.3	86.2
Protein	9.3	10.6	11.9	12.3	3.0
Fat	9.5	34.9	42.8	13.1	4.8
Sugar	3.0	0.9	0.0	1.9	5.3

Limbs used for digging and burrowing

Limbs used for flying

■ **Figure 30.9**
Left: The mole has powerful, short forelimbs that are adapted for digging and burrowing in the ground.
Right: The bat can fly with thin membranes spread between the elongated arm and hand bones.

Movement Mammals must find food, shelter, and escape from predators. They have evolved a variety of limb types that enable them to carry out these essential behaviors. Some mammals, such as coyotes and foxes, run. The fastest land mammal—the cheetah—can reach speeds as fast as 110 km/h.

Other mammals, such as kangaroos, leap. Some mammals, including dolphins, swim. Bats are the only mammals that fly. The structure of the skeletal and muscular systems in animals reflects the type of movement that an animal uses. Examine **Figure 30.9,** which shows the forelimbs of a mole and those of a bat. How does the structure of these limbs reflect the habitat and behavior of these animals?

Section 30.1 Assessment

Section Summary

- Mammals are successful in a wide variety of habitats.
- Mammals have specialized teeth.
- Respiratory, circulatory, and nervous systems have complex adaptations that enable mammals to have the extra energy they need and to maintain homeostasis.
- Mammals have internal fertilization and, in most mammals, offspring develop within the female uterus.

Understand Main Ideas

1. MAIN Idea **List** two characteristics unique to mammals.
2. **Explain** how mammals maintain a constant body temperature.
3. **Classify** the mammals that live in your area as herbivores, carnivores, omnivores, or insectivores.
4. **Summarize** how the respiratory and circulatory systems of mammals work together to enable mammals to have high energy levels.
5. **Compare and contrast** how respiration occurs in mammals to respiration in birds. Use **Figure 30.6** and **Figure 29.16** for reference.

Think Critically

6. **Hypothesize** A sperm whale's clicking sound is one of the loudest sounds produced by any animal. The larger the whale, the louder the sound it can make. Form a hypothesis that explains why the whale might produce this sound.

MATH in Biology

7. Suppose a hare spotted a coyote and tried to run away. Coyotes can run at a speed of 70 km/h. Hares can run at a speed of 65 km/h. How far could the hare run before the coyote catches up? Assume that the hare is 25 m from the coyote.

Biology Online **Self-Check Quiz** biologygmh.com

Section 30.2

Reading Preview

Objectives

- **Examine** the characteristics of mammals in each of the three subgroups of living mammals.
- **Distinguish** adaptations that contribute to the diversity of mammals and enable them to live in a variety of habitats.
- **Theorize** about the evolution of mammals.

Review Vocabulary

chromosome: cell structure that carries genetic material that is copied and passed from generation to generation of cells

New Vocabulary

monotreme
marsupial
placental mammal
therapsid

Diversity of Mammals

MAIN Idea Class Mammalia is divided into three subgroups based on reproductive methods.

Real-World Reading Link Think about the mammals you see every day, such as dogs or squirrels. They are only a small part of the 4500 living species of mammals found on Earth. Scientists have developed zoos and wild animal parks that offer opportunities to learn about and enjoy the great variety of mammal species found in the world today.

Mammal Classification

Class Mammalia is divided into three subgroups based on methods of reproduction—monotremes, marsupials, and placental mammals.

Monotremes The animal shown in **Figure 30.10** with its duck bill and webbed feet might not look like any mammal you have seen. However, it has hair and mammary glands, which makes it a mammal. The duck-billed platypus is a monotreme that lays eggs similar to those of reptiles. **Monotremes** are mammals that reproduce by laying eggs. The only other living monotremes besides the duck-billed platypus are echidnas. An adult echidna and an echidna egg are shown in **Figure 30.10.** Both the duck-billed platypus and the echidna only live in Australia, Tasmania, and New Guinea. Besides laying eggs, other unique features of these mammals include reptilian bone structure in the shoulder area, lower body temperature than most mammals, and a unique mix of normal-sized chromosomes, mammal-sized chromosomes, and small, reptilelike chromosomes.

Reading Check **Identify** how monotremes are different from other subgroups of mammals.

Figure 30.10 The echidna, like the duck-billed platypus, is an egg-laying mammal. Once an egg hatches, the offspring receive nourishment from the mother's mammary glands.

Duck-billed platypus

Echidna

Echidna hatching from egg

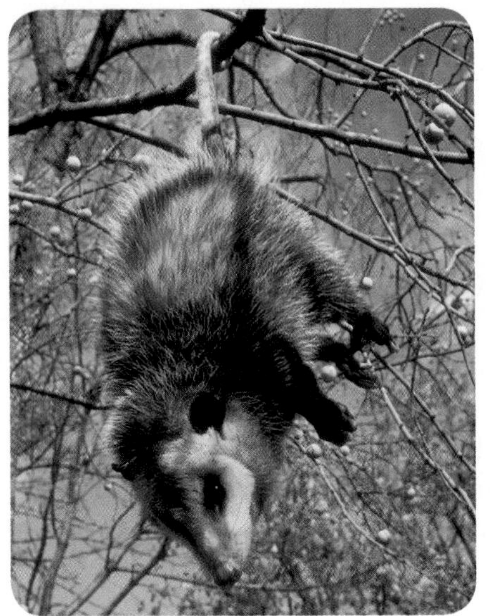

■ **Figure 30.11** Opossums are the only marsupials in North America. Most opossums spend much of their time in the trees.

Marsupials Pouched mammals that have a very short gestation period are **marsupials.** Immediately following birth, the offspring crawl into a pouch made of skin and hair on the outside of the mother's body. Within the pouch, the offspring continue development while being nourished by milk from the mother's mammary glands. In some species of marsupials, offspring are born and crawl into the mother's pouch only eight days after fertilization occurs.

The only North American marsupial is the opossum, shown hanging by its tail in **Figure 30.11.** Other marsupials include the koala, wallaby, kangaroo, and cuscus, some of which are shown in **Figure 30.12.** Australia and its nearby islands are home to most marsupials.

Connection to Earth Science Why marsupials are limited mostly to Australia is still a subject for debate among scientists. One theory, based on fossil evidence, is that marsupials originated in North America, then spread to South America and Europe while the continents were still connected as one giant landmass. From South America, marsupials moved across Africa to Antarctica then to Australia. Then, about 200 million years ago, the continents separated due to the movement of Earth's plates. This isolated the ancestors of today's marsupials to Australia and nearby islands.

Australian marsupials thrived because they were isolated from competing placental mammals. In North and South America, however, placental mammals had competitive adaptive advantages. For example, placental mammals evolved highly social behavior, had more variety in food sources, and evolved greater diversity in form and function than marsupials.

Marsupials in Australia and New Guinea fill the niches occupied by placental mammals elsewhere in the world. For example, kangaroos are the grazers in Australia. They fill the niche that deer, antelope, and buffalo fill in other areas of the world.

■ **Figure 30.12** The cuscus is a nocturnal marsupial that is found in northern Australia and New Guinea. The red kangaroo has a development period of only 33 days, after which the newborn begins nursing in the pouch.

Cuscus

Baby red kangaroo

Red kangaroo

Humpback whale

Pygmy shrew

■ **Figure 30.13**
The humpback whale, weighing 100,000 kg, is one of the largest mammals. The pygmy shrew, weighing 1.5 g, is one of the smallest mammals. Notice the size of the pygmy shrew by comparing it to the size of the earthworm it is eating.

Placental mammals Most mammals living today, including humans, are placental mammals. **Placental mammals** have a placenta, the organ that provides food and oxygen to and removes waste from developing young. Placental mammals give birth to young that do not need further development within a pouch.

Placental mammals are represented by 18 orders. Some orders are represented only by a few species. For example, there are only two species of flying lemurs in order Dermoptera. The aardvark, a termite-eating mammal found in Africa, is the only species in its order. Other orders, such as Rodentia, which includes squirrels and rats, have almost 2000 species. Sizes of placental mammals range from 1.5-g pygmy shrews to 100,000-kg whales, both of which are shown in **Figure 30.13.** Placental mammals range from marine dolphins with adaptations for swimming to moles adapted for subterranean life and bats equipped with wings and ultrasonic echolocation for flying in the dark. **Table 30.2** on page 894 describes some of the different orders of placental mammals.

Scientists have hypothesized several reasons why there are greater numbers and kinds of placental mammals compared to marsupials. One hypothesis is that marsupial young must cling to their mother's fur at birth. Limbs, therefore, are limited in their ability to evolve into structures such as the flippers and wings of some placental mammals. Another hypothesis that explains the success of placental mammals points out that the cerebral cortex of placental mammals is larger and more complex than that of marsupials. This might be due to the more stable, oxygen-rich environment they experience inside the uterus. According to this hypothesis, this might have enabled placental mammals to develop more complex social behaviors that led to their success.

 **Reading Check** **Identify** how placental mammals differ from marsupials.

FOLDABLES
Incorporate information from this section into your Foldable.

LAUNCH Lab
Review Based on what you've read about mammal classification, how would you now answer the analysis questions?

Order Insectivora—shrew

Order Chiroptera—flying fox

■ **Figure 30.14** Shrews are members of order Insectivora. The flying fox is a bat that is a member of order Chiroptera.

Order Insectivora As the name implies, these mammals' main food source is insects. The shrew, shown in **Figure 30.14,** is an insectivore, as are hedgehogs and moles. Members of order Insectivora usually are small and have pointed snouts that allow them to capture insects easily. Shrews include the smallest of all mammals. Shrews can be found in almost all parts of the world and spend most of their lives underground.

VOCABULARY

ACADEMIC VOCABULARY

Modify:
to make changes that result in a new purpose
The wings of a bat are modified forelimbs for flying.

Order Chiroptera There are about 925 species in order Chiroptera (ky RAHP ter uh)—all of which are bats. As mentioned previously, bats are the only mammals that truly can fly. Their wings are thin membranes supported by modified forelimbs. Bats feed on a variety of foods. Some eat insects, some eat fruit, and others feed on blood. The most common North American bat is the little brown bat which you might have seen swooping and darting at dusk to catch insects. The flying fox, shown in **Figure 30.14,** is the largest of all bats. It lives in tropical regions worldwide and feeds on fruit.

■ **Figure 30.15** Golden lion tamarins are omnivores that live in the coastal forests of Brazil. **Identify** *other animals that are members of order Primates.*

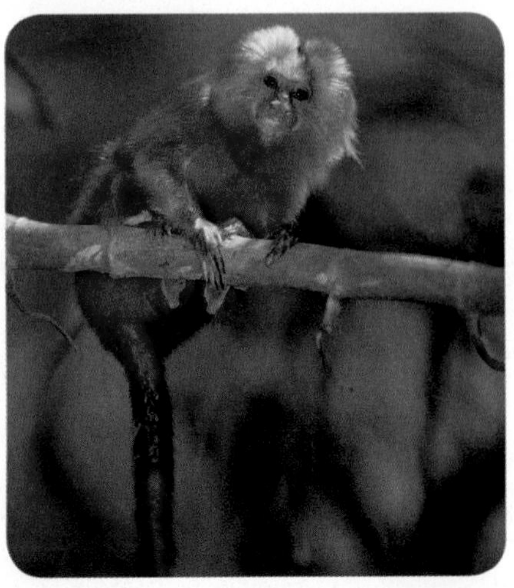

Order Primates Monkeys, apes, and humans are all examples of primates. Primates' brains, with large cerebral hemispheres, are the most developed brains of all mammals. Most primates are tree dwellers. This leads scientists to hypothesize that the need to perform complex movements while in trees, such as those involved in capturing food and avoiding enemies, led to advances in the brain structure of primates. Primate forelimbs often are adapted for grasping, and most have nails instead of claws. The golden lion tamarin in **Figure 30.15** is grasping with its hands the branch on which it is sitting.

Order Xenarthra Animals in order Xenarthra (zen AR thra) either have no teeth or have simple, peglike teeth. Anteaters, like the one shown in **Figure 30.16,** are toothless. Anteaters have a spiny tongue and sticky saliva that allow them to easily capture ants and termites in their nests. Sloths and armadillos both have peglike teeth. Sloths mostly feed on leaves, and armadillos feed on insects. Most mammals in this order live in Central America and South America with the exception of the armadillo, which also can be found in the southern United States.

Giant anteater

Beaver

Order Rodentia These gnawing mammals in order Rodentia, called rodents, include the beaver, which is shown in **Figure 30.16,** rats, woodchucks, marmots, squirrels, hamsters, and gerbils. Rodents make up nearly 40 percent of all mammalian species. Two pairs of razor-sharp incisor teeth continue to grow throughout the life of a rodent. They use their sharp teeth to gnaw through wood, seed pods, or shells to get food. The ability of rodents to invade all land habitats and their successful reproductive behavior have made them ecologically important in all terrestrial ecosystems.

■ **Figure 30.16** The giant anteater, the largest anteater, is found throughout Central and South America. The largest rodent is the beaver, weighing as much as 80 kg.
Describe ***the characteristics of members of order Xenarthra.***

Order Lagomorpha Like rodents, members of the the order Lagomorpha—rabbits, pikas, and hares—have long sharp incisors that continue to grow. Lagomorphs also have a pair of peglike incisors that grow behind the first pair. These mammals are herbivores that eat grasses, herbs, fruits, and seeds. The pika shown in **Figure 30.17** lives in high latitude or high altitude environments in which the ground is covered with snow for parts of the year. The grasseaters adapt to these conditions by harvesting grass during the warm months and storing it. The pikas then eat the grass during the winter when no fresh vegetation is available.

VOCABULARY
WORD ORIGIN
Lagomorpha
lago– from the Greek word *lagos,* meaning *hare.*
–morph, from the Greek word *morphe,* meaning *form.*

Order Carnivora You might have a pet dog or cat. It, along with wolves, bears, seals, walruses, coyotes, skunks, otters, minks, and weasels, belongs to the order Carnivora. All of these carnivores are predators with teeth adapted to tearing flesh. The lioness, shown in **Figure 30.17,** feeds on antelope, giraffes, and even crocodiles. After she captures her prey, she uses her incisors to tear off chunks of meat.

■ **Figure 30.17** The American pika can be found in alpine regions of the western U.S. and southwestern Canada. The lioness uses her canines to stab and pierce her prey.

Pika

Lioness

■ **Figure 30.18** The trunk of an elephant is called a proboscis. Trunks are unique to members of order Proboscidea.

Order Proboscidea Elephants are the largest living land mammals. They have flexible trunks adapted for gathering plants and taking in water. Two upper incisors are modified as tusks for digging up roots and tearing bark from trees. Some elephants are trained to help lift heavy objects. The elephant shown in **Figure 30.18** is helping remove debris that washed ashore during a tsunami in Indonesia on December 26, 2004. Ancient mastodons and mammoths are the extinct relatives of today's African and Asian elephants.

Order Sirenia Manatees and dugongs—members of the order Sirenia—are large, slow-moving mammals with big heads and no hind limbs. Their forelimbs are modified into flippers that aid in swimming. These animals are herbivores, feeding on seagrasses, algae, and other aquatic plants. Depending on their size, manatees can consume as much as 50 kg of vegetation per day. Sirenians can be found cruising the surface of warm tropical rivers and lagoons. Because they are so slow and prefer the surface of the water, they often are injured or killed by the propellers of speedboats. Notice the scars on the back of the manatee in **Figure 30.19.**

Order Perissodactyla These hoofed mammals include horses, zebras, and rhinoceroses. Members of this order have an odd number of toes, either one or three on each foot. These mammals are herbivores and have teeth that are adapted for grinding plant material. Perissodactyls can be found on all continents except Antarctica.

Reading Check **Compare** placental mammals using **Table 30.2.**

Table 30.2 Orders of Placental Mammals

Concepts In Motion — **Interactive Table** To explore more about placental mammals, visit biologygmh.com.

Order	Example	Characteristics
Insectivora	Shrews, hedgehogs, moles	Pointed snouts, smallest mammals, live underground, insect-eaters
Chiroptera	Bats	Nocturnal, use sonar, adapted for flight, fruit and insect-eaters
Primates	Monkeys, apes, humans	Binocular vision, large brains, most are tree-dwellers, opposable thumb
Xenarthra	Anteaters, sloths, armadillos	Toothless or peg-like teeth, insect-eaters
Rodentia	Beavers, rats, woodchucks, marmots, squirrels, hamsters, and gerbils	Sharp incisor teeth, plant-eaters
Lagomorpha	Rabbits, pikas, hares	Back legs longer than front legs, adapted to jumping, incisors that continually grow
Carnivora	Dogs, cats, wolves, bears, seals, walruses, coyotes, skunks, otters, minks, and weasels	Teeth adapted to tear flesh, meat-eaters
Proboscidea	Elephants	Long trunks, incisors become long tusks, largest land animal
Sirenia	Manatees and dugongs	Slow moving, big heads, no hind limbs
Perissodactyla	Horses, zebras, rhinoceroses	Hoofed, odd number of toes, plant-eaters
Artiodactyla	Deer, antelopes, cattle, sheep, pigs, goats, hippopotamuses	Hoofed, even number of toes, plant-eaters that chew cud
Cetacea	Whales, dolphins, porpoises	Front limbs that are flippers, no hind limbs, nostril forms a blowhole

Manatee

Humpback whale

■ **Figure 30.19** The West Indian manatee is endangered. Wildlife managers help rescue manatees that have been injured by boat propellers. The baleen of a whale is similar to a sieve.

Order Artiodactyla Members of the order Artiodactyla also are hoofed mammals. They differ from perissodactyls in that they have an even number of toes, either two or four, on each limb. Deer, antelopes, cattle, sheep, pigs, goats and hippopotamuses are all artiodactyls. Many cattle, sheep, and deer have horns or antlers. Mammals in this order are herbivorous and most chew their cud.

Order Cetacea Whales, dolphins, and porpoises have front limbs modified into flippers that aid in swimming. They have no hind limbs and the tail consists of fleshy flukes. Nostrils are modified into a single or double blowhole on top of the head. Except for a few muzzle hairs, their bodies are hairless. Some whales are predators. Others, like the blue whale, have a specialized structure inside their mouths called a baleen that is used to filter plankton for food. The baleen of a humpback whale is shown in **Figure 30.19.**

DATA ANALYSIS LAB 30.1

Based on Real Data*

Analyze and Conclude

How does boat noise affect whales? Killer whales might coordinate their cooperative hunting and other social behavior with certain calls that have meaning to the pod, or group, of whales with which they travel. The number of boats in the area of study increased by about five times from 1990–2000.

Data and Observations

Biologists examined the duration of whales' calls in three different pods for several years. Examine the graphs to the right.

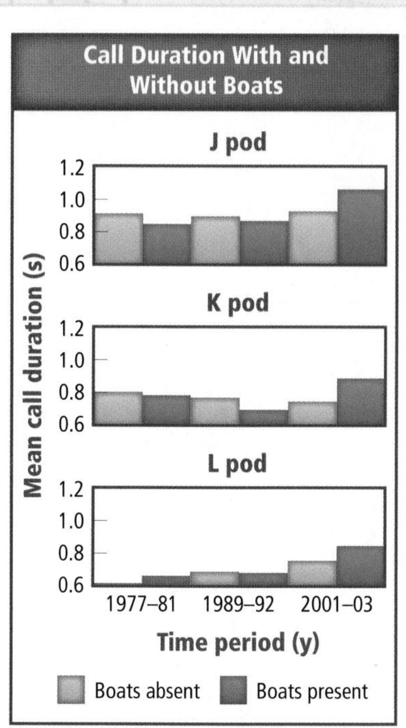

Think Critically

1. **Evaluate** the trend in call duration of whales in J, K, and L pods from 1977 to 2003. What might account for this trend?
2. **Hypothesis** Form a hypothesis that describes what the researchers were investigating in this study.

*Data obtained from: Foote, A., et al. 2004. Whale-call response to masking boat noise. *Nature* 428: 910.

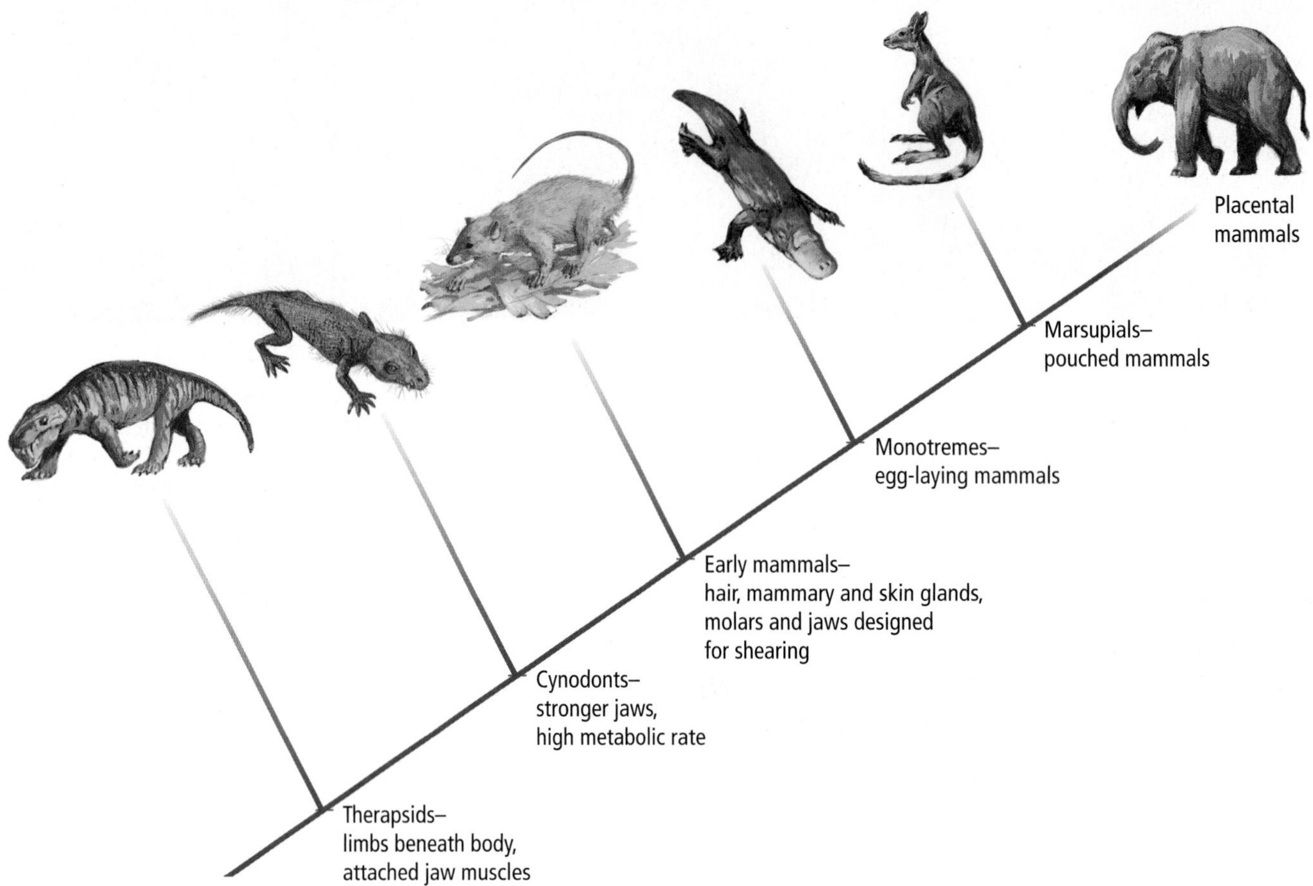

■ **Figure 30.20** Fossil evidence has enabled scientists to make this cladogram that reflects the phylogeny of the orders of living placental mammals.
Interpret ***Which present-day mammals have been on Earth longest?***

Evolution of Mammals

The first mammals probably evolved from reptiles in the mid-Triassic period about 220 million years ago. A few lived side by side with the dinosaurs, but mammals did not become common until the dinosaurs disappeared. The cladogram in **Figure 30.20** shows one interpretation of the evolution of mammals.

Therapsids Fossil evidence indicates that the first mammals probably arose from a group of mammal-like reptiles called therapsids. A **therapsid** is an extinct vertebrate with both mammalian and reptilian features. Therapsids had some characteristics of mammals today, including a pair of holes in the roof of the skull that allowed for the attachment of jaw muscles. Therapsids also had limbs positioned beneath their bodies that allowed for more efficient movement.

Evidence shows that therapsids might have been endotherms. They ate more food than their ancestors, which might have provided them with the energy to produce their own body heat. Being endothermic would have given therapsids an advantage over other ectothermic vertebrates in that they would have been able to be more active during the winter. Therapsids went extinct about 170 million years ago. One group of therapsids called cynodonts continued to evolve more mammalian characteristics, including a high metabolic rate, stronger jaws, and a structure in the mouth that allowed them to breathe while holding food or nursing. A cynodont is shown in **Figure 30.21.**

Cynodont

Eomaia

Figure 30.21 Cynodonts were animals that had some characteristics of mammals and were about the size of a weasel. *Eomaia* is the oldest placental mammal fossil discovered.

The Age of Mammals According to fossil evidence, the first placental mammals might have been mouse-sized animals such as *Eomaia,* shown in **Figure 30.21.** Recently unearthed fossils show some mammals were larger. One was 1m in length with a squat body and predatory teeth. Another had a beaverlike tail adapted to swim. When dinosaurs disappeared at the end of the Mesozoic Era, mammals underwent extraordinary adaptations to the environment. As flowering plants flourished, new sources of nutrition and new habitats became available. Mammals had new environments to fill. For example, fast-moving herbivores and their predators evolved to fill the niches in the drier prairies. The huge expansion in mammalian diversity and numbers led some scientists to call the Cenozoic Era the "Age of Mammals."

Section 30.2 Assessment

Section Summary

- Of the three subgroups of mammals, only the members of one lay eggs.
- The members of one of the mammalian subgroups have pouches in which the young spend most of their development time.
- Placental mammals have young that are nourished by the placenta as they develop in the uterus.
- Mammals might have evolved from reptilian ancestors called therapsids.
- There was a huge expansion in the diversity of mammals in the Cenozoic era.

Understand Main Ideas

1. MAIN Idea **Name** the three subgroups of mammals and describe their features.
2. **Identify** the order or orders to which the following mammal might belong and explain your reasoning: it has reddish-brown fur, two pairs of incisors in the upper jaw (one pair behind the other), claws, a body that is a little smaller than a basketball, and it can jump easily.
3. **Compare and contrast** the characteristics of mammals in order Perissodactyla to those in order Artiodactyla.
4. **Explain** how mammals might have evolved from reptiles.

Think Critically

5. **Hypothesize** The bill of a platypus can detect the electrical fields of muscle contraction of other animals. This is how the platypus searches for prey. Form a hypothesis about how this complicated adaptation developed in place of simply searching visually for prey.

WRITING in Biology

6. Some people have the misconception that marsupials are inferior to placental mammals. Analyze and explain the faulty reasoning of this idea.

Biology & Society

Canine Helpers

Picture Trixie, a mixed-breed dog, who was alone with her owner who just had a stroke. The man was paralyzed and unable to leave his bed. Trixie went to the door, opened the door, and barked on the porch to get attention for her owner. The neighbors heard Trixie bark, came over and saw the owner needed medical attention. Trixie had saved her owner's life.

Pack animals Dogs were domesticated approximately 14,000 years ago and began to diverge from the wolf about 135,000 years ago. Because wolves are pack animals and dogs evolved from wolves, dogs behave like pack animals. When a dog becomes a member of a human family, it often interprets the family as part of its pack. Helping, protecting, and saving members of the pack from harm are part of a dog's behavior.

The sense of smell A dog's sense of smell is much more sensitive than a human's sense of smell. A dog has 200 million scent receptors compared to 5 million in humans. Dogs routinely are used to help locate drugs, explosives, and missing people. Certified avalanche dogs are used to help locate people who have been buried by avalanches. The dogs can locate people buried in up to 5 m of snow. An avalanche dog can cover an area the size of a football field and up to 36 m of snow depth in 30 min. It would take five people with probes 15 h to cover the same volume of space.

Sensing cancer Dogs also are used to detect the presence of cancerous tumors in a person. In a recent research study, dogs were able to recognize the presence of bladder cancer by sniffing patients' urine. In this experiment, the dogs were trained to lie down when they detected cancerous cells in a urine sample.

This dog, a member of the Transportation Security Administration Explosive Detection Canine Team, is sniffing luggage for explosives.

There is some evidence that dogs also can detect skin cancer by detecting odors given off by cancerous moles. Currently, studies are under way in which dogs are being tested to see if they can detect lung cancer and prostate cancer. Dogs may be able to provide an early detection system that medical science has yet to achieve.

Sensing seizures Some dogs also can sense when a person might be about to have a seizure. Seizure alert dogs are used to help warn people who have a seizure disorder of an impending seizure anywhere from 15 minutes to 12 h before the seizure. This can allow people time to take anti-seizure medication, call for help, or move to a safe place. Currently, the belief is that these dogs sense a change in the facial expression or sense that something is different about the rhythm of the person's personality.

COMMUNITY SERVICE

Contact a retirement community near your school to see if it has a pet therapy program. Find out more about how the program works. Find out if your class could assist with the program to learn more about how pets assist senior citizens. For more information about pet therapy, visit biologygmh.com.

BIOLAB

INTERNET: HOW DO WE IDENTIFY MAMMALS?

Background: The physical characteristics that all mammals share, such as fur and mammary glands, have enabled them to adapt to nearly every ecosystem in the biosphere. Mammals thrive in rain forests, deserts, and polar regions, and they have adapted to the environment near your home or school as well.

Question: *What diversity of mammals can be found in your area?*

Materials

North American mammal identification field guide
binoculars
field journal

Safety Precautions

Procedure

1. Read and complete the lab safety form.
2. List the mammals you have observed in your area of the country.
3. Predict how these species of mammals would be classified.
4. Design and construct a data table for recording the species; physical characteristics, such as size, body shape, and unique features; and taxonomic classifications of the mammals you have observed.
5. Research the mammals to fill in information in your data table. Either observe the animals in their natural habitat in your local area, such as a park or wetlands, or visit the zoo. If you cannot observe the animals in their natural habitats, obtain information about local mammals from a guide book.
6. Record your observations in your field journal and transfer the information to your data table.
7. Post your results at biologygmh.com.

Analyze and Conclude

1. **Describe** basic characteristics shared by all mammals that you have observed.
2. **Compare and contrast** the mammals from your study to those of other students around the country.
3. **Compare and contrast** the physical characteristics scientists could use to separate the mammals into different taxonomic orders.
4. **Infer** how the mammals from your list have adapted to and survived in their environments.
5. **Describe** other observation strategies that could be used to conduct a more comprehensive mammal search of your chosen search area.
6. **Error Analysis** Compare your list of identified mammal species with the lists compiled by other students to determine possible identification errors.

POSTER SESSION

Make a Presentation Collect photographs of the mammals from another area of the country and create a poster to present to your class. Include information about the specific characteristics and adaptations of each mammal. To find out more about mammals, visit biologygmh.com.

Chapter 30 Study Guide

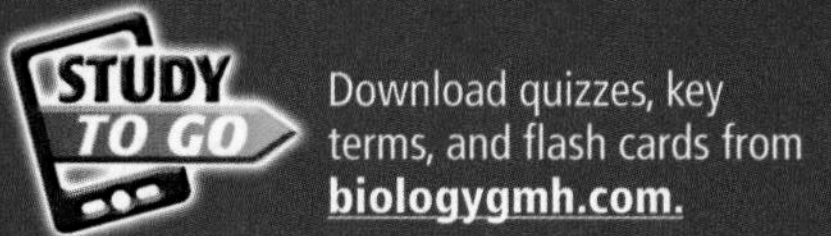

Download quizzes, key terms, and flash cards from biologygmh.com.

Foldables **Hypothesize** Only three species of monotremes are living today—one species of duck-billed platypus, and two species of echidna. Form a hypothesis that explains why this subgroup of mammals has low diversity compared to the diversity of marsupials and placental mammals.

Vocabulary	Key Concepts
Section 30.1 Mammalian Characteristics	
• cerebellum (p. 886) • cerebral cortex (p. 886) • diaphragm (p. 885) • gestation (p. 887) • gland (p. 887) • mammary gland (p. 880) • placenta (p. 887) • uterus (p. 887)	**MAIN Idea** Mammals have two distinct characteristics: hair and mammary glands. • Mammals are successful in a wide variety of habitats. • Mammals have specialized teeth. • Respiratory, circulatory, and nervous systems have complex adaptations that enable mammals to have the extra energy they need and to maintain homeostasis. • Mammals have internal fertilization and, in most mammals, offspring develop within the female uterus.
Section 30.2 Diversity of Mammals	
• marsupial (p. 890) • monotreme (p. 889) • placental mammal (p. 891) • therapsid (p. 896)	**MAIN Idea** Class Mammalia is divided into three subgroups based on reproductive methods. • Of the three subgroups of mammals, only the members of one lay eggs. • The members of one of the mammalian subgroups have pouches in which the young spend most of their development time. • Placental mammals have young that are nourished by the placenta as they develop in the uterus. • Mammals might have evolved from reptilian ancestors called therapsids. • There was a huge expansion in the diversity of mammals in the Cenozoic era.

Biology Online Vocabulary PuzzleMaker biologygmh.com

Assessment

Section 30.1

Vocabulary Review

In the analogies that follow, one of the words is missing. Complete each analogy by filling in the blank with a vocabulary term from the Study Guide page.

1. A yolk is to a bird as a __________ is to a mammal.
2. Incubation period is to a bird as a __________ period is to a mammal.
3. The nucleus is to the cell as the __________ is to the brain.

Understand Key Concepts

Use the diagram below to answer questions 4 and 5.

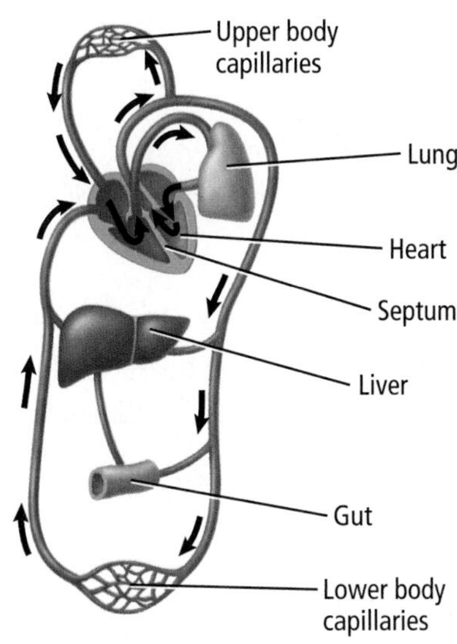

4. Which body system is illustrated in the diagram?
   - **A.** excretory system
   - **B.** skeletal system
   - **C.** circulatory system
   - **D.** reproductive system
5. Which best explains how this system supports endothermy in mammals?
   - **A.** Oxygenated blood is separated from deoxygenated blood.
   - **B.** The heart has three chambers and is able to pump more blood.
   - **C.** This system moves oxygenated blood to the lungs.
   - **D.** This system moves deoxygenated blood from the heart to the body.
6. Which is the least involved in maintaining homeostasis in mammals?
   - **A.** kidneys
   - **B.** heart
   - **C.** sweat glands
   - **D.** claws
7. Oil glands, sweat glands, and mammary glands are responsible for which functions?
   - **A.** hair and skin maintenance, temperature regulation, milk production
   - **B.** reproduction, hair and skin maintenance, temperature regulation
   - **C.** temperature regulation, milk production, reproduction
   - **D.** milk production, oxygen delivery, hair and skin maintenance

Use the diagram below to answer questions 8 and 9.

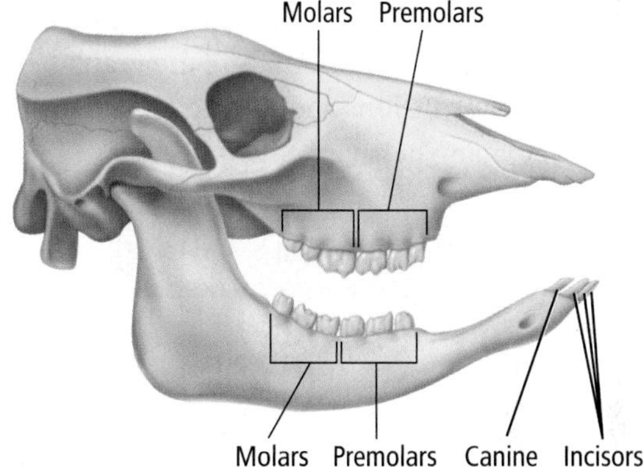

8. In what way did having a variety of tooth types contribute to the presence of mammals in all habitat types?
   - **A.** They could eat a variety of foods.
   - **B.** They could hunt effectively.
   - **C.** They could digest their food more easily.
   - **D.** Their digestive tracts were modified.
9. In which trophic category does this mammal belong?
   - **A.** herbivore
   - **B.** insectivore
   - **C.** carnivore
   - **D.** detritivore

Constructed Response

10. **Open Ended** Examine **Table 30.1** and form a hypothesis that explains why there are such big differences in the fat content of seal milk, dolphin milk, and milk of other mammals.

11. **Open Ended** Many animals that live in the Arctic have large bodies with short extremities, such as ears and legs. Explain how this adaptation might help them keep warm.

Think Critically

12. **Design an Experiment** Hippopotamuses secrete a fluid from glands deep in their skin that may function as sweat but can have other functions as well. Biologists hypothesize that this fluid might act as a sunscreen for the skin of the hippopotamus. Design an experiment using beads that absorb ultraviolet light that would test if the fluid on the skin of this mammal provides protection from the Sun.

13. **Analyze and Conclude** Biologists hypothesized that carnivores with large home range sizes, when in captivity in small spaces, had higher incidences of pacing behavior. They studied the arctic fox, the polar bear, and the lion. Analyze the graph below and make conclusions about the effect of confinement on pacing behavior.

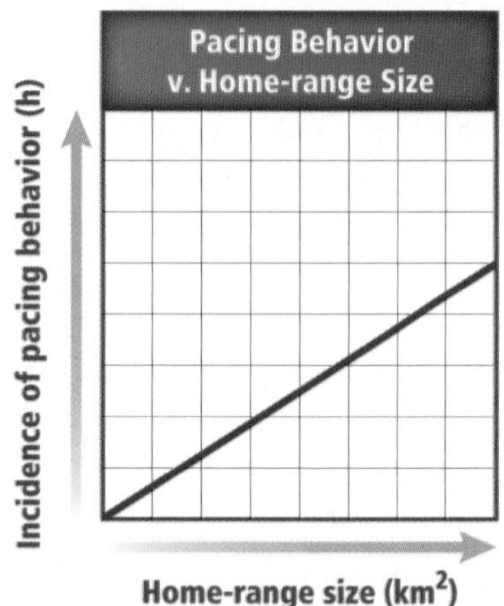

Section 30.2

Vocabulary Review

Each of the following sentences is false. Make the sentence true by replacing the italicized word with a vocabulary term from the Study Guide page.

14. An elephant is an example of a *marsupial.*

15. Mammals might have evolved from *monotremes.*

16. *Therapsids* are egg-laying mammals.

17. *Monotremes* are mammals that have a pouch.

Understand Key Concepts

18. Which mammal is a member of order Cetacea?
    - **A.** beaver
    - **B.** whale
    - **C.** zebra
    - **D.** manatee

19. Which is a benefit of the development of young within a uterus?
    - **A.** Young are born alive.
    - **B.** Predation of the young is less likely.
    - **C.** Predation of the young is more likely.
    - **D.** Young are more fully developed at birth.

20. Which mammal is not a marsupial?
    - **A.** opossum
    - **B.** kangaroo
    - **C.** echidna
    - **D.** wallaby

21. Which is not a characteristic of the duck-billed platypus?
    - **A.** webbed feet
    - **B.** egg-laying ability
    - **C.** three-chambered heart
    - **D.** small, reptilelike chromosomes

22. Examine **Figure 30.20.** Which mammal evolved first?
    - **A.** elephant
    - **B.** opossum
    - **C.** echidna
    - **D.** blue whale

Constructed Response

23. **Open Ended** Sketch and explain the ideal adaptations of a mammal that lives in 1-m deep marsh water, much underwater vegetation, and predatory snakes.

24. **Open Ended** Suggest reasons why you should study the orders of mammals.

25. **Open Ended** Arrange for a debate in your class about the use of animals for testing medicines and cosmetics. Visit biologygmh.com to do your research.

Think Critically

26. Infer Fossil evidence indicates that mammals lived at the same time as dinosaurs for many millions of years. During this time, mammals were very small compared to the dinosaurs. Infer why it might have been an advantage for mammals to remain so small when dinosaurs roamed Earth.

27. CAREERS IN BIOLOGY Find out what mammals are endangered in your area. Assume that you will be the zookeeper responsible for providing and maintaining a space for a new animal that is locally endangered and will be kept on exhibit at the zoo. Design a space, feeding routine, and other care instructions for maintaining this animal in your local zoo. Prepare a sign for the space that will alert people of the importance of protecting this endangered species and ways in which individuals can participate in conservation measures.

28. Research Select your favorite group of mammals. Make a map that shows its world distribution. Reflect on ecological factors that might currently be limiting its potential range or might affect the group in the future. Make recommendations about what should be done to insure the success of your favorite mammal group.

Use the table below to answer question 29.

Birth Weight and Protein Content of Milk

Mammal	Days Needed to Double Birth Weight	Protein Content of Milk (g/1000)
Human	180	12
Horse	60	26
Cow	47	33
Pig	18	37
Sheep	10	51
Cat	9	101

29. Analyze Data Explain the relationship between the number of days it takes to double birth weight and the protein content of milk. Make a graph of this table.

Additional Assessment

30. WRITING in Biology Visit biologygmh.com to research which mammal genomes have been sequenced. Write a summary paragraph describing what you learned.

DBQ Document-Based Questions

A specific type of ground squirrel was found to have the ability to produce ultrasonic calls that could not be heard by other mammals as well as calls that could be heard (audible). Biologists exposed ground squirrels to the ultrasonic call, background noise, a tone similar to the ultrasonic calls, and an audible call. Then they observed the portion of time the animals spent in vigilant behavior (looking for predators) during each sound. Use this graph to answer the questions below.

Data obtained from: Wilson, D. and Hare, F. 2004. Ground squirrel uses ultrasonic alarms. *Nature* 430: 523.

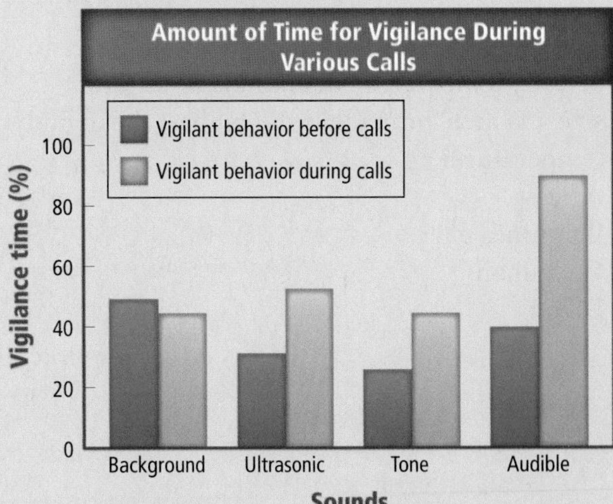

31. Under which conditions did ground squirrels exhibit the most vigilant behavior overall?

32. Under which conditions might an ultrasonic signal be more effective as a warning?

Cumulative Review

33. Distinguish between vascular and nonvascular plants. **(Chapter 21)**

34. Suggest ways to avoid becoming the host of a flatworm. **(Chapter 25)**

Standardized Test Practice

Cumulative

Multiple Choice

Use the graphs below to answer questions 1 and 2.

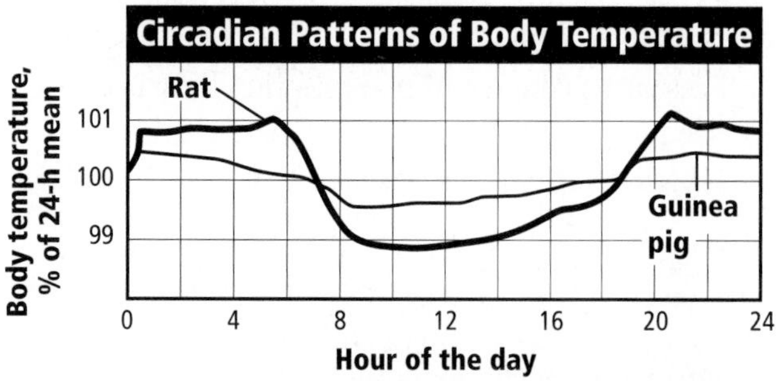

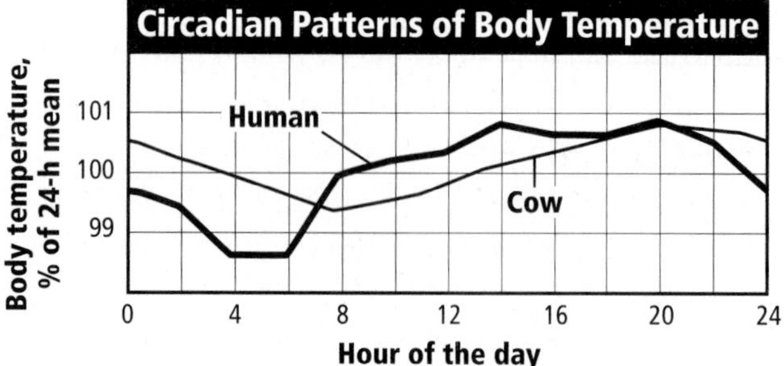

1. The graphs above show the circadian pattern of body temperature in animals of different sizes. Which animal has the highest mean body temperature?
   A. cow
   B. guinea pig
   C. human
   D. rat

2. The rat and guinea pig on the graph above are mainly nocturnal animals. What can you infer from this graph about the body temperatures of nocturnal animals?
   A. They have higher body temperatures than animals that are active during the day.
   B. They have more extreme temperature changes than animals that are active during the day.
   C. They have lower body temperatures than animals that are active during the day.
   D. They have less extreme temperature changes than animals that are active during the day.

3. Which statement describes the difference between invertebrate chordates and the rest of the phylum Chordata?
   A. Invertebrate chordates lack a backbone.
   B. Invertebrate chordates lack a notochord.
   C. Other members of the phylum lack a backbone.
   D. Other members of the phylum lack a notochord.

Use the table below to answer question 4.

Row	Group	Some Components of the Digestive System
1	Amphibians	Has gizzard, stomach, intestines
2	Reptiles	Has crop, large and small intestines
3	Birds	Has crop, gizzard, intestines
4	Fishes	Has swim bladder, stomach, intestines

4. Which row of information in the table contains correct information about the digestive system?
   A. 1
   B. 2
   C. 3
   D. 4

5. Pharyngeal pouches are defined as which of the following?
   A. cavities that hold food as it is digested
   B. sacs that hold the coiled digestive system in place
   C. structures that link the mouth cavity and the esophagus
   D. structures that regulate metabolism, growth, and development

6. How many pairs of jointed appendages do spiders have?
   A. 3
   B. 4
   C. 5
   D. 6

Biology Online Standardized Test Practice biologygmh.com

Short Answer

7. Describe four different characteristics and/or processes that enable mammals to maintain homeostatic temperature control.

8. Compare and contrast the two types of bird feathers.

9. What are two benefits of young of mammals receiving milk from their mothers?

Use the diagram below to answer question 10.

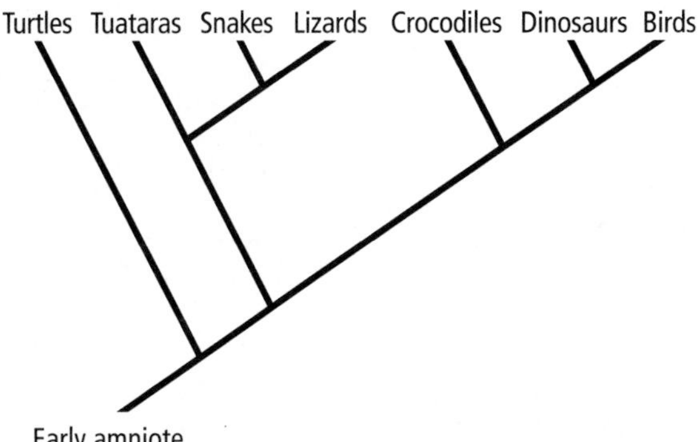

10. Assess which group of modern reptiles is most closely related to dinosaurs.

11. Hypothesize why there are so many different types of birds living today.

12. List two ways in which seedless vascular plants are better adapted than nonvascular plants to survive in a changing environment.

Extended Response

13. Sea cucumbers protect themselves by casting out their internal organs, but other echinoderms cannot. Hypothesize why this adaptation is found only in sea cucumbers.

14. Assess how the skeleton of a bird is adapted for flight.

15. Critique why an *r*-strategist population might be better suited to live in an unpredictable climate than a *k*-strategist population.

Essay Question

In most bird species, both parents care for the young. The parents come together during mating season to mate and raise their young. In some bird species, one parent builds the nest and then attracts a mate. In other bird species, the parents build the nest together. The parents often take turns guarding the eggs and keeping them warm. When the young hatch, the parents bring food that is similar to what young birds will eat as adults. This parental care continues until the chicks are ready to fly away. After they leave the nest, young birds are on their own and seldom have any more contact with their parents.

Using the information in the paragraph above, answer the following question in essay format.

16. Young birds usually are cared for by both parents. In mammals, the mother often raises the young by herself. Propose a hypothesis that explains why both bird parents care for the young while only the mammalian mother cares for her young. Discuss how the hypothesis could be tested.

NEED EXTRA HELP?																
If You Missed Question . . .	1	2	3	4	5	6	7	8	9	10	11	12	13	14	15	16
Review Section . . .	30.1	30.1	27.2	29.2	27.2	26.2	30.1	29.2	30.1	29.1	29.2	21.2	27.1	29.2	4.1	29.2

CHAPTER 31

Animal Behavior

BIG Idea **Many animal behaviors are influenced by both genetics and environmental experiences.**

Section 1
Basic Behaviors

MAIN Idea Animal behaviors are innate and learned behaviors that evolve by natural selection.

Section 2
Ecological Behaviors

MAIN Idea Animals that engage in complex behaviors might survive and reproduce because they have inherited more favorable behaviors.

BioFacts

- Emperor penguins, like those shown on the right, usually find a new mate each breeding season. It is the male penguin who incubates the egg.
- The longest migration made by a mammal is that of a gray whale—more than 19,000 km from the Arctic Ocean to Mexico and back.
- Some cocoon-building spiders make more than 6000 movements in an identical pattern each time they make a cocoon.

LAUNCH Lab

How do scientists observe animal behavior in the field?

Observing animals in their natural habitat is one way that scientists can study animal behavior. The photo on the left shows a colony of emperor penguins in Antarctica. Penguins exhibit behavior associated with courtship, care of their young, grooming, and defending territories. In this lab, you will watch a short video or view photos of bird behavior.

Procedure

1. Read and complete the lab safety form.
2. Record a description of all the different behaviors you observe in the **video** or **photos.**
3. Review your list and infer why the birds might have exhibited each type of behavior.

Analysis

1. **Explain** If you want to understand penguin behavior, you need to study many birds under various conditions. Why is this?
2. **Infer** Some of the behaviors you noticed might have been competitive behaviors. For what resources might animals compete? How could competitive behaviors benefit any animal?

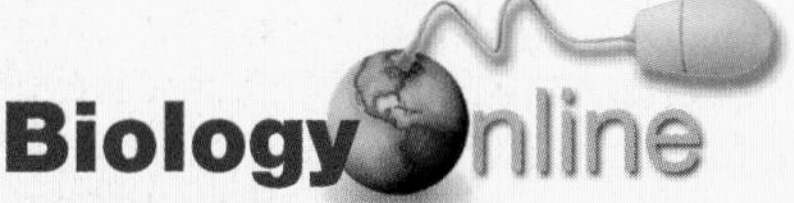

Visit biologygmh.com to:

- **study the entire chapter online**
- **explore the Interactive Time Line, Concepts in Motion, the Interactive Table, Virtual Labs, Microscopy Links, and links to virtual dissections**
- **access Web links for more information, projects, and activities**
- **review content online with the Interactive Tutor, and take Self-Check Quizzes**

Learned Behavior Make the following Foldable to help you organize information about the different types of learned behavior.

STEP 1 Fold a sheet of paper in half vertically.

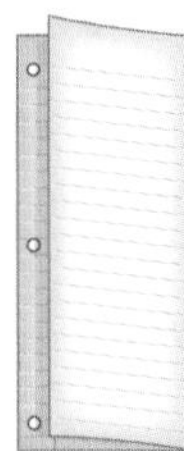

STEP 2 Cut five equal slits in one layer to form tabs.

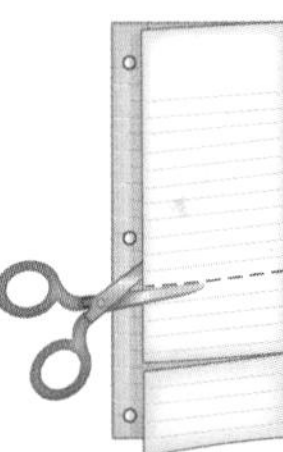

STEP 3 Label each tab with one of the five types of learned behavior presented in Section 31.1: habituation, classical conditioning, operant conditioning, imprinting, and cognitive behavior.

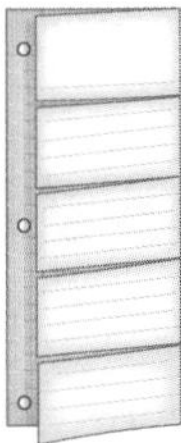

FOLDABLES **Use this Foldable with Section 31.1.** As you read the section, summarize information about the different types of learned behavior under the tabs.

Section 31.1

Reading Preview

Objectives

- **Relate** animal behaviors to evolution by natural selection.
- **Distinguish** between innate and learned behavior.
- **Identify** different types of animal behavior and provide examples of each.

Review Vocabulary

natural selection: population process by which heritable traits that result in the greatest number of offspring eventually become the most common traits in the population

New Vocabulary

behavior
innate behavior
fixed action pattern
learned behavior
habituation
classical conditioning
operant conditioning
imprinting
cognitive behavior

Basic Behaviors

MAIN Idea **Animal behaviors are innate and learned behaviors that evolve by natural selection.**

Real-World Reading Link Think about what happens when you smell your favorite food as you walk by a restaurant. Whether you are hungry or not, your mouth might start to water and you might start thinking about how good that food tastes. Other animals have similar behaviors.

Behavior

You might have seen a lizard lying on a rock in the sunlight. The lizard is regulating its body temperature through its behavior. In order to raise its body temperature, the lizard absorbs the Sun's heat. If the lizard's body temperature starts to get too high, it will move into the shade. This is an example of behavior. **Behavior** is the way an animal responds to a stimulus. A stimulus (STIHM yuh lus) is an environmental change that directly influences the activity of an organism.

Behavior can occur in response to an internal stimulus, which is a stimulus that comes from inside the body, as in the case of the lizard. Behavior can also be caused by an external stimulus—a stimulus that comes from outside the body. An external stimulus could be the smell of food, someone calling your name, or the sight of a predator.

Reading Check **Summarize** why a lizard might lie on a warm rock in the morning.

■ **Figure 31.1**
Studying Animal Behavior

The study of animal behavior only began about 100 years ago.

1898 Ivan Pavlov, a Russian physiologist, conditions a dog to salivate in response to the stimulus of a ringing bell.

1923 Austrian zoologist Karl von Frisch discovers that bees communicate by performing rhythmic dances.

1935 Konrad Lorenz describes and names the behavior of imprinting in baby ducks and goslings.

1971 British zoologist Jane Goodall first documents that chimpanzees use tools.

What influences behavior? For many years, scientists asked the question about whether behavior was genetically based or based on experiences. Studies have shown that some behavior is based solely on genetics and is not influenced by experience. Other behaviors, such as a finch learning the song of its species, are known to result from a combination of genetics and environmental influences. Today, many behaviors are considered to be the result of both genes and experience. In many cases, behavior results from the interaction of genetically based behaviors and behaviors based on experience. **Figure 31.1** shows some important discoveries about animal behavior.

The evolution of behavior Two general questions are asked when studying animal behavior. The first question focuses on what triggers an animal to react to specific stimuli. For example, what triggers a male bird, like the one shown in **Figure 31.2,** to sing during breeding season? The answer usually is found by studying the internal biology of an animal. Scientists now know that some male birds sing during breeding season in response to the internal stimulus of increased levels of the hormone testosterone.

The second question focuses on what advantages certain behaviors provide animals. The answers to this question are tied to the evolution of behavior through natural selection. What advantage does singing during breeding season provide the male bird? Perhaps the singing helps the male bird keep other male birds away. Perhaps the singing helps the male attract a mate.

You already have learned that animals with traits giving them a competitive advantage over other animals that do not possess those traits are more likely to reproduce and pass their genes on to future generations. In the past, birds that sang tended to have more offspring than birds that did not sing. Over a number of generations, birds that sang became the only birds contributing to the population's gene pool. The behavior has been naturally selected.

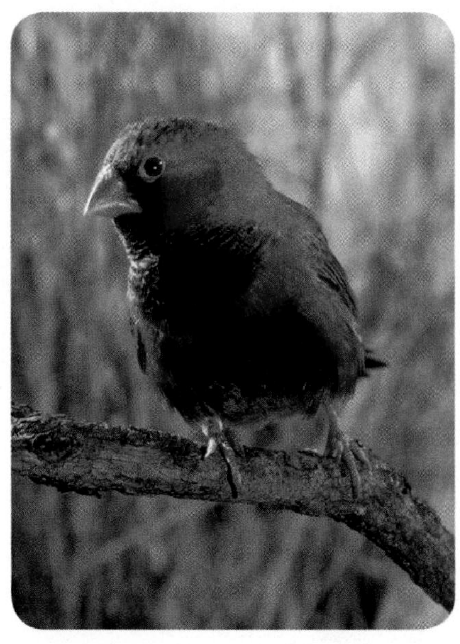

Figure 31.2 The male black-breasted zebra finch sings during mating season to attract a female.

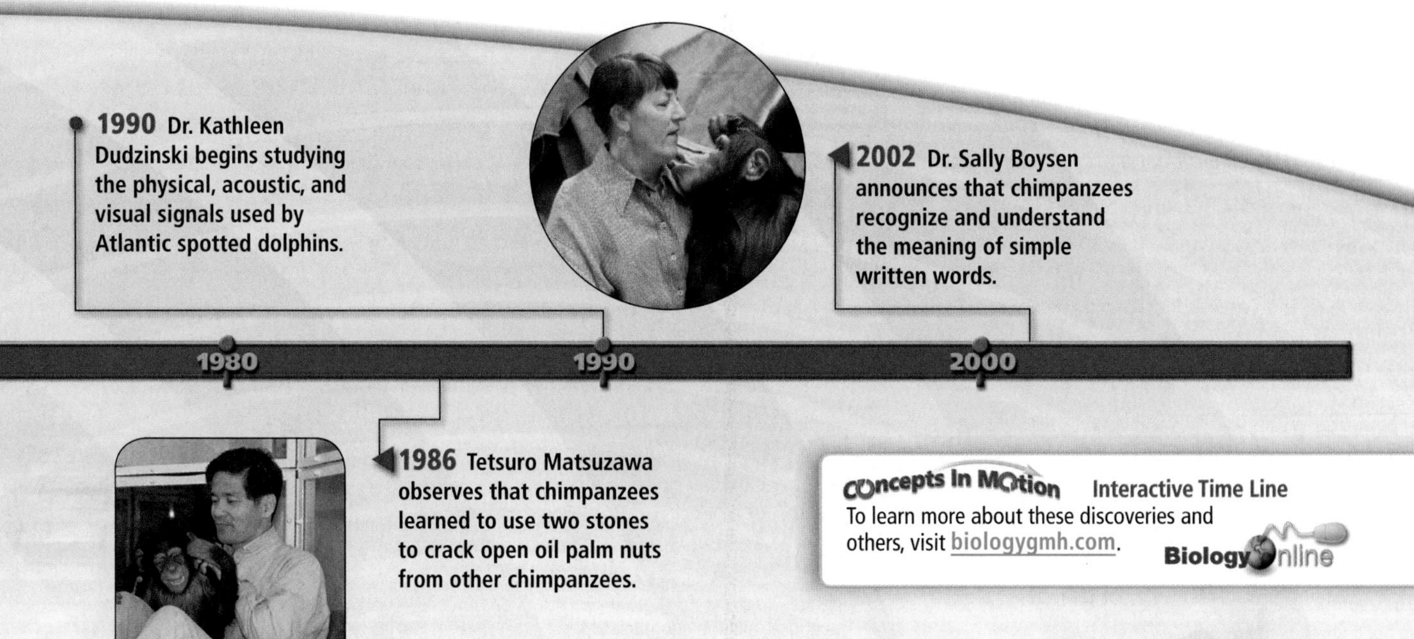

1990 Dr. Kathleen Dudzinski begins studying the physical, acoustic, and visual signals used by Atlantic spotted dolphins.
2002 Dr. Sally Boysen announces that chimpanzees recognize and understand the meaning of simple written words.
1980
1990
2000
1986 Tetsuro Matsuzawa observes that chimpanzees learned to use two stones to crack open oil palm nuts from other chimpanzees.

Concepts In Motion Interactive Time Line
To learn more about these discoveries and others, visit biologygmh.com.
Biology Online

CAREERS IN BIOLOGY

Animal Behaviorist Scientists that study the causes, development, and evolution of behavior are animal behaviorists. Animal behaviorists might work at zoos, aquariums, or museums, or they might teach or conduct research at a university. For more information on biology careers, visit biologygmh.com.

Innate Behavior

Behaviors that are genetically based and not linked to past experiences are called **innate** (ih NAYT) **behaviors.** However, you might say that all animal behaviors occur in and are influenced by the environment. Behaviors are referred to as innate, or instinct, when the same behavior commonly is observed among a large number of individuals within a population, even if the environments are different. For example, in some species, newly-hatched birds will make innate chirping sounds while opening their mouths in an upward direction when a parent lands in the nest. As part of an innate response to the chick's open mouth, the parent will feed the chirping bird. In addition, members of a particular group of mammals typically begin to walk at the same age, depending on their species. Therefore, walking generally is considered an innate behavior.

Fixed action patterns The goose in **Figure 31.3** is exhibiting innate behavior. When an animal carries out a specific set of actions in sequence, in response to a stimulus, it is called **fixed action pattern.** The goose is responding to the stimulus of an egg that is out of the nest. The set of actions carried out usually is the same and usually in the same order. The goose will extend its neck toward the egg and then stand up. It will then roll the egg back to the nest with a side-to-side motion of its neck with the egg held beneath its bill. The stimulus—finding that the egg is out of the nest—triggers the innate behavior, and the entire sequence of actions is carried out. Even if the egg is removed midway through the retrieval process, the goose will continue the behavior without the egg. This is the key to a fixed action pattern—the stimulus triggers an innate response that the animal does not control and is not directly influenced by environmental conditions or past experiences. Another example of a fixed action pattern is shown in **Figure 31.4.**

Reading Check **Explain** why a fixed action pattern is an example of innate behavior.

■ **Figure 31.3** The goose is carrying out a fixed action pattern.
Infer *What would happen if the egg were replaced with a similarly shaped object, such as a small rubber ball?*

A The goose responds to the stimulus of an egg out of the nest.

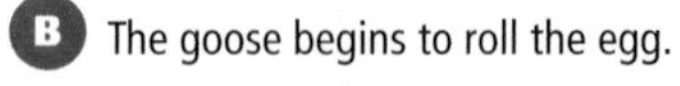

B The goose begins to roll the egg.

C The goose rolls the egg back to the nest with the underside of its bill.

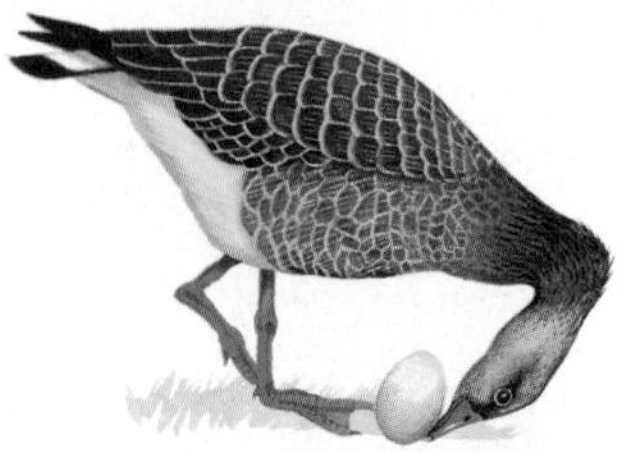

D The goose continues to roll the egg until it is in the nest.

Visualizing Types of Behavior

Figure 31.4
Animal behavior is innate or learned. Fixed action pattern behavior is innate because it is genetically based and is not linked to past experience. Habituation and operant conditioning are learned behaviors because each results from situations that the animal experiences.

Fixed Action Pattern This newly hatched cuckoo is carrying out a fixed action pattern. An adult female cuckoo lays her eggs in the nests of other bird species. When the baby cuckoo hatches, it ejects the other eggs from the nest before its eyes are even open. The process of ejection is a fixed action pattern.

Habituation These birds have become habituated to the scarecrow. Although they might have avoided it when it was first placed in the field, they learned that there were no positive or negative effects associated with it.

Operant Conditioning These ducks have learned to associate the presence of humans near the edge of the pond with the reward of food.

Concepts In Motion **Interactive Figure** To see an animation of animal behavior, visit biologygmh.com.

Biology Online

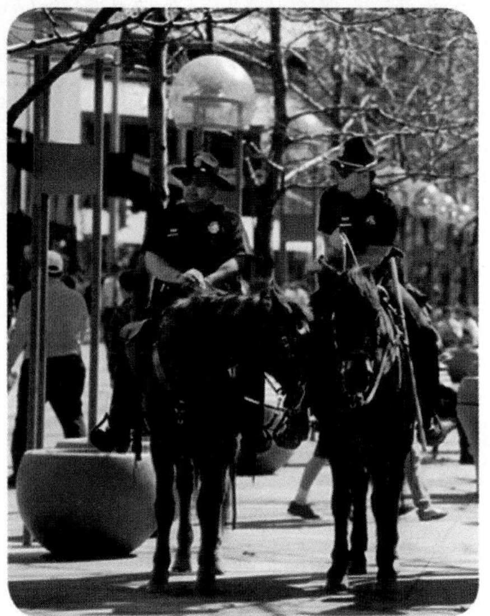

■ **Figure 31.5** Police horses become habituated to noise from crowds and traffic.
Recall ***Give an example of a time when you became habituated to a stimulus.***

FOLDABLES Incorporate information from this section into your Foldable.

Learned Behavior

Which activities do you enjoy—playing a sport, driving a car, playing video games or a musical instrument? These activities are examples of learned behaviors. **Learned behaviors** result from an interaction between innate behaviors and past experiences within a particular environment. Examples of learned behavior include habituation, conditioning, imprinting, and cognitive behavior.

Habituation Sometimes, animals learn over time that a potentially important stimulus deserves little or no attention. For example, baby birds in a nest see many types of objects moving overhead. At first, they might respond to these stimuli by crouching down and staying still. Some of the objects, such as falling leaves or members of their own species flying by, often are seen and have no positive or negative effects to the birds. Over time, the birds will stop responding to these stimuli. This is referred to as **habituation** (huh bit choo AY shun), which is a decrease in an animal's response after repeatedly being exposed to a stimulus that has no positive or negative effects.

The horses shown in **Figure 31.5** have become habituated to street and crowd noise. Habituation can be thought of as learning not to respond to a stimulus. Habituation is important to an animal's success because it allows an animal to ignore unimportant stimuli and focus on and respond to important stimuli, such as the presence of food, a mate, or a predator. Another example of habituation is shown in **Figure 31.4.** Birds often become habituated to a scarecrow because they learn that it has no positive or negative effects.

MiniLab 31.1

Explore Habituation

Does an earthworm habituate to touch? In this lab, you will observe whether an earthworm will learn that a stimulus can be ignored.

Procedure

WARNING: *Treat the earthworm in a humane manner at all times.*

1. Read and complete the lab safety form.
2. Line a small narrow **tray** with a **paper towel** moistened with **aged tap water.** Put on a pair of **gloves** and wet them with aged tap water.
3. Use your gloved hand to pick up an **earthworm** and gently transfer it to your tray. Allow the worm to rest for 1 min.
4. Determine which is the anterior end (head) of your worm. Lightly touch the anterior of the worm with the bristles of a small **paintbrush.**
5. After the worm recovers from its withdrawal reflex, touch it lightly again.

6. Repeat Step 5 five more times and record any changes in the worm's behavior.

Analysis

1. **Explain** Did the earthworm become habituated to the stimulus? How do you know?
2. **Think Critically** Why is the earthworm's withdrawal reflex likely an innate behavior? How does this behavior help the worm survive in its natural environment?

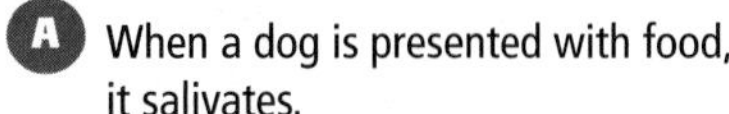

A When a dog is presented with food, it salivates.

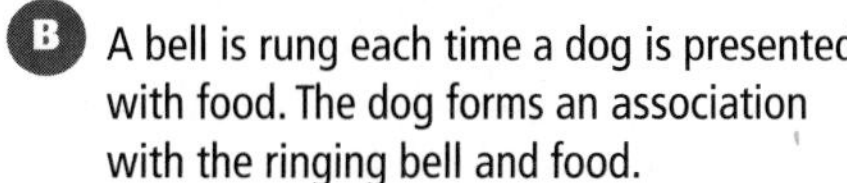

B A bell is rung each time a dog is presented with food. The dog forms an association with the ringing bell and food.

C Eventually, the dog will salivate to the sound of the bell alone. It has been conditioned to respond to the ringing bell.

■ **Figure 31.6** Through classical conditioning, the dog learns to associate the sound of a ringing bell with food.

Classical conditioning Ivan Pavlov, a Russian scientist who conducted experiments in the late 1890s and early 1900s, noticed that after he presented meat powder to a dog, the animal produced saliva. Later, Pavlov rang a bell each time he presented the meat powder. After repeated trials, the dog salivated when it heard the bell alone, without smelling or tasting meat powder.

Pavlov concluded that the dog related the sound of the ringing bell with the meat powder. Animal behaviorists refer to this type of learning as classical conditioning, which is illustrated in **Figure 31.6.** **Classical conditioning** occurs when an association is made between two different kinds of stimuli. In Pavlov's experiment, the dog learned to associate the sound of the bell with the unrelated stimulus of meat powder. The sound of the bell could produce the response of salivation.

Reading Check **Describe** an example of when you were conditioned by unrelated stimuli.

Operant conditioning B.F. Skinner, an American psychologist, carried out experiments on operant conditioning. In **operant conditioning,** an animal learns to associate its response to a stimulus with a reward or a punishment. In Skinner's experiment, a rat was placed in a box. As the rat explored the box, it accidentally would hit a lever, causing a food pellet to be released into the box. At first, the rat ignored the lever. It would eat the pellet and continue to move around the box. Eventually, the rat learned to associate pressing the lever with getting food. The animal was rewarded positively (receiving the food pellet) for its response (pressing the lever) to the stimulus (the lever).

In some cases, animals learn to associate their response with a negative reward. Monarch butterflies, which have a bright orange color pattern, are toxic to many predators. When a young blue jay eats a monarch butterfly for the first time, the bird becomes ill and vomits the butterfly. The bird quickly associates eating the butterfly with illness. In the future, the bird avoids eating monarch butterflies and other butterflies with a similar color pattern.

VOCABULARY

SCIENCE USAGE V. COMMON USAGE

Trial

Science usage: one of a number of repetitions in an experiment.
The biologist collected data for 50 trials during her study of behavior.

Common usage: a formal examination of a matter in a civil or criminal court.
The defendant was found innocent at the end of the trial.

Operant conditioning is a more powerful, long-lasting kind of learning that dominates much of everyday learning of humans and other vertebrates. For example, animals, including humans, learn ways of finding food by exploring a variety of locations. When certain locations prove to be a good source of food, animals are positively reinforced. Research shows that such animals are more likely to seek food the next time in the same location or locations that appear similar.

Imprinting Learning that can only occur within a specific time period in an animal's life and is permanent is called **imprinting.** The time during which an animal imprints is called the *sensitive period.* In some animals, the sensitive period occurs immediately after birth. Newborn offspring can form a strong bond with another animal, such as a parent, during this time. Some animals, such as whooping cranes, form a social attachment to the first object that they see after birth. Other animals, such as salmon, imprint on the chemical composition of the water in which they are hatched. The salmon use this imprint to return to this location when it is time for them to spawn.

Evidence of the influence of genetics on imprinting comes from experiments with newly hatched birds. In nature, the first object the offspring sees most likely will be its parent. This ensures that the offspring will have a higher chance of success by being nurtured by a parent. Experiments have shown that newly hatched birds will imprint on whatever object they see first, whether it is an animal of a different species, such as a human, or an inanimate object, such as a box.

VOCABULARY

ACADEMIC VOCABULARY

Migratory:
Characterized by moving from one location to another.
Migratory birds fly south for the winter.

Connection to History In 1999, only one flock of 180 migratory whooping cranes existed naturally. Scientists created a plan to introduce a second migratory flock of cranes to help ensure the species would not become extinct. Crane chicks were hatched in Wisconsin at the northernmost point of their migratory path. The chicks were imprinted using an ultralight plane like the one shown in **Figure 31.7.** Each year since 2001, a group of newly hatched chicks is imprinted by an ultralight. They follow it to the winter migration site in Florida and back to Wisconsin in the spring. In doing this, a second population of migratory cranes has been established successfully.

Figure 31.7 The first flock of whooping cranes to be imprinted using the ultralight arrived at their winter destination on December 3, 2001. Each year since then, a new flock has been imprinted, with all the cranes following the ultralight back to Wisconsin in the spring.
Infer ***what would happen if newly hatched cranes imprinted using a crane from the first flock.***

■ **Figure 31.8**
Left: The raven appears to be using problem-solving skills to reach the piece of food at the end of the string.
Right: The chimpanzee uses a stone to crack open nuts. Some scientists interpret this as cognitive behavior.

Cognitive behavior Thinking, reasoning, and processing information to understand complex concepts and solve problems are **cognitive behaviors.** Humans exhibit cognitive behaviors when they solve problems, make decisions, and plan for the future. Some experimental evidence supports the idea that other animals, such as chimpanzees and ravens, exhibit cognitive behavior. The raven shown in **Figure 31.8** appears to be using problem-solving skills to reach a piece of food.

Observations made by scientists of animals in their natural habitats also seem to show examples of cognitive behavior. Chimpanzees, like the one shown in **Figure 31.8,** have been observed using rocks to break open nuts. This behavior suggests that the chimpanzees are thinking and using tools to solve problems. Experiments are being conducted to find out if some primates purposely deceive, or lie to, other animals in their group—another sign of cognitive behavior.

Section 31.1 Assessment

Section Summary

- Behavior can be influenced by both genes and experience.
- Successful behaviors are those that give individuals an advantage for survival and reproduction.
- Behavior can be innate or learned.
- Learned behavior includes habituation, conditioning, and imprinting.
- Cognitive behavior involves thinking, reasoning, and problem solving.

Understand Main Ideas

1. MAIN Idea **Explain** how behavior could evolve.
2. **Explain** the difference between an internal stimulus and an external stimulus, and give an example of each.
3. **Compare and contrast** innate and learned behavior.
4. **Illustrate** specific examples of two types of learned behavior.

Think Critically

5. **Infer** A toad eats a bumblebee and receives a painful sting on its tongue. From then on, the toad avoids feeding on bumblebees or any other yellow and black insects. What kind of behavior is the toad exhibiting?

WRITING in **Biology**

6. **Explain,** using the terms *classical conditioning* and *operant conditioning,* how you would train an animal, such as a dog, to do tricks.

Section 31.2

Reading Preview

Objectives

- **Describe** different types of competitive behaviors and give examples of each.
- **Identify** types of communication, nurturing, and cooperative behaviors.
- **Analyze** the advantages and disadvantages of behavior in terms of survival and reproductive success.

Review Vocabulary

colony: a group of unicellular or multicellular organisms that live together in a close association

New Vocabulary

agonistic behavior
dominance hierarchy
territorial behavior
foraging behavior
migratory behavior
circadian rhythm
language
courting behavior
nurturing behavior
altruistic behavior

Ecological Behaviors

MAIN Idea **Animals that engage in complex behaviors might survive and reproduce because they have inherited more favorable behaviors.**

Real-World Reading Link Think about the advantages and disadvantages of owning a car. You would be able to drive yourself and your friends around town. However, you would also have to pay for gasoline, car insurance, and repairs. In a similar way, there are advantages and disadvantages to every type of animal behavior.

Types of Behaviors

All animal behaviors are somewhat ecologically based. Ecology is the study of the interactions of living things with each other and with their environment. These interactions can occur between members of the same species or between members of different species. Animals that engage in complex behaviors survive and reproduce because they have inherited genes that allow them to be successful in a particular environment.

Examine **Figure 31.9,** which shows two bighorn sheep fighting over a mate. Although it looks painful, the thick horns of the sheep protect them from injury when they butt heads. One of the sheep eventually will give up the contest, leaving the other the winner. What are the survival and reproductive advantages and disadvantages of this behavior? The winner is able to court and mate with a female without interference from the other male. The genes of the winner most likely will be passed on to future generations. Genes that provide adaptive advantages will increase in relative frequency according to the principles of evolution by natural selection. Genes that do not help an individual animal survive and produce offspring are likely to decrease in frequency in the gene pool of future generations. As you read about different types of behavior in this section, think about why a particular behavior might have evolved.

Figure 31.9 These bighorn sheep spar until one sheep gives in. The winner will be able to court a mate without interference from the other male. **Explain** *why this behavior favors natural selection.*

Figure 31.10 Polar bears engage in agonistic behavior. They spar until one bear leaves.
Infer ***What are some advantages of agonistic behavior?***

Competitive behaviors Competition for food, space, mates, and other resources occurs between individuals within a population. Competitive behaviors, like the example shown in **Figure 31.9,** allow individuals to establish dominance or control of an area or resource. Animals that are successful at competitive behaviors are more likely to obtain resources needed for survival and reproduction. Animals usually do not critically injure or kill one another when competing for food, mates, or other resources. Types of competitive behaviors include agonistic behavior, dominance hierarchies, and territorial behavior.

Agonistic behavior The polar bears in **Figure 31.10** are engaging in behavior in which one bear will be the winner and will have control over resources such as food or potential mates. This type of threatening or combative interaction between two individuals of the same species is called **agonistic** (ag oh NIHS tihk) **behavior.** Although the bears look as though they might hurt each other, agonistic behavior usually does not result in injury or death to either individual. The challenge will end when one animal eventually stops participating and leaves.

Dominance hierarchies A hierarchy is a grouping in which objects or individual animals are ranked in order from highest to lowest. Some animals living in groups develop **dominance hierarchies** (DAH muh nunts • HI rar keez) in which a top-ranked animal gets access to resources without conflict from other animals in the group. This ranking system helps reduce hostile behaviors among animals. These hostile behaviors would take time and energy away from finding food or a mate, or caring for offspring. Higher-ranked animals are more likely to get what they need to survive and reproduce. Female wolves, baboons, some song birds, and the chickens shown in **Figure 31.11,** establish dominance hierarchies.

Study Tip

Flashcards Make flashcards of the vocabulary terms in this section. Use the flashcards to review the terms with a partner or small group.

Figure 31.11 Female chickens, called hens, establish hierarchies in which one hen is dominant over the others. The dominant hen pecks other hens to maintain dominance.

■ **Figure 31.12** Gannets breed in large colonies. They establish a small area of territory in which to make a nest. Territorial behaviors include fighting, jabbing at each other, and biting each other's necks.

Territorial behaviors Many animals establish a territory. A territory is a specific area that contains resources, such as food or potential mates, that an individual continually defends against other individuals of the same species. The size of territories varies widely, depending on the animal and the particular environment. **Territorial behaviors** are attempts to adopt and control a physical area against the other animals of the same species. Territorial behaviors include verbal signals, such as the singing of birds or chattering of squirrels, as well as chemical signals, such as a male cheetah's urine. Birds, such as the North American gannets shown in **Figure 31.12,** that gather in large colonies to breed engage in territorial behavior by fighting and jabbing to maintain space in the nesting colony. Territories usually are defended by males in order to increase their chance of obtaining adequate food, mates, and places to rear their offspring.

Foraging behaviors Finding and eating food are examples of **foraging behaviors.** These behaviors have obvious advantages for animals. Foraging successfully means obtaining needed nutrients, while avoiding predators and poisonous foods. Foraging involves a trade-off between a food's energy content and the cost of finding, pursuing, and eating it. Scientists theorize that natural selection favors individual animals whose foraging behaviors use the least amount of energy to obtain the maximum amount of energy possible. These are the animals that will be most able to reproduce successfully and pass genes on to future generations.

Reading Check **List** some of the costs of foraging behaviors.

DATA ANALYSIS LAB 31.1

Based on Real Data*

Interpret the Data

Can the advantages of territorial behavior be observed? Surgeonfish are algae-eating fishes that vigorously defend their territory against other algae-eating fishes. They maintain a territory of about 2–3 m^2.

Data and Observations

The graph shows the results of a study that compared the feeding rates of territorial surgeonfish to those of nonterritorial surgeonfish.

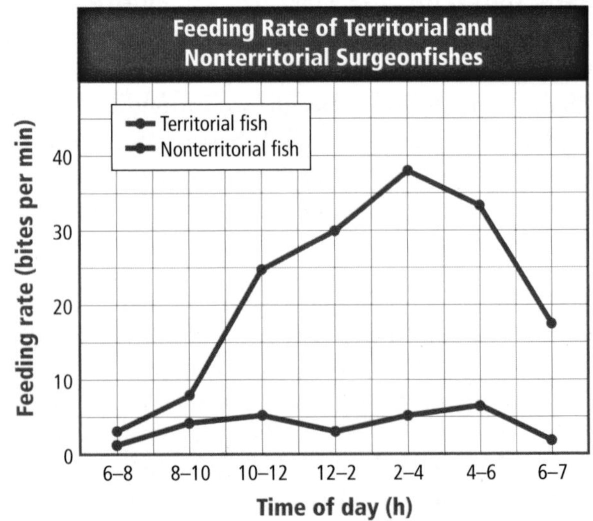

Think Critically

1. **Interpret** What is the meaning of each set of graphed data?
2. **Interpret** What is the advantage of the surgeonfishes' territorial behavior?
3. **Hypothesize** Form a hypothesis that explains why this behavior has evolved.

*Data obtained from: Craig, P. 1996. Intertidal territoriality and time-budget of the surgeonfish), *Acanthurus lineatus*, in American Samoa. *Environmental Biology* 46: 27–36.

Migratory behaviors Some animals, such as birds and grazing mammals, engage in **migratory behaviors**, moving long distances seasonally to new locations increasing their chances of survival. Land animals, like the wildebeest and zebra in East Africa, migrate almost continuously as different areas receive the rain needed for their food sources to grow. Each fall in North America, about two-thirds of bird species fly south to areas such as South America where food is available during the North American winter. The birds fly north in the spring to areas where they feed and breed during the summer.

How do the snow geese, shown in **Figure 31.13,** and other birds know which direction to fly? Sometimes migrations cover thousands of kilometers each year, with seemingly little navigational information. Recent studies show that the first migration of some birds is guided innately by both the position of the stars and Earth's magnetic field. Future migrations are influenced by external cues that the bird learns while flying that help it navigate more precisely.

Figure 31.13 Snow geese are one of the many bird species that migrate to find better conditions as seasons change.
Explain *why animals may engage in migratory behaviors.*

Biological rhythms Many animals, including humans, repeat behaviors in a rhythmic cycle. A **circadian** (sur KAY dee uhn) **rhythm** is a cycle, such as sleeping and waking, that occurs daily. Other biological cycles are seasonal or yearly. These cycles are influenced by environmental factors such as temperature changes, the increase or decrease of daylight hours, and the availability of food and water. These factors act as cues for animals to move into another phase of the cycle.

The daily cycle of sleeping and waking is influenced by external cues in animals. However, experiments have shown that many animals have an internal clock, often referred to as a biological clock, that maintains the daily rhythm of the sleep/wake cycle of about 24 hours. The graphs in **Figure 31.14** show the results of an experiment in which the activity level of nocturnal squirrels was monitored under two sets of conditions for 23 days—one in which a squirrel was exposed to a light cycle of 12 hours of light followed by 12 hours of darkness, and one in which a squirrel was kept in continual darkness. The biological clock of the squirrel maintained a sleep/wake cycle of 24 hours and 21 minutes in the absence of an external light and dark cycle. Controlled experiments show that the human biological clock has a cycle length of about 24 hours and 11 minutes.

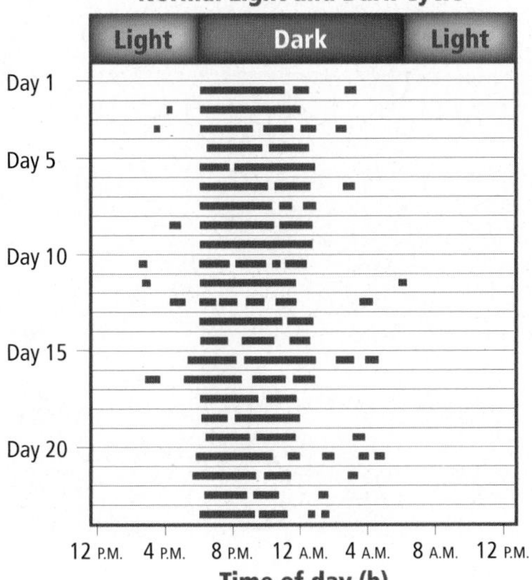

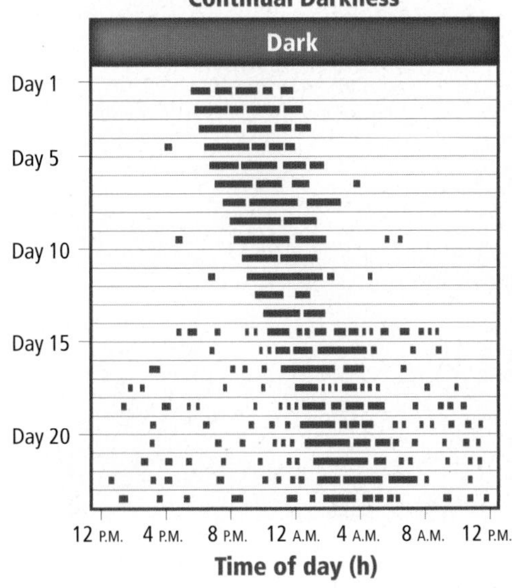

Figure 31.14 The green bars represent periods of the squirrels' activity, confirming they have a sleep/wake cycle of about 24 h.
Left: When exposed to a normal cycle of light and dark, the nocturnal squirrel was active when it was dark. It slept while it was light.
Right: When in the dark all the time, the squirrel maintained a sleep/wake cycle of 24 h and 21 min, instead of 24 h.

VOCABULARY

WORD ORIGIN

Auditory

audio– from Latin, meaning *relating to sound*

-ory suffix; from Latin, meaning *producing*

Communication Behaviors

Dogs bark, birds chirp, wolves howl, and lions growl and snarl. These are all examples of animal communication. Wolves howl to communicate information over long distances, including letting other wolves know their location, attracting mates, and signaling the presence of a predator. Such communication behaviors are critical to the survival and reproductive success of animals. Animals have several types of communication behaviors.

Pheromones Recall from Chapter 26 that some animals communicate by spreading highly specific chemicals called pheromones. These chemicals are specific to species, ensuring that individuals within a population receive important information. An advantage of species-specific pheromones is that predators cannot detect them, unlike other more noticeable communication behaviors, such as barks or howls. Pheromones often are used to relay messages between males and females about reproduction. For example, female silk moths produce a pheromone that is used to attract male moths for mating. Pheromones also can be used to relay messages of alarm in response to a predator attack. The cheetah in **Figure 31.15** is leaving its scent to communicate with other cheetahs.

Auditory communication If you ever have spent an evening outside in a park or a forest, you might have heard many animals using auditory communication. Howls, hoots, barks, and chirps are just a few of the sounds you might have heard. Auditory communication permits animals to send and receive sound messages that move faster than chemical messages. Male crickets, frogs, birds, and the howler monkey shown in **Figure 31.15** communicate information about mating, predators, and territory to others in the population using auditory communication. Humans use language to communicate complex information. **Language** is a form of auditory communication in which animals use vocal organs to produce groups of sounds that have shared meanings.

Figure 31.15 Some animals, like this cheetah, use pheromones to communicate and mark their territory. Male howler monkeys defend their territories with howls that can be heard over 4 km through dense forest.
Predict *Which communication behavior sends a message the farthest distance?*

Cheetah

Howler monkey

■ **Figure 31.16** Male frigate birds on the Galapagos Islands inflate and display red throat sacs to attract females during breeding season.

Courting and Nurturing Behaviors

Certain behaviors displayed by animals are directly related to the reproductive success of an individual animal. Attracting a mate and caring for offspring are important aspects of reproductive success.

Courting behaviors An animal engages in **courting behaviors** in order to attract a mate. An example of courting behavior is shown in **Figure 31.16.** The male frigate bird has inflated its bright red throat sac and is displaying it to attract the attention of female frigate birds. Courtship signals, whether they are a display of brightly colored feathers or a series of movements or sounds, are species specific. This is important in ensuring the reproductive success of a species. Courting behavior can last for minutes or months, depending on the species.

Selecting a male often is the female's role in the courtship process. Females often choose to mate with males that appear relatively larger and healthier than others. Thus, males with desired traits have a competitive advantage over other males and typically have a better chance of mating and successfully producing offspring.

Nurturing behaviors When parents provide care to their offspring in the early stages of development, they are engaging in **nurturing behaviors.** This includes providing food, protection, and skills needed for survival. Nurturing behaviors cost parents energy because of the extra work required to sustain offspring until they can take care of themselves. Animal species that spend time nurturing young often produce fewer offspring than animals that do not nurture. Reproductive energy can be spent producing millions of eggs, with very little if any energy spent on nurturing.

For example, a female cod can produce as many as nine million eggs during a single reproductive period. Only a small percentage of these eggs will survive. In contrast to the cod, animals that nurture, such as primates, produce far fewer eggs and offspring. A female orangutan, like the one shown in **Figure 31.17,** will give birth to one baby which she will nurse for up to three years. The baby will stay with the mother for five to seven years. In this case, more energy is spent nurturing young after birth to ensure they successfully reach a reproductive age. Although each of the reproductive strategies uses energy differently, each can be selected for because ultimately each results in there being at least one offspring that reaches maturity to successfully reproduce.

Reading Check Compare and contrast courting and nurturing behaviors.

■ **Figure 31.17** Nursing is an example of a nurturing behavior.
Expand ***What are some other nurturing behaviors?***

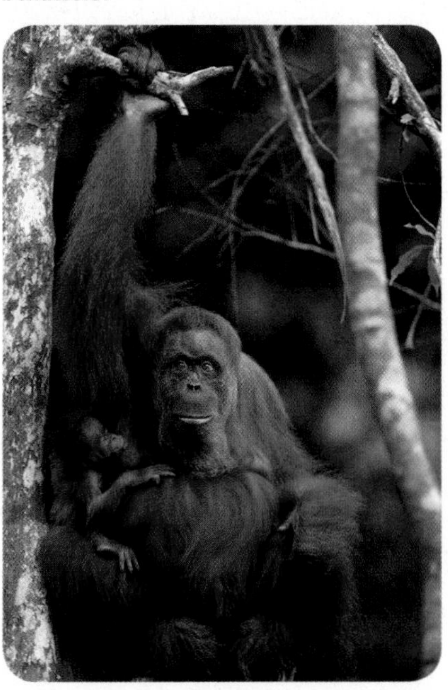

Cooperative Behaviors

Cooperative behaviors can exist in groups of same-species animals. Cooperative behaviors, such as those of the honeybee described in Chapter 26, can benefit all members of the group. However, some examples of cooperative behavior lead to an individual animal performing altruistic (al trew IHS tihk), or self-sacrificing, behaviors.

CAREERS IN BIOLOGY

Evolutionary Psychologist The social behavior of animals, including human social behavior and altruistic behavior, is studied by evolutionary psychologists. For more information on biology careers, visit biologygmh.com.

Altruistic behavior Sometimes an animal will perform an action that benefits another individual at a cost to itself. This type of behavior is called **altruistic behavior.** One example of altruistic behavior occurs in naked mole rats. Naked mole rats live underground in colonies. Each colony consists of one female that reproduces, called the queen, several males with whom the queen mates, called kings, and between 75–250 other males and females that do not reproduce. The nonreproductive members of the colony forage for food and care for and protect the queen, kings, and newborn offspring. **Figure 31.18** shows the nonreproductive individuals of a colony surrounding a queen and her offspring. By huddling around her, they are helping to keep the offspring warm.

Scientists have wondered what the advantage of altruistic behavior would be to an individual animal. Why should behavior that might hurt the individual animal ever be selected?

LAUNCH Lab

Review Based on what you have read about animal behavior, how would you now answer the analysis questions?

Kin selection One theory that has been presented to explain some types of altruistic behavior is kin selection. According to the idea of kin selection, altruistic behavior evolves because it increases the number of copies of a gene that is common to a population. It does not matter which individual passes the gene on to future generations. In the case of the mole rats, scientists have discovered through DNA analysis that all of the individuals in a colony of naked mole rats are closely related. The nonreproductive members of the colony will not pass on their genes. However, genes that are similar, if not identical, to their own will be passed on by the queen. As the nonreproductive members work to protect the queen and bring her food, they are ensuring that genes similar to their own will be passed on to future generations.

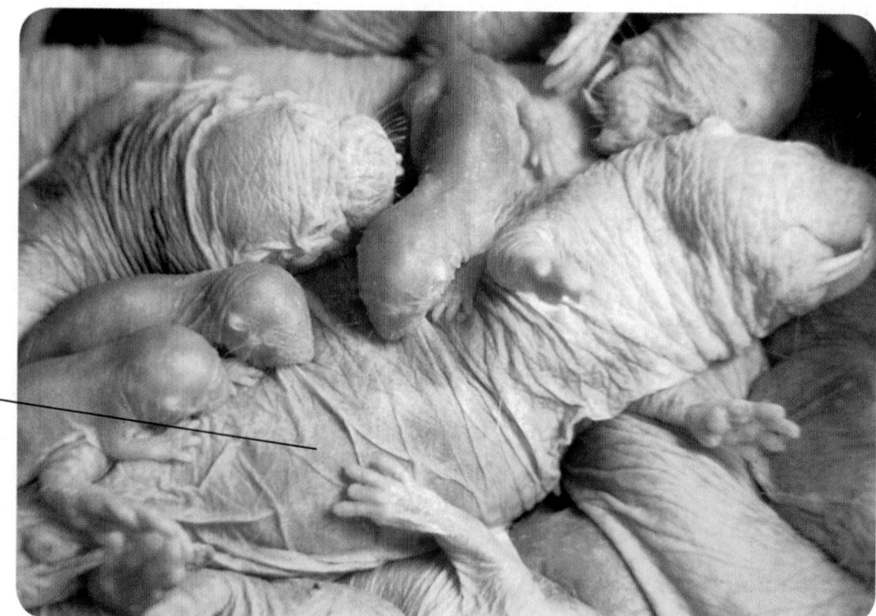

■ **Figure 31.18** The nonreproductive members of a colony of naked mole rats exhibit altruistic behavior. They forage for food, protect the queen, and huddle around her to provide warmth while she nurses her offspring.

CONCEPTS In MOtion

Interactive Table To explore more about animal behaviors, visit biologygmh.com.

Table 31.1 Effects of Behaviors

Behavior	Example	Advantage	Disadvantage
Migration		Animals that migrate increase their chance of survival by moving to a location that has better climate conditions and more food.	A large amount of energy is needed to move long distances and there is the possibility of increased predation while moving.
Pheromone communication		Pheromones provide a species-specific form of communication, which works without alerting predators.	Pheromones have a more limited range of communication than auditory or visual cues.
Nurturing		Nurturing increases an offspring's chance of survival. Genes of the parents continue to be present in future generations.	Parents spend increased amounts of energy on caring for offspring, possibly at the cost of the parents' health or safety.

Advantages and Disadvantages

Many behaviors have benefits and disadvantages related to survival and reproductive success. A cost-benefit analysis examines the advantages and disadvantages of a particular behavior in terms of survival and reproductive success. **Table 31.1** shows the cost-benefit analysis of some types of animal behavior.

Section 31.2 Assessment

Section Summary

- Behavior evolves when genes from successfully reproductive animals remain in a gene pool.
- Competitive behaviors allow animals to establish dominance without serious injury or death to other individuals.
- Communication behaviors are critical to the survival and reproductive success of animals.
- Certain behaviors, such as courting and nurturing, are directly related to the reproductive success of an individual animal.

Understand Main Ideas

1. MAIN Idea **Explain** how the behavior of an animal relates to its survival and reproductive success.
2. **Define** agonistic behavior and give one example of this type of behavior.
3. **Analyze** the advantages and disadvantages of nurturing behaviors.
4. **Describe** how animals can communicate using pheromones.
5. **Explain** why altruistic behavior is advantageous to an individual in a population.

Think Critically

6. **Infer** Extend **Table 31.1** by providing examples of the advantages and disadvantages of three other behaviors presented in this section.

MATH in Biology

7. The data in **Figure 31.14** show that the squirrel kept in continual darkness shifted the time of its activity slightly each day. After 23 days, the squirrel's activity cycle had shifted by eight hours. On average, how much, in minutes, did the activity cycle change each day?

BioDiscoveries

Eavesdropping on Elephants

Elephant ESP? Humans can hear many of an elephant's calls, from the loud, shrill trumpet to low moans and grumbles. However, people used to believe that elephants also used Extra Sensory Perception (ESP) to communicate with each other. ESP might include the ability to read other's minds or know their thoughts. ESP was used to explain how a male elephant, traveling for kilometers, avoids other male elephants but finds a female that is ready to mate, which occurs once every few years.

Solving the mystery Enter Katy Payne, a bioacoustics researcher at Cornell University. In 1984, she was visiting the elephant display at the Washington Park Zoo in Portland, Oregon, when she realized that the air throbbed near the elephants. Was something going on that people could not hear? She recorded "elephant talk" and found that the low rumbles that people could hear were only a small part of an elephant's way of communicating. The elephants were using infrasonic sound waves to communicate. Infrasonic sound is produced by sound waves that are below the range of human hearing. Those deep elephant sounds people could hear actually were the overtones of sounds so low and powerful they could travel without interference over long distances. In fact, these calls can be heard by other elephants and felt as vibrations in the ground many kilometers away.

Copy cat Not only do elephants use infrasonic sound to communicate, they also are capable of vocal learning and mimicry. Scientists hypothesize that vocal imitation is used within complex social groups to enhance bonds between individuals.

Most infrasonic calling occurs within family groups, and females with young tend to be the most vocal.

Just why exactly do elephants need to communicate? And why is it important to biology? The way animals communicate can reveal some evolutionary secrets, such as how "talking" to each other can increase the chances of survival of individuals in a species. The wide variety of communication methods that have evolved demonstrates the importance of communication among all creatures. Future research might enhance our understanding of animal communication, as well as uncover many more methods of communication.

WRITING in Biology

Time Line Visit biologygmh.com to research at least four scientists from the past and present who have made discoveries about animal communication. Create a time line with your results. Detail the research that they conducted, including their hypotheses, scientific methods, data, and results.

BIOLAB

Design Your Own

HOW DOES THE EXTERNAL STIMULUS OF LIGHT AFFECT BEHAVIOR?

Background: **A response to light can be an important part of an animal's ecological behavior in that it might help make an animal more successful in finding food, escaping predators, or maintaining homeostasis. In this lab, you will design a testing chamber and use it to test how isopods respond to light.**

Question: *How do isopods respond to light?*

Materials

clear plastic food wrap
forceps
plastic Petri dishes with lids
cardboard boxes or trays
small paper plates
aged tap water
black paper
isopods
scissors
light source
filter paper
paper towels
tape
graph paper

Safety Precautions

WARNING: *Be careful when working with a light source that can become hot. Treat the isopods in a humane manner at all times.*

Plan and Perform the Experiment

1. Read and complete the lab safety form.
2. Form a hypothesis about how the isopods will respond to light.
3. Plan how you will build a testing chamber, and then design an experiment to test your hypothesis. Keep in mind that isopods need to be kept moist at all times. Make sure your experiment has a control group of isopods. Identify the variables and ensure that your experiment tests only one variable at a time. What will you measure? How will you measure it?
4. Design and construct a data table you can use to record the data you collect concerning the behavior of the isopods in response to light.
5. Make sure your teacher approves your plan before you proceed.
6. Collect material needed for your experiment and construct your testing chamber. Handle isopods gently and carefully.
7. Carry out your experiment.
8. **Cleanup and Disposal** Return isopods to their classroom habitat. Disassemble any equipment you put together and return any reusable materials to their proper storage area. Be sure to wash your hands thoroughly.

Analyze and Conclude

1. **Organize Data** Create a graph that illustrates your findings.
2. **Explain** what your graph shows about the response of isopods to light.
3. **Draw Conclusions** Did the data you collected from your observations of the control and experimental groups of isopods support your hypothesis?
4. **Use Scientific Explanations** What types of complex ecological behaviors of isopods might involve their response to light?
5. **Think Critically** Isopods also respond to the stimulus of low moisture by crowding together. Predict how this behavior would maximize their fitness and success.
6. **Error Analysis** What variables in your experiment would affect your data if they were not well controlled?

APPLY YOUR SKILL

Field Investigation Look for isopods in their natural habitat. How do the data you collected in this lab help you select places to begin your search? Write a summary describing your observations of isopods in their natural habitat. To learn more about isopod behavior, visit Biolabs at biologygmh.com.

CHAPTER 31

Study Guide

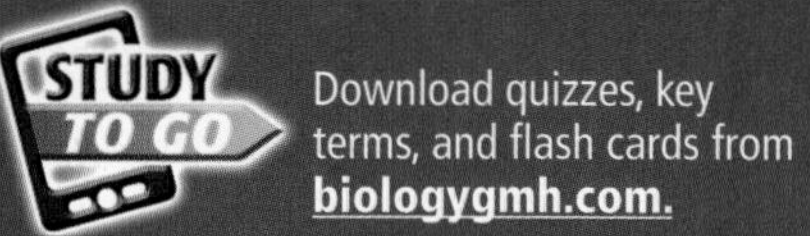

Download quizzes, key terms, and flash cards from biologygmh.com.

FOLDABLES **Illustrate** Use art and/or photos to illustrate an example of each type of learned behavior you described in the Foldable you made at the beginning of this chapter.

Vocabulary	Key Concepts
Section 31.1 Basic Behaviors	
• behavior (p. 908) • classical conditioning (p. 913) • cognitive behavior (p. 915) • fixed action pattern (p. 910) • habituation (p. 912) • imprinting (p. 914) • innate behavior (p. 910) • learned behavior (p. 912) • operant conditioning (p. 913)	MAIN Idea Animal behaviors are innate and learned behaviors that evolve by natural selection. • Behavior can be influenced by both genes and experience. • Successful behaviors are those that give individuals an advantage for survival and reproduction. • Behavior can be innate or learned. • Learned behavior includes habituation, conditioning, and imprinting. • Cognitive behavior involves thinking, reasoning, and problem solving. 
Section 31.2 Ecological Behaviors	
• agonistic behavior (p. 917) • altruistic behavior (p. 922) • circadian rhythm (p. 919) • courting behavior (p. 921) • dominance hierarchy (p. 917) • foraging behavior (p. 918) • language (p. 920) • migratory behavior (p. 919) • nurturing behavior (p. 921) • territorial behavior (p. 918)	MAIN Idea Animals that engage in complex behaviors might survive and reproduce because they have inherited more favorable behaviors. • Behavior evolves when genes from successfully reproductive animals remain in a gene pool. • Competitive behaviors allow animals to establish dominance without serious injury or death to other individuals. • Communication behaviors are critical to the survival and reproductive success of animals. • Certain behaviors, such as courting and nurturing, are directly related to the reproductive success of an individual animal. 

CHAPTER 31

Assessment

Section 31.1

Vocabulary Review

Use what you know about the vocabulary terms found on the Study Guide page to answer the following questions.

1. What type of behavior is carried out in a sequence of specific actions in response to a stimulus?
2. What type of behavior occurs when an association is made between two different kinds of stimuli?
3. What type of learning is permanent and occurs within a specific time period of an animal's life?
4. What type of behavior leads to a decrease in an animal's response after being exposed repeatedly to a stimulus that has no positive or negative consequences?
5. What type of behavior involves an animal associating its response with a reward or punishment?

Understand Key Concepts

6. Which behavior is genetically based and not linked to past experience?
   A. habituation **C.** fixed action pattern
   B. classical conditioning **D.** operant conditioning
7. Which is an example of imprinting?
   A. salmon returning to the water in which they hatched to spawn
   B. a rat learning to press a lever to get food
   C. a baby lion learning how to hunt
   D. baby birds getting used to seeing objects above them
8. An animal that solves a problem is engaging in what type of behavior?
   A. fixed action pattern
   B. cognitive behavior
   C. imprinting
   D. conditioning
9. Seasonal movement is an example of which type of behavior?
   A. migratory behavior
   B. classical conditioning
   C. cognitive behavior
   D. imprinting

Use the figure below to answer question 10.

10. Which type of behavior is shown above?
    A. imprinting **C.** habituation
    B. fixed action pattern **D.** operant conditioning
11. What is the time during which an animal imprints?
    A. nurturing period
    B. cognitive period
    C. sensitive period
    D. learning period

Constructed Response

12. **Short Answer** Compare and contrast classical conditioning and operant conditioning.
13. **Open Ended** What difficulties might scientists have when trying to determine if animals engage in cognitive behaviors?
14. **Open Ended** Describe an example of habituation. Do not use the examples given in this chapter.

Think Critically

15. **Hypothesize** why a behavior of an animal would cause it not to spend energy and time caring for its offspring.
16. **CAREERS IN BIOLOGY** Animal behaviorists observed that one species of lovebird carries nest-building materials in its beak. Another species of lovebird carries the material under its feathers. Hybrid offspring are produced by breeding these two species. The hybrids repeatedly shift the material between their beaks and their feathers while carrying it. What conclusion can be drawn about the influence of genetics on behavior from the results of this experiment?

Section 31.2

Vocabulary Review

Use the vocabulary terms found on the Study Guide page to answer the following questions.

17. What is a form of auditory communication in which animals use vocal organs to produce groups of sounds which have shared meanings?

18. In which situation does a top-ranked individual get access to resources without conflict from other individuals in the group?

19. What is a specific chemical spread by animals in order to communicate?

20. Which type of behavior results in an animal adopting and controlling a physical area against other animals of the same species?

21. Which type of behavior results in a threatening or combative interaction between two individuals of the same species?

Understand Key Concepts

22. Which behavior usually is concerned with finding and eating food?
- **A.** nurturing
- **B.** courting
- **C.** foraging
- **D.** migration

23. Which behavior is directly related to reproductive success within a species?
- **A.** altruism
- **B.** courting
- **C.** foraging
- **D.** migration

Use the figure below to answer question 24.

24. What is shown in the figure above?
- **A.** agonistic behavior
- **B.** migration
- **C.** dominance hierarchy
- **D.** nurturing behavior

25. What behavior is linked with pheromones?
- **A.** agonistic
- **B.** migration
- **C.** nurturing
- **D.** communication

26. Which is an example of a circadian rhythm?
- **A.** migration
- **B.** sleep/wake cycle
- **C.** hibernation
- **D.** reproductive cycle

27. Ensuring that offspring have an increased chance of survival is an example of which type of behavior?
- **A.** agonistic
- **B.** migration
- **C.** nurturing
- **D.** territorial

Constructed Response

28. Short Answer Differentiate between agonistic and territorial behaviors.

29. Short Answer Distinguish dominance hierarchy from territorial behaviors.

30. Open Ended Hypothesize what would happen if circadian rhythms disappeared.

Think Critically

31. Hypothesize the successful evolutionary advantages of animals sacrificing themselves for their sibling in a competitive battle with a predator.

Use the graph below to answer questions 32 and 33.

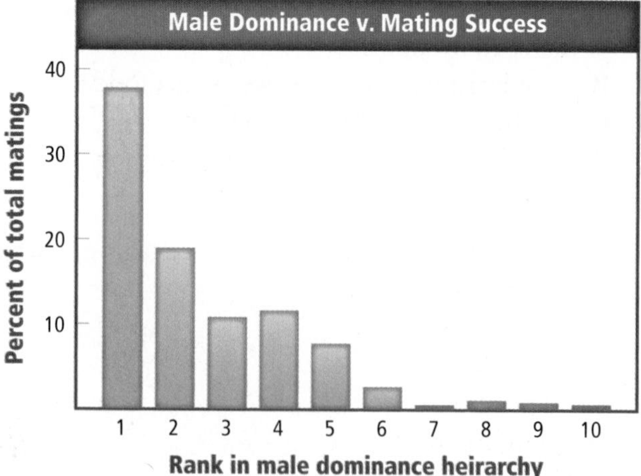

32. Draw Conclusions about the relationship between the order of male seals in the dominance hierarchy and their number of matings.

33. Hypothesize a reason for this behavior.

Biology Online Chapter Test biologygmh.com

34. **Infer** how an animal might starve if its parents failed to teach it competitive behaviors.

35. **Infer** If an individual animal no longer was able to learn, how might this condition affect its ability to engage in competitive behaviors in the near future?

36. **Compare and contrast** two strategies of spending reproductive energy—producing large numbers of eggs with little or no parental care, and producing a smaller number of eggs and engaging in nurturing behavior. Give an example of animals that use each strategy.

37. **Conclude** Of the three animals you have observed in this chapter—emperor penguins, earthworms, and isopods, which has the most complex ecological behavior? Based on what you learned in earlier chapters about these animals, why do you think this might be?

Use the figure below to answer questions 38 and 39.

A species of marine isopods lives in sponges in intertidal zones. The males of this species exist in three different sizes—alpha, beta, and gamma. Females of this species are similar in size to the beta males. Each size of male has a different strategy for mating.

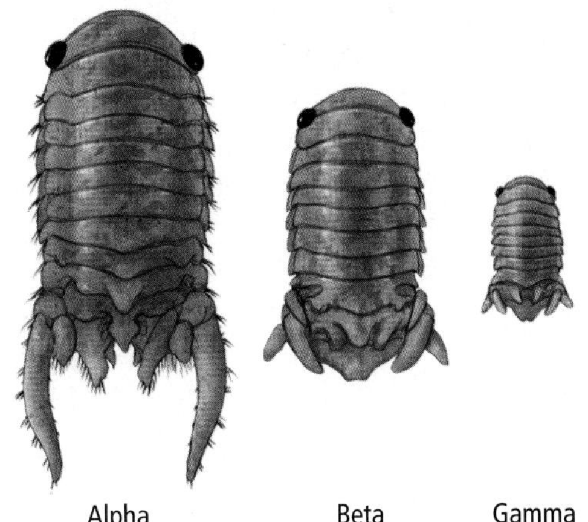

38. **Identify** which size of male would use the mating strategy that involves avoiding the alpha-sized males and hiding in a sponge to mate with a female. Explain your answer.

39. **Identify** which size of male would use the mating strategy that involves fighting with an alpha-sized male until one isopod wins. What is this type of behavior called?

Additional Assessment

40. WRITING in Biology Argue a case about why altruistic behavior by an individual animal might result in the animal's genes appearing in future generations.

DBQ Document-Based Questions

Oystercatchers are small shore birds that eat mussels as one of their primary foods. The birds must spend time and effort to hammer or stab the mussels to open them.

Use this graph to answer the questions below.

Data obtained from: Meire, P.M., and Ervynck, A. 1986. Are oystercatchers *(Haematopus ostralegus)* selecting the most profitable mussels *(Mytilus edulis)*? *Animal Behaviour* 34: 1427-1435.

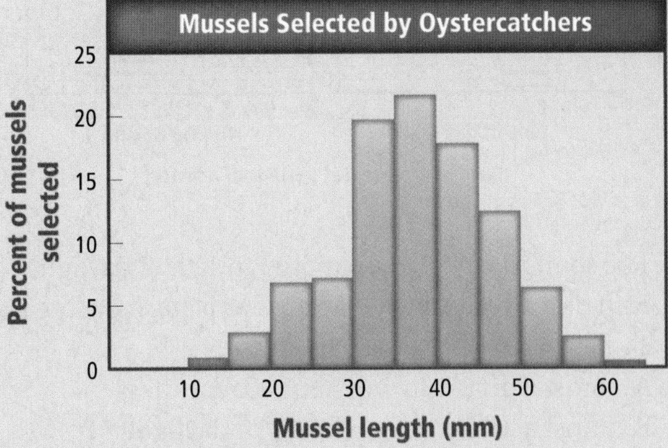

41. Which mussel size do the oystercatchers prefer?

42. The 10-mm mussels are the most abundant. Hypothesize why oystercatchers often do not forage for them.

43. Larger mussels provide many more calories than smaller mussels. The larger the mussel, the more it tends to be encrusted with barnacles that make it harder to open. Hypothesize why oystercatchers do not forage for the largest, most energy-rich mussels.

Cumulative Review

44. Calculate the amount of energy that would be available to the fifth level of an energy pyramid if 41,900 J represent 100 percent of the energy at the producer level. **(Chapter 2)**

Standardized Test Practice

Cumulative

Multiple Choice

1. Which describes a function of feathers?
   A. insulation
   B. nesting
   C. conserving water
   D. swimming

Use the diagram below to answer questions 2 and 3.

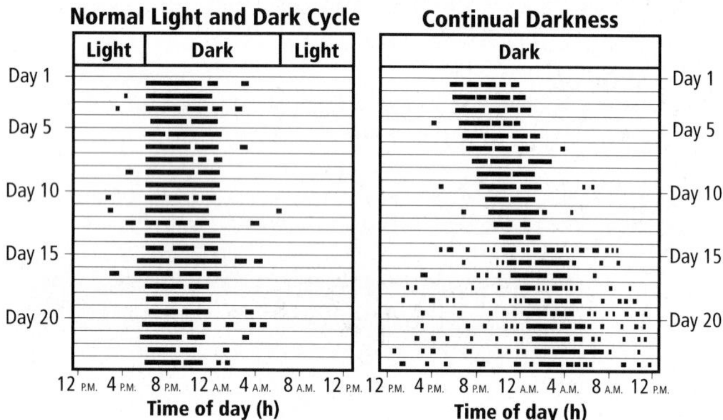

2. The squirrels that were exposed to 12 h of daylight each day displayed which behavior pattern during the 24-h cycles?
   A. most activity during hours of darkness
   B. most activity during hours of daylight
   C. constant sleeping
   D. continuous activity

3. Squirrels that were exposed to 24 h of darkness displayed which circadian rhythm?
   A. cycles of exactly 12 h
   B. cycles of less than 12 h
   C. cycles of exactly 24 h
   D. cycles of more than 24 h

4. How did the earliest fishes obtain their food?
   A. by grazing on phytoplankton at the water surface
   B. by living as parasites inside larger marine animals
   C. by sucking up organic matter off the ocean floor
   D. by using sharp teeth to break apart mollusks

Use the diagram below to answer question 5.

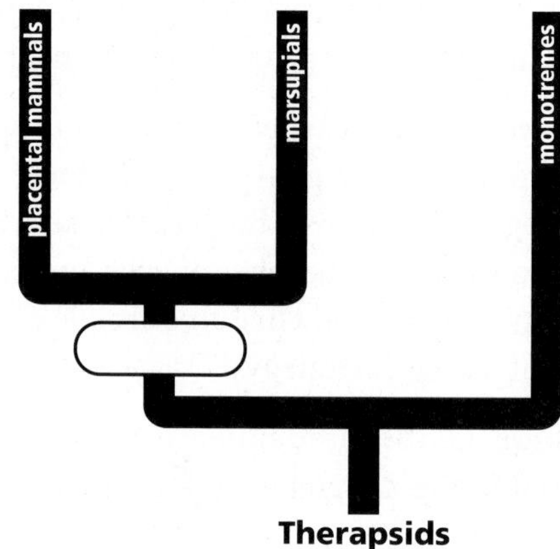

5. What information belongs in the bubble in the diagram?
   A. Adults give birth to live young.
   B. Adults lay eggs.
   C. Offspring live in mother's pouch after birth.
   D. Offspring receive milk from their mother.

6. Which structures are used by fishes to take in oxygen and transport it to body cells?
   A. gills and a closed circulatory system
   B. gills and an open circulatory system
   C. lungs and a closed circulatory system
   D. lungs and an open circulatory system

7. Echinoderms, such as sea stars, use their tube feet for locomotion and what else?
   A. reproduction
   B. respiration
   C. sensing gravity
   D. sensing light

8. Which characteristic is used to classify dinosaurs into two groups?
   A. structure of the hipbones
   B. structure of the skull and jaw
   C. whether they are ectotherms or endotherms
   D. whether they are herbivores or carnivores

Short Answer

Use the diagram below to answer question 9.

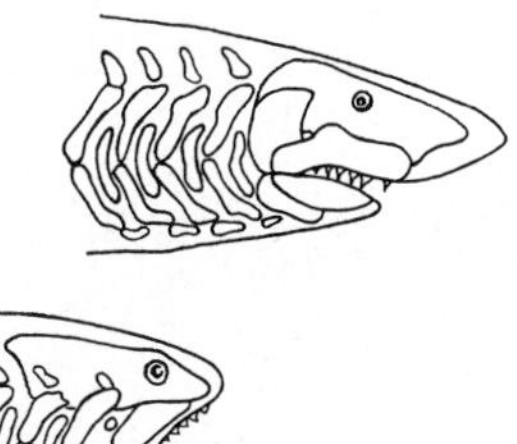

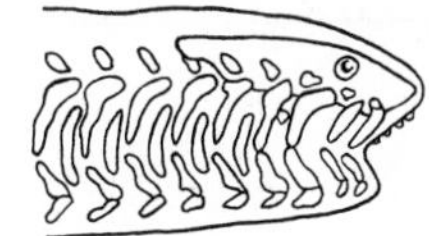

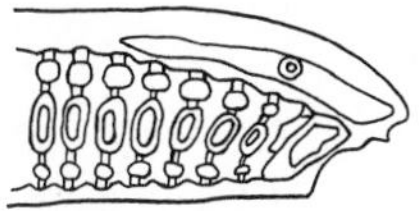

9. Describe the evolution of the jaw and explain how it was an important advancement for fishes.

10. Hypothesize why some birds migrate thousands of miles each year.

11. List three traits of mammals and explain why they are necessary for endotherms.

12. Compare and contrast an open circulatory system and a closed circulatory system.

13. Compare and contrast organisms in the order Rodentia with those in order Lagomorpha.

14. Hypothesize how an animal would benefit from a dominance hierarchy if it does not defend a territory.

Extended Response

15. Suppose a plant with adaptations for survival in a tropical rain forest is transplanted to a tropical desert. What adaptations in the rain forest plant could cause it to have trouble surviving in the new environment?

16. A certain type of insect uses pheromones to attract mates. The insect is most active during the day. Propose the advantages and disadvantages of this type of behavior for attracting mates.

Essay Question

The ring-tailed lemur is an herbivore. It eats a variety of plants and plant materials. Ring-trailed lemurs eat up to three dozen species of vegetation, but one of their favorites is the kily tree.

Groups of ring-tailed lemurs are led by a dominant female. A group usually contains between 15 and 30 lemurs. They can travel over a large area, some days more than 4 km. When the lemurs aren't eating, they often bathe in the Sun, groom each other, or play. Ring-tailed lemurs sleep under large trees. Settling down for the night is usually preceded by a loud whooplike call from all the lemurs.

Using the information in the paragraph above answer the following question in essay format.

17. The passage above describes the diet and behavior of ring-tailed lemurs. Suppose you want to do a study of lemur behavior. In an organized essay, explain what your research question would be and how you would study the behavior of ring-tailed lemurs.

NEED EXTRA HELP?

If You Missed Question . . .	1	2	3	4	5	6	7	8	9	10	11	12	13	14	15	16	17
Review Section . . .	29.2	31.1	31.2	28.2	30.2	28.1	27.1	29.1	28.2	31.2	30.1	25.3	30.2	31.2	21.1	31.2	30.2

UNIT 9 The Human Body

Chapter 32
Integumentary, Skeletal, and Muscular Systems
BIG Idea These systems work together to maintain homeostasis by protecting, supporting, and moving the body.

Chapter 33
Nervous System
BIG Idea The nervous system is essential for communication among cells, tissues, and organs.

Chapter 34
Circulatory, Respiratory, and Excretory Systems
BIG Idea These systems function together to maintain homeostasis by delivering important substances to the body's cells while removing wastes.

Chapter 35
Digestive and Endocrine Systems
BIG Idea The digestive system breaks down food to provide energy and nutrients for the body. Hormones regulate body functions.

Chapter 36
Human Reproduction and Development
BIG Idea Human reproduction involves the joining together of sperm and egg.

Chapter 37
Immune System
BIG Idea The immune system attempts to protect the body from contracting an infection through pathogens.

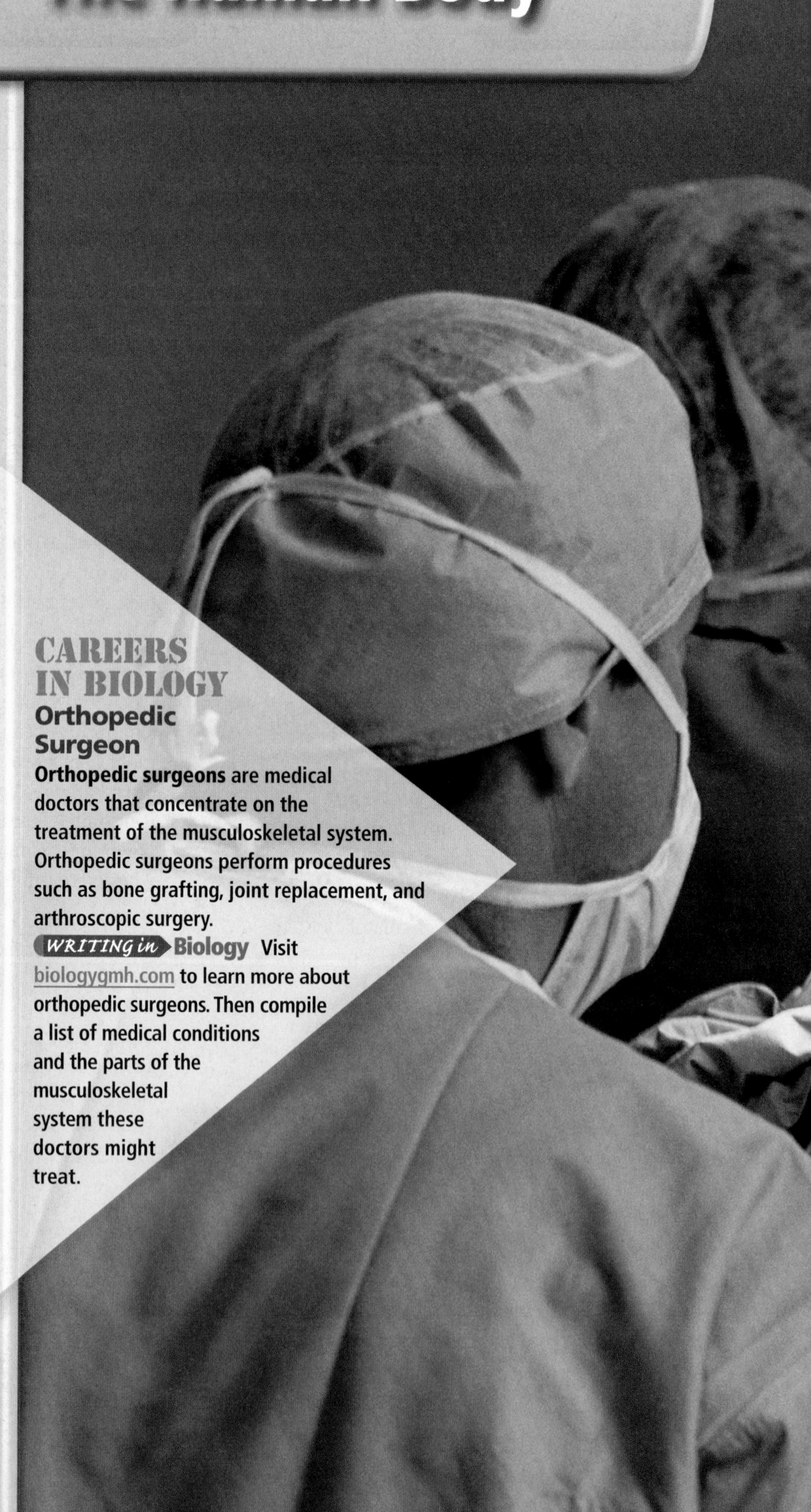

CAREERS IN BIOLOGY

Orthopedic Surgeon

Orthopedic surgeons are medical doctors that concentrate on the treatment of the musculoskeletal system. Orthopedic surgeons perform procedures such as bone grafting, joint replacement, and arthroscopic surgery.

WRITING in Biology Visit biologygmh.com to learn more about orthopedic surgeons. Then compile a list of medical conditions and the parts of the musculoskeletal system these doctors might treat.

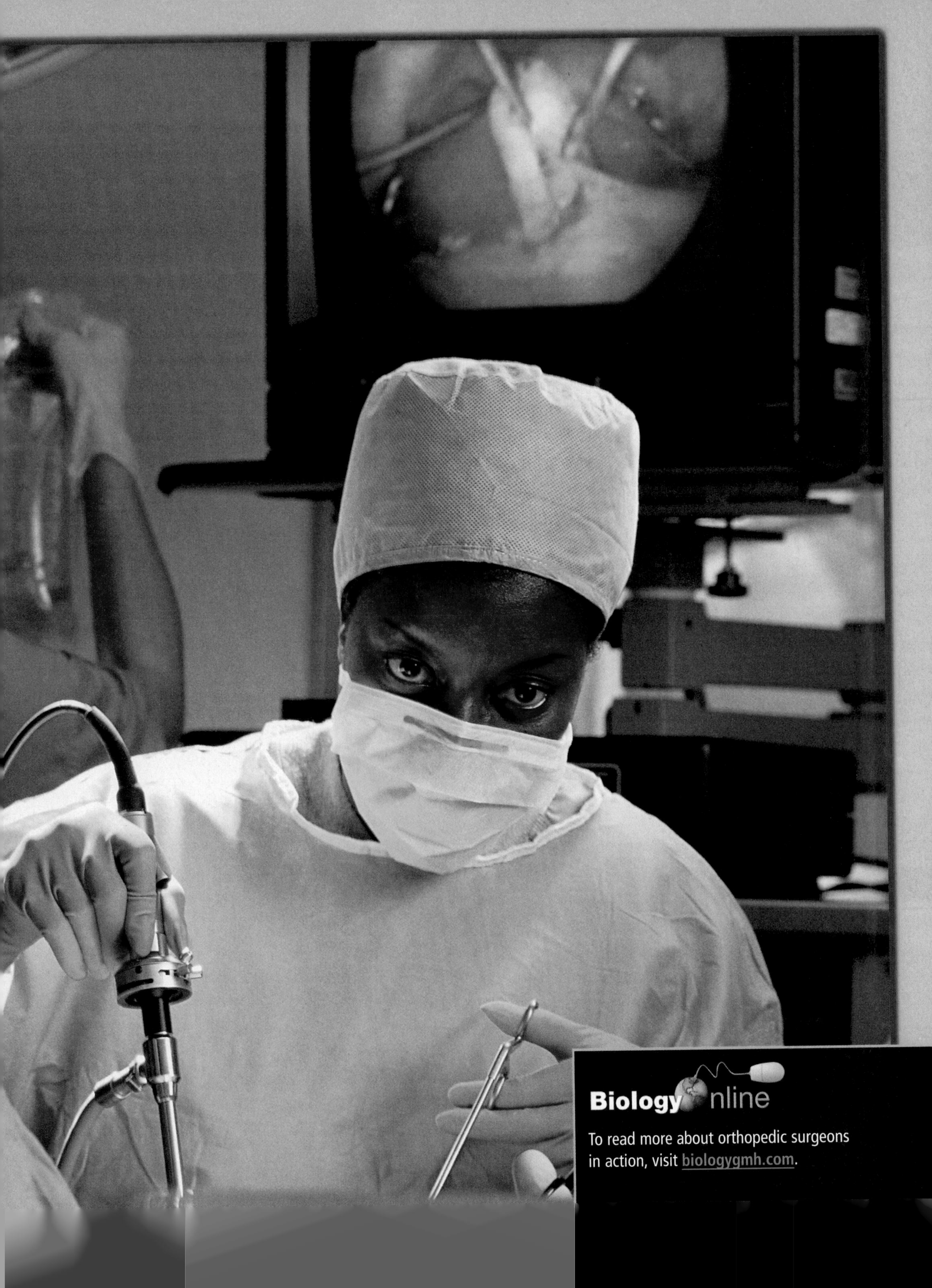
Biology Online
To read more about orthopedic surgeons in action, visit biologygmh.com.

CHAPTER 32

Integumentary, Skeletal, and Muscular Systems

BIG Idea **These systems work together to maintain homeostasis by protecting, supporting, and moving the body.**

Section 1
The Integumentary System
MAIN Idea Skin is a multi-layered organ that covers and protects the body.

Section 2
The Skeletal System
MAIN Idea The skeleton provides a structural framework for the body and protects internal organs such as the heart, lungs, and brain.

Section 3
The Muscular System
MAIN Idea The three major types of muscle tissue differ in structure and function.

BioFacts

- The skin of an adult can measure up to 18,580 cm^2 and can weigh as much as 3.5 kg.
- Adult humans have 206 bones in their bodies.
- Muscles work by contracting.

Start-Up Activities

LAUNCH Lab

How is a chicken's wing like your arm?

Chickens have structures similar to ours. You will examine a chicken wing and begin to explore it.

Procedure

1. Read and complete the lab safety form.
2. Obtain a **treated chicken wing** in a **self-sealing sandwich bag.** Observe the skin of the wing.
3. Without removing the wing from the bag, manipulate the wing to determine how it moves and where the joints are located.
4. Lay the bag on a flat surface and gently press and massage the wing to determine where bones and muscles are located.
5. Based on your observations, draw the wing as you imagine it might look if the skin was removed. Show the bones and muscles.

Analysis

1. **Label** your drawing to show which parts correspond to your upper arm, elbow, wrist, and hand.
2. **Differentiate** How are the parts that make up your arm different from the chicken wing?

Biology Online

Visit biologygmh.com to:

- study the entire chapter online
- explore Concepts in Motion, the Interactive Table, Microscopy Links, Virtual Labs, and links to virtual dissections
- access Web links for more information, projects, and activities
- review content online with the Interactive Tutor and take Self-Check Quizzes

Layers of Skin Make this foldable to help you understand skin as a multilayered organ of the body.

STEP 1 Place two sheets of notebook paper on top of each other with the top edges 1.5 cm apart.

STEP 2 Roll up the bottom edges, making all tabs 1.5 cm in size. Crease to form four tabs of equal size.

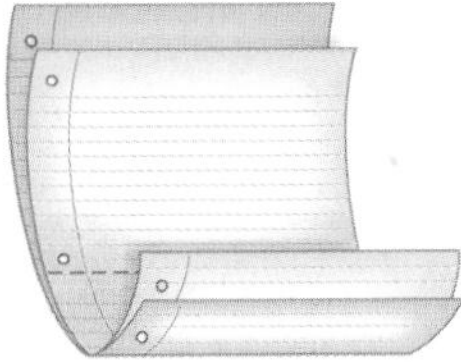

STEP 3 Staple along the folded edge to secure all sheets. With the stapled end on the bottom, label the tabs as illustrated.

Subcutaneous
Dermis
Epidermis
SKIN

FOLDABLES Use this Foldable with Section 32.1. As you study the section, record what you learn about each layer of tissue, and explain how the layers work together to perform specific functions.

Section 32.1

Reading Preview

Objectives

- **List** the four tissue types that are found in the integumentary system.
- **Explain** the functions of the integumentary system.
- **Describe** the composition of the two layers of skin.
- **Summarize** events that occur when skin is repaired.

Review Vocabulary

integument: an enveloping layer of an organism

New Vocabulary

epidermis
keratin
melanin
dermis
hair follicle
sebaceous gland

The Integumentary System

MAIN Idea Skin is a multilayered organ that covers and protects the body.

Real-World Reading Link The skin on the tips of fingers and toes is thick and is composed of curving ridges that form the basis of fingerprints. Fingerprints were first used in criminal investigations in 1860 by Henry Faulds, a Scottish medical missionary. Your skin is not just a simple covering that keeps your body together. It is complex and is essential for your survival. Your ridges are uniquely yours!

The Structure of Skin

The integumentary (ihn TEG yuh MEN tuh ree) system is the organ system that covers and protects the body. Skin is the main organ of the integumentary system and is composed of four types of tissues: epithelial tissue, connective tissue, muscle tissue, and nerve tissue. Epithelial tissue covers body surfaces, and connective tissue provides support and protection. Muscle tissue is involved in body movement. Nerve tissue forms the body's communication network. You will learn more about muscle tissue in Section 32.3, and you will learn about nerve tissue in Chapter 33.

FOLDABLES Incorporate information from this section into your Foldable.

The epidermis Refer to **Figure 32.1,** which illustrates the two main layers of skin as seen through a microscope. The outer superficial layer of skin is the **epidermis.** The epidermis consists of epithelial cells and is about 10 to 30 cells thick, or about as thick as this page. The outer layers of epidermal cells contain **keratin** (KER uh tun), a protein which waterproofs and protects the cells and tissues that lie underneath. These dead, outer cells are constantly shed. **Figure 32.2** shows that some of the dust in a house are dead skin cells. As much as an entire layer of skin cells can be lost each month.

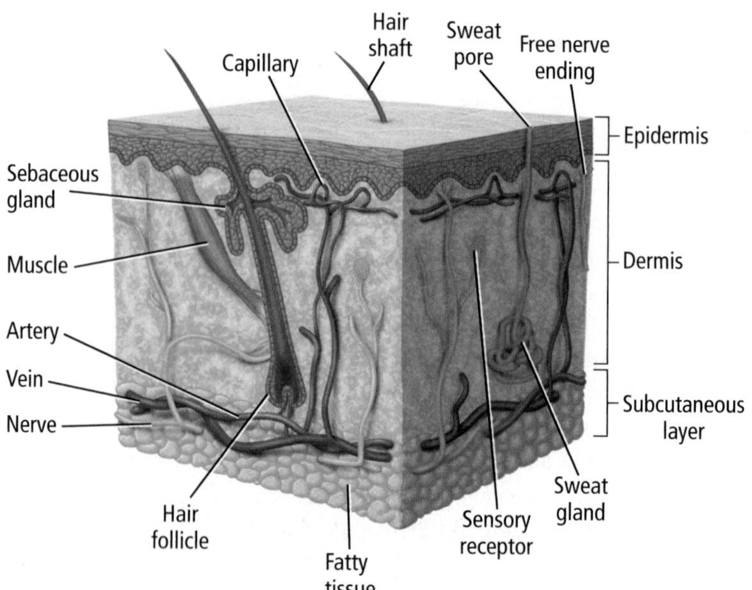

Figure 32.1 Skin is an organ because it consists of different types of tissues joined together for specific purposes.
Summarize *What types of tissues make up the skin?*

The inner layer of the epidermis contains cells that continually are dividing by mitosis to replace cells that are lost or die. Some cells in the inner layer of the epidermis provide protection from harmful ultraviolet radiation by making a pigment called melanin. **Melanin** is a pigment that absorbs light energy, which protects deeper cells from the damaging effects of ultraviolet rays of sunlight. The amount of melanin that is produced also influences the color of a person's skin. A suntan results when melanin is produced in response to exposure to the ultraviolet radiation in sunlight.

The dermis Directly beneath the epidermis is the **dermis,** the second layer of skin. The thickness of the dermis varies but usually is 15–40 times thicker than the epidermis. The dermis consists of connective tissue, a type of tissue that prevents the skin from tearing and also enables the skin to return to its normal state after being stretched. This layer contains other structures including nerve cells, muscle fibers, sweat glands, oil glands, and hair follicles. Beneath the dermis is the subcutaneous layer, a layer of connective tissue that stores fat and helps the body retain heat.

Hair and nails Hair, fingernails, and toenails also are parts of the integumentary system. Both hair and nails contain keratin and develop from epithelial cells. Hair cells grow out of narrow cavities in the dermis called **hair follicles.** Cells at the base of a hair follicle divide and push cells away from the follicle, causing hair to grow.

Hair follicles usually have sebaceous or oil glands associated with them, as shown in **Figure 32.3. Sebaceous glands** lubricate skin and hair. When glands produce too much oil, the follicles can become blocked. The blockage can close the opening of a follicle, causing a whitehead, blackhead, or acne—an inflammation of the sebaceous glands.

Reading Check **Summarize** the differences in structure and function of the epidermis and the dermis.

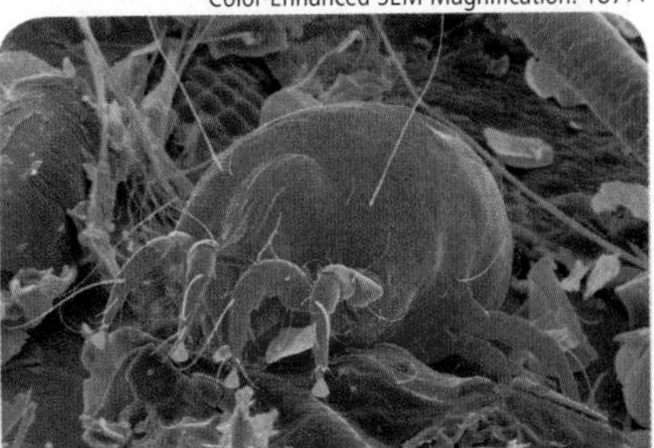

Figure 32.2 The dust mite pictured here is feeding on dead skin cells—a major component of dust.

Study Tip

Chart Make a chart with *Skin, Bones,* and *Muscles* as row labels, and *Components and structure* and *Function and purpose* as the column labels. Work in small groups to complete your chart as you review the text.

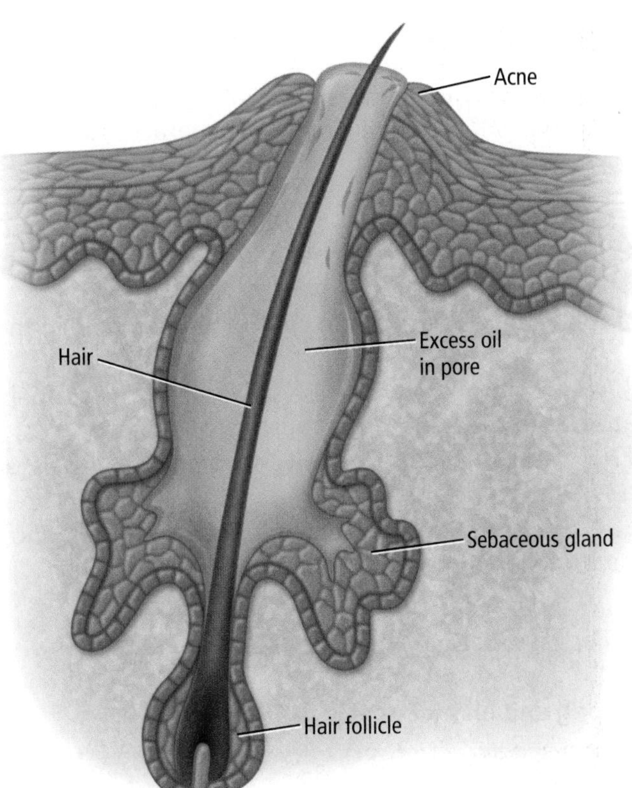

Figure 32.3 Oil, dirt, and bacteria can become trapped in follicles and erupt and spread to the surrounding area, causing localized inflammation.

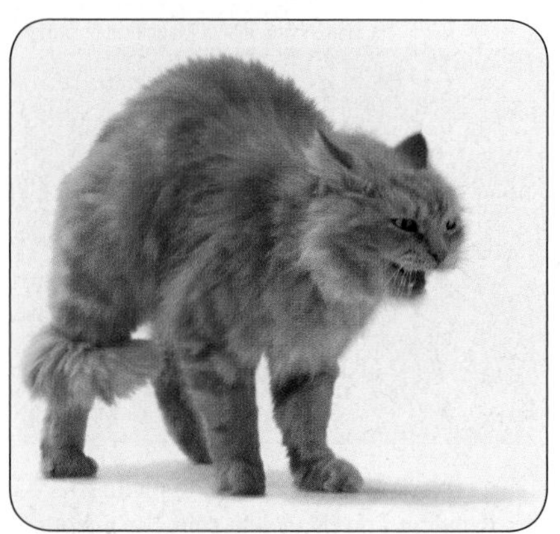

■ **Figure 32.4** Muscles in the skin cause the hair of some mammals to stand on end, and cause "goose bumps" on human skin.
Relate *What environmental changes produce "goose bumps"?*

Fingernails and toenails grow from specialized epithelial cells at the base of each nail. As cells at the base of a nail divide, older dead cells are compacted and pushed out. Nails grow about 0.5 to 1.2 mm per day. You might have heard that nails and hair continue to grow for several days after death. This is a myth; cells surrounding the nail and hair cells dehydrate causing the cells to shrink and pull away from nails and hair. This makes both appear longer.

Functions of the Integumentary System

Skin serves several important functions including regulation of body temperature, production of vitamin D, protection, and perception of one's surroundings.

Temperature regulation What happens when a person is working outside on a hot summer day? In order to regulate body temperature, the person sweats. As sweat evaporates it absorbs body heat, thereby cooling the body. What happens to skin when a person gets cold or frightened? "Goose bumps" are caused by the contraction of muscle cells in the dermis. In other mammals, when these muscles contract, the hair (fur) stands on end.

Notice the frightened cat in **Figure 32.4.** The cat appears larger, perhaps as a way to scare off enemies. This also is a mechanism for trapping air, which insulates or warms the mammal. Humans do not have as much hair as most other mammals, but "goose bumps" are caused by the same type of muscles that make a cat's fur stand on end. Humans rely on fat in the subcutaneous layer instead of hair to keep warm.

MiniLab 32.1

Examine Skin

How is chicken skin similar to human skin? The skin of chicken has characteristics similar to human skin. Using the chicken wing from the Launch Lab, you will further examine the characteristics of skin.

Procedure

1. Read and complete the lab safety form.
2. Wear disposable **lab gloves.** Remove the **chicken wing** from the **self-sealing bag** and place it in a **dissecting pan.**
3. Use a **dissecting kit** to remove the skin from the wing. Use **scissors** to carefully snip a hole in the skin that is loosely attached to the wing.
4. Make a cut about 6 cm in length. Pull the skin away from the wing. Use scissors and the **scalpel** to cut through the transparent membrane that attaches the skin to the muscles.
5. Try to remove the skin without making any more holes. Look for pockets of fat, blood vessels, and muscle fibers attached to the skin. Note the strength of the skin.
6. Dispose of the skin and used gloves as directed by your teacher. Clean your dissecting tools and dissecting pan with **warm, soapy water.** Save the skinned wing to use in the next MiniLab.

Analysis

1. **Think Critically** Human skin contains hair follicles. What type of follicles might you find on chicken skin?
2. **Explain** Why is it important for skin to be strong and elastic?

Vitamin production Skin also responds to exposure to ultraviolet light rays from the Sun by producing vitamin D. Vitamin D increases absorption of calcium into the bloodstream and is essential for proper bone formation. Many food products are now fortified with vitamin D.

Protection and senses Intact skin prevents the entry of microorganisms and other foreign substances. Skin helps maintain body temperature by preventing excessive water loss. Melanin in the skin protects against ultraviolet rays. Information about changes in the environment, such as pain, pressure, and temperature changes, is relayed to the brain.

VOCABULARY
ACADEMIC VOCABULARY
Function:
action, purpose.
One function of the skin is to protect the body.

Damage to the Skin

Skin has remarkable abilities to repair itself. Without a repair mechanism, the body would be subject to invasion by microbes through breaks in the skin.

Cuts and scrapes Sometimes, as in the case of a minor scrape, only the epidermis is injured. Cells deep in the epidermis divide to replace the lost or injured cells. When the injury is deep, blood vessels might be injured, resulting in bleeding. Blood flows out of the wound and a clot is formed. Blood clots form a scab to close the wound, and cells beneath the scab multiply and fill in the wound. At the same time, infection-fighting white blood cells will help get rid of any bacteria that might have entered the wound.

Effects of the Sun and burns As people age, the elasticity of their skin decreases and they start to get wrinkles. Exposure to ultraviolet rays from the Sun might accelerate this process and can result in burning of the skin.

Connection to Health Burns, whether caused by the Sun, heat, or chemicals, usually are classified according to their severity. The types of burns are summarized in **Table 32.1**. First-degree burns generally are mild and involve only cells in the epidermis. A burn that blisters or leaves a scar is a second-degree burn and involves damage to both the epidermis and dermis. Third-degree burns are the most severe. Muscle tissue and nerve cells in both the epidermis and dermis might be destroyed, and skin function is lost. Healthy skin might have to be transplanted from another place on the body in order to restore the protective layer of the body.

CAREERS IN BIOLOGY

Physical Therapist A physical therapist helps injured or disabled people to improve or regain physical functions using techniques such as exercise and massage. For more information on biology careers, visit biologygmh.com.

Table 32.1 Classification of Burns

Concepts In Motion Interactive Table To explore more about burns, visit biologygmh.com.

Severity of burn	Damage	Effect
First-degree	Cells in the epidermis are injured and may die.	• Redness and swelling • Mild pain
Second-degree	Cells deeper in the epidermis die. Cells in the dermis are injured and may die.	• Blisters • Pain
Third-degree	Cells in the epidermis and dermis die. Nerve cells and muscles cells are injured.	• Skin function lost • Healthy skin needs to be transplanted • No pain because of nerve cell damage

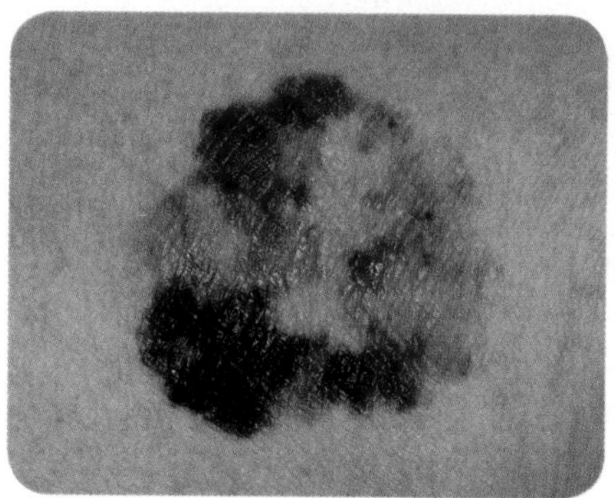

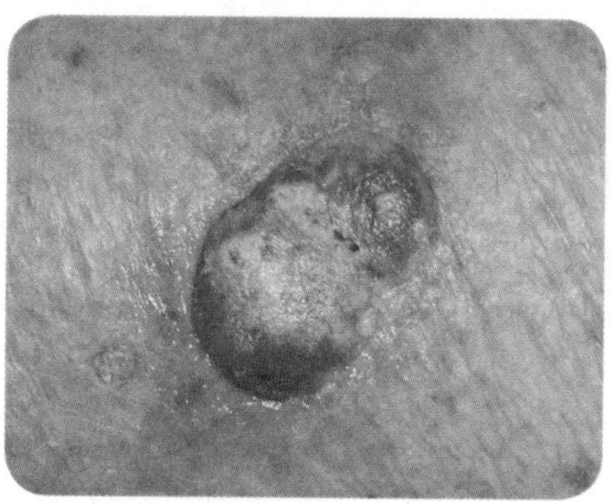

Figure 32.5 Warning signs of skin cancer include any obvious change in a wart or mole, or moles that are irregularly shaped, varied in color, or are larger than the diameter of a pencil.

Skin cancer Exposure to ultraviolet radiation, whether it is from the Sun or from artificial sources such as tanning beds, is recognized as an important risk factor for the development of skin cancer. Ultraviolet radiation can damage the DNA in skin cells, causing those cells to grow and divide uncontrollably. When this happens, skin cancer results. Refer to **Figure 32.5** to see some warning signs of skin cancer.

Skin cancer is the most common cancer in the United States. There are two main categories of skin cancer: melanoma and nonmelanoma. Melanoma begins in melanocytes, the cells that produce the pigment melanin. Melanoma is the deadliest form of skin cancer. Melanoma can spread to internal organs and the lymphatic system. It is estimated that one person dies from melanoma every hour in the United States. Teens are at greater risk for melanoma because as they grow, their skin cells divide more rapidly than they will when they reach adulthood.

Anyone can get skin cancer. However, individuals with light skin, light-colored eyes, light hair color, and a tendency to burn or freckle are at the greatest risk. Everyone should try to avoid prolonged exposure to the Sun, especially between 10 A.M. and 4 P.M. when the Sun's rays are the strongest. Other preventative measures include wearing protective clothing or sunscreen with a Sun Protection Factor (SPF) of at least 15.

Section 32.1 Assessment

Section Summary

- The skin is the major organ of the integumentary system.
- Maintaining homeostasis is one function of the integumentary system.
- There are four types of tissues in the integumentary system.
- Hair, fingernails, and toenails develop from epithelial cells.
- Burns are classified according to the severity of the damage to skin tissues.

Understand Main Ideas

1. MAIN Idea **Diagram** the two layers of the skin.
2. **Summarize** the types of tissues in the integumentary system and their functions.
3. **Generalize** different ways the integumentary system helps a human survive.
4. **Sequence** the process of skin repair in response to a cut.
5. **Compare** effects of first-degree, second-degree, and third-degree burns.

Think Critically

6. **Evaluate** the labels of two name-brand skin creams to compare how the two products claim to benefit the skin.

MATH in Biology

7. To determine how long an SPF will protect a person from burning in the Sun, multiply the amount of time the person can spend in the Sun before starting to burn by the SPF rating. If an individual who usually burns in 10 min uses a product with an SPF of 15, how long will the protection last?

Biology Online Self-Check Quiz biologygmh.com

Section 32.2

Reading Preview

Objectives

- **Distinguish** between the bones of the axial and appendicular skeletons.
- **Describe** how new bone is formed.
- **Summarize** the functions of the skeletal system.

Review Vocabulary

cartilage: tough, flexible connective tissue that forms the skeletons of embryos and later covers the surface of bones that move against each other in joints

New Vocabulary

axial skeleton
appendicular skeleton
compact bone
osteocyte
spongy bone
red bone marrow
yellow bone marrow
osteoblast
ossification
osteoclast
ligament

Personal Tutor

To learn about skeletal systems, visit biologygmh.com.

The Skeletal System

MAIN Idea **The skeleton provides a structural framework for the body and protects internal organs such as the heart, lungs, and brain.**

Real-World Reading Link Framing is an early stage of building a house. A person can walk through a house at that stage and know the plan of the house because of the framework. The skeletal system can be compared to the framework of a house. The framework provides structure and protection.

Structure of the Skeletal System

Notice all the bones in the adult skeleton pictured in **Figure 32.6.** If you counted them, you would find there are 206 bones. The human skeleton consists of two divisions—the axial skeleton and the appendicular skeleton. The **axial skeleton** includes the skull, the vertebral column, the ribs, and the sternum. The **appendicular skeleton** includes the bones of the shoulders, arms, hands, hips, legs, and feet.

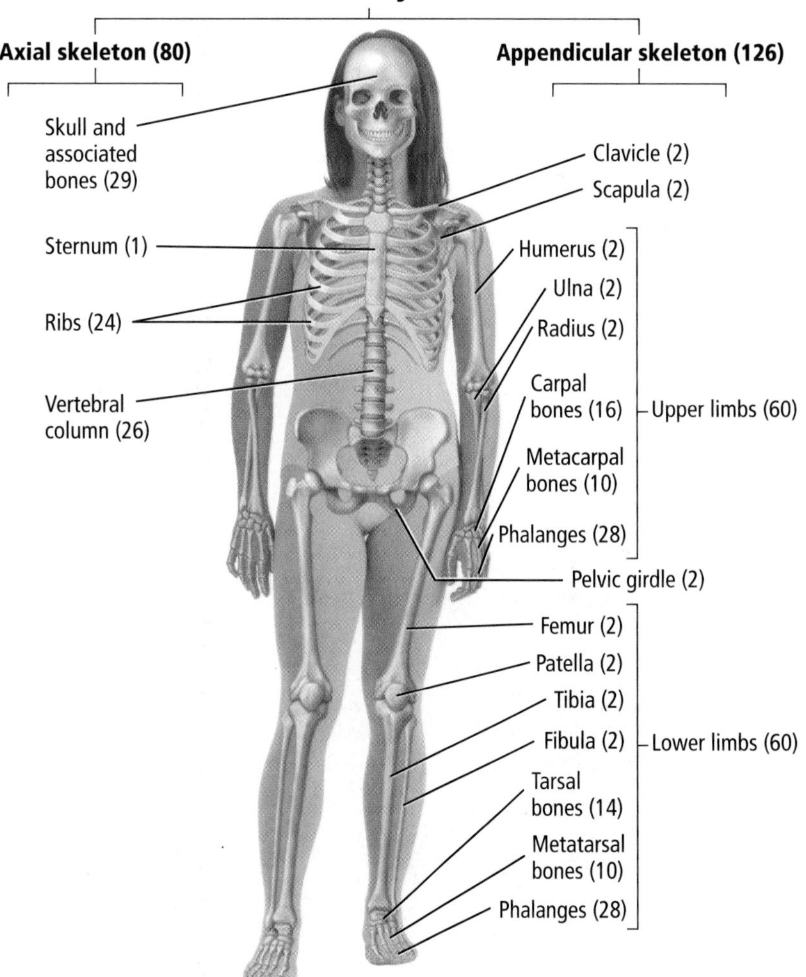

■ **Figure 32.6** The axial skeleton includes the bones of the head, back, and chest. Bones in the appendicular skeleton are related to movement of the limbs.

CAREERS IN BIOLOGY

Medical Illustrator A medical illustrator combines a talent in art with an interest in biology to illustrate procedures or to create teaching models. For more information on biology careers, visit biologygmh.com.

Compact and spongy bone Bone is a connective tissue that has many shapes and sizes. Bones are classified as long, short, flat, or irregular. Refer back to **Figure 32.6.** Arm and leg bones are examples of long bones, and wrist bones are examples of short bones. Flat bones make up the skull. Facial bones and vertebrae are irregular bones.

The outer layers of all bones are composed of compact bone. **Compact bone** is dense and strong; it provides strength and protection. Running the length of compact bones are tubelike structures called osteons, or Haversian systems, which contain blood vessels and nerves. The blood vessels provide oxygen and nutrients to **osteocytes**—living bone cells. The centers of bones can differ greatly, as illustrated in **Figure 32.7.**

As the name suggests, **spongy bone** is less dense and has many cavities that contain bone marrow. Spongy bone is found in the center of short or flat bones and at the end of long bones. Spongy bone is surrounded by compact bone and does not contain Haversian systems.

There are two types of bone marrow—red and yellow. Red and white blood cells and platelets are produced in **red bone marrow.** Red bone marrow is found in the humerus bone of the arm, the femur bone of the leg, the sternum and ribs, the vertebrae, and the pelvis. The cavities of an infant's bones are composed of red marrow. Children's bones have more red marrow than adult bones. **Yellow bone marrow,** found in many other bones, consists of stored fat. The body can convert yellow bone marrow to red bone marrow in cases of extreme blood loss or anemia.

Formation of bone The skeletons of embryos are composed of cartilage. During fetal development, cells in fetal cartilage develop into bone-forming cells called **osteoblasts.** The formation of bone from osteoblasts is called **ossification.** Except for the tip of the nose, outer ears, discs between vertebrae, and the lining of movable joints, the human adult skeleton is all bone. Osteoblasts also are the cells responsible for the growth and repair of bones.

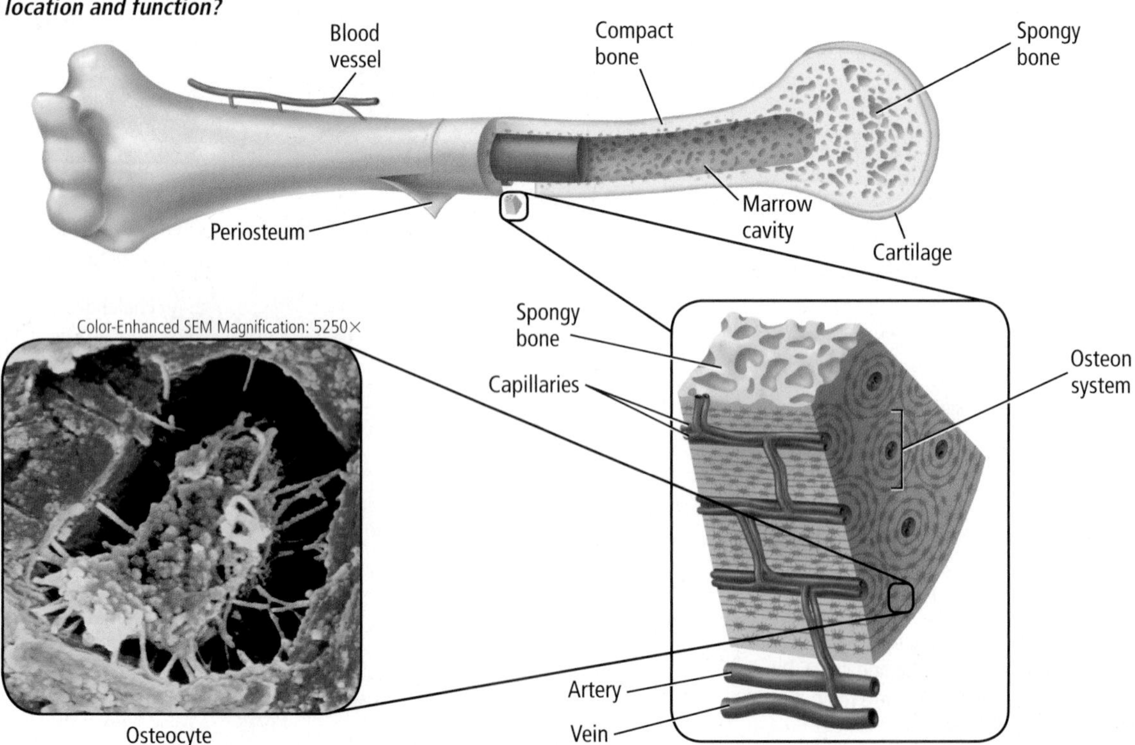

■ **Figure 32.7** Bone is either compact bone or spongy bone.
Classify ***How do spongy bone and compact bone differ in location and function?***

Remodeling of bone Bones constantly are being remodeled, which involves replacing old cells with new cells. This process is continual throughout life and is important in the growth of an individual. Cells called **osteoclasts** break down bone cells, which are then replaced by new bone tissue. Bone growth involves several factors, including nutrition and physical exercise. For example, a person with insufficient calcium can develop a condition known as osteoporosis that results in weak, fragile bones that break easily.

Reading Check **Compare** the roles of osteoblasts and osteoclasts.

Repair of bone Fractures are very common bone injuries. When a bone breaks but does not come through the skin, it is a simple fracture. A compound fracture is one in which the bone protrudes through the skin. A stress fracture is a thin crack in the bone. When a bone is fractured, repair begins immediately. Refer to **Figure 32.8,** which illustrates the steps in the repair of a broken bone.

Fracture Upon injury, endorphins, chemicals produced in the brain and sometimes called "the body's natural painkillers," flood the area of the injury to reduce the amount of pain temporarily. The injured area quickly becomes inflamed, or swollen. The swelling can last for two or three weeks.

Within about eight hours, a blood clot forms between the broken ends of the bone and new bone begins to form. First, a soft callus, or mass, of cartilage forms at the location of the break. This tissue is weak, so the broken bone must remain in place.

Callus formation About three weeks later, osteoblasts form a callus made of spongy bone that surrounds the fracture. The spongy bone is then replaced by compact bone. Osteoclasts remove the spongy bone while osteoblasts produce stronger, compact bone.

Splints, casts, and sometimes traction can ensure that the broken bone remains in place until new bone tissue has formed. Broken fingers often are kept in place by being taped to an adjacent finger.

Remodeling Bones require different amounts of time to heal. Age, nutrition, location, and severity of the break are all factors. A lack of calcium in a person's diet will slow down bone repair. Bones of younger people usually heal more quickly than bones of older people. For example, a fracture might take only four to six weeks to be repaired in a toddler, but it might take six months in an adult.

■ **Figure 32.8** Bone repair requires several steps. First, a mass of clotted blood forms in the space between the broken bones. Then connective tissue fills the space of the broken bone. Eventually, osteoblasts produce new bone tissue.

Fracture

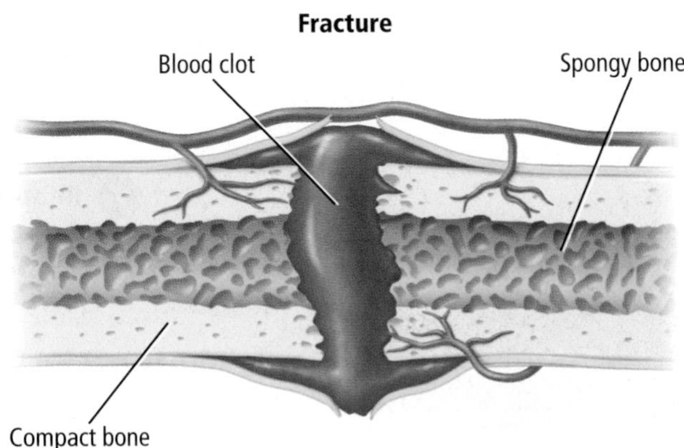

Callus Formation

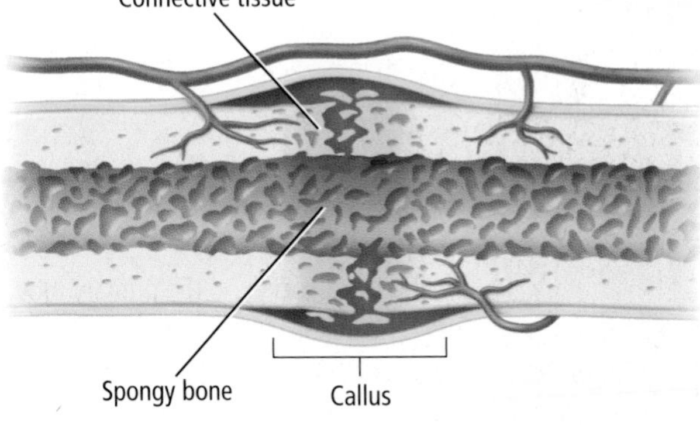

Remodeling

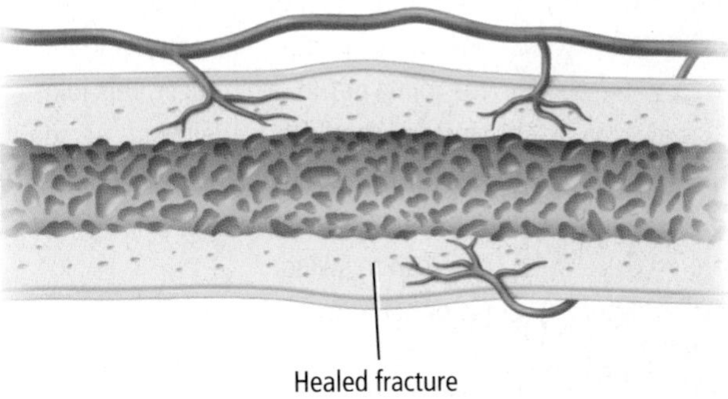

Joints

Joints occur where two or more bones meet. Except for the joints in the skull, they can be classified according to the movement they allow and the shapes of their parts. **Table 32.2** identifies five kinds of joints—ball-and-socket, pivot, hinge, gliding, and sutures. Study **Table 32.2** to identify the type of movement that each kind of joint allows and also the bones involved in each example.

Not all joints are movable. The joints between some skull bones are fixed. At birth, however, skull bones are not all fused together. They become fused by the time a baby is about three months old. Gliding joints, like those found in the hand, have limited movement. Other joints, such as the hinge joint of the elbow and the pivot joint in the lower arm, allow back-and-forth movement and twisting. The ball-and-socket joints of the hips and shoulders have the widest range of motion.

The bones of joints are held together by ligaments. **Ligaments** are tough bands of connective tissue that attach one bone to another. You will learn more about ligaments and tendons, which attach muscle to bone, in the following section.

LAUNCH Lab

Review Based on what you've read about joints, how would you now answer the analysis questions?

Reading Check **Review** the types of joints and how joints are classified.

Concepts In Motion Interactive Table To explore more about how joints move, visit biologygmh.com.

Table 32.2 Some Joints of the Skeletal System

Name of Joint	Ball-and-Socket	Pivot	Hinge	Gliding	Sutures
Example					
Description	In a ball-and-socket joint, the ball-like surface of one bone fits into a cuplike depression of another bone and allows the widest range of motion of any kind of joint. The joints of the hips and shoulders are ball and socket joints. They allow a person to swing his or her arms and legs.	The primary movement at a pivot joint is rotation. One example of a pivot joint is the elbow joint where two bones of the lower arm, the radius, and the ulna meet. This joint allows a person to twist the lower arm.	In a hinge joint, the convex surface of one bone fits into the concave surface of another bone. Elbows and knees are hinge joints. They allow back-and-forth movement like that of a door hinge.	Gliding joints allow side-to-side and back-and-forth movement. The joints in wrists and ankles are gliding joints. The joints of vertebrae also are gliding joints.	Sutures are joints in the skull that are not movable. There are 22 bones in an adult skull. All skull bones except the lower jaw bone are joined at sutures.

Osteoarthritis (ahs tee oh ar THRI tus) The ends of bones in movable joints, such as the knee, are covered by cartilage, which serves as a cushion and allows smooth movement of the joint. Osteoarthritis is a painful condition that affects joints and results from the deterioration of the cartilage. It is a very common condition in knees and hips and also affects the neck and back. Osteoarthritis affects about ten percent of Americans and the frequency increases with age. A young person who has a joint injury is at risk to develop osteoarthritis later in life.

Rheumatoid arthritis Rheumatoid (roo MAH toyd) arthritis is another form of arthritis that affects joints. Rheumatoid arthritis is not the result of cartilage deterioration or of wear and tear on the joint. Affected joints lose strength and function and are inflamed, swollen, and painful. Fingers can look deformed, as illustrated in **Figure 32.9.**

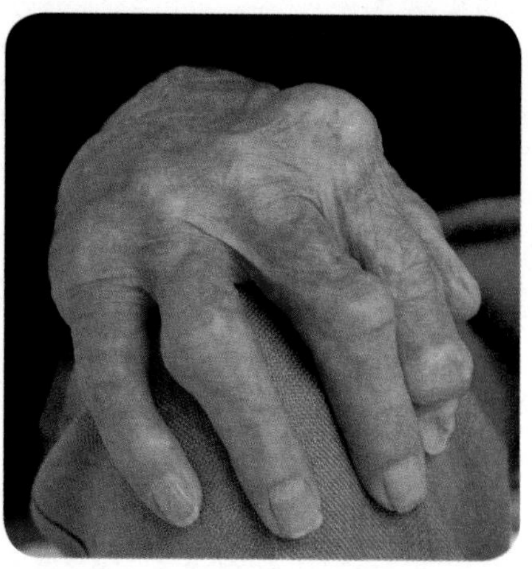

■ **Figure 32.9** Rheumatoid arthritis can cause loss of strength and function and involves severe pain.
Compare ***How does rheumatoid arthritis differ from the more common osteoarthritis?***

Bursitis Shoulders and knees also have fluid-filled sacs called bursae that surround these joints. Bursae decrease friction and act as a cushion between bones and tendons. Bursitis is an inflamation of the bursae and can reduce joint movement and cause pain and swelling. Perhaps you have heard of "tennis elbow" which is a form of bursitis. Treatment usually involves resting the joint involved.

Sprains A sprain involves damage to the ligaments that hold joints together. It is caused when a joint is twisted or overstretched and usually causes the joint to swell and be tender and painful.

MiniLab 32.2

Examine Bone Attachments

How are bones attached to muscles and other bones? Tendons attach muscle to bone, and ligaments attach bone to bone. You will examine these attachments using the skinned chicken wing from **MiniLab 32.1.**

Procedure

1. Read and complete the lab safety form.
2. Wear disposable **lab gloves.** Put the **skinned chicken wing** in a **dissection pan.**
3. Choose one muscle and use a pair of **dissection scissors** to cut the muscle away from the bone, leaving each end intact. Look for the long, white, tough tendons that connect the muscle to the bone.
4. Move the bones at the joint and notice how the tendon moves as the bones are pulled.
5. Carefully cut away all the muscles from the bones. The bones will still be attached to each other. Look for the white ligaments that hold them together. Examine the ends of each bone.
6. Draw a diagram of the wing without the muscles showing how the bones are attached to each other. Compare this drawing to the one you made in the Launch Lab.

Analysis

1. **Compare and Contrast** How is the drawing you made in the Launch Lab different from the drawing you made of the wing in this lab?
2. **Observe and Infer** Did you notice how a muscle is attached at one end to a bone and then how the ligament at the other end runs across a joint to attach that end of the muscle to the next bone? Explain why this is important. A diagram probably will help your explanation.
3. **Think Critically** At movable joints, what is the color of the ends of the bones? What do you think this material is?

Concepts In Motion
Interactive Table To explore more about the functions of the skeletal system, visit biologygmh.com.

Table 32.3 Functions of the Skeletal System

Function	Description
Support	• Legs, pelvis, and vertebral column hold up the body • Mandible supports the teeth • Almost all bones support muscles
Protection	• Skull protects the brain • Vertebrae protect the spinal column • Rib cage protects the heart, lungs, and other organs
Formation of blood cells	• Red bone marrow produces red blood cells, white blood cells, and platelets
Reservoir	• Stores calcium and phosphorus
Movement	• Attached muscles pull on bones of arms and legs • Diaphragm allows normal breathing

Functions of the Skeletal System

You might think that the only purpose of a skeleton is to serve as a framework to support the body. The bones of the legs, pelvis, and the vertebral column hold up the body. The mandible supports the teeth, and almost all bones support muscles. Many soft organs are directly or indirectly supported by nearby bones.

The skeletal system serves other functions besides support, as shown in **Table 32.3.** The skull protects the brain, vertebrae protect the spinal cord, and the rib cage protects the heart, lungs, and other organs.

The outer layers of bone tissue also protect the bone marrow found inside bones. In addition to forming red blood cells and white blood cells, red bone marrow forms platelets, which are involved in blood clotting. Red blood cells are produced at the rate of more than two million per second.

Until a person reaches about seven years of age, all bone marrow is red bone marrow. Then, fat tissue replaces some red marrow and gives the marrow a yellowish appearance, which gives it its name. Fat is an important source of energy.

Bones are reservoirs for the storage of minerals such as calcium and phosphorus. When blood calcium levels are too low, calcium is released from bones. When blood calcium levels are high, excess calcium is stored in bone tissue. In this way, the skeletal system helps to maintain homeostasis.

Bones that have muscles attached to them allow movement of the body. For example, as muscles pull on the bones of the arms and legs, they cause movement. Muscles that are attached to your ribs allow you to breathe normally.

Section 32.2 Assessment

Section Summary

- The human skeleton consists of two divisions.
- Most bones are composed of two different types of tissue.
- Bones are being remodeled constantly.
- Bones work in conjunction with muscles.
- The skeleton has several important functions.

Understand Main Ideas

1. MAIN Idea **List** and describe the functions of the axial skeleton and the appendicular skeleton.
2. **Compare** the compositions of red bone marrow and yellow bone marrow.
3. **Compare** the body's mechanism for repairing a fractured bone with the original development of bone.
4. **Construct** a classification scheme for all of the bones shown in **Figure 32.6.**

Think Critically

5. **Consider** What might be the result if osteoblast and osteoclast cells did not function properly both in a developing fetus and in an adult?
6. **Distinguish** between compact and spongy bone based on their appearance, location, and function.

Biology Online Self-Check Quiz biologygmh.com

Section 32.3

The Muscular System

Reading Preview

Objectives

- **Describe** the three types of muscle tissue.
- **Explain** at cellular and molecular levels the events involved in muscle contraction.
- **Distinguish** between slow-twitch and fast-twitch muscle fibers.

Review Vocabulary

anaerobic: chemical reactions that do not require the presence of oxygen

New Vocabulary

smooth muscle
involuntary muscle
cardiac muscle
skeletal muscle
voluntary muscle
tendon
myofibril
myosin
actin
sarcomere

MAIN Idea **The three major types of muscle tissue differ in structure and function.**

Real-World Reading Link Leonardo da Vinci contributed a great amount of knowledge to the scientific community. He studied the human body by examining cadavers. Da Vinci replaced muscles with string and learned that muscles shorten and pull on bones to make them move.

Three Types of Muscle

A muscle consists of groups of fibers or muscle cells that are bound together. When the word muscle is used, many people immediately think of skeletal muscle. Examine **Figure 32.10** to see that there are three types of muscle: smooth muscle, cardiac muscle and skeletal muscle. Muscles are classified according to their structure and function.

Smooth muscle Many hollow internal organs such as the stomach, intestines, bladder, and uterus are lined with **smooth muscle.** Smooth muscle is called **involuntary muscle** because it cannot be controlled consciously. For example, food moves through the digestive tract because of the action of smooth muscles that line the esophagus, stomach, and small and large intestines. Under a microscope, smooth muscle does not appear striated, or striped, and each cell has one nucleus.

Cardiac muscle The involuntary muscle present only in the heart is called **cardiac muscle.** Cardiac muscle cells are arranged in a network, or web, that allows the heart muscle to contract efficiently and rhythmically. This arrangement gives strength to the heart. Cardiac muscle is striped, or striated, with light and dark bands of cells with many nuclei. Cells usually have one nucleus and are connected by gap junctions.

■ **Figure 32.10** When magnified, differences in muscle shape and appearance can be seen. Smooth muscle fibers appear spindle-shaped; cardiac muscle appears striated or striped; skeletal muscle also appears striated.
Explain *In addition to their appearance, how else are muscles classified?*

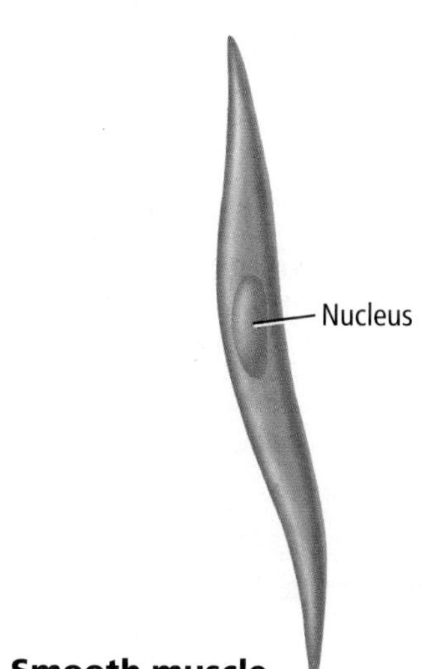

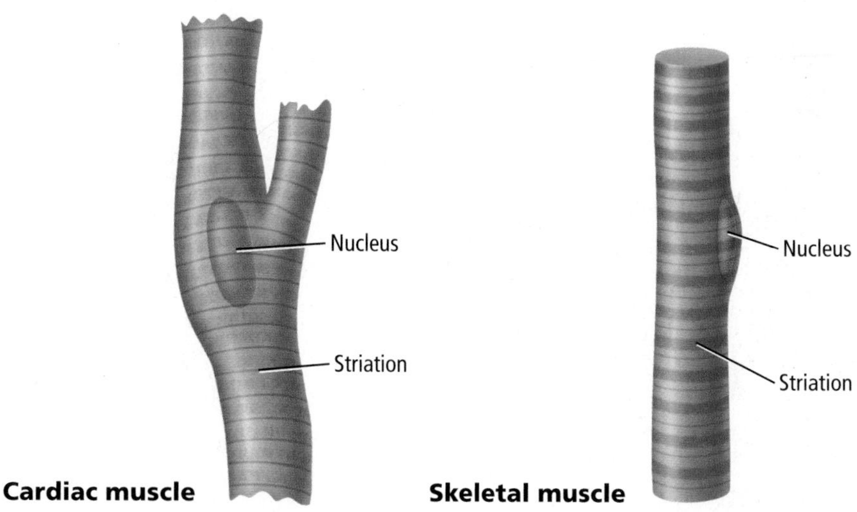

Vocabulary

Science usage v. Common usage

Contract

Science usage: to tighten or to shorten.

Muscles contract and cause movement.

Common usage: to become affected with.

If you are exposed to the flu, you may contract the illness.

Skeletal muscle Most of the muscles in the body are skeletal muscles. **Skeletal muscles** are muscles attached to bones by tendons and when tightened, or contracted, cause movement. Skeletal muscles are **voluntary muscles** that are consciously controlled to move bones. **Tendons,** which are tough bands of connective tissue, connect muscles to bones. Under a microscope, skeletal muscles also appear striated.

Skeletal Muscle Contraction

Most skeletal muscles are arranged in opposing, or antagonistic pairs. Refer to **Figure 32.11,** which illustrates muscles that you use to raise your arm and opposing muscles that you use to lower your arm. Skeletal muscle is arranged into fibers, which are fused muscle cells. Muscle fibers consist of many smaller units called **myofibrils.** Myofibrils consist of even smaller units, **myosin** and **actin,** which are protein filaments. Myofibrils are arranged in sections called sarcomeres. A **sarcomere** is the functional unit of a muscle and the part of the muscle that contracts as illustrated in **Figure 32.12.** The striations of skeletal muscles are a result of the sarcomeres, which run Z line to Z line. Z lines are where actin filaments attach within a myofibril. The overlap of actin and myosin filaments results in a dark band called the A band. The M line consists of only myosin filaments. The arrangement of the components of a sarcomere causes a muscle to shorten and then relax.

Sliding filament theory The sliding filament theory is also illustrated in **Figure 32.12**. This theory states that once a nerve signal reaches a muscle, the actin filaments slide toward one another, causing the muscle to contract. Notice that the myosin filaments do not move. There are many skeletal muscles involved in a simple motion, such as turning this page.

Connection to Chemistry When the nerve impulse reaches the muscle, calcium is released into the myofibrils. Calcium causes the myosin and actin to attach to each other. The actin filaments then are pulled toward the center of the sarcomere, which causes muscle contraction. ATP produced in the mitochondria is necessary for this step of muscle contraction. When the muscle relaxes, the filaments slide back into their original positions. You will learn more about nerve function in Chapter 33.

Figure 32.11 Muscles are arranged in antagonistic pairs.

Concepts In Motion

Interactive Figure To see an animation of muscles contracting, visit biologygmh.com.

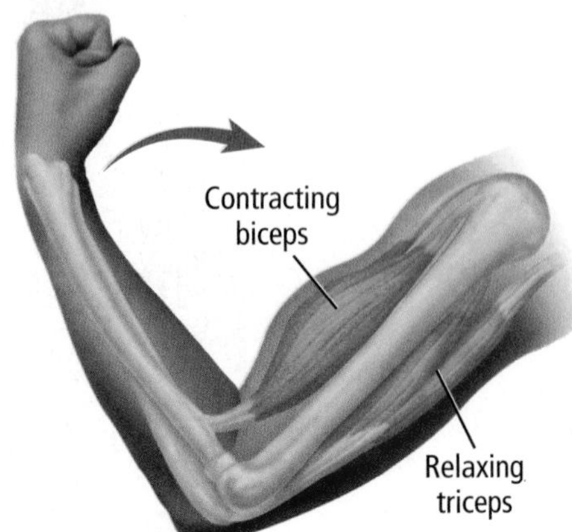

When the biceps muscle contracts, the lower arm is moved upward.

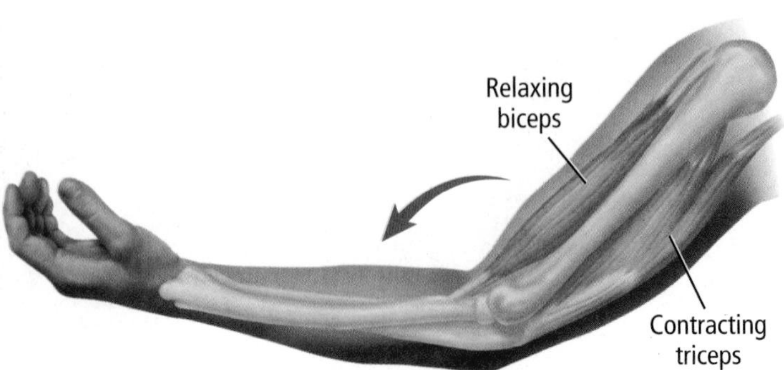

When the triceps muscle on the back of the upper arm contracts, the lower arm moves downward.

Visualizing Muscle Contraction

Figure 32.12
A muscle fiber is made of myofibrils. The protein filaments actin and myosin form myofibrils.

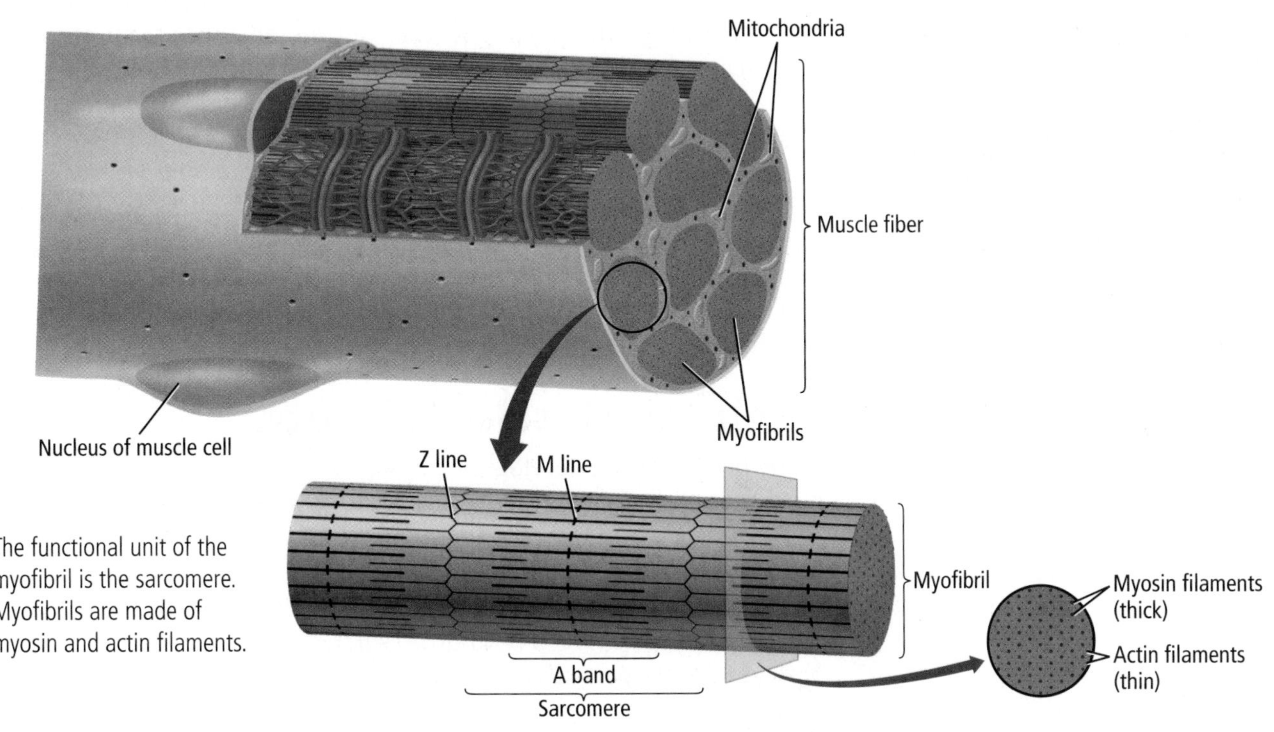

The functional unit of the myofibril is the sarcomere. Myofibrils are made of myosin and actin filaments.

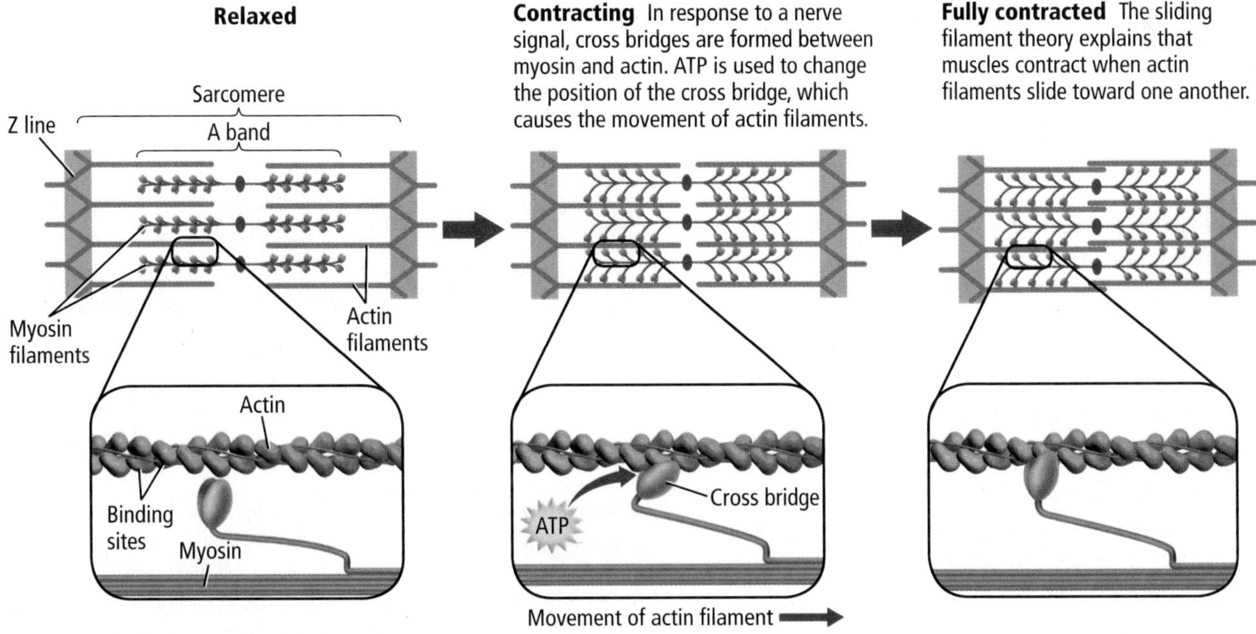

Concepts In Motion **Interactive Figure** To see an animation of muscle contraction, visit biologygmh.com.

Biology Online

■ **Figure 32.13** Crossing the finish line is a moment of intense energy.
Explain *How is normal breathing restored after intense exercise?*

Energy for muscle contraction All muscle cells metabolize aerobically and anaerobically. When sufficient oxygen is available, aerobic cellular respiration occurs in muscle cells.

Recall from Chapter 8 that cellular respiration process provides ATP for energy. After a period of intense exercise, muscles might not get enough oxygen to sustain cellular respiration, limiting the amount of ATP that is available. Muscles, like those of the athlete in **Figure 32.13,** then must rely on the anaerobic process of lactic acid fermentation for energy.

During exercise, lactic acid builds up in muscle cells, causing fatigue. Excess lactic acid enters the bloodstream and this stimulates rapid breathing. After resting for a short time, adequate amounts of oxygen are restored and lactic acid is broken down.

You probably have seen a dead animal along the side of the road. When an animal dies, rigor mortis sets in. Rigor mortis is a state of prolonged muscular contraction. ATP is required to pump the calcium back out of the myofibrils, which causes the muscles to relax. In rigor mortis, the dead animal cannot produce ATP, so the calcium remains in the myofibrils and the muscle remains contracted. After 24 h, cells and tissues begin degrading and the muscle fibers cannot remain contracted.

Skeletal Muscle Strength

Many people do not develop the physiques of champion bodybuilders, no matter how often they work out in the weight room. A person might be the fastest sprinter on the track team, but quickly becomes fatigued in a long-distance race. What might be the reason for these differences? The reason in both cases is the ratio of slow-twitch muscle fibers to fast-twitch muscle fibers. Both slow-twitch and fast-twitch fibers are present in every person's muscles.

DATA ANALYSIS LAB 32.1

Based on Real Data*

Interpret the Data

How is the percentage of slow-twitch muscle related to action of a muscle? The proportion of slow-twitch to fast-twitch muscle fibers can be determined by removing a small piece of a muscle and staining the cells with a dye called *ATPase stain*. Fast-twitch muscle fibers with a high amount of ATP activity stain dark brown.

Think Critically

1. **Hypothesize** why a muscle such as the soleus has more slow-twitch muscle fibers than a muscle such as the orbicularis oculi.
2. **Classify** muscles by giving examples of muscles that have a high proportion of fast-twitch muscle fibers.

Data and Observations

Muscle	Action	Percent Slow Twitch
Soleus (leg)	Elevates the foot	87
Biceps femoris (leg)	Flexes the leg	67
Deltoid (shoulder)	Lifts the arm	52
Sternocleidomastoideus (neck)	Moves the head	35
Orbicularis oculi (face)	Closes the eyelid	15

*Data adapted from: Lamb, D.R. 1984. *Physiology of Exercise* New York: Macmillan Co.

Slow-twitch muscles Muscles vary in the speeds at which they contract. Slow-twitch muscles contract more slowly than fast-twitch muscle fibers. Slow-twitch muscle fibers have more endurance than fast-twitch muscle fibers. The body of the triathlete in **Figure 32.14** has many slow-twitch fibers. These kinds of muscle fibers function well in long-distance running or swimming because they resist fatigue more than fast-twitch muscle fibers.

Slow-twitch muscle fibers have many mitochondria needed for cellular respiration. They also contain myoglobin, a respiratory molecule that stores oxygen and serves as an oxygen reserve. Myoglobin causes the muscles to have a dark appearance. Exercise increases the number of mitochondria in these fibers, but the overall increase in the size of the muscle is minimal.

Fast-twitch muscles Fast-twitch muscle fibers fatigue easily but provide great strength for rapid, short movements. Fast-twitch muscle fibers are adapted for strength. They function well in exercises requiring short bursts of energy such as sprinting or weightlifting, as illustrated in **Figure 32.14.**

Fast-twitch fibers are lighter in color because they lack myoglobin. Because they have fewer mitochondria, they rely on anaerobic metabolism, which causes a buildup of lactic acid. This causes these muscles to fatigue easily. Exercise increases the number of myofibrils in a muscle, thereby increasing the diameter of the entire muscle.

Most skeletal muscles contain a mixture of slow-twitch and fast-twitch muscle fibers. The ratio of these fibers is determined genetically. If there is a very high ratio of slow-twitch to fast-twitch, a person might be a champion cross-country runner. Champion sprinters have a high proportion of fast-twitch muscle fibers. Most people are somewhere in between.

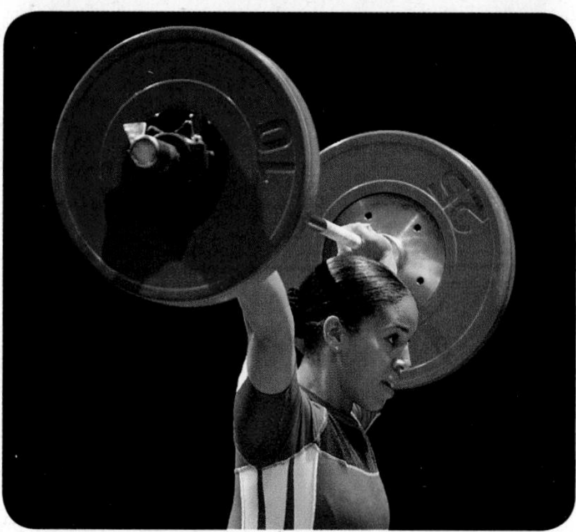

■ **Figure 32.14** Triathletes have a high proportion of slow-twitch muscle fibers. Weight lifters have a high proportion of fast-twitch muscle fibers.

Section 32.3 Assessment

Section Summary

- There are three types of muscle.
- Skeletal muscles are arranged in antagonistic pairs that work opposite to each other.
- Smooth muscles line many internal organs.
- Cardiac muscle is present only in the heart.
- All muscle cells metabolize both aerobically and anaerobically.

Understand Main Ideas

1. MAIN Idea **Construct** a chart that lists similarities and differences among the three types of muscles.
2. **Identify** which type of muscles are voluntary and which are involuntary.
3. **Explain** why aerobic respiration occurs before lactic acid fermentation in most muscles.
4. **Compare** the role of mitochondria in slow-twitch and fast-twitch muscle fibers.

Think Critically

5. **Infer** Wild turkeys have a higher ratio of dark meat (muscle) to white meat than farm-raised turkeys. Why does this allow wild turkeys to fly longer distances than domesticated turkeys?

WRITING in **Biology**

6. Write a short story that describes the sequence of events involved in skeletal muscle contraction. Tell your story from the point of view of a calcium ion.

CUTTING-EDGE BIOLOGY

Make Some Bones About It: Petri Dish Style

In futuristic movies, sometimes characters lose an arm or leg and these limbs are replaced with robotic limbs. This sounds like science fiction, but today's tissue engineers are working on making it a reality by working with tissues grown in Petri dishes.

How are tissues grown in the lab? Tissue engineering is the process of making human repairs by first starting at the cellular level. Tissue engineers are working to grow cartilage, nerves, bones, teeth, breast tissue, and arteries. Scientists use synthetic, biodegradable scaffolds that give cells an environment similar to the body. These scaffolds often are polymers, which are porous like a sponge and can house many cells that attach and grow. The polymer also allows for nutrients to be diffused. Eventually, the polymer degrades when the tissue develops its own matrix and does not need the scaffolding any longer.

It is important to determine how cells communicate with and respond to their environment, and to determine how cells move around. Mesenchymal stem cells in the body produce bone, cartilage, tendon, teeth, fat, and skin. These cells also are responsible for the connective tissue in the bone marrow. As cells naturally die in the body, mesenchymal stem cells receive signals and are mobilized to differentiate into the specific tissue needed. They also make excellent candidates for the tissue engineering activities of the scientists who primarily get them from the bone marrow.

Tissue engineering progress Although skin was the first tissue to be successfully engineered and made available to people, progress has been made with developing bone tissue.

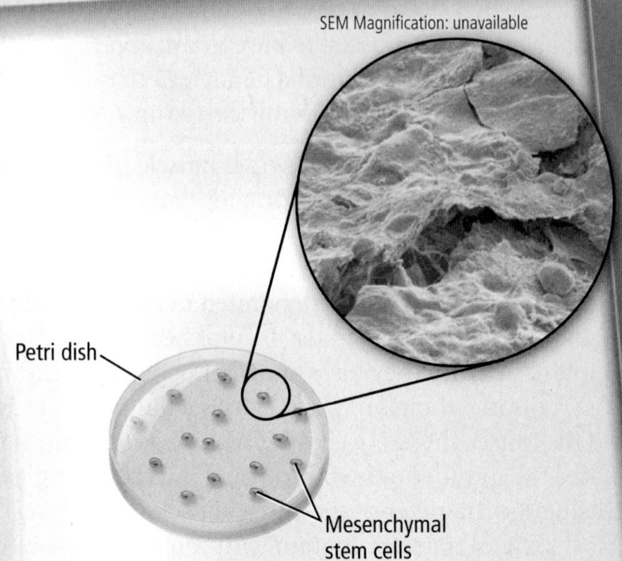

After eight weeks, mesenchymal stem cells have produced thick layers of bone cells.

Conventional titanium alloys used in hip and knee replacements are smooth. The body reacts to the smooth artificial surfaces, such as those of titanium, by covering them with fibrous tissue that interferes with proper functioning. Because natural bone and other tissues have bumps about 100 nm wide, biomedical engineers are working with bone cells that attach best to metals with nanometer-scale surface bumps. This offers hope for improved prosthetic hips, knees, and other implants. These new, bumpier parts could keep the body from rejecting the artificial parts and keep them functioning longer. The bone cells, grown in Petri dishes, will help researchers use nanotechnology to design implants that last longer and work better.

CAREERS in Biology

Research Visit biologygmh.com to research careers in tissue engineering or biomedical engineering related to the topics discussed in this feature. Design a brochure that informs the public about these professions including late-breaking work in the area, scientific research methodology, and necessary educational background. Include illustrations or photographs.

BIOLAB

FORENSICS: HOW CAN SKELETONS HELP YOU SOLVE A "CRIME"?

Background: **Imagine there is a National Museum of Domestic Chickens and it has been robbed. Several bones from the first chicken eaten in America are missing. Three dogs are suspects. Your job is to examine impressions of bones that were found in mud near the doghouse of each dog and to determine if any of the bones came from a chicken. You will be given a clue for each unknown bone.**

Question: *Can the structure and form of a bone tell you from which animal it came?*

Materials

impressions of three unknown bones
set of clues
various animal skeletons
hand lens
metric ruler
string

Safety Precautions

Procedure

1. Read and complete the lab safety form.
2. Collect materials you will use to measure and examine the skeletons. Determine what types of measurements you will make.
3. Obtain impressions of three bones and a set of clues from your teacher. Do not open the clues until you are told to do so.
4. Design a data table to record your measurements.
5. Examine the skeletons. Compare them to the impressions.
6. Make measurements and record the data.
7. Open the clues you were given and reexamine your data and answers.
8. **Cleanup and Disposal** Return any reusable materials to their proper storage areas.

Analyze and Conclude

1. **Analyze Data** Based on your observations and measurements, determine which one of the impressions came from a chicken.
2. **Interpret Data** How did you use information concerning the size and shape of each impression to help you determine from which animal it came?
3. **Evaluate** Did your conclusions change after you opened the clues? Explain your reasoning if your conclusions changed.
4. **Compare and Contrast** What similarities did you notice between each impression and bones in the human skeleton? What differences did you notice?
5. **Relate** Which skeletons seem to share the most characteristics with a human skeleton?
6. **Draw Conclusions** Which dog stole the chicken bones?

Poster Session Paleontologists are scientists who study fossils. Through their studies of fossil bones they have found evidence that birds had a dinosaur ancestor. Find out what type of evidence has been found and create a poster that shows what you learned. To learn more about fossil bones, visit BioLabs at biologygmh.com.

CHAPTER 32

Study Guide

Download quizzes, key terms, and flash cards from biologygmh.com.

FOLDABLES **Differentiate** Use what you have learned to differentiate between thick skin (like that found on the heel of the foot) and thin skin (like that found on the tip of a finger). How are the layers different? How are they similar? Why?

Vocabulary	Key Concepts
Section 32.1 The Integumentary System	
• dermis (p. 937) • epidermis (p. 936) • hair follicle (p. 937) • keratin (p. 936) • melanin (p. 937) • sebaceous gland (p. 937)	**MAIN Idea** Skin is a multilayered organ that covers and protects the body. • The skin is the major organ of the integumentary system. • Maintaining homeostasis is one function of the integumentary system. • There are four types of tissues in the integumentary system. • Hair, fingernails, and toenails develop from epithelial cells. • Burns are classified according to the severity of the damage to skin tissues.
Section 32.2 The Skeletal System	
• appendicular skeleton (p. 941) • axial skeleton (p. 941) • compact bone (p. 942) • ligament (p. 944) • ossification (p. 942) • osteoblast (p. 942) • osteoclast (p. 943) • osteocyte (p. 942) • red bone marrow (p. 942) • spongy bone (p. 942) • yellow bone marrow (p. 942)	**MAIN Idea** The skeleton provides a structural framework for the body and protects internal organs such as the heart, lungs, and brain. • The human skeleton consists of two divisions. • Most bones are composed of two different types of tissue. • Bones are being remodeled constantly. • Bones work in conjunction with muscles. • The skeleton has several important functions.
Section 32.3 The Muscular System	
• actin (p. 948) • cardiac muscle (p. 947) • involuntary muscle (p. 947) • myofibril (p. 948) • myosin (p. 948) • sarcomere (p. 948) • skeletal muscle (p. 948) • smooth muscle (p. 947) • tendon (p. 948) • voluntary muscle (p. 948)	**MAIN Idea** The three major types of muscle tissue differ in structure and function. • There are three types of muscle. • Skeletal muscles are arranged in antagonistic pairs that work opposite to each other. • Smooth muscles line many internal organs. • Cardiac muscle is present only in the heart. • All muscle cells metabolize both aerobically and anaerobically.

CHAPTER 32

Assessment

Section 32.1

Vocabulary Review

Explain the difference between the terms in each set.

1. epidermis, dermis
2. melanin, keratin
3. sebaceous glands, hair follicles

Understand Key Concepts

Use the diagram below to answer question 4.

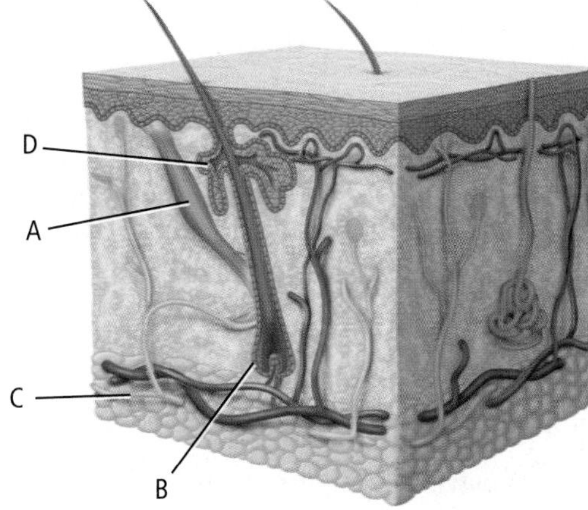

4. Which tissue type is responsible for the formation of "goose bumps"?
   A. A **C.** C
   B. B **D.** D

5. When are blackheads formed?
   A. when sebaceous glands become clogged
   B. when grooves in the epidermis gather dirt
   C. when hair follicles grow inward rather than outward
   D. when there is an excess of keratin produced

6. How does the skin regulate body temperature?
   A. by increasing sweat production
   B. by retaining water
   C. by producing vitamin D
   D. by regulating fat content in the epidermis

7. Which are not found in the dermis?
   A. muscles
   B. sweat and oil glands
   C. fat cells
   D. nerve cells

8. What could be inferred from suntans?
   A. Sunning for the purpose of tanning produces healthier skin.
   B. A tan might indicate sun damage to the skin.
   C. Tanning strengthens the elastic in the skin making the skin feel tight.
   D. Tanning promotes skin that has a more youthful appearance.

Constructed Response

9. **Open Ended** What possible effects on the body might there be if the epidermis was absent?
10. **Open Ended** What possible effects on the body might there be if the dermis was absent?
11. **Short Answer** Describe how the integumentary system contributes to homeostasis.

Think Critically

12. **Explain** why it does not hurt when you get a haircut.
13. **Assess** the reason why people with third degree burns might not feel pain at the site of the burn.

Section 32.2

Vocabulary Review

Explain the difference between the terms in each set.

14. spongy bone, compact bone
15. tendons, ligaments
16. osteoblasts, osteoclasts

Understand Key Concepts

Use the figure below to answer question 17.

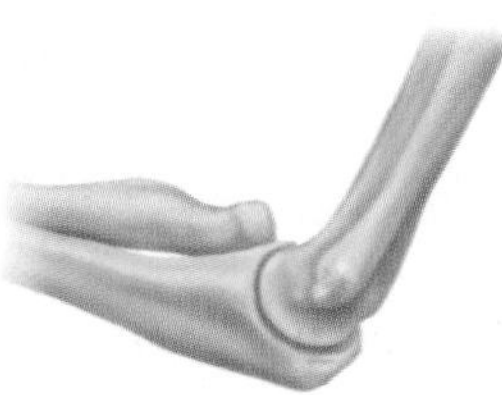

17. Where would you find the type of joint shown above?
    A. hip **C.** elbow
    B. vertebrae **D.** skull

18. Which is not a function of bone?
    A. production of vitamin D
    B. internal support
    C. protection of internal organs
    D. storage of calcium

Use the diagram below to answer question 19.

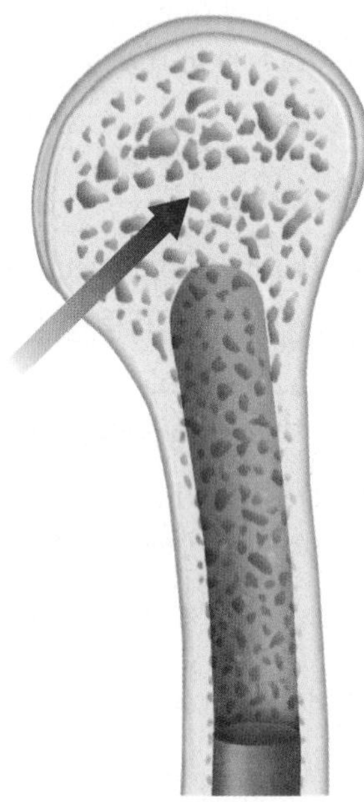

19. What is a characteristic of the portion of the bone indicated by the arrow?
    A. It contains no living cells.
    B. It contains bone marrow.
    C. It is the only type of bone tissue in long bones.
    D. It is made of overlapping osteon systems.

20. Which set of terms is mismatched?
    A. cranium—sutures
    B. wrist—pivot joint
    C. shoulder—ball-and-socket joint
    D. knee—hinge joint

21. What are the cells that remove old bone tissue called?
    A. osteoblasts
    B. osteocytes
    C. osteoclasts
    D. osteozymes

22. Which is not part of the axial skeleton?
    A. skull
    B. ribs
    C. hip bone
    D. vertebral column

23. Which is part of the appendicular skeleton?

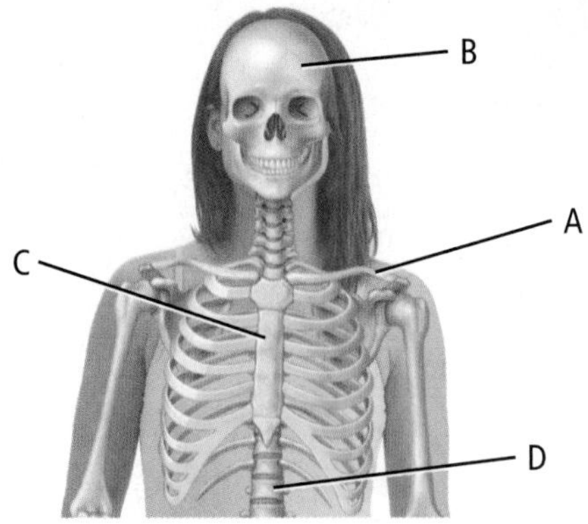

    A. A
    B. B
    C. C
    D. D

Constructed Response

24. **Open Ended** Describe potential consequences if all bone tissue in humans was comprised of spongy bone and there was no compact bone.

25. **Open Ended** Describe potential consequences if all bone tissue in humans was comprised of compact bone and there was no spongy bone.

26. **Short Answer** Compare the function of osteoclasts and osteoblasts.

Think Critically

27. **Analyze** the following scenario. A person enters the emergency room with an ankle injury. What structures of the patient's ankle need to be examined to determine the proper treatment?

28. **Hypothesize** what might happen to a woman's bones if she did not increase her intake of calcium during pregnancy?

Section 32.3

Vocabulary Review

For each set of terms below, choose the one term that does not belong and explain why it does not belong.

29. actin, melanin, myosin

30. cardiac muscle, smooth muscle, fast-twitch muscle

31. sarcomere, myofibril, myoglobin

Biology Online Chapter Test biologygmh.com

Understand Key Concepts

32. Which requires ATP?
- **A.** muscle contraction
- **B.** muscle relaxation
- **C.** both muscle contraction and relaxation
- **D.** neither muscle contraction nor relaxation

Use the diagram below to answer question 33.

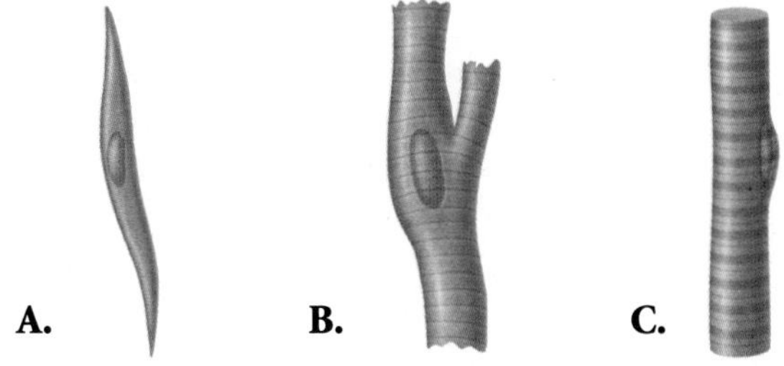

33. What muscles shown above are classified as voluntary muscles?
- **A.** the muscle type shown in A
- **B.** the muscle type shown in B
- **C.** the muscle type shown in C
- **D.** all muscles

34. Which is a characteristic of fast-twitch muscle fibers?
- **A.** They contain more myoglobin than slow-twitch fibers.
- **B.** They are resistant to fatigue.
- **C.** They have fewer mitochondria than slow-twitch fibers.
- **D.** They require high amounts of oxygen in order to function.

Constructed Response

35. Short Answer Compare and contrast the structure of skeletal, smooth, and cardiac muscle.

36. Short Answer Explain, based on the structure of the muscle fibers, why skeletal muscles can contract but not lengthen.

Think Critically

37. Predict any possible consequences if cardiac and smooth muscle had the same structure as skeletal muscle.

38. Infer why it is important that no muscle contains solely slow-twitch or fast-twitch fibers.

Additional Assessment

39. WRITING in Biology Imagine you are a writer for a health magazine. Write a brief article about the need for calcium for the skeletal and muscular systems.

Document-Based Questions

Athletes burn fat at a maximum rate when they exercise at an intensity near the lactate threshold—the point at which lactic acid starts to build up in the muscles. In addition, athletes who consume the greatest amounts of oxygen during intense exercise [VO_{2peak}] burn the most fat. Researchers compared the lactate threshold and oxygen consumption of overweight subjects who did not exercise to those of highly-trained athletes.

Data obtained from: Bircher, S. and Knechtle, B. 2004. Relationship between fat oxidation and lactate threshold in athletes and obese women and men. Journal of Sports Science and Medicine 3:174–181.

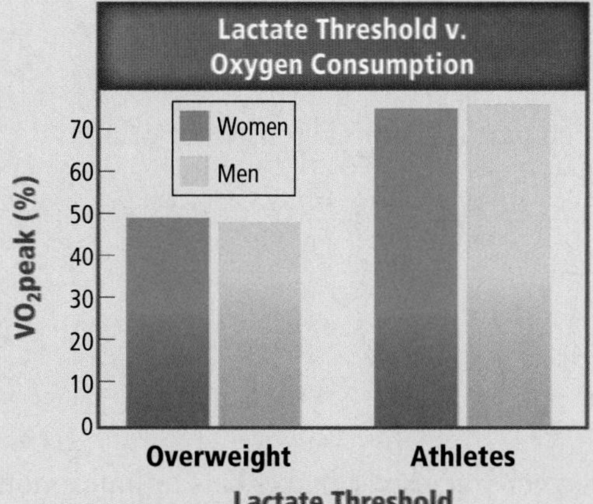

40. At what percent of $VO_{2\ peak}$ was the lactate threshold reached in overweight subjects?

41. How might an overweight person who does not exercise increase his or her $VO_{2\ peak}$ and, therefore, his or her lactate threshold?

Cumulative Review

42. Make sketches of the distinctive parts of mammals that distinguish them from all other groups. **(Chapter 30)**

Standardized Test Practice

Cumulative

Multiple Choice

1. Which describes the circulatory system of most reptiles?
   A. double loop, four-chambered heart
   B. double loop, three-chambered heart
   C. single loop, three-chambered heart
   D. single loop, two-chambered heart

Use the figure below to answer question 2.

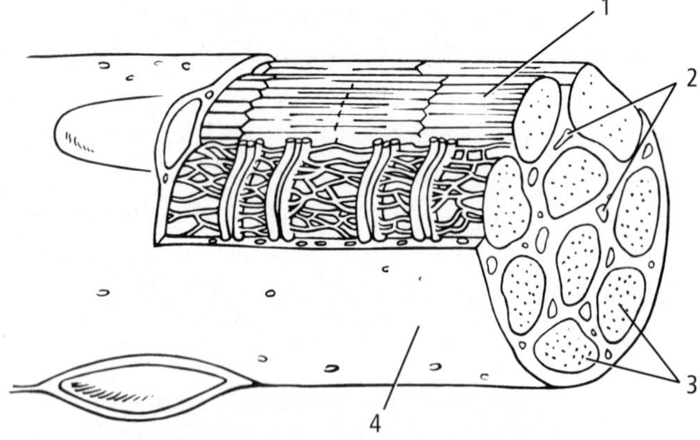

2. Which part of a muscle is used for cellular respiration?
   A. 1
   B. 2
   C. 3
   D. 4

3. Which characteristic makes bats unique among mammals?
   A. eyesight
   B. feathers
   C. flight
   D. teeth

4. Which learned behavior occurs only at a certain critical time in an animal's life?
   A. classical conditioning
   B. fixed action pattern
   C. habituation
   D. imprinting

Use the figure of the joint below to answer question 5.

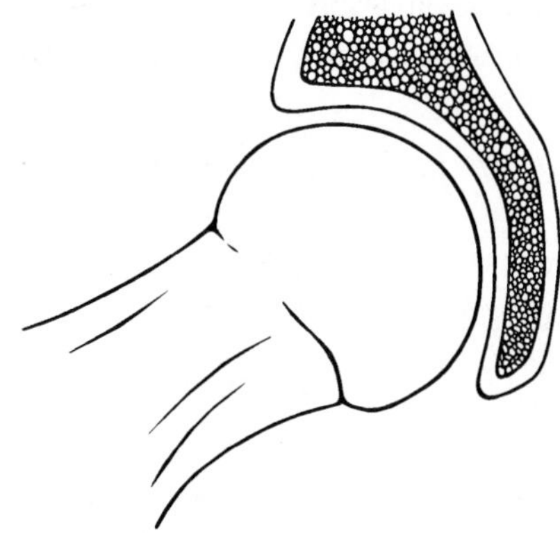

5. Where is the type of joint shown in the figure found?
   A. elbows and knees
   B. fingers and toes
   C. hips and shoulders
   D. wrists and ankles

6. Which describes the characteristics of a bird's brain?
   A. Birds have a large medulla to process their vision.
   B. Birds have a large cerebellum to control respiration and digestion.
   C. Birds have a large cerebrum to coordinate movement and balance.
   D. Birds have a large cerebral cortex to control flight.

7. Which type of bone is classified as irregular?
   A. leg bones
   B. skull
   C. vertebrae
   D. wrist bones

8. Which adaptation helps stop fishes from rolling side to side in the water?
   A. ctenoid scales
   B. paired fins
   C. placoid scales
   D. swim bladders

Short Answer

Use the diagram below to answer questions 9 and 10.

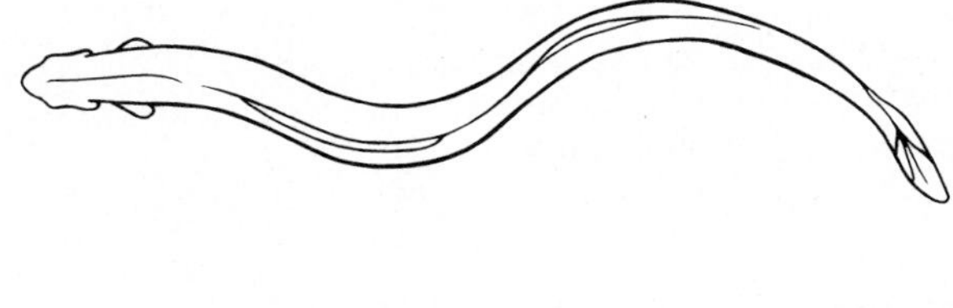

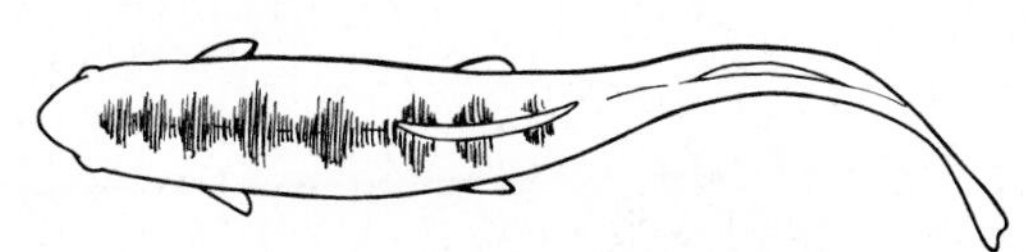

9. Describe the difference between how a fish with an S-shaped pattern swims and a fish that moves its tail only.

10. Decide where a fish with an S-shaped pattern would be likely found swimming.

11. Relate the key events in the life cycle of a butterfly to the key events in the life cycle of a grasshopper.

12. Howler monkeys are the loudest land animals. Their calls are heard for miles across the jungle. They use their calls to mark their territory. Assess this type of behavior.

13. Describe how fetal cartilage becomes bone.

14. A chimpanzee picks up a blade of grass and sticks it in an anthole. When it pulls the blade of grass out, it has ants on it. The chimpanzee eats the ants. The chimpanzee continues doing this because it is an easy way to get ants. Assess this activity as it relates to animal behaviors.

15. Describe two types of joint conditions.

Extended Response

Use the diagram to answer questions 16 and 17.

Pigeon

Eagle

16. Evaluate what the location of the eyes on these two birds reveals about their behavior.

17. Explain how the beaks of these two birds give evidence of what they eat.

Essay Question

Whooping cranes are an endangered species. One of the reasons for this is that they hatch in nesting areas and then migrate south for the winter. Humans can raise chicks, but teaching them to migrate is a different problem. Operation Migration solved this problem in 2001. Operation Migration used ultralight aircraft to lead a migration of human-raised whooping cranes on a 2000-km migration from Wisconsin to Florida. The whooping cranes followed the ultralight aircraft that used recorded calls to learn the migration route.

Using the information in the paragraph above, answer the following question in essay format.

18. Migratory behavior has been shown to be an innate behavior. Evaluate why it is necessary to use ultralight aircraft to guide the birds so they can learn the migration route.

NEED EXTRA HELP?

If You Missed Question . . .	1	2	3	4	5	6	7	8	9	10	11	12	13	14	15	16	17	18
Review Section . . .	29.1	32.3	30.1	31.1	32.2	29.2	32.2	28.1	28.1	28.1	26.3	31.2	32.2	31.1	32.2	29.2	29.2	31.2

CHAPTER 33

Nervous System

BIG Idea **The nervous system is essential for communication among cells, tissues, and organs.**

Section 1
Structure of the Nervous System
MAIN Idea Neurons conduct electrical impulses that allow cells, tissues, and organs to detect and respond to stimuli.

Section 2
Organization of the Nervous System
MAIN Idea The central nervous system and the peripheral nervous system are the two major divisions of the nervous system.

Section 3
The Senses
MAIN Idea Sensory receptors allow you to detect the world around you.

Section 4
Effects of Drugs
MAIN Idea Some drugs alter the function of the nervous system.

BioFacts

- A nerve impulse can travel as fast as 402 km/h.
- There are more than 100 billion neurons in the brain alone.
- A single neuron can connect with over 1000 other neurons.

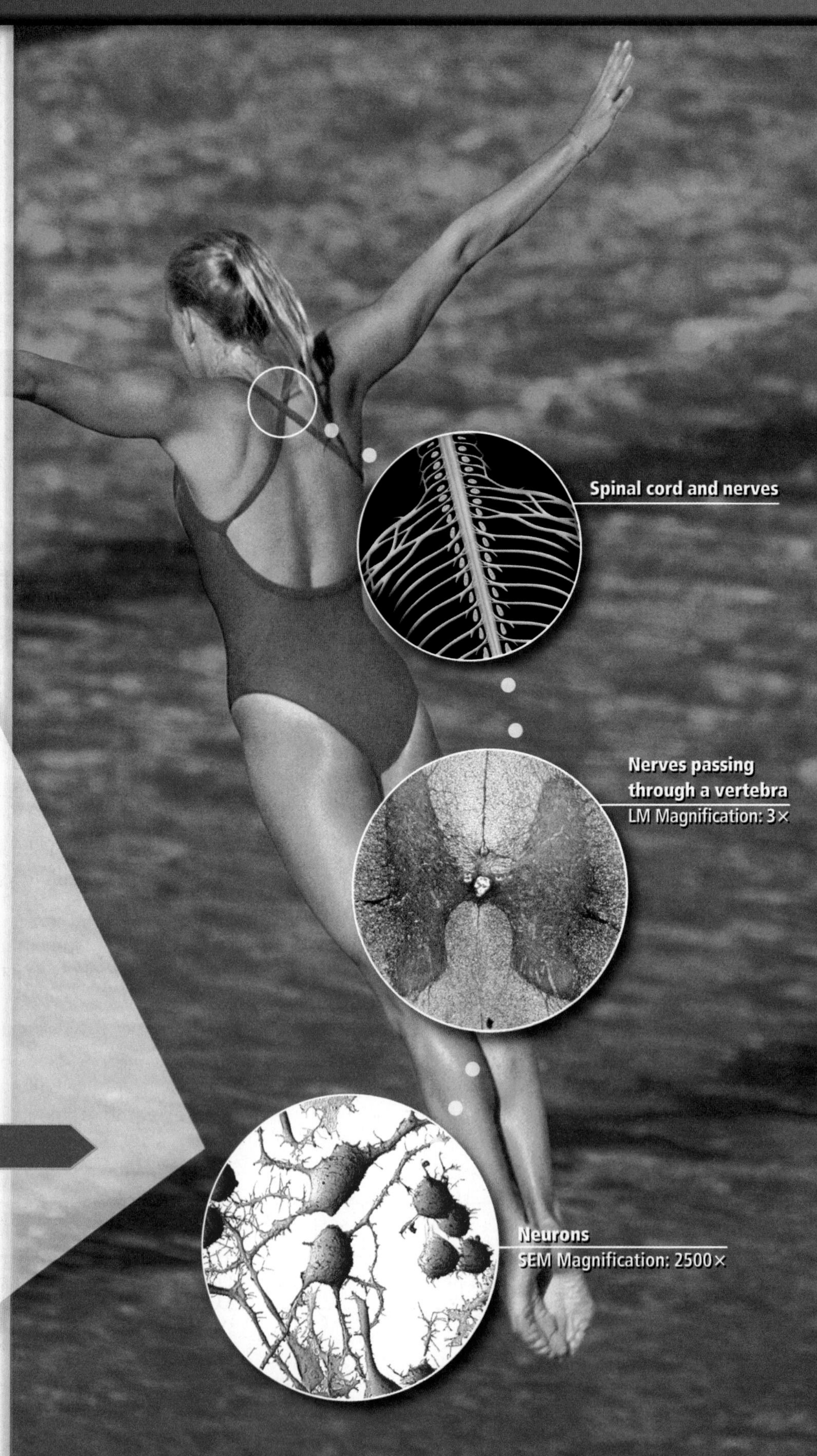

Spinal cord and nerves

Nerves passing through a vertebra
LM Magnification: 3×

Neurons
SEM Magnification: 2500×

LAUNCH Lab

How does information travel in the nervous system?

Your body is bombarded by sounds, odors, sights, tastes, and physical contact almost constantly. The nervous system makes sense of these stimuli, and reacts in ways that promote your survival. In this lab, you will model that communication process.

Procedure

1. Form groups of four and assign one student to each of the following roles: a sensor, a relayer, an interpreter, and an actor.
2. Brainstorm situations, such as touching a hot object, in which your senses receive information and you respond.
3. Model one situation. The sensor should describe what he or she senses to the relayer, who passes the information to the interpreter, who decides on a body response. The relayer then passes the response to the actor to act out the response.
4. Repeat Step 3 using different situations.

Analysis

Explain What factors could cause the situations you modeled to vary in speed?

Visit biologygmh.com to:

- study the entire chapter online
- explore the Interactive Time Line, Concepts in Motion, Interactive Tables, Microscopy Links, and links to virtual dissections
- access Web links for more information, projects, and activities
- review content online with the Interactive Tutor, and take Self-Check Quizzes

The Effects of Drugs Make this Foldable to help you understand the positive and negative effects of drugs.

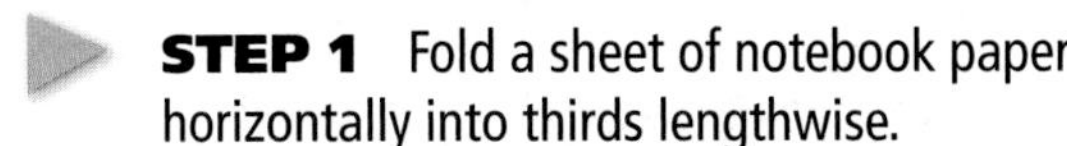

STEP 1 Fold a sheet of notebook paper horizontally into thirds lengthwise.

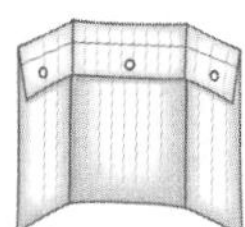

STEP 2 Open the folded paper and make a fold two inches from one of the long edges.

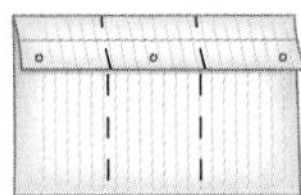

STEP 3 As illustrated, draw lines to create three columns. Label the columns as shown.

A. Increase | B. Block | C. Block

FOLDABLES Use this Foldable with Section 33.4. As you study the section, record what you learn about how drugs cause changes in the nervous system in the appropriate columns of your chart.

A column: Increase the rate at which neurotransmitters are synthesized and released.

B column: Block the transmitter from leaving the synapse.

C column: Block normal activity by mimicking other chemicals.

Reading Preview

Objectives

- **Identify** the major parts of a neuron and describe the function of each.
- **Explain** how a nerve impulse is similar to an electrical signal, and how it moves along a neuron.

Review Vocabulary

diffusion: random movement of particles from an area of higher concentration to an area of lower concentration resulting in even distribution

New Vocabulary

neuron
dendrite
cell body
axon
reflex arc
action potential
threshold
node
synapse
neurotransmitter

Structure of the Nervous System

MAIN Idea **Neurons conduct electrical impulses that allow cells, tissues, and organs to detect and respond to stimuli.**

Real-World Reading Link Imagine that you wake up in the middle of the night and get out of bed. On your way to the kitchen you stub your toe. You know right away what happened. Was it one second before you said "ouch?" Or was it less than that? How did your brain get the message so quickly?

Neurons

Electricity and chemistry were both involved as your brain received the message that you stubbed your toe. **Neurons** are specialized cells that help you gather information about your environment, interpret the information, and react to it. Neurons make up an enormous communication network in your body called the nervous system.

Figure 33.1 shows that a neuron consists of three main regions: the dendrites, a cell body, and an axon. **Dendrites** receive signals called impulses from other neurons and conduct the impulses to the cell body. Each neuron contains several dendrites. The nucleus of the neuron and many of the cell organelles are found in the **cell body.** Lastly, an **axon** carries the nerve impulse from the cell body to other neurons and muscles.

Reading Check **Relate** dendrites, axons, and cell bodies.

■ **Figure 33.1** There are three main parts of a neuron: the dendrites, a cell body, and an axon. Neurons are highly specialized cells that are organized to form complex networks.

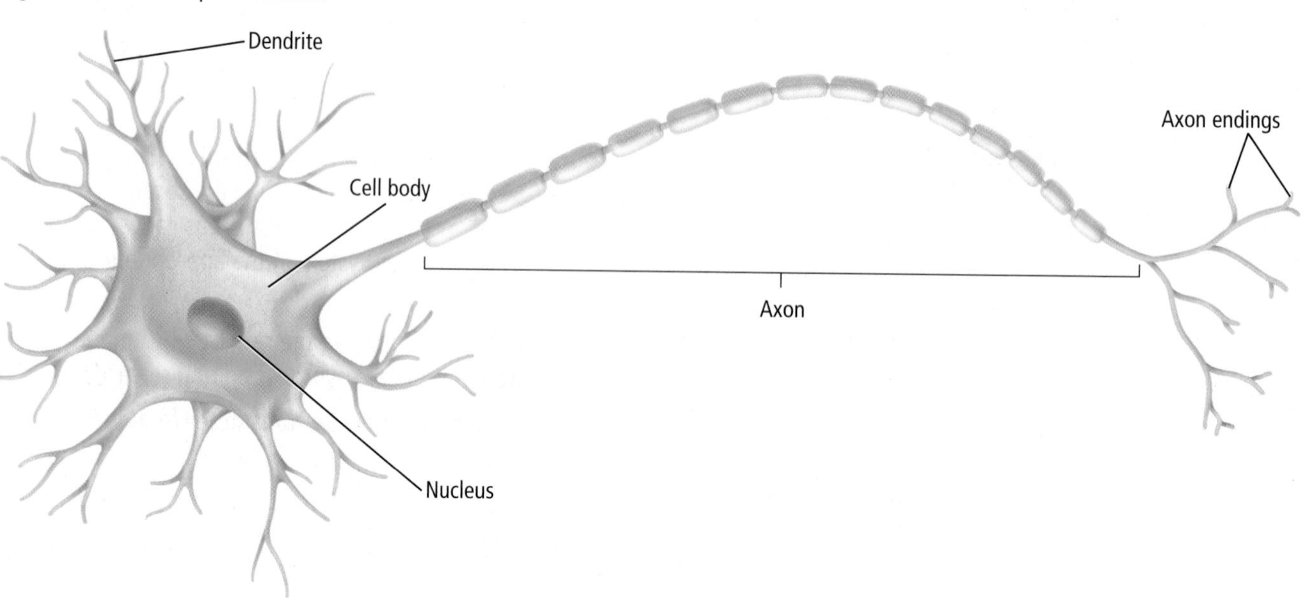

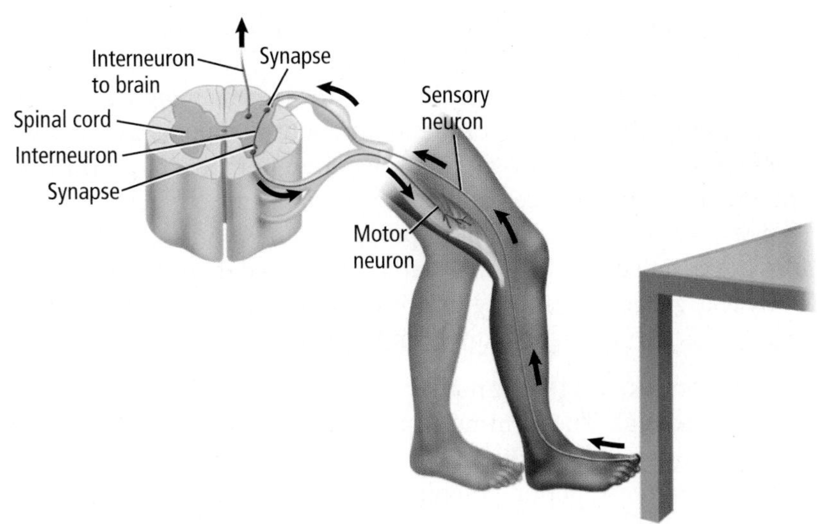

Figure 33.2 A simple reflex involves a sensory neuron, an interneuron, and a motor neuron. Interneurons can also carry impulses to the brain.
Explain ***How might a reflex be completed before the brain interprets the event?***

Concepts In Motion

Interactive Figure To see an animation of two examples of the reflex arc, visit biologygmh.com.

There are three kinds of neurons: sensory neurons, interneurons, and motor neurons. Sensory neurons send impulses from receptors in the skin and sense organs to the brain and spinal cord. Sensory neurons signal interneurons, which are found in the spinal cord and brain. Interneurons carry the impulse to motor neurons, which carry impulses away from the brain and spinal cord to a gland or muscle, which results in a response. Refer to **Figure 33.2** to follow the path of an impulse for a simple, involuntary reflex. The nerve impulse completes what is called a reflex arc. A **reflex arc** is a nerve pathway that consists of a sensory neuron, an interneuron, and a motor neuron. Notice that the brain is not involved. A reflex arc is a basic structure of the nervous system.

A Nerve Impulse

Connection to Physics A nerve impulse is an electrical charge traveling the length of a neuron. An impulse results from a stimulus, such as a touch or perhaps a loud bang that causes you to jump.

A neuron at rest When a neuron is at rest, as shown in **Figure 33.3,** it is not conducting an impulse. Notice that there are more sodium ions (Na^+) outside the cell than inside the cell. The reverse is true for potassium ions (K^+)—there are more potassium ions inside the cell than outside the cell.

Figure 33.3 The distribution of Na^+ and K^+ ions, and the presence of negatively charged protein molecules in the cytoplasm, keep the inside of the cell more negatively charged than the outside when a neuron is at rest.

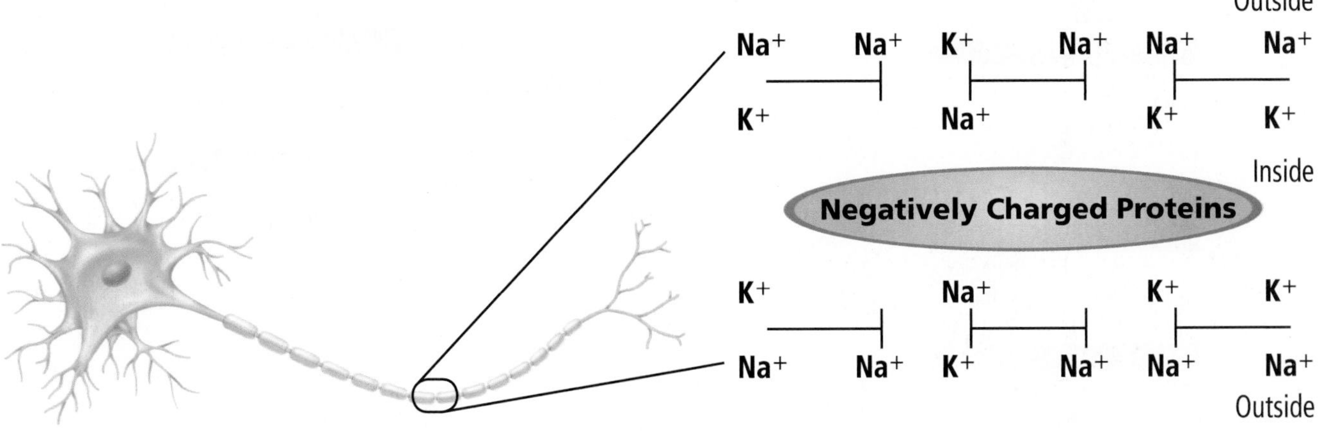

Vocabulary

Science usage v. Common usage

Channel

Science usage: a path along which information in the form of ions or molecules passes.

Nerve impulses move through neurons as channels open in the plasma membrane.

Common usage: the deeper part of a river, harbor, or strait.

Large ships move through a harbor channel.

Recall from Chapter 7 that ions tend to diffuse across the plasma membrane from an area of high concentration of ions to an area of low concentration of ions. Proteins found in the plasma membrane work to counteract the diffusion of the sodium ions and potassium ions. These proteins, called the sodium-potassium pump, actively transport sodium ions out of the cell and potassium ions into the cell.

For every two potassium ions pumped into a neuron, three sodium ions are pumped out. This maintains an unequal distribution of positively charged ions, resulting in a positive charge outside the neuron and a negatively-charged cytoplasm inside the neuron.

An action potential Another name for a nerve impulse is an **action potential.** The minimum stimulus to cause an action potential to be produced is a **threshold.** However, a stronger stimulus does not generate a stronger action potential. Action potentials are described as being "all or nothing," meaning a nerve impulse is either strong enough to travel along the neuron or it is not strong enough.

When a stimulus reaches threshold, channels in the plasma membrane open. Sodium ions rapidly move into the cytoplasm of the neuron through these channels, causing a temporary reversal in electrical charges. The inside of the cell now has a positive charge, which causes other channels to open. Potassium ions leave the cell through these channels, restoring a positive charge outside the cell. **Figure 33.4** shows that this change in charge moves like a wave along the length of the axon.

■ **Figure 33.4** Follow as an action potential moves along an axon from left to right. Notice what happens to the Na^+ and K^+ and how this changes the relative electrical charges inside and outside the neuron.

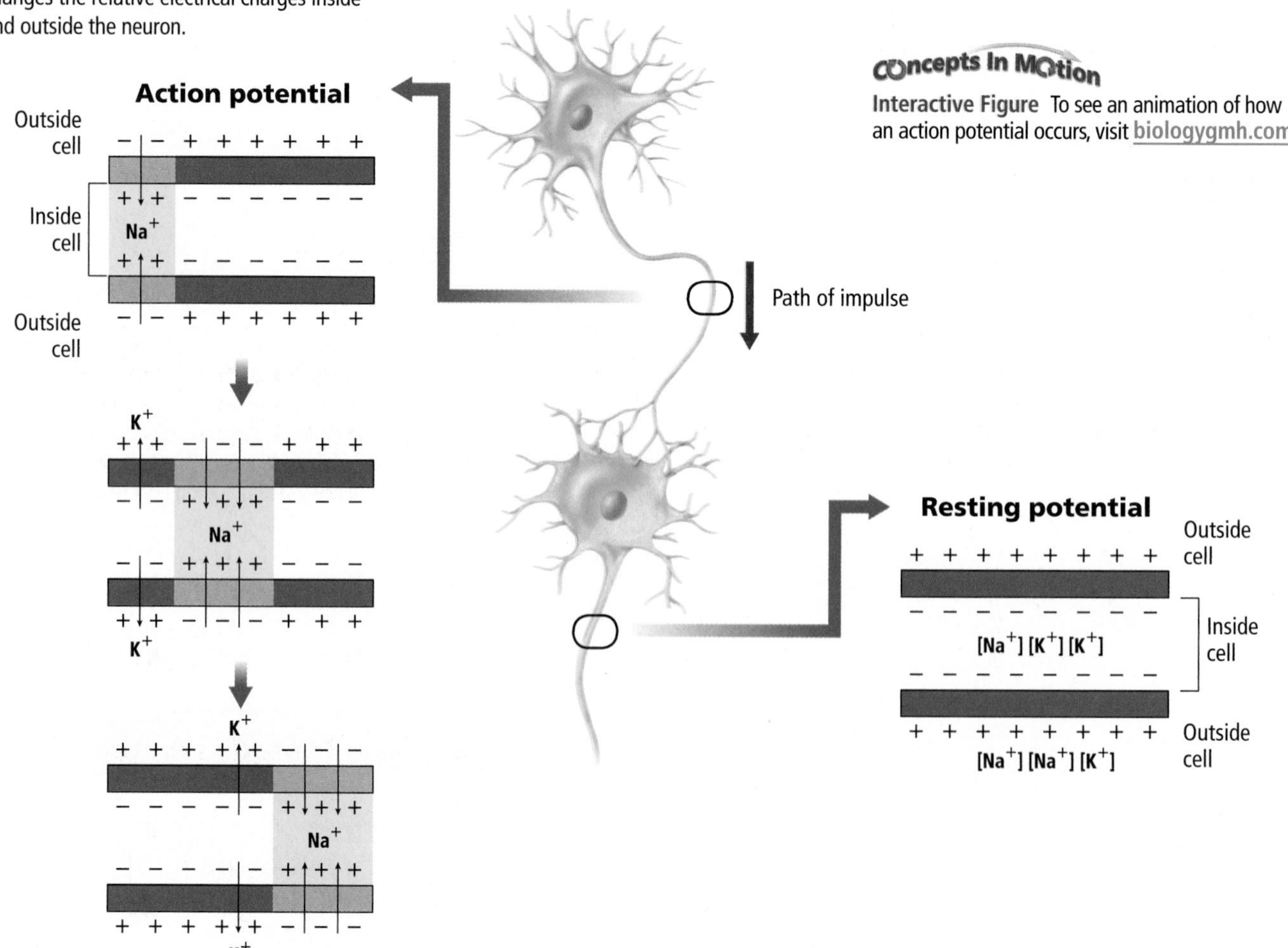

Concepts In Motion

Interactive Figure To see an animation of how an action potential occurs, visit biologygmh.com.

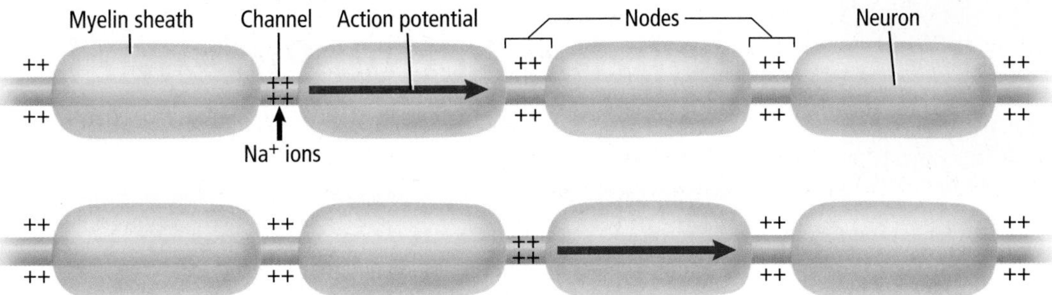

■ **Figure 33.5** A nerve impulse moves from node to node along myelinated axons.
Explain ***What happens at a node when an impulse moves along a myelinated axon?***

Speed of an action potential The speed of an action potential varies. Many axons have a covering of a lipid called myelin, which forms an insulating layer called a sheath around the axon. The myelin sheath has many gaps, called **nodes,** along the length of the axon, as shown in **Figure 33.5.** Sodium ions and potassium ions cannot diffuse through myelin, but they can reach the plasma membrane at these nodes. This allows the action potential to jump from node to node, greatly increasing the speed of the impulse as it travels the length of the axon.

In the human body, there are neurons that have myelin, and neurons that do not have myelin. Neurons with myelin carry impulses that are associated with sharp pain; neurons that lack myelin carry impulses associated with dull, throbbing pain. The action potentials in these neurons travel much more slowly than they do in neurons with myelin. When you stubbed your toe, which kind of neurons were involved?

 **Reading Check** **Explain** the relationship of a threshold to an action potential.

LAUNCH Lab

Review **Based on what you've read about action potentials, how would you now answer the analysis questions?**

MiniLab 33.1

Investigate the Blink Reflex

What factors affect the blink reflex? Have you ever been in a car when an object hit the windshield? You probably blinked. The blink reflex, in which the eye closes and opens again rapidly, is an involuntary response to stimuli the brain interprets as harmful. Nerve impulses associated with the blink reflex travel short, simple pathways in milliseconds, allowing for rapid reaction time that can prevent eye damage.

Procedure

1. Read and complete the lab safety form.
2. Form a group of three. One person, the subject, should sit behind a 1 m^2 piece of **acrylic.** A second person will monitor and record the subject's responses.
3. The third person should stand 1 m from the barrier and gently toss a **table tennis ball** so that it hits the barrier.
4. Repeat Step 3 and record the subject's response after each trial.
5. Brainstorm variables that might affect the subject's response. Predict the effect of each on the blink reflex.

Analysis

Interpret Data Did the subject perceive the stimuli in each trial the same way? Explain.

Visualizing Action Potential

Figure 33.6
To cause the voluntary contraction of a muscle, a signal from the brain creates an action potential in a motor neuron. This action potential travels along the motor neuron, which leads to the release of a neurotransmitter that signals the fibers of the muscle to contract.

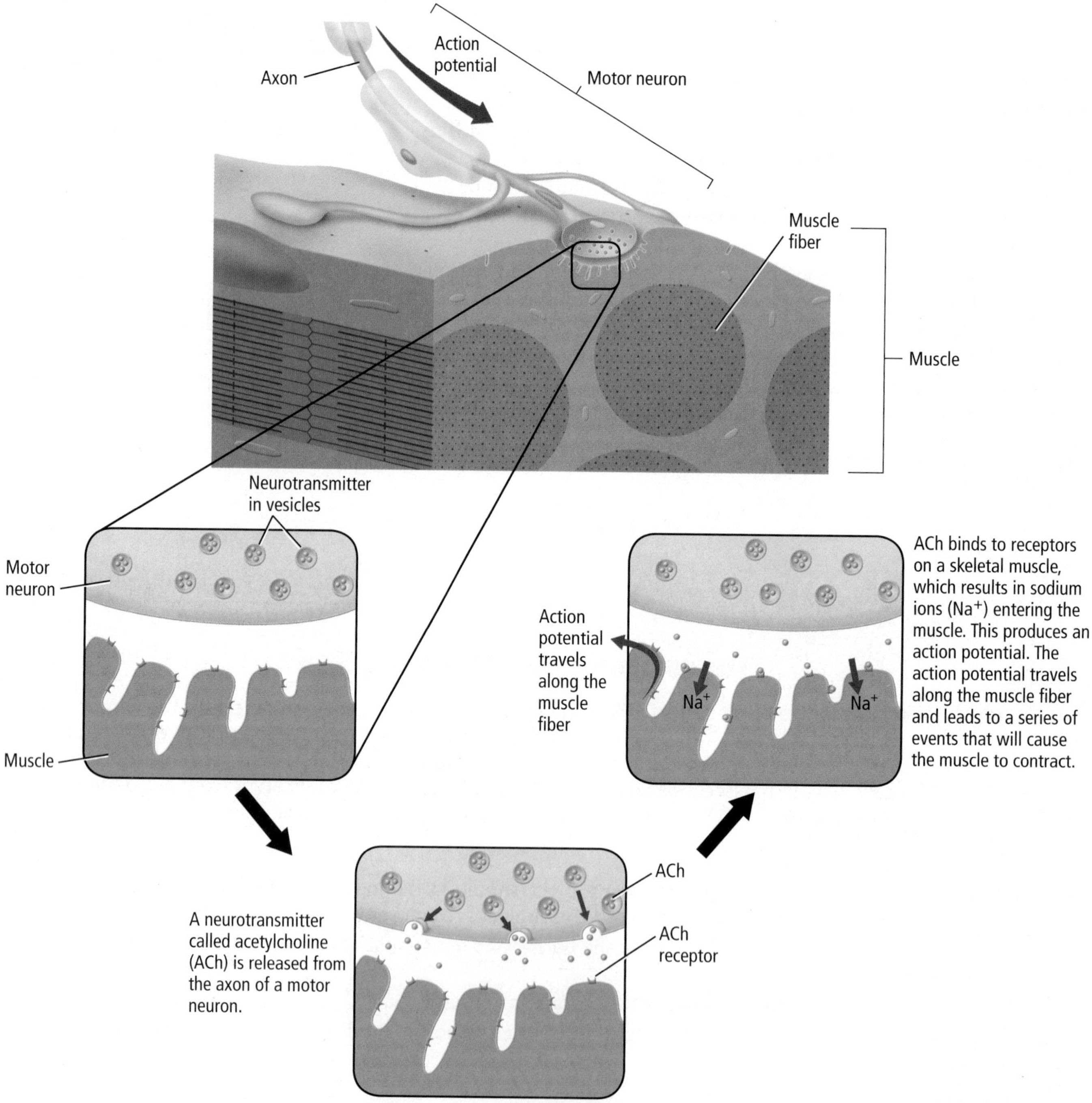

ACh binds to receptors on a skeletal muscle, which results in sodium ions (Na^+) entering the muscle. This produces an action potential. The action potential travels along the muscle fiber and leads to a series of events that will cause the muscle to contract.

A neurotransmitter called acetylcholine (ACh) is released from the axon of a motor neuron.

Concepts In Motion **Interactive Figure** To see an animation of action potentials, visit biologygmh.com.

Biology Online

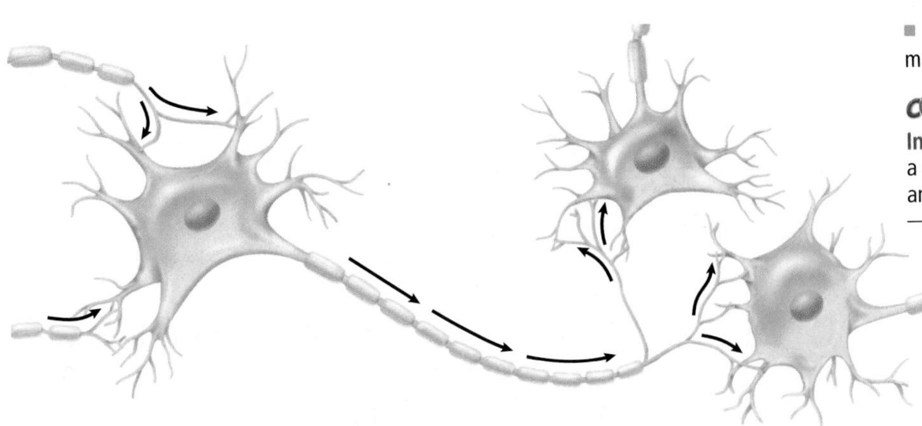

■ **Figure 33.7** A single neuron can have multiple connections with other neurons.

Concepts In Motion

Interactive Figure To see an animation of how a nerve impulse travels from one neuron to another neuron, visit biologygmh.com.

The synapse A small gap exists between the axon of one neuron and the dendrite of another neuron. This gap is called a **synapse** (SIH naps). When an action potential reaches the end of an axon, small sacs called vesicles carrying neurotransmitters fuse with the plasma membrane and release a neurotransmitter by exocytosis. When a motor neuron synapses with a muscle cell, as illustrated in **Figure 33.6,** the released neurotransmitter crosses the synapse and causes a muscle to contract.

Connection to Chemistry A **neurotransmitter** is a chemical that diffuses across a synapse and binds to receptors on the dendrite of a neighboring neuron. This causes channels to open on the neighboring cell and creates a new action potential.

There are more than 25 known neurotransmitters. Once a neurotransmitter has been released into a synapse, it does not remain there for long. Depending on the neurotransmitter, it might simply diffuse away from the synapse, or enzymes might break it down. Some neurotransmitters are recycled and used again. **Figure 33.7** shows that a single neuron can communicate with many other neurons.

Section 33.1 Assessment

Section Summary

- There are three major parts of a neuron.
- There are three basic types of neurons.
- A nerve impulse is an electric charge and is called an action potential.
- Neurons use chemicals and electricity to relay impulses.

Understand Main Ideas

1. MAIN Idea **Compare** How is the nervous system similar to the Internet as a communication network?
2. **Infer** why energy is necessary to counteract the diffusion of Na^+ and K^+ ions across the plasma membrane of a neuron.
3. **Predict** If the sensory nerves in a person's foot are nonfunctional, would the person feel pain if the foot was severely burned?

Think Critically

MATH in Biology

4. The sciatic nerve extends from the lower spinal cord to the foot. If a person's sciatic nerve is 0.914 m in length and the speed of an action potential is 107 m/s, how long will it take for a nerve impulse to travel the full distance of this nerve?
5. **Plan an experiment** that neurobiologists could use to determine that an action potential travels faster along a myelinated axon than along a nonmyelinated axon.

Section 33.2

Reading Preview

Objectives

- **Create** a flowchart that illustrates the major divisions of the nervous system.
- **Compare and contrast** the somatic nervous system with the autonomic nervous system.

Review Vocabulary

Sensory: conveying nerve impulses from the sense organs to the nerve centers

New Vocabulary

central nervous system
peripheral nervous system
cerebrum
medulla oblongata
pons
hypothalamus
somatic nervous system
autonomic nervous system
sympathetic nervous system
parasympathetic nervous system

Organization of the Nervous System

MAIN Idea The central nervous system and the peripheral nervous system are the two major divisions of the nervous system.

Real-World Reading Link Imagine you are taking a test. When you look at the first question, you are not sure how to answer it. You picture your notes. Your memory clicks and you answer the question. How does this happen?

The Central Nervous System

The nervous system consists of two major divisions. The interneurons of the brain and the spinal cord make up the **central nervous system** (CNS). The **peripheral nervous system** (PNS) consists of the sensory neurons and motor neurons that carry information to and from the CNS.

The function of the CNS is the coordination of all the body's activities. It relays messages, processes information, and analyzes responses. When sensory neurons carry information about the environment to the spinal cord, interneurons might respond via a reflex arc, or they might relay this information to the brain. Some brain interneurons send a message by way of the spinal cord to motor neurons, and the body responds. Other neurons in the brain might store the information.

Reading Check Describe the central nervous system.

Figure 33.8
Brainstorm

For thousands of years, scientists have studied the brain and investigated ways to treat neurological disease.

750 B.C. — 1800 — 1850

2000 B.C. Ancient surgeons use bronze tools to drill holes in the skull.

300 B.C. The first known human dissection is performed.

1818 Mary Wollstonecraft Shelley publishes *Frankenstein* as scientists begin to explore the connection between electricity and the nervous system.

1848 An iron rod pierces railroad worker Phineas Gage's frontal lobe. He survives, but his personality changes from quiet and hard-working to restless and aggressive.

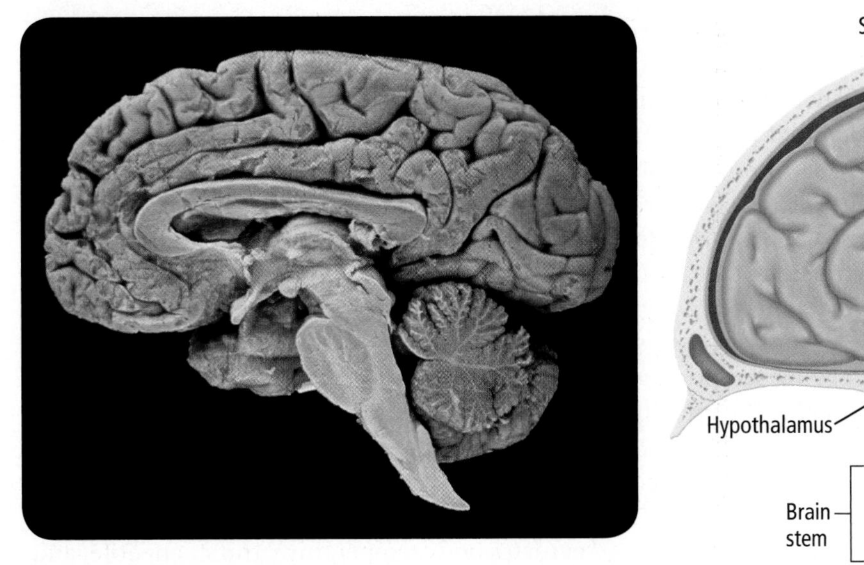

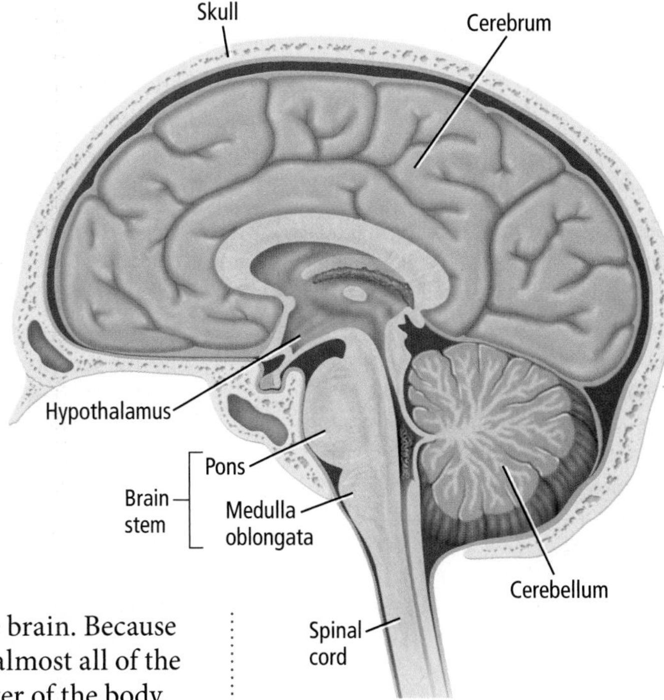

Figure 33.9
Left: A photograph of a human brain shows distinct sections.
Right: The major sections of the brain are the cerebrum, the cerebellum, and the brain stem.

The brain Over 100 billion neurons are found in the brain. Because the brain maintains homeostasis and is involved with almost all of the body's activities, it is sometimes called the control center of the body. Refer to **Figure 33.8** to learn about important events that have led to understanding of the functions of the brain.

Refer to **Figure 33.9.** The **cerebrum** (suh REE brum) is the largest part of the brain and is divided into two halves called hemispheres. The two hemispheres are not independent of each other—they are connected by a bundle of nerves. The cerebrum carries out thought processes involved with learning, memory, language, speech, voluntary body movements, and sensory perception. Most of these higher thought processes occur near the surface of the brain. The folds and grooves on the surface of the cerebrum, as shown in **Figure 33.9,** increase the surface area and allow more complicated thought processes.

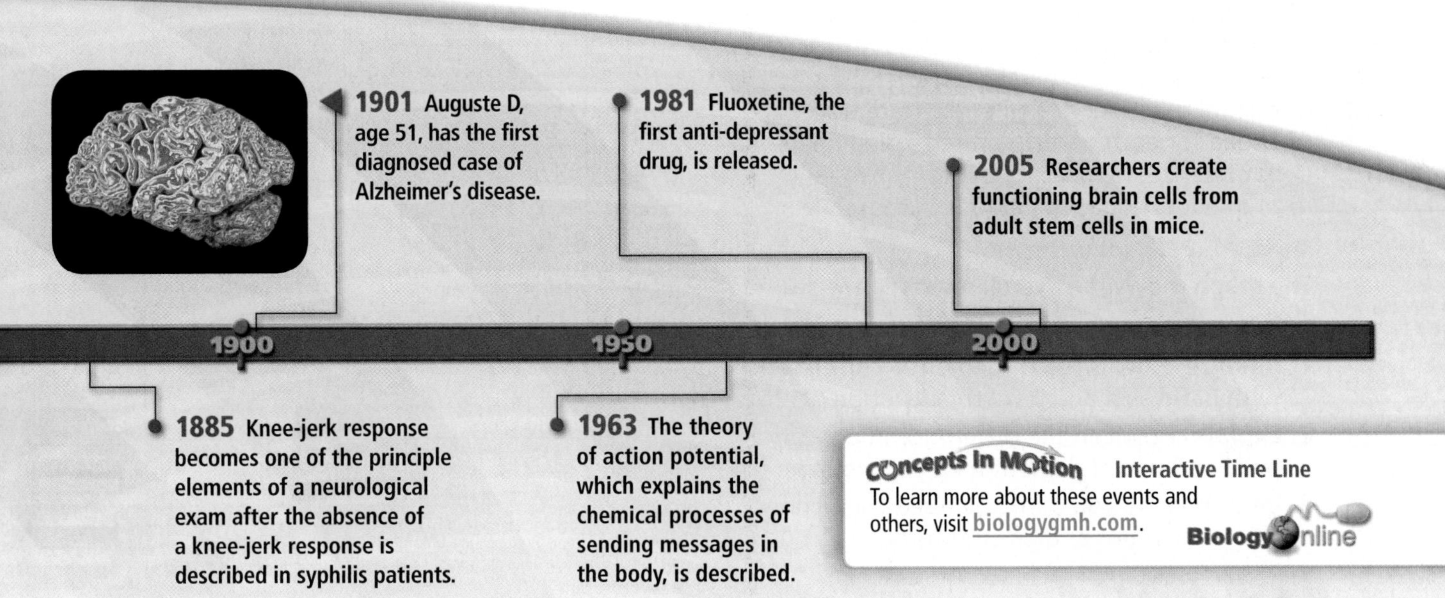

1901 Auguste D, age 51, has the first diagnosed case of Alzheimer's disease.
1981 Fluoxetine, the first anti-depressant drug, is released.
2005 Researchers create functioning brain cells from adult stem cells in mice.
1900
1950
2000
1885 Knee-jerk response becomes one of the principle elements of a neurological exam after the absence of a knee-jerk response is described in syphilis patients.
1963 The theory of action potential, which explains the chemical processes of sending messages in the body, is described.
Concepts In Motion Interactive Time Line
To learn more about these events and others, visit biologygmh.com.
Biology Online

The cerebellum controls balance, posture, and coordination, and is located at the back of the brain. The cerebellum is responsible for the smooth and coordinated movement of skeletal muscles and is also involved with motor skills, such as playing the piano or riding a bike.

The brain stem connects the brain to the spinal cord and is made up of two regions called the medulla oblongata and the pons. The **medulla oblongata** relays signals between the brain and the spinal cord. It also helps control breathing rate, heart rate, and blood pressure. The **pons** relays signals between the cerebrum and the cerebellum. The pons also helps control the rate of breathing. Have you ever felt a gagging sensation when your doctor put a tongue depressor in your mouth? The medulla oblongata contains the interneurons responsible for the swallowing, vomiting, coughing, and sneezing reflexes.

Located between the brain stem and the cerebrum, the hypothalamus is essential for maintaining homeostasis. The **hypothalamus** (hi poh THA luh mus) regulates body temperature, thirst, appetite, and water balance. It also partially regulates blood pressure, sleep, aggression, fear, and sexual behavior. It is about the size of a fingernail and performs more functions than any other brain structure of its size.

The spinal cord The spinal cord is a nerve column that extends from the brain to the lower back. It is protected by the vertebrae. Spinal nerves extend from the spinal cord to parts of the body and connect them to the central nervous system. Reflexes are processed in the spinal cord.

Reading Check **Review** the functions of the CNS.

CAREERS IN BIOLOGY

EEG Technologist EEG technologists operate electroencephalographs, instruments that record the brain's activity (brain waves). Hospitals and community colleges offer training for people who want to be eligible to work in hospitals and clinics. For more information on biology careers, visit biologygmh.com.

DATA ANALYSIS LAB 33.1

Based on Real Data*

Interpret the Data

Is there a correlation between head size, level of education, and the risk of developing dementia? In a ten-year study, 294 Catholic nuns were assessed annually for severe loss of mental function, or dementia. Data was recorded for each participant regarding head circumference—a measure of brain size—and level of education completed.

Data and Observations

The graph shows the overall results of the study.

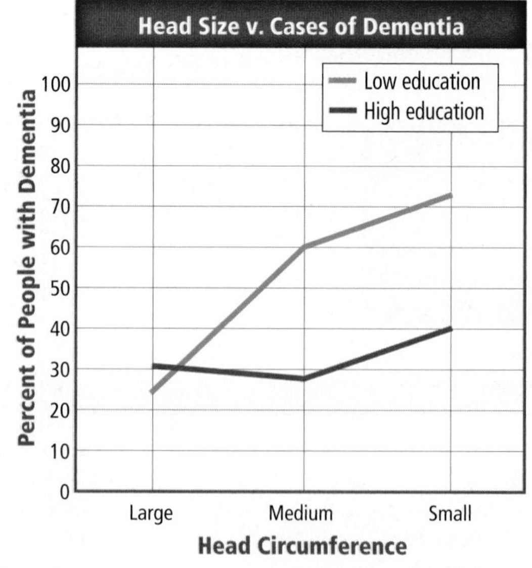

Think Critically

1. **Analyze** How is risk of dementia correlated with brain size and level of education?
2. **Explain** How can the difference in education level and risk of dementia be explained?
3. **Infer** Why do you think the researchers chose a group of nuns as their study group?

Data obtained from: Mortimer, James A., et al. 2003. Head circumference, education and risk of dementia: findings from the nun study. *Journal of Clinical & Experimental Neuropsychology* 25: 671–679.

The Peripheral Nervous System

When you hear the word nerve, you might initially think of a neuron. However, a nerve is a bundle of axons. Many nerves contain both sensory and motor neurons. For example, there are 12 cranial nerves that lead to and from the brain and 31 spinal nerves (and their branches) that lead to and from the spinal cord, as shown in **Figure 33.10.** You could think of nerves as two-way streets. Information travels to and from the brain through these sensory and motor neurons.

Refer to **Figure 33.11** as you read about the peripheral nervous system. This system includes all neurons that are not part of the central nervous system, including sensory neurons and motor neurons. Neurons in the peripheral nervous system can be classified further as being either part of the somatic nervous system or part of the autonomic nervous system.

The somatic nervous system Nerves in the **somatic nervous system** relay information from external sensory receptors to the central nervous system. Somatic motor nerves relay information from the central nervous system to skeletal muscles. Usually, this is voluntary. However, not all reactions of the central nervous system are voluntary. Some responses are the result of a reflex, which is a fast response to a change in the environment. Reflexes do not require conscious thought and are involuntary. Most signals in reflexes go only to the spinal cord and not to the brain. Remember the example of stubbing your toe? Refer back to **Figure 33.2** and note that the illustrated reflex is part of the somatic nervous system.

The autonomic nervous system Remember the last time you had a scary dream? You might have awakened and realized that your heart was pounding. This type of reaction is the result of the action of the autonomic nervous system. The **autonomic nervous system** carries impulses from the central nervous system to the heart and other internal organs. The body responds involuntarily, not under conscious control. The autonomic nervous system is important in two different kinds of situations. When you have a bad nightmare or perhaps find yourself in a scary situation, your body responds with what is known as a fight-or-flight response. When everything is calm, your body rests and digests.

Reading Check Compare and contrast voluntary responses and involuntary responses.

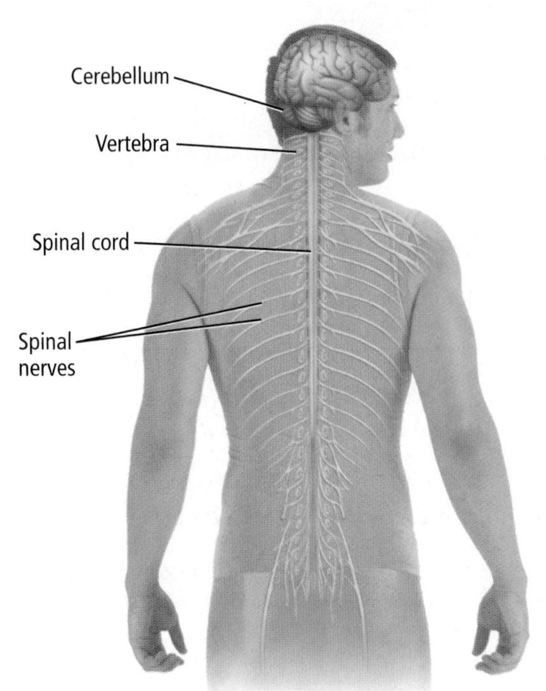

Figure 33.10 Thirty-one pairs of spinal nerves extend from the spinal cord.
Differentiate ***How is a neuron related to a nerve?***

Personal Tutor

To learn about the nervous system, visit biologygmh.com.

Figure 33.11 Each division of the nervous system functions in the control of the body and the communication within the body.

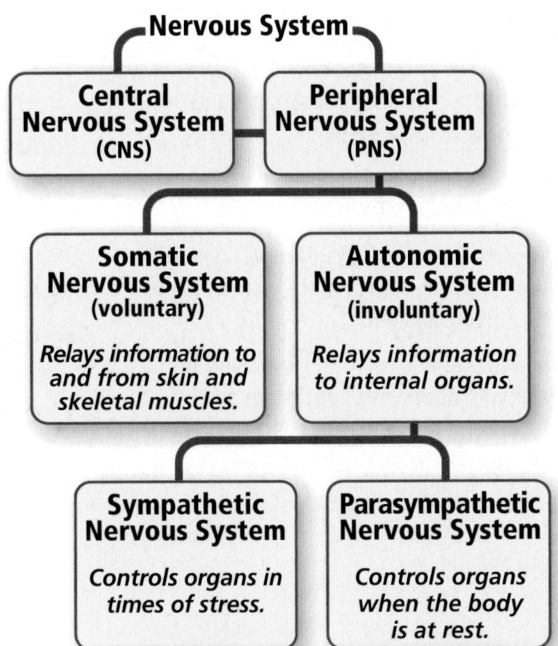

Concepts In Motion **Interactive Table** To explore more about the autonomic nervous system, visit biologygmh.com.

Table 33.1 The Autonomic Nervous System

Structure	Sympathetic Stimulation	Parasympathetic Stimulation
Iris (eye muscle)	Pupil dilation	Pupil constriction
Salivary Glands	Saliva production reduced	Saliva production increased
Oral/Nasal Mucosa	Mucus production reduced	Mucus production increased
Heart	Heart rate and force increased	Heart rate and force decreased
Lung	Bronchial muscle relaxed	Bronchial muscle contracted
Stomach	Muscle contractions reduced	Gastric juice secreted; motility increased
Small Intestine	Muscle contractions reduced	Digestion increased
Large Intestine	Muscle contractions reduced	Secretions and motility increased

Connection to Health There are two branches of the autonomic nervous system and they act together. The **sympathetic nervous system** is most active in times of emergency or stress when the heart rate and breathing rate increase. The **parasympathetic nervous system** is most active when the body is relaxed. It counterbalances the effects of the sympathetic system and restores the body to a resting state after a stressful experience. **Table 33.1** compares and contrasts the two systems. Both the sympathetic and parasympathetic systems relay impulses to the same organs, but the overall response depends on the intensities of the opposing signals.

Section 33.2 Assessment

Section Summary

- The nervous system has two major divisions—the central nervous system and the peripheral nervous system.
- The brain and spinal cord make up the central nervous system.
- The somatic nervous system and the autonomic nervous system make up the peripheral nervous system.
- The sympathetic nervous system and the parasympathetic nervous system are branches of the autonomic nervous system.

Understand Main Ideas

1. MAIN Idea **Compare** the structures of the central nervous system with the structures of the peripheral nervous system and explain their relationships.
2. **Assess** the similarities and differences between the somatic nervous system and the autonomic nervous system.
3. **Explain** Which part of the nervous system is involved in a fight-or-flight response? Why is such a response important?

Think Critically

4. **Hypothesize** What types of tests could a researcher perform to check whether different sections of the brain were functioning?
5. **Design an experiment** to demonstrate the actions of the sympathetic and parasympathetic nervous systems on the iris of the eye.

WRITING in **Biology**

6. Write a short story that describes a situation involving the heart when the sympathetic and parasympathetic nervous systems work together to maintain homeostasis.

Biology Online **Self-Check Quiz** biologygmh.com

Section 33.3

Reading Preview

Objectives

- **Identify** different sensory structures and what each is able to detect.
- **Compare** how each sense organ is able to transmit a nerve impulse.
- **Explain** the relationship between smell and taste.

Review Vocabulary

stimulus: anything in the internal or external environment that causes an organism to react

New Vocabulary

taste bud
lens
retina
rods
cones
cochlea
semicircular canal

The Senses

MAIN Idea **Sensory receptors allow you to detect the world around you.**

Real-World Reading Link Who can resist the smell of chocolate-chip cookies baking in the oven? When the aroma travels from the kitchen, you are responding to chemicals in the air. Your senses allow you to be aware of changes in your environment. You are interpreting the environment around you every second. You even were reacting to environmental stimuli before you were born.

Taste and Smell

Specialized neurons in your body called sensory receptors enable you to taste, smell, hear, see, and touch, and to detect motion and temperature.

The senses of taste and smell are stimulated by chemicals and often function together. Specialized receptors located high in the nose respond to chemicals in the air and send the information to the olfactory bulb in the brain. **Taste buds** are areas of specialized chemical receptors on the tongue that detect the tastes of sweet, sour, salty, and bitter. These receptors detect the different combinations of chemicals in food and send this information to another part of the brain.

The receptors associated with taste and smell are shown in **Figure 33.12.** Signals from these receptors work together to create a combined effect in the brain. Try eating while holding your nose. You will find that your food loses much of its flavor.

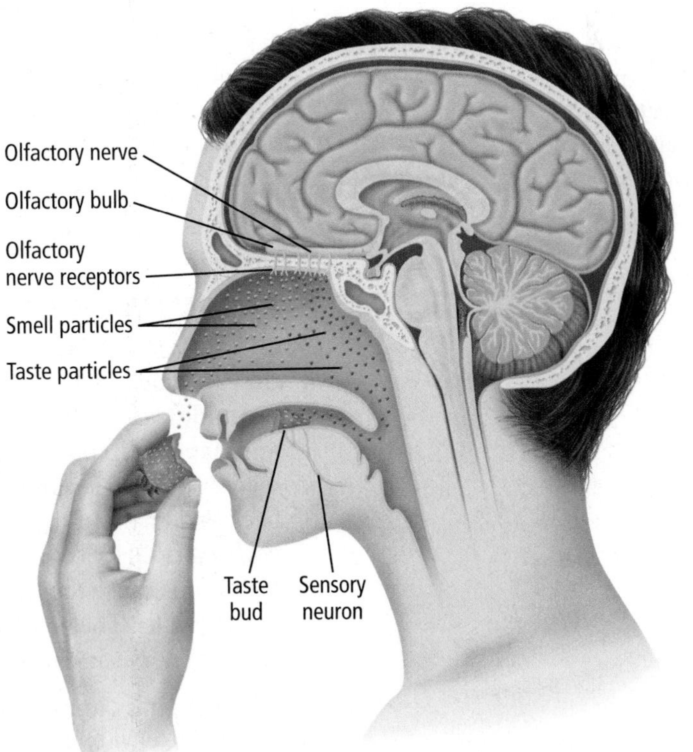

■ **Figure 33.12** The receptors of taste and smell function together and are stimulated in similar ways. Food is often smelled as it is tasted.

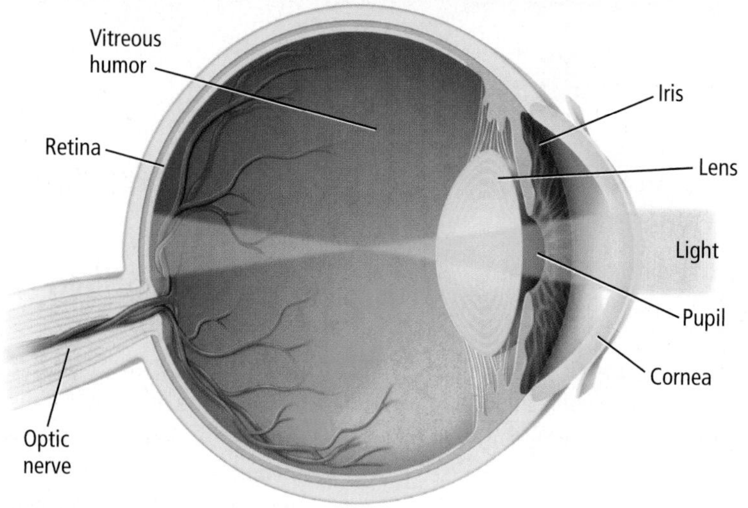

■ **Figure 33.13** Light travels through the cornea and the pupil to the lens, which focuses the image on the retina. Rods and cones in the retina send information to the brain through the optic nerve.

CAREERS IN BIOLOGY

Ophthalmologist An ophthalmologist is a medical specialist who deals with the structure, functions, and diseases of the eye. For more information on biology careers, visit biologygmh.com.

Sight

Figure 33.13 shows the path of light as it travels through the eye. Light first enters the eye through a transparent, yet durable, layer of cells called the cornea. The cornea helps to focus the light through an opening called the pupil. The size of the pupil is regulated by muscles in the iris—the colored part of the eye. Behind the iris is the **lens,** which inverts the image and projects it onto the retina. The image travels through the vitreous humor, which is a colorless, gelatinlike liquid between the lens and the retina. The **retina** contains numerous receptor cells called rods and cones. **Rods** are light-sensitive cells that are excited by low levels of light. **Cones** function in bright light and provide information about color to the brain. These receptors send action potentials to the brain via the neurons in the optic nerve. The brain then interprets the specific combination of signals received from the retina and forms a visual image.

VOCABULARY

ACADEMIC VOCABULARY

Interpret:
to explain or tell the meaning of.
Our senses help us interpret our environment.

Hearing and Balance

Hearing and balance are the two major functions of the ear. From a soft sound, like whispering, to a loud sound, such as a crowd cheering at a sporting event, specialized receptors in the ear can detect both the volume and the highness and lowness of sounds. Canals in the inner ear are responsible for your sense of balance, or equilibrium.

Hearing Vibrations called sound waves cause particles in the air to vibrate. **Figure 33.14** illustrates the path of sound waves as they travel through the ear.

Connection to Physics Sound waves enter the auditory, or ear, canal and cause a membrane, called the eardrum or tympanum, at the end of the ear canal to vibrate. These vibrations travel through three bones in the middle ear—the malleus (also called the hammer), the incus (anvil), and stapes (stirrup). As the stapes vibrates, it causes the oval window—a membrane that separates the middle ear from the inner ear—to move back and forth. In the inner ear, a snail-shaped structure called the **cochlea** (KOH klee uh) is filled with fluid and lined with tiny hair cells. Vibrations cause the fluid inside the cochlea to move like a wave against the hair cells. The hairs cells respond by generating nerve impulses in the auditory nerve and transmitting them to the brain.

Reading Check **Summarize** how each sense organ detects changes in the environment.

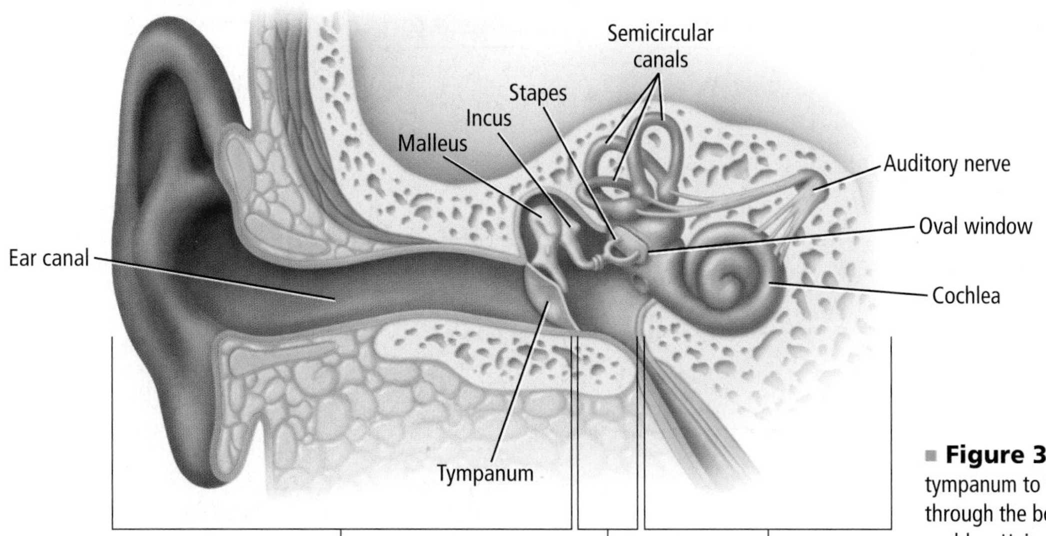

Figure 33.14 Sound waves cause the tympanum to vibrate, and the vibrations travel through the bones of the middle ear to the cochlea. Hair cells in the cochlea generate nerve impulses, which are sent to the brain through the auditory nerve.

Balance The inner ear also contains organs for balance, including three semicircular canals. **Semicircular canals** transmit information about body position and balance to the brain. The three canals are positioned at right angles to one another and, like the cochlea, they are fluid-filled and lined with hair cells. When the position of your head changes, fluid moves through the canals. This causes the hair cells to bend, which in turn sends nerve impulses to the brain. The brain then is able to determine your position and whether your body is still or in motion.

MiniLab 33.2

Investigate Adaptations to Darkness

How fast do light receptors in the retina adapt to low light conditions? The retina contains two types of receptor cells. Cones, adapted for vision in bright light, allow you to perceive color. Rods, adapted for vision in dim light, help you detect shape and movement. The brain combines and interprets nerve impulses received from these cells, making it possible for you to see in various light conditions.

1. Work with a partner. Using a **stopwatch,** time how long it takes to separate 30 **plastic bottle caps** into groups based on color.
2. Record the time, the number of caps in each group, and the percent accuracy of the grouping.
3. Predict changes in the data if the experiment is repeated in dim light.
4. Mix the caps into one group. Dim the lights. Immediately repeat Step 1.
5. Restore light conditions and record the data.
6. Discuss the data with your group. Predict changes in the data if the experiment is repeated after five minutes in dim light. Dim the lights.
7. Wait five minutes and repeat Step 1. Restore the light and record data.

Analysis

1. **Analyze** Graph the time required and the percent accuracy in each trial. How do these variables compare across trials?
2. **Think Critically** Based on the data, compare the action of the blink reflex (**MiniLab 33.1**) to the action of the eyes in adjusting to low light conditions.

■ **Figure 33.15** Many types of receptors are found in the skin. A person can tell if an object is hot or cold, sharp or smooth.

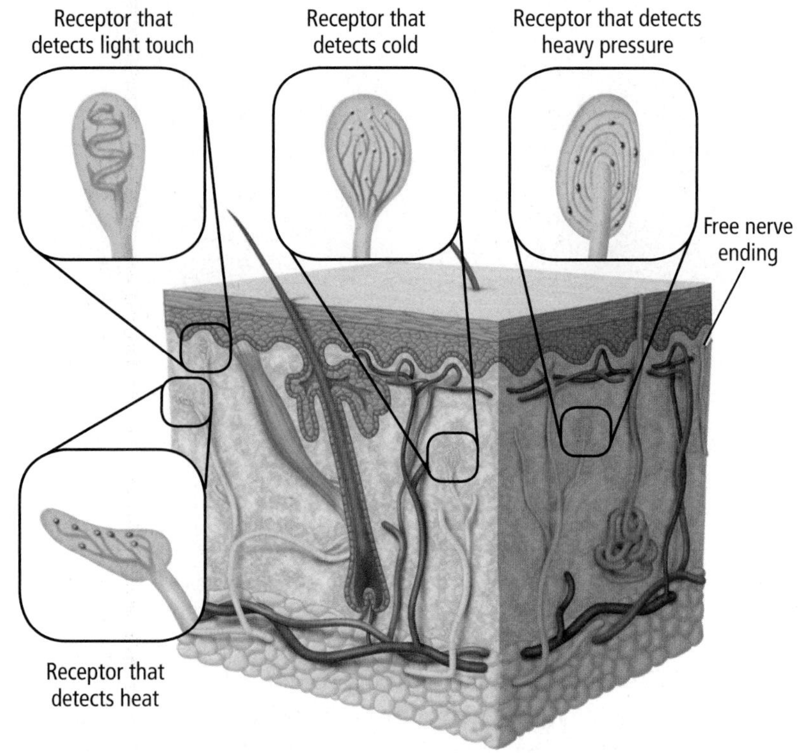

Touch

Many types of sensory receptors that respond to temperature, pressure, and pain are found in the epidermis and dermis layers of the skin. **Figure 33.15** illustrates the different types of receptors—some that respond to light touches, and others that respond to heavy pressure.

Distribution of receptors is not uniform in all areas of the body. The tips of the fingers have many receptors that detect light touch. The soles of the feet have many receptors that respond to heavy pressure. Pain receptors are simple, consisting of free nerve endings that are found in all tissues of the body except the brain. The brain constantly receives signals from these receptors and responds appropriately.

Section 33.3 Assessment

Section Summary

- The senses of taste and smell work together.
- The eye has two types of receptors.
- The ear is involved in both hearing and balance.
- The skin has many types of sensory receptors.
- Some sensory receptors are more complex than others.

Understand Main Ideas

1. MAIN Idea **Diagram** the route of a sound wave from the auditory canal until it causes a nerve impulse to be generated.
2. **Predict** what might be the result if the cornea was damaged.
3. **Analyze** the importance of the kind of receptors found in the fingers.
4. **Explain** why it might be difficult to taste when you have a cold and your nasal passages are clogged.

Think Critically

5. **Construct** an experiment to test the idea that certain areas of the tongue are taste-specific.
6. **Develop a hypothesis** People who have lost their sense of sight still experience sight occasionally. People who once could hear occasionally experience sound. Why might these phenomena occur?

Biology Online Self-Check Quiz biologygmh.com

Section 33.4

Reading Preview

Objectives

- **Identify** four ways drugs can affect the nervous system.
- **Describe** different ways drugs can harm the body or cause death.
- **Explain** how, at the cellular level, a person can become addicted to a drug.

Review Vocabulary

threshold: the minimum strength of a stimulus that causes an action potential to be generated

New Vocabulary

drug
dopamine
stimulant
depressant
tolerance
addiction

Effects of Drugs

MAIN Idea **Some drugs alter the function of the nervous system.**

Real-World Reading Link What is a drug? Some people think of substances like heroin or cocaine when they hear the term *drug*. However, some drugs are common, everyday substances. When you have a headache and take aspirin, you are taking a drug.

How Drugs Work

A **drug** is a substance, natural or artificial, that alters the function of the body. There are many types of drugs. Drugs range from prescriptions such as antibiotics, which fight bacterial infections, to over-the-counter pain relievers. There are also illegal drugs, such as cocaine and marijuana, which can cause addiction and death. Common substances such as caffeine, nicotine, and alcohol are also drugs.

Drugs affect a person's body in many different ways. Drugs that affect the nervous system work in one or more of the following ways:

- a drug can cause an increase in the amount of a neurotransmitter that is released into a synapse
- a drug can block a receptor site on a dendrite, preventing a neurotransmitter from binding
- a drug can prevent a neurotransmitter from leaving a synapse
- a drug can imitate a neurotransmitter

Concepts In Motion Interactive Table To explore more about common drugs, visit biologygmh.com.

Table 33.2 Some Common Drugs

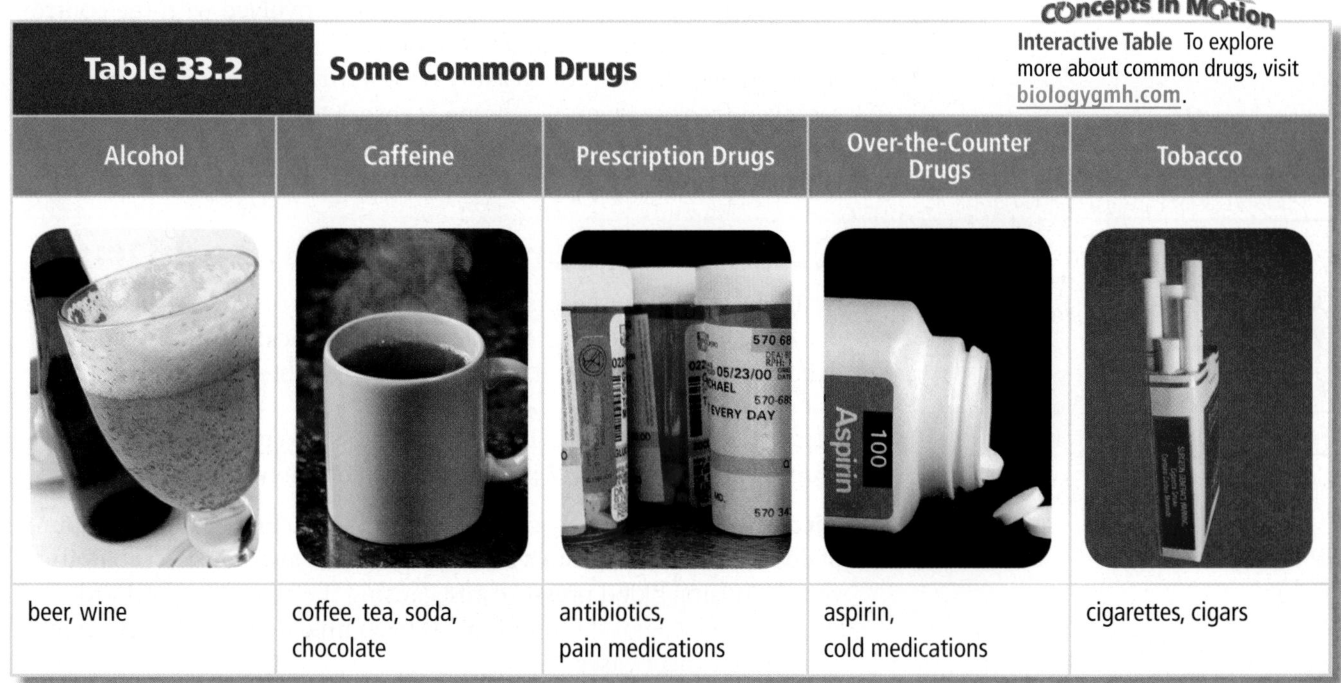

Alcohol	Caffeine	Prescription Drugs	Over-the-Counter Drugs	Tobacco
beer, wine	coffee, tea, soda, chocolate	antibiotics, pain medications	aspirin, cold medications	cigarettes, cigars

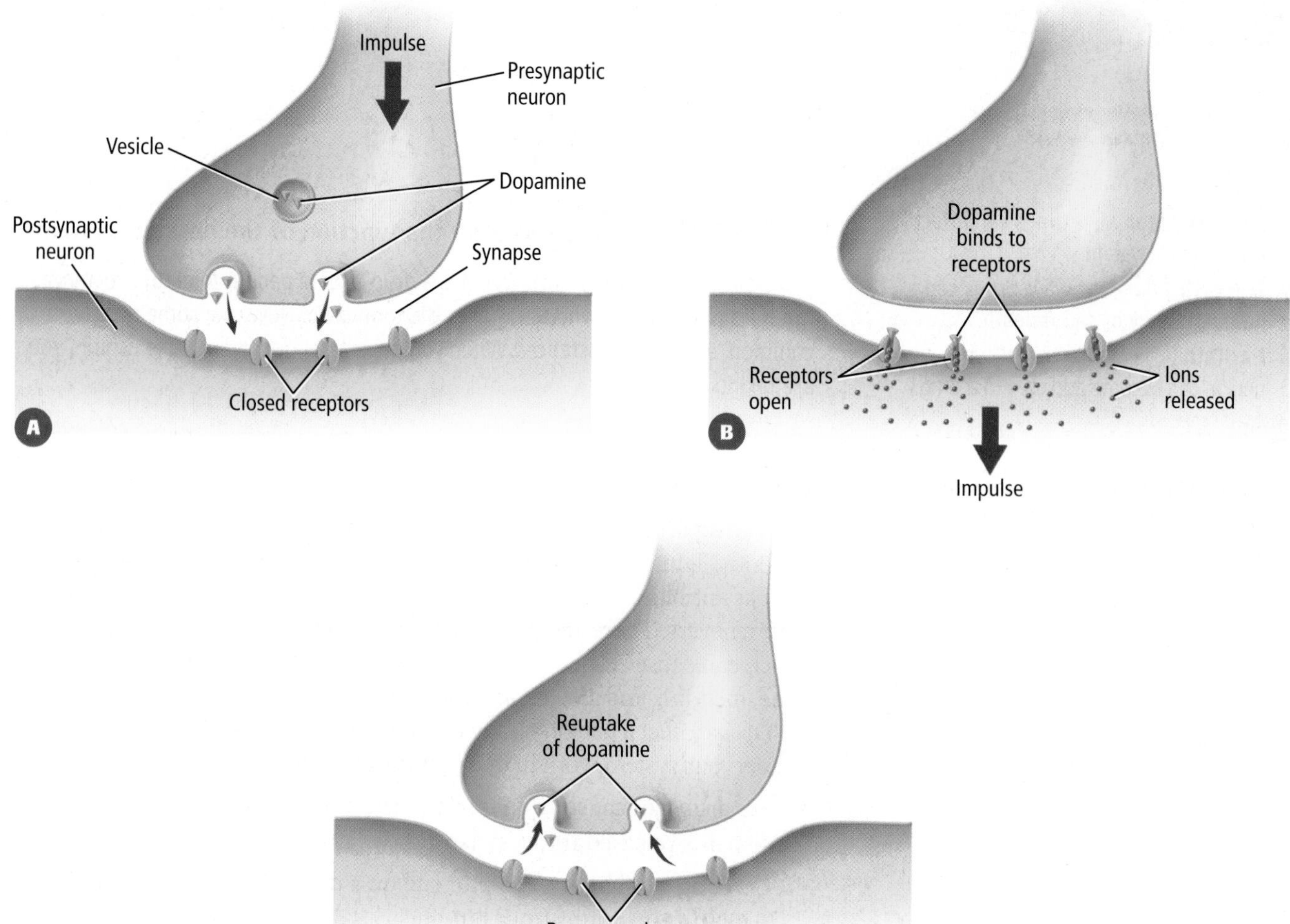

Figure 33.16 Dopamine crosses the gap from one neuron and binds to receiver sites, or receptors, on the membrane of another neuron. This occurs at a synapse.

Many drugs that affect the nervous system influence the level of a neurotransmitter called dopamine. **Dopamine** (DOH puh meen) is a neurotransmitter found in the brain that is involved with the control of body movements and other functions. Dopamine also is strongly involved with feelings of pleasure or reward. Dopamine normally is removed from a synapse by being reabsorbed by the neuron that released it, as illustrated in **Figure 33.16.**

VOCABULARY

WORD ORIGIN

Dopamine

dopa– refers to an amino acid

–amine refers to a derivative of ammonia

Classes of Commonly Abused Drugs

Drug abuse does not necessarily involve the use of illegal drugs. Any use of a drug for reasons other than legitimate medical purposes, whether deliberate or unintentional, can be considered abuse of that drug.

Stimulants Drugs that increase alertness and physical activity are **stimulants. Figure 33.17** indicates some common stimulants.

Nicotine Nicotine in cigarette and cigar smoke increases the amount of dopamine released into a synapse. Nicotine also constricts blood vessels, raising blood pressure and causing the heart to work harder than normal. Cigarette smoking has been linked to about 90 percent of all lung cancer cases.

Caffeine The most commonly used, and often abused, stimulant probably is caffeine. Caffeine is found in coffee, tea, some soft drinks, and even some foods like chocolate. Caffeine works by binding to adenosine receptors on neurons in the brain. Adenosine slows down neural activity, causing drowsiness. When caffeine binds to these receptors, it has the opposite effect. It makes users feel awake and alert. Caffeine also temporarily raises epinephrine (adrenaline) levels in the body, giving a quick burst of energy that soon wears off.

Study Tip

Chart Create a three-column chart titled *The Effects of Drugs.* Before reading this section, check the section heads, the boldfaced terms, and figures. Identify what you know about drugs in the first column. In the second column, list what you would like to know. In the third column, write what you have learned after studying this section.

Depressants Drugs that tend to slow down the central nervous system are **depressants.** These drugs can lower blood pressure, interrupt breathing, and slow the heart rate. Depressants can relieve anxiety, but they also can cause the noticeable effect of sedation.

Alcohol Alcohol is a depressant. It affects the central nervous system and is one of the most widely abused drugs in the world today. It is produced by the fermentation of grains and fruits. Alcohol is known to affect at least four different neurotransmitters, resulting in a feeling of relaxation and sluggishness. Short-term alcohol use impairs judgment, coordination, and reaction time. Long-term effects of alcohol abuse include a reduction in brain mass, liver damage, stomach and intestinal ulcers, and high blood pressure. Consumption of alcohol during pregnancy is the number-one cause of fetal alcohol syndrome, which can result in damage to a baby's brain and nervous system.

Inhalants Inhalants are chemical fumes that have an influence on the nervous system. Exposure to inhalants might be accidental due to poor ventilation. Inhalants generally work by acting as a depressant on the central nervous system. Inhalants might produce a short-term effect of intoxication, as well as nausea and vomiting. Death can occur. Long-term exposure to inhalants can cause memory loss, hearing loss, vision problems, peripheral nerve damage, and brain damage.

■ **Figure 33.17** There are many common stimulant drugs, such as coffee, tea, cocoa, and chocolate.

Illegal drugs Amphetamines and cocaine both increase dopamine levels and both prevent dopamine from being reabsorbed, so it remains in the synapse. This ultimately increases the levels of dopamine in the brain, which results in a feeling of pleasure and well-being.

The use of cocaine and amphetamines has short-term and long-term effects. Cocaine abuse might result in disturbances in heart rhythm, heart attacks, chest pain, respiratory failure, strokes, seizures, headaches, abdominal pain, and nausea. Abuse of amphetamines might result in rapid heart rate, irregular heartbeat, increased blood pressure, and irreversible, stroke-producing damage to small blood vessels in the brain. Elevated body temperature, called hyperthermia, and convulsions can result from an amphetamine or cocaine overdose, and if not treated immediately, can result in death. Abusers also can experience episodes of violent behavior, paranoia, anxiety, confusion, and insomnia. It can take a year or longer for users of methamphetamine—the strongest type of amphetamine—to recover after quitting the drug.

Marijuana is the most-used illegal drug in the United States. The active chemical in marijuana is tetrahydrocannabinol, or THC. Smoking marijuana quickly gets THC into the bloodstream where it is carried to the brain. THC binds to receptors on neurons in the brain, which produces the effect of intense pleasure. These receptors are found on neurons associated with many body activities. Short-term effects of marijuana use include problems with memory and learning, loss of coordination, increased heart rate, anxiety, paranoia, and panic attacks. Long-term smoking of marijuana might also cause lung cancer.

 **Reading Check** **Explain** the function of a neurotransmitter.

DATA ANALYSIS LAB 33.2

Based on Real Data*

Interpret the Data

Can the effects of alcohol use be observed? Two groups of students, ages 15–16, were given memory tasks to perform. Group 1 included heavy drinkers. Group 2 were nondrinkers. The images indicate typical results of comparing students from each group. The amount of the red-pink color indicates the amount of brain activity associated with performing the memory tasks.

Think Critically

1. **Describe** the difference between the brain activity of heavy drinkers and the brain activity of nondrinkers.
2. **Analyze** Based on these results, what long-term consequences might result from drinking as a teen?

Data and Observations

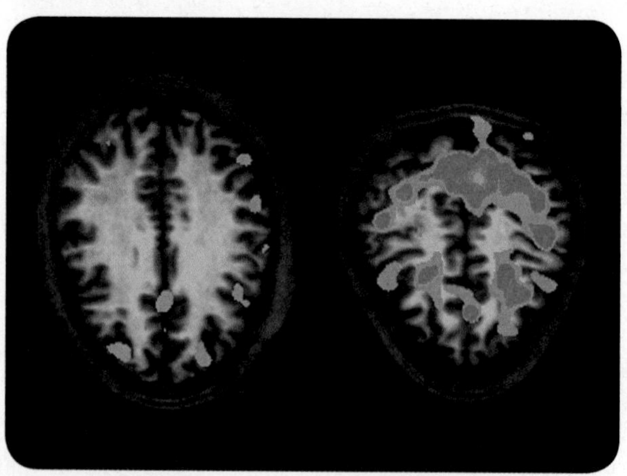

Group 1 Group 2

*Data obtained from: Brown, S.A., et al. 2000. Neurocognitive functioning of adolescents: effects of protracted alcohol use. *Alcoholism: Clinical and Experimental Research.* 24:164-171.

Tolerance and Addiction

Tolerance occurs when a person needs more and more of the same drug to get the same effect. The dosage increases because the body becomes less responsive to the drug. Drug tolerance can lead to addiction.

Addiction The psychological and physiological dependence on a drug is **addiction.** Current research suggests that the neurotransmitter dopamine is involved with most types of physiological addiction. Recall that dopamine normally is removed from a synapse as it is reabsorbed by the neuron that released it. However, certain drugs prevent that reabsorption, which results in an increase of dopamine in the brain. A person addicted to drugs derives pleasure from increased levels of dopamine and builds up a tolerance to the drug. As a result, the person takes more of the drug. When people who are addicted try to quit, the levels of dopamine decrease, making it difficult to resist going back to the drug.

Addictions can also be psychological. An individual with a psychological dependence on a drug such as marijuana has a strong desire to use the drug for emotional reasons. Both physiological and psychological dependence can affect emotional and physical health. Both types are strong, making it difficult to quit a drug.

Treatment People who are either psychologically or physiologically dependent on a drug experience serious withdrawal symptoms without it. It is very difficult for dependent users to quit on their own. They might be able to quit for short periods of time, but they are likely to use the drug again. Medical supervision is necessary when people who are psychologically and physiologically dependent on a drug try to quit.

The best way to avoid an addiction is never to use drugs in the first place, even when pressured to use them. Encourage people who abuse drugs to seek treatment for drug dependency. Physicians, nurses, counselors, clergy, and social workers are trained to direct people to the resources they need to get help, as illustrated in **Figure 33.18.**

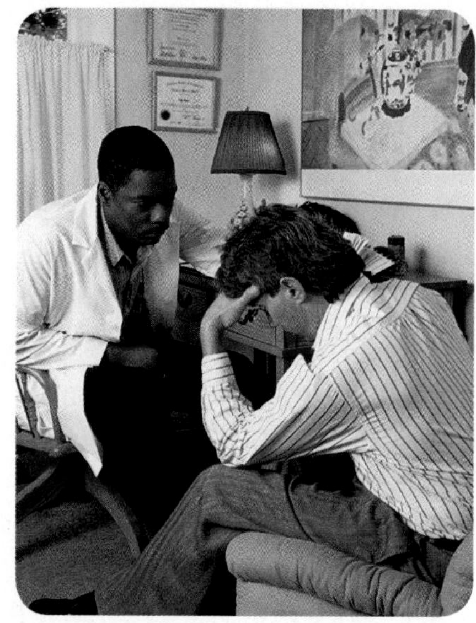

Figure 33.18 Counseling often is necessary to break an addiction.

Section 33.4 Assessment

Section Summary

- Drugs affect the nervous system in four different ways.
- Common substances such as caffeine and alcohol are considered drugs.
- Many addictive drugs increase levels of dopamine.
- Drug abuse has many negative consequences.
- A person can become psychologically and physiologically addicted to drugs.

Understand Main Ideas

1. MAIN Idea **Describe** four ways that drugs can influence the nervous system.
2. **Compare** the actions of cocaine, amphetamines, and nicotine.
3. **Explain** why the effects of stimulants and depressants do not necessarily counteract each other.
4. **Evaluate** why students who abuse amphetamines are likely to experience failing grades.

Think Critically

5. **Plan** possible treatments to help individuals overcome addiction by using your knowledge of neurotransmitters.
6. **Design an experiment** You might know that drugs affect people in different ways and at different rates. How would you design an experiment to determine the rate at which a drug is delivered to different body tissues?

CUTTING-EDGE BIOLOGY

BRAIN-CONTROLLED LIMBS: NO LONGER SCIENCE FICTION

For centuries, the only recourse for people who lose an arm or leg to accident or disease has been a prosthetic limb. These limbs help people regain some of the functions of a real arm or leg. However, their effectiveness is limited because the limbs are not controlled by the brain. Current scientific research is about to change all that.

What are brain-controlled prostheses?

Scientists currently are developing thought-controlled robotic arms with fully mobile shoulders and elbows. The hand is in the shape of a gripper that functions much like a real hand. Used primarily with monkeys and now humans in research, these arms are connected to the brain using implants.

How do the implants work? The implants are in the form of hundreds of electrodes that are as thin as a human hair. The electrodes are placed in the motor cortex of the brain 3 mm beneath the skull to pick up nerve signals in the brain. The implant transmits these signals to a computer. A mathematical procedure translates them into instructions for the arm. Within 30 milliseconds of the command, the arm can, for example, pick up food and bring it to the mouth. The arm is equipped with several motors, and moves in three dimensions just like a real arm. The arm responds and brings food to the patient when the patient thinks about the food.

During these experiments, the patient used its own arms to experiment with a joystick to get used to working with the robotic arm. Once the patient had practiced with the joystick, the scientists removed it and gently restrained the patient's own arm. To their amazement, the robotic arm began to move as a result of the patient's thoughts.

Scientists want to refine the technology so the system is completely wireless. One concern is that the current electrodes last only about six months.

How might these brain-controlled devices help society? Scientists plan to continue researching and using these devices with humans in the next few years. The hope is that these brain-computer interfaces (BCI) will help people who are parapalegic regain some movement or ability to communicate with others. Brain implants could also allow hand-free control of small robots that could perform everyday tasks. BCIs also might benefit people who are not paralyzed or who have not lost a limb. BCIs could also be used to perform tasks in dangerous environments or war zones.

WRITING in Biology

Newspaper article Based on the information from the feature and additional research at biologygmh.com, create a model of a device similar to that described in this feature. Use materials provided by your teacher or from your home. Write a 200-word description of your invention, how it works, and some benefits of this invention.

BIOLAB

HOW DO NEURAL PATHWAYS DEVELOP AND BECOME MORE EFFICIENT?

Background: **Imagine forging a narrow path through a wooded area. As the path is traveled over time, it becomes more defined and easier to follow. In a similar manner, neural pathways are developed in the brain when you learn something new. As you practice what you learned, connections between neurons strengthen, causing nerve impulses to pass more quickly and efficiently along the circuit.**

Question: *What effect do learning strategies have on the efficiency of a neural circuit?*

Materials

graph paper
paper
pencil
calculator

Procedure

1. Read and complete the lab safety form.
2. Work with one student in your group to write a list of 20 concrete words that describe specific physical objects. Assign a number, 1 to 20, to each word.
3. Read the list aloud to three other members of your group—the test subjects. Immediately, and without discussion, have them write down as many words as they can remember from the list.
4. Calculate and record the percent recall for each word: divide the number of subjects who recalled each word by the total number of subjects. Multiply by 100.
5. Graph the percent recall for each word. Note patterns in the data.
6. Calculate the average percent recall: add the percent recall for each word, divide by 20, and multiply by 100.
7. Brainstorm techniques to increase the average percent recall. Choose one technique. Predict how it will affect the average percent recall. Design an experiment to test the prediction.
8. Once your teacher approves the plan, implement it with the same test subjects, using another list of 20 concrete words that describe specific physical objects.
9. Repeat Steps 4–6 to evaluate changes in the average percent recall.

Analyze and Conclude

1. **Identify** patterns in the percent recall data after the list was read the first time. Which words were most likely to be remembered?
2. **Interpret Data** Describe the technique you used to increase the average percent recall. Compare the average percent recall before and after the technique was used.
3. **Analyze** Did the technique strengthen the neural circuits responsible for remembering the list of words as well as you predicted? Explain.
4. **Error Analysis** Identify factors, other than the technique you used, that might have affected the average percent recall.

APPLY YOUR SKILL

Design an experiment to determine if a specific learning strategy is equally effective with different test subjects. To learn more about learning strategies, visit BioLabs at biologygmh.com.

CHAPTER 33 Study Guide

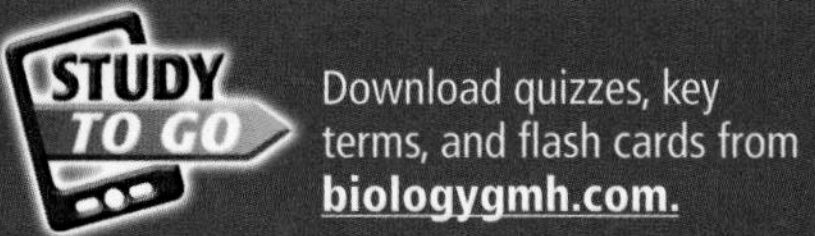

Download quizzes, key terms, and flash cards from biologygmh.com.

FOLDABLES Activity Pretend that you must develop a new drug. Explain how your drug works on the nervous system. How could you determine what side effects the drug might have?

Vocabulary	Key Concepts
Section 33.1 Structure of the Nervous System	
• action potential (p. 964) • axon (p. 962) • cell body (p. 962) • dendrite (p. 962) • neuron (p. 962) • neurotransmitter (p. 967) • node (p. 965) • reflex arc (p. 963) • synapse (p. 967) • threshold (p. 964)	**MAIN Idea** Neurons conduct electrical impulses that allow cells, tissues, and organs to detect and respond to stimuli. • There are three major parts of a neuron. • There are three basic types of neurons. • A nerve impulse is an electric charge and is called an action potential. • Neurons use chemicals and electricity to relay impulses.
Section 33.2 Organization of the Nervous System	
• autonomic nervous system (p. 971) • central nervous system (p. 968) • cerebrum (p. 969) • hypothalamus (p. 970) • medulla oblongata (p. 970) • parasympathetic nervous system (p. 972) • peripheral nervous system (p. 968) • pons (p. 970) • somatic nervous system (p. 971) • sympathetic nervous system (p. 972)	**MAIN Idea** The central nervous system and the peripheral nervous system are the two major divisions of the nervous system. • The nervous system has two major divisions—the central nervous system and the peripheral nervous system. • The brain and spinal cord make up the central nervous system. • The somatic nervous system and the autonomic nervous system make up the peripheral nervous system. • The sympathetic nervous system and the parasympathetic nervous system are branches of the autonomic nervous system.
Section 33.3 The Senses	
• cochlea (p. 974) • cone (p. 974) • lens (p. 974) • retina (p. 974) • rod (p. 974) • semicircular canal (p. 975) • taste bud (p. 973)	**MAIN Idea** Sensory receptors allow you to detect the world around you. • The senses of taste and smell work together. • The eye has two types of receptors. • The ear is involved in both hearing and balance. • The skin has many types of sensory receptors. • Some sensory receptors are more complex than others.
Section 33.4 Effects of Drugs	
• addiction (p. 981) • depressant (p. 979) • dopamine (p. 978) • drug (p. 977) • stimulant (p. 978) • tolerance (p. 981)	**MAIN Idea** Some drugs alter the function of the nervous system. • Drugs affect the nervous system in four different ways. • Common substances such as caffeine and alcohol are considered drugs. • Many addictive drugs increase levels of dopamine. • Drug abuse has many negative consequences. • A person can become psychologically and physiologically addicted to drugs.

Biology Online Vocabulary PuzzleMaker biologygmh.com

CHAPTER 33 Assessment

Section 33.1

Vocabulary Review

For each set of terms below, choose the one term that does not belong and explain why it does not belong.

1. axon—dendrite—reflex arc

2. cell body—synapse—neurotransmitter

3. myelin—node—threshold

Understand Key Concepts

Use the diagram below to answer question 4.

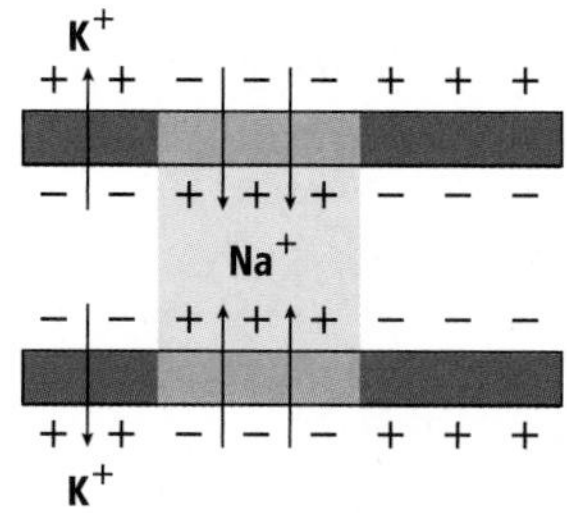

4. What is occurring in the diagram above?
A. K^+ ions are entering the neuron.
B. Negatively charged proteins are leaving the neuron.
C. Na^+ ions are entering the neuron.
D. The myelin coat has broken down, allowing ions to freely cross the plasma membrane.

5. Which is the correct path a nerve impulse will follow in a reflex arc?
A. motor neuron → interneuron → sensory neuron
B. interneuron → motor neuron → sensory neuron
C. motor neuron → sensory neuron → interneuron
D. sensory neuron → interneuron → motor neuron

Constructed Response

6. Short Answer Hypothesize why it takes more energy for a nerve impulse to travel an axon that lacks myelin as opposed to an axon that has myelin.

7. Short Answer Explain the following analogy: A neuron is like a one-way street, while a nerve is like a two-way street.

Think Critically

8. Infer In most animals, an action potential will travel only in one direction along a neuron. Infer what the result might be in humans if nerve impulses could travel in both directions on a single neuron.

Section 33.2

Vocabulary Review

For each set of terms below, choose the one term that does not belong and explain why it does not belong.

9. somatic system—parasympathetic system—sympathetic system

10. cerebrum—pons—medulla oblongata

11. autonomic nervous system—somatic nervous system—central nervous system

Understand Key Concepts

12. Which is characteristic of the sympathetic division of the autonomic system?
A. stimulates digestion
B. dilates the bronchi
C. slows the heart rate
D. converts glucose to glycogen

Use the diagram below to answer question 13.

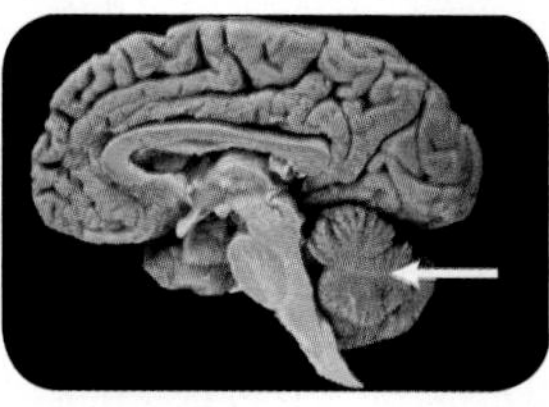

13. If the portion indicated by the arrow was damaged due to trauma, what effects would this person most likely experience?
A. partial or complete memory loss
B. body temperature fluctuations
C. trouble maintaining balance
D. rapid breathing

14. Which nervous system is the hypothalamus most involved in regulating?
A. voluntary
B. peripheral
C. sensory
D. autonomic

Constructed Response

15. Open Ended Suppose you are on the debate team at school. You must support the following statement: The autonomic nervous system is more involved with homeostasis than the somatic nervous system. Build your case.

Think Critically

16. Critique You might have heard the statement "humans use only ten percent of their brains." Use the Internet or other sources to compile evidence that either supports or refutes this idea.

17. Analyze The human cerebrum is disproportionately large compared to the cerebrum of other animals. What advantage does this give to humans?

Section 33.3

Vocabulary Review

Distinguish between the terms in each of the following sets:

18. rods—cones

19. cochlea—semicircular canals

20. retina—taste buds

Understand Key Concepts

21. If there were a power outage in a movie theater and only a few dim emergency lights were lit, which cells of the retina would be most important for seeing your way to the exit?
- **A.** rods
- **B.** cones
- **C.** Rods and cones are equally important.

22. Which represents the correct sequence as sound waves travel in the ear to trigger an impulse?
- **A.** cochlea, incus, stape, eardrum
- **B.** tympanum, bones in the middle ear, cochlea, hair cells
- **C.** auditory canal, tympanum, hair cells, cochlea
- **D.** hair cells, auditory canal, cochlea, malleus

23. With which sense are free nerve endings associated?
- **A.** taste
- **B.** hearing
- **C.** touch
- **D.** sight

Use the diagram below to answer question 24.

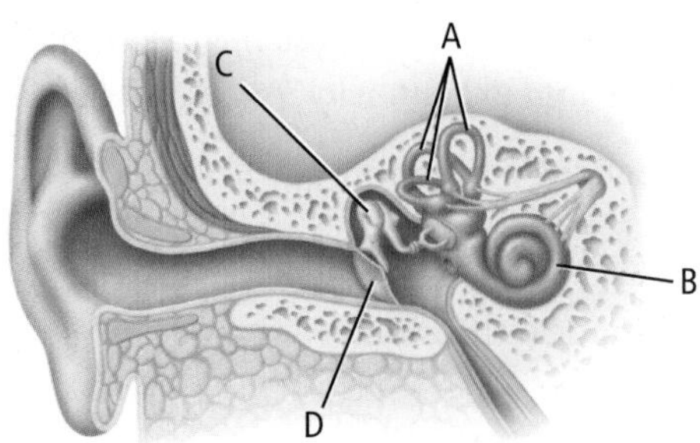

24. Some rides at amusement parks cause a person to become dizzy when the ride stops. Which structure in the diagram is most likely involved with the dizzy feeling?
- **A.** A
- **B.** B
- **C.** C
- **D.** D

Constructed Response

25. Open Ended A rare condition exists in which a person cannot feel pain. Is this desirable or undesirable? Explain your response.

Think Critically

26. Explain You have receptors for light (soft) touch all over your body. In terms of what you know about the nervous system, why are you not always conscious of things like wearing clothes or a wristwatch?

27. Categorize Rate the senses from 1 to 5 in order of importance (with 1 representing the most important.) Be prepared to debate this issue with other students in the class.

Section 33.4

Vocabulary Review

Explain the difference between the terms in each set. Then explain how the terms are related.

28. stimulants—depressants

29. tolerance—addiction

30. dopamine—drug

BiologyOnline Chapter Test biologygmh.com

Understand Key Concepts

31. Which decreases brain activity?
- **A.** nicotine
- **B.** amphetamines
- **C.** cocaine
- **D.** alcohol

32. What is the most likely function of amphetamines?
- **A.** to stimulate the sympathetic nervous system
- **B.** to stimulate the parasympathetic nervous system
- **C.** to stimulate the sympathetic and parasympathetic systems equally
- **D.** do not affect either the sympathetic or parasympathetic nervous system

Use the diagram below to answer question 33.

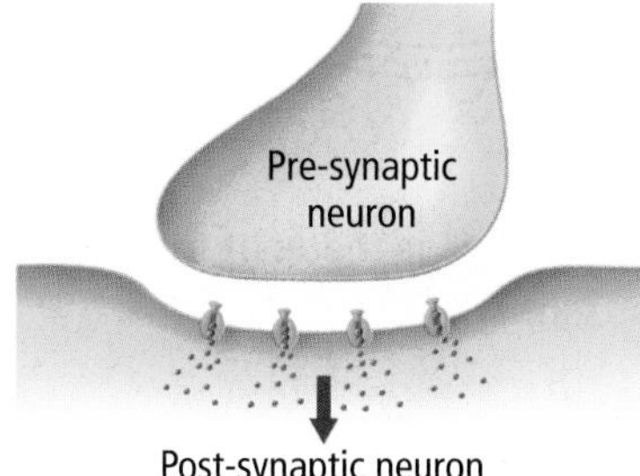

33. If a person is suffering from depression, which drug is one recommended treatment of the presynaptic neuron?
- **A.** one that increases the reuptake of dopamine
- **B.** one that increases the production of dopamine
- **C.** one that decreases the receptors for dopamine
- **D.** one that decreases the reuptake of dopamine

Constructed Response

34. Short Answer What does it mean when someone is addicted to a drug?

35. Open Ended Discuss what consequences might arise if a person's gene for the production of dopamine was defective.

Think Critically

36. Defend Form a conclusion about the following statement: "It is more difficult for someone to get addicted to drugs than it is to stop using drugs." Defend your position.

Additional Assessment

37.  WRITING in Biology Write a short story about a person who heard a loud noise and became afraid. Include in your story events that might occur in each division of the nervous system during such an experience.

DBQ Document-Based Questions

Data obtained from: Blinkov, S.M., and Glezer, I.I. 1968. *The human brain in figures and tables: a quantitative handbook*. New York: Plenum Press.
Nieuwenhuys, R., Ten Donkelaar, H.J., and Nicholson, C. 1998. *The central nervous system of vertebrates*. Vol. 3. Berlin: Springer.
Berta, A., et al. 1999. *Marine mammals: evolutionary biology*. San Diego: Academic Press.

Average Brain Mass (in grams)

Species	Mass (g)	Species	Mass (g)
Fin whale	6930	Dog (beagle)	72
Elephant	6000	Cat	30
Cow	425–458	Turtle	0.3–0.7
Adult human	1300–1400	Rat	2

38. Does there appear to be a correlation between body size and brain mass?

39. Discuss possible explanations (in terms of adaptations) that would account for your response to question 38.

Cumulative Review

40. Evaluate the role of fungi on Earth. **(Chapter 20)**

41. Examine the adaptations that have made arthropods the most evolutionarily successful animals. **(Chapter 26)**

42. Make an argument for or against the following statement: The skin should be considered an organ rather than a tissue. **(Chapter 32)**

Standardized Test Practice

Cumulative

Multiple Choice

1. Which characteristic is unique to mammals?
   A. hair
   B. endothermy
   C. four-chambered heart
   D. internal fertilization

Use the diagram below to answer questions 2 and 3.

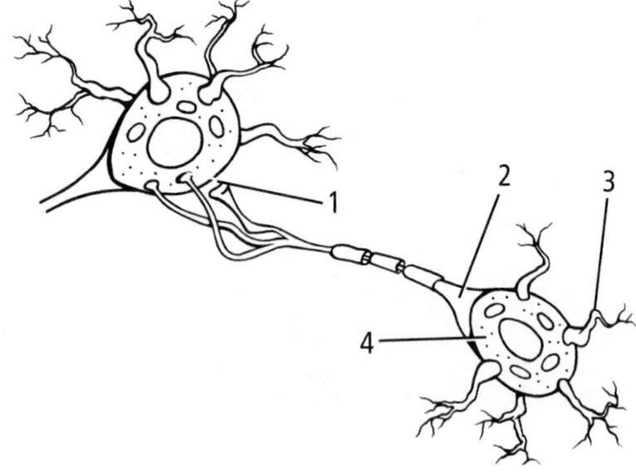

2. In which part of the diagram above would you expect to find myelin?
   A. 1
   B. 2
   C. 3
   D. 4

3. In which part of the diagram above would you expect to find neurotransmitters when an action potential reaches the end of the neuron?
   A. 1
   B. 2
   C. 3
   D. 4

4. What is the purpose of the epithelial tissue in the integumentary system?
   A. cover the body surface and protect its tissues
   B. move joints and bones
   C. provide a structural framework for the body
   D. transmit nerve signals

5. Which animal is a placental mammal?
   A. hummingbird
   B. kangaroo
   C. duck-billed platypus
   D. whale

Use the diagram below to answer questions 6 and 7.

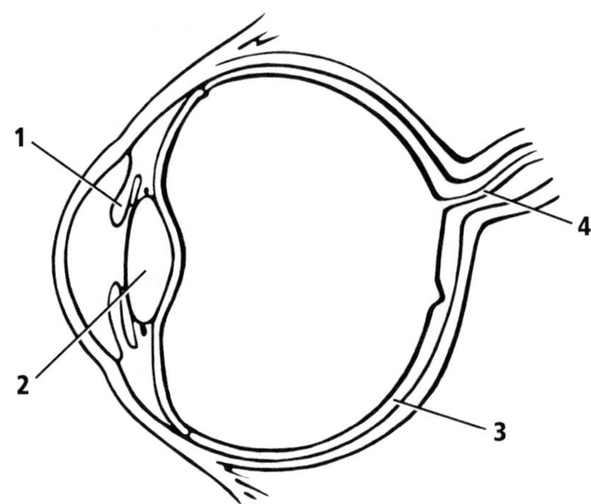

6. Which part of the eye is made of muscles that respond to stimuli?
   A. 1
   B. 2
   C. 3
   D. 4

7. If a person cannot see certain colors, what part of the eye might be damaged?
   A. 1
   B. 2
   C. 3
   D. 4

Use the graph below to answer question 8.

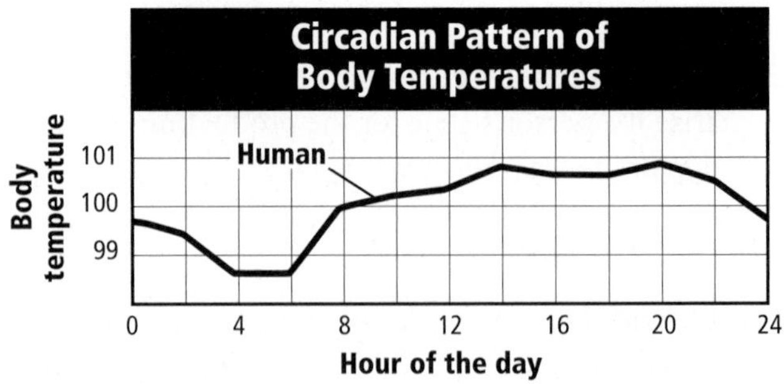

8. The graph above shows the circadian pattern of body temperature in humans. When does the body temperature of humans seem to be the lowest?
   A. after eating
   B. in the afternoon
   C. just before dawn
   D. late at night

Short Answer

Use the diagram below to answer questions 9 and 10.

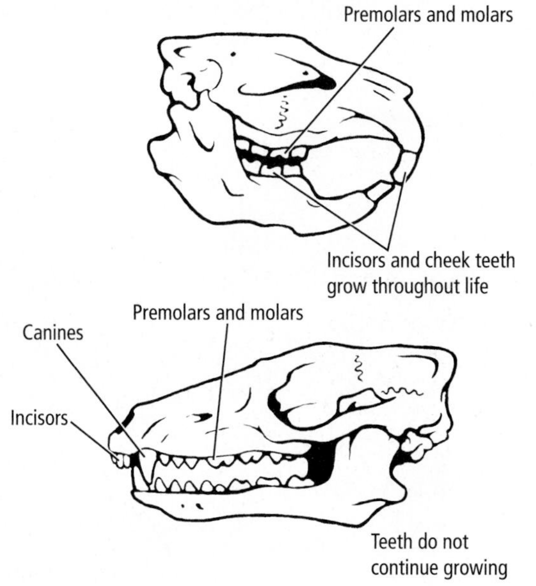

9. The figure above shows the teeth of two different types of mammals. From these teeth, what can you infer about the diets of these mammals?

10. Which animal's teeth most closely resemble those of humans? Explain your answer.

11. Explain how spiders predigest their food and compare this process to the digestion process of another animal with which you are familiar.

12. Suppose that a person who used to drink one cup of coffee to stay awake at night finds she needs to drink two cups. What is the name of this phenomenon and what causes it?

13. What is the role of the gametophyte generation in seed plants?

Extended Response

14. Two abandoned whooping crane chicks are found several days after they had hatched. A scientist wants to raise the chicks. To make the chicks feel comfortable, the scientist uses a hand puppet that looks like a whooping crane. The scientist offers the chicks mealworms but they will not take them. Formulate a hypothesis that gives a possible explanation of the actions of the chicks.

15. How are the actions of myosin and actin fibers related to the contraction of a muscle?

16. What is the main difference between segmented worms and other worms? What is the importance of this difference?

Essay Question

Each year doctors perform more than 450,000 joint repair and replacement surgeries. This surgery reduces pain and increases movement in the joints. Joint repair surgery involves removing any debris or excess bone growth from around the joint. This restores the functioning of the joint. Joint replacement surgery involves replacing the joint with a synthetic joint. The synthetic joint is made of polyethylene, ceramic, or metal. Joint replacement enables the joint to function in the same way as a natural joint. Joint replacements usually are performed on the knee, hip, or shoulder.

Using the information in the paragraph above, answer the following question in essay format.

17. Doctors usually only replace knee or hip joints on older patients who are less active than younger patients. Explain why doctors recommend this.

NEED EXTRA HELP?																	
If You Missed Question . . .	1	2	3	4	5	6	7	8	9	10	11	12	13	14	15	16	17
Review Section . . .	30.1	33.1	33.1	32.1	30.2	33.3	33.3	30.1	30.2	30.2	26.1	33.2	21.4	31.1	32.3	25.1	32.2

CHAPTER 34

Circulatory, Respiratory, and Excretory Systems

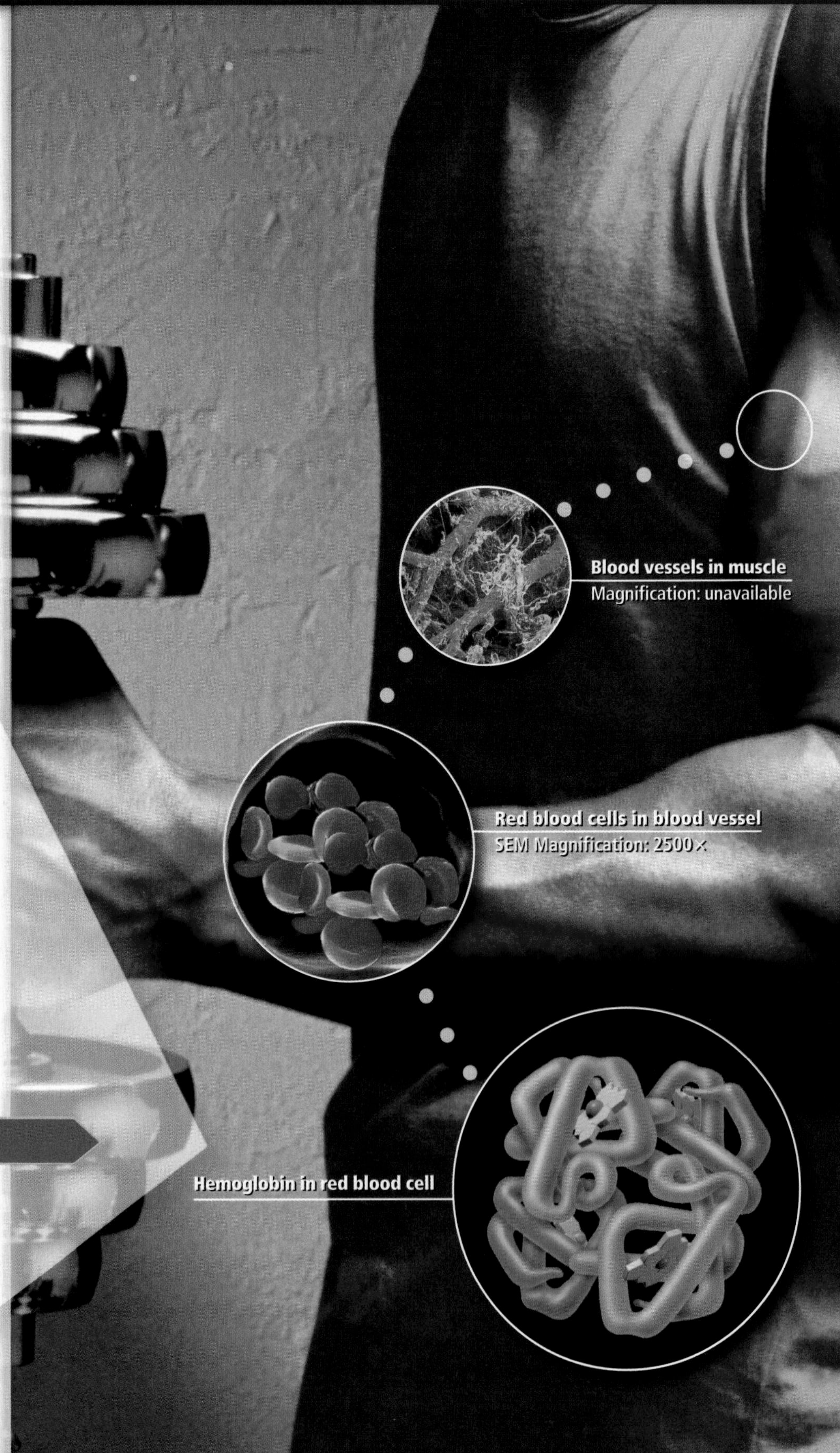

BIG Idea **These systems function together to maintain homeostasis by delivering important substances to the body's cells while removing wastes.**

Section 1
Circulatory System
MAIN Idea The circulatory system transports blood to deliver important substances, such as oxygen, to cells and to remove wastes, such as carbon dioxide.

Section 2
Respiratory System
MAIN Idea The function of the respiratory system is the exchange of oxygen and carbon dioxide between the atmosphere and the blood and between the blood and the body's cells.

Section 3
Excretory System
MAIN Idea The kidneys maintain homeostasis by removing wastes and excess water from the body and by maintaining the pH of blood.

BioFacts

- The only tissue in the human body that lacks blood vessels is the cornea of the eye.
- The surface area of the air sacs in your lungs could cover a tennis court.

LAUNCH Lab

What changes take place in the body during exercise?

Body systems, including the respiratory and circulatory systems, function together to meet the demands of exercise and to maintain homeostasis. For example, red blood cells circulate throughout the body to deliver oxygen to cells, where it is used to help produce the energy required for exercise. In this lab, you will investigate how body system responses to exercise might be related to each other.

Procedure

1. Read and complete the lab safety form.
2. Do a rhythmic exercise, such as jogging or marching in place, for two minutes. As you exercise, note how your body responds.
3. Make a list of the body system responses you identified as you exercised.

Analysis

1. **Create** a flowchart showing how these body responses might be related to each other.
2. **Analyze** Propose how one of the body system responses on your list helps regulate the body's internal environment.

Visit biologygmh.com to:

- study the entire chapter online
- explore the Interactive Time Line, Concepts in Motion, Microscopy Links, Virtual Labs, and links to virtual dissections
- access Web links for more information, projects, and activities
- review content online with the Interactive Tutor, and take Self-Check Quizzes

ABO Blood Types Make the following Foldable to help you identify the four ABO blood groups: A, B, AB, and O.

STEP 1 Fold a sheet of 11" × 17" paper into thirds lengthwise as shown.

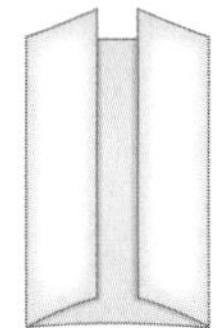

STEP 2 Fold your paper in half. Crease the fold well.

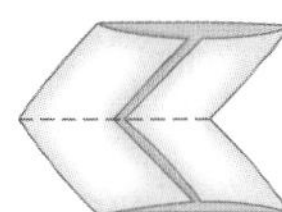

STEP 3 Open the two vertical tabs and cut along the creases.

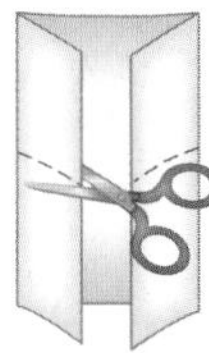

STEP 4 Label the four tabs as shown.

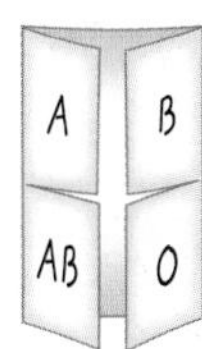

FOLDABLES Use this Foldable with Section 34.1. As you study the section, record what you learn about each of the four ABO blood groups.

Section 34.1

Reading Preview

Objectives

- **Identify** the main functions of the circulatory system.
- **Diagram** the flow of blood through the heart and body.
- **Compare and contrast** the major components of the blood.

Review Vocabulary

muscle contraction: muscle cells or fibers shorten in response to stimuli

New Vocabulary

artery
capillary
vein
valve
heart
pacemaker
plasma
red blood cell
platelet
white blood cell
atherosclerosis

Circulatory System

MAIN Idea The circulatory system transports blood to deliver important substances, such as oxygen, to cells and to remove wastes, such as carbon dioxide.

Real-World Reading Link Fast-moving highway traffic gets people to and from work quickly. Similarly, blood flowing in your body supplies nutrients and removes waste products quickly. When either traffic or blood flow is blocked, normal functions slow down or stop.

Functions of the Circulatory System

Cells must have oxygen and nutrients and must get rid of waste products. This exchange is done by the circulatory system—the body's transport system. The circulatory system consists of blood, the heart, blood vessels, and the lymphatic system. Blood carries important substances to all parts of the body. The heart pumps blood through a vast network of tubes inside your body called blood vessels. The lymphatic system is considered part of the circulatory and immune systems. You will learn about the lymphatic system in Chapter 37. All of these components work together to maintain homeostasis in the body.

The circulatory system transports many important substances, such as oxygen and nutrients. The blood also carries disease-fighting materials produced by the immune system. The blood contains cell fragments and proteins for blood clotting. Finally, the circulatory system distributes heat throughout the body to help regulate body temperature.

Figure 34.1
From Cadavers to Artificial Hearts

The human circulatory system has been studied for thousands of years, leading to great advances in medical technology.

350 B.C. Greek physician Praxagoras recognizes that veins and arteries are two different kinds of vessels.

1628 The first accurate description is made of the human heart—a pump that circulates blood in a one-way system.

1500 1600 1900

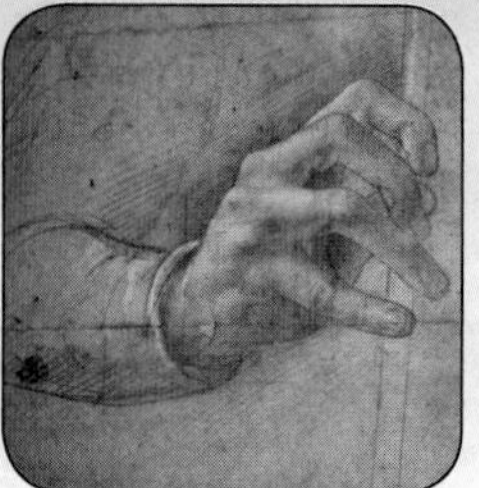

1452–1519 Leonardo da Vinci conducts extensive research on human cadavers. It is believed that he dissected about 30 corpses in his lifetime.

1903 The first electrocardiograph records the electrical activity of the heart.

Blood Vessels

Highways have lanes that separate traffic. They also have access ramps that take vehicles to and from roads. Similarly, the body also has a network of channels—the blood vessels. Blood vessels circulate blood throughout the body and help keep the blood flowing to and from the heart. The fact that there are different kinds of blood vessels was first observed by Greek physician Praxagoras, as noted in **Figure 34.1.** The three major blood vessels are arteries, capillaries, and veins, as illustrated in **Figure 34.2.**

Arteries Oxygen-rich blood, or oxygenated blood, is carried away from the heart in large blood vessels called **arteries.** These strong, thick-walled vessels are elastic and durable. They are capable of withstanding high pressures exerted by blood as it is pumped by the heart.

As shown in **Figure 34.2,** arteries are composed of three layers: an outer layer of connective tissue, a middle layer of smooth muscle, and an inner layer of endothelial tissue. The endothelial layer of the artery is thicker than that of the other blood vessels. The endothelial layer of arteries needs to be thicker because blood is under higher pressure when it is pumped from the heart into the arteries.

Capillaries Arteries branch through the body like the branches of a tree, becoming smaller in diameter as they grow farther away from the main vessel. The smallest branches are capillaries. **Capillaries** are microscopic blood vessels where the exchange of important substances and wastes occurs. Capillary walls are only one cell thick, as illustrated in **Figure 34.2.** This permits easy exchange of materials between the blood and body cells, through the process of diffusion. These tubes are so small that red blood cells move single-file through these vessels.

The diameter of blood vessels changes in response to the needs of the body. For example, when you are exercising, muscle capillaries will expand, or dilate. This increases blood flow to working muscles, which brings more oxygen to cells and removes extra wastes from cells.

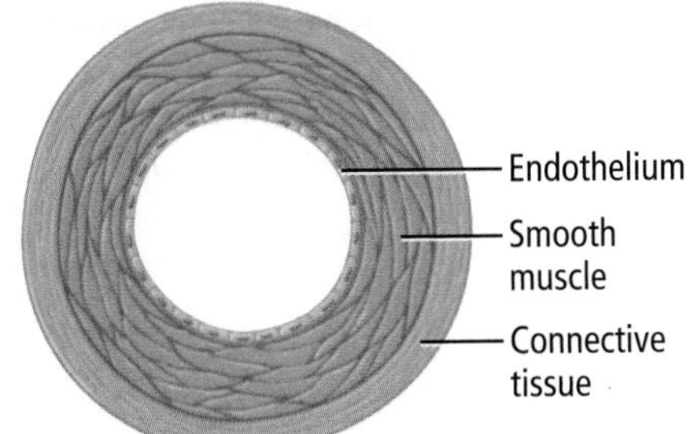

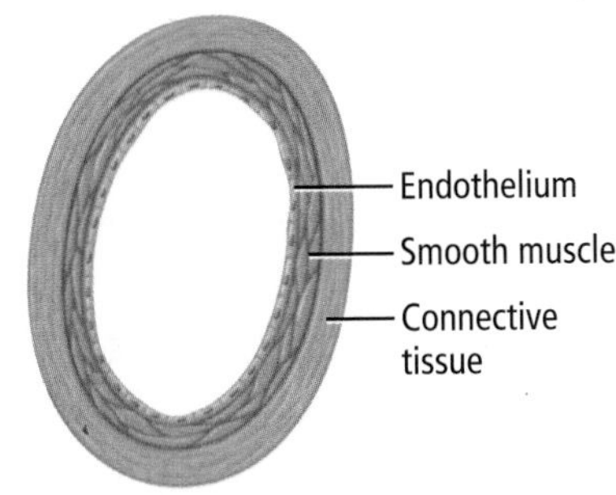

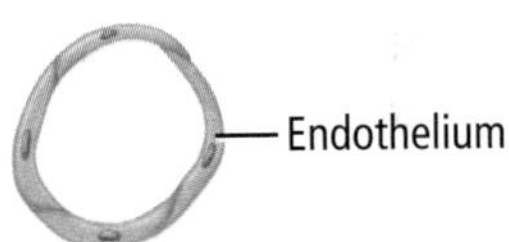

Figure 34.2 The three major blood vessels in the body are arteries, veins, and capillaries.
Predict ***By what process do you think materials cross the walls of capillaries?***

Concepts In Motion
Interactive Figure To see an animation of the structure of blood vessels, visit biologygmh.com.

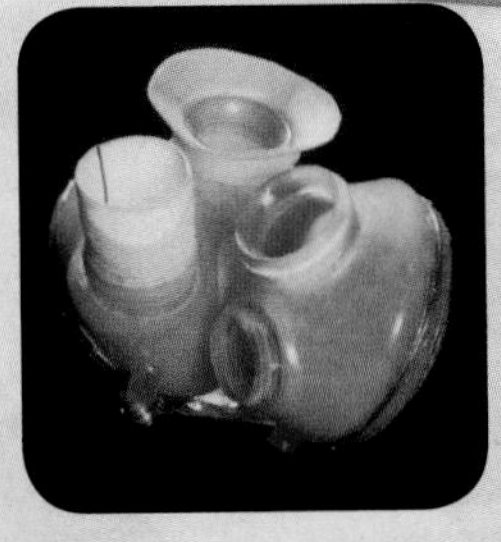

1982 The first artificial heart is implanted by William DeVries, a surgeon.

2004 Research shows that cardiac stem cells can generate new muscle cells. This opens up new treatment possibilities for treating heart failure.

1930 | 1965 | 2000

1940–1941 Dr. Charles R. Drew establishes the first blood banks for blood transfusions.

1967–1969 Surgeons perform the first heart transplant. An artificial heart keeps the patient alive until a donated heart replaces it.

Concepts In Motion Interactive Time Line
To learn more about these discoveries and others, visit biologygmh.com.
Biology Online

■ **Figure 34.3** Blood circulates throughout the body inside blood vessels.
Hypothesize ***How can body temperature be regulated by the diameter of blood vessels?***

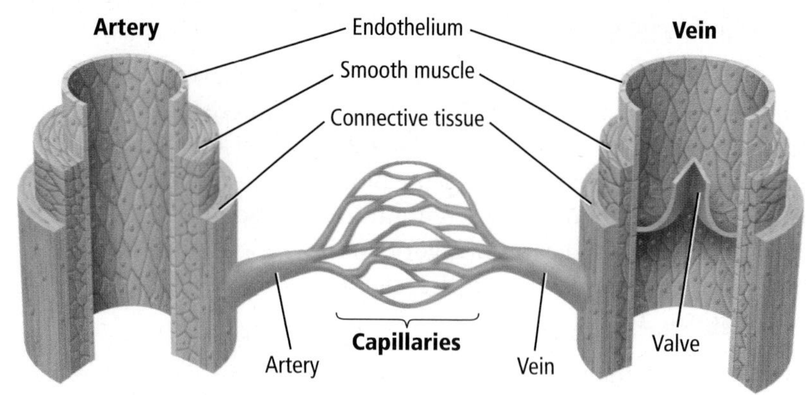

Veins After blood moves through the tiny capillaries, it enters the largest blood vessels, called veins. **Veins** carry oxygen-poor blood, or deoxygenated blood, back to the heart. The endothelial walls of veins are much thinner than the walls of arteries. Pressure of the blood decreases when the blood flows through capillaries before it enters the veins. By the time blood flows into the veins, the heart's original pushing force has less effect on making the blood move. So how does the blood keep moving? Because many veins are located near skeletal muscles, the contraction of these muscles helps keep the blood moving. Larger veins in your body also have flaps of tissue called **valves,** such as the one in **Figure 34.3,** that prevent blood from flowing backward. Lastly, breathing movements exert a squeezing pressure against veins in the chest, forcing blood back to the heart.

Reading Check **Describe** the differences in structure among arteries, capillaries, and veins.

The Heart

The **heart** is a muscular organ that is about as large as your fist and is located at the center of your chest. This hollow organ pumps blood throughout the body. The heart performs two pumping functions at the same time. The heart pumps oxygenated blood to the body, and it pumps deoxygenated blood to the lungs.

Structure of the heart Recall from Chapter 32 that the heart is made of cardiac muscle. It is capable of conducting electrical impulses for muscular contractions. The heart is divided into four compartments called chambers, as illustrated in **Figure 34.4.** The two chambers in the top half of the heart—the right atrium and the left atrium (plural, atria)—receive blood returning to the heart. Below the atria are the right and left ventricles, which pump blood away from the heart. A strong muscular wall separates the left side of the heart from the right side of the heart. The right and left atria have thinner muscular walls and do less work than the ventricles. Notice the valves in **Figure 34.4** that separate the atria from the ventricles and keep blood flowing in one direction. Valves also are found in between each ventricle and the large blood vessels that carry blood away from the heart, such as the aortic valve shown in a closed position in **Figure 34.4.**

CAREERS IN BIOLOGY

Exercise Physiologist Scientists who study the effects of exercise on the body are called exercise physiologists. They develop exercise programs and administer medical tests, such as stress tests. Their responsibilities might include monitoring heart activity and monitoring blood pressure levels. For more information on biology careers, visit biologygmh.com.

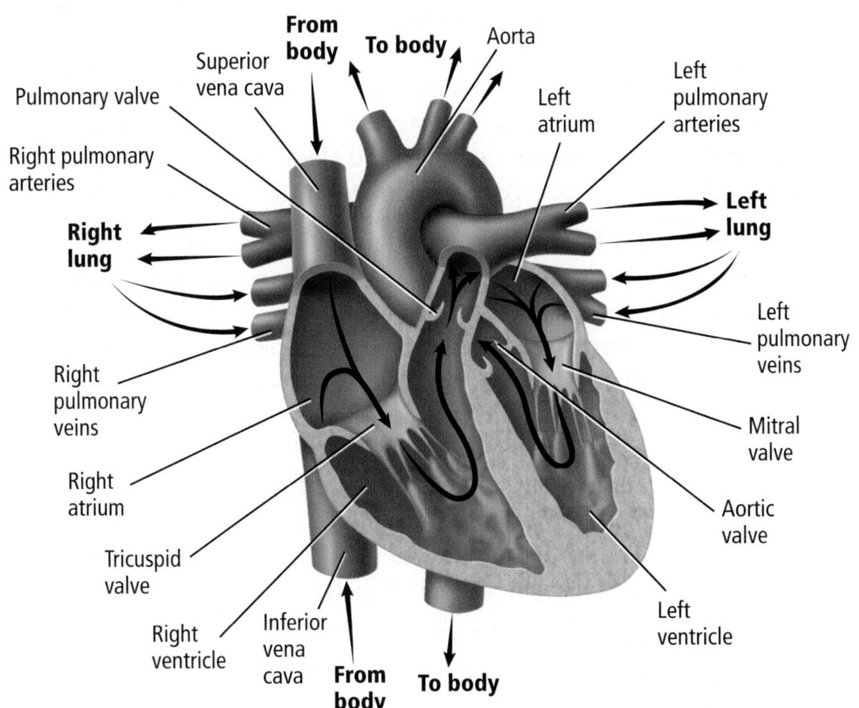

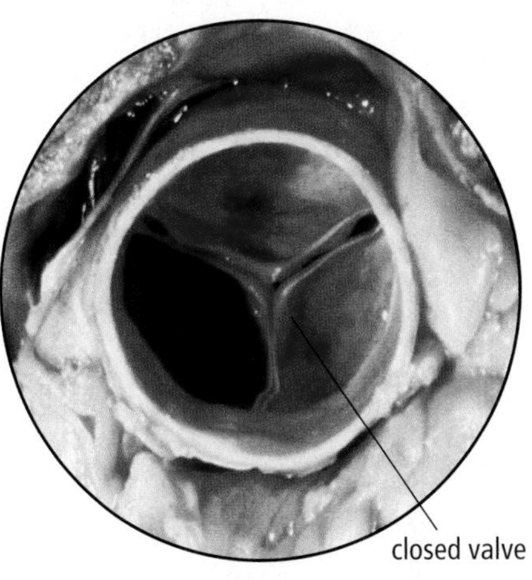

Aortic valve—closed position

■ **Figure 34.4** The arrows map out the path of blood as it circulates through the heart.
Diagram ***Trace the path of blood through the heart.***

How the heart beats The heart acts in two main phases. In the first phase, the atria fill with blood. The atria contract, filling the ventricles with blood. In the second phase, the ventricles contract to pump blood out of the heart, into the lungs, and forward into the body.

The heart works in a regular rhythm. A group of cells found in the right atrium, called the **pacemaker** or sinoatrial (SA) node, send out signals that tell the heart muscle to contract. The SA node receives internal stimuli about the body's oxygen needs, and then responds by adjusting the heart rate. The signal initiated by the SA node causes both atria to contract. This signal then travels to another area in the heart called the atrioventricular (AV) node, illustrated in **Figure 34.5.** This signal travels through fibers, causing both ventricles to contract. This two-step contraction makes up one complete heartbeat.

Pulse The heart pulses about 70 times each minute. If you touch the inside of your wrist just below your thumb, you can feel a pulse in the artery in your wrist rise and fall. This pulse is the alternating expansion and relaxation of the artery wall caused by contraction of the left ventricle. The number of times the artery pulses is the number of times your heart beats.

Blood pressure Blood pressure is a measure of how much pressure is exerted against the vessel walls by the blood. Blood-pressure readings can provide information about the condition of arteries. The contraction of the heart, or systole (SIS tuh lee), causes the blood pressure to rise to its highest point, and the relaxation of the heart, or diastole (di AS tuh lee) brings the pressure down to its lowest point. A normal blood-pressure reading for a healthy adult is a reading below 120 (systolic pressure)/80 (diastolic pressure).

■ **Figure 34.5** The SA node initiates the contraction of the heart, which spreads through both atria to the AV node. The AV node transmits the signal through excitable fibers that stimulate both ventricles.

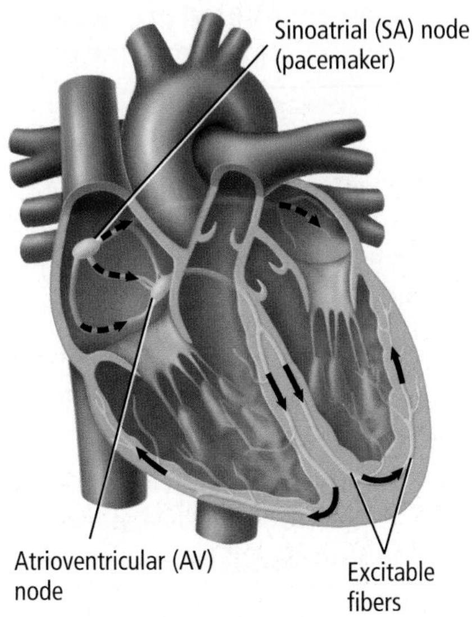

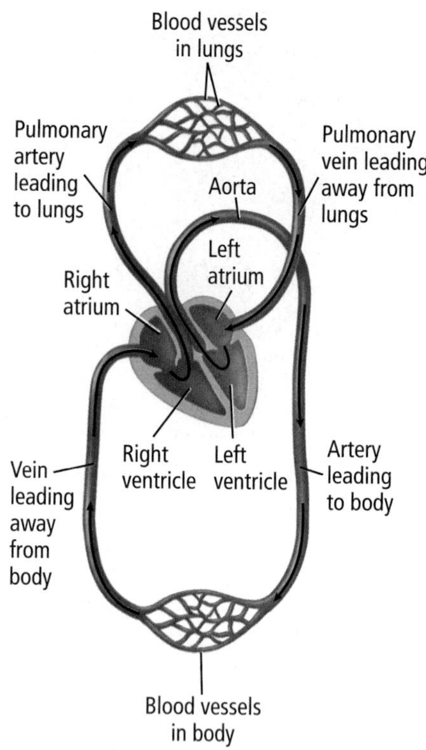

■ **Figure 34.6** Blood flow through the body consists of two different circulatory loops.

Concepts In Motion

Interactive Figure To see an animation of how blood flows through the circulatory system, visit biologygmh.com.

Blood flow in the body If you follow the flow of blood shown in **Figure 34.6,** you might notice that it flows in two loops. First, the blood travels from the heart to the lungs and back to the heart. Then, the blood is pumped in another loop from the heart through the body and back. The right side of the heart pumps deoxygenated blood to the lungs, and the left side of the heart pumps oxygenated blood to the rest of the body.

To the lungs and back When blood from the body flows into the right atrium, it has a low concentration of oxygen but a high concentration of carbon dioxide. This deoxygenated blood is dark red. The blood flows from the right atrium into the right ventricle and is pumped into the pulmonary arteries that lead to the lungs, as shown in **Figure 34.6.**

Eventually, blood flows into capillaries in the lungs that are in close contact with the air that comes into the lungs. The air in the lungs has a greater concentration of oxygen than the blood in the capillaries does, so oxygen diffuses from the lungs into the blood. At the same time, carbon dioxide diffuses in the opposite direction—from the blood into the air space in the lungs. Oxygenated blood, which is now bright red, flows to the left atrium of the heart to be pumped out to the body.

To the body and back The left atrium fills with oxygenated blood from the lungs, beginning the second loop of the figure eight. As shown in **Figure 34.6,** the blood then moves from the left atrium into the left ventricle. The left ventricle pumps the blood into the largest artery in the body called the aorta. Eventually, blood flows into the capillaries that branch throughout the body. Importantly, the capillaries are in close contact with body cells. Oxygen is released from the blood into the body cells by diffusion, and carbon dioxide moves from the cells to the blood by diffusion. The deoxygenated blood then flows back to the right atrium through veins.

MiniLab 34.1

Investigate Blood Pressure

How does blood pressure change in response to physical activity? Blood pressure changes from day to day, and throughout the day, and is affected by physical, psychological, behavioral, and inherited factors.

Procedure

1. Read and complete the lab safety form.
2. Watch the instructor demonstrate how to safely take a blood pressure. Practice using a **blood-pressure cuff** to measure a partner's blood pressure. Refer to a **blood-pressure chart** to interpret the reading.
3. Predict how exercise will affect systolic and diastolic blood pressure.
4. Take the resting blood-pressure reading of one member of your group.
5. Have the person whose blood pressure was taken do rhythmic exercise for one minute.
6. Take a second blood-pressure reading and compare it to the resting blood-pressure reading.

Analysis

1. **Identify** What are the experimental constants, the independent and dependent variables, and the control in your experiment?
2. **Conclude** Were your predictions supported? Explain.

Blood Components

Blood is the fluid of life because it transports important substances throughout the body. Blood is made up of a liquid medium called plasma, red blood cells, platelets, and white blood cells.

Plasma The clear, yellowish fluid portion of blood is the **plasma**. More than 50 percent of blood is plasma. Ninety percent of plasma is water and nearly 10 percent is dissolved materials. Plasma carries the breakdown products of digested food, such as glucose and fats. Plasma also transports vitamins, minerals, and chemical messengers such as hormones that signal body activities, such as the uptake of glucose by the cells. In addition, waste products from the cells are carried away by plasma.

There are three groups of plasma proteins that give plasma its yellow color. One group helps to regulate the amount of water in blood. The second group, produced by white blood cells, helps fight disease. The third group helps to form blood clots.

 **Reading Check** **Explain** the function of plasma.

Red blood cells The **red blood cells** carry oxygen to all of the body's cells. Red blood cells resemble discs with pinched-in centers, as shown in **Figure 34.7.** Recall that red blood cells develop in the marrow—the center portion of large bones. Red blood cells have no nuclei and only live for about 120 days.

Red blood cells mostly consist of an iron-containing protein called hemoglobin. Hemoglobin chemically binds with oxygen molecules and carries oxygen to the body's cells.

Platelets Have you ever cut your finger? You probably noticed that in a short while, the blood flowing from the cut slows and then stops as a blood clot forms a scab. **Platelets** are cell fragments, shown in **Figure 34.7,** that play an important part in forming blood clots.

When a blood vessel is cut, platelets collect and stick to the vessel at the site of the wound. The platelets then release chemicals that produce a protein called fibrin. Fibrin weaves a network of fibers across the cut that traps blood platelets and red blood cells, as shown in **Figure 34.8.** As more and more platelets and blood cells get trapped, a blood clot forms.

Color-Enhanced SEM Magnification: 1825×

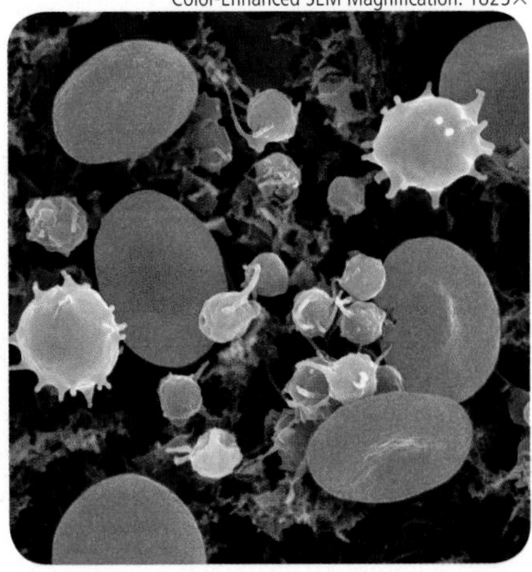

Figure 34.7 Blood is composed of liquid plasma, red blood cells (dimpled discs), white blood cells (irregularly shaped cells), and platelets (flat fragments).

Infer ***What might be occurring if there were too many white blood cells?***

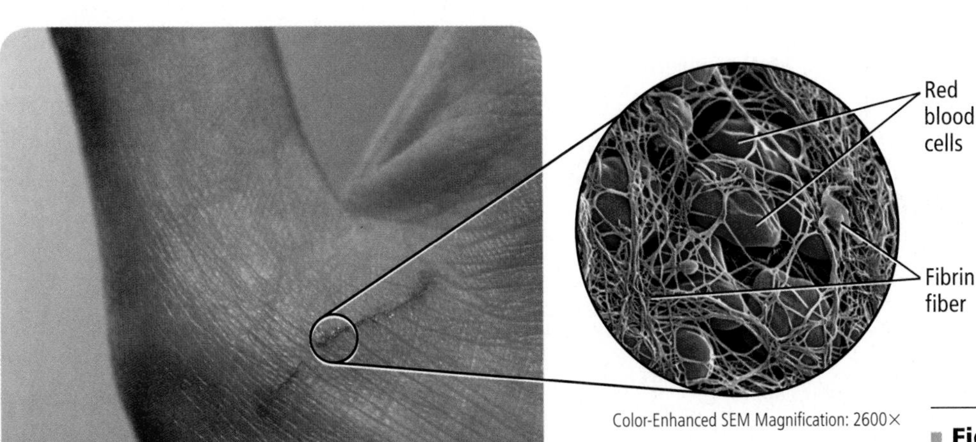

Color-Enhanced SEM Magnification: 2600×

Figure 34.8 A scab forms due to fibrin threads trapping blood cells and platelets.

White blood cells The body's disease fighters are the **white blood cells**. Like red blood cells, white blood cells are produced in bone marrow. Some white blood cells recognize disease-causing organisms, such as bacteria, and alert the body that it has been invaded. Other white blood cells produce chemicals to fight the invaders. Still, other white blood cells surround and kill the invaders. You will learn more about white blood cells in Chapter 37.

White blood cells are different from red blood cells in important ways. First, many white blood cells move from the marrow to other sites in the body to mature. Unlike red blood cells, there are fewer white blood cells—only about one white blood cell for every 500 to 1000 red blood cells. Also, white blood cells have nuclei. Finally, most white blood cells live for months or years.

Study Tip

Graphic Organizer Make a word map with the word *blood* in a large circle in the middle. Place the words *components, blood groups, circulation,* and *heart* in smaller circles around the large circle. Find information you learned in the chapter and add it to the appropriate smaller circles.

Blood Types

How do you know what type of blood you have? There are marker molecules attached to red blood cells. These markers determine blood type.

ABO blood groups There are four types of blood—A, B, AB, and O. If your blood type is A, you have A markers on your blood cells. If your blood type is B, you have B markers on your blood cells. People with blood type AB have both A and B markers. If your blood type is O, you do not have A or B markers.

FOLDABLES

Incorporate information from this section into your Foldable.

Importance of blood type If you ever need a blood transfusion, you only will be able to receive certain blood types, as shown in **Table 34.1.** This is because plasma contains proteins called antibodies that recognize red blood cells with foreign markers and cause those cells to clump together. For example, if you have blood type B, your blood contains antibodies that cause cells with A markers to clump. If you received a transfusion of type-A blood, your clumping proteins would make the type-A cells clump together. Clumping of blood cells can be dangerous because it can block blood flow.

Personal Tutor

To learn about blood groups, visit biologygmh.com.

Concepts In Motion

Interactive Table To explore more about blood groups, visit biologygmh.com.

Table 34.1 Blood Groups

Blood type	A	B	AB	O
Marker molecule and antibody	Marker molecule: A Antibody: anti-B	Marker molecules: B Antibody: anti-A	Marker molecules: AB Antibody: none	Marker molecules: none Antibodies: anti-A, anti-B
Example	A, A, A, A, A, A	B, B, B, B, B, B	A, B, B, A, A, B	
Can donate blood to:	A or AB	B or AB	AB	A, B, AB, or O
Can receive blood from:	A or O	B or O	A, B, AB, or O	O

Rh factor Another marker found on the surface of red blood cells is called the Rh factor. The Rh marker can cause a problem when an Rh-negative person, someone without the Rh factor, receives a transfusion of Rh-positive blood that has the Rh marker. This can result in clumping of red blood cells because Rh-negative blood contains Rh antibodies against Rh-positive cells.

The Rh factor can cause complications during some pregnancies. If the Rh-positive blood of a fetus mixes with the mother's Rh-negative blood, the mother will make anti-Rh antibodies. If the mother becomes pregnant again, these antibodies can cross the placenta and can destroy red blood cells if the fetus has Rh-positive blood. Rh-negative mothers are given a substance that prevents the production of Rh antibodies in the blood, so these problems can be avoided.

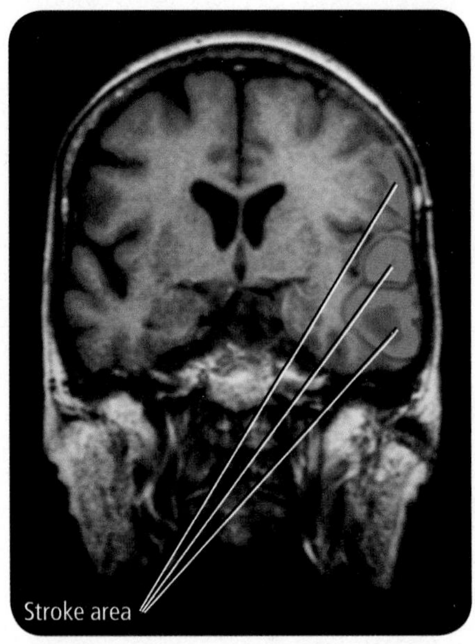

■ **Figure 34.9** Stroke is associated with ruptured blood vessels in the brain, as shown in red.

Circulatory System Disorders

Several disorders of the blood vessels, heart, and brain are associated with the circulatory system. Blood clots and other matter, such as fat deposits, can reduce the flow of oxygen-rich and nutrient-rich blood traveling through arteries. Physicians refer to the condition of blocked arteries as **atherosclerosis** (a thuh roh skluh ROH sus). The Greek word *sclerosis* means *hardening.* Signs of clogged arteries include high blood pressure and high cholesterol levels. When blood flow is reduced or blocked, the heart must work even harder to pump blood, and vessels can burst.

Atherosclerosis can lead to a heart attack or stroke. Heart attacks occur when blood does not reach the heart muscle. This can result in damage to the heart, or could even result in death if not treated. Strokes occur when clots form in blood vessels supplying oxygen to the brain. This can lead to ruptured blood vessels and internal bleeding, as shown in **Figure 34.9.** Parts of the brain die because brain cells are deprived of oxygen.

Section 34.1 Assessment

Section Summary

- Blood vessels transport important substances throughout the body.
- The top half of the heart is made up of two atria, and the bottom half is made up of two ventricles.
- The heart pumps deoxygenated blood to the lungs, and it pumps oxygenated blood to the body.
- Blood is made up of plasma, red blood cells, white blood cells, and platelets.
- Blood can be classified into the following four blood types: A, B, AB, and O.

Understand Main Ideas

1. MAIN Idea **Explain** the main functions of the circulatory system.
2. **Diagram** the path of blood through the heart and body.
3. **Compare and contrast** the structure of arteries with the structure of veins.
4. **Calculate** the average number of red blood cells for every 100 white blood cells in the human body.
5. **Summarize** the functions of the four components of blood.

Think Critically

6. **Cause and Effect** If a pacemaker received faulty signals from the brain, what would happen?
7. **Hypothesize** why exercise is a way to maintain a healthy heart.

MATH in Biology

8. Count the number of times your heart beats during 15 seconds. What is your heart rate per minute?

Section 34.2

Respiratory System

Reading Preview

Objectives

- **Distinguish** between internal and external respiration.
- **Summarize** the path of air through the respiratory system.
- **Identify** what changes occur in the body during breathing.

Review Vocabulary

ATP: biological molecule that provides the body's cells with chemical energy

New Vocabulary

breathing
external respiration
internal respiration
trachea
bronchus
lung
alveolus

MAIN Idea The function of the respiratory system is the exchange of oxygen and carbon dioxide between the atmosphere and the blood and between the blood and the body's cells.

Real-World Reading Link Air filters separate out dust and other particles from the air before they enter your car's engine. This prevents engine problems and helps ensure good air flow. Similarly, your respiratory system has features that ensure enough clean air gets into your lungs.

The Importance of Respiration

Your body's cells require oxygen. Recall from Chapter 8 that oxygen and glucose are used by cells to produce energy-rich ATP molecules needed to maintain cellular metabolism. This process is called cellular respiration. In addition to releasing energy, cellular respiration releases carbon dioxide and water.

Breathing and respiration The function of the respiratory system is to sustain cellular respiration by supplying oxygen to body cells and removing carbon dioxide waste from cells. The respiratory system can be divided into two processes: breathing and respiration. First, air must enter the body through breathing. **Breathing** is the mechanical movement of air into and out of your lungs. **Figure 34.10** illustrates air being released from the lungs into the air. Second, gases are exchanged in the body. **External respiration** is the exchange of gases between the atmosphere and the blood, which occurs in the lungs. **Internal respiration** is the exchange of gases between the blood and the body's cells.

Figure 34.10 Exhaled air from your lungs can be seen on a chilly evening.
Infer *How is the air that you inhale different than the air you exhale?*

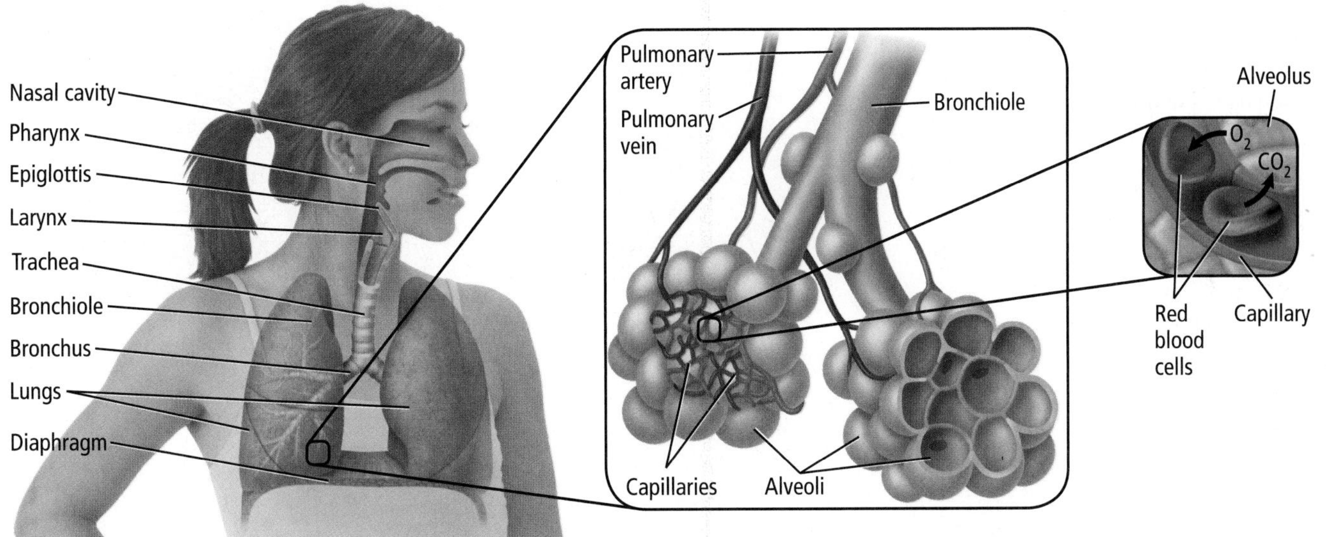

■ **Figure 34.11** Air travels into the alveoli of the lungs, where gases are exchanged across thin capillary walls.
Diagram ***Trace the path of oxygen from the atmosphere to the alveoli in the lungs.***

The Path of Air

The respiratory system is made up of the nasal passages, pharynx (FER ingks), larynx (LER ingks), epiglottis, trachea, lungs, bronchi, bronchioles, alveoli (al VEE uh li), and diaphragm. Air travels from the outside environment to the lungs where it passes through the alveoli, as shown in **Figure 34.11.**

First, air enters your mouth or nose. Hairs in the nose filter out dust and other large particles in the air. Hairlike structures called cilia, shown in **Figure 34.12,** also line the nasal passages, as well as other respiratory tubes. Cilia trap foreign particles from the air and sweep them toward the throat so that they do not enter the lungs. Mucous membranes beneath the cilia in the nasal passages, also shown in **Figure 34.12,** warm and moisten the air while trapping foreign materials.

Filtered air then passes through the upper throat called the pharynx. A flap of tissue called the epiglottis, which covers the opening to the larynx, prevents food particles from entering the respiratory tubes. The epiglottis allows air to pass from the larynx to a long tube in the chest cavity called the **trachea,** or windpipe. The trachea branches into two large tubes, called **bronchi** (BRAHN ki) (singular, bronchus), which lead to the lungs. The **lungs** are the largest organs in the respiratory system, and gas exchange takes place in the lungs. Each bronchus branches into smaller tubes called bronchioles (BRAHN kee ohlz). Each of these small tubes continues to branch into even smaller passageways, each of which ends in an individual air sac called an **alveolus** (plural, alveoli). Each alveolus has a thin wall—only one cell thick—and is surrounded by very thin capillaries.

Gas exchange in the lungs Air travels to individual alveoli where oxygen diffuses across the moist, thin walls into capillaries and then into red blood cells, as shown in **Figure 34.11.** The oxygen is then transported by the blood to be released to tissue cells in the body during internal respiration. Meanwhile, carbon dioxide moves in the opposite direction in the alveoli. Carbon dioxide in the blood crosses capillary walls, and then diffuses into the alveoli to be returned to the atmosphere during external respiration.

Reading Check **Infer** why gas exchange is effective in alveoli.

VOCABULARY
WORD ORIGIN
Alveolus
comes from the Latin word *alveus,* meaning *belly* or *hollow space.*

■ **Figure 34.12** Hairlike cilia line the mucous membranes of the nasal cavity.

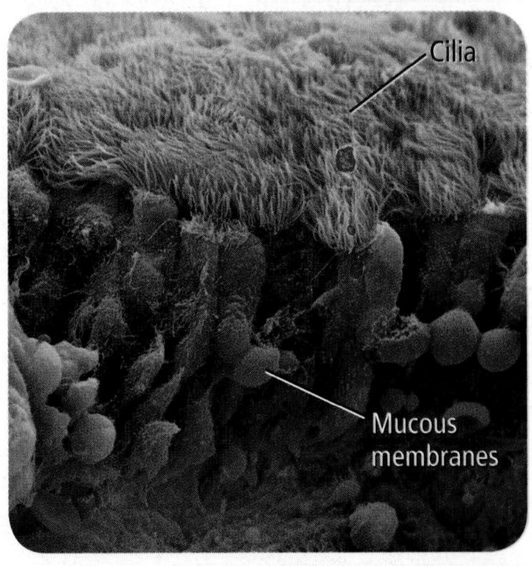

Color-Enhanced SEM Magnification: 2000×

■ **Figure 34.13** Rib and diaphragm muscles contract and relax during breathing.
Analyze ***How do you think air pressure is involved in breathing?***

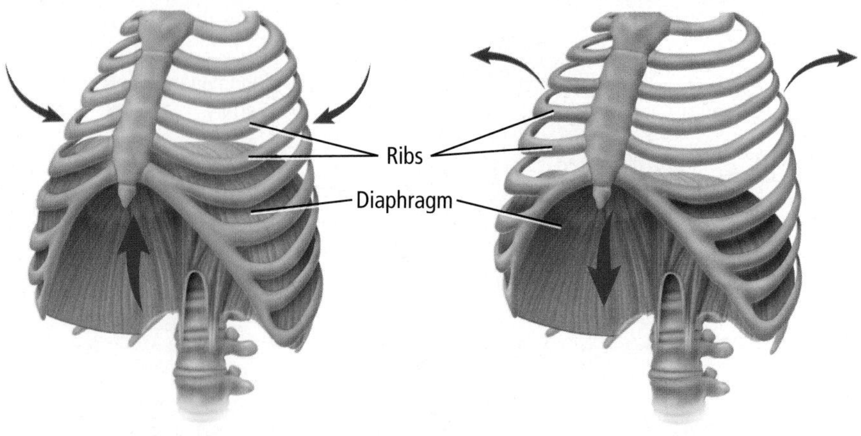

Breathing

The brain directs the rate of breathing by responding to internal stimuli that indicate how much oxygen the body needs. When the concentration of carbon dioxide in the blood is high, the breathing rate increases because cells need more oxygen.

Inhalation is the act of taking air into the lungs. During inhalation, as shown in **Figure 34.13,** the diaphragm contracts. This causes the chest cavity to expand as the diaphragm moves down, allowing air to move into the lungs. During exhalation, the diaphragm relaxes and returns to its normal resting position. This reduces the size of the chest cavity as the diaphragm moves up. Air naturally flows out from the greater pressure of the lungs. Follow **Figure 34.14** to learn how circulation and respiration work together to supply the needed oxygen and to get rid of carbon dioxide.

LAUNCH Lab

Review **Based on what you've read about respiration, how would you now answer the analysis questions?**

MiniLab 34.2

Recognize Cause and Effect

Does exercise affect metabolism? All of the chemical reactions that occur in your cells make up your metabolism. In this lab, you will explore how exercise affects the circulatory and respiratory systems and infer how this affects metabolism.

Procedure

1. Read and complete the lab safety form.
2. Record the number of heartbeats and number of breaths per minute for ten classmates.
3. Instruct the same students to walk in place for five minutes. At the end of that time, record each person's heartbeat per minute and number of breaths per minute.
4. After students have rested for five minutes, instruct them to jog or walk briskly in place for five minutes. Then, record each person's heartbeat per minute and number of breaths per minute.
5. Plot your results on **graph paper.** Each coordinate point should indicate breaths per minute on the horizontal axis and heartbeats per minute on the vertical axis.

Analysis

1. **Interpret** What is the relationship between the two dependent variables of your experiment—heart rate and breathing rate?
2. **Conclude** Does exercise affect metabolism? Why?
3. **Hypothesize** Why might students have different numbers of heartbeats per minute and breaths per minute even though they all walked or jogged for the same amount of time?

Visualizing Gas Exchange

Figure 34.14
Gases are exchanged in the lungs and in the tissue cells of the body.

In the lungs, oxygen (O_2) that is inhaled moves into capillaries and is transported to body cells. Carbon dioxide (CO_2) leaves the capillaries and is exhaled from the lungs.

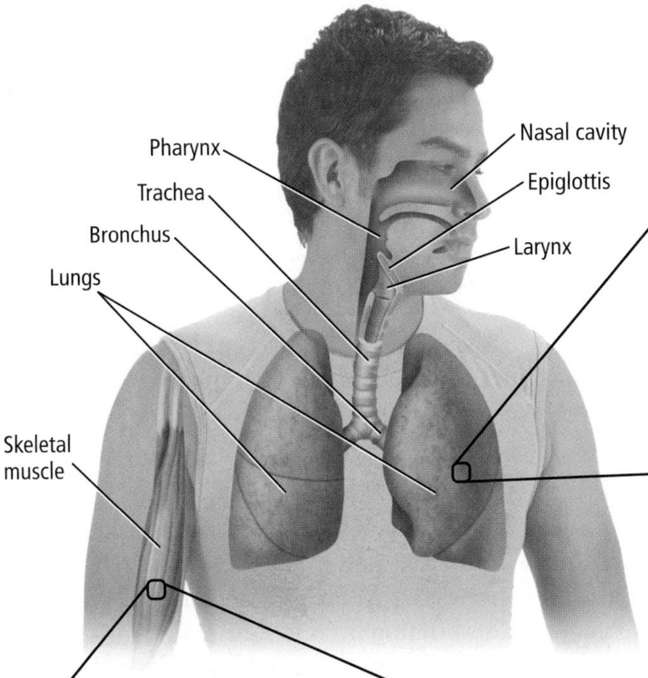

Color-Enhanced SEM Magnification: 300×

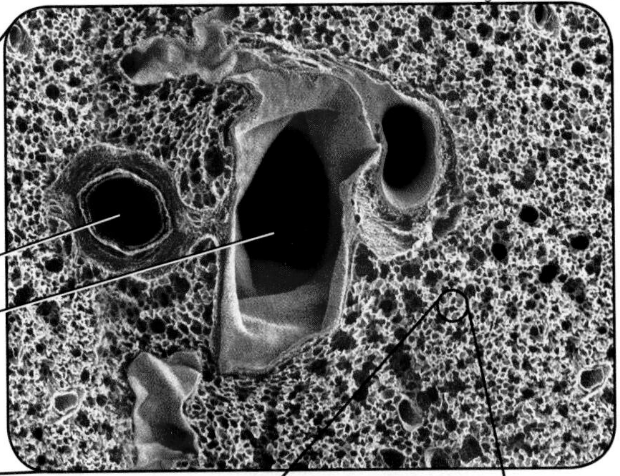

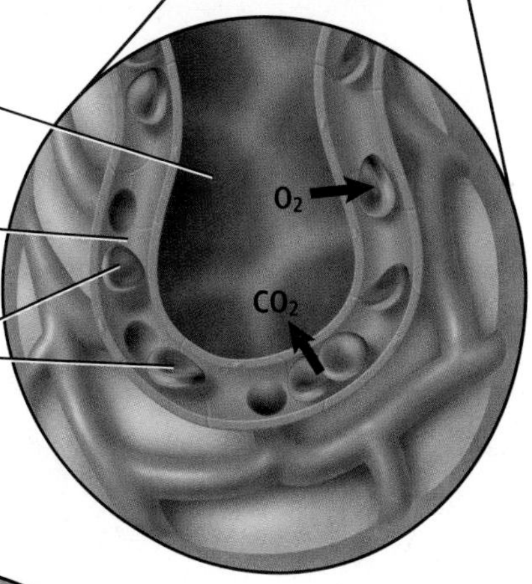

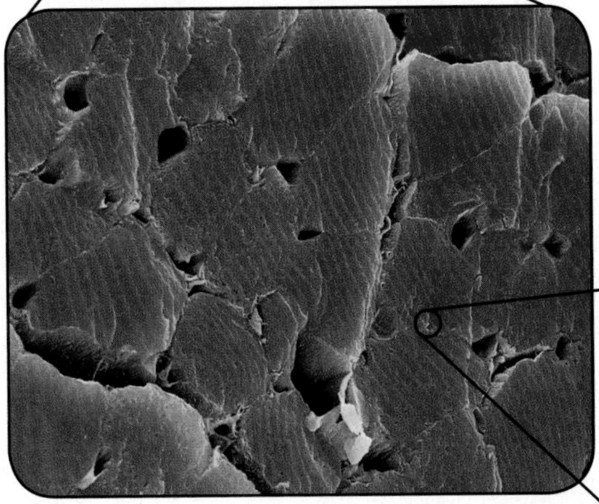

Color-Enhanced SEM Magnification: 1000×

In body tissues, such as muscle tissues, oxygen (O_2) moves from capillaries into tissue cells. Carbon dioxide (CO_2) produced by cellular respiration leaves tissue cells and moves into capillaries, and then is transported to the lungs.

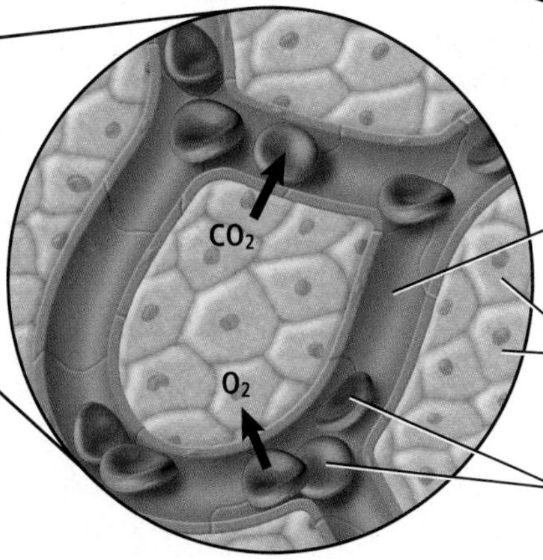

Concepts In Motion Interactive Figure To see an animation of gas exchange, visit biologygmh.com.

Biology Online

Table 34.2	Common Respiratory Disorders
Lung Disorder	Brief Description
Asthma	Respiratory pathways become irritated and bronchioles constrict.
Bronchitis	Respiratory pathways become infected, resulting in coughing and production of mucus.
Emphysema	Alveoli break down, resulting in reduced surface area needed for gas exchange with alveoli's blood capillaries.
Pneumonia	Infection of the lungs that causes alveoli to collect mucous material.
Pulmonary tuberculosis	A specific bacterium infects the lungs, resulting in less elasticity of the blood capillaries surrounding alveoli, thus decreasing effective gas exchange between the air and blood.
Lung cancer	Uncontrolled cell growth in lung tissue can lead to a persistent cough, shortness of breath, bronchitis, or pneumonia, and can lead to death.

Concepts In Motion **Interactive Table** To explore more about respiratory disorders, visit biologygmh.com.

Respiratory Disorders

Some diseases and disorders irritate, inflame, or infect the respiratory system, as described in **Table 34.2.** These disorders can produce tissue damage that reduces the effectiveness of the bronchi and alveoli. When these tissues become damaged, respiration becomes difficult. Smoking also causes chronic irritation to respiratory tissues and inhibits cellular metabolism. Finally, exposure to airborne materials, such as pollen, can produce respiratory problems in some people due to allergic reactions.

Section 34.2 Assessment

Section Summary

- Alveoli in the lungs are the sites of gas exchange between the respiratory and circulatory systems.
- The pathway of air starts with the mouth or nose and ends at the alveoli located in the lungs.
- Inhalation and exhalation are the processes of taking in and expelling air.
- The respiratory and circulatory systems work together to help maintain homeostasis.
- Respiratory disorders can inhibit respiration.

Understand Main Ideas

1. MAIN Idea **Identify** the main function of the respiratory system.
2. **Distinguish** between internal and external respiration.
3. **Sequence** the path of air from the nasal passages to the bloodstream.
4. **Describe** the mechanics of inhalation and exhalation.
5. **Infer** how the respiratory system would compensate for a circulatory disorder.
6. **Describe** three disorders of the respiratory system.

Think Critically

7. **Hypothesize** an advantage of heating and moisturizing air before it reaches alveoli.

MATH in Biology

8. The total surface area of the alveoli tissue in your lungs is approximately 70 m^2. This is more than 40 times the surface area of the skin. What is the surface area of your skin?

Reading Preview

Objectives

- **Summarize** the function of the kidneys in the body.
- **Sequence** the steps of the excretion of wastes from the Bowman's capsule to the urethra.
- **Distinguish** between filtration and reabsorption in the kidney.

Review Vocabulary

pH: measure of acidity and alkalinity of a solution

New Vocabulary

kidney
urea

Excretory System

MAIN Idea The kidneys maintain homeostasis by removing wastes and excess water from the body and by maintaining the pH of blood.

Real-World Reading Link Suppose that you cleaned your bedroom by first moving everything except large items into the hallway. You then return only the items you will keep to your bedroom and leave the items you don't want any longer in the hallway for later disposal. This is similar to how your kidneys filter materials in your blood.

Parts of the Excretory System

The body collects wastes, such as toxins, waste products, and carbon dioxide, that result from metabolism in the body. The excretory system removes these toxins and wastes from the body. In addition, the excretory system regulates the amount of fluid and salts in the body, and maintains the pH of the blood. All of these functions help to maintain homeostasis.

The components that make up the excretory system include the lungs, skin, and kidneys, as illustrated in **Figure 34.15.** The lungs primarily excrete carbon dioxide. The skin primarily excretes water and salts contained in sweat. The kidneys, however, are the major excretory organ in the body.

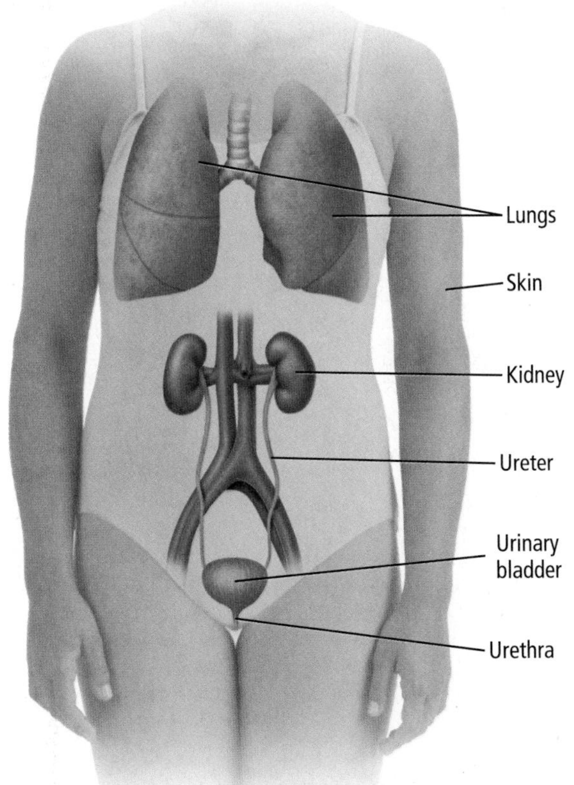

Figure 34.15 The organs of excretion work together to eliminate wastes from the body. They include the lungs, skin, and kidneys.

The Kidneys

As shown in **Figure 34.16,** the **kidneys** are bean-shaped organs that filter out wastes, water, and salts from the blood. The kidneys are divided into two distinct regions, also illustrated in **Figure 34.16.** The outer portion is called the renal cortex and the inner region is called the renal medulla. Each of these regions contains microscopic tubes and blood vessels. In the center of the kidney is a region called the renal pelvis, where urine collection occurs. Follow **Figure 34.16** as you read about how the kidneys function.

Nephron filtration Each kidney contains approximately one million filtering units called nephrons. Blood enters each nephron through a long tube that is surrounded by a ball of capillaries called the glomerulus (gluh MER uh lus) (plural, glomeruli). The glomerulus is surrounded by a structure called the Bowman's capsule.

The renal artery transports nutrients and wastes to the kidney and branches into smaller and smaller blood vessels, eventually reaching the tiny capillaries in the glomerulus. The walls of the capillaries are very thin and the blood is under great pressure. As a result, water and substances dissolved in the water, such as the nitrogenous waste product called **urea,** are pushed through the capillary walls into the Bowman's capsule. Larger molecules, such as red blood cells and proteins, remain in the bloodstream.

■ **Figure 34.16** Nephrons are the functional units of the kidney.
Sequence ***Summarize the path of urine as it is excreted from the body.***

Concepts In Motion

Interactive Figure To see an animation of how the kidney filters wastes, visit biologygmh.com.

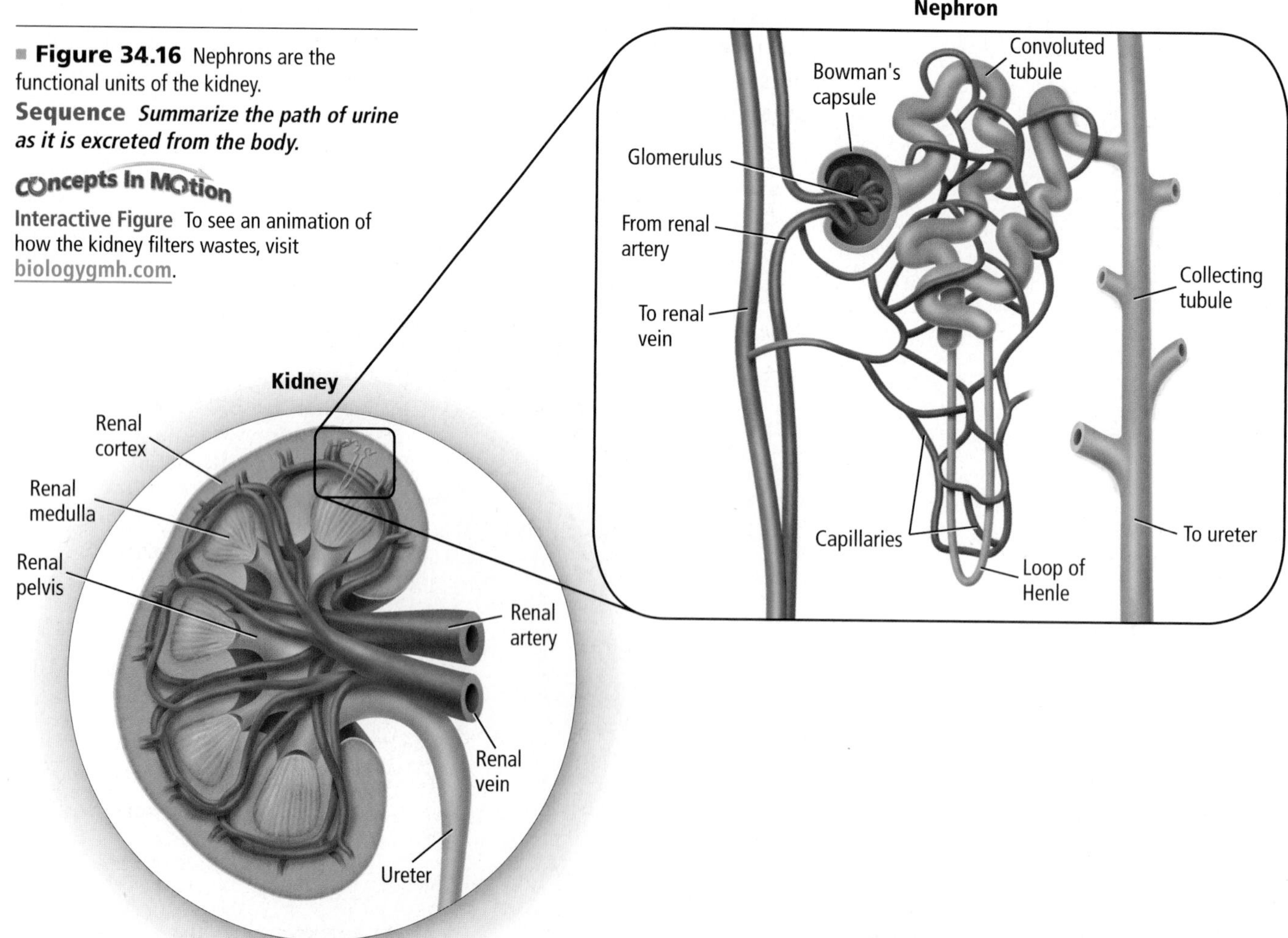

Reabsorption and the formation of urine The filtrate collected in the Bowman's capsule flows through the renal tubule that consists of the convoluted tubule, the loop of Henle, and the collecting tubule, as illustrated in **Figure 34.16.** Much of the lost water and useful substances, such as glucose and minerals, are reabsorbed back into the capillaries surrounding the renal tubule. This process is called reabsorption. At the same time, excess fluids and toxic substances in the capillaries are passed to the collecting tubules. This waste product is called urine. Urine leaves the kidney through ducts called the ureters (YOO ruh turz), shown in **Figure 34.16.** Urine is then stored in the urinary bladder and exits the body through the urethra.

The kidneys filter about 180 L of blood each day in adults, but produce only about 1.5 L of urine. The process of filtration and reabsorption from the blood requires large amounts of energy. Although kidneys account for only one percent of body weight, they use 20 to 25 percent of the body's oxygen intake for their internal energy requirements.

Connection to Chemistry The kidney can help maintain a normal pH in the blood by adjusting the acid-base balance. Recall from Chapter 6 that low pH results when there is an abundance of H^+. When the blood pH is too low, the kidney can increase pH levels in the body by excreting hydrogen (H^+) ions and ammonia into the renal tubules. The kidney can decrease pH levels by reabsorbing buffers such as bicarbonate (HCO_3^+) and sodium (Na^+) ions. Because biological processes normally require pH between 6.5 and 7.5, the kidneys help to maintain homeostasis by keeping pH levels within the normal range.

CAREERS IN BIOLOGY

Urologist A urologist is a medical doctor who has specialized knowledge about problems of the male and female urinary systems. Responsibilities can include examining patients, performing laboratory tests, interpreting test results and examination findings, and treating injuries and disorders. For more information on biology careers, visit biologygmh.com.

DATA ANALYSIS LAB 34.1

Based On Real Data*

Interpret the Data

How do extreme conditions affect the average daily loss of water in the human body? The body obtains water by absorbing it through the digestive tract. The body loses water primarily by excreting it in urine from the kidneys, through sweat, and through the lungs.

Think Critically

1. **Identify** What is the major source of water loss during normal weather?
2. **Hypothesize** Why is more water lost in sweat during rigorous exercise than in urine?
3. **Calculate** What is the percent of water loss for all three conditions?

Data and Observations

The table shows data collected in normal weather, hot weather, and during heavy exercise.

Average Daily Water Loss in Humans (in mL)

Source	Normal Temperatures	High Temperatures	Rigorous Exercise
Kidneys	1500	1400	750
Skin	450	1800	5000
Lungs	450	350	650

*Data obtained from: Beers, M. 2003. *The Merck Manual of Medical Information, Second Edition* West Point, PA.: Merck & Co. Inc.

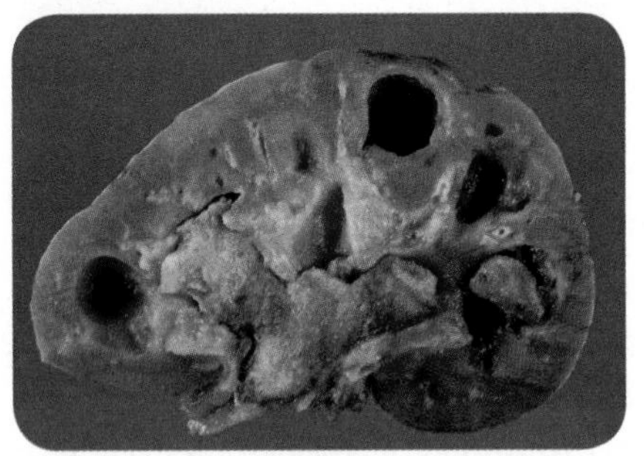

Figure 34.17 Kidney stones form as minerals such as calcium become solid masses.

Kidney Disorders

Sometimes kidney function can be inhibited or impaired by infections or disorders. When kidney function is impaired, the body cannot rid itself of wastes and homeostasis might be disrupted.

Infections Symptoms of a kidney infection include fever, chills, and mid- to low-back pain. Kidney infections often start as urinary bladder infections that spread to the kidney. Obstructions in the kidney also can cause an infection. If the infection is not treated, the kidneys can become scarred and their function might be permanently impaired. Antibiotics usually are effective in treating a bacterial infection.

Nephritis Another common kidney problem is nephritis (nih FRIH tus), which often is due to inflammation or painful swelling of some of the glomeruli, as listed in **Table 34.3.** This occurs for many reasons, such as when large particles in the bloodstream become lodged in some of the glomeruli. Symptoms of this condition include blood in the urine, swelling of body tissues, and protein in the urine. If this condition does not improve on its own, the patient may need a special diet or prescription drugs to treat the infection.

Kidney stones Kidney stones are another type of kidney disorder, as listed in **Table 34.3** and shown in **Figure 34.17.** A kidney stone is a crystallized solid, such as calcium compounds, that forms in the kidney. Small stones can pass out of the body in urine; this can be quite painful. Larger stones often are broken into small pieces by ultrasonic sound waves. The smaller stones then can pass out of the body. In some cases, surgery might be required to remove large stones.

Kidneys also can be damaged by other diseases present in the body. Diabetes and high blood pressure are the two most common reasons for reduced kidney function and kidney failure. In addition, kidneys can be damaged by prescription and illegal drug use.

VOCABULARY

ACADEMIC VOCABULARY

Inhibit:
To hold back, restrain, or block the action or function of something.
The concentration of the protein in the blood inhibited the organ from producing more of the same protein.

Concepts In Motion
Interactive Table To explore more about excretory disorders, visit biologygmh.com.

Table 34.3 Common Excretory Disorders

Excretory Disorder	Brief Description
Nephritis	Inflammation of the glomeruli can lead to inflammation of the entire kidneys. This disorder can lead to kidney failure if left untreated.
Kidney stones	Hard deposits form in the kidney that might pass out of the body in urine. Larger kidney stones can block urine flow or irritate the lining of the urinary tract, leading to possible infection.
Urinary tract blockage	Malformations present at birth can lead to blockage of the normal flow of urine. If untreated, this blockage can lead to permanent damage of the kidneys.
Polycystic (pah lee SIHS tihk) kidney disease	This is a genetic disorder distinguished by the growth of many fluid-filled cysts in the kidneys. This disorder can reduce kidney function and lead to kidney failure.
Kidney cancer	Uncontrolled cell growth often begins in the cells that line the tubules within the kidneys. This can lead to blood in the urine, a mass in the kidneys, or affect other organs due to the cancer spreading, which can lead to death.

Kidney Treatments

A large percentage of kidney function can be lost before kidney failure becomes apparent. If kidney problems are left untreated, the buildup of waste products in the body can lead to seizures, a comatose state, or death. However, modern medicine offers two possible treatments for reduced kidney function or complete kidney failure.

Dialysis Dialysis (di AH luh sus) is a procedure in which an artificial kidney machine filters out wastes and toxins from a patient's blood. There are two different types of dialysis, and one is illustrated in **Figure 34.18.** Blood is passed through a machine that temporarily filters and cleanses the blood. The filtered blood is then returned to the patient's body. The procedure lasts about three to four hours and requires three sessions per week.

In the second type of dialysis, the membrane lining the abdomen acts as an artificial kidney. The abdominal cavity is injected with a special fluid through a small tube attached to the body. The patient's fluid, that contains wastes from the blood, is drained. This procedure is performed on a daily schedule for 30 to 40 minutes.

Kidney transplant A kidney transplant is the surgical placement of a healthy kidney from another person, called a donor, into the patient's body. Kidney transplants have shown increasing success in recent years. However, there is a limited supply of donated kidneys. The number of patients waiting for kidney transplants far exceeds the organs available for transplant.

The major complication of a transplant is possible rejection of the donated organ. Rejection is prevented with medications such as steroids and cyclosporine. Cyclosporine is a drug given to transplant recipients to help prevent organ rejection. Many transplant patients also need blood pressure medication and other drugs to prevent infections.

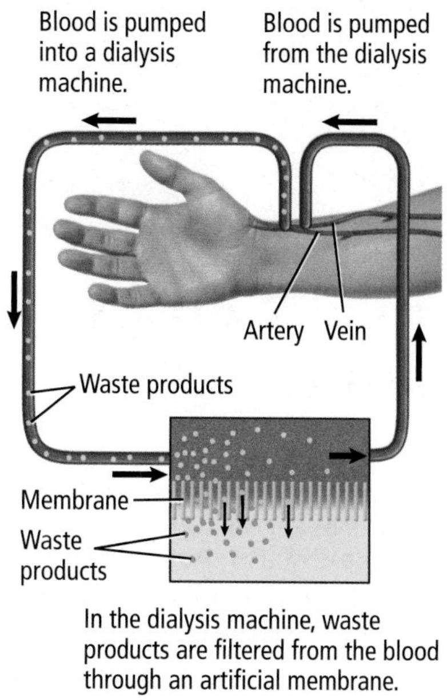

Blood is pumped into a dialysis machine.

Blood is pumped from the dialysis machine.

Artery

Vein

Waste products

Membrane

Waste products

In the dialysis machine, waste products are filtered from the blood through an artificial membrane.

■ **Figure 34.18** Dialysis is used to filter wastes and toxins from a patient's blood.

Personal Tutor

To learn about dialysis, visit biologygmh.com.

Section 34.3 Assessment

Section Summary

- The kidneys are the main excretory organ in the body.
- Nephrons are independent filtration units in the kidneys.
- Water and important substances are reabsorbed into the blood after filtration.
- The kidneys produce a waste product called urine.

Understand Main Ideas

1. MAIN Idea **Explain** how the kidneys help maintain homeostasis.
2. **Diagram** the excretion of waste from the Bowman's capsule to the urethra.
3. **Compare and contrast** filtration and reabsorption in a nephron.
4. **Identify** three types of kidney disorders.

Think Critically

5. **Hypothesize** why kidney failure without dialysis can result in death.

WRITING in **Biology**

6. Research the effects of a high-protein diet on the excretory system. Summarize your findings in a public service announcement.

MATH in Biology

7. Calculate the average amount of urine the body produces in a week.

Biology & Society

Mercury and the Environment

In the early 1950s, many residents in the area around Minamata Bay in Southwestern Japan contracted a disease that caused nerve damage, birth defects, and even some fatalities. Scientists found that the cause of this disease was mercury that was discharged into the bay by local industry. Many of the residents who ate fish contaminated by this mercury became ill.

Mercury sources Mercury is a metallic element that is liquid at room temperature. Mercury forms compounds that are highly toxic to humans. However, mercury has been a part of our environment for a very long time. Volcanoes and weathering of rocks naturally release mercury into the environment. Mercury also is used in many industrial processes.

Disposal of mercury-containing objects in landfills releases mercury into soil and water supplies. Burning mercury-containing objects, including industrial refuse and coal, releases mercury into the atmosphere. A coal-fired power plant might emit up to 50 metric tons of mercury into the air each year if it burns coal that contains mercury.

Mercury in the food chain The main source of human exposure to mercury occurs when it is concentrated in the food chain. Mercury enters the food chain when it is washed into surface water from the air by rain and when soil and rock particles enter surface water. Bacteria in the water then convert the mercury to an organic compound called methylmercury.

Methylmercury circulates in the body and is taken into the tissues of organisms easily. Once it is there, it is very difficult to eliminate through the kidneys. As a result, methylmercury accumulates in the tissues of fish and other aquatic organisms. Furthermore, this accumulation becomes greater in organisms that are long-lived or higher in the food chain.

Mercury and its effects Fish and shellfish are important to healthy diets because they contain many healthful proteins and other nutrients. However, fish and shellfish also contain mercury, as shown in the table. Which have the greatest and least average concentrations? Why do you think the concentration in sharks is so high?

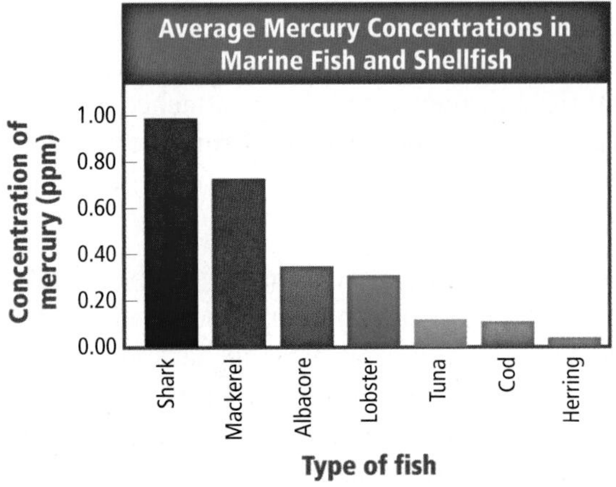

The FDA recommends that seafood selections during pregnancy and nursing should include varieties known to have lower than average methylmercury levels. Varieties known to contain higher levels should be consumed no more than twice per week.

Women can eat up to 340 grams of shrimp, canned light tuna, salmon, pollock, and catfish per week. Albacore tuna contains more mercury than canned light tuna, so women should not eat more than 170 grams of it per week. Young children should follow these same recommendations, but they should eat smaller portions.

WRITING in Biology

Community Service As a class, research local disposal programs for hazardous items such as thermometers and batteries. Collaborate to create a brochure about the programs. For more information about mercury, visit biologygmh.com.

BIOLAB

INTERNET: MAKE POSITIVE HEALTH CHOICES

Background: Both heredity and lifestyle choices affect overall health. Achieving optimal health involves making wise choices regarding exercise, nutrition, drugs and alcohol, stress management, and smoking. Because body systems function together to maintain homeostasis, changes in one system can impact overall health. In this lab, you will design a presentation that focuses on how specific health choices influence the functionality of body systems.

Question: *How do lifestyle choices affect the function of the circulatory, respiratory, and excretory systems?*

Materials

Choose materials that would be appropriate for creating the type of presentation you create.
Possible materials include:
resource materials about health choices from the school library or classroom

Procedure

1. Read and complete the lab safety form.
2. Develop an outline of information you would like to include in your presentation. Include information about how specific health choices affect the respiratory, circulatory, and excretory systems.
3. Use resources and data you collected in this chapter's labs to determine the effects of specific health choices on your body.
4. Choose a presentation medium. Ideas include a multimedia presentation, video, poster, or pamphlet. The medium you choose should appeal to a specific audience.
5. Share your presentation with your target audience. Post your research and presentation at biologygmh.com so others can benefit from what you have learned.

6. Use the evaluation information provided by your teacher to evaluate the effectiveness of the presentation.

Analyze and Conclude

1. **Describe** What is the intended audience for your presentation? How did you modify the information included to target this audience?
2. **Summarize** Identify the key points of your presentation.
3. **Explain** How do the health choices you described affect multiple body systems?
4. **Evaluate** Do you think your presentation will influence the health choices of your target audience? Explain.
5. **Critique your presentation** How could you increase the effectiveness of your presentation?

COMMUNITY INVOLVEMENT

CREATE Choose one or more health-promoting behaviors from your presentation. Design a survey to gather data about the choices that members of your target audience make regarding this health-promoting behavior. To learn more about health and body system function, visit BioLabs at biologygmh.com.

Study Guide

Download quizzes, key terms, and flash cards from biologygmh.com.

FOLDABLES **Draw a Conclusion** Determine which of the four blood types is characterized as the universal recipient. Explain your answer.

Vocabulary	Key Concepts
Section 34.1 Circulatory System	
• artery (p. 993) • atherosclerosis (p. 999) • capillary (p. 993) • heart (p. 994) • pacemaker (p. 995) • plasma (p. 997) • platelet (p. 997) • red blood cell (p. 997) • valve (p. 994) • vein (p. 994) • white blood cell (p. 998)	MAIN Idea The circulatory system transports blood to deliver important substances, such as oxygen, to cells and to remove wastes, such as carbon dioxide. • Blood vessels transport important substances throughout the body. • The top half of the heart is made up of two atria, and the bottom half is made up of two ventricles. • The heart pumps deoxygenated blood to the lungs, and it pumps oxygenated blood to the body. • Blood is made up of plasma, red blood cells, white blood cells, and platelets. • Blood can be classified into the following four blood types: A, B, AB, and O.
Section 34.2 Respiratory System	
• alveolus (p. 1001) • breathing (p. 1000) • bronchus (p. 1001) • external respiration (p. 1000) • internal respiration (p. 1000) • lung (p. 1001) • trachea (p. 1001)	MAIN Idea The function of the respiratory system is the exchange of oxygen and carbon dioxide between the atmosphere and the blood and between the blood and the body's cells. • Alveoli in the lungs are the sites of gas exchange between the respiratory and circulatory systems. • The pathway of air starts with the mouth or nose and ends at the alveoli located in the lungs. • Inhalation and exhalation are the processes of taking in and expelling air. • The respiratory and circulatory systems work together to help maintain homeostasis. • Respiratory disorders can inhibit respiration.
Section 34.3 Excretory System	
• kidney (p. 1006) • urea (p. 1006)	MAIN Idea The kidneys maintain homeostasis by removing wastes and excess water from the body and by maintaining the pH of blood. • The kidneys are the main excretory organ in the body. • Nephrons are independent filtration units in the kidneys. • Water and important substances are reabsorbed into the blood after filtration. • The kidneys produce a waste product called urine.

CHAPTER 34

Assessment

Section 34.1

Vocabulary Review

Match the following definitions with the correct vocabulary term from the Study Guide page.

1. a vessel carrying oxygen-rich blood
2. involved in blood vessel repair
3. stimulates the heart to contract

Understand Key Concepts

4. When blood leaves the heart, where does it exit?
   A. the aorta **C.** the lungs
   B. the capillaries **D.** the pulmonary vein

Use the diagram below to answer questions 5 and 6.

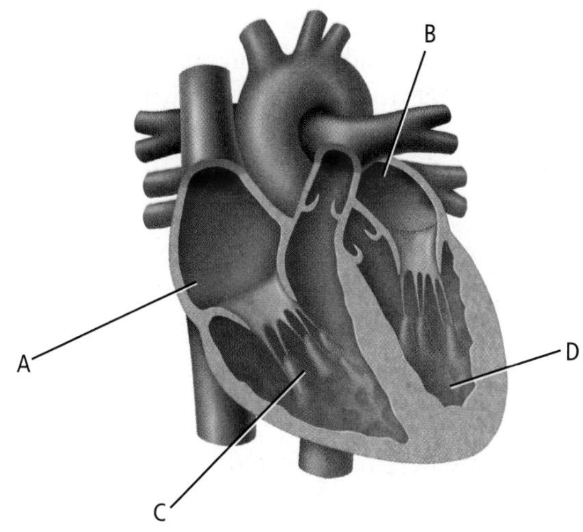

5. Which represents the right ventricle?
   A. A **C.** C
   B. B **D.** D
6. Into what part of the heart does oxygen-rich blood enter?
   A. A **C.** C
   B. B **D.** D
7. If a teenager with type A blood is injured in an auto accident and needs a blood transfusion, what type blood will he or she receive?
   A. only type A
   B. type A or type O
   C. only type AB
   D. only type O
8. Where are one-way valves in the circulatory system located?
   A. arteries **C.** veins
   B. capillaries **D.** white blood cells
9. When a small blood vessel in your hand is cut open, which plays an active defensive role against possible disease?
   A. plasma **C.** red blood cells
   B. platelets **D.** white blood cells

Constructed Response

10. **Short Answer** Differentiate between the function of the atria and the function of the ventricles.

Use the diagram to answer question 11.

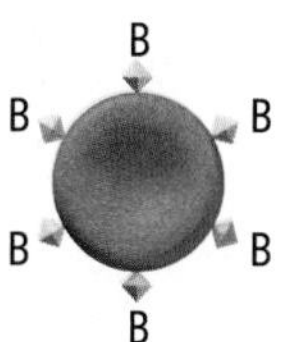

11. **Short Answer** A person has the blood type represented above. What type of blood can the person receive in a transfusion? Explain.

Think Critically

12. **Hypothesize** an advantage of your heart containing two pumping systems rather than one pumping system within the same organ.
13. **Deduce** which ABO blood type—A, B, AB or O—is the most valuable to medical personnel in an extreme emergency situation and explain why.

Section 34.2

Vocabulary Review

Use the vocabulary terms from the Study Guide page to answer the following questions.

14. In what structure does external respiration take place?
15. Which term defines the exchange of gases between the blood and the body's cells?
16. Which part of the air pathway branches off the trachea?

Understand Key Concepts

Use the diagram below to answer questions 17 and 18.

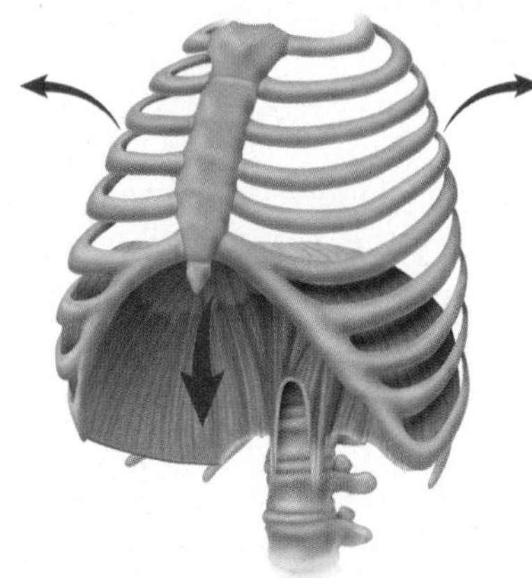

17. Which process is shown above?
    A. inhalation
    B. exhalation
    C. cellular respiration
    D. filtration

18. Which structure moves down as its muscles contract?
    A. trachea
    B. diaphragm
    C. pharynx
    D. ribs

19. Which process occurs inside the tissue cells in your legs?
    A. filtration
    B. breathing
    C. external respiration
    D. internal respiration

20. Which process causes the diaphragm to move back up?
    A. cellular respiration
    B. exhalation
    C. inspiration
    D. internal respiration

21. Which gas is needed by all cells?
    A. sulfur
    B. hydrogen
    C. carbon dioxide
    D. oxygen

22. How many breaths will a person take in one day if they take 12 breaths per minute?
    A. about 1000
    B. about 10,000
    C. about 17,000
    D. about 1,000,000

Constructed Response

23. **Short Answer** Differentiate among asthma, bronchitis, and emphysema.

Use the photo below to answer question 24.

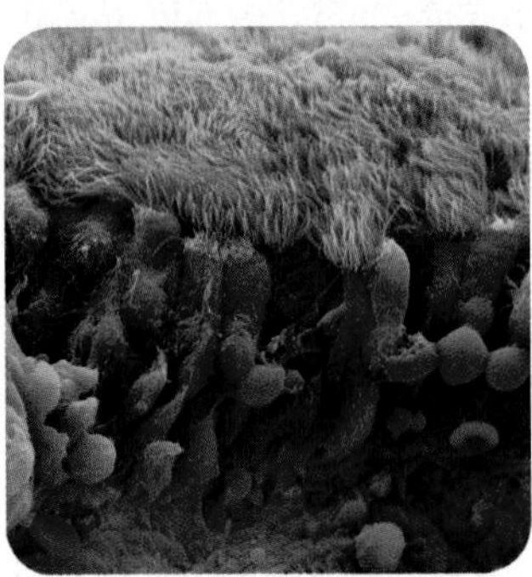

24. **Short Answer** Describe the function of the structures above. Where would these structures be found?

Think Critically

25. **Hypothesize** an advantage in breathing more deeply during exercise compared to another person engaged in similar exercise breathing at a normal rate.

Section 34.3

Vocabulary Review

Review the vocabulary terms found on the Study Guide page. Use the terms to answer the following questions.

26. Where are nephrons located?

27. Which waste product is found in urine?

Understand Key Concepts

28. Where is the loop of Henle?
    A. renal tubule
    B. glomerulus
    C. Bowman's capsule
    D. urethra

29. Which one of the kidney functions conserves water in the body?
    A. absorption
    B. filtration
    C. reabsorption
    D. breathing

30. Which process returns glucose to the blood?
    A. excretion
    B. filtration
    C. reabsorption
    D. exhalation

Biology Online Chapter Test biologygmh.com

Use the table below to answer questions 31, 32, and 33.

Reabsorption of Some Substances in the Kidneys

Chemical substance	Amount Filtered by Kidneys (g/day)	Amount Excreted by Kidneys (g/day)	Percent of Filtered Chemical Reabsorbed (per day)
Glucose	180	0	100
Urea	46.8	23.4	50
Protein	1.8	1.8	0

31. Based on the estimates from the table above, how much urea is absorbed by the kidneys?
- **A.** 0.50 g/min
- **B.** 23.4 g/day
- **C.** 46.8 g/day
- **D.** 50.0 g/day

32. Based on the table estimates above, what happens to glucose in the kidneys?
- **A.** It is reabsorbed into the blood.
- **B.** It is permanently filtered out of the blood.
- **C.** It is treated in the kidney like creatinine.
- **D.** It is treated in the kidney like urea.

33. Infer why proteins are not removed by nephrons.
- **A.** Collecting ducts are too small.
- **B.** Proteins cannot be filtered.
- **C.** Proteins never enter the nephron.
- **D.** Proteins are reabsorbed by nephrons.

Constructed Response

34. Short Answer How many liters of blood flow through your kidneys in one hour?

35. Short Answer Explain the differences between filtration and reabsorption in the kidney.

36. Open Ended Infer why kidneys require so much energy to function.

Think Critically

37. CAREERS IN BIOLOGY Formulate a list of questions one might ask a a urologist regarding urinary problems or keeping the male reproductive system healthy.

Additional Assessment

38. WRITING in Biology Construct an analogy about the circulatory system that is based on your local highway system in your town, city, or rural area.

DBQ Document-Based Questions

The following data compare the state of five subjects whose circulation was monitored. (The weight, age, and sex of all five subjects were the same.) All of Subject A's data were within normal limits; the other four were not.

Data obtained from: Macey, R. 1968. *Human Physiology*. Englewood Cliffs, NJ: Prentice Hall.

Subject	Hemoglobin (Hb) content of blood (Hb/100 mL blood)	Oxygen contents of blood in arteries (mL O_2/100 mL blood)	Oxygen content of blood in veins (mL O_2/100 mL blood)
A	15	19	15
B	15	15	12
C	8	9.5	6.5
D	16	20	13
E	15	19	18

39. Which subject might be suffering from a dietary iron deficiency? Explain your choice.

40. Which subject might have lived at a high altitude where the atmospheric oxygen is low? Explain your choice.

41. Which subject might have been poisoned by carbon monoxide that prevents tissue cells from using oxygen? Explain your choice.

Cumulative Review

42. Draw diagrams of a typical example from each of the three classes of mollusks and label the distinctive characteristics. **(Chapter 25)**

Standardized Test Practice

Cumulative

Multiple Choice

1. What happens to a skeletal muscle when the actin fibers are pulled toward the center of the sarcomeres?
   A. It contracts.
   B. It grows.
   C. It relaxes.
   D. It stretches.

Use the diagram to answer questions 2 and 3.

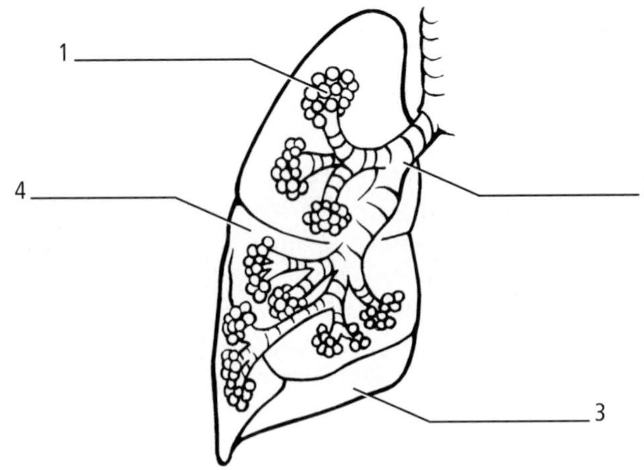

2. Which part of the respiratory system has hairs to filter particles from the air?
   A. 1
   B. 2
   C. 3
   D. 4

3. In which numbered location does gas exchange take place?
   A. 1
   B. 2
   C. 3
   D. 4

4. Which is an example of operant conditioning?
   A. A dog salivates when it hears a bell.
   B. A horse becomes accustomed to street noises.
   C. A newborn forms an attachment to the first animal seen after birth.
   D. A rat learns that it can get food by pulling a lever.

5. Which is an example of nurturing behavior?
   A. An animal in a colony spots a predator and warns the whole colony.
   B. A female chimpanzee takes care of her infant for three years.
   C. A male peacock displays its feathers in front of a female.
   D. A squirrel chatters at another squirrel to drive it away.

Use the table below to answer question 6.

Muscle Type	Function
Skeletal muscles	attached to bones and tighten when contracted causing movement
Smooth muscles	line the hollow internal organs such as stomach, intestines, bladder, and uterus
Cardiac muscles	

6. Where is the muscle type that is missing a description in the table located?
   A. in the heart
   B. in the kidneys
   C. lining the blood vessels
   D. lining the lymph vessels

7. Which answer choice is a result of parasympathetic stimulation?
   A. decreased heart rate
   B. decreased mucus production
   C. increased digestive activity
   D. increased pupil size

8. Which characteristic directly affects homeostatic temperature control in mammals?
   A. four-chambered heart
   B. high metabolic rate
   C. milk production
   D. signaling devices in fur

Biology Online Standardized Test Practice biologygmh.com

Short Answer

Use the diagram below to answer questions 9 and 10.

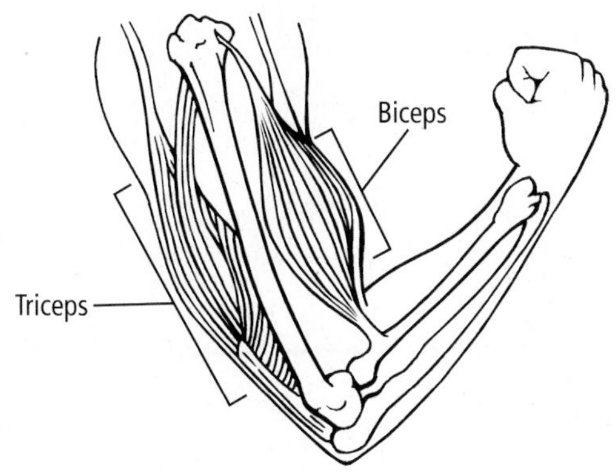

9. Describe how the biceps and triceps allow movement in the arm.

10. Explain why muscles are always in antagonistic pairs.

11. Some drugs cause an increased level of dopamine in nerve synapses. Name one of these drugs and relate the increased dopamine level to other effects that result from using the drug.

12. Use a table to organize information about the autonomic and somatic nervous systems. List the types of responses, systems affected, and include an example.

13. Monotremes are mammals that are similar to reptiles in some ways. Classify monotreme characteristics as similar to reptiles or similar to mammals.

14. A rare disease called amyotrophic lateral sclerosis (ALS) causes motor neurons in the body to lose myelin. What do you think would be the initial symptoms a person with ALS would have?

15. Explain how nephrons filter blood.

Extended Response

Use the illustration below to answer question 16.

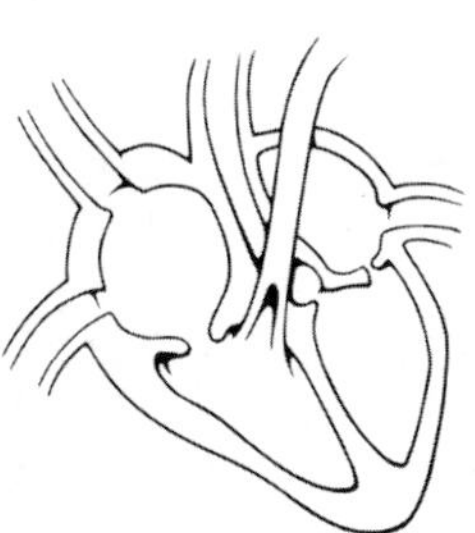

16. The illustration above shows a four-chambered mammalian heart. Write an explanation of the role of the four-chambered heart in circulating oxygenated blood throughout the body.

17. Compare and contrast apical meristems and lateral meristems in plants.

18. The invention of the microscope allowed scientists to discover hundreds of tiny living organisms that were never seen before. Distinguish, in a written statement, between an advance in technology and an advance in science using this historical example.

Essay Question

The human nervous system consists of a complex arrangement of voluntary and involuntary responses and activities. The presence of these different types of responses has evolved in humans to help with survival.

Using the information in the paragraph above, answer the following question in essay format.

19. From what you know about different nervous system responses, write a well-organized essay explaining how different types of involuntary response systems in humans are helpful for survival.

NEED EXTRA HELP?

If You Missed Question . . .	1	2	3	4	5	6	7	8	9	10	11	12	13	14	15	16	17	18	19
Review Section . . .	32.3	34.2	34.2	31.1	31.2	32.3	33.2	30.1	32.3	32.3	33.4	33.2	22.1	33.1	34.3	30.2	22.1	1.2	33.2

CHAPTER 35

Digestive and Endocrine Systems

BIG Idea **The digestive system breaks down food to provide energy and nutrients for the body. Hormones regulate body functions.**

Section 1
The Digestive System
MAIN Idea The digestive system breaks down food so nutrients can be absorbed by the body.

Section 2
Nutrition
MAIN Idea Certain nutrients are essential for the proper function of the body.

Section 3
The Endocrine System
MAIN Idea Systems of the human body are regulated by hormonal feedback mechanisms.

BioFacts

- A person's stomach lining is replaced every few days.
- A person secretes almost one liter of saliva every day.
- The small intestine is about 6 m long, and the large intestine is about 1.5 m long.

LAUNCH Lab

How does the enzyme pepsin aid digestion?

The acidic digestive juices in the stomach contain the enzyme pepsin. In this lab, you will investigate the role of pepsin in digestion.

Procedure

1. Read and complete the lab safety form.
2. Label and prepare **three test tubes.**
   A: 15 mL water
   B: 10 mL water, 5 mL **HCl solution**
   C: 5 mL each water, HCl solution, and **pepsin solution**
3. Cut **hard-boiled egg white** portions into pea-sized chunks with a **knife.**
4. Add equal amounts of egg white to each tube. Predict the relative amount of digestion in each test tube.
5. Place test tubes in an **incubator** overnight at 37°C. Record observations the next day.

Analysis

Evaluate Rank the test tubes based on the amount of digestion that occurred. Based on your results, describe the roles of pepsin and pH in digestion of proteins.

Biology Online

Visit biologygmh.com to:

- study the entire chapter online
- explore Concepts in Motion, Interactive Tables, Microscopy Links, and links to virtual dissections
- access Web links for more information, projects, and activities
- review content online with the Interactive Tutor and take Self-Check Quizzes

Negative Feedback System Make this Foldable to help you record what you learn about the role four hormones play in the negative feedback system.

STEP 1 Fold a 5-cm tab along the short side of a sheet of 11″ × 17″ paper.

STEP 2 Fold the same sheet of paper into fourths along the long axis to form a four-row chart.

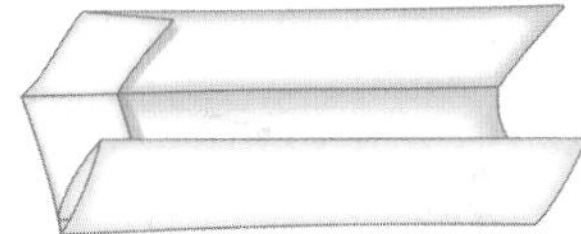

STEP 3 Draw lines along the creases.

STEP 4 Label the rows as follows: *Parathyroid hormone, Antidiuretic hormone,* and *Human growth hormone,* then choose another hormone to include in your chart.

FOLDABLES Use this Foldable with Section 35.3. As you study the section, record what you learn about the importance of the negative feedback system to the production of each of the hormones listed in your chart.

Section 35.1

Reading Preview

Objectives

- **Summarize** the three main functions of the digestive system.
- **Identify** structures of the digestive system and their functions.
- **Describe** the process of chemical digestion.

Review Vocabulary

nutrient: vital component of foods that provides energy and materials for growth and body functions

New Vocabulary

mechanical digestion
chemical digestion
amylase
esophagus
peristalsis
pepsin
small intestine
liver
villus
large intestine

The Digestive System

MAIN Idea **The digestive system breaks down food so nutrients can be absorbed by the body.**

Real-World Reading Link During an average lifespan, as much as 45 tons of food can pass through a person's digestive system. The food will travel almost 3 m through the digestive tract. What happens as food passes through this long tube?

Functions of the Digestive System

There are three main functions of the digestive system. The digestive system ingests food, breaks it down so nutrients can be absorbed, and eliminates what cannot be digested. Refer to **Figure 35.1** and **Figure 35.2** as you learn about the structure and function of the digestive system.

Digestion On Friday night, you and your friends go out for pizza. You bite into a slice and begin to chew. How does your body digest that pizza?

Mechanical digestion involves chewing food to break it down into smaller pieces. It also includes the action of smooth muscles in the stomach and small intestine that churn the food. **Chemical digestion** involves the breakdown of large molecules in food into smaller substances by enzymes. The smaller substances can be absorbed into the body's cells. Recall from Chapter 6 that enzymes are proteins that speed up biological reactions. When you chew the bites of pizza, **amylase,** an enzyme found in saliva, begins the process of chemical digestion by breaking down starches into sugars.

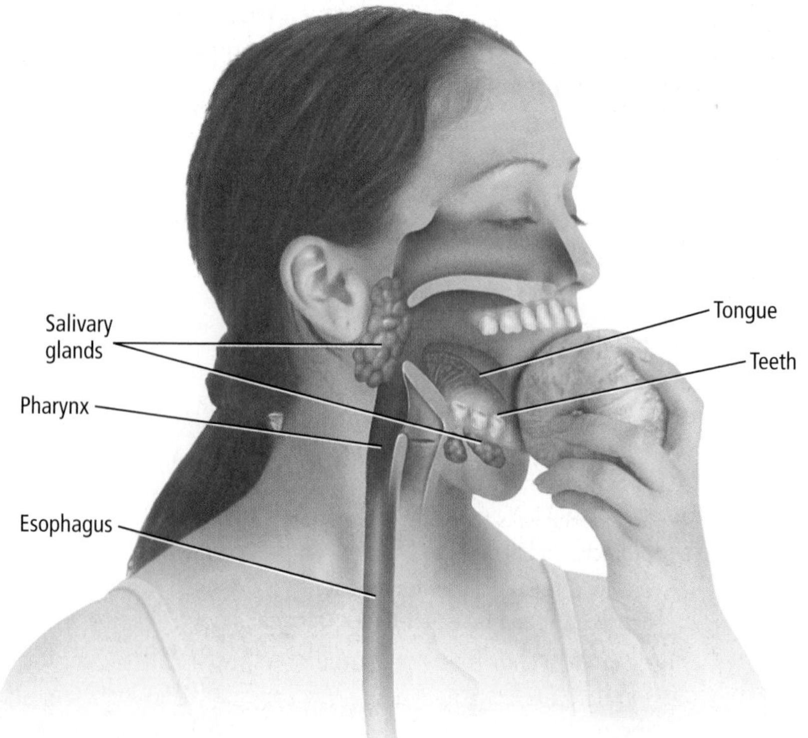

Figure 35.1 Mechanical digestion starts in the mouth. Secretions from the salivary glands keep food moist and begin the process of chemical digestion. Food moves through the pharynx into the esophagus.

Concepts In Motion

Interactive Figure To see an animation of food moving through the esophagus, visit biologygmh.com.

Esophagus When the tongue pushes chewed food to the back of the mouth, the swallowing reflex is stimulated. The food is forced by the action of the tongue into the upper portion of the esophagus. The **esophagus** (ih SAH fuh gus) is a muscular tube that connects the pharynx, or throat, to the stomach, illustrated in **Figure 35.2.** The wall of the esophagus is lined with smooth muscles that contract rhythmically to move the food through the digestive system in a process called **peristalsis** (per uh STAHL sus). Peristalsis continues throughout the digestive tract. Even if a person were upside down, food would still move toward the stomach.

When a person swallows, the small plate of cartilage called the epiglottis covers the trachea. If this opening is not closed, food can enter the trachea and cause a person to choke. The body responds to this by initiating the coughing reflex in an attempt to expel the food to keep the food from entering the lungs.

Stomach When food leaves the esophagus, it passes through a circular muscle called a sphincter, and then it moves into the stomach. The sphincter between the esophagus and stomach is the cardiac sphincter. The walls of the stomach are composed of three overlapping layers of smooth muscle that are involved with mechanical digestion. As the muscles contract, they further break down the food and mix it with the secretions of glands that line the inner wall of the stomach.

Connection to Chemistry Recall from Chapter 6 that pH is a measure of a solution's acidity. The environment inside the stomach is very acidic. Stomach glands, called gastric glands, secrete an acidic solution, which lowers the pH in the stomach to about 2. This is about the same acidity as lemon juice. If the sphincter in the upper portion of the stomach allows any leakage, some of this acid might move back into the esophagus, causing what is commonly known as heartburn.

The acidic environment in the stomach is favorable to the action of **pepsin,** an enzyme involved in the process of the chemical digestion of proteins. Cells in the lining of the stomach secrete mucus to help prevent damage from pepsin and the acidic environment. Although most absorption occurs in the small intestine, some substances, such as alcohol and aspirin, are absorbed by cells that line the stomach. While empty, the capacity of the stomach is about 50 mL. When full, it can expand to 2–4 L.

The muscular walls of the stomach contract and push food farther along the digestive tract. The consistency of the food resembles tomato soup as it passes through the pyloric sphincter at the lower end of the stomach into the small intestine. **Figure 35.3** illustrates peristalsis in the small intestine.

Reading Check **Compare** digestion in the mouth with digestion in the stomach.

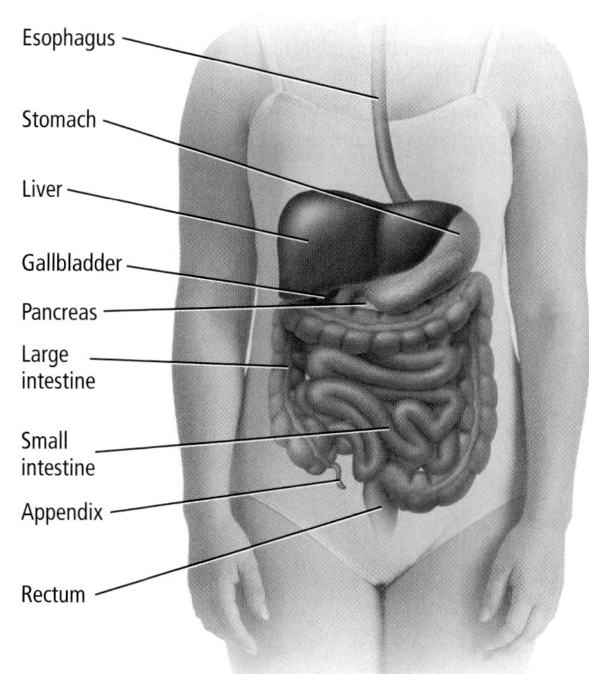

Figure 35.2 The esophagus extends from the pharynx to the stomach and is approximately 25 cm long.
Describe *Why are humans classified as coelomates?*

Figure 35.3 The smooth muscles in the walls of the digestive tract contract in the process of peristalsis.

Concepts In Motion
Interactive Figure To see an animation of peristalsis, visit biologygmh.com.

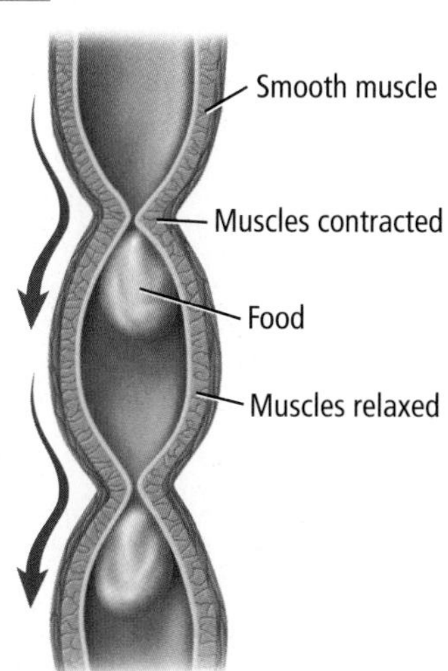

■ **Figure 35.4** Chemical digestion in the small intestine depends on the activities of the liver, pancreas, and gallbladder.
Discuss ***What is the importance of each of these organs in the process of chemical digestion?***

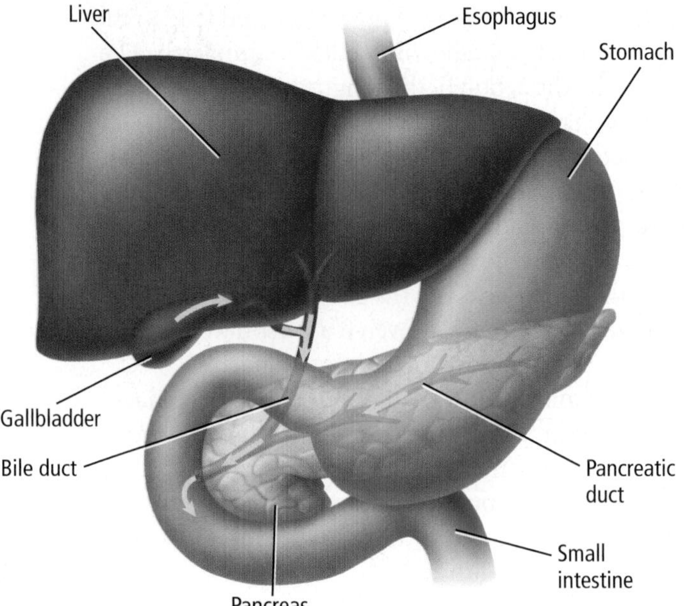

Small intestine The **small intestine** is approximately 7 m in length and is the longest part of the digestive tract. It is called small because it has a diameter of 2.5 cm compared to the 6.5 cm diameter of the large intestine. The smooth muscles in the wall of the small intestine continue the process of mechanical digestion and push the food farther through the digestive tract by peristalsis.

The completion of chemical digestion in the small intestine depends on three accessory organs—the pancreas, liver, and gallbladder, illustrated in **Figure 35.4.** The pancreas serves two main functions. One is to produce enzymes that digest carbohydrates, proteins, and fats. The other is to produce hormones, which will be discussed later in this chapter. The pancreas secretes an alkaline fluid to raise the pH in the small intestine to slightly above 7, which creates a favorable environment for the action of intestinal enzymes.

The **liver** is the largest internal organ of the body and produces bile, which helps to break down fats. About 1 L of bile is produced every day, and excess bile is stored in the gallbladder to be released into the small intestine when needed. **Figure 35.5** shows gallstones, which are cholesterol crystals that can form in the gallbladder.

Study Tip

Sequence and Order Using your notes, work with a partner to review the sequence of the organs in the digestive system. Then, practice retelling the sequence without your notes. Ask questions of one another for deeper learning.

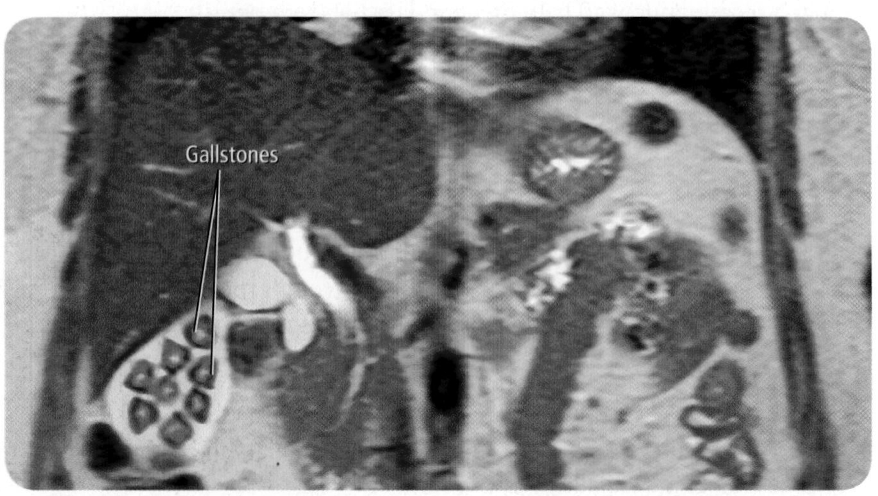

■ **Figure 35.5** Gallstones can obstruct the flow of bile from the gallbladder. Note the gallstones on this MRI film of a gallbladder.

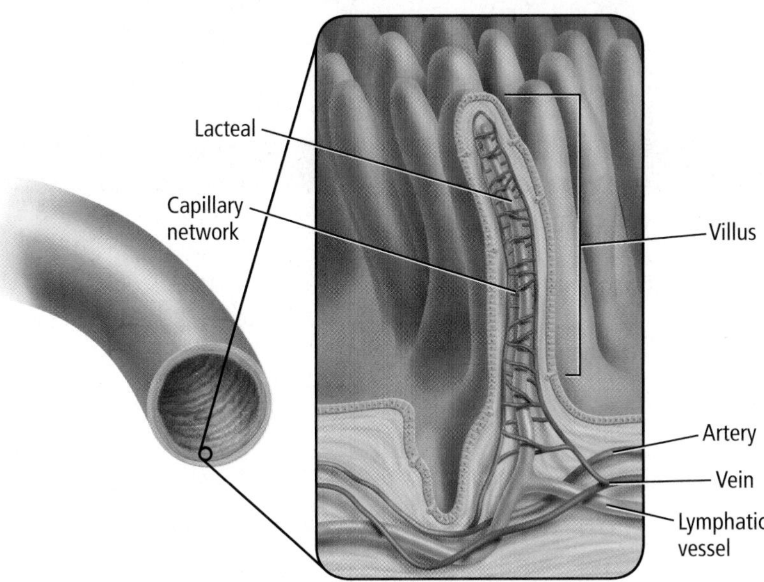

Figure 35.6 A villus is a fingerlike extension of the lining of the small intestine. Nutrients diffuse into capillaries in the villi and reach body cells by means of circulating blood.

Chemical digestion is completed and most of the nutrients from food are absorbed from the small intestine into the bloodstream through fingerlike structures called **villi** (VIH li) (singular, villus). Villi, illustrated in **Figure 35.6,** increase the surface area of the small intestine, giving the small intestine approximately the same surface area as a tennis court.

Refer again to **Figure 35.1** and **Figure 35.2** to follow the movement of digested food through the digestive system. Once digestion is complete, the remaining food, now in a semiliquid form called chyme (KIME), moves into the large intestine. Chyme is made up of materials that cannot be digested or absorbed by villi in the small intestine.

MiniLab 35.1

Investigate Digestion of Lipids

How do bile salts and pancreatic solution affect digestion? Lipids, or fats, are not water soluble. The body compensates by producing bile, a chemical that breaks apart fat and helps the molecules mix with the watery solution in the small intestine. In this lab, you will investigate the breakdown of lipids.

Procedure

1. Read and complete the lab safety form.
2. Study the lab procedure and construct a data chart.
3. Label **three test tubes.** Add 5 mL **vegetable oil** and 8-10 drops **phenolphthalein** to each. Shake well. If the color is not pink, add **NaOH solution** one drop at a time until the solution turns pink.
4. Add 125 mL **water** to a **250-mL beaker.** Warm to about 40°C.
5. Prepare the test tubes as follows, then seal each with a **stopper.**
   Test Tube A: 5 mL **distilled water,** pinch of **bile salt**
   Test Tube B: 5 mL **pancreatic solution,** pinch of bile salt
   Test Tube C: 5 mL pancreatic solution
6. Shake each tube to mix the contents and gently place in the beaker. Record your observations.
7. Dispose of test tube contents in the designated container.

Analysis

1. **Analyze** What did a color change inside a test tube indicate? What caused the change?
2. **Draw Conclusions** Based on your results, describe the roles of bile and pancreatic solution in digestion.

Concepts In Motion Interactive Table To explore more about digestion, visit biologygmh.com.

Table 35.1 Time for Digestion

Digestive Structure	Primary Function	Time Food in Structure
Mouth	Mechanical and chemical digestion	5–30 s
Esophagus	Transport (swallowing)	10 s
Stomach	Mechanical and chemical digestion	2–24 h
Small intestine	Mechanical and chemical digestion	3–4 h
Large intestine	Water absorption	18 h–2 days

Large intestine The **large intestine** is the end portion of the digestive tract. It is about 1.5 m long and includes the colon, the rectum, and a small saclike appendage called the appendix. Although the appendix has no known function, it can become inflamed and swollen, resulting in appendicitis. If inflamed, the appendix likely will have to be removed surgically.

Some kinds of beneficial bacteria are normal in the colon. These bacteria produce vitamin K and some B vitamins available to the body.

A primary function of the colon is to absorb water from the chyme. The indigestible material then becomes more solid and is called feces. Peristalsis continues to move feces toward the rectum, causing the walls of the rectum to stretch. This initiates a reflex that causes the final sphincter muscle to relax, and the feces are eliminated from the body through the anus. Refer to **Table 35.1** to review the primary function of each structure of the digestive system and how long food usually remains in each structure as it is being digested.

Section 35.1 Assessment

Section Summary

- The digestive system has three main functions.
- Digestion can be categorized as mechanical or chemical.
- Most nutrients are absorbed in the small intestine.
- Accessory organs provide enzymes and bile to aid digestion.
- Water is absorbed from chyme in the colon.

Understand Main Ideas

1. MAIN Idea **Describe** the process that breaks down food so nutrients can be absorbed by the body.
2. **Analyze** the difference between mechanical digestion and chemical digestion, and explain why chemical digestion is necessary for the body.
3. **Summarize** the three main functions of the digestive system.
4. **Analyze** what the consequence might be if the lining of the small intestine were completely smooth instead of having villi.

Think Critically

5. **Design** an experiment to gather data about the effect of pH on the digestion of different types of food.

MATH in Biology

6. A can of carbonated beverage typically holds about 354 mL of fluid. Compare this amount with the volume of an empty stomach. Give a ratio.
7. **Explain** The pH in the digestive system changes. Give examples and explain the importance of these changes.

 Biology Online Self-Check Quiz biologygmh.com

Reading Preview

Objectives

- **Correlate** activity level with the intake of Calories needed to maintain proper body weight.
- **Describe** how proteins, carbohydrates, and fats are broken down in the digestive tract.
- **Explain** the roles of vitamins and minerals in maintaining homeostasis.
- **Apply** the information in MyPyramid and on food labels as tools for establishing healthy eating habits.

Review Vocabulary

amino acid: basic building block of proteins

New Vocabulary

nutrition
Calorie
vitamin
mineral

Nutrition

MAIN Idea **Certain nutrients are essential for the proper function of the body.**

Real-World Reading Link There is a saying, "You are what you eat." What do you think that means? Much of the time you have freedom to choose what you will eat. However, your choices have consequences. What you eat can affect your health now and in the future.

Calories

Nutrition is the process by which a person takes in and uses food. Foods supply the building blocks and energy to maintain body mass. The daily input of energy from food should equal the amount of energy a person uses daily. A **Calorie** (with an uppercase C) is the unit used to measure the energy content of foods. A Calorie is equal to 1 kilocalorie, or 1000 calories (with a lowercase c). A calorie is the amount of heat needed to raise the temperature of 1 mL of water by 1°C.

The energy content of a food can be measured by burning the food and converting the stored energy to heat. Not all foods have the same energy content. The same mass of different foods does not always equal the same number of Calories. For example, one gram of carbohydrate or protein contains four Calories. One gram of fat contains nine Calories. To lose weight, more Calories must be used than consumed. The opposite is true to gain weight. In 2005, the United States Department of Agriculture released new guidelines for nutrition and suggested that people should become more active and use more Calories. **Table 35.2** compares average Calorie usage with different activities. The exact number of calories burned will vary depending on weight and gender.

Table 35.2 Activities and Average Calorie Usage

Concepts In Motion
Interactive Table To explore more about activities and Calories, visit biologygmh.com.

Activity	Calories Used Per Hour	Activity	Calories Used Per Hour
Baseball	282	Hiking and backpacking	564
Basketball	564	Hockey (field and ice)	546
Bicycling	240–410	Jogging	740–920
Cross-country skiing	700	Skating	300
Football	540	Soccer	540

■ **Figure 35.7** Your body needs carbohydrate-rich foods every day.
Analyze ***Which items in the photo are complex carbohydrates?***

VOCABULARY
SCIENCE USAGE V. COMMON USAGE
Consume
Science usage: to eat or drink.
We consume Calories when we eat food.

Common usage: to destroy.
The fire consumed several buildings.

Carbohydrates

Cereal, pasta, potatoes, strawberries, and rice all contain a high proportion of carbohydrates. Recall from Chapter 6 that sugars, such as glucose, fructose, and sucrose, are simple carbohydrates that are found in fruits, soda pop, and candy. Complex carbohydrates are macromolecules such as starches, which are long chains of sugars. Foods such as those shown in **Figure 35.7** have a high starch content, as do some vegetables.

Complex carbohydrates are broken down into simple sugars in the digestive tract. Simple sugars are absorbed through villi in the small intestine into blood capillaries and circulated throughout the body to provide energy for cells. Excess glucose is stored in the liver in the form of glycogen. Cellulose, sometimes called dietary fiber, is another complex carbohydrate found in plant foods. Although humans cannot digest fiber, it is important because fiber helps keep food moving through the digestive tract and helps with the elimination of wastes. Bran, whole-grain breads, and beans are good sources of fiber.

 **Reading Check** **Compare** simple and complex carbohydrates.

■ **Figure 35.8** Unprocessed fruits and vegetables have low fat content. The way in which naturally low-fat foods are cooked and served can increase fat content, such as when potatoes are fried in saturated fats.

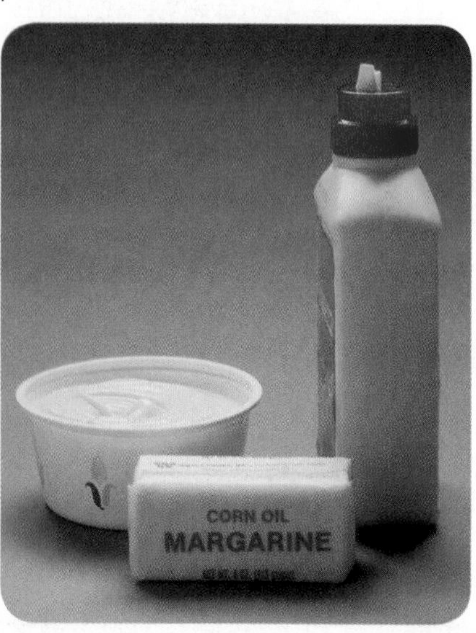

Fats

In proper amounts, fats are an essential part of a healthful diet. Fats are the most concentrated energy source available to the body, and they are building blocks for the body. Fats also protect some internal organs and help maintain homeostasis by providing energy, and storing and transporting certain vitamins. However, not all fats are beneficial.

Connection to Health Recall from Chapter 6 that fats are classified according to their chemical structure as saturated or unsaturated. Meats, cheeses, and other dairy products are sources of saturated fats. A diet high in saturated fats might result in high blood levels of cholesterol, which can lead to high blood pressure and other heart problems. Plants are the main source of unsaturated fats. They are not associated with heart disease, although excessive consumption of any type of fat can lead to weight gain.

A general rule is that saturated fats are solid and unsaturated fats are liquid at room temperature. The margarine in **Figure 35.8** contains less saturated fat than butter. Fats are digested in the small intestine and broken down into fatty acids and glycerol. Fatty acids can be absorbed through the villi and circulated in the blood throughout the body.

Proteins

You have learned that proteins are basic structural components of all cells, and that amino acids are the building blocks of proteins. Enzymes, hormones, neurotransmitters, and membrane receptors are just a few important proteins in the body.

During the process of digestion, proteins in foods are broken down to their subunit amino acids. The amino acids are absorbed into the bloodstream and carried to various body cells. These body cells, through the process of protein synthesis, assemble the amino acids into proteins needed for body structures and functions.

Humans require 20 different amino acids for protein synthesis. The human body can produce 12 of the 20 amino acids needed for cellular function. Essential amino acids are the eight amino acids that must be included in a person's diet. Animal products, such as meats, fish, poultry, eggs, and dairy products, are sources of all eight essential amino acids. Vegetables, fruits, and grains contain amino acids, but no single plant food source contains all eight essential amino acids. However, certain combinations, such as the beans and rice shown in **Figure 35.9,** provide all of the essential amino acids.

Figure 35.9 Beans and rice can be combined to provide all the essential amino acids.
Explain ***Why is it important to eat foods that contain the essential amino acids?***

Food Pyramid

In 2005, the United States Department of Agriculture published a new food pyramid, MyPyramid, which replaces the old pyramid that had been a symbol of good nutrition since 1992. **Figure 35.10** shows a version of the new pyramid. Notice that the orange and green sections are wider than the purple and yellow sections. The message of the pyramid is that a person needs more nutrients from grains and vegetables than from meats and oils.

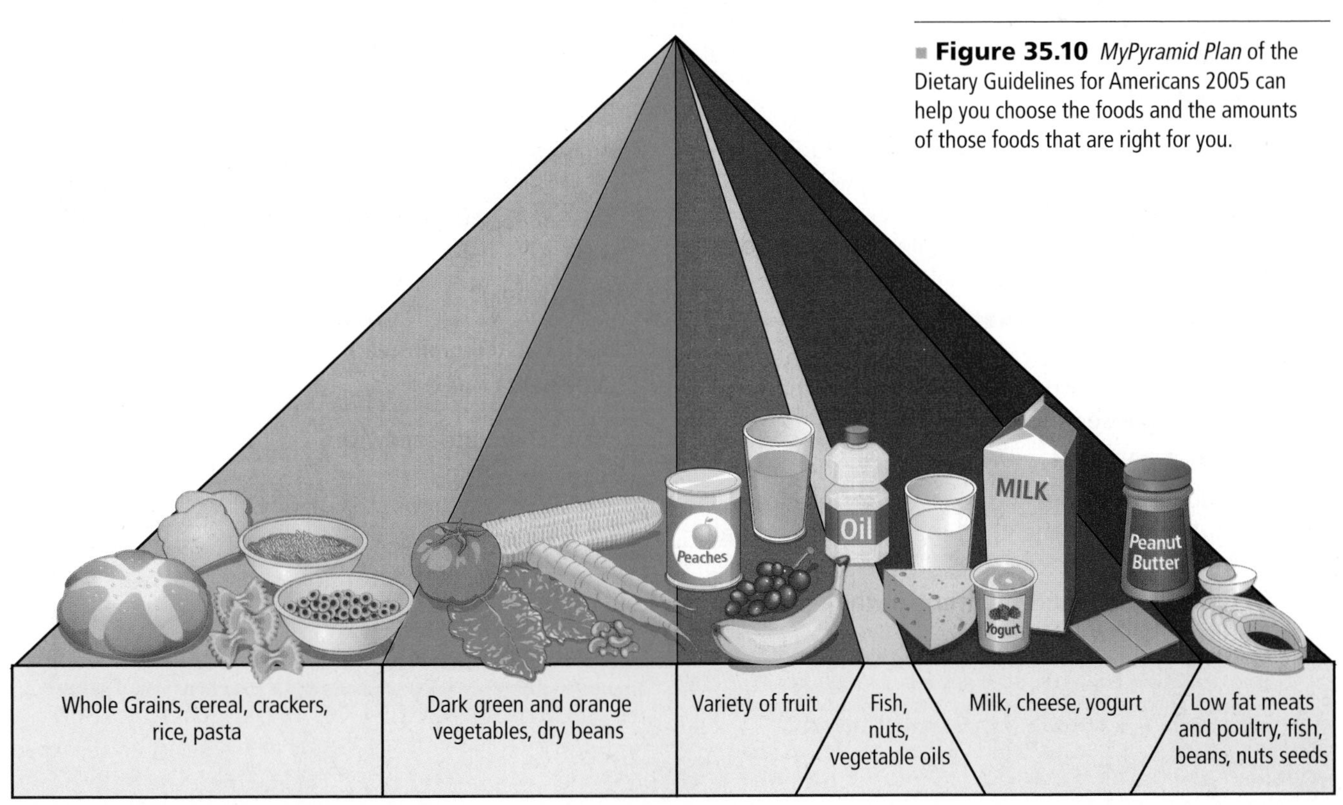

Figure 35.10 *MyPyramid Plan* of the Dietary Guidelines for Americans 2005 can help you choose the foods and the amounts of those foods that are right for you.

Vitamins and Minerals

In addition to carbohydrates, fats, and proteins, your body needs vitamins and minerals to function properly. **Vitamins** are organic compounds that are needed in small amounts for metabolic activities. Many vitamins help enzymes function properly. Some vitamins are produced within the body. Vitamin D is made by cells in your skin. Some B vitamins and vitamin K are produced by bacteria living in the large intestine. However, sufficient quantities of most vitamins cannot be made by the body, but a well-balanced diet can provide the vitamins that are needed. Some vitamins that are fat-soluble can be stored in small quantities in the liver and fatty tissues of the body. Other vitamins are water-soluble and cannot be stored in the body. Foods providing an adequate level of these vitamins should be included in a person's diet on a regular basis.

Minerals are inorganic compounds used by the body as building material, and they are involved with metabolic functions. For example, the mineral iron is needed to make hemoglobin. Recall from Chapter 34 that oxygen binds to hemoglobin in red blood cells and is delivered to body cells as blood circulates in the body. Calcium, another mineral, is an important component of bones.

Vitamins and minerals are essential parts of a healthy diet. **Table 35.3** on the next page lists some important vitamins and minerals, their benefits, and some food sources that can provide these necessary nutrients. Over-the-counter vitamins are also available. Taking more than the recommended daily allowance, however, can be dangerous and should not be done without consulting a doctor.

CAREERS IN BIOLOGY

Registered Dietician A registered dietician helps people address a variety of health issues by showing them how to make healthful decisions about their diets. For more information on biology careers, visit biologygmh.com.

DATA ANALYSIS LAB 35.1

Based on Real Data*

Compare Data

How reliable are food labels? In a study conducted at the U.S. Department of Agriculture Human Nutrition Research Center, scientists measured the mass of 99 single-serving food products.

Data and Observations

The table compares the mass listed on the food package label with the actual mass of the food in five single-serving packages.

Food (1 serving)	Label Mass (g)	Actual Mass (g)
Cereal, bran flakes with raisins (1 box)	39	54.2
Cereal, toasted grains with supplement (1 box)	23	39.6
Cookie, chocolate sandwich (1 pkg)	57	67.0
Mini danish, apple (1 per serving)	35	44.8
Mini donut, chocolate covered (4 per serving)	100	116.5

Think Critically

1. **Calculate** What is the percent difference in mass between the label mass and the actual mass of the cookies?
2. **Compare** What is the trend in the percent differences?

*Data obtained from: Conway, J.M., D.G. Rhodes, and W.V. Rumpler. 2004. Commercial portion-controlled foods in research studies: how accurate are label weights? *Journal of the American Dietetic Association* 104: 1420–1424.

Concepts In Motion

Interactive Table To explore more about vitamins and minerals, visit biologygmh.com.

Table 35.3 Major Roles of Some Vitamins and Minerals

Vitamin	Major Role in the Body	Possible Sources	Mineral	Major Role in the Body
A	• Vision • Health of skin and bones		Ca	• Strengthening of teeth and bone • Nerve conduction • Contraction of muscle
D	• Health of bones and teeth		P	• Strengthening of teeth and bone
E	• Strengthening of red blood cell membrane		Mg	• Synthesis of proteins
Riboflavin (B_2)	• Metabolism of energy		Fe	• Synthesis of hemoglobin
Folic Acid	• Formation of red blood cells • Formation of DNA and RNA		Cu	• Synthesis of hemoglobin
Thiamine	• Metabolism of carbohydrates		Zn	• Healing of wounds
Niacin (B_3)	• Metabolism of energy		Cl	• Balance of water
Pyridoxine (B_6)	• Metabolism of amino acids		I	• Synthesis of thyroid hormone
B_{12}	• Formation of red blood cells		Na	• Nerve conduction • Balance of pH
C	• Formation of collagen		K	• Nerve conduction • Contraction of muscle

■ **Figure 35.11** Notice how many servings are in each food container. The percent daily values are based on an individual serving, not the entire package.

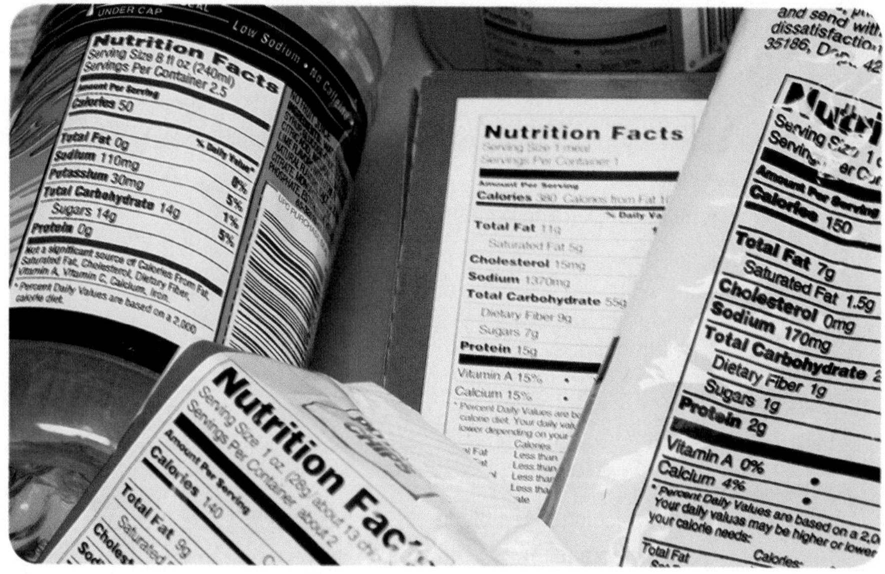

Nutrition Labels

Nutrition labels are provided on commercially packaged foods like those shown in **Figure 35.11.** These labels are based on a 2000-Calorie per day diet. Labels can be especially useful for monitoring fat and sodium intake—two nutrients that need to be consumed in moderation. The FDA requires that food labels list the following information:

- name of the food
- net weight or volume
- name and address of manufacturer, distributor, or packager
- ingredients
- nutrient content

Section 35.2 Assessment

Section Summary

- The energy content of food is measured in Calories.
- Carbohydrates, fats, and proteins are three major groups of nutrients.
- Carbohydrates are a major source of energy for the body.
- Fats and proteins are important building blocks for the body and also provide energy.
- Vitamins and minerals are essential for proper metabolic functioning.
- The *MyPyramid Plan* and food labels are tools you can use to eat healthfully.

Understand Main Ideas

1. MAIN Idea **Explain** why keeping a count of Calories consumed and Calories used is important in maintaining proper functioning of the body.
2. **Describe** how proteins, carbohydrates, and fats are changed in the process of digestion.
3. **Recommend** what nutrients a vegetarian should add to his or her diet.
4. **Explain** the roles of vitamins and minerals in the process of maintaining homeostasis.

Think Critically

5. **Summarize** how many Calories you consume during one day. Record all the foods and beverages you eat and drink in one day. If possible, do the same for total fat and saturated fat.

WRITING in **Biology**

6. Write a short article for your school newspaper describing what is needed for a well-balanced diet.

Section 35.3

The Endocrine System

Reading Preview

Objectives

- **Identify** and describe the function of glands that make up the endocrine system.
- **Explain** the role of the endocrine system in maintaining homeostasis.
- **Describe** feedback mechanisms that regulate hormone levels in the body.

Review Vocabulary

homeostasis: regulation of an organism's internal environment to maintain life

New Vocabulary

endocrine gland
hormone
pituitary gland
thyroxine
calcitonin
parathyroid hormone
insulin
glucagon
aldosterone
cortisol
antidiuretic hormone

MAIN Idea **Systems of the human body are regulated by hormonal feedback mechanisms.**

Real-World Reading Link When driving a car, everyone usually goes a similar speed. When cars go faster or slower than the accepted speed, the chance of an accident increases. Similarly, hormones must stay in the proper balance to maintain homeostasis in the body.

Action of Hormones

The endocrine system is composed of glands and functions as a communication system. **Endocrine glands** produce hormones, which are released into the bloodstream and distributed to body cells. A **hormone** is a substance that acts on certain target cells and tissues to produce a specific response. Hormones are classified as steroid hormones and non-steroid or amino acid hormones based on their structure and mechanism of action.

Steroid hormones Estrogen and testosterone are two examples of steroid hormones. You will learn how each of these affect the human reproductive systems in Chapter 36. All steroid hormones work by causing the target cells to initiate protein synthesis, as illustrated in **Figure 35.12.**

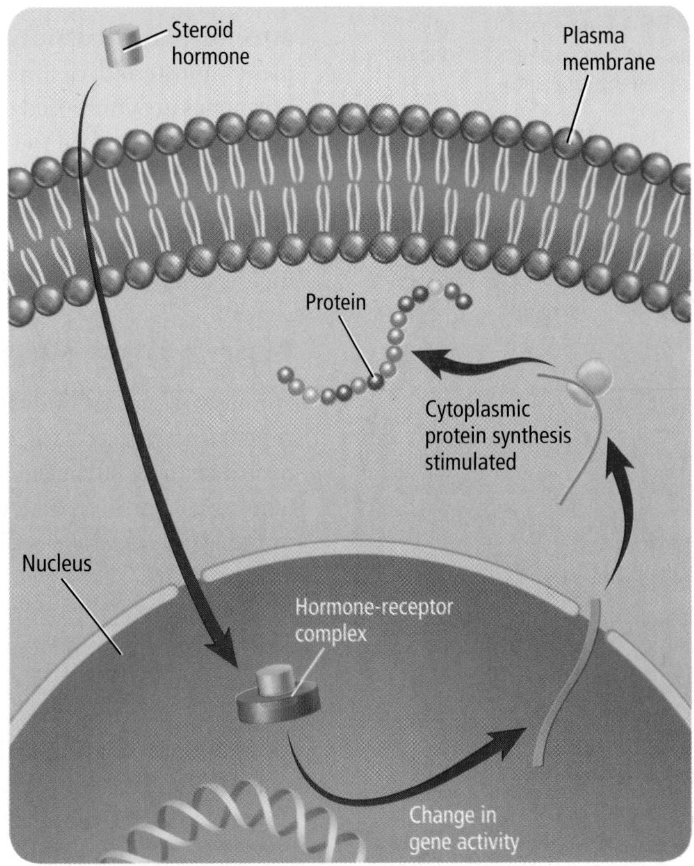

Figure 35.12 A steroid hormone passes through a cell membrane, binds to a receptor within the cell, and stimulates protein synthesis.

Concepts In Motion

Interactive Figure To see an animation of how steroid hormones lead to the production of proteins in a cell, visit biologygmh.com.

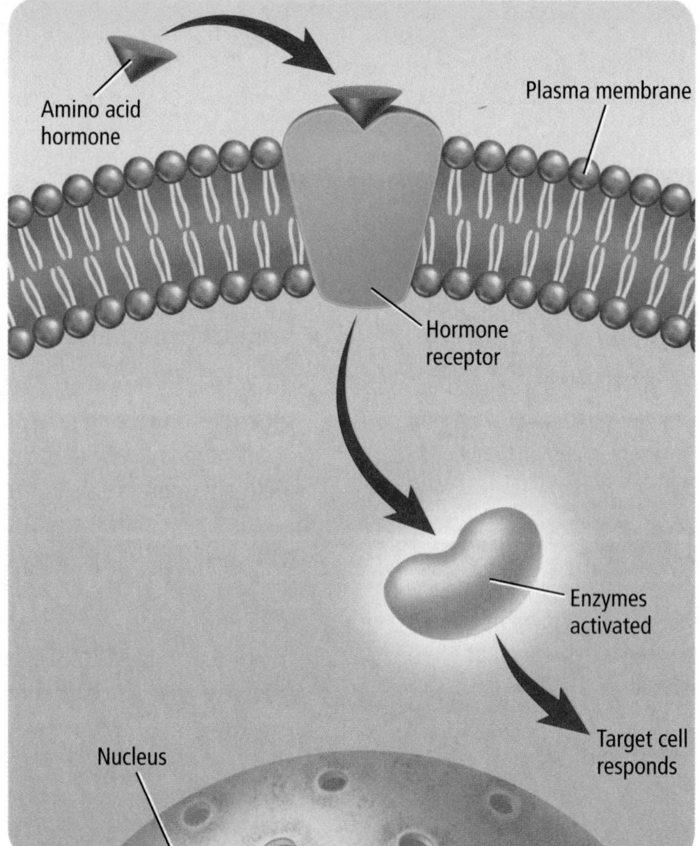

■ **Figure 35.13** An amino acid hormone binds to a receptor on the plasma membrane before entering the cell.
Explain ***What is the difference between amino acid hormones and steroid hormones?***

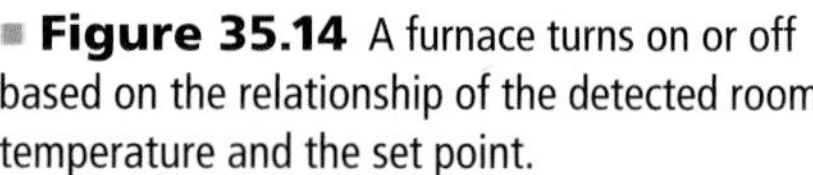

Interactive Figure To see an animation of the two ways an amino acid hormone can control what goes on in a target cell, visit biologygmh.com.

Steroid hormones are soluble in lipids and therefore can diffuse through the plasma membrane of a target cell. Once inside a target cell, they bind to a receptor in the cell. The hormone and the receptor that are bound together bind to DNA in the nucleus, which activates specific genes.

Amino acid hormones Insulin and growth hormones are two examples of nonsteroid, or amino acid, hormones. As the name implies, these hormones are composed of amino acids. Amino acid hormones must bind to receptors found on the plasma membrane of a target cell because they cannot diffuse through the plasma membrane. Once the hormone binds to the receptor, the receptor activates an enzyme found on the inside of the membrane. This usually initiates a biochemical pathway, eventually causing the cell to produce the desired response, as illustrated in **Figure 35.13.**

■ **Figure 35.14** A furnace turns on or off based on the relationship of the detected room temperature and the set point.

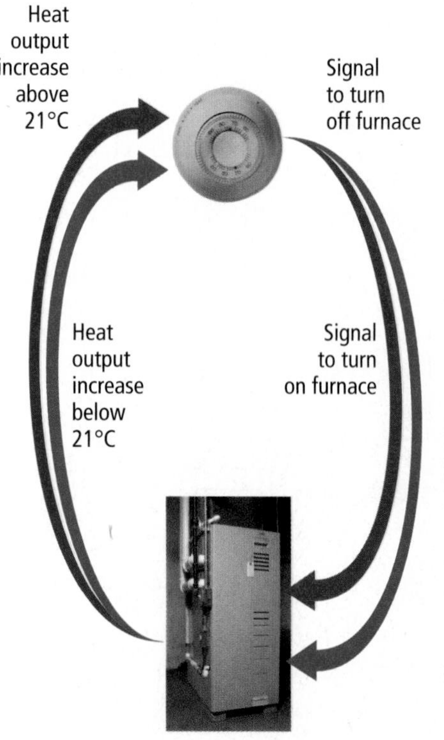

Negative Feedback

Homeostasis in the body is maintained by internal feedback mechanisms called negative feedback. Negative feedback returns a system to a set point once it deviates sufficiently from that set point. As a consequence, the system varies within a particular range. You already might be familiar with an example of a negative feedback system in your own home, illustrated in **Figure 35.14.**

For example, the temperature in a house might be maintained at 21°C. The thermostat in the house detects the temperature, and when the temperature drops below 21°C, the thermostat sends a signal to the heat source, which turns it on and produces more heat. Soon the temperature rises above 21°C, and the thermostat sends a signal to the heat source to shut off. The heat source will not turn on again until the room temperature drops below 21°C and is detected by the thermostat. Because this process can go on indefinitely, negative feedback often is described as a "loop."

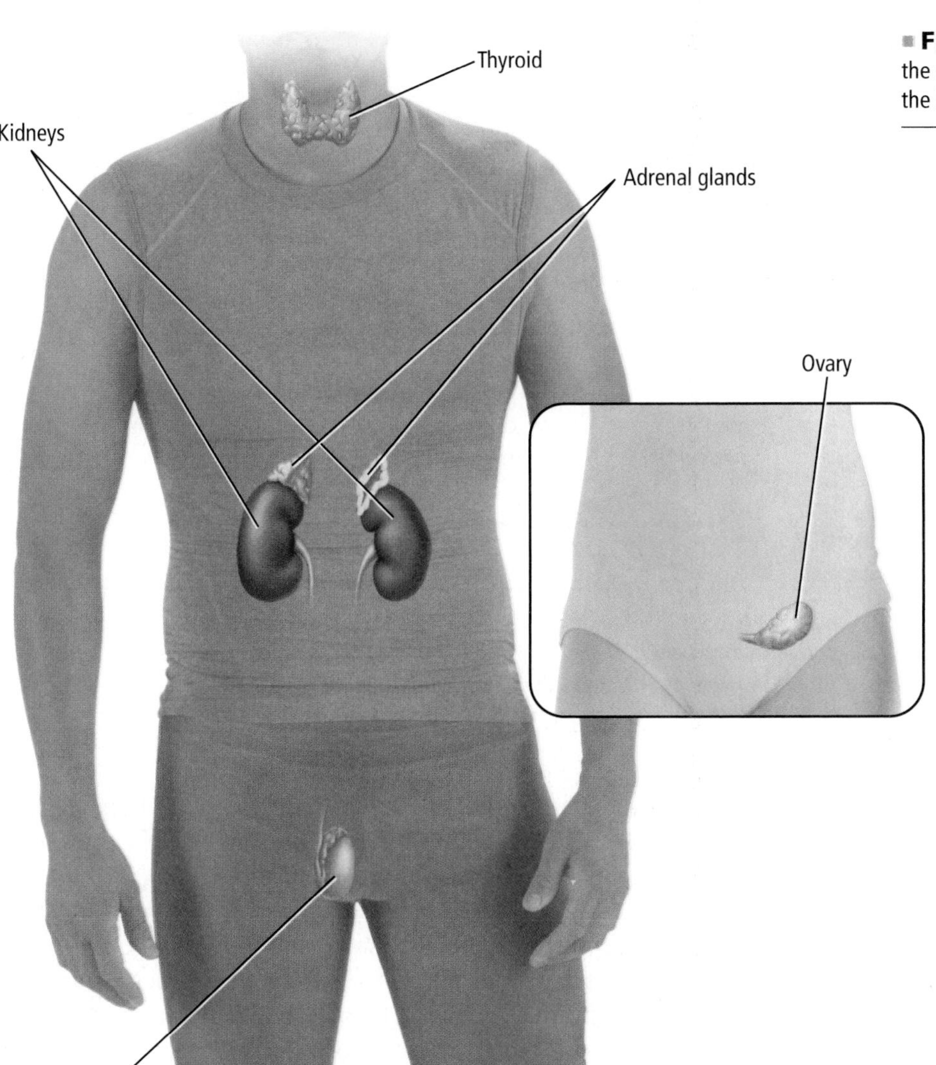

■ **Figure 35.15** The principal glands of the endocrine system are located throughout the body.

Endocrine Glands and Their Hormones

The endocrine system, shown in **Figure 35.15,** includes all the glands that secrete hormones—pituitary, thyroid, parathyroid, and adrenal glands, the pancreas, ovaries, testes, pineal gland, and the thymus gland.

Pituitary gland The pituitary gland is situated at the base of the brain as illustrated in **Figure 35.16.** This gland is sometimes called the "master gland" because it regulates so many body functions. Despite its size, it is the most important endocrine gland. The **pituitary gland** secretes hormones that not only regulate many body functions but also regulates other endocrine glands, such as the thyroid gland, the adrenal glands, the testes, and the ovaries.

A few pituitary hormones act on tissues rather than on specific organs. Human growth hormone (hGH) regulates the body's physical growth by stimulating cell division in muscle and bone tissue. This hormone is especially active during childhood and adolescence.

■ **Figure 35.16** The pituitary gland is located at the base of the brain. The gland has a diameter of approximately 1 cm and weighs 0.5–1 g.

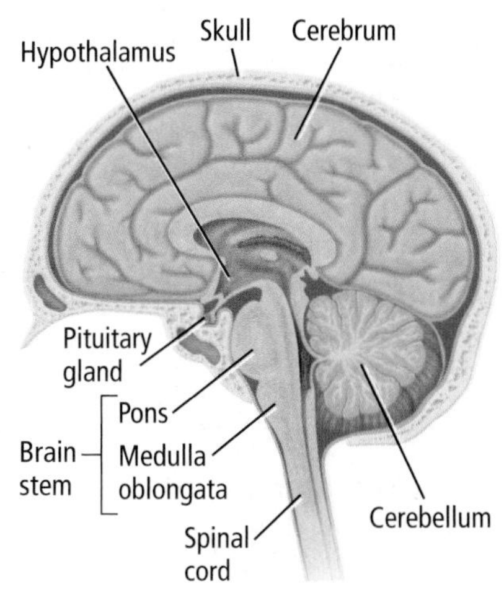

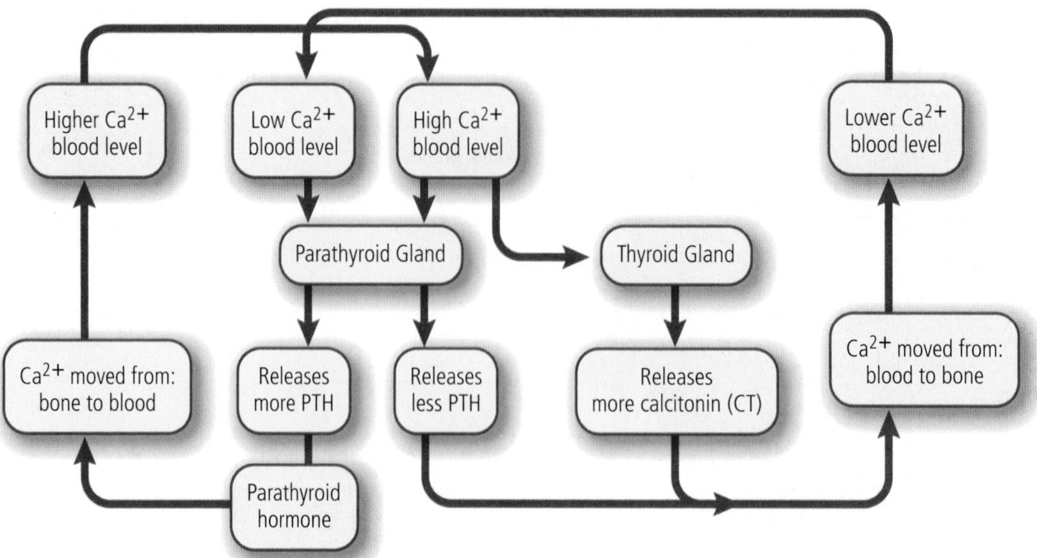

Figure 35.17 Parathyroid hormone (PTH) and calcitonin (CT) regulate the level of calcium in the blood.
Explain *How do PTH and CT illustrate negative feedback?*

Thyroid and parathyroid glands Identify the thyroid and parathyroid glands in **Figure 35.17.** One hormone produced by the thyroid gland is thyroxine. Like hGH, **thyroxine** does not act on specific organs; rather, it causes cells of the body to have a higher rate of metabolism. The thyroid gland also produces calcitonin. **Calcitonin** (kal suh TOH nun) is a hormone that is partly responsible for the regulation of calcium, an important mineral for bone formation, blood clotting, nerve function, and muscle contraction. Calcitonin lowers blood calcium levels by signaling bones to increase calcium absorption and also signaling the kidneys to excrete more calcium.

When blood calcium levels are too low, the parathyroid glands increase production of parathyroid hormone. **Parathyroid hormone** increases blood calcium levels by stimulating the bones to release calcium. The action of this hormone also causes the kidneys to reabsorb more calcium and the intestines to absorb more calcium from food. The thyroid and parathyroid glands have opposite effects on blood calcium levels. However, as they work together, they maintain homeostasis.

Reading Check **Explain** how negative feedback is important in maintaining homeostasis.

Personal Tutor

To learn about insulin feedback, visit biologygmh.com.

Pancreas As discussed in Section 1, the pancreas has a crucial role in the production of enzymes that digest carbohydrates, proteins, and fats. The pancreas also secretes the hormones insulin and glucagon, which work together to maintain homeostasis, as illustrated in **Figure 35.18.** When blood glucose levels are high, the pancreas releases insulin. **Insulin** signals body cells, especially liver and muscle cells, to accelerate the conversion of glucose to glycogen, which is stored in the liver. When blood glucose levels are low, glucagon is released from the pancreas. **Glucagon** (GLEW kuh gahn) binds to liver cells, signaling them to convert glycogen to glucose and release the glucose into the blood.

Figure 35.18 Glucagon and insulin work together to maintain the level of sugar in the blood.

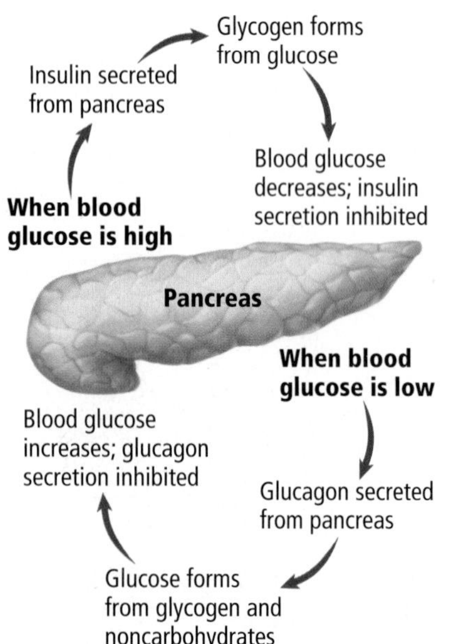

Diabetes is a disease that results from the body not producing enough insulin or not properly using insulin. Type 1 diabetes, which usually appears in people by the age of 20, occurs when the body cannot produce insulin. Type 2 diabetes occurs in 70–80 percent of people diagnosed with diabetes, and usually occurs after the age of 40. It results from the cells of the body becoming insensitive to insulin. Complications from diabetes include coronary heart disease, retinal and nerve damage, and acidosis, or low blood pH. In either type of diabetes, the blood glucose levels must be monitored and maintained to prevent complications from the disease.

Adrenal glands Refer again to **Figure 35.15.** The adrenal glands are located just above the kidneys. The outer part of the adrenals is called the cortex, which manufactures the steroid hormone aldosterone and a group of hormones called glucocorticoids. **Aldosterone** (al DAWS tuh rohn) primarily affects the kidneys and is important for reabsorbing sodium. **Cortisol,** another glucocorticoid, raises blood glucose levels and also reduces inflammation.

The body has different mechanisms for responding to stress, such as those discussed in Chapter 33 concerning the role of the nervous system and the "fight or flight response." The endocrine system also is involved with these types of responses. An "adrenaline rush" occurs when there seems to be a sudden burst of energy during a stressful situation. The inner portions of the adrenal glands secrete epinephrine (eh puh NEH frun), also called adrenaline, and norepinephrine. Together, these hormones increase heart rate, blood pressure, breathing rate, and blood sugar levels, all of which are important in increasing the activity of body cells.

CAREERS IN BIOLOGY

Endocrinologist An endocrinologist studies glands that secrete hormones and the diseases involving them. For more information on biology careers, visit biologygmh.com.

FOLDABLES
Incorporate information from this section into your Foldable.

MiniLab 35.2

Model the Endocrine System

How do hormones help the body maintain homeostasis? Activities like taking a test or running a race place demands on your body. Your body's responses to these demands cause changes in your body. Your endocrine and nervous systems work together to ensure a stable internal environment.

Procedure

1. Read and complete the lab safety form.
2. Identify a sport or activity. Brainstorm what body actions occur as you prepare for, take part in, and recover from the activity.
3. Imagine you are writing a computer program that your body will follow to complete the activity. Sequence the steps you brainstormed in Step 2.
4. Review your program. Insert steps where the endocrine system might secrete hormones to maintain homeostasis. Use your knowledge and available resources to identify the specific hormones involved. Include body responses to these hormones as separate steps.
5. Compare your program with those developed by other students.

Analysis

1. **Think Critically** Did some of the same hormones appear in most of the other programs you studied in Step 5? Why or why not?
2. **Draw Conclusions** List the major body systems represented in your program. What does this show about the range of body functions controlled by the endocrine system?

Visualizing the Endocrine System

Figure 35.19
The hypothalamus maintains homeostasis by serving as a link between the nervous system and the endocrine system. The pituitary releases growth hormone, ADH, and oxytocin as needed by the body. The pituitary gland also manufactures and secretes hormones that regulate the testes, ovaries, and the thyroid and adrenal glands.

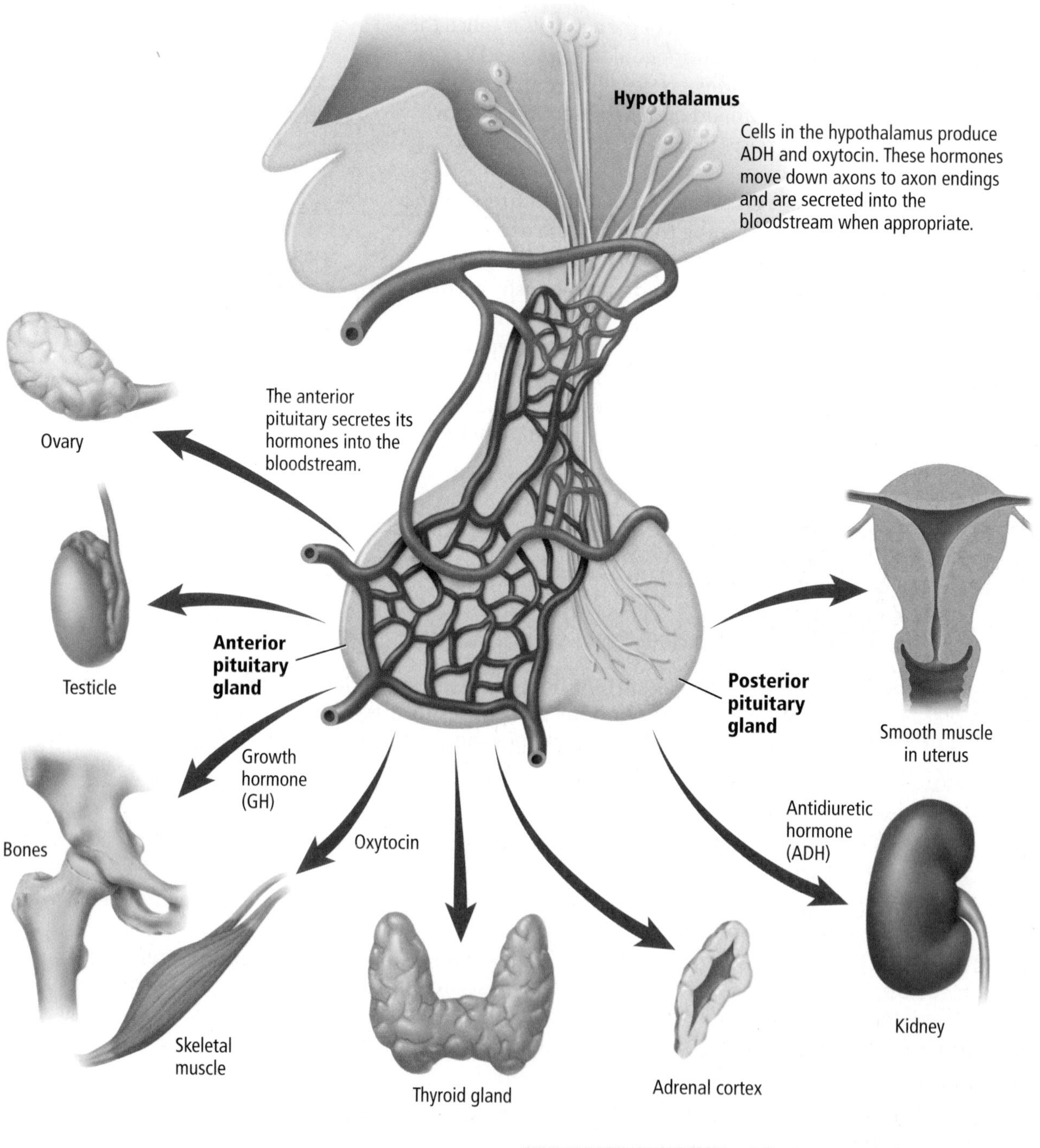

Hypothalamus

Cells in the hypothalamus produce ADH and oxytocin. These hormones move down axons to axon endings and are secreted into the bloodstream when appropriate.

The anterior pituitary secretes its hormones into the bloodstream.

Ovary

Testicle

Anterior pituitary gland

Posterior pituitary gland

Smooth muscle in uterus

Growth hormone (GH)

Bones

Oxytocin

Antidiuretic hormone (ADH)

Skeletal muscle

Thyroid gland

Adrenal cortex

Kidney

Concepts In Motion **Interactive Figure** To see an animation of the endocrine system, visit biologygmh.com.

Biology Online

Link to the Nervous System

The nervous and endocrine systems are similar in that they both are involved in regulating the activities of the body and maintaining homeostasis. Refer to **Figure 35.19** to study the role of the hypothalamus in homeostasis. Recall that this part of the brain is involved with many aspects of homeostasis. The hypothalamus produces two hormones, oxytocin (ahk sih TOH sun) and antidiuretic hormone (ADH). These hormones are transported through axons and stored in axon endings located in the pituitary gland. You will learn more about oxytocin in Chapter 36.

The **antidiuretic** (AN ti DY yuh REH tic) **hormone** (ADH) functions in homeostasis by regulating water balance. ADH affects portions of the kidneys called the collecting tubules. Think back to the last time you were working outside on a hot summer day. You produced a lot of sweat to help keep you cool and you might have become dehydrated. When this happens, cells in your hypothalamus detect that you are dehydrated—that the level of water in the blood is low—and respond by releasing ADH from axons in the pituitary gland that have been storing the hormone.

As illustrated in **Figure 35.20,** ADH travels in the blood to the kidney, where it binds to receptors on certain kidney cells. This causes the kidney to reabsorb more water and decrease the amount of water in the urine, increasing the water level in the blood. If there is too much water in a person's blood, the hypothalamus decreases ADH release, and the urine tends to be more dilute. ADH production is stimulated by nausea and vomiting, both of which cause dehydration. Blood loss of 15 or 20 percent by hemorrhage results in the release of ADH.

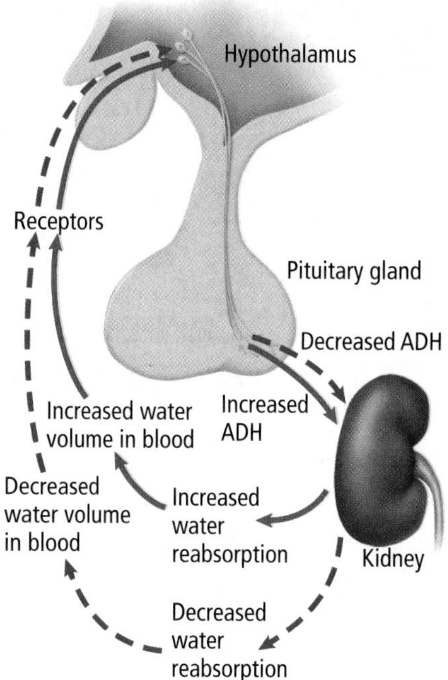

Figure 35.20 Antidiuretic hormone (ADH) helps to control the concentration of water in the blood.

Section 35.3 Assessment

Section Summary

- Endocrine glands produce substances called hormones.
- Hormones travel throughout the body in the bloodstream.
- Hormones are classified as steroid hormones or amino acid hormones.
- Hormone levels are influenced by feedback systems.
- The endocrine system helps to maintain homeostasis with signals from internal mechanisms called negative feedback.

Understand Main Ideas

1. MAIN Idea **Assess** the reasons why hormone feedback systems are referred to as "negative feedback."
2. **Predict** when you would expect to find high levels of insulin in a person's blood and when you would expect to find high levels of glucagon in a person's blood.
3. **Explain** how the endocrine and nervous systems work together to maintain homeostasis.
4. **Identify** and describe the functions of pituitary, thyroid, parathyroid, pancreas, and adrenal glands.

Think Critically

5. **Research** Iodine is essential for thyroid gland function. Fetal and childhood iodine deficiency is a major cause of mental retardation in the world, yet is easily preventable. Predict how iodine deficiency might lead to mental retardation or other health issues. Use your school library or the Internet to research what has been and is being done to alleviate this concern. Include information about possible sources of iodine in your response.
6. **Analyze** how a malfunction in a negative feedback mechanism can lead to the death of an organism.

In the Field

Careers: Forensic Pathologist and Forensic Toxicologist

Tools and Techniques of Forensic Pathology

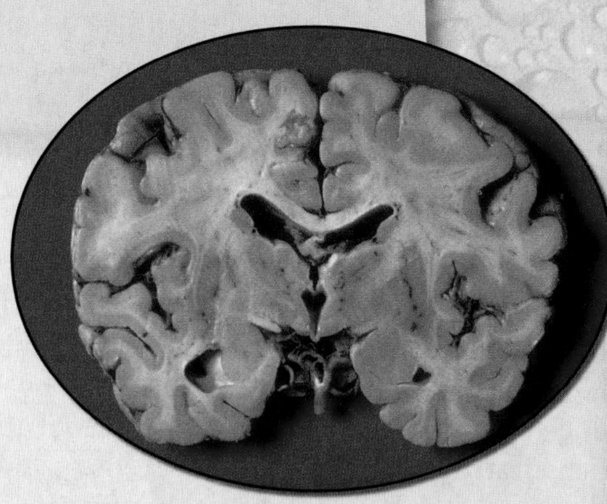
A brain slice might be used to determine a cause of death

Can a dead person talk? In a way, yes. The condition of a dead body can speak volumes about the circumstances surrounding the death. Forensic pathologists gather data from a body, then analyze it to determine when and how a person died. The tools, techniques, and scientific methods that forensic pathologists use help investigators plot the last hours of a person's life, as well as the events that led to death.

Clues from an autopsy The purpose of an autopsy is to make a permanent legal record of a body's characteristics. A forensic pathologist is trained to examine victims of sudden, unexpected, or violent deaths. During an autopsy, the pathologist examines and weighs the lungs, brain, heart, liver, and stomach. He or she uses a scalpel to slice thin sections of the organs, such as the brain slice shown at the right. The slices are chemically preserved to prevent further decay.

Digestion and time of death During the autopsy, the pathologist examines the victim's stomach contents. Why is this important? At the moment of death, digestion stops. The pathologist can use the condition of the stomach to estimate a time line. If the stomach is entirely empty, the victim probably died at least three hours after he or she last ate. If the small intestine also is empty, death likely occurred at least ten hours after the last meal.

Is it possible to identify the type of food in the stomach? In some cases, yes. A scanning electron microscope can be used to identify food particles. A stomach sample that matches the last known meal also can help investigators establish a time period.

Stomach contents can reveal poisoning Toxic substances such as household products, poisons, and drugs can be involved in a death. A forensic toxicologist, a specialist who can identify foreign chemicals that can lead to death, might be called.

While one piece of evidence rarely serves as conclusive proof, forensic pathologists are trained to note specific details. These details can add up and sometimes help tell the story of the final hours of a person's life.

WRITING in Biology

Want Ad Your city has a job opening for a forensic pathologist. Write an advertisement for the job. Be sure to include specific techniques and procedures with which applicants should be familiar, as well as general skills and characteristics applicants should have. For more information about forensic pathology, visit biologygmh.com.

Design Your Own

BIOLAB

HOW DOES THE RATE OF STARCH DIGESTION COMPARE AMONG CRACKERS?

Background: **Starch digestion begins in the mouth. The enzyme amylase, present in saliva, catalyzes the breakdown of starch into sugar molecules, the smallest of which is glucose, an important energy source. Foods, including crackers, vary in starch content. In this lab, you will compare how quickly starch is digested in several types of crackers to determine the relative amount in each.**

Question: *How does the amount of time required for starch digestion by amylase compare among various types of crackers?*

Possible Materials

variety of crackers
mortar and pestle
test tubes and test tube rack
filter paper
funnels
balance
beaker
Bunsen burner or hot plate
graduated cylinder
iodine solution
droppers
watch glasses
amylase solution
glass markers or wax pencil

Safety Precautions

WARNING: *Iodine can irritate and will stain skin.*

Plan and Perform the Experiment

1. Read and complete the lab safety form.
2. Examine three types of crackers. Design an experiment to compare the amount of time required to digest the starch in each. You will use the enzyme amylase to stimulate the digestion of starch. Iodine, a chemical indicator which turns blue-black when starch is present, will indicate when starch digestion is complete.
3. Construct a data chart to record your observations.
4. Consider these points with your group and modify the plan as necessary:
   - What factors will be held constant?
   - Have you established a control sample?
   - How will you know when starch digestion is complete in each sample?
   - How will you keep constant the amount of each type of cracker tested?
   - Will the chart accommodate your data?
5. Make sure your teacher approves your plan before you proceed.
6. Carry out your experiment.
7. **Cleanup and Disposal** Dispose of test tube contents as directed. Clean and return glassware and equipment. Wash hands thoroughly after handling chemicals and glassware.

Analyze and Conclude

1. **Analyze** How did the amylase affect the starch in the crackers?
2. **Observe and Infer** In which cracker was starch digested most quickly? What does this indicate about the amount of starch in this cracker compared to the others?
3. **Think Critically** What variations among human mouths might affect the action of amylase on starch? Explain.
4. **Error Analysis** Did any steps in your procedure introduce uncontrolled variables into the experiment? Explain how the procedure could be redesigned to make these factors constant.

APPLY YOUR SKILL

Redesign your experiment to determine how varying a condition like temperature or pH would affect the digestion of starch by amylase in one of the crackers. To learn more about amylase and digestion, visit BioLabs at biologygmh.com.

Chapter 35 Study Guide

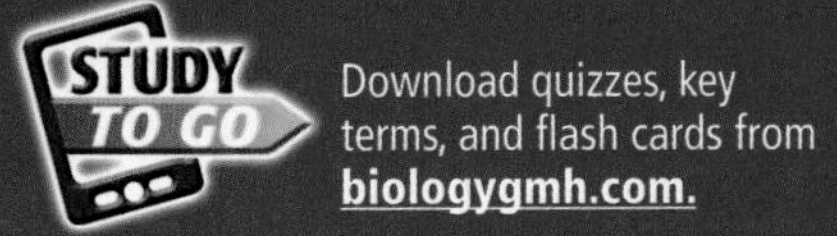

Download quizzes, key terms, and flash cards from biologygmh.com.

FOLDABLES **Predict** what would happen if an organ in the endocrine system did not produce a specific hormone, and the receptive feedback system was either not "switched on" or not "turned off."

Vocabulary	Key Concepts
Section 35.1 The Digestive System	
• amylase (p. 1020) • chemical digestion (p. 1020) • esophagus (p. 1021) • large intestine (p. 1024) • liver (p. 1022) • mechanical digestion (p. 1020) • pepsin (p. 1021) • peristalsis (p. 1021) • small intestine (p. 1022) • villus (p. 1023)	**MAIN Idea** The digestive system breaks down food so nutrients can be absorbed by the body. • The digestive system has three main functions. • Digestion can be categorized as mechanical or chemical. • Most nutrients are absorbed in the small intestine. • Accessory organs provide enzymes and bile to aid digestion. • Water is absorbed from chyme in the colon.
Section 35.2 Nutrition	
• Calorie (p. 1025) • mineral (p. 1028) • nutrition (p. 1025) • vitamin (p. 1028)	**MAIN Idea** Certain nutrients are essential for the proper function of the body. • The energy content of food is measured in Calories. • Carbohydrates, fats, and proteins are three major groups of nutrients. • Carbohydrates are a major source of energy for the body. • Fats and proteins are important building blocks for the body and also provide energy. • Vitamins and minerals are essential for proper metabolic functioning. • The *MyPyramid Plan* and food labels are tools for establishing healthy eating habits.
Section 35.3 The Endocrine System	
• aldosterone (p. 1035) • antidiuretic hormone (p. 1037) • calcitonin (p. 1034) • cortisol (p. 1035) • endocrine gland (p. 1031) • glucagon (p. 1034) • hormone (p. 1031) • insulin (p. 1034) • parathyroid hormone (p. 1034) • pituitary gland (p. 1033) • thyroxine (p. 1034)	**MAIN Idea** Systems of the human body are regulated by hormonal feedback mechanisms. • Endocrine glands produce substances called hormones. • Hormones travel throughout the body in the bloodstream. • Hormones are classified as steroid hormones or amino acid hormones. • Hormone levels are influenced by feedback systems. • The endocrine system helps to maintain homeostasis with signals from internal mechanisms called negative feedback.

CHAPTER 35 Assessment

Section 35.1

Vocabulary Review

For each set of terms, choose the one term that does not belong and explain why it does not belong.

1. esophagus—pancreas—large intestine
2. pepsin—glycogen—glucose
3. bile—amylase—peristalsis

Understand Key Concepts

4. Which action takes place in the stomach?
   A. Large fat molecules are digested into smaller molecules.
   B. Proteins are broken down.
   C. Amylase breaks down starches into smaller sugar molecules.
   D. Insulin is secreted for use in the small intestine.

5. Which row in the chart contains the words that best complete this statement? The (1) produces (2), which is secreted into the (3).

Row	1	2	3
A	liver	bile	small intestine
B	gallbladder	pepsin	stomach
C	pancreas	acid	large intestine
D	villi	amylase	mouth

   A. Row A
   B. Row B
   C. Row C
   D. Row D

6. A person complaining of digestion problems is determined to be not digesting fats well. Which explains this condition?
   A. The sphincter at the bottom of the stomach is not allowing bile to pass into the small intestine.
   B. The duct leading from the liver and gallbladder is blocked.
   C. The person is secreting excess bile.
   D. The stomach is not acidic enough for the digestion of fats.

Use the graph to answer question 7.

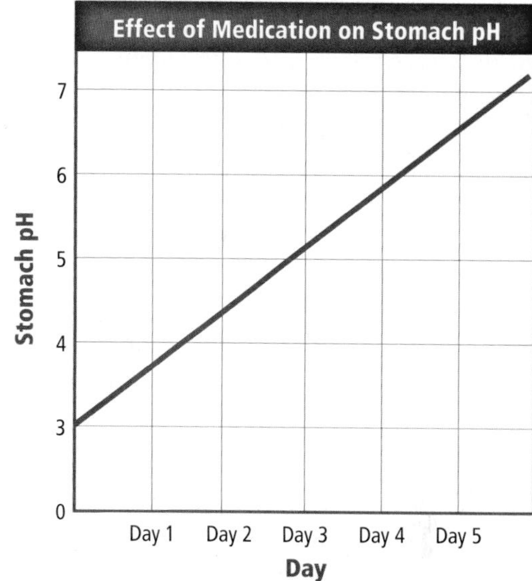

7. A person has been taking a medication for 5 days. Which of the following is likely to be a consequence of this medication?
   A. Pepsin would not be able to break down proteins.
   B. Amylase would not be able to break down starch.
   C. Bile would not be able to be produced.
   D. Enzymes secreted by the pancreas would not function well.

Constructed Response

8. **Short Answer** Explain why the term *heartburn* is an inaccurate description of this condition.

9. **Short Answer** Refer to **Table 35.1** to summarize the digestive processes that occur in the following structures: mouth, large intestine, stomach, small intestine, and esophagus.

10. **Open Ended** Why can a person live without a gall bladder? Assess the effects (if any) that this would have on the person's ability to digest food.

Think Critically

11. **Explain** why a drug manufacturer might add vitamin K to some antibiotics in tablet or pill form.

12. **Hypothesize** why humans have an appendix if it has no known useful function.

Section 35.2

Vocabulary Review

Distinguish between the terms in each pair.

13. saturated fats—unsaturated fats

14. micronutrients—macronutrients

15. vitamins—minerals

Understand Key Concepts

16. Which are characteristic of saturated fats?
- **A.** liquid at room temperature and found in vegetable oils
- **B.** mostly absorbed in the large intestine
- **C.** derived from animal sources and are solid at room temperature
- **D.** tend to lower blood cholesterol

17. Which carbohydrate is not digestible and provides fiber in your diet?
- **A.** sucrose
- **B.** starch
- **C.** glycogen
- **D.** cellulose

18. Which combinations in the stomach break down high-protein foods?
- **A.** a low pH and pepsin
- **B.** a high pH and bile
- **C.** a high pH and pepsin
- **D.** a low pH and bile

Use the image below to answer question 19.

19. If you ate the entire bag of chips, what percent of the recommended daily value of saturated fat would you consume?
- **A.** 14 percent
- **B.** 28 percent
- **C.** 5 percent
- **D.** 35 percent

Constructed Response

20. **CAREERS IN BIOLOGY** According to dieticians, low carbohydrate diets are usually high-fat, high-protein diets. Evaluate what health risks might be associated with long-term intake of foods high in fats and proteins.

21. **Open Ended** Point out what factors, besides not having enough food, might cause an individual to be malnourished.

Think Critically

22. **Explain** why a diet high in fiber might reduce the chance of colon cancer.

23. **Infer** the reasons why obesity rates in the United States have continued to rise steadily for at least the past 30 years.

Section 35.3

Vocabulary Review

Explain the difference between each term. Then explain how the terms are related.

24. insulin—glucagon

25. estrogen—growth hormone

26. cortisol—epinephrine

Understand Key Concepts

Use the graph below to answer question 27.

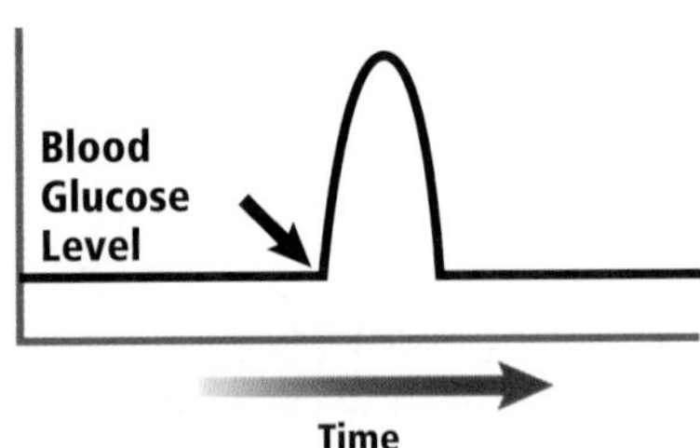

27. The graph shows blood glucose levels over a period of time. Which hormone might have caused a sudden surge as indicated by the arrow?
- **A.** antidiuretic hormone
- **B.** growth hormone
- **C.** glucagon
- **D.** insulin

28. Which hormones are released from nerve cells rather than endocrine glands?
 A. antidiuretic hormone and oxytocin
 B. growth hormone and thyroxine
 C. insulin and glucagon
 D. norepinephrine and epinephrine

29. Which pairs of hormones have opposite effects?
 A. calcitonin and parathyroid hormone
 B. epinephrine and norepinephrine
 C. growth hormone and thyroxine
 D. aldosterone and cortisol

Use the photo below to answer question 30.

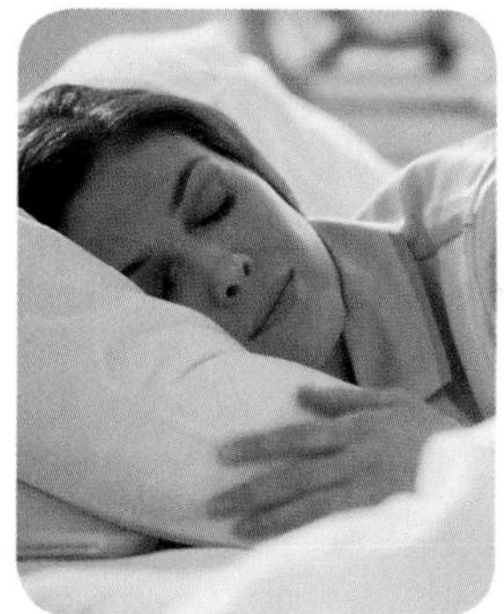
A.

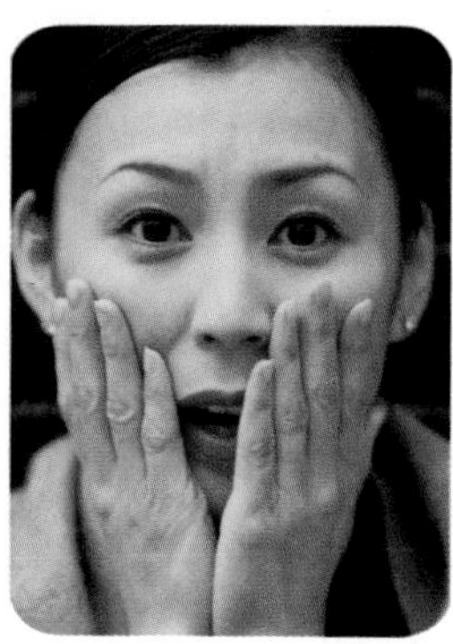
B.

30. Which person is likely to have high levels of epinephrine?
 A. person A
 B. person B
 C. both persons
 D. neither person

Constructed Response

31. **Open Ended** What would be the direct effect of overproduction of calcitonin? Analyze how this might disrupt homeostasis in systems other than the endocrine system.

32. **Short Answer** Assess how the long-term use of cortisol would impact one's ability to fight infection.

Think Critically

33. **Create** an analogy using a balance describing the relationship between calcitonin and parathyroid hormone.

34. **Hypothesize** Why is insulin usually injected instead of taken orally?

Additional Assessment

35. WRITING in Biology This chapter began with a situation in which you are eating a pizza. Write a short story describing the events that occur as the food moves through your digestive tract. *Hint: Be sure to include all major groups of nutrients.*

DBQ Document-Based Questions

Source: *Dietary Guidelines for America 2005*

Estimated Calorie Requirements in Gender and Age Groups

Gender	Age	Moderately Active	Active
Female	9–13	1600–2000	1800–2200
	14–18	2000	2400
	19–30	2000–2200	2400
	31–50	2000	2200
	51+	1800	2000–2200
Male	9–13	1800–2200	2000–2600
	14–18	2400–2800	2800–3200
	19–30	2600–2800	3000
	31–50	2400–2600	2800–3000
	51+	2400	2400–2800

36. According to the chart, which gender needs more Calories?

37. Describe the general trend regarding the number of Calories needed to maintain energy balance in relation to age.

38. Why do individuals in the 19–30-year-old group need the most Calories?

Cumulative Review

39. Describe the three different types of plant cells. **(Chapter 22)**

40. Apply your knowledge of radial symmetry to describe the feeding habits of cnidarians. **(Chapter 24)**

41. Sketch a typical arthropod and label its structures. **(Chapter 26)**

Standardized Test Practice

Cumulative

Multiple Choice

1. What is the function of melanin in the epidermis?
   A. to protect tissue from ultraviolet radiation
   B. to provide support for blood vessels
   C. to stimulate the growth of hair in the follicles
   D. to waterproof and protect the skin surface

Use the diagram below to answer questions 2 and 3.

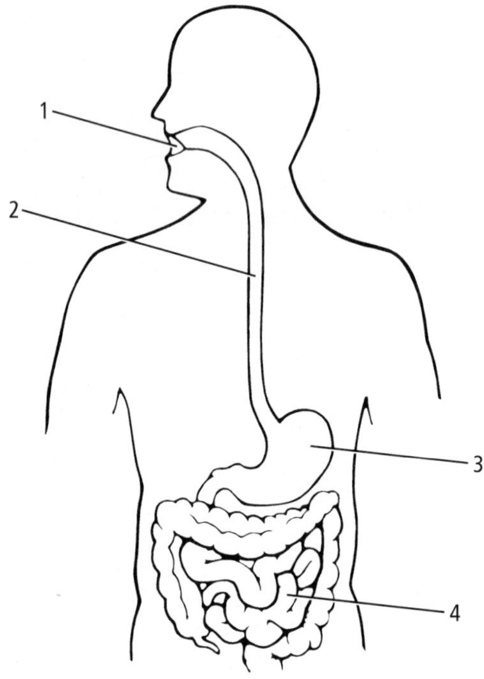

2. In which part of the digestive system does chemical and mechanical digestion first occur?
   A. 1
   B. 2
   C. 3
   D. 4

3. Which process happens first in a nerve cell when a stimulus reaches threshold?
   A. Potassium channels in the cell membrane open.
   B. Neurotransmitters are released into the synapse.
   C. Sodium ions move into the nerve cell.
   D. The cell becomes negatively charged.

4. Where would fat stored in bones be found?
   A. compact bone
   B. osteocytes
   C. red marrow
   D. yellow marrow

Use the diagram below to answer question 5.

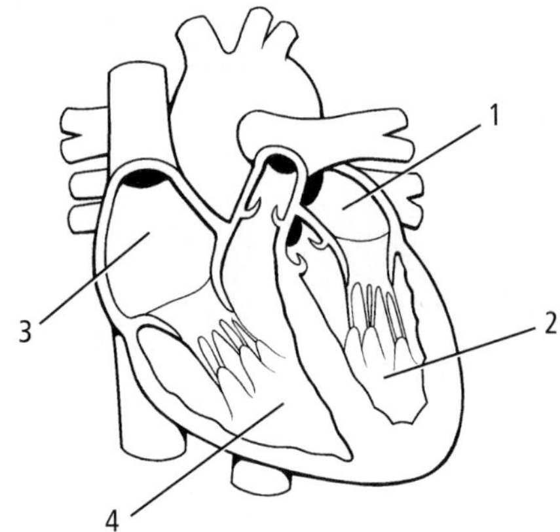

5. Which is the path that blood follows as it flows through the heart immediately after returning from the head and body?
   A. $1 \rightarrow 2$
   B. $2 \rightarrow 1$
   C. $3 \rightarrow 4$
   D. $4 \rightarrow 3$

6. Which describes how filtering occurs in the excretory system?
   A. Blood enters nephrons of the kidneys, and excess water and wastes are filtered from the blood.
   B. Urine leaves the kidneys through ureters.
   C. Water and nutrients are absorbed back into the blood.
   D. Water is added to excess nitrogenous wastes from the digestive system to form urine.

Short Answer

Use the graph below to answer questions 7 and 8.

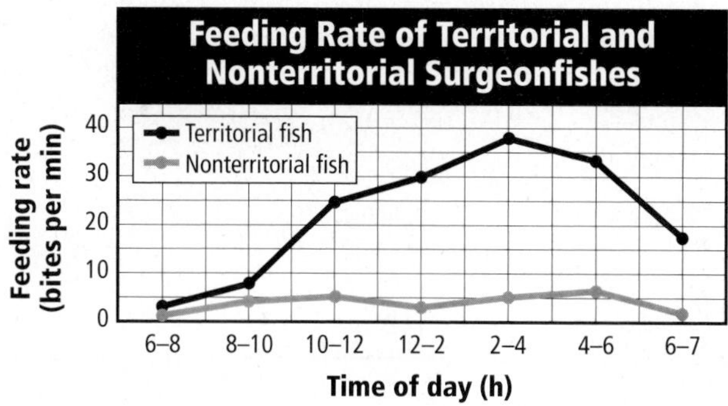

7. Compare and contrast the feeding behavior of the fishes shown in the graph.

8. Predict how the graph might appear if the territorial fish showed territorial behavior only during one season of the year.

9. Assess why a diet with no protein would be unhealthy.

10. What are two benefits to the young of mammals in receiving milk from their mothers?

11. Explain how the different body structures in roundworms and annelids enable them to move.

12. A person who exercises in extreme heat can lose salts that contain potassium and sodium through his or her sweat. What can you infer about the effect of overexertion on the nervous system?

13. Differentiate the three main vessels that blood flows through as it goes from the heart through the body, and returns to the heart.

Extended Response

14. Evaluate how a swim bladder helps a fish maintain its depth.

15. Evaluate how high blood pressure and kidney damage could be related.

16. Name three components of sympathetic stimulation, and assess how they could be helpful to a human's survival.

Essay Question

Humans need Vitamin C in their diet because it strengthens the function of the immune system and prevents a disease called scurvy. Vitamin C is water-soluble so it is not stored in the body. Vitamin C often is suggested for someone who is just getting sick or is already sick. Some people recommend taking very high doses of Vitamin C, sometimes even thousands of times higher than the recommended dose. Medical researchers disagree about the effectiveness of taking large doses of Vitamin C. Some researchers think that it does nothing while others think that it is helpful. Almost all medical researchers agree that taking large doses of Vitamin C for short periods of time is probably not harmful.

Using the information in the paragraph above, answer the following question in essay format.

17. Formulate a hypothesis about whether or not taking large doses of Vitamin C for a cold is helpful. Explain one way this hypothesis could be tested.

NEED EXTRA HELP?																	
If You Missed Question . . .	1	2	3	4	5	6	7	8	9	10	11	12	13	14	15	16	17
Review Section . . .	32.1	35.1	33.1	32.2	34.1	34.3	31.1	31.1	35.2	30.1	25.3	33.1	34.1	28.2	34.3	33.2	35.2

CHAPTER 36

Human Reproduction and Development

BIG Idea Human reproduction involves the joining together of sperm and egg.

Section 1
Reproductive Systems
MAIN Idea Hormones regulate human reproductive systems, including the production of gametes.

Section 2
Human Development Before Birth
MAIN Idea A human develops from a single fertilized cell into trillions of cells with specialized functions.

Section 3
Birth, Growth, and Aging
MAIN Idea Developmental changes continue throughout the stages of life.

BioFacts

- A human embryo increases in size 10,000 times in the first 30 days.
- The largest human newborn weighed 10.8 kg at birth.
- Approximately 97 percent of all live births in the United States are single births and 3 percent are multiple births.

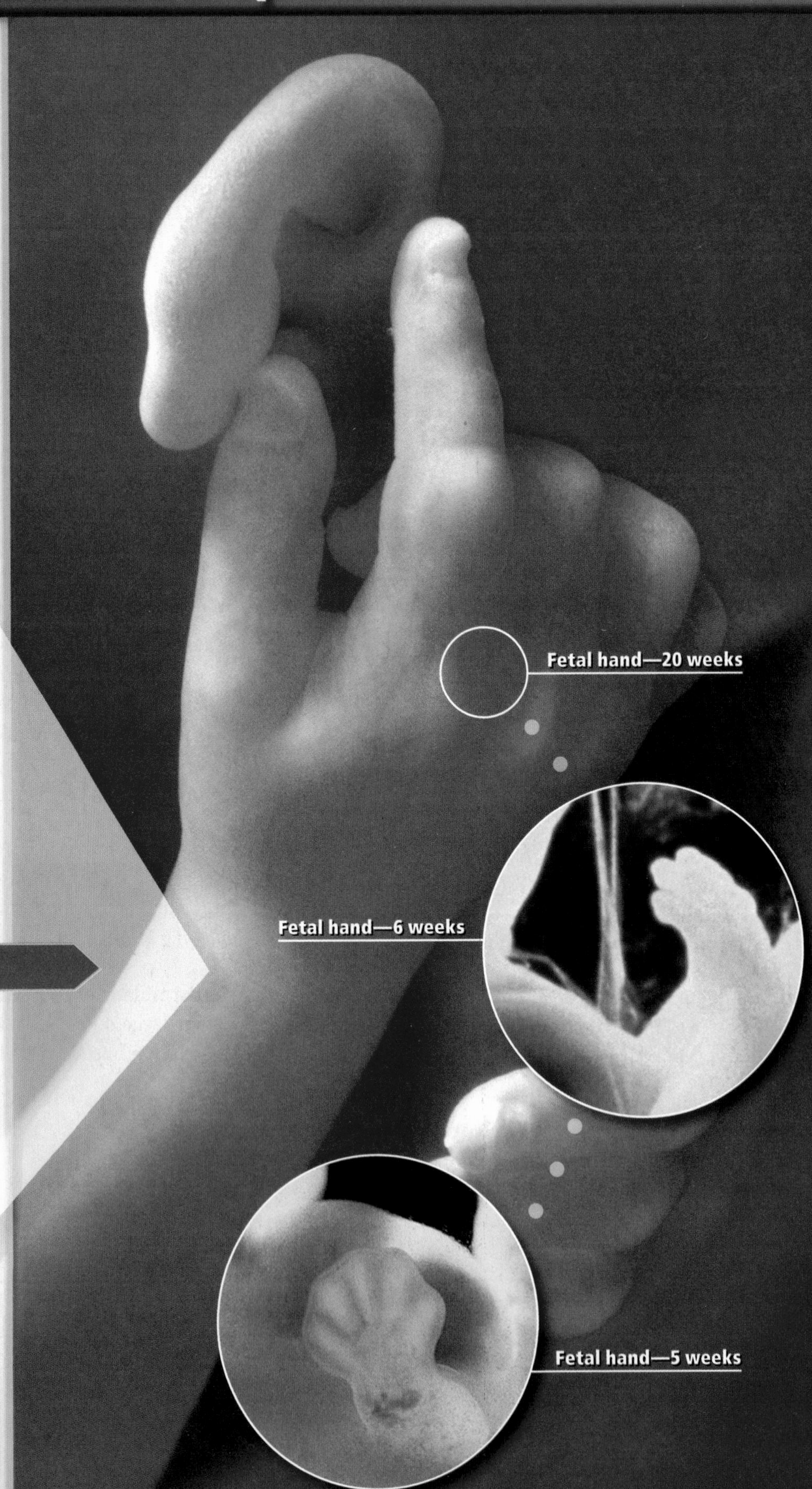

LAUNCH Lab

Sex Cell Characteristics

How are sex cells specialized for the formation of a zygote? Reproduction is a process which follows a predictable pattern. The production of sex cells is a crucial step in reproduction. Sperm and egg cells have specific characteristics that support their roles in reproduction. In this lab, you will investigate how the design of sex cells supports their function.

Procedure

1. Read and complete the lab safety form.
2. Observe the **slide of the egg cell** under the **microscope** and identify its characteristics. Make a sketch.
3. Observe the **slide of the sperm cell** under the microscope and identify its characteristics. Make a sketch.

Analysis

1. **Compare and contrast** the sperm and egg cells you studied. How do they differ?
2. **Identify** What structures and characteristics did you observe that might affect each cell's role in reproduction?

Visit biologygmh.com to:

- study the entire chapter online
- explore Concepts in Motion, Interactive Tables, Microscopy Links, and links to virtual dissections
- access Web links for more information, projects, and activities
- review content online with the Interactive Tutor, and take Self-Check Quizzes

Reproductive System Make this Foldable to help you compare and contrast the production of eggs and sperm.

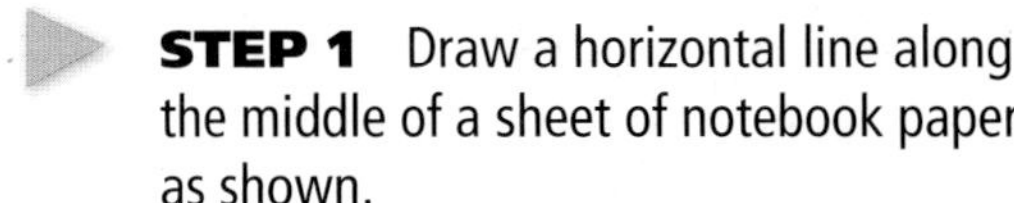

STEP 1 Draw a horizontal line along the middle of a sheet of notebook paper as shown.

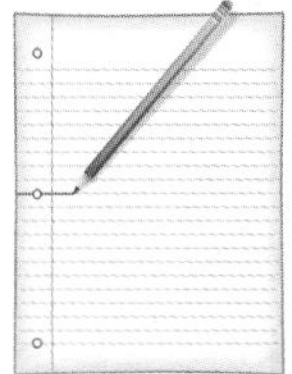

STEP 2 Fold the paper from the top and bottom so the edges meet at the center line.

STEP 3 Label the two tabs as shown.

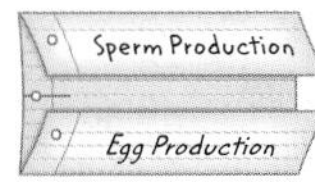

FOLDABLES **Use this Foldable with Section 36.1.** As you study the section, use the Foldable to record what you learn about the production of sperm in the testes and the production of eggs in the ovary.

Section 36.1

Reading Preview

Objectives

- **Summarize and discuss** the structures of the reproductive systems.
- **Explain** how hormones regulate the reproductive systems.
- **Sequence** the menstrual cycle.

Review Vocabulary

hypothalamus: portion of the brain that connects the endocrine and nervous systems, and controls the pituitary gland

New Vocabulary

seminiferous tubule
epididymis
vas deferens
urethra
semen
puberty
oocyte
oviduct
menstrual cycle
polar body

Reproductive Systems

MAIN Idea Hormones regulate human reproductive systems, including the production of gametes.

Real-World Reading Link You might have noticed how the temperature of a room affects the thermostat that controls furnace activity. If the room is warm, the thermostat will not allow the furnace to run. Similarly, male and female hormones in the human body have effects on body structures and influence human reproduction.

Human Male Reproductive System

Reproduction is necessary to ensure continuation of a species. The result of the human reproductive process is the union of an egg cell and a sperm cell, development of the fetus, and the birth of an infant. The organs, glands, and hormones of the male and female reproductive systems are instrumental in meeting this goal.

Figure 36.1 illustrates the male reproductive structures. The male reproductive glands are called the testes (tes TEEZ) (singular, testis) and are located outside of the body cavity in a pouch called the scrotum (SKROH tum). A temperature lower than 37°C—the average body temperature—is required for the development of sperm. Because the scrotum is located outside of the body cavity, it is several degrees cooler. This makes the environment suitable for the normal development of sperm.

Figure 36.1 The male reproductive system produces gametes called sperm in the testes.

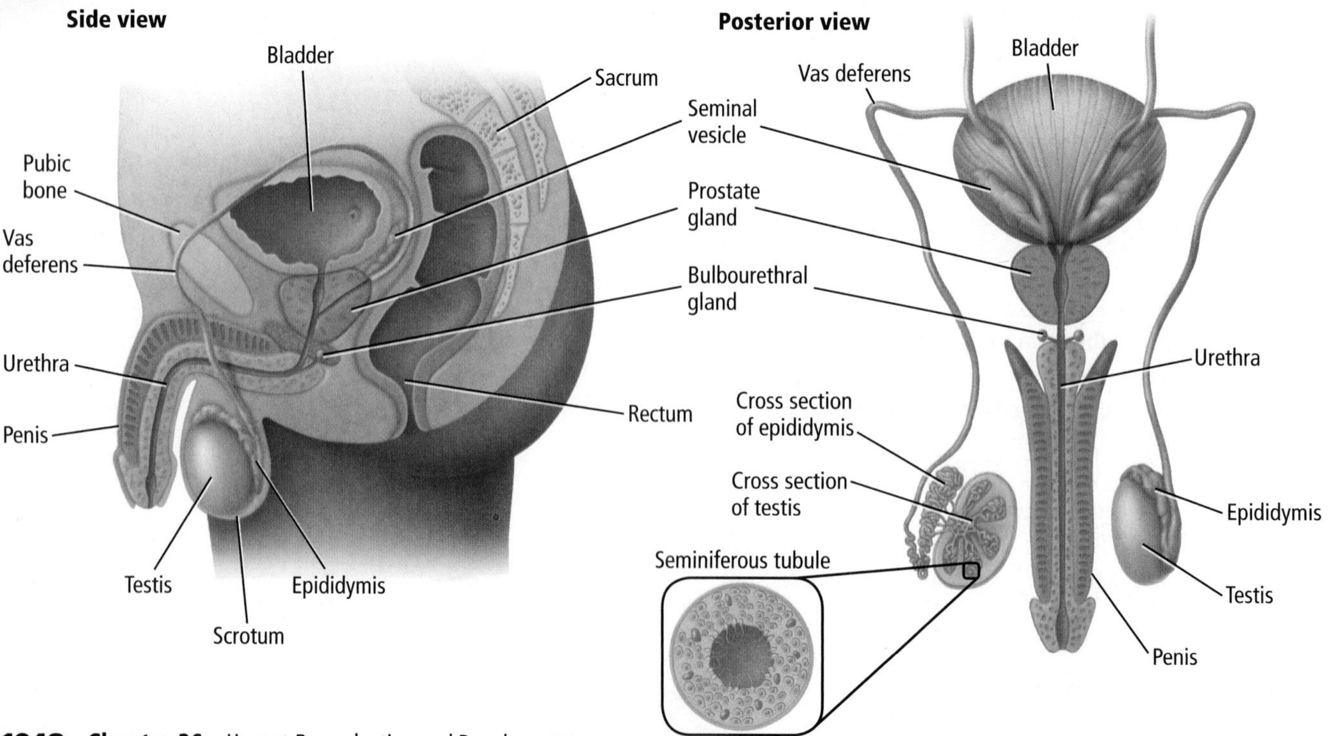

Sperm cells The male reproductive cells, called sperm cells, are produced in the testes. Follow the path that sperm travel in **Figure 36.1** as you read about the structures in the male reproductive system. Sperm, like the one shown in **Figure 36.2,** develop in the testes in the **seminiferous tubules** (se muh NIHF rus • TEW byulz). These tubules produce 100–200 million sperm each day. Next, sperm travel to the **epididymis** (eh puh DIH duh mus), a structure located on top of each testis where sperm mature and are stored. When the sperm are released from the body, they travel through the **vas deferens** (VAS • DEF uh runz), a duct leading away from the testis. There are two vas deferens, one leading away from each testis. The two vas deferens join together and enter the **urethra** (yoo REE thruh), the tube that carries both semen and urine outside of the body through the penis.

Sperm require a nourishing fluid to survive long enough to fertilize an egg. **Semen** (SEE mun) refers to the fluid that contains sperm, the nourishment, and other fluids from the male reproductive glands. The seminal vesicles contribute over half of the semen and secrete sugar into the fluid, which provides energy, other nutrients, proteins, and enzymes for the sperm. The prostate gland and bulbourethral glands contribute an alkaline solution to the fluid to neutralize acidic conditions sperm might encounter in the urethra and the female reproductive tract.

Male hormones Testosterone (tes TAHS tuh rohn), which is made in the testes, is a steroid hormone that is necessary for the production of sperm. It also influences the development of male secondary sex characteristics that begin to appear at **puberty,** the period of growth when sexual maturity is reached. These characteristics include hair on the face and chest, broad shoulders, increased muscle development, and a deeper voice. Recall from Chapter 34 that the larynx contains the vocal cords. Because the vocal cords are longer in males than in females, the male voice is deeper. Later in life, testosterone might lead to a receding hairline or baldness.

Three hormones influence testosterone production. **Figure 36.3** indicates that the hypothalamus, discussed in Chapter 33, produces gonadotropin-releasing hormone (GnRH) that acts on the anterior pituitary gland. GnRH increases the production of follicle-stimulating hormone (FSH) and luteinizing (LEW tee uh ni zing) hormone (LH). Both FSH and LH travel from the anterior pituitary gland through the bloodstream and to the testes. In the testes, FSH promotes the production of sperm, and LH stimulates the production and secretion of testosterone.

Levels of the male hormones are regulated by a negative feedback system that starts with the hypothalamus. Increased levels of testosterone in the blood are detected by cells in the hypothalamus and anterior pituitary, and the production of LH and FSH is decreased. When testosterone levels in the blood drop, the body responds by making more LH and FSH, as shown in **Figure 36.3.**

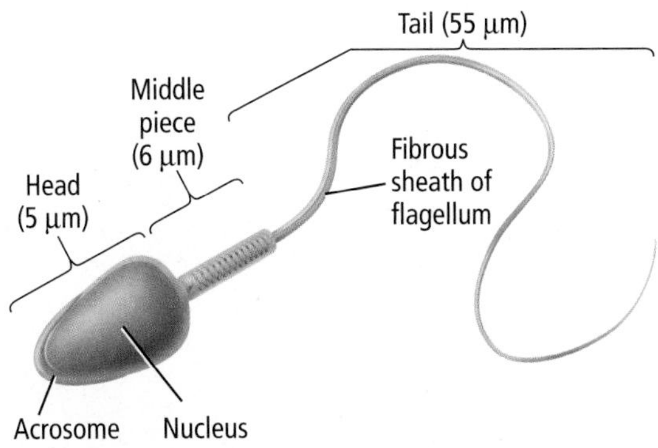

Figure 36.2 A sperm is a flagellated cell composed of a head, midpiece, and tail.
List, ***in correct sequence, the structures a sperm cell passes through or encounters as it makes its way out of the body.***

Figure 36.3 The hypothalamus produces the releasing hormone GnRH that travels to the pituitary gland. GnRH influences the rate of LH and FSH production. The levels of LH and FSH are regulated by a negative feedback pathway.

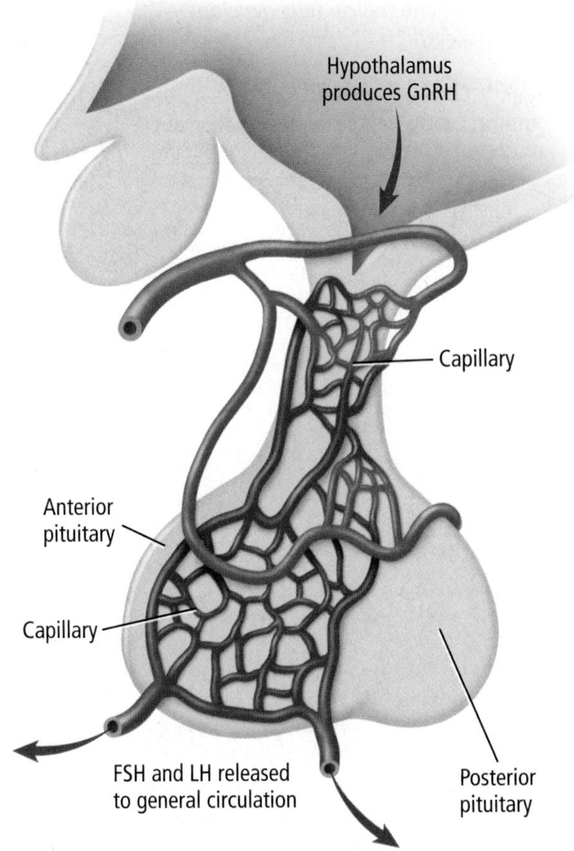

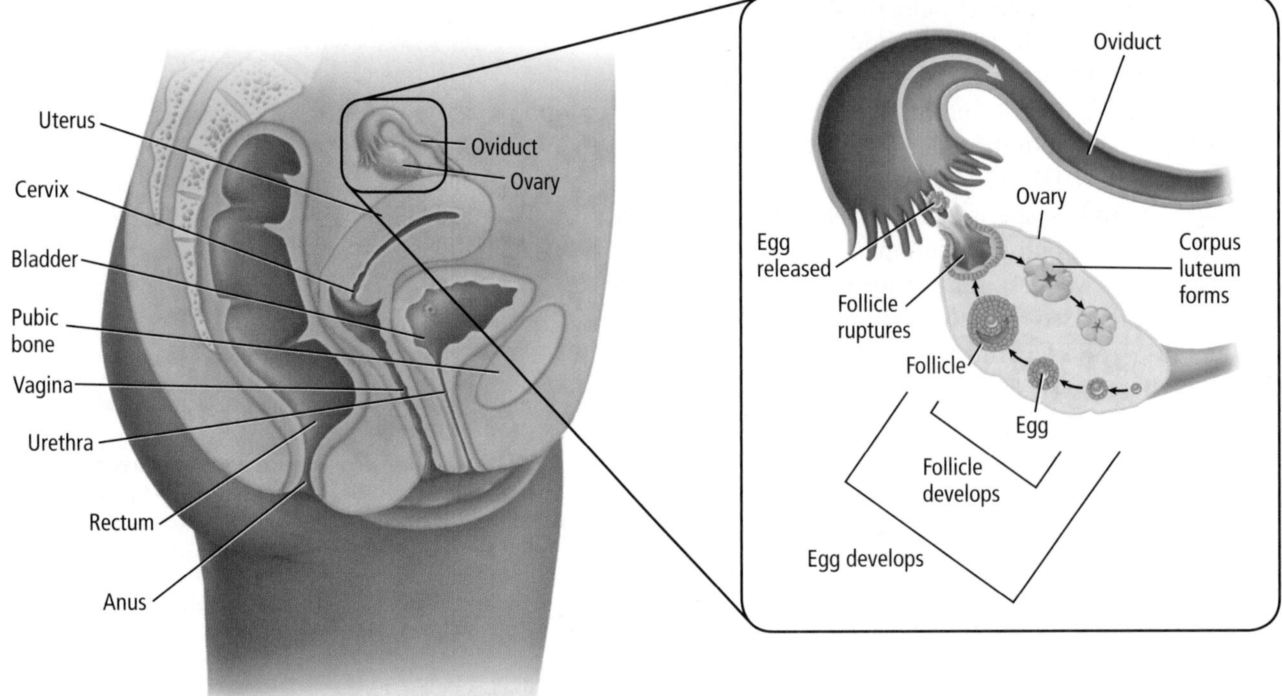

Figure 36.4
Left: The main structures of the female reproductive system are the vagina, uterus, and ovaries.
Right: During every menstrual cycle, one follicle fully matures and releases an egg. The follicle is then called the corpus luteum.
Predict ***What might be the result if more than one follicle fully develops during a cycle?***

Concepts In Motion

Interactive Figure To see an animation of the process of ovulation, visit biologygmh.com.

Human Female Reproductive System

A female's reproductive system is specialized to produce egg cells, receive sperm, and provide an environment that is right for fertilization of an egg and the development of an embryo. Refer to **Figure 36.4** as you read about the structures of the female reproductive system.

Egg cells The female reproductive cells, called egg cells, are produced in the ovaries, also illustrated in **Figure 36.4.** Each ovary is about the size of an almond. Inside each ovary are **oocytes** (OH uh sites), which are immature eggs. Approximately once every 28 days, oocyte development is stimulated and an egg, called an ovum, is formed. The ovum is surrounded by follicle cells that provide protection and nourishment.

After the egg is released from the ovary, it travels through an **oviduct** (OH vuh duct), a tube that connects to the uterus. The uterus, or womb, is about the size of an average human fist and is where a baby develops before birth. The cervix at the lower end of the uterus has a narrow opening into the vagina, which leads to the outside of the female's body.

Female hormones Estrogen and progesterone (proh JES tuh rohn) are steroid hormones made by cells in the ovaries. A female's anterior pituitary gland also produces LH and FSH, which influence estrogen and progesterone levels in a negative feedback loop. Effects of LH and FSH are different in males and females. During puberty, an increase in estrogen levels causes a female's breasts to develop, her hips to widen, and her amount of fat tissue to increase. During puberty, a female also will experience her first **menstrual** (MEN stroo ul) **cycle**—the events that take place each month in the human female to help prepare the female body for pregnancy.

Personal Tutor

To learn about sperm and egg formation, visit biologygmh.com.

Sex Cell Production

In Chapter 10, you learned that through meiosis, one cell in the male or female gonads—called testes and ovaries in humans—gives rise to four sex cells called gametes. In the human male, sperm are produced from primary spermatocytes daily beginning at puberty and continuing throughout a male's lifetime.

The production of eggs in the human female differs, as illustrated in **Figure 36.5.** A female is born with all of her eggs already beginning to develop. The genetic material has replicated in primary oocytes before birth, and the process of meiosis stops before the first meiotic division is completed. Then, once each menstrual cycle during the reproductive years, meiosis continues for a single developing oocyte. The resulting structures at the end of the first meiotic division of the oocyte are of unequal size. The smaller of the two structures is called a **polar body.** The chromosomes have segregated, but there is an unequal division of the cytoplasm. Most of the cytoplasm from the original cell goes to the cell that eventually will become the egg, and the polar body disintegrates.

During the second meiotic division, a similar process takes place. During metaphase of the second meiotic division, an egg ruptures through the ovary wall in a process called ovulation. The second meiotic division is completed only if fertilization takes place. Then, the zygote and the second polar body are formed, as shown in **Figure 36.5.** The second polar body also disintegrates.

Thus, the two meiotic divisions have yielded only one egg instead of four. If four eggs were formed and released midway through a female's menstrual cycle, more multiple births would be expected.

The Menstrual Cycle

The length of the menstrual cycle can vary from 23 to 35 days, but it typically lasts around 28 days. The entire menstrual cycle can be divided into three phases: the flow phase, the follicular phase, and the luteal phase.

Flow phase Day one of the menstrual cycle is when menstrual flow begins. Menstrual flow is the shedding of blood, tissue fluid, mucus, and epithelial cells from the endometrium—the tissue that lines the uterus. The endometrium is where the embryo will implant if fertilization of the egg occurs. Because an embryo will need oxygen and nutrients, the endometrium has a good supply of blood. During menstruation, bleeding occurs because the outer layers of the endometrium tear away, and blood vessels that supply the endometrium are ruptured. Around day five, repair of the endometrial lining begins, and it becomes thicker as the cycle continues.

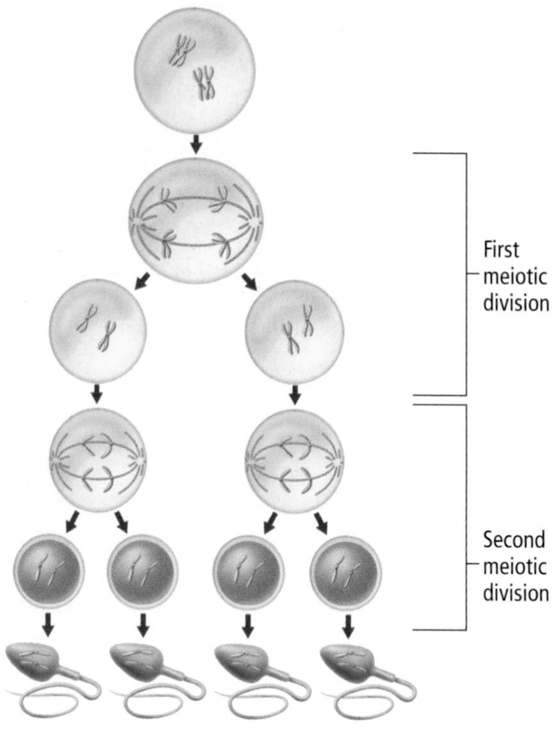

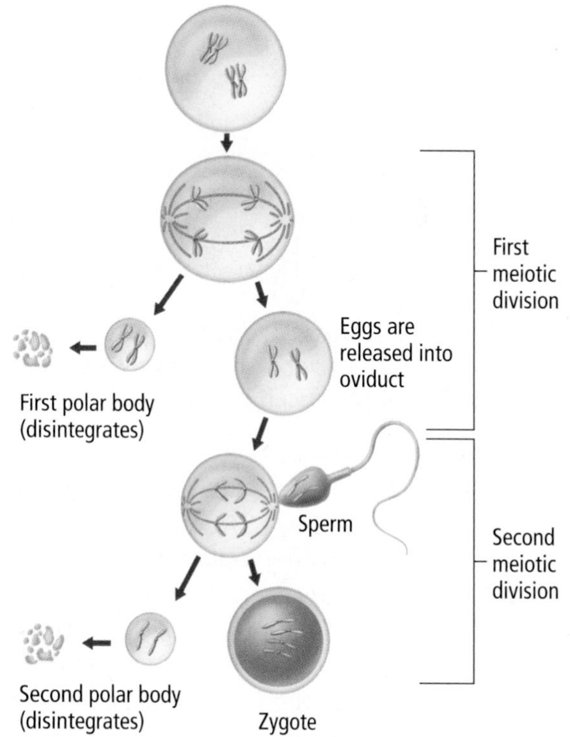

■ **Figure 36.5**
Top: The human male sex cell production follows the general pattern of meiosis and results in many sperms.
Bottom: Meiosis in the human female results in one egg. The second division in meiosis will not be completed in a human female unless the egg is fertilized.

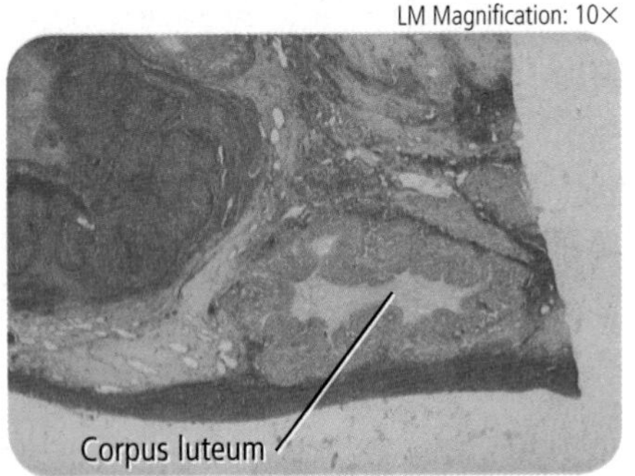

Figure 36.6 The corpus luteum produces progesterone and some estrogen.

Follicular phase During the menstrual cycle, changes also occur in the ovaries as a result of changing hormone levels, as illustrated in **Table 36.1.** At the beginning of a menstrual cycle, when estrogen levels are low, the anterior pituitary begins to increase production of LH and FSH. This stimulates a few follicles to begin to mature in the ovary. Cells in the follicles then begin to produce estrogen and a small amount of progesterone. Inside each follicle is an immature egg—the oocyte. After about a week, usually only one of the growing follicles remains. This remaining follicle continues to grow and secrete estrogen, which keeps levels of LH and FSH low—an example of negative feedback.

On day 12, the high level of estrogen causes the anterior pituitary gland to release a surge of LH. This rapid release of a large quantity of LH causes the follicle to rupture, and ovulation occurs.

Luteal phase After ovulation, the cells of the follicle change, and the follicle is transformed into a structure called the corpus luteum (KOR pus • LEW tee um), illustrated in **Figure 36.6.** The corpus luteum slowly degenerates as the menstrual cycle continues. The corpus luteum produces high amounts of progesterone and some estrogen, which keep levels of LH and FSH low through negative feedback. Recall that FSH and LH stimulate new follicles to develop, but when these hormones are kept at low levels, new follicles are temporarily prevented from maturing. Toward the end of the cycle, the corpus luteum breaks down, no longer producing progesterone and estrogen. This results in a rapid decrease in progesterone and estrogen levels. A rapid decrease in hormones triggers detachment of the endometrium, and the flow phase of a new menstrual cycle will begin.

VOCABULARY
WORD ORIGIN
Corpus luteum
corpus from Latin, meaning *body*
luteum from Latin, meaning *yellow.*

MiniLab 36.1

Model Sex Cell Production

Why does meiosis produce four sperm but only one egg? The difference in the division of cytoplasm is the major reason meiosis is different in human males and females. Use clay to model how the sex cells are produced during meiosis.

Procedure

1. Read and complete the lab safety form.
2. Choose **two lumps of clay,** each of a different color. Choose one to represent a primary spermatocyte and the other a primary oocyte.
3. Use the primary spermatocyte to simulate the meiotic divisions as they occur in males.
4. Simulate maturation of the sperm by removing about half of the clay from each sperm and using a small part of it to add a flagellum to each cell.
5. Next simulate the first meiotic division in females.
6. Use one of the sperm and mold it to one side of the large cell. Now simulate the second meiotic division.

Analysis

1. **Use Models** Make drawings of each step above and label the following: primary spermatocyte and oocyte, egg, sperm, first polar body, second polar body, fertilized egg, and zygote.
2. **Explain** What is the benefit of meiosis concentrating most of the cytoplasm into one egg?

Concepts In Motion

Interactive Table To explore more about the menstrual cycle, visit biologygmh.com.

Table 36.1 Menstrual Cycle Phases

	Flow Phase	Follicular Phase	Luteal Phase
Days	1–5	6–14	15–28
Ovarian activity			
Hormone levels (FSH, LH, Estrogen, Progesterone)			
Endometrium			

If the egg is fertilized, a different chain of events occurs, and a new menstrual cycle does not begin. The progesterone levels remain high and increase the blood supply to the endometrium. The corpus luteum does not degenerate and hormone levels do not drop. The endometrium accumulates lipids and begins secreting a fluid rich in nutrients for the developing embryo.

Section 36.1 Assessment

Section Summary

- Levels of male and female hormones are regulated by negative feedback systems.
- The human male produces millions of sperm cells every day.
- The number of sex cells resulting from meiosis differs in males and females.
- The human female has a reproductive cycle called the menstrual cycle.
- The menstrual cycle has three phases: the flow phase, the follicular phase, and the luteal phase.

Understand Main Ideas

1. MAIN Idea **Describe** how hormones regulate sperm and egg cells.
2. **Summarize** the structures of the reproductive systems and their functions.
3. **Describe** the origin and importance of substances found in semen.
4. **Explain** the major events that take place in the endometrium and in the ovary during the menstrual cycle.

Think Critically

5. **Infer** On day 12, estrogen levels cause a sharp increase in the amount of LH that is released. According to a negative feedback model, what would you expect to happen?

MATH in Biology

6. A woman began menstruating at age 12 and stopped menstruating at age 55. If she never became pregnant and her menstrual cycles averaged 28 days, how many eggs did she ovulate during her reproductive years?

Section 36.2

Reading Preview

Objectives

- **Discuss** the development that takes place during the first week following fertilization.
- **Describe** the major changes that occur during each trimester of development.
- **Explain** how female hormone levels are altered during pregnancy.

Review Vocabulary

lysosome: organelle that contains digestive enzymes

New Vocabulary

morula
blastocyst
amniotic fluid

Human Development Before Birth

MAIN Idea A human develops from a single fertilized cell into trillions of cells with specialized functions.

Real-World Reading Link Just as a single seed can grow into a plant with a beautiful flower, your complex body began as a single cell at the union of an egg and a sperm at fertilization.

Fertilization

Figure 36.7 shows the process of a sperm joining with an egg, which is called fertilization. Fertilization usually occurs in the upper portion of an oviduct near the ovary. In humans, sperm and eggs each are haploid, and each normally has 23 chromosomes. Fertilization brings these chromosomes together, restoring the diploid number of 46 chromosomes.

Sperm enter the vagina of the female's reproductive system when strong muscular contractions ejaculate semen from the male's penis during intercourse. Some sperm can exit through the penis before ejaculation without the male's knowledge. As a result, sexual activity that does not result in ejaculation can lead to the release of sperm, fertilization, and pregnancy.

Sperm can survive for 48 hours in the female reproductive tract, but an unfertilized egg can survive for only 24 hours. Fertilization can happen if intercourse occurs anytime from a few days before ovulation to a day after ovulation. Overall, there is a relatively short time when fertilization can occur successfully. But, it is important to remember that the length of the menstrual cycle can vary and ovulation can occur at any time.

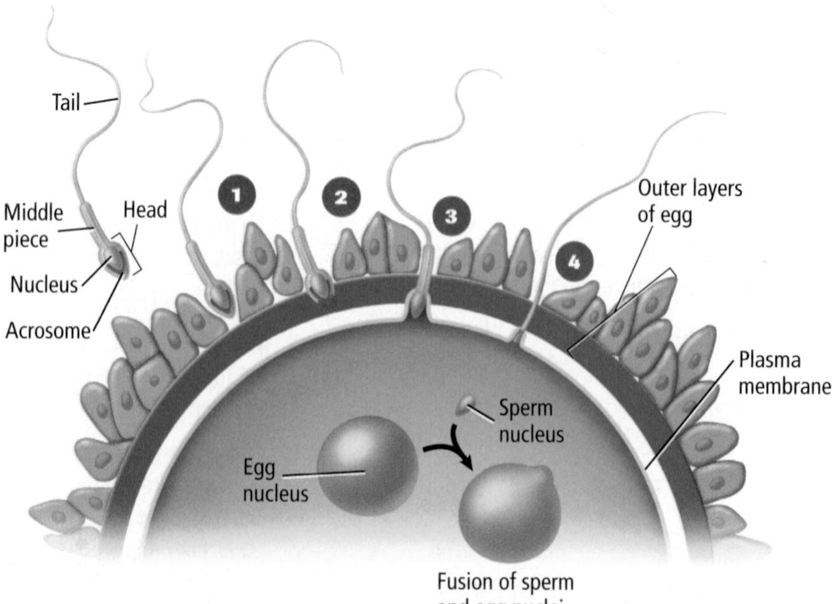

Figure 36.7 Although many are needed to weaken the barrier that surrounds the egg, only one sperm fertilizes an egg (steps 1-4). Fertilization is complete when the sperm nucleus fuses with the egg nucleus.

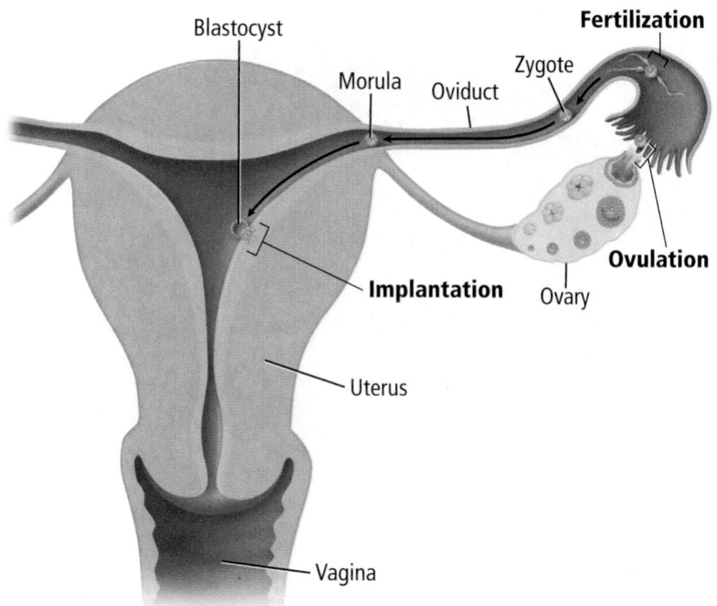

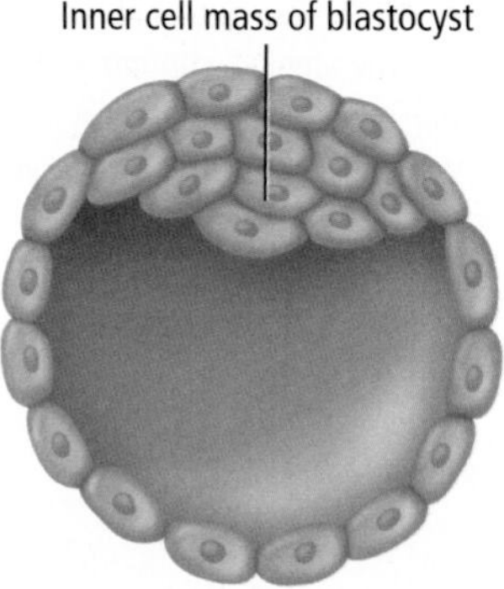

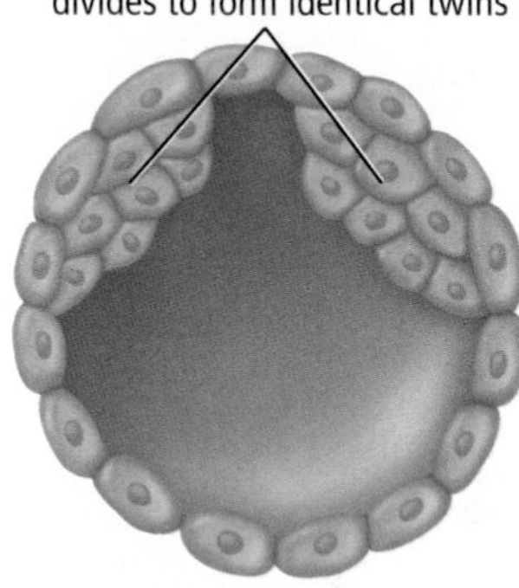

Figure 36.8
Left: During the first week of development, many developmental changes occur as the zygote travels through the oviduct.
Right: The inner cell mass of the blastocyst will develop into a fetus (top). If the inner cell mass divides, identical twins might form (bottom).

About 300 million sperm are released into the vagina during intercourse. Only several hundred of them will successfully complete the journey to the egg. Many never make it out of the vagina, some are attacked by white blood cells, and many simply die along the way. Only one sperm can fertilize an egg, but it takes several hundred to participate in the process.

Connection to Chemistry A single sperm cannot penetrate the plasma membrane that surrounds the human egg. Recall from Chapter 7 that lysosomes are organelles that contain digestive enzymes. Notice in **Figure 36.7** the tip of each sperm is a specialized lysosome called an acrosome. As each of several hundred sperm bombard the egg, the enzymes inside of the acrosome weaken the plasma membrane surrounding the egg. Eventually the plasma membrane becomes weak enough that one sperm can penetrate the egg. Immediately following this penetration, the egg forms a barrier to prevent other sperm from entering the now-fertilized egg.

 **Reading Check Explain** why hundreds of sperm are necessary for fertilization to take place.

Early Development

Figure 36.8 illustrates the first week of human development. The fertilized egg, which is called a zygote (ZI goht), moves through the oviduct propelled by involuntary smooth muscle contractions and by the cilia lining the oviduct. Around 30 hours after fertilization, the zygote undergoes its first mitosis and cell division. Cell division continues, and by the third day, the embryo leaves the oviduct and enters the uterus. At this point, the embryo is described as a **morula**—a solid ball of cells.

By the fifth day, the morula has developed into a **blastocyst,** which can be described as a hollow ball of cells. The blastocyst attaches to the endometrium around the sixth day and is fully implanted by Day 10. **Figure 36.8** shows that the blastocyst is not completely hollow. Inside the blastocyst is a group of cells called the inner cell mass. The inner cell mass eventually will become the embryo. Sometimes, the inner cell mass splits, and identical twins might form.

CAREERS IN BIOLOGY

Reproductive Endocrinologist Physicians who have advanced training in the treatment of infertility and disorders involving the reproductive hormones are called reproductive endocrinologists. A reproductive endocrinologist also might conduct research or train medical students. For more information about biology careers, visit biologygmh.com.

Figure 36.9 Four extraembryonic membranes—the amnion, chorion, yolk sac, and allantois—are important in development.
Identify *What is the role of the yolk sac in humans?*

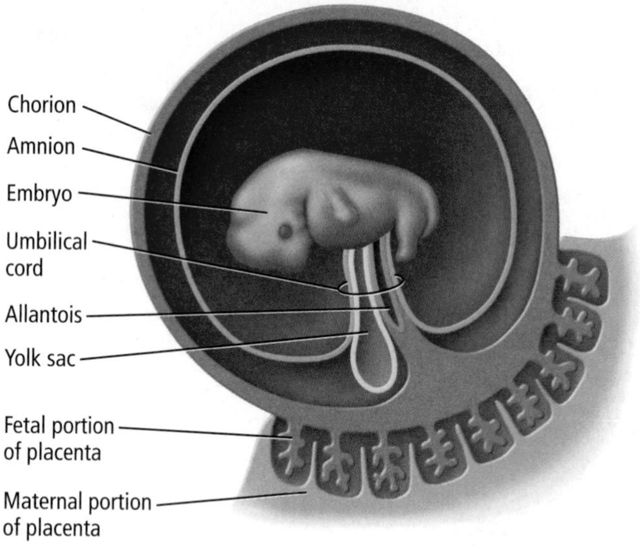

Chorion
Amnion
Embryo
Umbilical cord
Allantois
Yolk sac
Fetal portion of placenta
Maternal portion of placenta

Extraembryonic membranes In previous chapters, you learned the importance of the membranes that extend beyond an embryo called the extraembryonic membranes. You also learned about the development of the amniotic egg, and how this enabled animals to reproduce on land. Developing humans have these membranes, shown in **Figure 36.9,** but because humans and most other mammals develop inside the mother's body, these membranes have somewhat different functions.

Early in human development, four extraembryonic membranes form. These membranes are the amnion, the chorion (KOR ee ahn), the yolk sac, and the allantois (uh LAN tuh wus), as illustrated in **Figure 36.9.** The amnion is a thin layer that forms a sac around the embryo. Inside this sac is the **amniotic fluid** (am nee AH tihk • FLU id), which protects, cushions, and insulates the embryo. Outside of the amnion is the chorion, which, together with the allantois, contributes to the formation of the placenta. The yolk sac in humans does not contain any yolk but serves as the first site of red blood cell formation for the embryo.

VOCABULARY
ACADEMIC VOCABULARY
Enable:
To make able or feasible.
Amniotic eggs enable reproduction on land.

The placenta About two weeks after fertilization, tiny fingerlike projections of the chorion, called chorionic villi (VIH li), begin to grow into the wall of the uterus. The placenta (pluh SEN tuh), the organ that provides food and oxygen and removes waste, begins to form and is fully formed by the tenth week. The placenta has two surfaces—a fetal side that forms from the chorion and faces the fetus, and a maternal side that forms from uterine tissue. When completely formed, the placenta is 15–20 cm in diameter, 2.5 cm thick, and has a mass of about 0.45 kg. The umbilical cord, a tube containing blood vessels, serves as the connection between the fetus and the mother. **Figure 36.10** illustrates the connection between the mother and fetus.

The placenta regulates what passes from the mother to the fetus and from the fetus to the mother. Oxygen and nutrients can travel from the mother to the fetus. Alcohol, drugs, various other substances, and the human immunodeficiency virus (HIV) also can pass through the placenta to the developing fetus.

Metabolic waste products and carbon dioxide travel from the fetus to the mother. Because the mother and the fetus have their own separate circulatory systems, blood cells do not pass through the placenta. However, the mother's antibodies pass to the fetus and help protect the newborn until its immune system is functioning.

Study Tip

Time Line Create a time line showing the development of a human being from fertilization to adulthood. Use average ages for various stages of development. Include major characteristics of each stage of development.

Visualizing a Placenta

Figure 36.10
A growing fetus exchanges nutrients, oxygen, and wastes with the mother through the placenta. The placenta contains tissue from both mother and fetus.

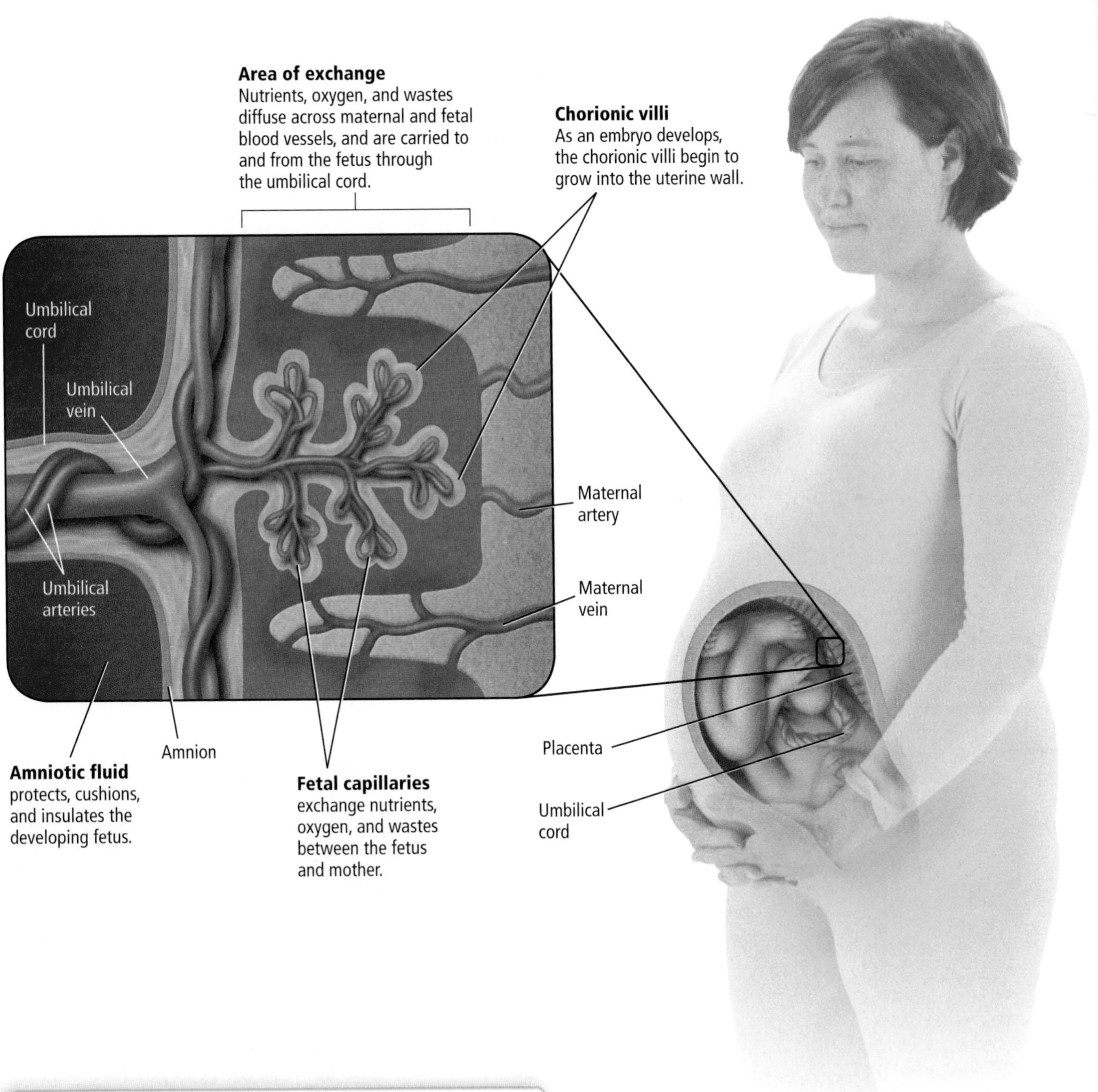

Area of exchange
Nutrients, oxygen, and wastes diffuse across maternal and fetal blood vessels, and are carried to and from the fetus through the umbilical cord.

Chorionic villi
As an embryo develops, the chorionic villi begin to grow into the uterine wall.

Umbilical cord

Umbilical vein

Umbilical arteries

Maternal artery

Maternal vein

Amnion

Amniotic fluid protects, cushions, and insulates the developing fetus.

Fetal capillaries exchange nutrients, oxygen, and wastes between the fetus and mother.

Placenta

Umbilical cord

Concepts In Motion Interactive Figure To see an animation of the placenta, visit biologygmh.com.

Biology Online

Hormonal regulation during pregnancy During the first week of development, the embryo begins to secrete a hormone, called human chorionic gonadotropin (hCG) (kor ee AH nihk • go na duh TROH pen), which keeps the corpus luteum from degenerating. If the corpus luteum remains active, progesterone levels, and to a lesser extent estrogen levels, remain high. Remember from the previous section that the decline of progesterone triggers a new menstrual cycle. If levels of these hormones remain high, a new menstrual cycle will not begin. Two to three months into development, the placenta secretes enough progesterone and estrogen to maintain the proper conditions for pregnancy.

Reading Check **Compare** two functions of the placenta.

Three Trimesters of Development

On average, human development takes around 266 days from fertilization to birth. This time span is divided into three trimesters, each around three months long. During this time, many events take place. The zygote grows from a single cell into a baby that has trillions of cells. These cells develop into tissues and organs with specialized functions. Follow **Figure 36.11,** which shows different stages of human development during the first trimester.

The first trimester In the first trimester, all tissues, organs, and organ systems begin to develop. During this time of development, the embryo is especially vulnerable to the effects of alcohol, tobacco, drugs, and other environmental influences, such as environmental pollutants. During the first two weeks of development, the mother might not realize she is pregnant because she has not missed a menstrual period yet. A lack of certain essential nutrients during this time might cause irreversible damage to the developing embryo. A few of the major causes of preventable birth defects are listed in **Table 36.2.**

At the end of eight weeks, the embryo is called a fetus. All of the organ systems have begun to form. By the end of the first trimester, the fetus can move its arms, fingers, and toes and make facial expressions. Fingerprints also are present.

Figure 36.11 The embryo develops into a fetus during the first trimester of pregnancy. By the end of the third month, the fetus can make small movements.

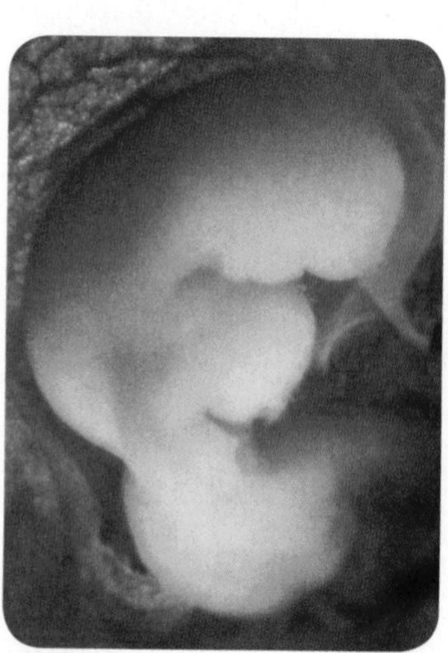

4 weeks

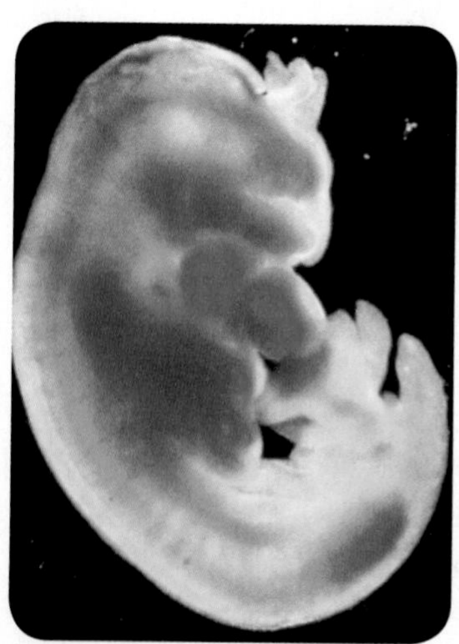

5–6 weeks

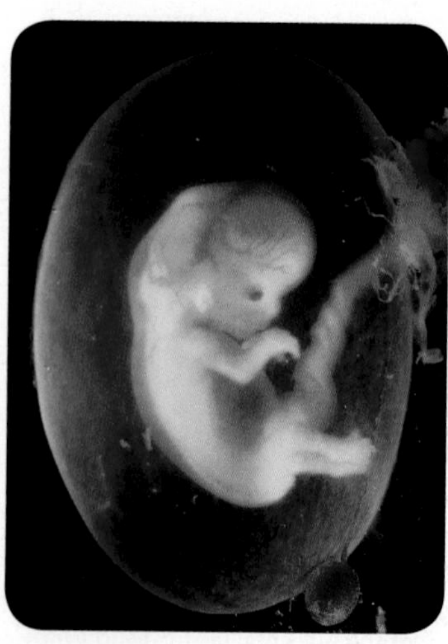

7–8 weeks

Table 36.2	Preventable Causes of Birth Defects
Cause	Defect
Alcohol consumption	• Mental retardation
Cigarette smoking	• Health problems related to premature births and underweight babies
Lack of folic acid in diet	• Anencephaly (head and brain do not completely form) • Spina bifida (nerve cells from the spinal cord are exposed, leading to paralysis)
Cocaine	• Low birth weight • Premature birth • Possible permanent brain damage and behavioral disorders
Methamphetamine	• Premature birth • Extreme irritability

Concepts In Motion

Interactive Table To explore more about birth defects, visit biologygmh.com.

The second trimester The second trimester primarily is a period of growth. Around 18 to 20 weeks, the fetal heartbeat might be heard using a stethoscope. The developing fetus is capable of sucking its thumb and can develop the hiccups. The mother might feel a fluttering sensation or might even feel light kicks. Hair usually forms, and the fetal eyes will open during this period. At the end of this trimester, the fetus might be able to survive outside the mother's uterus with the aid of medical intervention, but the chances for survival are not very high. If born this early, the baby cannot maintain a constant body temperature. The baby's lungs have not developed fully, so respiratory failure is a great risk. Also, the baby is very likely to become seriously ill because its immune system is not fully functional.

The third trimester During the third trimester, the fetus continues to grow at a rapid rate. Fat accumulates under the skin to provide insulation for the fetus once it is born. Adequate protein intake by the mother is important during this time. Protein is essential for the rapid amount of brain growth that occurs. New nerve cells in the brain are forming at a rate of 250,000 cells per minute. The fetus now might respond to sounds in the environment, such as music or the sound of its mother's voice.

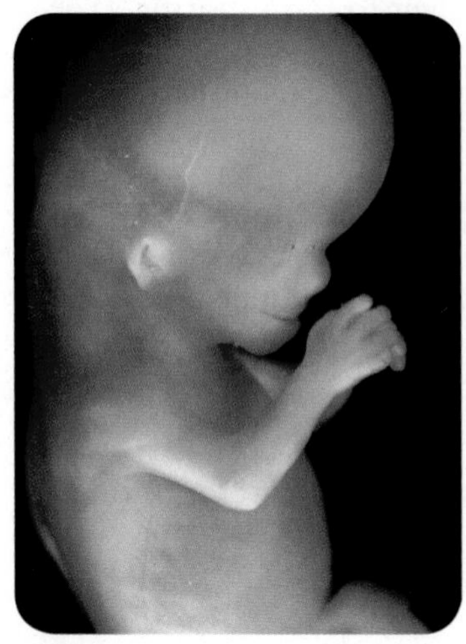

9–10 weeks

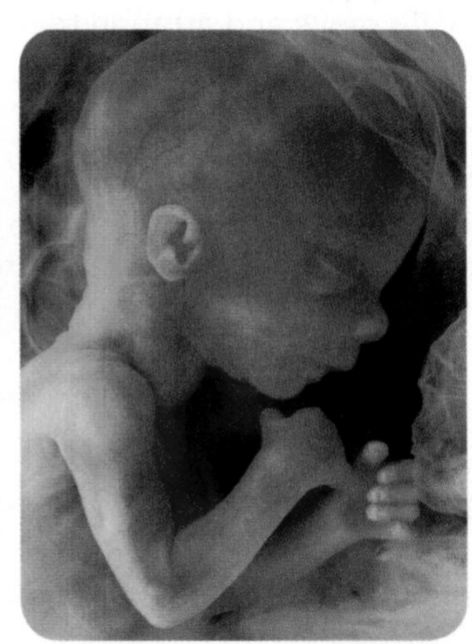

12 weeks

■ **Figure 36.12** In amniocentesis, fluid and cells lost from the fetus are removed from the amniotic fluid and analyzed.

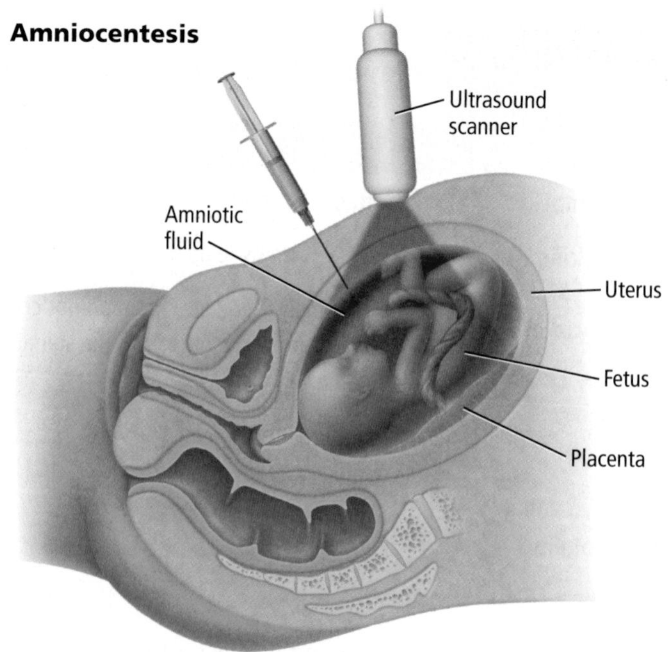

CAREERS IN BIOLOGY

Ultrasound Technician
Technical skills are needed in biology. An ultrasound technician obtains ultrasound images that are interpreted by a physician. For more information on biology careers, visit biologygmh.com.

Diagnosis in the Fetus

Many conditions can be diagnosed before a baby is born. Identifying certain conditions as early as possible increases the chance for proper medical treatment to help a newborn baby have the highest quality of life possible.

Ultrasound One way to identify conditions in the fetus is by using ultrasound, a procedure in which sound waves are bounced off the fetus. These sound waves are converted into light images that can be seen on a video monitor. Ultrasound can be used to determine if the fetus is growing properly, the position of the fetus in the uterus, and the gender of the fetus.

MiniLab 36.2

Sequence Early Human Development

What developmental changes occur during the first eight weeks of life? Fertilization begins when a sperm penetrates the egg. The zygote undergoes predictable developmental changes. Cell division produces increasing numbers of cells. Cells move and arrange themselves to form specific organs, making it possible for cells to perform specific functions.

Procedure

1. Visit biologygmh.com to see **images of embryos.**
2. Study the images and information provided for Stage 1 through Stage 23, the first ten weeks after fertilization. Choose one factor to track through this developmental period. Factors might include embryonic size, cell differentiation, overall structural changes, specific organ or organ system development, or others.
3. Chart the development of this factor along a time line through the ten-week period.

Analysis

1. **Analyze** the time line you created. Identify developmental milestones related to this factor during the ten-week period.
2. **Summarize** the level of development of the factor you examined by the end of Stage 23.

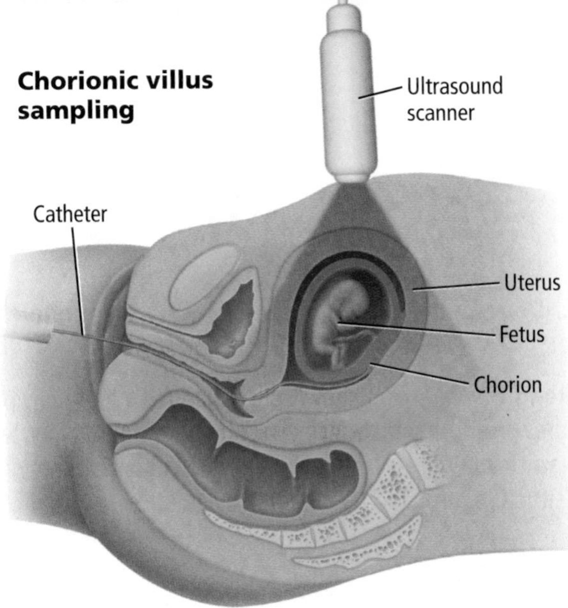

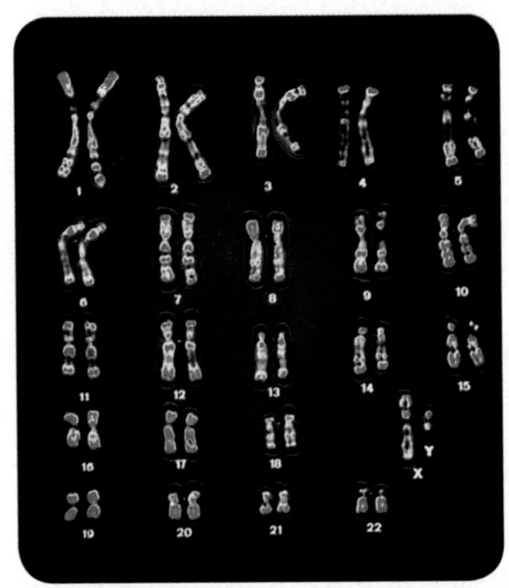

Karyotype

■ **Figure 36.13**
Left: Chorionic villus sampling involves removing cells from the chorion and analyzing them. Both procedures carry a small risk of a miscarriage.
Right: Karyotypes can be analyzed to help with diagnosis.

Amniocentesis and chorionic villus sampling Amniocentesis (am nee oh sen TEE sus) and chorionic villus sampling are prenatal tests. During amniocentesis—usually performed in the second trimester—a needle is inserted through the abdomen of the pregnant female, illustrated in **Figure 36.12.** Fluid from the amniotic sac is removed and analyzed. Tests that measure enzyme levels associated with certain conditions can be performed. Fetal cells can be examined by a karyotype or even by DNA analysis. Recall from Chapter 11 that a karyotype is a chart of chromosome pairs, shown in **Figure 36.13,** that is valuable in identifying unusual chromosome numbers or the sex of the fetus.

In chorionic villus sampling—usually performed during the first trimester—a small tube, called a catheter, is inserted through the vagina and cervix of the mother, illustrated in **Figure 36.13.** Cells from the chorion are removed and analyzed by karyotyping. The chromosomes in the cells of the chorion are identical to those of the cells in the fetus.

Section 36.2 Assessment

Section Summary

- Fertilization is the joining of egg and sperm.
- Four extraembryonic membranes are associated with a human embryo.
- The placenta regulates what substances can be exchanged between a fetus and its mother.
- Hormone regulation during pregnancy is different from hormone regulation during the menstrual cycle.
- Some medical conditions of a baby can be detected before it is born.

Understand Main Ideas

1. MAIN Idea **Describe** the changes that the zygote undergoes during the first week following fertilization.
2. **Describe** how defective acrosomes would affect the process of fertilization.
3. **Summarize** the development that occurs during each trimester.
4. **Compare and contrast** hormonal regulation during pregnancy with hormonal regulation during the menstrual cycle.

Think Critically

WRITING in Biology

5. Write a paragraph explaining the functions of the extraembryonic membrane in humans, and contrast those functions with the functions in other animals.

MATH in Biology

6. Determine the due date (predicted birth date) of the baby if the egg was fertilized on January 1.

Section 36.3

Reading Preview

Objectives

- **Discuss** the events that occur during the three stages of birth.
- **Describe** the stages of human development from infancy to adulthood.
- **Identify** hormones necessary for growth.

Review Vocabulary

growth: increase in the amount of living material and formation of new structures in an organism

New Vocabulary

labor
dilation
expulsion stage
placental stage
adolescence
infancy
adulthood

Birth, Growth, and Aging

MAIN Idea **Developmental changes continue throughout the stages of life.**

Real-World Reading Link You know from looking at your family photo album that you have grown and changed since you were born. Your bones, teeth, eyes, and muscles have changed. You can look forward to continued changes in your face and body structure throughout your life.

Birth

Birth occurs in three stages: dilation, expulsion, and the placental stage, as shown in **Figure 36.14.** Just before giving birth, the posterior pituitary gland releases the hormone oxytocin (ahk sih TOH sun), which stimulates involuntary muscles in the wall of the uterus to contract. This is the beginning of the birthing process called **labor.**

Another sign that the baby is going to be born is the **dilation** (di LAY shun), or opening, of the cervix. The cervix must open to allow the baby to leave the uterus. Contractions of the uterus become stronger and more frequent, and at some point the amniotic sac tears. The amniotic fluid flows out of the vagina, which is sometimes described as the "water breaking."

After a period of time that could be as short as a few hours or as long as a couple of days, the cervix fully dilates to around 10 cm. The uterine contractions are now very strong. The mother consciously will contract her abdominal muscles to help push the baby, usually head first, through the vagina in the **expulsion stage.** When the baby is out of the mother's body, the umbilical cord is clamped and cut. A small piece of the cord still attached to the baby soon will dry up and fall off, forming the navel, or belly button.

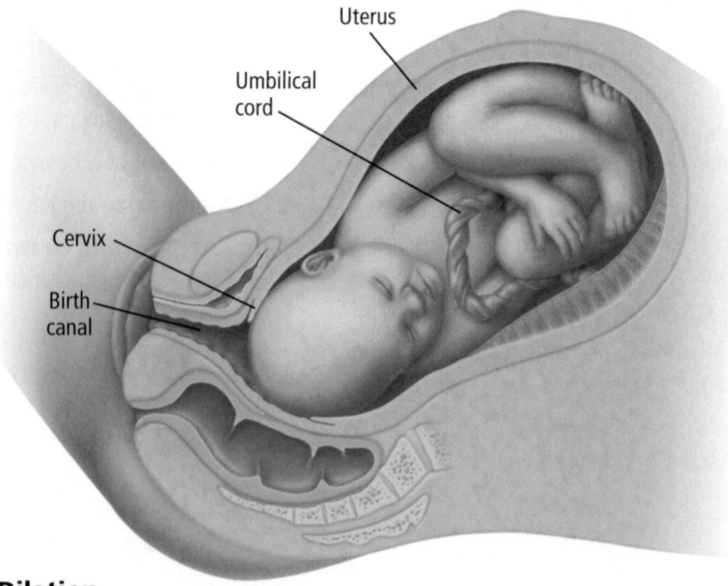

■ **Figure 36.14** Note the three stages of birth: Dilation stage: contractions open the cervix. Expulsion stage: The baby rotates as it moves through the birth canal, making expulsion easier. Placental stage: The placenta and umbilical cord are expelled.

Hypothesize ***What might happen if the placenta was not expelled quickly?***

Shortly after the baby is delivered, the placenta detaches from the uterus and leaves the mother's body along with extraembryonic membranes. This is the **placental stage** of the birthing process.

Sometimes, complications prevent the baby from being born through the vagina. In these cases, an incision is made through the mother's abdomen and uterus, and the baby is removed from the mother's body. This process is called a cesarean section.

During the first four weeks of life, the baby is called a newborn. Human newborns vary in size. However, on average, a newborn human baby has a mass of 3300 g and is 51 cm long.

 **Reading Check** **Describe** major events that occur during each stage of labor.

VOCABULARY

SCIENCE USAGE V. COMMON USAGE

Labor

Science usage: the process of giving birth.

There are three stages in labor: dilation, expulsion, and the placental stage.

Common usage: use of physical or mental effort.

Lifting the heavy bags of soil requires labor.

Growth and Aging

Humans go through many stages of growth during their lives. After you were born, you were in your infancy, but soon you will enter adulthood. You now are in a major development phase called **adolescence** (a dul ES unts) that began with puberty and ends at adulthood.

Hormones, such as human growth hormone, thyroxine, and steroids, influence growth. Human growth hormone stimulates most areas of the body to grow as cells replicate by the process of mitosis. This hormone works by increasing the rates of protein synthesis and breakdown of fats. Thyroxine from the thyroid increases the overall metabolic rate, and is essential for growth to occur. Steroid hormones, such as estrogen and testosterone, are also important for growth. Recall from Chapter 35 that testosterone and estrogen pass through the plasma membrane and into the nucleus of a target cell. The hormones activate certain genes that promote the formation of proteins. In this way, testosterone, and to a lesser extent, estrogen, cause an increase in the size of cells.

 **Reading Check** **Summarize** the roles of human growth hormone and thyroxine.

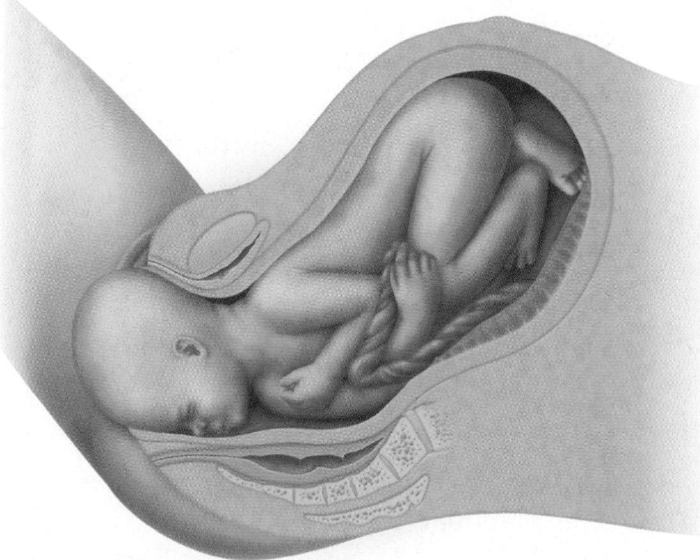

Expulsion stage

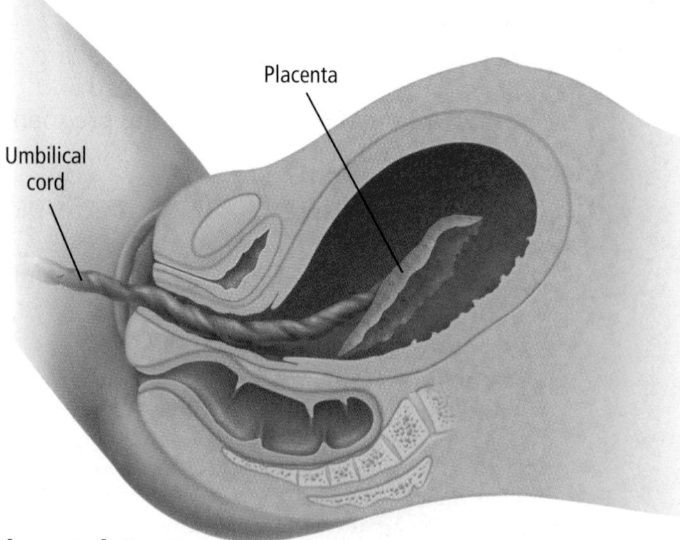

Placental stage

Infancy The first two years of life are known as **infancy.** Many changes take place during these years. An infant learns how to roll and crawl, grasp objects, and perform simple tasks. By the end of the first year, the infant likely is walking and might be uttering a few words. An enormous amount of mental development also occurs during these first two years.

In the first year, a baby typically grows about 25 cm in length and weighs three times more than when the baby was born. The child's growth slows during the second year; children grow at a rate of around 6 cm per year until the beginning of puberty.

Childhood and adolescence Childhood is the period of growth and development that extends from infancy to adolescence. The child's ability to reason and solve problems develops progressively during childhood. Puberty marks the beginning of adolescence, the period of growth between childhood and adulthood. Puberty usually begins between ages 8 to 13 in girls and ages 10 to 15 in boys.

In addition to the hormonal and sexual development that takes place during this time, other physical changes also take place. An adolescent experiences a growth spurt—girls grow approximately 6–11 cm and boys grow approximately 7–13 cm—in one year. In girls, the hips become wider and the waist might become narrower. In boys, the shoulders usually become broader. At the end of adolescence, physical growth is complete, marking the beginning of **adulthood.** The transition between adolescence and adulthood can be hard to define because of physical, emotional, and behavioral changes.

DATA ANALYSIS LAB 36.1

Based on Real Data*

Form a Conclusion

Is SIDS linked to smoking? In 1994, doctors began to recommend that infants sleep on their backs to reduce the risk of Sudden Infant Death Syndrome (SIDS).

Data and Observations

The table summarizes the annual SIDS rate per 1000 infants for mothers who smoked and mothers who did not smoke during pregnancy.

SIDS Deaths		
Year	Smoke Exposed	Unexposed
1989	3.21	1.33
1990	2.96	1.34
1991	3.32	1.72
1992	2.93	1.41
1993	3.28	1.17
1994	1.65	0.79
1995	2.19	0.65
1996	1.61	0.82
1997	3.21	0.64
1998	1.80	0.37

Think Critically

1. **Analyze** Did sleeping position affect SIDS? Explain.
2. **Calculate** the percentage of SIDS for babies born to smoking mothers and to those of nonsmokers each year.
3. **Conclude** How does this data show that some SIDS cases might be linked to smoking?

*Data obtained from: Anderson, M.E., et al. 2005. Sudden Infant Death Syndrome and prenatal maternal smoking: rising attributed risk in the Back to Sleep era. *BMC Medicine* 3: 4.

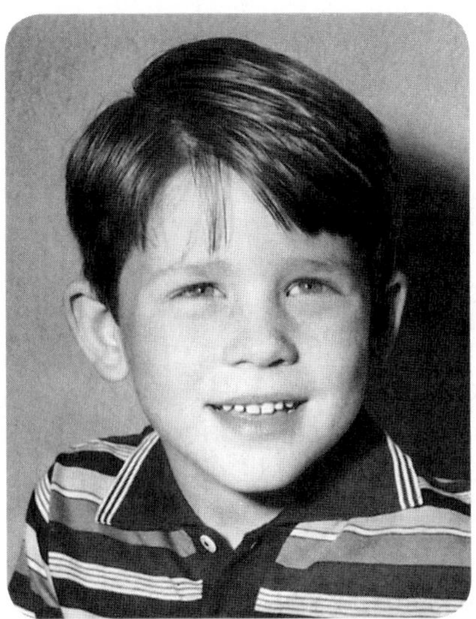

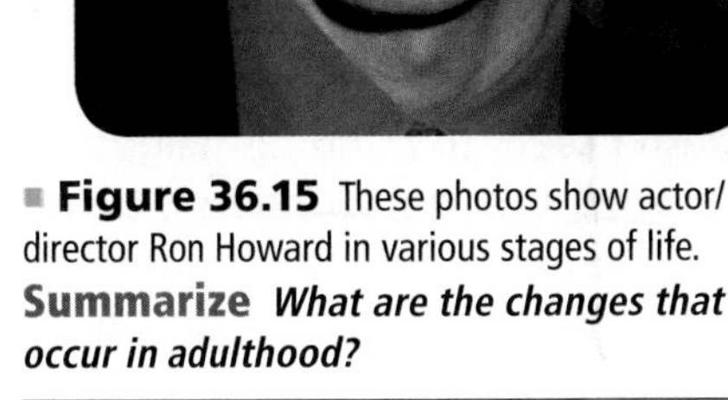

■ **Figure 36.15** These photos show actor/director Ron Howard in various stages of life.
Summarize ***What are the changes that occur in adulthood?***

Adulthood There are a number of theories on why people age, however, scientists do agree that the body goes through many changes as it ages, as shown in **Figure 36.15.** Physical changes perhaps are the most noticeable signs of aging, such as hair turning gray or white as pigment production declines. An individual might lose as much as two centimeters of height during the aging process because the discs between the vertebrae in the spine become flattened. Other changes include a decrease in muscle mass, a slowing of overall metabolism, and a decreased pumping ability of the heart. The skin loses its elasticity and sensory perceptions might diminish somewhat. In women, the ability to have children ends with menopause (MEN uh pawz), and sperm production decreases in men.

Despite all of the potential changes of aging, many people continue to be physically and mentally active as they grow older. Some older adults might even begin new careers. Anna Mary Robertson Moses, known as Grandma Moses, became a famous artist when she was in her late 70s.

Section 36.3 Assessment

Section Summary

- Humans go through many changes throughout the stages of life.
- There are three stages of the birthing process.
- Levels of several hormones influence human growth.
- The first year of life is a time of learning motor skills and of rapid growth.
- Puberty causes many changes in the body, and changes continue to occur as an adult grows older.

Understand Main Ideas

1. MAIN Idea **Construct** a chart that illustrates major changes that occur during the stages of human growth and aging.
2. **Identify** two signs that tell a pregnant woman she is almost ready to give birth.
3. **List** the events that occur during the three stages of birth.
4. **Describe** how the human growth hormone causes a person to grow.

Think Critically

5. **Hypothesize** Robert Wadlow—the tallest human being on record—was 272 cm tall and weighed 220 kg when he died at age 22. He was an average-size newborn, but developed a tumor in his anterior pituitary gland. Develop a hypothesis to explain how this tumor led to his height.
6. **Infer** How do you think scientists determine which specific substances cause birth defects?

Biology & Society

HGH: The Tall and Short of It

Terry is a 157.5-cm tall senior. He has not grown in the last two years. His dad is 190.5 cm and his other three brothers are 177.8 cm or taller. He doesn't seem to mind his height—or lack of it. His mom, though, wonders if he might have a disadvantage in sports because he is short. She keeps suggesting that he take growth hormones to become taller. She thinks it might help him be more successful in sports—and in life. What should he do?

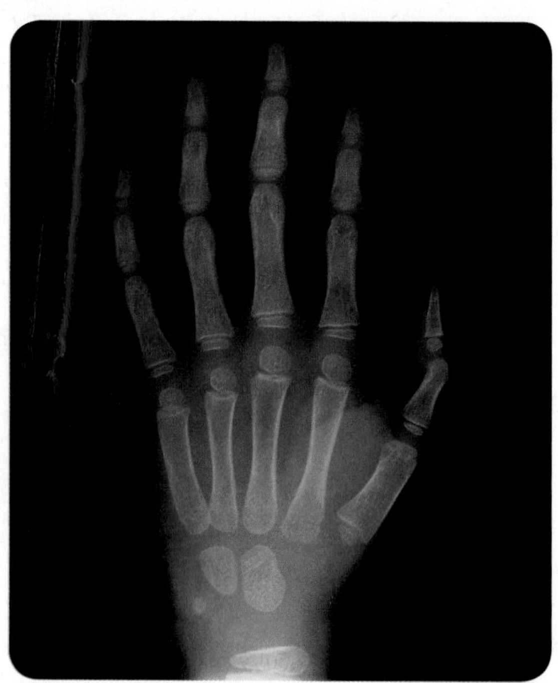

The oval bones are the growth plates, where bone growth occurs. If the growth plate is no longer visible, no more growth can occur.

What is human growth hormone? Human growth hormone (HGH) is a protein produced in the pituitary gland found in the brain. It is plentiful during the growth period of youth. Children with a lack of HGH are known as pituitary dwarfs and generally do not reach a height of over 135 cm.

What is HGH therapy? During adolescence, pituitary dwarfs can receive injections of synthetic HGH to increase their growth 10–12 cm during the first year. Less growth occurs during subsequent years. In 2003, the Food and Drug Administration approved HGH therapy for children who are otherwise healthy, but are predicted to reach adult heights of 160.0 cm or shorter for males and 149.9 cm or shorter for females. For these children, HGH therapy adds an average of 4–7 cm of height by adulthood. Using X-rays, a child's bone age—and therefore their growth potential—can be determined.

Therapy v. enhancement Sometimes HGH therapy can be used by physicians for individuals who are short and want to be taller or stronger athletes; however, this type of treatment is rare. There have been instances in which HGH drugs have been sold illegally and obtained by professional athletes to enhance their performance. These athletes were fined heavily or suspended because the drugs were in their systems.

HGH supplements sold in health food stores have less than one percent of HGH in them. Based on numerous scientific studies, they show no significant impact on human performance. Only HGH injections can improve growth and increase one's metabolism.

DEBATE in Biology

Debate Should HGH therapy be permissible when a teen is dissatisfied with his or her height for primarily cosmetic or athletic reasons? Consider Terry's situation in which he is feeling pressured to use HGH to become taller simply because the option exists. Conduct additional research about HGH and HGH therapy at biologygmh.com.

BIOLAB

INTERNET: HOW ARE ULTRASOUND IMAGES USED TO TRACK FETAL DEVELOPMENT?

Background: Ultrasound is a medical imaging technique that uses high frequency sound waves and their echoes to produce an image of something inside the body. While two-dimensional images are the current standard, technology capable of producing three-dimensional fetal images and four-dimensional, or moving images, is now available.

Question: *How are ultrasound images used to assess fetal characteristics and development?*

Materials

computer with Internet access
labeled ultrasound images showing embryos and fetuses at various developmental stages
ultrasound images showing embryos and fetuses at unknown stages of development

Procedure

1. Read and complete the lab safety form.
2. Visit biologygmh.com to examine fetal development from the second trimester through week 40. Use this information to complete the development time line you started in **MiniLab 36.2.**
3. Study the ultrasound images of fetuses during identified stages of development provided by your teacher. Compare these to your time line, and identify as many features as possible. As you study the images, choose a body structure that you would like to examine further.
4. Study the ultrasound images provided by your teacher of fetuses at an unknown stage of development. Use your time line and what you have learned to determine the approximate stage of fetal development. Look for clues based on development of the system you choose.

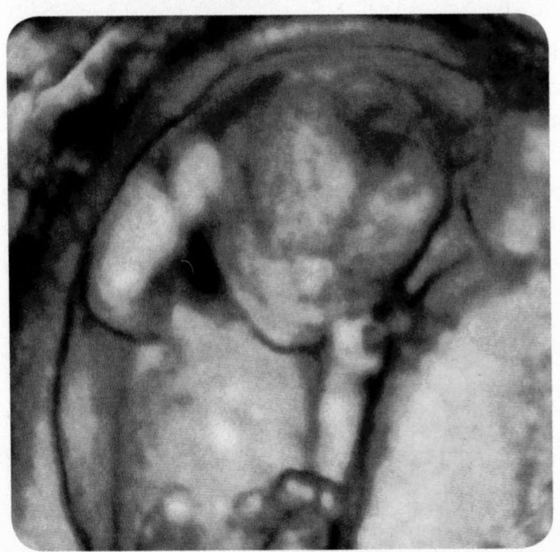

Analyze and Conclude

1. **Interpret Data** During which time period does the developing embryo or fetus change the most? Justify your answer.
2. **Analyze** What physical characteristics were most helpful in identifying the level of fetal development? Explain.
3. **Compare** two- and three-dimensional ultrasound images. Which are easiest to interpret?
4. **Think Critically** What advantages are provided by four-dimensional imaging?
5. **Error Analysis** How accurate were your estimates of fetal development? Explain how your estimates could have been improved.

WRITING in Biology

Poster Session Create a flow chart that illustrates the reproductive process. Begin with the creation of sex cells and end with a fetus at full term. To learn more about ultrasound, visit Biolabs at biologygmh.com.

CHAPTER 36

Study Guide

STUDY TO GO Download quizzes, key terms, and flash cards from biologygmh.com.

FOLDABLES **Research and assess** how hormones have a regulatory and/or stimulatory effect in each of the following: reproduction, metabolism, and human growth and development.

Vocabulary	Key Concepts
Section 36.1 Reproductive Systems	
• epididymis (p. 1049) • menstrual cycle (p. 1050) • oocyte (p. 1050) • oviduct (p. 1050) • polar body (p. 1051) • puberty (p. 1049) • semen (p. 1049) • seminiferous tubule (p. 1049) • urethra (p. 1049) • vas deferens (p. 1049)	**MAIN Idea** Hormones regulate human reproductive systems, including the production of gametes. • Levels of male and female hormones are regulated by negative feedback systems. • The human male produces millions of sperm cells every day. • The number of sex cells resulting from meiosis differs in males and females. • The human female has a reproductive cycle called the menstrual cycle. • The menstrual cycle has three phases: the flow phase, the follicular phase, and the luteal phase.
Section 36.2 Human Development Before Birth	
• amniotic fluid (p. 1056) • blastocyst (p. 1055) • morula (p. 1055)	**MAIN Idea** A human develops from a single fertilized cell into trillions of cells with specialized functions. • Fertilization is the joining of egg and sperm. • Four extraembryonic membranes are associated with a human embryo. • The placenta regulates what substances can be exchanged between a fetus and its mother. • Hormone regulation during pregnancy is different from hormone regulation during the menstrual cycle. • Some medical conditions of a baby can be detected before it is born.
Section 36.3 Birth, Growth, and Aging	
• adolescence (p. 1063) • adulthood (p. 1064) • dilation (p. 1062) • expulsion stage (p. 1062) • infancy (p. 1064) • labor (p. 1062) • placental stage (p.1063)	**MAIN Idea** Developmental changes continue throughout the stages of life. • Humans go through many changes throughout the stages of life. • There are three stages of the birthing process. • Levels of several hormones influence human growth. • The first year of life is a time of learning motor skills and of rapid growth. • Puberty causes many changes in the body, and changes continue to occur as an adult grows older.

CHAPTER 36 Assessment

Section 36.1

Vocabulary Review

Explain the difference between the terms in each set below, then explain how the terms are related.

1. urethra—semen
2. oocyte—oviduct
3. menstrual cycle—polar body

Understand Key Concepts

4. What would happen if the testes were located inside the body cavity?
   A. Sperm would not be produced because it is too warm.
   B. Testosterone levels would increase because of the warm temperature.
   C. The seminal vesicles would no longer be needed.
   D. Hormones from the testes would have difficulty entering the bloodstream.

Use the diagram below to answer questions 5 and 6.

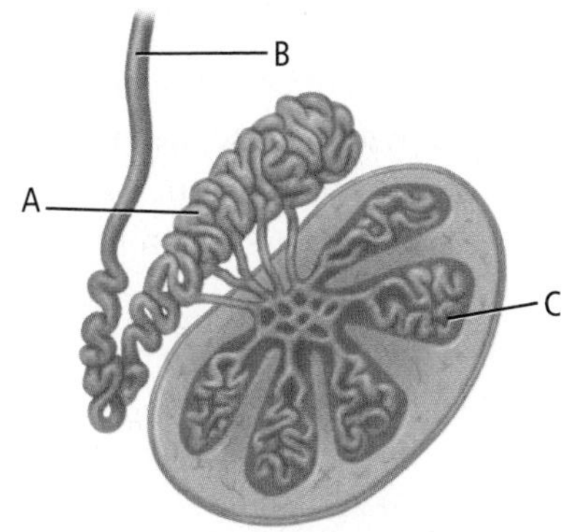

5. What occurs in the structure labeled *C* in the illustration?
   A. sperm cell storage and maturation
   B. sperm cell production
   C. secretion of sugar
   D. production of FSH

6. What is the function of the structure labeled *A* in the illustration?
   A. sperm cell storage and maturation
   B. sperm cell production
   C. secretion of sugar
   D. production of FSH

Constructed Response

7. **Short Answer** Why are the secretions of the male reproductive glands so important to sperm?
8. **Short Answer** Compare the actions of FSH and LH in the ovaries and testes.
9. **Short Answer** What advantages are there for the formation of one egg and polar bodies as compared to four eggs?

Think Critically

Use the diagram below to answer question 10.

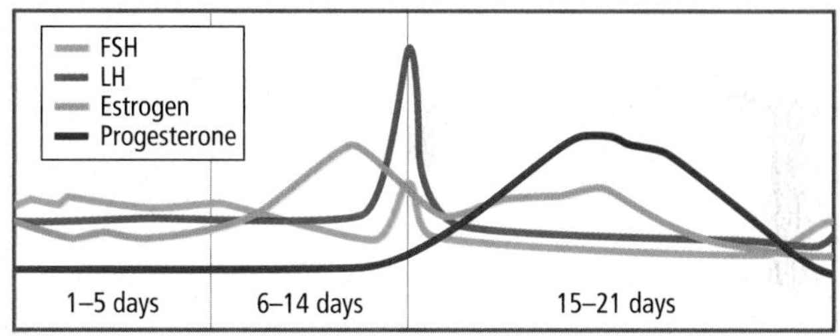

10. **Cause and Effect** Based on what you know about the hormonal control of a woman's reproductive cycle shown above, explain the hormonal basis of why a woman cannot get pregnant again while she is pregnant.
11. **Hypothesize** All of the reproductive hormones in a human male are present at birth. Develop a hypothesis to explain why the hormones have a greater influence on the body once puberty begins.

Section 36.2

Vocabulary Review

For questions 12–14, describe each of the following vocabulary terms.

12. morula
13. blastocyst
14. amniotic fluid

Understand Key Concepts

15. Where in the human female reproductive tract does fertilization usually occur?
    A. in the uterus
    B. in the vagina
    C. in the corpus luteum
    D. in an oviduct

16. Which of the following describes the proper sequence of development?
    A. zygote, blastocyst, morula
    B. morula, zygote, blastocyst
    C. zygote, morula, blastocyst
    D. morula, blastocyst, zygote

17. Which is produced by the placenta?
    A. human chorionic gonadotropin
    B. estrogen and progesterone
    C. oxytocin
    D. endometrial birth hormone

Use the diagram below to answer question 18.

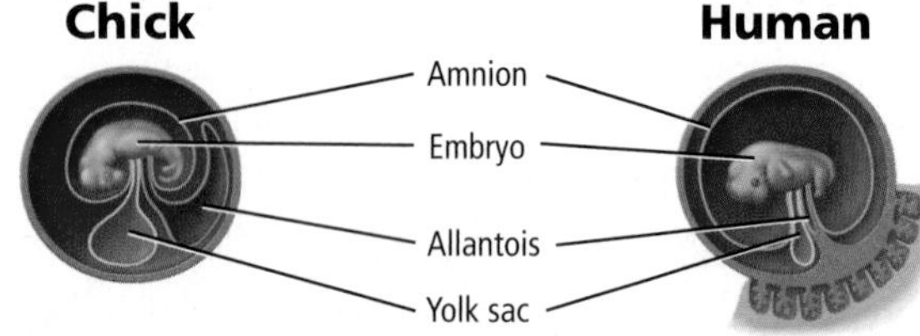

18. Why is the human yolk sac shown in the illustration smaller than that of the chick?
    A. The yolk in humans is converted into muscle.
    B. The yolk sac in chicks keeps the embryo warm.
    C. Developing humans get their nourishment from the placenta.
    D. The yolk sac serves no purpose for a developing human.

19. When can a pregnant woman first feel the movements of her fetus?
    A. in the first trimester
    B. in the second trimester
    C. in the third trimester
    D. in the last month only

Constructed Response

20. **Short Answer** Why is it important that the endometrium is refreshed each cycle?

21. **CAREERS IN BIOLOGY** Some couples consult with a reproductive endocrinologist because they are having difficulty conceiving a child. What are some possible biological reasons that could contribute to this difficulty?

22. **Open Ended** What do you think are the reasons that the greatest amount of harm to an embryo or fetus caused by alcohol or drugs occurs in the first trimester?

Think Critically

23. **Compare and contrast** the division of the inner cell mass during normal development and during the development of identical twins.

24. **Formulate a Model** A woman is carrying an embryo, but not enough hCG is getting into her system. Propose a possible treatment that might allow the embryo to be saved.

Section 36.3

Vocabulary Review

Explain the difference between the terms in each set below. Then explain how the terms are related.

25. labor—placental stage

26. dilation—expulsion

27. adolescence—adulthood

Understand Key Concepts

28. At which measurement is the cervix fully dilated?
    A. 10 mm
    B. 2 cm
    C. 10 cm
    D. 20 cm

29. When a pregnant woman tells her doctor that "her water broke," what does she mean?
    A. The amniotic sac has torn.
    B. There is a lot of pressure on her bladder.
    C. The yolk sac has torn.
    D. The placenta is leaking.

Use the diagram below to answer questions 30 and 31.

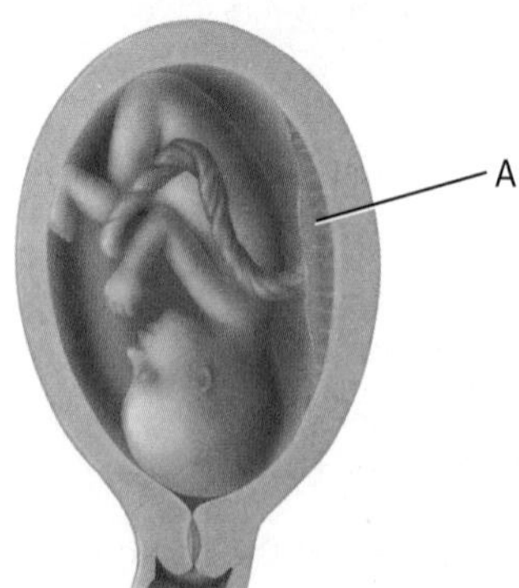

30. What is the name for the structure labeled A in the illustration?
    A. uterus
    B. placenta
    C. fetus
    D. cervix

Biology Online Chapter Test biologygmh.com

31. During which stage of birth does structure *A* leave the female's body?
    A. first
    B. second
    C. third
    D. fourth

32. During which year of a person's life does the most rapid rate of growth occur?
    A. the first year of infancy
    B. the first year of puberty
    C. the second year of puberty
    D. the first year of adulthood

Constructed Response

33. **Open Ended** What biological reasons can you think of to explain why women go through menopause and stop producing eggs while men can produce sperm all their lives?

34. **Short Answer** Compare puberty in females with puberty in males.

35. **CAREERS IN BIOLOGY** During rare occasions, a pediatrician examines a newborn baby who does not produce enough thyroxin. What are some possible results of this? Suggest a treatment for this condition.

Think Critically

Use the graph below to answer question 36.

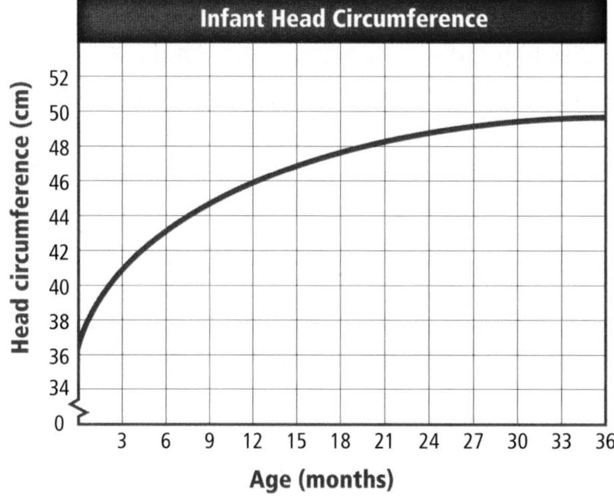

36. **Evaluate** During which period shown on the graph is the rate of change in head circumference greatest?

Additional Assessment

37. WRITING in Biology Prepare a pamphlet for pregnant women on health and lifestyle issues during pregnancy. Include a chart on major events of fetal development.

DBQ Document-Based Questions

To reduce the chances of brain and spine birth defects, the U.S. Public Health Service recommended in 1992 that women of childbearing age increase folic acid in their diets. The U.S. Food and Drug Administration required all cereal products be enriched with folic acid beginning in January 1998 (an optional period began in March 1996).

Below is a table showing the rate per 100,000 births of anencephaly—incomplete head and brain development—from 1991–2002.

Year	Rate	Year	Rate
1991	18.38	1997	12.51
1992	12.79	1998	9.92
1993	13.50	1999	10.81
1994	10.97	2000	10.33
1995	11.71	2001	9.42
1996	11.96	2002	9.55

Data obtained from: Mathews, T.J. Trends in Spina Bifida and Anencephalus in the United States, 1991–2002. National Center for Health Statistics/Centers for Disease Control and Prevention/Department of Health and Human Services.

38. Construct a graph to represent this data and describe the relationship between the variables that you observe.

39. Explain the overall trend in the number of cases of anencephaly during this time period.

Cumulative Review

40. How do the concepts of population growth, natality, and birth rate differ? **(Chapter 4)**

41. What are the three main differences between DNA and RNA? **(Chapter 12)**

Standardized Test Practice

Cumulative

Multiple Choice

1. Which is the role of arteries in the circulatory system?
   A. to carry blood away from the heart
   B. to carry blood back to the heart
   C. to provide individual cells with nutrients
   D. to prevent blood from flowing backward

Use the diagram below to answer question 2.

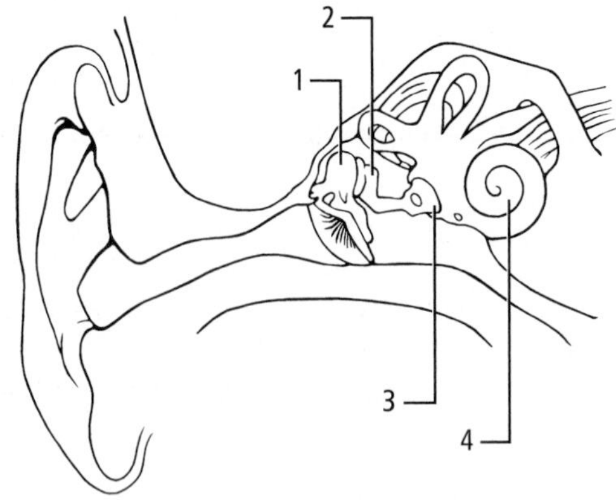

2. Which contains sensors for the auditory nerve?
   A. 1
   B. 2
   C. 3
   D. 4

3. What is the role of hormones in the body?
   A. They act as reaction catalysts.
   B. They control the breathing process.
   C. They help synthesize proteins.
   D. They regulate many body functions.

4. Which is the sequence of human development during the first week?
   A. egg → morula → blastocyst → zygote
   B. egg → zygote → morula → blastocyst
   C. morula → blastocyst → egg → zygote
   D. morula → egg → zygote → blastocyst

5. Which is the function of the kidneys?
   A. deplete carbon dioxide from the blood
   B. eliminate undigested foods from the body
   C. remove excess water and wastes from the blood
   D. rid excess proteins from the blood

Use the diagram below to answer question 6.

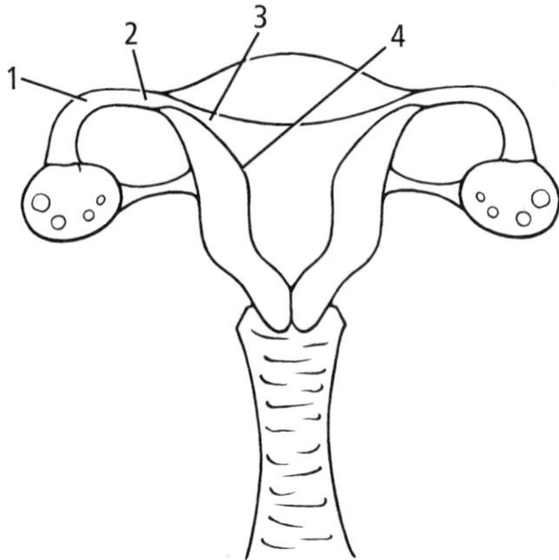

6. Where does fertilization take place?
   A. 1
   B. 2
   C. 3
   D. 4

7. When blood glucose levels are very high, what does the pancreas secrete?
   A. glycogen
   B. insulin
   C. insulin and glycogen
   D. neither insulin nor glycogen

8. Which describes the human circulatory system?
   A. four-chambered heart, one circulatory loop
   B. four-chambered heart, two circulatory loops
   C. two-chambered heart, one circulatory loop
   D. two-chambered heart, two circulatory loops

9. Which statement describes what happens during internal respiration?
   A. Carbon dioxide is used to derive energy from glucose.
   B. Gases are exchanged between the atmosphere and the blood.
   C. Gases are exchanged between the blood and the body's cells.
   D. Oxygen is used to derive energy from glucose.

Biology Online Standardized Test Practice biologygmh.com

Short Answer

Use the diagram below to answer questions 10 and 11.

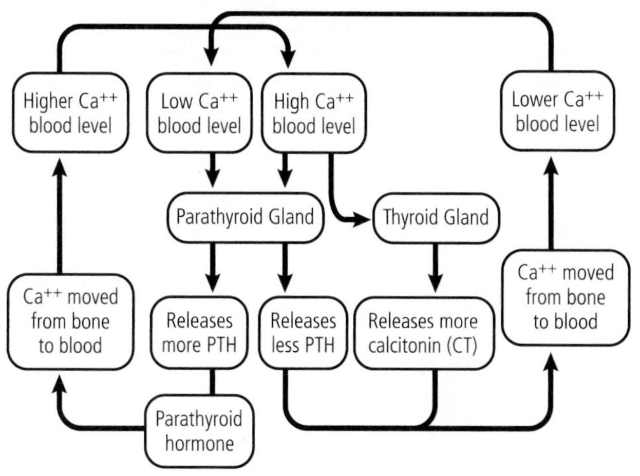

10. Assess how the parathyroid hormone affects bone tissue.

11. Evaluate how a person's blood calcium levels would be affected if his or her thyroid gland stopped working.

12. Analyze how Ivan Pavlov conditioned dogs to salivate when they heard a bell.

13. Assess how emphysema would cause difficulty for a person when climbing stairs.

14. Name and describe the two divisions of the human skeleton. Critique this division of the skeleton.

15. Think about the structure of the middle and inner ear. Infer why people might experience a temporary hearing loss after attending a loud concert.

16. Explain how the villi in the small intestine affect the rate of absorption.

Extended Response

17. A student did an experiment in a sunny room using unripe bananas. He found that bananas ripened faster in a paper bag than on top of a plate. Based on these results, what conclusion could the student make about the ripening of the bananas? Give one example of a way to improve the experiment.

18. Muscles in the legs tend to store large amounts of glycogen and fat. Muscles in the arms do not. When the muscles are used repeatedly, why will the muscles in the arm fatigue the quickest?

19. Different kinds of mammals have different digestive systems. Explain how the digestive systems of ruminant herbivores differ from other herbivores.

Essay Question

As elevation increases, air pressure decreases. At sea level, air pressure is about 760 mmHg. The percentage of oxygen in the atmosphere is about 21 percent. At 3200 m in elevation, the air pressure is 30 percent less than at sea level, however, the percentage of oxygen is the same. The difference in pressure occurs because the molecules of gas are spread farther apart. As altitude continues to increase, the pressure continues to decrease. Mountain climbers face the problems of decreased pressure when they climb a mountain. When climbers go to the summit of the highest mountains, they carry oxygen tanks with them to help them breathe.

Using the information in the paragraph above, answer the following question in essay format.

20. Evaluate why breathing oxygen would enable a mountain climber to reach a higher altitude.

NEED EXTRA HELP?																				
If You Missed Question . . .	1	2	3	4	5	6	7	8	9	10	11	12	13	14	15	16	17	18	19	20
Review Section . . .	34.1	33.3	35.1	36.2	34.3	36.2	35.3	34.3	34.2	35.3	35.3	31.1	34.2	32.2	33.3	35.1	1.3	32.3	30.2	34.2

CHAPTER 37

The Immune System

BIG Idea The immune system attempts to protect the body from contracting an infection through pathogens.

Section 1
Infectious Diseases
MAIN Idea Pathogens are dispersed by people, other animals, and objects.

Section 2
The Immune System
MAIN Idea The immune system has two main components: nonspecific immunity and specific immunity.

Section 3
Noninfectious Disorders
MAIN Idea Noninfectious disorders include genetic disorders, degenerative diseases, metabolic diseases, cancer, and inflammatory diseases.

BioFacts

- There are more than 600 lymph nodes, such as the tonsil, in the human body.
- Macrophages have cytoplasm that is in constant motion.
- Millions of viruses could fit on the head of a pin.

LAUNCH Lab

How do you track a cold?

Colds and many other illnesses are caused by pathogens that can pass from person to person. In this lab, you will trace the path of a cold.

Procedure

1. Read and complete the lab safety form.
2. Create a series of questions you can ask your classmates about the last time they had a cold: their symptoms, other family members and friends who had the same symptoms, and the hygiene precautions they used to avoid illnesses.
3. Interview your classmates using your list.
4. Design a concept map that organizes the data you have collected to trace the paths the colds in your classmates took as they passed from person to person.

Analysis

1. **Describe** how your concept map distinguishes between different cold symptoms present in your classmates.
2. **Infer** what paths the different colds might have taken as they passed from person to person among your classmates and their friends and family.

Visit biologygmh.com to:

- **study the entire chapter online**
- **explore the Interactive Time Line, Interactive Tables, Concepts in Motion, Microscopy Links, Virtual Labs, and links to virtual dissections**
- **access Web links for more information, projects, and activities**
- **review content online with the Interactive Tutor and take Self-Check Quizzes**

Describing Immunity Make the following Foldable to help you organize the ideas of immunity.

STEP 1 Stack three sheets of notebook paper, each 2.5 cm apart.

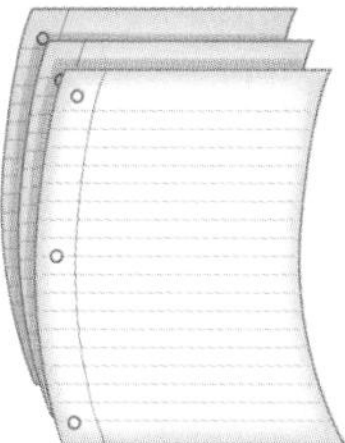

STEP 2 Fold the sheets in half so that all the layers are the same distance apart.

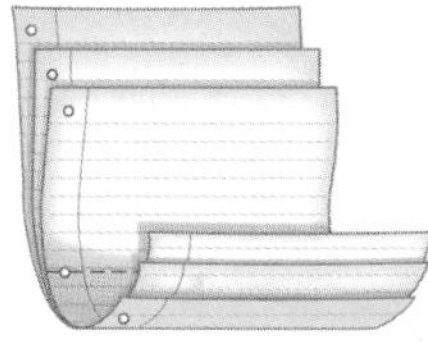

STEP 3 Staple the Foldable at the bottom and label each tab as shown with the following titles: *Immunity from Disease, Innate Immunity, Antibody Immunity, Cellular Immunity, Passive Immunity,* and *Acquired Immunity.*

FOLDABLES Use this Foldable with Section 37.2. As you study the section, describe each type of immunity on the page opposite the title. Use the Foldable to review what you have learned about immunity.

Reading Preview

Objectives

- **Construct** a flowchart demonstrating Koch's postulates.
- **Explain** how diseases are transmitted and how reservoirs play a role in disease dispersal.
- **Describe** symptoms of bacterial infectious disease.

Review Vocabulary

protozoan: unicellular, heterotrophic, animal-like protist

New Vocabulary

infectious disease
pathogen
Koch's postulates
reservoir
endemic disease
epidemic
pandemic
antibiotic

■ **Figure 37.1** These rodlike bacteria cause the disease anthrax.

Color-Enhanced SEM Magnification: 50×

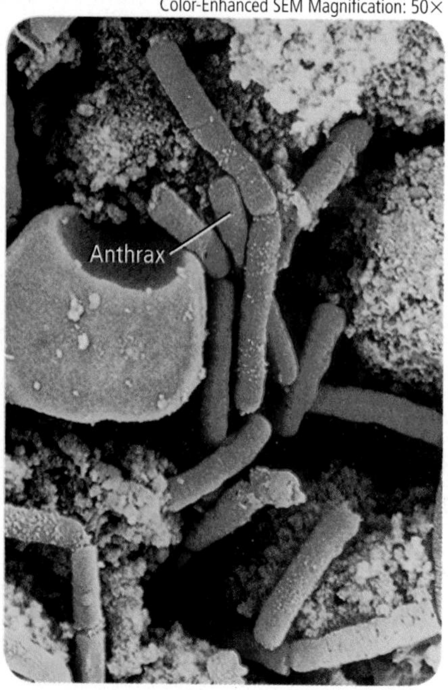

Infectious Diseases

MAIN Idea Pathogens are dispersed by people, other animals, and objects.

Real-World Reading Link Have you ever gotten something sticky on your hands? As you touched other objects, they too became sticky. In a similar manner, viruses transfer to objects that you touch. When these objects are touched by someone else, the virus can be picked up by another person.

Pathogens Cause Infectious Disease

What do a cold and athlete's foot have in common? They are both examples of an infectious disease. An **infectious disease** is a disease that is caused when a pathogen is passed from one organism to another, disrupting homeostasis in the organism's body. Agents called **pathogens** are the cause of infectious disease. Some but not all types of bacteria, viruses, protozoans, fungi, and parasites are pathogens.

Recall from Chapter 18 that many types of these organisms are present in the world around us without causing infectious diseases. Your body benefits from organisms, such as certain types of bacteria and protozoans, that normally live in your intestinal and reproductive tracts. Other bacteria live on your skin, especially in the shafts of your hair follicles. These organisms keep pathogens from thriving and multiplying on your body.

Germ Theory and Koch's Experiments

Before the invention of the microscope, people thought "something" passed from a sick person to a well person to cause an illness. Then, scientists discovered microorganisms and Louis Pasteur demonstrated that microorganisms from the air are able to grow in nutrient solutions, as discussed in Chapter 14. With the knowledge gained from these and other discoveries, doctors and scientists began to develop the germ theory. The germ theory states that some microorganisms are pathogens. However, scientists were not able to clearly demonstrate this theory until Robert Koch developed his postulates.

Identification of the first disease pathogen In the late 1800s, Robert Koch, a German physician, was studying anthrax (AN thraks)—a deadly disease that affects cattle and sheep and can also affect people. Koch isolated bacteria, like those in **Figure 37.1,** from the blood of cattle that had died from anthrax. After growing the bacteria in the laboratory, Koch injected the bacteria into healthy cattle. These animals developed the disease anthrax. He then isolated bacteria from the blood of newly infected cattle and grew the bacteria in the laboratory. The characteristics of the two sets of cultures were identical, indicating that the same type of bacteria caused the illness in both sets of cattle. Thus, Koch demonstrated that the bacteria he originally isolated were the cause of anthrax.

 **Reading Check** **Explain** how Koch proved the germ theory correct.

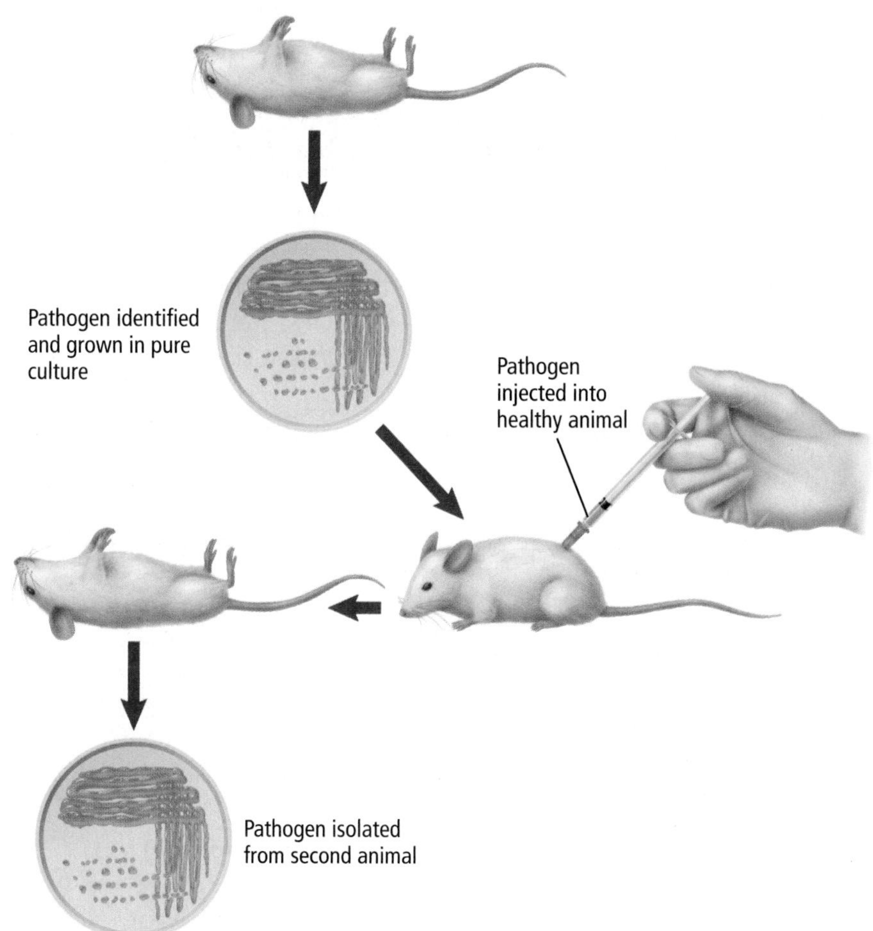

Postulate 1

The suspected pathogen must be isolated from the diseased host in every case of the disease.

Postulate 2

The suspected pathogen must be grown in pure culture on artificial media in the laboratory.

Postulate 3

The suspected pathogen from the pure culture must cause the same disease when placed in a healthy new host.

Postulate 4

The suspected pathogen must be isolated from the new host, grown again in pure culture, and shown to have the same characteristics as the original pathogen.

■ **Figure 37.2** Koch's postulates demonstrate that a specific pathogen causes a specific disease.

Think Critically ***What did Koch demonstrate when he isolated the same bacteria from the cattle the second time?***

Koch's postulates Koch established and published experimental steps known as **Koch's postulates,** which are rules for demonstrating that an organism causes a disease. These steps are followed today to identify a specific pathogen as the agent of a specific disease. Follow the steps in **Figure 37.2** as you read each of the four postulates.

Postulate 1: The suspected pathogen must be isolated from the diseased host in every case of the disease.

Postulate 2: The suspected pathogen must be grown in pure culture on artificial media in the laboratory. A pure culture is a culture that contains no other types of microorganisms—only the suspected pathogen.

Postulate 3: The suspected pathogen from the pure culture must cause the disease when placed in a healthy new host.

Postulate 4: The suspected pathogen must be isolated from the new host, grown again in pure culture, and shown to have the same characteristics as the original pathogen.

Some exceptions to Koch's postulates do exist. Some pathogens, such as the pathogen that is believed to cause syphilis (SIH fuh lus), cannot be grown in pure culture on artificial media. Artificial media are the nutrients that the bacteria need to survive and reproduce. Pathogens are grown on this media in the laboratory. Also, in the case of viruses, cultured cells are needed because viruses cannot be grown on artificial media.

Study Tip

Purposeful Reading Before reading, predict how the information you learn about diseases can be applied to your daily life. Scan the chapter and focus on the boldfaced headings to get an idea about what you will study. Record your ideas. Refer to the list as you study the chapter.

Concepts In Motion
Interactive Table To explore more about infectious diseases, visit biologygmh.com.

Table 37.1 Human Infectious Diseases

Disease	Cause	Affected Organ System	How Disease is Spread
Tetanus	Bacteria	Nervous system	Soil in deep puncture wound
Strep throat	Bacteria	Respiratory system	Droplets/direct contact
Meningitis	Bacteria	Nervous system	Droplets/direct contact
Lyme disease	Bacteria	Skeletal and nervous system	Vector (tick)
Chicken pox	Virus	Skin	Droplets/direct contact
Rabies	Virus	Nervous system	Animal bite
Common cold	Virus	Respiratory system	Droplets/direct contact
Influenza	Virus	Respiratory system	Droplets/direct contact
Hepatitis B	Virus	Liver	Direct contact with exchange of body fluids
West Nile	Virus	Nervous system	Vector (mosquito)
Giardia	Protozoan	Digestive tract	Contaminated water
Malaria	Protozoan	Blood and liver	Vector (mosquito)
Athlete's foot	Fungus	Skin	Direct contact or contaminated objects

Spread of Disease

Of the large number of microorganisms that coexist with humans, only a few cause disease. The pathogens vary as much as the diseases themselves. Some might cause mild diseases, such as the common cold. Others cause serious diseases, such as meningitis (men in JI tus), an infection of the coverings of the brain and spinal cord. **Table 37.1** lists some of the human infectious diseases you might know.

For a pathogen to spread, it must have both a reservoir and a way to spread. A disease **reservoir** is a source of the pathogen in the environment. Reservoirs might be animals, people, or inanimate objects, such as soil.

Human reservoirs Humans are the main reservoir for pathogens that affect humans. They might pass the pathogen directly or indirectly to other humans. Many pathogens might be passed on to other hosts before the person even knows he or she has the disease. An individual that is symptom-free but capable of passing the pathogen is called a carrier. Pathogens that cause colds, influenza (commonly referred to as the flu), and sexually transmitted diseases, such as human immunodeficiency (ih MYEWN noh dih fih shun see) virus (HIV), can be passed on without the person knowing he or she is infected.

VOCABULARY

SCIENCE USAGE V. COMMON USAGE

Carrier

Science usage: person who spreads germs while remaining well.
Typhoid fever was spread by a carrier known as "Typhoid Mary."

Common usage: a person or corporation in the transportation business.
Freight is shipped by carriers.

Animal reservoirs Other animals also are reservoirs of pathogens that can be passed to humans. Influenza and rabies are examples of human diseases listed in **Table 37.1** that are caused by pathogens passed to humans from other animals. Influenza can infect pigs and various types of birds. Rabies is found in domestic dogs and many wild animals, such as bats, foxes, skunks, and raccoons.

Other reservoirs Some bacteria normally found in the soil, such as tetanus bacteria, can cause disease in humans. The tetanus bacteria can cause a serious infection if it contaminates a deep wound in the body. Contamination of wounds by bacteria was a major cause of death during wars before the development of antibiotics and vaccinations.

Contaminated water or food is another reservoir of pathogens for human disease. One of the main purposes of sewage treatment plants is the safe disposal of human feces, which prevents contamination of the water supply by pathogens. Contaminated water used in growing or preparing food can transfer pathogens. Food also can become contaminated through contact with humans or insects such as flies.

Transmission of pathogens Pathogens mainly are transmitted to humans in four ways: direct contact, indirectly through the air, indirectly through touching contaminated objects, or by organisms called vectors that carry pathogens. Study **Figure 37.3,** which illustrates some of the ways pathogens can be transmitted to humans.

Direct Contact Direct contact with other humans is one of the major modes of transmission of pathogens. Diseases such as colds, infectious mononucleosis (mah noh new klee OH sus)(commonly referred to as mono, or the "kissing disease"), herpes (HUR peez), and sexually transmitted diseases are caused by pathogens passed through direct contact.

CAREERS IN BIOLOGY

Epidemiologist An epidemiologist studies disease patterns to help prevent and control the spread of diseases. For more information on biology careers, visit biologygmh.com.

■ **Figure 37.3** Diseases can be transmitted to humans in various ways.
Think Critically ***Identify ways to prevent contracting diseases if contact cannot be avoided.***

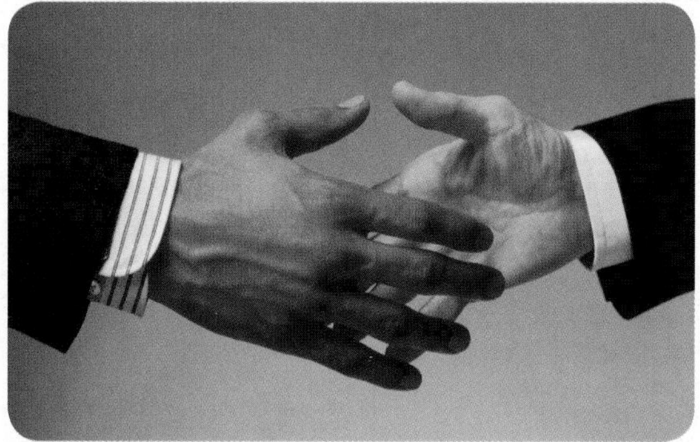

Direct contact

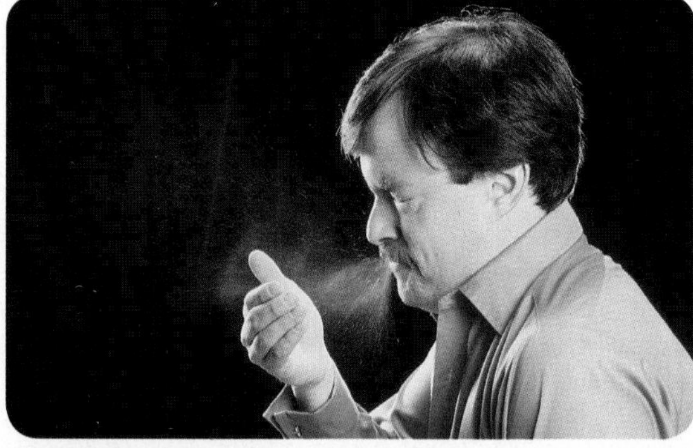

Indirect contact through air

Indirect contact by objects

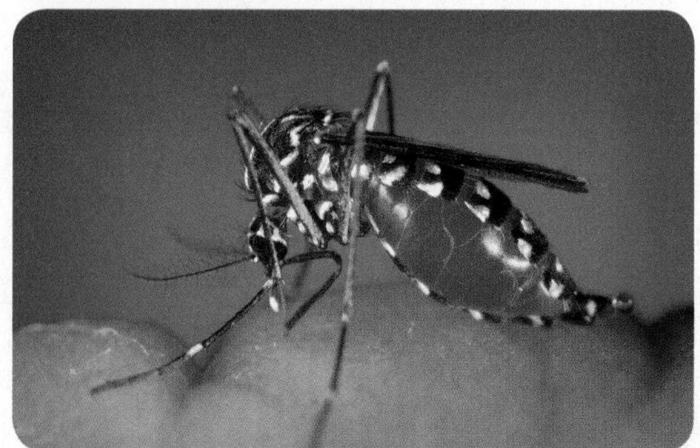

Vectors

Indirect contact Some pathogens can be passed through the air. When a person with an infectious disease sneezes or coughs, pathogens can be passed along with the tiny mucus droplets. These droplets then can spread pathogens to another person or to an object.

Many organisms can survive on objects handled by humans. Cleansing of dishes, utensils, and countertops with detergents as well as careful hand-washing help prevent the spread of diseases that are passed in this manner. As a result, there are various food rules that restaurants must abide by that are based on preventing the spread of disease.

Vectors Certain diseases can be transmitted by vectors. The most common vectors are arthropods, which include biting insects such as mosquitoes and ticks. Recall from **Table 37.1** that Lyme disease, malaria, and West Nile virus are diseases that are passed to humans by vectors. The West Nile virus, which is currently spreading across the United States, is transmitted from horses and other mammals to humans by mosquitoes. Flies can transmit pathogens by landing on infected materials, such as feces, and then landing on materials handled or eaten by humans.

LAUNCH Lab

Review Based on what you have read about spread of disease, how would you now answer the analysis questions?

 **Reading Check** **Describe** how diseases are spread to humans.

Symptoms of Disease

When you become ill with a disease such as the flu, why do you feel aches and pains, and why do you cough and sneeze? The pathogen, such as an influenza virus or bacteria, has invaded some of the cells of your body. The virus multiplies in the cells and leaves the cells either by exocytosis or by causing the cell to burst. Thus, the virus damages tissues and even kills some cells. When pathogenic bacteria invade the body, harmful chemicals or toxins might be produced. The toxins can be carried throughout the body via the bloodstream and damage various parts of the body.

Figure 37.4

Immunology Through Time

For centuries, scientists have struggled to learn about the human immune system. Today, scientists are working to stop a virus that has attacked the immune system of over 40 million people worldwide—HIV.

1800 1900 1970

1796 Edward Jenner discovers that a patient vaccinated with the cowpox virus is immune to smallpox.

1908 Elie Metchnikoff observes phagocytosis, and Paul Erlich describes antibodies. They share a Nobel Prize for their discoveries.

1975 César Milstein and his research team develop a technique to clone a specific antibody.

1981 The first clinical description of acquired immunodeficiency syndrome (AIDS) is established.

Toxins produced by pathogens can affect specific organ systems. The tetanus bacteria produce a potent toxin that causes spasms in the voluntary muscles. The disease botulism (BAH chuh lih zum) usually is caused when a person consumes food in which the botulism bacteria have grown and produced a toxin. This toxin paralyzes nerves. The toxin from the botulism bacteria can cause disease in humans even when no bacteria are present.

Some types of bacteria, some protozoans, and all viruses invade and live inside cells, causing damage. Because the cells are damaged, they might die, causing symptoms in the host. Some disease symptoms, such as coughing and sneezing, are triggered by the immune system, as discussed later in this chapter. For a closer look at research on the immune system, examine **Figure 37.4.**

Disease Patterns

As outbreaks of diseases spread, certain patterns are observed. Agencies such as the community health departments, the Centers for Disease Control and Prevention (CDC), and the World Health Organization (WHO) continually monitor disease patterns to help control the spread of diseases. The CDC, with headquarters in Atlanta, Georgia, receives information from doctors and medical clinics and publishes a weekly report about the incidence of specific diseases, as shown in **Figure 37.5.** The WHO similarly watches disease incidence throughout the world.

Some diseases, such as the common cold, are known as **endemic diseases** because they continually are found in small amounts within the population. Sometimes, a particular disease will have a large outbreak in an area and afflict many people, causing an **epidemic.** If an epidemic is widespread throughout a large region, such as a country, continent, or the entire globe, it is described as **pandemic.**

TABLE 2. Reported cases of notifiable diseases,* by geographic division and area — United States, 2003

Area	Total resident population (in thousands)	AIDS†
UNITED STATES	287,974	44,232**
NEW ENGLAND	14,134	1,697
Maine	1,295	52
N.H.	1,274	37
Vt.	616	16
Mass.	6,422	757
R.I.	1,068	102
Conn.	3,459	733
MID. ATLANTIC	40,038	10,142
Upstate N.Y.	11,385	1,589
N.Y. City	7,749	5,133
N.J.	8,575	1,514
Pa.	12,329	1,906
E.N. CENTRAL	45,635	3,875
Ohio	11,409	775
Ind.	6,157	506
Ill.	12,586	1,734
Mich.	10,043	676
Wis.	5,440	184
W.N. CENTRAL	19,464	844
Minn.	5,025	179
Iowa	2,936	75
Mo.	5,670	404
N. Dak.	634	2
S. Dak.	760	13
Nebr.	1,728	60
Kans.	2,712	111
S. ATLANTIC	53,564	12,191
Del.	806	216
Md.	5,451	1,572
D.C.	569	961

■ **Figure 37.5** The Centers for Disease Control and Prevention publish reports on the incidence of certain diseases.

Infer ***how these reports are helpful in understanding disease patterns.***

1985 Flossie Wong-Staal and her team clone HIV, enabling scientists to create a test to determine whether or not a person has HIV.

2004 HIV infection is pandemic in sub-Saharan Africa, where 10 percent of the world's population has 60 percent of the world's HIV infections.

1980 — 1990 — 2000

1984 Luc Montagnier and Robert Gallo independently announce the discovery of the virus that causes AIDS.

1999 Dr. Beatrice Hahn hypothesizes that humans most likely were exposed to HIV from a chimp species found in west equatorial Africa.

Concepts In Motion Interactive Time Line
To learn more about these discoveries and others, visit biologygmh.com. **Biology Online**

■ **Figure 37.6** Penicillin is secreted by the mold called *Penicillium,* shown growing on this orange.
Hypothesize ***Why are there so many antibiotics available?***

Treating and Fighting Diseases

A medical professional may prescribe a drug to help the body fight a disease. One type of prescription drug is an **antibiotic** (an ti bi AH tihk), which is a substance that can kill or inhibit the growth of other microorganisms. Recall from Chapter 20 that penicillin is secreted by the fungus *Penicillium,* which is shown in **Figure 37.6.** This fungus secretes the chemical penicillin to kill competing bacteria that grow on the fungal food source. Penicillin was isolated, purified, and first used in humans during World War II. Many other fungal secretions are used as antibiotics, such as erythromycin, neomycin, and gentamicin. Synthetic antibiotics also have been developed by pharmaceutical companies.

Chemical agents also are used in the treatment of protozoan and fungal diseases. Some antiviral drugs are used to treat herpes infections, influenza in the elderly, and HIV infections. Most viral diseases are handled by the body's built-in defense system—the immune system.

Connection to Health Over the last 60 years, the widespread use of antibiotics has caused many bacteria to become resistant to particular antibiotics. Recall from Chapter 15 that natural selection occurs when organisms with favorable variations survive, reproduce, and pass their variations to the next generation. Bacteria in a population might have a trait that enables them to survive when a particular antibiotic is present. These bacteria can reproduce quickly and pass on the variation. Because reproduction can occur so rapidly in bacteria, the number of antibiotic-resistant bacteria in a population can increase quickly, too.

MiniLab 37.1

Evaluate the Spread of Pathogens

How can you evaluate the spread of disease? Investigate what possible diseases might be transmitted by common items.

Procedure

1. Read and complete the lab safety form
2. Observe all the items given to you by your teacher.
3. Infer the types of diseases each item could pass on to a human (if any).
4. Evaluate the likelihood of each item transmitting a disease to a human and devise a scale for assessing each item's probability for transmitting an infectious disease.

Analysis

1. **Identify** the types of pathogens that might be transmitted by the items you were given and the methods of transmission of each pathogen.
2. **Infer** the items most likely to be disease reservoirs.
3. **Describe** possible disease patterns of each pathogen.
4. **Infer** how you could prevent getting diseases from these possible pathogens.

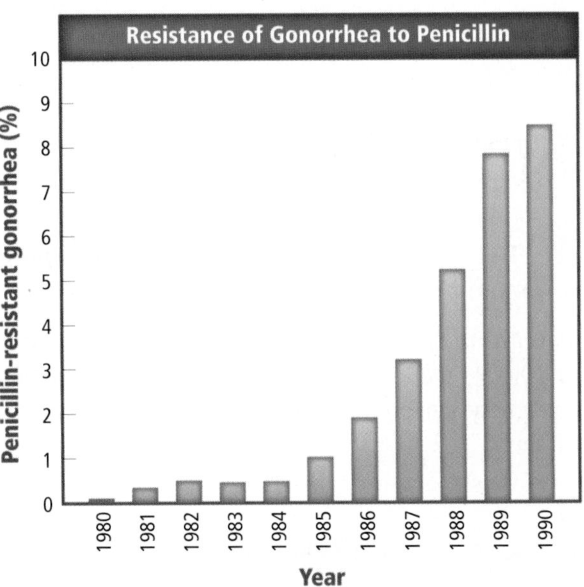

Figure 37.7 The graph shows the reported incidence of penicillin-resistant gonorrhea in the United States from 1980–1990.

Analyze ***What is the percentage increase from 1980 to 1990?***

Antibiotic resistance of bacteria has presented the medical community with some problems with treating certain diseases. For example, penicillin was used effectively for many years to treat gonorrhea (gah nuh REE uh), a sexually transmitted disease, but now most strains of gonorrhea bacteria are resistant to penicillin. As a result, new drug therapies are needed to treat gonorrhea. **Figure 37.7** shows the increase in gonorrhea resistance as the bacteria have gained resistance to treatment with antibiotics.

Another treatment problem is with staphylococcal disease—it is acquired in a hospital, which can result in skin infections, pneumonia (noo MOH nyuh), or meningitis. These staphylococci are often strains of bacteria that are resistant to many current antibiotics and can be difficult to treat.

Section 37.1 Assessment

Section Summary

- Pathogens, such as bacteria, viruses, protozoans, and fungi, cause infectious diseases.
- Koch's postulates demonstrate how to show that a particular pathogen causes a certain disease.
- Pathogens are found in disease reservoirs and are transmitted to humans by direct and indirect methods.
- The symptoms of disease are caused by invasion of the pathogen and the response of the host immune system.
- Treatment of infectious disease includes the use of antibiotics and antiviral drugs.

Understand Main Ideas

1. MAIN Idea **Compare** the mode of transmission of the common cold with that of malaria.
2. **Summarize** some symptoms of bacterial infectious disease.
3. **Define** *infectious disease* and give three examples of infectious diseases.
4. **Application** Draw a graphic organizer or concept map illustrating Koch's postulates for a bacterial infectious disease in a rabbit.
5. **Infer** why a person might be exposed to tetanus bacteria after stepping on a dirty nail.

Think Critically

6. **Evaluate** the following scenario: Two days after visiting a pet shop and observing green parrots in a display cage and fish in an aquarium, a student developed a fever, became ill, and was diagnosed with parrot fever. What might be the disease reservoir and possible transmission method?
7. **Evaluate** Animal feed is often medicated with a low level of antibiotics. Evaluate how this might play a role in the development of antibiotic-resistant bacteria.

Section 37.2

Reading Preview

Objectives

- **Compare and contrast** nonspecific and specific immunity.
- **Summarize** the structure and function of the lymphatic system.
- **Distinguish** between passive and active immunity.

Review Vocabulary

white blood cells: large, nucleated blood cells that play a major role in protecting the body from foreign substances and microorganisms

New Vocabulary

complement protein
interferon
lymphocyte
antibody
B cell
helper T cell
cytotoxic T cell
memory cell
immunization

The Immune System

MAIN Idea The immune system has two main components: nonspecific immunity and specific immunity.

Real-World Reading Link We live with a number of potential pathogens such as bacteria and viruses that can cause disease. Like a fort protecting a city from attack, the immune system protects the body against these and other disease-causing organisms.

Nonspecific Immunity

At the time of birth, the body has a number of defenses in the immune system that fight off pathogens. These defenses are nonspecific because they are not aimed at a specific pathogen. They protect the body from any pathogen that the body encounters.

The nonspecific immunity provided by the body helps to prevent disease. Nonspecific immunity also helps to slow the progression of the disease while the specific immunity begins to develop its defenses. Specific immunity is the most effective immune response, but nonspecific immunity is the first line of defense.

Barriers Like the strong walls of a fort, barriers are used by the body to protect against pathogens. These barriers are found in areas of the body where pathogens might enter.

Skin barrier One of the simplest ways that the body avoids infectious disease is by preventing foreign organisms from entering the body. This major line of defense is the unbroken skin and its secretions. Recall that the skin contains layers of living cells covered by many layers of dead skin cells. By forming a barrier, the layers of dead skin cells help protect against invasion by microorganisms. Many of the bacteria that live symbiotically on the skin digest skin oils to produce acids that inhibit many pathogens. **Figure 37.8** shows some normal bacteria found on the skin that protect the skin from attack.

■ **Figure 37.8** These bacteria normally are found on human skin.

Color-Enhanced SEM Magnification: 14,000×

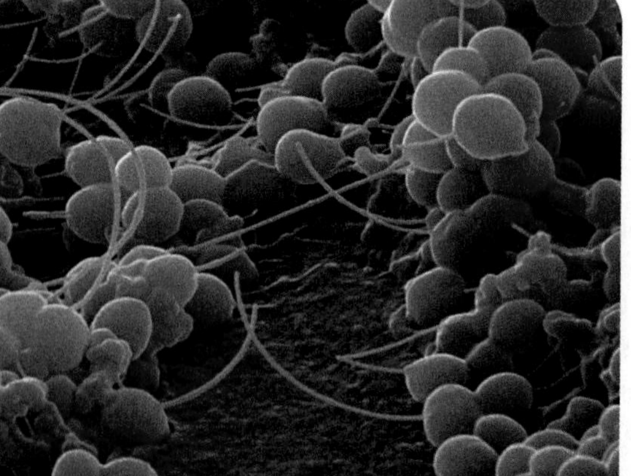

Chemical barriers Saliva, tears, and nasal secretions contain the enzyme lysozyme. Lysozyme breaks down bacterial cell walls, which kills pathogens.

Another chemical defense is mucus, which is secreted by many inner surfaces of the body. It acts as a protective barrier, blocking bacteria from sticking to the inner epithelial cells. Cilia, discussed in Chapter 7, also line the airway. Their beating motion sends any bacteria caught in the mucus away from the lungs. When the airway becomes infected, extra mucus is secreted, which triggers coughing and sneezing to help move the infected mucus out of the body.

A third chemical defense is the hydrochloric acid secreted in your stomach. In addition to its purpose in digestion, stomach acid kills many microorganisms found in food that could cause disease.

Nonspecific responses to invasion Even if an enemy gets through the walls of a town's fort, defense doesn't end. Similarly, the body has nonspecific immune responses to pathogens that get beyond its barriers.

Cellular defense If foreign microorganisms enter the body, the cells of the immune system, shown in **Table 37.2,** defend the body. One method of defense is phagocytosis. White blood cells, especially neutrophils and macrophages, are phagocytic. Recall from Chapter 7 and Chapter 34 that phagocytosis is the process by which phagocytic cells surround and internalize the foreign microorganisms. The phagocytes then release digestive enzymes and other harmful chemicals from their lysosomes, destroying the microorganism.

A series of about 20 proteins that are found in the blood plasma are involved in phagocytosis. These proteins are called complement proteins. **Complement proteins** enhance phagocytosis by helping the phagocytic cells bind better to pathogens, activating the phagocytes and enhancing the destruction of the pathogen's membrane, as illustrated in **Figure 37.9.** They are activated by materials that are in the cell wall of bacteria.

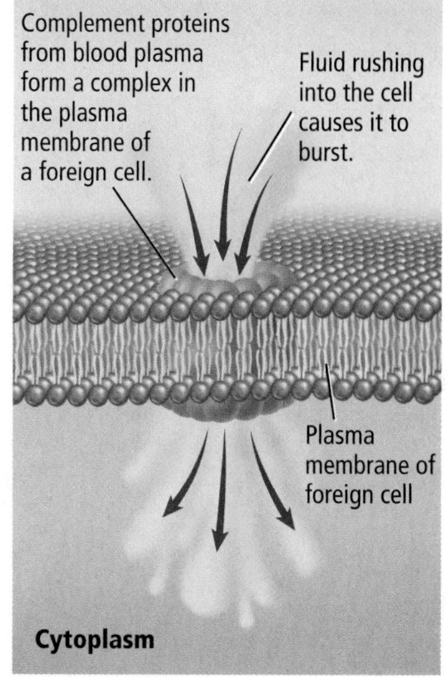

Figure 37.9 Complement proteins form a hole in the plasma membrane of the invading cell.

Concepts In Motion
Interactive Figure To see an animation of how the complement system can destroy a pathogen, visit biologygmh.com.

Interferon When a virus enters the body, another cellular defense helps prevent the virus from spreading. Virus-infected cells secrete a protein called **interferon.** Interferon binds to neighboring cells and stimulates these cells to produce antiviral proteins which can prevent viral replication in these cells.

Inflammatory response Another nonspecific response—the inflammatory response—is a complex series of events that involves many chemicals and immune cells that help enhance the overall immune response. When pathogens damage tissue, chemicals are released by both the invader and cells of the body. These chemicals attract phagocytes to the area, increase blood flow to the infected area, and make blood vessels more permeable to allow white blood cells to escape into the infected area. This response aids in the accumulation of white blood cells in the area. Some of the pain, heat, and redness experienced in an infectious disease are the result of the inflammatory response.

Table 37.2 Cells of the Immune System

Concepts In Motion
Interactive Table To explore more about cells of the immune system, visit biologygmh.com.

Type of Cell	Example	Function
Neutrophils	Stained LM Magnification: 2150×	Phagocytosis: blood cells that ingest bacteria
Macrophages	Stained LM Magnification: 380×	Phagocytosis: blood cells that ingest bacteria and remove dead neutrophils and other debris
Lymphocytes	Stained LM Magnification: 1800×	Specific immunity (antibodies and killing of pathogens): blood cells that produce antibodies and other chemicals

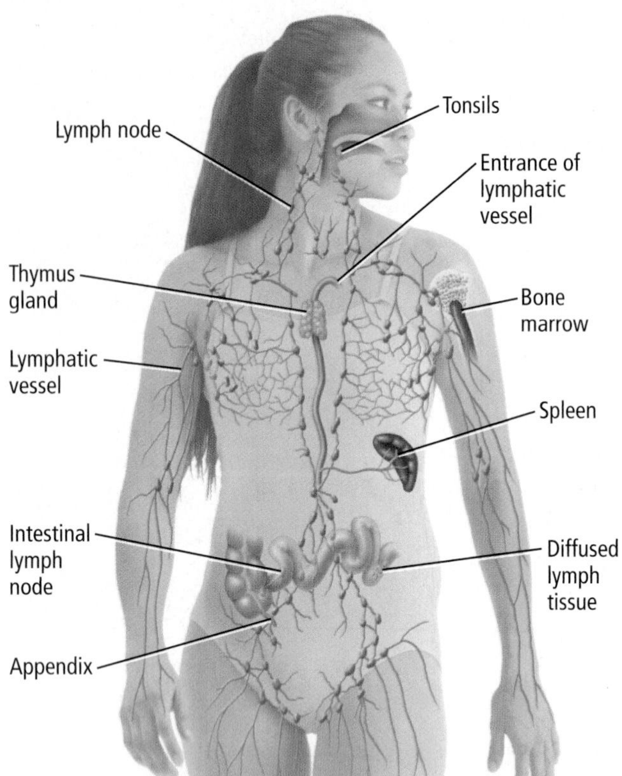

■ **Figure 37.10** The lymphatic system contains the organs involved in the specific immune response.
Identify ***Locate the lymphatic organ that is important for T cell development.***

VOCABULARY
WORD ORIGIN
Thymus
Comes from the Greek word *thymos,* meaning *warty excrescence.*

Specific Immunity

Pathogens sometimes get past the nonspecific defense mechanisms. The body has a second line of defense that attacks these pathogens. Specific immunity is more effective, but takes time to develop. This specific response involves the tissues and organs found in the lymphatic system.

Lymphatic system The lymphatic (lim FA tihk) system, illustrated in **Figure 37.10,** includes organs and cells that filter lymph and blood and destroy foreign microorganisms. The lymphatic system also absorbs fat. Lymph is the fluid that leaks out of capillaries to bathe body cells. This fluid circulates among the tissue cells, is collected by lymphatic vessels, and is returned to the veins near the heart.

Lymphatic organs The organs of the lymphatic system contain lymphatic tissue, lymphocytes, a few other cell types, and connective tissue. **Lymphocytes** are a type of white blood cell that is produced in red bone marrow. These lymphatic organs include the lymph nodes, tonsils, spleen, thymus (THI mus) gland, and diffused lymphatic tissue found in mucous membranes of the intestinal, respiratory, urinary, and genital tracts.

The lymph nodes filter the lymph and remove foreign materials from the lymph. The tonsils form a protective ring of lymphatic tissue between the nasal and oral cavities. This helps protect against bacteria and other harmful materials in the nose and mouth. The spleen stores blood and destroys damaged red blood cells. It also contains lymphatic tissue that responds to foreign substances in the blood. The thymus gland, which is located above the heart, plays a role in activating a special kind of lymphocyte called T cells. T cells are produced in the bone marrow, but they mature in the thymus gland.

B Cell Response

Antibodies are proteins produced by B lymphocytes that specifically react with a foreign antigen. An antigen is a substance foreign to the body that causes an imune response; it can bind to an antibody or T cell. B lymphocytes, often called **B cells,** are located in all lymphatic tissues and can be thought of as antibody factories. When a portion of a pathogen is presented by a macrophage, B cells produce antibodies. Follow along in **Figure 37.11,** as you learn about how B cells are activated to produce antibodies.

Visualizing Specific Immune Responses

Figure 37.11
Specific immune responses involve antigens, phagocytes, B cells, helper T cells, and cytotoxic T cells. The antibody-mediated response involves antibodies produced by B cells and memory B cells. The cytotoxic T cell response results in cytotoxic T cell activation.

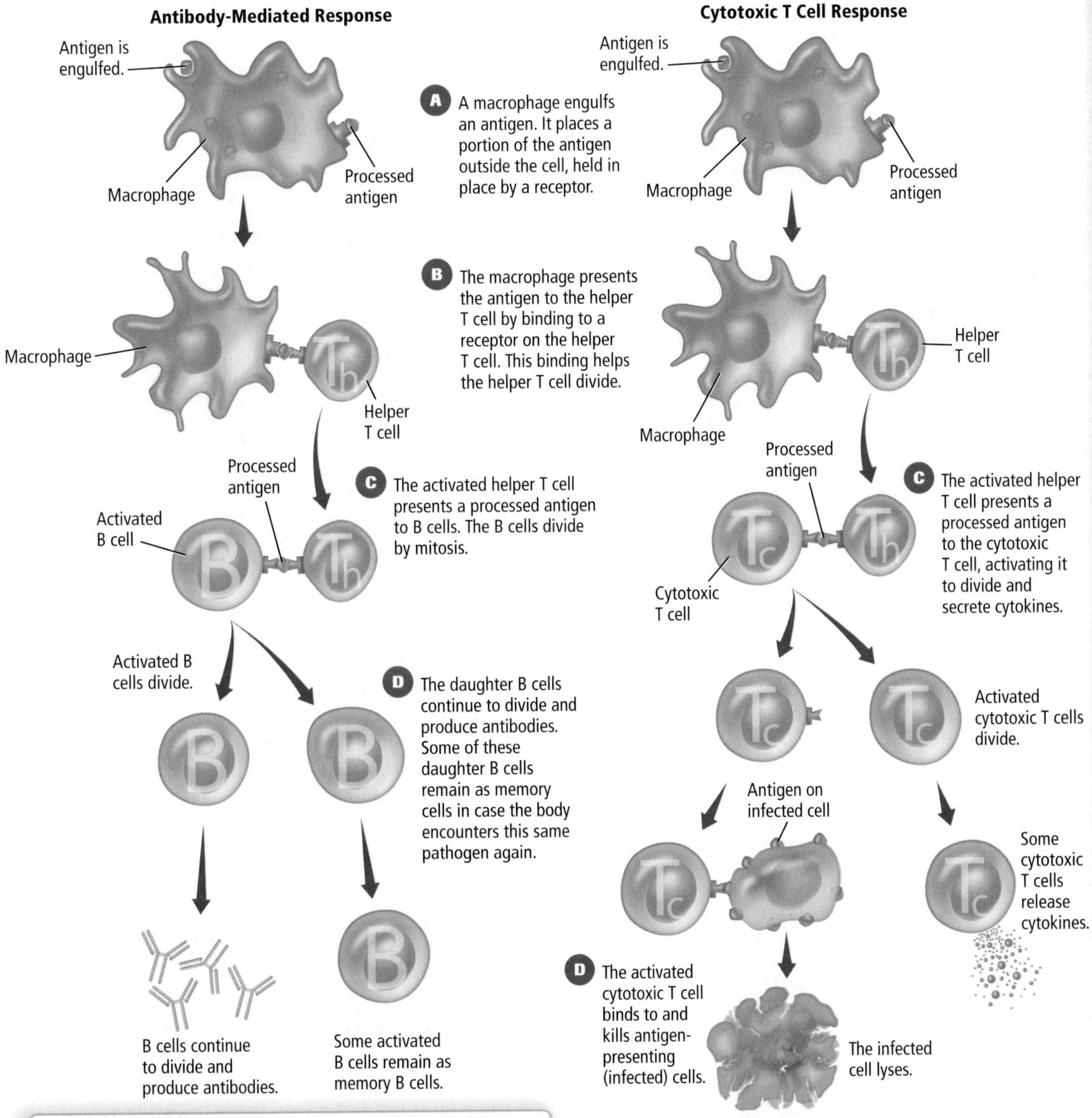

Antibody-Mediated Response

Antigen is engulfed.

Macrophage

Processed antigen

A A macrophage engulfs an antigen. It places a portion of the antigen outside the cell, held in place by a receptor.

Macrophage

Helper T cell

B The macrophage presents the antigen to the helper T cell by binding to a receptor on the helper T cell. This binding helps the helper T cell divide.

Processed antigen

Activated B cell

C The activated helper T cell presents a processed antigen to B cells. The B cells divide by mitosis.

Activated B cells divide.

D The daughter B cells continue to divide and produce antibodies. Some of these daughter B cells remain as memory cells in case the body encounters this same pathogen again.

B cells continue to divide and produce antibodies.

Some activated B cells remain as memory B cells.

Cytotoxic T Cell Response

Antigen is engulfed.

Macrophage

Processed antigen

Helper T cell

Macrophage

Processed antigen

Cytotoxic T cell

C The activated helper T cell presents a processed antigen to the cytotoxic T cell, activating it to divide and secrete cytokines.

Activated cytotoxic T cells divide.

Antigen on infected cell

Some cytotoxic T cells release cytokines.

D The activated cytotoxic T cell binds to and kills antigen-presenting (infected) cells.

The infected cell lyses.

Concepts In Motion Interactive Figure To see an animation of immune responses, visit biologygmh.com.

Biology Online

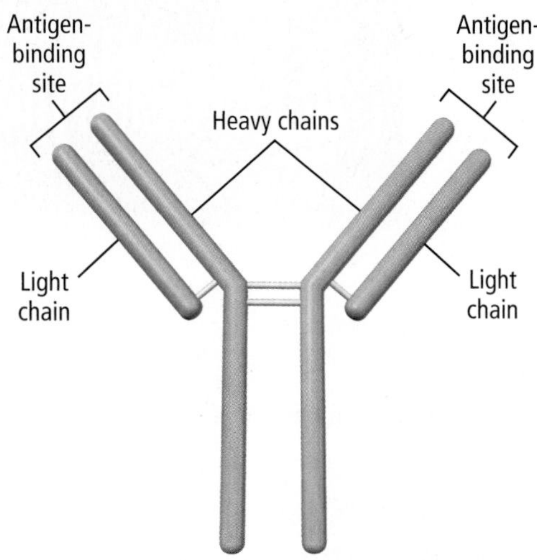

■ **Figure 37.12** Antibodies are made up of two types of protein chains—heavy and light chains.
Summarize *What cells produce antibodies?*

When a macrophage surrounds, internalizes, and digests a pathogen, it takes a piece of the pathogen, which is called a processed antigen, and displays it on its membrane, as illustrated in **Figure 37.11.** In the lymphatic tissues, such as the lymph nodes, the macrophage, with the processed antigen on its surface, binds to a type of lymphocyte called a **helper T cell.** This process activates the helper T cell. This lymphocyte is called a "helper" because it activates antibody secretion in B cells and another type of T cell, which will be discussed later, that aids in killing microorganisms:

- The activated helper T cell reproduces, binds processed antigens, and attaches to a B cell.
- The new helper T cells continue the process of binding antigens, attaching to B cells, and reproducing.
- Once an activated helper T cell binds to a B cell holding an antigen, the B cell begins to manufacture antibodies that specifically bind to the antigen.
- The antibodies can enhance the immune response by binding to microorganisms, making them more susceptible to phagocytosis and by initiating the inflammatory response, helping promote the nonspecific response.

B cells make many combinations of antibodies by using DNA that codes for the production of various heavy and light protein chains that make up antibodies as shown in **Figure 37.12.** Any heavy chain can combine with any light chain. If a B cell can make 16,000 different kinds of heavy chains and 1200 kinds of light chains, it can make 19,200,000 different types of antibodies (1200 × 16,000).

T Cell Response

Once helper T cells are activated by the presentation of an antigen by macrophages, helper T cells can also bind to and activate a group of lymphocytes called cytotoxic T cells. Activated **cytotoxic T cells** destroy pathogens and release chemicals called cytokines. Cytokines stimulate the cells of the immune system to divide and recruit immune cells to an area of infection. Cytotoxic T cells bind to pathogens, release a chemical attack, and destroy the pathogens. Multiple target cells can be destroyed by a single cytotoxic T cell. **Figure 37.11** summarizes the activation of cytotoxic T cells.

 **Reading Check** **Summarize** the role that lymphocytes play in immunity.

Passive and Active Immunity

The body's first response to an invasion by a pathogen is called the primary response. For example, if the viral pathogen that causes chicken pox enters the body, nonspecific and specific immune responses eventually defeat the foreign virus and the body is cleared of the pathogen.

One result of the specific immune response is the production of memory B and T cells. **Memory cells** are long-living cells that are exposed to the antigen during the primary immune response. These cells are ready to respond rapidly if the body encounters the same pathogen later. Memory cells protect the body by reducing the likelihood of developing the disease if exposed again to the same pathogen.

Passive immunity Sometimes temporary protection against an infectious disease is needed. This type of temporary protection occurs when antibodies are made by other people or animals and are transferred or injected into the body. For example, passive immunity occurs between a mother and her child. Antibodies produced by the mother are passed through the placenta to the developing fetus and through breast milk to the infant child. These antibodies can protect the child until the infant's immune system matures.

Antibodies developed in humans and animals that are already immune to a specific infectious disease are used to treat some infectious diseases in others. These antibodies are injected into people who have been exposed to that particular infectious disease. Passive immune therapy is available for people who have been exposed to hepatitis A and B, tetanus, and rabies. Antibodies also are available to inactivate snake and scorpion venoms.

Active immunity Active immunity occurs after the immune system is exposed to disease antigens and memory cells are produced. Active immunity can result from having an infectious disease or immunization. **Immunization,** also called vaccination, is the deliberate exposure of the body to an antigen so that a primary response and immune memory cells will develop. **Table 37.3** lists some of the common immunizations offered in the United States. Immunizations contain killed or weakened pathogens, which are incapable of causing the disease.

Most immunizations include more than one stimulus to the immune system, given after the first immunization. These booster shots increase the immune response, providing further protection from the disease-causing organism.

Vocabulary

Academic Vocabulary

Passive:
not active; acted upon.
The passive monkey stared lazily at the zoo visitors.

Table 37.3 Common Immunizations

Concepts In Motion
Interactive Table To explore more about immunizations, visit biologygmh.com.

Immunization	Disease	Contents
DPT	Diphtheria (D), tetanus (T), pertussis (P) (whooping cough)	D: inactivated toxin, T: inactivated toxin, P: inactivated bacteria
Inactivated polio	Poliomyelitis	Inactivated virus
MMR	Measles, mumps, rubella	All three inactivated viruses
Varicella	Chicken pox	Inactivated virus
HIB	Haemophilus influenzae (flu) type b	Portions of bacteria cell wall covering
HBV	Hepatitis B	Subunit of virus

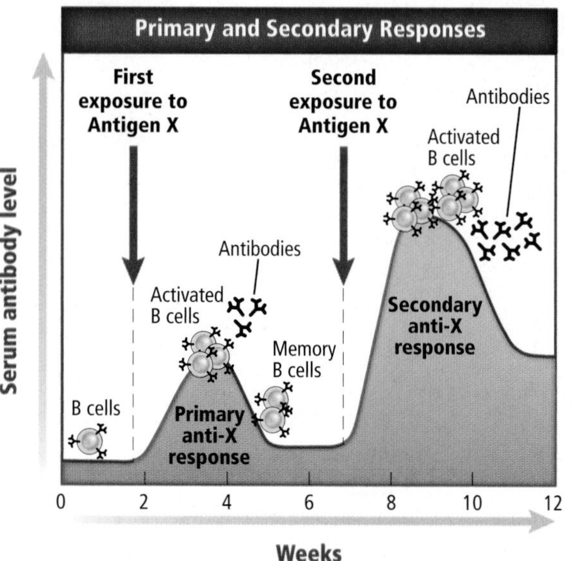

Figure 37.13 This graph shows the difference between the primary and secondary immune responses to exposure to an antigen.

Analyze ***What are the differences between the primary and secondary immune responses?***

Personal Tutor

To learn about primary and secondary responses, visit biologygmh.com.

Why are immunizations effective in preventing disease? The characteristics of the secondary immune response, which is the response to a second exposure to an antigen, enable immunizations to be effective in preventing disease. Study the graph in **Figure 37.13.** Note that the secondary response to the antigen has a number of different characteristics. First, the response is more rapid than the primary response, as shown by the greater steepness in the portion of the curves plotted in red. Second, the overall response, both B and T cell response, is greater during the second exposure. Lastly, the overall memory lasts longer after the second exposure.

Immune System Failure

Defects in the immune system can result in an increased likelihood of developing infectious diseases as well as certain types of cancers. Some diseases can affect the immune system's effectiveness. One such disease called acquired immunodeficiency syndrome (AIDS) results from an infection by human immunodeficiency virus (HIV). AIDS is a serious health problem worldwide.

In 2003, approximately 43,171 AIDS cases were diagnosed in the U.S. In 2003, 18,017 people died of AIDS in the U.S. In 2004, an estimated 40 million people globally were living with HIV infection.

DATA ANALYSIS LAB 37.1

Based on Real Data*

Draw a Conclusion

Is passive immune therapy effective for HIV infection? The standard treatment for a patient with an HIV infection is antiviral drug therapy. Unfortunately, the side effects and increasing prevalence of drug-resistant viruses create a need for additional therapies. One area being studied is passive immune therapy.

Data and Observations

The graph shows HIV patient responses to passive immune therapy. The number of viral copies/mL is a measure of the amount of virus in the patient's blood.

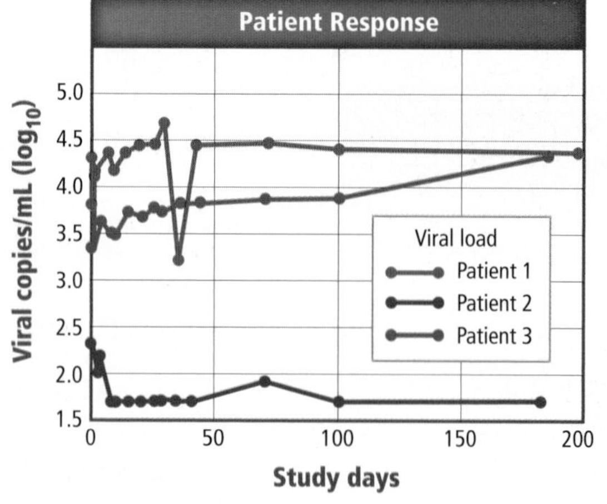

Think Critically

1. **Compare** the patient responses to passive immune therapy.
2. **Draw a conclusion** Can the researchers conclude passive immune therapy is effective? Explain.

*Data obtained from: Stiegler G., et al. 2002. Antiviral activity of the neutralizing antibodies 2FS and 2F12 in asymptomatic HIV-1-infected humans: a phase I evaluation. *AIDS* 16: 2019-2025.

Recall the important role that helper T cells play in specific immunity. HIV infects mainly helper T cells, also called $CD4^+$ cells because these cells have a receptor on the outside of their plasma membrane. This $CD4^+$ receptor is used by medical professionals to identify these cells, as illustrated in **Figure 37.14.**

In Chapter 18, you learned that HIV is an RNA virus that infects helper T cells. The helper T cells become HIV factories, producing new viruses that are released and infect other helper T cells. Over time, the number of helper T cells in an infected person decreases, making the person less able to fight disease. HIV infection usually has an early phase during the first six to twelve weeks while viruses are replicating in helper T cells.

The patient suffers symptoms such as night sweats and fever, but these symptoms are reduced after about eight to ten weeks. Then, the patient exhibits few symptoms for a period of time as long as ten years but is capable of passing the infection through sexual intercourse or blood products. HIV is a secondary immunodeficiency disease, which means that the immune system of a previously healthy person fails. Without antiviral drug therapy, the patient usually dies from a secondary infection from another pathogen after about ten years of being infected with HIV. Current antiviral drug therapy is aimed at controlling the replication of HIV in the body. The therapy is expensive and its long-term results are not known. Resistant strains of HIV are also becoming increasingly common.

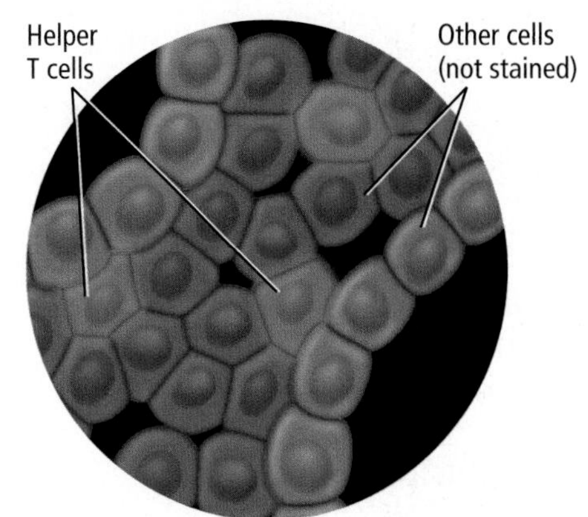

Figure 37.14 Helper T cells have receptors on the surface that are used to identify them in the laboratory.

Section 37.2 Assessment

Section Summary

- The nonspecific immune response includes the skin barrier, secreted chemicals, and cellular pathways that activate phagocytosis.
- The specific immune response involves the activation of B cells, which produce antibodies, and T cells, which include helper T cells and cytotoxic T cells.
- Passive immunity involves receiving antibodies against a disease.
- Active immunity results in immune memory against a disease.
- HIV attacks helper T cells, causing an immune system failure.

Understand Main Ideas

1. MAIN Idea **Compare** specific and nonspecific immune responses.
2. **Describe** the steps involved in activating an antibody response to an antigen.
3. **Make** an illustration demonstrating passive and active immunity.
4. **Describe** the structure and function of the lymphatic system.
5. **Infer** Why is the destruction of helper T cells in HIV infection so devastating to specific immunity?

Think Critically

6. **Hypothesize** what happens when an HIV strain mutates such that viral-replication drugs are no longer effective.
7. **Evaluate** In the disease called severe combined immune deficiency, a child is born without T cell immunity. Evaluate the effects of this disease.

MATH in Biology

8. Antibodies are made of two light protein chains and two heavy protein chains. If the molecular weight of a light chain is 25,000 and the molecular weight of a heavy chain is 50,000, what is the molecular weight of an antibody?

Section 37.3

Reading Preview

Objectives

- **Describe** five categories of noninfectious diseases.
- **Summarize** the role of allergens in allergies.
- **Distinguish** between allergies and anaphylactic shock.

Review Vocabulary

cancer: uncontrolled cell division that can be caused by environmental factors or changes in enzyme production in the cell cycle

New Vocabulary

degenerative disease
metabolic disease
allergy
anaphylactic shock

Noninfectious Disorders

MAIN Idea Noninfectious disorders include genetic disorders, degenerative diseases, metabolic diseases, cancer, and inflammatory diseases.

Real-World Reading Link Maybe you have heard your parents or grandparents complain about their arthritis, which causes achy bones and joints. Perhaps some of your relatives have survived cancer or have diabetes. You or a friend might have an allergy to dust, plant pollens, or other environmental substances. These disorders are different from infectious diseases caused by pathogens.

Genetic Disorders

Not all diseases or body disorders are caused by pathogens. In Chapter 11 you read about some diseases caused by the inheritance of genes that do not function properly in the body, such as albinism, sickle cell anemia, Huntington disease, and hemophilia. There are also chromosomal disorders that result from abnormal chromosome numbers, such as Down syndrome. Many diseases are complex and have both an environmental and a genetic cause.

Coronary artery disease (CAD) is an example of a condition with environmental and genetic origin. This cardiovascular disease can result in blockage of arteries, as shown in **Figure 37.15,** that deliver oxygenated blood to the heart muscle. There is a genetic component that increases a person's risk of developing CAD. Environmental factors such as diet contribute to the development of this complex disease. Families with a history of CAD have a two to seven times greater risk of having CAD than families without a history of CAD. The exact genetic factors, however, are not known.

Reading Check **Summarize** the factors that cause coronary artery disease.

Figure 37.15 When blood cannot flow through a coronary artery, such as the one shown here, a heart attack or sudden death can result.

Stained LM Magnification: 5×

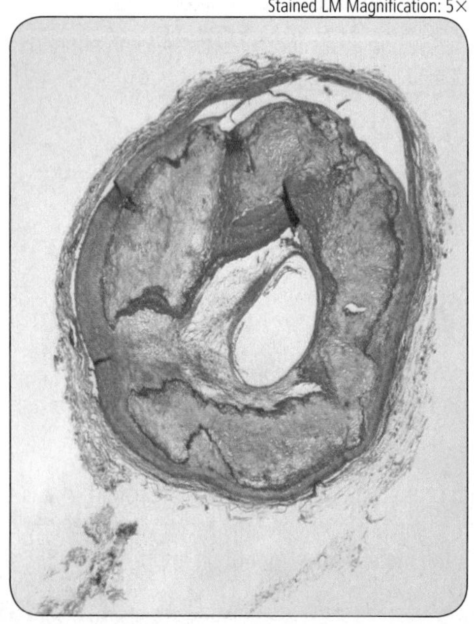

Degenerative Diseases

Some diseases called **degenerative** (dih JEH nuh ruh tihv) **diseases** are the result of a part of the body wearing out. This can be due to the natural aging process. However, a degenerative condition, such as degenerative arthritis, could occur sooner than would be expected if the person is genetically predisposed to the disease or if the person's joints have experienced an increased amount of wear and tear. Degenerative arthritis is common; most people have it by age 70. It is found in almost all vertebrate animals. Arteriosclerosis (ar tir ee oh skluh ROH sus), which is a hardening of the arteries, is another example of a degenerative disease. Because degenerative diseases also have a genetic component, some individuals might be more likely to develop a degenerative disease because of their genetic makeup.

Metabolic Diseases

Metabolic disease results from an error in a biochemical pathway. Some metabolic diseases result in the inability to digest specific amino acids or to regulate body processes. When the pancreas does not make the proper amount of insulin and glucose does not enter body cells normally, the condition known as Type 1 diabetes results. This results in high glucose levels in the bloodstream, which causes damage to many organs including the kidneys and the retinas of the eyes. Metabolic diseases can have a genetic component but also can involve environmental factors such as diet.

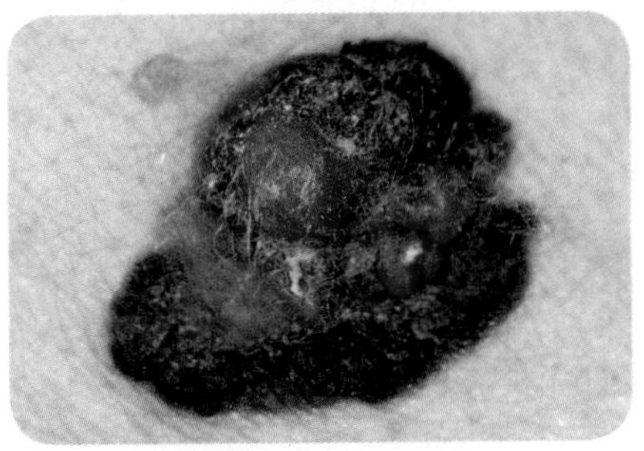

Figure 37.16 Cancer is due to an abnormal increase in cell division in the body, which results in a tumor such as this skin tumor.
Infer ***Why is this large growth so life-threatening?***

Cancer

In Chapter 9, you learned about cancer. Cancer is characterized by abnormal cell growth. Normally, certain regulatory molecules in the body control the beginning and end of the cell cycle. If this control is lost, abnormal cell growth results that could lead to various types of tumors, as shown in **Figure 37.16.** The abnormal cells can interfere with normal body functions and can travel throughout the body. Cancer can develop in any body tissue or organ, including the blood cells. Cancer in the blood cells is called leukemia. Both genetic and environmental factors have been shown to cause cancer.

Connection to History Cancer has been a disease that affects humans since ancient times. Egyptian mummies show evidence of bone cancer, and ancient Greek scientists described different kinds of cancer. Medieval manuscripts have reported details about cancer.

MiniLab 37.2

Compare Cancerous and Healthy Cells

How do cancerous cells and healthy cells differ in appearance? Observe and compare liver cells afflicted with this common noninfectious disease to healthy liver cells.

Procedure

1. Read and complete the lab safety form.
2. Place a prepared **slide of healthy human liver cells** on a **microscope.**

WARNING: ***Never touch broken microscope slides or other broken glass materials.***

3. Observe the healthy liver cells under several different magnifications.
4. Sketch a diagram of several healthy liver cells.
5. Repeat steps 2–4 with a prepared **slide of cancerous human liver cells.**

Analysis

1. **Compare and contrast** the features of healthy liver cells with those of cancerous liver cells.
2. **Infer** why it would not be dangerous to handle an object that was handled by a patient with liver cancer.
3. **Describe** how cancer disrupts the body's homeostasis.

LM Magnification: 50×

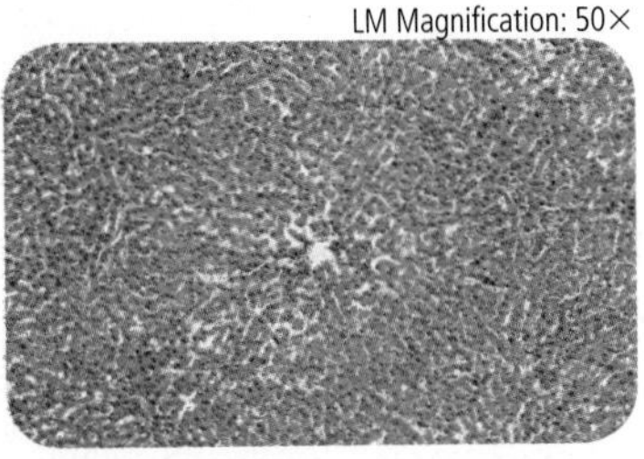

Healthy cells

LM Magnification: 50×

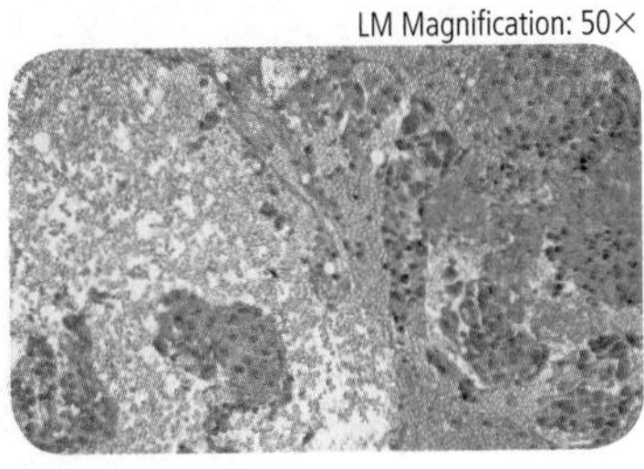

Cancerous cells

Inflammatory Diseases

Inflammatory diseases, such as allergies and autoimmunity, are diseases in which the body produces an inflammatory response to a common substance. Recall from Section 2 that infectious diseases also result in an inflammatory response. However, the inflammatory response in an infectious disease enhances the overall immune response. This inflammatory response is a result of the immune system removing bacteria or other microorganisms from the body. In inflammatory disease, the inflammatory response is not helpful to the body.

Allergies Certain individuals might have an abnormal reaction to environmental antigens. A response to environmental antigens is called an **allergy.** These antigens are called allergens and include things such as plant pollens, dust, dust mites, and various foods, as illustrated in **Table 37.4.** An individual becomes sensitized to the allergen and has localized inflammatory response with swollen itchy eyes, stuffy nose, sneezing, and sometimes a skin rash. These symptoms are the result of a chemical called histamine that is released by certain white blood cells. Antihistamine medications can help alleviate some of these symptoms.

Reading Check **Explain** how allergies are related to the immune system.

Table 37.4 **Common Allergens**

Concepts In Motion

Interactive Table To explore more about common allergens, visit biologygmh.com.

Allergen	Example	Description
Dust mite	Color-Enhanced SEM Magnification: 170×	Dust mites are found in mattresses, pillows, and carpets. Mites and mite feces are allergens.
Plant pollen	Color-Enhanced SEM Magnification: 2300×	Different parts of the country have very different pollen seasons; people can react to one or more pollens, and a person's pollen allergy season might be from early spring to late fall.
Animal dander	Color-Enhanced SEM Magnification: 80×	Dander is skin flakes; cat and dog allergies are the most common, but people also are allergic to pets such as birds, hamsters, rabbits, mice, and gerbils.
Peanut		Allergic reaction to peanuts can result in anaphylaxis. Peanut allergy is responsible for more fatalities than any other type of allergy.
Latex		Latex comes from the milky sap of the rubber tree, found in Africa and Southeast Asia; the exact cause of latex allergy is unknown.

Severe allergic reactions to particular allergens can result in **anaphylactic** (an uh fuh LAK tik) **shock,** which causes a massive release of histamine. In anaphylactic shock, the smooth muscles in the bronchioles contract, which restricts air flow into and out of the lungs.

Common allergens that cause severe allergic reactions are bee stings, penicillin, peanuts, and latex, which is used to make balloons and surgical gloves. People who are extremely sensitive to these allergens require prompt medical treatment if exposed to these agents because anaphylactic reactions are life-threatening. Allergies and anaphylactic reactions are known to have an inherited component.

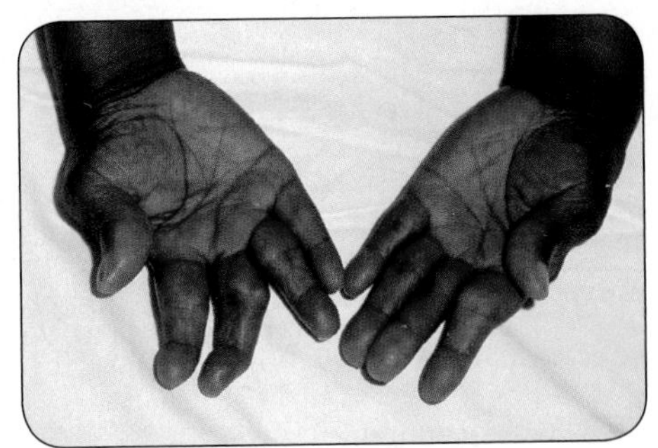

Figure 37.17 The large knobs on these fingers are due to rheumatoid arthritis—an autoimmune disease.

Autoimmunity During the development of the immune system, the immune system learns not to attack proteins produced by the body. However, some people develop autoimmunity (aw toh ih MYOON ih tee) and do form antibodies to their own proteins, which injures their cells.

Figure 37.17 shows the hands of a person with rheumatoid arthritis—a form of arthritis in which antibodies attack the joints. Degenerative arthritis, the form of arthritis that you read about earlier in the section on degenerative diseases, is not caused by autoimmunity.

Rheumatic fever and lupus (LEW pus) are other examples of autoimmune disorders. Rheumatic fever is an inflammation in which antibodies attack the valves of the heart. This can lead to damage to the heart valves and cause the valves to leak or not close properly as blood moves through the heart. Lupus is a disorder in which antibodies against cell nuclei, called antinuclear antibodies, are formed. As a result, many organs are vulnerable to attack by the body's own immune system.

Section 37.3 Assessment

Section Summary

- Noninfectious disorders often have both a genetic and an environmental component.
- The inflammatory response to an infectious disease enhances the immune response, but the inflammatory response to an inflammatory disease is not helpful to the body.
- Allergies are due to an overactive immune response to allergens found in the environment.
- Anaphylactic shock is a severe hypersensitivity to particular allergens.
- Autoimmunity results in an immune attack on body cells.

Understand Main Ideas

1. MAIN Idea **Identify** the type of noninfectious disease shown in **Figure 37.15.**
2. **Explain** the role of allergens in allergies.
3. **Sketch** a diagram demonstrating the process of anaphylactic shock.
4. **Categorize** the following diseases into the categories used in this section: sickle cell disease, diabetes, vertebral degeneration, autoimmunity, and leukemia.

Think Critically

5. **Hypothesize** several causes of chronic bronchitis (inflammation of the bronchioles) found in coal miners.
6. **Create a plan** A child is found to be allergic to cat dander. Create a plan that limits the child's exposure to the allergen.

WRITING in **Biology**

7. Create a pamphlet explaining the symptoms of allergies and listing common allergens.

CUTTING-EDGE BIOLOGY

Buckyballs: A Cure for Allergies?

If you've ever had a sneezing fit after smelling flowers or become sick after eating shellfish, you have had an allergic reaction. Many people have some type of allergy.

A Common Ailment Each year in the United States, consumers spend millions of dollars on nose sprays, pills, shots, doctor visits, and allergen avoidance in the fight against allergies. Common allergens include food, medications, animal venom and dander, and latex. Allergies, particularly food allergies, are increasingly common in the United States.

A New Study A recent study gives hope to allergy sufferers in the form of a tiny, carbon ball. Buckminsterfullerenes (nicknamed buckyballs) are spherical cages about 1–10 nanometers in size made up of 60 carbon atoms. They were discovered in 1985 by scientists who vaporized graphite with a laser.

In 2007, a study revealed that buckyballs prevent allergic responses in tissue cultures and in mice. Your immune system reacts to allergens by releasing histamines and other chemicals. It is thought that buckyballs can prevent allergens from activating the histamine response.

Some buckyballs were modified by adding chemical side groups to increase their solubility. Some human immune cells called mast cells grown in tissue cultures were treated with the buckyballs while others were not.

When exposed to allergenlike molecules, the buckyball-treated cultures released 50 times less histamine and inhibited 30 to 40 other chemicals involved in allergic reactions.

Soccer ball-shaped spheres of carbon atoms called buckyballs might provide relief for allergy sufferers in the future.

Mice injected with buckyballs also released less histamine when exposed to allergens. What scientists don't yet know is exactly what triggers mast cells to produce histamine and how buckyballs block that reaction.

WRITING in Biology

Research buckyballs and other new treatments for allergies. Work with a partner to develop a creative way to share your findings with the class. Ideas include a public service announcement, news article, poster, or presentation.

BIOLAB

FORENSICS: HOW DO YOU FIND PATIENT ZERO?

Background: Imagine that a new disease—"Cellphonitis"—has invaded your school. One of the symptoms of this disease is the urge to use a cell phone during class. Cellphonitis is easily transferred from person to person by direct contact and there is no natural immunity to the disease. A student in your class has the disease, and is Patient Zero. The disease is spreading in your class and you need to track the disease to prevent the spread of an epidemic.

Question: *Is it possible to track a disease and determine the identity of Patient Zero?*

Materials

Pasteur pipets (1 per group)
numbered test tubes of water, one infected with simulated "cellphonitis" (1 per group)
test tube racks (1 per group)
small paper cups (1 per group)
pencil and paper
testing indicator

Safety Precautions

Procedure

1. Read and complete the lab safety form.
2. Prepare a table to keep track of the contacts you make. Select a test tube and record the number of the test tube.
3. Use a Pasteur pipet and move a small amount of the fluid from the test tube to a paper cup.
4. Your teacher will divide your class into groups. When your group is called, you will simulate the sharing of saliva during drinking water by using your pipets to exchange the fluid in your test tubes with another member of your group.
5. Record who you exchanged with in your tables.
6. Roll the test tube gently between your hands to mix and repeat Step 4 every time your group is told to exchange. Be sure to pick someone different to exchange with each time.
7. When the exchanges are complete, your teacher will act as the epidemiologist and use the testing indicator to see who has the disease.
8. Share the information and work together as groups to see if you can determine the identity of Patient Zero.
9. Once each group has made their hypothesis, test the original fluid in each cup to see who really was Patient Zero.
10. Return the test tubes. Dispose of the other materials you used as instructed by your teacher.

Analyze and Conclude

1. **Analyze** Use your data and draw a diagram for each possible Patient Zero. Use arrows to show who should be infected with each possible Patient Zero.
2. **Compare and Contrast** How was the spread of "cellphonitis" in this simulation similar to the spread of disease in real life? How was it different?
3. **Think Critically** If this simulation were run in a large class, why might the disease not be passed in later exchanges?
4. **Error Analysis** What problems did you run into as you tried to determine the identity of Patient Zero?

COMMUNICATE

Newscast Use newspapers and other sources to learn about a current disease epidemic. Prepare a newscast of how epidemiologists are searching for the source of disease and present it to your class. To find out more about tracking disease, visit BioLabs at biologygmh.com.

CHAPTER 37

Study Guide

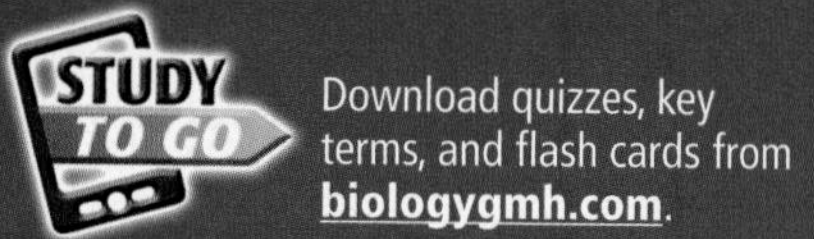

Download quizzes, key terms, and flash cards from biologygmh.com.

FOLDABLES **Infer** the conditions under which each type of immunity would be used to deter pathogens.

Vocabulary	Key Concepts
Section 37.1 Infectious Diseases	
• antibiotic (p. 1082) • endemic disease (p. 1081) • epidemic (p. 1081) • infectious disease (p. 1076) • Koch's postulates (p. 1077) • pandemic (p. 1081) • pathogen (p. 1076) • reservoir (p. 1078)	MAIN Idea Pathogens are dispersed by people, other animals, and objects. • Pathogens, such as bacteria, viruses, protozoans, and fungi, cause infectious diseases. • Koch's postulates demonstrate how to show that a particular pathogen causes a certain disease. • Pathogens are found in disease reservoirs and are transmitted to humans by direct and indirect methods. • The symptoms of disease are caused by invasion of the pathogen and the response of the host immune system. • Treatment of infectious disease includes the use of antibiotics and antiviral drugs.
Section 37.2 The Immune System	
• antibody (p. 1086) • B cell (p. 1086) • complement protein (p. 1085) • cytotoxic T cell (p. 1088) • helper T cell (p. 1088) • immunization (p. 1089) • interferon (p. 1085) • lymphocyte (p. 1086) • memory cell (p. 1089)	MAIN Idea The immune system has two main components: nonspecific immunity and specific immunity. • The nonspecific immune response includes the skin barrier, secreted chemicals, and cellular pathways that activate phagocytosis. • The specific immune response involves the activation of B cells, which produce antibodies, and T cells, which include helper T cells and cytotoxic T cells. • Passive immunity involves receiving antibodies against a disease. • Active immunity results in immune memory against a disease. • HIV attacks helper T cells, causing an immune system failure.
Section 37.3 Noninfectious Disorders	
• allergy (p. 1094) • anaphylactic shock (p. 1095) • degenerative disease (p. 1092) • metabolic disease (p. 1093)	MAIN Idea Noninfectious disorders include genetic disorders, degenerative diseases, metabolic diseases, cancer, and inflammatory diseases. • Noninfectious disorders often have both a genetic and an environmental component. • The inflammatory response to an infectious disease enhances the immune response, but the inflammatory response to an inflammatory disease is not helpful to the body. • Allergies are due to an overactive immune response to allergens found in the environment. • Anaphylactic shock is a severe hypersensitivity to particular allergens. • Autoimmunity results in an immune attack on body cells.

Chapter 37 Assessment

Section 37.1

Vocabulary Review

Match the definitions below with a vocabulary term from the Study Guide page.

1. A(n) __________ is an agent that causes an infectious disease.
2. When a disease becomes widespread in a particular area, it is called a/an __________.
3. A source of disease organisms is called a __________.

Understand Key Concepts

4. Which national organization tracks disease patterns in the United States?
   - **A.** The Centers for Disease Control and Prevention
   - **B.** National Disease Center
   - **C.** World Health Organization
   - **D.** United Nations
5. Which scientist established a method for determining whether a microorganism caused a specific disease?
   - **A.** Koch
   - **B.** Hooke
   - **C.** Sagan
   - **D.** Mendel
6. Which is the most common way that humans acquire an infectious disease?
   - **A.** contaminated water
   - **B.** mosquito bites
   - **C.** sick animals
   - **D.** infected humans

Use the photo below to answer question 7.

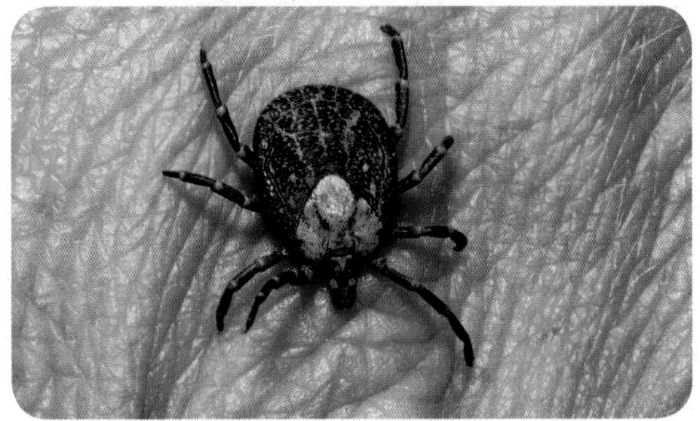

7. Which type of disease transmission is shown above?
   - **A.** direct contact
   - **B.** air transmission
   - **C.** object transmission
   - **D.** vector transmission

Use the photo below to answer question 8.

8. Which substance is secreted by the organism shown above?
   - **A.** anthrax
   - **B.** influenza
   - **C.** gentamicin
   - **D.** penicillin

Constructed Response

9. **Open Ended** Explain how you could prove that a particular bacteria was causing an infectious disease in a mouse population.
10. **Open Ended** Explain how the Centers for Disease Control and Prevention would be able to determine if an epidemic was occurring in your city.
11. **CAREERS IN BIOLOGY** Imagine you are the school nurse. Describe to students more than one way the cold virus could be transmitted from one person to another.

Think Critically

12. **Design** a feasible plan that could decrease the spread of infectious disease within your school.
13. **Evaluate** why growing viruses in cell cultures would be an exception to Koch's postulates.

Section 37.2

Vocabulary Review

For questions 14–16, match each definition with a vocabulary term from the Study Guide page.

14. a chemical produced by B cells in response to antigen stimulation
15. a cell that activates B cells and cytotoxic T cells
16. a type of white blood cell produced in the bone marrow that includes B and T cells

Understand Key Concepts

Use the diagram below to answer questions 17 and 18.

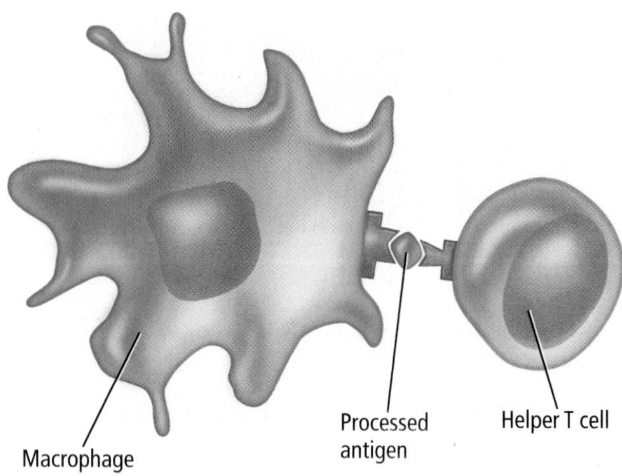

17. What kind of immune response is demonstrated in the diagram above?
    A. genetic
    B. nonspecific
    C. specific
    D. hormonal

18. To which does the activated helper T cell present its antigen?
    A. a pathogen
    B. bone marrow
    C. a B cell
    D. the thymus gland

19. Which is the first defense your body has against infectious disease?
    A. the helper T cell
    B. an antibody
    C. your skin
    D. phagocytosis

20. What is the role of complement proteins, found in the plasma, in the immune response?
    A. enhance phagocytosis
    B. activate phagocytes
    C. enhance destruction of a pathogen
    D. all of the above

21. Where are lymphocytes produced?
    A. bone marrow
    B. thymus gland
    C. spleen
    D. lymph nodes

Constructed Response

22. **Short Answer** Describe how the thymus is involved in the development of immunity.

23. **Open Ended** Evaluate why the body needs both a nonspecific and a specific immune response.

24. **Open Ended** Form a hypothesis as to why the proportion of unvaccinated Americans is increasing.

Think Critically

25. **Organize** the sequence of events that occur to activate an antibody response to tetanus bacteria.

26. **Compare** the role of helper T cells and cytotoxic T cells in the specific immune response.

Section 37.3

Vocabulary Review

Use a vocabulary term from the Study Guide page to answer questions 27–29.

27. What type of reaction is a hypersensitivity to an allergen such as a bee sting?

28. Which type of disease happens when people abnormally respond to environmental antigens?

29. Which type of disease is caused by a body part wearing out?

Understand Key Concepts

Use the photo below to answer question 30.

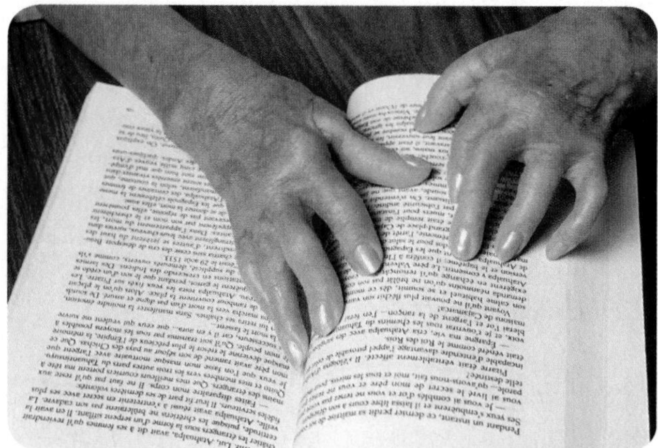

30. The above photo demonstrates which disease?
    A. tetanus
    B. sickle-cell disease
    C. rheumatoid arthritis
    D. allergy

31. Which type of noninfectious disease is defined as a problem in a biochemical pathway in the body?
    A. inflammatory disease
    B. metabolic disease
    C. degenerative disease
    D. cancer

BiologyOnline Chapter Test biologygmh.com

32. Which of the following substances is released in the body to cause most of the symptoms of allergies?
    A. insulin
    B. allergens
    C. histamine
    D. acetylcholine

33. Individuals can have a dangerous response to particular allergens, such as latex, and go into anaphylactic shock. What will be the result?
    A. breathing problems
    B. epileptic seizures
    C. atherosclerosis
    D. arthritis

34. In autoimmunity, which attacks the body's own proteins?
    A. antigens
    B. allergens
    C. antibodies
    D. antihistamines

Constructed Response

35. **Short Answer** Describe how an allergy differs from a common cold, considering that the symptoms are similar.

36. **Short Answer** Discuss the effects on the organs of the body when the smooth muscles in the bronchioles constrict, causing breathing to be difficult.

37. **Short Answer** Evaluate why lupus causes systemic problems in the body.

Think Critically

38. **Construct** a table listing each of the types of noninfectious disease and give an example of each type.

Use the graph below to answer question 39.

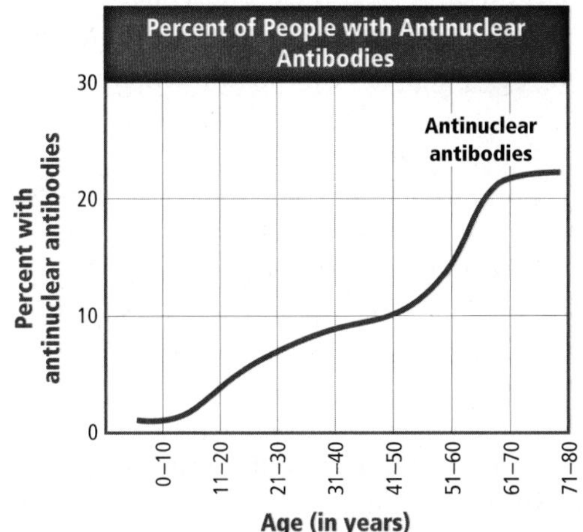

39. **Summarize** the relationship between antinuclear antibodies and age.

Additional Assessment

40. **WRITING in Biology** Construct an analogy comparing the immune system to a castle being attacked by invaders from a neighboring territory.

DBQ Document-Based Questions

The table below illustrates the effectiveness of using vaccinations to prevent the contraction of disease. There was a large decrease in cases of the diseases listed after the use of vaccinations.

Data obtained from: Mandell, G. L., et al. 1995. *Principles and Practice of Infectious Diseases*, 4th ed. Churchill Livingstone, and Centers for Disease Control and Prevention. 2000. *Morbidity and Mortality Weekly Report* 48: 1162-1192.

Disease	Maximum Number of Cases in a Year	Number of Cases in 1999 in U.S.	Percent Change
Measles	894,134	60	–99.99
Mumps	152,209	352	–99.77
Polio (paralytic)	21,269	0	–100.0
Tetanus	1560	33	–97.88
Hepatitis B	26,611	6495	–75.59

41. Which disease has shown the greatest change in occurrence since the year of its maximum number of cases?

42. Tetanus has shown a large decline since the United States started vaccinating. Explain why this disease will not be completely eradicated.

43. Create a bar graph showing the percent change in number of cases as a result of vaccination for each disease.

Cumulative Review

44. Describe the biome in your area, including the abiotic factors that help define the biome and at least five species that are prevalent. **(Chapter 3)**

45. Based on your knowledge of bird adaptations, design the perfect bird for a cold, marshy area inhabited by fish and other invertebrates. **(Chapter 29)**

Standardized Test Practice

Cumulative

Multiple Choice

1. In the digestive system, complex carbohydrates are broken down into which substance?
   A. amino acids
   B. fatty acids
   C. simple sugars
   D. starches

Use the diagram below to answer questions 2 and 3.

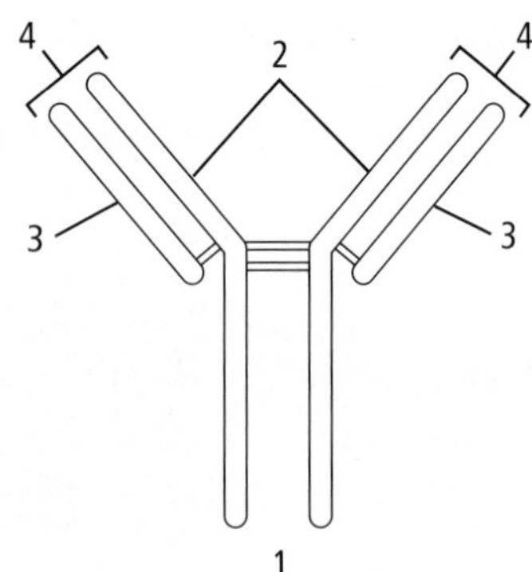

2. The diagram above shows the basic structure of an antibody. Which part of the diagram corresponds to the antigen binding site?
   A. 1
   B. 2
   C. 3
   D. 4

3. Why are parts 2 and 3 of the diagram above important for the formation of antibodies?
   A. They allow for an enormous number of possible antibodies to form.
   B. They are created by the T cells in the immune system.
   C. They help reduce the number of antibodies that form.
   D. They help stimulate the inflammatory response.

4. Which is the role of estrogen during puberty in females?
   A. It causes development of the female body.
   B. It causes eggs to begin to mature in the ovaries.
   C. It causes meiosis to start to produce an egg.
   D. It causes ovaries to release mature eggs.

5. Which is true of the appendix?
   A. It absorbs sodium hydrogen carbonate to neutralize acid.
   B. It has no known function in the digestive system.
   C. It helps break down fats.
   D. It secretes acids to help break down foods.

Use the diagram below to answer question 6.

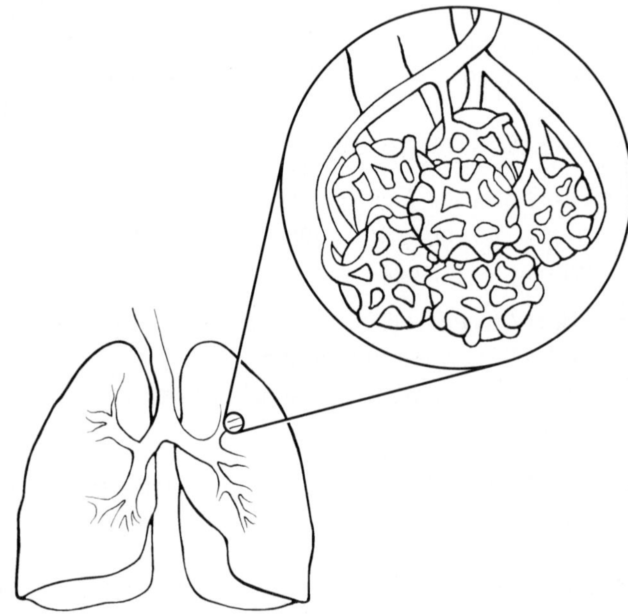

6. Which happens in the blood in these structures?
   A. Carbon dioxide and oxygen are exchanged.
   B. Carbon dioxide and oxygen remain constant.
   C. Nitrogen and carbon dioxide are exchanged.
   D. Nitrogen and carbon dioxide remain constant.

7. Puberty takes place during which transition in life?
   A. adolescence to adulthood
   B. childhood to adolescence
   C. fetus to infant
   D. zygote to fetus

8. What is the role of hormones in the body?
   A. to act as reaction catalysts
   B. to control breathing process
   C. to help synthesize proteins
   D. to regulate many body functions

Biology Online Standardized Test Practice biologygmh.com

Short Answer

Use the graph below to answer questions 9 and 10.

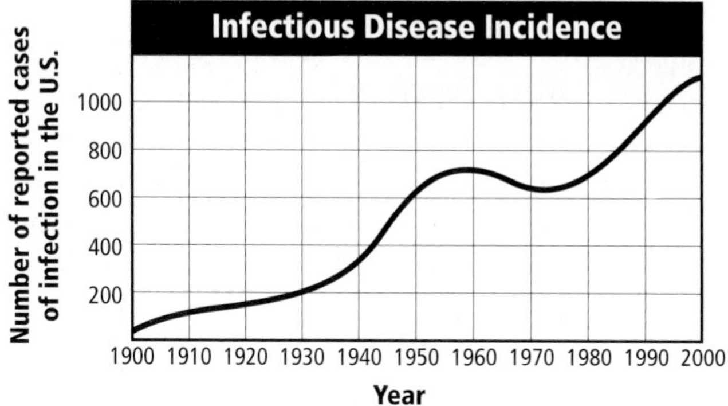

9. What is the overall trend shown in the above graph?

10. What are two possible explanations for the pattern in the above graph?

11. What characteristics are used to classify protists into three groups?

12. Describe the process of dilation during birthing. Assess why it is important.

13. Identify the function of the large intestine.

14. Assess how the respiratory system of most reptiles is adapted for life on land.

15. Free-living flatworms have some unique body structures: eyespots, a ganglion, and auricles that detect chemical stimuli. How are these body structures related to each other?

Extended Response

16. Arthropods first moved onto land about 400 million years ago and have survived several mass extinctions. Propose a hypothesis about why arthropods have been so successful.

17. Compare the production of sperm cells and egg cells during meiosis.

Essay Question

Scientist Mark Lappé wrote the following in 1981, in a book called *Germs That Won't Die:*

"Unfortunately, we played a trick on the natural world by seizing control of these [natural] chemicals, making them more perfect in a way that has changed the whole microbial constitution of the developing countries. We have organisms now proliferating that never existed before in nature. We have selected them. We have organisms that probably caused a tenth of a percent of human disease in the past that now cause twenty, thirty percent of the diseases that we're seeing. We have changed the whole face of the earth by the use of antibiotics."

Using the information in the paragraph above, answer the following question in essay format.

18. As Lappé predicted in 1981, many diseases have emerged in forms that are resistant to treatment by antibiotics and other powerful drugs. Have antibiotics "changed the whole face of the earth" for the better or for the worse? In an organized essay, discuss the pros and cons of antibiotics as they are used today.

NEED EXTRA HELP?

If You Missed Question . . .	1	2	3	4	5	6	7	8	9	10	11	12	13	14	15	16	17	18
Review Section . . .	35.2	37.2	37.2	36.2	35.1	34.2	36.2	35.2	37.1	37.1	19.1	36.2	35.1	29.1	25.1	26.1	36.1	37.1

Student Resources

For students and parents/guardians

This skillbuilder handbook helps you sharpen your problem-solving skills so you can get the most out of reading and understanding scientific writing and data. Improving skills such as making comparisons, analyzing information, reading time lines, and using graphic organizers also can help you boost your test scores. In addition, you'll find useful instructions on how to hold a debate and a review of math skills.

The reference handbook is another tool that will assist you. The classification tables, word origins, and the periodic table of the elements are resources that will help increase your comprehension.

Table of Contents

Make Comparisons

Why learn this skill?

Suppose you want to buy portable MP3 music player, and you must choose between three models. You would probably compare the characteristics of the three models, such as price, amount of memory, sound quality, and size to determine which model is best for you. In the study of biology, you often make comparisons between the structures and functions of organisms. You will also compare scientific discoveries or events from one time period with those from another period.

Learn the Skill

When making comparisons, you examine two or more items, groups, situations, events, or theories. You must first decide what will be compared and which characteristics you will use to compare them. Then identify any similarities and differences.

For example, comparisons can be made between the two illustrations on this page. The different structures of the animal cell can be compared to the different structures of the plant cell. By reading the labels, you can see that both types of cells have a nucleus.

Practice the Skill

Create a table with the heading *Animal and Plant Cells.* Make three columns. Label the first column *Cell Structures.* Label the second column *Animal Cells.* Label the third column *Plant Cells.* List all the cell structures in the first column. Place a check mark under either *Animal Cell* or *Plant Cell* or both if that structure is shown in the illustration. When you have finished the table, answer these questions.

1. What items are being compared? How are they being compared?
2. What structures do animal and plant cells have in common?
3. What structures are unique to animal cells? What structures are unique to plant cells?

Apply the Skill

Make Comparisons On pages 708 and 712 respectively, you will find illustrations for a sponge's life cycle and a jellyfish's life cycle. Compare the two illustrations carefully. Then, identify the similarities and the differences between the two cycles.

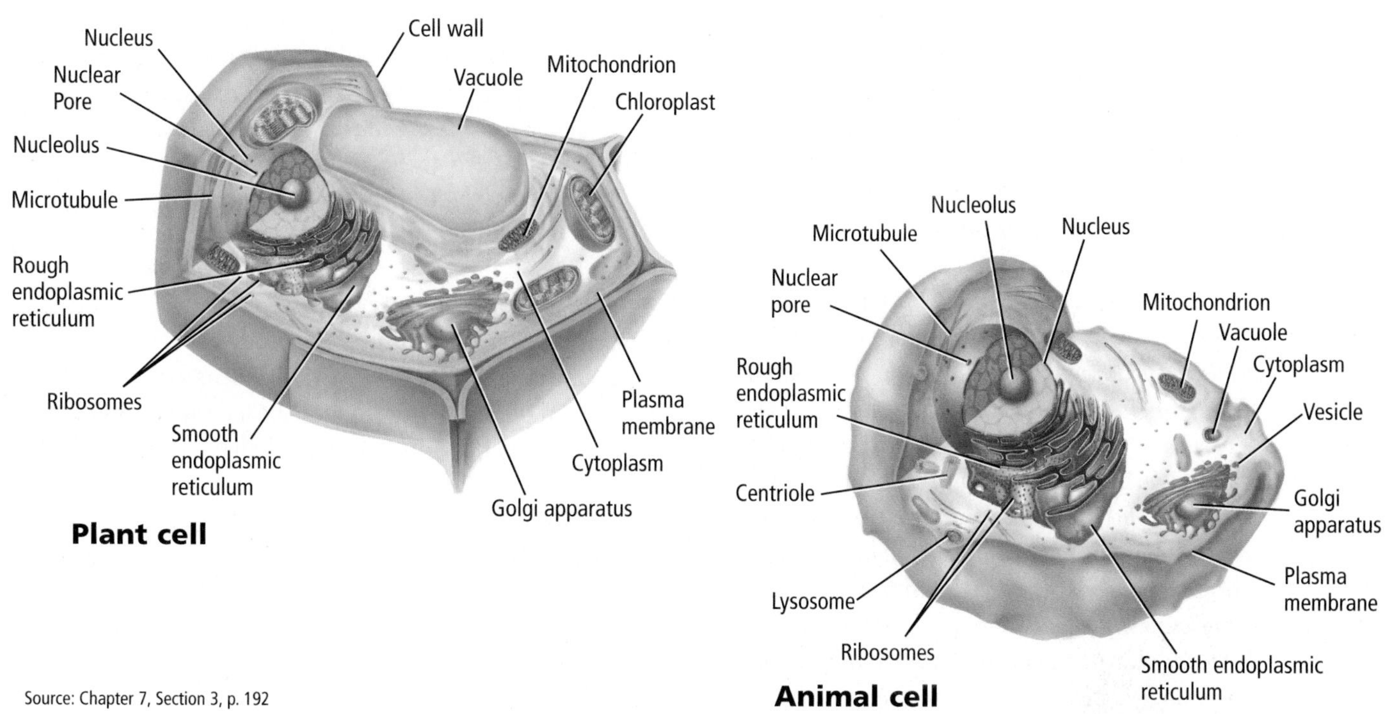

Plant cell

Animal cell

Source: Chapter 7, Section 3, p. 192

Analyze Information

Why learn this skill?

Analyzing, or looking at separate parts of something to understand the entire piece, is a way to think critically about written work. The ability to analyze information is important when determining which ideas are the most important.

Learn the Skill

To analyze information, use the following steps:

- identify the topic being discussed
- examine how the information is organized—identify the main points
- summarize the information in your own words, and then make a statement based on your understanding of the topic and what you already know

Practice the Skill

Read the following excerpt from *National Geographic*. Use the steps listed above to analyze the information and answer the questions that follow.

Like something straight out of a Jules Verne novel, an enormous tentacled creature looms out of the inky blackness of the deep Pacific waters. But this isn't science fiction. A set of extraordinary images captured by Japanese scientists marks the first-ever record of a live giant squid (Architeuthis) *in the wild.*

The animal—which measures roughly 8 meters long—was photographed 900 meters beneath the North Pacific Ocean. Japanese scientists attracted the squid toward cameras attached to a baited fishing line. The scientists say they snapped more than 500 images of the massive cephalopod before it broke free after snagging itself on a hook. They also recovered one of the giant squid's two longest tentacles, which severed during its struggle.

The photo sequence, taken off Japan's Ogasawara Islands in September 2004, shows the squid homing in on the baited line and enveloping it in "a ball of tentacles." Tsunemi Kubodera of the National Science Museum in Tokyo and Kyoichi Mori of the Ogasawara Whale Watching Association report their observations in the journal Proceedings of the Royal Society B.

"Architeuthis *appears to be a much more active predator than previously suspected, using its elongated feeding tentacles to strike and tangle prey," the researchers write. They add that the squid was found feeding at depths where no light penetrates even during the day.*

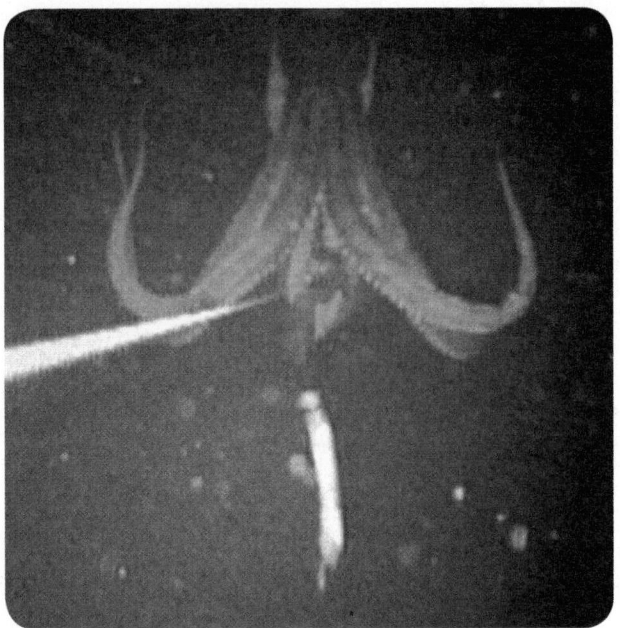

Giant squid on a fishing line

Squid expert Martin Collins of the British Antarctic Survey based in Cambridge, England is especially interested in clues the images might provide to the way giant squid swim and hunt in the deep ocean.

Collins says there were two competing schools of thought among giant squid experts. "One was the idea that [giant squid] were fairly inactive and just drifted around, dangling their tentacles below them like fishing lures to catch what came by," he said.

"The other theory was that they were actually quite active. This new evidence supports this, suggesting they are active predators which can move reasonably quickly. The efforts the squid went to untangle itself [from the baited fishing line] also shows they are capable of quite strong and rapid movement," he added.

1. What topic is being discussed?
2. What are the main points of the article?
3. Summarize the information in this article, and then provide your analysis based on this information and your own knowledge.

Apply the Skill

Analyze Information Analyze a short, informative article on a new scientific discovery or new science technology, such as the hybrid car. Summarize the information and make a statement of your own.

Synthesize Information

Why learn this skill?

The skill of synthesizing involves combining and analyzing information gathered from separate sources or at different times to make logical connections. Being able to synthesize information can be a useful skill for you as a student when you need to gather data from several sources for a report or a presentation.

Learn the Skill

Follow these steps to synthesize information:

- select important and relevant information
- analyze the information and build connections
- reinforce or modify the connections as you acquire new information

Suppose you need to write a research paper on endangered species. You would need to synthesize what you learn to inform others. You could begin by detailing the ideas and information you already have about endangered species. A table such as the one below could help you categorize the facts.

Table SH.1 Endangered Species Statistics

Group	October 2000		October 2005	
	U.S.	Foreign	U.S.	Foreign
Mammals	63	251	68	251
Birds	78	175	77	175
Reptiles	14	64	14	64
Amphibians	10	8	12	8
Fishes	69	11	71	11
Clams	61	2	62	2
Snails	20	1	24	1
Insects	30	4	35	4
Arachnids	6	0	12	0
Crustaceans	18	0	19	0

Source: U.S. Fish and Wildlife Service

Then you could select a passage about endangered species like the sample below, which is adapted from Chapter 5.

Stable ecosystems can be changed by the activity of other organisms, climate, or natural disasters. This natural process of extinctions is not what scientists are worried about. Many worry about a recent increase in the rate of extinction.

One of the factors that is increasing the current rate of extinction is the overexploitation, or excessive use, of species that have economic values. Historically, overexploitation was the primary cause of species extinction. However, the number one cause of species extinction today is the loss or destruction of habitat.

There are several ways that species can lose their habitats. If a habitat is destroyed or disrupted, the native species might have to relocate or die. For example, humans are clearing areas of tropical rain forests and are replacing the native plants with agricultural crops or grazing lands.

Practice the Skill

Use the table and the passage on this page to answer these questions.

1. What information is presented in the table?
2. What is the main idea of the passage? What information does the passage add to your knowledge about the topic?
3. By synthesizing the two sources and using your own knowledge, what conclusions can you draw about habitat conservation practices for endangered species?
4. Using what you learned in your studies and from this activity, contrast two types of habitat changes and their effects on the ecosystem.

Apply the Skill

Synthesize Information Find two sources of information on the same topic and write a short report. In your report, answer the following questions: What are the main ideas of each source? How does each source add to your understanding of the topic? Do the sources support or contradict each other? What conclusions can you draw from the sources?

Skillbuilder Handbook

Take Notes and Outline

Why learn this skill?

One of the best ways to remember something is to write it down. Taking notes—writing down information in a brief and orderly format—not only helps you remember, but also makes studying easier.

Learn the Skill

There are several styles of note taking, but all explain and put information in a logical order. As you read, identify and summarize the main ideas and details that support them and write them in your notes. Paraphrase, that is, state in your own words, the information rather then copying it directly from the text. Using note cards or developing a personal "shorthand"—using symbols to represent words—can help.

You might also find it helpful to create an outline when taking notes. When outlining material, first read the material to identify the main ideas. In textbooks, section headings provide clues to main topics. Identify the subheadings. Place supporting details under the appropriate heading. The basic pattern for outlines is as follows:

MAIN TOPIC
- I. FIRST IDEA OR ITEM
  - A. FIRST DETAIL
    - 1. SUBDETAIL
    - 2. SUBDETAIL
  - B. SECOND DETAIL
- II. SECOND IDEA OR ITEM
  - A. FIRST DETAIL
  - B. SECOND DETAIL
    - 1. SUBDETAIL
    - 2. SUBDETAIL
- III. THIRD IDEA OR ITEM

Practice the Skill

Read the following excerpt from *National Geographic*. Use the steps you just read about to take notes or create an outline. Then answer the questions that follow.

Mapping the three billion letters of the human genome has helped researchers better understand the 99.9 percent of DNA that is identical in all humans. Now a new project aims to map the 0.1 percent of DNA where differences occur. The International HapMap Project will look at variations that dictate susceptibility to genetic influences, such as environmental toxins and inherited diseases.

Researchers "read" DNA code by its structural units called nucleotides. These chemical building blocks are designated by the letters A (adenine), C (cytosine), G (guanine), and T (thymine). Single-letter variations in genes—called single nucleotide polymorphisms, or SNPs (pronounced "snips")—are often the culprits behind a wide range of genetic diseases. For example, changing an A to a T in the gene for the blood molecule hemoglobin causes sickle cell anemia.

But most diseases and disorders are not caused by a single gene. Instead they are caused by a complex combination of linked genetic variations at multiple sites on different chromosomes.

Haplotypes are sets of adjacent SNPs that are closely associated and are inherited as a group. Certain haplotypes are known to have a role in diseases, including Alzheimer's, deep vein thrombosis, type 2 diabetes, and age-related macular degeneration, a leading cause of blindness.

1. What is the main topic of the article?
2. What are the first, second, and third ideas?
3. Name one detail for each of the ideas.
4. Name one subdetail for each of the details.

Apply the Skill

Take Notes and Outline Go to Section 2.1 and take notes by paraphrasing and using shorthand or by creating an outline. Use the section title and headings to help you create your outline. Summarize the section using only your notes.

Understand Cause and Effect

Why learn this skill?

In order to understand an event, you should look for how that event or chain of events came about. When scientists are unsure of the cause for an event, they often design experiments. Although there might be an explanation, an experiment should be performed to be certain the cause created the event you observed. This process examines the causes and effects of events.

Learn the Skill

Every human body regulates its own temperature to maintain conditions suitable for survival. Exercise *causes* a body to heat up. The stimulated nerves in the skin are the *effect,* or result, of exercise. The figure below shows how one event—the **cause**—led to another—the **effect.**

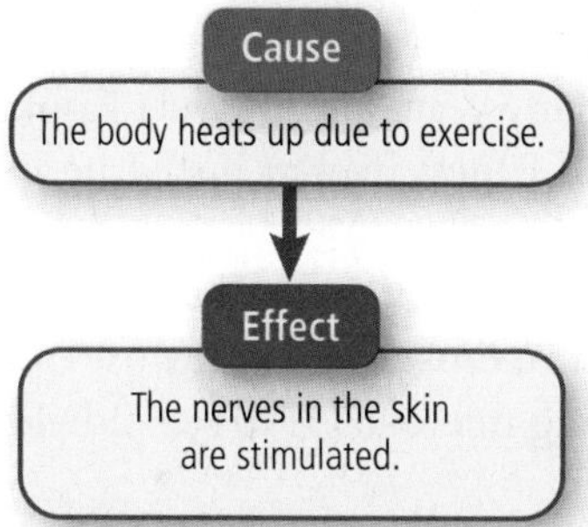

You can also identify cause-and-effect relationships in sentences from clue words such as:

because	**thus**
that is why	**due to**
led to	**for this reason**
so that	**produced**
consequently	**therefore**
as a result	**in order to**

Read the sample sentences below.

"A message is sent to sweat glands. As a result, perspiration occurs."

In the example above, the cause is a message being sent. The cause-and-effect clue words "as a result" tell you that the perspiration is the effect of the message.

In a chain of events, an effect often becomes the cause of other events. The next chart shows the complete chain of events that occur when exercise raises body temperature and the body returns to homeostasis.

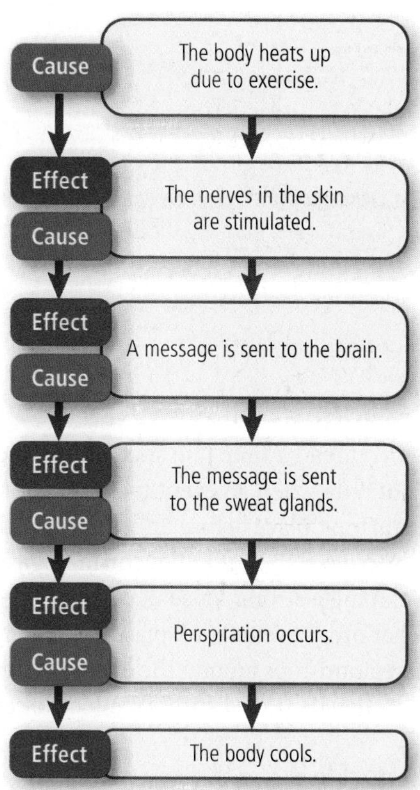

Practice the Skill

Make a chart, like the one above, showing which events are causes and which are effects using these sentences. Use Chapter 33 to help you.

1. The hair cells respond by generating nerve impulses in the auditory nerve and transmitting them to the brain.
2. As the stapes vibrates, it causes the oval window to move back and forth.
3. Sound waves enter the auditory canal and cause the eardrum to vibrate.
4. Vibrations cause the fluid inside the cochlea to move like a wave against the hair cells.
5. Vibrations travel through the malleus, the incus, and the stapes.

Apply the Skill

Understand Cause and Effect Read an account of a recent scientific event or discovery in a science article. Determine the causes and effects that lead to the event or discovery. Show the chain of events in a chart.

Read a Time Line

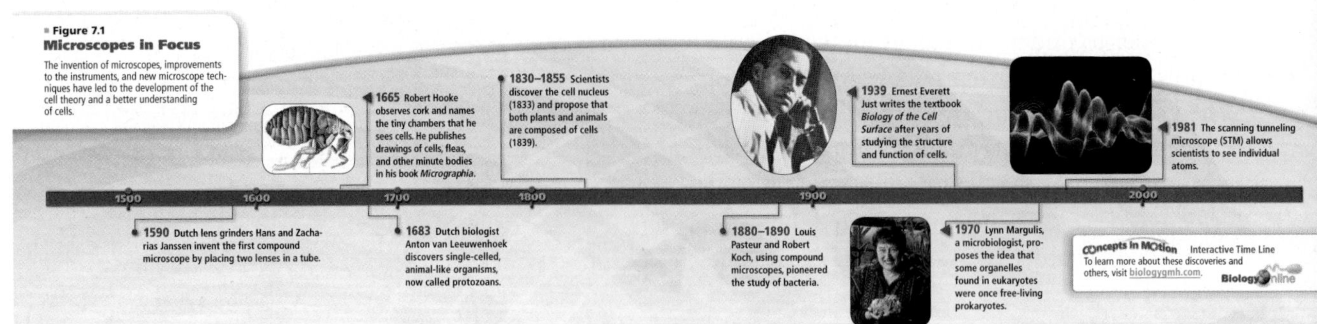

Figure 7.1
Microscopes in Focus
The invention of microscopes, improvements to the instruments, and new microscope techniques have led to the development of the cell theory and a better understanding of cells.
1500
1600
1700
1800
1900
2000
1590 Dutch lens grinders Hans and Zacharias Janssen invent the first compound microscope by placing two lenses in a tube.
1665 Robert Hooke observes cork and names the tiny chambers that he sees cells. He publishes drawings of cells, fleas, and other minute bodies in his book Micrographia.
1683 Dutch biologist Anton van Leeuwenhoek discovers single-celled, animal-like organisms, now called protozoans.
1830–1855 Scientists discover the cell nucleus (1833) and propose that both plants and animals are composed of cells (1839).
1880–1890 Louis Pasteur and Robert Koch, using compound microscopes, pioneered the study of bacteria.
1939 Ernest Everett Just writes the textbook Biology of the Cell Surface after years of studying the structure and function of cells.
1970 Lynn Margulis, a microbiologist, proposes the idea that some organelles found in eukaryotes were once free-living prokaryotes.
1981 The scanning tunneling microscope (STM) allows scientists to see individual atoms.
Concepts In Motion Interactive Time Line
To learn more about these discoveries and others, visit biologygmh.com.
Biology Online

Source: Chapter 7, Section 1 pp. 182–183

Why learn this skill?

When you read a time line such as the one above, you see not only when an event took place, but also what events took place before and after it. A time line can help you develop the skill of chronological thinking. Developing a strong sense of chronology—when and in what order events took place—will help you examine relationships among the events. It will also help you understand the causes or effects of events.

Learn the Skill

A time line is a linear chart that list events that occurred on specific dates. The number of years between dates at the begining and end of the time line is the time span. A time line that begins in 1910 and ends in 1920 has a ten-year time span. Some time lines span centuries. Examine the time lines below. What time spans do they cover?

Time lines are usually divided into smaller parts called time intervals. On the two time lines below, the first time line has a 300-year time span divided into 100-year time intervals. The second time line has a six-year time span divided into two-year time intervals.

Practice the Skill

Study the time line above and then answer these questions.

1. What time span and intervals appear on this time line?
2. Which scientist was the first to observe cells with a microscope?
3. How many years after Robert Hooke observed cork did Ernest Everett Just write *Biology of the Cell Surface?*
4. What was the time span between the creation of the first microscope and the use of the scanning tunneling microscope to see individual atoms.

Apply the Skill

Read a Time Line Sometimes a time line shows events that occurred during the same period, but related to two different subjects. The time line above shows events related to cells between 1500 and 2000. Copy the time line and events onto a piece of paper. Then use a different color to add events related to genetics during this same time span. Use the chapters in Unit 3 to help you.

Analyze Media Sources

Why learn this skill?

To stay informed, people use a variety of media sources, including print media, broadcast media, and electronic media. The Internet has become an especially valuable research tool. It is convenient to use, and the information contained on the Internet is plentiful. Whichever media source you use to gather information, it is important to analyze the source to determine its accuracy and reliability.

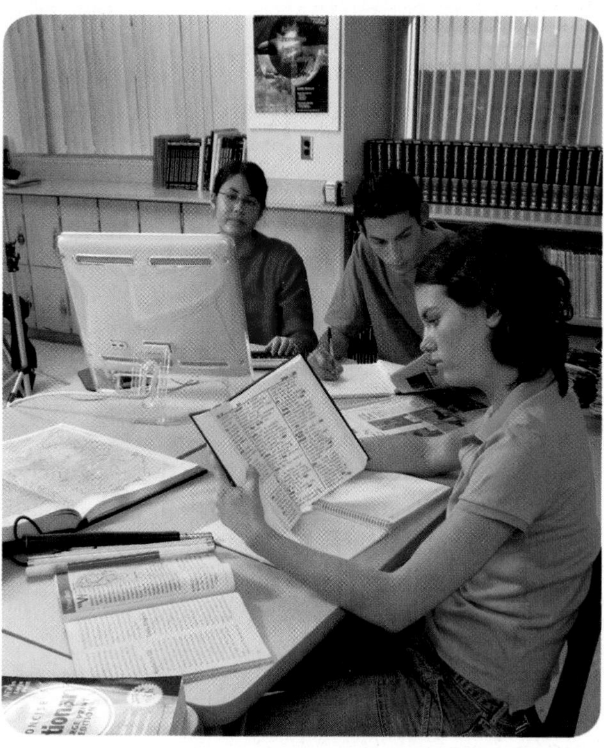

Learn the Skill

There are a number of issues to consider when analyzing a media source. Most important is to check the accuracy of the source and content. The author and publisher or sponsors should be credible and clearly indicated. To analyze print media or broadcast media, ask yourself the following questions:

- Is the information current?
- Are the resources revealed?
- Is more than one resource used?
- Is the information biased?
- Does the information represent both sides of an issue?
- Is the information reported firsthand or secondhand?

For electronic media, ask yourself these questions in addition to the ones above.

- Is the author credible and clearly identified? Web site addresses that end in .edu, .gov, and .org tend to be credible and contain reliable information.
- Are the facts on the Web site documented?
- Are the links within the Web site appropriate and current?
- Does the Web site contain links to other useful resources?

Practice the Skill

To analyze print media, choose two articles, one from a newspaper and the other from a newsmagazine, on an issue on which public opinion is divided. Then, answer these questions.

1. What points are the articles trying to make? Were the articles successful? Can the facts be verified?
2. Did either article reflect a bias toward one viewpoint or another? List any unsupported statements.
3. Was the information reported firsthand or secondhand? Do the articles seem to represent both sides fairly?
4. How many resources can you identify in the articles? List them.

To analyze electronic media, visit biologygmh.com and select Web Links. Choose one link from the list, read the information on that Web site, and then answer these questions.

1. Who is the author or sponsor of the Web site?
2. What links does the Web site contain? How are they appropriate to the topic?
3. What resources were used for the information on the Web site?

Apply the Skill

Analyze Sources of Information Think of an issue in the nation on which public opinion is divided. Use a variety of media resources to read about this issue. Which news source more fairly represents the issue? Which news source has the most reliable information? Can you identify any biases? Can you verify the credibility of the news source?

Use Graphic Organizers

Why learn this skill?

While you read this textbook, you will be looking for important ideas or concepts. One way to arrange these ideas is to create a graphic organizer. In addition to Foldables™, you will find various other graphic organizers throughout your book. Some organizers show a sequence, or flow, of events. Other organizers emphasize the relationship between concepts. Develop your own organizers to help you better understand and remember what you read.

Learn the Skill

An **events chain concept map** describes a sequence of events, such as stages of a process or procedure. When making an events-chain map, first identify the event that starts the sequence and add events in chronological order until you reach an outcome.

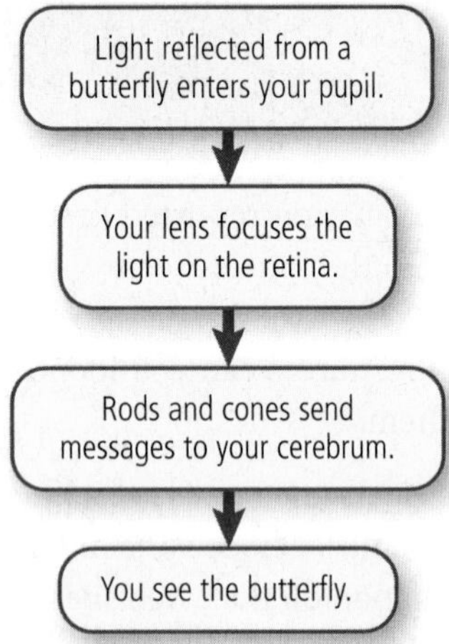

In a **cycle concept map,** the series of events do not produce a final outcome. The event that appears to be the final event relates back to the event that appears to be the initiating event. Therefore, the cycle repeats itself.

Blood Flow in the Body

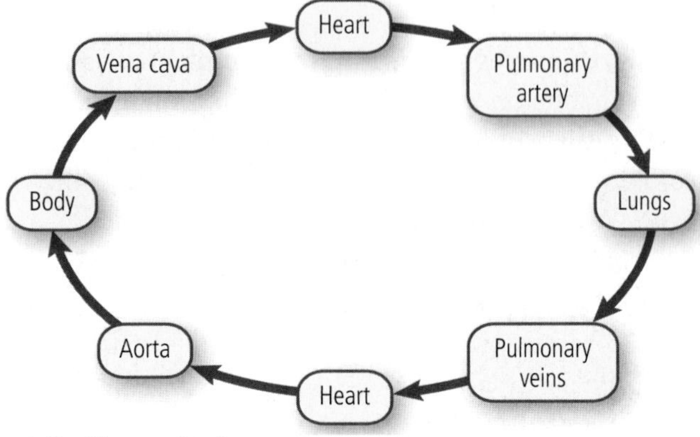

A **network tree concept map** shows the relationship among concepts, which are written in order from general to specific. The words written on the lines between the circles, called linking words, describe the relationships among the concepts; the concepts and the linking words can form a sentence.

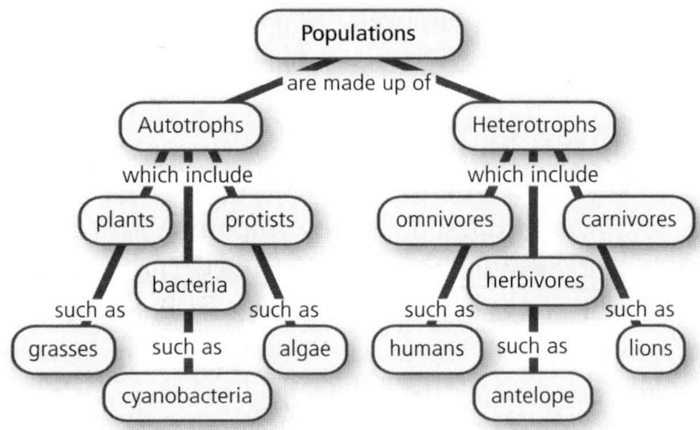

Practice the Skill

1. Create an events chain concept map that describes the process of hearing the ring of a bell. Begin with sound waves entering the outer ear. End with hearing the bell ring. Use Chapter 33 to help you.
2. Create a cycle concept map of human respiration. Make sure that the cycle is complete with the event that appears to be the final event relating back to the event that appears to be the starting event. Go to Chapter 34 for help.
3. Create a network tree concept map with these words: *Biomes, aquatic biomes, terrestrial biomes, marine biomes, estuary biomes, freshwater biomes, desert, grasslands, temperate forest, salt water, mixed waters, freshwater, sparse plant life, grasses,* and *broad-leaved trees.* Add linking words to describe the relationships between concepts. Refer to Chapter 3 for help.

Apply the Skill

Use Graphic Organizers Create an events chain concept map of succession using information from Chapter 3. Create a cycle concept map of the water cycle using information from Chapter 2. Create a network tree concept map of animals that includes invertebrates and vertebrates, characteristics of each type, and examples. Use Chapters 24–30 to help you.

Debate Skills

New research leads to new scientific information. There are often opposing points of view on how this research is conducted, how it is interpreted, and how it is communicated. The *Biology and Society* features in your book offer a chance to debate a current controversial topic. Here is an overview on how to conduct a debate.

Choose a Position and Research

First, choose a scientific issue that has at least two opposing viewpoints. The issue can come from current events, your textbook, or your teacher. These topics could include human cloning or environmental issues. Topics are stated as affirmative declarations, such as "Cloning human beings is beneficial to society."

One speaker will argue the viewpoint that agrees with the statement, called the positive position, and another speaker will argue the viewpoint that disagrees with the statement, called the negative position. Either individually or with a group, choose the position for which you will argue. The viewpoint that you choose does not have to reflect your personal belief. The purpose of debate is to create a strong argument supported by scientific evidence.

After choosing your position, conduct research to support your viewpoint. Use resources in your media center or library to find articles, or use your textbook to gather evidence to support your argument. A strong argument is supported by scientific evidence, expert opinions, and your own analysis of the issue. Research the opposing position also. Becoming aware of what points the other side might argue will help you to strengthen the evidence for your position.

Hold the Debate

You will have a specific amount of time, determined by your teacher, in which to present your argument. Organize your speech to fit within the time limit: explain the viewpoint that you will be arguing, present an analysis of your evidence, and conclude by summing up your most important points. Try to vary the elements of your argument. Your speech should not be a list of facts, a reading of a newspaper article, or a statement of your personal opinion, but an analysis of your evidence in an organized manner. It is also important to remember that you must never make personal attacks against your opponent. Argue the issue. You will be evaluated on your overall presentation, organization and development of ideas, and strength of support for your argument.

Additional Roles There are other roles that you or your classmates can play in a debate. You can act as the timekeeper. The timekeeper times the length of the debaters' speeches and gives quiet signals to the speaker when time is almost up (usually a hand signal).

You can also act as a judge. There are important elements to look for when judging a speech: an introduction that tells the audience what position the speaker will be arguing, strong evidence that supports the speaker's position, and organization. The speaker also must speak clearly and loudly enough for everyone to hear. It is helpful to take notes during the debate to summarize the main points of each side's argument. Then, decide which debater presented the strongest argument for his or her position. You can have a class discussion about the strengths and weaknesses of the debate and other viewpoints on this issue that could be argued.

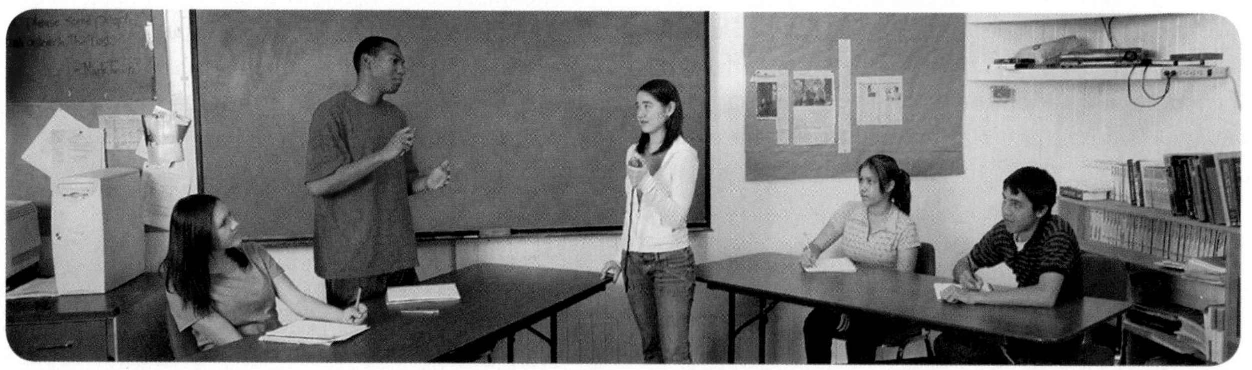

Math Skills

Experimental data is often quantitative and is expressed using numbers and units. The following sections provide an overview of the common system of units and some calculations involving units.

Measure in SI

The International System of Measurement, abbreviated SI, is accepted as the standard for measurement throughout most of the world. The SI system contains seven base units. All other units of measurement can be derived from these base units.

Table SH.2 SI Base Units

Measurement	Unit	Symbol
Length	Meter	m
Mass	Kilogram	kg
Time	Second	s
Electric current	Ampere	A
Temperature	Kelvin	K
Amount of substance	Mole	mol
Intensity of light	Candela	cd

Some units are derived by combining base units. For example, units for volume are derived from units of length. A liter (L) is a cubic decimeter ($dm^3 = dm \times dm \times dm$). Units of density (g/L) are derived from units of mass (g) and units of volume (L).

When units are multiplied by factors of ten, new units are created. For example, if a base unit is multiplied by 1000, the new unit has the prefix *kilo-*. Prefixes for some units are shown in **Table SH.3.**

To convert a given unit to a unit with a different factor of ten, multiply the unit by a conversion factor. A conversion factor is a ratio equal to one. The equivalents in **Table SH.3** can be used to make such a ratio. For example, 1 km = 1000 m. Two conversion factors can be made from this equivalent.

$$\frac{1000\text{ m}}{1\text{ km}} = 1 \quad \text{and} \quad \frac{1\text{ km}}{1000\text{ m}} = 1$$

To convert one unit to another factor of ten, choose the conversion factor that has the unit you are converting from in the denominator. For example, to convert one kilometer to meters, use the following equation.

$$1\ \cancel{\text{km}} \times \frac{1000\text{ m}}{1\ \cancel{\text{km}}} = 1000\text{ m}$$

A unit can be multiplied by several conversion factors to obtain the desired unit.

Table SH.3 Common SI Prefixes

Prefix	Symbol	Equivalents
mega	M	1×10^6 base units
kilo	k	1×10^3 base units
hecto	h	1×10^2 base units
deka	da	1×10^1 base units
deci	d	1×10^{-1} base units
centi	c	1×10^{-2} base units
milli	m	1×10^{-3} base units
micro	μ	1×10^{-6} base units
nano	n	1×10^{-9} base units
pico	p	1×10^{-12} base units

Practice Problem 1 How would you change 1000 micrometers to kilometers?

Convert Temperature

The following formulas can be used to convert between Fahrenheit and Celsius temperatures. Notice that each equation can be obtained by algebraically rearranging the other. Therefore, you only need to remember one of the equations.

Conversion of Fahrenheit to Celsius

$$°C = \frac{(°F) - 32}{1.8}$$

Conversion of Celsius to Fahrenheit

$$°F = 1.8(°C) + 32$$

Make and Use Tables

Tables help organize data so that it can be interpreted more easily. Tables are composed of several components—a title describing the contents of the table, columns and rows that separate and organize information, and headings that describe the information in each column or row.

Table SH.4 Effects of Exercise on Heart Rate

Pulse taken	Individual heart rate (Beats per min)	Class average (Beats per min)
At rest	73	72
After exercise	110	112
1 minute after exercise	94	90
5 minutes after exercise	76	75

Looking at this table, you should not only be able to pick out specific information, such as the class average heart rate after five minutes of exercise, but you should also notice trends.

Practice Problem 2 Did the exercise have an effect on the heart rate one minute after exercise? How can you tell? What can you conclude about the effects of exercise on heart rate during and after exercise?

Make and Use Graphs

After scientists organize data in tables, they often display the data in graphs. A graph is a diagram that shows relationships among variables. Graphs make interpretation and analysis of data easier. The three basic types of graphs used in science are the line graph, the bar graph, and the circle graph.

Line Graphs A line graph is used to show the relationship between two variables. The independent variable is plotted on the horizontal axis, called the x-axis. The dependent variable is plotted on the vertical axis, called the y-axis. The dependent variable (y) changes as a result of a change in the independent variable (x).

Suppose a school started a bird-watching group to observe the number of birds in the school courtyard. The number of birds in the courtyard was recorded each day for four months. The average number of birds per month was calculated. A table of the birds' visitations is shown below.

Table SH.5 Average Number of Birds Viewed

Time (days)	Average Number of Birds per Day
30	24
60	27
90	30
120	32

To make a graph of the average number of birds over a period of time, start by determining the dependent and independent variables. The average number of birds after each period of time is the dependent variable and is plotted on the y-axis. The independent variable, or the number of days, is plotted on the x-axis.

Plain or graph paper can be used to construct graphs. Draw a grid on your paper or a box around the squares that you intend to use on your graph paper. Give your graph a title and label each axis with a title and units. In this example, label the number of days on the x-axis. Because the lowest average of birds viewed was 24 and the highest was 32, you know that you will have to start numbers on the y-axis at least 24 and number to at least 32. You could decide to number 20–40 by intervals of two spaced at equal distances.

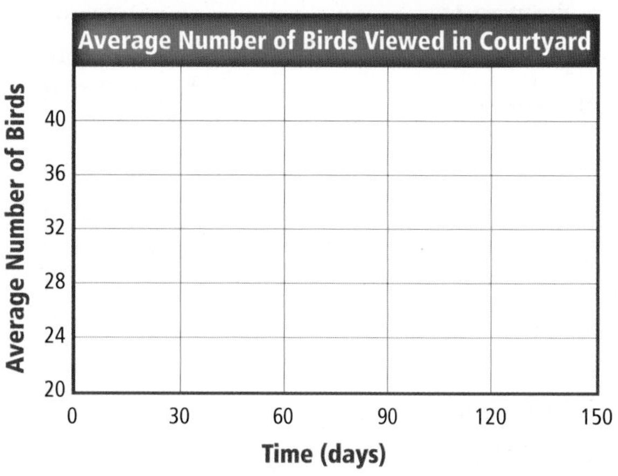

Begin plotting points by locating 30 days on the x-axis and 24 on the y-axis. Where an imaginary vertical line from the x-axis and an imaginary horizontal line from the y-axis meet, place the first data point. Place other data points using the same process. After all the points are plotted, draw a "best fit" straight line through all the points.

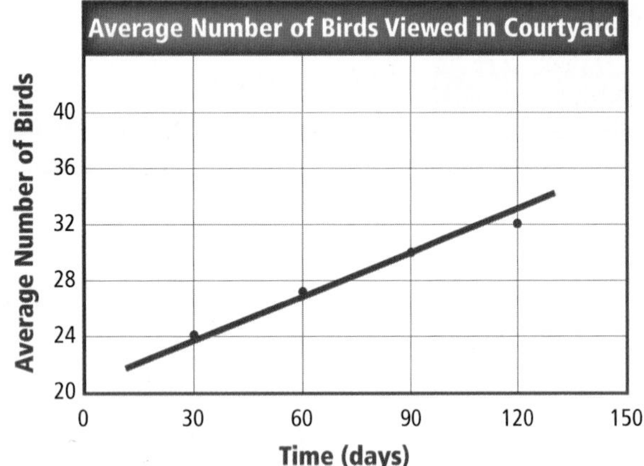

The bird-watching group also recorded the average number of brown-feathered birds they observed in the school courtyard. In the first month they averaged 21 brown-feathered birds per day. In the second month they averaged 24 brown-feathered birds. An average of 28 brown-feathered birds per day was observed in the third month. In the final month an average of 30 brown-feathered birds was observed.

What if you want to compare the average number of birds viewed with the average number of brown-feathered birds? The average brown-feathered bird data can be plotted on the same graph. Include a key with different lines indicating different sets of data.

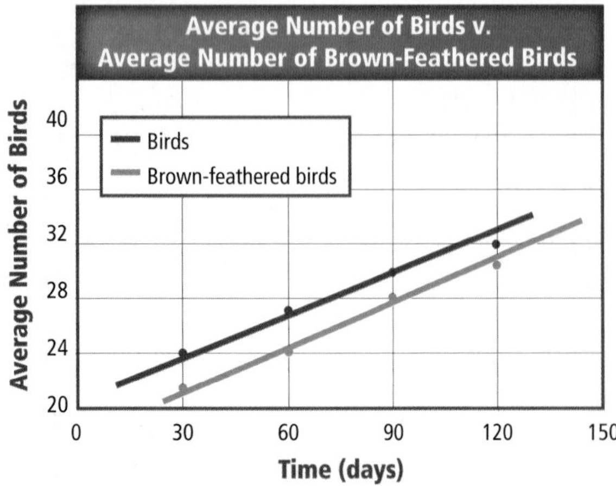

Practice Problem 3 Between 30 days and 120 days, what is the change in the average number of birds viewed?

Practice Problem 4 For the 120 days how did the average number of brown-feathered birds change as the average number of birds changed?

Slope of a Linear Graph The slope of a line is a number determined by any two points on the line. This number describes how steep the line is. The greater the absolute value of the slope, the steeper the line. Slope is the ratio of the change in the y-coordinates (rise) to the change in the x-coordinates (run) as you move from one point to the other.

The graph below shows a line that passes through (5, 4) and (9, 6).

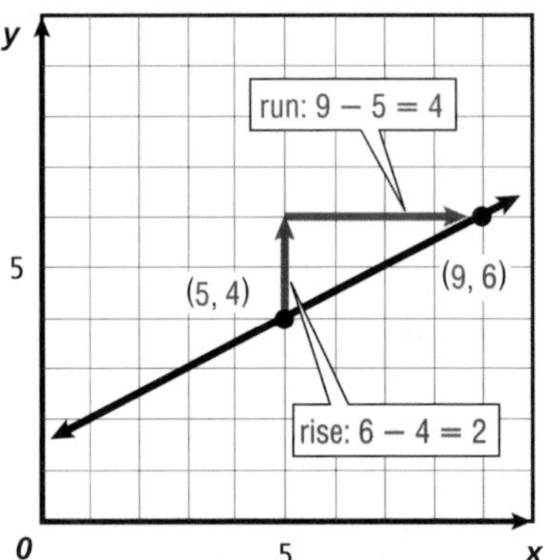

$$\text{Slope} = \frac{\text{rise}}{\text{run}}$$

$$= \frac{\text{change in } y\text{-coordinates}}{\text{change in } x\text{-coordinates}}$$

$$= \frac{6-4}{9-5}$$

$$= \frac{2}{4} \text{ or } \frac{1}{2}$$

So, the slope of the line is $\frac{1}{2}$.

A linear relationship can be translated into equation form. The equation for a straight line is

$$y = mx + b$$

where y represents the dependent variable, m is the slope of the line, x represents the independent variable, and b is the y-intercept, which is the point where the line crosses the y-axis.

Linear and Exponential Trends Two types of trends you are likely to see when you graph data in biology are linear trends and exponential trends. A linear trend has a constant increase or decrease in data values. In an exponential trend the values are increasing or decreasing more and more rapidly. The graphs below are examples of these two common trends.

In the graph below, there are two lines describing two frog species. Both lines show an increasing linear trend. As the temperature increases, so does the call pulse rates of the frogs. The rate of increase is constant.

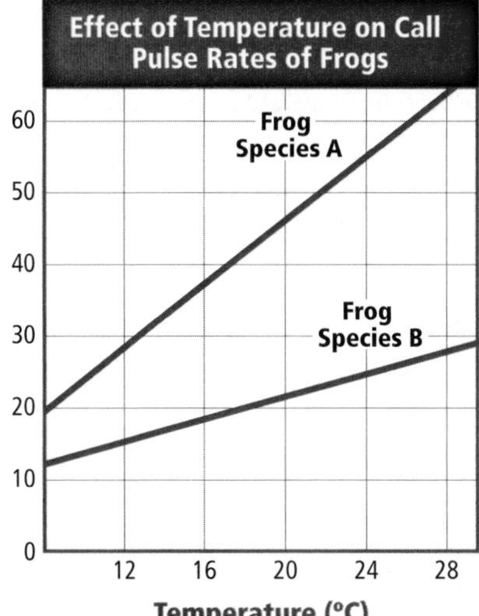

Source: Chapter 28, Section 3, p. 837

The example below shows how a mouse population would grow if the mice were allowed to reproduce unhindered. At first the population would grow slowly. The population growth rate soon accelerates because the total number of mice that are able to reproduce has increased. Notice that the portion of the graph where the population is increasing more and more rapidly is J-shaped. A J-shaped curve generally indicates exponential growth.

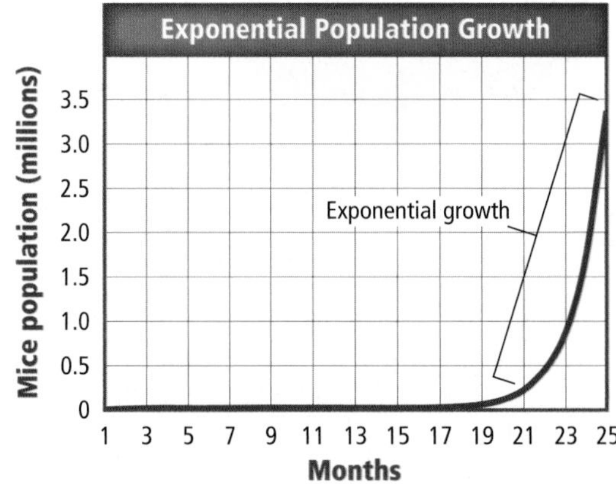

Source: Chapter 4, Section 1, p. 97

Bar Graphs A bar graph displays a comparison of different categories of data by representing each category with a bar. The length of the bar is related to the category's frequency. To make a bar graph, set up the x-axis and y-axis as you did for the line graph. Plot the data by drawing thick bars from the x-axis up to the y-axis point.

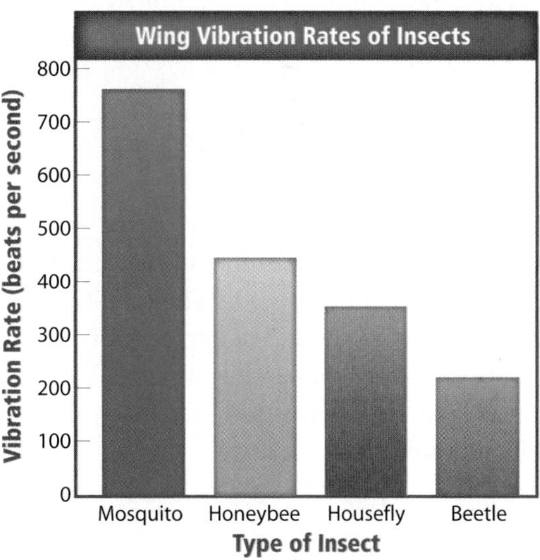

Look at the graph above. The independent variable is the type of insect. The dependent variable is the number of wing vibrations per second.

Bar graphs can also be used to display multiple sets of data in different categories at the same time. A bar graph that displays two sets of data is a called double bar graph. Double bar graphs have a legend to denote which bars represent each set of data. The graph below is an example of a double bar graph.

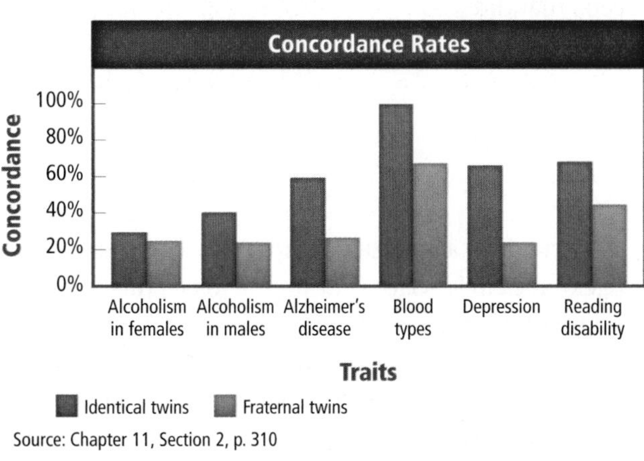

Source: Chapter 11, Section 2, p. 310

Practice Problem 5 Which type of insect has the highest number of wing vibrations per second? Is this more than twice as fast as the housefly? Explain.

Circle Graphs A circle graph consists of a circle divided into sections that represent parts of a whole. When all the sections are placed together, they equal 100 percent of the whole.

Suppose you want to make a circle graph to show the number of seeds that germinate in a package. You would first determine the total number of seeds and the numbers of seeds that germinate out of the total. You plant 143 seeds. Therefore, the whole circle represents this amount. You find that 129 seeds germinate. The seeds that germinate make up one section of the circle graph and the seeds that do not germinate make up another section.

To find out how much of the circle each section should cover, divide the number of seeds that germinate by the total number seeds. Then multiply the answer by 360, the number of degrees in a circle. Round your answer to the nearest whole number. The sum of all the segments of the circle graph should add up to 360°.

$$\text{Segment of circle for seeds that germinated} = \frac{\text{Seeds that germinated}}{\text{Total number of seeds}}$$

Divide $= \frac{129}{143}$

Multiply by number of degrees in a circle $= 0.902 \times 360°$

$= 324.72°$

Round to nearest whole number $= 325°$

Segment of circle for seeds that did not germinate $= 360° - 325°$

$= 35°$

To draw your circle graph, you will need a compass and a protractor. First, use the compass to draw a circle.

Then, draw a straight line from the center to the edge of the circle. Place your protractor on this line, and mark the point on the circle where 35° angle will intersect the circle. Draw a straight line from the center of the circle to the intersection point. This is the section for the seeds that did not germinate. The other section represents the group of seeds that did germinate.

Next, determine the percentages for each part of the whole. Calculate percentages by dividing the part by the total and multiplying by 100. Repeat this calculation for each part.

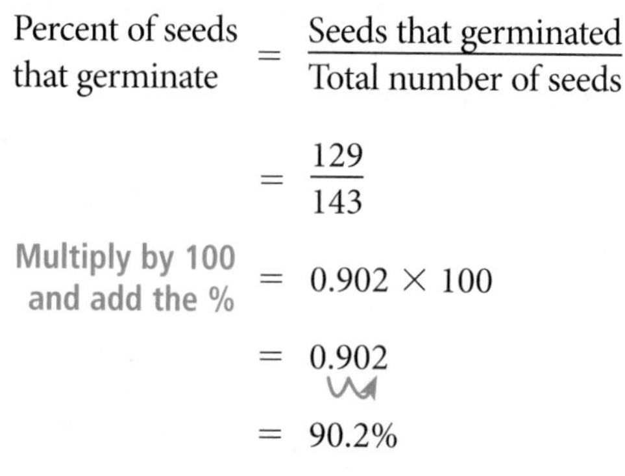

$$\text{Percent of seeds that germinate} = \frac{\text{Seeds that germinated}}{\text{Total number of seeds}}$$

$= \frac{129}{143}$

Multiply by 100 and add the % $= 0.902 \times 100$

$= 0.902$

$= 90.2\%$

Percent of seeds that do not germinate $= 100\% - 90.2\%$
$= 9.8\%$

Complete the graph by labeling the sections of the graph with percentages and giving the graph a title. Your completed graph should look similar to the one below.

If your circle graph has more than two sections, you will need to construct a segment for each entry. Place your protractor on the last line segment that you have drawn and mark off the appropriate angle. Draw a line segment from the center of the circle to the new mark on the circle. Continue this process until all of the segments have been drawn.

Percentage of Germinating and Non-Germinating Seeds

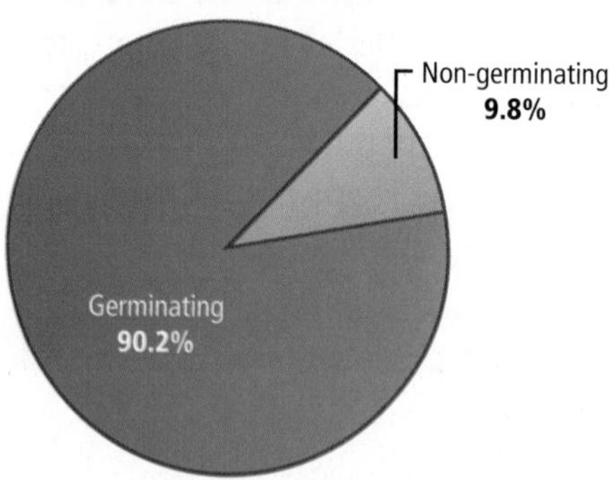

Practice Problem 6 There are 25 varieties of flowering plants growing around the high school. Construct a circle graph showing the percentage of each flowers' color. Two varieties have yellow blooms, five varieties have blue-purple blooms, eight varieties have white blooms, and ten varieties have red blooms.

Six-Kingdom Classification

The classification used in this text combines information gathered from the systems of many different fields of biology. For example, phycologists—biologists who study algae—have developed their own system of classification, as have mycologists—biologists who study fungi. The naming of animals and plants is controlled by two completely different sets of rules. The six-kingdom system, although not ideal for reflecting the phylogeny of all life, is useful for showing relationships. Taxonomy is an area of biology that evolves just the like species it studies. In **Table RH.1,** only the major phyla are listed, and one genus is named as an example. For more information about each taxon, refer to the chapter in the text in which the group is described.

Table RH.1 Six-Kingdom Classification

Concepts In Motion — **Interactive Table** To explore more about classification, visit biologygmh.com.

Kingdom	Phylum/Division* (Common Name)	Typical Example (Common Name)	Characteristics
Bacteria Salmonella Source: Chapter 18, Section 1, p. 519	Actinobacteria	*Mycobacterium*	• unicellular • most absorb food from surroundings • some are photosynthetic • some are chemosynthetic • many are parasites • many are round, spiral, or rod-shaped • some form colonies
	Omnibaceria	*Salmonella* (salmonella)	
	Spirochaetae (spirochetes)	*Treponema*	
	Chloroxybacteria	*Prochloron*	
	Cyanobacteria (blue green algae)	*Nostoc* (nostoc)	
Archaea Methanococcus jannaschii Source: Chapter 18, Section 1, p. 516	Aphragmabacteria	*Mycoplasma*	• unicellular • some absorb food from surroundings • some are photosynthetic • some are chemosynthetic • many are found in extremely harsh environments including salt ponds, hot springs, swamps, and deep-sea hydrothermal vents
	Halobacteria	*Halobacerium*	
	Methanocreatrices	*Methanobacillus*	
Protista Amoeba Source: Chapter 19, Section 1, p. 543	Sarcodina (amoeba)	*Amoeba* (amoeba)	• unicellular • take in food • free-living or parasitic • move by means of pseudopods
	Ciliophora (ciliates)	*Paramecium* (paramecium)	• unicellular • take in food • have large numbers of cilia
	Apicomplexa (apicomplexan)	*Plasmodium* (plasmodium)	• unicellular • take in food • no means of movement • are parasites in animals

*In the Kingdom Plantae the major phyla are referred to as "divisions."

Kingdom	Phylum/Division* (Common Name)	Typical Example (Common Name)	Characteristics
Protista *(continued)*	Zoomastigina (zooflagellates)	*Trypanosoma*	• unicellular • take in food • free-living or parasitic • have one or more flagella
	Euglenophyta (euglenoids)	*Euglena* (euglena)	• unicellular • photosynthetic or take in food • most have one flagellum
	Bacillariophyta (diatoms)	*Navicula*	• unicellular • photosynthetic • have unique double shells made of silica
	Pyrrophyta (dinoflagellates)	*Gonyaulax*	• unicellular • photosynthetic • contain red pigments • have two flagella
	Rhodophyta (red algae)	*Chondrus*	• most are multicellular • photosynthetic • contain red pigments • most live in deep, salt water
	Phaeophyta (brown algae)	*Laminaria*	• most are multicellular • photosynthetic • contain brown pigments • most live in salt water
	Chlorophyta (green algae)	*Ulva*	• unicellular, multicellular, or colonies • photosynthetic • contain chlorophyll • live on land, in freshwater, or salt water
	Acrasiomycota (cellular slime mold)	*Dictyostelium*	• unicellular or multicellular • absorb food • change form during life cycle • cellular and plasmodial slime molds
	Myxomycota (acellular slime mold)	*Physarum*	
	Oomycota (water mold/ downy mildew	*Phytophthora*	• multicellular • are either parasites or decomposers • live in freshwater or salt water

Diatom
Source: Chapter 19, Section 3, p. 554

Red algae
Source: Chapter 19, Section 3, p. 559

Slime mold
Source: Chapter 19, Section 4, p. 561

*In the Kingdom Plantae the major phyla are referred to as "divisions."

Reference Handbook

Kingdom	Phylum/Division* (Common Name)	Typical Example (Common Name)	Characteristics
Fungi	Zygomycota (common mold)	*Rhizopus* (bread mold)	• multicellular • absorb food • spores are produced in sporangia
	Ascomycota (sac fungi)	*Saccharomyces* (yeast)	•unicellular and multicellular • absorb food • spores produced in asci
	Basidiomycota (club fungi)	*Crucibulum* (bird's nest fungus)	• multicellular • absorb food • spores produced in basidia
Bread mold Source: Chapter 20, Section 2, p. 583	Deuteromycota (imperfect fungi)	*Penicillium* (penicillum)	• members with unknown reproductive structures • imperfect fungi
	Chytridiomycota	*Chytridium* (chrytid)	• some are saprobes • some parasitize protists, plants, and animals
Plantae	Hepaticophyta (liverworts)	*Pellia*	• multicellular nonvascular plants • reproduce by spores produced in capsules • green • grow in moist, land environments
	Anthocerophyta (hornworts)	*Anthoceros*	
	Bryophyta (moss)	*Polytrichum* (haircap moss)	
Liverwort Source: Chapter 21, Section 2, p. 612	Lycophyta (club moss)	*Lycopodium* (wolf's claw)	• multicellular vascular plants • spores are produced in cone-like structures • live on land • photosynthetic
	Arthrophyta	*Equisetum* (horsetails)	• vascular plants • ribbed and jointed stems • scale-like leaves • spores produced in cone-like structures
	Pterophyta (ferns)	*Polypodium* (ferns)	• vascular plants • leaves called fronds • spores produce in clusters or sporangia called sori • live on land or in water
Wood fern Source: Chapter 21, Section 3, p. 614	Ginkgophyta (ginko)	*Ginkgo* (ginko)	• deciduous trees • only one living species • have fan-shaped leaves with branching veins and fleshy cones with seeds

*In the Kingdom Plantae the major phyla are referred to as "divisions."

Kingdom	Phylum/Division* (Common Name)	Typical Example (Common Name)	Characteristics
Plantae *(continued)*	Cycadophyta (cycad)	*Cyas* (palm tree)	• palm-like plants • have large, feather-like leaves • produce seeds in cones
	Coniferophyta (conifer)	*Pinus* (pine tree)	• deciduous or evergreen • trees or shrubs • needle-like or scale-like leaves • seeds produced in cones
Welwitschia Source: Chapter 21, Section 4, p. 618	Gnetophyta (gnetophyte)	*Welwitschia* (welwitschia)	• shrubs or woody vines • seeds produced in cones • division contains only three genera
	Anthophyta (flowering plant)	*Rhododendron* (rhododendron)	• dominant group of plants • flowering plants • have fruit with seeds
Animalia	Porifera (sponges)	*Spongilla* (sponge)	• aquatic organisms that lack true tissues and organs • asymmetrical and sessile
	Cnidaria (cnidarians)	*Hydra* (hydra)	• radially symmetrical • digestive cavity with one opening • most have tentacles armed with stinging cells • live in aquatic environments singly or in colonies
Sponge Source: Chapter 24, Section 3, p. 707	Platyhelminthes (flatworms)	*Dugesia* (planaria)	• unsegmented, bilaterally symmetrical • no body cavity • digestive cavity, if present, has only one opening • parasitic and free-living species
	Nematoda (roundworms)	*Trichinella* (trichinella)	• pseudocoelomate, unseg-mented, bilaterally symmetrical • tubular digestive tract • without cilia • live in great numbers in soil and aquatic sediments
Abalone Source: Chapter 25, Section 3, p. 742	Mollusca (mollusks)	*Nautilus* (nautilus)	• soft-bodied coelomates • bodies are divided into three parts: head-foot, visceral mass, and mantle • many have shells • almost all have a radula • aquatic and terrestrial species

*In the Kingdom Plantae the major phyla are referred to as "divisions."

Kingdom	Phylum/Division* (Common Name)	Typical Example (Common Name)	Characteristics
Animalia *(continued)*	Annelida (segmented worms)	*Hirudo* (leech)	• coelomate, serially segmented, bilaterally symmetrical • complete digestive tract • most have setae on each segment that anchor them during crawling • terrestrial and aquatic species
	Arthropoda (arthopods)	*Colias* (butterflies)	• chitinous exoskeleton covering segmented bodies • paired, jointed appendages • many have wings • land and aquatic species
	Echinodermata (echinoderm)	*Cucumaria* (sea cucumber)	• marine organisms • have spiny or leathery skin and a water-vascular system with tube feet • radially symmetrical
	Chordata (chordates)		• segmented coelomates with a notochord • possess a dorsal nerve cord, pharyngeal slits, and a tail at some stage of life • most have paired appendages
	Chordata Subphylum: Urochordata	*Polycarpa* (sea squirt)	• young have all of the main chordate features; adults have only pharyngeal gill slits
	Chordata Subphylum: Cephalochordata	*Branchiostoma* (amphioxus)	• adults have all of the main features of chordates
	Chordata Subphylum: Vertebrata	*Panthera* (panther)	• the hallmark feature of all vertebrates is a spinal column

Sand dollar
Source: Chapter 27, Section 1, p. 798

Sea otter
Source: Chapter 30, Section 1, p. 880

*In the Kingdom Plantae the major phyla are referred to as "divisions."

Three-Domain Classification

Increasingly, biologists are classifying organisms into categories larger than kingdoms called domains. The three domains are: Domain Bacteria, Domain Archaea, and Domain Eukarya. With future discoveries, this classification system might change to incorporate new information.

DOMAIN	Bacteria	Archaea	Eukarya			
KINGDOM	Bacteria	Archaea	Protista	Fungi	Plantae	Animalia

Scientific Word Origins

This list of prefixes, suffixes, and roots is provided to help you understand science terms used throughout this biology textbook. The list identifies whether the prefix, suffix, or root is of Greek *(G)* or Latin *(L)* origin. Also listed is the meaning of the prefix, suffix, or root and a science word in which it is used.

Origin	Meaning	Example
A		
ad *(L)*	to, toward	adaxial
aero *(G)*	air	aerobic
an *(G)*	without	anaerobic
ana *(G)*	up	anaphase
andro *(G)*	male	androceium
angio *(G)*	of seed	angiosperm
anth/o *(G)*	flower	anthophyte
anti *(G)*	against	antibody
aqu/a *(L)*	of water	aquatic
archae *(G)*	ancient	archaeologist
arthro, artio *(G)*	jointed	arthropod
askos *(G)*	bag	ascospore
aster *(G)*	star	Asteroidea
autos *(G)*	self	autoimmune
B		
bi *(L)*	two	bipedal
bio *(G)*	life	biosphere
C		
carn *(L)*	flesh	carnivore
ceph *(G)*	head	cephalopod
chloros *(G)*	light green	chlorophyll
chroma *(G)*	pigmented	chromosome
cide *(L)*	to kill	insecticide
circ *(L)*	circular	circadian
cocc/coccus *(G)*	small and round	streptococcus
con *(L)*	together	convergent
cyte *(G)*	cell	cytoplasm
D		
de *(L)*	remove	decompose
dendron *(G)*	tree	dendrite
dent *(L)*	tooth	edentate
derm *(G)*	skin	epidermis
di *(G)*	two	disaccharide
dia *(G)*	apart	diaphragm
dorm *(L)*	sleep	dormancy

Origin	Meaning	Example
E		
echino *(G)*	spiny	echinoderm
ec *(G)*	outer	ecosystem
ella(e) *(L)*	small	organelle
endo *(G)*	within	endosperm
epi *(G)*	upon	epidermis
eu *(G)*	true	eukaryote
exo *(G)*	outside	exoskeleton
F		
fer *(L)*	to carry	conifer
G		
gastro *(G)*	stomach	gastropod
genesis *(G)*	to originate	oogenesis
gen/(e)(o) *(G)*	kind	genotype
gon *(G)*	reproductive	archegonium
gravi *(L)*	heavy	gravitropism
gymn/o *(G)*	naked	gymnosperm
gyn/e *(G)*	female	gynecium
H		
hal(o) *(G)*	salt	halophyte
hapl(o) *(G)*	single	haploid
hemi *(G)*	half	hemisphere
hem(o) *(G)*	blood	hemoglobin
herb/a(i) *(L)*	vegetation	herbivore
heter/o *(G)*	different	heterotrophic
hom(e)/o *(G)*	same	homeostasis
hom *(L)*	human	hominid
hydr/o *(G)*	water	hydrolysis
I		
inter *(L)*	between	internode
intra *(L)*	within	intracellular
is/o *(G)*	equal	isotonic
J		
jug *(L)*	to join	jugular

Origin	Meaning	Example
K		
kary *(G)*	nucleus	eukaryote
kera *(G)*	hornlike	keratin
L		
leuc/o *(G)*	white	leukocyte
logy *(G)*	study of	biology
lymph/o *(L)*	water	lymphocyte
lysis *(G)*	break up	dialysis
M		
macr/o *(G)*	large	macromolecule
meg/a *(G)*	great	megaspore
meso *(L)*	in the middle	mesophyll
meta *(G)*	after	metaphase
micr/o *(G)*	small	microscope
mon/o *(G)*	only one	monocotyledon
morph/o *(G)*	form	morphology
N		
nema *(G)*	a thread	nematode
neuro *(G)*	nerve	neuron
nod *(L)*	knot	nodule
nomy(e) *(G)*	system of laws	taxonomy
O		
olig/o *(G)*	small, few	oligochaete
omn *(L)*	all	omnivore
orni(s) *(G)*	bird	ornithology
oste/o *(G)*	bone formation	osteocyte
ov *(L)*	an egg	oviduct
P		
pal(a)e/o *(G)*	ancient	paleontology
para *(G)*	beside	parathyroid
path/o *(G)*	suffering	pathogen
ped *(L)*	foot	centipede
per *(L)*	through	permeable
peri *(G)*	around, about	peristalsis
phag/o *(G)*	eating	phagocyte
phot/o *(G)*	light	photosynthesis
phyl *(G)*	race, class	phylogeny
phyll *(G)*	leaf	chlorophyll
phyte *(G)*	plant	epiphyte
pinna *(L)*	feather	pinnate

Origin	Meaning	Example
P (continued)		
plasm/o *(G)*	to form	plasmodium
pod *(G)*	foot	gastropod
poly *(G)*	many	polymer
post *(L)*	after	posterior
pro *(G) (L)*	before	prokaryote
prot/o *(G)*	first	protocells
pseud/o *(G)*	false	pseudopodium
R		
re *(L)*	back to original	reproduce
rhiz/o *(L)*	root	rhizoid
S		
scope *(G)*	to look	microscope
some *(G)*	body	lysome
sperm *(G)*	seed	gymnosperm
stasis *(G)*	remain constant	homeostasis
stom *(G)*	mouthlike opening	stomata
syn *(G)*	together	synapse
T		
tel/o *(G)*	end	telophase
terr *(L)*	of Earth	terrestrial
therm *(G)*	heat	endotherm
thylak *(G)*	sack	thylakoid
trans *(L)*	across	transpiration
trich *(G)*	hair	trichome
trop/o *(G)*	a change	gravitropism
trophic *(G)*	nourishment	heterotrophic
U		
uni *(L)*	one	unicellular
V		
vacc/a *(L)*	cow	vaccine
vore *(L)*	eat greedily	omnivore
X		
xer/o *(G)*	dry	xerophye
Z		
zo/o *(G)*	living being	zoology
zygous *(G)*	two joined	homozygous

Reference Handbook

PERIODIC TABLE OF THE ELEMENTS

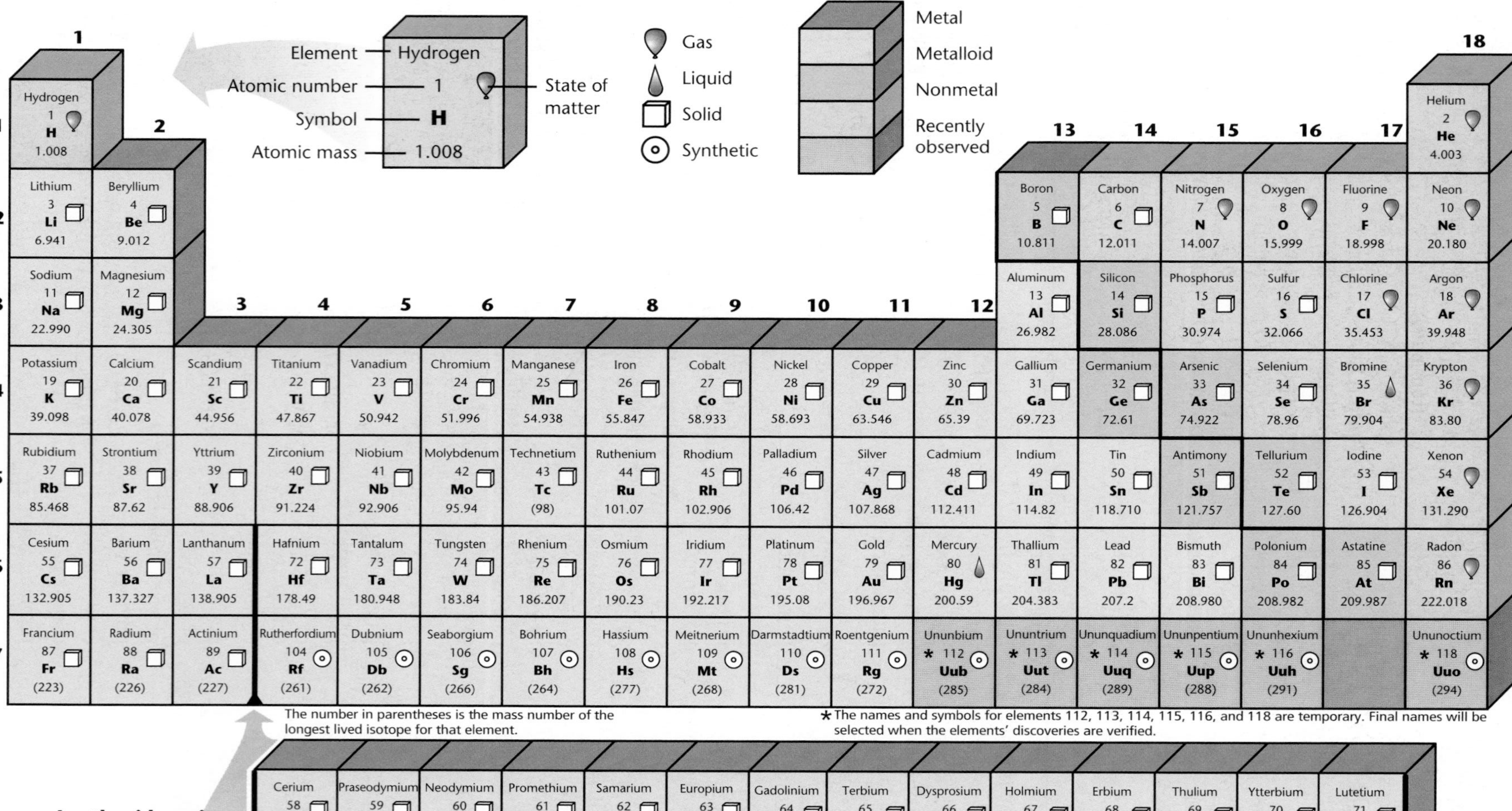

Period	1	2	3	4	5	6	7	8	9	10	11	12	13	14	15	16	17	18
1	Hydrogen 1 **H** 1.008																	Helium 2 **He** 4.003
2	Lithium 3 **Li** 6.941	Beryllium 4 **Be** 9.012											Boron 5 **B** 10.811	Carbon 6 **C** 12.011	Nitrogen 7 **N** 14.007	Oxygen 8 **O** 15.999	Fluorine 9 **F** 18.998	Neon 10 **Ne** 20.180
3	Sodium 11 **Na** 22.990	Magnesium 12 **Mg** 24.305											Aluminum 13 **Al** 26.982	Silicon 14 **Si** 28.086	Phosphorus 15 **P** 30.974	Sulfur 16 **S** 32.066	Chlorine 17 **Cl** 35.453	Argon 18 **Ar** 39.948
4	Potassium 19 **K** 39.098	Calcium 20 **Ca** 40.078	Scandium 21 **Sc** 44.956	Titanium 22 **Ti** 47.867	Vanadium 23 **V** 50.942	Chromium 24 **Cr** 51.996	Manganese 25 **Mn** 54.938	Iron 26 **Fe** 55.847	Cobalt 27 **Co** 58.933	Nickel 28 **Ni** 58.693	Copper 29 **Cu** 63.546	Zinc 30 **Zn** 65.39	Gallium 31 **Ga** 69.723	Germanium 32 **Ge** 72.61	Arsenic 33 **As** 74.922	Selenium 34 **Se** 78.96	Bromine 35 **Br** 79.904	Krypton 36 **Kr** 83.80
5	Rubidium 37 **Rb** 85.468	Strontium 38 **Sr** 87.62	Yttrium 39 **Y** 88.906	Zirconium 40 **Zr** 91.224	Niobium 41 **Nb** 92.906	Molybdenum 42 **Mo** 95.94	Technetium 43 **Tc** (98)	Ruthenium 44 **Ru** 101.07	Rhodium 45 **Rh** 102.906	Palladium 46 **Pd** 106.42	Silver 47 **Ag** 107.868	Cadmium 48 **Cd** 112.411	Indium 49 **In** 114.82	Tin 50 **Sn** 118.710	Antimony 51 **Sb** 121.757	Tellurium 52 **Te** 127.60	Iodine 53 **I** 126.904	Xenon 54 **Xe** 131.290
6	Cesium 55 **Cs** 132.905	Barium 56 **Ba** 137.327	Lanthanum 57 **La** 138.905	Hafnium 72 **Hf** 178.49	Tantalum 73 **Ta** 180.948	Tungsten 74 **W** 183.84	Rhenium 75 **Re** 186.207	Osmium 76 **Os** 190.23	Iridium 77 **Ir** 192.217	Platinum 78 **Pt** 195.08	Gold 79 **Au** 196.967	Mercury 80 **Hg** 200.59	Thallium 81 **Tl** 204.383	Lead 82 **Pb** 207.2	Bismuth 83 **Bi** 208.980	Polonium 84 **Po** 208.982	Astatine 85 **At** 209.987	Radon 86 **Rn** 222.018
7	Francium 87 **Fr** (223)	Radium 88 **Ra** (226)	Actinium 89 **Ac** (227)	Rutherfordium 104 **Rf** (261)	Dubnium 105 **Db** (262)	Seaborgium 106 **Sg** (266)	Bohrium 107 **Bh** (264)	Hassium 108 **Hs** (277)	Meitnerium 109 **Mt** (268)	Darmstadtium 110 **Ds** (281)	Roentgenium 111 **Rg** (272)	Ununbium * 112 **Uub** (285)	Ununtrium * 113 **Uut** (284)	Ununquadium * 114 **Uuq** (289)	Ununpentium * 115 **Uup** (288)	Ununhexium * 116 **Uuh** (291)		Ununoctium * 118 **Uuo** (294)

The number in parentheses is the mass number of the longest lived isotope for that element.

* The names and symbols for elements 112, 113, 114, 115, 116, and 118 are temporary. Final names will be selected when the elements' discoveries are verified.

Lanthanide series	Cerium 58 **Ce** 140.115	Praseodymium 59 **Pr** 140.908	Neodymium 60 **Nd** 144.242	Promethium 61 **Pm** (145)	Samarium 62 **Sm** 150.36	Europium 63 **Eu** 151.965	Gadolinium 64 **Gd** 157.25	Terbium 65 **Tb** 158.925	Dysprosium 66 **Dy** 162.50	Holmium 67 **Ho** 164.930	Erbium 68 **Er** 167.259	Thulium 69 **Tm** 168.934	Ytterbium 70 **Yb** 173.04	Lutetium 71 **Lu** 174.967
Actinide series	Thorium 90 **Th** 232.038	Protactinium 91 **Pa** 231.036	Uranium 92 **U** 238.029	Neptunium 93 **Np** (237)	Plutonium 94 **Pu** (244)	Americium 95 **Am** (243)	Curium 96 **Cm** (247)	Berkelium 97 **Bk** (247)	Californium 98 **Cf** (251)	Einsteinium 99 **Es** (252)	Fermium 100 **Fm** (257)	Mendelevium 101 **Md** (258)	Nobelium 102 **No** (259)	Lawrencium 103 **Lr** (262)

Glossary/Glosario

A multilingual science glossary at biologygmh.com includes Arabic, Bengali, Chinese, English, Haitian Creole, Hmong, Korean, Portuguese, Russian, Spanish, Tagalog, Urdu, and Vietnamese.

Pronunciation Key

Use the following key to help you sound out words in the glossary.

a	back (BAK)	**ew**	food (FEWD)
ay	day (DAY)	**yoo**	pure (PYOOR)
ah	father (FAH thur)	**yew**	few (FYEW)
ow	flower (FLOW ur)	**uh**	comma (CAHM uh)
ar	car (CAR)	**u** (+ con)	rub (RUB)
e	less (LES)	**sh**	shelf (SHELF)
ee	leaf (LEEF)	**ch**	nature (NAY chur)
ih	trip (TRIHP)	**g**	gift (GIHFT)
i (i + con + e)	idea, life (i DEE uh, life)	**j**	gem (JEM)
oh	go (GOH)	**ing**	sing (SING)
aw	soft (SAWFT)	**zh**	vision (VIHZH un)
or	orbit (OR but)	**k**	cake (KAYK)
oy	coin (COYN)	**s**	seed, cent (SEED, SENT)
oo	foot (FOOT)	**z**	zone, raise (ZOHN, RAYZ)

Cómo usar el glosario en español:
1. Busca el término en inglés que desees encontrar.
2. El término en español, junto con la definición, se encuentran en la columna de la derecha.

Glossary/ Glosario

A

English	Español
abdomen: (p. 763) in arthropods, posterior body region of an arthropod that contains fused segments, digestive structures, reproductive organs, and bears additional legs; in humans, part of body that is between the diaphragm and pelvis.	**abdomen: (pág. 763)** región posterior del cuerpo de un artrópodo, la que contiene segmentos fusionados, estructuras digestivas, los órganos reproductores y que sostiene patas adicionales.
abiotic (ay bi AH tihk) factor: (p. 35) any nonliving factor in an organism's environment, such as soil, water temperature, and light availability.	**factor abiótico: (pág. 35)** todo factor inanimado en el ambiente de un organismo, como el suelo, el agua, la temperatura del agua y la disponibilidad de luz.
abyssal zone: (p. 81) deepest, very cold region of the open ocean.	**zona abisal: (pág. 81)** la zona más profunda y más fría del océano.
acid: (p. 164) substance that releases hydrogen ions (H^+) when dissolved in water; an acidic solution has a pH less than 7.	**ácido: (pág. 164)** sustancia que libera iones hidrógeno (H+) cuando se halla disuelta en agua; una solución ácida contiene un pH menor que 7.
acoelomate (ay SEE lum ayt): (p. 701) animal with a solid body that lacks a fluid-filled body cavity between the gut and the body wall.	**acelomado: (pág. 701)** animal de cuerpo sólido que carece de una cavidad corporal llena de fluido entre los órganos internos y las paredes del cuerpo.
acrasin (uh KRA sun): (p. 563) chemical given off by starving amoeba-like cells that serves as a signal to the cells to form a sluglike colony.	**acrasina: (pág. 563)** sustancia química que liberan ciertas células ameboides, cuando tienen hambre, y que sirve de señal para que estas células formen una colonia viscosa.
actin: (p. 948) protein filament in muscle cells that functions with myosin in contraction.	**actina: (pág. 948)** filamento proteico de las células musculares que, junto con la miosina, participan en la contracción muscular.
action potential: (p. 964) nerve impulse.	**potencial de acción: (pág. 964)** un impulso nervioso.

activation energy: (p. 158) minimum amount of energy needed for reactants to form products in a chemical reaction.

active site: (p. 160) specific place where a substrate binds on an enzyme.

active transport: (p. 205) energy-requiring process by which substances move across the plasma membrane against a concentration gradient.

adaptation (a dap TAY shun): (p. 10) inherited characteristic of a species that develops over time in response to an environmental factor, enabling the species to survive.

adaptive radiation: (p. 439) diversification of a species into a number of different species, often over a relatively short time span.

addiction: (p. 981) psychological and/or physiological dependence on a drug.

adenosine triphosphate (uh DEN uh seen • tri FAHS fayt) (ATP): (p. 221) energy-carrying biological molecule, which, when broken down, drives cellular activities.

adolescence (a dul ES unts): (p. 1063) developmental phase that begins with puberty and ends at adulthood.

adulthood: (p. 1064) developmental phase that occurs at the end of adolescence, when physical growth is complete.

aerobic process: (p. 228) a metabolic process that requires oxygen.

aerobic respiration: (p. 228) metabolic process in which pyruvate is broken down and electron-carrier molecules are used to produce ATP through electron transport.

age structure: (p. 104) in any population, the number of individuals in their pre-reproductive, reproductive, and post-reproductive years.

agonistic (ag oh NIHS tihk) behavior: (p. 917) threatening or combative behavior between two members of the same species that usually does not result in injury or death.

air sac: (p. 863) in birds, the posterior and anterior structure used in respiration, resulting in only oxygenated air moving through the lungs.

aldosterone (al DAWS tuh rohn): (p. 1035) steroid hormone produced by the adrenal cortex that acts on the kidneys and is important for sodium reabsorption.

allele: (p. 278) alternative form that a single gene may have for a particular trait.

allergy: (p. 1094) overactive immune response to environmental antigens.

allopatric speciation: (p. 438) occurs when a population divided by a geographic barrier evolves into two or more populations unable to interbreed.

energía de activación: (pág. 158) cantidad mínima de energía que requieren los reactivos para formar productos durante una reacción química.

sitio activo: (pág. 160) lugar específico donde un sustrato se une a una enzima.

transporte activo: (pág. 205) proceso que requiere energía y que le permite a una sustancia atravesar la membrana plasmática contra un gradiente de concentración.

adaptación: (pág. 10) característica heredada de una especie; esta característica evoluciona a lo largo del tiempo en respuesta a un factor ambiental y le ayuda a la especie a sobrevivir.

radiación adaptativa: (pág. 439) diversificación de una especie en diferentes especies, a menudo en un período relativamente corto.

adicción: (pág. 981) dependencia psicológica o fisiológica a una droga.

trifosfato de adenosina (ATP): (pág. 221) molécula biológica que transporta energía, y que al desdoblarse, hace funcionar las actividades celulares.

adolescencia: (pág. 1063) fase del desarrollo que se inicia en la pubertad y termina al comenzar la edad adulta.

edad adulta: (pág. 1064) fase del desarrollo que empieza al terminar la adolescencia, cuando se completa el crecimiento físico.

proceso aeróbico: (pág. 228) proceso metabólico que requiere oxígeno.

respiración aeróbica: (pág. 228) proceso metabólico en que se desdobla el piruvato y las moléculas transportadoras de electrones ayudan a producir ATP mediante el transporte de electrones.

estructura etaria: (pág. 104) el número de individuos en edad prereproductora, reproductora y postreproductora en una población.

comportamiento agonístico: (pág. 917) comportamiento amenazador o combativo entre dos miembros de la misma especie y que normalmente no produce heridas o muerte.

sacos aéreos: (pág. 863) estructuras anteriores y posteriores de las aves que participan en la respiración celular y permiten sólo el paso de sangre oxigenada por los pulmones.

aldosterona: (pág. 1035) hormona esteroide producida por la corteza adrenal, la cual actúa sobre los riñones y es importante para la reabsorción de sodio.

alelo: (pág. 278) forma alternativa de un gene determinado para un rasgo dado.

alergia: (pág. 1094) acentuada respuesta inmune a un antígeno del ambiente.

especiación alopátrica: (pág. 438) sucede cuando una población separada en dos por una barrera geográfica, evoluciona y se convierte en dos poblaciones incapaces de entrecruzarse.

alternation of generations: **(p. 560)** reproductive life cycle that alternates between a diploid ($2n$) sporophyte generation and a haploid (n) gametophyte generation.

altruistic behavior: **(p. 922)** self-sacrificing behavior that benefits another individual.

alveolus: **(p. 1001)** in the lung, a thin-walled air sac surrounded by capillaries.

amino acid: **(p. 170)** carbon compound joined by peptide bonds; building block of proteins.

amnion (AM nee ahn): **(p. 852)** fluid-filled membrane that surrounds and protects a developing embryo.

amniotic egg: **(p. 853)** egg that provides a complete environment for the developing embryo with a yolk sac for nutrition, protective internal membranes and fluid, and a protective outer shell.

amniotic fluid (am nee AH tihk • FLU id): **(p. 1056)** amniotic sac fluid that cushions, insulates, and protects the embryo.

ampulla (AM pyew luh): **(p. 795)** in echinoderms, the muscular sac that contracts to force water into the tube foot, allowing it to extend.

amylase: **(p. 1020)** digestive enzyme in saliva that begins the process of chemical digestion in the mouth by breaking down starches into sugars.

anaerobic process: **(p. 228)** metabolic process that does not require oxygen.

analogous structure: **(p. 426)** structure that has the same function but different construction and was not inherited from a common ancestor.

anaphase: **(p. 251)** third stage of mitosis in which sister chromatids are pulled apart and microtubules, along with motor proteins, move the chromosomes to opposite poles of the cell.

anaphylactic (an uh fuh LAK tik) shock: **(p. 1095)** severe hypersensitivity to a specific antigen, causing a massive histamine release.

ancestral character: **(p. 495)** morphological or biochemical feature present in various groups within the line of descent.

ancestral trait: **(p. 424)** more-primitive characteristic that appeared in common ancestors.

annual: **(p. 620)** plant that completes its life span in one growing season or less.

anterior: **(p. 700)** head end of an animal with bilateral symmetry.

anthropoid: **(p. 455)** part of haplorhines; humanlike primates that include New World monkeys, Old World monkeys, and hominoids.

antibiotic (an ti bi AH tihk): **(p. 1082)** substance that is able to kill or inhibit the growth of some microorganisms.

alternancia de generaciones: **(pág. 560)** ciclo de vida reproductor que se alterna entre una generación esporofita diploide ($2n$) y una generación gametofita haploide (n).

comportamiento altruista: **(pág. 922)** comportamiento de autosacrificio que beneficia a otro individuo.

alvéolo: **(pág. 1001)** saco aéreo de paredes delgadas, localizado dentro de los pulmones, y que está rodeado por capilares.

aminoácido: **(pág. 170)** compuestos de carbono con enlaces peptídicos; son la unidad básica de las proteínas.

amnios: **(pág. 852)** membrana llena de fluido que rodea y protege al embrión en desarrollo.

huevo amniótico: **(pág. 853)** huevo que provee un ambiente completo para un embrión en desarrollo. Tiene un saco vitelino para la nutrición, membranas internas y fluidos protectores, y una cubierta protectora externa.

fluido amniótico: **(pág. 1056)** fluido del saco amniótico que acojina, aísla y protege al embrión.

ampolla: **(pág. 795)** saco muscular de los equinodermos que se contrae para impulsar el agua hacia las patas ambulacrales y provocar su extensión.

amilasa: **(pág. 1020)** enzima digestiva de la saliva que inicia el proceso de digestión química en la boca, al desdoblar almidones en azúcares.

aproceso anaeróbico: **(pág. 228)** proceso metabólico que no requiere oxígeno.

estructura análoga: **(pág. 426)** estructura que tiene la misma función, pero diferente construcción y que no se originó a partir de un antepasado común.

anafase: **(pág. 251)** tercera fase de la mitosis. En ella, las cromátides hermanas se separan y los microtúbulos, junto con proteínas motoras, mueven los cromosomas hacia polos opuestos de la célula.

choque anafiláctico: **(pág. 1095)** grave hipersensibilidad a un antígeno específico que causa una liberación masiva de histamina.

carácter ancestral: **(pág. 495)** característica morfológica o bioquímica presente en varios grupos dentro de un linaje.

rasgo ancestral: **(pág. 424)** característica más primitiva que aparecían en antepasados comunes.

anual: **(pág. 620)** planta que completa su ciclo vital en una temporada de crecimiento o menos.

anterior: **(pág. 700)** extremo delantero de un animal con simetría bilateral.

antropoide: **(pág. 455)** parte de los haplorrinos; primates parecidos a humanos.

antibiótico: **(pág. 1082)** sustancia que destruye algunos microorganismos o que inhibe su crecimiento.

antibody: (p. 1086) protein produced by B lymphocytes that specifically reacts with a foreign antigen.

anticuerpo: (pág. 1086) proteína producida por los linfocitos B que reacciona específicamente con un antígeno extraño.

antidiuretic (an ti di yuh REH tic) hormone: (p. 1037) functions in homeostasis by regulating water balance.

hormona antidiurética: (pág. 1037) hormona que ayuda en la homeostasis al regular el equilibrio del agua.

antigen: (p. 1086) a substance foreign to the body that causes an immune response; it can bind to an antibody or T cell.

antígeno: (pág. 1086) sustancia foránea al cuerpo que causa una reacción inmunológica; se puede enlazar a un anticuerpo o a una célula T.

aphotic zone: (p. 80) open-ocean zone through which sunlight cannot penetrate.

zona afótica: (pág. 80) zona del mar abierto a la que no llega la luz solar.

apoptosis (a pup TOH sus): (p. 256) programmed cell death.

apoptosis: (pág. 256) muerte celular programada.

appendage (uh PEN dihj): (p. 764) in arthropods, a structure such as a leg or an antenna that grows and extends from the outside body covering.

apéndice: (pág. 764) estructura de los artrópodos, como las patas o las antenas, que crece y se extiende desde la cubierta exterior del cuerpo.

appendicular skeleton: (p. 941) one of the two divisions of the human skeleton; includes the bones of the arms, legs, feet, hands, hips, and shoulders.

esqueleto apendicular: (pág. 941) una de las dos divisiones del esqueleto humano; incluye los huesos de brazos, piernas, manos, pies, caderas y hombros.

arboreal: (p. 455) tree dwelling—it is a characteristic of many primates.

arborícola: (pág. 455) que vive en los árboles: es una característica de muchos primates.

archaebacteria: (p. 500) prokaryotes whose cell walls do not contain peptidoglycan.

arquebacterias: (pág. 500) procariotas cuyas paredes celulares no contienen peptidoglucanos.

artery: (p. 993) elastic, thick-walled blood vessel that carries oxygenated blood away from the heart.

arteria: (pág. 993) vaso sanguíneo de paredes gruesas y elásticas que transporta sangre oxigenada desde el corazón hacia el resto del cuerpo.

artificial selection: (p. 419) Darwin's term for the selective breeding of organisms selected for certain traits in order to produce offspring having those traits.

selección artificial: (pág. 419) término empleado por Darwin para referirse a la cría de selección de organismos, la cual se efectúa para seleccionar y obtener una progenie con ciertos rasgos.

ascocarp: (p. 585) in sac fungi, the reproductive structure in which haploid nuclei fuse to form a zygote.

ascocarpo: (pág. 585) estructura reproductora de los ascomicetos; la estructura reproductora en que núcleos haploides se fusionan para formar un cigoto.

ascospore: (p. 585) spore produced by an ascus.

ascospora: (pág. 585) espora producida por un asco.

ascus: (p. 585) spore-producing saclike structure of sac fungi.

asco: (pág. 585) estructura en forma de saco que produce esporas en los ascomicetos.

atherosclerosis (a thuh roh skluh ROH sus): (p. 999) circulatory system disorder in which arteries are blocked, restricting blood flow.

aterosclerosis: (pág. 999) trastorno del sistema circulatorio en que las arterias se bloquean y restringen el flujo de sangre.

atom: (p. 148) building block of matter; contains subatomic particles—neutrons, protons, and electrons.

átomo: (pág. 148) unidad básica de la materia; contiene las siguientes partículas subatómicas: neutrones, protones y electrones.

atrium: (p. 824) heart chamber that receives blood from the body.

aurícula: (pág. 824) cavidad del corazón que recibe la sangre proveniente del cuerpo.

australopithecine: (p. 465) genus that lived in the east-central and southern part of Africa between 4.2 and 1 mya.

australopitecinos: (pág. 465) género que vivió en la parte centro oriental y meridional de África entre hace 4.2 y 1 millón de años.

autonomic nervous system: (p. 971) part of the peripheral nervous system that transmits impulses from the central nervous system to internal organs.

sistema nervioso autónomo: (pág. 971) parte del sistema nervioso periférico que transmite impulsos desde el sistema nervioso central hacia los órganos internos.

autosome: **(p. 305)** chromosome that is not a sex chromosome.

autotroph (AW tuh trohf): **(p. 41)** organism that captures energy from sunlight or inorganic substances to produce its own food; provides the foundation of the food supply for other organisms; also called a producer.

auxin: **(p. 648)** plant hormone that moves in only one direction away from the site where it was produced and can stimulate the elongation of cells.

axial skeleton: **(p. 941)** one of the two divisions of the human skeleton; includes the bones of the vetebral column, ribs, skull, and sternum.

axon: **(p. 962)** neuron structure that transmits nerve impulses from the cell body to other neurons and muscles.

autosoma: **(pág. 305)** cromosoma que no es un cromosoma sexual.

autótrofo: **(pág. 41)** organismo que captura energía del sol o de sustancias inorgánicas para producir sus propios alimentos; es la base para la alimentación de otros organismos; también llamado productor.

auxina: **(pág. 648)** hormona vegetal que es transportada en una sola dirección, alejándose del sitio donde fue producida, y que estimula la elongación celular.

esqueleto axial: **(pág. 941)** una de las dos divisiones del esqueleto humano e incluye los huesos de la columna vertebral, las costillas, el cráneo y el esternón.

axón: **(pág. 962)** estructura de la neurona que transmite impulsos nerviosos desde el cuerpo de la célula hacia los músculos y otras neuronas.

B

B cell: **(p. 1086)** antibody-producing B lymphocyte.

background extinction: **(p. 122)** gradual process of a species becoming extinct.

bacteria: **(p. 516)** microscopic prokaryotes most are beneficial to humans and to the environment, but a small percentage can cause disease.

base: **(p. 164)** substance that releases hydroxide ions (OH^-) when dissolved in water; a basic solution has a pH greater than 7.

basidiocarp (buh SIH dee oh karp): **(p. 586)** fruiting body of basidiomycetes.

basidiospore: **(p. 586)** haploid spore released by a basidium during reproduction.

basidium: **(p. 586)** club-shaped, spore-producing hypha of basidiomycetes.

behavior: **(p. 908)** the way in which an animal responds to an external or internal stimulus.

benthic zone: **(p. 80)** ocean-floor area consisting of sand, silt, and dead organisms.

biennial: **(p. 621)** plant with a two-year life span.

bilateral (bi LA tuh rul) symmetry: **(p. 700)** body plan that can be divided into mirror images along only one plane through the central axis.

binary fission: **(p. 520)** asexual form of reproduction used by some prokaryotes in which a cell divides into two genetically identical cells.

binocular vision: **(p. 452)** overlapping fields of vision as a result of eyes located on the front of the face—a characteristic of primates.

célula B: **(pág. 1086)** linfocito B productor de antígenos.

extinción tradicional o natural: **(pág. 122)** proceso paulatino de extinción de una especie.

bacterias: **(pág. 516)** procariota microscópico; la mayoría son benéficos a los humanos y al ambiente; sólo un pequeño porcentaje puede causar enfermedades.

base: **(pág. 164)** sustancia que libera iones hidróxido (OH^-) al disolverse en agua; una solución básica tiene un pH mayor que 7.

basidiocarpo: **(pág. 586)** órgano productor de esporas en los basidiomicetos.

basidiospora: **(pág. 586)** espora haploide liberada por un basidio durante la reproducción .

basidio: **(pág. 586)** hifa productora de esporas de los basidiomicetos; tiene forma de maza.

comportamiento: **(pág. 908)** manera en que un animal responde a un estímulo externo o interno.

zona béntica: **(pág. 80)** zona del fondo marino formada por arena, limo y organismos muertos.

bienal: **(pág. 621)** planta con un ciclo de vida de dos años.

simetría bilateral: **(pág. 700)** plan corporal que se puede dividir en imágenes especulares, a lo largo de un solo plano a través del eje central.

fisión binaria: **(pág. 520)** forma de reproducción asexual de algunos procariotas, en la cual la célula se divide en dos células genéticamente idénticas.

visión binocular: **(pág. 452)** campos de visión sobrepuestos como resultado de ojos ubicados enfrente de la cara: una característica de los primates.

binomial nomenclature (bi NOH mee ul • NOH mun klay chur): (p. 485) Linnaeus's system of naming organisms, which gives a scientific two-word Latin name to each species—the first part is the genus name and the second is the specific epithet.

biodiversity: (p. 116) number of different species living in a specific area.

biogeochemical cycle: (p. 45) exchange of matter through the biosphere involving living organisms, chemical processes, and geological processes.

biogeography: (p. 428) study of the distribution of plants and animals on Earth.

bioindicator: (p. 588) living organism that is sensitive to environmental conditions and is one of the first to respond to changes.

bioinformatics: (p. 375) field of study that creates and maintains databases of biological information, especially genomic data.

biological augmentation: (p. 135) technique of adding essential materials to a degraded ecosystem.

biological community: (p. 36) all the interacting populations of different species that live in the same geographic location at the same time.

biological magnification: (p. 126) increasing concentration of toxic substances, such as DDT, in organisms as trophic levels increase in food chains or food webs.

biology: (p. 4) science of life; examines how living things interact, how systems function, and how they function at a molecular level.

bioluminescent: (p. 555) able to emit light.

biomass: (p. 44) total mass of living matter at each trophic level.

biome: (p. 36) large group of ecosystems that share the same climate and have similar types of communities.

bioremediation: (p. 134) technique using living organisms to detoxify a polluted area.

biosphere (BI uh sfihr): (p. 34) relatively thin layer of Earth and its atmosphere that supports life.

biotic (by AH tihk) factor: (p. 35) any living factor in an organism's environment.

bipedal: (p. 463) walking upright on two legs.

blastocyst: (p. 1055) a modified blastula whose inner cell mass will develop into a fetus.

blastula (BLAS chuh luh): (p. 696) fluid-filled ball of cells formed by mitotic cell division of the embryo.

nomenclatura binaria: (pág. 485) sistema desarrollado por Linneo para nombrar los organismos, en que se otorga un nombre científico de dos palabras a cada especie, la primera palabra es el género y la segunda es la especie.

biodiversidad: (pág. 116) número de especies diferentes que viven en un área determinada.

ciclo biogeoquímico: (pág. 45) intercambio de material a través de la biosfera en que participan seres vivos, procesos químicos y proceso geológicos.

biogeografía: (pág. 428) estudio de la distribución de plantas y animales en la Tierra.

bioindicador: (pág. 588) organismo vivo que es sensible a las condiciones ambientales y que es uno de los primeros en responder a los cambios.

bioinformática: (pág. 375) campo de estudio en que se crean y mantienen bases de datos con información biológica, particularmente genética.

bioaumento: (pág. 135) técnica en que se agregan materiales esenciales a un ecosistema degradado.

comunidad biológica: (pág. 36) todas las poblaciones de diferentes especies que interactúan y viven en una misma zona geográfica, al mismo tiempo.

amplificación biológica: (pág. 126) aumento en la concentración de sustancias tóxicas (como el DDT) en los organismos, a medida que aumenta el nivel trófico de cadenas y redes alimenticias.

biología: (pág. 4) ciencia que estudia la vida; examina las interacciones entre los seres vivos, el funcionamiento de sus sistemas y el funcionamiento a nivel molecular.

bioluminiscencia: (pág. 555) capacidad de emitir luz.

biomasa: (pág. 44) masa total de materia viva en cada nivel trófico.

bioma: (pág. 36) gran grupo de ecosistemas que comparte un mismo clima y que posee comunidades similares.

biorremediación: (pág. 134) técnica en que se usan organismos vivos para descontaminar un área.

biosfera: (pág. 34) capa relativamente delgada de la Tierra y su atmósfera que mantiene la vida.

factor biótico: (pág. 35) todo factor vivo en el ambiente de un organismo.

bípedo: (pág. 463) que camina erguido sobre dos piernas.

blastocisto: (pág. 1055) blástula modificada; sus células internas se convierten más tarde en el feto.

blástula: (pág. 696) esfera llena de fluido que se forma a partir de células originadas por la división mitótica de las células del embrión.

book lung (p. 767) in spiders and some other arthropods, the respiratory structure with highly folded walls whose membranes look like book pages.

filotráquea: (pág. 767) estructura respiratoria de arañas y otros artrópodos; estructura respiratoria con paredes muy plegadas, cuyas membranas semejan las páginas de un libro.

boreal forest: (p. 68) biome south of the tundra with dense evergreen forests and long, cold, dry winters.

bosque boreal: (pág. 68) bioma situado al sur de la tundra, tiene densos bosques de siempreverdes e inviernos largos, fríos y secos.

bottleneck: (p. 433) process in which a large population declines in number, then rebounds.

cuello de botella: (pág. 433) proceso en que el número de miembros de una población declina y luego aumenta.

breathing: (p. 1000) mechanical movement of air into and out of the lungs.

respiración: (pág. 1000) movimiento mecánico de entrada y salida de aire de los pulmones.

bronchus (BRAHN kuhs): (p. 1001) one of the two large tubes that carries air to the lungs.

bronquio: (pág. 1001) uno de los dos grandes conductos por el que se introduce aire a los pulmones.

buffer: (p. 165) mixture that can react with an acid or a base to maintain the pH within a specific range.

amortiguador: (pág. 165) mezcla que puede reaccionar con un ácido o una base y mantener un pH dentro de cierto rango.

C

calcitonin (kal suh TOH nun): (p. 1034) thyroid hormone involved in regulation of blood calcium levels.

calcitonina: (pág. 1034) hormona tiroidea que participa en la regulación de los niveles de calcio en la sangre.

Calorie: (p. 1025) unit used to measure the energy content of food; 1 Calorie equals 1 kilocalorie, or 1000 calories.

Caloría: (pág. 1025) unidad de medida de la energía que contienen los alimentos; 1 Caloría equivale a 1 kilocaloría ó 1000 calorías.

Calvin cycle: (p. 226) light-independent reactions during phase two of photosynthesis in which energy is stored in organic molecules as glucose.

ciclo de Calvin: (pág. 226) reacciones independiente de la luz de la fase II de la fotosíntesis, en que la energía es almacenada como glucosa en moléculas orgánicas.

Cambrian explosion: (p. 398) rapid diversification of most major animal groups during the Paleozoic era.

explosión del Cámbrico: (pág. 398) rápida diversificación de la mayoría de los grupos animales que sucedió en la era Paleozoica.

camouflage (KA muh flahj): (p. 428) morphological adaptations that allow organisms to blend into their surroundings.

camuflaje: (pág. 428) adaptaciones morfológicas que permiten a los organismos disimularse en su ambiente.

cancer: (p. 254) uncontrolled growth and division of cells that can be caused by changes in control of the cell cycle and also may be caused by environmental factors.

cáncer: (pág. 254) crecimiento y división descontrolados de células que puede ser producido por cambios en ciclo celular o también puede ser causado por factores ambientales.

capillary: (p. 993) microscopic, one-cell-wall thick blood vessel where exchange of materials occurs between blood and body cells.

capilares: (pág. 993) vaso sanguíneo microscópico con paredes de una célula de grosor, en que sucede el intercambio de materiales entre la sangre y las células corporales.

capsid: (p. 526) outer protein layer that surrounds the genetic material of a virus.

cápsida: (pág. 526) capa externa proteica que envuelve el material genético de un virus.

capsule: (p. 518) polysaccharide layer secreted around the cell wall by some prokaryotes that prevents the cell from drying out and helps the cell attach to environmental surfaces.

cápsula: (pág. 518) capa de polisacáridos secretada alrededor de la pared celular de algunos procariotas que evita la deshidratación de la célula y que la ayuda a fijarse sobre superficies del ambiente.

carapace (KAR ah pays): (p. 857) dorsal part of a turtle's shell.

caparazón: (pág. 857) parte dorsal de la concha de una tortuga.

carbohydrate: (p. 168) organic compound containing carbon, hydrogen, and oxygen in a ratio of one oxygen and two hydrogen atoms for each carbon atom.

carbohidrato: (pág. 168) compuesto orgánico que contiene carbono, hidrógeno y oxígeno, en una razón de un átomo de oxígeno y dos átomos de hidrógeno, por cada átomo de carbono.

carcinogen (kar SIH nuh jun): (p. 254) cancer-causing substance.

carcinógeno: (pág. 254) sustancia que causa cáncer.

cardiac muscle: (p. 947) involuntary muscle found only in the heart.

músculo cardíaco: (pág. 947) músculo involuntario que sólo se halla en el corazón.

carnivore (KAR nuh vor): (p. 41) heterotroph that preys on other heterotrophs.

carnívoro: (pág. 41) heterótrofo que se alimenta de otros heterótrofos.

carrier: (p. 296) individual heterozygous for a recessive disorder such as cystic fibrosis or Tay-Sachs disease.

portador: (pág. 296) individuo heterocigoto para un trastorno recesivo como la fibrosis quística o la enfermedad de Tay-Sachs.

carrying capacity: (p. 98) largest number of individuals in a species that an environment can support long-term.

capacidad de carga: (pág. 98) el número mayor de individuos de una misma especie que un ambiente puede mantener a largo plazo.

cartilage (KAR tuh lihj): (p. 820) flexible, tough material that makes up vertebrate skeletons or parts of vertebrate skeletons.

cartílago: (pág. 820) material flexible y duro que forma todo o parte del esqueleto en los vertebrados.

caste: (p. 779) specialized group of individuals in an insect society that performs specific tasks.

casta: (pág. 779) grupo especializado de individuos en una sociedad de insectos que se encarga de realizar tareas específicas.

catalyst: (p. 159) substance that speeds up a chemical reaction by reducing the activation energy.

catalizador: (pág. 159) sustancia que acelera una reacción química al reducir energía de activación.

cell: (p. 182) basic unit of structure and organization of all living organisms.

célula: (pág. 182) unidad básica de estructura y organización de todos los seres vivos.

cell body: (p. 962) neuron structure that contains the nucleus and many organelles.

cuerpo celular: (pág. 962) estructura de la neurona que contiene el núcleo y muchos organelos.

cell cycle: (p. 246) process of cellular reproduction, occurring in three main stages—interphase (growth), mitosis (nuclear division), and cytokinesis (cytoplasm division).

ciclo celular: (pág. 246) proceso de reproducción celular; consta de tres fases principales: interfase (crecimiento), mitosis (división nuclear) y citoquinesis (división del citoplasma).

cell theory: (p. 183) states that (1) organisms are made of one or more cells; (2) cells are the basic unit of life; and (3) all cells come only from other cells.

teoría celular: (pág. 183) establece que (1) los organismos están formados por una o más células; (2) las células son la unidad básica de la vida y (3) todas las células provienen de otras células.

cell wall: (p. 198) in plants, the rigid barrier that surrounds the outside of the plasma membrane, is made of cellulose, and provides support and protection to the cell.

pared celular: (pág. 198) barrera rígida que rodea el exterior de la membrana plasmática de las plantas; está formada por celulosa y brinda soporte y protección a la célula.

cellular respiration: (p. 220) catabolic pathway in which organic molecules are broken down to release energy for use by the cell.

respiración celular: (pág. 220) vía catabólica en que se desdoblan moléculas orgánicas a fin de obtener energía para la célula.

central nervous system: (p. 968) consists of the brain and spinal cord and coordinates all of the body's activities.

sistema nervioso central: (pág. 968) está formado por el encéfalo y la medula espinal y coordina todas las actividades del cuerpo.

centriole: (p. 196) organelle that plays a role in cell division and is made of microtubules.

centríolo: (pág. 196) organelo formado por microtúbulos y que participa en la división celular.

centromere: (p. 248) cell structure that joins two sister chromatids.

centrómero: (pág. 248) estructura celular que une a dos cromátides hermanas.

cephalization (sef uh luh ZA shun): (p. 700) tendency to concentrate sensory organs and nervous tissue at an animal's anterior end.

cefalización: (pág. 700) tendencia a concentrar los órganos de los sentidos y el tejido nervioso en el extremo anterior del cuerpo del animal.

cephalothorax (sef uh luh THOR aks): (p. 763) in arthropods, the structure formed from the thorax region fused with the head.

cefalotórax: (pág. 763) estructura de los artrópodos formada por la fusión del tórax con la cabeza.

cerebellum: (p. 886) part of the brain responsible for balance and coordination.

cerebelo: (pág. 886) parte del encéfalo encargada del equilibrio y la coordinación.

cerebral cortex: (p. 886) highly folded outer layer of the cerebrum that is responsible for coordinating conscious activities, memory, and the ability to learn.

corteza cerebral: (pág. 886) capa exterior del cerebro que posee muchos pliegues y que se encarga de coordinar las actividades conscientes, la memoria y la capacidad de aprender.

cerebrum (suh REE brum): (p. 969) largest part of the brain; is divided into two hemispheres and carries out higher thought processes involved with language, learning, memory, and voluntary body movements.

cerebro: (pág. 969) órgano más grande del encéfalo; se divide en dos hemisferios y realiza los procesos más complejos de pensamiento relacionados con el lenguaje, el aprendizaje, la memoria y los movimientos voluntarios del cuerpo.

character: (p. 492) inherited morphological or biochemical feature that varies among species and can be used to determine patterns of descent.

carácter: (pág. 492) característica morfológica o bioquímica heredada que varía entre especies y que sirve para determinar patrones de la herencia.

chelicera (kih LIH suh ruh): (p. 771) one of a pair of arachnid appendages modified to function as fangs or pincers.

quelíceros: (pág. 771) uno de los pares de apéndices de los arácnidos que están modificados para funcionar como colmillos o tenazas.

cheliped: (p. 771) in most crustaceans, the first pair of legs, which has large claws to trap and crush food.

quelípedo: (pág. 771) el primer par de patas de los crustáceos y que consiste en grandes tenazas que sirven para atrapar y triturar los alimentos.

chemical digestion: (p. 1020) chemical breakdown of food by digestive enzymes such as amylase into smaller molecules that cells can absorb.

digestión química: (pág. 1020) desdoblamiento químico de los alimentos por enzimas digestivas, como la amilasa, en moléculas más pequeñas que las células puedan absorber.

chemical reaction: (p. 156) energy-requiring process by which atoms or groups of atoms in substances are changed into different substances.

reacción química: (pág. 156) proceso que requiere energía, en que los átomos o grupos de átomos se convierten en diferentes sustancias.

chemotaxis (KEE moh taks us): (p. 664) movement of a cell or organism towards or away from a particular chemical.

quimiotaxis: (pág. 664) en los musgos, es el movimiento de respuesta de un espermatozoide que es atraído por una sustancia producida por el arquegonio.

chitin (KI tun): (p. 577) tough, flexible polysaccharide in the exoskeletons of insects and crustaceans and in fungal cell walls.

quitina: (pág. 577) polisacárido fuerte y flexible del exoesqueleto de insectos y crustáceos y de la pared celular de los hongos.

chloroplast: (p. 197) double-membrane organelle that captures light energy and converts it to chemical energy through photosynthesis.

cloroplasto: (pág. 197) organelo de doble membrana que captura la energía de la luz y la convierte en energía química mediante la fotosíntesis.

chordate: (p. 803) animal of the phylum Chordata having a dorsal tubular nerve cord, a notochord, pharyngeal pouches, and a postanal tail at some point in its development.

cordado: (pág. 803) animal perteneciente al filo Chordata; presenta cordón tubular nervioso dorsal, notocordio, somitas y cola postanal, en algún momento de su desarrollo.

chromatin (KROH muh tun) (p. 247) relaxed form of DNA in the nucleus of a cell.

cromatina: (pág. 247) forma no condensada de DNA en el núcleo de una célula.

chromosome (KROH muh sohm) (p. 247) DNA-containing structure that carries genetic material from one generation to another.

cromosoma: (pág. 247) estructura que contiene DNA y que lleva el material genético de una generación a la siguiente.

cilium: (p. 198) short, hairlike projection that functions in cell movement.

cilio: (pág. 198) extensión corta y filiforme que funciona en la locomoción celular.

circadian (sur KAY dee uhn) rhythm: (p. 919) cycle that occurs daily, such as sleeping and waking.

ritmo circadiano: (pág. 919) ciclo que sucede diariamente, como dormir y despertar.

cladistics (kla DIHS tiks): (p. 495) taxonomic method that models evolutionary relationships based on shared derived characters and phylogenetic trees.

cladogram (KLA duh gram): (p. 496) diagram with branches that represents the hypothesized phylogeny or evolution of a species or group; uses bioinformatics, morphological studies, and information from DNA studies.

class: (p. 488) taxonomic group that contains one or more related orders.

classical conditioning: (p. 913) learned behavior that occurs when an association is made between two different kinds of stimuli.

classification: (p. 484) grouping of organisms or objects based on a set of criteria that helps organize, communicate, and retain information.

climate: (p. 66) average weather conditions in a specific area, determined by latitude, elevation, ocean currents, and other factors.

climax community: (p. 63) stable, mature ecological community with little change in the composition of species.

clitellum: (p. 748) thickened band of segments that produce a cocoon from which young earthworms hatch.

cloaca (kloh AY kuh): (p. 835) in amphibians, the chamber that receives digestive waste, urinary waste, and eggs or sperm before they leave the body.

cloning: (p. 367) process in which large numbers of identical recombinant DNA molecules are produced.

closed circulatory system: (p. 739) blood is confined to the vessels as it moves through the body.

cnidocyte (NI duh site): (p. 710) nematocyst-containing stinging cell on a cnidarian's tentacle.

cochlea (KOH klee uh): (p. 974) snail-shaped, sound-sensitive, inner ear structure filled with fluid and lined with hair cells; generates nerve impulses sent to the brain through the auditory nerve.

codominance: (p. 302) complex inheritance pattern that occurs when neither allele is dominant and both alleles are expressed.

codon: (p. 338) three-base code in DNA or RNA.

coelom (SEE lum): (p. 701) fluid-filled body cavity completely surrounded by mesoderm.

cognitive behavior: (p. 915) learned behavior that involves thinking, reasoning, and information processing.

cladística: (pág. 495) método taxonómico que modela las relaciones evolutivas basándose en caracteres derivados compartidos y árboles filogenéticos.

cladograma: (pág. 496) diagrama con ramas que representan la filogenia hipotética, o evolución, de una especie o grupo; utiliza la bioinformática, los estudios morfológicos y la información proveniente de estudios del DNA.

clase: (pág. 488) grupo taxonómico que contiene uno o más órdenes relacionados.

condicionamiento clásico: (pág. 913) comportamiento adquirido que sucede cuando se establece una asociación entre dos diferentes tipos de estímulo.

clasificación: (pág. 484) agrupamiento de organismos u objetos en base a una serie de criterios y que permite organizar, comunicar y retener información.

clima: (pág. 66) condiciones meteorológicas promedio en un área específica; son determinadas por la latitud, la elevación, las corrientes oceánicas y otros factores.

comunidad clímax: (pág. 63) comunidad ecológica madura y estable que presenta pocos cambios en el número de especies.

clitelo: (pág. 748) banda de segmentos engrosados que produce las cápsulas de las que eclosionan las nuevas lombrices de tierra.

cloaca: (pág. 835) en los anfibios, es la cavidad que recibe los desechos digestivos y urinarios, así como los huevos o el esperma, antes de que sean expulsados del cuerpo.

clonación: (pág. 367) proceso en que se producen grandes cantidades de moléculas idénticas de DNA recombinante.

sistema circulatorio cerrado: (pág. 739) sistema en que la sangre queda confinada en el interior de vasos cuando se desplaza a través del cuerpo.

cnidocito: (pág. 710) células urticantes, en los tentáculos de los cnidarios, que contienen nematocistos.

cóclea: (pág. 974) estructura del oído interno, con forma de caracol y sensible al sonido, que está llena de un fluido y revestida con células ciliadas; genera impulsos nerviosos que envía al encéfalo a través del nervio auditivo.

codominancia: (pág. 302) patrón hereditario complejo que sucede cuando ninguno de los alelos es dominante y ambos se expresan.

codón: (pág. 338) código de tres bases del DNA o el RNA.

celoma: (pág. 701) cavidad corporal llena de fluido y completamente rodeada por el mesodermo.

comportamiento cognitivo: (pág. 915) comportamiento adquirido que incluye razonamiento, pensamiento y procesamiento de información.

collenchyma cell: **(p. 633)** often elongated plant cell that provides flexibility for the plant, support for surrounding tissues, and functions in tissue repair and replacement.

célula colenquimatosa: **(pág. 633)** células vegetales, a menudo alargadas, que proveen flexibilidad a la planta, sostén a los tejidos que lo rodean y que funcionan como tejido para reparar y sustituir otros tejidos.

colony: **(p. 557)** group of cells or organisms that join together, forming a close association.

colonia: **(pág. 557)** grupo de células que se unen y establecen una asociación muy estrecha.

commensalism (kuh MEN suh lih zum): **(p. 40)** symbiotic relationship in which one organism benefits and the other organism is neither helped nor harmed.

comensalismo: **(pág. 40)** relación simbiótica en que un organismo se beneficia, mientras que el otro no obtiene beneficios pero tampoco es perjudicado.

community: **(p. 60)** group of interacting populations that live in the same geographic area at the same time.

comunidad: **(pág. 60)** grupo de poblaciones que interactúan y que viven en la misma región geográfica al mismo tiempo.

compact bone: **(p. 942)** strong, dense outer bone layer that contains Haversian systems.

hueso compacto: **(pág. 942)** capa de hueso externa, más fuerte y más densa, que contiene los canales de Havers.

companion cell: **(p. 638)** nucleated cell that helps the mature sieve tube member function in transporting dissolved substances in the phloem of vascular plants.

célula acompañante: **(pág. 638)** célula con núcleo que ayuda a los tubos cribosos maduros a realizar su función, en el transporte de sustancias disueltas, en el floema de las plantas vasculares.

complement protein: **(p. 1085)** protein in blood plasma that enhances phagocytosis.

complemento: **(pág. 1085)** proteínas del plasma de la sangre que estimulan la fagocitosis.

compound: **(p. 151)** pure substance with unique properties; formed when two or more different elements combine.

compuesto: **(pág. 151)** sustancia pura con propiedades particulares y que se forma cuando se combinan dos o más elementos.

cone: **(p. 618)** feature that contains male or female reproductive structures of cycads and other gymnosperms. **(p. 974)** a type of cell in the retina of the eye that is responsible for sharp vision in bright light and seeing color.

cono: **(pág. 618)** estructura que contiene estructuras masculinas o femeninas en las cicadáceas y otras gimnospermas. **(pág. 974)** tipo de célula en la retina del ojo responsable de la visión nítida en luz brillante y de la visión a color.

conidiophore (koh NIH dee uh for): **(p. 584)** spore-producing hypha of sac fungi.

conidióforo: **(pág. 584)** hifa de los ascomicetos que produce las esporas.

conjugation: **(p. 520)** form of reproduction used by some prokaryotes in which the prokaryotic cells attach to each other and exchange genetic material.

conjugación: **(pág. 520)** forma de reproducción de algunos procariotas en que dos células procariotas se conectan e intercambian material genético.

constant: **(p. 19)** a factor that remains fixed during an experiment while the independent and dependent variables change.

constante: **(pág. 19)** factor que permanece fijo durante un experimento mientras que las variables independiente y dependiente cambian.

contour feather: **(p. 862)** barbed feather that covers a bird's body, wings, and tail and forms the body contour.

pluma de contorno: **(pág. 862)** plumas barbadas que cubren el cuerpo, las alas y la cola de un ave y que dan contorno al cuerpo del ave.

contractile vacuole: **(p. 547)** organelle that collects excess water in the cytoplasm and expels it from the cell; maintains homeostasis in hypotonic environments.

vacuola contráctil: **(pág. 547)** organelo que recoge el exceso de agua en el citoplasma y lo expulsa de la célula; mantiene la homeostasis en ambientes hipotónicos.

control group: **(p. 19)** in a controlled experiment, the group not receiving the factor being tested.

grupo control: **(pág. 19)** en un experimento controlado, el grupo al que no se aplica el factor que se está probando.

cork cambium: **(p. 634)** meristematic tissue that produces cells with tough cell walls that form the protective outside layer on stems and roots.

cambio suberoso: **(pág. 634)** tejido meristemático que produce células con fuertes paredes celulares y que forma la capa protectora en el exterior de tallos y raíces.

cortex: **(p. 639)** layer composed of ground tissues between the epidermis and vascular tissue of a root.

corteza: **(pág. 639)** capa formada por tejido fundamental, situada entre la epidermis y el tejido vascular de la raíz.

cortisol: **(p. 1035)** a glucocorticoid that raises blood glucose levels, reduces inflammation, and is produced by the adrenal cortex.

cortisol: **(pág. 1035)** un glucocorticoide que eleva el nivel de glucosa en sangre, reduce la inflamación y que es producido por la corteza suprarrenal.

cotyledon (kah tuh LEE dun): (p. 617) seed structure that stores food or helps absorb food for the sporophyte of vascular seed plants.

cotiledón: (pág. 617) estructura de la semilla que almacena alimentos o que ayuda a absorber alimentos para el esporofito de la semilla de una planta vascular.

courting behavior: (p. 921) species-specific series of movements or sounds to attract a mate.

comportamiento de cortejo: (pág. 921) serie de movimientos o sonidos específicos de cada especie, los cuales sirven para atraer a una pareja.

covalent bond: (p. 152) type of chemical bond formed when atoms share electrons.

enlace covalente: (pág. 152) tipo de enlace químico que se forma cuando los átomos comparten electrones.

Cro-Magnon: (p. 473) a species also referred to as *Homo sapiens;* seem to have replaced Neanderthals.

Cromañón: (pág. 473) especie también conocida como *Homo sapiens sapiens;* parece haber reemplazado a los Neandertales.

crop: (p. 746) sac in which food and soil are stored until they pass to the earthworm's gizzard.

buche: (pág. 746) saco en que se almacenan los alimentos y el suelo hasta que pasan a la molleja de una lombriz de tierra.

crossing over: (p. 272) exchange of chromosomal segments between a pair of homologous chromosomes during prophase I of meisois.

entrecruzamiento: (pág. 272) intercambio de segmentos de cromosomas entre un par de cromosomas homólogos, el cual ocurre durante la profase I de la meiosis.

cyclin: (p. 253) one of the specific proteins that regulate the cell cycle.

ciclina: (pág. 253) una de las proteínas específicas que regulan el ciclo celular.

cyclin-dependent kinase: (p. 253) enzyme to which cyclin binds during interphase and mitosis, triggering and controlling activities during the cell cycle.

quinasa dependiente de la ciclina: (pág. 253) enzima a la que se une la ciclina durante la interfase y la mitosis, iniciando y controlando, de este modo, las actividades del ciclo celular.

cytokinesis (si toh kih NEE sis): (p. 246) third main stage of the cell cycle, during which the cell's cytoplasm divides, creating a new cell.

citoquinesis: (pág. 246) tercera etapa del ciclo celular; en esta etapa el citoplasma de la célula se divide y se origina una nueva célula.

cytokinin (si tuh KI nihn): (p. 650) plant hormone that promotes cell division by stimulating production of proteins required for mitosis and cytokinesis.

citoquinina: (pág. 650) hormona vegetal que promueve la división celular al estimular la producción de las proteínas que se requieren para la mitosis y la citoquinesis.

cytoplasm: (p. 191) semifluid material inside the cell's plasma membrane.

citoplasma: (pág. 191) material semifluido que está rodeado por la membrana plasmática de la célula.

cytoskeleton: (p. 191) supporting network of protein fibers that provide a framework for the cell within the cytoplasm.

citoesqueleto: (pág. 191) red de fibras proteicas de soporte que provee una estructura para la célula, dentro del citoplasma.

cytotoxic T cell: (p. 1088) lymphocyte that destroys pathogens and releases cytokines when activated.

linfocito T citotóxico: (pág. 1088) linfocito que al ser activado, destruye patógenos y libera citoquinas.

D

data: (p. 19) quantitative or qualitative information gained from scientific investigation.

datos: (pág. 19) información cualitativa o cuantitativa obtenida durante una investigación científica.

day-neutral plant: (p. 673) plant that flowers over a wide range in the number of hours of darkness.

planta de días neutros: (pág. 673) planta que florece bajo un amplio rango de horas de oscuridad.

degenerative (di JEH nuh ruh tihv) disease: (p. 1092) noninfectious disease, such as arthritis, that results from part of the body wearing out.

enfermedad degenerativa: (pág. 1092) trastorno no infeccioso, como la artritis, que resulta del desgaste de una parte del cuerpo.

demographic transition: (p. 102) population change from high birth rates and death rates to low birth rates and death rates.

transición demográfica: (pág. 102) cambio en una población que pasa de tener altas tasas de natalidad y de mortalidad, a tener bajas tasas de natalidad y de mortalidad.

demography: **(de MAH gra fee)** **(p. 100)** study of human populations based on size, density, movement, distribution, and birth and death rates.

dendrite: **(p. 962)** neuron structure that receives nerve impulses from other neurons and transmits them to the cell body.

denitrification: **(p. 48)** process in which fixed nitrogen compounds are converted back into nitrogen gas and returned to the atmosphere.

density-dependent factor: **(p. 95)** environmental factor, such as predation, disease, and competition, that depends on the number of members in a population per unit area.

density-independent factor: **(p. 94)** environmental factor, such as storms and extreme heat or cold, that affects populations regardless of their density.

dependent variable: **(p. 19)** factor being measured in a controlled experiment; its value changes because of changes to the independent variable.

depressant: **(p. 979)** substance/drug that slows down the central nervous system.

derived character: **(p. 495)** morphological or biochemical feature found in one group of a line but not in common ancestors.

derived trait: **(p. 424)** new feature that had not appeared in common ancestors.

dermis: **(p. 937)** skin layer beneath the epidermis; contains nerve cells, muscle fibers, sweat glands, oil glands, and hair follicles.

desert: **(p. 70)** area with low rainfall, whose annual rate of evaporation exceeds its annual rate of precipitation; can support cacti and some grasses and animal species such as snakes and lizards.

detritivore **(duh TRYD tuh vor):** **(p. 42)** heterotroph that decomposes organic material and returns the nutrients to soil, air, and water, making the nutrients available to other organisms.

deuterostome **(DEW tihr uh stohm):** **(p. 702)** coelomate animal whose anus develops from the opening in the gastrula.

development: **(p. 8)** changes an organism undergoes in its lifetime before reaching its adult form.

diaphragm: **(p. 885)** sheet of muscle beneath the lungs that separates the mammalian chest cavity from the abdominal cavity.

diffusion: **(p. 201)** net movement of particles from an area of higher concentration to an area of lower concentration.

dilation **(di LAY shun):** **(p. 1062)** the opening of the cervix during labor.

demografía: **(pág. 100)** estudio de las poblaciones humanas en base al tamaño, la densidad, el movimiento, la distribución y las tasas de natalidad y mortalidad de dichas poblaciones.

dendrita: **(pág. 962)** estructura de la neurona que recibe impulsos nerviosos de otras neuronas y que luego los transmite hacia el cuerpo de la neurona.

desnitrificación: **(pág. 48)** proceso en que los compuestos de nitrógeno fijado son convertidos a gas nitrógeno y devueltos a la atmósfera.

factor dependiente de la densidad: **(pág. 95)** factor ambiental, como la depredación, las enfermedades y la competencia, que depende del número de miembros de la población por unidad de área.

factor independiente de la densidad: **(pág. 94)** factor ambiental, como las tormentas y el calor o el frío extremos, que afectan a las poblaciones independientemente de su densidad.

variable dependiente: **(pág. 19)** factor que se mide en un experimento controlado; su valor cambia de acuerdo con los cambios en la variable independiente.

depresor: **(pág. 979)** sustancia o droga que disminuye la actividad del sistema nervioso central.

carácter derivado: **(pág. 495)** característica morfológica o bioquímica presente en un grupo de un linaje, pero no en los antepasados comunes.

rasgo derivado: **(pág. 424)** nueva característica que no aparece en antepasados comunes.

dermis: **(pág. 937)** capa de la piel situada bajo la epidermis; contiene células nerviosas, fibras musculares, glándulas sudoríparas y folículos pilosos.

desierto: **(pág. 70)** área con lluvias escasas y en que la tasa anual de evaporación excede la tasa anual de precipitación; es la morada de cactos, pastos y especies animales como serpientes y lagartijas.

detritívoro: **(pág. 42)** heterótrofo que descompone material orgánico y devuelve los nutrientes al suelo, al aire y al agua, poniendo los nutrientes a disposición de otros organismos.

deuterostomado: **(pág. 702)** animal celomado cuyo ano se desarrolla a partir de la apertura de la gástrula.

desarrollo: **(pág. 8)** cambios que sufre un organismo a lo largo de su vida, hasta alcanzar la vida adulta.

diafragma: **(pág. 885)** banda de músculos situada bajo los pulmones y que separa, en los mamíferos, el pecho de la cavidad abdominal.

difusión: **(pág. 201)** movimiento neto de partículas de una región de mayor concentración hacia una región de menor concentración.

dilatación: **(pág. 1062)** apertura del cuello uterino durante el parto.

diploid: **(p. 271)** having two copies of each chromosome (2*n*).

directional selection: **(p. 435)** shift of a population toward an extreme version of a beneficial trait.

dispersion: **(p. 92)** arrangement of a population in its environment.

disruptive selection: **(p. 436)** process in which individuals with average traits are removed, creating two populations with extreme traits.

diurnal: **(p. 452)** organisms that are active during the day.

division: **(p. 488)** taxonomic term used instead of *phylum* to group related classes of plants and bacteria.

DNA fingerprinting: **(p. 373)** separating an individual's unique sequence of DNA fragments to observe distinct banding patterns; can be used by forensic scientists to identify suspects and determine paternity.

DNA ligase: **(p. 366)** enzyme that chemically links DNA fragments together.

DNA microarray: **(p. 375)** silicon chips or microscope slides with DNA fragments that can allow many genes in a genome to be studied simultaneously.

DNA polymerase: **(p. 334)** enzyme that catalyzes synthesis of new DNA molecules.

domain: **(p. 488)** taxonomic group of one or more kingdoms.

dominance hierarchy **(DAH muh nunts • HI rar kee):** **(p. 917)** ranking system in which the top-ranked animal gets access to resources without conflict from others in the group.

dominant: **(p. 278)** Mendel's name for a specific trait that appeared in the F1 generation.

dopamine: **(p. 978)** neurotransmitter in the brain involved with feelings of pleasure, control of body movement, and other functions.

dormancy: **(p. 679)** period of little or no growth that varies from species to species; in plants, an adaptation that increases the survival rate of seeds in harsh environments.

dorsal **(DOR sul):** **(p. 700)** backside of an animal with bilateral symmetry.

dorsal tubular nerve cord: **(p. 803)** tube-shaped chordate nerve cord located above the digestive organs.

double helix: **(p. 330)** twisted-ladder shape of DNA, formed by two nucleotide strands twisted around each other.

down feather: **(p. 862)** soft feather beneath a bird's contour feathers that provides insulation by trapping air.

diploide: **(pág. 271)** células con dos copias de cada cromosoma (2*n*).

selección direccional: **(pág. 435)** cambio en una población hacia una versión extrema de un rasgo benéfico.

dispersión: **(pág. 92)** diseminación de una población en su ambiente.

selección disruptiva: **(pág. 436)** proceso en que los individuos con rasgos promedio son eliminados, creando dos poblaciones con rasgos extremos.

diurno: **(pág. 452)** organismos activos durante el día.

división: **(pág. 488)** término taxonómico que se usa en vez de *phylum* para agrupar clases relacionadas de plantas y bacterias.

huella genética: **(pág. 373)** separación de las secuencias de fragmentos de DNA propias de un individuo, para obtener su patrón único de bandas; se puede usar en estudios forenses para identificar a sospechosos o en estudios de paternidad.

DNA ligasa: **(pág. 366)** enzima que une químicamente entre sí, fragmentos de DNA.

micromatrices de DNA: **(pág. 375)** chips de silicio, o placas microscópicas con fragmentos de DNA que permiten el estudio simultáneo de todos los genes de un genoma.

DNA polimerasa: **(pág. 334)** enzima que cataliza la síntesis de nuevas moléculas de DNA.

dominio: **(pág. 488)** grupo taxonómico formado por uno o más reinos.

jerarquía de dominancia: **(pág. 917)** sistema de rango en que los animales de mayor jerarquía obtienen acceso a los recursos, sin conflictos con los otros miembros del grupo.

dominante: **(pág. 278)** nombre que dio Mendel a rasgos específicos que aparecían en la generación F1.

dopamina: **(pág. 978)** neurotransmisor cerebral presente en las sensaciones de placer, control de los movimientos del cuerpo y otras funciones.

latencia: **(pág. 679)** período en el cual ocurre muy poco o ningún crecimiento y que varía entre las especies; es una adaptación que aumenta la tasa de supervivencia de las semillas en ambientes hostiles.

dorsal: **(pág. 700)** parte trasera del cuerpo de un animal con simetría bilateral.

cordón nervioso tubular dorsal: **(pág. 803)** cordón nervioso de los cordados, de forma tubular, situado sobre los órganos digestivos.

doble hélice: **(pág. 330)** forma del DNA; semeja una escalera que se tuerce sobre sí misma y está constituida por dos cadenas enroscadas de nucleótidos.

plumón: **(pág. 862)** plumas suaves situadas bajo las plumas de contorno y que, al atrapar aire, proveen aislamiento al ave.

drug: **(p. 977)** natural or artificial substance that alters the body's function.

dynamic equilibrium: **(p. 202)** condition of continuous, random movement of particles but no overall change in concentration of materials.

droga: **(pág. 977)** sustancia natural o artificial que altera las funciones corporales.

equilibrio dinámico: **(pág. 202)** condición en que ocurre movimiento continuo y aleatorio de partículas, sin que haya un cambio general en la concentración de materiales.

E

ecological succession: **(p. 62)** process by which one community replaces another community because of changing abiotic and biotic factors.

ecology: **(p. 32)** scientific study of all the interrelationships between organisms and their environment.

ecosystem: **(p. 36)** biological community and all the nonliving factors that affect it.

ecosystem diversity: **(p. 118)** variety of ecosystems in the biosphere.

ectoderm: **(p. 697)** outer layer of cells in the gastrula that develops into nervous tissue and skin.

ectotherm: **(p. 837)** animal that cannot regulate its body temperature through its metabolism and obtains its body heat from the external environment.

edge effect: **(p. 126)** any different environmental condition occurring along an ecosystem's boundaries.

electron: **(p. 148)** negatively charged particle that occupies space around an atom's nucleus.

element: **(p. 149)** pure substance composed of only one type of atom; cannot be broken down into another substance by physical or chemical means.

embryo: **(p. 426)** organism's early prebirth stage of development.

emigration **(em uh GRAY shun):** **(p. 97)** movement of individuals away from a population.

endemic: **(p. 133)** found only in one specific geographic area.

endemic disease: **(p. 1081)** a disease found in only a few individuals within a population.

endocrine gland: **(p. 1031)** hormone-producing gland that releases its product into the bloodstream.

endocytosis: **(p. 207)** energy-requiring process by which large substances from the outside environment can enter a cell.

endoderm: **(p. 697)** inner layer of cells in the gastrula that develops into digestive organs and the digestive tract lining.

sucesión ecológica: **(pág. 62)** proceso en que una comunidad reemplaza a otra, debido a cambios en los factores bióticos y abióticos.

ecología: **(pág. 32)** ciencia que estudia todas las interrelaciones entre los organismos y su ambiente.

ecosistema: **(pág. 36)** comunidad biológica y todos los factores inanimados que la afectan.

diversidad de ecosistemas: **(pág. 118)** variedad de ecosistemas en la biosfera.

ectodermo: **(pág. 697)** capa exterior de células de la gástrula que origina el tejido nervioso y la piel.

poiquilotermo: **(pág. 837)** animal que no puede regular su temperatura corporal mediante su metabolismo y que obtiene el calor corporal a partir del ambiente externo.

efecto borde: **(pág. 126)** son todas las condiciones ambientales diferentes que suceden a lo largo de los límites de un ecosistema.

electrón: **(pág. 148)** partícula con carga negativa que gira alrededor del núcleo del átomo.

elemento: **(pág. 149)** sustancia pura compuesta por un solo tipo de átomo; no se puede descomponer en otra sustancia por medios físicos ni por medios químicos.

embrión: **(pág. 426)** etapa inicial del desarrollo de un organismo antes del nacimiento.

emigración: **(pág. 97)** salida de individuos de una población.

endémico: **(pág. 133)** que sólo se halla en una región geográfica determinada.

enfermedad endémica: **(pág. 1081)** enfermedad que sólo contraen unos cuantos individuos dentro de una población.

glándula endocrina: **(pág. 1031)** glándula productora de hormonas que libera su producto hacia el torrente sanguíneo.

endocitosis: **(pág. 207)** proceso que requiere energía y que permite la entrada de sustancias muy grandes a la célula.

endodermo: **(pág. 697)** capa interior de células de la gástrula; forma los órganos digestivos y el revestimiento del tracto digestivo.

endodermis: **(p. 640)** cell layer at the inner boundary of the cortex; regulates the material that enters the plant's vascular tissues.

endoplasmic reticulum **(en duh PLAZ mihk • rih TIHK yuh lum):** **(p. 194)** highly folded membrane system in eukaryotic cells that is the site for protein and lipid synthesis.

endoskeleton: **(p. 693)** internal skeleton that protects internal organs, provides support for the organism's body, and can provide an internal brace for muscles to pull against.

endosperm **(EN duh spurm):** **(p. 676)** tissue that provides nourishment to the developing embryo of flowering plants.

endospore: **(p. 521)** dormant bacterial cell able to survive for long periods of time during extreme environmental conditions.

endosymbiont theory: **(p. 406)** explains that eukaryotic cells may have evolved from prokaryotic cells.

endotherm: **(p. 861)** organism that generates its body heat internally by its own metabolism.

energy: **(p. 218)** ability to do work; energy cannot be created or destroyed, only transformed.

enzyme: **(p. 159)** protein that speeds up a biological reaction by lowering the activation energy needed to start the reaction.

epidemic: **(p. 1081)** large outbreak of a particular disease in a specific area.

epidermis: **(p. 636)** dermal tissue that makes up a plant's outer covering. **(p. 936)** in humans and some other animals, the outer superficial layer of skin made up of epithelial cells.

epididymis **(eh puh DIH duh mus):** **(p. 1049)** structure on top of each testis where sperm mature and are stored.

epiphyte: **(p. 614)** plant that lives anchored to an object or to another plant.

epistasis: **(p. 305)** interaction between alleles in which one allele hides the effects of another allele.

era: **(p. 396)** a large division of Earth's geologic time scale that is further divided into one or more periods.

esophagus **(ih SAH fuh gus):** **(p. 1021)** muscular tube that connects the pharynx to the stomach and moves food to the stomach by the process of peristalsis.

estuary **(ES chuh wer ee):** **(p. 78)** unique, transitional ecosystem that supports diverse species and is formed where freshwater and ocean water merge.

ethics: **(p. 15)** a set of values.

ethylene: **(p. 649)** gaseous plant hormone that affects the ripening of fruits.

endodermis: **(pág. 640)** capa de células situada en el límite interior de la corteza y que regula los materiales que entran al tejido vascular de la planta.

retículo endoplásmico: **(pág. 194)** sistema de membranas de las células eucariotas; presenta numerosos pliegues y es el sitio donde ocurre la síntesis de proteínas y lípidos.

endoesqueleto: **(pág. 693)** esqueleto interno que protege los órganos internos, provee soporte al cuerpo del organismo y sirve como punto de apoyo para la contracción de los músculos.

endosperma: **(pág. 676)** tejido que provee alimentos al embrión en desarrollo de las plantas con flores.

endospora: **(pág. 521)** célula bacteriana en estado latente que puede sobrevivir durante largos períodos, bajo condiciones ambientales extremas.

teoría endosimbiótica: **(pág. 406)** propone que las células eucarióticas evolucionaron a partir de células procariotas.

homeotermo: **(pág. 861)** organismo que genera su calor corporal internamente, debido a su metabolismo.

energía: **(pág. 218)** capacidad de realizar trabajo; la energía no se puede crear o destruir, sólo se puede transformar.

enzima: **(pág. 159)** proteína que acelera una reacción biológica, al disminuir energía de activación que se requiere para iniciar la reacción.

epidemia: **(pág. 1081)** diseminación amplia de una enfermedad dada, en un área específica.

epidermis: **(pág. 636)** tejido dérmico que forma la cubierta más externa de una planta. **(pág. 936)** en humanos y algunos otros animales, la capa superficial externa de la piel compuesta por células epiteliales.

epidídimo: **(pág. 1049)** estructura situada en la parte superior del testículo en que los espermatozoides maduran y se almacenan.

epifita: **(pág. 614)** planta que vive sujeta a un objeto o a otra planta.

epistasis: **(pág. 305)** interacción entre alelos en que un alelo oculta el efecto de otro.

era: **(pág. 396)** gran división de la escala del tiempo geológico de la Tierra que incluye uno o más períodos.

esófago: **(pág. 1021)** conducto muscular que conecta la faringe con el estómago; transporta los alimentos hacia el estómago mediante movimientos peristálticos.

estuario: **(pág. 78)** ecosistema único de transición que mantiene gran diversidad de especies y que se forma donde el agua dulce se mezcla con el agua de los mares.

ética: **(pág. 15)** conjunto de valores.

etileno: **(pág. 649)** hormona gaseosa de las plantas que afecta la maduración de los frutos.

eukaryotic cell: (p. 186) unicellular organism with membrane-bound nucleus and organelles; generally larger and more complex than a prokaryotic cell.

eutrophication (yoo troh fih KAY shun): (p. 127) water pollution from nitrogen-rich and phosphorus-rich substances flowing into waterways, causing algal overgrowth.

evolution: (p. 422) hereditary changes in groups of living organisms over time.

exocytosis: (p. 207) energy-requiring process by which a cell expels wastes and secretes substances at the plasma membrane.

exon: (p. 337) in RNA processing, the coding sequence that remains in the final mRNA.

exoskeleton: (p. 693) hard or tough outer covering of many invertebrates that provides support, protects body tissues, prevents water loss, and protects the organism from predation.

experiment: (p. 18) procedure performed in a controlled setting to test a hypothesis and collect precise data.

experimental group: (p. 19) in a controlled experiment, the group receiving the factor being tested.

expulsion stage: (p. 1062) birthing stage during which a baby travels through the birth canal and exits the mother's body.

external fertilization: (p. 695) type of fertilization that occurs when sperm and egg combine outside an animal's body.

external respiration: (p. 1000) gas exchange between the atmosphere and the blood, occurring in the lungs.

extinction: (p. 116) the disappearance of a species when the last of its members dies.

célula eucariota: (pág. 186) organismo unicelular con núcleo y organelos rodeados de membrana; generalmente son más grandes y complejas que las células procariotas.

eutroficación: (pág. 127) contaminación del agua causada por sustancias ricas en nitrógeno y fósforo que fluyen hacia masas de agua y que producen un crecimiento explosivo de algas.

evolución: (pág. 422) cambios hereditarios que sufren grupos de organismos a lo largo del tiempo.

exocitosis: (pág. 207) proceso que requiere energía y que permite a una célula expulsar desechos y secretar sustancias, a través de la membrana plasmática.

exón: (pág. 337) durante el procesamiento de RNA, la secuencia codificadora que queda en el mRNA final.

exoesqueleto: (pág. 693) cubierta exterior dura o de un material resistente que tienen muchos invertebrados; provee sostén, protege los tejidos del cuerpo, evita la pérdida de agua y los protege contra la depredación.

experimento: (pág. 18) procedimiento realizado bajo condiciones controladas, para recopilar datos precisos y probar una hipótesis.

grupo experimental: (pág. 19) grupo al que se aplica el factor que se está probando durante un experimento controlado.

etapa de expulsión: (pág. 1062) etapa del nacimiento durante la cual el bebé pasa a través del canal de parto y sale del cuerpo de la madre.

fecundación externa: (pág. 695) tipo de fecundación que ocurre cuando el esperma y el óvulo se unen fuera del cuerpo del animal.

respiración externa: (pág. 1000) intercambio de gases entre la atmósfera y la sangre, que ocurre en los pulmones.

extinción: (pág. 116) desaparición de una especie que ocurre cuando muere el último de sus miembros.

F

facilitated diffusion: (p. 202) passive transport of ions and small molecules across the plasma membrane by transport proteins.

family: (p. 487) taxonomic group of similar, related genera that is smaller than a genus and larger than an order.

feather: (p. 861) specialized outgrowth of the skin of birds used for flight and insulation.

fermentation: (p. 231) process in which NAD+ is regenerated, allowing cells to maintain glycolysis in the absence of oxygen.

difusión facilitada: (pág. 202) transporte pasivo de iones y moléculas pequeñas, a través de la membrana plasmática, por medio de proteínas transportadoras.

familia: (pág. 487) grupo taxonómico que contiene géneros similares relacionados, está por encima del género y por debajo del orden.

pluma: (pág. 861) estructura especializada que crece sobre la piel de las aves y que sirve para el vuelo y como aislamiento.

fermentación: (pág. 231) proceso de regeneración de NAD^+ que permite a las células realizar la glucólisis en ausencia de oxígeno.

fertilization: (p. 271) process by which haploid gametes combine, forming a diploid cell with $2n$ chromosomes, with n chromosomes from the female parent and n chromosomes from the male parent.

filter feeder: (p. 706) organism that filters small particles from water to get its food.

fin: (p. 822) paddle-shaped structure of a fish or other aquatic animal used for steering, balance, and propulsion.

fitness: (p. 428) measure of a trait's relative contribution to the following generation.

fixed action pattern: (p. 910) innate behavior that occurs in a sequence of specific actions in response to a stimulus.

flagellum: (p. 198) long, tail-like projection with a whiplike motion that helps a cell move through a watery environment.

flame cell: (p. 727) in flatworms, a cilia-lined, bulblike cell that moves water and certain substances into excretory tubules for elimination outside the body.

fluid mosaic model: (p. 190) a plasma membrane with components constantly in motion, sliding past one another within the lipid bilayer.

food chain: (p. 43) simplified model that shows a single path for energy flow through an ecosystem.

food web: (p. 43) model that shows many interconnected food chains and pathways in which energy and matter flow through an ecosystem.

foraging behavior: (p. 918) ecological behavior that involves finding and eating food.

forensics: (p. 15) the field of study that applies science to matters of legal interest and other areas such as archaeology.

fossil: (p. 393) preserved evidence of an organism, often found in sedimentary rock, that provides evidence of past life.

founder effect: (p. 433) random effect that can occur when a small population settles in an area separated from the rest of the population and interbreeds, producing unique allelic variations.

fruiting body: (p. 577) spore-producing fungal reproductive structure.

fungus: (p. 501) unicellular or multicellular eukaryote that is stationary, absorbs nutrients from organic materials in the environment, and has cell walls that contain chitin.

fecundación: (pág. 271) proceso de combinación de gametos haploides que origina una célula diploide con $2n$ cromosomas; n cromosomas provienen de la madre y n cromosoma provienen del padre.

animal filtrador: (pág. 706) organismo que filtra el agua para obtener partículas que le sirven de alimento.

aleta: (pág. 822) estructura de los peces u otros animales acuáticos, con forma de remo, que les sirve para dar dirección a sus movimientos, mantener el equilibrio y lograr propulsión.

aptitud: (pág. 428) medida de la contribución relativa de un rasgo a la siguiente generación.

pauta fija de acción: (pág. 910) comportamiento innato en respuesta a un estímulo que sucede como una secuencia de actos específicos.

flagelo: (pág. 198) filamento largo y móvil que, al sacudirse como un látigo, permite a una célula moverse en un medio acuático.

célula flamígera: (pág. 727) células de las planarias, con forma de bulbo y revestidas con cilios, que transportan agua y ciertas sustancias hacia conductos excretorios para su posterior eliminación del cuerpo.

modelo del mosaico fluido: (pág. 190) membrana plasmática cuyos componentes se encuentran en movimiento constante, deslizándose dentro de la capa doble de lípidos.

cadena alimenticia: (pág. 43) modelo simplificado que muestra una sola vía para el flujo de energía en un ecosistema.

red alimenticia: (pág. 43) modelo que muestra muchas cadenas alimenticias y vías interconectadas a través de las cuales fluyen la materia y la energía en un ecosistema.

comportamiento de forrajeo: (pág. 918) comportamiento ecológico relacionado con la búsqueda y el consumo de alimentos.

medicina forense: (pág. 15) campo de estudio que aplica la ciencia a asuntos de interés legal y otras áreas como la arqueología.

fósil: (pág. 393) pruebas preservadas de un organismo que a menudo se hallan en rocas sedimentarias y que aportan datos y hechos sobre la vida en el pasado.

efecto fundador: (pág. 433) efecto aleatorio que sucede cuando una población pequeña se establece y se entrecruza en una región, separada del resto de la población, produciendo variaciones alélicas únicas.

cuerpo fructífero: (pág. 577) estructura reproductora de los hongos que produce esporas.

hongo: (pág. 501) eucariota sésil, unicelular o multicelular, que absorbe nutrientes de la materia orgánica del ambiente y que tiene una pared celular de quitina.

G

gametangium (ga muh TAN jee um): (p. 583) reproductive hyphal structure of zygomycetes that contains a haploid nucleus.

gametangio: (pág. 583) estructura reproductora de las hifas de los cigomicetos: contiene un núcleo haploide.

gamete: (p. 271) a haploid sex cell, formed during meiosis, that can combine with another haploid sex cell and produce a diploid fertilized egg.

gameto: (pág. 271) célula sexual haploide, formada durante la meiosis, que se puede combinar con otra célula sexual haploide y producir un huevo diploide fecundado.

ganglion: (p. 728) group of nerve-cell bodies that coordinates incoming and outgoing nerve impulses.

ganglio: (pág. 728) conjunto de cuerpos celulares de neuronas que se encargan de coordinar la entrada y salida de impulsos nerviosos.

gastrovascular (gas troh VAS kyuh lur) cavity: (p. 711) in cnidarians, the space surrounded by an inner cell layer, where digestion take place.

cavidad gastrovascular: (pág. 711) en los cnidarios, espacio rodeado por una capa interna de células y en que ocurre la digestión.

gastrula (GAS truh luh): (p. 696) two-cell-layer sac with an opening at one end that forms from the blastula during embryonic development.

gástrula: (pág. 696) saco de dos células de espesor, con una apertura en uno de sus extremos, que se forma a partir de la blástula durante el desarrollo embrionario.

gel electrophoresis: (p. 365) process that involves using electric current to separate certain biological molecules by size.

electroforesis en gel: (pág. 365) proceso en que se usa corriente eléctrica para separar ciertas moléculas biológicas, según su tamaño.

gene: (p. 270) functional unit that controls inherited trait expression that is passed on from one generation to another generation.

gene: (pág. 270) unidad funcional que controla la expresión de un rasgo heredado y que se transmite de una generación a otra.

gene regulation: (p. 342) ability of an organism to control which genes are transcribed in response to the environment.

regulación génica: (pág. 342) capacidad de un organismo para controlar los genes que se transcriben en respuesta a un ambiente.

gene therapy: (p. 378) technique to correct mutated disease-causing genes.

terapia génica: (pág. 378) técnica para corregir genes con mutaciones que causan enfermedades.

genetic diversity: (p. 116) variety of inheritable characteristics or genes in an interbreeding population.

diversidad genética: (pág. 116) variedad de características o genes heredables en una población que se entrecruza.

genetic drift: (p. 433) random change in allelic frequencies in a population.

deriva genética: (pág. 433) cambio aleatorio de frecuencias alélicas en una población.

genetic engineering: (p. 363) technology used to manipulate an organism's DNA by inserting the DNA of another organism.

ingeniería genética: (pág. 363) tecnología que se aplica para manipular el DNA de un organismo, mediante la inserción del DNA de otro organismo.

genetic recombination: (p. 283) new combination of genes produced by crossing over and independent assortment.

recombinación genética: (pág. 283) nueva combinación de genes producida por el entrecruzamiento y la distribución independiente de genes.

genetics: (p. 277) science of heredity.

genética: (pág. 277) ciencia que estudia la herencia.

genome: (p. 364) total DNA in each cell nucleus of an organism.

genoma: (pág. 364) todo el DNA en el núcleo de cada célula de un organismo.

genomics: (p. 378) study of an organism's genome.

genómica: (pág. 378) estudio del genoma de un organismo.

genotype: (p. 279) an organism's allele pairs.

genotipo: (pág. 279) pares de alelos de un organismo.

genus: (p. 487) taxonomic group of closely related species with a common ancester.

género: (pág. 487) grupo taxonómico de especies estrechamente emparentadas que comparten un antepasado común.

geologic time scale: (p. 396) model showing major geological and biological events in Earth's history.

escala del tiempo geológico: (pág. 396) modelo que muestra los principales eventos geológicos y biológicos de la historia de la Tierra.

Glossary/
Glosario

germination: (p. 678) process in which a seed's embryo begins to grow.

gestation: (p. 887) species-specific amount of time during which the young develop in the uterus before they are born.

gibberellins: (p. 649) group of plant hormones that are transported in vascular tissue and that can affect seed growth, stimulate cell division, and cause cell elongation.

gill (p. 738) respiratory structure of most mollusks and aquatic arthropods.

gizzard: (p. 746) muscular sac in birds that contains hard particles that help grind soil and food before they pass into the intestine.

gland: (p. 887) an organ or group of cells that secretes a substance for use elsewhere in the body.

glucagon (GLEW kuh gahn): (p. 1034) hormone produced by the pancreas that signals liver cells to convert glycogen to glucose and release glucose into the blood.

glycolysis: (p. 229) anaerobic process; first stage of cellular respiration in which glucose is broken down into two molecules of pyruvate.

Golgi apparatus: (p. 195) flattened stack of tubular membranes that modifies, sorts, and packages proteins into vesicles and transports them to other organelles or out of the cell.

gradualism: (p. 440) theory that evolution occurs in small, gradual steps over time.

granum: (p. 223) one of the stacks of pigment-containing thylakoids in a plant's chloroplasts.

grassland: (p. 70) biome characterized by fertile soils with a thick cover of grasses.

ground tissue: (p. 638) plant tissue consisting of parenchyma, collenchyma, and sclerenchyma.

growth: (p. 9) process that results in mass being added to an organism; may include formation of new cells and new structures.

guard cell: (p. 636) one of a pair of cells that function in the opening and closing of a plant's stomata by changes in their shape.

germinación: (pág. 678) proceso que inicia el crecimiento del embrión de una semilla.

gestación: (pág. 887) período específico para cada especie, durante el cual las crías se desarrollan en el útero, antes de nacer.

giberelinas: (pág. 649) grupo de hormonas vegetales que son transportadas por el tejido vascular y que pueden afectar el crecimiento de las semillas, y estimular la división y la elongacion celular.

branquia: (pág. 738) estructura respiratoria de la mayoría de los moluscos.

molleja: (pág. 746) saco muscular que contiene partículas duras que ayudan a moler el suelo y los alimentos, antes de que pasen al intestino.

glándula: (pág. 887) grupo de células que secretan una sustancia a usarse en alguna otra parte del cuerpo.

glucagón: (pág. 1034) hormona producida por el páncreas; les indica a las células del hígado que conviertan glucógeno en glucosa y que liberen la glucosa hacia el torrente sanguíneo.

glucólisis: (pág. 229) proceso anaeróbico; primera etapa de la respiración celular, en la cual la glucosa se transforma en dos moléculas de piruvato.

aparato de Golgi: (pág. 195) conjunto de membranas tubulares aplanadas que modifica, acomoda y empaca proteínas en vesículas y luego las transporta hacia otros organelos o hacia afuera de la célula.

gradualismo: (pág. 440) teoría que señala que la evolución sucede gradualmente, en pasos pequeños, a lo largo del tiempo.

grana: (pág. 223) conjunto de tilacoides con pigmentos de los cloroplastos de una planta.

pradera: (pág. 70) bioma caracterizado por suelos fértiles con una espesa cubierta de pastos.

tejido fundamental: (pág. 638) tejido vegetal que consiste en parénquima, colénquima y esclerénquima.

crecimiento: (pág. 9) proceso que provoca el aumento de masa en un organismo; puede incluir la formación de células y estructuras nuevas.

célula guardiana: (pág. 636) una de las células del par de células cuya función es abrir y cerrar, mediante cambios en su forma, los estomas de la planta.

habitat: (p. 38) physical area in which an organism lives.

habitat fragmentation: (p. 127) habitat loss from separation of an ecosystem into small pieces of land.

hábitat: (pág. 38) área física en que vive un organismo.

fragmentación del hábitat: (pág. 127) pérdida de hábitat como resultado de la partición de un ecosistema en terrenos pequeños.

habituation (huh bit choo AY shun): (p. 912) decrease in an animal's response after it has been repeatedly exposed to a specific stimulus that has no positive or negative effects.

habituación: (pág. 912) disminución de la respuesta de un animal, luego de haber sido expuesto repetidamente a un estímulo determinado, que no tiene efectos positivos ni negativos.

hair follicle: (p. 937) narrow cavity in the dermis from which a hair grows.

folículo piloso: (pág. 937) cavidad estrecha de la dermis de la cual crece un cabello.

half-life: (p. 395) amount of time required for half of a radioactive isotope to decay.

media vida: (pág. 395) cantidad de tiempo que se requiere para que se desintegre la mitad de un isótopo radiactivo.

haploid: (p. 271) cell with half the number of chromosomes (n) as a diploid ($2n$) cell.

haploide: (pág. 271) célula con la mitad del número de cromosomas (n) que una célula diploide ($2n$).

haplotype: (p. 378) area of linked genetic variations in the human genome.

haplotipo: (pág. 378) área del genoma humano con variaciones genéticas ligadas.

Hardy-Weinberg principle: (p. 431) states that allelic frequencies in populations stay the same unless they are affected by a factor that causes change.

principio de Hardy-Weinberg: (pág. 431) establece que las frecuencias alélicas de una población permanecen inalterables, a menos que sean afectadas por un factor que produzca un cambio.

haustorium (haws toh REE um): (p. 578) specialized hypha of parasitic fungi that grows into a host's tissues and absorbs its nutrients.

haustorio: (pág. 578) hifa especializada de los hongos parásitos cuya función es invadir los tejidos del huésped para absorber sus nutrientes.

heart: (p. 994) hollow, muscular organ that pumps oxygenated blood to the body and deoxygenated blood to the lungs.

corazón: (pág. 994) órgano muscular y hueco que bombea sangre oxigenada hacia el cuerpo y sangre desoxigenada hacia los pulmones.

helper T cell: (p. 1088) lymphocyte that activates antibody secretion in B cells and cytotoxic T cells.

célula T ayudante: (pág. 1088) linfocito que activa los linfocitos B y los linfocitos T citotóxicos para que secreten anticuerpos.

herbivore (HUR buh vor): (p. 41) heterotroph that eats only plants.

herbívoro: (pág. 41) heterótrofo que sólo se alimenta de plantas.

hermaphrodite (hur MAF ruh dite): (p. 695) animal that produces both sperm and eggs in its body, generally at different times.

hermafrodita: (pág. 695) animal que produce óvulos y espermatozoides en su cuerpo, generalmente a diferentes tiempos.

heterosporous (he tuh roh SPOR uhs): (p. 665) able to produce two types of spores—megaspores and microspores—that develop into female or male gametophytes.

heterospóreo: (pág. 665) capaz de producir dos tipos de esporas (megasporas y microsporas) que, al desarrollarse, forman el gametofito masculino o el femenino.

heterotroph (HE tuh roh trohf): (p. 41) organism that cannot make its own food and gets its nutrients and energy requirements by feeding on other organisms; also called a consumer.

heterótrofo: (pág. 41) organismo que no puede producir su propio alimento y que obtiene los nutrientes y la energía que necesita, alimentándose de otros organismos; también se llaman consumidores.

heterozygous (heh tuh roh ZI gus): (p. 279) organism with two different alleles for a specific trait.

heterocigoto: (pág. 279) organismo con dos diferentes alelos para un rasgo específico.

homeostasis (hoh mee oh STAY sus): (p. 10) regulation of an organism's internal environment to maintain conditions needed for life.

homeostasis: (pág. 10) regulación del ambiente interno de un organismo a fin de mantener las condiciones necesarias para la vida.

hominin: (p. 458) humanlike primate that appears to be more closely related to present-day humans than to present-day chimpanzees and bonobos.

hominiano: (pág. 458) primate tipo humano que parece estar más estrechamente emparentado con los humanos actuales que los chimpancés y bonobos actuales.

hominoid: (p. 461) group that includes all nonmonkey anthropoids—the living and extinct gibbons, orangutans, chimpanzees, gorillas, and humans.

hominoide: (pág. 461) grupo que incluye todos los antropoides que no son monos: los gibones, orangutanes, chimpancés, gorilas y humanos vivos y extintos.

***Homo*: (p. 467)** genus that includes living and extinct humans.

***Homo*: (pág. 467)** género que incluye a los humanos vivos y extintos.

homologous chromosome: (p. 270) one of two paired chromosomes, one from each parent, that carries genes for a specific trait at the same location.

cromosoma homólogo: (pág. 270) uno de los cromosomas, de un par de cromosomas, que contienen en un mismo sitio los genes para un rasgo específico. Cada progenitor contribuye un cromosoma de cada par.

homologous structure: (p. 424) anatomically similar structure inherited from a common ancestor.

estructura homóloga: (pág. 424) estructura anatómicamente similar heredada de un antepasado común.

homozygous (ho muh ZI gus): (p. 279) organism with two of the same alleles for a specific trait.

homocigoto: (pág. 279) organismo con dos alelos iguales para un rasgo específico.

hormone: (p. 1031) substance, such as estrogen, that is produced by an endocrine gland and acts on target cells.

hormona: (pág. 1031) sustancia, como el estrógeno, que es producida por una glándula endocrina y que actúa sobre células blanco.

hybrid: (p. 279) organism heterozygous for a specific trait.

híbrido: (pág. 279) organismo heterocigoto para un rasgo específico.

hydrogen bond: (p. 161) weak electrostatic bond formed by the attraction of opposite charges between a hydrogen atom and an oxygen, fluorine, or nitrogen atom.

enlace de hidrógeno: (pág. 161) enlace electrostático débil, formado por la atracción de cargas opuestas entre un átomo de hidrógeno y un átomo de oxígeno, flúor o nitrógeno.

hydrostatic skeleton: (p. 732) the pseudocoelom in roundworms; the fluid within a closed space that gives rigid support for muscles to work against.

esqueleto hidrostático: (pág. 732) seudoceloma de los gusanos redondos; fluido dentro de un espacio cerrado que provee un punto rígido de apoyo a los músculos.

hypertonic solution: (p. 205) a solution that has a higher concentration of solute outside than inside a cell, causing water to leave the cell by osmosis.

solución hipertónica: (pág. 205) solución que tiene una mayor concentración de soluto que el interior de la célula, la cual se encoge y arruga debido a que el agua sale de su interior por osmosis.

hypha (HI fah): (p. 577) threadlike filament that makes up the basic structural unit of a multicellular fungus.

hifa: (pág. 577) estructura con forma de filamento que constituye la unidad básica estructural de un hongo multicelular.

hypocotyl: (p. 679) region of the stem nearest the seed.

hipocótilo: (pág. 679) región del tallo más cercana a la semilla.

hypothalamus (hi poh THA luh mus): (p. 970) part of the brain that regulates body temperature, appetite, thirst, and water balance.

hipotálamo: (pág. 970) parte del encéfalo que regula la temperatura del cuerpo, el apetito, la sed y el equilibrio del agua.

hypothesis (hi PAH thuh sus): (p. 16) testable explanation of a situation.

hipótesis: (pág. 16) explicación comprobable de una situación.

hypotonic solution: (p. 204) a solution that has a lower concentration of solute outside than inside the cell, causing water to flow into the cell by osmosis.

solución hipotónica: (pág. 204) solución que tiene una menor concentración de soluto; hay más agua fuera que dentro de la célula.

I

immigration (ih muh GRAY shun): (p. 97) movement of individuals into a population.

inmigración: (pág. 97) entrada de individuos a una población.

immunization: (p. 1089) vaccination; develops active immunity.

inmunización: (pág. 1089) vacunación; desarrolla inmunidad activa.

imprinting: (p. 914) permanent learning that occurs only within a specific period of time in an animal's life.

impronta: (pág. 914) aprendizaje permanente que sucede sólo en un período específico de la vida de un animal.

inbreeding: (p. 361) selective breeding of closely related organisms to produce desired traits and eliminate undesired traits, resulting in pure lines—however, harmful recessive traits can also be passed on.

endogamia: (pág. 361) cruce selectivo de organismos emparentados para obtener rasgos deseados y eliminar rasgos indeseados; permite obtener linajes puros, aunque también puede transmitir rasgos recesivos dañinos.

incomplete dominance: (p. 302) complex inheritance pattern in which the heterozygous phenotype is intermediate between those of the two homozygous parent organisms.

incubate: (p. 866) to maintain an egg or eggs at favorable environmental conditions for hatching.

independent variable: (p. 19) the one factor that can be changed in a controlled experiment; is the factor tested and affects the experiment outcome.

infancy: (p. 1064) first two years of human life.

infectious disease: (p. 1076) pathogen-caused disease passed from one organism to another organism.

inference: (p. 16) assumption based on prior experience.

innate (ih NAYT) behavior: (p. 910) genetically based behavior.

insulin: (p. 1034) hormone produced by the pancreas that works with glucagon to maintain the level of sugar in the blood.

interferon: (p. 1085) antiviral protein secreted by virus-infected cells.

intermediate-day plant: (p. 673) plant that flowers as long as the number of hours of darkness is neither too great nor too few.

internal fertilization: (p. 695) type of fertilization that occurs when sperm and egg combine inside an animal's body.

internal respiration: (p. 1000) gas exchange between the body's cells and the blood.

interphase: (p. 246) first stage of the cell cycle, during which a cell grows, matures, and replicates its DNA.

intertidal zone: (p. 79) narrow band of shoreline where the ocean and land meet that is alternately submerged and exposed and is home to constantly changing communities.

introduced species: (p. 128) nonnative species deliberately or accidentally introduced into a new habitat.

intron: (p. 338) in RNA processing, the intervening coding sequence missing from the final mRNA.

invertebrate: (p. 693) animal without a backbone; between 95 and 99 percent of animal species are invertebrates.

invertebrate chordate: (p. 803) chordate without a backbone.

involuntary muscle: (p. 947) smooth muscle, which cannot be controlled consciously.

ion: (p. 153) atom that is negatively or positively charged because it has lost or gained one or more electrons.

dominancia incompleta: (pág. 302) patrón complejo de herencia en que el fenotipo heterocigoto es intermedio entre los dos fenotipos de los progenitores homocigotos.

incubar: (pág. 866) mantener los huevos bajo condiciones ambientales favorables para que luego eclosionen.

variable independiente: (pág. 19) el único factor que se puede cambiar en un experimento controlado; es el factor que se está probando y afecta los resultados del experimento.

lactancia: (pág. 1064) los primeros dos años de vida en los humanos.

enfermedad infecciosa: (pág. 1076) enfermedad causada por un patógeno que se transmite de un organismo a otro.

inferencia: (pág. 16) supuesto basado en la experiencia previa.

comportamiento innato: (pág. 910) comportamiento basado en la genética.

insulina: (pág. 1034) hormona producida por el páncreas que, junto con el glucagón, mantiene los niveles adecuados de azúcar en la sangre.

interferón: (pág. 1085) proteína secretada por células infectadas por virus.

planta de días intermedios: (pág. 673) planta que florece mientras el número de horas de oscuridad no sea ni muy grande ni muy pequeño.

fecundación interna: (pág. 695) tipo de fecundación que sucede cuando el espermatozoide se une al óvulo dentro del cuerpo del animal.

respiración interna: (pág. 1000) intercambio de gases entre las células del cuerpo y la sangre.

interfase: (pág. 246) primera fase del ciclo celular; en esta fase, la célula crece, madura y replica su DNA.

zona intermareal: (pág. 79) franja estrecha de la costa donde se encuentran la tierra y el mar; es inundada periódicamente por las mareas y presenta un cambio constante en su comunidad.

especie introducida: (pág. 128) especie no nativa que es introducida deliberadamente o por accidente a un nuevo hábitat.

intrón: (pág. 338) en la transcripción del RNA, la secuencia de codones que se transcribe pero que no forma parte del mRNA final.

invertebrado: (pág. 693) animal sin columna vertebral; entre el 95 y el 99 por ciento de las especies animales son invertebrados.

cordado invertebrado: (pág. 803) cordado sin columna vertebral.

músculo involuntario: (pág. 947) músculo liso, no se puede controlar a voluntad.

ion: (pág. 153) átomo con carga positiva o negativa porque ha perdido o ganado uno o más electrones.

ionic bond: **(p. 153)** electrical attraction between two oppositely charged atoms or groups of atoms.

isotonic solution: **(p. 204)** a solution with the same concentration of water and solutes as inside a cell, resulting in the cell retaining its normal shape because there is no net movement of water.

isotope: **(p. 150)** two or more atoms of the same element having different numbers of neutrons.

enlace iónico: **(pág. 153)** atracción eléctrica entre dos átomos o grupos de átomos con carga opuesta.

solución isotónica: **(pág. 204)** solución con la misma concentración de agua y solutos que el interior de la célula; permite a la célula mantener su forma original porque no hay movimiento neto de agua.

isótopo: **(pág. 150)** dos o más átomos de un mismo elemento que tienen diferente número de neutrones.

J

Jacobson's organ: **(p. 855)** saclike, odor-sensing structure on the roof of a snake's mouth.

órgano de Jacobson: **(pág. 855)** órgano con forma de saco cuya función es detectar olores; está situado en el paladar de la boca de las serpientes.

karyotype (KER ee uh tipe): **(p. 311)** micrograph in which the pairs of homologous chromosomes are arranged in decreasing size.

keratin (KER uh tun): **(p. 936)** protein contained in the skin's outer epidermal cells that waterproofs and protects underlying cells and tissues.

kidney: **(p. 1006)** bean-shaped, excretory system organ that filters out wastes, water, and salts from the blood and maintains blood pH.

kingdom: **(p. 488)** taxonomic group of related phyla or divisions.

Koch's postulates: **(p. 1077)** rules for demonstrating that an organism causes a disease.

Krebs cycle: **(p. 229)** series of reactions in which pyruvate is broken down into carbon dioxide inside the mitochondria of cells; also called the tricarboxylic acid cycle and the citric acid cycle.

K-T boundary: **(p. 399)** layer of iridium-rich material betweeen rocks of the Cretaceous period and rocks of the Paleogene period that provides evidence of a meteorite impact.

cariotipo: **(pág. 311)** micrografía en que los pares de cromosomas homólogos aparecen ordenados en tamaño decreciente.

queratina: **(pág. 936)** proteína que contienen las células más externas de la epidermis; es impermeable al agua y protege las células y tejidos subyacentes.

riñón: **(pág. 1006)** órgano del sistema excretor con forma de frijol; mantiene el pH y elimina por filtración los desechos, el agua y las sales.

reino: **(pág. 488)** grupo taxonómico que incluye filos o divisiones relacionadas.

postulados de Koch: **(pág. 1077)** reglas para demostrar que un organismo causa una enfermedad.

ciclo de Krebs: **(pág. 229)** serie de reacciones en que el piruvato se desdobla en dióxido de carbono dentro de las mitocondrias de las células; también se llama ciclo del ácido tricarboxílico y ciclo del ácido cítrico.

límite KT: **(pág. 399)** capa de material rico en iridio, situada entre las rocas de los períodos Cretáceo y Paleoceno; provee pruebas del impacto de un meteorito.

labor: **(p. 1062)** three-stage birthing process that begins with uterine contractions and ends with expulsion of the placenta and umbilical cord.

language: **(p. 920)** auditory communication in which animals use their vocal organs to produce sounds with shared meanings.

parto: **(pág. 1062)** las tres etapas del proceso de alumbramiento; se inicia con las contracciones uterinas y termina con la expulsión de la placenta y el cordón umbilical.

lenguaje: **(pág. 920)** comunicación mediante sonidos en que el animal usa sus órganos vocales para producir sonidos con significados dados.

large intestine: (p. 1024) end portion of the digestive tract; involved primarily in water absorption.

intestino grueso: (pág. 1024) porción final del tracto digestivo; su función principal es la absorción de agua.

lateral line system: (p. 826) sensory receptors that enable fishes to detect vibrations, or sound waves, in water.

sistema de la línea lateral: (pág. 826) receptores sensoriales que permiten a los peces detectar vibraciones u ondas sonoras en el agua.

latitude: (p. 65) distance of a point on Earth's surface north or south of the equator.

latitud: (pág. 65) distancia de un punto sobre la superficie de la Tierra, hacia el norte o hacia el sur del ecuador.

law of independent assortment: (p. 280) Mendelian law stating that a random distribution of alleles occurs during the formation of gametes.

ley de la distribución independiente: (pág. 280) ley de Mendel que establece que la distribución independiente de alelos sucede durante la formación de los gametos.

law of segregation: (p. 279) Mendelian law stating that two alleles for each trait separate during meiosis.

ley de la segregación: (pág. 279) ley de Mendel que establece que los dos alelos para cada rasgo se separan durante la meiosis.

law of superposition: (p. 394) states that the oldest layers of rock are found at the bottom and the youngest layers of rock are found at the top of a formation if the rock layers have not been disturbed.

ley de superposición: (pág. 394) establece que, en una formación rocosa inalterada, los estratos rocosos más antiguos se hallan a mayor profundidad y los estratos más recientes se hallan más cerca de la superficie.

learned behavior: (p. 912) results from an interaction between innate behavior and past experience within a specific environment; includes habituation, conditioning, and imprinting.

comportamiento adquirido: (pág. 912) es resultado de la interacción entre el comportamiento innato y las experiencias previas en un ambiente específico; incluye la habituación, el acondicionamiento y la impronta.

lens: (p. 974) part of the eye behind the iris that inverts an image and focuses it on the retina.

cristalino: (pág. 974) parte del ojo situada detrás del iris; invierte la imagen y la enfoca sobre la retina.

lichen (LI ken): (p. 587) symbiotic relationship between a fungus (usually an ascomycete) and an alga or a photosynthetic partner.

liquen: (pág. 587) relación simbiótica entre un hongo (a menudo un ascomiceto) y un alga u otro organismo fotosintético.

ligament: (p. 944) tough connective tissue band that attaches bones to each other.

ligamento: (pág. 944) bandas fuertes de tejido conectivo que unen los huesos entre sí.

limiting factor: (p. 61) biotic or abiotic factor that restricts the number, distribution, or reproduction of a population within a community.

factor limitante: (pág. 61) factor biótico o abiótico que restringe el número, la distribución o la reproducción de una población en una comunidad.

limnetic zone: (p. 77) well-lit, open-water area of a lake or pond.

zona limnética: (pág. 77) área de agua abierta y bien iluminada de un lago o laguna.

lipid: (p. 169) hydrophobic biological molecule composed mostly of carbon and hydrogen; fats, oils, and waxes are lipids.

lípido: (pág. 169) molécula biológica hidrofóbica compuesta principalmente por carbono e hidrógeno; las grasas, los aceites y las ceras son lípidos.

littoral zone: (p. 76) area of a lake or pond closest to the shore.

zona litoral: (pág. 76) en lagos y lagunas, comprende la zona de agua poco profunda de la orilla y parte del fondo hasta donde penetra la luz solar.

liver: (p. 1022) largest internal organ of the body; produces bile.

hígado: (pág. 1022) es el órgano interno más grande del cuerpo; produce bilis.

long-day plant: (p. 672) plant that flowers in the summer, when there are fewer hours of darkness than the plant's critical period.

planta de días largos: (pág. 672) planta que florece en el verano, cuando hay menos horas de oscuridad que el período crítico de la planta.

lung: (p. 1001) largest respiratory system organ in which gas exchange takes place.

pulmón: (pág. 1001) órgano más grande del sistema respiratorio en que se lleva a cabo el intercambio de gases.

lymphocyte: (p. 1086) white blood cell involved in specific immunity; a B cell or a T cell.

linfocito: (pág. 1086) glóbulo blanco que participa en la inmunidad específica; linfocitos B o linfocitos T.

lysogenic cycle: (p. 528) viral replication process in which viral DNA inserts into the host cell's chromosome, may remain dormant and later activate and instruct the host cell to produce more viruses.

ciclo lisogénico: (pág. 528) proceso vírico de replicación en que el DNA vírico es insertado en el cromosoma de la célula huésped; puede permanecer latente, activarse más tarde y dar instrucciones a la célula huésped para que produzca más virus.

lysosome: (p. 196) vesicle that uses enzymes to digest excess or worn-out cellular substances.

lisosoma: (pág. 196) vesícula que usa enzimas para digerir sustancias celulares gastadas o que se hallan en número excesivo.

lytic cycle: (p. 528) viral replication process in which genetic material of the virus enters the host cell's cytoplasm, the cell replicates the viral DNA or RNA, and the host cell is instructed to manufacture capsids and assemble new viral particles which then leave the cell.

ciclo lítico: (pág. 528) proceso de replicación vírica en que el material genético del virus entra al citoplasma de la célula huésped. Después, la célula replica el DNA o RNA viral y recibe instrucciones para fabricar cápsides y ensamblar nuevos virus que luego salen de la célula.

M

macromolecule: (p. 167) large molecule formed by joining smaller organic molecules together.

macromolécula: (pág. 167) molécula de gran tamaño formada por la unión de moléculas orgánicas más pequeñas.

madreporite (MA druh pohr it): (p. 795) strainerlike opening through which water enters the water-vascular system in most echinoderms.

madreporita: (pág. 795) abertura con forma de colador, a través de la cual entra y sale el agua del sistema vascular acuático.

Malpighian (mal PIH gee un) tubule: (p. 767) in most arthropods, the waste-excreting structure that also helps maintain homeostatic water balance.

túbulo de Malpighi: (pág. 767) en la mayoría de los artrópodos, la estructura excretora que también ayuda a mantener el equilibrio homeostático del agua.

mammary gland: (p. 880) mammalian gland that produces and secretes milk to nourish developing young.

glándula mamaria: (pág. 880) glándula de los mamíferos que produce y secreta leche para alimentar a las crías.

mandible (MAN duh bul): (p. 765) in most arthropods, one of a pair of mouthparts adapted for biting and chewing food.

mandíbulas: (pág. 765) en la mayor parte de los artrópodos, las partes bucales adaptadas para morder y masticar el alimento.

mantle (MAN tuhl): (p. 737) membrane that surrounds a mollusk's internal organs.

manto: (pág. 737) membrana que rodea los órganos internos de los moluscos.

marsupial: (p. 890) pouched mammal whose offspring have a short period of development inside the uterus, then after birth have a longer period of development within the pouch.

marsupial: (pág. 890) mamífero con bolsa, cuyas crías se desarrollan en el útero durante un corto período de tiempo y completan su desarrollo dentro de la bolsa.

mass extinction: (p. 122) a large-scale dying out of a large percentage of all living organisms in an area within a short time.

extinción masiva: (pág. 122) desaparición a gran escala de un porcentaje grande de todos los organismos vivos de un área dada durante un corto tiempo.

matter: (p. 45) anything that takes up space and has mass.

materia: (pág. 45) cualquier cosa que ocupa lugar y tiene masa.

mechanical digestion: (p. 1020) physical breakdown of food that occurs when food is chewed into smaller pieces and then churned by the stomach and small intestine.

digestión mecánica: (pág. 1020) desintegración física del alimento que ocurre al masticarlo en trozos más pequeños y luego revolverlo en el estómago y el intestino delgado.

medulla oblongata: (p. 970) part of the brain stem that helps control blood pressure, heart rate, and breathing rate.

médula oblongada: (pág. 970) parte del bulbo raquídeo que ayuda a controlar la presión sanguínea y el ritmo cardíaco y el respiratorio.

medusa (mih DEW suh): (p. 712) umbrella-shaped, free-swimming body form of cnidarians.

medusa: (pág. 712) forma libre de los cnidarios que semeja un paraguas con tentáculos que cuelgan hacia abajo.

megaspore: (p. 665) spore that develops into a female gametophyte and is produced by a conifer's female cone.

megáspora: (pág. 665) espora que se desarrolla en un gametofito femenino y que produce el cono femenino de una conífera.

meiosis: (p. 271) reduction division process, occurring only in reproductive cells, in which one diploid (2n) cell produces four haploid (n) cells that are not genetically identical.

melanin: (p. 937) pigment in the inner layer of the epidermis that protects against harmful ultraviolet radiation and influences skin color.

memory cell: (p. 1089) long-lived lymphocyte produced during exposure to an antigen during the primary immune response; can function in future immune response to the same antigen.

menstrual (MEN strew ul) cycle: (p. 1050) monthly reproductive cycle that helps prepare the human female body for pregnancy; involves the shedding of blood, tissue fluid, mucus, and epithelial cells if an egg is not fertilized.

meristem: (p. 634) region of rapid cell division in plants; produces cells that can develop into many different types of plant cells.

mesoderm: (p. 697) layer of cells between the endoderm and the ectoderm that can become muscle tissue and tissue of the circulatory, respiratory, and excretory systems.

messenger RNA: (p. 336) type of RNA that carries genetic information from DNA in the nucleus to direct protein synthesis in the cytoplasm.

metabolic disease: (p. 1093) disease, such as type 1 diabetes, that results from an error in a biochemical pathway.

metabolism: (p. 220) all of the chemical reactions that occur within an organism.

metamorphosis: (p. 778) in most insects, the series of changes from a larval form to an adult form.

metaphase: (p. 250) second stage of mitosis in which motor proteins pull sister chromatids to the cell's equator.

metric system: (p. 14) measurement system whose divisions are powers of ten.

micropyle: (p. 666) opening of a conifer's ovule where a pollen grain can be trapped in a pollen drop.

microspore: (p. 666) spore that develops into a male gametophyte (a pollen grain) and is produced by a conifer's male cone.

microsporidium (mi kroh spo RIH dee um): (p. 544) microscopic protozoan parasite that infects insects and other organisms, causing disease.

migratory behavior: (p. 919) seasonal movement of a group of animals to a new location for feeding and breeding.

meiosis: (pág. 271) proceso divisorio de reducción que sólo ocurre en las células reproductoras, mediante el cual una célula diploide (2*n*) produce cuatro células haploides (*n*) no idénticas genéticamente.

melanina: (pág. 937) pigmento que se encuentra en la capa interna de la epidermis; protege a las células del daño causado por la radiación solar e influye en el color de la piel.

célula de memoria: (pág. 1089) linfocito de larga vida producido debido a la exposición a un antígeno durante la respuesta inmunológica primaria; capaz de funcionar en una futura respuesta inmunológica al mismo antígeno.

ciclo menstrual: (pág. 1050) ciclo reproductor mensual que ayuda a preparar el cuerpo de la hembra humana para el embarazo; comprende derrame de sangre, tejido líquido, mucosidad y células epiteliales si el óvulo no ha sido fecundado.

meristema: (pág. 634) región de rápida división celular vegetal; produce células capaces de desarrollarse en muchos tipos de células vegetales.

mesodermo: (pág. 697) capa celular entre el ectodermo y el endodermo que puede desarrollarse en tejido muscular y tejido de los sistemas circulatorio, respiratorio y excretor.

RNA mensajero: (pág. 336) tipo de RNA que transporta información desde el DNA en el núcleo hasta la síntesis directa de proteína en el citoplasma.

enfermedad metabólica: (pág. 1093) enfermedad, como la diabetes tipo 1, que resulta de un error en un trayecto bioquímico.

metabolismo: (pág. 220) todas las reacciones químicas que ocurren dentro de un organismo.

metamorfosis: (pág. 778) en la mayoría de los insectos, la serie de cambios desde una forma larval hasta una forma adulta.

metafase: (pág. 250) segunda fase de la mitosis en la cual las proteínas motores atraen a las cromátides hermanas hacia el ecuador de la célula.

sistema métrico: (pág. 14) sistema de medida cuyas divisiones son potencias de diez.

micrópilo: (pág. 666) abertura en el óvulo de una conífera donde puede atraparse un grano de polen en una descarga de polen.

micróspora: (pág. 666) espora que se convierte en el gametofito masculino (un grano de polen) y se forma por el cono masculino de una conífera.

microsporídeo: (pág. 544) protozoo microscópico que vive en las tripas de termitas y produce enzimas que digieren madera.

comportamiento migratorio: (pág. 919) movimiento estacional de un grupo de animales hacia una nueva localidad para alimentarse y reproducirse.

mimicry: **(p. 429)** morphological adaptation in which one species evolves to resemble another species for protection or other advantages.

mineral: **(p. 1028)** inorganic compound, such as calcium, that is used as building material by the body and is involved with metabolic functions.

mitochondrion **(mi tuh KAHN dree un): (p. 197)** membrane-bound organelle that converts fuel into energy that is available to the rest of the cell.

mitosis **(mi TOH sus): (p. 246)** second main stage of the cell cycle during which the cell's replicated DNA divides and two genetically identical diploid daughter cells are produced.

mixture: **(p. 163)** combination of two or more different substances in which each substance keeps its individual characteristics; can have a uniform composition (homogeneous) or have distinct areas of substances (heterogeneous).

molecular clock: **(p. 495)** model that uses comparisons of DNA sequences to estimate phylogeny and rate of evolutionary change.

molecule: **(p. 152)** compound whose atoms are held together by covalent bonds.

molting: **(p. 764)** in arthropods, the periodic shedding of the protective exoskeleton so their bodies can continue to grow.

monotreme: **(p. 889)** mammal that reproduces by laying eggs.

morula: **(p. 1055)** solid ball of embryonic cells that forms before the blastocyst.

multiple alleles: **(p. 304)** having more than two alleles for a specific trait.

mutagen **(MYEW tuh jun): (p. 348)** substance, such as a chemical, that causes mutations.

mutation: **(p. 345)** permanent change in a cell's DNA, ranging from changes in a single base pair to deletions of large sections of chromosomes.

mutualism **(MYEW chuh wuh lih zum): (p. 39)** symbiotic relationship in which both organisms benefit.

mycelium **(mi SEE lee um): (p. 577)** complex, netlike mass made up of branching hyphae.

mycorrhiza **(my kuh RHY zuh): (p. 589)** symbiotic relationship between a specialized fungus and plant roots fungal hyphae help plants obtain water and minerals and plants supply carbohydrates and amino acids to the fungus.

myofibril: **(p. 948)** small muscle fiber that functions in contraction and consists of myosin and actin protein filaments.

myosin: **(p. 948)** protein filament in muscle cells that functions with actin in contraction.

mimetismo: **(pág. 429)** adaptación morfológica en la cual una especie evoluciona para parecerse a otra a modo de protección u otras ventajas.

mineral: **(pág. 1028)** compuesto inorgánico, como el calcio, utilizado por el cuerpo como material de construcción, presente en las funciones metabólicas.

mitocondria: **(pág. 197)** organelo membranoso que transforman el combustible en energía disponible al resto de las célula.

mitosis: **(pág. 246)** segundo período principal del ciclo celular, durante el cual el DNA replicado de la célula se divide y se forman dos células hijas diploides idénticas.

mezcla: **(pág. 163)** combinación de dos o más sustancias diferentes en la cual cada una mantiene sus características individuales; pueden tener una composición uniforme (homogénea) o áreas distintivas de sustancias (heterogénea).

reloj molecular: **(pág. 495)** modelo que usa las comparaciones secuenciales de DNA para calcular la filogenia y la tasa de cambio evolutivo.

molécula: **(pág. 152)** compuesto cuyos átomos se mantienen unidos por medio de enlaces covalentes.

muda: **(pág. 764)** en los artrópodos, el cambio periódico del exoesqueleto protector de forma que sus cuerpos puedan seguir creciendo.

monotrema: **(pág. 889)** mamífero que se reproduce al poner huevos.

mórula: **(pág. 1055)** bola sólida de células embrionarias que se forma antes del blastocito.

alelos múltiples: **(pág. 304)** presencia de más de dos alelos para un rasgo genético.

mutágeno: **(pág. 348)** sustancia, como un químico, que causa mutaciones.

mutación: **(pág. 345)** cambio permanente en el DNA de una célula, desde cambios en un par de base simple a eliminaciones de grandes secciones de cromosomas.

mutualismo: **(pág. 39)** relación simbiótica en la cual ambos organismos se benefician.

micelio: **(pág. 577)** masa compleja compuesta por hifas ramificadas.

micorriza: **(pág. 589)** asociación simbiótica de un hongo especializado con las raíces de una planta; las hifas fúngicas ayudan a las plantas a obtener agua y minerales y las plantas proveen carbohidratos y aminoácidos al hongo.

miofibrilla: **(pág. 948)** fibra muscular pequeña que funciona por contracción, compuesta por los filamentos proteicos miosina y actina.

miosina: **(pág. 948)** filamento proteico en las células musculares que funciona con la actina en la contracción muscular.

N

NADP+: (p. 224) in photosynthesis, the major electron carrier involved in electron transport.

nastic response: (p. 650) reversible, responsive movement of a plant that occurs independent of the direction of the stimulus.

natural resource: (p. 123) any material or organism in the biosphere, including water, soil, fuel, and plants and animals.

natural selection: (p. 420) theory of evolution developed by Darwin, based on four ideas: excess reproduction, variations, inheritance, and the advantages of specific traits in an environment.

Neanderthal: (p. 470) a species also referred to as *Homo neanderthalensis* that evolved exclusively in Europe and Asia about 200,000 years ago.

nematocyst (nih MA tuh sihst): (p. 710) capsule whose threadlike tube contains poison and barbs and is discharged when prey touches a cnidarian.

nephridium (nih FRIH dee um): (p. 739) structure through which most mollusks eliminate metabolic wastes from cellular processes.

nephron: (p. 825) filtering unit of the kidney.

nerve net: (p. 711) cnidarian nervous system that conducts impulses to and from all parts of the body.

neural crest: (p. 821) group of cells that develops from the embryo's ectoderm and contributes to the development of many vertebrate structures.

neuron: (p. 962) cell that carries nerve impulses throughout the body and is composed of a cell body, an axon, and dendrites.

neurotransmitter: (p. 967) chemical that diffuses across a synapse and binds to receptors on a neighboring neuron's dendrite, causing channels to open on the neighboring cell and the creation of a new action potential.

neutron: (p. 148) particle without a charge in an atom's nucleus.

niche (NIHCH): (p. 38) role, or position, of an organism in its environment.

nictitating membrane: (p. 837) in amphibians, the transparent eyelid that moves across the eye to prevent it from drying out on land and to protect it under water.

nitrogen fixation: (p. 48) process in which nitrogen gas is captured and converted into a form plants can use.

nocturnal: (p. 452) organisms that are active at night.

node: (p. 965) gap in the myelin sheath along the length of an axon; nerve impulses move from node to node.

NADP+: (pág. 224) en la fotosíntesis, el principal portador de electrones presente en el transporte de electrones.

respuesta nástica: (pág. 650) movimiento reversible y sensible de una planta que ocurre independientemente de la dirección del estímulo.

recurso natural: (pág. 123) cualquier material u organismo en la biosfera, incluidos agua, suelo, combustible, plantas y animales.

selección natural: (pág. 420) teoría de la evolución desarrollada por Darwin, basada en cuatro ideas: reproducción excesiva, variaciones, herencia y las ventajas de cualidades específicas en un medioambiente.

Neanderthal: (pág. 470) especie también conocida como *Homo neanderthalensis,* la cual evolucionó exclusivamente en Europa y Asia hace unos 200,000 años.

nematocisto: (pág. 710) cápsula cuyo tubo filamentoso contiene veneno y bárbulas; se descarga cuando una presa toca un cnidario.

nefridios: (pág. 739) estructura por la cual la mayoría de los moluscos eliminan los desechos metabólicos de los procesos celulares.

nefrón: (pág. 825) unidad de filtración del riñón.

red nerviosa: (pág. 711) sistema nervioso cnidario que conduce los impulsos hacia y desde todas las partes del cuerpo.

cresta neural: (pág. 821) grupo de células que se desarrollan del ectodermo del embrión y contribuyen al desarrollo de muchas estructuras vertebradas.

neurona: (pág. 962) célula nerviosa que conduce los impulsos a través del cuerpo y se compone de un cuerpo celular, un axón y dendritas.

neurotransmisores: (pág. 967) químico que se difunde por la sinapsis y se enlaza a los receptores en la dendrita de una neurona vecina; causa la apertura de los canales en la célula vecina para crear un nuevo impulso.

neutrón: (pág. 148) partícula sin carga en el núcleo de un átomo.

nicho: (pág. 43) la función o posición de un organismo en su ambiente.

membrana nictitante: (pág. 837) en los anfibios, el párpado transparente que se mueve a lo largo del ojo para evitar que se seque al estar en tierra y para protegerlo bajo el agua.

nitrificación: (pág. 48) proceso mediante el cual el gas nitrógeno se captura y se convierte en una forma utilizable por las plantas.

nocturno: (pág. 452) organismos activos durante la noche.

nódulo: (pág. 965) brecha en la vaina de mielina a lo largo de un axón; los impulsos nerviosos se desplazan de nodo a nodo.

nondisjunction: (p. 313) cell division in which the sister chromatids do not separate correctly, resulting in gametes with an abnormal number of chromosomes.

no disyunción: (pág. 313) división celular en la cual las cromátides no se separan correctamente lo cual resulta en gametos con un número anormal de cromosomas.

nonrenewable resource: (p. 130) any natural resource available in limited amounts or replaced extremely slowly by natural processes.

recurso no renovable: (pág. 130) cualquier recurso natural disponible en cantidades limitadas o reemplazado en forma extremadamente lenta por los procesos naturales.

nonvascular plant: (p. 606) type of plant that lacks vascular tissues, moves substances slowly from cell to cell by osmosis and diffusion, and grows only in a damp environment.

plantas no vasculares: (pág. 606) tipo de planta que carece de tejidos vasculares, mueve sustancias lentamente de célula a célula mediante osmosis y difusión y sólo crece en un ambiente húmedo.

notochord (NOH tuh kord): (p. 803) flexible, rodlike structure extending the length of the chordate body, enabling the body to bend and make side-to-side movements.

notocordio: (pág. 803) estructura cilíndrica flexible que se extiende a lo largo del cuerpo de un cordado y le permite doblarse y realizar movimientos laterales.

nucleic acid: (p. 171) complex macromolecule that stores and communicates genetic information.

ácido nucleico: (pág. 171) macromolécula compleja que almacena y comunica información genética.

nucleoid: (p. 518) area in a prokaryotic cell that contains a large, circular chromosome.

nucleoide: (pág. 518) área de una célula procariota que contiene un cromosoma circular grande.

nucleolus: (p. 193) the site of ribosome production within the nucleus of eukaryotic cells.

nucléolo: (pág. 193) el sitio de producción de ribosomas dentro del núcleo de las células eucariotas.

nucleosome: (p. 332) repeating subunit of chromatin fibers, consisting of DNA coiled around histones.

nucleosoma: (pág. 332) subunidad repetitiva de fibras de cromatina que consiste en DNA enroscado alrededor de histonas.

nucleotide: (p. 171) a subunit of nucleic acid formed from a simple sugar, a phosphate group, and a nitrogenous base.

nucleótido: (pág. 171) subunidades de ácidos nucleicos formadas por un azúcar simple, un grupo fosfato y una base nitrogenada.

nucleus: (p. 148) center of an atom; contains neutrons and protons. **(p. 186)** in eukaryotic cells, the central membrane-bound organelle that manages cellular functions and contains DNA.

núcleo: (pág. 148) centro de un átomo; contiene neutrones y protones. **(pág. 186)** en las células eucariotas es el organelo membranoso central que se encarga de las funciones celulares y que contiene el DNA.

nurturing behavior: (p. 921) caretaking behavior that a parent provides to its offspring during the early stages of development.

comportamiento de crianza: (pág. 921) comportamiento de cuidado y formación que un progenitor provee a su cría durante las primeras etapas del desarrollo de ésta.

nutrient: (p. 45) chemical substance that living organisms obtain from the environment to carry out life processes and sustain life.

nutriente: (pág. 45) sustancia química que obtienen los organismos vivos del medioambiente para el desarrollo de los procesos vitales y el sustento de la vida.

nutrition: (p. 1025) process by which an individual takes in and uses food, which provides building blocks for growth and energy to maintain body mass.

nutrición: (pág. 1025) proceso mediante el cual un individuo consume y usa alimento, lo cual provee las bases para el crecimiento y la energía para mantener la masa corporal.

nymph (NIHMF): (p. 778) immature form of an insect during incomplete metamorphosis—the hatchling looks like a small adult insect and goes through several molts, eventually becoming a mature winged adult.

ninfa: (pág. 778) forma inmadura de un insecto durante la metamorfosis incompleta: el insecto recién salido del cascarón se parece a un insecto adulto pequeño y pasa por varias mudas hasta convertirse en un adulto alado maduro.

O

observation: (p. 16) orderly, direct information gathering about a natural phenomenon.

observación: (pág. 16) forma directa y ordenada de recopilar información sobre un fenómeno natural.

Okazaki fragment: (p. 334) short segment of DNA synthesized discontinuously in small segments in the 3' to 5' direction by DNA polymerase.

fragmento Okazaki: (pág. 334) segmento corto de DNA que la enzima polimerasa de DNA sintetiza discontinuamente en segmentos pequeños en la dirección de 3' a 5'.

omnivore (AHM nih vor): (p. 42) heterotroph that consumes both plants and animals.

omnívoro: (pág. 42) heterótrofo que consume tanto plantas como animales.

oocyte (OH uh site): (p. 1050) immature egg inside an ovary.

oocito: (pág. 1050) óvulo inmaduro dentro de un ovario.

open circulatory system: (p. 739) blood is pumped out of vessels into open spaces surrounding body organs.

sistema circulatorio abierto: (pág. 739) la sangre se bombea fuera de los vasos hacia los espacios abiertos que rodean los órganos corporales.

operant conditioning: (p. 913) learned behavior that occurs when an association is made between a response to a stimulus and a punishment or a reward.

condicionamiento operante: (pág. 913) comportamiento adquirido que ocurre al asociar una respuesta con un estímulo y un castigo o una recompensa.

operculum (oh PUR kyuh lum): (p. 824) movable, protective flap that covers a fish's gills and helps to pump water that enters the mouth and moves over the gills.

opérculo: (pág. 824) protector móvil que cubre las agallas de los peces y ayuda a bombear el agua que entra a la boca y se desplaza sobre las agallas.

operon: (p. 342) section of DNA containing genes for proteins required for a specific metabolic pathway—consists of an operator, promoter, regulatory gene, and genes coding for proteins.

operón: (pág. 342) sección de DNA que contiene los genes para las proteínas requeridas para un trayecto metabólico específico; consiste en un operador, un promotor, un gene regulador y un código de genes para las proteínas.

opposable first digit: (p. 452) a digit, either a thumb or a toe, that is set apart from the other digits and can be brought across the palm or foot so that it touches or nearly touches the other digits; this allows animals to grasp an object in a powerful grip.

primer dígito oponible: (pág. 452) dígito, ya sea un pulgar o un dígito del pie, que se diferencia del resto de los dígitos y el cual se puede cruzar a través de la palma de la mano o del pie y puede tocar o casi tocar los otros dígitos; esto les permite a los animales asir objetos fuertemente.

order: (p. 488) taxonomic group that contains related families.

orden: (pág. 488) agrupación taxonómica de familias relacionadas.

organelle: (p. 186) specialized internal cell structure that carries out specific cell functions such as protein synthesis and energy transformation.

organelo: (pág. 186) estructura celular especializada interna con funciones celulares específicas como la síntesis y la transformación de energía.

organism: (p. 6) anything that has or once had all the characteristics of life.

organismo: (pág. 6) cualquier cosa que tuvo o tiene todas las características de la vida.

organization: (p. 8) orderly structure shown by living things.

organización: (pág. 8) estructura ordenada de todos los seres vivos.

osmosis (ahs MOH sus): (p. 203) diffusion of water across a selectively permeable membrane.

osmosis: (pág. 203) difusión del agua a través de una membrana de permeabilidad selectiva.

ossification: (p. 942) formation of bone from osteoblasts.

osificación: (pág. 942) formación ósea a partir de los osteoblastos.

osteoblast: (p. 942) bone-forming cell.

osteoblasto: (pág. 942) célula formadora de hueso.

osteoclast: (p. 943) cell that breaks down bone cells.

osteoclasto: (pág. 943) célula que destruye las células óseas.

osteocyte: (p. 942) living bone cell.

osteocito: (pág. 942) célula ósea viva.

overexploitation: (p. 124) overuse of species with economic value—a factor in species extinction.

sobre-explotación: (pág. 124) uso excesivo de las especies con un valor económico; es un factor en la extinción de especies.

oviduct (OH vuh duct): (p. 1050) tube that transports an egg released from an ovary to the uterus.

oviducto: (pág. 1050) conducto que transporta un óvulo desde el ovario hasta el útero.

P

pacemaker: (p. 995) heart's sinoatrial node, which initiates contraction of the heart.

marcapaso: (pág. 995) nódulo atrioventricular del corazón que inicia la contracción cardíaca.

paleontologist (pay lee ahn TAH luh just): (p. 394) scientist who studies fossils.

paleontólogo: (pág. 394) científico que estudia los fósiles.

palisade mesophyll (mehz uh fihl): (p. 644) leaf-tissue layer that contains many chloroplasts and is the site where most photosynthesis takes place.

mesófilo en empalizada: (pág. 644) capa de tejido de la hoja que contiene muchos cloroplastos y donde se ubica la mayor parte de la fotosíntesis.

pandemic: (p. 1081) widespread epidemic.

pandemia: (pág. 1081) epidemia que se extiende a muchos países.

parasitism (PER uh suh tih zum): (p. 40) symbiotic relationship in which one organism benefits at the expense of another organism.

parasitismo: (pág. 40) relación simbiótica en la cual un organismo se beneficia a expensas de otro.

parasympathetic nervous system: (p. 972) branch of the autonomic nervous system that controls organs and is most active when the body is at rest.

sistema nervioso parasimpático (SNP): (pág. 972) división del sistema nervioso autónomo que controla los órganos y es más activo cuando el cuerpo está en reposo.

parathyroid hormone: (p. 1034) substance produced by the parathyroid gland that increases blood calcium levels by stimulating bones to release calcium.

hormona paratiroides: (pág. 1034) sustancia producida por la glándula tiroides que aumenta los niveles de calcio en la sangre al estimular la liberación de calcio en los huesos.

parenchyma (puh RENG kuh muh) cell: (p. 632) spherical, thin-walled cell found throughout most plants that can function in photosynthesis, gas exchange, protection, storage, and tissue repair and replacement.

célula de parénquima: (pág. 632) célula esférica con paredes delgadas que se encuentra en la mayoría de las plantas y que funciona en la fotosíntesis, el intercambio de gases, la protección, el almacenamiento y reparación o reemplazo de tejidos.

pathogen: (p. 1076) agent, such as a bacterium, virus, protozoan, or fungus, that causes infectious disease.

patógeno: (pág. 1076) agente, como las bacterias, los virus, los protozoarios o los hongos, causante de enfermedades infecciosas.

pedicellaria (peh dih sih LAHR ee uh): (p. 793) small pincher that helps echinoderms catch food and remove foreign materials from the skin.

pedicelarios: (pág. 793) pinza minúscula de los equinodermos que los ayuda a obtener alimento y eliminar objetos extraños de la piel.

pedigree: (p. 299) diagrammed family history that is used to study inheritance patterns of a trait through several generations and that can be used to predict disorders in future offspring.

pedigrí: (pág. 299) historia familiar diagramada que se emplea para el estudio de los patrones hereditarios de un rasgo a través de varias generaciones, capaz de predecir trastornos en la progenie futura.

pedipalp: (p. 772) one of a pair of arachnid appendages used for sensing and holding prey and in male spiders used for reproduction.

pedipalpo: (pág. 772) uno de un par de apéndices de los arácnidos utilizado para manipular la presa y, en las arañas macho, para la reproducción.

peer review: (p. 14) a process in which the procedures used during an experiment may be repeated and the results are evaluated by scientists who are in the same field or are conducting similar research.

evaluación de compañeros: (pág. 14) proceso en que los procedimientos que se usan durante un experimento pueden repetirse y otros científicos en el mismo campo de estudio o que realizan investigaciones similares pueden evaluar los resultados.

pellicle: (p. 547) membrane layer that encloses a paramecium and some other protists.

película: (pág. 547) capa membranosa que encierra un paramecio.

pepsin: (p. 1021) digestive enzyme involved in the stomach's chemical digestion of proteins.

pepsina: (pág. 1021) enzima digestiva presente en la digestión química de las proteínas.

perennial: (p. 621) plant that can live for several years.

perenne: (pág. 621) planta que vive por varios años.

pericycle: (p. 640) plant tissue that produces lateral roots.

periciclo: (pág. 640) tejido vegetal que produce raíces laterales.

period: **(p. 396)** subdivision of an era on the geologic time scale.

peripheral nervous system: **(p. 968)** consists of sensory and motor neurons that transmit information to and from the central nervous system.

peristalsis **(per uh STAHL sus):** **(p. 1021)** rhythmic, wavelike muscular contractions that move food throughout the digestive tract.

petal: **(p. 668)** colorful flower structure that attracts pollinators and provides them a landing place.

petiole **(PET ee ohl):** **(p. 644)** stalk that connects a plant's blade to the stem.

pH: **(p. 165)** measure of the concentration of hydrogen ions (H^+) in a solution.

pharmacogenomics **(far muh koh jeh NAH mihks):** **(p. 378)** study of how genetic inheritance affects the body's response to drugs in order to produce safer and more specific drug dosing.

pharyngeal pouch: **(p. 804)** in chordate embryos, one of the paired structures connecting the muscular tube lining the mouth cavity and the esophagus.

pharynx **(FER ingks):** **(p. 727)** in free-living flatworms, the tubelike muscular organ that can extend out of the mouth and suck food particles into the digestive tract.

phenotype: **(p. 279)** observable characteristic that is expressed as a result of an allele pair.

pheromone **(FER uh mohn):** **(p. 768)** chemical secreted by an animal species to influence the behavior of other members of the same species.

phloem **(FLOH em):** **(p. 638)** vascular plant tissue composed of sieve tube members and companion cells that conducts dissolved sugars and other organic compounds from the leaves and stems to the roots and from the roots to the leaves and stems.

phospholipid bilayer: **(p. 188)** plasma membrane layers composed of phospholipid molecules arranged with polar heads facing the outside and nonpolar tails facing the inside.

photic zone: **(p. 80)** open-ocean zone shallow enough for sunlight to penetrate.

photoperiodism **(foh toh PIHR ee uh dih zum):** **(p. 672)** flowering response of a plant based on the number of hours of darkness it is exposed to.

photosynthesis: **(p. 220)** two-phase anabolic pathway in which the Sun's light energy is converted to chemical energy for use by the cell.

phylogeny **(fy LAH juh nee):** **(p. 491)** evolutionary history of a species.

período: **(pág. 396)** subdivisión de una era en la escala geológica.

sistema nervioso periférico (SNP): **(pág. 968)** compuesto por neuronas sensoriales y motoras que transportan información desde y hacia el sistema nervioso central.

peristaltismo: **(pág. 1021)** serie de contracciones musculares rítmicas ondulantes que mueven el alimento por el esófago.

pétalo: **(pág. 668)** estructura floral colorida que atrae a los agentes polinizadores y les provee un lugar de aterrizaje.

pecíolo: **(pág. 644)** tallito de la hoja que une la lámina foliar con el tallo.

pH: **(pág. 165)** medida de la concentración de iones de hidrógeno (H^+) en una solución.

farmacogenética: **(pág. 378)** estudio de la influencia de la herencia genética en la respuesta corporal a los medicamentos a fin de producir posologías más seguras y específicas.

bolsa faríngea: **(pág. 804)** en los embriones de los cordados, una de las estructuras pareadas que conectan el conducto muscular que cubre la cavidad bucal con el esófago.

faringe: **(pág. 727)** en las planarias, el órgano muscular tubular que se extiende desde la boca y chupa las partículas de alimento hacia el tubo digestivo.

fenotipo: **(pág. 279)** apariencia externa que se expresa como resultado de un par de alelos.

feromona: **(pág. 768)** señal química que secreta una especie animal para influir en el comportamiento de otros miembros de la misma especie.

floema: **(pág. 638)** tejido vascular vegetal formado por los miembros del tubo criboso y células acompañantes que transporta azúcares disueltos y otros compuestos orgánicos de las hojas y tallos hacia las raíces; y de allí a las hojas y tallos.

bicapa fosfolípida: **(pág. 188)** capas membranosas del plasma compuestas por moléculas fosfolípidas cuyas cabezas polares miran hacia fuera y cuyas colas no polares miran hacia adentro.

zona fótica: **(pág. 80)** zona a mar abierto lo suficientemente baja para que penetre la luz solar.

fotoperiodicidad: **(pág. 672)** respuesta de floración de una planta al número de horas de oscuridad a la cual se expone.

fotosíntesis: **(pág. 220)** sendero anabólico bifásico por medio del cual la energía luminosa solar se transforma en energía química para uso de la célula.

filogenia: **(pág. 491)** historia evolutiva de una especie.

phylum (FI lum): (p. 488) taxonomic group of related classes.

pigment: (p. 223) light-absorbing colored molecule, such as chlorophyll and carotenoid, in the thylakoid membranes of chloroplasts.

pilus: (p. 518) hairlike, submicroscopic structure made of protein that can help a bacterial cell attach to environmental surfaces and act as a bridge between cells.

pistil: (p. 669) flower's female reproductive structure; it is usually composed of a stigma, a style, and an ovary.

pituitary gland: (p. 1033) endocrine gland located at the base of the brain; called the "master gland" because it regulates many body functions.

placenta: (p. 887) in most mammals, the specialized organ that provides food and oxygen to the developing young and removes their wastes.

placental mammal: (p. 891) mammal that has a placenta and gives birth to young that need no further development within a pouch.

placental stage: (p. 1063) birthing stage in which the placenta and umbilical cord are expelled from the mother's body.

plankton: (p. 77) tiny marine or freshwater photosynthetic, free-floating autotrophs that serve as a food source for many fish species.

plasma: (p. 997) clear, yellowish fluid portion of the blood.

plasma membrane: (p. 185) flexible, selectively permeable boundary that helps control what enters and leaves the cell.

plasmid: (p. 366) any of the small, circular, double-stranded DNA molecules that can be used as a vector.

plasmodium (plaz MOH dee um): (p. 562) feeding stage of a slime mold in which it is a mobile cytoplasmic mass with many diploid nuclei but no separate cells.

plastron (PLAS trahn): (p. 857) ventral part of a turtle's shell.

platelet: (p. 997) flat cell fragment that functions in blood clotting.

plate tectonics: (p. 400) geologic theory that Earth's surface is broken into several huge plates that move slowly on a partially molten rock layer.

polar body: (p. 1051) tiny cell that is produced and eventually disintegrates in the development of an oocyte.

polar molecule: (p. 161) molecule with oppositely charged regions.

filo: (pág. 488) agrupación taxonómica de clases que se relacionan.

pigmento: (pág. 223) molécula de color que absorbe la luz, como la clorofila y la carotenoide, en la membranas tilacoides de los cloroplastos.

pilus: (pág. 518) estructura submicroscópica filiforme compuesta por proteína que ayuda a una célula bacteriana a adherirse a las superficies ambientales y actuar como puente entre las células.

pistilo: (pág. 669) estructura reproductora femenina de la flor, compuesta generalmente por un estigma, un estilo y un ovario.

glándula pituitaria: (pág. 1033) glándula endocrina localizada en la base del cerebro, conocida como la "glándula maestra" puesto que regula muchas funciones corporales.

placenta: (pág. 887) en la mayoría de los mamíferos, el órgano especializado que provee alimento y oxígeno a la cría en desarrollo y elimina sus desechos.

mamífero placentario: (pág. 891) mamífero con placenta que pare a las crías que no requieren desarrollarse adicionalmente en una bolsa.

etapa placentaria: (pág. 1063) etapa de alumbramiento en la cual la placenta y el cordón umbilical se expulsan del cuerpo de la madre.

plancton: (pág. 77) diminutos autótrofos fotosintéticos marinos o de agua dulce, que flotan libremente y constituyen la fuente alimenticia de muchas especies de peces.

plasma: (pág. 997) porción fluida clara y amarillenta de la sangre.

membrana plasmática: (pág. 185) frontera flexible, selectivamente permeable, que ayuda a controlar lo que entra y sale de la célula.

plásmido: (pág. 366) cualquiera de las pequeñas moléculas de DNA circulares de filamento doble que pueden usarse como vector.

plasmodio: (pág. 562) etapa alimenticia de un hongo plasmódico en la cual es una masa de citoplasma móvil con muchos núcleos diploides pero sin membranas separadas.

plastrón: (pág. 857) parte ventral del caparazón de una tortuga.

plaqueta: (pág. 997) fragmentos celulares planos que funcionan en la coagulación de la sangre.

tectónica de placas: (pág. 400) teoría geológica que afirma que la corteza terrestre se divide en varias placas enormes que se mueven lentamente sobre una capa rocosa parcialmente fundida.

cuerpo polar: (pág. 1051) célula diminuta que se produce y posteriormente se desintegra en el desarrollo de un oocito.

molécula polar: (pág. 161) molécula con regiones cargadas opuestamente.

polar nuclei: (p. 674) in anthophytes, the two nuclei in the center of a megaspore.

polygenic trait: (p. 309) characteristic, such as eye color or skin color, that results from the interaction of multiple gene pairs.

polymer: (p. 167) large molecule formed from smaller repeating units of identical, or nearly identical, compounds linked by covalent bonds.

polymerase chain reaction (PCR): (p. 368) genetic engineering technique that can make copies of specific regions of a DNA fragment.

polyp (PAH lup): (p. 712) tube-shaped, sessile body form of cnidarians.

polyploidy: (p. 285) having one or more extra sets of all chromosomes, which, in polyploid plants, can often result in greater size and better growth and survival.

pons: (p. 970) part of the brain stem that helps control breathing rate.

population: (p. 36) group of organisms of the same species that occupy the same geographic place at the same time.

population density: (p. 92) number of organisms per unit of living area.

population growth rate: (p. 97) how fast a specific population grows.

postanal tail: (p. 803) chordate structure used primarily for locomotion.

posterior: (p. 700) tail end of an animal with bilateral symmetry.

postzygotic isolating mechanism: (p. 437) occurring after formation of a zygote.

predation (prih DAY shun): (p. 38) act of one organism feeding on another organism.

preen gland: (p. 862) oil-secreting gland located near the base of a bird's tail.

prehensile tail: (p. 456) functions like a fifth limb, provides the ability to grasp tree limbs or other objects and can support the body weight of some animals.

prezygotic isolating mechanism: (p. 437) occurring before breeding; produces a fertilized egg, or zygote.

primary succession: (p. 62) establishment of a community in an area of bare rock or bare sand, where no topsoil is present.

prion (PREE ahn): (p. 531) protein that can cause infection or disease.

product: (p. 157) substance formed by a chemical reaction; located on the right side of the arrow in a chemical equation.

profundal zone: (p. 77) deepest, coldest area of a large lake with little light and limited biodiversity.

núcleos polares: (pág. 674) en las antofitas, los dos núcleos en el centro de una megáspora.

rasgo poligénico: (pág. 309) característica, como el color de los ojos o de la piel, que resulta de la interacción de múltiples pares de genes.

polímero: (pág. 167) molécula gigante formada por unidades pequeñas, repetitivas, idénticas, o casi idénticas, de compuestos unidos por enlaces covalentes.

reacción en cadena de polimerasa (RCP): (pág. 368) técnica de ingeniería genética capaz de hacer copias de regiones específicas de un fragmento de DNA.

pólipo: (pág. 712) cuerpo sésil cilíndrico de los cnidarios.

poliploide: (pág. 285) que tiene uno o más grupos de todos los cromosomas que, en plantas poliploides, puede resultar en un mayor tamaño y mejor crecimiento y supervivencia.

pons: (pág. 970) parte del bulbo raquídeo que ayuda a controlar el ritmo de la respiración.

población: (pág. 36) grupo de organismos de la misma especie que viven en la misma localidad geográfica al mismo tiempo

densidad demográfica: (pág. 92) número de organismos por unidad de área o superficie habitable.

tasa de crecimiento demográfico: (pág. 97) el grado de rapidez con que crece una población específica.

cola postnatal: (pág. 803) estructura de los cordados que se usa principalmente para la locomoción.

posterior: (pág. 700) extremo de la cola de un animal con simetría bilateral.

mecanismo aislado postcigótico: (pág. 437) que ocurre después de la formación de un cigoto.

depredación: (pág. 38) modo de nutrición de un organismo al alimentarse de otro.

uropigio: (pág. 862) glándula secretora de aceite localizada cerca de la base de la cola de un ave.

cola prensil: (pág. 456) funciona como una quinta extremidad y permite que algunos animales se agarren de las ramas de los árboles u otros objetos y la cual puede sostener el peso de algunos animales.

mecanismo aislado precigótico: (pág. 437) que ocurre antes de la procreación; produce un óvulo fecundado o cigoto.

sucesión primaria: (pág. 62) colonización en un área de roca o arena desnudas, sin mantillo (capa vegetal superior).

prión: (pág. 531) proteína que puede causar infecciones o enfermedades.

producto: (pág. 157) sustancia formada por una reacción química; localizada en el lado derecho de la flecha en una ecuación química.

zona profunda: (pág. 77) el área más fría y profunda de un lago grande, con poca luz y una biodiversidad limitada.

proglottid (proh GLAH tihd): (p. 730) continuously formed, detachable section of a tapeworm that contains male and female reproductive organs, flame cells, muscles, and nerves; breaks off when its eggs are fertilized and passes out of the host's intestine.

prokaryotic cell: (p. 186) microscopic, unicellular organism without a nucleus or other membrane-bound organelles.

prophase: (p. 248) first stage of mitosis, during which the cell's chromatin condenses into chromosomes.

protein: (p. 170) organic compound made of amino acids joined by peptide bonds; primary building block of organisms.

proteomics: (p. 379) study of the structure and function of proteins in the human body.

prothallus (pro THA lus): (p. 665) heart-shaped, tiny fern gametophyte.

protist: (p. 501) unicellular, multicellular, or colonial eukaryote whose cell walls may contain cellulose; can be plantlike, animal-like, or funguslike.

proton: (p. 148) positively charged particle in an atom's nucleus.

protonema: (p. 664) small, threadlike structure produced by mosses that can develop into the gametophyte plant.

protostome (PROH tuh stohm): (p. 702) coelomate animal whose mouth develops from the opening in the gastrula.

protozoan (proh tuh ZOH un): (p. 542) heterotrophic, unicellular, animal-like protist.

pseudocoelom (soo duh SEE lum): (p. 701) fluid-filled body cavity between the mesoderm and the endoderm.

pseudopod (SEW duh pahd): (p. 550) temporary cytoplasmic extension that sarcodines use for feeding and movement.

puberty: (p. 1049) growth period during which sexual maturity is reached.

punctuated equilibrium: (p. 440) theory that evolution occurs with relatively sudden periods of speciation followed by long periods of stability.

pupa (PYEW puh): (p. 778) nonfeeding stage of complete metamorphosis in which the insect changes from the larval form to the adult form.

proglótido: (pág. 730) sección de una tenia que contiene músculos, nervios, bulbos ciliados y órganos reproductores; desprende cuando sus huevos son fecundados y sale por el intestino del huésped.

célula procariótica: (pág. 186) organismo unicelular, microscópico, sin núcleo u otros organelos limitados por membranas.

profase: (pág. 248) primera etapa de la mitosis, durante la cual la cromatina celular se condensa para formar cromosomas.

proteína: (pág. 170) compuesto orgánico formado por aminoácidos unidos por enlaces pépticos; piedra angular de los organismos.

proteómica: (pág. 379) estudio de la estructura y función de proteínas en el cuerpo humano.

prótalo: (pág. 665) gametofito diminuto de helecho, en forma de corazón.

protista: (pág. 501) eucariota unicelular, multicelular o colonial cuyas paredes celulares pueden contener celulosa; pueden tener forma vegetal, animal o fungosa.

protón: (pág. 148) particular cargada positivamente en el núcleo de un átomo.

protonema: (pág. 664) estructura filamentosa pequeña producida por musgos, capaz de desarrollarse en la planta gametofita.

protostomado: (pág. 702) animal celomado cuya boca se desarrolla de la abertura de la gástrula.

protozoario: (pág. 542) protista unicelular heterótrofo parecido a un animal.

seudoceloma: (pág. 701) cavidad corporal llena de fluido entre el mesodermo y el endodermo.

seudópodos: (pág. 550) extensión citoplásmica temporal que emplean los sarcodinos en la alimentación y locomoción.

pubertad: (pág. 1049) período durante el cual se llega a la madurez sexual.

equilibrio puntuado: (pág. 440) teoría que sostiene que la evolución ocurre con períodos relativamente súbitos de especiación, seguido de largos períodos de estabilidad.

pupa: (pág. 778) etapa no alimenticia de la metamorfosis completa de un insecto en la cual el insecto cambia de la forma larval a la adulta.

radial (RAY dee uhl) symmetry: (p. 700) body plan that can be divided along any plane, through a central axis, into roughly equal halves.

simetría radial: (pág. 700) plano corporal que, a través de un eje central a lo largo de cualquier plano, puede dividirse en casi dos partes iguales.

radicle: **(p. 679)** first part of the embryo to emerge from the seed and begin to absorb water and nutrients from the environment.

radiometric dating: **(p. 395)** method used to determine the age of rocks using the rate of decay of radioactive isotopes.

radula (RA juh luh): **(p. 738)** rasping tonguelike organ with rows of teeth that many mollusks use in feeding.

reactant: **(p. 157)** substance that exists before a chemical reaction starts; located on the left side of the arrow in a chemical equation.

recessive: **(p. 278)** Mendel's name for a specific trait hidden or masked in the F_1 generation.

recombinant DNA: **(p. 366)** newly generated DNA fragment containing exogenous DNA.

red blood cell: **(p. 997)** hemoglobin-containing, disc-shaped, short-lived blood cell that lacks a nucleus and that transports oxygen to all the body's cells.

red bone marrow: **(p. 942)** type of marrow that produces red and white blood cells and platelets.

reflex arc: **(p. 963)** nerve pathway consisting of a sensory neuron, an interneuron, and a motor neuron.

regeneration: **(p. 728)** ability to replace or regrow body parts missing due to predation or damage.

relative dating: **(p. 394)** method used to determine the age of rocks by comparing the rocks with younger and older rock layers.

renewable resource: **(p. 130)** any resource replaced by natural processes more quickly than it is consumed.

reproduction: **(p. 9)** production of offspring.

reservoir: **(p. 1078)** source of a pathogen in the environment.

response: **(p. 9)** organism's reaction to a stimulus.

restriction enzyme: **(p. 364)** bacterial protein that cuts DNA into fragments.

retina: **(p. 974)** innermost layer of the eye that contains rods and cones.

retrovirus: **(p. 530)** RNA virus, such as HIV, with reverse transcriptase in its core.

rhizoid (RIH zoyd): **(p. 583)** type of hypha formed by a mold that penetrates a food's surface.

rhizome: **(p. 615)** fern's thick underground stem that functions as a food-storage organ.

ribosomal RNA (rRNA): **(p. 336)** type of RNA that associates with proteins to form ribosomes.

radícula: **(pág. 679)** primera parte del embrión que emerge de la semilla y comienza a absorber agua y nutrientes del medio ambiente.

datación radiométrica: **(pág. 395)** método utilizado para determinar la edad de las rocas mediante la tasa de desintegración de los isótopos radioactivos.

rádula: **(pág. 738)** órgano raspador parecido a una lengua con hileras de dientes que emplean muchos moluscos para alimentarse.

reactivo: **(pág. 157)** sustancia que existe antes de empezar una reacción química; localizada al lado izquierdo de la flecha en una ecuación química.

recesivo: **(pág. 256)** nombre de Mendel para una rasgo específico oculto o encubierto en la generación F_1.

DNA recombinante: **(pág. 366)** fragmento de DNA recién generado que contiene DNA exógeno.

glóbulo rojo: **(pág. 997)** célula sanguínea de corta vida, esférica, anucleada, que contiene hemoglobina y que transporta oxígeno a todas las células del cuerpo.

médula roja: **(pág. 942)** tipo de médula que produce glóbulos rojos, glóbulos blancos y plaquetas.

arco reflejo: **(pág. 963)** trayecto nervioso que consiste en una neurona sensorial, una interneurona y una neurona motora.

regeneración: **(pág. 728)** capacidad de reemplazar o regenerar partes corporales perdidas debido a la depredación o daños.

datación relativa: **(pág. 394)** método empleado para determinar la edad de las rocas al compararlas con capas rocosas más recientes y más antiguas.

recurso renovable: **(pág. 130)** cualquier recurso reemplazable por procesos naturales de manera más rápida de lo que se consume.

reproducción: **(pág. 9)** producción de la progenie.

reservorio: **(pág. 1078)** fuente de un patógeno en el medio ambiente.

respuesta: **(pág. 9)** la reacción de un organismo a un estímulo.

enzima restrictiva: **(pág. 364)** proteína bacterial que corta el DNA en fragmentos.

retina: **(pág. 974)** capa más interna del ojo que contiene bastoncillos y conos.

retrovirus: **(pág. 530)** virus RNA, como el HIV, con transcriptasa inversa en su núcleo.

rizoide: **(pág. 583)** tipo de hifa formada por un musgo que penetra la superficie del alimento.

rizoma: **(pág. 615)** el tallo grueso subterráneo de un helecho que funciona como órgano de almacenamiento de alimento.

RNA ribosomal (rRNA): **(pág. 336)** tipo de RNA que se asocia con las proteínas para formar ribosomas.

ribosome: **(p. 193)** simple cell organelle that helps manufacture proteins.

ribosoma: **(pág. 193)** organelo celular simple que ayuda a elaborar proteínas.

RNA: **(p. 336)** ribonucleic acid; guides protein synthesis.

RNA: **(pág. 336)** ácido ribonucleico; guía la síntesis de proteínas.

RNA polymerase: **(p. 337)** enzyme that regulates RNA synthesis.

RNA polimerasa: **(pág. 337)** enzima que regula la síntesis de RNA.

rod: **(p. 974)** one of the light-sensitive cells in the retina that sends action potentials to the brain via neurons in the optic nerve.

bastoncillo: **(pág. 974)** una de las células de la retina que es sensible a la luz y que envía potenciales de acción al cerebro mediante las neuronas del nervio óptico.

root cap: **(p. 639)** layer of parenchyma cells that covers the root tip and helps protect root tissues during growth.

piloriza: **(pág. 639)** capa de células del parénquima que cubre la punta de las raíces y ayuda a proteger su tejido durante el crecimiento.

rubisco: **(p. 226)** enzyme that converts inorganic carbon dioxide molecules into organic molecules during the final step of the Calvin cycle.

rubisco: **(pág. 226)** enzima que convierte las moléculas inorgánicas de dióxido de carbono en moléculas orgánicas durante la etapa final del ciclo de Calvin.

S

safety symbol: **(p. 21)** logo representing a specific danger such as radioactivity, electrical or biological hazard, or irritants that may be present in a lab activity or field investigation.

símbolo de seguridad: **(pág. 21)** logotipo que advierte acerca de algún peligro, como radioactividad, agentes irritantes, riesgos eléctricos o biológicos, que pudieran presentarse en una actividad de laboratorio o investigación de campo.

sarcomere: **(p. 948)** in skeletal muscle, the functional unit that contracts and is composed of myofibrils.

sarcómero: **(pág. 948)** en el músculo esquelético, la unidad funcional que se contrae y se compone de miofibrilla muscular.

scale: **(p. 823)** small, flat, platelike structure near the surface of the skin of most fishes; can be ctenoid, cycloid, placoid, or ganoid.

escama: **(pág. 823)** estructura pequeña, plana y lameliforme, cerca de la superficie de la piel de la mayoría de los peces; puede ser serrada, cicloide, placoidea o ganoidea.

science: **(p. 11)** a body of knowledge based on the study of nature.

ciencia: **(pág. 11)** conjunto de conocimiento basado en el estudio de la naturaleza y su entorno físico.

scientific methods: **(p. 16)** a series of problem-solving procedures that might include observations, forming a hypothesis, experimenting, gathering and analyzing data, and drawing conclusions.

método científico: **(pág. 16)** una serie de procedimientos de solución de problemas que pueden incluir observaciones, formulación de una hipótesis, experimentación, recopilación y análisis de datos y sacar conclusiones.

sclerenchyma (skle RENG kuh muh) cell: **(p. 633)** plant cell that lacks cytoplasm and other living components when mature, leaving thick, rigid cell walls that provide support and function in transport of materials.

célula esclerénquima: **(pág. 633)** célula vegetal carente de citoplasma y de otros componentes vitales en su etapa madura, caracterizada por paredes celulares gruesas y rígidas que proveen soporte y funcionan en el transporte de materiales.

scolex (SKOH leks): **(p. 730)** parasitically adapted, knoblike anterior end of a tapeworm, having hooks and suckers that attach to the host's intestinal lining.

escólex: **(pág. 730)** extremidad adaptada parasitariamente con forma de perilla que poseen las tenias; posee ganchos y chupones que se adhieren a la cubierta intestinal del huésped.

sebaceous gland: **(p. 937)** oil-producing gland in the dermis that lubricates skin and hair.

glándula sebácea: **(pág. 937)** glándula productora de aceite en la dermis que lubrica la piel y el cabello.

secondary succession: **(p. 63)** orderly change that occurs in a place where soil remains after a community of organisms has been removed.

sucesión secundaria: **(pág. 63)** cambio ordenado que ocurre en el suelo de los lugares que experimentaron la expulsión de una comunidad de organismos.

sediment: **(p. 75)** material deposited by water, wind, or glaciers.

sedimento: **(pág. 75)** material depositado por el agua, el viento o los glaciares.

seed: (p. 607) adaptive reproductive structure of some vascular plants that contains an embryo, nutrients for the embryo, and is covered by a protective coat.

semilla: (pág. 607) estructura reproductora y adaptable de algunas plantas vasculares que contiene un embrión con su fuente de nutrientes y una capa protectora.

seed coat: (p. 677) protective tissue that forms from the hardening of the outside layers of the ovule.

tegumento: (pág. 677) tejido protector formado del endurecimiento de las capas externas del óvulo.

selective breeding: (p. 360) directed breeding to produce plants and animals with desired traits.

criaza selectiva: (pág. 360) crianza dirigida hacia la producción de plantas y animales con rasgos deseados.

selective permeability (pur mee uh BIH luh tee): (p. 187) property of the plasma membrane that allows it to control movement of substances into or out of the cell.

permeabilidad selectiva: (pág. 187) propiedad de la membrana plasmática que le permite controlar el movimiento de las sustancias dentro o fuera de la célula.

semen (SEE mun): (p. 1049) fluid that contains sperm, nourishment, and other fluids of the male reproductive system.

semen: (pág. 1049) fluido que contiene espermatozoides, nutrientes y otros fluidos del sistema reproductor masculino.

semicircular canal: (p. 975) inner-ear structure that transmits information about body position and balance to the brain.

canal semicircular: (pág. 975) estructura interna del oído que transmite al cerebro información relativa a la posición y equilibrio corporales.

semiconservative replication: (p. 333) method of DNA replication in which parental strands separate, act as templates, and produce molecules of DNA with one parental DNA strand and one new DNA strand.

replicación semiconservadora: (pág. 333) método de replicación del DNA mediante el cual los filamentos paternos se separan, actúan como plantillas y producen moléculas de DNA con un filamento paterno de DNA y otro nuevo de DNA.

seminiferous tubule (se muh NIHF rus • TEW byul): (p. 1049) tubule of the testis in which sperm develop.

túbulo seminífero: (pág. 1049) túbulo del teste donde se desarrollan los espermatozoides.

sepal: (p. 668) flower organ that protects the bud.

sépalo: (pág. 668) órgano de la flor que protege el botón.

septum: (p. 578) cross-wall that divides a hypha into cells.

septo: (pág. 578) tabique que divide una hifa en células.

serendipity: (p. 18) occurrence of accidental or unexpected but fortunate outcomes.

serendipia: (pág. 18) hecho accidental o inesperado con resultados favorables.

sessile (SEH sul): (p. 706) organism permanently attached to one place.

sésil: (pág. 706) organismo que permanece adherido a una superficie.

seta (SEE tuh): (p. 747) tiny bristle that digs into soil and anchors an earthworm as it moves forward.

seta: (pág. 747) pequeña cerda que penetra en el suelo y provee el soporte que requiere una lombriz al avanzar.

sex chromosome: (p. 305) X or Y chromosome; paired sex chromosomes determine an individual's gender—XX individuals are female and XY individuals are male.

cromosoma sexual: (pág. 305) cromosoma X o Z; los cromosomas sexuales pareados determinan el sexo del individuo: los individuos XX son femeninos y los XY, masculinos.

sex-linked trait: (p. 307) characteristic, such as red-green color blindness, controlled by genes on the X chromosome; also called an X-linked trait.

rasgo ligado al sexo: (pág. 307) característica, como el daltonismo, controlada por los genes en el cromosoma X; también denominado rasgo ligado a la X.

sexual selection: (p. 436) change in the frequency of a trait based on competition for a mate.

selección sexual: (pág. 436) cambio de la frecuencia de un rasgo basado en la rivalidad por una pareja.

short-day plant: (p. 672) plant that flowers in the winter, spring, or fall, when the number of hours of darkness is greater than the number of hours of light.

planta de días cortos: (pág. 672) planta que florece en el invierno, primavera u otoño cuando el número de horas de oscuridad es mayor que el número de horas diurnas.

SI: (p. 14) system of measurements used by scientists, abbreviation of the International System of Units.

SI: (pág. 14) sistema de medición que usan los científicos, abreviatura del Sistema Internacional de Unidades.

sieve tube member: (p. 638) nonnucleated, cytoplasmic cell of the phloem.

miembro de los tubos cribosos: (pág. 638) célula citoplasmática del floema que carece de núcleo.

single nucleotide polymorphism: (p. 376) variation in a DNA sequence occurring when a single nucleotide in a genome is altered.

polimorfismo de un nucleótido simple: (pág. 376) variación en una secuencia de DNA que ocurre al alterarse un solo nucleótido en un genoma.

siphon **(p. 741)** tubular organ through which octopuses and squids eject water, at times so rapidly that their movement appears jet-propelled.

sister chromatid: **(p. 248)** structure that contains identical DNA copies and is formed during DNA replication.

skeletal muscle: **(p. 948)** striated muscle that causes movement when contracted and is attached to bones by tendons.

small intestine: **(p. 1022)** longest part of the digestive tract; involved in mechanical and chemical digestion.

smooth muscle: **(p. 947)** muscle that lines many hollow internal organs, such as the stomach and uterus.

solute: **(p. 163)** substance dissolved in a solvent.

solution: **(p. 163)** homogeneous mixture formed when a substance (the solute) is dissolved in another substance (the solvent).

solvent: **(p. 163)** substance in which another substance is dissolved.

somatic nervous system: **(p. 971)** part of the peripheral nervous system that transmits impulses to and from skin and skeletal muscles.

sorus: **(p. 616)** fern structure formed by clusters of sporangia, usually on the undersides of a frond.

spawning: **(p. 826)** process by which male and female fishes release their gametes near each other in the water.

species: **(p. 9)** group of organisms that can interbreed and produce fertile offspring.

species diversity: **(p. 117)** in a biological community, the number and abundance of different species.

spindle apparatus: **(p. 250)** structure made of spindle fibers, centrioles, and aster fibers that is involved in moving and organizing chromosomes before the cell divides.

spinneret: **(p. 772)** in spiders, the structure that spins silk from a fluid protein secreted by their glands.

spiracle **(SPIHR ih kul):** **(p. 767)** opening in the arthropod body through which air enters and waste gases leave.

spongy bone: **(p. 942)** less dense inner-bone layer with many cavities that contain bone marrow.

spongy mesophyll: **(p. 644)** loosely packed, irregularly shaped cells with spaces around them located below the palisade mesophyll.

spontaneous generation: **(p. 401)** idea that life arises from nonliving things.

sporangium: **(p. 581)** sac or case in which fungal spores are produced.

sifón: **(pág. 741)** órgano tubular por el cual los pulpos y los calamares expulsan agua, a veces tan rápido, que asemeja una propulsión a chorro.

cromátides hermanas: **(pág. 248)** estructura formada durante la replicación del DNA, que contiene copias idénticas de DNA.

músculo óseo: **(pág. 948)** músculo estriado que causa movimiento al contraerse, adherido a los huesos por los tendones.

intestino delgado: **(pág. 1022)** parte más larga del tracto digestivo; presente en la digestión mecánica y química.

músculo liso: **(pág. 947)** músculo que recubre las paredes de muchos órganos internos, como el estómago y el útero.

soluto: **(pág. 163)** sustancia disuelta en un disolvente.

solución: **(pág. 163)** mezcla homogénea formada al disolverse una sustancia (el soluto) en otra sustancia (el disolvente).

disolvente: **(pág. 163)** sustancia en la cual se disuelve otra sustancia.

sistema nervioso somático: **(pág. 971)** porción del sistema nervioso periférico que transmite impulsos hacia y desde la piel a los músculos esqueléticos.

soro: **(pág. 616)** estructura de helecho formada por grupos de esporangios, ubicada generalmente en la super ficie inferior de una fronda.

desove: **(pág. 826)** proceso mediante el cual tanto los peces macho como las hembras liberan sus gametos cerca uno del otro en el agua.

especie: **(pág. 9)** grupo de organismos que pueden cruzarse y producir progenies fértiles.

diversidad de especies: **(pág. 117)** en una comunidad biológica, el número y la abundancia de diferentes especies.

huso: **(pág. 250)** estructura compuesta por fibras de microtúbulos, centriolos y áster encargada de movilizar y organizar los cromosomas antes de la división celular.

hileras: **(pág. 772)** en las arañas, la estructura productora de seda a partir del fluido proteico que segregan sus glándulas.

espiráculo: **(pág. 767)** abertura en el cuerpo de los artrópodos a través de la cual entra el aire y salen los gases de desecho.

hueso esponjoso: **(pág. 942)** capa de hueso interno menos densa con muchos orificios que contienen médula.

mesófilo esponjoso: **(pág. 644)** células irregulares, ligeramente empacadas, con espacios circundantes localizados bajo el mesófilo en empalizada.

generación espontánea: **(pág. 401)** idea de que la vida surge de la materia no viva.

esporangio: **(pág. 581)** en los hongos, un saco o envoltura donde se producen las esporas.

spore: (p. 580) reproductive haploid (*n*) cell with a hard outer shell that forms a new organism without the fusion of gametes and is produced in the asexual and sexual life cycles of most fungi and some other organisms.

stabilizing selection: (p. 434) most common form of natural selection in which organisms with extreme expressions of a trait are removed.

stamen: (p. 669) male reproductive organ of most flowers composed of a filament and an anther.

stem cell: (p. 256) unspecialized cell that can develop into a specialized cell under the right conditions.

sternum (STUR num): (p. 862) in birds, the large breastbone to which flight muscles are attached.

stimulant: (p. 978) substance/drug that increases alertness and physical activity.

stimulus: (p. 9) any change in an organism's internal or external environment that causes the organism to react.

stolon (STOH lun): (p. 583) type of hypha formed by a mold that spreads over a food's surface.

stomata: (p. 606) openings in the outer cell layer of leaf surfaces and some stems that allow the exchange of water, carbon dioxide, oxygen, and other gases between a plant and its environment.

strobilus (STROH bih lus): (p. 613) compact cluster of spore-bearing structures in some seedless vascular plant sporophytes.

stroma: (p. 223) fluid-filled space outside the grana in which light-dependent reactions take place.

substrate: (p. 160) reactant to which an enzyme binds.

sustainable use: (p. 130) use of resources at a rate that they can be replaced or recycled.

swim bladder: (p. 827) gas-filled internal space in bony fishes that allows them to regulate their buoyancy.

swimmeret: (p. 771) crustacean appendage used as a flipper during swimming.

symbiosis (sihm bee OH sus): (p. 39) close mutualistic, parasitic, or commensal association between two or more species that live together.

symmetry (SIH muh tree): (p. 700) balance or similarity in body structures of organisms.

sympathetic nervous system: (p. 972) branch of the autonomic nervous system that controls organs and is most active during emergencies or stress.

sympatric speciation: (p. 438) occurs when a species evolves into a new species in an area without a geographic barrier.

espora: (pág. 580) célula reproductora haploide (n) con una cubierta protectora dura capaz de formar un nuevo organismo sin la fusión de gametos; y se produce en los ciclos vitales sexuales y asexuales de la mayoría de los hongos.

selección estabilizadora: (pág. 434) la selección natural más común, mediante la cual se eliminan los organismos con expresiones extremas de un rasgo.

estambre: (pág. 669) órgano reproductor masculino de la mayoría de las flores, compuesto por un filamento y una antera.

célula madre: (pág. 256) célula no especializada capaz de desarrollarse en una célula especializada bajo las condiciones adecuadas.

esternón: (pág. 862) en las aves, hueso pectoral grande al cual se adhieren los músculos de vuelo.

estimulante: (pág. 978) sustancia / droga que aumenta la agudeza y actividad física.

estímulo: (pág. 9) cualquier cambio en el ambiente interno o externo de un organismo que ocasiona una reacción en el organismo.

estolón: (pág. 583) tipo de hifa formada por un hongo que se extiende sobre la superficie de los alimentos.

estomas: (pág. 606) aberturas en la capa celular externa superficial de las hojas y de algunos tallos que permiten el intercambio de agua, dióxido de carbono y otros gases entre una planta y su medioambiente.

estróbilo: (pág. 613) racimo compacto de estructuras que contienen esporas en algunos esporófitos de plantas vasculares sin semilla.

estroma: (pág. 223) espacio relleno de fluido fuera de las granas donde suceden reacciones lumino-dependientes.

sustrato: (pág. 160) reactivo al cual se adhiere una enzima.

uso sostenible: (pág. 130) uso de los recursos a una tasa tal que puedan reemplazarse o reciclarse.

vejiga natatoria: (pág. 827) espacio interno relleno de gas en los peces óseos que les ayuda a controlar su flotabilidad.

pleópodo: (pág. 771) apéndice crustáceo empleado como aleta durante la natación.

simbiosis: (pág. 39) asociación estrecha mutualista, parasítica o comensal entre dos o más especies que viven juntas.

simetría: (pág. 700) equilibrio o similitud en las estructuras corporales de los organismos.

sistema nervioso simpático: (pág. 972) rama del sistema nervioso autónomo que controla los órganos y es muy activo durante las emergencias o el estrés.

especiación simpátrica: (pág. 438) ocurre cuando una especie evoluciona en una especie nueva dentro de un área sin frontera geográfica.

synapse (SIH naps): (p. 967) gap between one neuron's axon and another nueron's dendrite.

sinapsis: (pág. 967) brecha entre el axón de una neurona y las dendritas de otra.

T

taste bud: (p. 973) one of a number of specialized chemical receptors on the tongue that can detect sweet, sour, salty, and bitter tastes.

papila gustativa: (pág. 973) una de un número de receptores químicos especializados de la lengua que detectan los sabores dulces, agrios, salados y amargos.

taxon: (p. 487) named group of organisms, such as a phylum, genus, or species.

taxón: (pág. 487) grupo nombrado de organismos, como un filo, un género o una especie.

taxonomy (tak SAH nuh mee): (p. 485) branch of biology that identifies, names, and classifies species based on their natural relationships.

taxonomía: (pág. 485) rama de la biología que identifica, nombra y clasifica las especies en base a su morfología y comportamiento.

technology (tek NAH luh jee): (p. 15) application of knowledge gained from scientific reasearch to solve society's needs and problems and improve the quality of life.

tecnología: (pág. 15) aplicación del conocimiento derivado de la investigación científica a fin de resolver los problemas y necesidades de la sociedad y mejorar la calidad de vida.

telomere: (p. 311) protective cap made of DNA that is found on the ends of a chromosome.

telómero: (pág. 311) capa protectora de DNA que se encuentra en los extremos de un cromosoma.

telophase: (p. 251) last stage of mitosis in which nucleoli reappear. Two new nuclear membranes begin to form, but the cell has not yet completely divided.

telofase: (pág. 251) fase final de la mitosis en que reaparecen los nucléolos. Comienzan a formarse dos nuevas membranas nucleares sin que la célula haya terminado de dividirse.

temperate forest: (p. 69) biome south of the boreal forest characterized by broad-leaved, deciduous trees, well-defined seasons, and average yearly precipitation of 75–150 cm.

bosque templado: (pág. 69) bioma al sur del bosque boreal compuesto por árboles caducifolios de hojas anchas, estaciones bien definidas y entre 70 y 150 cm de precipitación promedio anual.

tendon: (p. 948) tough connective-tissue band that connects muscle to bone.

tendón: (pág. 948) banda dura de tejido conectivo que adhieren los músculos a los huesos.

territorial behavior: (p. 918) competitive behavior in which an animal tries to adopt and defend a physical area against others of the same species.

comportamiento territorial: (pág. 918) comportamiento competitivo mediante el cual un animal trata de adoptar y defender un área física de otros animales de la misma especie.

test: (p. 550) hard, porous, shell-like covering of an amoeba.

testa: (pág. 550) cubierta dura, porosa, con forma de cáscara, de una ameba.

test cross: (p. 362) breeding that can be used to determine an organism's genotype.

cruzamiento de prueba: (pág. 362) cruce que puede ayudar a determinar el genotipo de un organismo.

tetrapod: (p. 830) four-footed animal with legs that have feet and toes with joints.

tetrápodo: (pág. 830) animal cuadrúpedo cuyas patas tienen pies y dedos con articulaciones.

thallose (THAL lohs): (p. 612) liverwort with a fleshy, lobed body shape.

talosa: (pág. 612) hepática con forma corporal carnosa lobulada.

theory: (p. 14) explanation of a natural phenomenon based on many observations and investigations over time.

teoría: (pág. 14) explicación de un fenómeno natural basado en muchas observaciones y experimentos con el correr del tiempo.

theory of biogenesis (bi oh JEN uh sus): (p. 402) states that only living organisms can produce other living organisms.

teoría de la biogénesis: (pág. 402) plantea que sólo los organismos vivos pueden producir otros organismos vivos.

therapsid: (p. 896) extinct mammal-like reptile from which the first mammals probably arose.

terápsido: (pág. 896) reptil extinto parecido a un mamífero del cual probablemente surgieron los primeros mamíferos.

thermodynamics: (p. 218) study of the flow and transformation of energy in the universe.

termodinámica: (pág. 218) estudio del flujo y transformación de la energía del universo.

thorax: (p. 763) middle body region of an arthropod consisting of three fused main segments that may bear legs and wings.

tórax: (pág. 763) región del cuerpo medio de un artrópodo compuesta por tres segmentos principales fusionados capaz de soportar patas y alas.

threshold: (p. 964) minimum stimulus needed to produce a nerve impulse.

umbral: (pág. 964) estímulo mínimo requerido para producir un impulso nervioso.

thylakoid: (p. 223) in choroplasts, one of the stacked, flattened, pigment-containing membranes in which light-dependent reactions occur.

tilacoide: (pág. 223) en los cloroplastos, una de las membranas apiladas y aplanadas que contienen pigmento donde ocurren las reacciones lumino-dependientes.

thyroxine: (p. 1034) thyroid hormone that increases the metabolic rate of cells.

tiroxina: (pág. 1034) hormona tiroidea que aumenta la tasa metabólica de las células.

tolerance: (p. 61) organism's ability to survive biotic and abiotic factors. **(p. 981)** as the body becomes less responsive to a drug, an individual needs larger and more frequent doses to achieve the same effect.

tolerancia: (pág. 61) capacidad de un organismo de sobrevivir factores bióticos y abióticos. **(pág. 981)** a medida que el cuerpo se vuelve menos sensible a una droga, un individuo necesita dosis más frecuentes y mayores para obtener el mismo efecto.

trachea: (p. 1001) tube that carries air from the larynx to the bronchi.

tráquea: (pág. 1001) conducto que lleva el aire desde la laringe hasta los bronquios.

tracheal (TRAY kee ul) tube: (p. 767) in most terrestrial arthropods, one of a system of tubes that branch into smaller tubules and carry oxygen throughout the body.

tubo traqueal: (pág. 767) en la mayoría de los artrópodos terrestres, uno entre un sistema de conductos que se ramifican en otros más pequeños y transportan el oxígeno por todo el cuerpo.

tracheid (TRAY key ihd): (p. 637) long, cylindrical plant cell in which water passes from cell to cell through pitted ends.

traqueida: (pág. 637) célula vegetal alargada y cilíndrica en la cual pasa el agua de célula a célula a través de extremos picados.

transcription (trans KRIHP shun): (p. 337) process in which mRNA is synthesized from the template DNA.

transcripción: (pág. 337) proceso en que el mRNA se sintetiza del patrón de DNA.

transfer RNA: (p. 336) type of RNA that transports amino acids to the ribosome.

RNA de transferencia: (pág. 336) tipo de RNA que transporta los aminoácidos a los ribosomas.

transformation: (p. 367) process in which bacterial cells take up recombinant plasmid DNA.

transformación: (pág. 367) proceso en el cual las células bacterianas recogen el DNA plásmido recombinante.

transgenic organism: (p. 370) organism that is genetically engineered by inserting a gene from another organism.

organismo transgénico: (pág. 370) organismo generado genéticamente al insertar el gene de un organismo distinto.

translation: (p. 338) process in which mRNA attaches to the ribosome and a protein is assembled.

traducción: (pág. 338) proceso mediante el cual el mRNA se adhiere al ribosoma y se sintetiza una proteína.

transpiration: (p. 645) process in which water evaporates from the inside of leaves to the outside through stomata.

transpiración: (pág. 645) proceso en el cual el agua se evapora de adentro hacia fuera de las hojas a través de los estomas.

transport protein: (p. 189) protein that moves substances or wastes through the plasma membrane.

proteína de transporte: (pág. 189) proteína que mueve sustancias o desechos a través de la membrana plasmática.

trichinosis (trih kuh NOH sus): (p. 733) disease caused by eating raw or undercooked meat, usually pork, infected with *Trichinella* larvae.

triquinosis: (pág. 733) enfermedad causada por la carne cruda o poco cocinada, generalmente de cerdo, infectada con larvas de la *Trichinella.*

trichocyst (TRIH kuh sihst): (p. 547) elongated, cylindrical structure that can discharge a spinelike structure that may function in defense, as an anchoring device, or to capture prey.

tricocisto: (pág. 547) estructura alargada y cilíndrica que puede descargar una estructura husiforme capaz de reaccionar como defensa, como sistema de anclaje o captura de presa.

trophic (TROH fihk) level: (p. 42) each step in a food chain or food web.

nivel trófico: (pág. 42) cada paso de una cadena o red alimenticia.

tropical rain forest: (p. 72) hot, wet biome with year-round humidity; contains Earth's most diverse species of plants and animals.

pluviselva tropical: (pág. 72) bioma caliente, lluvioso, con una humedad anual continua; contiene las especies más diversas de plantas y animales terrestres.

tropical savanna: (p. 71) biome characterized by grasses and scattered trees, and herd animals such as zebras and antelopes.

tropical seasonal forest: (p. 71) biome characterized by deciduous and evergreen trees, a dry season, and animal species that include monkeys, elephants, and Bengal tigers.

tropism (TROH pih zum): (p. 651) response to an external stimulus in a specific direction.

tube foot: (p. 795) one of the muscular, small, fluid-filled tubes with suction-cuplike ends that enable echinoderms to move and collect food.

tundra: (p. 68) treeless biome with permanently frozen soil under the surface and average yearly precipitation of 15–25 cm.

tympanic (tihm PA nihk) membrane: (p. 837) eardrum.

sabana tropical: (pág. 71) bioma caracterizado por hierbas, árboles dispersos y animales que se agrupan en manadas, como cebras y antílopes.

bosque estacional tropical: (pág. 71) bioma caracterizado por árboles caducifolios y siempreverdes, una estación seca y especies de animales que incluyen a los monos, los elefantes y los tigres de Bengala.

tropismo: (pág. 651) crecimiento de una planta en respuesta a estímulos externos proveniente de una dirección específica.

pie ambulacral: (pág. 795) uno de los conductos musculares pequeños rellenos de fluido y con ventosas de los equinodermos que posibilita el movimiento y la recolección de alimento.

tundra: (pág. 68) bioma carente de árboles, con suelo permanentemente congelado bajo la superficie; y una precipitación promedio anual de 15-25 cm.

membrana timpánica: (pág. 837) tímpano.

U

urea: (p. 1006) nitrogenous waste product of the excretory system.

urethra (yoo REE thruh): (p. 1049) tube that conducts semen and urine out of the body through the penis in males and transports urine out of the body in females.

uterus: (p. 887) saclike muscular female organ in which embryos develop.

urea: (pág. 1006) producto de desecho nitrogenado del sistema excretorio.

uretra: (pág. 1049) conducto que conduce el semen y la orina fuera del cuerpo a través del pene en los machos y transporta la orina fuera del cuerpo de las hembras.

útero: (pág. 887) órgano femenino muscular con forma de saco hueco donde se desarrollan los embriones.

V

vacuole: (p. 195) membrane-bound vesicle for temporary storage of materials such as food, enzymes, and wastes.

valve: (p. 994) one of the tissue flaps in veins that prevents backflow of blood.

van der Waals forces: (p. 155) attractive forces between molecules.

vascular cambium: (p. 634) thin cylinder of meristematic tissue that produces new transport cells.

vascular plant: (p. 606) type of plant with vascular tissues adapted to land environments; most widely distributed type of plant on Earth.

vascular tissue: (p. 606) specialized tissue that transports water, food, and other substances in vascular plants and can also provide structure and support.

vacuola: (pág. 195) espacio encerrado por una membrana para el almacenamiento temporal de materiales como alimento, enzimas y desechos.

válvula: (pág. 994) uno de los opérculos de los tejidos en las venas que evita que la sangre fluya hacia atrás.

fuerzas de van der Waals: (pág. 155) fuerzas de atracción entre las moléculas.

cámbium vascular: (pág. 634) cilindro delgado de tejido meristémico que produce células de transporte nuevas.

planta vascular: (pág. 606) tipo de planta con tejido vascular adaptada a ambientes terrestres; tipo de planta ampliamente distribuida en la Tierra.

tejido vascular: (pág. 606) tejido especializado que transporta agua, alimento y otras sustancias en las plantas vasculares y también proveen estructura y soporte.

vas deferens (VAS • DEF uh runz): (p. 1049) duct through which sperm move away from the testis and toward the urethra.

conducto deferente: (pág. 1049) ducto por el cual los espermatozoides se alejan de los testículos hacia la uretra.

vegetative reproduction: (p. 662) asexual reproduction in which new plants grow from parts of an existing plant.

reproducción vegetativa: (pág. 662) reproducción asexual en la cual crecen plantas nuevas de las partes de una planta existente.

vein: (p. 994) blood vessel that carries deoxygenated blood back to the heart.

vena: (pág. 994) vaso sanguíneo que devuelve la sangre desoxigenada al corazón.

ventral (VEN trul): (p. 700) underside or belly of an animal with bilateral symmetry.

ventral: (pág. 700) la superficie inferior o barriga de un animal con simetría bilateral.

ventricle: (p. 824) the heart chamber that pumps blood from the heart to the gills.

ventrículo: (pág. 894) la cavidad cardíaca que bombea la sangre del corazón a las agallas.

vertebrate: (p. 693) animal with an endoskeleton and a backbone.

vertebrado: (pág. 693) animal que posee endoesqueleto y columna vertebral.

vessel element: (p. 637) elongated, tubular plant cell that forms xylem strands (vessels) and conducts water and dissolved substances.

elemento vascular: (pág. 637) células vegetales alargadas, tubulares, que forman filamentos de xilema (vasos) y conducen agua y sustancias disueltas.

vestigial structure: (p. 425) reduced form of a functional structure that indicates shared ancestry.

estructura vestigial: (pág. 425) forma reducida de una estructura funcional que indica ascendencia compartida.

villus (VIH luhs): (p. 1023) fingerlike structure through which most nutrients are absorbed from the small intestine.

vellosidad: (pág. 1023) estructura en forma de dedos por la cual el intestino delgado absorbe la mayor parte de los nutrientes.

virus: (p. 525) nonliving strand of genetic material that cannot replicate on its own, has a nucleic acid core, a protein coat, and can invade cells and alter cellular function.

virus: (pág. 525) hebra sin vida, de material genético, incapaz de duplicarse por sí misma; tiene un núcleo de ácido nucleico, un revestimiento de proteína y puede invadir las células y alterar sus funciones.

vitamin: (p. 1028) fat-soluble or water-soluble organic compound needed in very small amounts for the body's metabolic activities.

vitamina: (pág. 1028) compuesto orgánico liposoluble o hidrosoluble, que se necesita en porciones muy pequeñas para las actividades metabólicas del cuerpo.

voluntary muscle: (p. 948) consciously controlled skeletal muscle.

músculo voluntario: (pág. 948) músculo esquelético controlado en forma consciente.

W

water-vascular system: (p. 795) system of fluid-filled, closed tubes that allow echinoderms to control movement and get food.

sistema vascular acuático: (pág. 795) sistema de conductos cerrados, rellenos de fluido, que permite a los equinodermos controlar el movimiento y obtener alimento.

weather: (p. 65) atmospheric conditions such as temperature and precipitation at a specific place and time.

tiempo: (pág. 65) condiciones atmosféricas, como la temperatura y la precipitación, en un lugar y tiempo específico.

wetland: (p. 78) water-saturated land area that supports aquatic plants.

humedal: (pág. 78) terreno saturado de agua que mantiene a las plantas acuáticas.

white blood cell: (p. 998) large, nucleated, disease-fighting blood cell produced in the bone marrow.

glóbulo blanco: (pág. 998) célula sanguínea gigante y nucleada que combate las enfermedades y se produce en la médula ósea.

woodland: (p. 69) biome characterized by small trees and mixed shrub communities.

zona boscosa: (pág. 69) bioma caracterizado por árboles pequeños y comunidades de arbustos mixtas.

X

xylem (ZI lum): (p. 637) vascular plant tissue that transports water and dissolved minerals away from the roots throughout the plant and is composed of vessel elements and tracheids.

xilema: (pág. 637) tejido vegetal vascular que transporta el agua y los minerales disueltos desde las raíces hacia el resto de la planta, compuesto por elementos de los vasos y traqueidas.

Y

yellow bone marrow: (p. 942) type of marrow that consists of stored fat.

médula ósea amarilla: (pág. 942) tipo de médula que consiste en grasas almacenadas.

Z

zero population growth (ZPG): (p. 104) occurs when the birthrate equals the death rate.

zygote (ZI goht): (p. 695) fertilized egg formed when a sperm cell penetrates an egg.

crecimiento demográfico nulo (CDN): (pág. 104) sucede cuando la tasa de natalidad es igual a la tasa de mortalidad.

cigoto: (pág. 695) óvulo fecundado que se forma cuando un espermatozoide fecunda un óvulo.

Index

Index Key

Italic page numbers = illustration/photo **Bold page numbers = vocabulary term**
Italic page numbers + act = activity

A

Index

B

Index

Index

Index

Index

Index

E

Index

G

Index

J

K

L

Index

Index

Index

S

Index

U

V

Index

Art Credits

Acknowledgements: Glencoe would like to acknowledge the artists and agencies who participated in illustrating this program: Alan Male, represented by American Artists Representatives, Inc.; Annette Lasker; Argosy; Articulate Graphics, represented by Deborah Wolfe Limited; Barbara Harmon; Barbara Higgins Bond, represented by American Artists Representatives, Inc.; Michael Rothman, represented by Melissa Turk & The Artist Network; Morgan Cain & Associates; Wendy Smith, represented by Melissa Turk & The Artist Network.

Photo Credits

Cover Steve Bloom/Stevebloom.com; **Frontsheet** Peter Griffith/Masterfile; **iv** (t) Alton Biggs, (tc) Whitney Hagins, (c) William G. Holliday, (bc) Dr. Chris Kapicka, (b) Linda Lundgren; **v** (t) Ann Haley MacKenzie, (tc) Dr. William Rogers, (bc) Dr. Marion Sewer, (b) Dinah Zike; **xxix** Michael Newman/PhotoEdit; **xxxii** (l) Charles D. Winters/Photo Researchers, (r) Dr. Bayard Brattstrom/Visuals Unlimited; **xxxix** (l) Robert Calentine/Visuals Unlimited, (r) Wolfgang Baumeister/SPL/Photo Researchers; **xli** Matt Meadows**; xxxviii** SCIMAT/Photo Researchers; **2** (t) Photo Library International/Photo Researchers, (c) CORBIS, (b) SPL/Photo Researchers, (bkgd) Myron Jay Dorf/CORBIS; **4** Karl Ammann/CORBIS; **5** (t) Reinhard Dirscherl/Visuals Unlimited, (b) Mike Derer/AP/Wide World Photos; **6** (t) Salk Institute for Biological Studies, (b) Dr. Fred Hossler/Visuals Unlimited; **7** (t to b) M.I. Walker/Photo Researchers, **(2)** Tom J. Ulrich/Visuals Unlimited, **(3 4)** Gary Meszaros/Visuals Unlimited, **(5)** Tom McHugh/Photo Researchers, **(6)** W. Wisniewski/zefa/CORBIS, **(7)** OSF/G.I. Bernard/Animals Animals, **(8)** Ron Fehling/Masterfile, **(9)** Stephen J. Krasemann/Photo Researchers; **8** (l) Alan & Sandy Carey/Peter Arnold, Inc., (c) Richard Nowitz/National Geographic Image Collection, (r) Joe McDonald/CORBIS; **9** (l) Hal Horwitz/CORBIS, (r) David M. Dennis/Animals Animals; **10** Gerry Ellis/Minden Pictures; **11** Jeremy Bishop/SPL/Photo Researchers; **12** (t) Mary Evans Picture Library/The Image Works, (c) Dr. Tim Evans/Photo Researchers, (b) SPL/Photo Researchers; **13** (t) John Reader/SPL/Photo Researchers, (b) Andrew Syred/Photo Researchers; **15** David Parker/SPL/Photo Researchers; **16** (l) Frans Lanting/Minden Pictures; (r) Reprinted by arrangement from the book *Field Guide to Birds of North America,* Fourth edition. Copyright © 1983, 1987, 1999, 2002 National Geographic Society; **18** George McCarthy/CORBIS; **20** Azure Computer & Photo Services/Animals Animals; **22** courtesy of California University; **23** Fred Habegger/Grant Heilman Photography; **25** Brandon Cole/Visuals Unlimited; **28–29** Ron Niebrugge; **30** (t) Galen Rowell/CORBIS, (c) Paul Souders/Getty Images, (b) Dan Suzio/Photo Researchers, (bkgd) CORBIS; **32** (l) CORBIS, (r) Yann Arthus-Bertran/CORBIS; **33** (t) Flip Nicklin/Minden Pictures, (bl) Nic Paget-Clark/In Motion Magazine, (br) Radu Sigheti/Reuters/CORBIS; **34** (t) NASA, (b) NASA Goddard Space Flight Center; **35** National Geographic/Getty Images; 37 Photo Library International/ESA/Photo Researchers; **38** (t) CORBIS, (b) Martin Harvey/Foto Natura/Minden Pictures; **39** Ken Lucas/Visuals Unlimited; **40** RC Hall/Custom Medical Stock Photo; **41** Jeffrey Lepore/Photo Researchers; **42** Michael P. Gadomski/Photo Researchers; **45** K. Oster/CORBIS; **47** Kevin Schafer/CORBIS; **50** Larry Lee/CORBIS; **51** David Young-Wolff/PhotoEdit; **53** William Manning/CORBIS; **58** (t) Peter Arnold, Inc., (c) ABPL/Stephanie Lamberti/Animals Animals, (b) Taxi/Getty Images, (bkgd) Gary Bell/Australian Picture Library/CORBIS; **60** Yann Arthus-Bertrand/CORBIS; **73** Fritz Polking/Peter Arnold, Inc.; **75** Ned Therrien/Visuals Unlimited; **78** David Sieren/Visuals Unlimited; **81** CORBIS; **83** Matt Meadows; **85** Getty Images; **90** (t) Eye of Science/SPL/Photo Researchers, (b) K. Kjeldsen/Photo Researchers, (bkgd) George McCarthy/CORBIS; **92** P. Kahl/Photo Researchers; **94** (l) Michael Ord/Photo Researchers, (r) Arco Images/Alamy Images; **95** (l) Tom Bean/CORBIS, (r) Charlie Ott/Photo Researchers; **96** Jim Cartier/Photo Researchers; **98** Pierre Holtz/Reuters/CORBIS; **99** Daryl Balfour/Photo Reseachers; **102** (l) Pete Oxford/Minden Pictures, (r) Bettmann/CORBIS; **103** (l) Photo by MPI/Getty Images, (r) John Griffin/The Image Works; **106** Joanna McCarthy/Getty Images; **108** (t) Bob Cranston/Animals Animals, (b) Bettman/CORBIS; **109** Gary W. Carter/CORBIS; **110** CORBIS; **111** Bob Cranston/Animals Animals; **114** (t) Terry Spivey/USDA Forest Service/www.insectimages.org, (c) Edward Kinsman/Photo Researchers, (b) Alexis Rosenfeld/Photo Researchers, (bkgd) Roland Gerth/zefa/CORBIS; **116** PSU Entomology/Photo Researchers; **117** OSF/R. Packwood/Animals Animals; **118** (l) Stefan Meyers/Animals Animals, (r) Michael and Patricia Fogden/Minden Pictures; **119** (tl) David Cavagnaro/Visuals Unlimited, (tr) David R. Frazier/Photo Researchers, (b) Nigel J. Dennis/Photo Researchers; **121** Adam Jones/Photo Researchers; **124** (tl) Michael Sewell/Peter Arnold,(tr) Adam Jones/Visuals Unlimited, (c) Frans Lanting/Minden Pictures, (b) Frans Lanting/Minden Pictures; **127** (t) Michael Gadomski/Earth Scenes, (b) Kirtley-Perkins/Visuals Unlimited; **130** (t) Dr. Marli Miller/Visuals Unlimited, (b) Gary Braasch/CORBIS; **133** Alan Sirulnikoff/SPL/Photo Researchers; **134** Robert Brook/Photo Researchers; **135** Anthony Bannister/Gallo Images/CORBIS; **136** Wendy Stone/CORBIS; **139** (l) B. Runk/S. Schoenberger/Grant Heilman Photography, (r) B. Runk/S. Schoenberger/Grant Heilman Photography; **140** Alan Sirulnikoff/SPL/Photo Researchers; **144–145** Karen Kasmauski/CORBIS; **146** (t) David M. Phillips/Photo Researchers, (b) BSIP/Photo Researchers, Inc., (bkgd) David Young-Wolff/PhotoEdit; **150** (l) Custom Medical Stock Photography, (r) Neil Borden/Photo Researchers; **151** (tl) Peter Bowater/Photo Researchers, (tr) Spencer Jones/Picture Arts/CORBIS, (c) W. Cody/CORBIS, (b) Charles D. Winters/Photo Researchers; **155** (inset) Dennis Kunkel/PhotoTake NYC, (bkgd) Zigmund Leszczynski/Animals Animals; **156** (l) Julian Calder/CORBIS, (r) Charles D. Winters/Photo Researchers; **157** David Young-Wolff/PhotoEdit; **158** (t) PhotoLink/Getty Images, (b) Matt Meadows; **161** Color-Pic/Animals Animals; **162** David Whitten/Index Stock Imagery; **163** (t) David Young-Wolff/PhotoEdit, (bl) Matt Meadows, (br) Martin Rotker/PhotoTake NYC; **166** Elizabeth Opalenik/CORBIS; **167** (t) Charles D. Winters/Photo Researchers, (c) E.S. Ross/Visuals Unlimited, (b) Digital Art/CORBIS; **168** (l) John Anderson/Animals Animals, (r) Dr. Dennis Kunkel/Visuals Unlimited; **172** Tom Stewart/CORBIS; **173** Horizons; **175** Daniel J. McCleery/Grant Heilman Photography; **176** Paul Katz/Index Stock Imagery; **180** Getty Images; **182** Bridgeman Art Library; **183** (t) Lester V. Bergman/CORBIS, (bl) SPL/Photo Researchers, (bc) Karen Kuehn Photography, (br) LBNL/SPL/Photo Researchers; **185** (t) Driscoll, Youngquist & Baldeschwieler, California Institute of Technology; Photo Researchers, (bl) Lester V. Bergman/CORBIS, (br) Biophoto Associates/Photo Researchers; **187** Jill Barton/AP/Wide World Photos; **193** Dr. Dennis Kunkel/Phototake NYC; **194** Dr. Dennis Kunkel/Phototake NYC; **195** (t) Dr. Dennis Kunkel/Visuals Unlimited, (b) Henry Aldrich/Visuals Unlimited; **196** (t) Gopal Murti/Phototake NYC, (b) Gopal Murti/Phototake NYC; **197** (t) Dr. Donald Fawcett/Visuals Unlimited, (b) George Chapman/Visuals Unlimited; **198** (t) Marilyn Schaller/Photo Researchers, (bl) Dr. David M. Phillips/Visuals Unlimited, (br) Dr. Linda Stannard-Uct/Photo Researchers, **204** (tl) Custom Medical Stock Photo, (tr) Carolina Biological Supply/Visuals Unlimited, (bl) Custom Medical Stock Photo, (br) Carolina Biological Supply/Visuals Unlimited; **205** (l) Custom Medical Stock Photo, (r) Dr. Linda Stannard-Uct/Photo Researchers; **208** SPL/Photo Researchers; **211** Lester V. Bergman/CORBIS; **213** A.M. Siegelman/Visuals Unlimited; **216** Jose Fuste Raga/CORBIS; **218** (t) Pr. S. Cinti, University of Ancona/PhotoTake NYC, (b) CMEABG-LYON-1/PhotoTake NYC; **219** Kevin Salemme/Merrimack College; **224** CORBIS; **227** Ed Reschke/Peter Arnold, Inc.; **234** Collection CNRI/PhotoTake NYC; **242** (t) M.I. Walker/SPL/Photo Researchers, (b) B. Runk/S. Schoenberger/Grant Heilman Photography, (bkgd) B. Runk/Grant Heilman Photography; **245** Dr. Gopal Murti/Visuals Unlimited; **247** Michael Abbey/Visuals Unlimited; **248** Andrew Syred/Photo Researchers; **249** Thomas Deerinck/Visuals Unlimited; **250** (t) Dr. Conley L. Rieder and Dr. Alexey Khodjakov/Visuals Unlimited, (b) Carolina Biological Supply Co./PhotoTake NYC; **251** Michael Abbey/Photo Researchers; **252** (l) RMF/Visuals Unlimited, (r) B. Runk/S. Schoenberger/Grant Heilman Photography; **257** P. Sorrentino-Eurelios/PhotoTake NYC; **262** Biodisc/Visuals Unlimited; **266–267** Louie Psihoyos/CORBIS; **268** (t) CNRI/Photo Researchers, (c) Manfred Kage/Peter Arnold, Inc, (b) Dr. David M. Phillips/Visuals Unlimited, (bkgd) Gallo Images/Heinrich van den Berg/Getty Images; **270** (l) Amie Meister/Custom Medical Stock Photography, (r) Davies & Davies/Getty Images; **277** Bettmann/CORBIS; **281** Robert Folz/Visuals Unlimited; **285** (l) Inga Spence/Getty Images, (r) Marc Moritsch/Getty Images; **286** Barry Runk/Stan Schoenberger/Grant Heilman Photography; **290** (l) Yann Arthus-Bertrand/CORBIS, (c) Ric Frazier/Masterfile, (r) Barbara Von Hoffmann/Animals Animals; **291** Glenn Oliver/Visuals Unlimited; **294** (t b) Addenbrookes Hospital/Photo Researchers, (bkgd) Anne Ackerman/Getty Images; **303** Dr. Stanley Flegler/Visuals Unlimited; **304** (tl) B. Runk/ S. Schoenberger/Grant Heilman Photography, (tr) Carolyn A. McKeone/Photo Researchers, **(bl)** Jane Burton/Photo Researchers, **(br)** James Strawser/Grant Heilman Photography; **305** (tl) Cheryl Ertelt/Visuals Unlimited, (tcl) Jacques Brun/Photo Researchers, (tcr) Dale C. Spartas/CORBIS, (tr) M. Claye/Photo Researchers, (b) Andrew Syred/Photo Researchers; **306** (t) Razi Searles/Bruce Coleman, Inc., (b) Dr. George Wilder/Getty Images; **307** (l) Charles Krebs/CORBIS, (r) Charles Krebs/CORBIS; **309** Carolyne A. McKeone/Photo Researchers; **311** CNRI/Photo Researchers; **313** (l) Lauren Shear/Photo Researchers, (r) CNRI/Photo Researchers; **316** Michael Newman/PhotoEdit; **317** Matt Meadows; **319** Jason Edwards/National Geographic/Getty Images; **320** (tl tc) Michael P. Gadomski/Photo Researchers, (tr) Wally Eberhart/Visuals Unlimited, (r) CNRI/Photo Researchers; **321** CNRI/Photo Researchers; **324** (bkgd) Jennifer Waters & Adrian Salic/Photo Researchers, (inset) Adrian T. Sumner/Getty Images; **326** (l) Dr. Jack M. Bostrack/Visuals Unlimited, (r) Dr. Frederick Skvara/Visuals Unlimited; **330** (t) Omikron/Photo Researchers, (b) A. Barrington Brown/Photo Researchers; **347** (t) Dr. Stanley Flegler/Visuals Unlimited, (b) Jackie Lewin, Royal Free Hospital/Photo Researchers; **350** SPL/Photo Researchers; **353** Omikron/Photo Researchers; **358** (t b) Jeffrey Plautz and Steve Kay, (bkgd) Dr. Steve Robinow/Dennis Kunkel Microscopy; **360** (lc) Yann Arthus-Bertrand/CORBIS, (r) GK Hart/Vikki Hart/Getty Images; **363** Carolyn A. McKeone/Photo Researchers; **365** (l) Klaus Guldbrandsen/Photo Researchers, (r) NOAA; **371** Jim Richardson/National Geographic Image Collection; **374** (t) Pascal Goetgheluck/Photo Researchers, (b) English Greg/SYGMA/CORBIS; **375** USDA; **380** Farina, Wyckoff, Condeelis and Segall/Albert Einstein College of Medicine; **388–389** Reuters/CORBIS; **390** Louie Psihoyos/CORBIS; **392** Roger Ressmeyer/CORBIS; **393** (l to r) John Reader/Photo Researchers, **(2)** Biophoto Associates/Photo Researchers, **(3)** Dominique Braud/Earth Scenes, **(4)** Bernard Edmaier/Photo Researchers, **(5)** Francois Gohier/Photo Researchers, **(6)** Francis Latreille/CORBIS; **398** (t) Georgette Douwma/Photo Researchers, (b) Ken Lucas/Visuals Unlimited; **405** (t) Dr. Ken MacDonald/Photo Researchers, (b) Francois Gohier/Photo Researchers; **408** O. Louis Mazzatenta/National Geographic Image Collection; **410** (t) Biophoto Associates/Photo Researchers, (b) Francois Gohier/Photo Researchers; **412** (l) Barbara Strnadova/Photo Researchers, (r) Roger Garwood & Trish Ainslie/CORBIS; **413** NPS Photo by Jim Peaco; **416** (t) Dennis Kunkel Microscopy, (c b) Michael and Patricia Fogden, (bkgd) Michael Fogden/Animals Animals; **418** AKG/Photo Researchers; **423** Phyllis Greenberg/Animals Animals; **425** (t) Oliver Strewe/Lonely Planet Images, (b) Tim Fuller; **426** (tl) Alan & Sandy Carey/Getty Images, (tr) Hans Pfletschinger/Peter Arnold, Inc., (bl) David M. Dennis/Animals Animals, (br) Clouds Hill Imaging Ltd./CORBIS; **427** (t) Tom Boyden/Lonely Planet Images, (b) Gerald & Buff Corsi/Visuals Unlimited; **428** James Watt/Animals Animals; **429** (l) Erwin Bud Nielsen/Index Stock Imagery, (r) Zig Leszczynski/Animals Animals; **430** Steve Raymer/National Geographic Image Collection; **435** (t) Peter Parks/OSF/Animals Animals, (b) B. Frederick/OSF/Animals Animals; **436** (tl) Ohio Department of Natural Resources, (tr) Suzanne L. & Joseph T. Collins/Photo Researchers, (b) William Weber/Visuals Unlimited; **437** Gerard Lacz/Animals Animals; **438** (l r)Thomas & Pat Leeson/Photo Researchers, (r)Enlightened Images/Earth Scenes; **439** Mitsuhiko Imamori/Minden Pictures; **442** (t) AP Images, (bl) Markus Botzek/CORBIS, (br) DEA Picture Library/Getty Images; **445** (cr) Alan & Sandy Carey/Getty Images, (r) Hans Pfletschinger/Peter Arnold, Inc., (l) Zig Leszczynski/Animals Animals; **446** A.N.T./Photo Researchers; **447** Patti Murray/Animals Animals; **450** (t) Robert Pickett/CORBIS, (c b) Theo Allofs/Visuals Unlimited, (bkgd) Frank Lukasseck/zefa/CORBIS; **452** Gary Retherford/Photo Researchers; **454** Jeff Cadge/Getty Images; **455** (l to r) Patti Murray/Animals Animals, **(2)** Studio Carlo Dani/Animals Animals, **(3)** OSF/David Haring/Animals Animals, **(4)** George Holton/Photo Researchers; **456** Andy Rouse/Stock Image/Getty Images, (b) Frans Lanting/CORBIS; **457** (t) Tom McHugh/Photo Researchers, (bl) ZSSD/Minden Pictures, (br) Renee Lynn/CORBIS; **458** Jorg & Petra Wegner/Animals Animals; **459** Barry Slaven/Visuals Unlimited; **462** (t) Pascal Goetgheluck/SPL/Photo Researchers, (b) Tom McHugh/Photo Researchers; **463** João Zilhão/Portuguese Institute of Archaeology; **465** (t) Pascal Goetgheluck/Photo Researchers, (bl br) John Reader/Photo Researchers; **468** John Gurche Illustrations; **469** (t) John Gurche Illustrations, (b) Ira Block/National Geographic Image Collection; **471 (1 2 4)** Pascal Goetgheluck/Photo Researchers, **(3 5)** KAREN/Sygma/CORBIS; **473** (l) Serge de Sazo/Photo Researchers, (r) John Gurche Illustrations; **474** Robert Sisson/National Geographic Image Collection; **475** Cyril Ruoso/JH Editorial/Minden Pictures; **477** Tom McHugh/Photo Researchers; **478** Pascal Goetgheluck/Photo Researchers; **482** (t) John Cancalosi/Peter Arnold, Inc., (c) Gary Braasch/CORBIS, (b) Leroy Simon/Visuals Unlimited, (bkgd) Altrendo Nature/Getty Images; **485** (l) Daniel A. Bedell/Animals Animals, (c) Martin Bruce/Animals Animals, (r) Azure Computer & Photo Services/Animals Animals; **486** Gay Bumgarner/Getty Images; **487** (l) Joe McDonald/Animals Animals, (c) Richard Sobol/Animals Animals, (r) Joseph H. Bailey/Getty Images; **489** Doug Wechsler/AG Pix; **490** Barbara Strnadova/Photo Researchers;

492 Frank Lane Picture Agency/CORBIS; **493** (l) Michael Nichols/National Geographic Image Collection, (c) Martin Harvey/Gallo Images/CORBIS, (r) Theo Allofs/zefa/CORBIS; **494** (l) D. Allen Photograhy/Animals Animals, (c) Peter Weimann/Animals Animals, (r) Dallas and John Heaton/Free Agents Limited/CORBIS; **497** Gerald & Buff Corsi/Visuals Unlimited; **498** Ann Ronan Picture Library/HIP/The Image Works; **499** (l) Dr. Kari Lounatmaa/Photo Researchers, (r) James Richardson/Visuals Unlimited; **500** Wolfgang Baumeister/SPL/Photo Researchers; **501** (l) G.I. Bernard/OSF/Animals Animals, (c) M.I. Walker/Photo Researchers, (r) Patti Murray/Earth Scenes; **502** (t) H. Zettl/CORBIS, (l to r) Eye of Science/Photo Researchers, **(2)** Dr. M. Rohde, GBF/SPL/Photo Researchers, **(3)** Eric V. Grave/Photo Researchers, **(4)** Biophoto Associates/Photo Researchers, **(5)** Norbert Rosing/National Geographic Image Collection, **(6)** Valerie Giles/Photo Researchers; **503** (l) Getty Images, (c) Norbert Wu/Minden Pictures, (r) Eric Wanders/Foto Natura/Minden Pictures; **504** (b) Tom Vezo/Minden Pictures, (bc) Tim Wright/CORBIS, (t) H. Zettl/zefa/CORBIS, (tc) B. Runk/S. Schoenberger/Grant Heilman Photography; **508** Frank Lane Picture Agency/CORBIS; **509** Flip Nicklin/Minden Pictures; **512–513** Colin Cuthbert/Newcastle University/SPL/Photo Researchers; **514** (l) ISM/Phototake NYC, (r) Dr. Richard Kessel/Visuals Unlimited, (bkgd) CORBIS; **516** (l) B. Boonyaratanakornkit & D.S. Clark, G. Vrdoljak/EM Lab, University of California at Berkley/Visuals Unlimited, (c) Dr. David M. Phillips/Visuals Unlimited, (r) Dr. Dennis Kunkel/Dennis Kunkel Microscopy; **517** (l) Jim Brandenburg/Minden Pictures, (r) Craig J. Brown/Index Stock Imagery; **518** Eye of Science/Photo Researchers; **519** (t) Eye of Science/Photo Researchers, (c) Dr. Gary Gaugler/Photo Researchers, (b) Science Source/Photo Researchers; **520** Dr. Dennis Kunkel/Phototake NYC; **522** Dr. Jeremy Burgess/SPL/Photo Researchers; **523** (l) Tim Fuller, (r) Eye of Science/Photo Researchers; **526** Archivo Iconografico, S.A./CORBIS; **527** (t) AP/Wide World Photos, (b) Bettmann/CORBIS; **531** Tim Fuller; **532** Eye of Science/Photo Researchers; **533** Gary Conner/Phototake NYC; **534** Dr. Gary Gaugler/Photo Researchers; **535** (t) Dr. Dennis Kunkel/Visuals Unlimited, (c) Dr. David M. Phillips/Visuals Unlimited, (b) Dr. Dennis Kunkel/Phototake NYC; **540** (t) Oliver Meckes/Nicole Ottawa/Photo Researchers, (c) Gerald and Buff Corsi/Visuals Unlimited, (b) Michael Abbey/Photo Researchers, (bkgd) Gerald and Buff Corsi/Visuals Unlimited; **542** Dr. Dennis Kunkel/Visuals Unlimited; **543** (l) Eric Grave/Photo Researchers, (c) Richard Herrmann/Visuals Unlimited, (r) Dwight Kuhn; **544** (l) Michel Gunter/Peter Arnold Inc., (r) Michael & Patricia Fogden; **546** (l) Eric V. Grave/Photo Researchers, (r) Wim van Egmond/Visuals Unlimited; **547** (l) Michael Abbey/Visuals Unlimited, (r) Sinclair Stammers/SPL/Photo Researchers; **550** Andrew Syred/Photo Researchers; **552** (t) Charles Melton/Visuals Unlimited, (b) Anthony Bannister/Getty Images; **553** (l) Astrid & Hanns-Frieder Michler/Photo Researchers, (r) Michael Abbey/Visuals Unlimited; **554** (t) Astrid & Hanns-Frieder Michler/Photo Researchers, (b) Dee Breger/Photo Researchers; **556** (l) Dr. David M. Phillips/Visuals Unlimited, (r) Bill Bachman/Photo Researchers; **557** (l) Michael Abbey/Visuals Unlimited, (r) Wim Van Egmond/Visuals Unlimited; **558** (l) M.I. Walker/Photo Researchers, (c) Brad Mogen/Visuals Unlimited, (r) M.I. Walker/Photo Researchers; **559** Robert De Goursey/Visuals Unlimited; **561** (l) Biology Media/Photo Researchers, (r) Bill Beatty/Visuals Unlimited; **563** (t) Carolina Biological Supply Co./Phototake NYC, (b) Dr. Richard Kessel & Dr. Gene Shih/Getty Images; **564** James Richardson/Visuals Unlimited; **565** (l) Astrid & Hanns-Freider Michler/Photo Researchers, (r) David Munns/Photo Researchers; **566** (t) Andrew Syred/Photo Researchers, (c) Steve Gschmeissner/Photo Researchers, (b) SPL/Photo Researchers; **567 569** M.I. Walker/Photo Researchers; **570** John Burbidge/Photo Researchers; **574** (t) Ed Reschke/Peter Arnold Inc., (c) Biodisc/Visuals Unlimited, (b) Darlyne A. Murawski/National Geographic Image Collection, (bkgd) Jeff Lepore/SPL/Photo Researchers; **576** (l) WSL Handout/Reuters/CORBIS, (r) Dr. David Phillips/Visuals Unlimited; **577** M. Kalab/Visuals Unlimited; **578** (l) Charles McRae/Visuals Unlimited, (c) Darlyne A. Murawski/Peter Arnold Inc., (r) M.F. Brown/Visuals Unlimited; **579** Loren Winters/Visuals Unlimited; **580** Dr. Dennis Kunkel/Visuals Unlimited; **581** Bill Keogh/Visuals Unlimited; **582** Joyce Longcore/University of Maine; **583** Vaughan Fleming/SPL/Photo Researchers; **584** Dr. David M. Phillips/Visuals Unlimited; **585** (1) Joyce Longcore/University of Maine, **(2)** Barry Runk/Stan Schoenberger/Grant Heilman Photography, **(3)** Ray Coleman/Visuals Unlimited, **(4)** Jeffrey Lepore/Photo Researchers, **(5)** Dr. Dennis Kunkel/Visuals Unlimited; **586** Gerald & Buff Corsi/Visuals Unlimited; **587** (l) Stephen Sharnoff/Visuals Unlimited, (r) V. Ahmadijan/Visuals Unlimited; **588** (t) Phyllis Greenberg/Earth Scenes, (b) age fotostock/SuperStock; **589** Dr. Gerald Van Dyke/Visuals Unlimited; **591** (l) Michael Busselle/CORBIS, (c) Vaughan Fleming/Photo Researchers, (r) Dr. Lucille K. Georg/Centers for Disease Control; **592** Graeme Teague Photography; **595** M. Kalab/Visuals Unlimited; **596** Stephen Sharnoff/Visuals Unlimited; **597** Dr. Gerald Van Dyke/Visuals Unlimited; **600–601** Jim Sugar/CORBIS; **602** (t) David Muench/CORBIS, (c) George H.H. Huey/CORBIS, (b) Theo Allofs/zefa/CORBIS, (bkgd) Photo Library International/ESA/Photo Researchers; **605** (l) John Lemker/Animals Animals, (r) Eye of Science/Photo Researchers; **606** (t) Barry Runk/Stan Schoenberger/Grant Heilman Photography, (b) Michael Boys/CORBIS; **607** (l) Rob Simpson/Visuals Unlimited, (r) SCIMAT/Photo Researchers; **610** (l) John Heseltin/SPL/Photo Researchers, (c) Brad Mogen/Visuals Unlimited, (r) Dr. R. Calentine/Visuals Unlimited; **611** (t) Geoff Butler, (b) Henry W. Robison/Visuals Unlimited; **612** (l) Hal Horwitz/CORBIS, (r) Dr. Ken Wagner/Visuals Unlimited; **613** Winifred Wisniewski/Frank Lane Picture Agency/CORBIS; **614** (t) Jane Grushow/Grant Heilman Photography, (cw from top) Dr. Jeremy Burgess/Photo Researchers, **(2)** Carlyn Iverson/Photo Researchers, **(3)** age fotostock/SuperStock, **(4)** Clay Perry/CORBIS; **615** (l) W.P. Armstrong, (r) J.B. Rapkins/Photo Researchers; **616** (l) W.P. Armstrong, (r) Barry Runk/Grant Heilman Photography; **617** (cw from top) William Harlow/Photo Researchers, **(2)** Barry Runk/Stan Schoenberger/Grant Heilman Photography, **(3)** Gail Jankus/Photo Researchers, **(4)** Inga Spence/Visuals Unlimited, **(5)** Michael Rose/CORBIS; **618** M. Philip Kahl/Photo Researchers; **619** (t) Adam Jones/Photo Researchers, (c) Dr. Daniel Nickrent and Dr. Kenneth Robertson, (bl) Dave Powell/USDA Forest Service/www.forestry-images.org, (bc) Gary Braasch/CORBIS, (br) Inga Spence/Visuals Unlimited; **621** (l) Randall G. Prostak/University of Massachusetts, (r) Randall G. Prostak/University of Massachusetts; **625** (cw from top) Barry Runk/Stan Schoenberger/Grant Heilman Photography, **(2)** Dr. Ken Wagner/Visuals Unlimited, **(3)** SCIMAT/Photo Researchers, **(4)** Eye of Science/Photo Researchers; **626** (tl) Barry Runk/Grant Heilman Photography, (r) Winifred Wisniewski/Frank Lane Picture Agency/CORBIS, (bl) Jane Grushow/Grant Heilman Photography, (br) W.P. Armstrong; **627** Dr. Daniel Nickrent and Dr. Kenneth Robertson; **630** (t) Dr. David Webb/University of Hawaii at Manoa, (b) Biodisc/Visuals Unlimited, (bkgd) Lowell Georgia/CORBIS; **633** (b) Frank H. McClung Museum/University of Tennessee, (cw from top) Brad Mogen/Visuals Unlimited, **(2)** Dr. Arnold Karpoff/Visuals Unlimited, **(3)** Ed Reschke/Peter Arnold, **(4)** Dr. Kent Simmons, **(5)** Jack Bostrack/Visuals Unlimited; **635** (t) Ken Wagner/Phototake NYC, (c) Sidney Moulds/SPL/Photo Researchers, (b) Eye of Science/Photo Researchers; **636** (t) Barry Runk/Stan Schoenberger/Grant Heilman Photography, (bl) Dr. Richard Kessel & Dr. Gene Shih/Visuals Unlimited, (br) Brad Mogen/Visuals Unlimited; **637** Wilfred Cote, SUNY College of Environmental Forestry; **639** Dr. Jeremy Burgess/Photo Researchers; **640** (l) Science VU/Visuals Unlimited, (r) Brad Mogen/Visuals Unlimited; **641** (tl) Antonia Reeve/Photo Researchers, (tc) Science VU/Visuals Unlimited, (tr) Oregon State University Jed Colquhoun Photo Collection, (bl) Geoff Tompkinson/Photo Researchers, (br) Beatrice Neff/Photo Researchers; **642** (t) Carolina Biological Supply/Phototake NYC, (b) Robert Calentine/Visuals Unlimited; **643** (tl) Bill Beatty/Visuals Unlimited, (tc) Science Pictures Limited/Photo Researchers, (tr) Barry Runk/Stan Schoenberger/Grant Heilman Photography, (bl br) Matt Meadows; **646** (l) Peter Marbach/Grant Heilman Photography, (r) The Picture Store/SPL/Photo Researchers; **647** (l) Darrell Gulin/CORBIS, (r) David Sieren/Visuals Unlimited; **649** (t c) Chris Althof, (b) Barry Runk/Stan Schoenberger/Grant Heilman Photography; **651** (t) Maryann Frazier/Photo Researchers, (c) David M. Dennis, (b) Adam Jones/Visuals Unlimited; **652** (t) Andrei Sourakov and Consuelo De Moraes; **655** (tl) Jack Bostrack/Visuals Unlimited, (tr) Dr. Jeremy Burgess/Photo Researchers, (bl) Barry Runk/Stan Schoenberger/Grant Heilman Photography, (br) Dr. Richard Kessel & Dr. Gene Shih/Visuals Unlimited; **656** (tl) Brad Mogen/Visuals Unlimited, (tr) Science VU/Visuals Unlimited, (cl cr) Carolina Biological Supply/Phototake NYC, (bl br) Chris Althof/The McGraw-Hill Companies; **660** (t) Dr. Jeremy Burgess/SPL/Photo Researchers, (b) M.F. Merlet/SPL/Photo Researchers, (bkgd) Roger Wilmshurst/Photo Researchers; **662** Dr. Gary Gaugler/Photo Researchers; **663** Rosenfeld Images Ltd./Photo Researchers; **669** (l) Dr. Nick Kurzenko/SPL/Photo Researchers, (r) Hal Horwitz/CORBIS; **671** (l) Patti Murray/Animals Animals, (r) Jerome Wexler/Visuals Unlimited; **677** (cw from top) Tom Stack & Associates, **(2)** Joy Spurr/Photo Researchers, **(3)** Joel W. Rogers/CORBIS, **(4)** Burke/Triolo Productions/Getty Images, **(5)** Chris Evans/www.forestryimages.org, **(6)** Paul Wray/Iowa State University/www.forestryimages.org, **(7)** Chris Hellier/CORBIS, **(8)** David Stuckel; **680** Cristina Pedrazzini/Photo Researchers; **681** Laura Sifferlin; **683** Rosenfeld Images Ltd./Photo Researchers; **688–689** Pierre Perrin/SYGMA/CORBIS; **690** (t) Alexis Rosenfeld/Photo Researchers, (b) Dr. Brad Amos/SPL/Photo Researchers, (bkgd) Taxi/Getty Images; **692** (l) E. Hanumantha Rao/Photo Researchers, (r) Darlyne A. Murawski/Peter Arnold, Inc.; **693** (t) Judy Davidson/Photo Researchers, (b) Rod Planck/Photo Researchers; **694** (t) Sheila Terry/Photo Researchers, (b) CORBIS; **695** (t) Mark Stouffer Enterprises/Animals Animals, (b) Spencer Platt/Getty Images; **696** (l) Dr. Richard Kessel & Dr. Gene Shih/Visuals Unlimited, (c) Dr. Richard Kessel & Dr. Gene Shih/Visuals Unlimited, (r) Dr. Richard Kessel & Dr. Gene Shih/Visuals Unlimited; **698** (l) O.S.F./Animals Animals, (c) D. Robert & Lori Franz/CORBIS, (r) Shah, Manjo/Animals Animals; **704** William Dow/CORBIS; **705** Therisa Stack/Tom Stack & Associates; **707** (tl) Lawson Wood/CORBIS, (tr) CORBIS, (bl) Tom Stack/Tom Stack & Associates, (br) Patti Murray/Animals Animals; **708** Azure Computer & Photo Services/Animals Animals; **709** (t) Chris McLaughlin/Animals Animals, (bl bc br) Harbor Branch Oceanographic Institution; **710** (l) Brandon D. Cole/CORBIS, (r) CORBIS; **713** (t) Georgette Douwma/Photo Researchers, (b) Jeff Rotman/Photo Researchers; **716** M. Amend, NOAA NMFS; **717** Matt Meadows; **720** Lawson Wood/CORBIS; **724** (t) SPL/Photo Researchers, (b) SPL/Photo Researchers, (bkgd) Scott W. Smith/Animals Animals; **726** (l) Arthur Seigelman/Visuals Unlimited, (r) James Robinson/Animals Animals; **728** Michael Abbey/Photo Researchers; **729** (t) Tom Adams/Visuals Unlimited, (b) E. R. Degginger/Photo Researchers; **730** (l) Barry Runk/Stan Schoenberger/Grant Heilman Photography, (r) Dr. Richard Kessel & Dr. Gene Shih/Visuals Unlimited; **731** (l) Barry Runk/Stan Schoenberger/Grant Heilman Photography, (r) Lauritz Jensen/Visuals Unlimited; **733** (t) Sinclair Stammers/Photo Researchers, (b) Dr. Robert Calentine/Getty Images; **734** (l) Centers for Disease Control, (c)Sinclair Stammers/SPL/Photo Researchers, (r) Lester V. Bergman/CORBIS; **735** (l) Michael P. Gadomski/Photo Researchers, (c) Nigel Cattlin/Photo Researchers, (r) Nigel Cattlin/Photo Researchers; **736** John Walsh/Photo Researchers; **737** (l) age fotostock/SuperStock, (r) OSF/D. Fleetham/Animals Animals; **742** (l) age fotostock/SuperStock, (r) Andrew J. Martinez/Photo Researchers; **743** Georgette Douwma/Photo Researchers; **744** Gary Meszaros/Photo Researchers; **745** (l) John Anderson/Animals Animals, (r) Marty Snyderman/Visuals Unlimited; **748** Barry Runk/Stan Schoenberger/Grant Heilman Photography; **749** (tl) Robert Yin/CORBIS, (tr) Amos Nachoum/CORBIS, (b) OSF/Kathy Atkinson/Animals Animals; **752** Eric Luse/San Francisco Chronicle; **760** (t) Sinclair Stammers/Photo Researchers, (c) Laguna Design/Photo Researchers, (b) Clouds Hill Imaging Ltd./CORBIS, (bkgd) Ralph A. Clevenger/CORBIS; **764** (t) Oliver Meckes/Nicole Ottawa/Photo Researchers, (b) Raymond Mendez/Animals Animals; **765** Mark Moffett/Minden Pictures; **768** (l) Harry Rogers/Photo Researchers, (c) Kjell Sandved/Visuals Unlimited, (r) Oliver Meckes/Nicole Ottawa/Photo Researchers; **770** (l) John Giustina/Getty Images, (c) CORBIS, (r) Barbara Gerlach/Visuals Unlimited; **771** Gregory G. Dimijian, M. D./Photo Researchers; **772** Niall Benvie/CORBIS; **773** (l) Brad Mogen/Visuals Unlimited, (c) Dennis Kunkel/Phototake NYC, (r) Fabio Colombini Medeiros/Animals Animals; **774** James Zipp/Photo Researchers; **777** (t) Milt & Patti Putnam/CORBIS, (b) Getty Images, (inset) Ralph A. Clevenger/CORBIS; **780** (t) Raymond Coleman/Visuals Unlimited, (bl) DK Limited/CORBIS, (br) David Aubrey/CORBIS; **781** (l) DK Limited/CORBIS, (r) Diane Nelson/Visuals Unlimited; **782** Clive Druett/Papilio/CORBIS; **790** (t b) Chris Newbert/Minden Pictures, (bkgd) Franklin Viola/Animals Animals; **792** (inset) Randy Morse/Animals Animals, (bkgd) Fred McConaughey/Photo Researchers; **793** (inset) Douglas P. Wilson/Frank Lane Picture Agency/CORBIS, (bkgd) Robert L. Dunne/Photo Researchers; **795** Nancy Rotenberg/Animals Animals; **796** Zigmund Leszczynski/Animals Animals; **797** (b) Stuart Westmorland/Photo Researchers, (l to r, t to b) Zigmund Leszczynski/Animals Animals, **(2)** Joyce & Frank Burek/Animals Animals, **(3)** C.C. Lockwood/Animals Animals, **(4)** Fred McConaughey/Photo Researchers, **(5)** Gustav Verderber/Visuals Unlimited, **(6)** Science VU/Alan Baker/Visuals Unlimited; **798** (t) Fred McConaughey/Photo Researchers, (bl) David Wrobel/Visuals Unlimited, (br) Scott Johnson/Animals Animals; **799** (t) Ken Lucas/Visuals Unlimited, (bl) James L. Amos/Photo Researchers, (br) Norbert Wu/Minden Pictures; **800** (t) David Steven Miller/Animals Animals, (b) Science VU/Alan Baker/Visuals Unlimited; **801** Norbert Wu/Minden Pictures; **802** (l) G.I. Bernard/OSF/Animals Animals, (r) Bruce Watkins/Animals Animals; **809** Andrew J. Martinez/Photo Researchers; **810** (t) Robert L. Dunne/Photo Researchers, (b) G.I. Bernard/OSF/Animals Animals; **816–817** China Photos/Getty Images; **818** (t) Jeffrey L. Rotman/CORBIS, (b) Michael Davidson/Molecular Expressions, (bkgd) Hal Beral/Grant Heilman Photography; **820** (l) Tom Brakefield/CORBIS, (r) C. Allan Morgan/Peter Arnold, Inc.; **823** CORBIS; **826** Ed Reschke/Peter Arnold, Inc.; **828** (l) Sinclair Stammers/SPL/Photo Researchers, (r) M.I. Walker/Photo Researchers; **829** (t) Jeffrey L. Rotman/CORBIS, (bl) Tim Davis/CORBIS, (br) Dr. Dennis Kunkel/Getty Images; **830** Jeff Rotman/Photo Researchers; **832** James L. Amos/National Geographic Image Collection; **833** Lee Rentz Photography; **835** Heidi & Hans-Jurgen Koch; **838** (l t) Fred Habegger/Grant Heilman Photography, (r) Joel Sartore/Grant Heilman Photography; **839** (tl) Fred Habegger/Grant

Heilman Photography, (tr) Derek Middleton/FLPA/Minden Pictures, (b) Juan Manuel Renjifo/Animals Animals; **841** Dr. Joseph Kiesecker/Pennsylvania State University; **843** Matt Meadows; **850** (t) Jim Merli/Visuals Unlimited, (b bkgd) Joe McDonald/Visuals Unlimited; **852** Gerold and Cynthia Merker/Visuals Unlimited; **853** Heidi & Hans-Jurgen Koch/Minden Pictures; **854** Michael & Patricia Fogden/Minden Pictures; **855** (l) Francesc Muntada/CORBIS, (r) Arthur Morris/Visuals Unlimited; **856** (l) Azure Computer & Photo Services/Animals Animals, (r) David M. Dennis/Animals Animals; **857** (tl) Zigmund Leszczynski/Animals Animals, (tr) Joe McDonald/CORBIS, (b) Doug Wechsler/Animals Animals; **860** Suzanne L. and Joseph T. Collins/Photo Researchers; **864** (t) Arthur Morris/CORBIS, (bl) Michael & Patricia Fogden/CORBIS, (bc) Hal Beral/Visuals Unlimited, (br) Tom Vezo/Peter Arnold, Inc.; **865** Anthony Mercieca/Photo Researchers; **866** Steve Maslowski/Visuals Unlimited; **869 871** Claus Meyer/Minden Pictures; **872** (t) Azure Computer & Photo Services/Animals Animals, (b) Michael & Patricia Fogden/CORBIS; **873** Breck P. Kent/Animals Animals; **875** James Zipp/Photo Researchers; **878** (t) Michael W. Davidson/Florida State University, (c) Susumu Nishinaga/Photo Researchers, (b) Ron Spoomer/Visuals Unlimited, (bkgd) Greg Stott/Masterfile; **880** Kevin Schafer/zefa/CORBIS; **881** (l) Tom & Pat Leeson/Photo Researchers, (r) Art Wolfe/Getty Images; **887** (l to r) Peter Steiner/CORBIS, (2) Marty Snyderman/Visuals Unlimited, (3) Kevin Schafer/CORBIS, (4) L. Rue/CORBIS, (5) A. & M. Shah/Animals Animals; **888** (l) Breck P. Kent/Animals Animals, (r) Stephen Dalton/Animals Animals; **889** (l) Tom McHugh/Photo Researchers, (c) J. & C. Sohns/Animals Animals, (r) McKelvey/Rismiller; **890** (t) Stouffer Productions/Animals Animals, (bl) Keith Scholey/Getty Images, (bc) Charles Philip Cangialosi/CORBIS, (br) Alan Root/OSF/Animals Animals; **891** (l) Tim Davis/CORBIS, (r) Tim Shepherd/OSF/Animals Animals; **892** (tl) Barry Runk/Stan Schoenberger/Grant Heilman Photography, (tr) David Lazenby/Animals Animals, (b) Tui De Roy/Minden Pictures; **893** (tl) Alan G. Nelson/Animals Animals, (tr) Alan & Sandy Carey/OSF/Animals Animals, (bl) Phyllis Greenberg/Animals Animals, (br) Panthera Productions/ABPL/Animals Animals; **894** Wally Santana/AP/Wide World Photos; **895** (l) Sirenia Project, U.S. Geological Survey, (r) Brandon D. Cole/CORBIS; **898** Alex Wong/Getty Images; **899** Zig Leszczynski/Animals Animals; 900 (t) Kevin Schafer/zefa/CORBIS, (b) Sirenia Project, U.S. Geological Survey; **906** (t) Tim Davis/CORBIS, (c) Tim Davis/Getty Images, (b) Frans Lanting/CORBIS, (bkgd) Joel Simon/Getty Images; **908** (t) Digital Vision/Getty Images, (b) Sinclair Stammers/Photo Researchers; **909** (t) B&B Wells/OSF/Animals Animals, (c) Jo McCulty/Ohio State University, (b) Dr. Tetsuro Matsuzawa; **912** (t) Ferrero/Labat/Peter Arnold, Inc., (b) Matt Meadows; **914** Scott Martin/AP/Wide World Photos; **915** (l) Courtesy of Dr. Bernd Heinrich, University of Vermont, (r) Clive Bromhall/OSF/Animals Animals; **916** Ray Richardson/Animals Animals; **917** (t) Norbert Rosing/Animals Animals, (b) H. Wiesenhofer/Photolink/Getty Images; **918** Wolfgang Kaehler/CORBIS; **919** Doug Wilson/CORBIS; **920** (l) Manoj Shah/Animals Animals, (r) Juergen & Christine Sohns/Animals Animals; **921** (t) Tui De Roy/Minden Pictures, (b) Manoj Shah/Animals Animals; **922** Jennifer Jarvis/Visuals Unlimited; **924** Sharna Balfour/Gallo Images/CORBIS; **926** (t) Digital Vision/Getty Images, (b) Norbert Rosing/Animals Animals; **932–933** Ted Horowitz/CORBIS; **934** (t) Anatomical Travelogue/Photo Researchers, (b) Dr. Ken Wagner/Visuals Unlimited, (bkgd) Paul Conklin/PhotoEdit; **937** Andrew Syred/Photo Researchers; **938** Peter Lavery/Masterfile; **940** (l) Dr. Ken Greer/Visuals Unlimited, (r) Dr. Ken Greer/Visuals Unlimited; **941** Tim Fuller; **942** Prof. P. Motta/Department of Anatomy/University, "La Sapienza", Rome/SPL/Photo; Researchers; **945** CORBIS; **950** Denis Paquin/AP/Wide World Photos; **951** (t) Matthew Impey/Colorsport/CORBIS, (b) Matthew Stockman/Getty Images; **952** Kacey Marra, Lee Weiss, Prashant Kumta and Jay Calvert/University of Pittsburgh; **953** 3B Scientific GmbH, Germany; **956** Tim Fuller; **960** (t) Biodisc/Visuals Unlimited, (b) Dr. David M. Phillips/Visuals Unlimited, (bkgd) Focus on Sport/Getty Images; **966** Dr. Fred Hossler/Visuals Unlimited; **968** (l) Hiram Bingham/National Geographic Image Collection, (r) Bettman/CORBIS; **969** (t) Science Pictures Limited/Photo Researchers, (b) Alfred Pasieka/SPL/Photo Researchers; **971** Tim Fuller; **977** (l to r) (1) Jochen Tack/Peter Arnold, Inc., (2) Don Farrall, (3) Vince Bucci/Stringer/Getty Images, (4) Tom Vezo/Peter Arnold, Inc., (5) Michael P. Gadomski/ Photo Researchers; **979** Laura Sifferlin; **980** Dr. Susan Tapert, University of California at San Diego.; **981** Michael Newman/PhotoEdit; **982** Duke University Photo Department; 983 Horizons; **985** Science Pictures Limited/Photo Researchers; **990** (t) Susumu Nishinaga/Photo Researchers, (b) Dr. Philippa Uwins, Whistler Research Photography/SPL/Photo Researchers, (bkgd) Lee White/CORBIS; **992** (l) Art Resource, (r) SSPL/The Image Works; **993** (t) Goldberg Diego/SYGMA/CORBIS, (b) SPL/Photo Researchers; **995** Biophoto Associates/Photo Researchers; **997** (t) Dr. Dennis Kunkel/Visuals Unlimited, (bl) Lauren Shear/Photo Researchers, (br) Dr. David M. Phillips/Visuals Unlimited; **999** Scott Camazine/Photo Researchers; **1000** David Wrobel/Visuals Unlimited; **1001** (t) Tim Fuller, (b) Eye of Science/Photo Researchers; **1003** (tl) Tim Fuller, (tr) Steve Gschmeissner/Photo Researchers, (b) Dr. Kessel & Dr. Kerdon/Tissues & Organs/Getty Images; **1005** Tim Fuller; **1008** Dr. E. Walker/Photo Researchers; **1011** Spencer Grant/PhotoEdit; **1014** Eye of Science/Photo Researchers; **1018** (bkgd) David Young-Wolff/PhotoEdit; **1020 1021** Tim Fuller; **1022** Simon Fraser/SPL/Photo Researchers; **1026** (t) Photolibrary.com/Index Stock Imagery, (b) Michael Newman/PhotoEdit; **1027** Jodi Cobb/National Geographic Image Collection; **1029** (t to b) Getty Images, (2) CORBIS, (3) Steven Mark Needham/Getty Images, (4) Foodcollection/Getty Images, (5) CORBIS, (6) Getty Images, (7) Lew Robertson/CORBIS, (8) Getty Images; **1030** C. Sherburne/PhotoLink/Getty Images; **1032** (t) Getty Images, (b) Sam A. Marshall; **1033** Tim Fuller; **1038** Lester V. Bergman/CORBIS; **1042** Jonathan Nourok/PhotoEdit; **1043** (l) CORBIS, (r) Yang Liu/CORBIS; **1046** (t) Biophoto Associates/Photo Researchers, (b) Lennart Nilsson/Bonnier-Alba, (bkgd) Neil Bromhall/Genesis Films/SPL/Photo Researchers; **1052** Lester V. Bergman/CORBIS; **1057** Tim Fuller; **1058** (l) Petit Format/Photo Researchers, (c) Claude Edelmann/Photo Researchers, (r) Dr. G. Moscoso/Photo Researchers; **1059** (l) Mediscan/CORBIS, (r) Steve Allen/Getty Images; **1061** Carolina Biological Supply Co./Visuals Unlimited; **1065** (l c) Hulton Archive/Stringer/Getty Images, (r) Mitchell Gerber/CORBIS; **1066** Getty Images; **1067** Zephyr/Photo Researchers; **1074** (t) SIU/Visuals Unlimited, (b) Susumu Nishinaga/SPL/Photo Researchers, (bkgd) Michael Newman/PhotoEdit; **1076** Scott Camazine/Photo Researchers; **1079** (tl) SuperStock, (tr) Grapes-Michaud/Photo Researchers, (bl) Bonnie Kamin/PhotoEdit; **1079** (br) James Gathany/Centers for Disease Control; **1080** (l) A.B. Dowsett/SPL/Photo Researchers, (r) Nick Sinclair/SPL/Photo Researchers; **1081** (t) James Gathany/Centers for Disease Control, (c) Immusol Inc., (b) Steve Wood/UAB Creative and Marketing; **1082** age fotostock/SuperStock; **1084** David Scharf/Photo Researchers; **1085** (t) Michael Ross/Photo Researchers, (c) Dennis Kunkel Microscopy, (b) Hossler/Custom Medical Stock Photo; **1086** Tim Fuller; **1092** Dr. Gladden Willis/Visuals Unlimited; **1093** (t) James Stevenson/Photo Researchers, (c b) Custom Medical Stock Photography; **1094** (1) Andrew Syred/Photo Researchers, (2) Eye of Science/Photo Researchers, (3) Dr. Dennis Kunkel/Visuals Unlimited, (4) C Squared Studios/Getty Images, (5) Brian G. Green/Getty Images; **1095** Sue Ford/Photo Researchers; **1096** Digital Art/CORBIS; **1099** (t) age footstock/Superstock, (b) Joe McDonald/CORBIS; **1100** Publiphoto/Photo Researchers; **1104** CORBIS; **1106** National Science Museum/HO/epa/Corbis; **1108** Glencoe; **1110** (tl) Bridgeman Art Library, (tc) SPL/Photo Researchers, (tr) LBNL/SPL/Photo Researchers, (b) Karen Kuehn Photography; **1111, 1113** Tim Fuller; **1119** (t) Hans Pfletschinger/Peter Arnold, Inc., (c) B. Boonyaratanakornkit & D.S. Clark, G. Vrdoljak/EM Lab, University of California at Berkley/Visuals Unlimited, (b) Eric Grave/Photo Researchers; **1120** (t) Dee Breger/Photo Researchers, (c) Robert De Goursey/Visuals Unlimited, (b) Biology Media/Photo Researchers; **1121** (t) Barry Runk/Stan Schoenberger/Grant Heilman Photography, (c) Hal Horwitz/CORBIS, (b) age fotostock/SuperStock; **1122** (t) M. Philip Kahl/Photo Researchers, (c) Lawson Wood/CORBIS, (b) age fotostock/SuperStock; **1123** (t) Scott Johnson/Animals Animals; **1123** (b) Kevin Schafer/zefa/CORBIS.

A Biologist's Guide To The Periodic Table

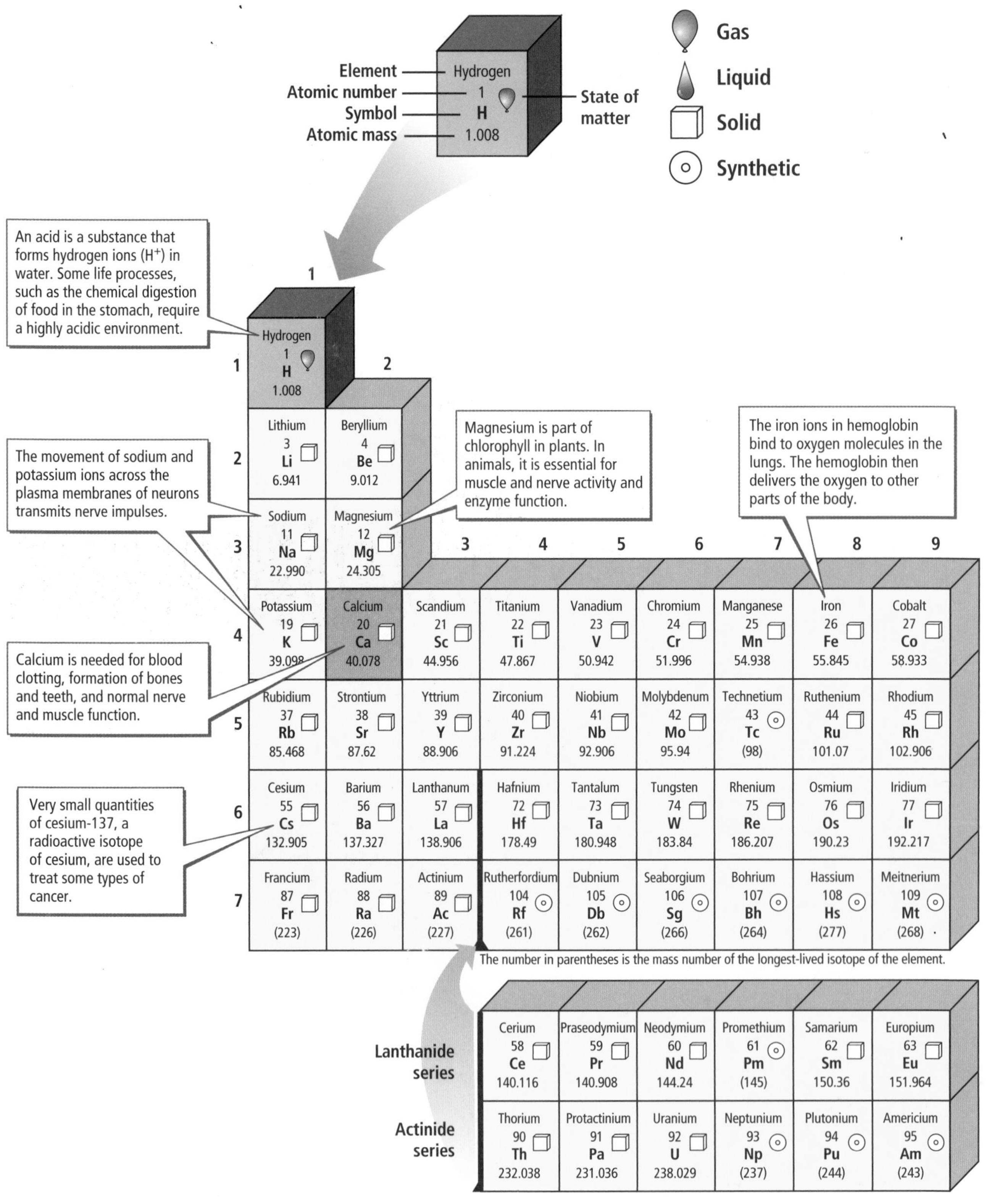

Element
Atomic number
Symbol
Atomic mass
Hydrogen
1
H
1.008
State of matter
Gas
Liquid
Solid
Synthetic
An acid is a substance that forms hydrogen ions (H^+) in water. Some life processes, such as the chemical digestion of food in the stomach, require a highly acidic environment.
The movement of sodium and potassium ions across the plasma membranes of neurons transmits nerve impulses.
Calcium is needed for blood clotting, formation of bones and teeth, and normal nerve and muscle function.
Very small quantities of cesium-137, a radioactive isotope of cesium, are used to treat some types of cancer.
Magnesium is part of chlorophyll in plants. In animals, it is essential for muscle and nerve activity and enzyme function.
The iron ions in hemoglobin bind to oxygen molecules in the lungs. The hemoglobin then delivers the oxygen to other parts of the body.
1 2 3 4 5 6 7 8 9
Hydrogen 1 H 1.008
Lithium 3 Li 6.941
Beryllium 4 Be 9.012
Sodium 11 Na 22.990
Magnesium 12 Mg 24.305
Potassium 19 K 39.098
Calcium 20 Ca 40.078
Scandium 21 Sc 44.956
Titanium 22 Ti 47.867
Vanadium 23 V 50.942
Chromium 24 Cr 51.996
Manganese 25 Mn 54.938
Iron 26 Fe 55.845
Cobalt 27 Co 58.933
Rubidium 37 Rb 85.468
Strontium 38 Sr 87.62
Yttrium 39 Y 88.906
Zirconium 40 Zr 91.224
Niobium 41 Nb 92.906
Molybdenum 42 Mo 95.94
Technetium 43 Tc (98)
Ruthenium 44 Ru 101.07
Rhodium 45 Rh 102.906
Cesium 55 Cs 132.905
Barium 56 Ba 137.327
Lanthanum 57 La 138.906
Hafnium 72 Hf 178.49
Tantalum 73 Ta 180.948
Tungsten 74 W 183.84
Rhenium 75 Re 186.207
Osmium 76 Os 190.23
Iridium 77 Ir 192.217
Francium 87 Fr (223)
Radium 88 Ra (226)
Actinium 89 Ac (227)
Rutherfordium 104 Rf (261)
Dubnium 105 Db (262)
Seaborgium 106 Sg (266)
Bohrium 107 Bh (264)
Hassium 108 Hs (277)
Meitnerium 109 Mt (268)
The number in parentheses is the mass number of the longest-lived isotope of the element.
Lanthanide series
Cerium 58 Ce 140.116
Praseodymium 59 Pr 140.908
Neodymium 60 Nd 144.24
Promethium 61 Pm (145)
Samarium 62 Sm 150.36
Europium 63 Eu 151.964
Actinide series
Thorium 90 Th 232.038
Protactinium 91 Pa 231.036
Uranium 92 U 238.029
Neptunium 93 Np (237)
Plutonium 94 Pu (244)
Americium 95 Am (243)